THE
OXFORD COMPANION TO
THE HISTORY OF MODERN
SCIENCE

THE
OXFORD COMPANION TO
THE HISTORY OF MODERN
SCIENCE

Editor in Chief

J. L. Heilbron

Editors

James Bartholomew   Jim Bennett   Frederic L. Holmes

Rachel Laudan   Giuliano Pancaldi

OXFORD
UNIVERSITY PRESS
2003

OXFORD
UNIVERSITY PRESS

Oxford   New York
Auckland   Bangkok   Buenos Aires   Cape Town   Chennai
Dar es Salaam   Delhi   Hong Kong   Istanbul   Karachi   Kolkata
Kuala Lumpur   Madrid   Melbourne   Mexico City   Mumbai   Nairobi
São Paulo   Shanghai   Singapore   Taipei   Tokyo   Toronto

Published by Oxford University Press, Inc.
198 Madison Avenue, New York, NY 10016
http://www.oup-usa.org

Oxford is a registered trademark of Oxford University Press

Library of Congress Cataloging-in-Publication Data

The Oxford companion to the history of modern science/editor in chief,
J. L. Heilbron; editors, James Bartholomew . . . [et al.].
p. cm.
Includes bibliographical references and index.
ISBN 0-19-511229-6 (acid-free paper)
1. Science–History–Encyclopedias. 2. Science–Social
aspects–Encyclopedias. I. Heilbron, J. L.
Q125 .O86 2003
509–dc21   2002153783

9 8 7 6 5 4 3 2 1

Printed in the United States of America
on acid-free paper

CONTENTS

PREFACE

"Eppure si muove"—"still it moves"—quipped Pope John Paul II as he hobbled on his newly repaired hip to preside over a synod of bishops in 1994. Everyone understood the unspoken reference to Galileo's apocryphal muttered defiance 331 years earlier during his forced recantation of his detestable opinion that the earth turns. The Pope joked that he too was a martyr to science. Our boon *Companion* prepares its readers for such unlikely references to the historical culture of science.

THE CHARACTER OF THE COMPANION

The *Oxford Companion to the History of Modern Science* reliably describes the development and ramification of the main branches and twigs of natural knowledge and their uses in industry, literature, religion, war, entertainment, and much else. In keeping with this serious side of its personality, it can be self-contemplative. It considers various approaches to the history of science including the feminist and Marxist; defines and historicizes terms of art like model and theory; and discusses the fluctuating influence of epistemological notions like hypothesis and proof. By turns the *Companion* is also playful, as in "Putti in Science," "Slogans from Science," and, in a Pickwickian way, "Quark"; expansive, as in "Asia" and "Latin America"; thoughtful, as in "Race" and "Clone"; and celebratory, as in "Noble Prize."

The *Companion* depicts its subject with a wider palette than other single-volume and most multi-volume reference works in the history of science and technology. It depicts for a general audience the process by which our colorful, vigorous, demanding, and sometimes off-putting science and its products have come to direct world history. According to George Sarton, the architect of the academic study of the history of science during the last century, "The history of science is the history of civilization." That is at least half true. Our *Companion* is more than an account of a specialized and to some tastes unpleasant human activity. It is also a *Companion* to world history, modern in coverage, proper in demeanor, generous in breadth, and cosmopolitan in scope.

Modernity. It may not be easy to say when modernity started or ended, if it is over, but few will dispute that Columbus sailed before it began or that it flourished during the late nineteenth and twentieth centuries, and that it saw the creation and exploitation of natural science as we know it. Hence we limit our coverage to the modern period. Apart from appropriate backward glances and biographies of early sixteenth-century pioneers like Nicholas Copernicus and Andreas Vesalius, the *Companion* picks up around 1550 and dwells on the seventeenth and eighteenth centuries before turning to the development of the modern scientific disciplines and their interactions with the societies that support them. The earlier period receives emphasis not to allow contributors to expatiate on the so-called Scientific Revolution, which has been the lynch-pin of the historiography of science, but to stress that modern science is a discovery as well as an invention. It was a discovery that nature generally acts regularly enough to be described by laws and even by mathematics; and it required invention to devise the techniques, abstractions, apparatus, and organization for exhibiting the regularities and securing their law-like descriptions.

The discovery of the regularities of planetary motions goes back to antiquity; but the discovery of principles that allowed for ever finer quantitative agreement between the prediction and observation of astronomical phenomena dates from the seventeenth century. The means by which the agreement was secured—the instruments, computational techniques, and physical models—suggested that the phenomena of physics and chemistry might also be (or with proper definitions and abstractions become) as law-like as astronomy. The systematic production of evidence favorable to the suggestion dates from the last few decades of the eighteenth century.

The redoubtable Immanuel Kant declared in 1786, while deprecating chemistry as no more than an art, that "in every special doctrine of nature only so much science proper can be found as there is mathematics in it." Chemistry went modern from Kant's point of view in the nineteenth century and several of the biological sciences in the twentieth. The trend was resisted by people who deplored the loss of the spontaneous human element from natural science and the coming of professionalization, with its specialization and standardization, and of fraternization with industry and the military. But their resistance did not arrest the trend, which only broadened and strengthened. Modern Western science provides much of the force and substance of globalization. This is not the fault of historians.

Answers to the questions when modern science began and how it grew express values about modernity. Some historians, put off by science's methods of aggrandizement, represent the secularly progressive path of science as a locally bumpy road along which scientists advance by fighting with one another over facts, interpretations, and authority. Others have located the fighting between forward-thinking philosophers of nature and reactionary powers of religion, state, and society. Still others couple the development of science tightly with practical applications, with pharmacy, medicine, mining, and manufactures.

The *Companion* speaks to all of these possibilities and more besides. Readers can choose among them or construct their own stories, aware that the choice matters; that placing the threshold of modern science in the work of Galileo, or the marginalization of women, or the second world war (to pick slight variations of standard candidates) implies and conveys a worldview.

Propriety. The word "scientist" was not seriously propounded before 1840 and not used widely until well into the twentieth century ("man of science" or "scientific man" being preferred). Nonetheless, historians and others who should know better refer readily to doubtful "scientists" of the Renaissance and mythical creatures like "geologists" of the Middle Ages and "biologists" of Antiquity. This usage can easily hide discontinuities in the subject matters of the sciences and in the circumstances of their cultivation. Anachronistic vocabulary distorts understanding of the nature of modern sciences and their place in the societies that support them.

The *Companion* avoids using the names of modern sciences in reference to the sixteenth, seventeenth, and eighteenth centuries, except for anatomy, astronomy, botany, and chemistry, which then covered phenomena that now fall to disciplines with the same names. Otherwise it recommends terms of the period, like "natural knowledge," "natural philosophy," and "natural history." It shrinks from labeling as a "scientist" any student of nature active before the middle of the nineteenth century, preferring "scholar," "savant," "man of letters," "academician," "professor," and, where relevant, "anatomist" or "astronomer"; and it prohibits altogether honorifica and horrifica like "the father [or mother] of biology." Earlier students of nature should be allowed the roles they played. Few were, or would have wished to be, professional researchers of today's type.

The *Companion* abates this prissiness when enforcing it would result in pedanticism, which it abhors even more than bad language. Nevertheless, to paraphrase the editor of the *Saggi* or *Essays* issued in 1667 by the Accademia del Cimento of Florence, perhaps the earliest of scientific societies, "if sometimes there shall be inserted any hints of Anachronism, we request that they may be taken always for the thoughts, and particular sense of some one of the Contributors, but not imputed to the whole of the Company."

Generosity. Today's policy analyst blurs the distinction between science and its technological applications: engineering, electronics, and industry now affect the biological sciences as strongly as they do the physical sciences. The *Companion* grants that science and high technology, research and development, have grown cozily close, and that governments now encourage a degree of intimacy between academia and industry previously experienced only in wartime. Consequently the *Companion* devotes much of the space it allots to the uses of science to practical or industrial applications. It makes special provision for scientific instruments and apparatus, which it esteems not only as implements of modernity, but also as products of cooperation between science and industry. However, these applications by no means account for all the important uses of science surveyed by the *Companion*. The interaction of science with philosophy gives rise to many articles. There are others on the reciprocal influence of science and ethics, government, law, the mass media, the military, music, philanthropy, polite conversation, religion, theology, and so on.

The programs of industrial, governmental, and charitable patrons insure that, no matter what scientists think, science no longer can be a disinterested pursuit of truth. Some philosophers and historians have arrived independently at this last proposition by proving that there is no truth and that, if there were, scientists could not recognize it. The *Companion* gives due weight to the view that modern science is but an engine for the creation and convenient arrangement of facts. It does not disfavor naive realism, however, and generously declines to decide whether existence is more truly predicated of electrons than of readers.

Cosmopolitanism. During the seventeenth and eighteenth centuries European centers operated a mercantilist system with respect to science. Information about the flora, fauna, and natural curiosities of colonies returned to the mother country for incorporation in herbaria, botanical gardens, zoos, and books. During the nineteenth century a few colonies acquired their own observatories and gardens, and countries that would challenge European hegemony in the twentieth century began to modernize—Japan, Russia, and the United States. By 1940, leadership had transferred to America, to be challenged in particular sectors after World War II by Europe, Japan, and the Soviet Union. American institutions profited from the migrations of trained people, not only the emigrés from the dictatorships of the interwar period but also freely mobile scientists and engineers who clogged the brain drain of the last decades of the twentieth century.

Meanwhile the former colonies strove with varying success to bring home relevant parts of Western science. Some countries in Latin America, notably Brazil, and in Asia, especially India and Pakistan, have managed to create a few institutions that have nourished investigators and prompted discoveries on a par with European and American models. China has followed its own route to the domestication of Western science and is beginning to be a major force in some advanced technologies. But scarcity of resources, lack of international recognition, doubts about the fitness and relevance of high-end science for developing countries, and the difficulty of insuring continuity of purpose in unstable political situations have hampered efforts to transfer Western standards and practices. Our cosmopolitan *Companion* offers accounts of the expansion of science and scientists from their traditional centers throughout the rest of the world. This spread in space parallels the growth of scientific specialties characteristic of the twentieth century. "Spread" and "growth" are key concepts for the *Companion*.

STRUCTURE AND USE

Classifications of science with explicit schemes of subordination, usually with physics on top, go back as far as Aristotle. Something of the same order survives today in the seniority of the Nobel Prizes, although the Swedish Academy of Sciences, Nobel, and Aristotle mean and meant different things by "physics." The *Companion* also has a hierarchical structure. It devotes a primary article to each comprehensive discipline (e.g., Astronomy, Botany, Chemistry) that indicates general historical development, appropriate terminology, and main subdivisions. These subdivisions form the subject matter of secondary articles (Non-optical astronomy, Horticulture, Atom and Molecule), which, in turn, spin off tertiary notices of people, discoveries, concepts, or instruments (Black Hole, Linnaeus, Quark).

Parallel to these disciplinary articles the *Companion* presents a set describing the institutions of science. Primary entries such as Academies and Learned Societies, Research Schools, and University sponsor secondary entries like Government, Industrial Laboratory, and Institute. It would not be companionable, however, to leave the institutional articles disconnected from the disciplinary ones. They are tied together through articles like Entropy, Evolution, and Materialism and Vitalism, which deal with concepts that figured in many different disciplines, and through overarching approaches to history of science and its uses, as set out in the article Historiography of Science.

Individuals, whether people or institutions, present a special problem and opportunity. On the one hand, the easy availability of the 18-volume *Dictionary of Scientific Biography* (Scribners) relieves the *Companion* of the need to rehearse the standard facts about careers and discoveries of most of the people who appear in its pages. On the other hand, the *Companion* cannot ask its readers to go elsewhere to learn where Newton or Madame Curie fit in time and space. It presents a hundred or so biographies, some of them comparative in the style of Plutarch. Those chosen were notable not only for their contributions to science and its uses, but also for the representativeness or interest of their lives.

The users and browsers we have in mind are the usual audiences for Oxford *Companions*: the casual and general reader, who will want to know where science is in society and how it got there; students preparing papers and (there are some) merely expanding their minds; all who teach, study, apply, or analyze any science whatsoever; every expert in need of breadth and refreshment; and anyone who likes to see a difficult job conscientiously tackled.

ACKNOWLEDGMENTS

The Editor expresses his heartfelt thanks to his associate editors and consultants, whose names appear on the verso of the title page, and to the 217 contributors, identified after their articles and listed, along with their institutions, on p. xiii. The contributors who have written over 20,000 words each deserve special thanks not only for their articles but also for helping to insure that the *Companion* has the character the editors desired. To create this character they assembled in New York City a lustrum ago at the invitation of Linda Halvorson Morse of Oxford University Press, who conceived the notion that the history of science might make a suitable subject for a *Companion*. That two-day meeting, in which James Miller, who helped to turn the plan into action, also participated, was the best of seminars. Even before the first article was assigned, the *Companion* had revealed to its editors many new facets of the business they had been engaged in for decades.

The major work at the Press has taken place in the Trade Reference Department headed by Nancy Toff. There Fritz McDonald and, since January 2001, Martin Coleman have been chiefly responsible for the *Companion*'s fate. Martin, on whom the major burden of dealing with contributors has fallen, deserves the thanks and gratitude of everyone associated with the book for his tact, resourcefulness, and unfailing dependability. To our picture researcher and clerks Jennifer Smith, Fatin Sabur, and Scott Kenemore, our copyeditors Ann Adelman, Ronald Cohen, Dan Geist, Margaret Hogan, and Linda Strange, our proofreaders Carol Holmes, Adams Holman, and Ryan Sullivan, our indexers Peter Brigaitis and Marie Nuchols, and our overseer Nancy Toff go warm thanks for their parts in giving the *Oxford Companion to the History of Modern Science* its comely shape and distinct personality.

EDITORS AND CONSULTANTS

DIRECTORY OF CONTRIBUTORS

Finn Aaserud, *Director, Niels Bohr Archive, Copenhagen, Denmark*
Bohr Institute; Bohr, Niels; Complementarity and Uncertainty

Pnina Abir-Am, *University of California, Berkeley*
Women in Science

Theodore Arabatzis, *Assistant Professor of History and Philosophy of Science, University of Athens*
Thermodynamics and Statistical Mechanics

Hervé Arribart, *Scientific Director, Saint-Gobain Recherche*
Glass

Jon Arrizabalaga, *Department of History of Science, Institución Milà I Fontanals, Consejo Superior de Investigaciones Científicas (CSIC), Barcelona, Spain*
Syphilis

Lawrence Badash, *Professor of History of Science, University of California, Santa Barbara*
Antinuclear Movement

James Bartholomew, *Professor of History, The Ohio State University*
Asia; English-Speaking World; Fukui, Ken'ichi; Internationalism and Nationalism; Minority Groups; Political Economy of Science

Jim Bennett, *Director, Museum of the History of Science, University of Oxford*
Astrolabe; Clock and Chronometer; Instruments and Instrument Making; Instruments, Astronomical Measuring; Instruments, Surveying; Longitude; Navigation; Sundial

Bernadette Bensaude-Vincent, *Professor of History and Philosophy of Science, Université de Paris X*
Glass; Materials Science; Plastics

Mario Bertolotti, *Professor of Physics, University of Rome*
Laser and Maser

Alan Beyerchen, *Professor of History, Ohio State University*
Ethics and Science; World War II and Cold War

Mario Biagioli, *Professor of History of Science, Harvard University*
Courts and Salons; Peer Review

Marvin Bolt, *Associate Curator, History of Astronomy Department, Adler Planetarium & Astronomy Museum*
Planetarium

Joanne Bourgeois, *Associate Professor of Geological Sciences, University of Washington, Seattle, Washington*
Gaia Hypothesis; Glaciology; Mohole Project and Mohorovicic Discontinuity; Planetary Science; Seismology

Jonathan Bowen, *Professor of Computing, South Bank University, London*
Computer Science

Brian Bowers, *Retired Senior Curator (Electrical Engineering), Science Museum, London*
Lighting

Bruce Bradley, *Librarian for History of Science, Linda Hall Library of Science, Engineering & Technology*
Library

Robert Brain, *Associate Professor of History of Science, Harvard University*
Exhibitions

Jeffrey C. Brautigam, *Assistant Professor of History, Hanover College, Indiana*
Biometrics

Paolo Brenni, *Istituto E Museo di Storia della Scienza, Florence*
Battery; Electrical Machine; Electrotechnology; Instruments, Electrical Measuring; Polarimeter

William H. Brock, *Professor Emeritus of History of Science, University of Leicester*
Dissertation; Nylon; Radioactive Tracer; Transuranic Elements; Urey, Harold

John H. Brooke, *Andreas Idreos Professor of Science and Religion, University of Oxford*
Natural Theology

Randall Brooks, *Curator, Physical Sciences and Space, Canada Science and Technology Museum*
Cybernetics; Feedback; Micrometer

Laurie M. Brown, *Professor of Physics and Astronomy, Emeritus, Northwestern University*
  Yukawa, Hideki

Janet Browne, *Reader in the History of Biology, Wellcome Trust Centre for the History of Medicine, University College, London*
  Botanical Garden; Botany; Classification in Science; Darwin, Charles; Entomology; Horticulture; Monsters; Natural History; Ovism and Animalculism; Photosynthesis; Respiration and Transpiration; Wardian Case

Jed Buchwald, *Dreyfuss Professor of History, California Institute of Technology*
  Carnot, Sadi, and Augustin Fresnel

Victor P. Budura, *Vice President for Space Support, East West Enterprises, Inc. , Huntsville, Alabama*
  Space Station

Joe D. Burchfield, *Associate Professor of History, Northern Illinois University*
  Earth, Age of the; Universe, Age and Size of the

Piers Bursill-Hall, *Department of Pure Mathematics and Mathematical Statistics, University of Cambridge*
  Mathematics (to 1800)

Johannes Büttner, *Univ. -Professor, Medizinische Hochschule Hannover*
  Clinical Chemistry

Helen Bynum, *Honorary Research Fellow, Wellcome Trust Centre for the History of Medicine, University College, London*
  Antibody; Early Humans

William Bynum, *Professor of the History of Medicine, Wellcome Trust Centre for the History of Medicine, University College, London*
  Endocrinology; Medicine; Mental Sciences

Regis Cabral, *International R&D Manager, Uminova Center, Umeå University, Umeå, Sweden*
  Development, Science and; International Organizations; Latin America; Lattes, Cesar, and Jose Leite Lopes; Multi-National Laboratories

David Cahan, *Professor of History, University of Nebraska-Lincoln*
  Helmholtz, Hermann von, and Heinrich Hertz

Geoffrey Cantor, *Professor of the History of Science, University of Leeds*
  Religion and Science; Young, Thomas

Tian Yu Cao, *Associate Professor of Physics, Boston University*
  Space and Time

Andrea Carlino, *Maître d'Enseignement et de Recherche, Institut d'Histoire de la Médecine et de la Santé, Université de Genève*
  Anatomical Theater

Kenneth J. Carpenter, *Professor Emeritus of Experimental Nutrition, University of California, Berkeley*
  Nutrition

David C. Cassidy, *Professor of Natural Science, Hofstra University*
  Heisenberg, Werner, and Wolfgang Pauli

Paul Ceruzzi, *Curator, Aerospace Electronics and Computing, Smithsonian Institution*
  Artificial Intelligence

Allan Chapman, *Member of the Faculty of Modern History, University of Oxford*
  Dividing Engine; Moon; Planet

Gail Charnley, *Healthrisk Strategies, Washington, D.C.*
  Risk Assessment/Scientific Expertise

Guido Cimino, *Professor of History of Science and Psychology, Università di Roma*
  Golgi, Camillo, and Santiago Ramon Cajal

Eugene Cittadino, *New York University*
  Ecology

William T. Clower, *Research Associate, Department of Neurobiology, University of Pittsburgh*
  Reflex

I. Bernard Cohen, *Victor S. Thomas Professor of the History of Science, Emeritus, Harvard University*
  Newton, Isaac

H. Floris Cohen, *Professor Emeritus of History of Science, University of Twente*
  Scientific Revolution; Scientific Revolutions

Roger Cooter, *Professor of History of Medicine, Wellcome Unit for the History of Medicine, University of East Anglia, Norwich*
  Pseudoscience and Quackery

Pietro Corsi, *Professor of the History of Science and Directeur d'Etudes, Université de Paris and Ecole des Hautes Etudes en Sciences Sociales*
  Cuvier, Georges, and Jean-Baptiste de Lamarck

Patrick Curry, *Associate Lecturer, Centre for the Study of Cultural Astronomy and Astrology, Bath Spa University College, London*
  Astrology

Olivier Darrigol, *CNRS, Paris*
  Hydrodynamics and Hydraulics

Edward B. Davis, *Distinguished Professor of the History of Science, Messiah College, Granthan, Pennsylvania*
Creationism

Alexis De Greiff, *Assistant Professor, Observatorio Astronómico Nacional, Universidad Nacional de Colombia at Santafé de Bogotá*
Third World Academy of Science

Margaret Deacon, *Southampton Oceanography Centre, University of Southampton*
Oceanographic Institutions

Suzanne Débarbat, *Astronome Titulaire Honoraire, Observatoire de Paris*
Carte du ciel

Robert J. Deltete, *Professor of Philosophy, Seattle University*
Energetics

Michael Dettelbach, *Associate Director, Corporate and Foundation Relations, Boston University*
Geography

Matthias Doerries, *Professor of the History of Science, Université Louis Pasteur, Strasbourg, France*
Long Fin-de-Siècle, The

François Duchesneau, *Professor of Philosophy, University of Montreal, Canada*
Schwann, Theodor

Jacalyn Duffin, *Hannah Chair, History of Medicine, Queen's University, Kingston, Ontario, Canada*
Stethoscope

William Eamon, *Professor of History, New Mexico State University*
Magic, Natural

Frank Egerton, *Professor of History, University of Wisconsin*
Phrenology and Physiognomy

Ralph Eshelman, *Principal, Eshelman and Associates, Lusby, Maryland*
Lighthouse

Henry Etzkowitz, *Director, Science Policy Institute, State University of New York at Purchase*
Patents

Isobel Falconer, *Research Associate, Open University, United Kingdom*
Electron; Mass Spectrograph; Thomson, Joseph John

Bernardino Fantini, *Director, Louis-Jeantet Institute for the History of Medicine, Geneva, Switzerland*
Jacob, François, and Jacques Lucien Monod

Patricia Fara, *Fellow of Clare College, University of Cambridge*
Analog/Digital Devices; Magnet and Compass; Turing, Alan

Theodore S. Feldman, *Boston, Massachusetts*
Baconianism; Barometer; Chemical and Biological Warfare; Climate; Climate Change and Global Warming; Graph; Hooke, Robert; Manhattan Project; Mathematization and Quantification; Meteorological Station; Meteorology; Military Institutions; Military-Industrial-Scientific Complex; Telegraph; Thermometer

Maurice A. Finocchiaro, *Distinguished Professor of Philosophy, University of Nevada, Las Vegas*
Galileo

Paul Forman, *Curator, Modern Physics Collection, National Museum of American History, Smithsonian Institution*
Schrödinger, Erwin

Robert Fox, *Professor of the History of Science, University of Oxford*
Heat Engine

Tore Frängsmyr, *Professor in History of Science, Uppsala University*
Linnaeus, Carl

Alan Gabbey, *Professor of Philosophy, Barnard College, Columbia University*
Mechanics

Elizabeth Garber, *Associate Professor of History, State University of New York at Stony Brook*
Conservation Laws; Gibbs, J. Willard

Thomas P. Gariepy, *Professor of the History of Science, Stonehill College, Easton, Massachusetts*
Antisepsis

Kostas Gavroglu, *Professor of History of Science, University of Athens, Greece*
Cold and Cryonics; Entropy; Solid State (Condensed Matter) Physics

Owen Gingerich, *Research Professor of Astronomy and History of Science, Harvard-Smithsonian Center for Astrophysics*
Brahe, Tycho; Copernicus, Nicholas

Tal Golan, *Assistant Professor of History, Ben Gurion University, Israel*
Law and Science

Gregory A. Good, *Associate Professor of History of Science, West Virginia University*
Atmospheric Electricity; Ionosphere; Lightning

Michael Gordin, *Assistant Professor of History, Princeton University*
Mendeleev, Dmitrii Ivanovich

Gennady Gorelik, *Research Fellow, Center for Philosophy and History of Science, Boston University*
Sakharov, Andrei, and Edward Teller

Penelope Gouk, *Senior Research Lecturer, Wellcome Unit for the History of Medicine, University of Manchester*
Music and Science

Anthony Grafton, *Henry Putnam University Professor of History, Princeton University*
Renaissance

Loren R. Graham, *Professor of History of Science, Massachusetts Institute of Technology/Harvard University*
Dialectical Materialism; Kapitsa, Pyotr; Lomonosov, Mikhail Vasilievich; Lysenko Affair

Ivor Grattan-Guinness, *Professor of the History of Mathematics and Logic Emeritus, Middlesex University*
Infinity; Mathematics (1800 to the present)

Jeremy Gray, *Senior Lecturer in Mathematics, The Open University*
Non-Euclidean Geometry

John L. Greenberg, *Paris*
Euler, Leonhard

Mott T. Greene, *John Magee Professor of Science and Values, University of Puget Sound*
Birth Order; Ozone Layer; Polar Science

Anita Guerrini, *Associate Professor of Environmental Studies and History, University of California, Santa Barbara*
Anatomy; Animal Care and Experimentation; Dissection

John F. Guilmartin, *Associate Professor of History, Ohio State University*
Sonar

W. D. Hackmann, *Emeritus Senior Assistant Keeper, Museum of the History of Science, Oxford*
Apparatus; Phonograph; Radio and Television; Transistor; Ultrasonics; Valve, Thermionic

Roger Hahn, *Professor of History, University of California, Berkeley*
Laplace, Pierre-Simon

Lesley A. Hall, *Archivist, Honorary Lecturer in History of Medicine, London Wellcome Library, University College, London*
Contraception; Sex

Karl Hall, *Assistant Professor, Central European University, Budapest*
Europe and Russia

Robert Dale Hall, *Rabat American School, Morocco*
Africa

Anne Hardy, *Reader in the History of Medicine, Wellcome Trust Centre for the History of Medicine, University College, London*
Germ

Oren Solomon Harman, *History of Science, Harvard University*
Population Genetics

P. M. Harman, *Professor of the History of Science, Lancaster University*
Maxwell, James Clerk

Takehiko Hashimoto, *Professor of History of Science and Technology, University of Tokyo*
Aeronautics

J. L. Heilbron, *Professor of History and the Vice Chancellor, Emeritus, University of California, Berkeley; Senior Research Fellow, Worcester College, Oxford*
Academies and Learned Societies; Atomic Structure; Cathedral; Cavendish, Henry, and Charles-Augustin Coulomb; Cosmic Rays; Curie, Marie, and Pierre Curie; Dynasty; Ether; Experimental Philosophy; Geodesy; History of Science; Imponderables; Leyden Jar and Electrophore; Lightning Conductor; Magneto-optics; Nobel Prize; Noble Gases; Physics; Planck, Max; Pneumatics; Putti in Science; Quantum Physics; Rainbow; Relativity; Röntgen, Wilhelm Conrad; Rutherford, Ernest; Scientist; Slogans from Science; Spontaneous Combustion; Standard Model; Sympathy and Occult Quality; Terminology; X Rays

Arne Hessenbruch, *Researcher in the Project on History of Recent Science and Technology on the Web, Dibner Institute, Massachusetts Institute of Technology*
Cathode Ray Tube; Metrology

Norriss S. Hetherington, *Director, Institute for the History of Astronomy, And Visiting Scholar, University of California, Berkeley*
Anthropic Principle; Astronomy; Astronomy, Non-Optical; Astrophysics; Black Hole; Chandrasekhar, Subrahmanyan; Cosmology; Extraterrestrial Life; Galaxy; Hawking, Stephen, and Carl Sagan; Hubble, Edwin; Light, Speed of; Milky Way; Nebula; Observatory; Parallax; Pulsars and Quasars; Steady-State Universe; Telescope

Frederic Lawrence Holmes, *Avalon Professor of the History of Medicine, Section of the History of Medicine, Yale University*
  Balance; Bernard, Claude; Berzelius, J. J.; Biochemistry; Chemical Compound; Chemistry; Distillation; Earths; Enzyme; Fermentation; Fire and Heat; Genetic Mapping; Homeostasis; Krebs, Hans Adolf; Laboratory, Chemical; Lavoisier, Antoine; Liebig, Justus von; Mass Action; Metabolism; Metals; Oxygen; Phlogiston; Physiology; Secretion

Gerald Holton, *Mallinckrodt Professor of Physics and Professor of History of Science, Emeritus, Harvard University*
  Einstein, Albert

R. W. Home, *Professor of History and Philosophy of Science, University of Melbourne*
  Aepinus, F. U. T; Cohesion; Electricity; Franklin, Benjamin; Magnetism

David W. Hughes, *Professor of Astronomy, University of Sheffield, Department of Physics and Astronomy*
  Aberration, Stellar; Eclipse; Orbit; Solar Physics; Star; Supernova

Bruce J. Hunt, *Associate Professor of History, University of Texas*
  Michelson, A. A.

Myles W. Jackson, *Assistant Professor of the History of Science, Willamette University*
  Bureaus of Standards; Optics and Vision

Margaret Jacob, *Professor of History, University of California, Los Angeles*
  Enlightenment and Industrial Revolution

Stephen Jacyna, *Wellcome Trust Centre for the History of Medicine, University College, London*
  Cell; Cytology

Frank A. J. L. James, *Reader in History of Science, The Royal Institution of Great Britain*
  Faraday, Michael; Field; Spectroscopy

William T. Johnson, *Mesa, Arizona*
  Non-Western Traditions

Jeffrey Allan Johnson, *Associate Professor of History, Villanova University*
  I. G. Farben

Stephen Johnston, *Assistant Keeper, Museum of the History of Science, University of Oxford*
  Slide Rule

Paul Josephson, *Associate Professor of History, Colby College, Waterville, Maine*
  Social Responsibility in Science

Lily Kay, deceased
  Crick, Francis, and James D. Watson

A. G. Keller, *Department of History, University of Leicester*
  Nuclear Bomb

Bettyann Holtzmann Kevles, *Lecturer, Department of History, Yale University*
  Primatology; Wilson, Edward O.

D. J. Kevles, *Stanley Woodward Professor of History, Yale University*
  AIDS; Biotechnology; Cancer Research; Chernobyl; Clone; Cold Fusion; Computer; Computerized Tomography; Environment; Eugenics; Genetics; Heredity; Human Genome Project; Intelligence Quotient; Microchip; Nuclear Power; Sociobiology; Sputnik

Peggy Kidwell, *Curator of Mathematics, National Museum of American History*
  Calculator

Rudolph Kippenhahn, *The Max Planck Institute of Astrophysics at Garching (retired)*
  Cryptography

Sally Gregory Kohlstedt, *Program of History of Science and Technology, University of Minnesota*
  Advancement of Science, National Associations for the; Zoological Garden

Alexei Kojevnikov, *Associate Professor of History, Department of History, University of Georgia, Athens*
  Vavilov, Nikolai, and Sergey Ivanovich Vavilov

Helge Kragh, *Professor of History of Science and Technology, University of Aarhus, Denmark*
  Telephone

Jan Lacki, *Assistant Professor of History and Philosophy of Physics, University of Geneva*
  Neumann, John von

Marcel C. LaFollette, *Washington, D.C.*
  Film, Television, and Science

L. R. Lagerstrom, *Senior Lecturer, Electrical and Computer Engineering, University of California, Davis*
  Constants, Fundamental; Standardization

Rachel Laudan, *Guanajuato, Mexico*
Cartography; Classification of Sciences; Crystallography; Earth Science; Empiricism; Fossil; Fraud; Geology; Geophysics; Hermeticism; Hutton, James; Hypothesis; Ice Age; Instrumentalism and Realism; International Geophysical Year; Isostasy; Lunar Society of Birmingham; Lyell, Charles; Method, Scientific; Mineralogy and Petrology; Mining Academy; Model; Neoplatonism; Neptunism and Plutonism; Observation and Experiment; Paleontology; Philosophy and Science; Plate Tectonics; Positivism and Scientism; Powell, John Wesley; Proof; Reductionism; Rhetoric in Science; Stratigraphy and Geochronology; Terrestrial Magnetism; Uniformitarianism and Catastrophism; Worldview

Peretz Lavie, *André Ballard Professor of Biological Psychiatry, The Technion Sleep Laboratory, Haifa, Israel*
Sleep

Susan E. Lederer, *Associate Professor, History of Medicine, Yale University*
Blood

John E. Lesch, *Professor of History, University of California, Berkeley*
Alkaloids

Bruce V. Lewenstein, *Associate Professor of Science Communication, Cornell University*
Popularization

Henry Lowood, *Curator for History of Science and Technology Collections, Stanford University Libraries*
Forestry; Journal; National Parks and Nature Reserves; Patriotic and Economic Societies; Printing House

A. J. Lustig, *Postdoctoral Fellow, Dibner Institute, Massachusetts Institute of Technology*
Cabinets and Collections; Jussieu, Antoine-Laurent de

Marjorie C. Malley, *Cary, North Carolina*
Radium

Joan Mark, *Research Associate in the History of Anthropology, Peabody Museum of Archaeology and Ethnology, Harvard University*
Mead, Margaret

Ulrich Marsch, *Max Planck Gesellschaft Munich, Germany*
Laboratory, Industrial

Ben Marsden, *Cultural History Programme Coordinator, School of History and History of Art, University of Aberdeen*
Engineer; Engineering Science

Pauline M. H. Mazumdar, *Professor Emeritus of the History of Medicine, University of Toronto*
Immunology

Renato G. Mazzolini, *Professor of History of Science, Università degli Studi di Trento, Italy*
Animal Machine; Developmental Mechanics; Embryology; Epigenesis and Preformation; Generation; Infusoria; Irritability; Metamorphosis; Polyp; Spontaneous Generation

Domenico Bertoloni Meli, *Professor of the History of Science, Indiana University*
Malpighi, Marcello

Andrew Mendelsohn, *Governor's Lecturer in History of Science and Medicine, Imperial College, University of London*
Bacteriology and Microbiology

David P. Miller, *Senior Lecturer, School of History and Philosophy of Science, University of New South Wales, Sydney*
Professional Society

John Mills, *Adjunct Professor of Psychology, University of Calgary, Calgary, Alberta, Canada*
Behaviorism

Philip Mirowski, *Carl Koch Professor of Economics and the History and Philosophy of Science, University of Notre Dame*
Operations Research

Susan Mossman, *Curator of Materials Science, The Science Museum, London*
Metallurgy

Shigeru Nakayama, *Professor Emeritus, Kanagawa University, Tokyo*
Diffusion in the East

Meera Nanda, *West Hartford, Connecticut*
Antiscience

Simon Naylor, *Lecturer in Human Geography, University of Bristol*
Exploration and Field Work

William R. Newman, *Professor, Department of History and Philosophy of Science, Indiana University*
Boyle, Robert

Thomas Nickles, *Professor and Chair, Philosophy Department, University of Nevada, Reno*
  Scientific Development, Theories of

Vivian Nutton, *Professor of the History of Medicine, University College, London*
  Galenism; Harvey, William; Vesalius, Andreas

Mary Jo Nye, *Thomas Hart and Mary Jones Horning Professor of the Humanities and Professor of History, Oregon State University*
  Cathode Rays and Gas Discharge; Chemical Nomenclature; Deoxyribonucleic Acid; Electrolysis; Elements, Chemical; Hodgkin, Dorothy; Ideal Gas; Pauling, Linus; Physical Chemistry; Protein; Protyle; Quantum Chemistry; Radiation Science; Radioactivity

Robert C. Olby, *Research Professor, Department of the History and Philosophy of Science, University of Pittsburgh*
  Chromosome; Double Helix; Franklin, Rosalind; Mendel, Gregor; Molecular Biology; Mutation

Kathryn Olesko, *Associate Professor, Georgetown University*
  Art and Science; Cameralism; Daily Life, Science and; Discipline(s); Error and the Personal Equation; Historiography of Science; Humboldt, Alexander von; Humboldtian Science; Institute; Kaiser-Wilhelm/Max-Planck-Gesellschaft; Marxism; Modernity and Postmodernity; Nazi Science; Precision and Accuracy; Probability and Chance; Schools, Research; Seminar; Textbook; University

Paolo Palladino, *Senior Lecturer in History of Science, Technology, and Medicine, Lancaster University*
  Breeding

Frank A. Palocsay, *Emeritus Professor, Department of Chemistry, James Madison University, Harrisonburg, Virginia*
  Inorganic Chemistry

Giuliano Pancaldi, *Professor of History of Science, University of Bologna*
  Biogenetic Law; Biology; Darwinism; Evolution; Extinction; Fact and Theory; Galvani, Luigi, and Alessandro Volta; Holism; In Vitro Fertilization; Instinct; Instruments, Biological; Lamarckism; Law of Science; Levi-Montalcini, Rita; Life; Museum; Naturalism and Physicalism; Nature; Network and Virtual College; Organism; Priority; Progress; Rational/Irrational; Tacit Knowledge; Teleology; Zoology

John Parascandola, *Public Health Service Historian, U. S. Department of Health and Human Services*
  Pharmacology

Manolis Patiniotis, *Lecturer, Department of History and Philosophy of Science, Athens University*
  Boltzmann, Ludwig

John Powers, *Instructor, New School University, New York*
  Boerhaave, Herman

Lewis Pyenson, *Research Professor, University of Louisiana at Lafayette*
  National Culture and Styles; Shift of Hegemony

Jessica Ratcliff, *Museum of the History of Science, University of Oxford*
  Electronic Media

Scott C. Ratzan, *Editor,* Journal of Health Communication: International Perspectives, *George Washington University School of Public Health*
  Mad Cow Disease

Philip Rehbock, *Professor, Department of History, University of Hawaii*
  Naturphilosophie; Oceanography

Richard Rice, *Florence, Montana*
  Inorganic Chemistry

Robert J. Richards, *Professor of History and Philosophy of Science, The University of Chicago*
  Race

Robert C. Richardson, *Professor of Philosophy, University of Cincinnati*
  Orthogenesis

Michael Riordan, *Adjunct Professor of Physics, University of California, Santa Cruz*
  Collider

Jessica Riskin, *Assistant Professor of History, Stanford University, California*
  Newtonianism

Guenter B. Risse, *Professor, Department of Anthropology, History, and Social Medicine, University of California, San Francisco*
  Hospital; Houssay, Bernardo Alberto

Alan J. Rocke, *Henry Eldridge Bourne Professor of History, Case Western Reserve University, Cleveland, Ohio*
  Alchemy; Analytical Chemistry; Atom and Molecule; Atomic Weight; Bunsen Burner; Carbohydrates and Fats; Chemical Bond and Valence; Chemical Equivalent; Dalton, John; Dyestuffs; Fischer, Emil; Food Preservation; Kekulé, August; Organic Chemistry; Periodic Table; pH; Polymer; Radical, Chemical; Stereochemistry; Woodward, Robert

Nicolaas A. Rupke, *Professor of the History of Science and Director of the Institute for the History of Science, Göttingen University*
Archetype; Gauss, Johann Carl Friedrich

Arturo Russo, *Associate Professor of History of Physics, University of Palermo*
Fermi, Enrico; Satellite

Rose-Mary Sargent, *Professor of Philosophy, Merrimack College, North Andover, Massachusetts*
Analysis and Synthesis; Aristotelianism; Causality; Matter; Mechanical Philosophy; Revolution, Restoration, and the Royal Society; Skepticism

Sara Schechner, *David P. Wheatland Curator, Collection of Historical Scientific Instruments, Harvard University*
Comets and Meteors

Jutta Schickore, *Research Fellow, Department of History and Philosophy of Science, University of Cambridge*
Microscope

Londa Schiebinger, *Edwin E. Sparks Professor of History of Science, Pennsylvania State University*
Gender and Science

Steven E. Schoenherr, *Professor, Department of History, University of San Diego*
Bell Labs

Silvan S. Schweber, *Professor of Physics and Richard Koret Professor in the History of Ideas, Brandeis University, Waltham, Massachusetts*
Bethe, Hans; Dirac, Paul Adrien Maurice; Elementary Particles; Feynman, Richard; Quantum Electrodynamics; Quantum Field Theory; Quark; Salam, Abdus; Tomonaga, Sin-Itirō

Robert Seidel, *Professor of History of Science and Technology, University of Minnesota*
Accelerator; Government; Grants; Klystron; RAND

Dennis Sepper, *Professor of Philosophy, University of Dallas*
Mind/Body Problems; Phenomenology

H. Otto Sibum, *Associate Professor, Research Director, Max Planck Institute for the History of Science, Berlin*
Joule, James, and Robert Mayer

Ruth Lewin Sime, *Professor Emeritus of Chemistry, Sacramento City College*
Meitner, Lise

Crosbie Smith, *Professor of History of Science, University of Kent at Canterbury*
Thomson, William, Lord Kelvin

Vassiliki B. Smocovitis, *Associate Professor of History, University of Florida*
Dobzhansky, Theodosius; Fisher, R. A.; Morgan, Thomas Hunt

Darwin Stapleton, *Executive Director, Rockefeller Archive Center*
Foundations; Philanthropy

F. Richard Stephenson, *Professorial Fellow in Astronomy, Department of Physics, University of Durham*
Eclipse Expeditions

Robert K. Stewart, *Sing Tao Professor of International Journalism, Ohio University*
Mission

James Strick, *Assistant Professor, Science, Technology, and Society Program, Franklin and Marshall College, Lancaster, Pennsylvania*
Pasteur, Louis

W. T. Sullivan, *Professor of Astronomy, University of Washington*
Radar

William C. Summers, *Professor of Therapeutic Radiology, Molecular Biophysics and Biochemistry, and Lecturer in History, Yale University*
Antibiotics; Vaccination; Virology

Abha Sur, *Lecturer, Department of Urban Studies and Planning, Massachusetts Institute of Technology*
Raman, C. V.

Liba Taub, *Curator and Director, Whipple Museum of the History of Science*
Orrery

Mary M. Thomas, *Department of History of Science and Technology, University of Minnesota*
Fortification

Phillip Thurtle, *Assistant Professor of Anthropology and Sociology, Carleton University, Ottawa, Ontario, Canada*
Electrophoresis

Daniel P. Todes, *Associate Professor, The Johns Hopkins University*
Pavlov, Ivan

Maria Trumpler, *Lecturer, Department of the History of Science, Harvard University*
Galvanism

Robert C. Ulin, *Professor of Anthropology, University of Western Michigan*
Anthropology

Anne van Helden, *Museum Boerhaave, Leiden, The Netherlands*
    Air Pump and Vacuum Pump

Denys Vaughan, *Research Fellow, Science Museum Library, London*
    Pendulum

Theo Verbeek, *Professor of Early Modern Philosophy, Utrecht University*
    Descartes, René

Christiane Vilain, *Assistant Professor of Physics and Epistemology of Physics, University Denis-Diderot, Paris*
    Huygens, Christiaan

James Voelkel, *Professor, Department of the History of Science, Johns Hopkins University*
    Kepler, Johannes

J. Samuel Walker, *Historian, Nuclear Regulatory Commission*
    Nuclear Diplomacy

Alice N. Walters, *Associate Professor of History, University of Massachusetts, Lowell*
    Grand Tour

Mike Ware, *Honorary Fellow in Chemistry, University of Manchester*
    Photography

Stephen J. Weininger, *Max-Planck-Institut für Wissenschaftsgeschichte*
    Acid and Base; Osmosis

Kathleen Wellman, *Professor of History, Southern Methodist University*
    Materialism and Vitalism

Petra Werner, *Berlin-Brandenburgian Academy of Science*
    Haber, Fritz; Warburg, Otto

Gary Westfahl, *Instructor, University of California, Riverside*
    Literature and Science; Science Fiction

Peter Westwick, *Senior Research Fellow in Humanities, California Institute of Technology*
    Acoustics and Hearing; Atoms for Peace; Basic and Applied Science; Blackett, Patrick M. S., and Ernest O. Lawrence; Brain Drains and Paperclip Operations; CERN; Chaos and Complexity; Cherenkov Radiation; Cloud and Bubble Chambers; Discovery; Electromagnetism; Geiger and Electronic Counters; High-Energy Physics; Institute for Scientific Information, Philadelphia; Kurchatov, Igor Vasilyevich, and J. Robert Oppenheimer; Lee, T. D., C.S. Wu, and C.N. Yang; Low-Temperature Physics; Nuclear Magnetic Resonance; Nuclear Physics and Nuclear Chemistry; Plasma Physics and Fusion; Science Wars; Secrecy in Science; Solvay Congresses and Institute; Strangeness; Theory of Everything; Vacuum

Curtis Wilson, *Tutor Emeritus, St. John's College, Annapolis, Maryland*
    Celestial Mechanics

Eric Winsberg, *Assistant Professor, Department of Philosophy, University of South Florida*
    Simulation

Alison Winter, *Associate Professor of History, University of Chicago*
    Mesmerism and Animal Magnetism

Roland Wittje, *Norwegian University of Science and Technology*
    Centrifuge and Ultracentrifuge

Richard Yeo, *Professor, History of Science, School of Humanities, Griffith University, Australia*
    Encyclopedias

A THEMATIC LISTING OF ENTRIES

HISTORIOGRAPHY OF SCIENCE

GENERAL CONCEPTS AND APPROACHES
Classification in Science
Discipline(s)
Gender and Science
Historiography of Science
History of Science
Modernity and Postmodernity
National Culture and Styles
Non-Western Traditions
Priority
Scientific Development, Theories of
Scientific Revolutions
Terminology

MAJOR PERIODS IN TIME
Enlightenment and Industrial Revolution
Long Fin-de-Siecle, the
Positivism and Scientism
Renaissance
Revolution, Restoration, and the Royal
Society
Scientific Revolution
Shift of Hegemony
World War II and Cold War

MAJOR DIVISIONS
Aristotelianism
Baconianism
Darwinism
Hermeticism
Humboldtian Science
Mechanical Philosophy
Naturphilosophie
Neoplatonism
Newtonianism

ORGANIZATION AND DIFFUSION OF SCIENCE

THE SCIENTIFIC PROFESSION
Engineer
Internationalism and Nationalism
Scientist

GENERALIZED INSTITUTIONS
Academies and Learned Societies
Advancement of Science, National Associa-
tions for the
Botanical Garden
Bureaus of Standards
Cabinets and Collections
Hospital
Institute
International Organizations
Laboratory, Chemical
Laboratory, Industrial
Library
Meteorological Station
Military Institutions
Mining Academy
Multi-National Laboratories
National Parks and Nature Reserves
Observatory
Oceanographic Institutions
Professional Society
Schools, Research
Seminar
University
Zoological Garden

INDIVIDUAL INSTITUTIONS
Bell Labs
Bohr Institute
CERN
I. G. Farben
Institute for Scientific Information,
Philadelphia
Kaiser-Wilhelm/Max-Planck-Gesellschaft
Lunar Society of Birmingham
Manhattan Project
RAND
Solvay Congresses and Institute
Third World Academy of Science

DIFFUSION (BEYOND SCIENCE)
Anatomical Theater
Encyclopedias
Exhibitions
Museum
Planetarium
Popularization
Printing House
Slogans from Science
Textbook

COMMUNICATION (WITHIN SCIENCE)
Diffusion in the East
Dissertation

THE
OXFORD COMPANION TO
THE HISTORY OF MODERN
SCIENCE

A

ABERRATION, STELLAR. Because the observer on the earth is often moving across the path of the incoming light from a star, the observed direction of the star deviates from its true direction. This deviation, known as aberration, depends on the velocity of the observer on the earth and on the velocity of light. The maximum deviation owing to the earth's moving around its orbit is 20.5 seconds of arc. The earth's spin produces an additional much smaller diurnal aberration.

James Bradley, England's third Astronomer Royal, discovered stellar aberration serendipitously. He was looking for evidence of stellar parallax, a concept at the heart of the heliocentric solar system. Since the diameter of the earth's orbit is 300 million km, nearby stars should move with respect to background stars as the earth orbits the sun. Since typical visual stars are twenty million times further away than the sun, the parallax angle usually is much less than a second of arc, and therefore extremely difficult to measure.

Robert *Hooke in London and Jacques Cassini in Paris attempted to measure parallax angles, but with little success. Bradley and his friend Samuel Molyneux decided to check Hooke's observations of 1669 of the star Gamma Draconis, which passed overhead at the latitude of London.

The *telescopes of the day were long, cumbersome, and suffered considerably from tube flexure. Molyneux commissioned George Graham to constructed a special vertical, 24-four-foot long, refracting telescope that could image Gamma Draconis once a day. Observations of stars at the zenith had the additional advantage that no correction for atmospheric refraction was required. Bradley and Molyneux observed deviations in the position of the star, but not of the sort expected for parallax. Other overhead stars shifted about in the manner of Gamma Draconis. After discounting the possibility that the axis of the earth might be changing direction, Bradley traced the phenomenon to the vector addition of the velocities of starlight and of the earth. He announced this result to the Royal Society in January 1729.

This observation confirmed the heliocentric *cosmology and helped to prompt Pope Benedetto XIV to remove the blanket proscription of Copernican treatises from the *Index of Prohibited Books*. Another removal—that of the effect of stellar aberration from recorded stellar positions—heralded a more accurate approach to positional astronomy. The fact that the aberration for all the stars in a specific direction had the same value, independent of the brightnesses (and thus the distances) of the stars, indicated the constancy of the velocity of the light. Bradley calculated that sunlight took an average of 8.2 minutes to reach the earth, about 0.1 minutes from the time accepted today. The precision of Bradley's instrument indicated that stellar parallax must be less than 1 second of arc.

Robert Grant, *History of Physical Astronomy* (1852). Colin A. Ronan, *Their Majesties Astronomers: A survey of astronomy in Britain between the two Elizabeths* (1967).

DAVID W. HUGHES

ACADEMIES AND LEARNED SOCIETIES. "Ci-gît un qui ne fut rien, pas même académicien" ("Here lies someone who was nobody, not even an academician"). This epitaph, on a tombstone now in the Musée des Beaux Arts in Dijon, suggests the scale and tone of the academic movement during the Enlightenment. Between 1750 and 1774, seventeen learned societies dedicated at least partly to natural science were founded in France alone. But as the epitaph insinuates, most of the academies of the great age of academies had more to do with society than with science. They drew their memberships primarily from the professions—lawyers, doctors, clerics, and military men—and from landed gentry. They discussed local improvements, maintained a library and collection of instruments, and performed, usually irregularly, the two essential offices of a learned society in the age of Enlightenment: giving prizes to winners of essay competitions and publishing papers presented by its members.

Although the various groups, clubs, and societies that occupied themselves with natural knowledge during the modern period elude exact classification, an academy may be defined sufficiently as a formal institution (however meager its statutes) with a restricted membership (never more than a few hundred before

Four seals from Academic societies. Clockwise from top left, they are those of the Accademia dei Lincei ("lynxes"), a small group of natural philosophers founded in 1603; the Kungliga Svenska Vetenskapsakademien, founded 1739; the Académie royale des sciences, Paris, founded 1666; the Accademia reale, Turin, founded 1783.

the twentieth century) devoted in substantial part to the advancement of all the natural sciences, particularly by the mutual encouragement of its members. A university, professional society, research council, museum, laboratory, or observatory is not an academy, although all of them exercise one or more of the same functions—publishing memoirs, encouraging research, funding expeditions, giving prizes, imposing standards, establishing solidarity, and promoting science among the wider society—first brought together in academies.

The earliest scientific academies of any importance appeared in Italy in the seventeenth century as variants of a common Renaissance type—a group of literary men maintained for a few years by a wealthy aristocrat or prelate. The Accademia dei Lincei (1603–1630), organized but not underwritten by a duke, did not meet together, but it gave its members—including Galileo Galilei, who used to style himself "linceo" (lynx) on the title pages of his books—a sense of common enterprise. Galileo's disciples

established a closer-knit body, the Accademia del Cimento (1657–1667), under higher patronage (the Grand Duke of Tuscany and his brother, a cardinal) and with better resources (salaries, instruments, and a meeting place). Inspired also by Francis Bacon's project of improving the arts and sciences through cooperative effort, the dozen or so members of the Cimento, "whose sole Design [was] to make Experiments, and Relate them," published an important collection of their "trials," *Saggi di naturali esperienze* (1667), as a corporate contribution, without attributing discoveries to individuals.

No other academy in seventeenth-century Italy had such lavish support as the Cimento, although several had even higher patronage. Examples are the Accademia Fisicomatematica of Rome, founded in 1677 by a cleric, Giovanni Giustino Ciampini, and encouraged by a cardinal (later elected pope) and by Christina, the converted former queen of Sweden; and the Accademia degli Inquieti (ca. 1690–1714), founded by professors at the University of Bologna, later (when endowed by a private citizen and the pope) the nucleus of the Accademia delle Scienze dell'Istituto di Bologna.

A similar transformation had occurred in France in 1666 when Louis XIV's chief minister decided that the king needed royal academicians to map the realm, perfect astronomy, improve navigation, and keep France current in natural knowledge. The Académie Royale des Sciences came into existence with a handful of salaried members including distinguished foreigners, notably Christiaan Huygens. Reorganized in 1699 and again definitively in 1716, the Paris Academy became the exemplar of a learned body in the service of a great monarch. Its forty-four regular members were distributed into six classes, three "mathematical" (geometry, astronomy, and mechanics) and three "physical" (chemistry, anatomy, and botany), and into three ranks (pensionnaire, associé, and adjoint). Its standing secretary had the duties of composing a history of the Academy's work each year as a preface to the annual volume of original *mémoires* presented obligatorily by members and, in due course, of composing their obituaries, which, taken together, became a role model for aspiring contributors to natural knowledge. The Paris Academy set prize competitions, advised the crown, and occasionally sponsored special projects like excursions to determine the shape of the earth (*see* GEODESY). Typically, however, its members did their work at their own expense and in their own rooms or with material and in space provided by other Parisian institutions.

Only a very few monarchs could copy the academy of the King of France. The first to try, the King of Prussia acting under pressure from Gottfried Leibniz, proved too stingy; the academy he established in 1700 had an uncertain and impoverished existence until refounded by Frederick the Great on Parisian principles. Its new statutes of 1744 provided for four classes (experimental philosophy, mathematics, speculative philosophy, and literature), members divided by rank, and, like the Paris Academy, a variable number of nonresident "correspondents." It published its histories and memoirs in French and, again like its model, engaged distinguished foreigners, notably Pierre de Maupertuis and Leonhard *Euler. The westernizing Peter the Great of Russia also had the ambition and wherewithal to create an academy of salaried savants, which he accomplished in St. Petersburg in 1724, without Russian members. Monarchs more modest than the rulers of France, Prussia, and Russia founded smaller academies on similar principles. In 1783 the King of Sardinia transformed a group that had existed since 1757 into the

Académie Royale of Turin, at the trifling cost of a building and a budget, and in 1757 the Elector of Bavaria established the Akademie der Wissenschaften of Munich with a meeting place, operating expenses, and half a dozen paid "professors." Closed or proprietary academies of the Paris type placed particular emphasis on mathematics and its applications, which they did much to advance.

Meanwhile, larger academies free to choose their members and able to salary only two or three functionaries flourished. The prototype was the Royal Society of London (founded in 1662), which had nothing from its patron, King Charles II, but a charter, a mace, and the right to call itself "royal." Like the Accademia del Cimento, it took much of its rhetoric and purpose from Bacon. Its *Philosophical Transactions* constituted, with the *Mémoires* of the Paris Academy, the most important and reliable periodical literature of natural knowledge during the eighteenth century. Like the Paris Academy, the Royal Society set up commissions to investigate debated matters: the calibration of thermometers, the shapes of lightning rods, and the merits of mesmerism. Since the Society depended on the dues and influence of its fellows, it began by enlisting any interested gentlemen or distinguished foreigners and made no distinction of class or rank. During the seventeenth and most of the eighteenth centuries, the number of fellows elected for their social status or professional affiliation exceeded in number those chosen for their contributions to science. Among the latter were many *instrument makers.

Several important academies formed on the pattern of the Royal Society in Britain during the eighteenth century: the Royal Society of Edinburgh (1783), a contribution to the consolidation of the United Kingdom as well as a recognition of the flourishing of natural philosophy in the north; the Manchester Literary and Philosophical Society (1781), which reflected the aspirations and discussed the technologies of the industrializing midlands; and the American Philosophical Society (1768), a venture in colonial Philadelphia that became a symbol of intellectual equality with Europe. Important European counterparts of the Royal Society included Kungl. Vetenskapsakademien of Stockholm (1739), which stressed technological problems of Swedish interest, and the Hollandsche Maatschappij der Wetenschappen of Haarlem (1752), which innovated by distinguishing "directors," who paid dues, from "scientific members," who did the work. Few of the many royal regional academies of France could do more than give out an occasional prize or publication. An exception, the Académie Royale des Sciences et Belles-Lettres of Bordeaux (1712), though it had fewer than twenty members at any time, had its own building, library, botanical garden, and instrument collection, all given to it by local philanthropists.

The division of labor in the republic of learning, according to which universities disseminated, and academies increased, knowledge, often created tensions between professors and academicians. Universities sometimes opposed the creation of academies, as the University of Leyden did that of the Hollandsche Maatschappij. The modus vivendi established by the Istituto di Bologna, whereby university and academy exchanged personnel, existed formally in only one other place during the eighteenth century. In 1752 the English Hannoverian government set up a Königliche Societät der Wissenschaften as an adjunct to the university it had established in Göttingen in 1737. A few of the professors ran the academy, which issued publications, including an influential review, the *Göttingische gelehrte Anzeigen*, read all over Europe.

The Bologna–Göttingen model did not become the route by which professors came to dominate the academies during the nineteenth century. Instead, professors added research to their duties. This development, which began at the end of the eighteenth century, accelerated under the transformation of higher education and academic life caused by the French Revolution.

The importance of academies in the advancement of natural knowledge declined as professors took them over. Two factors were chiefly responsible. For one, *journals independent of academies came to dominate scientific communication. Traditional academies published only a small fraction of the millions of papers carried in scientific journals between 1800 and 1900. Academies found an important niche in publishing, however, by printing papers rapidly, for example, in the weekly *Comptes rendus* (begun in 1835) of the Paris Academy and in the monthly *Proceedings* of the Royal Society (begun in 1856). Owing to the revolution in communications effected by railroads and the International Postal Union, a professor could send reprints of an article from the unread proceedings of his local academy directly to those who needed to know.

The second factor that diminished the scientific role of learned societies was the confirmation of the research activity of professors and provision for it in the higher schools and universities. With the invention of scientific *institutes and departments with their own laboratories and apparatus after the middle of the century, academies ceased to be the main locus for the presentation and criticism

of new results, for peer review, or for the financial and rhetorical support of research. This trend continued briskly in the twentieth century with the creation of governmental and industrial research laboratories and central funding agencies. The number of competent research scientists active in 2000 greatly exceeded the number of members in the national and regional academies; and even though more places have been added, academies have long since been unable to meet the expectation implied by the epitaph from Dijon, that they could admit all worthy contributors to science.

The nineteenth and twentieth centuries saw the spread of national academies of the Royal Society type throughout the world and the invention of an important new type, represented by the Akademiia Nauk, successor to the Petersburg Academy. The Soviet Union chose to funnel state support for research and advanced training in science through the Akademiia and its regional dependencies, which thereby captured functions that further west belonged to universities and government laboratories. A senior academician in the Soviet Union had useful perquisites and often enjoyed political influence or protection. Similar arrangements held elsewhere in the Soviet bloc.

Most academies now primarily serve organizational and honorific functions. They are nodes in national and international networks that promote research or outreach beyond their own memberships. The international coupling of major academies began in the nineteenth century as the International Association of Academies. Headquartered in Berlin, it oversaw such projects as the Carte du Ciel and the International Catalogue of Scientific Literature. During World War I, the Paris Academy, the Royal Society, and the U.S. National Academy of Sciences (through a subsidiary it set up called the National Research Council) organized science for military purposes. After the war the allies revived the cooperation of academies under an International Research Council, which at first excluded institutions belonging to the defeated Central Powers. The international network survived World War II in the form of the International Council of Scientific Unions (ICSU). A regional subset of these academies, the European Federation of National Academies of Sciences and Humanities (ALLEA), declares the objectives of its thirty-eight members to be to "promote excellence in science and scholarship... value and promote independence from ideological and political interests... [and] serve society with advice on science policy." They differ from their forerunners in asserting independence from the regimes that support them.

Although the national academies do advise governments when asked, and also award prizes and research grants (if any) and do what they can to improve public understanding of science, their main function now is honorific. Election to a major scientific academy is itself a distinction. The leading modern institute in the bestowal of prizes is an old academy, that of Stockholm, which exerts a worldwide influence far beyond the scientific work it fosters by giving the *Nobel awards in physics and chemistry. Carrying the tendency toward absurdity, "academies" with virtually no purpose but to elect members multiplied in the twentieth century. Examples are the Academia Europaea (founded in 1983) and the Académie International d'Histoire des Sciences (1928). These organizations differ little in function and notoriety from the Baseball Hall of Fame.

Roger Hahn, *The Anatomy of a Scientific Institution: The Paris Académie des Sciences (1666–1803)* (1971). Martha Ornstein, *The Role of Scientific Societies in the 17th Century*, 3d ed. (1938). Daniel Roche, *Le siècle des lumières en provence: Académies and académiciens provinciaux, 1680–1789* (2 vols., 1978). J. L. Heilbron, *Electricity in the 17th and 18th Centuries: A Study of Early Modern Physics* (1979; rev. ed. 1999). Charles Paul, *Science and Immortality: The Eloges of the Paris Academy of Sciences, 1699–1791* (1980). James E. McClellan II, *Science Reorganized: Scientific Societies in the 18th Century* (1985). http//www.allea.org.

J. L. HEILBRON

ACCELERATOR. During the twentieth century physicists developed increasingly powerful artificial means to produce very high-energy particles to transform or disintegrate atoms. At the end of World War I, Ernest *Rutherford used alpha particles from naturally occurring sources of radiation to transform nitrogen into oxygen. He called for the development of more energetic sources of charged particles for nuclear experiments. Two of his students, John D. Cockroft and Ernest T. S. Walton, completed the first successful particle accelerator at the Cavendish Laboratory in Cambridge in 1932. By accumulating a potential of hundreds of thousands of volts, Cockroft and Walton accelerated protons to energies sufficient to disintegrate the nuclei of light elements. Owing to the repulsion between nuclei and protons, and the difficulty of creating and maintaining high potentials, their machine could not transform heavier elements.

Physicists soon turned to other means to accelerate particles. In the United States, Ernest Lawrence's magnetic resonance accelerator (the cyclotron), which applied energy to protons or deuterons in successive small steps rather than, as in the Cockroft–Walton machine, in one large jump, provided bombardments able

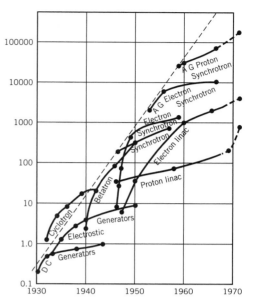

The increase in energy attainable by particle accelerators. The envelope of the rise is logarithmic.

to transform almost all nuclear species. His linear accelerator used the same principle of resonance acceleration to propel heavier nuclei. Another American, Robert Van de Graaff, returned to the one-jump method by a technique that allowed the accumulation of up to about ten million volts on a spherical conductor. The Van de Graaff, cyclotron, and linear accelerators were used at many universities and research institutes in the 1930s to explore the new field of nuclear physics. Financial support for the development of particle accelerators came largely from medical philanthropies. They hoped their high-voltage X rays, neutrons, and other particles as well as the artificially radioactive products of their interactions with other substances would be more effective against cancer and other diseases than the natural radiations from radium and other substances.

Lawrence was especially successful in generating support for his cyclotrons. The parameter most often employed to indicate their power—the diameter of the pole pieces of the magnet that retained the particles in their spiraling orbits as they accumulated energy—grew from a few inches to sixty. The increase in size gave a monumental increase in the energy with which the particles escaped from the magnet—from a few hundred thousand to thirty-two million electron volts. In 1939 Lawrence received the Nobel Prize in physics for his cyclotron and work done with it. The consequent enlargement of his prestige helped him to convince the

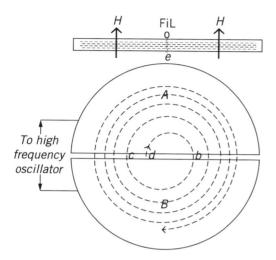

The cyclotron principle. Ions admitted at the center of the apparatus (at fil) are accelerated as they cross the gap (c, d, b) between the two D-shaped hollow cans (A, B) and describe circular orbits under a magnetic field when within the "dees;" the combination of circular arcs and horizontal accelerations produces the spiral paths shown.

Rockefeller Foundation to give the money to build a giant cyclotron with pole pieces 184 inches in diameter. Intended to be the last and largest of all particle accelerators and, in that way, the counterpart to the 200-inch telescope the foundation supported at Palomar, the 184-inch proved instead to be the first of a generation of much larger particle accelerators.

The cyclotron did not accelerate electrons because their loss in energy and increase in mass owing to acceleration made them unattractive for work at high energy. Nonetheless, in 1939 Donald William Kerst invented a "betatron," which produced electrons or energies useful in nuclear investigations. After the war linear electron accelerators became competitive with proton accelerators for some purposes, and physicists turned the intense radiation of electrons maintained in circular orbits to advantage in "synchrotrons" whose "light" could be used in *materials science and other applications.

The discovery of nuclear fission in uranium in 1939 provided a new role for particle accelerators and nuclear physicists. Cyclotrons at Berkeley and Los Alamos produced the first samples of plutonium, and cyclotroneering principles underlay the electromagnetic separation of isotopes in large banks of "calutrons" (after California University) that separated the fissile isotope of uranium for use in the first atomic bombs. Although other techniques eventually proved more efficient, these wartime successes opened the door to federal funding of nuclear physics by the Manhattan Engineer District and its successor, the Atomic Energy Commission.

The completion of the 184-inch cyclotron as a synchrocyclotron in 1946 was the first successful application of the synchrotron principle discovered by Edwin M. McMillan and Vladimir I. Veksler during the war. This principle enabled designers to accelerate particles in tight bunches by changing the frequency of the accelerating fields in step with relativistic changes in the mass of the particles. A series of accelerators built throughout the second half of the century, ranging from Brookhaven's 'Cosmotron' to the Tevatron, which produces trillions of electron-volts, incorporate the synchrotron principle and advances in magnetic technique that permit the confinement of the beam of accelerating particles to a narrow, evacuated pipe of fixed radius. The only link to the size of these machines is financial. The first version of CERN had a diameter of 200 meters; the one now under construction these extends to 27 kilometers. The largest current machine in the United States, at Fermilab, has a circumference of four miles, within which a herd of bison graze.

The linear accelerator also developed rapidly after World War II. At Berkeley, Luis Alvarez invented a type for protons using war surplus radar equipment. His student Wolfgang Panofsky applied the scheme to electrons at the Stanford Linear Accelerator (1962). Discoveries of new elements and particles by particle accelerators led to Nobel Prizes as well as increasingly larger accelerators in the United States, western Europe, and the Soviet Union. The prestige and power of these machines made them political as well as physical icons. Only after the end of the Cold War did the enormous cost of accelerators prompt the United States to withdraw from the competition by canceling the Superconducting Super Collider.

See also ELEMENTARY PARTICLES; NUCLEAR PHYSICS; PHILANTHROPY; WORLD WAR II AND COLD WAR.

Milton Stanley Livingston, *Particle Accelerators; a Brief History* (1969). Armin Hermann et al., *History of CERN*, 2 vols. (1987–1990). J. L. Heilbron and Robert W. Seidel, *Lawrence and His Laboratory: A History of the Lawrence Berkeley Laboratory* (1990). John Krige, ed., *History of CERN*, vol. 3 (1996).

ROBERT W. SEIDEL

ACCURACY. See PRECISION AND ACCURACY.

ACID AND BASE. To bring some order to the natural world, people have long classified substances according to taste and appearance. Among the oldest categories of substances are acids, with their sharp, sour taste; alkalis, with their bitter flavor; and salts, notable for being crystalline and water-soluble. However, these criteria have changed drastically over time in response to changing chemical techniques and theories.

The "antagonism" between acids and alkalis had been known in antiquity; only gradually did chemists recognize that the products of their "strife" were salts. In the eighteenth century the word "base" came to signify any substance, including metals, that reacted with acids to generate salts.

Among the oldest known alkaline substances are soda (sodium carbonate) and potash (potassium carbonate). Obtained from aqueous extracts of ashes used in soap and glass manufacture, these so-called fixed (nonvolatile) alkalis were contrasted with volatile alkali (ammonia) formed when urine decomposes. Chalk and limestone, naturally occurring forms of calcium carbonate, were classified as alkaline earths.

In addition to these substances, medieval alchemists had spirit of vinegar (acetic acid), which had been purified by their Arab predecessors, and the more powerful and corrosive inorganic acids—spirit of salt (hydrochloric), spirit of nitre (nitric), and spirit of vitriol (sulfuric)—which could dissolve many materials, including metals. Even gold, the noblest of them, succumbed to aqua regia (mixed hydrochloric and nitric acids). The solvent powers of these mineral acids made them valuable articles of commerce and potent symbols in the alchemical lexicon.

Contact between any of these acids and alkalis caused a vigorous, often violent reaction that produced much heat. If the alkali were a carbonate, a marked effervescence occurred as well owing to release of "air" (carbon dioxide). These manifestations fit in well with the anthropomorphic worldview of the alchemists, who thought in binary terms and saw acids and alkalis as mutually antagonistic.

Alchemists and chemists continuously speculated about the ultimate causes of acidity and basicity. A doctrine derived from Aristotle, which lasted into the eighteenth century, held that tangible matter itself had no intrinsic properties (see ARISTOTELIANISM). Rather, intangible essences or "principles"—such as the presumed principles of acidity, alkalinity, and salinity—united with ordinary matter. Such principles could not be isolated however, and acids and alkalis could only be defined circularly: a substance that reacted vigorously with the one was classified as the other. Furthermore, the fate of these principles after reaction presented a major conceptual problem, since the salts often differed substantially in properties from their progenitors.

Robert *Boyle discovered that certain plant infusions changed color when exposed to known acids or alkalis. For example, syrup of violets is blue when neutral; if treated with an acid, it turns red, with an alkali, green. (Litmus is the indicator now widely used, hence the term "litmus test.") Boyle's discovery permitted chemists to classify acids and alkalis independently of one another, and to show that specific acids and alkalis would always neutralize one another in the same weight ratio.

Extensive work on salts in the seventeenth and eighteenth centuries established that they consisted of an acid and a base. Since the components could sometimes be recovered, chemists inferred that they had remained intact in the salt. Moreover, the same salt could be synthesized by different routes. These results inclined chemists to turn away from principles and towards composition as the fundamental property of substances.

Nonetheless, *phlogiston, the principle of inflammability, played a dominant role through most of the eighteenth century. Its primary advocate, Georg Ernst Stahl, used it to explain acidity. He declared that all acids derived from vitriolic and differed only in their quantity of phlogiston. Antoine-Laurent *Lavoisier's ultimately successful attack on phlogiston required a new understanding of acids. When nonmetals (carbon, nitrogen, sulfur) burned, they produced acidic oxides; thus the indispensable generator of acidity was oxygen. By contrast, metal oxides were basic, and when they combined with acidic oxides the constituent oxygens bound the salt together.

A great many of Lavoisier's innovations have endured. Not so his theory of acidity. Around 1810 Humphry Davy demonstrated that hydrochloric acid (HCl) lacked oxygen. He intimated that hydrogen (H) must be the source of acidity, a view well accepted by midcentury. During the nineteenth century chemists gradually, and with many qualifications, accepted the idea that chemical compounds are constituted of submicroscopic molecules; that these molecules are held together by electrical forces between their atoms; that when dissolved in water many molecules disintegrate into charged fragments, called ions; and that many acids and bases and all salts exist as ions even in the undissolved state (such as table salt, Na^+Cl^-). From these investigations came the postulate that acids and bases generate, respectively, hydrogen (H^+) and hydroxyl (OH^-) ions in solution, whose interaction produces neutral water, H_2O.

The H^+/OH^- theory could not explain all acid/base reactions. In 1923 Gilbert N. Lewis proposed that acids are electron acceptors and bases electron donors; H^+ (electron deficient, hence positive) and OH^- (electron rich, hence negative) are just special cases. The Lewis theory of acids and bases still reigns as one of the most comprehensive principles in chemistry. One of its corollaries makes acidity and basicity relative properties that depend on particular conditions. These properties are not only of theoretical interest; acids, bases, and salts are among the most important commodity chemicals worldwide. In his *History of the Inductive Sciences* (1837), William Whewell asserted that "the whole fabric of chemistry rests ... upon the opposition of acids and bases." That is as true today as it was then.

See also ELECTROCHEMISTRY; ION; PHYSICAL CHEMISTRY.

Paul Walden, *Salts, Acids, and Bases: Electrolyes: Stereochemistry* (1929). William H. Brock, *The Norton History of Chemistry* (1993). Bernadette Bensaude-Vincent and Isabelle Stengers, *A History of Chemistry*, trans. Deborah van Dam (1996).

STEPHEN J. WEININGER

ACOUSTICS AND HEARING. Acoustics, the science of sound, falls at the intersection of several fields, including mechanics, hydrodynamics, thermodynamics, and electromagnetism. Scientific interest in sound also derives from human hearing, and hence involves physiology and psychology; and acoustics engages fields outside science such as music and architecture. A distinct field of acoustics, including but not limited to hearing, gradually emerged from this disparate background in the eighteenth and nineteenth centuries, aided by the use of quantitative apparatus to produce and detect sound and the mathematical analysis of the results.

Ancient and medieval philosophers of nature studied acoustics mainly as a means to understand music. Mathematical theories of music dated at least to the Pythagoreans, who identified musical intervals as ratios of whole numbers and related musical pitches to lengths of vibrating strings. Although a few ancient writers speculated on the wave nature of sound and the propagation of compressions, Arabic and European natural philosophers through the Middle Ages and Renaissance continued to study acoustics only as part of music theory, if at all. In the early modern period, natural philosophers began to undertake systematic experiments and to extend their investigations to sound in general. Experiments with vibrating strings led *Galileo Galilei to posit the relation between pitch and frequency, elucidated around the same time by Giovanni Benedetti and Isaac Beeckman, and also to suggest that pitch depended on the tension and diameter of the string as well as its length. Marin Mersenne in the early seventeenth century used very long vibrating strings, some over one hundred feet long, to arrive at a quantitative relation between pitch and frequency. Mersenne also measured the speed of sound, as did his contemporary Pierre Gassendi, who asserted that soft and loud sounds traveled at the same speed.

The scientific academies that sprang up later in the seventeenth century would make the speed of sound a prime program. Around the same time, experiments with improved air pumps, beginning with Otto von Guericke and extended by Robert *Boyle and Francis Hauksbee, convinced natural philosophers that sound did not travel in a vacuum. Experiments on the speed of sound tested in particular the theory of Isaac *Newton, whose work on fluid mechanics in the *Principia* included acoustics. But Newton arrived at a figure for the speed of sound, based on the pressure and density of the medium, at odds with contemporary estimates. Mathematicians in the eighteenth century, notably Leonhard *Euler, Jean d'Alembert, and Joseph Louis Lagrange, extended Newton's analytical mechanics and pneumatics to explain the discrepancy between theory and measurement, and Pierre-Simon, Marquis de Laplace, Jean-Baptiste Biot, and Siméon-Denis Poisson at the start of the nineteenth century succeeded by suggesting that the passing sound wave heated the medium.

The quantification of acoustics accelerated in the nineteenth century, driven by the use of laboratory apparatus such as tuning forks, vibrating plates, and sirens to produce standardized tones, and sounding boards and the stethoscope to detect them. The new instruments helped bring acoustics into the realm of precision physics and also indicated the increasing quantification of physiology. Hermann von *Helmholtz combined knowledge of the physiology and physics of sound in his synthetic treatise *On Sensations of Tone* of 1862, which confirmed that the ear analyzes periodic sound waves into Fourier sums of simple harmonics and predicted the existence of nonlinear summation tones. John Tyndall helped bring Helmholtz's work to English-speaking audiences, elaborating it with his own experiments and generalizing beyond Helmholtz's particular interest in music. Lord Rayleigh then provided a systematic mathematical analysis of sound in his *Theory of Sound* of 1877–78; although Helmholtz provided mathematical details in appendices to his works, he had

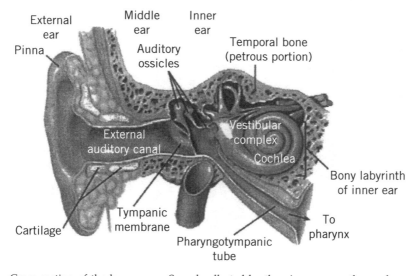

Cross section of the human ear. Sound collected by the pinna causes the eardrum (tympanic membrane) to vibrate; the vibrations are transmitted to the fluid-filled cochlea lined with tiny hairs (cilia) that generate the nerve impulses for input into the brain. The equalizing tube from the middle ear to the nose (phryngotympanic or Eustachian Tube) is named after the sixteenth century Paduan anatomist Bartolomeo Eustachi.

kept the main text nonmathematical. Rayleigh's two-volume work analyzed diverse phenomena of sound based on the vibration of air, liquids, and gases and of solid strings, plates, and rods under various perturbations; the books completed the edifice of classical acoustics.

New problems and programs were meanwhile emerging from the development of electromagnetism and its application to acoustics. Alexander Graham Bell, familiar with Helmholtz's research and inspired by his own work with deaf students, invented a means for transmitting speech over wires, called the telephone. Bell faced several coclaimants for the invention and strong competition for its development, including an improved system designed by Thomas Edison. Edison in the meantime invented the phonograph, the foundation for the recording industry. The development of radio further spurred the invention of microphones, loudspeakers, amplifiers, vacuum tubes, and oscillators by the burgeoning electrical industry. Industrial engineers also agreed on an international standard for the intensity of sound, the decibel. Magnetostriction, piezoelectricity, and acousto-optics, three more developments of the late nineteenth century, would also provide new ways to produce and detect sound. Electroacoustics opened up a rich new field for the study of sound in the twentieth century.

Electroacoustics also helped to transform architectural acoustics, the study of which dated to the ancients—the Roman architect Vitruvius wrote on the acoustics of theaters in the first century B.C.—and was extended in the seventeenth century by the Jesuit polymath Athanasius Kircher, among others. Around 1900 Wallace Sabine revived the subject with his work relating reverberation times to the volume and building materials of rooms. The introduction of electroacoustic technology, along with new sound-absorbing building materials, provided a means to active control of sound inside buildings, and in 1930 Carl Eyring revised Sabine's results to accommodate the new acoustic environments.

Sound does not just travel through air. The use of submarines in World War I spurred efforts to develop ways to detect them, which soon focused on sound. The subsequent development of sonar for submarine warfare provided much support of acoustics research and also new tools for marine biology and oceanography. Sound also travels through matter, and ultra-high frequency sound, or ultrasound, became an important probe for solid-state research and found application in industrial materials and in medicine in the second half of the twentieth century.

Frederick Vinton Hunt, *Origins in Acoustics: The Science of Sound from Antiquity to the Age of Newton* (1978). Emily Thompson, "Dead Rooms and Live Wires: Harvard, Hollywood, and the Deconstruction of Architectural Acoustics," *Isis* 88 (1997): 597–616. Robert T. Beyer, *Sounds of Our Times: Two Hundred Years of Acoustics* (1999).

PETER J. WESTWICK

**ACQUIRED IMMUNODEFICIENCY SYN-
DROME.** See AIDS.

**ADVANCEMENT OF SCIENCE, NATIONAL
ASSOCIATIONS FOR THE.** During the seventeenth century naturalists and philosophers emphasized scientific communication among themselves and sought venues for public demonstrations of scientific achievement. Francis Bacon shifted discussion toward what he identified as the "advancement of learning" through human design and proposed a hierarchical model of organization in his fictional work, *The New Atlantis* (1627). Advancing knowledge became associated not only with individual genius but with organizations like the Royal Society of London (1662), which sponsored research activities and publications to share new knowledge (*see* BACONIANISM). Local voluntary societies provided repositories for specimens, artifacts, and instruments important for advancing research, but these proved expensive to maintain and of benefit primarily to a local membership.

By the nineteenth century, transportation networks of canals and railroads facilitated travel while public interest in science was creating popular audiences and paying occupations for scientists just as cultivators of natural science began to feel a sense of community. German-speaking physicians and scientists, who also felt political isolation, were the first to organize on a national level. Lorenz Oken proposed the foundation of the Gesellschaft Deutscher Naturforscher ünd Arzte (GDNA), modeled loosely on a Swiss organization, in 1822. The GDNA grew quickly, met in a different city each year, and facilitated communication among its members.

Several British attendees at the German meetings, who perceived a decline of science in England, Scotland, and Ireland, rallied colleagues by letter and publication to attend a meeting in York, "the most centrical city for the three kingdoms," in 1831. The resulting peripatetic British Association for the Advancement of Science (BAAS) proved highly successful at attracting public audiences. It quickly became a forum where physical, natural, and social scientists could discuss matters of scientific content and policy, capacities that proved particularly attractive to the visiting North Americans who subsequently established the American Association for the Advancement of Science (AAAS), founded in Philadelphia in 1848. Positioned to enhance the professional ambitions of leaders and serve as advocates for their plans, the AAAS members conducted business at their annual meetings, published proceedings, and created committees to look into special topics from terrestrial magnetism to science education. As a

scientific forum the AAAS allowed disputants the opportunity to argue from evidence before peers and to wrestle with such complicated issues as chemical nomenclature and universal species taxonomy. Over the next half century, France, the Netherlands, and Belgium established similar organizations and the movement spread into increasingly independent colonies seeking industrial and government sponsorship for scientific and technological activities. The Italians reinvented a Riunione degli Scienziati Italiani (1839) under various titles over the next century as well.

The BAAS leaders extended membership privileges to British colonials and held five overseas meetings in colonies between 1884 and 1914. The resultant linkages advanced scientific dialogue but sustained international hierarchies as well. Public lectures by distinguished scientists provided stimulus to local science, and sometimes local scientists and visiting experts collaborated on research projects, but inevitably much of the scientific data, and credit, returned home at the conclusion of collaborations. That, at least, was the case for anthropology in Canada, where important early work on northwestern Native Americans and an ethnographic survey of tribes provided unprecedented, immediate opportunities for Canadian amateurs but left no national school in anthropology when the experts went back to England. Far from Britain, the Australia and New Zealand Association for the Advancement of Science (1888) began to link regional scientists and publicize their work. During the first quarter of the twentieth century similar initiatives came in India, South Africa, Hong Kong, and other colonial outposts.

In the twentieth century countries aspiring to democracy as well as technological and scientific progress initiated associations to advance science. Government sponsorship, rather than grassroots voluntarism, became common; for example, the Japan Society for the Promotion of Science (1932) became a channel for funds to large, expensive research facilities. After World War II the model seemed promising for aspiring nations like Brazil, where the Brazilian Society for the Advancement of Science grew from scientific initiatives that persuaded the government to provide financial support. Scientific members, however, found themselves politicized and, indeed, in conflict with policies of a military regime during turbulent times. Nationally based societies, regardless of their financial autonomy, inevitably worked within and through their home political systems, carefully balancing professional norms as they lobbied for scientific budgets, educational policies to support science, and independent peer review for evaluation of research.

National societies varied considerably as they pursued the goal of advancing science. In Britain the early success of the BAAS correlated with the status of scientific activity and high visibility of prominent scientists like Charles *Lyell, and was sustained by its outreach in provincial areas to enhance science education. In the United States, members relied on the AAAS to establish a forum for widely dispersed scientists, to maintain links among scientists across the increasingly distinct scientific specialties, and to advocate science to state and federal governments. In the twentieth century the AAAS published *Science* magazine, under editor James McKeen Cattell, the leading journal across the sciences with the largest international readership in the early twentieth century. In this respect the AAAS outdistanced the BAAS, which does not run the leading general British science journal, *Nature*.

The AAAS, with affiliates from most major professional societies, moved to rooms in the Smithsonian Institution until it built its own permanent headquarters in Washington, D.C., where it also reported on research and developments in federal agencies and budgets. By the late twentieth century the AAAS had an administrative and editorial infrastructure whose capacity to report on scientific developments was international, while specific programs in the organization concentrated on science education, promoted gender and racial diversity, and responded to ethical issues raised by advances in science, technology, and medicine. National societies produced a sense of identity even as they provided vehicles for international exchange of publications, visiting status to foreign attendees at meetings, and occasional collaborations.

Efforts to link these national organizations in a formal way have had mixed success. Shortly after World War II, in 1950, the United Nations Educational, Scientific, and Cultural Organization (UNESCO) sponsored an International Meeting for Associations for the Advancement of Science, but the effort faded from lack of support by the well-established organizations like the AAAS and BAAS. An International Federation of Associations for the Advancement of Science and Technology founded in the 1990s persisted, suggesting that the rootedness in particular settings has allowed many national associations to be highly successful but that emphasis on national needs and interests constrained systematic cooperation, at least before the advent of the Internet.

See also ACADEMIES AND LEARNED SOCIETIES; PROFESSIONAL SOCIETY.

R. von Gizycki, "The Association for the Advancement of Science: An International Comparative Study," *Zeitschrift für Soziologie* vol. 8, no. 1 (1979): 28–49. Jack Morrill and Arnold Thackray, *Gentlemen of Science: Early Years of the British Association for the Advancement of Science* (1981). Ivy G. Avrith, *Science at the Margins: The British Association and the Foundation of Canadian Anthropology, 1884–1910* (1987). Anna Maria Fernandes, *The Scientific Community and the State in Brazil: The Role of the Brazilian Society for the Advancement of Science, 1948–1980* (1987). Roy MacLeod, ed., *The Commonwealth of Science: ANZAAS and the Scientific Enterprise in Australasia, 1888–1988* (1988). Sally Gregory Kohlstedt, Michael M. Sokal, and Bruce V. Lewenstein, *The Establishment of Science in America: 150 Years of the American Association for the Advancement of Science* (1999).

SALLY GREGORY KOHLSTEDT

AEPINUS, Franz Ulrich Theodosius. (1724–1802), German natural philosopher.

Aepinus's work *Tentamen theoriae electricitatis et magnetismi* (1759) revolutionized the study of *electricity and *magnetism, adding new conceptual rigor and bringing these subjects within the reach of mathematical analysis. Aepinus greatly strengthened and improved Benjamin *Franklin's controversial new theory of electricity and also developed, in close analogy to Franklin's theory, a highly successful new theory of magnetism. In the process he provided a model for the mathematization of other fields of experimental natural philosophy, which until then had been entirely qualitative studies.

Aepinus was born in Rostock, Germany, where his father was a professor of theology. He studied at Rostock and Jena universities, then taught mathematics and experimental physics at Rostock for several years. He also began a program of astronomical observation that led to his appointment as astronomer at the Berlin Academy of Sciences in 1755.

In Berlin, Aepinus joined the circle surrounding the mathematician Leonhard *Euler, who became a powerful patron. Electricity was then a leading subject of inquiry, following the discovery of the electrical nature of *lightning. Aepinus became a highly skilled experimenter, working with his student Johan Carl Wilcke and Euler's son Johann Albrecht Euler. Their work soon carried them to the forefront of the field.

Aepinus and Wilcke adopted Franklin's idea that electrification consisted in the redistribution of a subtle "electric fluid" that supposedly pervaded all bodies. When a substance was rubbed—say, a glass rod with a cloth—some fluid transferred from one body to the other, leaving one with a surplus and the other with a deficiency of electric fluid. This implied two different states of electrification, "plus" and "minus," which in appropriate circumstances could neutralize each other. In their experiments, Aepinus and Wilcke systematically pursued this notion of two contrary electricities. Aepinus used this notion to make sense of the extraordinary behavior of tourmaline crystals, which, he showed, acquired opposite electrical charges on opposing faces when warmed. Struck by the analogy between an electrified tourmaline and a magnet, he conceived the idea that magnetization resulted from the redistribution within a piece of iron of a subtle magnetic fluid, different from but entirely analogous to Franklin's electric fluid, the poles of a magnet being regions of "plus" and "minus" magnetic charge. Aepinus pursued this idea in a brilliant series of studies, which he brought together in 1759 in his *Tentamen*.

Using Isaac *Newton's discussion of gravity as a model, Aepinus reduced all electrical effects (and, by analogy, all magnetic effects) to unexplained forces acting at a distance between particles of ordinary matter and particles of electric (or magnetic) fluid. He transformed Franklin's inspired but somewhat inchoate ideas into a coherent and mathematically consistent theory of electricity, which he then transferred in toto to a theory of magnetism. Most notably, he was able to explain in detail a wide range of puzzling induction effects—including the most pressing problem confronting theorists of electricity at the time, the shock delivered by the *Leyden jar—in which changing the electrical (or magnetic) condition of bodies induced changes in the electrical (or magnetic) condition of other, nearby bodies. The forces he invoked to explain this varied, he assumed, with distance. With the (inverse square) law of variation not yet established, however, Aepinus, being unwilling to base his work on speculation, could not progress beyond semiquantitative computations; his analyses never became fully mathematical. Nevertheless, his work constituted a dramatic advance toward a fully mathematized physics. A generation later, it provided the starting point for the work of both Charles-Augustin Coulomb and Alessandro Volta (*see* CAVENDISH, HENRY AND CHARLES-AUGUSTIN COULOMB AND GALVANI, LUIGI, AND ALESSANDRO VOLTA).

In 1757 Aepinus became professor of physics at the Academy of Sciences in Saint Petersburg. There, he caught the eye of the Grand Duchess, soon to become Russia's Empress Catherine the Great, and became her personal tutor in natural philosophy. For several years he published extensively on electricity and magnetism, mathematical questions, astronomy, and geography. However, further appointments—as director of studies at the Imperial Corps of Noble Cadets, tutor to Catherine's son (later the Emperor Paul),

and head of the cipher department in the Russian Foreign Ministry—took him away from natural philosophy. In the early 1780s, at Catherine's request, he prepared detailed proposals for a national school system, which were only partly implemented. In 1798 he retired to Dorpat (Tartu), Estonia, where he died several years later.

J. L. Heilbron, *Electricity in the 17th and 18th Centuries* (1979). R. W. Home, *Aepinus's Essay on the Theory of Electricity and Magnetism* (1979).

R. W. HOME

AERONAUTICS. Aerodynamic theories and experiments both in the air and wind tunnels provide the scientific basis for the applied science of aeronautics. Before the invention of the airplane by Orville and Wilbur Wright in 1903, some engineers made notable contributions to the development of aeronautical science. George Cayley in England was the first to study the lift and drag of a flat plate at different angles of attack using a whirling-arm device in the early nineteenth century. Later, Francis Wenham and Horatio Phillips in Britain constructed a wind tunnel, measured the lift and drag of a wing-shaped body, and demonstrated the better performance of cambered wings and the preference of a small angle of attack to obtain a better lift and drag ratio. The German Otto Lilienthal systematically measured the lift and drag of wings, and graphically represented the results for his own design of bird-shaped gliders, before falling to his death during an experimental flight. The Wright brothers utilized his data. They, too, devised a wind tunnel and other tools to obtain aerodynamic data until they finally made the world's first successful flight in a heavier-than-air machine.

Aeronautical pioneers at the turn of the century worried about stability and control of the airplane as well as its motive power. The Wrights emphasized control over stability in designing their airplane, which they modeled after a bicycle. Most subsequent designers, however, aimed at stable configurations, leading to the present form. A stability theory, formulated by George Bryan in 1911, consisted of a set of six equations corresponding to the six degrees of motions of the airplane. This theory could indicate the plane's stability given its basic dynamic characteristics. The National Physical Laboratory in England experimented with models to obtain the necessary data, which helped to produce highly stable biplanes. Many national governments promoted and financially assisted aeronautical research and development after the Wright brothers' demonstration in Europe in 1908.

In aeronautics, theories generally followed experiments and technologies. The Navier–Stokes

A scale model of an X-15 airplane in a wind tunnel at Langley Air Force Base, Virginia. The air is driven past the model at supersonic speeds, producing the shock waves shown.

equations, a set of equations that precisely described the flow of fluid taking into account its viscocity, refused to be solved analytically due to the non-linearity of the equations, and required alternative approximate equations to describe air flow around the airplane (*see* HYDRODYNAMICS AND HYDRAULICS). The boundary-layer theory proposed by Ludwig Prandtl in 1904 showed that the air around the airplane basically behaved as an inviscid and incompressible fluid outside the "boundary layer" covering the surface of a moving body: The flowing air flows viscously inside the layer and freely outside it.

Two theories emerged at the turn of the century to explain and estimate lift and drag produced by air flow around a wing. One—the discontinuous theory—assumed a discontinuity between the air flow and stagnating air behind the wing, causing a difference of pressure between the sides. But the unrealistic assumption of discontinuity could not produce a theory that gave accurate estimates, despite its subsequent mathematical sophistication. Another—the circulation theory—assumed the circulation of air around the airfoil. Together with the more remote steady air flow, the circulated air produced lift just as the air around a sliced tennis ball did. Martin William Kutta in Germany and Nikolai Joukowski in Russia developed the circulation theory independently; knowledge that Lilienthal's cambered wings gave a lift even at a zero angle of attack may have guided Kutta's thinking. The circulation theory, however, only dealt with two-dimensional air flow and disregarded the effects of wing tips. Prandtl took the matter further with the concept of vortices trailing from wing tips, and so generalized from two to three dimensions. The British engineer Frederick Lanchester anticipated trailing vortices, but his mode of presentation put off most of his readers.

Prandtl's aerodynamic theory was applied to wing design, wind-tunnel measurements, and the aerodynamics problems. The theory outran practice, however. It showed that an elliptical form would be most efficient in terms of lift per induced drag, but the insight could not be implemented until all-metal airplanes became practical. Prandtl's theory provided an important clue for estimating the effect of the wall of wind tunnels. Experimenters welcomed this "Prandtl correction" because it reduced discrepancies between full-scale and model experiments and between experiments in different wind tunnels; better agreement, however, awaited the creation of a variable density wind tunnel. Previously the discrepancy between full-scale and model experiments had been ascribed to the "scale effect" resulting from the difference of the Reynolds numbers. The Reynolds number was a non-dimensional number determined by the viscosity, the density, and the velocity of fluid as well as the size of a body in it, and was expected to characterize the behavior of fluid flow around the body such as an airplane or its model. The first variable density wind tunnel contained in an airtight tank was constructed in the United States in 1922.

In 1929 the British engineer Melvill Jones showed graphically the theoretical limit of the speed of an airplane when its total resistance consists only of skin friction. His diagram indicated a potentially multifold increase of speed by streamlining the airplane. In the 1930s, engineers sought to attain this goal through cowling, retractable wheels, flush riveting, and so on. The effort resulted in the successful DC–3 airplane.

Engineers tackled the problems of air flow near the speed of sound in the 1920s when the tips of propellers reached a speed close to that of sound. Experiments showed a dramatic increase of drag and decrease of lift in this transonic region. Engineers at the U.S. National Advisory Committee for Aeronautics, the forerunner of the National Aeronautics and Space Administration (NASA), demonstrated the separation of air flow at the so-called critical speed and the generation of shock wave above this speed. Supersonic effects had been studied in the nineteenth century by Ernst Mach, who first photographed the shock wave, and by Carl de Laval, who designed a high-speed wind tunnel to study supersonic aerodynamics. A journalist coined the phrase "sound barrier" to signify the sudden and dramatic increase of drag. The American pilot Chuck Yeager was the first man to fly through this "barrier" in his Bell–X-1 in 1949.

Scientists and engineers investigated supersonic and hypersonic aerodynamics intensively after World War II to develop supersonic airplanes, missiles, and rockets. Among the results are the area rule found by Richard Whitcomb, which prescribes that the cross-sectional area of the airplane body should not vary rapidly or discontinuously along its longitudinal axis; swept-back wings, developed in the 1960s; and a fairing of advance design at the top of a rocket, designed to protect against heating at hypersonic speed.

R. Giacomelli and E. Pistolesi, "Historical Sketch," in *Aerodynamic Theory*, vol. 1, ed. William F. Durand (1934): 305–394. Charles H. Gibbs-Smith, *Aviation: An Historical Survey from Its Origins to the End of World War II* (1970). Paul A. Hanle, *Bringing Aerodynamics to America* (1982). Peter Jakab, *Visions of a Flying Machine* (1990). Walter G. Vincenti, *What Engineers Know and How They Know It* (1990). John D. Anderson, *History of Aerodynamics* (1997).
TAKEHIKO HASHIMOTO

AETHER. See ETHER.

AFRICA. The history of science in the diverse continent of Africa is best dealt with under four headings: precolonial science in Islamic North Africa, precolonial science in animist Sub-Saharan Africa, European-based colonial science, and science in independent African nations.

Precolonial North Africa
From the eighth century on, North Africa had close ties, culturally and scientifically, with neighboring Islamic regions. From the ninth to the thirteenth century, scholars in Egypt looked to the Baghdad caliphates. In the Maghreb, stretching west of Egypt to Morocco, scholars had close links with Islamic Spain, which saw a flowering of culture and learning through the twelfth century.

For a time, mathematics, astronomy, and medicine flourished under the patronage of the sultans. However, by the fifteenth century, North African Islamic movements had nearly extinguished these intellectual inquiries. Legal scholars and theologians came to view rational study of nature as irrelevant and potentially heretical. Moreover, mystical Sufi orders became widely popular among the people. The role of patient empirical and rational investigation of nature as a way to learn about God was subordinated far below the revelations of the Quran and mysticism. Centers of Islamic learning, such as al-Azhar of Cairo, the Zaituna of Tunis, and the Qarawiyin of Fez no longer supported teachers of natural philosophy. Those who sought such teachers had to search for dwindling numbers of isolated scholars.

What remained was routine work, such as compilation of astronomical tables, which continued to the eighteenth century. Because it was needed to determine the *qibla* (the direction to Mecca) and the five daily times of prayer, astronomy had a privileged status. Even so, it became suspect because of its association with astrology, which was often practiced in the courts of the sultans. Increasingly, folk astronomical methods determined the *qibla* and times of call to prayer in place of the earlier complex, mathematical astronomy. Mathematical usage narrowed to practical matters such as the calculation of inheritances. Medicine too declined; only the vestiges of early Islamic medical science survived. In its place, a folk medicine, which continues to be practiced today, brought together Yunani (Greek) medicine derived from Galen and mystical prophetic medicine associated with the sayings of Mohammed. By the time of European colonization, innovative work in mathematics, medicine, and natural philosophy had long disappeared.

Nor did such work survive along the southern "shores" of the Sahara, the Sahel, a cultural outpost of North Africa. North Africans brought to the Sahel writing and formal education for teaching the Quran. By the sixteenth century, Timbuktu had emerged as a major center of Islamic learning, where mathematics of inheritance was taught, but not natural philosophy. By then other leading Islamic universities, such as the Qarawiyin and the Zaituna, had already jettisoned nonreligious studies.

Precolonial Sub-Saharan Africa
South of the Sahel, traditional medicine and various technologies, including sophisticated iron and bronze metallurgy, flourished prior to European colonization. They did little, though, to point the way to anything resembling modern science. In the African worldview, the natural world was a realm of spirits approachable by sorcery. In this context, analyzing nature according to the rational, disinterested ideal of Western science was unthinkable. Smelting iron required placation of spirits. Traditional healers interpreted the spiritual, psychological aspects of the ill and their communities using mediation with the spirits (although they also turned to herbalists for remedies that had proven effective). Moreover, lack of written languages impeded the progressive accumulation of natural knowledge beyond the limits of oral traditions. Even after the arrival of Christian missionaries, animistic worldviews persisted.

The incursions of the Europeans further discouraged the appearance of science. Importation of European goods, such as weapons, iron, cloth, and glass, undermined traditional technologies. The consequences of the slave trade and warfare in some regions exacerbated the loss of indigenous technology. Traditional knowledge and skills dwindled yet further, sending the societies into cultural decline.

Colonial Africa

Napoleon's invasion of Egypt (1798–1802), with an accompanying body of scientists who studied the riches of its ancient past, marked the beginning of European scientific interest in Africa. The French began colonizing in earnest in Algeria in the 1830s by sending a large medical corps. Under the direction of the French military, doctors investigated Algeria's social structure and environment during their off-duty hours, exploring its potential as a colony. Several meteorological stations were set up extending deep into the Sahara. Also, a small observatory, established in 1858 to provide mariners with accurate time, became the Observatory of Algiers. From these beginnings, French scientific societies took root in North Africa, some of which carried out notable scientific work. For example, research in the Pasteur Institutes in Algeria and Tunisia led to the discovery of the carriers of malaria, typhus, and bubonic plague. These successes were rewarded with Nobel Prizes for Charles Louis Alphonse Laveran in 1907 and Charles Nicolle in 1928. Despite these achievements, the mother institutes in Paris tightly controlled scientific work. Following World War I, the French government founded the Ecole Supérieure d'Agriculture Tropicale and Institut de Médecine Vétérinaire Exotique to train colonial personnel in tropical agriculture and veterinary medicine. However, coordination of research in the colonies had to wait until the Vichy government of World War II formed the Office de la Recherche Scientifique Coloniale, which set long-term research and training goals.

In British colonies, scientific initiative, usually applied to agriculture and medicine, rested with the colonial governments, although fundamental research and training remained in Britain. In particular, research in tropical medicine, most importantly work on parasitic diseases and their hosts, which initially fell to medical doctors in the field, centered in Britain with the founding of the London and Liverpool Schools of Tropical Medicine in 1899. After World War I, organizations covering different regions, such as the Forestry and Agricultural Organization of British East Africa and the West African Research Organization, promoted and coordinated research, largely in the agriculture of cash crops such as cocoa and palm oil. Scientific work shifted back to the colonies and regional centers with financing directly through the colonial governments.

Characteristically, European research in Africa was performed to benefit the colonialists or the mother country. Africans rarely advanced beyond assistants. Moreover, education for Africans imbued them with contempt of manual labor. Consequently they eschewed farming and generally did not apply their knowledge to make greatly needed improvements in agriculture.

Medicine had an important role in colonial Africa. The few doctors who went south of the Sahara were typically missionaries employing medicine as a proselytizing vehicle. They deprecated indigenous medical practice, promoting a decline in herbal medicine and spiritual healing. At the same time, the impersonal, antispiritual character of Western medicine and the disdain of doctors for local traditions alienated much of the population.

The only part of Africa to develop leading, independent schools of research was the Union of South Africa. The Royal Observatory of the Cape of Good Hope, established in 1820 to study the southern sky, became a respected observatory under the Scotsman Sir David Gill (director, 1897–1907), a leading proponent for using photography to accurately measure stellar positions. South Africa became the favored site for southern hemisphere observatories owing to its dry highlands and political stability. Sir John Herschel, a celebrated English astronomer, travelled there in the 1830s and made extensive observations of nebulae in the southern skies. In the 1920s, three American universities, Yale, Michigan, and Harvard, each set up astronomical stations in the interior of South Africa to complement their observatories in the northern hemisphere. In the 1920s and 1930s, South Africa moved to the forefront of geological and paleontological research as well, largely through the work of Alexander du Toit, a Huguenot descendent. He was one of the greatest field geologists of the twentieth century and an early supporter of the theory of continental drift, for which he accumulated an impressive amount of evidence. During the same period, research on ancient ancestors of humans began to flourish in Africa with the discoveries of fossils of *Australopithecus africanus* by the South Africans Raymond Dart and Robert Broom. Another field in which South Africa has made a mark is botany. Of the botanical gardens that exist in several African countries, the Kirstenbosch gardens of South Africa (established in 1913) are the largest and most ambitious, with extensive programs devoted to saving endangered plant species.

Independent Africa

Immediately after independence from the colonial powers, which had occurred for most African countries by the mid-1960s, the new governments typically considered support of scientific research unnecessary. Building a national research capability was often deemed to consume too much time and money for young nations with limited resources; besides, the argument went, any research needed could be acquired from the former colonizing countries. On the other side of the coin, the most scientifically advanced country on the continent, South Africa, exploited anthropology and psychology to justify apartheid.

By the late 1960s, African governments began to recognize the need to generate their own scientific research and technological development. Through several conferences held with African governments, such as the International Conference of the Organization of Research and Training in Africa in Relation to the Study, Conservation and Utilization of National Resources (1964), the United Nations played a major role in changing local attitudes toward science and technology. UNESCO has reinforced the idea with dozens of missions to African countries promoting science and technology.

The hurdles are high. Cash-strapped government efforts often lead to routine, uncoordinated, understaffed, and unproductive investigations. African countries face serious shortages of scientists. Many of the brightest students study overseas and remain there. The African universities, having few resources, cannot supply the needed body of scientists. Overwhelmed by increasing enrollments and decreasing funds, the universities do little research. The bulk of research takes place in institutes headed by African scientists and financed by international organizations or governments. For example, the Kainji Lake Research Institute in Nigeria searches for new viral outbreaks, such as Lassa and Ebola fevers.

Africa has a choice between adopting Western science exclusively to solve its problems and bridging the gap between Western and indigenous, traditional approaches. In some countries, Western and traditional medicine are merging into systems to care better for the people. Many Africans today regard traditional medicine as a positive force that can complement the advantages of Western practices. This fresh outlook may help encourage a science in Africa for solving African problems and building the African future.

See also Asia; Diffusion.

Seyyed Hossein Nasr, *Science and Civilization in Islam* (1968). *The Cambridge History of Africa* (8 volumes, 1975–86). Thomas Owen Eisemon, *The Scientific Profession in the Third World: Studies from India and Kenya* (1982). John W. Forje, *Science and Technology in Africa* (1989). Thomas A. Bass, *Camping with the Prince and Other Tales of Science in Africa* (1991). Steven Feierman and John M. Janzen, eds., *The Social Basis of Health and Healing in Africa* (1992). Lewis Pyenson, *Civilizing Mission: Exact Sciences and French Overseas Expansion, 1830–1940* (1993). Helaine Selin, ed., *Encyclopaedia of the History of Science, Technology, and Medicine in Non-Western Cultures* (1997). Saul Dubow, ed., *Science and Society in Southern Africa* (2000).

ROBERT DALE HALL

AGRICULTURE. See AGRICULTURAL CHEMISTRY; BREEDING.

AI. See ARTIFICIAL INTELLIGENCE.

AIDS. Acquired immune deficiency syndrome (AIDS) was recognized in 1981 as the condition underlying a cluster of several different diseases, notably a type of pneumonia and certain cancers, afflicting certain demographic groups—in the United States, homosexual men in particular. A group headed by Luc Montagnier at the Pasteur Institute in Paris and another under Robert Gallo at the National Cancer Institute in Bethesda, Maryland, sought the cause of the syndrome. Both suspected that the disease agent was a retrovirus similar to the two viruses then known to produce leukemia in human beings. The two groups cooperated. In 1984 both isolated the culprit retrovirus from the tissues of AIDS victims, and characterized it using both serological and molecular techniques. It is much disputed whether Gallo's laboratory succeeded independently of Montagnier's or whether it isolated the retrovirus from material provided by Montagnier's and then failed appropriately to acknowledge the source.

Wholly different from the two leukemic retroviruses, the AIDS agent was soon dubbed the human immunodeficiency virus (HIV) because it attacks the immune system. Scientists in many laboratories learned that HIV targets the white blood cells called "T cells" (because they come from the thymus), particularly the major fraction of them that bristle at their surfaces with a molecule termed CD4 and that direct the body's immune response. In AIDS patients, the number of T cells with CD4 molecules is sharply reduced, steadily compromising the body's ability to mobilize against one or more wasting diseases.

HIV enters the body mainly by transfusion with infected blood, intravenous drug injection with infected needles, or unprotected sexual intercourse with an infected person. The Montagnier and Gallo groups both quickly developed and patented different blood tests that detect the presence of the virus from antigens produced against its several parts by the body's immune system. A dispute between the governments of the United States and France over which group properly had rights in them was resolved by a compromise in 1987 that divided the revenues. Blood-screening tests greatly diminished the risk of infection to people who depended on blood transfusions, notably hemophiliacs.

Public officials, especially in the United States, were initially skittish towards dealing with AIDS because of its strong link to homosexuality. However, coming to recognize that it posed a major threat to public health, they substantially increased public funding of AIDS research and urged sexual abstinence, monogamy, and the use of condoms to halt the spread of the disease. The search for a vaccine against AIDS has so far proved fruitless. However, during the 1990s several biotechnology firms developed pharmaceuticals that inhibit the reproduction of HIV. Daily cocktails of these drugs have turned HIV infection into a chronic rather than a deadly condition for those who can afford the treatment. Between 1993 and 1997, the incidence of AIDS in the United States fell by almost 50 percent, to about 22 per 100,000 people. But many live in regions where preventative measures are unknown or ignored, and the cocktails unaffordable. Across the globe at the end of 2001, an estimated 40 million people were infected with HIV, and in sub-Saharan Africa AIDS was epidemic.

Mrko D. Grmek, A *History of AIDS: Emergence and Origin of a Modern Pandemic*, trans. Russell C. Maulitz and Jacalyn Duffin (1990). Robert Gallo, *Virus Hunting: AIDS, Cancer, and the Human Retrovirus: A Story of Scientific Discovery* (1991). Luc Montagnier, *Virus: The Co-Discoverer of Hiv Tracks Its Rampage and Charts the Future*, trans. Stephen Sartarelli (1999).

DANIEL J. KEVLES

AIRCRAFT. There are two types of aircraft: lighter than air, encompassing airships and balloons; and heavier than air, encompassing ornithopters, machines that use flapping wings for lift and propulsion, and airplanes. In addition to airplanes with fixed lifting surfaces, the term encompasses autogiros and helicopters, which derive lift, and, in the case of helicopters, power, from rotating wings. The earliest attempts at human flight involved muscle-powered ornithopters and led nowhere. Toys and models aside, the first aircraft were balloons, the French brothers Joseph-Michel and Jacques-Etienne Montgolfier successfully flying a man-carrying hot-air balloon in 1783. Hydrogen balloons soon followed, but, aside from the knowledge of chemistry needed to produce hydrogen, they did not depend on academic science. Powered piloted airships first flew in the final decades of the nineteenth century, but they, too, were designed empirically.

With the exception of balloons used to investigate the upper atmosphere, airplanes have been the focal point of aeronautical science since airplane development supplanted ballistics as the main driving force behind scientific aerodynamics during the twentieth century. The origins of the airplane lay in the work of Sir George Cayley, who invented the first successful configuration—fixed lifting surfaces, separate, movable control surfaces, and an independent source of propulsion—and tested lifting surfaces with a whirling arm device. Experiments with man-carrying hang gliders by Otto Lilienthal showed that controlled flight was possible, though his method—shifting the pilot's weight—proved inadequate. En route to inventing the man-carrying, powered airplane, the Wright brothers, Wilbur and Orville, conducted extensive tests on airfoils (the shape of a wing's transverse section) using a wind tunnel of their own design. Their most important contribution, however, was a means of continuously controlling the movement of an aircraft about all three axes, roll, pitch, and yaw, inspired by their systematic studies of birds in flight. They had help from Octave Chanute, whose hang gliders pioneered the Pratt-truss rigged biplane, the dominant airplane configuration until the mid-1930s. Though the Wrights were not the first to use a reciprocating gasoline engine to power an aircraft, they made important contributions in this area and effectively invented the aircraft propeller.

The mathematical basis of scientific aerodynamics predated the airplane. Seminal contributions include the ability to predict the transition from laminar to turbulent flow in a fluid by Osborne Reynolds; the reduction to a simple equation of Daniel Bernoulli's discovery that air pressure exerted upon a surface was inversely proportional to the velocity squared by Leonhard *Euler; the discovery and mathematical analysis of the shock waves created by an aerodynamic body in supersonic flight by Ernst Mach; and the ability to predict and describe the boundary between laminar and turbulent flow over an aerodynamic surface by Ludwig Prandtl. Prandtl's study of the so called boundary layer led directly to equations describing and predicting aerodynamic lift and drag, the force resisting the movement of a body through the air, still used today.

Aircraft design remained largely empirical well into the interwar period when aeronautical engineers began increasingly to apply science to their craft, particularly in the areas of structural design. Results included the replacement of externally-braced biplanes by unbraced, or monocoque, monoplanes; substantial increases in power through supercharging and improvements in cooling; knock-resistant aviation gasoline; drag reduction, featuring the extensive use of wind tunnel data; mathematical tools to describe and predict stability and control; and new materials, notably aluminum as the dominant structural material. The next major developments were the jet engine (1939), piloted supersonic flight (1947), and airplanes capable of exceeding Mach 2 (twice the speed of sound) from the late 1950s.

By the 1970s, aircraft design had reached a plateau in structural design, speed, and size. Subsequent developments have centered on efficiency, both aerodynamic and propulsive; new materials, particularly to withstand the elevated temperatures encountered at high Mach numbers; and—particularly—electronics. The most important electronic developments are fly-by-wire flight controls that dispense with mechanical linkages, avionics that achieve unprecedented reliability by means of miniaturized transistor circuitry, and computer-assisted design (CAD) using analytical computers of unprecedented power. In combination with novel radar-absorbent materials, CAD has made possible the development of "stealth" aircraft with almost no radar return.

Edward W. Constant II, *The Origins of the Turbojet Revolution* (1980). John D. Anderson, Jr., *Introduction to Flight*, 4th ed. (2000). James E. Tomayko, *Computers Take Flight: A History of NASA's Pioneering Digital Fly-by-wire Project* (2000).

JOHN GUILMARTIN

AIR PUMP AND VACUUM PUMP. Air pumps exist in two variants: vacuum pumps to take gas from a vessel to create a (near) *vacuum, and compressors to supply gas to create a high pressure. Compressors have been important since the mid-nineteenth century for work on the thermodynamics of gases and for the cooling apparatus for low-temperature research. From about the same time, the vacuum pump became significant for research on *cathode rays. At some points, the production of sufficiently low pressures required state-of-the-art scientific knowledge, but in the modern period the device has served merely as a technical tool. In the late seventeenth and early eighteenth centuries, however, the air pump played a crucial role at a more fundamental level.

The idea of pumping air was not new when the Magdeburg Burgomaster Otto von Guericke adapted a fire syringe around 1647 to remove air from a vessel. Compressors had been known since

antiquity. What was new was the application of a pump to a philosophical question (the possibility of a macroscopic vacuum) as well as the idea of evacuating rather than compressing. Guericke's work received attention through the dramatic demonstrations he gave of the weight and force sustainable by the pressure of the air. The vacuum pump caught on as a philosophical instrument only after Robert *Boyle published his pneumatic experiments in the 1660s. Boyle recognized the vacuum pump as an ideal tool for the experimental method he wished to apply. Rather than answering one specific philosophical question, Boyle wanted to map the properties of the vacuum and of air. His work met with serious criticism from people who did not believe in the experimental method, but within the circle of Baconian scholars Boyle's example reached an emblematic status. Pumps were depicted in frontispieces to works in natural philosophy, and savants had themselves portrayed with the instrument in order to express their commitment to the new science.

The frontispiece depicted here shows a philosopher between the tools he employs in his study of nature. A large air pump stands at his feet. This instrument, dominated by the inclined cylinder containing a piston, sucks air in from under the bell jar on the left during the outward stroke of the piston and expels it during the inward stroke. A tap between the cylinder and the vacuum prevents the air from flowing back. This specific model fit the emblematic role of the air pump perfectly because it focused attention on the pumping machinery. The earliest air pumps, as well as those of the later eighteenth century, had the vacuum chamber as the central element. With these pumps, the emphasis changed to the experiments taking place in this chamber.

Early air pumps leaked. This fault proved significant both in using pumps and securing their acceptance as a reliable tool of philosophy. Correcting it was essential, but far from easy—hence the small number of air pumps initially made. Before the instrument became available commercially in the 1670s, some fifteen savants or institutions managed to acquire one. These early pumps were extremely expensive. Even in the eighteenth century, the air pump remained an expensive piece of philosophical equipment.

During the eighteenth century, the technical improvement of the air pump became a problem in engineering without much philosophical importance. Craftsmen introduced mechanically operated valves and diminished the dead space at the bottom of the cylinder. Makers made extravagant claims for the degree of exhaustion their pumps could reach, but in fact the vapor pressure of the lubricant barred access to pressure below 10 millibars (0.01 atmospheres). Only when the lubricants were improved or eliminated could further progress be made. In 1858, the German glassblower J. H. W. Geissler introduced a vacuum pump that employed a piston of liquid mercury. Variants of the pump reached pressures sufficiently low to allow the discovery of *X rays in 1895. In the early twentieth century, a series of entirely new pump techniques, developed by Wolfgang Gaede, relied on the molecular behavior of the gas. Thanks to these techniques, pressures of 10^{-9} bars became available for many fields of experimental research. By then, the air pump had an important place in the background technology of science.

See also BACONIANISM; BAROMETER; ETHER; VACUUM.

Gerard L'E. Turner, *Nineteenth-Century Scientific Instruments* (1983). Steven Shapin and Simon Schaffer, *Leviathan and the Air-Pump, Hobbes, Boyle and the Experimental Life* (1985). Anne C. van Helden, "The Age of the Air-Pump," *Tractrix, Yearbook for the History of Science, Medicine, Technology and Mathematics* 3 (1991): 149–172. Willem D. Hackmann, *Museo di Storia della Scienza, Catalogue of Pneumatical, Magnetical and Electrical Instruments* (1995).

ANNE C. VAN HELDEN

ALCHEMY. Scholars can make solid arguments for the rise of alchemical ideas in ancient China or Persia, but most identify its birthplace in the Hellenistic cultural period, probably in Egypt under the Ptolemaic Kingdom and the early Roman Empire. The sort of alchemy practiced there was born from a mix of Aristotelian theories of matter; Hellenistic religious philosophies such as gnosticism, hermeticism, and the mystery cults; and various ancient metallurgical and chemical techniques. The first practitioners probably were a secretive group or sect, perhaps Hellenized Egyptians in Alexandria.

The earliest definitely known alchemical writer, Zosimus, lived in Alexandria about 300 A.D. and wrote a handbook or encyclopedia of his craft. The allegorical and allusive character of the writing has made its interpretation problematic. Zosimus's book forms a part of the Greek alchemical corpus, a collection of mostly anonymous or pseudonymous writings first assembled a hundred years or more after Zosimus, and known today from a handful of medieval Greek copies. Zosimus and other Greek alchemists aimed to produce valuable metals from less expensive materials, prepare pharmaceutical remedies, and undergo philosophical or religious purification. Important materials in the craft included sulfur, mercury, arsenic, and electrum (a gold-silver alloy).

Leading Islamic scholars, who derived the craft from the Greeks, practiced it, especially in the tenth and eleventh centuries. The number of modern chemical terms that derive from Arabic—alchemy, alkali, alcohol, elixir, and naphtha, for example—indicate the importance of this period to science. The most famous Arabic alchemist, Jabir ibn Mayyan, supposed to have lived in the early ninth century, seems in fact to have been a mystical Islamic sect that wrote pseudonymously a hundred years after Jabir's death. The celebrated philosophers al-Razi (Rhazes) and Avicenna (Ibn Sīnā) both pursued alchemy. A chief idea of medieval Arabic alchemy held that metals were formed of sulfur and mercury in various proportions, and that altering the proportions in a given metal could change its type, even from lead to gold. Not every alchemist, however, believed in the possibility of such transmutations.

During the twelfth century, scholars in the Christian West translated the ancient Greek philosophical and scientific corpus from Arabic into Latin; among these were works on alchemy elaborated by the Muslims. Albertus Magnus and Roger Bacon occupied themselves, among much else, with alchemical pursuits. A slightly later figure (fl. ca. 1300), perhaps the most important Western alchemist of the Middle Ages, called himself Geber. (Until about a century ago, scholars regarded "Geber" as the Latin form of "Jabir," and Geber's books as Latin translations of the *Suppositions Jabir*; they now regard Geber as a pseudonym of an unidentified Latin scholar.)

In Geber we first see unequivocal evidence for the use of concentrated mineral acids, with proof of distilled alcohol coming around this same period. These new substances, made possible by technological innovations, especially the more efficient cooling of distillates, dramatically altered the operational repertoire of chemists and alchemists. Three fourteenth-century Catalonians—Arnald of Villanova, Ramon Lull, and John of Rupescissa—incorporated these innovations into alchemy. Among other changes, the position of mercury in alchemical manipulations became more central. Despite this activity, interest in alchemy declined, only to reappear with renewed vigor in the Renaissance.

Alchemical ideas grew out of the cultures of which they were a part. Consequently, Greek, Arabic, late medieval, Renaissance, and early modern alchemical traditions each had distinctive characteristics. The craft also diffused within each cultural tradition. Some alchemists devoted themselves to practical tasks: pharmaceutical preparations; techniques of smelting, assaying, and metalworking; or manufacture of dyes and other chemical substances of commercial importance. Other alchemists sought to produce gold from base metals, or the "elixir of life" that would cure any disease. Still others regarded their discipline as an ethical or religious doctrine. Just as one could cure a metal of its imperfections to produce eternally incorruptible gold, or cure a human body of its imperfections to produce eternal life, so also, with the right discipline and approach, could one cure one's soul of its flaws and achieve salvation in heaven. In Renaissance Europe, the hermetic arts experienced a resurgence along with the humanist movement. Alchemy was regarded as one of these secret sciences, and many practitioners were powerfully influenced by Christian mysticism.

A leading early sixteenth-century alchemist was Theophrastus Bombast von Hohenheim, known as Paracelsus. Paracelsus's philosophy mixed gnosticism, cabalism, astrology, magic, and heterodox Christianity, and he gained significant fame (or notoriety) as a peripatetic iconoclastic physician. He championed the role of alchemically prepared mineral remedies, and certain of his treatments—mercury for the new scourge of syphilis, for instance—were doubtless efficacious. Paracelsus's emphasis on chemicals in pharmacy and medicine, called iatrochemistry or chemiatry, influenced later figures. Lively controversies over his approach still raged around 1600.

Other sixteenth- and seventeenth-century figures expressed doubts about the possibility of alchemical transmutations and cures, and gradually the hermetic influence declined in Europe. Two of the chief heroes of the *Scientific Revolution, Robert *Boyle and Isaac *Newton, were active in alchemical pursuits, which provides a measure of the staying power of these ideas. Through most of the seventeenth century, however, no one distinguished between scientific "chemistry" and "alchemy" (eventually viewed as pseudoscience); both words existed, but each indicated the same set of diverse activities, practical and mystical. Only during the 1680s did the popular chemical textbook of Niccolas Lewey open the first of a series of attacks on alchemy as fraudulent. Early Enlightenment thinkers then banished "alchemy" to the discredited category of occult doctrines (*see* SYMPATHY AND OCCULT QUALITY) while restricting the word "chemistry" to the science we know today.

Arthur John Hopkins, *Alchemy, Child of Greek Philosophy* (1934). E. John Holmyard, *Alchemy* (1957). Robert P. Multhauf, *The Origins of Chemistry* (1966). William Newman, *Gehennical Fire: The Lives of George Starkey* (1994). Pamela H. Smith, *The Business of Alchemy* (1994). Lawrence M. Principe, *The Aspiring Adept: Robert Boyle and His Alchemical Quest* (1998).

A. J. ROCKE

ALIEN LIFE. See EXTRATERRESTRIAL LIFE.

ALKALOIDS. The chemical category and name alkaloid were introduced in the 1810s by pharmacists and chemists who isolated and recognized the first organic bases capable of forming salts with acids, including morphine (1817), strychnine (1818), and quinine (1820), from plants. Thereafter the numbers of known alkaloids increased steadily. The approximate number stood at 30 in the 1830s, 80 in the 1870s, 200 in 1940, and over 12,000 by the end of the twentieth century. Long before they were isolated and subjected to chemical and physiological study, several of the alkaloids played significant roles in human societies in the form of plant extracts, as poisons or medicines. Since the 1810s alkaloids have taken a prominent part in pharmacological and physiological investigations, and some have important places in medicine. Studies of the biological sources and functions of alkaloids continue in a variety of disciplines.

In 1817 French chemist Joseph Louis Gay-Lussac and pharmacist Pierre-Jean Robiquet recognized the need for a new chemical category, that of alkaline substances of organic origin. The identification of morphine as a salifiable base in the same year by German pharmacist Friedrich Wilhelm Sertürner provoked this recognition. Gay-Lussac's and Robiquet's insight quickly led to a general research program for the isolation and characterization of new organic bases. This program was pursued especially strongly in France, where the practice of proximate organic analysis had developed since the eighteenth century in association with the professional ambitions of pharmacists and the Collège and later École de Pharmacie. Paris pharmacists Pierre-Joseph Pelletier and Joseph-Bienaimé Caventou soon isolated several more members of the new class of compounds, including strychnine and quinine. The term "alkaloid" was proposed in 1818 by German pharmacist K. F. W. Meissner.

Chemical isolation of alkaloids benefited from, and helped to provoke, early experimental studies of drug action such as those carried out by Paris physician and physiologist François Magendie from the 1800s to the 1840s. Magendie's advocacy and work, which included experimental studies of the action of strychnine, emetine, quinine, and other alkaloids, gave impetus to the dual goals of rationalization of drug therapy and the use of drugs to analyze human and animal function. Magendie's student Claude Bernard carried the latter program forward most notably in his analysis of the action of the alkaloid-containing plant poison curare.

Since Bernard's time alkaloids have figured in numerous physiological and pharmacological investigations. In the late nineteenth century Oswald Schmiedeberg demonstrated the parallelism of vagal nerve stimulation and the action of muscarine on the heart, and John Newport Langley used nicotine and pilocarpine to map and study the functions of the autonomic nervous system. Early in the twentieth century Henry Hallett Dale demonstrated the antagonism of the ergot alkaloid ergotoxine and adrenaline and compared the action of acetylcholine to that of muscarine and nicotine, while Otto Loewi used the alkaloids physostigmine and ergotamine to suggest the identity of acetylcholine and the humoral neurotransmitter substance. Alkaloids have also served as models for the chemical synthesis of analogs, related molecules with improved medicinal qualities.

B. Holmstedt and G. Liljestrand, eds., *Readings in Pharmacology* (1963). John E. Lesch, "Conceptual Change in an Empirical Science: The Discovery of the First Alkaloids," *Historical Studies in the Physical Sciences* 11, no. 2 (1981): 305–328.

JOHN E. LESCH

ANALOG/DIGITAL DEVICES. Analog devices model the world by using one continuous variable to represent another. Familiar examples include dials that measure a quantity such as speed, pressure, or current by the angle through which a needle turns; *thermometers and barometers, in which the length of a column of mercury corresponds to temperature or pressure; and telephones and recording equipment that convert sound into the vibration of a diaphragm and hence into an electric signal. Digital devices record quantities in discrete units: thus digital watches show time as a row of numbers that change abruptly from one to the other. The best known simple digital calculating device is the abacus, while all modern computers as well as many other electronic instruments operate digitally.

Digital devices can tackle a wide range of problems, and any desired degree of accuracy can be acquired by increasing the number of digits used. Analog devices are limited in their range of applications and cannot be made more accurate than their components, but respond immediately and can model entire systems. As microelectronic equipment has become faster, smaller, and cheaper, digital computers have eclipsed analog ones. Although digital calculating machines were invented in the seventeenth century, only in the mid-nineteenth century did developments in mathematics and mechanical equipment enable them to perform more quickly and accurately than analog ones.

Ancient civilizations invented several analog devices to tell the time, such as sundials, water clocks (clepsydras), and sand glasses, still in use as egg timers. More complex instruments relied on astronomical observations. *Astrolabes, which carry a map of the heavens and consist of complex rotatable flat plates, were in widespread use for making the accurate calculations essential for navigating, casting horoscopes, telling time, and finding the number of daylight hours. Still more sophisticated was the Greek Antikythera mechanism, made in the first century B.C.E. and discovered in 1901. Encased in a box that opened up like a book, its metal gears and pointers predicted the motion of the stars and planets.

During the early modern period, many traditional devices used for navigation, surveying, and other practical purposes developed into accurate analog instruments capable of performing a variety of calculations. For instance, gunners used sectors for the rapid solution of complicated arithmetical problems about cannon elevations and ammunition requirements. Consisting of two arms hinged together and engraved with several scales, sectors relied on dividers to measure distances between pairs of different scales, enabling volumes, circumferences, and so on to be read instantly. Sectors remained important, versatile calculators well into the nineteenth century, even after slide rules—based on John Napier's invention of logarithms—were introduced in the 1630s. Slide rules executed lengthy arithmetical and trigonometrical calculations by moving scales alongside one another. Because of their speed, accuracy, and portability, slide rules remained in use until the 1970s.

Maritime communities need good knowledge of the tides, and approximate analog tide predictors survive from the fifteenth century. The behavior of the tides is described mathematically by differential equations, solved from 1814 with the help of analog planimeters and linear integrators. In 1873, William *Thomson built a machine that could forecast tidal changes many years into the future. Based on a wheel-and-disc integrator invented by his brother, Thomson's tide predictor mechanically combined simple waves into a more complex pattern traced out by the pen of a chart recorder.

Thomson also outlined a general-purpose differential analyzer, but although more integrating devices continued to be developed, mechanical limitations hindered their adoption into automatic calculating machinery. In 1931, benefiting from technological improvements, Vannevar Bush constructed his first differential analyzer at Massachusetts Institute of Technology (MIT). Bush's analyzer was constructed from three basic units: integrators, gears for constant multiplication, and differential gears for addition and subtraction. Frustrated by the slowness and inaccuracy of mechanical components, he replaced some of them in later models with thermionic valves to store information. These machines were the first that could be set up to solve a range of problems involving differential equations, including modelling the sway of bridges in windy conditions and calculating the scattering of electrons by atoms. Nevertheless, they were huge, slow, and laborious to reset. Of the several copies made, the small meccano version at Manchester University is perhaps the most famous. During World War II, the military used Bush's analyzers extensively to compile ballistic firing tables and to calculate possible trajectories of V2 rockets. Bush also developed a hybrid analog and digital Comparator to crack enemy codes by rapidly comparing microfilms. As with the digital Colossus in England, further military applications will be revealed when secret papers become declassified.

After the war, electronic circuitry replaced mechanical components. Research was also underway into digital computers, which rely on electrical and magnetic devices that can be in either one of two possible states. In contrast, most electronic analog computers manipulate continuous voltage signals via an operational amplifier whose output current is proportional to its input potential difference. Digital computers did not immediately supplant analog computers. In 1954, analog computers still performed over half the weekly computer hours at MIT. In the 1990s, new analog electronic circuits were invented to imitate complex biological organisms such as the eye. The growth of digital computers mushroomed in the 1970s, but analog and hybrid computers (which incorporate both digital and analog components) remained important for dedicated real-time control and simulation applications, such as nuclear power plants, industrial chemical processes, and aircraft simulators.

See also CALCULATOR; COMPUTER.

Herman H. Goldstine, *The Computer from Pascal to von Neumann* (1972). Stan Augarten, *Bit by Bit: An Illustrated History of Computers* (1984). Michael R. Williams, *A History of Computing Technology* (1985). Anthony Turner, *Early Scientific Instruments: Europe 1400–1800* (1987). G. Pascal Zachary, *Endless Frontier: Vannevar Bush, Engineer of the American Century* (1997).

PATRICIA FARA

ANALYSIS AND SYNTHESIS. At the most elementary level, analysis concerns the separation of a whole into its component parts, whereas synthesis is the reverse process of combining parts to form a complex whole. In the history

of modern science, however, these methods have been formulated in many different ways. In mathematics, for example, synthesis referred to a method of proof described by René *Descartes in 1641 as that used by geometers to demonstrate a conclusion by a long series of definitions, postulates, and axioms. Mathematical analysis, which is concerned with the theory of functions, became important later in the century with the introduction of the calculus by Isaac *Newton and Gottfried Leibniz.

Analysis and synthesis have also been used to establish physical theory. In his *Discourse on Method* (1637), Descartes described how he had used conceptual analysis to arrive at theories concerning the existence of, and radical difference between, material and immaterial beings. Based upon his analysis of the essence of matter as extended substance, he then followed a synthetic method to explain the processes operative in the physical world in terms of the laws that governed the motion of the least parts of undifferentiated matter and gave rise to the various manifest qualities of macroscopic bodies. Descartes' metaphysical dualism set the stage for the development of sciences based solely on the physical properties of bodies, and his method of conceptual analysis and synthesis would be influential especially in areas of science where direct access to the subject of an investigation is impossible. In 1781, Immanuel Kant gave a linguistic formulation to the analytic/synthetic distinction. He defined analytic statements as ones known through reasoning alone, and synthetic statements as ones that refer to the physical world of sense experience. Although scientists did not make much use of Kant's distinction, twentieth-century philosophers who sought to understand the logical structure and empirical content of physical theory employed it extensively.

Analysis and synthesis are also used for practices employed in empirical inquiries, as in the chemical analysis of bodies and the chemical synthesis of new compounds. In 1620, Francis Bacon supported the idea of explaining physical processes in terms of the motion of the least parts of matter, but unlike Descartes, he insisted that natural philosophers had to dissect nature to find the real particles. In the next generation, Robert *Boyle followed Bacon's advice. His chemical experiments proved that all bodies could be analyzed neither into the Aristotelian four elements (earth, water, air, fire) nor into the Paracelsian three elements (salt, sulfur, mercury). Based upon these results, as well as those from his synthetic trials with nitrates, Boyle constructed experimental proofs for the new mechanical philosophy. Isaac

Newton too, in addition to his mathematical contributions, performed numerous experiments that depended upon analysis and synthesis, such as the manipulation of light rays by a series of prisms, described in his *Opticks* (1704).

Rom Harré, *Great Scientific Experiments* (1981). John Losee, *A Historical Introduction to the Philosophy of Science,* 3d ed. (1993).

ROSE-MARY SARGENT

ANALYTICAL CHEMISTRY. If practical chemistry is nearly as old as civilization, analytical chemistry is not much younger, for merchants and consumers have always sought means to be assured of the identity and purity of commodities in trade. Precious metals afford an obvious instance; ancient kings had every incentive to seek techniques by which they could be obtained, assayed, and regulated. In one such method, cupellation, a nodule of impure gold or silver held in a porous cup (cupel) was heated strongly in air; the cupel absorbed impurities and a very pure sample emerged from the fire. Comparing the sample's weight before and after the firing provided a precise measure of its purity. In the method of the touchstone, the color of a streak left on the stone by an unknown sample was compared with streaks left by test needles of varying known compositions; color comparisons could easily distinguish between, say, 16-karat and 20-karat gold. A third method, invented by Archimedes in the third century B.C., utilized specific gravity to assess purity.

Medieval alchemists were skilled analytical chemists, especially in assaying, *fire analysis, and *distillation (*see* ALCHEMY). In the sixteenth century, Georgius Agricola and Vannoccio Biringuccio wrote about the numerous sophisticated chemical methods of their day. In the following century, Robert *Boyle and Otto Tachenius established new essential features of qualitative analysis, including flame and spot tests, solvent actions, and color *indicators. Theories of *acids, bases, and salts provided a foundation for analytical investigations in the late seventeenth and eighteenth centuries.

Analytical chemistry underwent a fundamental change during the later eighteenth century. In the 1750s, the Scottish chemist Joseph Black carried out the first investigation in which weights of combining materials were carefully followed throughout a series of reactions. A generation later, the Frenchman Antoine-Laurent *Lavoisier used gravimetric techniques to establish a new oxygen-centered chemical theory based on presumed simple substances that became known as "elements." An essential feature of the chemical revolution was the new understanding that gases participate like solids and liquids in chemical reactions. The primary task of analysts thereafter was to determine both the proximate and the ultimate elemental composition of compounds and mixtures.

The modes of analysis possible at any given time depend on the available tools. Consequently, the history of analytical chemistry has been closely tied to the development of instruments and apparatus. The rise of blowpipe analysis towards the end of the eighteenth century, the invention of the *battery (1800), and the development of the spectroscope (1850), transformed inorganic analysis, and each of them prompted the discovery of several new elements. Martin Heinrich Klaproth, Jöns Jacob *Berzelius, Heinrich Rose, and Carl Remigius Fresenius gradually systematized analytical procedures, and in 1862 Fresenius founded a journal for analytical chemistry. By this time the balance, known in some form since antiquity, had been refined into a precision instrument found in every laboratory.

Volumetric methods of analysis appeared along with qualitative and gravimetric methods, particularly in France. In the 1780s and 1790s, Henri Descroizilles invented titration devices (burettes and pipettes). Early in the next century, J. L. Gay-Lussac and Jean-Baptiste-André Dumas developed additional volumetric methods with extraordinary ingenuity and skill. Somewhat later, in the 1840s and 1850s, the German chemist Robert Bunsen worked out elaborate methods of highly accurate gas analysis.

In the meantime, analytical methods for organic compounds had been invented. In 1831, German chemist Justus von *Liebig transformed what had been essentially a volumetric procedure into a gravimetric one. He captured the combustion products of a sample—gaseous carbon dioxide and water vapor—in solid form in lye-filled glass bulbs and in a calcium chloride tube, respectively. The increase in the weight of these devices provided a direct measure of the weight of the trapped combustion products, hence indirectly of the carbon and hydrogen content of the sample. A modification of earlier methods devised by Berzelius and Gay-Lussac, Liebig's quicker and simpler procedure for elemental organic analysis transformed the practice of organic chemistry. The field began a period of explosive growth, especially in Germany.

The classical era of chemical analysis culminated in the extraordinary accomplishments of two American chemists, Edward Morley and Theodore W. Richards. Working independently, each arrived at elemental atomic weights that became the definitive values around the world. They worked at the end of the nineteenth century, shortly before a great reformation in the practice of analytical chemistry. Instead of solutions, crystallizations, filtrations, titrations, test-tube reactions, and combustions—"wet" chemistry—the new approach would be typified by complex physical instruments whose results could be read from a detector, screen, or dial.

Instrumental analysis was by no means unknown in the nineteenth century. The *polarimeter, which measures the rotation of the plane of polarized light as it passes through a solution of an optically active substance, came in 1840, a generation before the cause of optical activity was understood. The spectroscope identified a new element in the sun, helium, before it had been detected on the earth. *Chromatography, which separates mixtures (originally of plant pigments) by selective adsorption, has roots deep in the nineteenth century. Shortly after 1900, Mikhail S. Tsvet transformed this curiosity into a valuable analytical method. The various substrates on which the components of the mixture were adsorbed dictated the species of chromatography: paper, thin-layer, ion-exchange, gas-solid and gas-liquid chromatography, and others. By the middle of the twentieth century, these methods had been fully instrumentalized, and in some cases automated, and were found in every chemical laboratory.

In 1935, Arnold Beckman began selling the first production pH meters. Colorimetry, aided by elaborate optics and electronics, rapidly evolved into ultraviolet, visible, and infrared spectrophotometry. A schematically simple device of J. J. Thomson was gradually transformed, first into a mass spectrograph, then into a production-model mass spectrometer. These innovations were commercialized mainly during the 1940s. Then, in the 1950s, nuclear magnetic resonance spectrometers began to reach maturity. By the 1960s, the sort of wet-chemical analysis characteristic of the previous generation had become rare, and the test tube an icon of ancient practices.

J. R. Partington, *A History of Chemistry*, vols. 3 and 4 (1962–1964). A. J. Ihde, *Development of Modern Chemistry* (1964). Ferenc Szabadváry, *History of Analytical Chemistry* (1966). W. H. Brock, *The Norton [Fontana] History of Chemistry* (1992). R. Bud and D. J. Warner, eds., *Instruments of Science: An Historical Encyclopedia* (1998).

ALAN J. ROCKE

ANATOMICAL THEATERS. Since antiquity *dissection has been performed on animals in two radically different ways: as a private practice of enquiry and as a public spectacle. In his *Anatomical Procedures*, written in the last quarter of the second century A.D. Galen distinguished between private anatomical explorations,

intended for rehearsal and investigation, and public dissections and vivisections, held before a large audience in the presence of religious, political, and intellectual authorities, and in a space that Galen himself consistently defined as a public one. The anatomical "exhibitions" publicized before prominent witnesses the truthfulness of Galen's own anatomical, physiological, and philosophical learning and distinguished his medical approach from that of other practitioners active in the medical marketplace (*see* GALENISM). These features of public anatomies, with their intentions of persuasion, propaganda, and professional distinction, became even more evident in the late Middle Ages and the Renaissance. Human dissection and public anatomy lessons were meticulously regulated by university statutes, framed in a theatrical space, and shaped as a theatrical event. These norms disciplined and structured a didactic experience; ritualized and hence domesticated a practice that might have been perceived as sacrilegious and inhumane; and solemnly celebrated the prestige and learning of the university and the medical establishment.

The first description of an anatomical theatre occurs in Alessandro Benedetti's *Historia corporis humani* (1502). Benedetti recommended that a temporary wooden theater with seats arranged in a circle be built in a large and well-ventilated space. Spectators should be seated by rank, a prefect appointed to manage and monitor the procedures, guards stationed to block access to undesirables, and two reliable stewards deputed to collect and disburse the necessary funds. Such wooden temporary theaters were erected in Padua and in Bologna for anatomy lessons. One of these theatres is represented in the title page of Andrea Vesalius' *De humani corporis fabrica* (Basle 1543). A crowd of 70 or 80 people is attending the public lesson, and some peer from behind the columns and from the windows above the theater. The public—in the woodcut as well as according to university regulations—consists of students and professors of the Faculty of Medicine and Philosophy, as well as by physicians, surgeons, and barbers. Other sources testify to the presence of clerics, artists, gentlemen, and even common people who came to attend the anatomical spectacles.

The engagement of Hyeronimus Fabricius ab Aquapendente—professor of anatomy and an influential nobleman—led to the erection of the first permanent anatomical theater in Padua in 1594. The theater, which still exists, could hold more than two hundred spectators. It had no seats: the standing audience leaned over the balustrade to see the dissecting table and the cadaver lighted only by two chandeliers and a few

Founded in the sixteenth century and later rebuilt, the beautiful anatomical theater of the University of Padua was the site of dissections of the human body both for the instruction of medical students and the edification of the aristocracy.

candles held by the students. The theater of Padua has hosted dissections for almost three centuries, attended and/or performed by such anatomists and physicians as Peter Pauw, William Harvey, Thomas Bartholin, Giovanni Battista Morgagni, and Antonio Scarpa.

When Pauw left Padua to become professor of anatomy in Leyden, he urged the construction of an anatomical theater in the recently founded Leyden University. The theater built in 1597 doubled as a *Wunderkammer* filled with human and animal skeletons. Likewise Bartholin, after his medical education in Padua, encouraged the erection of the anatomical theater in Copenhagen completed in 1643.

The spread of anatomical theaters across Europe during the seventeenth century reflected and assisted the incorporation of anatomy in medical education and culture. Anatomical theaters were built in Bologna (1649), Altdorf (1650), Uppsala (1662), and Prague (1688). London had at least two—one, probably designed by the architect Inigo Jones, in the Barber-Surgeons' Hall (1638), and another at the Royal College of Physicians. Similarly in Paris, a temporary anatomical theater used since 1604 gave way to a permanent structure at the Medical Faculty in 1620. Louis XIV provided a competing theater

for the surgeons. New anatomical theaters were inaugurated at the Paris medical school in 1744 and included in a building for the Royal Academy of Surgeons commissioned by Louis XVI in 1769.

See also ANATOMY.

Ruth Richardson, *Death, Dissection and the Destitute* (1987; 2000). Heinrich Von Staden, "Anatomy as Rhetoric: Galen on Dissection and Persuasion," *Journal of the History of Medicine and Allied Sciences* 50 (1995): 47–66. Roger French, *Dissection and Vivisection in the European Renaissance* (1999).

ANDREA CARLINO

ANATOMY. In modern parlance, anatomy is synonymous with the dissection of a human or animal body. In history, however, the term referred more broadly to a metaphor and a model, a method and a practice. In the early modern period, "anatomy" implied an analysis in the sense of taking something apart—not necessarily a body—to its fundamental components. In the early seventeenth century Francis Bacon advertised "an anatomy of the world," and in 1621 Robert Burton published his *Anatomy of Melancholy.*

Through the eighteenth century, anatomy referred both to the act of dissection and to its subject, the human or animal body. In this era, the act referred to both living and dead bodies, although later anatomy came to denote dissection of the dead. In addition, anatomical models, such as wax ones used for teaching purposes, as well as dried and preserved specimens, came under the rubric. Many anatomy cabinets existed to teach students basic structures before they witnessed an anatomy lesson and to reinforce their knowledge afterwards.

Modern anatomy began with the work of Andreas *Vesalius and his contemporaries in the sixteenth century. By means of human and animal dissection and animal vivisection, they explored and analyzed the structure of the human body. At almost the same time, however, some anatomists began to find the animal body of interest in itself, and comparative anatomy developed alongside human anatomy.

Anatomy and *physiology were not distinct disciplines until the nineteenth century, and early modern anatomists explored both. William *Harvey, for example, explained both the form and the function of the human heart and also compared the hearts of different species. Researchers in the seventeenth and eighteenth centuries dissected and vivisected many animals both to explore morphology and, especially, to answer questions about function. Because of the variety of animals they used, much of the work of men such as Marcello Malpighi was broadly comparative. A group of some of the first members of the Paris Academy of Sciences, including Claude Perrault and Joseph-Guichard Duverney, used the resources of the Jardin des Plantes and the menagerie at Versailles to compile their important *Mémoires pour servir à l'histoire naturelle des animaux* (1676), which notes morphological similarities but without an underlying theory of type.

Public anatomy, in which a human cadaver (and usually several live and dead animals) was dissected before an audience, flourished in the seventeenth and eighteenth centuries. While such exhibitions taught medical students about the form of the human body, they also attracted members of the general public who found the spectacle entertaining and the contemplation of mortality edifying (*see* ANATOMICAL THEATER). The decline of public anatomy at the end of the eighteenth century revealed changing sensibilities as well as a recognition that such dissection could no longer reveal anything new.

Municipal authorities usually allowed one or two cadavers of executed criminals per year for public dissection, although the number varied according to local custom. Those who gave multiple courses, or wished to do research,

usually resorted to grave robbing. Vesalius bragged about his skill pilfering corpses. Two centuries later Duverney bribed the grave-digger at the Hotel-Dieu in Paris to provide him with bodies. In England, the so-called Murder Act of 1752 gave the College of Surgeons first right to the bodies of executed criminals, but other researchers and lecturers continued to rely on "resurrection men." In Edinburgh in the late 1820s, an entrepreneurial pair named Burke and Hare murdered indigent people and sold their bodies to the medical school; this celebrated case helped lead to the passage of the Anatomy Act (1831), which specified that the bodies of the poor who died in workhouses or hospitals be made available for research and teaching.

Over the course of the eighteenth century, Giovanni Battista Morgagni's attention to autopsy indicated the value of revealing pathological conditions. At the end of the century, Marie-François-Xavier Bichat developed his concept of "general anatomy," which he defined as the study of the constituent tissues of the human body in health and disease. Bichat's tissue doctrine turned attention from the clinical description of pathological phenomena to physiological specificity and physical localization. Following Bichat, Karl von Rokitansky in Vienna brought pathological anatomy to its peak and made it an essential part of medical education in the nineteenth century. The *cell theory and improvements in the *microscope at midcentury led to Rudolf Virchow's emphasis on the cell as the fundamental unit of anatomy and the ultimate seat of disease, a concept he called his "anatomical idea."

The interest in systematics exemplified by Carl *Linnaeus in the eighteenth century also led to a new interest in comparative anatomy as morphological similarities came to be viewed as classificatory signposts. At the Jardin des Plantes, Georges-Louis LeClerc, Comte du Buffon and the anatomist Louis-Jean-Marie Daubenton collaborated on the massive *Histoire naturelle des animaux* (1749–1788), which developed an idea of species based on morphological similarities. In the early nineteenth century, Georges Cuvier continued Daubenton's work on comparative anatomy and combined it with the study of fossils and with Bichat's emphasis on the functionality of the parts of the body to develop his classification of animals. His skills in comparative anatomy and knowledge of the internal structures of animals led to a new classificatory scheme that shattered the old notion of a hierarchical chain of being.

Cuvier's *Règne animal* (1817) identified four principal types of animal form based on structural similarities and functional characteristics: the vertebrates, mollusks, articulates, and radiates. Cuvier's concept of form greatly influenced nineteenth-century work in *paleontology and comparative anatomy as well as the theory of *evolution. Cuvier's detailed work on fossils confirmed that fossil animals were now extinct, which, to him, did not imply a notion of transformism (see EXTINCTION).

His near contemporary, Étienne Geoffroy Saint-Hilaire, recognized homologous structures in organisms that implied the existence of a generalized type and so opened the door to a transformist, or evolutionary, theory. In the 1830s and 1840s, Richard Owen, who had worked exclusively in comparative anatomy, described an *archetype of the simplest vertebrate form. Owen accepted Karl Ernst von Baer's view that embryological development followed a branching, rather than a linear, plan and showed that fossils also show a radiating pattern toward individual specialization rather than a linear progression toward the human form. Charles *Darwin did little anatomical work himself. He relied on the investigations of Owen and others on homologies in his theory of evolution by natural selection, published in *The Origin of Species* (1859).

The skeleton, whether animal or human, fossil or new, was at the center of anatomical study. The skeleton appeared at both the beginning and the end of anatomical demonstrations, and took pride of place in anatomy collections. The human skeleton symbolized death and the end of all flesh, and the irreducible bedrock of organic form. In the eighteenth century, some illustrations of skeletons were used to emphasize, and even exaggerate, gender differences. In the nineteenth and early twentieth centuries, skeletons, and especially skulls, were used to adumbrate specious racial theories. The French anthropologist Pierre Paul Broca measured thousands of skulls to confirm his view that skull size (and therefore brain size) correlated with intelligence and with *race, gender, and criminality (see GENDER AND SCIENCE).

By the twenty-first century, anatomical determinism had long been discredited. Dissection, though still important in teaching medicine, had lost its value as a method of discovery. Anatomical atlases display the body in detail and virtual reality computer programs provide three-dimensional views of everything.

Knowledge of the structure of the living body is now gained in many different ways. *X rays, discovered at the end of the nineteenth century, were supplemented in the twentieth by ultrasound, the CT scan, magnetic resonance imaging, and the PET scan. Ultrasound employs high-frequency sound waves that form an image on a monitor. The CT (*computerized tomography) scan can be one hundred times more sensitive than conventional X rays. It allows a part of the body to be visualized in "slices" in two or three dimensions. Magnetic resonance imagery (MRI) allows visualization of soft tissue by means of magnetic fields and radio waves. Positron emission tomography (PET) shows the entire body at once. A small amount of radioactive glucose is injected into the bloodstream, after which the patient passes through a scanner like that used in CT scans. These new technologies have provided a cornucopia of information about the structure of the human body.

E. S. Russell, *Form and Function* (1916). F. J. Cole, *A History of Comparative Anatomy from Aristotle to the Eighteenth Century* (1949). K. E. Rothschuh, *History of Physiology*, trans. G. B. Risse (1973). Stephen Jay Gould, *The Mismeasure of Man* (1981). L. Schiebinger, "Skeletons in the Closet: the First Illustrations of the Female Skeleton in Eighteenth-Century Anatomy," in *The Making of the Modern Body*, ed. C. Gallagher and T. Laqueur (1987). Ruth Richardson, *Death, Dissection, and the Destitute* (1988). P. J. Bowler, *Evolution: the History of an Idea*, rev. ed. (1989). K. B. Roberts and J. D. W. Tomlinson, *The Fabric of the Body: European Traditions of Anatomical Illustration* (1992). Bettyann Kevles, *Naked to the Bone: Medical Imaging in the Twentieth Century* (1997).

ANITA GUERRINI

ANIMAL CARE AND EXPERIMENTATION. The use of animals as surrogates for the human body in research has a long history. While Aristotle (fourth century B.C.) probably only dissected dead animals, the Alexandrian physicians (third century B.C.) and the Roman physician Galen (second century A.D.) experimented with both live and dead animals. Modern animal experimentation began in the sixteenth century with Andreas *Vesalius and his colleagues in Italy. They used live animals, especially dogs and pigs, to demonstrate various functions of the human and animal body. The first to develop a systematic program of experimentation on live animals was the English physician William *Harvey. Harvey used hundreds of animals of different species to demonstrate the circulation of the blood. Many after 1650 followed his model of research, which included vivisection (surgical intervention on live animals).

René *Descartes argued in his *Discourse on Method* (1637) that animals did not have an intellect and therefore could not experience pain in the way that humans could. This assertion caused, and still causes, great controversy, but few in the seventeenth century used it to justify experimenting on live animals. Most experimenters believed that animals felt pain, and used earlier theological statements about the relative place of animals and humans in God's plan to justify causing pain for human benefit.

Seventeenth-century experiments focused on physiological problems brought to the fore by the theory of the circulation. Injection, inflation of lungs, various surgical interventions, and the vacuum pump were employed to demonstrate aspects of respiration and metabolism. Dogs and sheep were the animals most often used in this way, but the Italian physician Marcello *Malpighi used dozens of frogs to demonstrate the anastomoses between veins and arteries. Blood transfusions attempted between animals and between animals and humans usually did not succeed.

In the eighteenth century, notable experimenters included Stephen Hales, who measured blood pressure in horses and dogs, and Albrecht von Haller, who used almost two hundred different animals in a series of experiments on irritability in living tissues (the ability of tissue to respond to stimuli). He distinguished irritability, an unconscious response of the organism, from sensibility, a conscious response of tissues that have nerves. Pain exemplified sensibility. Haller's work presupposed that animals felt pain, since he measured reactions to various often painful stimuli, and he apologized for his use of animals in the preface to his essay on irritability and sensibility (1752). His apology indicates that a new sense of responsibility toward animals was developing, evidenced by several publications from the 1690s onward. In 1780, the utilitarian Jeremy Bentham argued that animals, like slaves, might be admitted to the moral community since the criterion for admission was not cognitive ability but the ability to suffer. In the same year Immanuel Kant stated that although he deplored excessive cruelty, animals existed for human benefit.

Animals were used in increasing numbers in the nineteenth century for research both in physiology and medicine. François Magendie was one of a generation of French scientists who established systematic experimentation on animals as the key feature of physiological research. Magendie's protégé Claude *Bernard confirmed the laboratory, rather than the clinic or the dissecting table, as the primary site for learning about the body, both animal and human, and its functions. Bernard's *Introduction to the Study of Experimental Medicine* (1865) was a manifesto for experimental physiology as well as a defense of it against a growing anti-vivisection movement.

Britain passed the first animal protection law in 1822, and the Society for the Prevention of Cruelty to Animals was founded in London two years later. By the 1840s, anti-vivisection sentiment began to emerge, much of it directed against British researcher Marshall Hall, one of the few in Britain to perform animal research on the French model. The introduction of anesthesia in the 1840s might have removed a major objection to experiments had it been used universally, but in neurological research it would nullify the results, and debate continued about the capacity of animals (as well as of differing classes and races of humans) to feel pain.

The British anti-vivisection advocate Frances Power Cobbe protested against animal experimentation in France and Italy in the 1860s; by the 1870s, however, many British researchers were beginning to emulate continental models. Cobbe and her circle, which had the support of Queen Victoria, published many articles and letters in the popular press. In 1876, Cobbe's agitation led to the introduction of a bill in Parliament to restrict animal experimentation; after much emendation, it was passed as the Cruelty to Animals Act of 1876, the first attempt by a national government to regulate experimentation. Although scientists retained control of the regulatory function, the legislation took an important first step.

At the same time, other experimenters used animals to develop the basic premises of the germ theory of disease and to begin the development of vaccines and therapies. These activities raised much less anti-vivisectionist attention. Robert Koch published the first proof of the germ theory of disease in 1877. By means of extensive testing on animals, mainly sheep and cows but also rabbits and mice, Koch demonstrated that the anthrax bacillus, which he had observed under the microscope, caused the disease in animals. He established "Koch's postulates" in 1878, which provided rules for bacteriological research. These stated that the microorganism suspected of causing disease had to be found in every case of the disease (in an experimental animal); that the microorganism had to be isolated and cultured outside of the animal; and that the culture when introduced into a healthy animal had to make it ill. Only if the organism was then recovered from this diseased animal could it be identified as the cause of the disease.

Louis *Pasteur developed a vaccine against the animal disease rabies in 1885. Because rabies was caused by a virus that could not be seen under the microscopes then in use, Koch's postulates could not be applied; however, the effectiveness of the vaccine helped to confirm the germ theory of disease. From the 1890s onward, several disease-causing microorganisms were identified and some vaccines and anti-toxins developed. This research used thousands of animals—mice and rabbits mostly, but also dogs, horses, and other mammals. Paul Ehrlich used thousands of mice to develop salvarsan, the first specific drug against *syphilis (1907), and many thousands more mice were used to develop the first antibiotic drugs in the 1930s and 1940s.

Experimental mice increasingly became the animal of choice in disease research; inbreeding to attain mice with certain characteristics began around 1910. By the 1930s, the Jackson Laboratories in the U.S. supplied researchers around the world with millions of specially-bred mice for research on a variety of topics, including cancer, infectious diseases, and physiology.

Research in psychology also consumed many animals, particularly in behavioral tests. The salivating dogs of Ivan *Pavlov are famous, and animals, especially rats, have been used in many experiments on behavior and the acquisition of cognitive functions. Experiments by Harry Harlow and others on the development of maternal bonding in monkeys investigated important psychological issues and also raised questions about the use of primates in research.

Primates were seldom used in research before the twentieth century. Although their anatomical similarity to humans made them (as Galen had argued) ideal anatomical proxies, this similarity also made researchers such as Claude Bernard uncomfortable. Because it was long thought that only primates (including humans) could contract the poliomyelitis virus, monkeys began to be used regularly in polio research around 1900. The development of polio vaccines in the 1950s relied on the use of millions of rhesus macaques; today, monkeys and chimpanzees figure in *AIDS research. Recent work revealing the cognitive abilities and emotional range of apes in particular has rendered the research use of primates problematic.

In the twentieth and twenty-first centuries, animals continue to be used in biomedical research in a number of fields. A majority of these are rats and mice bred and genetically modified for research purposes. Mice, rats, rabbits, dogs, cats, and other animals are also used to test the possible toxicity of cosmetics and household products.

After a period of quiescence following the successes of bacteriology, the anti-vivisection movement revived in the 1970s with a new philosophical basis in the notion of animal rights or animal liberation. Organized opposition to animal experimentation greatly increased. In part because of the attention activists directed toward research, most Western countries passed new laws in the 1980s governing the treatment of experimental animals. While details differ among nations, in general these laws strengthened protection for animals and set up a bureaucratic apparatus to ensure inspection of facilities and monitoring of research protocols. The use of animals for both scientific research and product testing has declined since the introduction of these laws, and alternatives to animal research (cell and tissue cultures, computer models, invertebrates rather than vertebrates) continue to be investigated. However, Western biomedical science is unlikely to reduce its heavy dependence on animal experimental subjects in the short term.

R. D. French, *Antivivisection and Medical Science in Victorian Society* (1975). H. Dowling, *Fighting Infection* (1977). R. G. Frank Jr., *Harvey and the Oxford Physiologists* (1980). J. E. Lesch, *Science and Medicine in France* (1984). Andrew Rowan, *Of Mice, Models, and Men* (1984). Nicolaas Rupke, ed., *Vivisection in Historical Perspective* (1987). F. Barbara Orlans, *In the Name of Science* (1993). Deborah Blum, *The Monkey Wars* (1994). Anita Guerrini, *Animal and Human Experimentation: A History* (2002).

ANITA GUERRINI

ANIMALCULISM. See OVISM AND ANIMALCULISM.

ANIMAL MACHINE. Analogies drawn between animals and both machines and human beings, on the one hand, and directly between machines and human beings, on the other, have been fruitful in the life sciences and theories of knowledge from antiquity to the present. Belief in the fundamental truth of these analogies implied a mechanistic view of nature and a deterministic view of life. The origins and cultural, social, and political significance of parallels between machines and animals or humans remain a matter of debate.

The notion of the animal-machine can be traced to comparisons made by Aristotle (384–322 B.C.) and to the mechanical contrivances of Hero of Alexandria (first century A.D.) in his *Automata*. During the late sixteenth and early seventeenth century, physicians and philosophers developed the notion in characteristic ways. Gómez Pereira, a medical doctor, sought to explain animal movements by comparing them to the mechanisms in use at the time, while another doctor, Santorio Santorio, maintained that the human's body mobility resembled that of a watch and depended on the number, position, and configuration of its constituent parts (*Methodi vitandarum errorum*, 1602). Some philosophers instead endeavored to explain sensations as an automatic process independent of the intentionality of reason. According to Marin Mersenne, animals automatically responded to sounds because they lacked reason. Rational creatures—human beings—could appraise the nature and properties of sounds and distinguish them from other objects (*Harmonie*

A mechanical drum player constructed for Marie-Antoinette around 1785. Many such automatons were built in the eighteenth century, not only to demonstrate the art of the maker and to entertain the wealthy, but also to indicate that animal bodies, including those of human beings, might be distinguished from windmills only by their greater complexity.

Universelle, 1636). Unlike animals, humans could construct things with their hands and reason.

René *Descartes formulated the clearest notion of the animal-machine in his posthumously published *De homine* (1662). Animal and human bodies functioned mechanically as automata; in addition, humans had reason or spirit, which endowed them with knowledge, free will, and language. All physiological functions like digestion and involuntary movements could be explained in terms of automatisms; only rational animals could control their movements. By attributing to human beings an immaterial soul (spirit/mind), Descartes freed them from Renaissance naturalistic animism and spiritualized them. And by considering all bodies as machines he paved the way for abolition of the distinction between natural and artificial bodies, thus extending the range of mechanistic explanations. The extraordinary debate provoked by Descartes's doctrine focused on the questions

"Do animals have souls and can they learn and communicate?", "How does the human mind interact with the body?" (the so-called mind-body problem), "Are there differences between living and nonliving matter?", and "Can one explain feelings, language, and thought in an entirely mechanistic way?"

Inspired by Jacques de Vaucanson's celebrated automata and by experiments on the *polyp and on *irritability, Julien de La Mettrie replied in the affirmative to the last question in his epoch-making book *L'Homme machine* (1748). By extending the animal-machine doctrine to human beings and excluding Cartesian dualism, he attributed motion, sensation, regeneration, and irritability to living matter in a materialistic manner. Though prohibited in France and burnt in The Hague, La Mettrie's work circulated widely if clandestinely.

During the first industrial revolution the steam engine provided a metaphor for the organism that suggested how heat could be converted into mechanical work. For most physiologists, organisms acted as energy-conversion devices similar to those analyzed by thermodynamics; Thomas Henry Huxley claimed that animal bodies were living machines in action (1874). By carefully measuring the production of animal heat in metabolism, Max Rubner showed that it followed the principle of energy conservation, while Étienne-Jules Marey applied methods of graphic recording to living organs and built artificial ones.

The crisis of determinism in the early twentieth century did not diminish the heuristic value of the notion of the animal-machine. Recast in new terms like feedback, input, output, control, and homeostasis, it reappeared in Norbert Wiener's *Cybernetics, or the Control and Communication in the Animal and the Machine* (1948). The self-regulatory features of information systems suggested that organisms and machines differed in organization more than in materials, and the aim of the cybernetic model was to mechanize intellectual processes. At the end of the twentieth century, analogies between mental processes and the operation of *computers had become a commonplace.

Leonora C. Rosenfield, *From Beast-Machine to Man-Machine* (1941). Heikki Kirkinen, *Les origines de la conception moderne de l'homme-machine* (1960). Julien de La Mettrie, *L'Homme Machine: A Study in the Origins of an Idea*, ed. Aram Vartanian (1960). Alan R. Anderson, ed., *Minds and Machines* (1964). Robert Lenoble, *Mersenne ou la naissance du mécanisme*, 2d ed. (1971). Steve J. Heims, *The Cybernetics Group* (1991).

RENATO G. MAZZOLINI

ANIMAL MAGNETISM. See MESMERISM AND ANIMAL MAGNETISM.

ANTHROPIC PRINCIPLE. The anthropic cosmological principle, in its weak form, states that the universe must be such as to admit and sustain life. From Descartes's "I think, therefore I am," we proceed to "I am, therefore the nature of the universe permits me to be." Although a tautology, this has interesting implications. Hydrogen and helium formed nearly instantaneously in the primordial inferno of the cosmological Big Bang, but the building blocks of life, including carbon, oxygen, and nitrogen, form in the interiors of stars over long times. Hence the universe we observe must be billions of years old, older than the first generation of stars to spew forth the building blocks of our life, and still young enough that our own sun has not expired, taking us with it. The physicist Robert Dicke at Princeton University in the 1950s noted this necessary connection between observers and the observed age of the universe. The universe is expanding. Not too slowly, or it would come to a halt and collapse. Nor too fast, lest it become too dilute for stars to coalesce. The rate of expansion depends on the density of the universe. Had the initial density differed by as little as one part in 10 to the 60th power, the result would be either a big crunch or a big chill, and no life. Slight changes in the values of physical constants, including the strengths of fundamental forces such as the gravitational and electromagnetic, and the masses and charges of subatomic particles, would also render life impossible. The remarkable set of coincidences apparently necessary for human life has prompted many grandiose inferences. The strong version of the anthropic principle, articulated in the 1970s, asserts that the universe was created and fine-tuned so that intelligent life could evolve in it. Here we have passed from science to religion.

John D. Barrow and Frank J. Tipler, *The Anthropic Cosmological Principle* (1986).

NORRISS S. HETHERINGTON

ANTHROPOLOGY. While anthropology has existed as an academic discipline only since the beginning of the twentieth century, the perspective it represents has a considerably longer history. Most anthropologists trace its history to the Anglo-French Enlightenment of the eighteenth century and the ensuing period of European global expansion. The Enlightenment is generally acknowledged for its challenge to "time-honored" traditions and for championing reason as the foundation of truth. The Enlightenment relativized customs and beliefs that had been understood as universal. Jean-Jacques Rousseau's essay, *Discourse on the Origin and Bases of the Inequality among Men* (1754), attributes inequality to the development of private property rather than

variable natural endowments or the vicissitudes of nature. Despite the criticism of Rousseau's romantic notion of the "Noble Savage," his sympathetic reference to indigenous peoples anticipated anthropology's critical turn 150 years later by emphasizing humanity's diversity and historical character.

The same historical theme is central to Giambattista Vico's *The New Science* (1725). Vico's book has been claimed as ancestral not only to anthropology but the human sciences more generally because it challenged essentialized notions of human nature. Vico argues like Rousseau against an unchanging human nature and for a vision of humanity as historically produced. For Vico, history presented a more suitable means for grasping the human condition than did the emerging natural sciences, a theme that resonates with contemporary anthropology.

The European expansion throughout the eighteenth and nineteenth centuries contributed significantly to the development of anthropology as an academic discipline. The nineteenth century witnessed the reconfigured political and economic landscapes of European and North American nation-states and indigenous cultures worldwide as colonial sources of labor, raw materials, and export commodities. In a manner that betrayed the critical insights of Rousseau, yet in part fulfilled the nature of Enlightenment reason, nineteenth-century colonialism was regarded in moral terms by many Western scholars, colonial administrators, and missionaries as bringing civilization to so-called "savage peoples." A linear and hierarchical mapping of human history consigned indigenous peoples to the dawn of humanity and place Western peoples at the pinnacle of human development.

The nineteenth century saw a critical tension between intellectual commitment to social evolution and opposing discourses that embraced evolution while supporting an alternative vision of indigenous peoples. The encounter with indigenous peoples could be an occasion for the celebration of difference and for exploring alternative possibilities for the human condition. The works of Lewis Henry Morgan, often regarded in anthropology as the founder of modern kinship studies, illustrate the point. In *League of the Iroquois* (1851) and *Ancient Society* (1877), Morgan showed the wide variability of kinship structures in relation to property and changing forms of human community. His work caught the attention of Karl Marx, who used it to show that many of the characteristics nineteenth-century scholars attributed to human nature were in fact particular to the history of capitalism. Morgan and Marx can be claimed as anthropological ancestors because they appreciated the unique historical nature of indigenous societies as a challenge to regnant ideas of unilineal development and progress that ultimately dismissed the humanity of indigenous peoples.

Anthropology Enters the Academy

The founding of American academic anthropology can be attributed to the German immigrant Franz Boas, while English and French anthropology owes much to the intellectual traditions established by Herbert Spencer and Emile Durkheim. Named Professor of Anthropology at Columbia University in 1896, Boas gained the institutional support that enabled him to launch Columbia as the outstanding American center for anthropological research and training of the early twentieth century.

Boas advocated local histories and strongly supported the American four-field approach, integrating cultural anthropology, archaeology, linguistics, and biological anthropology. He criticized nineteenth-century unilineal evolution and argued that ethnographic materials, practices, and culture should be understood from the context of their development. Boas brought linguistic, cultural, archaeological, and biological evidence to bear in studying the historical origins of cultural phenomena. However, his promotion of the American four-field approach came from his ethnographic research on America's indigenous population and his critique of the racist policies espoused by the American Immigration and Naturalization Service at Ellis Island. His numerous students—Margaret *Mead, Ruth Benedict, and others—left a lasting mark on American anthropology.

One would expect Edward Tylor's classic *Primitive Culture* (1871), rather than Herbert Spencer's *The Principles of Sociology* (1885) and Emile Durkheim's *Rules of the Sociological Method* (1895), to be the pervasive intellectual influence on British social anthropology given the enduring nature of Tylor's cultural concept of the "complex whole." However, Tylor embraced evolutionism and failed to establish a tradition of field work as the foundation of anthropological practice. Spencer favored an organic model of human society, while Durkheim introduced the notion of the "social fact" to explain the relation between the human individual and the determining cultural influences of groups or communities as the basic unit of social analysis. Spencer and Durkheim served as principal influences on British functionalism and structural-functionalism more generally, which rested on the pervasive idea that social beliefs and practices reinforce the identity and continuity of the social whole.

Bronisław Malinowski was primarily responsible for establishing rigorous field research as the hallmark of sociocultural anthropology. In contrast to much of late nineteenth- and early twentieth-century anthropology, in which fieldwork was not considered suitable for "gentlemen," Malinowski argued for immersion in the social round of indigenous societies over a sustained period of time. He also emphasized in *Argonauts of the Western Pacific* (1922) the importance of learning indigenous languages, of being able to "think in their symbols." He rejected evolutionary models and was skeptical toward historical approaches he considered too dependent on oral accounts. Malinowski argued for an organic vision of society as consisting of interdependent parts and emphasized individuals and the capacity of culture to satisfy their biological, psychological, and social needs.

Alfred Reginald Radcliffe-Brown's structural-functionalism was the primary competitor to Malinowskian functionalism. Radcliffe-Brown made his reputation through field research in the Andaman Islands and Australia, and, unlike Malinowski, placed a theoretical emphasis on social structure rather than the individual. For Radcliffe-Brown and Durkheim, individuals come and go while social structures endure. However, both Malinowski and Radcliffe-Brown conceived of societies as integrated wholes, and conflict as essentially dysfunctional. They and their students worked during the period of English colonial rule. While the connection between English social anthropology and colonialism was indirect, Malinowski and Radcliffe-Brown's rejection of history and de-emphasis of human agency except as a function of needs and structures precluded a critical understanding of the social, political, and economic circumstances that prefigured the anthropology of their times.

There is little doubt that Claude Lévi-Strauss remains the predominant figure in French anthropology. His work was heavily indebted to Durkheim, especially his *Elementary Structures of Kinship* (1949), as well as to the formalism of the Prague school of linguistics. Lévi-Strauss's most famous book, *Tristes Tropiques*, reported on his sole fieldwork experience, in the Amazon. This book is celebrated for its elegant prose and the questions it raises with respect to overcoming the cultural distance that separates the ethnographer from the indigenous "other." However, unlike Malinowski, Lévi-Strauss committed his scholarly career to anthropology from a distance through the formal analysis of myths worldwide. He argued that in spite of the variability of narratives or stories, myths should be understood as permutations of invariable human themes.

The structural method, the internal mechanics of the myth's primary oppositions, enabled Lévi-Strauss to transform the surface semantics of a myth to its underlying universal logic.

Structuralism and structural-functionalism continued to be influential right through the colonial era until it subsided in the wake of the critical anthropology of the 1970s. However, with the exception of sociology and political science, the influence of critical anthropology tended to be more pronounced in England and continental Europe than in the United States. Not all versions of structuralism reproduced the original models established by Malinowski, Radcliffe-Brown, and Lévi-Strauss. Edmund Leach introduced a more dynamic version of structuralism than his English predecessors, exhibiting a sensitivity to history and human agency in *Political Systems of Highland Burma* (1954). Although reproducing the functionalism of his mentor Malinowski with an ecological tone, E. E. Evans-Pritchard, like Leach, was more attuned to the importance of history to anthropology. Well versed in the hermeneutic philosophy of Wilhelm Dilthey, Evans-Pritchard identified anthropology with the human rather than natural sciences. The symbolic structuralism of Victor Turner's analysis of rituals and Mary Douglas's use of structure and anti-structure to articulate boundaries and matter out of place also influenced structuralism and functionalism. However, they lacked the deeper reflective and historical perspective necessary to grasp the intellectual contiguity between static theoretical models and the politics of European colonialism.

Mead and Benedict followed up Boas's work on the relation between culture and personality. Mead's books became known worldwide and posed a considerable challenge to dominant ideas of sex and gender. Edward Sapir advanced Boas's initiatives on language by arguing for the close association between language and worldview, a perspective now known as "linguistic relativity." Sapir evaluated the potentials that human cultures offered individuals in his important essay "Culture, Genuine and Spurious" (1924), which engaged Alfred Krober's notion of culture as "superorganic," suspended above the fray of daily practice. Sapir taught that cultures were genuine only if they enhanced the potentials of individuals, and so spoke positively of indigenous societies.

In both the specific traditions of French and British sociocultural anthropology and the American four-field approach, the culture concept served as the common bridge and point of contest among disparate theoretical traditions and practices. Even adamant defenders of the primacy of structure had to concede something to the "cultural other." Throughout the 1930s and 1940s there was wide agreement on fieldwork as central to the identity of anthropology, but "culture" eluded all efforts at operationalization. Perhaps out of resignation, Tylor's view of culture as the "complex whole" served as the minimalist position.

In the 1950s, American anthropology took a dramatic materialist turn through the theoretical contributions of Leslie White and Julian H. Steward. Following the influence of Karl Marx, both White and Steward argued against explaining culture in cultural terms. White embraced in *Evolution of Culture* (1959) a comprehensive theory of unilineal social evolution that had long been criticized and abandoned by the Boasian tradition. According to White, the capacity to harness energy was the primary force in societies' progress from one stage of technological complexity to another. On the other hand, Steward sought, in his *Theory of Culture Change* (1955), to distance himself from the untenable Eurocentric implications of nineteenth-century social evolution. Like Boas, Steward was sensitive to the historically particular in arguing that cultures could evolve in distinct patterns, depending on diverse environmental circumstances. He called this "multilinear evolution." Although Steward emphasized material processes as central to evolution, he did not dismiss the importance of culture. Rather, culture was central to his theoretical framework through the "culture core," the cultural activities most directly related to subsistence.

Contemporary Directions

The end of European colonialism during the 1960s and the political initiatives taken by newly independent states rendered problematic such static theories as structuralism and structural-functionalism, which dismissed history and eclipsed human agency. Anthropologists pursued numerous new directions. Marxist anthropology developed in the United States and France in the 1960s followed by the very important and influential appeal of Dell Hymes, who sought to refigure anthropology along critical lines in his edited *Reinventing Anthropology* (1969). In the United States, three of Steward's students—Stanley Diamond, Sidney Mintz, and Eric Wolf—elaborated Marxist anthropology. Diamond wrote romantically, as had Rousseau and Marx, about alternative potentials of human freedom based upon the reputed equality of indigenous societies. Mintz and Wolf advanced understanding of peasant societies and capitalist accumulation through their work on the Caribbean and Mexico, respectively. Hymes's book of 1969, which included contributions from

Diamond, Wolf, Scholte, and others less directly related to the Marxist tradition, called for the political reformulation of anthropology in a manner that acknowledged the close connections between anthropology and colonialism. According to the "reinventing" group, the practice of anthropology should not be divorced from political practice and should be sensitive to the social, political, and historical circumstances that concretely locate anthropology as a discipline.

French anthropology had its own version of Marxism represented in the works of Maurice Godelier and Claude Meillassoux. Godelier and Meillassoux wrote prolifically on the application of mode of production analysis to indigenous societies. They wrote, however, from a largely structuralist perspective that tended to grasp culture as shaped by the forces and relations of production. Godelier invoked the traditional Marxist categories associated with the mode of production and criticized Meillassoux for inventing a new mode of production for every variant in an indigenous social order. Godelier has recently moved away from the strict base-and-superstructure model of Marxism and has been a major figure in establishing a tradition of French anthropology distinct from that of Lévi-Strauss. Recent French anthropologists like Françoise Zonabend have clearly broken from both the colonial past and Lévi-Strauss by focusing on ethnographically detailed and historical accounts of rural France.

White's and Steward's materialism also influenced the development of American cultural ecology. Marvin Harris is perhaps the best known of the cultural materialists. Although hostile to American Marxist and critical anthropology for embracing dialectics and disavowing science, Harris borrowed such concepts as "mode of production," "etic behavioral mode of production and reproduction," and "emic superstructure" strangely reminiscent of Marx. Emics has traditionally referred to the informant's perspective, while etics entails the studied conclusions of the researcher. However, Harris regards culture as derivative of a techno-environmental determinism, a perspective quite remote from Marx and Marxist anthropology. The cultural ecology of Roy Rappaport presents a more subtle and complex rendering of human culture. In *Pigs for the Ancestors* (1967) Rappaport shows how the horticultural Maring of New Guinea use ritual and symbols to mediate the ecology of their gardens, thus avoiding the reductive tendencies associated with ecological castings of culture and daily practices.

Clifford Geertz's *The Interpretation of Culture* (1973), a book whose concept of "thick description" has influenced scholars from history to literary criticism, redefined the concept of culture as lived symbolic beliefs and practices having public meanings. Geertz argued that anthropology was not a science in search of laws but a hermeneutic one in search of meaning, a remark that essentially identifies one of the primary tensions in contemporary anthropology.

Other new directions also seek to bridge political economy and cultural theory. Recently, Marshall Sahlins has sought to reinvigorate structuralism by developing a dynamic, if not dialectical, relationship between structure and agency. George Marcus and James Clifford's *Writing Culture* (1986) contributed importantly to serious discussion of the rhetorical and political dimensions of ethnographic writing, a concern foreshadowed earlier in the writings of historian Hayden White. Marcus, Clifford, Fischer, and especially Tyler have come to be associated with anthropology's postmodern turn through challenging meta-narratives and privileging ethnography as local pastiche.

On another front, contributors to *Recapturing Anthropology* (1991), edited by Richard Fox, and *Anthropological Locations* (1997), edited by Akhil Gupta and James Ferguson, have discussed the reinventing of anthropology to address the problems that ensue from globalization and cultural diaspora. Authors in both volumes argue against the antiquated notion of a circumscribed field site and for what George Marcus calls multisited field research. Moreover, like John and Jean Comaroff and Johannes Fabian, Lila Abu-Lughod argues against univocal views of culture and for culture as contested practice.

Anthropology continues to challenge its own assumptions and to seek to respond to global dislocations and inequalities. Within American anthropology, some academic departments like Stanford's and Duke's have been torn apart by struggles over science and critical practice, while others seek new ways to bridge the subdisciplines. It seems clear that anthropology on a global basis will continue to be relevant to the challenges that face humanity because it is one of the few disciplines that through its critical practice embraces both diversity and interdisciplinary modes of knowing.

Marvin Harris, *The Rise of Anthropological Theory* (1968). Clifford Geertz, *The Interpretation of Cultures* (1973). Talal Asad, ed., *Anthropology and the Colonial Encounter* (1974). George W. Stocking, Jr., ed., *The Ethnographer's Magic and Other Essays in the History of Anthropology* (1992). Adam Kuper, *Anthropology and Anthropologists: The Modern British School* (1996). Akhil Gupta and James Ferguson, eds., *Anthropological Locations* (1997). George W. Stocking, Jr., ed., *Bones, Bodies, Behavior: Essays on Biological Anthropology* (1988). Michael

Herzfeld, *Anthropology* (2001). Thomas C. Patterson, *A Social History of Anthropology in the United States* (2001). Robert C. Ulin, *Understanding Cultures*, 2nd ed. (2001).
ROBERT C. ULIN

ANTIBIOTICS. Jean Antoine Villemin introduced the term "antibiosis" (French, *antibiose*) in 1889 to refer to "an organism [that] destroys the life of another organism to entertain its own life." "Antibiotic" has come to mean a substance elaborated by one organism that inhibits or kills another. The biological origin of antibiotics distinguishes these antimicrobial agents from other agents first produced in the laboratory, for example, sulfa drugs.

Microbial antagonism and its potential were recognized quite early. In 1885 Arnaldo Cantani described the use of one bacterium to cure another (tuberculosis) by "replacement therapy"; the same year, André-Victor Cornil and Victor Babés predicted therapeutic benefits from further study of the "reciprocal action of bacteria." In 1899 Rudolf Emmerich and Oscar Löw introduced the clinical use of pyocyanase, a bacterial preparation from aged cultures of *Pseudomonas pyocyanea*. Pyocyanase could lyse (disintegrate or dissolve) several pathogenic bacteria, but after a decade of enthusiastic use, this early antibiotic seems to have been abandoned.

The first antibiotic from a mold rather than a bacterium was isolated in 1896 by Bartolomeo Gosio, who described a crystalline material (mycophenolic acid) produced by *Penicillium* and other molds that could inhibit the growth of anthrax bacilli. Analysts of antibiotics used two tests—lysis of bacterial cultures and inhibition of growth on solid media. Early assays foundered because four mechanistically distinct phenomena occur in them: the so-called secondary metabolites that we call antibiotics; lytic enzymes such as lysozyme; bacterial viruses that kill by active parasitic infection; and bacteriocins, lethal proteins elaborated by one strain of bacteria to kill closely related but nonidentical strains.

Early in the twentieth century the search continued for antibiotics with more potent activity and fewer side effects. In 1922 Alexander Fleming discovered lysozyme, an enzyme that induced the complete and rapid lysis of certain test bacteria in human tears (Fleming's own). He became interested in the role of antibacterial agents in the body and in the natural defense against disease. This ecological viewpoint and his prior discovery of lysozyme helped Fleming interpret his famous finding of growth inhibition of *Staphylococcus* by fortuitous contamination with the mold *Penicillium notatum*. In 1929 Fleming reported the antibiotic action of culture fluids of *Penicillium* and termed his antibiotic "penicillin."

Given the history of antibiotics at the time, it is not surprising that penicillin provoked very little interest. Furthermore, focus on bacteriophage therapy as a potential cure for bacterial infections diverted attention from Fleming's work. By the late 1930s, however, enthusiasm for bacteriophage therapy began to wane.

René Dubos systematically searched soil organisms for secondary metabolites with antibiotic activity and in 1939 isolated tyrothricin (a mixture of tyrocidin and gramicidin) from *Bacillus brevis*. Although these agents were highly effective topically, they proved too toxic for parenteral use.

The need for better treatment of wound infections during World War II accelerated the work of Howard Walter Florey and Ernst Boris Chain, who were already engaged in purifying penicillin. With its remarkable antibacterial potency and incredibly low toxicity, it was a promising candidate for the first clinically useful antibiotic. Vigorous coordinated research on the purification, production, and testing of penicillin, a joint British–American war effort, led to its first clinical use in 1942.

Dubos's general approach soon resulted in the discovery of three important antibiotics: streptomycin, the first antibiotic active against tuberculosis, was isolated from *Streptomyces griseus* by Albert Schatz, E. Bugie, and Selman A. Waksman in 1944; chloramphenicol, a broad spectrum antibiotic from *S. venezuelae*, was isolated by Paul Burkholder and his group in 1947; and chlortetracycline was isolated from *S. aureofaciens* by Benjamin Duggar in 1948. Their immediate success in combating infectious diseases earned them a reputation as "wonder drugs." A broad search began for other such antibiotics.

Resistance to antibacterial agents was well known before the use of antibiotics in medicine, and with their widespread and indiscriminate use, antibiotic-resistant pathogens duly appeared. Initially, the employment of another antibiotic usually sufficed, but multiple-drug resistance soon appeared. First on theoretical and then on empirical grounds, multiple-agent chemotherapy was developed to minimize the appearance of drug-resistant variants. Much effort has gone into the search for new antibiotics to use when resistance develops, but the winner in the race between drug development and mutation to drug resistance has not been declared.

Selman A. Waksman, *Microbial Antagonisms and Antibiotic Substances* (1945). André Maurois, *The Life of Sir Alexander Fleming* (1959). Gladys L. Hobby, *Penicillin: Meeting the Challenge* (1985). Carol L. Moberg and Zanvil A. Cohen, eds., *Launching the Antibiotic Era: Personal Accounts of the Discovery and Use of the First Antibiotics* (1990).
WILLIAM C. SUMMERS

ANTIBODY. Antibodies are immunoglobulins (proteins) that circulate in the blood and lymphatic systems and function as part of the immune system by interacting with antigens (any substance, protein or polysaccharide, recognized as nonidiopathic, not belonging to the individual). Viruses, bacteria, parasites, and foreign transplant tissue all act as antigens. The antibody/antigen interaction is highly specific; the human body can produce over one million different antibodies.

The concept of antibodies developed in the late-nineteenth-century search for ways to treat and prevent infectious diseases. In 1890, following the discovery of bacterial toxins, Emil von Behring developed an experimental antidiphtheria serum. His gradual exposure of guinea pigs to diphtheria toxins resulted in animals whose blood serum, when injected into a previously unexposed guinea pig, prevented the disease. Paul Ehrlich described the unknown protective elements in the serum as "antibodies." Intense research in infection immunology followed.

Around 1900, Karl Landsteiner and his students isolated the four main human blood groups (A, B, AB, O) by virtue of their antigen/antibody interaction. Besides making blood transfusion safer, Landsteiner demonstrated that antibodies were not limited to combating infection. In a series of experiments beginning the following year, he attached simple organic chemicals to protein molecules to act as artificial antigens of a known structure. Analyzing the resulting antibodies demonstrated that their specificity was not a vague property but precisely defined in chemical terms—reminiscent of Ehrlich's earlier lock-and-key analogy for antibody/antigen interaction.

In 1939 Arne Tiselius and Elvin A. Kabat identified antibodies as protein molecules (later named immunoglobulins). Rodney Porter and Gerald Maurice Edelman used enzymes to split these molecules in order to elucidate their structure (high-resolution x-ray crystallography was later used for the purpose). Each immunoglobulin molecule consists of four protein chains: two duplicate long/heavy chains and two duplicate short/light chains arranged in a Y shape (the stem of the Y consists of paired long chains; the short chains paired with the long chains only along the arms). Particular sites on the chains have specific functions; the ends of both arms bind antigen. The long and short chains each consist of a constant and a highly variable portion. The flexible structure of the variable portion explains the ability of a single form to carry the huge range of antibody.

Research in the 1930s located antibody manufacture in the lymph nodes (they are also made in the spleen). Later research showed that B lymphocytes, arising in the bone marrow and moving to the lymphatic tissues, generate antibody-producing plasma cells by cell division and differentiation in response to the presence of antigen—the "clonal selection theory" of Frank Macfarlane Burnet (1957). The B cells do not work in isolation but require the presence of "helper" T cells (also produced in bone marrow but passing to the thymus gland) for production of antibody.

Antibodies have become valuable biomedical tools. In 1975 Cesar Milstein and Georges J. F. Köhler pioneered the production of monoclonal antibodies. Their fusing of a tumor cell and a B cell led to a new hybrid cell (hybridoma), which divides in vitro producing quantities of pure homogenous antibodies. These can be directed against preselected antigens. Monoclonal antibodies have become invaluable diagnostic reagents and assisted new therapies.

Debra Jan Bibel, *Milestones in Immunology* (1988). Arthur M. Silverstein, *A History of Immunology* (1989).

HELEN BYNUM

ANTINUCLEAR MOVEMENT. From their inception nuclear weapons were controversial, and this ambivalence extended to their civilian counterpart, nuclear reactors, because of the dangers associated with them.

Opposition to Nuclear Weapons.
The United States used nuclear weapons against Hiroshima and Nagasaki in August 1945 to hasten the end of World War II. Since then, despite calls for their use in the Korean and Vietnamese conflicts and concerns about their employment during the Cuban missile crisis, no nuclear weapon has been fired in anger.

Although the victors of World War II regarded the atomic bomb as the "winning weapon" and expected it to keep the peace, many people considered it an immoral weapon. The Soviet Union's explosion of its first device in 1949 made clear that a nuclear war would be catastrophic for its participants.

The first antinuclear protesters were the *Manhattan Project scientists themselves. The wartime Franck Report (named after its principal author, the emigré Nobel Prize–winning physicist James Franck) unsuccessfully urged the U.S. government to issue a warning before dropping a bomb on a city. After the war's end, many scientists joined journalists, clergy, and other citizens in advocating domestic legislation to ensure civilian control of nuclear research and development, and strong international control of weapons through the newly formed United Nations. The Federation

Linus *Pauling photographed in a protest against nuclear testing. The protest occurred outside the White House in 1961, on the day before Pauling was to dine there with other American winners of the Nobel Prize.

a Livable World, the Center for Defense Information, Physicians for Social Responsibility, and the American Physical Society, developed arguments that appealed to the intellect as well as the emotions. Composed in large part of scientists, physicians, and retired military officers, their expertise and patriotism could not be impugned.

Organized religion, surprisingly, was not an early and persistent critic of weapons of mass destruction; many churches supported the anticommunist sentiments of the period. During the 1980s, however, several panels of Catholic and Protestant bishops issued reports questioning the morality of nuclear deterrence. Filmmakers created visual images of the horrors of nuclear warfare in *The War Game*, *Threads*, *On the Beach*, and *The Day After*, and exposed the danger and absurdity of war initiated by a maniac (*Dr. Strangelove*) or by technological failure (*Fail Safe*).

The United States did not have a monopoly on antinuclear agitation. The Japanese people keep the memory of the bombings of 1945 alive in annual ceremonies in the Hiroshima Peace Park, which has a museum of artifacts and pictures. The government of India, acting as a leader of developing countries, criticized the superpowers' arsenals until it developed its own nuclear weapons. In the Soviet Union, physicist Andrei Sakharov first designed hydrogen bombs and then campaigned vigorously against their use. The Soviet government supported International Physicians for the Prevention of Nuclear War, a Nobel Peace Prize–winning organization (1985) founded by a Russian and an American. The Pugwash movement, a decades-long series of conferences by scientists from many countries, also received the Nobel Peace Prize (1995). The Green party and other organizations in Germany organized massive marches in the early 1980s to protest the introduction of intermediate-range ballistic missiles with atomic warheads in Europe. Australians and New Zealanders objected to tests by France in the South Pacific as late as the 1990s.

A number of international treaties have created nuclear weapon–free zones in Antarctica (1959), Latin America (1967), the South Pacific (1985), Southeast Asia (1995), and Africa (1996). These accords are more than mere gestures of compliance, but less than enforceable laws. Seeking an outright legal ban on nuclear weapons, a number of nongovernmental organizations requested a ruling from the World Court. The court issued an advisory opinion in 1996 stating that "the threat or use of nuclear weapons would generally be contrary to the rules of international law."

of American Scientists (FAS) led these efforts. A best-selling anthology of articles by leading scientists, *One World or None* (1946), helped mobilize public opinion.

During the mid-1950s, testing of newly developed hydrogen bombs revealed the danger of radioactive fallout circulating worldwide. A petition to ban testing, drawn up by the Nobel Prize–winning chemist Linus Pauling and signed by thousands of scientists, raised the level of international concern. The National Committee for a Sane Nuclear Policy (SANE) and the Campaign for Nuclear Disarmament (CND) sprang up in the United States and Britain, respectively. These and "ban the bomb" organizations in other countries were often slandered as procommunist. They obtained a following by appealing to the emotions, which allowed opponents to charge, with some accuracy, that they knew little about weapons and strategy. The Limited Test Ban Treaty of 1963, which permitted testing only underground and so removed fallout as an issue, deprived the antinuclear movement of much of its emotional steam. In the 1970s, however, a revitalized FAS, together with the Union of Concerned Scientists, the Council for

Opposition to Reactors and Radioactive Waste.
The United States drew a sharp separation between civilian reactors primarily producing electricity and military reactors primarily growing plutonium. American utilities that owned reactors did not want the albatross of militarism hung around their necks. In other countries, such as Great Britain, reactors served dual purposes. Since all reactors produce plutonium that could become fuel for breeder reactors as well as an explosive, many people feared that terrorists might hijack some of the metal while it was in transit.

During the 1960s and early 1970s, American utilities ordered the building of more than one hundred reactors. Grassroots opposition coalesced around fear of lethal doses of radioactivity released in a meltdown of the core (the theme of the popular film *The China Syndrome*), but other potential problems added to popular concerns: Lesser accidents might release dangerous levels of radioactivity beyond the site perimeter; plans for plants close to urban areas increased anxiety; water from the reactors' cooling system dumped in rivers, lakes, and oceans threatened fish and other animals; and the likely targeting of reactors in a nuclear war made them undesirable neighbors.

By protesting reactor plans at every stage, the opposition (often local citizen groups) slowed the building process so much that the utilities incurred enormous interest charges on the money they borrowed for construction. This often made the proposed reactor uncompetitive with fossil-fuel. By the early 1970s, some utilities had canceled their plans for reactors (not only because of the high costs but also owing to overestimates of the nation's electrical needs). A safety report by the U.S. Nuclear Regulatory Commission in 1975, which minimized the dangers of reactors, was widely criticized. Since 1974, not a single utility in the United States has ordered a reactor, although American manufacturers continue to construct them for other nations. About 100 aging reactors provided around 20 percent of domestic electricity in the United States in 2000. The corresponding numbers were considerably higher in France and Japan.

The U.S. Atomic Energy Commission (AEC) began to promote construction of civilian reactors in the 1950s but did not give sufficient attention to waste disposal. High-level radioactive wastes were planned to be stored "permanently" in abandoned salt mines in Kansas in the 1970s. When Kansas state geologists showed the presence of drill shafts that would allow the entrance of rainwater, the AEC abandoned the site.

No state then volunteered to accept a high-level waste repository. Spent fuel rods accumulated in the already crowded cooling ponds at each reactor location. In the 1980s, the U.S. Congress forced development of the Yucca Mountain site in Nevada, but even after the expenditure of more than four billion dollars by the end of the twentieth century, all the technical problems of containing radioactivity there for ten thousand years had not been solved.

The states of Nevada, Washington, and South Carolina accepted low-level radioactive wastes from universities, hospitals, and industry until the accident at Three Mile Island in Pennsylvania (1979) and the catastrophe at Chernobyl, Ukraine (1986). So-called NIMBY ("not in my backyard") opposition, often based on the fear that radioactivity would reach urban water supplies, then stifled progress in waste disposal. Having brought the nuclear reactor industry to a standstill in the 1970s, the antinuclear movement joined with state and local officials and increasingly sophisticated antinuclear and environmental organizations, such as the Sierra Club, Natural Resources Defense Council, and Union of Concerned Scientists, to challenge proposals for siting storage facilities.

Elsewhere, opposition to reactors has also been effective. In 2000 Germany decided to close all its nuclear power reactors within twenty years. The Green party, one of the few environmental groups to have members elected to a national parliament, turned its once-ridiculed position into national policy. Civilian antinuclear activities have also succeeded in opposing nuclear-powered merchant ships, which many ports now ban. Fears of radioactive contamination faced the American *Savannah* (launched in 1962), the German *Otto Hahn* (1964), the Japanese *Mutsu* (1972), and the Soviet *Sevmorput* (1989). Like reactors used to produce electricity, propulsion reactors changed from a once-promising technology into expensive, controversial, and feared objects. It would be premature to bury nuclear power, however. As other sources of energy fall behind the world's demand and air pollution from fossil-fuel plants reaches unacceptable levels, there may be no alternative but self-denial to the nuclear reactor.

Christopher P. Driver, *The Disarmers* (1964). Michio Kaku and Jennifer Trainer, eds., *Nuclear Power: Both Sides* (1982). Jonathan Schell, *The Fate of the Earth* (1982). Milton S. Katz, *Ban the Bomb: A History of SANE, the Committee for a Sane Nuclear Policy, 1957–1985* (1986). Lawrence S. Wittner, *One World or None: A History of the World Nuclear Disarmament Movement through 1953* (1993). Lawrence S. Wittner, *Resisting the Bomb: A History of the World Nuclear Disarmament Movement, 1954–1970* (1997).

LAWRENCE BADASH

ANTISCIENCE is a phantom phenomenon. There are no self-described anti-scientists or degrees in antiscience. Even the most anti-modernist movements oppose science in the name of a better science. Yet the specter of antiscience has haunted natural science from its beginnings into the twenty-first century.

No society can oppose science understood in its generic sense of "systematized knowledge derived from observation, study, and experimentation" (*Webster's New World Dictionary*), and survive long enough to discuss its opposition. What has been passionately resisted through history are those tendencies of the scientific enterprise that seek autonomy of nature from the supernatural and the independence of human reason from the dictates of the customs and commonsense of the age. Not science per se, but the attempt of science to differentiate itself from theology, philosophy, and cultural commonsense, has fueled the rage against science after every episode of its development and growth.

Antiscience sentiments and movements in the contemporary world are, to paraphrase Karl Popper, "shock waves" generated by the birth of the open society midwifed by the *Scientific Revolution, the *Enlightenment, and capitalism. However much they may differ in their modes of production and local customs, traditional or premodern societies everywhere were and are closed societies. Closed societies do not distinguish the nomos from the cosmos, or the conventional, societal laws from the divinely ordained laws of nature. Social norms are understood and experienced as dictated by the order of nature, created and sustained by a law-giver god, or by an immaterial, all-pervading spirit. For all its valiant attempts to accommodate the Newtonian worldview with the traditional theistic conception of god, the Scientific Revolution laid the groundwork for the Enlightenment, which demanded that theology justify itself by the methods of science. Gradually, the link between god, nature, and humanity was broken: God was not banished but denied the authority to explain the natural world. To the extent that traditional morality invoked the divinely ordained order of nature as a source of morality and ethics, traditional ideas of right and wrong, too, could not stand unchanged.

All antiscience and anti-Enlightenment movements—whether looking backward to the lost community of the *Volk*, or looking forward to future Utopias—display a strong urge to restore the lost unity between humanity, nature, and god. They see the disenchantment of nature as modern humanity's hubris against god, and

the claims of value-freedom and universality as modern science's hubris against centuries of wisdom of various traditions, big and small, around the world. Thus, calls for the re-enchantment of nature and re-contextualization of science recur repeatedly in the movements for "theistic science" (Creation, Islamic, or Vedic sciences) or "alternative sciences" (feminist, proletarian, or non-Western "ethno" sciences).

Totalitarian Antiscience Movements: Nazism and Communism

Nazism was the first organized antiscience movement that acquired state power and translated its worldview into a concrete social program—with disastrous results. Faced with the social ferment caused by the French Revolution and rapid but late industrialization, prominent nineteenth-century German Romantic intellectuals tended to ascribe these problems to the dualism created by modern science between a mechanical universe and a transcendent god. Some of them (notably Alfred Rosenberg, the chief ideologue of the Nazi party) further ascribed the dualist tendencies of modern scientific *Weltanschauung* to the influence of the Judaic conception of a radically transcendent god that had "corrupted" Christianity. The Nazi ideologues sought to overcome the dualist "Jewish" science with a holistic "Aryan" or "Nordic" science that could grasp the unseen but all-pervading divine life force that connected mind to matter and god to humanity and nature. The proponents of Aryan science, including a handful of prominent physicists and a large number of well-respected men of letters, believed that objectivity and autonomy in science were merely slogans invented by self-serving scientists. To quote Adolf Hitler, "There could only be the science of a particular type of humanity, and of a particular age. There is very likely a Nordic science and a National Socialist science, which are bound to be opposed to the Liberal–Jewish science." The vitalistic ontology and the holistic epistemology of "Nordic science" encouraged many experiments in the occult, like *phrenology, *astrology, and theosophy; food fads; pagan rituals; and also serious scientific programs in *biology, *ecology, and organic agriculture (*see* NAZI SCIENCE).

As a doctrine based on *dialectical materialism, *Marxism has no use for the vitalistic holism of Nordic science. The unity that Marxist regimes sought was between hand, heart, and mind, or between productive labor and scientific knowledge. Theorists made the major advances in modern science fit the fundamental Marxist assumptions that material existence shapes consciousness and that working classes have a consciousness with less interest in maintaining falsehoods and illusions than do the ruling classes. Thus, the Soviet Union under Stalin deemed Lysenko's *Lamarckism closer to reality than ordinary *genetics (*see* LYSENKO AFFAIR) because it reflected proletarian interests, while rejecting *quantum physics because it denied class struggle and the materiality of nature. Marxism denied the autonomy of modern science to set its own norms and standards of validity; science would be the handmaiden of the secular theology of Marxism.

The Postmodern Route to Antiscience

After a brief outbreak of hope for liberal modernity that followed the defeat of Nazism and the end of colonialism, the twin themes of the re-enchantment of nature and the re-contextualization of science have returned with a vengeance. This time, however, these ideas are not sponsored by totalitarian states (although postcolonial states make opportunistic use of them). Their chief sponsors are the populist, transnational new social movements that espouse ideals often associated with the political left: anticolonialism, opposition to global capitalism, feminism, protection of the cultural rights of minorities at home and "the other" elsewhere, and protection of the natural environment, to name just a few.

Like the cultural pessimists in fin-de-siècle Germany, recent postmodernist critics tend to blame the nature of scientific enterprise itself—especially its universality, value-freedom, and "reductionism"—for colonialism, patriarchy, racism, and the instrumental rationality of capitalism. Postmodernism slates scientific knowledge for serving Western power by defining reality in Eurocentric and patriarchal categories, thereby silencing the traditional sciences of non-Western people, women, and other historically oppressed social groups. Whereas modernist intellectuals in newly independent, ex-colonial countries, including those representing such oppressed people such as the untouchable castes in India, had embraced modern science in order to demystify traditional cosmologies that often justified highly non-egalitarian cultural traditions, postmodernist intellectuals have come to see the same traditional cosmologies as a source of emancipatory and ecologically sustainable knowledge. These postmodernists thus regard modern science, with its links to global post-industrial capitalism, as a greater evil than the customary, often religion-sanctioned, sources of oppression in closed societies.

The fundamental urge of all antiscience movements—to oppose the differentiation and relative autonomy of scientific knowledge from the *Geist*

or commonsense of the rest of the culture—is the motivating force behind social and cultural studies of science. These postmodernist approaches explicitly set out to bridge the gap between the prevailing social interests or cultural biases in a society and the content of what that society accepts as valid knowledge of nature. Thus the "Strong Programme" in sociology of scientific knowledge (SSK) offers to explain the "very content and nature of scientific knowledge. . .and not just its conditions of production" in sociological terms. SSK argues that all knowledge-systems are equally local and context-bound in their ways of justifying their beliefs. What gets accepted as logical and rational in any society is decided by the social values, goals, and metaphysical assumptions of the inquirers. A corollary is "epistemic charity" toward irrational beliefs: if modern science is a local construct like all other knowledge, we could abandon it without fear of irrationality, or at the minimum, we should feel no obligation to gauge the validity of other beliefs against scientifically justified beliefs in the same domain.

This "charitable" stance has become axiomatic among postmodern and multiculturalist critics of science. Any attempt to critically assess premodern understandings of nature—which typically do not distinguish workings of nature from the will of god, spirit, or other extra-sensory, untestable forces—against scientific knowledge is condemned as a colonial or Orientalist mindset. Feminist and postcolonial proponents of "alternative sciences" go an extra step and *reverse* the direction of critique: they hold up the methods and content of modern science for a critical examination from the vantage point of local knowledge of women and non-Western cultures, most of which have not yet undergone the process of disenchantment and secularization. The reversal is expressed in activist movements for ecology and feminism, which routinely invoke the goddess as a symbol of women's empowerment and ecological sustainability. These enthusiasts do not adequately understand the role that the goddesses and the idea of sacredness of nature have played (and still play) in perpetuating the oppression of actual women.

Far from being an isolated activity confined to the laboratory, science exerts a powerful influence on power, production, and belief. The danger of postmodern antiscience movements is that while they have failed in preventing postcolonial states from acquiring the most modern of technologies to enhance their power and production, they have succeeded in preventing science from entering—and altering—the belief systems of postcolonial societies. Thus we see a proliferation of postcolonial states that bolster their power and prestige with cutting-edge technologies, including nuclear weapons, while simultaneously reviving religious dogmas and traditional values as antidotes to the "western" ideas of secularism and individualism. The coming together of the instrumental rationality of modern technology with the rejection of the Enlightenment values of reason and secularism is the essence of "reactionary modernity," first seen in Nazi Germany.

State-sponsored antiscience movements flourish once again—this time in societies that have taken a turn to religious nationalism or fundamentalism. The Hindu nationalist government in India, for example, announced in March 2001 its decision to teach Vedic astrology as a part of the science curriculum in public sector colleges and universities. The Indian government is also sponsoring "research" on the practice of faith-healing, Vedic techniques of rain-making, air-purification and architecture, telepathy, parapsychology, and reincarnation. This growing obscurantism is openly defended as politically "progressive" (because "decolonizing the mind") and "scientific" by local standards set by a holistic Hindu metaphysics. At a more theoretical level, Hindu philosophers selectively interpret developments in quantum mechanics and ecology to declare Vedantic theosophy to be not just compatible with modern science but to be ahead of it. This selective reading of scientific evidence to support theological dogma is the mainstay of creation science in the West. Fundamentalist Islamic regimes, especially in Saudi Arabia and Pakistan, actively promote all kinds of miracles and superstitions as a part of Islamic science.

Whatever their doctrinal differences, all "religious sciences" justify themselves as healing the rift in the "sacred canopy" of closed societies inflicted by modern science, and as providing a fully rational, holistic science equal to, if not superior to, the reductionist science of the West. The new religio-political movements are fulfilling the postmodern agenda.

As the history of Fascism and Communism testifies, the coming together of state power with irrationalism portends grave danger. A strenuous defense of the rationality of science, the Enlightenment, and open societies is the only antidote to the dangers that lie ahead.

See also Anti-nuclear Movement; Outsiders and Minorities; Science Wars.

Karl Popper, *The Open Society and Its Enemies. Vol. 1: The Spell of Plato*, 5th ed. (1962). Jeffrey Herf, *Reactionary Modernism: Technology, Politics and Culture in Weimar and the Third Reich* (1984). Robert Pois, *National Socialism and the Religion of Nature* (1985). Partha Chatterjee, *Nationalist Thought and the Colonial World: A Derivative Discourse* (1986). Anna Bramwell, *Ecology in the Twentieth Century* (1989). Pervez Hoodbhoy, *Islam and Science: Religious Orthodoxy and the Battle for Rationality* (1991). Ernest Gellner, *Reason and Culture: Historic Role of Rationality and Rationalism* (1992). Gerald Holton, *Science and Antiscience* (1993). Anne Harrington, *Reenchanted Science: Holism in German Culture form Wilhelm II to Hitler* (1996). Paul Josephson, *Totalitarian Science and Technology* (1996). Swami Mukhyananda, *Vedanta in the Context of Modern Science: A Comparative Study* (1997). Sandra Harding, *Is Science Multicultural?* (1998). Lance E. Nelson, ed., *Purifying the Body of God: Religion and Ecology in Hindu India* (1998). Alf Hiltebeitel and Kathleen M. Erndl, eds., *Is the Goddess a Feminist? The Politics of South Asian Goddesses* (2000).

MEERA NANDA

ANTISEPSIS is commonly defined as the use of chemical agents to disinfect an operative area. Its theoretical origins lie in Louis *Pasteur's *germ theory. Through elegant experiments conducted during the 1850s and 1860s, he demonstrated that living microorganisms caused the chemical processes associated with fermentation and putrefaction. Pasteur's theory gradually replaced the then-common idea that miasmas or a decreased vital force created these phenomena.

Joseph Lister, a Glasgow surgeon, heard of Pasteur's work and recognized the connection between it and wound infection. He concluded that if he could kill the septic microorganisms he could reduce the high mortality rates associated with all but the simplest surgical procedures. He chose carbolic acid and in 1865 began to develop a rigorous method of killing or excluding microorganisms from the operative area. After a series of eleven operations on patients with compound leg fractures—nine successful, one leg requiring amputation due to gangrene, and one mortality—Lister began to publish his results in 1867. His method consisted of washing with a carbolic acid solution anything that came into contact with a patient during an operation and using carbolic acid–soaked catgut and dressings on the wound. In 1870 he introduced the carbolic acid spray, which soon became the method's icon.

Although Lister's antiseptic method dramatically lowered surgical mortality, it took almost twenty years before surgeons generally accepted it. It was exacting in detail and difficult to follow; the smallest distraction could allow infection. Other surgeons had proposed techniques that gave equally or almost equally favorable results, so Lister's complicated method faced lively competition. Surgeons found carbolic acid to be a noxious agent. Not all patients tolerated it well, and many surgeons experienced eye or skin irritation brought on by long exposure to it. Furthermore, not all surgeons were as easily persuaded as Lister had been that Pasteur's germ theory was correct. However, as evidence favoring the germ theory accumulated during the 1870s and 1880s, surgeons increasingly accepted antisepsis.

Beginning in the 1890s, asepsis replaced antisepsis. Asepsis emphasizes mechanical, as contrasted to chemical, means to create a germ-free operating space: autoclaves sterilize instruments with steam; ultraviolet light also destroys microorganisms. Robert Koch and Lister had independently discovered that airborne microorganisms are rarely as pathogenic as those found on surgeons or instruments. Consequently, Lister no longer insisted on the spray and lowered the concentration of the carbolic acid solutions. William Macewen's introduction of the surgical gown and William Stewart Halsted's of surgical rubber gloves further reduced the patient's risk of infection.

Antisepsis and asepsis are the bright line marking surgical epochs, for they both created modern surgery and shaped the development of surgery, hospitals, and the medical and surgical professions. Surgical anesthesia (1846) controlled pain and allowed surgeons more time to perform operations. Before antisepsis, surgeons confined themselves mostly to fixing superficial wounds; control over infection allowed them to explore other regions of the body as well. Theodor Billroth advanced operative techniques for stomach surgery, while Harvey Cushing almost single-handedly created neurosurgery.

Surgery's increased safety did not create a demand for elective surgery, but it did hasten the transfer of medical care away from a patient's home into a hospital. Safe surgery relied on instruments that surgeons could not carry from place to place, so *hospitals became the sites for operations. In turn, hospitals became increasingly more complex organizations, for they, not the surgeons, became responsible for not only the space but also the tools and ancillary nursing and technical support personnel. This shift increased the hospitals' financial burdens, so they had to learn how to finance charity and health care. Early on, surgeons had prepared their own antiseptic dressings; as antisepsis spread, industries grew that provided standardized chemicals, dressings, and instruments.

Better surgical techniques increased the prestige of the profession and its specialization. In a time when physicians could do little to alleviate illness, surgeons' bold advances turned them from medical pariahs into medical heroes.

Historians debate whether antisepsis and its underlying theoretical foundation made medicine more "scientific," but late-nineteenth-century and early-twentieth-century physicians and surgeons certainly claimed that antisepsis created a scientific and therefore more authoritative and privileged profession.

Leo M. Zimmerman and Ilza Veith, *Great Ideas in the History of Surgery*, 2d ed. (1967). Roy Porter, *The Greatest Benefit to Mankind* (1997). James Le Fanu, *The Rise and Fall of Modern Medicine* (1999). Harold Ellis, *A History of Surgery* (2001).

Thomas P. Gariepy

APPARATUS. The terms "apparatus" and *"instrument" have not been used consistently by scientific practitioners. At the root of "apparatus" is the notion of "preparing" or "making ready"; as the mechanical requisite for scientific experiments or investigations, "apparatus" only gained currency in the late seventeenth or early eighteenth century. "Instrument" referred to a tool, an implement, or a weapon, but in time it began to denote a device for more delicate work, such as making music or philosophical (scientific) experiments. As natural science developed, "instrument" came to refer to a specific device, and "apparatus" to a set of devices making up an experiment; thus, mathematical, optical, and philosophical "instruments," but electrical and chemical "apparatus."

The three broad categories of mathematical, optical, and philosophical instruments had evolved by the eighteenth century. Mathematical instruments consisted of devices used in *astronomy, time-telling, navigation, and surveying. These were angle-measuring instruments that originated from the most ancient, the *astrolabe. They were two-dimensional (apart from the armillary sphere) and constructed of a limited range of metals and using a limited range of techniques that had hardly changed for centuries. Optical instruments evolved next. Disregarding the ancient burning mirrors and the eye glasses of the late Middle Ages, these began with the invention of the *telescope in the late sixteenth century, probably in Middleburg in the Netherlands. The telescope opened up the instrument-making trade since it employed a variety of materials (vellum, wood, glass, metals) and required a wide set of craft skills in its manufacture. The third category, philosophical instruments, so named in the late seventeenth century, consisted of devices such as the air pump and the electrical machine. These instruments relied on a great variety of craftsmen—cabinet-makers, glassmakers and metalworkers—and stood at the forefront of the technologically possible. Several of these necessary skills came to prominence during the Industrial Revolution. For instance, increasing the vacuum of the air pump relied on improving the boring of the barrel and the manufacture of small parts to a high degree of accuracy, techniques developed in the manufacture of steam engines. Metallurgy also had to be improved to prevent engine parts from failing. The diverse skills required by instrument makers (casting, forging, engraving, silvering and gilding, leather and wood-working, and so on) meant that these craftsmen were not confined to one trade or guild. By the eighteenth century, England (and more specifically London) had become the center of the instrument-making trade, losing this dominance by the time of the Great Crystal Palace Exhibition of 1851.

In the last decades of the nineteenth century, science became increasingly professionalized. An important factor in this was the growth of teaching and industrial laboratories, placing new demands on the instrument-making trade. Manufacturing techniques of scientific apparatus remained predominantly pre-industrial, carried out by skilled craftsmen with little in the way of mass-production techniques. The early twentieth century saw the beginning of a general restructuring of the instrument-making trade from craft-based to industrial. Many of the best-known nineteenth-century firms amalgamated. World War I accelerated the process when apparatus for the scientific conduct of the war became strategically important for the first time.

Because of the demands of laboratory research, the earlier brass-and-glass apparatus rapidly became obsolete. It was replaced by items made of plastics, ceramics (for insulation), and metals such as aluminium. By 1930, functional design had become a major concern, and all traces of earlier forms of decoration had been eliminated. More important for the design of scientific apparatus between the wars was the rise of electronic technology. The incorporation of electronics—thermionic valves in the 1920s and transistors in the 1950s—led to the redesigning of traditional apparatus for control and measure and to the development of entirely new devices such as the oscilloscope. Integrated circuits appeared at the end of the 1950s, which led to the miniaturization of increasingly complicated electronic circuitry and the change from electromechanical analogue apparatus to digital ones connected to powerful computers.

G. L'E. Turner, *Nineteenth-Century Scientific Instruments* (1983). A. J. Turner, *Early Scientific Instruments: Europe, 1400–1800* (1987). A. J. Turner, *Mathematical Instruments in Antiquity and the Middle Ages: An Introduction* (1994). Paolo Brenni, "Physics Instruments in the Twentieth Century," in John Krige and Domenique Pestre, eds., *Science in the Twentieth Century* (1997), pp. 741–757.

W. D. HACKMANN

APPLIED SCIENCE. See BASIC AND APPLIED SCIENCE.

ARCHETYPE. The concept of an archetype (from the Greek *arkhe*, "original," and *tupos*, "imprinted image") has connotations that range from Plato's theory of ideas to Carl Jung's notion of pervasive cultural symbols in our collective unconscious. In the context of the natural sciences, the notion of archetype figured prominently in vertebrate morphology, during the "morphological period" in the history of biology (approximately 1800–1860). The vertebrate archetype represented the generalized and simplified skeleton of all backboned animals, to which the many parts of real skeletons of fishes, amphibians, reptiles, birds, and mammals could be reduced on the basis of their homological relations. The archetype construct gave expression to a belief in the fundamental relatedness, if not of all organisms, at least of all animals with endoskeletons.

The idea of a vertebrate archetype found its most authoritative and complete formulation in *On the Archetype and Homologies of the Vertebrate Skeleton* (1848) by the London biologist and conservator of the Hunterian Museum, Richard Owen. The first use in osteology of the term "archetype" was not Owen's, however; it occurs in a paper in *Lancet* in 1846 by the minor Scottish morphologist Joseph Maclise. Also, the famous archetype figure drawn by Owen was not original, but represented an improved version of a drawing by the German comparative anatomist and polymath Carl Gustav Carus in his *Von den Ur-Theilen des Knochen- und Schalengerüstes* (1828).

Owen initially interpreted his archetype as an Aristotelian, vaguely pantheistic entity, distinguishing it from a platonic idea and following the nature-philosophical speculations of Lorenz Oken. An organic body develops—Owen maintained—by the interaction of two antithetical forces, one the "special organizing force," the other the "polarizing force." The archetype was a manifestation of the second force, which produces repetition of parts and thus unity of organization. The first force was the Platonic idea, producing specific modifications and adaptations. In a later book, *On the Nature of Limbs* (1849), Owen redefined his vertebrate archetype as a Platonic idea—a blueprint in the mind of the Creator. This he did under pressure of leading Anglican patrons of his museum, who worried about the vaguely evolutionary pantheism of German nature philosophy. They regarded this as a threat to Paleyan natural theology (*see* *RELIGION AND SCIENCE). Owen placated his supporters by Platonizing the vertebrate archetype, thereby reintegrating animal morphology into a traditional, teleological epistemology (*see* TELEOLOGY).

*Darwinism absorbed much of Owen's homological research program as the archetype became an evolutionary ancestor and homological relationships were used to establish phylogenetic descent. Yet the vertebrate archetype was a more complex, richer model than that of a primordial ancestor. It functioned as a guide not only to the homological connections between different groups of vertebrates, but also to the homologies that occur within the skeleton of a single vertebrate, such as the serial repetition of vertebrae along the backbone. In recent decades, in the context of molecular genetics, Owen's archetype has received a renewed appreciation for this complexity.

Nicolaas A. Rupke, "Richard Owen's Vertebrate Archetype," *Isis* 84 (1993): 231–251. Alec L. Panchen, "Richard Owen and the Concept of Homology," in *Homology: The Hierarchical Basis of Comparative Biology*, ed. Brian K. Hall (1994): 21–62.

NICOLAAS A. RUPKE

ARISTOTELIANISM became the philosophy of the newly formed universities of Oxford and Paris and other higher schools during the thirteenth century. The Council of Trent (1545–1563) declared Thomistic Aristotelianism to be the official doctrine of the Roman Catholic church. Aristotelianism retained its cultural, theological, and intellectual dominance even in Protestant countries well into the seventeenth century although Aristotelian natural philosophy met with telling criticism during the *Scientific Revolution. Elements of Aristotle's philosophy continued to influence the development of science well into the nineteenth century.

In 1586 the Jesuits introduced the *Ratio studiorum* (revised in 1599 and 1616) that established the standard texts used for teaching Aristotelian philosophy at their colleges, including La Flèche (where René *Descartes was educated) and their central teaching establishment, the Collegio Romano. Many Christian non-Jesuitical institutions adopted the *Ratio* in one form or another. In general, the texts favored by the Jesuits described a physical world consisting of four elements (earth, air, water, and fire) that in varying combinations composed all earthly bodies. The texts portrayed a geocentric universe and maintained that the Copernican heliocentric theory of 1543 was at best a mathematical tool for calculating the positions of celestial bodies. Traditionalists like the Jesuit cardinal Robert Bellarmine of the Collegio Romano and the authors of the commentaries on Aristotle's *Libri naturales* issued by the University of Coimbra around 1600 showed how to reconcile new observations with Aristotelian theory. Philosophers like *Galileo's friend and

colleague at the University of Padua, Cesare Cremonini, defended these accommodations because they preserved the systematic explanations of Aristotle against Galileo's piecemeal, unsystematic approach. The close relation between Catholic dogma and Aristotelian doctrine also encouraged continuing with the older philosophy.

Increasingly, the alterations produced by individuals in an attempt to keep Aristotelianism coherent in the face of new observations in astronomy and elsewhere resulted in the erosion of agreement on specific philosophical details. Despite these variations, however, Aristotelianism continued to offer a framework for constructing plausible explanatory systems. The framework considered natural bodies in terms of matter, privation, and form, which gave individual portions of matter the qualities that characterize particular bodies. Four different causes elucidated the generation, construction, or alteration of a body: the material cause out of which the change comes; the formal cause that provides the specifying or defining factor; the agent or efficient cause that initiates the change; and the final cause that provides the purpose (*telos*) for which a thing came into existence.

The *mechanical philosophy of the seventeenth century subverted the framework. In his *Discourse on Method* (1637), Descartes rejected the notion of final causality and the theory of four qualititively different elements in favor of one undifferentiated matter. In place of the Aristotelian qualitative approach to explanation, Descartes defended, though he seldom practiced, a quantitative and mathematical method that referred the manifest properties of bodies to the effects of bits of matter in motion. Despite these disagreements of principle, Descartes wrote in a systematic, deductive style agreeable to the Aristotelians, and he presented his works to his former teachers, the Jesuits at La Flèche and Paris, explicitly asking for, although never receiving, their approbation.

In England, Francis Bacon directed his criticism of the Aristotelian framework primarily against its use of syllogistic reasoning to construct elaborate theoretical systems. In *The New Organon* (1620), he argued that the syllogism should serve only as a method of demonstration. It could not be used for the discovery of new knowledge or for the production of new works. In place of the syllogism, Bacon advocated an inductive method based upon the careful compilation and analysis of observational and experimental data. Bacon also rejected the notion of the four elements. He retained the notion of forms, but he described them in a mechanical fashion as the laws that governed the motions of bodies. Although Descartes and Bacon both rejected the Aristotelian conception of final causality, not all mechanical philosophers did. Robert *Boyle, who brought together the Baconian and mechanical programs in the next generation, maintained that teleological inferences, such as those used by William *Harvey in elucidating his discoveries concerning the circulation of the blood (1628), were a necessary adjunct to mechanical hypotheses in biological investigations.

Parts of the works of Galileo, Descartes, Bacon, and Boyle were placed on the church's *Index of Prohibited Books* and opposed by the Jesuits. Aristotelians such as Antoine Goudin, Jean-Marie-Constant Duhamel, and Jacques Grandamy continued to write well into the eighteenth century, often setting the agenda and parameters for philosophical debate. By the nineteenth century Aristotelianism was no longer viable as a natural philosophy. Even so, remnants of it continued to be influential, primarily in the study of living matter. Thomas Aquinas had used Aristotle's natural philosophy to construct a teleological proof for the existence of God. In the seventeenth century John Ray argued for the necessity of a designer God based upon considerations drawn from observations of the complexity and organization exhibited in living systems. Although belief in the purposefulness of nature opposed strict mechanistic laws, at the end of the seventeenth century Richard Bentley constructed teleological arguments based upon considerations drawn from Newtonian mechanics. William Paley would continue to popularize this type of argument in his *Natural theology* (1802). By the mid-nineteenth century, however, defenders of Darwinian *evolution rejected the Aristotelian idea of purposes in nature, and the stage was once again set for a fierce battle between *religion and science.

Stillman Drake, *Galileo at Work* (1978). James G. Lennox, "Robert Boyle's Defense of Teleological Inference in Experimental Science," *Isis* 74 (1983): 38–52. C. B. Schmitt, *Aristotle and the Renaissance* (1983). Norma E. Emerton, *The Scientific Reinterpretation of Form* (1984). Dennis des Chene, *Physiologia. Natural Philosophy in Late Aristotelian and Cartesian Thought* (1996). Markku Peltonen, ed., *The Cambridge Companion to Bacon* (1996). Roger Ariew, *Descartes and the Last Scholastics* (1999).

ROSE-MARY SARGENT

ART AND SCIENCE have been interdependent in highly creative ways since the fifteenth century. Artists have utilized scientific and mathematical principles since the popularization of perspective in Leon Battista Alberti's *On Painting* (1435–36), which rested on geometrical principles and classical optics. Combined with Giotto's development of chiaroscuro—the juxtaposition of light and dark colors to create volumetric form—in the fourteenth century, perspective gave the illusion

of three-dimensionality on a two-dimensional surface and so fostered the belief that the artistic representation of nature was an accurate depiction of its reality. Perspective machines, including the perspectograph (Ludovico Cardi da Cigoli) and pantograph (Christoph Scheiner) invented in the seventeenth century, suggest how far perspectival representation was thought to capture nature itself. Isaac *Newton argued in his *Optics* (1704) that color was a sensation produced by light on the eye (rather than a property of surfaces) and white light a composite of rays "differently refrangible" (rather than a homogeneous absence of all color). Although challenged by David Brewster, Johann Wolfgang von Goethe, Joseph Priestley, Benjamin Thompson, Thomas Young, and others in the late eighteenth and early nineteenth centuries, Newton's theory of color, especially his color circle, had a profound effect on the development of color theory in the nineteenth century.

Color was a key concern in the textile industry, a leading sector of European industrialization, and by the early nineteenth century physiologists, chemists, and physicists made color, including its applications to industry and design, a strategic area of investigation. In 1839 the French chemist Michel-Eugène Chevreul, the director of dyeing at a tapestry factory, discovered the law of simultaneous contrast: one color takes on the hue complementary to its neighboring color. His discovery rested in part on the study of a modified Newtonian color wheel. Eugène Delacroix applied Chevreul's law in his paintings, producing images striking for their emotional impact, such as his *Death of Sardanapalus* (1827). Fascinated by theories of light and color developed in the wake of Newton, Joseph Mallord William Turner abandoned the Euclidean box for a space created by luminosity, as in *Keelman Heaving Coals by Moonlight* (1835) or, more strikingly, *Light and Color (Goethe's Theory)—The Morning after the Deluge* (1843). But it was Delacroix's experiments with color that influenced a new generation of painters who experimented with the creation of form through the juxtaposition of minute amounts of complementary colors, including impressionists like Claude Monet or neo-impressionists like the pointillist Georges Seurat. Empiricist studies of vision, of which a leading proponent was Hermann von *Helmholtz, also strongly influenced the development of impressionism, including the work of Seurat. Seurat's *Sunday Afternoon on the Island of La Grande Jatte* (1884–1885) unites several different features of nineteenth-century color theory, including the notion that the retina, not the painted surface, creates the images as well as the effects of light and color. Later the development of photographic techniques, which

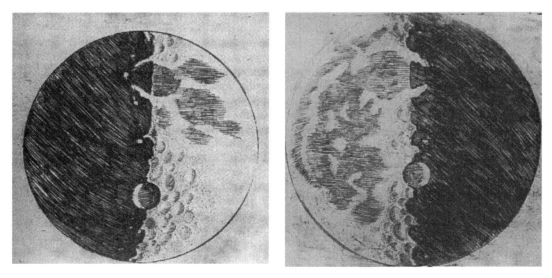

Created by an engraver from Galileo's wash drawings for *The Starry Messenger* (1610), these two images of the moon deploy the Italian techniques of perspective and chiaroscuro to accentuate the ring mountain of the crater Albategnius. Galileo, a skilled draftsman, used the geometry of shadows to depict the moon's earth-like topography.

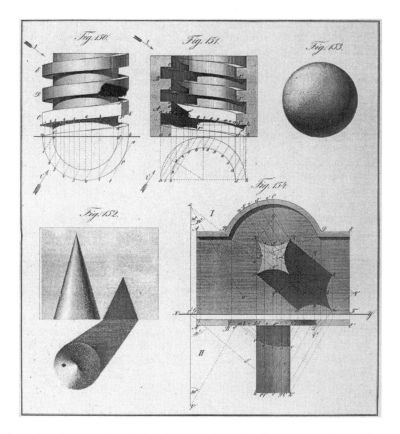

Meno Burg's geometrical drawing blended descriptive geometry with the arts of perspective, illumination, and rendering of shadows. These figures, modeled on architectural elements, were used for the manufacture of precision interchangeable parts.

allowed for the production of frames closely separated in time, influenced artistic representations of motion, including Marcel Duchamp's *Nude Descending a Staircase* (1912).

Since the *Scientific Revolution, words and numbers have not always been sufficient to represent either knowledge about nature or the protocols used to attain it. Although the printing press did not immediately change the custom of using traditional images in scientific and medical textbooks, by the time of Andreas *Vesalius's *De humanini corporis fabrica* (1543) illustrations based on empirical observations (and perspective and chiaroscuro) had begun to replace them. The Reformation divided supporters of images: Catholics, and especially the Jesuits, pro, Protestants contra. After the Peace of Westphalia in 1648, which was supposed to end religious wars in Europe, painting and image construction in the Protestant world revived and new visual cultures appeared. Illustrations became especially important in rendering instrumental findings of the seventeenth century. A master of perspective, *Galileo used it and chiaroscuro in *Starry Messenger* (1610) to present his telescopic discoveries—the surface irregularities of the moon, the moon's light, and the phases and positions of the new moons he found around Jupiter—which challenged views of the heavens held since Aristotle. Aristotelians in Rome, shaken, debated to what degree seeing could be regarded as knowing. Robert *Hooke's *Micrographia* (1665) popularized the role of the microscope in constructing images below the threshold of everyday sight. In some cases artistic representation rendered discoveries without words that could describe them. Maria Merian's *Metamorphosis of the Insects of Surinam* (1705) depicted insects and amphibians in stages of development alongside plant vestiges of their natural habitats, but she did not explain the meaning of her ecologically sensitive style of representation. Natural philosophers have used images not only simply to render nature, but also to help them see in the particularities of the natural world the universal principles upon which scientific knowledge is based.

The development of technical drawing added another artistic dimension to science. Beginning in the eighteenth century, illustrations of scientific instruments filled textbooks in the physical sciences. Typical for the time was Jean Antoine Nollet's *Leçons de physique expérimentale* (six volumes, 1749–1764). In England, Joseph Priestley pointed out in 1780 that words, written or verbal, were insufficient for rendering instructions suited to replicate experiments because they could not accurately convey the quantitative dimensions of real objects that perspective, which he regarded as a mathematical science, could. Illustrations of instruments and apparatus, objects of investigation, and experimental techniques as drawn by Priestley and others helped to standardize scientific protocols, aiding not only scientific communication, but also training in laboratory work. In the first half of the nineteenth century, perspective merged with projective geometry and the early modern engineer's cutaway, transparent, and exploded views (as were found in Agricola's *De re metallica liber XII* [1556]) to create geometrical or mechanical drawing. First in France at the École polytechnique and then at German military schools, especially in Berlin under Meno Burg, geometrical drawing became the foundation of objects manufactured within standards of tolerance set by gauges, including scientific instruments.

In some scientific disciplines important epistemological, ideological, and even ethical questions have been raised about the relationship between artistic representation and reality. Around 1800 French artists and scientists, reviving the issues raised by Galileo's critics, questioned whether perspective rendered visual truth, while French thinkers valued reason over the senses in interpreting visual reality. The proliferation of mechanical means of rendering accuracy in the nineteenth century—when the camera obscura, camera lucida, magic lantern, *microscope, *telescope, and instruments for accurate perspectival drawing were joined not only by the diorama, panorama, ophthalmoscope, photometers, and stereoscope, but also by lithography, *photography, and the *X ray—seemed to eliminate the aesthetic and subjective in representation. But what is "true to nature" and objective? The individual with its particularities and imperfections or the "type" that depicted the universal features of the object of investigation? William Hooker's fruits in *Pomona Londinensis* (1813–1815), drawn for the Horticultural Society of London, or Pierre-Joseph Redouté's *Les Roses* (1817) or *Les Lilacées* (1802–1816) represent the typical—all three volumes have been used by scientists to identify species—but are the images real? Conversely, photographic images, though far more accurate than ideal types, are hard to use to identify species. Francis Galton turned the photograph's liability into an asset when by mechanical means he created from the facial features of criminals a composite photograph of the "typical" criminal. Perhaps no mechanical reproduction is absolutely real or entirely free of social, cultural, ideological, and other biases.

Yet in the sciences that depend on visualization more than others—including *anatomy, *biology, *botany, *geology, *mineralogy, *natural history, *paleontology, *spectroscopy, and *zoology, to name a few—objects "drawn from nature," ideologically tainted or not, serve crucial roles in the creation of universals. The massive production during the nineteenth and twentieth centuries of atlases in these visually-based sciences testifies to the crucial epistemological role that images play not only in identification, but also in the construction of objects of investigation and in the training of the scientific eye to recognize optical consistency. Scientific illustration accomplishes by visual representation what measurement does in the quantitative sciences: it averages the features of the natural world. Only in those sciences directed at understanding culture, such as archaeology, does the particular take precedence over the universal in artistic representations, for particularity is essential to rooting the object in a specific time and place.

From idealization to deception and illusion is not a big step. Paintings portray miracles. Science reflects them. To dispel supernatural and miraculous origins of the Virgin Mary's "tears" in a painting, Johann Nepomuk Fischer in 1781 took recourse to early photometric techniques, especially as developed by Johann Heinrich Lambert; he successfully separated optical illusion from reality. In the reverse direction, devices used to create phantasmagoria and optical illusions in the nineteenth century were successfully modified to solve scientific problems. Photometric techniques depended upon nineteenth-century projection culture, including the magic lantern (an early form of the slide projector), which offered ways of displaying images so that luminosity and hue could be compared and quantified. Karl Zöllner's projection apparatus was crucial to the standardization of colors used to identify and classify stars for astronomical catalogues.

Objectivity has been compromised in representations of the human body, especially in images of race, ethnicity, and gender. Samuel Thomas von Soemmerring's *Tabula sceleti feminini* (1796) depicted the female skeleton as a composite drawn from artists' renditions (including classical statues) and skeletal artifacts in order to create a universal woman with perfect proportions corresponding to contemporary notions of femininity. Max Brödel, a product of the University of Leipzig school of medical illustrators and holder of the first chair in the newly formed Department of Art as Applied to Medicine at The Johns Hopkins University (1911), believed that the draftsman's mind had to exclude the distractions found in photographs—the particular, the individual, the extraneous detail—to favor the technical precision necessary to aid surgery and train the eye.

Yet objectivity could also be compromised in more sinister ways. Ethnographic representations of races from the early modern period onward have shaped western understanding of the "other" as well as false ideas of western superiority. These representations have served as the "scientific" foundations of filtering processes including the extermination of peoples identified by visual features alone, as was done in the Third Reich. Eduard Pernkopf's *Atlas of Topographical and Applied Human Anatomy* (1944 and several subsequent editions), based on the corpses of a selection of fourteen hundred Austrian prisoners including children executed by the Gestapo, has since the 1990s been the subject of debate among physicians worldwide. Pernkopf's illustrations, which surgeons still consult, originally included prisoners' tattooed camp numbers and swastikas next to his signature. His images of the brain were based on the organs of deformed and retarded children whose remains the Austrian government did not bury until 2002. Only in historical and cultural terms can the tainted objectivity of these and other scientific illustrations be understood.

The relationship between science and art includes iconographical, metaphorical, or allegorical representations of scientists, their instruments, and sometimes even their theories. Such simulacra have helped to create both the public image of science and the social and ideological significance of "objective" knowledge since the early modern period. Hans Holbein's *The Ambassadors* (1533) is an early symbolic representation of scientific instruments and visual illusion. The two shelves in the painting separate terrestrial and astronomical instruments of the day, but together they emphasize the importance of direct experience in the acquisition of knowledge. Between the two men, on the floor, is an anamorphic image of a human skull, which, because not immediately recognizable, serves as a sign of the limitation of human vision. The observer requires a knowledge of the distortions possible with perspective to interpret it. Less symbolic but no less striking are Johannes Vermeer's portraits of *The Geographer* and *The Astronomer* from the seventeenth century: both are not only engaged in visual activities, but their gaze through the window is a sign of their preoccupation with visual information. Likewise Joseph Wright of Derby's *Philosopher Lecturing by a Planetarium* (1766) features Newton's universe as one that can only be viewed by the light of the sun, which emanates from the center of the painting. This type of portraiture, which rose exponentially with the growth of modern science, created images of heroism and genius that have become a part of its popular understanding. Portraiture was crucial in establishing the cultural role of art in science and of science in art.

For the modern period the genres of visual information have multiplied with no less effect on the sciences. For its central role in representing science and scientists, film also must be counted among the artworks that have rendered for modern audiences what Holbein rendered for his. Classics include Stanley Kubrick's *Dr. Strangelove* (1964), for its Cold War mad scientist surrounded with the weapons of nuclear war; Ridley Scott's *Blade Runner* (1982), for its nightmarish vision of the impact of advanced technology on human beings; Andrew Niccol's *Gattaca* (1997), for its scientific and bureaucratic regime of DNA–driven biological determinism; and Steven Spielberg's *AI: Artificial Intelligence* (2001), for its scientist-entrepreneurs oblivious to the heartbreak at the interface between machine and emotion. From Holbein to Kubrick, artists have woven the sciences into the history of politics, society, economics, and culture.   •

See also FILM, TELEVISION, AND SCIENCE; LITERATURE AND SCIENCE.

Joseph Priestley, *A Familiar Introduction to the Theory and Practice of Perspective* (1780). Svetlana Alpers, *The Art of Describing: Dutch Art in the Seventeenth Century* (1983). Barbara Stafford, *Voyage into Substance: Art, Science, Nature, and the Illustrated Travel Account* (1984). Martin Rudwick, *The Great Devonian Controversy* (1985). Ludmilla Jordanova, *Sexual Visions: Images of Gender in Science and Medicine* (1989). Martin Kemp, *The Science of Art: Optical Themes in Western Art from Brunelleschi to Seurat* (1990). Samuel Y. Edgerton, Jr., *The Heritage of Giotto's Geometry: Art and Science on the Eve of the Scientific Revolution* (1991). Barbara Stafford, *Body Criticism: Imaging the Unseen in Enlightenment Art and Medicine* (1991). Jonathan Crary, *Techniques of the Observer: On Vision and Modernity in the Nineteenth Century* (1992). Lorraine Daston and Peter Galison, "The Image of Objectivity," *Representations* 40 (1992): 81–128. John Douard, ed., *Visualization in the Sciences* (1992). Martin Jay, *Downcast Eyes: The Denigration of Vision in Twentieth-Century French Thought* (1993). Hubert Damisch, *The Origin of Perspective* (1994). Barbara Stafford, *Artful Science: Enlightenment, Entertainment, and the Eclipse of Visual Education* (1994). Caroline Jones, et al., eds., *Picturing Science, Producing Art* (1998). Jonathan Crary, *Suspensions of Perception: Attention, Spectacle, and Modern Culture* (1999). Klaus Hentschel and Axel D. Whitmann, eds., *The Role of Visual Representations in Astronomy* (2000). Ludmilla Jordanova, *Defining Features: Scientific and Medical Portraits, 1660–2000* (2000). Klaus B. Staubermann, "Making Stars: Projection Culture in Nineteenth-Century German Astronomy," *British Journal for the History of Science* 34 (2001): 439–451.

KATHRYN OLESKO

ARTIFICIAL INTELLIGENCE. The first machines simply augmented human or animal muscle power, but later pre-industrial inventors introduced devices that also augmented or replaced the human intellect: for example, windmills with sails that adjusted for the intensity of the wind and float valves that maintained water in a tank at a desired level. These devices used the technique of *feedback, whereby information about a machine's status or behavior—say the height of water in a tank—is fed back to the input stages—the valve supplying water to the tank. By the nineteenth century, the development of information-processing machines reached a sufficient level to generate speculation about the similarities between self-regulating mechanisms and human thought processes. The most famous publication on this theme was Mary Shelley's *Frankenstein* (1818), whose main character (Frankenstein's creation) has entered popular consciousness not as a person but as anything that becomes dangerous to its creator. Further philosophical speculation on this topic followed. In 1889 Mark Twain discussed the intelligence of the Paige typesetting machine, whose mechanical complexity outdistanced its day and almost bankrupted its inventor. Specifically, Twain believed that the Paige machine's ability to right-justify type automatically—something that he knew required human skill—was an obvious indication of mechanical intelligence.

The invention of the digital computer in the mid-twentieth century once again raised these philosophical questions. In 1948, Norbert Wiener, a mathematician at Massachusetts Institute of Technology, coined the term *"cybernetics" in a book that discussed not only the mathematical nature but also the philosophical implications of some machines. These machines could regulate themselves to adjust to changes in their physical environment, select alternative courses of action based on the results of previous actions, and remember what actions they had taken in the past—and their consequences. In short, they could learn from experience, adapt to new conditions, and interact with their human creators.

In England in 1951 Alan *Turing published a paper that described a form of parlor game in which players guessed whether they were communicating (through a teletype or similar impersonal device) with a person or a machine (a computer). If players could not distinguish between human and computer, Turing argued, they would have to concede that the computer is "intelligent." The Turing Test for machine "intelligence," a concise, simple, and witty definition of a seemingly intractable conundrum, remains valid. No machine has ever passed the test, although in some restricted domains computers come close.

Similar interests in machine intelligence arose in the United States. At a seminal conference at Dartmouth College in 1956, John McCarthy coined the term "artificial intelligence" (later shortened to AI) to describe ways of making computers act "intelligently." (Some accounts of the Dartmouth conference mention that the group considered Wiener's term "cybernetics," but rejected it so as to avoid introducing a philosophical dimension to their work. Wiener's term resurfaced in the 1990s as "cyberspace," meaning a world existing entirely inside networks of computers.) One of McCarthy's colleagues, Marvin Minsky, defined AI as "the science of making machines do things that would require intelligence if done by men." The term has been criticized for being vague and imprecise; yet it has persisted, perhaps because it captures the imprecision and frustration encountered by humans when they try to understand in an objective way intelligent behavior—what most humans, even children, do effortlessly.

McCarthy developed the programming language LISP (List Processing), which simplified the writing of programs that handled symbolic and logical problems as opposed to numerical problems, which other programming languages were designed to solve. For years LISP remained the preferred language for AI researchers. For nearly all of AI's history, research consisted mainly of writing programs in LISP for ordinary, general-purpose electronic computers. A famous example, ELIZA, mimicked a question-and-answer session between psychoanalyst and patient. AI has focused primarily on computer programming, using text almost exclusively as input and output in keeping with the Turing Test for intelligence. AI has little connection with robotics, the art of building machines that replicate activities carried out by intelligent creatures. Occasionally AI researchers have attempted to couple their programs to microphones, loudspeakers, cameras, electric motors, and so forth, but these projects were exceptions. AI researchers argue that understanding basic cognition gives them enough to do. Also, the U.S. Defense Department's Advanced Research Projects Agency (ARPA), which provided most of the funding for AI research, supported the restricted definition of AI's domain. Since about 1985, AI and robotics have moved much closer together, in part because ARPA's support for AI was drastically cut in that year and because computers, cameras, sensors, motors, and so forth have become much smaller and more capable of putting AI ideas into practice.

Attempts to get computers to perform intelligent tasks have led to new insights into human psychology, and AI has given new life to philosophy as well. But AI research has not led to easier methods of programming computers or to techniques of writing computer programs without errors—both very desirable goals for which the marketplace would pay well. Nor does it have a cupboardful of exemplary programs. As soon as AI perfects a program that can do what was once considered a task requiring human intelligence, computer scientists no longer see it as important and it is demoted from its classification as AI. In the early 1950s computer scientists thought that getting a computer to play a good game of chess would penetrate to the basis of human intelligence. By 1996 a chess program had defeated the human World Champion, and pocket chess computers that played a respectable game were selling for $30. Consequently, AI researchers no longer consider chess such an important topic. Pocket computers that recognize handwriting have captured another frontier of AI research.

The branch of AI called expert systems has had some commercial success. Expert systems consist of large structured pools of data on a specific topic—say, medical diagnosis—plus a set of rules acquired by interviewing specialists in the field—in this case, skilled physicians. The resultant system performs as well as and in some cases better than humans. The most famous fictional computer, HAL, of the 1968 Stanley Kubrick film *2001: A Space Odyssey*, could diagnose and suggest repairs for the spacecraft on which it traveled. HAL's voice recognition—really voice understanding—has not been approached in reality, although some progress has occurred with limited vocabularies. And HAL's emotional breakdown suggests a mentality that no computer today comes close to achieving. Researchers have reached no agreement on how to program so sensitive a computer or whether such a program is even possible.

Artificial intelligence continues to defy easy definition. It has not produced any dramatic breakthroughs—something that people seem to demand elsewhere in computing—but over the years it has made genuine progress. Its principal contribution to society may lie more in telling human beings about themselves than in providing us with the more prosaic benefits of computer technology.

Edward A. Feigenbaum and Julian Feldman, eds., *Computers and Thought* (1963). Hubert L. Dreyfus, *What Computers Can't Do: A Critique of Artificial Intelligence* (1972; rev. ed., 1979). Pamela McCorduck, *Machines Who Think* (1979). Arthur Norberg and Judy O'Neill, "The Search for Intelligent Systems," in *Transforming*

Computer Technology: Information Processing for the Pentagon 1962–1986 (1996): chapter 5. Ray Kurzweil, *The Age of Spiritual Machines* (1999).

PAUL E. CERUZZI

ASIA. As the world's largest and most culturally diverse region, modern Asia displays an exceedingly wide range of scientific achievements and institutions, with historical patterns to match. Japan lies at one end of the spectrum, Bhutan or Laos at the other. Two in the middle, India and China, are among the world's oldest recorded civilizations; both participated in modern science during the twentieth century, and to some extent before. And each can boast of important scientific achievements from the pre-modern era. The fourteen *Nobel Prizes in science awarded to investigators from Asia constitute an imperfect indicator of the region's standing in modern science; and it is telling that only half of the prizes went for work done in Asian countries. This record serves as a stark reminder of Asia's strengths in, and contributions to, modern science, on the one hand, and of the weaknesses and continuing difficulties that confront science in the region, on the other.

Origins and Early Growth
Modern science in Asia traces its origins to the arrival of European scientists interested in the region's flora, fauna, minerals, topography, and so on. India, China, the Philippines, and Japan felt the European *Scientific Revolution through the activities of Jesuit missionaries in the late sixteenth and early seventeenth centuries, although the Philippines had a European style college by the early seventeenth century. Spain, as the dominant power in the country, long remained marginal to the scientific enterprise. The sustained development of modern science in Asia actually began with the arrival of British scientists in India shortly before 1800. Other European investigators followed, to China, Southeast Asia, Japan, and elsewhere; but developments in India proceeded more rapidly because of the subcontinent's continuing, ever-more intimate association with Britain. British scientists carried out astronomical studies in Bengal and a Trigonometric Survey throughout the whole of India (including today's Pakistan and Bangladesh) beginning in 1800. These and other early investigations simultaneously satisfied curiosity and promoted British commercial and political interests in the region. Direct Crown Rule, however, was not established until 1858.

India's attraction for the British led to a much earlier founding of scientific societies and professional journals than in any other Asian country. British colonial authorities in India

provided Western-style education there long before other European powers did so in the regions under their control. In 1813, an Indian reformer with British encouragement established an English-language school for Hindu boys; this school evolved into Presidency College, Calcutta (later the University of Calcutta), and became one of India's leading sites for the growth of modern science. Two other universities, founded at Bombay and Madras by 1857, were modeled on the University of London, and maintained the same academic standards. In 1898, a wealthy Indian business leader, J. N. Tata, dispatched an associate to Europe and the United States to study the operations of universities. After many delays, but with British encouragement, his initiative led in 1911 to the founding of the Indian Institute of Science at Bangalore. It would later constitute the nucleus of the Indian Academy of Sciences. In 1859, the Jesuits established a modern university with a scientific curriculum in the Philippines, the Ateneo de Manila, and added a medical school in 1872. Similarly, Japan and China, which never came formally under European rule, managed to establish universities with programs in science at Tokyo and Peking (Beijing) in 1877 and 1898, respectively. But only later were Indian developments as a whole matched elsewhere in Asia.

Important as the legacy of European scientific activity was for the development of modern science in Asia, no sustained growth occurred before indigenous leaders assumed long-term control. Conditions varied widely from one country to another in this respect. Under the reformist Meiji regime, Japan from 1868 began to lay the foundations required for modern science in a comprehensive manner. An ultra-modern engineering school opened in 1869 under the directorship of Henry Dyer, a Scot. After consulting leading scientific authorities in Europe, Japan's government began hiring young European and American scientists, engineers, and physicians at very high salaries to staff several institutions, including Tokyo University. Most importantly, a systematic program to send young Japanese to Europe and the United States for scientific and other modern studies began in 1872. After 1878, scientific societies and academic journals were established in rapid succession. By contrast, China's non-native Manchu regime only grudgingly developed a program for technical translation and modern armaments manufacture at the Kiangnan Arsenal in 1860, but scarcely went beyond it until 1898. Other needed developments—research laboratories, journals, professional societies—only appeared in the twentieth century. Thailand (then Siam) under King Chulalongkorn began to establish French-style *grand écoles* in the 1890s, but had no university until 1916, and only a modest number of Thai nationals studied in Europe. Local populations in Vietnam and the East Indies (today's Indonesia) had no prospects for studying science, to say nothing of conducting research, until even later.

Parameters of Development

To what extent did the formation of a local scientific community draw effectively on traditional intellectual culture and the social structure already extant before Europeans arrived? India and Japan enjoyed relatively positive experiences in this respect. Early recruits to scientific and technical studies in India such as Chandrasekhara Raman, India's first Nobel laureate in science, and Srinivasa Ramanujan, the eminent mathematician, came from the traditional Brahmin priestly caste or the Kyasthas scholarly community. In Japan, the first generation of modern scientists (before 1920) arose predominantly from the educated samurai population inherited from the old regime (6 percent of all Japanese); Hideki *Yukawa, the first Japanese Nobel laureate, came from a lineage of samurai Confucian scholars. In China, however, the first generation of recruits to modern science had tenuous links to traditional intellectual culture and none at all in significant numbers. This was not conducive to the establishment of modern science there.

Political independence, as opposed to colonial dependency or subjugation, was another important variable. Japan's political independence allowed it to define its own agenda, and facilitated rapid progress overall. A government of educated leaders chose which nations, institutions, and laboratories would receive the students sent to the more developed countries of Europe or (less frequently) North America. For example, they could dispatch aspiring Japanese clinicians and medical researchers to German universities at a time when Germany was preeminent in medicine. India did not have this freedom, probably one of the reasons that its modern medicine lagged behind other fields. The lack of political independence was particularly harmful in Southeast Asia. In Vietnam, the Pasteur Institute maintained a significant local research institute for tropical medicine and microbiology at Nha Trang from the 1880s; but it had almost no local educational infrastructure. Well-informed authorities such as Gaston Darboux, permanent secretary of the Académie des sciences in Paris, tried to encourage more forward-looking policies, but the retrograde views of French colonial authorities in Vietnam prevented both the permanent establishment of a university (at Hanoi) and

even a permanent scientific mission of French scientists until 1917. Vietnamese had no opportunity to study modern science before 1920; some historians say that as late as 1930, the medical school had the only "serious" scientific program at the University of Hanoi. Dutch colonial authorities in the East Indies at best duplicated French science policy concerning the local population. There was no university at all in this period, only the Institute of Technology at Bandung, which opened in 1920. In 1930, Dutch nationals constituted 60 percent of its student body, Indonesians only a third.

China was neither truly independent nor formally a colony. Taking advantage of the national government's obvious weaknesses under the alien Manchu dynasty, various foreign powers began to establish protected enclaves in several parts of China, and pursued developmental agendas tailored to their own objectives. Some of their strategies favored the growth of science at a time when the Manchu regime was, at best, ambivalent and, at worst, antagonistic to it. The brief spasm of reformism that allowed the establishment of Peking University in 1898 soon spent itself; the anti-foreign Boxer Rebellion, encouraged by reactionary elements of the Manchu elite, broke out in 1900. Although China was forced to pay reparations for damage inflicted on foreign properties by the Boxers, some of this "Boxer Indemnity" was rebated (especially by the United States) to China in support of overseas study by Chinese nationals at (mostly) American universities. This program made possible the formation in 1914 of China's first professional scientific organization, the Science Society of China, a diverse organization of young aspiring scientists then enrolled at Cornell University; in 1918 they returned to China and began promoting the growth of scientific institutions and research projects by Chinese citizens.

Mass education—its presence or absence—has been important for the growth of modern science in Asia. Japan was the first Asian country to provide elementary and mid-level schooling for everyone, and its twentieth-century record in science is the continent's most impressive. India underscores the point. Mass education was only sporadically available in the country during the period of Crown Rule before 1947. Har Gobind Khorana, born in 1922 in a region now part of Pakistan, and one of only two scientists from Asia to win a *Nobel Prize in medicine, had the good fortune to come from the only family in his village whose members could read and write. Yet mass education's importance should not be overstated. Partly under U.S. leadership after 1901, the Philippines achieved universal primary education not long after Japan did so; and in recent years, China appears to have followed the Philippines. Yet in technical fields during the first half of the twentieth century, the Philippines achieved only modest distinction, and that only in medicine and meteorology. The consequences of mass education for science in China cannot be evaluated yet.

The degree of political stability and the ease with which private and public interests could be deployed on behalf of scientific ventures have profoundly affected modern scientific achievement in Asia. India, Japan, and Taiwan enjoyed far greater relative stability during the twentieth century than all the other countries in the region. Stability allowed both governments and private patrons to support scientific education and research in the expectation of a positive return on their investments. The British authorities, who stabilized India, did little to support modern science in India directly, but encouraged private individuals and organizations to do so. Thus, J. N. Tata could take a significant part in founding the Indian Academy of Science, with later support from the Maharajah of Mysore, the head of one of India's princely states. Similarly, a trio of wealthy Indian lawyers raised the funds to endow chairs in physics at the University of Calcutta that supported the work of Raman and Meghnad Saha, both Nobel nominees, and one a laureate.

Japan's record in this respect parallels India's. Despite being under American military occupation from 1945 to 1952, following its defeat in World War II, Japan retained control of its own internal affairs, and generally adhered to policies favorable to modern science. Thus were universities founded, laboratories established, journals published, conferences held, and research conducted on a regular basis. Japan had its Imperial Academy of Sciences (founded 1906) and seven comprehensive state ("imperial") universities by 1940, as well as a host of private institutions. Taiwan and Korea, under Japanese rule from 1895 and 1910, respectively, had one imperial university each in addition to private schools. All of the state institutions and many of the private ones were active in research and teaching. Moreover, contrary to some assertions, private interests in Japan supported scientific activities at state as well as at private institutions. Two financial services executives from Osaka endowed the world's first chair of genetics, at Tokyo University, in 1918; the country's leading scientific institution of the inter-war period, the Research Institute for Physics and Chemistry (Riken), was founded in 1917 through substantial private contributions as well as state funding.

While the growth of modern science greatly benefited Japan, its military and related applications did considerable harm to the rest of Asia,

with the exception of India. In Manchuria, from the time Japan seized it from China in September 1931 until the end of World War II, the Japanese army conducted research in biochemical warfare, which employed human subjects. Several thousand victims were systematically killed in the process. The U.S. Army later found the data generated sufficiently useful to justify exonerating the perpetrators. Several years before the loss of Manchuria, and after the turmoil of 1911–1927, China experienced an effective central government, which made its capital at Nanking and launched a series of reforms much like those of Japan in the late nineteenth century. During this so-called "Nanking Decade," 1928–1937, the new regime founded a science academy, the Academia Sinica, with more than a dozen research institutes, as well as a number of universities, mostly in coastal cities.

Much of this investment appeared to be lost when Japan invaded China in 1937. Many professors and university students fled to the interior to escape the conflict, taking books, journals, and research equipment with them. Some resumed their activities, at Chunking and Kunming, especially those associated with the newly constituted National Southwest University, created by merging the faculties of the transported coastal schools. Harmful as the Japanese invasion was for modern Chinese science in general, the regrouping that followed had some positive effects. Two young physicists, C. N. Yang and T. D. Lee, who met in Kunming, moved to the University of Chicago to study for their Ph.D.'s in theoretical physics after the war. Working together after completing their studies, Yang and Lee surprised the international physics community by challenging the law of the conservation of parity in certain atomic reactions. For this achievement they became in 1957 the first Chinese scientists to receive the Nobel Prize. By providing them with a strong basic education in physics, China's institutional advances in the Nanking Decade and wartime years facilitated their later scientific work. What else might have been accomplished if China's political stability had matched that of India or Japan?

Contributions of Japanese and Indians to world science before 1945 include Shibasaburo Kitasato's collaboration (in 1900) with Emil von Behring on the discovery of natural immunity. During World War I, Katsusaburo Yamagiwa developed the world's first efficient means of creating cancer in the laboratory; many believe he ought at least to have shared the Nobel Prize in Physiology or Medicine for 1926. In 1920, Meghnad Saha, a professor of physics at the University of Calcutta, published his theory of thermal ionization, a landmark achievement in modern astrophysics. Another Indian colleague, S. N. Bose, then teaching at the University of Dacca (in today's Bangladesh), made what some consider the greatest contribution of an Indian physicist to science when in 1924 he took the first major step in the formulation of quantum statistical mechanics. While Bose never won a Nobel Prize, nor was even nominated, one of the two grand classes of particles in modern physics, the bosons, is named for him. Shortly thereafter, Chandresekhara Vankata Raman published the discovery that did win India's first Nobel in science. This so-called "Raman effect" in quantum optics found important application in spectroscopy. Published in 1928, Raman's achievement, was one of those rare instances in which an important discovery was almost immediately recognized (in this case in 1930) by a Nobel Prize.

More typical were the delays experienced by Hideki *Yukawa and Shin'ichiro Tomonaga. Yukawa postulated the existence of a new particle to help explain nuclear forces in 1935 (see ELEMENTARY PARTICLE), for which he received the Nobel Prize in 1949; Tomonaga solved a difficult problem in *quantum electrodynamics in 1944, and shared the prize for 1964. Their candidacies developed rapidly compared with that of Raman's nephew, Subrahmanyan *Chandrasekhar. Between 1930 and 1933, working in India and Cambridge, "Chandra" developed the astrophysical theory of white dwarf stars that brought him the Nobel Prize in 1983.

World War II and the Impact of Big Science

The effects of the war on science were felt as strongly in Asia as anywhere else. Seizing power in 1949, China's new Communist regime claimed to be based on scientific principles, and in its first decade appropriated Soviet models that promised a massive expansion of scientific infrastructure and research activity. Academia Sinica merged with another institution to create the powerful Chinese Academy of Sciences in 1950. New institutes and universities were founded, and emigré scientists enticed to return home. Intellectual constraints and Communist dogma offended some; surging nationalism and the dynamic expansion of the scientific enterprise excited others. During a period of relative calm in 1965, a team of biochemists in Shanghai achieved perhaps the most important modern scientific advance ever carried out in China by synthesizing bovine insulin. Seen as a harbinger of future success, the insulin achievement was soon revealed as an epitaph; China descended into the maelstrom of the Cultural Revolution the following year. Between 1966 and 1974, all the nation's universities and most of its laboratories were closed as political chaos engulfed the country.

India, Pakistan, and other Asian nations also sought progress through science after gaining independence. Several countries managed to expand the institutional base of science by building laboratories, universities, or even science academies. But few had the personnel with graduate degrees in sufficient numbers to staff these institutions at an appropriate level. Widespread poverty, illiteracy, and political instability hampered progress. The arrival of big science in the United States and Europe proved a singular spur and challenge. World War II had raised the bar for significant achievement in science, especially in physics. Raman in India wanted to undertake research in nuclear physics after the war, but had to return to optics because of a lack of resources. Abdus Salam had hoped to practice physics in his native Pakistan after completing his doctorate in 1951, but fearing intellectual stagnation, he returned to Cambridge after three years at Lahore. His Nobel Prize came in 1979. In China, Mao had largely shielded the physicists from the destructive forces of the Cultural Revolution; partly at the urging of C. N. Yang and T. D. Lee, Mao and his successors made high-energy physics a target for investment. In 1975, the Chinese leadership endorsed a proposal from the physicists to build a 50 Ge V proton accelerator. Comparable to some earlier accelerators in Europe and the United States, the facility would have brought China welcome prestige. However, the cost of the project and disagreements among likely users led to its termination in the early 1980s.

Because of earlier progress and relative stability, India and Pakistan developed faster than most other Asian countries. Jawaharlal Nehru's prime ministership, 1947–1964, gave India both a rhetorical and a substantive commitment to science that fully matched the new regime of Mao in China but without the cross-currents of revolutionary politics. Nehru's commitment to science expressed itself partly in the foundation of the campuses of the Indian Institute of Technology in five Indian cities with support from Indian and other governments. Focusing primarily on undergraduate education but with some research programs, these schools took MIT as a model, and were widely seen to produce graduates of comparable quality. In a major private initiative of the period, one of Nehru's leading scientific advisers, the physicist Homi Bhabha, persuaded the Tata family to establish the Tata Institute for Fundamental Research at Bombay.

Big science stretched even Japan's resources. Its economy began to recover from the war in late 1950, but the country spent at only half the level of Western Europe on science for the next two decades. Leo Esaki's career in this period illustrates the situation. In 1958, he discovered the tunneling effect in semiconductors while working as a graduate student for the firm that later became Sony. A shared Nobel Prize came to him for this work in 1973, but he worked in the United States after 1960. Susumu Tonegawa's research career followed a similar pattern. After earning a Ph.D. in molecular biology at the University of California, San Diego, he moved to Switzerland, and there investigated the genetic origins of antibody diversity that in 1987 made him the first Japanese recipient of a Nobel Prize in medicine. He was at the time, and has remained, a professor at MIT. Ken'ichi Fukui, however, stayed in Japan, studying chemical reaction theory at Kyoto University. By applying quantum theory to knowledge of hydrocarbons gleaned from wartime work on synthetic fuels, he became in 1981 the first scientist in Asia to win the chemistry Nobel Prize (shared with Roald Hoffmann). In 1984, Y. T. Lee, from Japan's former colony of Taiwan, also won in chemistry, for work done at the University of California, Berkeley.

Only after 1970 did Japan begin to extend the level of support for science long provided by the affluent nations. The founding of Tsukuba University and the Tsukuba Science City in the early 1970s exemplified the new commitment by providing equipment and facilities well beyond those of Japan's other institutions. After 1980, it appeared that Japan intended to raise its level of accomplishment in science, and after 1990 it made explicit declarations to that effect. Convinced by these declarations, Leo Esaki accepted the presidency of Tsukuba University in 1992. Japan's business establishment added weight to expectations by publicly advocating huge increases in funding for research in academic institutions. For Esaki and the Japanese establishment, securing more Nobel Prizes in science became both a public aspiration and a symbolic obsession. Specific initiatives such as the annual Japanese Forum of Nobel Prize Recipients were launched to fulfill this objective. In 2000, a retired professor from Tsukuba University, Hideki Shirakawa, received a share of the Nobel Prize in chemistry for his work on conducting polymers. The following year, another Japanese chemist, Ryoji Noyori, also won in chemistry.

Japan's accomplishments and greater wealth in the region have elicited significant responses on behalf of science in other East Asian nations. South Korea, Taiwan, and Singapore in particular have built substantial new research facilities in recent years, although most have emphasized applied rather than basic science. There is a strong conviction in Asia that greater investment in science will promote more rapid economic growth, as well as cultural prestige. Several nations have launched formal programs to induce emigré nationals to return. In 1957, Chinese authorities tried to persuade C. N. Yang to leave the United States and return to China while he was in Stockholm receiving his Nobel Prize. But few of the efforts succeeded before the early 1990s, when Taiwan successfully invited Y. T. Lee to leave Berkeley to lead Academia Sinica (1993) and Japan enticed Esaki to take the presidency at Tsukuba (1992). Although wealth does not in itself lead directly to distinguished achievement in science, the era of big science makes wealth essential. The major economic reforms set in train by India, China, and other Asian countries in recent years should further raise their collective profile in science.

See also AFRICA; ENGLISH-SPEAKING WORLD.

UNESCO, *National Science Policy and Organization of Research in the Philippines* (1970). Jack N. Behrman, *Industry Ties with Science and Technology Policies in Developing Countries* (1980). Leo A. Orleans, ed., *Science in Contemporary China* (1980). O. P. Jaggi, *Science in Modern India* (1984). G. Venkataraman, *Journey Into Light: Life and Science of C. V. Raman* (1988). James R. Bartholomew, *The Formation of Science in Japan: Building a Research Tradition* (1989). Frank N. Magill, ed., *The Nobel Prize Winners: Physics, Chemistry, Medicine,* 9 vols. (1989). Lewis Pyenson, *Empire of Reason: Exact Sciences in Indonesia, 1840–1940* (1989). James Reardon-Anderson, *The Study of Change: Chemistry in China, 1840–1940* (1991). Lewis Pyenson, *Civilizing Mission: Exact Sciences and French Overseas Expansion, 1830–1940* (1993). Yeu-Farn Wang, *China's Science and Technology Policy: 1949–1989* (1993). Jacques Gaillard, V. V. Krishna, and Roland Waast, eds., *Scientific Communities in the Developing World* (1997).

JAMES R. BARTHOLOMEW

ASSOCIATIONS. See ACADEMIES AND LEARNED SOCIETIES; ADVANCEMENT OF SCIENCE, NATIONAL ASSOCIATIONS FOR THE; PROFESSIONAL SOCIETY.

ASTROLABE. The astrolabe is an astronomical and astrological instrument used for calculations involving the positions of the stars and the sun. Although based on the assumption of a spherical sky in daily rotation around a central earth, almost all astrolabes are flat.

The traditional form of astrolabe has a circular, planispheric map of the stars known as the "rete" (Latin for 'net'). This star-chart usually takes the form of a brass plate with most of the metal cut away by fretwork, leaving a network of bands supporting pointers indicating the positions of the stars, and a circular band for the annual path of the

An analogue computer of the motions of the sun and stars made in Toledo by Ibrahim ibn Sa'id al-Sahli in 1068. The fine lines inscribed on the back plate (tympan) indicate the observer's grid of azimuths and altitudes. The cut-away upper plate (the "rete") represents the ecliptic circles and (via the stylized pointers) the positions of prominent stars. The rete can be rotated over the tympan to simulate the diurnal motion of the heavens. The astrolabe remained in use in Western Europe for calculations, observations (via a sight on its verso), and astrological determinations into the seventeenth century.

sun through the stars, known as the ecliptic. The plane of projection coincides with the equator, while the rectilinear projection lines converge at the south celestial pole. The same projection renders the local horizon of the user, and circles of altitude and azimuth, on a solid brass disc the same size as the rete. These projected lines form the characteristic pattern of the "latitude plate," where circles and arcs cluster around the point that indicates the local zenith, while arcs of azimuth radiate out from it. Both projections extend from the north celestial pole, usually to the tropic of Capricorn. The rete, pivoted at the pole, rotates above the stationary latitude plate; adjusted to any orientation of the sky it will indicate the positions of the stars or the sun above the horizon. Since the position of the sun in its daily rotation measures the time, an hour scale at the edge of the instrument can give the time at any hour of day and night: a rule pivoted at the center carries the sun's position (indicated by a date or zodiac scale on the ecliptic ring) to the hour scale. On the back of the astrolabe is an alidade, or sighting rule, with an altitude scale, for measuring the altitude of the sun or one of the stars included on the rete. By this means the rete can be set to the current position of the sky and

the time found. Other scales on the front and back of the instrument allow astrolabes to perform a range of different functions.

Ptolemy described the geometrical basis of the traditional instrument. The earliest surviving astrolabes come from ninth-century Baghdad and Iran. The earliest dated European ones come from the fourteenth century, though astrolabes from several centuries earlier have survived. In the traditional form, separate projections are given for different latitudes, so instruments often have a number of alternative latitude plates. Muslim astronomers found several ways of making a "universal" astrolabe—one that would serve in any latitude—and Europeans devised or rediscovered similar designs in the sixteenth century. The European instrument was in rapid decline by the early seventeenth century, but fine astrolabes continued to be made in parts of the Islamic world through the nineteenth century. Astrolabes have always been treasured and admired, and they remain one of the most desirable instruments for collectors today.

The mariner's astrolabe, by contrast with the complex and ornate instruments for astronomy and astrology, had a simple and robust design. Combining an altitude ring and an alidade, they were used for finding latitude at sea mainly in the sixteenth and early seventeenth centuries.

The Planispheric Astrolabe, published by the National Maritime Museum (1974). Harold Saunders, *All the Astrolabes* (1984). A. J. Turner, *The Time Museum: Astrolabes* (1985).

JIM BENNETT

ASTROLOGY. Astrology is best defined as the set of theories and practices interpreting the positions of the heavenly bodies in terms of human and terrestrial implications. (The positions have variously been considered signs and, more controversially, causes.) The subject—and therefore its study—is fascinating, difficult, and often paradoxical. Although inextricably entangled with what are now demarcated as science, magic, religion, politics, psychology, and so on, astrology cannot be reduced to any of these. The historical longevity and cultural diversity of astrology are far too great for it to have been precisely the same thing in all times and places, yet it has always managed to reconstitute itself as much the same thing in the minds of its practitioners, public, and opponents alike. These points have particular relevance in relation to historians of science, who until recent decades predominantly analyzed astrology anachronistically as a "pseudo-science," the human meanings of which could largely be derived from its lack of epistemological credentials.

Western astrology originated as Mesopotamian astral divination. The planets and prominent stars, identified with gods in ways that have since changed remarkably little, were considered celestial omens in which the divine messages, largely answering royal concerns, could be discerned. The origins of many key elements of the astrological tradition—not only the planetary deities, zodiacal signs, risings, and settings, but also the effort to systematize divination through what we would now consider astronomical and empirical observations—developed between its apparent origins around 2000 B.C. and the fifth century B.C., when natal astrology first appeared. The latter, following Alexander's conquest of Persia, was absorbed and transformed by Greek geometric and kinetic models, which added the aspects, or angles of separation between planets and points, and emphasized the importance of the *horoscopos* or Ascendent, the degree of the zodiacal sign rising on the Eastern horizon. (The first known horoscope dates from 4 B.C.) Astrology was also fruitfully married to Aristotelian cosmology and Greek medicine in the form of Hippocratic humors, and, slightly later, Galenic temperaments. Astrologers tended to develop increasingly flexible interpretive schemes, of which the most famous and

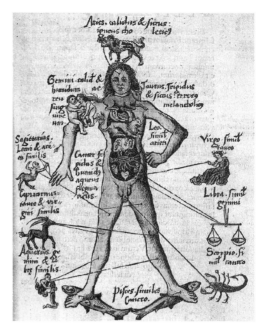

A "Zodiac man" from Georg Reich, *Margarita philosophica* (1503), one of the first printed encyclopedias. The figure indicates associations of the zodiacal signs with the body parts, beginning with the head (Aries) and proceeding to the feet (Pisces) via the "privates" (Scorpio). According to medical lore, no part of the body should be bled when the moon stands in the corresponding zodiacal sign.

influential was formulated by Ptolemy (c. A.D. 100–170) in his *Tetrabiblos*.

Astrology played an important role in Roman life, for the most part in crudely populist and overtly political contexts. A more fruitful course followed in the wake of Alexander the Great's conquests, as Greek astrology spread to Persia and throughout Eastern Asia as far as India. In this way Greek astrology eventually became incorporated into, and benefited from, the learning of the Arabic world. It was introduced into medieval Europe in Latin translations, notably, from the mid-twelfth century onwards, of works by Abu Ma'shar (787–886). These supplied a philosophical basis (largely Aristotelian) for astrology and popularized the idea that conjunctions of Jupiter and Saturn ("grand conjunctions") in particular regions of the heavens signify changes of political rulership. A complete revolution of the conjunctions around the zodiac, which takes 960 years, indicated changes in the fortunes of entire religions. Pierre d'Ailly and Roger Bacon took up this astral historiography.

In the late fifteenth century, a series of influential translations by Marsilio Ficino made available rediscovered Greek texts, including much of Plato, Plotinus, and Iamblichus and the *Corpus Hermeticum*. These placed a renewed magical and/or mystical astrology at the heart of the *Renaissance revival of neo-Platonism and hermeticism. Typically, it managed to evade Giovanni Pico della Mirandola's powerful critique of astrology in his *Disputationes* (1494) by finding shelter elsewhere in the set of ideas that had inspired him (for example, occult *sympathy and antipathy).

Astrology was controversial within the Christian church. It survived the condemnations of St. Augustine and the early church fathers, who saw it as pagan (and in particular polytheistic) and a transgression of both human free will and divine omnipotence. Augustine did not deny that astrologers could speak truthfully, only that when they did so it was with the help of, and in the service of, demons.

At both popular and elite levels, however, astrology in one form or another remained entrenched. It fell to Thomas Aquinas in the late thirteenth century to arrange a compromise that secured for it a long-lived, if limited, niche. His synthesis of Christian theology and Aristotelian natural philosophy permitted "natural astrology" to influence physical and collective phenomena but not human souls directly; the individual judgments (and in particular predictions) of "judicial astrology" were therefore illicit. Since Aquinas admitted that most people followed the promptings of their bodies, which felt the influence of the stars, he gave a tacit legitimation of astrology in practice. But the Reformation presented a serious new challenge in the sixteenth century. Luther and Calvin objected violently to astrology's idolatry, as they saw it, which they stigmatized as "superstition."

The seventeenth century was pivotal in the history of astrology. Contrary to the argument of Keith Thomas's influential *Religion and the Decline of Magic* (1971), the historical puzzle is not why so many intelligent people then believed in astrology (at a time when most people did), but why did they cease to believe in it?

Strong social and political forces abetted its fall from favor. In the English Revolution the pamphlets and almanacs of astrologers on both sides—but especially those of William Lilly for Parliament—played a major, and highly visible, role. In the late seventeenth and eighteenth centuries the new patrician and commercial alliance sought to put sectarian strife and upheaval behind it, and astrology became firmly identified as vulgar plebeian (rather than religious) superstition, to be contrasted with the spirit of rationalism and realism. This

perception was now most often articulated by a new set of opponents: the metropolitan literati. Jonathan Swift's issue of a mock almanac in 1707 predicting the death of the prominent astrologer John Partridge, followed by another putatively confirming its fulfilment, epitomized the attack. Partridge became a laughing-stock in coffee-house circles, although his almanacs continued to sell. Benjamin Franklin later employed the same tactic in the American colonies.

At the same time, increasing political centralization in France made astrologers' unlicensed prophecies unwelcome there too. After a short period of ambivalence, most prominent European natural philosophers also started to close ranks against astrology, ignoring or criticizing it as part of the old Aristotelian order, and/or as (plebeian or Platonic) magic. Isaac *Newton's success set the seal on this development. He borrowed the old idea of attraction at a distance, but substituted a single and quantifiable force for an astrological sine qua non: the planets as a qualitative plurality. Natural philosophy quietly absorbed natural astrology (including the moon's effects on tides), but judicial astrology, as a symbolic rather than mathematical system addressing merely "secondary" qualities and "subjective" concerns, had no place in a newly disenchanted (and commodified) world.

In this context the charge against astrology of "superstition" began to acquire its present meaning as a cognate of stupidity or ignorance. To begin with, natural philosophers made common cause with the guardians of religious orthodoxy; but as natural philosophy moved in the direction of modern science, the hostility to astrology increasingly became a secular opposition.

The early modern period has too often been described, by those mistaking contemporary rhetoric for reality, as the time of the death of astrology. It did decline seriously as it was pushed into largely (but not entirely) rural strongholds dominated by farmers' almanacs, and into a relatively simple and magical set of beliefs. But early in the nineteenth century, as the middle classes grew in power and began to break away from patrician hegemony, a new urban astrology appeared that still remains. More individualistic than before, it succeeded in adapting to consumer capitalist society. And in the early twentieth century, through the work of Alan Leo and his commercially canny Theosophy, astrology secured a firm footing in both the popular press and the thriving middle-class market for psychology-cum-spirituality. At present it seems to meet a demand for (re-)enchantment that no amount of technical, technological, or purely theoretical progress can serve.

Thus astrology has managed to adapt to, and even exploit, every challenge history has thrown it. There is no reason to expect it will ever fail to do so, despite the outraged denunciations it continues to attract from contemporary guardians of scientific probity.

Eugenio Garin, *Astrology in the Renaissance* (1983). Patrick Curry, ed., *Astrology, Science and Society* (1987). Jim Tester, *A History of Western Astrology* (1987). Patrick Curry, *Prophecy and Power: Astrology in Early Modern England* (1989). Tamsyn Barton, *Ancient Astrology* (1994). Nicholas Campion, *The Great Year* (1994).

PATRICK CURRY

ASTRONOMICAL MEASURING INSTRU-MENTS. See INSTRUMENTS, ASTRONOMICAL MEAS-URING.

ASTRONOMY. Astronomy, unlike modern sciences formed during the Scientific Revolution of the seventeenth century, has an ancient pedigree. The goals of astronomers, however—their theories, their instruments and techniques, their training, their places of work, and their sources of patronage—have undergone changes as revolutionary as those experienced by other sciences over the past four centuries.

For two millennia before the seventeenth century, a primary problem for astronomy in the Western world was to discover the true system of uniform circular motions believed to underlie the observed and seemingly irregular motions of the planets, Sun, and Moon. Astronomers observed and recorded a few planetary, solar, and lunar positions; attempted to fit geometrical models to the observations; and constructed tables of positions. In addition to the effort for its own sake, there were practical offshoots, including personal horoscopes, more general warnings of man-made and natural catastrophes, and calendars foretelling times of religious celebrations and the agricultural seasons. Solar and stellar navigation became important only later; fourteenth- and fifteenth-century explorers still found their way in sight of land. The rest of the universe scarcely existed for ancient and medieval astronomers other than as the limiting outer sphere of the stars. Nor did they concern themselves with the physical composition of the universe. Beyond the region of the Earth lay one unchanging element.

The Copernican Revolution, begun in the sixteenth century, switched the places of the Earth and the Sun in a geometrical model still composed of uniform circular motions. In other important ways, however, *Copernicus radically redefined the astronomical agenda. After new observations refuted the ancient assumption of circular motion, the physical nature and cause of orbital motion, previously outside the province of astronomers, became crucial. Interest in the physical composition of the heavens increased as well once the Earth left the center of the universe, and especially after *Galileo's revolutionary new telescopic observations.

In 1609 rumor from Holland of a device using pieces of curved glass to make distant objects on the Earth appear near reached Galileo in Italy. He constructed his own *telescope and turned it on the heavens. Among his discoveries were four moons circling Jupiter. These Medicean stars, as he named them, secured him a position at the court of the Grand Duke of Tuscany. Galileo also observed mountains on the Moon and sunspots. Henceforth the telescopic discovery of hitherto unknown planets, moons, asteroids, comets, and nebulae, as well as examination of their more prominent activities, was a standard feature of astronomers. Amateurs wealthy enough to procure relatively large telescopes could excel.

The *telescope became an instrument of precise measurement through a happy accident. Noticing that a spider's web spun in the focal plane of his telescope was superimposed on the telescopic image, the Englishman William Gascoigne realized that crosshairs or wires could help center telescopes on objects and also help measure angles between them. By 1700, after some resistance, astronomers had accepted the telescope as the primary instrument of astronomical measurement. The cost of larger telescopes, their operation, and analysis of data speeded the transition from individual observers to organized *observatories under government patronage. The Paris Observatory was founded as part of the new Paris Academy of Sciences in 1666, and the Greenwich Observatory began operations a decade later. In a new age of exploration, the task of the Greenwich Observatory, as stated in a royal warrant, was to rectify "the tables of the motions of the heavens, and the places of the fixed stars, so as to find out the so much-desired longitude of places for the perfecting the art of navigation."

The last great achievement of pre-telescopic observations was Tycho *Brahe's body of positional measurements at the end of the sixteenth century. He enjoyed inherited wealth and also royal patronage, given in exchange for the glory his discoveries and fame cast over his patrons. Early in the seventeenth century, Johannes *Kepler used Brahe's positions to destroy faith in uniform circular motion, showing instead that ellipses more accurately describe planetary motions around the Sun. Aristotelian physics, explaining all motion around a central Earth, did not work in a Sun-centered universe. An explanation of why the planets retrace their

The wealthy seventeenth-century Danzig brewer and astronomer Johannes Hevelius (right). The instrument, a brass sextant six feet in radius, was modeled on Tycho Brahe's instrument of a century earlier.

elliptical paths around the Sun became a central problem of astronomy.

Near the end of the seventeenth century, Isaac *Newton treated celestial motions as problems in mechanics governed by the same laws that determined terrestrial motions. Bodies remain at rest or move uniformly in straight lines unless external forces alter their state. A force of attraction toward the Sun continually draws the planets away from rectilinear paths and holds them in their orbits. Newton showed mathematically that, on his mechanical principles,

Kepler's elliptical orbits result from a universal inverse-square law of gravity. The working out of details left undone by Newton focused the energies of many mathematical astronomers during the eighteenth century. Newtonians added quantitative success upon success, though not in the form Newton himself had used. Newton had employed geometry, the accepted medium of mathematical proof, in his demonstrations. His successors used new, more powerful algebraic methods. This change may help explain why Newton's loyal followers in England made little progress compared to scientists on the continent. Royal academies with generous support for astronomers, particularly at Paris, Berlin, and St. Petersburg, also made a difference. London's Royal Society, in contrast, neither received nor paid out royal emoluments, and the astronomers royal at Greenwich had to provide some of their own instruments.

Exact orbital calculations incorporate the influence of small perturbation effects. The Sun alters the Moon's motion around the Earth, and Jupiter and Saturn modify the motions of each other about the Sun. The Swiss-born mathematician Leonhard *Euler, at St. Petersburg and later at the Berlin Academy of Sciences, helped develop mathematical techniques to compute perturbation effects. He applied them first to the Moon, and in 1748 to Jupiter and Saturn, whose motions were the subject of that year's prize topic of the Paris Academy of Sciences. Euler, and Joseph Louis Lagrange and Pierre-Simon *Laplace, both members of the Paris Academy, applied the newly all-powerful calculus (see MATHEMATICS) to the perturbations of planets and satellites, the motions of comets, the shape of the Earth, precession (a slow conical motion of the Earth's axis of rotation caused primarily by the gravitational pull of the Sun and the Moon on the Earth's equatorial bulge), and nutation (a smaller wobble superimposed on the precessional motion of the Earth's axis).

Orbital calculations now lie outside mainstream astronomy. NASA scientists use computers to calculate trajectories for their spacecraft. Astronomers would be willing, for additional government funding, to calculate whether various passing asteroids will safely miss the Earth.

The business of astronomy has at times been a family profession. Gian Domenico Cassini, recruited from Italy to set up the Paris Observatory in 1669, was succeeded by a son, a grandson, and a great-grandson, before the French Revolution drove the family from the observatory in 1793. Friedrich Struve, who helped Czar Nicholas I chart his vast empire and also recorded the positions of double stars at the Pulkovo Observatory, opened in 1839 outside St. Petersburg, founded an astronomical dynasty spanning four generations. A son succeeded him at Pulkovo, a grandson directed the Königsberg Observatory, and a great-grandson, having fled Russia in 1921 after the revolution, directed the Yerkes Observatory of the University of Chicago and then, in the 1950s, the astronomy department of the University of California (see DYNASTY).

In the middle of the nineteenth century, the Pulkovo Observatory shared the honor of possessing the largest refracting telescope in the world. Refractors bend to a focus light passing through curved glass lenses. As late as the end of the eighteenth century, glass of the quality necessary for optical instruments could be cast only in small pieces, up to two or three inches in diameter. The English duty on manufacturing limited production and made further experimentation too costly. Progress occurred on the continent, but even there the largest lens achieved by 1824 was a 9.5-inch disc for the Dorpat Observatory (in what is now Tartu, Estonia). In 1847 the 15-inch refractors at Pulkovo and the Harvard College Observatory were the largest in the world.

British industry made possible the first large reflecting telescope (1780 to 1860), which used metal mirrors to reflect light to a focus. William Herschel built a 48-inch metal mirror in 1789; in 1845 William Parsons, the Earl of Rosse, completed in Ireland his "leviathan," which had a metal mirror 72 inches in diameter. Large reflectors with their tremendous light-gathering power yielded remarkable observations of distant stellar conglomerations. But difficulty in aiming tons of metal and rapid tarnishing of mirrors rendered early reflectors unsuitable for observatories and professional astronomers, who could not harness consistently the instrument's raw power. Something new arrived with the twentieth century: a 60-inch reflecting telescope built and installed at the new Mount Wilson Observatory in 1908. The telescope had a reflective silver coating on a glass disc ground to bring incoming light to a focus, and a mounting system and drive capable of keeping the multiton instrument fixed on a celestial object while the Earth turned beneath it. The mountain observatory, funded by Andrew Carnegie's philanthropic Carnegie Institution, was one of the first located above most of the Earth's obscuring atmosphere. Its reflecting telescope, specifically designed for photographic work, completed the revolution in astronomical practice—which had required the tedious drawing by hand of features seen through telescopes—begun with the invention of *photography.

With professional astronomers by definition already employed in research projects, amateur astronomers pioneered in applying photography. The American John Draper took the first known photograph of a celestial object, the Moon, in the 1840s. In 1851 a new process using plates exposed in a wet condition made possible a few photographs of the brightest stars, but not until the introduction of more sensitive dry plates after 1878 did *photography become common in astronomical studies. Draper's son Henry took the first photograph of a nebula, the Orion Nebula, in 1880. In England long exposures taken by Andrew Common and Isaac Roberts brought out details too faint for the eye to see. Photography facilitated comparisons over time, produced permanent recordings of positions suitable for more precise measurement, and was essential for exploitation of the other new astronomical tool of the nineteenth century, *spectroscopy.

The development of spectroscopy and the subsequent rise of the new science of *astrophysics created new activities for astronomers. When attached to telescopes, prisms splitting light into spectra opened to investigation the physical and chemical nature of stars. In 1859 Gustav Kirchhoff, professor of physics at Heidelberg, showed that each element produces its own pattern of spectral lines. In the first qualitative chemical analysis of a celestial body, he compared the sun's spectrum to laboratory spectra. The English amateur astronomer William Huggins seized on news of Kirchhoff's work. By 1870 he had identified several elements in spectra of stars and nebulae. He also measured motions of stars revealed by slight shifts of spectral lines. Early in the twentieth century Vesto M. Slipher at the Lowell Observatory in Arizona was the first to measure Doppler shifts in spectra of faint spiral nebulae, whose receding motions revealed the expansion of the universe. It required an extended photographic exposure over three nights to capture enough light for the measurement.

Astronomical entrepreneurship late in the nineteenth century saw the construction of new and larger instruments and a shift of the center of spectroscopic research from England to the United States. Charles Yerkes and James Lick put up the funds for their eponymous observatories, which came under the direction, respectively, of the University of Chicago and the University of California. Percival Lowell directed his own observatory in Arizona. All three observatories were far removed from cities; the latter two sit on mountain peaks.

A scientific education became necessary for professional astronomers in the later nineteenth century, as astrophysics came to predominate and the concerns of professionals and amateurs diverged. As late as the 1870s and 1880s the self-educated American astronomer Edward Emerson Barnard, an observaholic with indefatigable energy and sharp eyes, could earn a place for himself at the Lick and Yerkes observatories with his visual observations of planetary details and his discovery of comets and the fifth satellite of Jupiter, but he was an exception and an anachronism. A project begun in 1886 at the Harvard College Observatory and continued well into the twentieth century to obtain photographs and catalog stellar spectra furthered another social shift in astronomy. It employed women, for lower wages than men would have received, but at least made space for them in a male profession. Annie Jump Cannon was largely responsible for the Henry Draper Catalogue, published between 1918 and 1924, which gave spectral type and magnitude for some 225,000 stars. Also, she rearranged the previous order of spectra into one with progressive changes in the appearance of the spectral lines. Although she developed her spectral sequence without any theory in mind, astronomers quickly realized that changes in the strength of hydrogen lines indicated decreasing surface temperature.

Spectral class became even more useful when, early in the twentieth century, it was related to luminosity. The relationship could be used to estimate distances. Students of stellar evolution asked what in the constitution of stars gave rise to dwarfs and giants and why brightness increased systematically with spectral type. The source of stellar energy, and with it the constitution of stars, became better known after the discovery of *radioactivity at the beginning of the twentieth century. As late as 1920, however, the English astrophysicist Arthur Eddington complained that the inertia of tradition was delaying acceptance of the most likely source of stellar energy: the fusion of hydrogen into helium. Eddington calculated how fast pressure increases downward into a star and how fast temperature increases to withstand the pressure. He used qualitative physical laws regarding the ionization of elements developed in the 1920s by Meghnad Saha, an Indian nuclear physicist. Saha's work also provided a theoretical basis for relating the spectral classes of stars to surface temperatures. World War II produced a deeper understanding of nuclear physics and more powerful computational techniques. The practice of astrophysics has moved from observatories to scientific laboratories to giant computers running simulations.

*Cosmology, the study of the structure and evolution of the universe, only belatedly insinuated itself into modern mainstream astronomy.

An inability to measure great distances limited cosmology for centuries to philosophical speculations, often focused on the nature of nebulae and the possible existence of island universes similar to our galaxy. Cosmology achieved an observational foundation early in the twentieth century, when Harlow Shapley and Edwin *Hubble at the Mount Wilson Observatory made the observations that revealed the size of our *galaxy, the existence of other galaxies, and the expansion of the universe.

At first these observations made little connection with Einstein's relativity theory. Astronomers, especially in the United States, possessed only limited mathematical knowledge, and were largely content to produce observations while leaving theory to theoreticians. In England, in contrast, interest in relativity theory relatively flourished. But the work was strictly mathematical, without observational input. In the 1930s Hubble attempted to bridge the gulf between observation and theory. Cooperation, he wrote, featured prominently in nebular research at the Mount Wilson Observatory, and he struck up a close collaboration with colleagues at the nearby California Institute of Technology. Not until the 1960s, however, after the discovery of quasars (*see* PULSARS AND QUASARS) and the intense theoretical effort to find a new energy source to explain them, did relativity theory secure a place in mainstream astronomy.

During the 1970s and 1980s scientists realized that important cosmological features could be explained as natural and inevitable consequences of new theories of *elementary particle physics, and particle physics now increasingly drives cosmology. Also, particle physicists, having exhausted the limits of particle *accelerators and public funding for yet larger instruments, now turn to cosmology for information regarding the behavior of matter under extreme conditions, such as prevailed in the early universe.

The greatest change in astronomy during the twentieth century in understanding the universe and also in the backgrounds of astronomers and their activities followed from observations beyond visual light. In the 1930s the American radio engineer Karl Jansky and the radio amateur Grote Reber pioneered detection of radio emissions from celestial phenomena, and radar research during World War II helped develop *radio astronomy, especially in England. X-ray astronomy also took off after the war, first aboard captured German V-2 rockets carrying detectors above the Earth's absorbing atmosphere. NASA subsequently funded a rocket survey program and then satellites to detect X rays. Most of the new X-ray astronomers came over from experimental physics with expertise in designing and building instruments to detect high-energy particles. Their discoveries followed from technological innovations. Gamma rays and infrared and ultraviolet light provided further means of non-optical discoveries in the space age. Unlike the relatively quiescent universe open to Earth-bound astronomers' visual observations, the universe newly revealed to engineers and physicists observing at other wavelengths from satellites is violently energetic.

See also ASTRONOMY, NON-OPTICAL; ASTROPHYSICS; COSMOLOGY.

Otto Struve and Velta Zebergs, *Astronomy of the 20th Century* (1962). Martin Harwit, *Cosmic Discovery: The Search, Scope, and Heritage of Astronomy* (1981). Michael Hoskin, *Stellar Astronomy: Historical Studies* (1982). Dieter B. Hermann, *The History of Astronomy from Herschel to Hertzsprung* (1984). John North, *The Norton History of Astronomy and Cosmology* (1994). John Gribbin, *Companion to the Cosmos* (1996). *American Astronomy: Community, Careers, and Power, 1859–1940* (1997). Michael Hoskin, *The Cambridge Illustrated History of Astronomy* (1997). John Lankford, ed., *History of Astronomy: An Encyclopedia* (1997).

NORRISS S. HETHERINGTON

ASTRONOMY, NON-OPTICAL. Many recent astronomical discoveries have been made without using visible light. The newly revealed universe is fascinating in its activity. Stars are born, galaxies collide, neutron stars collapse into black holes, quasars vary in hours through luminosities greater than that of our entire galaxy, and pulsars rotate gigantic radio beams in fractions of seconds.

Radio astronomy is the oldest branch of the new astronomy. In 1932 Karl Jansky, an American radio engineer with the Bell Telephone Company, detected electrical emissions from the center of our galaxy while studying sources of radio noise. Neither optical astronomers nor Jansky's practical-minded supervisors cared. Grote Reber, an ardent radio amateur and distance-communication addict, took an interest, and built for a few thousand dollars a pointable radio antenna 31 feet in diameter in his backyard in Wheaton, Illinois. In 1940 he reported the intensity of radio sources at different positions in the sky.

Fundamental knowledge underlying radio astronomy techniques increased during World War II, especially with research on radar, and especially in England. After the war, research programs at Cambridge, at Manchester, and at Sydney, Australia, dominated radio astronomy for the next decade.

Manchester's large steerable radio telescope at Jodrell Bank was rescued from financial disaster in 1958 by its ability to track *Sputnik* (*see* SATELLITE). Other uses floated had included mobilizing the telescope as part of a ballistic missile tracking system and for long-range bomber navigation. As an astronomical instrument it detected radar signals bounced off ionized meteor trails in the earth's atmosphere. Other scientists bounced radar signals off the Moon and developed planetary radar astronomy to map the surfaces of planets.

The program at Cambridge was led by Martin Ryle, who received a Nobel Prize in 1974 for his overall contributions to radio astronomy. He completed a survey of almost two thousand radio sources, most of them extragalactic, in 1955. These sources had a bearing on Fred Hoyle's steady-state cosmological theory. Within a few years radio data and their interpretation were more certain, and argued convincingly against Hoyle's *cosmology, which allowed for fewer faint radio sources than were detected.

The major blow to Hoyle's theory came in 1965 with discovery of the cosmic microwave (short radio wave) background radiation. It had been predicted in 1948 by nuclear physicists exploring the consequences of the cosmological big bang, but no astronomer looked for it. In 1963 Arno Penzias and Robert Wilson with the Bell Telephone Laboratories detected what the first regarded as noise in an antenna, but soon realized must be excess radiation of cosmic origin. In 1965 Robert Dicke at Princeton interpreted the work for which Penzias and Wilson received a Nobel Prize as a remote consequence of the origin of the universe.

Radio sources were identified in the early 1960s with star-like objects, now called quasi-stellar objects, or quasars. They have luminosities a thousand times that of our entire galaxy. The only known source for so much energy from such a small volume is a *black hole. Quasars also have extremely large red shifts, probably a manifestation of an expanding universe. Ryle shared the Nobel Prize with his Cambridge colleague Anthony Hewish, who was credited with discovering pulsating radio sources, although his student Jocelyn Bell made the actual observation, in 1967 (*see* PULSARS AND QUASARS).

Radio astronomy is increasingly threatened by modern society's growing use of garage-door openers, microwave ovens, wireless telephones, and other sources of interference with radio signals from astronomical objects. The commercial use of radio frequencies is light pollution's invisible cousin.

As did radio astronomy, X-ray astronomy also took off after World War II, first aboard captured German V-2 rockets that took detectors above the earth's absorbing atmosphere. Herbert Friedman at the U.S. Naval Research Laboratory found that the Sun weakly emits X rays, as expected. Astronomers did not expect to find strong X-ray sources, and doubted that brief and expensive rocket-borne experiments could contribute much to observational astronomy. In the immediate post-*Sputnik* period, however, more government money existed for astronomical research than imaginative scientists could spend.

Help came from Italian-born Riccardo Giacconi, who worked at the private company American Science and Engineering and then at the Harvard–Smithsonian Observatory before becoming director of the Space Telescope Science Institute. In 1960, with funding from the Air Force Cambridge Research Laboratories, Giacconi discovered the first cosmic X-ray source, Sco X-1 (the strongest X-ray source in the constellation Scorpius). In 1963 Friedman's group detected a second strong X-ray source, associated with the Crab nebula, the remnant of a supernova explosion.

NASA now funded a rocket survey program and a small satellite devoted exclusively to X-ray astronomy. Launched into an equatorial orbit from Kenya on its independence day in 1970, the satellite was nicknamed *Uhuru*, Swahili for freedom. Its detectors discovered binary X-ray pulsars, neutron stars whose energies arise from infalling matter from companion stars. The Einstein X-ray telescope, launched in 1978, revealed that individual sources account for much of the X-ray background radiation. Giacconi had wanted to name this satellite *Pequod*, from Melville's *Moby Dick*, as a reminder of Massachusetts and the American Indian. The inevitable comparisons of Giacconi and the egomaniacal Captain Ahab indicate the qualities needed to drive a large and complex scientific project to completion. NASA declined to associate its satellite with a white whale.

A single scientific paper was published on X-ray astronomy in 1962, and 311 a decade later. Only 4 out of 507 American astronomers participated in X-ray astronomy in 1962, compared to 170 out of 1,518 in 1972. Over 80 percent of the participants came from experimental physics with expertise in designing and building instruments to detect high-energy particles.

More energetic than X rays, gamma rays arise from nuclear reactions, including those that create elements in stars and bombs. A satellite watching for nuclear tests in the atmosphere made the first observation of cosmic gamma ray sources, in 1973. The Compton Gamma Ray Observatory, deployed from a U.S. space shuttle in 1991, has recorded bursts of gamma radiation, about one a day, perhaps from mergers of extremely distant neutron stars into black holes. Gravitational waves from these events may be detectable by the Laser Interferometric Gravitationwave Observatory, constructed in 1999. The final seconds of such an event would be more luminous than a million galaxies. If it occurred in our galaxy, the burst of gamma radiation would destroy the earth's ozone layer, kill all life on Earth, and leave the earth's surface radioactive for thousands of years.

The ultraviolet is relatively quiet. The Extreme Ultraviolet Explorer Satellite, launched in 1992, has observed hot plasma flares in the outer regions of a few nearby stars. Much of the interstellar medium, however, stops ultraviolet light. The Hubble Space Telescope spans the spectral region from far-ultraviolet through visible to near-infrared.

Cool stars, planets, grains of dust in interstellar space, and also animals radiate most of their energy at infrared frequencies. Frank Low at Texas Instruments developed an infrared detector in 1961 so sensitive that under the impossible ideal condition of no other interfering infrared source, and hooked up to the world's largest telescope, it could have detected the body heat of a mouse on the Moon. Carl Sagan wanted to use it to beat the Russians in the race to detect life on Mars. In 1967 Low, now at the Kitt Peak National Observatory in Arizona, discovered a giant cloud of gas and dust in the constellation of Orion invisible at optical wavelengths but emitting enormous energy in the infrared. The cloud might be an interstellar nursery in which stars are born and grow in their violent way. The Infrared Astronomical Satellite during ten months in 1983 before all its cooling helium evaporated found more extensive dust tails for comets than are visible optically, rings of dust in our solar system, and an extensive ring or shell of gas and dust around the star Vega, perhaps left over from the original cloud of gas and dust from which Vega condensed.

The recent revolution in non-optical astronomy has important policy lessons. Should bureaucrats follow Thomas Kuhn's theory of the structure of scientific revolutions and act as keepers of the paradigm, directing research funds into observations intended to extend and consolidate mature theories? Or should they favor the early stages of development of new fields, when practitioners fumble along without any fixed conceptual framework? Non-optical astronomical discoveries have followed largely from technological innovations, with little advance prediction or justification for their search. Is it, then, nobler of mind and more cost-efficient to serve old paradigms or new technologies?

See also ASTRONOMY; RELATIVITY.

Sir Bernard Lovell, *The Story of Jodrell Bank* (1968). David O. Edge and Michael J. Mulkay, *Astronomy Transformed—The Emergence of Radio Astronomy in Britain* (1976). Martin Harwit, *Cosmic Discovery: The Search, Scope, and Heritage of Astronomy* (1981). Richard F. Hirsch, *Glimpsing an Invisible Universe: The Emergence of X-Ray Astronomy* (1983). Woodruff T. Sullivan, *The Early Years of Radio Astronomy: Reflections Fifty Years after Jansky's Discovery* (1984). Wallace Tucker and Riccardo Giacconi, *The X-Ray Universe* (1985). Wallace Tucker and Karen Tucker, *The Cosmic Inquirers: Modern Telescopes and Their Makers* (1986). David H. DeVorkin, *Race to the Stratosphere: Manned Scientific Ballooning in America* (1989). Andrew J. Butrica, *To See the Unseen: A History of Planetary Radar Astronomy* (1996).

NORRISS S. HETHERINGTON

ASTROPHYSICS. Telescopes equipped with prisms that split starlight into rainbow-like spectra made possible investigations of the physical and chemical nature of astronomical objects. In 1802 the English chemist William Wollaston found several dark lines in the solar spectrum, and in 1814 the German optician Joseph Fraunhofer observed and catalogued hundreds of solar lines. In 1859 Gustav Kirchhoff, professor of physics at the University of Heidelberg, showed that each element produces its own pattern of spectral lines. In the first qualitative chemical analysis of a celestial body, Kirchhoff compared laboratory spectra from thirty elements to the Sun's spectrum and found matches for iron, calcium, magnesium, sodium, nickel, and chromium.

The English astronomer William Huggins, tired of making drawings of planets and timing the meridian passage of stars, likened news of Kirchoff's work to "coming upon a spring of water in a dry and thirsty land." Huggins suggested to his friend and neighbor, William Allen Miller, a professor of chemistry at Kings College, London, that they commence observations of stellar spectra. Initially Miller doubted the wisdom of applying Kirchhoff's methods to the stars, because of their faint light; but by 1870 Huggins and Miller had identified several elements in spectra of stars and nebulae.

Spectroscopic techniques were also employed to measure motion in the line of sight. In 1842, Johann Christian Doppler, an Austrian physicist, argued that motion of a source of light should shift the lines in its spectrum. A more correct explanation of the principle involved was presented by the French physicist Armand-Hippolyte-Louis Fizeau in a paper read in 1841 but not published until 1848. Not all scientists accepted the theory. In 1868, however, Huggins found what appeared to be a slight shift for a hydrogen line in the spectrum of the bright star Sirius, and by 1872 he had more conclusive evidence of the motion of Sirius and several other stars. Early in the twentieth century Vesto M. Slipher at the Lowell Observatory in Arizona measured Doppler shifts in spectra of faint spiral nebulae, whose receding motions revealed the expansion of the universe (*see* COSMOLOGY).

Instrumental limitations prevented Huggins from extending his spectroscopic investigations to other galaxies. Astronomical entrepreneurship in America's gilded age saw the construction of new and larger instruments and a shift of the center of astronomical spectroscopic research from England to the United States. Also, a scientific education became necessary for astronomers, as astrophysics predominated and the concerns of professional researchers and amateurs like Huggins diverged.

George Ellery Hale, a leader in founding the *Astrophysical Journal* in 1895, the American Astronomical and Astrophysical Society in 1899, the Mount Wilson Observatory in 1904, and the International Astronomical Union in 1919, was a prototype of the high-pressure, heavy-hardware, big-spending, team-organized scientific entrepreneur. While an undergraduate at the Massachusetts Institute of Technology in 1889, he invented a device to photograph outbursts of gas at the Sun's limb. He continued studying the Sun at his home observatory and then at the Yerkes Observatory of the University of Chicago. In 1902 Andrew Carnegie established the Carnegie Institution of Washington with a $10,000,000 endowment for research, exceeding the total of endowed funds for research of all American colleges combined. With grants from the Carnegie Institution, Hale built the Mount Wilson Observatory with a 60-inch reflecting telescope in 1908 and a 100-inch completed in 1917. In his own research, Hale in 1908 detected the magnetic splitting of spectral lines from sunspots (*see* MAGNETO-OPTICS).

Stellar spectra were obtained at the Harvard College Observatory and catalogued in a project begun in 1886 and continued well into the twentieth century. Women did much of the tedious computing work. Edward Charles Pickering, the newly appointed director of the Harvard College Observatory in 1881 and an advocate of advanced study for women, declared that even his maid could do a better job of copying and computing than his incompetent male assistant. And so she did. And so did some twenty more women over the next several decades, recruited for their steadiness, adaptability, acuteness of vision, and willingness to work for low wages. Initially the stars were catalogued alphabetically, beginning with A, on the strength of their hydrogen lines,

but Annie Jump Cannon found a continuous sequence of gradual changes among them: O, B, A, F, G, K, M, R, N, S (mnemonically: Oh, be a fine girl; kiss me right now, Sweet.). Antonia Maury added a second dimension to the classification system by noting that spectral lines were narrower in more luminous, giant stars of the same spectral class. Cecilia Payne-Gaposkin, a graduate student at Harvard, determined from spectra relative abundances of elements in stellar atmospheres. Her Ph.D. thesis of 1925, published as *Stellar Atmospheres*, has been lauded as the most brilliant thesis written in astronomy. She received her degree from Radcliffe College, however—Harvard did not then grant degrees to women—and later when employed at Harvard she was initially budgeted as "equipment."

Ejnar Hertzsprung in Copenhagen and Henry Norris Russell at Princeton University independently, in 1911 and 1913 respectively, related spectral class to luminosity. The Hertzsprung–Russell diagram was used to estimate distances by determining spectral class, reading absolute brightness off the diagram, and comparing that to the observed brightness diminished by distance. Once the source of stellar energy became better known, evolutionary tracks of stars would be drawn on the H–R diagram.

By the middle of the nineteenth century, discussions of the source of solar energy had rejected chemical combustion, which would have burnt away a mass as large as the Sun in only 8,000 years. From 1860 on, William *Thomson (Lord Kelvin) switched from an influx of meteoritic matter as the Sun's energy source to gravitational contraction. Thomson estimated the age of the Sun at twenty million years, less than a tenth that required by *Darwin's theory of *evolution. In 1903 Pierre and Marie *Curie measured the heat given off by a gram of radium. This hitherto unknown source of energy opened up vast spans of time for geological and biological evolution, though astronomers were slow to abandon gravitational contraction (*see* EARTH, AGE OF).

In 1938 the German-born Hans Bethe (b. 1906), then at Cornell University, proposed a plausible mechanism for energy production in stars. He knew much about atoms but little about stellar interiors when he attended a conference in April 1938 reviewing the problem of thermonuclear sources in stars. Shortly thereafter, Bethe worked out the carbon cycle. It begins with a carbon-12 nucleus; adds four protons in stages, converting the carbon to nitrogen and then to oxygen; and ends in nuclei of carbon-12 and helium-4 plus energy. A bit later Bethe also envisioned a proton-proton reaction: two protons (hydrogen nuclei) form a nucleus of deuterium (one proton becoming one neutron); the addition of a third proton creates a helium-3 nucleus (two protons, one neutron), two of which collide to form helium-4 (two protons, two neutrons) while ejecting two protons. The net result in both cases is the fusion of four hydrogen nuclei into one helium nucleus plus a release of energy. The proton-proton reaction provides the main source of energy in stars about the mass of our Sun, 70 percent of which is hydrogen and 28 percent helium. In more massive and hotter stars with more heavy elements, the carbon cycle is more important. Bethe received the Nobel Prize in 1967 for his work on the mechanisms of energy production in stars.

See also ASTRONOMY; ASTRONOMY, NON-OPTICAL; COSMOLOGY.

Cecilia Payne, *Stellar Atmospheres* (1925). A. S. Eddington, *Stars and Atoms, Fourth Impression* (1929). Otto Struve and Velta Zebergs, *Astronomy of the 20th Century* (1962). Helen Wright, Joan N. Warnow, and Charles Weiner, *The Legacy of George Ellery Hale: Evolution of Astronomy and Scientific Institutions, in Pictures and Documents* (1972). J. B. Hearnshaw, *The Analysis of Starlight: One Hundred and Fifty Years of Astronomical Spectroscopy* (1986). John Gribbin, *Companion to the Cosmos* (1996).
NORRISS S. HETHERINGTON

ATMOSPHERIC ELECTRICITY. In the early eighteenth century, noticing the similarity between lightning and static electric discharges produced in the laboratory, Francis Hauksbee and William Wall suggested that there was electricity in the atmosphere. Lightning, aurora borealis, the odor of ozone on a mountain ridge, and St. Elmo's fire had not yet been associated with electricity.

In 1752, following a suggestion of Benjamin Franklin, the Comte du Buffon arranged for a test of the electrical character of *lightning. That same year, astronomer Pierre Charles Lemonnier discovered that the clear atmosphere exhibits an electrical charge. This was a transforming moment: air as well as clouds could be electrified. In 1757 Giambattista Beccaria, Franklin's first defender in Italy, began systematic research into atmospheric electricity using rockets and other probes. He published the first extended treatise on the subject in 1775. Charles Augustin *Coulomb investigated the electrical conductivity of the atmosphere. He showed that inadequate insulation could not account for most of the charge lost by an electrified metal plate and that losses owing to atmospheric conductivity increased with the charge on the test body.

The early nineteenth century saw natural historical studies of atmospheric electricity and speculations about its processes. Around 1812 Gustav Scöbler studied the diurnal period of atmospheric electricity, a phenomenon later investigated by François Arago and Alexander von *Humboldt. In the early 1840s Jean Charles Athanase Peltier discovered that the earth has a negative charge. Theorists pondered how processes on the earth's surface such as evaporation and chemical changes might produce atmospheric electricity.

The new physics of the 1890s, particularly the new understanding of the atom, *electron, radiation, and *electromagnetic theory, reinvigorated studies of atmospheric electricity. Since Coulomb, scientists had believed that dust or moisture in the atmosphere, and not the air itself, was responsible for dissipating charges. Early-nineteenth-century chemists Jöns Jacob Berzelius and Humphry Davy, among others, suggested that chemical compounds are held together by electrical force. After F. Linss discovered in 1887 that dry air can conduct electricity, scientists began to apply the theory of ions, introduced by Svante Arrhenius the same year to explain electrolytic and other phenomena in liquids, to gases. Joseph John Thomson's discovery of the electron in the late 1890s strengthened this understanding. Around 1900, Julius Elster and Hans Geitel, gymnasium teachers in Germany, explained charge dissipation as a result of the movement of ions and electrons. The investigation of ionized gas flow formed a central part of atmospheric research throughout the twentieth century.

Because the ground is usually charged negatively with respect to the air, a potential gradient (or voltage) amounting to at least hundreds and often thousands of volts per meter exists between the two. During the late nineteenth and early twentieth centuries, scientists measured its daily, seasonal, and other regularities at land observatories, at high altitude using balloons, and at sea, especially on the research vessel *Carnegie* between 1909 and 1929. They developed ingenious instruments for capturing values of the potential gradient: William Thomson's water-drop electroscope and flame electrometer, Franz Exner's gold-leaf electroscope with attached oil lamp, and Hans Benndorf's device that used *radium to bring itself to the ambient potential of the air at that spot. Victor Hess summarized the research in *Electrical Conductivity of the Atmosphere* (1926 in German, 1928 in English). *Radioactivity, discovered in 1896, the physics of various "rays" (*cathode and *X rays, for example), and the discovery by Henri Becquerel and Ernest *Rutherford that rays from radioactive substances ionize gases suggested exploration of the effects of naturally occurring radioactivity on the atmosphere. Julius Elster and Hans Geitel, C. T. R. Wilson, John Joly, H. Gerdien, Victor

Hess, and John Satterly all investigated radioactivity in the atmosphere long before nuclear fallout brought it to public attention. Their work resulted in the discovery of *cosmic rays. From the 1920s on, Victor Hess, Werner Kolhörster, Robert Millikan, Arthur Holly Compton, Scott Forbush, and Oliver Gish linked investigations of gamma radiation with atmospheric electricity through the effects of altitude on the ionization of gases in closed vessels.

In the later twentieth century, as some scientists concentrated on modeling the global electrical circuit, others examined the physics of lightning discharge closely, including the electrical fields produced, and discovered a wide assortment of new electrical phenomena, including red sprites (extensive, brief light flashes above large thunderstorms), trolls (Transient Red Optical Luminous Lineament—red discharges occurring just above cloud tops after an especially strong sprite), blue jets (conical discharges traveling upward from cloud tops to 45 km at 100 km/sec), and elves (Emissions of Light and Very low frequency perturbations from Electromagnetic pulse Sources—brief, expanding disks of red light at about 100 km altitude). Many of the newly discovered events occur in the stratosphere and above it.

Park Benjamin, *A History of Electricity (The Intellectual Rise in Electricity) from Antiquity to the Days of Benjamin Franklin* (1898). Paul Fleury Mottelay, *Bibliographical History of Electricity and Magnetism* (1922). W. J. Humphreys, *Physics of the Air*, 3d ed. (1964). Peter E. Viemeister, *The Lightning Book* (1972). J. L. Heilbron, *Elements of Early Modern Physics* (1982). Donald R. MacGorman and W. David Rust, *The Electrical Nature of Storms* (1998).

GREGORY A. GOOD

ATOM AND MOLECULE. Greek philosophers in the pre-Socratic era first expressed the belief that matter is not infinitely divisible, and that there exist invisibly small ultimate particles called "atoms" (Greek for "unsplittable") that constitute all perceptible matter. The Stoic philosophers and to some degree Aristotle and his followers opposed this opinion, and the debate continued sporadically into the early modern era. The idea of atoms became increasingly popular in the seventeenth and eighteenth centuries, but natural philosophers could do little more than speculate about them.

Chemists began vicariously to explore the world of ultimate particles soon after the death of Antoine-Laurent *Lavoisier in 1794. Lavoisier's definition of a chemical "element"—a chemically irreducible species of matter—provided for the first time a basis for an empirical approach to atoms, for each element might be thought to consist of a characteristic kind of irreducible

particle. If the atoms of the various elements had truly distinguishable characteristics, such as weight, for example, then that fact ought to be discernible in the combinations of the various elements with each other. And so it was, for the laws of "stoichiometry" (regularities in the weight proportions in which elements combine with each other) were soon discovered, by Joseph Louis Proust, Jeremias Richter, and others.

These laws could only have been discerned by starting with the assumption of atoms. An example will make this point clear. Carbon and oxygen combine in two different ways to make carbonic oxide and carbonic acid gas. The first has 43 percent carbon and 57 percent oxygen; the second has 27 percent carbon and 73 percent oxygen by weight. These numbers seemed to suggest nothing about integral ratios associated with presumed integral atoms. The Englishman John Dalton in 1803 set out under the assumption that the substance that is less rich in oxygen consists of collections of identical molecules, each molecule consisting of one atom of carbon united to one atom of oxygen. In such a case, the carbon atom would have to weigh 12 to the oxygen atom's 16 units (or 6 to 8, or 3 to 4, or any other equal ratio), for 12 is to 16 as 43 is to 57. But if that were true, the second substance could easily be understood as consisting of molecules containing one carbon atom united to *two* of oxygen, for 12 is to 32 (16 + 16) as 27 is to 73. In this fashion, Dalton used the regularities of elemental weight relations to justify the theory of chemical atoms.

In this reasoning, assumptions about the composition of the molecules help to determine the calculation of relative weights of carbon and oxygen atoms. For example, eleven years after Dalton's calculation, Joseph Louis Gay-Lussac chose to believe that the more highly oxygenated gas consisted of a "volume" of carbon united with a "volume" of oxygen, with relative weights of 6 and 16 respectively (reflecting the 27/73 percentage composition). Then the other compound would have two carbons and one oxygen, for 12 (6 + 6) is to 16 as 43 is to 57. Thus, even if stoichiometry implied chemical atoms, it did not dictate *unambiguous* relative weight relations among them, for those calculations derive from assumptions regarding molecular formulas. (The terms "atom" and "molecule" are used here as they are defined today, definitions not uniformly adopted until the 1860s.)

Gay-Lussac preferred to reason in terms of "volumes" because he had discovered in 1808 that chemical reactions between gases take place in integral volumes. In 1811 and 1814, respectively, Amedeo Avogadro and Andre-Marie Ampère independently concluded from this law of combining volumes that the ultimate particles of gases must all have the same volume, so that equal volumes of any two gases under the same temperature and pressure must contain equal numbers of particles. "Avogadro's hypothesis," as it came to be known, suggested a simple physical method of determining relative atomic weights, namely by measuring relative vapor densities. But Avogadro's hypothesis did not seem to be reconcilable with certain facts, such as that water vapor was less dense than one of its constituents, oxygen. To solve this problem, Avogadro suggested that oxygen particles split in two in combining with hydrogen, each reactant oxygen particle thus providing the basis for two product molecules of water. This seemed both ad hoc and highly improbable to most of Avogadro's contemporaries.

Another disputed point was whether all atomic weights were integral multiples of that of the lightest element, hydrogen—a thesis defended as early as 1815 by the English physician William Prout. So many atomic weights were close to integers (when, by convention, hydrogen is set equal to one) that Prout's hypothesis seemed probable to some prominent atomists. The hypothesis led to the notion that there might be a fundamental particle whose aggregation accounted for the various elemental atoms. But contrary cases such as chlorine (35.5 times that of hydrogen) gradually drew most chemists away from Prout's hypothesis.

Chemists working their way through the puzzles of atomic theory distinguished between atoms and molecules by thinking of the former as the smallest integral particle of an element, and the latter as the smallest integral particle of a compound; the first was homogeneous and unsplittable, the second a heterogeneous compound particle that could be formed, transformed, or dismembered. Every chemist applied his own distinct terminology to these concepts. Avogadro complicated the scheme, for his theory required a distinction between atoms and molecules *of elements*. Since most chemists supposed that the force uniting atoms into molecules was electrical, the idea that identical atoms of an element could attract each other seemed as absurd as mutual attraction of two identical electrostatic charges.

The rejection of Avogadro's theory, though reasonable in the context, involved chemists in difficulties of a different order. Gradually there developed other physical approaches to understanding the atomic-molecular level: in addition to Gay-Lussac's law of combining volumes, there was Alexis Petit and Pierre Dulong's law of atomic heats, and Eilhard Mitscherlich's study of isomorphic crystalline compounds. The great Swedish chemist Jöns Jacob *Berzelius gave the fullest early development of atomic theory between 1813 and 1826. Berzelius used all of the physical methods known to him, as well as sensitive and broad-ranging chemical data and analogies, to construct a system of atomic weights and formulas that most European chemists accepted from the late 1820s through the early 1840s.

That did not stop new proposals for deriving atomic weights and formulas, each system implying its own conception of the atoms and molecules it represented. By 1850 four versions including Berzelius's competed for European dominance. The two leading variants were a system of chemical "equivalents" whose claimed advantage was being purely empirical, hence permanently defensible and unalterable, and a revised system of atomic weights advertised as being ontologically true, even though theoretically derived.

The latter had been developed by the Frenchmen Auguste Laurent and Charles Gerhardt. The Laurent-Gerhardt system resembled Berzelius's, but used Avogadro's hypothesis more consistently. Laurent proposed definitions of atom and molecule in 1846 that were subsequently adopted into all of the major European languages and are current today: the atom is the smallest *chemically* active unit of an element, while the molecule is the smallest freely existing and *physically* active unit of an element or compound.

During the 1850s a consensus quietly developed among leading German and British theorists in favor of the Laurent–Gerhardt system. The evidence driving this shift came primarily from ingeniously designed (and interpreted) chemical reactions, not from evidence from physics. Chemists had long puzzled over the phenomenon called isomerism (first named by Berzelius in 1830): the word applied to cases of two or more unequivocally distinct substances that contained the same elements combined in the same proportions. Many had conjectured, reasonably but vaguely, that the differences in properties might be traced to differing arrangements of the atoms in the molecules of isomeric compounds. By the 1850s chemists had begun to get purchase on the question of intramolecular atomic arrangements, and by the end of the decade August Kekulé's theory of "chemical structure" (as his Russian rival Aleksandr Butlerov called it) was beginning to gain adherents. These novelties were made possible only by manipulating the newer formulas and atomic weights.

To gain fuller support for the new system, a small coterie of reformers organized an international chemical symposium, the first of its kind, which took place in Karlsruhe, Germany, in September 1860. Although the meeting itself had little drama, it was successful in the long term in helping to widen the consensus; here the persuasive exposition of the Italian theorist Stanislao Cannizzaro played a leading role. By the 1860s, most chemists had adopted the Laurent–Gerhardt system, which is nearly identical to that used today. Only in France were equivalents preferred by the majority of chemists, until they, too, adopted atomic weights in the 1890s. The capstone of the chemical atomic theory was the periodic system of the elements (*see* PERIODIC TABLE), beginning with the work of Lothar Meyer and especially Dmitrii *Mendeleev in the late 1860s, and elaborated by many others during ensuing decades.

In the late 1850s physicists were also penetrating the invisible world of atoms and molecules. The early development of the kinetic theory of gases by Rudolf Clausius and James Clerk Maxwell led to a reaffirmation of Avogadro's hypothesis. As the kinetic theory gained in stature and success, Avogadro's hypothesis took on the status of an accepted fact of nature. This sense was reinforced when a number of estimates of the sizes of molecules, inferred from experimental physical data, emerged in the late 1860s and early 1870s, all fairly consistent among themselves.

Still, a current of skepticism toward the physical reality of atoms and molecules can be followed throughout the nineteenth century. This skepticism was finally and permanently removed at the beginning of the twentieth century, as the result of the discovery of radioactivity, subatomic particles, light quanta, and a new interpretation of Brownian motion.

A. G. van Melsen, *From Atomos to Atom* (1952). Frank Greenaway, *John Dalton and the Atom* (1966). W. H. Brock, *The Atomic Debates* (1967). David M. Knight, *Atoms and Elements* (1967). Stephen G. Brush, *The Kind of Motion We Call Heat* (1976). Mary Jo Nye, *The Question of the Atom* (1984). Alan J. Rocke, *Chemical Atomism in the Nineteenth Century* (1984).

A. J. ROCKE

ATOM BOMB. See NUCLEAR BOMB.

ATOMIC STRUCTURE. By 1890, much evidence had accumulated that the atom of chemistry and the molecule of physics must have parts. The chemical evidence included analogies between the behavior of dilute solutions and of gases, as developed in the ionic theory of Svante Arrhenius (Nobel Prize in chemistry, 1903), and the implication from the *periodic table that the elements must have some ingredient in common. The physical evidence included the emission of characteristic spectra, which was likened to the ringing of a bell, and the formation of ions in gas discharges.

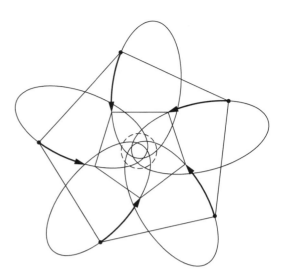

Sommerfeld's "Ellipsenverein" showing the linked motions of five electrons describing similar precessing elliptical orbits; at any instant the electrons occupy the vertices of a rotating, pulsating pentagon.

Study of the rays emanating from the cathode in these discharges prompted *Joseph John *Thomson (Nobel Prize in physics, 1906) to assert in 1897 that the rays consisted of tiny charged corpuscles, which made up chemical atoms and constituted their common bond, and also emitted spectral lines and proffered the key to ionization. This hazardous extrapolation quickly received confirmation. Thomson showed that the *ions liberated from metals by ultraviolet light had the same low mass-to-charge ratio (m/e) as corpuscles, around one one-thousandth the corresponding ratio for hydrogen atoms in *electrolysis. Thomson's student Ernest *Rutherford (Nobel Prize in chemistry, 1908) determined that the rays from radioactive substances consisted of two sorts, one, alpha, unbendable, the other, beta, bendable, by a magnetic field; and he and the discoverer of *radioactivity, Henri Becquerel (Nobel Prize in physics, 1903), showed by the degree of bending that beta rays also had a very small m/e.

Indication of the presence of corpuscles within atoms came from the magnetic splitting of spectral lines accomplished by Pieter Zeeman (Nobel Prize in physics, 1902) as elucidated by Hendrik Antoon Lorentz (Nobel Prize in physics, 1902). Lorentz traced spectra to the vibrations of "ions" whose m/e, as determined from Zeeman's magneto-optical splitting, was very close to those of the cathode-ray, photo-electric, and beta-ray particles. The omnipresent corpuscle became the *electron around 1900 when experiments at Thomson's laboratory at Cambridge indicated that the charge on corpuscles was about the same as that on a hydrogen ion.

Thomson was a leader of the British school, largely Cambridge trained, that who, to paraphrase another of its lights, William *Thomson, Lord Kelvin, understood a problem best after constructing a physical model of it. Thomson supposed that the corpuscle-electrons constituting an atom circulated within a spherical space that acted as if filled with a homogeneous, unresisting, diffuse, weightless positive charge. On this last assumption, an atom of mass A would contain around 1,000A electrons. In 1904 Thomson exhibited analogies between a swarm of electrons circulating in concentric rings and the chemical properties of the elements. He thus introduced the important idea that one element differed from another only in its electronic structure. In 1906, having calculated the capacity of his model atoms to scatter *X rays and beta rays, he deduced that the number of scattering centers was more nearly 3A than 1000A, thus giving weight and reality to the positive charge.

In 1910, Rutherford and Hans Geiger showed that alpha particles could be scattered through large angles. Rutherford inferred that Thomson's latest estimate was still too high. The scattering experiments indicated that the charges making up an individual scatterer acted as if assembled together at one point. Such an assemblage needed a smaller total charge to deflect an alpha particle through a given angle than a diffuse scatterer. Thus the atomic nucleus, which in Rutherford's original formulation could be either positive or negative, entered physics. He soon chose a positive center carrying the entire mass A and a charge of around $Ae/2$. Rutherford's student Henry G. J. Moseley then (in 1913–1914) deduced from his examination of characteristic x-ray spectra that the nucleus could be characterized by a whole number Z, beginning at 1 with hydrogen and increasing by a unit thereafter through the periodic table. Rutherford's school associated Z (the "atomic number") with the positive charge on the nucleus and recognized it to be a more reliable indicator of chemical nature than atomic weight. Classification by Z allowed an explanation of the few places in the periodic table where chemical properties did not follow weight and provided space for the accommodation of isotopes according to the ideas put forward by Rutherford's former colleagues Frederick Soddy (Nobel Prize in chemistry, 1921), Georg von György Hevesy (Nobel Prize in chemistry, 1943), Kasimir Fajans, and others around 1913.

During a study trip to Rutherford's laboratory in 1911, Niels *Bohr (Nobel Prize in physics, 1922) convinced himself that the nuclear atom was the route to a theory of atomic structure much better than Thomson's. One cause of his conviction was that ordinary electrodynamics required that a nuclear hydrogen atom with one electron could destroy itself instantly by radiation, whereas atoms stuffed with electrons could last a very long time. By presenting the problem of the existence of atoms in its strongest form, the nuclear hydrogen atom appealed to Bohr's dialectical mind. He solved the problem by fiat, by declaring that electrons could circulate in atoms only on orbits restricting their angular momenta to integral multiples of the *quantum of action. With this condition Bohr derived the wavelengths of the spectra of hydrogen and ionized helium and gave indications of the nature of molecular bonding in (1913–1914). Arnold Sommerfeld extended the scheme to spectra containing doublets and triplets, and also to the x- ray spectra investigated by Moseley and Manne Siegbahn (Nobel Prize in physics, 1924).

Physicists returning from World War I found that Bohr (who as a Dane had been neutral), Sommerfeld (who was too old to fight), and their coworkers had made Bohr's combination of ordinary mechanics and quantum conditions into a hybrid, inconsistent, vigorous theory where none had existed before. Reasoning about electrons orbiting nuclei in quantized orbits, they could explain many features of atomic spectra, ionization and electron impact, and molecular bonding. The high point of the hybrid approach, known subsequently as "the old quantum theory," came in 1922, when Bohr claimed to be able to derive the lengths of the periods of the table of elements. But by then quantitative mismatches between theory and data on atoms more complicated than hydrogen had begun to undermine confidence in the internally structured mechanical atom. During the years between 1925 and to 1927, it was superseded by ascribing non-mechanical spin, unsociability (Pauli Exclusion), and exchange forces to the electron, and by replacing the Bohr-Sommerfeld equations with matrix and wave mechanics (see *QUANTUM MECHANICS). Physicists declared that they knew everything about the atom "in principle" and moved their model-making into the recesses of the nucleus.

J. L. Heilbron, *Historical Studies in the Theory of Atomic Structure* (1981). Abraham Pais, *Inward Bbound* (1986).

J. L. HEILBRON

ATOMIC WEIGHT. In his work of the 1770s and 1780s, Antoine-Laurent *Lavoisier took weight to be the central measure of amount of material; when he defined a chemical element as a chemically undecomposable substance, he took loss of weight as his criterion for decomposition. In the generation after Lavoisier's death, chemists developed ideas about chemical atoms whose single measurable feature was their weight. Since the absolute weight of an object as small as an atom could not be measured, atomic weights have always been expressed in units relative to a conventional standard.

Determining the relative weights of the atoms of each of the elements involved three pragmatic issues: first, choosing that conventional standard; second, determining the formulas for the compounds used to derive the atomic weights; and third, achieving the greatest possible accuracy in measurement. Among the first generation of chemical atomists, John Dalton, Humphry Davy, and William Prout chose the fiducial standard that the atom of hydrogen = 1 exactly; on the other hand, Thomas Thomson chose oxygen = 1, William Wollaston chose oxygen = 10, and Jöns Jacob Berzelius chose oxygen = 100. By the middle of the nineteenth century, nearly every chemist had adopted the hydrogen standard, which is still used today in modified form (namely, carbon-12 = 12 exactly, which places hydrogen very close to 1).

Throughout the nineteenth century, the chemist had to determine (or to *assume*, if empirical determination proved impossible) the formulas for any chemical compounds used to derive atomic weights. For example, Dalton's data (1810) indicated that water consists of 87.5 percent oxygen and 12.5 percent hydrogen. He assumed that molecules of water consist of an atom of hydrogen united to an atom of oxygen; consequently oxygen weighed 7 using the standard of hydrogen = 1. By contrast, Davy and Berzelius (in 1812 and 1814, respectively) took H_2O rather than HO as the formula for water. In this case, *two* atoms of hydrogen must make up the same quantity that one did in the previous case, so hydrogen must be 0.5 on the scale of oxygen = 7, or if H = 1, then O = 14. Actually, Berzelius's more precise chemical analysis indicated that water consisted of 88.8 percent oxygen by weight, making the weight of an oxygen atom 16 if calculated on the scale H = 1.

Since no physical methods for the determination of chemical formulas existed at the beginning of the nineteenth century, the varying assumptions about formulas led to conflicting systems of atomic weight, many of which gave values for the same atom that differed by small integral multiples (such as O = 8 versus O = 16). Only gradually did chemists learn how to determine formulas in a way that could command a consensus in the specialist community (*see* ATOM AND MOLECULE). By the 1860s, most European chemists had come to agreement on a single system of atomic weights and formulas, nearly identical with that used today. Still, many fine chemists in France continued to accept HO as the formula of water, until they too converted to the majority viewpoint in the 1890s.

The dispute between Dalton and Berzelius over hydrogen, oxygen, and water also illustrates the significance of accuracy in gravimetric analyses of compounds to arrive at precise atomic weights. Since many atomic weights appeared to be close to integers on the scale of H = 1, Prout and Thomson both believed that all atomic weights should be integral. If true, all atoms might be built up from subatomic particles that represented the integral weight units—perhaps hydrogen atoms themselves. "Prout's hypothesis," as it came to be known, was formally proposed in 1815.

Berzelius rejected the hypothesis on empirical grounds. Thomson and others, he said, were letting their predilection damage their objectivity. An extended discussion between Berzelius, Justus von *Liebig, Jean-Baptiste-André Dumas, Charles Marignac, and others between 1838 and 1849 over the relative atomic weights for the crucial elements hydrogen, carbon, and oxygen established that Berzelius's nonintegral weight for carbon was nearly 2 percent too high. The revised weights (1, 12, and 16 almost exactly) gave another boost to Prout's hypothesis. Subsequent new determinations by the Belgian chemist J. S. Stas established that many elements had nonintegral atomic weights and drew opinion against Prout once more.

Edward Morley and T. W. Richards, who won a Nobel Prize for his work, made the finest atomic-weight determinations using purely chemical means at the end of the century. After the development of the mass spectrometer, atomic weights could be determined with great accuracy and directness. The history of the mass spectrometer is intimately connected with investigations of *atomic structure, including the discovery of protons, neutrons, and isotopes. All atomic nuclei consist of integral numbers of unit masses of protons and neutrons, the elemental atomic weight being nothing more than a weighted average of all the naturally occurring isotopes of the element. There was some truth to Prout's hypothesis after all.

W. H. Brock, *The Atomic Debates* (1967). David M. Knight, *Atoms and Elements* (1967). Mary Jo Nye, *The Question of the Atom* (1984). Alan J. Rocke, *Chemical Atomism in the Nineteenth Century* (1984).

A. J. ROCKE

ATOMS FOR PEACE. The application of modern science to warfare reached new heights, or depths, with the development of nuclear weapons during World War II. The destructiveness of nuclear bombs spurred scientists and politicians to try to ban them after the war through international agreements, but the onset of the Cold War derailed their efforts. The Soviet Union developed a nuclear arsenal, the United States pushed ahead with its own, and by 1953 both nations had tested versions of H-bombs much more destructive than the bombs used against Japan. The advent of the H-bomb rekindled American efforts for arms control. One of these, a study led by J. Robert Oppenheimer (*see* KURCHATOV AND OPPENHEIMER) called Operation Candor, urged a new approach to disarmament, including the release of information about nuclear weapons to encourage public understanding. Stalin's death, peace in Korea, and the Candor report inspired President Eisenhower to develop a new idea.

On 8 December 1953 Eisenhower proposed to the United Nations that nuclear nations contribute uranium and fissionable materials to an International Atomic Energy Agency (IAEA), which would use its endowment to develop peaceful applications of nuclear technology such as power generators. The plan, called "Atoms for Peace," would not only divert fissionable material from the production of nuclear weapons, but would also provide plentiful sources of electrical power for economic development worldwide. Atoms for Peace reflected Eisenhower's genuine belief that diversion of resources from weapons to power reactors could lead the world to peace instead of war. At the same time, Eisenhower and his advisors recognized the scheme's propaganda benefits and its relative advantage to the United States, since the U.S. possessed a steady supply of uranium and a larger stockpile of weapons than the Soviets.

The redeeming features of nuclear energy had figured in many public pronouncements since Hiroshima promoting the peaceful side of the atom. American policy, however, had emphasized weapons and had focused reactor research on military propulsion instead of civilian atomic power. In the early 1950s the fruits of these research programs, combined with fears of an international race for civilian power, renewed American interest in civilian applications. The domestic push for nuclear power in the United States coincided with and reinforced Eisenhower's scheme to promote nuclear power internationally.

Atoms for Peace met immediate general acclaim outside the Soviet bloc, and eventually resulted in the creation of the IAEA under the United Nations in 1956. But the original plan to pool material in the international agency bogged down in Soviet resistance and diplomatic squabbles. Atoms for Peace instead came to focus on bilateral agreements with individual countries, under which the U.S. helped to establish research and power reactors abroad, and with Euratom, a European regional association. By mid-1955 the United States had negotiated research bilaterals with two dozen nations. The atomic energy labs in the U.S. hosted foreign students and researchers, and sent their own scientists abroad with blueprints for reactors and research centers.

Atoms for Peace foundered on a basic problem: reactors for civilian power could also produce and hence proliferate weapons material. The proposal thus failed to advance arms control, although it did aid the spread of nuclear power. It also gave impetus to the International Conferences on the Peaceful Uses of Atomic Energy, held in Geneva in 1955 and 1958 under the auspices of the United Nations. These conferences gave scientists a chance to discuss their research with colleagues abroad and gave nations an opportunity to maneuver for Cold War prestige and propaganda. Internationalist ideals would continue to butt up against nationalist realities, as the military stand-off produced by the arms race brought Cold War competition to a wider socio-political arena. Despite the hopes of Atoms for Peace, science as well as technology entered what Eisenhower in 1958 called the "total Cold War."

See also WORLD WAR II AND COLD WAR.

Arnold Kramish, *The Peaceful Atom in Foreign Policy* (1963). Richard G. Hewlett and Jack M. Holl, *Atoms for Peace and War 1953–1961: Eisenhower and the Atomic Energy Commission* (1989).

PETER J. WESTWICK

B

BACONIANISM. Francis Bacon, Lord Chancellor of England, aimed to reform his country's legal system and humankind's method of acquiring natural knowledge. The latter program, the *Instauratio magna*, was to have influence in the history of science well into the nineteenth century. Bacon exposed the weaknesses of existing forms of knowledge: the "frivolous disputations" of scholasticism and the "blind experiments and auricular traditions" of the "mechanical arts." He proposed to replace them with an approach that joined reason and experience, contemplation and action. Because the truth of a thing is at the same time its utility, this method would put knowledge to the service of humanity: "human Knowledge and human Power do really meet in one."

Bacon presented his method in *New Organon* (1620), which he intended to replace Aristotle's logic books known as the *Organon*. The method is induction. Bacon argued that the deductive Aristotelian syllogism was dry and barren. To use it, one had first to jump immediately from casual experience to general principles. General principles, Bacon countered, themselves must derive from an exhaustive survey of experience via induction. Baconian induction is not simplistic but involves multiple stages, each requiring abstraction and invention, until one arrives at the truth.

Because the material for induction is so vast, Baconian research requires collaboration. This theme appears throughout Bacon's writings, but most clearly in his posthumous *New Atlantis*, a utopia where state-supported research is carried on at "Salomon's House"—an institution also responsible for the country's religious and moral culture.

It is a commonplace that no student of nature has ever followed the Baconian method. Nonetheless, Bacon's ideas resonated strongly in early-seventeenth-century England, where natural philosophers and educational reformers—particularly the circle around Samuel Hartlib—invoked his name and claimed to follow his precepts. Salomon's House was called "the prophetic scheme" for the Royal Society of London. The *philosophies* of the French Enlightenment venerated him; British scientists of the early Victorian period considered him the "supreme legislator of the republic of science." The Frankfurt School of philosophy found in his work the seeds of modern materialism, the mercantilization of culture, and the destruction of human values. All saw Bacon through the prism of their own desire. It is also common, though not useful, to label certain disciplines depending heavily on data collection as "Baconian." Such a generalization ignores both the details of Bacon's method and the variety of methodological problems involved in sciences such as meteorology, botany, geology, and natural history. The term "Baconianism" is as little likely to have significant application outside its proper historical context as "Romanticism."

Finally, some scholars have criticized Bacon for failing to appreciate the role of mathematics in "the emergence of modern physical science." In his time, however, mathematics and physics were distinct disciplines that few if any thought to unite. Astronomy and mechanics—the glory of seventeenth-century science—belonged to mathematics and had been treated quantitatively since ancient times; electricity, heat, and other parts of physics would not develop a mathematical methodology until the late eighteenth century. This criticism of Bacon, like so much written about him, neglects the historical context in which he lived, worked, and wrote.

Theodore Brown, "The Rise of Baconianism in Seventeenth-Century England," in *Science and History: Studies in Honor of Edward Rosen*, ed. Erna Hilfstein, Pawel Czartoryski, and Frank D. Grande (1978): 501–522. Markuu Peltonen, ed., *The Cambridge Companion to Bacon* (1996).

THEODORE S. FELDMAN

BACTERIOLOGY AND MICROBIOLOGY. The science known as bacteriology or microbiology came into existence when the French chemist Louis *Pasteur, the German physician Robert Koch, and their associates established the role of specific microorganisms in causing specific fermentations and diseases. Yet the full significance of the new science they wrought extends well beyond providing the evidence to support and the tools to apply what their medical contemporaries called "germ theory." Bacteriology was less a new discipline with its own set of research problems than a new kind of laboratory or experimental system in which myriad physiological, chemical, and biological problems could be addressed and from which new sciences could emerge, notably immunology. Moreover, the development of bacteriology was nearly inseparable from the role it played in wider material, social, and conceptual transformations that unfolded in and beyond the industrializing world in the century after 1860.

Ever since the seventeenth century, the learned and the curious had bent over *microscopes to peer at a hitherto unseen new world that seemed to teem with *life. These visions inspired everything from philosophical systems by Gottfried Wilhelm Leibniz to comical verse by Jonathan Swift to natural history of "animalcules" and *"infusoria" and controversy over the origin of life. By the mid-nineteenth century, field observation and innovation by men like the physician Ignaz Semmelweis in the maternity clinic, the civil servant and amateur agronomist Agostino Maria Bassi on the silk plantation, and the surgeon Joseph Lister were yielding the first testable conjectures concerning the role of microscopic life in processes of putrefaction and disease.

Decisive in the transition from this field experimental work to systematic laboratory study of a wide range of specific microorganisms grown in artificial media under controlled conditions were, first, application of chemistry and experimental physiology to the study of apparently "germ"-related processes and, second, radical innovation in methods and materials. The first of these phases, roughly 1835–1875, began in the Berlin physiology laboratory of Johannes Peter Müller. His student Theodor *Schwann designed an experimental apparatus allowing him to observe that putrefaction and alcoholic fermentation occurred only in the presence of air that had not been heated to destroy any microscopic life it contained.

Schwann published one important paper and then moved on to other research; twenty years later Louis Pasteur extended Schwann's work to other fermentations and to disease, publishing hundreds of papers and numerous books, training students and gathering associates into his laboratory and thus establishing a school of research on microscopic organisms. Whereas his forerunners and contemporaries were unable to observe microorganisms in isolation from complex natural media such as grape juice, beer, and blood, the chemist Pasteur created a general experimental system for cultivating microorganisms in artificial media.

These media were chemically calibrated and their environments controlled so as to favor the growth of any given species over others, thus allowing each to be studied in isolation, its chemical input and output subjected to exact analysis.

The second phase in the emergence of bacteriology, roughly 1875–1890, is associated especially with Robert Koch. The methods and instruments of chemistry and physiology were supplemented and often replaced by novel ways of cultivating, manipulating, and representing microorganisms. Thus the delicate blown-glass globe taken from chemistry by Schwann and Pasteur for use in growing microbial cultures—tricky to manage and precarious to transport, filled with infusions difficult to maintain in biological purity and impossible to scan with the microscope—became the simple, transportable, scannable glass dish named after Koch's associate Richard Petri, now ubiquitous in biomedicine, with its conveniently immobile growing-surface of gelatine or agar-agar, upon which a mixed culture could be sown and then divided into colonies or individual bacteria to be "isolated" and replated as "pure cultures." Whereas Pasteur's difficult methods had been adopted by a mere handful of investigators outside his laboratory, now the new bacteriological technique spread throughout the world.

Koch's fundamental contributions are threefold. First, his studies on anthrax joined those of Pasteur on silkworm diseases in showing the direct relevance of the laboratory to understanding events in the field. Koch watched the seemingly static "little rods" (bacilli) observed by other microscopists in anthrax blood multiply by division and, in a characteristic life-cycle, change form into highly resistant spores. The spore stage explained the ability of the disease to persist in abandoned pastures. Second, Koch showed that the welter of microscopic life that others had observed in infected wounds could be resolved into a limited number of species, each correlated with a distinct set of pathological effects. Finally, in the work that made him virtually a household name, Koch used his new method of solid-media culture to isolate a microscopic species responsible for the leading cause of death in the Western world, pulmonary consumption or tuberculosis, as well as for the dread disease cholera, while his students Friedrich Loeffler and Georg Gaffky did the same for diphtheria and typhoid fever.

Though these and similar demonstrations are often thought of as the end of etiological research, they were in fact its beginning. By what mechanisms and under what circumstances did microorganisms cause disease and epidemics? Why did some people exposed to infection become ill while others did not? Some early bacteriologists avoided these difficult questions. Others explored them through work on variable virulence, secreted toxins, and, above all, natural and acquired immunity.

In 1880 Pasteur's invention of artificial vaccines launched the study of immunity and immunization. But it also made bacteriology something it had not been before, namely, experimental biology in the strict sense: experimental study of species, inheritance, variation, and the Darwinian evolutionary mechanism of natural selection. Pasteur's vaccines were weakened cultures of pathogenic species, and many bacteriologists understood this weakening or "attenuation" to be biological variation. In the twentieth century, the study of microbial variation and of the associated phenomenon of infection of bacteria by virus (bacteriophage) came to have momentous theoretical consequences for the life sciences through its central role in the origins of *molecular biology and *genetics.

From its earliest, dramatic achievements, bacteriology was an icon of the triumph of technical skill and experimental discipline over speculation and superstition. Yet as some bacteriologists themselves pointed out, bacteriological research has also been shaped by myths and beliefs concerning purity and danger, life and death: disease-causing germs as demons or enemies, the seeds of disease and the soil of the body, the body as an armed citadel. The scope of meanings of microbiological research has continued ever since to range from the mundane (the quality of beer and the safety of milk) to the cosmic (the origin of life and the cycle of matter) to the political and social (relative importance of "seed" and "soil," or germs and social conditions, in disease causation).

Alongside these wider meanings, the rise of bacteriology involved a conceptual transformation in *medicine: from defining diseases by their circumstances, symptoms, and pathologies to defining them by their causes; from clinical and pathological to etiological definitions of disease. Bacteriology gave medicine and hygiene powerful reasons for focusing on the identification and control of necessary, specific causes.

With the conceptual transformations came practical ones. Through bacteriology, not only could diseases be defined and understood in etiological terms, but they could be routinely diagnosed and sometimes even treated according to their causes rather than their symptoms. Bacteriological and serological testing put *laboratories, for the first time, at the very center of everyday medical and public health practice. The emergence of routine bacteriological diagnosis in hospitals and municipal health departments may be the single most important event in the wider and longer history by which the means and authority of diagnosis, which once lay solely in the skilled and experienced hands of the physician, was increasingly placed in laboratories and machines.

Beyond the laboratory revolution in medicine, bacteriological methods such as testing of bodies and waters and wastes, products such as vaccines, and technologies such as pasteurization remove the fabric of everyday human existence, consumption, and material production in and beyond the industrializing world. The immediacy of this impact has been unparalleled among the life sciences and comparable to the changes in civilization associated with electromagnetism and synthetic chemistry.

With its array of products and services, the bacteriological laboratory ushered in an era in the history of science in which experiment is often synonymous with invention, inquiry with application. The intimate and constant interaction of Pasteur, Koch, Paul Ehrlich, and others with government and business became a compelling model for the organization of scientific life in the twentieth century. The establishment of the Pasteur Institute in Paris (1888), Koch's institute in Berlin (1891), the Lister Institute in London (1891), Ehrlich's institute in Frankfurt (1899), and the Rockefeller Institute in New York (1901) inaugurated a new kind of large, national, nonuniversity research institution and an expansion in the scale and social role of the sciences of health and life, their growth into an omnipresent complex involving industry, commerce, philanthropy, the state, and the consumer.

Ludwik Fleck, *Genesis and Development of a Scientific Fact*, ed. T. J. Trenn and R. K. Merton, trans. F. Bradley and T. J. Trenn, foreword by T. S. Kuhn (1935; 1979). William Bulloch, *The History of Bacteriology* (1938; 1979). Bruno Latour, *The Pasteurization of France*, trans. A. Sheridan and J. Law (1988). Paul Weindling, *Health, Race and German Politics between National Unification and Nazism, 1870–1945* (1989). Thomas D. Brock, *The Emergence of Bacterial Genetics* (1990). Andrew Cunningham and Perry Williams, eds., *The Laboratory Revolution in Medicine* (1992). Pauline M. H. Mazumdar, *Species and Specificity: An Interpretation of the History of Immunology* (1995). Michael Worboys, *Spreading Germs: Disease Theories and Medical Practice in Britain, 1865–1900* (2000). Jean-Paul Gaudillière and Ilana Löwy, eds., *Heredity and Infection: The History of Disease Transmission* (2001).

ANDREW MENDELSOHN

BALANCE. The need to determine the weights of objects is at least as old as recorded history. Commercial transactions, from the determination of the value of the metal in coins to selling fresh fish, required measurements of their weight. The method most commonly used, the equal arm balance, placed the matter to be weighed on a pan hanging from one arm of the balance and added known weights to the other pan until the beam from which they were both suspended remained horizontal. Many such balances appear in drawings dating from ancient Egypt, Greece, and Rome, but the balance also had independent origins in places such as Peru. The simplest balances had wooden beams hung from a cord, but quite early their accuracy was improved by providing them with a fixed fulcrum around which the beam rotated. Later beams were made of metal, often brass, and pans suspended by cords. Lengthening the arms of the beam increased the sensitivity.

Weight measurements were important in early modern metallurgy and pharmacy, and became so also in early modern chemistry. Seventeenth-century chemical texts do not describe balances as they do furnaces (*see* DISTILLATION) and other vessels, perhaps because the balances on which they relied differed little from those used in commerce. Nevertheless, some seventeenth-century chemists, such as Nicolas Lemery, attained quantitative analytical results of impressive accuracy. The early chemists in the Academy of Sciences went to great lengths to carry out their distillation procedures so as to avoid losses, and to ensure that the weights of the products they obtained equaled that of their starting materials.

Summing up such procedures in the mid-eighteenth century, Pierre Joseph Macquer wrote in his *Dictionnaire de chymie* in 1766:

Two quantities of matter each suspended at one of the extremities of the lever or beam of the balance are regarded as equal in weight when the beam stays in a perfectly horizontal direction. This is the most accurate and best method of all to determine the quantities of matter employed. It is much used in commerce and elsewhere; it is also the only method to use in chemical operations requiring exactness.

For reasons not yet fully understood, however, chemists of the mid-eighteenth century placed less emphasis on the "absolute weights" of the materials on which they operated than had their predecessors of the late seventeenth. They were more interested in the specific weight, or density, of materials, a property thought to be characteristic of a particular substance, and, consequently, helpful in its qualitative identification.

When Antoine-Laurent *Lavoisier began the chemical investigations that restored the measurement of absolute weights to a primary analytical criterion, he employed balances that were unusually well crafted, but similar in design to those in ordinary use. As his need for ever more

accurate results increased, however, he had balances of unprecedented precision constructed for him by the leading instrument makers of France. In addition to very long beams, these balances incorporated knife-edge fulcrums to reduce friction and magnifying lenses to read off the scale on which the indicator measured the deviations of the beam from horizontal. Lavoisier's most sensitive Fortin balance was accurate to 1 part in 400,000.

The sensitivity of Lavoisier's later balances probably exceeded his requirements. Impurities, leaks in his apparatus, and other factors beyond his control set the limits of accuracy of his results. By the early nineteenth century, however, improved methods and criteria for purity enabled chemists to reduce their analytical errors to less than 0.1 percent in the determination of combining proportions. For such purposes accurate and reliable balances became essential to any well-equipped chemical laboratory and symbolic of the quantitatively precise science that chemistry had become.

Pierre Joseph Macquer, *Dictionnaire de chymie* (1766). Ernest Child, *The Tools of the Chemist* (1940). Maurice Daumas, *Lavoisier: Théoricien er expérimenteur* (1955). Bruno Kisch, *Scales and Weights: a Historical Outline* (1965). John Stock, *Development of the Chemical Balance* (1969). Hans Jenemann, *Die Waage des Chemikers* (1979). Bernadette Bensaude-Vincent, *Lavoisier: Mémoires d'une révolution* (1993).

FREDERIC LAWRENCE HOLMES

BALLISTICS. The science of ballistics encompasses attempts to describe and predict the behavior of projectiles, particularly those fired from guns, logically extended to include unguided aerial bombs and missile warheads. The subject has three main branches: interior ballistics, addressing the behavior of propellant and projectile inside the gun; exterior ballistics, addressing the behavior of the projectile in flight; and terminal ballistics, addressing the interaction between projectile and target. Niccolò Tartaglia's *Nuovo Scientia* (Venice, 1537), the first comprehensive treatment, was based on Aristotelian mechanics and had no experimental foundation. *Galileo laid the groundwork for exterior ballistics by deducing that a projectile fired in a vacuum would follow a parabolic arc; however, since he did not take atmospheric resistance into account, his insight had practical value only for high trajectory mortar projectiles fired at low velocities.

The birth of ballistics as a science dates to the publication of Benjamin Robins's *New Principles of Gunnery* (London, 1742) and his invention of the ballistic pendulum. By permitting calculation of the impact velocity of a projectile through conservation of momentum, the ballistic pendulum enabled direct measurement of the effectiveness of gunpowder, leading to improvements in powder manufacture. Of greater theoretical importance, atmospheric resistance, or drag, could be measured by varying the distance between gun and pendulum. Robins demonstrated that aerodynamic drag exceeded earlier estimates and increased sharply at about the speed of sound. His development of a whirling arm apparatus for measuring drag confirmed, as Christiaan *Huygens and Isaac *Newton had foreseen, that subsonic drag varies as the velocity squared. Although transonic drag defied precise mathematical analysis, Robins's discoveries rendered mathematical modeling of ballistic trajectories feasible, a feat accomplished for subsonic velocities by Leonhard *Euler in 1753 using nonlinear differential equations. Interior ballistics proved less tractable, despite Robins's success in mathematically modeling the behavior of propellant gases and in demonstrating that the area beneath the pressure-volume curve, describing the pressure exerted on the projectile as a function of distance traveled, equaled the projectile's kinetic energy at the muzzle. Extended by Joseph Louis Lagrange, circa 1793, to reflect the velocity of propellant gases as proportional to the distance traveled, Robins's equations remain fundamental.

The late 1850s saw the birth of experimental interior ballistics when Thomas Jefferson Rodman in the United States perfected a gauge that directly measured peak pressures within a gun's chamber by means of a piston that drove a knife-edge against a copper cylinder, the pressure being proportional to the depth of the indentation. Though Rodman's apparatus yielded inconsistent results—because of refracted shock waves as we now know—he determined the internal distribution of pressure within the barrel and designed guns accordingly. Of equal importance, Rodman demonstrated that peak internal pressures could be reduced for the same propellant force by pressing the gunpowder into larger grains to reduce the initial burning rate. Andrew Noble in Britain advanced internal ballistics further by developing a means of measuring the acceleration of the projectile within the bore via spark circuits cut by the advancing projectile. He set the open ends of the circuits on opposite sides of a series of rapidly whirling paper disks on a single shaft; a tiny hole burned through the paper as each successive circuit broke; the angular distance between holes measured the elapsed time, and acceleration could be calculated directly. With the projectile mass and acceleration known, Noble could compute propellant force and pressure as a function of time. Pizeoelectric gauges now permit the direct and instantaneous measurement of internal pressures, and small accelerometers within experimental projectiles allow measurements of velocity in near real time. Basic theory, however, has remained essentially unchanged.

Scientific terminal ballistics emerged from the 1870s as a product of the growing strategic importance of armored warships. Engineers pursued improvements in armor and armor-piercing ammunition through extensive tests, but the fiendish complexity of the interaction between projectile and target defied all but the most basic mathematical modeling. At the same time, black powder—the classic mixture of saltpeter, charcoal, and sulfur—gave way to nitrocellulose-based propellants, rendering the practice, if not the theory, of interior ballistics obsolete overnight since the burning rates of the new propellants, unlike that of black powder, varied as a function of temperature and pressure.

From the late nineteenth century to the mid-twentieth, ballistic theory advanced steadily but incrementally through progressive incorporation of empirical data. The only major conceptual breakthrough came in the 1880s through supersonic aerodynamics associated with Ernst Mach and the use of spark photography to capture shock wave patterns in wind tunnels. By the early 1900s, gunners could predict the fall of shot at long ranges with useful accuracy, a matter of particular importance in naval warfare, by taking into account air temperature, the temperature of charge and gun, wind velocity, Coriolis effect (caused by the earth's rotation beneath the projectile), and air density and its diminution as a function of altitude. The two world wars forced steady improvements in ballistic practice and refinements in theory, particularly through the analytical use of electronic computers from the late 1930s, applied to aerial bombs and ballistic missiles as well as gun projectiles. Mathematical analysis of transonic aerodynamics in the late 1960s finally made it possible to model mathematically a high-speed projectile's entire trajectory.

During the Cold War, the development of long-range ballistic missiles forced refinements in the theory and practice of ballistics. The precise accuracies demanded of intercontinental ballistic missiles, in particular, demanded not only more accurate modeling of trajectories, but major advances in the theory and practice of geodesics, the precise measurement of the earth and its gravitational and magnetic fields. More recently, high-speed digital computers have enabled ballisticians to solve problems previously considered not practically susceptible to solution, even as the underlying theory remains the same.

Benjamin Robins, *New Principles of Gunnery* (1742). J. G. Benton, *A Course of Instruction in Ordnance and Gunnery Compiled for the Use of the Cadets of the U.S.M.A.* (1862). A. G. Greenhill, "Ballistics," *Encyclopedia Britannica*, 11th ed. (1911): vol. III, 270–279. C. K. Thornhill, *A New Special Solution to the Complete Problem of Internal Ballistics in Guns*, North Atlantic Treaty Organization Advisory Group for Aerospace Research and Development, Pamphlet No. 550 (1966).

JOHN GUILMARTIN

BAROMETER. The barometer grew out of practical hydrostatics in the early seventeenth century. Italian mining and hydro-engineers had noticed that pumps would not raise water more than about thirty feet. *Galileo proposed that "the force of the vacuum" could hold up a column of water only so tall in a pump before it broke, as if the *vacuum were a rope holding up a weight. Isaac Beeckman, Giovanni Baliani, and others argued that the weight of the air outside the pump balanced the water column. Around 1641 Gasparo Berti attached a forty-foot lead pipe to the side of his house, filled it with water, sealed the top, and opened a cock at the bottom, which stood in a large vessel of water. Ten feet of water flowed out, leaving a column suspended some thirty feet high and a space above it that posed a difficult puzzle, since the reigning Aristotelian physics held the vacuum to be an impossibility.

Galileo's disciples repeated Berti's experiment with different liquids until, in 1644, Evangelista Torricelli filled a glass tube with mercury, inverted it in a bowl of the same liquid, and watched the silver liquid fall. Torricelli reasoned from the equality of the ratios of the heights of the mercury and water columns and their specific weights that the atmosphere indeed balanced the standing column. He suggested that his instrument "might show the changes of the air," but the meteorological possibilities of the instrument were largely ignored during two decades' debate over the nature of the space above the mercury and the balancing act of the air. The "Torricellian experiment" remained an experiment for the demonstration and investigation of the vacuum. Blaise Pascal pushed the experiment further towards a measuring instrument by fitting the tube with a paper scale and watching the mercury travel "up or down according as the weather is more or less overcast." This arrangement he designated a "'continuous experiment,' because one may observe, if he wishes, continually." The word "barometer" appeared in 1663 in the work of Robert *Boyle, who with Robert Hooke had set a tube in a window for weather observation.

Soon many new types of barometers appeared: siphon instruments, in which a recurved lower

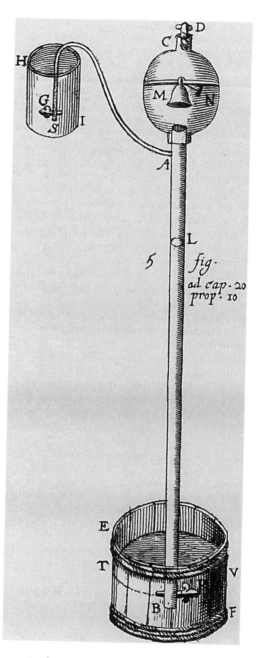

The water barometer standing over thirty feet high set up by Galileo's disciple Gasparo Berti in Rome around 1640; a hammer moved by a magnet outside the experimental space above the water struck a bell to determine whether sound could propagate in the space.

end replaced the barometer's cistern; double and triple barometers, in which successive liquids magnified the motions of the column; diagonal barometers, whose inclination had the identical function; and others. Many designs aimed at a growing market for philosophical and mathematical instruments among the well-to-do. None increased the barometer's accuracy, which did not concern natural philosophers of the seventeenth and early eighteenth centuries.

Precision became important after the Seven Years' War. European states inaugurated national cartographic projects to provide accurate topographic information for military campaigns, taxation, agricultural reform, and other programs of the late Enlightenment. Barometers provided a convenient, if not the most authoritative, method for measuring heights on these surveys, and a demand arose for their precision. The instruments also proved useful to scientific travelers and for investigations into the properties of the air at a time when philosophers were discovering the different gases constituting it and puzzling over water vapor and its pressure in the atmosphere. The Genevan natural philosopher Jean André Deluc, an avid Alpine explorer, designed the first barometer capable of precise measurement. Deluc's exhaustive *Recherches sur les modifications de l'atmosphère* (1772) included such reforms as boiling the mercury to remove dissolved air, methods for leveling the barometer and for reading the mercury meniscus, and corrections for the expansion with temperature of mercury, glass, and the ambient air. Equally important, Deluc showed how to deploy the instrument in repeated series of exhaustive measurements—a revolutionary technique. Deluc's rule for converting barometric measurements to heights was quickly adopted in England, where a national geodetic survey was underway.

In England instrument makers, moving away from craft-based organization, implemented industrial techniques such as division of labor, machine manufacture, and research and development financed out of profits from government trade. By 1770 Jesse Ramsden, the most advanced among them, was manufacturing barometers accurate to one-thousandth of an inch. In France the guilds hobbled efforts by the enlightened monarchy to modernize the instrument trade, though a few excellent instrument makers worked there towards the end of the ancien régime. The dissolution of the guilds during the Revolution and the need for instruments for military purposes and for metrification strengthened the instrument trade in France. Jean Nicolas Fortin's barometer, accurate to two-thousandths of an inch, became the standard model in France (*see* INSTRUMENTS AND INSTRUMENT MAKING).

In the nineteenth century the popularity of scientific travel and the increasing reach of European imperialism favored the development of the aneroid barometer. At the end of the seventeenth century Leibniz had suggested an instrument that would balance the pressure of the atmosphere against a spring-loaded, flexible bellows or box. In 1844 Lucien Vidie developed a practical barometer based on this principle. The aneroid became popular among mountaineers and mariners; in the twentieth century aviation pushed the development of highly accurate aneroid instruments.

The requirements of networks of weather observers organized by national governments in the late nineteenth century helped create a new class of barometers: the high-precision "primary" barometer, against which observers' instruments were calibrated. Manufacturers employed sophisticated methods, including the Sprengel mercury vacuum pump, to clean, fill, and evacuate these instruments (*see* AIR PUMP AND VACUUM PUMP).

W. K. Middleton, *History of the Barometer* (1964). Anthony J. Turner, *Early Scientific Instruments, Europe 1400–1800* (1987). Theodore S. Feldman, "Late Enlightenment Meteorology," in *The Quantifying Spirit in the 18th Century*, ed. Tore Frängsmyr, J. L. Heilbron, and Robin E. Rider (1990): 143–178. Jan Golinski, "Barometers of Change: Meteorological Instruments as Machines of Enlightenment," in *The Sciences in Enlightened Europe*, ed. William Clark, Jan Golinski, and Simon Schaffer (1999): 69–93.

THEODORE S. FELDMAN

BASE. See ACID AND BASE.

BASIC AND APPLIED SCIENCE. There is a common distinction in modern science between two types of research. Basic science is supposed to aim for new knowledge; it is also known as fundamental or pure science, suggesting that it is uncontaminated by such worldly concerns as practicality, patents, or profits. Applied science, or mission-oriented research, instead aims to produce technologies for social use, such as for industry or the military. The distinction began appearing in the late nineteenth century as a reaction to the increasing integration of science and technology, manifested in the science-based industries of the second industrial revolution. Academic scientists began referring to their work as pure science, in order to distinguish it from technology. The insistence on science for its own sake, especially strong in Germany, received forceful expression from Henry Rowland in the United States in a famous address of 1883 entitled "A Plea for Pure Science."

Scientists and their sponsors have often posited a directional flow from basic to applied science; basic science, the argument runs, furnishes the foundation for applications. The concept, deriving ultimately from Francis Bacon's utilitarianism, spurred industrial corporations at the end of the nineteenth century to integrate knowledge-making into their firms through the creation of research laboratories. Subsequent successes in the electrical, chemical, and pharmaceutical industries validated basic research as a source of new technologies, as did the spectacular military applications of World War II. In 1945 Vannevar Bush endorsed the concept in his *Science: The Endless Frontier*, which helped elicit vast infusions of federal funds into American science to ensure future military and industrial security. Later studies by the federal government on this investment in basic research, including Projects Hindsight and Trace, reached no consensus on the returns, but the perception persisted that basic science led to applications.

But influence also flows in the opposite direction. Many important developments in basic science have stemmed from applied research. The problem of blackbody radiation that spurred the emergence of quantum theory derived from data relevant to light bulbs for the then-new electric lighting industry; and Albert Einstein's definition of simultaneity in his theory of special relativity may have drawn inspiration from his work at the Swiss patent office on coordinated clocks for railroad systems.

Such examples underscore the difficulty of distinguishing between basic and applied science. Were the experiments on blackbody radiation, conducted in Germany's industrial standards laboratory, basic or applied science? Is research into the properties of semiconductor materials basic or applied? One may locate basic research in universities and applied research in industrial or government labs, but such institutional distinctions overlook the research subjects and techniques held in common, as well as the social goal of pedagogy inherent in academic research. One may also try to perceive a rough dividing line based on the motivation of individual scientists and whether they seek to develop new knowledge or new technology. Scientists may tailor their motivations to fit the audience, however, by telling colleagues or sponsors what they want to hear. High-energy physicists in the postwar United States portrayed their field as the most basic and fundamental of the sciences, but in appearances before political representatives they stressed the social relevance of their work in the form of potential military technologies or industrial spin-offs.

Locating the blurry boundary between basic and applied research is not a purely academic exercise. Scientific institutions—in academia, industry, and government—make a hard distinction in the apportionment of scientific support and in determining whether to keep research results secret, and their reward systems likewise often distinguish between basic and applied science.

Ronald Kline, "Construing 'Technology' as 'Applied Science': Public Rhetoric of Scientists and Engineers in the United States, 1880–1945," *Isis* 86 (1995): 194–221.

PETER J. WESTWICK

BATTERY. In 1800 Alessandro Volta (*see* GALVANI, LUIGI, AND ALESSANDRO VOLTA) announced to the Royal Society of London the invention of the electric cell, the first device capable of generating an electric current thanks to an electrochemical process. The invention of the battery led to the discovery of phenomena that pointed the way to electrochemistry (*electrolysis, *electroplating) and to electromagnetism (action of currents on magnets). For a few decades the electric cell had no important practical application; it was too expensive, weak, and inefficient for lighting or operating electric engines. Batteries found their first large-scale application in electric telegraphy, whose spectacular development started in the 1840s. From the 1880s electric cells and storage batteries found wide use as electricity sources for telephones.

The simple voltaic battery is subject to polarization. The hydrogen bubbles produced tend to isolate the copper electrode and decrease the electromotive force. During the nineteenth century scientists and inventors struggled to develop more reliable and longer-lasting batteries by trying different types of electrodes and electrolytes. In 1836 the English chemist John F. Daniell invented a nonpolarizing cell, but the forerunner of the modern dry cell did not appear until 1886. The invention of the French chemist Georges Leclanché, it had zinc and carbon as electrodes and a paste of ammonium chloride as electrolyte. A mixture of manganese dioxide and carbon powder prevented polarization. Mass production of modified and improved dry cells started at the beginning of the twentieth century. Today, thanks to the increasing miniaturization of electronic components developed in the last decades of the twentieth century, dry cells of very different sizes and types (zinc-carbon, alkaline, lithium, mercury) supply electricity for an extremely wide range of compact apparatus. Nevertheless, scientists still search for a cheap, nonpolluting, long-lasting, and powerful "ideal cell."

In the late 1850s the French scientist Gaston Planté invented the lead-acid storage (or secondary) battery, which amounted to reversible, and thus rechargeable, electrochemical cells. These proved to be very successful (for example in the automobile industry) and they are the most common ones in use today. Commercialization of nickel-iron and nickel-cadmium batteries began just after 1900. Although more expensive than lead batteries, they have found many applications in portable devices.

Another sort of battery descends from the discovery of the thermoelectric effect by the German nature philosopher Thomas Seebeck in 1822. It led to the invention of thermopiles, which generate electricity by maintaining the junctions of an array of couples of dissimilar metals at different temperatures. Thermoelectric piles developed in the 1820s by the Italian physicist Leopoldo Nobili played an important role in studies concerning infrared radiation. Various devices based on the thermoelectric effect are used to measure temperatures and generate electricity.

The fuel cell invented in 1829 by the English chemist and physicist William Grove produces electricity by slowly combining hydrogen and oxygen. Rediscovered in the 1960s, modern fuel cells are used today in spacecraft and may eventually be exploited commercially.

Bell Laboratories developed silicon photovoltaic cells, which transform solar light into electricity, in the 1950s. These expensive, non-polluting cells, which are generally connected to storage batteries, require little maintenance and have a long working life, but are still expensive. Therefore they find applications in remote areas unconnected to an electric grid and in isolated devices (such as telephonic and radio apparatus) and spacecraft and satellites.

Nuclear cells produce electricity by collecting negative charges generated by the radiation emitted by radioactive substances. Very expensive and requiring special handling, they are used in space vehicles.

George W. Vinal, *Primary Battery* (1950). James W. King, "The Development of Electrical Technology in the 19th Century. 1. The Electrochemical Cell and the Electromagnet," *Contribution from the Museum of History and Technology United States National Museum* 228 (1962): 233–271. Richard H. Schallenberg, *Bottled Energy: Electrical Engineering and the Evolution of Chemical Energy Storage* (1982). Colin A. Vincent and Bruno Scrosati, *Modern Batteries: An Introduction to Electrochemical Power Sources*, 2d ed. (1998).

PAOLO BRENNI

BEHAVIORISM. Psychological behaviorists believe that mental contents should be identified with their physical expression. Methodological behaviorists adopt that view as a working stance, not as a metaphysical doctrine, while radical behaviorists believe that all mental contents can be exhaustively analyzed physically. Psychological behaviorism is intimately linked to American social pragmatism, making its first appearance in American sociology in the 1890s and, a little later, in American neorealist philosophy. Max Meyer

published the first American behaviorist text, *The Fundamental Laws of Human Behavior*, in 1911. At about the same time, John Watson switched from comparative psychology to abnormal and developmental psychology, believing that emotions were construable as actions. In 1913, he published "Psychology as the Behaviorist Views It," a short article in which he suggested extending the same analysis to all mental contents.

In 1920, Watson abandoned academia and his research but became an influential popular writer and broadcaster, disseminating a social technology aimed at comprehensive human betterment. Within academia, Jacob Kantor, Walter Hunter, Stevenson Smith, and Max Meyer's student Albert Weiss made behaviorism a flourishing and diverse doctrine in the 1920s. With the exception of Hunter, however, none engaged in research, so that behaviorism remained a speculative doctrine. By the end of the decade several writers proclaimed its demise.

In the early 1930s, behaviorism led a subterranean existence, nourished by the writings of Edwin Guthrie and by B. F. Skinner's and Edward Tolman's animal research. It regained its influence as neobehaviorism due to an alliance between a statistical procedure, analysis of variance, animal research, and the doctrine of operationism. Analysis of variance permitted experimenters to assess the joint effects of several variables simultaneously. Capitalizing on decades of previous experience and the existence of specially bred populations, neobehaviorists carried out thousands of animal experiments. Operationism was their crucial interpretive tool. Operationists believe that all concepts, including all mental contents and attributes, are definable in terms of the techniques for measuring their degree of expression. It follows that animals' states of mind can be equated with humans', so that animal experimentation can be extrapolated directly into the human realm.

Clark Hull and Skinner dominated neobehaviorism, Hull as the leader of an energetic group of theorists and researchers and Skinner as the founder of the extraordinarily pervasive technique of behavior modification. In Hull's circle, O. H. Mowrer operationized numerous psychoanalytic concepts, thereby extending neobehaviorism's scope, while Kenneth Spence advanced its cause by training many influential doctoral students. In behavior modification, behaviorism closed its historical cycle by regaining its pragmatic origins. Whatever behavior modification's theoretical shortcomings, it is a powerful tool in such areas as clinical psychology, mental deficiency (especially autism), attention deficit syndrome, and obesity.

Laurence D. Smith, *Behaviorism and Logical Positivism: A Reassessment of the Alliance* (1986). John A. Mills, *Control: A History of Behavioral Psychology* (1998).

JOHN A. MILLS

BELL LABS, founded as a consequence of the *telephone experiments of Alexander Graham Bell and Elisha Gray, evolved to become one of the largest corporate research laboratories in the modern world. The official organization of Bell Telephone Laboratories in 1925 consolidated several earlier organizations. Bell's company expanded after winning a patent fight in 1876, bought out Gray's Western Electric equipment company in 1881, and created in Boston a mechanical department in 1884 responsible for research and development within the new AT&T that emerged in 1885. Theodore Newton Vail became AT&T's president in 1907 and transferred the research labs from Boston to the new Western Electric Engineering Department at 463 West Street in New York City that would become the headquarters of Bell Labs. These early departments began the pioneering research that produced the amplifying vacuum tube, condenser microphone, loudspeakers, public address system, and the hearing aid.

In 1925, the research departments of Western Electric and AT&T were consolidated into the Bell Telephone Laboratories under John Joseph Carty. One of the first products to emerge from this new Bell Labs was Henry Harrison's electrical recording system, which replaced the old hand-cranked Victrola with higher-fidelity radiolas and orthophonic *phonographs. By 1926, Bell Labs had perfected a system for recording and playing back talking motion pictures, and created a licensing-supply-training company called Electrical Research Products, Inc. (ERPI) to provide sound equipment and technicians to Hollywood studios and theaters. Arthur Keller continued research into sound recording in the 1930s, working with Leopold Stokowski to develop high-fidelity stereophonic records by 1932 and a multiple channel recording system that Walt Disney would use for *Fantasia* in 1940.

Bell Labs sponsored research in its widest possible meaning, from pure science to practical application. Clinton Davisson was the first of its scientists to win a Nobel Prize (in 1937 for his detection of electron waves). From similar research would come the transistor in 1947 (which also brought Nobel Prizes), the solar cell in 1954, and the laser in 1958 (another Nobel Prize–winning invention).

Theoretical breakthroughs became the foundation for improving the telephone system that the Labs had been created to serve. The vacuum tube developed by Harold Arnold in 1913 made possible long-distance and international telephone transmission. The transistor propelled the digital revolution, making possible the T1 digital multiplex network and the pager and the Telstar communications satellite by 1962, the touch-tone phone in 1963, the Unix-based computer network in 1969, and the single-chip DSP signal processor in 1979. The laser laid the foundation for fiber optic communication systems, from the first network installed in Chicago in 1977 to the transatlantic network of 1988. When AT&T lost its regional operating companies in 1984, Bell Telephone Laboratories became AT&T Bell Laboratories. When AT&T again restructured in 1995, the name was changed to AT&T Laboratories. Throughout the history of the Bell Labs, it has remained a pioneer in research that has touched almost every aspect of daily life.

M. D. Fagen, ed., *A History of Engineering and Science in the Bell System*, 2 vols. (1975). Prescott C. Mabon, *Mission Communications: The Story of Bell Laboratories* (1975).

STEVEN E. SCHOENHERR

BERNARD, Claude (1813–1878), experimental physiologist.

Among his peers Bernard was distinguished by the prolific discoveries of his middle years; the unexpectedness of several of the most important of them; the extraordinary surgical skill he deployed in his vivisection experiments; his imaginative, independent, sometimes idiosyncratic approach to the problems of his day; the limited range of his knowledge of the chemical methods of his time relevant to the physiological problems he studied; and a confidence in his own methods and insights that sometimes caused him to underestimate the importance of the contributions of his contemporaries. He thought more deeply about the principles underlying his experimental practice than any other physiologist of his or any other time. He became a French scientific hero in an era of increasing German scientific domination. Warm, poised, and gracious in public, a charismatic teacher who attracted not only medical students and aspiring young physiologists, but the fashionable world of Paris to his lectures, and who surrounded himself in his later years with a small band of devoted "disciples," Bernard was in private a lonely, often tormented man, who sought in a passionately intellectual correspondence with the wife of a Russian diplomat to offset the aftermath of a cold, arranged marriage that had ended in a legal separation and isolation from his two surviving children.

Born in Villefranche, a small village near Lyons, to two vineyard workers, Bernard was apprenticed to an apothecary but aspired to be a playwright. After some local successes, he came to Paris in 1834 to continue his writing, but instead took up medicine. Entering François Magendie's small laboratory at the Collège de France in 1841 as his *préparateur*, Bernard began there to carry out experiments closely related to Magendie's ongoing investigations.

One of Bernard's earliest independent research lines was animal nutrition. From the beginning he contended with the efforts of prominent chemists, especially Justus *Liebig in Germany and Jean-Baptiste-André Dumas, to infer the chemical processes taking place in organisms from their burgeoning knowledge of the properties and reactions of biologically important organic compounds such as carbohydrates, fats, *proteins, and organic acids. Bernard distrusted conclusions drawn from chemical reasoning alone, and looked for opportunities to demonstrate the necessity for physiological tests of chemical changes through vivisection.

For more than five years, Bernard and his sometime collaborator, the chemist Charles-Louis Barreswil, proposed several theories of digestion that they subsequently had to abandon. At the same time Bernard struggled to survive without a secure position, and at one point nearly resigned himself to returning to Villefranche as a village doctor—a fate from which his arranged marriage saved him.

Along the way toward his untenable theories of digestion, he had made some interesting subsidiary discoveries, such as that the turbid, alkaline urine of herbivorous animals turns clear and acidic when they fast; but in his main endeavor to follow the process of nutrition through the digestive tract and on into the bloodstream, he bogged down.

Suddenly, in 1848, his luck turned. Almost by chance he found that pancreatic juice exerts a special emulsifying action on fats. A few months later, while seeking unsuccessfully to locate the site in the circulation where sugar ingested in foodstuffs is destroyed, he discovered instead that blood in the portal vein contains sugar even when the animal has received none in its food. Through a set of experiments he traced the source of this sugar to the liver. This was not only a great discovery but, as he realized immediately, the chance he had long awaited to demonstrate that chemistry must not "adventure alone" in physiology. Dumas had taught that animals receive all of their main body constituents ready-formed in their food, and merely oxidize them by stages to produce heat and work. Chemical methods could supply only the tools that physiologists needed to determine what really happens in the organism.

Major and minor discoveries followed one another in unremitting succession for the next decade. In 1851 Bernard found that when he sectioned the sympathetic nerve on one side of a rabbit, the innervated portions of its body did not become cooler, as he had anticipated, but warmer and engorged with blood. This observation can be regarded as the first step in the discovery of the action of the vasomotor nerves that control the flow of blood into local regions of the body, and so help regulate body temperature. Following up his discovery of the production of sugar in the liver, in 1857 Bernard isolated from the liver an insoluble compound similar to starch, which he named glycogen, and showed that it was the immediate source of the sugar that the liver secretes into the blood. During the 1850s he elucidated the action of the South American poison curare, showing that it paralyzes the action of motor nerves, while leaving the sensory nerves intact.

In 1854 Bernard finally acquired an academic position commensurate with the reputation he had long attained, a chair in general physiology was created for him at the Sorbonne. In 1855, Bernard succeeded also to Magendie's chair in experimental medicine at the Collège de France. He gradually developed broader perspectives on the basic principles of physiology, including the idea of the *milieu intérieur*, now regarded as his most original and important biological generalization.

J. M. D. Olmsted and E. Harris Olmsted, *Claude Bernard and the Experimental Method* (1952). Claude Bernard, *An Introduction to the Study of Experimental Medicine*, trans. H. C. Greene (1957). Frederic L. Holmes, *Claude Bernard and Animal Chemistry* (1974). Mirko D. Grmek, *Le legs de Claude Bernard* (1997).

FREDERIC LAWRENCE HOLMES

BERZELIUS, Jöns Jacob (1779–1848), Swedish chemist, the dominant authority in European chemistry in the early nineteenth century.

Born in a small Swedish town in 1779, Jacob Berzelius was orphaned at an early age and raised by a stepfather. He maintained himself as a medical student at the University of Uppsala on a scholarship. He learned chemistry from recent textbooks, adopted the chemistry of Antoine-Laurent *Lavoisier when the university still favored the phlogiston theory, and—fanatic that he was—carried out experiments in his room.

With his friend and supporter Wilhelm Hisinger, Berzelius was among the first to exploit the Voltaic *battery for chemical analysis. By 1803, they had found that the electrolytic current applied to a neutral salt in solution caused its acid to accumulate around the positive pole and its base around the negative pole. In 1807 and 1808, the English chemist Humphry Davy decomposed soda and potash, and soon afterward, the alkaline earths. From these results, Berzelius constructed what came to be known as the "dualistic theory." All of the "bases" that combined with oxygen, as well as the oxides themselves, formed a continuous series from oxygen, the most electronegative, to potassium, the most electropositive. When oxides, or first-order compounds, as Berzelius called them, combined to form salts, or second-order compounds, the more electronegative played the role of the acid, the more electropositive the role of the base.

Berzelius was stimulated also by a work by a little-known German chemist, Jeremias Benjamin Richter, entitled *Foundations of Stoïchiometry*, which claimed, among other regularities, that if two different bases saturate the same acid in some proportion by weight, they would saturate any other acid in the same proportion. Finding that Richter's experimental results, as well as those of other chemists, did not fully support these generalizations, Berzelius resolved to test them "with the most scrupulous precision." In the small, simply equipped laboratory where he worked after becoming Professor of Medicine and Pharmacy at the school of surgery in Stockholm in 1807, Berzelius reached a new standard of quantitative accuracy in chemical research. Where his predecessors had been content with round numbers, Berzelius and his imitators expected to obtain results agreeing to a few parts in a thousand.

While engaged in these analyses, Berzelius learned indirectly about John *Dalton's ideas about combination in simple proportions (*see* ATOM AND MOLECULE). Dalton's clarifying perspective seemed to Berzelius one of the "greatest steps that chemistry has made toward its perfection as a science." Berzelius did not fully adopt Dalton's atomic theory, however, and unlike Dalton he did accept the experimental results of Joseph Gay-Lussac showing that gases combine in simple integral ratios. From his own accumulating analytical results, Berzelius developed rules to calculate the number of "volumes" or of atoms that could combine with one another in oxides and in neutral salts formed by the further combinations of acid and basic oxides.

"When we endeavor to express chemical proportions," Berzelius wrote in 1814, "we find the necessity of chemical signs." Earlier chemists had devised special chemical symbols derived from alchemy; Berzelius rejected them as useless because they were difficult to draw or reproduce in a printed text. Instead, he used ordinary letters, taken from the first letter or, in some cases, the first two letters of the Latin name for each elementary body. The chemical sign represented one volume of the substance, and the number of volumes

contained in a compound was indicated by a number. Berzelius wrote water as $2H + O$. With many modifications in detail, Berzelius's signs soon became the universal language of chemistry.

Berzelius first doubted whether the principles of composition that he was establishing for inorganic chemistry applied also to organic compounds. Improving the method of elementary analysis by combustion of the organic compounds, however, he attained a degree of accuracy comparable to that for inorganic analyses, and confirmed that organic compounds followed the same basic rules of combining proportions. During the 1820s, he extended his dualistic theory to explain organic composition.

In 1822, Berzelius introduced the literary innovation of a "Yearly Report of Progress in the Physical Sciences," in which he summarized and evaluated the most important publications of the previous year in physics and chemistry. For the next fifteen years, the *Jahres-Bericht* consolidated Berzelius's authority. After he resigned his professorship in 1832 and relinquished his laboratory, Berzelius exercised his authority through these annual reports and a multivolume advanced textbook that he tried to keep current.

Like other chemists of his era, Berzelius held a realistic, pragmatic view of the role of theory in science. "Every theory," he wrote, "is nothing more than a manner of representing the interior of phenomena." As new facts appear, the mode of representing the phenomena must change, "perhaps without ever finding the true" way. In practice, Berzelius found it difficult to give up his own theories in the face of new facts that challenged their adequacy. During the 1830s, a new generation of ambitious chemists came to prominence. They admired what Berzelius had contributed, but increasingly sought to supercede his authority. Disapproving of their more aggressive style, unable to compete experimentally with the larger laboratories and cohorts of students they commanded, Berzelius resisted his gradual loss of leadership and the changes the new men wished to introduce into chemical theory. By the time of his death in 1848, he had become rather embittered by trends he could not stop. The fundamental achievements of his younger days, however, far outlasted his time.

See also ATOM AND MOLECULE.

Evan Melhado, *Jacob Berzelius: The Emergence of his Chemical System* (1981). Alan J. Rocke, *Chemical Atomism in the Nineteenth Century* (1984).

FREDERIC LAWRENCE HOLMES

BETHE, Hans Albrecht, theoretical physicist. Born 2 July 1906 in Strasbourg, then a part of

Germany. His father, a respected physiologist, transferred to the newly founded Frankfurt University in 1915. His mother had been raised in Strasbourg, where her father had been a professor of medicine. Bethe's father was Protestant. His mother, born Jewish, had become a Lutheran before she met Hans's father. Bethe grew up in a Protestant household in which religion did not play an important role.

Bethe attended a traditional Humanistisches Gymnasium in Frankfurt that emphasized Greek and Latin. Although instruction in mathematics and the sciences at gymnasia was in general poorer than in the humanities, several of the mathematics instructors at the Goethe Gymnasium that Bethe attended stimulated his interest. He left school with a fair amount of mathematics, a good deal of science, much Latin and Greek, and a knowledge of the German classics—Kant, Goethe, and Schiller—and the French and English languages.

By the time Bethe had finished Gymnasium, he knew he wanted to be a scientist, and his poor manual dexterity steered him first into mathematics and then into theoretical physics. After completing two years at the University in Frankfurt, he went to Arnold Sommerfeld's seminar in Munich. He obtained his doctorate in 1928 summa cum laude. After a brief stay in Stuttgart as Paul Ewald's assistant, he returned to Munich to habilitate under Sommerfeld. During the academic year 1930–1931, Bethe was a Rockefeller fellow at the Cavendish Laboratory in Cambridge, then directed by Ernest *Rutherford, and in Rome in Enrico *Fermi's Institute. In 1932, he again spent six months in Rome working with Fermi. By 1933, he was recognized as one of the outstanding theorists of his generation. His book-length articles in the *Handbuch der Physik* on the quantum theory of one- and two-electron systems and on the quantum theory of solids immediately became classics. In April 1933, after Hitler's accession to power, he was removed from his position in Tübigen because he had two Jewish grandparents. He went to England, and in the fall of 1934 accepted a position at Cornell University. He arrived there in February 1935, and has been there ever since.

Bethe's explanation of the mechanism for energy generation in stars grew out of his participation in the third Washington conference on theoretical physics in April 1938. These conferences, organized by George Gamow, Edward Teller, at the time one of Bethe's closest friends, (*see* SAKHAROV, ANDREI AND EDWARD TELLER), and Merle Tuve, had become annual events. Most of the participants at that conference conflated the problem of nucleosynthesis with the problem of

energy generation. Bethe returned to Cornell and separated the two problems. He advanced two sets of reactions—the proton-proton and the carbon cycle—to account for energy production in stars such as the sun. The carbon cycle depended on the presence of carbon in the star. At that time, physicists knew no mechanism to account for the abundance of carbon in stars. However, the presence of carbon in stars had been corroborated by their spectral lines in stellar atmospheres. Bethe accepted this fact, and proceeded to compute the characteristics of stars nourished by the two cycles. He found that the carbon-nitrogen cycle gave about the correct energy production in the sun.

During World War II, Bethe worked on armor penetration, radar, and atomic weaponry. After a stint at the Radiation Laboratory at M.I.T, he joined Robert Oppenheimer (*see* KURCHATOV, IGOR VASILIEVICH, AND J. ROBERT OPPENHEIMER) at Los Alamos in 1943 as the head of the theoretical division. His ability to translate his understanding of the microscopic world to the design of macroscopic devices made his services invaluable at Los Alamos. Bethe, Fermi, and others converted their knowledge of the interaction of neutrons with nuclei into diffusion equations, and the solutions of the equations into reactors and bombs.

After the war, Bethe became deeply involved in the peaceful applications of *nuclear power, as well as in the investigation of the feasibility of developing fusion bombs and of ballistic missiles, and helping to design them. He served on the President's Science Advisory Committee (PSAC) and other government committees. In 1967, he won the Nobel Prize for his theoretical investigations in 1938 explaining the mechanism of energy production in stars.

Bethe's life can be divided into well-delineated stages. Between 1906 and 1933, German culture and German institutions molded him. From 1934 until 1940, Cornell was his haven. He then published *Bethe's Bible* (in the *Reviews of Modern Physics*, 1935–37), a synthesis and concise presentation of all the extant knowledge of nuclear structure and reactions, and his solution of the problem of energy generation in stars. During the third period, World War II, he acquired new sorts of powers at M.I.T. and Los Alamos. Capitalizing on them, he played major national and international roles in the first postwar decade. He was at the center of developments in *quantum electrodynamics and meson theory, helped Cornell become one of the outstanding universities in the world, consulted with private industries trying to develop atomic energy for peaceful purposes, and exerted great influence on national security. His membership in PSAC and its subcommittees, consultancies with Avco, GE, and other industrial firms, and crises within the home help explain the routine character of his scientific production between 1955 and 1970. From the mid 1970s, in collaboration with Gerald E. Brown, Bethe returned to creative and productive work. The life cycle of supernovas and the properties of the neutrinos involved in solar fusion particularly engaged his attention.

J. Bernstein, *Hans Bethe: Prophet of Energy* (1980). *Hans Bethe: A Life in Science* [videorecording: 9 videocassettes (VHS NTSC)] (1997). S. S. Schweber, *In the Shadow of the Bomb. Bethe, Oppenheimer and the Moral Responsibility of the Scientist* (2000).

SILVAN. S. SCHWEBER

BIG BANG. See SPACE AND TIME.

BIOCHEMISTRY is the name most commonly used during the twentieth century to designate the branch of science that investigates the chemical constituents of living matter, the substances produced by organisms, the functions

and transformations of these substances within organisms, and the energetic changes associated with them. Many of these topics had been studied under such names as iatrochemistry, animal and plant chemistry, and physiological chemistry.

From the time that *chemistry itself emerged as a distinct investigative activity during the seventeenth century, chemical knowledge was regularly applied to the study or explanation of processes occurring in living organisms. The successive stages of digestion and the formation of *blood came to be interpreted as a series of interactions of acids and alkalis when their study became prominent in the middle of the century. Alcoholic fermentation, which was often characterized as an "intestine motion," also became a model for the explanation of internal vital processes. Van Helmont was the first of many to suggest that "ferments" caused digestion and other transformations. During the seventeenth and eighteenth centuries, chemists divided their field into three parts according to the three kingdoms of nature. Substances obtained from the animal kingdom were characterized

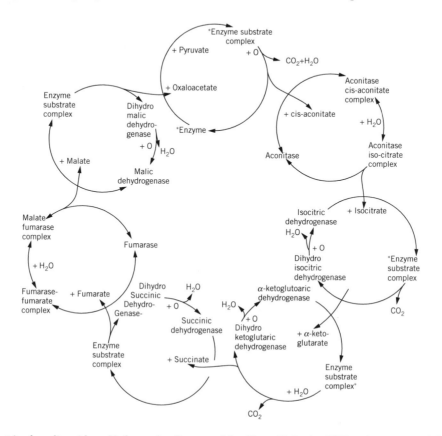

The tricarboxylic-acid or Krebs cycle, discovered by Hans Krebs in 1937 and represented here as a chain of enzyme cycles. It is the final common pathway for degradation of foodstuffs in the organism.

by "volatile alkali," an alkali that sublimed or passed over into the receiver in the distillation procedures central to early chemical analysis. Plant material contained, instead, predominantly fixed alkali. This distinction led to various concepts of *nutrition during the eighteenth century, such as that proposed by Herman *Boerhaave, who inferred that food entered the body in an "acidescent" condition, gradually becoming "alkalescent" as it transformed into blood, tissues, and eventually excretory products.

The changes in chemistry associated with the chemical revolution at the end of the eighteenth century also marked a transformation in the study of the chemical processes of life. Antoine-Laurent *Lavoisier made three central contributions. Through the extension of his studies of combustion to several representative plant substances, especially sugar and oils, he showed that plant matter comprises carbon, hydrogen, and *oxygen, individual substances being distinguished by the proportions in which they contain these elements. Drawing on the demonstration by his colleague, Claude Louis Berthollet, that "volatile alkali" is composed of nitrogen and hydrogen, Lavoisier concluded that animal substances contain four elements—carbon, hydrogen, oxygen, and nitrogen. By this time, chemists understood that "animal substances" did not derive from the animal kingdom alone, but constituted a class of substances, including fibrin, albumin, casein, and gluten, some of which came from plant matter.

Adapting his quantitative methods to the study of alcoholic fermentation, Lavoisier showed that the chemical change involved is the conversion of sugar to alcohol and carbonic acid. Determining the elementary composition of each of the substances involved, he produced the first balanced chemical equation for a biologically important process.

Through experiments carried out on birds and guinea pigs breathing in closed chambers and in a calorimeter (the latter experiment in collaboration with Pierre Simon de *Laplace), Lavoisier identified respiration as a slow combustion in which carbon converts to carbonic acid and hydrogen to water, and heat and work are derived. To maintain its equilibrium, an animal or human must ingest in its food quantities of carbon and hydrogen equal to those consumed in respiration. Lavoisier could not specify in detail the nature of the substances containing the carbon and hydrogen, or the internal processes connecting the input to the final products of respiration, but his theory has provided ever since the boundaries within which the material exchanges between the organism and its surroundings and the dynamic

chemical processes within the organism have been studied.

During the first four decades of the nineteenth century, the rapid development of chemistry provided increasingly powerful methods to study the composition of plant and animal matter and the transformations they undergo. By the 1830s, the three most important general classes of foodstuffs, as well as the major constituents of the organism, had been identified as carbohydrates, fats, and "albuminous" matter (later renamed proteins). The process of chemical change in the organism most open to direct investigation was digestion. A series of investigators, including Friedrich Tiedemann and Leopold Gmelin, Johann Eberle, Theodor Schwann, Louis Mialhe, Claude Bernard, and Willy Kühne, followed the changes that the three classes of foodstuff undergo in their passage through the stomach and intestines. In 1836, Schwann identified pepsin as a "ferment" that induced digestive changes in albuminous foodstuffs. The further characterization of pepsin has occupied investigators ever since. Pepsin became the prototype for the general concept that "ferments," acting in minute quantities as catalysts, direct many key processes in organisms. In 1876, Willy Kühne proposed the word "enzyme" to denote the class of ferments, including pepsin, that can "proceed without the presence of organisms." Later, his term gradually replaced the term ferment to denote the agents of both intracellular and intercellular chemical transformations.

During the 1840s, chemists and physiologists sometimes competed to determine the conditions under which the chemical processes of life could be established, chemists inferring their nature from the chemical properties of the substances involved, physiologists such as Claude *Bernard insisting on the need to follow the chemical changes into the living body itself. Among the chemists, Justus *Liebig and Jean-Baptiste Dumas showed how the new knowledge of the chemistry of proteins, carbohydrates, fats, and their decomposition products could be used to give more comprehensive meaning to Lavoisier's theory of respiration. Thus they took the first speculative steps toward an account of the intermediate stages connecting foodstuffs with final decomposition products and an ascription to these processes of the sources of animal heat and muscular work. Hermann von *Helmholtz gave further meaning to these processes by linking them with his formulation of the general principle of the conservation of energy. Bernard demonstrated the fruitfulness of his approach by discovering through vivisection the special action of the pancreatic juice on fats and the glycogen function of the liver. Others saw the need to establish a new scientific specialty devoted to the study of the chemical processes of life. Carl Lehmann became a professor of physiological chemistry at Leipzig in 1847, and Felix Hoppe-Seyler professor of the same at Strasbourg in 1872. Both argued for the necessity of an independent discipline, but their example did not spread, and the subject continued to be pursued mainly in Institutes of Physiology in Germany, though a series of textbooks of physiological chemistry represented the subject as a coherent and growing body of knowledge and field of investigation.

During the second half of the nineteenth century, organic and physiological chemists identified a number of the amino acids obtainable by decomposing proteins. At the end of the century, Emil Fischer and Franz Hofmeister independently proposed that amino acids linked together through "peptide" ($-CO-NH-$) bonds compose proteins. Fischer also established the stereospecific formulas for a number of biologically important sugars. By this time it had become clear that starch, glycogen, and other "polysaccharides" were made of many simple sugars linked together. That fats can be decomposed into glycerol and fatty acids had been established early in the century by Michel Chevreul. The general appreciation that each of the three major foodstuffs could be broken down into smaller molecules, and growing evidence that this happened during digestion, led to the "building block" concept. This assigned the creation of proteins, fats, and carbohydrates of the organism to the reconstruction of similar materials, assimilated from the food, and transferred the central problem of "metabolism" from (in the words of Frederick Gowland Hopkins) "complex substances which elude ordinary chemical methods" to "simple substances undergoing comprehensible reactions." This shift in perspective energized the effort during the first decades of the twentieth century to detect the nature of the intermediary steps by new experimental methods. The discovery by Eduard Buchner in 1902 that the archetypical "ferment" (the agent of alcoholic fermentation) could be separated from yeast cells and produce "cell-free" fermentation further energized the search.

At the turn of the century, biochemical processes were studied in many institutional settings ranging from departments of physiology and organic chemistry to internal medicine, pathology, and brewery laboratories. In 1914, Hopkins acquired the title of Professor of Biochemistry, in a newly established Department of Biochemistry at Cambridge. Departments of biochemistry sprung up in the United States at leading medical schools, finding an institutional niche in the service role of teaching biochemistry to medical students. Although the discipline in Germany remained subordinate to physiology or organic chemistry, German scientists like Otto *Warburg, Otto Meyerhof, Carl Neuberg, Franz Knoop, Heinrich Wieland, and Gustav Embden continued to dominate research in the field until the Nazi takeover in 1933 dispersed much of the German research community. Émigrés from Germany played a substantial role in the rise to prominence of biochemical research in American universities during the 1930s and the postwar period.

While some biochemists worked out metabolic pathways involving small molecules, others studied the structure of the largest biological molecules, the proteins. During the first two decades of the century, they ascribed special properties of proteins to their "colloidal state." After John Northrop crystallized pepsin in 1930, opinion shifted back to the view that proteins were large molecules of definite composition. Frederick Sanger established the complete sequence of amino acids in insulin between 1945 and 1955. But the earlier hope of finding larger patterns in the number and ordering of amino in proteins was not realized.

The crystallization of pepsin settled a long debate over whether enzymes were proteins or whether, as some prominent biochemists such as Richard Willstätter had maintained, proteins only acted as carriers for catalysts of small molecular weight. The study of the specificity and kinetics of enzyme reactions was one of the central focuses of biochemical research during the first half of the twentieth century.

In the 1920s, Warburg and Wieland proposed competing theories about the nature of biological oxidations. Warburg's theory that an iron-containing *Atmungsferment* transferred atmospheric oxygen to substrates undergoing oxidation seemed to be in conflict with Wieland's claim that cellular oxidations were dehydrogenation reactions catalyzed by "dehydrogenases." By the 1930s, these theories could be seen as complementary. Through this and associated research, a further class of smaller molecules essential to enzymatic oxidations and reductions, and named "co-enzymes," was discovered.

By the mid-twentieth century, biochemistry appeared to be a flourishing, mature science dealing with a broad range of problems among the chemical phenomena of life. Rather suddenly, there arose a new discipline, named by its founders *molecular biology, aimed at explaining biological phenomena at the molecular level. The new field crystallized particularly around the helical structure proposed in 1953 by James Watson and Francis Crick (*see* CRICK, FRANCIS, AND

JAMES WATSON) for the molecule of deoxyribonucleic acid, and the genetic implications inherent in it. Biochemistry and its predecessor fields had been involved with the investigation of vital phenomena at the level of molecules ever since chemical molecules had been defined in the early nineteenth century. Why then the need for a new field?

The answer is a mixed intellectual and social one. That the structure of DNA immediately offered a possible explanation of the molecular basis of the classical *gene prompted a convergence between structural chemistry and *genetics that had not been part of the agenda of biochemistry. Molecular biology hybridized two earlier fields. Also, some of the pioneers of the field of molecular biology, especially those who had moved into biological research from physics, had little training in biochemistry, regarded it as uninteresting, and hoped that they could solve the fundamental problems of life without it. Some biochemists, in turn, regarded the newcomers to molecular problems as interlopers, and resisted their assimilation. The institutional separation that resulted has persisted, even though the barrier was artificial. Much of the investigation pursued in biochemical laboratories during the last decades of the twentieth century was indistinguishable from that carried out by molecular biologists, who have long since learned that they, too, need to incorporate biochemistry into their work and thought.

Dorothy M. Needham, *Machina Carnis: the Biochemistry of Muscular Contraction in its Historical Development* (1971). Marcel Florkin, *A History of Biochemistry* (Section VI of *Comprehensive Biochemistry*) (1977). Robert E. Kohler, *From Medical Chemistry to Biochemistry; the Making of a Biomedical Discipline* (1982). Frederic Lawrence Holmes, *Between Biology and Medicine: the Formation of Intermediary Metabolism* (1992). Mikulas Teich with Dorothy Needham, *A Documentary History of Biochemistry, 1770–1940* (1992). Joseph S. Fruton, *Proteins, Enzymes, Genes: the Interplay of Chemistry and Biology* (1999). Graeme Hunter, *Vital Forces: the Discovery of the Molecular Basis of Life* (2000).

FREDERIC LAWRENCE HOLMES

BIOGENETIC LAW is the thesis or principle according to which the development of an individual (ontogeny) recapitulates the evolutionary history of its species (phylogeny). Announced several times since 1800 and expressed in a number of ways, the principle received its classical formulation in Ernst Haeckel's *Generelle Morphologie* (1866). For Haeckel, the "biogenetic law" (his phrase) was the fundamental law of *evolution. It asserted that phylogeny was the mechanical cause of ontogeny, and acted by means of ordinary physical and chemical processes. Despite the reductionist language (*see*

*REDUCTIONISM), Haeckel's law owed something to the naturalists of the Romantic age and the earliest evolutionists. While favored by the contemporaries of Charles *Darwin, the fortunes of the biogenetic law declined with the advent of experimental *embryology, *genetics, and neo-Darwinism during the first half of the twentieth century (*see* *DARWINISM).

During the early nineteenth century many naturalists influenced by the idealistic philosophy of nature, including Lorenz Oken, Gottfried Reinhold Treviranus, Friedrich Tiedemann, Johann Friedrich Meckel, Étienne Geoffroy Saint-Hilaire, and Étienne R. A. Serres, favored the idea that the embryo recapitulated the morphological types or forms of the species below it. Their theories on recapitulation were criticized by Karl Ernst von Baer, who in 1828 maintained that living forms belonged to four distinct archetypes, or fundamental arrangements of organs, that could not be placed in a single line of development, and thus offered evidence against the very possibility of evolution. Von Baer, however, admitted that in their earliest stages the embryos of different classes of animals could be similar to one another because embryos develop the general traits of their type first, and then the particular traits of their species. Charles Darwin cautiously endorsed recapitulation in the sixth edition (1872) of the *Origin of Species* by saying that the embryo often shows more or less plainly the structure of the progenitor of the group.

The reasons for the demise of the biogenetic law in the twentieth century have been debated. Attacks on recapitulation came from Thomas Hunt *Morgan in 1916 and 1932, Walter Garstang in 1922, and Gavin de Beer in 1930. Three explanations of their effectiveness have been suggested: scientific developments made recapitulation unfashionable in approach and untenable in theory; Morgan proclaimed genetics as the new, chief body of knowledge and practices dealing with evolution, and rejected the claims of the old morphology; finally, biologists and historians alike have tried to keep the histories of Darwinian evolutionism and the biogenetic law separate, perhaps from a desire to assert the independence of present-day neo-Darwinism from the embarrassing heritage of Romantic biology, Haeckelian evolutionism, and social Darwinism.

Stephen Jay Gould, *Ontogeny and Phylogeny* (1977). Robert J. Richards, *The Meaning of Evolution. The Morphological Construction and Ideological Reconstruction of Darwin's Theory* (1992).

GIULIANO PANCALDI

BIOLOGICAL INSTRUMENTS. See INSTRUMENTS, BIOLOGICAL.

BIOLOGICAL WARFARE. See Chemical and
Biological Warfare.

BIOLOGY. The word "biology," from Greek
"life discourse," was introduced around 1800
by several authors to denote "the science of
life" (Gottfried Reinhold Treviranus, 1802), or
"everything that pertains to living bodies" (Jean-
Baptiste *Lamarck, 1802). The word conveyed the
new, central role that the notions of *life and
*organism were then acquiring, and a dissatisfac-
tion with the emphasis on the mere classification
of natural objects typical of naturalists of earlier
generations (*see* *Classification in Science). However, since Charles *Darwin's *Origin of Species*
could still do without the word "biology" in 1859,
neither the word nor its introduction should be
taken as rigid points of departure. Also, the Nobel
Prize for "physiology or medicine," first awarded
in 1901, did not and does not include the kind of
natural history practiced by Darwin.

What we now call biology or the life sciences
comprises a rich variety of research traditions and
practices. They include, beside the general pursuit
of knowledge, *medicine, surgery, midwifery,
pharmaceutics, agriculture, *forestry, hunting,
fishing, *breeding, collecting, voyages of exploration, veterinary practice, *food preservation and
industries, *biotechnology, and *environment
preservation. The places devoted to the study of
the living world since the Renaissance also point
to diversity. The medical school, *university,
*anatomical theater, *botanical garden, *hospital,
doctor's room, school of agriculture, veterinary
school, natural history *museum, fieldwork, and
the *laboratory—each with its own set of goals,
instruments, and habits—have all contributed to
the shaping of what we now call biology.

The historical development of the life sciences
fits the transition from natural knowledge to
modern science set out in the article *history of
science, which derives primarily from the history
of natural philosophy, but with some distinctive
traits and contrasts. One such trait is that old
research traditions seem to have coexisted with
newer traditions more constantly and pervasively
than in the physical sciences (*see* Physics).

Because of these distinctive traits, the life sciences have received mixed reviews in appraisals
of the *Scientific Revolution, which until recently
focused typically on the physical sciences.
Because of these same distinctive traits, the history of biology offers scope for reflections aimed
at resisting the recurrent temptation, on the part
of scientists, historians, and philosophers alike, to
force on science a simplified, unilinear pattern of
historical development.

Thus, for example, the adoption of experimental techniques took on different forms in the life

sciences depending on the field. *Dissection—the
main experimental technique revived in anatomy
and medicine in the late sixteenth and early seventeenth centuries—stemmed mainly from within
medicine itself, and implied no necessary revolt
against the Aristotelian tradition, of the kind
then frequent among astronomers and natural
philosophers. The works of Andreas Vesalius
on *anatomy, William *Harvey on the circulation of the *blood, and Marcello *Malpighi on
development (*see* *Embryology) exemplify the shift
toward experiment stemming from within the
medical tradition.

During the seventeenth and eighteenth centuries a growing number of new concepts and
experimental techniques developed within natural philosophy and *chemistry were introduced
successfully into the life sciences (*see* Mechanical
Philosophy). The works of René *Descartes
on *physiology, John Mayow on *respiration,
*Hermann Boerhaave on animal heat, Stephen
Hales on perspiration, Albrecht von Haller on
*irritability, Lazzaro Spallanzani on *generation,
and Antoine-Laurent *Lavoisier on respiration
indicate how intense and creative the interaction
between the life sciences and natural philosophy could be.

Meanwhile, much new information on the
animate and inanimate objects of the globe—
gathered through the voyages of exploration and
the imperial networks set up by European powers from the fifteenth century onwards—was
collected, named, and classified along lines suggested by the ancient natural history tradition revived by Renaissance humanists with
their interest in erudition, antiquities, and collections. Largely unaffected by the mechanical philosophy of the seventeenth century, this
tradition—cultivated by authors like Konrad
Gesner, Ulisse Aldrovandi, John Ray, and Carl
*Linnaeus—continued to bear fruit well into
the eighteenth century, especially in the fields
of *botany, *zoology, systematics, the *earth sciences, and *paleontology.

Repeatedly in the early development of the
life sciences, interests and practices that could
be regarded as of merely specialist interest
impinged upon broad philosophical and religious
issues. For example, during the eighteenth
century, in the controversy over *generation,
the concerns of naturalists and physicians
merged and occasionally clashed with issues of
interest to philosophers, theologians, and public
authorities, producing a recurrent battleground
for the ideological strands that characterized
Enlightenment Europe. Enlightened curiosity
and secularization (*see* *Religion and Science)
made the egg, the animalcules supposed to

be found in sperm, regenerating polyps, the microscopic animals found in vegetable and meat soups (called *infusoria), *monsters, fossils, as well as topics like the age of the earth, *spontaneous generation, and *extraterrestrial life subjects of enduring concern to philosophers and cultivated elites. Meanwhile, herborizing, *entomology, and an amateur interest in natural history became fashionable among the leisured classes of European countries and their colonies. Similar interests continued to be popular well into the twentieth century and beyond, favored by the creation of natural history museums with their impressive dioramas and dinosaur displays, and by the entertainment and movie industries (see *POPULARIZATION).

A combination of the traditional concerns and practices of the naturalist with philosophical, secularized theory building is found also in the development of *evolution theories and Darwinism throughout the nineteenth century. The history of evolution theories—a major component of the life sciences over the past two hundred years—reveals the extraordinary power that a blend of the empirical and the philosophical, the scientific and the popular, played in the history of biology. Here the works of Lamarck and Darwin, with their considerable impact on the general public, are emblematic.

A new experimental turn within the life sciences developed after 1839 with the introduction of *cell theories and the new microscopic techniques used to explore the fine texture and development of living bodies. The works of Matthias Jakob Schleiden on plant cells and Theodor *Schwann on animal cells promoted the idea that the cell was the fundamental unit of life, provided a powerful generalization for biology, and stimulated microscopical researches. Together, they contributed to shaping the age of the (compound) *microscope: a long, tremendously productive season in the history of laboratory biology and the medical sciences, which lasted until the introduction of the electron microscope in the 1930s.

Early experimental developments in the life sciences were seldom accompanied by the sort of *mathematization and quantification that the physical sciences underwent in the same period. Systematic, successful attempts at quantification first materialized in the central decades of the nineteenth century. These had their origins in the gradual affirmation of laboratory biology, the studies of Gregor *Mendel on *heredity, and *biometrics.

The introduction of new instruments and the associated development of new laboratory techniques marked dramatic changes in the

life sciences as in other fields of natural science. Chemical analysis and improved microscopes in the eighteenth and nineteenth centuries, X-ray *crystallography and *electrophoresis in the 1930s, the *centrifuge and ultracentrifuge, *chromatography and *nuclear magnetic resonance from the middle decades of the twentieth century, the introduction of new monitoring techniques throughout the century, and the development of computational biology in the 1990s changed the outlook and perspectives of the life sciences repeatedly. The cumulative effect of these changes in laboratory technique, which had begun to accelerate with the earlier work of Justus *Liebig, Claude *Bernard, and Louis *Pasteur has been the enrollment of laboratory biology and medicine as major branches of the experimental sciences, and the growth of industries based on biological knowledge and techniques as new, powerful forces in society.

Despite this now recognizable trend, the panorama of the life sciences around 1900 still looked deeply fragmented. The unifying concepts offered by Darwin's theory of evolution in the 1860s and 1870s had only occasionally elicited the commitment of scientists working in other fields, like experimental *physiology. The growing number of biologists trained in the laboratory and concentrating on such disciplines as *cytology, *embryology, or *bacteriology felt uneasy about the controversy surrounding evolution theories. Biologists devoted to the empirical work of botany and zoology shared the malaise, while the continuation of Darwin's research tradition in the form of the biometric program developed by Francis Galton, Karl Pearson, and the adepts of *eugenics often failed to appeal either to laboratory or field biologists.

The reemergence of Mendelian *genetics in 1900, thanks to the works of Hugo de Vries, Karl Correns, and Erich Tschermak von Seysenegg, took place in this fragmented panorama, and for a quarter century seemed unable to contribute to its unification. Only after 1930 did some consensus on the mechanisms of evolution develop leading to the "evolutionary synthesis" or neo-Darwinian theory of evolution (see DARWINISM, EVOLUTION). Contemporary and later accounts of developments between 1920 and 1940 suggest, however, that the degree of unification achieved in biology through neo-Darwinism occurred via a federative effort rather than as a movement dominated by only one of the strands making up the field. Current historiography also testifies to the varied impact the synthesis produced in different disciplinary and national contexts.

Meanwhile, the consolidation of the research universities and the growing support given to the biomedical sciences by public and private bodies (*see* POLITICAL ECONOMY OF SCIENCE, GOVERNMENT, FOUNDATION) helped to launch powerful research schools that set new standards in training and research, and favored teamwork that occasionally transcended national borders. The work of Thomas Hunt *Morgan and his school on the genetics of the fruit fly *Drosophila* during the 1910s and 1920s may be taken as prototypical. In the fly room—combining the old, simple tools of the naturalists and the sophisticated laboratory apparatus of the cytologists—biologists learned how to combine the laws of Mendelian genetics with laboratory work on the physical basis of the *gene. A similar research school developed later in molecular genetics around the study of the viruses called *Bacteriophage*.

Meanwhile, evolution theories underwent important changes. Around 1930, R. A. *Fisher, J. B. S. Haldane, and Sewall Wright developed mathematical methods that resulted in *population genetics, demonstrating that a theory of natural selection could be based on (rather than, as some believed at the beginning of the twentieth century, refuted by) Mendelian genetics. Population genetics also showed that statistics and mathematical modeling could be basic tools for evolutionary biologists, as they were for physicists.

From the 1940s onwards, two other fronts potentially controversial for evolution theories—systematics and paleontology—fell into line in the works of Theodosius Dobzhansky, George Gaylord Simpson, Ernst Mayr, and their pupils. The evolutionary synthesis could be presented as proof that biology had at last some unification and a widespread consensus around the heritage left by Charles Darwin. Biologists and historians of biology thus represented the synthesis during the centennial celebrations in 1982 commemorating Darwin's death.

The neo-Darwinian synthesis had (and retains) several features appealing to the era of the cold war, when expert and lay audiences alike looking for broad generalizations thought to answer the challenges, and tame the uncertainties, associated with the deep ideological commitments and recurrent conflicts of a divided world order (*see* LYSENKO AFFAIR, ANTISCIENCE). However, in the early 1980s, just when the synthesis was being celebrated as the major development of twentieth-century biology, it was losing its central position in the life sciences. It no longer reflected the increasingly experimental and instrumentalist ethos attracting biologists and their public.

Since the early 1950s, *molecular biology had been shifting the careful balance pursued within the synthesis between the naturalists' and the experimenters' traditions of the life sciences in

favor of the experimenters. The shift had been prepared during the 1930s by incursions of leading physicists like Niels *Bohr into the terrain of biology, followed by the pathbreaking work of biologists with a background in physics like Max Delbrück, and by books such as Erwin *Schrödinger's *What Is Life?* (1942). The new shift of biology toward experiment, the laboratory, and the techniques and metaphors of the new information technologies born of World War II (*see* COMPUTER, ARTIFICIAL INTELLIGENCE), was later celebrated by biologists in books addressed to a wide public, like James Watson's *The Double Helix* (1968), Jacques Monod's *Chance and Necessity* (1970), and François Jacob's *The Logic of Life* (1970).

The identification of DNA as the carrier of heredity in the 1940s, the discovery of the *double-helix structure of DNA by Francis *Crick and James *Watson in 1953, the breaking of the genetic code in the mid-1960s, and the development of recombinant DNA techniques leading to the emergence of biotechnology industries in the 1970s all can be depicted as the products of a new wave of instrumentalism in biology. So can initiatives like the publicly funded, international *Human Genome Project, launched in 1988, and the publication in 2001 of two draft sequences of the human genome (*see* GENE) produced by biologists participating in the project and by the firm Celera Genomics. Instrumentalism fueled by computer-based techniques and inspired by the conceptual tools of the information sciences seems to constitute the main connecting thread among these developments, though perhaps not the compass likely to orient biologists toward the long-sought unity of the life sciences.

The detailed recent histories of the people, research schools, laboratory techniques, national institutions, foundations, private industries, and transnational groups of scholars involved in the development of the biomedical sciences during the second half of the twentieth century indicate a complex story. The wealth of stimuli generated by the diverse interests, training, disciplinary traditions, and cultural backgrounds of a growing international population of researchers seems no less important than the overall instrumentalist strategy adopted by the leading actors and programs. A portrait of the life sciences at the turn of the twenty-first century would emphasize the tensions as well as the synergisms among the major, different threads that continue to make up biology. Given the rapid emergence and disappearance of specialties, fields, and subfields, the major threads can best be identified through a few catchwords used by experts, science policy makers, science journalists, and the public when trying to capture the current focus of biology: "life sciences," suggesting the multiplicity of research traditions and practices that continue to be deployed by humans in order to understand and bend to their own advantage life and its complexity; "evolution," referring to the persistent expectation, associated with the life sciences generally and anthropology in particular, that biology can tell us something about our place in nature; "neurosciences," alluding to the widespread feeling that the brain is the next citadel to be addressed; and "biotechnologies" and "bioethics," currently the most popular of the catchwords, as indicating the powerful, manipulative skills through which biologists have at last joined other scientists and technologists in supplying high-tech goods for the planetary market economy of the turn of the century, and as evoking the mixture of expectations and fears that their new power is generating in the public.

Jacques Roger, *The Life Sciences in Eighteenth-Century French Thought* (1963; Engl. transl. 1997). Frederic L. Holmes, *Claude Bernard and Animal Chemistry: The Emergence of a Scientist* (1974). Robert C. Olby, *The Path to the Double Helix* (1974). David Elliston Allen, *The Naturalist in Britain* (1976). Garland E. Allen, *Thomas Hunt Morgan: The Man and His Science* (1978). Robert G. Frank, Jr., *Harvey and the Oxford Physiologists: Scientific Ideas and Social Interaction* (1980). Ernst Mayr, *The Growth of Biological Thought* (1982). David Kohn, ed., *The Darwinian Heritage* (1985). William Coleman and Frederic L. Holmes, eds., *The Investigative Enterprise: Experimental Physiology in Nineteenth-Century Medicine* (1988). Paula Findlen, *Possessing Nature: Museums, Collecting, and Scientific Culture in Early Modern Italy* (1994). Gerald L. Geison, *The Private Science of Louis Pasteur* (1995). Michel Morange, *A History of Molecular Biology* (1998).

GIULIANO PANCALDI

BIOMETRICS, or biometry, is the statistical study of populations of living organisms and their variations. Biometrics had its origin in the collaboration between W. F. R. Weldon and Karl Pearson at University College London (UCL) from 1891 to 1906. As a morphologist, Weldon was interested in the functional relationships between the physical characteristics of living organisms. But Darwin's theory of evolution by natural selection had changed the context in which morphologists thought about those relationships. His theory challenged future generations of researchers to answer two fundamental questions: how were physical characteristics passed on to offspring, and could the process of natural selection cause the physical character of a population to deviate enough to make it a new and separate species? Investigating the latter question, Weldon realized that before a determination whether natural selection could cause significant deviation in the physical character of a population could be made,

ways for measuring the influence of heredity and for establishing the normal degree of variation for a given species had to be established. Weldon concluded, therefore, that the question of the efficacy of natural selection was a statistical question. To make the necessary measurements and calculations, he borrowed the concepts and techniques of correlation and regression developed by Francis Galton in his *Natural Inheritance* (1889). Soon, however, Weldon confronted his own inadequacies as a statistician and turned to Pearson.

Pearson, a Cambridge-educated mathematician, was a polymathic dabbler. Living in London in the 1880s, he made his mark as a freethinking socialist and as an advocate of scientific rationalism. Appointed to the chair of applied mathematics and mechanics at University College in 1884, Pearson found a focus for his mathematical work once he began to collaborate with Weldon, who came to UCL in 1891 as the new Jodrell Professor of zoology.

Pearson also had been impressed with Galton's *Natural Inheritance*, both for its statistical approach and for its eugenic concerns. Under Weldon's influence, Pearson began work on ways to adapt and extend the concepts of correlation and regression to statistical problems presented by the study of dynamic populations. The work done by Pearson between 1891 and 1896 laid many of the foundations of modern statistical theory. By 1900, researchers who applied statistical theory to the study of populations of living organisms were identified as the "biometric school."

The school faced severe criticism of its methods from researchers who, bolstered by the "rediscovery" of Gregor *Mendel's theory of *heredity, argued that the study of genetics held the key to understanding evolution. Faced with this challenge, Pearson and Weldon launched *Biometrika* (1901), a journal dedicated to "the statistical study of biological problems." The controversy between the biometricians and the Mendelians raged for more than two decades.

Biometrics continued to be controversial, both for its statistical approach and for its ties to *eugenics, until the 1930s, when the final elements of a synthesis of biometrics and Mendelism were put in place. One of the contributors to the synthesis, R. A. *Fisher, another Cambridge-trained mathematician, succeeded Pearson in 1933 as Galton Professor of eugenics at UCL. Fisher became interested in evolution by reading Pearson's articles entitled "Mathematical Contributions to the Theory of Evolution." Fisher sought to demonstrate the compatibility between the effects of Darwinian natural selection as described by biometrics and Mendelian genetics. His paper of 1918, "The Correlation Between Relatives on the Supposition of Mendelian Inheritance," did much to establish that compatibility. By the late 1930s, work in a similar vein by Sewall Wright and J. B. S. Haldane had completed the synthesis and laid the foundations for the science of population genetics.

Meanwhile, Pearson's successor as editor of *Biometrika*, his son, Egon Pearson, took advantage of the cessation of hostilities between the biometric and Mendelism camps. No longer needing to defend the legitimacy of the statistical study of evolution, he began to purge *Biometrika* of its antigenetics rhetoric and of its eugenic concerns, transforming biometrics into the production of statistical tools and data for use by biologists of all persuasions.

William B. Provine, *The Origins of Theoretical Population Genetics* (1971). Theodore M. Porter, *The Rise of Statistical Thinking, 1820–1900* (1986). Stephen M. Stigler, *The History of Statistics: The Measurement of Uncertainty before 1900* (1986).

JEFFREY C. BRAUTIGAM

BIOTECHNOLOGY. The Hungarian scientist Karl Erky coined the term "biotechnology" in the title of a treatise that he published in 1919 calling for an industrialized agriculture for the production of meat, fat, and milk. Erky's program built on processes that involved fermentation to produce goods like beer. The growing understanding that microbiological processes are chemical ones turned Erky's branch of biotechnology toward the exploitation of biological organisms to work on an industrial scale. After World War II industrial biotechnology expanded into the production of pharmaceuticals such as penicillin and, in the 1960s and 1970s, into attempts to coax single-cell organisms to transform hydrocarbons into edible proteins and barnyard waste into gasohol. Japan, Germany, Britain, and the European Commission encouraged these programs because they promised to yield better medicine and more abundant food and materials, and to bolster international competitiveness in the face of declines in traditional manufacturing.

A new biotechnology, which depended on molecular genetics, particularly the technique of recombinant DNA, was invented in 1973 by Stanley Cohen, at Stanford University, and Herbert Boyer, at the University of California San Francisco Medical School. Their technique allowed biologists to snip out a gene from the genome of one species and insert it into the genome of another, where it might replicate and express itself. The method was first used to insert foreign genes into bacteria to study their function, but molecular biologists recognized that it could be employed to modify virtually any organism, including plants, animals, and perhaps even human beings.

To many scientists, the fact that humans could now easily create combinations of genetic material unknown in nature was more worrisome than welcome. They feared that, while many DNA hybrids would no doubt prove to be innocuous, some—for example, bacteria containing genes suspected to cause cancer—might be hazardous to human health. In response to that fear, 140 biologists from the United States and Europe convened in February 1975 at the Asilomar Conference Center, on the Monterey Peninsula, in Pacific Grove, California, to probe the promise and hazards of joining DNA across species. The conference agreed that recombinant research could and should be done under guidelines that ensured its safety. Variants of the guidelines were adopted by the National Institutes of Health in the United States and its counterpart research agencies in Europe.

From the mid-1970s onward, recombinant research proceeded with increasing momentum, spurred ahead not only by its scientific but also its commercial possibilities. Molecular biologists predicted that it would revolutionize medicine, by replacing disease-causing genes with normal ones; pharmaceuticals, by turning bacteria into factories for the production of drugs to order; and agriculture, by equipping plants to fix their own nitrogen from the air.

Some of the biologists, including Herbert Boyer, took their case to investment bankers. Together with a young venture capitalist, Boyer scraped up $1,000 to found a biotechnology company to exploit recombinant *DNA commercially. The company was formally incorporated (with additional capital including $100,000 from Thomas Perkins of the Kleiner and Perkins firm) as Genentech ("genetic engineering technology").

Boyer and Swanson determined that Genentech's first project should be the production by recombinant techniques of human insulin. The world's supply of therapeutic insulin came from the pancreases of slaughtered cows and pigs. Projections at Eli Lilly & Co., which accounted for at least 80 percent of all insulin sales in the United States, indicated that the insulin needs of the American diabetic population might eventually outrun the animal supply. In early September 1978, at a crowded press conference, Genentech announced that it had bioengineered human insulin and that, about two weeks earlier, it had entered into an agreement whereby Eli Lilly & Co. would manufacture and market the hormone. When in mid-October 1980, Genentech—assigned the stock symbol "GENE"—went public, investors snapped up its shares at more than twice the offering price of $35, astonishing Wall Street observers.

The financial markets' strong interest in biotechnology was matched by the attention it now received among federal policymakers in the United States. Biotechnological products promised to increase the country's international trade surplus in high-technology goods, which since the mid-1970s had been offsetting a sizable trade deficit in other types of manufactures. In 1980, the government granted the biotechnology industry a triple boost: the National Institutes of Health virtually ended its restrictions on recombinant research; Congress passed the Bayh-Dole Act, which explicitly encouraged universities to patent and privatize the results of federally sponsored high-technology research; and in June, the United States Supreme Court ruled in *Diamond v. Chakrabarty* that a patent could be issued on a genetically modified living organism, holding, over the legal and moral objections of critics, that whether an invention was living or dead was irrelevant to its qualification for intellectual-property protection.

Building on the Chakrabarty case, the United States Patent Office issued its first patent on a genetically engineered plant in 1985 and the first patent on an animal in the history of the world in 1988 (a mouse genetically modified at Harvard to be supersusceptible to cancer). Biotechnology soon received similar encouragement from the European Community. In 1991, the European Patent Office granted Harvard a patent on its mouse, and in 1998 the European Commission, the executive arm of the Community, issued a sweeping directive authorizing patents on a wide range of biotechnological inventions.

By the 1990s, the biotechnology industry had established itself in the United States and Europe and had branches in Asia and Latin America. In 1999, it generated $20 billion in revenues in the United States alone. It comprised start-up firms, many of them spun off from academic laboratories, as well as major pharmaceutical companies and several oil and chemical giants that supported research programs of their own or purchased one or more of the fledglings. Some 80 to 90 percent of the biotechnology companies produced pharmaceuticals and health-service tests.

Agricultural biotechnology accounted for about 10 percent of the industry with plants engineered to resist pests and pesticides or to remain fresh longer after harvest. Some firms in the agricultural area attempted to modify animals genetically with the goal of making sheep that produced more and better wool or cows and pigs that provided more and leaner meat. A few pioneers in "molecular farming" tried to turn common animals genetically into factories for the production of valuable human proteins otherwise difficult and expensive to obtain, if they could be obtained at all.

Yet the new biotechnology had been creating enemies since its inception. Scientists and laypeople both questioned the wisdom of leaping into research with recombinant DNA, holding that the resulting organisms might threaten delicate ecological balances and that, in any case, creating them would be an act of hubris, an assumption of the powers of God. As the biotechnology industry developed, the attention of the critics turned to the patenting of living organisms, the best purchase they could obtain on the advancing biotechnological juggernaut. Some critics declared that such patents would foster monopoly in vital areas such as the food industry; others insisted that the genetic engineering of animals would lead to their suffering; still others contended that patenting them was sacrilegious, turning God's creatures into commodities. In the 1990s, when the *Human Genome Project fostered the rapid identification and patenting of genes, one dissident warned the United States Congress that "we are right in the middle of an ethical struggle on the ownership of the gene pool."

The critics made no significant headway until the late 1990s, when genetically modified foods began coming to market. In the United States, these foods, those derived from corn, soy, canola, cotton, and milk, were too widespread to avoid easily. They were also making inroads abroad. Europeans greeted their arrival with protests and boycotts. Critics called them "Frankenfoods" and claimed that they posed hazards to human health because they contained proteins that did not occur in the natural varieties of food crops. The outcry spread to the United States and led a number of food markets to refuse to stock genetically modified products.

Most of the genetically modified food crops available in 2000 had been designed to assist agricultural producers. A notable example, the Monsanto Corporation's Round-Up Ready soybean, was engineered to withstand the company's Round-Up pesticide. Analysts suggested that once the agricultural biotechnology industry produced foods directly beneficial to consumers, the opposition would diminish. But whatever the future may bring to agricultural biotechnology, the biomedical sector of the industry continues to flourish and enjoy the strong support of the public, including investors, because of its promise for health care.

Martin Kenney, *Biotechnology: The University-Industrial Complex* (1986). Stephen Hall, *Invisible Frontiers: The Race to Synthesize a Human Gene* (1987). Jack Ralph

Kloppenburg, Jr., *First the Seed: The Political Economy of Plant Biotechnology, 1492–2000* (1988). Sheldon Krimsky, *Biotechnics and Society: The Rise of Industrial Genetics* (1991). Robert Bud, *The Uses of Life: A History of Biotechnology* (1993). Susan Wright, *Molecular Politics: Developing American and British Regulatory Policy for Genetic Engineering, 1972–1982* (1994). Herbert Gottweiss, *Governing Molecules: The Discursive Politics of Genetic Engineering in Europe and the United States* (1998). Arnold Thackray, ed. *Private Science: Biotechnology and the Rise of the Molecular Sciences* (1999).

D. J. KEVLES

BIRTH CONTROL. See CONTRACEPTION.

BIRTH ORDER. The study of birth order as a determinant of intellectual outlook has only recently entered the history of science. Historians of science have long pondered the question why, during *scientific revolutions, do some people readily discard their old ways of thinking while others cling fast to the status quo? Common answers attend to such factors as education, age, class, nationality, political allegiance, and so on. Recent work by Frank Sulloway indicates that the answer may lie elsewhere: that "openness to innovation" and its opposite, strong identification with existing power and authority, are behaviors learned early in life as a part of sibling competition for the attention of parents.

Frank Sulloway suggests that one's position (serial rank) in the birth order of a family exercises a powerful influence over the strategy one adopts to engage the attention of one's parents. Firstborn children tend to side with and identify with their parents in order to win their attention and approval, while laterborn children, finding this behavioral niche already filled, must find another strategy to win attention and favor.

This Darwinian claim gains strength from the finding (a standard of personality theory and research) that siblings raised together have personalities hardly more like one another than those of people they meet on the street. Moreover, birth order is a proxy for differences in age, size, power, and status in the family. Younger, smaller, weaker, and lower-status laterborns begin life at odds with the status quo and may in time adopt a "revolutionary personality." Gender, temperament, parent-offspring conflict, availability of parental resources, and family size provide "interaction effects" that shape sibling survival strategies, but these other variables seem only to modify the fundamental influence of birth order without supplanting or neutralizing it.

Applied to the history of science, a "birth-order effect" suggests that in the making of scientific judgments there is, in addition to our usual notion of "influences," an appreciable component of involuntary behavior, a residue of childhood strategies for gaining attention, favor, and psychic security. Sulloway undertakes to provide an empirical foundation for this hypothesis by examining the role of several thousand scientists actively involved in twenty-eight major scientific revolutions.

What Sulloway's data show are that the more socially radical an innovation, the more likely it will be led by laterborns and opposed by firstborns; the more socially conservative a theoretical innovation, the more likely it is to be led and supported by firstborns; there are no historical examples of radical innovations by firstborns and no examples of conservative innovations led by laterborns. The samples are large, the correlations are robust, and their statistical significance is demonstrated using several different measures.

Frank J. Sulloway, *Born to Rebel: Birth Order, Family Dynamics, and Creative Lives* (1996).

MOTT T. GREENE

BLACKETT, Patrick M. S. (1897–1974), British experimental physicist, co-detector of pair production of positron, and opponent of nuclear weapons, and **Ernest O. LAWRENCE** (1901–1958), American experimental physicist, inventor of the cyclotron, and proponent of nuclear weapons.

Blackett and Lawrence were among the leading experimental physicists of their time and worked in the preeminent physics labs, Blackett at the Cavendish Laboratory in Cambridge, Lawrence at the Radiation Laboratory he founded at the University of California at Berkeley. Both won Nobel Prizes for their work in nuclear physics, as much for their development of instruments as for the discoveries they made with them. And both applied physics to military problems in World War II, although afterward Blackett spoke out against nuclear weapons while Lawrence strongly supported their development.

Blackett was born and raised in cosmopolitan London, Lawrence in the American Midwest. As youths both tinkered with radio sets, thus gaining valuable familiarity with electronics. Blackett trained in naval schools and saw action in the British Navy in World War I. After the war he went to Cambridge, where he took the mathematical tripos (honors) and then in 1921 joined the Cavendish under Ernest *Rutherford. Rutherford had just achieved the disintegration of a nitrogen nucleus by alpha particles, and Blackett undertook to observe the disintegration in a *cloud chamber. In 1924, after culling some 23,000 cloud chamber photographs of particle tracks, he identified eight that showed disintegration; the famous photos would adorn generations of nuclear physics textbooks.

Blackett then turned the cloud chamber to the study of *cosmic rays, incorporating a coincidence circuit using *Geiger counters to trigger the expansion of the chamber when cosmic rays passed through it. In 1932 Blackett and Giuseppe P. S. Occhialini obtained photos of positive electrons in cosmic ray showers and explained their appearance by pair production. Blackett won the Nobel Prize in physics in 1948, "for his development of the Wilson [cloud chamber] method and his discoveries, made by this method, in nuclear physics and on cosmic radiation."

Lawrence exhibited energy, enthusiasm, lack of sophistication, and dedication to science. Unlike Blackett's education in elite naval schools, Lawrence studied in small midwestern colleges before finishing his Ph.D. in physics at Yale in 1924 with a thesis on the photoelectric effect. He stayed on at Yale until Berkeley hired him away in 1928. He took up the problem of accelerating subatomic particles to energies comparable to those of particles produced by natural *radioactivity, such as the alphas used by Rutherford and Blackett in their disintegration experiments.

Instead of trying to maintain a very high voltage, a difficult task with available electrical technology, Lawrence in 1929 thought to turn a particle through a lower voltage many times. By bending particles into a circular path using a magnetic field and sending them through an alternating electric field provided by a radio-frequency oscillator, Lawrence could give them successive pushes and pulls to reach high voltages. The device, dubbed the cyclotron, promised ever-higher energies, which Lawrence pursued through the 1930s in a series of ever-larger machines and an ever-widening circle of funding sources. The cyclotrons provided copious sources of high-energy particles for nuclear physics and new radioactive isotopes for chemistry, biology, and medicine. Lawrence won the Nobel Prize in physics in 1939, "for the invention and development of the cyclotron and for results attained with it, especially with regard to artificial radioactive elements."

Lawrence pursued a pragmatic approach to the art of *accelerator building, often resorting to cut-and-try instead of theory to achieve higher energies. The Cavendish cultivated a similar tradition of string-and-sealing-wax experiment. But both labs increasingly adopted characteristics of big science, including large collaborative teams of engineers and scientists, complex organization, and connections to industrial equipment-makers.

World War II diverted both men, like many of their colleagues, to war work. By 1935 Blackett was involved in the British project to develop radar for air defense; after the outbreak of war he started work on bomb-sights and joined the so-called MAUD committee on the British atomic bomb project. He spent most of the war itself engaged in operations research, a new field that applied quantitative methods to military tactics, such as the optimum placement of antiaircraft batteries and the coordination of antisubmarine warfare. Lawrence lent several of his top scientists to the American radar project, but his main contribution was to the Manhattan Project for the production of atomic bombs, where he turned his cyclotrons into giant mass spectrographs to separate uranium isotopes. As one of the scientific leaders of the project, Lawrence helped advise the government on the use of atomic bombs.

Blackett became involved in left-leaning politics in the 1930s and began to explore relations between science and society. During the war he had opposed in vain Britain's strategic bombing of German cities, both on moral grounds and on the practical military consideration, backed up by operations research, that bombers would be more useful in antisubmarine warfare. His opposition to strategic bombing underlay his postwar criticism of atomic weapons, again on both moral and military grounds: he discounted the effectiveness of atomic weapons and urged Britain not to develop them and to stay neutral in the deepening cold war. Blackett publicized his views in a widely read book, *Military and Political Consequences of Atomic Energy* (1948; published in 1949 in the United States as *Fear, War, and the Bomb*), which also introduced the argument that the United States had used atomic bombs not so much as a military weapon against Japan but as a political weapon against the Soviet Union. His activism earned him an undeserved reputation as a communist fellow-traveler, but he continued to speak out against nuclear weapons and the arms race.

The pragmatic Lawrence professed to keep politics out of science and vice versa. His ascent into influential circles sharpened his political views. He displayed no moral scruples after Hiroshima; he felt atomic bombs had ended the war and further bloodshed, and might prevent future wars. He became a staunch advocate of nuclear-weapons development by the United States, arguing in favor of a crash program to build a hydrogen bomb in 1949 and mobilizing his laboratory to aid the effort. A consequence of his advocacy was the establishment of an offshoot of the Berkeley lab that became the Livermore weapons laboratory. Although Lawrence did not testify in person at the security hearing of J. Robert Oppenheimer in 1954, he opposed the continued influence of Oppenheimer, whom he thought insufficiently hawkish on nuclear policy (*see* KURCHATOV, IGOR VASILIEVICH, AND J. ROBERT OPPENHEIMER).

The scientific careers of Lawrence and Blackett also diverged. Lawrence was the exemplar of the entrepreneurial lab director, his attention after the war devoted more to administration than to his own research. Blackett turned down the directorship of the Cavendish in 1953. He remained immersed in research but left nuclear physics for problems in geomagnetism, where he exercised his instrumental dexterity free of the trappings of big science.

Some parallels persisted. Both Blackett and Lawrence propagated their opinions in the postwar decade not from official advisory positions, but as physicists of high reputation with personal connections to political and scientific leaders. Their respective governments honored their achievements. The two national laboratories Lawrence established in Berkeley and Livermore would bear his name. In the 1960s Blackett re-entered high-level advising under the Labor government, served as president of the Royal Society, and was elevated to the peerage as Baron Blackett of Chelsea.

Herbert Childs, *An American Genius: The Life of Ernest Orlando Lawrence* (1968). Bernard Lovell, *P. M. S. Blackett: A Biographical Memoir* (1976). J. L. Heilbron and Robert W. Seidel, *Lawrence and His Laboratory: A History of the Lawrence Berkeley Laboratory* (1989). Mary Jo Nye, "A Physicist in the Corridors of Power: P. M. S. Blackett's Opposition to Atomic Weapons Following the War," *Physics in Perspective* 1 (1999): 136–156.

PETER J. WESTWICK

BLACK HOLE. A black hole in space has a gravitational force so strong that nothing can escape from it, not even light. Hence no one can see a black hole, but astronomers are convinced they exist. An early intellectual precursor to the concept of black holes may be found in John Michell's dark stars. Michell noted in 1783 that if bodies with densities not less than the Sun's, and hundreds of times greater in diameter, really existed, gravity would prevent their light particles from reaching us. Their existence might be inferred from the motions of luminous bodies around them. Michell's idea was ignored even before wave theory, in which gravity does not act on light, overthrew the particle theory from which he reasoned.

Collapsed stars make another intellectual precursor to black holes. After 1915 and Einstein's relativity theory, gravity again could act on light. In the 1930s, Subrahmanyan *Chandrasekhar, recently arrived from India to study in England, modeled stellar structures. He found that stars of less than 1.4 solar mass shrink until they become white dwarfs, but more massive stars continue contracting. The British astrophysicist Arthur

Eddington noted that at high compression, gravity would be so great that radiation could not escape, a situation he regarded as absurd. Others accepted Chandrasekhar's mathematics but believed that continuous or catastrophic mass ejection would act as a universal regulating mechanism to bring stars below the critical mass. Also, massive stars might evolve into stars composed of neutrons. In 1939, however, the American physicist J. Robert Oppenheimer established a mass limit for neutron stars (*see* KURCHATOV, IGOR VASILIEVICH, AND J. ROBERT OPPENHEIMER). When it has exhausted thermonuclear sources of energy, a sufficiently heavy star will collapse indefinitely, unless it can reduce its mass.

World War II and then the hydrogen bomb project diverted research away from stellar structure. Nuclear weapons programs, though, did develop a deeper understanding of physics and more powerful computational techniques, and in the 1960s, computer programs that simulated bomb explosions were modified to simulate implosions of stars. A renewed theoretical assault followed on "black holes," as they were named by the American nuclear physicist John Wheeler in 1967. In contrast to collapsed or frozen stars, black holes are now known to be dynamic, evolving, energy-storing, and energy-releasing objects.

Because no light escapes from black holes, detection of them requires observing manifestations of their gravitational attraction. From a companion star, a black hole captures and heats gas to millions of degrees, hot enough to emit X rays. Because the Earth's atmosphere absorbs X rays, devices to detect them must be lofted on rockets or *satellites. A few black holes probably have been found, although other explanations of the observational data are possible.

In another predicted manifestation, two black holes spiral together, gyrate wildly while coalescing, and then become steady. Outward ripples of curvature of spacetime, also called gravitational waves, would carry an unequivocal black-hole signature. Gravitational waves should propagate through matter, diminishing in intensity with distance. On Earth, they should create tides the size of an atom's nucleus, in contrast to lunar tides of about a meter. Gravitational-wave detectors may be operational early in the twenty-first century.

Meanwhile, without benefit of prediction and intent, we may already have observed manifestations of black holes. Extraordinarily strong radio emissions from both the centers of *galaxies and from quasars (compact, highly luminous objects) may be powered by the rotational energy of gigantic black holes, either coalesced from many stars or from the implosion of a single supermassive rotating star a hundred million times heavier than our sun. Other possible explanations for radio galaxies and *quasars, however, do not require black holes.

S. W. Hawking and W. Israel, eds., *Three Hundred Years of Gravitation* (1987). Kip S. Thorne, *Black Holes and Time Warps: Einstein's Outrageous Legacy* (1994).

NORRISS S. HETHERINGTON

BLOOD is a complex physiological fluid essential to a wide array of life-supporting functions in the body. Through the circulatory system of arteries, veins, and capillaries, blood transports oxygen from the lungs to cells, and removes waste products like carbon dioxide and other cellular metabolic by-products, delivering them to the lungs, kidneys, and skin for excretion. In addition, blood is the primary vehicle of immunity. White blood cells, or leukocytes, constitute the cellular components of the immune system, responsible for warding off infections and infiltration of foreign proteins.

Since antiquity, blood has been regarded as integral to life. Recognizing that the heart and blood vessels were connected in a unified system, Aristotle defined the blood as a nutritive fluid that sustained the vitality of the solid parts of the body. In the Hippocratic and Galenic traditions, blood was identified as one of the four humors that comprised the human body. In this tradition, health and disease were interpreted as the result of the interaction of these four constituents. Imbalance or disequilibrium in the humors offered a powerful rationale for bloodletting or phlebotomy, which remained an important medical therapy for more than two millennia.

Initial interest in the blood focused on its physical movement in the animal and human body. In the early seventeenth century, English physician William *Harvey provided statistical and experimental support in *De Motu Cordis* (1628) for his claims that blood circulated in the bodies of animals and humans. At this time, physicians made little differentiation between the blood of animals and humans, as the earliest efforts in transfusion illustrate. In 1667 French physician Jean Baptiste Denis performed the first human transfusion when he transfused blood from sheep and calves into several patients. At the Royal Society in London, English physician Richard Lower paid a clergyman the sum of one guinea for his willingness to "suffer the experiment of transfusion" with blood from a lamb.

The advent of the *microscope revealed that blood was not a simple, uniform fluid. The Dutch lens maker Antoni van Leeuwenhoek, examining his own blood under a primitive microscope, first identified the red blood cells, the small, red globules "swimming in a liquor, called by physicians, the serum." Jan Swammerdam and Marcello *Malpighi, who identified the role of the capillaries in the circulatory system, also noted these corpuscles. English physician William Hewson was among the first to describe white blood cells, or leukocytes, and demonstrated some of the essential features of blood clotting or coagulation. In 1851 German physiologist Otto Funke identified hemoglobin, the reddish pigment in red blood cells. Investigations by Felix Hoppe-Seyler established that hemoglobin was responsible for taking up and discharging oxygen. Using new chemical dyes and improved microscopes, Paul Ehrlich profoundly influenced the development of hematology. Ehrlich identified several new types of white blood cells, described differences between healthy and diseased blood cells, and distinguished the various forms of anemias.

In the early twentieth century, attention focused on the role of blood in the immune system. In 1900 Karl Landsteiner discovered that human blood could be divided into three groups when he combined red blood cells of individuals with the sera taken from others. (The fourth, rarer blood group was identified by two of his associates, who tested a larger group of people.) Although surgeons attached little importance to this work before World War I, the importance of blood grouping to avert transfusion reactions intensified when Landsteiner, with Alexander Wiener, discovered in 1940 the Rh blood group, a major cause of fetal and infant death. In the 1950s serologists sought to define the antigens on white blood cells. In 1958 Jean-Baptiste Dausset reported his discovery of a gene complex (human leukocyte A complex) that accounted for different immunological responses in blood transfusions. The identification of other human leukocyte antigens (HLA) followed; these formed the basis of the HLA system for typing tissues and organs to reduce the threat of organ and tissue graft rejection in transplantation.

During World War II, the biochemist Edwin Cohn pioneered a new method for plasma fractionation of human blood, fostering renewed interest in blood biochemistry. Cohn's method for separating the protein fraction from plasma and the production of such pharmaceutical products as gamma globulin (used in the treatment of infectious disease) and antihemophilic factor (which altered the lives of thousands of hemophilia patients) ushered in a new era of blood-based research. Extending some of the work on blood fractionation, chemist Linus Pauling studied the physical chemistry of hemoglobin variants and their role in sickle cell disease. In 1949 *Pauling concluded that the inherited disorder resulted from a flaw in the molecular composition of hemoglobin, and proclaimed sickle cell anemia to be the first molecular disease. Although significant for the emergence of molecular medicine, these insights did not foster immediate clinical benefits for sickle cell patients.

Over the course of the twentieth century, the therapeutic advances in blood transfusion and blood components provided an important and life-saving medical intervention. At the same time, the therapeutic use of blood fostered new dangers; medically administered blood became the vehicle for the transmission of disease. During World War II, the infection of military personnel with hepatitis raised serious concerns that disease might spread through the blood supply. In the 1980s blood-borne spread of a hitherto unknown virus, the human immunodeficiency virus (HIV), was linked to a newly identified and lethal disease, *AIDS. Although heat treatment and viral antibody testing have decreased the threat of HIV transmission, concern about the transmission of such diseases as *Mad Cow Disease continue to trouble blood bankers and health policy makers.

Blood is a potent biological substance studied by physicians and medical researchers; blood is also a complex cultural entity with social meanings that vary considerably with time and place. Ideas about blood purity extend far beyond the physical properties of a biological fluid to encompass notions about social networks and contamination—both moral and physical. In twentieth-century Japan, many believed that a person's blood type determines personality and character. In the American *eugenics movement, blood and heredity were often closely linked in ideas about lineage or bloodlines. As Karl Landsteiner had predicted, the science of blood came to play an important role in forensic science and in cases of disputed paternity. By the 1930s, German, Austrian, and Danish courts all accepted blood tests for establishing paternity. In the United States, courts remained reluctant for decades to rely on serological tests to establish paternity, and, in the case of several highly publicized cases in which infants were mistakenly given to the wrong parents, maternity. Serological tests to determine guilt and innocence in criminal proceedings came to play an important role in criminal prosecutions on both sides of the Atlantic. Even as DNA testing has come to dominate such proceedings, blood remains a powerful signifier.

Richard Titmuss, *The Gift Relationship: From Human Blood to Social Policy* (1971). Maxwell M. Wintrobe, ed., *Blood Pure and Eloquent* (1980). Keith Wailoo, *Drawing Blood* (1997). Douglass Starr, *Blood: An Epic History of Medicine and Commerce* (1998). Angela Creager, "'What Blood Told Dr. Cohn': World War II, Plasma Fractionation, and the Growth of Human Blood Research," *Studies in History and Philosophy of Biology and Biomedical Sciences* 30C no. 3 (1999): 377–405.

SUSAN E. LEDERER

BOERHAAVE, Hermann (1668–1738), physician, chemist, botanist, and educator.

Boerhaave was born in the village of Voorhout, outside Leyden, in the Dutch Republic on 31 December 1668, to Jacob Boerhaave, a Calvinist minister, and Hagar Daalders, daughter of a nautical-instrument maker. After his mother died, his father married Eva de Bois, a minister's daughter from Leyden who raised Hermann and his eight younger siblings.

He received his early education in the Bible and the classics from his father, who groomed him for the ministry. After attending the Latin school affiliated with Leyden University, he matriculated into the Arts Faculty in 1684, where he absorbed the new Cartesian, experimental, and mathematical approaches of Burchard de Volder. Boerhaave then matriculated into the Theology Faculty and privately undertook the study of medicine, a pursuit greatly enhanced by an appointment at the university library, where he read the library's holdings in medicine and chemistry. In July 1693 Boerhaave obtained his medical degree at Harderwijk to avoid the higher graduation fees and requirements in Leyden. Boerhaave's hopes for a career in the ministry were dashed after an argument on a canal boat led to the circulation of rumors that he was a follower of Baruch Spinoza. He settled in Leyden and opened a private medical practice.

In 1701, Boerhaave was appointed a lecturer in medicine at Leyden University, and the next year he received permission to offer private lectures in anatomy and chemistry. His courses were extremely popular. In 1709, Boerhaave became professor of botany and medicine although he had no formal training or experience in botany. Eventually, he became one of the premier botanists in Europe, reorganizing and expanding the university's botanical garden and publishing two botanical catalogs, *Index plantarum* (1710) and *Index alter plantarum* (1720). In 1718, he was appointed professor of chemistry. For the next decade he lectured for up to five hours each morning and saw patients every afternoon. He resigned from his professorships in botany and chemistry in 1729 but continued to give courses in medicine and experiment in the chemical laboratory, the fruit of which was a series of three papers on the alchemical properties of mercury that he published in the Royal Society of London's *Philosophical Transactions* (1733–1736).

He ceased his lecturing in April 1738 due to failing health.

Boerhaave's renown as a physician derived from his pedagogical "method," which incorporated the discoveries of the previous two centuries and espoused a medical philosophy based on experiment and mechanical reasoning. Boerhaave asserted that physicians must be skilled in mathematics, mechanics, and chemistry, and must apply these arts to understand the human body in terms of its structure, internal motions, and observable properties. Boerhaave presented his medical method through a series of frequently reprinted textbooks. His *Institutes medicinae* (1708) outlined basic medical theory and physiology; his *Aphorisms* (1709) presented medical diagnoses and treatments in an organized fashion; and his *Libellus de materiemedica* (1719), a pharmacopeia, was cross-referenced with his *Aphorisms*.

Boerhaave's most notable achievement, his *Elementa chemiae* (1732), moved away from the traditional textbook emphasis on chemical recipes and focused on the experimental investigation and demonstration of theoretical principles. He devised novel approaches to establish these principles, such as the systematic use of Daniel Fahrenheit's thermometers to study heat and the air-pump to study the chemical properties of air. The *Elementa* became the model for academic or "philosophical" chemistry in the mid-eighteenth century.

Albrecht von Haller called Boerhaave "*communis Europae praeceptor*" (the general teacher of Europe). More than 1,900 students matriculated into the Leyden Medical Faculty during Boerhaave's tenure, and 178 of these earned medical degrees under his guidance. His students established curricula on the Leyden model in Edinburgh, Vienna, Göttingen, Moscow, and Philadelphia. Among his most prominent students were Haller, Carl *Linnaeus, Julien de La Mettrie, Johann Theodor Eller von Brockhausen, Petrus van Musschenbroek, and forty members of the Royal Society of London, including four secretaries and one president.

F. W. Gibbs, "Boerhaave and the Botanists," *Annals of Science* 6 (1957): 47–61. Lester King, *The Medical World of the Eighteenth Century* (1958). G. A. Lindeboom, *Herman Boerhaave: The Man and His Work* (1968). E. Kegel-Brinkgreve and A. M. Luyendijk-Elshout, eds., *Boerhaave's Orations* (1983). Andrew Cunningham, "Medicine to Calm the Mind: Boerhaave's Medical System and Why It Was Adopted in Edinburgh," in *The Medical Enlightenment of the Eighteenth Century*, eds. Andrew Cunningham and Roger French (1990): 40–66. John C. Powers, *Herman Boerhaave and the Pedagogical Reform of Eighteenth-Century Chemistry* (Ph.D. diss., Indiana University, 2001).

JOHN C. POWERS

BOHR INSTITUTE. The Niels Bohr Institute, given its present name in 1965 in commemoration of the eightieth birthday of the Danish physicist Niels *Bohr, was inaugurated in 1921 as Copenhagen University's Institute for Theoretical Physics. Like Bohr's professorship created in 1916, its establishment was promoted by a group of influential friends both at home and abroad. In his inaugural speech Bohr explained that the purposes of the new institute were to test experimentally the predictions of the new field of atomic physics and to accommodate young scientists from all over the world for short periods. At that time, so soon after World War I, neutral Denmark was ideally situated to attract and accept visitors from any country.

Bohr's goals were fulfilled beyond all expectations. He began with one building (which following tradition included the private residence of the professor and his family), a small staff, and four or five foreign visitors. Bohr soon obtained substantial funding for a new building, improved experimental facilities, and more staff and visitors. The Danish state, as well as funding from Danish foundations and the International Education Board (the newly established, Rockefeller–supported philanthropical organization), underwrote the development.

Simultaneously with Bohr's receipt of the Nobel Prize in 1922, physicists at the institute established experimentally that element number 72 (hafnium) was not a rare earth, as most chemists had thought, but a homologue of zirconium, as Bohr had predicted. Yet the main scientific contributions from the institute were theoretical. Werner *Heisenberg developed quantum mechanics in close contact with Copenhagen physicists and subsequently worked out his uncertainty principle while staying at Bohr's institute (*see* COMPLEMENTARITY AND UNCERTAINTY). The first informal international conference at the institute, an annual event subsequently emulated by several other places of research, took place in 1929.

In the early 1930s, Bohr recognized that the central concerns of theoretical physics were moving from the atom to the nucleus. Arguing that the same experimental equipment was required for research in physics and biology, Bohr applied successfully for funding from the Rockefeller Foundation's newly instituted program in experimental biology. The cyclotron at Bohr's institute was one of the first in Europe. Bohr and his collaborators contributed substantially to theoretical nuclear physics, not least by explaining and analyzing the fission process.

With the German occupation of Denmark, international connections were severed. For a month in early 1944, after Bohr had escaped Denmark for Britain and the United States, German forces took over the institute.

The years immediately following the war saw a boom in institute activities, with financial support from the Ford Foundation among others. It persisted as a center for theoretical and experimental nuclear physics, moving part of its experimental activities to new quarters outside Copenhagen. And it continued to be an exemplar for international cooperation as the only physics institute in the Western world to receive Soviet and Chinese visitors during the most trying periods of the cold war. In 1975, Bohr's son Aage Bohr (then institute director) and his collaborator Ben Mottelson (professor at the Nordic Institute for Atomic Physics, established in 1957 next to Bohr's institute) were awarded the Nobel Prize for their theory of nuclear structure.

Like other physics institutes of comparable size, the Copenhagen facility was compelled to move its large-scale experimental activities to specialized laboratories, such as *CERN in Geneva. Although recently hit by severe budget cuts, it remains a world center for theoretical physics.

Peter Robertson, *The Early Years: The Neils Bohr Institute 1921–1930* (1979). Finn Aaserud, *Redirecting Science: Niels Bohr, Philanthropy, and the Rise of Nuclear Physics* (1990).

FINN AASERUD

BOHR, Niels (1885–1962), physicist, inventor of the quantum theory of the atom.

Niels Bohr was born in Copenhagen on 7 October 1885. His mother Ellen, née Adler, belonged to a flourishing Danish–Jewish banking family. His father, Christian Bohr, was an internationally renowned physiologist at the University of Copenhagen. Around the turn of the century, Christian Bohr hosted an informal discussion group with three other prominent Copenhagen professors that young Niels was allowed to attend. What he heard helped shape his mind.

At the University of Copenhagen, Bohr's teacher was Christian Christiansen, the university's only physics professor and a member of Christian Bohr's discussion group. Niels Bohr's first major contribution to science—which was to guide the development of physics for years to come—derived from a postdoctoral stay at Manchester, where Ernest *Rutherford and his international group of collaborators had just established that the mass of the atom was concentrated in a small nucleus with electrons swirling around it at relatively large distances. According to classical physics, such a system would be unstable. As a remedy, Bohr postulated, in a trilogy of papers published in 1913, that the quantum of action introduced by Max *Planck in 1900 set a condition on the number of orbits that an electron can occupy. This seemingly arbitrary postulate proved to have impressive predictive power, accounting for hitherto inexplicable spectroscopic data as well as providing a theoretical basis for the *periodic table of the elements. Bohr received the Nobel Prize for this work in 1922.

At this time Bohr had secured himself a solid institutional base at the University of Copenhagen, where a new Institute for Theoretical Physics was inaugurated in 1921 (*see* BOHR

Bohr lecturing at Columbia, 1937.

INSTITUTE). The very best of the younger theoretical physicists from all over the world flocked to Copenhagen. Bohr's celebrated need for a "helper" in developing his ideas contributed to the institute's unique atmosphere, subsequently described as the "Copenhagen spirit."

After the formulation of quantum mechanics in 1925, which completed the development of quantum theory, Bohr resumed the philosophical interest of his youth, taking a leading role in interpreting the new theory and pondering its philosophical implications outside the field of physics. In 1927, he introduced the complementarity argument, which he continued to refine and promote for the rest of his life and which still constitutes the basis for the "Copenhagen interpretation" of quantum mechanics (*see* COMPLEMENTARITY AND UNCERTAINTY). Yet his emphasis on an experimental basis for theoretical work did not subside, and beginning in the early 1930s he changed the object of experimental and theoretical research at the institute from the atom as a whole to its nucleus. In so doing, Bohr kept his institute in the forefront of contemporary international physics research.

In 1931 the distinguished Danish philosopher Harald Høffding, one of Christian Bohr's discussion circle, died. Høffding had been the first occupant of the honorary residence at Carlsberg, which was conferred for life by the Royal Danish Academy of Sciences and Letters on the most prominent intellectual in Danish society. The academy chose Bohr as the second occupant. In 1932 he settled in with his wife, Margrethe, and their six sons.

Because several of the institute's guests came from the Soviet Union and Germany, Bohr learned early on about the lack of openness of Soviet society and Hitler's persecution of Jews. Virtually overnight Bohr's institute became a sanctuary for young German physicists unable to return to their homeland—until Bohr was able to find permanent placement for them, most often in the United States. At the end of 1938, physicists at the Copenhagen Institute provided a theory for the recent discovery of fission based on Bohr's liquid drop model of the nucleus, and in 1939, during a stay of several months in the United States, Bohr contributed important insights about the mechanism of the fission process. Yet, as he announced publicly in several lectures and publications, he did not believe in the feasibility of an atomic bomb within the foreseeable future.

Bohr held this view until he arrived in England in October 1943 after escaping from Nazi-occupied Denmark. Once acquainted with the Allied efforts, he came to consider the atomic bomb project feasible and took part in it for the rest of the war. At the same time he carried on his own personal crusade to convince Churchill and Roosevelt that Stalin needed to be informed about the project in order to retain mutual confidence among nations after the war as well as to avoid a nuclear arms race. Bohr's continued insistence on an "open world" among nations, the necessity of which he brought to the attention of statesmen at every opportunity, came to public expression in 1950 in an Open Letter to the United Nations.

Bohr served as a mentor and guide to several generations of theoretical physicists during a particularly important and exciting period for the field. He died peacefully at his home in Copenhagen on 18 November 1962.

Stefan Rozental, ed., *Niels Bohr: His Life and Work as Seen by His Friends and Colleagues* (1967). Abraham Pais, *Niels Bohr's Times, in Physics, Philosophy, and Polity* (1991).

FINN AASERUD

BOLTZMANN, Ludwig (1844–1906), theoretical physicist.

Born in Vienna to a comfortable middle-class family, Boltzmann studied at the university there from 1863 to 1866, when he received his doctorate on the kinetic theory of gases. In 1869 he became professor of mathematical physics at the University of Graz, where he remained until 1873. During this period he spent some months with Robert Bunsen and Leo Königsberg in Heidelberg and with Gustav Kirchhoff and Hermann von *Helmholtz in Berlin. In 1873 he returned to Vienna to the chair of mathematics at the university, which he held for the next three years. In 1876 he relocated again to Graz as professor of experimental physics and began his acquaintance with his future friend but persistent opponent in scientific matters, Wilhelm Ostwald.

The peripatetic Boltzmann went to the University of Munich as professor of theoretical physics in 1890 and then, four years later, to the University of Vienna in the same capacity, to take the chair vacated by the death of his teacher Joseph Stephan. At about the same time, Ernst Mach, who would become both a philosophical and personal adversary, became professor of history and theory of the inductive sciences at Vienna. The friction between the two prompted Boltzmann to accept the offer of Ostwald to move to Leipzig. The hoped-for heaven turned into a hell, owing to disagreements with Ostwald over atomism. Upon Mach's retirement in 1901, Boltzmann returned to his previous post. Emperor Francis Joseph asked him to give his word of honor that he would never accept a position outside the Empire again.

During much of the nineteenth century the belief in the strictly deterministic character of the physical laws was the cornerstone of the physicists' worldview. Boltzmann's work seriously undermined it. In 1877, he published a famous paper, "On the relation between the second law of the mechanical theory of heat and the probability calculus with respect to the theorems on thermal equilibrium," which ascribed only a probabilistic value to the second law of *thermodynamics. A system tends to the state of thermodynamic equilibrium as the most probable, but by no means the only, state the system can reach.

Having realized that the second law could not be interpreted via mechanical principles alone, Boltzmann studied James Clerk *Maxwell's approach to the kinetic theory of gases. In a paper on thermal equilibrium, Boltzmann extended Maxwell's theory of distribution of energy among colliding gas particles, treating the case when external forces are present. He deduced that the average energy of a molecule was roughly equal to kT (where k is "Boltzmann's constant" and T the absolute temperature); larger or smaller energies could occur, but with proportionately lower probability.

Boltzmann next turned his attention to nonequilibrium systems. How could kinetic theory account for the process through which a gas tends towards an equilibrium state? In 1872 he formulated "Boltzmann's H-theorem," which states that H (as the negative *entropy) always decreases, except when the distribution of molecular velocities complies with Maxwell's law. Boltzmann was the first to show that the increase of entropy corresponds to an increasing randomness of molecular motion, as required by Maxwell's distribution law.

These results gave rise to a paradox. If Newtonian mechanics held on the molecular level, interactions between particles had to be reversible, whereas thermodynamic changes on the macroscopic level were irreversible. The answer to this "reversibility paradox" lay in the statistical character of the second law. Nevertheless, the community of physicists became apprehensive about the statistical approach. The reversibility paradox—which was initially pointed out by William *Thomson—formed the basis of a controversy between Boltzmann and his friend and colleague Josef Loschmidt.

The main difficulty lay in accepting the implications of the use of the theory of probability in the formulation of a fundamental law of physics. Boltzmann took on the task of persuading his colleagues that the statistical approach could account legitimately for the macroscopic phenomena of the real world. In developing his position he reached one of his major results, $S = k \log W$, which connected the entropy S of a system in a given state with the probability W of the state. The formula connects a thermodynamic or macroscopic quantity, the entropy, with a statistical or microscopic quantity, probability.

Boltzmann aggressively defended his belief in the atomic structure of matter and tried to reconcile this perspective with the statistical description of macroscopic phenomena. In 1903, he started offering a university course on "Methods and General Theory of the Natural Sciences." It appeared that, at last, he could defend his views in a wider setting. But ill health and recurrence of the depression that sometimes plagued him caused him to take his own life in October 1906, while on vacation with his family at the Bay of Duino, a resort near Trieste.

Engelbert Broda, *Ludwig Boltzmann: Man, Physicist, Philosopher* (1983). John Blackmore, ed., *Ludwig Boltzmann, His Later Life and Philosophy, 1900–1906*, 2 vols. (1995). Carlo Cercignani, *Ludwig Boltzmann: The Man Who Trusted Atoms* (1998). David Lindley, *Boltzmann's Atom: The Great Debate That Launched a Revolution in Physics* (2001).

MANOLIS PATINIOTIS

BOTANICAL GARDEN. A botanical garden differs from pleasure gardens or horticultural establishments in that it exists primarily for scientific research and education. Nonetheless the first gardens combined several functions, especially medicinal and recreational. Religious allegory also played an important role in gardens within the Islamic and Judeo-Christian traditions, drawing on notions of earthly paradises or a former golden age.

The great era of European botanical gardens followed the discovery of the New World. There was a close relationship between the foundation of the six preeminent European gardens in Padua, Leiden, and Montpellier (sixteenth century) and Oxford, Paris, and Uppsala (seventeenth century) and the rise of modern science. Before then, however, many important gardens existed in Mediterranean countries, Arabia, and other parts of the globe in which religious and contemplative functions played a significant role. In the west, the ancient Greeks distinguished between Arcadia, a rustic paradise, and Elysium, the land of the dead where the gods lived. In the Jewish rabbinical tradition the Garden was the blessed part of Sheol where the just awaited resurrection. Early Christians held that humankind was created in an earthly paradise, believed to be the Garden of Eden, and at death would be received into a comparable heavenly paradise where they would dwell with Christ. The Old Testament Song of Solomon and the Mosaic account of creation consequently led in general to the equation of paradise with an enclosed garden.

In the New Testament, the risen Christ appeared to Mary Magdalene as a gardener. Medieval manuscripts were therefore often illuminated with allegorical scenes set in an enclosed garden, where individual flowers and figures represented specific church doctrine. A typical symbolic design was a square enclosure divided into four quarters with a fountain at the center. The straight, raked paths reflected the conduct expected of a Christian.

For Christians, ways of thinking about botanical gardens shifted in tempo with the rise and fall of biblical literalism. Many scholars came to believe in a literal Garden of Eden that had been swept away by the biblical flood. This Edenic garden, it was supposed, could be re-created by bringing back together all the species of the known world. Fragmentary archives indicate that the earliest European gardens were indeed planted to represent the four continents. Entering these would be like visiting the divisions of the earth in succession. This motif of gathering the whole world in one area more or less continues to the present day. By bringing all the plants of the earth together one can study and name them, and by knowing their properties scholars can display power over nature or other lands and peoples.

Most early botanical gardens had a connection to a university and medical school, and hence were useful as well as decorative. An illustration of the Leiden garden in 1601 shows students being taught there. The Dutch East India Company established a garden at the Cape of Good Hope in South Africa around 1686 that supplied fruit and vegetables to sailors. Some gardens displayed curiosities, such as stones and minerals. The one at the University of Uppsala had a menagerie. Plants might be arranged according to their medicinal, morphological (shrub, herb, etc.), or classificatory properties. In Paris citrus trees were arranged in rectangular beds according to their fruit, and during Carl *Linnaeus's lifetime, the garden at Uppsala was planted according to his classification scheme.

During the eighteenth century voyages of exploration to Australia, South Africa, and the South Seas brought many exotic collections to Europe. Living plants could be seen in England at Kew Gardens (royal property until 1841), stately homes such as Chatsworth or Syon Park near Chiswick, and in Paris in the great Jardin des Plantes, which were opened to the public on arranged occasions. Enterprising private groups like the Royal Horticultural Society (1804) sponsored collecting trips abroad and ran public gardens and competitions for their members. In India, the British East India Company took the lead in establishing botanical gardens in Saharanpore and Calcutta. Kew Gardens, under Sir Joseph Banks, became a center of proto-imperial science.

Botanical gardens became highly significant in the nineteenth century. Under the autocratic rule of Banks, followed by the botanical dynasty of William Jackson Hooker, Joseph Dalton Hooker, and William Thiselton-Dyer, Kew Gardens directed a network of colonial gardens, some large, some small, whose curators and botanists Kew appointed, distributed, and rewarded. Their primary function was to supply interesting research plants to London and act as subservient intellectual outposts: a striking example of the center-and-periphery model of colonial activity beloved by Victorian administrators. The textbooks, classification schemes, and colonial floras issued by men like Joseph Hooker and George Bentham depended entirely on the flow of specimens from the colonies. Through their display of trophy specimens and ethos of comprehensiveness, botanical gardens readily supported scientific imperialism. Curators were key figures in developing the plantation economy of the British empire and in moving cash crops from one area to another. Seeds and information were exchanged between gardens and with the local population. By the end of the century, some overseas gardens had become awe-inspiring places, with dramatic palm avenues and lakes glistening with giant water lilies.

Botanical gardens lost much of their sense of purpose in the early years of the twentieth century when the colonial model broke down and the social, practical, and intellectual functions of science shifted. They recovered legitimacy only toward the end of the twentieth century with renewed interest in ecological and environmental movements, and the rise of urgent conservation issues. Kew Gardens, for example, holds a seed bank, and most gardens contribute markedly to research and conservation policy, especially defense against tropical deforestation.

See also CABINETS AND COLLECTIONS.

Lucille Brockway, *Science and Colonial Expansion* (1979). John Prest, *The Garden of Eden: The Botanic Garden and the Re-Creation of Paradise* (1981). Henry Hobhouse, *Seeds of Change* (1985). Donald P. McCracken, *Gardens of Empire: Botanical Institutions of the Victorian British Empire* (1997). Richard Drayton, *Nature's Government: Science, Imperial Britain, and the "Improvement" of the World* (2000).

JANET BROWNE

BOTANY. Much of the earliest botany was concerned with materia medica. An exception, the work of the ancient Greek philosopher Theophrastus, classified plants in the Aristotelian tradition. He included much information on the agricultural techniques of his day. But by far

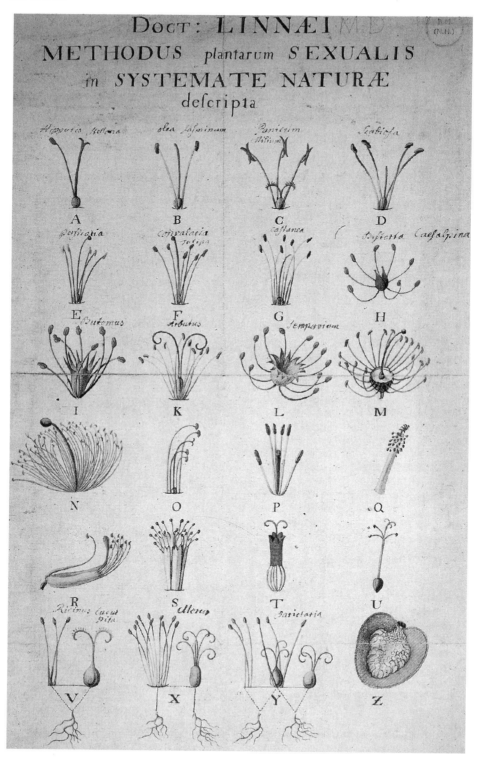

A plate from Carl *Linnaeus' *Systema naturae* (1732). It indicates one of the distinctive features of his system of classification, the distribution of genera into higher taxa by the numbers of stamens and pistils in the flowering parts.

the most famous ancient text was Dioscorides's herbal, a plant encyclopedia written in Greek for physicians. Translated subsequently into Latin and Arabic, and copied and recopied, often with additions and manuscript illustrations because herbalists and apothecaries had to be sure of the plants' identities, Discorides's work underwrote the first printed herbals, particularly those of Otto Brunfels and Leonhart Fuchs. These were expensive books, often hand-colored and annotated, which circulated mainly among physicians, theologians, university dons, and courts.

During the late sixteenth century practitioners increasingly noticed that Arabic or ancient texts did not describe the plants of western Europe and started adding to traditional lists. This enlargement coincided with the colonial expansion of Europe, most notably into the East Indies and the New World, and changing systems of medical education, itself partly a consequence of the new texts available. Plants, remedies, and herbs were introduced to Europe from overseas, encouraging the development of the physic gardens already attached to the main universities, enriching the gardens of the royal courts, and posing questions about geographical distribution. By the end of the century, Rembert Dodoens and William Turner were producing works not so much to supplement but to replace ancient ones. Herbals and botanical compendia continued to be published in large numbers in relatively stereotyped format. Nicholas Culpeper's *English Physician Enlarged; or, The Herbal* (1653) was typical. Through these volumes, countless people learned a utilitarian and domestic combination of medical and botanical science.

During the seventeenth century the study of botany broadened to come into contact with other disciplines besides medicine and classification. The use of the microscope in Britain, Italy, and the Netherlands revealed the unsuspected world of the small. Robert *Hooke saw "pores" in cork, which he called "cells." Nehemiah Grew and Marcello *Malpighi independently identified the principal tissues of the plant body, including stem, root, and leaf. Plant functions were mostly interpreted by analogy with animals. For Grew the xylem vessels resembled animal tracheae; Stephen Hales suggested that plant sap circulated like blood. Grew also identified stamens as the male reproductive organs and supposed that plants possessed two sexes, again like animals. Proof came in the work of Rudolph Jakob Camerarius in 1694, although the sexuality of plants was contested for decades afterwards.

The Swedish doctor Carl *Linnaeus transformed botany in the eighteenth century by devising an "artificial" classification system whereby plants were sorted into higher taxa by the numbers of stamens and pistils they possessed. Linnaeus trained and sent plant collectors all over the increasingly accessible globe, to Japan, South Africa, the Carolinas, Asia, and central Spain. They found many specimens of plants unknown in Europe. Linnaeus also set out philosophical principles for the science. He introduced binomial names, the first denoting the genus, the second the specific character of an individual. Furthermore, Linnaeus helped botanical knowledge move out of the elite sphere of universities, museums, and physic gardens to a broader constituency. Many gentlefolk, women, and working men encountered botany through popularizations of Linnaeus's writings. His devout definition of plant species as stable entities, fixed since their creation by God, stimulated studies of hybridization, which even Linnaeus conceded must occur sometimes. Johann Wolfgang von Goethe proposed that all floral parts had been modified from leaves. Later, Augustin-Pyramus de Candolle formulated useful concepts of symmetry, abortion, modification, and fusion.

During the nineteenth century naturalists turned to the functional anatomy of plants. Robert Brown made important investigations into fertilization before making his better known discoveries of the *cell nucleus and Brownian motion. Jacob Mathias Schleiden laid the groundwork for the cell theory of Theodor *Schwann. Hugo von Mohl summarized advances in cell anatomy in 1851. Meanwhile fundamental observations, including identification of free swimming spermatozoids, were made on the nature of fertilization in algae. René-Joachim-Henri Dutrochet studied osmosis. J. B. J. D. Boussingault elucidated the nitrogen cycle. The essential steps in carbon fixation were recognized when Julius von Sachs showed that the starch present in green cells came from the carbon dioxide absorbed. Gregor *Mendel and Charles Naudin did essential experimental work on plant hybridization. Modern perceptions of the overwhelming importance of evolutionary theory have obscured the significance of these physiological, experimental, and microscopic investigations.

Although evolutionary theory gave the geography and morphology of plants new legitimacy, physiology led the way. Sachs's *Lehrbuch der Botanik* (1868) inspired every plant physiologist with the hope of spending time in Sachs's laboratory in Wurzburg. Sachs made significant studies on plant hormones, growth mechanisms, and tropisms like the effect of gravity or light on roots and shoots. The emphasis of research shifted so much toward the physiological, notably in enzyme studies and fermentation, that Joseph Dalton Hooker in Kew Gardens, England, felt it necessary to reassert the value of taxonomy as an academic discipline in the *Index Kewensis* and his and George Bentham's *Genera Plantarum*. Botany soon became the prime area for genetic research, as in pure-line experiments, *cytology, fertilization, and cell division.

In the twentieth century increasing investigations into fossil plants led to debate over the origin of flowering plants (angiosperms) in the Cretaceous period. This also involved controversy over what should properly be considered primitive. Plant paleontology became an important part of science with the rise of quarternary studies, the study of the geology and environment of the recent archaeological past.

See also BOTANICAL GARDEN.

Julius von Sachs, *History of Botany, 1530–1860* (1890). J. R. Green, *A History of Botany, 1860–1900, Being a Continuation of Sachs' History of Botany 1530–1860* (1909). A. G. Morton, *History of Botanical Science* (1981).

JANET BROWNE

BOVINE SPONGIFORM ENCEPHALOPATHY. See MAD COW DISEASE.

BOYLE, Robert (1627–1691), chemist, natural philosopher.

Robert Boyle was born in Lismore, Ireland, the fourteenth child of Richard Boyle, the Earl of Cork. As the son of the immensely wealthy "Great Earl," Boyle had a privileged upbringing. He was initially educated at home by a tutor, and attended Eton from 1635 to 1638. Soon thereafter he came under the tutorship of Isaac Marcombes, a native of Auvergne, with whom Robert and his brother Francis toured parts of Europe from 1639 until 1642. When the Great Earl experienced a sudden reversal of fortune owing to the Irish Rebellion, Robert went to live with Marcombes in Geneva until 1644, when he returned to England.

Ensconced in the family estate at Stalbridge in Dorset, Boyle bent himself to devotional writing. He composed early versions of *Seraphic Love, The Martyrdom of Theodora*, and other pious reveries. During the second half of the 1640s, Boyle made contact with several members of the loosely organized group of technical and utopian writers inspired by Francis Bacon and clustering around the expatriate "intelligencer" Samuel Hartlib. Although much attention has been focused on the role that Benjamin Worsley, a member of the group, played in Boyle's scientific formation, his first exposure to systematic experimentation occurred at the hands of George Starkey, an emigré from New England who wrote immensely popular chrysopoetic treatises under the pseudonym Eirenaeus Philalethes. From 1650 until the middle of the decade, Boyle acquired from Starkey a full experimental knowledge of Helmontian "chymistry," a discipline that fused mundane chemical pursuits with the quest for such "grand arcana" as the universal dissolvent or alkahest and the philosophers' stone.

After an interlude in Ireland, Boyle moved in 1655–1656 to Oxford, where he came into contact with a group of physicians and natural philosophers who encouraged his pursuit of natural philosophy. Before or around this time, Boyle became a corpuscularian thinker, committed to the idea that matter is composed of discrete particles rather than making up an infinitely divisible continuum. Already in his treatise *Of the Atomicall Philosophy* (c. 1654–1655) Boyle attempted to ground his corpuscular doctrine on the phenomena of the laboratory, an approach that would develop into a lifelong quest. Although Boyle's matter theory pitted him against the mainstream of school philosophy, his most important early source for the experimental verification of corpuscles at the microlevel was Daniel Sennert, a scholastic medical professor and iatrochemist at the University of Wittenberg (*see* IATROCHEMISTRY).

In later works such as *The Sceptical Chymist* (1661) and *The Origine of Formes and Qualities* (1666–1667), Boyle would fuse his experimentally based corpuscular theory with continental versions of mechanism to arrive at "the *mechanical philosophy*"—a programmatic attempt to reduce sensible phenomena to the two "catholick principles" matter and motion. In these and other works, Boyle constructed highly effective experimental means of debunking such scholastic entities as substantial forms and "real qualities" in favor of the interaction of unseen particles. This Baconian program would exercise a major influence on subsequent philosophers and scientists as diverse as John Locke and Isaac *Newton.

During the early 1660s Boyle began publishing a succession of experiments with the *air-pump designed by Robert Hooke at his request. These experiments, which led to the formulation of Boyle's law on the inverse proportionality of volume and pressure in gases, became an important venue for the justification of the Royal Society's experimentalist program against the aprioristic rationalist stance of Thomas Hobbes. Boyle's pneumatic experiments by no means represent the major thrust of his continuing research, however, which focused rather on the justification of the mechanical philosophy by means of experiment.

Boyle's pursuit of the philosophers' stone seems to have reached a sort of climax in the late 1670s, and was probably related to his increasing discomfort with the uneasy relationship between his religion and the mechanical philosophy. Other "chymical" pursuits also continued to engage him, such as the use of indicator tests to divide salts into the three "tribes" of acid, "urinous" (ammonia and its compounds), and "alcalizate" (fixed alkalies such as potassium carbonate). This Helmontian project, mediated by Starkey, led Boyle to some of his most fruitful discoveries in the realm of chemistry, and may have helped set the stage for the increasing emphasis on the chemistry of salts in the subsequent century.

R. E. W. Maddison, *The Life of the Honourable Robert Boyle, F.R.S.* (1969). Steven Shapin and Simon Schaffer, *Leviathan and the Air Pump: Hobbes, Boyle, and the Experimental Life* (1985). Michael Hunter, ed., *Robert Boyle Reconsidered* (1994). Rose-Mary Sargent, *The Diffident Naturalist: Robert Boyle and the Philosophy of Experiment* (1995). Lawrence M. Principe, *The Aspiring Adept: Robert Boyle and his Alchemical Quest* (1998). William R. Newman and Lawrence M. Principe, *Tried in the Fire: Starkey, Boyle, and the Fate of Helmontian Chymistry* (forthcoming).

BILL NEWMAN

BRAHE, Tycho (1546–1601), astronomer.

Tycho once boasted that his observatory on the island of Hven had cost King Frederick II of Denmark more than a ton of gold. Born into a noble Danish family, Tycho was accustomed to walk in corridors of wealth and power, yet he chose a commoner wife and devoted his energies to the construction of an unrivaled scientific establishment. A volume published in 1667 dramatically illustrates the scope of his achievement: in an attempt to catalog all known astronomical observations of planetary and stellar positions, the book devotes 92 pages to pre-Tychonic observations and 65 pages to three decades of post-Tychonic data, compared to an overwhelming 912 pages to Tycho's own measurements.

As a teenager Tycho was impressed by the ability of astronomers to predict an eclipse accurately, but dismayed by their errors in predicting the time of the great conjunction of Jupiter and Saturn in 1563. The appearance of a brilliant new star in 1572 gave him an occasion to make his mark in astronomy. Besides giving the usual astrological interpretations, he demonstrated that the nova lay beyond the moon, that is, in the realm of the eternal and incorruptible stars and planets. His lectures and small booklet on the star, *De nova stella* (1573), procured for him several fiefdoms, including Hven, a small island in the strait north of Copenhagen. There, as he was beginning to build his observatory, he observed the conspicuous comet of 1577, and again, was able to show that it lay in the supralunary sphere.

On Hven Tycho built Uraniborg, his castle of the stars, containing observing decks for his elaborately graduated naked-eye instruments as well as a cellar full of alchemical furnaces. His extensive observations of the sun, moon, planets, and stars led to revised and improved solar and lunar theories as well as to a star catalogue that would finally supersede Ptolemy's. Flushed with his success in establishing large minimum distances to the nova and comet, Tycho undertook an even more ambitious project: to determine the distance to Mars by using evening and morning observations, that is, with a triangulation baseline approximately equal to the diameter of the earth. In hindsight it is clear that the parallactic angle is too tiny for naked-eye detection, but Tycho and his predecessors assumed a distance twenty times too small and therefore a parallactic angle twenty times too large, so he had good reason to believe that his project was feasible. After an initial failure at the close approach of Mars in 1582, he began a major overhaul of his instruments. He built a subterranean observatory, Stjerneborg, on Hven near Uraniborg to give his devices greater stability and protection from the wind.

While making his Martian observations, Tycho discovered the critical role of refraction for observations made with altitudes under 30 degrees. Correcting for refraction with an erroneous table, in 1587 Tycho convinced himself that he had finally found a parallax for Mars. The calculated distance ruled out the traditional earth-centered arrangement of the planets, but rather than adopting the Copernican system, which, according to Tycho, violated both physics and sacred scripture, he postulated an alternative geo-heliocentric system in which the moon and sun circled a fixed central earth while the sun in turn carried a retinue of planets in orbit around it. This he published in his *De mundi aetherei recentioribus phaenomenis* (1588). Subsequently he realized that his refraction correction was erroneous, but he never withdrew his unsubstantiated claim about the Mars parallax or his cosmological Tychonic system. Although he kept silent about his failed campaign to detect the distance to Mars, his treasure trove of detailed observations would provide the precious database for Johannes *Kepler's study of the Martian orbit.

In the mid-1590s Tycho grew increasingly concerned about the future of his observatory, since by Danish law his children by his commoner wife could not inherit his property. Appeals to the young new king Christian IV not only failed, but Tycho lost much of the lavish support to which he had grown accustomed. Deciding to leave Denmark for the Prague court of Emperor Rudolf II, where the inheritance laws were more liberal, in 1597 Tycho packed up his instruments, library, and printing press for the journey to Bohemia. He paused outside the Danish border, but when no call came for his return, he printed up a well-illustrated book, *Astronomiae instauratae mechanica* (*Instruments for the Reform of Astronomy*, 1598), as his calling card for the nobility of the Holy Roman Empire. Welcomed in Prague by Rudolf II, he began setting up his great instruments at Benatky Castle, some distance north of the capital, and there he was joined by Kepler, a young, precariously employed apprentice. A renewed observational program had barely begun when it was interrupted by Rudolf's call to relocate in Prague. Tycho did not last long in Prague, his life possibly shortened by mercurial elixirs brewed in his alchemical laboratory. His partially printed study of the 1572 supernova, *Astronomiae instauratae progymnasmata* (*Exercises for the Reform of Astronomy*), was completed by Kepler and published posthumously in 1602.

J. L. E. Dreyer, *Tycho Brahe* (1890). Victor Thoren, *Lord of Uraniborg* (1990). Owen Gingerich and James Voelkel, "Tycho Brahe's Copernican Campaign," *Journal for the History of Astronomy* 29 (1998): 1–34. John Robert Christianson, *On Tycho's Island: Tycho Brahe and His Assistants, 1570–1601* (2000).

OWEN GINGERICH

BRAIN DRAINS AND PAPERCLIP OPERA-TIONS. Like other forms of labor and capital crucial to economic development in the modern world, scientists and their intellectual capital have moved from nation to nation in the global economy of science. The phenomenon appeared in the early modern period, when proliferating scientific academies under state sponsorship recruited leading scientists from abroad: the Academy of Berlin, for example, in the early eighteenth century lured Pierre de Maupertuis from Paris and Leonhard *Euler from St. Petersburg; and the St. Petersburg Academy drew most of its initial members from Germany and Switzerland.

In the twentieth century flows of intellectual capital increased from trickles to torrents, measured not just in individuals but in dozens and hundreds of scientists. Some migrated to take advantage of professional opportunity, for instance abandoning the backlog of academic jobs in Germany for the growing academic and industrial research system in the United States early in the century. But many scientists were uprooted either as victims of political persecution or as spoils of war claimed by victorious nations. The rise of fascism in the 1930s drove hundreds of scientists from Germany, Austria, Hungary, and Italy, including many of the leading lights of European science. Over thirty nations took in émigrés, but most went to Britain or the United States. Their colleagues tried to find academic jobs for them, whether out of obligation or opportunity, and often succeeded despite the Great Depression and anti-Semitism.

The victors of World War II engaged in a form of intellectual reparations. The United States pursued Project Paperclip, which aimed to bring leading German scientists with their families to America and put them to work on problems of military or economic importance. Paperclip would also deny America's competitors, especially the Soviet Union, the services of the Germans it recruited. Between 1945 and 1952, the United States imported over six hundred technical specialists under Paperclip. Similarly, Great Britain undertook Operation Matchbox to spirit away scientists from the Soviet sphere in Germany; the Soviets meanwhile took Germans east. Some of these postwar expatriates went voluntarily, either out of political conviction or pragmatic recognition that working conditions would be better in the host country than in their ruined homelands.

The prewar brain drain and postwar Paperclip operations helped shift the scientific center of gravity away from central Europe. European émigrés contributed to many important scientific developments, from the electronic computer to molecular biology. Émigrés helped the Allies develop the atomic bomb in World War II, and expatriated scientists after the war brought German rocket technology to the United States and Soviet Union.

Another brain drain flowed from Asia to the United States. The Nationalist Chinese government sent promising students to the United States for graduate training in the 1930s and 1940s; some of them stayed there to take advantage of greater resources or to escape the world war and ensuing civil war in China. The rapprochement between the United States and China in the early 1970s led to another influx of Chinese scientists; by the 1990s some eight hundred scientists and engineers from China had settled in America. The flow of scientific capital from developing countries to the developed world continued through the end of the twentieth century in the labor market for high-tech industry, which relied heavily on foreign workers, especially from South and East Asia.

Donald Fleming and Bernard Bailyn, eds., *The Intellectual Migration: Europe and America, 1930–1960* (1969). Clarence G. Lasby, *Project Paperclip: German Scientists and the Cold War* (1971).

PETER J. WESTWICK

BREEDING. Before the "agricultural revolution," the term "breeding" was associated with aristocratic notions that an individual's worth depended on descent from the founders of the political community. Worth was inborn and guaranteed by divine right. In the eighteenth century, however, agricultural "improvers" began to discuss the variable productivity of individual plants and animals and to improve the technique of "selective" reproduction of superior individuals. During the following century, these practices moved beyond their original agricultural context to the modification of domestic plants and animals, most notably roses, dogs, and pigeons, for aesthetic purposes. They were also predicated ever more explicitly on the assumption that superiority was an intrinsic, but alterable property. The named descendants of Robert Bakewell and Patrick Shireff's bulls and wheats could be crossed with inferior individuals to transfer their remarkable superiority. Worth could be made. Yet, the terms "pedigree" and "inheritance," used by growing numbers of breeders' associations as technical terms, betrayed how the aristocratic connotations of "breeding" underlay these breeders' increasingly formalised practices.

Charles *Darwin famously naturalized pigeon breeders' transforming practices, but he never explained how the random mating of superior and inferior individuals would avoid reversion to the norm. During the late nineteenth century, however, August *Weismann introduced a radical disjunction between the apparent, historically variable characteristics of an organism, and essential ones transmitted unchanged from one generation to the next. This speculative disjunction was taken up by Wilhelm Johannsen and integrated with commercial breeders' practices of "line" breeding—the inbreeding of outstanding plants or animals to fix their characteristics for faithful reproduction in farmers' crops and herds. He then demonstrated experimentally how apparent, historically variable characteristics could be mapped onto combinations of discrete, but hidden and unchanging units of inheritance, which he named "genes."

During the early twentieth century, the emerging science of *genetics provided the ideal tool for the management of the modern nation-state. *Eugenic and agricultural reformers now shared a mutually reinforcing language for the propagation of the "elite," which, significantly, became a technical term in plant and animal breeding. Eugenics was sufficiently controversial that its institutionalization was the exception rather than the norm. Plant and animal breeding, however, enjoyed massive public investment, predicated on the assumption that their potential to improve national agricultural output depended on the systematic application of genetic principles. They

became forms of "applied genetics" and the gene became a networking technology, integrating dispersed farmers into the global flow of industrially processed foods and other biological materials.

Not everyone accepted the transformation of breeding into applied genetics. By focusing on hybridization, the crossing of different elite lines to produce more productive crops and herds, renowned commercial breeders such as Edwin Sloper Beaven argued that genetics ignored the inextricable connection between the characteristics of an organism and its historically specific environment. Eventually, geneticists who maintained the initial close associations with breeders and persisted in focusing on the whole organism took up this line of argument. During the second half of the twentieth century, however, the investigation of the physiological mechanisms linking the gene, the organism, and its environment led to increasing attention to the material constitution of the gene itself. The "gene" was then further uprooted, this time from its historically specific lineage. Plant and animal breeding came to be understood as a form of engineering, giving rise to what Ralph Riley, the director of the Plant Breeding Institute in Cambridge, England, called "genetic engineering."

The economic importance of the movement of increasingly autonomous genes, manifest throughout the twentieth century as breeders sought to extend patent laws to their own "technological" artifacts, vastly increased with the emergence of genetic engineering. Leading multinational corporations such as Monsanto and Unilever acquired the institutional network built around the practice of breeding as it was privatized during the last two decades of the century. The contemporary furor over genetically modified crops, however, is the response to the metropolitan articulation of the global "green revolution" once celebrated by awarding the Nobel Prize for peace to the plant breeder Norman Borlaug and, indirectly, to the Consultative Group on International Agricultural Research. The parallel debates over "designer babies" are symptomatic of growing concern as the welfare state seeks to optimize its expenditures by encouraging the wider adoption of prenatal diagnostic tools emerging from the more recent medical applications of genetic engineering, often in institutions once associated with animal breeding.

The different temporalities of human and agricultural applications of genetic engineering mark the differences that the discourse of genealogical relationship has sought to erase through terms such as "coding," "copying," and "cloning." The notions of lineage and worth on which the original aristocratic connotations of "breeding" were once predicated no longer have any meaning.

Harriet Ritvo, *The Animal Estate* (1987). Jack Kloppenburg, *First the Seed* (1988). Christophe Bonneuil, *Des Savants pour l'Empire* (1991). Sarah Franklin, Celia Lury, and Jackie Stacey, *Global Nature, Global Culture* (2000). Paolo Palladino, *Plants, Patients and the Historian* (2002).

PAOLO PALLADINO

BSE. See MAD COW DISEASE.

BUBBLE CHAMBER. See CLOUD AND BUBBLE CHAMBER.

BUNSEN BURNER. Robert Bunsen (1811–1899) was one of the finest and most versatile chemists of the nineteenth century. When hired by the University of Heidelberg in 1852, he was promised a new laboratory building, which was ready for occupancy in the spring of 1855. Coincidentally, Heidelberg had just begun to light city streets by coal gas, and Bunsen specified that his laboratory should be similarly equipped. At this time, chemists used a variety of fuels for heating: "spirit" (alcohol) lamps, oils, coal, and charcoal. Coal gas had been tried as well, but incomplete combustion produced a flame more notable for its luminosity than its heat.

In the fall of 1854, Bunsen suggested to the university's mechanic, Peter Desaga, a way to obtain a very hot, sootless, nonluminous flame by mixing the gas with air in a controlled fashion before combustion. The flow of the gas could pull in air through apertures at the bottom of a cylindrical burner, the flame igniting at the top. Given this concept, Desaga developed a workable design and had produced fifty burners by the time the new lab opened for business. Bunsen published a description of the burner two years later. It was rapidly and widely adopted. He never sought a patent, in effect donating this important invention to the world of science.

The Bunsen burner, simple, inexpensive, and effective, immediately displaced its predecessors. The easily adjusted flame burned hot and clean, and was perfectly suited to laboratory operations. The present form of the Bunsen burner, familiar to every science student today, has scarcely changed from the original of 1855.

G. Lockemann, "The Centenary of the Bunsen Burner," *Journal of Chemical Education* 33 (1956): 20–21.

A. J. ROCKE

BUREAUS OF STANDARDS. The origin of national and international bureaus of standards lies in the interest in metrology during the late eighteenth and early nineteenth centuries. On the eve of the French Revolution, France had over a thousand legal units of measurement with approximately 250,000 local variations. Spurred on by the work of French and foreign natural philosophers, the French government began to revamp its metrological system. French experimental natural philosophers argued that all of the earlier systems of the ancien régime needed to be disposed of, and a new standard based on precision and simplicity needed to be implemented. In 1790 the French Academy approved the decimal metric system. On 22 June 1799 platinum standards of the meter and the kilogram were placed in the Archives in la République de Paris. The leading German physicist and mathematician of the early nineteenth century, Carl Friedrich *Gauss, promoted the kilogram and meter, as well as the second, as the units of the physical sciences. His measurements of the earth's magnetic force pioneered the use of these three units and their derivatives. Subsequent research by Gauss and Wilhelm Weber on electrical phenomena also employed this set of units.

James Clerk *Maxwell and Joseph John *Thomson furthered Gauss and Weber's research under the aegis of the British Association for the Advancement of Science (BAAS). In 1874 the BAAS announced the CGS system, a unit system based on the mechanical units of the centimeter, gram, and second and adding a set of standardized prefixes, from micro to mega, indicating levels of magnitude. During the 1880s the BAAS and the International Electrical Congress approved a new set of practical units for electricity and magnetism: the ohm for electrical resistance, the volt for the electromotive force, and the ampere for electric current. The schematic origins of Germany's Physikalisch-Technische Reichsanstalt (PTR)—the Imperial Physical Technical Institute for the Experimental Advancement of the Exact Sciences and Precision Technology—can be traced to an essay written by Karl Schellbach, "On the Foundations of a Museum for the Exact Sciences," published in 1872, a year after German unification. The renowned industrialist and engineer Ernst Werner von Siemens was responsible for the actualization of Schellbach's plan. In 1882 Siemens donated the building site that was to house the PTR. The German Reichstag officially approved the PTR on 28 March 1887. It originally comprised physics and technical departments. The physics department was composed of research laboratories for heat, electricity, and optics, while the technical department housed laboratories for precision metrology, heat, and pressure, as well as a second optical laboratory. The first president of the PTR was Germany's premier physicist, Hermann von *Helmholtz.

The PTR not only served as the nation's leading metrological laboratory; its governing committee sought to forge links among scientists, engineers, instrument makers, private entrepreneurs, and government officials. In addition to Siemens and Helmholtz, other leading scientists on the advisory board of the PTR included Ernst Abbe, Rudolf Clausius, and Wilhelm Conrad Röntgen. From the very beginning the objectives of the PTR were fourfold: the execution of scientific experiments that necessitated the collaboration of teams of scientists and engineers; the testing and verification of the properties of materials used in precision-measuring devices; the testing and verification of the uniformity of the components used to produce precision-measuring devices; and the testing and verification of measuring tools.

From 1898 until 1945, the PTR was empowered by the Reichstag to provide the legal units of electricity and to monitor instruments for measuring electric current. In 1923 the Imperial Institute for Weights and Measures was integrated into the PTR, rendering the PTR the sole legal arbiter for monitoring calibration and testing offices. The PTR sustained heavy damage during World War II, and was rebuilt in 1946 as the office for the West Berlin Senate. Built in Brunswick in 1950, the Physikalisch-Technische Bundesanstalt (PTB) became the PTR's successor. The PTB, like the PTR, is totally state supported.

The PTR served as the model for the American National Bureau of Standards (NBS, now called the National Institute of Standards and Technology) founded in 1901 and the British National Physical Laboratory (NPL) founded in 1905. The NBS was the first physical science research laboratory funded by the U.S. government. Its directors have included some of the nation's leading physicists and engineers, including Samuel Wesley Stratton, Lyman J. Briggs, and Lewis M. Branscomb. Throughout its history, the NBS has undertaken research on a broad range of topics including electrical standards, aviation technology, uranium calibration research,

atomic clocks, electronics, and parity. The NPL was created by the Cavendish physicist Richard Tetley Glazebrook. Starting out with a total of twenty-four physicists, the NPL had sixty-three by the outbreak of World War I. Enjoying less federal support than Germany's standards bureaus, the NPL and NBS have historically relied on a mix of government grants, testing fees, and (particularly in the case of the NPL) gifts from individuals and companies.

Perhaps the most important office of international weights and measures is the Bureau International des Poids et Mesures in France (BIMP). This agency ensures the international uniformity of standards. Its authority is granted by the Convention of the Meter, signed in Paris in 1895, which now represents the collaborative work of fifty-one member nations, including all industrialized countries. The BIPM, operating under the exclusive supervision of the International Committee for Weights and Measures, comprises committees whose members belong to the national meteorological laboratories of these member nations. The BIPM currently conducts research in seven principle areas: laser, wavelength, and frequency standards; mass; time; electricity; radiometry and photometry; ionizing radiation; and chemistry. It also undertakes research in thermometry, pressure, and humidity. During the eleventh Conférence Générale des Poids et Mesures in 1960, the International System of Units (SI) was officially recommended as the basis of scientific measurement. Its seven fundamental units are the meter, kilogram, second, ampere, kelvin, mole, and candela.

H. J. Griffin and Erwin Reifler, *A Comparative History of Metrology* (1984). David Cahan, *An Institute for an Empire: The Physikalisch-Technische Reichsanstalt, 1871–1918* (1989). Ronald Edward Zupko, *Revolution in Measurement: Western European Weights and Measures since the Age of Science* (1990). Jed Z. Buchwald, *Scientific Credibility and Technical Standards in 19th and Early 20th Century Germany and Britain* (1996).

MYLES W. JACKSON

C

CABINETS AND COLLECTIONS. Natural history cabinets and collections took shape in two forms during the Renaissance: the cabinet of curiosities and the botanical garden. The cabinet of curiosities, or *Wunderkammer*, was a repository of objects noteworthy to the Renaissance philosopher: rarities, exotica (from Asia or the New World), monstrosities, artifacts, natural objects, and items of historical interest. The great Renaissance collectors included Ulisse Aldrovandi, whose museum (which he bequeathed to the city of Bologna), or *teatro* or *microcosmo*, as he variously termed it, contained a dragon; and the German Jesuit Athanasius Kircher, whose collection in Rome reflected not only his interests in natural philosophy but also in Egyptian hieroglyphics. Owners restricted access to a discriminating (and carefully chosen) public of fellow scholars and aristocratic or clerical patrons. Aldrovandi kept a ledger of visitors, among them one pope, six princes, twenty-one professors, eighty-seven doctors, six painters, and a notary. Although a few noblewomen visited his museum, they were not enumerated in the roll.

The physical composition and arrangement of an early modern cabinet of curiosities was an idiosyncrasy of its owner. In general, no clear differentiation could be made between objects displayed and the means of display, as cupboards, shelves, drawers, mountings, and frames—the ceiling also served for and as display—were splendid objects in themselves, fully integrated with their contents. Within the room or rooms, objects followed one another according to the three kingdoms of nature, animal, vegetable, and mineral; the importance to collectors of objects like cameos and worked baroque pearls made the boundaries between the artificial and the natural relatively porous and unmeaning.

Besides the splendid display of material objects, the cabinet had another dimension: the catalog, which ordered and described its contents in a form portable beyond its walls. The Dutch merchant and manufacturer Levinus Vincent, whose cabinet had been visited by Peter the Great and Carlos III of Spain, published in 1719 a splendid catalog depicting and describing his holdings. The collection of the Italian antiquary Cassiano dal Pozzo consisted entirely of representations rather than objects: over seven thousand drawings of classical and medieval antiquities, fossils, and botanical, zoological, and horticultural subjects. Several theoreticians of collecting, including the German Samuel Quiccheberg, in his *Inscriptiones vel tituli theatri amplissimi* (1565), discussed the relation of the material cabinet and the idealized catalog.

Botanical gardens began in sixteenth-century Italy as adjuncts to the teaching of medicine and evolved into collections not just of plants useful in *materia medica* but of botanical diversity in general. Over time they developed into repositories of botanical diversity for its own sake. By the mid-eighteenth century the garden was paired with the herbarium, another of the great genres of modern collecting. Botanical gardens through the eighteenth century were laid out almost exclusively along medical or taxonomic lines (according, for example, to the system of Joseph Pitton de Tournefort, Bernard de Jussieu, or Carl *Linnaeus), with little concession to horticultural (rather than architectural) aesthetics. All the earliest gardens (in Padua, Pisa, Leyden, Oxford, and Montpellier) and most later ones were associated with universities, although governments maintained some (the Jardin du Roi, later Jardin des Plantes, Paris) and private individuals others (George Clifford, Holland). In all cases, they restricted admission, generally to students and well-to-do patrons.

From the early eighteenth century, the contents of collections and their administration became ever more standardized. To return to botany, herbarium sheets of the seventeenth and early eighteenth centuries containing pressed and dried plant specimens were often ornamented—dried plants spilling from engraved flowerpots, for example—and had no standard size. Each collector's system of preparation, labeling, sorting, and storage was unique. The growth of networks of botanical collectors and systematists and of widely shared taxonomic systems fostered standardization in the second half of the eighteenth century. The most important botanical authority,

The "museum" of Ferrante Imperato, from his *Dell'historia naturale* (1672). The illustration presents a relatively coherent collection; many of the period included instruments and precious objects besides books and natural history specimens.

Linnaeus, succeeded in imposing his standard size folio herbarium sheet on most of his correspondents, and he ruthlessly cut down sheets he received to the size of his own cupboards, heedless of ornament or aesthetics. At the same time, administrative techniques drawn from commerce and government regulated the contents of collections and kept track of exchanges and purchases among them. The eighteenth century also saw the purchase and transfer of whole collections, like Sir Hans Sloane's assortment of natural objects and antiquarian and ethnological material, bought by the British government in 1753, and Linnaeus's herbarium, sold to J. E. Smith of London in 1784. The first collection formed the core of the British Museum; the second ended up in the Linnean Society, founded in 1788. This new genre of collection, systematic rather than aesthetic, transferable, and increasingly standardized, had almost completely supplanted the cabinet of curiosities by the turn of the nineteenth century. Even amateur collectors of seashells, plants, birds' eggs, and other hobbyists came

to order their collections along scientific lines, labeled according to the latest systematics and stored out of sight in boxes, cases, and cupboards analogous to those used in scientific institutions and university collections.

The nineteenth century opened a new era in the typology of collections, still recognizable today and marked by differentiation into three genres: systematic, applied, and aesthetic and didactic. The work of classifying the world's plants and animals begun in the Renaissance went on, but it did so behind the scenes, in herbaria and zoological museums open only to accredited scholars and invisible to the public. The institutions housing this work—Kew Gardens near London, the Jardin des Plantes and Muséum d'Histoire Naturelle in Paris, the Schönbrunn Gardens in Vienna, and later their colonial satellites in India, Java, Mauritius, Jamaica, and elsewhere—also carried out imperial schemes of acclimatization, hybridization, and plant and animal transplantation. This work was linked to surveys important for systematics and provided revenues both for the gardens and their overseeing governments. Finally, as part of larger nineteenth-century movements for public health, education, and recreation, these institutions came to serve as teaching establishments and places of entertainment for an ever-widening general public.

Parallel to the enclosure of the herbarium as a de-aestheticized, private, and academic space was the opening of the botanical garden as a public park. This theoretical reconception accompanied a shift from the rigidly rectilinear garden plans of the seventeenth and eighteenth centuries taking taxonomy as their planting principle, to more openly aesthetic designs drawing on the picturesque eighteenth- and nineteenth-century English landscape style and organized by geographical or climatic constellations of flora. Guidebooks, identifying labels, and public lectures furthered the education of visitors. Likewise, modern zoological gardens, with their combination of the spectacular and the educational, taking their cue from the zoo founded in Regent's Park, London, in 1828, served an important didactic and recreational function in major European cities. Within doors, the nineteenth century's great museums likewise split their functions between public spectacle and instruction—competing in this in the later part of the century with the great exhibitions—and restricted scholarly collections.

In the twentieth century, this tripartite division continued to hold despite profound changes: transformations in biological theory and systematics (particularly the theory of evolution); political shifts (especially the collapse of the European empires, which altered the context of agricultural interventions); and changing fashions in pedagogical theory and museum display. The modern herbarium, zoo, and science museum would be quite recognizable to a visitor from the mid-nineteenth century, whereas the early modern collector of curiosities would be baffled in spaces epistemologically so foreign. Nevertheless, some new features and types of collections came into existence in the twentieth century. The rise of ecology and conservation biology have led to a return to the collection of biological (and ethnological) diversity for its own sake, reminiscent of the catchall collecting of the sixteenth and seventeenth centuries. The computer has increasingly made information the object of collection, in public media and forums like the World Wide Web. These collections include the genomes of fruit flies, *Arabidopsis*, and human beings, and the widely accessible electronic catalogs of great libraries, themselves collected into metacatalogs like the Karlsruher Virtueller Katalog ⟨http://www.ubka.uni-karlsruhe.de/hylib/virtueller_katalog.html⟩.

André Malraux, *Museum without Walls*, trans. Stuart Gilbert and Francis Price (1967). Oliver Impey and Arthur MacGregor, eds., *The Origins of Museums: The Cabinet of Curiosities in Sixteenth and Seventeenth-Century Europe* (1985). Susan Sheets-Pyenson, *Cathedrals of Science: The Development of Colonial Natural History Museums during the Late Nineteenth Century* (1988). Eilean Hooper-Greenhill, *Museums and the Shaping of Knowledge* (1992). Paula Findlen, *Possessing Nature: Museums, Collecting, and Scientific Culture in Early Modern Italy* (1994). Horst Bredekamp, *The Lure of Antiquity and the Cult of the Machine: The Kunstkammer and the Evolution of Nature, Art, and Technology*, trans. Allison Brown (1995).

A. J. LUSTIG

CAJAL, Santiago Ramon. See GOLGI, CAMILLO, AND SANTIAGO RAMON CAJAL.

CALCULATOR. Although calculation had long been a part of the daily routine of commercial life, it was two natural philosophers of the seventeenth century, Blaise Pascal and Gottfried Leibniz, who invented the first machines that could carry out arithmetic processes automatically. Pascal's could add and subtract; Leibniz's did well at addition and multiplication. Although considered mechanical marvels, these devices had little immediate impact on the way people did arithmetic. Of greater practical importance was John Napier's discovery of logarithms, which made it possible to multiply or divide by addition or subtraction. The properties of logarithms were embodied in the slide rule, which soon found use in calculations relating to taxation and, during the nineteenth century, in all sorts of problems considered by scientists and engineers.

In the second half of the nineteenth century, machines that could carry out arithmetic operations came into use in commerce. The most successful of them did addition best. A few—notably the Swiss-made Millionaire—could multiply directly rather than by repeated addition, and were of particular use to scientists. Observatories and laboratories hired people called computers to do routine calculations by hand, using published mathematical tables, and by machine.

Nineteenth-century scientists and mathematicians had little to do with the invention of calculating machines. They took a more active role in the development of difference engines, machines that computed values of functions using the method of finite differences. Charles Babbage envisioned and Georg and Edvard Scheutz successfully built a difference engine. One Scheutz difference engine was used briefly at the Dudley Observatory in Albany, New York.

Instruments for carrying out more complex mathematical operations also received the attention of scholars. In 1855 James Clerk *Maxwell invented a form of planimeter, a device for finding the area bounded by a closed curve. Planimeters had applications to many scientific and engineering problems, for example, finding the work done by a steam engine from an indicator diagram. In 1876 William *Thomson, Lord Kelvin, proposed an instrument for representing a curve as the sum of terms in a harmonic series. The British long used Kelvin's harmonic analyzer and its descendants to predict the tides. Kelvin's instrument inspired several imitators, such as the predictor, designed by William Ferrel of the U.S. Coast and Geodetic Survey in 1880. Larger versions of this machine operated in the U.S. into the 1960s.

In the 1930s, Vannevar Bush at the Massachusetts Institute of Technology developed a much more complex, room-sized mechanical device for solving differential equations. To use Bush's differential analyzer, as well as similar instruments built in Great Britain and Norway, the operator had first to assemble an array of gears, shafts, disk integrators, and electric motors and to enter data by tracing a curve on an input table. The result appeared as a curve on an output table.

During the 1920s and 1930s, punched card equipment, initially developed for census tabulation and then applied to accounting work, came into the social sciences and astronomy. Centers devoted to computation developed at government agencies and universities, particularly Columbia in New York City and Iowa State in Ames. In the late 1930s, a physics graduate student at Harvard University, Howard Aiken, conceived a calculator that could be programmed to carry out calculations. With the help of Harvard faculty, he persuaded International Business Machines (IBM), the most successful maker of punched card equipment, to build the calculator and donate it to Harvard. The machine, called the Automatic Sequence Controlled Calculator, or Mark I, calculated for the U.S. Navy when completed during World War II. As with IBM accounting equipment, calculations were carried out using electromechanical relays. Germany also built relay calculators. During the same period, physicist John Mauchly and engineer J. Presper Eckert designed and built the first general purpose electronic computer, the ENIAC, at the University of Pennsylvania. All of these machines filled a room.

By the 1960s, the introduction of smaller, more reliable electronic components made it possible to make electronic calculators that fit on a desktop. Those introduced by the British firm of Sumlock

Comptometer and sold under the name "Anita" used vacuum tubes. In 1964, Sony exhibited a prototype transistorized calculator at the New York World's Fair. Sony and other companies, such as Sharp and Wang Laboratories, soon were selling desktop electronic devices. They cost several thousand dollars, but numerous scientists used them.

In the 1970s, the introduction of the integrated circuit made it possible to build still smaller, hand-held electronic calculators. These initially cost several hundred dollars, but the price quickly fell. The hand-held calculator soon displaced the slide rule and many printed tables. Some early electronic calculators could only do business arithmetic. Others, intended for scientists and engineers, could compute trigonometric functions, exponents, and statistical functions. Early makers of hand-held calculators included the Japanese firm of Busicom and such American companies as Texas Instruments and Hewlett Packard.

As integrated circuits became more powerful, so did calculators. In the mid-1980s the Japanese firm Casio introduced a hand-held calculator that could graph simple functions. Hewlett Packard, Texas Instruments, and other companies soon followed suit; graphing calculators are now widely used in the classroom as well as the laboratory. The more expensive forms of these machines can do symbolic manipulations—not only simple algebra, but also calculus. The programs used in these calculators rely heavily on the results of university mathematicians and scientists, but the devices themselves, like nineteenth-century calculating machines, are very much commercial products.

See also ANALOG/DIGITAL DEVICES; COMPUTER.

William Aspray, ed., *Computing before Computers* (1990). Mary Croarken, *Early Scientific Computing in Britain* (1990). James W. Cortada, *Before the Computer: IBM, NCR, Burroughs, and Remington Rand and the Industry They Created 1865–1956* (1993). Michael R. Williams, *A History of Computing Technology* (1997).

PEGGY ALDRICH KIDWELL

CALORIMETER. See INSTRUMENTS, HEAT AND LIGHT.

CAMERALISM was the science of managing the state's resources in early modern German principalities. Originally concerned only with public finance, cameralists broadened their purview at the end of the Thirty Years' War in 1648 in order to restore order in both state and society. By 1750 cameral science embraced the administration of all the state's resources—water, agriculture, forests, buildings, roads, population, mines, coinage, weights and measures, the cadaster, and the commons—as well as the management, protection, and disciplining of the population in daily life. To create administrative policy, cameralists drew upon knowledge of chemistry, forestry, geography, natural history, physics, technology, mining, medicine, mathematics, and other sciences.

Taught at universities from 1727, cameral science developed theories of the common good and of the use of technical skills and material resources for the happiness of all. One of its practical aims was the resolution of conflicts between royal prerogative and social demands. By late in the century, cameral science inspired the formation of local physicoeconomic societies. These grassroots organizations reflected the public's growing interest in using the natural sciences, medicine, and technology to manage resources. They also served as seedbeds for generating patriotic sentiment.

Several major cameralists wrote on scientific, technical, or mathematical matters. Johann Becher published on chemistry, medicine, and other sciences. Christian von Wolff was known for his mathematical textbooks and compendia. Johann Heinrich Gottlob von Justi published on manufactories, mining, and agriculture as well as chemistry, mineralogy, and natural history.

Embracing both Baroque classificatory schemes for organizing knowledge and the emancipatory goals of Enlightenment reason, cameralists integrated natural knowledge into daily life and state practice. They promoted the study of the chemical, mechanical, and natural sciences among manufacturers and artisans and viewed technology as a basis for reform, repair, and the reinstatement of social order. They drew parallels between the certainty of mathematics and the laws of nature, on the one hand, and, on the other, the certainty they thought could exist in the state's laws, especially those based on experience. Interested in measurement, especially the exactitude of weights and measures, cameralists helped to shape an ethos of precision in German society.

By the end of the eighteenth century, as the sciences became more specialized and an ideal unity of knowledge less sustainable, subjects that formerly united under the umbrella of cameralist science moved into one of two camps: either the sciences of state (law, economy, politics, and history) or the natural sciences and engineering technologies, especially agriculture and mining. Vestiges of the cameralist tradition of using natural knowledge for the common good can be seen in the public service work of leading astronomers in the early nineteenth century, such as that of

Carl Friedrich *Gauss, Christian Ludwig Gerling, and Friedrich Wilhelm Bessel on weights and measures and cadastral reform.

Henry Lowood, *Patriotism, Profit, and the Promotion of Science in the German Enlightenment: The Economic and Scientific Societies, 1760–1815* (1991). David Lindenfeld, *The Practical Imagination: The German Sciences of State in the Nineteenth Century* (1997).

KATHRYN OLESKO

CANCER RESEARCH. At the end of the nineteenth century, little was known about the causes of cancer, but the prospects for learning more appeared to be brightening with the emergence of evidence that at least some cancers might be produced by infectious microorganisms. Scientists in many laboratories searched for such cancer-causing agents without success until 1909, when a farmer brought Peyton Rous, a young biologist on the staff of the Rockefeller Institute for Medical Research, a chicken with a large sarcoma-type tumor protruding from its right breast. Rous obtained a cell-free extract from the tumor by mincing and then filtering the tissue. He then injected a solution of the extract into healthy chickens of the same breed and found that they, too, developed sarcomas. Rous contended that the tumors might have been stimulated by a "minute parasitic organism" carried in the extract—perhaps a virus.

Other scientists using Rous's methods failed to induce tumors in mice, rats, rabbits, or dogs. By the 1930s the theory that viruses cause cancers had fallen into deep disrepute. John Bittner, a biologist at the Jackson Laboratory in Bar Harbor, Maine, found that a tendency to breast cancer in certain strains of mice could be transmitted from mothers to foster children—that is, independently of the animals' genetic makeup. Bittner suspected that the mothers transmitted a virus to the infant mice when they suckled. Reluctant to challenge the prevailing antiviral paradigm of oncogenesis, he called the agent involved a "milk factor." His suspicion was later validated by recognition of the mouse mammary tumor virus (MMTV).

In the scientific community at large, theories of viral oncogenesis were revived by Ludwik Gross, a refugee from Poland who joined the staff of the Veterans Hospital in the Bronx. In 1944, Gross began injecting healthy mice with an extract obtained from ground-up organs of adult leukemic ones. He failed to induce leukemia in the recipient mice until, probably in 1949, he tried injecting the extract into newborn mice. They all developed the disease within two weeks. In the 1950s, scientists in America and Europe obtained filtrates from a large variety of tumors, injected them into newborn mice, and isolated an abundance of viruses that provoked tumors in many species, including hamsters, rats, apes, and cats. By the early 1960s, animal tumor virology had become a major branch of basic medical and biological science. In 1966, at the age of eighty-five, Rous shared the Nobel Prize in physiology or medicine.

Knowledge that viruses are DNA or RNA wrapped in a protein coat allowed tumor virologists to begin reaching into the mystery of what occurs when a virus transforms a normal cell into a cancerous one. In the early 1960s, Renato Dulbecco, at the Salk Institute in La Jolla, California, demonstrated that when a virus invades a cell, its genetic material is incorporated into the cell's native DNA, with the result that it perverts the cell's machinery of regulated growth, and causes it to multiply malignantly. Dulbecco worked with DNA viruses. But Howard Temin, a former student of Dulbecco now at the University of Wisconsin, ran into difficulties pursuing a similar line of research using the Rous sarcoma virus, whose genetic material is RNA. RNA is synthesized using the code that DNA contains, but according to the then-prevailing central dogma in molecular biology, it could not generate DNA and, hence, could not integrate its genetic information into the DNA of a cell.

Temin contended that the central dogma must at least in part be wrong, that Rous viral RNA somehow generated DNA complementary to itself that could integrate into the DNA of the cell. Although ridiculed for his claim, Temin pursued it experimentally through the 1960s, focusing on showing that such integration occurred. In 1969 he turned to a consideration of how the virus's RNA could be physically made into complementary DNA. In 1970, he found the answer—an enzyme that catalyzes the synthesis of DNA from RNA.

The same discovery was made simultaneously and independently by David Baltimore, a member of the MIT faculty who since his graduate-school days had been working on the genetic systems of RNA viruses. In 1970, Temin and Baltimore reported, in separate, back-to-back articles published in *Nature*, the discovery of the enzyme, promptly dubbed "reverse transcriptase" in recognition of its ability to transcribe RNA back into DNA. At the end of his paper announcing the enzyme, Temin noted that the light it cast on how retroviruses reproduce raised "strong implications for theories of viral carcinogenesis."

No virus had been shown to cause cancer in human beings, but the rapid development of tumor virology had convinced a coalition of scientists and influential laypeople to agitate for a stepped-up federal cancer program. In 1971, the agitation yielded the National Cancer Act, inaugurating a "War on Cancer" that would more than triple the budget of the National Cancer Institute by 1976. Biomedical scientists managed to divert resources from the war to research into the interaction of tumor viruses with the cell. They argued that victory over cancer would come not with searching for human tumor viruses but from understanding how the integration of a virus's genetic information into a cell's DNA transformed the cell into a source of malignancy.

In 1969, the biologists Robert J. Huebner and George J. Todaro, addressing the transformation puzzle, theorized that the cells of many, if not all vertebrates must naturally contain two kinds of cancer-related DNA: "virogenes," to create RNA viruses, and "oncogenes," a term they coined to denote genes with the power to transform normal cells into tumor cells. They speculated that both genetic substances lie latent in the cell, having entered it by some ancient infection, until activated by natural causes or environmental carcinogens.

In the early 1970s, J. Michael Bishop and Harold E. Varmus, both young faculty members of the medical school of the University of California in San Francisco, began investigating the virogene/oncogene hypothesis. They sought to determine whether any stretch of DNA in a normal chicken cell resembles the Rous virus's transforming gene, a tentative identification of which had recently been established by several other scientists in the San Francisco area. Bishop and Varmus dubbed the transforming viral RNA the "sarc" gene and sought to determine whether the viral "sarc" gene has a cousin—scientists call it a homologue—in the DNA of a normal chicken.

Between 1976 and 1978, the Bishop–Varmus group found that the viral "sarc" fragment is homologous not only to a region in the DNA of chickens but also to regions in the DNA of quail, turkeys, ducks, emus, calves, mice, and salmon. They even had detected evidence of them in human DNA. "Sarc" seemed to have cousins in DNA everywhere. The "sarc" homologues were obviously not the oncogenes—the latent oncogenic DNA—that Huebner and Todaro had originally proposed. Evolutionary reasoning helped identify them. The major groups of species—birds, mammals, and fish—that carry the "sarc" cousins had separated at least 400 million years earlier. To Bishop and Varmus, the plain evidence that the "sarc" homologues had been conserved through so much time and speciation indicated that they might be involved in some critical cellular function such as growth and development. They appeared to be normal genes that can be turned into oncogenes by viral action or by a physical or chemical agent.

In 1979, the laboratory of Robert Allan Weinberg at MIT reported that normal cellular DNA could indeed be transformed into oncogenic DNA with a chemical carcinogen. Soon, experiments in other laboratories detected transformed DNA—DNA changed from that of a normal cellular gene—in a variety of cancer cells, including carcinomas taken from rabbits, rats, mice, and people. Most of these cellular genes seem to exist all over the tree of animal evolution, just like the cousins to the viral "sarc" gene (now designated *src*, in conformity with the classical rules of three-letter genetic nomenclature). Like the *src* gene, these other cellular genes probably have a role in the fundamental cellular processes of growth, regulation, and differentiation. They are thus normal cellular genes that can be turned into oncogenes by chance processes within the cell as well as by environmental carcinogens, including tobacco smoke. These "proto-oncogenes"—potential agents of cancer—comprise, as Bishop has said, a kind of "enemy within."

The multiple steps that led to the discovery of oncogenes revolutionized cancer research and widened opportunities for the study of normal cell growth and regulation. Parts of the work earned Nobel Prizes, including one, in 1975, for Dulbecco, Baltimore, and Temin and another, in 1989, for Bishop and Varmus. Coupled with increasing knowledge of the oncogenic role of environmental carcinogens, the oncogene revolution has shown that cancer can arise from genes alone or from interactions between our genes and what we ingest, inhale, or encounter.

James Patterson, *The Dread Disease: Cancer and Modern American Culture* (1987). Natalie Angier, *Natural Obsessions: Striving to Unlock the Deepest Secrets of the Cancer Cell* (1988). Daniel J. Kevles, "Pursuing the Unpopular: A History of Courage, Viruses, and Cancer," in *Hidden Histories of Science*, ed. Robert B. Silvers (1995): 69–112. Robert Proctor, *Cancer Wars: How Politics Shapes What We Know and Don't Know About Cancer* (1995). Robert A. Weinberg, *Racing to the Beginning of the Road: The Search for the Origin of Cancer* (1996). Michel Morange, *A History of Molecular Biology*, trans. Matthew Cobb (1998).

D. J. KEVLES

CARBOHYDRATES AND FATS.

Sugar, starch, fats, and oils have been prepared, consumed, and traded since the birth of civilization. The initial steps toward a chemical understanding of these materials awaited Antoine-Laurent *Lavoiser's chemical revolution, John *Dalton's atomic theory, and the earliest methods of elemental organic analysis.

During the first thirty years of the nineteenth century, chemists isolated and characterized sucrose (table sugar), glucose, fructose, starch, cellulose, gum arabic, and many other "carbohydrates." The name derived from the observation by Jospeh Louis Gay-Lussac and Louis Jacques Thenard in 1810–1811 that the formulas of all of the members of this class of organic compounds could be reduced to the generic formula $C_n(H_2O)_n$. In 1812, Konstantin Sigizmundovic Kirchhoff made the shrewd (and industrially important) observation that starch treated with dilute sulfuric acid yields glucose.

Over the next decade, Michel Eugène Chevreul investigated fats and oils. When he saponified these substances (converted them into soap by treatment with potash or soda), he found that the fat split into glycerin and a mixture of complex organic acids. Fats, therefore, must be a kind of natural salt formed between glycerin and various "fatty acids," and soapmaking represents a replacement of the glycerin by potash or soda. A generation later (in the 1850s), Marcellin Berthelot developed a more detailed understanding of this relationship. Fats and oils, he showed, are triglycerides—triesters of fatty acids with glycerin, a trialcohol—and saponification is simple hydrolysis (splitting apart) of these esters. Berthelot's work opened the door to the preparation of a huge number of natural and artificial triglycerides, as well as mono- and diglycerides. Some of these new compounds later became commercially important.

Berthelot also studied sugars. He found that, like glycerin, they had multiple hydroxyl groups, but that they also had aldehyde or ketone reactions. Several important simple sugars—such as glucose, fructose, and galactose—had the same molecular formula. With the rise of *stereochemistry in the late 1870s and 1880s, it became clear that these relatively simple organic molecules had no fewer than four centers of asymmetry, so that this one single structural formula corresponded to sixteen possible stereoisomers, each of which would have distinct chemical and physical properties.

During the 1880s and 1890s, Emil *Fischer demonstrated the spatial configurations of several of these simple sugars, including glucose. No one had ever before carried out such a complex series of structural analyses, and Fischer's investigation remains one of the most impressive accomplishments in the history of the science. Fischer also showed that polysaccharides such as cellulose and starch consisted of glucose units linked together to form a natural carbohydrate polymer.

The structures and chemistry of the sugars turned out to be extraordinarily complex. For example, glucose exists normally in a ring structure, and the precise character of the linkage of glucose and fructose to form sucrose proved to be exceedingly difficult to unravel. A full understanding came only over several additional decades and from the efforts of many organic chemists.

A. J. Ihde, *Development of Modern Chemistry* (1964). J. R. Partington, *A History of Chemistry*, vol. 4 (1964).

A. J. ROCKE

CAREERS IN SCIENCE. See SCIENTIST.

CARNOT, Sadi (1796–1832), and **Augustin FRESNEL** (1788–1827), physicists.

On 10 May 1788, the year before the French Revolution began, Fresnel was born in Broglie in Normandy. His father, an architect, and his mother, a daughter of the manager of the local chateau, had religious and Royalist sympathies. Revolutionary turmoil soon forced them to move. On 1 June 1796 Carnot was born at the Palais de Luxembourg to Lazare Carnot, one of the post-Thermidorean directors, a surviving member of the Committee on Public Safety, and an organizer of the successful military campaign in 1793 against Austrian and Dutch forces.

Fresnel entered the new Central School at Caen in 1799, where he was well trained in mathematics and the sciences. Sadi Carnot was educated at home by his father until he was sixteen. Both Fresnel (in 1804) and Carnot (in 1812) did well in the admissions examinations for the École Polytechnique; Fresnel placed seventeenth and Carnot twenty-fourth, each out of nearly two hundred applicants. Both entered the engineering corps after graduation; Fresnel joined Ponts et Chausées (Bridges and Roads), and Carnot (after two years at the Artillery and Engineering School at Metz) the Royal Corps of Engineers. True to family sympathies (and like many others at Ponts et Chausées), Fresnel resisted the return of Napoleon from Elba in 1815 and was sent into home exile. Both men spent much of their careers in Paris and both died young: Fresnel, always physically weak, of tuberculosis, and Carnot probably of cholera.

Each man produced novel though unequally influential physical theories. Beginning in 1816 and extending over nearly seven years, Fresnel developed the mathematical, theoretical, and experimental foundations of wave optics. By the early 1830s his principles and devices had spread widely. Carnot's book, *Reflections on the Motive Power of Fire* (1824), provided for the first time a general theoretical system for understanding the conditions under which the efficiency of heat engines could be maximized. His scheme

lay fallow for over two decades, until William *Thomson, during a sojourn at Victor Regnault's laboratory in Paris, learned of it in 1845. Carnot's theory, suitably adjusted, became central to the development of thermodynamics by Thomson and Rudolf Clausius.

When Fresnel began his innovative work in 1815, several *physiciens* had been exploring novel optical phenomena for about half a decade as a result of Étienne Louis Malus's discovery of polarization in 1809. François Arago and Jean-Baptiste Biot (like Malus, graduates of the École Polytechnique) had exchanged bitter words over priority issues concerning the colors produced by polarized light that passes through thin crystals. This topic (chromatic polarization) lay at the center of Parisian interest in optics. Fresnel knew nothing about the dispute or the details of polarization. His interest in optics had been stimulated by difficulties he perceived in the contemporary French understanding of heat and light as fluids consisting of minute particles subject to various forces.

Fresnel probably met Arago at the École Polytechnique, but his first substantial encounter with him took place in Paris at a dinner party given by Fresnel's maternal uncle, Léonor Merimée. At home in the countryside, and apparently ignorant of previous work in the area by Christiaan Huygens and Thomas Young, Fresnel developed a theory of light as a wave in a universally present medium (or *ether), according to which striped patterns would form when light passes through narrow slits or near edges. This phenomenon of diffraction, known since the seventeenth century and discussed by Isaac *Newton in the *Opticks*, inspired Fresnel to create a mathematical system tightly connected with persuasive experiments that eventually embraced the influential topic of polarization.

Arago acted as Fresnel's patron and collaborator, and, through his access to the centers of power, stimulated ongoing interest in Fresnel's work, which culminated in 1818 in Fresnel's winning an Academy of Sciences prize contest on diffraction. Supported by impressive new experiments, quantitative, bound to a subject (polarization) of intense contemporary interest to powerful people like Pierre-Simon, Marquis de *Laplace, and pushed forward by the aggressively political Arago, Fresnel's new wave optics spread rapidly in French circles and soon thereafter throughout Europe.

Carnot's theory of heat engines had a different fate. Despite or perhaps because of the fame and power his father had wielded before Napoleon, Carnot did not have the access to the centers of Parisian science that Arago (despite his republican sympathies) and Merimée gave Fresnel. Moreover, Carnot's subject did not connect with the main interests of his contemporaries. His system was not based on any novel physical concept of heat (which might have excited curiosity), but rather on an obscure analogy between the transfer of heat by means of an engine from a hot to a cold source and the fall of water from a high to a low place. Carnot's father had considered the requirements that ordinary machines must satisfy to produce the greatest possible effect under given circumstances and had clarified how that effect should be estimated. His son appropriated several of Lazare Carnot's notions and produced a theory based on the principal axioms that, for maximum effect, no heat must flow directly from hot to cold but must rather be transferred through an engine; and that the total quantity of heat is always conserved. Carnot doubted the truth of this second requirement, but could never restructure his theory to do without it.

The principles of Fresnel's wave optics endure essentially unchanged, despite the photon, because many optical instruments are based upon them. Carnot's system has no comparable legacy, except indirectly through the principles of thermodynamics, which stabilized during the late 1850s and 1860s. This difference may be explained as follows. Unlike Fresnel, Carnot had neither a patron nor the skill to turn his scientific knowledge to practical success. (Fresnel invented the layered lens and associated lighting elements that transformed coastal illumination in France and elsewhere.) Carnot did not develop his system to a high mathematical level, perhaps because he wished to reach an audience of working engineers, who, however, gleaned little new from it. Nonetheless Carnot, like Fresnel, was a signal contributor to the creation of *physics during Napoleonic and Restoration France.

D. S. L. Cardwell, *From Watt to Clausius* (1971). Sadi Carnot, *Reflexions on the Motive Power of Fire: A Critical Edition with the Surviving Manuscripts*, ed. Robert Fox (1986). Jed Buchwald, *The Rise of the Wave Theory of Light* (1989).

JED Z. BUCHWALD

CARTE DU CIEL. Astronomical photography, for which François Arago foresaw a promising future in January 1839, was soon afterwards inaugurated by Henry Draper, Armand-Hippolyte-Louis Fizeau, and Léon Foucault. At the Cape of Good Hope in 1882, David Gill succeeded in photographing a very bright comet with many stars clearly visible in the field of view. He sent his photograph to Ernest Barthélémy Mouchez, director of the Paris Observatory, who, when presenting it to the Académie des Sciences, proposed a scheme for the photographic production of celestial charts, an idea that had been advanced twenty-five years previously by Warren De la Rue.

At the Paris Observatory, the brothers Paul and Prosper Henry installed a successful astrograph in 1885. Mouchez, with the help of Gill, organized an international congress in 1887 to establish a project for a general photographic chart of the heavens. Sixteen observatories established the proposal; the number rose to eighteen in 1889. In 1896 a further agreement was reached on a unified system of astronomical constants.

The work of the Carte du Ciel (the French name being used by all the participants) aimed at establishing standard observations for future generations and began with two sets of plates of the sky around 1900: "catalogue" plates including stars up to magnitude 11, and "carte" plates up to magnitude 14. In the 1950s new plates were taken to determine proper motions by comparisons with the original photographs, but the comparisons did not prove satisfactory, and the whole operation ceased in 1970.

The success of the astrometric satellite *Hipparchos* between 1989 and 1992 led to the production of the Hipparchos and Tycho catalogues of stars and led astronomers at the United States Naval Observatory to reconsider the original project. By comparison with the satellite observations, they assigned new values to the proper motions of the stars of the general catalogue of the Carte du Ciel within 0.002 s. Two CD-ROMs were issued from data based on an interval of about one hundred years, an appropriate outcome for those who, at the end of the nineteenth century, proposed a photographic map of the heavens for the benefit of their successors.

Suzanne De'barbat et al., *Mapping the Sky—Past Heritage and Future Directions* (1988).

See also PHOTOGRAPHY.

SUZANNE DÉBARBAT

CARTESIANISM. See COPERNICUS; DESCARTES, RENÉ; MECHANICAL PHILOSOPHY.

CARTOGRAPHY, the art and science of making representations of areas of the earth and other spatially extended objects, has been connected with the history of modern science both as a field of inquiry and as a tool. As a field, it has supplemented and stimulated geodesy and, later, *planetary science. As a tool, it has been indispensable to *geology, *geophysics, *oceanography, *meteorology, and *biology, all of which rely on thematic maps to represent and analyze spatially distributed phenomena such as

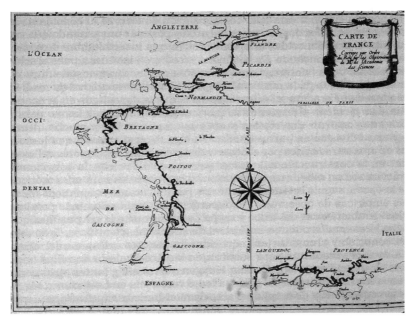

The result of a geodetic survey of France begun by the Académie royale des sciences in the 1660s; the dotted and solid lines indicate the contours as accepted before and after the measurements, respectively.

magnetic variation, rock outcrops, or warm and cold fronts.

Three major events around 1500 triggered an explosion of cartographic activity in Europe. First, cartographers assimilated the lessons of Claudius Ptolemy's *Geographia*, the greatest cartographic achievement of the ancient world. Brought to Italy from Constantinople, the complete text was translated into Latin early in the fifteenth century. By the century's end, at least seven different editions had been published. From Ptolemy, cartographers learned to arrange geographical knowledge according to a coordinate system of parallels (latitudes) and meridians (longitudes) and to project the system onto flat surfaces. Second, thanks to the invention of printing and advances in engraving, cartographers found ways to reproduce maps, first as wood engravings and, in the sixteenth century, by copper engravings that made much clearer maps. Third, the voyages of discovery and increased international trade that followed produced a demand for more accurate maps of the world. Thus the chief focus of cartographers became the production of atlases.

The Low Countries were the center of atlas-making in the sixteenth century. There, Gerardus Mercator invented the famous projection that showed lines of constant direction (rhumb lines) as straight lines, just as the portolan charts of the late Middle Ages had done earlier. Mercator used this projection in the marine chart he published

in 1569. In 1599, Edward Wright used the newly invented logarithms to lay out the mathematical basis of this projection. In 1570, Abraham Ortelius published one of the most comprehensive early atlases, the *Teatrum Orbis Terrarum*, with 53 maps by various authors. Numerous other atlases followed. By the eighteenth century, France had become the center of cartography. Guillaume Delisle published an atlas of 98 maps, and Jean-Baptiste Bourguignon d'Anville included 211 maps in his *Atlas général* (1737–1789).

Part of France's lead in cartography came from its research in geodesy, which led not only to a determination of the earth's shape and the length of a degree, but also to the first detailed topographical map of any nation, the *Carte géométrique de la France* (or *Carte de Cassini*). Its 182 sheets were finally completed and published by 1793, the culmination of decades of work.

Across Europe, nations founded surveys, mostly run by the military: the Ordnance Survey of Great Britain in 1791, the Institut Géographique National in France, the Landestopographie in Switzerland. The Spanish set up their topographic survey in the eighteenth century, the Austro-Hungarians, Swiss, and Americans in the nineteenth century. The Germans combined the existing state surveys into a national survey following unification in 1871. In 1888, Japan instituted its Imperial Land Survey. European nations also surveyed their overseas territories. The most ambitious of these surveys, the Great Trigono-metrical Survey of India, established early in the nineteenth century, not only mapped the sub-continent but raised gravimetric problems that quickly led to the theory of *isostasy.

Already in the seventeenth century, scientists had designed maps for specific scientific pur-poses. Between 1698 and 1700, Edmund Halley sailed the Atlantic, measured the variations in magnetic declination, and charted them on a pio-neering map that appeared in different editions between 1701 and 1703. By drawing lines between points of equal variation, he pioneered the tech-nique of isogonic lines that would be used often in later thematic maps. In the early nineteenth century, *geologists invented the stratigraphic map that represented each kind of rock that out-cropped at the earth's surface by a different color. Constructing such maps became the major goal of newly instituted national geological surveys. So important was stratigraphic mapping, so urgent its need for good underlying base maps, that in the United States the government gave the Geological Survey, established in 1878, the responsibility for producing the prerequisite topographic maps.

As national topographic surveys designed to promote national military and commercial inter-ests began to publish their results, entrepreneurs founded private companies to exploit the infor-mation in maps for the use of the general public. Among the more important of the companies they founded were Bartholemew in Great Britain, Jus-tus Perthes in Germany (an enterprise begun in the late eighteenth century), and Rand McNally in the United States.

From the 1880s to World War I, cartographers, like many other scientists, created international institutions. In 1875, the Convention of the Meter, attended by twenty nations, accepted the metric system, initially proposed in France in 1791, as the universal system of measurement. This aided the systematization of map scales. In 1884, the International Meridian Conference in Washington, D.C., decided on Greenwich, London, as the site of the prime meridian; the choice was universally accepted by World War I. At the International Geographical Congress of 1891, the German geographer Albrecht Penck proposed an International Map of the World on a scale of 1:1,000,000 in 1,500 sheets. The proposal was accepted but the work proceeded slowly. The most significant accomplishment of this still incomplete project was the 107-sheet *Map of Hispanic America* published by the American Geographical Society in 1945.

World War I brought a halt to international cooperation and ushered in the most productive century ever in cartography. In succession, the airplane, aerial photography, seismic techniques, echo sounders and sonar, radar, satellites, and computers made possible the mapping of features of the land, ocean bottoms, and extraterrestrial objects with an accuracy, ease, and variety hitherto undreamed of.

Without a growth industry in aeronautical charts, aviation would have foundered. Con-versely, airplanes made aerial photography and reconnaissance mapping possible once cartogra-phers had developed techniques to translate pho-tographs into maps. Meteorologists responded to the needs of the aviation industry, and benefited from it as they developed theories of high- and low-pressure systems and hot and cold fronts, all of them represented on maps. Today, meteo-rology is a map-intensive enterprise. Seismology, the echo sounder, and sonar displaced sound-ing as methods of mapping underwater. The unexpected topographic features detected with these new tools helped bring about the *plate-tectonic revolution.

In the 1960s, the already important remote-sensing systems came into their own. Electronic measuring helped extend continental surveys to oceanic areas. Soon thereafter, satellite tri-angulation became possible, helping to tie the triangulation of the continents together into a sin-gle system. Innovations multiplied. For example, scientists developed SLAR (side-looking airborne radar) in the 1970s. After SLAR had been tested in Ecuador and Nicaragua, RADAM (Radar Ama-zon Commission) of Brazil and Aero Service of Philadelphia mapped the hitherto inacces-sible Amazon Basin. Wide-angle cameras and improved film allowed the mapping of the earth and the moon and planets.

Computers were quickly harnessed for the purposes of mapping by institutions such as the Laboratory for Computer Graphics and Spatial Analysis at Harvard. They made the previously tedious business of laying out projections routine. After experiments in the 1970s, computers by the 1980s could translate aerial photographs into topographic maps complete with contours, allowing rapid and accurate contour mapping for the first time. Statistical mapping also became much easier.

The greatest change brought about by com-puters was a shift from considering the map as a completed object to the map as a manipulable tool. In the 1960s, geographers developed computer-ized information systems. Instead of entering data on a two-dimensional surface, they digitalized it. This meant they could easily superimpose differ-ent sets of information and, even more important, draw conclusions based on the superimposition. With this new tool in hand, cartographers were better equipped to prepare specialized maps for business, law-enforcement, natural-disaster prediction, or other specialized interests.

Today, cartography flourishes as never before. The military remains heavily involved, and the private companies multiply. Most nations have their surveys. Non-governmental groups such as the American Geographical Society and the National Geographic Society promote cartography. Professional societies such as the American Congress on Surveying and Mapping, the American Society of Photogrammetry, and the American Society of Civil Engineers (focusing here on the United States) have multiplied. They publish journals such as the *Manual of Photogrammetry, Photogrametria*, and *Surveying and Mapping*. Finally, a variety of international organizations attempt to coordinate mapping worldwide. Among them are the U.N. Office of Cartography, the Inter-American Geodetic Survey, the Pan-American Institute, and the International Hydrographic Organization in Monaco.

During most of the twentieth century, historians of cartography described how their discipline had achieved increasing scope and precision. At the end of the century, the critical cartography movement led by J. B. Harley and other British cartographers emphasized instead that, however exact maps may be, they are far from neutral. Cartographers' choice of projections, scales, symbols, and units reflect their own and their patrons' interests. According to these historians, maps depend not only on scientific knowledge but also on political interests.

See also NAVIGATION.

Arthur Robinson, *Early Thematic Mapping in the History of Cartography* (1982). Leo Bagrow, rev. R. A. Skelton, *History of Cartography*, 2d ed. (1985). J. B. Harley and David Woodward, eds., *History of Cartography*, 4 vols. (1987–1999). Timothy Foresman, ed., *The History of Geographic Information Systems: Perspectives from the Pioneers* (1998). Mark Monmonier, *Air Apparent* (1999). J. B. Harley, Paul Laxton, and J. H. Andrews, eds., *The New Nature of Maps* (2001).

RACHEL LAUDAN

CATASTROPHISM. See UNIFORMITARIANISM AND CATASTROPHISM.

CATHEDRAL. The long unimpeded drops in large churches from lantern to pavement made the best early-modern test bed for measuring free fall under gravity. Isaac *Newton confirmed his mainly mistaken notions of air resistance by observing the descent of inflated pigs' bladders from the top of Saint Paul's Cathedral in London in 1718, and, but for the death of the cardinal sponsoring the experiment, the earth's rotation would have been confirmed by dropping weights from the cupola to the crypt of Saint Peter's in Rome seventy years later.

The most significant scientific work done in cathedrals concerned two fundamental constants of the solar theory: the eccentricity of the sun's (or the earth's) orbit and the obliquity of the ecliptic, the inclination of the earth's axis to the plane of its (or the sun's) annual motion. During the seventeenth century, the eccentricity of the orbit played an important part in discriminating between the Ptolemaic theory and Johannes *Kepler's version of Nicholas *Copernicus's theory. A large camera obscura was required. The great church of San Petronio in Bologna served admirably after Gian Domenico Cassini made a hole in its roof and embedded a rod running north-south (a meridian line) in its pavement to catch the image of the midday sun. This "heliometer," as Cassini called it, found unequivocally for Kepler. Notably, Cassini accomplished this in the capital of the Papal States two generations after the papal condemnation of *Galileo Galilei.

Similar cathedral observatories were built in Rome (at Santa Maria degli Angeli) just after 1700, in Paris (at Saint Sulpice) in the 1740s, and in Florence (at Santa Maria del Fiore) in 1755. Observers at these instruments helped to establish that the obliquity changes in time, by very little to be sure, but still another reason to prefer Copernicus to Ptolemy.

While installing the meridian in Florence, its builder, Leonardo Ximenes, expanded the necessary measurements and operations into an investigation in experimental physics. He measured the lengthening of a long metal chain under its weight (while determining the exact height of the hole), the rate of evaporation of water as a function of temperature (while leveling the meridian line using a water-filled trough), and the decline of air pressure with height (while climbing the 90 meters [295 ft.] to the lantern). For a short time, Ximenes turned the vast cathedral of Florence, which he fenced off for the purpose, into a physical laboratory as well as a solar observatory, literally practicing science in the church.

J. L. Heilbron, *The Sun in the Church: Cathedrals as Solar Observatories* (1999).

J. L. HEILBRON

CATHODE RAYS AND GAS DISCHARGE. The study of cathode rays and gas discharge laid the groundwork for the elucidation of the properties of the electron and its role in determining physical and chemical properties of matter.

As early as 1833 Michael *Faraday investigated glows produced by electrical discharge through

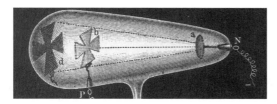

Demonstration by William Crookes that the invisible cathode rays follow straight lines (because they throw sharp shadows).

gases. Subsequently the invention of a mercury pump that could reduce gas pressures in glass tubes to as low as 10^{-6} atmosphere multiplied investigations of the phenomena that Eugen Goldstein in 1876 called "Kathodenstrahlen" since the radiations appeared to flow from the negative, or cathode, pole of the vacuum tube. William Crookes, editor of *Chemical News*, found evidence in the 1870s and 1880s that the "rays" are negatively charged particles of matter. He proposed that these particles constituted a "fourth state" of matter that was neither solid, liquid, nor gas. Heinrich Hertz in 1883 and Philipp Lenard in the early 1890s reported results inconsistent with the particle hypothesis and supporting an electromagnetic-wave interpretation of the cathode radiation. There appeared to be an English interpretation and a German interpretation of the agent of the Kathodenstrahlen.

After assuming the directorship of the Cavendish Laboratory in Cambridge in 1884, Joseph John Thomson directed much of the laboratory's research toward studies of ionization and electric gas discharge and, later, to *X rays and *radioactivity. Working on the hypothesis that the cathode radiation consists of fast-moving charged particles that produce pulses of X rays when they hit glass or metal in the vacuum tube, Thomson studied the velocity of the suppositious particles and their response to electrical and magnetic fields. Here he drew on recent work by Jean Baptiste Perrin in Paris to help counter the evidence that the rays could not be charged.

In 1897 Thomson calculated the mass-to-charge ratio for the cathode-ray particles and made a case that the cathode radiation comprises negatively charged corpuscles a thousand times smaller in mass than the hydrogen *ion. Thomson proposed a model of the atom containing large numbers of these particles arranged spherically to account for physical and chemical properties of chemical elements. X-ray scattering experiments begun in 1903 by Charles Glover Barkla and C. A. Sadler at Liverpool suggested that the number of electrons in an atom is approximately twice the relative atomic weight (taking hydrogen's as one), which led to the hypothesis that light atoms contain tens of electrons rather than thousands of corpuscles.

A beam of negative particles in the electric discharge tube must be associated with an oppositely directed beam of positive particles. Physicists obtained evidence that these counterparts existed in the late 1880s. Wilhelm Wien reported investigations in 1902 on a positive radiation that he could draw through holes or canals bored in the cathode ("Kanalstrahlen"). Canal rays consist of gaseous ions with a mass-to-charge ratio identical to that of the ions made of the residual gas in the electric discharge tube. J. J. Thomson and his assistant Francis Aston devised ways of manipulating the various constituents of canal rays that led to the physical separation of isotopes and, in the 1920s, to the rapid deployment of *mass spectrometry.

See also ATOMIC STRUCTURE; ION.

E. A. Davis and I. J. Falconer, *J. J. Thomson and the Discovery of the Electron* (1997). Mary Jo Nye, *Before Big Science: The Pursuit of Modern Chemistry and Physics, 1800–1940* (1999).

MARY JO NYE

CATHODE RAY TUBE. A cathode ray tube is a partially evacuated glass tube with two or more electrodes inserted. When the vacuum is sufficiently high and the voltage across the electrodes sufficiently great, rays issue from the cathode.

Electrical discharges through gases were studied throughout the second half of the nineteenth century in such vacuum tubes. All colors of the rainbow could be produced in the residual gases, making striking and widely admired effects. Investigators analyzed the regions of light and darkness within a tube and the spectroscopic composition of the light in order to illuminate the relationship between *electricity, matter, and the *ether. Julius Plücker (1801–1868), William Crookes (1832–1919), and Eugen Goldstein (1850–1930) contributed to the instrumentation and analysis. With the achievement of higher vacua, the electrical discharge receded, and the presence of the ray issuing from the cathode could be inferred from the fluorescence it caused in the tube wall. In 1876, Goldstein termed this agent "cathode ray." In the fourth quarter of the nineteenth century, determining the nature of this ray became a significant research project. Vacuum techniques continued to improve, largely because of the demand in the quickly growing electrical *lighting industry. By the end of the century, electrical, *vacuum, and glassblowing techniques made possible the generation of X rays, where the cathode rays impinged on the walls of the tube or on a special target in the tube.

Much of early twentieth-century physics was based on research performed with the cathode ray tube. With results obtained by its use, Wilhelm Conrad *Röntgen (1845–1923), Philipp Lenard (1862–1947), Joseph John *Thomson (1856–1940), and Ferdinand Braun (1850–1918) all won Nobel Prizes.

In the first half of the twentieth century, the X-ray tube was by far the most important type of cathode ray tube as a result of the explosive growth of radiology. The hot-cathode tube became particularly important in radiology from the 1920s onward because the spectrum of X rays produced became easier to control. There were two other important areas of application. The ENIAC computer built in 1946 was one of the largest vacuum systems ever assembled; it suffered from the fragility, size, and slowness of the tubes, which were gradually displaced by the semiconductor transistor. The oscilloscope provided the second significant application. Other applications continued to be explored, notably for television, and from the 1930s onward much research took place in the commercial sphere. Prominent companies included RCA, DuMont, Cosser, Telefunken, Sylvania, and Tektronix (and after World War II, Japanese companies such as Sony). In 1939, 50,000 (non-X-ray) cathode ray tubes were sold worldwide. During World War II, demand grew explosively in military electronics and in radar. By 1944, sales amounted to two million units. After 1945, growth continued primarily in televisions and computer displays. By 1987, sales topped 100 million units. Today, cathode ray tubes serve in many scientific instruments, both as indicators and for data display.

During the 1990s, the flat-panel display using liquid crystals began to challenge the cathode ray tube. This display is much smaller than the cathode ray tube, but as yet cannot compete with it in price.

Peter A. Keller, *The Cathode-Ray Tube—Technology, History and Applications* (1991). Arne Hessenbruch, "Geissler Tube," in *Instruments of Science—An Historical Encyclopedia*, eds. Robert Bud and Deborah Jean Warner (1998): 279–281.

ARNE HESSENBRUCH

CAT SCAN. See COMPUTERIZED TOMOGRAPHY.

CAUSALITY refers to the power or propensity that an object or event has to produce a change in itself or in another object or event. In the history of modern science, however, there has been no agreement about the concept, or even the existence, of causality. During the late Medieval and Renaissance periods, Aristotelians argued that four distinct types of causes all had to be invoked in order to explain change in any object (*see* ARISTOTELIANISM).

Many natural philosophers of the seventeenth century rejected the Aristotelian conception of causality. Francis Bacon and René *Descartes dismissed final causes and argued for a mechanical version of efficient causality whereby all physical processes would be explained by contact action. According to the mechanists, laws of motion governed the least parts of matter and gave rise in a deterministic manner to the causal powers possessed by bodies to produce effects in other bodies. Not all mechanical philosophers rejected final causes, however. Robert Boyle, for example, argued that physiological research could be advanced by inquiry into the purposes of the parts of living bodies.

Throughout the seventeenth and eighteenth centuries another group of thinkers defended the ancient empirical belief that humans can know only the sensible appearance of things but not the underlying causal processes that bring the appearances about. Empirical doubt about causality received its most vigorous and detailed explication from the philosopher David Hume. In his *Enquiry Concerning the Principles of Human Understanding* (1748), Hume argued that all we can ever know is the constant conjunction between two events. We call the event that comes first in the series the "cause," but we can never see a necessary causal connection with its "effect." Partly as a result of Hume's work, Immanuel Kant concluded that causality is a category that the mind imposes upon nature.

These philosophical considerations did not have an immediate impact on the practice of the sciences. During the early 1800s chemists such as Humphry Davy and John *Dalton continued to experiment and theorize about the underlying causal properties of matter. As the century progressed, however, skepticism about causality became more intrusive. Around 1900, for example, Ernst Mach and other physicists insisted that the real aim of science was to provide accurate descriptions of the phenomena and not speculative causal explanations. Further researches into the subatomic recesses of nature have also raised problems for the general principle of causality—that a cause must precede its effect. Despite such problems and ambiguities, however, scientists find it difficult to suspend causal inquiry.

See also MECHANICAL PHILOSOPHY; POSITIVISM.

Larry Laudan, *Science and Hypothesis* (1981). Ernan McMullin, "Conceptions of Science in the Scientific Revolution," in *Reappraisals of the Scientific Revolution*, ed. David C. Lindberg and Robert S. Westman (1990): 27–92.

ROSE-MARY SARGENT

CAVENDISH, Henry (1731–1810), and **Charles Augustin COULOMB** (1736–1806), the two most important contributors to the quantification of physical science during the eighteenth century. Both derived their major scientific questions, and the keys to answering these questions, from Isaac *Newton's conception of distance *forces. Otherwise, the two men were altogether different.

Cavendish, a grandson of dukes on both sides of his family, became one of the wealthiest men in England. Educated at an upper-class school and at Cambridge, and lacking no personal comfort, he nonetheless developed a painful shyness from which his solitary scientific pursuits provided a refuge. He spent his entire life in England and did not marry. Coulomb also came from a wealthy family and received his early education in Parisian colleges. But his father's business speculations eventually ended in bankruptcy and Coulomb had to make his way in the world. He became a military engineer. Neither shy nor home-bound, he spent eight years directing the building of a huge fortress in Martinique. Following his return to France in 1764 he wrote papers on mechanics and magnetic compasses, which won prizes from the Académie des Sciences of Paris (*see* ACADEMIES AND LEARNED SOCIETIES). On the strength of these works, the academy

admitted Coulomb as a member in 1781, at the age of forty-five. In contrast, Cavendish was elected a Fellow of the Royal Society before he was thirty, not because of prize-winning work but because his father, a leading light of the society, wished it. Academy membership allowed Coulomb to withdraw from active service and to raise a family. Membership in the Royal Society gave Cavendish a social life.

Cavendish's first important investigation, which he did not publish, aimed at producing a Newtonian theory of heat based on corpuscular motions and distance forces. While pursuing these ideas he came naturally to the study of airs, the expansion of which provided a simple and convenient model for the interaction of heat and matter (*see* PNEUMATICS). In 1762 Cavendish discovered that on treatment with acids, metals released an inflammable air (H_2). He was one of several experimenters who reached the implausible conclusion (in his case in 1784) that inflammable air combined with "eminently respirable air" (so named by Joseph Priestley; *see* OXYGEN) to make water. In Cavendish's explanation, which preserved the old chemistry against the innovations of Antoine-Laurent *Lavoisier, inflammable and eminently respirable air were water plus phlogiston and water minus phlogiston, respectively.

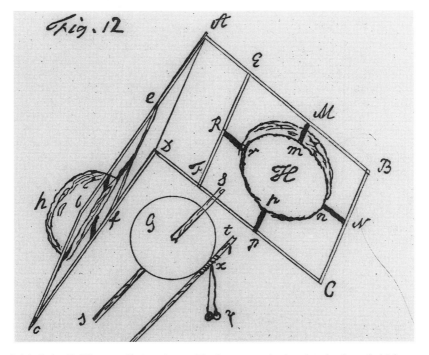

Cavendish's Rube-Goldberg null experiment. The large exterior insulated sphere (which opens like a book) is shut to enclose the previously charged insulated internal sphere; a wire briefly connects the two internally, whereupon the outer sphere opens and the experimenter shows that no charge remains on the inner.

Cavendish's deepest work concerned electricity. Conceiving electrical matter to be compressible in the manner of a pneumatic fluid (a gas), and supposing further that the particles of electrical matter acted on one another and on particles of common matter by a distance force diminishing as some unknown power of the distance between them, he described the electrical states of systems of charged and uncharged conductors exposed to one another's influence. His long and difficult memoir of 1771 introduced the concepts of electrical capacity and, as represented by the compression of the gas-like electrical matter, electrical potential. Cavendish's interpretation of the results of electrical induction in terms of potential was well beyond the grasp of almost all his contemporaries. Unfortunately, his thoroughness, combined with the profligacy of nature, defeated his intention of producing a *Principia electricitatis*. As he learned more about the range of inductive capacities and conductivities of bodies, he came to doubt the possibility of describing all these phenomena with the aid of only mathematics and a single undifferentiated electric fluid.

Cavendish's unpublished papers on electricity contain many results that would have been important if released when he obtained them. One was a demonstration of the inverse-square repulsion between particles of electric fluid, based on a mathematical consequence of the law of squares—that, given a perfect conductor, any charge must migrate to its surface. He managed this demonstration with the rickety device pictured here and enriched it by estimating how far the exponent in the force law could differ from −2 and remain consistent with his measurements (answer: by <0.02). This quantitative judgment of possible error in determining an empirical "law" was probably unique in its time. Another important investigation concerned the division of current in parallel circuits. Cavendish made his results public and palpable in 1776 by constructing a model of a torpedo (an electric fish) that made clear to his colleagues how an animal living in salt water (a passable conductor) could store electricity and appear to direct it to numb its prey.

In his last significant published work Cavendish returned to old Newtonian problems. In 1798 he used a torsion balance to detect the gravitational force between masses in the laboratory. He did not derive a value of the universal constant of gravitation from the tiny deflection of the balance, but rather estimated from the measurements a likely value for the density of the earth—that too being a Newtonian problem.

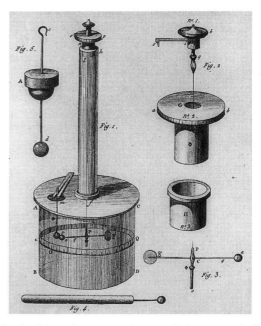

Coulomb's well-engineered torsion balance, in which the repulsive force between two charged pith balls (one fixed to the housing and the other to an end of the mobile horizontal insulated arm) is balanced by the twist in the vertical wire suspending the arm.

Coulomb's most important contribution to physics also concerned the forces between particles of electric fluids, of which, in contrast to Cavendish, he admitted two. Like Cavendish, in designing his method of measurement he made use of a Newtonian theorem: that a gravitating spherical shell acts on bodies outside it as if its entire matter resided at its center. If the particles of each electric fluid repelled one another by an inverse-square force, a charged conducting sphere should act on bodies outside it as if its charge were concentrated at its center. Coulomb adopted a torsion balance he had invented to test the strength of wires and to investigate magnetism to measure "the force of electricity." The apparatus with which he claimed to demonstrate the law of squares (Coulomb's law; 1785) has become an icon in physics. It is a carefully engineered device for obtaining numbers (the degree of twist of a wire), in striking contrast to Cavendish's homemade rig for registering zeros (the lack of a charge on a previously charged sphere). Coulomb's contemporaries had trouble reproducing his measurements (keeping the balance rod horizontal is difficult), and modern attempts have succeeded only by using precautions that Coulomb did not specify. Nonetheless, Coulomb's academic colleagues accepted these results and his similar results for magnetism, no doubt because they had already enlisted electricity among Newtonian *imponderable fluids.

Coulomb followed up his measurements of electrical forces with determinations of the distribution of electricity on the surfaces of charged, touching, insulated spheres. Powerful mathematicians, notably Simeon-Denis Poisson and Carl Friedrich *Gauss, confirmed Coulomb's determinations, and in the early nineteenth century they built up a fully quantitative electrostatics and magnetostatics.

The difficulty historians have found in repeating Coulomb's experiments has given rise to doubts about the veracity of his reports and to meditations on the meaning of "replication" in science. One answer to the doubters is that Coulomb perfected his technique over many years and that the coordination of eye and hand acquired from his long engineering practice is not a common talent of historians.

A. J. Berry, *Henry Cavendish* (1960). C. Stewart Gillmor, *Coulomb and the Evolution of Physics and Engineering in 18th Century France* (1971). J. L. Heilbron, *Electricity in the 17th and 18th Centuries* (1979; 2d ed. 1999). Christine Blondel and Matthias Dörries, eds., *Restaging Coulomb* (1994). Christa Jungnickel and Russell McCormmach, *Cavendish* (1996).

J. L. HEILBRON

CELESTIAL MECHANICS. The study that Pierre-Simon Laplace named celestial mechanics originated in Isaac *Newton's *Principia* (1687). Newton assembled impressive empirical evidence for the inverse-square law of universal gravity and gave closed solutions for the one- and two-body problems (to determine the motion of a body under gravitational attraction towards a fixed center, and to determine the motions of two bodies attracting each other gravitationally). The three-body problem (to determine the motions of as many as three mutually attracting bodies) admits of no such solution, and Newton's geometrical approach to it did not provide for systematic refinement. Celestial mechanics emerged as an algebraic-style attack on these problems, first treating the bodies as points, and then taking into account their extendedness in space. Thus the Moon's attraction for the Earth's equatorial bulge causes a precessional motion of the Earth's axis ("the precession of the equinoxes"); Newton's derivation of this motion was badly flawed, posing for the new science one of its major challenges.

Advance came through formulating these problems as differential equations and solving them with trigonometric functions. Having systematized the calculus of these functions in 1739, Leonhard *Euler in 1748 introduced trigonometric series and statistical procedures for the differential correction of orbital elements. He also obtained the differential variation of the inclination and node. In 1749 Jean Le Rond d'Alembert carried out the first valid derivation of the precession of the equinoxes. Euler, thus prodded, then completed a formal theory of the dynamics of rigid bodies. Alexis-Claude Clairaut in 1752 gave the first satisfactory derivation of the moon's apsidal motion. In 1758–1759, by carrying out the first extended numerical integration, he achieved a prediction of the return of Halley's Comet accurate to a month.

Lunar theories—both Newton's and the newer analytic ones—remained insufficiently accurate to permit determination of longitude at sea to within one degree. To achieve the necessary accuracy, Tobias Mayer combined analytic theory with statistical correction of coefficients, thus providing the basis for the British *Nautical Almanac* (published from 1767 onwards). Semiempirical lunar theories would remain the basis of the national almanacs until 1862.

Further eighteenth-century advances were due to Joseph Louis Lagrange and Laplace. Lagrange introduced the "perturbing function." Assume a planet or satellite moving in an ellipse about its primary in accordance with the solution of the two-body problem, while the attraction of a third body disturbs the motion. Lagrange's new function, when differentiated with respect to any coordinate, yields the force disturbing the motion in the direction of the coordinate. Lagrange also completed and systematized the method of variation of orbital parameters. Laplace in 1773 showed that, to a good approximation, the planetary mean motions are immune from secular variation; an observed deceleration of Saturn and acceleration of Jupiter remained unexplained. In 1785 he accounted for these anomalies by a 900-year periodic variation proportional to the cubes and three-dimensional products of the orbital eccentricities and inclinations. In 1787 he provided an apparently cogent gravitational explanation for the moon's observed secular acceleration (though in the 1850s his account was shown to require fundamental revision). Laplace's *Mécanique céleste* remained the prime textbook of celestial mechanics until the 1890s; its treatment of second- and higher-order approximations relative to the perturbing forces, however, was unsystematic.

Resolutions of this difficulty came from the method of variation of orbital parameters as developed by Simeon-Denis Poisson and Lagrange in 1808–1809. Peter Andreas Hansen made all perturbations additive to the mean motion; his lunar theory was used for the national ephemerides from 1862 to 1922. Charles-Eugène

Delaunay successively removed perturbations from the perturbing function, incorporating them in progressively refined canonical variables. The slow convergence of his method led George William Hill to found lunar theory anew, starting from periodic solutions of a restricted form of the three-body problem. Hill's work stimulated the fundamental topological researches of Henri Poincaré into the stability of dynamical systems.

René Taton and Curtis Wilson, eds., *Planetary Astronomy from the Renaissance to the Rise of Astrophysics, Part B: The Eighteenth and Nineteenth Centuries* (1995): chs. 20–22, 28. Martin C. Gutzwiller, "Moon-Earth-Sun: The Oldest Three-Body Problem," *Reviews of Modern Physics* 70 (1998): 589–639.

CURTIS WILSON

CELL. The term "cell" was coined by the seventeenth-century British natural philosopher Robert *Hooke to describe the walled cavities he discerned through the *microscope in vegetable tissue. These spaces reminded him of the cells of a monastery. To confuse matters, eighteenth- and nineteenth-century anatomists sometimes used the term "cellular tissue" to refer to connective tissue visible to the naked eye.

The origins of modern cell theory lie in the nineteenth century and in developments in microscopy. The wide availability of achromatic microscopes after 1830 obviated some of the optical problems that had beset earlier attempts to elucidate the minute structure of living tissue. Still more important, however, was the growth of communities of microscopic workers in the middle decades of the nineteenth century. The emergence of schools of histology in Berlin, Breslau, Edinburgh, and London made possible a concerted and extended investigation into questions that had previously received only cursory attention.

Credit for first enunciating the cell theory goes to two German histologists, Jakob Matthias Schleiden (1804–1881) and Theodor *Schwann (1810–1882). Schleiden was a botanist who in 1838 announced that all the constituent parts of plants were composed of microscopic cellular structures. Schwann discerned a similar pattern of organization in animal tissue. On the basis of this analogy he promulgated the fundamental tenet that all living tissue consisted of cells or cell derivatives. He set out this claim in his *Microscopic Researches into the Accordance in the Structure and Growth of Animals and Plants*. As the title suggests, Schwann was particularly interested in the developmental role of cells. He also suggested that cells performed an important, indeed central, functional role.

The cell theory as promulgated by Schwann met with opposition, and his doctrine of cytogenesis was eventually deemed to be fallacious. However, the claim that a single structure underlay the diversity of the organic world answered an evident need in nineteenth-century biological thought for a unifying principle around which various programs of research all tending to a common end could be organized.

Schleiden and Schwann maintained that cells arose originally from a structureless substance or "blastema." This accorded with the doctrines of such speculative *Naturphilosophen* as Lorenz Oken, who had argued that the primitive spherical vesicles from which living beings had developed emerged from a primordial fluid (*see* NATURPHILOSOPHIE). In the course of the nineteenth century several scientists challenged the doctrine of the spontaneous generation of cells. Embryologists such as Robert Remak showed that even in the earliest stages of embryogenesis, cells originated solely through the division of pre-existing cells. This critique of the blastema culminated in 1855 when the German pathologist Rudolf Virchow enunciated the doctrine *omnis cellula e cellula*—all cells come from cells. This stark claim met considerable skepticism; it was, however, eventually to become scientific orthodoxy. Virchow also drew attention to the role of cells in various morbid processes, thus laying the foundations for cellular pathology.

The cell theory identified the basal structural and functional unit of the organism. But, although in this sense "primitive," the cell was by no means a simple entity. From the outset, histologists recognized that it consisted of several distinct parts and disputed which of its elements had the greatest functional significance. Some thought the cell wall the most active part; others settled on the cell content, or "protoplasm." In time, however, the focus of investigation shifted to the cell nucleus.

In the 1870s microscopists identified the presence of two distinct nuclei, derived from each of the parents, in the fertilized ovum. In the same decade accounts appeared of the process of mitosis, the mode in which cells divided in the adult organism. Developments in staining methods permitted histologists to discern thread-like structures within the nucleus that appeared to be central both to mitosis and to meiosis, the division of reproductive cells. In 1888, these strands of nuclear matter were given the name *"chromosomes." By the end of the nineteenth century there was general agreement that these chromosomes provided the material basis for continuity within the organism and for the transmission of characteristics between generations.

The early twentieth century saw the amalgamation of histological and embryological findings with Mendelian notions of heredity. Despite some protests, by 1910 informed scientists concurred that the "factors" postulated by Gregor Mendel as the transmission agents of hereditary characteristics consisted of elements of the chromosome. American scientists, notably Thomas Hunt Morgan and his associates at Columbia University, effected a union between cell biology and breeding experiments. Morgan maintained that it was possible to map on the chromosome the distribution of the Mendelian factors, or "*genes," responsible for particular characteristics.

Among the other cell elements, mitochondria have been identified as crucial to metabolic functions, and cell membranes have been ascribed complex biochemical functions. Following ideas of the nineteenth-century French physiologist Claude *Bernard, scientists regard cells not as isolated units, but as partners in a dynamic interaction with their milieu. Research has vindicated the chief tenet of the theory expounded by Schwann in 1838: the cell is itself an organism, and all living beings, including the most complex, perform the sum of their functions by means of their constituent cells.

J. Walter Wilson, "Virchow's Contribution to the Cell Theory," *Journal of the History of Medicine*, vol. 2 (1947). John R. Baker, "The Cell Theory: A Restatement, History and Critique," *Quarterly Journal of Microscopical Science*, vol. 89 (1948). Frederick Lawrence Holmes, "The Milieu Interieur and the Cell Theory," *Bulletin of the History of Medicine*, vol. 37 (1963). Henry Harris, *The Birth of the Cell* (1999).

STEPHEN JACYNA

CENTRIFUGE AND ULTRACENTRIFUGE. The primary use of centrifuges in research laboratories is to sediment substances in a solution, taking advantage of density differences. Centrifuges work by driving a solution in a container spun at high speeds toward the periphery of the container. Machines have been used since the eighteenth century to demonstrate centrifugal force, a basic concept of mechanics well known since *Newton's time. Centrifuges used for separating and precipitating different substances came into general use in chemical, medical, and biological laboratories in the middle of the nineteenth century. They transferred into the laboratory after their widespread application in sugar refinery and cream separation (1878). An early laboratory application was the hematocrit for the separation of the constituents of *blood. Around the turn of the twentieth century, amphibian and fish eggs were centrifuged to observe their development under conditions of changed gravity.

Today, "ultracentrifuge" signifies all centrifuges that spin faster than 20,000 rpm. The first machine so called, developed by the Swedish colloid chemist The Svedberg in the 1920s, was a device used solely to determine the size of colloid particles. Svedberg and his collaborators developed two different methods: velocity sedimentation (the observation of the sedimentation process in time) and equilibrium sedimentation (the observation of the distribution of particles at equilibrium between acceleration by the centrifugal field and diffusion). Concentrations in the solution were measured by spectroscopic methods, usually absorption spectroscopy, which employs a light beam passed through the glass windows of the sedimentation cell ("analytic ultracentrifuge"). Experiments to determine the size of proteins such as hemoglobin required large gravitational fields of the order of 100,000 times the force of gravity, which Svedberg achieved in 1926 with a rotor spinning up to 45,000 rpm. Svedberg received the Nobel Prize for chemistry in 1926, ensuring him the necessary support for the further development of his ultracentrifuge, but the device was too expensive to spread to other laboratories.

In the 1930s, Jesse Beams and Edward Pickles developed ultracentrifuges based on the principle of a spinning top (a small, cone-shaped top, driven by air jets, rotating in a close-fitting recess). A rotor spun in a vacuum to prevent heating caused by air friction was connected to a spinning top drive by means of steel wire. The simpler drive and a compact design made this ultracentrifuge cheaper than Svedberg's. In the 1950s, it became commercially available standard equipment in many laboratories. Besides the analytic ultracentrifuge, Beams and Pickles also developed purely preparative high-speed centrifuges. The *Manhattan Project investigated the use of gas centrifuges for the enrichment of uranium, but abandoned them in favor of gaseous diffusion. The rejected process was later developed as an industrial technology.

Both preparative and analytic ultracentrifuges played a crucial role in the emergence of *molecular biology and *genetics. In 1958, Matthew Meselson and J. F. Stahl centrifuged DNA, and thereby gave evidence for James Watson's and Francis Crick's proposed duplication mechanism of the DNA molecule. Differential methods—separating cell fragments into different fractions by high-speed centrifugation—developed between the 1930s and the 1960s, with electron microscopy and biochemical analysis, revolutionized cell biology.

Boelie Elzen, "The Failure of a Successful Artifact: The Svedberg Ultracentrifuge," in *Center on the Periphery: Historical Aspects of 20th-Century Swedish*

Physics, ed. Svante Lindqvist (1993): 347–377. Hans-Jörg Rheinberger, "From Microsomes to Ribosomes: 'Strategies' of 'Representation,' " in *Journal of the History of Biology* 28 (1995): 49–89.

ROLAND WITTJE

CERN. In the early 1950s a dozen European nations banded together to build a laboratory for high energy *accelerators, known by its acronym as CERN. The first steps toward CERN came in late 1949, when Raoul Dautry, administrator general of the French Atomic Energy Commission, raised the possibility of a collaborative European institute for nuclear science. The following year I. I. Rabi proposed to UNESCO that European states create regional laboratories, including one in nuclear science, similar to the one Rabi had helped establish at Brookhaven National Laboratory in the United States. European nuclear physicists and science administrators took up the proposals and in 1952 established the European Council (later Organization) for Nuclear Research (CERN).

The original member nations were Britain, Switzerland, Denmark, the Netherlands, Greece, Sweden, Belgium, France, West Germany, Norway, Yugoslavia, and Italy; membership would eventually swell to twenty nations. The organizers decided to concentrate their efforts in a new laboratory in Geneva, which opened in 1954, and planned as its centerpiece a large particle accelerator. They focused on high-energy physics because it shared in the glamour of nuclear physics without apparent military applications that might preclude cooperation, and also because several of the founding scientists worked in that field and could not hope to compete with Americans with the resources available in their own countries. Though CERN's founders proscribed military research at the lab, they recognized that CERN would serve as a training ground for technical experts who might then work on military or industrial applications of nuclear energy.

CERN emerged from a confluence of diverse national and scientific interests. Rabi's role in the creation of CERN indicates American aims to foster European science as a bulwark against Communism in the Cold War. The lab drew on increasing pan-European sentiment in the early 1950s; the choice of facilities nevertheless displayed the usual rivalries between individuals, institutions, and nations within Europe, evident in wrangling over the award of CERN contracts to industrial firms. And although CERN meant to pool the resources of European countries in high-energy physics in order to compete with the United States, individual nations at first did not forsake domestic accelerator programs; some British scientists in particular feared a

commitment to CERN would siphon funds from their own work.

Several generations of machines mark the history of CERN. In 1954 the lab and the member nations decided to pursue a 25–30-BeV proton synchrotron (PS), which boasted the highest energy in the world when completed in 1959. In 1965 CERN for its next machine chose Intersecting Storage Rings (ISR), a new type of accelerator that provided greater efficiencies through head-on collisions rather than impact of the beam on a stationary target. Physicists at the time preferred a higher-energy, 300-BeV proton synchrotron, but engineers backed the ISR and national administrators preferred its lower price. In 1971 the physicists won approval of the deferred Super Proton Synchrotron or SPS, which the lab in the late 1970s converted to a proton-antiproton collider. In the 1980s CERN built the Large Electron-Positron Collider, or LEP, to explore the carriers of the weak force. The LEP ran particles around a ring nine kilometers across, dwarfing the two-kilometer diameter of the SPS. In the 1990s the lab sought to enter the tevatron energy range with a new Large Hadron Collider, which would share the underground tunnel of the LEP but use superconducting magnets. CERN also supported the construction of detectors such as bubble chambers, whose increasing size and complexity overtaxed the capabilities of individual researchers or laboratories. Despite its impressive facilities and accumulating data, CERN in the 1960s lost out on major scientific discoveries to American accelerator laboratories, where experimental research and engineering were integrated in a single organization. CERN reflected European traditions of separating science from engineering, and theory from experimental practice, as the debate between physicists and engineers over the ISR made plain. Only in the 1970s did CERN researchers begin to master the new mode of big science pioneered in the United States. The detection of weak neutral currents by the PS in 1973 and of the W and Z particles in the SPS collider in 1983, both of which provided important evidence for the unified theory of the electromagnetic and weak forces, signaled CERN's maturity.

As CERN grew and outstripped national facilities, it felt increased pressure for access to its unique machines from national physics communities. Expanding budgets also invited increased attention from national governments, which sought evidence of return on their investments. Member states contributed in proportion to their net national income, with Germany, Britain, France, and Italy each usually contributing from 10 to 25 percent of the total annual budget, which had reached about one billion

Swiss francs, or about $600 million, by the end of the century. The increasing complexity and expense of its facilities forced CERN to heed the demands of national governments and physics communities for greater accountability, in order for it to procure the human and material resources necessary to sustain it as one of the leading labs in the world for high-energy physics.

See also ELEMENTARY PARTICLES; NUCLEAR PHYSICS.

Arminn Hermann, John Krige, Ulrike Mersits, and Dominique Pestre, *History of CERN*, vol. 1 (1987) and vol. 2 (1990). John Krige, ed., *History of CERN*, vol. 3 (1996).

PETER J. WESTWICK

CHANCE. See PROBABILITY AND CHANCE.

CHANDRASEKHAR, Subrahmanyan (1910–1995), mathematical astrophysicist.

Chandrasekhar showed that stars up to a particular mass limit evolve into white dwarfs, implying that more massive stars collapse to form black holes. He wrote thorough mathematical monographs inspired by a variety of astrophysical problems, including *relativity and *black holes, and elevated the *Astrophysical Journal* to world-class status during his nineteen-year editorship.

In 1930 on a boat from India to England, where he would pursue graduate studies in stellar structure, Chandrasekhar modified stellar models for relativistic and quantum effects and found a limit—the Chandrasekhar limit—to the mass of a star (approximately 1.4 solar masses) that can evolve into a white dwarf (an extremely dense star). This result was startling, but won little attention. In 1934 Chandrasekhar was advised that to persuade astronomers, he would have to compute the masses of a representative sample of white dwarfs and demonstrate that all fell below the limit. Chandrasekhar presented his results before the Royal Astronomical Society in 1935. The famous English astrophysicist Arthur Eddington pointed out that in Chandrasekhar's mathematical model heavier stars would keep contracting and gravity would become strong enough to hold in radiation, which, in Eddington's opinion, was an absurd way for stars to behave. Devastated, Chandrasekhar abandoned this line of research, resuming it again only in the 1960s with a mathematical study of relativity and black holes. In 1983 he received a Nobel Prize for the work.

In 1937 Chandrasekhar took up a position at the University of Chicago's Yerkes Observatory, where his theoretical work could be valuably combined with investigations by observers. Yerkes quickly built an outstanding graduate program in astronomy and astrophysics, with Chandrasekhar as its most active faculty member. In 1952, however, the astronomy department made an abrupt change in emphasis, from theoretical to observational courses, and Chandrasekhar shifted much of his teaching to the physics department.

Also in 1952 he became managing editor of the *Astrophysical Journal*, a position he exercised autocratically and energetically for nineteen years while transforming the journal from the private publication of a university into the leading astrophysical journal in the world.

After the debacle with Eddington in 1935, Chandrasekhar, a classical applied mathematician in the Cambridge University tradition, chose mathematically beautiful problems usually neglected because of their difficulty, solved fundamental equations about a problem, and eventually published a formal, logical, rigorous treatise about it. Then he would move on to a new problem. Stellar dynamics was followed in the 1940s by radiative transfer, particularly with regard to stellar and planetary atmospheres. Next Chandrasekhar explored hydrodynamic and hydromagnetic stability, and then ellipsoidal figures in equilibrium, particularly the stability of rotating and vibrating stars within the framework of general relativity. This work, culminating in 1969, turned out to be relevant to pulsars, rapidly rotating neutron stars discovered in 1967.

In the 1960s Chandrasekhar finally felt free to unleash his early interest in relativity theory, a field he had long perceived as a graveyard for theoretical astronomers. Astronomy had gradually abandoned interest in relativity, except for cosmologists, who were not considered real astronomers. In 1963, however, a relativistic astrophysics revolution began with the discovery of the first *quasar (quasi-stellar object). Its tremendous energy output exceeded that obtainable from thermonuclear combustion of all its mass. In the intense theoretical effort to find a new energy source, gravitational collapse was the most likely mechanism, and relativity theorists suddenly were respectable. Perhaps also a factor freeing Chandrasekhar to pursue research in a field he earlier had shied away from as too risky was his now firmly established scientific reputation and professional position. He had been elected to the Royal Society in 1944 and received its Gold Medal in 1952, and would receive the U.S. National Medal of Science in 1967. The Chandra X-ray Observatory, launched into space in 1999, was named in his honor. Chandrasekhar's work on relativity and black holes remained the consuming scientific passion of the remaining three decades of his life. Black holes were, for

Chandrasekhar, the perfect object, their construction dependent solely on mathematical concepts of space and time. The beauty of the equations drew him to the mathematics, irrespective of any astrophysical relevance.

Chandrasekhar's intellectual legacy includes lectures in 1983 in memory of Arthur Eddington; a collection of essays published in 1987 on aesthetics and motivations in science, especially the quest for beauty in mathematical equations, which he likened in 1992 to a sequence of paintings by Claude Monet; and an analysis of Isaac *Newton's *Principia* in 1995.

Kameshwar C. Wali, *Chandra: A Biography of S. Chandrasekhar* (1991). G. Srinivasan, ed., *White Dwarfs to Black Holes: The Legacy of S. Chandrasekhar* (1999). Web Site: http://www.phys-astro.sonoma.edu/BruceMedalists.
NORRISS S. HETHERINGTON

CHAOS AND COMPLEXITY. Nature is complicated. Scientists trying to understand it have to simplify and approximate in order to discern regularity in phenomena and describe it mathematically. In the late twentieth century a new field called chaos theory emerged that instead embraced complexity and the nonlinear mathematical equations that expressed it.

Scientists and mathematicians had previously addressed the topic of complexity and nonlinear equations, notably Henri Poincaré, who worked on the theory of differential equations and dynamical systems. In 1908 Poincaré pointed out that small differences in the initial conditions of a system could result in large changes in their long-term evolution, and noted as an example the unpredictability of the weather. But Poincaré's work did not immediately spark a new line of research; physicists at the time were fruitfully exploiting linear differential equations in the development of relativity and quantum theory.

Chaos theory first emerged from the increasing use of computers in meteorology after World War II, when computer scientists viewed the complex, nonlinear problems of meteorology as a testing ground and meteorologists used computers as a way to handle their accumulations of data and to model weather systems. In 1961 Edward Lorenz, a meteorologist at the Massachusetts Institute of Technology (MIT), was running simplified atmospheric models through his computer. He decided to retrace a run, but instead of starting at the beginning he started halfway through, typing in the numbers for the initial conditions from the printout for the first run. The printout had rounded off the six decimal places used by the computer to just three, but Lorenz assumed a difference of one part in a thousand would be inconsequential. Instead he found that

the second run, from almost the same initial conditions, diverged wildly from the first. He first thought he had blown a vacuum tube, but then recognized the importance of the small difference in initial conditions.

Using a still simpler system of three nonlinear equations modeling convection, Lorenz demonstrated sensitive dependence on initial conditions, as the phenomenon was called, and cast doubt on the prospects of long-range weather forecasts. But he also revealed a sort of abstract order within the disorderly behavior that resulted: a plot of the results in three dimensions traced a complex double spiral, nonintersecting and nonrepeating yet with distinctive boundaries and structure. The image, later called a Lorenz attractor, appeared with Lorenz's results in 1963, in a paper entitled "Deterministic Nonperiodic Flow" in the *Journal of the Atmospheric Sciences*. The title asserted the persistence of determinism; the avenue of publication indicated the source of Lorenz's interest in the problem and ensured that most physicists and mathematicians would miss its initial appearance.

In the meantime Stephen Smale and several other mathematicians at the University of California at Berkeley were developing ways to model dynamical systems through topology, folding and stretching plots in phase space to reproduce the unpredictable histories of nonlinear systems; two points on the plot could be close together or far apart depending on the sequence of folds, thus exhibiting sensitivity to initial conditions. In the early 1970s mathematician James Yorke came across Lorenz's paper and publicized it in a mathematics journal, in which he applied the term "chaos" to the subject. Yorke drew on the work of Robert May, a mathematical physicist who had turned to population biology. May had found that nonlinear equations describing cyclic changes in populations could begin doubling rapidly in period before giving way to apparently random fluctuations; but within the random behavior stable cycles with different periods would reappear, then start doubling again toward randomness. Yorke explained May's results with chaos theory. A review article in *Nature* in May 1976 brought chaos to a still wider audience.

Yorke learned that Soviet mathematicians and physicists had been pursuing similar lines of research, starting with the work of A. N. Kolmogorov in the 1950s and extended by A. N. Sarkovskii, who arrived at the same conclusions as Yorke, and Yakov Grigorevich Sinai, who developed the theory in the framework of thermodynamics. Physicists in the Soviet Union, the United States, and Europe saw in chaos a way to tackle long-standing problems in fluid dynamics, especially regarding turbulence and phase transitions. The appearance of periodic order within longer-term disorder found visual expression in the geometry developed by Benoit Mandelbrot and other mathematicians in the 1970s. Mandelbrot coined the term "fractal" to describe the new class of irregular shapes that seemed to duplicate their irregularity when viewed at different scales and dimensions. In 1976 Mitchell Feigenbaum found that a single constant described the scaling or convergence rate—that is, the rate at which cycles doubled in period on the way to chaos—no matter the type of physical system or mathematical function (quadratic or trigonometric). Shortly after Feigenbaum announced the single universal scaling law, the first conference on chaos convened in Como, Italy. That was in 1977, fifty years after another conference of physicists there had considered the competing interpretations of quantum mechanics.

Chaos theory emerged from diverse disciplinary and institutional origins: from Lorenz, an academic meteorologist; to Mandelbrot, working at International Business Machines (IBM) Corp. on mathematical economics; to Feigenbaum, a theoretical physicist at Los Alamos National Laboratory. Los Alamos eventually created a Center for Nonlinear Studies, and other centers for chaos theory emerged at the University of California at Santa Cruz, at Gorky in the Soviet Union, and elsewhere. Chaos theory served to bridge disparate disciplines dealing with apparent disorder: biology, ecology, economics, meteorology, and physics. Digital electronic computers were central to all of the work. Chaos also connected abstract mathematics with real-world problems. The theory provided tools for astronomers studying the red spot on Jupiter and galactic structure, population biologists modeling the fluctuations of species, epidemiologists charting the cycles of disease, physiologists investigating cardiac fibrillations, and urban engineers tracking traffic flows. A few enterprising chaos theorists sought to predict the stock market and make investors, and themselves, rich. Some practitioners viewed chaos theory as a subset of a wider field called complexity, which studied neural nets, cellular automata, spin glasses, and other exotic systems exhibiting complex interconnections among individual components.

Chaos theory emerged in the 1960s and 1970s, a time of general cultural ferment often manifested in antiscientism. Chaos theory itself seemed to reject reductionism and determinism in favor of a holistic embrace of complexity and flux, even if it still found rules and regularity buried deeper in disorder, and it thus resonated with critics of deterministic science. Several chaos pioneers themselves drew inspiration from the romanticism of Goethe. The emergence of centers for the field in Santa Cruz and Santa Fe, towns with New Age reputations, suggest the countercultural component in the chaos community.

Benoit Mandelbrot, *The Fractal Geometry of Nature* (1983). James Gleick, *Chaos: The Making of a New Science* (1987). David Ruelle, *Chance and Chaos* (1991). George A. Cowan, David Pines, and David Meltzer, eds., *Complexity: Metaphors, Models, and Reality* (1994). Ilya Prigogine, *The End of Certainty: Time, Chaos, and the New Laws of Nature* (1997).

PETER J. WESTWICK

CHEMICAL AND BIOLOGICAL WARFARE. "In no future war will the military be able to ignore poison gas. It is a higher form of killing."—Fritz Haber, on receiving the Nobel Prize for chemistry, 1919.

The age of modern chemical warfare opened on 22 April 1915. At the urging of Fritz *Haber, Germany's leading industrial chemist, the German Army mounted a chlorine attack at Ypres in Flanders that killed five thousand men and opened a four-mile hole in the front. The new weapon immediately launched a war of measure and countermeasure. New gases included phosgene, undetectable and twenty times more powerful than chlorine, and mustard gas. British facilities on 7,000 acres at Porton Down bred test animals; Edgewood Arsenal in Maryland, unrivalled in size until the Manhattan Project, produced 200,000 gas shells and bombs per day. We remember a war of machine guns, heavy artillery, and barbed wire; but gas also became a constant presence in the Great War. By 1918 half the shells fired contained gas; mustard gas accounted for more than a quarter of American casualties. Nevertheless, armies failed to make gas a decisive weapon. It generated constant fear among the soldiers and forced them to wear heavy protective equipment, but its strategic value was primarily psychological and defensive.

During World War II gas might have provided a decisive tactical advantage, for example against the German blitzkrieg invasions of France and the Soviet Union, where rapid mobility was required; a gas counterattack, requiring the Germans to carry heavy protective equipment and to spend time decontaminating gassed sites, could have hobbled the invasions. In both cases the defenders were too quickly disorganized to mount an effective attack. The German chemical company I. G. Farben had discovered the organophosphorous compound "tabun" in 1936 during research on insecticides. This so-called nerve gas, a cholinesterase inhibitor, was far more dangerous than earlier gases. With it the Germans could have slowed the Allied landings in Normandy significantly. The Germans calculated however that Allied retaliation in kind would so burden war industries with the production of protective gear and decontaminants that the Germans would lose more than their opponents. All the major belligerents nevertheless prepared extensively for chemical warfare; the United States' Chemical Warfare Service employed 20,000 in 1942 and built thirteen new production plants between 1942 and 1945.

A biological arms race began in the 1930s. The Japanese Pingfan Institute near Harbin, under the direction of General Shiro Ishii, weaponized half a dozen different diseases, carried out human experiments on a large scale, and deployed biological agents against Chinese towns. After World War II the United States protected Ishii from prosecution in return for the information he held. The United States initiated its own program in 1942, involving twenty-eight universities, concentrating on anthrax and botulism, and manufacturing probably 250,000 anthrax bombs in 1944 for strategic bombing of German cities. Porton Down produced five million anthrax-filled cakes; botulinal toxin from a Porton grenade likely killed German security chief Reinhard Heydrich. Both the United States and Great Britain continued to develop biological weapons during the Cold War; the United States used an herbicide, code-named Agent Orange, in a massive campaign to defoliate the rainforests of Vietnam in the 1960s. Small amounts of dioxin in the herbicide accumulated, causing stillbirths and birth defects among the Vietnamese. The United States also used CS gas, an irritant similar to tear gas, against Vietnamese soldiers. The Soviet Union in the 1970s and 1980s created the largest biological weapons program in the world and probably deployed poison gases, including a mycotoxin agent known as "yellow rain," in Southeast Asia and in its war in Afghanistan. During the 1970s the United States developed binary weapons, in which two relatively harmless compounds combine to form a nerve gas inside a shell upon firing.

Unlike nuclear weapons, which require substantial industrial resources, chemical and biological weapons lie within reach of smaller nations and subnational groups. Recent uses of chemical weapons have all occurred in developing countries, where the weapons exert great effect against armies heavily weighted towards infantry. As was true at the beginning of the story, chemical weapons favor defense over offense, and they offer substantial leverage to subnational groups and smaller powers against much larger nations.

Victor A. Utgoff, *The Challenge of Chemical Weapons. An American Perspective* (1980). Robert Harris and Jeremy Paxman, *A Higher Form of Killing. The Secret Story of Gas and Germ Warfare* (1982). L. F. Haber, *The Poisonous Cloud. Gas Warfare in World War I* (1986). Ken Alibek, *Biohazard* (1999).

THEODORE S. FELDMAN

CHEMICAL BOND AND VALENCE. In the eighteenth century and earlier, chemical combination was thought to be ruled by the laws of "chemical affinity" as measured (among other ways) by the energy and proportions in which acids and bases combine to form salts. When atomic theory and electrochemistry began to be developed almost simultaneously shortly after 1800, most theorists thought that the atoms that form chemical compounds must be held together by polar electrostatic forces. The electronegativity or electropositivity of a substance could be measured by its behavior in an electrochemical cell; the sign and intensity varied with the substance, and was a primary characteristic of each element. Developed primarily by Jacob *Berzelius in the 1810s and 1820s, the theory of electrochemical dualism worked well for inorganic compounds, but by the 1830s it ran into increasing difficulties in *organic chemistry.

Certain elements seemed to be able to substitute for one another in compounds irrespective of their electrochemical character. A study of inorganic and organometallic compounds led Edward Frankland in 1852 to point out that atoms of nitrogen, phosphorus, antimony, and arsenic seemed always to combine with either three or five other atoms, regardless of electrochemistry. In the late 1850s Frankland's friend Hermann Kolbe began applying this concept to organic compounds. Even before Frankland's paper of 1852, Alexander Williamson had pointed out the "bibasic" nature of certain organic radicals, and soon he applied this concept to the oxygen atom as well. Following Williamson's lead, Charles-Adolphe Wurtz, William Odling, and August Kekulé also explored the same concept in organic chemistry, with regard to hydrogen, oxygen, sulfur, nitrogen, and ultimately carbon.

Kekulé summarized this emerging theme in two articles published in 1857–1858. He stated that atoms of each element appear to have a certain fixed capacity to combine with atoms of the same or of other elements; he called these components of combination capacity "affinity units." Hydrogen and chlorine atoms had one such unit; oxygen and sulfur two; nitrogen, phosphorus, and antimony three; and carbon four. About a decade later this concept was renamed "valence." Kekulé wrote that the formula for water, H_2O, signifies two monovalent hydrogen atoms combined with one divalent oxygen atom; ethane, H_3CCH_3, holds together because each methyl group (CH_3) has one unused valence unit of the tetravalent carbon atom, and the two valences satisfy each other in hooking together; and so on with other formulas.

This theory of atomic valence seemed to be supported by abundant evidence, and most leading chemists rapidly accepted it. Kekulé used valence concepts to develop a theory of "chemical structure." Carbon atoms, he wrote, could use some of their valences to bond together to create carbon "chains," forming the "skeleton" of a molecule. Following the valence rules, atoms of other elements, such as hydrogen, oxygen, and nitrogen, could add onto (or into) this skeleton, to form molecules of (potentially) all the organic compounds then known.

The gradual formulation of accessory assumptions, such as that of multiple bonds (ethylene as $H_2C{=}CH_2$, formaldehyde as $H_2C{=}O$, and so on), demonstrated that the structure theory could cover a wide range of organic formulas. In 1865 Kekulé showed how carbon tetravalence could be used to derive a promising candidate formula for the crucially important benzene molecule, C_6H_6, the prototype of all so-called aromatic compounds. This formula suggested a closed ring of six carbon atoms with alternating single and double bonds, and with a hydrogen atom attached to each carbon. Chemists soon accepted Kekulé's benzene theory.

Despite their important roles in its initial formulation, Frankland and Kolbe rejected the full implications of valence and structure theory, though Frankland's resistance collapsed quickly in the early 1860s. One problem they identified was the number of additional assumptions necessary to make the theory work well. Moreover, it was not clear whether valence was a constant property, characteristic of each element, or variable with chemical circumstances, as Frankland had initially proposed. Disputes about the fixity of valence continued for decades.

Some chemists objected that the valence connections between atoms resembled crude "hooks" or "glue" rather than an isotropic (spatially uniform) natural force such as gravitational or electrical attraction. Frankland referred to valence connections as chemical "bonds," but demurred at describing them further. Gravitation seemed to be excluded as the cause of valence owing to the complexity and stability of chemical compounds; electrical attraction seemed impossible because atoms of a single element could bond to each other. Therefore, accepting the idea of valence appeared to require renouncing the possibility of understanding chemical affinity as a familiar macroscopic physical force. It was a new and uniquely chemical concept.

The residual anomalies associated with Kekulé's highly successful benzene theory led to new proposals by Lothar Meyer, Henry Edward Armstrong, and F. K. Johannes Thiele that invoked partial or center-directed valence bonds. In 1874 J. H. van't Hoff and Joseph Le Bel began to explore valence bonds in three dimensions (see STEREOCHEMISTRY). In the 1890s Alfred Werner used a modified valence theory successfully to represent the molecular composition of certain complex inorganic substances ("coordination compounds"). In this endeavor he made use of the recently formulated theory of ionization of Svante Arrhenius.

When J. J. Thomson demonstrated the existence of the electron (1897), some physicists and chemists immediately began to inquire whether this offered a new way to understand valence. In 1904 Thomson developed an atomic model in which electrons were supposed to circulate in exterior shells. In the same year in Berlin, Richard Abegg formulated a more explicit "rule of eight," corresponding to periodic valence regularities.

During and just after World War I, Walther Kossel and Gilbert N. Lewis began independently to develop electronic theories of the chemical bond, a concept fruitfully extended shortly thereafter by Irving Langmuir. Neutral atoms have as many electrons outside the nucleus as protons within (see ATOMIC STRUCTURE). In the new theory, the second and third periods of the *periodic table each have eight members, the last of which has a stable nonbonding "octet" of electrons in a shell. Beyond the octet shells are the odd electrons in the outer shell, the "valence electrons," which can be shared with adjacent atoms to form chemical bonds. For instance, aluminum, with atomic number 13, is eleven places past helium in the periodic table, or three places past the first stable octet. The aluminum atom therefore has three valence electrons available to share with other atoms—a valence of three.

Langmuir thus distinguished an "ionic" bond, where an electron transferred from an electropositive to an electronegative atom, from what he called a "covalent" bond, where the two electrons, one from each atom forming the bond, were shared more or less equally. In the first case, coulombic attraction held the atoms together; in the second case, electricity was involved in a manner not understood until the advent of *quantum mechanics.

Neils *Bohr and his collaborators began to develop their theories of electronic structure by examining atomic spectra with reference to quantum principles and the basic patterns revealed by the periodic table. Meanwhile, work on wave mechanics by Louis de Broglie and Erwin *Schrödinger began to provide theoretical explanations for the spatially directed nature of valence bonds. In the late 1920s and early 1930s, Walter Heitler, Fritz London, John Slater, Linus *Pauling, and others developed "valence-bond theory" as an application of the new quantum mechanics of Erwin Schrödinger and Werner *Heisenberg. This involved constructing wave functions to represent the older Lewis-style electron pairs of a covalent bond.

At about the same time, Robert Mullikan developed an alternative quantum-mechanical technique for understanding chemical bonding based on what he called "molecular orbitals." Erich Hückel applied both valence-bond and molecular-orbital methods to the problem of aromatic compounds and found Mullikan's method superior. The molecular-orbital approach seemed to many to provide a cleaner and more satisfying model, and after World War II it gradually displaced the valence-bond model championed by Pauling.

Both valence-bond and molecular-orbital approaches had led to the idea of "resonance": the bonds between the carbon atoms of the benzene ring could not be considered either as single or as double bonds, but rather as uniform "resonance hybrids" halfway between the two states. Theories of resonance have successfully accounted for the curiously passive character of aromatic compounds toward addition to the double bonds of the ring, and the concept of "aromaticity" has broadened to include nonbenzenoid compounds.

After welcoming and then abandoning electrical theories of chemical bonding in the course of the nineteenth century, chemists in the twentieth century, armed with physical theories of the atom and quantum-mechanical principles, embraced them once more, and with considerable success. Chemists still have a long way to go, however, in developing theoretical models for the observed chemical behavior of molecules.

Colin A. Russell, *The History of Valency* (1971). Mary Jo Nye, *The Question of the Atom* (1984). A. J. Rocke, *Chemical Atomism in the Nineteenth Century* (1984).

A. J. Rocke

CHEMICAL COMPOUND. The term "mixt" that seventeenth-century chemists used to designate composed bodies made little distinction between chemical combinations and physical mixtures. During the eighteenth century, however, a clear separation occurred. In his *Dictionnaire de chymie* (1766), for example, Pierre Joseph Macquer distinguished between a "*chemical combination or composition...* in which there must be

in addition a mutual adherence between the substances that combine," to avoid "*mixts or mixtion*, by which one might mean only a simple mixture, a simple interposition of parts."

Georg Ernst Stahl had introduced the idea of orders of composition, ranging from the union of simple elements or principles upward, and gave the successive levels the names "*mixtes, composés, décomposés*, and *surdécomposés*." Macquer preferred *composés* of the first, second, third, and fourth orders. English-language chemical texts generally translated *composé* as *compound*.

Eighteenth-century definitions of compounds did not include the criterion that their constituent parts must be combined in fixed proportions. Nevertheless, it was common knowledge that the class of compounds most often studied, the neutral salts, were each composed of an acid and a base that had to be mixed in particular proportions to saturate one another (to create a compound that did not turn litmus either red or blue). In 1754 Guillaume-François Rouelle described a new class of "neutral" salts that contained an excess of acid, formed crystals, and turned litmus red. The additional acid required to form such a salt turned out also to be in a fixed proportion to the quantity necessary to form the neutral salt, as Torbern Bergman noted in 1775, when he included acid salts in his table of affinities.

Antoine-Laurent *Lavoisier devoted great effort to establishing the proportions by weight of the carbon and oxygen combined in "fixed air" (later carbonic acid gas), and of the inflammable air (later hydrogen) and oxygen that formed water. After demonstrating that metals and nonmetals absorb oxygen when calcined or burned, Lavoisier also determined their combining proportions, and recognized that some of them exhibited up to four degrees of oxygenation. Lavoisier had no strong theoretical reasons for assuming that the constituents of these compounds combine in constant proportions; he determined the proportions to implement his "balance sheet" style of chemical reasoning. At the end of the eighteenth century, no precise ideas about the nature of chemical compounds could explain the empirical evidence about constant proportions. Claude Louis Berthollet commented in 1803 that the notion that compounds consist of ingredients in fixed proportions was a mere "hypothesis" based on an unwarranted distinction between solutions and combinations.

Berthollet believed that substances acted on one another chemically with a force proportional to their respective affinities and their masses. In solution particles combined in whatever proportions they were present. If, however, a particular proportion produced a combination having maximum volatility or maximum cohesion, the combination separated out as a substance characterized by that proportion. Gases displayed "more uniform proportions than other combinations" because when their constituents combine they undergo much greater contraction. Thus Berthollet gave, for the first time, explanations derived from a general theory of chemical combination that accounted for "the dispositions and circumstances that determine the fixed proportions in certain combinations." In doing so, however, he asserted that fixed proportions resulted from special circumstances that "interrupt" the normal chemical action of particles on one another.

In a debate famous for its civility and the competence of both parties, as well as the importance of the issues it raised, Joseph Louis Proust challenged both Berthollet's general conception and the experimental evidence on which he had based his claims that various salts and oxides can form in a continuous range of proportions. Proust made a fundamental distinction between solutions, alloys, and glasses, on the one hand, and "true compounds" on the other. A compound, he asserted, "is a privileged product to which nature assigns fixed proportions....The characters of true compounds are as invariable as the ratio of their elements." Though he provided extensive experimental evidence for the fixed proportions of substances among the contested cases, Proust could find no precise definition of a true compound to differentiate it from a solution or other homogenous substances whose proportions are variable.

By identifying the relative weights of the elements as their defining property, John Dalton transformed the meaning of combining weights. Their determination was no longer an end in itself, but the means by which to find out the "number of simple elementary particles which constitute one compound particle, and the number of less compound particles which enter into the formation of one more compound particle." The doctrine that compound bodies consist of small whole numbers of atoms made their constant proportions no longer the outcome of analytical experience, but essential to their definition.

John Dalton, *A New System of Chemical Philosophy, Part I* (1803). J. R. Partington, *A History of Chemistry*, vol. 3 (1962). Torbern Bergman, *Dissertation on Elective Attractions* (1968).

FREDERIC LAWRENCE HOLMES

CHEMICAL EQUIVALENT. William Wollaston created the expression and concept of "chemical equivalent" or "equivalent weight" in 1814. He wanted to avoid the hypothetical part of John Dalton's atomic theory while addressing the pragmatic need for a set of elemental weights to calculate stoichiometry and manipulate formulas (*see* ATOM AND MOLECULE; ATOMIC WEIGHT). In the most general sense, Wollaston defined an "equivalent" as the relative amount of one element that combines with or replaces another chemically. Since in water one part by weight of hydrogen combines with eight parts of oxygen, they were regarded as equivalent quantities.

Given the central importance of oxygen and hydrogen for chemistry, a more specific version of Wollaston's definition soon became well established: the quantity of any element that combines with eight parts of oxygen or one part of hydrogen. The relative proportions of potassium and sodium in their respective oxides provided chemical equivalents for them; analysis of potassium chloride gave a value for chlorine, which could also be derived by examining hydrogen chloride. For hydrogen = 1, chlorine = 35.5, sodium = 23, and potassium = 39, since 23 parts of sodium combine with 35.5 of chlorine or 8 of oxygen, 1 of hydrogen with 35.5 of chlorine, and so on.

The concept of chemical equivalent proved popular among chemists during the nineteenth century. It served as a rallying point for opposition to atomic theory, since it seemed to provide a route to molecular formulas and reaction equations without hypotheses. Unfortunately, the definition had difficulties and its application in concrete cases was not always secure. Multiple proportions presented the most intractable problem. For example, the chemical equivalent of oxygen is eight in water but sixteen in hydrogen peroxide. Similarly, the chemical equivalent of carbon in the various hydrocarbons had almost as many different values as there were known hydrocarbons.

The principal architect of atomic theory in the first half of the century, J. Jacob *Berzelius, opposed replacing atomic weights with chemical equivalents. He understood that the system of equivalent weights as it actually developed rested ultimately upon assumptions and conventions as much as chemical atomism rested upon theoretical constructs. Nevertheless, in the early 1840s there was a Europe-wide movement to adopt equivalent weights in preference to atomic weights. This consensus survived until the development of theories of valence and structure in the 1850s and 1860s. At first, structuralists salvaged the concept of equivalent weight by defining it as "atomic weight divided by valence." This formulation applies properly, however, only to simple hydrides and halides, and moreover valence is itself a variable property.

During the 1860s most chemists agreed to use reformed atomic weights nearly identical to those used today. In France, however, equivalent weights in formulas survived until the 1890s; many leading chemists there regarded them as more empirical and positive than atomic weights. Since then, the concept of chemical equivalent has rarely been taught or used.

Alan J. Rocke, *Chemical Atomism in the Nineteenth Century* (1984).

A. J. ROCKE

CHEMICAL LABORATORY. See LABORATORY, CHEMICAL.

CHEMICAL NOMENCLATURE. Before the mid-eighteenth century the language of chemistry was often haphazard and ambiguous. In the 1780s Antoine-Laurent *Lavoisier and his colleagues created a new, systematic nomenclature that became the adopted standard, providing a basis upon which subsequent additions and amendments arose in the nineteenth and twentieth centuries. By the beginning of the twentieth century, codification of nomenclature had come to lie in the hands of international committees of chemists rather than in the preferences of local groups.

The French nomenclature of 1787 was the work of Louis-Bernard Guyton de Morveau (who began the project in 1782) and Antoine François de Fourcroy, Claude Louis Berthollet, and Lavoisier. Theirs was not the first binomial nomenclature. Precedents occur in Oswald Croll's 1609 *Basilica chymica* and in Torbern Bergman's mid-eighteenth-century efforts to establish a Latin nomenclature in chemistry analogous to the natural history taxonomy of his Swedish colleague Carl *Linnaeus. In 1746 the Royal College of Physicians published a dictionary, which influenced Pierre-Joseph Macquer's dictionary of chemistry (1766), using the principle that names of substances should reflect their constituents rather than their observable properties or geographical origins.

Lavoisier provided philosophical legitimation for the new nomenclature by grounding it in Étienne B. de Condillac's philosophy of language and in the argument that a language constructs a science ("une langue bien faite est une science bien faite"). The new nomenclature eliminated phlogiston from scientific vocabulary and from scientific theory. It organized thirty-three simple substances into four categories and named a chemical compound by the two elements or radicals (roots) supposed to compose it. The

system subordinated the traditional languages of pharmacy, metallurgy, and textiles to a new dualistic logic. White of lead became lead oxide and stinking air became sulfuretted hydrogen gas. On logical grounds, Lavoisier's "oxygen" (acidifying principle) should have been renamed once Humphry Davy established that muriatic acid (hydrochloric acid) does not contain *oxygen. But the name oxygen and the French system were here to stay.

German chemists accepted the system, if not all the French names. For Germans, oxygen remains the ill-named "Sauerstoff" (acid stuff); hydrogen is "Wasserstoff," carbon "Kohlenstoff." In the long run, the system gave way to some traditional naming practices: observable properties provided the naming principle for the new elements chlorine (Greek *chloros*, "green") and bromine (*bromos*, "stink"), just as national pride determined the naming of gallium, germanium, scandium, and polonium (*see* ELEMENT).

With the establishment of the structural theory of organic chemistry in the 1860s, single hydrocarbon chains (or backbones) became the basis for naming organic substances, with branching chains given the names methyl, ethyl, and so forth, and number-prefixes indicating the position on the chain of substituents. As early as 1865 August Wilhelm Hofmann suggested the still-used endings of "ene" for hydrocarbons with one double bond, "diene" for two double bonds, and "ine" or "yne" for triple bonds. The presence of functional groups was indicated by the suffixes "ol" for alcohol, "al" for aldehydes, "one" for ketones, and "oic acid" for acids.

The International Conference on Chemical Nomenclature, chaired by the French organic chemist Charles Friedel, met in Geneva in 1892 and standardized these and other conventions in a series of sixty-two resolutions. Still, some commonplace names remained in standard use, for example, lactic acid rather than alpha-hydroxypropanoic acid and glycine instead of 2-aminoethanoic acid. The International Union of Pure and Applied Chemistry, created in 1919, excluded German scientists, used French as its official language, and submitted increasingly to influence from American chemists. It reorganized after World War II with English as the official language.

As noted by historians of chemistry, the rules and codes of language adopted at international nomenclature meetings, for example, those at Liège in 1930 and Rome in 1938, largely conformed in practice to the customs of naming in the standard reference works of *Chemical Abstracts*, published in the United States, and *Beilstein's Handbuch der Organischen Chemie*, published in Germany. Thus, the eighteenth-century ideal of a rigorous system coexisted in the twentieth century with pragmatic and simple names common in everyday laboratory use.

Aaron J. Ihde, *The Development of Modern Chemistry* (1964). William H. Brock, *The Norton History of Chemistry* (1992). Bernadette Bensaude-Vincent and Isabelle Stengers, *A History of Chemistry* (1996). Bernadette Bensaude-Vincent, "Languages in Science: Chemistry," in *Modern Physical and Mathematical Sciences*, ed. Mary Jo Nye, vol. 5, *The Cambridge History of Science*, 8 vols. (in press).

MARY JO NYE

CHEMISTRY. The term chemistry first appeared in references to a practice consolidated in Alexandria at the beginning of the Christian era, and known later as the "Egyptian" art. Historians of chemistry have characteristically distinguished chemistry from *alchemy, a product of the Arabic and European Middle Ages and later depicted as a secrecy-laden search for methods to produce noble metals by transmutation and to obtain the mysterious philosophers' stone. According to this story, chemistry emerged during the late sixteenth and seventeenth centuries. Drawing on practices of alchemy, metallurgy, and herbalist distillers, chemistry distinguished itself from its earlier roots both by its openness and its goal:

the determination of the composition of substances drawn from the plant, animal, and mineral kingdoms by separating them into their elementary constituents. Recent historians, however, believe that until well into the seventeenth century alchemy and chemistry were synonymous terms that denoted a broad range of inquiries, and that only after practitioners of chemistry wished to distance themselves from the seekers of gold and the philosophers' stone did they narrow the definition of alchemy to activities they discountenanced.

The remarkable stability of the identity of chemistry despite deep mutations in its aims and conceptual frameworks rests on the nature of the place in which it has been practiced, from the earliest times of which we have records to the present. The original meaning of *"laboratory" was the space in which chemists "elaborated" chemical and medicinal substances. Chemists traditionally assembled in their laboratories a characteristic set of apparatus capable of a well-defined repertoire of operations. With furnaces and *distillation apparatus constructed from components made of earth, metal, or glass, they attempted through the agency of *fire to separate substances into their volatile and fixed components. Filtration, solution, precipitation,

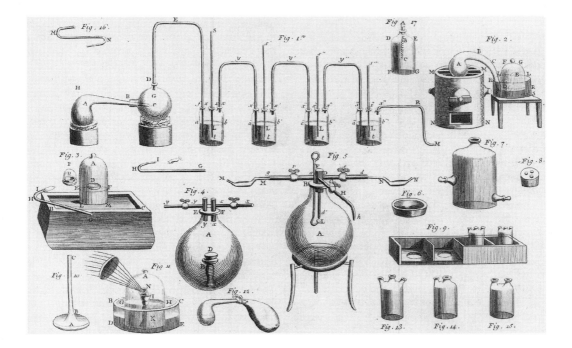

Lavoisier's apparatus: Figure 1, for collecting gases released in chemical operations; Figure 2, for demonstrating the fixation and production of oxygen by calcining mercury and reducing the calx; Figure 5, for synthesizing water.

maceration, solvent extraction, and other operations were usually subsidiary to the central procedures of distillation and sublimation.

During the seventeenth century, a series of textbooks, of which the prototype was Andreas Libavius's *Alchemie* (1597), organized knowledge of chemical substances and operations into teachable form. They culminated in the popular *Cours de Chymie* of Nicolas Lemery, first published in 1673. He and other chemists in France gave chemical lectures and demonstrations at the Jardin des plantes and in their apothecary shops. In the German principalities, chemical instruction entered the universities during the seventeenth century, an early example being the teaching of chemical medicine at the University of Marburg (1609). By the end of the century, chemistry had become a sufficiently distinct branch of natural knowledge to form, together with *physics, *anatomy, and *botany, the physical section of the French Academy of Sciences. In England, Robert *Boyle led those who sought to make chemistry part of the broadly forming new natural philosophy.

Seventeenth-century chemists interpreted their separations by theories of matter through hybrids drawn in part from the three "principles" (salt, sulfur, and mercury) adapted by Paracelsus from alchemical views, and in part from the four-element theory of Aristotle (*see* ELEMENTS, CHEMICAL). The most common compromise identified the products of a distillation as oils (characterized by the combustible principle, sulfur); "spirituous" liquids (mercury); insipid liquids (water); substances soluble in water (salt); and fixed insoluble substances (earth). At a more pragmatic level, there gradually emerged a scheme in which *metals, alkalis, and "absorbent earths" could be separated from and recombined with the three known mineral acids and a "vegetable" acid. Boyle's systematic application of color indicators able to detect the presence of acids and alkalis facilitated these identifications.

The incorporation of the mechanical philosophy into chemistry during the second half of the seventeenth century abetted the interpretation of chemical processes as consequences of the particular shapes of the ultimate participating particles. Sharp-pointed acid particles embedded themselves in pores in alkali particles, resulting in a salt that displayed neither the properties of the acid nor the alkali, but from which both could later be recovered.

During the first half of the eighteenth century, chemists gradually subordinated early theories of matter to the more pragmatic concept of neutral salts formed of an acid and a base. Elaborate mechanistic explanations gave place to the more generic particulate picture. In 1718, Etienne-François Geoffroy systemized the well-defined chemical changes then known in a Table of Rapports, the columns of which indicated the order in which substances would replace one another in combination with the substance shown at the top of each column. During the following decades, chemists adopted *Newton's idea that differential short-range forces of attraction, or *"affinities," drove these replacements.

Chemists continued to support their activities largely from apothecary shops. The teaching of chemistry in German medical faculties expanded. A group of chemists in the reformed Paris Academy of Sciences became a leading forum for the pursuit of experimental chemistry in the first three decades of the eighteenth century. At the Prussian Academy of Sciences, reformed by Frederick II, Andreas Marggraf and Johann Pott became prominent after 1750 in the investigative expansion of the chemistry of salts and in their application to both plant and mineral analysis. The growing sense of a chemical community in Germany gave rise in 1778 to the first specialized chemical periodical, *Chemisches Journal*, edited by Lorenz Crell.

Because fire figured so prominently in their operations and distinctions, chemists were much concerned to explain combustion. Chemists long associated combustibility with the sulfur principle because common sulfur conspicuously burned. The dominant form of this approach in the eighteenth century was Georg Ernst Stahl's theory of *phlogiston. Stahl recognized that the calcination of metals and the reduction of metallic ores to their metallic form, as well as the relationship between sulfur and vitriolic acid, could all be explained as exchanges of the same principle. Phlogiston served to link an extensive series of the most important chemical changes known at the time into a coherent system.

Until the late eighteenth century, chemists operated mainly on solids and liquids. Since vapors could not be handled by ordinary techniques, they were usually neglected in accounting for the substances that entered or left during a chemical change. Atmospheric air was regarded as an elementary, homogeneous, elastic fluid, which could contain various impurities. In 1756, following up earlier work by Stephen Hales (*see* PNEUMATICS), Joseph Black showed through quantitative balance experiments that ordinary alkalis contain a distinct species of air, which he named "fixed air." During the next two decades, several British natural philosophers discovered other "airs." In 1775, the most prolific of these discoverers, Joseph Priestley, discovered (by reducing mercury without charcoal), a "purer and more respirable" air than the atmosphere. Extending the phlogiston theory to explain the properties of the various new airs, Priestley named his discovery "dephlogisticated air."

In 1772, Antoine Laurent *Lavoisier took up the problem of the absorption and release of air. His discoveries that phosphorus and sulfur gain weight when burned, and metallic calces lose weight when reduced, led him to a distinctive quantitative style of experiment, the so-called "balance-sheet method," which many historians identify as the basis of the modern science of chemistry. In 1777, Lavoisier proposed a new general theory of combustion, dispensing with phlogiston, in which "pure" air combined with metals and combustible bodies, releasing the material principle of heat. Soon he renamed this principle "caloric" and pure air *"oxygen," which, as its Greek root implies, is regarded as the principle of acidity.

By 1785, Lavoisier's demonstration of the decomposition and synthesis of water consolidated his theoretical structure and converted several influential French chemists to his side. They lobbied chemists who visited Paris, founded a new journal, and collaborated to reform *CHEMICAL NOMENCLATURE. Although Priestley never acceded to the "French chemistry," other British chemists, as well as the Germans who remained loyal to the Stahlian viewpoint until 1790, rapidly adopted the essential features of Lavoisier's reform during the last decade of the century.

To help consolidate what supporters and opponents alike were calling a revolution in chemistry, Lavoisier published a *Traité élémentaire de chimie* (1789) that summarized the experiments and theories on which he had based his movement, incorporated into the new framework earlier knowledge of the chemistry of salts, and presented recent applications of the new theory of combustion to organic substances. In addition, the *Traité* replaced the traditional chemical elements, which Lavoisier regarded as "metaphysical," with a pragmatic definition of elements as substances that "we have not been able to decompose by any method." His list of elements provided the basis from which the modern *periodic table of elements has grown.

In principle, eighteenth-century chemists recognized a hierarchy of levels of composition, from elements to various orders of "mixtes," but the only levels generally established through analysis and synthesis were salts and their constituent acids, metals, alkalis, and alkaline earths. Chemists often assumed that these constituents were present in fixed proportions because a specific quantity of an acid was required to neutralize a particular quantity of a base. At the end of the century, Claude-Louis Berthollet challenged this assumption. From the principle that substances combine because of general laws of attraction, Berthollet inferred that compounds could form in an indefinite series of proportions, depending on the relative masses of the constituents placed together. Joseph Proust argued that true compounds occurred only in fixed proportions. Their debate ended indecisively, but contributed to a clearer distinction between *chemical compounds and solutions, mixtures, or alloys.

Along with his attack on fixed proportions, Berthollet broke with eighteenth-century affinity theory, according to which a substance with greater affinity for one of two substances combined together will completely displace the substance of lesser affinity. Instead, according to Berthollet, an equilibrium sets in that depends on the relative masses of each of the substances present. His idea took hold gradually, culminating in the 1860s to quantitative laws of *mass action.

Considerations about the composition of the atmosphere led John *Dalton during the first years of the nineteenth century to identify chemical atoms with characteristic *atomic weights. By assuming that when only one compound of any two elements was known it consisted of one atom of each, that when two were known one of them consisted of one atom of each, the second of two of one and one of the other, and so on, Dalton calculated, from the measured proportions by weight of the elements, the relative weights of their atoms. In 1808, he published atomic weights for thirty-six elements relative to hydrogen = 1. Dalton conceived of his atoms as hard, spherical bodies, and depicted their assumed spatial arrangements in the compounds they formed.

Several prominent English chemists, including Thomas Thomson and William Hyde Wollaston, adopted Dalton's atomic theory after obtaining independent evidence for multiple combining proportions. To avoid commitment to the reality of the atoms, Wollaston referred to "equivalent" rather than atomic weights, but he calculated them in Dalton's manner. In Sweden, Jöns Jacob *Berzelius accepted the general principle of the atomic theory, but established more rigorous standards of experimental precision for determining atomic weights. Those that he established during the first two decades of the century were generally accurate to within 1 percent. In place of Dalton's "law of simplicity," Berzelius devised rules and analogies by which to decide on the empirical formula for a given compound. Since chemists operated with several sets of rules, they produced atomic weights for some elements that differed by small integral multiples.

The invention of the *battery (pile) by Alessandro *Volta in 1800 influenced chemistry powerfully. Anthony Carlisle and William Nicholson found even before Volta's paper was published that the current generated by the pile can decompose water into its elements. During the next decade, Humphry Davy passed a current through alkaline solutions to isolate the metals potassium and sodium. Later he recovered magnesium, strontium, calcium, and barium from amalgams of the alkaline earths. Because the electric current could overcome the force of the affinities supposed to hold substances together in compounds, Davy inferred that affinities were electric forces. Berzelius, who also experimented extensively with Voltaic currents, developed a general theory of chemical composition based on the same assumption. According to Berzelius's "dualistic theory," oxides formed between highly negative oxygen and positive, or less negative, atoms. The remaining net positive or negative charges of these oxides allowed them to combine into neutral salts and other more complex compounds.

The discovery in 1810 by Joseph Louis Gay-Lussac that gases combine in ratios of small whole numbers by volume provided a potential resolution of uncertainties over atomic weights. One needed only the additional assumption that a given volume of any gas contains the same number of atoms (or molecules). Some, including most famously Amedeo Avogadro and André Marie Ampère, devised theories based on this conclusion, but Dalton rejected the generalization that all gases contained the same number of particles in the same volume. The consequence that individual molecules of several elementary gases must then be composed of multiple identical atoms caused other prominent chemists to reject or restrict the generalization.

By the second decade of the nineteenth century, chemists had established fundamental principles sufficient to support rapidly growing investigative and educational activities. New institutional arrangements strengthened the role of chemistry in higher education. In France, the École Polytechnique, set up during the Revolution, included lectures and laboratories in chemistry. Leading chemists, such as Gay-Lussac and Louis Jacques Thenard, became professors there, at the Collège de France, or at the University of Paris, and gave lectures that attracted large audiences. Aspiring foreign chemists came to Paris to learn from these masters of rigorously logical exposition and experimental demonstration. In London, Davy achieved great popularity through his chemical lectures at the Royal Institution. Chemical instruction took place in Germany in several universities, and especially in small proprietary pharmacy schools. One of those who began his education in Germany, but continued it in Paris, was Justus *Liebig. After working with Gay-Lussac, Liebig returned to his native state of Hesse in 1824 with an appointment to teach chemistry at the small University of Giessen. There he developed a teaching laboratory, intended at first to train pharmacists toward instruction in general chemistry. By the mid 1830s, Liebig's laboratory was training dozens of chemists and pharmacists in a systematic program that moved from elementary exercises to original research projects.

The laboratories in which these activities took place resembled earlier ones but now were also centers for innovations in apparatus and procedures. Lavoisier had initiated the break from the traditional array of material objects that had long characterized the places where chemists worked. Sometimes adapting and combining customary retorts and pneumatic troughs, sometimes designing, for particular purposes, apparatus that was quite unfamiliar, Lavoisier both increased the range and the cost of chemical equipment dramatically. Some of his designs, such as the Baroque gazometer depicted in his *Traité*, his successors simplified. But by the 1820s, chemists so regularly designed and redesigned apparatus for more special operations and for better precision that publications began to appear to keep them informed of the latest developments in laboratory organization and equipment.

As modern chemistry emerged in the nineteenth century, the traditional small-scale commercial producers of various chemicals grew into a heavy chemical industry. By the 1830s, factories rather than shops produced sulfuric acid and soda. Historians often treat this early stage of industrialization as independent of the *Scientific Revolution. Chemical crafts and deliberate chemical experimentation with the intention to advance the science were closely interwoven by the early eighteenth century. What each owed to the other differed from case to case.

By the mid-nineteenth century, chemists comprised the most prominent class of the new genre of *scientists. With roots in long-established practices, driven forward first by the chemical revolution identified with Lavoisier, and then by the atomic theory associated with Dalton, chemists led the way into what fin-de-siècle spokesmen called the "century of science." The laboratories that had been their peculiar abodes for several centuries now became the prototypes for the places in which other fields, such as physiology, began to establish themselves (*see*

INSTITUTE). Whatever may have been the source of the earlier development of industrial chemistry, chemists were confident that further advances depended on the application of chemical science to factory production. By the 1840s, Justus Liebig could proclaim in his *Familiar Letters on Chemistry* that the wealth of a nation could be measured by its chemical production.

No sooner had chemistry crystallized as a modern scientific discipline than it began to split into subfields. First, *organic chemistry separated from *inorganic chemistry. Lavoisier had already applied his theories and methods of combustion analysis to determine the elementary composition of plant and animal matters. A generation later, several leading chemists took up the problem of obtaining the quantitative precision that had eluded him. By the 1820s, they had shown that organic compounds conformed to the laws of combining proportions. The application of principles of composition derived from the study of inorganic compounds ran into severe difficulties. The discovery of compounds identical in elementary composition, but different in properties, showed that empirical formulas did not fully characterize them, and set off decades of controversy. A series of "radical" theories proposed during the 1830s posited groups of elements that remained constantly associated while taking part in the reactions that transformed the organic compounds of which they formed constituent parts. The substitution theory asserted, on the other hand, that one element within a radical or compound could be exchanged for another, such as chlorine for hydrogen, without deeply altering the properties of the molecule.

Meanwhile, as the focus in organic chemistry shifted from the analysis of substances derived from plants or animals to the derivation of carbon compounds from them by partial decomposition, substitution, and later by synthesis, another field, most often called physiological chemistry, arose to take up the problems lost in the shift. By the early twentieth century, physiological chemistry had evolved into *biochemistry. Liebig's efforts, beginning in 1840 to apply the new chemical knowledge to improve agriculture, had already become a focal point for the formation of agricultural chemistry.

Around 1850, organic chemists still used competing theories of composition, different systems of atomic weights, and several versions of the letter symbols that Berzelius had introduced to represent atoms. In addition to using empirical formulas, chemists used various "rational" formulas that grouped the atoms within a molecule into several "types." Generally they regarded these types not as literal descriptions

of the spatial arrangement of the atoms, but as means to classify organic compounds, to understand their relationships to one another and their reactions. From these efforts emerged generalizations about the proportions in which elements and radicals combined with one another that gradually produced theories of chemical combination applicable to both organic and inorganic compounds. Put in their most general form by August *Kekulé beginning in 1857, these new views underwrote the classical theory of *valence and the basis for structural formulas of organic compounds. Organic chemistry enjoyed a rapid expansion, remaining the dominant subfield of chemistry for the rest of the century.

Debates over the constitution of organic compounds during the 1850s ended in reforms of the system of atomic weights that achieved consistent interpretations of the relations between compounds. Among the most influential proposals were those of Charles Gerhardt, based in part on the views of Auguste Laurent. These efforts culminated at the first international Congress of chemists, which met in Karlsruhe in 1860 to standardize atomic weights and chemical notation. There, Stanislaus Cannizzaro circulated a proposal that although not immediately accepted, afterward quickly resolved the outstanding differences.

From the time it first coalesced during the seventeenth century, chemistry shared a wide borderland with other branches of physical science. Its theories of matter borrowed from Aristotelian natural philosophy and then from the mechanical philosophy. Even within their own operational domain, chemists of the eighteenth century distinguished between the "physical" and the "chemical" properties of the substances they studied. At the end of the eighteenth century, the exploration of the properties of heat (*see* FIRE AND HEAT) and of gases again blurred the boundaries between chemistry and what was by then emerging as a discipline of experimental physics. The rapid growth of chemical investigations oriented around accurate combining proportions, the atomic theory, and the isolation and analysis of organic compounds seemed to distance chemistry from a physics defined around a different set of problems and methods. Nevertheless, the investigation of electricity (*see* ELECTROLYSIS) and other phenomena common to both fields often pierced the boundaries. During the second half of the nineteenth century, a new subfield arose, deliberately situated at this persistent interface.

The discipline of *physical chemistry was formed by Jacobus Henricus van't Hoff, Svante Arrhenius, and Wilhelm Ostwald. They concentrated on the application of thermodynamics to chemical processes (*see* THERMODYNAMICS AND STATISTICAL MECHANICS), the causes of chemical affinity, and the properties of solutions. The central importance of Arrhenius's theory of ionization to the new field earned them and their followers the sobriquet "ionists." Begun in Germany, the field spread most rapidly after 1900 in the United States under the leadership of a generation trained in the laboratories of Ostwald and Van't Hoff.

Until nearly the end of the nineteenth century, the atom as conceived and employed by chemists and physicists had little in common. The physical atom, usually called "molecule," was defined mainly by the kinetic theory of gases, while chemical atom functioned as a unit of composition defined by its atomic weight. The situation changed after physicists identified the *electron and the nucleus as the principal sub-units of the atom. During the first decades of the twentieth century, several chemists proposed theories of chemical affinity and valence based on the view that electrons in the outer shells of atoms form chemical bonds. The most influential of these theories, published by Gilbert N. Lewis in 1916, connected the electronic structure of the atom with the properties of the elements defined by their place in the periodic table. Reviving in a new form the electrochemical theory of Berzelius, Lewis postulated that covalent bonds consist of an electron pair shared by two atoms, whereas an ionic bond results from the transfer of an electron from one atom to another atom to form a pair that holds the molecule together electrostatically.

The development of physical chemistry and the elucidation of the electronic structure of the atom gave rise to a school of chemists who sought to describe the reaction mechanisms of organic compounds in terms of the positive and negative regions of molecules, the displacements of electrons, and the formation of transient intermediate compounds. Systematized by Christopher Ingold during the 1930s, these efforts culminated in the formation of the subfield of physical organic chemistry.

Meanwhile, *quantum mechanics declared that the properties of the chemical elements and their combinations could be explained by the basic laws of physics. It referred the chemical bond (*see* CHEMICAL BOND AND VALENCE) to the distribution of the electrons believed to form both ionic and covalent bonds. In the event, however, the quantum mechanical equation could not be solved for complex molecules. Chemists and physicists had recourse to approximate solutions, combined with other knowledge of the properties of compounds and the nature of their reactions. The leading figure in this development was Linus *Pauling.

Chemical laboratories and chemical industries continued to grow in size and complexity during the second half of the nineteenth century. The synthetic dye industry became possible through the growing capabilities of chemists both to synthesize naturally occurring organic compounds and to produce previously unknown compounds by modification of the natural ones. German industry led the way in this development. At first drawing on the knowledge of academic chemists, these industries soon began to hire chemists to work directly on the discovery of new dyes, creating the first industrial research laboratories (*see* LABORATORY, INDUSTRIAL RESEARCH). The special problems and expertise required to scale up laboratory operations for volume production gave rise to the profession of chemical engineering.

During the early twentieth century, new classes of chemical industry emerged. The increasing worldwide demand for fertilizer led to the invention of catalytic processes requiring very high temperatures and pressures to obtain nitrogen from the atmosphere. The best known of these was the Haber-Bosch process for ammonia synthesis (*see* HABER, FRITZ). World War I deprived France, Great Britain, and the United States of the supplies of chemicals previously acquired from Germany and stimulated the growth of chemical industries that afterward became strong competitors with Germany. World War I also saw a great expansion of chemical plants devoted to munitions and poison gas.

After World War II, the traditional character of the chemical laboratory began to change. Physical instrumentation became more prominent, and automated analytical methods replaced some of the chemist's traditional skills. To some observers, chemistry lost its status as a fundamental science. *Quantum chemistry appeared to be merely an application of physics to the particular phenomena that chemists studied. As the applications of chemistry to many other fields of science and technology multiplied, concern whether it retained any core unity intensified. It had come to be, in the words of one general history of the subject, a "field without a territory," existing both "everywhere and nowhere." Others have argued, however, that even though its most fundamental principles may be borrowed from physics, extension of these principles to the wide range of chemical phenomena still requires methods of reasoning and investigation that have long been characteristic of chemistry.

See also ELECTROLYSIS; INSTITUTE; PNEUMATICS.

James R. Partington, *A History of Chemistry* (1961–70). Aaron J. Ihde, *Development of Modern Chemistry* (1964). Robert P. Multhauf, *The Origins of Chemistry* (1966). Owen Hannaway, *The Chemists and the Word: The*

Didactic Origins of Chemistry (1975). Frederic L. Holmes, *Lavoisier and the Chemistry of Life: An Exploration of Scientific Creativity* (1985). John Servos, *Physical Chemistry from Ostwald to Pauling* (1990). William Brock, *The Norton History of Chemistry* (1992). David Knight and Helge Kragh, eds., *The Making of the Chemistry: The Social History of Chemistry in Europe, 1789–1914* (1998). Lawrence Principe, *The Aspiring Adept: Robert Boyle and His Alchemical Quest* (1998). Trevor H. Levere, *Transforming Matter: A History of Chemistry from Alchemy to the Buckyball* (2001).

FREDERIC LAWRENCE HOLMES

CHERENKOV RADIATION. Scientists working on radioactivity around 1900 noticed a faint bluish light emanating from transparent substances, such as liquids, near their sources. The strange phenomenon seemed to pose less of a mystery than radioactivity itself, and inspired comment but not further investigation at the time. The effect got its name from Pavel Cherenkov, a Soviet physicist who in 1934, while investigating luminescence, found that clear liquids excited by gamma rays from a radium source emitted faint blue light. An intense examination of the spectrum, lifetime, and polarization of the radiation and the characteristics of the source convinced Cherenkov that the phenomenon was not luminescence but a new type of radiation. His colleague Sergey Ivanovich Vavilov suggested that electrons knocked from atoms by gamma rays caused the radiation, which led Cherenkov to more experiments showing that most of the radiation was emitted in the direction of the forward path of the particle that stimulated it. Igor Tamm and Ilya Frank explained the phenomenon in 1937 as a consequence of classical electrodynamics: a charged particle, such as an electron, traveling through a transparent medium faster than the phase velocity of light in that medium will emit radiation in a well-defined direction, analogous to the sonic boom produced by a plane flying faster than the speed of sound. Cherenkov, Frank, and Tamm shared the Nobel Prize in physics for 1958 for the work; Vavilov had died by that time and thus was ineligible.

Cherenkov radiation later reached popular audiences in images of nuclear reactors, where radiation from fuel rods excited a blue glow in pools of cooling water. It found scientific application in detectors for high-energy experiments at particle accelerators. The Cherenkov effect converted particles to light and thus offered a way to detect and count them via photomultipliers in combination with electronic logic circuits. Furthermore, since the radiation emerged sharply collimated, and since the emission angle depended only on the incident particle's velocity and the refractive index of the medium, Cherenkov radiation also provided the speed and direction of high-energy particles. Cherenkov counters would figure in several prominent discoveries in postwar particle physics, including the antiproton.

J. V. Jelley, *Cherenkov Radiation and Its Applications* (1958). V. P. Zrelov, *Cherenkov Radiation in High-Energy Physics*, trans. Y. Oren, 2 vols. (1970).

PETER J. WESTWICK

CHERNOBYL. After midnight on 26 April 1986, several tests were underway during the shutdown of the number 4 reactor of the Chernobyl nuclear power station in the eastern region of the Ukraine, then a part of the Soviet Union. The reactor, fueled by uranium and moderated by graphite, included a number of mechanisms designed to ensure its safety by preventing the chain reaction that powered it from running out of control and, if it did, by ensuring that the reactor would be promptly cooled and its nuclear reactions halted. For the sake of efficiency, however, the test program called for the disabling of several of these mechanisms, including the reactor's emergency core cooling system.

Shortly before 1 A.M., the reactor started behaving abnormally. Two of the operators wanted to stabilize it at a low level of power, but their superior instructed them to increase the power level so the tests could proceed. The reactor rapidly overheated, and with the emergency cooling system unavailable, a series of mechanical breakdowns and chemical reactions generated an explosive mixture of hydrogen and oxygen. At 1:23:58 A.M., the mixture detonated, destroying the reactor building, setting its uranium core ablaze, and jeopardizing the entire reactor complex. The explosion spewed forth red-hot pieces of nuclear fuel and radioactive graphite, more than ten times as much by weight as had been delivered by the atomic bomb that destroyed Hiroshima. More than thirty of the reactor operators, guards, and local firefighters who fought to contain the damage and extinguish the fire died from radiation poisoning. Much of the radioactive debris rose high into the atmosphere. Winds carried it to the northwest across Byelorussia and the Baltic Republics, then into western Europe. The fallout contaminated the soil and water across thousands of miles of territory.

Soviet spokespeople initially tried to conceal the scope of the disaster. A. M. Petrosyants, the chairman of the State Committee on the Use of Nuclear Energy, had previously assured the public that nuclear reactors were perfectly safe and posed no threat to local populations. Now he regarded the Chernobyl disaster as a valuable learning experience. "Science requires victims," he declared at a press conference. But the growing demands for *glasnost* (openness) in Soviet society soon forced the authorities to reveal the extent of the calamity and the hazards arising in its wake. Official inquiries blamed the operators alone for the catastrophe, covering up the share of responsibility that later assessments assigned to the reactor designers and to policies that, in the phrase of one report, discouraged a "safety culture" in nuclear plants. Powerful indictments were leveled at the Soviet propensity for secrecy, which had kept previous but smaller nuclear accidents from receiving the kind of evaluations that might have helped prevent the catastrophe at Chernobyl. Coming seven years after Three Mile Island, the disaster boosted the movement against nuclear power in the West. But in the end, both Russian and foreign analysts interpreted the Chernobyl nuclear plant as a technological system mismatched to the idiosyncrasies of human behavior. As such, it was a metaphor for the failures of the Soviet system.

Grigori Medvedev, *The Truth about Chernobyl* (1991). Piers Paul Read, *Ablaze: The Story of the Heroes and Victims of Chernobyl* (1993).

DANIEL J. KEVLES

CHROMOSOME. In 1833 the renowned botanist, Robert Brown, described the invariable presence in the cellular tissues of plants of a granular body he called the nucleus. Over the following half-century improvements in the microscope and in staining techniques made possible the resolution of the granular content of this body into a collection of so-called "nuclear threads" formed it was thought by fragmentation of a continuous thread or "spireme" as the cell prepared to divide. These threads in the form of V-shaped loops then migrated from the "equator" of the cell to opposite ends or "poles" where, it was believed, they each reformed into a fresh spireme. The German histologist Walther Flemming described this process in detail and called it "mitosis" in 1882. Its remarkable character suggested the importance of these threads in the transmission of hereditary information from mother cell to daughter cells, but the name given to them by Wilhelm Waldeyer, "chromosomes," referred only to their staining capacity.

The German zoologist August Weismann pursued the question of the functions served by mitosis and the nuclear threads. He placed the chromosomes in the theater of evolution by identifying their substance as the material determinants of the hereditary characters and by attributing the production of variations to the varied commingling of the chromosomes in sexual reproduction. The remarkable dance of the threads in mitosis ensured the faithful transmission of the determinants from one cell generation to the next, whereas the special form of cell division that gave rise to the germ cells had the role of halving the number of hereditary determinants, hence his term "reduction division" for what we know as "meiosis." Weismann maintained that reduction division also brought about a diversity in the chromosomal contents of the germ cells, and thus yielded variations among the offspring upon which natural selection could act.

Before the turn of the century a number of Weismann's interpretations of the cytological evidence had to be revised or withdrawn. Most important was the establishment of the doctrine of the "individuality of the chromosomes," first enunciated by Carl Rabl in 1885 and established by the classic experiments of Theodor Boveri in 1902. The same chromosomes persist from one mitosis to another, their loss of identity following cell division being only apparent, and each pair of chromosomes is constituted of a maternal and a paternal member, each pair differing from the others.

With the rediscovery of Mendel's laws in 1900 came the possibility of a relation between Mendelian and chromosomal phenomena. In 1902 Walter Stanborough Sutton pointed to the parallel between the association of maternal and paternal chromosomes and their subsequent separation in meiosis with character pairs in Mendelian heredity and their subsequent segregation in germ cell formation. But many Mendelian factors would have to lie on the same chromosome, thus overly restricting their independence. Thomas Hunt *Morgan overcame this difficulty in 1911 by suggesting that chromosomes can exchange segments by "crossing over." Two years later his student Alfred Henry Sturtevant, realizing that the differing degrees of linkage between different pairs of *genes offered a measure of their physical separation on the chromosome, produced the first genetic map. Covering six genes, Sturtevant's method gave birth to chromosome mapping in terms of degree of linkage. The remainder of the century has seen the continuing refinement and extension of genetic mapping, to which have been added measures of the physical map (linear distances) and the sequence map (sequence of the bases in DNA).

Bruce R. Voeller, ed., *The Chromosome Theory of Inheritance. Classic Papers in Development and Heredity* (1968). Alice Baxter and John Farley, "Mendel and Meiosis," *Journal of the History of Biology* 12 (1979): 137–173.

ROBERT OLBY

CHRONOMETER. See CLOCK AND CHRONOMETER.

CINEMA. See FILM, TELEVISION, AND SCIENCE.

CLASSIFICATION IN SCIENCE. Attempts to order and classify natural objects and phenomena have been primary aims of human thought for centuries. In the Western philosophical tradition, Aristotle's system enshrined various classifications that were taken for granted down to the seventeenth century. Celestial bodies consisted of materials different from the earthly *elements; different sorts of matter had different kinds of motion; and animals came either viviparous or oviparous, sanguineous or bloodless. Aristotle also recognized a hierarchy of natural beings that he arranged into a "scala naturae" or ladder of life, stretching, he said in *Historia Animalium*, little by little from things lifeless to animal life. Frequently referred to as the great chain of being, Aristotle's ladder remained a key concept in the West. Theophrastus divided plants into a logical succession of genera and species to yield a definition that applied to one form alone. The Hippocratic corpus did much the same for medicine, characterizing the human body by four humors.

The ancient schemes of classification dominated Western thought for years. They assisted in arranging animals in bestiaries according to their presumed characters and moral meanings, and plants according to their medicinal properties and astrological connotations. Occasionally, however, as in Edward Topsell's *The Historie of Four-Footed Beasts* (1607), animals were considered in alphabetical, that is, artificial, order.

The botanical treatises of Gaspard Bauhin (1590) and the French botanist Charles de L'Écluse or Clusius (1618) opened the great age of botanical taxonomy. Each attempted his own solution to the problem of the weight to assign to various criteria, and each searched for a "natural" arrangement of kinds—seeking a classification scheme that would reflect morphological relationships (or "affinities") as seen in the number and relative position of petals, sepals, and reproductive organs, the shape of leaves, and means of reproduction, whether by bulbs or seeds. Andrea Cesalpino returned to Aristotle's and Theophrastus's logical hierarchies and gave a series of "essential" characters whose successive subdivisions would not violate intuitive perceptions of affinity, rather in the way that animal classification schemes used differences in the heart, nerves, or respiratory system. Cesalpino's scheme underlay most of the plant classifications published afterwards. John Ray found that flowering plants fall into two natural groups based on the number of seed-leaves (cotyledons), the monocotyledons and dicotyledons; and nomenclature was much advanced by Joseph Pitton de Tournefort when he defined and named nearly seven hundred genera in his *Elements de Botanique*

of 1694. In general, most seventeenth-century authors sought to arrange nature according to its surface features, using anatomy and "habitus" (external physiognomy, as in herb, shrub, or tree) to assist them.

The definitions for individual animals and plants became longer and longer. In his *Systema naturae* (1735) Carl *Linnaeus proposed an "artificial system" for plants whereby the numbers of stamens and pistils (male and female reproductive organs) allocated plants into groups without further ado. Plants with five stamens would be placed in a new category called "Pentandria" (Greek for five male organs). This system, often called the numerical or sexual system, was understood as a method of classifying plants purely for human convenience, for it grouped plants together on the basis of numbers alone. Georges Louis Buffon, who favoured a naturalistic scheme, critiqued it harshly. Linnaeus also introduced the binomial nomenclature, the first word of which signified the genus, and the second the specific name, of the species. In his *Genera Plantarum* (1737) and *Species Plantarum* (1753) he defined all the known plants of the world. By convention, Linnaeus's names have been taken retrospectively as defining the type. The popularity of the artificial system percolated through other disciplines. Boissier de Sauvages classified diseases in this manner and Antoine-Laurent *Lavoisier did the same for chemistry. In 1869, Dmitrii *Mendeleev became known as the Linnaeus of chemistry for ordering chemical elements in a structured series of relationships, the *periodic table.

Linnean botanists achieved remarkable results. Nevertheless alternatives to Linnaeus's rigidly logical scheme emerged as a number of botanists searched for ways to acknowledge natural relationships. Michel Adanson in *Famille des Plantes* (1763) described purely empirical relationships by including a large number of characters. Through his work the natural system came to be understood as a means of classification in which several morphological characters, not just one, determined relationships. Georges Cuvier famously created a comparable scheme for animals based on four basic "types" or "plans" in which various characteristics were subordinated to others. Analogies and homologies became significant phenomena to be explained, and a number of proposals, especially from philosophically sophisticated scholars like Wolfgang Goethe, highlighted such resemblances as the material reflection of an idealized plan or archetype, often discussed in theological terms as the divine plan for natural beings. Cuvier subsequently argued bitterly with Geoffroy Saint Hilaire over the presumed order and stability of nature.

Taxonomic relationships became more complex by the continued discovery of unknown or intermediary fossil types. Early-nineteenth-century naturalists proposed various unusual schemes for animals, the most notable being the Quinarian system of William Macleay, which grouped genera and families into sets of five. Most classification schemes necessarily assumed that species were fixed or relatively stable entities. If organisms changed over time, they could have no "essence" or definition. Charles *Darwin found it difficult to counter these views in his *Origin of Species* (1859) but personally regarded the evolutionary origins of anatomical relationships as one of the great advantages of accepting his proposals. Even though classification schemes did not change much in practice after the acceptance of his and Alfred Russel Wallace's, the issue of the classificatory relationship between humans and apes played a key part in the Darwinian controversies. Thomas Henry Huxley used the structural similarities between apes and mankind to defend Darwin's theory at the same time as Richard Owen used them to attack it. The same controversies emerged when naturalists attempted to insert early hominid remains in pre-existing classification schemes.

Many of the metaphysical issues surrounding classification are still hotly debated. Purely numerical systems have been introduced to avoid the need for individual judgment. Cladistics provides an evolutionary scheme that links related groups in time.

David Knight, *Ordering the World: A History of Classifying Man* (1981). Peter F. Stevens, *The Development of Biological Systematics: Antoine Laurent de Jussieu, Nature and the Natural System* (1994). Harriet Ritvo, *The Platypus and the Mermaid and Other Figments of the Classifying Imagination* (1997).

JANET BROWNE

CLASSIFICATION OF THE SCIENCES. From the *Renaissance to the end of the nineteenth century, some of Europe's keenest minds struggled to classify knowledge (*scientia*). The roots of the problem were ancient. In antiquity, Aristotle had discussed classification. In the Medieval period, universities had divided the liberal arts—which students studied before beginning course work in law, medicine, or theology—into the trivium of grammar, dialectic, and logic, and the quadrivium of arithmetic, geometry, astronomy, and music.

Francis Bacon set the agenda for the modern discussion. He wrote about classification, not always consistently, in a number of works, but most importantly in the *Advancement of Learning* (1605). He wrote on knowledge in general, not the sciences in our more restricted modern sense.

He divided knowledge according to the faculty of the mind chiefly exercised in its creation. This gave him three great classes: history, created by the faculty of memory; philosophy, by that of reason; and poesy, by that of imagination. What we call science was partly to be found in history and partly in philosophy. Within history, natural phenomena fell to *natural history, while civil affairs, church affairs, and the written word belonged to civil, ecclesiastical, and literary histories. Within philosophy, natural philosophy treated the natural world, and other branches dealt with the divine and human realms. Mathematics lay outside the classification. *Natural history (zoology, botany, and *mineralogy) and natural philosophy (which included *mechanics, *optics, *pneumatics, etc.) continued to be the chief categories of knowledge through the eighteenth century. But Christian von Wolff modified Bacon's classification by adding mathematics, and Jean Le Rond d'Alembert, in the *Preliminary Discourse* (1751) to the *Encyclopédie* of Denis Diderot, pruned the theological branches of the tree of knowledge to tiny stubs, while allowing the branches representing knowledge of the natural world and practical techniques to enlarge and ramify.

In the first half of the nineteenth century, classifications flourished as never before. Scientists and philosophers for the first time dealt with science in much the terms we now use, without including literature, theology, and history. They also abandoned the use of mental faculties as organizing principles. One alternative, adopted by William Whewell, scientific polymath and Master of Trinity College, Cambridge, was to use the fundamental ideas around which, he supposed, each science developed. Thus, in his *Philosophy of the Inductive Sciences* (1840), his categories included geometry (with space as the fundamental idea), mechanics (cause, force, and matter), and *chemistry (affinity and substance). In France, André-Marie Ampère constructed a dichotomous classification of the sciences into 128 divisions in his *Essai sur la philosophie des sciences* (1834). Thus he divided general physics into elementary general physics and mathematical physics and these in turn into experimental physics, chemistry, stereonomy, and atomology. The latter pair of names illustrates the many neologisms Ampère introduced, which probably did little to encourage others to accept his classification. Neither Ampère's scheme nor Whewell's found favor.

What did catch the public imagination were classifications based on the supposed development of human thought. Auguste Comte invented the earliest and most influential of these developmental classifications, and set it out in his *Cours de philosophie positive* (1830). There he pronounced the "law" by which all branches of knowledge passed successively through stages—theological, metaphysical, and positivistic (scientific) stages, thus relegating many previously respectable branches of knowledge, theology in particular, to prescientific status. Comte identified five main positive or empirical sciences: *astronomy, physics, chemistry, *physiology or *biology, and social physics or sociology. Each had its appropriate method and each had to be mastered before much progress could be made with its successor. In his time, he thought, sociology teetered on the brink of achieving scientific status. Mathematics once more stood outside the classification as an achievement in its own right and a necessary tool for science.

Other scholars, among them the physiologist Wilhelm Wundt, the chemist Wilhelm Ostwald, and the statistician Karl Pearson followed Comte's lead and produced variations in all of which the sciences appeared in essentially the Comtean order. Even Herbert Spencer who, in his *Essays: Scientific, Political, and Speculative* (1875), produced what he trumpeted as an alternative to Comte's classification, accepted its basics.

With the success of the Comtean scheme, the classification of the sciences ceased to be of interest except to librarians. The major histories of classification were written not by philosophers or scientists, but by the Librarian of Princeton University, Ernest Cushing Richardson, and the Professor of Divinity at Edinburgh University, Robert Flint. Composed at the beginning of the twentieth century, their books have not been superseded, a fact that suggests that the debate about the order, connection, and relative importance of different branches of knowledge, including the sciences, has been resolved.

Were these classifications mere playthings, the idle ruminations of scholars in their studies? Even in the encyclopedias to which they were so often companions, classifications featured chiefly in the prefaces. The editors then usually proceeded to arrange the articles in alphabetical order, ignoring the maps and trees of knowledge that they had elaborated in their preliminary materials.

The repeated attempts to classify knowledge and the sciences from the seventeenth to the twentieth century indicate rapid or periodic changes in the organization of knowledge as new forms, particularly the natural sciences, challenged others such as philosophy and theology that had enjoyed high prestige for centuries. With the changes went competition for control of school and university curricula, for prestige, and for the patronage of the powerful.

The context that made the classificatory enterprise so absorbing to leading scholars throughout the three centuries that saw the formation of modern science rewards study.

Ernest Cushing Richardson, *Classification, Theoretical and Practical* (1901). Robert Flint, *Philosophy as Scientia Scientiarum and a History of the Classification of the Sciences* (1904). Robert Dolby, "Classification of the Sciences: The Nineteenth-Century Tradition," in R. F. Ellen and D. Reason, eds., *Classifications in Their Social Context* (1979): 167–93. Robert Darnton, "Philosophers Trim the Tree of Knowledge: The Epistemological Strategy of the Encyclopédie," in Darnton, *The Great Cat Massacre and Other Episodes in French Cultural History* (1984).

RACHEL LAUDAN

CLIMATE. The mid-eighteenth-century *Encyclopèdie* of Denis Diderot and Jean d'Alembert offered three definitions of climate. A climate, first, is a latitude-band around the earth, of such a width that the longest day along its polar circle exceeds that along its equatorial circle by some set amount, say one half-hour. Second, climate denotes a region characterized by its seasons, the quality of the soil, "or even the manners of the inhabitants." Third, climates are synonymous with the temperatures or "degrees of heat" proper to them. Climate, according to these definitions, is a region rather than a pattern of typical weather; the term belonged to geography rather than meteorology.

Geographers had discussed climate and its relationship to culture since ancient times. The source of this tradition, the Hippocratic treatise *Airs, Waters, and Places*, attributed a population's character to the winds ("airs"), water sources ("waters"), and soil and orientation ("places") of its locale, as well as to diet, sanitation, occupational patterns, and so on. So much did the Hippocratic tradition flourish in the Enlightenment that the *Encyclopèdie* took Montesquieu to task for his famous discussion of climate in *The Spirit of the Laws* which, they claimed, added nothing to "such familiar topics."

Before the late eighteenth century the geographic tradition had little to do with meteorology. Geographers' pronouncements on climate remained general and qualitative, benefiting little from the increasing availability of good meteorological observations. Meteorologists for their part showed little interest in climate. Their chief goal was to discover recurring patterns they could use to predict the weather and its influence on agriculture and health. In the late eighteenth century, using precise instruments then being developed and organized by learned societies, observers began to collect a significant body of reliable weather data. But they did not construct from it an understanding of climates: they did not integrate observations of the weather at many locales into perceptions of the unity of the weather over periods of time and extents of space. Kant's criticism of contemporary natural history applies well to this approach, which placed objects "merely beside each other and ordered in sequence one after another," rather than integrating them into a "whole out of which the manifold character of things is derived."

Medical topography came closest to uniting meteorology, precise measurement, and geography in the late eighteenth century. The largest undertaking occurred in France, where the Royal Society of Medicine dispatched physicians to report on environmental conditions throughout the realm in Hippocratic fashion. The members gathered weather observations, descriptions of "airs, waters, and places," and information on local populations in order to publish a "medical and topographical map" of France. Before their efforts could be integrated in this way into what would have been a climatology of the nation, the Revolution closed the Society.

A true climatology required a new vision of nature. This vision emerged in the first decades of the nineteenth century, integrating geography with meteorology, the Romantic vision of the unity of nature with late Enlightenment methods of precise instrumentation and measurement. Alexander von *Humboldt is the best known representative of this synthesis. Schooled in late-eighteenth-century experimental natural philosophy, Humboldt also enjoyed close relations with the leaders of the German Romantic movement. Drawing on available observations, he constructed a unity among all aspects of nature, seeking in a region's geology, climate, flora, fauna, and human culture a coherent, interacting whole, which he called "physique gènèrale." Climate, in Humboldt's formulation, grew out of the manifold relations of physical geography: the size and orientation of landforms, their height and geological constitution, the relations of land and water, plant and snow cover, and so on. Suddenly Humboldt and his contemporaries discovered regional and continental breadth in meteorological patterns that their predecessors had perceived as merely local. Humboldt's famous isotherms, lines of equal mean temperature drawn on a map, were a visual representation of one aspect of this breadth, uniting global observations of temperature into a coherent whole.

Climatologists of the age of Humboldt studied the distribution of climates over the globe, applied climatic considerations to biogeography, and speculated on the climates of historical and geological time periods, drawing on the evidence of written and fossil records. After the 1850s imperialists and racists increasingly applied the old Hippocratic arguments to buttress claims of European racial superiority and to justify African and Asian conquests. Following Hippocrates' dictum that "races are the daughters of climate," geographers and anthropologists ascribed Victorian virtues of intelligence, industry, sobriety, and more to the influence of temperate European and American climates. The enervating climates of colonial regions, on the other hand, had given birth to feeble races fit only to be ruled. Tropical medicine addressed closely related questions of the climatological basis of diseases that threatened Europeans' ability to govern their possessions and the physical and moral degeneration of those resident in the colonies. Among the more notorious climatological racists was Ellsworth Huntington, who also investigated the extent to which workers' performance depended on their climatological environment, particularly temperature. Experiments inspired by Huntington and carried out in the 1920s by the National Research Council of the United States included one in which workers were subjected to rectal measurements of body temperature at ten-minute intervals while laboring under heavy loads at environmental temperatures of up to 115° F. Eugenicists and Nazi ideologues continued to propagate theories of racial climatology as late as the mid-twentieth century.

The more empirical and practical branches of climatology in the nineteenth century focused particularly on biogeographical and agricultural problems. Typical questions included the influence of climatological elements such as temperature, humidity, and precipitation, on the life-cycles of food crops and on human settlement. Climatologists developed classifications of climates, integrating both meteorological and biogeographical elements in their schemes. The best-known of these was presented in 1884 by Wladimir Köppen and refined over a half century. In the last third of the nineteenth century, so-called "classical climatology" devoted much energy to standardizing measurements and insuring their reliability.

Particularly in the universities and schools, the dominant role of climatology led many to understand meteorology as part of geography; the association reflected Europeans' pursuit of knowledge of exotic regions in an age of imperial expansion. After the turn of the century, as meteorologists like Vilhelm Bjerknes pushed to ally their field with physics, climatology receded from center stage while developing in new directions. Especially for the amphibious landings of World War II, military planners drew on climatology for knowledge of times and seasons offering best conditions for operations. Earlier in the century agricultural climatology had developed sophisticated criteria, such as "critical periods," during which plants particularly need moisture, but it was not until after World War II that meteorologists applied detailed climatological and microclimatological studies to agriculture, precisely quantifying the climatic environment and integrating climatology and plant physiology. Since the middle of the twentieth century, issues of climate change—present in climatological discourse since the Enlightenment—have become prominent.

See also CLIMATE CHANGE AND GLOBAL WARMING; METEOROLOGY.

Aleksandr Khristoforovivh Khrgian, *Meteorology. A Historical Survey*, 2d ed., vol. 1, ed. Kh. P. Pogosyan (1959). Theodore S. Feldman, "Late Enlightenment Meteorology," in *The Quantifying Spirit in the Eighteenth Century*, eds. Tore Frängsmyr, John L. Heilbron, and Robin E. Rider (1990): 143–178. David N. Livingstone, "The Moral Discourse Of Climate: Historical Considerations On Race, Place, And Virtue," *Journal Of Historical Geography* 17 (1991): 413–434. Richard Grove, "The East India Company, the Raj and the El Niño: The Critical Role Played by Colonial Scientists in Establishing the Mechanisms of Global Climate Teleconnections, 1770–1930," in *Nature and the Orient: The Environmental History of South and Southeast Asia* (1998): 301–23. David Livingstone, "Tropical Climate and Moral Hygiene: the Anatomy of a Victorian Debate," *British Journal for the History of Science* 32 (1999): 93–110.

THEODORE S. FELDMAN

CLIMATE CHANGE AND GLOBAL WARMING. Theories of climate change date from very early times; Theophrastus (371–287 B.C.) wrote on desiccation wrought by deforestation. His work, revived in the *Renaissance, helped fuel concern over deforestation in European colonies, and from the late eighteenth century onward, colonial governments established forest reserves that were among the earliest measures of environmental conservation. Meanwhile, Enlightenment students of the classical world uncovered literary evidence for climate change since ancient times, while Americans such as Thomas Jefferson argued that deforestation from European settlement had moderated the climate, rendering it fit for civilization. Nineteenth-century climatologists, applying more exacting historical and scientific analysis to these questions, found no convincing evidence of climate change in historical times.

Climate change on astronomical and geological time-scales on the other hand was more nearly certain. From Buffon, Kant, and *Laplace to Lord Kelvin and beyond, cosmologists speculated about the long-term cooling of the earth from its origin as a molten ball and about the longer-term cooling of the sun. Nineteenth-century geologists found indisputable evidence in *fossils of a warmer climate in ancient times, and around the middle of the century, discovered the ice ages. In order to reconcile these discoveries with his belief in a uniform state of the earth, Charles Lyell developed a theory of cyclical climate change, according to which "all … changes are to happen in future again, and iguanodons … must as surely live in the latitude of Cuckfield as they have done so."

Several eighteenth- and early nineteenth-century natural philosophers and mathematicians, including Horace Bénédict de Saussure, Joseph Fourier, and Claude Pouillet, had noted the atmosphere's selective transmission of heat; Fourier compared the atmosphere to the glass of a greenhouse. Beginning in 1859, John Tyndall began experiments on the radiative properties of atmospheric gases, and speculated that variations in their amounts might have altered the earth's climate on the geological time scale. In 1895, Svante Arrhenius, trying to explain the *Ice Ages, calculated temperature changes from variations in carbon dioxide. Two decades later, he predicted that industrial generation of CO_2 would protect the globe from recurring ice ages and allow increased food production for a larger world population. Arrhenius's ideas were welcomed by the American geologist T. C. Chamberlin, who, in the first three decades of the twentieth century, developed a theory of the atmosphere as a large-scale geological agent based on a carbon cycle. Crustal uplifts expose large surface areas to weathering, a process that absorbs CO_2; global cooling and glaciation follow. The cycle turns about when mountain ranges reduce to nearly base levels.

By the 1950s, alteration in insolation—solar radiation received by the earth—orbital changes, mountain building, and volcanism had all been identified as agents of climate change. Water vapor eclipsed CO_2 as an agent of global warming until G. S. Callendar published a series of articles from 1938 to 1961 emphasizing anthropogenic influences on the amount of carbon dioxide in the atmosphere. Political tension feeds fears of climate degradation, and the Cold War intensified anxiety over both global cooling and global warming. Proposals were floated for large-scale interventions such as damming the Bering Strait, orbiting fleets of mirrors in space, and spraying sulfur dioxide into the upper atmosphere. Today, the scientific community has reached a consensus that CO_2 levels have increased owing to industrial activity and that warming is taking place on a global scale. The relation between these two phenomena remains in doubt, as well as the role of other factors and the future course of climate change.

See also OZONE HOLE; POLLUTION.

Spencer Weart, "The Discover of the Risk of Global Warming," *Physics Today* (January 1997): 34–40. James Rodger Fleming, *Historical Perspectives on Climate Change* (1998).

THEODORE S. FELDMAN

CLINICAL CHEMISTRY. "Clinical chemistry" is the term used today for the application of chemical, molecular, and cellular methods to the understanding and assessment of health and diseases in human beings. The roots of this specialist field go back to early modern times, when Paracelsus introduced methods of alchemy and chemistry into medicine to obtain chemical remedies, as well as to find chemical explanations for body functions and pathological processes. Analytical methods for recognizing diseases in the investigation of human and animal body parts were first devised near the end of the eighteenth century by, among others, Antoine François de Fourcroy, Jöns Jakob *Berzelius, William Prout, and Justus von *Liebig. In the nineteenth century, microscopy was introduced into medical laboratories, in particular for investigating *blood once suitable coloring materials had become available (Paul Ehrlich and others). From the middle of the nineteenth century, the development of new chemical methods caused clinicians and physicians to take an increasing interest in chemical investigations. Laboratories were set up in hospitals, clinics, and even in some medical practices. Universities created professorships in the new field of science. As early as 1795, Fourcroy founded an Éstablissement chimico-clinique at the Paris École de Pharmacie. In 1842, Johann Joseph Scherer, a student of Liebig, was appointed to the professorial chair for clinical chemistry (organic chemistry) at the University of Würzburg, the first such professorship in the world. By the end of the century, clinical chemistry and microscopy had become a regular part of the curriculum of medical students in many countries and practical teaching had entered some clinical laboratories.

The development of clinical chemistry in the twentieth century was closely linked to the fundamental subjects of analytical chemistry and biochemistry. At first methods devised in biochemical or chemical laboratories dominated. After World War I, however, specialized industrial companies increasingly took over development. They worked out investigative methods for medical diagnostics, including the detection and determination of substances that occur in very low concentrations in the body. The new methods emphasized rapid performance of urgent analyses and minimization of the material required.

Colorimetric (photometric) methods to measure dyes formed by chemical reaction proved particularly suitable. First carried out with apparatus requiring visual measurement by the analyst (Otto Folin and others), colorimetry came to rely on electronic readouts from photocells. For treating acute diseases, apparatus for analyzing gases (blood and respiratory gases—Donald D. Van Slyke and Poul Astrup) became of increasing importance. After World War II, analysts developed techniques requiring only a few microliters of material and using mechanized systems like the AutoAnalyzer of Leonard T. Skeggs, which greatly reduced the human work required. The output of automatic analyzers when run through computers resulted in diagnostic reports suitable for use by the physician.

By the 1980s, the initially predominant chemical analyses increasingly shared the field with molecular-biological methods, which allowed investigation of very complex biological substances (proteins, nucleic acids, etc.) and further reduced the amount of substance required for detection. Today, the results of clinical chemical investigations are widely used in practical medicine. Medical diagnostics has grown with the development of preventive medicine, early disease detection, disease monitoring, and intensive-care procedures. Frequently, the clinical chemical laboratories control the therapeutic steps by continuously monitoring the concentration of drugs administered. These important functions require extensive experimental research: basic research on chemical changes in pathological states (pathophysiology) and applied research for developing and controlling analytical methods. Such research is usually carried out today in the clinical chemical institutes of universities. Companies making the apparatus and reagents also develop new methods. Scientists working in the field of clinical chemistry generally have advanced degrees in medicine, chemistry, biochemistry, pharmaceutics, or biology.

Noel G. Coley, *From Animal Chemistry to Biochemistry* (1973). Stanley Joel Reiser, *Medicine and the Reign of Technology* (1978). Robert E. Kohler, *From Medical Chemistry to Biochemistry. The Making of a Biomedical Discipline* (1982). Johannes Büttner, ed., *History of Clinical Chemistry* (1983). Johannes Büttner and Christa Habrich, *Roots of Clinical Chemistry* (1987). Louis Rosenfeld, *Four Centuries of Clinical Chemistry* (1999).

JOHANNES BÜTTNER

CLOCK AND CHRONOMETER. Ambitions of astronomers and navigators, and lately of physicists, have driven many of the most important developments in the history of mechanical, electrical, and electronic horology. Through these developments science has had some of its most important and least remarked impacts on social life. Before the advent of reliable clocks and apart from a few exceptional activities such as monastic prayer, there was little demand for accurate timekeeping to regulate social or working practices. Clocks created possibilities for regulation, and these were put to use in ways that had profound effects on the lives of almost everyone.

The earliest mechanical clocks of which we have detailed records, beginning in the fourteenth century, originated in an astronomical milieu and might better be considered as clockwork *astrolabes or equatoria than as timekeepers. Celebrated examples are the early-fourteenth-century *horologium astronomicum* of Richard of Wallingford, abbott of St. Albans in England, and the slightly later *astrarium* of astronomer Giovanni Dondi in Pavia, Italy. The ten-foot diameter face of Richard's clock had the appearance of an astrolabe, while the astrarium has separate faces for each planet, echoing the treatment of the planets in Ptolemy's *Almagest*. These clocks were considered mechanical marvels. The astronomer Johannes Regiomontanus reported of his visit to Dondi's machine in 1463 that prelates and princes came to see it as though to witness a miracle.

Close links existed between astronomy and the production of clocks at a more modest level as well: astrolabe makers like Jean Fusoris in the early fifteenth century in France and Georg Hartmann in Germany in the sixteenth also made clocks, and famous clockmakers like Joost Bürgi in Prague and Girolamo della Volpaia in Florence also made astronomical instruments. The founding director of the Royal Observatory at Greenwich in the late seventeenth century, John Flamsteed, turned to the leading English clockmaker Thomas Tompion for the manufacture of large astronomical instruments, while Tompion's sometime partner in business, the clock- and watchmaker George Graham, designed and built a set of instruments for Greenwich in the early eighteenth century that became standard equipment throughout Europe.

The most famous representation of clocks in the service of early-modern astronomy is the image in Tycho *Brahe's *Astronomiae instauratae Mechanica* (1598) of his mural quadrant attended by three assistants—one taking the observation, one noting the time given by two clocks, and the third writing down the results. Tycho observed in the accompanying text that the clocks could seldom give the time accurately to seconds. With an instrument set in the meridian, the most convenient orientation for measurement in declination (the angular distance of an object from the celestial equator), noting the passage of time between the occurrence of stars on the meridian would be a measure of the other positional coordinate, right ascension (angular distance along the equator). Thus, by the simple addition of a reliable and sufficiently accurate clock, one instrument could supply all the needs of positional astronomy. This goal was to be one of the principal stimuli for the development of clocks.

Before the seventeenth century no mechanical clock had a natural oscillator controlling its motion and therefore its timekeeping. By a series of cogwheels a descending weight might be arranged to move pointers across dials, or to activate other movements or representations, and through an "escapement" mechanism the to-and-fro motion of a pivoted bar or balance could be used to restrain the weight's descent, but on its own the balance had no oscillatory motion that could regulate the whole machine. The first device to do so was the pendulum.

*Galileo designed a form of pendulum control for a clock, but it was the Dutch mathematician and astronomer Christiaan *Huygens who devised the first form of pendulum clock to go into production—Salomon Coster of The Hague began to make clocks to Huygens's design in 1657. Huygens published his idea in his *Horologium* (1658) and more extensively in his classic of mechanical horology, *Horologium oscillatorium* (1673). Huygens's clock retained the traditional "verge" escapement in which two pallets on a staff alternately engage and disengage the teeth of a crown-wheel but, instead of attaching this staff to a balance bar, linked it to the motion of an independently suspended pendulum. The force transmitted by the wheel from the descending weight served to maintain the motion of the pendulum. It would be important in the future development of pendulum clocks to arrange for the maintaining action of the escapement to interfere as little as possible with the natural oscillation of the pendulum.

In *Horologium oscillatorium* Huygens identified one source of error in pendulum clocks and offered a solution. The period of a pendulum swinging in a vertical circular arc is not completely independent of amplitude of swing. Huygens demonstrated that the fully isochronous curve is a cycloid (the line traced by a point on the circumference of a circle rolling along a straight line) and showed further that a pendulum bob could be made to describe a cycloid if the suspending cord swings between and winds along a pair of cycloidal "cheeks."

A branch of pure geometry dealing with the evolutes and involutes of curves derives from this problem in practical horology. Coster's clocks have cycloidal cheeks, but for the most part subsequent horologists found it sufficient to keep the amplitude of swing relatively small and, by delivering a steady impulse to the pendulum, as constant as possible.

With Huygens we encounter the second major influence on mechanical horology: navigation. By using a clock to keep a standard time and by comparing its readings with local time found astronomically, a solution might be found to the great practical problem of the age—finding longitude at sea. Huygens designed a seagoing version of his clock and sea trials of it had some success, but a pendulum would never provide the oscillator in a reliable marine chronometer.

An improved escapement for a pendulum clock, the "anchor" escapement, emerged in the late seventeenth century. A pair of pallets successively engage and disengage teeth of the escape wheel, but these contacts are accompanied by recoil of the wheel, which is detrimental to timekeeping. By the 1720s George Graham had shaped pallets and teeth to engage without recoil. This "dead-beat" escapement together with Graham's compensation pendulum finally made the pendulum clock (or "astronomical regulator" as it could now be called) sufficiently accurate for measurement in astronomy. His pendulum, introduced in 1715, used a jar of mercury as the bob, so arranged that as the rod lengthened in heat, the mercury level moved in the opposite direction to retain the center of mass at the same distance from the point of suspension.

The longitude prize offered by the British government in 1714—as much as £20,000 for a method accurate to half a degree—powerfully stimulated horological ingenuity and application. John Harrison submitted his first marine timekeeper for a sea test in 1736. For some twenty-five years he received generous support from the Board of Longitude, while producing only three more marine timekeepers—a rate of manufacture that discouraged support for the chronometer as a solution to the longitude problem. Harrison added a number of inventions to mechanical horology such as the gridiron pendulum, in which sets of brass and steel rods expand in opposite directions when heated so as to maintain the center of mass unchanged. Eventually Harrison's fourth timekeeper, which took the form of a large watch, performed in trials to the most exacting standards set by the Longitude Act, and a protracted argument ensued with the board over whether the reward should be paid in full.

Other chronometer makers, such as Thomas Mudge in England and Pierre Le Roy and Ferdinand Berthoud in France, advanced the art, Le Roy in particular with an escapement and the idea of incorporating a temperature compensation into the arms of the balance. In a spring-regulated timekeeper such as a chronometer, the most pronounced effect of temperature change arose from the changing elasticity of the balance spring. Harrison had used a bimetallic strip to alter the effective length of the spring. Other chronometer makers, such as John Arnold and Thomas Earnshaw, who both contributed to the evolution of a standard instrument and to its manufacture in large numbers, made balance arms of bimetallic strips, altering the moment of inertia and thus the period of the balance with changing temperature to counteract the effects on the balance spring.

By the early nineteenth century the chronometer had arrived at a stable design, though much effort would still be devoted to shaping the spring for an isochronous motion and to improving the compensation by different designs of balance. Clocks, however, continued their radical development. On the mechanical side, constant force escapements using a released spring or the action of gravity were developed. Zinc and steel compensation pendulums came in the nineteenth century. In 1895 the Swiss physicist Charles-Edouard Guillaume, director of the International Bureau of Weights and Measures, produced an alloy of nickel and steel—"invar"—with a zero coefficient of expansion. In 1899 Sigmund Riefler of Munich patented an escapement in which impulse to the pendulum arrives through the suspension spring. Riefler placed his mercury-in-steel pendulum in a constant-pressure vessel and made clocks accurate to a few seconds a year.

From 1840 on there were serious attempts to apply electricity to clockwork both by maintaining the motion of a pendulum by electromagnetic force and by sending currents as signals to distant slave clocks so as to synchronize their timekeeping with a master clock. Pioneers in the field included Alexander Bain and Charles Wheatstone in Britain and Matthäus Hipp in Switzerland. Charles Sheppard supplied an electric clock of historical importance to the Royal Observatory in Greenwich in 1852; its empire of slave displays spread through the observatory and beyond, and eventually time signals carried by the telegraph wires of the expanding railway system provided the basis of a national time service based on measurements at Greenwich.

In a free-pendulum clock, the pendulum swings almost entirely without interference while its impulse is controlled by an auxiliary or slave clock regulated by signals from the free pendulum. In the very successful system by William Hamilton Shortt, dating from the early 1920s, a synchronizing signal from an invar pendulum contained in a partially evacuated cylinder is sent to a separate slave pendulum clock that in turn controls the invar pendulum by a gravity arm. Timekeeping to within a second a year can be achieved by the Shortt free pendulum, an accuracy exceeded only by quartz clocks.

As Pierre *Curie and his brother Jacques-Paul had discovered, the elastic vibrations of quartz crystals are accompanied by small electrical potentials. These can be amplified and maintained, and used to control circuits linked to counters or dials. The Bell Telephone Laboratories contrived to use a crystal for time measurement in the late 1920s. While quartz clocks improved accuracy of timekeeping over the best pendulum clocks by a factor of ten, oscillations at the atomic level offered even greater precision. Work at the National Physical Laboratory in England produced clocks based on magnetically induced oscillations of cesium atoms. By 1959 accuracies to within one second in a thousand years had been achieved, which was better than the best astronomical measurements. Clocks had begun by keeping time regulated by astronomical phenomena. They have become more accurate than the heavens and now set a new definition of time independent of sun and stars.

See also STANDARDIZATION.

Frank Hope Jones, *Electrical Timekeeping* (1949). Rubert T. Gould, *The Marine Chronometer: Its History and Development* (1960). Cedric Jagger, *The World's Great Clocks and Watches* (1977). Derek Howse, *Greenwich Time and the Discovery of Longitude* (1980). William J. H. Andrewes, ed., *The Quest for Longitude* (1993).

JIM BENNETT

CLONE. Clones, genetically identical copies of living organisms, were first produced with certain plants by the cultivation of a stem or branch of the original. The process originated in ancient times and is widely used by horticulturalists and home gardeners to obtain reproductions of fruit trees, vines, and roses. The biologist J. B. S. Haldane took the word "clone," which he coined in 1963, from the Greek word for "twig." Scientists applied the term to the replication of genes themselves beginning in the 1970s, when the new techniques of recombinant *DNA permitted them to snip out individual genes from an organism's genome and insert them into bacteria, where they would multiply with the bugs. The chemist Kary B. Mullis's invention of the polymerase chain reaction (PCR) in 1983 enabled scientists to produce a billion clones of a gene in a test tube.

In the late 1920s, the German embryologist Hans Spemann had achieved a type of animal cloning by inserting the nucleus from a salamander embryo cell into a denucleated egg that then developed into an independent salamander. In 1938, by then a Nobel laureate, Spemann proposed that it might be possible similarly to clone an adult animal. But he described the experiment as "fantastical," partly because he did not know how to do it, partly because he did not know whether the nucleus from an adult animal's fully differentiated cells possessed the capacity to direct the full development of the organism from the egg.

In 1951 Americans Robert Briggs and Thomas J. King, biologists at the National Institutes of Health, cloned embryonic frog cells but failed in attempts to clone more differentiated ones. Their results, along with those from other laboratories, indicated that the older the cells from which the nuclei came, the less likely the clones would develop. In 1962, John Gurdon, a developmental biologist at Oxford University, reported that he had been able to produce fully developed frogs by cloning putatively differentiated cells from the intestinal linings of tadpoles. Gurdon argued that his experiment, although successful only 2 percent of the time, confirmed that fully differentiated cells retained the genetic capacity to direct development. Other embryologists could not replicate his results. Biologists working on farm animals were able to clone horses, pigs, rabbits, and goats from early embryo cells, but most considered cloning from adult cells impossible.

In the mid-1990s, however, Ian Wilmut, a biologist at the Roslin Research Institute in Scotland, resolved to try to improve the efficiency of "molecular farming," the genetic engineering of animals to produce valuable human proteins such as blood clotting factors. Animals were then being engineered for the purpose by injecting the gene for the protein into newly fertilized eggs that a surrogate mother brought to term. The gene generated the desired protein in only about 5 percent of the resulting animals. But if those animals could be cloned, the 95-percent failure rate could be turned into a 100-percent success rate. That prospect brought Wilmut support for his research from PPL Therapeutics, a biotechnology firm.

Wilmut thought that adult cells might be clonable if the nucleus was taken from cells in the right condition. Keith Campbell, an embryologist whom he hired to assist him in the research, suggested that cells in the G0 condition—a state of quiescence they enter when near starvation—might do. They confirmed this possibility in March 1996 when two lambs were born that they had cloned, using the G0 approach, from differentiated embryo cells. They then attempted to clone a sheep from adult udder cells. The effort produced 277 failures and one success—a sheep, born in July, that they named Dolly, in honor of the singer and actress Dolly Parton. "No one could think of a more impressive set of mammary glands than Dolly Parton's," Wilmut explained.

The announcement of Dolly's birth the following February—it was delayed to give PPL Therapeutics time to file a patent on the cloning technology—prompted immediate international debate about the application of cloning to human beings. Opponents demanded that it be legally banned. They prevailed in several countries, but not absolutely, since the biotechnology industry and many biomedical scientists backed human cloning for research purposes. Proponents see in cloning opportunities to study cellular differentiation and development and to obtain human stem cells for medical purposes.

See also ETHICS IN SCIENCE; GENE AND GENETICS.

Gina Kolata, *Clone: The Road to Dolly, and the Path Ahead* (1998). Ian Wilmut, Keith Campbell, and Colin Tudge, *The Second Creation: Dolly and the Age of Biological Control* (2000).

D. J. KEVLES

CLOUD AND BUBBLE CHAMBERS. C. T. R. Wilson built the first cloud chamber in 1895 to satisfy his interest in the weather. Previous work by John Aitken had examined the role of dust as a nucleating agent for water vapor in air, and hence as a source of fog and clouds. Like Aitken, Wilson built a chamber to reproduce in the laboratory the condensation of clouds: sudden expansion of the volume of a closed vessel containing saturated air produced a temperature drop and supersaturated the air. Unlike Aitken, Wilson brought a background in physics and the program of the Cavendish Laboratory at Cambridge University, especially in ion physics and discharge tubes, to bear on the subject. Instead of dust, Wilson focused on the role of ions as nuclei for water droplets. He also began photographing the formation of drops. Then in 1910 Wilson thought to use the device as a detector of charged particles, whose passage through the chamber would leave a trail of ions and hence water droplets. The next year Wilson obtained his first photographs of the tracks left by alpha and beta rays, as well as evidence of X and gamma rays through the beta rays they produced, and thus provided compelling visual evidence of individual atoms and electrons and their interactions.

The cloud chamber then became a popular tool for the study of particles given off by radioactivity, nuclear physics, and cosmic ray physics. Commercial firms helped propagate the device by providing affordable, experiment-ready Wilson chambers. P. M. S. Blackett, also at the Cavendish, used a Wilson chamber in the early 1920s to confirm Ernest Rutherford's transmutation of nitrogen into oxygen. In the early 1930s Blackett and Giuseppe P. S. Occhialini built a countercontrolled chamber, with a Geiger counter wired to trigger the expansion of the chamber whenever a cosmic ray passed through it. They used the chamber in 1932 to confirm the existence of the positron, which Carl Anderson of the California Institute of Technology (Caltech) had just detected in his own cloud chamber.

The Wilson cloud chamber suffered from a slow cycle time, and its diffuse gas offered few chances for interaction with incoming particles. The solid film of nuclear emulsions yielded more interactions and hence grew in popularity in the 1930s; but emulsions also constantly recorded tracks and hence complicated the resolution of occurrence times. In 1952 Donald Glaser, a physicist at the University of Michigan, tried to solve these problems by turning to a liquid analogue of the cloud chamber: instead of supersaturated gas, Glaser used a superheated liquid. He filled a small glass bulb, a couple of centimeters wide, with liquid diethyl ether held under pressure above its boiling point, then suddenly released the pressure, superheating the liquid. A charged particle passing through the chamber further heated the liquid and left a line of vapor bubbles in its wake. A high-speed camera filmed the first tracks in late 1952, and Glaser announced his results in 1953.

Glaser's bubble chamber provided sufficient density for numerous interactions and a fast cycle time. He intended to use his bubble chamber to detect *cosmic rays and hence tried various means of triggering the expansion as Blackett had done with the cloud chamber (*see* ELEMENTARY PARTICLES). The bubble chamber instead became an important particle detector for high-energy accelerators, whose predictable output dispensed with the need for countercontrolled expansion. A particular implication of the bubble chamber intrigued accelerator physicists: the possibility of using liquid hydrogen, which, because of its simple nucleus, already served as a target for interaction experiments at accelerators. A hydrogen bubble chamber would combine target and detector in one device.

Accelerator laboratories soon began building bubble chambers, including a group at the University of California at Berkeley under Luis Alvarez. The Berkeley group produced the first tracks in a hydrogen chamber in late 1953 and soon scaled up to larger chambers, culminating in a 180-cm. (72-in.) version to accommodate the greater interaction lengths of hyperons and other strange particles. The massive chamber, completed in 1959, presented new problems in cryogenics, optical systems, and computerized data analysis. The coordination of the physicists, engineers, and technicians building the chamber produced a complex, corporate organizational structure.

The Berkeley bubble chamber signaled the growing importance accorded the detector in accelerator experiments. It also marked a decisive shift from the table-top device of Glaser, built for cosmic-ray physics, to the big science of high-energy physics. Bubble chambers at several accelerator laboratories paid dividends in evidence of new particles and resonances that supported the SU(3) theory for classifying strange particles in the early 1960s and of neutrino interactions that helped confirm the electroweak theory in the early 1970s. Both Glaser and Alvarez won Nobel Prizes in physics for their work with bubble chambers, and both eventually drifted away from the field, disillusioned with the routinization, specialization, and automation that the bubble chamber had brought to particle detection.

R. P. Shutt, ed., *Bubble and Spark Chambers: Principles and Use* (1967). Peter Galison, *Image and Logic: A Material Culture of Microphysics* (1997). Peter Westwick, "Chamber, Bubble," and Jeff Hughes, "Chamber, Cloud," in *Instruments of Science: An Historical Encyclopedia*, eds. Robert Bud and Deborah Jean Warner (1998): 98–100, 100–102.

PETER J. WESTWICK

CODES. See CRYPTOGRAPHY.

COHESION, the property of sticking together, became a pressing theoretical issue in the seventeenth century, for with the rise of "corpuscular" or "mechanical" philosophy, it became the key to understanding what distinguished solid from fluid matter.

Corpuscular philosophers, who held that nature consisted solely of material particles of different shapes and sizes variously moved, faced a challenge in explaining how particles hold together. There could be no glue, since an adhesive, too, could consist only of particles in motion; nor could the philosopher invoke an inherent tendency among particles to cohere, since this would amount to the same empty verbalism for which corpuscular philosophers derided their Aristotelian opponents, and which

they prided themselves on having transcended (*see* SYMPATHY AND OCCULT QUALITY).

One line of thinking linked cohesion with the traditional notion, itself unacceptable to doctrinaire mechanists, that nature abhors a vacuum. Galileo *Galilei adopted this view in his *Discourse on Two New Sciences* (1638). He argued, however, that because a siphon could not raise water more than 18 *braccia* (about 10 m [33 ft]) before the water column broke under its own weight, the "attraction of the vacuum" could not exceed the weight of the breaking column. The cohesion of solid bodies resulted, he supposed, from their being composed of a multitude of minute particles separated by minute vacua, each of which exerted its attraction.

Pierre Gassendi, the great reviver of atomism, supposed that atoms that cohered did so because they possessed various cavities and protuberances that could catch together like the hooks and eyes used by dressmakers. René *Descartes, on the other hand, attributed cohesion to a state of relative rest between adjacent particles. Isaac *Newton in his *Opticks* dismissed both suggestions out of hand. Invoking hooked atoms was begging the question, he said, while saying that bodies adhere by rest made them stick together by nothing at all. As part of his wider challenge to the mechanical philosophy, Newton suggested that the particles of hard bodies attract one another with a force that is exceedingly strong at contact, but weakens very rapidly with separation. He claimed that experiments performed by Francis Hauksbee involving drops of oil creeping between sheets of glass slightly inclined to each other demonstrated the existence in nature of powerful, short-range forces such as Newton invoked.

With the widespread acceptance of the Newtonian program during the eighteenth century, natural philosophers concurred in explaining cohesion on the basis of a short-range interatomic force. They came to draw a distinction between the force of cohesion, now restricted to an attraction between corpuscles of the same kind, and the forces of affinity binding corpuscles of different chemical species together in compounds. In the late nineteenth century, molecular theorists invoked such an attraction, acting between molecules of a gas, to explain the deviation of real gases from ideal gas behavior. The twentieth century identified the force in question as ultimately electrical in nature, deriving from the electronic charge distributions within neighboring atoms.

Richard S. Westfall, *The Construction of Modern Science: Mechanisms and Mechanics* (1971).

R. W. HOME

COLD AND CRYONICS. Artificially produced cold has always been welcome. From the time ice was brought from the mountains so that the Roman emperors could enjoy their wine chilled in the summer to our days of high temperature superconductivity, the wonders of low temperatures have been publicly displayed and their benefits privately pursued. From unexpected phenomena of the very cold to dramatic changes in food production and eating habits, the production and harnessing of cold has probed nature and shaped society.

The first systematic researches on cold were reported by Robert *Boyle in his *New Experiments Touching Cold* (1665)—the subject that the master of experimenters found "the most difficult" of all. The ingenious experiments of Joseph Black to determine the latent heat and specific heat of water involved ordinary ice (*see* FIRE AND HEAT). Cryogenics received a boost from Michael *Faraday's liquefaction of nearly all the gases known in the 1820s. Among the gases he could not liquefy were oxygen, nitrogen, and hydrogen. By the end of the nineteenth century all gases except helium had been liquefied. Raoul-Pierre Pictet in Geneva and Louis Paul Cailletet in Paris first obtained small droplets of oxygen and nitrogen in 1877. Zygmunt Florenty von Wróblewski and Karol Stanisław Olszewski liquefied oxygen in appreciable quantities in 1883. Carl von Linde and William Hampson made significant improvements to the apparatus for reaching low temperatures. James Dewar liquefied hydrogen in 1898 at the Royal Institution in London. Heike Kamerlingh Onnes managed to liquefy helium on 10 July 1908 at the Physical Laboratory of the University of Leiden. Using the regeneration method and starting from liquid hydrogen temperatures, he made liquid helium and found its boiling point to be $4.25°K$ and its critical temperature $5°K$.

The development of *thermodynamics, especially James Prescott Joule's and William *Thomson's proofs that the temperature of a gas dropped when it expanded very quickly, provided the necessary background for the investigation and the understanding of the properties of matter in the very cold. Thomas Andrews's experiments determining the critical point—the temperature at which a gas whose pressure is increased at constant volume liquefies—and Johannes Diderik van der Waals's discussion of the continuity of the gaseous and liquid states brought further insight into the characteristics of very cold fluids.

The nineteenth century saw remarkable developments in the large-scale production of cold, especially through the development of the vapor compression process that led to different types of refrigerating machines and refrigeration processes. The plentiful availability of artificial cold transformed the preservation, circulation, and consumption of food. By the end of the nineteenth century the Linde Company had sold about 2600 gas liquefiers: 1406 were used in breweries, 403 for cooling land stores for meat and provisions, 204 for cooling ships' holds for transportation of meat and food, 220 for ice making, 73 in dairies for butter making, 64 in chemical factories, 15 in sugar refining, 15 in candle making, the rest for other purposes.

In 1911 the Institut International du Froid was founded in Paris to regulate the industry and to formulate directions of further research on cold. The preservation and transport of agrarian, fish, and dairy products, the standardization of the specifications for home refrigerators, the construction of trains and ships with large refrigerators, the installation of special refrigerators in mortuaries and slaughterhouses, the building of new hotels with air cooling systems, the design of breweries, the manufacture of transparent ice, and the possibilities of medical benefits from cold were some of the issues on which the national delegates who founded the Institut reached consensus.

In 1911, Kamerlingh Onnes observed that certain metals become superconductors, losing all resistance to electrical current, below $4°K$. In recent decades materials have been made that reach the superconducting state at much higher temperatures. Another bizarre bit of cold behavior, which came to light in the 1930s, is the superfluidity that liquid helium acquires below $2.19°K$ in virtue of which it does not display any of the features of classical fluid. These two phenomena turned out to be explicable only on the principles of *quantum mechanics. The explanation forced quantum mechanics to negate one of its basic methodological and historical tenets—that it made a difference only in the microscopic world. Superconductivity and superfluidity showed that macroscopic quantum phenomena exist.

With the availability of liquefied gases, the variation of the electrical resistance of metals with temperature was persistently studied. Dewar and John Ambrose Fleming made the first systematic measurements in 1896. Their results derived at liquid oxygen temperatures suggested that electrical resistance would become zero at absolute zero. But the same measurements at liquid hydrogen temperatures showed that the resistances after reaching a minimum started increasing again. In 1911 Kamerlingh Onnes measured the resistance of platinum and that of pure mercury at helium temperatures. At 4.19°K the value of the resistance dropped abruptly and became 0.0001 times that of solid mercury at 0°C. Impurities did not affect the superconductivity of mercury, but a high magnetic field could destroy it.

The first successful quantum mechanical theory of electrical conduction, proposed by Felix Bloch (1928), predicted that superconductivity was impossible. This theory considers the electrons in a metal as uncoupled, though it calculates the field in which any one electron moves by averaging over the other electrons. If the metal was at absolute zero, its immobile lattice determined a periodic potential field for the electronic motions and offered no electrical resistance. Bloch used an analogy to ferromagnetism to try to understand superconductivity. He found that the most stable state of a conductor, in the absence of an external magnetic field, had no currents. But since superconductivity was a stable state displaying persistent currents without external fields, his theory did not explain how superconductivity could come about in the first place.

At the beginning of November 1933 there appeared a short letter in *Naturwissenschaften* by Walther Meissner and Robert Ochsenfeld. It presented strong evidence that, contrary to every expectation and belief of the previous twenty years, a superconductor expelled the magnetic field after the transition to the superconducting state and the magnetic flux became zero (the Meissner effect). Superconductors were found to be diamagnetic and, hence, superconductivity a reversible phenomenon, thus allowing the application of thermodynamics.

In 1934 Fritz London and his brother Heinz, on the assumption that diamagnetism must be an intrinsic property of an ideal superconductor, and not merely a consequence of perfect conductivity, proposed that superconductivity involved a connection not with the electric, but with the magnetic field. Their assumption led to the electrodynamics of a superconductor consistent both with the zero resistance and the Meissner effect. Fritz London, in his discussion of superconductivity in 1936, formulated for the first time the notion of a macroscopic quantum phenomenon.

Because ionic masses are so much larger than the electron mass, physicists doubted that ions played an important role in the establishment of the superconducting state. Herbert Frohlich in 1950 asserted the opposite and found that the interaction of the electrons in a metal with the lattice vibrations would lead to an attraction between the electrons. Experiments confirmed his assertion. The mass became an important parameter when the motion of the ions was involved, and this, in turn, suggested that superconductivity could be derived from an interaction between the electrons and zero-point vibrations of the lattice. Soon after learning about these results, John Bardeen showed that superconductivity might arise from a new attraction between the electrons and the phonons resulting from lattice vibrations, thus laying the foundations for the electron pair theory. In 1956 Leon Cooper argued that such an interaction could provide what was needed. Based on these ideas, in 1957 Bardeen, Cooper, and John R. Schrieffer worked out the details of a microscopic theory of superconductivity, and shared the Nobel Prize in physics of 1972 for their successful explanation of this elusive phenomenon.

All liquids solidify under their own pressure at low enough temperatures. Helium can only be solidified under a pressure of 26 atmospheres. The densities and specific heats of all liquids follow a continuous change and increase as the temperature goes down. In the case of helium, however, these parameters display a maximum at 2.19°K and then decrease. The two methods for measuring the viscosity of any liquid—rotating a disk in it or forcing it through very small capillaries—give identical results. Not so for liquid helium below 2°K. The first of these methods gives a value a million times larger than the second. Finally, all liquids can be deposited in open containers, kept in containers with extremely small holes through which they cannot flow, and remain at rest when exposed to light. Liquid helium does not tolerate any such constraints. It goes over open containers, leaks through the smallest capillaries, and springs up in a fountain when light falls on it. Below 2.19°K liquid helium becomes a superfluid.

In 1938, Fritz London proposed that the transition to the superfluid state can be understood in terms of the Bose–Einstein condensation mechanism, first discussed by Albert *Einstein in 1924. For an ideal Bose–Einstein gas, the condensation phenomenon represented a discontinuity in the derivative of the specific heat. London argued that the sudden changes in the properties of helium at 2.19°K could result from such discontinuities. Below a certain temperature and depending on the mass and density of the particles, a finite fraction of them begins to collect in the energy state of zero momentum. The remaining particles fly about as individuals, like the molecules in a normal gas. Laszlo Tisza proposed to regard superfluid helium as a mixture of a normal and a superfluid. These two components had different hydrodynamical behaviors as well as different heat contents. At absolute zero, the entire liquid became a superfluid consisting of condensed atoms, while at the transition temperature the superfluid component vanished.

Excluding some applied fields, low-temperature physics became the high point of Soviet physics, especially during World War II. After receiving his doctorate under the supervision of Ernest Rutherford, Pyotr Kapitsa served as an Assistant Director of magnetic research at the Cavendish Laboratory, before becoming the director of the Mond Laboratory in Cambridge. There he liquefied helium in 1934. Though he was not allowed to return to England by the Soviet authorities after a trip in 1934, he was, by 1935, appointed as director of a new Institute of Physical Problems within the Academy of Sciences in Moscow. That is where he conducted his experiments with liquid helium in 1941 and coined the term "superfluid" when he discovered the remarkable characteristics of its viscosity. In 1941 Lev Landau developed a quantized hydrodynamics that explained the transition to the superfluid state in terms of rotons and phonons. The ground states and the excitations played the roles of the superfluid and normal state, respectively. The excitations were the normal state because they could be scattered and reflected and, hence, exhibit viscosity. The ground state described the superfluid because it could not absorb a phonon from the walls of the tube or a roton unless it met some conditions on the velocity. Landau's formalism led to two different equations for the velocity of sound. One was related to the usual velocity of compression, while the other depended strongly on temperature. Landau named it "second sound." Victor Peshkov demonstrated the existence of standing thermal waves in 1944 for the first time, and in 1949 the experiments of Maurer and Herlin settled the issue of the temperature dependence of the second sound velocity below 1°K. By 1956 Richard Feynman could show that some of Landau's assumptions could be justified quantum mechanically and that the rotons were a quantum mechanical analog of a microscopic vortex ring.

Oscar Edward Anderson, Jr., *Refrigeration in America. A History of a New Technology and Its Impact* (1953). Kurt Mendelssohn, *The Quest of Absolute Zero* (1977). Roger Thevenot, *A History of Refrigeration Throughout the World* (1980). Kostas Gavroglu and Yorgos Goudaroulis, *Methodological Aspects of the Development of Low Temperature Physics 1881–1956. Concepts out of Context(s)* (1989). Kostas Gavroglu and Yorgos Goudaroulis, eds., *Through Measurement to Knowledge. The Selected Papers of Heike Kamerlingh Onnes 1853–1926* (1991). Per Fridtjof Dahl, *Superconductivity. Its Historical Roots and Development from Mercury to the Ceramic Oxides* (1992). Ralph Scurlock, ed., *History and Origins of Cryogenics* (1992). Kostas Gavroglu, *Fritz London, A Scientific Biography* (1995). Tom Shachtman, *Absolute Zero and the Conquest of Cold* (1999).

KOSTAS GAVROGLU

COLD FUSION. On 23 March 1989 in a press conference in Salt Lake City, B. Stanley Pons, a professor of chemistry at the University of Utah, and Martin Fleischmann, his collaborator from the University of Southampton, proclaimed that they had achieved the fusion of deuterium nuclei—the type of reaction that fuels the sun—in a laboratory experiment. They reported that when they passed an electrical current through palladium metal immersed in a beaker of heavy water with a bit of lithium, the cell produced an excess of heat—enough at one point to melt a cube of the metal and far more than could be accounted for by current running through the electrode or by ordinary chemistry. The heat, they said, had to be a product of cold fusion, so-called because it had been achieved at room temperature.

Their scientific miracle won wide coverage in the world press. It promised cheap, clean energy, which, in the wake of the fuel shortages of the 1970s, governments had been spending hundreds of millions of dollars to find. It captured attention in the White House, the U.S. Congress, and the government of Utah, which promptly appropriated five million dollars for a new National Cold Fusion Institute. Chase Peterson, the president of the university, enthused that the breakthrough "ranks right up there with fire, with cultivation of plants and with electricity."

Many scientists, however, greeted cold fusion with considerable skepticism. It defied the known laws of physics. Moreover, Pons and Fleischman revealed few details of the experimental apparatus and methods that produced their astonishing result and persistently ducked requests from scientists for more information. Some scientists wryly observed that fusion cells running as hot as Pons and Fleischmann's should have been producing enough neutron radiation to kill them.

Laboratories in the United States, Europe, and Japan geared up to witness the miracle firsthand or debunk it. Theoretical calculations demonstrated that the fusion of deuterium was impossible with Pons and Fleischman's apparatus, and experimental tests in various laboratories showed that the heat generated with it did not arise from fusion. On 1 May 1989, at the meetings of the American Physical Society in Baltimore, Maryland, the evidence for cold fusion was authoritatively reviewed and found wanting at a special session that drew two thousand people.

Some observers wondered whether Pons and Fleischman had attempted to perpetrate a scientific fraud. At the least, one physicist remarked, cold fusion was a result of their "incompetence and perhaps delusion." Cold fusion indelibly tarred Fleischmann's reputation and forced Pons's departure from the University of Utah in the fall of 1989. Inspired by odd bits of seemingly encouraging evidence and more than a dash of hope, research into cold fusion continued at several laboratories in the United States, Japan, and Italy at least until the mid-1990s, some of it funded by the Japanese government, but none of the efforts yielded compelling results.

David L. Goodstein, "Pariah Science—Whatever Happened to Cold Fusion?" *The American Scholar* 63 (Fall 1994): 527–541. Gary Taubes, *Bad Science: The Short Life and Weird Times of Cold Fusion* (1993).

DANIEL J. KEVLES

COLD WAR. See MANHATTAN PROJECT; NUCLEAR DIPLOMACY; WORLD WAR II AND COLD WAR.

COLLECTIONS. See CABINETS AND COLLECTIONS.

COLLIDER. During the final quarter of the twentieth century, particle colliders emerged as the preferred instruments in high-energy physics. Their defining characteristic, besides their great size—the largest are measured in kilometers—is their manner of generating collisions. They accelerate two beams of subatomic particles and bring them together at interaction points, where a particle in one beam can collide with a particle in the other. Surrounding each interaction point, a particle detector records the tracks, energies, and other characteristics of particles emanating from these collisions, allowing physicists to analyze what transpired.

The great advantage of colliders over conventional "fixed-target" machines (such as cyclotrons), in which particle beams strike stationary objects, is that essentially all the energy of the individual colliding entities can be used to create new subatomic particles. The available "center-of-mass" energy, or total collision energy, grows in proportion to the beam energy rather than to its square root, as in fixed-target experiments. All discoveries of massive new subatomic particles since 1975 have been made using colliders, while fixed-target experiments have excelled at examining the structures of known particles such as protons. To permit meaningful experiments, colliders must attain sufficient luminosity, a key measure of the rates of interaction between particles in the opposing beams.

The idea of particle colliders occurred to Rolf Wideröe and Donald Kerst in the mid-1950s, but the first significant work on developing such an instrument began at Stanford University in 1958. Led by Gerard O'Neill of Princeton University, a small group of physicists built two evacuated "storage rings" in a figure-eight configuration. Beams of electrons circulated in opposite directions within these rings at energies of up to 500 million electron volts (MeV); collisions occurred in the shared segment where the rings touched.

In parallel with this effort, physicists at the Frascati National Laboratory in Italy, led by Bruno Touschek, built a single-ring collider in which electrons circulated one way and positrons (their antimatter opposites) the other. Following the success of this prototype, the Italian physicists developed a full-scale electron-positron collider called ADONE, with beam energies of up to 1,500 MeV, or 1.5 billion electron volts (GeV). Experiments using this instrument began in 1968, recording electron-positron annihilations that usually created other subatomic particles.

Physicists at the European Center for Nuclear Physics (*CERN), led by Kjell Johnsen, pioneered proton-proton colliders. In 1971 they successfully operated the Intersecting Storage Rings, in which beams of protons circulated at energies of up to 28 GeV. Collisions occurred at six interaction points where the interlaced rings crossed.

The most productive electron-positron collider was the SPEAR facility built at the Stanford Linear Accelerator Center (SLAC) under the direction of Burton Richter. Completed in 1972, SPEAR generated collisions at combined energies of up to 8 GeV. It yielded the discoveries of the massive psi particles and tau lepton, and Nobel Prizes for Richter (in 1976) and SLAC physicist Martin Perl (in 1995).

Following these advances, physicists built colliders at all leading high-energy physics laboratories. Especially noteworthy was a proton-antiproton collider built at CERN as an upgrade of its existing Super Proton Synchroton, stimulated by ideas and inventions of Peter McIntyre, Carlo Rubbia, and Simon Van der Meer. By observing proton-antiproton collisions at total energies of up to 540 GeV in 1982–1983, two teams of physicists discovered the massive W and Z bosons, the mediators of weak interactions and key elements of the *standard model.

These significant discoveries and the development of superconducting magnets for the Tevatron proton-antiproton collider at the Fermi National Accelerator Laboratory (Fermilab) encouraged U.S. physicists to design the Superconducting Super Collider (SSC), a 40,000 GeV proton-proton collider that was to have a circumference of 86 km (54 miles), several interaction points, and a cost of $5.9 billion. Those were its parameters in 1989 when construction began south of Dallas, Texas. Congress terminated the project in 1993 owing to cost overruns, lack of major participation from other countries, and a concern to reduce budget deficits after the Cold War.

Since the SSC's demise, the development of particle colliders has continued largely through upgrades of existing instruments at CERN, Cornell University, Frascati, Fermilab, and SLAC, and at national laboratories in China and Japan. A prime example was the conversion of the Stanford Linear Accelerator into a linear electron-positron collider. The new machine accelerated individual "bunches" of the two types of particle and brought them together after a single pass through the linear accelerator. This approach contrasts with that of storage-ring colliders, in which the bunches of particles circulate continuously in fixed, intersecting orbits.

Burton Richter, "The Rise of Colliding Beams," in *The Rise of the Standard Model*, ed. L. Hoddeson et al. (1997): 261–284. Michael Riordan, "The Demise of the Superconducting Super Collider," *Physics in Perspective* 2, no. 4 (2000): 411–425.

MICHAEL RIORDAN

COMETS AND METEORS. Until the seventeenth century, comets and meteors were classified as related natural phenomena and heavenly wonders that heralded calamity. The roots of these views reached back to antiquity. Aristotle saw comets as a type of fiery meteor that formed when terrestrial exhalations ascended into the upper atmosphere, below the moon's sphere, and began to burn. Other fiery meteors included shooting stars, fireballs, and the aurora borealis. Comets and meteors augured windy weather, drought, tidal waves, earthquakes, and stones falling from the sky because both the meteors and the portended disasters derived from hot, dry exhalations that had escaped from the earth. Romans came to view comets and showy meteors as monsters, contrary to nature.

Medieval chronicles recorded meteorological apparitions that heralded the death of holy men and kings, and augured wars of religion and civil strife. According to some early church fathers and later theologians, these heavenly signs demarcated critical periods in the history of the world and religion. Thus Origen and John of Damascus thought that the Star of Bethlehem had been a comet, whereas Saint Jerome, Thomas Aquinas, Martin Luther, and Thomas Burnet expected comets and fiery meteors to precede the Day of Judgment and consummation of all things.

Medieval and Renaissance natural philosophers agreed that comets and meteors prefigured calamity. John of Legnano and Johannes *Kepler looked for causal connections. Many more wrote guides, both in Latin and in vernacular languages, to interpret the meaning of these celestial hieroglyphs. Their tracts served astrologers and propagandists who used fiery meteors to legitimate political authority and to fortify conspirators up through the eighteenth century.

In the Renaissance, observations of the parallax, tails, and motion of comets by Tycho *Brahe and others convinced astronomers that comets were not sublunar meteorological phenomena, but celestial bodies traveling through interplanetary regions. The separation of comets from meteors was completed by the end of the seventeenth century when astronomers agreed with Isaac *Newton and Edmond Halley that comets traveled in elliptical orbits around the sun. In the mid-eighteenth century, philosophers began to consider extraterrestrial origins for meteors such as the aurora and shooting stars.

How far these theoretical developments caused the decline of divination from comets and meteors has been much debated. Whatever the answer, in the late seventeenth century the learned of England and France (followed later by those in central and eastern Europe) rejected as vulgar the notion that comets and meteors were miraculous signs sent by God to rebuke infidels. They no longer saw them as causes of murder, rebellion, drought, flood, or plague. Nevertheless, neither the celestial locus nor the periodic orbits of comets required believers to give up their faith in the eschatological or prophetic functions of comets.

Newton suggested that comets transported life-sustaining materials to the earth and fuel to the sun. He, Halley, and William Whiston argued that comets had key roles to play in the earth's creation, Noachian deluge, and ultimate destruction. The final conflagration would be ignited by a comet, many theologians believed, and natural philosophers concurred that a blazing star could serve this function by immersing the earth in its fiery tail, by dropping into the sun and causing a solar flare, or by kicking the earth out of its orbit and transforming it into a comet. Forced to travel in a much more elongated circuit around the sun, the old earth would be scorched and frozen in turns, a very plausible model of hell. The new periodic theory of comets did not destroy the belief in comets as agents of upheaval or renewal, nor as tools God might use to punish the wicked or save the elect.

In the eighteenth and early nineteenth centuries, Georges-Louis LeClerc, Comte de Buffon, William Herschel, and Pierre-Simon *Laplace, continued to connect comets to the creation and dissolution of planets, but separated astrotheology from celestial mechanics. Unlike their predecessors, they neither hoped nor expected to find the moral order reflected in the natural world. When catastrophism (*see* UNIFORMITARIANISM) went out of style in the mid-nineteenth century, comets appeared to pose little risk or benefit to the earth. In recent years, however, the tide has turned, and the stage may be set for a new theological interpretation of comets and meteors. Most scientists now believe that comets (and their meteoric debris) may have been both the agents of death (most notably of the dinosaurs) and the conveyors of life's building blocks.

Roberta J. M. Olson, *Fire and Ice: A History of Comets in Art* (1985). John G. Burke, *Cosmic Debris: Meteorites in History* (1986). Donald K. Yeomans, *Comets: A Chronological History of Observation, Science, Myth, and Folklore* (1991). Sara J. Schechner, *Comets, Popular Culture, and the Birth of Modern Cosmology* (1997).

SARA J. SCHECHNER

COMPASS. See MAGNET AND COMPASS.

COMPLEMENTARITY AND UNCERTAINTY constitute the foundation of the "Copenhagen interpretation" of quantum physics, an acausal understanding of physics that remains predominant today. Both concepts were introduced in 1927, by Niels *Bohr and Werner *Heisenberg, respectively. They arose as part of the development of quantum physics when the field was in tremendous flux. Several institutions—notably the *Bohr Institute for Theoretical Physics and the physics institute at the University of Göttingen—and many individual physicists had placed high stakes on establishing an acceptable theory as well as its interpretation.

Bohr had been a leader in the development of quantum physics since he published his revolutionary atomic model in 1913 (*see* ATOMIC STRUCTURE). In Bohr's model, atomic electrons could exist only in orbits determined by the quantum of action (*see* QUANTUM PHYSICS) and emit electromagnetic radiation only when jumping from one orbit to another. During the "old quantum theory" (1913–1925), theorists invoked a mixture of Bohr's "correspondence principle" (which specified a numerical connection between quantum and classical physics) and arguments based on the quantum of action. As one of a long line of increasingly radical attempts to arrive at an overarching theory, in 1924 Bohr, his assistant Hendrik Kramers, and an American postdoctoral researcher, John Slater, published a paper based on a denial of the well-established principle of energy conservation (*see* CONSERVATION LAWS). In their view, the principle held only statistically and not for individual atomic processes.

Experimental results showing energy conservation in collisions between individual photons and atomic electrons in the Compton effect (*see* X RAYS) quickly forced the abandonment of this view, and others took the lead in seeking to formulate a quantum theory. In the fall of 1925, Heisenberg, then working as Bohr's assistant, devised a means to calculate spectral data without explicit appeal to the correspondence principle. Whereas Heisenberg's severely operationalist, as well as particle-oriented, theory involved complicated matrix calculations, Erwin *Schrödinger's wave-oriented version of "quantum mechanics," published in the fall of 1926, involved mathematics with which the average physicist felt more comfortable. In spite of this, and although Schrödinger's "wave mechanics" seemed at first to allow the visualization of atomic processes by emphasizing continuity and retaining causality, Schrödinger and others soon showed that his approach was mathematically equivalent to Heisenberg's.

In September 1926 Schrödinger paid a now famous visit to Copenhagen. Bohr stubbornly sought to convince him of the reality of quantum jumps. In this tense environment Heisenberg wrote the article containing his "uncertainty principle," which stated that in the atomic domain the quantum of action set a limit to the precision with which two conjugate variables, such as a particle's position and momentum, or the time and energy of an interaction, could be measured. Since the present therefore cannot be fully specified, Heisenberg argued, neither can the future. By explaining the indeterminism of quantum physics in this way, Heisenberg tried to make his presentation more visualizable (*anschaulich*) and hence more acceptable to his fellow physicists.

Heisenberg overstepped common practice by submitting his article from Copenhagen without asking Bohr's permission. It turned out that Bohr disagreed so strongly with Heisenberg's presentation of the quantum that Heisenberg felt compelled to add a correction in the proofs that allowed a greater role for the wave picture. Bohr was then perfecting his own formulation of the foundations of quantum theory. At the end of a lecture surveying the general situation given in Como in September 1927, he proposed his notion of complementarity for the first time. The new notion provided an understanding of quantum mechanics in general and of Heisenberg's uncertainty principle in particular. Bohr maintained that, unlike in classical theory, a description of processes in space-time and a strictly causal account (by which Bohr meant an account recognizing conservation laws) of physical processes excluded one another. This meant in practice that the investigator could choose which aspect of microphysical reality he wished to see expressed by his choice of experimental setup. Although the setup required for realizing one aspect excluded the realization of the other—for example, an apparatus for exhibiting light with particulate properties cannot also show it as a wave—both sets of properties had to be invoked to obtain a complete description of the microphysical reality. While presented as a direct result of quantum mechanics, Bohr's interpretation and his subsequent elaborations of it resonated with philosophical views with which he had struggled in his youth. Only with Bohr's complementarity of 1927 did his work and that of Heisenberg, Max Born, Wolfgang Pauli, Pascual Jordan, and others begin to converge into what came to be seen as the unified "Copenhagen interpretation" of quantum mechanics.

The group surrounding Bohr soon came to perceive complementarity and uncertainty as so closely intertwined that in 1928 Heisenberg gave Bohr's concept precedence over his own. In 1935 Albert *Einstein and two collaborators challenged Bohr's interpretation for being inherently incomplete; they thought that they could obtain more information by experiment than complementarity allowed. Bohr repelled their attack by widening the divide between classical and quantum ideas. In the larger physics community, however, the uncertainty principle became inseparable from any presentation of quantum mechanics, while complementarity figured little in the teaching of the new physics. It tended to be regarded as overly philosophical, vague, and irrelevant.

In recompense, complementarity took on a life beyond physics. Bohr sought to generalize its application, first to psychology, then to biology, and ultimately beyond the scope of natural science. Although he did not complete the book on the topic that he had hoped to write, Bohr conceived complementarity as a general epistemological argument of great import for humanity. It constituted a guiding principle for his own activities, inside and outside physics. At the same time, Bohr's disciples sought to spread their understanding of Bohr's word, sometimes—as in the case of Jordan, who tried to use it to save the freedom of the will—to Bohr's embarrassment. Variations of complementarity became part of severe ideological struggles in Nazi Germany and the Soviet Union.

While extremely devoted, the audience for Bohr's philosophical statements was never large, and today consists largely of a specialized set of philosophers. Nevertheless, complementarity played an important role in providing a conceptual basis for the early work on quantum mechanics.

Max Jammer, *The Conceptual Development of Quantum Mechanics* (1966). Jørgen Kalckar, ed., *Niels Bohr Collected Works*, vols. 6 and 7 (1985, 1996). Mara Beller, *Quantum Dialogue: The Making of a Revolution* (1999). David Favrholdt, ed., *Niels Bohr Collected Works*, vol. 10 (1999).
FINN AASERUD

COMPLEXITY. See CHAOS AND COMPLEXITY.

COMPUTER. The modern electronic computer has theoretical antecedents in the English scientist Charles Babbage's conception in the 1820s of an "analytical engine" that would solve any mathematical problem put to it; and operating forebears in the electromechanical machines that the engineer Herman Hollerith devised for the 1890 American census and that processed data entered on punch cards. The digital electronic computer itself was pioneered during World War II by J. Presper Eckert, an electronics engineer, and John Mauchly, a physicist, in response to the military's need for fast and efficient calculation of artillery firing tables. Their machine, called ENIAC (for electronic numerical integrator and computer) and completed near the end of 1945, used eighteen thousand vacuum tubes and could perform five thousand operations per second. In a report published in June that year, the mathematical physicist John von *Neumann laid out what came to be the basic constituents of a fully capable electronic computer (which ENIAC was not): units for processing, program, input, and output.

During the next few years, computers containing the von Neumann elements were developed in several projects driven by the agencies of national security and then spun out into the civilian sector. For example, in the Air Force–sponsored Project Whirlwind at MIT, the engineer Jay Forrester devised a memory constructed of small magnetizable cores, each capable of rapidly storing and returning coded information. In 1955, the International Business Machines (IBM) company produced the first commercial computer with magnetic core memory, a behemoth weighing 250 tons and occupying a large room. By the end of the 1950s, digital electronic computers were becoming commonplace in government, business, and universities.

In the 1970s, hobbyists developed the much smaller desktop computers using semiconductor chips etched with integrated circuits. In 1974, a small firm that three hobbyists had founded in Albuquerque, New Mexico, started marketing a

The "Pilot ACE," the oldest complete general purpose electronic computer in Britain as it appeared at its inauguration at the National Physical Laboratory in 1950. The Pilot incorporated plans for the larger ACE (Automatic Computing Engine) drawn up by Alan *Turing.

personal computer kit called the Altair. Bill Gates, a 20-year-old Harvard student, and his high-school friend Paul Allen, 22, wrote a software program for it. Gates dropped out of Harvard to develop the Microsoft Corporation, the software firm he and Allen founded in 1975 for the Altair venture. In 1976, Steve Wozniak, 25, and Steve Jobs, 20, began marketing a personal computer, the Apple, the first model of which they built in the home garage of Jobs's parents.

In 1977, Jobs and Wozniak brought out the Apple II, which sported a keyboard, a monitor, and a floppy disk drive for storage. A later version, introduced in 1983, operated with a mouse and pull-down menus, both of which had been developed under contracts from the defense department and NASA. In 1981, IBM entered the PC market, enlisting Microsoft to provide the operating software for its machines. In response, Microsoft devised MS-DOS (for Microsoft disk operating system), which became an industry standard. The PC caught on so fast that two years later, *Time* magazine designated the personal computer as its "Man of the Year."

See also CALCULATOR.

Kenneth Flamm, *Creating the Computer: Government, Industry, and High Technology* (1988). Martin Campbell-Kelly and William Aspray, *Computer: A History of the Information Machine* (1996).

DANIEL J. KEVLES

COMPUTER SCIENCE. Computer science is the study of the principles and the use of devices for the processing and storing of usually digital data using instructions in the form of a program. Before the existence of modern computers, people who performed calculations manually were known as "computers." The term "computer science," signifying a particular combination of applied mathematics (particularly logic and set theory), and engineering (normally electronic) first occurred as the name of a university department at Purdue (U.S.) in 1962. The two key areas of development have been hardware (the computers themselves) and more recently software (the intangible programs that run on them).

Mechanical computing devices long preceded electronic ones. The earliest known mechanical adding machine, the creation of the German inventor Wilhelm Schickard, dates from 1621. The mechanical calculators created by Blaise Pascal (1623–1662) and Gottfried Leibniz (1646–1716) received wider attention than Schickard's machine, which fell into oblivion, and formed part of the intellectual inheritance of Charles Babbage (1792–1871), often celebrated as a pioneer of computing. Alarmed by the number of mistakes in hand-computed mathematical tables, he invented the Difference Engine and subsequently the Analytical Engine with many of the features of a modern computer. The "mill" (the gears and wheels that performed the arithmetical operations) corresponded to a modern central processing unit (CPU) for computation and the "store" was a mechanical memory for reading and writing numerical values. Ada Lovelace (1815–1852), the daughter of Lord Byron, provided the earliest comprehensive description of this first programmable computer partially based on notes by the Italian Luigi Menabrea (1809–1896). Babbage never completed the Analytical Engine, which would have stretched cogwheel machinery to its limits at the time.

Leibniz was the first mathematician thoroughly to study the binary system, upon which all modern digital computers are based. George Boole presented what became known as Boolean algebra or logic in his masterwork of 1854, *An Investigation of the Laws of Thought.* Boole's laws can be used to formalize binary computer circuits. Later, David Hilbert (1862–1943) argued that in an axiomatic logical system all propositions could be proved or disproved, but Kurt Gödel (1906–1978) demonstrated otherwise, with important implications for the theory of computability. Propositional and predicate logic, together with set theory, as formulated by Ernst Zermelo (1871–1953) and Adolf Fraenkel (1891–1965) among others, provide important underpinnings for computer science.

Analog computers use continuous rather than discrete digital values. They enjoyed some success before digital technology became established for systems of related variables in equational form. Vannevar Bush devised the successful Differential Analyzer for solving differential equations at the Massachusetts Institute of Technology during the 1930s. Later he wrote a seminal article, "As We May Think" (1945), that predicted some of the features of the World Wide Web, illustrating a very broad appreciation of computer science.

In 1936, the English mathematician Alan *Turing, influenced by Gödel, devised a theoretical "universal machine," later known as a Turing machine, that helped to define the limits of possible computations on any computing machine. Turing had a practical as well as a theoretical bent. He played a significant part in the building of Colossus, at Bletchley Park, which made possible the breaking of German codes during World War II (*see* CRYPTOGRAPHY). Although it has a claim to be the first modern digital computer, Colossus had little influence since it remained secret for several decades. Turing subsequently worked on the design of the Pilot ACE computer at the National Physical Laboratory and the programming of the Manchester Mark I at the University of Manchester, both successful postwar British computers. Turing's standing in modern computing is indicated by the Turing Award, computer science's highest distinction. It has been given

annually since 1966 by the Association for Computing Machinery (ACM), the subject's foremost professional body, founded in 1947.

John Atanasoff built what may have been the first electronic digital computer, the Atanasoff-Berry, prototyped in 1939 and made functional in 1940. It influenced John Mauchly, who with J. Presper Eckert constructed the famous ENIAC at the Moore School of Electrical Engineering in Philadelphia between 1943 and 1945. EDVAC, the first U.S.-built stored-program computer, followed in 1951. Maurice Wilkes attended a summer program at the Moore School in 1946, returned to the University of Cambridge in England, and completed the EDSAC in 1949. Its run on 6 May 1949 made it the world's first practical electronic stored-program computer. The Lyons company copied much of EDSAC to produce the first commercial data processing computer, the LEO (Lyons Electronic Office), in 1951. Other important early computer pioneers include Konrad Zuse, who worked separately on mechanical relay machines, including floating-point numbers, producing the Z1-Z4 models in Berlin between 1936 and 1945. Significant U.S. engineers include Howard Aiken, who developed the electromechanical calculator Harvard Mark I, launched in August 1944, and George Stibitz, who illustrated remote job entry in September 1940 by communicating between Dartmouth College in New Hampshire and his Model 1, first operational in 1939, located in New York. Aiken established programming courses at Harvard long before the university computer science courses of the 1960s.

Programming facilities for early computers initially operated at the binary level of zeros and ones. Assembler programs allowed the input of instructions in mnemonic form, but still matching the machine code very closely. Higher-level programming less dependent on machine language requires a compiler program run on a computer. Noam Chomsky provided influential formal characterizations of grammars for languages, including programming languages, in the late 1950s and early 1960s.

Early high-level programming languages for scientific and engineering applications included Fortran, developed between 1954 and 1957 by John Backus and others at IBM in New York City. Backus also devised Backus Normal Form (BNF) for the formal description of the syntax of programming languages. Successive versions of Fortran have kept it in use. COBOL was another important programming language, developed for business applications in the late 1950s. U.S. Navy Captain Grace Hopper played a key part in its creation. ALGOL, the first programming language described using BNF, included in its

1960 version important new features such as block structuring, parameter passing by name or value, and recursive procedures that greatly influenced subsequent programming languages.

Pascal, designed by Nichlaus Wirth in Zurich between 1968 and 1970, embodied the concepts of structured programming espoused by Edsger Dijkstra and C. A. R. Hoare. Its simplicity suited it for educational purposes as well for practical commercial use. Wirth went on to develop Modula-2 and Oberon and is widely considered as the world's foremost designer of programming languages. Ada was developed in the 1970s for U.S. military applications. It proved to be the opposite of Pascal in the scale of complexity.

Dennis Ritchie created "C" as a general-purpose procedural language. It served as a basis for the highly influential Unix operating system, developed by Ritchie and Kenneth Lee Thompson at Bell Laboratories in New Jersey and refined still further there into C++. C++ encourages information hiding, as suggested by David Parnas, or encapsulation within "objects" considered as instances of classes, a technique first used in the SIMULA language, produced by Ole-Johan Dahl and others in 1967. The highly successful language Java, designed as a portable object-oriented programming language for distributed applications, dates from the early 1990s.

The languages so far considered follow an order of instructions. Some higher-level languages, such as LISP (late 1950s and early 1960s, at MIT), widely used in artificial intelligence, express computations in the form of mathematical functions, and so reduce the importance of the ordering of execution. Logic programming, as in Prolog (1970s), is a relational approach admitting nondeterministic answers. An extension, constraint logic programming, allows the convenient inclusion of extra conditions on variables.

A von *Neumann machine, similar but not identical to a theoretical Turing machine, refers to the standard arrangement of early sequential computers with CPU and memory still widely used. However, parallel architectures have become increasingly important, as computer-processing power presses against physical limits. Architecture has evolved through valves or tubes, solid-state transistors, and integrated circuits of increasing complexity.

Theoretical underpinnings for computer science include the definition of computability incorporated in the Turing machine, the λ-calculus of Alonso Church (1903–1995), and recursive functions. Complexity theory aids reasoning about the efficiency of computation. Other theoretical computer science subdisciplines include automata theory, computational geometry (for computer graphics), graph theory, and formal languages.

Software engineering encompasses the process of producing programs from requirements and specifications via a design process. Dijkstra from Holland has been a major contributor to the field. His influential paper *GO TO Statement Considered Harmful* (1968) led to the acceptance of structured programming, a term he coined, in the 1970s, in which abstraction is encouraged in the design process and program constructs are limited to make reasoning about the program easier. Dijkstra, Dahl, and Hoare wrote the widely read *Structured Programming* (1972). Hoare has also made important contributions to formal reasoning about programs using assertions, sorting algorithms, and the formalization of concurrency. Donald Knuth of Stanford University has been a major innovator in computer algorithms. His multivolume and still unfinished magnum opus, *The Art of Computer Programming*, is one of the best-known and influential books in computer science. He has contributed especially to parsing, reasoning, and searching algorithms, all important computer science techniques. Like all good computer scientists, he has expertise in both theory and practice. As well as major theoretical contributions, he has produced the TeX document preparation system, still widely used by computer scientists internationally for the production of books and papers.

Artificial intelligence (AI) has held out huge promises that have been slower to mature than expected. Major contributions have been made by John McCarthy, latterly at Stanford University, and Marvin Minsky at MIT. Important aspects of AI include automated reasoning, computer vision, decision making, expert systems, machine learning, natural language processing, pattern recognition, planning, problem solving, and robot control. A successful outcome of the Turing test, where the responses of a human are essentially indistinguishable from those of a computer, has proved elusive in practice unless the knowledge domain is very limited. Connectionism, using massively parallel systems, has opened up newer interesting areas for machine learning such as neural networks (similar to the workings of the brain) and also genetic algorithms (inspired by Darwin's theory of evolution). Databases are an important method of storing, organizing, and retrieving information. The Briton Edgar Codd created the relational model for databases in the late 1960s and early 1970s at the IBM Research Laboratory in San Jose, California. The two important categories of database objects are "entities" (items to be modeled) and "relationships" (connections between the entities) for which a good underlying theory has been established. Communication has become as significant as computation in computing. Claude Shannon provided an important theoretical approach in his paper of 1948, *A Mathematical Theory of Communication*. He contributed to both network theory and data compression. Donald Davies of NPL and others developed packet switching in the 1960s, a precursor to the Internet, originally established in 1969 and known as the ARPAnet for many years. More recently, the expansion of the Internet has made possible the proliferation of the World Wide Web (WWW), a distributed information system devised in the early 1990s by the British scientist Tim Berners-Lee at *CERN in Switzerland. His unique insight combined a number of key principles: a standard network-wide naming convention for use by hyperlinks in traversing information; a simple but extensible markup language to record the information; and an efficient transfer protocol for the transmission of this information between the server and a client user.

Computer science has had an enormous social impact in recent years. Yet it is still a relativity young and perhaps immature science. Quantum computers offer the possibility of removing some of the stumbling blocks encountered today and could theoretically render useless many of the data security mechanisms currently in place. The future of computer science looks even more interesting than its past.

J. A. N. Lee, ed., *International Biographical Dictionary of Computer Pioneers* (1995). Valerie Illingworth, ed., *A Dictionary of Computing*, 4th ed. (1996). Martin Campbell-Kelly and William Aspray, *Computer: A History of the Information Machine* (1997). David Harel, *Computers Ltd.: What They Really Can't Do* (2000). Mark W. Greenia, *History of Computing: An Encyclopedia of the People and Machines that Made Computer History*, CD-ROM (2001). Raúl Rojas, ed., *Encyclopedia of Computers and Computer History*, 2 vols. (2001). *IEEE Annals of the History of Computing*, available online at www.computer.org/annals. The Virtual Museum of Computing, available online at vmoc.museophile.com.
JONATHAN P. BOWEN

COMPUTERIZED TOMOGRAPHY. In the 1920s, innovators in several different countries devised methods of using X rays to obtain an image of a planar cross-section (P) in the interior of the body. The rays were directed obliquely through P to a film plate placed perpendicular to the body's vertical axis and at a height such that the rays through the center of the plane would strike the center of the plate. The body was rotated around its vertical axis and the film plate rotated in synchronization with it around a parallel axis but in the opposite sense. The rotation would blur or cut off part of the image of any plane above or below P, while the image of P would remain fixed in place on the film and be visible through the blur. The first imaging machines came to be called tomographs, from the Greek *tomo*, 'slice.' Tomographs, produced commercially by the 1940s, provided images of the chest, the gastrointestinal tract, ears, sinuses, noses, kidneys, and urethral canals. The blurring of the surrounding bone and tissue, however, made the images less than perfect.

The deficiencies were largely overcome with the arrival, in 1971, of scans using computer-assisted tomography, originally called CAT scans but now commonly known as CT scans. In CT, an x-ray tube rotates around the body's vertical axis while sending a beam of parallel rays at regular intervals through it. Detectors that rotate with the tube register the beam's emergent energy, which depends on the density of tissue and bone it has encountered. A computer analyzes and integrates the data to construct images of successive cross-sectional planes in the bodily region of interest.

CT had many pioneers. Among them was William Oldendorf, a neurologist at UCLA, who proposed in the early 1960s to obtain an image of the brain by rotating a beam of X rays through it and who constructed a small working model of the device. About the same time, Allan McLead Cormack, a nuclear physicist who had dealt with radiation safety at hospitals in South Africa and Boston, worked out mathematical methods for calculating from x-ray absorption data the intensity of the radiation at planar points in heterogeneous tissues. Then, around 1970, Godfrey N. Hounsfield, a computer engineer at the laboratory of the London record company EMI, realized that the analysis of x-ray scanning data could yield images of soft-tissue organs of previously unseen accuracy.

Though Hounsfield's and Cormack's modes of analysis differed, both required the solution of thousands of equations. The computers of the day were up to the task. EMI, flush with revenues from Beatles recordings, paid for part of the development costs; Hounsfield persuaded the British Department of Health and Social Services to fund the rest by arguing that his method of scanning would permit physicians to peer into the brain. In 1971, he produced the first CT scanner, which, put to the test, detected a woman's brain tumor. In 1979, Hounsfield and Cormack shared the Nobel Prize in physiology or medicine, a distinction that surprised some people aware of the number of scientists who contributed to the development of CT.

CT scanning was rapidly adopted in the United States, where Robert Ledley, a scientist at Georgetown University Medical School, developed a whole-body scanner. With CT, physicians could spot tumors and rapidly diagnose organ trauma, fractures, internal bleeding, and swelling. The scanners did away with a great deal of exploratory surgery. CT images also found their way into the courtroom, where defense lawyers used brain scans to argue, often successfully, that their clients committed crimes because they suffered from diminished mental capacities.

Charles Suskind, "The Invention of Computed Tomography," in *History of Technology*, eds., A. Rupert Hall and Norman Smitz (vol. 6; 1981). Bettyann Holtzmann Kevles, *Naked to the Bone: Medical Imaging in the Twentieth Century* (1997).

DANIEL J. KEVLES

CONDENSED MATTER PHYSICS. See SOLID STATE (CONDENSED MATTER) PHYSICS.

CONSERVATION LAWS. Isolated physical and chemical systems possess certain unchanging properties, for example, mass and energy, and, if in thermal equilibrium, temperature. Conservation laws refer to a subset of these properties conserved when these systems interact (conservation of energy, conservation of mass). Natural philosophers first explicitly set out such rules in the eighteenth century. Conservation laws have guided theory in the physical sciences ever since. Many instructive conflicts have erupted over the identity of the property conserved and the conditions of its conservation.

The first of these conflicts, fought out in the early eighteenth century, concerned the "force" of a particle or set of particles. "Force" could mean a particle's mass m multiplied by its velocity v (momentum), mv^2, or $mv^2/2$ (vis viva). Colin Maclaurin, Gottfried Leibniz, and Johann Bernoulli I put forward conflicting claims for the conservation of "force" based in metaphysical principles logically developed and eventually expressed mathematically. Experimental data was incorporated into various metaphysical schemes. These arguments intensified with the 1724 prize competition of the Paris Royal Academy of Sciences. Other laws and controversies followed including conservation of angular momentum (Jean d'Alembert, 1749). These quarrels died with their adherents after d'Alembert rooted rational mechanics in virtual velocity rather than conservation. The physical conditions governing momentum, vis viva, energy, force, and so on were disentangled only in the nineteenth century.

Another conserved quantity of the eighteenth century was weight, which became an important

guide to chemical theory with the discovery and identification of the several sorts of air (*see* PNEUMATICS). Conservation of weight became a foundation of the reformed chemistry of Antoine-Laurent *Lavoisier. A third conservation law developed during the eighteenth century had to do with static electricity. Benjamin *Franklin's theory of positive and negative electricity (1747) explicitly conserved charge and, moreover, made good use of the law in explaining the operation of the *leyden jar. Later natural philosophers, who used two electrical fluids where Franklin had made do with one, also practiced, if they did not make explicit, the conservation of electricity. Most theories of caloric, the weightless matter supposed to cause the phenomena of heat (*see* IMPONDERABLES) also supposed its conservation. Conservation laws in physical sciences were thus well established by 1800.

During the nineteenth century, conservation became a tool for discovery. In 1824 the military engineer Sadi Carnot (1824) applied the principle of conservation to caloric considered as the fuel of steam engines. The work extracted from the engine came from the cooling of the caloric from the temperature of the boiler to that of the environment just as the fall of water works a mill. Carnot's analysis, which resulted in the important insight that no engine more efficient than a reversible one can exist, was put into mathematical form by Benoit-Pierre-Émile Clapeyron in 1837 and largely ignored. Meanwhile, Michael *Faraday, William Grove, and others explored the conservation of force, including *electricity and *magnetism.

In the 1840s this work changed direction and several men from various backgrounds became "discoverers" of the conservation of energy. William *Thomson, Lord Kelvin, developed Clapeyron's work, which led him to the definition of absolute temperature. In 1847 Thomson heard James Joule present an account of his measurements of the heat produced by an electrical current (Joule's law) and by mechanical motion. Joule had concluded that the forces of nature were not conserved but transformable one into another in accordance with an exact calculus. A certain amount of heat will always generate the same amount of mechanical work (mechanical equivalent of heat).

In *The Conservation of Force* (1847), Hermann von *Helmholtz announced a general principle of nature that he extracted from a representation of matter as a collection of atoms held together by central forces. He equated the change in vis viva of a particle moving under the influence of a center of force to the change in the "intensity of the force." He identified the latter with the potential function introduced earlier by Carl Friedrich *Gauss. Helmholtz showed how the results of experiments, like Joule's measurements of the production of heat in current-carrying wires, supported his principle of the interconvertibility of force.

In 1850 Rudolf Clausius put forward the clearest statement of the conservation of energy. He redid Carnot's analysis, replacing the conservation of caloric by the conservation of the "energy" of the perfect gas he assumed as the working substance of his heat engine and gave a mathematical expression for the conservation of energy, the first law of thermodynamics. Later he presented heat as the vis viva of gas molecules and the raising of a weight by the engine as the transformation of one type of mechanical energy (kinetic) into another (potential). In the second half of the nineteenth century the conservation of energy became a mainstay of the physical sciences.

The conservation of energy had no prominent place in James Clerk *Maxwell's kinetic theory or statistical mechanics but was central to Ludwig *Boltzmann's work on both mechanics and thermodynamics. J. Willard *Gibbs extended thermodynamics from physics into chemistry and developed, along with Helmholtz, other conservation laws (enthalpy and free energy) useful in physical chemistry. Energy conservation underwrote a new philosophical approach to physics, developed by Wilhelm Ostwald and Georg Helm (see ENERGETICS). Also, Maxwell reworked his theory of electromagnetism and light within an energy framework using Thomson and Peter Guthrie Tait's *A Treatise on Natural Philosophy* (1867) as a guide.

Despite these substantial acquisitions, physicists had trouble making energy conservation fit certain phenomena of heat and radiation, and its applicability to *radioactivity at first appeared doubtful. The bleak situation, which Thomson described as clouds over the otherwise sunny landscape of physics, was saved by the quantum theory of Max *Planck and the demonstration of the conservation of weight and energy in radioactive decay through measurements by Marie Sklodowska *Curie, Ernest *Rutherford, and others and the mass-energy law ($E = mc^2$) of relativity.

Conservation of energy was an integral part of Niels Bohr's theory of *atomic structure (1913). As the problems of his quantized atoms mounted in the early 1920s, however, he limited conservation of energy to the average of all the interactions of atoms with the electromagnetic field, and freed individual interactions from the

necessity of obeying the first law of thermodynamics (1924). This *lèse majesté* played a part in Werner *Heisenberg's quantum mechanics (1925), which changed the place of conservation laws in physics. Conservation laws now sprang from the mathematical symmetries inherent in the expressions for the matrices representing the operations that take a physical system from one state to another. Symmetry here required that under geometrical change or time reversal the mathematical form remains the same: a rotation in space implied conservation of angular momentum; time reversal, conservation of energy; and linear translation, conservation of momentum. In addition, there was a nonclassical symmetry associated with the intrinsic angular momentum (the spin) of a particle at rest. In the 1930s an associated concept, isospin, was introduced and developed into a method of classifying the known nuclear particles. During the 1950s, isospin helped in classifying and predicting antiparticles and in generating a new conservation law, the conservation of nucleons.

To explore the nucleus physicists had to incorporate light into their theories of the atom and nucleus. The simplest problem, the interaction of the electron and the electromagnetic field (see QUANTUM ELECTRODYNAMICS), included a synthesis of quantum mechanics and the special theory of relativity. Techniques developed to bring convergence (renormalization) often forced changes in the conception of the nucleus and its constituents. P. A. M. Dirac's derivation of the wave equation for the electron (1928) implied the existence of negative energy states, which he interpreted as the domain of an "antiparticle" with the same mass as the electron but with positive charge. In Dirac's theory the two sorts of electrons could be created and annihilated together and contradicted an implicit assumption in quantum mechanics—the conservation of particles. Dirac's interpretation gained credence through Carl David A. Anderson's observation in 1931 of the positron (positive electron) in tracks made in his cloud chamber by *cosmic rays.

The problem of beta decay further undermined assumptions as basic as the conservation of energy and momentum. Wolfgang Pauli suggested in 1931–1932 that an undetected particle of zero mass and electrical charge, named the neutrino by Enrico *Fermi, carried away the missing energy and momentum. The neutrino first revealed itself directly in experiments done by Frederick Reines and Clyde L. Cowan in 1956.

The development of particle *accelerators resulted in the discovery or manufacture of more and more "fundamental" particles and graver and graver problems for conservationists. Novelties included particles produced in associated pairs under circumstances so improbable that physicists gave them a quantum number named strangeness. Scientists postulated a new force within the nucleus, the weak force. The neutrino became a left-handed particle and the antineutrino right-handed. Conservation laws again needed revision, including those springing from the assumption of parity (P) conservation in weak interactions. (Parity requires that a device and its mirror image, if made of the same materials, function in the same way.) Nonconservation of parity led to investigations of other symmetry relations and their conservation laws, including charge conjugation (C), in which all the charges entering an equation become their opposites, and time reversal (T), in which the time variable t is changed to $-t$. The strongest result that physicists could produce was the conservation of CPT, in which the transformations C, P, and T simultaneously take place, with the corollary that if T conserves the relations, so does CP.

Experiments to test nonconservation under P demonstrated that some particles are intrinsically right-, and others intrinsically left-handed. By the 1970s a generally accepted model for particle behavior emerged, the so-called *Standard Model, whose fundamental building blocks, the quarks, have fractional charges $1/3, 2/3, -1/3$ of the electron's. In this model, however, all hadrons (protons, neutrons, etc.) should have the same mass, which they do not. The theory "breaks" this unwanted symmetry by introducing different sorts or "flavors" of quarks. Further symmetries and conservation laws emerged from the requirement that quarks be confined by the strong force. This led to further symmetries and new conservation laws. The dependence of physical interpretation on difficult mathematics seemed justified by the experimental identification of the different quarks in the 1970s and 1980s.

Thus conservation laws, at first intuitive expressions of physical regularities and lately of esoteric mathematical symmetries, have guided physics over the past 250 years.

J. L. Heilbron, *Electricity in the Seventeenth and Eighteenth Centuries* (1979). Abraham Pais, *Inward Bound: Of Matter and Forces in the Physical World* (1986). Necia Grant Cooper and Geoffrey B. West, *Particle Physics, A Los Alamos Primer* (1988). Silvan S. Schweber, *QED and the Men Who Made It: Dyson, Feynman, Schwinger, and Tomonaga* (1994). Crosbie Smith, *The Science of Energy: A Cultural History of Energy Physics in Victorian Britain* (1998).

ELIZABETH GARBER

CONSTANTS, FUNDAMENTAL. The belief that numbers constitute the essence of the universe runs deep in the human experience. The patterns of the sky and seasons provided the first opportunity for discovering nature's numbers, and the agricultural, commercial, and religious needs of early civilizations provided the motivation for inscribing them in calendars. The Greeks in particular anchored thought in number, both in the metaphorical, as in the speculations of the Pythagoreans and the Platonists, and in the practical, as in the exact geometrical astronomy of Hipparchus and Ptolemy. These two approaches came together from time to time, notably in the work of Johannes *Kepler, who combined numerological beliefs in the literal harmonies of the celestial spheres with Tycho *Brahe's precision measurements to discover that the cube of a planet's average orbital diameter divided by the square of its orbital period was the same value, no matter the planet.

*Galileo Galilei confined his numerical endeavors to terrestrial bodies. He measured the rate of fall of various weights and found that the ratio of the distance traveled and the time of fall squared was the same for each body examined. Galileo's falling bodies and Kepler's fruitful numerology came together in Isaac *Newton's theory of universal gravitation. Newton's theory implied the existence of a fundamental constant (later labeled G) that specified the force of attraction not only between a planet and the Sun but also between a falling object and Earth.

Not much attention was paid to these constants of gravity. The mathematical methods of the time, which focused on the form of ratios between quantities and not their proportionality constants, veiled the importance of the constants themselves. Even in the late eighteenth century, Henry *Cavendish devised his famous torsion balance experiment not to measure the force between two weights and thus what modern physicists call the gravitational constant, but to measure the density of the earth. A significant exception to the lack of interest in the natural constants prevalent in early modern science was the measurement of the speed of light by Ole Rømer in 1675. A possible explanation of this exception is that, in contrast to gravitational acceleration, speed was conceptually familiar and, in the case of light, would be determined by astronomical phenomena—Rømer used eclipses of Jupiter's satellites—frequently subjected to measurement.

Not until the middle of the nineteenth century did the modern interest in fundamental constants in physics evolve. A broad-based quantifying spirit that had arisen during the eighteenth century supplied the general motivation, and the burgeoning telegraph industry, in dire need of well-defined electrical units and standards, supplied the immediate requirement. In 1851, Wilhelm Weber proposed a system of electrical units founded on the metric system. A decade later, the British Association for the Advancement of Science, under the leadership of William *Thomson, Lord Kelvin and James Clerk *Maxwell, took up the challenge of promulgating an international system of electrical units and standards that could meet the needs of both science and industry. In such a system, certain fundamental quantities of nature played a key role, such as the magnetic permeability and electrical permittivity of the *ether. The work also raised the possibility of defining a "natural," non-arbitrary system of units, perhaps based on the wavelength, mass, and period of vibration of the hypothesized atoms.

Meanwhile, certain key numbers were appearing in pathbreaking physical theories. James *Joule demonstrated the mechanical equivalent of heat and calculated its value. Ludwig *Boltzmann reframed thermodynamics in terms of statistics and an important constant, later called k, relating a molecule's average energy to temperature. Max *Planck introduced another key constant related to molecular energy, h, in his blackbody radiation law. Maxwell's electromagnetic theory emphasized that the speed of light c was actually the speed of electromagnetic radiation in general. Albert *Einstein's theory of special relativity and his mass-energy equivalence, $E = mc^2$, further established the fundamental status of c. And Joseph John *Thomson's discovery of the electron introduced its mass and charge (m_e and e) as candidates for the fundamental quantities of matter and electricity.

As the recognized physical constants multiplied into the twentieth century, they raised the question, how fundamental? Although they could be categorized by type, such as properties of objects or factors in physical laws, and some clearly possessed deeper and broader significance than others, it became apparent that many of them were interrelated and that the term "fundamental" always hid some arbitrariness.

A second question was, how precise? The 1920s and 1930s saw an informal international effort to identify not only the best extant value of each fundamental constant but also its precision. The technologies of World War II, and offspring like the laser and atomic clock, assisted by enabling revolutionary increases in decimal places. In the 1960s the project gained a formal footing with the establishment by the International Council of Science of a Committee on Data for Science and Technology (CODATA) and its Task Force on Fundamental Constants. It surveyed the literature and produced a set of "best" values for fundamental constants in 1973, 1986, and 1998.

The hard-eyed quest for the next decimal place did not eradicate interest in numerology, however. Certain combinations of e, h, and c seemed to contain deeper magic. The dimensionless constant e^2/hc, for example, was revealed by quantum electrodynamics to be the constant that defined the strength of the electromagnetic force. Dimensionless constants held great allure because their value did not depend on the system of units chosen, but seemed to be pure numbers of the universe. Moreover, some simple combinations of fundamental constants yielded very large dimensionless numbers all on the order of 10^{40}. In his "large number hypothesis," P. A. M. *Dirac proposed that this coincidence hinted at an undiscovered law of the universe. The large number hypothesis also raised the question, how constant?, as it implied that some fundamental constants, such as G, might vary as the universe evolves. On some cosmological theories, small changes in the values of the constants can trigger large consequences for the development of the universe. Only for values close to those observed could complex life develop. Thus a consideration of the fundamental constants renewed interest in the anthropic principle and raised hopes that nature's numbers could be derived from the fact of human existence.

B. W. Petley, *The Fundamental Physical Constants and the Frontier of Measurement* (1985). John D. Barrow and Frank J. Tipler, *The Anthropic Cosmological Principle* (1986). M. Norton Wise, ed., *The Values of Precision* (1995). National Institute of Standards and Technology, *The NIST Reference on Constants, Units, and Uncertainty* (2000), online at http://physics.nist.gov/cuu.

LARRY R. LAGERSTROM

CONTRACEPTION. There have been significant developments in the means of controlling conception in the modern era, both improvements in traditional methods and the creation of new ones on the basis of changed scientific understanding of the process of conception. A relatively effective basis for the allegedly natural expedient of restricting sexual intercourse to the woman's "safe period" was only reliably established in 1929 by Kyusaku Ogino of Japan and Hermann Knausof Vienna, working independently of one another, following several decades of scientific research on the female reproductive cycle. Devices using hormonal analysis of female urine now enable precise pinpointing of the fertile period.

Barrier methods locally applied to the genitals have a long history. For centuries the manufacture and purveyance of contraceptives was a furtive and unrespectable trade, even when not illegal, which adversely affected research and quality control. Linen sheaths were invented in the sixteenth century as prophylactics against venereal disease, and animal-gut versions (condoms) had appeared by the mid-seventeenth century. The extent to which these devices were then employed for contraception is debatable, but they were certainly being used for that purpose by the early nineteenth century, if not earlier. Vulcanization of rubber in the 1840s lowered mass production costs and sheaths became increasingly popular. Female occlusive caps were also introduced around this time. The early, custom-made versions were superseded in the 1870s when German physician Wilhelm Mensinga devised the "Dutch cap" or diaphragm, a domed rubber cap with a metallic spring rim. Thinner and more flexible condoms followed the development of latex in the 1930s. Sponges for birth control date back probably to the eighteenth century and have enjoyed a late-twentieth-century comeback.

The nineteenth century saw the commercial development of chemical contraceptives, such as W. J. Rendell's vaginal pessaries (London, around 1880). They contained quinine as a spermicide but probably owed their efficacy to the greasy binding medium that hindered sperm motility. From the late 1920s the British Birth Control Investigation Committee conducted research into spermicides, although continuing stigmatization of the topic, even within the scientific community, created problems. Besides pessaries, these now come as creams or foams, ideally used with a barrier method for increased reliability.

The intrauterine device was devised by German gynecologists, notably Ernst Gräfenburg, before World War I. First made of gold or silver, IUDs are now plastic or copper. They work by preventing implantation of the fertilized ovum. Sterilization of women, formerly a major abdominal operation, has been performed since the 1970s using a laparoscope via a small incision and clips on the fallopian tubes. Vasectomy, a much less invasive procedure, has been available since the 1890s.

Anxieties about the quantity and quality of population stimulated interest in developing easy-to-use reliable contraceptives. In spite of continuing taboos, research was pursued in both Europe and North America during the mid-twentieth century. Investigations of sex hormones and steroids on an international scale led to the development of the birth control pill during the 1950s, an extremely reliable method that does not involve action at the time of intercourse. While many scientific and clinical disciplines have been involved in ensuring the safety and efficacy of the pill, periodic scares over its health risks have led to sudden rises in unplanned pregnancies. The implications for sexually transmitted diseases of using a nonbarrier method of contraception have also dimmed initial enthusiasm, but the pill remains one of the most widely used methods of birth control. Hormonal contraceptives are also given as implants or injections with long-term efficacy. Related developments include the "morning after" pill, a postcoital contraceptive, and the increasing likelihood of an effective and acceptable "male pill." An ideal method has still not been discovered, and research continues.

Historians and demographers continue to debate the relationship between the development of birth control methods and population decline. Many now consider that abstention, rather than technological developments, played a significant role in the industrialized West's declining birthrate from the later nineteenth century, and that improvements in contraceptive technology did not create the demand for reproductive control so much as respond to it.

Norman Himes, *A Medical History of Contraception* (1936). Angus McLaren, *A History of Contraception: From Antiquity to the Present Day* (1990). Lara Marks, *Sexual Chemistry: The History of the Contraceptive Pill* (2001).

LESLEY A. HALL

COPERNICUS, Nicholas (1473–1543), astronomer.

Copernicus stands at the crossroads to modern science because his radically new plan for the cosmos, a heliocentric system to replace the time-honored, earth-centered cosmology, required a radically new physics. Only with a sun-centered arrangement does a causal scheme of gravitational physics follow, but in Copernicus's own work such a consequence was at best simply in an embryonic form. He had no empirical evidence to support the motion of the earth; his sixteenth-century successors for the most part suspended judgement on his cosmology.

Copernicus was born 19 February 1473 in Torun, a Hanseatic town that had shortly before transferred its allegiance to the Polish monarchy. His education at the University of Cracow was underwritten by his maternal uncle, Lucas Watzenrode, who, after becoming bishop of the northernmost Catholic diocese in Poland, provided a canonry for his nephew. Copernicus continued his education in Italy, in canon law at the University of Bologna (1496–1500) and then in medicine at Padua (1501–1503). He used these skills as a personal secretary and physician to his uncle and as a cathedral administrator.

Copernicus had begun to develop the heliocentric hypothesis by 1514, by which time a

Nicholas Copernicus, after a self-portrait drawn during his student days.

manuscript of his so-called *Commentariolus* had reached a scholar in Cracow. In this small work Copernicus outlined the postulates behind his new theory and expressed his doubt about Ptolemy's use of the equant, a point within an orbit but outside its center from which the orbiting body appears to move with constant angular velocity. Copernicus went to considerable effort to eliminate the equant, which he felt violated the fundamental cosmological principle of representing celestial appearances by uniform circular motions. A major part of his final work, the replacement of the equant with combinations of circles, was greatly admired by his immediate successors. However, as some of his Islamic predecessors had already demonstrated, the equant can be eliminated without supposing a heliocentric framework.

In examining the arrangement of the major circles in Ptolemy's epicyclic theory (the deferent or carrying circle and the epicycle carried on it), Copernicus must have noticed that the directional line to the planet was preserved when the circles were interchanged. If he did not discover this for himself, he could have found it in Johannes Regiomontanus's *Epitome of the Almagest* (1496). This could have led Copernicus to a rescaling of the circles to allow one to be placed heliocentrically for each planet. Although

very few working notes survive to show the progress of his thinking, the preserved fragments document this step.

The heliocentric arrangement led to a beautiful discovery: the swiftest planet, Mercury, took up the orbit closest to the Sun, whereas the slowest planet, Saturn, automatically fell farthest from the Sun. "In this arrangement, therefore, we discover a marvelous commensurability of the universe and a sure harmonious connection between the motion of the spheres and their size, such as can be found in no other way," Copernicus proclaimed in his great treatise, *De revolutionibus orbium coelestium*. It also explained naturally why the retrogression (or occasional apparent westward motion) of Mars, Jupiter, and Saturn always occurs when these three superior planets are opposite the sun in the sky, and why the retrogression of Mars exceeds Jupiter's, and Jupiter, Saturn's. Consideration of harmony and unity apparently drove Copernicus to his new vision of the cosmos.

Copernicus carried out occasional planetary observations to confirm the parameters used by Ptolemy. In order to check slow-moving Saturn, he made these observations over fourteen years, between 1514 and 1527. But because his final scheme amounted to a geometrical transformation of Ptolemy's models with essentially the same numerical parameters, his system had little quantitative advantage over the earlier ones. Both the Ptolemaic and Copernican tables give maximum errors exceeding four degrees in the worst cases for the planet Mars.

After 1530, with the basic observations in hand, Copernicus worked slowly to make the numerical work of his book consistent. Whether he would have succeeded under his own steam in getting the work ready for publication seems unlikely. The necessary catalyst, a young Lutheran astronomy professor from Wittenberg, George Joachim Rheticus, came to stay with Copernicus in Poland in 1539. The twenty-six-year-old Rheticus was so enthusiastic about the novel cosmology that Copernicus gave him permission to publish a short "first report" or *Narratio prima* (1540). When no explosion ensued, Copernicus allowed Rheticus to take a copy of the manuscript to Johannes Petreius in Nuremberg, who published *De revolutionibus* in 1543. Although Copernicus must have proofread finished pages as they arrived from Nuremberg, he received the completed book only on his deathbed, in May 1543.

While the great majority of readers understood the heliocentric cosmology simply as an astronomical hypothesis and not as physical reality, all perceived Copernicus's book as an important text. Additional printings, in Basel in 1566 and Amsterdam in 1617, supplied the continuing demand for this classic work.

Edward Rosen, *Three Copernican Treatises: The* Commentariolus *of Copernicus, The Letter against Werner, The* Narratio prima *of Rheticus*, 3d ed. (1971). N. M. Swerdlow and O. Neugebauer, *Mathematical Astronomy in Copernicus's* De revolutionibus (1984). Nicholas Copernicus, *On the Revolutions*, trans. Edward Rosen (1992). Owen Gingerich, *The Eye of Heaven: Ptolemy, Copernicus, Kepler* (1993).

OWEN GINGERICH

COSMIC RAYS. The first explorers of *radioactivity found it in air and water as well as in the earth. Shielded electroscopes placed out of doors lost their charges as if they were exposed to penetrating radiation. Since leak diminished with height, physicists assigned its cause to rays emanating from the earth. As they mounted ever higher, however, from church steeples to the Eiffel Tower to manned balloons, the leak leveled off or even increased. In 1912–1913, Victor Hess of the Radium Institute of Vienna ascertained that the ionization causing the leak declined during the first 1,000 m (3,280 ft) of ascent, but then began to rise, to reach double that at the earth's surface at 5,000 m (16,400 ft). Hess found further, by flying his balloon at night and during a solar eclipse, that the ionizing radiation did not come from the sun. He made the good guess—it brought him the Nobel Prize in physics in 1936—that the radiation came from the great beyond.

The need to know the meteorological state of the upper atmosphere for directing artillery during the Great War improved balloon technology. Robert Millikan, who would receive the Nobel Prize in 1923 for his measurement of the charge on the electron, was a powerful organizer of American science for war. His observations made with Army balloons seemed to show that the ionization

in the atmosphere declines continually from the earth's surface. Experiments in lakes at different heights showed him his error and he became the champion of what he called "cosmic rays." He regarded them as the "birth cries of infant atoms" since, by his calculations, the relativistic conversion of mass into energy during the formation of light elements from hydrogen would produce high-frequency radiation (photons) of the penetrating power of Hess's rays.

The advance of electronics transformed the study of cosmic rays from a guessing game to an exact, expensive, and productive science. In 1929 Werner Kolhörster, who had confirmed Hess's measurements, and Walther Bothe placed two Geiger counters one above the other, separated them by a lead block, and arranged a circuit to register only when both counters fired simultaneously. They detected too many coincidences for photons to produce and they inferred that cosmic-ray primaries must be charged particles with at least a thousand times more energy than the hardest rays from radioactive substances. After learning the coincidence method from Bothe (who received a Nobel Prize for it in 1954), Bruno Rossi, one of the Italian pioneers in the field, set three counters in a triangular array and deduced the existence of showers of particles produced by stopping the primaries (1932). A new particle, the positive electron (positron), was found among the secondaries in cloud chambers by Carl David Anderson, who worked with Millikan at the California Institute of Technology (Caltech), and by P. M. S. Blackett and Giuseppe P. S. Occhialini, who used Rossi's electronics. Anderson and Blackett received Nobel Prizes in 1936 (sharing with Hess) and in 1948, respectively.

Cosmic rays became big science when Arthur Holly Compton (Nobel Prize, 1927) undertook to annihilate Millikan's primary photons. Compton designed a method to show that the sea-level intensity of cosmic rays at the poles exceeds that at the equator. If the primaries were charged

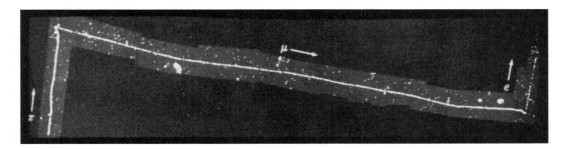

Tracks of a pi-meson decaying in a photographic emulsion into an electron and a mu-meson, which then disintegrates into another electron and an invisible neutrino, obtained in 1949 by a group at the University of Bristol under Cecil F. Powell.

particles, they would be deflected by the earth's magnetic field more strongly the lower the magnetic latitude. Compton's project mobilized sixty investigators who carried expensive standardized apparatus on eight expeditions. Government agencies used to funding geophysical work and the Carnegie Institution of Washington paid the bills. By 1934 Compton's company had confirmed the latitude effect and silenced the birth cries of infant atoms.

The victory at first appeared to have a heavy cost, however, since *quantum electrodynamics required that the charged primaries, if electrons or protons, be absorbed more quickly by the atmosphere than was compatible with their intensity at the surface. Anderson's cloud chamber soon disclosed the existence of a secondary particle (now called the μ meson or muon) whose intervention would slow the absorption of cosmic rays. At first identified with the particle Hideki *Yukawa (Nobel Prize,1949) had postulated as the carrier of nuclear force (now called the π meson or pion), the muon has played an important role in the systematics of the *standard model. It also afforded, through the difference in time between its decay in flight and at rest, the first experimental confirmation of the time dilation required by the theory of relativity.

The last major contributions of students of cosmic rays to fundamental physics occurred just after World War II. With the help of photographic emulsions developed for wartime use, Cecil Powell (Nobel Prize, 1950) and his colleagues at the University of Bristol (including Cesar *Lattes) caught a pion as it turned into a muon, confirming the growing realization that Anderson's meson was not Yukawa's. In the same year, 1947, two other British physicists, George D. Rochester and Clifford C. Butler, found evidence in cloud chamber tracks of the decay of unknown neutral particles (now called hyperons) into neutrons, protons, and pions. Two years later, Powell, immersed again in emulsions, found a particle (the K meson) that decayed into three mesons.

The data needed to unravel the relations among *elementary particles did not come from cosmic rays, however, but from *accelerators. Already in 1947 the Berkeley synchrotron was making μ and π mesons in greater quantity than cosmic rays furnished. Whereas the synchrotron confirmed and extended the discoveries of mesons made through the study of cosmic rays, cosmic-ray physicists could find the antiproton only by scanning emulsions exposed to the beam of the Berkeley Bevatron, where it was first made and detected in 1955. In compensation, cosmic-ray physicists came to agree that the primary radiation striking the atmosphere consists almost entirely of protons, a substantial sprinkling of alpha particles and other nuclei, and a few electrons.

The invention of artificial *satellites gave cosmic-ray physicists a needed fillip. _Sputnik I_ and the U.S. runners-up, _Explorer I_ and _Explorer III_, all carried counters to measure cosmic-ray intensity. The _Explorers'_ counters stopped working high above the earth. The man in charge of the instrumentation, James A. Van Allen of the University of Iowa, interpreted the silence as evidence that the satellite had passed through a region so full of charged particles that the counters jammed. The region, now known as the Van Allen Belt(s), consists of cosmic-ray and solar particles trapped in the earth's magnetic field. Hess's assertion that the Sun does not contribute to cosmic radiation succumbed to technological advances that replaced the balloons of the years around World War I with the rockets of the Sputnik era.

Bruno Rossi, _Cosmic Rays_ (1966). Laurie M. Braun and Lillian Hoddeson, eds., _The Birth of Particle Physics_ (1983). Michael W. Friedlander, _Cosmic Rays_ (1989).

J. L. HEILBRON

COSMOLOGY. For two millennia the *Aristotelian cosmos of rotating spheres carrying the Moon, Sun, planets, and stars around the central earth permeated Western thought. Then in 1543 Nicholas *Copernicus switched the positions of the earth and Sun. To account for the daily motion of the heavens, his scheme had the earth rotating on its axis. Revolutions in science, as in politics, often go further than their initiators anticipate. In Copernicus's system, the outer sphere of the stars was not needed to move them. Human imagination soon distributed the stars throughout a perhaps infinite space.

The earth lost more than its uniqueness in space in the Copernican system. *Galileo's telescopic observations of the Moon's surface emphatically demanded the revolutionary conclusion that the Moon was not a smooth sphere, as Aristotelians had maintained, but resembled the earth, with mountains and valleys. His discovery of four satellites of Jupiter, similar to the earth's single moon, forged another bond between the earth and the other planets.

In late Aristotelian cosmology, solid crystalline spheres carried the planets around the earth and also provided the physical structure of the universe. With his observation of comets coursing through the solar system, Tycho *Brahe shattered the crystalline spheres. Belief in uniform circular motion, fundamental to both Aristotelian and Copernican cosmology, died early in the seventeenth century. Johannes *Kepler showed that the earth and the other planets all travel around the Sun in elliptical orbits.

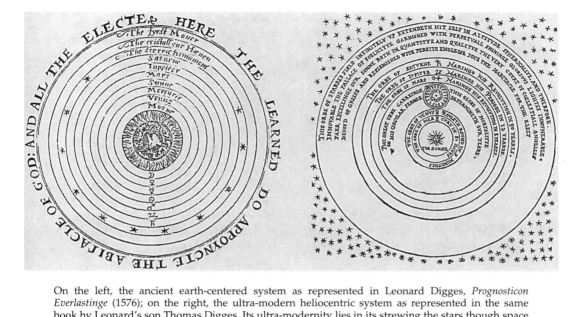

On the left, the ancient earth-centered system as represented in Leonard Digges, *Prognosticon Everlastinge* (1576); on the right, the ultra-modern heliocentric system as represented in the same book by Leonard's son Thomas Digges. Its ultra-modernity lies in its strewing the stars though space rather than, as in Copernicus' original scheme, placing them on the surface of a sphere.

Aristotelian physics no longer worked in a Copernican universe. A new explanation of how the planets retrace the same paths forever around the Sun became a central problem of cosmology. Isaac *Newton explained how a force of attraction or gravitation toward the Sun continually draws the planets away from straight-line motion and holds them in Kepler's elliptical orbits. Gravity would collapse the Sun and the fixed stars together unless the stars moved in orbits around the center of a system, as do the planets around the Sun. In 1718 the English astronomer Edmond Halley reported his discovery that three bright stars no longer occupied the positions determined by ancient observations. Formerly fixed, stars were now freed to roam.

Before the observations of William Herschel (1738–1822), astronomers represented the heavens as the concave surface of a sphere surrounding the observer in the center. Herschel's large *telescopes forced them to treat the heavens as an expanded firmament of three dimensions. Herschel observed that most stars seemed to lie between two parallel planes. He concluded that the Milky Way is the appearance of the stars in the stratum to an observer on the earth.

At the beginning of the twentieth century astronomers reckoned the diameter of our stellar system at a few tens of thousands of light years across. (A light year, nearly six trillion miles, is the distance light travels in a year in a vacuum.) Also, they generally assumed the Sun to be at or near the center of the universe.

By 1920, however, the American astronomer Harlow Shapley had persuaded his peers that the system of stars was ten or even a hundred times larger than previous estimates and that the Sun stood many tens of thousands of light years away from the center. Shapley supposed erroneously that the galactic universe made a single, enormous, all-inclusive unit (*see* GALAXY). In 1925, the American astronomer Edwin *Hubble demonstrated conclusively that spiral nebulae are independent entities far beyond the boundary of our own galaxy. Just as Shapley had used the period-luminosity relation for certain double stars (Cepheids) to determine distances to globular clusters in our galaxy, Hubble used the relation to determine the distance to a Cepheid variable he discovered in a spiral nebula. It worked out to be several times more distant than Shapley's estimate of the outer limits of our galaxy. In 1926 Hubble published the now-standard classification system for the thousands of extragalactic *nebulae then known or within reach of existing telescopes.

At the time, most scientists, including Albert *Einstein, assumed that the universe was static. The Dutch astronomer Willem de Sitter formulated in 1916 a static model of the universe with apparent redshifts in the spectra of nebulae greater for nebulae at greater distances. In 1929 Hubble tested and confirmed de Sitter's prediction. He interpreted the redshifts not in the context of de Sitter's static model of the universe, but as Doppler shifts indicative of real velocities of recession. The universe was not static, but expanding.

Hubble's velocity-distance relation made possible greater distance determinations because redshifts can be measured even when individual stars and galaxies cannot be distinguished, out to billions of light years. In 1927 the Belgian priest and astrophysicist Georges Lemaître published a theory of a universe expanding from what he called a "cosmic egg." Initially ignored, it attracted favorable attention in 1931 after Hubble had demonstrated his velocity-distance relation. De Sitter hailed Lemaître's brilliant discovery of the expanding universe and Einstein confirmed that Lemaître's theoretical investigations fit well into the general theory of *relativity.

Advances in nuclear physics helped transform speculations about an expanding universe. Beginning in 1946, the Ukrainian-born American physicist George Gamow set for cosmology the problem of explaining with laws of physics the cosmic distribution of the chemical elements as the result of thermonuclear reactions in an early, extremely hot, dense phase of an expanding universe. The British astronomer Fred Hoyle in 1950 derisively called Gamow's process the "big bang," and the term stuck. After astronomers agreed that elements heavier than hydrogen and helium could not have been formed during the dense hot origin of the universe, Hoyle explained the later creation of heavy elements in stellar interiors. In opposition to the big bang theory, Hoyle championed steady-state creation, in which the universe expands but does not change in density because of the continuous appearance of new matter. The crucial observation in the cosmological debate came from radio astronomy (*see* ASTRONOMY NON-OPTICAL). In 1963 Arno Penzias and Robert Wilson at the Bell Telephone Laboratories measured a faint cosmic background radiation, which was interpreted by Robert Dicke at Princeton University as a remnant of the big bang.

A major conceptual change in cosmology during the 1970s and 1980s focused attention on the question why the universe is the way it is. A young American particle physicist, Alan Guth, proposed in 1979 that important cosmological features can be explained as natural and inevitable consequences of new theories of particle physics. Guth's theory of inflation states that in the first minuscule fraction of a second of the universe's evolution a huge inflation occurred. After that, the inflationary universe theory merges with the standard big-bang theory. If the mass density of the universe exceeds a certain critical value, gravity eventually will reverse the expansion of the universe and reunite everything in a "big crunch." With a density less than critical, the universe will expand forever, resulting in a "big chill." Neither case is indefinitely hospitable

to life as we know it. At the critical density, the resulting "flat" universe will continue to expand, but at an ever slower rate. Why the density of the universe one second after the big bang came very close to the theoretical critical density is called the "flatness problem." The standard big bang theory offers no solution. In the inflationary theory, a brief burst of exponential expansion automatically drives the density, whatever its initial value, incredibly close to the critical density. Another success of inflationary theory involves the "horizon problem." Exchanges of energy cannot be transferred farther than the "horizon distance," the maximum distance that light can have traveled since the beginning of the big bang universe. But the cosmic background radiation is uniform over a space much greater than the horizon distance. The inflationary theory makes the universe far smaller during its initial phase than the standard big bang theory does, smaller than the horizon distance, thus enabling the universe to come to a uniform temperature before the process of inflation switched on. If inflation created a uniform universe, how did the small primordial non-uniformities of matter necessary to begin the process of gravitational clumping come about to seed the evolution of cosmic structure? Answer: inflation could have stretched quantum mechanical fluctuations enormously, and the resulting wrinkles in space-time could have caused large-scale cosmic structures, such as a galaxy or even a cluster of galaxies, to coalesce.

Cosmological theories live or die on the basis of their predictions. When NASA in 1982 approved funding for the Cosmic Background Explorer Satellite, another young American particle physicist, George Smoot, proposed to seek in the cosmic background radiation the tiny differences from uniformity predicted by the inflationary theory. Eventually the experiment involved more than a thousand people, at a cost of over $160 million. After a decade of work, Smoot announced in 1992 what Stephen Hawking called "the scientific discovery of the century, if not of all time." Wrinkles in the fabric of space and time showed that matter was not uniformly distributed, but contained the seeds out of which a complex universe could grow. Furthermore, the fifteen-billion-year-old wrinkles in the background radiation had the size expected.

Astronomers have seen only about 10 percent of the universe; the remaining 90 percent is so far revealed only by its gravitational effects. Weakly interacting massive particles (WIMPs), predicted from theoretical physics, make up most of the missing matter, though they have not been detected in the laboratory. Some of the invisible universe may consist of stars of very low luminosity, such as brown dwarfs, and also primordial black holes. Other forms of dark matter may be topological defects remaining from phase changes in the early universe. The question of large-scale structure and galaxy formation has become central; cosmic strings may provide an answer. Also in question is the expansion of the universe. Many astronomers assumed that either the expansion will proceed indefinitely at a constant rate, or will be slowed and ultimately reversed by gravity. In 1997, however, astronomers observed several gigantic exploding stars or *supernovae at immense distances, seemingly even farther than a constant expansion would have flung them. A runaway universe, its expansion continuously accelerated by some mysterious form of energy, suddenly became a likelihood.

See also ASTRONOMY, NON-OPTICAL; STEADY-STATE UNIVERSE.

Edwin Hubble, *The Realm of the Nebulae* (1936). Richard Berendzen, Richard Hart, and Daniel Seeley, *Man Discovers the Galaxies* (1976). Robert W. Smith, *The Expanding Universe: Astronomy's "Great Debate" 1900–1931* (1982). Michael J. Crowe, *Theories of the World from Antiquity to the Copernican Revolution* (1990); Norriss S. Hetherington, ed., *Encyclopedia of Cosmology: Historical, Philosophical, and Scientific Foundations of Modern Cosmology* (1993). Michael J. Crowe, *Modern Theories of the Universe from Herschel to Hubble* (1994). Norriss S. Hetherington, *Hubble's Cosmology: A Guided Study of Selected Texts* (1996). Helge Kragh, *Cosmology and Controversy: The Historical Development of Two Theories of the Universe* (1996). Alan H. Guth, *The Inflationary Universe: The Quest for a New Theory of Cosmic Origins* (1997).

NORRISS S. HETHERINGTON

COULOMB, Charles-Augustin. See CAVENDISH, HENRY, AND CHARLES-AUGUSTIN COULOMB.

COURTS AND SALONS. Since the middle ages, European princely courts were populated by practitioners of natural philosophy, mostly physicians and astrologers catering directly to the sovereign's body and peace of mind. Their number increased in the early modern period as courts became larger and wealthier, and their activities diversified. Natural history, alchemy, natural magic, clocks, automata, lodestones, anamorphic devices, mirrors, telescopes, microscopes, and, eventually, experiments took their places at court.

These subjects did not share the same courtly spaces or visibility. Some were incorporated in court pageants, played out in court disputations, or displayed in botanical gardens, cabinets of curiosity and, later, galleries and museums. Others were confined to the prince's study or to the workshops of court jewelers, glassmakers, or apothecaries. Princes and courtiers had an eye for intriguing objects and intriguing philosophers. They did not care where the objects fell in classifications of knowledge or academic disciplines. As shown by the remarkable variety of objects included in early modern aristocratic cabinets of curiosity, the line between *naturalia* (natural objects) and *mirabilia* (spectacular human artifacts) often blurred. Similarly, the courtly interest in artistic representations of botanical and zoological specimens was rooted in their location at the intersection of natural history and the visual arts.

Spectacle was serious business at court. The power of the prince and the relative positions of his aristocratic subjects were continuously staged through a multitude of rituals, some of them quite spectacular. Natural philosophy moved from serving the prince's personal and bodily needs to strengthening and legitimizing his political role and authority through the production of novelties and curiosities—objects that would help to cast him as a unique person entitled to a unique, exalted political role. The personal demeanor and the writing and argumentative style of the natural philosopher also adapted to fit courtly standards of elegance, wit, and nonchalance.

The court employed natural philosophers, paid them well, and, most importantly, provided them with a social status they could attain nowhere else. The implications of a court position extended well beyond material rewards or personal status to the authority of the natural philosopher and his discipline. Disciplinary hierarchies (quite rigid in the university and the craft guilds) could be easily redrawn if the prince wished it so. Artisans could become artists and (here *Galileo is the exemplar) mathematicians could become philosophers.

On the other hand, court life demanded a certain level of social and linguistic skill and only topics that were novel, spectacular, and nontechnical were likely to be approved. Moreover, the court favored objects, not theories or long-term projects. Although princes appreciated and often promoted philosophical disputations, they worried more about the quality of the performance than the truth-value of the debate. Ultimately, natural philosophers at court, being courtiers, existed at the whim of the prince. Reaction to these disadvantages of the court system was a central impetus in the establishment of scientific academies—institutions often connected to courts but with relative stability and some tools for self-governance.

Salons were gatherings in the urban homes of nobles or patricians. Their size, schedule, discussion topics, exclusivity, or internal structure varied greatly, as did the role of women in them. Usually organized and managed directly by the host, they relied in some cases on a litterateur or philosopher (the "secretary") to run the salon's activities. As an institution, the salon resembled both the court and the academy. In mid-seventeenth-century Paris, some salons amounted to small full-fledged courts; others were the direct ancestors of royal academies of science. In turn, in the eighteenth century many provincial academies resembled seventeenth-century urban salons in the scope of their interests and the nature of their membership.

Like courts, salons played an important role in the development of the scientific community and, more generally, the so-called republic of letters. They provided forums for discussion of the members' work or, more often, of philosophical literature and news. Their polite conversation helped to prepare non-noble savants for careers at court or in academies. But conversation could turn contentious, disrupting the sociability that underpinned the salons, and cast them as places where natural philosophy was consumed, not produced.

The salons supported and constrained natural philosophy the way the court did. The court could be more effective in providing social legitimation and financial support. The salons offered social training, more leeway in discussion topics, somewhat greater accessibility, and contacts that might make a career—Bernard de Fontenelle and Jean d'Alembert's careers were made in this way. Unlike the court, the salon came with schedules for meetings and activities but, like the court, might come to an end with the death of its patron. Also like courts, salons tended to blend natural philosophical interests with literature, poetry, and moral questions.

Salons have been celebrated as hotbeds of Enlightenment thought and as crucial sites for women natural philosophers—an association that can be traced back to the seventeenth century. But even as they reached an apex of cultural prominence, salons came to be associated with scientific marginality. By the eighteenth century, academies had taken over as the main scientific institutions, relegating the salons to the role of launching pads for future academicians or merely occasions for the nonprofessional consumption of academic science.

See also CABINETS AND COLLECTIONS.

Vivian Nutton, ed., *Medicine at the Courts of Europe, 1500–1837* (1990). Bruce Moran, ed., *Patronage and Institutions* (1991). Mario Biagioli, *Galileo Courtier* (1993). Paula Findlen, *Possessing Nature* (1994). Pamela Smith, *The Business of Alchemy* (1994). Mary Terrall, "Gendered Spaces, Gendered Audiences: Inside and Outside the Paris Academy of Sciences," *Configurations* 2 (1995): 207–232. Mario Biagioli, "Etiquette, Interdependence, and Sociability in Seventeenth-Century Science," *Critical Inquiry* 22 (1996): 193–238.

MARIO BIAGIOLI

CREATIONISM is a popular movement arising from deep-seated objections to perceived religious and social consequences of the acceptance of *evolution and from the conviction that evolution has never been satisfactorily demonstrated. These objections predate Charles *Darwin. "If the book be true," Cambridge geologist Adam Sedgwick wrote of the evolutionary speculations published (anonymously) by Robert Chambers in *Vestiges of the Natural History of Creation* (1844), then "the labours of sober induction are in vain; religion is a lie; human law is a mass of folly, and a base injustice; morality is moonshine; our labours for the black people of Africa were works of madmen; and man and woman are only better beasts."

Modern creationists raise similar concerns, linking evolution with racism, sexual promiscuity, totalitarianism, nihilism, and various forms of irreligion. Many philosophical and empirical objections to evolution voiced today—the existence of significant "gaps" in the *fossil record (which Darwin admitted and modern paleontologists continue to debate), questions about the limits of variability in organisms, the difficulty of explaining the origin of complex organs such as the eye (also admitted by Darwin), and the denial that the earth has existed long enough for natural selection to produce all forms of life—were likewise expressed in the mid-nineteenth century, for example in the review of the fourth edition of Darwin's *On the Origin of Species* by Scottish engineer Fleeming Jenkin (*North British Review*, 1867). Political and social objections—including the argument (detailed by biologist Vernon Lyman Kellogg in *Headquarters Nights* [1917]) that there were close links between the teaching of evolution and German militarism—drove prominent American politician and lawyer William Jennings Bryan to head efforts to outlaw the teaching of evolution in publicly funded American schools, culminating in the staged trial of John Scopes in Dayton, Tennessee (1925).

Although modern creationists still try sometimes to prevent the teaching of evolution, more often they push for "equal time" to teach their ideas alongside evolution in public schools. They challenge the ways in which evolution is presented in standard textbooks, calling for more attention to its perceived difficulties and clearer statements about the nature of evolution as a "theory" rather than a fact. Creationists have also developed their own alternative textbooks; most are written for fundamentalist religious schools though some are intended for public schools, which do not use them, as a rule.

Where most earlier antievolutionists, including Bryan, accepted evidence for an old earth and universe and did not see the biblical flood as a major geological event, the self-styled "scientific creationists" of today insist on a young earth (roughly the traditional biblical age of six thousand years) and embrace "flood geology," the claim that most fossiliferous rocks are relics of the flood and therefore not evidence for evolution. These ideas are central to *The Genesis Flood* (1961) by engineer Henry Morris and theologian John C. Whitcomb, Jr., the single most important creationist book since the 1920s. This work, which popularized flood geology for a wide audience of conservative Protestants, rests on an idea that had circulated for many years within the thoroughly creationist Seventh-day Adventist religious tradition—an idea ultimately derived from prophetess Ellen Gould White but directly taken from Canadian schoolteacher George McCready Price, author of *The New Geology* (1923) and many other works. Morris helped found two organizations, the Creation Research Society and the Institute for Creation Research (San Diego), that are leading disseminators of creationism. Their reach extends beyond the United States to a number of other countries, though creationism—as has been true for most of its history—remains largely an American phenomenon.

Ronald L. Numbers, *The Creationists* (1992). Edward J. Larson, *Summer for the Gods* (1997).

EDWARD B. DAVIS

CRICK, Francis (b. 1916), and **James D. WATSON** (b. 1928), molecular biologists, collaborators in pioneering research on DNA.

The discoverers of the structure of the DNA molecule standing under their model of the double helix.

In 1953 Watson and Crick published two joint articles in *Nature* announcing the structure of *deoxyribonucleic acid (DNA) and its genetic significance. This double-helix structure, often termed the W-C model, has since become a scientific and cultural icon, signifying the power of *molecular biology. The intertwining strands of Watson's and Crick's careers—like the double helix before its splitting—produced one of the most potent bonds in science.

Prior to their convergence, Watson's and Crick's trajectories exemplified different trends in twentieth-century science. James Dewey Watson, born in Chicago, Illinois, received a B.Sc. in zoology in 1947 from the University of Chicago. His fascination with genetics won over ornithology and brought him to Indiana University, Bloomington, for graduate work. His Ph.D. thesis in 1950, on the effects of X rays on bacteriophage multiplication, signified his membership in the expanding phage group (led by his mentors Salvador Luria and Max Delbrück), placing him at the vanguard of molecular biology. Dominated by physiochemical approaches to bacterial and viral genetics, the young field was shifting from a view of heredity based on proteins to one based on DNA. Watson spent his first postdoctoral year in Copenhagen studying DNA turnover in phage. But after meeting Maurice Wilkins, who worked in King's College, London, on x-ray crystallography of DNA, Watson changed his project. Supported by a fellowship from the National Foundation of Infantile Paralysis (1951–1953), he came to the Cavendish Laboratory in Cambridge, ostensibly to work on the tobacco mosaic virus (TMV). He studied x-ray diffraction of proteins but his interest lay in DNA structure. He shared an office with Crick.

Francis Harry Compton Crick, born on June 8, 1916, at Northampton, England, received a B.Sc. in physics from University College, London, in 1937. His doctoral work on the viscosity of pressurized water at high temperature was interrupted by World War II. Like most scientists, Crick engaged in war projects. He left the British Admiralty and physics in 1947 to study life science, thus contributing to the migratory trend from physics to molecular biology in the postwar decade. Supported by a studentship from the Medical Research Council (MRC), Crick studied the physical properties of cytoplasm in chick fibroblast cells at the Strangeway Laboratory in Cambridge. In 1950 he joined the MRC Unit at the Cavendish Laboratory for doctoral studies of protein structure, but by the time he met Watson his attention had already turned to DNA structure and its genetic significance.

Watson and Crick's commitment to DNA and the challenges of unraveling its structure were reinforced by friendship and a synergetic partnership. They shared intense intellectual energies, youthful arrogance, and complementary scientific strengths: Crick's physiochemical and theoretical bent and Watson's experimental and biological insights. They relied heavily on data from Wilkins's laboratory, where Rosalind Franklin was studying x-ray diffraction of DNA. Her work suggested a helical structure, similar to the protein alpha-helix elucidated by Linus Pauling (1951). Deploying Pauling's model-building and racing against him, Watson and Crick displayed DNA's architecture and the interactions among its molecular groups.

It was known by 1950 that the DNA chain consisted of linked deoxyribose-phosphate groups (sugar-phosphate backbone) and four bases: adenine (A), guanine (G), cytosine (C), and thymine (T). Though Erwin Chargaff had shown in 1950 that the ratios A/T and G/C were always 1:1, it was not until his meeting with Watson and Crick in 1952 that they grasped the structural significance of this feature. The challenge was to fit all these pieces together.

Watson's and Crick's feat was threefold. They established that DNA consisted of two anti-parallel complementary helical chains—two sugar-phosphate backbones—held together by hydrogen bonds that linked consecutive base pairs attached to the chain: A with T, and G with C (Chargaff's rule). They also showed how this structure explained genetic replication: The chains would separate—each a template for the synthesis of a new one—the old and new helical backbones held together through complementary base-pairing. Equally important, they identified DNA as the code that carries genetic information, thus prompting research on the genetic code with a new information paradigm. The explanatory power of their DNA model was recognized immediately. In 1954, when Crick obtained his Ph.D. from Caius College, Cambridge, for studies of "X-ray diffraction: polypeptides and proteins," their collaboration was already transforming molecular biology.

After 1953 Crick, based at the MRC's Laboratory of Molecular Biology, worked on the genetic code. Following George Gamow's lead and collaborating with two dozen physical and life scientists, including Watson, Crick focused his energies on the problem of how genetic information in the four DNA bases specified the assembly of the twenty amino acids into proteins. Though he did not "break the code" (this was accomplished in 1961 by Marshall Nirenberg and Heinrich Matthaei), Crick contributed original, as well as synthetic, insights to its solution. Watson, first at the California Institute of Technology then back in Cambridge, worked on RNA structure in TMV. After 1956, as a professor at Harvard University, he led a research group that in 1961 (concurrently with another team) identified messenger RNA in phage. In 1962 Watson, Crick, and Wilkins shared the Nobel Prize.

By the mid-1960s Watson's and Crick's paths diverged personally and professionally. Watson's notorious book *The Double Helix* (1968), recounting their research as a race to the Nobel Prize, led to a permanent rift between Crick and Watson. Crick remained in research, his 1962 appointment at the new Salk Institute in La Jolla, California, becoming permanent in 1976. Like many molecular biologists, he moved to neuroscience, making significant theoretical contributions on this new frontier. Watson devoted his energies to the institutional promotion of molecular biology. His textbook, *The Molecular Biology of the Gene* (seven editions since 1965), has shaped the field, as did his directorship of Cold Spring Harbor. In 1989 Watson became the first director of the Human Genome Project, in a sense fulfilling the potential of the Watson-Crick collaboration.

James D. Watson, *The Molecular Biology of the Gene* (1965). James D. Watson, *The Double Helix: A Personal Account of the Discovery of the Structure of DNA* (1968). Francis Crick, *What Mad Pursuit: A Personal View of Scientific Discovery* (1988). Francis Crick, *The Astonishing Hypothesis: The Scientific Search for the Soul* (1994). James D. Watson, *A passion for DNA: Genes, Genomes, and Society* (2000).

LILY E. KAY

CRYONICS. See COLD AND CRYONICS.

CRYPTOGRAPHY. Early cryptography, the science of encoding and decoding secret messages, used two distinct methods. In one, the words to be encoded (the plain text) were replaced by other words or groups of symbols, thus producing a secret, or encrypted, text. In the other, symbols replaced letters. The first method was widely used during the latter half of the nineteenth century and up to the early months of World War I. Code books contained the original words and their replacement terms opposite each other like entries in a dictionary. The weakness of the code books became evident when the Allied Powers captured the German code book and deciphered the infamous Zimmerman Telegram sent by Germany's War Ministry to the German ambassador in Mexico. (The telegram promised Mexico territorial gains at U.S. expense if it entered the war on Germany's side.)

After World War I, cryptographers preferred the method of replacing the original individual letters with coded symbols. This method had a long tradition. Julius Caesar encoded reports by replacing the original letters with letters three positions away in the alphabet. Thus A became D, X became A, and so on. Replacing original letters with letters three positions away in the alphabet allowed a total of twenty-five different "shifts." In Caesar's system, each original letter had the same coded symbol; the scheme can thus be described as a mono-alphabetic shift. Not only can the twenty-six original letters be transposed into others, they can be randomly mixed. The number of such combinations has twenty-seven digits, and each provides the opportunity for a mono-alphabetic encryption.

Mono-alphabetic encoding is easy to crack because the frequency of the letters in the original text recurs at the same frequency as the letters in the coded text. The code becomes harder to break when each letter is encoded with another "Caesar shift." This leads to poly-alphabetic encryption, which was already known in the fifteenth century. It culminated in the various encryption machines developed during the two world wars and actively employed in World War II. Germany's Enigma machine, with four or five cascaded encryptions, presented a formidable challenge. Polish and British mathematicians broke Enigma, and thereby influenced the progress of the war in Europe.

The Japanese during World War II used a machine that the Americans nicknamed Purple. It worked like Enigma, with Japanese words typed in the Latin alphabet. Although Purple was more difficult to break than Enigma, a group of Americans, including the famous William Friedman, perhaps the best cryptologist ever, succeeded in cracking it in several months of hard work.

A new type of mathematical encoding arose at the same time as Enigma and other encryption machines. Numbers replaced letters, such as 01 for A, 02 for B, and so on. In this way, a plain text could be converted to an encoded numerical text by various mathematical calculations. Reversing the calculations re-created the original text. The three-position Caesar shifts can be seen as such an operation. From the original letters, A becomes numerical 1; by adding 3, it becomes 4, that is, E; Y becomes $25 + 3 - 26 = 2 = B$.

The substantially refined method of the so-called "one-time pads" falls into the same category. Sender and receiver each possess a table with the same random digits, such as 390763692880. The encoder converts his original text, such as the word "rose," into an original numerical text—for example, 18151905. He then adds to this the pairs of random numbers from the beginning, while not carrying the 1s. Thus, 18 + 39, 15 + 07, 19 + 63, and 05 + 69 yields the secret text 47127264. The decoder removes the pairs of random digits without carrying the 1s, and by doing so recovers the original text. Soviet spies used this method after World War II. They wrote the random numbers on stamp-sized pieces of notebook paper and discarded each page after use so that no number combinations would be used more than once.

With these procedures, the sender and the receiver had to exchange a key, something like a code book, the first setting of Enigma, or a table of random numbers. Transmission of the code keys always presented a special risk, since they could fall into the wrong hands. However, in the 1970s, a new mathematical procedure presented cryptography with a new twist, the public key method. The best known, the so-called "RSA" method, uses three convenient numbers, usually designated N, E, and D. N is the product of two large prime numbers of at least 100 digits each. It does not matter which two prime numbers create N. From N, two key numbers, D and E, are determined. N and E allow the conversion of an original numerical text into a secret numerical text. Decoding is then possible only with the help of the numbers N and D. Thus, if Alice makes N and E public, but not D, anyone can send her encoded reports with N and E. But only she can read the messages. Today, banks and the Internet use the public key method. It also allows documents to be signed electronically, guaranteeing the authenticity of the signatures.

See also TURING, ALAN.

David Kahn, *The Codebreakers: The Story of Secret Writing* 2d ed. (1996). F. L. Bauer, *Decrypted Secrets: Methods and Maxims of Cryptology* (1997). Rudolf Kippenhahn, *Code Breaking: A History and Exploration*, trans. Ewald Osers and Rudolf Kippenhahn (1999). Simon Singh, *The Code Book: The Science of Secrecy from Ancient Egypt to Quantum Cryptography* (2000)

RUDOLF KIPPENHAHN

CRYSTALLOGRAPHY. Crystals have attracted attention because of their striking and often beautiful forms since early in human history. As an object of systematic study, they have attracted the attention of very different investigators: natural philosophers, mineralogists, chemists, physicists, mathematicians, metallurgists, and biologists. Only in the twentieth century did crystallography become an institutionalized scientific discipline. Perhaps for this reason a comprehensive history of the study of crystals is yet to be written.

In the sixteenth and seventeenth centuries, natural historians thought that crystals, like snowflakes and fossils, bridged the conventional categories of the material world. Like living things, their symmetric form indicated organization. It was even possible to see them growing. Yet they did not seem to be fully alive. Some observers speculated that the variety of crystalline appearance indicated astrological influence, others that it evidenced nature's ability to impose form on

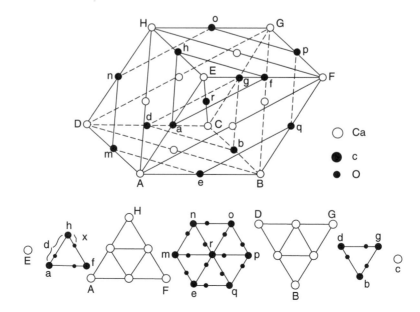

Arrangement of atoms in a crystal of calcium carbonate as determined by x-ray diffraction; the lower diagrams show the arrangement of atoms in special planes.

matter. Crystals were thought by some to grow from seeds in the earth; by others, from circulating fluids. Mechanical philosophers such as Robert *Boyle and Robert Hooke saw in the regular form of crystals a reflection of the underlying arrangement of corpuscles or atoms. Nicolaus Steno, in his *Produmus to a Dissertation on Solids Naturally Contained within Solids* (1669), explored possible manners of growth and stated the principle that crystals of the same kind have the same angles between adjacent faces.

In the eighteenth century, the study of crystals advanced on a number of fronts. Carl *Linnaeus proposed that minerals could be classified by counting their faces, a suggestion of limited utility at the time since most minerals do not appear crystalline to the naked eye and since the polarizing microscope was yet to be invented (*see* Mineralogy and Petrology). René-Just Haüy, professor of mineralogy at the Muséum d'Histoire Natural in Paris, followed earlier work by Romé de l'Isle on the relation of visible crystal forms to the units composing them. In his *Traité de cristallographie* (1822), Haüy suggested that cleaving a crystal divided it into one of six basic polyhedral units, themselves formed of smaller units that might or might not have other polyhedral forms. William Wollaston's invention of a reflecting goniometer that used light rays to measure crystal angles made possible much more precise measurements of crystal forms.

Haüy assumed a univocal relation between crystal form and chemical composition. The chemist Eilhard Mitscherlich dissented. He observed that substances with different chemical compositions could crystallize in the same form, a property he called isomorphism. This bothered the young Louis Pasteur who took advantage of the peculiar optical properties of crystals to investigate the matter further. He carefully separated tartrate crystals that differed just slightly in their facial angles. Polarized light passed through a solution of one set of crystals rotated in one direction; when passed through a solution of a second set, in the reverse—indicating, Pasteur argued, crystalline asymmetry at the molecular level.

The optical properties of minerals proved a reliable aid in identifying them. Henry Clifton Sorby's polarizing microscope enabled mineralogists to examine rocks in thin section and identify their previously invisible crystal constituents. This became a basic technique of petrography. In the second half of the nineteenth century, physicists studied the elasticity, density, and electrical properties of minerals. In 1880, Pierre *Curie and his brother Jacques-Paul discovered that pressure exerted at the right point on a crystal produced an electric field, a phenomenon known

as piezoelectricity. Pierre later incorporated this effect in an electrometer used in the detection of *radioactivity. Other scientists worked out the mathematical possibilities for crystal structure: Auguste Bravais, professor of physics at the École Polytechnique of Paris, showed in his *Études cristallographiques* (1866) that only fourteen possible arrangements of points in space lattices were possible.

The discovery of *X rays revolutionized crystallography. In 1912 Max von Laue and his group at the University of Munich photographed the diffraction pattern produced by a copper sulphate crystal, showed that X rays passed through a crystal scattered and deflected at regular angles. William Henry Bragg, professor of physics at the University of Leeds, and his son William Lawrence Bragg, then a student at Cambridge, realized that the pattern depended on the atomic structure of the crystal and succeeded in analyzing the crystal structure of the mineral halite (sodium chloride). This in turn led to the invention of the X-ray powder diffractometer. X-ray crystallographers began organizing themselves into formal bodies shortly after World War I. In 1925, they held an informal meeting in Germany to restore disrupted relations. At the Royal Institution in London and the University of Manchester, the Braggs trained students from all over the world in the techniques of x-ray crystallography. Societies were founded in Germany in 1929, the United States in 1941, and the United Kingdom in 1943.

Zeitschrift für Kristallographie became the leading journal in the field in 1927 when it started accepting papers in French and English as well as in German. International cooperation in crystallography revived immediately after World War II. The International Union of Crystallography was founded and quickly joined the International Council of Scientific Unions, the intermediary for UNESCO funding. Soon after came the debut of *Acta crystallographica*, quickly to become the premier journal in the field. X-ray crystallography proved crucial to the rapidly expanding field of *molecular biology, itself the result of the coalescence of biochemistry and crystal structure analysis. After the discovery that *proteins could be crystallized, W. T. Astbury began exploring their structure with X rays. In the late 1950s the double-helix structure of nucleic acid, predicted by Francis *Crick and James D. Watson, was confirmed by Maurice Wilkins and Rosalind Franklin; in the same period Max Perutz and John Kendrew determined the structure of hemoglobin and myoglobin.

In the 1960s, x-ray crystallographers began adding computers to their apparatus, increasing the speed and precision of analysis. X-ray crystallography remains the most powerful, accurate tool for determining the structure of single crystals. It is widely used in disparate fields including mineralogy, metallurgy, and biology.

Paul Ewald, ed., *Fifty Years of X-ray Diffraction* (1962). John Burke, *Origins of the Science of Crystals* (1966). Seymour Mauskopf, "Crystals and Compounds: Molecular Structure and Composition in Nineteenth-Century French Science," *Transactions of the American Philosophical Society* (1976). Dan McLachlan, Jr., and Jenny P. Glusker, eds., *Crystallography in North America* (1983).

RACHEL LAUDAN

CURIE, Marie Sklodowska (1867–1934), and **Pierre CURIE** (1859–1906), co-discoverers of radium.

This most famous couple in the history of science met in 1894 in Paris where Marie (then Maria) had come from her native Warsaw to study mathematics. With great determination, she came first in her class in physics in 1894 and second in mathematics in 1895.

Pierre was as sunk in his work—teaching physics and studying magnetism—as Maria was in hers. They had other things in common as well. He had been teaching at a new, non-elite technical school (the Ecole Municipale de Physique et Chimie Industrielle) since its foundation in 1882. This attachment and his inherited republicanism gave him the same humanitarian ideals that Maria, who had tried to educate peasants in Poland, not only held but practiced. They also shared a somber love of nature. They married in 1895.

Marie chose for her dissertation the examination of pitchblende, a complex uranium-bearing ore. She hoped that some of its constituents might give off rays of the kind that Henri Becquerel had found in a crystal containing uranium (*see* RADIOACTIVITY). In conducting her tests, she used several instruments that Pierre had invented. One was a balance that measured weight by electricity. Pierre and his brother Jacques-Paul had discovered its principle, piezoelectricity (the capacity of some crystals to become charged when strained), in 1880. When Jacques-Paul left Paris for a provincial professorship in 1883, Pierre turned from crystallography to magnetism, about which he discovered several notable things: that all substances are diamagnetic (a property overcome in para- and ferromagnetic materials), that magnets become paramagnetic when heated beyond a certain temperature (the Curie point), and that paramagnetism diminishes in strength in inverse proportion to the absolute temperature (Curie's

law). Among the instruments he invented during these investigations was a sensitive electroscope with magnetic damping. With this instrument Marie found that pitchblende was more strongly radioactive than the uranium it contained.

Pierre joined her in the laborious task of isolating the source of pitchblende's activity. By July 1898 they had a bismuth sample four hundred times more active than uranium. They called the hypothetical agent it bore "polonium." They then examined the radioactive barium fractions. Radium was the result, their Christmas present to the world. To obtain it in its barium carrier they had processed a few hundred kilograms of pitchblende. That was nothing. Marie turned Pierre's school, at which they did their chemical separations, into a toxic dump. She began with ten tons of pitchblende tailings (the ore minus the uranium). She worked in a large, unheated shed once used for dissections, a cold, smelly, dangerous, eerie place, stirring her cauldrons with a pole as big as herself and filling the endless vessels in which radium was concentrated. Within a year she had a preparation 100,000 times as powerful as uranium. It glowed in the dark. It took three more years to obtain enough moderately pure radium to weigh on Pierre's balance.

This singular intellectual and athletic achievement brought Marie not only her doctorate but also a share, with Pierre and Henri Becquerel, of the Nobel Prize in physics for 1903.

The Curies played almost no part in elucidating radioactive lineages or in devising the concept of isotope with which to capture the fact that different radioelements can have identical chemical properties. Their poor showing resulted from fidelity to a peculiarly French notion of scientific theorizing that put them at a disadvantage in competition with Ernest *Rutherford. The Curies believed that scientists should express their ideas in the most general way consistent with the known facts. Only as experiment progressed should options be closed. In contrast, Rutherford followed the British procedure of modeling the phenomena in clear, visualizable images.

Pierre Curie's positivism agreed with the practice of the French establishment that belatedly admitted him to its midst in 1904, as professor at the Sorbonne. Marie Curie held a similar concept of science strengthened by the political activism of her youth. She believed that to throw off the Russian yoke ordinary people would have to learn practical, efficacious, technical, and useful science, with nothing of the speculative or romantic about it.

In 1906 Pierre Curie was killed by a runaway truck. At one stroke Marie lost her husband, best friend, and closest collaborator. The Sorbonne gave her the chair of physics they had created two years before for Pierre, and she began her first lecture where he had left off. She described the disintegration theory as the most useful interpretation of the known facts. True to herself and Pierre, she added, "It seems to me useful not to lose sight of the other explanations of radioactivity that can be proposed." In 1911, Stockholm called again, this time with the chemistry prize, for her studies of radium.

Between the announcement and the ceremony, Marie received advice from the chairman of the committee that had proposed her that she should renounce the prize. Stolen letters had revealed that she and Pierre's former student Paul Langevin were lovers. The revelation fueled attacks that linked antifeminism and xenophobia with the defense of the family and old-fashioned morality. The journalistic savaging resulted in no fewer than five duels.

Marie insisted that her prize-winning achievements had nothing to do with her private life. She went to Stockholm. The emotional strain of the year overtaxed even her physical capacity and will power, however, and she never fully recovered her strength. That did not stop her from driving around the battlefields of World War I in one of the ambulances she had equipped with medical x-ray machines; from visiting the United States in a triumphal tour in 1921 to receive a gram of radium from the women of America; or from establishing her Institut du Radium as a premier laboratory for the study of radioactivity and nuclear physics, and the nurturing of women scientists. There in the early 1930s her daughter Irène Joliot-Curie and son-in-law Frédéric Joliot found that bodies could be made artificially radioactive by irradiating them with neutrons. They shared the Nobel Prize in chemistry for 1935. Marie Curie did not live to see this family triumph. She died in 1934 after suffering for some years from the results of her intimate association with radium.

Marie Curie, *Pierre Curie* (1923; reprint 1963). Anna Hurwic, *Pierre Curie* (1995). Susan Quinn, *Marie Curie* (1996).

J. L. Heilbron

CUVIER, Georges (1769–1832), and **Jean-Baptiste de LAMARCK** (1744–1829), protagonists of French natural history divided by a quarter of a century, conceptions of science, social aptitudes, and political abilities.

Although Lamarck's first published work was the *Flore Françoise* (1779), botany appears to have been for him no more than an intensely pursued pastime. His lifelong work and passion was the establishment of a thoroughly materialistic natural philosophy, inspired by the post-Newtonian dream to reduce the endless variety of phenomena to a handful of physical laws derived from basic properties of matter. Even movement, according to Lamarck, would one day be explained in terms of arrangements of matter. In his *Recherches sur les principaux faits physiques* (1794) and in subsequent works he violently criticized Antoine-Laurent *Lavoisier's claim that there existed in nature several chemical elements endowed with specific characteristics as a treason of Newtonian precepts: contrary to the view defended by Lavoisier and his followers, Lamarck maintained that the goal of chemistry should be the derivation of all chemical compounds and of their behavior from the properties of the four basic material principles, fire, air, water and earth.

To Lamarck, *fire was the most active form of matter. Highly compressible, it could regain its original volume by generating movement from the explosion of gunpowder and lightning to the burning of a candle or the deployment of the nervous fluids in animals. His physicochemical speculations led him to abandon the fixism of his youth and to investigate life's properties and progressive complication in terms of a fluid dynamics he called "biology." Life and the progress of structural complexity were to him the result of strictly material processes, which could even account for man's intellectual abilities as well as for his moral propensities and social aptitudes. In a constantly changing environment (which he studied in his *Hydrogéologie*, 1802), exposed to atmospheric influences (which were the subject of his meteorological investigations), the internal fluids—blood, lymph, electric-like nervous fluids, and so forth—modified their pathways and distribution, slowly changing the internal and the external configuration of the living body.

Twenty-five years Lamarck's junior, Cuvier took Lavoisier's chemistry and Pierre-Simon *Laplace's physics as models for the reform of natural history. Naturalists should cultivate careful observation and refrain from all-encompassing speculation. In the life sciences, reasoning from first principles could not rectify the lack of basic knowledge of the main organic systems, and of the functions they supported and made possible.

Cuvier's first encounter with Lamarck was not pleasant. The young naturalist, then tutor to an aristocratic family of Normandy, had distinguished himself with pioneering work on the anatomy of marine invertebrates and his proposal to take the internal organization of the animal, and not its shell, as the basis for classification. In December 1794 Cuvier applied for a junior post at the newly established Natural History Museum in Paris; Lamarck advised his colleagues not to give the job to so overqualified a candidate. Lamarck had made ample use of Cuvier's anatomical discoveries in his lectures as professor of invertebrate zoology at the museum, as well as in his publications: a borrowing that Cuvier later eagerly publicized.

Supported by the followers of Lavoisier and by Laplace, Cuvier was admitted to the First Class of the Institut, the reconstituted Académie des Sciences, in 1795, and rose to the position of perpetual secretary of the class in 1803. Yet, his career in the Natural History Museum was not easy. He bribed the professor of human anatomy Jean-Claude Mertrud into giving him the post of assistant to the chair, and in this capacity he gave his famous lectures on comparative anatomy, published between 1800 and 1802. Only when his scientific authority and administrative power had been established was he elected to a full chair in the museum.

Cuvier regarded living organisms as machines that could not be taken apart (*Leçons d'anatomie comparée*, 1800). Complex interactions between systems of organs created and sustained vital functions: any derangement in any part of the complex functional equilibrium between internal organs would inevitably lead to death. Limited variation might occur at the surface of the animal, involving a slight change of color or skin thickness. Environmental change could only produce extinction, never transformation of form or type, as Lamarck and other contemporaries taught.

Cuvier found further success, and a powerful weapon against evolutionists, in paleontology. His awe-inspiring reconstructions of extinct, gigantic species fired popular imagination. The study of the geology of the Paris basin he conducted with Alexandre Brongniart proved to him that earth history had been marked by a series of catastrophic events, each having caused the extinction of several forms of life. Evolution was not only theoretically impossible, but also historically untenable.

During the Consulate and the Napoleonic Empire, mistrust of flights of imagination in science became an expression of political prudence. Speculations, especially of a materialistic bent, were regarded as potentially subversive. After 1802 and the publication of his *Recherches sur l'organisation des corps vivans* (the first version of his evolutionary doctrines, afterwards elaborated in the much celebrated and little read *Philosophie zoologique*, 1809) Lamarck refrained from airing his views and wrote several scores of purely technical memoirs in invertebrate zoology.

In 1809 Napoleon forced him to terminate his series of *Annuaires météorologiques*, which enjoyed considerable success in the provinces, but were loathed by Cuvier and Laplace.

Cuvier's *Recherches sur les ossemens fossiles* (1812), the summa of his paleontological discoveries, was perhaps the last original work he composed. His time was increasingly occupied with administrative and political functions as he consolidated his grip on scientific activity in France. His political power increased during the Restoration, after the fall of Napoleon. Younger naturalists came to regard him as a negative influence on French science. They hailed Lamarck's last major editorial effort, the seven-volume *Histoire naturelle des animaux sans vertèbres*, as the work of the new, French Linnaeus. They admired the revamped formulation of his evolutionary ideas, in the first volume of the work, for its philosophical daring, and as an advertisement for a natural history different from the heaping up of facts in the manner of Cuvier.

During the 1820s, Lamarck, blind and old, abandoned active work, and Cuvier came under attack from his former colleague and friend Étienne Geoffroy Saint-Hilaire and younger followers of Lamarck. The stage figures of Lamarck as the destitute, isolated yet prophetic figure of French natural science and a plump and powerful Cuvier, a man ready to serve any master, made their appearance.

Cuvier wrote the history of science through his *Reports* as perpetual secretary of the Institut and his necrologies of deceased academicians. Historians have only recently started to question the accuracy of his reconstructions and to unveil the political agenda behind them. Cuvier has had the last word against Lamarck. Indeed, some of the very last words he wrote were directed against his old adversary. In his necrology of Lamarck (1832), Cuvier rightly stressed the outdated physicochemical ideas that filled Lamarck's works, though he wrongly asserted that Lamarck's evolutionary ideas had no followers and were an object of public derision. He sketched a portrait of a bitter, isolated, self-absorbed man, whose only claim to fame was the remarkable work on invertebrate classification made possible, he added, thanks only to the fundamental contributions that he himself, Cuvier, had made to comparative anatomy.

Dorinda Outram, *Georges Cuvier: Vocation, Science and Authority in Post-Revolutionary France* (1984). Pietro Corsi, *The Age of Lamarck. Evolutionary Theories in France* (1988; expanded edition, *Lamarck. Genèse et enjeux du transformisme, 1770–1830* [2000]). Pietro Corsi, ed., *Jean Baptiste Lamarck: Works and Heritage*, www.lamarck.net.
PIETRO CORSI

CYBERNETICS is the discipline that studies communication and control in living beings and machines or the art of managing and directing highly complex systems. Concepts investigated by cyberneticists include systems (animal or machine), communication between systems, and their regulation or self-regulation.

The MIT mathematician Norbert Wiener invented the term in 1947 (first used in print, 1948) from a Greek term for "steersman" or "governor." The origins of cybernetics lie in the development of feedback mechanisms such as the spring governors for steam engines invented by James Watt and Matthew Boulton (1788), James Clerk *Maxwell's consideration of ships' steering engines (1868), and Weiner's involvement in developing automated range finders (1941), which led to the construction of the ILLIAC *computer.

Wiener joined MIT in 1919 as a professor of mathematics. Together with MIT neurophysiologist Arturo Rosenblueth he established small interdisciplinary teams to investigate unexplored links between established sciences. Working with the engineer Julian Bigelow, Wiener developed automatic range finders for antiaircraft guns able to predict an aircraft's course by taking into account the elements of past trajectories. Wiener and Bigelow observed the seemingly "intelligent" behavior of these machines and the "diseases" that could affect them. The servomechanisms appeared to exhibit "intelligent" behavior because they dealt with "experience" (recording of past events) and predictions of the future. They also observed a strange defect in performance. With too little friction, the system entered a series of uncontrollable oscillations. Rosenblueth pointed out that humans exhibited similar behavior and Wiener inferred that in order to control a finalized action (an action with a purpose) the circulation of information needed for control must form a closed loop, allowing the evaluation of the effects of previous actions and the adaptation of future conduct based on them. Wiener and Bigelow thus discovered the negative feedback loop.

Rosenblueth's multidisciplinary teams approached the study of living organisms from the viewpoint of a servomechanism engineer and considered servomechanisms from the perspective of the physiologist. Rosenblueth and neurophysiologist Warren McCulloch (then director of the Neuropsychiatric Institute at the University of Illinois) brought together mathematicians, physiologists, and mechanical and electrical engineers in 1942 at a seminar at Princeton's Institute for Advanced Study. After ten more meetings, two seminal publications resulted: Wiener's *Cybernetics, or Control and Communication in the Animal and the Machine* (1948) and Claude Shannon and Warren Weaver's *The Mathematical Theory of Communication* (also 1948), which established information theory. Although it had an interdisciplinary orientation, early cybernetic studies employed an engineering approach focusing on feedback loops and control systems and on constructing intelligent machines. Influenced by the changing and challenging perspectives, McCulloch moved from neurophysiology to mathematics and then to engineering.

The new disciplines inspired biologist Ludwig von Bertalanffy to found the Society for General Systems Research. The society included disciplines far removed from engineering: sociology, political science, and psychiatry. The excitement attracted researchers such as mathematician Anatole Rapoport, biologist W. Ross Ashby, biophysicist Nicolas Rashevsky, economists Kenneth Boulding and Oskar Morgenstern, and anthropologist Margaret Mead. Mead urged Wiener to extend his ideas to society as a whole. The Society's *Yearbooks* first appeared in 1954 and exerted a profound influence on those interested in applying the cybernetic approach to social systems and to industrial problems.

Early interest in cybernetics in the USSR was stifled on ideological grounds but in 1955 Aleksandr Mikhailovich Lyapunov, founder of Soviet cybernetics and programming, along with Sergei Sobolev and Anatoly Kitov, published the first permitted paper on cybernetics in *Voprosy Filisofii* (Problems of Philosophy). Soon thereafter, V. M. Glushkov established the Institute of Cybernetics in the Ukraine though its independent work does not appear to have had much impact on western studies.

Cybernetics was closely related to the development of ENIAC (1946), EDVAC or EDSAC (1947), Whirlwind II (1951), and other early *computers. The latter used a superfast magnetic memory invented by an electronics engineer from MIT's Lincoln Laboratory, Jay Forrester. Beginning in 1952, he coordinated the SAGE (Semi-Automatic Ground Environment) alert and defense system for the U.S. Air Force. It combined radar and computers for the first time to detect and prevent attacks by enemy rockets. Forrester designed SAGE to be capable of making vital decisions as information arrived while interacting in real time with humans inputting data and with other humans making decisions on appropriate reactions.

Ten years later, back at MIT, Forrester created industrial dynamics, which regards all industries as cybernetic systems in order to simulate and predict their behavior. In 1964, confronted with the problems of the growth and decay of cities, he extended industrial dynamics to urban systems (urban dynamics), generalized his theories as system dynamics, and published its definitive text, *World Dynamics* (1971).

Heinz von Foerster coined the term "second-order cybernetics" in 1970. First-order cybernetics concerned observed systems, and second-order cybernetics observing systems. Biologists like von Bertalanffy stayed at the first level. Second-order cybernetics deals with living systems and not with developing control systems for inanimate technological devices and explicitly includes the observer(s) in the systems under study. These living systems range from simple cells to human beings. From the 1970s "cybernetics" has served as an umbrella term for several related disciplines: general systems theory, information theory, system dynamics, dynamic systems theory (including catastrophe theory), and *chaos theory, among others. Cybernetics and systems science now constitute an academic domain encompassing recently developing "sciences of complexity," including artificial intelligence, neural networks, dynamical systems, chaos, and complex adaptive systems such as those used in flexible or multi-mirrored astronomical telescopes. Problems that appear beyond solution now may one day succumb to quantum computers able to mimic more closely the analytical powers of the human brain.

See also COMPUTER SCIENCE.

Norbert Wiener, *Cybernetics, or Control and Communication in the Animal and the Machine* (1948). Heinz von Foerster, "Ethics and Second-Order Cybernetics," *Cybernetics & Human Knowing*, (1992) vol. 1, no. 1, pp. 9–20, Aalborg, Denmark. Francis Heylighen, Cliff Joslyn, and Valentin Turchin, "What Are Cybernetics and Systems Science?" Principia Cybernetica Project (1999) at: http://pespmc1.vub.ac.be/CYBSWHAT.html. Joel de Rosnay, "History of Cybernetics and Systems Science," Principia Cybernetica Project (2000) at: http://pespmc1.vub.ac.be/CYBSHIST.html.

RANDALL C. BROOKS

CYTOLOGY. The term "cytology," denoting the study of plant and animal cells, has been in use from the 1880s, but it only acquired widespread currency in the early decades of the twentieth century. The seventeenth-century British natural philosopher and first Curator of the Royal Society of London, Robert *Hooke, may, however, be regarded as the first cytologist. He observed, named, and figured vegetable cells using the recently invented compound *microscope. Nehemiah Grew, using the same

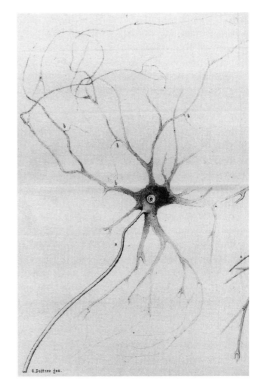

A large ganglion cell from a spinal cord magnified 300 to 400 times. This famous illustration showed that nervous tissue consists of isolatable cells, not of a continuous net.

sort of microscope as Hooke, published many illustrations of vegetable cells. During the last quarter of the seventeenth century, the Dutch draper Antoni van Leeuwenhoek reported observations of structures including what would later be identified as protozoa and blood globules.

These were the efforts of a handful of individuals and did not amount to a concerted program of research into the minute structure of living tissue. Only in the mid-nineteenth century did schools of histological inquiry began to develop. These histologists made use of new achromatic microscopes that corrected some of the optical defects of earlier microscopes. Among the most important figures in this first generation of microscopic anatomists were Jan Evangelista Purkyně , Professor of Physiology at Breslau, and Johannes Peter Müller, Professor of Physiology at Berlin. Müller in particular produced a school of microscopic observers who went on to produce important original research. One of these, Theodor *Schwann, was to make a fundamental theoretical contribution to cytology.

In 1831 the Scottish botanist Robert Brown identified a nucleus in vegetable cells. Jacob

Matthias Schleiden came to the conclusion in 1838 that cells were a ubiquitous feature of plant organization and that the nucleus played a crucial role in the growth of plant cells. When Schwann became acquainted with the results of Schleiden's researches, he recognized an analogy between these nucleated cells and certain structures he had himself discerned in animal tissue. In 1839 Schwann argued that the cell was the common unit of structure, function, and development in both plants and animals. This "cell theory" became a fundamental tenet of biology.

In 1846 Schleiden's former associate, Carl Wilhelm von Naegeli, concluded that in almost all cases new plant cells arose through the division of preexisting ones. Embryological studies made by Robert Remak supported the view that the same held for animal cells. The pathologist Rudolf Virchow was at first a proponent of free cell generation. By 1855, however, on the basis of his microscopic investigation into disease processes, Virchow rejected this concept of cytogenesis.

Later developments in cytology were closely associated with the increasingly refined techniques for fixing and staining tissue that became available to investigators. The German anatomist Walther Flemming elaborated techniques to reveal the internal structure of the cell. He was the first to describe the presence of thread-like bodies, later named chromosomes, in the nucleus, and to describe the role they played in cell division. Flemming coined the term "mitosis" for this process in 1879. In the 1890s, the mode of division characteristic of reproductive cells ("meiosis") was distinguished.

Oscar Hertwig had reported in 1876 that the fertilized ovum contained nuclear matter from both parents. Edouard van Beneden expanded upon these observations in 1883, giving a definitive account of the processes involved in the production of the new nucleus in the fertilized egg. The full significance of these cellular phenomena for the understanding of heredity came in the early twentieth century after the rediscovery of Gregor *Mendel's ideas (see GENETICS).

During this period cytology and genetics became increasingly intertwined. The work of the American biologist Edmund Beecher Wilson summed up what had been done in cytology in the previous half century and made a bridge to later developments: his *The Cell in Development and Inheritance* (1896) synthesized what was known about the structure and functions of the cell. In his later researches on the role of chromosomes in reproduction, he supplied the cytological basis for the researches of his associate, Thomas Hunt Morgan, on the mechanisms of heredity.

Although attention thus focused on the nucleus, the cell contents, or "cytoplasm," also received intensive study. Several observers discerned a reticular structure in the cytoplasm. By the end of the century, however, these appearances had been dismissed as artifacts produced by the fixatives employed. A considerable debate broke out during the mid-nineteenth century over whether the animal cell, like the vegetable, in all instances possessed a true "wall" or dividing membrane. Ernst Wilhelm von Brücke and Max Schultze argued against the traditional view of the cell as a bounded cavity. The essential parts of the cell were the nucleus and its surrounding "protoplasm."

Around the turn of the twentieth century, Carl Benda employed sophisticated staining methods to identify cytoplasmic inclusions to which he gave the name "mitochondria." Microscopy offered little insight into the functions these bodies might perform, although Benda speculated that they might play a role in heredity. Later biochemical analysis revealed that they participated in cell respiration. The structure of the mitochondria was elucidated after World War II by the use of the electron microscope.

In the second half of the twentieth century cytology became increasingly detached from its early reliance upon the light microscope. At the same time its links with cognate disciplines like microbiology, genetics, *biochemistry, and *molecular biology have become ever more intimate. Using techniques drawn from these sciences, cytologists have been able to provide fuller accounts of how the different parts of the cell perform their metabolic and reproductive functions.

Arthur Hughes, *A History of Cytology* (1959). H. Grunze and A. I. Spriggs, *History of Clinical Cytology* (1983). G. E. Palade, "A Short History of Cell Biology," *Noesis*, no. 9 (1984). Henry Harriss, *The Cells of the Body: A History of Somatic Cell Genetics* (1995). Nicolas Ramussen, "Mitochondrial Structure and the Practice of Cell Biology in the 1950s," *Journal of the History of Biology*, vol. 66 (1995): 151–166.

STEPHEN JACYNA

D

DAILY LIFE, SCIENCE AND. As an academic subject, "science and daily life" studies how the forms of rationality present in everyday life interact with scientific ways of thinking and knowing. It assumes that what is taken to be everyday rationality (including common sense) changes over time, as does the relationship between everyday rationality and sensibilities, emotions, the senses, and social behavior. Christa Wolf's book *Accident* (1989), which logs her thoughts on the day of the Chernobyl disaster in 1986 (also the day on which her brother had brain surgery), captures the problematic of the study of science and daily life. Listening to radio reports of meltdown and radioactivity as the brain surgeon describes her brother's condition and chances for recovery over the telephone, Wolf wonders how life has become "a continuous acclimatization to ever harsher lighting, sharper insights, increasing matter-of-factness." Thus conceived, the study of science and daily life examines one strand of the historical process of rationalization outlined by the nineteenth century sociologist Max Weber.

The philosopher Alfred Schutz raised the question of the relationship between rationality and the everyday in the 1930s. He believed that the closer rationality in the everyday life of a society imitated that in the natural sciences, the more rational the society was. Scholars have dropped Schutz's rigid measure of everyday rationality but have retained his goal of understanding the nature of rationality in daily life. Also in the 1930s French historians created the Annales paradigm, a part of which captured the "mentality" of an age by exposing the rational principles at work in ordinary language and speech. Since the 1930s in the English-speaking world the theoretical study of the everyday has fallen to ethnomethodologists—including Harold Garfinkel, Edward T. Hall, and Erving Goffmann—and sociologists, especially Peter Berger and Thomas Luckmann, who reshaped the foundation that had been laid by Schutz. They were especially active during the Cold War when the study of the everyday had strategic importance for intelligence activities. In the German-speaking world neither ethnomethodology nor the Annales paradigm gathered much attention until the end of the twentieth century. Scholars in the field of *Alltagsgeschichte* (the history of daily life) have for the most part not considered the relationship between science, forms of rationality, and everyday existence. By contrast, German philosophers like Jürgen Habermas, who have studied the problem of rational principles in the everyday with only tacit acknowledgment of ethnomethodological principles, have nonetheless contributed significantly to the understanding of science and daily life. Far from being only a matter of concern in the industrialized world, the forms of rationality present in the localized, indigenous knowledge present in the everyday life of more traditional cultures have become major scholarly and political concerns as a result of globalization. Ethnobiologists, for instance, argue that indigenous forms of knowledge offer means superior to the natural sciences to preserve biological diversity and ecological balance precisely because indigenous knowledge agrees with the thinking and habits of everyday life and because it preserves the harmony and balance of the natural and social worlds.

The study of science and daily life includes the history of the senses, forms of scientific argumentation, the entry of probabilistic reasoning into the everyday, and the functions of technical rationality. In a groundbreaking study on French history, Robert Mandrou argued that hearing was the principal sense until displaced by vision during the Enlightenment (no matter the appearance of the *telescope and *microscope in the seventeenth century). Before about 1700, he contends, the imagination responded primarily to sound, and information was conveyed primarily by speech, not by reading. Mandrou attributed the shift from sound to sight to the spread of Cartesian thinking in daily life when rational principles superseded emotional ones in guiding everyday thinking. Steven Shapin connected scientific and everyday reasoning in the work of Robert Boyle during the *Scientific Revolution. Guided especially by sociologists like Peter Berger and Thomas Luckmann who argued that the natural sciences embody forms of everyday reasoning,

Shapin demonstrated how gentlemanly discourse in the seventeenth century shaped the atmosphere of trust upon which scientific agreement rested, thus contributing significantly to the affirmation and spread of scientific truth. The history of probability and statistics in the eighteenth and nineteenth centuries incorporated much from everyday reasoning and experience about life insurance, widow's funds, the lottery, and other games of chance, while the spread of probabilistic reasoning from science to the everyday has conditioned views on issues ranging from weather predictions to life expectancy. Near the end of the Cold War, Habermas feared the continued infiltration of technical forms of rationality into the everyday. In his view technical rationality narrowed choices in decision-making and threatened the democratic process by imposing uniformities. Rather than dispense with rationality altogether, Habermas—who believed in the Enlightenment promise of emancipation through reason—sought instead to redefine rationality in the everyday so as to resurrect the creative possibilities of ordinary language, common sense, and social communication.

See also HISTORIOGRAPHY OF SCIENCE.

Peter Berger and Thomas Luckmann, *The Social Construction of Reality* (1966). Harold Garfinkel, *Studies in Ethnomethodology* (1967). Alfred Schutz, *The Phenomenon of the Social World* (1967). Jürgen Habermas, *Toward a Rational Society* (1970). Robert Mandrou, *Introduction to Modern France, 1500–1640: An Essay in Historical Psychology* (1976). Gerd Gigerenzer et al., eds., *The Empire of Chance: How Probability Changed Science and Everyday Life* (1989). Steven Shapin, *A Social History of Truth: Civility and Science in Seventeenth Century England* (1994).

KATHRYN OLESKO

DALTON, John (1766–1844), English natural philosopher, proposed the modern atomic theory.

Son of a modest Quaker weaver, Dalton grew up near Kendal, in the Lake District of England. Mentored by a local polymath, Dalton showed early promise in mathematics and natural philosophy, and became a schoolmaster at the age of nineteen. A few years later he was hired at the New College, a "dissenters' academy" in Manchester. He resigned his post in 1800, preferring to support himself by private teaching for the rest of his life. Dalton was an active and respected member of the flourishing scientific scene in Manchester. He never married.

In 1794, a month after his election to the Manchester Literary and Philosophical Society, Dalton presented his first paper there, an investigation of color-blindness, with which he himself was afflicted (it is still often referred to as "Daltonism"). The preceding year he had published his first book, *Meteorological Observations and Essays*. These publications reveal a deep and abiding interest in physiology and natural history, in addition to a commitment to physics and chemistry. He remained particularly engaged with theories of gases and solubilities, especially the solution of water vapor in air, and of gases in water.

In 1801, Dalton published a new theory of mixed gases applicable to the atmosphere. He supposed that in a gaseous mixture, each component exerts pressure independently of the other gases. What later became known as "Dalton's law of partial pressures" suggested that the interactions between gases are purely physical and not chemical, and he applied this thesis to the solution of gases in water. Although it solved some anomalies, the theory created further difficulties. While defending and elaborating his physical theory of gases, Dalton made his greatest contribution to science, his chemical atomic theory.

Influenced by Antoine-Laurent *Lavoisier's new chemistry and by the popular Newtonianism then prevalent in Britain, Dalton began to investigate the relative weights of the (then-hypothetical) atoms of elements. His method was simplicity itself. According to Lavoisier's analysis, water was 85 percent oxygen and 15 percent hydrogen. If water consists of a combination of the ultimate particles of oxygen and hydrogen, and if each "compound atom" of water consists of one "simple atom" of each element, then the atom of oxygen must weigh about 5.7 times (85/15) as much as the atom of hydrogen. If, then, hydrogen is conventionally assigned the dimensionless atomic weight of 1, the relative weight of an oxygen atom is 5.7. To take another example, carbonic oxide was 45 percent carbon and 55 percent oxygen. Again assuming a "binary" formula for the carbonic oxide molecule and with the weight of an oxygen atom now known to be 5.7, carbon must then be about 4.5. The latter weight matched the calculated weight for the carbon atom from the analysis of "olefiant gas" (ethylene), assuming a binary formula for the latter.

Such atomic and molecular weights appear in an entry in Dalton's notebook of 6 September 1803, apparently in a presentation read the following month to the Manchester "Lit and Phil" and in the revised paper as published two years later. Dalton did not reveal here his method of calculation, however. The first published discussion of Dalton's new atomic theory occurred in the pages of the textbook of his friend, Thomas Thomson, in 1807. Thomson rightly stressed that these calculations could proceed only by assuming molecular formulas for the simple compounds under study. Dalton's axiom was to assume the simplest possible formulas. The procedure was arbitrary (or at least not fully empirical). The finest and most productive new ideas often require such leaps.

In 1808 and 1810 Dalton published the first two volumes of *A New System of Chemical Philosophy*, in which he developed his atomic theory in detail for the first time. The book, although much criticized for the simplicity of assumptions it invoked, marked an epoch in the history of chemistry. Subsequent elaborators of chemical atomism, such as Jöns Jacob Berzelius, Joseph Louis Gay-Lussac, Amedeo Avogadro, Justus von *Liebig, and Jean-Baptiste-André Dumas, began with Dalton's atomistic vision even if they rejected or modified certain details within it.

Gradually, Dalton's axioms for formula assignments were replaced by more systematic and empirical methods, and the chemical atomic theory gained steadily in stature. From its birth in the early nineteenth century to the present day, John Dalton's creation has formed the cornerstone of the science of chemistry.

Frank Greenaway, *John Dalton and the Atom* (1966). Elizabeth Patterson, *John Dalton and the Atomic Theory* (1970). Arnold Thackray, *John Dalton* (1972). A. Rocke, *Chemical Atomism in the Nineteenth Century* (1984).

A. J. ROCKE

DARWIN, Charles (1809–1882), naturalist and proponent of the theory of evolution by natural selection.

Darwin was born in Shrewsbury, England, the son of a medical doctor, and educated at Shrewsbury School. In 1826 his father sent him to Edinburgh University to study medicine. Darwin pursued natural history rather than the medical curriculum and in 1827, recoiling from surgery, he asked to be withdrawn. During this period he encountered much that was new and exciting in science. Robert Grant explained Jean Baptiste de Lamarck's transformist theory to him, he attended Robert Jameson's natural history lectures, and studied the evolutionary writings of his own grandfather, Erasmus Darwin. He went to Cambridge in January 1828 with a view to studying to become a parson. He did not fulfil this ambition. After graduation he joined Captain Robert Fitzroy on the H.M.S. *Beagle* voyage (December 1831 to October 1836), the invitation indirectly coming through John Stevens Henslow, botany professor at Cambridge, Darwin's friend and mentor.

The voyage, the key event in Darwin's life, greatly expanded the way he thought about the natural world. The ship surveyed the

The monkey sage caricatured in the *Hornet* (1871), which drew on the popular misconception that Darwin taught that man descended directly from apes.

southern coastlines of South America and also visited Brazil, Tahiti, the Galapagos Islands, New Zealand, Australia, and various ports in the Indian and Atlantic oceans. Darwin spent several years in South America traveling on land in Chile, Patagonia, and Tierra del Fuego. He collected extensively. His account of his journey, *Journal of Researches into the Geology and Natural History of the Various Countries Visited by H.M.S Beagle* (1839) appeared in many successive editions.

During the voyage Darwin learned from Charles Lyell's *Principles of Geology* (1831–1833) that the environment is constantly changing. He applied this idea fruitfully to the origin of mountain ranges and coral reefs, and to other questions of natural history. Throughout the voyage, his mind evidently teemed with questions and ambitions to become an expert naturalist. As the journey drew to a close, he pondered the question of the origin of species and the wide variety of adaptations that organisms possess. Phenomena he observed on the Galapagos Islands, particularly the differences among the finches, may have stimulated his thinking but did not produce a "eureka" moment.

Darwin became converted to *evolution by applying to living beings Lyell's ideas of gradualism and constant small changes. He took the idea of competitive struggle and survival rates from Thomas Malthus's *Essay on the Principle of Population* (1798). By the end of 1838 he had set out the basic ideas that would underlie the *Origin of Species* some twenty years later. During the 1840s and 1850s, Darwin expanded his theories with a wide range of evidence supplemented by extensive reading, practical inquiries, and animal and plant breeding experiments. He spent several years investigating barnacles, living and fossil, which gave him valuable skills in anatomy and classification. His reputation as a talented geologist and naturalist soared. Even so, he kept his evolutionary ideas secret except for discussions with a few close friends. In 1856 Lyell urged him to publish.

While writing up his theories for publication, Darwin received a short essay containing near-identical ideas from Alfred Russel Wallace, who was in Malaysia collecting specimens at the time. Friends hurriedly arranged a joint announcement at the Linnean Society of London in July 1858, and Darwin published *On the Origin of Species* in November 1859. In this he claimed that every animal and plant is variable. Individuals best adapted to surrounding conditions will survive and reproduce. Over eons of time, and under gradually changing conditions, organisms evolve. Although not the first to propose evolution, Darwin aroused intense controversy. Victorians found it hard to accept gradual organic change and equally hard to remove God from the creative process. The *Origin of Species* was often perceived as a dangerously atheistic tract although Darwin did not deliberately defy either Christianity or the religious beliefs of his friends and relatives. In the book he scarcely referred to human beings and avoided confrontation over the biblical account of human creation. The book was vigorously defended by T. H. Huxley, "Darwin's bulldog." Darwin later expanded and confirmed his ideas in the *Descent of Man* (1871) and *Expression of the Emotions* (1872). His work on plants and earthworms, though customarily ignored, were significant elements in this program of justification. In later life, Darwin was revered as a grand old man of science. He died in April 1882, at Down House, near Bromley, Kent, where he had lived and worked for forty years. He was buried in Westminster Abbey.

See also CREATIONISM.

David Kohn, ed., *The Darwinian Heritage* (1985). Robert M. Young, *Darwin's Metaphor: Nature's Place in Victorian Culture* (1985). A. J. Desmond and J. R. Moore, *Darwin* (1991). Ernst Mayr, *One Long Argument: Charles Darwin and the Genesis of Modern Evolutionary Thought* (1991). Janet Browne, *Charles Darwin: Voyaging* (1995).

JANET BROWNE

DARWINISM, a term introduced by zoologist T. H. Huxley in his review of Charles Darwin's *Origin of Species* (in April 1860). By "Darwinism," Huxley meant Darwin's theory of natural selection, which asserted that individuals having any advantage over others in the struggle for life would have the best chance of surviving and procreating. Hence, if favorable, small, individual inheritable variations would slowly cause the species to change. In his review, however, Huxley endorsed the general notion of a common descent of living beings (humankind included) and evolutionary *naturalism, rather than the particular theory Darwin used to explain evolutionary change. Furthermore, he hailed Darwin's book as a gun in the hands of liberalism. Since then, "Darwinism" has oscillated between the narrower meaning (Darwin's theory of evolutionary change) and the broader one (common descent and evolutionary naturalism), as it has been appropriated, debated, or rejected by authors ranging from professional biologists to sociologists, philosophers to religious leaders, journalists to politicians.

Until the establishment of the history of biology as a specialty within the history of science in the mid-twentieth century, historical assessments of Darwinism were closely intertwined with the needs and fortunes of evolutionism, and Darwin's role in it. Thus Darwin himself, while asserting his theory of natural selection in the first edition of the *Origin* (1859), passed over earlier evolutionary theories almost in silence. The historical sketch that he added to the *Origin* in 1861 emphasized discontinuity from the several authors (notably Jean Baptiste de Lamarck) who since the mid-eighteenth century had advocated one form or another of transformism, as evolutionary naturalism was then called. Darwin was concerned in his historical sketch to stress that until the *Origin*, most naturalists had believed that species were immutable, separately created productions (*see* CREATIONISM).

Darwin's success during the 1860s and 1870s in convincing the majority of biologists and large sectors of public opinion in several countries to adopt an evolutionary perspective did not carry a univocal interpretation of Darwinism with it. Even among Darwin's closest supporters, strict adhesion to the theory of natural selection remained the exception rather than the rule. As for Darwin himself, in the two decades following 1859 he adjusted his own explanation of evolutionary change by introducing a number of other mechanisms. In the sixth edition of the *Origin* (1872, the most widely read), the mechanisms included the use and disuse of parts (an old favorite of Lamarck's), sexual selection, directed variation, correlated variation, spontaneous variations, and family selection, in addition to natural selection.

As to the broader philosophical, religious, and political implications involved in the adoption of an evolutionary perspective in biology, Darwin's supporters divided into several separate camps. These ran from the secularized naturalism of T. H. Huxley in Britain through the freethinking of Fritz Müller, a German expatriate in Brazil, to the theistic evolutionism of Asa Gray in America. The many outside the circle of Darwin's close allies who had some reason for aligning themselves with Darwin and basking in his enormous prestige developed an even greater variety of positions. These included overall evolutionary worldviews embracing the whole cosmos as well as human societies, like that of the British philosopher Herbert Spencer, who had a large following in continental Europe and America, and the monistic philosophy supported by the zoologist Ernst Haeckel, whose books were read by several generations of Germans. The special brand of Darwinism called social Darwinism, which gained recurrent, considerable followings in political arenas in several countries well into the twentieth century, typically did not distinguish between the different strands of Darwinism.

While preaching evolution, as Darwin did, and addressing the same audiences that venerated him toward the end of his life, neither Spencer nor Haeckel had a role for natural selection: both adopted one form or another of Lamarckian evolutionism. Darwin accepted them among his supporters, preferring to support the common cause of evolutionism rather than divide the camp that had been constructed around his name. Darwin pursued this strategy in his publications and by circulating cautious signs of approval or dissent through his worldwide network of correspondents.

The posthumous appropriation of the Darwinian heritage began immediately after his death in 1882. An early example is *Darwinism*, published in 1886 by Alfred Russel Wallace, the co-discoverer of natural selection. In the name of "pure Darwinism," Wallace reasserted Darwin's earlier, strictly selectionist and adaptationist explanation of evolutionary change against the later attempts by Darwin himself and George John Romanes to make room for other agencies in the production of new species. In 1888, Romanes coined the term "neo-Darwinism" to denote the position of those who regarded selection as the all-sufficient factor in species formation, conceived as a gradual process. The term applied to the German zoologist August Weismann, who devoted considerable energy to

disproving the inheritability of acquired characters admitted by Lamarck, by Darwin himself, by many early Darwinians, and by turn-of-the-century neo-Lamarckians, who combined it with their preferred notion of a direct action of the environment on variation (*see* LAMARCKISM).

In the decades around 1900 the neo-Darwinian view suffered a sharp decline under the attacks of the early geneticists Hugo de Vries and William Bateson, who maintained that evolution proceeded by large, discontinuous variations. By the early 1920s, disagreement on the mechanisms of evolution was the norm among biologists. The situation began to change with the publication in 1930 of R. A. *Fisher's *The Genetical Theory of Natural Selection*. Fisher combined a high regard for Darwin's theory of natural selection with a new, quantitative approach to the study of variation in natural populations. This proved to be one of the key factors that, in the mid-twentieth century, led to the theory known as the evolutionary (or neo-Darwinian) synthesis, in which the previously scattered efforts of experimental geneticists, systematists, and paleontologists aiming to explain evolutionary change merged. With some significant oscillations and controversy, this synthesis is the evolutionary theory adopted by the majority of biologists today (*see* EVOLUTION).

Peter J. Bowler, *The Non-Darwinian Revolution: Reinterpreting a Historical Myth* (1988). Thomas F. Glick, ed., *The Comparative Reception of Darwinism* (1988). Adrian Desmond and James Moore, *Darwin* (1991). Adrian Desmond, *Huxley: The Devil's Disciple* (1994). Ernst Mayr and William B. Provine, eds., *The Evolutionary Synthesis. Perspectives in the Unification of Biology* (1998). Ronald L. Numbers, *Darwinism Comes to America* (1998).

GIULIANO PANCALDI

DEOXYRIBONUCLEIC ACID, universally known by its acronym DNA, became one of the glamour chemicals of the twentieth century following the publication in 1953 of a proposed helical structure that offered an immediate explanation for its genetic role by Francis *Crick and James D. Watson. Before 1950, nucleic acids stood as poor cousins to *proteins, which were then regarded as the best candidates for the material secret of life.

Felix Hoppe-Seyler, founder of the *Zeitschrift für physiologische Chemie* and professor of chemistry at Tübingen and Strassburg, first isolated nucleic acid from pus-forming bacteria in 1869. Albrecht Kossel, who received the Nobel Prize in physiology or medicine in 1910, and Phoebus Aaron Levene independently established the presence in nucleic acids of purines (adenine and guanine) and pyrimidines (cytosine, thymine, and uracil) in a repeating unit containing phosphoric acid and pentose sugars (ribose and deoxyribose).

In the 1930s and 1940s, interest increased among biologists, bacteriologists, biochemists, and physiological chemists in identifying the chemical nature of the gene; chemists favored protein because of the size, complexity, and diversity of its substructure of amino acids. Bacteriologists Oswald T. Avery, Colin M. MacLeod, and Maclyn McCarty of the Rockefeller Institute in New York City produced evidence from 1942 to 1944 that the substance transforming the nonvirulent pneumococcus bacteria into the virulent form is purely deoxyribonucleic acid and not protein. They proposed nucleic acid as the hereditary material of living organisms.

Columbia University biochemist Erwin Chargaff began studying nucleic acids as a result of the Avery group's work. He found that the ratios of adenine to thymine, and of guanine to cytosine in DNA came very close to 1:1 (the "Chargaff rules"). In the meantime, Rockefeller Institute biochemist Alfred Mirsky rejected the Avery interpretation on the ground that nucleic acids in DNA are too simple (just four base pairs) to provide the blueprint for the gene.

Linus *Pauling had begun collaborating with Mirsky on protein structure when Pauling turned to the study of biological molecules in the 1930s. After attending a biology seminar at Caltech in 1953 in which Robley Williams of Berkeley showed electron-microscope photographs of molecules of sodium ribonucleate, Pauling calculated a structure for DNA. Unable to get recent x-ray diffraction photographs from the laboratory of Maurice Wilkins at King's College in London, Pauling based his model-building on photographs taken earlier by William Thomas Astbury. These turned out to be photographs of a mixture of two forms of DNA, so that Pauling was using data which did not correspond to any real single structure.

Pauling's structure, published in early 1953, modeled three intertwined chains, with the phosphates near the fiber axis and the bases on the outside. Using x-ray photographs belonging to Maurice Wilkins and Rosalind Franklin, noting the Chargaff rules, believing DNA to be the hereditary material, and considering that reproduction occurs in pairs, Watson and Crick published their double-helix model in 1953 just after Pauling's triple-helix model. The double-helix model excited instant admiration for the American-British team as well as long-standing puzzlement over how Pauling had missed the structure. Watson and Crick used the idea of complementary of base pairs with the two helices on the outside and the bases on the inside.

Wilkins and Franklin followed up with further confirmation from diffraction data. The double-helix structure was remarkable for the simplicity and elegance with which it revealed how genes are duplicated in heredity. For this revolutionary work Crick, Watson, and Wilkins shared the 1963 Nobel Prize in physiology or medicine. Franklin had died of cancer in 1958.

Robert C. Olby, *The Path to the Double Helix* (1974). James D. Watson, *The Double Helix: A Norton Critical Edition*, ed. Gunther S. Stent (1980). Maclyn McCarty, *The Transforming Principle* (1985). Joseph S. Fruton, *Proteins, Enzymes, Genes: The Interplay of Chemistry and Biology* (1999).

MARY JO NYE

DESCARTES, René (1596–1650), philosopher and mathematician.

Born in La Haye, France, Descartes was educated at the Jesuit College of La Flèche and at the University of Poitiers. In 1618 he journeyed to the Netherlands and Germany, returning in 1622. In the Netherlands he met the natural philosopher

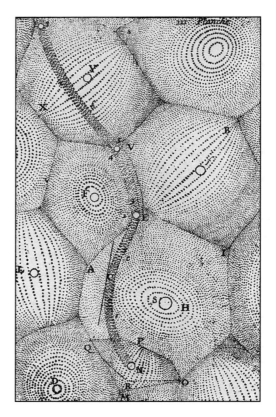

The celestial regions according to Descartes. Stars sit at the centers of large vortices of subtle matter, which sweep around dead captured suns (planets) with their captured suns (moons), while other suns that have been stripped of their vortices roam freely as comets. From René Descartes, *Principia philosophiae* (1644).

Isaac Beeckman, who probably gave him the idea of a universal "mathematics" as the basis of a science of nature. Descartes worked out this idea in Paris between 1625 and 1628; he then worked on analytical geometry and optics, undertaking a general treatise on method (*Regulae ad directionem ingenii*) that was never finished. In 1629 he returned to the Low Countries where he remained until the end of 1649. He died the following year in Stockholm, in the service of Queen Christina.

Descartes was fascinated by the problems of ascertaining natural knowledge. His metaphysics can be seen as an attempt to make a mathematical physics possible while paying tribute to traditional metaphysical and theological concerns like the existence of God and the immateriality of the soul. Later generations regarded his physics, published in its fullest form in his *Principia philosophia* (1644), as a purely hypothetical construction. What lay behind it, however, and what appeared more clearly in Descartes's early work *Le monde* (1633), withdrawn in compliance with the Roman Inquisition's Condemnation of Copernicanism in 1633, was the revolutionary idea that nature (as the object of natural science) is not something given but is whatever can be seen as the interpretation of a theoretical model. Descartes expected his approach to provoke a revolution in the general theory of the universe and also in practical disciplines like medicine. He supposed that medicine could be put on a more solid basis by adopting the ideas that animal bodies are machines and that the explanation of any biological function (breathing, sleeping, digesting, etc.) consists in imagining a mechanical device capable of performing it. Descartes linked this idea with the notion that the most fundamental biological process is the circulation of the blood (which, contrary to William *Harvey, he explained as a process based on a quality of the blood), because it is self-regulatory and the basis of all other functions. His followers took heuristic advantage of the concept, leaving it to anatomy to discover the actual machine. Descartes's natural philosophy at first spread most rapidly among physicians, especially in the Low Countries, then among philosophers.

Descartes's merits are least contested in mathematics, which fascinated him from his days at La Flèche and which may have provided the model for his method. His mathematical work was based on the idea that algebra, to which he gave a simplified notation still in use, made it possible to bring together problems that in their geometrical formulation appear unrelated. He contributed to optics, demonstrating the theoretical advantages of hyperbolic over spherical lenses in telescopes. He

did not manage, however, to develop a controlled, industrial method for producing such lenses. He also gave a demonstration of the law of refraction based on a particulate model of light and an indication of the possibility of mathematical physics in his theory of the *rainbow.

J. F. Scott, *The Scientific Work of René Descartes (1596–1650)* (1952). William R. Shea, *The Magic of Numbers and Motion: The Scientific Career of René Descartes* (1991). Stephen Gaukroger, John Schuster, and John Sutton, eds., *Descartes' Natural Philosophy* (2000). Henk J. M. Bos, *Redefining Geometrical Exactness: Descartes' Transformation of the Early Modern Concept of Construction* (2001).

THEO VERBEEK

DEVELOPMENT. See SCIENTIFIC DEVELOPMENT, THEORIES OF.

DEVELOPMENT, SCIENCE AND. Development, like its older sister progress, has often eluded historians. They may have refuted Boris Hessen's position that science was a necessary component of the flourishing of the bourgeoisie and dismissed as insufficient Susantha Goonatilake's argument that Europe and the United States aborted Third World science. They have been puzzled that Islamic and Chinese knowledge did not raise the standard of living for the people, and they have not agreed about why the outstanding science of the Soviet Union and Russia did not translate into improvements, that is, development. While historians have linked development to science education and practice, the case of Russia demonstrates that these links do not suffice. India has tremendous research and development activity; by 1990 it had over 440 scientific journals; but its people were (and are) poor.

Only recently has it been understood that innovation is a necessary linkage between science and development. An innovation is an element that, when introduced in a social network, reduces the transaction costs between some or all network nodes. The great challenge for China is to link its scientific and technical system to its productive sector through innovation. From this perspective, problems such as the retaining of trained people and the difficulty of establishing traditions of scientific excellence can be understood if not resolved.

According to UNESCO and the World Bank, Third World countries have very low scientific budgets and, consequently, few scientific workers and low productivity even in fields like agriculture, public health, and environment, where innovations are easiest to see. A survey of 1990 indicated that the research productivity of the 441 (full-time equivalent) Senegalese researchers had steadily declined, in agriculture as in other areas.

A country like Nigeria, a giant in African science, has not been able to create fifty scientists per million population. Even so, its scientific community is underutilized.

When not engaged in innovations of local interest, science assumes a role outside the region's productive sphere. Its practitioners tend to link their research results to external networks that legitimize the practitioners in their regional network. They do "developed" science, presented as international and neutral. The elite nature of universities and research centers in developing regions is the institutional reflection of this problem.

When scientists from developing regions encounter barriers to their scientific plans or increased transaction costs, they engage in other activities or move out (*see* BRAIN DRAINS AND PAPERCLIP OPERATIONS). International recognition is hard to achieve for those remaining at home, as revealed by the very small number of Nobel Prizes in science and technical fields awarded to developing countries (*see* ASIA). Those who do win recognition abroad gain an important resource for institutional development back home, as illustrated by the cases of Cesar *Lattes in Brazil, Manuel Sandoval Vallarta in Mexico, and Abdus *Salam in Pakistan. Development costs can be reduced by global recognition, and research continuity improved. The Rockefeller Foundation (*see* PHILANTHROPY) has been instrumental in providing such recognition and funding, as illustrated by the career of Argentina's Bernardo Alberto Houssay, a Nobel laureate.

Politics can increase transaction costs, forcing scientists to depart, as occurred in Russia, China, Chile, Argentina, and Algeria. Developing regions plagued by financial stringency and political unrest cannot retain their scientists. No tradition of scientific practice can be established under these conditions, except in the canonical realms of economic interest—public health, agriculture, and environment. Utilitarianism dominates investment decisions in science. Thus Brazil has public health research institutions that go back a century. The Fundação Oswaldo Cruz in Rio de Janeiro has a long history of vaccine production. The utilitarian nature of the health sector may combine with important local factors to motivate medical research. Diseases can be regionally specific, creating an obvious research niche. Such was the case with Chagas disease in Brazil and Carrión disease in Peru. International medicine did not outshine the efforts of Latin Americans in these areas.

In another example of regional advantage, the hundred-year-old Campinas agronomic institute had its origins in coffee research (1887–1897) done by F. W. Dafert. Export requirements also have stimulated scientific research. Again in Brazil the EMBRAPA (Empresa Brasileira de Pesguisa Agropecuária—the Brazilian Company for Agriculture and Food Research) is one of the world's leading networks in its field. Supported by a network of university agricultural research institutes, EMBRAPA has been responsible for major increases in productivity in the Brazilian food industry. Military and transportation requirements have made the aerospace sector a major contributor to Brazilian exports. Here an important ideological factor came into play: for Brazilians, the inventor of the airplane was Alberto Santos-Dumont and aerospace research and development amount to recovery of a technological tradition. Other developing countries show a similar preference for practical science. India's *Journal of Genetics*, founded in 1910, is among the oldest of its research journals. The *Transactions* of the Bose Research Institute, focused on plant physiology, followed in 1917. Two advanced research centers in molecular biology, TIFR in Bombay and AIIMS in Delhi, were established in 1960.

Environmental issues have recently led to the establishment of research institutions in agriculture and forestry—for instance the Indian Institute of Forest Management founded in Bhopal in 1982 and the Tropical Forest Research Institute, founded in 1988 in Jabalpur—that may provide the infrastructure for broader environmental research. Food production is closely tied to environmental issues. Moreover, the environmental field benefits both from international interests and from concerns of the local political elite. No doubt power relations affect research focus: whereas much substantial research in developing regions addresses rain forests and desertification, little is done on marine pollution, in particular on the destruction of oceanic plankton.

Lack of investment in scientific research in developing regions of the world increases the gap between the haves and have-nots. As the global economy and the information society, both based on science, assert themselves, the divide further widens. Without a science base, some regions, notably African, will confront transaction costs that will forbid innovation. Without science they run the risk of becoming underdeveloped ghettos. The lack of proper attention to environmental and medical sciences has betrayed development in most of the world.

Richard B. Norgaard, *Development Betrayed* (1994). Juan José Saldaña, *Historia Social de las Ciencias en América Latina* (1996). Jacques Gaillard, V. V. Krishna, and Roland Waast, *Scientific Communities in the*

Developing World (1997). Gilbert Rist, The History of Development. From Western Origins to Global Faith (1997). Regis Cabral, "Refining the Cabral-Dahab Science Park Management Paradigm," International Journal of Technology Management 16 (1998): 813–818. Hebe Vessuri, La investigaciòn y desarrollo (I+D) en las universidades de América Latina (1998). Marcos Cueto, The Return of Epidemics: Health and Society in Peru during the Twentieth Century (2001).

REGIS CABRAL

DEVELOPMENTAL MECHANICS (Entwicklungsmechanik) is the name given in the 1890s by the German anatomist Wilhelm Roux to his experimental program for causal analysis of the transformation of fertilized eggs into multicellular organisms. More generally it also denotes the discipline of experimental embryology that flourished from the 1880s to the 1950s.

In France, Laurent Chabry was the first to apply the methods of experimental teratology—the mutilation of an embryo and comparison of its development to that of a normal embryo—to ascidian fertilized eggs. He built his own microtools to observe, manipulate, and kill a single blastomere (one of the segments formed by the division of the ovum) at the two-cell or four-cell stage. In his experiments he obtained half individuals or larvae with anomalies and concluded that each blastomere contained potential parts that were lost if destroyed. In Germany, Wilhelm His, who much improved research techniques by introducing serial microscopic sections, accepted descriptive embryology as essential for understanding the development of individual organs, but criticized its inability to reveal the factors responsible for differentiation. Chabry, His, and their colleagues criticized Ernst Haeckel for studying embryos only for the sake of establishing phylogenetic relationships, and mistrusted his so-called biogenetic law (ontogeny recapitulates phylogeny—the development of the individual follows that of the species).

Eduard Pflüger, professor of physiology at Breslau University, tested whether factors additional to ones internal to the fertilized egg directed development. In 1883 he analyzed the effects of gravitation on the cleavage of fertilized frog eggs and concluded that their development depended on external factors. Gustav Born and Roux, both then also working in Breslau, contested this interpretation. Roux refined Pflüger's experimental techniques and maintained that embryos developed normally even in altered environmental conditions, and that after the second cell division they were self-differentiating.

In 1888 Roux performed his celebrated experiment of puncturing one of the blastomeres of the frog's egg at the two-cell stage with a heated needle. Since in some cases the surviving blastomere developed into a well-formed hemi-embryo, Roux concluded that the experiment confirmed his theory of self-differentiation. According to Roux, every cell of a developing embryo has a capacity for self-determination that resided in the structure of the nucleus. He suggested that each cell division separated structurally differentiated nuclear material. He described the process as the production of a mosaic-work (mosaicism) in which each cell gives rise to different tissues and organs.

By shaking water containing sea urchin eggs, Hans Driesch separated the blastomeres from each other and demonstrated, in 1891, that isolated blastomeres at the two-cell and four-cell stages could give rise to complete but dwarf embryos. Since these results contrasted with Roux's theory of self-differentiation by nuclear division, Driesch argued that the isolated blastomeres could produce all the parts of the organism by responding to the needs of the whole. Conflicting views over the interpretation of these experiments provoked a debate reminiscent of that between preformation and *epigenesis in the eighteenth century. Roux's embryonic development was mechanistic and neo-preformationist, Driesch's vitalistic and neo-epigenetic.

Roux's research program acquired a commanding position in experimental embryology, which by innovating manipulative operations with extirpation, transplantation, and explantation techniques obtained astounding results usually published either in the Archiv für Entwicklungsmechanik der Organismen, a journal founded by Roux in 1893, or in the Journal of Experimental Zoology. The leading figures studying heredity and development at the cellular level were Theodor Boveri, Hans Spemann, Kurt Herbst, and Johannes Holtfreter in Germany; Charles Otis Whitman, Edmund Beecher Wilson, Thomas Hunt Morgan, Edwin Grant Conklin, Charles Manning Child, and Ross Granville Harrison in the United States; and Jean Eugène Bataillon and Étienne Wolff in France. Although experimental results did not give rise to a comprehensive theory concerning the transformation of a fertilized egg into a multicellular organism, they did clarify the external conditions necessary for development and showed that the factors guiding differentiation reside in the nucleus. Transplantation experiments in early embryos have shown the existence of organ-forming areas exhibiting various characteristics (Harrison's morphogenetic fields) and the capacity for integrated inductions of the upper lip of the blastopore (Spemann's organizer).

Frederick Barton Churchill, "From Machine-Theory to Entelechy. Two Studies in Developmental Teleology," Journal of the History of Biology 2 (1969): 165–185. Jane Marion Oppenheimer, "Hans Driesch and the Theory and the Practice of Embryonic Transplantation," *Bulletin of the History of Medicine* 44 (1970): 378–382. Jean-Louis Fischer, "Experimental Embryology in France (1887–1936)," *The International Journal of Developmental Biology* 34 (1990): 11–23. Klaus Sander, *Landmarks in Developmental Biology 1883–1924* (1997).

RENATO G. MAZZOLINI

DIALECTICAL MATERIALISM is a form of traditional philosophical materialism that postulates that all nature can be explained in terms of matter and energy. As a philosophy of science, it posits an objective reality, obedient to natural laws. Knowledge derives from the influence of the material world on the knowing subject, who is also a material being. Dialectical materialism denies divine influence on nature and the existence of a deity. It also opposes the view that there exist any forces or phenomena inaccessible, in principle, to scientific explanation.

In these characteristics dialectical materialism resembles traditional materialism as espoused by the classical Greek atomists or nineteenth-century scientific materialists. However, dialectical materialists differ from most of these earlier materialists in their sharp criticism of reductionism—the belief that all phenomena in nature, including human behavior, can be explained in terms of the simplest interactions of matter. Greek atomists such as Democritus and Leucippus and nineteenth-century materialists such as Ludwig Büchner and Jacob Moleschott were strict reductionists. In contrast, dialectical materialists argue for differentiated "levels of being," such as the physical, the biological, and the social. Phenomena on a higher level of being, for example the social, cannot be exhaustively explained in terms of laws pertaining to a lower level, for example the physical.

The basic assumptions of dialectical materialism are:

• The world is material, and consists of what contemporary science defines as matter-energy.
• The material world forms an interconnected whole.
• Human knowledge is derived from objectively existing reality, both natural and social; being determines consciousness.
• There are no truly static entities in the world.
• The changes in matter-energy occur in accordance with certain overall regularities or laws.
• The laws of the development of matter exist on different levels corresponding to the different subject matters of the sciences; in some cases complex biological and psychological phenomena may not be explicable in terms of elementary physicochemical laws.
• Matter is infinite in its properties; human knowledge will never be complete.

• The motion present in the world requires no external mover.
• Human knowledge grows with time, as illustrated by the increasing success of applied science and by the accumulation of relative—not absolute—truths.

Dialectical materialism was most fully developed and dogmatically applied in the Soviet Union. Some Soviet ideologists used dialectical materialism to attack *relativity physics and *quantum mechanics, while other Soviet scientists, notably physicist Vladimir Fok, used it to defend those theories. Some scientists in the United States and Great Britain, including John Desmond Bernal and Hermann Joseph Muller, viewed dialectical materialism favorably, but the majority of scientists outside the Soviet bloc either criticized or ignored it.

See also CAUSALITY; VAVILOV, NIKOLAI IVANOVICH, AND SERGEI IVANOVICH VAVILOV.

Loren R. Graham, *Science and Philosophy in the Soviet Union* (1972). Helena Sheehan, *Marxism and the Philosophy of Science: A Critical History* (1993). Ernst Mayr, "Roots of Dialectical Materialism," in *Na Perelome: Sovetskaia Biologiaa v 20–30kh Godakh*, ed. E. I. Kolchinskii (1997): 12–17.

LOREN R. GRAHAM

DIET. See FOOD PRESERVATION; NUTRITION.

DIFFUSION IN THE EAST. The diffusion of Western science to East Asia historically centers on Japan. Diffusion at the individual level was paramount during the Edo period (1600–1867); at the systemic level, during the Meiji era (1868–1912).

The Edo Period: Diffusion at the Individual Level

During the Edo period, systemic religious prohibitions intended as a safeguard against Christianity severely restricted access to books in Japan. What knowledge of European science the Japanese obtained in the seventeenth century came from books translated into Chinese by the Jesuits, who used the excellence of European astronomy to persuade Chinese audiences of the superiority of the Christian religion that lay behind it. These books, beginning with those of Matteo Ricci, promoted both Christianity and European astronomy. Japanese on the receiving end could not easily distinguish between the two. One work, the *Huan'youquan* (An Explanation of the World), imported in the late seventeenth century, was considered to be so permeated by both religion and astronomy that it prompted the prohibition of all writings by Jesuits.

The shogunal house tried to monopolize natural knowledge. Yoshimune Tokugawa, the eighth shogun, had his court mathematician, Genkei Nakane, consult prohibited works by Jesuits. The shogun was surprised to learn that Chinese calendrical astronomy had already made European calendrical astronomy part of its tradition. Since Chinese astronomy, the source of inspiration for the Japanese, had already moved in a Western direction, Japan would have to do the same. Yoshimune relaxed the book prohibition and directed that subordinates study European science.

This relaxed policy lasted to the end of Tokugawa Shogunate's regime in the mid-nineteenth century, but the abandonment of restrictions on books was never announced publicly. Shogunal astronomers could consult Jesuit writings on astronomy; but because their calendrical reforms were not publicized, their knowledge never reached beyond the limited circle of government specialists.

The loosening of restrictions extended to language studies. Trade with the Dutch led to a specialized translation industry centered in Nagasaki. The private nature of this business permitted translation not only of documents concerned with practical, trade-related matters such as navigation but also with topics in natural philosophy. In the latter half of the eighteenth century the work of the Nagasaki translators facilitated the introduction of European science, including Copernican and Newtonian thought, to Japan.

Ryoei Motoki first came into contact with the Copernican theory in an article about the trial of Galileo featured in a history of astronomy attached to a compendium of maps. Omitting the section related to the forbidden teachings of Christianity, he still hesitated over introducing Copernican theory before moving ahead. Since translation did not involve publication, he distributed a manuscript copy among a small group of scholars. The manuscript, never printed, was later distributed by Kōkan Shiba, an itinerant painter, man of letters, and author of books on Copernican theory. Except among a few Buddhist scholars, no opposition developed to the heliocentric, or—as it became known in Japan—"moving earth," theory.

Tadao Shizuki, an aspiring philosopher, resigned his position as an official translator to devote himself to the translation of books, one of which, *Rekisho Shinsho*, introduced Newtonian mechanics. Struggling with such problems as the absence in East Asian thought of the notion of atoms, Shizuki tried to explain Newtonian ideas on the basis of East Asian natural philosophy, especially theories of divination or yin-yang. He conceived of gravitation not only in terms of

Newton's inverse square law but also on the basis of an inverse-cube or inverse fourth-power law; he tried to explain chemical affinity and biological phenomena in the same way.

There also appeared in the late eighteenth century a group of physicians who studied Dutch and presented European science beginning with anatomy. Because of the shogunal restrictions, they generally avoided spreading Western knowledge publicly, but they traveled freely exchanging information with each other. By the end of the century some of these physicians and scholars had concluded that European science was superior to traditional Chinese science.

In the early 1840s the Opium War between Great Britain and China demonstrated the superiority of European military technology to that of East Asia. The arrival of Matthew Perry's fleet in 1853 in the fortified Uraga harbor effectively impressed this fact on Japan's ruling samurai class. The focus of Dutch studies shifted from medicine to military science and interest broadened to works in English, French, and other Western languages. Where Western science had previously been studied at the individual level, military science brought attempts at systematization that bore fruit after the Meiji Restoration of 1868.

The Meiji Period: Diffusion at the Systemic Level

The Meiji government recognized that industrial technology underpinned European military superiority. From the 1870s the new government set forth a policy of Westernization or modernization. In the Edo period, the Japanese had lacked the notion of "progress," a characteristic feature of modern Western science. However, promoted by Yukichi Fukuzawa and other Western-leaning journalists impressed by qualitative differences between Japanese and Western military technologies, this idea gradually permeated all of Japanese society.

In 1872 following the adoption of a new administrative structure, an organized effort to import Western science on a national level began under the slogans of *fukoku kyohei* (enrich the nation, strengthen the military) and *shokusan kogyo* (enhance commerce and build up industry). Foreigners came in as teachers and technicians, while Japanese went to Europe and the United States as students; upon returning home, these students replaced the foreigners. At the systemic level, this importation of Western science was completed about 1886. Until then, every work site and government office had pursued its own strategy; subsequently, most importation activities took place under the sponsorship of the Ministry of Education and the imperial universities.

In the Edo period Western science had been referred to as *kyuri* (investigation of principles), emphasizing an aspect of Western scientific thought lacking in the East Asian tradition. In Meiji, at a time when *kyuri* referred only to specialized branches of physics, Western science came to be known as *kagaku*, meaning the study of learning divided into its specialized parts. Opinion makers seized on a characteristic feature of late-nineteenth-century Western science, presenting it not as natural philosophy but as a set of specialized studies. Young men from samurai families who had lost their stipends and privileges under Meiji rule often chose careers in government works rooted in Western science, bringing new social status to the science and engineering fields.

What Japanese of the Meiji period wanted more than Western science, however, was modern Western technology, especially military technology. Consequently, when the Imperial University was founded in 1886 it included a faculty of engineering with departments of ordnance, explosives, and, at the Navy's insistence, shipbuilding. These departments, reflecting a commitment to strengthening the military, had no precedent in Europe or the United States. The label "academic discipline" was affixed to all the important Western technologies, indicating the high esteem in which they were held; the English term "strength of materials" was reconceived as "materials science." Thus was the concept of engineering established in Japan.

Technology Transfer

A Westernization scheme based on blueprints prepared by the Meiji government working from the top down could not be implemented without the involvement of the private sector. Nearly all of the government's so-called model factories failed and, in the 1880s, came under the control of private entrepreneurs. Within the private sphere, artisans took the lead in implementing and diffusing Western science and technology. During the early years of Meiji, the government often sponsored exhibitions designed to stimulate domestic industry and promote inventions. Tokimune Gaun's 1876 invention of a simple spinning machine made him famous; but it had no patent protection. Although the government had some concern that a patent system would hinder the diffusion of science and technology, the Ministry of Agriculture and Commerce showed a great sympathy toward penniless inventors by creating patent protection. In 1884, the International Bureau of Exhibitions, established to prevent the theft of inventions exhibited at multilateral events, invited Japan to join the International Convention for the Protection of Industrial Property. Membership offered little advantage to a late-developing country like Japan; but adherence to the convention became a quid pro quo for securing revision of certain unequal treaties between Japan and the Western powers. However, the treaties were not revised at that time; and the Bureau of Patents that Japan had set up in 1885 as a step toward membership in the international organization protected only domestic inventions. Sakichi Toyoda, who founded Toyoda Motors, exploited the gap in the world system by obtaining a Japanese patent on foreign technology. For a decade before the revision of the treaties, Japan spread the fruits of Western science and technology within the private sector and carried out an industrial revolution without being troubled by international patent law.

An Era of Greater Independence

While the diffusion of science was primarily an individual matter in the Edo period and a systemic phenomenon in Meiji, in the twentieth century science spread significantly at both levels. Government provided a stable infrastructure and predictable, if parsimonious, funding. The regime had created just two so-called imperial universities before 1900—Tokyo (1886) and Kyoto (1897). In the first two decades of the new century it established three more—Tohoku (1906), Kyushu (1911), and Hokkaido (1918). Private sector initiatives and those of individual scientists, however, remained the motive force for change, just as in other developed nations. The physical chemist Kotaro Honda introduced Gustav Tammann's Göttingen paradigm of research on metals at Tohoku University; Katsusaburo Yamagiwa imported Rudolf Virchow's Berlin research program of cancer studies to Tokyo University.

With the onset of World War I, Japanese scientists could no longer study in Germany, now an adversary. Many switched to Great Britain and a smaller contingent went to the United States. Though Germany again attracted some Japanese after the end of hostilities, the shift to English-speaking nations proved to be long-term. The war also forced greater independence on the Japanese: industrialists had to manufacture products—synthetic dyes, electric turbines, certain pharmaceuticals—that had previously been imported; the elites greatly increased their investments in research infrastructure, leading to the establishment of the Research Institute for Physics and Chemistry, the Aeronautics Research Institute, and other facilities. Rising militarism in the 1930s intensified these effects. In 1932 the government founded the Japan Society for the Promotion of Science. In the same decade new imperial universities were established at Osaka and Nagoya;

Kyushu University got a faculty of science separate from its faculty of engineering.

Contributions from individual scientists in the inter-war years also showed a greater degree of independence. In February 1935 Hideki Yukawa, who had never studied or even traveled abroad, published the meson particle theory that would win him Japan's first Nobel Prize. The country's second Nobel laureate, Shin'ichirō Tomonaga, studied with Werner Heisenberg at Leipzig for two years in the late 1930s; but the third and fourth laureates, Reona Esaki and Ken'ichi Fukui, followed the Yukawa model of working strictly in Japan. (After completing his prize-winning work, Esaki moved to the United States in 1960).

China and East and Southeast Asia

The diffusion of science encountered far more obstacles in nineteenth-century China and the adjacent nations of East and Southeast Asia than it did in Japan. Despite a common legacy of Confucianism, which invested formal education with a high moral value, educational opportunities were very limited under indigenous leadership. Moreover, the arrival or reassertion of European influence around the region after the Opium War (1839–1842) did less to change this in the short term than is generally recognized. The Europeans did introduce modern science and technology; but the support they received in this effort, prior to the end of the nineteenth century, was far less generous than it was in Japan. China's Qing Dynasty (1644–1911) was controlled by a non-native ethnic group, the Manchus, and with the exception of a few officials was either indifferent or actively hostile to Western knowledge. After the Opium War, the prevailing attitude became one of passive toleration but rarely took the form of official activism. Official indifference and passivity meant that much of the work of diffusing modern science fell, far more than in Japan, to Western missionaries and other lay teachers. John Fryer, an English layman, was largely responsible for introducing modern chemistry to China in the late nineteenth century and coining appropriate Chinese vocabulary for it. In 1860 the Qing government did establish several facilities, including the Kiangnan Arsenal, through which Western science could penetrate China. There was no further movement by the regime until 1898 when Peking University was founded during a brief flurry of reformist activity. Much of the work of diffusing scientific knowledge to and within China thus remained a task of the twentieth century (see ASIA).

The situation in Korea, Taiwan, and Vietnam was profoundly influenced by the presence of colonial regimes. Because France was a leading scientific power, French influence in Vietnam should in principle have produced better conditions there than in the other countries of the region. However, local French colonial authorities lacked both interest and imagination and the consolidation of French rule after 1885 produced few real improvements. Modern-style education for Vietnamese above the elementary level was severely limited. As late as 1940 Vietnam had only fourteen secondary schools, and there was just one university (founded in 1917). Only after 1930 did Vietnamese nationals as a practical matter have an opportunity to study modern science. Notwithstanding the oppressions of Japanese rule in Taiwan (from 1895) and Korea (from 1910), conditions for the diffusion of science were much better in those countries. In the early 1920s Japan established in each an imperial university with a full range of programs in technical and nontechnical fields; by following the appropriate courses of study, Koreans and Taiwan Chinese could and did pursue careers in science domestically, in Japan, and elsewhere.

Recent Developments

With military defeat in World War II and subsequent disarmament, Japanese science and technology—once dominated by military concerns—shifted to market-driven priorities. During the war, Japan had been isolated from the rest of world and had fallen behind in research; in the 1950s it compensated by importing technology. Long-term purchases of foreign, principally American, technology pushed development faster than independent local efforts would have. At first, low-cost labor allowed Japan to seize markets from the United States in textiles, shipbuilding, home appliances, and steel; high-quality production methods based on QC (quality control) circles later accomplished the same thing. These trends attracted world attention in the 1980s. In the 1970s the newly industrializing countries (NICs) of South Korea, Taiwan, Singapore, and Hong Kong had begun importing Japanese technology on a broad scale. Beginning in the 1980s, Southeast Asian countries did the same, joined by China in the 1990s. These technology transfers led to rising incomes around the region but movement of the locus of production away from Japan led to a technological hollowing out there. Pursued as they are by other Asian countries, Japan and the United States both recognized the key status of scientific and technical research and development. The Japanese government has called for *kagaku gijutsu rikkoku* (national reliance on science and technology). Since the early 1990s' collapse of the Cold War and its big-budget military technology demands, the scientific and technological competition between the United States—with its superior infrastructure—and Japan has intensified in certain respects as they proceed along more similar paths. Other nations in East and Southeast Asia may eventually play a significant role in this competition as well.

Adrian Arthur Bennett, *John Fryer: The Introduction of Western Science and Technology into Nineteenth-Century China* (1967). Shigeru Nakayama, *A History of Japanese Astronomy: Chinese Background and Western Impact* (1969). Shigeru Nakayama, "Japanese Scientific Thought," in *Dictionary of Scientific Biography 15*, supplement 1, ed. C. C. Gillespie (1978): 728–758. Shigeru Nakayama, *Academic and Scientific Traditions in China, Japan, and the West*, trans. Jerry Dusenbury (1984). James R. Bartholomew, *The Formation of Science in Japan: Building a Research Tradition* (1989). James Reardon-Anderson, *The Study of Change: Chemistry in China, 1840–1949* (1991).

SHIGERU NAKAYAMA

DIGITAL DEVICES. See ANALOG / DIGITAL DEVICES.

DIRAC, Paul Adrien Maurice (1902–1984), British physicist, one of the pioneers of relativistic *quantum physics, *quantum electrodynamics, and *quantum field theory.

All the major developments in quantum field theory in the 1930s and 1940s have as their point of departure some work of Dirac's. Werner *Heisenberg characterized Dirac's postulation of antimatter in 1931 as "the most decisive discovery in connection with the properties or the nature of elementary particles." And all this does not take into account Dirac's famous exposition *The Principles of Quantum Mechanics* (1930), the guide to several generations of physicists.

Dirac was born on 8 August 1902 in Bristol, England, the son of a Swiss father, Charles Adrien Ladislas Dirac, and an English mother, Florence Hannah Holten. Charles Dirac emigrated to England in 1888 and in 1896 obtained a position teaching French in the Merchant Venturers' Technical College (M.V.) at Bristol, Paul's mother's home town. Paul had an older brother and a younger sister. Dirac's father was a strong-willed, dominating personality. He demanded that his children speak to him in French, threatening to punish them for grammatical errors. Dirac traced his legendary taciturnity to this linguistic program. Because the rest of the family could not meet Charles's standards, they ate dinner in the kitchen, while Paul dined with his father—Paul never saw his parents eat a meal together, nor could he recall anyone ever making a social call at the house. Paul's brother Reginald wanted to become a physician, but Charles forced him to study mechanical engineering at Bristol. He obtained a third class degree, accepted a position as a draftsman with an engineering firm, and

committed suicide at 24. Thereafter Paul had very little interaction with his father.

Dirac began his secondary education in 1914 at the M.V. His schoolmates recalled him as introverted, reticent, and aloof. He followed Reginald's footsteps to Bristol University and electrical engineering. Although he preferred mathematics, he thought that the only way to earn a living as a mathematician was as a school teacher, a prospect that did not appeal to him. He graduated from Bristol with first class honors in 1921. Unable to find a suitable engineering position owing to the economic recession that then gripped England, Dirac accepted free tuition at Bristol to study mathematics. An 1851 Exhibition Studentship and a grant from the Department of Scientific and Industrial Research led him to Cambridge as a research student in 1923.

At Cambridge, Dirac came under the influence of Ralph Howard Fowler and Arthur Eddington. He attended Fowler's lectures on the old quantum theory and became his research student. Within six months of his arrival, Dirac wrote two papers on statistical mechanics and in May 1924 he submitted his first paper on quantum problems.

Upon first reading, Dirac failed to see the significance of Heisenberg's first paper on quantum mechanics, which he saw in September 1925 in the page proof that Heisenberg had sent Fowler. A week later Dirac realized that the non-commuting quantities that bothered him made up the essence of Heisenberg's new approach. It occurred to him to try to connect the commutator of two non-commuting dynamical variables with their Poisson brackets (a form of the equations of classical mechanics that resembled the commutator). His success convinced him that the new quantum mechanics represented an extension of classical physics rather than, as Heisenberg had argued, a break with it. He stated his position succinctly in the paper he submitted to the *Proceedings of the Royal Society* early in 1926: "Only one basic assumption of classical theory is false ... the laws of classical mechanics must be generalized when applied to atomic systems, the generalization being that the commutative law of multiplication, as applied to dynamical variables, is to be replaced by certain quantum conditions."

Dirac's contributions to the development of quantum mechanics were immediately acknowledged as central and by 1927 he had become a member of the core set that judged theoretical advances. In 1930 he was elected a fellow of the Royal Society, and in 1932 Lucasian Professor of Mathematics in Cambridge. In 1933 he and Erwin Schrödinger shared the Nobel Prize in physics for their contributions to the developments of quantum mechanics. In 1937 Dirac married Margit Balasz, the sister of Eugene Wigner, another Nobel Prize–winning mathematical physicist.

During World War II Dirac taught at Cambridge while carrying out research on uranium isotope separation and on atomic bomb design. After the war he was often a visitor at the Institute for Advanced Study in Princeton. In 1969 he accepted a research professorship at Florida State University in Tallahassee, which he held until his death on October 20, 1984.

R. H. Dalitz and R. Peierls, "Paul Adrien Maurice Dirac" in *Biographical Memoirs of Fellows of the Royal Society* 32 (1986): 139–185. H. S. Kragh, *Dirac. A Scientific Biography* (1990).

SILVAN S. SCHWEBER

DISCIPLINES. Discipline refers to a demarcated body of knowledge identified as a separate science (physics, chemistry, biology, mathematics, etc.) with its own methods, problems, and practitioners. The disciplines represent one stage in the historical evolution of organizational schemes for knowledge and the social systems that support it. The modern disciplines, which date from the early nineteenth century, possess institutionalized systems of training, recruitment, and professional behavior as well as specialized forms of association and communication (professional societies and journals). Disciplinary training cultivates the skill-based activities and critical perspectives central to professionalization. Members of different disciplines compete among each other for funding and academic turf. To exist disciplines require considerable political, economic, social, and cultural resources.

Until the seventeenth century the scholastic curriculum organized knowledge into the seven liberal arts (astronomy, music, geometry, arithmetic, rhetoric, grammar, and dialectic) and philosophy (logic, ethics, and physics). Early scientific societies, the Royal Society of London and the Academie des Sciénces in Paris, eschewed scholastic divisions in favor of a broad and sometimes eclectic organization of natural knowledge; they also elevated the status of practical knowledge, which was considered inferior in the scholastic framework. Although specialized textbooks proliferated in the eighteenth century, natural philosophy, comprehensive encyclopedias, and other broad ways of assembling knowledge about the natural world persisted. By the end of the eighteenth century, institutional questions about the organization and presentation of knowledge, including the status of the "lower" or philosophical faculty in comparison to the "higher" faculties of law, theology, and medicine, began to weaken traditional classifications of the

sciences, especially in the German universities where debates ensued over which branches of knowledge had the right to certify truth. In the context of intense reform of the school and the university, disciplines emerged, and with them differentiated roles for scholars, soon to be called scientists, took shape. The modern disciplines appeared first in the natural sciences, followed by the evolution of moral philosophy into the social sciences; the humanities became everything that remained. Disciplinary knowledge continued to evolve and divide over the course of the nineteenth century. By 1900, interdisciplinary combinations and subdisciplinary specialties had emerged in response to stagnation, enlargement, or difficulties of the classical scientific disciplines.

The widespread success of the disciplines spawned criticism of the restrictive ways of thinking disciplinary knowledge imposed. Disciplinary knowledge not only mechanized creativity, but continued to exist and thrive in large part because it could be transmitted in instruction. The German philosopher Georg Wilhelm Friedrich Hegel was among the first to challenge the rigidity of disciplinary scientific knowledge in the preface to his *Phenomenology of the Spirit* (1807), where he criticized its conceptual ossification and advocated more fluid intellectual practices. By midcentury disciplinary knowledge was the subject of satire in works like Gustave Flaubert's *Bouvard and Pécuchet* and his *Dictionary of Received Ideas*. Flaubert cast a more positive light upon ways of thinking associated with the arts, including fiction. At the end of the century Friedrich Nietzsche found the intellectual practices of the disciplines so abusive that, in essays like *On the Advantage and Disadvantage of History for Life*, he likened disciplinary knowledge to death. He preferred metaphorical ways of thinking that released concepts from their hardened shells. These examples vividly illustrate the historical wedge that disciplinary knowledge placed between science and art. No wonder that in the nineteenth century the appellation "genius" became more closely associated with art (and eccentricity) than with science (and conformity).

Although Émile Durkheim had addressed the social meaning of classification systems in 1912, it was not until the 1930s that sociologists of knowledge like Karl Mannheim pointed out that disciplines, like other organizational systems, represent ideological interests. In the 1970s the French philosopher Michel Foucault exposed this nearly hidden trait of disciplines by demonstrating the ways in which all academic disciplines controlled discursive practices and hence contributed to the political constraints of modernity. Building on Durkheim and others,

Rudolf Stichweh has considered disciplines not only as "knowledge assembled to the taught," intimately tied to educational systems and their mechanisms of transmission and training, but also as knowledge classification systems responsive to social change. Disciplinary differentiation around 1800, he has claimed, occurred not only alongside social differentiation (the emergence of the middle class), but also because of it.

Since 1968 postmodern thinkers have drawn attention to accelerated shifts in disciplinary knowledge. Advanced systems of communication, especially the personal computer, have made information in principle available to everyone. According to Jean-François Lyotard, the former privileged position of disciplines and those who practiced them are challenged by the democratization of access to them. He cites the popularity of interdisciplinarity as a response to the diluted status of the disciplines. Because interdisciplinarity cultivates forms of knowledge and ways of thinking that place a premium upon imaginative combinations of research techniques generally shunned by disciplinary practitioners, the postmodern production of knowledge emphasizes "performativity"—preoccupation with creating novelty in the techniques of research—over the achievement of specific goals. Interdisciplinarity thus attacks the balkanization of knowledge sustained and jealously guarded by the disciplines, especially in academic departments. Unlike earlier disciplines, which were intracompetitive but inherently stable, interdisciplinary forms of knowledge promote and feed on instabilities. Lyotard's interpretation resonates with broad contemporary movements that stem from the growing pressure for multiculturalism (a form of interdisciplinarity) in traditional centers of learning concurrent with globalization and new patterns in demographic migration. Just as the modern disciplines emerged during a period of profound social change, so interdisciplinarity seems to be taking hold during a period of similar change, albeit on a global scale.

While a convenient way of approaching science's past, disciplinary history can fall short of capturing science's role in broader historical processes. Disciplinary history remains superior, however, in rendering the intellectual and social activities of a science.

See also SCHOOLS, RESEARCH; TEXTBOOK; UNIVERSITY.

Owen Hannaway, *The Chemists and the Word: The Didactic Origins of Chemistry* (1975). Michel Foucault, *Discipline and Punish* (1978). Jan Goldstein, "Foucault among the Sociologists: The 'Disciplines' and the History of the Professions," *History and Theory* 23 (1984): 170–192. Rudolf Stichweh, *Zur Entstehung des modernen Systems wissenchaftlicher Disziplinen: Physik in Deutschland, 1740–1890* (1984). David R. Shumway and Ellen Messer-Davidow, "Disciplinarity: An Introduction," *Poetics Today* 12 (Summer 1991): 201–225. Timothy Lenoir, *Instituting Science: The Cultural Production of Scientific Disciplines* (1997). Peter Burke, *A Social History of Knowledge* (2000).

KATHRYN OLESKO

DISCOVERY. Scientists base their enterprise on the notion of discovery. Their work aims at the production of new knowledge about nature, which may consist of recognition or identification of a new phenomenon—a new chemical element, such as oxygen, or a new sort of radiation, such as X rays or radioactivity—or a new conceptual approach to explain existing phenomena, such as quantum theory or relativity. The accumulation of discoveries over time, by this account, gives scientists a more complete and accurate understanding of the natural world. The value and reward systems of science reflect this conception of discovery: the goal of every scientist is to make an important discovery, which will elicit the respect of peers and corresponding compensation in career advancement.

The centrality of discoveries makes them a prime subject for the history of science. The study of particular discoveries, however, demonstrates that they are often extended in place and time and difficult to attribute to a single person. The history of modern science thus poses a deep challenge to the concept of unitary discovery that underlies the scientific enterprise.

The discovery of oxygen in the late eighteenth century exemplifies the difficulties with the concept of discovery. There are four primary candidates for credit for the discovery. In early 1774 Pierre Bayen, an apothecary in the French army, found that red precipitate of mercury (what we would designate HgO) yielded a gas when heated, which he identified as the fixed air (CO_2) familiar from the work of Joseph Black. Soon afterward Joseph Priestley repeated Bayen's experiment, probably independently, saw that the emitted air supported combustion, and thus identified it as nitrous air (N_2O). Antoine-Laurent Lavoisier refined Priestley's experiment and in early 1775 concluded that the product was regular atmospheric air. Priestley then determined that the gas was in fact better than common air: candles burned in it "with an amazing strength" and a mouse stuck in a bell jar of it not only survived for an hour but "was taken out quite vigorous." In line with the prevailing phlogiston theory of combustion, Priestley called the gas "dephlogisticated air." Lavoisier then returned to the experiment and within a year decided that the gas was a separate component, "eminently respirable," of atmospheric air, which had previously been considered homogeneous. Lavoisier believed that the gas carried the principle of acidity, so he called it "oxygen," or "acid former"; he went on to systematize chemistry, including the introduction of modern nomenclature. In the meantime, Carl Scheele in Sweden had also isolated a gas that he called "fire air" for its influence on candle flames, and postulated that fire air was a part of the atmospheric air.

So, who discovered oxygen and when? What constituted the discovery? Bayen first isolated it but thought it was something else; Priestley's experiments identified its chemical significance; Lavoisier incorporated the gas in a new theory of combustion and gave it its modern name, but mistook it for the source of acidity and hence misnamed it; and Scheele pursued a similar line of reasoning but published his results late. One might also attribute the discovery to succeeding generations of chemists, who identified oxygen's atomic weight, its atomic and nuclear structure, and the molecular formula and structure of the gas.

Similar complexities emerge in accounts of the discoveries of the planet Uranus, X rays, and quantum theory, all of which are usually attributed to an individual at a particular point in time, respectively, William Herschel in 1781, Wilhelm Conrad Röntgen in 1896, and Max Planck in 1900. These examples share a characteristic: all were unexpected within prevailing theories. The historian Thomas S. Kuhn has distinguished between such anomalies that challenge conceptual frameworks, and discoveries that fit into and were predicted by existing theories. New concepts that challenge prevailing views require more time and more work to accommodate in a new framework and to convince scientists to relinquish the old one; the product of this drawn-out process, according to Kuhn, is a scientific revolution. The second case, what Kuhn called normal science, consists of puzzle-solving within the existing framework: deducing theoretical implications from or detecting experimental phenomena predicted by current theory. These puzzles, such as the isolation of chemical elements at expected places in the periodic table, or detection of elementary particles predicted by quantum field theory, do not challenge the underlying conceptual framework and thus, according to Kuhn, may be localized in time and place.

Even in these cases of expected discoveries, however, scientists (or their historians) may not agree on what constitutes convincing evidence for a discovery. Normal science still suffers from priority disputes, those contentious debates

that pinpoint the problem of discovery—for instance, the discovery of element number 72, eventually called hafnium, in 1922, or of elements 99 and 100, einsteinium and fermium, in the early 1950s. Similar episodes in the history of biomedical research, exemplified by the discoveries of insulin and sulfa drugs, indicate that such distended discoveries are not limited to a particular discipline or an institutional setting. The emergence of new scientific institutions has further complicated the definition of discovery, especially in the cooperative research programs in industrial and national laboratories of the twentieth century, where scientists from diverse disciplines collaborate on particular problems and whose publications may list tens or hundreds of authors.

The difficulty faced by Nobel Prize committees in apportioning awards, and the sometimes bitter disputes that follow their decisions, highlights the difficulty in reconciling the concept of unitary discovery with the complicating details of history. To take just one example, the discovery of the antiproton at the University of California at Berkeley in 1955 led to a Nobel Prize shared by two physicists, although several others made crucial contributions to the work and one would later file a legal suit, unsuccessfully, for a portion of the award. The Nobel Prizes also illuminate a distinction between discovery and invention. The Nobel statutes allowed prizes for "discovery or invention in the field of physics," and Nobel committees have on occasion awarded prizes as much for invention as for discovery, as in the cases of wireless telegraphy, color photography, the cloud chamber, the cyclotron, and holography. But the Nobel Prize in physics has usually gone not to inventors of devices but creators of new phenomena or theories, reflecting the notion that an invention is an artificial construct, but what one discovers was already present in nature. The Nobel Prize in physiology or medicine, and, to a lesser degree, the Nobel Prize in chemistry, have rewarded inventions more frequently, reflecting their proximity to application.

Thomas S. Kuhn, "The Historical Structure of Scientific Discovery," in *The Essential Tension: Selected Studies in Scientific Tradition and Change*, ed. Kuhn (1977): 165–177. Thomas Nickles, ed., *Scientific Discovery: Case Studies* (1980). Michael Bliss, *The Discovery of Insulin* (1982). Peter Galison, *How Experiments End* (1987).

PETER J. WESTWICK

DISSECTION. Dissection is the separation by surgical means of the parts of an animal or human. Although it encompasses vivisection, dissection usually is confined to the division of the dead.

As a method of research, dissection goes back at least as far as Aristotle in the fourth century

B.C. The Roman physician Galen in the second century A.D. was the most skilled practitioner before the modern era, and his dissection manual, *On Anatomical Procedures*, influenced many Renaissance anatomists. Autopsies for forensic purposes began in Europe at the end of the thirteenth century.

While dissection of animals and of an infrequent human cadaver became a standard feature of European medical schools by 1500, modern dissection as a method of research began with the work of Andreas *Vesalius in the 1530s. His *On the Fabric of the Human Body* (1543) decisively established dissection of the human body (supplemented by dissection and vivisection of animals) as the best method of determining the structure, and to some extent the function, of the body. Dissection retained a major role in anatomical, physiological, forensic, and disease research until the twentieth century.

William *Harvey made extensive use of animal and human dissection and animal vivisection in his discovery of the circulation of the *blood, published in 1628, and many researchers in the seventeenth century followed his techniques. However, the systematic use of human autopsy to diagnose and identify diseases only occurred in the work of Giovanni Battista Morgagni a century later. By the end of the eighteenth century, the autopsy was recognized as a central feature of medical research, with the work in particular of Matthew Baillie and John Hunter in Britain and Jean-Nicolas Corvisart and Marie-François-Xavier Bichat in France.

Pathological anatomy became institutionalized in Vienna under the leadership of Karl von Rokitansky between 1830 and the 1870s. In Berlin in the second half of the century, Rudolf Virchow combined dissection with cell theory (*see* CYTOLOGY) to create a modern medical science focused on the autopsy, and his model was adopted in North America by the influential physician William Osler. While pathology thus became closely linked with the autopsy, physiological research abandoned work on the human cadaver in favor of live animals.

The autopsy maintained its importance in the classification and diagnosis of disease between 1850 and 1950, but has since declined precipitously. In the United States, one out of every two hospital deaths attracted an autopsy in 1950; by 1990 the number was one in seven. Most practitioners believe that modern technologies have superseded the autopsy, which now finds its almost exclusive use in forensic laboratories and classrooms.

See also ANIMAL CARE AND EXPERIMENTATION.

K. E. Rothschuh, *History of Physiology*, trans. G. B. Risse (1973). R. B. Hill and R. E. Anderson, *The Autopsy: Medical Practice and Public Policy* (1988).

ANITA GUERRINI

DISSERTATION. The academic dissertation in the United Kingdom and France is usually known as a "thesis"—an Aristotelian term signifying an intellectual position maintained in verbal debate or in writing. Medieval philosophers in the Latin West used the term *dissertare*, "to debate." Later the sense extended to printed books, as seen in Johannes Kepler's opinion of Galileo's telescopic observations, *Dissertatio cum nuncio sidereo* (1610), and in the three important dissertations on the progress of philosophy, mathematics, and the physical sciences that accompanied the fourth edition of the *Encyclopaedia Britannica* (1816–1824). In European universities from the twelfth to the sixteenth centuries the academic thesis was a scholarly initiation rite, in which an outline of arguments in Latin for and against received and new opinions was presented—particularly those of the professor promoting the student. With the introduction of printing, and as original observations and experiments grew more common, eight- to sixteen-page dissertations were often abstracted and collected together from different European universities and became an important element in developing the scientific periodical, as in Albrecht von Haller's *Auserlesene Chirurgische Disputationen* (1777–1787).

As the secondary school curriculum broadened in the eighteenth century, the *Abitur* and *baccalauréat* effectively absorbed the bachelor's degree and allowed the doctorate of philosophy to become a mark of graduation in continental universities. German students usually had the option of either writing a thesis or paying a fine for this degree; the former was usually chosen by students who aspired to postgraduate studies and academic careers. Until the 1800s, the higher examined, or by thesis, doctorates (D.Phil.) offered by Oxford and Cambridge could be taken only by their own graduates. For this reason, and because of religious tests, British students (particularly chemists) in search of research experience tended to spend a year at German universities. For such foreign students the Ph.D. awarded for a short dissertation on an experimental or theoretical subject became a postgraduate qualification.

By the 1880s a doctoral degree that included writing a dissertation, formerly only common among physicians, was the expected qualification for anyone aspiring to an academic career in science. Evidence of original research was demanded in the higher doctorate required for university teaching purposes in France and Germany (*Doctorat, Habilitationsschrift*) and this encouraged the development of graduate research schools. American students in search of European enlightenment also spent time in Europe after graduation before the chemist Ira Remsen copied the German model of a graduate school at Johns Hopkins University in 1876. The British scientific community strongly advocated research doctoral degrees on the German dissertation model, but the innovation spread only slowly. Finally the economic and political necessity of colonial cohesion, and the British government's concern to wean British and American students from German universities after World War I, persuaded British universities to introduce the Ph.D/D.Phil. degrees. Since the 1920s, the dissertation has become the central feature of postgraduate education in the sciences and humanities worldwide. In all cases it involves a written thesis based upon original research that is submitted to a university as a requirement for a higher degree.

Owen Hannaway, "The German Model of Chemical Education in America: Ira Remsen at John Hopkins," *Ambix* 23 (1976): 145–164. Renate Simpson, *How the PhD Came to Britain. A Century of Struggle for Postgraduate Education* (1983).

WILLIAM H. BROCK

A whiskey still at work in the nineteenth century. The beaked head is an old design.

DISTILLATION. According to Nicolas Lemery, distillation was "an operation for separating and collecting, via an appropriate degree of heat, the fluid and volatile principles of bodies." Believing that *fire was the most powerful agent of material change, but that unchecked fire only destroyed the bodies whose principles they sought to separate, chemists viewed and employed distillation procedures as a carefully controlled means to apply fire to achieve their ends. Distillation methods go back to antiquity, but no artifacts have survived to reveal the forms of the apparatus used. Seventeenth- and eighteenth-century textbooks, however, provide verbal descriptions and drawings that allow detailed reconstruction of the procedures that their users followed.

The distillation apparatus comprised a vessel containing the materials to be heated; a head to which the volatile materials "flew up," fitted with a collecting gutter and downward slanting spout for cooling and condensing them; and a receiver to collect the condensed liquids. To obtain very volatile products, such as ardent spirits, essential oils, and the *spiritus rector* thought to contain the essence of the "virtues" of particular plant matters, the chemist used an alembic with a high vertical neck in order that the less volatile constituents would flow into the vessel rather than the receiver. To prevent

the vessel from overheating, chemists placed a water or sand bath between it and the furnace fire. For less volatile substances, such as heavier oils, and the mineral acids, which required a great heat, the vessel of choice was the retort, a one-piece spherical vessel and spout, which allowed the products to leave the distilled matter "laterally." Retorts could be exposed directly to the open fire.

By the seventeenth century chemists sometimes surrounded the head of the distillation apparatus with a *refrigeratory*, an open metal tub filled with cold water to aid in the condensation of the volatile products. A refinement of the refrigeratory, a barrel through which the spout carrying the vapors downward passed, ensured that the condensed products would flow only into the receiver. Sometimes the path of the condensing vapors through the refrigeratory was extended by making the tube into a spiral.

Having no direct measure of the degree of heat applied to the materials they distilled, seventeenth- and eighteenth-century chemists cultivated a highly skilled art that depended on sensitive attention to the size of the charcoal fire employed, the type of furnace used, and the nature and rate of the products obtained. Gross control of the fire was attained by the number of coals used, finer regulation by opening or closing ventilating holes. Often chemists controlled the distillation by observing the rate at which the condensed product dripped into the receiver. In a distillation aimed at complete separation, they gradually raised the heat, in a process sometimes lasting for many days or weeks, changing the receiver often to obtain separately the products of differing volatility. Because the process often separated the several products only incompletely, each fraction obtained might be redistilled, a procedure known as rectification.

During the eighteenth century distillation gradually lost its place as the dominant operation in a chemical laboratory. Chemists learned that the constituents of plant or animal matter could be separated with less risk of destruction or alteration by means of solvents such as water, alcohol, and ether. They still relied on distillations, however, for further analysis.

Chemists were slow to adopt *thermometers to control distillation procedures. By the early nineteenth century, however, distillations at carefully controlled temperatures gave new life to this traditional procedure. During the 1830s Théophile-Jules Pelouze thus separated a series of partial decomposition products from well-defined organic compounds. Better glass, rubber stoppers, ground glass connections, and *Bunsen burners all contributed to the further refinement of distillation apparatus and procedures, which have remained indispensable to the chemical laboratory.

Nicolas Lemery, *Cours de chymie*, 4th ed. (1681): 45–71. Pierre-Joseph Macquer, *Dictionnaire de chymie*, vol. 1 (1766): 64–67, 285–286, 354–357. R. J. Forbes, *A Short History of the Art of Distillation from the Beginnings up to the Death of Cellier Blumenthal* (1948). Robert Anderson, "Distillation," in *Instruments of Science: An Historical Encyclopedia*, eds. R. Bud and D. Warner (1998). Robert Anderson, "The Archeology of Chemistry," in *Instruments and Experimentation in the History of Chemistry*, ed. F. L. Holmes and Trevor Levere (2000): 5–34.

FREDERIC LAWRENCE HOLMES

DIVIDING ENGINE is a machine employed to generate exact divisions on scientific scales. These can be linear scales, such as those on slide rules, though the greatest energy in mechanical division was devoted to the generation of angular scales for astronomical purposes. In 1674, Robert Hooke made reference to a "quadrantal dividing plate of 10 feet radius" to copy-divide mathematical instruments; however, the earliest surviving true dividing engines were those used by clockmakers to manufacture horological gears. One of the earliest such machines, dating from around 1670 and now in the Science Museum, London, has a circular plate of about 30 centimeters (12 in)

in diameter within which small holes divide concentric circles into equal spaces. A metal peg inserted into each successive hole divides the rotation of a central arbor that carries the blank clock wheel to a rotary cutter.

The crucial step in the development of the dividing engine came with the application of an accurate tangent worm screw to the edge of a circular engine plate, making it possible to equate full and part turns of the same screw with exact angular intervals in the circular plate. In origin, however, this use of a worm screw on a denticulated arc to delineate angles probably goes back to the 2-meter (7-ft) radius sextant (1676) and mural arch (1688) of John Flamsteed at Greenwich. Flamsteed's instruments, in turn, developed from William Gascoigne's micrometer (1639), in which an accurate screw of forty threads to an inch allowed measurements of single arc-seconds from fractions of a full turn as shown upon a graduated dial.

The first person to bring these components together into a precision dividing engine was Henry Hindley, a York clockmaker, around 1739. Jesse Ramsden in London developed Hindley's worm screw and denticulated wheel to produce a prototype for all subsequent engines in 1766 and then again in 1775. Ramsden's engine of 1775 with its 9-meter (3-ft) wheel incised with 2,160 teeth won an award of £615 from the Board of Longitude. Its treadle action and "Hindley" graduation cutter made it semiautomatic. It influenced all subsequent dividing engines, especially those of John and Edward Troughton (1778 and 1793) and William Simms (1846), down to the present.

The pressure to perfect the dividing engine in the eighteenth century came from the need to find the longitude at sea, for in competition with the better-known chronometer method of John Harrison, the superior lunar tables of Johann Tobias Mayer (1754) made it possible to find longitude from the moon's position. Crucial to the success of the lunar method, however, was an abundant supply of cheap yet critically accurate sextants for shipboard use. Dividing engines produced the scales of these sextants and of other mathematical instruments. In 1850 Simms's automatic dividing engine generated the graduations on George Biddell Airy's transit circle, which defined the Greenwich meridian. Dividing engines remain in widespread use for the rapid yet definitive graduation of instrument scales.

Jesse Ramsden, *Description of an Engine for Dividing Mathematical Instruments* (1777). A. Chapman, *Dividing the Circle: The Development of Critical Angular Measurement in Astronomy, 1500–1850* (1990, 1995): ch. 8.

ALLAN CHAPMAN

DNA. See Deoxyribonucleic Acid; Double Helix.

DOBZHANSKY, Theodosius (1900–1975), founder of evolutionary genetics.

The only child of Sophia Voinarsky and Grigory Dobrzhansky, a high school mathematics teacher, Theodosius Dobzhansky's family history included a long line of Russian Orthodox priests. From Nemirov, where Theodosius was born, the family moved to the outskirts of Kiev in 1910.

Dobzhansky's lifelong interest in natural history and evolution manifested itself early in insect collecting around Kiev. He loved outdoors activity, horseback riding in particular. As a high-school student he resolved to become a biologist. He graduated with a major in biology from the University of Kiev in 1921 and took a position at the Polytechnic Institute of Kiev on the faculty of agriculture. There his interests turned from beetles to genetics under the influence of the plant geneticist Gregory Levitsky.

Levitsky was part of the sphere of the geneticist and reformer Nikolai I. Vavilov, who was bringing the pioneering genetical researches of Thomas Hunt *Morgan to the Soviet Union. Dobzhansky began research that concentrated on newer areas of genetics and in 1922 undertook experiments on the genetics of the fruit-fly *Drosophila melanogaster*. In 1924, he moved to Leningrad where he lectured in genetics until 1927 under the wing of Yuri Filipchenko, who also followed Morgan's pioneering work. While there, Dobzhansky led an expedition to Central Asia to study the genetics of horses and cattle. In 1924 he married a fellow geneticist, Natalie ("Natasha") Siverstev; they had one daughter, Sophie.

Dobzhansky's life changed in 1927 when he received a fellowship from the Rockefeller Foundation to study genetics with Morgan at Columbia University. He left the Soviet Union just before Trofim Denisovich Lysenko's purging of geneticists, and never returned (*see* Lysenko Affair). At Columbia, he trained in the classical genetics of the Morgan "fly-group," working closely with Calvin Blackman Bridges and Alfred Henry Sturtevant. In 1928 Dobzhansky followed Morgan to the California Institute of Technology, where he became assistant professor in genetics in 1929. In 1933 he made a significant breakthrough when he switched his model organism to *Drosophila pseudoobscura*. The study of the genetics of natural populations of this species formed the backbone of Dobzhansky's pioneering researches into evolutionary genetics and geographic races and varieties of this diverse organism. He used the species' giant salivary chromosomes, which permitted the reconstruction of the phylogenetic history of *D. pseudoobscura* and its closest relatives from visible chromosome markers.

This work underpinned his *Genetics and the Origin of Species* (1937), which became the best known and most influential book on evolutionary biology of the twentieth century. A synthesis between Darwinian selection and the newer genetics, it served as a textbook and catalyst for future researches in evolution. It drew especially on Dobzhansky's Russian background and also on the theoretical insights of R. A. Fisher, J. B. S. Haldane, and especially Sewall Wright, with whom he collaborated closely. Dobzhansky is regarded as one of the prime architects of the modern synthetic theory of evolution.

In 1940 Dobzhansky left the Morgan group to return to Columbia University. There he continued his research into *D. pseudoobscura* and created a lively community of researchers around him. During the 1950s Dobzhansky traveled extensively in South America studying speciation patterns in tropical species of *Drosophila*. In 1962 he moved to Rockefeller University where he remained until his retirement in 1970.

Dobzhansky was deeply interested in the application of genetics and evolution to the understanding of human beings, writing extensively on anthropological and philosophical themes. In 1962 he published *Mankind Evolving*, another synthesis that drew on evolution and cultural anthropology. In 1967 he confronted the broader philosophical implications of evolution in *The Biology of Ultimate Concern*. He campaigned actively against Lysenko and the destruction of Soviet genetics.

In 1970, after retiring from Rockefeller University, Dobzhansky moved to the University of California at Davis where he again supervised an active group of geneticists. He died five years later after a long battle with leukemia. Despite his enthusiastic promotion of evolutionary biology, Dobzhansky remained a fundamentally religious person and a member of the Russian Orthodox church.

Howard Levene, Lee Ehrman, and Rollin Richmond, "Theodosius Dobzhansky Up to Now," in *Essays in Evolution and Genetics in Honor of Theodosius Dobzhansky*, ed. Max K. Hecht and William C. Steere (1970): 1–41. B. Land, *Evolution of a Scientist: The Two Worlds of Theodosius Dobzhansky* (1973). R. C. Lewontin, John A. Moore, William B. Provine, and Bruce Wallace, *Dobzhansky's Genetics of Natural Populations I-XLIII.* (1981). Mark. B. Adams, *The Evolution of Theodosius Dobzhansky* (1994). Louis Levine, ed., *Genetics of Natural Populations. The Continuing Importance of Theodosius Dobzhansky* (1995). Francisco J. Ayala, "Theodosius Dobzhansky," in *Biographical Memoirs: National Academy of Sciences* 55 (1985): 163–213.

Vassiliki Betty Smocovitis

DOUBLE HELIX. The double helix is the name associated with the molecular structure of *deoxyribonucleic acid (DNA). Although isolated in the nineteenth century and identified chemically by 1909, its structure was not clarified until 1953 when Francis *Crick and James D. Watson published their now famous model. The world authority on the nucleic acids, Phoebus Aaron Levene, had formulated the now infamous tetranucleotide structure, according to which the DNA molecule comprised four bases—adenine, guanine, cytosine, and thymine—attached to each other by a sugar-phosphate backbone. So small and boring a structure did not seem able to serve as a repository for the subtle and diverse specificities of the genetic material, yet DNA was long known to be the chief constituent of the chromosomes, the material bearers of our genes (*see* GENETICS). Hence arose the suggestion that it acted as a "midwife" molecule assisting protein genes to duplicate. When in the 1940s it became clear that DNA was a macromolecule implicated in the transfer of *genetic material in bacteria and bacterial viruses, DNA became a contender for the role of the genetic material.

Mounting an attack on the structure called for the skills of the x-ray crystallographer, good knowledge of chemistry, and much patience. The physical chemist Rosalind *Franklin had all of these. When she took over the work on DNA at King's College London, she distinguished the crystalline A form and the wet B form, established the crystal parameters of the former, assigned it to its correct crystalline group (C2 monoclinic), and determined the change in length of the fibers in the transition from one form to the other. Her diffraction photograph of the B form in 1952 was stunning in its clarity and simplicity.

James D. Watson and Francis Crick were not officially working on DNA, but were so convinced of the importance of solving the structure and concerned that Franklin was going in the wrong direction, that they made two attempts on it themselves. Their first attempt failed. Their second, aided by knowledge of the contents of Franklin and Raymond Gosling's report to the Medical Research Council of December 1952, led to their famous double helical model. Unlike Franklin they had long been convinced that the DNA molecule is helical. Both also understood the potentially great biological significance of the structure.

Franklin told Watson and Crick that the sugar-phosphate backbone had to be on the outside. Her data suggested more than one helical strand. Crick realized that the symmetry of the crystalline A form indicated that the sugar-phosphate chains should run in opposite directions (anti-parallel).

This proved to be a fundamentally important point in solving the structure. Watson solved the problem of putting the bases inside the helical cylinder of the two backbones by his innovation of the matching of the large purine bases (adenine and guanine) with the smaller pyrimidine bases (thymine and cytosine), so that adenine pairs with thymine and guanine with cytosine—another fundamental feature.

As refinements to the crystallographic data appeared, the double helix enjoyed growing support. Details were modified, but the basic structure has remained. Admittedly it was a shock when in 1959 the first single-crystal pictures of mixed nucleotides gave like-with-like pairing of the bases. Another seventeen years passed before data from single-crystal pictures supported the Watson-Crick scheme of pairing. A year later the non-helical "zipper" model was introduced as an alternative to the double helix. Not being a double helix, it could open out more easily for replication and transcription. But as we now know, nature long ago solved the problems associated with opening up the double helix.

J. D. Watson, ed. Gunther S. Stent, *The Double Helix. A Personal Account of the Discovery of the Structure of DNA.* Norton Critical Edition (1980). Robert Olby, *The Path to the Double Helix* (1994). Horace Judson, *The Eighth Day of Creation* (1996).

ROBERT OLBY

DYESTUFFS. Early dyes were prepared from natural organic materials derived from plants, insects, or shellfish, or from inorganic salts. Well-known examples are indigo (from a plant native to India), alizarin (from the madder plant), and Tyrian purple (from a kind of snail).

Born just before 1860, the synthetic dye industry expanded explosively over the following decades. Coal tar, a noxious by-product of the manufacture of two valuable industrial commodities, coke and coal gas, provided the material from which all the early synthetic dyes derived. The earliest competent analysis of coal tar was the doctoral work of August Wilhelm von Hofmann, who would later become the most famous chemist of his generation. In 1845 Hofmann emigrated from Bonn to London, where he directed the new Royal College of Chemistry.

An English student of Hofmann's, William Henry Perkin, experimented with impure aniline derived from coal tar on Easter holiday 1856. He isolated an unknown purple substance that seemed to have good properties as a dye. With the assistance of his father and brother (and against Hofmann's advice), Perkin began to produce "aniline purple" commercially. When early in

1859 the color became fashionable in Paris as "mauve," Perkin's fortune was made.

Mauve ushered in a commercial revolution. In 1859 a French dye firm near Lyon began marketing a new aniline-derived color named "fuchsine" in France and "magenta" elsewhere. Proving wildly popular, magenta soon displaced mauve. Many European firms that became leading chemical companies began their existence by producing magenta. By 1862, just five years after the first mauve emerged from Perkin's factory, twenty-nine European firms, mostly British and French, were offering more than a dozen coal-tar colors. Prices often plummeted for existing dyes as production became more efficient or fashion changed, providing constant incentive to producers for innovation.

Around 1870 the industry entered a new phase as German firms began to predominate over British and French rivals. Among the reasons for this shift was the growth in Germany of successful theories of chemical composition. A better understanding of materials that had previously been manipulated purely empirically proved heuristically powerful; in particular, August Kekulé's theory of aromatic compounds (1865) provided excellent guidance to innovation in the industry. One consequence was that the Bayer and BASF companies established the first industrial research laboratories.

In 1869 Karl Graebe and Carl Lieberman synthesized the first important natural dye, alizarin, in Adolf von Baeyer's Berlin laboratory. Firms began to produce huge quantities of alizarin, spelling ruin for the extensive madder plantations of Provence and Turkey. In academic-commercial cooperation as well as in the world of theory the Germans proved highly effective. The nation's victory in the Franco-Prussian War and the subsequent founding of the German Empire provided the right psychological, political, and economic conditions for industrial expansion, especially in the economically overheated years of the early 1870s, and secured German domination of the synthetic dye field.

By this time dye producers could set their sights on the most important remaining natural dye, indigo. Adolf von Baeyer achieved a laboratory synthesis in 1880 and BASF acquired the rights. The industrialization of the synthesis proved difficult, but was achieved in 1897, after a research expenditure of more than three million marks. Three years later synthetic production reached the equivalent of a quarter million acres of cultivation of indigo plants; ten years later natural production in India had virtually ceased.

The synthetic dye industry, which reached maturity in the last third of the nineteenth century, was the basis and nucleus for the entire fine chemicals industry. Many of the largest dye companies diversified early into other products, such as pharmaceuticals, photochemicals, food chemicals, pesticides, organic intermediates, heavy chemicals, plastics, and all the other products of chemical industry. The dye industry also claims pride of place in the origin of *industrial research laboratories.

The German ascendancy in fine chemicals, established in the international dye industry of the 1870s, developed into a near-monopoly in the early years of the twentieth century. Consequently, the Allies had great difficulty in obtaining many strategic chemicals during World War I. German monopolistic practices were fully dismantled only after the end of World War II.

John J. Beer, *The Emergence of the German Dye Industry* (1959). Anthony S. Travis, *The Rainbow Makers: The Origins of the Synthetic Dyestuffs Industry in Western Europe* (1993).

A. J. ROCKE

DYNASTY. Those who think that scientists are marked out by a special cast of mind can cite the examples of Isaac *Newton, René *Descartes, or Albert *Einstein, who neither came from, nor left, families of natural philosophers. The concept of the call to science fit well with the transfer of the ideas of genius, originality, and creativity from the arts to the sciences toward the end of the nineteenth century. Nonetheless, science has been promoted largely by families of savants who have handed down the cultivation of science as if they were lines of politicians or shoemakers. Many examples exist of two or three successive generations of professors and academicians, father to son, in continental universities of the seventeenth and eighteenth centuries: Bernoullis, Snels, Mayers, and so on, and, above all, the Cassinis, who headed the Paris Observatory for four generations.

The French system of *cumul* (the simultaneous tenure of several posts by the same individual) in the nineteenth century allowed the creation and transmission of astonishing amounts of academic capital. Here the Becquerels were exemplary. Henri, who discovered *radioactivity while experimenting with a family heirloom (a crystal containing uranium), was simultaneously professor of physics at the Muséum d'Histoire Naturelle, the Conservatoire des Arts et Métiers, and the École Polytechnique. His father had held the same posts at the Muséum and the Conservatoire; and his grandfather had begun the accumulation with a professorship at the Muséum. Henri's son Jean and nephew Paul Becquerel also made careers in science. Other

French families in which the scientific gene has been well and frequently expressed include the *Curies, the Langevins, and the Perrins, to speak only of physics and chemistry. The English have provided more Darwins and Huxleys than can easily be counted. Nobel Prize winners sometimes come in generational doublets, mother-daughter (the Curies) and father-son (*Bohrs, Braggs, Siegbahns, *Thomsons). And there are several pairs of distinguished scientists in which either the father or the son, or one of two brothers, had the prize, for example, the Alvarezes, Borns, De Broglies, Comptons, Huxleys, Perrins, Rayleighs, and Warburgs.

These few indications will perhaps suffice to ground the generalization that science has been a family business like law, medicine, and butchery, often carried on with great success, and not only the domain of sporadic geniuses.

J. L. HEILBRON

E

EARLY HUMANS. The modern study of early humans depended on two major intellectual developments in the second half of the nineteenth century: rejection of the belief that human beings had existed only for the six thousand years of biblical time, and acceptance of the biological principle of human *evolution. The major proponents of the antiquity of humans came from geology (Charles *Lyell), *paleontology (Hugh Falconer), and archaeology (John Lubbock and John Evans). Chance finds of flint tools predated the discovery of Brixham cave (1858), but the human-made artifacts and fossilized animal bones uncovered during its meticulous excavation proved conclusively that humans coexisted with extinct animals like mammoths of the late Pleistocene. The field methods of geology and relative dating using the fossil index had become a means to study our ancestors.

The comparative anatomy of great apes and humans provided evidence for human evolution. A committed Darwinian, T. H. Huxley argued in *Evidence as to Man's Place in Nature* (1863) for a common ancestor of humans and simians, and suggested searching for it in the fossil record. Some fossil discoveries, at Engis Cave, Belgium (1829), and Neander Valley, Germany (1856), predated the finds at Brixham. But the robustly built Neanderthal Man, with its prominent brow ridges, was often dismissed as a pathological aberration of a modern skeleton.

Gabriel de Mortillet organized the artifacts known in 1872 into a developmental sequence, an important improvement on the earlier systems of naming an object after the location where the "type specimen" was found. Mortillet's sequence overlay Édouard Lartet's for the Paleolithic based on supposedly predominant animal species. Edward B. Taylor and other anthropologists added a developmental perspective to studies of primitive societies, seeking observations of the "living Stone Age" in remote tribes (*see* ANTHROPOLOGY). Further discoveries of Neanderthals in Europe, however, including almost complete skeletons at Spy, Belgium (1886), encouraged the search for fossils. Human paleontology (later paleoanthropology)

thus became the dominant discipline in modern studies of early humans. While at first this work consisted of description and naming, the use of anthropomorphic measuring techniques (Aleš Hdrlička) and later complex statistical analysis (William H. Howell) changed the search for type specimens into a much more dynamic endeavor.

Ernst Haeckel championed the idea that fossils would provide evidence for the "missing link" between humans and other animals. He also believed that Asia had been the cradle of humankind. Inspired by his writings, the young Dutch doctor Eugène Dubois went to search in Java (now Indonesia) where, in 1891, he found a primitive skullcap. Further finds indicated that this hominid had walked upright. As with the Neanderthals, debate raged over the place of *Pithecanthropus erectus* in human evolution, but the notion of a large-brained, bipedal hominid ancestor received a strong fillip. Authors speculated for many years about the order in which the four main events purported to be essential for "humanness"—bipedalism, a large brain, descent from the trees, and tool-use/sociability/language—had occurred. Speculation had dangers, though. Grafton Elliot Smith, Arthur Smith Woodward, and Arthur Keith's belief in the primacy of a large brain led to their acceptance of the Piltdown fossil (1908), famously proved a hoax in 1953.

Marcellin Boule's analysis of the Neanderthal skeleton from La Chapelle-aux-Saints as brutish, shambling, and semiupright (1911–1913) unequivocally restored to this group its fossil status, but sidelined its ancestral role. The archaic modern fossil skeletons found at Cro-Magnon, France (1868), apparently provided a more acceptable forerunner for *Homo sapiens*. Boule championed the "pre-sapiens" theory, according to which a large-brained human ancestor with modern body proportions had been present as far back as the Pliocene. The ramifications of this idea resonated throughout the twentieth century. In South Africa (1924) the "Taung baby" (*Australopithecus africanus*)—a juvenile face and fossilized brain, about two million years old (myo)—was rejected by all save Raymond Dart

Cro-Magnon I, the first early-human remains discovered (in 1864). It is about 30,000 years old.

and Robert Broom. It looked too apelike to belong in the human family. Moreover Asia was still the favored continent for human evolution despite Broom's finds of adult *Australopithecine* skulls in South African caves at Sterkfontein, Swartkrans, and Kromdraai. The Taung child was not accepted until the 1950s, after the exposure of Piltdown.

American vertebrate paleontologist Henry Fairfield Osborn continued to promote an Asian origin of the human family. He sponsored a series of expeditions to Central Asia. Davidson Black's later work (1927–1929) at Chou Kou Tien near Beijing, China, unearthed "Peking Man" (*Homo erectus*). Africa too had champions. In the 1920s, Louis Leakey returned to his native Kenya from Cambridge committed to the pre-sapiens theory, but with an African origin. Together with his second wife Mary, also an archaeologist, Leakey established the significance of Africa. Their son Richard took up the mantle with spectacular

success in the latter half of the twentieth century, becoming with Donald Johanson one of the most celebrated "fossil hunters"—academics with a media profile, useful in funding increasingly complex and costly expeditions. The East African fossil finds—including "Zinj" (*Australopithecus boisei*, 1959), 1.7 myo; "1470" (*Homo habilis*, 1972), 1.9 myo; "Lucy" (*A. afarensis*, 1974), 3 myo; "the first family" (*A. afarensis*, 1975), 3 myo; "Homo erectus boy" (1984), 1.6 myo; "the black skull" (*A. aethiopicus*, 1985), 2.6 myo; and the preserved footprints of an upright hominid (1978), 3.5 to 3.7 myo—have extended our lineage back in time and suggested the coexistence of more than one species of hominids in Africa at intervals in the past. Charles Loring Brace and Milford Wolpoff in rehabilitating the Neanderthals argued that the same may be true in Europe and the Near East. The importance for paleoanthropology of thinking of populations as biological units, according to the new evolutionary synthesis launched by Ernst Mayr, Theodosius Dobzhansky, and George Gaylord Simpson in 1947, has become clear.

Increasingly sophisticated radiometric dating techniques have helped determine the absolute age of fossils. Computerized tomography scanners now allow paleontologists to see within fossils. Molecular biologists have suggested divergence dates based on similarities among human *deoxyribonucleic acid (DNA) and mitochondrial DNA. An intellectual overhaul of archaeology, helped by Lewis Binford's view that we should not seek current behavior patterns in the historical record, turned attention away from the past search for "humanness." Allied with the rise of ecological thinking and general principles such as the effects of climate as an evolutionary lever, fossil hominids have been freed from the legacy of humans' once unique position in the natural order, to take their place in the changing environment of the planet's evolving ecosystem.

Peter Bowler, *Theories of Human Evolution: A Century of Debate, 1844–1944* (1986). Roger Lewin, *Bones of Contention* (1987). Eric Trinkaus and Pat Shipman, *The Neanderthals* (1993). A. Bowdoin Van Riper, *Men among the Mammoths* (1993).

HELEN BYNUM

EARTH, AGE OF THE. Before the mid-eighteenth century, few scholars or scientists in the Christian West questioned the adequacy of the chronologies derived from the Mosaic narrative. They believed that the earth was little older than the few thousand years of recorded human history. Beginning in the second half of the eighteenth century, however, investigations of the earth's strata and fossils suggested that the earth's crust had undergone innumerable cycles of formation and decay and that it had supported an ever-changing sequence of living beings long before the first appearance of humans. For geologists such as James Hutton and Charles Lyell, the earth's age seemed too vast for human comprehension or measurement. By 1840, geologists had identified most of the major subdivisions of the stratigraphic column and arranged them in a chronological sequence, but it was a chronology without a measurable scale of duration, a history of the earth without dates.

In 1859, Charles *Darwin attempted to determine one geological date by estimating how long it would take to erode a measured thickness of the earth's strata. His conclusion that 300 million years had been required to denude even the relatively recent strata of the Weald, a district of southern England, brought an immediate reaction. In 1860, the geologist John Phillips rebutted with an estimate that the composite thickness of the whole stratigraphic column could be denuded in only 100 million years. Soon afterward, the physicist William Thomson, Lord Kelvin calculated that 100 million years would be sufficient for the earth to cool from an assumed primordial molten condition to its present temperature. Kelvin's conclusions, based on the widely held hypothesis that the earth had been formed from the scattered particles of a primordial nebula, and supported by the latest theories of thermodynamics, carried great weight during the remainder of the century. Subsequent estimates of rates of erosion and sedimentation, of solar radiation and cooling, and of the time of formation of the Moon and the oceans converged on his figure of 100 million years.

All this was changed by the discovery in 1903 that radioactive elements constantly emit heat. A year later, Ernest *Rutherford suggested that the ratio of the abundance of radioactive elements to their decay products provide a way to measure the ages of the rocks and minerals containing them. Robert John Strutt and his student, the geologist Arthur Holmes, pursued Rutherford's idea. By 1911, Holmes had used uranium/lead ratios to estimate the ages of several rocks from the ancient Precambrian period. One appeared to be 1,600 million years old. Many geologists were initially skeptical, but by 1930, largely as a result of the work of Holmes, most accepted radioactive dating as the only reliable means to determine the ages of rocks and of the earth itself. The discovery of isotopes in 1913, and the development of the modern mass spectrometer of the 1930s, greatly facilitated radioactive dating. By the late 1940s, the method produced an estimate of between 4,000 and 5,000 million years for the age of the earth. In 1956, the American geochemist Clare Cameron Patterson compared the isotopes of the earth's crust with those of five meteorites. On this basis, he decided that the earth and its meteorites had an age of about 4,550 million years. All subsequent estimates of the age of the earth have tended to confirm Patterson's conclusion.

Joe D. Burchfield, *Lord Kelvin and the Age of the Earth* (1975, 1990). G. Brent Dalrymple, *The Age of the Earth* (1991). C. L. E. Lewis and S. J. Knell, eds. *The Age of the Earth: from 4004 B.C. to A.D. 2002* (2001).

JOE D. BURCHFIELD

EARTHS. During the sixteenth and early seventeenth centuries the three principles of Paracelsus (salt, sulfur, and mercury) competed with the four Aristotelian elements (air, earth, fire, and water) in the definition of chemical composition. A common compromise accepted the Paracelsian principles as the "active" constituents of matter, and two of the Aristotelian elements, water and earth, as the "passive" principles. These five principles were loosely identified with the several types of product thought to be obtained by the analysis of any composed, or "mixed" body into its simplest parts, but were most easily identified in the distillation of plant matter. The earth was what remained at the end in the distillation vessel and could not be dissolved in water. Because distillers sought "active" principles, especially those with medical "virtues," they ordinarily had little interest in the earth. As Nicolas Lemery put it in describing the distillation of Gayac wood, which yielded medically useful spirit, "The earth, called *Caput mortuum*, cannot serve for anything."

Throughout this period chemists argued whether the principles were truly simple and unchangeable and whether multiple specific salts, earths, and sulfurs (or oils) existed. A particular class of earths derived from stones as well as coral and seashells, dissolved in acids, and when strongly heated formed a calx that absorbed water. These were sometimes called *absorbent earths*.

By the early eighteenth century the substances that dissolved in acids and formed neutral salts with them were divided into three classes—alkalis, metals, and earths—collectively known as *bases*. The absorbent earths derived from corals, shells, and limestone were by then recognized as identical, and known as *calcareous earth*. They formed with each known acid a salt recognized by its special properties: with vitriolic acid, for example, the scarcely soluble *selenite*.

Increasing discrimination in the identification of salts of calcareous earth led to the identification of other earths, such as magnesia and barytes, whose salts were distinct from those of calcareous earth. This emerging class of bases became known by the late eighteenth century as alkaline earths.

Several French chemists, including Guillaume-François Rouelle, Pierre Joseph Macquer, and L. B. Guyton de Morveau revived in the mid-eighteenth century the idea that the simple elements earth, air, fire, and water composed all matter. They did so not by philosophical arguments resembling those Aristotle used two thousand years earlier, however, but on the grounds that the four elements were the simplest constituents that could be revealed through chemical analyses. Macquer discussed at length the question whether a single earth underlay all tangible substances known as earths that had not been isolated in pure form, or whether different fundamental kinds of earth existed. Antoine-Laurent Lavoisier pragmatically dismissed all of the earlier philosophical definitions of elements or principles. That the class of bases previously regarded as forms of earth continued to be called

"alkaline earths" left, however, a permanent trace of ideas descended from antiquity in the language of modern chemistry.

Nicolas Lemery, *Cours de chymie*, 4th ed. (1681): 370–409. Pierre Joseph Macquer, *A Dictionary of Chemistry, Containing the Theory and Practice of that Science*, trans. anon, 1 (1777): unpaginated, entry "Earths." R. Hooykas, "Die Elementenlehre deer Iatrochemiker," *Janus* 41 (1937): 1–28.

FREDERIC LAWRENCE HOLMES

EARTH SCIENCE. The discipline of earth science was invented during the 1960s. It replaced *geology as the major institutional framework for studying the earth just as geology had replaced *mineralogy in the early nineteenth century. As with the simultaneous invention of *planetary science, geologists had additional reasons for rethinking the status quo. Geology had lacked focus and exciting new ideas and techniques for the previous half century. Oceanographers and geophysicists (*see* OCEANOGRAPHY; GEOPHYSICS) had new instruments and techniques, particularly in submarine gravity studies, submarine seismology, deep ocean drilling and paleomagnetism.

The needs of submarine warfare during World War II and the Cold War gave a strong boost to seismology and studies of submarine gravity. Before World War II, the Dutch physicist Felix Andries Vening-Meinesz had developed a pendulum apparatus (now named after him) that made it possible to measure gravity at sea as accurately as on land. Using this apparatus, Vening-Meinesz and others mapped the gravity anomalies of the ocean floors and discovered the areas of downbuckling frequently associated with island arcs. Following the War, Maurice Ewing developed sea-floor seismic equipment, measured velocities in the sediments of the deep ocean, and contributed importantly to the understanding and detection of long range sound transmission in oceans.

In the late 1950s, the *International Geophysical Year (IGY), which included gravity, geomagnetism, oceanography and seismology in its areas of concentration, further encouraged these lines of research. Deep ocean drilling that brought up cores from the sediments on the ocean floor began in the 1960s with the establishment of JOIDES (Joint Oceanographic Institutions for Deep Sea Drilling). At the same time, scientists at the United States Geological Survey, the University of Newcastle, and the Australian National University were racing to produce time scales of global magnetic reversals and theories of the motion of the magnetic north pole.

The promotion of the nascent discipline of earth science owed much to a few pioneering individuals and institutions. Among the individuals, in addition to those already mentioned, were J. Turzo Wilson, who led the Canadian contribution to the IGY and made major contributions to plate tectonics, and W. H. (Bill) Menard of Scripps Institution of Oceanography, later Director of the United States Geological Survey. The institutions that saw some of the important early work were spread across the English speaking world: Princeton University, Cambridge University, the University of Toronto, and the Australian National University, together with Lamont Doherty Geological Observatory (founded by Columbia University as a result of Ewing's successes) and Scripps Institution of Oceanography.

By the early 1970s, geophysicists, oceanographers, and geologists had pieced together the theory of *plate tectonics. By the end of the decade, the theory had won worldwide acceptance, except in the Soviet Union. The theory vindicated the new approach to the science of the earth. It showed that expensive oceanographic surveys were necessary because the geology of the ocean floor turned out to be unexpectedly different from that of the land surface. It demonstrated that theorizing on a global scale led to the detection of patterns that would never have emerged from the laborious mapping of one square mile of the earth after another. And it showed once and for all that sophisticated instruments and high levels of funding need not be a waste of public money.

Simultaneous developments in other areas supported these conclusions. Geologists showed that the earth had suffered multiple impacts from meteors, asteroids, and other extraterrestrial bodies, ending the century-long convention that extraterrestrial phenomena were irrelevant to the study of the earth. When geologists found evidence supporting the thesis, put forward by Walter Alvarez, that one of these impacts precipitated the widespread extinctions at the Cretaceous–Tertiary boundary, the move to integrate earth and planetary science gained further momentum. The growing environmental movement and the popular belief that in some way all earth systems were interconnected doubtless also helped accelerate the shift to earth science.

During the 1970s and 1980s, established geology departments changed their names to earth science, or earth and ocean science, or earth and planetary science. They hired professors trained in physics and chemistry as well as in geology, and overhauled their curricula. Geological surveys rethought their missions, broadened their scope, and began using new instruments. The movement spread beyond research science. In 1982, a group of scientists founded the History of the Earth Sciences Society,

and shortly thereafter began publishing *Earth Sciences History*, now the major journal for the history of geology and earth science. In 1983, the National Earth Science Teachers Association was chartered in the United States and began publishing *The Earth Scientist* to promote the teaching of earth science in the school system. Since the 1970s, instruments have continually improved. Magnetometers for measuring fossil magnetism which in the 1950s and 1960s were so new and intricate that many scientists believed the instruments created the effects observed by "paleomagicians" became standard, everyday, off-the-shelf tools. Theorizing has continued apace. Plate tectonic theory, which at first swept all before it, has been refined, and to some extent reformulated.

For all the success of earth science, traces of its roots in the formerly distinct areas of geology, geophysics, and oceanography remain. Major societies founded before the disciplinary transition, such as the American Association of Petroleum Geologists, the Geological Society of London, and the American Geophysical Union, still exist and continue to publish journals specializing in their traditional interests. And although histories of plate tectonics and biographies of its creators abound, as yet we have no comprehensive history of the disciplinary change to earth science.

H. W. Menard, *Science: Growth and Change* (1971). Robert Muir Wood, *The Dark Side of the Earth* (1985). H. W. Menard, *Ocean of Truth: A Personal History of Global Tectonics* (1986). Walter Lenz and Margaret Deacon, eds., "Ocean Sciences: Their History and Relation to Man," *Proceedings of the Fourth International Congress on the History of Oceanography*, Hamburg, September 1987 (1990).

RACHEL LAUDAN

EARTH, SIZE OF THE. See GEODESY.

ECLIPSE. The angular diameters of the Sun and the Moon happen to be nearly the same. Occasionally the two bodies line up and the Sun is eclipsed by the Moon. The first known record of a solar eclipse occurs in a Chinese report of 2136 B.C.; the first recorded British eclipse dates from A.D. 538.

The shadow cast by the Moon only just intersects the surface of Earth, so each total solar eclipse can be observed only over a narrow path. Totality lasts for seven minutes at most. The application of Newtonian gravitational physics to calculate the orbital motion of the Moon enabled Edmond Halley to produce accurate predictions of the path of the solar eclipse of 1715.

A total solar eclipse provides the opportunity to observe the fine features of the Sun. By the middle of the nineteenth century, astronomers had associated the corona and prominences visible during totality with the outer solar atmosphere. Chromospheric flash spectra (*see* SPECTROSCOPY) were taken in 1870, and by the 1930s the highly ionized state of the corona, and its mean temperature of 2,000,000°C, had been established. Study of solar eclipses has also yielded information about matters beyond the Sun. Detailed analysis of eyewitness records of the positions of ancient eclipse paths indicate that the Moon is gradually receding from the Earth, causing the day to lengthen by about two milliseconds per century.

Arthur Eddington used the total solar eclipse of 29 May 1919 to confirm Albert *Einstein's general theory of *relativity. The Sun's gravitational field bent the light beams that passed tangentially to the solar surface by something like the predicted 1.75 seconds of arc.

Lunar eclipses occur when the Moon passes through the earth's shadow cone. Accurate timing of the different phases of a lunar eclipse and knowledge of the angular diameter of the Moon and the earth's radius enabled the Greek astronomer Aristarchus of Samos (c. 310–c. 230 B.C.) to deduce that the Moon lies around 60 terrestrial radii from the earth.

Eclipses elsewhere in the solar system have also been significant in the history of astronomy. The seventeenth-century Danish astronomer Ole Rømer timed the eclipses of Jupiter's moons behind the planet when Earth was at different positions around its orbit. That provided enough information to calculate the velocity of light (*see* LIGHT, SPEED OF). *Galileo, who had discovered the satellites in 1609, proposed that these eclipses, together with optical coincidences between the satellites, could provide the basis of a standard clock, visible in different parts of the earth, for finding differences of longitude. The technique could be used on land but a telescope necessary for determining the moments of coincidence could not be managed at sea. More recently (1985 to 1991) the eclipses of Charon by Pluto have led to accurate calculations of the sizes and colors of these two bodies.

The importance of stellar eclipses was recognized by Edward Pigott and John Goodricke. Observing the variations in brightness of the star Algol (Beta Persei) in the 1780s, they suggested that the alterations might be caused by a planet (half Algol's size) revolving round the star and sometimes eclipsing it. Between 1843 and about 1870 Friedrich Wilhelm August Argelander produced ever more accurate values for variations in brightness throughout stellar eclipses. In 1919 Gustav Müller and Ernst Hartwig produced a

catalogue of 131 eclipsing binaries. Joel Stebbins introduced a selenium photometer in 1910 and with it discovered the secondary minimum in Algol's light curve, indicating that its companion was a faint star and not a planet. Analysis of light from eclipsing binaries allows calculation of the sizes and luminosities of the two components. They often show evidence of limb darkening and deviations from sphericity.

Frank Dyson and R. v.d. R. Woolley, *Eclipses of the Sun and Moon* (1937). J. B. Zirker, *Total Eclipses of the Sun* (1995).

DAVID W. HUGHES

ECLIPSE EXPEDITIONS. When the Moon completely covers the Sun, the sudden onset of darkness, visibility of stars by day, and the appearance of the solar corona are an enthralling sight. Total eclipses also reveal much information of scientific importance about the nature of the Sun. Although these events occur some sixty-five times in a typical century, they are extremely rare at any given location—only three or so in an average millennium. Hence travel, often to remote regions of the globe, has become an essential requirement for avid eclipse observers. They must be requited quickly: the total phase never lasts more than seven and a half minutes. Both the path of the Moon's umbral shadow across the earth's surface and the instant when the Sun will disappear and reappear can be predicted with high accuracy. As a result, modern eclipse expeditions can be planned in detail years in advance. In fact, once a particular eclipse is over, astronomers are often already considering their next venture.

The history of total eclipse observations can be traced back to about 700 B.C. However, until about three hundred years ago, predictions were too crude to allow the anticipation of a total eclipse. Newtonian celestial mechanics made eclipse predictions sufficiently accurate to justify eclipse expeditions.

What may well be described as the first true eclipse expedition (from Paris to London) took place in 1715. Some time before the event, Edmond Halley circulated his predictions of the track of totality over England. Accordingly, the French astronomer Joseph Liouville traveled to London with the express intent of observing totality. Joining Halley, he was rewarded with a fine view of the solar corona, as well as several prominences. Over the next 120 years or so, total eclipse expeditions occurred rarely, partly because of the infrequent occurrence of these events in the more accessible areas of Europe. The transatlantic eclipse travels of the Spanish astronomer Jose Joaquin de Ferrer, who journeyed to Cuba in 1803 and to New York State in 1806, in each case successfully observing a total eclipse, were exceptional.

The year 1842 signaled the beginning of the era of large-scale expeditions. Many European astronomers, probably none of whom had witnessed a total eclipse before, traveled to southern France and Northern Italy to observe totality. These observers had a wide variety of instruments. Although they photographed the partially eclipsed Sun, not until 1851 was the first successful photograph taken during totality—by Berkowski, a Prussian professional photographer. This picture clearly shows the corona, as well as several prominences (see PHOTOGRAPHY).

Regular intercontinental eclipse expeditions were soon followed. Both European and American astronomers enjoyed the adventure of travel to remote places: Peru in 1858, Labrador in 1860, and Aden, India, and what was then Siam (Thailand) in 1868. At the Siam expedition, astronomers made the first spectroscopic observations of the Sun at a total eclipse. These and later observations yielded much information about the nature and chemical composition of prominences, the chromosphere, and the corona. By 1894, eclipse travel had begun to assume such proportions in England that the Royal Society and Royal Astronomical Society set up the Joint Permanent Eclipse Committee (JPEC) to supervise expeditions. It disbanded in 1970, on the ground that "few problems [now] seem urgently in need of eclipse observations."

Probably the most remarkable JPEC venture took place in 1919, soon after the end of World War I. At the instigation of Frank Dyson (then Astronomer Royal) and Arthur Eddington, total eclipse observations were made to test Einstein's general theory of *relativity. Separate expeditions went to Brazil and West Africa where Eddington himself observed. Their photographs of the deflection of starlight demonstrated that Einstein's prediction was accurate within a margin of 10 percent.

Recently, eclipse expeditions have become the province of amateur astronomers, although they still have much to offer professionals. Problems of interest today include the structure of the chromosphere and corona, and variations in the solar diameter over the eleven-year solar cycle. With today's relative ease of long-distance travel, eclipse expeditions will remain high on the agenda of adventuresome astronomers far into the future.

Mabel Loomis Todd, *Total Eclipses of the Sun* (1894). F. Richard Stephenson, *Historical Eclipses and Earth's Rotation* (1997). Pierre Guillermier and Serge Koutchmy, *Total Eclipses: Science, Observations, Myths and Legends* (1999). Mark Littmann, Ken Wilcox, and Fred Espenak, *Totality: Eclipses of the Sun*, 2d ed. (1999). Serge Brunier and Jean-Pierre Luminet, *Glorious Eclipses: Their Past, Present and Future* (2000).

F. RICHARD STEPHENSON

ECOLOGY. In his groundbreaking 1927 textbook, *Animal Ecology*, Charles Elton characterized "ecology" as a new name for an old subject, or simply "scientific natural history." Much of ecology does concern phenomena that have been observed and pondered since antiquity, such as patterns in the geographical distribution

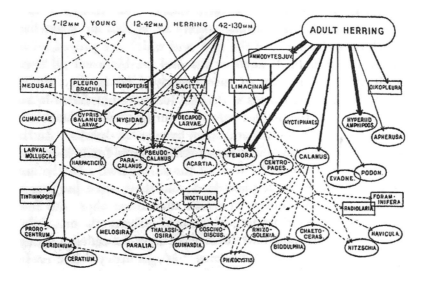

A chart showing the intricate relations between North Sea herring and the plankton on which they feed. The fish vary their diet according to their size.

of organisms, population fluctuations, predator-prey interactions, food chains, life cycles, and cycles of materials. Before the twentieth century, most of these phenomena fell within the domain of natural history. However, experts took up the many pieces of the subject in different studies in which the individual organisms, not their interrelationships, were the chief objects of study. The process by which the interrelationships themselves became the objects of study is the process by which ecology became a science.

The name came earlier than the science. In 1866 Ernst Haeckel coined the word "ecology," with more or less its current meaning, in a long treatise on animal morphology that owed much of its inspiration to Charles *Darwin's theory of evolution. Haeckel wanted a name for a new branch of science that examined the collection of phenomena Darwin characterized as the "conditions of existence," that is, everything in the physical and biological environment of an organism that affects its survival in the broadest sense. Haeckel coined the term but did not make much use of it. The first to do so were botanists working during the 1890s mainly on the geographical distribution of plants. Although this line of research owed something to Darwin, it had deeper roots.

The leader of this older line of research, and one of Darwin's major sources of inspiration, was the German polymath Alexander von *Humboldt, whose treatise on plant geography (or phytogeography), based upon a turn-of-the-century expedition to South America, became a model for the entire field. Humboldt's central contribution was to identify particular assemblages of plants, characterized by vegetation type, with particular environmental regimes. He divided the globe into discrete vegetation zones and made the observation, not unique to him, that the zones traversed in ascending high mountains, such as those he had climbed in the Andes, paralleled the zones passed through in traveling from the equator to the poles.

Over the course of the century, Humboldt's line of inquiry would be expanded, elaborated upon, and modified by a succession of European researchers. By the 1890s a rich collection of observations and descriptions existed alongside a growing vocabulary of phytogeographical terms. This geographical tradition then meshed with a laboratory tradition, born in the revitalized and now thriving German university system, to produce detailed studies of plant adaptation under natural conditions—studies aided considerably by opportunities for travel opened up for European scientists as a result of intense colonization efforts in Asia and Africa and continued economic ventures in South America. In the United States,

the maturation of the university system, combined with the expansion of the nation into the western regions of the North American continent, afforded similar opportunities for botanists. The science of ecology emerged out of this combined functional and geographical emphasis in botany around the turn of the century. North American botanists quietly adopted Haeckel's new term in 1893, and the Danish botanist Eugenius Warming published the first textbook to employ both the term and the concept in 1895.

With his knowledge of the laboratory methods taught in the German universities and his extensive travel experiences in South America and Europe, Warming developed his own classification scheme for plant communities based mainly upon available soil moisture in the habitats in which the characteristic plant groups are generally found. He also offered a discussion of ecological succession, the process by which different communities replace one another in the occupation of a particular site, such as a recently exposed rocky slope, a sand dune, or a burnt-over patch of forest. Warming's counterpart in Germany, Andreas Franz Wilhelm Schimper, in 1898 produced a similar textbook oriented toward plant geography, although with less emphasis on community structure. Schimper, to whom we owe the term "tropical rainforest," came from a school of laboratory-trained botanical field researchers who found more inspiration in Darwin's natural selection theory than did most early plant ecologists. Like many of his colleagues, Schimper traveled widely in the American tropics as well as in Germany's newly acquired Pacific island territories and the island of Java, where the Dutch colonial authorities maintained a botanical laboratory and garden. His textbook, like Warming's, provided a wealth of examples drawn from nature and became a source of inspiration for many budding ecologists in Europe and North America.

Meanwhile, young American botanists found ample opportunities for applying the latest insights and practices from European science to the study of North America's forests and grasslands, fast disappearing under the ax and the plow. A dynamic approach to plant ecology took shape, with emphasis on the process of succession and its inevitable culmination in the climatically determined "climax" community. This conception received its most thorough and persuasive treatment in the hands of Nebraska-born Frederic E. Clements, who spent most of his career in the western United States as a special ecological researcher for the Carnegie Institution. Although Clements's notion of "climax" came under attack at home and abroad, it nevertheless served as a productive focal point for the study of plant communities throughout the first half of the twentieth century. Clements's ideas found modest support among British plant ecologists, but numerous rival schools of plant ecology emerged in Germany, France, Switzerland, Scandinavia, and Russia, most of them promoting versions of what came to be known as phytosociology, or plant sociology—essentially the identification, often by elaborate sampling techniques and statistical methods, of natural plant groups associated with particular sets of environmental conditions. Whereas Clements and his followers focused on the climax community as the single, natural endpoint of a successional sequence, as well as a self-regulating, integrated whole, most European phytosociologists recognized multiple outcomes of successional sequences and did not necessarily emphasize the plant community as a stable, self-sustaining entity.

Studies of animal distribution began in the nineteenth century, but the formal development of animal ecology did not occur until the 1920s. British zoologist Charles Elton, whose field research emphasized the study of populations in the wild, was perhaps the most influential figure. Elton's work, often involving northern fur-bearing animals of commercial value, made a number of concepts part of the naturalist's vocabulary, including the ecological niche, the food chain, and the pyramid of numbers, that is, the decrease in numbers of individual organisms, or total quantity (weight) of organisms, at each successive stage in a food chain, from plants and plant-eating animals at the bottom to large carnivores at the top. Just as with plant ecology, diverse schools of animal ecology emerged in Europe and the United States during the first half of the twentieth century. Some schools, like Elton's, focused on empirical studies of predator-prey interactions and population fluctuations, others focused on animal community organization, still others on broader patterns of distribution and abundance.

Although some of the early work in animal ecology, particularly in the United States, attempted to model itself on plant ecology, by the 1930s animal ecology had emerged as an independent field. There was little overlap or interaction between the work of animal and plant ecologists. Effective impetus for an integrated perspective in ecology came from work in aquatic biology, best exemplified in the late nineteenth century by Karl Möbius's studies of the depleted oyster bank off Germany's north coast and the pioneering limnological (freshwater) studies of François Alphonse Forel on Swiss lakes. This work was continued and refined in the early twentieth century by many researchers, including August Thienemann in Germany and Einar Naumann in Sweden. Möbius's concept of the "biocenosis," the integrated community consisting of all living beings associated with a given habitat or a particular set of environmental conditions, was adopted widely by German and Russian ecologists in the 1920s and 1930s. An integrative perspective also emerged in soil science, as in Sergei Winogradsky's turn-of-the-century studies of soil microbiology, and in studies of biogeochemical cycles, as in the work of Russian geochemist Vladímir Vernadsky, who introduced the term "biosphere" in 1914. However, the integrative concept that had the broadest appeal and played a central role in bringing together the many different strands of ecological science was that of the "ecosystem," introduced by British botanist Arthur G. Tansley in 1935 but first used effectively in an aquatic setting.

Tansley was Britain's foremost plant ecologist and the founder in 1913 of the British Ecological Society, the first such national organization, formed two years earlier than its American counterpart. A pioneer in vegetation surveys, a critic of Clements's idea of the climax community, a passionate conservationist, and a student of Sigmund Freud, Tansley brought his broad experience and erudition to bear on the problem of identifying the ideal ecological unit of study. He suggested that the term "ecosystem" captured this concept best without implying any mysterious vital properties. The new term received its fullest early treatment in a seminal paper published in 1942 by a young American limnologist, Raymond Lindeman. Making use of the concept of ecological succession, Elton's pyramid of numbers and food chains, earlier studies of energy flow in aquatic systems, and Clements's notion of the stable climax community, Lindeman traced the flow of energy through the different trophic (feeding) levels (producers, primary consumers, secondary consumers) in a small Minnesota pond as a way to map its structure as an ecosystem and to demonstrate its progress in development toward a stable, equilibrium state.

World War II proved to be a watershed for ecology. Although earlier preoccupations with community classification and structure, population dynamics, and patterns of distribution continued in the postwar years, newer methodologies, practices, and conceptual schemes took hold, and ecology as a science and a profession grew in size, status, and organization. In the postwar period, Lindeman's groundbreaking work on ecosystem ecology found a home among biologists funded by the U.S. Atomic Energy Commission, who used radionuclides to trace the flow of materials and energy through natural ecosytems. Ecosystem research soon expanded from its base in the Atomic Energy Commission. It also prospered among a small group of Tansley's followers at the new Nature Conservancy in Britain. It became an essential feature of modern ecological science, a message conveyed to several generations of students worldwide through the successive editions of Eugene P. Odum's *Introduction to Ecology*, first published in 1953. Meanwhile, the prewar synthesis of Darwinian natural selection theory with Mendelian genetics resulted in the gradual postwar emergence of a more strongly Darwinian perspective in population and community ecology.

The postwar years also saw a shift toward quantitative aspects of ecology. Mathematical techniques developed in the United States, Europe, and the Soviet Union during the interwar period joined with war-born techniques involving information systems and *cybernetics to produce a movement toward mathematical modeling and computer simulation of populations, communities, and ecosystems. Much of this modeling and its techniques came under attack during the last decades of the twentieth century. Some ecologists abandoned model building for empirical studies, others worked on refining and improving the models, and many called into question the underlying notions of stability and equilibrium upon which most of the models were based.

The devastation brought by World War II also contributed to greater postwar interest in the conservation of natural resources, protection of wildlife, and preservation of natural environments, a trend that, when linked in the 1960s with social criticism, blossomed into an international environmental movement that drew heavily upon concepts and theories of ecology. As had occurred before the war in a more limited way among a few visionaries, ecology now came to be widely viewed not only as the source of remedies for environmental ills but also as the scientific underpinning for a new social order. This proved to be a mixed blessing for ecologists. On the one hand, funding for ecological research increased considerably, and many more people were drawn into the field. On the other hand, the theoretical framework of ecological science, being neither unified nor consistent, could not provide easy, unambiguous solutions to environmental problems, let alone unified and consistent social visions. Toward the end of the twentieth century, this disagreement and uncertainty among ecologists was used as fuel in legislative and legal debates arguing against the protection of endangered species and the maintenance of pristine nature reserves. This situation encouraged the further refinement and integration of ecological science toward the incorporation of human disturbance and the notion of managed ecosystems.

Robert P. McIntosh, *The Background of Ecology* (1985). Eugene Cittadino, *Nature as the Laboratory: Darwinian Plant Ecology in the German Empire, 1880–1900* (1990). Leslie A. Real and James H. Brown, eds., *Foundations of Ecology* (1991). Gregg Mitman, *The State of Nature: Ecology, Community, and American Social Thought, 1900–1950* (1992). Michael Shortland, ed., *Science and Nature: Essays in the History of the Environmental Sciences* (1993). Donald Worster, *Nature's Economy: A History of Ecological Ideas*, 2d ed. (1994). Sharon Kingsland, *Modeling Nature: Episodes in the History of Population Ecology*, 2d ed. (1995). Stephen Bocking, *Ecologists and Environmental Politics: A History of Contemporary Ecology* (1997). Pascal Acot, ed., *The European Origins of Scientific Ecology (1800–1901)*, 2 vols. (1998).

EUGENE CITTADINO

ECONOMIC SOCIETIES. See PATRIOTIC AND ECONOMIC SOCIETIES.

EINSTEIN, Albert (1879–1955), theoretical physicist, developer of the special and general theories of relativity.

Einstein transformed and advanced science as only Isaac *Newton and Charles *Darwin had done. His most celebrated contributions were the theories of *relativity: (1) the special theory (SRT), which revised the notions of *space and time, brought together under one view *electricity, *magnetism, and *mechanics, dismissed the concept of the *ether, and revealed as a by-product the equivalence of mass and energy ($E = mc^2$); and the general theory (GRT), which reinterpreted gravitation as the effect of the curvature of space-time and opened the way to the development of a so-far unachieved unified field theory that would also geometrize electromagnetic fields.

From GRT, Einstein deduced the stability of a spatially bounded universe by adding a "cosmological constant" (which he later retracted); gravitational waves; and the imaging of remoter objects using the gravitational field of nearer galaxies. GRT's early successes included the prediction of the degree of deflection of starlight passing close to the sun (observed in 1919 during an *eclipse); the red shift of light traversing a gravitational field; and the explanation of the precession of the planet Mercury. Einstein worked out most of SRT in 1905. The creation of GRT required intense labor from 1907 to 1915/16.

Einstein responded to his successes with expressions of humorous self-derogation. He once said that his greatest gifts were stubbornness and taking seriously the sorts of questions only children ask. His personal behavior and opinions often alarmed his more conventional colleagues. He had bohemian tendencies in dress

Einstein discussing atomic energy with the *Pittsburgh Post-Gazette* in December 1934. At the time, four years almost to the day before the discovery of uranium fission, the leading physicists doubted that useful energy could be derived from the atom.

and demeanor; urged pacifism during World War I and arms control after World War II; decried nationalism and undemocratic, hierarchical rules; supported Zionism (with provision for the Arabs in Palestine); and opposed religious establishments in favor of a cosmic religion in the spirit of Spinoza. Einstein's fame and charisma made him the target of attacks, chiefly by anti-Semites, and also flooded him with adoring or opportunistic appeals. An example of the latter was his acting as intermediary between three of his fellow émigré colleagues, who in August 1939, on the eve of World War II, wanted to alert President Roosevelt to the possibilities of the German war machine's use of nuclear energy. Einstein had been living in Princeton since 1933 and was thought to have the necessary influence.

In 1914, Einstein moved to Berlin as professor of physics at the university, among other flattering appointments. The move was the culmination of a long series of events. Born in Ulm, Germany, he had spent his childhood in Munich, where his father, an electrical manufacturer and ultimately unsuccessful entrepreneur, hoped to improve his business. In 1888, Einstein entered the Luitpold Gymnasium, but, disliking its military style of instruction, resorted to self-education. He read geometry, popular science, Kant's *Critique of Pure Reason*, and classic works of literature.

In 1894, Einstein tried, and failed, the entrance examination for the Swiss Polytechnic in Zurich. After two years of further preparation in a secondary school, which repaired his deficiencies

in foreign language and brought him to the usual age of entry, he gained admission to the Polytechnic to study to be a schoolteacher. He fell in love with one of his classmates, Mileva Marić, with whom he had a daughter, and after their marriage in 1903, two sons. After a long separation they divorced in 1919, whereupon Einstein married his cousin Elsa Löwenthal.

Failing to find a teaching position or a university assistantship after graduating from the Polytechnic, Einstein moved to Bern in 1904 to become a clerk in the Swiss patent office. His work examining applications for patents on electromagnetic devices may have helped him form ideas used in SRT, described in one of his several groundbreaking publications of 1905 in *Annalen der Physik*. The others—the statistical account of Brownian motion and the postulation of the light quantum—did not have obvious ties to patentable machines but were major theoretical advances. As Einstein's extraordinary talents became known, universities vied for him. He became professor at the University of Prague (1911) and the Swiss Polytechnic (1912) before going to Berlin.

Einstein's great creative work of 1905 stemmed from his preoccupation with fluctuation phenomena. He did not start with a crisis brought on by puzzling new experimental facts, but by stating his dissatisfaction with an asymmetry or lack of generality in a theory that others would dismiss as an "aesthetic blotch." He then proposed several principles, and showed how their consequences removed his original dissatisfaction. All three papers endeavored to bring together and unify apparent opposites, and ended with brief proposals for a series of confirming experiments. Similarly, GRT and the unified field theory arose from his dissatisfaction with the incompleteness of SRT. As he once said, he was driven by his "need to generalize."

In generalizing, he relied on what he called "schemes of thought, the selection of which is, in principle, entirely open to us, and whose qualification can only be judged by the degree to which its use contributes to making the totality of consciousness intelligible." Among his guiding presuppositions, or "themata," were the primacy of formal explanations, unity on a cosmological scale, logical parsimony and necessity, symmetry, simplicity, completeness, continuity, constancy and invariance, and causality. In contrast, the concepts of probabilism and indeterminacy rather than causality and completeness, fundamental in the quantum mechanics of Niels *Bohr and his school, were abhorrent to Einstein.

Of all Einstein's thematic presuppositions, the concept of unity, or, as he described it, the longing to behold the pre-established harmony beyond the harshness and dreariness of everyday life, made him an enduring icon for science.

Paul Arthur Schilpp, ed., *Albert Einstein: Philosopher-Scientist* (1949). Albert Einstein, *Ideas and Opinions* (1954). Abraham Pais, *Subtle Is the Lord...: The Science and the Life of Albert Einstein* (1982). (Various editors), *The Collected Papers of Albert Einstein* (1987–). Gerald Holton, *Thematic Origins of Scientific Thought: Kepler to Einstein*, rev. ed. (1988). Albrecht Fölsing, *Albert Einstein* (1998).

GERALD HOLTON

ELECTRICAL MACHINE generally designates a high-tension electrostatic generator. In friction machines, charges are produced by rubbing a moving insulator (usually made of glass, but also of sulfur, resins, etc.) and accumulated on a metallic conductor. At the beginning of the eighteenth century Francis Hauksbee invented a friction machine with a rotating evacuated glass globe rubbed by hand for studying glows in vacuum. The development of electrical studies and the invention of the *Leyden jar led to the first electrical machines around midcentury. These machines had rotating globes, disks, or cylinders made of glass and were rubbed most often with stuffed leather cushions. Large "prime conductors" (usually insulated metal bars) collected the charges through a set of metallic points or combs placed close to the glass. The largest machines could generate a few hundred thousand volts and produce long sparks. The greatest of these machines, built by the English instrument maker John Cuthbertson

for the Dutch natural philosopher Martin van Marum in the 1780s, had glass disks 65 inches in diameter. Smaller and portable machines became popular for electrotherapy.

In 1775 the Italian physicist Alessandro Volta (see GALVANI, LUIGI, AND ALESSANDRO VOLTA) invented the electrophorus, a device based on electrostatic induction, which could produce an almost endless series of electrical discharges. In 1788 William Nicholson mechanized the routine of charging the electrophorus in "doublers" that raised to detectable magnitude charges otherwise too small to measure. Other natural philosophers of the period produced similar devices. A similar technique underlies the powerful induction generators of the nineteenth century. Introduced by the Italian physicist Giuseppe Belli in the 1830s, they completely superseded friction generators in the 1860s, when the German physicists August Toepler and Wilhelm Holtz invented plate induction machines. In the early 1880s English electrical engineer James Wimshurst introduced particularly efficient and popular induction machines that played an important role in several branches of physics. These machines were used for producing sparks, electrical discharges in vacuum, and *X rays, as well as for electrotherapy.

From the beginning of the twentieth century better types of high tension generators and transformers began to replace electrical machines. One, the endless-belt electrostatic generator invented in the late 1920s by the American physicist Robert Van de Graaff, could produce tensions up to several million volts. These machines were used as particle accelerators and for the production of highly penetrating X rays. Today, high-energy research laboratories and several industries (such as neutron and X-ray production, polymerization, ion implantation) employ large, improved Van de Graaff accelerators.

See also ACCELERATORS.

John Gray, *Electrical Influence Machines. A Full Account of Their Historical Development*, 2d ed. (1890). Willem D. Hackmann, *Electricity from Glass. The History of the Frictional Electrical Machine* (1978).

PAOLO BRENNI

ELECTRICAL MEASURING INSTRUMENTS. See INSTRUMENTS, ELECTRICAL MEASURING.

ELECTRICITY was originally the term used to describe the power acquired by certain substances, notably amber, when rubbed, to attract nearby small objects. This power, known to the Greeks, was carefully distinguished by William Gilbert (1600) from magnetism. Seventeenth-century philosophers were

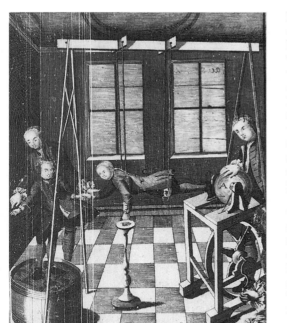

A human capacitor, a standard piece of apparatus in electrostatic experiments around the middle of the eighteenth century. The harness insulates the boy; the electrical machine charges him.

unanimous in believing that the rubbing agitated a subtle matter associated with ordinary matter, causing it to be ejected into the surrounding air, where it (or perhaps air rushing in to fill the space left empty) swept up any light objects in the way.

Francis Hauksbee (c. 1666–1713) obtained more powerful effects by mounting a glass globe on a spindle and rubbing it as it rotated. He showed that electrification was linked to the emission of light—indeed, in Hauksbee's experiments it seemed that the subtle matter streaming from electrified bodies could be felt, seen, and heard. As elaborated by Jean-Antoine Nollet in the 1740s, the theory envisaged fiery matter streaming from electrified bodies, while other streams flowed in to replace the matter that had left.

In 1731, Stephen Gray discovered that the electrical attracting power could be transmitted over great distances, provided that the conducting line was made of an appropriate material and suitably supported. This led to a distinction between "electrics"—substances electrifiable by friction but poor conductors—and "non-electrics"—substances not electrifiable by friction but good conductors.

In 1746, Petrus van Musschenbroek announced a sensational discovery, the "Leyden experiment," in which a bottle filled with water and electrified

by means of a conductor leading from a generating machine into the water delivered a terrible blow when contact was made simultaneously with the conductor dipping into the water and the outer surface of the bottle. "I thought it was all up with me," Musschenbroek reported (*see* LEYDEN JAR; ELECTROPHORE). Enthusiasts everywhere rushed to repeat the experiment. Nollet delighted the French king by discharging a bottle through a line of monks, making them leap into the air simultaneously. In America, Benjamin *Franklin and his friends amused themselves with electrical party tricks.

Franklin also devised a new theory of electricity that in time became generally accepted. He, too, supposed that ordinary matter was suffused with subtle fluid. For Franklin, however, electrification consisted in the redistribution of this fluid between rubber and rubbed, one finishing up with more than its natural quantity, the other with less. The notion of electrical "charge" thus acquired a meaning, the bodies becoming electrified "plus" and "minus" respectively. Noting the ability of pointed conductors to discharge nearby charged objects, Franklin conceived an experiment, first successfully performed near Paris in May 1752, to demonstrate that thunderclouds were electrified and lightning was an electrical discharge. His conclusion that erecting pointed conductors (*lightning rods) on buildings could protect them from lightning strikes was hailed as a triumph of reason over nature.

Franklin's theory worked well in explaining when a shock would be experienced in a variety of situations. He did not provide a coherent dynamical basis for his theory. In 1759, Franz Ulrich Theodosius *Aepinus published a fully consistent version of Franklin's theory, based explicitly on forces acting at a distance between particles, whether of fluid or of matter. By adding the forces acting in various situations, he gave satisfactory explanations for a wide range of effects, and successfully predicted others. His ideas were taken up by leading investigators of the next generation such as Alessandro Volta (*see* GALVANI, LUIGI, AND ALESSANDRO VOLTA) and Charles Augustin Coulomb (*see* CAVENDISH, HENRY, AND CHARLES AUGUSTIN COULOMB), who showed in 1785 that the forces involved obeyed an inverse-square law with respect to distance. On this basis, Simeon-Denis Poisson developed a full mathematical theory of electrostatics in 1812.

To make his theory consistent with elementary observation, Aepinus found it necessary to assume that particles of ordinary matter repelled each other. Many people rejected this idea, supposing instead the existence of two electric fluids that normally neutralized each other but that became separated in electrification. This theory attributed to the second electric fluid the additional repulsive force that Aepinus invoked. Operationally, the one-fluid and two-fluid theories could not be distinguished, and each found adherents.

In the 1790s, controversy erupted between Alessandro Volta and Luigi Galvani over Galvani's experiments in which frogs' legs jerked spasmodically when a conducting circuit was completed between the crural nerve and the leg muscle. Galvani attributed the effect to the discharge of "animal electricity" accumulated in the muscle, which he saw as analogous to a Leyden jar. Volta believed the cause was ordinary electricity, and concluded that dissimilar conductors in contact generate an electromotive force. He built a "pile" comprising pairs of silver and zinc disks separated by pieces of moist cardboard, the electrical force of which he managed to detect with a sensitive electroscope.

Volta's device was the first source of continuous electric current. Its operation was accompanied by chemical dissociations within the moist conductor and in any other conducting solutions forming part of the electrical circuit. These became a focus of research, leading to Humphry Davy's successful isolation of potassium and sodium in 1807 by electrolyzing molten potash and soda, respectively, and eventually to large-scale industrial applications of electrochemical processes.

In 1820, Hans Christian Oersted discovered that a wire carrying an electric current deflects a magnetic needle. André-Marie Ampère quickly showed that a current-carrying loop or solenoid was equivalent to a magnet, and proposed that all magnetism arose from solenoidal electric currents in molecules of iron (see MAGNETISM). Ampère's discovery led to the construction of electromagnets—iron-cored solenoids carrying ever-larger currents—that produced magnetic effects far more powerful than any previously known. Then, in 1831, Michael *Faraday discovered electromagnetic induction: when an electrical conductor cuts across lines of force, an electromotive force is generated in the conductor. These discoveries underpinned the development, first, of the electric telegraph and, later in the nineteenth century, the electrical power industry, which provided a continuous supply of electrical power for industrial or domestic use from generators in which coils of conducting wire placed between the poles of a powerful electromagnet were rotated by steam or water pressure. In the twentieth century, the increasing availability of such power, and the rapid proliferation of appliances to exploit it, transformed civilized life.

But what is electricity? For most nineteenth-century physicists, sources of electromotive force literally drove a current of the electric fluid (or fluids) around a conducting circuit. In an electrolytic cell, the electrodes acted as poles, attracting the constituent parts of the solute into which they were dipped. For Faraday, however, and later for James Clerk *Maxwell, the energy resided not in electrified bodies but in the medium surrounding them, which was thrown into a state of strain by the presence of a source of electromotive force (see FIELD). If the medium were a conductor, the tension would collapse, only to be immediately restored; the product of this continuous repetition, the "current," was a shock wave propagated down the conductor. Static electric charges represented the ends of lines of unrelieved electric tension.

These ideas were widely influential in late-nineteenth-century physics, even though they left the relationship between electricity and matter far from clear. Following the discovery of the *electron by Joseph John *Thomson in 1897, however, and the recognition that it was a universal constituent of matter, electric currents again came to be conceived in terms of a flow of "subtle fluid"—now clouds of electrons semi-detached from their parent atoms, driven along by an electromotive force. No longer, however, was "charge" defined in terms of accumulations or deficits of this fluid. Rather, charge became a primitive term, a quality attributed to the fundamental constituents of matter that was itself left unexplained. Some kinds of fundamental particles, including electrons, carry negative charge, whatever that might be, whereas others carry an equally mysterious positive charge: in combination, these two kinds of charge negate each other. After Robert Millikan's experiments (1913), most physicists accepted the notion that there is a natural unit of electricity, equal to the charge on the electron. The current *standard model in elementary particle physics, however, assumes the existence of sub-nucleonic constituents ("quarks") bearing either one-third or two-thirds of the unit charge. Neither these quarks nor their fractional charges have been detected in a free state.

E. T. Whittaker, A History of the Theories of Aether and Electricity (2 vols., 1951). René Taton, History of Science, trans. A. J. Pomerans, vol. 3, ch. 4, "Electricity and Magnetism (1790–1895)," 178–234; vol. 4, ch. 8, "Electricity and Electronics," 152–202 (4 vols., 1964–1966). L. Pearce Williams, The Origins of Field Theory (1966). R. W. Home, Aepinus's Essay on the Theory of Electricity and Magnetism (1979). J. L. Heilbron, Electricity in the 17th and 18th Centuries (1979). Bruce J. Hunt, The Maxwellians (1991).

R. W. HOME

ELECTRICITY, ATMOSPHERIC. See ATMOSPHERIC ELECTRICITY.

ELECTROLYSIS. Alessandro Volta's publication of his invention of the *battery in 1800 opened an era in science. Learning of Volta's work from Sir Joseph Banks before the publication of Volta's paper, William Nicholson and Anthony Carlisle immediately constructed Voltaic piles of half-crown silver disks, pieces of zinc, and pasteboard soaked in salt water. When they inserted platinum wires from the ends of their pile into a dish of water, hydrogen gas was generated at one wire and oxygen gas at the other. Nicholson announced the results in his journal (*Nicholson's Journal*) before the world learned about the battery from its inventor.

In 1803, Jöns Jacob *Berzelius in Stockholm observed that chemical acids formed around the positive pole and chemical bases around the negative pole of a Voltaic pile. Three years later, Humphry Davy summarized his observations of hydrogen, metals, metallic oxides, and alkalies around the negative pole during electrolysis and oxygen or acids around the positive pole. In 1807, Davy succeeded in obtaining the new metals potassium and sodium from electrolysis of their dry fused alkalis.

Both Davy and Berzelius drew the obvious conclusion that electrical attractions and chemical affinities are identical. About 1809, Amedeo Avogadro had devised a scale of oxygenicity or acidity for the affinity of oxygen with other elements. Berzelius transformed Avogadro's concept into a scale of electronegativity of the elements in 1818. The scale supported Berzelius's theory that chemical combination results from the union of atoms containing unequal amounts of positive and negative electrical fluid combining together with the release of heat (caloric fluid). Berzelius suggested that each chemical compound retains a small excess of either negative or positive fluid, so that acids, for example, with an excess of negative fluid, combine with bases, carrying an excess of positive fluid, to form salts.

Berzelius's dualistic theory enjoyed considerable support until it began to seem irreconcilable with new results in organic chemistry in the 1840s. In the meantime, Michael *Faraday, in collaboration with Whitlock Nicholl, developed the terms 'electrode,' 'electrolysis,' and 'electrolyte,' and with William Whewell, 'ion,' 'cation,' 'anion,' 'cathode,' and 'anode' for the new electrochemical science. Assuming a quantitative relationship between the amount of chemical substance decomposed and the quantity of current that passes through a solution, Faraday devised a "volta-electrometer" to measure the quantity of current. With it, he demonstrated that the quantity of electricity that frees 1 of hydrogen gas liberates other elementary gases in the amount of their chemical equivalent weights.

As early as 1805, Theodor von Grotthuss contrived a model to explain the appearance of electrolysis products at the electrodes, implying that ions in solution move with equal speeds in different directions. Further work with electrochemical cells suggested that cations move faster than anions, an idea developed in 1854 by Wilhelm Hittorf into a quantitative measure, the "transport number" on fraction of the current carried by a particular ion. In 1857, Rudolf Clausius suggested that some of the molecules in solution are already ionized even before the current is applied. Hittorf and François Raoult accepted the suggestion, which became the underlying principle of Svante Recte Arrhenius's ionic theory.

In his doctoral thesis of 1884, Arrhenius proposed (unclearly) that some "complex molecules" exist in dilute solutions as potential carriers of current before the flow of current begins. By 1887, under the influence of Wilhelm Ostwald and Jacobus Henricus van't Hoff, Arrhenius moved to the truly novel interpretation that a substantial number of charged ions already exist in dilute electrolytes. Ostwald, Van't Hoff, and Arrhenius promulgated this new ionic theory in the new journal *Zeitschrift für physikalische Chemie*, founded in 1887.

Arrhenius's argument that a shifting equilibrium holds between ions and undissociated molecules in strong electrolytes was refuted by the Danish physical chemist Niels J. Bjerrum in 1909 over Arrhenius's persistent objections. A complete and detailed treatment of strong electrolytes was developed by Peter J. W. Debye and Erich Hückel in 1923, with improvements by Lars Onsager in 1926 that took into account the Brownian motion of ions surrounded by groups of oppositely charged ions (the ionic atmosphere).

James *Joule made the first attempt to relate the electromotive force of an electrochemical cell to the thermodynamics of chemical reactions in 1840. This problem remained an important theme in thermodynamics, notably in the work of Hermann von *Helmholtz, J. Willard *Gibbs, and Walther Nernst. Merle Randall and Gilbert N. Lewis, who had worked with Ostwald and Nernst, codified the practical application of these studies their book *Thermodynamics and the Free Energy of Chemical Substances* (1924). The ionist theory and the development of electron theory around 1900 revived interest in electrochemical explanations of the chemical bond, developments to which Lewis decisively contributed, along with Irving Langmuir, between 1916 and 1919.

J. R. Partington, *A History of Chemistry, Vol. 4* (1970). John W. Servos, *Physical Chemistry from Ostwald to Pauling: The Making of a Science in America* (1990). William H. Brock, *The Norton History of Chemistry* (1992).
MARY JO NYE

ELECTROMAGNETISM. The study of electricity and magnetism has alternated between theories that represented the two as manifestations of a single effect and the view that they were separate phenomena, with general inflection points around 1600 (when the two fields were divided) and 1820 (when they reunited).

Ancient and medieval philosophers did not distinguish between the ability of amber to attract objects and the action of the lodestone on iron. In 1600 William Gilbert insisted on a distinction between the two, coining the term *"electricity" to describe the effect of amber and attributing it to many other substances; the lodestone remained for Gilbert the sole source of magnetism, around whose force he constructed an entire cosmology. Gilbert's successors spent the seventeenth century searching for the source of electrical and magnetic phenomena and generally finding it in material emanations, such as René *Descartes's theory of magnetic effluvia that swarmed around iron and accounted for magnetic action.

Over the course of the eighteenth century natural philosophers incorporated the distinct fields of electricity and magnetism within a new quantitative, experimental discipline of physics, aided by new instruments to produce phenomena—electrostatic machines like Francis Hauksbee's spinning glass globe, and the *Leyden jar, an early form of capacitor—and to measure them, such as the electroscope, which indicated electrical force by the displacement of threads, straws, or gold leaf. Quantification allowed mathematicians, notably Henry *Cavendish and Charles Augustin Coulomb, to follow Isaac *Newton's example for gravitation and reduce electro- and magnetostatics to distance forces. Philosophers around 1800 explained their measurements in terms of a system of imponderable fluids, with one (or two) weightless, elastic fluid(s) each for electrical and magnetic forces. But despite similar mathematical and conceptual approaches to electricity and magnetism, the phenomena still appeared unrelated. Electricity produced violent action like lightning or sparks, unlike milder magnetic effects, and did not have the same polar behavior.

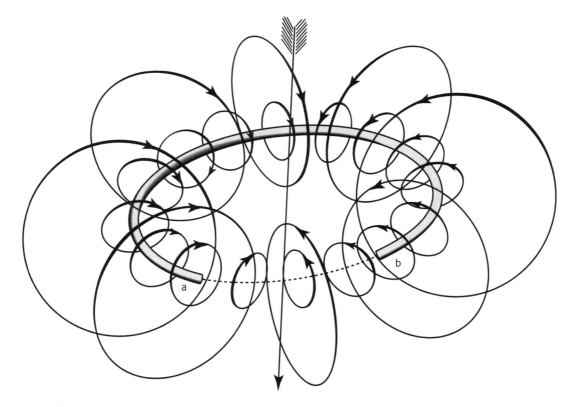

Diagram showing the lines of magnetic force and the direction of the magnetic field of a current-carrying wire.

One of the new instruments, Alessandro Volta's *battery, produced a flow of current electricity. It thus enabled the unification, or reunification, of electricity and magnetism, whose interactions stemmed from dynamic electric and magnetic effects. In 1820 Hans Christian Oersted noticed that a current-carrying wire displaced a nearby compass needle. Oersted's response reflected his adherence to *Naturphilosophie, a metaphysical system advanced at the time by German philosophers who sought single polar forces underlying various domains of natural phenomena. Oersted thought his experiment gave evidence of such a unifying force and proposed a connection between electricity and magnetism. André-Marie Ampère soon mathematized Oersted's experiment, providing a force law for currents in wires. In 1831 Michael Faraday, who may also have dabbled in Naturphilosophie, explored the reverse effect, the ability of a moving magnet to generate electric currents. Oersted, Ampère, and *Faraday thus established electromagnetism, although all three noted a difference between the linear action of electricity and the circular action of magnetism.

Faraday formulated the concept of fields to explain the action of electric and magnetic forces at a distance. Instead of material, if weightless, fluids carrying the force, Faraday attributed the source of electromagnetic action to lines of force in the medium between currents and magnets: lines of force ran from one magnetic pole to the opposite pole, and between opposite electrical charges; magnets or conductors experienced pushes or pulls as their motion intersected these lines of force. James Clerk *Maxwell in the 1860s and 1870s systematized the field concept and the interaction of electric and magnetic forces in a set of differential equations, later simplified into four basic equations. Maxwell's theory, which interpreted light as very rapidly oscillating electric and magnetic *fields, received experimental confirmation by Heinrich Hertz, who starting in 1887 demonstrated the propagation of electromagnetic waves through space.

Over the course of the nineteenth century experimental work on electromagnetism, especially in Germany, spurred the emergence of precision physics, which in turn promoted the establishment of academic research institutes. Electromagnetism provided one of the two main lines of development of nineteenth-century physics, along with *thermodynamics and the kinetic theory. Whereas thermodynamics derived from the first industrial revolution of the late eighteenth century, specifically the program to describe the working of the steam engine, electromagnetism stimulated the second industrial revolution of the late nineteenth century.

Electromagnetism gave rise to industries around the *telegraph, a result of Oersted's discovery of mechanical motion produced by electric current; electric power generation and transmission, with the dynamo deriving from Faraday's work on the transformation of mechanical motion into electric current; and *radio, a consequence of Hertz's experiments on the free propagation of electromagnetic radiation. Together with chemistry and the chemical industry, electromagnetism and its applications exemplified the science-based industry of the second industrial revolution. But the electrical industry did not just apply scientific theories developed in isolation from commercial concerns; rather, prominent scientists, such as William *Thomson (later Lord Kelvin) and Oliver Heaviside tackled practical problems, especially in telegraphy, and thus advanced the state of electromagnetic theory.

Since Maxwell's theory established a conception of light as a form of electromagnetic waves, electromagnetism merged with theoretical optics. At the end of the century some physicists sought to extend the domain of electromagnetism to all physical phenomena, including mechanics and gravity, in what was called the electromagnetic worldview. Many electromagnetic theories relied on an all-pervasive *ether as the medium for electromagnetic waves. Albert *Einstein's theory of *relativity at the outset of the twentieth century enshrined the interchangeability of electrical and magnetic forces and banished the ether, leaving electromagnetic waves to propagate through empty space at the speed of light. The concurrent development of quantum theory, which itself arose from the application of thermodynamics to electromagnetic waves (in the form of heat radiation), led to the wave-particle duality, in which electromagnetic radiation at high frequencies may behave either as a wave or a particle. The elaboration of quantum electrodynamics starting around 1930 provided a *quantum theory of electromagnetism and a description of the duality.

Physicists later in the twentieth century came to view electromagnetism as one of four fundamental forces, the others being gravity and the strong and weak nuclear forces. Particle physicists hoped to show that each force was a particular manifestation of a more general force; they thus joined the unified electromagnetism with the weak (in the so-called electroweak force) and then the strong force (in quantum chromodynamics), and sought in vain for final unification with gravity in a so-called *theory of everything.

E. T. Whittaker, *A History of the Theories of Aether and Electricity,* 2 vols. (1951–1953). J. L. Heilbron, *Electricity in the 17th and 18th Centuries: A Study in Early Modern Physics* (1979). Jed Z. Buchwald, *From*

Maxwell to Microphysics: Aspects of Electromagnetic Theory in the Last Quarter of the Nineteenth Century (1985). Silvan S. Schweber, *QED and the Men Who Made It: Dyson, Feynman, Schwinger, and Tomonoga* (1994). Olivier Darrigol, *Electrodynamics from Ampère to Einstein* (2000).

PETER J. WESTWICK

ELECTRON. By 1890, many chemists and physicists believed that atoms of different elements might represent arrangements of a more fundamental unit, a belief that stemmed primarily from the conviction that nature was essentially simple. However, models devised by physicists on the basis of kinetic theory disagreed with the chemists' results of spectral analysis.

Electrical theories of the atom offered a possible solution. These followed from James Clerk *Maxwell's electromagnetic theory and from Michael *Faraday's work on *electrolysis. Maxwell proposed that the vibrations of light were not mechanical, as previously thought, but electromagnetic. At the same time, the laws of electrolysis implied that electricity existed in discrete units with a charge equal to that on the hydrogen ion. Might the atom contain such units, whose oscillations would explain the emission of line spectra? In 1891, George Johnstone Stoney named these units of charge "electrons" and attempted to find out how big they were by reconciling the spectroscopic and kinetic data.

Simultaneously, Hendrik Antoon Lorentz and Joseph Larmor were trying to accommodate discrete charges within Maxwell's theory, which was expressed in terms of the continuous *ether. They proposed models of charges as vortices or strain centers in the ether and Larmor adopted the term "electron" for his charge. In 1896, the discovery of the magnetic splitting of spectral lines by Lorentz's student Pieter Zeeman lent support to these theories. For Lorentz and Larmor the electron was embedded within the atom, but played no role in determining its chemical nature. This view changed in the years 1895–1905 following the discovery of *X rays and *radioactivity, and investigations of *cathode rays.

Cathode rays were discovered by Julius Plücker in 1858. They are found when an electric potential is applied across a gas at low pressure, but initially they seemed peripheral to mainstream physics. The discovery of X rays in 1895 revived interest in the cathode rays that caused them. Recognition that cathode rays were negatively charged particles, about 2,000 times lighter than atoms, depended primarily on the work of four men: Philipp Lenard, who showed that cathode rays traveled much farther than expected through gases and that their absorption depended on the molecular weight of the gas; and Emil Wiechert, Walter Kaufmann, and Joseph John *Thomson, who measured the charge to mass ratio of the rays by various means. Of these, Thomson went the furthest theoretically, proposing on 30 April 1897 that cathode rays were subatomic, negatively charged particles from which all atoms were built up: they provided the essential mass of the atom and its chemical constitution. To mark the distinction from electrons, Thomson called the particles "corpuscles."

Thomson's contemporaries at first found his suggestion difficult to accept: it sounded like alchemy. They preferred George Francis FitzGerald's alternative proposal that cathode rays were free Larmor-type "electrons." Thus, the name electron became firmly attached to the particles several years before realization that they were indeed essential constituents of the chemical atom. Chief among the corroborating evidence was that work on radioactivity that demonstrated the identity of beta and cathode ray particles and showed that atoms could and did split up and change their chemical nature.

FitzGerald's suggestion ensured that Thomson's cathode-ray particles attracted wide attention by tying them to the attempt by Lorentz, Henri Poincaré, Kaufmann, and others to formulate an entirely electromagnetic theory of matter; an attempt that fostered, but proved incompatible with Einstein's relativity theory. But it was Thomson's concept, as interpreted by Ernest *Rutherford and Niels *Bohr, that made the electron fundamental for new theories of atoms and chemical bonding, and together with relativity and quantum mechanics, to ideas of the nature of matter. Development of these ideas has been very largely the elucidation of the properties of electrons, starting with its charge, first measured by Thomson's students John S. Townsend and H. A. Wilson (1899 and 1903), but firmly established by Robert A. Millikan's experiments beginning in 1907. Electron spin, proposed by Samuel Goudsmit and George Uhlenbeck in 1925 to complete the explanation of the fine structure of spectra, was the last stage in the development of the old quantum theory; the wave nature of the electron, demonstrated by Clinton Davisson and Lester Germer, and George P. Thomson in 1927, and the relativistic electron, formulated by P. A. M. *Dirac in 1928, became cornerstones of the new quantum mechanics.

Quantum electrodynamics was developed from the late 1920s by Dirac, Werner *Heisenberg, Ernst Pascual Jordan, and Wolfgang Pauli to describe the interactions between electromagnetic radiation and charged particles such as electrons. It was completed in the early 1950s by Freeman Dyson, Richard Feynman, Julian Schwinger, and Shin'ichirō Tomonaga, as "renormalization" became accepted. This provided a way of handling the infinite values for calculated quantities that arose in any attempt to make the electron's (or other particle's) equation of state relativistically invariant.

An unforeseen consequence of Dirac's equation for the relativistic electron was the existence of negative energy states, interpreted as antimatter, which spelled the end of an era of subatomic physics predicated on the assumption of conservation of particles. Initially identified as protons, by 1930 physicists began to realize that the anti-particles of electrons must have the same mass as the electron, an idea confirmed by Carl David A. Anderson's identification of the "positron" in 1932.

In the 1950s, accelerated bombardment by electrons suggested that protons and neutrons had a complex structure, the number of elementary particles known had proliferated, and physicists sought further simplification. In 1964, based largely on observed symmetries, George Zweig and Murray Gell-Mann independently proposed that many particles were themselves composed of "quarks" with charge one- or two-thirds that of an electron. The *"Standard Model" of elementary particles synthesizes research suggesting that quarks form a family of six members subject to the "strong interaction" and do not appear independently, being bound together in pairs to form more familiar particles such as the proton and neutron, as described by quantum chronodynamics. A corresponding family of six "leptons" is not sensitive to the strong interaction and has, so far, proved structureless. The electron is the lightest and most stable of the three negatively charged particles that, together with their associated neutrinos, make up the leptons. Any proof that the electron has a finer structure would undermine the edifice of modern particle physics.

J. L. Heilbron, *Historical Studies in the Theory of Atomic Structure* (1981). Arthur Miller, *Albert Einstein's Special Theory of Relativity: Emergence (1905) and Early Interpretation (1905–1922)* (1981). Jed Buchwald, *From Maxwell to Microphysics* (1985). Silvan S. Schwever, *QED and the Men Who Made It* (1994). Laurie Brown et al., eds., *The Rise of the Standard Model* (1996). Per F. Dahl, *Flash of the Cathode Rays* (1997). Edward Davies and Isobel Falconer, *J. J. Thomson and the Discovery of the Electron* (1997). Jed Buchwald and Andrew Warwick, eds., *Histories of the Electron* (2001).

ISOBEL FALCONER

ELECTRONIC COUNTERS. See GEIGER AND ELECTRONIC COUNTERS.

ELECTRONIC MEDIA. Communication within science has changed as electronic modes of communication—e-mail, bulletin boards, websites, multimedia presentation technology, and wireless networks—have developed over the last forty years. Especially relevant to science is the electronic journal. Scientists have long regarded journals as their most important medium of publication. Since the 1960s, as the number and cost of journals have exploded, criticism has mounted against the print-journal system. Some proponents of electronic publishing and archiving systems express hope that electronic journals will replace the print system. But as print journals increasingly include data sets, video, and images on CD-ROM or online, electronic media are beginning to supplement rather than replace existing publications.

New publication formats are also developing on the Internet, for example the public archive hosted by Los Alamos National Laboratory at http://arxiv.org. Created in 1991, it had roughly 170,000 submissions by February 2001 and expected another 35,000 by the end of the year. Specializing in physics, mathematics, and computer science, it is a completely automated publication archive, with no peer review. At the UNESCO conference of 2001 on electronic publishing in science, Paul Ginsparg, the administrator of arxiv.org, estimated the cost associated with each article to be under five dollars compared to thousands of dollars per printed article. Advantage in pricing is not decisive, however. The same UNESCO conference highlighted three central outstanding difficulties with electronic publication: ensuring that the data or text in an online article remains fixed for purposes of referencing and establishing priority; guaranteeing that a given article will remain available indefinitely; and building an electronic equivalent of peer review to weed out irrelevant, inaccurate, or incomplete submissions.

One of the most frequently touted benefits of electronic publication is making scientific knowledge more widely available to researchers in remote or developing regions. So far, visitors to websites like arxiv.org come overwhelmingly from developed Western countries. However, other forms of electronic communication provided significant sharing of information with scientists in developing countries. Healthnet.org maintains a low earth-orbit satellite system, to which over 100,000 people currently subscribe, that provides health workers in remote regions of the developing world with access to one another, to the rest of the medical community, and to data banks. Healthnet.org alleviates isolation through discussion groups, publications, and an e-mail tool that retrieves Web content. Scientists who conduct fieldwork in remote areas, such as geologists and ecologists at the McMurdo research and observation station in Antarctica, also depend on satellite communications.

Electronic media provide data as well as communication. The biological sciences in particular utilize globally available databases as sources of, and repositories for, experimental data. The Genome Database of the Human Genome Project is continually updated with data from numerous laboratories. Rather than technical problems, private interests, concerns about priority, and lack of funding constrain the free sharing of data electronically.

Sharing also assists computing. Massive computing power can be brought to bear through parallel distributed processing, or grid computing, that divides the work across networked computers. Electronic media often couple a scientific instrument to several teams of researchers. Well-known examples include remote-sensing instruments like the Hubble Space Telescope and the Mars Global Surveyor, which researchers communicate with through wireless networks.

The extent and character of electronic usage vary across scientific disciplines. A study by John P. Walsh and Todd Bayma (1996) related the different social structures and work organizations in a field to the ways it uses computer-mediated communication. Mathematicians, who tend to work alone, employ e-mail extensively to maintain contact with their small, dispersed, and highly specialized communities. In contrast, particle physicists require e-mail to coordinate and administer the research of large teams, typically assembled around expensive government-owned equipment. Theoretical chemistry and experimental biology share data with external colleagues according to the commercial value of the data.

All of these developments have taken place over the last fifteen years or so, since the widespread availability of desktop computers in the late 1980s and routine access to the Internet from the early 1990s. A revolution in science communication is under way. Its development has been rapid, and its full outcome is far from clear.

John P. Walsh and Todd Bayma, "Computer Networks and Scientific Work," *Social Studies of Science* 26 (1996): 661–703. American Association for the Advancement of Science website on e-publishing at http://www.aaas.org/spp/dspp/SFRL/projects/epub.htm; in particular, AAAS/UNESCO/ICSU Workshop on Developing Standards in Electronic Publishing in Science, October 12–14 1998, *Report*, at http://xserver.aaas.org/spp/dspp/SFRL/projects/epub/report.htm. Second Joint ICSU Press—UNESCO Conference on Electronic Publishing in Science, *Proceedings*, posted in full at http://associnst.ox.ac.uk/~icsuinfo/confer01.htm. The arxiv project at http://arxiv.org.; in particular, Paul Ginsparg, "Creating a Global Knowledge Network," presented at the second UNESCO Conference on Electronic Publishing in Science, at http://arxiv.org/blurb/pgo1unesco.html.

JESSICA RATCLIFF

ELECTROPHORE. See LEYDEN JAR AND ELECTROPHORE.

ELECTROPHORESIS. Electrophoresis is used to characterize and prepare molecules of biological and chemical interest. When an electric field is applied to a solution, molecules separate on the basis of their electrical charge. The Russian, Alexander Reuss, performed what is retrospectively regarded as the first electrophoretic experiment in 1807. He placed two pieces of glass tubing filled with water in a slab of wet clay, carefully layered sand at the bottom of each tube and then connected the tubes to a battery. Water at the end of the tube connected to the positive pole of the battery became milky as clay particles migrated through the sand. Hermann von *Helmholtz generalized experimental observations into an equation of electrophoretic principles in 1879. Since electrophoresis was relatively non-disruptive, researchers began using electrophoretic separation for large molecules of biological interest. In 1900, W. B. Hardy quantitatively determined the mobility of separated proteins, and in 1908, Karl Landsteiner separated protein from blood sera.

In the 1920s, The Svedburg, inventor of the ultracentrifuge, encouraged his student Arne Tiselius to improve electrophoretic techniques. Over the next few decades, Tiselius transformed the electrophoresis apparatus into a powerful analytical instrument. He created a cooling chamber that reduced aberrations due to buffer convection currents, incorporated a means to visualize the migration patterns of colorless substances utilizing their differential refraction of light, and developed a buffer counter-flow to separate substances that would otherwise migrate to the end of the tube. Tiselius used this apparatus to describe the moving boundaries corresponding to $\alpha-$, $\beta-$, and $\gamma-$globulin in blood serum. He received the Nobel Prize in chemistry in 1948 for this work.

By then, zonal electrophoresis, which came onto the scene in 1940, was challenging the Tiselius apparatus in popularity. In zonal electrophoresis, molecules migrated through a solid matrix, such as paper, infused with a buffer. This sharpened the migration boundaries of the substances by further reducing the effects of convection currents. The low adsorption of polyacrylamide gel, first used by Samuel Raymond and Lester Weintraub in 1959, further increased resolution. In 1961, Stellan Hjérten used agarose, created by eliminating the charged sulfur-containing component from agar, as a support.

In capillary electrophoresis, researchers reduced convection currents even further through the use of small-diameter electrophoretic chambers. Small tubes dissipate heat more efficiently, providing a more uniform temperature throughout the electrophoretic chamber. Sensitive detection techniques allow researchers to see the small amount of sample used in capillary electrophoresis.

Electrophoresis has evolved into one of the most adaptable of laboratory procedures. Modern researchers, for example, use agarose supports to separate whole chromosomes or polyacrylamide to separate nucleic acids a few nucleotides in length. Sensitive visualization techniques allow researchers to identify single copies of genes, while the robustness of solid supports allows for the separation of large amounts of raw material. Many of the procedures now used in biomolecular science utilize some form of electrophoresis. This adaptability has ensured that almost all laboratories interested in studying proteins and nucleic acids have several types of electrophoresis apparatus.

Curt Stern, "Method for Studying Electrophoresis," *Annals of the New York Academy of Sciences* 39 (1939): 147–186. C. J. O. R. Morris and P. Morris, "Experimental Methods of Electrophoresis," in Morris and Morris, *Separation Methods in Biochemistry* (1963): 664–770. C. Shafer-Nielson, "Steady-State Gel Electrophoresis Systems," in *Gel Electrophoresis of Proteins* Michael J. Dunn, ed. (1986). Lilly Kay, "Laboratory Technology and Biological Knowledge: The Tiselius Electrophoresis Apparatus, 1930–1945," *History and Philosophy of the Life Sciences* 10 (1988): 51–72. Olof Vesterberg, "A Short History of Electrophoretic Methods," *Electrophoresis* 14 (1993): 1243–1249.

PHILLIP THURTLE

ELECTROTECHNOLOGY. The first practical (and very questionable) application of electricity occurred in the eighteenth century. Using electrostatic machines and *Leyden jars, natural philosophers, doctors, and various charlatans administered sparks to paralytics and otherwise tried to repair the body by shocks and effluvia. Another early example of electrical technology, the *lightning conductor, was comparatively reliable. Alessandro Volta's invention of the electrochemical cell or battery grandly enlarged the range of application of electricity (*see* GALVANI, LUIGI, AND ALESSANDRO VOLTA). The battery's current could produce electrolysis, generate a bright light across a gap between two carbons, and act upon a magnet. Michael Faraday's discovery that the relative movement of wires and magnets

induce electric currents made motors possible (*see* ELECTROMAGNETISM). But the very first electromagnets, electric motors, generators, and lamps were little more than scientific toys, at home only in laboratories and in lecture rooms. Also, despite countless improvements introduced after Volta, the battery remained an inefficient and expensive source of energy. The first electric motors could not compete with steam engines. Electric arc lights, too bright for domestic application and requiring a complex regulating apparatus were confined to lighthouses until technical improvements suited them for public buildings and street illumination in the later decades of the nineteenth century.

The first successful practical applications of electricity were electroplating and electric telegraphy in the 1840s. Around the beginning of the decade an electrochemical process was discovered for depositing a thin layer of gold or silver on a non-noble metal (or a conductive mould made of graphite-covered gutta-percha). Electroplating and electroforming permitted the bulk production of inexpensive "artistic objects" for the growing middle class. Decoratively wrought metal objects (cutlery, vases, statuettes, candlesticks), previously manufactured only in limited numbers by skilful craftsmen, could be multiplied cheaply and accurately.

Telegraphy also boomed in the 1840s, the development of transmission lines accompanying the growth of national railroad systems. In the second half of the nineteenth century entrepreneurs pressed the creation of an international telegraphic network that demanded an unprecedented effort from physicists, engineers, and electrical manufacturers. New materials (e.g., for insulating cables) had to be developed, and measuring instruments refined. New technologies for making and laying underwater cable had to be invented. Universal electrical standards were needed to allow interconnection of apparatus of diverse origins into a world network; new apparatus was required for detecting the weak signals traveling over thousands of miles. The scientific and technical community, industrialists, tycoons, and politicians joined their efforts for the achievement of this global enterprise. The first transatlantic cables were laid in the 1860s with physicist William *Thomson, Lord Kelvin, in charge of testing the cable as it paid out.

The telegraphic industry supported many spin-offs and trained many inventors of electrical devices. Pantelegraphy, invented by the Italian Giovanni Caselli in the late 1850s, a primitive form of fax, allowed transmission of images by telegraph. It was deployed in France during the Second Empire, but the market was too small to sustain it after 1870. At the turn of the century, however, electric transmission of static images returned (with a better technology) to supply newspapers and magazines with photographs from around the world. Thomas Edison got his start as a telegraphist; his contributions to the field, such as the stock ticker, laid the groundwork for a number of his later inventions, including the phonograph.

In the late 1860s, a number of inventors and scientists with telegraphic experience, notably Charles Wheatstone and Ernst Werner von Siemens, introduced the first self-excited generators, far more powerful and efficient than earlier models. Another major step, taken by the Italian physicist Antonio Pacinotti in the 1860s, was a reversible dynamo machine, an ideal prototype for a direct current electric motor as well as generator. Pacinotti did not find a favorable industrial environment in Italy and tried to commercialize his apparatus abroad. In 1869 the Belgian Zénobe-Théophile Gramme, a skilled mechanician, transformed "Pacinotti's ring" into an industrial product. With better generators and motors electricity could be used not only for arc lighting, which gained ground in the 1870s, but also for mechanical power.

During the last decades of the nineteenth century several inventors proposed types of incandescent lamps to challenge gas and oil illumination. The new lamps required a complex system of generators, cables, switches, electricity meters, fuses, and lamps. Edison developed the first commercially successful system in the early 1880s—a triumph based not only on technology, but clever advertising and lobbying as well.

Supplying electricity was one thing, metering it another. The standard laboratory instruments were too delicate and temperamental for extended use in factories and power plants. The American electrical engineer Edward Weston patented in the 1880s a series of solid, reliable voltmeters and ammeters easily for unskilled workers to use. Weston's measuring instruments proved to be so well designed that they were used with few modifications for more than a century.

In 1885 both the Italian physicist and engineer Galileo Ferraris and the Serbian-born American electrician Nikola Tesla discovered the principle of the rotating magnetic field and the induction motor. Ferraris, a university professor, did not care to develop his discovery industrially and built only a few pieces of demonstration apparatus. Tesla, an inventor, patented various types of motors and generators and laid down the principles of modern polyphase electrical systems, which the industrialist George Westinghouse exploited. The alternating-current system met with fierce opposition from the supporters of the existing direct-current technology. The subsequent "battle of the currents" turned more on economic and industrial strategies than on scientific arguments.

In 1888 physics professor Heinrich Hertz demonstrated the existence of the electromagnetic waves deducible from James Clerk *Maxwell's equations. For several years these radio waves remained a lecture demonstration. Pioneers of wireless such as Oliver Joseph Lodge and Guglielmo Marconi faced the problem of detecting the waves over distances wider than lecture halls and transforming successful technology into a reliable commercial system. At the end of the nineteenth century, wave propagation was still mysterious; no one understood fully how antennas and detectors worked and the invention and improvement of wave detectors depended more on empirical trial and error, practical savoir-faire, and technical skill than on scientific research. Wireless experimenters advanced in an almost virgin field, or rather forest, and their work opened new perspectives to scientific research.

In the twentieth century the booming electrical (and later electronic) industry increased demand for systematic research and development. The growing complexity, variety, and cost of apparatus, components, networks, and systems, necessitated rational planning. Industries developed their own research and development units. The boundary between laboratory research and workshop activity became even vaguer. Science tended increasingly toward industrialization and industry toward subordination to science.

See also LABORATORY, INDUSTRIAL; LIGHTING; STANDARDIZATION; TELEGRAPH.

Sanford P. Bordeau, *Volts to Hertz ... the Rise of Electricity* (1982). Brian Bowers, *A History of Electric Light and Power* (1982). Thomas P. Hughes, *Network of Power. Electrification in Western Society, 1880–1930* (1983). W. A. Atherton, *From Compass to Computer. A History of Electrical and Electronics Engineering* (1984). David E. Nye, *Electrifying America. Social Meanings of a New Technology* (1990). Graeme J. Gooday, "The Morals of Energy Metering: Constructing and Deconstructing the Precision of the Victorian Electrical Engineers Ammeter and Voltmeter," in *The Values of Precision* (1995), ed. M. Norton Wise: 239–282. Joseph F. Keithley, *The Story of Electrical and Magnetic Measurement from 500 B.C. to the 1940s* (1999).

PAOLO BRENNI

ELEMENTARY PARTICLES. Modern particle physics began with the end of World War II. Peace and the Cold War ushered in an era of new *accelerators of ever increasing energy and intensity able to produce the particles that populate the subnuclear world. Simultaneously, particle detectors of ever increasing complexity and sensitivity

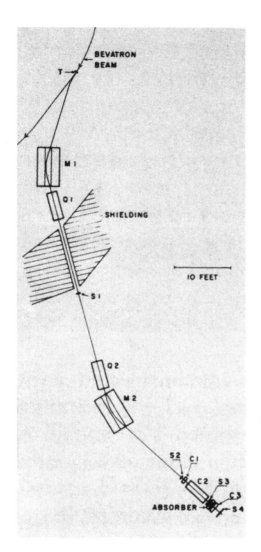

Experimental layout for the detection of the anti-proton: a mixed beam of negative particles from the *accelerator (the Berkeley Bevatron) is bent and focused by the magnets M and Q and analyzed by the counters S and C to distinguish the anti-protons from the much more plentiful pions.

recorded the imprints of high-energy subnuclear collisions. Challenges, opportunities, and resources attracted practitioners: the number of "high energy" physicists worldwide grew from a few hundred after World War II to some 8,000 in the early 1990s.

Developments in 1947 shaped the further evolution of particle physics. Experimental results regarding the decay of mesons observed at sea level presented to the Shelter Island conference led Robert Marshak to suggest that there existed two kinds of mesons. He identified the heavier one, the π meson, with the meson

copiously produced in the upper atmosphere in nuclear collisions of *cosmic-ray particles with atmospheric atoms and with the *Yukawa particle responsible for nuclear forces. The lighter one, the μ meson observed at sea level, was the decay product of a π meson and interacted but weakly with matter. A similar suggestion had been made earlier by Shoichi Sakata in Japan. Within a year, Wilson Powell identified tracks showing the decay of a π into a μ meson in a nuclear emulsion sent aloft in a high altitude balloon. During the early 1950s the data obtained from particles produced in accelerators led to the rapid determination of the characteristic properties of the three varieties of π mesons.

The two-meson hypothesis suggested amendments to the list of particles. Some particles ("leptons")—the electron, muon, and neutrino—do not experience the strong nuclear forces. Others ("hadrons")—the neutron, proton, and the π-mesons do interact strongly with one another. It proved useful to split the hadrons into baryons and mesons. Baryons, of which the proton and neutron are the lightest representatives, have odd-half integer spin and (except for the proton) are unstable, one of the decay products always being a proton. Mesons have integer spin and when free ultimately decay into leptons or photons.

In January 1949, Jack Steinberger gave evidence that the μ-meson decays into an electron and two neutrinos, and shortly thereafter several theorists indicated that the process could be described in the same manner as an ordinary β-decay. Moreover, they pointed out that the coupling constant for this interaction had roughly the same magnitude as the constant in nuclear β-decay. Attempts then were made to extend an idea Oskar Klein had put forth in 1938, that a spin 1 particle, the "W boson," mediated the weak interactions and that the weakness of the β-decay interaction could be explained by making the W mesons sufficiently heavy.

During the first half of the 1950s theoretical attempts to explain pion-nucleon scattering and the nuclear forces were based on *field theoretical models emulating *quantum electrodymics (QED). The success of QED rested on the validity of perturbative expansions in powers of the coupling constant, $e^2/hc = 1/137$. However, for the meson theory of the pion-nucleon interaction the coupling constant had to be large—around 15—to yield nuclear potentials that would bind the deuteron. No one found a valid method to deal with such strong couplings. By the end of the 1950s *quantum field theory (QFT) faced a crisis because of its inability to describe the strong interactions and the impossibility of solving any of the realistic models that had been proposed to explain the dynamics of hadrons. Theorists abandoned efforts to develop a theory of the strong interactions along the model of QED, although Chen Ning Yang and Robert L. Mills advanced a local gauge theory of isotopic spin symmetry in 1954 that proved influential later on. Local gauge invariance, however, implies that the gauge bosons are massless. This is not the case for the pion and thus Yang and Mills's theory was considered an interesting model but not relevant for understanding the strong interactions.

The crisis in theoretical particle physics at the end of the 1950s inspired several responses. It led to the explorations of the generic properties of QFT when only such general principles as causality, the conservation of probability (unitarity), and relativistic invariance figured and no specific assumptions regarding the form of the interactions were made. Geoffrey Chew's S-matrix program that rejected QFT and attempted to formulate a theory that made use only of observables was more radical.

Another response to the crisis made symmetry concepts central. First applied to the weak and the electromagnetic interactions of the hadrons, symmetry considerations were later extended to encompass low-energy strong interactions. Symmetry became one of the fundamental concepts of modern particle physics, used as a classificatory and organizing tool and as a foundational principle to describe dynamics. Interest in field theories, and in particular in gauge theories, revived after theorists had appreciated the notion of spontaneous symmetry breaking (SSB). SSB allows a field theory to have a much richer underlying symmetry than that observed. Usually a symmetry expresses itself in such a way that the vacuum state of the theory is invariant under the symmetry that leaves the description of the dynamics (the Lagrangian) invariant. In the early 1960s Julian Schwinger, Jeffrey Goldstone, Yoichiro Nambu, Steven Weinberg, Abdus Salam, and others noted that in quantum field theories symmetries could be realized differently: the Lagrangian could be invariant under some symmetry, without the symmetry applying to the ground state of the theory. Such symmetries are called "spontaneously broken" (SBS).

In 1967 Weinberg, and in 1968 Salam, independently proposed a gauge theory of the weak interactions that unified the electromagnetic and the weak interactions and made use of the Higgs mechanism, which generates the masses of the particles associated with the gauge theory. Their model incorporated previous suggestions by Sheldon Glashow (1961) for formulating a gauge theory in which the gauge bosons mediated weak forces. The renormalizability of such theories—the existence of consistent algorithms for extracting finite contributions from every order of perturbation theory—was proved by Gerard 't Hooft in his dissertation in 1972. The status of the Glashow-Weinberg-Salam theory changed dramatically in consequence. As Sidney Coleman noted in his article in *Science* describing the award of the Nobel Prize to Glashow, Salam, and Weinberg in 1978, "'t Hooft's kiss transformed Weinberg's frog into an enchanted prince."

As presently described, a common mechanism underlies the strong, weak, and electromagnetic interactions. Each is mediated by the exchange of a spin 1 gauge boson. The gauge bosons of the strong interactions are called gluons, those of the weak interactions, $W^{\pm}$ and Z bosons, and those of electromagnetism, photons. The charges are often called "colors": QED, the paradigmatic gauge theory, works with a single gauge boson, the photon, coupled to a single "color," namely the electric charge. The gauge bosons of the strong interactions carry a three-valued color, those of the weak interactions carry a "two-dimensional" weak color charge. Weak gauge bosons interact with quarks and leptons and some of them, when emitted or absorbed, can transform one kind of quark or lepton into another. When these gauge bosons are exchanged between leptons and quarks they are responsible for the force between them. They can also be emitted as radiation when the quarks or leptons accelerate.

Quantum chromodynamics (QCD) describes the strong interactions between the six quarks: the up and the down, the charmed and the strange, and the top and the bottom. They are usually denoted by u, d, c, s, t and b. Evidence for the top quark was advanced in the fall of 1994 and confirmed in the spring of 1995. Quarks carry electrical charge and in addition a "three dimensional" strong color charge. QCD is a gauge theory with three colors, eight massless gluons, and color-carrying gauge bosons, six that alter color and two that merely react to it. Gluons do not carry color in the same way as quarks do; they carry a color-anticolor, which enables them to interact with one another.

In QFT the vacuum is a dynamic entity. Within any small volume of space-time the root mean square values of the field strengths (electric and magnetic in QED, color-gluon field in QCD) averaged over the volume do not vanish. Virtual particle-antiparticle pairs are constantly being created, and as demanded by the energy-time uncertainty relations, particle and anti-particle annihilate one another shortly thereafter without traveling very far. These virtual pairs can be polarized in much the same way as molecules in a dielectric solid. Thus in QED the presence of an electric charge e_o polarizes the "vacuum," and the charge that is observed at a large distance differs from e_o and is given by $e = e_o/\epsilon$, with ϵ the dielectric constant of the vacuum. The dielectric constant depends on the distance (or equivalently, in a relativistic setting on energy) and in this way the notion of a "running charge" varying with the distance being probed, or equivalently varying with the energy scale, is introduced. Virtual dielectric screening tends to make the effective charge smaller at large distances. Similarly virtual quarks and leptons tend to screen the color charge they carry.

It turns out however that non-Abelian gauge theories like QCD have the property that virtual gluons "antiscreen" any color charge placed in the vacuum (and in fact overcome the screening due to the quarks). This means that a color charge that is observed to be big at large distances originates in a charge that is weaker at short distances, and in fact vanishingly small as $r \to 0$. This phenomenon has been called asymptotic freedom. The discovery of the antiscreening in spin 1 non Abelian gauge theories was made independently by 't Hooft in 1972, and by David Politzer and by David Gross and Frank Wilczek in 1973: non-Abelian gauge theories behave at short distances approximately as a free (non-interacting)theory. This behavior, called *asymptotic* freedom, could explain in a natural way the SLAC experiments on deep inelastic scattering of electrons by protons. Some physicists speculate that in non-Abelian gauge theories the complement to asymptotic freedom at short distances is confinement at large distances. This would explain the non-observability of free quarks. In other words, even though the forces among quarks become vanishingly small at short distances, the force between them increases very strongly at large distances. Although to this day confinement has not been proved in a rigorous fashion, non-perturbative calculations point to the correctness of the assumption.

The past two decades have seen many successful explanations of high-energy phenomena using QCD. The detection and identification of the $W^\pm$ and of the Z_0 in 1983 by Carlo Rubbia and coworkers at *CERN gave further confirmation. Similarly, the empirical data obtained in lepton and photon deep inelastic scattering, and in the study of jets in high energy collisions, can be accounted for quantitatively by QCD. Furthermore, computer simulations have presented convincing evidence that QCD confines quarks and gluons inside hadrons. Frank Wilczek, one of the important contributors to the field, remarked

at a conference in 1992 devoted to an assessment of QCD that it had become mature enough to be placed in its "conceptual universe with appropriate perspective."

L. Hoddeson, L. M. Brown, M. Dresden, and M. Riordan, eds., *The Rise of the Standard Model: Particle Physics in the 1960s and 1970s* (1997). Gerard 't Hooft, *In Search of the Ultimate Building Blocks* (1997).

SILVAN S. SCHWEBER

ELEMENTS, CHEMICAL. Antoine-Laurent *Lavoisier's *Traité élémentaire de chimie* (1789) introduced the modern definition of elements. Lavoisier explicitly rejected the obsolete four-element theory of matter, dating to Empedocles and Aristotle, in which everything was believed to be composed of earth, air, fire, and water, combined in varying proportions. The supposed four elements conveyed different pairs of essential qualities: earth the cold and dry; water, cold and wet; fire, hot and dry; air, hot and wet. One prediction from the four-element theory that Lavoisier specifically refuted was the transmutability of water into earth.

Lavoisier proposed that the term "elements," or principles of bodies, should refer only to the endpoint of observational analysis—those substances into which bodies have been reduced by decomposition. This made experiment the final arbiter and so improved on Robert Boyle's earlier more metaphysical definition of elements as perfectly unmixed bodies. Lavoisier drew up a list of thirty-three elements, or simple substances, including metallic and nonmetallic solids; earthy substances; the gases oxygen, nitrogen (azote), and hydrogen; and light and heat (caloric). Louis Bernard Guyton de Morveau, Antoine François de Fourcroy, and Claude Louis Berthollet were among Lavoisier's collaborators who also contributed to what became known as an eighteenth-century revolution in chemistry.

The new chemistry regarded an element as a simple substance with observable properties; compound substances were made up of one, two, or more simple substances. In his *New System of Chemical Philosophy* (1808–1810), the natural philosopher John Dalton identified each of the simple substances or elements with indivisible atomic particles and characteristic combining weights. Dalton devised combinatorial rules and visual images to describe the composition of ordinary bodies by fixed numbers of atom elements, taking hydrogen to be the smallest and lightest element, with an arbitrary atomic weight of 1 unit.

Between 1790 and 1844, thirty-one new elements were discovered, although in some cases without separation from their oxides. Chemists identified these elements and their properties by traditional analytical techniques (*see* CHEMISTRY), supplemented in the early nineteenth century by electrochemical decomposition and replacement by potassium. No further elements were identified until 1860, when Robert Bunsen and Gustav Kirchhoff noted unusual blue spectral lines in the spectrum of a salt (*see* SPECTROSCOPY). They gave the name "cesium" (from Latin *caesius*, "blue of the firmament") to the supposed emitter. The following year they discovered rubidium from its dark red spectral lines in certain alkaline compounds. The spectroscope figured in the discovery of thallium by William Crookes in 1861, and of indium in 1863, gallium, the rare earths, and the *noble gases.

With the proliferation of elements beyond Lavoisier's table of thirty-three, many natural philosophers and chemists speculated that there must be an underlying basic material in all simple substances. In 1815 William Prout proposed hydrogen as the basic building block, citing the experimental result that gas densities appeared to be exact multiples of the density of hydrogen. Tests of Prout's hypothesis continued over the next decades, even as chemists disagreed whether to take hydrogen or oxygen as the most effective standard for calculating relative combining weights of chemical elements, or atoms. By 1860, chemists were convinced that the careful measurements of Jean-Servais Stas demonstrated that atomic weights could not be multiples of 1 or 0.5 or 0.25 as a fundamental protyle.

In 1860 approximately 140 chemists convened at an international chemistry congress in Karlsruhe to discuss standardization of conventions for atomic weights and molecular formulas. Charles Frédéric Gerhardt's system, in which water has the composition H_2O (H = 1, C = 12, and O = 16), was widely adopted. During the era, scientists including Johann Wolfgang Döbereiner, Alexandre-Émile Béguyer de Chancourtois, and John Newlands attempted systematic groupings of the elements. During the 1860s, Dmitrii *Mendeleev, professor of technical chemistry in St. Petersburg, used Gerhardt's formula convention, along with combining values, or valences, and the analysis of other properties to develop what he called a natural system of the elements. By early 1869 Mendeleev arrived at a law relating atomic weights to periodicity of properties. His *periodic table left blank spaces for unknown elements. The idea of natural families also informed the table of elements published in 1864 by Lothar Meyer. In 1870 Meyer first used increasing atomic weights as the basis of vertical arrangement, complemented by horizontal arrangement of families, and a separate graphical figure plotting atomic weights against atomic volumes. Meyer's short paper of March 1870 brought wide attention to Mendeleev's publication of the previous year.

In 1875 Paul Émile Lecoq de Boisbaudran discovered the element gallium, which fit neatly into Mendeleev's blank space below aluminum, an important confirmation of his law of periodicity. Other predicted elements followed in 1879 (scandium) and 1886 (germanium). The known rare earths doubled in number from 1869 to 1886 and proved difficult to classify. In 1913 Henry G. J. Moseley demonstrated the existence of a constant relationship between the frequency of the shortest x-ray line emitted by an element and what Moseley, following A. Van den Broek, termed atomic number, beginning with 1 for hydrogen (*see* ATOMIC STRUCTURE). Moseley correctly predicted that there must be ninety-two natural elements up to and including uranium. Atomic number replaced atomic weight as the organizing principle for the periodic table of the elements.

Radioactivity produced elements possessing the same atomic numbers and chemical properties as well-known chemical elements, but with different atomic weights. Frederick Soddy coined the term "isotope" to signify any of these chemically identical "elements." In 1912 Joseph John Thomson obtained results suggesting that the inert gas neon (atomic number 10) is a mixture of neon atoms weighing 20 and 22. After World War I, Francis Aston designed a mass spectrograph that sorted out ions by weight and determined that isotopes can be found generally among the chemical elements. Thus a chemical element had a unique atomic number, but an average atomic weight determined by the relative abundance of its isotopes.

Following James Chadwick's discovery of the neutron in 1932, physicists and chemists systematically irradiated elements of the periodic table. Enrico Fermi and his collaborators found that neutrons that had been slowed down were more effective than fast neutrons in producing radioactive isotopes. Fermi's group believed that it had created elements heavier than uranium (atomic weight 238 and atomic number 92) when, in 1934, their irradiations of uranium produced new activities. Similar work by Irène Joliot-Curie and Frédéric Joliot and by Lise Meitner, Otto Hahn, and Fritz Strassmann, resulted in the discovery of uranium fission (*see* NUCLEAR PHYSICS AND NUCLEAR CHEMISTRY).

In 1940 Edwin M. McMillan and Philip H. Abelson produced the transuranium element 93 (neptunium) by bombarding uranium with neutrons in a Berkeley cyclotron. Glenn T. Seaborg and his colleagues produced element 94 (plutonium) in the same way in 1941.

At the end of the twentieth century, scientists recognized 112 elements. Of these, 90 occur in nature either free or in combination with other elements; three (atomic numbers 110–112) had not been named formally by the end of 2000. Since the introduction of Mendeleev's and Meyer's tables of 1869 and 1870, the form for the classification of the elements by means of a periodic system has changed remarkably little. The current asymmetrical rectangular table, in which the lathanide series (numbers 57–70) and the actinide series (numbers 89–102) fall outside the main body of the table was largely the design of Seaborg. A pyramidal periodic table, a form originally favored by Niels *Bohr, has been proposed but has not come into general use despite its more symmetrical appearance. The elements are today recognized not as simple substances in the physical meaning of undecomposable primary matter (see ELEMENTARY PARTICLES; QUARK), but as basic substances in the chemical sense of fundamental matter that exists freely or virtually in all known bodies.

See also RADIOACTIVITY.

Antoine-Laurent Lavoisier, *Elements of Chemistry*, trans. Robert Kerr (1790; trans. 1965). Aaron J. Ihde, *The Development of Modern Chemistry* (1964). William H. Brock, *The Norton History of Chemistry* (1992). Mary Jo Nye, *Before Big Science: The Pursuit of Modern Chemistry and Physics 1800–1940* (1996). Eric Scerri, "Realism, Reduction, and the 'Intermediate Position,'" in *Of Minds and Molecules: New Philosophical Perspectives on Chemistry*, ed. Nalini Bhushan and Stuart Rosenfeld (2000): 51–72.

MARY JO NYE

EMBRYOLOGY. The embryo and its development have been investigated since antiquity and, with few exceptions until the early nineteenth century, under the heading "generation." During the mid-1900s, the term "embryology" denoted the branch of anatomy or physiology concerned with the development of the individual before birth. At the end of the nineteenth century the perception of embryology as a distinct discipline encouraged some universities to establish chairs in "embryology and histology." Given the sensitive character of their subject matter, embryological investigations often reflected deep religious, philosophical, and gender beliefs and influenced legal and social issues.

By breaking open day after day eggs hatched by a hen, Aristotle recorded a sequence of observations on the development of the chick. On the third day he observed a palpitating heart and later distinguished a head with prominent

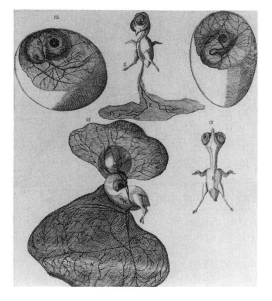

The development of a chick according to William *Harvey's teacher, Girolamo Fabrici, from Fabrici's posthumous *De formatione ovi et pulli* (1621).

eyes clearly separate from the rest of the body. Aristotle held that the first organ to develop in the embryo was the heart, which he considered to be the control core of animal life and the source of vital heat. He also maintained that death occurred when the heart stopped beating. During the Renaissance Aristotle's observations on the developing chick, a subsidiary part of his theory of generation, were replicated and examined by Ulisse Aldrovandi, Volcher Coiter, and Girolamo Fabrici, who investigated many viviparous animals using a comparative approach (*De formatione ovi et pulli*, 1621).

Drawing an analogy with shipbuilding, Fabrici believed that the embryo would build itself up from a bony framework, a view harshly criticized by his pupil William Harvey in his *De generatione animalium* (1651). Harvey based his theories on numerous observations and dissections of domestic fowl and deer carried out over many years. Although he never saw the ova of vivipera, he postulated that all female animals produced them and that their development followed the pattern of ovipera. Because his dissections of the uteri of fowl and deer after mating never revealed conception in the form of a mixture of male and female semen, he rejected ancient doctrines of conception. Instead, he held that females produced eggs endowed with a vitality that, with no material contribution, could be stimulated to develop by the male's semen. By focusing on a little scar (Fabrici's *cicatricula*, i.e., the blastoderm) on the surface of the hen's fertilized egg, Harvey made his most important discovery, namely that after the first day it started germinating. He also described how the chick developed from the *cicatricula* by budding and subdivision in sequence. He named this process *"'epigenesis." Contrary to Aristotle, Harvey maintained that the blood and not the heart formed first.

During the second half of the seventeenth and the eighteenth centuries, many scholars opposed Harvey's epigenesis with the alternative theory that development consisted in the unfolding and growth of all the parts of the adult organism, which pre-existed miniaturized in the egg. In his *Dissertatio epistolica de formatione pulli in ovo* (1673), Marcello Malpighi compared the development of the chick to that of plants, considered the intake of food from the yolk and albumen, and described how the parts of the embryo change shape and position before acquiring resemblance to their adult form. His description of embryogenesis and his discoveries (e.g., of the cardiac tube, the neural folds, and the neural tube) had a profound impact upon embryological research.

During the eighteenth century the opposition between these two main views of embryogenesis culminated in the controversy between Albrecht von Haller's mechanistic preformationism and Caspar Friedrich Wolff's vitalistic epigenesis. Wolff maintained that, just as plants formed from the structureless substance of the vegetational bud, so the chick emerged from a homogeneous primordium through the secretion and solidification of fluids regulated by a *vis essentialis* (essential force) present in living matter (*Theorie von der Generation*, 1764). Despite Wolff's denial that organs pre-existed in a latent form and his insistence that preformation was a chimera, epigenesis acquired preeminence only during the nineteenth century.

The period 1820 to 1880, the age of classical descriptive embryology, saw the seminal works of Christian Heinrich Pander, Karl Ernst von Baer, Martin Heinrich Rathke, Rudolf Albert von Koelliker, Robert Remak, Ernst Haeckel, Oscar Hertwig, Richard Hertwig, Francis Maitland Balfour, Aleksandr Kovalevsky, and others. Improvements of the *microscope first, and then the combined introduction of the microtome and staining techniques, helped overcome some of the technical barriers that had frustrated previous investigators in their observations. Study of the embryo was much influenced by morphology in the 1820s and 1830s, by *cell theory in the 1840s and 1850s, and by the theory of *evolution from the 1860s onwards. It was also influenced by the ideal, typical of the century, of furnishing historical explanations. Applied to the embryo, this ideal postulated that knowledge of the organization of living entities could be achieved only by investigating their gradual development. It also suggested a parallelism between the stages of development of the individual organism and the long-term transformations of the entire animal series.

One major contribution of early nineteenth century embryology was the doctrine of germ layers. In 1817 Pander described how the chick's blastoderm (a term he introduced) developed into three separate layers, which he considered to be the antecedents of later structures (*Beiträge zur Entwicklungsgeschichte des Hühnchens im Eye*). In the 1820s von Baer maintained that the germ-layer concept applied to other vertebrates as well, and Rathke extended it to invertebrates. In papers published between 1850 and 1855, Remak demonstrated that germ layers consisted of cells. Thus the goal of much embryological research of the time became that of tracing the origin of a single organ to a specific germ layer. In 1867–1871 Kovalevsky reported compelling observations that evidenced the universality and specificity of the germ layers. "Mesoderm" was coined in 1871, and "ectoderm" and "endoderm" in 1872, to denote the three germ layers. The doctrine of the absolute specificity of the germ layers has been disputed by more recent embryological research, which has emphasized the interactions among the layers as they develop.

Perhaps the most significant contribution of embryology during the early eighteenth century was the clarification of the fertilization mechanism. In papers published in 1824–1825 Jean-Louis Prevost and Jean-Baptiste-André Dumas described filtering experiments that showed that spermatozoa played an essential role in fertilization, and in 1827 von Baer announced the discovery of the egg, first in dogs and then in other mammals (*De ovi mammalium et hominis genesi*). In 1841 Koelliker argued for the cellular origin of spermatozoa and in 1844 suggested that the ovum is a cell. In the early 1850s Remak proved the cellular nature of the egg and argued that the original fertilized egg with its nuclear content underwent a process of division (cleavage) until it formed the germ layers. At that time, however, the prevailing theory of fertilization still denied that spermatozoa made any material contribution to the embryo. A host of investigations of animals and plants conducted by many scholars between 1875 and 1880 led to the conclusion that fertilization consisted in the union of a part of one spermatozoon or pollen with an egg or ovule. This fusion of two cells produced the unicellular fertilized egg that only then began dividing to form the embryo. This theory aroused great excitement because it explained the continuity of life by establishing a bond between generations through the transfer of some material substance from the parents to the new individual.

Nineteenth-century embryology sought laws of development. Von Baer enumerated four laws, which account for his endorsement of epigenesis and the existence of different types of embryonic development, and therefore the organisations that he had distinguished (radiates, articulates, mollusks, and vertebrates). In post-Darwinian thought the notion of evolutionary change eroded the type concept. A major problem, therefore, was envisaging a law that considered the results obtained by descriptive embryology and simultaneously accounted for individual and ancestral development. A law, later known as the biogenetic law, formulated by Haeckel in 1866 stated that ontogeny (the development of the individual) briefly recapitulates phylogeny (the development of the organic lineage to which it belongs) determined by heredity and adaptation (*Generelle Morphologie der Organismen*, 1866). This law had considerable impact upon nonspecialists as well and acquired dogma-like status. The next generation of embryologists raised doubts about recapitulation and, in order to gain better understanding of the mechanisms of development, turned once again to problems of causation, this time however by experimenting directly on the embryo.

See also CHROMOSOME; DEVELOPMENTAL MECHANICS; HEREDITY.

Howard B. Adelmann, *Marcello Malpighi and the Evolution of Embryology*, 5 vols. (1966). Jane M. Oppenheimer, *Essays in the History of Embryology and Biology* (1967). William Coleman, *Biology in the Nineteenth Century: Problems of Form, Function and Transformation* (1971). Shirley A. Roe, *Matter, Life and Generation: Eighteenth-Century Embryology and the Haller-Wolff Debate* (1981). T. J. Horder, J. A. Witkowski, and C. C. Wylie, eds., *A History of Embryology* (1986). Scott F. Gilbert, ed., *A Conceptual History of Modern Embryology* (1991).
RENATO G. MAZZOLINI

EMPIRICISM. Empiricism is the belief that all our knowledge derives from the experience of our senses (or from extensions of our senses in instruments). In the history of modern philosophy, debates about the merits of empiricism have generated a huge literature. Philosophers have seen empiricism as being opposed to rationalism, the belief that reason alone can arrive at basic truths about the world, a position frequently connected with a belief in innate ideas. Not surprisingly, given that modern science had roots in the philosophical tradition, these debates have spilled over into the history of science. By and large, those natural philosophers and scientists who have considered geometry the model for inquiry into the natural world, and who have advocated the use of hypotheses, have been inclined to rationalism, while those who favor the use of experiment and induction have inclined to empiricism. Pinning the labels empiricism and rationalism on natural philosophers and scientists, though, is apt to oversimplify their carefully considered positions.

Two of the most important natural philosophers, René *Descartes and G. W. Leibniz, are usually placed in the rationalist camp. Empiricism in science as well as in philosophy was strongly associated with the English. Francis *Bacon, whom the founders of the Royal Society of London and the editors of the great French *Encyclopédie* celebrated as a luminary, set the tone. Isaac *Newton frequently spoke as though he were mounting an empiricist answer to speculative systems. Certainly eighteenth-century England saw him in that light. John Locke laid out a systematic empiricist philosophy in his *Essay Concerning Human Understanding* (1690), and his work led to that of Bishop George Berkeley and David Hume. Even so, some British natural philosophers rejected empiricism, one example being James Hutton. On the Continent, Immanuel Kant attempted a reconciliation of rationalism and empiricism.

From the early nineteenth century through World War II, most scientists and philosophers of science subscribed to both empiricism and inductivism. It was a natural stance for Georges Cuvier and others who wanted to assert the scientific standing of nonmathematical sciences such as biology, chemistry, and geology. In England, John Herschel and John Stuart Mill spoke strongly for empiricism. Mill even made the controversial (and largely rejected) claim that even the truths of mathematics derived ultimately from experience. All the proponents of positivism in its varied forms, including Auguste Comte, Ernst Mach, and the logical positivists in the 1930s were empiricists. Positivist histories of science, such as George Sarton's *History of Science* (1952), naturally stressed the role of observation and experiment in the formation of modern science.

Pure empiricism did not completely win the day, though. In the nineteenth century, the mathematician, mineralogist, and philosopher William Whewell used his *History of the Inductive Sciences* (1837) and *Philosophy of the Inductive Sciences* (1840) to make the case that important as experience was, ideas had an equal role to play in science. Other philosophers and historians of science followed suit, including Ernst Cassirer in a whole series of books, most prominently *Das Erkenntnisproblem in der Philosophie und*

Wissenschaft (1906–1907), Edwin A. Burtt in his *Metaphysical Foundations of Modern Physical Science* (1932), and Alexandre Koyré in a variety of works including his *Études Galileénnes* (1939).

Since World War II, debate about science and empiricism has shifted to science studies. No longer educated as philosophers, most scientists give little thought to epistemology, though most would consider themselves empiricists. The postpositivist turn in philosophy of science has meant that many, though not all, scholars trained in history and philosophy of science have shifted away from empiricism.

RACHEL LAUDAN

ENCYCLOPEDIAS. We owe the word "encyclopaedia" to Quintilian's Latinized version of the Greek term denoting a circle of study or learning. This concept came to inform the notion of the seven liberal arts that passed into the medieval university curriculum. This set of favored subjects included geometry and some natural knowledge and appears in all major medieval and Renaissance encyclopedias, for example Gregor Reisch's *Margarita Philosophica* (1496) and Johann Heinrich Alsted's *Encyclopaedia* (4 vols., 1620).

During the early eighteenth century, a new encyclopedic genre, the dictionary of arts and sciences, made its appearance. It included information on the arts and crafts that had previously been excluded from liberal sciences and the universities. Antoine Furetière's *Dictionnaire Universel* (3 vols., 1690) and John Harris's *Lexicon Technicum* (2 vols., 1704, 1710) were the first examples. These works differed from the historical dictionaries of the time, such as Louis Moréri's *Grand Dictionnaire Historique* (2 vols., 1674) and Pierre Bayle's *Dictionnaire Historique et Critique* (2 vols., 1697), which covered history, geography, and biography rather than the arts and sciences. Knowledge was becoming specialized. The *Grosses vollständiges Universal Lexicon*, begun in 1732 by the Leipzig publisher Johann Zedler, which has entries on scientific topics as well as on history, theology, philosophy, and biography, reached sixty-four folio volumes by 1750. The much slimmer dictionaries of arts and sciences did not have biographical entries and treated history and geography only insofar as relevant to the account of technical terms.

These so-called "scientific dictionaries" focused on natural, mathematical, and craft or technical knowledge. The strength of Harris's *Lexicon* in mixed-mathematical subjects and their application in areas such as navigation, architecture, fortification, gunnery, and shipbuilding earned him the nickname "technical Harris." A member of the Royal Society (serving as its secretary in 1710), Harris incorporated substantial quotations from Isaac *Newton's *Opticks* (1704), which had just come out in English translation, and inserted an unpublished paper by Newton on acids in the second volume. Ephraim Chambers's *Cyclopaedia* (2 vols., 1728) built more widely on Harris's foundations. It claimed to contain a well-rounded course of ancient and modern learning, thus justifying its use of "cyclopaedia." Like Harris, Chambers covered subjects still categorized as scientia, such as law, grammar, music, and theology. He gave more attention to crafts and trades that fell outside the area of practical mathematics treated by Harris: paper, soap- and glassmaking, mining, forging, weaving, bleaching, dying, tanning, and the manufacture of cloth and pins. But Chambers distinguished between "Mechanical Arts...wherein the Hand, and Body are more concern'd than the Mind" and the mechanical sciences. In keeping with the implied preference, the *Cyclopaedia* was stronger on the "sciences" than the "arts," although later editions promised to improve the coverage of trades and manufacturing arts.

The content of these dictionaries of arts and sciences gives a clue to their commercial success. They provided information on subjects such as the mechanical arts excluded from university education together with detailed entries on the major mathematical and physical sciences: the dictionaries of Harris and Chambers amounted to practical manuals and Newtonian textbooks. Both appeared on study guides at Cambridge, and students were directed to read particular entries. Harris and Chambers treated chemistry, medicine, and natural history less fully, possibly because specialist lexicons for these subjects already existed. They met a need not only in England but also in Europe. Chambers's *Cyclopaedia* appeared in two Italian translations (Venice, 1748–1749, and Naples, 1747–1754), and inspired the creation of the greatest of all the eighteenth-century compendia of knowledge, the *Encyclopedie, ou Dictionnaire raisonné des arts et sciences* of Denis Diderot and Jean Le Rond d'Alembert (17 vols. of text and 11 vols. of plates, 1751–1772).

Eighteenth-century encyclopedias were published and sold by subscription. This method, introduced by English booksellers early in the seventeenth century, tested the market for large and expensive works. A prospectus announcing the work gave the names of subscribers and helped recruit additional ones. The range of occupations represented in these lists—from bishops and physicians to watchmakers and printers—indicate the breadth of the market. The

Frontispiece of Denis Diderot's *Encyclopédie* (1751). At the top, Reason and Philosophy peel the veil from truth; Geometry, Astronomy, and Physics occupy the center, above Optics, Botany, Chemistry, and Agriculture.

first volume of Harris's *Lexicon* cost twenty-five shillings, and Chambers's two folios cost four guineas, both expensive compared with the price of monthly magazines, about six pence an issue in 1750. Subscription and serialization placed the early English scientific dictionaries within the reach of a wider group of readers than the first edition of the *Encyclopédie*, which initially cost 280 livres (around 11 guineas) and rose to 980 livres (40 guineas) by the 1770s.

Three significant changes in encyclopedias may be discerned over the course of the eighteenth century. First, the scope and importance of the nonscientific content increased: although the *Encyclopédie* did not admit biographical entries, it boasted large and sometimes controversial essays on topics in history, literature, music, art, politics, and philosophy. Perhaps its most distinguished feature, apart from its anticlericalism, was its comprehensive documentation of the arts, crafts, and trades, which it illustrated with some 2,500 engravings. In this aspect of their work, the encyclopedists claimed inspiration from the philosophy of Francis Bacon (*see* BACONIANISM).

Second, the format shifted from relatively compressed entries on terms (the style of both Harris and Chambers) to longer essays, still arranged alphabetically. The *Encyclopaedia Britannica*, issued in one hundred installments from 1768 and published in Edinburgh in three volumes in 1771, departed from the format of the earlier dictionaries of arts and sciences, presenting the sciences as "systems" in separate treatises of at least twenty-five pages each. The *Encyclopédie Méthodique* (66 vols., 1782–1832)—the successor to the *Encyclopédie*—amounts to a set of specialist treatises, in which, in the words of a contemporary reviewer, "every science will have its dictionary, or system, apart."

The third development was the recruitment of specialists to write the articles. The *Encyclopédie* engaged some of the most distinguished natural philosophers and academicians of France, especially Gabriel François Venel in chemistry, Louis-Jean-Marie Daubenton in anatomy and zoology, Nicolas Desmarest in geology, and, of course, d'Alembert in mathematics. The third edition of the *Britannica* (10 vols., 1788–1797) followed suit. Whereas William Smellie (the main compiler of its first edition) wrote many of its treatises by collating from various books, the third edition brought in experts. For its six-volume *Supplement*, its editor, Macvey Napier, commissioned leading men of science to update articles, such as those in natural philosophy formerly done by the Edinburgh professor John Robison.

British competitors of the *Britannica* such as Abraham Rees's *New Cyclopaedia* (45 vols.,

1802–1819), David Brewster's *Edinburgh Encyclopaedia* (18 vols., 1809–1830), and the *Encyclopaedia Metropolitana* (28 vols., 1829–1845) all gave science high priority and sought out leading contributors. The *Metropolitana* boasted contributions from Charles Babbage and John Herschel, who wrote extensive treatises on astronomy, light, and sound. The ninth edition of the *Britannica* (1875–1889) responded with "Physical Sciences" by James Clerk *Maxwell and "Evolution in Biology" by Thomas Henry Huxley.

All but one of these nineteenth-century works abandoned the maps of knowledge that delineated the relationships between the branches of science. The exception, the *Metropolitana*, was arranged not alphabetically, but rather on a classification published in 1817 by Samuel Taylor Coleridge that placed subjects in logical or systematic order. All branches of mathematics preceded Herschel's articles on light and sound because these subjects assumed prior mathematical knowledge. The *Metropolitana* did follow the nineteenth-century practice of assigning detailed articles on scientific disciplines to experts, which raised concerns about the role of encyclopedias as a medium for the public communication of knowledge. The *Britannica* eventually found a solution. Its great eleventh edition of 1911 marked the high tide of the expert article. It then floated from Cambridge University Press to the United States, at first to Sears, Roebuck and Company and later to the University of Chicago Press. From the 1930s it was sold by door-to-door salesmen as an indispensable aid to social success. The revised fifteenth edition (from 1974) of the *Encyclopaedia Britannica*, which is now published by an independent corporation, came in two versions: a "micropaedia" for quick reference and a "macropaedia" for detailed specialist articles. This division resolved the tension between the interests of most people and the scholarly imperative.

During the nineteenth and twentieth centuries, European countries and the Soviet Union created encyclopedias emphasizing their national cultures within the international circle of knowledge. The most important of these works for the history of science is the *Enciclopedia Italiana* with its collateral publications. They are the products of the Istituto della Enciclopedia Italiana, founded in 1927 by the industrialist Giovanni Treccani and directed in its scientific program by the philosopher Giovanni Gentile. The *Enciclopedia* itself, complete in thirty-six massive volumes (1929–1939), is among the world's best. When supplemented by the Istituto's *Dizionario biografico degli Italiani* (now to the letter "G" in 55 volumes, 1960 to the present) and its specialized series in art, architecture, and so on, it represents the grandest attainment of the encyclopedia as a cultural resource. The Istituto is currently publishing a *Storia della scienza*, which will extend encyclopedism with an encyclopedia of the history of science twenty times the size of this Companion.

Philip Shorr, *Science and Superstition in the Eighteenth Century: A Study of the Treatment of Science in Two Encyclopedias of 1725–1750* (1932). Arthur Hughes, "Science in English Encyclopaedias 1704–1875," *Annals of Science* 7 (1951): 340–370; 8 (1952): 323–367; 9 (1953): 233–264; 11 (1955): 74–92. Robert Collison, *Encyclopedias: Their History Throughout the Ages* (1964). Robert Darnton, *The Business of Enlightenment: A Publishing History of the Encyclopédie, 1775–1800* (1979). Frank A. Kafker, ed., *Notable Encyclopedias of the Seventeenth and Eighteenth Centuries: Nine Predecessors of the Encyclopédie*, in *Studies on Voltaire and the Eighteenth Century* 194 (1981). Anna S. Arnar, ed., *Encyclopedism from Pliny to Borges in Memory of Robert Rosenthal* (1990). Richard Yeo, "Reading Encyclopedias: Science and the Organisation of Knowledge in British Dictionaries of Arts and Sciences, 1730–1850," *Isis* 82 (1991): 24–49. Frank A. Kafker, ed., 'Notable Encyclopedias of the Late Eighteenth Century: Eleven Successors of the Encyclopédie,' in *Studies on Voltaire and the Eighteenth Century* 315 (1994). Richard Yeo, *Encyclopaedic Visions: Scientific Dictionaries and Enlightenment Culture* (2001).

RICHARD YEO

ENDOCRINOLOGY. The clinical specialty of endocrinology is less than a century old, but knowledge of endocrine diseases and the structures and functions of the endocrine glands has a much longer history. The physicians of antiquity described the effects of castration and diseases such as diabetes and treated simple enlargement of the thyroid (goiter) with burnt sponge or seaweed, now known to contain iodine. The anatomy of the testes, ovaries, thyroid, pituitary gland, and adrenals was long known, although as separate structures, rather than a loosely connected endocrine system. In the eighteenth century, Albrecht von Haller noted the existence of "glands without ducts," which he distinguished from ducted ones such as the salivary and sweat glands. A century later Claude *Bernard crystallized the notion of "internal secretion" when he distinguished between the "external" secretion of bile by the liver and its internal one of sugar.

Despite these and many other anatomical, physiological, and pathological insights, endocrinology as a coherent body of knowledge did not emerge until the early twentieth century. It then carried with it the newsworthy but ambiguous legacy of Charles-Édouard Brown-Séquard, a serious scientific clinician who introduced in the 1880s testicular extracts as a sensational agent of rejuvenation. Brown-Séquard's death soon after his self-experimentation deflated the notion of frolicking octogenarians, but a second potent endocrine extract, from the thyroid gland, had demonstrable physiological effects. By the 1890s, the cluster of clinical conditions variously described as cretinism, myxoedema, and cachexia strumipriva had been referred to a failure of the thyroid gland. In 1891 George Murray reported the successful treatment of myxoedema with thyroid extract. That the mixture also acted as a stimulant gave it wider appeal, and it was frequently prescribed for lethargy, obesity, and general malaise.

The hormone concept appeared in 1902, when the British physiologists William Bayliss and Ernest Starling identified in the mucosa of the duodenum a substance they called "secretin." It could stimulate secretion by the pancreas even when the neurological connections were severed. This action pointed towards a chemical stimulus; three years later Starling called this class of substances "hormones," from the Greek for "to excite." The subsequent coining of the word "endocrine," from Greek words for "within" and "separate," codified the notion that hormones flow directly into the bloodstream and act on other organs or cells without the intermediating functions of the nervous system. Edward Sharpey-Schafer's monograph *The Endocrine Organs* (1916) helped define the field. He had earlier (with George Oliver) isolated a blood-pressure raising hormone of the adrenals, adrenaline. Although it took some time for the clinical specialty to rid itself of the enthusiastic claims of an earlier generation, Sharpey-Schafer's solidly scientific synthesis firmly established its experimental roots.

The dramatic therapeutic potentials of the discipline were realized in 1921 when Frederick Grant Banting and Charles Herbert Best isolated insulin, one of the active endocrine products of the pancreas. The Nobel Prize two years later went to Banting and J. J. R. Macleod, in whose lab the work took place, but Banting shared his prize with Best and Macleod his with James Bertram Collip, the biochemist who had assisted in its purification. The relative contributions of the four men have been much debated, but insulin itself stood out as a major therapeutic breakthrough in the treatment of diabetes. The interwar period proved to be fertile for endocrinology, with new biochemical and bioassay techniques to identify and purify many active hormones from the ovaries, testes, adrenals, pituitary, thyroid, and parathyroids. In 1936, Edward Doisy (who shared the 1943 Nobel Prize for his work on Vitamin K) defined four criteria by which hormones could be identified. These were: 1: a gland must be identified as producing an internal secretion; 2: the substance must be detectable; 3: it must be capable of being purified; and 4: the pure substance needs to be isolated, purified, and studied chemically. This followed his research on the role of ovarian hormones on the estrus cycle, which in turn laid the foundation for the development of hormonal contraceptives as well as agents to treat menstrual and other gynecological disorders.

Research on the several endocrine glands clarified that hormones belong to various classes of bioactive substances. These include steroids (the gonads and adrenal cortex), catecholamines (adrenal medulla), iodinated amino acids (thyroid), and proteins and active peptides (anterior and posterior pituitary, pancreas, gut, parathyroids, and thyroid). Work on the pituitary by Pierre Marie, Harvey Cushing, and many others proved to be especially significant. Not only does the pituitary synthesize many of the central hormones that regulate peripheral production (through a subtle system of negative feedback), it is in intimate contact with the hypothalamus, an area of the brain that also has important controlling functions on the nervous system, especially the autonomic system. The earlier notion that the endocrine system stands functionally apart from the nervous system has thus been fundamentally modified.

Research during the past half-century has been aimed primarily at clarifying the chemical structures, synthetic pathways, and molecular modes of actions of the diverse group of substances called hormones. Since these are involved in many fundamental physiological processes such as metabolism, digestion, growth, reproduction, salt-and-water maintenance, and the interaction of the organism with its environment, endocrinology has maintained its close ties with basic science, especially molecular biology. The complexity of the system allows many ways for it to go awry; the clinical discipline is now sub-specialized, with diabetes, gynecological endocrinology, and the thyroid each having its own group of specialists. Endocrinological surgery is now a recognized specialty, especially important in the treatment of tumors of the endocrine organs.

Humphry Davy Rolleston, *The Endocrine Organs in Health and Disease* (1936). Joseph Meites, B. T. Donovan, and Samuel McDonald McCann, *Pioneers in Neuroendocrinology*, 2 vols. (1975–78). Michael Bliss, *The Discovery of Insulin* (1982). V. C. Medvei, *A History of Endocrinology* (1982). Samuel McDonald McCann, ed., *Endocrinology: People and Ideas* (1988).

W. F. BYNUM

ENERGETICS. The great unsettled question of late-nineteenth-century physics was the status of the mechanical worldview. For more than two hundred years—from René *Descartes, Christiaan *Huygens, and Isaac *Newton in the seventeenth century to Hermann von *Helmholtz, Heinrich Hertz, and Ludwig *Boltzmann at the end of the nineteenth—physicists had generally sought mechanical explanations for natural phenomena. As the nineteenth century drew to a close, Hertz reaffirmed the classical goal of physical theory: "All physicists agree," he wrote in the preface to his *Principles of Mechanics* (1894), "that the problem of physics consists in tracing the phenomena of nature back to the simple laws of mechanics." But when these words were published, physicists were no longer in general agreement about the nature of their project. Many doubted, and some explicitly denied, that mechanics was the most basic science. Other candidates contended for the honor—thermodynamics and electromagnetic theory, in particular, and several comprehensive alternatives to the mechanical worldview were proposed and vigorously debated throughout the 1890s and early 1900s.

Energetics was one of the alternatives. Tracing its origins to the founders of the law of energy conservation, especially Robert Mayer, and to the thermodynamic writings of Rudolf Clausius, William *Thomson (Lord Kelvin), and Josiah Willard *Gibbs, energetics attempted to unify all of natural science through the concept of energy and by laws describing energy in its various forms. The energeticists believed that scientists should abandon their efforts to understand the natural world in mechanical terms and should give up atomism as well in favor of a new worldview based entirely on relations among quantities of energy.

Energetics as a scientific project of the late 1880s and 1890s took place largely in Germany. (A prominent exception was the work of the French physicist Pierre Duhem.) Its main German proponents were Georg Helm, a Dresden mathematician and physicist, and Wilhelm Ostwald, the professor of physical chemistry at Leipzig. Helm first urged the formulation of a "general energetics" in his *Theory of Energy* (1887), which proposed an "energy principle" (a law more general than the law of energy conservation) as its basis. An essay in 1890 sought to reduce mechanics to energetics by means of this energy principle, and another in 1892 was intended to do the same for electricity and magnetism. In 1894 Helm wrote a book on the energetic development of physical chemistry. These publications elicited an invitation to address the German Association of Scientists and Physicians at their meeting in Lübeck in 1895 on "the current state of energetics."

Ostwald's interest in energy stemmed from his reading, in mid-1886, of Dutch chemist Jacobus van't Hoff's studies in chemical dynamics and from his own efforts, in the late 1880s, to understand the thermodynamic writings of Gibbs, which Ostwald published in German translation in 1892. He was soon converted to the way of "pure energetics," the theory of which he developed in two essays published in 1891 and 1892. He then refined his theory and applied it to a variety of problems in general and physical chemistry in 1893–1894. Always the enthusiast, Ostwald traveled to the 1895 meeting in Lübeck, where he was also on the program, to demonstrate the demise of the mechanical worldview and to promote energetics as its proper replacement.

The heated debate at Lübeck turned out to be a disaster for energetics. The negative reactions of Boltzmann and Max *Planck to the energeticists were taken as definitive by younger physicists such as Arnold Sommerfeld and Albert *Einstein. Helm and Ostwald later replied to these criticisms, only to be rebutted again by Boltzmann (1896–1898). Ostwald published his *History of Electrochemistry* in 1896; Ernst Mach likely hurried his (incomplete) *Theory of Heat* into print in the same year to support the anti-mechanist cause; and Helm, in his history of energetics of 1898, tried to defend his own work. But the damage had been done. Ostwald continued to uphold energetics after 1900, but increasingly as a monistic worldview, not as a scientific project.

The scientific proposals of the energeticists were flawed, but the attention they received undermines the common assertion that the physical scientists of the late nineteenth century were satisfied with the state of their science. The long tradition of mechanical explanation in the natural sciences was coming to an end. The debate over energetics as a viable replacement for the mechanical worldview reflected the difficulties inherent in the mechanical view.

Robert J. Deltete, "Gibbs and the Energeticists," in *No Truth Except in the Details: Essays in Honor of Martin J. Klein* (1995): 135–169. Robert J. Deltete, "Helm and Boltzmann: Energetics at the Lübeck Naturforscherversammlung," *Synthèse* 119 (1999): 45–68. Georg Helm, *The Historical Development of Energetics*, trans. and intro. Robert J. Deltete (2000).

ROBERT J. DELTETE

ENGINEER. The engineer emerged in the second half of the eighteenth century as a social and professional type distinct from the skilled craftsman (carpenter, blacksmith, stonemason, or millwright), the architect, and the military engineer (responsible primarily for weapons and fortifications). The socially elevated, status-conscious, and (either academically or institutionally) accredited civil engineer oversaw the design, construction, and maintenance of the public works associated with industrialization: roads, bridges, canals, harbors and drainage schemes, lighthouses, factory machinery, railways, and so forth.

Within this basic pattern there were considerable variations concerning the role of science, forms of training, and the involvement of both the state and professional bodies. In Great Britain, John Smeaton's Society of Civil Engineers (founded 1771) celebrated gentlemanly collegial interaction and patronage rather than formal accreditation. Thomas Tredgold's charter (1828) of the London-based Institution of Civil Engineers (founded 1818) saw the engineer applying natural philosophy to direct "the great sources of power in Nature for the use and convenience of mankind." Yet for Samuel Smiles the engineering archetype was a self-made individualist, not a trained corporate player. In Great Britain, as in the United States, the autodidact doyens of (nonmilitary) engineering pointed to the achievements of empirical shop culture to justify the system of apprenticeship, rather than academic scientific training. From the 1830s, educational providers looking to foreign models promoted a mix of science, modern languages, and political economy as the foundation for a properly learned profession of engineering; and from the second half of the nineteenth century, college-trained engineers were prominent as managers, public servants, and empire builders seen as uniquely able to provide social leadership (and solutions) in response to industrial change. In the United States, similarly, only late in the century was there a consolidation of a school culture of engineering institutions construed as places in which the organized practices of engineering, different to those of science, were inculcated.

The *Encyclopédie* of Denis Diderot defined the engineer as a man at work in any of the state corps of engineering (military, naval, or bridges and roads), and possessed of the best mathematical, mechanical, and hydraulic knowledge. By the end of the ancien régime, entrants to these corps came almost exclusively from the associated professional schools, a subset of the network of *grandes écoles*. Among the most prestigious were the École de Ponts-et-Chaussées (founded 1747), the École de Mines (founded 1783), and, especially, the École Polytechnique (founded 1794) (a model for the Military Academy [founded 1802] at West Point, New York, garrison of the American Corps of Engineers). From the early nineteenth century the École Polytechnique was the first port of call for any state engineer; *polytechniciens* were "finished" for the engineering corps at specialized *écoles d'application*. Despite its meritocratic recruitment ideology, most socioprofessional groups did not send their sons to the École Polytechnique; the scientific training there, with its emphasis on solid deductive knowledge, produced a technocratic managing class better suited to safeguarding authority than to entrepreneurship. For the new and, by the twentieth century, dominant breed of (to use Claude Louis Berthollet's term) "industrial engineers"—technical innovators and managers attuned to industrial processes, responsive to corporate capital rather than bureaucracy or professional aims—there were institutions lower in the hierarchy, notably the École Centrale des Arts et Manufactures (founded 1829). Humbler still, the *écoles d'arts et métiers* recruited from the lower-middle classes and generated graduates (*gadzarts*) for mechanical industry.

In the German states, where the universities were reluctant to provide technical education, engineers fit for civil service were produced in polytechnic institutions established in Dresden (1851), Berlin (1866), Karlsruhe (1865), and elsewhere. As places for the cultivation of technological learning, these *Technische Hochschulen* initially stressed theoretical "knowledge for its own sake," according to the German bureaucratic ethos; but, especially from the 1870s, practical laboratory culture, supported by industry and state, was increasingly incorporated within the academy. Nonetheless, the *Technische Hochschulen* successfully campaigned against the universities for the right (granted around 1900) to give doctor's degrees. The institutions and their alumni attracted international admiration and imitation for their perceived role in stimulating economic vibrancy, as academics learned to present technical education as a safeguard against national industrial decline.

In the early twentieth century, specialized engineering sub-disciplines with their related institutions serving specific technological contexts (in telegraphy, electrical power, sanitation, and aeronautics) proliferated. Later still, *big science and technoscience redrew the boundaries between engineer and scientist, reconfiguring their relative social status.

Robert Fox and George Weisz, eds., *The Organization of Science and Technology in France 1808–1914* (1980). R. A. Buchanan, *The Engineers: A History of the Engineering Profession in Britain 1750–1914* (1989).
BEN MARSDEN

ENGINEERING SCIENCE. Although the term "engineering science" first came into widespread use in the early twentieth century, "the science of the engineer" and equivalent phrases had been employed much earlier to signify parts of professional engineering transcending untutored practice, parts of science that could usefully inform innovative practice, and canonical theoretical knowledge qualified to enhance the engineer's professional status. From the mid-nineteenth century, individual engineering sciences became integrated into a systematic field of study, modeled on the physical sciences, but with a content specific to some particular engineering practice.

Recently, commentators seeking to develop an independent field of the history of technology have redefined engineering science as a "mode of knowledge" distinct, in content and style, from any physical science (David Channell, *The History of Engineering Science* [1989]). John M. Staudenmaier, S. J., writes of "engineering theory" as a "body of knowledge using experimental methods to construct a formal and mathematically structured intellectual system" (*Technology's Storytellers: Reweaving the Human Fabric* [1985]) to explain the behavior of a particular class of (idealized) artifact or of artifact-related materials. Its experimental methods involve models, testing machines, towing tanks, and wind tunnels; it is structured by the demands of practice, and thus develops ways of comparing models with full-scale apparatus, often relying on pragmatic approximation rather than rigor.

According to this model, engineering science has provided a common language through which a community of status-conscious practitioners has articulated its increasingly specialized concerns. Furthermore, a network of professional societies, schools, and laboratories, and a technical literature, existing especially in Europe and the United States from the late nineteenth century, catered for the creation, inculcation, and critical evaluation of engineering science. Those institutions acted as a cohesive and mediating force, orchestrating a transition from the scattered skills and knowledge associated with local problem-solving to the standardized, universal, abstract theories, designs, and practices issuing from the schools.

Historical investigation reveals variations and tensions within this schematic account. From the late seventeenth century, the Royal Society of London and the Academy of Sciences in Paris gave a central role to natural philosophy in perfecting practical arts for the public good. In the aftermath of the French Revolution, plans for a reformed and centralized system of engineering training, designed to foster economic health, were implemented. The École Polytechnique (1794–1795) in

Paris exemplified the militaristic production of technocrats. Elite savants furnished bright student engineers with a core curriculum concentrating on mathematics and theoretical sciences in readiness for the traditional branches of engineering (bridges, roads, artillery, mining) taught at the *écoles d'application*. While the École Polytechnique tended to eschew practical concerns, the École Centrale celebrated its distinctive *science industrielle* as a means of producing neither savants nor artisans but men who were at once scientists, generalists, and technological problem-solvers for the new industries.

The École Polytechnique's first director, Gaspard Monge, did, however, create a science of descriptive geometry (1795) for engineering drawing; founder Gaspard de Prony published the influential *Architecture hydraulique* (1790–1796); Lazare Carnot generalized the study of machine efficiency. Subsequent engineering theorists, many of them alumni of the school, included Charles Burdin (turbines); J. V. Poncelet (author of the *Mécanique industrielle* [1829]); J. N. P. Hachette (*Traité élémentaire des machines* [1811]); and C. L. M. H. Navier (who revised Belidor's *Architecture hydraulique* [1819]). They replaced the abstraction and microscopic model-building of the physics of Pierre-Simon *Laplace with a focus on macroscopic phenomena and (especially in the case of Jean Poncelet and Gaspard Coriolis) transformed engineering mechanics from an offshoot of rational mechanics into a new science of work.

If the stereotypical European engineer was scientifically schooled but industrially ineffectual, his British equivalent was economically potent and scientifically illiterate. Practical apprenticeship, according to received wisdom, made the British engineer. Yet he too had at his fingertips a miscellany of "modern improvements" in handbooks, encyclopedias (such as those of Abraham Rees and, later, Andrew Ure), parliamentary reports, works of mathematical practice by men such as Charles Hutton and Olinthus Gregory (associated with the military colleges), and Thomas Tredgold's classics on carpentry (1820) and cast iron (1822). This scattered literature sat beside scientific transactions in the libraries of new engineering associations which, like the Institution of Civil Engineers (1818) in London, produced their own publications.

In Britain, the consolidation of the science of the engineer coincided with the establishment of university-based engineering education, with one eye on European models (especially the Freiberg School of Mines) and the other on an agenda of professionalization. From the 1820s, the Mechanics Institutes (such as the Franklin Institute in the United States) targeted artisans with popular science. From the late 1830s, educators at King's College and University College in London offered mathematics, chemistry, geology, and natural philosophy to student engineers, and the Edinburgh natural philosopher James David Forbes examined "academical engineers" at the new Durham University. Low-status lecturers or demonstrators gave practical tuition (for example, in surveying), but the scientific engineer had still to complete his training with a practical apprenticeship.

From the 1840s, professors of engineering in Britain and the United States gradually articulated a corpus of unified engineering theory. Their strategies varied. Lewis Gordon in Glasgow assembled the best of collective contemporary engineering experience in a textbook keenly attuned to recent European developments (especially Benoit Fourneyron on turbines, and Gordon's mentor at Freiberg, Julius Weisbach, for hydraulics, the mechanics of machinery, and "mechanical effect"). Charles Blacker Vignoles codified railway construction in lectures reproduced in London-based newspapers. Eaton Hodgkinson collaborated with industrialist William Fairbairn and the British Association for the Advancement of Science (founded in 1831) in developing experimental regimes for the study of new engineering materials (wrought and cast iron) and innovative structures (notably the Britannia Bridge).

At the other extreme, Robert Willis's *Principles of Mechanism* (1841) mimicked the forms of deductive geometry in its kinematics, or the classification of modes of communicating motion by machinery independent of force. A standard university text, its techniques were superseded only by the work of Franz Reuleaux at the end of the century. Henry Moseley's *Mechanical Principles of Engineering and Architecture* (1843) and especially William Whewell's *Mechanics of Engineering* (1841) borrowed from French theorists of structures and work while aiming, like Willis, to place sanitized and "progressive" engineering sciences within an English liberal education.

In systematizing engineering knowledge, these author-professors acted as mediators between disembodied recent science and actual (or potential and lucrative) industrial concerns; as translators, directing the application of science to practice; or as organizers of disorganized craft practices. Such roles are consistent with Eugene Ferguson's claim that the province of the engineering sciences lay between "pure physical science" and the "empirical and intuitive knowledge of the engineer" (*Bibliography of the History of Technology* [1968]). Despite Auguste Comte's insistence that the engineer, although responsible for organizing the mediation between science and practice, was not a man of science, the École Centrale claimed that "industrial science" readied its students to mediate between a complex body of scientific knowledge and its applications to industry; and the *Engineer* insisted (1856) that the application of science to practice was itself a science.

W. J. M. Rankine's practically oriented work in thermodynamics and good relations with local industrialists secured him a chair of engineering in Glasgow (1855). From there he argued that the academically trained engineer bridged the gap between distinct worlds of natural philosophical questions (what are we to think?) and practical questions (what are we to do?). The scientific engineer worked without waste, husbanding human and natural resources, economically achieving the practical aims precisely delimited by exact theoretical science, bearing in mind the quantifiable constraints of the market. With Rankine's Certificate of Proficiency in Engineering Science (from 1862), an engineer could plan with certainty and innovate with confidence.

A forum for topical scientific engineering discourse existed in technical periodicals, including *Engineering* (1866) and the publications of the Institution of Naval Architects, the American Society of Mechanical Engineers, and many other specialist professional engineering associations. Rankine's monumental and long-lasting textbooks, especially *Applied Mechanics* (1858), defined the bedrock of the scientific engineer. The basic repertoire offered theoretical engineering sciences, including soil mechanics, hydraulics, structures and frameworks, and elasticity, but many found Rankine's works inscrutable and, ironically, divorced from practice.

Thus, Isambard Kingdom Brunel objected to the anti-progressive standardization of engineering science and the "best practices" deduced by government commissions set up to learn from railway disasters, collapsing bridges, or naval catastrophes. These commissions harbored many professorial engineering experts and academicians, such as the electrical engineer Fleeming Jenkin (Edinburgh) and the heat-engine theorist Osborne Reynolds (Manchester), who lobbied for the training of scientific engineers to ensure public safety and economic prosperity. They looked covetously to well-funded institutions, academies, and polytechnics in Europe and the United States, but generally choked at the idea of mass-produced engineers, preferring to nurture a scientifically trained, gentlemanly elite of professional leaders.

In addition to having a theoretical base, the engineering sciences developed distinctive experimental practices. For science-intensive electrical engineering companies (such as Siemens), laboratories produced reliable electrical measures in a context of international *standardization. For German *Technische Hochschulen*, the research laboratory linked college technical practice with industrial production. From 1868, the Polytechnic Institute of Munich had a materials-testing laboratory. Toward the end of the nineteenth century, Robert Thurston at Cornell, Alexander Kennedy in London, James Alfred Ewing in Cambridge, and college engineering professors generally argued that purpose-built laboratories, long essentials for chemistry and physics, were now vital to engineering teaching and research, particularly as venues for precision measurement. Trinity College, Oxford, had an engineering laboratory from 1886, and the University followed suit in 1914 with its own lab—well away, however, from the "Science Area."

The engineering science laboratory also modeled practice. Schools of engineering accumulated both demonstration apparatus and experimental models. Exactly what the behavior of a small-scale experimental model or theoretical simulation revealed about its full-sized counterpart was crucial—and unclear. Benjamin Isherwood insisted that only from full-scale experimental researches like his, in steam engineering for the U.S. Navy in the 1860s, could valid general "engineering laws" emerge; the British Association accumulated vast quantities of (unreducible) data relating engine power, ship shapes, and speeds on a similar understanding (voiced by naval architect C. W. Merrifield). Economic pressures forced engineers to learn how to "scale up." In hydrodynamics, John Scott Russell's "wavelines" and Rankine's "streamlines" pointed plausibly to low-resistance hull shapes; but from the late 1860s, William Froude turned, additionally, to experimental tanks. In aeronautical engineering, a growth area especially after World War I, wind tunnels could give workable design solutions where fluid dynamics failed. Eventually, computer simulation would complement, and in some respects supersede, those modeling techniques.

At the beginning of the twentieth century, engineering science began to appear as a named academic discipline. Charles Frewen Jenkin, son of the Edinburgh academician and himself an expert in aeronautical materials, entitled his professorial address at Oxford "Engineering Science" (1908). The following decades saw a flurry of research publications in the aeronautical, mechanical, and marine fields—now deemed branches of engineering science; textbooks in engineering science began to appear; Macmillan's

Engineering Science Series (1922) included works on electrical engineering and telephony.

R. V. Southwell, Jenkin's successor at Oxford, launched a prestigious Oxford Engineering Science Series in 1932. Southwell's career neatly illustrates the ironies of engineering science in practice. In an anti-industrial academic environment, he offered a small group of students "essential scientific equipment" in the form of systematic theoretical knowledge of idealized engineering systems. Worried that more and more research took place in government labs or large firms, he wanted to enhance the fragile reputation of academic engineering. He taught that engineering science was not the key to industrial success; it used mathematics and physics, but unlike them, considered practical material constraints, approximate (and, increasingly, computable) solutions, matter in bulk, and visualizable models. In Oxford, at least, engineering science was a research end in itself.

E. T. Layton, Jr., "American Ideologies of Science and Engineering," *Technology and Culture* 17 (1976): 688–701. John Hubbel Weiss, *The Making of Technological Man: The Social Origins of French Engineering Education* (1982). David F. Channell, *The History of Engineering Science: An Annotated Bibliography* (1989). Ben Marsden, "Engineering Science in Glasgow: Economy, Efficiency and Measurement as Prime Movers in the Differentiation of an Academic Discipline," *British Journal for the History of Science* 25 (1992): 319–346. Thomas Wright, "Scale, Models, Similitude and Dimensions: Aspects of Mid-Nineteenth-Century Engineering Science," *Annals of Science* 49 (1992): 233–254. Jack Morrell, *Science at Oxford 1914–1939: Transforming an Arts University* (1997).

BEN MARSDEN

ENGLISH-SPEAKING WORLD. Science has had several principal languages over the centuries—Greek, Latin, Italian, Arabic, Chinese, French, German, and English. During the eighteenth century, French dominated discourse about natural knowledge. During the late nineteenth century, German became the principal scientific language for a large area including, besides Germany itself, Austria-Hungary, Sweden, Denmark, the Netherlands, and parts of Switzerland. Japanese who wished to pursue a career in medicine had to know German; so did citizens of the United States and Imperial Russia. English supplanted German to become the worldwide means of communication in commerce and travel as well as in science. The so-called "English-speaking world"—Britain, the United States, Australia, New Zealand, Ireland, most of Canada, and large enclaves elsewhere—has the great advantage of possessing this universal language as its mother tongue.

Britain's Legacy

Although during the nineteenth century, the British Isles produced extraordinary achievements in science—as indicated by the names Charles *Darwin, Charles *Lyell, James Clerk *Maxwell, and William *Thomson—and natural science had established a secure foothold in the other English-speaking countries, Germany was the leader in world science around 1900. Thanks in part to the work of Justus von *Liebig at Giessen from the late 1820s, the Germans had become the unchallenged leaders in chemistry, both in the academy and in industrial applications. Britain's precocious William H. Perkin founded the aniline dye industry in the late 1860s, but Germany's stronger institutional base in applied science allowed it to capture the manufacture of all synthetic organic dyes. German higher education seemed equally strong. The twenty-eight German-speaking universities in central Europe, mostly located in Germany itself, had no parallel in the world in 1900. With their many distinguished professors, excellent laboratories, and easy accessibility, the German institutions drew students from all over the world, including significant numbers from the United States and Britain.

Against this array, Britain had the ancient universities of Cambridge and Oxford, the old Scottish universities, Trinity College, Dublin, and a rising number of municipal or "red brick" institutions, notably the universities of London, Manchester, and Liverpool. McGill University, the University of Toronto, three small institutions of higher learning in New Zealand, and four universities of modest size in Australia represented the higher education available in the English-speaking parts of the British Empire. The United States had begun copying the German academic model with the establishment of Johns Hopkins in the 1870s, the University of Chicago in the 1890s, and the importation of an embryonic research ethic into the older East Coast institutions. The Land Grant Act of the Civil War years had established universities that would become research centers, notably in the Middle West and California. MIT was a modest engineering school; Cal Tech did not exist.

Partly in response to the German ascendancy, England debated the mission of its established universities, the role of research, and the place of experimental science in them. Cambridge had long excelled in mathematics, but until 1851, no course of study leading to a degree in chemistry or physics existed at Cambridge or Oxford. That year, Cambridge created the Natural Science Tripos (examination), which, however, long had a second-class standing in the university. Dirtying their hands with experimental work still grated on the sensibilities of Oxford dons, who regarded chemistry as "stinks." External pressures for change from political leaders, industrialists, and British scientists with German academic degrees had aroused strong opposition. The situation for natural science was much more favorable at the red-brick universities, especially the University of London, whose BSc degree could be obtained by examination even by non-resident students.

A concerted movement for sweeping changes in the ancient universities gathered momentum after 1870. The Devonshire Commission, led by William Cavendish, Duke of Devonshire, undertook a six-year investigation (1870–1876) into the state of British science. Scottish institutions came off well; the members of the Commission praised William Thomson's program in physics at the University of Glasgow, despite his master-apprentice approach to instruction. But the Commission harshly criticized conditions in England. It recommended the enhancement of existing resources and several new initiatives: eleven new chairs for science at Oxford, a National Ministry of Science and Education, a new astronomical observatory, a redirection of collegiate fellowships away from the Classics and toward the natural sciences, and a program of grants for research by established scholars to be administered by the Royal Society.

For the rise of the English-speaking countries in science, a particularly important initiative from the 1870s was the founding of the Cavendish Laboratory at Cambridge. The Duke of Devonshire served as Chancellor of Cambridge while chairing the reform commission. A descendant of two leading physical scientists (Henry *Cavendish and Robert *Boyle), Devonshire had excelled in mathematics as a student and had considerable experience in the iron and steel business. His financial contribution, more than sufficient to build and equip the laboratory, set a new standard of support for science in an academic setting. The first Cavendish Professor of Experimental Physics, James Clerk Maxwell, started a tradition of excellence that under his successors made the Cavendish for half a century what Niels *Bohr called the center of physics.

Lord Rayleigh, successor to Maxwell as Cavendish director, received a Nobel Prize in physics in 1904 for his discovery of argon. In 1906, Rayleigh's successor, J. J. *Thomson, received a Nobel for his discovery of the *electron. Between 1895 and 1898, the New Zealand–born Ernest *Rutherford worked with Thomson at the Cavendish on x-ray induced conductivity in gases. The work for which Rutherford received the Nobel Prize for chemistry in 1908—his

development of a modern theory of radioactivity—took place at McGill University in Canada. Rutherford succeeded Thomson as Cavendish Professor and Laboratory Director in 1919.

Impact of World War I

When war broke out in August 1914, science and engineering students in Germany numbered about 16,000; in Britain, 4,000 at most. A 1910 estimate put the number of working British industrial chemists at one-third the number working in Germany. Germany was spending three times as much on its universities as Britain. Britain depended on Germany for imports of dyestuffs as well as tungsten for making steel, pharmaceuticals, magnetos, certain kinds of optical glass, and, to some degree, even explosives.

In the United States, surging immigration, industrialization, and economic expansion had created a favorable environment for the growth of science at the beginning of the century. John D. Rockefeller's fortune made possible the establishment of the University of Chicago in 1891; unlike the East Coast institutions, it emphasized graduate education from the start. In 1901, the United States Congress created the National Bureau of Standards and authorized it to conduct research deemed necessary to establish appropriate standards for industry; a prominent physicist was named its director (see BUREAU OF STANDARDS). An unprecedented gift of $10 million from Andrew Carnegie resulted in the creation of the Carnegie Institution of Washington. Other industrialists founded Stanford University, Vanderbilt University, and the Rockefeller Institute for Medical Research in New York.

World War I initiated the rise to preeminence of the English-speaking countries in science, partly because the war ruined Germany's economy, but mainly because it accelerated institutional developments that might otherwise have been delayed. In 1914, Britain created the British Dyestuffs Corporation, and the following year set up what became the Department of Scientific and Industrial Research (DSIR) and encouraged the formation of Research Associations to offset deficiencies in applied science made all too apparent by the war. The DSIR survived the war to continue government support to science. In 1916, a Committee on the Neglect of Science, chaired by Lord Rayleigh, launched major initiatives directed at the creation of a scientifically more literate public. Although the United States did not enter the war until 1917, it had moved in the same direction, creating the National Research Council (NRC) in 1915, letting research contracts to twenty-one university laboratories, and undertaking research on optical glass and chemical weapons. The performance of applied science during the war caused many firms in Britain and the United States to set up or expand industrial research laboratories after the war.

Southern California first became prominent in science just after the war when George Ellery Hale, one of the creators of the NRC and director of the Mt. Wilson Observatory, persuaded local industrialists that the region should have a major research university with forward-looking programs in science. Hale's recruitment of several distinguished faculty members, together with the large endowment pledged by his business supporters, made possible the conversion of the small Throop Polytechnic Institute (founded 1891) into the California Institute of Technology in 1921, with R. A. Millikan as its founding president. Science and engineering in Canada, already strong at McGill from private benefactors, gained ground and reputation at the University of Toronto. There, with the active collaboration of Charles Best and help from J. J. R. Macleod and James Collip, Frederick Banting isolated insulin from the pancreas and demonstrated its effectiveness in treating diabetes. In 1923, Banting became the first Canadian to receive a Nobel Prize in physiology or medicine.

Just before the war ended, the Allies took steps to isolate German science. Delegates meeting in London in the fall of 1918 voted to dissolve the International Association of Academies, founded in the 1890s and headquartered in Berlin, and create a new International Research Council (IRC) from which citizens of the former Central Powers were excluded. The IRC finally admitted Germany in 1926. In the interim, Germany had not ceded its leadership. Despite the policy of the IRC, able young Americans such as Linus *Pauling and J. Robert Oppenheimer went to Munich and Göttingen to study quantum theory.

The English-speaking countries had three advantages that would gain them eventual leadership across the sciences: numbers, resources, and a language that enabled them to function as a single, large community of scholars. The point may best be made by pointing to migration within the community. British-style education throughout the Empire had long allowed colonials to study or work in Britain. Thus did the eminent medical clinician and researcher, William Osler, born and educated in Canada, later teach in Britain (1904–1919). Similarly, Rutherford, after taking his degree at Canterbury College in New Zealand (1894), went to study at Cambridge. The travel went both ways. In 1885, William Henry Bragg, a protégé of J. J. Thomson's at Cambridge, moved to Australia and taught physics at Adelaide until 1909, later returning to Leeds in England. His son, William Lawrence Bragg, born and educated in Australia, became Cavendish Professor of Physics at Cambridge in 1938.

The pace of these movements began to increase during and immediately following World War I, as a result of improved transportation and more opportunities in research. Born and educated at Capetown in South Africa, Max Theiler studied medicine at the London School of Tropical Medicine from 1916 to 1922; he then worked in the United States at Harvard, the Rockefeller Institute, and Yale. Also during World War I, India's great mathematician Srinivasa Ramanujan from Madras visited Britain and became a Fellow of Trinity College, Cambridge, before returning home. Frank McFarlane Burnet, born in Australia in 1899 and educated at the University of Melbourne, studied at the Lister Institute in London, then returned home. And in the 1930s, both Subramanyan *Chandrasekhar from India and John Eccles of Australia studied in England, at Cambridge and Oxford, respectively. Chandrasekhar became professor of *astrophysics at the University of Chicago after his years in England.

Eccles's career shows particularly well the career possibilities available to a scientist in the English-speaking world. Born in Melbourne in 1903, he studied and taught at Oxford from 1925 to 1937, directed a research institute in Sydney from 1937 to 1943, then moved to New Zealand's University of Otago. From 1952 to 1966, Eccles served as professor of neurophysiology at the Australian National University, Canberra, after which he moved to Chicago, and later Buffalo, New York. Other colonial systems—those of the French, the Dutch, the Spanish, and the Americans—also produced patterns of this kind, though none on anything like the scale of the British.

The Migration From Germany

Adolf Hitler's accession to power in Germany in 1933 completed what World War I had begun, preparing the way for the supremacy of the English-speaking countries in science. Following the enactment of the Nuremberg Laws later that year, most Jewish scientists were forbidden to work in Germany's universities. Those who could leave gradually left Germany, more often than not for Britain (Max Born) or the United States (James Franck). Albert *Einstein had left Germany in 1930; he and several other distinguished European scientists became members of the Institute for Advanced Study in Princeton, a private institution founded earlier that year by the family of Louis Bamberger, a department store magnate.

Science in the United States was deeply enriched by the European immigrants. The physicist Maria Goeppert Mayer, a protégé of Max Born's at Göttingen, came in 1930 as the wife of an American chemist, Joseph Mayer. Naturalized in 1933, Goeppert Mayer taught physics at several institutions, often without compensation. One was the University of Chicago, a leading center for physics after the arrival of Enrico *Fermi and Edward Teller, themselves exiles from Fascism and Nazism. Goeppert Mayer's presence at Chicago led to a position at the nearby Argonne National Laboratory, where she noted the existence of periodic properties for nuclear isotopes and their resemblance to electron shells in atoms. This work culminated in her shell model of the nucleus, for which she shared a Nobel Prize in 1963 with Hans Jensen and Eugene Wigner. In 1935, Hans Bethe arrived from Germany by way of a two-year position at Bristol in England. From his position at Cornell as professor of physics, Bethe moved to Los Alamos, New Mexico, where he played a leading role in the Manhattan Project to develop the atomic bomb. A consummate statesman of science, as well as a brilliant researcher, Bethe received an unshared Nobel in 1967, partly for his studies of energy production in the sun and other stars.

Nor was it only physics that benefited from the influx of European scientists. Though trained originally in theoretical physics, Max Delbrück—a great-grandson of Justus von *Liebig—came to the United States in 1937 on a Rockefeller Foundation fellowship, switched to viral genetics, and investigated bacteriophages at several institutions. He shared a Nobel Prize in physiology or medicine in 1969. Fritz Lipmann, a biochemist, arrived in 1939. Another Rockefeller Foundation fellow, he held positions at Cornell, Harvard, and the Rockefeller Institute (renamed Rockefeller University). Lipmann became interested in metabolism and the enzymes that aid in digestion. He shared a Nobel Prize in 1953 with another German émigré, Hans *Krebs. Krebs is notable as one of a significant but smaller number of scientists fleeing Nazi persecution who found opportunity in Britain, as opposed to the United States. Others—Otto Frisch, Rudolf Peierls, and Franz Simon—made suggestions that were instrumental in committing Britain and the United States to the atomic bomb project.

Postwar Science

Despite its much smaller scientific community and more modest funding base for research, Britain continued to attract talented foreign and domestic investigators in the postwar period. In 1951, the young American biologist James. D. *Watson arrived at the Cavendish Laboratory after completing his Ph.D. at Indiana University. At Cambridge, he met and began working with a somewhat older British colleague, Francis *Crick. In pursuing their classic work on the structure of *DNA, they took full advantage of opportunities and information produced in the English-speaking world. In 1944, Oswald Avery of Columbia University had argued that DNA was the genetic material of bacteria. Essential knowledge of amino acids came from Erwin Chargaff of Columbia, whom Watson and Crick met in Cambridge; Rosalind Franklin, Maurice Wilkins, and Raymond Gosling at the University of London supplied x-ray diffraction photographs of DNA; Linus Pauling of Cal Tech suggested a clever but mistaken model of the DNA molecule that Watson and Crick improved on. The foundation of *molecular biology, like the cooperation on the atomic bomb and many other episodes in science and technology in the twentieth century, was a product of interactions in the English-speaking world.

The United States, with its large land base and growing wealth, in some ways after 1900 duplicated the institutions of the British Empire (later British Commonwealth) within its own borders. In the pursuit of science, as in the acquisition of wealth, the American academic system, and to some degree the Canadian, were founts of opportunity. These nations created effective systems of universities and government and industrial laboratories, eager for competent staff irrespective of their national origins, readily open to one another, yet intensely competitive in a national context. Innovations historically successful in one institution—especially universities—were usually copied in others. In this sense, the fifty leading research universities of North America have something in common with the German universities of 1900. The industrial research laboratories have their parallels in the old dyestuffs industries, and the national laboratories a pale antecedent in the famous German bureau of standards, the Physikalische-Technische Reichsanstalt.

Ironically, the factor that underlay the hegemony of the English-speaking countries in science—a common language—has created a more even playing field for the rest of the world. Nearly all scientists now communicate in English, except with native speakers of their own language. That makes possible large transnational collaborations and migrations for study and research. An ever more tightly knit Europe, whose scientists communicate in English and sometimes enjoy support approaching American levels, has challenged the supremacy of the English-speaking nations, or rather of the United States, since Britain sometimes belongs to Europe, and Europe sometimes speaks English. Asia should not be left out of the equation. To take one straw in the wind, in the

1990s, Ken'ichi *Fukui, Japan's first Nobel laureate in chemistry, was able to attract postdoctoral fellows from the United States and other countries because all shared the common language of English.

See also WORLD WAR II AND COLD WAR.

D. S. L. Cardwell, *The Organisation of Science in England* (1967). Donald Fleming and Bernard Bailyn, eds., *The Intellectual Migration: Europe and America, 1930–1960* (1969). Daniel J. Kevles, *The Physicists: The History of a Scientific Community in Modern America* (1971). G. Bruce Doern, *Science and Politics in Canada* (1972). Roy M. MacLeod, "Resources of Science in Victorian England: The Endowment of Science Movement, 1868–1900," in Peter Mathias, ed., *Science and Society, 1600–1900* (1972): 111–166. J. G. Crowther, *The Cavendish Laboratory, 1874–1974* (1974). David Wilson, *Rutherford: Simple Genius* (1983). R. W. Home, ed., *Australian Science in the Making* (1988). Frank N. Magill, ed., *The Nobel Prize Winners: Physics, Chemistry, Medicine,* 9 vols. (1989). Jarlath Ronayne and Campbell Boag, *Science and Technology in Australasia, Antarctica and the Pacific Islands* (1989). Arthur M. Silverstein, *A History of Immunology* (1989). Thomas Hager, *Force of Nature: The Life of Linus Pauling* (1995).

JAMES BARTHOLOMEW

ENLIGHTENMENT AND INDUSTRIAL REVOLUTION.

The European Enlightenment and the First Industrial Revolution, not often seen as closely related, were in fact intertwined profoundly. Many people active in promoting enlightened causes, freedom of religion, the abolition of slavery, and (more in Catholic than in Protestant Europe) anticlericalism were also active in industrial development. For example, Joseph Priestley was minister in Birmingham's Unitarian chapel, where early industrialists and skilled workers worshiped. James Watt, perfecter of the steam engine, was among his closest friends. Both spoke out against slavery. Priestley championed the American Revolution and the early stages of the French Revolution, and his sermons addressed the issues raised for the Christian by worldly success and prosperity. In Manchester the Unitarians supported liberal causes and the first mechanizers of cotton like James M'Connel and John Kennedy associated with the chapel or its intellectual offshoots. In both places we can justifiably talk about an enlightened industrialism. At the same moment the Scottish Enlightenment provided the first and perhaps the greatest theorist of industry and capitalism, Adam Smith. He examined the inner workings of both in *The Wealth of Nations* (1776).

On a more general level the freedom of thought that was the hallmark of enlightened opinion bore close relation to the scientific and intellectual innovation we associate with industrial circles. To illustrate: Jean Antoine Chaptal, with a medical and chemical background, established one of the most successful factories in the Montpellier region, and in 1789 wrote on behalf of the French Revolution. He believed, as did many of the revolutionaries, that the clergy had inhibited French industry by not allowing the education system to address the applied and the practical. In the 1790s French education was turned on its head, new men and women replaced the priests and nuns, and science focused on application became a key to the new curriculum. Chaptal and his friends formed the intelligentsia of the revolution, barely managed to avoid death during the Terror, and went on to secularize education and to promote industrial development. They did so in conscious imitation of the British. After 1800 wherever in Western Europe industry was being encouraged, enlightened reformers looked to the French and British models.

Newtonian science provided an important bridge between the Enlightenment and the Industrial Revolution. *Newton's masterworks in natural philosophy demonstrated that a powerful mind free to follow its own way could discover the principles of the world system and overthrow ideas about light and matter that had misled lesser humans for centuries. Newton also thought for himself in matters of religion, further enhancing his standing among philosophers who identified freedom of thought with progress in science. The most influential of these free thinkers was Voltaire, who lived a few years in England, extolled English thought and institutions in a little book, *Lettres philosophiques* (1733), banned in France, and brought out the first French popularization of Newton's science, *Eléments de la philosophie de Newton* (1738). But Newton was no deist or atheist as were many of the leaders of the French Enlightenment. His private religiosity was intense and millenarian and should not be associated with the worldliness of enlightened pundits and industrialists.

Newtonian mechanics possessed a deeply practical side, particularly as developed by his assistant Jean Theophilus Desaguliers, the son of a Huguenot refugee. Desaguliers showed the application of mechanical principles to the design of machines, mills, pumps, and so forth. In that guise, Newtonianism fit in with the program identified a century earlier with Francis Bacon, the cultivation of science for the benefit of ordinary life. This program also informed the great midcentury engine of Enlightenment, the *Encyclopédie* of Denis Diderot and Jean d'Alembert. Besides its wide coverage of natural knowledge and pronounced anticlericalism, it displayed, in eleven volumes of plates, the processes of arts and manufacturers as carried on in France. The information served not only readers curious to know how commodities were made, but also society as a whole by inspiring natural philosophers to apply their science to improving artisanal practices. Inhibited by guild and other restrictions, however, French manufacturers lagged behind their English counterparts, notably in scientific instruments and technically advanced machinery, like Watt's steam engine. French educational institutions were two generations behind their British counterparts in adopting Newton's mechanics as the centerpiece of their science curriculum.

We may take James Watt as an example of what an eager young man in midcentury London could learn about science. Like many practical men, Watt read widely in science and in the enlightened writings of the time. He took a tutor in mathematics and worked out of the Newtonian textbook by the Dutch physicist Willem Jacob 'sGravesande. Once prosperous thanks to Watt's genius at invention, the family had little interest in sermon literature but bought books that would become the classics of the Enlightenment, works by Priestley, Montesquieu, and Smith, and even the occasional piece of pornography. They saw themselves as meritorious and deserving of advancement, and preferred systems of government closer to the republican than the monarchical. Science, liberal government, and freedom of thought worked to promote the interests of early industrialists better than clerical education and absolutist systems for controlling production. Early industrialists needed the government to build canals and turnpikes, but they wanted as little interference as possible when it came to the quality of their goods or the way they treated their workers. They gravitated naturally toward enlightened ideas although they were deeply suspicious of democracy for anyone but themselves. The radicalism of the Enlightenment—best symbolized by the democratic writings of Jean-Jacques Rousseau and the extravagant materialism of Baron d'Holbach—made early industrialists nervous. They approved of Voltaire but not Marat. The Enlightenment was not the Industrial Revolution, but the one is ultimately unthinkable without the other. Together they laid the foundations for modern systems of economy, government, and science.

John Gascoigne, *Joseph Banks and the English Enlightenment. Useful Knowledge and Polite Culture* (1994). Margaret C. Jacob, *Scientific Culture and the Making of the Industrial West* (1997). Maxine Berg and Kristine Bruland, eds., *Technological Revolutions in Europe. Historical Perspectives* (1998). Margaret Jacob and David Reid, "Technical Knowledge and the Mental Universe of Manchester's Early Cotton Manufacturers," *Canadian Journal of History* 36 (August 2001): 1–23.

MARGARET JACOB

ENTOMOLOGY. Entomology (along with ornithology) was one of the first fields of natural knowledge to professionalize. Men with specialist knowledge in the area were among the earliest of scientific employees in museums. Enthusiastic input from collectors, illustrators, travelers, and specialist taxonomists buoyed interest. During the seventeenth and early eighteenth centuries, insects featured in microscopical researches like those of Marcello Malpighi and Robert Hooke. Theologians and naturalists alike praised the beauty and complexity of insects, many seeing them as evidence for the wisdom and perfect design of God's works. Religious symbolism and entomological description joined in other ways: Jan Swammerdam explained insect life cycles in *The Natural History of Insects* (1737) as signifying Christ's resurrection. At the same time, the social insects, such as ants, bees, and wasps, provided civic metaphors, as in Bernard Mandeville's *Fable of the Bees* (1714), which proposes an idealized structure for human society. Notions about the efficiency of the hierarchical castes in ants' nests or beehives, and the concept of many individuals working for the general good, appeared in many philosophical tracts and utopian fictions. Moral lessons had long been found in locust plagues and the like. Insects also had economic value, as exemplified by the cochineal beetle, which, for several centuries, provided the primary source for red dye.

Many important entomological collections were founded and expanded during the eighteenth century. New techniques and devices for catching specimens, including nets, traps, and lamps, and conventions for adequate preservation and display, were developed. Insect collections, like those of shells or minerals, amused the wealthy, intrigued the learned, and provided financial opportunities for specimen hunters, shopkeepers, and book publishers. The publication of illustrated manuals boomed, complemented by catalogues and lists of identifying names. Specialist classification schemes—especially divisions of day fliers from night fliers, and beetles and bugs from butterflies—were introduced. Carl *Linnaeus identified seven insect orders (today, twenty-nine are generally recognized). Insects played a key role in systems of physicotheology, such as Linnaeus's scheme of natural economy, which adumbrated the concept of a food chain. The relative number of organisms and the balance, or harmony, between them depended on insects as an essential source of food for birds. William Paley, in his discussion of the "polity of nature" in *Natural Theology or evidences of the existence and attributes of the deity* (1807), argued that the usefulness and beauty of insects counterbalanced their stings and bites. The curious reproductive patterns of some insects such as aphids aroused great scientific and philosophical interest. For much of the century the word "insect" was applied indiscriminately to most small organisms.

In Great Britain, an Entomological Society was founded in 1833 by a small group of enthusiasts—associated with an entomological club and periodical—who broke from the Zoological Society. The new society rapidly became a locus for expert taxonomic work, publishing scientific *Transactions* from 1834. In Paris and elsewhere specialist taxonomists were also producing many detailed studies of individual genera or families.

A number of unusual classification schemes appeared in the early nineteenth century, most notably William Macleay's quinarian system (based on grouping genera and families in fives)—a significant attempt to reveal meaningful affinities and relationships between insects. Macleay's system was developed further by William Swainson. Parasitic insects were recognized by him as either highly complex or very reduced in structure when compared to the basic type to which they were related. This recognition allowed him to classify many organisms previously difficult to place. Identification of the various stages of insect life was also a focus of research throughout the century. Concepts of host organisms for different stages of the cycle, and insects as vectors of disease, were introduced, although they were not codified until much later.

The *spontaneous generation controversy—centered around the debate between Louis *Pasteur and Félix-Archimède Pouchet—demonstrated the importance of distinguishing between asexual (parthenogenetic) forms of reproduction, as displayed by aphids and other organisms, and sexual reproduction; and of establishing the mechanisms that underlay fertilization and cell multiplication. In Great Britain, the natural historical approach predominated. Sir John Lubbock, the Victorian politician and banker, began his career as a talented entomologist. Aided by advances in histological techniques and microscope optics, he described several key stages of the reproductive process in insects, tracing the "germ cells" through the generations. He later observed social behavior, perhaps inventing the glass observation hive in the process, and wrote the best-selling *Ants, Bees and Wasps* (1882). In the

early twentieth century Austrian biologist Karl von Frisch went further, investigating the routes taken by bees in search of food and the "dance" by which they communicated information to others in the hive.

Charles *Darwin devoted much time to studying the mutual dependency between insects and flowers, the one acquiring food or nectar and the other achieving cross-fertilization. Darwin's results on the respective adaptations developed by both sets of organisms appeared primarily in *The Effects of Cross and Self fertilisation* (1876) although he had included much original matter on the subject in his *Orchids* (1861). He based his notion of the mutual dependency of insects and flowers on a tract published by Christian Konrad Sprengel in 1793. Notwithstanding the fanfare about apes and angels following publication of Darwin's *Origin of Species* (1859), evolution was best demonstrated by the insect world. In the 1860s and 1870s, Henry Walter Bates's and Fritz Müller's demonstrations of various forms of mimicry, in which palatable insects mimic unpalatable ones or inanimate objects like leaves in order to survive, gave strong support to evolutionary theory. Henry Bernard Davis Kettlewell's study of industrial melanism in moths and their differential survival rates ultimately substantiated modern evolutionary theory. Insects also proved crucial for genetic research, especially into mutation and recombination of chromosomes. Thomas Hunt *Morgan deliberately chose the fruit fly for his laboratory work because of its rapid breeding, convenient maintenance, adaptable external characteristics, and large, easily observable chromosomes. In a few years the species was acknowledged as a highly suitable organism for experimental work. Much of Morgan's achievement, in fact, rested on having chosen an appropriate organism for the job. Other insects, particularly weevils, were similarly important in early population genetics and ecological modeling in the laboratory, as in Charles Elton's seminal work *Animal Ecology* (1927).

Insects are a significant factor in the medical health field, particularly epidemiology. Alexandre Yersin established fleas as the causative agent in plague in the 1880s. Alphonse Laveran saw the malarial parasite in human blood while working in Algeria in 1880; Patrick Manson identified the mosquito as the parasite's customary host, or vector (1894); and Ronald Ross disclosed the malarial life cycle in 1897. Throughout the twentieth century pest control—for both medical and agricultural purposes—focused on programs of insect extermination, often with powerful insecticides like DDT.

J. Michael Chalmers-Hunt, *Entomological Bygones or Historical Entomological Collecting Equipment and Associated Memorabilia, Archives of Natural History*, vol. 21, part 3 (1994): 357–378. Paul W. Riegert, *From Arsenic to DDT: A History of Entomology in Western Canada* (1980). Robert E. Kohler, *Lords of the Fly. Drosophila Genetics and the Experimental Life* (1994). W. Conner Sorensen, *Brethren of the Net: American Entomology, 1840–1880* (1995). Paolo Palladino, *Entomology, Ecology and Agriculture: The Making of Scientific Careers in North America, 1885–1985* (1996). Michael A. Salmon, *The Aurelian Legacy: British Butterflies and Their Collectors* (2001).

JANET BROWNE

ENTROPY. Many physicists and chemists quip that the second law of thermodynamics has as many formulations as there are physicists and chemists. Perhaps the most intriguing expression of the law is Ludwig *Boltzmann's paraphrase of Willard *Gibbs: "The impossibility of an uncompensated decrease in entropy seems to be reduced to improbability."

Entropy owes its birth to a paradox first pointed out by William *Thomson in 1847: energy cannot be destroyed or created, yet heat energy loses its capacity to do work (for example, to raise a weight) when it is transferred from a warm body to a cold one. In 1852 he suggested that in processes like heat conduction energy is not lost but becomes "dissipated" or unavailable. Furthermore, the dissipation, according to Thomson, amounts to a general law of nature, expressing the "directionality" of natural processes. The Scottish engineer Macquorn Rankine and Rudolf Clausius proposed a new concept, which represented the same tendency of energy towards dissipation. Initially called "thermodynamic function" by Rankine and "disgregation" by Clausius, in 1865 the latter gave the concept its definitive name, "entropy," after the Greek word for transformation. Every process that takes place in an isolated system increases the system's entropy. Clausius thus formulated the first and second laws of thermodynamics in his statement "The energy of the universe is constant, its entropy tends to a maximum." Hence, all large-scale matter will eventually reach a uniform temperature, there will be no available energy to do work, and the universe will suffer a slow "heat death."

In 1871 James Clerk *Maxwell published a thought-experiment attempting to show that heat need not always flow from a warmer to a colder body. A microscopic agent ("Maxwell's demon," as Thomson latter dubbed it), controlling a diaphragm on a wall separating a hot and a cold gas, could choose to let through only molecules of the cold gas moving faster than the average speed of the molecules of the hot gas. In that way, heat would flow from the cold to the hot gas. This thought-experiment indicated that the "dissipation" of energy was not inherent in nature, but arose from human inability to control microscopic processes. The second law of thermodynamics has only statistical validity—in macroscopic regions entropy *almost* always increases.

Boltzmann attempted to resolve a serious problem pointed out by his colleague Joseph Loschmidt in 1876, and by Thomson two years earlier, that undermined the mechanical interpretation of thermodynamics and of the second law. This law suggests that an asymmetry in times dominates natural processes; the passage of time results in an irreversible change, the increase of entropy. However, if the laws of mechanics govern the constituents of thermodynamic systems, their evolution should be reversible, since the laws of mechanics are the same whether time flows forward or backward: Newton's laws retrodict the moon's position a thousand years ago as readily as they predict its position a thousand years from now. Prima facie, there seems to be no mechanical counterpart to the second law of thermodynamics. In 1877 Boltzmann found a way out of this difficulty by interpreting the second law in the sense of Maxwell's demon. According to Boltzmann's calculus, to each macroscopic state of a system correspond many microstates (particular distributions of energy among the molecules of the system) that Boltzmann considered to be equally probable. Accordingly, the probability of a macroscopic state was determined by the number of microstates corresponding to it. Boltzmann then identified the entropy of a system with a logarithmic function of the probability of its macroscopic state. On that interpretation, the second law asserted that thermodynamic systems have the tendency to evolve toward more probable states. A decrease of entropy was unlikely, but not impossible.

In 1906 Walther Nernst formulated his heat theorem, which stated that if a chemical change took place between pure crystalline solids at absolute zero, there would be no change in entropy. Its more general formulation is accepted as the third law of thermodynamics: the maximum work obtainable from a process can be calculated from the heat evolved at temperatures close to absolute zero. More commonly the third law states that it is impossible to cool a body to absolute zero by any finite process and that at absolute zero all bodies tend to have the same constant entropy, which could be arbitrarily set to zero.

S. G. Brush, *The Kind of Motion We Call Heat: A History of the Kinetic Theory of Gases in the 19th Century*, 2 vols. (1976). Penha Maria Cardoso Dias, "The Conceptual Import of Carnot's Theorem to the Discovery of Entropy," in *Archive for History of Exact Sciences* 49, ed. Penha Maria Cardoso Dias, Simone Pinheiro Pinto, and Deisemar Hollanda Cassiano (1995): 135–161. Robert Locqueneux, *Prehistoire and Histoire de la Thermodynamique Classique* (1996). Crosbie Smith, *The Science of Energy* (1998).

THEODORE ARABATZIS AND KOSTAS GAVROGLU

ENVIRONMENT. In the nineteenth century, human exploitation of the natural environment came to be recognized in the industrializing West as a threat to human welfare. Romantics deplored the destruction of forests as diminishing the world's esthetic and spiritual reserves, and scientists warned that forest depletion entailed losses in watershed and habitats for numerous plant and animal species. The outcry led the governments of Germany and France to embark on programs of forest restoration and preservation. In the United States, the concern for forests joined with the realization that the frontier was closing to create a "conservation" movement, which flourished from the 1890s through the 1910s and made its overarching goal the maintenance of the nation's natural heritage. Championed by President Theodore Roosevelt, the movement won the creation of national parks, forest preserves, and laws for the protection of wildlife.

The American environmental movement lost force during the turn towards probusiness conservatism in the 1920s, but concern for the preservation of nature revived under the reformist leadership of President Franklin Roosevelt in the 1930s and drew renewed support after World War II. Spreading affluence permitted people the time and means to visit the national parks and to find in the unspoiled beauty of nature spiritual relief from the sameness of the suburbs, television, and fast-food restaurants.

The environmental agenda was significantly enlarged when, in 1962, Rachel Carson, a trained biologist and gifted writer, published *Silent Spring*, a powerful dissection of the intricate and myriad ways that herbicides and pesticides, particularly DDT, were poisoning man and nature. Carson called the chemicals of weed and insect control "elixirs of death," explained that they killed wildlife, especially birds, as they accumulated in the wild food chain, and stressed that they threatened human health. In 1972, the federal government banned the use of DDT. In the meantime, *Silent Spring*, which was widely translated, had helped to stimulate a worldwide environmental movement with the goals of protecting nature and health against the threats of poisonous pollutants entering the air, earth, and water.

In the late 1980s, scientists warned that environmental dangers global in scope had arrived. The growing commercial and industrial use of chlorofluorocarbons (CFCs) was depleting the upper atmosphere's ozone layer, which blocks the passage of cancer-causing ultraviolet light from the sun. The burning of fossil fuels was releasing enough carbon dioxide to create a

Los Angeles in the 1960s, with and without smog.

greenhouse-like effect that raised average temperatures around the globe. The clearing of the tropical rain forests was destroying large fractions of the world's species of insects, birds, and animals.

In the Montreal Protocol (1987) and its toughened revision (1990), the industrial nations agreed to limit the use of ozone-depleting gases. At Rio de Janeiro in 1992, they reached an accord in principle to reduce the production of greenhouse gases below the levels of 1990; and in Kyoto in 1997, they devised tentative mechanisms for achieving that goal. However, in 2001 the United States, which produces one-quarter of the world's greenhouse gases, withdrew from the Kyoto Protocol, declaring that the required limits on the burning of fossil fuels would injure its economy. A comparable commitment to indigenous economic growth in Latin America has interfered with attempts to slow the destruction of the rain forests.

Bill McKibben, *The End of Nature* (1989). Robert Gottlieb, *Forcing the Spring: The Transformation of the American Environmental Movement* (1993).

DANIEL J. KEVLES

ENZYME. The German physiologist Willy Kühne proposed the name *enzyme* in 1876 to denote one of the two classes into which the agents of chemical change in organisms known as *ferments* had been divided. Eventually, however, his term replaced *ferment* for both classes. Ferments had been known since antiquity as the agents that caused bread to rise and transformed grape juice into wine and malt extracts into beer. The description by Antoine-Laurent *Lavoisier of alcoholic fermentation as a chemical equation gave the activating ferment the special status of an agent, rather than a constituent in the process.

During the 1830s chemists became aware of a number of other reactions similarly activated by very small amounts of a substance distinct from the compounds that were transformed. Anselme Payen and Jean-François Persoz extracted from germinating seedlings a substance able to convert two thousand times its own weight of starch into sugar. They named the substance *diastase*. Eilhard Mitscherlich showed that the sulfuric acid whose action converts alcohol to ether is not consumed in the process; he pointed to many other such processes, which he called "decomposition and combination through contact." In 1835 Jöns Jacob *Berzelius named the general phenomenon *catalysis*. Berzelius predicted that many other organic catalytic agents like diastase would be found and that thousands of catalytic processes must occur between the tissues and fluids of plants and animals.

One year later, Theodor *Schwann identified in gastric juice a substance that acted in very small quantities, in the presence of acid, to digest albumin and other nitrogenous nutrient substances. Although unable to isolate the substance, which he named *pepsin*, he showed that its chemical properties distinguished it from any other known nitrogenous animal matter. Carefully comparing the properties of pepsin with the alcoholic ferment, Schwann concluded that they acted similarly enough to be considered members of a general class of organic ferments. Schwann's analysis of the action of pepsin opened the way to search for other digestive ferments.

When Louis *Pasteur asserted that the ferments for alcohol, lactic acid, and butyric acid required the presence of living organisms, the definition of ferments that Schwann had created by comparing pepsin and the alcoholic ferment seemed again sundered since the digestive ferments could act outside the body. To distinguish them from the "formed ferments" such as yeast that seemed inseparable from the organisms themselves, Pasteur's "unformed ferments" were so designated until Willy Kühne called them "enzymes."

After Eduard Buchner's demonstration of cell-free fermentation in 1900, the distinction became unnecessary. The preferred term in English was "enzyme," probably because "ferment" seemed too closely associated with the traditional fermentations to apply to all of the organic catalytic processes. In German, *Ferment* was preferred nearly until World War II as there was another word for fermentation— *Gärung*.

The reagents Schwann applied to characterize pepsin suggested that it fit within the general class of "albuminoid" matters later called *proteins. The extremely small quantities in which pepsin and other enzymes acted, however, made them extraordinarily difficult to isolate, and eluded the chemical tests normally applied to identify substances as proteins. As late as the 1920s the eminent biochemist Richard Willstätter maintained that the proteins associated with enzymes were inactive "carriers" for small, reactive molecules that conferred specificity on individual enzymes. Only after John Northrop crystallized pepsin in 1930 did opinion swing decisively to the belief that enzymes are proteins. Much research since then has been devoted to demonstrating how the structure of these immensely complex molecules can itself explain the exquisite specificity of the actions of enzymes.

Frederic Lawrence Holmes, *Claude Bernard and Animal Chemistry* (1974). Joseph S. Fruton, *Contrasts in Scientific Style: Research Groups in the Chemical and Biochemical Sciences* (1990). Joseph S. Fruton, *Proteins, Enzymes, Genes: The Interplay of Chemistry and Biology* (1999).

FREDERIC LAWRENCE HOLMES

EPIGENESIS AND PREFORMATION. Epigenesis and preformation are the names given to two sets of theories of generation debated during the second half of the seventeenth and the eighteenth centuries. Epigenesis taught that the germ comes into existence by fecundation and that the parts of the embryo arise one after the other by successive accretions. According to preformation, the germ contains the rudiments of all the parts of the future organism, which then unfold in embryonic development. Preformation theories were usually called theories of evolution or *emboîtement* (encasement). To avoid the ambiguity of the term "evolution," *emboîtement* became "preformation" or "pre-existence" during the nineteenth century. According to the time and context, supporters of epigenesis usually held animistic and later vitalistic as well as materialistic conceptions of organic nature. Those who instead embraced preformation theories upheld atomistic and mechanistic ones.

Preformation theories can be traced back to ancient Greek atomism and to its notions that the embryo resulted from a combination of miniaturized atoms deriving from the seminal fluids of both parents. Epigenesis, instead, may be traced back to Aristotle's (384–322 B.C.) theory of generation and, more specifically, to his observations and interpretations of the developing chick. A follower of Aristotle, William Harvey (1578–1657), first introduced the term epigenesis and defined the process as "the addition of parts budding out from one another" (*Exercitationes de generatione animalium*, 1651). According to Harvey the egg was a new formation and the embryo developed from the homogeneity of the primordium to the heterogeneity of the adult form through budding and subdivision.

Giuseppe degli Aromatari maintained that the future plant could be perceived in seeds and bulbs, and conjectured that the chick might be present in the egg before fecundation (*Epistola de generatione plantarum ex seminibus*, 1625). Given that plants and oviparous animals originated from seeds and eggs, authors suggested that an analogous rule must be at work among insects and viviparous animals. Nicolaus Steno's (1638–1686) recognition of the homology of the mammalian ovary with that of oviparous animals, the observations of Francesco Redi (1626–1697/98) and Jan Swammerdam (1637–1680) of insect eggs, and Regnier de Graaf's (1641–1673) discovery of the ovarian follicles, which he erroneously took to be the mammalian egg, gave rise to a special version of preformation called ovism. On the other hand, the discovery of spermatozoa favored a further version called animalculism. In both versions the future organism pre-existed, being miniaturized either in the egg or in the spermatozoa.

During the eighteenth century ovism dominated the life sciences. It was supported by the sophisticated theory of Charles Bonnet, the observations of Albrecht von Haller, and the experiments of Lazzaro Spallanzani. Bonnet argued contrary to epigenesis that the parts of an organism depended so intimately on one another that the existence of some presupposed the existence of the others. All the parts of the future organism must therefore exist in the embryo even if they were not visible. Only with growth and development, with unfolding, did they become apparent.

Caspar Friedrich Wolff put forward a new theory of epigenesis in his *Theorie von der Generation* (1764): both plants and animals started from a homogeneous primordium formed gradually through the secretion and solidification of fluids regulated by a *vis essentialis* (essential force) that was unique to living matter. This force acted on the fluids by attraction and repulsion of substances. With the rise in the 1820s of typological thinking in comparative embryology and with the demonstrations of the role played by spermatozoa in fertilization and of the existence of the mammalian egg, preformation theories declined and epigenesis established itself as the leading theory in the nineteenth century. However, the contrast between preformation and epigenesis reemerged in *developmental mechanics through the theories of Wilhelm Roux and Hans Driesch. Both epigenesis and preformation had merit, the former in considering the egg as undifferentiated at its primordium, the latter in envisaging the existence of something preformed, which is now recognized to be the genetic code contained in the nucleus of the fertilized egg.

Walter Pagel, *William Harvey's Biological Ideas* (1967). Shirley A. Roe, *Matter, Life, and Generation: Eighteenth-century Embryology and the Haller-Wolff Debate* (1981). Olivier Rieppel, "Atomism, Epigenesis, Preformation and Pre-existence: A Clarification of Terms and Consequences," *Biological Journal of the Linnean Society* 28 (1986): 331–334. Jacques Roger, *The Life Sciences in Eighteenth-century French Thought* (1998).

RENATO G. MAZZOLINI

ERROR AND THE PERSONAL EQUATION. Since Greek times astronomers have recognized that observations were afflicted by errors, that results based on them might only be approximate, and that the quality of data varied. Astronomers in early modern Europe took the first steps toward giving reliable estimates of those errors. Johannes *Kepler, who used Tycho *Brahe's observations to derive the elliptical shape of planetary orbits, was probably the first to construct a correction term that assigned a magnitude to error, and among the first to give a theory of an instrument (the Galilean telescope) for purposes of improving the accuracy of measurements taken with it.

During the eighteenth century steps were taken toward standardizing the analysis of measurements and understanding the conditions under which different sets of measurements could be combined. Analysts identified two types of errors: constant (affecting the instruments or the conditions of measurement) and accidental (randomly affecting the quality of the measurements themselves). Control over instrumental errors was achieved at first by codifying the behavior and demeanor of the observer, by taking into account the limitations of the human senses (especially vision), by examining how outside sources contaminate experiments, by perfecting the construction of instruments, and by developing methods for instrument calibration.

The second type of error, the random, relates to classical probability theory. Initially the criteria for the selection of good measurements rested mainly on the notion that the median or the mean of measurements reduced the effect of errors in any of them. In 1756 the mathematician Thomas Simpson countered reports that a single well-taken measurement sufficed by demonstrating the superiority of the mean; his presentation to the Royal Society of London included a discussion of the equal probability of positive and negative errors and an argument that the mean lies closer to the true value than any random measurement. But no consensus existed about the selection or combination of measurements. The first firm parameters of an error theory emerged from the consideration of observations of the Moon's motion, especially its libration; from secular inequalities in the motions of Jupiter and Saturn; and from measurements of the shape of the earth. During the second half of the eighteenth century, Johann Tobias Mayer, Leonhard *Euler, Rudjer J. Bošković, and Johann Heinrich Lambert developed ad hoc, limited, varied, but effective procedures for combining measurements made under different conditions. In 1774 Pierre-Simon *Laplace deduced a rule for the combination of measurements using probability theory.

The meridian measurements made during the French Revolution to determine the new standard of length, the meter, gave the occasion to devise the first general method for establishing an equilibrium among errors of observation by determining their "center of gravity." This method, the method of least squares, was so employed in 1805 by Adrien Marie Legendre. In 1806 Carl Friedrich *Gauss acknowledged Legendre's work but only

to say that he had been using the method for years. A priority dispute ensued. Three years later Gauss published the first rigorous proof of the method of least squares; he demonstrated that if the mean is the most probable value, then the errors of measurement form a bell curve (Gaussian) distribution. The true value (which has the smallest error) lies at the center of the distribution, while the width of the curve determines the precision of the measurement. (Application of the method assumes the absence of constant or systematic errors.) From astronomy the method spread to chemistry, physics, mineralogy, and geodesy. It was also applied to practical projects including the reform of weights and measures, longitude determinations, triangulations, the U.S. Coastal Survey, and cadasters, where it set the boundaries within which dispute could take place. The method of least squares made large-scale projects, like Gauss's magnetic map of the earth, manageable by providing a means to combine and assess data from geographically dispersed locations. Over the course of the nineteenth century, the method dominated error analysis (especially in the German-speaking world) and shaped the development of probability theory.

The power of the method of least squares seemed to eradicate the subjective element in the treatment of measurements. But Gauss and his colleague the astronomer Wilhelm Olbers acknowledged in 1827 that sometimes measurements displayed large deviations from the mean. When was the deviation large enough to justify ignoring a measurement? Gauss could not provide an objective answer and recommended reliance on intuition. In 1852 Benjamin Peirce developed a rigorous method for rejecting outliers.

Deviations of another sort led Friedrich Wilhelm Bessel to develop the personal equation. In 1823 he noted a constant difference in the measurements taken by the former British Astronomer Royal Nevil Maskelyne and his assistant, which Bessel referred to physiological differences. With the "personal equation" Bessel calculated the average difference between two observers; he could then combine measurements taken by several observers. The personal equation created the factory-like atmosphere of nineteenth-century astronomical observatories where teams of observers were calibrated according to its principles. It was seldom used elsewhere except in psychology. Especially in Wilhelm Wundt's physiological institute at Leipzig, the personal equation became the foundation of a research program in the determination of human reaction times.

The history of error and the personal equation embraces far more than the history of rules and methods. The determination of error is always an estimate; were the true error known, perfectly accurate results could be attained. Because reliable estimates of error generate confidence in results, the history of error theory also sheds light on how trust is established in a scientific community and beyond. In the teaching laboratory, the method of least squares indicated how well student investigators performed an experiment, and thus the level of expertise they had attained. In practical fields like surveying, the introduction of the method aided professionalization. Finally, the method of least squares shaped the moral economy of the sciences by promoting honesty in the execution and reduction of observations. Proper application of the method became a sign of the investigator's integrity.

See also PRECISION AND ACCURACY; PROBABILITY AND CHANCE.

Mansfield Merriman, *A List of Writings Relating to the Method of Least Squares, with Historical and Critical Notes* (1877). G. E. R. Lloyd, "Observational Error in Later Greek Science," in *Science and Speculation*, ed. Jonathan Barnes, et al. (1982). Stephen Stigler, *The History of Statistics* (1986). Giora Hon, "H. Hertz: 'The Electrostatic and Electromagnetic Properties of Cathode Rays Are Either Nil or Very Feeble,' (1883): A Case-Study of Experimental Error," *Studies in History and Philosophy of Science* 18 (1987): 367–382. Giora Hon, "On Kepler's Awareness of the Problem of Experimental Error," *Annals of Science* 44 (1987): 545–591. Giora Hon, "Towards a Typology of Experimental Errors: An Epistemological View," *Studies in History and Philosophy of Science* 20 (1989): 469–504. Gerd Gigerenzer, et al., eds., *The Empire of Chance* (1990).

KATHRYN OLESKO

ETHER, a possibly nonexistent entity invoked from time to time to fill otherwise empty spaces in the world and in natural philosophy. Descended from the Aristotelian quintessence, which occupied the realms through which the planets wandered, and the Stoics' pneuma, which held the world together, ether characterizes theories opposed to atomism, which admits spaces void of matter. Both sorts of theories—plenary and atomistic—enjoyed vigorous revivals during the *Scientific Revolution. With the invention of the *barometer and *air pump in the middle decades of the seventeenth century, void and ether became objects of experiment and, in practice, very much the same (no)thing.

The experimental investigation of void began above the mercury in the barometer tube. This space had the property of transmitting light and magnetic virtue, but not sound, and of allowing the free passage of bodies through it. It seemed infinitely compressible or, rather, was so subtle that it could pass right through

glass. Were these the properties of a space void of all matter or of one filled with a substance different from ordinary matter? A third way, preferred by René *Descartes, made the special substance and ordinary matter the same thing, except for the size and shape of their constituent parts. Isaac *Newton countered with a solution as hard to grasp as ether itself. In his world system, the planets move through resistanceless spaces replete with the presence of God and perhaps also with springy ethers that mediated gravitational attraction, chemical behavior, *electricity, *magnetism, and the interaction of light and ordinary matter. These ethers, unlike the Stoics' pneuma and Descartes' plenum, admitted voids among their particles.

With the acceptance of the wave theory of light in the nineteenth century, physicists felt obliged to suppose the existence of a subtle, imponderable medium whose undulations constituted the disturbance perceived as light. The first mathematicians to attempt a detailed picture of this "luminiferous ether" modeled it as a mechanical substance with rigidity and inertia. They managed thus to represent most of the properties of light—reflection, refraction, interference, and polarization. As William *Thomson (Lord Kelvin) and the Cambridge mathematician George Gabriel Stokes explained it to other model-makers, the luminiferous ether had to combine the properties of shoemakers' wax, which allows slow bodies to pass through it under steady pressure but shatters when struck a sharp blow, with those of rigid steel, which can support transverse vibrations without suffering permanent distortion.

The ether soon became so familiar that mathematicians assimilated it to ordinary matter, or vice-versa. As latter-day Cartesians, they pictured atoms and molecules as permanent vortex rings in an all pervasive ether (Kelvin, following a hydrodynamical theory of his friend Hermann von *Helmholtz, and Joesph John *Thomson) or as knots or twists in it (Joseph Larmor). This sort of modeling was a specialty of physicists who had passed through the honors course in mathematics at Cambridge (the mathematical tripos). Continental physicists, especially the French, regarded it with a mixture of puzzlement and distaste. Nonetheless, James Clerk *Maxwell devised an ether model for the mediation of electrical and magnetic forces that suggested that light was an electromagnetic phenomenon.

Maxwell's theory charged the ether with accounting for electricity and magnetism as well as for light. None of the several models with mechanical properties proposed to effect it succeeded. Hendrik Antoon Lorentz and others then introduced space-filling media that had nonmechanical properties in order to underpin an adequate electrodynamics of bodies moving through the suppositious ether (*relativity). In 1905, Albert *Einstein showed the value of discarding the ether as a substrate and reference frame for electrodynamic phenomena.

The acceptance of relativity theory, however, did not destroy the ether. Einstein himself, in his application of relativity principles to the gravitational theory (1915), supposed that a gravitating body distorts nearby space, and that these distortions determine the trajectory of a passing ponderable body. An entity that can distort its shape, deflect light, and propagate electric and magnetic disturbances can be called a void only by discourtesy. More recently, *quantum electrodynamics has filled the void with a vacuum that undergoes energy fluctuations and acts as a theater for the creation and annihilation of virtual particles. One such fluctuation is said to have given rise to the present universe (see COSMOLOGY). Physicists appear to need an ether on which to load all the properties of the physical world they cannot otherwise explain. Ether, alias the vacuum, exists. Void is anything but nothing.

E. T. Whittaker, *A History of the Theories of Aether and Electricity*, 2 vols. (1951). E. J. Aiton, *The Vortex Theory of Planetary Motions* (1972).

J. L. HEILBRON

ETHICS AND SCIENCE. Even if it is possible to distinguish, it is no longer possible to dissociate science from its practical applications, which often raise ethical questions. The interaction of ethics and science has always been reciprocal, but until the end of the nineteenth century, science mostly posed challenges for ethics, while in the twentieth century ethics posed new challenges for science.

"Ethics" derives from the Greek word for "character," and the Greeks and Romans assumed that if we knew the human good or could model our behavior on that of a virtuous person, ethical conduct would be natural. Christian medieval ethics added aristocratic mutual obligation and contract, the divine inspiration of natural law, and the moral fallibility of all human beings. By the seventeenth century, new knowledge was generating new forms of power, as recognized by Francis Bacon, which meant new potential for wealth, abuses, and corruption. The mechanistic division between mind and matter in the work of René *Descartes, which implied irrelevance of ethical issues to natural knowledge, was denied by Baruch Spinoza's equally rationalist *Ethics*. The various scientific academies established by the early eighteenth century stressed an

ethics of mutual reliability among their members as free-acting observers.

Echoing the Greco-Roman model, the story of *Galileo's conflict with the church, his prosecution and recantation under duress, and the ultimate vindication of his ideas has often been taken by scientists as a parable of the need for integrity and autonomy in scientific research. Scientists have had no direct analog to the Hippocratic Oath's injunction to "do no harm," but have taken as normative Galileo's insistence on following and finding the truth about this world in the face of convention and orthodoxy. The arguments of scientists have thus been suffused with a moral subtext of struggle against entrenched authority, constructing new knowledge not only in their own interests but in the interest of humanity.

During the eighteenth and nineteenth centuries, idealizations of science became entwined with the Enlightenment promise of human progress and the norms of reason. Impressed by the success of natural philosophy, Immanuel Kant laid out ethical claims revolving around an autonomous, rational agent. His (agent-based) "categorical imperative" to treat the humanity of others not simply as a means but as an end in itself stemmed from his understanding, in contrast to his skeptical contemporary David Hume, of logically consistent thinking as essential to our human nature. In (consequence-based) utilitarian ethics, science was modeled differently, becoming the root of cost-benefit analyses that offered the greatest good for the greatest number. Since science increasingly was associated with new developments in medicine, engineering, commerce, and warfare, the ethical systems represented by Kant and the utilitarians played ever greater roles in thought about how humans should treat each other. The nineteenth-century social sciences emerged in this climate, striving to treat human relationships as objects of study in order to better them. But when George Edward Moore published his *Principia Ethica* in 1903, he cordoned off reason from moral claims on the grounds that the "good" was fundamentally indefinable in natural terms. His contemporary Max Weber also stressed the distinction between fact and value, reinforcing Hume's argument that no bridge spanned the gulf between "is" and "ought." The impact of science on ethical thought had reached an impasse.

During the twentieth century, influence shifted to the impact of ethical thought on science, particularly physics, chemistry, and the life sciences, as the sciences themselves became entrenched authority for many people. Amid the moral chaos engendered by the trauma of World War I and the world economic depression of the

1930s, scientists made themselves increasingly important for military, economic, and medical advances. Robert K. Merton noted that this fact produced among scientists renewed calls for the autonomy, integrity, and insulation of science, but at the same time brought ever more anxious and insistent demands that science incorporate the ethic of social responsibility. The demands sometimes came from scientists themselves, sometimes from citizens galvanized by specific technological developments, and sometimes from governments responding to their publics. Following World War II, demands for responsibility coalesced into movements and various professional codes of behavior. As the country with the greatest resources and influence in science, the United States was often a barometer of these developments.

With the advent of the Cold War and the nuclear arms race, physicists and other scientists in every militarily powerful nation faced tensions between pursuing new knowledge and limiting its dangers. Some scientists who helped develop nuclear weapons led efforts to limit their further use, founding journals, creating federations, informing public opinion, and advising or critiquing governments and policies. These tensions and perplexities animated the Pugwash Conferences begun in 1957 in the Nova Scotia village by that name, at which scientists from around the world met as private individuals to discuss ways to limit international strains caused by the multiplication and increasing deadliness of armaments. As the philosopher Hans Jonas has observed, the first ethical responsibility in our era is to help insure that the possibility of ethics persists, that our creations do not destroy our world.

The importance of chemistry for new materials, agriculture, pharmaceuticals, energy, and many other factors in an elevated standard of living has raised a host of ethical concerns related to social justice and the environment. The same chemical production processes that give us new materials and energy generate pollution. New herbicides and pesticides that increase food production also have negative run-off effects on the environment. Drugs such as antibiotics have undesired evolutionary impact on microbes. All of these involve financial costs beyond the reach of much of the world's population, at a human price these people often bear. Soon after the publication of Rachel Carson's book *Silent Spring* (1962), the American Chemical Society enunciated the "Chemist's Creed" (1965), pledging an ethic of responsibility to the public, science, chemical profession, employer, self, employees, clients, students, and associates. The Green Revolution

in agriculture produced a backlash against monocultures in favor of biodiversity, especially in the non-Western world, and the environmental ethics movement has modified or redirected significant scientific resources and programs. Many people have argued that ethical systems may not be universally valid, but ethical concerns must be global.

World War II was also a watershed in the life and social sciences, as Nazi eugenics policies and experiments on captive human beings led to postwar trials in Nuremberg. The ten principles known as the Nuremberg Code (1947), which emphasize the informed, voluntary participation of human subjects and safeguards for their welfare, became the basis for a sequence of declarations by the World Medical Assembly and most ethical discussions of human experimentation. By 1966 in the United States, written informed consent and harm/benefit assessments had to be submitted to formal ethical review before research on human subjects could be conducted with funding from the National Institutes of Health. These considerations spread to other agencies, created a demand for a new specialty of bioethics, and altered research in the social sciences.

In the late 1960s climate of distrust of the role, direction, and authority of science in human affairs, geneticists and other biologists considered whether they were confronting ethical issues analogous to those facing physicists at the dawn of the nuclear age. After a brief, voluntary moratorium on certain lines of work, biologists agreed at the Asilimar Conference of 1975 upon a set of biological and physical safeguards in research on recombinant DNA and a set of ethical guidelines for informing staff of the possible hazards involved. Soon afterward, Peter Singer's book *Animal Rights* (1975) sparked a movement to extend to animals ethical protections similar to those afforded human subjects in research, some of which the American government put into place by the end of the 1980s. Anthropologists faced intensified disputes over the norms for study and disposition of human remains.

The growth of international corporate financial interest in genetic research in the 1980s and U.S. government funding for the Human Genome Project (1991) raised new ethical issues of ownership and privacy of genetic information, as well as new clinical, public health, and civil rights concerns. The HGP therefore decided to devote some of its funding to an Ethical, Legal and Social Implications Program. At the same time, the convoluted and politically controversial charges of scientific fraud in the decade-long David Baltimore case indicated the heavy-handedness of government policing of scientific integrity.

Cloning technologies, stem-cell research, defensive research on biological weapons, and programs in robotics and artificial intelligence (which has given rise to a machine-rights movement) indicate that new bioengineering and bioethical issues will continue to proliferate.

Change fostered by science has been occurring too rapidly for many people and institutions to assimilate. Science and its technologies are implicated in the tensions between Western and non-Western values. Science played its role in the colonial expansion of Europe, and is closely associated with technological and medical practices that conflict with indigenous ethical and cultural assumptions. The ethical implications of the global impact of nuclear threat, environmental degradation, and bioengineering have raised the question of whether scientists can—or should—remain individual, free-acting agents of change.

See also ANTI-NUCLEAR MOVEMENT; DEOXYRIBO-NUCLEIC ACID; ENVIRONMENT; GENETICS.

Loren Graham, *Between Science and Values* (1980). Sheldon Krimsky, *Genetic Alchemy: The Social History of the Recombinant DNA Controversy* (1982). Luciano Caglioti, *The Two Faces of Chemistry* (1983; Italian orig. 1979). Hans Jonas, *The Imperative of Responsibility* (1984; German orig. 1979). George J. Annas and Michael A. Grodin, eds., *The Nazi Doctors and the Nuremberg Code: Human Rights in Human Experimentation* (1992). Paul Durbin, *Social Responsibility in Science, Technology and Medicine* (1992). Ian Barbour, *Ethics in an Age of Technology* (1993). Vandana Shiva, *Monocultures of the Mind: Perspectives on Biodiversity and Biotechnology* (1993). Kristin Schrader-Frechette, *Ethics of Scientific Research* (1994). Nicholas Low, ed., *Global Ethics and Environment* (1999). S. S. Schweber, *In the Shadow of the Bomb: Bethe, Oppenheimer and the Moral Responsibility of the Scientist* (2000). Texts of professional codes available at the Center for the Study of Ethics in the Professions website at: http://www.iit.edu/departments/csep.

ALAN BEYERCHEN

EUGENICS. Francis Galton, a British scientist and a cousin of Charles *Darwin, coined the word "eugenics" in 1883, drawing on the Greek root meaning good in birth or noble in heredity. The term expressed his idea of improving the human race biologically by manipulating its hereditary essence—by, as he put it, getting rid of the "undesirables," multiplying the "desirables." Eugenics became popular after the rediscovery in 1900 of Mendel's theory that the biological

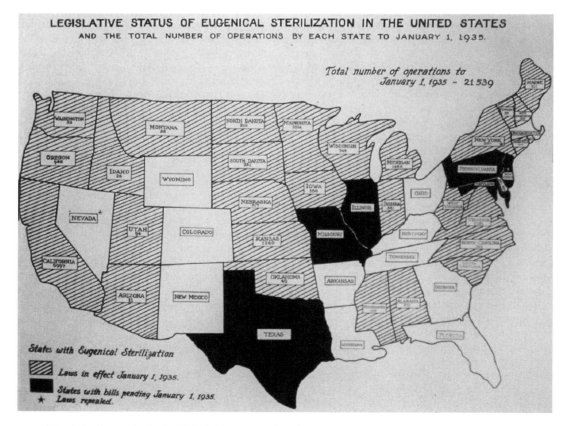

The state of eugenics in the United States around 1935.

makeup of organisms is determined by certain "factors," later identified with genes. Eugenics movements blossomed in the United States, Canada, Britain, Germany, Scandinavia, and elsewhere in Continental Europe and parts of Latin America and Asia.

Eugenicists insisted that genes made people prone to poverty, criminality, alcoholism, and prostitution, and that people carrying socially deleterious genes were proliferating at a threatening rate. Eugenicists on both sides of the Atlantic argued for a two-pronged program to increase the frequency of good genes in the population and decrease that of the bad variety. One prong was "positive" eugenics, encouraging the birth of "superior" people. The other was "negative" eugenics, eliminating or excluding biologically inferior people from the population.

Many important people on both sides of the Atlantic endorsed positive eugenics. The American president Theodore Roosevelt, worried by the declining birth rate among women of his class, urged them to bear more children for the good of the race. The social and sexual radical Bernard Shaw declared that in the interest of producing more superior children, society should allow able women to conceive children by able men whom they might never see again.

Negative eugenics flourished between the world wars through eugenic sterilization laws. Two dozen American states enacted them, although only a few, notably California, enforced them. Similar measures were passed in the Canadian provinces of British Columbia and Alberta, in Sweden, and in Germany. Sterilization rates climbed with the onset of the worldwide economic depression in 1929. Sterilization acquired broadened support because it promised to reduce the cost of institutional care and of poor relief. In Germany, within three years after the passage of enabling legislation in 1934, the Nazis had sterilized some 225,000 people, almost ten times the number so treated in the previous thirty years in the United States.

Little was attempted in the way of positive eugenics except in Germany, where the Nazi government provided subsidized loans to biologically sound couples whose fecundity would be a credit to the *Volk*. To foster the breeding of an Aryan elite, Heinrich Himmler urged members of the S.S. to father numerous children with racially preferred women, and in 1936 he instituted the *Lebensborn*—spa-like homes where S.S. mothers, married and unmarried, might receive the best medical care during their confinements.

In the early Hitler years, Nazi eugenics and Nazi racial polices operated independently of each other, but they increasingly merged after the Nuremberg Laws of 1935. In 1939, the Third Reich moved beyond sterilization to perform euthanasia on certain classes of the mentally diseased or disabled in German asylums. Among the classes were all Jews, no matter what the state of their mental health.

Beginning in the 1930s, eugenic doctrines were increasingly criticized on scientific grounds and for their class and racial bias. The critics showed that most human behaviors, including the deviant variety, are shaped by environment at least as much as by biological heredity, if genes play a role there at all. Roman Catholics and civil libertarians also vigorously resisted sterilization. After World War II, the revelations of the Holocaust made eugenics a dirty word.

The development of molecular genetics since the 1950s has prompted some biologists to call for a new eugenics, one that is free of racial and class bias, that will rid the human race of genetic disease, and that will, if possible, genetically engineer improved human beings. Whether such a eugenics becomes a matter of public policy remains to be seen, but it appears that prospective parents are already using contemporary knowledge to practice a private eugenics, deciding with their physicians what genetic type of children they will bear.

See also GENETICS.

Robert Proctor, *Racial Hygiene: Medicine Under the Nazis* (1988). William H. Schneider, *Quality and Quantity: The Quest for Biological Regeneration in Twentieth-Century France* (1990). Daniel J. Kevles, *In the Name of Eugenics: Genetics and the Uses of Human Heredity* (1995). Edward J. Larson, *Sex, Race, and Science: Eugenics in the Deep South* (1995). Gunnar Broberg and Nils Roll-Hansen, eds., *Eugenics and the Welfare State: Sterilization Policy in Denmark, Sweden, Norway, and Finland* (1996). Philip Kitcher, *The Lives to Come: The Genetic Revolution and Human Possibilities* (1996).

DANIEL J. KEVLES

EULER, Leonhard (1707–1783), mathematician, specialist in mechanics. Originally from Basel, Switzerland, Euler was the most productive mathematician of all time. The measure is not the number of papers, for which the current record holder is the Hungarian Paul Erdös, but the number of published pages. Yet productivity was perhaps the least important of Euler's claims to mathematical distinction. One of his great contributions was his clarity, in contrast to French mathematicians of the time, who rarely expressed themselves so lucidly. The polishing that the savants of the previous century carried to extremes was almost wholly abandoned in the prolific eighteenth century.

Eighteenth-century mathematicians often did not cite their sources of insight. Euler for one appropriated without acknowledgment fundamental ideas of others gleaned from their letters as well as from their works. For example, he borrowed from Jean d'Alembert ideas in fluid dynamics, which d'Alembert had first introduced in his essay on the resistance of fluids, submitted to the Berlin Academy for its prize of 1749. D'Alembert did not win the contest (of which Euler was a judge), but vice did lead to virtue, as Euler expressed limpidly the ideas he pirated.

Euler appropriated freely from other French writers besides d'Alembert. He clarified and developed their ideas well beyond the points to which their originators had brought them. His textbooks incorporate his and their insights. Used for generations, many of these books became classics, for example, *Introductio ad analysin infinitorum* (1748), *Institutiones calculi differentialis* (1755), and *Institutiones calculi integralis* (1768–1770).

Euler's clarity was not the only important factor in establishing his widespread influence. He contributed to every branch of mathematics of his day except probability. He achieved much in the realm of number theory. He arguably founded graph theory and combinatorics when he solved the Königsberg Bridge problem in 1736. These four topics also were among Erdös's specialties. But in addition Euler contributed to ordinary and partial differential equations, the calculus of variations, and differential geometry, which Erdös did not touch.

Moreover, Euler made major contributions to every branch of *mechanics. The motion of mass points, celestial mechanics, the mechanics of continuous media (mechanics of solids and nonviscous fluids, theories of materials, *hydrodynamics, hydraulics, elasticity theory, the motion of a vibrating string, and rigid-body kinematics and dynamics), *ballistics, *acoustics, vibration theory, *optics, and ship theory all received something important from him.

If Beethoven did not need to hear to compose music, Euler did not need to see to create mathematics. He began to go blind in one eye in 1738 and became totally blind thirty years later. This only increased his productivity, since total blindness relieved him of academic chores like proofreading and eliminated unwanted visual distractions. Euler did not miss eyes for another reason; he had a prodigious memory. He could recite the *Iliad* by heart, and he reveled in the most difficult mental computations.

Euler began his academic career in 1727 as a member of the Academy of Sciences of Saint Petersburg. In 1734 he married Katharina Gsell, who also belonged to Saint Petersburg's Swiss colony. She bore him thirteen children, of whom only three sons and two daughters survived

early childhood. Unfortunately, the xenophobic Russian nobles resented the foreign members of the Academy, and the Russian Orthodox censors did not tolerate new sciences like Copernican astronomy. Furthermore, the Alsatian bureaucrat Johann Schumacher, who presided over the Academy, acted like a despotic rug merchant. Euler had to haggle with him to get his salary raised.

Despite these adversities, Euler cheerfully undertook projects of the Saint Petersburg Academy like Russian mapmaking, which involved painstaking work correcting land maps and reducing astronomical observations. This toilsome and stressful activity took him away from his mathematics and probably affected his sight and health.

In 1741 the new Prussian monarch Frederick II offered Euler a position in the new Berlin Academy. Euler accepted it because of political tensions and heightened hostility towards foreigners in Saint Petersburg. The move did not bring joy. Neither man pleased the other. Frederick was totally uninterested in Euler's mathematics. Euler resented the increasing French influence in the Berlin Academy, whose members, at Frederick's insistence, had to speak and write in French.

Euler came to feel that *belles lettres* had gained too much ground in Berlin at the expense of mathematics and that Frederick, who referred to him as his "one-eyed geometer" or "octopus," was becoming ever less supportive. In 1766 Euler returned to Saint Petersburg, where he remained until he died.

Euler set the trends in research in the mathematical sciences for most of the eighteenth century. The same sort of dominance is not possible for twentieth-century mathematics. The mathematics community of the eighteenth century was small; in the twentieth century it is huge, as is the number of specialized areas of mathematical research. Even Erdös, with his gigantic and pathbreaking output, has not had as much influence on the large-scale development of contemporary mathematics.

Clifford A Truesdell, "Rational Fluid Mechanics, 1687–1765," in Leonhard Euler, *Opera omnia*, series 2, vol. XII (1954): ix–cxxv. Clifford A Truesdell, "The Rational Mechanics of Flexible or Elastic Bodies, 1638–1788," in ibid, series 2, vol. XI (1960): 7–435. Thomas L. Hankins, *Jean D'Alembert: Science and the Enlightenment* (1970). Ronald Calinger, "Leonhard Euler: The First St. Petersburg Years (1727–1741)," *Historia Mathematica* 23, no. 2 (1996): 121–166.

JOHN L. GREENBERG

EUROPEAN COUNCIL FOR NUCLEAR RESEARCH. See CERN.

EUROPE AND RUSSIA. The main countries of Europe (including Great Britain and European Russia from time to time) have followed a similar pattern in developing the sciences since the second half of the nineteenth century. The attitudes of the dominant churches and the rate of commercial and industrial development have variously affected the pace and intensity of the pursuit of natural knowledge in the several countries. But the ready exchange of information, people, and inventions; the similarity of institutions for the study and spread of science; and commercial competition, warfare, and colonial adventures coupled development in the main European countries. The surprising finding that in 1900 the number of academic physicists per unit of population, and the amount of investment per unit of national income, were virtually the same in Britain, France, and Germany suggests the tightness of the coupling. This equalization occurred despite considerable differences in the methods of funding and the operations of the universities in the different countries.

During the eighteenth century the chief institutions concerned with natural knowledge divided their work in the same way throughout Europe: the universities taught established science, the academies sought new knowledge. The larger observatories and botanical gardens, often associated with both universities and academies, taught, preserved, and developed their subject matters in the same way in Paris and Saint Petersburg. A flourishing book trade, expanding university libraries, and many review journals, some, like the *Göttingische gelehrte Anzeigen*, associated with universities, kept everyone interested abreast of the latest advances. European scholars saw themselves as inhabitants of a "republic of letters." The use of Latin and French as the languages of scholarly interchange, travel when war did not preclude it, the mobility of academicians (the moves of Leonhard *Euler from Switzerland to Saint Petersburg to Berlin to Saint Petersburg, and of Pierre de Maupertuis from Paris to Berlin and back, exemplify the traffic), and the institution of "corresponding members" of the learned societies further enforced the notion of a pan-European communion of philosophers. Immanuel Kant's *Idea for a universal history with a cosmopolitan purpose* (1784) made a "universal cosmopolitan existence ... the highest purpose of nature." Ludwig Wachler, the author of a *History of history and art from the renewal of culture in Europe* (1812), taught that the cultivation of the sciences was a peculiarly European activity.

The sciences literally helped put Europe on the map. More exacting techniques for surveying and *geodesy were developed in national academies, and academy expeditions helped establish the topographic contours of Europe. European governments learned to appreciate that precision measurement increased state revenues, and they set about the laborious task of establishing common units recognized beyond provincial boundaries. Enlightened states supported natural sciences for their utility; not only technical subjects like astronomy and geodesy with obvious applications, but also natural philosophy, geography, meteorology, mining, forestry, and agriculture, whose study would prepare better civil servants and effective policing (*see* CAMERALISM).

Science unified more than knowledge and bureaucratic process. It helped constitute modern European states. Beginning in the early 1820s the Versammlung Deutscher Naturforscher und Ärzte brought together German-speaking physicians and naturalists from the patchwork of principalities, free cities, and palatinates that made up the region. The renowned explorer Alexander von *Humboldt called its meeting in Berlin in 1828 "a noble manifestation of scientific union in Germany; it presents the spectacle of a nation divided in politics and religion, revealing its nationality in the realm of intellectual progress." The German institution spread—the European system was above all one of parallel institutional forms. In 1831 "gentlemen of science" founded the British Association for the Advancement of Science, which, like its German model, had a wide membership and moved to meetings around the country. The French counterpart came into existence just after the Franco-Prussian war exposed the error of not staying closer to the practices of its rivals.

By spreading French science, which already gave the tone to European science, the French Revolution marked the acme of the Republic of Letters; by having its way by force of arms, the Revolution simultaneously disseminated a republicanism that many scholars judged to be inimical to the old cooperative scholarship. Auguste Comte may have envisioned the approach of a "scientific" stage of human civilization in which the intellectual elites of Europe would form ties across political boundaries and initiate a peaceful "European revolution" (*Cours de philosophie positive*, 1830–1842). But he was too late—and too early. From the Napoleonic wars to the unification of Germany, rivalry rather than cooperation dominated the development of European scientific institutions. The main growth occurred in the universities and higher schools. The foundation of the University of Berlin in 1810 and the restructuring of elementary and secondary curricula in Prussia gave a strong impetus to the entire educational system. Research gradually became a responsibility of the professoriate and of their students who aimed to be secondary school teachers. The rationale became ideology: only those who had contributed to knowledge, however small the contribution might be, could transmit to others the right mixture of information and enthusiasm. The Germans invented the teaching laboratory (*see* LIEBIG, JUSTUS VON), the research *seminar, the disciplinary *institute, and the Ph.D. After the middle of the nineteenth century, foreign students began to flock to the German universities, to profit from their professors and facilities and—helping to integrate the European system of science—to bring back home what they found useful and transportable. After their defeat in 1870–1871, the French sent a *mission to discover what made German universities so strong. In consequence, the French strengthened their provincial faculties of science and set up new higher technical schools like the École Supérieure d'Electricité (1884) and the École Supérieure de Physique et de Chimie (1882).

The strength of the French educational system since the Revolution had been its technical schools. The École Polytechnique trained scientists and engineers in a great deal of higher mathematics combined with physics and chemistry. Its students, selected by competition, graduated to enter specialized engineering schools—for mining, civil engineering, artillery, and so on. They then practiced as state employees, civil or military. Along with the centralized École Polytechnique the Revolution created an École Normale Supérieure, which supplied the teachers for the new system of state secondary schools (the *lycées*). The excellence of the Parisian technical schools and the scientific culture they supported made France the mecca of European natural philosophers and mathematicians until the 1830s and 1840s. As late as 1845 William *Thomson chose Paris as the best place to complete his education in physics.

Thomson had been educated in old British universities (Glasgow and Cambridge). By the time he went to Paris, a new sort of university, which emphasized modern languages and science, had begun to grow. The University of London was put together from University College and King's College, founded in 1826 and 1831, respectively; later other colleges were added. Provincial manufacturing centers supported the creation of "redbrick" universities, which trained practical scientists of the second industrial revolution. Beginning in the 1850s committees of Parliament forcibly brought Oxford and Cambridge into the nineteenth century by establishing professorships in science and reducing the power of the humanistic dons.

As rivalry and war propagated the stronger institutional forms from one country to another (the Germans borrowed from the École Polytechnique for their *Technische Hochschulen*), pressures from science and its applications forced the invention of new means of cooperation. Some, like the famous meeting in Karlsruhe in 1860 where chemists came to settle their differences over atomic weights, were fleeting. Others, like the International Bureau of Weights and Measures, established near Paris in 1875, have endured. The need for agreement over measures made an irresistible force for internationalism.

The push toward standardization, the increasing mobility of students, the ever more efficient distribution of scientific journals, the remarkable expansion of the applied sciences that marked the second industrial revolution—all of these factors contributed to a pan-European identity for science even as imperial rivalries reached their peak. In *A History of European Thought in the Nineteenth Century* (1896–1914), John Theodore Merz, a German with a doctorate on Hegel who ran a chemical factory in Britain, gave voice to the growing conviction that "in the course of our century Science at least has become international.... [W]e can speak now of European thought, when at one time we should have had to distinguish between French, German, and English thought."

Around 1900 the reconstruction of the Republic of Letters was cemented by the foundation of the International Association of Academies, with headquarters in Berlin; the International Catalogue of Scientific Literature, a retrospective inventory of periodical literature run from the Royal Society of London; and, outlasting both, the Nobel Institution and its prizes. On the nationalist level, the *Kaiser-Wilhem Gesellschaft (founded 1911) copied in its own way the Royal Institution of Great Britain (founded 1799) and the Carnegie Institution of Washington (founded 1902), both examples of the use of private money for scientific research, then contrary to German practice. Alarmed British scientists pointed to Germany's expenditures on higher education to try to obtain, without much luck, greater resources from a stingy government. German scientists pointed to British trade schools and to the large expenditures on big research institutions made by the United States. French scientists pointed everywhere.

World War I shattered the growing internationalism. It replaced the International Association of Academies, headquartered in Berlin, with the International Research Council (IRC), dominated by the Belgians, French, British, and Americans, which did not admit the former Central Powers until 1926. The war also led to the creation of the Soviet Union, which was to spend most of the twentieth century outside the Western concert of science, and, indirectly, Nazi Germany, which would soon be ostracized for its hounding of Jewish academics. Still, World War I consolidated parallels among the belligerents: closer cooperation between science and the military and industry in the various countries, and, among the Allies, creation of equivalent institutions for the channeling of government money into academic science. The international *Rockefeller philanthropies set up to support the exchange of researchers and to help build scientific institutes in Europe (France, Germany, Denmark, and Sweden were among the recipients) also helped the advanced scientific nations progress together.

After the opening of the IRC to the former Central Powers and the foundation of various international unions for pure and applied science, cosmopolitanism had a brief renewal. Just before the Nazi takeover, for example, the staff of sixty chemists at Fritz *Haber's Kaiser Wilhelm Institut für Physikalische Chemie included seven Hungarians, four Austrians, three Russians, two Czechs, two Canadians, and one each from the United States, England, France, Poland, Ireland, Lithuania, Mexico, and Japan. During the late 1920s, international meetings increased in frequency. The Nazis and fascists then promoted internationalism in their special way by forcing some of the greatest European scientists to flee, particularly to Great Britain or the United States. At the same time, the Soviet Union drew in on itself. It had recruited left-leaning European scientists regardless of passport to help build up scientific research institutes in the People's Commissariat of Heavy Industry. The successive political purges of 1936–1938 expelled many visiting scientists, and the remainder left at the earliest opportunity.

World War II damaged the material infrastructure of European science and, together with the emigrations of the 1930s, made the United States by far the world's dominant scientific power. Soviet scientists who until the mid-1930s had played a lively part in multilateral European exchanges were sorely disappointed to find themselves further isolated by Cold War politics. Science remained an engine of prestige for the Soviet social experiment, however, and the growth in the Soviet Academy of Sciences in particular reflected its members' ability to turn the geopolitical insecurities of Soviet leaders into massive infusions of support for scientific research.

Two differences between Soviet and western European scientific institutions are especially notable. In Russia funding for individual programs of research did not come through formal independent peer review, but rather from large block grants to their host institutions. This practice exaggerated the discretionary powers of institute directors, who seldom resisted the temptation to cultivate huge patronage networks as the addition of classified research swelled the staffs of some institutes to a thousand or more. The Soviet Academy of Sciences also far outstripped its European counterparts in the control of institutional resources. In collective terms, it could thus dispense both resources and status, a function usually performed by separate institutions elsewhere.

Cold War rivalries and incipient European integration aided the rise of *multinational laboratories. High-energy physics, with its unprecedented concentration of material and human resources on the search for the ultimate constituents of matter, took the lead. If the popular rationales for these expenditures often made reference to bilateral geopolitics, each new generation of *accelerators still fostered increasingly complex multilateral collaborations. The best known institution in the western half of the continent was the European Center for Nuclear Research (*CERN in its French abbreviation), founded outside Geneva in 1954 after arduous negotiations involving twelve sponsor nations. On the other side of the Iron Curtain, members of the Warsaw Pact nations joined Soviet physicists in building a rival accelerator at Dubna, outside Moscow, at an analogous institute—the Joint Institute for Nuclear Research.

Large-scale collaborations came to the life sciences with the founding of the European Molecular Biology Laboratory (EMBL) in the 1970s. Based in Heidelberg, it boasted four affiliated facilities elsewhere in Europe, and more than a dozen member nations. Where the lengthy lead times for particle experiments dictated a large permanent staff along with a steady stream of visiting researchers at CERN, EMBL had few permanent staff and visiting appointments lasting several years at most before the researcher returned to a home institution. It aimed not so much to transcend national boundaries by means of a single institution as to ensure steady cross-fertilization among the scientific communities of its member nations. Other initiatives like the European Synchrotron Radiation Facility (Grenoble), EURATOM, and the European Space Agency have been launched as well, with varying degrees of success.

The collapse of the Soviet Union and the increasing federalization of the European Union have reshaped the playing field for science in Europe. Scientists from EU nations are among the primary beneficiaries of employment mobility across borders, and continuing economic disparities have also made it attractive for Russian,

Czech, or Hungarian scientific elites to pursue careers further west on the continent. Science in Russia, though financially impoverished in general, shows modest signs of stabilizing after the "brain drain" of the early 1990s, with a pronounced shift toward grant-based research funded by private (both foreign and domestic) and government foundations.

France and Germany continue to fund large systems of institutes for pure research with few rivals elsewhere in the world, but more and more of the money for academic research passes through Brussels. Since proposals seem to have a better chance of success the larger the spectrum of collaborators, EU grants make a powerful force for the integration of European science. Nonetheless, the entire enterprise may be regarded as underfunded, particularly in Britain. Both Japan and the United States spend half again as large a percentage of their gross domestic products on research and development compared to the average EU member. The increasing autonomy of European Council members vis-à-vis their national state bureaucracies, however, offers further opportunities for the scientists of Europe to cement broader institutional alliances that constrain national policy makers. European political union remains anything but certain, yet most scientists continue to claim "Europe" as one of the surest means (whether directly or indirectly) for the local advancement of science on a global stage.

See also ACADEMICS AND LEARNED SOCIETIES; ASSOCIATIONS FOR THE ADVANCEMENT OF SCIENCE; NOBEL PRIZE; STANDARDIZATION; UNIVERSITY; WORLD WAR II AND COLD WAR.

Alexander Vucinich, *Science in Russian Culture* (1963). Paul Forman, J. L. Heilbron, and Spencer Weart, "Physics around 1900: Personnel, Funding, and Productivity of the Academic Establishments," *Historical Studies in the Physical And Biological Sciences* 5 (1975): 1–185. Brigitte Schröder-Gudehus, *Scientifiques et la paix. La communauté scientifique internationale au cours des années* 20 (1978). Spencer Weart, *Scientists in Power* (1979). Charles Paul, *Science and Immortality: The Éloges of the Paris Academy of Sciences, 1699–1791* (1980). Alexander Vucinich, *Empire of Knowledge: The Soviet Academy of Sciences 1917–1970* (1984). James E. McClellan III, *Science Reorganized. Scientific Societies in the 18th Century* (1985). Harry Paul, *From Knowledge to Power. The Rise of the Science Empire in France, 1860–1939* (1985). Christa Jungnickel and Russell McCormmach, *Intellectual Mastery of Nature: Theoretical Physics from Ohm to Einstein*, 2 vols. (1986). Daniel J. Kevles. *The Physicists, The History of a Scientific Community in Modern America*, rev. ed. (1995). John Krige and Dominique Pestre, eds., *Science in the 20th Century* (1997).

KARL HALL

EVOLUTION. What we now call evolution in biology—the notion that organisms are related by descent—was first debated within the life sciences during the second half of the eighteenth century. The word "evolution," however, had been used in the seventeenth and eighteenth centuries to denote individual, embryonic development, not descent. "Evolution" (from Latin *evolvere*, "to unroll") then meant typically the unfolding of parts preexisting in the embryo, as conceived by the supporters of "preformation" in *embryology. Occasionally, supporters of epigenesis used evolution to denote what they regarded instead as the successive addition of new parts in individual development (*see* EPIGENESIS AND PREFORMATION).

From the mid-eighteenth century—both jointly with earlier embryological speculations (*see* BIOGENETIC LAW) and independently of them—a number of natural philosophers formulated hypotheses implying a dynamic conception of the history of the universe, the earth, and life, as opposed to a static conception of nature. Stasis was then increasingly regarded, and occasionally attacked, as typical of the major western religious traditions (*see* RELIGION AND SCIENCE). The Enlightenment produced new or renewed dynamic conceptions of nature: in astronomy, by attempts aimed at extending Newtonian concepts to explain the history of the planetary system as well as its functioning (Georges-Louis de Buffon, Immanuel Kant, Pierre-Simon de *Laplace); in the earth sciences, by evidence pointing at a formerly unthought of antiquity of the earth (Buffon, James *Hutton); and in the life sciences, by speculations on a possible temporalization of the traditionally static system of classification of living beings (Charles Bonnet, Jean-Baptiste Robinet), by attempts at formulating materialistic explanations of the origin of life, *generation, heredity, development, and change of organic structures (Benoît de Maillet, Pierre de Maupertuis, Erasmus Darwin), by the occasional observation of variability in species (Carl *Linnaeus), and by the transposition of the notion of embryonic development to the entire history of life on Earth (Carl Friedrich Kielmeyer).

Between 1802 and 1820, Jean Baptiste *Lamarck combined several of these themes to produce the first systematic, if not always clear, theory of organic change (in those years "evolution" in its present meaning is documented only in the works of Julien-Joseph Virey). Around 1830, Étienne and Isidore Geoffroy Saint-Hilaire developed and circulated Lamarck's notions further. Charles *Lyell discussed Lamarck's views from a critical perspective in his authoritative geological work of the early 1830s, where he used "evolution" for the first time in its present sense in English. Lamarckian notions, combined with speculations on embryonic development and the nebular hypothesis in astronomy (*see* NEBULA), figured also in the first popular book to spread an evolutionary worldview in the English-speaking countries: Robert Chambers's anonymously published *Vestiges of the Natural History of Creation* (1844).

From 1859, Charles *Darwin's *Origin of Species* attracted enormous attention to the issue of the natural derivation of all species from one, or few, original living forms. In scientific as well as popular circles the question at issue was "evolution." Darwin avoided the word, however, preferring the expression "descent with modification." The philosopher Herbert Spencer and many of Darwin's own followers preferred "evolution," because of its broad implications for a view of nature—often secularized, and embracing humankind and society as well as the cosmos—that emphasized gradual, progressive change, achieved through competition.

Darwin's own explanation of evolutionary change, the theory of natural selection (*Darwinism), faced strong opposition even among knowledgeable scientists. The late nineteenth century saw many attempts to fill up the gaps ("missing links") of the *fossil records documenting ancestral histories, and little agreement on the causes of evolution. Around 1900, neo-Darwinism, neo-Lamarckism, and *orthogenesis competed with one another and other theories of evolutionary change.

After the rediscovery of Gregor *Mendel's laws in 1900, it took about four decades for biologists from various specialties to build consensus around the so-called modern synthesis, or neo-Darwinian theory of evolution, combining Darwin's notion of natural selection with the science of *genetics. Since the 1940s, that has been the evolutionary theory adopted by the majority of biologists. From the mid-1960s, however, new approaches have emerged to the study of biological evolution, deploying the tools and concepts developed in the meantime by evolutionary *ecology, *paleontology, and, especially, *molecular biology. The study of evolutionary change at the molecular level, in particular, has become a major research field, and the results achieved have led to conclusions sometimes diverging from mainstream neo-Darwinian concepts.

For example, the circumstance that genetic variation (now measured with techniques like gel *electrophoresis of enzymes) is occasionally uncorrelated with reproductive success and adaptive evolution—and thus seems "neutral"

with regard to natural selection—has led some biologists, notably Motoo Kimura, to develop, beginning in 1964, the neutral theory of molecular evolution. The theory asserts that much of the evolutionary change observed at the molecular level occurs via random genetic drift, unaffected by natural selection.

On another, connected front, privileging the study of fossil evidence and the process of speciation, Niles Eldredge and Stephen Jay Gould began in 1972 to develop the hypothesis of punctuated equilibria, which departs from the neo-Darwinian notion of evolutionary change as gradual and continuous. The hypothesis asserts that the history of many fossil lineages shows long periods of little morphological change (stasis) alternating with brief periods of rapid change associated with speciation.

Since the 1960s, the issue of human evolution has also aroused renewed scientific interest. The frequent harvest of ancient fossil specimens, especially from Africa; the study of mitochondrial *DNA, allowing tentative dating of the more recent branches of the human family tree; and the statistical assessments of genetic variation, favoring a comprehension of how biological and cultural traits may interact in human populations through migrations; have made and continue to make dramatic news in popular as well as scientific circles.

Despite the opposition called up repeatedly by phenomena like social Darwinism and its several twentieth-century derivatives, and because of the recurrent conflicts over the religious and ideological implications of our understanding of evolution, evolutionary biology continues to carry with it today all the excitement that accompanied the publication of Darwin's *Origin* in 1859.

Ernst Mayr, *Evolution and the Diversity of Life* (1976; 1997). Peter J. Bowler, *Evolution, The History of an Idea* (1984; 1992). Niles Eldredge, *Time Frames: The Rethinking of Darwinian Evolution and the Theory of Punctuated Equilibria* (1985). Stephen Jay Gould, *Wonderful Life* (1989; 2000). Simon Conway Morris, *The Crucible of Creation* (1998). James A. Secord, *Victorian Sensation. The Extraordinary Publication, Reception, and Secret Authorship of Vestiges of the Natural History of Creation* (2000).

GIULIANO PANCALDI

EXHIBITIONS have shaped much of the cultural landscape of the modern world. Architecture and planning, mass entertainment and consumption, colonialism and imperialism, and not least, the cultures of science and technology owe much of their form and substance to the temples of modernity erected episodically throughout the world since the middle of the nineteenth century. Science suffused the very idea of exhibitions,

The machinery court at the Crystal Palace Exhibition, London, 1851.

their ideologies of progress, and the philosophies of classification that ordered displays. Exhibitions, moreover, diffused scientific ideas and introduced millions of people to scientific discoveries—radium and X rays, for example—and technological innovations—electric lighting, the telephone, plastics, and satellites. Scientists from different countries used exhibitions as a venue to meet and share their work and as a marketplace for instruments and technologies.

International exhibitions, also known as great exhibitions, universal expositions, and world's fairs, began with London's 1851 Crystal Palace Exhibition. The Great Exhibition of 1851, the brainchild of Albert, the Prince Consort, established the basic patterns of organization and display that would endure for decades without drastic modification. It built upon and to some extent supplanted traditions of local fairs and several decades of industrial displays at mechanic's institutes and national exhibitions in France and other countries. The success of the Crystal Palace Exhibition spurred other nations to follow suit, launching a veritable parade of international extravaganzas that Gustave Flaubert called "the delirium of the nineteenth century." By World

War I nearly one hundred international exhibitions had been held around the world. The largest late-nineteenth-century exhibitions frequently received visitors numbering in the tens of millions, culminating in the Paris Universal Exposition of 1900, attended by some fifty million people. In the twentieth century, international exhibitions disappeared during both world wars, only to be revived afterwards, each time with diminished status and importance.

Prince Albert promoted the Great Exhibition of 1851 as a carousel of industrial progress, a liberal and free trade antidote to the climate of unrest that haunted Europe in the years around 1848. The exhibition would provide an illustration of how "the great principle of division of labor, which may be called the moving power of civilization, is being extended to all branches of science, industry, and art." The exhibition would become a "scientific experiment," Albert announced, an attempt to reveal human progress as an unfolding of God's natural laws and to further "a naturalist's insight into trades." For Albert, a disciple of the Belgian statistician Lambert-Adolphe-Jacques Quetelet, the experimental results would show themselves in the system of classification of sciences, arts, and industries. Many scientists were consulted for the classification, including the biologist Richard Owen, the German chemist Justus von *Liebig, and especially, the British chemist Lyon Playfair, who presided over the final system of classes and stipulated the conditions of selection and the procedure of juries.

*Classification remained an emblem of the primacy of science and a central theme of debate in major exhibitions held around the world. Several major exhibitions experimented with radically different methods of classification. For the Paris Universal Exposition of 1867, for example, the French engineer and sociologist Frédéric Le Play devised a philosophical classification system that encompassed not only industry but the whole of human activity in a didactic arrangement linked with the architecture of the exhibition palace. Also noteworthy was the attempt of Harvard psychologist Hugo Münsterberg to deepen the philosophical content of the exhibition by subsuming the classification of the world's fair held in St. Louis in 1904 to the disciplinary principles of a vast international scientific congress held at the exhibition (*see* CLASSIFICATION OF SCIENCES).

Juries remained the key institution of international exhibitions. Composed of international experts, frequently scientists, their judgments set the standard for every field of endeavor. Both the selection and judgments of juries were often contested and the debates contributed greatly to the emerging concept of the scientific or technological

expert. The British scientist Charles Babbage, for example, suggested that the 1851 exhibitions jury be regarded as the prototype of a professional class of scientists. In practice, juries endlessly disputed the basic principles of evaluation. As a result, exhibition juries promoted the creation of modern international institutions of standard weights and measures (*see* STANDARDIZATION).

At the Paris Universal Exposition of 1855, jury members, beset with headaches over how to test, compare, and classify objects of diverse national origin, signed a document imploring the authorities of all countries to participate in international negotiations to establish coordinated meteorological standards. Several countries signaled their support, and in 1867 an international commission was established to debate and implement the metric system as the international standard for science, technology, and commerce. In the decades that followed, numerous international scientific commissions used exhibition sites to hammer out standards agreements. The most famous of these, the international agreement on standards of electrical resistance, was reached by several delegations of scientists, including William *Thomson, Lord Rayleigh of Britain, and Hermann von *Helmholtz, Werner von Siemens, and Rudolf Clausius of Germany, at the International Electrical Exhibition of Paris in 1881.

Several features of exhibitions attracted scientific pilgrims to the throng. Exhibitions played host to international scientific congresses, bringing scientists into direct contact with colleagues from other countries. At these meetings, as with similar gatherings of artists, industrialists, and politicians, scientists were exposed to national differences in method, apparatus, and philosophy. Exhibitions also furnished scientists with the opportunity to see and buy scientific instruments from around the world and to display and sell instruments of their own design. As in the industrial trades, early exhibition promoters expressed the hope that science would benefit from the increased contacts between instrument makers and eminent scientists. This desideratum seems to have been met frequently; for example, the Astronomer Royal George Biddell Airy visited the 1851 exhibition to examine Charles Shepard's electrically driven master clock controlling slave dials elsewhere in the building.

The optimism about this type of exchange reached its apogee in 1876, when an international exhibition devoted exclusively to scientific instruments took place at London's South Kensington museum. Scientists and instrument makers (and their apparatus) from many countries mingled profitably. After surveying what was probably the most complete spectrum of scientific instruments ever assembled, British physicist James Clerk *Maxwell composed a taxonomy of scientific apparatus according to their functions and principles of construction. Arraying the different components of physical and chemical apparatus in close analytical detail, Maxwell's "General considerations concerning scientific apparatus" was a perfect example of the "naturalist's insight into trades" to which exhibitions aspired but did not always produce.

By contrast, the broader attempt at science education that was part of every exhibition's mission delivered mixed results. The early exhibitions drew on their origins in mechanic's institutes and worker education movements to promote mass education and diffusion of science and technology. The early Victorians had become adept at "learning by seeing," at understanding the principles of steam engines and related machinery, tracing with the eye the visible processes of driveshafts, gears, belts, valves, pulleys, cutoffs, and levers that drove and transmitted power. Aiming at a similar clarity and directness, exhibition displays of various scientific disciplines including the physical sciences, as well as *geology, *biology, and *anthropology, generally eschewed the grotesque imagery of earlier curiosity *cabinets in favor of lucid, analytical depictions of objects and artifacts. Statistics featured prominently in exhibits of many kinds, offering a seemingly transparent representation of the hidden facts of nature and society.

Over the course of the nineteenth century, exhibition organizers and participants gradually pulled back from the more ambitious attempts at scientific content and created exhibits that used the results of science to dazzle, inspire, and impress. The visiting public turned out to be less avid and equipped for education than promoters hoped, and new kinds of edifying entertainment gradually supplanted rigorous instruction in the majority of exhibits. Historians have attributed this transition in part to the changing nature of science and technology themselves. Late-nineteenth-century science was increasingly dominated by electricity and other phenomena (such as *X rays) whose operations were invisible and whose principles proved difficult to communicate. After 1876, when the wonders of the phonograph, telephone, electric lighting, loudspeakers, and electric railways prevailed at exhibitions, entrepreneurs like Thomas Edison exploited the dazzling character of the new technologies to enhance the prestige of their business enterprises. Such developments, coupled with a growing tendency of elite professional scientists to distance themselves from applied science or technology, contributed to a growing bifurcation between pure and applied science at exhibitions. While attempts to communicate elements of applied science remained, exhibitions witnessed an increasing renunciation of attempts to communicate the content of pure science, opting instead for aestheticized displays of general convictions.

Because the diffusion of science at exhibitions overlapped with other forms of popularization, it is impossible to precisely gauge their broader cultural impact. But there can be no doubt that the ideologies and aesthetics of exhibition displays have left a legacy in many of the characteristic modes of scientific popularization, including fascination with scales of magnitude, emphasis on visual communication, and promotion of the material bounty of scientific progress.

Robert W. Rydell, *All the World's a Fair: A Vision of Empire at America's International Exhibitions, 1876–1916* (1984). Paul Greenhalgh, *Ephemeral Vistas: The Expositions Universelles, Great Exhibitions, and World's Fairs, 1851–1939* (1988). John E. Findling and Kimberly D. Pelle, eds., *Historical Dictionary of World's Fairs and Exhibitions, 1851–1983* (1990). Robert W. Rydell, *The Books of the Fairs: Materials about World's Fairs, 1834–1916* (1992). Robert M. Brain, *Going to the Fair: Readings in the Culture of Nineteenth-Century Exhibitions* (1993). Robert W. Rydell, *World of Fairs: The Century of Progress Exhibitions* (1993). David E. Nye, "Electrifying Exhibitions: 1880–1930," in *Fair Representations: World's Fairs and the Modern World (European Contributions to American Studies XXVII)*, ed. Robert W. Rydell and Nancy E. Gwinn (1994): 140–156. Tony Bennett, *The Birth of the Museum. History, Theory, Politics* (1995).
                                                    ROBERT M. BRAIN

EXPERIMENT. See OBSERVATION AND EXPERIMENT.

EXPERIMENTAL PHILOSOPHY. "The business of experimental philosophy," says the *Encyclopedia Britannica* in 1771, "is to inquire into, and to investigate the reasons and causes of, the various appearances and phenomena of nature; and to make the truth and probability thereof obvious and evident to the senses, by plain, undeniable, and adequate experiments." Experimental philosophy was thus a method, a practice, and a slogan. It signified to the eighteenth century the means and goal of the suppression of the philosophy of the schools begun in the seventeenth century (see ARISTOTELIANISM) and, more generally, of any entrenched "system" immune from the challenge of experience. The complacent natural philosophers of the Enlightenment ascribed the method to Francis Bacon (see BACONISM), and the first examples of its successful practice to *Galileo and the Accademia del Cimento (see ACADEMIES AND LEARNED SOCIETIES); next (according to the *philosophes*) the Royal Society of London demonstrated the superiority of the experimental over all other ways of philosophizing, with which the Académie royale des sciences of Paris concurred, once it had rid itself of the system of René *Descartes.

Isaac *Newton's *Opticks* (1704) demonstrated that, and how, the experimental philosopher could move from exact observations and careful manipulations to a new theory of wide application and great importance. Among the few who managed to achieve something similar to the *Opticks* during the eighteenth century were Benjamin *Franklin (*electricity), Henry Cavendish (electricity), and Antoine-Laurent *Lavoisier (*chemistry). Most other "experimental philosophers" repeated demonstrations or found new phenomena that they related to one or another of the dominant natural philosophies. Thus the influential Dutch professors Willem Jacob 'sGravesande and Petrus van Musschenbroek, who gave the first sustained university courses in natural philosophy illustrated with experiments, generally followed Newton (*see* NEWTONIANISM); the abbé Jean-Antoine Nollet, who from the 1730s offered courses similar in purpose though wider in coverage outside the universities, generally followed Descartes.

By 1750 professors everywhere were introducing experiments into their courses, particularly in the many small German universities and in the larger Jesuit colleges in France and Italy, in order to help capture the interest (and where possible, the fees) of students. Throughout the second half of the eighteenth century, the German universities combated a shrinking enrollment and the Jesuits faced increasing competition from other teaching orders and a general animosity that led to their suppression in 1773. Meanwhile the niche cut out by Nollet and his predecessors in England—the curators of the Royal Society Francis Hauksbee and John Theophilus Desaguliers—widened to nourish all sorts of lecturers who entertained paying audiences with the instruments of science. For many of these men and most of their auditors, "experimental philosophy" was a misnomer. They showed, looked, and talked, but neither experimented nor philosophized.

Experimental philosophy, as opposed to philosophical entertainment, rested on two principles of unequal weight. The lesser admitted Bacon's criticism of hasty theorizing and his remedies. The greater held that the bigger and more precise the experimental apparatus, the better, even if the size and sophistication served no immediate philosophical purpose. Ever larger electrostatic generators, for example, culminated in the gigantic machine, which could throw a spark a quill thick for two feet or more, built for the Teyler Foundation in Haarlem in the Netherlands (1785).

The reason, as given by the experimental philosopher who commissioned it, Martin van Marum: "I took it as certain that, if one could acquire a much greater electrical force than hitherto in use, it could lead to new discoveries." "Twist the lion's tail," Bacon had admonished, and wait for the results. No doubt the method had its successes, including, in electricity alone, the creation of the *Leyden jar, the assemblage of the jars into an electric *battery, the multiplication of the contact electricity of zinc and silver into the Voltaic pile (*see* GALVANI, LUIGI, AND ALESSANDRO VOLTA), and the connection of many piles into powerful engines of physical and chemical research.

Other examples of enlargement in the service of philosophy were the long, thin magnetic needles with well-defined poles used by Charles-Augustin Coulomb (*see* CAVENDISH, HENRY, AND CHARLES-AUGUSTIN COULOMB) to establish the law of magnetic force (1785), Lavoisier's apparatus for his experiments on gases, respiration, and heat (1770s–1780s), and Cavendish's balance for weighing the world in the laboratory (1798). Many significant instrumental improvements, however, did not require enlargement. The English instrument maker Edward Nairne advertised an air pump in 1777 that, he said, reduced the pressure to as little as 1/600 of normal atmospheric (at.) in only six minutes; it did not differ much in size from the standard machine of the time (which gave 1/165 at.) or from 'sGravesande's of the 1720s (at best 1/50 at.).

The most significant instrumental improvements for most experimental philosophers had to do with the *microscope and the meteorological instruments. As the microscope became easier to use, Sunday philosophers applied it more readily to increase their wonder at the works of the Creator, especially after Abraham Trembley spied the marvelous self-regenerating freshwater polyp in a drop of dirty ditch water (1740). The measurement of the weather occupied many people wanting to do something useful for themselves and for science. They could measure to many more places of decimals in 1780 than in 1730. At the earlier date, observers did not correct their *barometers for temperature because the correction would have been less than the error of the instrument. After the main cause of error (unwanted air) had been removed, serious philosophers, such as the Genevan Jean André Deluc, corrected for temperature, capillarity, and the curve of the meniscus, and claimed to be able to read their barometers to a few hundredths of a millimeter.

In 1730 the usually meticulous René-Antoine Ferchault de Réaumur rejected the suggestion that he use brass scales rather than paper ones on his *thermometers; that, he said, would be to carry precision to ludicrous lengths. As late as 1777 the Royal Society found wide variations in the boiling points of their thermometers. A committee chaired by Cavendish established procedures that removed the discrepancies. Meanwhile thermometers marked to a fifth or a tenth of a degree came on the market. Lavoisier's instruments, standardized by methods like those of the Cavendish committee and divided according to the latest technology, could be read reliably to perhaps a hundredth of a degree. The hygrometer, a device for determining humidity by the motion of a piece of animal matter (hair, gut, or ivory) attained similar perfection in the hands of Deluc.

The improvements paid off in several ways. Barometric measurement after suitable corrections could be used to measure heights of steeples and mountains to satisfactory accuracy. The improved thermometers made possible the detailed explorations of heat (*see* FIRE AND HEAT), especially of the laws of dilation of gases, made by the group of *physiciens* centered on Pierre-Simon, Marquis de *Laplace. And the hygrometer had the honor of helping Deluc to gain the odd post of natural philosopher to the Queen of England. The reasons for his preferment may make an apt symbol for much experimental philosophy in the Age of Reason. Deluc came to England recommended by his grand ideas about the history of the earth, which, as he labored to show, agreed perfectly with the account in Genesis, properly understood. He developed this project with the same keenness for observation and analysis he deployed in perfecting his instruments. Although they had little bearing on his grand philosophy, they propelled him to his royal roost. During negotiations he presented the King with a super-precise barometer for measuring heights during military campaigns and the Queen with the latest hygrometer for determining over-exactly the humidity of her greenhouses.

See also INSTRUMENTS AND INSTRUMENT MAKERS; MATHEMATIZATION; NEWTONIANISM; QUANTIFICATION.

J. L. Heilbron, *Electricity in the Seventeenth and Eighteenth Centuries: A Study of Early Modern Physics* (1979, 1999). Steven Shapin and Simon Schaffer, *Leviathan and the Air Pump: Hobbes, Boyle and the Experimental Life* (1985). Tore Frängsmyr et al., eds., *The Quantifying Spirit in the Eighteenth Century* (1990). Jan Golinski, *Science as Public Culture: Chemistry and Enlightenment in Britain, 1760–1820* (1992). Thomas Hankins and Robert J. Silverman, *Instruments and the Imagination* (1995). Stephen Gaukroger, *Francis Bacon and the Transformation of Early-Modern Philosophy* (2001).

J. L. HEILBRON

EXPERIMENTATION, ANIMAL. See ANIMAL CARE AND EXPERIMENTATION.

EXPERTISE, SCIENTIFIC. See RISK ASSESSMENT/ SCIENTIFIC EXPERTISE.

EXPLORATION AND FIELDWORK. The histories and geographies of fieldwork and exploration are closely bound together in the development of modern science. During the fifteenth, sixteenth, and seventeenth centuries, both the logic of exploration and the consequences of its discoveries helped force a reconceptualization of the fundamental principles upon which scientific authority rested. Through the eighteenth, nineteenth, and twentieth centuries, exploration and fieldwork continued to be powerful forces in the shaping of science, although their main contributions had less to do with redefining principles of method than with developing substantive scientific knowledge of the world.

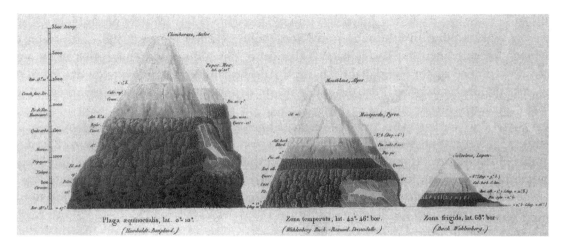

Plot of Andean peaks, showing variation of plants by height and latitude.

During the Age of Discovery (1450–1700), European nations, inspired by religious fervor and mercantile aspirations, developed a new understanding of the world beyond their own continent. Exploration in the Renaissance challenged established academic authority in three ways. First, the project of discovery, with its emphasis on the importance of experience, observation, and veracity, helped lay a new foundation for the production of knowledge about the world. This preeminently empirical form of knowledge acquisition threw into question the school philosophy with its emphasis on a priori reasoning, religious belief, and sedentary debate. However, this is not to say that Prince Henry the Navigator should be raised to the level of Nicholas *Copernicus or *Galileo Galilei as a hero of the *Scientific Revolution.

Second, the objects collected during voyages of discovery and the testimonies of respected observers on board ships revealed New World civilizations seemingly incompatible with Mosaic chronicles of creation. The unfamiliar environments of the Americas also challenged the established natural histories of the Old World. The representative *natural history of the sixteenth century—for example, Konrad Gesner's *Historia animalium* (1551–1558)—promoted an associative understanding of animal life in which descriptions of animals included their mythological, etymological, cosmological, and iconographic relations, as well as their habits and characteristics. Not only did the creatures of the New World undermine the supposedly comprehensive Renaissance catalogues, but attempts at their inclusion within an emblematic natural history brought existing formats and classifications into question. How could one write a natural history of an animal knowing nothing about its name, iconographic associations, roles in human culture, relations to other creatures, or place in myths and legends? By the mid-seventeenth century, natural histories included the animals of the New World and dispensed with previous classical conventions.

Thirdly, Renaissance exploration challenged classical scholarship by its emphasis on application. For instance, Prince Henry the Navigator sought geographical knowledge. However, his voyages were also commercial, colonial, and religious in motivation. As new protoscientific techniques helped merchants and rulers establish empires of trade and dominion, so Europe's more radical natural philosophers argued that science should help improve the lot of humankind. The labors of some natural historians, mathematicians, chemists, and astronomers had a direct bearing on the endeavors of those involved in oceanic exploration. For instance, John Evelyn, a founding member of the Royal Society of London, advised the English Navy about the supply of timber for their ships. Science and exploration also combined in the development of a set of technological devices that facilitated the efficient movement of people and things around the world. Practical charts of maritime navigation, the quadrant, accurate and robust timepieces, and new sorts of ships helped produce a science of exploration (*see* Clock and Chronometer; Longitude).

By 1650 most of the world's coastlines had been mapped, at a rudimentary level at least, and concerted efforts were underway to explore and map continental interiors. Although the majority of accounts came in the form of a topographical compendium, their geographical focus varied widely, ranging from descriptions of local areas up to continents and even the entire globe. In an attempt to make these texts useful to others, the Royal Society of London devised an agenda and method toward the standardization of exploration reports and the geographical scale of their enquiry. It preferred "regional surveys." Samuel Hartlib, the Prussian natural philosopher, promoted the use of regional surveys in the production of a "political anatomy" that organized geographical, economic, and social data. Meanwhile, Robert *Boyle prepared a treatise to aid the regional surveyor, a eulogy to the inductive method, entitled *General Heads for the Natural History of a Country, Great or Small; Drawn Out for the Use of Travellers and Navigators* (1692). Boyle's text might be seen as a very early fieldwork manual.

The work of Carl *Linnaeus gave a boost to international field studies, which both imitated his example and employed his classifications. Linnaeus botanized in Lapland, elsewhere in Scandinavia, Holland, France, and England. His students, or disciples as he called them, applied his binomial nomenclature to the world beyond Europe. From 1745 to 1792 nineteen of them left on voyages of discovery to, among other places, Australia and the Pacific, Siberia, Senegal, the Americas, the Arabian peninsula, the Ottoman Empire, and Africa. A series of other significant scientific expeditions of continental interiors took place during the eighteenth century. Mark Catesby and later John Bartram and his son William explored the American East Coast. Like Catesby before him, John Bartram made a good trade in supplying European natural history enthusiasts with North American specimens. Catesby and William and John Bartram, the latter acting as the King's Botanist in North America, also wrote and published influential natural histories of the Carolinas, Georgia, Florida, Pennsylvania, and the Bahamas. Many others continued to work in this tradition, culminating in the famous western territorial expedition of Meriwether Lewis and William Clark in the first decade of the nineteenth century.

Under French leadership, two international scientific teams set out in the 1730s to resolve the debate over the shape of the globe. One team, led by Pierre de Maupertuis and including Anders Celsuis, went to Lapland, the second team, led by Louis Godin and Charles Marie de Lacondamine, to Peru, to measure the length of a degree along the meridian. Their extensive collection of a wide range of measurements contributed to international scientific knowledge of the regions. At the very end of the century Alexander von *Humboldt and his companion Aimé-Jacques-Alexandre Bonpland traveled through South America, accumulating enough material for thirty published volumes. Using more than four dozen measuring instruments, including chronometers, an achromatic telescope, sextants, compasses, barometers, thermometers, rain gauges, and theodolites, the two men set a new standard for regional scientific study. In particular, they pioneered the "isomap," the cartographic method for the delineation of comparative natural features.

Another landmark voyage of exploration was that of the HMS *Endeavour*, captained by James Cook. From 1768 to 1771 Cook traveled to the Pacific accompanied by a group of naturalists including Joseph Banks, the founder of the Royal Botanic Gardens at Kew, and Daniel Carl Solander, one of Linnaeus's foremost students. The first voyage's main objective was to observe from Tahiti the transit of Venus across the face of the sun. Although part of a broader international astronomical enterprise, the voyages played an important role in the introduction to Europe of the natural histories of New Zealand and Australia.

Cook's orders included taking possession of any new lands that might benefit the British crown. Banks and Solander's involvement in Cook's conquest of Australasia therefore enrolled naturalists as agents of the European imperial endeavor. It became commonplace for expeditions organized by the military to include men of science to assess the value of newly discovered lands and collect new species of plants and animals. These and others involved in the development of European colonial rule in the eighteenth and nineteenth centuries brought a wide array of foreign specimens back to their respective countries, feeding the growth of botanical and zoological gardens, natural history museums, and the trade in exotic plants. This association also benefited the development of the sciences. For instance, Charles *Darwin's participation in the British Admiralty's hydrological survey of the waters around South America on the HMS *Beagle* was fundamental to the development of his theory of natural selection.

The study of ethnology also grew out of the close relations between science and empire in the nineteenth century. Although European traders had come into contact with different cultures for centuries, colonial expansion brought with it a heightened impetus to study aboriginal peoples, before their eradication by "civilization." The emerging sciences of *anthropology and ethnology developed standardized methods for the measurement of the human body, termed "anthropometry." Nineteenth-century advances in *photography made further important contributions to this effort. British colonial governmental officials carried out comprehensive photographic surveys of the peoples of India in the 1860s, and in Africa in the 1890s.

The university was an important site for the furthering of fieldwork as an integral aspect of science from at least the sixteenth century. From around 1525, university physic gardens introduced medical students to herbs as living plants rather than only as dried herbarium specimens. The lack of such a garden in Bologna forced the University's instructor in *medicina practica*, Luca Ghini, to take his charges into the countryside in search of herbs. Students of Ghini later led extended expeditions during university vacations that transformed into more general lectures on natural history. This practice spread to France in the 1550s and to Sweden, Denmark, England, and Scotland by 1650. London's Society of Apothecaries was the most persistent promoter of the field excursion as a necessary aspect of medical training. Its expeditions into the English countryside occurred for 214 years without a break, stopping in the nineteenth century only because it took so long to get beyond the sprawling metropolis.

Education in field study for medical students carried over into other scientific disciplines in the eighteenth and nineteenth centuries. It became increasingly common for botany and geology students at the European universities to undertake fieldwork as part of their study. This tradition was introduced to the United States by Louis Agassiz and Spencer Fullerton Baird, who pioneered formal field teaching at Harvard University and Dickinson College, Pennsylvania, respectively. Students benefited from the proliferation of inexpensive field equipment: guidebooks with directions for taking notes, collecting and preserving specimens, and formulating a collection; tools like specially designed and reasonably priced hammers, geological maps, and "botany boxes," with tightly fitting lids and carrying straps; nets for entomologists; and guns and cameras (though these required a heavier outlay) for zoologists. Easy access to field equipment contributed greatly to the widespread enthusiasm for field study in Europe and North America in the nineteenth century. Women, children, and artisans, as much as university-educated middle- and upper-class gentleman, participated.

The establishment of anthropology and *geography as discrete university disciplines at the very end of the nineteenth century extended the close connection between a scientific education and field study. Practitioners in both fields placed in situ study at the center of their disciplines. More generally, academic science began to fragment into increasingly specialized subdisciplines, and even the largest expeditions started to limit themselves to particular tasks. For instance, while Felix Andries Vening Meinesz surveyed marine gravity on world-ranging submarines, Jacob Clay and Robert Millikan circled the globe measuring cosmic rays at various altitudes, latitudes, and depths; while Jean Abraham Chrétien Oudemans produced geographical maps of the Dutch East Indies, others made magnetic measurements and took soundings of coastal waters.

Exploration and fieldwork declined in importance during the twentieth century. Automated data-collection devices such as satellites, self-recording weather stations, and seismometers reduced the role of the human observer. Meanwhile, exploration became increasingly driven by the agendas of multinational corporations scouting for new reserves of natural resources such as oil and gas. Notable exceptions to this rule were the mapping and scientific exploration of Antarctica and the sea floor (*see* OCEANOGRAPHY).

The high latitudes of Antarctica made it of special interest for geophysical studies. Cooperative "Polar Years" were organized by geophysicists in 1882–1883 and 1932–1933. Mapping of the continent continued after World War II by nations interested in staking claims on the landmass. However, in 1957–1958 the International Council of Scientific Unions implemented the third Polar Year, formalized as the International Geophysical Year (*IGY). The success of the IGY prompted the creation of the Special (changed later to Scientific) Committee on Antarctic Research (SCAR) in 1958. This led indirectly to the signing of the 1959 Antarctic Treaty, setting the continent aside for peaceful, international, and scientific

purposes. However, despite the best efforts of SCAR, Antarctica remains the one continental landmass not yet comprehensively mapped by fieldworkers working on the ground.

See also ANTHROPOLOGY; BOTANY; CABINETS AND COLLECTIONS; CARTOGRAPHY; GEOLOGY; HUMBOLDTIAN SCIENCE; PALEONTOLOGY.

Gordon Fogg, *A History of Antarctic Science* (1992). David Livingstone, *The Geographical Tradition* (1992). Mary Louise Pratt, *Imperial Eyes: Travel Writing and Transculturation* (1992). Henrika Kuklick and Robert Kohler, eds., *Science in the Field* (1996). Anne Larsen, "Equipment for the Field," in *Cultures of Natural History*, ed. Nicolas Jardine, James Secord, and Emma Spary (1996): 358–377. Lewis Pyenson and Susan Sheets-Pyenson, *Servants of Nature* (1998). Tony Rice, *Voyages of Discovery: Three Centuries of Natural History Exploration* (2000). David E. Allen, *Naturalists and Society* (2001). David Livingstone, *Spaces of Science* (2002).

SIMON NAYLOR

EXTINCTION. That some species found as *fossils might have become extinct in the course of time was first debated during the second half of the seventeenth century. What we now call fossils were then being separated from a broader class of objects that included inanimate as well as formerly animate objects. The notion of extinction, on the other hand, conflicted with a well-established philosophical idea that proclaimed the plenitude of nature and with the religious notion of the perfection of Creation. Some writers, however, notably Robert *Hooke, endorsed both extinction and the perfection of Creation by assuming that new species had been formed as others became extinct.

Throughout the eighteenth century, extinction continued to be a controversial issue, on a par with the still not universally admitted organic origin of fossils. At the end of the century, however, Georges *Cuvier brought compelling new evidence for the reality of extinction based on the case of large fossil mammals of the kinds he called *Megatherium* and *Mastodon*. Using the tools of comparative *anatomy, Cuvier argued that those animals were distinct from any living species, and, given their size and the systematic exploration of the landed part of the globe, their descendants were unlikely to be found alive in remote areas. Cuvier explained extinction by the notion, already circulated by Jean-André Deluc, that the earth had been subjected to a series of "revolutions" that had drastically altered living conditions. For Deluc (though not for Cuvier) the latest of these revolutions was Noah's flood. Other geologists, like Giovanni Battista Brocchi, followed by Charles Lyell, suggested instead that species might have a fixed duration, just as individuals of the same species have similar life spans.

Within Cuvier's perspective, the reality of extinction ruled out the very possibility of the transformation of old species into new ones, as advocated by naturalists like Jean-Baptiste *Lamarck. That changed with the introduction of Charles *Darwin's notion of "descent with modification." According to Darwin's *Origin of Species* (1859), extinction was one of the two possible outcomes of evolution by natural selection, the other being modification (*see* *DARWINISM).

Until the 1970s, emphasis on evolution—the positive side of the evolutionary process—tended to distract attention from extinction, the negative side of life on earth. That changed in 1980 with the introduction of the Alvarez hypothesis. Formulated by physicist Luis Alvarez, his son the geologist Walter Alvarez, and others, the hypothesis concerned a possible, catastrophic collision between Earth and a comet (or an asteroid or swarm of asteroids) that caused the extinction of the dinosaurs sixty-five million years ago. Together with a growing awareness about the role of the human species itself as cause of the extinction of many species (*see* *ECOLOGY), the Alvarez hypothesis contributed to a renewed interest in the dynamics of extinction. Currently, paleontologists admit six major global extinction events in the history of life on Earth, and extinction has become an integral part in the study of the patterns of evolution, or macroevolution.

Martin J. S. Rudwick, *The Meaning of Fossils. Episodes in the History of Palaeontology* (1972; 2d ed. 1976). Niles Eldredge, *Fossils. The Evolution and Extinction of Species* (1991).

GIULIANO PANCALDI

EXTRATERRESTRIAL LIFE. Displaced from the center of Aristotelian cosmology, the earth became one of many planets in the Copernican worldview. Galileo Galilei's telescopic observations of earthlike mountains on our moon, and of moons circling Jupiter, emphasized this displacement. The principle of plenitude, which interpreted any unrealized potential in nature as a restriction of the Creator's power, argued for inhabitants on other worlds. Although there was no evidence of lunar inhabitants, why else, asked the English clergyman John Wilkins in his 1638 *The Discovery of a World in the Moon*, would Providence have furnished the moon with all the conveniences of habitation enjoyed by the earth?

Social critics seized on lunar inhabitants either as members of a perfect society or as exemplars of all earth's vices. This literary convention furnished some defense in attacks against the establishment and helped spread the idea of a plurality of worlds. So did persistent rumors that England intended to colonize the moon.

Life spread beyond the moon in the French astronomer Bernard de Fontenelle's *Entretiens sur la pluralitè des mondes* (1686). During their evening promenades, the conversation of a beautiful marquise and her tutor turned to astronomy. On the second evening they spoke of an inhabited moon, on the third of life on the planets, and by the fifth night they had progressed to the idea of fixed stars as other suns, giving light to their own worlds. More conventional astronomy textbooks repeated these views.

Belief in extraterrestrial life permeated much of eighteenth- and nineteenth-century thought. It allowed an easy attack on Christianity, whose teachings about Adam, Eve, and Christ might appear ridiculous if the earth were not the whole of the habitable creation. Similar concerns regarding the immensity of the universe and the corresponding insignificance of humans appeared in novels.

The gullibility of a public raised on pluralist writings is illustrated by the widespread acceptance of reports in the *New York Sun* in 1835, purportedly from Sir John Herschel at the Cape of Good Hope, detailing his observations of winged quadrupeds on the moon. *The New York Times* judged these reports to be probable.

Many aspects of the extraterrestrial life debate appeared in Percival Lowell's Martian hypothesis and in reactions to it. Too modest to believe humankind the sole intelligence in the universe, Lowell, a wealthy Boston investor, announced in 1894 his intention to establish an observatory in the Arizona Territory and search for signs of intelligent life on Mars. There was already in America a lively interest in Mars, attributed by cynics to public imbecility and journalistic enterprise. Others hoped that the discovery of intelligent life elsewhere would increase reverence for the Creator. Lowell reported an amazing network of straight lines, which he interpreted as canals, and concluded that Mars was inhabited. Astronomers criticized Lowell for seeing only the evidence that supported his beliefs. Many readers, however, were persuaded by Lowell's literary skill. They also applauded the social arrangements of Mars, as elucidated by Lowell, particularly the abolition of war.

Changes in scientific knowledge in the twentieth century strengthened belief in extraterrestrial life. Larger telescopes expanded the observable universe to millions of galaxies, each containing millions of stars, all rendering it increasingly improbable that our earth alone shelters life. In 1953, Stanley Miller and Harold Urey at the University of Chicago synthesized amino acids,

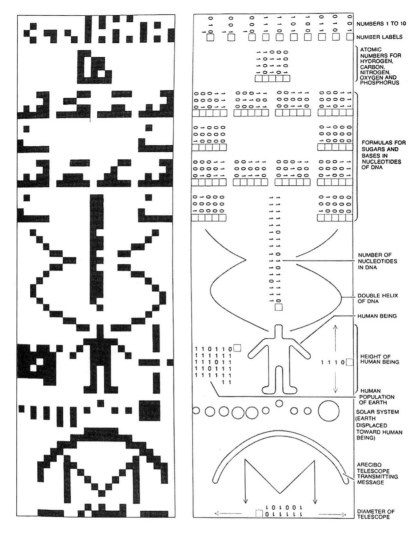

A calling card from planet earth (1975). On receiving the 1679 bits indicated, bright extraterrestrials will rearrange them into 73 columns of 23 bits each to produce the pattern on the left. Maybe.

the building blocks of life, from a mixture of methane, ammonia, water, and hydrogen, the supposed ingredients of our primitive earth's atmosphere. Although the late astronomer Fred Hoyle attributed both the origin of life on earth and much of subsequence evolution to showers of microorganisms from space, his was a minority view, though recently resurrected. Most scientists suppose that life occurs inevitably on earthlike planets, of which they estimate that millions exist in the universe. *Star Trek*, the television series, visits some of these planets every week.

Skeptics object that, if intelligent life is inevitable and has had billions of years to evolve and travel through the universe, it should long since have reached our earth. That extraterrestrials are not known argues against their existence. Believers in UFOs (unidentified flying objects) attribute the absence of evidence of extraterrestrials to a government cover-up.

The chemical theory of the origin of life coincided with the space age and did not long remain earthbound. In 1976, in one of the greatest exploratory adventures of the twentieth century, the National Aeronautics and Space Administration (NASA) landed two *Viking* spacecraft on the surface of the planet Mars, at the cost of over one billion dollars. Experiments detected metabolic activity, but probably from chemical rather than biological processes.

Post-*Viking*, interest shifted from microorganisms to direct communication with interstellar intelligence. Several early radio pioneers, including Guglielmo Marconi in 1920, thought they detected radio signals from Mars. The most comprehensive interstellar communication program was NASA's Search for Extraterrestrial Intelligence (SETI). From a small and inexpensive research and development project during the 1980s, SETI emerged in the early 1990s as a hundred-million-dollar program. A targeted search for radio signals focused on some thousand nearby stars, while a second element of SETI surveyed the entire sky. Ridiculed as "The Great Martian Chase," even after changing its name to "High Resolution Microwave Survey" in a vain attempt to highlight its potential for basic discoveries in astronomy, SETI lost its government funding in 1993. A scaled-back version of the original targeted search has been continued with private funding.

Would intercourse with extraterrestrials be beneficial? Suppose that they were a cancer of purposeless technological exploitation intent on enslaving us, rather than benign philosopher-kings willing to share their wisdom? Even if they should be helpful and benign, superior beings would be menacing. Anthropological studies of primitive societies confident of their place in the universe find them disintegrating upon contact with an advanced society pursuing different values and ways of life.

Steven J. Dick, *Plurality of Worlds: The Extraterrestrial Life Debate from Democritus to Kant* (1982). Michael J. Crowe, *The Extraterrestrial Life Debate 1750–1900: The Idea of a Plurality of Worlds from Kant to Lowell* (1986). Donald Goldsmith and Tobias Owen, *The Search for Life in the University*, 2d ed. (1992). Walter S. Sullivan, *We Are Not Alone: The Continuing Search for Extraterrestrial Intelligence*, rev. ed. (1993). Steven J. Dick, *The Biological Universe: The Twentieth-Century Extraterrestrial Life Debate and the Limits of Science* (1996). NASA maintains a website on the history of the search for extraterrestrial intelligence, which also links to the official SETI site: http://www.history.nasa.gov/seti.html.

NORRISS S. HETHERINGTON

F

FACT AND THEORY. Appeals to "fact," or reality, as distinguished from and often opposed to "theory," first appeared in the writings of natural philosophers during the seventeenth century. Before that, "fact"—in the sense of a fact accomplished, a deed, or a crime—belonged to the language of law. "Theory," on the other hand, meant a systematic view or speculation (hence also "sight," or "spectacle") as in the medieval and early modern usage of the Latin words *theoria* and *theorica*, and it already carried an implicit distinction from, or opposition to, "practice" with it.

Among the authors responsible for the new, combined use of "fact" and "theory," Francis Bacon and Galileo *Galilei figured prominently, together with several contributors to the periodical publications of institutions such as the Royal Society of London and the Paris Academy of Sciences (*see* ACADEMY; BACONIANISM). The success of discourses of the fact/theory kind in the scientific literature produced during the subsequent centuries has been interpreted as pointing to something constitutive of modern science itself.

In the twentieth century philosophers like Alfred Tarski and Karl Popper have claimed that to find theories corresponding to the facts was the regulative ideal of the scientific tradition—a claim most of those who feel they belong to that tradition would agree with today, as their predecessors did over the past four hundred years. Yet historically that claim has rarely gone unchallenged.

Robert *Boyle in the 1660s made a plea for natural knowledge to be based on "matters of fact" that had to be supported mainly by the individual beliefs of the experimenters. Boyle's plea was challenged by Thomas Hobbes, who maintained—together with a long, previous and subsequent philosophical tradition—that knowledge based on belief could never attain the kind of certainty that he expected from natural philosophy, which he wanted modeled on geometry.

Denis Diderot and Jean d'Alembert, leaders in conveying the ambitious program of modern natural philosophers to a wider public during the eighteenth century, were extremely cautious about the factual nature of experimental knowledge. They made their positions clear in their respective articles "Fait" (Fact), and "Expérimental" (Experimental) in their famous *Encyclopédie*.

During the central decades of the nineteenth century, discourse of the fact/theory kind probably reached the peak of its popularity in scientific literature. Yet texts emblematic of the period like Charles *Darwin's books conveyed all the possible tensions between "fact" and "theory," tensions that the author and the informed contemporary readers knew well. Already in the 1840s, the Cambridge polymath William Whewell had argued and perhaps demonstrated that "fact and theory pass into each other by insensible degrees."

If we judge from the vast literature that the physician and historian Ludwik Fleck could avail himself of when writing his posthumously famous *Genesis and Development of a Scientific Fact* (1935), awareness of the scientific, anthropological, epistemological, psychological, and sociological issues implied in fact/theory discourse had reached a climax in the 1930s. Forgetfulness or neglect of this literature must be invoked to explain the sometimes overdone claims of, and sometimes overharsh attacks against, otherwise valuable literature on "facts as social constructs" produced during the 1980s and 1990s.

See also CONSTRUCTIVISM; EMPIRICISM; INSTRUMENTALISM AND REALISM; MODEL AND WORLDVIEW; POSITIVISM AND SCIENTISM; SCIENCE WARS.

Ludwik Fleck, *Genesis and Development of a Scientific Fact* (1935; new ed. 1979). Steven Shapin and Simon Schaffer, *Leviathan and the Air-pump* (1985).

GIULIANO PANCALDI

FAMILY. See BIRTH ORDER; DYNASTY.

FARADAY, Michael (1791–1867), English natural philosopher and public man of science. Faraday was born on 22 September 1791 in Newington Butts, Surrey. His father was a blacksmith and a member of the Sandemanian Church, to which Faraday had a lifelong commitment. In many ways, Faraday's work can be viewed as his seeking the laws of nature that he believed God had written into the universe at the Creation.

Faraday attended a day school before being apprenticed as a bookbinder, 1805–1812. During his apprenticeship he developed an overwhelming interest in science, which he cultivated by attending various scientific lectures including those by Humphry Davy in the Royal Institution. Faraday was appointed chemical assistant there in 1813, but within a few months he was accompanying Davy on a tour of the Continent. On their return to England in 1815, Faraday was reappointed in the Royal Institution, where he rose to be superintendent of the house (1821), director of the laboratory (1825), and Fullerian professor of chemistry (1833).

Following Hans Christian Oersted's announcement of electromagnetism, Faraday discovered electromagnetic rotations (1821), the principle behind the electric motor. Ten years later he discovered electromagnetic induction and commenced a remarkable decade of work in which, among other things, he rewrote the theory of electrochemistry (coining in the process words such as electrode, anode, cathode, and ion and establishing his laws of *electrolysis). He built the Faraday cage (1836), which showed that measurements of electric charge depended on the electrical state of the observer. This observation led Faraday to develop his theory that electricity was the result of induction between contiguous particles rather than the action of a special fluid. In the 1840s he extended his skepticism of scientific theories by arguing against the existence both of chemical atoms and the luminiferous *ether. His skepticism about these theories was strengthened by his discovery of the magneto-optical effect and diamagnetism in 1845 (*see* MAGNETO-OPTICS), and culminated in his establishment of the field theory of *electromagnetism, which, when mathematized by William *Thomson (Lord Kelvin) and James Clerk *Maxwell, became one of the cornerstones of physics.

Faraday also made a number of purely chemical discoveries: In the 1820s he identified several new carbon chloride compounds, he liquefied gases, and he detected what became known as benzene. In the 1850s he carried out an extensive investigation of colloidal suspensions by passing light through them.

Faraday not only undertook research but was frequently invited to provide practical scientific advice. He helped Davy with the miners' safety lamp in 1815, in the 1820s with the electrochemical protection of the copper bottoms of ships, and, though unsuccessfully, with the improvement of optical glass. In his own right he was scientific adviser to Trinity House (the English and Welsh lighthouse authority) from 1836 to 1865 and oversaw the program to electrify lighthouses. He gave the government scientific advice on wide-ranging problems, including the conservation of pictures, the prevention of explosions in coal mines, the installation of lightning conductors, and the best way of attacking Kronstadt. For twenty years he taught chemistry to the cadets of the Royal Artillery and Royal Engineers at the Royal Military Academy in Woolwich.

Faraday was one of the most popular scientific lecturers of his day. In the mid-1820s he established the Christmas Lectures for children and the Friday Evening Discourses for members of the Royal Institution; both series continue to this day. He used his lectures to criticize passing fashions, for instance, of table-turning in the early 1850s, which offended both his religious beliefs and his sense of scientific propriety.

With his success in scientific research, his value as a government adviser, and his popularity as a lecturer, Faraday became one of the most famous men of the period. He was painted, sculpted, and photographed by leading artists, and was the friend of the scientifically minded Prince Albert. At Albert's suggestion, Queen Victoria gave Faraday in 1858 a "Grace and Favour House" at Hampton Court. Faraday spent an increasing amount of time there. He died on 25 August 1867 and was buried in the Sandemanian plot in Highgate Cemetery.

David Gooding and Frank A. J. L. James, *Faraday Rediscovered: Essays on the Life and Work of Michael Faraday, 1791–1867* (1985). David Gooding, *Experiment and the Making of Meaning: Human Agency in Scientific Observation and Experiment* (1990). Geoffrey Cantor, David Gooding, and Frank A. J. L. James, *Faraday* (1991). Geoffrey Cantor, *Michael Faraday: Sandemanian and Scientist. A Study of Science and Religion in the Nineteenth Century* (1991). Frank A. J. L. James, *The Correspondence of Michael Faraday*, 4 vols. to date (1991–).
                                            FRANK JAMES

FATS. See CARBOHYDRATES AND FATS.

FEEDBACK. Norbert Wiener defined feedback as "the property of being able to adjust future conduct by past performance" (1954). Feedback requires a sensing unit and a means of reacting to the value sensed, and may be applied either positively or negatively. In modern technology, feedback devices control speed, timing, pressure, and temperature.

The earliest recorded example of feedback dates to ca. 250–300 B.C. when Ctesibius of Alexandria used a float to control levels in a water clock to improve its accuracy. The first modern feedback devices (ca. 1620) were thermostats in a chemical laboratory furnace for incubating chickens made for James I of England by Cornelius Drebbel of Alkmaar, Holland.

Regulating the speed of moving components has been important in scientific instruments. The centrifugal governor for controlling windmill speeds patented in England (1787) by Thomas Mead was followed by William Bolton's suggestion to James Watt (1788) resulting in a device widely adopted in steam engines. This flyball governor was adapted (ca. 1845) to clock drives for telescopes and chronographs by Troughton and Simms, followed by the American firms Fauth and Company and Warner and Swasey. These devices vary inertia or friction on the rotating shaft by raising or lowering a pair of weights as the shaft spins. Recording instruments, such as meteographs and seismographs driven by falling weights or springs, also incorporated flyball governors.

George Biddell Airy and James Clerk *Maxwell developed the first mathematical theory of automatic control (1868), and the advent of electrical devices soon called for voltage and current regulation. Charles Brush, Elihu Thomson, and E. J. Houston developed a new generation of feedback regulators for arc lamps and battery chargers in the United States beginning around 1878. Max Kohl in Germany patented a speed controller for motors that led to the regulation of one motor by another (servomotors). World War II saw rapid progress with the use of feedback loops in radar, autopilots, missiles, and so on. Scientists and engineers transferred these devices into manufacturing and cybernetics. George Philbrick's analogue computer of 1946 incorporated a pneumatic three-response controller for computational work but was immediately challenged by digital computers utilizing electronic feedback.

Norbert Wiener, *The Human Use of Human Beings: Cybernetics and Society*, 2d ed. (1954). Otto Mayr, *The Origins of Feedback Control* (1970).

RANDALL BROOKS

FERMENTATION. According to the popular seventeenth-century textbook of Nicolas Lemery, "fermentation is an ebullition caused by spirits which are seeking to leave a body, and encountering the grosser terrestrial parts which oppose their passage, they swell and rarify the material until they detach it....The spirits divide, subtilize, and separate the principles, in such a way that they change the nature of the material from what it had been before." By Lemery's time it was known that the fermentation of grapes produced a strong "spirit of wine" separable by distillation, but that if allowed to continue too long, it produced an acetous acid, or "acid of vinegar," distinct from the common mineral acids.

These properties of fermentation distinguished it sharply from the chemical operations that chemists normally performed on substances derived from the plant or animal kingdom. The standard distillations employed fire to separate the plant or animal matter analyzed into simple principles. In fermentation the active agent, in addition to the ferment, was water; the slower and gentler process required only a mild heat, under which the principles composing the plant matter separated and recombined differently. Fermentation seemed the closest of all processes that chemists could observe to what they imagined to take place in digestion and nutrition.

Georg Ernst Stahl studied fermentation in great detail both to improve the arts of wine-making and brewing in Germany and to work out a theory of the process. He concluded that the constituent particles of fermentable plant material must include salt, oil, and earth, the standard products of distillation analyses that chemists of the time took to be the "principles" of all plant matters. But Stahl specified that the salt of fermentable substances was sharp and acidic, the oil light and volatile. Substances that contained an alkaline volatile salt and a thick oil (mainly animal matters) underwent putrefaction instead. Stahl stressed that fermentations required an agent to set the individual particles of the fermentable substance in motion. He identified the principle agent as water. The ferments added to further activate fermentations did so because their particles communicated the motion they easily took up to the particles of the fermentable matter.

Stahl's treatise on the art of fermentation carried the experimental investigation and theoretical interpretation to the limits of the chemistry of his time. When Antoine-Laurent *Lavoisier turned during the 1780s to what he called "one of the most striking and extraordinary [operations] that chemistry presents to us," he did so as the climactic test of his "new chemical theory." To simplify the situation as far as possible, he studied the fermentation of sugar, the only fermentable substance whose composition he thought he could establish with confidence. Borrowing from the new *pneumatic chemistry, he wrote, "must of grape [i.e. grape sugar] = carbonic acid + alcohol," the first representation of a chemical change in the form of an algebraic equation. He did not succeed in determining the proportions of carbon, hydrogen, and oxygen in each of the substances taking part. By 1815 Joseph Gay-Lussac and Louis Jacques Thenard had achieved accurate results for the elementary composition of sugar. Because sugar and water had the proportion of hydrogen to oxygen, they gave the class of substances to which sugar belongs the name "carbohydrate." During the next decades it became clear that complex carbohydrates, such as starch, convert to sugar before they ferment, and that fermentation is, fundamentally, a process by which simple sugars, such as glucose, decompose.

The association between fermentation and the processes of life could now be made far more specific than in the traditional definitions. Theodor Schwann showed during the 1830s that the process is associated with the growth of microscopic organisms. When in 1839 he defined cells as the basis of the organization of all plants and animals and as the locus of the "metabolic power" that alters the substances brought to them in the surrounding fluid medium, he used alcoholic fermentation as the model for the type of chemical transformations that take place within the cells. During the next three decades other investigators found that sugar can undergo other fermentations to produce such substances as lactic and butyric acid instead of alcohol. During the 1860s Louis *Pasteur showed that a specific microorganism stimulated each of these fermentations only under anaerobic conditions, and he consequently defined fermentation as "life without air." Pasteur believed that "proper" fermentations took place only in the presence of living cells. Although, as Eduard Buchner showed, the beginning of the alcoholic fermentation occurred outside the cell, Pasteur was correct in the sense that fermentation was not a mere chemical reaction, but an expression of the nutritional processes of living organisms. Buchner's technical triumph enabled biochemists of the twentieth century to study more effectively, outside of the living organism, the intermediate stages of a complex metabolic fundamental to life.

See also BIOCHEMISTRY; ENZYME; WARBURG.

Robert E. Kohler, "The Background to Eduard Buchner's Discovery of Cell-Free Fermentation," *Journal of the History of Biology* 4 (1971): 35–62. Robert E. Kohler, "The Reception of Eduard Buchner's Discovery of Cell-Free Fermentation," *Journal of the History of Biology* 5 (1972): 327–353. Frederic Lawrence Holmes, *Lavoisier and the Chemistry of Life* (1984). Joseph S. Fruton, *Proteins, Enzymes, Genes: The Interplay of Chemistry and Biology* (1999).

FREDERIC LAWRENCE HOLMES

FERMI, Enrico (1901–1954), nuclear physicist, designer of the first nuclear reactor.

Fermi had the widest scope of all the founders of quantum physics. As a theorist, he contributed decisively to quantum mechanics (Fermi–Dirac statistics) and nuclear physics (theory of beta decay). As an experimentalist, he introduced the technique of neutron bombardment to study artificial radioactivity, opening the way to the discovery of nuclear fission. He established a famous school of nuclear physics in Rome, but left fascist Italy because of anti-Jewish legislation. He settled in the United States, where he contributed to the atomic bomb program and served as an influential scientific advisor in postwar American nuclear policy. His name is honored in the unit of length for nuclear dimension (10^{-13} cm), in the transuranic element of atomic number 100 (fermium), in a class of elementary particles (fermions), and in one of the most important particle physics laboratories in the world (Fermilab, near Chicago).

Born in Rome on 28 November 1901, Fermi studied physics at the University of Pisa, as a fellow of the prestigious Scuola Normale Superiore. Soon after his graduation in 1922, the influential director of the Physics Institute of Rome University, Orso Mario Corbino, a former Minister of National Education and eventually (1923–1924) of National Economy, realized his promise and promoted his career. Corbino sent Fermi to pursue his studies in Göttingen (1923) and Leyden (1924), and in 1926 obtained for him a chair of theoretical physics (the first in Italy) in the institute in Rome. That year Fermi published a seminal paper on the quantization of the monatomic ideal gas, proposing a new quantum statistics for particles with half-integral spin (fermions).

With Corbino's support, Fermi set up a brilliant research group in nuclear physics that included Edoardo Amaldi, Franco Rasetti, Emilio Segrè, Oscar D'Agostino, Bruno Pontecorvo, and Ettore Majorana. The group followed up the then-new phenomenon of artificial *radioactivity by means of neutron bombardment of the *elements of the periodic table. Between March and July 1934, they "discovered" (that is, made and detected) about fifty new radionuclides. A few of these, which they misinterpreted as transuranic elements, turned out to be fission products of uranium, as Otto Hahn and Fritz Strassmann discovered four years later. Also in 1934, Fermi found that neutrons slowed down in passing through light elements and that, when suitably retarded, they became extremely effective in provoking nuclear transmutations. No less important was Fermi's theoretical analysis of nuclear beta decay (1934), which invigorated the study of weak interactions.

The Fermi group itself soon decayed. Most members left Rome, and Corbino died in January 1937, thereby depriving Fermi of important institutional support. The fascist racial legislation of 1938 hit Fermi's Jewish wife, Laura Capon, and political boycott added to scientific frustration. Fermi finally decided to leave Italy. After he collected his Nobel Prize in Stockholm in December 1938 for his neutron work, he and his family sailed to New York. He became a U.S. citizen in 1944.

Initially established at Columbia University, Fermi moved to Chicago in 1942 to work in the Manhattan Project. He led the construction of the first nuclear reactor, which went critical on 2 December, demonstrating the chain reaction and the feasibility of producing plutonium for an atomic bomb. He then collaborated with the Los Alamos teams involved in the construction of the bomb and attended the Trinity Test in Alamogordo, New Mexico, on 16 July 1945. After the war he served on the Atomic Energy Commission's General Advisory Committee, contributing to the definition of the American nuclear strategy and research policy. He also spent time in Los Alamos in 1950 to work on the H-bomb program launched by President Harry Truman that January. But Fermi's main base was the University of Chicago, where he inaugurated an important research program in particle physics centered on a new 450 MeV synchrocyclotron. His team studied pion-nucleon interactions (1952–1953), confirming the conservation law of isotopic spin in strong interactions and observing the first pion-nucleon resonance.

In 1952 Fermi was elected president of the American Physical Society. He had to cope with the drama unleashed in the American scientific community by the anticommunist campaign of Senator Joseph McCarthy and the ensuing trial of J. Robert Oppenheimer before the Atomic Energy Commission's Security Board. He testified in April 1954 in support of the former scientific leader of Los Alamos. In the summer of 1954, Fermi fell ill with a stomach cancer. After useless surgery, he died in Chicago on 29 November.

Laura Fermi, *Atoms in the Family* (1954). Emilio Segrè, *Enrico Fermi, Physicist* (1970). Edoardo Amaldi, "Personal Notes on Neutron Work in Rome in the '30s and Post-War European Collaboration in High Energy Physics," in *History of Twentieth Century Physics*, ed. Charles Weiner (1977): 293–351. Gerald Holton, "Fermi's Group and the Recapture of Italy's Place in Physics," in *The Scientific Imagination. Case Studies*, ed. Holton (1978): 155–198. Arturo Russo, "Science and Industry in Italy between the Two World Wars," *Historical Studies in the Physical and Biological Sciences* 16, no. 2 (1986): 281–320.

ARTURO RUSSO

FEYNMAN, Richard (1918–1988).

One of the greatest and most original physicists of the twentieth century, Feynman was born in Far Rockaway, New York. His father, a Russian emigré, grew up in Patchogue, Long Island; Feynman's mother came from a well-to-do family, and had attended the Ethical Culture

Feynman lecturing to a class at Caltech in 1963. The blackboard to the right presents quantum mechanics in P.A.M. *Dirac's "bra" and "ket" formulation.

School in New York. After public schools in Far Rockaway, where Feynman had excellent teachers in chemistry and mathematics, he entered the Massachusetts Institute of Technology (MIT) in 1935. Admitted to Princeton University as a graduate student in physics in 1939, he became assistant to the newly arrived John Wheeler, then twenty-seven and full of bold and original ideas. In the spring of 1942 Feynman obtained his Ph.D. and immediately started working on problems related to the development of an atomic bomb. In 1943 he was one of the first physicists to go to Los Alamos. Hans Bethe, the head of the theoretical division, and J. Robert Oppenheimer, the director of the laboratory, quickly recognized Feynman as one of the most valuable members of the theoretical division—versatile, imaginative, ingenious, and energetic. In 1944 he was put in charge of the computations for the theoretical division. He introduced punch card computers to Los Alamos and began a lifelong interest in computing. In 1945 Feynman joined the physics department at Cornell University as an assistant professor. He left in 1951 for the California Institute of Technology, where he remained until his death.

Feynman's first major contribution, the content of his doctoral dissertation and of his article in the *Reviews of Modern Physics* for 1948, was the path-integral formulation of nonrelativistic *quantum mechanics, which helped clarify the assumptions that underlay the usual quantum mechanical description of microscopic entities. In Feynman's approach, a particle going from the spatial point x_1 at time t_1 to the spatial point x_2 at time t_2 can take any path, each of which has a certain probability amplitude. The dynamic that results is the outcome of summing over all paths with their respective probability amplitude—a formulation that may well be Feynman's most profound and enduring contribution. It has deepened understanding of quantum mechanics and significantly extended the range of systems that can be quantized. Feynman's path integral has also enriched mathematics and provided new insights into spaces of infinite dimensions.

Feynman was awarded the Nobel Prize for physics in 1965 for his work on *quantum electrodynamics (QED). In 1948, simultaneously with Julian Schwinger and Shin'ichirō *Tomonaga, he showed that the infinite results that plagued QED could be removed by a redefinition of the parameters that describe the mass and charge of the electron, a process called "renormalization." Schwinger and Tomonaga had built on the existing formulation of the theory. Feynman invented a completely new diagrammatic approach that allowed the visualization of space-time processes,

clarified concepts, and simplified calculations. Using Feynman's methods it became possible to compute QED processes to amazing precision. The magnetic moments of the electron and muon have been calculated to an accuracy of one part in 10^{12} and found to be in agreement with experiment.

In 1953 Feynman developed a quantum mechanical explanation of liquid helium that justified the earlier phenomenological theories of Lev Landau and Laslo Tisza. Because a helium atom has zero total spin angular momentum, the wave function (the quantum-mechanical formulation) of a large collection of helium atoms is unchanged under the exchange of any two of them. When in this state the system behaves as one unit. Hence helium near absolute zero acts as if it had no viscosity.

In the late 1960s experiments at the Stanford Linear Accelerator indicated an unexpectedly large probability that high-energy electrons underwent large angle scattering in striking protons. Feynman found that he could explain the data by assuming that the proton consisted of small, pointlike entities. He called these subnuclear entities "partons," soon identified with the *quarks of Murray Gell-Mann and George Zweig. Feynman devoted much of his research during the 1980s to studying quarks and their interactions and explaining their confinement inside nucleons and mesons.

Feynman demonstrated his uncanny ability to get to the heart of a problem—whether in physics, applied physics, mathematics, or biology—on prime-time television when he served on the presidential commission that investigated the crash of the *Challenger* space shuttle. He pinpointed the central problem by dropping a rubber O-ring into a glass of ice water to illustrate the cause of the rocket's failure.

In his physics Feynman always stayed close to experiments and showed little interest in theories that could not be tested experimentally. He imparted these views to undergraduate students in his justly famous *Feynman Lectures on Physics* and to graduate students through the widely disseminated notes of his graduate courses. His writings on physics for the interested general public, *The Character of Physical Laws* and *QED*, convey the same message.

Richard Feynman, with R. B. Leighton and M. Sands, *The Feynman Lectures on Physics*, 3 vols. (1963). Richard Feynman, *The Character of Physical Laws* (1965). Richard Feynman, *QED: The Strange Theory of Light and Matter* (1985). James Gleick, *Genius. The Life and Science of Richard Feynman* (1992). Richard Feynman, *The Meaning of It All: Thoughts of a Citizen Scientist* (1998).

SILVAN S. SCHWEBER

FICTION. See LITERATURE AND SCIENCE; SCIENCE FICTION.

FIELD. The field, one of the most important concepts in modern physics, denotes the manner in which magnetic, electrical, and gravitational forces act through space. The field concept alleviated the difficulties many scientists found in assuming the existence of *forces acting at a distance without the intervention of some material entity. Fields thus serve many of the functions of *ether theories, which received their fullest development during the first half of the nineteenth century following the establishment of the wave theory of light by Augustin-Jean *Fresnel.

Michael *Faraday introduced the term "field" into natural philosophy on 7 November 1845, following his discovery of the magneto-optical effect and diamagnetism. He used the term operationally, in analogy to a field of stars seen through a telescope. During the next decade he developed the concept into a powerful explanatory framework for electromagnetic phenomena. This embodied much of his earlier thinking about the nature of magnetic action, especially his use of curved lines of force, which he had employed since the early 1830s to account for phenomena such as electromagnetic induction. It also embodied his opposition to conceptions such as atoms and the ether. As he wrote in 1846, he sought "to dismiss the aether, but not the vibrations."

Initially Faraday's contemporaries ignored his field concept, since it did not have the mathematical precision of action-at-a-distance theories such as André-Marie Ampère's *electrodynamics. William *Thomson (Lord Kelvin) reacted with contempt to Faraday's "way of speaking of the phenomena." However, Faraday's field theory had the great merit that in treating electrical events it took account not only of the wire carrying the electric current but also of the insulation of the wire, the surrounding medium, and so on. The notion of field made a good basis for developing a theory of long-distance *telegraph signaling, which became a pressing problem in the mid-1850s with the intended construction of the transatlantic cable. Action-at-a-distance theories could not cope with this problem; Thomson, using Faraday's field concept, solved it.

This practical success prompted the adoption of the field concept in Britain. In the hands of Thomson and of James Clerk *Maxwell it became a mathematical theory, much to the bemusement of the nonmathematical Faraday. However, unlike Faraday, who wished to abolish the ether from natural philosophy, both Thomson and Maxwell sought to interpret the field in terms of elaborate mechanical models (involving ethereal vortices) in an endeavor to retain the ether (*see* STANDARD MODEL). This project ultimately failed. Relativity theorists such as Hendrik Antoon Lorentz and Albert *Einstein replaced the ether by the nonmechanical field, a space capable of propagating forces, which is a cornerstone of modern physics.

William Berkson, *Fields of Force: The Development of a World View from Faraday to Einstein* (1974). David Gooding, "Faraday, Thomson, and the Concept of the Magnetic Field," *British Journal for the History of Science* 13 (1980): 91–120. Nancy J. Nersessian, *Faraday to Einstein: Constructing Meaning in Scientific Theories* (1984). Bruce J. Hunt, "Michael Faraday, Cable Telegraphy and the Rise of Field Theory," *History of Technology* 13 (1991): 1–19.

FRANK A. J. L. JAMES

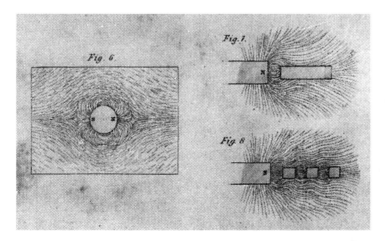

Distribution of iron filings around magnets. These pictures, from a paper by *Faraday of 1852, indicated the direction and intensity of the "physical lines of force" that he supposed to lie behind electric and magnetic phenomena.

FIELD WORK. See Exploration and Field Work.

FILM, TELEVISION, AND SCIENCE. Communication about science via film and television derives from two often contradictory impulses, to educate and entertain. Although nineteenth-century scientists helped to develop the cinematographic equipment and techniques later exploited by entertainers, the organized scientific community today plays only a minor role in sponsorship, control, and production of film and television programs about science for public and classroom use. Individual initiatives and commercial interests dominate.

Pioneers in the development of research film techniques like biologist Eadweard Muybridge and astronomer P. J. C. Janssen sought to record animal locomotion or astronomical events as means of gathering data for professional use and classroom demonstrations. Other innovators like Thomas Edison recognized film's potential for public education and entertainment. In the 1910s, for example, Edison made melodramatic "health propaganda" films for the Tuberculosis Association, which were shown in commercial movie houses, schools, factories, and churches to encourage better hygiene and preventative care. Science films and videos still are used regularly for health, engineering, and other practical education, as well as classroom instruction at all levels.

In the early twentieth century, even the sparsest research films seemed exotic to audiences unaccustomed to motion pictures. Anthropologists and biologists who made films to record their research found new audiences for the edited versions, demonstrating what Gregg Mitman calls the permeability of boundaries between scientific culture and popular communication. Films like Robert Flaherty's *Nanook of the North* (1922) exposed urban audiences to new places, peoples, and ways of life. And scientifically trained filmmakers like Jean Painlevé developed underwater photographic techniques to reveal the miniature worlds of sea horses and sea urchins, establishing a genre of authentic but entertaining nature films.

Natural history museums pioneered in using film for both research and public presentation, helping to underwrite and promote productions like Martin and Osa Johnson's *Trailing African Wild Animals* (1923), one of the first commercially successful feature-length animal pictures. The practice of editing films of scientific expeditions and "re-creating" their scenes of nature to enhance their entertainment value contributed to acceptance of a merging of fact and dramatization that became common in popular science

entertainment. Natural history subjects could also be layered with moral lessons, making them seem suitable for family audiences.

Hollywood's dramatic treatments of scientists and scientific knowledge in the 1930s and 1940s followed no single pattern. In horror movies based on Mary Shelley's *Frankenstein* or Robert Louis Stevenson's *Dr. Jekyll and Mr. Hyde*, fictional scientists used their expertise for immoral or criminal purposes, but in "science fiction" movies (like *The Things to Come*, 1936, and the Flash Gordon serials) science was represented as the route to the future, enabling space travel and eventually a better life. Biographical films like *The Story of Louis Pasteur* (1936) and *Madame Curie* (1943) portrayed scientists as creative heroes, while dramas like *Arrowsmith* (1931) emphasized that research can also contain personal, ethical, and economic challenges.

Science and health comprised about one percent of newsreel segments shown in U.S. theaters from 1939 to 1948. *The March of Time* offered an interview with Albert *Einstein, film from new stroboscopic cameras, and reports on atomic power and the Manhattan Project.

From the 1940s, Hollywood movies began to favor themes related to space and nuclear weapons. Both types accomplished a considerable incidental diffusion of scientific information, of varying accuracy, as well as lessons about scientific authority, scientists' loyalty to governments, and the usefulness of basic research. In science fiction films like *Destination Moon* (1950), *When Worlds Collide* (1951), and *The Day the Earth Stood Still* (1951), scientists figured primarily as politically naive dupes or calm heroes rather than villains; but science itself was presented as a force for both good and evil.

This pragmatic acceptance of scientific ability extended to early television, which gave serious attention to astronomy, medicine, and natural history. One of the first primetime science series in the United States, *The Nature of Things* (1948–1954), featured astronomer Roy Kenneth Marshall, Director of the Fels Planetarium in Philadelphia. Zoos and museums took to television early. Programs at the Chicago Museum of Science and Industry and the Bronx Zoo aired in 1948 and 1950, and the American Museum of Natural History in New York cooperated in the Columbia Broadcasting System (CBS) series *Adventure* (1953–1956). The series originally focused on researchers' own accounts of their work; by 1955, however, most presenters were professional entertainers or public figures, signaling a trend away from using practicing scientists to explain science on television. Another innovator in science television was Lynn Poole,

Public Relations Director for The Johns Hopkins University, who created *The Johns Hopkins Science Review* and *Johns Hopkins File 7*, which ran on U.S. commercial networks from 1948 to 1953, and in syndication from 1956 to 1958.

Science television programs aimed at children consisted primarily of science fiction (for example, *Captain Video*, 1949–1956), and puppet shows and animation (*Science Circus*, 1949, and *Mr. I Magination*, 1949–1952). The premier show of this type, *Watch Mr. Wizard* (1951–1965) and its sequel *Mr. Wizard's World* (1983–1991), both starring Don Herbert, perfected the format in which a "scientist" host leads a child actor through a series of experiments or adventures.

The Bell Telephone System underwrote an innovative series in the 1950s that remains in continuous use in classrooms. These films (which included *Our Mr. Sun*, 1956, and *Hemo the Magnificent*, 1965) combined live action, actors, animation, and actual scientific film to explain such things as weather, the cardiovascular system, and measurement of time and space, often with religious or philosophical interpretations.

Television quickly adapted the models that had been popularized in early nature documentaries, weaving film of animals in their natural habitats into little melodramas about survival. For over thirty years Franklin Park's zookeeper Marlin Perkins exploited this theme in his syndicated series *Zoo Parade* (1950–1957) and *Wild Kingdom* (1962–1988). Simplistic views of nature combined with calming bits of scientific information helped prime a commercial market for Walt Disney's *True-Life Adventures* series (e.g., *The Living Desert*, 1953), beautifully photographed feature films shown in movie theaters and on television.

The television audience's love of natural history films turned a few scientists into celebrities. One enduring series, the *National Geographic Specials*, first broadcast in 1964, helped to propel the primatologist Jane Goodall to fame. French oceanographer Jacques Cousteau had been known for spectacular undersea photography in films such as *Le Monde du Silence* (1956), but it was his series of television adventure specials that made him a household name worldwide.

Television offers more sobering images of science within its news and public affairs programming, addressing issues that have ranged from environmental pollution and genetic engineering to the development and containment of nuclear weapons and bioterrorism. Although scientists represent only a minority of guests on news interview shows, television nevertheless has helped to make popular celebrities of J. Robert *Oppenheimer, Edward *Teller, Linus *Pauling, and Carl *Sagan, all of whom have sought public platforms for their political and social messages. From the 1970s, both Hollywood movies and television series have favored scientist characters who are brilliant and possibly foolhardy but rarely villainous (such as *Andromeda Strain*, 1971; *Medicine Man*, 1992; and *Jurassic Park*, 1993), although film images of older "mad scientist" stereotypes survive on videotape and cable networks. Television news coverage of science has offered generally favorable portraits of scientific accomplishments, and a number of exceptional television specials and series like *The Ascent of Man* (1973, 1975) and *Cosmos* (1980) have pondered the social and ethical aspects of science. Continuing series like *Nova* (since 1973), *Nature* (1982), and *Scientific American Frontiers* (1990); the specials hosted by British naturalist David F. Attenborough; and American commercial cable ventures like the Discovery Channel and the Learning Channel, offer a range of well-produced and well-regarded science programming today.

Anthony R. Michaelis, *Research Films in Biology, Anthropology, Psychology, and Medicine* (1955). Marcel C. LaFollette, "Science on Television: Influences and Strategies," *Daedalus* 111 (1982): 183–198. John C. Burnham, *How Superstition Won and Science Lost: Popularizing Science and Health in the United States* (1987). Robert Lambourne, Michael Shallis, and Michael Shortland, *Close Encounters? Science and Science Fiction* (1990). Rima D. Apple and Michael W. Apple, "Screening Science," *Isis* 84, no. 4 (1993): 750–754. James Burkhart Gilbert, *Redeeming Culture: American Religion in an Age of Science* (1997), chs. 9 and 10. Vivian Carol Sobchack, *Screening Space: The American Science Fiction Film*, 2d ed. (1997). Gregg Mitman, *Reel Nature: America's Romance with Wildlife on Film* (1999). Andy Masaki Bellows and Marina McDougall, with Brigitte Berg, eds., *Science Is Fiction: The Films of Jean Painlevé* (2000).

MARCEL C. LAFOLLETTE

FIRE AND HEAT. The hidden natures of those most obvious of physical phenomena, fire and heat, puzzled natural philosophers until the middle of the nineteenth century. Whether the agent of heat, which could work without glowing, differed from that of fire, to which the ancients had accorded elemental status, became an insistent question toward the end of the eighteenth century.

When seventeenth-century chemists eliminated fire from their list of the five simple principles they believed to constitute all matter, they retained the idea that fire was the most powerful agent available for altering or reducing matter. Their refined distillation procedures enabled them to apply fire with subtle control. Although a few chemists and natural philosophers, including Francis Bacon, regarded fire and

The instrument with which *Lavoisier and *Laplace determined specific heat capacities. A specimen at a standard temperature was placed in the wire basket and packed with ice in equilibrium with ice-cold water. The amount of water from melted ice in the pot F measured the capacity of the specimen, though not very exactly.

heat as an expression of motion of particles of ordinary matter, most believed it to be a special substance. In Newtonian philosophy, this substance was thought to consist of particles that repelled one another but were attracted by ordinary matter. Consequently they spread throughout bodies, reaching an equilibrium in which all bodies in contact with one another contained the same degree of heat as determined by a *thermometer.

Increasingly precise thermometers enabled philosophers to measure even what they did not understand. The mixing of two samples of water at different initial temperatures and the subsequent measurement of the temperature of the combination addressed a question of long-standing interest to physicians: If two items possess the same quality in different degrees, what is the degree of the quality when they are mixed? According to measurements done in 1747–1748 by Georg Wilhelm Richmann, the Saint Petersburg academician later killed attempting Benjamin *Franklin's experiment with *lightning, the mixture's temperature $T = (m_1T_1 + m_2T_2)/(m_1 + m_2)$, where the subscripts indicate the temperatures and amounts of the two samples before mixing. The formula intimates the conservation of heat and measures it by the product of mass and temperature.

Daniel Gabriel Fahrenheit gave the first inkling that the mixing business was more interesting than Richmann's formula allowed. Fahrenheit observed that supercooled water when shaken converted to ice while its temperature rose to 32 degrees on his peculiar scale; and also that mercury had a smaller effect in the mixing experiments than an equal weight of water. Joseph Black followed up these observations in the 1750s and concluded that water at the freezing point must contain heat concealed from the thermometer. He confirmed this insight by measuring the time required to melt snow, which he took to be a measure of the amount of heat hidden or latent in ice-cold water. As for the different effectiveness of mercury and water, Black ascribed it to a difference in their capacity for heat. He measured the "specific heat capacity" of a substance by comparing the times required to heat the same weights of water and of the substance through the same interval of temperature (the shorter the time, the smaller the capacity). Black did not publish these measurements, most of which he made in a brewery to secure a steady warmth, nor did he declare his opinion about the true nature of heat.

Johan Carl Wilcke, professor of natural philosophy at the Royal Swedish Academy of Sciences, arrived in 1772 at the concept of latent heat by observing that hot water melted less snow than Richmann's formula called for. Ten years later Wilcke also conceived of specific heat, perhaps independently of Black, whose views had been circulated by his students and colleagues. Wilcke used Richmann's formula to calculate specific heats: if the mixture of a weight W of (say) mercury at temperature T with an equal weight of ice-cold water produces a final temperature Q, and w is the amount of water at temperature T that, if mixed with ice-cold water, also results in the temperature Q, then $Q = (wT + Q \times 0)/(w + W)$, where 0 represents freezing on the temperature scale of Wilcke's countryman Anders Celsius. From the last expression, w/W = specific heat of mercury = $Q/(T - Q)$. Unlike Black, Wilcke did not keep his concept of heat latent. In several publications, he discussed heat as an elastic fluid just like electricity (*see* IMPONDERABLE).

Natural philosophers quickly accepted the concepts of specific and latent heats, and chemists introduced a new species of combination to explain the presence of bound heat that did not act on a thermometer. Antoine-Laurent *Lavoisier drew on these ideas to define a "matter of fire" that was released in combustion and calcination and gave rise to the liquid and gaseous states. Lavoisier and the mathematician Pierre-Simon *Laplace devised a method in the early 1780s to measure the quantities of fire matter released in physical and chemical processes by the quantity of ice melted when the process took place in a closed environment maintained at the temperature of melting ice. Later they named this apparatus a "calorimeter" and renamed the matter of fire and heat "caloric." The new nomenclature only lightly veiled the descent of their more rigorously defined principle from earlier ideas about the nature of fire.

The question now centered on the relation between the caloric material and the phenomena of fire, light, heat, and (after 1800) radiant heat. Even after Benjamin Thompson, Count Rumford's famous demonstrations in the late 1790s of the improbably huge quantity of heat procurable from a gun barrel by grinding it, most chemists and natural philosophers still believed heat to be material and assimilated it to caloric. Perhaps fire was caloric plus particles of light? With the replacement of the material by the wave theory of *light in the 1820s and 1830s, and the acceptance of the kinetic theory of gases in the 1840s and 1850s (*see* THERMODYNAMICS), the question resolved itself. Chemists and physicists declared that the matters of fire, heat, and light were all the same in the sense that none of them existed.

Douglas McKie and N. H. de V. Heathcote, *The Discovery of Specific and Latent Heats* (1935). Antoine-Laurent Lavoisier, *Elements of Chemistry* (1965; French orig. 1789). Robert Fox, *The Caloric Theory of Gases: From Lavoisier to Regnault* (1971). J. L. Heilbron, *Weighing Imponderables* (1993).

J. L. HEILBRON

FISCHER, Emil (1852–1919), organic chemist.

The son of a merchant in Euskirchen (near Bonn), Fischer studied with August Kekulé and Adolf Baeyer at Bonn and Strasbourg, respectively. After earning his doctorate with Baeyer in 1874, he taught successively at the Universities of Munich, Erlangen, and Würzburg before being called in 1892 to the University of Berlin, as the successor of A. W. von Hofmann. There he remained until his death in 1919.

Fischer's doctoral research was on dye chemistry. In 1878 he completed an impressive investigation (carried out jointly with his cousin Otto Fischer) of the important artificial dye magenta, demonstrating structure and stimulating production of new synthetic dyes based on it.

When he turned to the chemistry of carbohydrates, chemists knew only seven sugars of which they understood the general structures but not their three-dimensional isomeric relationships or *stereochemistry. Because sugars tend to form gums and syrups rather than well-formed, sharply melting crystals, they are difficult to study and interpret. Fischer used two crucial aids in this work: a reaction sequence developed by Heinrich Kiliani that added one more carbon atom onto any sugar, and a series of compounds that he himself had discovered, hydrazine derivatives. Fischer found that phenylhydrazine reacted with sugars to form derivatives with beautiful crystals and sharp melting points; moreover, these transformations provided new information on the stereochemistry of the molecules.

Using Kiliani syntheses, phenylhydrazine derivatizations, and a variety of other chemical manipulations, Fischer gradually lifted the veil on the simple sugars. By 1891 he had worked out the structural details of all sixteen possible isomers of glucose, in the process discovering a number of new artificial sugars. This extraordinary accomplishment led directly to his call to Berlin. Fischer continued to work productively on carbohydrate chemistry the rest of his life, elucidating the structures of ever more complex molecules.

Another area that captured Fischer's interest early on and continued throughout his career was a group of related organic bases including uric acid, caffeine, and theobromine, which had frustrated the efforts of some of the finest chemists of the century. After many difficulties, in 1898 Fischer succeeded in synthesizing the parent substance for the family, which he named "purine," and he proposed formulas for the principal members of the group.

Fischer's research on the purines, of great significance to the pharmaceutical industry, simultaneously laid the basis for understanding a prime component of nucleic acids. He himself prepared the first nucleotide in the laboratory (a nucleotide—a component of what became known as *DNA—is the combination of a purine or pyrimidine base with a sugar molecule and a phosphate). Fischer received the Nobel Prize in 1902 for his research on carbohydrates and purines.

In 1899 Fischer began working on amino acids, then known to be components of all *proteins. He succeeded in adding several new amino acids to the thirteen already recognized, worked out many structures, and synthesized several new ones. He then began to prepare what he called "polypeptides" (linkages of several amino acids) and indicated exactly how the amino acids connected together chemically. By 1916 he had prepared no fewer than a hundred polypeptides containing up to eighteen amino acids. Chemical similarities between his large polypeptides and natural proteins led him to believe that the two were closely related.

Fischer also studied the action of "ferments" or *enzymes. Struck by the highly selective activity of all enzymes, he depicted them as "keys" that could unlock the chemical reactivity of a particular substrate. This topological "lock-and-key" image for biochemical reactions has been influential to the present day.

Fischer was a fine and highly productive teacher, directing about two hundred doctoral dissertations. In the public realm, he acted as the leading lobbyist for the formation of the *Kaiser Wilhelm Society, which set up its first two institutes to study aspects of chemistry. Just before World War I, he held a position of such authority in the Society as to be what one historian has called the "de facto president of science" in Germany. But losing two of his three sons and much of his personal wealth in the war embittered his last years.

As Richard Willstätter wrote, Fischer was "the unmatched classicist" of organic chemistry and a "princely" human being. His work was extraordinarily broad as well as deep. More than anyone else, Emil Fischer established the organic-chemical details of the science of *biochemistry.

M. O. Forster, "Emil Fischer Memorial Lecture," *Journal of the Chemical Society* 117 (1920): 1157–1201. Kurt Hoesch, *Emil Fischer, sein Leben und sein Werk* (1921).

Burckhardt Helferich, "Emil Fischer," in *Great Chemists*, ed. Eduard Farber (1961): 981–995. Eduard Farber, "Emil Fischer," in *Dictionary of Scientific Biography*, vol. 5 (1972): 1–5.

ALAN J. ROCKE

FISHER, R. A. (1890–1962), statistician and evolutionary theorist.

Sir Ronald Aylmer Fisher was born in East Finchley, London, the son of an auctioneer. He was the youngest of eight children, his twin brother having died in infancy. He displayed marked mathematical ability as a small boy, a natural talent probably reinforced by accommodating to his poor eyesight. Unable to read or write on his own, he performed mathematical functions in his head. He took an early interest in astronomy. He obtained a scholarship to Harrow, the well-known British public school; another scholarship followed his admission to Gonville and Caius College at Cambridge, where he studied mathematics, theoretical physics, and astronomy. For six years after leaving Cambridge in 1913, he worked on a farm in Canada, calculated for an investment house, and taught at Rugby School. His poor eyesight barred him from military service during World War I. In 1919 he obtained a position as a statistician at Rothamsted Experimental Station, in Hertfordshire, England, where he at last found a worthy outlet for his formidable mathematical skills.

Most of Fisher's scientific work falls into two related areas: the development of statistical methods for the design of experiments, especially in biology and agriculture, and the application of statistics to evolutionary theory. In 1925 he published his first important book in the former area of research, *Statistical Methods for Research Workers*, whose tables, expanded in collaboration with Frances Yates, were republished in 1938. Then came *The Design of Experiments* (1935) and *Statistical Methods and Scientific Inference* (1956). This work set the foundations of modern statistical analysis by introducing such concepts as random sampling and developing the analysis of variance.

Fisher's key work in his second research line, *The Genetical Theory of Natural Selection* (1930), demonstrated how new insights from Mendelian genetics could illuminate Darwinian natural selection (*see* DARWINISM). The book placed the cornerstone of the mathematical foundations of modern evolutionary theory and raised a new structure in mathematical or theoretical population genetics. In arguing for the efficacy and supremacy of natural selection, Fisher restored Darwinian evolutionary theory to its preeminent position preceding the rediscovery of Mendelian genetics. For his contributions to evolutionary theory, Fisher ranks as a founder of "neo-Darwinism," also known as the synthetic theory of evolution or the evolutionary synthesis, alongside mathematical theorists like the American Sewall Wright and the British polymath J. B. S. Haldane.

The Genetical Theory included extensive discussion on evolution and *eugenics and their use in improving western "civilization." These interests, which Fisher had developed during his Cambridge undergraduate period, informed a significant part of his subsequent work. His reputation as a eugenicist brought him appointment as Galton Professor of Eugenics at University College in London in 1933. Many of his views on class and gender would be unpopular if not offensive today. In 1943 Fisher accepted a position as Balfour Professor of Genetics at Cambridge University, where he lived until his retirement. In 1959, though officially retired, he left Cambridge for Adelaide, Australia, where he worked in the Division of Mathematical Statistics of the Commonwealth Scientific and Industrial Research Organization until his death.

Fisher married Ruth Eileen Guiness in 1917; they had eight children. Their close family centered on reading and conversation. Fisher and his wife eventually separated, however, and Fisher lived alone in his rooms in Cambridge. His book on evolutionary theory is still read and discussed by evolutionists, and his pioneering statistical work is recognized by statisticians the world over.

J. H. Bennett, ed., *Collected Papers of R. A. Fisher*, vol. 1 (1971). Norman T. Gridgeman, "R. A. Fisher," in *Dictionary of Scientific Biography*, ed. C. C. Gillispie (1972): 7–11. Joan Fisher Box, *R. A. Fisher, The Life of a Scientist* (1978).

VASSILIKI B. SMOCOVITIS

FLIGHT. See AERONAUTICS; AIRCRAFT.

FOOD. See FOOD PRESERVATION; NUTRITION.

FOOD PRESERVATION. The earliest technique of food preparation was probably cooking, which not only extended the range of edible matter, but also increased the usable lifetime of the food. The long list of traditional preservation techniques can be classified into the categories of chemical treatments, biological processes, and the use of physical barriers. Chemical treatments include salting, smoking, and adding sugar, vinegar, or saltpeter, together with drying. Biological processes include fermentation or related techniques (used in making beer, wine, cheese, butter, yogurt, sauerkraut, and other products), and application of spices, many of which not only disguise spoilage but actually retard it. Physical barriers include storage of the food in oil, fat, wax, brine, or vinegar, or in a tightly sealed pot. Another effective technique, used since time immemorial (and not just by humans), is burial in soil. Food kept cool in a cellar, well, or stream lasts longer as well.

Some preservation techniques have become so intimately connected with culinary arts that the former purposes have been concealed by the latter. The sharpest (and most actively antibacterial) spices have been used preferentially in warm climates from earliest times. Bacon and ham are still consumed with their smoky, salty, nitrated flavors, independent of any fear of spoilage. Few know that carbonation was introduced into soft drinks principally for its preservative action on the sweet syrup that otherwise would rapidly sour or ferment. Yeast, alcohol, carbonation, and flavorings such as hops preserve the sweet solution of malted grain that becomes beer. Even bread making can be considered a technique that extends the palatable lifetime of the grain pastes and porridges consumed in most traditional cultures.

Industrialization offered both new opportunities and new requirements for food technologies. On the one hand, the greater efficiency of food distribution offered some insurance against the cycles of feast and famine that characterize traditional agricultural societies. On the other hand, urbanization and faster means of transportation also meant that distances between the producers and consumers of foodstuffs gradually increased; foods had to survive the trip, as well as subsequent temporary storage in food shops. In 1810, the French confectioner Nicolas Appert published a technique for sealing heated food in containers; his glass jars, soon replaced by handmade tinned-steel canisters, started the canning industry. In 1869 Appert's countryman Hippolyte Mège Mouriès developed a manufacturing process for a butter substitute with good keeping properties, made from animal products and named "margarine." Both Appert and Mège Mouriès knew of the French army's interest in such inventions; Napoleon I promoted canning, and Napoleon III margarine, for explicitly military purposes.

In the mid-nineteenth century, efforts to maximize profits from the large-scale agriculture of the Americas exercised the ingenuity of inventors. Salt pork and whisky were regarded as complementary methods of "concentrating" the food value of American grain in a stably preserved fashion. The Texas dairyman Gail Borden developed techniques to prepare evaporated and sweetened condensed milk, and then (in the 1850s) a powdered skimmed milk that could keep almost indefinitely. The German chemist and entrepreneur Justus von *Liebig invented a process for "extract of beef" that was industrialized in cities neighboring the Argentinian pampas. Meat packers in Cincinnati and later Chicago developed efficient methods to butcher hogs and then cattle by a sort of mass-production dis-assembly line. The huge yields of grain from the Great Plains increased still further under an aggressively inventive mechanization of plowing, planting, and harvesting, and newly industrialized milling and distribution networks. Roller milling of wheat and bleaching of flour dramatically lowered the price of white bread, which not only had higher status but also better keeping properties than whole-wheat bread.

The Union army in the American Civil War exploited all of these advances in food technology. After the war, accelerating urbanization and industrialization, and the spread of railroads and steamboats, promoted the same advances throughout the world. Refrigeration, first by use of natural ice and after about 1880 also by mechanical means, provided a means to expand further the distribution of fresh foodstuffs. Food processors also increased their use of chemical additives, now supplied by a chemical industry developing rapidly in Europe. Sulfites, borax, salicylic and benzoic acids all found application in nineteenth-century processing, increasingly from about 1870. Some of these chemical preservatives were suspected of being hazardous to human health. A controversy erupted over "embalmed meat" during the Spanish-American War—the highly poisonous formaldehyde had been used to stabilize military rations—and gave impetus to the movement to regulate additives and food adulteration.

The bacteriology of Louis *Pasteur and others provided a rational basis for understanding food spoilage, previously dealt with on an empirical basis. It was now understood that most decay processes involved the action of microorganisms and that both traditional and later chemical techniques of preservation either destroyed these organisms or inhibited their growth. Pasteur's studies led to the heat treatment known as pasteurization, applied as early as the 1860s to beer and wine, and by the turn of the century to milk as well.

Food can be preserved by low temperatures. There was a thriving frozen meat trade in the nineteenth century, but mass marketing of a broad range of frozen food awaited the innovations of Clarence Birdseye, who developed a quick-freezing system in the 1920s. After home refrigerators became common in the following decade, consumers eagerly embraced frozen foods. Not all preservation techniques have been successful, however. Gamma-ray irradiation of foods using radioactive sources, introduced in the 1960s, has never enjoyed commercial success. The procedure kills microorganisms that cause decay, but consumers associate the process with dangers of radioactivity.

The food processing and additives industries expanded explosively after World War II, partly owing to wartime innovations and partly as the result of commercial and social developments. Processed foods—TV dinners, frozen vegetables, ready-to-eat products, and other "convenience foods"—became increasingly popular; many of these required the heavy use of chemical additives and other processing techniques. Large corporations mass marketed much of the world's food supply in the second half of the century. Many consumers live in environments in which automobiles, suburban communities, supermarkets, home refrigerators and freezers, and personal habits (of weekly rather than daily food shopping, for example) indirectly dictate the sort of foods required. The food processing industry has become an integral part of these patterns.

See also FERMENTATION.

Reay Tannahill, *Food in History* (1973). *The Cambridge World History of Food* (2000). Sue Shephard, *Pickled, Potted, and Canned: How the Art and Science of Food Preservation Changed the World* (2001).

ALAN J. ROCKE

FORESTRY. The medieval and early modern response to the need for long-term preservation of wooded lands was to refine the forest legally. Royal edicts and custodians restricted access to forests, but primarily for the protection of game for hunting. Game wardens and rangers also were responsible for the supply of wood and other forest products. As a result, the technical skills and experience associated with managing woodlands came by the seventeenth century to be concentrated in the state, especially in Europe.

Rational forest management addressed a growing concern in early modern Europe that centuries of mismanagement had jeopardized the supply of wood. Military needs, industrialization, and population growth all stimulated demand and strained supply, thus focusing attention on forests not as hunting grounds but as natural resources providing lumber, wood fuel, strong masts and planks for warships, charcoal, potash, and tar. Naval needs in particular stimulated the publication of John Evelyn's *Sylva; or a Discourse on Forest-Trees* (1664), which established a tradition of complaints about the damage done to European woodlands. In England, the annals recording deforestation and unsound forestry continued through the systematic surveys of forests and agriculture by the Board of Agriculture over a century later. In France, ordinances issued by Jean-Baptiste Colbert in 1669 concentrated responsibility for forests in the officers of the Crown and began the reform of forest law, a project continued by edicts and publications issued throughout the eighteenth century.

Scientific investigations also contributed to the emerging forestry literature of the century. René de Réaumur summarized the problem of growing demand for wood and called for the improvement of forestry in a paper presented to the Paris Academy of Sciences in 1721. George-Louis LeClerc (Comte du Buffon), Henri-Louis Duhamel du Monceau, and other influential academicians published on the growth of trees, physical properties of wood, forest botany, and naval architecture; but their fundamental investigations failed to lead directly to schemes of improved forestry.

Intensifying attention to preserving the wood supply dovetailed with confidence in the mathematics and science fields in the emergence of rational systems of forest management that first took hold in Germany. The last year of the Seven Years' War (1763) brought the first forestry school, founded by H. D. von Zanthier in the Harz Forest; the first book with the words "Forestry Science" in its title, Johann Gottlieb Beckmann's *Beiträge zur Verbesserung der Forstwissenschaft*; and the first journal devoted exclusively to forestry, J. F. Stahl's *Allgemeines oekonomisches Forstmagazin*. Administrative officials schooled in the curriculum of "cameral science," by then established throughout German-speaking central Europe, responded favorably to analysis of the forest and its management based on economic reason and quantitative measures.

Foresters of the old school generally set annual cuttings based on areal divisions of the forest. Equal annual yields meant roughly that equal areas were harvested. By the 1760s, new methods of forest economy tracking the mass or volume of wood began to replace area-based systems. Sophisticated methods for calculating the quantity of standing wood in the forest followed, also offering prediction of its future growth and cuttings. Mathematically inclined foresters, including Carl Christoph Oettelt, Johann Hossfeld, and Johann Vierenklee, relied on geometry, arithmetic, and algebra to convert the forest into an equivalent quantity of wood mass. The rational synthesis of calculation and cameralism required appropriate terms of analysis; first formulated in mathematical terms by Vierenklee in 1767, sustained yield (*Nachhaltigkeit*) filled this role, becoming the conceptual cornerstone of the new system.

Between 1760 and the early 1800s, forestry based on the sustained yield concept elaborated steps for determining the quantity of wood mass over ever longer periods of time. Chief foresters followed instructions together known as forestry assessment (*Forsttaxation*), methods firmly established by the middle of the nineteenth century in the teaching and publications of the "classical writers" of German forestry: Heinrich Cotta, Georg Ludwig Hartig, Johann Christian Hundeshagen, and Carl Gustav Heyer. Cotta's notion of the "regulated forest" and sustained yield delivered the classical framework for a sound system of forest economy. By the end of the nineteenth century the German forest symbolized the replacement of disorderly nature with neatly arranged constructs of science. Foresters in other nations—France, England, India, the United States—learned these methods through professional training in Germany or the direct influence of German foresters, though as in the case of Gifford Pinchot's first plan for the management of Biltmore Forest in western North Carolina, such methods were often modified by local or national circumstances.

In the United States, the appointment of Pinchot as director of the Division of Forestry of the Department of the Interior in 1898, Theodore Roosevelt's presidency beginning in 1901, and the transfer of the Division to the Department of Agriculture in 1905, stimulated the professionalization of public forest management and reinforced principles of resource and wildlife conservation in its practices. These influences guided the spectacular growth in federally managed forested lands during the first two decades of the twentieth century. Methods of forest inventory continued to rest on the science of the German School, supplemented through the 1950s by advances in sampling, statistical, and surveying techniques. The "Multiple-Use Sustained Yield Act" of 1960 printed the stamp of congressional approval on the management of forests according to sustained-yield and multiple-use forestry, but the introduction of practices such as clear-cutting to meet rising demand has prompted a series of public and scientific debates addressing concern over biodiversity and environmentally sustainable forest development.

Franz Heske, *German Forestry* (1938). Heinrich Rubner, *Forstgeschichte im Zeitalter der Industriellen Revolution* (1967). Michael Williams, *Americans and Their Forests: A Historical Geography* (1989). Roland Bechmann, *Trees and Man: The Forest in the Middle Ages* (1990).

HENRY LOWOOD

FORTIFICATION. In response to fifteenth-century improvements in large guns, the tall castle walls of medieval construction gave way to low, masonry ramparts. Reflecting the Renaissance attention to proportion, symmetry, and harmony, military engineers employed geometry as the language of fortification design. The angled bastion, a projecting section of the walls and a key element in the plan of fortifications, exemplified this trend. Engineers such as Francesco di Giorgio Martini, Antonio da Sangallo the Younger, and Michelangelo used these geometrical configurations to design defenses in which every part could be seen and defended from some other part.

Changes in the relationship between fortifications and science can be discerned from at least the early 1600s when *Galileo and others developed instrumentation to aid in the design and construction of fortifications. Galileo's work as a mathematical practitioner and instrument maker facilitated fortification design, while his work on mechanics as a natural philosopher opened new doors for exploring the nature of fortification material and form. During the seventeenth century the locus of progress in fortification switched from Italy to France, owing primarily to the work of the French military engineer Sébastien le Prestre de Vauban. Although Vauban advocated a flexible approach to the design, attack, and defense of fortified places, his successors often broke his work into a series of systems. These systems became textbook fundamentals and important teaching tools for military art and science, and continued to fuel polite dabbling in fortification as a mathematical entertainment.

Vauban and other prominent fortification engineers, including Menno van Coehoorn, promoted the development of formal education and professionalism. The curricula, faculty, and textbooks of eighteenth-century schools such as France's École du Génie at Mézières, established in 1749, demonstrate increasing interest in applying scientific information to fortifications. Bernard Forest de Bélidor's treatise *La science des ingenieurs dans la conduite des travaux de fortification et d'architecture civile* (1729), incorporating up-to-date knowledge of mechanics and the properties of materials, became a standard work in European military schools. The U.S. Military Academy at West Point, established in 1802 and modeled after the French military engineering schools, further encouraged engineers to approach forts as objects of scientific investigation.

The design and construction of fortifications provided support for wider scientific studies. Galileo and other natural philosophers of the Renaissance who taught the designing of fortifications used the resultant income in their scientific investigations. Later, Charles Augustin Coulomb (*see* CAVENDISH, HENRY, AND CHARLES AUGUSTIN COULOMB) and others used their education and practical experience in fortifications as the basis for fundamental studies in materials, experimental physics, and hydraulics. Posting at a fortification also created opportunities for observations on natural history, and astronomical and magnetic phenomena. Often military engineers, particularly in Britain and the United States, were active members in scientific societies and institutions, including the Royal Society and the National Academy of Sciences.

The interaction of science and fortifications ebbed with the arrival of rifled guns and mobile defense units. Rapid progress in the development of heavy ordnance and armoring at the end of the nineteenth century caused the replacement of large fortifications by gun batteries. The twentieth-century arrival of aircraft, aircraft carriers, and rocketry have placed the relationship between fortifications and science firmly in the historian's province.

Monte D. Wright and Lawrence J. Paszek, eds., *Science, Technology, and Warfare: The Proceedings of the Third Military History Symposium, United States Air Force Academy, 8–9 May 1969* (1970). Jim Bennett and Stephen Johnston, *The Geometry of War, 1500–1750* (1996).

MARY M. THOMAS

FOSSIL. Until the end of the eighteenth century, scholars could not decide whether fossils were the traces of dead creatures. For every fossil that clearly resembled a living animal, they found others unlike anything currently living. When minerals were believed to grow in the earth, and when spontaneous generation remained a serious hypothesis, many people assumed that fossils had grown in situ. In 1726, in a notorious example, Johann Beringer, a member of the faculty of medicine at the University of Würzburg, published a book full of illustrations of curiously shaped stones found in the region. He described them as the handiwork of God. They turned out to be the handiwork of jealous colleagues who had planted them as a hoax.

In the early nineteenth century, *paleontology began to emerge as a subfield of *geology. From the point of view of geologists, fossils were chiefly useful for identifying strata (*see* STRATIGRAPHY). Geologists concentrated on the fossils plentiful enough to serve this function, and so spent most of their time classifying marine invertebrate species. Questions that thrilled a wider public about the meaning of fossils and the history of life on earth were of limited interest to paleontologists. They have, however, loomed large in the public image of geology.

When Georges Cuvier and his followers excavated bones and reconstructed giant beasts

Unearthing of the jaw of a large fossil reptile, later christened "mosasaur" by Georges Cuvier, as depicted in Barthélemy Faujas de St Fond's *Histoire de la montagne. . .de Maestricht* (1799).

such as the mammoth or the ichthyosaurus, they showed to almost everyone's satisfaction that some species had become extinct. While fossil marine bivalves might be thought to have living representatives in some distant ocean, no one expected to find an ichthyosaurus in a remote part of the world. How and why extinction has occurred was to generate much speculation in both scientific and religious circles.

Meanwhile, the search for fossils of extinct vertebrates had started. In 1824, the English geologist William Buckland published the first paper describing a dinosaur. In 1842, another Englishman, Richard Owen, coined the term for these creatures. Almost immediately, dinosaurs captured the place in the public imagination they retain to this day. Professional fossil collectors mounted expeditions to likely sites. At the beginning of the twentieth century, the steel magnate Andrew Carnegie, reading of a discovery of a giant dinosaur in Wyoming, dispatched a team to secure it for his new museum in Pittsburgh. After various adventures, the bones of *Diplodocus carnegii* arrived. So spectacular was the creature when reconstructed that national museums throughout the world considered themselves incomplete without one. Plaster casts of Carnegie's dinosaur can now be seen in England, France, Austria, Russia, Germany, Spain, Argentina, and Mexico.

Charles *Darwin admitted in his *Origin of Species* (1859) that the fossil record with its many gaps posed a severe problem for his theory. He resolved the problem by asserting that the record was incomplete. Nonetheless, his followers, including T. H. Huxley and O. C. Marsh, professor of vertebrate paleontology at Yale University, thought it important to try to trace ancestries for at least some species. They were particularly successful with the horse. For the public, though, the more important question remained the "missing link" between humans and their simian ancestors. The remains of Neanderthal man came to light in 1856, followed by Peking Man (1903) and Piltdown Man (1912), exposed as a hoax in 1953 (*see* EARLY MAN).

The ability of fossils to stimulate the imagination has not declined in recent years. In the late 1960s, a new debate started when two Americans, Robert Bakker and John Ostrom, suggested that dinosaurs were hot blooded and thus lively, not the slow-moving creatures that had been imagined. Just a little over a decade later, a group of American scientists led by Walter Alvarez suggested that dinosaurs had been killed off by an extraterrestrial object hitting the earth about sixty five million years ago. Most scientists now seem to agree. Another round of interest in fossils, sparked by the discovery that traces of DNA can be found in some fossils, inspired Stephen Spielberg's film *Jurassic Park*.

See also EARLY MAN; STRATIGRAPHY.

Martin Rudwick, *The Meaning of Fossils* (1972). Adrian Desmond, *The Hot Blooded Dinosaurs* (1975). Peter Bowler, *Fossils and Progress* (1976). David Raup, *The Nemesis Affair* (1986).

RACHEL LAUDAN

FOUNDATIONS. During the twentieth century, scientists' need for expensive experimental apparatus, large research teams, and international travel required diligent searches for funds that have been integral to setting research agendas. As Robert Kohler described the situation, "Assembling the material and human resources for doing research is no less a part of the creative process than doing experiments." Throughout the century, but particularly in the first half, foundations were central to that process.

The perpetual charitable trust as a legal entity intended to deliver a specific public benefit dates from the Elizabethan era. The general-purpose philanthropic foundation awaited the late-nineteenth-century elaboration of the nonprofit organization to give it a vehicle, and the capitalist accumulation of wealth to make it substantial. Philanthropically created foundations that support science are primarily a recent American phenomenon.

The first large foundation to support fundamental research was the Carnegie Institution of Washington, initially endowed by Andrew Carnegie with $10 million and enlarged with later gifts. The Institution hired scientists and created laboratories of its own and made numerous small grants to individual scientists and large ones to institutions, such as the Mt. Wilson Observatory and the National Research Council. Over the next three decades, American science benefited from other foundations as well: Robert Goddard's rocket experiments had support from the Carnegie Institution and the Guggenheim Foundation; Albert *Einstein's visit to the United States in 1931 to collaborate with Robert Millikan was funded by the Oberlaender Trust; the China Foundation's fellowships sent Chinese geneticists and biologists abroad for advanced training. The Sauberán Foundation in Argentina helped to establish Nobel laureate Bernardo Houssay's Institute of Biology and Experimental Medicine in Buenos Aires in 1944. In the 1950s, the Nutrition Foundation provided fellowships for Central American food chemists to study in the United States.

The largest amount of money distributed in the first half of the twentieth century came from the cluster of philanthropies created by the Rockefeller fortune. Led by the Rockefeller Foundation, these philanthropies initiated programmatic giving that focused funding on particular institutions, disciplines, or problem areas, and relied heavily on knowledgeable program officers to make decisions about allocations. Rockefeller institution-building created or supported major research centers, including the Peking Union Medical College (founded in 1918), the *Bohr Institute in Copenhagen (1921), the Mathematical Institute at Göttingen University (1926), the Palomar Observatory (1948), and public health institutes in Europe, the United States, and South America.

In specific fields, Rockefeller funding had significant effects. Support of physicists in the 1920s and 1930s aided in shifting the center of physics from Europe to the United States. The Rockefeller Foundation helped to create molecular biology, contributing approximately $25 million to the field from the 1930s through the 1950s.

Rockefeller philanthropies greatly expanded the role of the fellowship, both resident and traveling, as a means of shaping or even redirecting careers. In 1919, the Rockefeller Foundation began a twenty-year program run by the National Research Council, which at first supported physicists and chemists, but soon expanded to biologists and psychologists. The International Education Board (IEB), created in 1923, was the first Rockefeller philanthropy to focus on fellowships as a means of promoting science. These fellowships enabled young researchers to devote their time to their specialties during what was generally their period of greatest productivity.

After the IEB merged with the Rockefeller Foundation in 1928, the fellowship program became global, and assisted many scientists and organizations in less-developed regions. Alberto Hurtado became research director of Peru's Institute of Andean Biology in 1934 after holding a postdoctoral research fellowship from the Rockefeller Foundation at the University of Rochester. Over the next twenty years, the Foundation granted ten fellowships to Hurtado's staff to study abroad; combined with Rockefeller grants for instruments, this program made the Institute a global leader in physiology studies. Extended throughout Latin America, the fellowships of the Rockefeller Foundation, and similar fellowships offered by the John Simon Guggenheim Foundation, helped to reorient Latin American science from Europe to North America.

After World War II, the rapidly increasing scale of government-funded research induced the largest foundations, including the Rockefeller Foundation, to withdraw from general support of the natural sciences. Both the older and newer foundations turned to more specific and applied areas such as agriculture, population studies, and medicine. Particularly wealthy foundations such as the Wellcome Trust (1936) and the Nuffield Foundation (1943) in Britain, and the Howard Hughes Medical Institute (1953) in the United States, supported medicine.

Because foundations can focus on specific research areas and dispose of funds flexibly, their institutional form was adopted throughout the world in the latter part of the twentieth century, often by government initiative. The Merieux Foundation (1967) supports research in biology and medicine in France; the Mario Negri Institute (1961) supports biomedical research in Italy; the interrelated Fundación Andes, Fundación Antorchas, and Vitae (Apoio à Cultura, Educação e Promoção Social) in Latin America (1985) and the Volkswagen Foundation (1961) in Germany give to higher education science research. A major example of government-created foundations that fund research is the Deutsche Bundesstiftung Umwelt, the largest foundation in Europe in the 1990s, established by the German government with the proceeds of the sale of a publically owned corporation. In the twentieth century, foundations became an integral element of the scientific enterprise, responsible for the founding or expansion of institutions, providing crucial support for the development of scientific careers, and aiding in the creation of new fields of science.

Mary Brown Bullock, *An American Transplant: The Rockefeller Foundation and Peking Union Medical College* (1980). Howard S. Berliner, *A System of Scientific Medicine: Philanthropic Foundations in the Flexner Era* (1985). Robert E. Kohler, *Partners in Science: Foundations and Natural Scientists, 1900–1940* (1991). Marcos Cueto, ed., *Missionaries of Science: The Rockefeller Foundation and Latin America* (1994). Jean-Francois Picard, *Fondation Rockefeller et la Récherche Medicale* (1999).

DARWIN STAPLETON

FRANKLIN, Benjamin (1706–1790), natural philosopher, diplomat, and inventor of the lightning rod.

A key figure in the struggle of Britain's American colonists for independence, Franklin served as colonial agent in London (1757–1762, 1765–1775), as a member of the Second Continental Congress, and as an author of the Declaration of Independence. Subsequently American plenipotentiary in Paris, he helped to negotiate the 1783 peace treaty with Britain and, later, to draw up the American constitution.

Franklin's reputation as a natural philosopher preceded his arrival in Europe as a politician. In the 1740s he developed a revolutionary new theory of *electricity that was eventually adopted almost universally. He also devised an experiment to prove the electrical nature of *lightning. Performed successfully in France in 1752, the demonstration caused a sensation, for it offered the prospect of human reason controlling nature's power, of conducting rods projecting above the roofline protecting buildings from lightning strikes.

Born in Boston, Massachusetts, to a tallow-chandler, at the age of 12 Franklin apprenticed with one of his older brothers, a printer. He had little formal schooling but read voraciously and also taught himself to write, modeling his style on Joseph Addison's. In 1723, he ran away to Philadelphia, where in time he built a successful printing business and became a public figure. He published the *Pennsylvania Gazette* and the enormously popular *Poor Richard's Almanack*. He also served as postmaster of Philadelphia from 1737 to 1753. Filled with ingenious ideas, he invented, among other things, the rocking chair, bifocal glasses, and the Pennsylvania fireplace. By 1748, he could retire from business and devote himself to public affairs.

By the mid-1740s, when he took up the study of electricity, Franklin was well-read in experimental natural philosophy, familiar with Robert *Boyle's work, Isaac *Newton's *Opticks*, and the writings of Herman *Boerhaave, Willem Jacob 'sGravesande, John Theophilus Desaguliers, and Stephen Hales, among others. Like most of his contemporaries, he attributed electrical effects to the action of a subtle fluid that supposedly pervaded all bodies. However, while others believed that electrification consisted in agitation of this fluid, Franklin saw it as a redistribution. Moreover, for him, the fluid was specific to electricity. When a person rubbed a substance like glass, he said, there was an exchange of fluid between the rubber and the body rubbed, leaving one with a surplus, the other with a deficiency. Hence two kinds of electrical "charge" existed, "plus" and "minus," that could in appropriate circumstances neutralize each other. Later, when Franklin became familiar with Charles-François Dufay's work, he identified these with the two different modes of electrification, "vitreous" and "resinous," that Dufay had discovered some years earlier.

Franklin's way of viewing things enabled him to give a coherent account not only of Dufay's distinction but of a range of electrical phenomena that existing theories had not handled well—phenomena that, in Franklin's terms, involved charging or discharging of bodies. Above all, he managed to explain the most famous electrical phenomenon of the age, the *Leyden jar experiment, something existing theories could not do. His observation that pointed conductors drew charge off bodies better than rounded ones did was the key to his idea of the lightning rod. These successes won his ideas many adherents. Other phenomena Franklin dealt with less successfully. Initially, he paid little attention to the traditional starting point of electrical inquiry, the power of rubbed bodies to attract light objects. Later, he suggested that when a body became charged, the additional electric fluid formed an atmosphere around it, and that contact between this atmosphere and other bodies brought about the attraction. However, his discussion of how it did so, and generally of the role such atmospheres might play in electrical attractions and repulsions, left many questions unanswered. In 1759, Franz Ulrich Theodosius *Aepinus brought new clarity to the subject by eliminating the atmospheres and converting Franklin's theory into one based on unexplained forces acting at a distance between particles of electric fluid and particles of ordinary matter.

Although Franklin continued to take an interest in science and to promote the erection of lightning rods on public buildings, his major contributions to electrical understanding were over by 1755. By then, however, he had established a reputation as a natural philosopher that opened doors for his diplomacy in both London and Paris, and enhanced the standing of the American Philosophical Society, which he helped to found and over which in later life he long presided. He died at Philadelphia on 17 April 1790.

Carl Van Doren, *Benjamin Franklin* (1938). I. Bernard Cohen, *Franklin and Newton* (1956). J. L. Heilbron, *Electricity in the 17th and 18th Centuries* (1979). R. W. Home, *Aepinus's Essay on the Theory of Electricity and Magnetism* (1979). I. Bernard Cohen, *Benjamin Franklin's Science* (1990).

R. W. HOME

FRANKLIN, Rosalind (1920–1958), geneticist and physical chemist.

Born into an intellectual family in London, Rosalind Franklin won scholarships to St. Paul's Girls' School and then to Newnham College, Cambridge. After graduating and carrying out research in gas-phase chromatography, she left Cambridge in 1942 to work at the British Coal Utilization Research Association. During the war years she built a reputation for her studies of the micellar structure of coals. She continued this research when in 1947 she moved to the Central Laboratory of the Chemical Services of the State in Paris. Here she acquired unusual skill in *x-ray diffraction techniques, applying them to the study of the graphitization of carbons.

The subject of the physical chemistry of carbon brought Franklin into contact with Charles Coulson, through whom she was introduced to Sir John Randall, director of the Medical Research Council Biophysics Unit at King's College, London. There, on a Turner Newall Fellowship, she intended to study proteins in solution, but Randall, learning of the promising results obtained by Maurice Wilkins and Raymond Gosling in their exploratory studies of deoxyribonucleic

acid (DNA), asked her to study *DNA with Gosling instead.

Franklin went on to distinguish two forms of DNA: at low humidity the crystalline A form and at high humidity the partially crystalline B form. As the water content increased, the threads of the sodium salt of DNA extended, and the packing of the molecules within them became less ordered. Franklin realized that the chain molecules of the A form were held together by the attraction between positive sodium ions and the negatively charged phosphate groups of the DNA, and that water molecules penetrating between the chains in the transition to B reduced their orientation to one another. The sugar-phosphate backbones of the chains must therefore be on the outside. Being crystalline, the diffraction pattern from the A form offered more discrete data than did that from the B pattern, so Franklin and Gosling concentrated on the former.

Franklin hoped to solve the structure in an inductive manner using a standard method to derive the spatial distribution of the atoms that dominate the diffraction pattern of the A form. The method did not, however, reveal the bonding between atoms. The diffraction pattern of the B form, on the other hand, though containing less data, was strikingly simple. To those familiar with the kind of pattern produced by a helix, the B-form diffraction pattern gave the clearest example yet of such a structure: a Maltese cross, the intensities moving out from the center or equator leaving an "absence" in the meridian before closing in again. Initially (1951) Franklin had favored a helical conformation for both forms of DNA. But in 1952 her more detailed studies—especially of the A form—led her to look for other possibilities, notwithstanding the remarkable B pattern she had obtained in May 1952. Not until February of the following year did she return to her interpretation of this B pattern. She was then close to solving the structure, but her earlier antihelical phase had slowed her down. At this juncture James D. *Watson and Francis Crick solved the structure. Without the data she and Gosling had produced, however, their feat would not have been possible. In announcing their discovery in 1953 they remarked that they had been "stimulated by a knowledge of the general nature of the unpublished experimental results and ideas" of the King's College group. This aspect of the discovery has led to debate about both the extent of their reliance on Franklin's work and the ethics of so doing unbeknownst to her.

In 1953 Franklin moved to J. D. Bernal's Crystallography Laboratory in Birkbeck College, London. There she worked with Aaron Klug, D. L. D. Casper, and K. C. Holmes on the structure of the tobacco mosaic virus (TMV), a nucleoprotein containing ribonucleic acid (RNA). They established the hollow nature of the molecule, which contains protein sub-units packed in a helical fashion around the cylinder and attached to the continuous RNA chain embedded within them. She correctly predicted that the RNA was present as a single chain.

Franklin was an excellent experimentalist, a patient and conscientious researcher who adopted a thoroughly professional approach to her work. In her work with Gosling on DNA, she seems to have been isolated. Hence she lacked the advantages of the frank and challenging interchanges generated by open collaboration. How much this isolation was of her own making, and how much due to others, has been much debated. But it would be a mistake to cast her as a feminist who spoke her mind and paid the penalty of ostracism. No such climate developed when she moved to Birkbeck College, and when she died of cancer at the age of thirty-seven she had already won an international reputation for her study of the structure of TMV. With TMV, as with DNA and coal, she had worked where she liked to—at the "edge" of crystallinity among the paracrystalline substances.

Rosalind Franklin, D. L. D. Casper, and Aaron Klug, "The Structure of Viruses as Determined by X-ray Diffraction," in *Plant Pathology: Problems and Progress, 1908–1958*, ed. C. S. Holton (1959): 447–461. Max F. Perutz, M. H. F. Wilkins, and J. D. Watson, "Three Letters to the Editor," *Science*, June 27 (1969): 1537–1538. Aaron Klug, "Rosalind Franklin and the Discovery of the Structure of DNA," *Nature* 219 (1968): 808–810, 843–844; corrigenda, 879, 1192; correspondence, 880. Brenda Maddox, *Rosalind Franklin. The Dark Lady of DNA* (2001).

ROBERT OLBY

FRAUD. In 1981 then-Representative Albert Gore, Jr., began the hearings of the Investigations and Oversight Subcommittee of the House Science and Technology Committee. Cases of scientific misconduct had been disclosed at four major research centers in the 1980s; doubts about the integrity of American scientific research circulated. Gore wondered publicly if these cases were not "just the tip of the iceberg." Answering in the affirmative in 1982, William Broad and Nicholas Wade, well-regarded science journalists who had worked for both the *New York Times* and *Science*, published in their *Betrayers of the Truth* that scientific fraud was much commoner than anyone had believed.

Fraud is a particularly serious sin in science, whose whole purpose is supposed to be the pursuit of truth or knowledge. In the biomedical sciences in particular, a distressingly common

pattern of misconduct seemed to be emerging. John Darsee, a cardiologist at Harvard Medical School, published over a hundred articles and abstracts between 1979 and 1981. After charges of fraud led to investigations in 1981, he retracted fifteen out of the twenty articles questioned. Other cases surfaced at the University of California at San Diego, the University of Pittsburgh, the Baylor College of Medicine, and the Massachusetts Institute of Technology. The latter became known as the Baltimore case because the senior author of the article allegedly containing a junior author's fraudulent data was David Baltimore, Nobel Prize winner and by then president of Rockefeller University. Although Baltimore himself was not suspected, his support of the junior author seemed ill advised. Eventually in 1991 she was cleared of all wrongdoing, but by then the case had caused Baltimore's resignation from Rockefeller and widespread public concern.

The furor over the Baltimore case reflected growing suspicion about the practices of science. Broad and Wade had claimed that fraud was endemic to the scientific enterprise. The supposed perpetrators listed in their book included some of science's most lustrous names. Ptolemy, they charged, invented astronomical measurements; *Galileo reported results too perfect to be credible; Isaac *Newton fudged data; John *Dalton described experiments he probably never performed; Gregor *Mendel produced statistical results too good to be true; and Robert Millikan, awarded the Nobel Prize in 1923 for measuring the charge on the electron, had selected only the readings that suited him.

Far from being disinterested seekers after truth, Broad and Wade argued, scientists were as ambitious and anxious for recognition as any other professionals. They rarely replicated experiments, supposedly one of the best checks on fraud, because they received no credit for doing so. And in the face of charges of fraud or threats of regulation from outside, the scientific community drew together instead of subjecting their colleagues to investigation.

Scientists, government officials, and historians undertook serious studies of the problem. They quickly discovered the difficulty of defining fraud. Faking data with the intent to deceive others (the motive, according to most scientists, behind the planting of the Piltdown skull) was clearly wrong. But what to make of the practices identified in the nineteenth century by the computer pioneer Charles Babbage: forging (reporting observations that had never been made), trimming (working with data to make them look better), and cooking (choosing the data that best fit the theory under test)? On occasion depending on the choice of the appropriate statistical methods, trimming or cooking might be appropriate. What about thought experiments, experiments that took place only in the scientist's head? And preparing a streamlined and cleaned-up version of research for publication, although it could involve fraud, was also in most cases unproblematic scientific practice. Historians of science believed that many of the historical cases adduced by Broad and Wade fell in a benign category.

In any case, the American scientific community had not been prepared to deal with fraud. Techniques for detecting it were rudimentary. "Whistle blowers," who were as likely to suffer as the scientists they charged, brought most cases to light. Procedures for handling cases once detected did not exist. In 1985 Congress passed the Health Research Extension Act. It included important provisions concerning fraud: institutions receiving federal money had to establish administrative processes to review reports of fraud; and the National Institutes of Health had to establish a process to respond to their findings. It took five years to draft and discuss guidelines. In 1989 these were published in the Federal Register. To avoid emphasizing questions of motive, the term "fraud," defined in law as intent to deceive, did not appear. Instead misconduct was defined as "fabrication, falsification, plagiarism, or other practices that seriously deviate from those that are commonly accepted within the scientific community for proposing, conducting and reporting research."

The guidelines created considerable controversy. Many scientists wanted the ruling confined to fabrication, falsification, and plagiarism rather than opening the door to the policing of other practices that, although dubious morally (e.g., sexual harassment), were not specific to the scientific enterprise. Efforts to set up a system to deal with misconduct continued. In 1992 the Public Health Service set up the Office of Research Integrity. Over the next five years it received approximately a thousand allegations of misconduct. It investigated 150 of these and found against the researchers in about half of them. Some of its conclusions, most spectacularly in the Baltimore case, were reversed.

Although the problem of fraud emerged in the United States and seems to have been most intense there, other countries with substantial programs of scientific research, including Great Britain, Japan, Germany, France, and Australia, have all put in place procedures to detect and regulate misconduct in science. In an enterprise funded by public money, it is not surprising that governments set up systems for assuring accountability, but that they did without them for so long.

Gerald Holton, *The Scientific Imagination* (1978). William Broad and Nicholas Wade, *Betrayers of the Truth* (1982). Alexander Kohn, *False Prophets* (1986). Robert Bell, *Impure Science: Fraud, Compromise, and Political Influence in Scientific Research* (1992). Marcel C. La Follette, *Stealing into Print: Fraud, Plagiarism, and Misconduct in Scientific Publishing* (1992). *Integrity and Misconduct in Research: Report of the Commission on Research Integrity* (1995). Daniel J. Kevles, *The Baltimore Case: A Trial of Politics, Science, and Character* (1998). Robert L. Park, *Voodoo Science: The Road from Foolishness to Fraud* (2000).

RACHEL LAUDAN

FRESNEL, AUGUSTIN. See CARNOT, SADI, AND AUGUSTIN FRESNEL.

FUKUI, Ken'ichi (1918–1988). One of the twentieth century's most eminent theoretical chemists, Ken'ichi Fukui was born in Nara, Japan, to an upper-middle-class family on 4 October 1918. In 1981, he became Japan's first Nobel laureate in chemistry, sharing the award with Roald Hoffmann of Cornell University for their contributions to chemical reaction theory. The work, which found broad application in industry and medicine, made it possible to predict whether the introduction of a chemical compound to the body might cause cancer.

Fukui's father, Ryokichi, graduated from the predecessor of today's Hitotsubashi University and worked for a time as the Japanese representative of a British export-import firm and later as a factory manager. Fukui's mother, Chie (Sugisawa) Fukui, graduated from the Nara Women's High School. Both parents took an active part in the education of Fukui and his two brothers. As a youth, Fukui had diverse interests ranging from history and literature to natural history and mathematics. His favorite reading included *National Geographic*, the novels of Soseki Natsume, and the writings of the nineteenth-century French entomologist and chemist Jean Henri Fabre. Formal schooling took a back seat to sports and outdoor activities; only Fukui's talent for mathematics and a general interest in nature foreshadowed his later career.

An older cousin, Dr. Gen'itsu Kita, professor of engineering at nearby Kyoto University, advised Fukui to study in the university's Faculty of Engineering and major in chemistry, to pursue his interest in mathematics, and to concentrate as much as possible on basic rather than applied science. Fukui followed Dr. Kita's advice, attended lectures on theoretical physics in the Faculty of Science, studied quantum mechanics, and read books on the subject by Paul Dirac and

Hideki Yukawa, sometimes alone and sometimes with friends.

Upon graduation in May 1941, Fukui entered the graduate program, and followed faculty advice to accept a temporary commission in the Japanese Army. For the next four years he divided his time between Kyoto and the Army Fuels Research Institute at Fuchu, in suburban Tokyo. At Fuchu, he served as part of a team assigned to research on synthetic fuels dedicated to the needs of aviation. This wartime research not only allowed him time for independent reading but deepened his knowledge of hydrocarbon chemistry, especially the analysis of iso-paraffin. When the war ended, he returned to the university full-time, and in 1951 he achieved the rank of full professor at the age of thirty-three. His duties as professor of fuel chemistry required him to take an interest in chemical applications and teach courses with a strong applied slant. As a senior faculty member, however, he was gradually able to shift the emphasis in his section away from purely applied issues and toward the theoretical aspects of chemistry, for which his reading in quantum mechanics and theoretical physics had well prepared him.

Fukui's first major publication dealing with chemical reaction theory appeared in 1952 in the *Journal of Chemical Physics*, the leading publication for scientists interested in quantum chemistry. Beginning with this paper, and relying on analyses of hydrocarbons, Fukui and his collaborators showed over the next few years that the site and rate of a chemical reaction depends on the geometries and relative energies of the highest occupied molecular orbital (HOMO) of one reactant and the lowest unoccupied molecular orbital (LUMO) of the other. Building on the concept of orbital symmetry control developed by Roald Hoffmann and Robert B. Woodward, Fukui applied perturbation theory to his original qualitative notions. Perturbations distort the HOMO of one reactant and the LUMO of the other and thus affect the energetics of various reaction pathways.

Fukui worked as a major theoretician in a strongly applied academic setting. He never studied abroad as a young chemist, and most of his overseas travel came after he was well known, rather than before. In 1962, however, he received the Japan Academy Prize for the "frontier orbitals" theory. After six months in 1970 as a visiting professor at the Illinois Institute of Technology in Chicago, he became better known in the United States. In 1981, he was elected an overseas member of the National Academy of Sciences. Japan conferred the Order of Cultural Merit on him in 1981, and in 1985 he became a member of the Pontifical Academy of Sciences in Rome.

After retiring from Kyoto University in 1982, Fukui was named director of the (Kyoto) Institute for Fundamental Chemistry, a private foundation created with a $30 million gift from the Kao Soap Company. The Nobel Prize brought him extraordinary public attention. On 9 January 1998, Fukui died in the city where he had spent most of his life.

Scott A. Davis, "Kenichi Fukui, 1981," in Frank N. Magill, ed. *The Nobel Prize Winners: Chemistry*, vol. 3. (1990): 1061–1066. William B. Jensen, "1981 Nobel Laureate: Kenichi Fukui," in Laylin K. James, ed. *Nobel Laureates in Chemistry*, 1901–1992 (1993): 639–647. Roald Hoffmann, "Obituary: Kenichi Fukui (1918–1998)," *Nature* vol. 391, 19 February (1998): 750.

JAMES BARTHOLOMEW

FUNDAMENTAL CONSTANTS. See CONSTANTS, FUNDAMENTAL.

FUSION. See PLASMA PHYSICS AND FUSION.

G

GAIA HYPOTHESIS. Independent scientist James Lovelock published his first book on the Gaia hypothesis in 1979. He gives cell biologist Lynn Margulis credit for copartnering it, and has traced roots of the idea to James *Hutton, Russian geochemist Vladímir Vernadsky, and Swedish chemist Lars Gunnar Sillen. The Gaia hypothesis states that the earth as a planet is a living and evolving organism, maintaining physical and chemical conditions suitable for life. The hypothesis can also be simply stated, "The Earth is homeostatic." Lovelock refers to the science of Gaia as "geophysiology." The Gaia hypothesis (which Lovelock now calls the Gaia theory) should not be confused with the anthropic principle.

Lovelock formulated the Gaia hypothesis by asking, How can we determine from a distance whether or not a planet has life? In comparing the earth to its neighbors, he concluded that it differs in that its atmosphere contains unstable compounds (e.g., O_2, methane) that are present because living organisms modulate the atmosphere. Oxygen content itself is not a prerequisite for life; in fact, oxygen is toxic to many important life forms such as some bacteria.

Before Lovelock's hypothesis, geoscientists had already recognized that the earth has maintained a temperate surface for more than three billion years. They had also described complex interactions among the physical, chemical, and biological systems on Earth that sustain this temperature. For example, Earth has not suffered a runaway greenhouse effect, as has Venus, in part because organisms remove carbon dioxide from the atmosphere, and organic carbon is subsequently stored in the earth's crust. Earth systems have a set of checks and balances. Thus, if more carbon dioxide is emitted from volcanoes, the climate becomes warmer, which in turn favors more weathering and photosynthesis, both of which consume carbon dioxide, driving temperatures back down.

Lovelock named his hypothesis after the Greek goddess of the earth even though he intended the hypothesis to be scientific. Partly because his early treatment included poetic and metaphorical language, some people responded to its possible religious or philosophical meanings, while many scientists were skeptical or dismissive. Now, however, the Gaian concept has entered the mainstream scientific literature. The convening of a Chapman Conference (a focused research symposium) by the American Geophysical Union in 1988 and publication of the proceedings, entitled *Scientists on Gaia*, in 1991 were landmarks in this process.

The Gaia hypothesis continues to cause debate. Is it more than a description of what is and has been? Is it more than metaphorical? Is it testable or falsifiable? Or, as a theory, like *plate tectonics or *evolution, can Gaia be accepted by scientists as paradigmatic? Moreover, because it refers to the earth as a living organism, the Gaia hypothesis leads to discussion of the grand question, What is life?

James Lovelock, *The Ages of Gaia. A Biography of Our Living Earth* (1988). Tyler Volk, *Gaia's Body: Toward a Physiology of the Earth* (1998).

JOANNE BOURGEOIS

GALAXY. In 1918 the American astronomer Harlow Shapley argued convincingly that our galaxy is a hundred times larger than previously believed and that the sun is not at its center. Shapley also concluded, erroneously, that the galactic universe made a single, enormous, all-comprehending unit.

A fifteen-year-old crime reporter in a tough Kansas oil town, Shapley discovered a Carnegie library and began to read, and in 1907 he entered the University of Missouri. He earned bachelor's and master's degrees in astronomy at Missouri, and in 1913, a Ph.D. at Princeton University. From there he went to the Mount Wilson Observatory. Using its 60-inch telescope, then the best in the world, Shapley conducted his revolutionary research on the scale of the galaxy. With his quick mind and a complete absence of humility, Shapley followed his own results while ignoring others.

The key to measuring distances was the period-luminosity relation for Cepheid-type variable stars, so named after the constellation Cephus, in which a typical Cepheid star is located. Cepheids

The famous great spiral nebula in the constellation Andromeda and its two small companions, as photographed at the Lick Observatory by N.U. Myall.

are giant stars visible for great distances. The relationship between their periods and luminosities, noted by the American astronomer Henrietta Leavitt in 1908, states that the longer the period, or time, from maximum brightness to minimum and back to maximum, the greater the luminosity of the star. (The observed, or apparent, brightness of a star decreases by the square of its distance from the observer. The absolute magnitude, a measure of brightness, is defined as the magnitude the star would have at a standard distance, approximately 32 light-years.) Shapley collected all the available data on Cepheid stars whose distances and magnitudes he knew, and plotted period against absolute magnitude. Next, he made the reasonable assumption that Cepheids in distant globular clusters resemble the nearby Cepheids upon which he based his period-magnitude curve. Globular clusters, 150 of which exist in our galaxy, are densely packed balls of hundreds of thousands of stars. Shapley observed the periods of Cepheids in globular clusters, read off their presumed absolute magnitudes from his graph of period against magnitude, and compared that absolute magnitude with the observed apparent magnitude. This calculation gave him the distance to the Cepheid variable star and to the globular cluster in which it resided.

Making the assumption that globular clusters constitute a galactic skeleton, Shapley determined the galactic outline, its size, and the place of the solar system within it. He put the Sun far toward one edge of the galactic plane, and made the size of the galaxy far larger than any previous estimate. He concluded that "globular clusters, though extensive and massive structures, are but subordinate items in the immensely greater organization which is dimly outlined by their positions. From the new point of view our galactic universe appears as a single, enormous, all-comprehending unit, the extent and form of which seem to be indicated through the dimensions of the widely extended assemblage of globular clusters." Shapley's big galaxy model argued against the existence of other galaxies. Even the most remote globular cluster should be inside our galaxy. Spiral-shaped *nebulae (concentrations of stars and dust) might lie outside our newly enlarged galaxy, but if so, their size would be implausibly large. Shapley made this point in the so-called "Great Debate" before the National Academy of Sciences on 26 April 1920.

In 1921, Shapley left Mount Wilson for the Harvard College Observatory, leaving Edwin *Hubble with the new 100-inch telescope, completed in 1918, to hunt for Cepheids in spiral nebulae. On 19 February 1924, Hubble wrote to Shapley that he had found a Cepheid in the Andromeda Nebula, which (by the period-distance relationship) he reckoned to be roughly a million light years away, far beyond Shapley's limit for our galaxy. Reading Hubble's letter, Shapley remarked to a colleague who happened to be in his office: "Here is the letter that has destroyed my universe." Shapley had enlarged our galaxy a hundred times. Showing that spiral nebulae are galaxies far outside our galaxy, Hubble made the universe larger yet, and shattered Shapley's all-comprehending unity.

Harlow Shapley, *Galaxies*, rev. ed. (1967). Richard Berendzen, Richard Hart, and Daniel Seeley, *Man Discovers the Galaxies* (1976). Robert W. Smith, *The Expanding Universe: Astronomy's "Great Debate" 1900–1931* (1982). Michael J. Crowe, *Modern Theories of the Universe from Herschel to Hubble* (1994). Norriss S. Hetherington, *Hubble's Cosmology: A Guided Study of Selected Texts* (1996).

NORRISS S. HETHERINGTON

GALENISM. Few people living in western Europe between 500 and 1100 had heard of Galen (129–c.199), but the Byzantine and Muslim worlds of that period based their formal medicine on his ideas, not least because his monotheism and teleology made his writings acceptable within a new religious environment. Although he forcefully paraded his commitment to a

Hippocratic view of illness (*see* MEDICINE) as resulting from a disturbance in the balance of an individual's four basic humors, the sheer size of his oeuvre and his tendency to enunciate partially developed, and at times contradictory, ideas presented problems for successors who stood in awe of his achievement. The difficulty of forming a coherent Galenic system grew ever greater from the fourth century onward as medical treatises written from an alternative standpoint were either gradually subsumed under his name or discarded as superfluous.

In late antique Alexandria (by 600 A.D.) there already existed a syllabus of selected works of Galen, which was then transferred, through translations, to the Syriac world and, in the ninth century, to the Islamic world. Here, leading physicians, notably Ibn Sīnā (known in the West as Avicenna), produced increasingly systematized and logically organized compendia with an emphasis on their theoretical content. Ibn Sīnā made the three bodily systems, centered on the brain, heart, and liver, parallel in every respect. Islamic physicians extended Galen's comparisons of the strengths of drugs from a few hundred "simples," simple herbal or mineral drugs, to all aspects of pharmacology. Galen also became the model of the ideal physician—learned, logical, and a marvel of a diagnostician. Rather than criticize his theories and results, scholars strove to reconcile apparently divergent notions. His system lives on as Yunani ("Ionian") medicine in many parts of the Muslim world today.

From the twelfth century onward, this Arabized Galenism passed to western Europe, where it dominated the medical schools in the new universities. The rediscovery, albeit in Latin translation, of Galen's own texts gave new direction to medical thought in the early fourteenth century, including the (sporadic) revival of dissection as a tool for understanding the body. But most of Galen's writings were not available in Latin, and the interpretation of those that were was heavily influenced by the Arabs.

The medical humanists of the late fifteenth and early sixteenth centuries demanded a return to Galen's original Greek. The Aldine *Editio princeps* (1525) and the subsequent flood of new Latin translations made accessible texts that had long been forgotten, especially in anatomy and philosophy—for example, *On the Opinions of Hippocrates and Plato; Exhortation to Medicine*. By going back beyond the Arabs to the Greek Galen, Renaissance doctors could maintain the traditional authority of their medical doctrines while improving upon their medieval predecessors. They could also see, for the first time for centuries, Galen's emphasis on anatomical dissection and his insistence on the unity of thought and experience. Many, like the Englishman John Caius, supported the introduction of regular dissections into university courses on the Galenic model.

But the rediscovery of Galen also revealed dissonances and errors within his writings, not all of them the fault of translators or copyists. Andreas Vesalius, in his *De humani corporis fabrica* (1543), although following a Galenic program of anatomical research, concluded that Galen's results were flawed because he dissected only animals. In physiology, William *Harvey used Galenic and Aristotelian ideas to prove the non-Galenic concept of the circulation of the blood (1628). By 1700 Galenist doctrines flourished only in hygiene and semiotics, but even there Galen's categories, such as temperament, were reinterpreted along different lines. Galenism came to represent everything bad in ancient medicine: theory divorced from observation, crude teleology, and a preference for logical pedantry over bedside medicine. Galen could be safely left to classicists and historians.

O. Temkin, *Galenism: Rise and Decline of a Medical Philosophy* (1973). V. Nutton, ed., *Galen: Problems and Prospects* (1981). A. Wear, R. K. French, and I. M. Lonie, eds., *The Medical Renaissance of the Sixteenth Century* (1985).

VIVIAN NUTTON

GALILEO [GALILEI, Galileo] (1564–1642), natural philosopher and astronomer, discovered the laws of falling bodies and telescopic evidence for Copernicanism.

Galileo was born in Pisa and died near Florence. In 1581 he enrolled at the University of Pisa, studying mostly mathematics but leaving in 1585 without a degree. For several years he did private teaching and independent research. In 1589 he was appointed professor of mathematics at Pisa, and then from 1592 to 1610 at Padua.

In his Pisan and Paduan years, he studied motion, especially that of falling bodies; opposed Aristotelian physics and followed an Archimedean approach; and pioneered experimentation as a procedure combining empirical observation with quantitative mathematization and conceptual theorizing. By this procedure he discovered an approximation to the law of inertia, the composition of motion into component elements, the laws that in free fall the distance fallen increases as the square of the time elapsed and the velocity acquired is directly proportional to the time, the isochronism of the pendulum, and the parabolic path of projectiles. He did not publish any of these results then, however, and a systematic account appeared only in the *Two New Sciences* (1638).

The frontispiece to the first collected edition of Galileo's works (1656), which omitted the *Dialogues* that got him into trouble. It shows him presenting his telescope to the muses and pointing to the sun surrounded by the planets. The crown-like band above the upper central planet represents the orbits of Jupiter's moons, whose eclipses Galileo tried to adapt to finding longitude at sea (hence the ship on the left). The cannon indicates the analysis of shell trajectories in the *Discourses on Two New Sciences* (1638).

A main reason for this delay was that in 1609 Galileo became actively involved in astronomy. Until then, although he appreciated the merits of Nicholas *Copernicus's theory, he was more impressed by the evidence against it. He was especially troubled by the astronomical consequences that could not be observed, such as the similarity between terrestrial and heavenly bodies, Venus's phases, and annual stellar parallax. In 1609 he perfected the telescope enough to make several discoveries, which he immediately published in a little book, *The Sidereal Messenger* (1610): lunar mountains, stars invisible to the naked eye, the stellar composition of the Milky Way and of nebulas, and Jupiter's moons. He became a celebrity. He resigned his professorship, became Philosopher and Chief Mathematician to the grand duke of Tuscany, and moved to Florence. Soon thereafter, he discovered Venus's phases and sunspots.

Galileo judged that his telescopic discoveries strengthened Copernicanism, but not conclusively: there was still astronomical counterevidence (no annual stellar parallax), and the objections drawn from Aristotelian physics had not yet been answered. He conceived a work treating all aspects of the question, but he did not finish it for two decades. It appeared in 1632 as a *Dialogue on the Two Chief World Systems*.

Galileo delayed because he came under attack from conservative philosophers and clergymen, who charged him with heresy for holding a belief contrary to Scripture. They argued that the earth cannot move because many scriptural passages state or imply that it stands still and Scripture cannot err. He defended himself by counterarguing that Scripture is not a scientific authority and does not really favor geostaticism over geokineticism. Responding to a complaint, the Inquisition of the Roman Catholic church launched an investigation, which yielded two results in 1616. Galileo received a private warning forbidding him to hold or defend the earth's motion. Furthermore, the church published a decree declaring the earth's motion to be physically false and contrary to Scripture, condemning attempts to show otherwise, and banning Copernicus's book of 1543 until revised. Issued in 1620, these revisions eliminated suggestions that the earth's motion is physically true and compatible with Scripture and conveyed the impression that Copernicus treated the earth's motion merely as a hypothesis useful for astronomical calculations.

Galileo kept quiet until 1623, when an old admirer became Pope Urban VIII. This encouraged Galileo to write the *Dialogue*, which showed that the pro-Copernican arguments were much stronger that the progeostatic ones. However, his enemies charged that the book defended the earth's motion, something he had been expressly forbidden to do. He was summoned to Rome to stand trial, which began in April 1633. At the first interrogation, Galileo denied that his book defended the earth's motion, claiming instead that it showed the geokinetic arguments to be inconclusive. There followed an out-of-court meeting during which he was persuaded to plead guilty to this charge in exchange for leniency. At the next deposition he admitted having defended the earth's motion but insisted that he did so unintentionally. The trial concluded with a sentence that did not exhibit the promised leniency: the cardinal-inquisitors, instructed by the Pope, found Galileo guilty of "vehement suspicion of heresy," an intermediate category of theological crime; forced him to recite an abjuration, expressing sorrow and cursing his errors; sentenced him to indefinite house arrest; and banned the *Dialogue*.

The interpretation and evaluation of Galileo's work are highly controversial. Nevertheless, his contributions to physics, astronomy, and methodology were epoch-making. Add to them the mythological status into which he was catapulted by the tragic trial, and he may properly be regarded as the "father of modern science."

Maurice Clavelin, *The Natural Philosophy of Galileo* (1974). Stillman Drake, *Galileo at Work* (1978). Alexandre Koyré, *Galileo Studies* (1978). Maurice A. Finocchiaro, *Galileo and the Art of Reasoning* (1980). William A. Wallace, *Galileo and His Sources* (1984). Mario Biagioli, *Galileo Courtier* (1993).

MAURICE A. FINOCCHIARO

GALVANI, Luigi (1737–1797), physician, natural philosopher, and investigator of animal electricity, and **Alessandro VOLTA** (1745–1827), natural philosopher and inventor of the battery.

While belonging to very different cultural milieus, Galvani and Volta shared an interest in both electricity and physiology, not uncommon during the second half of the eighteenth century. After the publication of Galvani's treatise *On the forces of electricity in muscular motion* (1791), their area of common interest commanded the attention of expert "electricians," physicians, and a wide public throughout Europe, because of their controversy over animal electricity or "galvanism." The controversy magnified their contemporary fame and has since nurtured arguments between physicists (siding with Volta) and biologists (siding with Galvani) as well as historians, many of whom have overemphasized the conflicting traits of their sciences and personalities.

Luigi Galvani was born in Bologna, then part of the Papal States, the son of Domenico, a goldsmith, and Barbara Caterina Foschi. From 1755 he studied medicine and philosophy at the University of Bologna. He also attended the courses in physics, chemistry, and natural history offered by the Bologna Institute of Sciences. He graduated in 1759 and began to practice medicine, including surgery, as he would do throughout his life. In 1762 he began an academic career, which enabled him to become lecturer of practical anatomy at the Institute in 1766.

In 1768 Galvani performed as the anatomist in charge of the "carnival anatomy," an annual celebration during which the body of a recently executed criminal was dissected to the bones in sixteen steps. Besides the professorate and the students, the event's audience included the local elite. Galvani performed the public anatomy three more times, increasing his fame as well as furthering his career: in 1771 he became a member of the powerful College of Physicians, in 1772 he was made president of the Bologna Academy of Sciences, and from 1775 he occupied the university chair in medicine that had been held by his former teacher who—after Galvani had married Lucia Galeazzi in 1762—was also his father-in-law. Lucia often assisted her husband in his experiments.

Deep religious feelings inform the private papers illustrating Galvani's uneventful but productive life. His religious views supported, if they did not prompt, his decision, late in life, not to swear loyalty to the Cisalpine Republic created in the region with the support of the French, led by General Bonaparte. Galvani died in 1797 deprived of his chair and offices.

Galvani's laboratory notes indicate an interest in electricity from 1780. A likely identity of the electrical and nervous fluids inspired his investigations. By exciting with (static) electricity the spinal medulla of frogs freshly killed, and seeing their legs jump as a consequence, he realized that frogs were the most sensitive detectors of electricity available. He devoted the following seventeen years to exploring the tricky interactions between what he regarded as an electricity internal to animals, in the form of their nervous fluid, and the electricity external to them generated by atmospheric electricity or by instruments like an electrostatic machine, a *Leyden jar, or a Franklin square.

Alessandro Volta was born in the commercial town of Como, northern Italy, then part of Austrian Lombardy, a younger son from a family of the lesser nobility. He did not attend university, but as a pupil of the Jesuits developed a lifelong interest in natural philosophy, combining it with a commitment to Enlightenment culture then fashionable among the educated classes and the public administrators of Lombardy. Volta began his international correspondence on scientific and literary topics at the age of eighteen. At twenty-four he published an ambitious treatise "on the attractive force of the electric fire." At thirty he embarked on a career as a civil servant and a teacher in the recently reformed educational institutions of Lombardy. Appointed professor of experimental physics at the University of Pavia in 1778, he held the position until 1820 and managed to build there one of the finest physics laboratories of his time.

Unlike Galvani, who communicated only irregularly with his peers outside Bologna, Volta traveled and corresponded extensively, sharing his enthusiasm for natural philosophy with colleagues in many countries. After his intention to marry an opera singer had brought him close to rupture with his family (and had got him a reprimand from no less than the Emperor of Austria), in 1794 he married the cultured and well-to-do Teresa Peregrini, nineteen years his junior. They had three children, to whose education they attended personally.

By the 1780s Volta had won European fame as an electrician, particularly as an inventor of electrical machines and as a chemist specializing in the chemistry of airs. Before the controversy with Galvani began, Volta had invented the electrophorus, eudiometer, electric pistol, *condensatore* (a device that made weak electricity detectable), and straw electrometer. Volta's contributions to the science of electricity included the notions of tension, capacity, and actuation (an ancestor of electrostatic induction). Thanks to painstaking measurement techniques refined throughout his life, Volta managed to combine these notions into simple quantitative laws. By the 1780s some regarded him as "the Newton of electricity."

Volta was intrigued by Galvani's electrophysiological treatise: it touched on his long-cultivated interest in the Enlightenment theme of the "mind of animals." After repeating and varying Galvani's experiments, Volta rejected a special electricity in animals and developed a theory explaining galvanic phenomena as a consequence of ordinary electricity set in motion by the contact of different metals. The controversy that ensued stirred a crescendo of investigative and theoretical ingenuity on both sides. In 1794 Galvani managed to make his frogs jump without recourse to metals. In 1796 Volta decided to do without frogs as measuring devices. In 1797 Galvani provoked his frogs to jump via a circuit in which only nerves intervened. Towards the close of 1799, Volta managed to build a mostly metallic device—the first electric *battery—that produced a steady flow of electricity while also imitating the electric organs of an animal, the torpedo fish.

Volta's apparatus and its chemical performances (*see* ELECTROLYSIS) made headline news in the European daily press. Bonaparte brought Volta to Paris and rewarded him generously before Europe's cultured elites, then worried by French ambitions. Already in 1794 the Royal Society of London had awarded the Copley Medal to Volta, the first foreigner to achieve the honor. Volta died in 1827 in Como, retaining the title of count conferred on him by Bonaparte, whose star had in the meantime fallen.

Through avenues conspicuously unforeseen by their initiators, Galvani had set in motion what

would become electrophysiology, while Volta had laid the foundations of electrochemistry, *electromagnetism, and a new industrial era.

Marcello Pera, *The Ambiguous Frog: The Galvani-Volta Controversy on Animal Electricity* (1992). John L. Heilbron, *Electricity in the 17th and 18th Centuries* (1999). Giuliano Pancaldi, *Volta: Science and Culture in the Age of Enlightenment* (2003).

GIULIANO PANCALDI

GALVANISM. In 1791, Italian anatomist Luigi Galvani published experimental evidence of an "animal electricity" stored in the muscles of frogs' legs. The main evidence was that joining the leg and the appropriate nerve of a freshly flayed frog by an arc made of two different metals made the remains of the animal jump. Soon investigators all over Europe were making dead frogs dance and debating the cause of the phenomenon.

Alexander von *Humboldt showed that a circuit of nerve and muscle could produce the effect, and suggested that the source lay in the organic elements; Alessandro *Volta identified the causal agent as ordinary electricity generated by the contact of dissimilar materials.

Before 1800, the term galvanism primarily referred to a subtle fluid that flowed through living matter. After Volta's invention of the electric *battery, which supplied a constant source of galvanism (1800), the term came to refer to the phenomena it produced. Since the galvanism of Volta's pile had all of the essential characteristics of the *electricity produced by electrostatic generators, including producing sparks and powerful effects on humans, the two were gradually identified as one. The vestiges of the term galvanism can still be seen in the name galvanometer, an instrument developed over the first decades of the nineteenth century to measure the effects of the pile, and used today to measure electric current. By the 1880s, galvanism referred exclusively to the technology of inexpensively coating objects with a thin metallic layer by passing electricity through a solution of metallic salt. Only the colloquial use of the verb "galvanize," to jolt into action, recalls the origin of current electricity in experiments on frogs.

A variety of quackish medical applications grew out of the early galvanic experiments. In 1793, Carl Creve proposed to use a galvanic arc to detect whether a body was genuinely dead or only apparently so. In the United States, Frances Perkins patented a galvanic arc, which he called a "tractor," for healing a host of ailments. Mary Shelley's Dr. Frankenstein (1818) displays a thorough knowledge of galvanic effects.

The original conception of galvanism as a powerful and mysterious force that seemed to

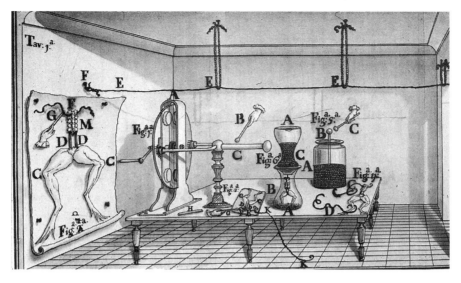

Diagram of the apparatus with which Galvani studied the electrical stimulation of muscular action. This illustration served as the basis for the better-known but cruder engravings in Galvani's memoir of 1791.

have an important role in both organic and inorganic phenomena owed much to German *Naturphilosophie*, which emphasized qualitative and mysterious phenomena and the interconnection of physical forces.

Marcello Pera, *The Ambiguous Frog: The Galvani-Volta Controversy on Animal Electricity* (1992).

MARIA TRUMPLER

GAS DISCHARGE. See CATHODE RAYS AND GAS DISCHARGE.

GAUSS, Carl Friedrich (1777–1855), German mathematician, astronomer, and physicist.

Gauss was born in Brunswick; his father was a manual laborer and, before her marriage, his semiliterate mother had worked as a maid. Against the wishes of his father, who wanted Gauss to learn a proper trade, the mathematically precocious boy (he could calculate before he could speak) was encouraged by his teachers to study. At the age of fourteen he was presented to the Duke of Brunswick, Karl Wilhelm Ferdinand, who provided him with financial support until 1806. From 1792 to 1795 Gauss used the ducal subvention to study in the Collegium Carolinum, a preparatory school for university study, where he developed a variety of brilliant mathematical ideas. Between 1795 and 1798 he attended the University of Göttingen, where he read the mathematical classics and found that many of his discoveries were not new.

Gauss was a gifted linguist, and he did not know at first what path to follow. He decided on mathematics when in 1796 he

solved the ancient problem of constructing a regular polygon with seventeen sides using only compasses and a straight edge. In 1798 he returned to Brunswick, where he worked as a private scholar and produced his finest mathematics, which was in number theory. He obtained his doctorate in absentia in 1799, nominally under Johann Friedrich Pfaff at the duchy's University of Helmstedt, after submitting some earlier work on the first of his four proofs of the fundamental theorem of algebra. With publication of his great work, *Disquisitiones arithmeticae* (1801), Gauss became, according to his older contemporary Joseph-Louis Lagrange, the leading mathematician of Europe.

Gauss next turned his attention to astronomy, following the discovery of the minor planet Ceres by Palermo astronomer Giuseppe Piazzi on New Year's Day 1801. Piazzi soon lost track of his planetoid, but from the few observational data he had provided, Gauss determined the orbit and, when Piazzi found Ceres again, it was in the position Gauss predicted. For this extraordinary feat Gauss used an improved orbit theory as well as the method of least squares, which he had previously developed to assist in analyzing geodetic measurements. This work led to his second remarkable book, *Theoria motus corporum coelestium* (1809), which showed him to be one of Europe's finest mathematical astronomers.

With the death of his patron in 1806 following defeat by Napoleon's troops, Gauss had to look for a new source of income. In 1807 he accepted a position at Göttingen University as professor

The portrait shows Gauss in middle age in a costume not unusual for a savant of the time.

1831 until 1837, he established a magnetic observatory. In 1834 he organized the Magnetic Association, the first major scientific project based on international cooperation. This association, set up to conduct Europe-wide geomagnetic observations, was subsequently expanded by Humboldt into a worldwide network. Gauss summarized the results in a joint publication with Weber, *Allgemeine Theorie des Erdmagnetismus* (1839), followed by an *Atlas* (1840). Gauss also applied himself to the improvement and invention of scientific instruments, developing the heliotrope (a sextant-like instrument that uses reflected sunlight for geodetic measurements), the bifilar magnetometer (in which two silk threads are used for measuring geomagnetic force), and in 1833, together with Weber, the first electrical telegraph.

Gauss's motto was *pauca sed matura*, yet his many (323) "mature" publications were "few" only in relation to the multitude of his ideas. (According to an exact historian of mathematics, Gauss had precisely 404 ideas, 178 of which he discussed in print.) He has been compared to Archimedes and Isaac *Newton, and in the ranking game—which mathematicians play with much greater accord than do other groups of academics—Gauss has been classed as a credible candidate for *princeps mathematicorum*. A preoccupation with his brilliance has deflected many of his biographers from properly historicizing him. The historical conditions under which his talent found a niche, and his work found its initial acceptance, must be sought in the world of German *Kleinstaaterei* (the politics of patchwork-quilt principalities before the 1871 unification of Germany). He remained loyal to the enlightened despots and university officials who supported him. Although not a churchman, Gauss became increasingly preoccupied with the metaphysical issue of immortality and believed staunchly in life after death. He was married twice; the first marriage (1805–1809; one son and one daughter) was happy, the second (1810–1831; two sons and a daughter) less so.

of astronomy and director of the Observatory. By this time he had already produced his main mathematical work and had toyed with non-Euclidean geometry, but he kept the results by and large to himself. Much of his time and energy went into equipping and running the Observatory and into teaching. Gauss did not enjoy teaching, yet among his pupils were such outstanding mathematicians as Moritz Benedikt Cantor, Richard Dedekind, and Bernhard Riemann.

Gauss applied his unequalled facility with numbers to various scientific projects, most notably in his geodetic survey of the Kingdom of Hanover, to which Göttingen belonged. From 1818 until 1825, he personally led the triangulation fieldwork and subsequently supervised the data conversion, making more than a million calculations. Among the resulting theoretical advances was his method of overcoming the difficulties of mapping the terrestrial ellipsoid on a sphere (conformal mapping), as described in his *Disquisitiones generales circa superficies curvas* (1828).

In 1828, at the request of Alexander von *Humboldt, Gauss took part in the only scientific convention he ever attended, a meeting of the Gesellschaft deutscher Naturwissenschaftler und Ärtze in Berlin. It inspired him to develop his long-standing interest in earth magnetism. Together with the physicist Wilhelm Eduard Weber, who was professor at Göttingen from

Wolfgang Sartorius von Waltershausen, *Gauss, a Memorial* (1856; trans. 1966). G. Waldo Dunnington, *Carl Friedrich Gauss: Titan of Science* (1955). Kenneth O. May, "Gauss, Carl Friedrich," in *Dictionary of Scientific Biography*, vol. 5 (1972): 298–315. Karin Reich, *Carl Friedrich Gauss, 1777–1977* (1977). Walter K. Bühler, *Gauss: A Biographical Study* (1981). Horst Michling, *Carl Friedrich Gauss: Aus dem Leben des Princeps Mathematicorum*, 2d ed. (1982).

NICOLAAS A. RUPKE

GEIGER AND ELECTRONIC COUNTERS.

Twentieth-century physicists first built electronic counters to detect charged particles from *radioactivity, then turned them to *cosmic ray research, and later combined them in arrays for high-energy particle experiments. In 1908 Hans Geiger built a cylindrical capacitor to help Ernest *Rutherford determine the charge on alpha particles. A thin wire with a high voltage ran down the center of a brass cylinder filled with carbon dioxide. A charged particle passing through the chamber would ionize some of the gas in proportion to its charge, setting up a current of ions traveling toward the wall of the tube and electrons toward the center wire; an electrometer needle registered the resultant voltage step. Increasing the voltage on the capacitor increased its sensitivity by creating an avalanche of electrons. The quantity of these secondary electrons depended on the speed and nature of the primary particles. Geiger's initial counter could count up to ten particles per minute and confirmed the double charge of the alpha particle.

Geiger subsequently refined the device by using a sharp needle instead of a coaxial wire for an anode; the additional sensitivity could detect electrons as well as alpha particles, and even photons via secondary electrons knocked out of atoms. But the Geiger counter also registered random, spontaneous counts, even with no nearby radioactive source and especially with maximum sensitivity. In 1928 Walther Müller, a postdoctoral assistant to Geiger at the University of Kiel, designed a still more sensitive counter and found that the spontaneous discharges were caused not by contamination of the gas or metal, but by cosmic rays passing through the chamber. Müller's work turned an apparent defect of the Geiger counter into an advantage, and Geiger counters (or, as they were sometimes called, Geiger-Müller counters) became a prime tool in the fruitful program of cosmic ray physics in the 1930s.

Walther Bothe soon incorporated Geiger counters into coincidence circuits, two counters wired together so as to register a particle when both counters fired together. Bothe and Werner Kolhörster at first measured counts optically, using the electronic output to nudge a mirror and photographic film to record the reflected light. Bruno Rossi, an Italian physicist visiting Bothe's institute, dispensed with the film. A simple logic circuit designed by Rossi ensured that if and only if two counters fired, then a voltage difference on the circuit would result. An experimenter could expand Rossi's circuit to more intricate arrays of counters, firing in coincidence or anticoincidence, with logic gates in the circuit to sort out the particles produced in cosmic ray showers. For example, by measuring the counting rate with and without lead blocks inserted between three

Geiger counters, one could determine whether particles originated from the sky or the lead block.

The radar and atomic bomb projects in the United States in World War II developed new amplifiers, pulse-height analyzers, and other electronic devices that would find use in electronic detectors. Rossi worked at Los Alamos on fast timing circuits to measure nuclear processes, such as the time between the emission of prompt and delayed neutrons from nuclear fission. Nuclear energy also provided a market for Geiger counters as radiation detectors. Following the development of vacuum tube amplifiers by the radio industry, Geiger counters were connected to audio speakers to register particle counts as clicks. A popular image of the nuclear age featured the crackle of Geiger counters betraying the presence of nuclear radiation.

After the war, electronic counters became both high art and big science in detectors for high-energy physics. Particle physicists wired new sorts of counters into electronic logic circuits: first scintillation and *Cherenkov counters, which converted flashes of light into electronic pulses, and later spark chambers, a sort of flattened Geiger counter with a parallel-plate configuration, and drift chambers and multiwire proportional counters, both of which worked according to the same principle as the Geiger counter. All of these counters generally recorded electronic data instead of the visual output of *cloud and bubble chambers, and used high particle counts to provide statistical evidence of phenomena instead of snapshots of individual events. By the 1970s, physicists were using computers to reconstruct particle tracks from complex collections of electronic counters.

Thaddeus J. Trenn, "The Geiger-Müller Counter of 1928," *Annals of Science* 43 (1986): 111–135. Peter Galison, *Image and Logic: A Material Culture of Microphysics* (1997). Jeff Hughes, "Geiger and Geiger-Müller Counters," in *Instruments of Science: An Historical Encyclopedia*, eds. Robert Bud and Deborah Jean Warner (1998).

PETER J. WESTWICK

GENDER AND SCIENCE. Gender studies of science began perhaps with Christine de Pizan's *Book of the City of Ladies* (1405), but took off in the 1980s with the appearance of Carolyn Merchant's *Death of Nature* (1980) and Evelyn Fox Keller's *Reflections on Gender and Science* (1985). This field of study, closely allied with the gender studies of medicine and technology, typically includes the history of women in science, the science of woman (or sexual science), gender in scientific cultures, and gender in the content of the sciences.

The history of women in science examines social trends and academic structures that have encouraged or discouraged women's full participation. Scholars have analyzed women's place in mainstream institutions of modern science, such as universities and scientific societies. Women were generally admitted to these institutions in the late nineteenth century, to graduate education in the early twentieth century, and to professorships in growing numbers only after 1975. Because women have not flourished in traditional scientific institutions, scholars have identified other ways that women have insinuated themselves into scientific work, including noble networks and artisanal workshops; creative couples; and women's colleges. In addition, definitions of science have been challenged to include activities in which women have historically played a dominant role but which are not usually considered science, such as midwifery. The history of women in science also pays close attention to how political trends and social movements—World War II and the women's movement, for example—have influenced women's participation.

The science of woman (or sexual science) analyzes how science has studied, especially, women's intellect and their potential for creative endeavor. When women's exclusion from modern science was formalized, new justifications, based on scientific, not Biblical, authorities arose. The eighteenth century saw the rise of both scientific sexism and racism—the rigorous study of sexual (and racial) differences—whose results (women's narrow skulls, for instance) were used to argue for women's natural incapacity for scientific pursuits. Feminists have challenged scientific sexism (strong in the 1970s as sociobiology and in the 1990s in the form of evolutionary psychology), often by fighting science with science.

Studies of gender in scientific cultures declared, in the 1980s, that science is "masculine," not only in the person of its practitioners but in its ethos and institutions. A major stumbling block along the road to women's equality has been the presumption that women should assimilate—that they should enter science on its terms, checking talents, traits, and styles not compatible with its cultures at the laboratory door. Scholars have focused on different aspects of the historical conflict between cultural ideals of femininity and of science. Many have traced this conflict to the deep gulf between the public realm of science, presumably bristling with masculine reason, impartiality, and intellectual virility, and the private sphere of domesticity, radiating with feminine warmth, tender feeling, and quiet intuition. Others have sought the origins of the masculine grip on scientific culture in the homosocial bonding said to fire male creativity. Still others have viewed the masculine character of the culture of science as an outgrowth of the early scientific societies and their economy of civility that required men of independent economic and social standing as guarantors of truth. While the notion that science is masculine alienated many female scientists in the 1980s, science institutions in the 1990s implemented reforms to overcome the historical tendency to model scientific culture (and professional culture more generally) on male heads-of-households. These reforms included partner hiring, flexibility in the tenure track, parental leave, the use of inclusive language, lactation rooms, expanded day care, and textbook revisions to include women and their accomplishments.

Finally, feminists have uncovered the way gender has molded particular aspects of specific sciences. They have demonstrated that gender inequalities, built into the institutions of science, are often found also in the knowledge issuing from those institutions. The 1980s saw efforts to develop a "feminist science," typically a science built on neglected feminine values variously styled as cooperation, a feeling for the organism, or reflexivity. The 1980s and 1990s also saw efforts among scientists to identify "how women do science differently," based on the notion that previously excluded groups (women and minorities) bring culturally wrought unique standpoints with them into science. This, it was thought, would foster scientific creativity. Today, a number of scholars have shifted the focus to how feminists—both men and women—can change science by mainstreaming gender analysis into normal science.

The Women's Health Movement, along with the development of gender analytics, for example, enabled scientists to understand how females have largely been excluded from drug testing in the United States. Documenting the dangers that arise from not knowing how particular medications metabolize in females was one factor leading to the founding in 1990 of the Office of Research on Women's Health at the National Institutes of Health. Between 1990 and 1994, the U.S. Congress enacted no fewer than twenty-five pieces of legislation to improve the health of American women. The right to have public funds used fairly for developing medicine that will benefit both men and women is now guaranteed by federal law.

Gender analysis has not, however, yielded results uniformly across the sciences. Instances of gender can be documented in the humanities, social sciences, medical and life sciences, where research objects are sexed or easily imagined to have sex and gender. The physical sciences have by and large resisted gender analysis, perhaps because of the extremely small number of people trained in a physical science and in gender studies. It may be premature to conclude, as some do, that there simply are no gender dimensions to uncover. Rather, the question of gender in the physical sciences remains open and will require sustained and careful research. A critical understanding of how gender operates in science acts as yet another filter to remove bias from science and has the potential to introduce new questions and directions into science.

David Noble, *A World Without Women: The Christian Clerical Culture of Western Science* (1992). Elizabeth Fee and Nancy Krieger, eds., *Women's Health, Politics, and Power* (1994). Margaret Rossiter, *Women Scientists in America: Before Affirmative Action, 1940–1972* (1995). Evelyn Fox Keller and Helen Longino, eds., *Feminism and Science* (1996). Helena Pycior, Nancy Slack, and Pnina Abir-Am, eds., *Creative Couples in the Sciences* (1996). Londa Schiebinger, *Has Feminism Changed Science?* (1999). Angela Creager, Elizabeth Lunbeck, and Londa Schiebinger, eds., *Feminism in Twentieth-Century Science, Technology, and Medicine* (2001).

LONDA SCHIEBINGER

GENERATION. Until the early nineteenth century the term "generation" indicated the production or coming into existence of living beings, their growth and the phenomena of regeneration. Aristotle (384–322 B.C.) coupled the notion of generation with that of corruption to describe the cycle by which earthly things—plants, animals, minerals—come into being, acquire form (which they subsequently lose), and then disintegrate. He postulated that a vegetative soul regulated the generation, nutrition, and growth of all living things. He also described the generation of plants from seeds, recorded the breeding habits of animals, investigated the embryonic development of the chick, distinguished between oviparous and viviparous animals, and accepted the idea that some insects arose spontaneously from rotting matter. He maintained that, in the generation of human beings, the male provided the seed and the female the menstruum. In keeping with his conception of the higher rank of the male, Aristotle rated the seed as the efficient cause of generation, and the menstruum as the material cause. Because of its metaphysical foundation and rich empirical analysis, Aristotle's theory of generation exerted enormous influence until the mid-seventeenth century, when misgivings over the ancient doctrine of fecundation, conceived as the merging of the two seminal fluids, and controversies over which of the fluids determined the form of the future organism inspired alternative theories of generation in which the role of fecundation remained profoundly enigmatic.

In 1651 William *Harvey stated that no mass of semen, menstrual clot, or their mixture could be detected in the dissected uteruses of mammals after intercourse. He thus disproved the Aristotelian and Galenic doctrines of fecundation on an observational basis and postulated that all animals are produced from eggs that originate independently in the female and carry a vegetative soul. The male's contribution, Harvey contended, consisted in imparting a stimulus, the animal soul, that enabled the egg to develop into a new individual endowed with motion and sensation. Harvey's observations and ideas, purged of Aristotelian terminology, were used by supporters of both *epigenesis and preformation, the two most widespread theories of generation of the seventeenth and eighteenth centuries.

By the end of the seventeenth century, most natural philosophers agreed that all plants originate from seeds and all animals from eggs. In animals, however, the specific contribution of the male to the generating process remained enigmatic. When Antoni van Leeuwenhoek reported in 1677 the discovery of animalcules (spermatozoa) in semen, these were interpreted by some authors as miniaturized individuals (animalculism), but mostly as parasites of the testis. Carl *Linnaeus opposed preformation theories and in his systematic analogies between the sexual organs of plants and animals equated pollen, or fecundating dust, to male sperm. He argued that generation should be understood as a sexual process and as the universal method by which animals and plants propagate. The universality of sexual generation was contested by scholars, for instance Charles Bonnet, who discovered parthenogenesis in aphids in 1740, and several French botanists of the late eighteenth and early nineteenth centuries, who claimed that cryptogamic plants did not reproduce sexually.

In 1749 Georges-Louis Buffon introduced the term "reproduction" as an alternative to "generation." He argued that what distinguished living from nonliving matter was its organization, which had the unique property of reproducing itself. All living beings, he maintained, had the power of reproduction by copulation and thus perpetuated the species to which they belonged. He insisted that, although reproduction was a fact that could not be explained, its modes of action could be studied. He set out a general mechanistic explanation based on the notions of primitive "living organic molecules" and an "internal mold" with which he accounted for most of the phenomena then under discussion, like growth, fecundation, embryonic development, hybrids, monsters, heredity, and regeneration of the *polyp. During the first half of the nineteenth century, "reproduction" replaced "generation" as the central concept of the life sciences after Jean-Louis Prevost and Jean-Baptiste-André Dumas had shown in 1824–1825 the essential role of sperm fertilization and Karl Ernst von Baer had demonstrated in 1827 the mammalian ovum.

Elizabeth B. Gasking, *Investigations into Generation 1651–1828* (1967). John Farley, *Gametes & Spores: Ideas about Sexual Reproduction 1750–1914* (1982). Jacques Roger, *Buffon: A Life in Natural History* (1997). Jacques Roger, *The Life Sciences in Eighteenth-century French Thought* (1998).

RENATO G. MAZZOLINI

GENERATOR, ELECTRIC. See ELECTROTECHNOLOGY.

GENETIC MAPPING. The several scientists who rediscovered Gregor Mendel's experiments on plant hybridization around 1900 believed that he had established the independence of the various traits of plants. "[The] behavior of any two of the differentiating marks in hybrid combination is independent of the other distinguishing marks of the two strains of the plant." Breeding experiments on other organisms than those Mendel studied soon showed, however, that not all pairs of distinguishing characters segregated completely independently of other pairs. In some cases combinations represented in the first hybrid generation remained together in subsequent generations; in other cases they did not always stay together, but did so more frequently than would occur by chance. Moreover, when Walter Stanborough Sutton and Theodor Boveri independently drew attention to the parallelism in the behavior of chromosomes observed in cell division and the Mendelian factors in breeding experiments, each noticed that there were far fewer chromosomes to assort independently than factors assumed to segregate independently.

In 1911 Thomas Hunt *Morgan, who had recently organized a small group of students at Columbia University to conduct genetic experiments on the fruit fly *Drosophila*, suggested that these discrepancies could be reconciled if the Mendelian factors lined up along the chromosomes. Drawing on a hypothesis proposed two years earlier by F. A. Janssens, that homologous chromosomes twist around each other during the reduction divisions of meiosis and may thus exchange material, Morgan proposed that the different degrees of "coupling" observed between the factors could be explained by their relative separations on the same chromosome.

Morgan showed that two factors known to be on the same *Drosophila* chromosome could be recombined by crossing flies containing one or the other of each of the character pairs.

In a conversation with Morgan late in 1911, one of his young assistants, Alfred Henry Sturtevant, realized that the proportion of "cross-overs" between any two factors could serve as an index of the distance between them. Working through the night, Sturtevant calculated the proportions of cross-overs between six sex-linked factors previously identified in the laboratory and showed that they could be represented on a linear diagram. For three factors relatively close together, the proportion of cross-overs between the outer two equaled the sum of the proportions between each of these and the intermediate factor. For greater distances, the proportion of cross-overs for the more distant two fell under the sum. Sturtevant explained the shortfall as a consequence of "double crossovers" in which the middle factor was exchanged, but the outer two remained on the same chromosome.

During the next two years Morgan, Sturtevant, and Calvin Blackman Bridges produced similar linear diagrams of the factors on two more of the four *Drosophila* chromosomes, and another member of the "fly-room" group, Hermann Joseph Muller, did the fourth. Morgan and Sturtevant believed that the linear character of these diagrams indicated that the Mendelian factors are physically located along the chromosomes, although they stressed that the linear

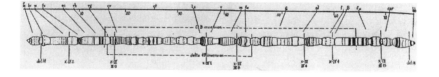

A drawing published in 1934 by the American geneticist T.S. Painter representing the appearance of genes along chromosomes taken from the salivary gland of a fruit fly. These chromosomes, a hundred times larger than those in the germ cells, could be mapped in detail under the microscope.

"linkage" distances might not be proportional to their physical spacing. The more theoretically oriented Muller set out to provide stronger demonstrations. In 1916 he showed that the total linkage distances of each of the four groups in *Drosophila* that segregated independently from the other groups were proportional to the lengths of the four chromosomes. In an elaborate set of experiments for which he constructed a *Drosophila* strain in which the female contained twenty-two different recessive mutations located on two different chromosomes, he determined that all factors on either side of the cross-over segregated together, as would be expected if they constituted segments of the chromosomes exchanged intact. By then the Morgan group was confident of their interpretation of the factors as a "series of beads" of which "whole sections will come to lie, now on one side, now on the other side, in the double chromosome." When the chromosomes separate, the series may break apart "between the beads at the crossing point." Although not all geneticists accepted this conception at first, it soon prevailed.

In *The Mechanism of Mendelian Heredity* (1915) Morgan, Sturtevant, Muller, and Bridges presented the first comprehensive summary of "classical genetics." They placed the linear diagrams or maps showing the spatial locations of the factors of all four of the linkage groups of *Drosophila* side-by-side as a frontispiece to the book. The renaming of what Sturtevant in his original paper had merely called a "diagram" as a "chromosome map" has had long-lasting consequences for the language of *genetics. Calling linear representations of the cross-over frequencies of Mendelian factors "maps" not only shaped the relationship perceived between the linkage diagram and the physical chromosome by analogy to the relation of an ordinary map with a portion of the surface of the earth. Although both the methods of mapping and the conception of the factors mapped (soon renamed "genes") have changed substantially since the early days of classical genetics, the phrase "genetic mapping" has remained firmly in place.

A. H. Sturtevant, "The Linear Arrangement of Six Sex-Linked Factors in *Drosophila*, as Shown by their Mode of Association," *Journal of Experimental Zoology* 14 (1913): 43–59. T. H. Morgan, A. H. Sturtevant, H. J. Muller, and C. B. Bridges, *The Mechanism of Mendelian Heredity* (1915; 1972). A. H. Sturtevant, *A History of Genetics* (1965). Elof Axel Carlson, *The Gene: a Critical History* (1966). Robert E. Kohler, *Lords of the Fly* (1994).

FREDERIC LAWRENCE HOLMES

GENETICS is grounded in the papers on inheritance in peas by the Austrian monk Gregor *Mendel that he published in 1865. It emerged as a field of biology after Mendel's long-ignored work was rediscovered in 1900 by scientists concerned with what might be learned about the mechanism of *evolution from hybridization experiments in plants. In the United States and England, Mendel's laws of biological inheritance were immediately embraced by a number of evolutionary biologists and plant breeders, including William Bateson at Cambridge University, who coined the term "genetics" in 1903. Yet the theory also ran into a good deal of skepticism. The mathematics of Mendelian inheritance seemed to conflict with the one-to-one male-female ratio of sexually reproducing species. Then, too, many characters varied with transmission from one generation to the next and many expressed themselves not as alternates—e.g., tall or short—but in a blended fashion, intermediate between the characters of parents.

However, in 1905, Edmund Beecher Wilson, at Columbia University, and Nettie M. Stevens, at Bryn Mawr, concluded independently that the determination of sex, including the one-to-one sex ratio, was caused in Mendelian fashion by the segregation and reunion of the X and Y chromosomes. In 1909, the Danish biologist Wilhelm Johannsen drew on his studies of heredity in selected lines of plants to propose that variation arises from two sources: one is the influence of environment on the developing organism; the other is slight variation in Mendel's transmitted factors of heredity, which Johannsen called "genes." Johannsen's delineation of these shaping differences gave rise to the concepts of, respectively, "genotype" (the sum of all an organism's genes) and "phenotype" (the characters it displays, which are a product of both its genes and the environment it experiences).

Like Mendel's rediscoverers, Thomas Hunt *Morgan, a biologist at Columbia University, came to the study of heredity through concern with the problem of evolution. Initially skeptical of Mendelism, he became a convert to it after 1910, when he observed a heritable mutation—the appearance of a white-eye—in the fruit fly *Drosophila melanogaster*, whose eyes were normally red. Morgan and a team of three graduate students—Calvin Bridges, Hermann Muller, and Alfred Sturtevant—proceeded to scrutinize the offspring of innumerable generations of fruit flies, finding the creatures convenient for the study of heredity because they reproduce very rapidly and in great abundance and because they are cheap to maintain, requiring only jars for housing and rotting bananas for feed.

The Morgan group demonstrated that a number of characters—not only eye color but, for example, wing shape—are transmitted by genes. They also showed that the genes reside on the flies' chromosomes and that occasionally two members of a chromosome pair exchange parts with each other. Called "crossing over," the phenomenon allows characters from genes on the same chromosome to be inherited independently. The Morgan group recognized that the frequency with which such independent inheritance occurs provides a measure of the relative physical closeness—termed "linkage"—of genes to each other. Using frequency data, they were able to draw the first genetic maps—that is, linear pictures of where in linkage units different genes are located along the chromosome. The Morgan group summarized their findings in a highly influential book, *The Mechanism of Mendelian Heredity* (1915). By this time, studies of inheritance in a number of organisms strongly indicated that many traits, including those of an apparent blending nature, are the product of combinations of genes. In 1933 Morgan was awarded the Nobel Prize in physiology or medicine for his pioneering work.

The increasing success of Mendelism, however, was accompanied by debates among biologists about the nature of the gene. Was it a physical entity or some kind of dynamic organizing principle? Convincing evidence that genes are material entities came in 1927, when Herman Muller, working with fruit flies at the University of Texas, showed that exposing them to X rays would induce genetic mutations. In Berlin in 1935, Max Delbrück, a physicist then turning to genetics, collaborated with two other scientists to try to ferret out information about the actual structure of the gene. Using X rays, they combined experimental data and quantum mechanical theory to account for mutations in terms of the chemical and physical behavior of atoms. Their findings indicated that genes are relatively stable macromolecules susceptible to analysis by physical and chemical methods.

In 1937, at the California Institute of Technology, where he had gone on a fellowship, Delbrück embarked on research with bacteriophage—a virus that infects bacteria and multiplies rapidly inside its host. He saw in this interaction a simple system for the study of genetics, a kind of hydrogen atom for biology. Remaining in the United States because of World War II, Delbrück worked on bacteriophage at Cold Spring Harbor, on Long Island, New York, during the summers with the Italian refugee scientist, Salvador Luria, and Alfred Hershey, a chemist on the faculty of the Washington University Medical School, in St. Louis, Missouri. Their collaboration gave rise to what came to be known as the Phage Group, a small, informal network of viral and bacterial geneticists, and it eventually brought

them a Nobel Prize. The Phage Group produced relatively few experimental results of high significance in microbiological genetics, but under Delbrück's guiding influence it focused attention on the problem of self-replication in living organisms, helped recruit a number of physicists into the field, and fostered the use of physical, chemical, and statistical reasoning in microbiology.

In the early 1940s, while the Phage Group was forming, further evidence that genes are some sort of macromolecule engaged in biochemical functions came from the Nobel Prize–winning research of the Americans George Beadle, a geneticist, and Edward Tatum, a biochemist. As the result of previous work in Paris with the biologist Boris Ephrussi, Beadle thought that genes somehow shape the biochemical pathways that produce different eye colors in fruit flies. At each step along the pathway, an assist is given by an enzyme, an organic catalyst essential to the biochemical transformation in which it is involved. At Stanford in 1940, Beadle and Tatum began to pursue the hypothesis with Neurospora, an ordinary bread mold. They triggered genetic mutations in the mold with X rays and analyzed the resulting metabolic variations. They found that, with a specific gene bred into it, the mold could metabolize a given substance, while with the gene bred out, it could not—in short, that the absence of the gene forced the mold into a metabolic error. In 1945, Beadle spelled out the striking import of their research and recent related work by others: "...that to every gene it is possible to assign one primary action and that, conversely, every enzymatically controlled chemical transformation is under the immediate supervision of one gene, and in general only one." That idea was soon distilled into an apothem—the "one gene-one enzyme hypothesis," which became a guiding principle for the emerging field of biochemical genetics.

All the while, many biologists supposed that genes must be proteins, which were known to make up much of the cell and to catalyze its remarkable range of chemical synthesis. The supposition was reinforced by Beadle and Tatum's work, since it associated genes with enzymes, which are proteins. But evidence to the contrary came from a group at the Rockefeller Institute for Medical Research that was headed by Oswald T. Avery. Avery and his associates had worked for a decade to understand why an infectious form of pneumococcus could transform an uninfectious type into a similarly infectious one. In 1944 they reported that the change was caused by a transforming factor in the infectious bacteria that they identified as a *deoxyribonucleic acid (DNA).

The Avery group's results suggested to some biologists that genetic material might be composed of DNA, but Avery, an older man and a retiring personality, did not push the idea and most biologists remained attached to the belief that genes are proteins. However, members of the Phage Group determined that phage consists of DNA wrapped in a protein coat. In 1952 Hershey, in collaboration with Martha Chase, showed that, when a phage infects a bacterium, it injects its DNA into the bacterial cell, leaving the protein coat on the cell surface. The clear implication of the experiment was that the injected DNA was responsible for the multiplication of the virus inside the cell. At least for members of the Phage Group, the Hershey-Chase experiment strongly argued that the phage's genetic material must be DNA.

In 1951, James D. Watson, a member of the Phage Group, arrived at Cambridge University on a postdoctoral fellowship. Convinced that DNA is the material of genes, he began collaborative work on its structure with Francis Crick, a young physicist attempting a doctorate in biology (*see* CRICK, FRANCIS, AND JAMES D. WATSON). Their partnership soon expanded to include Maurice Wilkins, another former physicist who was investigating the physical properties of DNA at Kings College, in London. A key member of his group was Rosalind *Franklin, an expert in X-ray diffraction studies of biological molecules. Watson and Crick did no experiments. Rather, they devised tinker-toy models of DNA drawing on chemical and physical data about the molecule, especially some of Franklin's diffraction results with DNA. In 1953, they succeeded in determining that DNA comprises a double helix joined at regular intervals across the distance between them by one of two complementary nucleotide base pairs—adenine with thymine, or cytosine with guanine.

Watson, Crick, and Wilkins held that the structure settled the question that DNA is the genetic material. The structure immediately suggested to them how genes can replicate themselves. The two strands of the double helix would separate, each with its single string of complementary nucleotides; then each would form a template for the creation of a new double helix identical to the first. The physicist George Gamow suggested that the sequence of nucleotides must contain an organism's genetic information in the form of a code. By 1964, biologists had demonstrated experimentally that specific sequences of three base pairs code for specific amino acids. Through a complicated biochemical mechanism, a series of such triplets

is translated at a cellular site into a chain of amino acids, which enfold themselves into a specific protein—for example, a constituent of the eye—involved in the organism's structure or, as in the case of an enzyme, figuring in one of its processes, like metabolism.

The working out of the genetic code enabled biologists to begin studying gene function at the molecular level. Their task was greatly facilitated by several experimental and technological innovations, including in the 1970s the technique of recombinant DNA, which allowed the removal of single genes from one organism and their insertion into another; in the 1980s, table-top machines that would determine the sequence of base pairs in a strand of DNA; and, in the 1990s, the creation of computerized databases that made available on the World Wide Web the DNA sequences in the genomes of human beings of laboratory organisms such as yeast and mice. A century after the rediscovery of Mendel's papers, genetics was a highly populated, multidisciplinary field, wealthy in resources, and rich in promise for understanding inheritance, development, and disease.

See also DEOXYRIBONUCLEIC ACID; GENETIC MAPPING; MOLECULAR BIOLOGY; POPULATION GENETICS.

Elof Axel Carlson, *The Gene: A Critical History* (1966). James D. Watson, *The Double Helix* (1968). Robert Olby, *The Path to the Double Helix* (1974). Garland E. Allen, *Thomas Hunt Morgan: The Man and His Science* (1978). Horace Freeland Judson, *The Eighth Day of Creation: Makers of the Revolution in Biology* (1979). Francis Crick, *What Mad Pursuit* (1988). Lily E. Kay, *Who Wrote the Book of Life: A History of the Genetic Code* (2000). Evelyn Fox Keller, *The Century of the Gene* (2000). Michel Morange, *The Misunderstood Gene*, trans. Matthew Cobb (2001).

DANIEL J. KEVLES

GEOCHRONOLOGY. See STRATIGRAPHY AND GEOCHRONOLOGY.

GEODESY. The librarian of Alexandria, Eratosthenes, called "beta" in antiquity because he was second best at everything, made the first significant measurement of the size of the earth. His technique required the determination of the terrestrial distance between two points on the same meridian and the difference in latitude between them. The figure shows Eratosthenes's stations at noon on the day of measurement, when the sun stood at the zenith in Aswān and a little over seven degrees from it at Alexandria. Since the sun's rays strike the earth almost parallel, the difference in latitude $\Delta\phi$ equals the measurable zenith distance z; the station separation Alexandria-Aswān is the same fraction of the earth's circumference C as $\Delta\phi$ is of 360°. Eratosthenes's value of C may have been close to correct

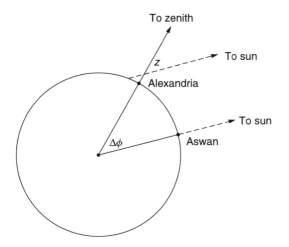

The relation between the difference in latitude between Alexandria and Aswan ($\Delta\phi$) and the distance of the sun from the zenith at Alexandria (z) when it stands overhead at Aswan.

(40,000 km [24,800 miles])—the size of his ruler is not known exactly—an extraordinary coincidence since he used gross approximations.

Knowledge of C had a practical value to people planning long trips. The Arabs, who traveled far across deserts, devised several geodetic methods, and the Caliph al-Ma'mūn (reigned 813–833), a great patron of mathematicians, sent out expeditions to deploy them. The principal technique was to take the latitude of a place in a flat region and walk due north or south, measuring the distance, until the latitude had changed by a degree.

Columbus chose the values of the circumference from Greek and Arab sources that made the earth small enough that his project to reach Asia by crossing the great Western ocean might seem feasible. His landing in America revealed the world to be twice the size he had calculated. Renewed interest in geodetic measurement, stimulated by the European voyages of discovery, returned mathematicians to Eratosthenes's technique with the significant difference, however, that henceforth they would determine distance by trigonometry rather than by walking or guessing.

Although Tycho *Brahe carried out a trigonometric survey of parts of Denmark, the first thorough and accurate geodetic measurement, which became the early-modern exemplar, took place in Holland in 1616. The surveyor, the professor of mathematics at the University of Leyden Willebrord Snel, gave the book in which he published his results the appropriately boastful title, *Eratosthenes batavus* (*The Dutch Eratosthenes*, 1617).

He found for C the equivalent of 38,640 km (24,015 miles). He determined the distance between his two stations (around 130 km [80 miles]) by setting out a series of virtual triangles with church steeples and town towers at their vertices, measuring the angles at each vertex, and obtaining the lengths of the sides of the triangles by linking them (by angles taken from the steeples and towers) to base lines laid out by a surveyor's chain.

Snel's technique was first applied on a large scale by Louis XIV's astronomers, who, as members of the new Paris Académie Royale des Sciences (*see* ACADEMIES AND LEARNED SOCIETIES), had the chore and challenge of making an exact survey of France. Beginning in the 1660s under the direction of Jean Picard, they measured increasingly long arcs of the meridian through Paris. Picard improved the instrumentation by replacing open sights with lenses and provided a second baseline to check results obtained with the first. He made $C = 40,042$ km (24,886 miles). The effective head of the Paris Observatory, Gian Domenico Cassini, and then his son and grandson, extended Picard's arc, crisscrossed France with virtual triangles, and delivered the news that the conventional maps of the kingdom ascribed to it lands that in fact lay in the English Channel and the Bay of Biscay. France may have lost more territory to its astronomers than to its neighbors.

The conversion of the Cassinis' measurements to a value for the earth's circumference presented

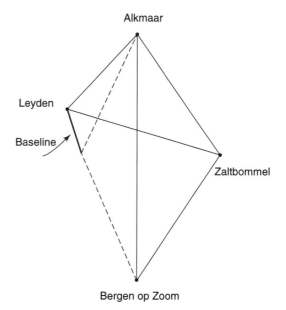

Scheme for Snel's trigonometric determination of the distance between Alkmaar and Bergen op Zoom.

a new problem for geodesy. On a mission for the Académie to the island of Cayenne, Picard found that a pendulum clock went more slowly near the equator than it had done in Paris. Isaac *Newton explained this tropical lethargy with his theory of gravity, which suggested that if the earth had cooled from a spinning fluid ball, its equatorial diameter would exceed the distance between the poles by an amount he could calculate. Consequently geodecists should try to fit their measurements to an ellipse. Cassini's son Jacques reported the first such attempt in his *De la grandeur et de la figure de la terre* (1720). The ellipse that fit his facts had the shape of an American football instead of a Newtonian pumpkin.

Confrontation between the Cartesians, who dominated the Académie, and the slowly growing Newtonian band of younger members came to center on the shape of the earth. In the 1730s the Newtonians, led by Pierre de Maupertuis, convinced the crown to finance a definitive test, in the interest of navigation as well as science. Two expeditions went out, one to Peru, the other, under the direction of Maupertuis, to Lapland. Both braved harsh weather, primitive conditions, uncomprehending natives, and vicious mosquitoes, all to learn whether a degree of the meridian was longer at the pole than at the equator. The outcome "flattened the poles and the Cassinis," according to Newton's popularizer Voltaire, although, in fact, Maupertuis's measurement was so far out that had it been in the opposite direction the expeditions would have found for a pointed earth.

During the second half of the eighteenth century the leading states of Europe undertook trigonometric surveys of their domains, primarily for military and economic purposes, which gave abundant data for fixing the ellipticity of the Newtonian earth. The more the data, the less plausible the ellipsoid. Geodecists assigned the discrepancies to inhomogeneities in the density of the earth and to mountains that drew aside plumb bobs and falsified the verticals necessary for finding latitudes. They came to agree that no fiddling would fit their knowledge to an ellipsoid; the shape of the earth was just that, the earth's shape, a "geoid," defined by Carl Friedrich *Gauss and other leading geodecists as a surface everywhere perpendicular to the direction of gravity.

Beginning in 1787 with the linkage of the meridians through Greenwich and Paris via a series of triangles that crossed the channel, Europe gradually approached unification, trigonometrically speaking. The French furthered unification by running surveys in territories conquered by Napoléon and his generals and by imposing where they could the metric system of weights and measures, itself based on a remeasurement of the Paris meridian from Dunkirk to Perpignan. Introduced during the first years of the French Revolution to replace the welter of provincial units that obstructed commerce and victimized the consumer, the standard meter was defined in theory as one ten-millionth of the distance from the pole to the equator and in practice as the ten-millionth part of ninety times the average length of a degree along the Paris meridian. The meter, its decimal multiples, its progeny of grams and liters, and its imposition on Europe are conspicuous, recurrent reminders of the prestige of geodesy in the eighteenth century, the rationalism of the Enlightenment, and the power of the French Revolution.

After 1800 trigonometrical surveys were pursued outside of Europe notably by the British in India and the new government of the United States. The American Coast and Geodetic Survey became a principal route for the little support the U.S. federal government gave science during the nineteenth century. Warfare, and especially the cold war, again pushed geodesy to the front in the twentieth century. Plans made from precisely established points played a major part in the artillery duels of World War I and corrected many errors in maps previously regarded as accurate. Development of aerial photography and radio during the war gave geodesy powerful new tools for postwar mapping. Similarly the invention of the laser after World War II enhanced the means, and the creation of intercontinental ballistic missiles increased the importance, of knowing with precision how far one point lies from another on the earth's surface. The emergence of geophysics as a major science during the last fifty years owed much to military interest in the geoid as a firing range.

Hunter Dupree, *Science and the Federal Government* (1957). N. D. Haasbroek, *Gemma Frisius, Tycho Brahe and Snellius and their Triangulations* (1968). Henri Lacombe and Pierre Costabel, eds., *La figure de la terre du xviiie siècle à l'ère spatiale* (1988). J. L. Heilbron, *Weighing Imponderables and Other Quantitative Science around 1800* (1990).

J. L. HEILBRON

GEOGRAPHY has its roots in ancient efforts to describe the surface of the earth, the form and extent of its lands and seas, and all that is observed in them. The peculiarly modern, European art of geography, as it emerged in the *Renaissance, combined mathematical delineation and historical description with varying emphases. For Renaissance scholars, geography was both a body of knowledge recovered from the classical past and a newly discovered art of systematically ordering knowledge of the earth. The transmission of the *Geographia* of Claudius Ptolemy (c. 90–168) from Constantinople to Italy and its translation into Latin early in the fifteenth century was central to making the mathematical description of the earth's lands and seas the core of the new art of geography. It gave instructions for arranging geographical knowledge according to a coordinate system of parallels (latitudes) and meridians (longitudes) and for projecting those lines onto flat surfaces. Renaissance geography was closely tied to the production of maps of the earth's surface and its separate divisions, culminating in the first "atlases" of the Low-country cartographers Gerhard Mercator and Abraham Ortelius in the mid-sixteenth century.

This Renaissance "invention" of geography was sponsored by centralizing states making ever more universal claims to legitimacy. By advertising the usefulness of their art for administering and defending principalities and cities, geographers secured more or less official roles in European states from the sixteenth century. As these states began to explore and trade overseas, geographers used words and pictures to help interpret, manage, and celebrate novel discoveries and new possessions for kings, princes, and churches. Geography also carried all the prestige of humanism and the mathematical arts. Sebastian Münster, author of the *Cosmographica universalis* (Basel, 1544), was typical in advertising geography as the essential study of the active Christian prince. How could anyone understand Homer and Virgil, the campaigns of Caesar and Cato, the acts of God, the wanderings of the Jews, and the missions of the Apostles, without geography?

The Protestant Reformation gave added impetus to official patronage. As reformed and counterreformed princes and cities turned to the foundation of gymnasia and universities, geographical systems assumed a pedagogical and doctrinal function, exhibiting God's wise design of the world. The mathematical basis of geography was progressively refined as a theater for God's providence in the *Systema geographicum* (Hanover, 1611) of the theologian Bartholomäus Keckermann and the *Geographia generalis* (Amsterdam, 1650) of the Dutch physician Bernhard Varenius. Varenius's book was one of the most influential geographical systems of the seventeenth and eighteenth centuries. Isaac *Newton edited the second and third editions despite Varenius's Cartesian leanings. In the work of the Jesuit Athanasius Kircher or the globe-maker Vincenzo Coronelli, who served Louis XIV at Versailles, mathematical techniques imbued geography with a renewed sense of cosmographical mission under the sponsorship of Roman Catholic Counter-Reformation and royal absolutism.

In the eighteenth century, geography flourished as a descriptive science with a mathematical foundation, often with an appeal to divine design. Enlightenment geography aligned itself programmatically, with the "systematic spirit" and against the "spirit of system," often restricting itself to positive description and tying itself ever more firmly to a technical symbolic language. Although it relied upon the observations of travelers and field surveyors, geography was essentially a product of the cabinet. While maintaining its focus on devising critical and systematic descriptions of the earth's surface, geography expressed its cosmographical ambitions increasingly by attempts to standardize and universalize the language of geographical description—to establish uniform topographical languages and cartographic symbolism (including map scale).

These efforts reached their zenith in Enlightenment France with Cassini's trigonometric charts of France, the collection of maps for military purposes at the Dépôt de la Guerre (founded 1688), and the closely associated cabinet of J.-B.-B. d'Anville. Here, geography began to converge with and restrict itself to cartographical description, while occupying the center of a web of skills including the work of field surveyors, instrument makers, and engravers. Geographers such as d'Anville distinguished themselves from lesser workers by their ability to sift and compile geographical information critically and express it in ever more precise and standardized languages, especially its graphic and mathematical expression on a chart. In practice, however, the lines between positive description and philosophical speculation were ill defined. Geographers such as Philippe Buache and Nicolas Desmarest entered into debates over the nature of continents, seas, earthquakes, volcanoes, river basins, and mountain ranges—topics often labeled "physical geography" and "theory of the earth" to distinguish them from "geography" proper.

In Protestant Europe, where geography remained tied to natural philosophy and natural theology, geographers discussed such topics as the causes of the tides, the heat of the tropics, seasonal rains, the decrease of temperature with elevation, and so on. Political geography, human geography, mineral geography, plant geography, and zoogeography emerged as distinct fields, justified principally through their service to natural theology. All exposed the wise hand of Providence by revealing the diversity and distribution of created things, and the balances and economies of nature (language eschewed by French geographers). In Germany and Scotland, at Hamburg, Edinburgh, and Göttingen, in works such as A. F. Büsching's *Neue Erdbeschreibung* (11 vols., 1754–1792) and Christoph Ebeling's *Erdbeschreibung und Geschichte von Amerika* (7 vols., 1793–1816), geography took the form of theologically informed gazetteering aimed at developing industrious, purposeful citizens of the world such as great merchants and administrative officials. At Göttingen, under the auspices of Gottfried Achenwall and A. L. Schlözer, geography was closely allied with statistics and political history (*Staatenkunde*). Immanuel Kant, who spent most of his forty years at Königsberg lecturing on "physical geography" (including physical anthropology), promoted geography's "extensive utility," supplying "the purposeful arrangement of our knowledge, our own enjoyment, and plentiful material for sociable conversation." Johann Reinhold Forster and his son Georg Forster, who together accompanied James Cook on his second expedition around the world from 1772 to 1776, similarly used geographical exploration to demonstrate the providential design of the world and to make its audience into active witnesses and instruments of providence. In his German translations of contemporary voyages and in his briefer, essayistic works, the nomadic younger Forster decisively focused geography on active exploration and the gradual emergence of underlying dynamic laws and processes.

Toward the end of the eighteenth century, the natural-historical concern with systematic collection, classification, and exposition gave way to philosophical concern with dynamic processes and physical causes. Taking inspiration from the Forsters' "philanthropic" approach to understanding the reciprocal influences of nature and human civilization, the Prussian mining official and naturalist Alexander von *Humboldt sought physical and historical laws in systematic geographical investigations of everything from climate to language and art. Humboldt shunned the term "geography" and referred to his science as "physics of the earth." As the most renowned traveler and naturalist of the early nineteenth century, Humboldt put the systematic spatial analysis of phenomena usually left to physicists and natural historians—rocks, terrestrial magnetism, atmospheric temperature and chemistry, plant and animal species, anything open to calibrated and standardized perception—at the center of geography. He also emphasized the aesthetic and emotional responses of the traveler. And he stressed that records of medieval travel and conquest, including the works of Christopher Columbus, be preserved and rehabilitated as important geographical sources. Although

Humboldt omitted God from his unfinished *Kosmos* (1845–1862, subtitled "sketch of a physical description of the world"), his dual revelation of the lawfulness of nature and the progress and limits of human knowledge of nature was implicitly congenial to a recognition of divine design.

Humboldt's contemporary Carl Ritter presented geography as a comparative study of the world's "terrestrial units" that supplied the key to a developmental understanding of the history of civilization, expressly overseen by divine providence. As delivered for three decades at the newly founded Berlin University and at the Berlin Military Academy and in his massive but unfinished *Erdkunde* (1817), Ritter's demonstration that civilization had migrated from East to West influenced generations of statesmen, soldiers, and scholars. Among them was Arnold Guyot, whose Protestant geographical theodicy found a home at Princeton University after 1848. In Britain, Mary Somerville discovered a similar providence at work in geography: both objectively, in the interaction of physical forces over the surface of the globe to produce geographical laws (*Physical Geography*, 1848), and subjectively, in the progressive interaction of the different branches of the physical sciences in geographical science (*On the Connexion of the Physical Sciences*, 1834).

Geographical societies began emerging in the late eighteenth century, imbued with a civilizing mission and informed by this sense of divine lawfulness The Association for Promoting the Discovery of the Interior Parts of Africa, founded in London in 1788, was followed by groups in Paris (1821), Berlin (1828), London (1830), Saint Petersburg (1845), and New York (1851). These metropolitan societies brought together army and navy officials, statesmen, and scholars to promote exploration and publish maps and accounts of voyages. Although membership waned in mid-century, it revived in the 1860s and 1870s with the recognition of the centrality of geography to imperial expansion and the cultivation of national identities. Between 1870 and 1890, the number of geographical societies in Europe quadrupled to over eighty, driven by the establishment of provincial geographical societies. In 1875, these societies began convening quadrennial International Congresses of the Geographical Sciences. They also encouraged the proliferation of popular geographical magazines and literature. Taking their cue from Germany, where geography was established in natural science faculties, and where academic geography and geographical societies had close ties, geographical societies urged that geography be taught in universities and secondary schools. In 1871, after the Prussian victory over France, the Paris society and the ministry of education established a Committee on the Teaching of Geography and a number of professorships at French universities. In the 1880s, the Royal Geographical Society of London succeeded in establishing chairs of geography at Oxford and Cambridge.

In the universities, geography was dominated by the prevalent enthusiasm for evolutionary theories. Friedrich Ratzel, who trained with Ernst Haeckel in Jena before settling at Munich, Halford Mackinder at Oxford, Alfred Hettner at Tübingen, and William Morris Davis at Harvard all argued that the geographical diversity of humans resulted from the variation of physical conditions over time and space. The Russian exile Peter Kropotkin and the Scottish social critic Patrick Geddes stressed the independent laws of organisms in interpreting responses to inorganic conditions. Paul Vidal de la Blache and Otto Schlüter insisted that human modes of social life (lifestyles, *Kulturformen, genres de vie*) shaped the landscape.

Regardless of their understanding of evolution, academic geographers took the region or the landscape as the fundamental unit of analysis. A few modern geographers explicitly abjured questions of *evolution, and claimed that the task of the geographer was to comprehend the unique character of regions or landscapes, without attributing causality to any particular factor. For all their sometimes bitter disagreement, Carl Sauer at Berkeley and Richard Hartshorne at the University of Wisconsin agreed that regional description lay at the heart of scientific geography.

Between the 1950s and the 1970s, geographers enthusiastic about the so-called "quantitative revolution" rejected Hartshorne's strictures, and argued that by using quantification and statistics, geographers could discern the laws of social dynamics. More recently, geographers have analyzed the subjective experience of space and produced "behavioral geographies" and "mental maps."

The very ubiquity of geographical knowledge made it difficult to define the field. Since geography became an academic discipline in the late nineteenth century, its diffuseness has fueled a great deal of historical writing by academic geographers concerned with defining and defending the scientific territory and prestige of their field. This literature, rife with talk of "disciplinary crisis," is understandably preoccupied with discovering the essence of geography in past founders and precursors. Hartshorne's *The Nature of Geography* (1939) drew up a historical lineage and philosophical foundation for the discipline that established geography as a descriptive science of regional differentiation. It remains an influential interpretation of the history and philosophy of geography. Other geographers have adopted philosophies from phenomenology and logical positivism to *Marxism, structuralism, and postmodernism in the search for philosophical foundations. In the 1990s, the emphasis began to shift to studying the institutional, political, and social contexts, interests, and languages of earlier geographers. The bread and butter of geography, though, has remained the training of school teachers and the preparation of regional descriptions applicable to a variety of policy needs.

See also GEODESY; HUMBOLDTIAN SCIENCE; GEOPHYSICS; ETHNOLOGY.

International Geographical Union, *Geographers: Bio-bibliographical Studies* (1977–), 3 vols. so far. Margarita J. Bowen, *Empiricism and Geographical Thought from Bacon to Humboldt* (1981). David Stoddart, ed., *Geography, Ideology, and Social Concern* (1981). David Livingstone, *The Geographical Tradition* (1992). Morag Bell, Robin Butlin, and Michael Heffernan, eds., *Geography and Imperialism, 1820–1940* (1995). Anne Godlewska, *Geography Unbound: French Geographic Science from Cassini to Humboldt* (1999).

MICHAEL DETTELBACH

GEOLOGY. The word geology as a general term for the study of the earth was popularized in the late eighteenth century by the Swiss naturalists Jean André De Luc, who made his career in England, and Horace Bénédict de Saussure, famous for his voyages in the Alps. Abraham Gottlob Werner disliked the term as being too suggestive of theorizing, and promoted the alternative "geognosy." But when a group of Englishmen decided in 1807 to found a society for the study of the earth that eschewed the practical, utilitarian goals of the Continental *mining schools, they called it the Geological Society of London. Within two or three decades, geology had its own specialists, societies, textbooks, and journals, the Geological Society's Transactions being the first. Worldwide, the word geology (or its cognates) rapidly became the preferred term for the study of the earth.

Geologists generally divided their specialty into two parts: historical geology, which reconstructed earth history using *stratigraphy and *paleontology, and physical geology, which investigated the earth's structure and causal processes. The latter consisted of *mineralogy, diminished from the overarching category for the study of the earth to a mere subdiscipline, petrology, and structural geology.

The successes of historical geology were manifest in the outlines of the stratigraphic column as worked out by the mid-nineteenth century. Field work with a hammer and a

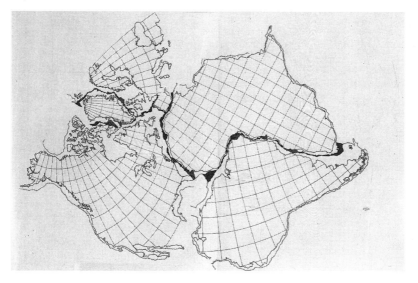

The Atlantic continents snuggled up at 500-fathoms in the influential diagram by
Edward C. Bullard et al. The dark patches indicate gaps and overlaps.

map was the preferred way of investigating, and the geological map an ingenious way of summarizing the results. Georges Cuvier and Alexandre Brongniart produced a pioneering map of the Paris area in 1812. The first national geological map, of England, was prepared by William Smith in 1815; the Geological Society's map followed in 1819. Although these maps were expensive and time-consuming to produce, state governments set up geological surveys to prepare them because of their economic importance for the extractive industries and agriculture. The French began fieldwork for a national geological map in 1825. The British Geological Survey was founded in 1835, followed by the Canadian (1842), the Irish (1845), and the Indian (1854). In Europe, the Austro-Hungarian Empire and Spain created surveys in 1849, Sweden and Norway in 1858, Switzerland in 1859, Prussia and Italy in 1873, and Russia in 1882. The American states founded their surveys early: Massachusetts and Tennessee in 1831 and Maryland in 1833, but a national survey was not created until 1879.

Though the work of mineralogists and petrologists attracted less public attention than that of the historical geologists, their field rapidly moved ahead in the nineteenth century. And although they too carried out field studies, they retained their ties to chemistry and the laboratory.

Historical and physical geology supposedly worked in concert, but the results obtained in the one frequently held little interest for the other. Nonetheless, geologists sought ways of presenting a coherent picture of the earth. One possible strategy, tried by Charles *Lyell in his *Principles of Geology* (1830–1833), made the study of present geological processes a necessary key to understanding earth history (*see* UNIFORMITARIANISM). Although rejected by most geologists, it provoked a useful methodological discussion.

A second strategy looked for an overarching causal process that could elucidate the details of the stratigraphic record. For most nineteenth-century geologists, the preferred hypothesis was that the earth had cooled and contracted. The nebular hypothesis proposed by Pierre Simon de *Laplace and William Herschel, and the work on rates of cooling by Jean Baptiste Fourier, supported their model. In 1831, the French geologist Élie de Beaumont suggested that as the earth had cooled from a molten body, the crust at intervals had buckled under the strain, throwing up mountain chains and exterminating whole genera in the great floods that coursed down their sides. Variants of this theory, and criticisms of it, flourished for the rest of the century and culminated in the four-volume *Face of the Earth* (1883–1904) by Eduard Suess, Professor of Geology at the University of Vienna. According to Suess, the molten center of the earth had once been covered with a thin, solid crust. As the earth cooled, portions of the crust collapsed, creating ocean basins. Later, the remaining higher areas became unstable, and collapsed in turn, forming new ocean basins and leaving former ocean beds exposed as new continents.

It was a triumphant moment for geology. Two great geologists summed up the history of their science during the nineteenth century.

Sir Archibald Geikie, head of the Geological Survey of Great Britain, published his *Founders of Geology* in 1905. He celebrated the achievements of British geologists, especially the Scots James Hutton and Charles Lyell. He arranged his history around battles between Neptunists and Plutonists (*see* NEPTUNISM) and Uniformitarians and Catastrophists (*see* UNIFORMITARIANISM AND CATASTROPHISM). Karl Zittel, a professor at the University of Munich and a renowned paleontologist, published a *History of Geology and Paleontology* in 1901. He gave more weight to mineralogy, petrology, and theories of mountain elevation, and praised Suess as having achieved "almost general recognition for the contraction theory."

This celebration of geology's progress was soon to seem inappropriate. The cooling earth with its foundering continents did not survive as a synthesis for more than a decade. The discovery of *radioactivity revealed a heat source within the earth that counteracted the cooling from some original molten state. The discovery of *isostasy made it highly improbable that continents could have foundered. Detailed studies of the Alps made it clear that simple up-down forces acting on the earth's crust could not explain the tens or even hundreds of miles of foreshortening revealed by their folded strata.

The resultant theoretical vacuum led to a proliferation of alternatives in the 1910s and 1920s. Some scientists, such as Harold Jeffreys in England and Hans Stille in Germany, attempted to revamp contraction theory in light of the criticisms. Others preferred more radical alternatives: the planetismal hypothesis advanced by Thomas Chamberlin, the radiogenic by John Joly, and the theory of continental drift of Alfred Wegener. None of these theories succeeded in garnering enough evidential support to win widespread acceptance.

For the next fifty years, geologists hunkered down and continued their map-making and surveying. Mineralogy and petrology made important advances. Underwater exploration revealed interesting gravity anomaly patterns around island arcs (*see* OCEANOGRAPHY). Geologists found new work in the oil industry, which joined geological surveys and mining as the main sources of employment outside the universities. The American Association of Petroleum Geologists, founded in 1917, had become the world's largest professional geological society by 2000, with over 30,000 members in more than 100 countries. By this time, however, geology was no longer the umbrella discipline for the study of the earth, but just one branch of the *earth sciences.

See also NEPTUNISM; OCEANOGRAPHY; UNIFORMITARIANISM.

Mott Greene, *Geology in the Nineteenth Century* (1982). Rachel Laudan, *From Mineralogy to Geology* (1987). Gabriel Gohau, *History of Geology* (1990). David Oldroyd, *Thinking About the Earth* (1996).

RACHEL LAUDAN

GEOMETRY, NON-EUCLIDEAN. See NON-EUCLIDEAN GEOMETRY.

GEOPHYSICS. Until the late nineteenth century no institutionalized discipline of geophysics, the physics of the earth, existed. Most geophysical efforts responded to specific national interests, relying heavily on government funding.

In the eighteenth century, although many natural philosophers still produced speculative cosmogonies based on Newtonian or Cartesian physics, quantitative, empirical geophysical work gradually displaced speculation. In particular, physicists and cartographers alike sought to determine the figure of the earth (*see* GEODESY).

In the nineteenth and early twentieth centuries, researchers turned to problems of *terrestrial magnetism, regional variations in the earth's gravitational field (gravity anomalies), and seismology in addition to geodetics. In the 1830s, scientists in Germany and England including Alexander von *Humboldt, Carl Friedrich *Gauss, Wilhelm Weber, and Edward Sabine proposed and initiated far-flung surveys of magnetic declination and dip (*see* TERRESTRIAL MAGNETISM). In the 1850s, surveyors for the Great Trignometrical Survey of India, established early in the century, found that their plumb bobs were not deflected by the gravitational force of the Himalayas to the extent they had predicted. This led to theories of *isostasy that preoccupied physicists and geologists in the first half of the twentieth century. In the last decades of the nineteenth century, the invention of reasonably accurate and compact seismographs spurred the development of *seismology into a quantitative discipline.

By the early twentieth century, geophysicists clustered in their own societies and institutes. In 1880 John Milne had founded the Seismological Society of Japan, and the Japanese quickly joined the Germans and the Americans as leaders in geophysics. The geophysical laboratory of the Carnegie Institution of Washington, founded in 1902, became a major center for American geophysics. The American Geophysical Union, established in 1919 as an affiliate of the National Academy of Sciences by the National Research Council of the United States, quickly assumed world leadership in geophysics.

World War I and World War II provided new applications (and research funding) for geophysics—seismology, geomagnetism, and submarine cartography, for instance, aided submarine warfare. The history of the military's role in the development of geophysics has yet to be written as many of the relevant documents have only recently been declassified. The *International Geophysical Year that began in 1957 gave a further boost to geophysics. The U.S. government established the National Aeronautical and Space Administration (NASA) in 1958 and the National Oceanic and Atmospheric Administration (NOAA) in 1970.

The prestige of geophysics rose with spectacular discoveries such as *cosmic rays, the core and mantle of the earth, sea floor spreading, and El Niño. Applied geophysics boomed as geophysicists turned their attention to problems such as global warming, ozone depletion, water supply and quality, and earthquake prediction. At the end of the twentieth century, the largest international organization of geophysicists was the American Geophysical Union, with over 35,000 members in more than 100 countries. Although geophysics retained a separate identity, by the late twentieth century it was much more tightly linked to geology and oceanography than ever before. Through most of its history, geophysics was practiced by physicists chiefly interested in problems arising from physics itself. Geology, by contrast, was from its start in the early nineteenth century the province of professional geologists. Their primary goal was to tell the history of the earth. Although they were interested in geological causes, they inferred these from current geological processes and past geological monuments, not from physical first principles. Insofar as geologists and geophysicists interacted, the encounters were often tense, as in the debate between William *Thomson, Lord Kelvin, and the geological community over the age of the earth (see EARTH, AGE OF THE). Oceanography had its origins in the mapping of the oceans and in exploration rather than in the physical sciences. With the creation of the overarching discipline of *earth science and the broad acceptance of the theory of *plate tectonics in the 1960s and 1970s, geophysicists, geologists, and oceanographers began working on related problems using the same techniques and implements. Simultaneously the creation of the related umbrella category of planetary science meant that geophysicists, geologists, and oceanographers extended the scope of their investigations beyond the earth itself to the domain formerly dominated by astronomers, the solar system. Though traces of the independent historical trajectories of these disciplines remain, their merging in earth and planetary sciences, although not always easy, has led to rapid scientific advances.

American Geophysical Union, *History of Geophysics* (1984–ongoing). Robert Muir Wood, *The Dark Side of the Earth* (1985). Gregory Good, ed., *The Earth, the Heavens and the Carnegie Institution of Washington* (1994). John Leonard Greenberg, *The Problem of the Earth's Shape from Newton to Clairaut* (1995).

RACHEL LAUDAN

GERM. The word "germ" derives from the Latin for "seed"; it was first used in its modern sense of pathogenic microbe in the nineteenth century. The theoretical association of "seed" with disease dates from antiquity but in modern times is associated with the *De contagione* (1546) of Girolamo Fracastoro. The development of the microscope in the early seventeenth century made possible the discovery of formerly invisible living creatures. Pioneering observations by the Dutch microscopist Antoni van Leeuwenhoek later in the century disclosed identifiable "animalcules" in various waters and on human teeth. The imperfections of most available optical instruments hindered further significant observations until critical improvements in the early 1830s. In 1835, Agostino Maria Bassi showed that a minute fungus caused muscardine disease in silkworms. Microbes soon were associated with many other diseases: Casimir Joseph Davaine, notably, repeatedly observed rod-shaped structures, which he named "bacteridia," in the blood of creatures dead of anthrax. The relationship between such structures and disease, whether cause or result, remained a matter of debate.

From the 1830s, European scientists interested in the processes of fermentation and putrefaction moved steadily towards the development of a germ theory of disease. In 1840, Jacob Henle set out the theoretical framework for such a theory, while Theodor Schwann demonstrated that not the air itself, but something in the air initiated putrefaction in organic substances. Louis *Pasteur attracted popular and scientific attention when he announced his own germ theory of disease causation in 1864. The issue was widely debated in the following decade, until Robert Koch identified the causative organism of anthrax (1876), and later those of tuberculosis and cholera. Several recent historians have shown that several different germ theories of disease circulated in the later nineteenth century, most in a continual process of modification. Consensus was reached only around the turn of the century.

The contribution of Koch and his associates to the establishment of the germ theory extended beyond the identification of organisms. Koch drew up the program known as "Koch's postulates," which, through several subsequent revisions, continues in use as the basic tool for confirmation of an association between specific microbes and diseases. Koch and his colleagues made important contributions to the identification of bacteria by developing staining methods based on the new industrial dyes of the period, and by 1900 the bacterial causes of several dozen diseases had been established. At this time many physicians believed that every disease must have its causal bacterium. Increasingly too, they identified diseases not by symptom as in the past, but by cause, in the laboratory.

Improvements in bacterial filtration processes after 1880 led to the realization that there must be yet smaller disease-causing agents, and in 1896 Martinus Willem Beijerinck launched his controversial but valid concept of the filterable virus—possibly water-soluble microbes small enough to pass through filters and invisible to contemporary microscopes. Among the first pathogens identified as such were the agents of tobacco mosaic virus, foot and mouth disease, and yellow fever. *Virology evolved rapidly as a science in the early twentieth century. The bacteriophage phenomenon was independently identified by Frederick William Twort and Félix d'Hérelle during World War I. Only with the invention of the electron microscope in the early 1930s, however, did it become possible to obtain visual impressions of viruses.

In the twentieth century, the idea of germs as disease agents became firmly established in the popular consciousness, partly through the use of the concept in advertising for domestic cleansing products. Virology remained a field of dynamic scientific interest; new or emergent virus diseases attracted attention across the twentieth century. The expansion of air travel, international trade, and global tourism alerted the medical community around 1970 to the possibilities of the diffusion of previously unrecognized and dangerous viruses.

At a more mundane level, Western countries became increasingly uneasy at an apparent upsurge in food-borne infections such as salmonellosis and listeriosis, which suggested the continuing threat to physical well-being of the endemic germs encountered in everyday life. In the last decades of the twentieth century, the discovery of prions—infectious dead proteins—forced a reassessment of the classic germ theory of disease, which was based on the belief in living disease organisms.

See also HOSPITALS.

William Bulloch, *A History of Bacteriology* (1935). Anthony P. Waterson and Lise Wilkinson, *An Introduction to the History of Virology* (1978). Frank Fenner, and A. Adrian Gibbs, eds., *Portraits of Viruses. A History of Virology* (1988). Gerald Geison, *The Private Science of Louis Pasteur* (1995). Nancy Tomes, *The Gospel of Germs: Men, Women and the Microbe in American Life* (1998). Michael Worboys, *Spreading Germs. Disease Theories and Medical Practice in Britain 1865–1900* (2000).

ANNE HARDY

GIBBS, Josiah Willard (1839–1903), mathematical physicist.

The only son of the Professor of Sacred Literature at Yale College, Gibbs seldom ventured beyond New Haven. He suffered constantly from ill health, and after the death of his parents, continued to live in the family house with his two sisters and one brother-in-law. His ordered existence centered on Yale, as a student (1854–1863), tutor (1863–1866), and professor (of mathematical physics, 1871–1903). Gibbs did his graduate work at Yale's advanced engineering school, from which he emerged as one of the very first PhDs in the United States. His thesis on gear design scarcely suggested the direction of the researches in fundamental physics that made his reputation. In 1866, Gibbs tore himself away from New Haven for a three-year stay in Europe, during which he attended courses in mathematics and experimental physics at the Collège de France in Paris and the universities of Berlin and Heidelberg, and read the works of European mathematicians extensively. Gibbs left no repository of personal papers for historians trying to understand the enigma of the mind behind the papers and the man behind the work.

After his return to the family house, Gibbs taught and worked gratis at Yale until 1879, when the university gave him a salary to counter an offer that Johns Hopkins had made on the basis of Gibbs's growing reputation in Europe. He had developed this reputation by staying away from meetings and sending copies of his work to European and American physicists. He was elected to the major physical and mathematical societies of Europe, and did not achieve the same reputation in the United States, where empiricism dominated the sciences. He seems to have thought deeply for a long time before committing his ideas to paper, publishing only a few, crucial works. Yet, after some years working on his pivotal and complex 1879 thermodynamics paper, he judged it to be too verbose.

Gibbs's research interests encompassed engineering, *mechanics, *thermodynamics, the electromagnetic theory of light, vector analysis, and statistical mechanics. He always worked from general principles, basing his mechanics on Hamilton's equations, his thermodynamics on the first and second laws, and his statistical mechanics on ensembles of particles that obeyed

Hamiltonian dynamics. He explored the implications of these basic laws at the most general level, moving from the simplest to the most general case. He sought physically significant, rather than mathematically elegant, results, and avoided speculations about molecules.

Gibbs transformed thermodynamics. Whereas physicists agreed in their understanding of the law of conservation of energy, they held three different interpretations of the second law, and many notions of the relationship between energy and entropy. In 1873, Gibbs restated both laws of thermodynamics distinctly, established entropy as a function of state; investigated the thermodynamic properties of a homogeneous substance, and developed new thermodynamic diagrams to display these properties; investigated the solid, liquid, and gaseous states of a substance; showed the conditions for their coexistence; and identified the critical points. By 1878, he had developed a theory of chemical mixtures and arrived at his phase rule, which set both the conditions under which a particular chemical mixture was stable, and if unstable, the direction of change to reach a new equilibrium.

It was James Clerk *Maxwell who first drew chemists' attention to the significance of Gibbs's work, which was then drawn on by the energeticists of the 1880s and 1890s (see *ENERGETICS).

Reading Maxwell led Gibbs to develop vector analysis, based on William Rowan Hamilton's quarternions and William Kingdom Clifford's mechanics. As a result, he drew the ire of the Edinburgh professor of *physics Peter Guthrie Tait, a fiery defender of British priority in science, who insisted on keeping quarternions pure and undefiled. Gibbs countered by demonstrating the use of vectors in *astronomy and *electromagnetism.

In his last publication, Gibbs in 1903 created a statistical mechanics more general than those of Maxwell or Ludwig *Boltzmann. Initially he considered the behavior of ensembles of mechanical systems, obeying Hamilton's mechanics, sharing the same total energy but having different positions and velocities. He moved from these "microcanonical ensembles" to "canonical ensembles," whose only common quantity was the same number of particles, and demonstrated that such ensembles behaved as thermodynamic systems. His "grand canonical system" was an ensemble of systems with different numbers of particles.

Gibbs was well aware of the major problems within his system, particularly its incorporation of the equipartition theorem (that the average energy for every degree of freedom of motion was equal) and the ergodic hypothesis (that all the systems in an ensemble passed through every possible configuration consistent with its total energy). He was also dissatisfied with his understanding of irreversibility, and did not see his work as a description of real gases.

Gibbs's approach ended in paradox. The entropy of the Grand Ensemble depended on whether the particles in the systems were distinguishable or indistinguishable from those in other systems of the ensemble. For Gibbs, this was a technical problem; for his critics, it addressed the properties of real molecules.

Gibbs's paradox was not so much solved, as bypassed, by the development of quantum physics. Max *Planck and Albert *Einstein postulated statistical distributions of energy for blackbody radiation (photons) and the photoelectric effect (electrons), respectively, for indistinguishable particles, avoiding the conundrum of deriving them from mechanical principles.

Josiah Willard Gibbs, *Elementary Principles of Statistical Mechanics* (1903). Josiah Willard Gibbs, *The Scientific Papers of J. Willard Gibbs*, Henry Andrews Bumstead and Ralph Gibbs Van Name, eds., 2 vols. (1906). Lynde Phelps Wheeler, *Josiah Willard Gibbs: The History of a Great Mind* (1952). Martin J. Klein, "The Physics of J. Willard Gibbs in His Time," in *Proceedings of the Gibbs Symposium*, Yale 1989, D. G. Caldi and G. D. Mostow, eds. (1990).

ELIZABETH GARBER

GLACIOLOGY is the discipline that examines how glaciers and ice sheets behave. It studies their origin and accumulation, deformation and movement, sublimation and melting, as well as how glacial ice interacts with climate. A subdiscipline of *geophysics, glaciology also has strong ties to the atmospheric sciences, particularly climatology, and to glacial geology, which analyzes the history and geological effects of glaciers, glaciation, and ice ages. The study of mountain glaciers and polar ice caps has required mountaineering skills and other high-risk tactics, so the history of glacier exploration and examination is also a history of adventure and, in some cases, tragedy.

In the late eighteenth century, Horace Bénédict de Saussure and other naturalists began studying mountain glaciers in the Alps, descriptive work that was summed up by the Swiss naturalist Louis Agassiz in his *Études sur les Glaciers* (1840). In the mid-nineteenth century, James David Forbes began systematic observations of glacial flow, using physical theory in an attempt to understand the phenomenon. In the latter part of the century, John Tyndall's publications on glaciers and glaciation influenced both scientists and the public.

The first International Polar Year (1881–1883) was largely responsible for the first studies of high-latitude glaciers and ice sheets. Expeditions to Greenland directed by Fridtjof Nansen in the

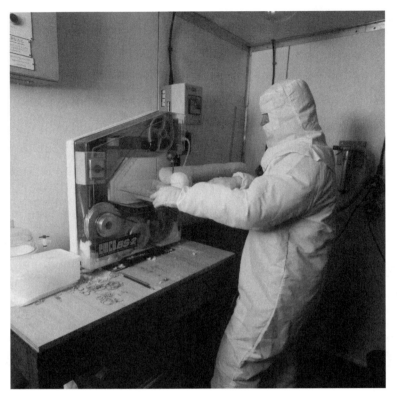

A glaciologist shown sawing through a section of an ice core from Antarctica. He works in a refrigerated clean room in Cambridge, England, to minimize melting and contamination.

1890s and by Alfred Wegener in the early 1900s surveyed the ice sheet. Robert Scott's Antarctic expedition of 1901–1904 conducted the first studies of the southern ice sheet. The renowned valley glaciers of southeast Alaska were first examined by Grove Karl Gilbert while on the Harriman Alaska expedition in 1899.

In the early twentieth century, R. M. Deeley and P. H. Parr made the first successful mathematical models of glaciers as viscous fluids. The first professional society for glaciology was founded in the 1930. In the 1950s—recognized within the field as the beginning of its modern era—John Nye led in putting glaciology on a sound physical footing. In the same decade the International Geophysical Year (IGY) (1957–1958) sparked further studies of high-latitude phenomena.

Current glaciological investigations are interdisciplinary and international. Ice sheets preserve one of the finest records of climate change over the last hundred thousand years or more. Among the notable discoveries of recent decades is that the ice record of atmospheric carbon dioxide content shows a marked positive correlation with fluctuations in global temperature. Fears of

global warming have impelled scientists also to examine short-term behavior of these large ice sheets, which hold much of the world's water and hence have an important influence on global changes in sea level. Programs are in place for drilling and examining cores from the Greenland and Antarctic ice sheets. Other glaciologists are studying the rapid surging and retreat of ice shelves and of glaciers, the basal boundary condition of glaciers, and the interaction of glaciers with their substrates. The study of ice on other planetary bodies, notably on Jupiter's moons Europa and Ganymede, on Saturn's icy satellites, and on Triton, is an emerging field. Icy outer crusts on these bodies exhibit phenomena uncommon to Earth—impact cratering preserved in ice, and ice volcanism—eruption of liquid (probably water) through the icy crust.

"The Histories of the International Polar Years and the Inception and Development of the International Geophysical Year," *Annals of the IGY*, vol. 1 (1959). G. K. C. Clarke, "A Short History of Scientific Investigations on Glaciers," *Journal of Glaciology*, special issue commemorating the fiftieth anniversary of the International Glaciological Society (1987): 4–24.

JOANNE BOURGEOIS

GLASS. Although glass is one of the most ancient synthetic materials, not until the last decades of the twentieth century did scientific understanding of its structure and the conditions of its formation play a major role in its manufacture and improvement. The science of glass still lags behind the science of metals and basic concepts remain unclear.

The empirical knowledge accumulated since the origin of glassmaking was first compiled and organized by Antonio Neri in *L'Arte Vetraria* (1612). Apart from local variations, the composition of glass—with soda, lime, and silica as the main components—remained remarkably stable over the centuries until 1674, when George Ravenscroft of London invented lead crystal. This new glass marked the end of Venetian leadership in the glass trade. Towards the end of the seventeenth century two methods for making flat glass joined glass blowing (an invention of the first century B.C.) as the main means of production of glass objects. Denis Diderot and Jean d'Alembert's *Encyclopédie* offers a detailed description of the state of the art in the mid-eighteenth century. Although the international exhibition in London in 1851 tried to promote the modernity of glass with the Crystal Palace, the mechanization of glassmaking proved slower than that of steel. Continuous production did not start until 1867, when Friedrich Siemens invented a continuous crossfired furnace equipped with regenerators. Philip Arbogast invented the "press-and-blow" process for making bottles in 1882, but its full automation was only effected in the United States in the 1910s. Emile Fourcault in Belgium and Irving Wrightman Colburn in the United States introduced continuous flat-glass machines in the early twentieth century. The most significant advance in flat-glass manufacture was the development of the float process by William and Richard Pilkington in 1959. A molten glass ribbon floated on molten metallic tin gives a very good surface finish, avoiding expensive and time-consuming polishing.

While industrial companies improved processes, the development of optical instruments, in particular by Joseph Fraunhofer, prompted changes in chemical composition. The need for special glasses for microscopes prompted the collaboration of Ernst Abbe, a physicist, Otto Friedrich Schott, a chemist, and Carl Zeiss, the instrument maker. In the 1880s, Jena Glass Works became the world leader in optical glasses. During World War I, optical glasses became strategic materials. The Allied governments funded research on and production of optical glass. The creation of a Department of Glass Technology at Sheffield, England, in 1921, was the first landmark in the formation of a science of glass. William E. S.

Turner launched a campaign to measure systematically the physical properties of a great variety of glasses. There gradually emerged a general notion of the "glassy state," which designates not conventional glasses but a state of solid matter characterized by nonequilibrium and disorder. In the 1930s, W. H. Zachariasen defined glass as a three-dimensional network of atoms forming a solid that lacks periodicity. This negative definition highlights the difficulty in understanding the relations between microstructure and properties necessary to provide the foundations for a science of glass. Despite recent advances in the exploration of glass transition and structure, the factors that determine whether a liquid will "freeze" in a glassy state during cooling are not fully understood. The familiar brittle material known as glass still retains part of its mystery. This difficulty did not prevent the technological development of a large array of new glasses, such as borosilicate pyrex, in the 1920s; or more recently, ultrapure silica fibers for long-range communication; chalcogenide glasses and fluoride glasses, transparent to infrared radiation; and optically nonlinear rare-earth glasses for lasers and active optical devices.

See also SOLID STATE PHYSICS.

Chloé Zerwick, *A Short History of Glass* (1980). Don Klein and Ward L. Boyd, *The History of Glass* (1991).

HERVÉ ARRIBART AND BERNADETTE BENSAUDE-VINCENT

GLOBAL WARMING. See CLIMATE CHANGE AND GLOBAL WARMING; OZONE LAYER.

GOLGI, Camillo (1843–1926), Italian neurohistologist, inventor of the black reaction staining technique, and discoverer of the Golgi apparatus; and **Santiago RAMÓN Y CAJAL** (1852–1934), Spanish neurohistologist, inventor of the neuron theory.

The work of Camillo Golgi and Santiago Ramón y Cajal—joint recipients of the Nobel Prize for Physiology or Medicine in 1906—is responsible for two of the major scientific achievements of the nineteenth century: the histomorphology of the nervous system and the neuron theory.

Golgi was born in Corteno near Brescia. Following in the footsteps of his father, a physician, he graduated in medicine from the University of Pavia in 1865, with a thesis on mental illness under Cesare Lombroso. He soon abandoned psychiatry, however, and turned to the histology of the nervous system. In 1873, after numerous trials, he developed the black reaction microscopic method for staining nervous tissues. This milestone technique allowed microscopic observation of the nerve cell structures with a clear visualization of the development and ramification of both the axons and the dendrites.

Thanks to the systematic use of the black reaction method, Golgi offered a new image of the fine anatomy of the cerebrospinal axis. He also offered a new description of the nerve cell, distinguishing clearly (from a morphological perspective) between the nervous prolongation and the protoplasmic prolongations, and demonstrating that the axon could emit collateral branches. He maintained that gray matter contained an extremely intricate, fine, dense network (a "diffuse nervous network," as he called it) made up of the plexus of branches coming from the axons of different cell layers. In the second half of the nineteenth century, reticular theories like Golgi's partly overturned the atomistic-reductionist assumption underlying cell theory, favoring a holistic approach that regarded the cerebrospinal axis as a continuous reticular structure and its functions as the result of collective action.

In 1886 Golgi collected his neurohistological works in a volume, *On the Fine Anatomy of the Central Organs of the Nervous System*, that was one of the main contributions to nineteenth-century neuroscience. Professor of histology at the University of Pavia from 1876 and of general pathology from 1881, he also directed the pathology laboratory, which became a leading center of medical and biological research. He was also rector of his university and a senator of the Kingdom of Italy.

Among Golgi's numerous research endeavors his work on malaria (he discovered the relation between the biological cycle of the parasites and the intermittence of the fever, and he identified different species of plasmodia for the different forms of malaria) and on cytoplasm (in 1898 he identified a new organelle in the form of a network later known as the Golgi apparatus) deserve special mention. In 1887 Ramón y Cajal learned of Golgi's staining method, which was a turning point of his career.

Ramón y Cajal was born in Petilla de Aragón. After a restless youth, he was convinced by his surgeon father to study medicine. He graduated from the University of Zaragoza in 1873, obtained his doctorate from the University of Madrid in 1877, and in 1883 won the chair of general anatomy at the University of Valencia. In 1887 he became professor of histology and pathological anatomy at the University of Barcelona, where he remained until 1892, when he was called to the same chair at the University of Madrid. During his years in Barcelona he made his main discoveries and formulated the theory of the neuron. Ramón y Cajal improved Golgi's staining techniques with the method of "double impregnation," which he used on small mammals and on embryos in early stages of development, when the nerve structure is simpler. He focused on a series of histological data that seemed to contradict the theory of the "diffuse nervous network." Beginning around 1888, he began to elaborate a new theoretical synthesis anticipated by Auguste-Henri Forel, Wilhelm His, and Fridtjof Nansen (all proponents of the idea that the nerve cell was an independent unit). In 1891 Wilhelm von Waldeyer-Hartz labeled this synthesis the "neuron theory."

The neuron theory asserts the morphological and functional "individuality" of each nerve cell. It maintains that the entire unit comprising the cell body, the protoplasmic prolongations, and the nervous prolongation constitutes a single cell, anatomically separate from other cells. According to the theory there are properly no nervous networks: relations between neurons are established by contact (Charles Sherrington's "synapse") between the extremities of a cell's axon and the dendrites or cell body of another cell and not, as Golgi thought, by anastomosis or continuity between the nervous prolongations of different cells. Ramón y Cajal also asserted that, according to the "law of dynamic polarization," which he had elaborated with A. van Gehuchten, each neuron could be regarded as a physiological unit composed of a receptor apparatus (the cell body and dendrites), a conduction apparatus (the nervous prolongation and its ramifications), and an emission apparatus or discharge organ (the terminal ramifications of the axon).

In the following years Ramón y Cajal worked on neurogenesis and neuropathology, showing the "embryological, regenerative, and reactive individuality" of the neuron. He collected most of his results in his epoch-making *Texture of the Nervous System of Man and Vertebrates* (1897–1904). After receiving the Nobel Prize, Ramón y Cajal held a position of great prestige and influence in Spanish society, and wrote books on several subjects, including a very successful autobiography (*Recollections of My Life*, 1901–1917).

Around 1900, the neuron theory was attacked both by Golgi and the "neo-reticular" neurohistologists (Stephan Apáthy, Hans Albrecht Bethe, Franz Nissl, and H. Held), supporters of the "neurofibrillar theories." A controversy arose between Golgi and Ramón y Cajal that apparently concerned histological data (microscopic observations presented a degree of uncertainty, and no decisive facts existed in favor of the dominant neuron theory), but in effect revolved around a different conception of the anatomo-functional architecture of the nervous system. Golgi remained tied to a holistic model of the structure and activity of the cerebrospinal axis (the same underlying the theories of the adversaries of cerebral localization); a model that constituted for him a sort of "ideal lens" through which to observe the nervous tissues and interpret histological data. Without Golgi, however, it would be difficult to understand and explain Ramón y Cajal's work and the invention of the neuron theory.

Camillo Golgi, *Opera omnia*, 4 vols. (1903, 1929). José M. López Piñero, *Ramón y Cajal* (1985). Javier DeFelipe and Edward G. Jones, eds., *Cajal on Cerebral Cortex. An Annotated Translation of the Complete Writings* (1988). Gordon M. Shepherd, *Foundations of the Neuron Doctrine* (1991). Paolo Mazzarello, *The Hidden Structure: A Scientific Biography of Camillo Golgi* (1999). Guido Cimino, "Reticular Theory versus Neuron Theory in the Work of Camillo Golgi," *Physis* 36 (1999): 431–472.

GUIDO CIMINO

GOVERNMENT. In antiquity, Archimedes and Aristotle both enjoyed the patronage of monarchs. In modern times, royal and aristocratic patronage of science has been replaced by more democratic support. Motivated by policy rather than patronage, nation-states pursue their military, political, and economic goals through subsidy of science. Early modern states extended patronage to corporate bodies like the Paris Academy of Sciences, which, while primarily controlled by the academicians themselves, were conceived as serving the technological needs of the state. This form of patronage survived the age of revolution. Governments in capitalist and socialist systems alike reconciled support of science with its relative autonomy in order to harvest the technological fruits of research and the educational benefits of technical training. The application of science to warfare, a by-product of this process, stimulated vast increases in support for science during and after World War II. In totalitarian states, however, support was accompanied by control that throttled scientific initiative and distorted scientific knowledge to support political ideologies.

Although Francis Bacon proposed the subsidy of science by the state early in the seventeenth century, it remained for Jean-Baptiste Colbert to articulate a mechanism for it in the Paris Academy of Sciences during the later eighteenth century. The academy's members included state-salaried savants responsible for advice on technical matters. French revolutionary governments sought to abolish it and other academies, and even executed one of the most distinguished academicians, Antoine-Laurent *Lavoisier. The École Polytechnique, created as an alternative, was recast as the flagship of the elitist system of French education by Napoléon, who also restored the Academy of Sciences and the *grandes écoles* and created the Université de France. This scientific establishment enjoyed a level of patronage that was the envy of English and German scientists.

As Charles Babbage complained in his *Reflections on the Decline of Science in England* (1827),

France offered "emoluments for science" that far exceeded those available in England. The British Association for the Advancement of Science (BAAS) sought to redress the imbalance after its formation in 1830. Its presidents included Prince Albert and Lyon Playfair, both of whom pursued careers in science and government. They perceived a threat to national power in the ascendancy of German science and industry in the later nineteenth century. Word War I demonstrated the scientific superiority of Germany and caused the Entente nations to build up their own science.

In the wake of their defeats by Napoléon at the beginning of the nineteenth century, Prussia and other German states had reformed their educational systems. Competition between state-run German universities ensured both rewards for research and *Lernfreiheit* (academic freedom), at least for professors heading research institutes. Since German scientists were civil servants, the advancement of science through research became a state activity through their entrepreneurship. After German unification, the growing chemical, electrical, and metallurgical industries supplied demand for research and scientific expertise.

The relationship between science and the federal government in the United States was forged over a century and a half. The founders included men of science like Thomas Jefferson and Benjamin Franklin. President George Washington urged government support for the increase and diffusion of knowledge. In addition to protecting intellectual property in the form of copyrights and patents on inventions, the government supported science in its military academy at West Point, where a French model of military training was adopted. Like the École Polytechnique, the West Point academy created a cadre of technical talent who found employment in engineering the canals and railroads that knit the new nation together. Federal explorations encouraged national expansion. The Coast Survey encouraged maritime commerce and was the leading antebellum federal patron of science. Franklin's great-grandson, Alexander Dallas Bache, its second superintendent, used his position to stimulate reforms in American science. Joseph Henry, the first secretary of the semi-governmental Smithsonian Institution, joined him in forming the American Association for the Advancement of Science (AAAS) in 1848. During the Civil War the federal government chartered its own National Academy of Sciences and offered land grants for colleges to train farmers and mechanics in support of mechanized agriculture.

In 1901, the National Bureau of Standards was established as the first federal laboratory in the physical sciences. Private demand for chemists and engineers led American universities such as Johns Hopkins and the University of Chicago to adapt European models—particularly those of the German universities—for scientific education. By 1910 the scientific establishment in the United States rivaled its European counterparts in size, if not in accomplishments.

The mobilization of scientific talent in Western Europe and the United States in World War I, particularly in chemistry, made a lasting impression on their governments. Thus, while Britain, France, and the United States began by attempting to harvest technological innovation from the public at large, all of them ended by relying more upon trained scientific and engineering talent, giving rise to an ideology of national science in the United States and analogous drives for elitist research institutions in France and other modernizing countries. The new Soviet Union forged associations of scientists to support the nation's rapid industrialization. Germany used its Russian connections to help rebuild its own scientific establishment, damaged by the war and postwar inflation and boycotted by the Allied scientific establishments. The totalitarian policies of Hitler and Stalin, in particular the Nazis' anti-Semitic measures, required alignment of scientific concepts with political ideology. While many scientists emigrated as a conseqence, the continuing strength of German science became a source of alarm to the West when Hitler and Stalin joined forces to dismember Poland in 1939. In particular, the discovery of fission in Berlin moved emigré scientists in the United States and Britain to encourage their governments to explore the development of nuclear energy.

The mobilization of science in a democracy required the creation of extraordinary scientific institutions. Vannevar Bush, president of the Carnegie Institution, organized American scientists in a unique way, contracting with universities and other research organizations to do the kind of research previously performed in military laboratories under military direction. The National Defense Research Committee consolidated scientific research in government bureaus and universities to prepare for war; it was succeeded by the Office of Scientific Research and Development, which mobilized American scientists at university-based laboratories like the MIT Radiation Laboratory. The effort to develop nuclear energy was eventually transferred to the War Department, and its Manhattan Engineer District (MED) covertly established the world's largest scientific and technical enterprises to build nuclear weapons. Following the first use of the atomic bomb, Congress sought to naturalize nuclear energy through the creation of an Atomic Energy Commission (AEC) and science through a National Science Foundation (NSF). Protracted debates over the composition, administration, and purview of the AEC and the NSF left science in the hands of the military in the immediate postwar period. The MED created a system of national laboratories at Brookhaven, Argonne, and Oak Ridge, and funded academic laboratories like the University of California's Radiation Laboratory; the military services promoted the continuing development of radar and other tools of war with investments in academic research. Although the civilian Atomic Energy Commission took over the MED laboratory system, the new Department of Defense became the most generous patron of science in the history of the world. With the National Science Foundation dormant between 1945 and 1950, the AEC and the Office of Naval Research (ONR) played central roles in funding basic scientific research in universities. Even as the NSF was funded again in the early 1950s, the Army and the new Air Force followed the ONR pattern of extramural research support. While other, nonmilitary government bureaus (e.g., the National Institutes of Health and the National Bureau of Standards) also expanded support for research and development in the postwar era, the Department of Defense continued to be the primary U.S. patron of science and technology through the rest of the century.

By the 1950s, Russia's system of scientific support had reached sufficient maturity to challenge the United States for leadership in national security fields such as nuclear weapons and missile development. The launch of *Sputnik* by the Soviet Union in 1957 inspired the United States government to create the National Air and Space Administration (NASA) and the Defense Advanced Research Projects Agency (DARPA) and to strengthen science education programs with the National Defense Education Act.

Western Europe sought to compete with both of the superpowers by creating an accelerator laboratory of surpassing size at *CERN. A European space program was initiated, though with goals more modest than the manned lunar visit John F. Kennedy declared as the object of U.S. space efforts in the 1960s.

After Apollo fulfilled Kennedy's dream, federal support for science in the United States declined as discontent with scientific policy grew. The President's Science Advisory Panel, created in 1957 by Dwight Eisenhower, was abolished by Richard Nixon in 1969. Although the election of Ronald Reagan to the presidency in 1980 revived interest in nuclear weapons and countermeasures, the demise of the Soviet Union under his successor all but terminated it.

There has been a steady increase in U.S. federal support for basic research over the last half-century. However, while the federal government was once the nation's main research and development funder—accounting for as much as 67 percent of all such funding in 1964—its share of support fell below 50 percent for the first time in 1979, remained between 45 and 47 percent from 1980 to 1988, and then fell steadily to 29.5 percent in 1998, the lowest share ever recorded in the National Science Foundation's (NSF) data series (which began in 1953). These patterns are similar to those in western European nations and Japan. In 1997, for example, roughly one-third of all research and development funds derived from government sources—down considerably from the 45 percent share reported sixteen years earlier, in part due to the privatization of government research in a number of countries. In Russia, the collapse of the Soviet Union also saw a collapse of government funding for research and development, which fell nearly 75 percent in the 1990s, reducing the nation's scientific and technical workforce by half.

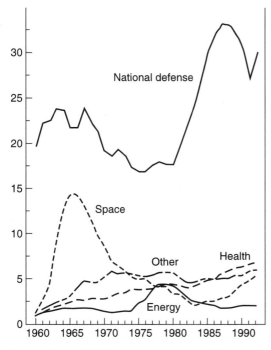

Comparison of U.S. government expenditures for research and development in several fields from 1960 to 1990 in constant dollars. The outline of the defense curve follows temperature trends in the Cold War. The hump in space records reaction to Sputnik. From National Science Foundation, *Science and Engineering Indicators* (1991).

These trends in government financial support of science in part reflect growing political controversy over nuclear power, genetic engineering, technological impact on the environment, and a variety of other science-related issues. Nuclear power, in particular, has become unpopular in most nations since the accidents at Three Mile Island in Pennsylvania and Chernobyl in the Ukraine. Scientists as well as activists have questioned the safety and the long-term effects of science-based technologies, as well as some of the more ambitious schemes for military applications such as the Strategic Defense Initiative in the United States. While for two decades after World War II scientists enjoyed relative immunity from the political process, this has faded along with the memories of wartime scientific accomplishments.

The role of the state in the support of science has become vital to its continued growth and survival in the last sixty years. Historians have recently begun to analyze the role of government support in shaping the subject matter of scientific investigations. In particular, the elite nature of scientific activity financed by government and the attractions of military, industrial, and medical applications have come under critical scrutiny by those who adhere to various theories of social construction of science and postmodernist historians who regard traditional analytical categories, derived from the sciences, as inadequate to evaluate scientific activity.

A. Hunter Dupree, *Science in the Federal Government: A History of Policies and Practices* (1986). Margaret C. Jacob, *The Politics of Western Science, 1640–1990* (1990). Hugh Slotten, *Patronage, Practice, and the Culture of American Science* (1994). Lewis Pyenson and Susan Sheets-Pyenson, *Servants of Nature* (1999).

ROBERT W. SEIDEL

GRAMOPHONE. See PHONOGRAPH.

GRAND TOUR usually refers to British travelers' explorations of Continental Europe, especially Paris and Italy, between the English Revolution and the late nineteenth century. When considering the impact of tourism on the development of science, however, it is useful to include non-British tourists traveling for pleasure and enlightenment throughout Europe and the United Kingdom. The grand tour afforded an opportunity to meet, and exchange ideas with, natural philosophers living in other countries; to witness natural scenes and phenomena not found at home; to visit *cabinets of curiosities and other collections of unique objects scattered throughout Europe; and to acquire "philosophical" objects of personal interest.

The youthful experiences of Robert *Boyle provide an early example of the role of the

Grand Tour in the education of a natural philosopher. Sent on an extended tour of Europe at the age of twelve, Boyle learned French and Italian and studied natural phenomena, particularly in the Alps, that he subsequently mentioned in philosophical works he wrote as an adult. Wintering in Florence in 1641–1642, Boyle became familiar with the works of the city's most famous celebrity, *Galileo. Visits to Tuscan palaces acquainted him with collectors' items like giant lodestones and branches of coral.

Tourism exploded during the eighteenth century. Travelers complained of the "swarms" of their fellow countrymen they encountered in their journeys. The increase in numbers reflected both the participation of more Europeans in travel, and the inclusion of more travelers from the lower-aristocracy and the upper-middle classes in the mix. Paris and Italy retained their status as the places most frequently visited, but other destinations acquired appeal for the tourist—the Low Countries, Germany, and especially Switzerland, with its unsurpassed scenery and Alpine wonders.

Whereas many English travelers went to the Continent to experience nature and antiquity, Continental travelers in England interested in natural philosophy focused on the institutions that fostered the philosophical spirit of the age. While staying in London in the early eighteenth century, the antiquarian collector Zacharias Conrad von Uffenbach recorded his visits to the headquarters of the Royal Society at Gresham College, the lecture demonstrations of Francis Hauksbee, and the instrument makers of the City. Browsing in the shops of the London instrument trade was often a highlight of visits to England. The astronomers Johann (Jean) Bernoulli III and Thomas Bugge recorded their respective pleasures in visiting the workshops of the English instrument makers and seeing their products at the observatories they served.

War did not discourage the determined scientific tourist. Ferdinand Rudolph Hassler, the Swiss surveyor who became the first superintendent of the U.S. Coast Survey, acquired much of his early scientific education while touring France and Germany in the 1790s. In Paris in 1793 to observe a solar eclipse with his teacher and companion John George Tralles, Hassler was not disturbed, according to his biographer, Emil Zschokke, "by the bloody reign of terror" during the French Revolution. Although clouds obscured the eclipse, Hassler's stay in Paris gave him the opportunity to meet Joseph-Jérôme-Le Français de Lalande, Jean-Charles Borda, and Antoine-Laurent *Lavoisier. Hassler proceeded to Göttingen to study at the university, stopping off on his way to visit the Observatory at Gotha. After returning to Paris in 1796, he spent some time in formal study before returning to Switzerland.

In the nineteenth century, Americans increasingly adopted the tradition of the grand tour and with it its role in science education. Americans traveled to Europe to earn medical degrees or to see the new wonders of industrialization; even respected American scientists could benefit from the experiences and prestige attached to a sojourn in Europe. Thus, although Maria Mitchell had already established herself as an important contributor to American astronomy by the 1850s, she seized the opportunity provided in 1857 by a Chicago banker to chaperone his daughter in her travels through Europe. Armed with letters of introduction from the president of Harvard, Edward Everett, and the secretary of the Smithsonian, Joseph Henry, Mitchell toured the observatories of England and the Continent and met with important European scientists. Her American reputation enhanced by her European reception, she subsequently became the first professor of astronomy at Vassar College, a Fellow of the American Philosophical Society, and a tireless fighter for improved educational opportunities for women.

R. E. W. Maddison, "Studies in the Life of Robert Boyle, F. R. S. Part VII: The Grand Tour," *Royal Society Notes and Records* 20 (1965): 51–77. G. LE. Turner, "The London Trade in Scientific Instrument-Making in the 18th Century," *Vistas in Astronomy* 20 (1976): 173–182. Darwin H. Stapleton, *Accounts of European Science, Technology, and Medicine Written by American Travelers Abroad, 1735–1860, in the Collections of the American Philosophical Society* (1985). Jeremy Black, *The British Abroad: The Grand Tour in the Eighteenth Century* (1992).

ALICE N. WALTERS

GRANTS make possible most of basic scientific research conducted in universities and colleges. In the United States and Britain, the role of private philanthropies has been prominent in developing the system of research grants. In other countries the government has played the leading role.

Many forms of patronage of science and technology can be regarded as grant-making. For example, during its existence from 1714 to 1828, the British Board of Longitude gave over £100,000 for efforts to make nautical clocks accurate enough to determine longitude at sea during. The Royal Society of London and the French Académie des Sciences funded some research support to their members, although the totals did not amount to much before the twentieth century. The U.S. National Academy of Sciences (NAS) had a total endowment for such purposes of under $100,000 in 1900 (*see* ACADEMIES).

GRAPH 351

Shortly thereafter, philanthropists Andrew Carnegie and John D. Rockefeller created several organizations to promote science and learning, and Henry Ford, Simon Guggenheim, and many others sought deductions allowed by the federal income tax by setting up their own foundations after 1916. World War I, which saw the mobilization of academic and industrial scientists by the National Research Council (NRC) of the NAS, catalyzed philanthropic interest in the physical sciences; after the war, the NRC dispersed Carnegie and Rockefeller Foundation funds for basic research in physics, chemistry, and biology, provided peer review of the research proposals and, as advisors and officers of these organizations, made large grants for institutional support of research to universities and research institutions. The NRC provided the managerial skills to interpret progressive foundation values into scientific programs and provided a shield against unsolicited proposals. The International Education Board of the Rockefeller Foundation extended its support of scientific research to war-torn Europe, where it helped to sponsor both the creation and the diffusion of quantum mechanics to the rest of the world, particularly the United States. It also impressed its values on distinguished European scientists like Niels *Bohr, who redirected the efforts of his Institute for Theoretical Physics (*see* BOHR INSTITUTE) toward nuclear physics with their support.

Ernst Solvay, Alfred Nobel, and Henry Wellcome had sought to endow the sciences that they believed to be the foundations of their fortunes, thereby providing support beyond the meager offering of universities and colleges to those among their faculties who merited it. The governments of several European nations provided research support, though initially not much, through institutions like the British Department of Scientific and Industrial Research (DSIR) (founded 1916), the Caisse des Recherches Scientifique (founded 1901) and its successor the Caisse Nationale des Sciences (founded 1930) in France, and the Consiglio Nazionale delle Ricerche (founded 1923) in Italy. Germany, where the public-private Kaiser Wilhelm Society provided the bulk of national patronage for scientific research, the formation of the Notgemeinschaft der Deutschen Wissenschaft in 1920 responded to the postwar exclusion of Germany from the international scientific community and succeeded in winning philanthropic support from the International Education Board of the Rockefeller Foundation, which also provided grants for individuals, as did the Rockefeller-funded National Research Council.

World War II revolutionized government research funding. The president of the Carnegie Institute of Washington, Vannevar Bush, organized the National Defense Research Committee and the Office of Scientific Research and Development to support war-related research during World War II, primarily through research contracts awarded by similar mechanisms. Although they were disbanded following the war, the military services and the U.S. Congress provided for a much larger system of research grants through the organization of the Office of Naval Research (ONR) and the Atomic Energy Commission (AEC). The question of political control of grants for basic research delayed the creation of Bush's proposed National Research Foundation, however, until 1950, when the National Science Foundation was created and placed under a National Science Board appointed by President Truman. By this time, most academic concerns about federal control and educational policy-making had been swept away and by 1953 the federal government was providing over half of the nation's research and development funding. While the federal government has come to be the leading extramural patron of research, the level of support offered has been affected primarily by considerations relating to national security or prestige, so that the Korean War, Russia's orbiting of the first artificial satellite, *Sputnik*, in 1957, and the escalation of American activity in Vietnam were more important factors in determining federal expenditure patterns than the internal needs of science. Foundation support has increased primarily to biomedical research.

In Europe, Japan, and mainland Asia, support for scientific research increased after World War II in response to the damage inflicted by the war as well as in support of postcolonial economic development. These investments helped to slow the brain drain of technical talent to the United States, while at the same time expanding national and international scientific institutions like the European Center for Nuclear Research (CERN). These expenditures have been approximately equal in size to those of the United States, and have followed similar trends.

Howard Miller, *Dollars for Research: Science and its Patrons in Nineteenth-Century America* (1970). Robert Kohler, *Partners in Science Foundations and Natural Scientists, 1900–1945* (1991). National Science Board, *Science and Engineering Indicators—2000* (2000).

ROBERT W. SEIDEL

GRAPH. The graph dates back at least to the mid-seventeenth century, when Christopher Wren and Robert Hooke invented a weather clock, the mechanism of which recorded rainfall amounts, wind direction, and temperature as pen-drawn lines on

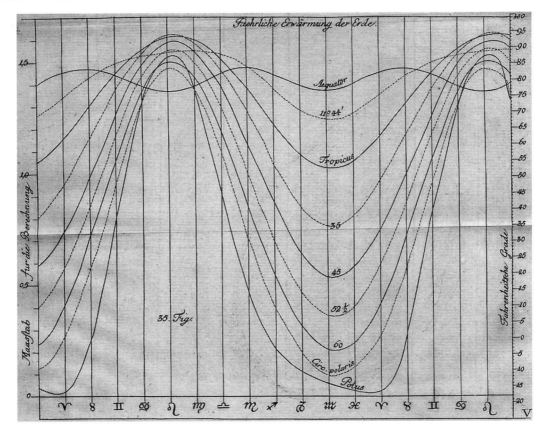

J. H. Lambert's graph of the seasonal variation of soil temperature at various depths.

paper. Scarcely a handful of such graphs were devised before 1800. Even as weather and tidal clocks became more common in the first third of the nineteenth century, natural philosophers continued mostly to prefer their data in tabular form. They understood the advantage of automatic recording devices to be in their functioning independently of human intervention, not in their graphical presentation of data. In two eighteenth-century examples, the phenomenon itself presents a graph. In 1712 Brook Taylor and Francis Hauksbee presented an image of the capillary rise of water between two glass plates; the water's profile is a segment of a hyperbola. Around 1796 John Southern invented the indicator diagram, drawn by a pen attached to a pressure gauge along the horizontal axis, while the piston in the cylinder of a steam engine contributed motion along the vertical axis. The result was a graph of pressure versus volume in the engine.

The Swiss philosopher, natural philosopher, and mathematician Johann Heinrich Lambert alone in the eighteenth century published numerous graphs, used them for data analysis, and theorized about their usefulness and limitations.

Lambert published graphs representing annual temperature variation at different soil depths, magnetic variation over several centuries, rates of evaporation of water with temperature, and other experimental data. He used graphs to test Petrus van Musschenbroek's assertion that the expansion of a metal rod with temperature is linear, showed how to determine a periodic variation from a graph, applied differentiation to graphical representations to determine rates of change, and discussed curve fitting. These techniques were far ahead of their time and Lambert's contemporaries knew little of his work.

The early decades of the nineteenth century saw several further isolated examples of graphs: William Herschel's comparison of the temperatures generated by visible light and by radiant heat; John Robison's relation of the breaking load on a wire to its length; and John Dalton's graph of the dependence of the boiling points of several compounds on their concentration. Graphs of meteorological phenomena, magnetic variation, and tides became common in the 1820s; among these were Alexander von Humboldt's representation of isotherms over the globe and many displays by Carl Friedrich *Gauss of the variation of magnetic intensity and declination over time. By 1820 graphical representation of data had become common enough that it did not require special explanation. For the most part however natural philosophers used graphs for display rather than analysis. In 1832 John Herschel presented a graphical method for determining the orbits of binary stars. Plotting the angular position of one of the pair against time, Herschel fitted a curve to the points by eye, derived an ellipse from the curve, then used the ellipse as the true orbit of the star "to the exclusion of all the original observations" if it matched the original data within the limits of observational error. Herschel's contemporary, the historian and philosopher of science William Whewell agreed that this "method of curves" eliminated errors of observation and could "correct these observations, so as to obtain data more true than the observed facts themselves."

Beginning in the 1830s and until his death in 1864 James Forbes, professor of natural philosophy at the University of Edinburgh, published dozens of data graphs, particularly in the areas of meteorology and heat. Although, like Lambert, Forbes was considerably more comfortable with graphical representation than were most of his contemporaries, he did not march alone. By the mid-nineteenth century, the representation of data in graphs had become common, if not widespread, and practitioners had developed some sophistication in curve fitting, treatment of errors, and graphical analysis.

Laura Tilling, "Early Experimental Graphs," *British Journal for the History of Science* 8 (1975): 193–213. Edward Tufte, *The Visual Display of Quantitative Information* (1983).

THEODORE S. FELDMAN

GREENHOUSE EFFECT, THE. See CLIMATE CHANGE AND GLOBAL WARMING; OZONE LAYER.

H

HABER, Fritz (1868–1934), German chemist, scientific administrator, statesman; the "father of gas warfare" and the discoverer of ammonia synthesis. Haber is one of the most controversial scientific personalities of the last century—as perhaps best expressed by his motto that a scientist should serve mankind in times of peace but the fatherland in war.

Haber was born in Breslau (Wrocław). His father was a dye merchant and chemical trader. Between 1886 and 1891 Haber studied in Berlin and Heidelberg. After obtaining his degree, he worked for several years for various chemical manufacturers. He began his scientific work in 1894 at the Technische Hochschule in Karlsruhe, first on electrochemical reactions, then on thermal gas reactions. In the latter field he was the first to synthesize ammonia, for which he later became famous.

Ammonia was of great economic importance as an essential ingredient in fertilizers and munitions. In about 1900 Haber began to investigate the nitrogen-hydrogen system, and by 1905 he had found that the formation of ammonia required a very high temperature (300°C) and a catalyst; in 1908 he used uranium and osmium as catalysts. After he had tested the prototype of an ammonia synthesizer, producing 100 cm^3/hour, the Badische Anilin- und Sodafabriken (BASF) put the synthesizer into large-scale production.

Haber was not only a talented chemist but also a brilliant organizer. In 1911 he became the first director of the Kaiser-Wilhelm Institute for Physical and Electrochemical Chemistry at Berlin-Dahlem. At the beginning of World War I he served as a chief and section director in the Ministry of War. There he proposed the use of chlorine gas as a chemical weapon and led the first mass introduction of this gas in April 1915 at Ypres. Haber also introduced other chemical weapons, including phosgene (an asphyxiating gas) and chlorarsenic.

Some of Haber's biographers interpret the suicide of his first wife, Clara Immerwahr, as being directly connected to Haber's role in gas warfare, but other factors undoubtedly contributed. That marriage to Haber was difficult is evident in the recollections of his second wife, from whom he separated in 1927, and is implied in his correspondence with his fellow chemist Richard Willstätter. In his letters to Willstätter, Haber observed that until the age of fifty the lives of men are dominated by women, then afterward by digestion. He compared himself to bread baked in a fire, describing himself as crusty, moody, and mistrustful. He blamed himself for being impatient and for a lack of purposefulness.

The Allies placed Haber on the list of war criminals because of his role in chemical warfare. Almost simultaneously, in 1919, he received the Nobel Prize in chemistry for his synthesis of ammonia as the starting point for the production of chemical fertilizer. Haber valued this decision of the Swedish Academy—which also honored two other Germans that year: Max *Planck (retrospectively awarded the physics prize for 1918) and Johannes Stark (the 1919 physics prize)—as an important initiative toward reconciliation. Haber urged that distinguished scholars keep in mind their role as models for society, and he developed conceptions of an elite made up of people of achievement and merit. He engaged himself after the war in efforts to advance chemistry and science as a whole in Germany, materially assisted talented investigators with original ideas, and tried to keep them from leaving the impoverished country. But in another mood, he lamented his tendency to enlarge his scientific activities; he pretended to envy people who spent their lives playing cards and indulging in harmless family conversation instead of devoting themselves to the great questions of science and its management. Perhaps this sourness resulted from the failure of his bold attempt to obtain gold from sea water to help Germany pay its reparation costs.

Haber was born into a Jewish family but assimilated into the Christian community through baptism. When Jews were dismissed from public service following enactment of the law for reform of the civil service (1 April 1933), which affected many of his closest co-workers, Haber sent in his own resignation. The experience created a deep personal crisis for him, as his correspondence with Albert *Einstein documents. Haber sought another post, traveled through Great Britain, the Netherlands, and France, restored old contacts in Switzerland, and died early in his exile in a hotel in Basel, an outcast from his German homeland.

L. F. Haber, *The Chemical Industry 1900–1930: International Growth and Technological Change* (1971). L. F. Haber, *The Poisonous Cloud. Chemical Warfare in the First World War* (1986). Morris Gorman, *The Story of Fritz Haber* (1987). Dietrich Stolzenberg, *Fritz Haber: Chemiker, Nobelpreisträger, Deutscher, Jude* (1994). Petra Werner in collaboration with Angelika Irmscher, *Fritz Haber: Briefe an Richard Willstätter 1910–1934* (1995).

PETRA WERNER

HARVEY, William (1578–1657), English physician and anatomist, discoverer of the circulation of the *blood. William Harvey was educated at King's School Canterbury and Gonville and Caius College, Cambridge. He took his B.A. in 1597 and remained in Cambridge studying medicine until 1600, when he migrated to Padua. There he came under the influence of Girolamo Fabrici, the leading anatomist of his day, who was carrying out comparative researches into the organs of animals. Harvey learned about Fabrici's discovery in 1574 of the valves in the veins, which would be the key to his own major discovery. Fabrici did not publish this until 1603, and did not draw any major conclusions from it.

Graduating M.D. in 1602 (and, according to the Padua copy of his degree certificate, swearing fealty to the pope), Harvey returned to London, setting up in practice and marrying, in 1604, Elizabeth Browne, daughter of a royal physician. Appointed physician to St. Bartholomew's Hospital in 1609, he became a member of the College of Physicians early in 1610. He remained devoted to the college for the rest of his life, giving benefactions and helping to enforce its Galenist views (*see* GALENISM).

For twenty eight years, beginning in 1616, Harvey gave the Lumleian Lectures on anatomy and surgery at the college. His London practice prospered. He became a royal physician in 1617 or 1618 and served the Crown loyally. He took part in embassies and, during the Civil War, acted as warden of Merton College, Oxford. Harvey died at his brother's house in London or Roehampton, on 3 June 1657.

Harvey's interest and expertise in anatomy, already clear in Cambridge, was sharpened by his period at Padua. His lecture notes for 1616 show that, like many of the younger anatomists, he had by then accepted the theory of Realdo Colombo that the active movement of the heart was its systole, when it expelled the blood into the arteries, and that blood passed from one side of the heart to the other via the lungs. Harvey came to believe in the total circulation of the blood around the body sometime after 1616 or perhaps 1619, a date implied by him in *Exercitatio de motu cordis et sanguinis in animalibus* (*On the Movement of the Heart and Blood in Animals*, 1628).

This little book, an anatomical exercise, is notable for its combination of descriptive anatomy with a strong logical reasoning to prove the circulation of the blood. Harvey pondered the results of his dissections in the light of a Galenic argument about the size of the vessels and the quantity of blood involved. His belief in the greater circulation was strengthened by an Aristotelian belief in the primacy of the heart and of the perfect circle of the macrocosm, with the warm Sun at its center. He offered a range of proofs, including a consideration of the role of the valves in the veins.

Harvey's contemporaries did not immediately accept his circulatory system, and when they did, not always for the reasons Harvey would have preferred. He extended his ideas in response to criticisms, most importantly in his public correspondence with Jean Riolan, Jr. (1649). He was preternaturally concerned with effective proof, as can be seen in his manuscript notes on Galen, in the British Library, and he may have delayed publication of *De Motu Cordis* for some years until he felt certain of its invulnerability to argument.

The same methodology of dissection and rigorous logic in support of Aristotelian theories can be seen in Harvey's longest book, *Exercitationes de generatione animalium* (*On Animal Generation*, 1651), the result of experiments carried out over more than twenty years. Following Fabrici, Harvey attempted to show that every living animal derives from an egg. He denied the Galenic two-seed theory of conception. In his view, the male seed upon entering the vagina transferred a vaporous spirit to the womb, and there transformed itself into an embryo. Although Harvey made many interesting observations, his anatomical information was soon outdated.

A conservative in many respects, strongly Galenist in his therapeutics and Aristotelian in his philosophical outlook, William Harvey made a revolutionary discovery that changed ideas on human physiology forever.

Walter Pagel, *William Harvey's Biological Ideas* (1967). Geoffrey Keynes, *The Life of William Harvey* (1978). William Harvey, *The Circulation of the Blood and Other Writings*, trans. Kenneth Franklin (1990). Don G. Bates, "Harvey's Account of His 'Discovery,'" *Medical History* 36 (1992): 361–378. Roger French, *William Harvey's Natural Philosophy* (1994).

VIVIAN NUTTON

HAWKING, Stephen (1942–), and **Carl SAGAN** (1934–1996), icons of science who starred on television and wrote popular books, but whose scientific reputations are very different.

Carl Sagan's 1980 thirteen-part television series *Cosmos* has been seen by an estimated 400 million people. *Time* magazine featured on its cover the entrancing visionary with the strangely halting yet melodic voice. Sagan was "the Prince of Popularizers," "the Showman of Science," and "the Nation's Scientific Mentor to the Masses." Many of his peers, however, never accepted Sagan as a serious scientist. He made connections and identified goals, but had a short attention span and often failed to follow through on details. Abrasive, arrogant, and egomaniacal, Sagan made few friends and many enemies. Envy no doubt also played a part.

Sagan began as an undergraduate at the University of Chicago in 1951; he finished a controversial doctoral dissertation, including sections on possible life on the Moon and a greenhouse model for the atmosphere of Venus, in 1960. His showmanship and chutzpah were evident early. A campus science lecture series he organized on the creation of life in the universe had several distinguished faculty and one young graduate student as lecturers.

Sagan won a postdoctoral fellowship to the University of California at Berkeley that year and went on to a junior appointment at Harvard University in 1963. His book *Intelligent Life in the Universe* (1966), co-authored with the Russian scientist I. S. Shklovskii, gave Sagan a taste of fame. But Harvard was not impressed, and Sagan, denied tenure, left for Cornell University. There he developed into an admired superstar, flitting like a butterfly from one scientific topic to another while becoming the preeminent voice of American space science and a national celebrity. For the 1972 *Pioneer 10* spacecraft, Sagan and his wife designed a plaque, a potential message to other civilizations, showing the location of Earth and naked male and female figures (*see* EXTRATERRESTRIAL LIFE). His book *The Cosmic Connection*, on the origin of *life, *extraterrestrial life, and space travel, came out a year later.

In 1985 he published a science fiction novel, *Contact*. The heroine's successful attempt to communicate with alien intelligences helped win in the real world a temporary reprieve from congressional budgetcutters for the SETI (Search for Extraterrestrial Intelligence) program. The novel was later made into a movie.

Sagan used his fame to champion the "nuclear winter" hypothesis: that a nuclear exchange might trigger urban fires, cloud the earth's atmosphere, block sunlight, lower temperatures to freezing, prevent crops from growing, and destroy civilization. This hypothesis remains contested. Its conclusion that a nuclear war is not winnable does not seem to have had much effect on nuclear strategy.

Sagan was rejected for membership in the National Academy of Sciences (NAS) in 1992, but later the NAS awarded him its Public Welfare Medal for his popularization of science. In 1994, he was diagnosed with myelodysplasia, a blood disorder leading to acute leukemia. A bone-marrow transplant prolonged his life until 20 December 1996.

In 1962, Stephen Hawking was at Cambridge studying for a doctorate in *cosmology. He had only worked about an hour a day as an Oxford undergraduate, physics and mathematics problems having presented little challenge to the young genius, and his laziness caused him to flounder at Cambridge. That year he was diagnosed with amyotrophic lateral sclerosis (motor neuron disease, also known as Lou Gehrig's disease). Thought and memory are not affected, but paralysis progresses. The disease rapidly worsened, then slowed. Now he realized that there were many worthwhile things he could do if reprieved. He completed his dissertation in 1965 and obtained a research position at Gonville and Caius College.

Einstein's general relativity theory predicted superdense neutron stars, black holes, and points of infinite density in the universe, but their existence seemed highly improbable. That changed with the discoveries of the *pulsar and quasar in the 1960s. Astronomers soon ascribed the power of quasars to the rotational energy of gigantic black holes and identified pulsars as rotating neutron stars.

Meanwhile, Hawking, in collaboration with Roger Penrose, a mathematician at the University of London, was developing a new mathematical technique for analyzing the relation of points in space-time, and proved mathematically that *space and time would have a beginning in a *Big Bang and end in black holes (*see* BLACK HOLE). Undertaken as an esoteric mathematical study, Hawking's work was suddenly transformed into a major contribution to the most exciting scientific topic of the decade. Just a few years out of school, Hawking already had the reputation of a new Einstein.

In the early 1970s, Hawking, unable physically to manipulate mathematical equations on paper, combined relativity theory and *quantum physics to produce a new understanding of black holes. He became a fellow of the Royal Society (1974) at the unusually early age of thirty-two. Scientific audiences gathered to see the sticklike figure hunched in a wheelchair mumbling almost incomprehensibly. In 1985, Hawking caught pneumonia and had a tracheostomy. Since then he has used a computer and a speech synthesizer to communicate.

Public fame intensified after Hawking was profiled on a 1983 BBC television program. In need of money for his children's schooling, his own nursing, and his family's support after his death, Hawking obtained a large advance for a popular book. Millions of copies, in twenty different languages, of *A Brief History of Time: From the Big Bang to Black Holes*, have been purchased, if not read, since 1988. The film version appeared in 1992. The film version appeared three years later, followed by a six-part television series, *Stephen Hawking's Universe*, in 1997.

Implications drawn from Einstein's general theory of relativity won fame and respect for Hawking after the discovery of quasars and pulsars demanded explanations that his work provided. In his person, Hawking is an inspiration for the physically handicapped. Sagan inspired searches for extraterrestrial life. It is perhaps possible that early fame and the duties of a popularizer short-circuited Sagan's potential for fundamental scientific advance and the type of respect that Hawking earned.

Michael White and John Gribbin, *Stephen Hawking: A Life in Science* (1992). Stephen Hawking, *Black Holes and Baby Universes and Other Essays* (1993). Keay Davidson, *Carl Sagan: A Life* (1999). William Poundstone, *Carl Sagan: A Life in the Cosmos* (1999).
Web Sites: http://www. Hawking.org.uk
http://www.sciam.com/explorations/010697sagan/010697explorations.html.

NORRISS S. HETHERINGTON

HEARING. See ACOUSTICS AND HEARING.

HEAT. See FIRE AND HEAT.

HEAT ENGINE. In the later seventeenth century, it was recognized that the pressure of the atmosphere or of steam might be used as a source of power. Early attempts to exploit the insight—notably by Denis Papin in France and Thomas Savery in England—had limited success. But early in the eighteenth century, the English engineer, Thomas Newcomen began erecting reliable and effective engines, the first of them at Dudley Castle in Worcestershire in 1712, which were soon in use throughout Britain and even in continental Europe and the Americas.

The Newcomen engine consisted of a large vertical cylinder, open at the top, in which a piston, suspended by a chain from one end of a centrally pivoted horizontal beam, moved up and down. The motion was generated by passing steam from a boiler into the cylinder below the piston and then condensing the steam by the injection of cold water. Acting against the resulting partial vacuum, the pressure of the atmosphere drove the piston down. A further injection of steam then raised the piston again and the process repeated, creating a vertical motion that could be harnessed by connecting the other extremity of the beam to a suitable machine, typically a pump for draining mines.

Outstanding among the eighteenth-century engineers who worked to improve the Newcomen engine was James Watt, instrument maker and natural philosopher, whose numerous innovations included a separate condenser to avoid the injection of condensing water into the cylinder. Watt used steam, rather than the atmosphere, to push down the piston and incorporated a phase in which steam raised the piston, giving power on the upward as well as the downward stroke. This innovation, which entailed the replacement of the chain of the Newcomen engine with a rod connecting the piston to the beam, greatly increased the potential of steam as a source of power.

The ending of Watt's patents in the early years of the nineteenth century undermined the virtual monopoly that he and his business partner Matthew Boulton had exercised in steampower technology for over two decades. The way was now open for the use of pressures above that of the atmosphere (of which Watt had made little use) and for the age of steam locomotion, in which Richard Trevithick and subsequently George Stephenson and his son Robert were the main innovators. These new departures, which led to significant improvements in economy (conventionally measured as the amount of work produced for a given weight of coal consumed) inspired scientists as well as engineers. Sadi *Carnot's *Reflections on the Motive Power of Fire* (Paris, 1824) offered the most penetrating analysis of the way in which the heat of the boiler yielded mechanical work. Carnot conceived the heat engine as a device to produce power by the "fall" of heat from the high temperature of the boiler to the low temperature of the condenser. The analogy referred to a waterfall: the bigger the fall, the greater the power produced. In the mid-century, James *Joule, William *Thomson, Lord Kelvin, and Rudolf Clausius showed that some of the "falling" heat converted into work, the ratio between heat consumed and the work produced being a constant, the mechanical equivalent of heat. Nevertheless, the core of Carnot's ideas survived in what became the second law of thermodynamics.

The possible use of working substances other than steam also attracted attention. But despite attempts to use heated air, and even liquids and solids, steam continued to hold sway. During the second half of the nineteenth century, improvements leading to enhanced economy and greater power and compactness gave the steam engine a new flexibility. By 1900, horizontal engines had largely ousted the slower and more cumbersome beam engine, and internal combustion engines were emerging as important alternatives for certain uses. Following the work of Charles Parsons and August Camille Edmond Rateau, the more efficient steam turbines too had won widespread favor. They retained a dominant position in the generation of electricity and, until the mid-twentieth century, in the powering of textile and other machinery.

N.-L.-S. Carnot, *Reflexions on the Motive Power of Fire* (1824; English trans., ed. Robert Fox, 1986). H. W. Dickinson, *A Short History of the Steam Engine* (1938; 1963). D. S. L. Cardwell, *From Watt to Clausius. The Rise of Thermodynamics in the Early Industrial Age* (1971). R. L. Hills, *Power from Steam. A History of the Stationary Steam Engine* (1989).

ROBERT FOX

HEGELIANISM. See DIALECTICAL MATERIALISM.

HEISENBERG, Werner (1901–1976), and **Wolfgang PAULI** (1900–1958), theoretical physicists.

Heisenberg and Pauli were key members of the small group of theorists who invented quantum mechanics, the physics of atomic and molecular processes, during the 1920s. Their close collaboration continued thereafter as they refined the theory, applied it to new phenomena, and trained new generations in its uses. Pauli was called "the conscience of physics" for his mastery of theoretical physics and his critical assessment of ongoing work. His review articles and books on relativity theory, statistical mechanics, and quantum physics are masterpieces of physical insight. Heisenberg also worked on hydrodynamics. His decisions to remain in Germany after Hitler's rise to power and to work on applications of nuclear fission during the war still evoke controversy.

Heisenberg was quiet and friendly, at once retiring and daring, virtuous and athletic. Pauli to the contrary was outspoken and aggressive, systematic yet inclined to mysticism and risqué nightlife. Like most quantum physicists of the 1920s, Heisenberg and Pauli were products of the European upper-middle-class cultural elite. Both of their fathers were professors: Pauli's, a professor of colloid chemistry at the University of Vienna; Heisenberg's, a professor of Byzantine philology at the University of Munich. Both sons attended outstanding humanistic gymnasia (high schools), which emphasized classical languages

and literature. Both were attracted to physics by the excitement attending Albert Einstein's theory of relativity, and both earned doctorates in theoretical physics with Arnold Sommerfeld in Munich. Pauli arrived at the University of Munich in 1918, two years ahead of Heisenberg. Heisenberg's early encounter with Pauli, then in his last semester, helped to turn Heisenberg toward atomic theory. Their semester together marked the beginning of the lifelong collaboration that encouraged their most significant contributions to quantum physics. Their voluminous technical correspondence offers a rich resource for the history of twentieth-century physics.

Pauli and Heisenberg each pursued postdoctoral work in quantum physics with Max Born in Göttingen and Niels Bohr in Copenhagen. Although still in their twenties, both assumed full professorships in theoretical physics in 1928: Heisenberg at the University of Leipzig, Pauli at the Swiss Federal Institute of Technology in Zurich. There they trained students from the world over until well after the outbreak of war. In 1942 Heisenberg transferred to the Kaiser Wilhelm Institute for Physics in Berlin, where he headed German fission research, and a year later assumed a professorship at the University of Berlin. Fearing a German invasion of Switzerland, Pauli moved to the Institute for Advanced Study in Princeton. He returned to Zurich after the war, while Heisenberg remained head of his institute, renamed for Max Planck, in Göttingen and Munich. He was heavily involved in West German science and cultural policy.

Heisenberg and Pauli entered their profession at a time of great ferment. New data and analyses indicated the inadequacy of the planetary quantum model of the atom developed earlier by Bohr and Sommerfeld. At the same time, turmoil surrounding the German defeat in World War I, hyperinflation, and a boycott of German science made professional advancement difficult for young physicists. Special scholarships and foreign philanthropy helped maintain the financial stability of German physics, while social disengagement and intense work by its practitioners fostered rapid developments.

Together with other physicists—P. A. M. *Dirac, Ernst Pascual Jordan, Hendrik Kramers, Alfred Landé, and Erwin *Schrödinger—and their mentors—Bohr, Born, and Sommerfeld—Heisenberg and Pauli probed the boundaries of the quantum theory in its applications to spectroscopy, atomic models, and the interactions of light with atoms. Pauli's emphasis on empirical data and his constant criticism of Heisenberg's inconsistencies helped push Heisenberg to achieve the first breakthrough to quantum mechanics in 1925. Heisenberg's often radical approach helped to encourage Pauli's well-known "exclusion principle" in the same year, a principle limiting the number of electrons in each quantum state of atoms and molecules. After further correspondence with Pauli in 1926 and 1927, Heisenberg, then in Copenhagen, proposed his well-known principle of uncertainty, or indeterminacy, one of the pillars of the current Copenhagen Interpretation of quantum mechanics.

The relativistic formulation of quantum mechanics and its extension to the electromagnetic field during the late 1920s formed the background to the only joint Heisenberg-Pauli publication, a two-part paper in 1929 and 1930 presenting a general quantum theory of fields. It provided the foundation for subsequent work. But their paper also confirmed the existence of infinite results in the calculation of observed finite quantities, a problem that plagued field theories for years. Two other collaborations were influential but unpublished at the time: a report for the canceled 1939 Solvay Congress, and a proposed unified theory from which Pauli resigned shortly before his death.

New data raised additional challenges for quantum mechanics during the 1930s. Among these were the discovery of the neutron, the energy distribution of electrons emitted by neutrons in beta decay, and the mysterious showers of particles created when a cosmic ray smashes into matter. Correspondence, conference meetings, and collaborations between their institutes enabled Heisenberg, Pauli, and their coworkers to achieve fundamental new results. Heisenberg presented the first neutron-proton model of the nucleus in 1932. A year later Pauli proposed a solution to the energy puzzle in beta decay by postulating the existence of what became the neutrino, an elementary particle later associated with the weak force, one of the four forces of nature (two others being gravitational and electromagnetic forces). Utilizing new data on cosmic rays and their interactions with matter, Heisenberg and Pauli produced influential studies of another of the four forces, the proposed strong force, and its elementary particle, the meson, which challenged the adequacy of then-current quantum mechanics.

As accelerators replaced cosmic rays after the war as sources of elementary-particle data, Heisenberg and Pauli remained at the forefront of work on high-energy interactions and the search for a unified theory of the four forces (or fields). Nevertheless, they were skeptical that the infinities of field theory could be effectively eliminated through mere subtraction of the offending terms, a process known as "renormalization" that is now widely accepted. They argued instead for a new transformation in *quantum physics.

Max Jammer, *The Conceptual Development of Quantum Mechanics* (1966). Markus Fierz, "Pauli, Wolfgang," in *Dictionary of Scientific Biography*, vol. 10, ed. Charles Coulston Gillispie (1974): 422–425. Wolfgang Pauli, *Wissenschaftlicher Briefwechsel mit Bohr, Einstein, Heisenberg u. a.*, eds. Armin Hermann, Karl von Meyenn, Victor Weisskopf (1979–). Abraham Pais, *Inward Bound: Of Matter and Forces in the Physical World* (1986). David C. Cassidy, *Uncertainty: The Life and Science of Werner Heisenberg* (1992).

DAVID C. CASSIDY

HELMHOLTZ, Hermann von (1821–1894), physiologist, physicist, philosopher, and statesman of science; and **Heinrich HERTZ** (1857–1894), physicist

Hermann Ludwig Ferdinand von Helmholtz was born in Potsdam, Germany, on 31 August 1821. He came from a lower-middle-class family that stressed education and culture. As a youth, he wanted to become a physicist, but financial considerations forced him instead to become a medical doctor. In 1838, he enrolled in the Friedrich-Wilhelms-Institut in Berlin, the Prussian military's medical-training institution.

While serving as a Prussian military doctor in Potsdam (1843–1848), Helmholtz published articles on heat and muscle physiology. In 1847, he also published his epoch-making memoir, "On the Conservation of Force," which provided the conceptually clearest and most general presentation to that time of what became known as the law of conservation of energy.

Helmholtz left military service for academic life in 1848. He taught for a year at the Art Academy and served as an assistant at the Anatomical Museum in Berlin. In 1849, he became an associate professor of physiology at the University of Königsberg. He soon announced a major invention, the ophthalmoscope, and a major discovery: nervous impulses are propagated at a finite, measurable velocity, hence a short time intervenes between human thought and bodily reaction. He soon became a full professor (1851). Before leaving Königsberg for Bonn in 1855, Helmholtz contributed important studies of color, human perception, electric currents within the human body, and the eye's accommodation.

Helmholtz remained at Bonn for only three years, during which time he published the first part of his magisterial *Handbook of Physiological Optics* (1856). He left for Heidelberg in 1858, and there completed the second (1860) and third (1867) parts of the *Handbook*, which synthesized the empirical results of physiological optics and propounded Helmholtz's own theory of perception. During the same period, he revolutionized musicology with his *On the Sensations of Tone as a Physiological Basis for the Theory of Music* (1863). Along the way, he independently discovered non-Euclidean geometry and gave popular lectures on science to lay and professional audiences. These last were published as *Popular Scientific Lectures* (1865, 1867, 1876).

At Heidelberg, Helmholtz's intellectual interests had shifted from physiology to physics, especially hydrodynamics, acoustics, electric currents, and electrodynamics. With the founding of the German empire in 1871, Helmholtz was called to Berlin as professor of physics. Now widely perceived as Germany's foremost man of science, he was received into the Prussian nobility. From 1887 to 1894, he served as the founding president of the Imperial Institute of Physics and Technology in Berlin, an institute devoted to pure and applied physics, the setting of physical standards, and testing.

Helmholtz's most gifted student was Heinrich Hertz, who was born in Hamburg on 22 February 1857 to a well-situated and highly cultured family. Intending to become an engineer, Hertz studied at the Dresden Polytechnic and the Munich Technische Hochschule, but in 1877 he switched to physics and the University of Munich. A year later, he transferred to Berlin to study with Helmholtz.

In the 1860s and 1870s, Helmholtz was much concerned with evaluating competing theories of *electrodynamics. Hertz became Helmholtz's disciple in his visionary program to establish firm foundations for electrodynamics. Under his direction, Hertz first worked on a topic on electric currents. When Helmholtz suggested in 1879 that Hertz test experimentally the assumptions underlying James Clerk *Maxwell's theory of electromagnetism, Hertz demurred, in part because he thought it too difficult. Instead, he completed a doctoral dissertation on electromagnetic induction.

During the next three years (1880–1883), Hertz served as Helmholtz's assistant, conducted research in his institute, and occasionally dined with his family. In 1883, he left Berlin for a lectureship at the University of Kiel. There he produced a penetrating analysis of Maxwell's electrodynamic equations, showing their essential validity even as he doubted Maxwell's contiguous-action interpretation of them.

In 1885, Hertz became professor of physics at the Karlsruhe Polytechnic. During the next four years, he confronted the problem that Helmholtz had posed in 1879: He tested Maxwell's theory, demonstrating experimentally the existence of electromagnetic waves propagated at finite velocity in air. (During the course of his work, he also discovered, in 1887, the photoelectric effect.) Helmholtz arranged for the quick publication of Hertz's results. These years marked his greatest achievements in physics.

In 1889, Hertz moved to the University of Bonn. He further analyzed Maxwell's theory, now adopting Maxwell's contiguous-action interpretation of his equations, which he (and others) put into their canonical form. He also took up the new problem of the electrodynamics of moving bodies. He had clarified Maxwell's theory and thus fulfilled Helmholtz's program of clarifying electrodynamics, while simultaneously achieving world fame for himself.

During the 1880s Helmholtz had sought to reformulate mechanics and to provide a mechanical version of the second law of *thermodynamics (*see* ENTROPHY). Hertz now undertook to develop this program. In *The Principles of Mechanics* (1894), Hertz eliminated the concept of force and developed an ether-based physics. The *Principles* represented one of the last statements of the mechanical view of nature. The book was received with much skepticism, even by Helmholtz, who wrote a preface to it.

Hertz's death, in Bonn on 1 January 1894, and Helmholtz's in Berlin on 8 September 1894, marked the end of classical physics and its mechanical worldview. Helmholtz sought to unify physics, if not all the sciences, and indeed hoped for the ultimate unification of all culture. Hertz, by contrast, worked within the physics that Helmholtz had outlined. Together they cleared the ground in electrodynamics and *mechanics, and so paved the way for Max *Planck, Albert *Einstein, and others at the turn of the century. Furthermore, Hertz's results in electrodynamics proved as seminal for technology as for physical theory, for they set the stage for revolution in wireless communication.

Leo Koenigsberger, *Hermann von Helmholtz* (1906; reprint 1965). David Cahan, ed., *Hermann von Helmholtz and the Foundations of Nineteenth-Century Science* (1993). Jed Z. Buchwald, *The Creation of Scientific Effects: Heinrich Hertz and Electric Waves* (1994). Davis Baird, R. I. G. Hughes, and Alfred Nordmann, eds., *Heinrich Hertz: Classical Physicist, Modern Philosopher* (1998).

DAVID CAHAN

HEREDITY. Theories of heredity—the tendency of like to produce like—were long subordinate to the broader biological issues of gestation and development. For example, the popular theory of pre-formation, suffused with developmental speculations, held that organisms contained their successors in miniature, packed like tiny Chinese dolls, each ready to be enlarged. However, hybridization in plants and animals indicated

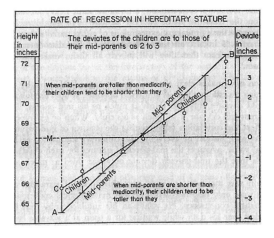

Francis Galton's depiction of his concept of regression toward the mean (1885). The height of the midparent is the average of the height of the father and 1.08 times the height of the mother.

that organisms could be modified. During the nineteenth century, this and other evidence helped to discredit pre-formationism and to add weight to the ancient idea that characters are transmitted from one generation to the next by the union of generative material from the male and the female.

Theories of heredity came to be embedded in the theories of *evolution that emerged in the late eighteenth century. In the 1760s, Carl *Linnaeus claimed that hybridization occurs in nature and leads to the creation of new species. Data from breeders did suggest that some "pre-potent" characters in long-standing lines of plants and animals were transmitted faithfully from one generation to the next. Botanists, later including Gregor *Mendel, conducted hybridization experiments to test Linnaeus's theory, but most succeeded only in demonstrating what breeders knew—that it was difficult to fix the characters of hybrid varieties. The varieties often showed the phenomenon of "reversion," their characters disappearing for many generations then suddenly reappearing. Charles *Darwin noted that a descendant in a line of hybrid pigeons selected to be brown-feathered might be born with the white-edged plumage of a lost founding contributor.

The reversion in the breeders' pigeons was one among many such instances that Darwin noted in his Variations in Plants and Animals under Domestication (1868), a comprehensive work on breeding and heredity. Darwin came to confront the problem of inheritance precisely because he wanted to resolve difficulties in his theory of evolution by natural selection. Scientists critical of the theory faulted it in part because it did

not account for the postulated objects of natural selection—advantageous traits that would be transmitted to future generations. His critics wondered how such traits arose, and why, even if they did, they would not blend in with the rest of the population and eventually disappear.

Darwin attempted to resolve the issue by resorting to the common view that organisms could acquire inheritable characters. Breeders and biologists held that such characters might be gained or lost in response to environmental influences or specific life experiences. In the mid-1790s, Darwin's grandfather Erasmus Darwin, an early evolutionist, proposed in his Zoonomia that all life had originated in "one living filament ... with the power of acquiring new parts ... and of delivering down those improvements to its posterity." In the Variations, Charles Darwin reported a number of instances of heritable gain and depletion—for example, a cow that had lost a horn from an accident and then produced three calves that were hornless on the same side of the head.

In the same work, Darwin also postulated a theory of pangenesis, a mechanism whereby acquired characters were made heritable. According to the theory, particles called "gemmules" circulated in the body's fluids, altered under environmental or life experiences, and then entered the body's reproductive system. There they made possible the transmission of the acquired character to the next generation. Powerful gemmules accounted for the phenomenon of pre-potency and latent ones for reversion.

Darwin's theory of pangenesis failed to attract much scientific support. In 1869–1870, his cousin Francis Galton, doubting that gemmules were responsible for the acquisition of inheritable characters, injected blood from one variety of rabbit into rabbits of another type to see whether the descendants of the recipients would take on any of the donors' characteristics. They did not. In 1871, Galton reported to the Royal Society that "the doctrine of pangenesis ... is incorrect."

Sidestepping unobservables such as gemmules, Galton devised a quantitative approach to the problem of heredity that depended on the measurement of an organism's observable characters. In 1876, he showed that in seeds from the sweet pea, a group of parental seeds of the same weight produced a family of daughter seeds in which the weights were distributed around a mean in a Gaussian fashion. Moreover, the mean weight of every daughter family fell closer to the mean of the total population than did that of its parent group. Galton interpreted this finding to suggest that the characteristics of offspring were products not only of the immediate parent but also of numerous forebears, a view consistent with his belief in the material nature of heredity. He argued that the effect of ancestry caused the progeny of one generation to revert toward the center of the population, and he dubbed the measure of that tendency the "coefficient of reversion."

Galton, a pioneer eugenicist, worked with pea seeds, he later said, "only as a means of throwing light on heredity in Man." In the mid-1880s, he gathered data on features such as height of parents and their grown children. He found that the intergenerational data were related by the coefficient of reversion, but recognizing that the variable expressed a general statistical property independent of heredity, he renamed it the "coefficient of regression." Further analysis of his anthropometric data led him to realize that its disparate parts might be mathematically connected through a "coefficient of correlation." In 1889, Galton summarized his statistical studies in *Natural Inheritance*, a work that gave heredity its first sharp definition: the quantitative, hence measurable, relationship between generations for given characters.

All during this period, Darwin's postulation of gemmules had helped prompt consideration of material mechanisms of inheritance. August Weismann, a biologist at the University of Freiburg, became convinced in 1868 from his studies of *cytology that "the direction of development"—his term for the expressed force of heredity—was transferred from parents to offspring through "germ plasm" contained in the sperm cell and the egg cell. Although an enthusiast of Darwin's theory of evolution, he concluded that acquired characters were probably not heritable because the germ plasm was continuous between generations, unaffected by environmental influences. Testing this idea in the 1880s, he cut off the tails of hundreds of mice, and found that the mutilation failed to produce hereditary effects.

By the 1890s, heredity had become a branch of biology in its own right. After 1900, Mendel's rediscovered papers gave the particulate, material theory of inheritance an enormous boost. Mendelism resolved the issues of pre-potency and reversion, and together with Galton's statistical approach to inheritance, threw a bright light on the mechanism of evolution by natural selection. It also raised a flood of new questions, with which scientists are still preoccupied, about how the material mechanisms of heredity work.

Robert Olby, *The Origins of Mendelism* (1967). Frederick B. Churchill, "August Weismann and A Break from Tradition," *Journal of the History of Biology*, 1 (1968): 91–112. Ruth Schwartz Cowan, "Francis Galton's Contribution to Genetics," *Journal of the History of Biology*, 5 (Fall 1972): 389–412. Gloria Robinson, *A Prelude to Genetics* (1976). Ruth Schwartz Cowan, "Nature and Nurture: The Interplay of Biology and Politics in the Work of Francis Galton," *Studies in the History of Biology*, 1 (1977): 133–207. N. Russell, *Like Engend'ring Like: Heredity and Animal Breeding in Early Modern England* (1986).

DANIEL J. KEVLES

HERMETICISM. In 1462–1463, the Florentine Renaissance humanist Marsilio Ficino translated into Latin a Byzantine collection of treatises known collectively as the Hermetic Corpus. His translation, for which he interrupted his life work of translating Plato and Plotinus, introduced hermeticism, hitherto known only by a few fragmentary texts and references, to European philosophers. For two centuries, Hermetic Corpus and associated texts fascinated humanists and philosophers. They believed them to be of great antiquity, written by a younger contemporary of Moses, the Egyptian sage known as Hermes Trismegistus, an alias for the Egyptian god of wisdom and knowledge, Thoth. In 1614 the scholar Isaac Casaubon threw cold water on this reverent attitude by concluding from textual analysis that the Hermetic Corpus, far from being one of the earliest texts of revealed religion, was a much later compilation. His conclusions held up. By the end of the seventeenth century hermeticism was in precipitous decline. Today scholars believe that members of eclectic, gnostic religious groups in Egypt wrote the texts in the first and second century A.D.

The Hermetic Corpus dealt primarily with religion. It inspired the hope in Renaissance scholars that it might be a guide to the Prisca Theologica, the ancient theology, and would thus serve, depending on the point of view of the scholar, to replace Christianity or to strengthen it by extending the repertoire of revealed religion. Some texts, however, considered how the wise could come to understand and control the correspondences between the macrocosm and the microcosm. They discussed astrology, alchemy, and the signatures of plants.

Marsilio Ficino and his associate Giovanni Pico della Mirandola assimilated hermeticism to two other intellectual traditions, *Neoplatonism, and natural magic. Magic has been practiced in many human societies and in broad outline everywhere depends on similar assumptions. Magicians assume that the powers by which one thing in the world affects another are hidden or occult (*see* SYMPATHY AND OCCULT QUALITY). They can be discovered, and thus controlled, only with difficulty, usually by the magician who has special insights because of his preparation, as much spiritual as intellectual or practical.

Two kinds of magic can be distinguished: spiritual and natural. In spiritual magic, the magician prevails upon the spirits, good or bad, white or black, to set in motion these occult powers. In natural magic, the magician relies on detecting correspondences and signatures in the natural world. Renaissance humanists and philosophers distanced themselves from spiritual magic, a touchy matter that quickly led to trouble with the church. They pursued instead a dignified version of natural magic. Since both magic and Neoplatonism had contributed to the synthesis of the Hermetic Corpus, Ficino and Pico did not have far to go to assimilate it to the versions of magic and Neoplatonism current in their own time. Giambattista della Porta, the most famous exponent of natural magic, published his *Magia naturalis* (Natural Magic) in 1558. Much of it dealt with the "magic" of mechanical gadgets, secret writing, and cosmetics.

Positivist historians of science naturally viewed this amalgam of hermeticism, natural magic, and Neoplatonism as inimical to the growth of modern science. Those not wedded to that tradition thought otherwise. The distinguished American historian Lynn Thorndike wrote his monumental *History of Magic and Experimental Science* (8 vols., 1923–1958) to demonstrate that magical tradition, with its utilitarian attitude to the world that contrasted strongly with the contemplative attitude of philosophers, spawned the method of experiment. The German physician and historian Walter Pagel, in a series of scholarly studies from the 1930s through the 1960s, showed how the hermetic traditions shaped the work of chemist-physicians such as Paracelsus (c. 1493–1541) and Jean Baptiste van Helmont (1579–1644). From there, certain hermetic and Neoplatonic ideas passed into chemistry and mineralogy and were revived in the late eighteenth century by the founders of *Naturphilosophie.

In 1964 the historian Frances Yates made strong claims that hermeticism had affected not just certain methods or particular chemists but was one of the basic causes of the Scientific *Revolution. Her thesis appeared when historians of science sought new interpretations of the formative period of modern science. Those who followed up her suggestions found that hermeticism in particular (as opposed to Neoplatonism) had been most influential in the theory of matter—whether in chemistry or in natural philosophy. Some historians of science continued the Pagel project. Others argued that Isaac *Newton, who labored over alchemical and hermetic works for years, may have found in them support for his ideas of attraction and repulsion, traditionally regarded as occult properties

(*see* ALCHEMY). Later it was suggested that music mediated the domains of magic and experimental science and that musical practices informed natural philosophy (*see* MUSIC). As the features of music thought to be occult were understood through experimentation, natural magic began to yield to science. Today few historians of science would doubt that hermeticism had an important role to play in the genesis of modern science though just what role and just how important a role have not been decided.

Walter Pagel, *Paracelsus* (1958). D. P. Walker, *Spiritual and Demonic Magic* (1958). Frances Yates, *Giordano Bruno and the Hermetic Tradition* (1964). Paolo Rossi, *Francis Bacon: From Magic to Science* (1968). B. J. T. Dobbs, *The Foundations of Newton's Alchemy* (1975). Charles Webster, *From Paracelsus to Newton: Magic and the Making of Modern Science* (1982). Alan Debus, *The French Paracelsians* (1991). Penelope Gouk, *Music, Science, and Natural Magic in Seventeenth-Century England* (1999).

RACHEL LAUDAN

HERTZ, HEINRICH. See HELMHOLTZ, HERMANN VON, AND HEINRICH HERTZ.

HIGH-ENERGY PHYSICS. The field of high-energy physics emerged after World War II out of research in *nuclear physics and *cosmic ray physics. Its name referred to the energies of the nuclear and subnuclear particles that physicists sought to study, and hence the field was also sometimes called elementary particle physics.

Physicists in the first decades of the twentieth century used particles emitted by *radioactivity as a probe of the atomic nucleus. But many of the charged alpha and beta particles emitted in natural radioactive decay had insufficient energy to overcome the electrical barrier of the nucleus, and exist in insufficient quantities to provoke enough reactions for convenient study. Starting in the 1920s nuclear physicists sought to increase the energy of the particles by accelerating them. Early accelerators passed electrically charged particles through a single large voltage drop (as in the Cockcroft-Walton and Van de Graaff accelerators) or multiple, smaller drops provided by high-frequency oscillators (as in the cyclotron and linear accelerator, or linac) to kick particles to higher speeds. Technical refinements and industrial-size apparatus allowed particle *accelerators to produce ever higher energies through the 1930s, beyond ten million electron volts.

In the meantime, starting in about 1930, physicists began to exploit a natural source of high-energy particles in cosmic rays. A thriving field of cosmic ray research focused on sorting the nature of cosmic radiation as well as new phenomena and new particles, including the positron (detected in 1932 by Carl David A.

Anderson) and a mysterious new particle with mass midway between that of the electron and proton, which was hence called the meson. Whereas nuclear physics in the 1930s emphasized the particle accelerator, cosmic ray research relied on particle detectors, especially *cloud chambers and *Geiger counters.

World War II interrupted the pursuit of high-energy particles but provided new resources, both technical and political, to exploit after the war. Technical resources included advances in microwave electronics and new designs for accelerators, both of which allowed physicists to push into energies of billions of electron volts. By about 1950 physicists could perceive some separation between nuclear physics and a new subfield of particle physics, which became known as high-energy physics (*see* WORLD WAR II AND COLD WAR).

New political resources stemmed from the contributions of physicists to military technology during World War II and their continued mobilization during the Cold War. National governments supported high-energy physics as a way to train new scientists and engineers who might then work on problems of national interest; as a means to keep talented scientists on tap in national laboratories in case of military emergency; and as a hedge against scientific discoveries that might have military or industrial applications. High-energy physics also provided a surrogate arena for international competition, with American, Soviet, and European labs jockeying for the claim to high-energy hegemony. High-energy physicists accepted and encouraged such justifications of state support, in exchange for the opportunity to pursue interesting and challenging scientific problems at higher energies.

Postwar high-energy physics combined detectors and accelerators in laboratories that typified a new type of big science, characterized by large and expensive equipment, state support, cooperative team research, collaboration of scientists and engineers, and industrial-style management. American laboratories led the postwar development of big-science accelerator programs, notably at the University of California at Berkeley and Brookhaven in New York, although the Soviet Union and a new European lab at *CERN soon offered strong competition. The United States in the 1960s would choose to centralize its largest accelerators in a new lab at Fermilab near Chicago, which would thereafter challenge CERN for the claim to the highest energies.

High-energy physics had theoretical and experimental components. Experimentalists pushed toward higher energies, lured by the possibility of producing previously unseen particles

with masses equivalent to the energy of particle collisions. In 1955 Berkeley scientists achieved the exemplary discovery of the antiproton using the first generation of postwar accelerators. Higher energies required ever larger magnets and vacuum chambers and bigger budgets, all of which constrained the ambitions of accelerator builders. The invention of strong focusing in 1952 provided a way around the problem of magnets and vacuum volumes, both of which contributed to costs. The principle used an alternating gradient in the magnetic field to compress the beam of particles, and quickly led to proposals from Brookhaven, where the idea originated, and CERN, whose design problems had inspired the Brookhaven work, for machines in the range of 25 billion electron volts.

Physicists still needed ways to study the output of accelerators, and particle detectors grew in importance and size in the 1950s and beyond. They fell into two general categories. Image detectors, such as cloud and bubble chambers and nuclear emulsions, produced a snapshot of the tracks left by individual particles at a particular time and place. Logic devices, such as electronic counters, accumulated statistical counts of large numbers of particles. By the 1970s physicists were combining the two types in new, even larger detectors.

Experiment led theory in high-energy physics through the 1950s, as theorists struggled to explain new resonances and particles emerging from accelerators. To make sense of the proliferating particles theorists classified them according to such properties as strangeness and isospin; then, in the 1960s, theorists developed the *quark model of matter to resimplify the concept of fundamental particles and began to seek a unified theory of the four fundamental forces that would incorporate the quark model. By the 1970s theory was driving experiment, as particle physicists directed their research to find particles predicted by the new theories. Their efforts paid off in the detection of the J/psi and W and Z particles, weak neutral currents, and several of the postulated quarks, thus confirming the so-called *standard model of the theorists and encouraging speculation about a grand unified theory, or *theory of everything.

The spiraling energies available in accelerators changed the definition of high energy. Accelerators that had pushed back the high-energy frontier in the late 1940s were considered medium-energy machines by the mid-1960s; succeeding generations of devices might last only a decade or so at the cutting edge. Theorists, in the meantime, extended their equations far beyond energies available in the laboratory or in nature, and particle physics began to merge with cosmology; the phenomena postulated by theorists existed only in the milliseconds after the Big Bang and disappeared as the universe expanded and cooled.

Scientific and popular attention to the quest of particle physicists made theirs the most glamorous field of physics in particular, if not science in general, in the second half of the twentieth century. High-energy physicists claimed to engage in the most fundamental research, from which all other science was derived, since they dealt with the elementary constituents of matter. Physicists in other fields, such as solid-state physics, questioned the pretensions of particle physicists and sought to divert some of their substantial government funding. The perceived fundamental nature of high-energy physics alone could not ensure continued state support, especially after the end of the Cold War. American physicists in the 1980s proposed a massive new accelerator, the Superconducting Super Collider (SSC), to pursue the Higgs boson and other exotic phenomena predicted by theory in the neighborhood of 40 trillion volts. In 1993 the U.S. Congress voted to end the multi-billion dollar project, unconvinced that the scientific results justified the social investment.

Laurie Brown and Lillian Hoddeson, eds., *The Birth of Particle Physics* (1983). Andrew Pickering, *Constructing Quarks: A Sociological History of Particle Physics* (1984). Laurie Brown, Lillian Hoddeson, and Max Dresden, eds., *Pions to Quarks: Particle Physics in the 1950s* (1989). Peter Galison, *Image and Logic: A Material Culture of Microphysics* (1997). Lillian Hoddeson, Laurie Brown, Max Dresden, and Michael Riordan, eds., *The Rise of the Standard Model: Particle Physics in the 1960s and 1970s* (1997).

PETER J. WESTWICK

HISTORIOGRAPHY OF SCIENCE is the history of the treatment of science as a subject of historical investigation. Historiography in the narrow sense means the history of historical writing; more broadly conceived, it also embraces historical methodology, including the values and principles guiding the selection of subjects, the construction of categories of historical analysis, and the demarcation of historical periodization.

Although historical scholarship was regarded with skepticism in the seventeenth century, the debate between the "ancients" and the "moderns" introduced into historical thinking the narrative of progress punctuated by inventions such as the compass, gunpowder, and the printing press, thus linking science and technology to the advancement of civilization. Comprehensive historical narratives that integrated science and technology into the story of civilization were rare. The

magisterial *Encyclopedia* of Denis Diderot and Jean d'Alembert (17 volumes of text and 11 of plates, 1751–1772), identified the sciences, arts, and trades as essential components of a grand story of enlightenment and progress. D'Alembert's *Preliminary Discourse to the Encyclopedia of Diderot* (1751) laid out the first narrative of the *Scientific Revolution from *Galileo to *Newton and Locke, depicted as the story of heroic efforts against the dark and repressive forces of ignorance, as in Galileo's struggles with the Inquisition. Published in the same year, Voltaire's *History of Louis XIV* was one of the few Enlightenment works to incorporate the natural sciences, albeit cursorily, into modern history. Despite criticisms of the Enlightenment's positive view of science—most notably by Jean Jacques Rousseau—the approach and tone of d'Alembert's narrative shaped subsequent historical studies of science as stories viewed from within.

Eighteenth century textbooks and *encyclopedias about the natural world both organized extant natural knowledge and created simple chronologies that mapped the temporal unfolding of science's principal ideas and methods of investigation. Jean Étienne Montucla's *History of Mathematics* (two volumes, 1758) was an early example of a type of history of the sciences that flourished at the end of the eighteenth century and was particularly popular in the German states. Perhaps the most ambitious and extensive example of this genre was the *History of the Arts and the Sciences since their Renewal* (around the year 1200) written by leading scholars and published in Göttingen from 1796 onward. This series included histories of mathematics (in four volumes by Abraham Gotthelf Kästner, 1796–1800), physics (in nine volumes, one by Friedrich Wilhelm August Murhard, 1798–1799, and eight by Johann Carl Fischer, 1801–1808), chemistry (in three volumes by Johann Friedrich Gmelin, 1797–1799), and technology (two volumes on the technologies of war by Johann Gottfried Hoyer, 1797–1800, and three on other technologies by Johann Heinrich Moritz Poppe, 1807–1811). The collective effort of these authors, all practitioners in the sciences and technology, resulted in positivist chronologies suited as rational overviews of their fields. Kästner limited the history of a science to how theories were discovered, became known, and later corrected, expanded, and applied. These histories furthermore separated science and its practitioners from the public: as Kästner explained, no one could read them who could not also write them.

This positivist and rationalist history of science guided the production and selection of personas and topics for the chronologies, biographies, bibliographies, and translations and editions of scientific works that appeared in the nineteenth century, especially at its close. As in the eighteenth century, these histories were written less in service of history than of science; they contributed, inter alia, to the intellectual definition, constitution, and boundary of disciplines. Guided by chronology rather than narrative, these histories took the study of nature out of its historical context and ordered results, investigations, and families of problems according to their relative success in maintaining a position in the corpus of scientific knowledge. Value judgments condemning certain traditions, like *Naturphilosophie, to the historical trashcan were therefore inevitable. Written not by trained historians but by scientists, philosophers, and others whose avocation was history, these histories fell outside the range of even history's historiography and methodology. Rarely did they employ historical categories of analysis more complex than the "views of nature" used by John Theodore Merz in *A History of European Thought in the Nineteenth Century* (four volumes, 1903–1914) to parse scientific discoveries into like-minded groups.

Histories of the sciences written not only in service of science but of the nation appeared in greater numbers after the unifications of Italy (1866–1870) and Germany (1871). Under the auspices of national agencies, the French published the works of Antoine-Laurent *Lavoisier (six volumes, 1864–1893); the Dutch, of Christiaan *Huygens (twenty-two volumes, 1888–1950); the Italians, of *Galileo Galilei (nineteen volumes, 1890–1909); and the Swiss, of Leonhard *Euler (thirty-two volumes, 1911–1998). These collections exhibited every bit of the nationalistic sentiment found in other cultural productions of the period like national operas, museums, theaters, and monuments that aided the construction of national identity. Here the motivation for elaborately documenting the history of great scientists dovetailed with that behind the grand state-supported historical projects—for example, the *Monumenta Germaniae Historica* (1819–present)—that coalesced the past in search of an identity for the present. Criticisms of this type of history were and still are vocal. Near the century's close, Friedrich Nietzsche condemned the deadening fetish for excess history, indigestible knowledge, and bibliographical quisquilia. At the end of the twentieth, Georges Canguilhem noted that this kind of history—based on sources, discoveries, influences, priorities and successions—was a victim of the classificatory scheme that made it possible (the disciplines) and so was doomed to failure. Nonetheless, these collections and similar

ones that followed provided a firm foundation for what later became known as "internal" history of science, the genre dominant at the field's inauguration early in the twentieth century and upon which professional growth in the history of science rested after World War II.

Nineteenth-century historical narratives of science organized science's past around biographies, ideas, theories, and investigative techniques. At the same time, institutional histories also became prominent. The result of the phenomenal growth of institutions dedicated to scientific practice and education across the century, these in-house histories, although often uncritically laudatory, partially exposed for historical examination the social and material network of support necessary for scientific practice. For Germany, where the production of historical scholarship was often a state supported enterprise, these histories ranged from Friedrich Paulsen's comprehensive history of the German university system (three volumes, 1896–1897; English translation, 1906) and Carl Gustav Adolf von Harnack's history of the Prussian Academy of Sciences (three volumes, 1900), to Max Lenz's history of the Friedrich-Wilhelm-University in Berlin (four volumes, 1910–1918) or articles by scientists on their institutes, for example, Eduard Riecke, Woldemar Voigt, and Franz Kopp on Göttingen's physical institute, published in a scientific journal, *Physikalische Zeitschrift* (1905). Also in the nineteenth century, philosophers—some of them practicing scientists—analyzed branches of scientific knowledge from an epistemological perspective. William Whewell's *History of the Inductive Sciences* (1837) introduced a method for rationally reconstructing a scientific discovery, which Whewell illustrated through his analysis of Isaac Newton's work on gravity. Ernst Mach's equally imaginative *The Development of Mechanics* (1883) influenced Albert *Einstein.

By 1900 the historical study of science included categories of analysis drawn largely from the sciences or philosophy. Most analysts applied them to the evolution of ideas or techniques within one discipline or a set of closely related disciplines, rather than integrating science and its activities into broader historical movements. For example, Albert Lange's influential *History of Materialism* (1866) grew out of his interest in the history of workers' movements. Nonetheless, he separated the history of science from social concerns. These disciplinary histories escaped the types of criticism that historians had by then developed to assure the production of objective and innovative historical knowledge, as in Karl Lamprecht's pathbreaking and comprehensive cultural history. With only a few exceptions—such as the sociologist Max Weber's narrative of the negative effects of rationalization in *The Protestant Ethic and the Spirit of Capitalism* (1904–1905) and the mathematician Wilhelm Ahrens's objection to the omission of politics in Leo Koenigsberger's biography (1904) of the mathematician Carl Gustav Jacobi—no attempt was made in the long nineteenth century, not even by a historian as innovative as Lamprecht, to link the history of science to broader historical trends. On the eve of World War I the typical history of science resembled Emanuel Rádl's *History of Biological Theories* (1913), which covered a single scientific discipline in terms of its major practitioners, ideas, theories, and methods. Rádl wrote from a philosophical perspective whose goal was to understand scientific ideas, individual genius, or more generally the progress of the human mind.

With the rise of the social sciences and the reform of historical scholarship after World War I, especially in France, consideration of the place of scientific knowledge in history began to move away from an association with progress, rationality, and individual achievement toward a more complex assessment of historical and social contingencies at work in the formation of scientific knowledge. Two founders of the *Annales* school of historians, Marc Bloch and Lucien Febvre, made a strong case for locally situated ways of knowing and the near impossibility of transcending the framework of mental tools available during a particular historical era. Their approach to knowledge and thinking could later be found in Michel Foucault's *The Order of Things* (1970), a history of the social sciences that sought to explain the "creation" of the human subject as an object of scientific investigation. In the 1930s French thinkers also turned to Sigmund Freud's psychoanalysis as a tool for understanding the history of science, as in Gaston Bachelard's *Formation of the Scientific Spirit* (1938). In Great Britain the influence of the social sciences upon the historical understanding of science came by way of *Marxism, while in the United States, the sociologist Robert K. Merton adopted Max Weber's sociological methods in his pioneering *Science, Technology, and Society in Seventeenth Century England* (1938). Marxist and Weberian categories of analysis, including the consideration of social and religious issues, seemed to portend a historiography of science that was more inclusive in its analysis of contextual factors. Despite the interdisciplinary richness of these approaches, however, they did not have much impact upon the historical analysis of science. Far more influential was the work of Alexandre Koyré, a French historian interested primarily in the rational reconstruction of scientific ideas.

His *Etudes galiléennes* (1939), a rich textual analysis of Galileo's work on motion, shaped the first generation of professional historians of science in the United States who became active after World War II, including Henry Guerlac, I. Bernard Cohen, and Thomas S. Kuhn.

The best known historian of science when Koyré appeared on the scene was George Sarton, who in 1912 founded *Isis*, the leading professional journal for history of science. Sarton's work, especially his *Introduction to the History of Science* (five volumes, 1927–1948), drew its inspiration from the positivism of Auguste Comte. Several results of World War II—including the international political ascendancy of the United States, the role of science and technology in the war, and the pedagogical importance of courses on the values of western civilization—popularized the Enlightenment metanarrative of the role of positive and rational science in western civilization. Works shorter than Sarton's—including Edwin A. Burtt's *Metaphysical Foundations of Modern Science* (1950) and Herbert Butterfield's *Origins of Modern Science* (1950)—became the principal means through which history of science entered history courses in the English-speaking world or, at some institutions, became required reading in stand-alone courses on the history of seventeenth century science. Both from within and in reaction to this tradition of the grand metanarrative, Thomas S. Kuhn wrote *The Copernican Revolution* (1957), which challenged the assumption that science can be understood on the basis of rational developments alone.

Since the early 1960s the historiography of science revolved largely around the relationship between science and its multiple contexts, including the community of scientists itself. For most of the Cold War, this debate raged on between "internal" and "external" history of science, a division some practitioners regarded as schizophrenic. Strong arguments for considering science from within the context of culture emerged from the Warburg Institute in London, where Frances Yates's *Giordano Bruno and the Hermetic Tradition* (1964) and *The Rosicrucian Enlightenment* (1972) set a new generation of scholars in quest of understanding the "irrational" amidst the "rational." The sheer weight of her evidence was enough to destroy a historiography of science guided by positivist, rational, or philosophical principles. The first work not only to challenge the autonomy of science successfully but also to offer new categories for analyzing its contextual dependencies was Kuhn's *Structure of Scientific Revolution* (1962). Historians of science appreciated Kuhn's arguments for the local character of scientific practice and values; for the importance of educational institutions, science pedagogy, and models for teaching problem-solving in science; and for the importance of disciplines in maintaining consensus. According to this point of view, science was shaped by its social institutions. (Ludwik Fleck's *Genesis and Development of a Scientific Fact* [1935, English translation 1979], argued for similar conclusions, but Kuhn claimed not to have known of Fleck's work when writing his own.)

Although some, like Georges Canguilhem, felt that Kuhn's categories of "normal science" and "paradigm" invoked compulsion, and hence did not divorce him so cleanly from positivist and rationalistic history, Kuhn's work was in most quarters considered novel. Kuhn's identification of controversy as revelatory of tacit practices and of crisis as the moment when cultural factors were most likely to influence the content and practice of science provided historians with two readily observable phenomena to analyze. Paul Forman's classic study of quantum theory and Weimar culture is an innovative early study of crisis. In a manner similar to Kuhn, Jerome R. Ravetz challenged the assumption that science was special among other forms of culture. He developed the idea of science as "craftsmen's work," likening scientific practice to ordinary life. These and other similar challenges to the autonomy and special status of science—including Paolo Rossi's study of science and the arts in the early modern era—enriched science historiography considerably.

Yet questions concerning the relationship between science and social factors remained. At Edinburgh in the mid-1970s emerged the so-called "Strong Programme" in the sociology of scientific knowledge (SSK) around Barry Barnes and other sociologists. Although the relationship between SSK and varieties of twentieth century Marxism has never been carefully examined, SSK shared with Marxism a belief that social classes and interests mattered in the construction of knowledge. And like Marxism, SSK posited that truth was not universal but historical. SSK found an able critic in the 1980s in the sociologist of science Bruno Latour who dissolved what remained of the division between the "inside" and "outside" of science. On first reading some of Latour's claims sound paradoxical, if not outright nonsensical. Latour has examined the transformations science has brought to modern society—transformations that make it impossible, he claims, to separate science and society into two mutually exclusive categories of analysis. Anthropologists and ethnographers rather than historians took up Latour's challenge: Karin Knorr-Cetina, Michael Lynch, and Sharon

Traweek have studied the interaction of small groups of researchers in their everyday activity.

By the last decade of the twentieth century historians of science responded to the changes that had been accruing since the 1960s in studies that Jan Golinski has labeled "constructivist." Constructivism assumes that scientific knowledge is locally created, produced, and situated. The "local" in scientific knowledge and the processes by which it becomes universally accepted are the two central historical issues in constructivist historiography. Constructivists view scientific knowledge not as revealed, but rather as "made" using methods, tools, and materials available in culture. As in SSK, in constructivism truth does not figure; perceptions of the strengths and weaknesses of the epistemic foundation of knowledge do. Despite their radical rejection of the rationality of science, historians working in the constructivist vein maintain a tight focus on science in their investigations, using categories of analysis like identity, discipline, pedagogy, rhetoric, laboratory, lecture, seminar, and representation. Leading practitioners of constructivism include Simon Schaffer and Steven Shapin. Dominated by local studies, constructivist historiography has marginalized studies of the "big picture" of universally accepted and acquired scientific knowledge that dominated the historiography of science in the 1950s. Shapin, for instance, has challenged prevailing traditions about a reigning grand narrative, that of the Scientific Revolution.

Not all writing in the history of science since 1960 can be neatly grouped together as "post-Kuhnian practices." Although all journals in the field have both historical and historiographical orientations, very few subscribe entirely to a single methodology, including constructivism. Historians of science use a wide variety of approaches, ranging from those found in the technical articles of *Archive for the History of the Exact Sciences* to the occasional studies that mediate history and the history of science in *American Historical Review*, *Past & Present*, *Annales*, or *Geschichte und Gesellschaft*. J. L. Heilbron's insightful reading of the early modern church as an observatory demonstrates that one need not take recourse to theory to understand the interaction of science and culture. Although the grand metanarrative has all but disappeared from the historiography of science, detailed narrative studies are still the only way to understand the complex interaction of writing, memory, and practice in scientific creativity, as Frederic L. Holmes has shown. The study of non-western science, which has burgeoned since decolonization, offers challenges for the historical understanding of science that strain present historiographical practices, as Gyan Prakash has shown in his refreshing examination of science in modern India. A genre that has been particularly popular on best-seller lists is the little book on a closely focused topic in the history of science, of which the bellwether was Dava Sobel's *Longitude* (1995). And at the fringes of historical writing still linger the alluring temptations of fiction where the reconstruction of the past can contain imaginative elements but still ring true, as Denis Guedj has admirably demonstrated in his novel on the French meter.

See also DISCIPLINES; GENDER AND SCIENCE; MODERNITY AND POSTMODERNITY; REVOLUTIONS AND SCIENCE; SCHOOLS; SCIENCE AND DAILY LIFE; THEORIES OF SCIENTIFIC DEVELOPMENT.

Wilhelm Ahrens, *C. G. J. Jacobi als Politiker* (1907). Thomas S. Kuhn, *The Structure of Scientific Revolutions* (1962). Paolo Rossi, *Science, Technology, and the Arts in the Early Modern Era* (1970). Paul Forman, "Weimar Culture, Causality, and Quantum Theory, 1918–1927: Adaptation by German Physicists and Mathematicians to a Hostile Intellectual Milieu," *Historical Studies in the Physical Sciences* 3 (1971): 1–115. Jerome R. Ravetz, *Scientific Knowledge and its Social Problems* (1971). Barry Barnes, *Scientific Knowledge and Sociological Theory* (1974). Steven Shapin and Simon Schaffer, *Leviathan and the Air Pump* (1985). Bruno Latour, *Science in Action* (1987). Georges Canguilhem, *Ideology and Rationality in the History of the Life Sciences* (1988). Dava Sobel, *Longitude* (1995). Steven Shapin, *The Scientific Revolution* (1996). Jan Golinski, *Making Natural Knowledge: Constructivism and the History of Science* (1998). J. L. Heilbron, *The Sun in the Church: Cathedrals as Solar Observatories* (1999). Gyan Prakash, *Another Reason: Science and the Imagination in Modern India* (1999). Denis Guedj, *The Measure of the World* (2001). Frederic L. Holmes, *Meselson, Stahl, and the Replication of DNA* (2001). J. L. Heilbron, "Records of Right Thinking" (ms. 2002).

KATHRYN OLESKO

HISTORY OF SCIENCE. According to the theory of preformation, the seeds or eggs of all subsequent generations were contained, one within the other, in the original ancestors of all living organisms, and these in turn contain the seeds and eggs of all creatures that are to be, to the end of time. This theory, long dead in science, still exerts an influence on the writing of its history. Today's molecular biology may be portrayed as descending from Gregor *Mendel's play with peas, or Albert *Einstein's relativity as arising from Isaac *Newton's gravity, in a preordained progression. From time to time the unfolding may have been hindered by unfriendly governments, hostile religions, prejudice, and ignorance, but, the impediments once set aside, science resumed its natural course.

Recent approaches, represented by this *Companion*, reject finalist and preformationist historiography. The *Companion* takes for granted that in

Allegory of the history of astronomy. The work of Hipparchus and Ptolemy, restarted by Copernicus, is completed by Tycho and Kepler with the help of gold coins falling from a two-headed eagle representing the Holy Roman Emperor, Rudolf.

the study of nature, as in other spheres of human activity, development depends on the interaction of individual personalities, local circumstances, and large-scale social forces, and that, had conditions been different than they were, our science would not be as it is. That is not to say that natural science is indefinitely labile; once people agreed that the purpose of their science was to develop accounts of the natural world in closer and closer accord with contrived observation and experiment, constraints imposed by success, as well as by the phenomena, became more and more limiting. Hence, despite the contingency, or owing to similar challenges prompting similar responses, the sciences have developed along similar trajectories that may be divided into corresponding periods.

Terms

In the late seventeenth century, John Ray, an English naturalist and classifier, extolled the theory of preformation as "one of the most beautiful discoveries of modern physics." Today's scientists would accuse him of being doubly mistaken, in accepting an erroneous theory and in foisting it on *physics. Historians who labor to understand natural knowledge in its context do not worry whether Ray was right or wrong; they know that all past theories are defective from today's viewpoint, which, in due course, also will be modified. But they do or should worry about his assignment of preformation to physics. Although his usage was standard in his time—"physics" then did include the entire natural world—it is obsolete today. Under what rubric should historians describe preformation? If they choose Ray's, they would put it where few interested readers will find it. If they put it under biology or the life sciences, they will be using categories that did not exist in the seventeenth century and that exclude associations presupposed by Ray's contemporaries. The constituents of a prehistory of biology, or of any other modern science, can only be specified retrospectively and arbitrarily.

Because the body of knowledge has been cut up in different ways at different times historians have difficulty dealing with extended periods. The problem is compounded by the fact that the cutting often preserves the names of the subdivisions while changing their content, as in Ray's "physics" and ours. The *Companion*'s solution to this problem is to use the names of modern sciences in descriptions of the past only when the practices and scope of the older subjects approximate those of the modern ones. For earlier times it uses period phrases like "*natural knowledge" or "*experimental philosophy," which have no modern misleading namesakes. The *Companion*'s solution to the specific problem presented by Ray's categorization is that preformation is neither physics nor biology, but natural knowledge. This solution is a luxury that an individual historian, working independently on some piece of science, seldom can afford. To describe the transformations of natural knowledge into its many modern disciplinary subdivisions requires many hands working together under a general conception of when and how the transformations took place. The *Companion* can offer a schema of the long-term growth of science more faithful to its history than would a series of disciplinary histories produced by specialists laboring alone.

What was said about "physics" applies also to "science." Since the seventeenth century the word has narrowed its meaning, especially in English, out of all recognition. Its root, "scientia," signified all systematic knowledge, as in German "Wissenschaft"; but in English "science" has come to mean natural science, as in the British Association for the Advancement of Science, founded in 1831 and divided into classes for the mathematical, physical, and what we would call life sciences. The association provided the opportunity for the invention of the word "*scientist" to refer to a person professionally interested in the subjects covered in its meetings. Although the French word "science" can have the meaning of "Wissenschaft," it has been used widely since the seventeenth century in the English sense. The prestigious Paris Academy of Sciences, founded in 1665, did not and does not admit representatives of the social and human sciences. The first steps toward restricting the meaning of "science" to natural science began before the narrowing of the meaning of "physics." From the end of the eighteenth century "science" in its restricted sense had the same scope that "physics" had had a century earlier.

The word puzzles so far considered have been drawn exclusively from European languages. This is not an oversight. The subject of the *Companion* is the growth and spread of Western science and its uses. Western science is a uniquely powerful method of representing and exploiting the natural world. These activities do not exhaust the interests of humankind. It is not a compliment to other sorts of knowledge and belief to enroll them as precursors or approximations to Western science; on the contrary, their strength and value must be judged by the extent to which they fulfill the purposes for which they were designed. This is not to say that Western science owes nothing to non-European cultures. Quite the reverse is the case. Many ingredients of Western science have come from other cultures and regions (e.g., India, Islamic regions, and colonial territories; *see* ASIA, NEAR EAST, AFRICA, RELIGION AND SCIENCE) and indigenous systems of natural knowledge have shaped the ways in which non-Western societies take up Western science when exposed to it. More recently, Western science has been enriched greatly by the direct contributions of people who are not of European descent. Much Western science is now done in Asia (especially Japan). The *Companion* gives due weight to all these contributions, ingredients, and influences without losing its focus on Western science, that is, using our terminology, on science *tout court*.

Periodization

*Classifications of knowledge and the names of its divisions are important demarcators of the periods into which the history of science falls. So too are institutions for the support, pursuit, and diffusion of science. Simultaneous and significant changes in institutional and intellectual factors make good indicators of a new period or subperiod. Typically modern institutions, like professional disciplinary societies and university research *laboratories, appeared around the time that the names of many sciences—*biology, *geology, physics, and psychology, for example—were first used in something like their present meaning. From these and other considerations the *Companion* sets the epoch of modern science around 1830. It regards the thirty years on either side of 1800 as a transition from an earlier period, usually called the *scientific revolution, marked by the rejection and creation of elaborate systems of natural philosophy, the beginning of sustained instrumental exploration of nature, and the foundations of institutions that encouraged the pursuit of natural knowledge. For symmetry (the date at which the revolution commenced is arbitrary and agitated) the *Companion* starts the *scientific revolution in 1600. To make clear what the revolution was about, it considers a preliminary period commencing, for convenience, in 1500. This rough periodization is further refined as follows.

The preliminary period, 1500–1600, may be characterized as an *Overloaded System*. It saw challenges to the *Renaissance cosmos from the new knowledge brought by the European discoveries of Africa and Asia, the reassertion of ancient philosophies opposed to Aristotle's, and the creation of new technological needs. It is covered in thematic articles like *Navigation, *Aristotelianism, and *Hermeticism; in biographies of the refounders of the anomalously precocious sciences of anatomy and astronomy (Nicholas *Copernicus, Tycho *Brahe, Andreas *Vesalius); and as background in many articles on the scientific revolution.

The scientific revolution is divided into two sets of overlapping subperiods. With *New Formulations and Responses*, 1600–1660, the older, bookish culture tried to absorb impulses from heliocentrists, atomists, mathematicians, and methodologists, and the results brought by the new world-opening instruments, the *telescope and *microscope. During *Attempted Reconstructions*, 1640–1750, the challenge was met by new world systems, encyclopedias, eclectic natural philosophies, rampant and random experimentation, and novel applications of mathematics, and by the founding of *academies, *observatories, *libraries, gardens, and *journals of natural knowledge. With *Effective Dismemberment*, 1720–1770, the fatiguing fight between disciples of Newton, René *Descartes, Gottfried Leibniz, and Saint Ignatius of Loyola ended with most combatants dropping their philosophical scruples against *instrumentalism and abstraction, and the *universities and academies confirming their division of knowledge, the first taking responsibility for its diffusion, the second for its increase. Through the entire period the dominant system of patronage and *communication was the so-called "republic of letters," a set of networks connecting people interested in science, literature, collecting, and politics, and maintained by correspondence, travel, and, after the middle of the seventeenth century, the *Journal des Sçavans* and its many imitators. The citizens of this republic included wealthy and well-placed people both lay and clerical, their clients and protégés, members of learned societies and universities, professional men, and writers of all kinds. They patronized the instrument makers and lecturers on natural knowledge who became prominent toward the end of the period.

The transition period is marked by the liberation of the *Quantitative Spirit*. Between 1770 and 1830 new types of experts, who relied upon special training and met much higher standards of knowledge and performance than most savants of the republic of letters, quantified parts of physics, systematized *chemistry, made a comparative *natural history, spurned romantic science (*Naturphilosophie), deployed the *lightning rod and other science-based appliances, and contributed to an ever-widening range of professional journals. During the transition period *professional societies for individual scientific disciplines first appeared; so did the model of later institutes of technology (such as the École Polytechnique), the metric system, the University of Berlin (the fountainhead of the German research tradition), and the first explicit, widely based scientific lobbies like the British Association for the Advancement of Science.

The *Companion* divides modern science into three subperiods. The first and longest, the *Classical Science*, 1830–1915, introduced the great substantive syntheses in the physical and life sciences (the laws of *thermodynamics and *electromagnetism, *evolution by natural selection, relativity, the *electron, the *cell, and the *gene); traced the history of life on earth; and multiplied university research institutes, engineering and medical schools, and bureaus of *standards. It saw the establishment of disciplinary science, the creation of effective popularization to reach a wider public left further and further behind the research frontier, and the creation of research laboratories.

During the remaining subperiods, which were shaped by the two *world wars, science sprang up everywhere. *New Sinews*, 1915–1945, indicates new funding mechanisms, the radical reconstruction of classical physics, the partial reduction of chemistry to a branch of physics and physics to a branch of *engineering, new injections from mathematics, the teachings of the fruit fly, the rise of biomedical sciences, the multiplication of industrial research laboratories and high-tech inventions, the intrusion of natural sciences into social sciences and humanities, and the spread of European science to new centers and new groups in old centers. The *Globalization of Western Science*, 1945–2000, saw first the Americanization and then, after Europe, the Soviet Union, and Japan had recovered from World War II, the *globalization of science; also the hegemony of *high-energy physics, the physicalization of the biological sciences (*see* MOLECULAR BIOLOGY), the synthesis of physics and geology in *plate tectonics, the introduction of the *computer, travel to the moon (*see* SPACE SCIENCE), a vast expansion of industrial and *military research, the continuing geographical spread of Western science, the widening of opportunity to become a scientist in Western societies, and the theology of the Big Bang.

Modulations by Region and Subject

The periodization just delivered is normative, not procrustean. All parts of science everywhere did not advance in lockstep. For example, Italy preceded and then trailed France and England during the scientific revolution and metropolitan centers were almost always in advance (that is, further along the normal periodization) of provincial and peripheral areas. The causes of advance or delay range from the chance appearance of a gifted leader to long-term historical processes. The gifted leader was particularly effective before the twentieth century. When the cultivation of natural knowledge and the practice of science does not involve many people, a single, singular individual more able than his peers can be disproportionately consequential. As the scientific enterprise grows, however, the peculiar talent and personality of an individual, however decisive on a local level, cannot thwart or deviate general trends.

Science, like most other social activities, does better when encouraged. Until the middle of the nineteenth century and with the exception of *astronomy, encouragement did not require much money. If a society set some prestige or reward on the study of science, science had students. The retardation of Italy and the advancement of England after 1660 may be assigned to differential encouragement. In Italy, the Roman Catholic church (*see* CHURCH, THE), though a sponsor of learning, inhibited the development of modern subjects construed as inimical to faith, whereas in England the same subjects were promoted as useful to the arts or helpful to religion. The differential encouragement in this case is explained by differences in the organization, doctrine, and power of established churches (*see* RELIGION AND SCIENCE).

Another type of encouragement is competition. During the nineteenth century the German states continually improved their science institutes as part of their efforts to obtain and retain the best professors. A man like Hermann von *Helmholtz, who served several universities, could leave behind him a string of improved facilities. In contrast, France had a central administration that decided where professors would teach. They had no bargaining chips for leveraging institutional improvements. People went to the provinces when sent and, if ambitious, angled to be called to Paris, where, if they were successful, they would accumulate professorships and other academic posts (*see* DYNASTY). The practice of centralization, which made Paris the leading place for science during the transition period, was not an effective strategy against the competitive system of Germany and, later, of Britain and the United States.

The political form of a regime does not seem to have as much power to influence science as might be supposed. Science has both flourished and stultified under absolute monarchies, empires, military dictatorships, totalitarian regimes, and democracies. Under Louis XIV, Napoléon I, and Kaiser Wilhelm II, during the first years of the Weimar Republic, in the United States in the twentieth century, and, usually, in the Soviet Union, it has done comparatively well; under the Russian czars and the Nazis, and in the United States before 1900, it did comparatively badly. What is important is not the government's form or ideology but its attitude toward science.

In our century, money has had much to do with success in science. Private means have long since ceased to keep research at the frontier. The United States, which lagged behind Europe in science in the nineteenth century, has outdistanced it in the twentieth primarily by bringing to bear the resources of foundations (*see* PHILANTHROPY), industry, and government. Bilateral and multinational combinations, however, have blunted American leadership in many areas and even in the prestigious subfield of high-energy physics the United States is no longer even the first among equals (*see* CERN).

The normative periodization fits physics better than other sciences, which follow its development the more faithfully the nearer to it they are in subject matter. For example, chemistry was influenced by its ties with practice much earlier than physics. Astronomy and human *anatomy, which deal with well-defined and restricted phenomena, acquired a high level of exact detail long before physics did. Geology's development paralleled physics' in time although it had different antecedents in natural knowledge and strong regional modulation owing to the location of exposed strata and other revealing phenomena.

The study of plants and animals, their distribution, relationships, and functioning, satisfy the normal periodization in gross, especially as to the introduction of effective schemes of classification and the powerful systematic generalization of evolution; but quantification came only in the twentieth century, well over a hundred years after it had begun in physics. The adoption of the methods of the physical sciences by some branches of psychology, *anthropology, and archaeology is also an innovation of our century. The *Companion* makes clear how and why the development of these several sciences follows or deviates from the normal timing in articles under their names.

J. L. HEILBRON

HIV. See AIDS.

HODGKIN, Dorothy Crowfoot (1910–1994), English chemist, pioneer in x-ray crystallography, winner of the Nobel Prize in chemistry in 1964 for her determination of the structures of biologically important molecules, including penicillin, vitamin B_{12}, and insulin. Her research career, which began in the 1930s, spanned more than forty years.

Dorothy Crowfoot was born in Cairo, where her father supervised Egyptian schools and ancient monuments for the British government. Her mother was a self-taught expert in botany and ancient weaving. Crowfoot and her two sisters lived mostly in England from 1914, while their parents stayed mostly in the Middle East. Crowfoot's interest in chemistry and in crystals began in 1923, when a family friend encouraged her to search for minerals in the garden. Crowfoot collected samples of minerals and analyzed their chemical identity. For a sixteenth-birthday gift, her mother gave Crowfoot a copy of William Henry Bragg's *Concerning the Nature of Crystals*.

Crowfoot entered Somerville College at Oxford in 1928, where she read chemistry and physics, and while a student she chemically analyzed the colored-glass tesserae that her mother had found in Palestine. On a recommendation to a friend from the physical chemist T. S. Lowry, Crowfoot began working with John Desmond Bernal at Cambridge in 1932. Like Bragg and his son William Lawrence Bragg, Bernal was unusually open-minded in the encouragement and employment of female colleagues.

In 1934 Crowfoot assisted Bernal in an analysis of the first x-ray photograph of a protein crystal, pepsin (an enzyme that aids digestion in the stomach). Somerville College offered her a two-year research fellowship, the first year of which was to be spent in Bernal's group at Cambridge. She received her doctorate from Cambridge in 1937 with a thesis on sterols, focusing on the three-dimensional structure of cholesterol. She did much of the research in her own tiny damp laboratory in the basement of the Oxford University Museum, which she equipped from some small grant monies and, later, with Bernal's x-ray equipment, which he gave to her when he started war research. In 1937 she married Thomas L. Hodgkin, a historian and educator. They had three children.

World War II created a demand for penicillin, the antibiotic mold discovered by Alexander Fleming in 1928, which had attracted little concerted attention following the introduction of sulfa drugs. Under the pressure of wartime conditions, Howard Walter Florey and Ernst Boris Chain isolated penicillin powder in 1942, and Florey persuaded several U.S. corporations (Merck, Pfizer, and Squibb) to begin industrial production by a fermentation process—even while the structure of penicillin remained in dispute among organic chemists.

Oxford chemist Robert Robinson, who along with Florey headed a large penicillin research group, opposed the hypothesis of Chain and Edward P. Abraham that the penicillin molecule includes a beta-lactam group (a very reactive carbon-nitrogen ring structure a hypothesis proven correct by Hodgkin and her student Barbara Rogers Low. Hodgkin and Low made use of one of the world's first computers, an early IBM analog machine using punch cards, which permitted much faster calculation of electron-density maps than had been possible in Hodgkin's earlier x-ray crystallography work. After the war Hodgkin's penicillin model was used to develop semisynthetic penicillins, until John C. Sheehan and K. R. Henery-Logan accomplished a complete synthesis of penicillin in 1957.

Hodgkin turned her attention in 1948 to the structure of vitamin B_{12}, when Lester Smith of the Glaxo drug company presented her with crystal samples. This cobalt-containing heavy compound (molecular weight 1,355) was used in the treatment of pernicious anemia. The mathematical work on this project was begun with punch-card computers, then in 1953 was assisted by Kenneth Trueblood's volunteering the services of a new SWAC (Standards Western Automatic Computer) machine at the University of California, Los Angeles. In 1956 Hodgkin and her research group arrived at the structural formula for the B_{12} molecule, the largest then completely determined: $C_{63}H_{88}N_{14}O_{14}PCo$.

Hodgkin's third major achievement came in 1969 with the announcement of the structure of the 777-atom molecule of insulin, for which Frederick Sanger in Cambridge had mapped the sequence of amino acids. By this time Hodgkin had moved up in university rank from lecturer and demonstrator (1946) to university reader (1957) to a Royal Society–appointed Wolfson Research Professorship (1960). She also had won a Nobel Prize, the first English woman so honored. Her laboratory, which she restricted to a research group of about ten people, many of them women, had moved to new quarters in 1958. Among her most famous students was the future prime minister Margaret Thatcher, with whom Hodgkin came to have considerable political differences but with whom she occasionally lunched in the 1980s.

Norman W. Hunt, "Dorothy Hodgkin," in *Nobel Laureates in Chemistry 1901–1992*, ed. Laylin K. James (1993): 456–461. Sharon Bertsch McGrayne, "Dorothy Crowfoot Hodgkin," in *Nobel Prize Women in Science: Their Lives, Struggles and Momentous Discoveries* (1993): 225–254. "Dorothy Hodgkin, 1910–1994, English Chemist and Physicist," *Physics Today* 48 (1995): 80–81. Georgina Ferry, *Dorothy Hodgkin: A Life* (1998).

MARY JO NYE

HOLISM. "Holism" (from Greek *olos*, whole) signifying that the whole exceeds the sum of its parts, was coined in 1926 by Jan Christian Smuts, the South African statesman and amateur botanist, in a book entitled *Holism and Evolutionism*. The book supported an "organic" view of the universe as against the mechanistic view that Smuts regarded as prevailing in contemporary scientific circles and culture at large. He claimed that nature displayed a tendency to form from particular wholes new units that acted as higher wholes, unlike anything a mechanism could achieve.

When Smuts wrote, ontological holistic views were widespread in Europe and the United States. Philosophers like Henri Louis Bergson in France and Alfred North Whitehead in Britain, embryologists Charles Manning Child and Charles Judson Herrick, and psychologist Karl Lashley in the States, biologists Jakob Johann von Uexküll and Hans Driesch, the neurologist Constantin von Monakow, Gestalt psychologist Max Wertheimer, and the neuropsychiatrist Kurt Goldstein in Germany all taught one or another form of holism.

A common thread uniting these otherwise diverse personalities was a dissatisfaction with (or a reaction against) the successful strategy carried out by an earlier generation of scientists who, from the mid-nineteenth century, had aimed to explain living phenomena with the ordinary concepts of physics and chemistry (*see* REDUCTIONISM). To early-twentieth-century eyes, some of the traumas and tensions generated by fast industrialization and urbanization, and released by World War I, could be blamed on scientific mechanism. To those same eyes, a return to a Romantic approach to the natural sciences represented by authors like Goethe, with their organicist and teleological themes, seemed an appropriate remedy. In Germany similar concerns merged with aspirations to assert the fundamental tenets of a national culture. The ideologies that accompanied the dramatic events of interwar politics and World War II were among the results. After the war, ontological holism found supporters in fields such as *ecology and evolutionary biology, as well as in popular movements such as the New Age.

Forms of epistemological holism have been adopted by several authors in fields such as the philosophy of science and the philosophy of language. Since the 1950s, Willard Van Orman Quine developed the view that experience counts for or against our beliefs in an holistic manner. According to Quine, little can be said about particular, everyday as well as scientific propositions except by adopting a background language that conveys our entire body of beliefs. Advocates of epistemological, semantic holism have further argued that only whole languages, theories or belief systems have meanings. Smaller units such as words, sentences, and hypotheses signify only in a derivative sense.

See also GAIA HYPOTHESIS.

Jerry Fodor and Ernest Lepore, *Holism. A Shopper's Guide* (1992). Anne Harrington, *Reenchanted Science. Holism in German Culture from Wilhelm II to Hitler* (1996).

GIULIANO PANCALDI

HOMEOSTASIS. The American physiologist Walter Cannon coined the word "homeostasis" in 1926 to designate the coordinated physiological reactions that maintain steady states in the body. He believed that a special term was necessary to differentiate the complex arrangements in living beings, involving the integrated coordination of a wide range of organs, from the relatively simple physico-chemical closed systems in which a balance of forces maintains an equilibrium. "Changes in the surroundings," Cannon wrote, "excite reactions in [the open system that constitutes a living being], or affect it directly, so that internal disturbances of the system are produced. Such disturbances are normally kept within narrow limits, because automatic adjustments within the system are brought into action, and thereby wide oscillations are prevented and the internal conditions are held fairly constant."

Cannon illustrated his concept by describing a variety of mechanisms that maintain constant conditions in the fluid matrix, or "internal environment," of higher animals. These included materials such as glucose and oxygen in the blood, as well as the fluid matrix's temperature, osmotic pressure, and hydrogen-ion concentration. The knowledge of the mechanisms that Cannon showed to be involved in these reactions came largely from his own previous experiments on the role of the autonomic nervous system and the adrenal secretions. The close association he established between homeostatic mechanisms and the preservation of conditions in the internal environment, however, derived in large part from the inspiration of the nineteenth-century French physiologist Claude Bernard, whom Cannon acknowledged as the first to give a "more precise analysis" to general ideas about the stability of organisms. Cannon quoted particularly Bernard's "pregnant sentence" that "It is the fixity of the 'milieu interieur' which is the condition of free and independent life."

Originally a physiological principle, homeostasis took on broader meanings after World War II, with the recognition of the similarity of homeostatic mechanisms to feedback controls in servomechanisms. Biologists applied the concept at all levels—cellular, organ system, individual, and social systems. The maintenance of steady concentrations of the intermediates of a metabolic pathway despite a constant flux of matter and energy through the pathway became an example of homeostatic regulation. Cannon himself had asked in 1932 in his popular book *The Wisdom of the Body*, in an epilogue entitled "Relations of Biological and Social Homeostasis," whether it might not "be useful to examine other forms of organization—industrial, domestic, or social—in the light of the organization of the body?" His suggestion has been followed, and homeostasis in its widest sense now means the "maintenance of a dynamically stable state within a system by means of internal regulatory processes that counteract external disturbances of the equilibrium."

Walter Cannon, "Organization for Physiological Homeostasis," *Physiological Reviews*, 9 (1929): 399–427. Claude Bernard, *Lectures on the Phenomena of Life Common to Animals and Plants* (1974).

FREDERIC LAWRENCE HOLMES

HOOKE, Robert (1635–1702), natural philosopher and inventor.

Hooke was born in Freshwater, Isle of Wight, in 1635 to John Hooke, a minister. He showed a talent for invention as a young child, making

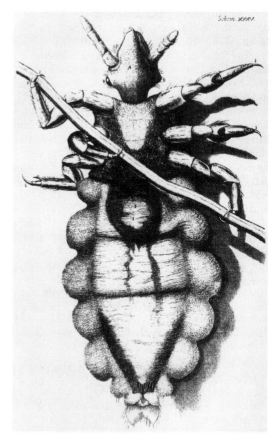

Hooke's louse as drawn by him from observations through his microscope. From Hooke's *Micrographia* (1665), which contains, besides many other beautifully executed drawings of microscopic objects, a description of the interference phenomenon known as Newton's rings.

sundials, a wooden clock, a model man-of-war that could sail and fire its guns, and other mechanical toys. Sent to London in 1648, he was taken into the home of Richard Busby, master of Westminster School, where he devoured classical languages and mathematics (the first six books of Euclid in a week) and continued his career of invention.

Hooke matriculated at Oxford in 1653 as a chorister, and by 1655 had found his way into the brilliant circle there that was fashioning the new mechanical philosophy. He became assistant to Robert *Boyle, who perceived Hooke's genius for devising simple yet conclusive experiments. Hooke built an air pump for Boyle around 1657, and collaborated in Boyle's investigations with the pump and related instruments. The experiments confirmed what we know as Boyle's law. Hooke's diary and manuscripts mention many techniques and devices for flying. He invented important improvements to the clock, including spring-controlled escapements, and probably the spiral spring for watch balances and the anchor escapement for pendulum clocks. His work with springs led by 1678 to his law of elasticity (Hooke's law), "*Ut tensio sic vis*" or "The power (sic) of any springy body is in the same proportion with the extension."

With the restoration of the English monarchy in 1660, the Oxford group moved to London and founded the Royal Society. In 1662, the Society appointed Hooke curator of experiments, charging him to furnish "three or four considerable experiments" for each weekly meeting. This seemingly impossible goal unleashed in him a flurry of creativity. From drawings for the Society of microscopical observations came the *Micrographia* (1665), one of the great works of the century, filled with images that still excite admiration today. The *Micrographia* included speculations on light, which Hooke regarded as a rapid vibration of small amplitude. He discovered and explained the interference phenomena occurring between two lenses, later known as "Newton's rings." He introduced the idea of a wave front, explained its deflection at the boundary between media, and described diffraction. He invented telescopic sights, crosshairs, and a clock-driven telescope and the conical pendulum that drove it, and carried out experiments showing how the Moon's features might have been caused by impacting bodies or volcanic action. The universal joint is also his invention.

Hooke has been called the first scientific meteorologist, organizing cooperative weather observations and designing a standard thermometer, graduated from a single fixed point, which modern analysis has proved remarkably accurate. He also invented a hygrometer, a wind gauge, numerous types of barometers, and a weather clock that recorded the readings of all of them. Hooke had an abiding interest in the earth's history, clearly recognizing the organic origin of fossils and suggesting that earthquakes, crustal movements, and other changes had cast marine fossils high upon dry land.

Hooke was also an accomplished architect. Appointed surveyor to the City of London after the Great Fire of 1666, he resurveyed most of the City; his buildings—the Royal College of Physicians and Bedlam Hospital among them—compare in quality to those of Christopher Wren, his close friend and collaborator.

Many of Hooke's inventions involved him in controversy. As always, mathematicians and natural philosophers pursued closely related problems, and betrayal was common. One of Hooke's backers probably provided Christiaan *Huygens with information on the spiral spring, which Huygens then claimed to have invented; considerable evidence also points to treachery on the part of Henry Oldenburg, secretary of the Royal Society, against Hooke and others. Hooke felt most bitterly toward Isaac *Newton. Hooke had been first to propose that the planets' orbital motions are compounded of a straight-line motion along the tangent and a centripetal gravitational motion; he postulated the inverse-square behavior of gravity, and communicated these ideas in 1679–1680 to Newton. Newton, nursing a hatred against him from an earlier controversy over their theories of light, proved these hypotheses mathematically, but refused to acknowledge Hooke in the *Principia*. Hooke, for his part, believed that Newton had stolen his ideas about gravity. Evidence suggests that Newton destroyed records of Hooke's achievement and planted forgeries to establish his own priority in several areas.

The historiographical tradition makes Hooke a difficult, jealous, and ungenerous man. The primary sources belie this portrayal. His diary shows him pursuing an active social life, always at coffeehouses and taverns with his many friends, enjoying deep and loving relationships. Historians have also faulted Hooke for taking on much but finishing little. On the contrary, he certainly completed his many scientific instruments, mechanical devices, surveys, and buildings. The truth seems to be that he had a practical rather than a theoretical mind, and that this has hurt his reputation at the hands of a historiography that values theory over practice.

Margaret Espinasse, *Robert Hooke* (1962). Michael Hunter and Simon Schaffer, eds., *Robert Hooke: New Studies* (1989). Ellen Tan Drake, *Restless Genius. Robert Hooke and His Early Thoughts* (1996).

THEODORE S. FELDMAN

HORTICULTURE. From the earliest times plants have been subject to utilitarian and commercial enterprise. Herbalists and apothecaries regulated the supply of medicinally useful plants to customers, while sailors and merchants took considerable risks to import valuable spices and exotics. Horticulture, however, implies the development and marketing of desirable plant commodities and as such essentially dates from the Renaissance. There are few clearer-cut examples of the long-term manipulation of natural organisms for fashion and changing human tastes.

The diversification of the eight classic florists' flowers—the auricula, polyanthus, hyacinth, anemone, ranunculus, tulip, pink, and carnation—dates from the sixteenth century. Of these, the tulip was the most important. Bulbs reached Vienna from Turkey in 1554, although they probably circulated well before that date. The plant was rapidly subjected to improvement, especially in the "breaking" or striping of the petals. Many different varieties appear in Dutch and Flemish still lifes of the seventeenth century. The production of tulip bulbs and their commerce became highly professional; monasteries provided most of the bulbs sold across Europe. In the 1630s the Dutch suffered tulipomania, called by them the "Wind trade." The market crashed in 1637.

During the eighteenth century the rise of *botanical gardens and interest in landscape gardening and the embellishment of the great estates of Europe led to a marked increase in horticultural activities. Glasshouses and hothouses of various kinds came into use. But the real growth of commercial horticulture came in the early nineteenth century as a consequence of the industrial revolution, with easier and cheaper glass production, the repeal of glass taxes in Britain, new distribution networks like the railways, and the development of coal-fired stoves, flues, and ventilation systems. Suddenly, nursery workers could contemplate mass-production of plants. There was a marked surge in commercial nursery firms and seed producers during the nineteenth century, two of the most important being the Vilmorin Company in Paris and James Veitch in England. Garden flowers, vegetables, vines, and fruits became highly diversified. Charles *Darwin said that he grew fifty-two varieties of gooseberry alone. Public gardens began to include serried ranks of identical, brightly colored flowers laid out in formal bedding schemes. John Claudius Loudon and his wife, Jane, fuelled popular interest with gardening manuals and the first garden magazine. Flower shows and societies also played a part, such as the Horticultural Society of London (founded in 1805, later the Royal Horticultural Society) with grounds in Chiswick, London. This society began the practice of awarding medals and certificates for choice specimens at their annual show. It especially encouraged the commercial development of orchid breeding.

Diversification continued rapidly through the early twentieth century. Practical expertise intermeshed increasingly with plant genetics, as at the John Innes Institute for Plant Breeding in Norwich, which played a central role in advancing the field. Toward the end of the century technological advances allowed highly sophisticated propagation techniques. The botanist William Stearn introduced the word "cultivar" to distinguish plants of human-made origin from real varieties.

H. R. Fletcher, *The Story of the Royal Horticultural Society of London* (1969). Ruth Duthie, *Florists' Flowers and Societies* (1988).

JANET BROWNE

HOSPITAL. The traditional charitable role of the hospital was to provide shelter and food as well as spiritual salvation and bodily recovery in times of famines and epidemics. Reflecting a more positive vision of health and the new mercantile economy, young urban workers, their livelihood threatened by illness, sought hospital care during the Renaissance. By the early 1500s, this shift prompted a regular medical presence, exemplified by conditions at the Santa Maria Nuova Hospital in Florence. Physicians medicalized hospitals they visited, experimenting on patients with established and new remedies and preserving their newly gained experience in casebooks. They also created disease classifications, occasionally instructed medical students, and subjected deceased and unclaimed inmates to anatomical dissection. By the 1730s, European hospitals were primarily places for physical restoration of the military and civilian labor force. Providing Enlightenment physicians with greater access to wide sectors of the population, hospitals became ideal settings for an expanding medical presence. They provided the necessary context for the construction of a new medical science and improved clinical skills, nurseries capable of training better medical professionals. Some hospitals offered programs of institutional apprenticeship for surgeons (at the Hôtel Dieu of Paris) and for physicians (at British voluntary institutions such as St. George's Hospital in London and the Royal Infirmary of Edinburgh). John Aikin, who considered the hospitalized sick poor as ideally suited for "experimental practice," and John Howard, a widely traveled prison and hospital reformer, were early leaders of this hospital movement.

The conversion of hospitals into instruments of medicine acquired greater momentum in the nineteenth century. European hospitals linked to local universities such as Guys Hospital in London, the Hôpital de la Charité in Paris, and the Allgemeines Krankenhaus in Vienna became sites for comprehensive programs of education and research. Notable early proponents of hospitals as medical institutions were the French physicians Pierre J. G. Cabanis and Jacques-René Tenon. With their numerous halls filled with the sick poor, large hospitals were veritable museums of pathology. In these controlled environments, physicians took advantage of the diversity of illnesses and funneled interesting cases into special teaching units for detailed studies and experimental management.

Clinical knowledge obtained in the wards was focused overwhelmingly on acute, complex, and life-threatening conditions. This institutional environment came to shape the character of Western medicine: dramatic, disease oriented, and interventionist, in effect removing sick individuals from medical cosmology and replacing them with depersonalized disease carriers. By contrast, the largely patient-dominated context of private practice continued to shape knowledge and treatments related to more common health problems and chronic conditions. In the end, the hospital context was decisive in grounding and framing biomedical medical care.

Changes in welfare schemes and professional power relationships, educational expediency, and innovation all contributed to the reform of the traditional hospital. In the early 1800s, novel ways of practicing medicine in Paris attempted to solve the dual problems of large numbers of institutionalized sick and severe manpower shortages among attending physicians. These new techniques, pioneered by French physicians including Philippe Pinel and T.-R.-H. Laennec, codified the disease manifestations of individual suffering inmates, including bodily clinical signs and anatomical changes. This ''medicine of observation'' thus became the main tool for acquiring professional knowledge: systematic study and classification of diseases and postmortem dissection of deceased patients. Diagnostic specificity and outcome were achieved through the establishment of clinical-anatomical correlations. A new paradigm localized all bodily suffering in organ systems. As the product of social and political factors linked to the French Revolution and its wars, this new anatomical-pathological knowledge came to characterize the Paris Medical School. Eventually it spread throughout the world. The pathological paradigm—disease must have an identifiable anatomical seat—continues to rule biomedicine by consensus of its practitioners, and remains largely responsible for the professional solidarity among scientifically trained physicians, especially by contrast with practitioners of alternative medicine.

Sanitary considerations and medical knowledge, in turn, profoundly influenced hospital architecture and organization, the role of caregivers, and patient management. Hospital space was divided into separate pavilions and provided with improved heating and ventilation, practices pioneered in the 1880s by John Shaw Billings during the construction of the Johns Hopkins Hospital in Baltimore. Also in the United States, Edward S. Stevens and Isadore Rosenfield were among influential twentieth-century architects whose designs facilitated internal hospital circulation and greater patient privacy.

By the late nineteenth century, the strict separation between administration and caregivers, including physicians and nurses, remained. Physicians had the power to admit and discharge patients, substituting criteria based on medical needs for previous religious and charitable yardsticks. The work of nurses, professionally trained on the model established by Florence Nightingale, became an able extension of medical management. Total immersion, peer control, and behavioral guidelines were demanded of all institutional caregivers, forging stronger professional bonds through the creation of hospital instruction programs such as internships and residencies.

For the past two centuries, hospitals have remained ideal locations for the most advanced clinical research, teaching, and patient management of acute medical conditions. Hospitals also function as testing grounds for new technological devices, starting with instruments such as the stethoscope, ophthalmoscope, and laryngoscope, and continuing with forms of imaging from X rays to body scans. Beginning in the 1820s, many of the clinical characteristics observed in particular groups of patients underwent quantification and statistical manipulation, thereby exposing the probabilistic nature of clinical diagnosis and prognosis. The growing body of knowledge of patient outcomes became an increasingly valuable tool for judging the efficacy of medical treatments, especially after the employment of random clinical trials following World War II. And most of the therapeutic revolution of the last half-century took place in hospital settings affiliated with academic medical centers.

Guenter B. Risse, *Hospital Life in Enlightenment Scotland: Care and Teaching at the Royal Infirmary of Edinburgh* (1986). Charles Rosenberg, *The Care of Strangers: The Rise of America's Hospital System* (1987). Colin Jones, *The Charitable Imperative: Hospitals and Nursing in Ancién Regime and Revolutionary France* (1989). Lindsay Granshaw and Roy Porter, eds., *The Hospital in History* (1989). Rosemary Stevens, *In Sickness and in Wealth: American Hospitals in the Twentieth Century* (1989). Guenter B. Risse, *Mending Bodies—Saving Souls: A History of Hospitals* (1999).

GUENTER B. RISSE

HOUSSAY, Bernardo Alberto (1887–1971), physiologist, experimental endocrinologist, and academic leader.

One of the most prominent and influential Latin American scientists of the twentieth century, Houssay was the son of a French lawyer. He received his baccalaureate degree with honors from the Colegio Nacional de Buenos Aires at the early age of thirteen, then entered the School of Pharmacy of the University of Buenos Aires in 1901, graduating four years later at the head of his class. Not satisfied with a pharmaceutical career, Houssay went on to study medicine between 1904 and 1910 at the University of Buenos Aires during a critical period in the evolution of academic medicine in Argentina. With the conservative Academy of Medicine controlling the appointment of new professors, more than seven hundred medical students including Houssay successfully protested in 1905 against the clinically oriented status quo. By 1908 he had received an assistantship in the Department of Physiology and during the following year, while still a student, he was selected to fill the chair of physiology at the School of Veterinary Science. He graduated with a doctoral dissertation about the physiological actions of the pituitary gland, a work that won him high honors and official publication.

Houssay began a busy career that combined medical practice and academic teaching. While attending private patients, he also held a post in a municipal hospital and taught at both the medical and veterinary schools. After 1915 he took on an additional post at the National Public Health Laboratories as chief of experimental pathology, studying the action of snake and insect venom on blood coagulation and developing effective antidotes against them. Then, in 1919, Houssay acquired the chair of physiology at the School of Medicine of the University of Buenos Aires, and rapidly converted this academic unit into an active center for experimental investigations and research training. Until the end of World War II, the prestigious Institute of Physiology—supported in part since the 1920s by funds from the Rockefeller Foundation—received worldwide recognition as a center for biomedical research.

A contextual, socially grounded analysis of Houssay's research topics and style discloses a distinctive experimental approach ideally suited to limited local budgets and technology. Influenced by the career and work of Claude *Bernard, Houssay strongly supported the application of the scientific method to study of the human body's internal regulatory mechanisms. After his dissertation, he had selected *endocrinology as an area of great promise. Among his interests was the physiological action of the hypophysis, a gland easily procured from animal carcasses being processed by local slaughterhouses. Lacking specialized laboratory personnel, apparatus, and an adequate payroll, Houssay carried out his work with the support of student volunteers, who assisted him in an assembly-like scheme. Houssay sought to turn technological deficiencies into

an asset; he proclaimed that excessive equipment could hamper the scientist's intellectual powers.

Following the discovery of insulin in 1921, it became clear that ductless glands directly releasing hormones into the bloodstream were central in human *metabolism. Houssay and his collaborators focused on the diabetogenic effects of extracts from the anterior lobe of the hypophysis. They discovered that hypophysectomized dogs became sensitive to the hypoglycemic action of insulin. During the 1930s, the research led to the discovery of several hormonal feedback mechanisms involving the thyroid, adrenals, and gonads. In recognition of this work on carbohydrate metabolism, Houssay shared the Nobel Prize in physiology and medicine in 1947 with Gerty Cori and Carl Ferdinand Cori.

Author of several books and over six hundred scientific papers, Houssay was venerated in Argentina as the key spokesman for modern biomedicine and defender of academic freedom. His prominence and liberal political leanings brought him into conflict with the military dictatorship that ruled Argentina after 1943. Dismissed from his university posts, he continued his research in a privately funded laboratory until the fall of General Perón's regime in 1955, when he was officially reinstated. He spent his last years as head of the Argentine National Council for Scientific and Technical Research, a governmental organization he had conceived and founded in 1957. Appealing to their patriotism, Houssay vainly tried to stop the growing exodus of young scientists, a brain drain caused by the deteriorating political and economic conditions of Argentina. During his long and successful career, Houssay belonged to the Academy of Medicine, founded the Argentine Society of Biology and the Association for the Advancement of Science, held foreign memberships in many scientific societies in the United States, Britain, Germany, France, Italy, and Spain, and garnered honorary degrees from Paris, Oxford, Cambridge, and Harvard.

Bernardo A. Houssay, "History of Hypophysial Diabetes," in Essays in Biology in Honor of Herbert M. Evans (1943): 247–256. Virgilio G. Foglia, "The History of Bernardo Houssay's Research Laboratory, Instituto de Biologia y Medicina Experimental: The First Twenty Years, 1944–1963," Journal of the History of Medicine 35 (1980): 380–397. Virgilio G. Foglia and Venancio Deulofeu, eds., Bernardo A. Houssay. Su Vida y Obra (1887–1971), 1981. Marcos Cueto, "Laboratory Styles in Argentine Physiology," Isis 85 (1994): 228–246. Marcos Cueto, "Science under Adversity: Latin American Medical Research and American Private Philanthropy 1920–1960," Minerva 35 (1997): 233–245.

GUENTER B. RISSE

HUBBLE, Edwin (1889–1953), astronomer, showed that spiral nebulae are independent island universes beyond our galaxy and that the universe is expanding.

Hubble was born on 20 November 1889 in Marshfield, Missouri. Attending the University of Chicago on an academic scholarship, he played on the undefeated 1909 basketball team and won letters in track. In 1910 he went to Oxford University as a Rhodes scholar. There he studied law, participated vigorously in sports, and traveled on the Continent. Back in the United States in 1913, Hubble taught high school physics and Spanish, and coached basketball. He joined the Kentucky bar, but never practiced law. Seemingly, science would be his metier. In 1914 Hubble returned to the University of Chicago and its Yerkes Observatory. When the United States declared war on Germany in April 1917, Hubble rushed through his doctoral dissertation, took his final examination, and reported to the army three days later. He served in France, reaching the rank of major. Discharged in 1919, Hubble joined the Mount Wilson Observatory, just east of Los Angeles.

Hubble's doctoral dissertation was a photographic investigation of nebulae, cloudy patches of light in the sky, of which little then was known. A few are nearby agglomerations of gas and dust, but most are distant clusters of stars. About 17,000 small, faint nebulous objects resolvable into groupings of stars had been cataloged, and some 150,000 were within reach of existing telescopes. A classification system was needed, and Hubble developed the now-standard classification system for galaxies during the 1920s.

At Mount Wilson, Harlow Shapley's determination of distances to globular clusters had suggested by 1918 that the universe was a single, enormous, all-comprehending unit. Soon Hubble destroyed Shapley's big galaxy model. Using the new 100-inch telescope at Mount Wilson, Hubble discovered Cepheid variable stars in several spiral-shaped nebulae. This type of star has a period-luminosity relation: the longer the period (time for its brightness to change from maximum to minimum and back to maximum), the brighter the star. From the observed period, Hubble could estimate the intrinsic, absolute brightness. Comparing this value to the observed apparent brightness diminished by distance, Hubble calculated the distance of the Cepheid and its nebula. Centuries of speculation over the possible existence of island universes similar to our galaxy came to an abrupt end in 1925 when Hubble announced his discovery of Cepheid variables in spiral nebulae and distance calculations placing them decisively beyond the boundaries of our galaxy. This multiplied the known size of the universe by about a factor of ten.

Early in the twentieth century most scientists, including Albert *Einstein, assumed that the universe is static. The Dutch astronomer Willem de Sitter formulated in 1916 a static model of the universe with apparent (but not real) redshifts in the spectra of nebulae greater for nebulae at greater distances. Normally, a shift of a spectral line toward the red end of the spectrum indicates that the object is moving away from the observer. In 1929 Hubble tested de Sitter's prediction. He had distances to forty-six extragalactic nebulae and redshifts measured by Milton Humason at Mount Wilson under Hubble's direction. The redshifts were greater at greater distances. De Sitter's static model of the universe, however, was abandoned, with the redshifts generally interpreted as real Doppler shifts indicative of real velocities of recession. The empirical relationship between redshift and distance (actually, between redshift and size or brightness, unambiguous indications of distance) became generally understood as a correlation between velocity and distance. Hubble's famous velocity-distance relation showed that more distant nebulae are receding from us at greater speeds, that the universe is not static but expanding.

The velocity-distance relation made possible determinations of greater distances because redshifts could be measured even when individual stars and galaxies were too faint to distinguish. Hubble increased the size of the known universe by yet another factor of ten.

Hubble traveled to Europe several times to give lectures and receive important honors. He read widely in the history of science and was a trustee of the Huntington Library. He also partied in nearby Hollywood with movie stars, including Charlie Chaplin and Greta Garbo, and with famous writers, including Aldous Huxley, Christopher Isherwood, and Anita Loos. He fished in Scotland and the Colorado Rockies.

World War II interrupted Hubble's work on cosmology. He served as chief of ballistics and director of the Supersonic Wind Tunnels Laboratory at the Army Proving Grounds in Maryland, and was awarded the Medal of Merit for his wartime work. Hubble died in San Marino, California, of a heart attack on 28 September 1953, soon after completion of the giant 200-inch telescope on Palomar Mountain, south of Los Angeles, and too soon for conclusive answers from the research program Hubble had planned for the new telescope.

See also GALAXY; RELATIVITY.

Edwin Hubble, *The Realm of the Nebulae* (1936). Richard Berendzen, Richard Hart, and Daniel Seeley, *Man Discovers the Galaxies* (1976). Robert Jastrow, *God and the Astronomers* (1978). Robert Smith, *The Expanding Universe: Astronomy's "Great Debate" 1900–1931* (1982). Gale E. Christianson, *Edwin Hubble: Mariner of the Nebulae* (1995). Norriss S. Hetherington, *Hubble's Cosmology: A Guided Study of Selected Texts* (1996).

NORRISS S. HETHERINGTON

HUMAN GENOME PROJECT. The human genome project originated in the latter half of the1980s in the United States. Its goal was to map the location of the 80,000 to 100,000 genes that human beings were thought to possess and to sequence the three billion base pairs that human DNA was estimated to contain. Recent technical developments had encouraged biologists to consider the mapping and sequencing effort a practical possibility. Human DNA had been found to include numerous restriction fragment length polymorphisms (RFLPs), so called because they were snipped from DNA by restriction enzymes and varied in length. Seemingly ubiquitous in the genome, they could provide numerous points of reference for gene mapping. Moreover, techniques for rapidly sequencing DNA had been incorporated into commercially available machines. At the then-current sequencing price of about one dollar per base pair, the cost of sequencing the entire human genome would come to three billion dollars. Enthusiasts of the human genome project needed government funding.

In the United States, they found it initially in the Department of Energy (DOE), which had a long-standing interest in the mutational impact of nuclear radiation on genomes. In 1987, DOE initiated an ambitious five-year human genome program that would comprise among other activities the development of automated high-speed sequencing technologies and research into the computer analysis of sequence data. The National Institutes of Health (NIH) quickly joined the genome game, if only to take principal control of it

away from the big science–oriented Department of Energy. However, the increasing NIH commitment to the genome project stimulated opposition within the biomedical scientific community. The dissenters feared that the genome project would be a three-billion-dollar big science crash program, built around a few large bureaucratized centers that would be given over to DNA sequencing. The project would be tedious, routinized, and likely to sap resources from meritorious areas of biology.

A coalition of genome enthusiasts and skeptics nevertheless found common ground in a genome project that would be spread over fifteen years and would serve a broad biological interest by sponsoring genomic investigations of nonhuman organisms such as mice and yeast. The project would speed the search for genes related to disease (a type of research that many biologists wanted to pursue anyway). Part of the money would be invested in the development of technologies that would make sequencing rapid and cheap enough to be accomplished in many ordinary-sized laboratories rather than in just a few large facilities.

Biomedical scientists and industrial representatives stressed to Congress that the project promised high medical payoffs and would be essential to national prowess in world biotechnology, especially if the United States expected to remain competitive with the Japanese. In October 1988, James D. Watson, the codiscoverer of the structure of DNA, was appointed head of what soon came to be called the National Center for Human Genome Research in NIH. The move effectively decided in favor of NIH the nagging issue of the lead federal agency in the biological side of the project; DOE would deal primarily with technology.

Following the United States, Britain, France, Italy, several other nations, and the European

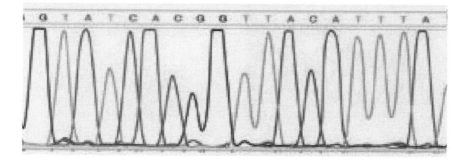

A minuscule portion of the printout of the results of sequencing the human genome. The letters A, C, G, T stand for the four bases that make up the staircase of the double helix.

Community established projects to map and sequence at least parts of the human genome. In 1988, the Human Genome Organization was established as an international body—a "U.N. for the genome," one biologist remarked—intended to foster collaborative efforts and the exchange of information. Both the NIH, at Watson's initiative, and the European Community, at the behest of the European Parliament, incorporated into their respective genome projects investigations of the ethical, legal, and social issues in human genetics—for example, eugenics and the privacy of genetic information. The project was the first in the history of science with a mandate to deal with such matters.

Through the 1990s, genomic data poured out of laboratories on both sides of the Atlantic and into centralized databases, including one at the European Molecular Biology Laboratory, another at the Los Alamos National Laboratory, and still another at NIH's genome center in Washington. Late in the decade, the databases were made easily and freely accessible on the new World Wide Web.

Profit-making competition entered genomics in 1992 when Craig Venter, a specialist in gene sequencing at NIH, left to head a new private center called The Institute for Genomic Research (TIGR). Although TIGR would be nonprofit, it was funded by a venture capital group that established Human Genome Sciences Inc. to develop and market products resulting from TIGR's research. Venter predicted that TIGR would track down one thousand genes daily and would identify the majority of human genes within three to five years. In 1998, Venter moved to a new, for-profit company called Celera that aimed to sequence the entire human genome by 2001 using rapid new automated machines supplied by its principal owner, the Perkin-Elmer Corporation. Goaded by Celera, the NIH genome center picked up its sequencing pace. In June 2000, at a White House ceremony presided over by President Bill Clinton, Venter and Francis Collins, the head of the NIH project, announced that they had both completed a full draft of the human genome.

Despite the triumph, the entrance of profit-making companies into wholesale gene-sequencing worried leading biomedical scientists. The companies, led by Celera, have proposed to patent large fractions of the human genome without knowing much more about them than their base-pair sequences. Their strategy stimulated a forceful statement in 2000 by the presidents of the Royal Society of London and the National Academy of Sciences in the United States stressing that "the human genome itself must be freely available to all humankind." Offsetting the drive to privatize the genome are the freely accessible databases to which the public projects and their grantees in the United States and Europe are steadily contributing, but how much of the genome will be locked up by private corporations depends on the policies of the world's patent offices.

Steve Jones, *The Language of Genes: Biology, History, and the Evolutionary Future* (1993). Daniel J. Kevles and Leroy Hood, eds., *The Code of Codes: Scientific and Social Issues in the Human Genome Project* (1993). Robert Cook-Deegan, *The Gene Wars: Science, Politics, and the Human Genome* (1995). Philip Kitcher, *The Lives to Come: The Genetic Revolution and Human Possibilities* (1996). Kevin Davies, *Cracking the Genome: Inside the Race to Unlock Human DNA* (2001).

DANIEL J. KEVLES

HUMAN IMMUNODEFICIENCY SYNDROME. See AIDS.

HUMANISM. See RENAISSANCE.

HUMBOLDT, Alexander von (1769–1859), naturalist, geographer, administrator, social critic, scientific popularizer, expeditionary, and correspondent.

Born in Berlin the son of a Prussian officer and his Huguenot wife, Alexander von Humboldt and his brother Wilhelm became by 1810 internationally known figures in science, education, and administration. Initially educated at home, Humboldt's formal education encapsulated older traditions in cameralism (the management of the state's natural, financial, and population resources) and exposed him to novel changes in administration and management, scientific and philosophical thinking, and observational practices. Between 1787 and 1792 he studied at the Universities of Frankfurt-an-der-Oder and Göttingen, the Hamburg Commercial Academy, and the Freiberg Academy of Mines. He also studied engineering in Berlin.

He published initially on volcanism (1790), muscle and nerve stimulation (1790), plants in relationship to the environment (1793), and galvanism (1797). Later he took photometric measurements of the southern stars and explained why sound is amplified at night (the Humboldt effect). He spent his early professional years as an administrator or traveler. His best known scientific expedition was to Latin America with the French botanist Aimé-Jacques-Alexandre Bonpland in 1799, returning to Paris in 1804. In 1826 the Prussian king called him back to Berlin to be privy councilor and court tutor to the crown prince; he also gave a popular university course on physical geography. He helped to organize the first international scientific conference in Berlin in 1828. Thereafter, he divided his time between Paris and Potsdam until his death in 1859.

Humboldt was an anti-Kantian who rejected innate ideas. Antoine-Laurent *Lavoisier's *Elements of Chemistry* (1789) convinced him to favor knowledge based on precise measurements, not imprecise metaphysical constructions. He eventually learned to use nearly every measuring instrument available at the time. By the 1790s he had committed himself to a "terrestrial physics" that substituted precise measurements and detailed natural drawings for verbal description and symbolic representations. The geographer Georg Forster in 1790 urged him to assemble observations into a meaningful whole; Jena philosophers in 1794 suggested he view nature aesthetically. With the assistance of leading French scientists from 1805 to 1826 Humboldt reduced his Latin American measurements, obtained estimates of their errors, and interpreted the results of his travels.

His terrestrial physics is expressed best by his "Physical Profile of the Andes and Surrounding Country" (1807), a chart of twenty-one columns presenting a grand combination of measurement-based results and natural observations. The "Physical Profile"—a representation of Mount Chimborazo, which Humboldt had climbed to within 365 meters (1,200 ft) of its peak—was to accompany his "Essay on the Geography of Plants" (1805) that argued for the dependency of plant life on geographical location. He was the first to use isotherms and isobars to depict average thermometric and barometric conditions. Averages represented to Humboldt lawful expressions of nature's equilibrium, the balance of all forces. His integration of his results into a tabular and pictorial whole was a monumental achievement. He tried to apply his method of averages to the geographical distribution in plants (1817). Humboldt's precision never captured all observational errors, however, and his averages, sometimes difficult to define and use, were often secured only by acts of faith. Others adopted the overall contours of *Humboldtian science—its focus on averages and the integration of nature's features—with some success.

Humboldt's *Views of Nature* (1808, 1826, 1849), partly based on results reported to him by others, popularized his idea that order could be achieved by finding the equilibrium points of nature's forces. His magisterial multivolume *Cosmos* (1845–1862) was an attempt to create for the entire universe the conceptual and historical unity he had already achieved for Latin America. A history beginning with the universe's primeval origins and ending with the creation of the physical landscape and life forms, *Cosmos* wove together Humboldt's prodigious correspondence with leading scientists of his day. Although often maligned by experts, *Cosmos* secured Humboldt's reputation as Germany's first widely read popularizer of science.

Humboldt's views of nature and politics intersected. Civilizations were to him morally developed to the degree that they overcame nature's challenges. Committed to a constitutional monarchy after the French Revolution, Humboldt looked upon civil disturbances as geological faults: pressures seeking equilibrium. He considered silver mines in the New World of value to European countries even though he was critical of slavery and political domination. Believing average conditions to be politically desirable, he advocated to colonial powers that they adopt his scientific observational and measuring methods as tools for the efficient management of local resources. Yet the European revolutions of 1848–1852 shocked him. Ironically in the reactionary 1850s, *Cosmos* contributed to the perception that Humboldt stood for impiety, republicanism, and revolution.

See also CAMERALISM.

Douglas Botting, *Humboldt and the Cosmos*, 2d ed. (1973). Margarita Bowen, *Empiricism and Geographical Thought: From Francis Bacon to Alexander von Humboldt* (1981). Kurt-Reinhold Biermann, *Alexander von Humboldt*, 3d ed. (1987). Alexander von Humboldt, *Cosmos: A Sketch of the Physical Description of the Universe*, 2 vols. (1996). Alexander von Humboldt, *Personal Narrative* (1996). Alexander von Humboldt, *Island of Cuba: A Political Essay* (2001).

KATHRYN OLESKO

HUMBOLDTIAN SCIENCE. Historians use the term "Humboldtian science" to describe a type of scientific practice during the nineteenth century that resembled the work of Alexander von *Humboldt, whether or not it resulted from Humboldt's direct influence. Susan Faye Cannon coined the term in 1978 to signify a scientific style that conducted observations with the latest instruments, corrected measurements for errors, and linked these to mathematical laws; constructed maps of isolines connecting points with the same average values; identified large, even global, units of investigation; and used nature rather than the laboratory as a site of investigation. The term, as applied to nineteenth-century science, has since acquired other connotations, including connecting different types of large-scale phenomena, demonstrating their interdependencies, seeking a universal science of nature, and using large-scale international organizational

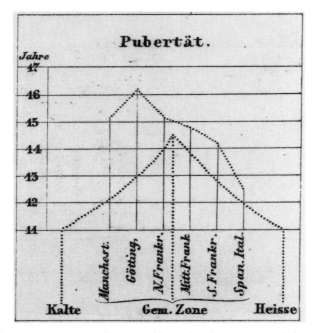

Onset of sexual activity as a function of latitude. The primitive inhabitants of hot and cold climates reach puberty early; the refined citizens of the university town of Göttingen follow five years later.

structures to execute local readings as part of a global effort.

Scholars since Cannon have been careful to differentiate what Humboldt did from Humboldtian science. Humboldt deliberately avoided speculation and description as was found in natural history and natural philosophy. The major theme of his master science of terrestrial physics was the equilibrium of the earth's forces. For Humboldt isolines represented not merely average values, but also a natural aesthetic order (as they did in patterns of maximum areas of concentration in the regional distribution of plants) and even political stability. This aesthetic sensibility rarely appears in Humboldtian science despite its importance to Humboldt, who in many ways incorporated the Romantic emphasis on aesthetics, the imagination, and the picturesque in image and word. Nature was to Humboldt not only the assemblage of averages and the balance of forces, but also an aesthetic composition. The language of his travel books projected strange worlds in living color, a literary quality with great public appeal. He also wanted to retain in nature study the morally didactic qualities that eighteenth-century aesthetic theory had valued. In this sense, Humboldt's science, in contrast to Humboldtian science, was not merely knowledge to be learned, but to be lived. Hence some of the social values associated with the "European tour" in the nineteenth century came from their association with Humboldt's science.

Humboldtian science, as it has been used, also does not include two diametrically opposed directions in which Humboldt's work was taken: popular science and disciplinary specialization. The aesthetic image of the unity and balance of nature's forces particularly appealed to the public. The theme of harmony in Humboldt's work acquired political, social, and religious connotations, and helped to make nature study a part of liberal culture in Germany. The image of nature's order became an antidote to social and political disarray after the European revolutions of 1848. Immediately following Humboldt's death in 1859, Humboldt Associations, and later Humboldt Festivals, were established throughout the German states. They promoted interest in nature through public participation in nature walks and specimen collecting. More focused disciplinary uses of Humboldt's work appeared after industrialization exuded pollution, damaged forests, and in other ways highlighted the interconnectedness of environmental conditions. August Grisebach published in 1872 the first comprehensive classification of the earth's vegetation according to climatic conditions in which he adopted Humboldt's techniques in plant geography, especially Humboldt's notion of "social plants" (plant species in a regional environment forming special communities, such as heaths, savannahs, and bogs, that excluded the germination of other species).

Humboldtian science differed markedly from the institutionalized, disciplinary-based sciences that took shape in the nineteenth century. Although Humboldtian science shared the methodological rigor of nineteenth-century scholarship (*Wissenschaft*), it eschewed the intellectual specialization of, and sharp divisions between, scientific disciplines. For example, Humboldt and his Berlin colleague the physicist Heinrich Wilhelm Dove took meteorological measurements, but Dove's meteorology never reached out to other scientific disciplines like botany and never included naturalistic drawings in color. Dove also conducted some of his experiments in a laboratory; a Humboldtian scientist worked only in nature. The laboratory scientist posed detailed, particular questions in a contrived environment where certain variables could be held constant. Practitioners of Humboldtian science viewed nature as an ensemble, an organic whole whose interrelatedness could be captured in a geography broadly defined and keenly sensitive to large-scale issues. British natural philosophers especially viewed Humboldtian science as a rigorous counterbalance to the specialization and professionalization then shaping the content and practice of science.

Despite its marginal institutional position, Humboldtian science had a powerful effect on certain areas of science in the nineteenth century, especially geomagnetism, as well as on the development of certain scientific communities. Historians have found Humboldtian science more in evidence in Anglo-Saxon regions than elsewhere. Humboldtian science appeared most frequently in Victorian Britain with its vast empire over which scientists could collect data on the scale advocated by Humboldt. The best example of Humboldtian science is the British Magnetic Crusade, an effort dedicated to measuring the magnetic features of the earth in the British empire and beyond. Humboldt himself had proposed to the Royal Society of London that its members undertake global geomagnetic observations; in the 1830s the British government donated over £100,000 to the enterprise. Directed by Humphrey Lloyd of Trinity College, Dublin, the Magnetic Crusade consisted of a chain of fixed magnetic observatories with standardized instruments whose measurements—magnetic declination and the horizontal and vertical components of magnetic intensity—were sent to England where Edward Sabine reduced them. Methods of observation were printed on instruction sheets and distributed; adherence to the rules of observation

was upheld by naval officers who conducted many of the observations. Through these observations Sabine could correlate deviations in Earth's magnetic phenomena to the action of sunspots and demonstrate the eleven-year cycle in sunspots.

The British viewed Humboldt as writing within their tradition of providential design in nature, and were annoyed that he did not specify the design's spiritual agency. Yet they believed that Humboldt's measuring methods were suited for shaping character and tempering laziness in young men. Humboldtian science defined the work of two leading British scientific organizations, the Royal Society of London during the 1820s and the British Association for the Advancement of Science in the 1830s. British imperial activity also bore the mark of Humboldtian science. Charles *Darwin carried a copy of Humboldt's *Personal Narrative* aboard the voyage of HMS *Beagle* from 1831 to 1836. Humboldt's aesthetic image of nature recurs in the paragraphs on the tangled bank near the end of Darwin's *Origin of Species* (1859). In Australia magnetic observations at the Rossbank Magnetic Observatory between 1840 and 1854 and astronomical ones at Melbourne's Flagstaff Observatory (founded 1857) were further examples of Humboldtian science. The Rossbank data became Australia's first project in physics, inspiring a younger generation, including Georg Neumayer, the German scientist who founded the Flagstaff Observatory. In characteristic Humboldtian fashion, Neumayer gathered data from remote locations for map construction by means of the telegraph. British colonial administrators believed Humboldtian science was a part of their "civilizing mission." Even British imperial literature captured its importance. Rudyard Kipling's *Kim* (1901) not only immortalized the empire's penchant for gathering data on natural and human activity (Colonel Creighton was an ethnographer, Kim a surveyor), but also took place during the period of political unrest when natural and social data about India became a form of surveillance deployed to hold the empire together.

Elsewhere, Humboldt directly influenced the Berlin geographer Carl Ritter, but he, like other Humboldt-inspired German scientists, directed no large-scale projects. When Humboldt proposed to the British that they inaugurate global geomagnetic observations, he did not know that Carl Friedrich Gauss had already published a mathematical theory of magnetic intensity in 1833. Working with the physicist Wilhelm Weber, Gauss nonetheless made the Magnetic Observatory at the University of Göttingen (then a part of the British empire) the center of the British

Magnetic Crusade. Gauss and Weber designed the instruments for the project, all chronometers were calculated, and all observations were reduced, according to Göttingen mean time. Neither *Gauss nor Weber, however, held up Humboldt's work as a model; both continued to work within the framework of laboratory- or observatory-based disciplinary sciences.

Many other scientists, even among the British, felt that Humboldt's science led to writing in outmoded traditions of travel literature and that the integration it claimed was rarely achieved. In the context of discipline-building and professionalization in Germany, Humboldt's science came to be viewed as amateurish, a throwback to an earlier era. Humboldtian science of a sort did promote scientific internationalism, but to what degree is still disputed. Although a concept constructed by historians rather than by historical actors to explain their own actions, Humboldtian science seems to capture the sciences of empire.

See also DISCIPLINE(S).

John Cawood, "Terrestrial Magnetism and the Development of International Collaboration in the Early Nineteenth Century," *Annals of Science* 34 (1977): 551–587. Susan Faye Cannon, *Science in Culture: The Early Victorian Period* (1978). Jack Morrell and Arnold Thackray, *Gentleman of Science* (1981). Malcolm Nicolson, "Alexander von Humboldt, Humboldtian Science and the Origins of the Study of Vegetation," *History of Science* 25 (1987): 167–194. Mary Pratt, *Imperial Eyes: Travel Writing and Transculturation* (1992). W. H. Brock, "Humboldt and the British: A Note on the Character of British Science," *Annals of Science* 50 (1993): 365–372. R. J. Home, "Humboldtian Science Revisited: An Australian Case Study," *History of Science* 33 (1995): 1–22. Michael Dettelbach, "Humboldtian Science," in *Cultures of Natural History*, eds. N. Jardine, J. A. Secord, and E. C. Sparry (1996): 287–304. David Philip Miller and Peter Hans Reill, eds., *Visions of Empire* (1996).

KATHRYN OLESKO

HUTTON, James (1726–1797), Scottish natural philosopher.

James Hutton was born in Edinburgh of merchant stock. He went to Edinburgh University, where he attended courses on mathematics and natural philosophy given by the famous Newtonian exponent Colin Maclaurin. Hutton then dithered about his future, trying law and medicine, spending four years in Paris, and receiving a medical degree from the University of Leiden in 1749 before returning to Scotland to improve a small farm inherited from his father. In 1768, he moved to Edinburgh, settled with his unmarried sisters, and threw himself into the brilliant intellectual life of the city at a time now described as the Scottish Enlightenment. Apart

from a tour of England and some excursions in Scotland, he spent the rest of his life in Edinburgh.

Hutton numbered the chemist Joseph Black, the political economist Adam Smith, the philosopher and historian Adam Ferguson, the inventor James Watt, and the mathematician (and later disciple) John Playfair among his friends. Stimulated by their company, he refined his intellectual commitments. A deist in religion, he believed that a powerful and benevolent deity governed the universe, and dismissed the Biblical miracles as fables. A Newtonian in natural philosophy, Hutton took the laws of mechanics as the basis of a system of the earth—when supplemented by Black's principles of heat and chemistry.

In 1783, an informal philosophical society to which Hutton belonged was reconstituted as the Royal Society of Edinburgh. Two years later, he outlined his theory of the earth for the members. "This globe of the earth," he declared, "is evidently made for man.... [It is] a habitable world; and on its fitness for this purpose, our sense of mission in its formation must depend." So that the globe would remain indefinitely habitable, the deity had designed a cycle of continuous decay and renewal. Soil, washed into the ocean by rain and rivers, sank to the bottom, where it consolidated into rock and was elevated by heat to form new habitable land. Hutton took Black's theory of specific and latent heat for granted. Solar heat in latent form consolidated rocks; as specific heat, it elevated them.

In 1788, a version of Hutton's presentation of 1785 to the Royal Society appeared in the Society's *Transactions*. Most of the British and Continental natural philosophers, chemists, and mineralogists who read it found it incomprehensible. Many argued that the soil, far from constantly disappearing, was a mantle that protected the earth. Although they agreed that fossils in calcareous rocks indicated that these strata had once been underwater, they ridiculed the idea that rocks had been consolidated and elevated by heat. Everyone knew that heat caused limestone to disintegrate to quicklime. Hutton's readers were equally baffled by his claim that granite had consolidated from a melt. Melts of materials similar to granite (the sand that made glass for example) cooled to glasses, not to crystalline rocks.

To counter these criticisms, Hutton set about writing his two-volume *Theory of the Earth* (1795). Simultaneously, he worked on other books: *Dissertations on ... Natural Philosophy*, mainly meteorology, chemistry, and matter theory (1792), *Philosophy of Light, Heat and Fire* and *Investigation of the Principles of Knowledge and the Progress of Reason* (both 1794), and a 1,000 page manuscript on the principles of agriculture unpublished at his death.

Hutton's Scottish admirers, notably John Playfair and James Hall, developed his ideas (*see* NEPTUNISM AND PLUTONISM). Charles Lyell made Hutton the hero of the influential introduction to his *Principles of Geology* (1830) and Archibald Geikie, in his *Founders of Geology* (1901), treated Hutton as a founding father of the discipline. Hutton thus came to be seen as the founder of modern geology, a Plutonist, a *uniformitarian, and a hardheaded, no-nonsense scientist reporting what he saw in the field, untouched by religious pre-conceptions. In the mid twentieth century, historians of science have tried to put Hutton back into his intellectual context in the Scottish Enlightenment.

See also NEPTUNISM AND PLUTONISM.

E. B. Bailey, *James Hutton: The Founder of Modern Geology* (1967). Roy Porter, *The Making of Geology* (1977). Rachel Laudan, *From Mineralogy to Geology* (1987). G. Y. Craig and J. H. Hull, eds., *James Hutton* (1999).

RACHEL LAUDAN

HUYGENS, Christiaan (1629–1695), mathematician, astronomer, and physicist.

Huygens was the second son of Constantijn Huygens, an important secretary of the independent Dutch States and a clever humanist. Christiaan's mother died when he was eight years old, but his father took great care with his sons' education. Interested in everything new, even optics and mechanics, Constantijn became a good friend of René *Descartes, who had moved to Holland in 1629. Christiaan was thus one of the first important natural philosophers to begin with Descartes's ideas rather than Aristotle's. Huygens learned Cartesian geometry privately from Frans Van Shoote before attending the University of Leyden. Although he accepted the new natural philosophy and geometry, he nevertheless studied carefully the treatises of Archimedes and *Galileo.

At the age of eighteen, Huygens was already corresponding with Marin Mersenne about the main problems in geometry and mechanics then under discussion in Paris. He traveled to London with his father and to Paris several times until he was employed officially as a foundation member of the Paris Royal Academy of Sciences in 1666 (*see* ACADEMIES AND LEARNED SOCIETIES). The main reason for his call to Paris was his work on instruments—telescopes and lenses, whose power he demonstrated by resolving the usual blurred image of Saturn into a spherical body surrounded by a ring, and the pendulum clock, the first reliable timekeeper for astronomical and navigational use.

Huygens remained a French academician for fifteen years, pursuing his own work and collaborating with Gian Domenico Cassini, the effective head of the then-new Observatory of Paris. Huygens's work covered most of the questions of interest to mathematicians and natural philosophers of his time: quadratures and rectifications of curves, practical optics and theoretical models for light, motion and space, gravity, and all mechanical devices related to clocks, telescopes, air pumps, boats, coaches, windmills, and so on. He returned to Holland in 1681 for health reasons. He remained there after his recovery because an increasing intolerance toward Protestants in Paris made his return impossible. After the death of his father in 1687, Christiaan became something of a recluse in the old family country home of Voorburg, where he died.

Huygens's most famous publication, *Horologium oscillatorium* (1660), goes far beyond the practice and theory of clocks (*see* CLOCK AND CHRONOMETER). It contains his main results about mechanics: free fall, the compound pendulum, and centrifugal force. Some geometrical results, about evolutes and involutes for instance, come in when necessary for physics. But he omitted a very important mechanical problem, although he found its solution in 1656: the demonstration of the classical rules for elastic collision, which he deduced from Galilean invariance (the principle that the velocity of a body measured by an observer in motion is the difference between the body's velocity and that of the observer, with respect to a third object supposed to be at rest). The invariance of physical laws with regard to any uniform motion, as determined by observers in uniform motion among themselves, such as that illustrated by Galileo's famous example of a falling weight in a moving boat, led him to the idea of an indefinite empty space where motion can only be relative. He held this idea strongly but confidentially, as he debated it only with Gottfried Leibniz.

Huygens's second most important book, the *Traité de la lumière* (1690), is the source of his reputation as a modern physicist and prescient genius. He was the first to describe the propagation of light as a wave of motion and to deduce from it a construction for the laws of refraction that can still be found in schoolbooks.

Less famous than Galileo, Descartes, Isaac *Newton, or Leibniz, Huygens nevertheless belongs among them for his exceptional mathematical and experimental skills and, more importantly, his deep intuitions about motion and light. He developed the best ideas of Cartesian mechanism with the help of Galilean principles and classical geometry. This syncretism did not result from conflict but from Huygens's personal sense of intelligibility. But since his convictions were intuitive and scarcely changed from his youth to his death, he had insuperable difficulties in arguing and publishing them. Thus, though his contemporaries admired his technical results and manual dexterity, his methods and principles remained almost unknown.

H. J. M. Bos et al., *Studies on Christiaan Huygens* (1980). Joella G. Yoder, *Unrolling Time: Christiaan Huygens and the Mathematization of Nature* (1988). Fokko Jan Dijksterhuis, *Lenses and Waves: Christiaan Huygens and the Mathematical Science of Optics in the Seventeenth Century* (1999).

CHRISTIANE VILAIN

HYDRAULICS. See HYDRODYNAMICS AND HYDRAULICS.

HYDRODYNAMICS AND HYDRAULICS. From at least the third millennium B.C., canals, dams, and reservoirs were used for irrigation in Egypt, in Mesopotamia, and probably also in Asia. The building of these constructions and contemporary techniques of navigation imply some practical understanding of fluid motion. A greater theoretical knowledge and more elaborate apparatus appeared in Alexandria in the third and second centuries B.C. Archimedes established the laws of equilibrium of immersed solids, and Hero discussed pneumatic and hydraulic devices and the connection between efflux rate and water head (depth of the mouth). The word "hydraulic" was originally used to describe an Alexandrian invention, the water-powered organ.

The tremendous growth of hydraulic construction during the Roman Empire and the medieval period led to important innovations, such as aqueducts and waterwheels, but added little to the Greeks' understanding of fluid equilibrium and motion. A Renaissance man, Leonardo da Vinci, brought late-medieval mechanics and the emerging experimental trend to bear on these questions. His insights into the pressure-head relation, eddy formation, flux conservation, and open-channel dynamics probably aided *Galileo's disciple Evangelista Torricelli, who in 1644 established the proportionality between the efflux velocity and the square root of the water head. Torricelli also explained the principle of the Florentine barometric tube by the balance between the weight of supported mercury and atmospheric pressure.

In seventeenth-century France, Blaise Pascal formulated the law of isotropic pressure, and persuaded his brother-in-law to verify the altitude-dependence of barometric pressure, which in Pascal's view excluded the Aristotelian *horror vacui.* Hydrostatics thus reached maturity. Fluid motion still challenged the new mechanical

The first systematic experimentation on the output of water wheels (1759). William Smeaton, an engineer and a fellow of the Royal Society, measured the rate at which wheels with variously shaped vanes raised weights as a function of the pressure head and other variables.

philosophy. In his *Principia mathematica* of 1687, Isaac Newton discussed fluid resistance in order to show, contra Aristotle and René Descartes, that matter could not fill interplanetary space. By theory and experiment, he established that the resistance of a fluid to motion through it was proportional to the cross section of the moving object, to the fluid density, and to the squared velocity.

Newton's reasoning used the balance between the momentum lost by the object and that acquired by the fluid and a drastic simplification of the flow pattern. In contrast, the Swiss geometer Daniel Bernoulli based his *Hydrodynamica* of 1738 on Leibnizian *vis viva* (kinetic energy) conservation, thus obtaining the relation between wall pressure, velocity, and height (Bernoulli's law). His word "hydrodynamica" expressed the synthesis between conceptions of hydrostatics and hydraulics. Only after suitable extensions of Newtonian dynamics and differential calculus did "hydrodynamics" come to mean a general theory of fluid motion. In 1744 and 1752, Jean Le Rond d'Alembert published two treatises in which he applied his general principle of dynamics to fluid motion and established the paradoxical lack of resistance to the motion of a solid through a perfect fluid. Probably motivated by this breakthrough, in 1755 the Swiss geometer Leonhard *Euler obtained the partial-differential equation for the motion of a perfect fluid by equating the forces acting on a fluid element to the product of its acceleration and mass. He also showed how to derive Bernoulli's law from this equation. In his *Méchanique analitique* of 1788, Joseph Louis Lagrange solved Euler's equation for simple cases of two-dimensional fluid motion and proved a few important theorems.

D'Alembert's paradox and the nonlinear structure of Euler's equation deterred geometers and engineers from applying the new fluid mechanics to concrete problems. The French masters of late eighteenth-century hydraulics, Jean-Charles Borda, Charles Bossut, and Pierre-Louis-Georges Du Buat, combined experiment, global balance of momentum or *vis viva*, and physical intuition. Borda and Du Buat corrected Newton's misconceptions about fluid resistance; Borda completed Bernoulli's ideas on efflux and on head loss in a suddenly expanding pipe; Bossut and Du Buat established the proportionality between the loss of head in a long pipe or channel and the squared velocity; Du Buat formulated the general condition of permanent (constant velocity) flow as the balance between wall friction, pressure gradient, and weight.

The scope and accuracy of this semi-empirical approach grew considerably in the nineteenth century with the work of French, British, and German engineers. In the 1830s, the increased interest in canal building and river navigability led Jean-Baptiste Bélanger and Gaspard Gustave de Coriolis to compute the backwater caused by weirs. Between 1850 and 1870, Henri-Philibert-Gaspard Darcy and Henri-Émile Bazin made extensive measurements of flow in pipes and channels. In Britain, John Scott Russell, William Rankine, and William Froude studied how wave formation, streamlining, and a vessel's skin friction affected ship resistance. Most influential were Froude's model-towing experiments and his formulation of the laws that relate small-scale data to true-scale resistance.

Fundamental hydrodynamics also progressed. The mathematical physicists Simeon-Denis Poisson, Augustin-Louis Cauchy, George Biddell Airy, George Gabriel Stokes, William *Thomson (Lord Kelvin), Joseph Boussinesq, and John William Strutt (Lord Rayleigh) provided solutions of Euler's equation for ocean waves, ship waves, canal waves, and solitary waves. In 1858, while studying the aerial motion in organ pipes, Hermann *Helmholtz discovered that rotational motion in a perfect, incompressible fluid obeyed remarkably simple conservation laws. Ten years later, Thomson exploited the resultant steadiness of annular vortices to represent atoms of matter. Meanwhile, von Helmholtz and Rayleigh argued that the formation of highly unstable vortex sheets (thin layers of uniformly rotating

fluid) behind solid obstacles provided a solution to d'Alembert's paradox.

In 1822, the French engineer-mathematician Claude-Louis Navier inserted a viscosity-dependent term in Euler's equation. As Stokes demonstrated in his memoir on pendulums of 1850, the equation correctly described the regular flows observed for small characteristic lengths and velocities (e.g., the diameter of the bulb and its velocity in the pendulum case), but it seemed useless for the irregular flows observed in hydraulic cases. In the 1840s, Navier's disciple Adhémar Barré de Saint-Venant suggested that a variable effective viscosity (viscosity depending on local agitation) could be used to describe the average large-scale motion in pipes and channels. Saint-Venant's protégé Boussinesq successfully implemented this approach in the 1870s.

Whereas the French separately studied the two kinds of flow—laminar and turbulent, in Thomson's parlance—in 1883 the British engineer Osborne Reynolds studied the transition between them. He found it to occur very suddenly (as had been observed by Gotthilf Hagen in 1839) for a given value of the number LUD/η, where L and U are the characteristic length and velocity of the flow, D the fluid's density, and η the fluid's viscosity. By astute experiments and by analogy with the kinetic theory of gases, Reynolds shed light on the implied instability. Thomson and Rayleigh then pioneered the mathematical study of this question.

Despite the practical orientation of some of its theorists, nineteenth-century hydrodynamics failed to meet hydraulic and other engineering needs. It did produce, however, some of the key concepts that permitted the success of applied fluid mechanics in the twentieth century. Helmholtz's theorems on vortex motion and his concept of surfaces of discontinuity (or vortex sheets) served Ludwig Prandtl's and Theodore von Kármán's theories of fluid resistance, Frederick Lanchester's, Martin William Kutta's, and Nikolai Joukowski's theories of the airfoil, and Vilhelm Bjerknes's theory of meteorological fronts. Reynolds's and Boussinesq's theories inspired Geoffrey Taylor's, Johannes Burgers's, Prandtl's, and Kármán's statistical approaches to turbulence.

The newer fluid mechanics bridged fields as diverse as hydraulics, marine architecture, meteorology, and aeronautics. Large laboratories were built to combine model measurements, theoretical analysis, and technical forecast. Prandtl's Göttingen institute set the trend early in the twentieth century; similar institutes were created in other industrializing countries. The United

States and the Soviet Union became leaders in theoretical and applied fluid mechanics.

René Dugas, *Histoire de la mécanique* (1950). Hunter Rouse and Simon Ince, *History of Hydraulics* (1957). G. A. Tokaty, *A History and Philosophy of Fluid Mechanics* (1971). Paul Hanle, *Bringing Aerodynamics to America* (1982). Thomas Wright, *Ship Hydrodynamics*, 1770–1880 (Ph.D. diss., London Science Museum, 1983).

OLIVIER DARRIGOL

HYPOTHESIS. A hypothesis is a guess, hunch, or supposition taken as a basis for reasoning. The method of hypothesis, one of the major scientific methods, has had a career coincident with, and tangled up in, that of modern science.

Rene *Descartes brought the method of hypothesis to the fore in his *Principles of Philosophy* (1644). He argued that the world presents itself as the face of a clock whose internal workings are hidden. To talk about the workings, we must resort to hypothesis or conjecture. Similarly, while we can observe the reflection and refraction of light, if we want to explain them in terms of corpuscles or atoms, we have no option but to use hypotheses. Descartes's argument was developed by many devoted to the *mechanical philosophy, the most prominent of them being Robert *Boyle.

As anyone acquainted with logic knew at the time, to accept the method of hypothesis was to commit the fallacy of affirming the consequent. Even if we give a plausible explanation of refraction as caused by atoms, the success of the explanation does not show that atoms actually exist. Following the method of hypothesis thus fell disastrously short of achieving the Aristotelian goal of certain knowledge. Isaac *Newton, horrified by the sloppiness of the method of hypothesis (although not above resorting to it in his own thinking), declared in the *Principia mathematica philosophiae naturalis* (1687): "I do not feign hypotheses."

Newton's prestige brought hypothesis into contempt for 150 years, in such works as Condillac's *Traité des systèmes* (1741), Jean d'Alembert's "Discours préliminaires" (1751) to the Encyclopédie, and Thomas Reid's *Inquiry into the Human Mind* (1764). With the resurgence of Francis *Bacon's reputation in the first half of the nineteenth century, the alternative methods of enumerative and eliminative induction became even more popular.

Nonetheless, from the mid eighteenth century on, a minority of natural philosophers and scientists argued that science could not avoid the method of hypothesis. Natural philosophers who used *imponderable fluids or *ethers to explain gravitational phenomena, such as George Le Sage, neurophysiological phenomena, such as David Hartley, or matter theory, such as Roger Boscovich, realized that to repudiate hypotheses was to be guilty of self deception. Attitudes toward the method of hypothesis played an important role in the debates about the wave theory of light at the beginning of the nineteenth century. Those who thought the method of hypothesis unacceptable, such as John Stuart Mill, rejected the wave theory. Supporters of the method of hypothesis worked to shore up it up, for example, by adding that the wider the range of phenomena a hypothesis could explain, the more likely it was to be true.

The method gradually gained credence as figures such as William Whewell in his *Philosophy of the Inductive Sciences* (1840) and Claude Bernard in his *Introduction to the Study of Experimental Medicine* (1865) argued for its use. Whewell changed the terms of the discussion by suggesting that acceptable hypotheses had to account for more than they had been invented to explain. They also had to predict new phenomena. Even so, many scientists continued to distrust the method of hypothesis. The influential physicist and epistemologist Ernst Mach had deep reservations about its use as anything more than a heuristic aid, reservations that contributed to his misgivings about accepting the reality of atoms. Many scientists and philosophers, especially positivists, shared his views.

However, given that theories about the deep structure of the world, usually not directly accessible to observation, remain central to science, hypothesis has become ever more, not less, important. Today, the method of hypothesis, now often called hypothetico deductivism, or the method of conjectures and refutations, is the most popular of the scientific methods among both scientists and philosophers of science.

Larry Laudan, *Science and Hypothesis* (1981).

RACHEL LAUDAN

I

ICE AGE. The *Etudes sur les glaciers* by the Swiss botanist and geologist Louis Agassiz (1840) opened the eyes of geologists to the possibility that a great ice age had occurred in the recent geological past. The evidence for the hypothesis was already well known. In northwestern Europe, the location of almost all early geological research, a thick layer of boulder clay covered the bedrock, huge stones (erratics) turned up far from their mother strata, bare rocks showed long, parallel scratches, and the remains of earlier beaches appeared well above existing sea levels. Until Agassiz, though, these manifestations had been put down to a vast flood. Since most naturalists still believed in the relative youth of the earth, they relied on written testimony as much as on field evidence. An ancient flood figured prominently in Greek authors such as Ovid as well as in Genesis. Moreover, from the early nineteenth century, most geologists held that the earth had cooled from an originally hot state, making it difficult to contemplate a past ice age (*see* GEOLOGY).

Louis Agassiz drew on work by earlier Swiss geologists to reinterpret the phenomena and argue that most of Europe had been covered by an ice sheet. By the early 1860s, even the most reluctant geologists, such as Charles *Lyell, had grudgingly come to agree. What caused the ice age, Agassiz never explained. In the nineteenth century, a Scottish autodidact, James Croll, set forth the most satisfactory attempt in his book, *Climate and Time* (1875). He suggested that as the shape of the earth's orbit slowly changed as a result of gravitational interactions with other planets, the variations in ellipticity caused occasional ice ages. One of the consequences of his theory was that glacial conditions in one hemisphere would be opposed by interglacials in the other hemisphere.

Scientists soon invoked ice ages to solve other problems. They explained the well-known changes in sea level around the Baltic by the locking up of sea water as ice or by the depression of the crust under the weight of the ice. They suggested that humans had reached North America by crossing from Eurasia on a bridge of ice. They told the story of much of human pre-history as a series of adaptations to life on the southern edge of the Eurasian ice cap.

By the 1920s, using fieldwork by the United States Geological Survey and studies on gravel river terraces by German geologists, scientists had decided that Agassiz's great ice age had in fact consisted of four different stages of advance and retreat. Since these stages occurred in the southern hemisphere as well as in the northern, geologists dropped Croll's theory. Following suggestions that radiation was the decisive cause of glaciation, Milutin Milankovich, a Serbian mathematician, calculated the radiation received in the two hemispheres at various times in the last half million years. At the time, these calculations did not seem to offer the needed support for the four-stage theory.

Following World War II, scientists abandoned the four stage theory as inconsistent with their findings from deep-sea cores. Correlations between the ages of the cores, their paleomagnetism, and their temperature at the time of formation (using oxygen isotopic ratios) revealed a more complicated story that seemed to fit better with information about radiation cycles. As regards the cause of ice ages, though, the verdict is still out.

See also GEOLOGY.

John Imbrie and Katherine Palmer Imbrie, *Ice Ages: Solving the Mystery* (1979).

RACHEL LAUDAN

IDEAL GAS. An ideal gas consists of a vast number of elastic spheres in rapid rectilinear motion with no forces acting between them. Alternatively, it is a fluid that satisfies the relation $pv = nRT$ between its pressure p, volume v, and temperature T; R is the "gas constant," and n the molar amount of substance, or the number of gas volumes containing Avogadro's number of molecules, in the gas sample. The ideal gas law rests on *Boyle's law, which Robert Boyle was not the first to state, but the first to publish. In 1661 he heard of a quantitative relation between volume and pressure of an enclosed sample of air proposed by Richard Towneley and Henry Power. Boyle's assistant

Robert *Hooke confirmed that he had made observations consistent with the statement that the product of the pressure and volume of air is a constant.

In 1718 the Swiss mathematician Daniel Bernoulli correlated air pressure with particle motion on the assumption that pressure arose from the impact of confined particles against the walls of the container. Bernoulli derived the relation $p = nmv^2/3s^3$, where n is the number of molecules, m the mass of a single particle, v the mean velocity of the molecules, and s the length of a side of the cubic container. Since s^3 represents volume, the relation can be restated as Boyle's law with $nmv^2/3$ expressed as a constant. In the early 1800s publications of Joseph Louis Gay-Lussac on the expansion of gases, along with unpublished work by Jacques-Alexandre-César Charles, established the linear relationship between the volume and temperature of gases.

Working within the tradition of molecular physics, Rudolf Clausius in 1857 deduced that the average speed of gas molecules is several hundred meters per second. He also worked out an equation for the ratio of specific heat at constant pressure to specific heat at constant volume for a gas particle. Clausius, James Clerk *Maxwell, and Ludwig *Boltzmann further developed the kinetic theory of gases using statistical and probabilistic laws (see STATISTICAL MECHANICS). Experimental information suggested to Clausius that some of the heat absorbed by a diatomic gas molecule may cause its atoms to rotate around a common center or to vibrate along their common axis rather than to increase the velocity of the molecule. Experimental results also called into question the equipartition of energy among the different modes of motion, sowing doubts by the end of the nineteenth century about the validity of the kinetic theory of gases. In the early 1900s the *quantum theory accounted for the apparently anomalous fact that rotation about the axis of symmetry does not contribute to specific heat.

Nineteenth-century researchers found that all gases deviate from the ideal gas laws at high pressures and low temperatures. The Irish physical chemist Thomas Andrews particularly investigated the conditions under which a gas can be liquefied by the application of pressure in his studies of the critical temperature above which the gas cannot be liquefied. In 1873 the Dutch physicist Johannes Diderik van der Waals modified the familiar ideal gas law to take into account the weak attractive forces between molecules of a gas and the size of the molecules. His equation $(p + a/V^2)(V - b) = nRT$, accounted for reduced pressure owing to attractive force (the term a/V^2), and for the excluded volume of the container owing to the finite size of the molecules (b). The intermolecular attractive forces or "Van der Waal forces" received an explanation in Fritz London's application of quantum mechanics to the problem of dispersion forces in 1931.

Stephen G. Brush, *The Kind of Motion We Call Heat* (1976).
Keith J. Laidler, *The World of Physical Chemistry* (1993).
MARY JO NYE

I. G. FARBEN. I. G. Farbenindustrie AG (Farben), perhaps the most controversial corporation in history, came into existence in 1925 through the merger of Germany's leading dye firms. By far the largest chemical enterprise in Germany, Farben had the largest staff and international sales in the world. Its formative leaders Carl Duisberg and Carl Bosch were self-made entrepreneurs, former laboratory chemists whose organizational and technical talents, respectively, had brought them into the leadership of the two biggest firms in the field, Bayer and BASF. Bosch, who shared the Nobel Prize for Chemistry in 1931 for the technical development of the Haber-Bosch ammonia synthesis, chaired Farben's board of directors until 1935, then took Duisberg's place on the supervisory board.

Under Bosch's leadership, Farben pursued a small amount of basic research and extensive applied research on fronts ranging from the traditional areas of synthetic organic dyes and pharmaceuticals to fertilizers and pesticides, photochemicals and films, polymers, synthetic fibers, and plastics. Kurt Heinrich Meyer and Herman Mark conducted especially significant high-polymer studies at Ludwigshafen until 1932. Gerhard Domagk was awarded the Nobel Prize for Medicine in 1939 for his work at Elberfeld on the chemotherapeutic effects of sulfa compounds. Farben also subsidized academic chemistry with postdoctoral fellowships and support for journals, and paid most of the operating budget of the *Kaiser Wilhelm Institute for Chemistry.

Farben's two most expensive projects were the catalytic high-pressure hydrogenation of coal products to produce synthetic gasoline, and the production of "buna" synthetic rubber from butadiene derived from acetylene. Technical and scientific leaders for the former were Carl Krauch and Matthias Pier; for the latter, Otto Ambros and Walter Reppe. The economic crisis of the early 1930s severely limited Farben's basic research and support for science. To subsidize its synthetic fuel against foreign petroleum it reluctantly concluded an agreement with the National Socialist regime that took power in 1933. The resulting ill-fated collaboration deepened with the Four Year Plan of 1936 (under which Farben built its first plants to produce "Buna S" rubber for tires), which fostered German economic autarky in preparation for World War II.

During the war, Farben took advantage of German conquests to expand its holdings as it supplied critical synthetics to the war effort. It never completed its fourth and last buna plant, whose construction involved the lethal exploitation of slave labor drawn from the concentration camp at Auschwitz. Farben's subsidiary Degesch (co-owned with Degussa) produced the pesticide Zyklon B, used for mass gassings of prisoners. Another Farben product, fortunately not used during the war, was sarin nerve gas. The victorious Allies dissolved the concern after World War II, and tried its executives for war crimes at Nuremberg in 1947–1948. Most were acquitted; some received light prison sentences, and all were freed by 1951. Out of Farben's dissolution emerged the principal German chemical firms of the postwar era.

Anthony S. Travis et al., eds., *Determinants in the Evolution of the European Chemical Industry, 1900–1939* (1998). John E. Lesch, ed., *The German Chemical Industry in the Twentieth Century* (2000). Peter Hayes, *Industry and Ideology: IG Farben in the Nazi Era*, 2d ed. (2001).

JEFFREY ALLAN JOHNSON

IGY. See INTERNATIONAL GEOPHYSICAL YEAR.

IMMUNOLOGY as science dates from the 1880s, although inoculation against smallpox goes back many centuries. The history of immunology falls into two periods, before and after World War II. It begins with serology: identification of bacteria, clinical application of vaccines and sera to infectious diseases, and the chemical problem of specificity. After the war, transplantation and grafting rather than infectious disease led immunological work. Unlike other biosciences, immunology was not reductionist: newer work concentrated on the activities of cells rather than on chemistry. The field grew rapidly in the 1960s and 1970s as the new cellular immunology developed; many of those who participated wrote accounts of the period. The historiography of the subject developed in the 1990s and deals with social, scientific, and business history. French writing on the subject has an epistemological twist.

Smallpox inoculation originated in Asia and the Middle East, came west in the 1700s, and may be responsible for the decline in smallpox deaths by about 1790. Vaccinia (cowpox) as inoculum was suggested several times in the late 1700s, most famously by Edward Jenner, a country doctor (though it is unclear if the material he used for vaccination was actually *Vaccinia*). Production

was unregulated, and the operation painful and sometimes ineffective; nonetheless health authorities enforced inoculation through such measures as the British Compulsory Vaccination Act of 1853. Compulsion led to worldwide antivaccination movements with strong political and anticolonial overtones.

The Age of Serology, 1890 to 1950

In the 1880s, evidence for germ theory began to mount. Louis *Pasteur announced his rabies vaccine and the Russian zoologist Elie Metchnikoff suggested that white blood cells defended the body against invaders. Pasteur liked Metchnikoff's idea and invited him to Paris. But the Franco-Prussian war of 1870 was still being waged on the intellectual front and a rush of publications from Robert Koch and his Berlin colleagues overtook Metchnikoff's work. As bacteriologists, the German scientists were more interested in immune serum, which could be used to identify bacteria, and ignored cells. Animals could be immunized not only with bacteria, but also with bacterial toxins, and immunity transferred passively to another animal via serum. In 1891, antiserum against the toxin of diphtheria proved to be effective in treating the disease. Serum manufacture using horses began at the Institut Pasteur; a global network of serum institutes followed. In Germany, the state guaranteed production and standardization. On the battlefields of World War I, tetanus antitoxin strikingly reduced the incidence of tetanus. At war's end, the victorious Allies through the League of Nations and its Health Organisation set up their own standardization project at the Statens Seruminstitut Copenhagen, by-passing German laboratories, but still using German techniques.

Standardization was central to serology. Paul Ehrlich's standardization in 1894 of diphtheria antitoxin provided a starting point for theory and research. According to his side-chain theory, cells had receptors (the side-chains) that normally functioned to capture nutrients. The receptors could be blocked by toxin, which caused the cell to repair itself by producing an excess of new side-chains, which were shed into the serum and formed serum antitoxin. Toxin and antitoxin bound irreversibly on contact. Specificity was absolute. Ehrlich called this a pluralistic view: receptors for every possible antigen were already present in the normal animal. His critics objected that the theory required far too many specific substances.

The alternative, propounded by Karl Landsteiner of Vienna, gained currency after 1900. Specificity became approximate, a matter of the closeness of fit, determined by the charge outline of antigens. The antigen-antibody reaction became reversible as in colloid chemistry. On this view, antibody might be formed by assembling protein molecules on antigen as template, as the Prague chemist Felix Haurowitz suggested in 1928; unlike Ehrlich's theory, Landsteiner's required no preformed antibody. But Ehrlich's vocabulary and his cartoon-like pictures of the union of antigen and antibody outlived his chemistry. A century later, we still speak of receptors. In the mid-1920s, Swedish chemists began differentiating serum proteins, first by molecular size in the ultracentrifuge, then by charge and then by both together. Antibody activity was found to lie in the globulin fraction of the serum protein. This work led to Rodney Porter's elucidation of the four-chain structure of antibody globulin (1962), the definition of a family of different immunoglobulins each with a different immunological function, and to the sequencing of the amino acids making up the chains.

Cellular Immunology, 1950s to 1980s

Antibiotics came in with World War II and although new vaccines still appeared, notably Jonas Salk's polio vaccine (along with antivaccinationism), serological treatment of disease declined. Thinking became more biological, dealing with reactions in animals and cells rather than serological chemistry. The key problem was graft rejection, important in wartime along with blood transfusion. Discussion centered on tolerance: Peter B. Medawar in London, arguing on the model of the blood groups, suggested that graft intolerance was immunological, while Frank Macfarlane Burnet in Australia taught that tolerance of self developed in fetal life. Various theories appeared bringing antibody-producing cells to the fore: selection theories of antibody production reappeared. Burnet's fruitful clonal selection theory proposed that immunocompetent cells carried a range of natural antibodies as receptors. Antigen triggered multiplication of a clone of cells, and each clone produced one specific antibody. Clones that recognized foreign (that is, nonidiopathic) material survived the fetal period, self-clones being eliminated. This view overtook the template theory about 1967. Burnet linked immunology to modern biology through Darwinian selection and the population genetics of clones, and to current thinking on protein genetics.

An avalanche of new work focused on populations of lymphocytes—then cells with no known function, now seen as mediating the immune response, including immune recognition and memory. Immunological experimentation turned to the reactions of animals, usually inbred mice. Old cells-versus-serum controversies were resolved as it appeared T lymphocytes, developing in the neonatal thymus, mediate cellular immunity and interact with B lymphocytes from the bone marrow, producers of serum antibody; the cells cooperate via messenger molecules named lymphokines. At this point, Anne-Marie Moulin suggests, researchers began to recognize an "immune system." Clinical applications of cellular immunology included autoimmunity, hypersensitivity, and transplantation surgery. Pharmaceutical companies patented immunosuppressants and, later, perfectly specific monoclonal antibodies originally derived from cultured myeloma, a cancer of lymphocytes.

In the 1970s, with increased funding especially in the United States, the profession expanded: journals proliferated, congresses national and international were initiated, symposia and courses organized.

Acquired immune deficiency syndrome (AIDS) appeared in 1982; human immunodeficiency virus (HIV), isolated in 1984, was found to affect a particular subpopulation of lymphocytes, the T4 cells, key to the immune response. T4 cells became part of popular discourse and historians too became interested in immunology.

See also AIDS.

William D. Foster, *A History of Bacteriology and Immunology*. (1970). Anne-Marie Moulin, *Le dernier langage de la médecine: histoire de l'immunologie de Pasteur au SIDA* (1989). Virginia Berridge and Philip Strong, eds., *AIDS and Contemporary History* (1993). Emily Martin, *Flexible Bodies: Tracking Immunity in American Culture from the Days of Polio to the Age of Aids* (1994). Alberto Cambrosio and Peter Keating, *Exquisite Specificity: The Monoclonal Antibody Revolution* (1995). Pauline M. H. Mazumdar, *Species and Specificity: An Interpretation of the History of Immunology* (1995). Christopher Sexton, *Burnet: A Life*, 2d ed. (1999).

PAULINE M. H. MAZUMDAR

IMPONDERABLES. Around 1800 physical science enjoyed a fleeting unification under a scheme that developed from study of the phenomena of heat and *electricity (see FIRE AND HEAT). The discovery of the conduction of electricity down damp threads by Stephen Gray in 1729 prompted the assimilation of the agent of electrical attraction to water running through a pipe, an analogy strengthened by Benjamin *Franklin's comparison (1751) of the machines used for generating electricity (globes or cylinders of glass spun against the hand) to pumps, and *Leyden jars (condensers) to reservoirs. For those who accepted Robert Symmer's version of Franklin's theory (1759), which made negative charge as real as positive, electricity was served by two fluids, which, since ordinary bodies appeared to weigh no more when electrified than when not, were taken to be imponderable.

In order to explain the most evident electrical phenomena, natural philosophers ascribed repulsive and attractive forces to the droplets of the electrical fluids: repulsive between droplets of the same fluid, attractive between those of different fluids and between the fluid(s) and ordinary matter. To account for the differences in the degree of electrification or tension exhibited by insulated conductors of different shapes and sizes electrified in the same way by the same machine, philosophers ascribed a pressure to the electrical fluid(s) and specific electrical capacities to the conductors. Johan Carl Wilcke and Alessandro Volta developed these concepts in the 1770s (*see* GALVANI, LUIGI, AND ALESSANDRO VOLTA).

Most chemists and natural philosophers of the eighteenth century traced the action of heat to a special substance, which, like electricity, was understood to be an expansive fluid because of its spontaneous "flow" from hot to cold bodies. Also like electricity, its parts were taken to be self-repellent in order to explain the expansion of bodies when heated. Within the Newtonian philosophy the self-repellency of heat arose from a repulsive force acting between the particles of the heat fluid. With the discoveries of latent and specific heats by Wilcke and Joseph Black in the 1770s and 1780s, the parallels between the heat fluid and the agent of electricity broadened: latency could be regarded as a bonding between heat and matter; specificity indicated an analogy between temperature and heat capacity, on the one hand, and tension and electrical capacity, on the other.

The standard representations of magnetism and visible light easily fit the imponderable model. By analogy to electricity, *magnetism came to be regarded as a distance force arising from magnetic fluid(s) whose particles obeyed the same rules of attraction and repulsion that regulated the traffic of the electrical fluids. The main distinction—that nothing comparable to conductors of electricity existed for magnetism—was regarded as a question of degree not kind. The magnetic fluid(s) stayed in magnetic substances as electrical fluid(s) did in strong insulators. Franz *Aepinus worked out these parallels in detail in 1759. As for *light, its particulate nature was assured by the widely held optical theories of Isaac *Newton, which endowed light particles with short-range forces by which they interacted with matter to produce the phenomena of reflection, refraction, and inflection (diffraction). The capstone of the arch of imponderables was the discovery by William Herschel in 1800 of radiant heat beyond the red end of the visible spectrum. Infrared light connected heat and ordinary light and, via the analogies between heat and

electricity, light with magnetism. More speculative philosophers added fire, flame, phlogiston, and what-not to the generally accepted five (or seven) imponderables.

The scheme, which functioned as a *standard model for physical science around 1800, had two important assets. For one, it immediately explained the existence of the phenomena it covered by the mere presence of the relevant agent. For another, it lent itself to the fashion of science of the time, quantification. In 1785, Charles Augustin Coulomb (*see* CAVENDISH, HENRY, AND CHARLES AUGUSTIN COULOMB) established to the satisfaction of the members of the Paris Académie des Sciences (and few others) that the interfluid forces in electricity and magnetism declined, as did the force of gravity, with the square of the distance between interacting elements. Pierre-Simon *Laplace and his school pursued the program of quantifying the distance forces that are supposed to act between elements of the heat fluid (which they called caloric) and between light particles and matter. Laplace and Jean-Baptiste Biot managed to give detailed accounts of refraction, both single and double; polarization; and other optical phenomena in these terms. By taking literally the concept of heat as a conserved fluid, Laplace created a brilliant theory of adiabatic processes that resolved the long-standing and scandalous discrepancy between theoretical treatments and measurements of the speed of sound in air (*see* ACOUSTICS AND HEARING). Although it did not appeal to the notion of distance forces, the adiabatic theory encouraged belief in the existence of caloric.

A serious fault with the scheme of imponderables, apart from its multiplication of weightless fluids at a time when chemistry was learning to live strictly by the balance, was the ontological independence of the several fluids. The unification the scheme brought rested on parallel treatment of diverse phenomena, not on connections among their agents. This weakness was partially overcome by the linking of electricity and magnetism beginning with the discovery of the action of a current-carrying wire on a magnet by Hans Christian Oersted in 1819. But the replacement of the particulate by the undulatory theory of light in the early nineteenth century, and the annihilation of caloric by the kinetic theory of heat in its middle decades, destroyed the old standard model. A new synthesis seemed imminent and immanent in James Clerk *Maxwell's unification of electricity, magnetism, light, and radiant heat, and the program pursued around 1900 to reduce ponderable matter to *electromagnetism. That program failed, leaving the *electron, the electric current, and the flow of heat as residues and reminders of the first standard model in physics.

Tore Frängsmyr et al., *The Quantifying Spirit in the 18th Century* (1990). J. L. Heilbron, *Weighing Imponderables and Other Quantitative Science around 1800* (1993).

J. L. HEILBRON

INDUSTRIAL LABORATORY. See LABORATORY, INDUSTRIAL.

INDUSTRIAL REVOLUTION. See ENLIGHTENMENT AND INDUSTRIAL REVOLUTION.

INFINITY. A network of questions surrounding infinite magnitudes and collections of things has taunted mankind since antiquity, especially when linked to continuity and to theories of space and time. Zeno's famous paradoxes, like that of Achilles unable to catch the tortoise, either confront matter with rest, or present clashes of finite with infinite.

Connections with the claimed infinitude of space brought clashes with both Aristotle (who gave several arguments against the possibility of an infinite universe) and certain religious beliefs. Mathematicians too eschewed reliance on infinity. For example, Euclid's parallel axiom seemed to be unintuitive in allowing coplanar parallel lines to be extendible "for ever" without meeting; indeed, he had stated it in the complementary form of two nonparallel lines meeting at one finitely distant point in the plane. In geometry innovations were limited to a point or line at infinity, introduced to understand perspective in art and then given a more general role in projective geometry.

Continuity is a major notion in geometry, the science of space. Body presents another enigma, well exemplified for physics by theories of matter. Is a continuous solid body really unbroken throughout its extension, or is it composed of minute elementary particles at tiny distances apart? If the latter, what if anything fills the gaps between them? Many answers have been advocated since antiquity, often accepting or rejecting Aristotle's views of the abhorrence of a vacuum: some thinkers assumed the existence of a continuous universal *ether, eliminating vacua; others welcomed void spaces and populated them with *atoms.

Analogous differences arose analyzing motion. An important example, Newton's second law in his *Principia* (1687), is usually stated as force = mass × acceleration. Newton often employed it this way himself; but he stated it in a manner analogous to atomism: micro-increment on impulse = mass × micro-increment on velocity.

To bridge the gap (as it were) between continuity and atomism mathematicians posited infinitesimals, variable quantities smaller than all ordinary ones and yet not equal to zero. Despite their paradoxical credentials, they have been crucial in several branches of mathematics, especially the differential and integral calculus. In particular, Leibniz's key notion was the infinitesimal increment dx, of the same dimension as x; itself variable, it admitted its own increment ddx, and so on. Leibniz also allowed for infinitely large variable quantities $\int x$, $\int\int x$, ... and presented the integral "$\int y \cdot dx$" as the sum ("$\int$") of rectangles y high and dx wide. Despite their superb utility in models of phenomena involving continuous phenomena, infinitesimals received a nervous press: even Leibniz called them "convenient fictions" and hinted at justification by limit processes. However, limit theory also posed difficulties: Newton gave them prime status in his rival "fluxional" calculus, but did not handle them well. Only in the early nineteenth century did Augustin-Louis Cauchy put forward a robust theory.

Cauchy subscribed to the traditional view of infinity as a quantity beyond the indefinitely large numbers, but possessing weird properties, such as the whole not being greater than the part because there are as many even integers $(2, 4, 6, \ldots)$ as integers $(1, 2, 3, \ldots)$. Occasionally mathematicians speculated that infinities might come in different sizes; but not until Georg Cantor did this idea bear fruit. Tackling a particular problem in mathematical analysis in the early 1870s, he was led to consider infinite sets of points on the unending straight line. He discovered that the manners of their distribution were far more complicated than had been thought. He defined different kinds of sets (such as "closed" and "dense") and studied the relationships between them, thus creating "point-set topology." He also produced a refined definition of continuity. One main consequence was a general theory of integration, which became known as "measure theory" when developed in the early 1900s by Henri Lebesgue and William Henry Young; it not only played an important role in mathematical analysis itself but also came into probability theory, statistical and quantum mechanics, and thence to other sciences.

Cantor found that some sets had an infinitude of members greater than that of the integers $1, 2, 3, \ldots$; for example, the set of real numbers, or of points on the straight line. He also found that there were "no fewer" points on a line L of finite length than in the square, or cube, or constructed upon L. Cantor was shocked by this refutation of the intuitive understanding of dimension in space, which was to inspire profound new results in topology from the 1920s.

The main notion in Cantor's theory was that of a limit point of a set P; for example, if P is the set of rationals {1/2, 1/3, ...}, then its (only) limit point is 0. He called the set of such limit points the "derived set" of P, itself a set and maybe with its own derived set; and he came to allow for an "infinitieth" derived set I of P, which he defined as the intersection set of all its predecessors. I may not be empty; for example, the set of all numbers between −4 and 7 inclusive always has itself as its derived sets. If so, then I had its own derived set, thus launching a new sequence and so on for infinitely ever.

Thereby Cantor rejected the traditional construal of infinity as the "last" of the finite integers, and instead took it to be the first of a new sequence of numbers, with no predecessor. He created a theory of "transfinite" numbers, cardinal and ordinal, closely related to point set topology but with a life of its own in containing an exact theory of inequalities between different orders of infinitude. The sequences have no end. If one posits the largest of such numbers (cardinal or ordinal) N, then the paradox arose that both $N = N$ and $N > N$. No definitive resolution of these (and other) paradoxes was found. However, important advances occurred in the axiomatization of set theory (initially by Ernst Zermelo) and in mathematical logic (initially by Alfred North Whitehead and Bertrand Russell), and thereby philosophy.

Cantor did not form the inverses of his transfinite cardinals to define infinitesimals; on the contrary, he declared them to be impossible. His opinion has been refuted in recent decades, especially in Abraham Robinson's theory of nonstandard analysis.

Abraham Fraenkel, *Abstract Set Theory* (1953). L. L. Whyte, *Essay on Atomism. From Democritus to 1960* (1961). Joseph Dauben, *Georg Cantor* (1979; 1990). Joseph Dauben, *Abraham Robinson* (1995).

IVOR GRATTAN-GUINNESS

INFORMATION TECHNOLOGY. See COMPUTER; COMPUTER SCIENCE.

INFUSORIA, the name given to the small creatures seen in infusions of decaying vegetable or animal matter during the century after 1760 and now a class of Protozoa.

In a letter dated 7 September 1674 to the Secretary of the Royal Society, the Dutch microscopist Antoni van Leeuwenhoek described observations, made with high-powered lenses of his own making, of "very many little animalcules" found in the water taken from an inland sea. Some of them "were roundish, while others, a bit bigger, consisted of an oval. On these last I saw two little legs near the head, and two little fins at the hindmost end of the body. Others were somewhat longer than an oval, and these were very low a-moving, and few in number." Some of the "animalcules" seen by Leeuwenhoek were Protozoa, probably ciliates. In subsequent reports he described more of them, both in infusions of decaying vegetable matter like pepper and cloves and in pond and rainwater. Microscopists of the late seventeenth and early eighteenth centuries gave many names to these creatures. In 1760 Martin Frobenius Ledermüller named them "little animals of infusions," and in 1765 Heinrich August Wrisberg invented "infusoria," a term with much the same meaning as "microorganisms" today. In the same year Lazzaro Spallanzani conducted experiments that showed that infusoria did not come about by spontaneous generation (*Dissertazioni due*, 1765), while Horace Bénédict de Saussure discovered that some of them could reproduce by constant division. In his taxonomic and posthumous work *Animalcula infusoria fluviatilia et marina* (1786), the Danish naturalist Otto Frederik Müller divided the order of Infusoria into several genera.

During the first half of the nineteenth century the structure, function, and classification of infusoria were studied vigorously. Building on Spallanzani's suggestion that the contracting vacuoles to be seen in the body of certain infusoria were respiratory organs (*Opuscoli di fisica animale e vegetable*, 1776), Christian Gottfried Ehrenberg maintained that infusoria amounted to complete organisms in miniature (*Die Infusionsthierchen als vollkommene Organismen*, 1838). He interpreted these vacuoles, however, as stomachs and therefore created a new class of animals, the Polygastrica, which included most species of infusoria. Conversely Félix Dujardin held that they were droplets of environing fluid present in the structureless substance (*sarcode*, later called protoplasm) constituting the bulk of the microscopic animals' bodies (*Histoire naturelles des zoophytes*, 1841).

During the second half of the nineteenth century, Carl Theodor Ernst von Siebold and Otto Bütschli found that most infusoria were unicellular and therefore classified them as Protozoa. Around 1900 investigators determined that some were parasites of man and animals. The old infusoria proved to be responsible for diseases like sleeping sickness, amoebic dysentery, and malaria.

Clifford Dobell, *Antony van Leeuwenhoek and His "Little Animals"* (1932). Marc J. Ratcliff, "Temporality, Sequential Iconography and Linearity in Figures: The Impact of the Discovery of Division in Infusoria," *History and Philosophy of the Life Sciences* 21 (1999): 255–292.

RENATO G. MAZZOLINI

INORGANIC CHEMISTRY. Traditionally, chemical substances were classified on the basis of their origins, the most general distinctions being between the mineral kingdom and the animal or vegetable kingdoms. After the demonstration in 1828 that urea, a compound produced by living things, could be synthesized from inorganic chemicals, the discipline of organic chemistry developed as a distinct area of study—the chemistry of compounds of carbon and the elements with which it readily combines (hydrogen, oxygen, nitrogen). Inorganic chemistry then became the chemistry of compounds of all the elements other than carbon. This early distinction based on origin accounts for the peculiarity that a few carbon compounds—such as the oxides, carbonates, and cyanides—are still generally regarded as inorganic. As a recognized area of study, inorganic chemistry dates from the nineteenth century, but as a practical interest in substances from inanimate sources, it is rooted in alchemy and it stretches back centuries to mining, metallurgy, medicine, dyeing, glassmaking, and gunpowder production.

Among the numerous problems facing nineteenth-century chemists, many revolved around the lack of agreement over standards, such as chemical notation and an atomic-weight scale. Two chemistries seemed to exist, organic and inorganic, each with its own set of atomic weights. At the International Chemical Congress at Karlsruhe, Germany, in 1860, the Italian chemist Stanislao Cannizzaro circulated a short paper on the problems associated with atomic weights and ways of resolving them. During the next decade chemists succeeded in clearly distinguishing between atoms and molecules, reforming and standardizing atomic weights, and reuniting the two chemistries with a single set of conventions.

While catalyzing reform, Cannizzaro's paper also had a profound effect on the thinking of many chemists, including Dmitrii *Mendeleev and Lothar Meyer, both of whom organized the elements on the basis of increasing atomic weight (Mendeleev in 1869 and Meyer in 1870). The modern *periodic table, the eventual outcome of this organization, is more closely associated with Mendeleev, who was by far the bolder of the two. He not only organized the elements into a table based on his periodic law, but also predicted the existence of several elements that were unknown at that time, as well as their properties. These represent one of the first instances of successful prediction in chemistry. Mendeleev's table assumed its modern form in 1914 when the English physicist Henry G. J. Moseley organized the elements on the basis of increasing charge in the atomic nucleus, i.e., increasing number of protons (see ATOMIC STRUCTURE).

In 1858 the German chemist August Kekulé advanced his theory that the atoms in molecules are arranged in specific ways relative to one another. These structural formulas, along with the idea that the valence or "combining power" of an atom is constant, worked very well for organic compounds and fueled the dramatic growth in this area of chemistry after 1860. Kekulé tried to extend these ideas to inorganic compounds, but they failed even in simple cases. More complicated molecules—like those of the coordination compounds first synthesized during the nineteenth century—were particularly problematic.

While an assistant to Robert Bunsen in Marburg, Germany, Frederick Augustus Genth synthesized several ammonia-cobalt compounds in 1847. He published this work in 1851 after emigrating to the United States. The following year the American chemist Wolcott Gibbs also began investigating these compounds, and in 1856 he and Genth jointly described the synthesis and analysis of thirty-five ammonia-cobalt compounds. Their report of the properties and striking colors of these unusual compounds presented a challenge to theorists.

In 1869 the Swedish chemist Christian Wilhelm Blomstrand made the first attempt to describe the structures of the compounds synthesized by Genth and Gibbs. Making use of the variable valence of nitrogen, he proposed a chain theory, which the Danish chemist Sophus Jørgensen subsequently adopted and modified. This model linked ammonia molecules (NH_3) in an ammonia-cobalt complex together in a chain like the CH_2 groups in an organic hydrocarbon compound. More importantly, Jørgensen prepared and examined many new complex compounds, including some with chromium, rhodium, and platinum, as well as cobalt. He also determined many of their physical properties. The vast amount of experimental data collected by Jørgensen, Gibbs, and Genth was essential to the subsequent development of the revolutionary coordination theory.

The theory, a foundation-pillar of inorganic chemistry proposed by Alfred Werner, represented a new way of describing inorganic compounds and was not merely a modification of organic structural theory. Werner obtained his doctorate from the University of Zurich in 1890 with a dissertation on the spatial arrangement of atoms in molecules containing nitrogen. In 1893, the year he began teaching at the University of Zurich, he published the first of his papers on his theory of variable valence. He described molecular inorganic compounds as having a central (metal) atom surrounded by a definite number of other atoms, molecules, ions, or radicals in a three-dimensional arrangement. The number of such groups around the central atom was termed the coordination number. The ammonia-cobalt compounds have a coordination number of six, the six ammonia molecules being arranged at the vertices of an octahedron (8 faces defined by 6 vertices—two square pyramids sharing a common square base) centered at the cobalt. Werner's coordination theory provided a tool for understanding not only inorganic compounds and reactions, but also many organic and biochemical reactions, such as those involving Grignard reagents in which a central magnesium ion is coordinated with both halide ions and organic groups. Werner received the Nobel Prize in 1913.

Since many chemists regarded Werner's monumental work as having completed inorganic chemistry, the subject generated little interest during the early part of the twentieth century. The American Chemical Society further marginalized the field by making it an area within physical chemistry. Not until the *Manhattan Project in the early 1940s and the need to separate uranium isotopes did the field of inorganic chemistry began to flourish again. A group of chemists organized a separate Division of Inorganic Chemistry within the American Chemical Society and began the journal *Inorganic Chemistry*.

The 1950s mark the reemergence of inorganic chemistry. Although the first organometallic compound was synthesized in 1827, chemists could not devise an adequate description of its structure and bonding until the early 1950s when they prepared and studied many other such compounds. The structure of one of these novel molecules, ferrocene, was elucidated by two Harvard chemists and subsequent Nobel laureates, Robert Woodward and Geoffrey Wilkinson. The ferrocene "sandwich" has an iron(II) ion as the bonding center between two parallel, aromatic, five-member carbon rings. The development of *quantum mechanics and significantly improved understanding of atomic structure led to new bonding theories that can interpret these novel structures satisfactorily.

Another pivotal figure in inorganic chemistry of the second half of the twentieth century is F. Albert Cotton, a former student of Wilkinson's. His textbook, *Advanced Inorganic Chemistry*, written with Wilkinson and currently in its sixth edition (1999), has trained several generations of chemists since its first edition in 1962. Cotton's 1963 textbook, *Chemical Applications of Group Theory*, which appeared in its third edition in 1994, has provided chemists with an extremely powerful tool for understanding, explaining, and modeling the structure and reactivity of inorganic compounds.

The resurgent inorganic chemistry originating in Werner's "completed masterpiece" is an essential component of *materials science, an interdisciplinary field that blends chemistry and engineering. Thin films, industrial diamonds, ceramics, superconductors, semiconductors (both traditional and organic/organometallic), LEDs (light-emitting diodes), and lasers are a few of the materials and devices that will continue to be improved through advances in inorganic chemistry. Current trends that should have significant results in the twenty-first century include greater theoretical understanding of the formation of ceramics, leading to improved materials for engineering; synthesis, development, and manufacture of organometallic LEDs; synthesis of diamond and other commercially important films; detailed computer models of every element in the periodic table and three-dimensional visualization tools accelerating growth in all aspects of inorganic chemistry; and a more powerful understanding of quantum mechanics, leading to improved communication and computational devices.

See also CHEMISTRY; ORGANIC CHEMISTRY.

Walter Hückel, *Structural Chemistry of Inorganic Compounds*, 2 vols., trans. L. H. Long (1950). J. R. Partington, *A History of Chemistry*, 4 vols. (1961–1972). George B. Kauffman, *Inorganic Coordination Compounds* (1981). C. A. Russell, ed., *Recent Developments in the History of Chemistry* (1985). Allen G. Debus, *Chemistry, Alchemy, and the New Philosophy* (1987). Arthur L. Donovan, ed., *The Chemical Revolution* (1988). John Hudson, *The History of Chemistry* (1992). William H. Brock, *The Chemical Tree: A History of Chemistry* (2000). Trevor H. Levere, *Transforming Matter: A History of Chemistry from Alchemy to the Buckyball* (2001).

RICHARD E. RICE AND FRANK A. PALOCSAY

INSTINCT. Early reflections on the role of instinct in animal and human behavior are rooted in the works of Aristotle, the Stoics, and Galen. Relying on these authors—as well as on the works of the Persian physician Avicenna (Ibn Sina) and on religious traditions—medieval and early modern authors emphasized the distinction between the invariability of animal behavior, pertaining to "nature," and the variability of human actions, pertaining to "reason." Aristotelians, however, granted animals at least a lesser kind of reason.

In the seventeenth century, with the diffusion of René *Descartes's mechanistic approach to the study of animal bodies, philosophers traced instinctive behavior to bodily organization. They regarded instincts as innate behavioral patterns proper to different species, and deduced that individuals having the same bodily organization displayed identical instincts. Empiricists, on the other hand, by privileging the role of sensory experience, often acknowledged a similar ability to learn from and elaborate sensory images both to animals and humans. Within a similar perspective, eighteenth-century authors like Abbé de Condillac and Erasmus Darwin regarded animal instincts and human reason as products of the same fundamental ability to learn from experience.

With Jean-Baptiste de *Lamarck, reflections on instincts became an important component of the argumentation aimed at establishing the causes able to explain evolutionary change. According to Lamarck and most of his followers, instincts as well as organs and functions arose through habits generated in living beings by their need to adjust to the environment, and to varying living conditions.

Late eighteenth- and early nineteenth-century literature on instincts—including that produced within the framework of *natural theology—played a role in Charles *Darwin's early reflections on evolutionary mechanisms. Later, a full chapter of Darwin's *Origin of Species* (1859) was devoted to showing that instincts, like complex organs and functions, could be explained with the gradual transformation in time brought about by natural selection (*see* *DARWINISM). Here, and more fully in his *Descent of Man* (1871), Darwin argued that the intellectual faculties of the human species had probably been perfected through natural selection, like animal instincts.

During the first half of the twentieth century, the study of fixed action patterns in animal behavior became central to the discipline of ethology, especially as it developed in Austria, Germany, and Holland. Karl von Frisch, Konrad *Lorenz, Nikolaas Tinbergen, and others accumulated a great quantity of observational data and conceptualization concerning instincts within the disciplinary framework of ethology.

Skepticism toward an extended use of the notion of instinct, on the other hand, was already being expressed in the 1920s by the American tradition of psychology known as *behaviorism. By emphasizing the processes of habit formation and conditioning, behaviorists like John Watson drastically restricted the role of instinctive or unlearned behavior in animal and human psychology.

Since the 1970s, several authors—especially those adhering to sociobiology, such as Edward O. Wilson—have developed approaches to the study of unlearned behavioral patterns that emphasize genetic and neurobiological analyses. Within the new perspective, the notion of instinct seems to have lost many of the functions it displayed in the Darwinian tradition and in ethology.

W. H. Thorpe, *The Origins and Rise of Ethology* (1979).
                                        GIULIANO PANCALDI

INSTITUTE. The university-based institute was the instrument and symbol of Germany's dominance of natural science at the end of the nineteenth century. In its spatial isolation from the rest of its *university, the institute represented and enforced the division of the sciences by *discipline and of the professoriate by research accomplishment.

Institutes evolved from two university practices. The earliest, dating from the eighteenth century, was the cabinet for instruments or other materials used for demonstration in teaching and to a limited degree, for research. Their maintenance required workrooms; their use for instructional demonstrations, lecture rooms. Initially privately owned, cabinets were placed on the university (state) budget by 1850. The evolution of cabinet to institute depended upon the establishment of university laboratories and the inauguration of the professorial research ethos in the early nineteenth century. Institutes were thus the principal venue through which the manual practices associated with experimental research became a regular part of university instruction in the sciences. An early example is Justus von *Liebig's chemistry laboratory at Giessen, founded in 1826.

The second university practice that gave rise to institutes dates from the early nineteenth century. The emphasis German elites placed on *Bildung* (the formation of mind and character)—based in study of ancient languages, philosophy, history, and later theoretical natural science and taken as the key to advancement at higher levels of government service—was a central impetus in the development of new methods of instruction based on learning the principles of research. To prepare secondary teachers capable of imparting proper *Bildung*, professors of philology created special *seminars in which students were systematically taught disciplinary knowledge and research methods that could in principle lead to the production of original research. This model entered the natural sciences in 1825 in the Bonn seminar for the natural sciences, followed in 1835–1836 by the establishment of the Königsberg seminar for physics and mathematics, directed by Carl Gustav Jacobi, a classicist turned mathematician, and the mineralogist turned physicist Franz Ernst Neumann. Like the university cabinet, the seminar required a dedicated budget, director, and space. Before 1850, when laboratories began to appear alongside cabinets, they also were found

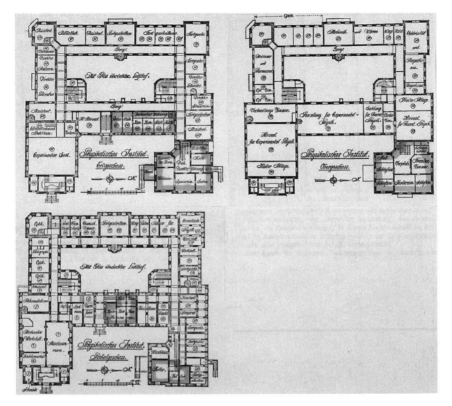

Floor plan of the physics institute built at the University of Leipzig in 1904. Its many small rooms indicate the arrival of the advanced independent researcher. The areas outlined made up the living quarters of the professor-director, "the baron of the institute."

associated with seminars in either the natural sciences or physics and mathematics combined. Managing and maintaining the cabinet, laboratory, and seminar required increasingly greater financial resources and support services, adding to the administrative complexity of these units. As the professorial research ethos took hold and instructional techniques for transmitting it to students were developed, the nature of instruction became ever more complex.

In the second half of the nineteenth century the institute emerged as the principal site in the university of original research, both professorial and student. The institute by then was an amalgamation of physical space, financial resources, material supplies and equipment, student clientele, support staff, and professorial leadership. The long-standing German tradition of having a single full professor (*Ordinarius*) per discipline stamped the organizational structure of the institute as one run by a director who had near total control over all persons in the institute's sphere, including assistant and associate professors (*Privatdozenten* and *Ausserordinarien*).

The success of the instructional mission of institutes beyond the production of scientific practitioners (where success was uneven) can be measured by the demographics of their student clientele. Chemical institutes, like Liebig's, included students who wished to become physicians, pharmacists, agriculturalists, and future secondary teachers. Similarly and simultaneously physics institutes furthered the careers of physicians and secondary teachers. Seminars in all the natural sciences, which generally had poorer resources than full-fledged institutes, catered in addition to geographers, technicians, astronomers, businessmen, and civil servants. Students interested in technology and engineering found places in institutes in the Technical High Schools (Technische Hochschulen), which began to multiply after 1870. Institutes thus spread through the educated male elite in Germany an outlook not only familiar with the techniques of research, but one highly appreciative of its social and cultural values as well.

The institute phenomenon eventually took over the entire university system in Germany.

Resources devoted to teacher (and research) training grew faster than resources devoted exclusively to service teaching. For example, in 1873 the University of Leipzig replaced an old physics cabinet, built in 1835, with an institute four times as large. This, in turn, was replaced by one three times larger still in 1904. Laboratory facilities occupied 12 percent of the cabinet, 46 percent of the first institute (1873), and 60 percent of the second—if the thirteen rooms and veranda of its baronial quarters are deducted, 75 percent. Most of the increase came in small rooms for individual research rather than in teaching laboratories and lecture halls.

The system was in principle continuously renewed by the "call"—the invitation of a professor in one state to take over an institute in another. The invitee would negotiate with both, for equipment and space as well as salary. A man in demand might leave a string of new buildings in his wake. Hermann von *Helmholtz received a new institute of physiology as a dowry when he moved from Bonn to Heidelberg in 1858, and a new physics institute when he moved on to Berlin in 1871.

The institute spread throughout Europe, including Britain, and, with an important exception, the United States. Although American university departments provided disciplinary separation, teaching facilities, and, in time, a research ethic, their more democratic organization thwarted any pretensions of their chairs to papal authority and perquisites, as was the custom in Germany. The long-term historical consequences of the fluidly democratic structure of the department versus the rigidly hierarchical structure of the institute were crucial for the future development of the scientific community in both countries. Whereas American universities have been receptive to interdisciplinary cooperation, German universities have been less so due to the persistence of the rigid administrative practices of the institute. Consequently at the end of the twentieth century, Germany's educational ministers favored interdisciplinarity by encouraging innovation outside the outmoded institute structure and by offering pathways for career advancement for younger scholars that did not rest on permanent appointments in institutes.

By the beginning of the twentieth century competition among the European nations in natural science and its applications prompted the formation of institutes with few if any teaching functions. Among them, the Carnegie Institution of Washington (founded in 1902 with a gift larger than Harvard's endowment), the research departments of the national *bureaus of science, and the *Nobel institutes for physics and chemistry are prominent international examples, alongside the many institutes of the *Kaiser Wilhelm Gesellschaft and the Soviet Academy of Sciences.

R. S. Turner, "The Growth of Professorial Research in Prussia, 1818 to 1848," *Historical Studies in the Physical Sciences* 3 (1971), 137–182. Paul Forman et al., "Physics circa 1900: Personnel, Funding, and Productivity of the Academic Establishments," *Historical Studies in the Physical Sciences* 5 (1975): 1–185. R. S. Turner, "Justus von Liebig versus Prussian Chemistry," *Historical Studies in the Physical Sciences* 13: 1 (1982): 129–162. David Cahan, "The Institutional Revolution in German Physics, 1865–1914," *Historical Studies in the Physical Sciences* 15: 2 (1985): 1–65. Frederic L. Holmes, "The Complementarity of Teaching and Research in Liebig's Laboratory," *Osiris* 5 (1989), 121–164. Kathryn Olesko, *Physics as a Calling. Discipline and Practice in the Königsberg Seminar for Physics* (1991).

KATHRYN OLESKO

INSTITUTE FOR SCIENTIFIC INFORMATION, PHILADELPHIA.

The historic growth in the number of scientists has been matched by an increase in their output. By the mid-twentieth-century the profusion of scientific publications made it impossible for an individual scientist to keep up with the literature in a particular field. The problem of information overload began to trouble government advisors in the United States in the late 1950s and led to proposals for ways to condense or distill literature for easier consumption.

The Institute for Scientific Information in Philadelphia was a private initiative to address the problem, albeit with government support. In the early 1950s Eugene Garfield, a chemist working on a medical indexing project at Johns Hopkins, had the idea to use calculators to generate standardized indexes of scientific literature automatically, thus cutting out much painstaking human labor and allowing the production of comprehensive indexes fast enough to keep up with current literature. Lawyers had long used citation indexes to look up legal cases. In 1955 Garfield wrote an article in *Science* proposing scientific citation indexes, as he called them, which would include not only each published article but also all other articles that referred to a given article. He soon began publishing a compilation of the tables of contents of about two hundred biomedical journals, which he sold by subscription as *Current Contents*. In 1960 he incorporated his business as the Institute for Scientific Information.

In 1961 Garfield, with the help of geneticist Joshua Lederberg, obtained funding from the National Institutes of Health to produce a sample citation index for the field of genetics. Garfield expanded the coverage beyond genetics into a general *Science Citation Index*, a 5-volume compilation of 1.4 million citations covering over 600 journals. The Institute for Scientific Information began regular publication of the *Science Citation Index* in 1964 and in the 1970s added indexes for the social sciences, arts, and humanities literature. The indexes became an important tool for researchers to locate relevant work among the ever-increasing mass of published literature. They also provided quantitative data on publication and citation patterns for sociologists and historians of science, and a new field of scientometrics emerged around the analysis of citation indexes. The indexes became a sort of scorecard for the performance of journals and authors, identifying oft-cited articles as well as those that had little apparent influence, and figuring in promotion and tenure reviews. The indexes perhaps contributed to specialization by allowing scientists to forego wide reading in favor of narrow directed searches within their particular field.

Dorothy B. Lilley and Ronald W. Trice, *A History of Information Science, 1945–1985* (1989).

PETER J. WESTWICK

INSTRUMENTALISM AND REALISM. Realists believe that the objects scientists study exist independently of the scientists. They also believe that science should aim at finding the truth, that the entities scientists have identified do actually exist in the world, and that the theories of modern science are at least approximately true. Some instrumentalists agree with the first of these propositions. All disagree with the second. Instrumentalists regard the objects of knowledge as tools for human purposes, and believe that the aim of science is empirical adequacy—explanation, prediction, and control—but not truth.

The modern dispute between realists and instrumentalists began in the late nineteenth century. It was sparked by disagreements about how to interpret atoms and molecules. Many physicists and chemists, such as Ludwig Boltzmann, believed that the atoms and molecules that figured in their theories really existed in the world. Other notable scientists, particularly Ernst Mach, Pierre Duhem, and Henri Poincaré disagreed. According to them, atoms and molecules were no more than convenient fictions. Their reasons for believing so varied, but their distrust of the method of hypothesis used to supposedly prove the existence of atoms loomed large.

Pierre Duhem decided to use historical examples as well as philosophical arguments to support his instrumentalist position. In *To Save the Phenomena* (1908) he advanced an enormously influential interpretation of the history of astronomy. Plato had set the agenda for astronomy in antiquity with his dictum that astronomers need do no more than "save the phenomena"—that is, produce an astronomical theory that fit the facts. They did not have to worry about whether the theory was true as long as it worked. According to Duhem, this became the predominant tradition in astronomy. On this interpretation Andreas Osiander thus drew on mainstream ideas when he declared in his preface to the *De revolutionibus* of Nicholas *Copernicus that readers need not read the book as asserting truths about the world. Duhem and many subsequent historians of astronomy went on to argue the empirical equivalence of geocentric and heliocentric planetary systems.

Max *Planck and Albert *Einstein opposed the trend away from realism. So did those who believed that if atoms could be split, they must exist. But the problem did not go away. In the mid-1920s, physicists debated whether the wave theory or the particle theory of subatomic structure or the *quantum theory was true. Niels *Bohr and Werner *Heisenberg championed an instrumentalist interpretation that was accepted by most physicists, who were unable to give a realist interpretation to subatomic theorizing. The logical positivists took the lessons of quantum physics to heart. The question of the reality of theoretical entities, they decided, was a metaphysical, not a scientific, question.

In the 1960s, scientific realism enjoyed a revival. Equipped with new arguments from philosophers of science such as Karl Popper, J. J. Smart, Hilary Putnam, and Wilfred Sellars, it became the orthodoxy in philosophy of science for a couple of decades. Many found particularly compelling the so-called miracles argument of Hilary Putnam, according to which the success of science would be a miracle if the objects it postulated did not actually exist and its theories were not at least approximately true. Soon, though, further challenges to realism were mounted both from within philosophy and from a new anti-realist movement, constructivism.

Philosophers pointed out that the success of science did not in fact mean that science followed the track to truth. Since the history of science is full of once highly successful theories that invoke entities that, like caloric and phlogiston, no contemporary scientists believe to exist, Larry Laudan made the pessimistic meta-induction that there is no reason to believe in the truth of contemporary scientific theories either. Other philosophers returned to the Duhem argument that there may be empirically equivalent theories between which no evidence can decide. Responding to these criticisms, Arthur Fine, Ian Hacking, Philip Kitcher, and Jarrett Leplin have all attempted to shore up the realist position.

Meanwhile, in the sociology of science, constructivists such as Karin Knorr-Cetina and Andrew Pickering produced detailed studies of historical and contemporary science to show that scientific knowledge was manufactured or constructed and had nothing to do with underlying physical reality or truth. A significant number of historians of science, notably Simon Schaffer and Steven Shapin, joined forces with these sociologists. This radical relativist position was rejected by realist and anti-realist philosophers of science and many historians of science, and greeted with horror by many scientists. The debate was one of the most heated in late twentieth-century science studies.

Mary Jo Nye, *Molecular Reality* (1972). Ian Hacking, *Representing and Intervening* (1983). Jarrett Leplin, ed., *Scientific Realism* (1984). Arthur Fine, *The Shaky Game: Einstein, Realism and the Quantum Theory* (1986). Karin Knorr-Cetina, *The Manufacture of Nature* (1980). Andrew Pickering, *Constructing Quarks* (1984).

RACHEL LAUDAN

INSTRUMENTS AND INSTRUMENT MAKING. Although the concept of a scientific instrument may seem clear, the historian is bound to find it problematic, not least because the term did not come into common use until the second half of the nineteenth century. The present usage, in its application to the past, comes partly from a projection of current scientific practice onto earlier activities and partly from choices made by collectors, curators, and dealers as they constructed a specialist interest in material culture that would guide their activities in museums and salesrooms.

An important function of contemporary scientific instruments is to investigate the natural world, to discover new truths about nature. But until the beginning of the seventeenth century instruments had no such role. In the terminology of the time, some "mathematical instruments" played an integral part in certain mathematical arts or sciences, but had no place in the science that dealt with causes and explanations in the material world, *natural philosophy. Practitioners used these to solve problems and produce practical results, such as casting a horoscope, telling the time, finding the latitude, or drawing a map.

Astronomy provides the earliest record of the use of mathematical instruments and through astronomy many of the technical characteristics of mathematical instruments developed. This precedence might be seen as the outcome of a Platonic belief in a geometrical cosmos. But

A trade card from the mathematical instrument maker Thomas Wright (1718). The card gives pride of place to the orrery, a clockwork planetarium, shown under the globe (upper center), armillary ring (left), and theodolite (right).

an alternative view is just as plausible. The regulation of time and calendar by the changing appearance of the sky required measurement and invited instrumentation: a Platonic concept of geometrical astronomy may have been a theoretical formulation of an existing operational practice in observing the heavens.

Astronomical instruments fixed in observatories were generally made by bringing together local resources such as woodwork, metalwork, and masonry, whereas personal portable instruments like *astrolabes and *sundials came from mathematical instrument makers, and specialist workshops in recognized centers of production. One of the earliest makers on whom we have much biographical information was Jean Fusoris, university educated and a church canon, who set up a workshop in Paris and produced astrolabes and clocks in the early fifteenth century.

Other early makers or founders of workshops were leading mathematicians and astronomers. Johannes Regiomontanus established a workshop in Nuremberg in the 1470s; several surviving instruments are attributed to him. Scholarly mathematicians from the sixteenth century with a strong commitment to the development of instrumentation include Peter Apian in Ingolstadt and Gemma Reines (called Frisius) in Louvain. Gemma worked in association first with Gerardus Mercator and later with his nephew Gualterus Arsenius to produce a great many instruments and a variety of new designs. An international trade developed. The Louvain workshop benefited from an understanding with Christophe Plantin whose printing house in Antwerp was used for their ordering and distribution. This arrangement represented a wider association between instrument production and other mathematical commerce, such as mapmaking and book publishing.

The sixteenth century saw a remarkable development in instrumentation as part of a general flourishing of practical mathematics. The development of navigation and commerce and changes in the conduct of warfare inspired new ways of harnessing geometry to more effective action. This activity tended to take place at court or in the city rather than in the university. Mathematics occupied an inferior position in the university hierarchy to medicine, law, and natural philosophy, but at court it could be used as a tool for political and territorial advance, and instruments were used for persuasion as well as action. The appearance of many surviving instruments from this period indicate that their role was partly rhetorical.

Among many new designs from the sixteenth century were different types of sundials, quadrants, and nocturnals for finding the time, universal astrolabes, theodolites, and other surveying instruments, the cross-staff and backstaff for astronomical navigation, and the sector for a wide range of calculations. The publication of many books on instrumentation and the spread of centers of production, notably to Florence, Venice, Nuremburg, Augsburg, Ingolstadt, Louvain, Paris, Antwerp, London, and Prague, accompanied and supported the development in the range of designs.

Because these mathematical instruments did not engage with natural philosophy, they did not have to respect the received account of the natural world. Terrestrial globes rotated on polar axes in advance of the publication of the Copernican theory simply for convenience. Different projections of the celestial sphere could be used on the two sides of a single astrolabe according to their intended applications, a freedom that reflected the variety of projections used in contemporary cartography. In both fields, convenience and efficiency were the criteria of success, not fidelity to nature. At the same time, the status enjoyed by practical mathematicians in nonacademic contexts for work, and their freedom from the disciplinary restrictions of the university, gave them a relative confidence and autonomy that would eventually facilitate the application of their practices to the reform of a demoralized natural philosophy.

The application of instruments to discovering the truths of nature began in the late sixteenth and early seventeenth centuries, most significantly in the use of the *telescope in astronomy. Instrumentation had always been part of astronomical practice, of course, but it had been applied to mathematical astronomy, not the natural philosophy of the heavens, which was treated as a separate discipline. *Galileo insisted that his telescopic discoveries from 1609 onwards gave evidence for the Copernican cosmology and so thrust the telescope into the forefront of a dispute in natural philosophy, where its reliability as a tool of discovery would be a critical issue.

The telescope and the *microscope created a new domain of instrumentation separate from the established trade of mathematical instruments. A different category of artisan, the more able and enterprising among the spectacle makers, produced the new optical instruments. Like the telescope, the microscope first arose in a commercial rather than a learned context. It was an optical toy with no agenda for use in natural philosophy until the midseventeenth century. Then an increasing interest in explaining natural phenomena through the interaction of invisible particles acting as tiny machines made the microscope a likely arbitrator of the claims of the *mechanical philosophy. Through the development of the microscope and telescope a new trade was born. By the late seventeenth century, although they included spectacles among their stock, some specialists had become "optical instrument makers."

The natural philosophers involved themselves closely in this development. Johannes *Kepler and René *Descartes had been concerned with the true form of an aplanatic lens, one that did not suffer from spherical aberration. Christopher Wren designed an unsuccessful machine for grinding hyperbolic lenses to remove the defect, while other early fellows of the Royal Society ground telescope objectives, as did Christiaan *Huygens in the Netherlands. In the case of microscopes, Robert *Hooke associated with the London makers and frequented their workshops, while in Delft Antoni van Leeuwenhoek made his own extraordinary microscopes with their single, tiny, spherical lenses. The best optical glass came from Italy, where Eustachio Divini in Rome and Giuseppe Campani in Bologna led the field. Italian natural philosophers, such as Gian Domenico Cassini for the telescope or Marcello Malpighi for the microscope, could rely on the products of the best commercial workshops. Leeuwenhoek had to make everything himself.

By the late seventeenth century, the commercial trade in optical instruments was particularly vigorous in London, where visitors to the shops of Christopher Cox, John Yarwell, or Richard Reeves might expect to buy a fine telescope or a microscope equivalent to the one illustrated in Hooke's *Micrographia* (1665). But the early promise of microscopes as arbitrator of philosophy proved hollow. If fleas and other tiny things were as complicated as they appeared to be, the fundamental mechanical corpuscles lay far beyond the instrument's reach. Microscopy declined in natural philosophy. Nonetheless Hooke's astonishing illustrations had made their mark: through much of the eighteenth century a widespread interest in natural history would supply the makers with a ready clientele for microscopes.

A third category of instrument with which the natural philosophers had an even stronger engagement than with optical instruments emerged in the later seventeenth century. These "instruments of natural philosophy," unlike the mathematical and optical ones, had no location of their own within the trade. Natural philosophers themselves designed the instruments and contracted assembly to artisans. Philosophical instruments included air pumps and electrical machines, which, unlike the passive telescopes and microscopes for observing, intervened and interfered with nature. They literally implemented the collaborative, public, and institutionalized experimental philosophy practiced in the Royal Society of London and other societies that cultivated natural knowledge. Experimental demonstrations were to be performed in public before witnesses and repeated at will: they created a need for instruments of natural philosophy.

Practical applications continued to drive improvements in mathematical instruments even while optical and philosophical instruments began to capture attention. Edmund Gunter in Gresham College, London devised a quadrant for telling the time and performing other astronomical calculations, a sector for navigating by the Mercator chart, and a rule for achieving the same end with logarithmic scales. The ubiquity and longevity of the "sliding Gunter" or logarithmic slide rule testify to the sophistication of mathematical instrumentation in the early-modern period.

As instruments changed over the seventeenth century so did their provenance and markets. London grew into a major center for instrument making and, in the eighteenth century, dominated the trade. Makers in London could belong to any guild or company—they could nominally be grocers, or haberdashers, or fishmongers—and the companies did not restrict the production methods, designs, or materials. This freedom, which contrasted with the centralized and regimented situation in Paris, suited a trade that needed to combine disparate materials, adopt new designs, adapt working practices, and merge artisanal skills. New configurations in the trade began to emerge, as certain London makers at the turn of the century, notably Edmund Culpeper and John Rowley, traded across the traditional boundaries by offering both mathematical and optical instruments.

During the eighteenth century the most ambitious makers dealt in "mathematical, optical and natural philosophical instruments." Demonstration apparatus became fully commercial under the stimulus of subscription courses in experimental natural philosophy, such as those given by Francis Hauksbee in London and the abbé Jean-Antoine Nollet in Paris. Makers offered books (often written by themselves, in the cases of George Adams or Benjamin Martin), demonstrations, and courses of lectures in addition to instruments. Shops presented their wares within the context of the regular trade in luxury goods, intriguing foreign visitors. The growth of material consumption within the middle classes benefited the makers and natural philosophy had a fashionable following, encouraged by entertaining lecture courses or domestic demonstration from itinerant lecturers. The formation of instrument collections spread from institutions, universities, and the aristocracy to the homes of the bourgeoisie—a development encouraged by entrepreneurial traders in a buoyant market.

The rise of a consumer market directed the production of instruments toward the elegant, such as barometers, globes, and orreries, and the spectacular, such as air pumps and electrical machines. Telescopes, particularly the Gregorian reflector, and microscopes multiplied in the same context, and their designs reflected their intended station in a library or a drawing room. The solar microscope was developed to project large images of microscopic subjects onto a wall to entertain a group. At the same time, however, London manufacturers consolidated a leading position in the most exacting part of the trade: measuring instruments for astronomers, navigators, and surveyors.

A succession of outstanding makers of precision instruments in eighteenth-century London, beginning with George Graham and continuing through Jonathan and Jeremiah Sisson, John Bird, Jesse Ramsden, and John and Edward Troughton, raised the status of makers among the community of mathematicians and natural philosophers to an unprecedented level. These makers produced observatory instruments for fundamental measurement in astronomy—at first mural quadrants, transit instrument, and zenith and equatorial sectors, and later meridian circles—and sextants, theodolites, and other precision measuring instruments for everyday professional use. Their work was complemented by that of several outstanding optical instrument makers, John and Peter Dollond for lenses and James Short for telescope mirrors.

While individual skills must figure in the explanation of this development, certain institutional factors also played an important part—commissions from the Royal Observatory and later the Ordnance Survey, and the liberality of the Royal Society, which elected the leading makers as fellows, awarded them medals, and published their papers in the *Philosophical Transactions*. The activities of the Board of Longitude charged with administering the longitude prize established in 1714 were also influential. One of the contending methods, that of lunar distances, demanded exact and robust instruments. The board publicized methods it rewarded, for example, Bird's prescriptions for making quadrants.

At the end of the eighteenth century makers began to move away from a concentration on handwork in small workshops. Jesse Ramsden employed some fifty men in his premises in Piccadilly: the Board of Longitude published his description of the dividing engine he built for the mechanical graduation of scales on sextants and other instruments. Hand division had previously been the most prized skill in the precision trade; it was now mechanized and, comparatively speaking, deskilled. Subcontracting, buying in parts, even buying whole instruments and adding the retailer's name, became common in the eighteenth century as the trade grew and its organization became more complex. These trends accelerated during the nineteenth century.

The acceleration proved costly to the London workshops. They lost their dominance, partly through complacency, partly through loss of status in the community of learning, partly through the vigorous rise of other centers of innovation. The downgrading of manual skill in the social changes brought by the industrial revolution may have been a factor in the loss of status, but the makers as individuals never regained their positions of respect. Even in astronomy, scientists with mechanical flair like George Biddell Airy designed the major instruments and commissioned components from different makers, adopting the division of labor from contemporary industry. Germany and France became increasingly competitive.

Munich was the center of the resurgent German industry. The able makers there included Georg von Reichenbach, Joseph Liebherr, Joseph von Utzschneider, Joseph Fraunhofer, Traugott Lebrecht Ertel, Georg Merz and Carl August, Ritter von Steinheil. They focused on precision instrumentation, which they pursued in partnerships of opticians and mechanicians. Thus they benefited from research-based improvements in glass quality as well as from innovative designs in structures and mountings. The workshop of the Repsold family formed another center in Hamburg, while Karl Philipp Heinrich Pistor and Johann August Daniel Oertling were active in Berlin from 1813 and 1826, respectively. Observatories, other than British ones, equipped in the nineteenth century usually had German instruments, whereas the many eighteenth-century foundations, including French ones, had been supplied from London.

The French Revolution swept away the old restrictive practices of the guilds and put in place reforms, such as the metric system of weights and measures, that would create work for instrument makers and encourage innovation. Étienne Lenoir, Jean Nicolas Fortin, and François-Antoine Jecker seized the new opportunities, supplied standards of length, weight, and capacity throughout the country, and met the renewed demand for portable instruments from mathematical professionals. Prominent and successful workshops in nineteenth-century France included Gambey, Lerebours and Secretan, Gautier, Morin, and others. The French developed a particular expertise in physical optics, led by Jules Duboscq and later by the aptly named Jean-Baptiste-François Soleil.

From 1851 onwards international *exhibitions furthered the international character of the instrument trade. The exhibitions' reports give a good indication of the relative strengths of the contributing nations. The British were taken aback at the success of their rivals at the Great Exhibition in London (1851). Microscopes, largely through the introduction of the achromatic objective, had again become serious tools of scientific research, and here London makers continued to shine; but elsewhere they had lost much ground. By the end of the century, the German workshops of Zeiss and Leitz had seized the lead in microscopy.

By 1900 the "scientific instrument" in the modern sense of the term had arrived. As makers sought to respond flexibly to rapid innovation, the old characterizations and distinctions became irrelevant. *Spectroscopy opened up a vast new area of instrumentation for chemistry and astronomy. Industrial instruments greatly expanded the market open to instrument entrepreneurs, while the coming of the electrical industry and the spread of power supply, of the electric *telegraph, and then of radio, opened up a large field for collaboration between scientists and manufacturers. Techniques of detection and measurement had not only to work on the laboratory bench, they had to be standardized and made sufficiently robust to travel successfully to distant stations.

The twentieth century was characterized by ever larger manufacturing units, close liaison with research laboratories in universities, in institutions, or in-house, a bewildering array of new techniques and instruments, and the growing irrelevance of regional contexts other than as economic determinants. This flexibility has been particularly marked as electronic technologies have increasingly displaced mechanical ones. Instrument making has become difficult to isolate from science itself. That may always have been the case. The ubiquity of instrumentation in today's science makes the relation obvious.

A further feature of the twentieth century was the rise of collecting, both institutional and private. Museums now have large collections of instruments—transferred to them from societies, universities, manufacturers, industries, collectors, dealers, and salesrooms. Historians have not realized the potential of this resource fully. Instruments have been integral to the story of science. Although material evidence may be more intractable and awkward to use than written sources, historians of science can scarcely afford to neglect it.

See also INDUSTRIAL LABORATORY; OBSERVATORY; STANDARDIZATION.

Maurice Daumas, *Scientific Instruments of the Seventeenth and Eighteenth Centuries and Their Makers*, trans. Mary Holbrook (1972). Jim Bennett, *The Divided Circle: A History of Instruments for Astronomy, Navigation and Surveying* (1987). Anthony Turner, *Early Scientific Instruments, Europe 1400–1800* (1987). Gerard L'E. Turner, *Nineteenth-Century Scientific Instruments* (1983). Robert Bud and Deborah Jean Warner, eds., *Instruments of Science: An Historical Encyclopedia* (1998).

JIM BENNETT

INSTRUMENTS, ASTRONOMICAL MEASURING. Ptolemy's *Almagest* provides the earliest written account of observatory instruments for measurement. They include rings set for measuring angles in a single coordinate, a quadrant marked on a plane surface in the meridian for taking the altitude of the sun, and a set of three jointed rules for finding the zenith distance and, over a series of observations, the parallax of the moon. One instrument, the armillary sphere, differed from the others in its ambition and complexity. A series of nested rings moved around an axis pointing to the celestial pole and so produced motions parallel to those of the heavens. This motion included that of a second axis set at an angle to the first equal to the obliquity of the ecliptic (the angle between the ecliptic or annual path of the sun and the equator) that could follow the daily motion of the ecliptic pole. Direct measurements thus could be taken in a coordinate system with the ecliptic as the fundamental reference circle, as required by the planetary geometry of the *Almagest*.

The *Almagest* thus presented a tension that runs through the history of measuring instruments in astronomy. Stable instruments that move as little as possible take the most reliable measurements; instruments that can be moved into alignment with the circles astronomers project onto the heavens take the most convenient ones. In addition, complexity and ingenuity, often prized in themselves, result in higher charges from the instrument maker. Ptolemy's armillary sphere could hardly have been used for worthwhile measurement, but it would certainly have been a valuable aid to teaching Ptolemaic astronomy. Measurement by armillary seldom occurred in practice. In this particular, as in much else, Tycho *Brahe was an exception.

Tycho used Ptolemaic and medieval precedents in equipping his observatory on the Danish island of Hven, but his total of some two dozen instruments, described in his *Astronomiae instauratae mechanica* of 1598, included many innovations and experiments. Some of his instruments have equatorial or even ecliptic motions. The most successful, however, were those with the greatest stability, in particular his great mural quadrant two meters in radius. Observing stars in the meridian immediately gave their declination (angular distance from the celestial equator). Tycho measured the complementary coordinate, right ascension, by timing his observations using mechanical clocks. Although the later technique was unsuccessful, this general method was to dominate astronomical measurement for centuries, as astronomers became accustomed to the nightly routine of "grinding the meridian."

Tycho achieved an accuracy for naked-eye measurements close to their theoretical limit, the resolving power of the human eye. Only with the aid of telescopic sights could astronomical accuracy surpass Tycho's fractions of a minute of arc. The earliest use of a telescopic sight moving across a divided arc, as well as of an eyepiece micrometer for measuring small angles without moving the *telescope, came with the English astronomer William Gascoigne in about 1640, and both techniques were established in the observatories of Paris and Greenwich in the late seventeenth century.

Telescopes with the new sights were fastened to iron frameworks supporting scales divided on brass and mounted on meridian walls. Astronomers hoped that these instruments would provide for both coordinates, in the manner proposed by Tycho, but they tended to bend out of the plane of the meridian under their own weight and could not be realigned. This mattered little for declination but significantly impaired accuracy in right ascension. Consequently a specialized "transit" instrument, first used by Ole Rømer who worked in Paris and Copenhagen in the later seventeenth century, became the standard companion to the mural quadrant. The transit instrument was essentially a telescopic sight mounted on a horizontal axis with bearings supported by piers on either side. Free to turn in the vertical plane and used with a clock, it could be readjusted to the meridian as necessary.

George Graham provided Edmond Halley at Greenwich with a set of instruments that became standard observatory equipment in the eighteenth century. Central to these were his mural quadrant of 1725, his transit instrument, and his regulator clock with its dead-beat escapement and mercury compensation pendulum. With Graham's clocks the goal of a timepiece sufficiently accurate for astronomical measurement had been achieved. His quadrant design became the basis for many such instruments made in London workshops and sent to observatories across Europe.

Confident of their reputations and abilities, London makers tried once again to introduce observatory equatorials late in the century. Both Jesse Ramsden and brothers John and Edward Troughton managed to install large equatorials in observatories, but they did not work well and the old lesson about keeping motion to a minimum had to be relearned. More significantly, the same two workshops replaced quadrants by full circles, which had significant advantages in stability, accuracy, and error reduction. Eventually the Greenwich quadrants gave way to a mural circle designed by Edward Troughton, but once again, despite contrary expectations, it had to be complemented by a transit instrument for the

separate measurement of right ascension. Despite this flaw, such was Troughton's reputation that observatories in Britain and in its overseas colonies adopted his arrangement.

German workshops developed an alternative to Troughton's combination of a mural circle and a transit instrument, and the German pattern was widely adopted, eventually even in Britain. This single instrument combined a full vertical circle for declination with the basic transit instrument and thus could take both coordinates at once with greater efficiency and less opportunity for confusion. A second influential German design from the early nineteenth century, an equatorial telescope with a divided object glass micrometer, allowed two halves of a lens to separate along their common diameter and be moved by micrometer screws. As the halves separated, the relative movement of the resulting pair of identical images could be measured, as, for example, when two partners in a double star were brought into coincidence. With such an instrument, built by Joseph Fraunhofer, Friedrich Wilhelm Bessel in Königsberg made the first measurement of stellar parallax in 1838.

The equatorial had at last joined the ranks of instruments for fundamental measurement, though not through the traditional direct reading of graduated arcs. However, equatorial mountings did carry the instruments that replaced the traditional ones when astrometry moved away from graduated arcs and adopted photography, interferometry, photometry, and other measuring techniques no longer concerned simply with position in the pattern of the heavens.

See also CLOCK AND CHRONOMETER; INSTRUMENTS AND INSTRUMENT-MAKING; OBSERVATORY; TELESCOPE.

Jim Bennett, *The Divided Circle: A History of Instruments for Astronomy, Navigation and Surveying* (1987). Allen Chapman, *Dividing the Circle: The Development of Critical Angular Measurement in Astronomy 1500–1850* (1990).

JIM BENNETT

INSTRUMENTS, BIOLOGICAL. The fine woodcut representing a few dozen anatomy instruments—from the most delicate lancet to rough saw and hammer—included in Andreas Vesalius's epoch-making *On the Fabric of the Human Body* (1543) is an icon of the then new regard for instruments in the pursuit of natural knowledge. So too were the *balance for measuring variations in weight owing to ingestion, excretion, and perspiration, discussed in Santorio Santorio's *On Static Medicine* (1614), and the diagrams describing simple loop and palpation experiments performed on the human body in William Harvey's *Movement of the Heart and Blood in Animals* (1628).

Many of the instruments represented by Vesalius, Santorio, and Harvey derived from established techniques in medicine, surgery, natural philosophy, alchemy, or chemistry. The same applied to several instruments used by early modern naturalists, which came from pharmacy, agriculture, hunting, breeding, and the multifaceted know-how of merchants and travelers. The emphasis on instruments, however, was new, and it affected the study of living things just as it did astronomy and other sciences infused with the "new philosophy." The positive evaluation of artisanal skills extended to cooperation with printers to employ the latest advances in the art to represent and spread knowledge and achievements—a trend from which anatomy, natural history, and printing itself benefited in turn.

From the 1620s a genuinely new instrument became an icon of what later would be called the life or biological sciences. This was the optical *microscope, put to good use after 1660 by Robert *Hooke, Marcello Malpighi, and Nehemiah Grew. Meanwhile the "mechanical philosophy" penetrated the study of plant and animal life. The *air pump played a role when Robert *Boyle showed that plants produce air (1680–1682). John Mayow and others used simple cupping glasses and siphon arrangements for similar purposes. But perhaps no one better than Stephen Hales, with his *Vegetable Staticks* (1728) and *Haemastaticks* (1733), conveyed Europe-wide the notion that organisms could be included in experimental apparatus that could measure the movements and pressure of blood or sap, and the quantification of the airs inspired, expired, and transpired.

The study of the chemistry of life throughout the eighteenth century continued to demand new instrumentation. Apparatus developed by chemists became basic tools for investigators exploring animal heat and respiration. The *thermometer and the *calorimeter have remained fundamental for physiology to the present. The *stethoscope came in 1816.

Electrical machines were adapted to the investigation and possible improvement of living organisms after the invention of the *Leyden jar. With the rise of *Galvanism and the invention of the voltaic battery in 1799, electric currents became available. From the 1840s, galvanometers—based on the deflection of a compass needle by a wire carrying a current, and developed mainly in connection with telegraphy—allowed scientists like Emil du Bois-Reymond to carry out new measurements of the small quantities of electricity involved in physiological processes.

The family of electrical measuring devices used in physiology expanded quickly after 1850. In 1855 Rudolf Albert von Koelliker and Heinrich Müller argued that electrical activity must accompany the functioning of the vertebrate heart. In 1874, physicist Gabriel Jonas Lippmann introduced the capillary electrometer based on the response of the surface tension and thus the shape of the meniscus of mercury to a change in electromotive force of as little as 0.0001 volt. Étienne-Jules Marey applied the instrument to the study of muscles, where the quick reactions of the new electrometer made it possible to distinguish individual muscle-action potentials, which the galvanometer did not. Photographic recording devices adapted to the electrometer allowed Lippmann and Marey, in 1876, to record the changes of the electromotive force in the contracting heart of a tortoise and a frog, producing the first electrocardiograms. Further improvements enabled Augustus D. Waller, in 1887, to obtain the first human electrocardiogram.

In the 1840s, another new family of apparatus had been introduced—pneumatic and mechanical instruments used to study the combined action of respiration and the systolic and diastolic rhythms of the heart. The first such device, Carl Ludwig's kymograph (from the Greek *kyma*, "wave"), combined a manometer connected to the artery and clockwork moving a recording cylinder (1846). The self-registering apparatus was inspired by the indicators that James Watt used to chart pressure variations in a steam engine. The myograph (for taking tracings of muscular contractions and relaxation), the sphygmograph (for recording the movements of the pulse), and the cardiograph followed soon.

Important improvements introduced after 1830 in microscopes (correction of spherical aberration), the preparation of specimens (better microtomes and staining techniques), and microsurgery facilitated an unprecedented flow of microscopic observations that molded the content, laboratory set-up, and teaching habits in cytology, histology, embryology, and neurology through the mid-twentieth century. Similar effects on research, training, and manipulative skills were produced after the mid-nineteenth century in *botany, *entomology, *zoology, *paleontology, and *anthropology by the creation of natural history *museums, agricultural experiment stations, and *oceanographic institutes.

Physiology and clinical medicine found new opportunities for instrumental developments after 1900 with the introduction of new electromagnetic apparatus. Willem Einthoven's string galvanometer (1903) made possible the first complete electrocardiograph, produced by Cambridge Scientific Instruments in 1908. Einthoven's machine, filling two rooms, evidently was *big science. Soon, however, the introduction of vacuum-tube amplifiers developed for radio communications allowed the production of more practical, and less expensive electrophysiological and medical apparatus based on electronic technology, setting a pattern of exchange that has continued to the present.

Also in the 1920s, the ultracentrifuge—in which an oil-turbine generated gravitational fields reaching 400,000 g by the 1940s—was developed by The Svedberg (*see* CENTRIFUGE). Meant to determine the size of colloidal particles, the ultracentrifuge proved essential for the study of hemoglobin and proteins. Later versions have become popular instruments in biochemical laboratories. So have *electrophoresis machines, beginning with those produced by Arne Tiselius in the 1930s. Since World War II these machines—together with x-ray diffraction installations, electron microscopes, electronic cell counters, flow cytometers, fluorescent-activated cell sorters, radioactive tracers, and peptide synthesizers—have reshaped life science laboratories. The new physical apparatus and some old and new laboratory animals, such as the mouse (*Mus musculus*), the fruit fly (*Drosophila melanogaster*), the bacterium *Escherichia coli*, and the fungus *Neurospora crassa*, have changed the sociology as well as the layout of the laboratory.

The development of *biotechnologies since the 1960s and genomics in the 1990s have brought an impressive host of new instruments: protein sequencers, peptide synthesizers, gene or DNA sequencers, biolistic apparatus (or "gene guns," using microprojectiles to inject DNA into cells), patch clamp amplifiers (measuring minute membrane cell currents), and polymerase chain reaction machines (to identify and reproduce a gene or a segment of DNA). Concomitantly, information technology has come to play an increasingly important role in biology. Magnetic resonance imaging (*see* NUCLEAR MAGNETIC RESONANCE), and *computerized tomography scanners have reshaped hospital, clinical, and patient practice.

With U.S. industry analysts expecting a 12 percent per year increase in the market for life science instrumentation in universities and other schools between 1998 and 2003, and with the related market in hospital and clinical research instrumentation expected to grow at 48 percent per year, biological instruments are more than ever industrial, as well as research, assets.

Robert J. Frank, Jr., *Harvey and the Oxford Physiologists* (1980). William Coleman and Frederic L. Holmes, eds., *The Investigative Enterprise. Experimental Physiology in Nineteenth-Century Medicine* (1988). Adele E. Clarke and Joan H. Fujimura, eds., *The Right Tools for the Job. At Work in Twentieth-Century Life Sciences* (1992). Robert Bud, *The Uses of Life. A History of Biotechnology* (1993). Nathan Rosenberg, ed., *Sources of Medical Technology: University and Industry* (1995). Bettyann Holtzmann Kevles, *Naked to the Bone. Medical Imaging in the Twentieth Century* (1997). Nicolas Rasmussen, *Picture Control. The Electron Microscope and the Transformation of Biology in America, 1940–1960* (1997). Robert Bud and Deborah Jean Warner, eds., *Instruments of Science* (1998). Jacalyn Duffin, *To See with a Better Eye. A Life of R. T. H. Laennec* (1998).

GIULIANO PANCALDI

INSTRUMENTS, ELECTRICAL MEASURING.

Electrometers

Until the invention of the electric cell by Alessandro Volta in 1800 (*see* GALVANI, LUIGI, AND ALESSANDRO VOLTA), studies of electricity concerned only electrostatics. The first electrical measuring devices quantified the amount of electricity produced by a frictional machine or stored in a Leyden jar. Although various electroscopes and electrometers had been in use since the 1740s, electricians (as they called themselves) lacked fundamental concepts of charge, tension, and capacity, and had not defined their units of measurement. The first instruments divide into three main groups in accordance with their principle of operation: electrostatic attraction and repulsion, spark length, and heating effects of discharges. The first group includes all the electroscopes with suspended balls, metallic strips, or movable pointers. Of the many types proposed during the later eighteenth century, those of John Canton (pith balls), Abraham Bennet (gold leaves), and Volta (straws) deserve special mention. These instruments were neither comparable nor absolute. The other groups include Timothy Lane's discharging electroscope (spark length) and Ebenezer Kinnersley's "electrical thermometer" (heating effect).

In the last quarter of the eighteenth century electrostatics became increasingly quantitative. Volta managed to make his straw electrometers comparable to one another and almost proportional to the intensity measured. By adding a condenser to his electroscopes Volta increased their sensitivity and detected the weak contact potential between two metals. It was a fundamental step leading to his discovery of the electrochemical pile.

Between the end of the eighteenth century and the first decades of the nineteenth century electrostatics became highly mathematized by Simeon-Denis Poisson and others who built on the laws of force demonstrated by Charles Augustin Coulomb (*see* CAVENDISH, HENRY, AND CHARLES AUGUSTIN COULOMB). New types of electroscopes were introduced. The most important later models derived from the absolute galvanometer (and its reduced form, the portable electrometer) developed by William *Thomson, Lord Kelvin, after 1850. It used the attraction of two metallic disks. Thomson also designed a quadrant electrometer in which a large suspended aluminum needle moved between four insulated brass quadrants. This instrument, often modified and improved, remained in use well into the twentieth century.

Galvanometers

In 1820 the Danish natural philosopher Hans Christian Oersted published the description of his experiments demonstrating that an electric current can deflect a magnetic needle. Johann S. Schweigger, Johann Poggendorff, and James Cumming soon introduced current multipliers, essentially copper wires coiled around a magnetic needle. The coil increased the action of the current on the magnet. These simple galvanoscopes were followed by "astatic" ones designed by Leopoldo Nobili around 1823. Nobili suspended a pair of identical and parallel needles with like poles pointing in opposite directions. He inserted the lower needle into the multiplying coil and used the upper one as an indicating pointer on a circular scale. Thanks to this system, Nobili's galvanometers did not respond to the earth's magnetism. They underwent many important improvements during the nineteenth century.

In 1837 the French physicist Claude Pouillet invented the tangent and sine galvanometers, the first devices for measuring absolute current. Here the action of a current flowing in a circular coil around a magnetic needle counterbalances the action of the earth's magnetism. Owing to their simple geometry, Pouillet's galvanometers allowed the measurement of a current in terms of the dimensions of the coil, the intensity of the earth's magnetic field, and the deflection of the needle. The absolute electrodynamometer of Wilhelm Weber (1845) had a movable coil in the center of a larger, fixed one. This instrument does not require one to know the value of the earth's magnetism and can also measure alternating currents.

The development of telegraphy, the laying of the first transatlantic cable in 1858, and the need to define electric standards produced great advances in galvanometry. Detecting the weak and damped signal transmitted by the cable engineers needed a very sensitive, quick, and robust instrument. William *Thomson designed a moving magnet galvanometer that met all these requirements. He suspended a tiny magnet with a small mirror attached and magnified its deflections by reflecting a ray of light onto a scale.

Thomson subsequently introduced a multiple magnet astatic version of his instrument.

Moving coil galvanometers had been in use since the 1820s, but Thomson's telegraphic siphon recorder of 1867 was the first commercial instrument of this type. In 1885 the French physicist and physiologist Arsène d'Arsonval together with the electrician Marcel Deprez, designed a galvanometer of the same type that found widespread use. It employed a light rectangular coil hanging between the poles of a powerful permanent magnet as the current detector.

Another popular late revival was the string galvanometer with a light, current-carrying metallic wire as the detector. First proposed in 1827, greatly improved by Willem Einthoven around 1900, and used in the first electrocardiographs, the string galvanometer, duly modified, figured in the oscillographs invented by William du Bois Duddell at the end of the nineteenth century.

Sophisticated current balances, based on electrodynamic action between movable and fixed coils, were used for high precision laboratory measurements. The indefatigable Thomson devised a series of such instruments in the 1880s for use as secondary standards for calibrating others' instruments. Additional clever nineteenth-century instruments included the "Wheatstone bridge," invented in 1843 by the English physicist Charles Wheatstone, who developed the idea from Samuel H. Christie. This bridge amounts to a circuit including a sensitive galvanometer. It enables one to determine the value of unknown resistances by balancing currents to sire a zero reading on the galvanometer.

Industrial Instruments: Ammeters, Voltmeters, and Supply Meters

The rise of the electrical industry required new and robust measurement instruments that would be easy to operate. In the early 1880s the English physicists William Ayrton and John Perry produced a moving iron ammeter, which sucked a piece of iron, connected to a helical spring, into a current-carrying solenoid. When the iron was attracted in the solenoid, the spring uncoiled and moved a pointer which was fixed to it. Similar apparatus with a much higher internal resistance absorbed a very small amount of current and served as voltmeters. In 1888 instruments patented by the British engineer Edward Weston began to supersede these instruments. Weston's ammeters and voltmeters had a pivoted moving coil between the poles of a strong permanent magnet. The instrument maker Robert Paul at the beginning of the twentieth century introduced the low-friction moving coil unipivot meter, which did not require accurate leveling. Although these instruments only measured direct current (DC), they continued to be widely used in the twentieth century.

For alternating current (AC), the German firm Simens and Halske, and then Weston and others, introduced compact electrodynamometers. A small moveable coil was inserted in a larger fixed coil. The coils were connected in series, and a current flowing through them produced a deflection of the moveable one. Thermal (hot wire) instruments became essential for measuring high-frequency currents, when solenoids and coils present high inductances. Philip Cardew patented the first practical apparatus of this kind in 1883; Hartmann and Braun produced a more compact and successful design in the 1890s. Voltage could be measured with a galvanometric voltmeter (fundamentally an ammeter with a very high internal resistance) or with an electrostatic voltmeter of the types invented by Thomson. The use of shunts made possible measurement of currents of widely differing magnitudes.

The introduction of electric light in the 1880s and the distribution of electricity to private customers required apparatus for measuring and recording the amount of electric energy supplied to the users. Thomas Edison patented the first electrolytic supply meter and Hermann Aaron invented an electricity-driven pendulum clock system, while others introduced electric motor meters operating against an electromagnetic brake. With the increasing use of AC, the induction-motor meters became standard for measuring domestic electricity consumption. Electronic supply meters, introduced in the 1980s, are now used for industrial applications.

The invention of the thermoionic valve revolutionized the technology of electrical measurement. In the 1920s the first meters using thermoionic elements were introduced. Electronic oscilloscopes replaced electromechanical models, DC amplifiers increased sensitivity, and copper oxide rectifiers and diodes adjusted DC instruments for AC. In the second half of the twentieth century solid-state transistors and miniaturized circuits created the possibility of a new generation of instruments. Voltage could be converted into time intervals and measured with a quartz-controlled clock. Today cheap electronic multimeters with digital displays can measure voltages, resistances, capacities, and so on in DC or AC.

K. Edgcumbe, *Industrial Electrical Measuring Instruments* (1908). H. Cobden Turner and E. H. W. Banner, *Electrical Measurements in Principle and Practice* (1940). William D. Cooper, *Electronic Instrumentation and Measurement Techniques*, 2d ed. (1978). Willem D. Hackmann, "Eighteenth Century Electrostatic Measuring Devices," *Annali dell'Istituto e Museo della Storia della scienza di Firenze* 3 (1978): 3–58. John T. Stock and Vaughan D. Denys, *The Development of Instruments to Measure Electric Current* (1983). Graeme J. Gooday, "The Morals of Energy Metering: Constructing and Deconstructing the Precision of the Victorian Electrical Engineer's Ammeter and Voltmeter," in M. Norton Wise, ed., *The Values of Precision* (1995): 239–282. Joseph F. Keithley, *The Story of Electrical and Magnetic Measurement from 500 B.C. to the 1940s* (1999). John Webster, *The Measurement, Instrumentation and Sensors Handbook* (1999).

PAOLO BRENNI

INSTRUMENTS, SURVEYING. Some ancient designs of surveying instrument are known— Vitruvius from the first century B.C. and Hero of Alexandria in the following century describe levels, angular measuring instruments, and a form of mechanical hodometer—but until the development of practical mathematics in the Renaissance, much surveying, insofar as it involved measurement, was confined to simple linear measure and its instrumentation to ropes and poles. Surveyors had other responsibilities in making assessments and valuations, and an early survey would be more likely to produce lists than maps. But from the sixteenth century on, scaled maps became valued as an efficient record from which information could be derived, including information not originally measured or envisaged by the surveyor. Such maps could be produced more readily through the application of mathematical instruments.

As happened in navigation, instruments for surveying originated in adaptations of portable astronomical instruments—the *astrolabe, quadrant, and cross-staff—to the less demanding requirements of land measurement. This indicates that the refounding of surveying practice on the mathematical science of geometry came from practical mathematicians rather than from working surveyors (though the groups did overlap) and participated in a wider effort to extend the application of geometry into a growing number of mathematical arts.

The astrolabe had a degree circle and alidade on the back that could be extracted to form an instrument more specifically designed for surveying. The resultant "theodolite" of the sixteenth century had only a single circle and while some could be suspended so as to measure vertical angles or altitudes, the main use was for angles in the horizontal—bearings or azimuths. Because "theodolite" later came to refer to an instrument for measuring simultaneously in the horizontal and vertical planes, and because this has confused historians, these early instruments should be called "simple theodolites."

Altazimuth instruments, combining horizontal and vertical measurement, as in the

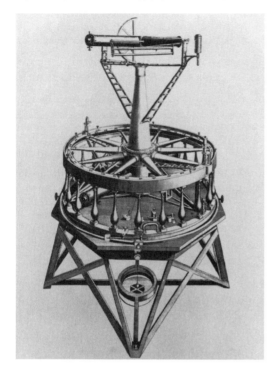

The grand theodolite made by Jesse Ramsden for the Royal Society in 1787 for geodetic measurements. It weighed 400 pounds fully equipped and could be read with a precision of one second of arc.

"topographical instrument" of Leonard Digges, were proposed in the sixteenth century, but they were too complicated to come into general use. Other complex and exotic instruments, their designers and promoters claiming all manner of uses and applications, made brief appearances at the same time, many of them as range finders for use in warfare. Much of the enthusiasm for these instruments derived from the introduction of triangulation, first described by Gemma Reines (called Frisius) in 1533. The possibility of plotting all visible stations by taking angles from either end of a single measured length was presented as an extraordinary advance to surveyors accustomed to measuring every required distance individually. If the application required mastery of unfamiliar techniques involving angles and trigonometry, so much the better for the mathematicians, the keepers of such specialist knowledge.

Surveyors made their contributions to the new geometrical practice also. The surveying compass or circumferentor, a straightforward instrument based on the magnetic compass, and in particular, the plane table, seemed to threaten the carefully fostered new importance of the mathematicians. The user of the plane table could mark base lines and lines of sight directly on to a sheet of paper, using a detached sighting rule or alidade, and so construct a map as the survey progressed, with no need to record angle measurements in degrees. The instrument became popular despite the anxiety of the geometers that it was all too simple.

Geodetic investigations of the eighteenth century, notably French expeditions to determine the shape of the earth by measuring the lengths of a degree along the meridian in different latitudes, and a project to link the observatories of Paris and Greenwich by triangulation, challenged established techniques and instrumentation. The French favored portable repeating circles. The Borda circle extended the reflecting principle of an octant or sextant to 360 degrees, while the English preferred large theodolites and portable zenith sectors. On the smaller scale, makers continued attempts to convince surveyors of the value of the altazimuth theodolite, but on the whole the azimuth-only instrument prevailed. A characteristic instrument of the period, called "the common theodolite" by the London maker George Adams, combined the simple theodolite with the circumferentor by mounting a magnetic compass on the alidade so that its scale rotated with the sights: bearings could be taken either by noting the position of the alidade on the fixed circle of the theodolite or by reading the compass scale beneath the stationary needle. Theodolites with telescopic sights were designed and made but were not widely adopted.

In the nineteenth century the altazimuth theodolite with a telescopic sight at last became the instrument of choice for everyday surveying. In its most common form, the "transit theodolite," the telescope could rotate through 360 degrees in the vertical plane. The "American transit," with a large compass and telescopic sight, developed from the circumferentor. In the 'Y' level, introduced in the eighteenth century, a telescopic sight was mounted with a spirit or bubble level and could be reversed in adjustable bearings, but with the coming of the railway boom rapid results were required from less skilled operators and a new "dumpy" level was easier to use and harder to derange.

The theodolite followed the same route in the twentieth century, with more and more mechanical and optical components safely enclosed from the operator. Micrometer microscopes improved the accuracy of scale reading while illuminated scales etched on glass for vertical and horizontal angles could be viewed together in a single eyepiece. While photogrammetry from aerial photographs was increasingly used for large-scale projects, electronic distance measurement transformed operations on the ground. From the 1970s digital results could be provided by theodolites in which glass encoder discs replaced divided circles and photodiodes replaced eyesight. The use of lasers for establishing alignments and gyro attachments for direction, together with microprocessors for computing and solid-state memory for recording, substituted the "complete positioning system" for the traditional theodolite. When radios could receive signals from navigational satellite systems and computers could give instant calculations of precise positions, the idea of a separate class of surveying instruments rapidly became obsolete.

See also ELECTROMAGNETISM.

Jim Bennett, *The Divided Circle: A History of Instruments for Astronomy, Navigation and Surveying* (1987).
JIM BENNETT

INTELLIGENCE QUOTIENT. The term "intelligence quotient" (I.Q.) arose from the attempts of psychologists early in the twentieth century to devise quantitative tests of human intelligence. The most influential of these tests was devised in France in 1904 by Alfred Binet and his collaborator Théodore Simon. They classified each test taker according to his "mental age"—that is, the age of the chronologically uniform group of children whose average test score his matched. The concept of intelligence quotient, invented in 1912 by the German psychologist William Louis Stern, received its initials from Lewis Madison Terman at Stanford University, who in 1916 accomplished the most important revision of the Binet-Simon tests. I.Q. expresses the ratio of a child's mental age to his chronological age, multiplied by one hundred. If the ratio is 1, the child's I.Q. will be 100; if nine-tenths, 90; and so on.

At the time of Stern and Terman's work, many psychologists claimed that the tests did not depend on the personal circumstances, especially education, of the examinees, and that I.Q. thus indicated native intelligence. During the 1920s, analysis of performance on such tests by U.S. army draftees during World War I concluded that white and black Americans had, on average, mental ages of, respectively, thirteen and ten years, and that immigrants from Eastern and Southern Europe were intellectually inferior to native whites from northern Europe. In Britain, the psychologist Cyril Lodowic Burt found that lower-income whites similarly possessed less native intelligence than members of the middle and upper classes. Such results were used in both countries to rationalize the existing social order, and in the United States to justify the passage of restrictions on immigration from Eastern and Southern Europe.

A growing number of critics, however, disparaged the mental-age interpretations drawn from the army tests as ludicrous. They contested the notion that human beings have some concrete, invariant entity called intelligence that can be measured unambiguously. In 1929, after reviewing the existing data, the American psychologist Carl Campbell Brigham concluded that verbal, mathematical, and behavioral tests for intelligence measured only how well the examinee did on a particular examination. To say that the scores, taken together, indicated something called general intelligence, he added, was to indulge in "psychophrenology"—to confuse the test name (e.g., "verbal") with the reality of the trait, and to misidentify the summed traits with intelligence.

By the early 1930s, psychologists were coming to recognize that performance on I.Q. tests reflected cultural biases in the tests themselves and differences in the social environment and educational preparation of the test takers. Confidence that the tests measured native intelligence was further undermined in the 1970s when investigators revealed that Burt's data appeared to be fraudulent. Some well-credentialed psychologists continue to hold that I.Q. indicates native intelligence. But their claims, often advanced to argue that minority groups of color have not achieved as much in society as whites because on average they are less intelligent, probably have no more merit than those of their predecessors.

See also HEREDITY; EUGENICS.

Daniel J. Kevles, *In the Name of Eugenics: Genetics and the Uses of Human Heredity* (1995). Stephen Jay Gould, *The Mismeasure of Man* (rev. ed., 1996).
DANIEL J. KEVLES

INTERNATIONAL GEOPHYSICAL YEAR. The International Geophysical Year (IGY) was an ambitious international scientific project that ran from July 1957 to December 1958. In 1950 American geophysicists had proposed a Third International Polar Year that would make significant advances on the earlier ones of 1882–1883 and 1932–1933 by using the rocketry, information processing, and other instrumentation developed during World War II. The project quickly widened to geophysics as a whole. Sanctioned by the International Council of Scientific Unions—the parent body of international scientific organizations—and implemented by national committees in participating countries, it involved some eight thousand scientists from about sixty different nations. It was timed to coincide with the twenty-fifth anniversary of the Second International Polar Year and with a peak in the sunspot cycle.

The IGY Special Committee decided to concentrate on the topics most likely to benefit from a global approach: aurora and airglow, cosmic

rays, geomagnetism, glaciology, gravity, ionospheric physics, longitude and latitude determinations, meteorology, oceanography, seismology,
and solar activity. The project produced an unparalleled database as well as a number of major
discoveries. Oceanographers confirmed the existence of a continuous worldwide system of submarine midocean ridges (actually huge mountain
chains), one piece of evidence that contributed
to plate tectonics in the mid–1970s. Satellites
launched by the United States detected belts of
radiation around the earth (named the Van Allen
belts) and the influx of charged solar particles
believed to be responsible for the auroras (*see*
Cosmic Rays). Scientists in Antarctica determined
the size and shape of the land mass underlying the ice and discovered a jet stream circling
the continent.

The IGY was a prime example of "big science." Governments contributed major funding.
Scientists participated in much larger numbers
than had been normal in peacetime. International
cooperation led to the invention and dissemination of new, intricate, and very expensive instruments. Scientists became more involved in and
gained new standing in politics and international
law. IGY's success encouraged the United States
to commit to Skylab; it led to more intense Antarctic exploration; and it prompted further international cooperation in the International Year of the
Quiet Sun (1964–1965), the International Hydrological Decade (1965–1975), and the International
Decade of Ocean Exploration (1970–1980).

The two-China problem, competition in
Antarctica, and the space race exemplify relations between politics and the IGY. In 1955, the
People's Republic of China agreed to join the
IGY. Two years later, on learning that the Nationalist Chinese of Taiwan had belatedly signed on,
the communist government backed out. Scientists
on the mainland decided to go ahead with the
research even though they could not do so under
the umbrella of the IGY. In the case of Antarctica,
seven countries that had been asserting territorial claims agreed to set them aside for the
duration of the IGY. Eleven nations cooperated
in establishing fifty-five research stations. The
nations shared logistical support, which their mil-
itaries often provided. Although these complex
arrangements caused some tensions, the experiment succeeded sufficiently that in December
1959 the Antarctic Treaty was signed preserving the continent for the purposes of peaceful
scientific research.

The space race dates from 4 October 1957
when the Soviets successfully launched *Sputnik,
the first artificial satellite to circle the Earth,
and part of the IGY. From then on scientists

used data gathered by both Soviet and U.S.
satellites, supposedly freely circulated through
three centers, one in Moscow, one in Washington,
and one divided among several other nations.
The Soviets, however, who used military rockets
to launch satellites, did not release data about
their launch vehicles, and the U.S. government
suppressed information about their bomb tests,
even as IGY scientists detected the effects. In
China, Antarctica, and during the space race,
scientists had to learn to cope with a world
in which science was no ivory tower activity
but deeply enmeshed in politics. The history
of the IGY is ripe with implications for the
politics and sociology of science as well as
information about the postwar explosion in earth
and planetary sciences.

Walter Sullivan, *Assault on the Unknown* (1961). J. Tuzo
Wilson, *The Year of the New Moons* (1961). Karl Hufbauer,
Exploring the Sun (1991). G. E. Fogg, *A History of Antarctic
Science* (1992).

RACHEL LAUDAN

INTERNATIONALISM AND NATIONALISM.
Bolstered by the relative weakness of secular
authority and the use of Latin as a common
language, science in late medieval and early
modern times probably mirrored the Platonic
ideal of scholarly internationalism better than
it did at any time in the past two centuries.
Nationalism in science, already apparent in the
competition between Catholic and Protestant
states after the Reformation, took on a new
intensity with the French Revolution. Rallying
to save their revolution from attack by other
European states, the French under Napoléon
Bonaparte forcibly extended their influence to
the rest of the continent and aroused nationalistic
responses everywhere they went. Prussia in 1810
founded the University of Berlin as an instrument
of nationalistic cultural renewal against the
French: by 1880 Berlin was the foremost center of
science and learning in a newly unified Germany.

Internationalism did not disappear as a coun-
tervailing force. The relative peace that prevailed
in Europe during the century between Napoléon's
demise and the Great War of 1914–1918 offered
many opportunities for international cooperation
in astronomy, geodetic science, seismology, meteorology, and other fields. During the same period
dramatic improvements in transportation and
communications—railways, steamships, telegraphy, the telephone, and postal service—greatly
facilitated long distance travel and exchanges of
information by scientists, and created demands
for international convention and standards. Universities in Germany, France, Britain, and other
nations attracted students from all over the world.

Aufforderung.

Durch den Krieg werden die Beziehungen der wissenschaftlichen physikalischen Kreise zum feindlichen Ausland eine Neuregelung erfahren. Sie wird sich besonders auf unser Verhältnis zu England beziehen, nachdem die deutschfeindliche, ohne jedes Verständnis für deutsches Wesen abgefasste Erklärung der englischen Gelehrten auch von acht bekannten Physikern unterschrieben ist (Bragg, Crookes, Fleming, Lamb, Lodge, Ramsay, Rayleigh, J. J. Thomson).

Es ist hierdurch erwiesen, daß die langjährigen Versuche, mit den Engländern zu einem bessern gegenseitigen Verständnis zu gelangen, gescheitert sind und für absehbare Zeit nicht wieder aufgenommen werden können. Die Rücksichten, die wir im Interesse einer Annäherung der wissenschaftlichen Kreise beider Völker genommen haben, sind nicht mehr gerechtfertigt. Daher ist es auch geboten, daß der unberechtigte englische Einfluß, der in die deutsche Physik eingedrungen ist, wieder beseitigt wird.

Es kann sich selbstverständlich nicht darum handeln, die englischen wissenschaftlichen Ideen und Anregungen abzulehnen. Aber die so oft getadelte Ausländerei der Deutschen hat sich auch in unserer Wissenschaft so bemerkbar gemacht, daß es nötig erscheint darauf hinzuweisen.

Nach diesem Hinweis beschränken wir uns zunächst darauf vorzuschlagen, daß alle Physiker dahin wirken:

1. daß bei der Erwähnung der Literatur die Engländer nicht mehr wie es vielfach vorgekommen ist, eine stärkere Berücksichtigung finden als wie unsere Landsleute;
2. daß die deutschen Physiker ihre Abhandlungen nicht in englischen Zeitschriften veröffentlichen, abgesehen von den Fällen, in denen es sich um Erwiderungen handelt;
3. daß die Verleger nur in deutscher Sprache geschriebene wissenschaftliche Werke und Übersetzungen nur dann aufnehmen, falls es sich nach fachmännischem Urteil um ganz bedeutende literarische Leistungen handelt;
4. daß Staatsgelder auf Übersetzungen nicht verwendet werden.

E. Dorn.   F. Exner.   W. Hallwachs.   F. Himstedt.   W. König.
E. Lecher.   O. Lummer.   G. Mie.   F. Richarz.   E. Riecke.
E. v. Schweidler.   A. Sommerfeld.   J. Stark.   M. Wien.   W. Wien.
O. Wiener.

Manifesto drawn up by Wilhelm Wien against British physics (1915). Its few signatories included the most fanatical German physicists.

European and American scientists began to teach and conduct research in the new scientific institutions of Japan, India, China, the Turkish Ottoman Empire, and South America. International scientific congresses were first convened in the latter half of the century; the International Association of Academies of Science (IAA) made its appearance at Berlin in the 1890s. The Nobel prizes, founded by the last testament of Alfred Nobel in 1896, symbolized the international ideals of the later nineteenth century. Nobel stated expressly that the prizes be awarded without regard to the nationality of the recipients.

World War I did considerable damage to internationalism in science. The burning of the library at Louvain University and other atrocities committed by German troops in their invasion of Belgium aroused neutral and Allied nations. German scientists and academics responded with the famous "Manifesto of the 93 Intellectuals" declaring solidarity with the German Army and repudiating the charges against it. Among the many prominent signers of the document were Max *Planck and Paul Ehrlich. Fritz *Haber's prominent role in German chemical weaponry caused equal or greater resentment among scientists in the Allied countries, not only during the war but immediately following. The award to Haber of the 1918 Nobel Prize in Chemistry for his nitrogen fixation process (which had helped Germany prolong the war) damaged the reputation of the prizes.

Because of the Belgian invasion, the Manifesto of the 93, the German initiation of gas warfare, and a putatively greater number of wartime atrocities committed by Germans, the Allied nations decided not to renew prewar scientific collaboration with the Central Powers. Even before the Armistice, the allies planned to replace the IAA with a new international body from which German, Austrian, Hungarian, and Bulgarian scientists would be excluded. The first of three meetings to implement the plan convened in early October 1918 at the Royal Society in London. France, Britain, Italy, the United States, Belgium, Japan, Brazil, and Serbia attended. Neutral nations like Sweden and the Netherlands were not invited to London, but several attended a second meeting in Paris six weeks later. Delegates at the London and Paris meetings voted to dissolve the IAA in favor of an International Research Council (IRC); states not already represented (but in no case the former Central Powers) could be voted in with a three-quarters majority of the IRC's member states. The boycott against the Central Powers ended in 1926. Although Germany urged its scientists to participate, many refused to accept as a concession what they regarded as a right. In 1931 the IRC changed its name—to the International Council of Scientific Unions (ICSU)—in part to remove the negative connotations of the war-born name.

In the early 1920s students from the former Allied nations once again studied in German universities and vice versa with the help of the Rockefeller Foundation's International Education Board. International conferences increased. Nominations of candidates for the Nobel Prize took on a less nationalistic tone. Some prominent scientists—in Britain (John Desmond Bernal), France (Frédéric Joliot and Irène Joliot-Curie), and the United States (J. Robert Oppenheimer)—during the depression years of the 1930s admired the Soviet Union for its support of science and putatively internationalist principles. But the rise of fascism and World War II once again reduced internationalism to allegiance to one set of countries against another.

During the Cold War many scientists remained attached to weapons programs while others became advocates for international control of the same weapons. This issue was sharply posed in the United States in Oppenheimer's security hearings of 1954. Relations with Japan presented another example of the same ambivalence. Having helped to design the atomic bombs that devastated Hiroshima and Nagasaki, American physicists protested vigorously when the U.S.

occupation authorities destroyed two Japanese cyclotrons in late November 1945.

Although intensifying rivalry between the United States and the Soviet Union substantially dampened prospects for international cooperation in science, it did not destroy them entirely. The Atoms for Peace proposal by President Dwight Eisenhower in 1953–1954 called for peaceful uses of atomic energy to serve civilian needs and recommended relaxation in the Atomic Energy Act to allow for sharing of nuclear material. Beginning in the mid-1950s, the annual Pugwash Conferences held in Nova Scotia brought together disarmament-minded scientists. Participation in them by scientists from formerly hostile nations—such as Oppenheimer and Japan's Hideki Yukawa—suggested that the Republic of Science still had life in it.

See also INTERNATIONAL ORGANIZATIONS; NOBEL PRIZES; STANDARDIZATION.

Roger Hahn, *The Anatomy of a Scientific Institution: The Paris Academy of Sciences, 1666–1803* (1971). J. L. Heilbron, *The Dilemmas of an Upright Man: Max Planck as Spokesman for German Science* (1986). James R. Bartholomew, *The Formation of Science in Japan: Building a Research Tradition* (1989). Constance Holden, "Chauvinism in Nobel Nominations," *Science* 243: 4890 (27 January 1989): 471. Elisabeth Crawford, *Nationalism and Internationalism in Science, 1880–1939* (1992). Arthur Donovan, *Antoine Lavoisier: Science, Administration and Revolution* (1993). Robert Marc Friedman, *The Politics of Excellence: Behind the Nobel Prize in Science* (2001).

JAMES R. BARTHOLOMEW

INTERNATIONAL ORGANIZATIONS. International science organizations are ideologically motivated by belief in the unity of nature, the universe, and scientific laws. Despite criticisms from social constructivists, universalism motivates internationalism. While coexisting with other knowledge forms, modern science was founded on ideas coming from several cultures and traditions, being thus international at its inception. Early examples include the international meteorological project started by James Jurin, secretary of the Royal Society of London in 1723. Observations of the transits of Venus in 1761 and 1769, coordinated by the Frenchman Joseph Delisle, involved Italians, Portuguese, Spaniards, Russians, Danes, and Swedes as well as the French and English, who happened then to be at war.

During the nineteenth century science took on a more national cast, with attendant glories and declines. National scientific associations emerged. The British Association for the Advancement of Science was founded in 1831, its German counterpoint in 1822, its American in 1848, and the Indian Association of Cultivation of Science in 1876. Scientific rivalries mirrored diplomatic relations. Internationalist impulses balanced these nationalist trends as even national efforts could contribute to international science. The metric system, invented and promoted by the French, prevailed throughout Europe. The Conférence Diplomatique du Mètre (1875) resulted in the creation of the International Bureau of Weights and Measures located at Sèvres.

International meetings sought to transcend national borders. International congresses of chemists met in 1860, of geologists in 1875, and of physicists in 1900. Universal exhibitions, such as Paris 1889 and 1900, included scientific sessions. International committees deliberated standards and nomenclature (for example, electrical units and atomic weights). Reacting to the growing nationalist sentiments of the time, many scientists adopted internationalist ideals by the end of the nineteenth century, culminating in the foundation of the International Association of Academies (IAA) in 1899. However, most of these early international associations succumbed in World War I. Science became a war instrument, and scientists national propagandists, as in the "Appeal of the 93 Intellectuals" defending German war aims (October 1914) and counter-manifestos issued by the Allies. Similarly, the Allied position against Germany in 1918 led in 1919 to the formation of the International Research Council (IRC) from which the former Central Powers were excluded until 1926. The coming to power of the Nazis in 1933 and rearmament snuffed out the brief interwar interlude of renewed internationalist impulses.

Twentieth-century scientific internationalism, however, worked to counterbalance the destructive association of science and the state. The number of international scientific meetings is revealing: 20 between 1870 and 1900, 14 between 1901 and 1914, 7 during the war, 15 between 1918 and 1923, and 45 in the early 1930s. Similarly, international organizations proliferated after 1918, fostering research and the free mobility of scientists. The International Union of Pure and Applied Chemistry (IUPAC), the International Union of Biological Sciences (IUBS), and the International Astronomical Union (IAU) were founded in 1919; the physicists followed with the International Union of Pure and Applied Physics (IUPAP) in 1922. The International Council of Scientific Unions (ICSU) replaced the IRC in 1931; its founders insisted that science was nonpolitical and universal. Scientists set the basis for the internationalism of UNESCO and other UN organizations after World War II. The experiences of UNESCO and the United Nations

Atomic Energy Commission (UNAEC) indicate the opportunities and perils of international organizations during the later twentieth century.

The UN General Assembly created the UNAEC as its exclusive agent in the field, with a mandate to create mechanisms for research and control of atomic energy and for the elimination of nuclear armaments. UNAEC's initiatives, however, were curtailed by the Security Council's veto power and it became a venue for nationalism. At the first UNAEC meetings, held in June 1946, the U.S. representative, Bernard Baruch, insisted on the veto, controls, and sanctions. The Soviets asked for the destruction of nuclear weapons. The French argued that it was necessary to determine the limits of the scientifically possible.

To complicate UNAEC activities, in July 1946 the United States tested improved nuclear devices at Bikini and in August passed the Atomic Energy Act imposing rigid controls on nuclear material. Despite a growing air of futility, the commission announced that international controls were still possible given enough safeguards. In November, the UNAEC presented a progress report to the Security Council. Despite French protests, armaments debate focussed on atomic weapons. The General Assembly demanded an atomic control convention from the Security Council. In December the UNAEC approved Baruch's proposal calling for a system to outlaw atomic weapons and instruments to enforce the system. The report was approved, Poland and the USSR abstaining. The commission submitted Baruch's recommendations to the Security Council.

Considering his job done, Baruch resigned in January 1947. His plan got nowhere. In February Andrey Gromyko proposed a convention prohibiting atomic weapons, the destruction of existing stockpiles, an international control system, and an international organization for atomic energy research. The Cold War gave little chance for an agreement. The UNAEC assembled for the last time on 29 July 1949. On 23 August the Soviet Union detonated its first atomic bomb.

The short lived and nearly forgotten UNAEC was superseded by the International Atomic Energy Agency. However, the United Nations now operates on atomic energy matters along the lines UNAEC discussed. No "illegal" atomic operations are allowed, and no atomic wars have occurred. By denying their national affiliation and working directly within the structure of the UNAEC, scientists reinforced the highest internationalist ideals and achieved long-term results. Despite this success, the story of UNAEC-IAEC shows that science can easily become a theater for international conflict. The history of UNESCO illustrates the point from a different perspective.

UNESCO was preceded by the International Committee of Intellectual Co-operation, founded in 1922 and based in Geneva, and its executive agency, the International Institute of Intellectual Co-operation, founded in 1925 and based in Paris. Both existed until 1946. UNESCO had another precursor in the International Bureau of Education (IBE) founded in 1925 and based in Geneva. IBE became part of UNESCO in 1969. In November 1945, the London Conference of Allied Ministers of Education approved UNESCO's constitution. It entered into effect in November 1946. One of UNESCO's five main functions is the advancement, transfer, and sharing of knowledge.

UNESCO comprises three bodies. The General Conference, composed of member states, approves the biannual program and budget. The Executive Board, composed of fifty-eight representatives, meets twice a year and functions as an administrative council, supervising the program's execution. The Secretariat is appointed every six years. UNESCO is supported by national commissions, nearly seven thousand schools, and over six thousand clubs and associations.

UNESCO's impressive achievements include a vast publishing program, over ten thousand titles between 1946 and 2000, among them the *World Science Report* and the *UNESCO Courier*, issued in twenty-seven languages. Membership increased steadily until 1980, when a crisis of credibility hit. The United States, Great Britain, and Singapore left in 1984 and 1985 (membership nonetheless rose from 153 in 1980 to 159 in 1990). When the crisis ended, numbers again increased rapidly, reaching 188 in 2000.

The Americans and the British quit UNESCO because they thought that its programs threatened their national interests. For them, UNESCO did not use its resources neutrally. Unable to convince the majority of the UNESCO assembly, they withdrew, thereby creating a stronger and more internationalist organization. Politically motivated projects declined, efficiency improved, and hiring processes became visible as other nations moved in to keep the dream of international science and education alive. Federico Mayor (Spain) returned UNESCO to political neutrality during his term (1987–1999). Improvements continued, made easier by the dissolution of the Soviet Union.

In the early twenty-first century the excessive emphasis on applications and financial gains may be more threatening to internationalism than nationalism. The problem obtrudes in the two great human adventures that survived the Cold War, polar research and the International Space Station. Polar research must be international because of the nature of the problems and demands on resources. Nationalistic positions are irrelevant in conflicts between different disciplines over the uses of research vessels, logistics, and instruments. Like polar research, national competition marked the early history of space exploration. Space research has always been very close to military applications. Nevertheless, most regard Yuri Gagarin and Neil Armstrong as individual astronauts, not as flag carriers. They have been taken out of their nationalist contexts to become international symbols. The International Space Station (ISS) may open a new chapter in the history of humanity. By confronting research issues above the planet, ISS reinforces the unity of human society and raises interesting possibilities for a reinterpretation of history.

See also NATIONAL ASSOCIATIONS FOR THE ADVANCEMENT OF SCIENCE; NAZI SCIENCE; STANDARDIZATION.

Erik Baark, Regis Cabral, and Andrew Jamison, *Science and Technology for Development in the United Nations System* (1988). Birgitte Schroeder-Gudehus, "Nationalism and Internationalism," in *Companion to the History of Modern Science*, ed. Geoffrey Cantor, Robert Olby, et al. (1990): 909–919. Tore Frängsmyr, ed., *Solomon's House Revisited. The Organisation and Institutionalisation of Science* (1990). Elisabeth Crawford, *Nationalism and Internationalism in Science, 1880–1939. Four Studies of the Nobel Population* (1992). Regis Cabral, "The United Nations Atomic Energy Commission: Science for International or National Security? 1945–1949," in *Debating the Nuclear*, ed. Regis Cabral (1994): 11–56. Aant Elzinga, ed., *Changing Trends in Antarctic Research* (1993). Michel Conil Lacoste, *The Story of a Grand Design: UNESCO 1946–1993* (1994). Frank Greenaway, *Science International: A History of the International Council of Scientific Unions* (1996). Human Space Flight at http://spaceflight.nasa.gov/

REGIS CABRAL

INTERNET. See COMPUTER; COMPUTER SCIENCE; ELECTRONIC MEDIA; NETWORK AND VIRTUAL COLLEGE.

IN VITRO FERTILIZATION (IVF) designates the practices of assisted human conception in which fertilization is achieved outside the organism, typically in a test tube (Latin *in vitro*, "in glass").

Experiments realizing some form of artificial insemination of animals date back to the eighteenth century. Experiments on humans, however, are documented only from the 1960s. The first baby conceived with IVF was born in Britain on 25 July 1978, thanks to techniques developed during the previous decade through the joint work of Robert G. Edwards, an endocrinologist and an expert in mammalian egg maturation, and Patrick Steptoe, a gynecologist and a leading expert in laparoscopy. Steptoe's specialty

was a technique for observing the abdominal cavity. It was used to harvest the human oocytes subsequently fertilized in vitro (with techniques that circumvented several causes of female and male infertility), and then placed back inside the mothers.

Since the 1980s, IVF has become a widespread medical practice. Centers devoted to the treatment of infertility and semen preservation multiplied in the wealthy industrial nations. The number of IVF babies numbered in the hundred thousands by the year 2000. International IVF research and practice have benefited from a cluster of specialties and interests similar to those that produced the oral contraceptive pill, a combination of steroid hormones inhibiting ovulation first approved by U.S. authorities in 1960. The standard account of the history of IVF, written by one of the protagonists, Robert G. Edwards, acknowledges the contributions of Gregory Pincus, the American endocrinologist expert in the comparative behavior of mammalian eggs in vivo and in vitro, who, with gynecologist John Rock, developed the first oral contraceptive. However, the original impulse sustaining research on human IVF in Great Britain, the United States, France, and Australia came from the medical profession and from patients themselves. The research community, public authorities, and medical industries initially expressed considerable skepticism owing to political considerations and the controversial heritage of *eugenics.

Since the 1970s these preoccupations have produced a wide literature on the ethical, religious, philosophical, and legal implications of assisted human conception. Such implications figured prominently in the public discussions generated by the first parliamentary document on human fertilization, issued in Britain in 1987 and derived from the report of a committee chaired by philosopher Mary Warnock. Public perceptions of the broad implications of IVF have merged since with the hopes and fears generated by what had been announced from the 1960s as the imminent (and often postponed) revolution of genetic engineering.

Albertus Th. Alberda, Raoul Arif Gan, and Hendricus Maria Vemer, eds., *Pioneers in in Vitro Fertilization*, 1995. Robert G. Edwards, "The History of Assisted Human Conception with Special Reference to Endocrinology," *Experimental and Clinical Endocrinology and Diabetes*, 104 (1996): 183–204.

GIULIANO PANCALDI

ION. William Whewell gave the world the English term "scientist" at a meeting of the British Association for the Advancement of Science in 1833. In that same year Whewell and Michael *Faraday collaborated on the creation of a new vocabulary to describe the results of Faraday's investigations in electrochemistry. In 1834 Faraday published an article in the *Philosophical Transactions* of the Royal Society of London in which he systematically introduced a new vocabulary for electrochemistry, including the terms "ion," "anion," and "cation" ("cathion").

Like Michael Faraday and Humphry Davy in London, Jöns Jacob *Berzelius in Stockholm used the battery to decompose compound substances by electrical current, depositing the separated components at the electrical poles or electrodes. Around 1812 Berzelius came to identify the force of chemical affinity with the force of electrical attraction, assuming that charged atoms or groups of atoms are held together in a molecule by slight opposite electrical charges or forces. This theory came to be called a dualistic or electrochemical theory of chemical composition and decomposition. Berzelius's electrochemical theory was gradually undermined by the recognition that molecules such as hydrogen contain like atoms, as well as by the discovery that (negatively charged) chlorine atoms can substitute for (positively charged) hydrogen atoms in hydrocarbons. Berzelius's molecular constituents anticipated ions joined in polar chemical bonds.

Svante Arrhenius's proposal in 1887 that molecules of electrolytes break up into charged ions in dilute solution, whether or not electric current is present, furthered the conviction that ions exist and participate in many chemical and physical processes. Scientists studied ionized particles with increased fervor in the late nineteenth century following investigations of the nature of *cathode rays and canal rays in the 1880s and 1890s, both of which consist of charged particles.

*X rays and *radioactivity became detectable by their capacity to produce charged ions and (in the case of alpha and beta particles) by their identification as charged ions. The improvement of electrometers allowed the detection of small amounts of ionization, as did the invention in the late 1920s of the automatic electrical counter dubbed the *"Geiger counter" after its invention by Hans Geiger and Walther Müller. *Cosmic radiation also could be detected through electrical effects and ionized particles could be seen as vapor trails in the *cloud chamber invented by C. T. R. Wilson in 1899.

The electrochemical hypothesis that "ions" exist fleetingly within activated chemical molecules gained adherents in the early 1900s when the electron theory of valence was being developed by Joseph John *Thomson, Walther Kossel, Gilbert N. Lewis and Irving Langmuir.

Lewis, and then Langmuir, identified the electron as the essential constituent of polar (ionic) and nonpolar (covalent) chemical valence bonds between 1916 and 1919. In the 1920s and 1930s Arthur Lapworth, Robert Robinson, and especially Christopher Ingold revolutionized the understanding of reaction mechanisms in organic chemistry by applying theories of electrons and ions to aromatic and aliphatic compounds. The history of modern chemistry is inseparable from the history of ions.

See also ATOMIC STRUCTURE; ELECTRON; PHYSICAL CHEMISTRY; VALENCE.

William H. Brock, *The Norton History of Chemistry* (1992). Elisabeth Crawford, *Arrhenius: From Ionic Theory to the Greenhouse Effect* (1996).

MARY JO NYE

IONOSPHERE. Beginning with the discovery in the mid-eighteenth century that air can be electrically charged, some investigators speculated about electrical conditions at higher elevations. Erik Pontoppidan suggested in 1752 that the aurora borealis is electrical. Jean-Baptiste Biot and Joseph Louis Gay-Lussac carried electrometers in a balloon ascent in 1804 that reached an elevation of several kilometers, far short of the ionosphere. By the 1830s, scientists such as John Herschel concluded that, somehow, diurnal variation of geomagnetism depended on changes in the electric state of the atmosphere.

Balfour Stewart first suggested the existence of electrical currents in a high, conductive layer of the atmosphere in 1884 to explain geomagnetic variations. Arthur Schuster in 1889 supported this supposition with an application of the spherical harmonic analysis of Carl Friedrich *Gauss. Studies of atmospheric conductivity and ionization by Schuster, Hans Linns, Franz Exner, and Svante Arrhenius at this time indicated that gas ions give rise to electrical processes in the atmosphere. In 1902, Arthur E. Kennelly in the United States and Oliver Heaviside in England explained Guglielmo Marconi's trans-Atlantic radio transmission of 1901 as arising from reflection from this layer.

Although many scientists still doubted the existence of an ionized layer, W. H. Eccles termed it the Heaviside layer in 1912, and soon others called it the Kennelly-Heaviside layer. By the 1920s researchers determined that there must exist several ionized layers. Merle A. Tuve and Gregory Breit in the United States and Edward Victor Appleton in England used radio waves purposefully to probe these conductive layers. Tuve and Breit determined a height of 230 km (140 miles) for one layer (the Appleton layer, now the F-layer), and Appleton 90 km (55 miles) for another (the Kennelly-Heaviside layer, now the E-layer). Appleton's sustained investigation of physical processes in these layers earned him the Nobel Prize for physics in 1947.

In 1926, specialists meeting at conferences and in committees named this region the ionosphere, in analogy with the troposphere and stratosphere, two terms then gaining currency. Robert Watson-Watt first used the name in print that year, and Hans Plendl next used its German cognate in 1931. The term became widely accepted during the Second International Polar Year of the 1930s. E. O. Hulburt and Sydney Chapman proposed the first detailed theories of the ionosphere at this time. Extreme solar ultraviolet radiation was found to cause much of the ionization.

Detailed examination of accumulated geomagnetic observatory data by Chapman in the 1940s indicated the existence of a tightly defined eastward electric current along the magnetic equator. Chapman called this current the Equatorial Electrojet. The electrical conductivity of the ionosphere came under study by sounding rockets beginning in the late 1940s in the United States and in the 1950s in Europe. Rocket-borne electrical research, along with investigations of cosmic rays and high-altitude chemical composition, greatly multiplied during the International Geophysical Year. The Van Allen Radiation Belts were discovered in 1958, and Thomas Gold termed the region of near space around Earth the magnetosphere in 1959.

Rockets and satellites carrying magnetometers, mass spectrometers, and electrostatic probes greatly increased knowledge of the ionosphere from the 1960s on. The vertical distributions of oxygen, ozone, helium, and hydrogen ions have been studied, as have electron densities. Likewise, investigators used these new capabilities to study the ionospheric effects of solar and interplanetary phenomena.

See also COSMIC RAYS.

C. S. Gillmor, "The History of the Term 'Ionosphere'," *Nature* 262 (1976): 347–348. Marco Ciardi, "Atmosphere, Structure of," in *Sciences of the Earth: An Encyclopedia of Events, People, and Phenomena*, ed. Gregory A. Good (1998): 47–50.

GREGORY A. GOOD

IQ. See INTELLIGENCE QUOTIENT.

IRRATIONAL. See RATIONAL/IRRATIONAL.

IRRITABILITY. Physiologists and biologists have used the term "irritability" to denote the property of organs, tissues, living matter, and protoplasm that allows their stimulation or excitation. According to ancient physiological

doctrines, the irritated part must be able to recognize being irritated in order to respond. Hence Galen (129/130–199/200) assigned a kind of "perception" not only to the nervous system, but also to organs like the stomach, uterus, and bladder, and why the notion of irritability was often associated with that of sensation, perception, and sensibility during the seventeenth and eighteenth centuries.

Building upon ancient doctrines and contemporary experiments, both William Harvey and Frances Glisson put forward doctrines of irritation. In his early work *Anatomia hepatis* (1654), Glisson maintained that the stomach and intestines expelled harmful agents because these organs were "capable of irritation." In his *Tractatus de ventriculo et intestinis* (1672) he shifted from a doctrine of irritation to one of irritability by assigning to fibers a "natural perception" independent of nerves, so that they could be irritated and respond to such irritation. This property he called irritability.

Glisson's doctrine lapsed into oblivion, until its partial revival by Albrecht von Haller in a paper published in Göttingen in 1753 ("De partibus corporis humani sensilibus et irritabilibus"). Haller suggested "a new division of the parts of the human body" based on experiments carried out mainly on living animals and consisting in diversified stimulation of different parts of the animal body. Haller named irritable any part that contracted when touched, and sensible any part that evidenced signs of pain. Since all parts displaying irritability were muscular, and those displaying sensibility were nervous, Haller stated that both properties depended "on the original fabric of the parts." In so doing he correlated a specific physiological behavior with a specific tissue structure.

Haller's doctrine aroused blazing controversy for over twenty years. Robert Whytt contested Haller's clear-cut distinction between irritability and sensibility in his *Physiological Essays* (1755), maintaining that muscular contractions arise from their sensibility and "are no more than an effort of nature to throw off what is hurtful." Felice Fontana investigated the laws of irritability, or muscular contraction, and discovered that the heart loses irritability during contraction, a phenomenon later named the refractory period of the heart (*De irritabilitatis legibus*, 1767).

By the early nineteenth century the notions of irritability and sensibility had been incorporated into most physiological systems. Some authors applied them also to plant physiology. However, while contractility replaced irritability in Haller's sense, biologists like Jean Baptiste *Lamarck considered irritability to be the fundamental property that distinguished animals from vegetables. During the second half of the nineteenth century, irritability was assigned to the protoplasm of cells. By the early twentieth century it concerned the response of all living cells and organisms to change in their environment, and was increasingly replaced by the term "excitability."

Owsei Temkin, "The Classical Roots of Glisson's Doctrine of Irritation," *Bulletin of the History of Medicine* 38 (1964): 297–328. Renato G. Mazzolini, *The Iris in Eighteenth-century Physiology* (1980).

RENATO G. MAZZOLINI

ISOMERISM AND MESOMERISM. See STEREO-CHEMISTRY.

ISOSTASY, the idea that different parts of the earth's crust are in gravitational balance and by implication floating on a semifluid substrate, had its origins in the mid-nineteenth century in mapping and geodetics. It played a crucial role in debates about earth tectonics in the first half of the twentieth century.

In the 1850s, surveyors working for the Great Trigonometrical Survey of India ran into a problem. They knew that the gravitational force of mountains should deflect their plumb bobs. The Himalayas, though, deflected their bobs less than anticipated. For the surveyors, this underperformance created cartographic problems. For scientists, it invited speculations about the structure of the earth. George Biddell Airy, the British Astronomer Royal, proposed that high mountains have low-density crustal roots extending to greater depth than the surrounding crust. John Henry Pratt, a mathematician as well as Archdeacon of Calcutta, proposed instead that the density of crustal columns varies inversely with their height. Formally these suggestions were equivalent. In either case, at some finite depth the load on the substrate would be everywhere the same.

Geologists were intrigued. The American geologist Clarence Dutton named the phenomenon isostasy and speculated about its tectonic consequences. European geologists saw in it an explanation for the changing levels of land and sea in Scandinavia. They proposed that during the Ice Age the weight of the ice sheets had depressed land that was now rebounding. Further developments in *geophysics and *geodesy changed the

status of isostasy from an exciting idea to a widely accepted truth. Geophysicists, notably the Finn Weikko Heiskanen, explored the implications of Airy's mechanism. Geodesists preferred Pratt's mechanism for convenience. Here the leader was John Hayford, Inspector of Geodetic Work and head of the U.S. Coast and Geodetic Survey. Geodesists knew only too well that determinations of latitude and longitude made by triangulation frequently differed from those made by astronomical observations. They attributed these discrepancies to deflections of plumb bobs. Hayford undertook a heroic series of calculations using Pratt's formulation of isostasy. He showed that plumb-bob deflections varied systematically. Boundaries between land and sea caused much larger variations than local topography. These results in hand, Hayford announced in 1909 a widely acclaimed new model for the figure of the earth. Furthermore he confirmed what some geologists had long suspected, namely that material of the continents was less dense than that of the ocean floors.

Geologists quickly realized that isostasy menaced the theory of the cooling contracting earth that had underpinned mainstream thinking about tectonics. If continents were lighter than ocean floors, then the proposal that continents foundered to form new ocean basins, put forward by Eduard Suess in his magisterial *Face of the Earth* (1883–1904), could not be true. Geologists and physicists scrambled to produce alternative tectonic theories. None of them, though, succeeded in dealing with the problems caused by isostasy. Continental drift, for example, seemed impossible: how could less dense continents move through more dense ocean floors? Only in the 1960s, when earth scientists developed the theory of *plate tectonics, was a satisfactory alternative found that sidestepped the problems of isostasy. Placing the definitive boundaries between plates of equivalent thickness, not between continents and oceans, made horizontal plate movement possible.

Mott Greene, *Geology in the Nineteenth Century* (1982). Gabriel Gohau, *History of Geology* (1990). David Oldroyd, *Thinking About the Earth* (1996). Naomi Oreskes, *The Rejection of Continental Drift* (1999).

RACHEL LAUDAN

J

JACOB, François (b. 1920), and **Jacques Lucien MONOD** (1910–1976), molecular biologists.

In the late 1950s, two research groups at the same scientific center, the Pasteur Institute in Paris, suddenly discovered that the problems on which they were working, and that everybody believed to be substantially different, depended on the same fundamental genetic and biochemical mechanism that controlled the expression of genetic information. One research program, led by Jacques Monod, focused on enzymatic adaptation; the other program, under the guidance of André Lwoff, allowed François Jacob and Elie Wollman to elucidate the phenomenon of lysogeny and bacterial sexuality.

Monod had thought to become a professional musician, before choosing biology as a profession. After obtaining a degree from the Faculty of Science in Paris, he received additional training in both Strasbourg and Paris, and in 1936 joined Boris Ephrussi on a visit to Thomas Hunt *Morgan's laboratory at the California Institute of Technology. Here Monod not only learned *genetics but also a new scientific style. He returned to the Sorbonne to prepare a doctoral dissertation (defended in 1941) on bacterial growth and to continue his activity as an amateur musician. In 1938 he married Odette Bruhl, an archaeologist. During World War II, Monod joined the Franc-Tireurs partisans and the French Communist Party, and he helped organize the general strike that led to the liberation of Paris. In 1945, however, he resigned from the Communist Party over a policy disagreement.

In late 1945 Monod joined André Lwoff's department at the Pasteur Institute. In collaboration with Alice Audureau, he continued his research on bacterial growth and enzyme adaptation. In 1947 he produced a general report on enzymatic adaptation, assessing the respective roles of the hereditary and environmental factors (the substratum) in enzyme synthesis. Between 1948 and 1953, the American immunologist Melvin Cohn made fundamental contributions to the implementation of Monod's research program on enzyme adaptation. Linear kinetics ("Monod's plot"), established through experiments and the study of the incorporation of the radioactive isotope sulfur-35 into a protein in the course of its induced synthesis, showed that the formation of β-galactosidase corresponded to the total synthesis of the protein, without the formation of precursors or intermediaries. In 1953 Monod became director of a new department of cellular *biochemistry, concentrating on the investigation of bacterial growth. In cooperation with George Cohen, he introduced the innovative idea of bacterial permease, an enzyme responsible for the permeability of the cellular membrane to metabolites.

Jacob entered the University of Paris intending to become a surgeon. After the invasion of France by Germany in 1940, he fled to London, where he joined the French Free Forces. For his service in the war he earned the Croix de la Liberation. He had suffered severe injuries, which dimmed his hopes of becoming a surgeon: after earning a medical degree in 1947, he decided to become a researcher. In the same year he married Lise Bloch, a pianist. In 1950 Jacob became an assistant at the Pasteur Institute, working under Lwoff on lysogenic bacteria, in which bacteriophages can exist in a noninfectious stage (the prophage). Lwoff demonstrated that the expression of the prophage and cytolysis could be stimulated by ultraviolet light. Jacob defended his doctoral dissertation on lysogenic bacteria and their prophage in 1954.

From 1954 to 1958 Jacob studied bacterial sexuality with Ellie Wollman. They described the mechanism of gene transfer between bacteria and developed techniques to interrupt the genetic exchange as they wished, which proved a powerful tool for localizing genes on the bacterial chromosome. Furthermore, the simultaneous presence in the same cytoplasm of two different chromosomes allowed the application of the methods of genetic analysis to bacteria. Both lysogeny and enzymatic induction (the induction by a metabolite of a specific biosynthetic pathway) proved to be inducible systems under genetic control.

Jacob and Monod decided to exploit the potentialities of this circumstance, using in one field the methods, ideas, and discoveries of

427

the other to clarify the mechanisms of gene expression. What Francis Crick called the "great collaboration" produced three theoretical models that proved fundamental for the development of molecular biology: the operon, messenger RNA, and allosteric interactions. Together, Jacob and Monod conceived the famous Pa-Ja-Mo experiment, conducted in collaboration with Arthur Pardee, which showed the existence of a double genetic determinism in protein synthesis. Two distinct sets of genes intervened, one determining the structure of the synthesized molecule, the other controlling the expression of the first. Each set, or "operon," was a unit of coordinated expression made up of an operator and the group of structural genes that it coordinated. The operon model opened the way to three further avenues of research: the nature of the repressor (the protein that blocks the expression of a given operon); the mechanism of the repressor's chemical action, with its relation to the target and the inductor (that is the chemical product that links to the repressor and allows the expression of the operon); and the molecular mechanism of the transfer of genetic information for protein synthesis, that is, messenger RNA.

In 1961 Jacob produced, with Sydney Brenner, evidence for the existence of messenger RNA. In the same year Jacob and Monod generalized the concept of "allosteric transition," a chemical interaction allowing complete freedom in the selection of chemical mechanisms. Monod later devised a more formal model of allostery in collaboration with Jean-Pierre Changeux and Jeffries Wyman.

In 1965 Jacob, Monod, and Lwoff received the Nobel Prize for their discoveries of cellular regulation mechanisms. After years of intense collaboration, the activities of Jacob and Monod then diverged. Monod became involved in the management of the Pasteur Institute, which he directed from 1971 until his death. Jacob changed his research field, introducing the concept of "replicon," a unit of independent replication, and studied gene expression during the early differentiation stages of the mouse embryo. Both Jacob and Monod wrote widely circulated books and articles on the philosophical aspects of the life sciences.

Jacques Monod, *Le hasard et la nécessité. Essai sur la philosophie naturelle de la biologie moderne* (1970), trans. A. Wainhouse as *Chance and Necessity* (1971). François Jacob, *The Logic of Life: A History of Heredity* (1973). André Lwoff and Agnès Ullmann, eds., *Origins of Molecular Biology—A Tribute to Jacques Monod* (1979). François Jacob, *The Statue Within* (1988). François Jacob, *Of Flies, Mice, and Men* (1998).

BERNARDINO FANTINI

JOULE, James (1818–1889), scientific brewer, and **Robert MAYER** (1814–1878), physician, natural philosophers commonly known as "co-discoverers" of the principle of energy conservation.

Joule was educated at home and by the chemist John *Dalton. As the son of the wealthiest brewing family in Manchester, England, he had the opportunity to choose his profession freely. But he actively participated in the brewing business while planning to become a natural philosopher. The brewery and Manchester industry in general made an ideal environment for studying the most current problems in science and technology. The new forces of *electricity and *magnetism then enjoyed the attention of most people interested in natural philosophy. Joule's first major research project was his investigation into the electric motor as a possible alternative to the steam engine. His observations of conversion processes displayed by the engine led him to estimate its duty (efficiency) and to ponder the ontological status of the forces involved.

Expressing phenomena numerically had become a habit of Joule's when he worked in the brewing world. Producing beer under a tax regime taught him to work by, and to trust in, numbers: every minute process of brewing was controlled and measured by the tax authorities, and expressed numerically. Joule successfully applied this moral economy and its related skills to natural philosophical research. He hoped that by quantifying his electrical research he could gain deeper insights into the relationship between *electricity and heat. He succeeded in deriving a quantitative law of heat production by a voltaic current: heat is proportional to the product of the resistance and the square of the intensity of the current ("Joule heat"). Joule hoped to reduce chemistry to absolute measures and elucidate the concept of latent heat (*see* FIRE AND HEAT).

Joule's experiments to determine the mechanical equivalent of heat (the ratio of the mechanical work performed to the heat produced) were part of a series of investigations. The experiments drew on the thermometric skills he had acquired in the brewery, and invoked a close collaboration with the local instrument maker and natural philosopher John Benjamin Dancer. Joule and Dancer produced the most precise working mercury thermometer available at the time. Their contemporaries still used air thermometers, although the constancy of the expansion coefficient for all gases had been doubted. Joule recognized the need for a more sensitive mercury thermometer in heat research.

Within the local Manchester network of practitioners, Joule found everything he needed to challenge the caloric theory of heat. In the manuscript of his paper "On the mechanical equivalent of heat" (Royal Society of London, *Philosophical Transactions*, 1850), he discussed his "paddle wheel" experiments (in which stirring water raises its temperature) and concluded that friction consisted in the conversion of mechanical force into heat. He based his claim on precision. But the scientific community (represented by the Royal Society) doubted the scientific brewer's claim. His conclusion about the conversion of stirring into heat does not appear in the printed paper. Joule was characterized as a "gentleman specialist" for having established the mechanical equivalent of heat through exact measurement, but his related reflections on the dynamical nature of heat and its significance for *thermodynamics carried little weight before he began his collaboration with William *Thomson. Thomson made "Joule's constant "(the ratio of mechanical work to heat) the building block of the science of energy. As a result of his new status as the master of precision measurement, Joule became a core member of the British Association for the Advancement of Science's program to establish absolute units based on the new science of energy.

Mayer, the youngest son of an apothecary from Heilbronn, Germany, devoted his free time to chemical and physical experiments, trained in the medical faculty of the University of Tübingen, and passed his medical examination in 1838 with a dissertation, *Über das Santonin.* He visited Paris in 1839–1840, and in 1840–1841 worked as the doctor aboard a Dutch vessel sailing to the East Indies. In 1842, he returned to Heilbronn. There he investigated the mechanics of heat and developed an overarching theory of force. Like Joule, Mayer was an outsider to the rising community of physical science. Through his practice as a physician, his upbringing in an apothecary household, and his childhood play with mechanisms, Mayer posited that neither a physical machine nor the human body could generate motion out of nothing: thus, production is proportional to consumption.

He made the curious observation that the blood drawn from Europeans in Java differed in color from blood taken in Europe. He ascribed the difference to a disparity in the heat requirements of the body in the two regions. He wrote: "For the maintenance of a uniform temperature of the human body the heat *produced* in it must necessarily stand in a quantitative relationship to its heat *loss*, also to the temperature of the surrounding medium." His subsequent scientific contributions focused on challenging the then-dominant materialistic notion of forces (*see*

IMPONDERABLE). For Mayer, force was distinct from matter, indestructible, transformable or mutable, and immaterial, but converted in invariable quantitative relations from one form to another. He insisted on the mechanical equivalence (not equivalent) of heat, which could be measured as a constant relationship between motion (expressed as mechanical work) and heat. Although he urged that physical sciences be grounded in exact measurements, he based his own claim of equivalency on disputed gas expansion coefficients. For those like Joule, who favored precision measurement, Mayer's knowledge claim—implying a strong antimaterialism—was mere speculation. Mayer managed neither to persuade his contemporaries (such as the editor of *Annalen der Physik und Chemie*, Johann Christian Poggendorff, who rejected his major article of 1841), or to forge an alliance with Justus von *Liebig and Hermann von *Helmholtz, respectively the dominant chemist and physicist of Germany at the time. Mayer suffered immensely both mentally and physically from being marginalized in the newly professionalized scientific community.

Joule and Mayer both participated actively in the rising culture of the exact sciences, which they wished to see anchored in quantitative measurement. Although neither of them coined or used a phrase equivalent to "conservation of energy," their investigations of the mechanics of heat became seeds for the nineteenth-century science of energy. However, theirs is not a case of simultaneous discovery. They did not announce the same thing at the same time, and they did not consciously seek to uncover a new principle hidden in nature. Instead, their work belonged to a long process that made energy conservation a key representation of nature. Their attempts to enhance their credibility in the rapidly expanding scientific community by claiming priority for the establishment of the equivalence between heat and mechanical work did much to construct the image of them as codiscoverers.

Donald S. Cardwell and James Joule, *A Biography* (1989). Kenneth L. Caneva, *Robert Mayer and the Conservation of Energy* (1993). H. Otto Sibum, "Reworking the Mechanical Value of Heat. Instruments of Precision and Gestures of Accuracy in Early Victorian England," *Studies in History and Philosophy of Science* 26, 1 (1995): 73–106. H. Otto Sibum, "Les gestes de la mesure. Joule, les pratiques de la brasserie et la science," *Annales: Histoire, Science Sociale* 4–5 (1998): 745–774. Crosbie Smith, *The Science of Energy. A Cultural History of Energy Physics in Victorian Britain* (1998).

H. OTTO SIBUM

JOURNAL. The scientific journal and its essential companion, the journal article, have continuously

Number of Journals

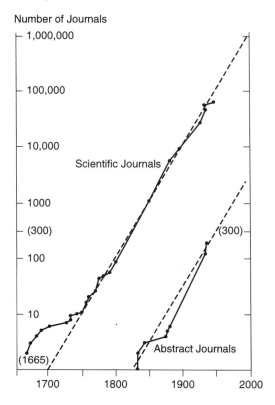

The number of scientific journals published per year, 1700–2000.

grown in importance since their debut in the mid-seventeenth century. As the word "journal"—and its substitute, "periodical"—suggest, this medium is characterized by currency and regularity of publication. Correspondence, appendices or revisions of books, and even manuscripts predated the journal as methods for communicating scientific news, but none of these formats could match the stability of the journal as a repository for the rapidly growing scientific literature.

The first regularly published periodical to include scientific essays and reviews was the *Journal des Sçavans*, which began publication on 5 January 1665. It was followed later that year by the first journal to specialize in experimental science and natural philosophy, the *Philosophical Transactions* of the Royal Society of London. Its articles generally reported briefly on experiments or observations, fresh and accessible material appropriate for prompt publication. Founded in 1682, the *Acta Eruditorum* of the Academia Naturae Curiosorum, by publishing in Latin, attracted articles in academic fields such as mathematics.

During the eighteenth century the founding of hundreds of new scientific academies and

societies paced the proliferation of scientific journals and concentrated the control of the scientific community over its burgeoning literature. One of the most tangible steps taken by these institutions was to provide opportunities for the dissemination of knowledge. The Paris Academy of Sciences, after reorganizing in 1699, established a record of its accomplishments through the *Histoire*, an annual publication required by the Crown, and the *Mémoires* written by individual members, reviewed by the assembly, and issued with the *Histoire* beginning in 1701. Societies sought to stimulate discoveries, inventions, and improvements through open communication of results; the successful ones created a readership of correspondents, professors, officials, agronomists, physicians, and others who consumed and contributed to their programs. Most smaller societies and academies in the eighteenth century, especially those in the German states, adhered more closely to the Royal Society's model of prompt publication than the Paris Academy's deliberate transactions, but not without borrowing from the Paris model the practice of explicit institutional affiliation with their journals.

This same period witnessed two more trends in journal publishing: the proliferation of privately published journals and specialization. The two trends were linked, if only because academies and societies with their learned memberships usually supported journals covering the whole of science, while independent publishers often followed more narrowly defined interests. Journals setting the latter trend were the Abbé Rozier's *Introduction aux observations sur la physique* (later the *Journal de physique*), founded in 1771; *Der Naturforscher*, founded in 1774 and edited by J. E. I. Walch; and William Nicholson's *Journal of Natural Philosophy, Chemistry, and the Arts*, founded in 1797 and more commonly known as "Nicholson's Journal." While these viewed all of experimental science as their domain, other journals in chemistry, forestry, natural history, mining, medicine, and additional fields sharpened the focus more narrowly. Private journals covering general science reached their high point by the mid-nineteenth century, then were overwhelmed by institutionally sponsored and single-discipline journals. The *Philosophical Magazine*, founded by Alexander Tilloch in 1798, exemplified this pattern by reducing its coverage gradually to physics research and by swallowing Nicholson's Journal. While the publications of academies continued to cover the sciences generally, a parallel trend toward the division of transactions along disciplinary lines emerged, led by the voluminous *Comptes Rendus* of the Paris Academy. Conferences became an important means of semiformal communication among scientists, and their proceedings joined journals in the publication of contributions. By the early twentieth century, the article and conference paper dominated formal scientific communication.

The growth of the journal literature encouraged innovations in bibliographic control of these publications. By the end of the eighteenth century, the expense and number of periodicals made it difficult for any individual to survey the literature, let alone acquire it. Annual and multiyear indexes for periodicals and review journals such as Lorenz Crell's *Chemisches Journal* (1778–1781) covered the literature and followed the example of the *Allgemeine Deutsche Bibliothek* (1765–1792) by providing intellectual access in the form of indexes. By the end of the century, review journals and analytical bibliographies of periodical literature began to appear, culminating in the *Allgemeines Repertorium der Literatur* of Johann Samuel Ersch, published between 1793 and 1807 to cover publications from 1785 to 1800. Indexes, bibliographies, and biobibliographies edited by Ersch, Johann Poggendorff, Georg Christoph Hamberger, and others prepared the way for comprehensive indexes to the scientific literature such as the *Royal Society Catalogue of Scientific Papers*, covering the entire nineteenth century; its successor, the *International Catalogue of Scientific Literature*, continuing to the beginning of World War I; *Science Abstracts*, which began publication in 1898; *Chemical Abstracts*, founded in 1907; and dozens of others. By 1960, *Science Abstracts* alone generated about 21,000 citations with abstracts annually, and *Chemical Abstracts* occupied roughly 3,000 abstractors. With the rise of bibliometrics, new forms of bibliographic control such as citation indexes, offered in Eugene Garfield's *Science Citation Index* since 1961, navigated a vast sea of citations. As the literature grew, abstracting and bibliographic services such as the INSPEC (Information Service in Physics, Electrotechnology, and Control) database (*Science Abstracts*) began to depend on computers to deliver searchable bibliographic databases. By the beginning of the twentieth century, printed books had yielded their central position in scientific communication to the journal, conference paper, and preprint. Issues of quality control, economics, and technology shaped the journal in this century. The system of peer review that evolved over the course of the century gave intellectual control of journal contents to editors and reviewers representing the ranks of scientists. The challenge of managing the expense of publication proved more difficult as the size of the published literature grew. One solution, represented by the founding of the American Institute of Physics in 1932, was to concentrate nonprofit publishing by member societies. But since the 1950s consolidation has favored private publishers such as Reed Elsevier and Springer Verlag, who are able to purchase stables of specialized journals and to profit from expensive subscriptions placed by research institutions and libraries, rather than individuals. In the 1980s and 1990s the reduction of library budgets and ceaseless growth of the printed journal literature precipitated a crisis. Computer technology and the Internet led to innovations such as online refereed publications, introduced by the *Online Journal of Current Clinical Trials* in 1992, or informal preprint servers and ftp sites; they offer the possibility of accelerating the pace of scientific publication while challenging the role of for-profit publishers by reducing the cost of disseminating results.

Harcourt Brown, *Scientific Organizations in Seventeenth Century France* (1967). Robert Mortimer Gascoigne, *A Historical Catalogue of Scientific Periodicals, 1665–1900, with a Survey of Their Development* (1985). Charles Bazerman, *Shaping Written Knowledge: The Genre and Activity of the Experimental Article in Science* (1988).

HENRY LOWOOD

JUSSIEU, Antoine-Laurent de (1748–1836), botanist.

Jussieu was a member of the prominent eighteenth-century natural history dynasty associated with the Paris Académie des Sciences and the Jardin du Roi. His primary instruction in botany came from his uncle Bernard de Jussieu, a systematist of profound though diffuse influence, who published little but designed the botanical garden of the Trianon and expanded the Jardin du Roi.

Antoine-Laurent de Jussieu took a doctorate of medicine in 1770 with a thesis on analogies between the animal and vegetable economies. His botanical career began in earnest in 1773, when he presented his paper on the buttercup family to the Académie and was elected adjunct botanist. During the revolutionary reorganization of the Jardin du Roi into the Muséum d'Histoire Naturelle in 1793, Jussieu was appointed professor of field botany; he became its second director in 1800.

As a director, among his most important institutional improvements was the establishment of a herbarium for the Muséum, comprised at its inception largely of captured Dutch, Belgian, and Italian herbaria and of monastic and private collections confiscated domestically. Both the Muséum's and Jussieu's own herbarium were further enriched through his connection to a web of botanists and collectors around the world. His primary work, supplemented by many subsequent articles, mostly in the *Annales du Muséum d'histoire naturelle*, was the *Genera plantarum* of

1789. Jussieu retired in 1826 owing to failing eyesight, a family affliction, and was succeeded in the chair of botany by his son Adrien.

Jussieu's great importance to the history of biology lies in his development of a natural system of classification for plants using the Linnaean binomial nomenclature and hierarchy of taxa. Carl *Linnaeus's own artificial, "sexual" system of plant classification, based solely upon the parts of fructification (in flowering plants, the pistils and stamens), had considerable virtues, being easy both to learn and employ; but it also suffered from severe drawbacks for the serious botanist. Plants out of flower could not be reliably classified, and the sexual system occasionally made hash of what were generally felt to be certain "natural" affiliations. (In several places Linnaeus himself had fudged the sexual system to accommodate nature.)

Jussieu's natural system of classification was not limited to the enumeration of flower parts. Like Linnaeus, he believed reproduction to be the most important biological function and therefore privileged it in classification. However, Jussieu extended the characters used for classification beyond the stamens and pistils: of primary importance for him were the number of cotyledons (seed leaves), their position in the seed, and the relative positions of attachment of the calyx (sepals), corolla (petals), stamens, and pistils to one another. These provided the highest levels of classification, into classes and orders; other morphological characters drawn from elsewhere on the plant further differentiated groups into families, genera, and species. Again like Linnaeus, however, Jussieu was willing to fudge the system where necessary; only by the exercise of experience and critical judgment, only by "assiduous meditation on truly natural genera," could botanists properly weight characteristics in assigning a taxonomic rank.

Unlike some of his taxonomic contemporaries and successors, particularly the Swiss botanist Augustin-Pyramus de Candolle, Jussieu believed firmly in a plenitude of nature. Properly understood, he maintained, species, genera, and families abutted one another, without morphological gaps between them—apparent gaps were artifacts of undiscovered species—so that the delineation of taxa was an essentially arbitrary affair. Accordingly, whether for pragmatic or aesthetic reasons, Jussieu avoided groups that were either too big (containing more than a hundred subtaxa, too many to retain in memory) or too small. The *Genera plantarum* has no monogeneric families and it broke up "natural" groups like the Compositae, which Jussieu split into three families of around a hundred genera each.

Jussieu's methodology exercised a profound influence on systematics and through systematics on natural history and biology. In providing the solid framework of a natural system, Jussieu fostered debates on the true shape of nature and on the continuity of its creatures that persisted (often in very confused form) into the period when evolutionary biology transformed ideas of species' meanings and relationships to one another. In purely pragmatic terms, the *Genera plantarum* remains one of the cornerstones of the systematic indexing of species without which biologists would be lost.

Peter F. Stevens, *The Development of Biological Systematics: Antoine-Laurent de Jussieu, Nature, and the Natural System* (1994). E. C. Spary, *Utopia's Garden: French Natural History from Old Regime to Revolution* (2000).

A. J. LUSTIG

K

KAISER WILHELM/MAX PLANCK GESELL-SCHAFT.

Established in 1911, the Kaiser Wilhelm Gesellschaft (KWG) was until 1945 the umbrella administration of a set of research institutes dedicated primarily to strategic areas of natural science and technology. The Kaiser Wilhelm Institutes (KWIs) existed outside universities, had no instructional responsibilities, and enjoyed funding from private, industrial, and state sources. Often perceived as novel because they were intended to overcome the stagnation of the *Humboldtian ideal of the unity of teaching and research, the KWIs were preceded in the nineteenth century by other privately funded research institutes, mostly for historical scholarship, intended to supplement university research institutes. The initial instigation for extra-university institutes came from chemists who wanted research facilities similar to those found in the Imperial Institute for Physics and Technology, established in 1887 as Germany's national *bureau of standards. International competition, especially the recently developed American practice of private funding for research led by the Carnegie Institution and the Rockefeller Foundation, as well as the decline of state support for research in the financial crisis of 1907–1910, also spurred the inauguration of the KWG. The number of KWIs grew from barely a handful before World War I to over thirty in Nazi Germany. After World War II, the Kaiser Wilhelm Gesellschaft became the Max *Planck Gesellschaft (a name chosen to honor the man who had begun the quantum revolution in physics and presided over the KWG from 1933 to 1938), although the historical continuity between the two organizations is disputed.

The KWG consisted of members and senators drawn from the industrial, financial, civil service, and academic sectors of society; each had its own obligatory levels of monetary contribution. In principal KWI directors set scientific research agendas, but support levels, set by donors, delimited research projects. Through the KWG, business and industry became accustomed to the idea of supporting research institutes outside industrial and educational contexts. The first KWIs investigated areas of direct interest to the state and its industrialized economy: chemistry, coal research, experimental therapy, labor physiology, and biology. Between 1914 and 1918 the KWIs developed new chemical weapons, synthetic fuels, and artificial substitutes for natural resources.

During the Weimar Republic (1919–1932), the KWG kept its distance from the new democracy but grew nonetheless, partly as a result of Rockefeller Foundation funding for physics and cell physiology (*see* WARBURG). By 1933 the KWG had attained international distinction and several of its directors had won Nobel Prizes. Soon, however, many KWIs aligned with the Third Reich. The KWI for Anthropology, Human Heredity, and Eugenics provided racial training and crafted racial policy for Nazi officials. Other KWIs provided ideological and practical foundations for the planned German resettlement of Eastern Europe or conducted politically strategic research on wind tunnels for aerodynamics, nuclear weapons, plant and animal genetics, chemical weapons, and twin research.

The Max Planck Gesellschaft (MPG), set up by officials in the British occupation zone of West Germany between 1946 and 1948, continued the personnel, form, and organization of the KWG. New Max Planck Institutes (MPIs) were designed for fields of research not covered elsewhere or for cultivating local historical strengths, such as precision instrument construction in Lower Saxony. The number of KWIs in 1939 was 28; of MPIs in 1946, 13; and of MPIs in 1990, when reunification of Germany reorganized the research landscape, 77. Since reunification the former East Germany has received most of the new Max Planck Institutes. These new MPIs have committed themselves to novel interdisciplinary fields and have forged a closer relationship with the educational system, especially through new predoctoral training programs. The Max Planck Institute for the History of Science, established in 1995, focuses on historical epistemology,

The KWG's first institute, dedicated to chemistry, Berlin-Dahlem, 1913.

although its visiting scholars represent many varieties of history of science.

See also INSTITUTES; NAZI SCIENCE.

Jeffrey Johnson, *The Kaiser's Chemists: Science and Modernization in Imperial Germany* (1990). Kristie Macrakis, *Surviving the Swastika: Scientific Research in Nazi Germany* (1993).

KATHRYN OLESKO

KAPITSA, Pyotr (1894–1984), one of the most famous physicists in the Soviet Union, a pioneering investigator of superconductivity.

Born on the island of Kronshtadt near St. Petersburg to Leonid and Olga Stebnitskii Kapitsa, in 1919 Pyotr Kapitsa graduated from the Petrograd Polytechnic Institute where he studied under A. F. Ioffe. For two years Kapitsa taught in this same institute and then transferred to Cambridge University in England, where he received a Ph.D. in 1923. During the 1920s and early 1930s Kapitsa remained in Cambridge, where he enjoyed the full support of Ernest *Rutherford. With financial support procured by Rutherford, Kapitsa built the Mond Laboratory within the Cavendish Laboratory at Cambridge as a center for the study of very high magnetic fields. Nobody expected Kapitsa to return to Russia, especially at a time when Joseph Stalin's ruthless policies were becoming more and more apparent. While he was visiting friends in Russia in 1934, however, the police detained Kapitsa on Stalin's orders and refused to permit him to return to England.

Kapitsa complained that he did not have the resources in Moscow that he had had in Cambridge. Stalin decided to give Kapitsa everything that he wanted materially. The Soviet government purchased the equipment in the Mond Laboratory and brought it to Moscow. Named the Institute of Physical Problems and located in a small woods near the Moscow River, Kapitsa's new institute could pursue delicate experiments under quiet conditions. Stalin also gave Kapitsa a Cambridge-style home in Moscow next to his institute. Although at first depressed about his captivity, Kapitsa soon responded scientifically to the favored conditions Moscow offered. During the first three years of his detainment, at the peak of Stalin's purges, Kapitsa accomplished the most important work of his scientific career, research on the superconductivity of liquid helium at temperatures near absolute zero. In 1976 he received a Nobel Prize for this work.

Kapitsa married twice, first to Nadezhda Tschernovsvitova and then to Anna Krylova, by whom he had two sons. He received many honors, both in the Soviet Union and abroad, and held a U.S. patent on a turbine for the production of liquid air.

In his scientific work, Kapitsa emphasized the importance of combining experimental and theoretical physics. He retained a remarkable independence of spirit. When the secret police arrested several other prominent Soviet physicists, including Lev Landau and Vladimir Fok, Kapitsa played an important role in gaining their release. During World War II Stalin asked him to work on the atomic bomb project. Kapitsa refused, not because of objections to the bomb itself but because he could not tolerate Lavrenty Beria, the head of the Soviet secret police who was also in charge of the bomb project. As a result of this refusal, Kapitsa was removed from the directorship of his institute and kept under house arrest at his country dacha, where he attempted to continue his experimental work. After the death of Stalin in 1953 Kapitsa returned to his former position.

During the 1960s and 1970s Kapitsa provided places in his institute for young physicists to do important research and also for artists to display paintings that did not follow official cultural policies. When the Soviet Academy of Sciences tried to expel the dissident physicist Andrei Sakharov, Kapitsa and others rose to his defense and the effort failed. When Kapitsa died in 1984, he was the only member of the presidium of the Soviet Academy of Sciences not also a member of the Communist party.

Only after Kapitsa's death did even his close friends learn that all during the Soviet period he wrote letters of protest and complaint to government leaders. The archives contain dozens of letters to Stalin, Vyacheslav Molotov, Georgy Malenkov, and Nikita Khrushchev in which Kapitsa criticized Soviet policies toward science, excoriated the secret police for arresting scientists, and practically challenged the police to detain him. We now know that Beria intended to arrest Kapitsa but that Stalin intervened, saying, "I'll take care of him personally. Don't you touch him." Stalin might have been motivated by respect for Kapitsa's opinions and appreciation that Kapitsa never made his protests public.

Peter Kapitsa, *Experiment, Theory, Practice* (1980). Lawrence Badash, ed., *Kapitza, Rutherford, and the Kremlin* (1985). "Peter Kapitsa: The Scientist Who Talked Back to Stalin," *Bulletin of Atomic Scientists*, vol. 46, no. 3 (April 1990): 26–33.

LOREN R. GRAHAM

KEKULÉ, August (1829–1896), German chemist, the founder of structural chemistry.

Kekulé was born to a bourgeois family in Darmstadt, the capital of the Grand Duchy of Hesse. Initially intending to become an architect, Kekulé studied at the nearby University of Giessen, but he was charmed by the lectures of Justus von *Liebig and decided to become a chemist. Following his doctorate, he spent four years on study trips in Switzerland, Paris, and London before becoming Privatdozent (lecturer) at the University of Heidelberg (1856). Two years later he was hired as professor at the University of Ghent, where he lectured and published his research in French. He returned to his homeland in 1867, when the University of Bonn called him to fill the position left vacant by August Wilhelm von Hofmann's decision to move to Berlin. There he remained until his death in 1896.

From contact with Charles Gerhardt in Paris and Alexander Williamson in London, by 1854 Kekulé was won over to a reformed system of atomic and molecular weights then being urged against the system of "equivalents" advocated by many older chemists. The concept of atomic *valence was just then being approached by several chemists simultaneously, including Kekulé. But Kekulé was the first to see that the idea of valence could be used to dissect molecules right down to the individual atoms that constitute them. By regarding hydrogen and chlorine atoms as capable of bonding to one atom only, oxygen and sulfur to two each, nitrogen and phosphorus to three, and carbon to four, the chemist could draw up structural plans of molecules that matched the various known possibilities. The theory could then be used as a guide for chemical analysis or synthesis (*see* ANALYSIS AND SYNTHESIS).

According to his later reminiscences, Kekulé developed this "theory of atomicity of the elements" (soon thereafter called the "theory of chemical structure") in London, probably in the summer of 1855; he published a preliminary version of it in two articles in 1857 and 1858. Crucial to this theory was the claim that carbon atoms could use their "affinity units" (valence bonds) to link up with each other, producing a carbon "chain" or "skeleton" that formed the backbone of the organic molecule. Coincidentally, the Scottish chemist Archibald Couper published a substantially similar theory, entirely independently and virtually simultaneously; unfortunately, his work had little influence on his contemporaries.

Expounded in detail in the pages of Kekulé's extraordinarily popular *Lehrbuch der Organischen Chemie* (published in installments beginning in 1859), Kekulé's structure theory had enormous success among the rising generation of organic chemists. Such leading contemporaries as Aleksandr Butlerov, Alexander Crum Brown, Emil Erlenmeyer, Adolf von Baeyer, Hofmann, Charles-Adolphe Wurtz, and Edward Frankland developed the theory further. The Karlsruhe Congress of 1860, the brainchild of Kekulé, helped win further support for the new atomic weights and formulas that had formed the starting point for structural considerations.

Emboldened by his success, Kekulé tackled the important but problematical family of organic compounds conventionally called "aromatic," which are all based on the molecule of benzene (C_6H_6). In 1865, he tentatively suggested that a cyclical structure for this molecule—the six carbons linked together in a hexagonal "ring" with alternating single and double bonds—could resolve most of the outstanding theoretical anomalies of aromatic compounds. He developed this theory more definitely and confidently in 1866 and in his textbook.

Kekulé's benzene theory was one of the most brilliant ideas of the century. Although it left important problems unresolved, it cast such an unexpectedly clear light on aromatic substances that most practicing chemists soon adopted the theory. Most technologically interesting organic compounds (dyes and drugs, for example) were aromatic, and Kekulé's theory became the master key to productive industrial research. So many "organikers" turned to aromatic chemistry as a specialty that by the end of the century the large majority of all known organic compounds were aromatic.

Kekulé was one of the most imaginative and innovative chemists of the nineteenth century. His private life was not happy (his beloved young wife died delivering their first child, and his second marriage was unfortunate), but his energetic, humorous, self-confident yet magnanimous character won him many good friends. Fluent in French, English, German, and Italian, he was an ardent internationalist and had nearly as much influence abroad as in Germany. His own countrymen benefited most from his research, however, since German supremacy in organic chemistry was largely founded on his ideas, which still form the bedrock of organic chemistry.

Richard Anschütz, *August Kekulé*, 2 vols. (1929). O. T. Benfey, *Kekulé Centennial* (1966). Alan J. Rocke, *The Quiet Revolution* (1993).

A. J. ROCKE

KEPLER, Johannes (1571–1630), astronomer and mathematician.

Kepler was born on 27 December 1571 in the imperial free city of Weil der Stadt, where his grandfather was mayor. His frequently absent father was a ne'er-do-well and sometime mercenary. Kepler's education in the fine school system of the Duchy of Württemberg culminated in a scholarship to study theology at the University of Tübingen. He passed his B.A. by examination in 1588, proceeding to Tübingen in 1589. The theologian Jacob Heerbrand and the astronomer Michael Mästlin, both adherents of Philipp Melanchthon in their view of the natural world, were among his teachers. After receiving his M.A. in 1591, Kepler continued advanced theological studies until he was assigned to teach at the Protestant school in Graz, Styria. He held the position of mathematics teacher and district mathematician from 1594 until the expulsion of the Protestants from Graz in 1600. He then became one of Tycho *Brahe's assistants in Prague. Upon Tycho's death in 1601, Kepler became

imperial mathematician to Holy Roman Emperor Rudolf II, a post he retained under emperors Matthias and Ferdinand II until his death. He moved to Linz in 1612 to become mathematician to the estates of upper Austria. When civil war and the expulsion of Protestants made his position untenable, he became mathematician to Albrecht Wallenstein in Sagan in 1628. He died in Regensburg on 15 November 1630.

Kepler's work was distinguished most of all by his fervent, realist defense of Nicholas *Copernicus's heliocentric system, and by his insistence that astronomy be physical, which ran against a contemporary division of astronomy into physical (cosmological) and mathematical (theoretical) parts. Kepler's conception of physical reasoning was broadly Aristotelian (*see* ARISTOTELIANISM). It included the notion of formal cause deriving from Kepler's belief in God's providential design. Thus his first publication, the *Mysterium cosmographicum* (1596), argued for Copernicus from the physical premise that God had constructed the solar system using the perfect Platonic solids as archetypes, determining the planets' unique commensurable spacing through the inscribing and circumscribing of their spheres around the solids.

When his work under Tycho Brahe shunted him into detailed work on planetary theory, Kepler pursued a controversial physical approach, analyzing Mars's motion in terms of a force coming from the sun that moved the planets around their orbits. From this he was able to derive the first two of what are now called his laws of planetary motion—that the orbits of the planets are elliptical with the sun at one focus, and that the planet-sun radius sweeps out equal distances in equal times. He published the long justificatory narrative of this research in his *Astronomia nova* (1609).

He returned to cosmological writing in his *Harmonice mundi* (1619), taking up especially the relationship of the planets' distances and periods. Almost accidentally, he discovered the third law of planetary motion—that the ratio of the square of a planet's period to the cube of its distance was constant for all of the planets. This finding fit only awkwardly within the book's broader argument that God had painstakingly arranged the planet's distances and periods so that their angular motions viewed from the sun would embody all musical harmonies.

Kepler inherited from Tycho the prestigious responsibility for the *Tabulae Rudolphinae* (1627). These were astronomical tables based on Tycho's unparalleled observations using Keplerian theory. The tables—the first ever using logarithms—were abstruse by contemporary standards, and their immense superiority was only made manifest by Kepler's prediction of the first observed transit of Mercury in 1631. In an effort to smooth the acceptance of his unorthodox astronomy, Kepler also published the *Epitome astronomiae Copernicanae* (3 vols., 1618–1621), a detailed exposition of Keplerian astronomy in textbook form.

Kepler was a productive polymath. In addition to his astronomical work, he made important contributions to optics and mathematics. His *Astronomia pars optica* (1604) was the foundational work of seventeenth-century optics. It included the inverse-square diminution of light and the first recognizably modern description of the formation of the retinal image. His *Dioptrice* (1611), the first theoretical analysis of the telescope, included an improved telescope design. In mathematics, his *Nova stereometria doliorum vinariorum* (1615), concerning the volumes of wine casks, was a pioneering work of precalculus, and his *Chilias logarithmorum* (1624), an important early work on logarithms. He also published chronological works and balanced defenses of astrology.

Kepler's mathematical genius and his important conviction that astronomical theory must be derived from physics places him without doubt among the greatest early modern applied mathematicians, although the technical difficulty of his works and an unwarranted reputation for mysticism impeded his recognition by historians.

Max Caspar, *Kepler*, trans. C. Doris Hellman (1959; reprinted 1993). Alexandre Koyré, *The Astronomical Revolution*, trans. R. E. W. Maddison (1973; reprinted 1992). Bruce Stephenson, *Kepler's Physical Astronomy* (1987). Owen Gingerich, *The Eye of Heaven: Ptolemy, Copernicus, Kepler* (1993). Bruce Stephenson, *The Music of the Spheres: Kepler's Harmonic Astronomy* (1994). James R. Voelkel, *The Composition of Kepler's* Astronomia nova (2001).

JAMES R. VOELKEL

KEPLER'S LAWS. See ORBIT.

KLYSTRON. William Hansen and Russell and Sigurd Varian invented a pioneering microwave generator at Stanford University in the late 1930s to accelerate electrons for atomic research and to serve as the heart of a guidance system for aircraft. They named this device the klystron, from the Greek word for "waves," because the bunching of electrons that took place in the generator reminded the inventors of water crashing on a beach. A similar device was independently invented by N. D. Devyatkov at Scientific Research Institute Number 9 in the Soviet Union. Ernest Lawrence brought the klystron to the attention of Alfred Loomis and Edward L. Bowles at the Massachusetts

Institute of Technology, and they interested Sperry Gyroscope in the invention. Sperry and the Radiation Laboratory at M.I.T. developed the klystron as a radar receiver during World War II, under Hansen's leadership.

The klystron used the principle of amplification that Hansen had developed with his "rhumbatron" resonant cavity of 1935. The resonant cavity, like a tuning fork, continues to vibrate at one frequency for a long time. A highly focused beam of electrons traversed a series of microwave-resonant cavities. A radio-frequency signal caused the electrons to bunch at its frequency in one cavity, the "buncher," and the bunched electrons then traveled through a drift tube in which they were shielded from electromagnetic forces and induced a subsequent cavity, the "catcher," to emit an amplified signal at the same frequency. Beam currents of up to 25 kilovolts could be manipulated in this way. The klystron was a high-power amplifier with considerable stability but limited frequency.

In England, wartime work in Marcus Oliphant's laboratory in Birmingham developed the klystron's radar potential. Two of his nuclear physicists, John Randall and Harry Boot, transformed the device into the multi-cavity magnetron, which reached higher frequencies and became the basis for most radar in the microwave range.

In addition to military applications, the klystron is used in ultra-high-frequency television broadcasting, satellite communications, and industrial heating. Its most significant scientific uses have been in particle *accelerators. After the war, Hansen returned to Stanford and worked with electrical engineer Edward L. Ginzton in the Microwave Laboratory to perfect the use of the klystron for particle acceleration, with the support of Sperry and the Office of Naval Research. This research led to the high-power klystrons used in the electron linear accelerators, from the Mark III to the two-mile-long Stanford Linear Accelerator built in the 1960s. Robert Hofstadter used the Mark III in his studies on the structure of the neutron, for which he shared the Nobel Prize in physics in 1961. Burton Richter and his colleagues used the Stanford Linear Accelerator in the discovery of the J/psi particle, for which he shared the Nobel Prize in physics (1976). Klystrons have become the workhorses of high-energy electron accelerators around the world.

Although a number of large electrical manufacturers acquired klystron technology to make *radar apparatus during World War II, it was Hansen, Ginzton, and the Varian brothers who led its development after the war. The Varian Corporation became one of the centers around which the Stanford Industrial Park, and later Silicon Valley, developed. More powerful klystrons enabled the armed services to develop accurate missile-tracking radar and other advanced defense technologies.

Anne Ginzton Cottrell and Leonard Cottrell, eds., *Times to Remember: The Life of Edward L. Ginzton* (1995).
ROBERT W. SEIDEL

KREBS, Hans Adolf (1900–1981), discoverer of the metabolic pathway commonly known as the "Krebs cycle," and one of the principal architects of the subfield of intermediary metabolism.

The oldest of three children of Georg Krebs, a respected otolaryngologist in the Hanoverian town of Hildesheim, and Alma Krebs, who came from a large family that had been settled there for many generations, Hans was born in 1900 into secure surroundings. His secondary school education in a classical Gymnasium included more Greek and Latin than science. Reaching the age of eighteen toward the end of World War I, he had only completed basic training when the shooting stopped. That brief experience, however, reinforced a sense of discipline and responsibility that he carried through his later life. After the war, Krebs studied medicine at the University of Göttingen and other universities before obtaining his M.D. in Munich in 1923. Along the way, the example set by some of the eminent teachers whose lectures he attended inspired in him the ambition to combine clinical practice with experimental investigation in a field such as internal medicine. An unexpected opportunity to become an assistant to Otto *Warburg in 1926 brought a decisive change in the direction of Krebs's aspirations. In Warburg's laboratory, he learned to use a powerful method for studying the respiratory exchanges of isolated tissues by placing thin slices in a medium in which they could survive for several hours. The oxygen the tissues absorbed and the carbon dioxide they emitted could be measured precisely and continuously in a micromanometer of Warburg's design.

Three years later, told by Warburg that he must leave the laboratory, and believing that his mentor thought him incapable of independent research, Krebs found a position in 1930 in a hospital near Hanover in which he could combine clinical duties with some experimental work. There he began to carry out a plan he had not been allowed to pursue in Warburg's laboratory, to apply Warburg's methods to the study of intermediate metabolic reactions. Moving the next year to Freiburg, Krebs took up the problem of how urea is synthesized in animals. Within nine months he had discovered the "ornithine cycle" of urea

synthesis, which quickly marked him as a rising star in his field.

None too soon. One year later, along with thousands of "non-Aryans," Krebs was dismissed from his post. Having already achieved an international reputation, he obtained a Rockefeller Foundation fellowship and entry into the well-known biochemical laboratory of Frederick Gowland Hopkins in Cambridge, England, where he resumed his experimental research after an interruption of less than two months. Although the ethos he found in English laboratories and in English life was more casual than the tightly disciplined life he had experienced in Germany, Krebs immediately felt at home. Despite his reticence and a degree of punctuality that appeared unusual to his hosts, his warmth, directness, and quiet scientific leadership won him ready acceptance among his colleagues. When he was appointed demonstrator in biochemistry in 1934, he became the first among the dozens of German refugees by then in England to attain a regular academic post.

In Cambridge, Krebs resumed his plan to study the processes by which foodstuff decompose in the energy-yielding reactions of animal tissues. Anticipating that he would not be able to stay in Cambridge permanently, he accepted an invitation in the summer of 1935 to move to the Department of Pharmacology at the University of Sheffield, where he had enough laboratory space to begin training younger biochemists to do research in his field. Continuing his own studies of oxidative carbohydrate metabolism, Krebs benefited from investigations by Albert Szent-Gyorgyi of the catalytic activity of various organic acids, and from a proposal by Carl Martius of the decomposition pathway of citric acid, to envision a cyclic process through which "triose" would be degraded in a series of steps to yield carbon dioxide and water. Through a well-conceived set of experiments, Krebs persuaded himself and the field that the "citric-acid cycle," as he named it, really takes place in animal tissues. Viewed at first as the pathway of carbohydrate metabolism, the citric-acid cycle eventually came to be seen as the "final common pathway" for the decomposition of all classes of foodstuff, and the centerpiece in the network of metabolic pathways that Krebs and many other biochemists traced out during the following decades (*see* BIOCHEMISTRY).

At the end of World War II, Krebs's laboratory received funding from the Medical Research Council to establish a Metabolic Research Unit, where he built an internationally renowned biochemical school. There, he and a small team of assistants, students, and postdoctoral fellows continued the lines of investigation that he had begun with such auspicious results before the war. Awarded the *Nobel Prize in 1953 for the discovery of the citric-acid cycle, Krebs soon transferred himself and his Metabolic Research Unit to the Department of Biochemistry at Oxford.

In the mid-1950s, Krebs turned his attention to the problem of how the pathways of intermediary metabolism are regulated. He pursued that problem productively for twenty-five years. Forced to retire from Oxford at the age of sixty-seven, Krebs feared that his scientific career would be ended, but he made arrangements to continue at the Radcliffe Infirmary, where he retained several members of his long-standing team and attracted students from many countries. He maintained his rigorously disciplined schedule, appearing and departing from his laboratory as punctually as ever, five and a half days a week, until a few weeks before his death in his eighty-first year.

Hans Krebs, *Reminiscences and Reflections* (1981). Frederic L. Holmes, *Hans Krebs* (2 vols., 1991– 1993).

FREDERIC LAWRENCE HOLMES

KURCHATOV, Igor Vasilyevich (1903–1960), Soviet nuclear physicist and director of the Soviet nuclear weapons project; and **J. Robert OPPENHEIMER** (1904–1967), American theoretical physicist, director of the Los Alamos laboratory for design of nuclear weapons in World War II, and government advisor.

Oppenheimer was born in 1904 to a well-to-do family in New York. Nominally Jewish, Oppenheimer obtained his early education at the assimilationist Ethical Culture School. He entered Harvard in 1922 and majored in chemistry, but studied widely in physics, math, philosophy, literature, and languages. His undergraduate work under Percy Bridgman convinced him to switch to physics, in which he displayed remarkable analytical ability but little experimental dexterity. To pursue physics he went to the Cavendish Laboratory at Cambridge, where he stuck to theory, mastered quantum mechanics, and overcame his emotional insecurities. His first papers from Cambridge in 1926 won him an invitation to study under Max Born at the University of Göttingen, where he obtained his Ph.D. in 1927 with a dissertation on the quantum theory of continuous spectra. The sixteen papers Oppenheimer published by 1929 marked him as a rising theoretical physicist.

After several stints as a postdoctoral researcher in Europe, Oppenheimer accepted joint appointments at the California Institute of Technology (Caltech) and the University of California at Berkeley. His research extended into *quantum electrodynamics, *cosmic rays and *nuclear physics, and astrophysics, including the first theoretical suggestion of *black holes. He presided over the main American school of theoretical physics in the 1930s, prolific in publications and students. Although he could be arrogant and impatient, Oppenheimer exuded aesthetic style, wit, and charisma, and students adopted his mannerisms, if not his tendency for sloppy calculation. The depression and the rise of fascism in the 1930s steered him to left-wing politics, although he shed formal political affiliations as World War II approached.

Kurchatov was born in 1903 in the southern Urals; his family moved to the Crimea in 1912, where Kurchatov attended the Gymnasium through the hardship of the first world war and then studied physics at the local state university. Unlike many Soviet physicists of his generation, and unlike Oppenheimer, Kurchatov did not study in Europe. In 1925 a former classmate recommended him to A. F. Ioffe, the director of the Leningrad Physical Technical Institute. Ioffe's institute, known as the Fiztekh, was a leading center for physics in the Soviet Union. At the Fiztekh Kurchatov worked on dielectrics and pioneered the study of ferroelectricity. He also demonstrated his abundant reservoirs of energy, enthusiasm, and confidence, and, unlike Oppenheimer, his dexterity in the lab.

In 1932 Kurchatov shifted to the emerging field of nuclear physics and Ioffe named him, at the age of twenty-nine, to lead the Fiztekh's nuclear physics program. Kurchatov helped make the Fiztekh a center for Soviet nuclear physics. His leadership and administrative abilities won him the nickname "the General." Although his father suffered through internal political exile, Kurchatov survived the terror of the 1930s, when Joseph Stalin's totalitarian regime imprisoned and exterminated millions of Soviets, including many leading scientists and engineers.

Kurchatov organized the first Soviet study of nuclear fission in 1939, which concluded that a self-sustaining chain reaction was possible. A world away, in Berkeley, Oppenheimer came to similar conclusions. A program to pursue an atomic bomb gradually coalesced in the United States, and in 1942 Oppenheimer joined it as head of research on fast neutrons, which was then underway at several sites. When the program came under the direction of the U.S. Army that summer, Oppenheimer suggested to General Leslie Groves that the design of a bomb be centralized in a new laboratory, in order to overcome the effects of secrecy and distance on communication. Groves accepted Oppenheimer's advice as well as his suggestion of a site, a remote mesa called Los Alamos in New Mexico, and named Oppenheimer to direct the lab. It was a surprising choice—first, since Oppenheimer, a theorist, would be leading a largely experimental program, and second, because he lacked a Nobel Prize and hence the stature of other possible candidates. Still, it was an inspired one, as Oppenheimer added his charisma and talent for identifying fruitful lines of research to the sense of purpose of the enterprise. In June 1945 Los Alamos exploded the first nuclear weapon in a test code-named Trinity; although Oppenheimer and some scientists who witnessed the test professed unease at the implications of their work, Oppenheimer and other project leaders concurred in the use of atomic bombs in combat against Japan.

Kurchatov's laboratory helped lead the exploration of fission in the Soviet Union, including the important discovery in 1940, to which Kurchatov contributed, of spontaneous fission. Germany's invasion of the Soviet Union diverted scientists to more immediate problems, and Kurchatov worked on the defense of ships against magnetic mines. After Soviet espionage in 1942 brought news of Britain's pursuit of an atomic bomb, the Soviet government revived fission research and selected Kurchatov as scientific director of the project. Like Oppenheimer, Kurchatov lacked prestige within the physics community—a nomination to the Soviet Academy of Sciences in 1938 had failed to win him election—but he made up for it with charisma and leadership.

Unlike the United States, the Soviets pursued only a small bomb project during the war. Continuous spy reports from Britain and the United States guided the Soviet work. At the end of the war Kurchatov's group was still a long way from an atomic weapon, but the awesome effects of the American bombs on Japan jolted Stalin and he immediately launched a crash project. Kurchatov and his colleagues did not display any moral scruples about atomic weapons, as did Oppenheimer and some Americans after Trinity and Hiroshima; they viewed atomic weapons at the time as necessary for the defense of their homeland and worked on them with patriotic dedication. Kurchatov continued to direct the work, displaying remarkable ability to navigate the treacherous political waters swirling around Stalin. In 1949 the Soviets tested their first atomic bomb, based on intelligence from the United States but also reflecting Soviet talent and commitment. Kurchatov was named a Hero of Socialist Labor after the test. The Soviets followed that with a test of a thermonuclear device in 1953, based on original work in Kurchatov's project. The Soviet thermonuclear test affected Kurchatov similarly to the way Trinity and Hiroshima affected Oppenheimer. The stressful preparations for the test itself took a toll—even if failure perhaps no longer meant the firing squad after Stalin's death in 1953—and then a tour of the destruction at ground zero brought home the power of the weapon. Kurchatov afterward increasingly turned to development of fission and fusion reactors for peaceful atomic energy.

Oppenheimer would also leave weapons work, although involuntarily and without the honors bestowed on Kurchatov. At the end of the war Oppenheimer resigned from Los Alamos and accepted appointment as director of the Institute for Advanced Study at Princeton, which during his tenure became a thriving center for theoretical physics. He continued to exercise substantial influence on nuclear policy, however, as chair of the General Advisory Committee to the Atomic Energy Commission from 1947 to 1952, during which time the United States accumulated a stockpile of nuclear weapons. Oppenheimer led the committee's opposition to development of the hydrogen bomb, not only on moral grounds but because fusion did not appear technically feasible, and because a crash program would divert scarce resources from new fission weapons. President Harry Truman instead heeded the advice of the Air Force and hawkish anticommunists and approved a crash program for the H-bomb. Oppenheimer's opponents in the H-bomb debate would lead an effort a few years later to revoke his security clearance, based as much on his advice on the H-bomb as on his mendacity in reporting contacts with communists during the war. The hearings on the matter in 1954 divided the scientific community and ended with a judgment that Oppenheimer constituted a security risk. After a dozen years as a leader of the American nuclear program, Oppenheimer found himself cut off from inside circles of nuclear policy. He continued to direct the Institute at Princeton, frustrated at his isolation, until 1966.

Alice Kimball Smith and Charles Weiner, eds., *Robert Oppenheimer: Letters and Recollections* (1980). Richard Rhodes, *The Making of the Atomic Bomb* (1986). David Holloway, *Stalin and the Bomb: The Soviet Union and Atomic Energy 1939–1956* (1994). Richard Rhodes, *Dark Sun: The Making of the Hydrogen Bomb* (1995). S. S. Schweber, *In the Shadow of the Bomb: Oppenheimer, Bethe, and the Moral Responsibility of the Scientist* (2000).

PETER J. WESTWICK

L

LABORATORY, CHEMICAL. A prominent text-book of chemistry published in 1777 defines a laboratory as "the place intended for chemical operations." The origin of the laboratory was intimately connected with the origin of chemistry. The connection appears neatly in the seventeenth-century word "elaboratory": the place where a chemical substance is "elaborated"—produced from its elements, fashioned, or transmuted.

In 1766 the French chemist Pierre Joseph Macquer explained the need for a laboratory of chemistry: "because chemistry is a science founded entirely on experiments." Chemists used a great many different agents impossible to describe in writing but "perfectly recognizable to the senses."

The identity of chemistry as an essential but distinct "part of *physics," as Macquer put it, was shaped by a particular repertoire of apparatus arranged in a characteristic way that chemists had inherited from alchemists, metallurgists, and herbal distillers. The arrangement persisted from the time chemistry emerged as a pedagogical *discipline in the early seventeenth century until early in the nineteenth century. Its most prominent features were a set of furnaces of several sizes and shapes, capable of delivering various degrees of heat. For greatest heat, chemists resorted to the reverberatory furnace, designed to fit around a vessel so that the flames closely surrounded it until they reached a narrow chimney. The most prominent operations carried out with the powerful agent of fire were *distillations, which required also a wide assortment of vessels to meet special conditions. Because the charcoal fires were sooty, and the distillation products sometimes included noxious vapors, operations took place under a large, well-ventilated chimney, the ancestor of the modern hood. Because a plentiful supply of water was necessary, both for doing chemistry and washing apparatus, and for ease of delivery of supplies, some chemists believed that laboratories should be on the ground floor. Here, however, the humidity often caused reagents to absorb moisture and labels to fall off bottles, so others advocated upper, drier, airier, lighter rooms.

Operations not requiring the heat of a furnace took place on a large table, the forerunner of the chemical "bench." A typical bench held a wide array of glass, copper, and earthen vessels, including funnels for filtrations, mortars and pestles, and stirring rods. Shelves contained reagents in neat rows, carefully labeled to avoid confusion. An eighteenth-century chemist would have on hand at least four acids, three alkalis, a dozen neutral salts, solvents such as alcohol and ether, various essential oils, and color indicators such as litmus and gall-nuts. The many other known neutral salts he could make up as needed.

Chemical laboratories also included balances. Contrary to a common assumption of historians of chemistry, chemists in the late seventeenth and early eighteenth century regularly weighed the substances they used, and often labored to carry out their distillation and other procedures so as to avoid losses of material. Some of them attained impressive quantitative accuracy.

Many early chemical laboratories were propri-etary. Some chemists made their living as apothe-caries; in their laboratories pursuit of chemical investigations mingled with the preparation of pharmaceuticals. During the seventeenth century, however, a few universities set up laboratories to demonstrate chemical and pharmaceutical oper-ations. At one of the earliest, the University of Marburg, Johannes Hartmann began in 1609 not only to demonstrate preparations, but also to invite students into the laboratory to learn to make them for themselves. By the eighteenth cen-tury, chemistry had become a regular part of the medical curriculum. German universities usually possessed instructional chemical laboratories, as did Glasgow and Edinburgh. The earliest labora-tory intended primarily for research rather than for education or for commercial and medical pur-poses was established in 1670 at the Academy of Sciences in Paris. The Berlin Academy of Sci-ences acquired a similar laboratory in 1753, where Andreas Marggraf and Johann Pott did important experimental work.

By the mid-eighteenth century the chemical laboratory began to stock special apparatus for *pneumatic chemistry—the trough invented by

The first practical teaching laboratory in chemistry, set up by Liebig at the University of Giessen in 1827. The drawing pictures the well-dressed experimenters around 1840.

Stephen Hales and, by the 1770s, glass vessels of various sizes. Antoine-Laurent *Lavoisier adapted existing pneumatic apparatus. When he took up the study of airs in 1773, he used retorts made of metal rather than clay or glass to withstand higher temperatures and pressures. Increasingly, however, his "balance-sheet method" induced him to design special apparatus for particular operations. Having ample means, he had them constructed by leading instrument makers in Paris. Some of the new instruments, such as the ice calorimeter, gasometers to measure the quantity of a gas led into a combustion process, the large glass vessel with connecting brass tubes in which he synthesized water, or the precision balances he had constructed as his need for quantitative accuracy grew, were very expensive. Some contemporaries, notably Joseph Priestley, complained that he was making chemical experimentation too costly for ordinary chemists. By the end of his career, Lavoisier had transformed the nature of chemical apparatus and the operations performed with them as deeply as his theories had transformed the structure of chemical thought. Nonetheless, much old equipment remained. Lavoisier depended as much as his predecessors on the charcoal furnaces that dominated the space of a laboratory.

In the early nineteenth century, as chemical activity burgeoned, the laboratories in which chemists worked became rapidly more complex. By the 1820s catalogues began to appear to acquaint chemists with the latest apparatus. The new availability of rubber tubing and stoppers gave greater flexibility in the design of apparatus and liberated chemists from the tedious task of

luting joints that had to be air-tight. By the 1850s the handy and compact *Bunsen burner freed chemical laboratories from the dominance of the traditional furnaces and moved distillations and other processes requiring modest heat to the laboratory benches that now lined the working space in neat rows.

Meanwhile chemistry laboratories began to grow in size, not only because of the increasing variety and complexity of equipment, but also because of the growing opportunities for chemical careers. The pioneer in this enterprise, Justus *Liebig, expanded his laboratory in proportion to his success as a teacher and a leading experimentalist. By 1840 he was organizing the advanced students in his laboratory into small research groups that collaborated in investigating problems he judged specially significant. His example spread to other universities, so that by the 1870s chemical laboratories had become teaching-research institutes with multiple rooms, elaborate equipment, special professional living and working quarters, and several assistants to share the responsibility of teaching the next generation how to do chemistry.

Antoine-Laurent Lavoisier, *Elements of Chemistry*, trans. Robert Kerr (1965). Jon Eklund, *The Incompleat Chymist: Being an Essay on the Eighteenth Century Chemist in His Laboratory* (1975). Frank A. J. L. James, ed., *The Development of the Laboratory* (1989). Bruce T. Moran, *Chemical Pharmacy Enters the University: Johannes Hartmann and the Didactic Care of Chymiatria in the Early Seventeenth Century* (1991). Alan J. Rocke, *Nationalizing Science: Adolphe Wurtz and the Battle for French Chemistry* (2001).

FREDERIC LAWRENCE HOLMES

LABORATORY, INDUSTRIAL RESEARCH.

The industrial research laboratory first appeared in commercial enterprises around 1860. Previously, research related to industrial processes took place in universities, academies, and private laboratories. Sometimes inventors set up their own enterprises with which they associated their laboratories. But after 1860, companies began establishing in-house research facilities.

The industrial research laboratory as an organizational entity has several typical features. It is generally separated from production facilities and not subject to immediate demands from production or production managers. Industrial research is a continuous activity; scientific investigations and their extension to new products become routine operations of the firm. Industrial research also features cooperation of scientists from different disciplines, engineers, technicians, and craftsmen, working in an interdisciplinary manner. Since interdisciplinary activities of people with different backgrounds and tasks make coordination and hierarchy necessary, a coordinating institution at the top of the research structure is often created for the purpose—either a central research laboratory or a responsible member of a board of directors. Those conditions of routine activity, novelty, interdisciplinarity, and coordinated hierarchy confirm a company's engagement in industrial research. The same features characterize the work carried out in industrial research laboratories.

Beginning in the late 1850s a considerable number of chemists began working in industry for iron foundries. After Henry Bessemer had announced his revolutionary process for converting iron into steel in 1856, German foundries quickly implemented his invention and employed chemists to develop and refine it, and to look for new alloys. Afred Krupp in 1863 established what was probably Germany's first industrial research laboratory at his crucible steel foundry in Essen; and in 1883 he created a second laboratory with the special mission of finding new alloys, improving existing steels, and investigating the properties of iron and steel. About one hundred foundries around Sheffield in England also set up chemical testing sites by 1900. In the United States, the Pennsylvania Railroad Company established a chemical laboratory in 1876, probably becoming the first American enterprise to do so in pursuit of technological innovations.

The emerging organic-chemical and electrical industries gave rise to a great number of industrial research laboratories from the 1870s onwards. Following the discovery in 1856 of synthetic aniline dyes derived from coal tar by William Henry Perkin and August Wilhelm von Hofmann in London, many firms were established in Britain, France, Switzerland, and Germany to develop it. These firms, synthesizing great numbers of dyes, grew rapidly. By 1880, however, advancing scientific progress, increasingly complex production processes, lawsuits over patents, and fierce competition had driven many European companies out of business. Apart from a handful of British enterprises, some ten German firms continued to grow and provided 80 to 90 percent of the world production of synthetic dyes by 1913. All the surviving companies had developed particularly strong industrial research laboratories.

The firms first sought enduring connections with university researchers to supplement and later replace the scientific knowledge of the aging company founders. Moreover, the first all-German Patent Law of 1877 made copying of other companies' inventions and production processes illegal. Companies had now to file their own patents, check their claims of novelty, screen competitors' patents, and evaluate inventions and discoveries offered to them from university professors. All these were the duties of the newly established industrial research laboratories, which emerged between 1877 and 1886 in the leading German dye companies. With rapid scientific progress in new fields like pharmaceuticals, photographic chemicals, artificial fibers, and synthetic rubber, and with a continuous stream of young chemists leaving German universities, the biggest German chemical companies—BASF, Hoechst, and Bayer—each employed around 300 chemists in their large research laboratories and testing stations by 1912.

In the United States, DuPont de Nemours started the investment in industrial research by creating research laboratories from 1902 on, after it had been threatened with antitrust suits for misusing its market power in explosives. Here, industrial research, embodied in industrial research laboratories, had the function of deliberately broadening and widening the enterprise's activities.

In the electrical industry, after the introduction of the *telegraph in the 1840s and 1850s, a number of firms in the United States, Germany, Italy, and Britain, and some independent inventors, notably Thomas Edison, started to work on the *telephone, high voltage systems, and—in the 1870s—the incandescent light bulb. Rapid scientific and technical progress, competition for market domination influenced by national patent laws and corporate acquisitions, and, in the case of Siemens, the death of the company founder and main inventor (Ernst Werner von Siemens), prompted

Siemens and Halske (Germany), General Electric (United States), and Philips (Netherlands) to set up research laboratories around 1900. They were to safeguard a constant stream of marketable new inventions and technologies. Their main activity in the first years centered on the incandescent light and lighting technologies. But with the arrival of the *telephone, *X rays, and new theories in physics, the scope of the research laboratories in the electrical industries widened to include *vacuum tubes and x-ray machinery, and then, more generally, to encompass theoretical physics, physical chemistry, chemistry, acoustics, and mechanical technologies. Inspired by the success of others, the American Telegraph and Telephone Co. (AT&T) and Eastman Kodak (photographic equipment) also began to pursue industrial research in newly created laboratories. Despite different aims and business strategies, at the onset of World War I most large industrial enterprises in the United States, Great Britain, and Germany had departments for pursuing industrial research. Usually a centralized main laboratory directed several dispersed laboratories and testing stations while maintaining close long-term relationships with nearby universities.

Drawing on war profits, a return to peacetime conditions, and a postwar economic boom, the years from 1919 to 1926 saw unprecedented growth in the number of new industrial research laboratories and the expansion of existing ones. Also, as companies grew during the war years and mergers became common, huge industrial complexes with several thousands of workers and large research facilities were established. By 1925, AT&T employed about 3,600 staff members in their "Bell Laboratories," a company it set up for the sole purpose of pursuing industrial research. The same year DuPont had a research staff of around 300 in the same year, which rose to about 700 by 1930. In automobiles, General Motors created probably the largest infrastructure for research and development. In Europe, the biggest spender on research and development was I. G. Farben, a chemical trust stemming from the merger of six German dye companies (including Bayer, BASF, and Hoechst) in 1925; it employed about 1,100 research chemists in at least sixty laboratories by 1930. Imperial Chemical Industries (ICI) in Britain, which also resulted from a merger of six chemical companies, had a research staff of around 400 chemists. In other sectors such as textiles and branded goods, British companies saw great investment in industrial research in the interwar years. In the electrical industries, Siemens and Halske employed roughly 900

chemists, physicists, and engineers in more than seventy laboratories. By 1934 Philips had a corps of 370 researchers. In steel, Germany was the home of probably the biggest industrial enterprise of the interwar years, the Vereinigte Stahlwerke AG (United Steel); it combined the research efforts of several firms, with a research staff of at least 400 people.

During World War II, each belligerent nation made use of the knowledge, technologies, and manpower of giant industrial enterprises for their war efforts. In the United States, DuPont was heavily involved in the *Manhattan Project leading to the atomic bomb; in Britain, ICI also participated in a nuclear power project. In Germany, I. G. Farben supplied synthetic fuel and rubber for the war effort. All required substantial investment in science and technology to produce nuclear power, *radar, antibiotics, jet engines, and early computers. The Cold War resulted in a continuation of large research efforts by both government and industry. Many companies established new industrial research laboratories.

The organization and management of industrial research laboratories have changed over time. Their number, location, administration, and coordination altered frequently as companies grew, diversified, closed, or sold production lines, or merged with other firms. Some companies found it useful to centralize all research efforts in one big department for research and development; others continued to decentralize research along product lines.

A typical modern industrial research laboratory has a managing director, group leaders, team heads, and scientists. The director reports to the company's board. An experienced scientist supervises the laboratory work of interdisciplinary groups of scientists and engineers. A large and diversified company may maintain many testing and investigating stations organized along the companies' different lines. But only the centralized, main industrial research laboratory with direct access to a company's top management enjoys the status of the scientific and technological heart of the enterprise.

Companies have tried various means of financing their industrial research laboratories. Until World War II, the divisions at Siemens had to pay for scientific investigations and tests done by its main industrial research laboratory, while DuPont allowed their laboratories to spend substantial amounts before evaluating their output. In contrast, Bayer and BASF for many decades permitted their laboratories, which produced new inventions and discoveries regularly, to function without regular evaluation. Researchers in testing stations could not choose their projects, whereas outstanding scientists in the centralized main laboratory might enjoy great freedom in deciding their agendas. Sometimes, large enterprises distanced their creative scientists from the immediate concerns of manufacturing by founding new companies or divisions only for the purpose of conducting industrial research, as AT&T did with Bell Laboratories in 1925 and Vereinigte Stahlwerke AG did with a new subsidiary in 1926. IBM located research divisions in a foreign country (Zurich, Switzerland) to safeguard academic freedom and benefit from the stimulation of a famous nearby technical university. Some companies, like DuPont during the 1930s, allowed fundamental research in their laboratories but abandoned the initiative when it gave no greater returns than earlier approaches.

Despite secrecy's importance in industrial research, especially for patentable discoveries, industrial scientists need intellectual exchange with colleagues and ways to discuss their theories and inventions. Beginning in the 1920s, after decades of secrecy, many enterprises set up their own journals to publish recent results or allowed their researchers to publish freely in academic journals. This relaxation aided recruitment of young scientists since they did not have to abandon academic standards and ambitions on entering industrial research laboratories. They could thus enhance their own and their company's reputations, build a record for transfer to a university, and earn a chance at a *Nobel Prize. Winners include Irving Langmuir and Horst Störmer at Bell Laboratories in 1932 and 1998, respectively, and Gerhard Domagk at Bayer/IG Farben in 1939.

Michael Aaron Dennis, "Accounting for Research: New Histories of Corporate Laboratories and the Social History of American Science," *Social Studies of Science* 17 (1987): 479–518. David A. Hounshell, *Science and Corporate Strategy. DuPont R&D, 1902–1980* (1988). Ernst Homburg, "The Emergence of Research Laboratories in the Dyestuffs Industry, 1870–1900," *British Journal for the History of Science* 25 (1992): 91–111. David Edgerton and Sally Horrocks, "British Industrial Research and Development before 1945," *Economic History Review* 48 (1994): 213–238. Ulrich Marsch, "Strategies for Success: Research Organization in German Chemical Companies and IG Farben until 1936," *History and Technology* 12 (1994): 23–77. Paul Erker, "The Choice between Competition and Cooperation: Research and Development in the Electrical Industry in Germany and the Netherlands, 1920–1936," in *Innovations in the European Economy between the Wars*, ed. François Caron, Paul Erker, and Wolfram Fischer (1995): 229–254. David Edgerton, *Science, Technology and the British Industrial "Decline"* (1996). David A. Hounshell, "The Evolution of Industrial Research in the United States", in *Engines of Innovation. U.S. Industrial Research at the End of an Era*, ed. Richard S. Rosenbloom and William J. Spencer (1996): 13–86. Ulrich Marsch, *Zwischen Wissenschaft und Wirtschaft. Industrieforschung in Deutschland und Grossbritannien 1880–1936* (2000).

ULRICH MARSCH

LABORATORY, INTERNATIONAL. See MULTI-NATIONAL LABORATORIES; CERN.

LAMARCK, Jean-Baptiste De. See CUVIER, GEORGES, AND JEAN-BAPTISTE DE LAMARCK.

LAMARCKISM became a critical term in biology in the 1880s and 1890s within the debate over evolution theories that developed during the temporary eclipse of *Darwinism.

Given the time and context, Lamarckism, "neo-Lamarckism" (introduced by the American entomologist Alpheus Packard in 1884), and "neo-Lamarckian" (attached to the doctrines of some evolutionists active around 1900) had little to do with the theory of evolution advanced by Jean Baptiste de Lamarck almost a century earlier.

In the later context, Lamarckism designated above all the doctrines that characteristics acquired through use during the life of an organism could be transmitted to its offspring and that the environment played a direct role in the modification of living beings. Many regarded the "Lamarckian" inheritance of acquired characters as an appropriate complement, or even a necessary alternative, to Charles *Darwin's notion of natural selection, which seemed unable to sustain the broad evolutionary theory and worldview that Darwin himself had advocated. The experimental evidence brought to support the inheritance of acquired characters by authors like Charles-Édouard Brown-Séquard, Paul Kammerer, Ivan *Pavlov, and William McDougall, however, proved insubstantial, and the notion fell by the wayside after the severe criticism of August Weismann.

"Lamarckism" as used in biology today retains its late-nineteenth-century significance. It also evokes memories of the *Lysenko affair, the dramatic confrontation in the 1930s and 1940s between a Soviet version of the doctrine of the inheritance of acquired characters, supported by Stalin's head biologist and agriculturalist Trofim Denisovich Lysenko, and Western, neo-Darwinian geneticists.

In recent decades historians of biology have unearthed evidence revealing other important, but relatively unknown chapters in the history of Lamarckism. It is now agreed that Lamarck's own works and ideas circulated more widely than previously suspected in several European countries before the publication of Darwin's *Origin of Species* in 1859. Thus, Lamarck's important legacy, and

the controversial reception accorded to it in scientific and popular circles from the 1830s through
the1850s, merged in depth with the assessments of
Darwin's own work. Darwin engaged in frequent
dialogue with Lamarck's views and concepts
while refining his own theory and presentation
strategies, and subscribed to the then-popular
notion of use-inheritance. Many who joined in
support of evolution after 1859, including prominent defenders of Darwin's like Thomas Henry
Huxley and Ernst Haeckel, and successful propagandists of evolutionism like Herbert Spencer
endorsed Lamarckian, rather than properly Darwinian views. The histories of Lamarckism and
Darwinism appear now more closely intertwined
than was allowed by mid-twentieth-century biologists in the wake of the triumphs of the neo-
Darwinian theory of evolution.

Peter J. Bowler, *The Eclipse of Darwinism* (1983).
Mario A. Di Gregorio, *T. H. Huxley's Place in Natural
Science* (1984).

GIULIANO PANCALDI

LAPLACE, Pierre-Simon de (1749–1827), mathematician, astronomer, and physicist.

Known as the "Newton of France," Pierre-
Simon de Laplace transformed the study of
mathematical astronomy with his five-volume
Traité de mécanique céleste (1799–1825). Using the
language of calculus and differential equations,
he expressed Newton's *Principia* in its modern
language, solved many of the issues left open by
his predecessor, and demonstrated the stability of
the solar system. At the time of his death, almost
exactly a century after Isaac *Newton, Laplace
had won every honor available. He was dean of
the Académie des Sciences, decorated with titles
and medals, a member of the Académie Française,
and a marquis serving in the Chamber of Peers.

Laplace began life as the second of three sons
of an average Norman farmer and later mayor
of Beaumont-en-Auge. Destined for the clergy,
he attended a local Benedictine school and then
the nearby University of Caen, where he found
his vocation as a mathematically adept natural
philosopher. As a student, he took a special interest in current discussions over the validity of the
Newtonian system of the world. During his stay
at Caen, he rejected traditional Catholic theology
and remained opposed to it for the rest of his life.

At the age of twenty, Laplace set out for
Paris with a recommendation to Jean Le Rond
d'Alembert, who sponsored him for a post as a
teacher of mathematics at the École Militaire in
Paris. The young instructor submitted a battery
of original papers on differential equations and
astronomy to the Paris Academy, where he gained
election in 1773. He rose rapidly in its ranks to

a senior position. Laplace participated actively in
the academy's transformation during the French
Revolution, when it became part of the Institut
de France, and he assumed a preponderant role
in all its affairs. During the Revolution, he
taught mathematics for a few months at the
École Normale de l'an III and served on the
board of directors of the École Polytechnique.
Laplace was a central figure in the creation of the
French Bureau of Longitude, whose activities he
dominated for thirty years. Although he served
briefly as minister of interior under Bonaparte
and as vice chancellor of the emperor's Senate,
his political activities were unremarkable.

Laplace's philosophical conviction in favor
of determinism led him to explore probability
theory and to generate its classical synthesis in
the *Théorie analytique des probabilités* (1812). His
most notable innovation was the establishment of
Bayesian inductive probabilities for calculating
the likelihood that a particular explanatory
hypothesis is the cause for a set of known
phenomena. He applied such a calculation in
proposing the nebular hypothesis for the origins
of the solar system that held sway for much of
the nineteenth century. Starting in 1781, Laplace
collaborated with Antoine-Laurent *Lavoisier
on several problems, notably the behavior and
nature of heat, for which he invented an ice
calorimeter to measure heat transfers, and the
systematization of the concepts and nomenclature
of chemistry. Lavoisier inspired their colleague
Claude Louis Berthollet to write an *Essai de statique
chimique* (1803) explaining chemical phenomena
as consequences of Newtonian laws of attraction.
These were the laws used by Laplace to explain
the phenomena of capillarity, optical refraction,
the cohesion of solids, and tidal motions. He
expected that all aspects of the inorganic world
could be elucidated in terms of Newton's laws.

With his neighbor Berthollet, Laplace organized a gathering of young scientists in 1806
named the Société d'Arcueil to encourage them to
pursue research in their Newtonian style. Many
of these disciples became the leaders of French
science in the next generation. Jean-Baptiste Biot
and Simeon-Denis Poisson furthered his program productively, but others, like François
Arago, Étienne Louis Malus, and Joseph Louis
Gay-Lussac, who attended the Arcueil meetings,
struck out on their own.

Laplace wrote two popular works without
equations that have been translated and reprinted
often. During the Terror, he had time for an
Exposition du système du monde (1796), which
contains a chapter on the history of astronomy
since antiquity. In 1814, he published his *Essai
philosophique sur les probabilités*, developed from

the introduction to his earlier mathematical treatise. It was here that he inserted his famous assertion about the determinism of the universe.

Maurice P. Crosland, *The Society of Arcueil* (1967). Roger Hahn, *Laplace as a Newtonian Scientist* (1967). Charles C. Gillispie, *Pierre-Simon Laplace, 1749–1827. A Life in Exact Science* (1997).

ROGER HAHN

LASER AND MASER. MASERs (microwave amplification by stimulated emission of radiation) produce coherent microwave radiation with very low noise. LASERs (light amplification by stimulated emission of radiation) emit infrared-visible radiation of very high coherence. Although the idea of using stimulated emission to produce coherent radiation dates from the 1930s, the first maser, built by Charles Hard Townes at Columbia University, did not materialize until 1954. The basic idea was that if more particles of a medium can be placed in an excited rather than in an unexcited state, then a radiation interacting with the particles is amplified rather than absorbed because stimulated emission prevails over the absorption. The presence of a resonant cavity enhances the interaction of the particles with the radiation and allows the formation of an electromagnetic vibratory mode.

In his first maser, Townes used a microwave transition at 23.870 GHz (= 1.25 cm) in ammonia molecules. He employed a microwave cavity with an inverted population obtained by using a quadrupole electric field to separate the excited ammonia molecules from the unexcited ammonia molecules in a beam emerging from an oven. The excited ammonia molecules enter a cavity capable of resonating at around 1.25 cm. Here they may amplify a low signal at the resonant frequency or, if sufficiently plentiful, may generate a signal. Nicolai Gennadievich Basov and Aleksandr Mikhailovich Prokhorov in Moscow arrived at the same idea as Townes at about the same time. They wanted to improve the characteristics of microwave spectroscopes. They had noticed that a medium with an inverted population had a stronger signal to noise ratio than a normal medium.

The first ammonia maser produced very monochromatic radiation with extremely low noise. In 1956, Nicolaas Bloembergen pointed out that solid state masers might be tuned and suggested a system whose energy levels depend on the strength of an imposed magnetic field, so as to permit tuning over a considerable frequency range. The system was soon put into operation by H. E. Derrick Scovil, George Feher, and Harold Seidel at Bell Telephone Laboratories.

Although the maser represented the first application of stimulated emission, and a revolutionary way to consider microwave sources, it did not have a dramatic technological application because less expensive, smaller, and simpler semiconductor devices also could emit or amplify microwave radiation with very low noise. However, Arno Penzias and Robert Wilson used a maser when they discovered the black-body cosmic background (*see* BIG BANG).

Lasers extend the use of stimulated emission into the visible and infrared. Townes and Arthur Schawlow gave the first thorough discussion of their operation and properties in 1958. Two years later Theodore Harold Maiman developed the first working laser by placing a rod of ruby in the center of a helical flash lamp. The two end faces of the rod were parallel to each other and silvered so as to create an optical cavity. Pulsed radiation at 6943 Å (0.69 μm) traveling along the axis of the rod was obtained by pumping with light from the flash lamp.

Soon Ali Javan and his associates at Bell Labs developed the first continuous gas laser utilizing a mixture of helium and neon. They obtained several inverted levels in neon by collision with helium atoms excited in a radiofrequency discharge. This laser emitted continuously at 3.95 μm, 1.1 μm, or 0.63 μm. It appeared that almost every substance can lase (serve as the medium of a laser). Helium-neon and ruby lasers were joined by neodymium lasers, which emit usually pulsed radiation of moderate coherence and high power at 1.06 μm, tunable lasers operable both continuously or with very short pulses, and tiny diode lasers stimulated directly by an electric current. The market now offers a hundred different lasers, some of which emit pulses as short as a few femtoseconds.

Even before lasers became practical, their possible applications were discussed. The military dreamed of secure communication channels, radar with great resolution, death rays to destroy enemy targets, and so on. Many applications of lasers have been found in *medicine, *chemistry, and *spectroscopy. Today, lasers read and write compact discs, control mechanical processing and measurements, and send signals down optical fibers. Nonlinear optics was born with lasers. And in physics itself, lasers have been used to demonstrate Bose-Einstein condensation, the movement of molecules during chemical reactions, and other exotic phenomena. Lasers also had a profound influence on *quantum mechanics with the development of quantum optics and the creation of coherent and nonclassical states.

Mario Bertolotti, *Maser and Lasers: An Historical Approach* (1983). Joan Lisa Bromberg, *The Laser in America* (1991).

MARIO BERTOLOTTI

LATIN AMERICA. The European invasion of the Americas produced a knowledge holocaust. We know little or nothing about the local institutions and artifacts that were destroyed. We eat tomatoes, corn, beans, and potatoes without understanding how they were developed. We have limited knowledge about the ways in which the indigenous peoples of what is now referred to as Latin America came to have zero, decimal positional numerical systems, books, accurate calendars, and an accounting for the motions of Venus. Although the pre-Columbian world possessed a rich network of knowledge-producing institutions, it would be anachronistic to call this knowledge science in the modern sense. The conquest, which brought with it the Inquisition and later legalized slavery, also brought conquerors who rejected "unscientific" knowledge.

The colonial period (roughly 1500 to 1800) saw some investigations of natural knowledge and some relevant institutional development. Gonzalo Fernández de Oviedo wrote his *Historia Natural y General de las Indias* describing the flora, fauna, and peoples known to the Spanish at that time. The Spanish founded institutions such as Cordoba University in Argentina in 1613. (The Portuguese did not create institutions at this level.) The region continued to draw European naturalists. Pehr Löfling, a disciple of Linnaeus, described the Orinoco flora. José Celestino Bruno Mutis, a Spaniard established in Colombia, continued Löfling's work. Mutis mobilized intellectuals in Nueva Granada in an Expedición Botanica. His disciples, Clement Ruiz Pabon and Juan José d'Elhuyar, visited Sweden in 1781.

José Bonifácio de Andrada e Silva, a Brazilian elected to the Portuguese Academy of Sciences, worked in Europe from 1790 to 1799. Andrada e Silva carried Antoine-Laurent *Lavoisier's ideas to Scandinavia. There he collected some 3500 minerals and fossils, which today are housed in the Brazilian Imperial Collection, Petrópolis. For a while he held the chair of metallurgy at Coimbra in Portugal. After Brazil became independent from Portugal in 1821, Andrada e Silva struggled to end slavery. Failing, he moved to France in 1823 and returned to science. Like Andrade e Silva, the naturalist Alexander von *Humboldt had an interest in mineralogy. He stayed in Caracas in 1799 and 1800 and collected materials in Cuba, Mexico, Bolivia, Colombia, Ecuador, and Peru. He expanded the horizons of European scholars with his descriptions of the New World, in which he did not always fully acknowledge Mutis's contributions.

In the half-century following the independence of Latin American nations from Spain and Portugal (roughly 1820 to 1870), several important Europeans and North American naturalists visited the region. Alfred Russel Wallace went to the Amazon; Charles *Darwin visited many parts of South America including Argentina and the Galapagos Islands; Eugenio Warming and Peter Wilhelm Lund carried out biological and paleontological studies in Lagoa Santa, central Brazil. Some of the stream of visitors stayed and contributed to the scientific and institutional growth of Latin America. An outstanding example, Fritz Müller, a Ph.D. from the University of Berlin and an M.D. from the University of Greifswald, moved to Blumenau in Brazil in 1852. He corresponded with Louis Agassiz, Max Schultze, Ernst Haeckel, and Charles Darwin. His book *Für Darwin* appeared in Leipzig in 1864 and, in an augmented English version, *Facts and Arguments for Darwin*, in London in 1869. His work on the evolution of Brazilian crustaceans provided solid support for

Aerial view of the Laboratório Nacional de Luz Síncroton, in Campinas, Brazil.

Darwin's theory and won Müller an honorary doctorate from the University of Bonn. Nearly all of Müller's 248 published scientific works were written in Brazil.

While positivism did not receive support from European natural scientists, it had a profound impact in Latin America (*see* Positivism and Scientism). Auguste *Comte, positivism's founder, maintained that humanity evolved according to a law of the three stages. The most primitive of these, the theological stage, was characterized by a belief in supernatural forces. Positivists regarded Catholic Latin American countries as theological states dependent intellectually on Europe. In Argentina, President Domingo Faustino Sarmiento, who greatly contributed to the country's development, believed—under the inspiration of Comte's theory of stages modified by a Romantic evolutionary theory—that civilization was urban and European. The doctrine made it difficult for Argentinian scientists to obtain local patronage. The director of the Buenos Aires Natural History Museum, zoologist Hermann C. C. Burmeister, refused to support Florentino Ameghino, a local genius whose biological and paleontological work was known in Europe. Europeans and North Americans occupied many important posts; Benjamin A. Gould at the Cordoba Astronomical Observatory and Emil Bose at the La Plata physics institute are examples.

In Brazil, Emperor Pedro II promoted science and letters. Nevertheless, the establishment in 1889 of a positivist republic by the military left the country without universities. Military schools taught from positivist writings and fostered applied research. The Campinas Agricultural Institute, the Butantã Biological Institute, and the Institute for Technological Research (IPT) exemplify the sorts of institutions then founded. The IPT, a test center for industry and civil engineering, was founded in 1899 by Antonio Francisco Paula Souza, one of the first ministers of the Brazilian positivist republic. Medical science made outstanding advances. Oswaldo Cruz, founder of Brazilian experimental medicine, eradicated yellow fever, bubonic plague, and smallpox from Rio de Janeiro despite positivist opposition. One of the major biomedical research institutions of Latin America, the Oswaldo Cruz Foundation, bears his name.

After 1914, the influence of positivism declined throughout Latin America. This decline had many benefits. Distorting Comte's work and accepting it as absolute truth, the Latin American positivists not only had opposed the teaching of most post-Newtonian physics, including the concept of electromagnetic *field and non-Euclidean geometry, but also governmental engagement in the institutional development of science and education, including the creation of universities. Their position in technical and military schools made it all the more difficult for scientific advance.

As a new way of thinking emerged, Argentine science flourished. With a per capita income greater than that of Italy or Sweden, Argentina attracted the European powers. Both the French and the Spanish promoted their own institutes and lectures in Argentina. The Germans supported physics at Emil Bose's La Plata physics institute. Bose hired Einstein's former assistant Johann Laub, the engineer Konrad Simons, and some brilliant Argentineans. Richard Gans, who had studied with Ferdinand Braun at Strasbourg, succeeded Bose in 1911. Niels *Bohr noticed Gans's application of H. A. Lorentz's theory of magnetic fields to free electrons. Between 1925 and 1945 Gans taught in Germany, which gave Argentineans, including Gans's student Enrique Gaviola, who received his doctorate from Berlin in 1926, opportunities to study there. Unfortunately this flowering of Argentinean science did not last.

In Brazil, the antipositivist reaction resulted in the foundation of the Brazilian Academy of Sciences (ABC) in May 1916, supported by the Associação Brasileira de Educação (Brazilian Association for Education) and the Instituto Franco-Brasileiro de Alta Cultura (French-Brazilian Institute for Advanced Studies). ABC had its own periodical, renamed in 1929 the *Anais da Academia Brasileira de Ciências*. Its leaders included naval officer Álvaro Alberto da Mota e Silva, its president from 1935 to 1937. The opposition to positivism became public during Einstein's visit in 1925 on his way to Argentina. At Sobral, in the northeast of Brazil, the expedition coordinated by Arthur Eddington photographed the eclipse of 1919, validating the general theory of *relativity. The Brazilian positivists had opposed Einstein's ideas in newspaper articles but suffered ridicule after the eclipse findings became known.

Brazilian and American agriculture prospered from international connections. In 1925 Iwar Beckman from Sweden developed the wheat hybrid Frontana at the Alfredo Chaves experimental station in Vereanópolis, Rio Grande do Sul. This variety saved North American wheat growers after the problems caused by leaf rust in the 1940s and 1950s.

In the 1930s, after a failed revolution against the central government, the São Paulo elite created the Universidade de São Paulo (USP) as an instrument for social and economic development. A number of European researchers, including Fernand Braudel and Claude Lévi-Strauss, became USP professors. Physicist Gleb Wataghin, a Ph.D.

from Turin University, also joined USP. He discovered *cosmic-ray showers and developed the theory of the multiple production of cosmic rays. He trained a number of students and his group actively cooperated with Arthur Holly Compton during his cosmic-ray expedition in the 1930s.

Mexicans also participated in Compton's project. Manuel Sandoval Vallarta, who graduated from MIT in 1921 and joined its faculty, became interested in Compton's discovery of charged particles in cosmic rays. Vallarta demonstrated experimentally Oliver Heaviside's formulas for the electromagnetic propagation in conductors as well as his operational calculus.

Vallarta and his colleague at MIT, Georges Lemaître, followed up Comptons's experimental results on the effect of the earth's magnetic field on cosmic rays. Lemaître had suggested that the universe originated from one primitive atom or "cosmic egg." His work with Vallarta, which concluded that charged particles made one component of primary cosmic rays, supported the cosmic-egg theory. In 1943 Vallarta became professor of physics at the National Autonomous University of Mexico (UNAM). He had a profound influence on Mexican physics. His students include Alfredo Baños, Carlos Graef Fernandez, Luis Enrique Erro, and Marcos Moshinsky. Outside Mexico, Vallarta's most famous student was Richard *Feynman.

Between 1945 and 1990, science and the state had a love-hate relationship in Latin America. On several occasions, states exiled their brightest while trying to foster science that might aid development. Under the regime of President Juan Perón, Argentina missed many opportunities to build. Immediately after World War II, Gaviola planned an atomic research institute and invited Werner *Heisenberg, who accepted but never came. The international press, fearing atomic weapons in the hands of a totalitarian regime, attacked the plan. When the United States announced at the United Nations Atomic Energy Commission (UNAEC) (see INTERNATIONAL ORGANIZATIONS) that it was ready to retaliate against what it considered "illegal" atomic programs, and a U.S. Air Force B-29 squadron flew over Montevideo, the Perón regime dropped Gaviola's project.

This incident, combined with conflicts between Perón and the universities, made the Richter adventure possible. Austrian chemist Ronald Richter arrived in Argentina in August 1948 and convinced Perón to build laboratories at Bariloche. In March 1951, Perón announced that Richter had discovered how to control nuclear fusion. Ridiculed by the scientific community, Perón ended the project in November 1952.

The nuclear field, under the leadership of Gaviola, Jorge Sabato, and José Antonio Balseiro, flourished. The Atucha 1 nuclear reactor went critical in 1967. Argentina has sold nuclear knowledge to Peru, Algeria, Turkey, Iran, and Australia.

The sort of support given nuclear physics did not recur in other areas of knowledge. Normally scientists and professors worked under pressure and even persecution. Expelled from his university, the Argentinian physiologist Bernardo Alberto Houssay went on to receive a *Nobel Prize in 1947. By the mid-1970s, the universities were completely dysfunctional, 547,000 Argentineans had left the country, and an unknown number of students had disappeared.

In Brazil, constant tension existed between the government and the scientists even though both sides believed that science should contribute to development. Gradually, thanks to the Brazilian National Research Council (CNPq), research centers such as the Brazilian Center for Physics Research (CBPF) were founded according to a plan outlined at the ABC meetings in 1945. Alvaro Alberto described the requirements for the atomic age: research, mineral prospecting, and proper industrial facilities for isotope production and assembly of instruments. The ABC requested that the government invest in research centers, send Brazilians abroad for training, and invite foreign scientists to visit Brazil. Nuclear relations between Brazil and the United States turned on Brazil's possession of monazite, a major source of thorium, cerium, and other rare earth metals and compounds. Brazil wanted to exchange monazite for nuclear technology. The United States responded with a systematic smuggling of the ore.

Getúlio Vargas, elected president in 1951, supported the Brazilian program of science for development. Like President Truman, Vargas considered nuclear energy an important component of industrialization. Nevertheless, the United States opposed the Brazilian program. Brazil started an independent program that included a secret agreement with occupied Germany. It collapsed in 1955, but the institutions it created became the basis of Brazilian scientific growth. In 1975, a new agreement was reached between Brazil and Germany, but Germany failed at transferring nuclear technology to Brazil.

After 1955 international cooperation in physics grew. One example is the partnership between Nordita, the Nordic Institute for Theoretical Physics headed by Bohr, and the Porto Alegre Physics Institute. In the late 1950s and early 1960s, Theodor A. J. Maris and Gerhard Jacob (president of CNPq in the 1990s) visited Nordita. Their leadership transformed Porto Alegre into a center for research on the Mössbauer effect and gamma ray correlation, two major research methods in nuclear physics. The application of these methods to solid state physics catapulted graduates from Porto Alegre into leading positions elsewhere. Other Latin Americans visited Nordita, for instance Argentinean solid-state physicist Leo Falicov, a link between Latin America and Berkeley. His best-known Porto Alegre student, Cylon E. T. Gonçalves da Silva, founded the Brazilian synchrocyclotron accelerator at Campinas.

In Mexico, President Aleman heeded the warning of Vallarta, Graef Fernandez, and others. In August 1945 he made science-for-development one of his main policies. In 1949, Vallarta moved to Mexico as full professor. Thanks to many like him, Mexico developed a unique nuclear policy that reflected views popular at the time. Many Mexicans believed that the United States's atomic monopoly increased injustice, poverty, and violation of rights. This perception led to the Tlatelolco Treaty of 1994, which renounced all military nuclear programs. Science-for-development became the enduring center of Mexican research, notably in agriculture. Supported by the Rockefeller Foundation and the Mexican Ministry of Agriculture, CIMMYT, the International Maize and Wheat Improvement Center, produced high-yield corn varieties. The Green Revolution, despite downplaying local varieties and increasing petroleum dependency, did achieve its goal of increasing global food production. Institutions modeled after CIMMYT were created in other parts of the world.

The difficulties of practicing science on the periphery may be indicated by the nominations of Latin American scientists for Nobel Prizes. In the 1920s and 1930s the Brazilian Carlos Chagas was nominated four times for the prize in medicine. In 1909 he discovered Tripanossomiase americana, identifying the disease parasite and vector. Known today as Chagas disease, it plagues over twenty million persons in at least eighteen countries. International recognition of his achievement would have translated into visibility and power at home. Envious, Afrânio Peixoto, rector of Rio de Janeiro University, mobilized attacks on Chagas. At the National Academy of Medicine, Chagas faced criticism from influential persons, and his work was not incorporated into medical education. In 1912, he received the Schaudinn Prize, awarded by the Hamburg Institute for Tropical Diseases. Yet when the Nobel Prize secretariat approached Brazilian authorities about the nomination, neither the medical community nor the politicians recommended Chagas. Other potential

Brazilian Nobel candidates were Leite Lopes and Cesar Lattes (see LATTES, CESAR, AND JOSE LEITE LOPES).

In 1947, Bernardo Houssay was awarded the Nobel Prize for medicine (along with Carl Ferdinand and Gerty Cori). As a student, Houssay had developed a method to investigate pituitary gland hormones. He became professor of physiology in 1909, a year before receiving his degree with an award-winning dissertation. In 1943 the Peronists, in a crackdown of potential opposition, evicted him from University of Buenos Aires. Supported by the Rockefeller Foundation and a local entrepreneur, Juan Bautista Sauberan, Houssay founded the Institute of Experimental Biology and Medicine. He continued his research on the pituitary, insulin, and diabetes; trained over 250 doctoral students; and published over two thousand scientific papers.

In 1970, Luis Federico Leloir from the Institute for Biochemical Research, Buenos Aires, received a Nobel Prize for his research on sugar nucleotides. Leloir took his M.D. in Buenos Aires in 1932 and worked in Houssay's institute until interrupted by Peronism. He moved to the Corises' biochemical laboratory at Washington University and then to David Green's experimental medicine institute at Columbia. On his return to Argentina, Leloir, like Houssay before him, founded his own institution, the Fundación Campomar (1947). The award of the Nobel Prize sheltered him from the vicissitudes of politics and military persecution. Leloir is a role model for Argentinean scientists.

In 1984, Niels Jerne, Georges Köhler, and Cesar Milstein shared the Nobel Prize for medicine or Physiology for their work on the immune system and the production of monoclonal antibodies. Born at Bahia Blanca, Argentina, Milstein was a graduate in chemistry from the University of Buenos Aires. In 1952, he completed a Ph.D. on aldehyde dehydrogenase at the Institute of Biological Chemistry. With a fellowship from the British Council, he received a second doctorate for work on enzyme active sites with Frederick Sanger at Cambridge in 1960. In 1961, Milstein became the head of the Division of Molecular Biology at the National Institute of Microbiology, Buenos Aires. In 1963 he resigned when the military began to persecute intellectuals and scientists. Milstein returned to Cambridge and to a Nobel Prize trajectory.

As can be seen from these cases, the lack of recognition for Latin American science can have as much to do with internal factors as with external ones. Historians of science, most but not all of them from Latin America, are trying to record the rich history of which only a few examples from Mexico, Brazil, and Argentina have been presented here. Further information about these and many other cases may be found in *QUIPU*, the main Latin American journal for the history of science, founded in 1984.

Nancy Stepan, *Beginnings of Brazilian Science: Oswaldo Cruz, Medical Research and Policy, 1890–1920* (1976). Simon Schwartzman, *Formação da Comunidade Científica no Brasil* (1979). Mario Mariscotti, *El Secreto Atomico de Huemul. Cronica del Origen de la Energia Atomica en la Argentina* (1985). Lewis Pyenson, *Cultural Imperialism and Exact Sciences: German Expansion Overseas, 1900–1930* (1985). Regis Cabral, "The Latin American Nuclear Debate," *Science Studies* 4 (1991): 53–60. Regis Cabral, *Knowledge Flows Between Scandinavia and Ibero America: A Network Approach* (1992). Regis Cabral, "Biotechnology, Wheat Production, and the Brazilian Company for Agricultural and Livestock Research (EMBRAPA), 1970–1990," *Science and Public Policy* 21 (1994): 147–156. J. J. Saldaña, *Historia Social de la Ciencia en America Latina* (1996). Hebe Vessuri, "Bitter Harvest: The Growth of a Scientific Community in Argentina," in *Scientific Communities in the Developing World*, ed. Jacques Gaillard, V. V. Krishna, and Roland Waast (1997): 307–335.

REGIS CABRAL

LATTES, Cesar (b. 1924), physicist, codiscoverer of the pi meson, and **Jose LEITE LOPES** (b. 1918), physicist, pioneer of "science for development," who predicted the Z° boson.

Brazilians place Cesar Lattes and Jose Leite Lopes among the builders of the twentieth century. Born in Recife, Leite Lopes studied at the Recife Chemistry School until he shifted to physics and specialized at the University of São Paulo (USP) in 1944. In 1946 he completed his Ph.D. at Princeton, working with Wolfgang Pauli and writing articles on meson theory and electron radiation. Leite Lopes became professor of theoretical physics in Rio de Janeiro, a career that became unsafe after the military came to power in 1964. The military suspected him for his views on development and forbade him from appearing in public. Under these circumstances, Leite Lopes accepted a position at the Orsay Faculty of Sciences, Paris, returning to Brazil in 1967. In April 1969, he and other researchers who favored an independent Brazilian nuclear program were banned from university activities. In September 1969, Leite Lopes left Brazil for Carnegie-Mellon University, Pittsburgh, moving from there to the Université Louis Pasteur, Strasbourg. He returned to Brazil in the 1980s.

In 1949 Leite Lopes cofounded the Centro Brasileiro de Pesquisas Físicas (CBPF), the Brazilian Center for Physical Research, which he directed from 1960 to 1964. He coordinated the center's theoretical work, while Lattes oversaw its experimental research. Until 1954, CBPF was a cornerstone in the efforts of Brazilian President Getúlio Vargas and National Research Council President Alvaro Alberto to apply nuclear science to development. A combination of American and Brazilian conservative forces stopped the plan. With Juan José Giambiagi (Argentina), Marcos Moshinsky (Mexico), and Lattes, Leite Lopes founded the Escola Latinoamericana de Física (LASP, the Latin American School for Physics) in 1959. LASP has been fundamental in improving the quality of Latin American research.

In 1958, Leite Lopes published his important paper on the unification of electromagnetic and weak forces, in which he predicted the boson Z°. The boson Z° is a subatomic particle with integral spin and mass 91 GeV. It is the carrier of the weak, but fundamental, force of nature. Confirmation of the prediction at *CERN by a team led by Carlo Rubbia and Simon Van der Meer brought them the *Nobel Prize. For his part Leite Lopes received the 1993 Mexican Prize for Science. It is widely believed in Latin America that he should have shared the Nobel Prize with Rubia and Van der Meer.

Leite Lopes originated much of the accepted wisdom on science for development. His *Ciência e Desenvolvimento* (*Science and Development*, 1964) shaped a generation of researchers and practitioners. It represented science as a liberating force, a necessary component of development, sovereignty, and democracy. A liberated Latin America required investment in science and in new generations of scientific workers. France recognized Leite Lopes for his contributions to science and science policy with the Academic Palm and the National Order of Merit (1989). In 1999 he received the UNESCO Science Prize for his application of science to development.

Lattes, born in Curitiba, completed the physics course at USP in 1943, working with Gleb Wataghin and G. P. S. Occhialini on *cosmic rays. After the war, he joined Occhialini and Cecil Powell, who were trying to detect in photographic emulsions particles produced by Powell's cyclotron at the University of Bristol. Lattes suggested that they expose the plates to *cosmic rays. Occhialini took plates prepared by Lattes to the Pyrenees, and Lattes exposed some himself at the Chacaltaya meteorological station in the Bolivian Andes. Lattes's addition of boron to the emulsions made possible the detection of the pi meson (predicted by Hideki *Yukawa in 1934) and demonstrated its decay into a mu meson and an electron. Powell received the Nobel Prize of 1950 for the discovery, an award that Latin Americans feel should have gone to Lattes. A letter from Niels *Bohr, to be opened in 2012, supposedly explains Lattes's exclusion.

In 1948, Lattes, still interested in the production of mesons by accelerators, visited the Radiation Laboratory (now the Lawrence Berkeley Laboratory) of the University of California at Berkeley. Working with Eugene Gardner, he discovered that the 470-centimeter (184-inch) cyclotron had been producing mesons for over a year, which Gardner had not been able to detect. Within two weeks, Gardner and Lattes could announce their results. With prestige unprecedented for a Brazilian physicist, Lattes returned to cofound CBPF and the Brazilian National Research Council. His work transformed the Chacaltaya station into an international center for cosmic-ray research. Lattes also organized high-energy research at USP and, in 1962, at Campinas University (UNICAMP). Today, UNICAMP's Gleb Wataghin Physics Institute, which ranks among Latin America's best, has a worldwide reputation.

Lattes was elected to the Brazilian Academy of Sciences and to the physics societies of Brazil, the United States, Germany, Italy, and Japan, among others. He received Brazil's Einstein Prize and Santos Dumont Medal. The Organization of American States and the Third World Academy of Sciences also honored him. Hundreds of streets and libraries have been named after him, a popular recognition of his efforts for human development.

Lattes and Leite Lopes represent the best in Latin American science and serve as role models for scientists. Both have contributed greatly to science for development and participated in the making of institutions in Brazil and *Latin America. The fact that others received Nobel Prizes based in part on their work epitomizes a central issue in development: knowledge producers from developing countries are rarely accepted on equal terms by their European and North American counterparts. Both Lattes and Leite Lopes have pointed out that, as long as this continues, scientists will be as much to blame for underdevelopment as financial institutions and power grabbers.

Jose Leite Lopes, *Ciência e Desenvolvimento* (1964). Jose Leite Lopes and Michel Paty, eds., *Quantum Mechanics, Half a Century Later* (1977). Simon Schwartzman, *Formação da Comunidade Científica no Brasil* (1979). Regis Cabral, *The Brazilian Nuclear Debate, 1945–1955* (1994). Ana Maria Ribeiro de Andrade, *Físicos, Mésons e Política: A Dinâmica da Ciência na Sociedade* (1998). A. M. R. Andrade, *Cesar Lattes, a Descoberta do Méson Pi e Outras Histórias* (1999).

REGIS CABRAL

LAVOISIER, Antoine-Laurent (1743–1794), founder of modern chemistry, discoverer of oxygen's role in chemical reactions, and public servant.

Born in 1743 in Paris to Jean-Antoine Lavoisier, solicitor to the Parlement of Paris, and to Émilie Punctis, the daughter of a wealthy attorney, Lavoisier was raised, after the early death of his mother, by a maiden aunt. His family life was warm and affectionate. He enrolled in 1754 at the Collège Mazarin, where he received an outstanding education that included both classical and literary training, science, and mathematics. Although he finished his education in the Faculty of Law in 1763, he quickly acquired such a consuming interest in the natural sciences that he never practiced law. With Jean-Étienne Guettard, who was planning a geological and mineralogical atlas of France, Lavoisier undertook several extensive expeditions through various regions of France, mapping, examining rocks, and observing the ordering of strata. In describing these formations, Lavoisier introduced the innovative practice of drawing vertical cross sections to represent the stratigraphical order. Recognizing the close relationship between field mineralogy and the chemical analysis of minerals, Lavoisier attended the popular chemical lectures of Guillaume-François Rouelle, and set up a laboratory in his own home.

Talented and ambitious, Lavoisier undertook various projects that he hoped would win him enough recognition that, with the help of his father's connections, would gain him election to the Academy of Sciences. He attained that goal in 1768 after presenting a paper on an improved hydrometer with which he claimed he would reform the analysis of mineral water. To secure an assured income, Lavoisier at the same time joined the Ferme Génerale, a private consortium that collected taxes and customs duties for the French government. Although this position, whose responsibilities he took seriously, caused him to spend much time away from Paris on inspection trips, it also enabled him to afford the expensive instruments and apparatus that his innovative programs of experimental investigation required. In 1771, he married Marie-Anne-Pierrette Paulze.

During 1772, Lavoisier became interested in a problem then attracting attention among French chemists: the processes by which air is fixed in or released from other bodies. In the fall of that year, he discovered that phosphorus and sulfur gain weight when they are burned and showed that a lead ore reduced in the presence of charcoal released an air. Convinced that this last discovery was "one of the most interesting . . . since the time of Stahl," he decided to read everything published on airs from the time of Stephen Hales in 1728 and Joseph Black in 1756 through the identification of several new "airs" by Joseph Priestley. Lavoisier

An experiment in Lavoisier's laboratory drawn by his wife (who sits recording the results). The apparatus, more elaborate than typical for the time, collects gases respired by Lavoisier's seated assistant for later analysis.

judged that this work had set the stage for a "revolution" in physics and in chemistry.

His own early experiments in the field were only marginally successful. His first publicly announced theory, that Black's "fixed air" was absorbed in combustion and the calcination of metals and released in the reduction of metallic calces, collapsed during his effort to support it experimentally; but in the process, he devised novel techniques for measuring the quantities of materials—the airs as well as solids and liquids—that took part in a chemical "operation," and to reason about composition in the terms that have become known as his "balance sheet" method. His growing mastery of these techniques and his persistence in overcoming the many pitfalls arising in their use, rather than his advocacy of the generally accepted principle that matter is neither created nor destroyed in chemical changes, most distinguished Lavoisier from his contemporaries. These talents gave him the critical foundation for theoretical reasoning that enabled him to surpass other chemists of his time and impose his views upon them.

By 1777, aided in part by Priestley's discovery that the air released by a mercury calx reduced without charcoal is "better" than common air, Lavoisier had identified the substance that takes part in the changes he had been studying as a particular portion of the atmosphere that he called at first "eminently respirable air," because only this portion supported the life of animals and humans. Because he had also found that several common acids contain the "base" of this air, he renamed the base the "acidifying principle," or "oxygen." By this time, he could also conclude that "fixed air" was a combination of oxygen and "charcoal." Thereupon, Lavoisier offered a new "general theory of combustion," which he claimed could explain the phenomena better and more simply than could Stahl's phlogiston theory.

It required ten more years, the discovery of the decomposition of water, and an extensive campaign to convert the leading chemists of Europe to adopt the new theory. The climax of the campaign came in 1787 when Lavoisier and his French followers, Guyton de Morveau, Claude Louis Berthollet, and Antoine Fourcroy, proposed a reform of the chemical nomenclature that expressed rationally the composition of compound bodies and embodied the new theories into the language of chemistry.

Despite his monumental achievements in chemistry, Lavoisier's work in that field constituted only one of his many activities. In addition to his duties in the Tax Farm, Lavoisier took on important administrative functions in the Academy of Sciences, developed an experimental farm at a country estate he had purchased outside Paris, made improvements in the manufacture of gunpowder, wrote important papers on economics, and became deeply involved in reformist political and administrative roles during the early stages of the French Revolution. As a member of the Temporary Commission on Weights and Measures from 1791 to 1793, he played an important part in the planning for the metric system. He devised plans for the reform of public instruction and the finances of the revolutionary government. In his efforts to deal rationally with the turbulent events of the Revolution, however, Lavoisier repeatedly underestimated the power of public passion and political manipulation, and despite his generally progressive views and impulses, he was guillotined, along with twenty-eight other Tax Farmers, as an enemy of the people at the height of the Reign of Terror in 1794.

Henry Guerlac, *Antoine-Laurent Lavoisier: Chemist and Revolutionary* (1975). Bernadette Bensaude-Vincent, *Lavoisier: Mémoires d'une revolution* (1993). Arthur Donovan, *Antoine Lavoisier: Science, Administration, Revolution* (1993). Jean-Pierre Poirier, *Lavoisier: Chemist, Biologist, Economist* (1996).

FREDERIC LAWRENCE HOLMES

LAW AND SCIENCE are central establishments of modern Western culture. Law directs our action in the world; science our knowledge of the world. Scientific knowledge and techniques have played an ever-increasing role in the spreading of justice in modern society. Institutions of the law have helped to clarify the character of legitimate scientific knowledge and practices, and to readjust the social and institutional relations that their application required.

The law was a major patron of nineteenth-century science. Men of science served as representatives, consultants, and witnesses for the courts and the interested parties in the rising tide of litigation on matters of energy (first gas and oil, and later electricity), environment (pollution and contamination), public health (food and drug adulteration, water supply, sewage treatment), transportation, communication, mining, industry, agriculture, insurance, patents, and, of course, forensics. At the beginning of the nineteenth century, forensic experts employed only basic microscopy and toxicology; by its end, they detected stains and forgeries using infrared and ultraviolet light, differentiated human from animal blood, identified people by body shape, traced minute quantities of inorganic substances by their line spectra, reconstructed important characteristics of a corpse from partial clusters of its bones, and photographed the insides of items with X rays.

Early in the twentieth century, experts learned to identify people by their fingerprints and firearms by their ballistic prints, and to determine genetic relations by blood groups. By the end of the century, toxicology broadened to evaluate occupational exposures, public health hazards, and toxic wastes. Epidemiology grew from the study of disease transmission to accidents, birth defects, and mental illnesses. *X-ray technology was joined by other medical imaging techniques, such as *computerized tomography (CT), positron emission tomography (PET), and magnetic resonance imaging (MRI). Electron microscopes detected the tiniest clues of crime, and *molecular biology identified people and determined their genetic relations. The twentieth century also saw a growing involvement of social scientists expert on the qualities of parenting, the causes of violence, and the validity of eyewitness testimony and repressed memories. Social scientists likewise provided statistical and methodological arguments vital for the resolution of disputes such as antitrust litigation and employment discrimination.

The criminal justice system was but one of the modes of interaction of science and law. During the nineteenth century patent law became a major mediator between the producers of scientific knowledge and those who adapted it to the various wants of society. Late in the century, the legal domain of regulation evolved to control the risks scientific knowledge and its technological products created for public safety and the environment. The twentieth century saw the continual intensification of this regulatory effort and the growth of governmental agencies to protect public health, the environment, workers, and consumers. Meanwhile, private tort law evolved to provide individuals with a way to seek compensation for scientific-technological breakdowns. In recent decades, massive legal controversies in which hundreds of thousands of claimants sought compensation for industrial accidents, polluted environments, workplace hazards, and defective products, have clogged the courts.

Twentieth-century developments in medical research, genetics, molecular biology, and their associated biotechnologies presented new legal challenges. Expanded options for contraception and abortion created one set of legal conflicts within changing public expectations about women's sexuality and liberty and the legal rights of the human fetus. Reproductive technologies such as artificial insemination, in vitro fertilization, and embryo implantation created a second set of legal conflicts for family and parenthood. Dramatic advances in surgery, organ transplantation, drug therapy, and resuscitation created a third set of legal conflicts around the questions of patients' rights and the responsibilities of the state and the medical community. Public fears and distrust of biotechnology created a fourth set of legal disputes, which have resulted more often than not in decisions favoring the interests of science and industry. The most notable instances were the patentability of genetically engineered life forms in the 1980s and their release into the environment.

The law typically refrained from intruding upon the practice, management, and dissemination of basic science. The processes of peer review, funding, teaching, publication, and laboratory and research administration ordinarily remained outside the purview of the courts. Still, in the late twentieth century courts were asked to clarify the boundaries of acceptable behavior by scientists and scientific institutions, and to adjudicate claims of fraud, misrepresentation, and misappropriation of research results; bias in peer review; and mistreatment of experimental subjects. A second avenue by which the courts became involved in the activities of basic science was litigation arising from moral and religious opposition to the purposes of modern science and challenges to the privileged status the scientific enterprise enjoys in modern culture. The attacks on nuclear energy and biotechnology by social critics are a case in point. So are the assaults on science by animal rights activists, conservationists, antiabortion groups, and creationists. In some of these cases the law cast its regulatory net over scientific practices. On the whole, though, the law continued to defer to science and has been reluctant to participate in assaults upon the scientific enterprise, or to adjudicate between science and other belief systems in ways that could obstruct the advance of science and its values.

Steven Goldberg, *Culture Clash: Law and Science in America* (1994). Sheila Jasanoff, *Science at the Bar: Law, Science, and Technology in America* (1995). Kenneth Foster and Peter Huber, *Judging Science: Scientific Knowledge and the Federal Courts* (1997). Tal Golan, ed., *Science and Law, special volume of Science in Context* (1999). Simon Cole, *Suspect Identities: A History of Fingerprinting and Criminal Identification* (2001).

TAL GOLAN

LAW OF SCIENCE. The statements, typically in mathematical form, with which scientists aim to describe nature are called laws of science. The search for quantitative laws of natural phenomena—in conjunction with, or independently of, the search for causes able to explain them—has been a prominent goal of natural philosophers and scientists since the early modern period. The nature and epistemological status of these laws, however, have been the subject of frequent debate.

Early examples of the use of "law" to denote the regularities observed in natural phenomena are found in the writings of Roger Bacon on optics. Its usage became established through the works of Johannes Regiomontanus, Nicholas *Copernicus, and Johannes *Kepler, who wrote about the laws of motion of the planets and of mathematics, as well as of optics.

At the beginning of the eighteenth century, debate over nature's laws centered on the controversy between the followers of Isaac *Newton and those of René *Descartes. Faced with the objections raised against the philosophical legitimacy of Newton's notion of attraction by those who accepted Descartes's criteria for the science of mechanics, Newtonians asserted that the true business of philosophers was to enquire into the laws of nature while remaining agnostic about causes. Early discourse about nature's laws was thus encouraged indirectly by disagreement over causes and the proper goals of explanation (*see* INSTRUMENTALISM; MATHEMATIZATION AND QUANTIFICATION).

Meanwhile, the analysis of the notion of causation carried out by philosophers such as John Locke and, above all, David Hume, added epistemological profundity, and uncertainty, to the issue. Locke claimed that the so-called necessary connections in nature could not be known. Hume argued that, because our knowledge derives uniquely from sense experience, the constant conjunction between some of our impressions is the only legitimate source of the notion of causation.

Undaunted by these difficulties, and following the example set by emblematic achievements like Kepler's laws for planetary motions, Newton's laws of mechanics, and Robert *Boyle's law of gases, subsequent generations of scientists devoted formidable energy to establishing general quantitative relations in every field of natural knowledge. By the end of the eighteenth century, the sciences of heat, *fire and heat *electricity, *magnetism, and chemistry could each boast a number of laws accepted by competent judges, providing proof that the strategy indicated by the earlier examples had paid rich dividends.

These achievements led Immanuel Kant to inquire further into the philosophical bases of the laws of science. According to Kant, sense experience alone, as postulated by Hume, could not explain the degree of generality ascribed to scientific laws. The generality we recognize in them was for Kant the product of an a priori, law-giving activity of the mind that merged with sense experience to produce the kind of judgments typical of the natural sciences.

Kant's reflections inaugurated a line of thought cultivated throughout the nineteenth century and beyond by William Whewell, Ernst Cassirer, and many others. It clashed with the position of those who continued to subscribe, in one form or another, to Hume's appraisal of causation, for example, John Stuart Mill, Ernst Mach, Bertrand Russell, Ernest Nagel, and Carl Gustav Hempel.

A reassessment of the role of laws in scientific practice began around 1980. D. M. Armstrong has argued that laws involve more than Hume's constant conjunction and, contrary to Kant, they are not a priori judgments. According to Armstrong, the laws of science (including probabilistic laws) concern states of affairs that are simultaneously universals, like mass, expressing relations of necessitation holding between these universals. Nancy Cartwright, on the other hand, has suggested an alternative approach to the traditional distinction between theoretical or fundamental laws, assumed to concern reality, and phenomenological laws concerning appearances. The received view regards theoretical laws as closer to truth than phenomenological laws, thus justifying the explanatory and predictive power we credit to theoretical laws. According to Cartwright, however, theoretical laws are true only of the models that we build to go from theory to reality, while phenomenological laws can be true of the objects in reality. Cartwright believes that her concept of the laws of science is compatible with a realist conception of the causal factors (including theoretical entities) involved in laws.

D. M. Armstrong, *What Is a Law of Nature?* (1983). Nancy Cartwright, *How the Laws of Physics Lie* (1983). Friedel Weinert, ed., *Laws of Nature. Essays on the Philosophical, Scientific and Historical Dimensions* (1995).

GIULIANO PANCALDI

LAWRENCE, Ernest O. See BLACKETT, PATRICK M. S., AND ERNEST O. LAWRENCE.

LEE, T. D. (b. 1926); **C. S. WU** (1912–1997); and **C. N. YANG** (b. 1922); Chinese-American physicists.

Chien-Shiung Wu was born in Liuhe, a small town near Shanghai, China, months after the downfall of the Qing (Manchu) dynasty. Girls had few educational prospects in China at the time, but Wu's father, an educator, encouraged her to study in Shanghai with a prominent scholar; her evident intellect led her to the National Central University in Nanjing for undergraduate work in physics. She moved to the United States in 1936 and earned her Ph.D. in physics at the University of California at Berkeley in 1940, with a dissertation on the fission products of uranium. She spent the war in various teaching positions

and then in work on radiation detectors at Columbia University for the *Manhattan Project. She stayed at Columbia after the war, in the end remaining for thirty-five years. Her postwar research focused on beta decay, the spontaneous, radioactive disintegration of an atom through emission of an electron. Her precise experimental technique confirmed Enrico Fermi's theory of beta decay and established her as an expert on the phenomenon.

Tsung-Dao Lee was born in Shanghai in 1926 and came of age during the war with Japan. Lee studied at the National Chekiang University and then the National Southwest Associated University, a combination of Tsing Hua University and other leading Chinese schools that relocated to the interior away from the combat. After graduating in 1946 he came to the United States to study with Fermi at Chicago. Fermi steered him towards astrophysics, and Lee's Ph.D. thesis of 1950 examined white dwarf stars. After stints at Berkeley and Princeton, he won appointment at Columbia in 1953 and stayed there for the rest of his career. He thus came into contact with Wu, a connection of consequence for physics. His research dealt with the theory of turbulence, statistical mechanics, and quantum field theory.

Chen Ning Yang grew up around the Tsing Hua University in prewar Beijing, where his father, a doctor of mathematics from Chicago, was a professor. Yang also attended Southwest Associated University, a few years before Lee. After graduate work at Tsing Hua he won a fellowship for further study in the United States and arrived there in the fall of 1945. Like Lee he chose Chicago, and obtained his Ph.D. there in 1949, studying with Edward Teller, Fermi, and Samuel Allison; his dissertation, directed by Teller, applied group theory to nuclear reactions. He then went to the Institute of Advanced Study in Princeton, where he would stay for seventeen years, working in quantum field theory and theoretical particle physics. In 1954, while spending a year at Brookhaven National Laboratory, he developed with R. L. Mills a nonlinear gauge field theory, later known as the Yang-Mills theory.

In the 1950s physicists faced a proliferation of new particles produced by high-energy *accelerators. Theoretical physicists struggled to compose a taxonomy for the denizens of the particle zoo, based on mass, charge, lifetime, spin, and other observed properties. Lee and Yang had started to collaborate in 1951 at Princeton on the theory of elementary particles. Starting in 1955 they tackled the so-called θ-τ (theta-tau) puzzle: experimental evidence conflicted over whether the theta and tau particles were the same animal or two different ones. To solve the puzzle, Lee and Yang in 1956 proposed that parity, or right-left symmetry in space, is not conserved in weak interactions—that is, that two weak interactions that are mirror images but otherwise identical do not behave the same way. Theta and tau were in effect the same particle in different parity states, one right-handed and the other left-handed. Lee and Yang's idea challenged current theories of beta decay in particular and fundamental assumptions about the symmetry of natural laws in general.

Lee and Yang proposed several experiments to test the hypothesis, including ones for beta decay, a prominent example of a weak interaction. Wu, the beta-decay expert, discussed experimental approaches with Lee at Columbia and became convinced of the importance of the problem and the possibility of its demonstration. She suggested an experiment using cobalt-60, an isotope that undergoes beta decay, with the cobalt atoms cooled to a fraction of a degree above absolute zero and their spins aligned by an electromagnetic field. Wu collaborated with several scientists at the National Bureau of Standards to take advantage of their expertise in cryogenics. The experiment demonstrated that in beta decay the cobalt atoms gave off electrons in one direction more than the other, against the requirements of symmetry. The confirmation of parity nonconservation led to challenges to other long-held principles of symmetry, especially with respect to electric charge and time. It also won Lee and Yang, but not Wu, the Nobel Prize in physics for 1957, the quickness of the reward for their work indicating the perceived importance of symmetry violation for physics.

All three continued with distinguished careers in physics, although Lee and Yang broke off their collaboration in 1962. Lee and Yang were the first Nobel laureates of Chinese descent, and thus a source of ethnic pride. Wu, Lee, and Yang were part of the large Chinese-American scientific diaspora in the second half of the twentieth century (see BRAIN DRAINS AND PAPERCLIP OPERATIONS). The diplomatic rapprochement between the United States and China in the early 1970s renewed scientific exchange between the two countries, and Lee and Yang played leading roles in the exchange. Their international stature as scientists also gave them a voice in Chinese science policy. Their advice sometimes diverged: Lee advocated vigorous government support of basic research in China, including expensive accelerators, as a way for China to stay at the cutting edge of science; Yang, although he supported basic research, opposed high-energy physics in China as a waste of money given the backward state of the economy. Lee and Yang would also provide strong responses to the Tiananmen Square massacre in 1989, both arguing for continued exchange with Chinese colleagues instead of sanctions, though Yang again favored economic development before political reform. Although Wu did not engage in foreign policy to the same degree as Lee and Yang, she enjoyed prominent status in the American community: she was elected president of the American Physical Society in 1975, the first Chinese and the first female to hold the post, and won the Wolf Prize the following year.

Laurie M. Brown et al., eds., *Pions to Quarks: Particle Physics in the 1950s* (1985). Robert Novick, ed., *Thirty Years since Parity Nonconservation: A Symposium for T. D. Lee* (1988). C. S. Liu and S.-T. Yau, eds., *Chen Ning Yang: A Great Physicist of the Twentieth Century* (1995). Zuoyue Wang, "U.S.-China Scientific Exchange: A Case Study of State-Sponsored Scientific Internationalism in the Cold War and Beyond," *Historical Studies in the Physical and Biological Sciences* 30, no. 1 (1999): 249–277.
                                             PETER J. WESTWICK

LEVI-MONTALCINI, Rita (b. 1909), neurobiologist, discoverer of the nerve growth factor, recipient in 1986, with Stanley Cohen, of the Nobel Prize in physiology or medicine.

Born in Turin, Italy, the youngest (together with twin sister Paola, who became a well-known painter) of the four children of Adamo Levi, engineer and entrepreneur, and Adele Montalcini, both Jewish, Rita Levi-Montalcini had not been slated for a professional career. At twenty, however, she enrolled in medicine at the University of Turin. She graduated in 1936 with a dissertation on the formation of reticular fibers, carried out under the direction of Giuseppe Levi, a prominent histologist who at that time was tutoring two other future Nobel prizewinners, Salvador E. Luria and Renato Dulbecco.

After specializing in neurology and psychiatry, Levi-Montalcini ran afoul of the 1938 law banning Jews from public office, the professions, and a university career. Having spent part of 1939 in Brussels, she returned to Turin just before the German invasion of Belgium. Prevented from continuing her research at the university, she set up a small laboratory at home. Inspired by an article of Viktor Hamburger, a pupil of Hans Spemann, the discoverer of the phenomenon now known as embryonic induction (see EMBRYOLOGY), she studied the interaction of genetic and environmental factors in controlling the differentiation processes that take place during the early formation of the nervous system in chickens. These studies further improved her expertise in silver staining and micro-surgical techniques.

After the collapse of the fascist regime and the invasion of northern Italy by German troops in September 1943, Levi-Montalcini and other members of her family found refuge in Florence. Using counterfeit documents, they managed to survive until the liberation of the city. Back in Turin, she received an invitation from Hamburger to spend a semester at Washington University in St. Louis. She left Italy on 19 September 1947 on the same ship that brought Dulbecco to the United States (he had been invited by Luria, then at Indiana University). The planned semester became a thirty-year stay, which from 1961 she combined with periods spent in Rome, where the U.S. National Science Foundation helped her set up a laboratory. In 2001, she was made one of the five life members of the Italian Senate. She never married.

The work for which Levi-Montalcini is known began during her early years in St. Louis, when she adopted a population model for interpreting the phenomena observed during the formation of the peripheral nervous system of the chick embryo. According to that model, the differentiation processes observed arose from mass migration and degeneration of the different populations of cells involved. She developed the model further by studying the acceleration of the differentiation processes caused by the transplantation of mouse-tumor cells into chick embryos. The procedure consisted of grafting small pieces of mouse sarcomas at the base of the limb buds of chick embryos two or three days old. The pieces invaded adjacent tissues. At seven days of incubation, large nerve bundles invaded the tumors; at eleven days, the density of the nerves in the sarcomas exceeded that of any normal tissue at any embryonic stage. By 1951, Levi-Montalcini regarded these phenomena as showing the existence of specific agents promoting nerve growth. The nature of the agents remained unknown.

Additional research carried out in 1952 in cooperation with Hertha Meyer (another former collaborator of Giuseppe Levi in Turin, then at the Universidade do Brasil, Rio de Janeiro) and Viktor Hamburger showed that effects similar to those studied *in vivo* also took place *in vitro*. The technique consisted of using a tissue-culture method that excluded influences of the organism as such on the nerve growth observed, and in producing the mysterious "diffusible agent" under investigation.

Subsequent work, carried out between 1954 and 1959 in cooperation with biochemist Stanley Cohen and others, established that a growth-stimulating agent also existed in snake venoms and in mouse salivary glands. Like the cause of the effects produced by mouse sarcomas, this agent was isolated and described as a protein or a substance bound to protein. Named Nerve Growth Factor (NGF), the agent is one of a class of endogenous products, identified in proteins collectively designated as Polypeptide Growth Factors, and of considerable interest for both basic and clinical research.

See also EMBRYOLOGY; MENTAL SCIENCES.

Rita Levi-Montalcini, *In Praise of Imperfection: My Life and Work*, trans. Luigi Attardi (1988). Rita Levi-Montalcini, *The Saga of the Nerve Growth Factor* (1997).

GIULIANO PANCALDI

LEYDEN JAR AND ELECTROPHORE. The Leyden jar, a bottle partially filled with water and electrified via a wire from an electrostatic machine, made a contribution to natural knowledge of the first importance. When it was invented in 1745, natural philosophers supposed that electrical attraction and repulsion arose from a vapor or effluvium given off by an electrified body. This effluvium drove away light objects in its path (repulsion), and, according to Europe's leading electrician of the time, the Parisian academician J. A. Nollet, stimulated neighboring bodies to emit effluvia, which caused what appeared to be attraction. These effluvia ran through conductors but were stopped in insulators like glass. Since objects responded to an electrified body across a thin glass screen, it was thought that glass had to be thick to insulate effectively.

When, following this theory, the professor of natural philosophy at the University of Leyden, Petrus van Musschenbroek, or one of his associates thought to collect electrical effluvia in a bottle he was in for a shock. He supported the bottle with a bare hand while charging the water. On disconnecting the charging wire with the other hand, he received a smart blow, which Musschenbroek told Nollet he would not repeat for the entire kingdom of France. Nollet bravely discovered that the stroke was greater the thinner the glass, when, according to his theory, it should have leaked the most.

Benjamin *Franklin secured a favorable audience in Europe for his theory of positive and negative electricity via his analysis of anomalies in the behavior of the Leyden jar. According to him, effluvia accumulating in the bottle repelled fluid from the exterior surface through the hand to the ground. The thinner the bottle the greater the repulsive force and the greater the effect—provided the glass was absolutely impermeable to the effluvia. When the experimenter joined the two surfaces externally, the stockpiled fluid within rushed through him to fill the void without. Thus the shock.

The first depiction of the Leyden experiment, shown beneath a diagram of the material system of electricity (the theory of effluences and affluences) that the experiment would destroy.

Franklin reduced the cumbersome bottle to a pane of glass armed on each side with a metal coating (the "Franklin square"). Franz *Aepinus and Johan Carl Wilcke then eliminated the glass. Their "air condenser" could be charged like a Leyden jar. Otherwise it appeared to act at a distance. Apparently the electrical fluid did not drift through the air except when it could be plainly seen, in the form of sparks. The subsequent grounding of electrical theory on Newtonian forces was an important step toward the system of *imponderables.

The Franklin square had more to teach. By gradually moving the pane's metallic coatings away, Wilcke examined the effects of electrical induction as a function of distance (1762). In 1775, Alessandro Volta (*see* GALVANI, LUIGI, AND ALESSANDRO VOLTA) incorporated the same idea in a useful machine. Replacing the glass with a special resinous mixture (the "cake") and giving the upper coating an insulated handle (the "shield"), he made an "elettroforo perpetuo," a perpetual purveyor of electricity. He charged the cake by friction, placed the shield on it by the handle, and touched the shield's conducting surface, thus making a condenser. When raised, the shield could be discharged at will and returned to the electrified cake to repeat the process. Analysis of

the electrophore and dissectible condenser gave electricians the fundamental concepts of electrical induction, tension (a forerunner of potential), and capacity.

J. L. Heilbron, *Electricity in the 17th and 18th Centuries* (1979; reprinted 2000).

J. L. HEILBRON

LIBRARY. Libraries have proved to be essential for both the practice and preservation of scientific research. Research that originates in the study of earlier work relies on the record available record in a library. By focusing on aspects of science and technology, some libraries have purposefully provided the scientific researchers of their community of users with two essential services: the collection, organization, and provision of access to a repository of recorded scientific information in various formats, including books, periodical publications, microforms, films, and electronic media, and the provision of reference assistance by helping library users find information.

Access to a library's collection and services usually is limited, although the community of users may be national or even international in scope, as in the case of the British Library. The earliest libraries to support scientific research specifically, however, belonged to individuals. For example, Conrad Gesner, the sixteenth-century Swiss naturalist, had an extensive library used by himself and, perhaps, a close circle of associates. Many personal scientific libraries, even into the nineteenth century, approached institutional size. Sir Joseph Banks collected 25,000 volumes for his library, and Alexander von *Humboldt 11,000. Many institutional libraries originated with the gift or purchase of a personal library. Sir Hans Sloan's library of 50,000 volumes formed the nucleus of the British Museum Library founded in 1753.

Learned societies, professional organizations, and research institutes created institutional libraries for scientific research. One of the first acts of many scientific societies was to set up a library. Even before Charles II granted it a charter of incorporation in 1662, the Royal Society of London, for example, had already created a library in 1660. The American Academy of Arts and Sciences started a library in 1780, the year of its establishment. These libraries collected not only the publications of the society and its membership, but also other scientific publications, especially scientific periodicals, often by exchange.

Scientific periodicals trace their origin to the seventeenth century; both the *Philosophical Transactions* of the Royal Society and the *Journal des sçavans* started publishing in 1665. Scientific periodicals are intended to communicate current results of scientific research quickly. Their inexorable growth in number and size challenged librarians to provide adequate shelf space and adequate financial resources.

Indexes were recognized as essential for the effective use of scientific periodicals from the time the earliest periodicals appeared. The first volume of the *Philosophical Transactions* has an index. To make it possible for researchers to review a large body of literature, however, libraries have emphasized the acquisition and even the creation of collective indexes to a series of periodicals.

Jeremias David Reuss, who ended his career as librarian at the University of Göttingen, edited the first periodical index devoted to the proceedings of scientific societies. His *Repertorium commentationum a societatibus litterariis editarum* (Göttingen, 1801–1821) is a classified subject index to the publications of learned societies issued before 1800. A continuation, the Royal Society's *Catalogue of Scientific Papers*, issued in nineteen volumes, purported to list by author the titles of all scientific articles published in all languages and fields in the nineteenth century. The Society managed to create a subject index of mathematics (38,748 entries), mechanics, and part of physics before running out of steam.

Other specialized indexes to the scientific literature were created throughout the nineteenth and twentieth centuries. The *British Engineering Index* began in the 1880s and soon relied heavily on a cooperative partnership with the Engineering Societies Library in New York, which did not consider the *Engineering Index* to have broad enough coverage and so began its own local index of periodicals in the late 1920s. Most libraries have relied on published collective indexes, *Science Abstracts* (1898–) for physics, *Chemical Abstracts* (1907–), and *Biological Abstracts* (1926–).

With the advent of online bibliographical databases in the 1960s, libraries began to offer a specialized search service. The earliest electronic databases of bibliographical information covered the sciences and medicine. QUOBIRD, an experimental system at Queen's University of Belfast, retrieved references on atomic and molecular physics in the early 1970s. The National Library of Medicine initiated AIM-TWX, an Abridged *Index Medicus*, in 1970 for experimental online bibliographical retrieval. In 1974, there were fewer than one hundred such databases. Ten years later, vendors or search services offered almost three thousand online databases, most of them in the pure and applied sciences. Specially trained reference librarians learned how to use the vendors' services and acted as intermediaries for the

endusers of the information. The cost of providing access to electronic indexes of scientific literature remains a major budgetary consideration for science libraries.

Another financial concern for libraries is the costly service of classification and cataloging of collections. Classification schemes such as the Universal Decimal Classification and the Library of Congress Classification organize collections by subject. Catalogs allow users to find out what is in a library and where it is located. Some scientific and technical libraries have created their own cross-referenced catalogs by taking advantage of the detailed subject analysis offered by the major classification schemes. The John Crerar Library in Chicago created a card catalog for its collection of science and technology in 1896 that, when appropriate, gave a book several designators within the Universal Decimal Classification. The Department of Technology at the Carnegie Library of Pittsburgh adopted a similar catalog in 1910, and the Engineering Societies Library in New York did so too in 1919. This system allowed a researcher to search the catalogue without browsing the shelves and took advantage of a classification scheme with which the library's users were familiar. Many online public access catalogs (OPACs) continue to provide a similar function by allowing sophisticated searching by keyword, classification number, publisher, and many other options.

At the end of the twentieth century, many scientists began to use repositories of scientific information by electronic access instead of traveling to a library building. A science library traditionally has provided a resource and the tools, services, and a place to use it. For many users of scientific information, the "place" may now be less important than the "function" a library provides by collecting, organizing, and providing access to a repository of information sources regardless of their location or format.

J. Christian Bay, *The John Crerar Library 1895–1944* (1945). David A. Kronick, *A History of Scientific and Technical Periodicals* (1962). Elmer D. Johnson, *History of Libraries in the Western World*, 2d ed. (1970). Ellis Mount, *Ahead of its Time: The Engineering Societies Library, 1913–80* (1982). D. W. Krummel, *Fiat lux, fiat laterbra: A Celebration of Historical Library Functions* (1999). John L. Thornton, *Thornton and Tully's Scientific Books Libraries and Collectors: A Study of Bibliography and the Book Trade in Relation to the History of Science*, 4th ed., ed. Andrew Hunter (2000).

BRUCE BRADLEY

LIEBIG, Justus von (1803–1873), one of the dominant figures in the emergence of organic chemistry, originator of the modern teaching research laboratory.

Born in 1803 in Darmstadt to Johann Liebig, an inventive hardware merchant, and Maria Caroline Moser, Liebig was the second son in a large family. Leaving the local classical Gymnasium at age fourteen without his *Abitur*, Justus was apprenticed for a time to an apothecary, but soon returned home, where he helped in his father's business, read old chemistry books borrowed from the local Ducal library, and carried out some of the laboratory operations described in them in his father's shop. Aspiring to become a pharmaceutical chemist, he studied at the University of Bonn and then at Erlangen under the direction of Carl Wilhelm Kastner, who enthusiastically supported Liebig's newly emerging aim to establish a school for pharmacists. To prepare him for this goal, the Duke of Hesse-Darmstadt sent him to Paris with a grant. There Liebig encountered a more rigorous, quantitative style of chemical reasoning and experimentation than he had known in Germany. He impressed Joseph Gay-Lussac and Louis-Jacques Thenard sufficiently to gain access to their laboratories, where he collaborated with them and expanded his ambition beyond merely teaching chemistry to becoming a leading researcher in the field. With the further support of Alexander von *Humboldt, Liebig received an appointment in 1824 as Extraordinary Professor at the small provincial University of Giessen in Hesse-Darmstadt.

At Giessen in 1826, Liebig and three other members of the faculty established a private pharmaceutical-chemical institute similar to several others in Germany whose primary purpose was to train pharmacists. Over the next two decades, as Liebig assumed sole leadership of the institute, it evolved into a chemical research laboratory in which students were systematically taught, in growing numbers, not only to perform analyses, but to carry out chemical investigations leading to original results (*see* LABORATORY, CHEMICAL). At its peak during the 1840s, more than fifty students worked in it at a time. Although the majority of them returned to apothecary shops or went into industry, some developed into leading research scientists. Liebig's laboratory became both a symbol of the larger scale science of an industrializing world and a model for other laboratories, in chemistry, physiology, and other fields. The productivity of the Giessen laboratory was greatly enhanced by Liebig's invention in 1830 of an improved method for the combustion analysis of organic compounds that enabled even beginning students to obtain accurate, reliable results.

Meanwhile, Liebig himself made important experimental contributions to the vigorously expanding field of organic chemistry. Together with his close friend, the chemist Friedrich Wöhler, he identified in 1832 the "benzoyl radical" that remained intact through various chemical transformations. He discovered new compounds such as acetal and aldehyde, and in 1838 wrote a landmark paper on organic acids. With Wöhler again, he investigated the many transformations that uric acid can undergo.

While Liebig and other leading chemists of the time, such as Jean-Baptiste Dumas and Jöns Jacob *Berzelius, expanded the number of known organic substances and their reactions, they argued, sometimes acrimoniously, about the nature of the compounds they made. The recognition that compounds of identical combining proportions differed in their properties, which they explained by making the assumption that the atoms were arranged differently within their molecules, led to various theories about the "radicals" that were supposed to be unchanging constituents of a series of related compounds. Dumas, Berzelius, and Liebig each proposed his own radical scheme.

Unable to separate intellectual disagreements from personal attacks, Liebig was often at odds with his contemporaries. An impulsive person who sometimes drove himself to exhaustion, he was generous and warm-hearted with friends and charismatic to his students, but a relentless foe of his enemies. Liebig became so weary of controversy that in 1840 he abruptly abandoned organic chemistry and turned to the application of chemistry in agriculture and physiology. There he produced works of interpretation and speculation that exerted strong influences on both fields. Of course, his work immersed him in intense controversy.

In this period of expansion in the research universities in the German-speaking states, scientists who achieved reputations as leading investigators regularly received calls to move to another university, with inducements such as increased salaries and enlarged or better equipped laboratories. Liebig turned down several such offers in return for improvements in his position at Giessen. These included an extensive enlargement of his laboratory in 1839. By 1852, when he received an invitation from the King of Bavaria to come to Munich, he was tired of the pace he had maintained as director of his teaching laboratory, and he accepted the King's invitation. In Munich he had greater resources than before, but he no longer trained students and only sporadically carried out experiments. He continued to revise his agricultural chemistry, write extensively, and promote industrial and other applications of chemistry.

William H. Brock, *Justus von Liebig: The Chemical Gatekeeper* (1997). Joseph S. Fruton, *Contrasts in Scientific Style: Research Groups in the Chemical and Biochemical Sciences* (1990).

FREDERIC LAWRENCE HOLMES

LIFE. The Aristotelian notion of change as, indifferently, physical movement and animated development, blurred a convenient later distinction between living and nonliving objects. Relying on the Aristotelian tradition, early modern naturalists did not have recourse to a distinct conception of life in their descriptions of the variegated objects included in their "natural histories" (*see* NATURAL HISTORY; CLASSIFICATION IN SCIENCE). Medieval and Renaissance medical traditions, on the other hand, availed themselves of notions of the human body that extended to animals and plants only with qualification. Alchemists and Paracelsians, for their part, found "vital" flames also in the minerals they treated in their furnaces.

The emergence during the second half of the seventeenth century of approaches to the phenomena of living beings consistent with the "new philosophy" of the moderns resulted from several different strands. Physicians like William *Harvey adopted experimental techniques. René *Descartes developed conceptual strategies inspired by mechanistic principles and arrived at new comparisons between humans, animals, plants, and machines (*see* ANIMAL MACHINE). The systematic use of the *microscope to explore the traditional kingdoms of nature made a third strand; a fourth was the need—felt strongly by Leibniz, for example—to counteract the influence of the mechanical philosophy by pointing to life and organization wherever mechanists saw just corpuscles and extension (*see* ORGANISM).

The combined effect of these developments after 1650 was to set natural philosophers, physicians, and microscopists on the track of new maps of the animated part of the world. Throughout the eighteenth century, the search focused mostly on the notion of *generation and its implications for the classification of animals and plants. In the meantime, Stephen Hales and Antoine-Laurent *Lavoisier paid growing attention to the physics and chemistry of plants and animals, and established important experimental traditions in the study of animal heat and respiration. Beginning in the 1790s naturalists and philosophers debated the possibility of doing for organic bodies what *Newton and the chemists had accomplished for inert matter. The debate favored the introduction of the word *"biology" and nurtured a long series of definitions, which carried religious and political implications, of what should count as life.

Concomitantly with the adoption of an evolutionary perspective, in 1802 Jean-Baptiste de *Lamarck defined life as a "physical fact." Marie François-Xavier Bichat, Georges Cuvier, Johannes Peter Müller, and Justus von *Liebig insisted instead that life was a special force that resisted until death the forces of physics and chemistry. From 1859, Charles *Darwin circulated a view of living organisms as the contingent product of an evolutionary process in which internal and external circumstances molded the "machinery of life" (his phrase) without a hint of purpose or design. Claude *Bernard saw directedness at least in individual development. Relying on evidence provided by laboratory biology, physiologists like Carl Ludwig, Emil du Bois-Reymond, Ernest von Brücke, and Hermann von *Helmholtz claimed that physical and chemical forces could explain living organisms fully. The search for the proper localization of life inside organisms led to similarly contrasting views; Bichat pointed to tissues, others to *cells. In 1869 Thomas Henry Huxley pointed to "protoplasm," the mucilaginous substance then observed inside each cell, as the "physical basis of life."

In hindsight, the program of physicochemical reduction led to the extraordinary development of the life sciences during the twentieth century. In fact, the development took place and continues to unfold within the fragmented, multifaceted framework provided by fields like *genetics, *molecular biology, and *biotechnology that avoid tackling the big question of the nature of life. The fact, however, that the question has been raised at key turning points and with seminal consequences by scientists of the stature of Niels *Bohr (1932), Erwin *Schrödinger (1944), Jacques *Monod (1970), and François *Jacob (1970) suggests that it continues to figure high on the agenda, despite the elusive or controversial answers it draws.

The authors of one of the two draft sequences of the *human genome published early in 2001 thought fit to convey amazement at some of their findings—for example, that the human genome has only 31,000 protein-encoding *genes against a worm's 18,000 and a plant's 26,000—by quoting a verse by T. S. Eliot: "We shall not cease from exploration/And the end of all our exploring/Will be to arrive where we started/And know the place for the first time."

Erwin Schrödinger, *What Is Life?* (1944). Jacques Monod, *Chance and Necessity* (1972). Stephen J. Gould, *Wonderful Life* (1989). Catherine Wilson, *The Invisible World: Early Modern Philosophy and the Invention of the Microscope* (1995). Simon Conway Morris, *The Crucible of Creation* (1998). Lily Kay, *Who Wrote the Book of Life?* (2000).

GIULIANO PANCALDI

LIGHTHOUSE. Samuel Johnson defined a lighthouse as "a high building at the top of which lights are hung to guide ships at sea." Lighthouses, however, are not restricted to guiding ships at sea, but may stand on any body of navigable water where they can be useful. But they are seldom found inland; the one that stood at Pica, Chile, at the turn of the nineteenth century, to guide travelers across the Atacama Desert, was an exception. Distinctive color patterns on some lighthouses serve as "day marks" to aid navigation in daylight. Many of the earliest lighthouses used smoke for this purpose.

The first purpose-built lighthouse, the Pharos of Alexandria, one of the seven wonders of the ancient world, dates from around 280 B.C. At 100 to 140 meters (350 to 450 feet) high, it was the tallest lighthouse ever made. The Roman emperors built about thirty lighthouses, the tower at Dover, England, being an extant example.

The earliest lights burned wood, coal, dried animal dung, candles, and various vegetable and animal oils. Glazed lanterns with air vents located below the fuel allowed upward drafts that increased the effectiveness of the burning and the brightness of flames. Candles and oil lamps, especially when grouped, were easier to use and control, and brighter than fires. The first major technological advance added reflectors to increase light intensity. A plano-convex lens set in front of a light source was found to have the same concentrating characteristic as the parabolic reflector. By 1757 Jonas Norberg had made a practical test of a parabolic reflector. In the same year he devised an oscillating light by turning the reflectors to and fro horizontally. In 1781 Norberg introduced the world's first revolving light, produced by revolving the reflector or lens around a stationary illuminate.

About 1782 Aimé Argand invented a hollow wick that allowed air to be pulled up inside and outside the flame, which produced a cleaner, brighter, and steadier flame. Argand also introduced a screw to alter the level of the wick. By about 1815 an Argand lamp gave off a light equal to seven to eight candles.

In 1822 Augustin Fresnel (*see* CARNOT, SADI, AND AUGUSTIN FRESNEL) invented a lenticular glass lens assemblage. Fresnel understood that light could be concentrated by reflection and refraction: a glass prism may refract or reflect light depending on its shape and the angle at which light strikes it. Fresnel assembled spherical lenses as refractors and prisms as reflectors in a geometry that concentrated as much light as possible from the illuminate and emitted it outward from the light source. Lens and prism assemblies similar to those still in use today did not become available until after 1850. The typical illuminant with Fresnel lenses was first mantle lamps, later replaced by electric lamps (*see* LIGHTING). The parabolic reflector fitted with an electric arc-light was introduced in 1902. The sealed beam electric lamp with xenon discharge, available from 1965, did not prove popular as the narrow horizontal light beam did not have much "loom" and thus could not be detected in the sky before the flash could be seen when far from shore. When conditions make the visibility of the light poor, lighthouses often employ sound devices (fog signals) such as bells, whistles, and horns.

The invention of radar, LORAN (long range navigation using land-based radio transmitters and receivers), and global positioning systems have made lighthouses secondary navigational aids. Most lighthouses in the world have been automated; others have been abandoned or demolished.

John Naish, *Seamarks: Their History and Development* (1985).

RALPH E. ESHELMAN

LIGHTING. Little changed in the light sources inherited from antiquity until the late eighteenth century, when combustion began to receive serious scientific attention. In 1784 the Swiss Aimé Argand patented an oil lamp with a circular wick that spread out the flame and allowed air to the inside as well as the outside of the wick. These measures increased brightness, because the fuel burned more completely, giving a larger flame.

From the late eighteenth century into the nineteenth century, several people experimented with light from gas flames and a company supplying gas was established in London in 1812. Metal burners, designed to spread the flame as widely as possible, inadvertently cooled it, a problem corrected in the 1850s by William Sugg, who obtained a hotter and brighter flame by using burners of steatite or soapstone, poor conductors of heat. Some large gas lights had extra metal ducting to bring in the air and carry away the exhaust gases. About 1860 Friedrich Siemens increased the efficiency by using the heat of the exhaust to warm the incoming air.

Flames give light because they contain hot solid particles. John William Draper, a professor of chemistry in New York, established the relation between the temperature of a very hot solid and the light it produces in 1847. Several oxides, especially of the rare earth elements, give light at relatively low temperatures. The first of these "selective emitters" exploited was lime (calcium oxide). Goldsworthy Gurney discovered the principle of limelight in 1820 when trying to decompose lime with an oxyhydrogen blowpipe.

Limelight never became an important source of light except in theaters.

Gas mantles use selective emitters. The Austrian chemist Carl Auer, later Freiherr von Welsbach, discovered the idea in 1872 when he spilled rare metal salts on an asbestos mat over a flame. As the salts dried they became luminous. Auer found that he could impregnate a cotton web with the salts, burn off the cotton, and leave a fine web of oxide. The first mantles were held above the flame on a fire-clay support. After 1905 the mantle was inverted and the flame pointed down into it, but the flame had to fit the mantle shape to heat it fully. Gas mantles competed seriously with early electric lamps as did Walther Nernst's lamp made of a rod of selective emitter material heated electrically, which first appeared commercially in 1898.

The continuous electric current made available by Alessandro Volta's pile opened a new road to illumination (see GALVANI, LUIGI, AND ALESSANDRO VOLTA). Humphry Davy showed that the current could produce light both by heating a thin wire to incandescence and by producing a bright spark, or "arc," between pieces of carbon. The electric arc did not see much use until the advent of electric generators in the 1870s. Since the ends of the carbons burn away, regulating arrangements must be provided to move them together. Electromechanical regulators monitored variations in either the current through the arc or the voltage across it as the arc lengthened. They were the first automatic electric control devices.

The incandescent filament lamp needed three things: a material that could be raised to white heat and cooled repeatedly, a method of sealing it in a glass vessel, and an adequate vacuum pump. Several inventors succeeded after the invention of the mercury vacuum pump in the mid-1870s. Best known were the Englishmen Joseph Swan and St. George Lane Fox and the Americans Thomas Edison and Hiram Maxim. All used carbon filaments: Swan's made from cotton, Lane Fox's from grass fibers, Edison's from bamboo fibers, and Maxim's from strips of paper. Each baked his raw material in an airless furnace, driving off all but the carbon atoms, leaving a filament of pure carbon. Later Swan, a chemist, dissolved the cotton and extruded a very smooth fiber of pure cellulose for his filaments. Carbon filament lamps, mostly made by Swan's or Edison's methods, provided domestic electric lighting until superseded by metal filament.

The efficiency of a filament lamp depends on its temperature. Osmium, tantalum, and tungsten have very high melting points but are difficult to draw into fine wires. Since their resistivity is much less than carbon's, lamps using them have very long filaments. Osmium lamps first appeared in 1902, soon followed by tantalum and, with the highest melting point of all, tungsten. The first tungsten lamps had the filament made by sintering the metal powder. Research by William Coolidge in the United States in 1911 made drawing of tungsten possible. Soon all lamp filaments were tungsten. The coiled filament of tungsten wire wound into a tight spiral made it easier to support the filament and reduced loss of energy by cooling.

A further improvement in efficiency came by introducing a little halogen gas into the bulb. A lamp fails because tungsten evaporates from the filament. In the cooler parts, away from the filament, tungsten vapor and halogen gas combine, but the compound breaks up when it drifts close to the filament and tungsten is deposited there. This self-repairing mechanism enables the lamp to run hotter and more efficiently.

An electric discharge through many gases at low pressure will ionize the gas and produce light. The first gas so used, mercury, gives a very blue light. From 1932 mercury lamps run at atmospheric pressure gave a more acceptable light suitable for street lighting. Meanwhile low-pressure sodium lamps proved to be the most efficient of all, but their monochromatic yellow light limited application. Also they were difficult to manufacture because sodium vapor attacks most glasses and metals. Subsequent development of discharge lamps relates to improved glasses and metals for the electrodes. In many lamps the envelope containing the discharge is now made of quartz rather than glass: for high-pressure sodium lamps sintered alumina is used. Discharge lamps cannot be connected directly to the electricity supply but require special control circuits that limit the current and, in some cases, provide a high voltage to start the discharge.

The fluorescent lamp, introduced in 1938, was not widely adopted until after World War II. Most of the light in a low-pressure mercury discharge comes at a single ultraviolet wavelength. A coating of fluorescent powder inside the tube converts the ultraviolet into visible light. Much research has gone into fluorescent materials. Modern compact fluorescent lamps incorporate the control gear in the cap and so can be plugged into an ordinary lampholder.

Light-emitting diodes are semiconductor devices that produce light efficiently at low levels. Now widely used for signs and indicators, they may in the future be made bright enough for general lighting.

Brian Bowers, *Lengthening the Day* (1998).

BRIAN BOWERS

LIGHTNING. Until the seventeenth century most ideas about lightning in the Western world were based on Aristotle's *Meteorologica*, which ascribes lightning and thunder to a dry, warm exhalation "forcibly ejected" from one cloud onto another. The impact produced the sound. As for the light, it came from a wind burning "with a fine and gentle fire."

In spite of his espousal of the mechanical philosophy, René *Descartes's account in his *Meteorology* (1637) resembled Aristotle's. Thunder occurs when a high cloud condenses and falls against a lower one. When the air between these clouds contains "very fine, highly inflammable exhalations," it usually produces "a light flame which is instantaneously dissipated." Others in the seventeenth century attributed lightning to combustion of "sulphurous vapors." In the early eighteenth century Francis Hauksbee and William Wall conjectured that lightning was related to "electric fire." As late as 1746 John Freke argued that both electricity and lightning arose from "a great quantity of the elementary fire driven together."

This echo of Aristotle, generally ignored by historians, underlay Benjamin *Franklin's famous identification of lightning and static electric discharge. Likewise, natural philosophers explored the similarity between lightning and electricity in the 1740s. Franklin suggested in 1750 that an iron rod extending 9 meters (30 ft) above a church steeple could be used to collect lightning (*see* LIGHTNING CONDUCTOR), and a few years later Thomas François Dalibard successfully performed the experiment in France. Franklin soon after tried his famous kite experiment and survived, though Georg Wilhelm Richmann died in 1753 when he personally conducted lightning from rod to ground. Franklin concluded that clouds are electrified, the bottom usually negatively, while John Canton discovered that clouds may be either positive or negative. In the 1770s Giambattista Beccaria investigated the frequency of lightning strikes.

Little more was learned until the twentieth century when photography, spectroscopes, and electromagnetic devices provided new tools of investigation. In the 1930s, T. E. Allibon and J. M. Meek concluded that sparks move from cloud to ground, ending in terminal "brushes," increasing gradually in length until the atmosphere is ionized. Meek found that a "pilot" discharge travels from the negative cloud bottom, followed by other discharges. A positive current follows the completed discharge path to the cloud.

Many scientists tried to explain how clouds become charged. G. C. Simpson hypothesized in 1927 that as raindrops separate the droplets take opposite charges. About the same time C. T. R. Wilson, Julius Elster, and Hans Geitel suggested that the charge separation intensifies as induction polarizes drops in a cloud. Bigger drops, falling, pick up electrons, making the bottom of the cloud increasingly negative. Others proposed that up-drafting air carries charged ions. In 1960 C. B. Moore and Bernard Vonnegut cast doubt on the idea that the ionization of droplets must occur to produce lightning since electrical fields in and around clouds can precede precipitation.

Although lightning scientists still use ground-based detection networks, in the second half of the twentieth century they have relied increasingly on high-altitude airplanes, sounding rockets, satellites, and the space shuttle. In 1990 this technique resulted in the dramatic discovery of "red sprites"—lightning that goes from cloud tops to as high as the ionosphere—and other stratospheric discharges.

Park Benjamin, *A History of Electricity (The Intellectual Rise in Electricity) from Antiquity to the Days of Benjamin Franklin* (1898). Peter E. Vermiester, *The Lightning Book* (1972).

GREGORY A. GOOD

LIGHTNING CONDUCTOR. The similarity between electrical discharge and lightning occurred to several minds after the invention of the electrical machine and the *Leyden jar in the late 1740s made impressive sparks and shocks available in the laboratory. Even with this modest equipment, Jean-Antoine Nollet made a line of insulated monks jump by using them to short-circuit a condenser. Nollet was the first to list the analogies between electricity and lightning (in 1748), and another Frenchman, Jacques de Romas, probably was the first to propose to bring atmospheric charge to earth via a kite (1750 or 1751). Benjamin *Franklin, with his characteristic boldness and optimism, first proposed an experiment to show that lightning could produce some of the effects of artificial electricity and offered measures to protect buildings from lightning strokes.

The experiment, first performed in France in 1752 under the instructions of the Comte Georges-Louis de Buffon, who expected, rightly, that a positive outcome would undermine the position of Nollet as Europe's leading electrician, employed a tall pointed insulated iron rod exposed to the elements. During a storm a dispensable old soldier approached his knuckle to the rod, drew a spark, and tried to confirm the theory. It worked. The soldier survived because the rod detected fluctuations in the electrical state of the lower atmosphere. Had he received

A method, not recommended, for protection against a lightning stroke–the parapluie-paratonnerre ('umbrella-lightning rod') invented by Jacques Barbeu Dubourg, Benjamin *Franklin's French translator.

a lightning stroke, the veteran would not have reported his success. The first and last natural philosopher who succeeded with Franklin's experiment was Georg Wilhelm Richmann.

After Richmann's electrocution in 1753, Franklinist electricians, particularly Giambattista Beccaria and Alessandro *Volta, monitored atmospheric electricity remotely. Others developed the art of protecting buildings from thunderbolts. Deployment progressed slowly partly because Franklin's theories suggested that a sharply pointed high metal pole, perfectly grounded, could despoil a cloud of its electricity silently as well as channel the stroke if it came. There arose a squabble over whether pointed rods might entice lightning that otherwise would strike somewhere else. The last significant holdouts, Nollet (died 1770) and Benjamin Wilson (died 1788), rejected as silly and arrogant the claim that puny man could disarm a thundercloud. Nollet recommended sitting in a grounded metal cage. Wilson preferred conductors that ended in balls or knobs so as

not to attract the attention of passing clouds. If a stroke came anyway, the blunt rod would be at least as effective as the pointed one in disposing of the lightning. Nollet's cage, though inconvenient, would have worked. Wilson's obtuse ends would have been no better than Franklin's points.

During the nineteenth century, many public and private buildings were armed with lightning rods installed so as to connect all the metal parts to the same grounded system. Failures in protection could almost always be traced to imperfect grounding. Thus electrical science saved property and lives, especially of people who rang church bells in an effort to break up storm clouds, and made its first significant contribution to technology.

J. L. Heilbron, *Electricity in the 17th and 18th Centuries* (1979; reprinted 2000). I. Bernard Cohen, "Prejudice against the Introduction of Lightning Rods," Franklin Institute *Journal* 253 (1952): 393–440.

J. L. HEILBRON

LIGHT, SPEED OF. Several investigators tried to measure the speed of light in the seventeenth and eighteenth centuries. Galileo, or perhaps later his followers, flashed light to an assistant, but the speed was too great to measure. In 1675 the Danish astronomer Ole Rømer noticed that intervals between eclipses of Jupiter's moons are less when Jupiter and the Earth approach each other; he correctly attributed the phenomenon to the time it takes light from Jupiter to reach the Earth. Using contemporary estimates of satellite periods and distances, Roemer calculated a velocity of 214,000 kilometers per second (k/s). James Bradley's discovery of the *aberration of starlight provided a second means of estimating the speed of light, since the aberration angle depends upon the ratio of the speeds of the observer and of light. If extended to the speed of light, Bradley's calculations would have produced about 264,000 k/s.

In the nineteenth century, French physicists Armand-Hippolyte-Louis Fizeau and Jean Bernard Léon Foucault made terrestrial measurements of the speed of light by passing a beam through the gaps in a rapidly spinning toothed wheel. They obtained values of 315,300 and 298,000 k/s. The American physicist Albert A. Michelson improved on their experiments by measuring the interference fringes produced by a light beam split up in his "interferometer" and made to traverse slightly different paths through it. That raised the light velocity to 299,910 k/s (1879). Next Michelson attempted to detect the expected change in the speed of light caused by its motion through the hypothetical luminiferous *ether. The surprising null result of the 1887 ether-drift experiment suggested that the speed of light is constant, independent of the speed of the emitting body. This result became a crucial element in the pedagogy, and perhaps in the discovery, of Einstein's *relativity theory. Subsequent measurements of the speed of light over precisely measured distances and timed electrically ended in the spurious agreement of several sets of pre-1941 measures at 299,776 k/s. Holding Michelson in awe, followers ended their searches for flaws in their experimental apparatus when their measures agreed with his. New technologies later increased the accepted speed by nearly 17 k/s, more than four times Michelson's purported 4 k/s margin of error, to 299,792.5 k/s. Newer technology, including radiotelemetry, also improved the determination of planetary distances and the astronomical unit.

Dorothy Michelson Livingstone, *The Master of Light: A Biography of Albert A. Michelson* (1973). Albert Van Helden, *Measuring the Universe: Cosmic Dimensions from Aristarchus to Halley* (1985).

NORRISS S. HETHERINGTON

LINNAEUS, Carl (1707–1778), founder of the modern system of botanical classification.

Carl Linnaeus, who was born 23 May 1707 at Södra Råshult and died 10 January 1778 at Uppsala, was ennobled in 1762, taking the name von Linné. His parents were the assistant vicar (later vicar) Nils Linnaeus and his wife, Christina. Through his father, Linnaeus acquired an early interest in botany, which was further encouraged by his teacher Johan Rothman at the grammar school at Växjö. Here he had to learn Boerhaave's medical and Tournefort's botanical systems. In 1727 he began to study medicine (of which botany was a part) at Lund University, but only a year later he went on to the larger university of Uppsala. The elderly professors of medicine, Lars Roberg and Olof Rudbeck, Jr., were not vigorous. Linnaeus soon took over instruction in botany. As early as 1729, he produced a short pamphlet on the sexuality of plants, "Praeludia sponsaliorum plantarum," in which he set out the basis of his system. He also compiled a flora of the Uppsala area in manuscript form, "Hortys Uplandicus" (1730). In the summer of 1732 he undertook a journey to Lapland to study its plant life: however, his "Iter Lapponicum" remained unpublished until 1889. In Linnaeus's time a medical student needed to travel to a foreign university before he could obtain his doctor's degree. Linnaeus began a three-year period abroad in 1735. At Harderwijk in Holland he obtained his medical degree with a thesis on ague. After that he was engaged by the merchant Georg Clifford to care for his garden and library.

While serving Clifford, Linnaeus published *Systema Naturae* (1735), in which he formulated his botanical sexual system. Botany then was confused and complex. Whether one followed Aristotle or the later systematists Andrea Cesalpino, Caspar Bauhin, or Jospeh Pitton de Tournefort, there were problems. They differed over whether to classify plants according to color, size, the corolla, or the fruit. From Rudolf Jakob Camerarius and Sebastien Vaillant, Linnaeus had learned that plants could be regarded as possessing sexuality, an insight he now developed as the basis of a new classification. By counting the stamens in the flower and noting how they were arranged, he divided them into twenty-four classes, and by counting the styles on the pistil, he divided them into orders, from which followed genera and species. This work appeared in twelve editions during Linnaeus's lifetime (the last in 1766–1768).

During his three years in Holland, Linnaeus published seven substantial works besides the *Systema Naturae*: *Hortus Cliffortianus* (1737), a catalogue of Clifford's garden; *Fundamenta Botanica* (1736), the basis of his method; *Critica Botanica* (1737), his rules for naming genera and species; *Genera Plantarum* (1737), a list of all the known genera of plants, arranged according to his system; *Flora Lapponica* (1737), an application of the system to the plants of Lapland; *Classes Plantarum* (1738), the classes of plants; and a historical survey of all botanical systems from that of Cesalpino to Tournefort's and his own. All this work rested on the presentation and application of the new sexual system.

Linnaeus's system was rapidly accepted by the leading botanists, although not without criticism. Among the critics were Georges-Louis, Comte de Buffon, and other French botanists who protested against the artificial nature of Linnaeus's systematization. Linnaeus himself believed at first that he had devised a natural system, but he soon realized that it was artificial. Also at first he believed that species had been constant since the Creation, but he began to doubt this too after he found that the plant *Peloria* was a hybrid—a cross between two species.

The sexual system of plant classification was Linnaeus's first great achievement. The second, the *Species Plantarum* (1753), listed every plant species then known in the world, about 8,000 in all. Here he presented the binary nomenclature, where each plant had a name of two words. Previously botanists had indicated the species by a generic name plus a long description; Linnaeus now prescribed that the first name should indicate the genus and the second the species. His third great contribution was his clear descriptions of species and his terminology for the parts of a plant essential to its identification. He had a keen eye for distinguishing features and gave short and concise descriptions. Linnaeus gave botanists a common scientific language.

After returning from Holland, Linnaeus worked as a doctor in Stockholm. In 1741, he was professor of medicine at Uppsala University, where he concentrated on botany but also lectured on dietetics. He was one of the founders of the Royal Swedish Academy of Sciences (1739), physician-in-ordinary to the royal family (from 1747), custodian of the royal cabinets, and member of many foreign academies. On social questions, he shared the mercantilism and economic thinking that prevailed in Sweden at that time. He always proclaimed the value of natural science and natural history. In the 1740s he visited various provinces of Sweden on commission from the Estates to discover useful plants and natural resources. Linnaeus regarded himself as a man of the Enlightenment and as a traditional Christian. He had a rational view of the economic utility of science, but at the same time he nurtured an almost religious feeling for the beauty of nature and the magnificence of Creation.

Linnaeus was the first to place man among the mammals and coined the name *Homo sapiens*, but he also believed in men with tails and cavemen (troglodytes) and other semihumans (anthropomorpha). After his journey to Holland, during which he also paid short visits to London, Oxford, and Paris, he never went abroad again, although he sent his pupils off to virtually every corner of the world—to Asia and America, to Iceland and Australia. In turn they sent him exotic plants and accounts of their travels. Many foreign students came to Uppsala to hear his lectures. Linnaeus was the first Swedish natural philosopher to achieve international fame. That helped him in his efforts to give science some status in Sweden.

William Blunt, *The Complete Naturalist: A Life of Linnaeus* (1971). James R. Larson, *Reason and Experience: The Representation of Natural Order in the Work of Carl von Linné* (1971). Tore Frängsmyr, ed., *Linnaeus, the Man and His Work* (1983, 1994). Lisbet Koerner, *Linnaeus: Nature and Nation* (1999).

TORE FRÄNGSMYR

LIPIDS. See CARBOHYDRATES AND FATS.

LITERATURE AND SCIENCE. By today's conventional wisdom, literature and science might be regarded as the opposite ends of a spectrum of disciplines. The writing and analysis of literature is humanity's most personal and subjective activity, science our most impersonal and objective activity, while fields like history and the social sciences fall between

By the glimmer of the half-extinguished light, I saw the dull, yellow eye of the creature open, it breathed hard, and a convulsive motion agitated its limbs.
. . . I rushed out of the room.

Victor Frankenstein beholds his creation for the first time. From Mary Shelley, *Frankenstein, or, The Modern Prometheus* (1831), frontispiece by Theodore von Holst.

these extremes. However, scholars approaching literature as a phenomenon amenable to scientific analysis and critics characterizing scientific research as a process analogous to writing narratives have attacked this viewpoint for over a century. The often-ignored literature of *science fiction has also presented itself as a meeting place for disparate literary and scientific interests.

These efforts to blend or blur the lines between literature and science would have seemed unnecessary centuries ago, when modern science emerged as a recognizable discipline. To learned people of the seventeenth and eighteenth centuries, recalling the ideal of the "Renaissance man," an appreciation for literature and attentiveness to new natural knowledge were equally important and respectable aspects of the universal wisdom to which they aspired. Isaac *Newton may have devoted as much effort to crafting phrases and sentences as he did to solving equations, and his *Opticks* (1704) can be considered excellent literature as well as a pioneering scientific text. Poets like John

Milton and Alexander Pope expressed interest in and admiration for new scientific ideas. To be sure, the natural philosopher, mathematician, and alchemical adept sometimes figured as fools in literature, unsociable misfits preoccupied with silly research while oblivious to the real world. Characters in Molière's *The Learned Ladies* (1672), Thomas Shadwell's *The Virtuoso* (1676), and Aphra Behn's *The Emperor of the Moon* (1687) come to mind. Yet such portrayals, which endure in stereotypes of the "absent-minded professor," were more good-natured than vindictive.

Attitudes changed in the nineteenth century, when applied science and technology was visibly transforming society, and new findings in geology and biology challenged traditional views of human development. A fear of scientific progress surfaced in early works of science fiction that harkened back to the old stories of Prometheus and Faust to represent scientists as dangerous overreachers, as in Mary Shelley's prototypical *Frankenstein* (1818) and Honoré de Balzac's *In Search of the Absolute* (1834). Yet Johann Wolfgang von Goethe's *Faust* (1808, 1832) provided a more positive picture of the boundless quest for knowledge, and other important writers expressed enthusiasm about science and technology—in poems like William Wordsworth's "Steamboats, Viaducts, and Railways" (1833); in positive portrayals of scientists such as Daniel Doyce of Charles Dickens's *Little Dorrit* (1857); and in novels by George Eliot and Thomas Hardy that displayed a keen interest in evolution. The emerging figure of the scientific detective, exemplified by Arthur Conan Doyle's Sherlock Holmes, also indicated admiration for science. In turn, scientists sometimes approached their work in ways reflecting familiarity with literature, as in the psychological theories of Sigmund Freud and the "thought-experiments" of Albert Einstein, which employed narrative patterns to garner insights into reason and *relativity. Alfred Nobel, in establishing his prizes, regarded superior achievements in literature along with successful pursuits of peace as the two non-scientific activities that most merited annual commemoration.

In the twentieth century, "literature" came to include not only the works of celebrated authors but also the products of a community of sophisticated writers and university-based critics who were generally baffled or bored by revolutionary new developments in mathematics, physics, and astronomy. Instead, they looked backward, newly attentive to traces of ancient myths in forms of contemporary expression, a critical approach paradoxically given a scientific veneer by structuralists like Claude Lévi-Strauss, who claimed that such mythic patterns could be detected and studied in an objective, unambiguous manner. Literary attention to scientists was rare—with significant exceptions including Sinclair Lewis's *Arrowsmith* (1925), Bertolt Brecht's *The Life of Galileo* (1943), and Friedrich Durrenmatt's *The Physicists* (1962)—though the idiosyncratic fables of Jorge Luis Borges and Italo Calvino at times provocatively played with scientific ideas. The term "science" now referred not to a loose association of gifted amateurs but rather to an institutionalized enterprise where educated professionals communicated with each other through narrowly focused and rigidly formatted papers. To outsiders, science was losing any sense of individuality or poetry. In 1959, novelist and former physicist C. P. Snow lamented in a famous speech the gap between the separated "two cultures" of science and the humanities.

Snow's controversial speech provoked some valuable reactions. In literature, alongside growing numbers of science fiction writers, authors like Kobo Abe, Aleksandr Solzhenitsyn, Thomas Pynchon, Don DeLillo, and Umberto Eco have dealt with contemporary science in realistic novels. Some scientists call for improvements in scientific writing, and a few, notably Stephen Hawking and Stephen Jay Gould, communicate scientific ideas to a wide general audience in polished works of nonfiction.

Another conspicuous response to Snow was a new field of literary criticism, "Literature and Science," devoted not only to explicating literary works in the context of scientific concepts but also to analyzing science itself as a sort of literature, considering scientific theories and texts in terms of flawed imagery, unacknowledged biases, and cultural conditioning. Scientists reading these studies may feel that their scientific underpinnings are often shallow, inaccurate, or misapplied, as Paul Gross and Norman Levitt argued in their *Higher Superstition* (1994). While critic Andrew Ross was preparing a collection of essays, *Science Wars* (1996), in response, the physicist Alan Sokal submitted an essay filled with what he regarded as a hodgepodge of nonsensical pseudoscientific jargon to a journal edited by Ross. Sokal then heralded the acceptance and publication of his article as evidence of the movement's intellectual bankruptcy in Sokal and Jean Bricmont's *Fashionable Nonsense* (1998). Although much that is interesting and significant has emerged from scholars in this area, working scientists proud of the predictive accuracy and practical applications of their theories might understandably regard arguments from literary critics about the purportedly shaky foundations of the scientific enterprise as irrelevant if not risible.

Martin Burgess Green, *Science and the Shabby Curate of Poetry* (1965). N. Katherine Hayles, *Chaos Bound* (1990). Stuart Peterfreund, ed., *Literature and Science* (1990). Joseph W. Slade and Judith Yaross, eds., *Beyond the Two Cultures* (1990). C. P. Snow, *The Two Cultures* (1993). Donald Bruce and Anthony Purdy, eds., *Literature and Science* (1994). Roslynn D. Haynes, *From Faust to Strangelove* (1994).

GARY WESTFAHL

LOMONOSOV, Mikhail Vasilievich (1711–1765), Russia's first eminent natural philosopher, who still occupies an important symbolic place in the Russian world of scholarship. Moscow University and a number of other institutions and academic prizes are named after him.

Lomonosov was born in the small village of Mishaninskaia near the far north of European Russia, on the White Sea. Although legally peasants, Lomonosov and his family enjoyed a freedom not known to the serfs on the estates of central Russia. Lomonosov's father, an active merchant, owned several fishing and cargo ships, and his mother was the daughter of a deacon. As a child Lomonosov learned to read and write, in both Russian and Church Slavonic.

Desiring further education, Lomonosov in 1730 applied for admittance to the Slavic-Greek-Latin Academy in Moscow, the best higher educational institution in Russia at that time, although devoted to theology and the preparation of clergy. Since peasants could not attend the academy, Lomonosov concealed his origins; he claimed to be the son of a priest, which his knowledge of Church Slavonic supported. By the time his superiors learned the truth, Lomonosov had so impressed them by catching up with his classmates (most of whom were younger than he) in Latin and surpassing them in other subjects that they permitted him to stay.

In 1735 the Imperial Academy of Sciences in St. Petersburg, then just ten years old, requested monasteries and ecclesiastical academies to send students to its fledgling university to study under foreign academicians. Lomonosov was chosen. In St. Petersburg he learned mathematics and physics. A year later the academy sent him to western Europe to study chemistry and mining. For almost five years Lomonosov studied natural science at the universities of Marburg and Freiburg. The professors who made the deepest impression upon him were Christian Wolff at Marburg and Johann Friedrich Henkel at Freiburg.

After returning to St. Petersburg in 1741 with his German wife, Elizabeth Zilch, Lomonosov was made an adjunct of the Academy of Sciences in physics and later professor of chemistry. In 1748 he opened the first scientific chemical laboratory in Russia, equipped with *balances and other equipment similar to what he had seen in Europe. Later he became head of the geographical department of the Academy of Sciences.

Lomonosov's scientific activity can be divided into three phases. From 1740 to 1748 he concerned himself with speculative physics, particularly the corpuscular philosophy, the nature of heat and cold, and the elasticity of air. He compiled a syllabus in physics and in 1746 delivered the first public lecture on the subject ever presented in the Russian language. Most of his scientific writings, however, were in Latin. From 1748 to 1757, after the construction of his chemical laboratory, Lomonosov worked on the characteristics of saltpeter, the nature of chemical affinity, the production of glass and mosaics, the freezing of liquids, and the nature of mixed bodies. From 1757 to his death in 1765 Lomonosov devoted his time to scientific administration, exploration, mining, metallurgy, and navigation. Throughout his professional life he wrote poetry and promoted the Russian language and Russian history.

Lomonosov's most significant work in natural philosophy was his extension of the corpuscular or *mechanical philosophy common in the seventeenth and early eighteenth centuries to a wide variety of phenomena. He liked to describe nature in concrete pictures and mechanical models, and to reason by analogy from them. This approach, applied literally and speculatively, sometimes led Lomonosov to concepts that seem prescient to modern readers, such as the lowest possible temperature occurring when all particles are motionless. However, much Soviet literature on Lomonosov contains exaggerated claims about his achievements. No truly definitive study of his scientific work exists.

Lomonosov symbolized emerging Russian scholarship, talented but still only partially developed. He had brilliant ideas but lacked discipline and scattered his efforts over too broad a front. Nonetheless, he was an unprecedented phenomenon, a native Russian champion of science and learning who, in important areas, stood at the forefront of knowledge. He would serve as a model for young Russian scientists for generations.

Henry M. Leicester, ed., *Mikhail Vasil'evich Lomonosov and His Corpuscular Theory* (1970). Galina E. Pavlova and Aleksandr S. Federov, *Mikhail Vasil'evich Lomonosov: His Life and Work*, trans. Arthur Aksenov (1984).

LOREN R. GRAHAM

LONG FIN-DE-SIÈCLE, THE. The term "fin-de-siècle," originally the title of a French play of 1888, has become a catchword for the turn from the nineteenth to the twentieth century. Although originating in a literary and artistic reaction against a worldview increasingly dominated by science, the term has also proved useful in discussing new and unexpected developments in the sciences of the period. The fin-de-siècle, which took quite different forms in different countries, describes a sensibility more than a set of ideas or doctrines. The works of Henri Louis Bergson and especially of Friedrich Nietzsche, which were read and vigorously debated from the 1890s, gave voice to a widespread feeling of decadence: their work, with its antihistoricism, *Lebensphilosophie*, use of myth, antihumanism, and elitism, was sharply opposed to the dominant current of unconditional celebration of progress. This optimistic expectation of a civilized, democratic society, and invocation of science as a solution to all problems of humanity, is well represented in the works of Herbert Spencer. Artistic reactions included inwardness (the retreat from public progress); solipsism (the denial of an objective world); sensuality (in Art Nouveau); the portrayal of feelings of decay, decadence, and disorientation; and a radical break with academic conventions and traditions (in cubism and atonal music). The fin-de-siècle thus meant both withdrawal and avant-garde, the celebration of scientific and technological progress (as in futurism) and the fear of its moral, social, and political consequences.

There was talk in the 1890s of science's bankruptcy for having failed to provide orientation and moral standards. Conversely, science itself could be invoked to express the fear of decadence: biology (like books on degeneration and the warnings of eugenicists) and *thermodynamics (as Hermann von *Helmholtz's thermodynamic theory of the heat-death of the earth) seemed to confirm fears of the future. The pace of scientific and technological development led to all kinds of new ailments, observers claimed, such as neurosis. The fin-de-siècle world consisted of rapid industrialization, growing cities, professionalization, social differentiation, and national competition. These developments showed in the exponential increase of scientists, universities, international congresses, exhibitions, and practical and metrical standardization; and in technological innovations like electricity, the tram, the *phonograph, and the *telephone.

The opinion that science might have exhausted itself and that all the universe's fundamental laws had been described was flatly contradicted by a series of startling new discoveries in the last decade of the nineteenth century, including Wilhelm Conrad Röntgen's *X rays (1896), Henri Becquerel's radioactive uranium

(1896), and Joseph John *Thomson's discovery of *electrons (1897); and by the development of new substances like aspirin (1899), cellophane (1908), Bakelite (1908), and synthetic ammonia (1910). This swift pace of discoveries and theories seemed to undercut a search for unity and coherence; the best example here is German physicist Max *Planck, whose *quantum physics of 1900 upset traditional beliefs, including his own, in the unity of science. He and others only reluctantly accepted the consequences of his results. Albert *Einstein's papers on *relativity from the first decade of the century were to shake the core of Newtonian physics, long considered unshakable. The mechanical worldview that dominated the physical sciences in the second half of the nineteenth century lay in shambles. The French physicist Henri Poincaré, overwhelmed by the latest developments, was a prominent advocate of this new demand for flexibility and practicability; his "conventionalism" advocated tolerance and diversity of approaches, ranking truth claims secondary.

*Biology was also riven by theoretical debates during the fin-de-siècle. Experts and others argued over mechanistic and vitalistic interpretations of the basic nature of life and development, and over the applicability of biology to human beings, whether in *eugenics, *anthropology, or sociology. Probably the single most influential text by a biologist of the period was Ernst Haeckel's *Die Welträtsel* (*The Riddle of the Universe*, 1899), which synthesized a mechanistic, antireligious "monistic" worldview from the creation of the cosmos and solar system to the evolution of human beings and society.

*Cytology, the avant-garde life science of the last three decades of the nineteenth century, was increasingly troubled by its inability to frame manageable questions and then answer them. The nature of life itself, and its manifestation in the development of organisms, was murky. Biologists split roughly along two lines: those who, like Hans Driesch, insisted on living matter's ability to organize itself, and those, like August Weismann, who insisted that biology must be in the end a mechanical elaboration of physics and chemistry. In 1900 Gregor *Mendel's work of the 1860s on particulate inheritance was simultaneously rediscovered by three different sets of people; from the 1910s the new experimental science of genetics came increasingly to eclipse cytology.

Furthermore, thanks to the unprecedented freedom of Western societies at the turn of the century, individuals explored unknown territories and pursued alternatives to established science, in numerous countercultures. These ranged from feminism, psychoanalysis, and vegetarianism to movements against academic medicine and new quasi-religious creeds like anthroposophy and theosophy. Prominent scientists studied mysterious phenomena like spiritualism and miracles and challenged the relation between religion and science anew. Some established their own scientific creeds, like the monism of the evolutionary biologist Haeckel and the *energetics of the chemist Wilhelm Ostwald; others, like Pierre Duhem, as ardent a Catholic as a physicist, endeavored to find ways for science and religion to coexist peacefully. All these currents attempted to respond to science's perceived deficiency: that it had exiled mystery, spirit, poetry, and morality from the world.

Two catastrophes ended the long fin-de-siècle: the sinking of the *Titanic* in 1912, a confirmation to those who warned against technological hubris, and the First World War, a demonstration beyond the worst fears of those who thought that science brought more swords than pruning hooks into the world.

Carl E. Schorske, *Fin-de-Siècle Vienna* (1981). Carl-Gustav Bernhard, Elisabeth Crawford, and Per Sèrböm, eds., *Science, Technology and Society in the Time of Alfred Nobel* (1982). Mikulás Teich and Roy Porter, eds., *Fin de Siècle and Its Legacy* (1990). Anson Rabinbach, *The Human Motor: Energy, Fatigue, and the Origins of Modernity* (1992).
MATTHIAS DÖRRIES

LONGITUDE is the angular distance east or west of a standard meridian, the complementary coordinate to latitude (the angular distance north or south of the equator), both used in expressing geographical location (*see* GEOGRAPHY). It has been important in science because of its significance for astronomical observation and because of the influence of attempts to devise a method for finding longitude when at sea.

Various standard meridians have been determined by the position of an important observatory, a large port, or a significant geographical feature such as the Canary Islands, as used by Gerardus Mercator. The international agreement of 1884 to accept the line of longitude through Greenwich as the world's prime meridian owed much to the diligent work of astronomers at the Royal Observatory, but also reflected Britain's imperial dominance and the quality of its official sea charts.

Difference in longitude amounts to difference in local time (places distant by fifteen degrees differ by one hour). The time at a distant meridian can be found in a number of ways and compared with local time determined astronomically. A lunar eclipse, for example, might be timed, and its occurrence compared with the time given in ephemerides calculated for a standard meridian, making allowance for a variety of factors such as parallax. *Galileo suggested that Jupiter's four moons, which he discovered, offered a handy celestial clock, with numerous moments of coincidence with the limb of the planet and between the moons themselves. Simultaneous timing of explosions of rockets, or direct connection by geodetic survey, provided other methods in favorable circumstances.

Methods viable on land were rarely useful at sea, where the problem of finding longitude, made acute by voyages of discovery into the Atlantic and Indian Oceans, became a synonym for the impossible. Three influential methods were proposed in the sixteenth century. The chronometer method involved the apparently simple expedient of carrying a portable watch set to standard time. Suggested by Gemma Reiner (called Frisius) in 1530, much mechanical ingenuity would be required to bring timekeepers up to the level of seagoing robustness and accuracy necessary to make this method useful. The lunar-distance method, proposed by Johann Werner in 1514, involved the measurement of the moon's position with respect to certain stars, and the calculation of standard time from the measurement via tables prepared by astronomers. At this stage, the lunar theory, the stellar positions, and the seagoing instrument to make the measurement did not exist. A third method did not depend on timekeeping but on global accounts of the distribution of magnetic variation with latitude (which could be measured directly) and longitude (which could not). Martin Cortes maintained the currency of this idea through his much-translated and republished textbook, *The Arte of Navigation* (1561).

Attempts to effect these methods led to many significant developments—from the beginnings of state-sponsored *observatories in Europe to the invention of the spring-regulated watch. The most famous of the rewards offered for a solution to the longitude problem—the £20,000 prize established in Britain in 1714 to be administered by a Board of Longitude—resulted in enormous interest in the problem in the eighteenth century, and eventually the completion of both the chronometer and lunar-distance methods, seen as rivals throughout their development but used in complementary ways in the navigational practices of the nineteenth century. Both Johann Tobias Meyer (posthumously), for lunar work, and John Harrison, for a chronometer, received part rewards from the Board, Harrison's being later made up to the maximum amount by the British Parliament.

See also GEOGRAPHY.

William J. H. Andrewes, ed., *The Quest for Longitude* (1996).
JIM BENNETT

LOPES, Jose Leite. See LATTES, CESAR, AND JOSE LEITE LOPES.

LOW-TEMPERATURE PHYSICS. The field of low-temperature physics, or cryogenics, emerged in the late nineteenth century. In 1877, two researchers succeeded in liquefying oxygen independently within days of each other. Raoul-Pierre Pictet, a Swiss physicist, and Louis Paul Cailletet, a French mining engineer, both cooled oxygen gas under pressure, then rapidly expanded the volume to condense the gas. Cailletet soon reproduced the feat with nitrogen. In 1895, Carl von Linde in Germany and William Hampson in England developed a method for industrial-scale production of liquid air, which aided subsequent cryogenics research. The three leading low-temperature laboratories—at the University of Leiden (under Heike Kamerlingh Onnes), the University in Cracow (Karol Olszewski), and the Royal Institution in London (James Dewar)—engaged in a race to lower temperatures and the liquefaction of hydrogen and, after its discovery on Earth in 1895, helium. In 1898, Dewar won the race for hydrogen, using a variation on Linde's technique to reach about 20° Kelvin. In 1908, Kamerlingh Onnes used liquid hydrogen to cool helium enough to condense it at low pressure, at around 5° K. The Leiden laboratory thereafter enjoyed a fifteen-year monopoly in the production of liquid helium.

The early history of cryogenics illustrates the proliferation of academic research laboratories in the late nineteenth century, and their move into fields requiring such relatively expensive apparatus. The work involved chemists, engineers, skilled glassblowers, and instrument makers as well as physicists. Kamerlingh Onnes's success with liquid helium owed as much to his ability to form a team with physical, mechanical, and chemical expertise as to his high standards of experimental precision and keen grasp of thermodynamic and electromagnetic theory.

Cryogenics demonstrates the interpenetration of industry and science in the second industrial revolution—in particular, the budding refrigeration industry, which had emerged as a rival to natural ice in the late nineteenth century, especially for brewing lager beer and shipping meat to Europe from Argentina and New Zealand. Industrial uses of liquid air and its components, such as liquid oxygen for oxyacetylene blowtorches, spurred the formation of firms such as Linde Air (founded by von Linde), British Oxygen Company, and L'Air Liquide. Dewar's development of a silvered, double-walled flask with an intervening vacuum led quickly to a commercial market in thermos bottles (although Dewar

Heike Kamerlingh Onnes, professor of physics at the University of Leyden, seated in front of his elaborate equipment for low-temperature work. He and his colleagues succeeded in liquefying helium and in discovering superconductivity in 1908 and 1911, respectively.

failed to patent his flask). Kamerlingh Onnes had ties to Dutch industry. The industrial relevance of low-temperature research induced national governments to sponsor it in their standards laboratories such as the U.S. National Bureau of Standards and the German Physikalisch-Technische Reichsanstalt (PTR).

Academic physicists instigated their own low-temperature programs. The elucidation of specific heats at low temperatures, explored by Dewar and extended by Walther Nernst and F. A. Lindemann, provided important evidence for the fledgling quantum theory before World War I and helped establish low-temperature research as a fruitful field of physics. Pyotr Kapitsa's investigation of magnetic effects at low temperatures in Cambridge, which he continued in Moscow, and William Giauque's work on magnetic cooling and specific heats in Berkeley illustrate the spread of low-temperature physics in the 1930s. Cryogenic techniques would find application after World War II in the production of the first thermonuclear fusion weapons, in rocket propellants, and in bubble chambers for particle physics experiments. The volatile liquids and high pressures of *cryogenics required elaborate safety precautions, although they did not always prevent catastrophic explosions.

Two important lines of research emerged from peculiar phenomena observed at low temperatures. In 1911, Kamerlingh Onnes and his collaborators found that electrical resistance in mercury suddenly vanished at 4° K. This "superconductivity," as Kamerlingh Onnes called it, puzzled theorists for decades. In 1933, Walther Meissner and Robert Ochsenfeld at the PTR in Berlin found that magnetic induction as well as electrical resistance disappears in a superconductor, and hence showed that superconductivity included more than its name indicated. In 1957, John Bardeen, Leon Cooper, and John Schrieffer of the University of Illinois produced a satisfactory microscopic explanation of superconductivity, based on quantum mechanical coupling of electrons with opposite spin.

Superconductivity promised spectacular technological applications in low-loss electrical power transmission and high-power superconducting electromagnets. But known materials that exhibited superconductivity proved too fragile for power transmission; and, like heat, high magnetic fields, as in electromagnets, destroy the superconductive state. Hopes for new technologies rekindled with the discovery of so-called type II superconductivity in 1961, which persevered in the presence of magnetic fields, and then flared anew with the announcement in the mid-1980s of a class of ceramic materials that stayed superconductive at temperatures up to 100° K.

The second peculiar phenomenon that stemmed from observations by Kamerlingh Onnes was a drop of density in liquid helium below about 2° K. With the spread of low-temperature physics, unexpected results with liquid helium began to accumulate, suggesting two different states of helium: normal helium I and low-temperature helium II. In 1937 and 1938 several physicists established that the viscosity as well as the density of helium II seemed to vanish, and that it could form a thin film that swept up the sides of vessels; Kapitza termed the effect "superfluidity." In 1955 Richard *Feynman arrived at a microscopic theory of superfluidity based on quantization of vortices in the fluid. The theories of both superconductivity and superfluidity rely on interactions among individual particles or atoms; Pauli's exclusion principle does not apply, and Bose-Einstein statistics instead of Fermi-Dirac statistics govern the behavior. In other words, superconductivity and superfluidity are macroscopic quantum phenomena; hence their

novelty and interest, and the difficulty in accommodating them within physical theory.

Kurt Mendelssohn, *The Quest for Absolute Zero* (1977). Kostas Gavroglu and Yorgos Goudaroulis, *Methodological Aspects of the Development of Low Temperature Physics 1881–1956* (1989). Per F. Dahl, *Superconductivity: Its Historical Roots and Development from Mercury to the Ceramic Oxides* (1992). Ralph G. Scurlock, ed., *History and Origins of Cryogenics* (1992).

PETER J. WESTWICK

LUNAR SOCIETY OF BIRMINGHAM. The Lunar Society was an informal association, most active between the late 1770s and the early 1790s. Never numbering more than fourteen members, without rules or formal charter, and centered on Birmingham, not London, the Society would have been unremarkable were it not for the fact that it both represented and indeed crystallized many of the forces that were transforming Britain from a rural society with power vested in lands and rents to an urban industrialized society. The Royal Society of London, centered in the metropolis, could no longer meet the country's needs and interests in natural philosophy, *chemistry, and the "arts" or technology. In London, the Society of Arts was established in 1754 to promote "improvements"; outside London, Literary and Philosophical Societies sprang up in Bristol, Bath, Manchester, Derby, and Newcastle.

Matthew Boulton, the Birmingham industrialist, sought out people interested in encouraging natural philosophy, industry, and social reform. Among his early associates were Erasmus Darwin, soon to become the leading doctor in provincial England, and John Whitehead, a clock and instrument maker who also published works on *mineralogy. Other important members were James Watt, Bolton's partner and the inventor of the separate condenser for the steam engine; Richard Lovell Edgeworth, who designed and constructed conveyances; James Keir, who experimented on alkalis and translated French chemical works; Josiah Wedgwood, founder of the famous china company; and the banker Samuel Galton. In 1781, Joseph Priestley, then one of Europe's leading chemists as well as a Nonconformist minister, completed the group.

Between them, the members, most of whom were members of the Royal Society, made contributions to almost every aspect of Britain's industrialization: the building of roads and canals, the replacement of water mills by the rotary steam engine, the design and running of factories, the understanding of chemical processes, the reform of agriculture, the transformation of everyday life by new consumer goods such as china tableware, and public education.

The Lunar Society (so-called because it met on the Monday nearest the full moon to assist travel at night) started to decline in the early 1790s. Priestley, whose house had been burned by a mob who distrusted his radical ideas, moved to the United States in 1794. The other members, who had not recruited a younger generation, met less frequently as their successful businesses demanded more of their time. By 1800, the society had vanished. The members had inculcated their beliefs in their descendants, many of whom became associated with leading Whig institutions such as the *Edinburgh Review*. Edgeworth's daughter Maria became a famous novelist, Francis Galton won a reputation for his work on heredity, and his cousin, Charles *Darwin, a descendant of Erasmus Darwin, had one Wedgwood for an uncle and another for a wife.

Robert E. Schofield, *The Lunar Society of Birmingham: A Social History of Provincial Science and Industry in Eighteenth-Century England* (1963).

RACHEL LAUDAN

LYELL, Charles (1797–1875), originator of the doctrine of uniformitarianism in *geology.

Lyell came from a Scottish family with a long tradition of serving in the English navy. Although he was educated and spent most of his life in England, he remained attached to the intellectual traditions of the Scottish Enlightenment. His father intended him to be a barrister. While studying at Oxford, he developed problems with his eyesight that made reading difficult and the prospect of a legal career unappealing. The geological lectures of William Buckland fascinated Lyell, and his ambitions began to shift. In 1819, he joined the Geological Society of London. He took several field trips to the Continent, some of them with another budding geologist, Robert Murchison. Sicily, in particular, impressed him, with its evidence of major geological changes in the recent past.

In the 1820s, Lyell read Jean Baptiste de *Lamarck's *Philosophie Zoologique* (1809) and John Playfair's *Illustrations of the Huttonian Theory of the Earth* (1802). Lamarck's suggestion that one species could transmute into another appalled Lyell as speculative and irreligious. Playfair's exposition of Hutton's theory of an indefinitely habitable earth, on the other hand, attracted Lyell, a deist like his fellow Scots James Hutton and John Playfair. Lyell thought that the deity would have approved of Hutton's plan. Playfair's use of the vera causa method also attracted him. Lyell began planning an ambitious book to bring Huttonianism up to date, establish the principles of reasoning in geology, and use them to discredit the theories of the transmutation of species and the cooling, contracting earth. He also hoped that the book would make his reputation and bring in enough money so that he could live comfortably and help his unmarried sisters.

In 1830, Lyell published the first of the three volumes of his *Principles of Geology*. He introduced the volume with a cautionary history to make plain the perils of an ill-chosen methodology, described the presently observable geological causes that would have been adequate to produce past geological effects, and proposed a theory of climate change based on alternating positions of land and sea that made the supposition of a cooling earth unnecessary. In the second volume (1832), he tackled the vexing question of the *fossil record, maintaining that from the first appearance of fossils, although individual species had become extinct and been replaced by newly created ones, the same broad classes had always existed. In the third volume (1833), he responded to his critics and introduced a new classification of the Tertiary strata.

The *Principles* succeeded brilliantly. It was a master work, rigorous in its argument yet accessible to a wide, educated public. It won respectful (though deeply critical) reviews from George Poulet Scrope and William Whewell. Lyell spent much of the rest of his life revising, reworking, and abstracting the *Principles*, which went into a twelfth edition after his death. In 1832, he married Mary Horner, an intelligent woman who acted as his secretary and became expert in the fossil shells so important to Lyell's classification of the Tertiary. In 1831–1832, he taught geology for a brief period at King's College London but did not enjoy it, and never again took a university position. From 1834 to 1836, he was president of the Geological Society. He and Charles *Darwin started a lifelong friendship when Darwin returned from the *Beagle* voyage in 1836.

In 1841–1842, the Lyells traveled in the United States, where Charles lectured at the Boston Atheneum. They found much to admire, and returned to the United States in 1845–1846 and 1852–1854. Lyell's *Travels in North America* (1845) became a small classic. In 1848, he was knighted. Soon, he began worrying once again about whether one species could have evolved into another. In 1863, he published the *The Antiquity of Man*, assembling a broad range of evidence for man's great age and descent from the lower animals, but leaving his readers to draw their own conclusions. In 1864, he announced at the Royal Society that Darwin's argument had finally persuaded him, and between 1865 and 1868 he rewrote the *Principles* for the tenth edition to reflect his change in belief.

Lyell died in 1875. Although his theory had not won over most geologists, it had sharpened their thinking and been a decisive influence on Charles Darwin. Lyell's reputation declined after his death, only to be restored by the assessment of Sir Archibald Geikie, head of the Geological Survey of Great Britain, in his *Founders of Geology* of 1905.

Leonard Wilson, *Charles Lyell, The Years to 1841* (1972). Charles Lyell, *Principles of Geology*, ed. Martin Rudwick (3 vols., 1991). Leonard Wilson, *Lyell in America: Transatlantic Geology, 1841–1853* (1998). Derek J. Blundell and Andrew C. Scott, eds., *Lyell: The Past is Key to the Present* (1998).

RACHEL LAUDAN

LYSENKO AFFAIR. Lysenkoism was a doctrine of heredity espoused by the Soviet agronomist Trofim Denisovich Lysenko that contradicted modern genetics. Lysenko began propagating his views in the 1930s, gradually winning the support of many journalists, educators, administrators, and officials of the secret police. In 1940 Lysenko's most prominent opponent, the internationally known geneticist Nikolai *Vavilov, was arrested; he died three years later in prison. In 1948 Lysenko won a complete victory by gaining the backing of Stalin and the Soviet government. Departments of genetics in universities and research institutes were forced to follow the new official view. Lysenko became the autocrat of Soviet biology, often casting his critics as traitors to the Soviet Union. Several thousand Soviet biologists opposing him were arrested. Not until 1965, after the fall of his supporter Nikita Khrushchev, was Lysenko finally overthrown.

Commentators in the West commonly explain Lysenko's influence by a consonance between his support for the doctrine of the inheritance of acquired characteristics and Soviet ideological desires to "create a new Soviet man." If people can inherit improvements acquired from the social environment, so the argument goes, then revolutionary changes in society can quickly result in the improvement of human beings. This explanation cannot stand. Lysenko believed in the inheritance of acquired characteristics in plants and animals, but he opposed applying his views to humans; he regarded attempts to alter human heredity as examples of bourgeois influence on science. All of Lysenko's work concerned plants and animals of agricultural value, including wheat, tomatoes, potatoes, corn,

chickens, and dairy cows. His rise to prominence was linked to his efforts to aid Soviet agriculture at a time when forced collectivization had lowered productivity. None of his nostrums led to genuine improvement of agriculture. Rather, his monopoly of Soviet biology for several decades prevented the Soviet Union from receiving the agricultural benefits of modern genetics.

Lysenko's unsophisticated biological views were embodied in a vague "theory of nutrients" that assigned primary importance to the environment in the determination of heredity. He denied the existence of genes and disputed the importance of DNA. He regarded heredity as a property of the "entire organism." He claimed to have changed several species of plants into new ones by manipulating environmental conditions, claims that many biologists at the time recognized as false. Most of his methods, however, were less ambitious; for example, his attempt to accelerate the maturation of crops by soaking seeds before planting. Soviet administrators liked this approach because, if successful, it might allow harvesting before the fall frosts that plagued Soviet agriculture. Lysenko's most costly, even disastrous, biological experiments came toward the end of his career, when he caused severe damage to the dairy industry by breeding pedigreed cows with less valuable bulls, canceling the results of generations of careful animal husbandry.

China developed a Lysenkoist movement in the early phases of Mao Tse-Tung's rule. Between 1949 and 1956, a period in which Chinese leaders copied Soviet policies, Lysenkoism was the only officially sanctioned approach to genetics. Led by Luo Tianyu, a plant breeder in the Beijing area, Chinese Lysenkoism stressed populist and nativist themes, ascendant since the 1949 revolution, and won political favor more for these reasons that for any imagined contribution to agricultural technology. Mao's brief "Hundred Flowers" campaign of relative intellectual openness in 1956 sent Lysenkoism into permanent eclipse. The cost of Chinese Lysenkoism was delay in the deployment of orthodox genetics in agriculture rather than destruction of the sort wrought by the parent movement in the USSR.

David Joravksy, *The Lysenko Affair* (1970). Loren Graham, *Science, Philosophy and Human Behavior in the Soviet Union* (1987). Valerii Soifer, *Lysenko and the Tragedy of Soviet Science* (1994).

LOREN R. GRAHAM

M

MAD COW DISEASE. In 1986, Bovine Spongiform Encephalopathy (BSE) was described in the United Kingdom and dubbed "Mad Cow Disease" since the infectious agent caused brain tissue to become sponge-like and make the animal "mad." Ten years later, after thousands of cows had been infected and prematurely slaughtered so that they would not enter the human food chain, a new human disease appeared. Fifteen young U.K. residents were diagnosed with a similar brain-wasting condition, variant Creutzfeldt-Jakob disease (vCJD). CJD had been known for some time.

During its first five years of named existence, vCJD had struck ninety-six people in the United Kingdom as of June 2001, and perhaps two or three elsewhere. Ninety percent of the world's 180,900 cases of BSE had occurred in the UK. That prompted the scientific community to conclude that "the most likely way people got vCJD was from eating tainted beef."

BSE aroused scientific and policy debate for some years. In 1996, some warned that the Mad Cow crisis would be the *AIDS epidemic the UK never had. In 2000, when the disease migrated to continental Europe, the *Frankfurter Allgemeine* compared it to Europe's Black Death of the fourteenth century.

Mad Cow Disease became a continuing problem not only because of the threat to animal health—a real threat that claimed the lives of thousands of cattle—but also because of the scientific uncertainty of the length of incubation, the presence of the infective agent, and the actual vector of transmission. Scientists still do not know how people get vCJD in the first place. A recent report of the House of Lords again pointed to eating beef as the most probable assumption.

United Nations agencies responded to the challenge through the Office of International Epizootes (OIE), the Food and Agriculture Organization (FAO), and the World Health Organization (WHO), which created policies to eliminate the threat to animal and human health. Since the consensus held that the disease originated from feeding infectious animal products to cattle, the practice was stopped, surveillance implemented, and certain beef and beef products banned.

The infectious agent responsible for the disease could be a self-replicating protein called a prion. Or, what many believe more likely, it could be viral-like, with nucleic acids carrying genetic information. The agent cannot be identified easily in live animals or humans since it resides mainly in the central nervous system. Science has hedged its bets about the risk of BSE to humans: "theoretical," "hypothetical," "negligible," "remote," and "incalculably small" were the terms used to describe the risk to humans. But the loss to agriculture, trade, and other industries has been almost incalculably large, some $20 billion according to recent estimates. Some believe the response has extended well beyond the threat to human health.

The case of Mad Cow Disease illustrates how modern society blurs facts and fears. Fears generated by the BSE crisis spilled over to genetically modified foods, biomedical research, and other scientific programs. The episode also highlights some of the challenges of globalization. Multilateral organizations and multinational companies provide the resources for protecting public health, but public fear and emotion often impede reason and appropriate policy responses.

Scott Ratzan, ed., *The Mad Cow Crisis: Health and the Public Good* (1998).

Scott C. Ratzan

MAGIC, NATURAL. Natural magic may be defined as the use of hidden (or occult) forces to manipulate nature and produce marvelous effects. Although the proponents of natural magic tirelessly argued that the hidden powers they invoked were natural, not supernatural, opponents insisted that they lay outside the normal course of nature, and were demonic.

The intellectual origins of natural magic in the West have been traced back to the so-called *Hermetica*, a body of writings, mainly from the second and third centuries, that purported to contain the secret doctrine of Hermes Trismegistus, venerated by the Egyptians as the god Thoth and thought to have lived just

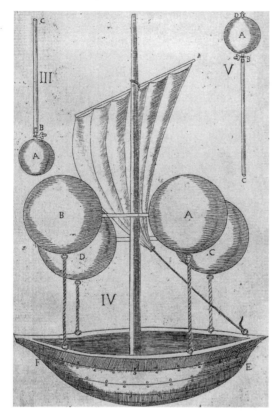

A seventeenth-century airplane supported by four large vessels emptied of air in the way a barometer was made. From Francesco Lana-Terzi, *Prodromo all'arte maestro* (1670).

but the survey of the whole course of nature....[I]t opens up unto us the properties and qualities of hidden things, and the knowledge of the whole course of nature....the works of magic are nothing else but the works of nature, whose dutiful handmaid magic is." As della Porta's famous work, *Magia naturalis* (1558), demonstrates, natural magic contrasted with traditional Aristotelian natural philosophy, whose goal was to explain normal, everyday occurrences by aiming to produce exceptional, unusual, and "marvelous" phenomena. For della Porta, natural magic was both the "consummation of natural philosophy" and its "practical part." The *Magia naturalis* combined the unusual and the mundane: techniques to produce animals of pre-determined colors, fruits that ripen out of season, steel that cuts through other metals, optical glasses to create strange illusions, ointments to improve the complexion, and methods of secret. The combination of practical improvements and theatrics made natural magic especially appealing to the Renaissance courts.

By the seventeenth century, natural magic had largely jettisoned its overriding concern with occult causes and had shifted its focus toward the "wonders" that might be produced with machines, hydraulics, and optical devices. Characteristic of this movement were the works of the Jesuit natural philosophers Athanasius Kircher and Gaspar Schott. Kircher, a polymath whose interests embraced mathematics, *acoustics, *optics, *astrology, *medicine, numerology, ciphers, philology, archaeology, *geology, and theology, wrote more than thirty encyclopedic volumes to exemplify his vision of *pansophia* (universal wisdom). Among his heterodox interests was *magnetism, long a province of natural magic. Kircher's dramatic demonstrations of magnetic and electrical phenomena at his laboratory/museum in Rome impressed visitors and revitalized interest in these phenomena.

Kircher never published his planned work on "Universal Magic." Instead, he handed over his notes to his disciple Schott, who published a four-volume compendium on natural magic, *Magia universalis* (1657–1659). This extraordinary work dealt with the production of wonders by means of optics, acoustics, mechanics, and physics. The entire work concerned natural effects that astound and amaze because they result from the manipulation of hidden forces, springs, levers, siphons, and so on. These works detailed a vast amount of data that bordered on, and was sometimes indistinguishable from, experimental science.

Discredited and ignored today, natural magic had an important impact on the *Scientific Revolution of the sixteenth and seventeenth

after the time of Moses. Despite its fraudulent attribution (not proven until 1614), the *Corpus hermeticum* exerted a profound influence on Renaissance natural philosophy. In her seminal work, *Giordano Bruno and the Hermetic Tradition* (1964), Frances A. Yates argued that the hermetic tradition inspired a new image of the natural philosopher as a powerful magus that intrigued Renaissance intellectuals with the possibility of understanding and controlling natural forces through magic. Long after the hermetic corpus's false pedigree was revealed, natural philosophers continued to cultivate the magical ideas that sprang from it. The religious and philosophical implications of hermeticism attracted no less a figure than Isaac *Newton.

Although Hellenistic in intellectual origin, natural magic was preeminently a *Renaissance science. The Italian natural philosopher Giambattista della Porta, Renaissance Europe's foremost exponent of natural magic, defined it as "nothing else centuries. Renaissance magic contributed to the emergence of a new conception of the scientific enterprise: the idea of science as a hunt for nature's secrets. Not satisfied with understanding nature on the basis of external appearances, the new philosophers insisted on penetrating into its hidden recesses and uncovering the occult causes of phenomena. Natural magic also played an important role in promoting a new concern for the pragmatic benefits of scientific knowledge. Finally, magic's emphasis on experiment was in accord with the program of the new philosophy.

The rise of the *mechanical philosophy in the seventeenth century dealt a nearly fatal blow to magic, as far as its relevance to science was concerned. In theory, at least, the mechanical philosophy banished occult qualities from natural philosophy by reducing explanations of phenomena to mechanical causes. Nevertheless, because the mechanical philosophy could not offer a plausible and comprehensive view of the physical world, the status of occult qualities continued to be debated.

See also SYMPATHY AND OCCULT QUALITY.

Lynn Thorndike, *History of Magic and Experimental Science*, 8 vols. (1923–1958). Paolo Rossi, *Francis Bacon. From Magic to Science*, trans. S. Rabinovitch (1968). Keith Thomas, *Religion and the Decline of Magic* (1971). Keith Hutchison, "What Happened to Occult Qualities in the Scientific Revolution?" *Isis* 73 (1982): 233–253. Brian Copenhaver, "Natural Magic, Hermetism, and Occultism in Early Modern Science," in *Reappraisals of the Scientific Revolution*, eds. David C. Lindberg and Robert S. Westman (1990): 261–302. William Eamon, *Science and the Secrets of Nature: Books of Secrets in Medieval and Early Modern Culture* (1994).

WILLIAM EAMON

MAGNET AND COMPASS. During the twelfth century Western Europeans learned about a magnet's tendency to point towards the geographic north. As compasses became important navigational aids during the sixteenth century, natural philosophers increasingly focused on the properties of lodestone, naturally occurring magnetite of varying quality. Mounted in silver, large carved lodestones became valuable collectors' items; lower-grade pieces, often armed with iron to conserve their strength, were used for entertainment and to remagnetize or "touch" compass needles.

Early expertise on magnetism belonged mainly to the maritime community, although compasses were also used for exploring mines and unknown territory. The first comprehensive and influential book on the topic, *De magnete* (1600) by the physician William Gilbert, described experiments with lodestones and iron filings, discussed techniques for magnetizing iron, and pictured instruments remodeled from navigational compasses. Gilbert

regarded the earth itself as a huge magnet and proposed an animistic magnetic cosmology. René *Descartes, rejecting *sympathies and occult powers, made the mechanical explanation of magnets central to his corpuscular view of the universe.

Stimulated by the need for more reliable compasses, and benefiting from improved steel manufacture, natural philosophers of the eighteenth century developed methods for making artificial magnets by aligning steel bars and repeatedly stroking them with magnets. Horseshoe-shaped magnets, stronger compound magnets, and powder magnets were also developed. Because they were more consistent, long-lasting, and convenient than lodestones, these artificial magnets facilitated systematic research and accurate measurements, and helped to replace magnetic theories based on *ethers or subtle fluids with ones based on distance forces. In 1785, Charles *Coulomb used long thin steel magnets to show that magnetic poles obey a Newtonian inverse square law.

After Hans Christian Ørsted demonstrated that electricity and magnetism are linked, Michael *Faraday carried out his pioneering research into *electromagnetism and *field theory. In 1850, William *Thomson introduced the quantitative concepts flux density and permeability to describe the behavior of magnetic material. Magnets became crucial for the swelling electrical industry, and scientists developed new magnetic alloys with different properties. Electromagnets, consisting of a metallic core inside a current-carrying coil, became vital for many new inventions, including generators, transformers, and switches (*see* ELECTROMAGNETIC INSTRUMENTS).

Faraday showed in 1845 that nonferrous substances also display magnetic effects, and by the end of the nineteenth century materials were classified as diamagnetic, paramagnetic, or ferromagnetic. In the early 1930s Louis Néel demonstrated the existence of elementary magnetic domains and identified antiferromagnetism, the property of crystals to divide into interlaced lattices with opposite magnetic. Magnets (including antiferromagnets) became increasingly important for sound- and video-recording devices, and for *computer memories and processors. Following a long period of eclipse, magnets were the subject of intensive research by the end of the twentieth century.

Research into magnets has always been closely tied to practical goals, particularly navigation. Although Francis Bacon and others acclaimed compasses along with printing and gunpowder as the momentous inventions of a new age, they were neither reliable nor accurate. They generally had a soft-iron lozenge-shaped needle, mounted

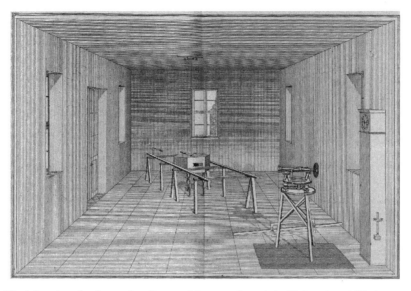

The laboratory for the study of terrestrial magnetism at the University of Göttingen.
The telescope measures displacement of a magnetic needle in the earth's field.

beneath a card divided into thirty-two points, all balanced on an upright pin and contained in a wooden box. In the mid-eighteenth century, the natural philosopher Gowin Knight introduced substantial changes, including a fine steel needle, a scale accurate to half a degree, a delicate suspension system, and a water-resistant brass case. Navigators complained that this philosophical instrument was too sensitive for use at sea. Instrument makers continued to try to meet navigational needs, especially after the foundation of the Admiralty Compass Committee in 1837.

The hulls of iron ships perturb compasses. From early in the nineteenth century, compasses carried iron bars and magnets to compensate for the perturbation, then called deviation. Towards the end of the century, liquid-filled compasses improved stability. However, since World War I magnetic navigation compasses have largely been displaced by small robust gyrocompasses. Special compasses have also been designed for airplanes.

Mariners had long recognized the need to adjust their compass readings to allow for declination (or variation)—the variation of angle between the magnetic and geographic north, which had been known about since at least the fifteenth century. To measure this variation accurately for different places and times, Knight and others developed sensitive variation compasses with long needles that could only swing through a very limited range. The angle of dip from the horizontal was discovered later. During the eighteenth century, instrument makers redesigned Gilbert's inclination compass,

a vertical needle rotating in front of a dial, to measure dip more accurately.

During the nineteenth century, further specialized magnetic compasses catered to the requirements of different communities—notably geomagnetists, navigators, and surveyors—and obtained still greater accuracy with devices such as Vernier scales and optical magnifiers. Scientists interested in terrestrial magnetic intensity developed magnetometers that measured the rate of oscillation of a suspended needle. In the 1830s, Carl Friedrich Gauss and Wilhelm Weber introduced more accurate variometers designed to record absolute intensity by the rotation of a magnetic bar.

Alfred Hine, *Magnetic Compasses and Magnetometers* (1968). W. E. May, *A History of Marine Navigation* (1973). Roderick Home, "Introduction," in Roderick Home and P. J. Connor, *Æpinus's Essay on the Theory of Electricity and Magnetism* (1979). Anthony Fanning, *Steady as She Goes: A History of the Compass Department of the Admiralty* (1986). Patricia Fara, *Sympathetic Attractions: Magnetic Practices, Beliefs, and Symbolism in Eighteenth-Century England* (1996). James Livingston, *Driving Force: The Natural Magic of Magnets* (1996).

Patricia Fara

MAGNETISM was known to the Greeks as the power of certain stones rich in iron to attract other, similar stones or pieces of iron. Later, it was recognized that when freely suspended, these "magnets" orientated themselves approximately north-south. This property, transferred to a floating iron needle by "touching" it with a magnet, became the basis of the mariner's

Illustrations of the development of the chick embryo published by Karl Ernst von Baer in 1828. Two years earlier he had made the long-sought discovery of the mammalian egg—not in a chicken, but in a colleague's pet dog. Baer studied not only the successive appearance of the various parts of the embryo, but also, as indicated in the figures, the development and function of the extra-embryonic membranes in the chick and the mammal.

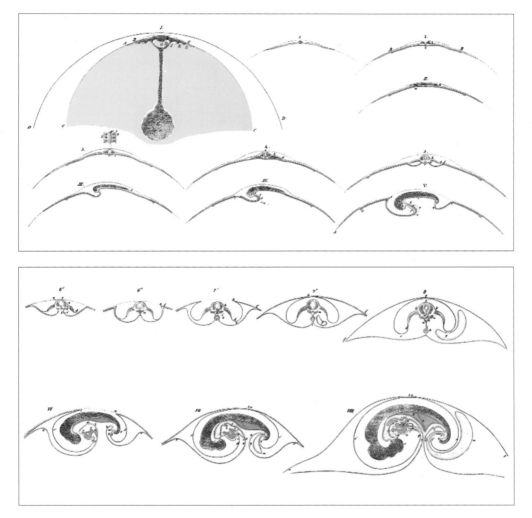

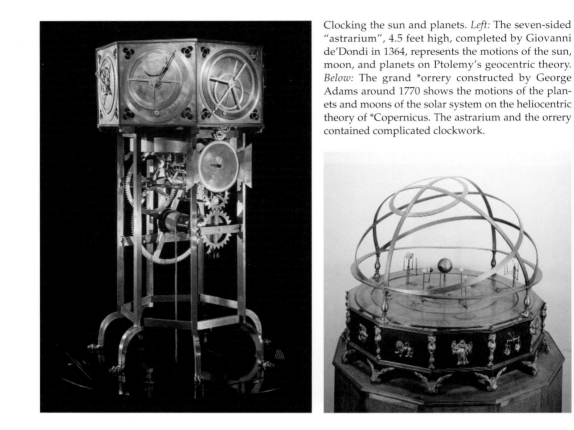

Clocking the sun and planets. *Left:* The seven-sided "astrarium", 4.5 feet high, completed by Giovanni de'Dondi in 1364, represents the motions of the sun, moon, and planets on Ptolemy's geocentric theory. *Below:* The grand *orrery constructed by George Adams around 1770 shows the motions of the planets and moons of the solar system on the heliocentric theory of *Copernicus. The astrarium and the orrery contained complicated clockwork.

Some of the most beautiful and useful astronomical instruments have no moving parts. *Below left:* The *sundial with style pointing to the north pole of the heavens, an invention of the fifteenth century, can be used in any orientation. The pair of vertical dials painted on the Johanniskirche in Ansbach in 1777 face south (the dial on the left) and east. *Below right:* The sun's noon crossings of the meridian line in the *church of San Petronio in Bologna gave important astronomical information during the late seventeenth and early eighteenth centuries.

The complete unraveling of nature's great mystery of light and color, the *rainbow, by the icon of reason, Isaac *Newton, may symbolize the power of rational thought and controlled experiment. The double bow circles the farm house in Lincolnshire where Newton was born.

Left: The rich *cabinet in which Laurens Theodosius kept his collections and instruments and instructed his children illustrates the penetration of science into ordinary well-to-do households during the eighteenth century. *Below:* The thunder-house, part of the standard equipment of the public lecturer on natural philosophy, indicates the uses of science in entertainment and technology. When struck by a spark from an electrical machine, the house remains unscathed or collapses, depending on whether or not its "*lightning rod" is properly earthed.

A serious elegant young lady interrupted during her reading of a book on Newtonian physics, very likely the French translation of Colin Maclaurin's *An Account of Sir Isaac Newton's Philosophical Discoveries* (1748). Maurice Quentin de la Tour, *Portrait of Mlle Ferrand* (1753).

The installations for the taking, study, and dissemination of astronomical observations built by Tycho *Brahe *(left)* on the island of Hven at the expense of the King of Denmark were unique in size and purpose. The central palace or observatory, Uraniborg *(below)*, begun in 1576, sat at the center of a square walled garden, 300 feet on a side, containing an orchard, accommodation for servants, and a printing house. The palace had deep basements for alchemical and other experiments.

The grandest instrument was the mural quadrant *(left)* over six feet in radius, five inches wide, cast in solid brass and fixed in the plane of the meridian. One of Tycho's assistants observed a star cross the meridian, another called out the time, and a third wrote it all down as the painted image of Tycho and his dog looked on. The mural also has pictures of other astronomical instruments, more assistants at work, and a basement laboratory.

After World War II, particle *accelerators grew to the size of Tycho's island. *Above left:* three acceler-ator rings are superposed on a photograph of Washington, D.C., taken by a Landsat *satellite. The outermost ring, about equal to the Beltway in diameter, is that of the ill-fated Superconducting Super Collider; the middle one, that of the Large Electron Positron Collider (LEP) at *CERN; the smallest is the Fermilab ring, which is shown in life-size in the photograph to the right.

The Fermilab ring has a circumference of four miles, within which lives a herd of bison native to the American mid-west. The particles undergoing acceleration may begin their journey in the octopus-shaped Van de Graaff machine *(below)* before being injected into a narrow vacuum tube in the underground tunnel. The tunnel *(left)* contains utility pipes as well as the string of exactly placed square-shaped magnets that control the orbits of the particles traveling at close to the speed of light within the vacuum tube.

Opposite page, left: China and Japan first received European astronomy—specifically the system of Tycho *Brahe—from the Jesuits during the late Ming (late sixteenth to early seventeenth centuries). This portrait of an elegant astronomer, sometimes said to be the Jesuit Ferdinand Verbiest, in fact depicts the bandit-sage Wu Yong in a Japanese version of the famous Chinese novel *Shuihu Zhuan* ("water margin"). The picture, by the Japanese woodblock artist Kuniyoshi (1797–1861) shows the sage counting the stars on his fingers and consulting a celestial globe while plotting strategy for his Liangshan rebels.

Opposite page, right: Astronomers working in or around the observatory established by Taqī al-Dīn in Istanbul around 1575. The *astrolabe, quadrant, clock, and desk should be compared with the contemporary Uraniborg on color-insert page four.

Below: Electrical phenomena had attracted the attention of several Japanese by at least 1765, and electrostatic generators were used for entertainment by the turn of the century. Edison's success with the incandescent bulb was noted almost as quickly in Japan as in the United States and Britain, and in March 1883 a group of private Japanese investors introduced electrical lighting to the Ginza district in Tokyo. A wood-block print by the noted artist Sadakichi Nozawa commemorates this event.

The invention of *spectroscopy and the spectroscope *(left)* in the mid-nineteenth century gave chemists a new way to identify terrestrial *elements and astronomers their only way to determine the makeup of *stars. The second and third of the stellar spectra shown below come from the double star β-Cygni. The upper spectrum appears orange to the naked eye, the lower one blue. This page is taken from Henry E. Roscoe's *Spectrum Analysis* (1885).

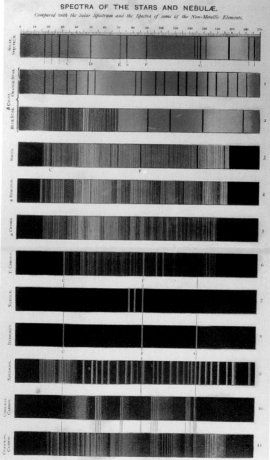

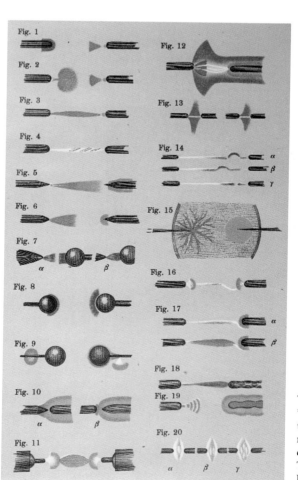

The glows around the electrodes in a gas discharge, as in a *cathode-ray tube, have shapes and colors that depend on the form and sign of the electrode, and on the pressure and nature of the residual gas. Experiments with gas discharges led to the discoveries of *X rays and the *electron. The glowing electrodes on the left come from Otto Lehman's *Die elektrischen Lichterscheinungen* (1898).

compass. Magnetism's mysterious powers led scholastic philosophers to consider it as the paradigm of an occult power at work in the world (*see* SYMPATHY AND OCCULT QUALITY).

William Gilbert's *De magnete* (1600) became the starting point for subsequent magnetic research. For Gilbert, magnets were characterized by their north and south "poles," the opposite ends of the axis according to which freely suspended magnets aligned themselves with the earth. Magnets brought together with like poles facing would flee each other, but if unlike poles were facing, they came together. The earth itself, Gilbert concluded, was a large magnet, this being what caused compass needles to orientate themselves. He described magnetic attraction, in animistic terms, as coition, two magnets coming together in mutual harmony.

Magnetism challenged seventeenth-century mechanists who proclaimed that all natural powers arose from matter and motion alone. *Descartes's success in providing a mechanistic explanation within the framework of his overall world picture was a triumph for the new philosophy. He envisaged distinctively shaped corpuscles of subtle matter passing through a magnet along channels suitable to receive them, then returning through the external air to form a loop of matter in motion. Such channels were peculiar to iron, magnetization bringing randomly oriented corpuscles into alignment. The patterns formed by iron filings around a magnet showed the streamlines of flow of the subtle matter through the air. Pressure from the streaming matter caused other, nearby magnets to align themselves as they did. While later authors often altered details, Descartes's depiction was widely accepted for more than a century. Even *Newton, who tried measuring the force between two magnets, adopted it.

In 1759, Franz Ulrich Theodosius *Aepinus proposed a new theory based on forces acting at a distance, analogous to his improved version of Benjamin *Franklin's theory of *electricity. Magnetization, he suggested, involved not currents of subtle matter but redistributions within samples of iron of a subtle fluid specific to magnetism, that left one part of a sample with a surplus of fluid and another with a deficit. These regions corresponded to the poles. Particles of fluid repelled each other but were attracted to particles of iron, which likewise repelled each other; other kinds of matter did not act on the fluid. Magnetic poles were therefore centers of force, not merely entry or exit points for streams of subtle matter. By adding the forces acting in various situations, Aepinus accounted semi-quantitatively for the known phenomena, and predicted new effects,

particularly in improving compass needles, that were quickly confirmed.

A modified form of Aepinus's theory became widely adopted following Charles Augustin *Coulomb's advocacy in the 1780s. Unhappy with Aepinus's notion that particles of iron repelled each other, Coulomb invoked a second subtle fluid as the carrier of this force. He also developed a molecular theory of magnetization, restricting the redistributions Aepinus discussed to individual molecules; the alignment of polarized molecules of iron made macroscopic magnets. In 1785, Coulomb reported that the forces between magnetic poles followed an inverse-square law. This opened the way to Simeon Denis Poisson's fully mathematized version of Coulomb's theory.

From Gilbert on, magnetism and electricity had been regarded as unconnected (if analogous) phenomena. However, in 1820, Hans Christian Ørsted discovered that a wire carrying an electric current affected a compass needle. André-Marie Ampère quickly showed that a current-carrying loop or solenoid was equivalent to a magnet. He argued from this that the magnetic fluids were a fiction and that magnetization derived from tiny solenoidal electric currents in molecules of iron. In the manner of Laplace, he showed how the observed forces between current-carrying wires could be compounded from elementary forces exerted by electrical charges in motion. Meanwhile, electromagnets—iron-cored solenoids carrying ever-larger currents—generated magnetic effects far more powerful that any previously obtained. In 1831, Michael *Faraday discovered the reverse of Ørsted's effect, electromagnetic induction: when an electrical conductor cuts across magnetic lines of force, an electromotive force appears in it. This effect became the basis of the electrical power industry, power being generated by rotating coils of wire in a magnetic field.

Dissatisfied with theories based on action at a distance, Faraday shifted attention to the *field surrounding a magnet, devising beautiful experiments to support his unorthodox ideas. He regarded magnetic lines of force not as mere geometrical constructs showing the direction of a compass needle at any point, but as lines of strain in space. Poles as centers of force did not exist—matter interacted with lines of force, conducting them well, as in iron, or poorly, as in diamagnets such as bismuth, the anomalous magnetic behavior of which was another of Faraday's discoveries. James Clerk *Maxwell started with Faraday's ideas when in the 1860s he developed his dynamical theory of the electromagnetic *field, which represented magnetic lines of force as lines of rotational strain in the *ether.

In the twentieth century, the equations of Maxwell's theory survived the abandonment of the ether that had initially given them physical meaning, and continued to provide the basis for understanding magnetic interactions. Meanwhile, modern theories of atomic structure gave Ampère's ideas on the origins of magnetism new currency, the motions of electrons in atoms constituting precisely the kind of elementary current loops he envisaged. New and much more powerful permanently magnetic materials—ferrites and, from the 1970s, various rare-earth alloys—became indispensable to much of late twentieth-century technology. In the 1970s, magnets exploiting the phenomenon of superconductivity at low temperatures (*see* LOW-TEMPERATURE SCIENCE) also became widely available; these provided stronger and more homogeneous fields that found widespread application in new forms of laboratory apparatus.

E. T. Whittaker, *A History of Theories of the Aether and Electricity* (2 vols., 1951). René Taton, *History of Science*, trans. A. J. Pomerans (4 vols., 1964–1966), vol. 3, ch. 4, "Electricity and Magnetism (1790–1895)," 178–234; vol. 4, ch. 7, "Magnetism," 142–151. L. Pearce Williams, *The Origins of Field Theory* (1966). R. W. Home, *Aepinus's Essay on the Theory of Electricity and Magnetism* (1979).

R. W. HOME

MAGNETO-OPTICS. The effects of magnetic fields on light have had an importance for physical theory far beyond their significance in nature since their first detection in 1845. That occurred because William *Thomson, Lord Kelvin, inferred from Michael *Faraday's ideas about electricity (*see* FIELD) and his own ideas about light that glass stressed by an electric field should rotate the plane of polarized light passing through it. Faraday looked, found nothing, and substituted a magnetic field, which worked. Success prompted a characteristic leap: Faraday inferred that magnetism could be concentrated, as it had been in the experiment, by materials other than ferromagnetics. This insight, the first spin-off from a magneto-optical effect, led him to the discoveries of para- and diamagnetism.

In 1862 Faraday sought to change the spectrum of a light source by placing it in a magnetic field. No luck. But the fact that he had tried encouraged a repetition some thirty years later. Pieter Zeeman had just obtained his doctorate from the University of Leyden with a prize-winning thesis on a second magneto-optic effect, the change in polarization of plane-polarized light reflected from an electromagnet. In 1896 Zeeman saw the bright yellow lines of sodium broaden when he placed their source in a magnetic field. He brought this news to his professor, Hendrik Antoon Lorentz. From his own model for the emission of light, in which the radiator is an "ion" of unknown charge e and mass m, Lorentz predicted that the broadened lines should be resolvable into triplets polarized in certain ways and separated by a distance proportional to e/m. Zeeman confirmed the prediction and deduced the value of e/m. It came out close to the number that Joseph John *Thomson had found for the ratio of charge to mass of *cathode-ray particles. The "Zeeman effect" played a major part in the establishment of the *electron as a building block of matter. Zeeman and Lorentz shared the Nobel Prize for physics in 1902.

Most Zeeman patterns differ from Lorentz's triplet. Explaining quartets, quintets, and so on proved too much for both classical and early quantum theories of *atomic structure and spectral emission. After World War I demobilizing physicists found the "anomalous Zeeman effect," which most of them had ignored, high among the outstanding problems of atomic physics as described in Arnold Sommerfeld's comprehensive survey, *Atombau and Spektrallinien* (1919).

Lorentz's theory derived the magnitude of the line splitting from g, the ratio of the magnetic moment to the orbital angular momentum of the radiating electron. Alfred Landé, a young Jewish theorist at the University of Frankfurt am Main, managed to refer the refractory splitting to an anomalous value of g and to accepted rules for quantum transitions. Attempts to derive the anomalous g from an atomic model failed until the introduction of electron spin in 1925 by Samuel Goudsmit and George Uhlenbeck, young Dutch physicists who, like Landé, made their careers in the United States. Meanwhile, analysis of the systematics of quantum transitions in the anomalous effect helped to guide Wolfgang Pauli to the most striking of all the discoveries prompted by magneto-optical phenomena: the "Exclusion Principle," which ascribes four quantum numbers to each electron in an atom and prohibits any two electrons from having the same values for all their quantum numbers. Pauli's fourth quantum number was soon associated with Goudsmit and Uhlenbeck's spin. Particles that, like electrons, are exclusive, divide the universe with particles that, like photons, are gregarious. Both sorts have revealed much when under the influence of a magnetic field.

Paul Forman, "Alfred Landé and the Anomalous Zeeman Effect," *Historical Studies in the Physical Sciences* 2 (1970): 153–261. J. L. Heilbron, "The Origins of the Exclusion Principle," *Historical Studies in the Physical Sciences* 13 (1982): 261–310. Theodore Arabatzis, "The Discovery of the Zeeman Effect," *Studies in the History and Philosophy of Science* 23 (1992): 365–388.

J. L. HEILBRON

MALPIGHI, Marcello (1628–1694), anatomist and microscopist.

Born in Crevalcore, near Bologna, Malpighi studied at the University of Bologna, graduating in medicine and philosophy in 1653. Among his instructors was Bartolomeo Massari, who, in addition to his regular classes, taught anatomy to a few select students—Malpighi among them—at his house, practicing dissections and vivisections and exploring issues of current interest, such as William *Harvey's views on the circulation of the blood.

After graduating, Malpighi spent three years at Bologna practicing medicine. In 1656 he accepted the chair of logic at the school, then almost immediately left to teach theoretical medicine at Pisa. During his three years in Pisa, Malpighi collaborated closely with Giovanni Alfonso Borelli, who was instrumental in converting him from an enlightened Aristotelian to a follower of the new corpuscular philosophy. In all likelihood, Malpighi learned to use the *microscope at Pisa and was influenced by the experimental philosophy practiced under Medicean patronage by the Accademia del Cimento.

Between 1659 and 1662 Malpighi taught theoretical medicine at Bologna and published his first two important works, *Epistolae de pulmonibus* (*Letters on Lungs*, 1661), where he announced his discovery of the microstructure of the lungs. By masterfully using the microscope and various observational techniques, Malpighi made the first significant anatomical finding about the microstructure of an organ. All his later publications relied on the microscope. He discovered air sacs (alveoli) and a minute network of blood vessels in the lungs of frogs, showing that blood always flows inside vessels and that arteries and veins are joined by anastomoses.

Between 1662 and 1666 Malpighi was primary professor of medicine at Messina, a position he owed to Borelli's influence. During those years he pursued his anatomical investigations and published *De lingua*, *De cerebro*, and *De omento* (*On the Tongue, On the Brain, On the Omentum,* 1665) and *De externo tactus organo* (*On the External Organ of Touch*, 1665). Sense perception (notably taste, vision, and touch) featured in most of his investigations.

In 1666 Malpighi returned to Bologna as professor of theoretical medicine. Soon after his arrival he published *De viscerum structura* (*On the Structure of the Viscera*, 1666), consisting of four treatises—*De hepate* (*On the Liver*), *De cerebri cortice* (*On the Cerebral Cortex*), *De renibus* (*On the Kidney*), and *De liene* (*On the Spleen*)—and an appendix, *De polypo cordis* (*On the Heart Polyp*). In the four treatises Malpighi asserted that the main organs of the body consist of microscopic glands, which he understood as minute mechanisms filtering different fluids from the blood. In his work on the heart polyp Malpighi used the deformations induced by a pathological state to shed light on anatomical features of the healthy body, notably the microstructure and composition of *blood.

In 1668 Malpighi established contacts with the Royal Society of London and was elected a Fellow in 1669 following his dedication to the society of a remarkable treatise, *De bombyce* (*On the Silkworm,* 1669), revealing the silkworm's microstructure. Subsequently, Malpighi devoted himself to the study of generation in the hope of uncovering the most minute structures of organs in their formative process or in simpler organisms; his would be the most important contribution to the study of generation in the century after Harvey. *De formatione pulli in ovo* (*On the Formation of the Chick in the Egg*, 1673) was followed by *De ovo incubato* (*On the Incubated Egg*), published together with the extensive first part of the *Anatome plantarum* (*Anatomy of Plants*, 1675). The second part of that work, *Anatome plantarum pars altera*, appeared four years later.

Malpighi published two further works in his lifetime, in 1684 a collection of anatomical observations in the *Philosophical Transactions* of the Royal Society and the short treatise *De structura glandularum conglobatarum* (*On the Structure of Conglobate Glands*, 1689), his crowning achievement on the subject.

In 1691 Malpighi moved to Rome as *archiater* (physician) to Pope Innocent XII. His move testified to his reputation as a physician as well as an anatomist. He died in Rome three years later, having ordered his disciples to wait thirty hours before dissecting him. He wanted to be anatomized, but not until it was certain that he was well and truly dead. Several years earlier, he had arranged for a number of his works to be published posthumously. They include his massive *Vita* and replies to his antagonists Michele Lipari, dating from his stay at Messina, and Professor Gerolamo Sbaraglia of Bologna. All three works are combative in tone, a marked contrast to the politeness of Malpighi's previous publications. Many of his medical consultations were printed in the first half of the eighteenth century.

Howard B. Adelmann, *Marcello Malpighi and the Evolution of Embryology*, 5 vols. (1966). Theodore M. Brown, *The Mechanical Philosophy and the "Animal Economy"* (1981). Domenico Bertoloni Meli, ed., *Marcello Malpighi, Anatomist and Physician* (1997).

DOMENICO BERTOLONI MELI

MANHATTAN PROJECT. Development of the first atomic bomb began with efforts by European emigré physicists to alert the government of the United States to the military potential of uranium fission. Leo Szilard, Eugene Wigner, and Edward Teller approached Albert *Einstein to use his connections to inform President Franklin Delano Roosevelt. The president turned the matter over to the U.S. Bureau of Standards for investigation. Most physicists still considered a practical bomb impossible. Calculations indicated that several tons of uranium would be required. It also appeared that any explosion that could be started would scatter the fissile material before the chain reaction had proceeded very far. But calculations done by German refugees Otto Robert Frisch and Rudolf Peierls in Britain revised the estimate to 10 kgs (22 lbs) or so of the light isotope of uranium (U-235). Another refugee, Franz Simon, estimated that enough of this isotope could be separated from natural uranium by gaseous diffusion.

When this information reached the United States in mid-1941, the uranium project had come under the civilian organization set up under Vannevar Bush to mobilize U.S. scientists for war. Roosevelt decided on an all-out effort. In June 1942 the U.S. Army took over the project, which had quickly grown beyond the resources of Bush's Office of Scientific Research and Development. General Leslie Groves, a specialist in the construction of large complexes (he had built the Pentagon), directed the Manhattan Engineer District, the code name given the project as part of a fruitless attempt at secrecy.

The chief challenge facing the uranium project was isolation of the explosive U-235 isotope. Because isotopes of the same element are chemically identical, atoms of U-235 could be separated from the much more prevalent atoms of U-238 only by exploiting the weight difference of less than 1/100 between them. Manhattan scientists developed two methods. Electromagnetic separation employed modified cyclotrons (called "calutrons" after their mother state) to hurl uranium atoms around semicircular orbits; the lighter atoms, following a tighter path, collected on a special target. Gaseous diffusion allowed gaseous uranium hexafluoride to diffuse through a barrier; the lighter molecules passed through more quickly and could be harvested. Both methods required cascading thousands of units to separate significant amounts of U-235.

Niels *Bohr had predicted that uranium separation would require turning the United States "into one huge factory." For the largest industrial operation in history the Army built three complexes in remote areas of the United States. Oak Ridge, Tennessee's gaseous diffusion

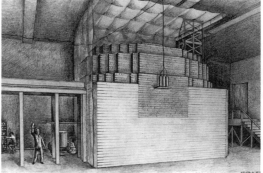

On the left, the last attempt of Werner *Heisenberg's group to provoke a self-sustaining chain reaction (1945) as compared with Enrico *Fermi's first critical pile (December 1942) on the right.

plant, was the largest concrete building in the world. Five cascades of ninety-six calutrons each, held an unprecedented amount of evacuated space. At Los Alamos, New Mexico, Robert Oppenheimer directed bond design. The third great complex, raised in Hanford, Washington, to have access to the Columbia River as a coolant, was devoted not to uranium separation but to the creation of a human-made nuclear explosive, plutonium. A nucleus of U-238 that absorbs a neutron can decay in two steps to plutonium, which can then be removed from its parent uranium by chemical means. The reactors at Hanford scaled up the first self-sustaining chain reaction (for the continuous production of plutonium) achieved by the emigré Italian physicist Enrico *Fermi in Chicago in 1942. The chemical separation of the plutonium built on processes developed by Glenn T. Seaborg and his associates using tiny amounts of the substance made at a cyclotron in Berkeley, California.

The design of the bombs took place at Los Alamos, New Mexico, under the general direction of J. Robert Oppenheimer. A shotgun assembly exploded the uranium bomb: two subcritical masses (lumps of uranium too small to sustain the reaction needed for explosion) were propelled together to generate a critical mass. The plutonium bomb required a more difficult design: high explosives imploded a spherical plutonium shell to critical density. Only this design required testing. On 16 July 1945, at the "Trinity" site in New Mexico, a test bomb exploded with the force of 20,000 tons of TNT. By early August two bombs were ready: "Little Boy" and "Fat Man," so-called from the long, narrow shape of the uranium gun assembly and from the spherical configuration of the plutonium bomb.

The army had originally planned its complexes as military installations staffed by scientists in uniform. Scientists, however, feared that military control would stifle them, and Oppenheimer insisted on a civilian establishment. Groves nevertheless maintained a tight and galling security. His strict compartmentalization of the project's divisions struck at what scientists saw as a foundation of their discipline: free communication. Their discontent took political form in Chicago, where a lull in activity followed the success of a pilot nuclear reactor. The Franck Report, named for James Franck and issued in June 1945, argued that the United States would be unable to maintain its monopoly on atomic weapons and that an atomic arms race would ensue if the United States initiated a surprise nuclear attack.

This insightful position ran counter to U.S. policy. As early as 1942 Roosevelt planned to use a U.S. atomic monopoly to counter Soviet power, dreaming of creating "a reservoir of force so powerful that no aggressor would dare to challenge it." After Roosevelt's death in April 1945, President Harry S. Truman used the U.S. monopoly as a diplomatic "big stick" against the Soviets, little dreaming that his adversaries would explode their own nuclear device four years later. Meanwhile, the Manhattan Project's scientific leadership (Fermi, Oppenheimer, Ernest Lawrence from Berkeley, and Arthur Holly Compton from Chicago), advising the War Department on atomic policy, showed no such hesitation as the Chicago physicists, but rather recommended immediate deployment of the weapon and consequent development on all fronts. On 6 August the United States exploded the uranium bomb over Hiroshima; three days later the plutonium bomb destroyed Nagasaki.

The Soviet Union's first atomic bomb, a faithful copy of Fat Man, was helped by espionage of British and U.S. atomic research. Many scientists, Oppenheimer among them, had been Soviet sympathizers in the 1930s. U.S. wartime relations with the Soviets favored friendship and openness. Under these conditions espionage flourished. Although Oppenheimer elicited the distrust of some of his colleagues, especially Teller and Lawrence, for his opposition to the hydrogen bomb, and lost his security clearance in 1954, he did not abet espionage. The most damaging of the atomic spies, a German emigré named Klaus Fuchs, came to the Manhattan Project through the British.

Richard G. Hewlett et al., *A History of the United States Atomic Energy Commission*, 2 vols. (1962–1969). Richard Rhodes, *The Making of the Atomic Bomb* (1986). Martin Sherwin, *A World Destroyed: The Atomic Bomb and the Grand Alliance* (1987). Larry Badash, *Scientists and the Development of Nuclear Weapons: From Fission to the Limited Test Ban Treaty, 1939–1963* (1995). Daniel J. Kevles, *The Physicists: The History of a Scientific Community in Modern America* (1995).

THEODORE S. FELDMAN

MAPS. See CARTOGRAPHY; GEOGRAPHY; NAVIGATION.

MARXISM is a blanket term covering the various theories and ideas associated with or based on the work of Karl Marx (1818–1883). Marx thought of himself as the *Newton of the social realm. He sought the "economic law of motion" of industrial society that would explain all cultural development, including the production of scientific ideas and technological artifacts. There are three prominent varieties of Marxism: historical materialism, communism, and humanistic Marxism. All three impose specific frameworks on the contemporary practice or historical understanding of science and technology.

Historical materialism, the earliest variety of Marxism drawn from Marx's writings from the late 1840s onward, argues that stages in social and cultural evolution depend on changes in the economic substructure. The laws of social evolution in historical materialism were developed contemporaneously with, and often as a challenge to, the laws of human evolution in *Darwinism. Deterministic in outlook, historical materialists argued that scientific ideas and technological artifacts are rooted in economic processes controlled by the bourgeoisie. At a now-famous session of the International Congress of the History of Science and Technology held in London in 1931, Boris Hessen and Nikolay Bukharin presented radical Marxist interpretations of science and technology. Hessen argued that the theoretical foundation of Newton's *Principia* was to be found not in problems of motion left unsolved by *Copernicus, *Kepler, and *Galileo, but in English economic and technological development, especially navigational problems. Hessen's interpretation of science encouraged the British socialist John Desmond Bernal to argue that science could best contribute to the material benefit of humanity if scientists were in charge of social policy as they were in the Soviet Union. By contrast in the United States, he thought, capitalism obstructed the development of science and technology by instituting competitive funding practices. Around the same time, Dirk Struik and Henry Sigerist applied Marxist social or economic theory to the interpretation of American science, while the sociologist Robert Merton demonstrated that the social and economic foundation of science and technology could also be understood by applying Max Weber's theories of modernization, in particular his study of the Protestant ethic which Merton applied to the rise of modern science in Puritan England. Since the 1930s the most prominent example of Marxist historiography of science has been in the work of Joseph Needham, whose magisterial study of Chinese science remains a standard.

Communism, as developed by V. I. Lenin and later Joseph Stalin, is a more rigid and less inclusive variety of Marxism in which orthodox dogma applies everywhere, including theoretical developments in the natural and social sciences. Maintaining that the state can accelerate historical change, communists impose strict political control over science and technology. Although the East German constitution, for instance, stipulated that science should protect and enrich life, communist economic policy dictated that scientific developments would lead to technological innovation with little regard for the quality of life. Although relaxed in 1963–1964 to allow for greater autonomy in science, East German Marxist dogma still imposed restrictions on basic research which failed to produce commercially viable products. Communist regimes raised questions about the ethical responsibility of scientists to truth and scientific freedom, as British scientists recognized in the 1930s when the Soviet Union pressured its scientists in the name of social responsibility and political conformity. The conflict concerned more than opposition between centralized planning and academic freedom, as the case of T. D. Lysenko later confirmed: Lysenko used Marxist dogma to destroy an entire school of Soviet genetics. Since the 1960s, the relations between communist Marxism and the natural sciences has been uneasy.

Humanistic Marxism, based principally on Marx's economic and philosophic manuscripts of 1844 discovered in 1932, focuses on the plight of human alienation and the possible venues for human development, especially for emancipation from the restrictions imposed by alienation. In humanistic Marxism, the unfolding of humanity's relationship to the natural world is a more important consideration than either the historical development of science and technology or the laws of social evolution. Humanistic Marxism has influenced the historical understanding of science and technology only recently, and in subtle ways. From within the framework of a socialist feminism, Donna Haraway has sought to uncover the symbols and discourses that suppress or mask human expression and development and to use that revelation for emancipation.

See also Historiography of Science; Lysenko Affair; Vavilov.

J. D. Bernal, *The Social Function of Science* (1939). Joseph Needham et al., *Science and Civilization in China*, 7 vols. (1954–2000). Boris Hessen, "The Social and Economic Roots of Newton's *Principia*," in *Science at the Crossroads*, ed. P. G. Wersky (1971). Donna Haraway, "A Cyborg Manifesto: Science, Technology, and Socialist Feminism in the Late Twentieth Century," in *Simians, Cyborgs and Women: The Reinvention of Nature* (1991). R. Olwell, "'Condemned to the Footnotes': Marxist Scholarship in History of Science," *Science and Society* 60 (1996): 7–26. *Science under Socialism: East Germany in Comparative Perspective*, ed. Kristie Macrakis and Dieter Hoffmann (1999).

KATHRYN OLESKO

MASER. See Laser and Maser.

MASS ACTION. The law of mass action expresses the mathematical relationship between the quantities of the substances present at the beginning of a chemical reaction in which they participate and the quantities of both reactants and products present when the reaction has reached an equilibrium. Although formulated in essentially its present form as an equation during the late nineteenth century, its underlying concept goes back to around 1800 and the French chemist Claude Louis Berthollet (*see* COMPOUND).

Eighteenth-century chemists organized much of their cumulative knowledge of the chemical operations they could carry out in their laboratories in the form of tables of "rapports" or "affinities," which they also sometimes called "elective attractions." These tables expressed in highly compressed form the order in which substances would replace each other in combination with a third substance. In the case of interactions or "double decompositions" between two neutral *salts, the affinity tables expressed the direction in which this exchange took place. The Swedish chemist Torbern Bergman published the most complete affinity table of the century in 1778. He performed thousands of experiments to gather the information arranged in it. In some cases the exchange seemed to go in the opposite direction from the expected order. Heating the substances, for example, disrupted the order established by operations carried out in solution at room temperature; but Bergman believed that more careful control and further experimentation would suffice to eliminate these apparent inconsistencies.

Berthollet encountered similar anomalies, some of which referred to the insolubility or volatility of a product, which removed it from the solution. Moreover, the normal direction of a chemical change could often be overcome by adding much greater amounts of one of the products. Berthollet concluded that chemical changes do not go to completion except when one or more of the products is removed; otherwise the proportion between the initial and final products depends on their relative masses as well as their affinities.

Berthollet connected these views with a belief that chemical compounds exist in indefinite proportions in solution. When the acceptance of John Dalton's atomic theory discredited this belief, some of Berthollet's contemporaries inferred that his views on mass action were untenable; others, however, including the influential Jöns Jacob *Berzelius, could separate these ideas. For several decades chemists regarded Berthollet's mass action as a competitor to Bergman's affinities. The issue appeared chemically insoluble, because the relative quantities of each substance present in a solution could not be determined without applying reagents that themselves disturbed the equilibrium.

In 1862 Marcellin Berthelot and Péan de Saint-Gilles devised an effective way to test Berthollet's concept of mass action by studying the very slow chemical reaction between an alcohol and an organic acid to form an ester and water. At intervals they removed small portions of the solution and titrated the acid with a base without affecting the reaction sensibly. Their results strongly confirmed Berthollet's mass action and stimulated two Norwegian chemists, Cato Guldberg and Peter Waage, to make similar determinations using particularly sluggish inorganic reactions. Guldberg and Waage obtained results analogous to those of the French team and in 1864 published a mathematical law of mass action, which they subsequently simplified. By 1879 they had arrived at the form of the law that has been used ever since.

Frederic Lawrence Holmes, "From Elective Affinities to Chemical Equilibria: Berthollet's Law of Mass Action," *Chymia* 8 (1962): 105–145. J. R. Partington, *A History of Chemistry*, vol. 4 (1964).

FREDERIC LAWRENCE HOLMES

MASS SPECTROGRAPH. The mass spectrograph is an electromagnetic instrument for separating ions on the basis of their charge to mass ratio (e/m), and hence for studying their mass and chemical nature. In 1912 Joseph John *Thomson and his assistant Francis Aston, analyzing positive rays (ions that stream through a hole in the cathode of a gas-discharge tube), discovered an ion closely associated with that of neon, atomic mass 20, but corresponding to mass 22. For several years identification of this ion as a compound, a new element, or an isotope of neon remained uncertain, but from 1913 Frederick Soddy actively promoted it as evidence of isotopes in nonradioactive substances.

The subsequent development of the mass spectrograph is intimately connected with Soddy's concept of isotopes and the Rutherford-Bohr atom (*see* ATOMIC STRUCTURE). The first instruments, invented by Arthur Dempster in Chicago in 1918 and Aston in Cambridge in 1919, both attempted to separate isotopes unambiguously using variations on Thomson's positive-ray apparatus. But it was Aston, working in the Cavendish Laboratory at Cambridge under Ernest *Rutherford, whose name became linked with the mass spectrograph.

The two instruments relied on different focusing techniques and produced different types of results. In Aston's mass spectrograph perpendicular electric and magnetic fields focused ions with different masses at different points on a photographic plate. Aston identified and measured the atomic weights of a large number of isotopes; established the "whole number rule" for atomic weights (that isotopic masses are integral multiples of that of hydrogen, then known only as a single isotope of mass 1); and, with his second instrument of 1925, measured deviations from this rule, the "packing fraction" of nuclei (a measure of the mass equivalent of the energy binding the constituents of a nucleus together). He received the Nobel Prize for chemistry in 1922 for his work. Mass spectrographs are used extensively for accurate atomic weight determination.

Dempster's instrument, which Aston refused to call a "mass spectrograph" because it provided a momentum rather than mass spectrum, established a tradition of "mass spectrometers." In this design, a magnetic field perpendicular to the plane of a beam of ions causes them to move in a circle whose radius depends on the mass and velocity of the particles. In half a turn around a narrow vacuum chamber, ions of the same mass and velocity can be caught in a cup, the others having ended in the walls of the chamber. Dempster used a quadrant electrometer (see INSTRUMENTS, ELECTRICAL MEASURING) to detect them in the cup quantitatively. By varying the accelerating potential he selected different ions and measured their relative abundance. Although he failed to distinguish unambiguously between the hypothetical isotopes of magnesium and chlorine, Dempster opened the way for accurate abundance determinations and subsequently discovered many isotopes.

From the mid-1920s on several different types of mass spectrograph have been developed to meet the needs of spectroscopists, atomic-weight chemists, and radioactivists. The most important of these instruments were those of Kenneth Bainbridge (1933), who provided the first experimental proof of the Einstein mass-energy relationship, and Alfred Neir, who introduced a 60 degree (rather than 180 degree) analyzer in 1940. In the 1930s and 1940s physicists invented other electromagnetic means of separating ions: time-of-flight, radio-frequency, and cyclotron-resonance instruments in particular. By this time isotopes had become fundamental to many physical sciences, commercial instruments were available, and the earlier distinction between mass spectrographs and spectrometers had been lost as mass spectrometry became established as a central technique in an era of increasing reliance on instrumentation.

Francis Aston, *Mass Spectra and Isotopes* (1933). H. E. Duckworth, R. C. Barber, and V. S. Venkatasubriamanian, eds., *Mass Spectroscopy*, 2d ed. (1986). J. A. Hughes, *The Radioactivists: Community, Controversy and the Use of Nuclear Physics* (Ph.D. diss., Cambridge University, 1994).

ISOBEL FALCONER

MATERIALISM AND VITALISM. Materialists make the ultimate principles matter and motion;

vitalists, the soul or an irreducible life force. Because both believed that they could answer the question "What is life?" their doctrines played a major role in *botany, *zoology, and *physiology in the late eighteenth and early nineteenth centuries. They also featured in political and religious debates. Critics of materialism, seizing on its irreconcilability with Christianity, identified it with attacks on all traditional political values, while connecting vitalism with traditional religious views and political orders.

Both theories go back to antiquity. Early Greek philosophers, such as Thales, Democritus, and Epicurus, outlined materialist positions, emphasizing the primacy of matter usually at the expense of the immaterial or the soul. They described the world as composed of matter and space, and change as the result of combinations or movements of atoms. From late antiquity through the Renaissance, the hold of the Catholic church and *Aristotelianism on intellectual life limited the role of materialism. It was revived in the seventeenth century as part of a challenge to Aristotle. Mechanical philosophers, like Greek materialists, explained phenomena by configurations of matter in motion, and thus many proponents of the new science argued for materialism, saving, of course, the human soul. Thomas Hobbes constructed a consistent but religiously unorthodox version. Pierre Gassendi reconciled materialism and religion by making atoms finite and by postulating an immortal intellect in man. René *Descartes though, by emphasizing the immaterial, immortal soul of man, made materialism less significant in the *Scientific Revolution than it might otherwise have been.

In the eighteenth century, philosophers turned to materialism hoping that it would increase knowledge of the physical world and of man. They believed that the properties of living beings arose from the complexity of their organization and rejected sharp distinctions between the living and the nonliving. Nineteenth-century materialists, like Hermann von *Helmholtz, Ludwig Büchner, and Jacob Moleschott, attempted to confute the prominence vitalism enjoyed in medicine. Their studies suggested to them that what had been explained as vital forces could be better understood as purely physical (that is, material) phenomena. The nervous system, for instance, could adequately explain mental activities like reason, memory, or the emotions, that the behavior vitalists were inclined to associate with the soul or life forces. Modern biology and biochemistry are basically materialist, although less dogmatically so than their early modern predecessors.

The term "vitalism" first occurs in the seventeenth century to designate the doctrine formulated, often with reliance on classical sources, by Paracelsus, Robert Fludd, J. B. Van Helmont, and others who posited life-bearing seeds or spirits. The term was also applied retroactively to much older philosophical works, like those of Aristotle and Plato, that fostered the broad notion that the universe is filled with life, that matter is endowed with life, and that many natural processes are life processes. The revival of ancient Stoicism contributed to vitalism by emphasizing a universal soul or *pneuma*. Defined most clearly by the German physician Georg Ernst Stahl, whose work had a European following, vitalism became prominent among physicians in the late seventeenth and early eighteenth century. They emphasized the soul as the critical determinant of human physiology and function, although they had different responses to questions of the specific, nonmechanical entity of life, its effects, and its connection to the body in terms of cause and effect. From the late eighteenth century, the vitalists' adamant rejection of materialism and staunch advocacy of life forces led to an alliance with the Romantics and with conservative political and intellectual movements. Late nineteenth- and early twentieth-century vitalists, like Hans Driesch and Henri Louis Bergson, insisted that mechanism could not adequately explain the development of the embryo or the spiritual dimension of human life and the vital energy of the mind. Insofar as there is a contemporary immaterialist position, it focuses specifically on functions of mind rather than on life in general.

Friedrich Lange, *The History of Materialism: And Criticism of Its Present Importance*, 3d ed. (1950). John Yolton, *Thinking Matter: Materialism in Eighteenth-Century Britain* (1983). Stephano Poggi and Maurizio Bossi, eds., *Romanticism in Science: Science in Europe, 1790–1840* (1994). Elizabeth Williams, *The Physical and the Moral: Anthropology, Physiology, and Philosophical Medicine, 1750–1850* (1994). Jacques Roger, *The Life Sciences in Eighteenth-Century French Thought* (1997).

KATHLEEN WELLMAN

MATERIALS SCIENCE. Although materials have long been objects of scientific inquiry— *Galileo discussed the strength of beams in his *Two New Sciences* (1638)—the academic discipline of materials science appeared only recently. The mechanics of elastic bodies, which developed as a kind of mixed science, became an integral part of the program of experimental philosophy developed by Robert *Hooke at the Royal Society, by Edmé Mariotte in France, and by Jakob Bernoulli in Switzerland. Only in retrospect, however, can we identify these studies of the mechanical properties of wood or iron as a protoscience of materials. The generic notion of materials is a fairly recent invention. The new discipline emerged around 1960 when the departments of metallurgy of a number of academic institutions were renamed "metallurgy and materials science" and a few years later materials science emerged as an autonomous entity. This linguistic change reflected an inner evolution of metallurgy toward the determination of crystalline structures.

Beginning with the study of crystals by *X-ray diffraction (1913), the determination of microstructure became the prime concern of physical metallurgy. The notions of crystal lattices, dislocation, and defect proved key to understanding the macroscopic behavior of metals. The connections between microstructure and mechanical properties were probed, and the models and theories elaborated by physicists put to work for designing new materials.

X-ray diffraction techniques gave precise atomic pictures of solids, and *quantum mechanics provided the theoretical foundations for their further description. The solid state became an object of investigation in itself. Solid-state physicists discriminated between properties depending on the idealized crystal pattern and properties dependent on "accidents" of the inner arrangement or of the surface of the solid. This stress on structure-sensitive properties in the study of crystals turned the subject toward materials science.

A solid is not a material, however. The notion of materials combines physical and chemical properties with social needs and industrial or military interests. This hybrid concept implies that knowing and producing cannot be separated. Significantly, the discipline that emerged in the 1960s has been named "materials science and engineering" (MSE).

The generic notion of materials first appeared in the language of policy makers and referred to a bottleneck holding up advances in space and military technologies. The idea that all materials were strategic emerged during the *Cold War in the United States in response to *Sputnik. Designers sought high-performance materials with previously unknown properties for use in extreme circumstances. The Department of Defense's Advanced Research Project Agency (ARPA) invested heavily in academic research on materials and created interdisciplinary laboratories equipped with expensive instruments to prompt joint research involving metallurgists; chemical, electrical, and mechanical engineers; chemists; solid-state physicists; and electronics specialists. This program created the research field of MSE, at this stage mainly an American science. In Europe, a number of materials science centers grew out of former metallurgy departments, but materials did not become a political concern and solid-state chemistry emerged more prominently.

In the 1970s, new social priorities reoriented academic research from military to civilian goals. Environmental and safety legislation, together with economic competition with Japan, put new demands on material scientists. Materials science was now conceived as a response to "man's needs." While the U.S. federal budget for research and development stagnated, industrial companies became more engaged and research became more concerned with process. The interrelation of structures and properties with functions and processes, visually represented by a tetrahedron in many textbooks, provided the conceptual framework for MSE and helped to make it an established, teachable discipline. Courses in MSE proliferated in engineering schools, an annual review of materials science started (1971), and a Materials Research Society (1973) and a European Materials Research Society (1983) appeared.

At the same time the leadership that physicists and metallurgists had exerted passed to chemists. Development of composite materials made of a matrix reinforced with fibers acted as a driving force in the formation of the discipline. Unlike conventional mass-produced *plastics, composite materials are shaped to achieve specific functions and respond to specific demands. They offer the paradigm of "materials by design," a product of science and technology working together. Composites technology reinforced thinking in terms of four variables—structure, properties, performance, and processes; changes made in any of them can have a significant effect on the performance of the whole system and require a rethinking of the entire device. The traditional linear approach—given a set of functions find the properties required and design the structure combining them—gave way to a systems approach aided by computer simulation. The synergy between the four composite material variables called for a synergy between various specialists and a new organization of labor in project teams.

In the 1990s, academic research regained dominance. Following a drastic reduction of technical staff at most industrial companies, many industrial researchers joined university laboratories. MSE continued to diversify. Part of the materials science community shifted from microscale to nanoscale analysis. Instruments, which had already played a decisive role in the emergence of MSE with x-ray diffraction in the

1920s and the transmission electron microscope in the 1950s, once again proved crucial: the scanning tunneling microscope and the atomic force microscope allow not only the visualization but also the manipulation of individual atoms. With nanotechnology, materials are no longer carved like a statue out of a block of marble but by bonding atoms or groups of atoms. The key step becomes the assembly of building blocks.

In the 1990s, materials scientists suddenly became interested in mollusk shells, insect cuticles, algae, and spider silk. These composite structures, associating the hard and the soft, combining inorganic and organic components, capable of high performance, appeared as models for human technology in three respects. They are models of functional materials, optimally performing several functions including growth, repair, and recycling. They are models of structure: the remarkable properties of bulk materials such as bone or wood result from a complex arrangement at different levels, each controlling the next; the hierarchy of structures from the molecular to the macroscopic appears as a mark of the superiority of nature's design over human engineering. Finally nature assembles components at low energy cost, teaching lessons in processing. Biomimetics thus became a fashionable topic covering projects ranging from the design of a shield inspired by the layered structure of the abalone shell to the development of genetically engineered materials such as artificial spider silk. More importantly, biomaterials inspired the emergence of a new style of chemistry, *chimie douce*, concerned with reactions at ambient temperature in open reactors resembling reactions occurring in biological systems.

Nanotechnology and biomimetics disrupt rather than reinforce the fragile coherence of the mixed discipline of MSE. The field is imploding. Its territory is dismembered into many research areas concerned with different scales and kinds of structures. The frontiers of MSE border bioengineering and artificial intelligence. Materials scientists follow divergent epistemic practices. Some advocate rational design by computer simulation. They want to build up a material by computer calculations *ab initio*, starting with the most fundamental information about the atoms and using the most basic rules of physics. Others apply a combinatorial approach, developed initially in the pharmaceutical industry. It calls for the simultaneous synthesis of a large array of compounds at once, screened to detect and select interesting structures. Despite their differences, both strategies look exceedingly systematic when compared with the semiempirical methods chosen by scientists who encourage new materials to assemble themselves. Again two different pathways may be distinguished. Some scientists engineer bacteria, while others try the chemical route. Whether materials science will split into branches integrated into established disciplines or become a discipline of its own, pioneering a new style of science, the future will tell. Meanwhile MSE functions as a laboratory to test new frontiers in science.

Cyril Stanley Smith, "The Development of Ideas on the Structure of Metals," in *Critical Problems in the History of Science*, ed. Marshall Clagett (1959): 467–498. National Academy of Science (COSMAT), *Materials and Man's Needs*, 3 vols. (1975). Spencer R. Weart, "The Solid Community," in *Out of the Crystal Maze. Chapters from the History of Solid State Physics*, ed. L. Hoddeson, E. Braun, J. Teichman, and Spencer Weart (1992): 617–666. Ivan Amato, *Stuff. The Materials the World is Made of* (1997). Bernadette Bensaude-Vincent, *Eloge du mixte. Matériaux nouveaux et philosophie ancienne* (1998). Merton C. Flemings and Robert W. Cahn, "Organization and Trends in Materials Science and Engineering Education in the U.S. and Europe," *Acta metallurgica* 48 (2000): 371–383. Robert W. Cahn, *The Coming of Materials Science* (2001).

BERNADETTE BENSAUDE-VINCENT

MATHEMATICS

Mathematics to 1800

From the eleventh to the fifteenth centuries mathematics was an adjunct to astronomy and astrology, subjects important for calendrical, judicial, and medical applications. The calculational prowess of surveyors, clock and calendar makers, masons and machine makers, and above all merchants and accountants far outstripped those of previous centuries. The introduction of Hindu-Arabic arithmetic and calculation methods in the abacus schools of Italy in the thirteenth to fifteenth centuries was to change the face of calculational mathematics.

Over the first half of the fifteenth century Italian traders, collectors, academics, and enthusiasts engaged in a brief but crucial trade in importing Byzantine copies of Ancient Latin and Greek texts. Early *Renaissance scholars, lacking the necessary technical skills, did not treat the mathematical texts that they found. Only in the last decades of the fifteenth century was significant attention paid to Ptolemy's mathematical model of the heavens (*see* COPERNICUS, NICHOLAS). Humanist scholars turned to the extant works of Archimedes in the middle of the sixteenth century, and to some parts of Apollonius only at its very end. These humanist-mathematicians were not engaged in mere historical reconstruction, however; they intended to continue mathematics in the style of the ancients. Developing Archimedes led to the study of the geometry of curves and to

the development of mathematical mechanics, statics, hydrostatics, and the mathematics of the simple machines.

Many Renaissance mathematicians followed what they thought to be Platonic doctrines that mathematics was the key to unlocking the secrets of the universe. Some regarded this form of knowledge as mystical and symbolic; others considered that it provided a knowledge of the world that could be sure and certain without engaging in the difficult, sometimes obscure, and seldom definitive argumentation of the natural philosophers.

The status of mathematics changed during the Renaissance in part because of the success of applications of mathematics in spectacular engineering feats, from Brunelleschi's dome of the Cathedral of Florence early in the fifteenth century to the raising of the Vatican obelisk in the last years of the sixteenth century. Engineers sometimes adopted a classical style so as to appear learned or gentlemanly, as opposed to craft practitioners; the association of their work with Archimedes' interest in practical matters helped to legitimize their efforts. This late-sixteenth-century development influenced the thinking of principal figures of the *Scientific Revolution such as *Galileo, Marin Mersenne, and René *Descartes.

The Renaissance saw the development of algebra from a commercial problem-solving art to something like the science of equations. The practical study of Hindu-Arabic calculation methods developed into a larger problem-solving art treating questions well beyond the scope of normal commercial needs; some of these (still hand-written) practical texts used abbreviations and symbols to ease the difficulties of calculation. By the early sixteenth century this field, sometimes called "the analytical art," started to look like what we call algebra. Printed versions of these calculational books appeared from the early sixteenth century. From the 1520s through the 1540s a small group of Italian mathematical practitioners competing with each other developed the general method of solving cubic equations. From this point the mathematics of equations—algebra—became a new and clear object of mathematical inquiry.

The general solution to the cubic required the manipulation and calculation with an entity whose square was a negative number. The conceptual difficulty in dealing with this entity was much greater than the challenge of negative numbers (already a problem for some mathematical thinkers). There was no obvious real model for the square root of a negative number in the way that debt could represent a negative quantity.

Contemporary mathematicians had to decide whether to believe the formalism that produced such oddities, or their intuitions. The suppression of common-sense intuitions has marked the development of mathematics ever since.

During the Scientific Revolution the study of curves broke deeply with the classical model that the Renaissance had tried to recreate, and by the end of the century mathematicians had developed a new and extraordinarily powerful tool for the study of curves: the infinitesimal calculus. More important still was the discovery of the link between algebra and geometry. In the 1620s Pierre de Fermat and Descartes independently applied the new formalism of algebra to the relationships that represented curves, and discovered that equations could be represented by curves and curves by equations: algebra and geometry were different languages for describing the same entity. Descartes published the explicit discovery of the relationship between algebra and geometry in his *La Géométrie* of 1637.

The mathematical work of Isaac *Newton and Gottfried Leibniz took place within the context of the intense study of the general properties of curves. Separately and probably independently each invented new and very general methods for treating the properties of curves, each with a different notation and slightly different conceptual tools, and each with different ways of obfuscating—or avoiding—the use of infinitesimals. Newton's discovery dates to the mid-1660s, Leibniz's to 1672–1675. Although we describe their achievement as the discovery of the calculus, in some ways it would be better to consider the discoveries of the 1660s and 1670s as general theories of the algebraic geometry of curves, and to reserve for the work of the 1720s through the 1740s the effective invention of the subject we now call the calculus.

Newton's mathematics was deeper than Leibniz's. However, from what Newton published during his lifetime others could scarcely learn his calculus; not until the 1730s did other mathematicians catch up with what he had accomplished in what we now call integration. Leibniz, on the other hand, developed simple, clear, and easy-to-use algorithmic methods for determining integrals and differentials, using a simple and efficient new notation. On the Continent advocates of the new calculus occupied strong institutional positions in the new scientific societies and some universities. Its rapid dissemination among European mathematicians coincided with (but was not linked to) the dissemination of Newton's theories of universal gravity, and of light and colors.

The Scientific Revolution gave mathematical ways of knowing about the physical world a new status. From the 1650s on, the idea of a mathematical mechanical physics remained a constant and intellectually stable thread in natural philosophy, Newton's *Principia* being the paradigmatic case. Mathematically minded natural philosophers hoped to derive the mathematical mechanics of nature without recourse to hypotheses, following in the style of Newton's theory of gravity.

The eighteenth century saw explosive growth in the scope of mathematics and its scientific applications. The rhetoric of the utility of the mathematical sciences finally began to have extensive and concrete justification, and the spread of the calculus through the many textbooks published on the Continent brought out both mathematical ability and mathematical ambitions. Access to mathematical instruction was widespread by the middle of the eighteenth century—in France, for example, in military and engineering schools, Jesuit colleges, and universities, among other places.

However, mathematicians recognized that the infinitesimal foundations of the new methods were far from secure. Jean le Rond d'Alembert best summed up the way that mathematicians of the time dealt with the problem in practice when he wrote to a pupil, "Carry on, and faith will follow." Bishop Berkeley's attack on the calculus in 1734 compared its foundations unfavorably to those of theology. There ensued a noisy squabble that ended with Colin Maclaurin's *Treatise on Fluxions* (1742), in which he tried to base the Newtonian fluxional calculus on synthetic Archimidean methods. Continental mathematicians accepted Maclaurin's arguments as proof that the calculus gave results confirmable by synthetic geometry; the English used them to bury their mathematical studies further into synthetic, geometric, ultimately dead-end work. On the Continent the best mathematicians, as well as popularizers and teachers, continued vainly to try to provide sound foundations for the infinitesimal calculus.

After the union of Newtonian mathematical physics and the Leibnizian version of the calculus by Leonhard *Euler and others, the marriage, or continuum, between physics, rational *mechanics, and higher mathematics (the differential and integral calculus) became complete. The problems that stimulated much mathematical study were generally related to physics and so only very rarely pushed the mathematics beyond safe limits. This kept the paradoxical nature of the foundations of the calculus tolerable.

The study of both ordinary and partial differential equations developed in tandem with the first great success of the new calculus:

the analysis of compressible and incompressible fluids (gases and liquids, respectively), elasticity, and the gravitational dynamics of the planets. The development of gas and *hydrodynamics from the very late 1730s, along with the analysis of elastic phenomena, brought a vast new range of phenomena within the range of mathematical science: tides, a mechanical theory of heat, sound, harmonics and *acoustics, fluid flow and the movement of bodies in resistive media, optimization of hull design, water flow in tubes and jets, the pressures of air on sails, color and the optics of lenses, and the mechanics of static electric charges. Later in the century the application of elementary statistics and probability theory to mortality tables produced an "arithmetic of life." Annuities and life insurance became much more precise, and man more clearly the subject of quantitative social laws.

The imperialist project of rational mechanics reached its apex in the work of Pierre-Simon de *Laplace and Joseph Louis Lagrange at the end of the eighteenth century. In his *Système du monde* (1796) and *Mécanique céleste* (1799–1825), Laplace gave a complete mathematical theory of the dynamics of the planetary system and outlined the project of a deterministic rational mechanical theory of the physical world. Lagrange's work was less politically dramatic, but more scientifically prescient. In his *Mécanique analytique* (1788) he recast mechanics, reducing it to pure analysis without the benefit of an intuitive basis in geometry: the book is famously without diagrams.

The most significant innovation of eighteenth-century mathematics was the concept of function, formulated by Euler as any reasonably continuous relationship between variables. Euler allowed functions to be perfectly general, although in practice he limited their extent to geometric or physical relations. During the second half of the century the definition of function became increasingly general as mathematicians discovered that many arbitrary curves or expressions could be treated as objects of analysis. It was around the concept of function that mathematical thinking began to relax its reliance on geometric and physical intuitions.

PIERS BURSILL-HALL

Mathematics 1800 to the Present

The initial changes in mathematics in the nineteenth century were consequences of the French Revolution. The École Polytechnique, founded in 1794 as the preparatory school for a range of engineering colleges, gained a worldwide reputation. A new university system established in 1808 led to further professional opportunities. The dominant research area, under the leadership of Joseph Louis Lagrange and Pierre-Simon de *Laplace, continued to be the calculus and its applications to mechanics, but a new generation would soon both challenge and build on their methods.

The two great figures in the vanguard of this new generation were Joseph Fourier, with his mathematization of heat diffusion and methods of solution by Fourier series and Fourier integrals, and Augustin-Louis Cauchy, who introduced a new and ultimately dominating approach to the calculus based upon limits, the inauguration of complex-variable analysis, and major extensions of elasticity theory and *hydrodynamics. Fourier's work joined the electrostatics of Simeon-Denis Poisson as the first major mathematizations of phenomena outside mechanics. The optics of Augustin Fresnel and the *electromagnetism of Andre-Marie Ampère soon followed, enlarging *mechanics into mathematical physics.

From the 1820s mathematics became more international, with the rise of Great Britain and Italy and especially of Germany, which rivaled France by the 1860s. One inspiration for the upsurge was the eventual recognition of several achievements of Carl Friedrich *Gauss, a great but rather isolated mathematician for much of his life in Göttingen. His most important follower there, Bernhard Riemann, proposed a new version of complex-variable analysis and facilitated the recognition of non-Euclidean geometries; he also inspired the set theory of Georg Cantor.

Göttingen maintained a strong interest in applied mathematics, in contrast to the preference for purer mathematics that emerged in the other major center, Berlin. There the lecture courses of Karl Weierstrass, especially on real- and (his own version of) complex-variable analysis, exerted a huge influence; he stressed the role of rigor. The Berlin purist attitude spread, even to France; thus the greatest French mathematician of the time, Henri Poincaré, was virtually alone in his prosecution of mathematical physics.

As usual, Britain remained distant from many Continental developments; but mathematical physics prospered, especially from the time of George Biddell Airy and George Gabriel Stokes. One reason was the Mathematical Tripos degree at Cambridge University, good for learning skills though notoriously free from attention to genuine problems. A major topic at Cambridge was potential theory, inspired by George Green's marvelous theorem relating the state of a phenomenon inside a body to that on its surface (1828). The self-taught Green published his work obscurely, but it gained massive attention after republication by William *Thomson (Lord Kelvin) and development by him, Stokes, and many other mathematicians, including some at Göttingen. But Berlin rose again: in 1870 Weierstrass showed that its "Dirichlet principle," named after the German mathematician Gustav Peter Lejeune Dirichlet but known also to Green, was untrustworthy; methods were profoundly affected.

Britain also played a major role in the introduction of new algebras, which handled objects other than numbers and magnitudes. The theories included matrices, determinants, invariants, quaternions, vectors, differential operators, groups, algebraic number theory, probability theory, and algebraic logic.

Throughout the century mathematics became steadily more professionalized everywhere, with ever more universities and military colleges. Textbooks and treatises came out in profusion; and journals proliferated, many inspired by the German *Journal für die Reine und angewandte Mathematik* (founded 1826). From the 1860s onwards national societies for mathematicians and/or mathematics schoolteachers were formed. In 1897 a series of International Congresses of Mathematicians was initiated. At that time the Göttingen mathematician Felix Klein launched a vast encyclopedic survey of mathematics, the *Encyklopädie der mathematischen Wissenschaften*. It was completed in the mid-1930s.

The scale of work achieved during the nineteenth century far exceeded that in preceding times, but it was surpassed during the following half-century, and then again in the next half. As Hermann Weyl wrote in the mid-twentieth century, mathematics was "like the Nile delta, its waters fanning out in all directions." All the branches and topics mentioned earlier continued in greatly expanded forms, and new branches and applications emerged. The growth in topology was especially marked, partly thanks to remarkable contributions from Poincaré. The ever-increasing variety of work, both internally and in applications, during the twentieth century makes summary impossible. A few movements that started around 1900 will be outlined.

After a slow start, Cantor's set theory became widely developed, both in its revolutionary handling of *infinity and for its technical methods; topology was one beneficiary. Set theory became a lingua franca among mathematicians in many fields, and from around 1950 was quite widely taught, even (and unfortunately) in elementary schools. However, the neighboring subject of logic remained relatively obscure. It has increased its role fitfully with the development of computing science as well as cryptography, especially after the wartime decoding achievements of Alan *Turing and others.

This breach from logic appears surprising because of the emphasis on axiomatization of theories, or at least on formal presentations, in many branches of mathematics. David Hilbert exerted a particularly strong influence here, though even he could not sell logic widely. Indeed, he was challenged by Luitzen Egbertus Jan Brouwer, who proposed "intuitionist" mathematics without recourse to the normal true-or-false logic. His idiosyncratic philosophy inspired few followers (Hermann Weyl was one), but he heralded various forays into "constructive" mathematics.

The advent of mechanical and then electrical calculators, and finally computers, gave new prominence to constructions (especially algorithms). Numerical mathematics also benefited: many methods envisioned in the past now became feasible—interest increased in nonlinear mathematics, for instance, lessening the dominance of linear theories (such as Fourier series).

One disappointment of nineteenth-century mathematics was the slow development of mathematical statistics. From around 1900, however, several of its main concepts and theories were developed and popularized by Karl Pearson and his followers in Britain, among others. The subject grew massively: applications to life sciences and medicine became especially wide, especially after World War II. Statisticians form a massive community in its own right, markedly different in character from that of the mathematicians.

On a national level, the United States advanced rapidly from a minor status in mathematics in 1900 to one close to supremacy by midcentury. Curiously, Americans preferred pure mathematics over applications. The Soviet Union also became notable for its mathematics. Germany retained its importance up to Hitler's time. After Word War II France enhanced its status still further with a group of pure mathematicians writing under the collective name Bourbaki—they favored structure in mathematics and formal presentations, and disdained the study of logic. Elsewhere, Hungary produced a disproportionate number of major figures in the field. The most eminent of all was John von *Neumann, who spent most of his career in the United States.

See also ASTRONOMY; PHYSICS; STATISTICS.

IVOR GRATTAN-GUINNESS

Morris Klein, *Mathematics in Western Culture* (1954 et sec). Morris Klein, *Mathematical Thought from Ancient to Modern Times* (1972). Joseph W. Dauben, ed., *Bibliography of the History of Mathematics* (1985). Andrei Kolmgorov and Adolph P. Yushkevich, eds., *Mathematics in the 19th Century*, 3 vols. (1992–1994). Ivor Grattan-Guinness, ed., *Companion Encyclopedia of the History and Philosophy of the Mathematical Sciences*, 2 vols. (1994). Jean-Paul

Pier, ed., *Development of Mathematics 1900–1950* (1994). Ivor Grattan-Guinness, *The Fontana History of the Mathematical Sciences* (1997). Victor J. Katz, *A History of Mathematics: An Introduction*, 2d ed. (1998). Albert C. Lewis, ed., *The History of Mathematics; a Selective Bibliography*, rev. ed., CD-ROM (2000). Jean-Paul Pier, ed., *Development of Mathematics 1950–2000* (2002).

MATHEMATIZATION AND QUANTIFICATION. To understand the quantification of natural philosophy in the early modern period, one must begin with the constellation of the subject and its relation to *mathematics. Early modern mathematics drew on a tradition dating to antiquity and included all practical subjects involving extensive calculation: *astronomy, navigation, *mechanics, civil and military engineering, surveying and cartography, geometrical optics, *acoustics, perspective, and music. The field had low status, as the pejorative use of "mechanic" suggests.

*Natural philosophy directed the eye of wisdom towards the natural, as distinguished from the divine, realm. Synonymous with "physics" in the early modern period, the field included subjects that today would be classified with *biology, *geology, and psychology, as well as *chemistry and *physics. From the beginnings of the *Scientific Revolution well into the eighteenth century, natural philosophy remained a literary discipline, as nonquantitative as it was broad.

From the beginnings of the Scientific Revolution, some natural philosophers, notably *Galileo, *Descartes, *Huygens, and *Newton, aimed to bring their discipline under the sway of mathematics. Mathematicians working in established mathematical disciplines sometimes appropriated parts of natural philosophy. For example, over the course of the sixteenth and seventeenth centuries astronomy took over from natural philosophy the concept of the cause of planetary motion, and mathematized it. Newton's triumph applied calculus to that cause: a force that, inherent in infinitesimal parts of matter, draws the planets infinitesimally into their orbits during infinitesimal moments of time (*see* FORCE; GRAVITATION).

Experimental physics resisted quantification until the late eighteenth century (*see* EXPERIMENTAL PHILOSOPHY). Even Newton, attacking *electricity, employed a kind of pictorial explanation often labeled "Cartesian" and popularized by successive generations of experimental natural philosophers (*see* MECHANICAL PHILOSOPHY). Newton's followers—Francis Hauksbee and Petrus van Musschenbroek among them—missed Newton's application of infinitesimals and groped after a mathematical law of magnetism based on the gross characteristics of the *magnet. The problem lay partly in the subject matter: astronomers had studied the comparatively simple regularities of the heavenly bodies over millennia; phenomena from magnetism to *meteorology presented a greater challenge to mathematization. The orientation of natural philosophy contributed to the challenge: the discipline aimed at edification and entertainment rather than measurement and calculation. Nor were scientific (or, as they were then called, philosophical) *instruments suitable for quantitative investigations. Mathematical instruments, intended for astronomy, surveying, engineering, and the other parts of mathematics, were of little enough accuracy. Philosophical instruments—aids to natural philosophy such as the *thermometer and *barometer—offered at best a qualitative indication, despite the numbers penned on their paper scales. Indications such as "blood heat" on thermometer scales reflect the anthropomorphic coloring of early modern measurement. Measures were also local: barometer readings taken in local inches, for example, frustrated attempts to coordinate meteorological observations across Europe.

Around 1760 a quantifying spirit swept over natural philosophy. A typical early result was Joseph Black's discovery of the latent heat of evaporation (*see* FIRE AND HEAT). Black calculated the rate at which a can of water, set on a stove with the fire "pretty regular," heated from room temperature to boiling; this quantity, multiplied by the time taken to boil off the water, gave the "degrees of heat...contained in [its] vapor." The mathematics, no higher than and perhaps borrowed from bookkeeping, characterized other early successes as well. Benjamin *Franklin applied similar reasoning to quantities of charge collected by a person drawing electricity from an electrostatic generator, as did Antoine-Laurent *Lavoisier to the quantities of oxygen fixed during combustion.

Mathematical instruments played a defining role in quantification. In the age of the emergence of the national state, governments and their armies and navies required accurate navigational and cartographic information; gentleman farmers draining fens and businessmen financing canals needed good surveys; engineers wanted precisely machined parts for the engines of the Industrial Revolution. The mathematical instrument trade responded, developing sophisticated precision instruments as it moved from a craft-based to a protocapitalist form of organization. Philosophical instruments required for survey like the thermometer and barometer benefited, but natural philosophers, increasingly involved in research, also demanded precision in instruments like the electrometer that had no practical application. Equally important, they developed a methodology of exact measurement, standardizing instrument scales, correcting for disturbing factors and repeating extensive series of readings.

In the last third of the eighteenth and the first years of the nineteenth century, enlightened governments organized precise measurement in many fields, including cartography, meteorology, national population census, natural resources survey, and the calculation of mathematical constants. The tabular format was perfected for the presentation of large amounts of information. The quantifying spirit penetrated even to fields like natural history, where the Linnaean method applied the simplest part of arithmetic, counting, to the classification of species into higher taxa. Calculation came to be regarded as the highest form of intellectual activity. Large quantities of reliable data provided the basis for later fruitful theorizing in the nineteenth century; the late *Enlightenment was not given to synthesis of insight from information.

In natural philosophy, nevertheless, quantification yielded important exemplars in *electricity, *magnetism, and the theory of heat. The military engineer Charles Augustin Coulomb (*see* CAVENDISH, HENRY, AND CHARLES-AUGUSTIN COULOMB) applied the techniques of engineering to the construction of his famous magnetic torsion balance, with which he demonstrated the inverse-square law of magnetic and electric forces. Coulomb represents an important late-eighteenth-century trend: the cross-fertilization of *engineering and experimental physics. Through the crossover of personnel and with the increasing availability of mathematical training at institutions like the École du Génie at Mézières and the École Polytechnique, mathematicians and natural philosophers, and their respective fields, became steadily less distinguishable. The process was one of the strongest threads in the quantification of natural philosophy.

J. L. Heilbron, *Electricity in the 17th and 18th Centuries* (1979). Theodore S. Feldman, "Applied Mathematics and the Quantification of Experimental Physics: The Example of Barometric Hypsometry," *Historical Studies in the Physical Sciences* 15 (1985): 127–197. Tore Frängsmyr, J. L. Heilbron, and Robin Rider, eds., *The Quantifying Spirit in the Eighteenth Century* (1990). J. L. Heilbron, *Weighing Imponderables and Other Quantitative Science around 1800* (1993).

THEODORE S. FELDMAN

MATTER. The concept of matter in modern science arose in response to theories originally formulated in ancient Greece. The Greeks thought of matter as the underlying stuff out of which all things are made. The earliest philosophers sought a single material substance, such as fire or water, that would be responsible for the appearance of all physical phenomena. For Plato matter became the recalcitrant stuff that received forms imperfectly. For Aristotle it was the primary substrate that possessed no qualities of its own but provided the limiting conditions with which the other three causes of change (form, agency, and purpose) had to work. Democritus described matter as undifferentiated as well, but also as being composed of indivisible particles (atoms) moving in a void. During the *Renaissance the Aristotelian conception of matter came under fire from several sides, notably in the sixteenth century by the physician Paracelsus, who defended an account of bodies as composed of three material elements (salt, sulfur, and mercury).

In the seventeenth century, Paracelsian chemistry gave way to the chemical theories of Jean Baptiste van Helmont, who attributed active powers of *sympathy and antipathy to matter. By far the dominant conception, however, was that of the *mechanical philosophy. In his *Discourse on Method* (1637), René *Descartes argued for the fundamental distinction between matter and mind. He made matter pure extension; the only quality Cartesian matter possesses is that it takes up space. All apparent qualities of bodies, such as color and texture, are merely appearances that result from the motions of the particles of bodies (corpuscles) impinging upon our nerves and exciting sensations in us.

In the next generation, Robert *Boyle defended an experimental version of corpuscularianism heavily influenced by his interest in chemistry. He agreed with Descartes that the diversity of bodies arose from the configuration of one universal matter. In his *Origin of forms and qualities* (1666), however, he argued that some configurations (primary concretions) remained relatively permanent because he found that they could not be broken down by chemical analysis. John Locke, in his *Essay concerning human understanding* (1690), elaborated upon the distinction between the primary qualities of matter, which refer to such quantitative aspects as size and shape, and the secondary qualities of matter, which refer to the powers that bodies have because of their particular configurations to produce sensations in human observers. But, Locke added, we can only know matter through the powers of these particular configurations and thus the general idea of a "substance" as that which possesses powers is a mere name signifying an unknown support for the qualities we experience.

Isaac *Newton went a long way toward eliminating the concept of matter entirely. According to his *Principia mathematica* (1687), matter is what resists change of motion and causes change of motion in other bodies. He distinguished between material objects, which have weight because of mass, and forces, which measure the interaction of material objects. An object's response to force depends upon its quantity of matter (mass). By focusing upon mass as a quantifiable aspect of matter, Newton could calculate and predict the motions of bodies. This understanding of matter differed significantly from the earlier conception of matter as substance.

After Newton, investigation into the underlying substrate became an empirical question. In the late eighteenth and nineteenth centuries, Joseph Priestley, Joseph Black, Antoine-Laurent *Lavoisier, Humphry Davy, and John *Dalton pursued the idea that chemical as well as physical phenomena could be explained by assuming that all material substances possessed mass and were composed of atoms. In the early twentieth century advocates of electromagnetic conceptions of basic physical phenomena questioned the need to postulate the existence of any underlying material substratum. In their approach, which had a brief vogue, matter became a label for objects studied by classical mechanics and no longer represented a scientific explanatory category.

See also ARISTOTELIANISM; ATOM AND MOLECULE; ELEMENT.

Stephen Toulmin and June Goodfield, *The Architecture of Matter* (1963). P. M. Heimann and J. E. McGuire, "Newtonian Forces and Lockean Powers: Concepts of Matter in Eighteenth-Century Thought," *Historical Studies in the Physical Sciences* 3 (1971): 233–306. Ernan McMullin, *Newton on Matter and Activity* (1978). Ernan McMullin, ed., *The Concept of Matter in Modern Philosophy* (1978). P. M. Harman, *Energy, Force, and Matter* (1982). Leon Lederman with Dick Teresi, *The God Particle* (1993).

ROSE-MARY SARGENT

MAX-PLANCK-GESELLSCHAFT. See KAISER-WILHELM/MAX-PLANCK-GESELLSCHAFT.

MAXWELL, James Clerk (1831–1879), physicist, creator of the electromagnetic theory of light and the statistical theory of gases.

Maxwell was born in Edinburgh, his father John Clerk having taken the name Maxwell as heir to estates in Galloway in Scotland. The young man's abiding concern with philosophical principles was established while a student at Edinburgh University (1847–1850). Writing substantial papers on the geometry of rolling curves and the theory of elastic solids, he became interested in color vision. In the 1850s, building on work by Thomas Young, Hermann von Helmholtz, and Hermann Grassmann, he used red, green, and blue primaries to form color

combinations in experiments with tinted papers and on the mixture of spectral colors. Awarded the Royal Society's Rumford medal in 1860, he projected the first trichromatic color photograph in May 1861.

Admitted to Peterhouse and then Trinity College, Cambridge, Maxwell became a pupil of the mathematics coach William Hopkins, graduating second wrangler in 1854. Elected a fellow of Trinity in 1855, he was appointed professor of natural philosophy at Marischal College, Aberdeen, in 1856 and at King's College London in 1860. Resigning this post in 1865, he was elected to the new professorship of experimental physics at Cambridge (1871), where he founded the Cavendish Laboratory. Building upon his own work on establishing a standard unit of electrical resistance, Maxwell directed experiments on precision measurements in electricity and edited Henry Cavendish's *Electrical Researches*, published shortly before his death from cancer in November 1879.

Maxwell's most sustained achievement was in formulating the theory of the electromagnetic *field, developing work by Michael *Faraday and William *Thomson (Lord Kelvin). Guided by Thomson, he illustrated Faraday's "lines of force" by the analogy of streamlines in a fluid, establishing a geometry of field relations. Seeking physical foundations, he proceeded in a deliberately hypothetical style, imagining a mechanical *ether model where rotating vortices representing magnetism were separated by particles whose motion represented the flow of an electric current. Extending the model, Maxwell obtained an unexpected result: the close agreement between the velocity of transverse waves in an electromagnetic ether and the measured velocity of light. His electromagnetic theory of light, first proposed in 1862, unified optics and *electromagnetism. He subsequently discarded the ether model, placing emphasis on the transmission of energy in the field, and stated equations of the electromagnetic field that later became codified as the four "Maxwell equations." Maxwell expounded his theory in his *Treatise on Electricity and Magnetism* (1873), deploying vectors, integral theorems, topology, and analytical dynamics, a style that joined geometry to dynamics and freed physical quantities from representation by a mechanical model. Heinrich Hertz's production of electromagnetic waves in 1887 led to the general acceptance of Maxwell's field theory.

In his 1857 Adams Prize essay at Cambridge on the stability of the motion of Saturn's rings, Maxwell had concluded that the system consists of concentric rings of particles. Alerted to problems of describing particle collisions, in spring 1859 he noticed a paper by Rudolf Clausius on the kinetic theory of gases, and was intrigued by his use of a probabilistic argument to calculate the motions of gas molecules. Maxwell introduced a statistical function, identical in form to the distribution formula in the theory of errors, to calculate the distribution of velocities among molecules. He applied this model to obtain results for gaseous diffusion, viscosity, and thermal conductivity. The theory was provisional and Clausius was able to point out some of its deficiencies; and Maxwell's own experimental study of the viscosity of gases, by observing the decay in the torsional oscillation of discs, led him to reconstruct his argument. In his mature paper of 1867 he revised his theory of gas molecules in a form consonant with his findings on viscosity. He provided a rigorous derivation of the distribution law, a formulation soon enlarged by Ludwig *Boltzmann in seeking statistical foundations for thermodynamics.

But in a famous argument, and in characteristic style, Maxwell ingeniously expanded his reasoning. According to the second law of thermodynamics recently developed by Clausius and Thomson, heat flows from hot to cold bodies. But because of the statistical distribution of molecular velocities in a gas, there will be fluctuations of individual molecules that take heat from a cold body to a hotter one. The intervention of Maxwell's "finite being" (termed "demon" by Thomson) would be needed to select molecules so as to make this process detectable; but the process occurs spontaneously at the molecular level. The second law of thermodynamics, Maxwell concluded, is a statistical law, which applies to systems of molecules, not to individual fluctuations.

Lewis Campbell and William Garnett, *The Life of James Clerk Maxwell* (1882; repr. 1969). C. W. F. Everitt, *James Clerk Maxwell: Physicist and Natural Philosopher* (1975). Stephen G. Brush, *The Kind of Motion We Call Heat: A History of the Kinetic Theory of Gases in the Nineteenth Century* (1976). P. M. Harman, *Energy, Force, and Matter: The Conceptual Development of Nineteenth-Century Physics* (1982). P. M. Harman, ed., *The Scientific Letters and Papers of James Clerk Maxwell*, 2 vols. to date (1990, 1995). Daniel M. Siegel, *Innovation in Maxwell's Electromagnetic Theory: Molecular Vortices, Displacement Current and Light* (1991). P. M. Harman, *The Natural Philosophy of James Clerk Maxwell* (1998). Olivier Darrigol, *Electrodynamics from Ampère to Einstein* (2000).

P. M. HARMAN

MAYER, ROBERT. See JOULE, JAMES, AND ROBERT MAYER.

MEAD, Margaret (1901–1978), American cultural anthropologist, best known for her studies of South Pacific peoples. Mead's theoretical interests included child development, gender and sex roles, and cultural change. She was the foremost popularizer of anthropology in her generation.

Mead was born in Philadelphia to Edward Sherwood Mead, an economist, and Emily Fogg Mead, a sociologist and feminist. The eldest of four children, her earliest formal schooling was irregular, largely carried out at home. After one year at DePauw University she transferred to Barnard College in New York City, where she received a B.A. in 1923 and an M.A. in 1924, both in psychology.

During her senior year at Barnard, Mead met two people who would transform her career: Franz Boas, professor of anthropology at Columbia University, and his teaching assistant, Ruth Benedict. From them she absorbed a vision of anthropology as a science in which one could pursue a career almost with the intensity of a religious vocation. The challenge was to document the life-ways of small societies before they disappeared, to study how these cultural systems affected the development of the individuals growing up within them, and to apply the knowledge thus gained to contemporary social problems in her own society.

Mead's first fieldwork was conducted during nine months spent in American Samoa in 1925–1926. In *Coming of Age in Samoa* (1928), she suggested that adolescent Samoan girls made a more peaceful transition to adulthood than did American teenagers because Samoans had a more casual attitude toward work and sex and the girls had fewer difficult choices to make. The book was a best-seller and remains a classic, despite an attack on Mead's work after her death by Australian anthropologist Derek Freeman (*Margaret Mead and Samoa*, 1983).

Mead returned from Samoa to complete her Ph.D. in anthropology at Columbia (finishing in 1929) and to begin work at the American Museum of Natural History in New York, where she remained for the rest of her life as assistant curator of ethnology (1926–1942), associate curator (1942–1964), curator (1964–1969), and curator emerita. Mead's first marriage, to Luther Cressman, a ministry student turned social scientist, ended in divorce in 1928. Mead and her second husband, New Zealand psychologist and anthropologist Reo Fortune (to whom she was married from 1928 to 1935), did fieldwork on Manus in the Admiralty Islands; in New Guinea among the Arapesh, the Mundugumor, and Tchambuli (later called the Chambri); and in Nebraska studying the Omaha Indians. During this time Mead wrote *Growing Up in New Guinea* (1930), *The Changing Culture of an Indian Tribe* (1932), and *Sex and Temperament in Three Primitive*

Societies (1935). With her third husband, the British anthropologist Gregory Bateson, to whom she was married from 1936 to 1950, she carried out a three-year study on Bali (1936–1939) that pioneered the use of film and photography to study dance, trance, ritual, and mother-child interactions. They published their results in *Balinese Character: A Photographic Analysis* (1942).

When World War II began, Mead went to Washington as executive secretary of the Committee on Food Habits of the National Research Council, working on food distribution and problems of morale. She wrote *And Keep Your Powder Dry* (1942), a deliberately upbeat analysis of American character, and with Ruth Benedict developed methods for the study of cultures at a distance, which in the postwar years became a subfield of ethnology called culture and personality.

Mead's only child, Mary Catherine Bateson, was born in 1939. Motherhood was important to Mead. In *Male and Female* (1949) she discussed biologically determined sex roles in addition to the culturally determined roles she had previously emphasized. In 1953 she took up fieldwork again, returning with two younger colleagues to the island of Manus. Manus had been a landing station for the U.S. Army during the war, and Mead was surprised to find how quickly the Manus people had adopted Western civilization. Sudden rapid change is easier for small societies than slow change, she decided, and this became the theme of *New Lives for Old* (1956).

Mead's postwar career was one of ever-increasing activity and influence as she lectured at colleges and universities around the United States, continued her interest in psychiatry, and took part in interdisciplinary conferences and organizations, including the World Council of Churches. She published a vivid autobiography, *Blackberry Winter: My Earlier Years* (1972), and collected *Letters from the Field, 1925–65* (1977). In the mid-1970s she was elected to the National Academy of Sciences and to the presidency of the American Association for the Advancement of Science. After her death (in New York City) she was awarded the Presidential Medal of Freedom, her nation's highest civilian honor.

Mead's emphasis on culture rather than genes as a determinant of personal behavior gave powerful impetus to cultural relativism and the acceptance of diversity in American life.

Mary Catherine Bateson, *With a Daughter's Eye* (1984). Jane Howard, *Margaret Mead* (1984). Lowell Holmes, *Quest for the Real Samoa* (1986). Michaela di Leonardo, *Exotics at Home* (1998). Joan Mark, *Margaret Mead* (1999).

JOAN MARK

MECHANICAL PHILOSOPHY. Many seventeenth-century natural philosophers sought to explain all physical properties and processes in terms of the motion of the least parts of matter of which physical bodies are composed. They usually referred to these least parts as corpuscles so as not to confuse the mechanical position with the type of ancient atomism that Pierre Gassendi had tried to revive early in the century. Although the mechanical philosophers (or corpuscularians) agreed in rejecting Aristotelian philosophy (*see* ARISTOTELIANISM) and most of the mystical elements associated with Renaissance naturalism, they divided over the positive formulation of their position. "Mechanical philosophy" is a cover term for a continuum of positions from a pure kinetic theory of motion on one end to a robust matter theory on the other. These variations can be seen in the works of René *Descartes, Francis Bacon, *Galileo, Robert *Boyle, and Isaac *Newton.

Descartes maintained in his *Discourse on Method* (1637) that matter is pure extension, from which it followed that there could be no empty space; that all motion must result from direct contact; and that all change must be change of place (local motion). Descartes concluded that any property possessed by a material body had to arise from the motion imposed on the matter it contained by an external source. All physical processes, therefore, were to be explained by the laws of motion that the least parts of matter obey. Only the human soul escaped mechanical explanation. The world was a vast machine made up of smaller machines (including human and animal bodies) consisting of inert particles moved by physical necessity. Although Descartes located the origin of motion in God, his principles of inertia and the plenum allowed him to describe a deterministic system where, on impact, motion is transferred but not destroyed.

Cartesianism dominated corpuscularism throughout the seventeenth and into the eighteenth centuries, but it had several strong competitors. In his *Novum organum* (1620), Bacon advocated explanations in terms of the motion of matter. In his investigation of heat, for example, he concluded that bodies feel warm when the particles of matter that compose them move rapidly. Unlike the later Cartesians, however, Bacon insisted that experimental and observational techniques had to be developed to discover the true nature of the particles responsible for such qualities. Galileo's *Assayer* (1623) offered a similar account of motion as the cause of heat, and his *Discourse on the Two New Sciences* (1638) presented detailed experimental studies of mechanical subjects. Both Bacon and Galileo brought the practices of craftsmen and mechanics to bear on natural philosophical issues. In this tradition, the mechanical philosophy helped to elevate the intellectual, social, and economic status of the technical arts.

Many natural philosophers in England in the generation after Bacon took his works as their model. Boyle followed Bacon's experimental program and believed in the Baconian ideal of useful knowledge. Writing in the 1660s, Boyle was also influenced by the works of Descartes, Gassendi, and Galileo, and introduced the term "mechanical philosophy" in 1674 to refer to all explanations of physical phenomena in terms of matter and motion. Unlike philosophers before him, however, Boyle tried to use chemical analysis to turn the mechanical philosophy into an experimentally based theory of matter. He also elaborated upon the distinction, first introduced by Galileo, of the primary and secondary qualities of bodies. In *The Origin of Forms and Qualities* (1666), Boyle maintained that quantifiable properties, such as size and shape, are primary because all material bodies possess them. Other qualities, such as color or texture, arise in us in consequence of the particular configurations of corpuscles in the bodies that we see or touch.

Newtonian or classical mechanics is often taken as the paradigm of mechanical explanation. Yet in his *Principia mathematica* (1687), Newton upset the mechanical philosophy by introducing the concept of force. Unlike Descartes and Boyle, for whom force amounted to the pressure of one body on another, Newton's force was the measure of the change in motion of a moving body. Thus he added a third element to the original principles of matter and motion. At first mechanical philosophers, especially Cartesians, rejected Newtonian forces as a throwback to sympathies and antipathies. But his scheme gradually gained acceptance as mathematicians succeeded in deploying gravitational force to ever finer phenomena and other examples of distance forces turned up in *electricity and *magnetism.

Following Newton's achievements, mechanical conceptions came to be applied to all areas of learning, not always with advantage. Ernst Mayr, for example, in *The Growth of Biological Thought* (1982), argued that reliance upon mechanics advanced physical sciences but led to the neglect of the biological ones. This assessment can of course be extended to the human sciences as well. Sociologists and psychologists in the nineteenth and twentieth centuries often attempted to find deterministic laws covering the behavior of groups and of individuals.

At a more global level, some have argued that ecological disaster can be attributed, at least in part, to mechanistic ideas of nature. In *The Death of Nature* (1980), Carolyn Merchant put forward the still controversial thesis that with the mechanical philosophy scientific inquiry became a masculine activity imposed upon a passive, feminine nature. According to Merchant, this attitude set the stage for, and ultimately justified, the "rape" of nature.

E. J. Dijksterhuis, *The Mechanization of the World Picture* (1961; 1989). Marie Boas Hall, *Robert Boyle on Natural Philosophy* (1965). Richard S. Westfall, *The Construction of Modern Science* (1977). Peter Alexander, *Ideas, Qualities and Corpuscles: Locke and Boyle on the External World* (1985). Daniel Garber, *Descartes' Metaphysical Physics* (1992). Michael Hunter, ed., *Robert Boyle Reconsidered* (1994).

ROSE-MARY SARGENT

MECHANICS. If bodies move, forces move them; if they are in equilibrium, opposing forces hold them at rest; for mechanical devices to work, there must be forces of both sorts. Motion, forces, and machines have been the subject of two distinct sciences: the science of motion, and mechanics. In the *Aristotelian tradition, the science of motion belonged to physics, or natural philosophy, the science of natural bodies "insofar as they are natural." Aristotle opposed natural motion (a falling rock) to violent motion (a weight raised by a pulley). Violent motions and the machines that created them were the concern of the mechanical arts—acting *against* nature for practical ends—and of mechanics, or the science of weights. The division of mechanics into manual or practical, and rational or theoretical, goes back to Pappus of Alexandria in the early fourth century A.D. and was current thirteen hundred years later.

The *Renaissance inherited many important achievements in rational mechanics, including the parallelogram rule for the composition and resolution of motions, early forms of the principles of virtual work and of virtual velocities, demonstrations of the law of the straight and angular lever, an embryonic notion of moment (torque), and determination of centers of gravity. Equally important was the understanding that geometrical demonstration was essential to mechanical theory. This legacy underpinned the evolution of mechanics in the early modern period. Simon Stevin and *Galileo simplified Archimedes' proof of the law of the lever; Christiaan *Huygens devised a more rigorous proof (1693). Galileo used the angular lever and the notion of moment to determine equilibrium conditions on the inclined plane (*Le meccaniche*, c. 1593).

The parallelogram rule for motions uses geometrical displacements straightforwardly. Not so the corresponding rule for forces, because it is not obvious what their "composition" and "resolution" mean. Following the work of Stevin and others, Pierre Varignon based his theory of equilibrium on the rule for forces (*Projet de la nouvelle mécanique*, 1687). Isaac *Newton recognized its indispensability, and cannily prepared for his own proof of the rule in the *Philosophiae naturalis principia mathematica* (1687) by stipulating in his second Law of Motion that every change in motion caused by an impressed force takes place "along the straight line in which the force is impressed." This ensured *by definition* the geometrical equivalence of the rules for forces and motions. As for the principle of work, Newton claimed that it depends on the equality of action and reaction (Law III). In René *Descartes's formulation of the principle of mechanical work, the same force that raises one hundred pounds through two feet will raise two hundred pounds through one foot (and so on). This he used effectively in short treatises on machines he sent in 1637 to Marin Mersenne and Constantijn Huygens.

These examples belong to "rational mechanics" as understood by Pappus. They remind us that the major figures, better known for their contributions to the science of motion, also contributed to mechanics, though their thinking in these areas revealed the conceptual fluidity characteristic of pivotal transformations in the development of science. Two contrasting signs of this fluidity were the creative coupling of principles from both sciences, and indecisiveness about the relations between mechanics and physics. Galileo corroborated his law of free fall through experiments on the inclined plane, and the equilibrium conditions on the inclined plane played key roles in the formal demonstration of the law in the Third Day (or part) of his *Discourses concerning Two New Sciences* (1638). In the fourth part, Galileo used the law of fall and the composition rule to demonstrate the parabolic path of projectiles, a result he obtained and confirmed experimentally around 1608. Apart from certain medieval innovations in the geometrization of natural motion, natural philosophy had not been mathematical, whereas mechanics had never been anything else. Galileo decisively blurred that dichotomy by showing that natural motions could be given mathematical descriptions in accord with experiment and mechanical principles. Still, he would have agreed that *Le meccaniche* belonged in a different disciplinary pigeonhole from the third and fourth part of his *Discourses*.

In *Principia philosophiae* (1644), Descartes set out his pioneering three "Laws of Nature," according to which a bodily state persists until forced to change by external causes, a moving body endeavors to move always in a straight line, and exchanges of motion (size × speed) between colliding bodies are determined by the contests between their forces of persistence and by Descartes's conservation law. Descartes claimed that his laws and the collision theory derived from them could explain the whole physical universe, including machines. Yet there is nothing on mechanics in his *Principia*, nothing that explains the work principle of 1637. Christiaan *Huygens's masterly solution to the problem of centre of oscillation (*Horologium oscillatorium*, 1673) required an insight from Galileo's *Discourses* (third part) enunciated as a principle by Evangelista Torricelli (1644): a system of heavy bodies cannot move of its own accord unless the common center of gravity descends. Torricelli's principle also played a crucial role in Christiaan Huygens's collision theory, out of which tumbled the result that in perfectly elastic collisions the quantity mv^2 remains constant. Huygens regarded this result as a notable corollary of his collision rules; to Gottfried Wilhelm von Leibniz it suggested the universal conservation of *vis viva*, a force measured by mv^2, to add to the already known conservation of "directed motion." This confirmation of force as a metaphysical reality led Leibniz to the creation in 1691 of a new science of force, which he baptized *dynamics*. It led in turn to a protracted argument in the early decades of the eighteenth century about whether motive force should be mv, the Cartesian and Newtonian measure, or mv^2, the Leibnizian measure.

Newton transformed Descartes's "Laws of Nature" into three "Axioms" or "Laws of Motion" according to which a body persists in its state of rest or straight-line motion (Law I), the force impressed on a body is proportional to its change of motion (Law II), and action and reaction are equal and opposite (Law III). From these laws Newton demonstrated the mutual dependence of his inverse-square law of universal gravitation and Johannes *Kepler's first and third laws of planetary motion—the second law being a consequence of inertial motion under any central force. The basic problem—to determine the central force given the planet's deviations from inertial motion summed as an orbit, and conversely—was quite foreign to traditional mechanics. Although in the *Principia* Newton links Law III to the work principle and proves the parallelogram rule for forces (which is "abundantly confirmed from mechanics"), the *Principia* is not a treatise on mechanics, but on "the mathematical principles of natural philosophy."

However, Newton's *Principia* was a major exercise in rational mechanics in a new sense that emerged in the work of Isaac Barrow and John Wallis. Barrow had argued (1664–1666) that geometrical theorems apply to all of physics, so that the principles of mechanics and of physics become identical. For Wallis, mechanics was "the part of geometry that deals with motion, and investigates, apodictically and using geometrical reasoning, the force with which such and such a motion takes place" (*Mechanica*, 1670–1671). Similarly, for Newton rational mechanics was "the science, set out in exact propositions and demonstrations, of the motions that result from any forces whatever and of the forces that are required for any motions whatever" (*Principia*, "Preface"), and natural philosophy was basically the problem of "finding the forces of nature from the phenomena of motions and then to demonstrate other phenomena from these forces." The same ideas were to inform Leonhard *Euler's *Mechanica* (1736), significantly subtitled *The Science of Motion Expounded Analytically*.

"Rational" or "theoretical" mechanics in the older sense should be distinguished from the post-Newtonian sense of "rational mechanics," which comprised dynamics and statics. Mechanics in its golden age (the eighteenth century) was not merely a set of variations on the principles and methods of Newton's *Principia*. Among additional ingredients were the concepts and symbolism of Leibnizian differential and integral calculus, which became standard in treatises on analytical mechanics of the period, and the new mathematics, particularly the calculus of variations, which made it possible to formulate new principles that solved new problems.

To take some notable examples, a general theory of rigid-body motion became a desideratum following the work on centers of oscillation of Huygens, Jakob Bernoulli (1703), and Jean d'Alembert (*Traité de dynamique*, 1743), and the researches on lunar libration and equinoctial precession by Newton (*Principia*), d'Alembert, and Joseph Louis Lagrange. Here the principal figure was Leonhard *Euler (memoirs of 1750, 1758), whose *Theoria motus corporum solidorum seu rigidorum* (1760) provided a general theory of rigid-body motion. Euler's researches depended on a "new principle of mechanics" (1750), his recasting of Newton's Law II in the analytic form $mdv_{x,y,z} \propto f_{x,y,z}\, dt$. Euler introduced moment of inertia (1749) and principal axes of rotation (suggested by the rolling of ships about three orthogonal axes), which he applied in his theory of the spinning top, an exceptionally difficult problem that had not even been recognized as a problem since the early seventeenth century.

Suppose in a system of bodies in mutual constraint the motion applied to each body a_i resolves into the motion actually acquired v_i and another motion V_i. That is the same as if the v_i and V_i had initially been communicated together, so that the system would have been in equilibrium had the V_i alone been present. That is "d'Alembert's Principle," the centrepiece of his *Traité de dynamique*, which allowed the methodological reduction of dynamics to statics. Lagrange reformulated the principle and coupled it with the principle of virtual velocities to obtain the first formulation of what became "Lagrange's Equations" (*Mécanique analytique*, 1788). Lagrange showed that the conservation of linear and angular momentum, and of *vis viva*, and the principle of least action, follow from his equations, rather than being foundational principles in their own right.

Pierre de Maupertuis's principle of least action had sounded a new note. Reflecting on the controversy of the 1660s over Fermat's least-time optical principle, Maupertuis argued (1744) that in all bodily changes, the "action" (Σ mass × speed × distance) is the least possible, a principle that for Maupertuis and Euler—though not for d'Alembert and Lagrange—pointed to the governance of all things by a Supreme Being. (Σ signifies a sum over all particles in the system under consideration.) Given the principle of virtual velocities, Σ force × ds = 0, which means, by the principles of the integral calculus, that the integral of this sum is a maximum or minimum. Euler developed the least action principle clearly and rigorously for a single particle (1744), and, in a memoir on lunar libration (1763), Lagrange extended Euler's result to an arbitrary system of bodies and derived a general procedure for solving dynamical problems.

The nineteenth century saw new departures in the application of principles established in the preceding two centuries. The relativity of motion, a central theme since the work of Galileo and Huygens, became the object of further study in the work of Alexis-Claude Clairaut, who asked how a system of moving bodies would behave if the system moved along noninertial curves (1742). Gaspard Gustave de Coriolis showed (1835) that the Newtonian laws of motion apply in a rotating reference frame if the equations of motion include a "Coriolis acceleration" in a plane perpendicular to the axis of rotation, a kinematic acceleration which Coriolis interpreted as an extra force (the "Coriolis force") and which became important in ballistics and meteorology. An important step in the formalization of mechanics was Heinrich Hertz's attempt to remove inconsistencies arising from the assumption within traditional classical mechanics that forces are ontologically prior to the motions they cause. He treated forces as "sleeping partners" in a formalized mechanics that depended on the operationally understood notions of time, space, and mass, and when necessary on linkages to hidden masses with hidden motions with respect to hidden coordinates (*Die Prinzipien der Mechanik*, 1894).

Maupertuis's variational principle enjoyed an improved mathematical treatment by William Rowan Hamilton (1834, 1835), whose transformation of Lagrange's equations was modified and generalized by Carl Gustav Jacobi in the form now known as the Hamilton-Jacobi Equation (1837). In turn, the Hamilton-Jacobi Equation found fruitful application in the establishment of the *quantum mechanics of Louis de Broglie (1923) and Erwin Schrödinger (1926).

See also RELATIVITY.

René Dugas, *Mechanics in the Seventeenth Century* (1958). Clifford Truesdell, *Essays in the History of Mechanics* (1968). Stillman Drake and I. E. Drabkin, eds., *Mechanics in Sixteenth-century Italy. Selections from Tartaglia, Benedetti, Guido Ubaldo, and Galileo* (1969). Richard S. Westfall, *Force in Newton's Physics: The Science of Dynamics in the Seventeenth Century* (1971). Ernst Mach, *The Science of Mechanics: A Critical and Historical Account of its Development* (1974). Pierre Duhem, *The Evolution of Mechanics* (1980). René Dugas, *A History of Mechanics* (1988). H. J. M. Bos, "Mathematics and Rational Mechanics," in *The Ferment of Knowledge: Studies in the Historiography of Eighteenth-century Science*, ed. G. S. Rousseau and Roy Porter (1980): 327–355. Ivor Grattan-Guinness, "The Varieties of Mechanics by 1800," *Historia Mathematica* 17 (1990): 313–338. Alan Gabbey, "Newton's Mathematical Principles of Natural Philosophy: a Treatise on 'Mechanics'?" in *The Investigation of Difficult Things: Essays on Newton and the History of the Exact Sciences*, ed. P. M. Harman and Alan Shapiro (1992): 305–322.

ALAN GABBEY

MEDICINE. The Hippocratics, the group of doctors who derived inspiration from Hippocrates of Cos, established the Western medical tradition. Their writings, produced over a period of more than two centuries, covered the whole of medicine, surgery, hygiene, and therapeutics. They based their ideas on careful bedside observation and accumulated their knowledge by transmission from practitioner to practitioner. Although they were observers, not experimentalists, the Hippocratics posited a naturalistic cause of disease, based on imbalance of the four humors (blood, yellow bile, black bile, and phlegm) whose perfect mix constituted health.

Since the Hippocratics, Western medicine has presented itself as "scientific," although what that constitutes has changed frequently. In any case, medicine as a social institution and medical practitioners both individually and collectively have been more concerned with the exigencies of disease and suffering (and the maintenance of professional boundaries and income) than with the disinterested furtherance of scientific knowledge. Even productive physicians such as Thomas Sydenham saw medicine as essentially an observational science, best learned by bedside practice and furthered by experience.

Sydenham conceptualized disease within a framework of Hippocratic humoralism, but his lasting theoretical legacy was his suggestion that diseases could be classified (nosology) in exactly the same way as plants, animals, and other species of being. Sydenham's nosological enterprise inspired many eighteenth-century doctors, whose own symptom-based nosologies became increasingly elaborate. At the same time, humoralism began to lose its hold as doctors turned to physiological theory and autopsy findings for confirmation of their systems. Early-eighteenth-century doctors stressed the cardiovascular system; from midcentury, the nervous system commanded center stage as the regulator of all life processes and the ultimate source of disease.

The medicine that emerged after 1794 in revolutionary Paris represented a break with the past, the sharpness of which is still a subject of historical debate. The medical schools reopened after their brief closure in the early days of the Revolution with a curriculum centered on the *hospital, and gradually incorporated a number of central ideas. These included an emphasis on the organs and (through Marie-François-Xavier Bichat's work) the tissues as the vital seats of disease; new methods of diagnosing disease, including the more systematic use of percussion and use of the stethoscope, invented by T. R. H. Laennec, which made mediate auscultation a routine medical exercise; and the systematic correlation of the signs and symptoms observed during life with the findings observable on autopsy (clinicopathological correlation). To these pillars of French hospital medicine was added the *méthode numérique* of Pierre Louis, the beginnings of clinical evaluation of therapy, but also an attitude of mind that sought truth through study of large numbers of cases. The hospital offered above all an expanded field of vision.

Large numbers were also central to a second important development in nineteenth-century medicine. This concerned public health and the prevention of disease, especially acute epidemic ones, of which cholera made the most visible impact. Asiatic cholera was a new disease in Europe and North America, and its six pandemics during the century generated much discussion about its cause, contagiousness, and prevention.

The progress of epidemic diseases (and all sorts of other social parameters) could be monitored more closely through the development of both the concepts and apparatus of social statistics: a crucial hallmark of the modern state. Measuring such things as population size, birth and death rates, illegitimacy and abortion rates, causes of death on a national basis, and many other parameters had its difficulties—interpreting the results inevitably raised contentions.

In most European and North American countries, rudimentary structures for both social statistics and public health developed in the first half of the nineteenth century. During the second half of the century, a third estate within medicine, the laboratory scientist, gradually acquired power. The philosophy of experimental medicine was eloquently articulated by one of the most fertile physiologists of the period, Claude *Bernard. Rudolf Virchow systematically applied the *cell theory to the understanding of disease processes. Equally influential on both clinical medicine and public health was the work of microbiologists and bacteriologists such as Louis *Pasteur and Robert Koch. Modern historians have demonstrated that many *germ theories competed for acceptance, and that the process of establishing a bacteriological orthodoxy was much more complicated than an earlier heroic historiography had assumed. Nevertheless, *bacteriology was undoubtedly the most influential of the several experimental life sciences—*physiology, animal (or bio-) chemistry, *nutrition, experimental pathology, and immunology—established before the outbreak of World War I. "The future belongs to science," wrote Sir William Osler, in many ways a traditional clinician, but also one imbued with the excitement of the scientific prospects for clinical medicine.

In many ways, Osler's prediction has been fulfilled: the medicine of the past century became increasingly science based, although modern science and technology are so inextricably linked that much of what passes for science in modern medicine might better be called technology. The application of *X rays to medical diagnosis shortly after Wilhelm Conrad *Röntgen announced their discovery in 1895 marked a significant moment within medicine. During the first half of the century, electrocardiograms, electroencephalograms, incubators for newborns, and a number of other diagnostic and therapeutic machines were introduced. Since World War II, technology has penetrated even further into medical practice: modern surgery depends on it, imaging equipment has gone far beyond what X rays could achieve, and mechanized blood tests make it possible for physicians to keep track of a wide range of diagnostic markers.

For roughly a half century from the 1920s, medicine enjoyed a golden period in Western society. Declining death rates and a widespread belief that science and technology could solve whatever problems lay ahead gave medicine a special status. The problems of the past few decades seem more intractable. The treatment of acute disease has given way to the less satisfactory management of chronic illnesses, especially those of the elderly. New diseases such as *AIDS, the emergence of microorganisms resistant to antibiotics and other chemotherapeutic agents, and the ethical issues surrounding molecular *genetics, rationing of scarce resources, definitions of death, and other trappings of modern medicine have created a new set of issues. Medicine, politics, ethics, and economics have become inextricably entwined, and there are no easy solutions.

W. F. Bynum, *Science and the Practice of Medicine in the Nineteenth Century* (1994). David Weatherall, *Science and the Quiet Art. Medical Research and Patient Care* (1995).

W. F. BYNUM

MEITNER, Lise (1878–1968), atomic physicist, co-discoverer of nuclear fission.

Lise Meitner was born in Vienna on 7 November 1878 and died in Cambridge, England, on 27 October 1968. The third of eight children of Philipp (a lawyer) and Hedwig (née Skovran) Meitner, Lise grew up in a family that was intellectual and socially progressive. Both parents were of Jewish origin, but the religion played no part in the children's upbringing and all were baptized as adults.

Meitner entered the University of Vienna in 1901, four years after the university first admitted women; earned a doctorate in physics in 1906; and worked in Berlin from 1907 until her forced emigration from Germany in 1938. Primarily an experimentalist, Meitner made important contributions to *radioactivity, *nuclear physics, and the discovery of nuclear fission. Her most formative teachers, however, were the theoretical physicists Ludwig *Boltzmann and Max *Planck, and her lifelong interest in theory informed and guided her work.

In Berlin, Meitner began an interdisciplinary collaboration with Otto Hahn, a chemist her age, that helped establish the field of radiochemistry. Together they identified several new radioactive species, discovered and developed the physical separation method known as radioactive recoil, and were the first to use photographic methods for studying magnetic beta spectra. Their radiochemical expertise permitted them to search for a rare "missing" element, and in 1918 they reported the discovery of a long-lived isotope of element 91, protactinium (Pa).

Meitner's career was a series of firsts for the inclusion of women into German science. She held no position whatsoever until Planck appointed her his assistant in 1912. In 1913 she was given a position and salary comparable to Hahn's in the *Kaiser-Wilhelm Institute for Chemistry (KWI). After volunteering as an x-ray nurse in the Austrian army during World War I, she returned to the KWI to head her own physics section in 1917, acquired the title of professor in 1920, and served as an adjunct professor at the University of Berlin from 1926 until she was dismissed by the Nazi regime in 1933. After fleeing Germany in July 1938, she held positions at the Nobel Institute for Experimental Physics and the Royal Institute of Technology in Stockholm. She was recognized by many international scientific societies and repeatedly nominated for a Nobel Prize.

In the 1920s, Meitner achieved exceptional prominence for her pioneering studies of the nucleus. Using magnetic beta-gamma spectra, she was the first to describe radiationless orbital electron transitions in 1923, an effect now named for Pierre Auger, and in 1924 she proved that gamma radiation follows particle emission in radioactive decay. Convinced that quantization must extend to the nucleus, she was dismayed when Charles D. Ellis proved the existence of the continuous primary beta spectrum in 1927. After Meitner confirmed the result in 1929, Wolfgang Pauli proposed a new nuclear particle, the neutrino, which was quickly incorporated into nuclear theory. Always close to theory, Meitner measured the Compton scattering of high-energy gamma radiation, verifying the formula of Oskar Klein and Yoshio Nishina that was based on the relativistic electron theory of P. A. M. *Dirac. In 1932, Meitner and her coworkers were the first to observe positrons from a noncosmic source and to observe the formation of electron-positron pairs in a cloud chamber.

In 1934, Meitner recruited Hahn and the chemist Fritz Strassmann for the "uranium project" that culminated in the discovery of nuclear fission. Believing that they were synthesizing artificial elements beyond uranium, the team spent four years disentangling a complex mixture of radioactive species, nearly all of which were later found to be smaller nuclei produced by fission. Only in December 1938, when barium was identified among the uranium products, was it recognized that the uranium nucleus had split. Although Meitner was in Stockholm during the final experiments, she collaborated closely with Hahn through their correspondence. She and her nephew, Otto Robert Frisch, provided the first theoretical interpretation of fission and calculated the energy released in the process. Their interpretation was regarded as seminal, but Meitner's contribution to the discovery itself was obscured and the 1944 Nobel Prize in chemistry was awarded only to Hahn. With recent historical correctives, the discovery is now more generally understood as the result of an interdisciplinary collaboration in which Meitner and nuclear physics played an essential role.

Meitner was invited to join the *Manhattan Project in Los Alamos in 1943, but she was unwilling to work on a nuclear weapon and remained in Sweden. She retired in 1954 and in 1960 moved to England, to be near Frisch and his family in Cambridge.

Lise Meitner, "Looking Back," *Bulletin of the Atomic Scientists* 20, no. 11 (1964): 2–7. O. R. Frisch, "Lise Meitner 1878–1968," *Biographical Memoirs of the Fellows of the Royal Society, London* 16 (1970): 405–420. O. R. Frisch, *What Little I Remember* (1979). Ruth Lewin Sime, *Lise Meitner: A Life in Physics* (1996). Elisabeth Crawford, Ruth Lewin Sime, and Mark Walker, "A Postwar Tale of Nobel Injustice," *Physics Today* 50, no. 9 (1997): 26–32. Ruth Lewin Sime, "Lise Meitner and the Discovery of Nuclear Fission," *Scientific American* 298, no. 1 (1998): 80–85.

RUTH LEWIN SIME

MENDEL, Gregor Johann (1822–1884), prelate, hybridist, and naturalist.

The son of a peasant farmer, Mendel was raised in German-speaking Silesia in the village of Heinzendorf. An excellent student, he gained admission to the University of Olmütz in 1841 where he studied philosophy, mathematics, and physics. But the stress of examinations and the inadequacy of his financial resources caused him in 1843 to seek the security of the Augustinian monastery of Saint Thomas in Brünn, Moravia. The Augustinians taught in the local schools and at the city's Philosophical Institute; Mendel, finding the duty of attending the sick and the dying deeply disturbing, preferred to teach Latin, Greek and mathematics at the Znaim gymnasium. His high school teaching career stretched from 1849 to 1868 when he was elected abbot. Although he proved to be an excellent educator, Mendel failed to qualify as a permanent teacher at the examination in 1850 and again in 1855 despite two years' remedial work at the University of Vienna between 1851 and 1853. Exam stress was his chief problem. Meanwhile he taught physics and natural history with much success at the new Brünn modern school.

While Mendel taught in Moravia, efforts were being made to develop the economy of the country. That involved meeting the threat to the Moravian sheep industry posed by Australia. The authorities considered plant and animal breeding to be the most promising instruments for improving and diversifying Moravian agriculture. The dynamic abbot Cyril Napp, concerned about the dependence of his monastery on the productivity of its many farms, encouraged Mendel, who wanted to explore the nature and results of hybridization both for practical and for theoretical reasons. An experimental approach, such as Mendel had learned in his physics courses in Vienna, might yield results that would guide the practical breeder and clarify the nature and origin of species for the naturalist.

The origin of species was discussed in central Europe long before *Darwin's theory reached there in connection with the question whether hybridization yielded offspring able to reproduce their like with the constancy of species. If so, as Carl *Linnaeus had suggested, might not the majority of species have arisen by hybridization between a much smaller number of created forms? The possibility had its attractions to a scientifically minded member of the Roman Catholic church.

Mendel planned a systematic investigation to answer this question. Treating the organism as composed of a number of independent hereditary traits, he designed a series of experiments to substantiate their independence and reveal the patterns governing their combinations. After testing thirty-four varieties of the edible pea for constancy of type (1854–1856), he embarked on his now famous hybridizations (1856–1863). To achieve statistically significant results he grew large numbers of plants. To avoid problems in the sorting of the progeny he studied only clearly distinguishable traits. One expression of each of the traits *dominated* in the hybrid so that another expression (the *recessive* trait) did not reappear until the following generation. The progeny of these hybrids showed the dominant characters three times as often as the recessive. The next generation revealed that the dominants consisted of one pure form for every two hybrids. Hence the ratio of pure dominants:hybrids:pure recessives was 1:2:1, like the coefficient in the expansion of the binomial equation $(A + a)^2 = A^2 + 2Aa + a^2$.

Mendel knew that in fertilization one pollen cell fertilizes one egg cell. Hence if these cells in the hybrid were of two kinds represented by the letters A and a, self-fertilizations would yield all combinations of germs cells for a and for A, for example: $AA + Aa + aA + aa$ [$A + 2Aa + a$].

Therefore he postulated that the differing potentials—A and a—brought together in the hybrid, separated in the formation of the germ cells. This is Mendel's principle of the *purity of the germ cells*, and it accounted for the reversion of hybrid offspring to their originating species. He accounted for the variability that followed from hybridization by the possibilities for recombination between differing germ cells. He thought that some hybrids did not separate, but remained constant in their progeny. But his subjection of some so-called constant hybrids to the test of experiment showed that they followed the same sort of segregation as the edible pea.

Mendel continued these researches until his duties as abbot became too heavy. He announced his results in three lectures to the Brünn Society of Naturalists, two papers on hybridization in *Pisum* and one on *Hieracium* in their *Proceedings*, and described his work in correspondence with the famous botanist Carl Wilhelm von Naegeli. Not until 1900, however, were his papers discovered and his work repeated. Then the science that would become *genetics was born.

Hugo Iltis, *Life of Mendel* (1932). Curt Stern and Eva Sherwood, eds., *The Origin of Genetics. A Mendel Sourcebook* (1966). Robert Olby. *Origins of Mendelism*, 2d ed. (1985). Franz Weiling, "Historical Study: Johann Gregor Mendel, 1822–1884," *American Journal of Medical Genetics*, 40 (1991): 1–25. Vítězslav Orel, *Gregor Mendel. The First Geneticist* (1996). Franz Weiling, "Gregor Mendel," *American Journal of Medical Genetics*. Robin Marantz Henig, *The Monk in the Garden* (2000).

ROBERT OLBY

MENDELEEV, Dmitrii Ivanovich (1834–1907), chemist, formulator of the periodic system of chemical elements.

Growing up in Tobol'sk, Siberia, Mendeleev had a typical childhood in the backwaters of the Russian empire. His mother determined that a child with his good grades should attend university. Having failed to matriculate at Moscow University, Russia's oldest educational institution, he was sent on to St. Petersburg, where he enrolled in the Chief Pedagogical Institute, the same place at which his father had trained. The Institute shared a faculty with the more prestigious St. Petersburg University, and there Mendeleev began to study the natural sciences, especially *chemistry, under Aleksandr A. Voskresenskii. In 1858, he was granted permission to study abroad for two years. After some deliberation, Mendeleev settled in Heidelberg. Rather than working with Robert Bunsen, as he had intended, Mendeleev established a laboratory in his apartment and began to study the capillarity of organic solutions. He served as the social center for the several Russian chemists in Heidelberg. In September 1860, he traveled with three Russian companions to the chemical congress at Karlsruhe, which proved seminal in standardizing various chemical conventions, including the crucial revival of Avogadro's hypothesis for the determination of consistent atomic weight values. In February 1861, a few days before Czar Alexander II abolished serfdom, Mendeleev returned to St. Petersburg to try to establish a career in the capital during the turbulent years of Imperial reform.

At first, these efforts proved abortive. While seeking a permanent position, Mendeleev wrote a textbook, *Organic Chemistry* (1861), which won him the Demidov Prize of the Petersburg Academy of Sciences. In 1864, Mendeleev obtained a position at the St. Petersburg Technological Institute. During the 1860s, he was one of the principal architects of the Russian Chemical Society, one of the first truly organized scientific societies in Russia. In 1867, he succeeded Voskresenskii at St. Petersburg University, and he began to teach inorganic chemistry. While preparing a textbook for this course, *The Principles of Chemistry* (first edition, 1869–1871), Mendeleev made his most important chemical discovery.

Today, Mendeleev's name is most closely identified with his formulation of the periodic system of chemical elements, an ordering of the sixty-three then-known elements by order of increasing atomic weight. Although seven other chemists formulated similar periodic classifications before Mendeleev in the 1860s, his generally recognized priority stems from his novel prediction of three elements to fill empty spaces in the *periodic table. These elements, eventually named gallium, germanium, and scandium, were discovered by other European chemists from 1876 to the mid-1880s, and Mendeleev's detailed forecasting of their chemical properties, from atomic weight to specific gravity, earned him an international reputation. The periodic system, originally conceived as a convenient pedagogical classification, is today almost universally employed as a teaching tool throughout the world.

Mendeleev did not pursue work on the periodic system after finalizing his predictions in 1871. Although he continued to teach inorganic chemistry, he began large-scale experimental research into the laws of gas expansion in an attempt to locate the substance of the luminiferous *ether. This effort, financed by the Russian Technical Society, collapsed in 1881, a few months after Mendeleev failed by one vote to win election to the Imperial Academy of Sciences. This event, which sparked a lively public debate about the role of ethnic Russians in Imperial institutions, along with Mendeleev's public attacks on spiritualists in St. Petersburg in 1876, helped forge a new public role for the natural scientist in Russian culture.

After his rebuff at the Academy of Sciences, Mendeleev retreated to his pedagogical duties while continuing to consult for the Ministry of Finances on oil exploitation in Baku and conducting research on chemical solutions. In 1890, during a dispute about a petition for students' rights at the University, Mendeleev resigned from his post at St. Petersburg University, and soon moved to the Navy, where he developed a form of smokeless gunpowder. In 1893 Mendeleev was appointed first director of the Chief Bureau of Weights and Measures and in 1899 he initiated and oversaw the partial introduction of the metric system into the Russian empire. In the final years of his life, he maintained his role as a public intellectual while reviving his ether speculations of the 1870s in a theoretical chemical work. He died as the most highly decorated chemist in Russian history.

O. Pisarzhevskii, *Dmitry Ivanovich Mendeleyev: His Life and Work* (1954). J. W. van Spronsen, *The Periodic System of Chemical Elements: A History of the First Hundred Years* (1969). Bernadette Bensaude-Vincent, "Mendeleev's Periodic System of Chemical Elements," *British Journal for the History of Science* 19 (1986): 3–17. Bernadette Bensaude-Vincent, "Mendeleev: Beyond the Periodic Table," *Ambix* 45 (1998), special issue.
MICHAEL D. GORDIN

MENTAL SCIENCES. The various dimensions of consciousness have long been the subjects of speculation and observation. The modern word "psychology" dates from the seventeenth century, but Greek doctors and philosophers established a framework that guided Western thinking on the subject. Plato and the Hippocratics believed that the soul and hence mental activity resided in the brain, whereas Aristotle located these functions in the heart. Galen, the most influential doctor of antiquity, sided with Hippocrates and Plato, and medieval physicians placed the important mental functions of reason, memory, and imagination along with sensation in the ventricles of the brain.

René *Descartes created a framework for a scientific physiology but at the expense of human psychology. His strict dualism, in which animals were merely complicated machines, potentially understandable in material terms, and human beings similar machines, with soul added, was a neat but awkward solution to long-standing problems. If animals were machines, then sensation, movement, and the apparent expression of emotions, for example in response to painful stimuli, should be explicable mechanically. Consciousness, free will, language, reason, and other "higher" aspects of mental life were functions of the soul (or mind, the French word being the same for both) and hence approached through the traditional philosophical exercise of introspection.

Descartes's doctrine of the *animal machine was more innovative than his dualism, which did not differ much from Plato's. It stimulated a mechanistic approach to physiology, which the physician Julien de La Mettrie took to a logical conclusion in the mid-eighteenth century. La Mettrie reasoned that human beings and animals resembled one another structurally and functionally and thus described what he called the "man-machine." La Mettrie's model of physiological and mental functioning made human thought and consciousness the direct consequence of the nuanced structures of the human body and brain.

In the late seventeenth century, John Locke, working within the philosophical tradition, had argued that at birth the human mind is a blank tablet (*tabula rasa*) and that individual personalities and the experiences and sensations imprinted on this tablet mold values. The stark cultural relativism of Locke's enterprise escaped many later thinkers but he also briefly elaborated a mental mechanism that was echoed by many subsequent accounts of the way the mind works: the association of ideas. This mechanism formed the basis of David Hartley's account of human psychology of 1749. Hartley became a rigorous materialist (though also a religious one) under the influence of the theory of the human mind proposed by the Unitarian chemist and clergyman Joseph Priestley.

Locke influenced many eighteenth-century philosophical expositions of mental functioning, including the so-called common sense philosophies of the Scottish Enlightenment, such as Thomas Reid's. The common-sense philosophers reinserted innate, God-given capacities into the mind, and developed a faculty psychology that resonated to some degree with the older faculty psychology of medieval churchmen. This faculty psychology was incorporated in turn into the physiognomic (or phrenological) system of the Austrian neuroanatomist Franz Joseph Gall. Gall and his disciple Johann Christoph Spurzheim developed a new science of mind (phrenology) that seemed to many early-nineteenth-century doctors and reformers to offer an objective, measurable method to relate mind, brain, and character (*see* PHRENOLOGY AND PHYSIOGNOMY). Gall the neuroanatomist elevated the white matter in the cerebral cortex to a place of prominence; Gall the physiognomist noticed as a schoolboy that his comrades with prominent eyes had good memories. His practical system posited a number of innate faculties located in the white matter of the cerebrum and cerebellum that directly influenced the shape of the overlying skull.

By the second half of the nineteenth century, phrenology had degenerated into a popular science of reading character from the bumps on the head, but for almost half a century it had offered the promise of a science of mind based on biological, objective principles. Many alienists believed that it explained madness, criminality, and social deviance; educational reformers believed that it offered a measure of innate ability and therefore a means of building on talent or counteracting undesirable traits.

Phrenology's lasting scientific legacy was its systematic exposition of the notion of cerebral localization (that different parts of the brain served different mental and neurological functions), its insistence on a degree of continuity between the functions of human and animal minds, and its privileging the white matter of the brain. Cerebral localization began to acquire more specific clinical and experimental evidence from the mid-nineteenth century. The French surgeon, anatomist, and anthropologist Pierre Paul Broca associated a lesion in the left inferior frontal gyrus of the cerebral context (since known as Broca's gyrus) with the loss of speech (aphasia). From the 1870s, the British neurologist David Ferrier and the German physiologists Eduard Hitzig and Gustav Theodor Fritsch independently produced experimental evidence in animals of the existence of motor and sensory strips in the cortex, ablation of which led to motor or sensory deprivation, while electrical stimulation produced movement or evidence of sensation.

Two decades earlier, the evolutionary philosopher Herbert Spencer reinforced the notion of psychological continuity between humans and animals in his *Principles of Psychology* (1855). He offered a model of mental functioning that drew on earlier experimental work on spinal reflexes and made use of the association of ideas as a fundamental principle of mental life. Charles *Darwin's evolutionary work helped to establish the biological basis for a comparative psychology. Darwin had a strong interest in the psychological development of infants (he observed his own children carefully), the adaptive functions and biological bases of *instincts, and the physical expression of emotion, the subject of one of his books (1872). He left many of his notes on psychology to his disciple George John Romanes, who published three books on aspects of comparative psychology beginning with *Animal Intelligence* (1882). In the next decade Charles Lloyd Morgan brought an additional rigor to the study of the field.

Paralleling these developments (and occasionally interacting with them), nineteenth-century psychiatrists dealt with "mental disease" (as opposed to madness, insanity, and so on). From the 1830s, major synthetic monographs in France, Germany, Great Britain, and the United States dealt with the causes, symptoms, and classification of mental disorders. Psychiatry as a discipline became rather marginalized from mainstream medicine, as massive psychiatric hospitals ("asylums") housed mostly chronic patients. Nerve specialists, increasingly called neurologists, dealt with what came to be called the neuroses. The German psychiatrist Emil Kraepelin (an almost exact contemporary of Sigmund Freud's) divided mental disorders into two grand groups, the psychoses and the neuroses. At the same time, Kraepelin attempted to integrate psychiatry with experimental psychology, especially as elaborated by his onetime colleague Wilhelm Wundt. In the United States, William James's *Principles of Psychology* (1890) offered an influential synthesis of psychology. In addition, James developed a notion of the "psychosomatic."

Psychiatrists liked to correlate the major mental disturbances with physical disease. General paralysis of the insane (GPI), a common condition in nineteenth-century asylums, became a paradigm. Its intriguing mix of neurological and psychiatric symptoms was shown, early in the twentieth century, to be the result of end-stage syphilis. At the same time, Freud's psychoanalysis (and the psychoanalytically inspired systems of others) sought the origin of adult neuroses in infant and early childhood experiences, real or imagined. Although Freud himself remained a closet psychological materialist and determinist, in practice his ideas encouraged an emphasis on psychological analyses and psychotherapy. Jean Piaget (1896–1980) greatly extended the developmental approaches of psychoanalysis as he investigated the origins of the child's conception of time, space, causality, and objectivity.

Behaviorism, a psychological movement founded by the American John Watson and extended by others including B. F. Skinner, denied any central role to emotions or feelings, attempting instead to understand human (and animal) behavior as a result of reflex actions, habits, and learned responses to certain stimuli. The Russian physiologist Ivan *Pavlov's famous work with dogs on what he called "conditioned reflexes" provided some scientific underpinning for behaviorism. Experimental psychology has spawned a multiplicity of other techniques and schools, such as the educational psychology of Alfred Binet and others concerned with measuring "intelligence" and *"Intelligence Quotient," and

the integrative, "Gestalt" approach of Wolfgang Köhler, noted especially for his study of the behavior of primates.

More recently, neuroscience has emerged as a major scientific specialty; it seeks to integrate many older disciplines devoted to understanding the structures and functions of the nervous system. In the 1880s scientists recognized the nerve cell (neurone) as an important functional unit of neurological activity. In 1897 Charles Sherrington called the connections between nerve cells (and between nerve endings and muscles) synapses; Henry Hallett Dale, Otto Loewi, and others showed that chemical neurotransmitters mediated the interaction of neurons. Histological studies elucidated the complexity of the brain and spinal cord, making possible a more precise appreciation of localization of function and the recognition of a subtle communication between different parts of the brain. A number of neurotransmitters, such as 5-Hydroxytryptamine and dopamine as well as acetylcholine and noradrenaline, the first two identified, have been shown to have important functions within the brain, and the use of modern imaging techniques, radioactive-tracer molecules, and other sophisticated tools has revealed much about the physical operations of the central nervous system. As the Hippocratics insisted, consciousness and mental functions reside in the brain. Unpacking exactly what that means has taken some time.

Edwin G. Boring, *A History of Experimental Psychology* (1950). Edwin Clarke and C. D. O'Malley, eds., *The Human Brain and Spinal Cord* (1968). Robert Maxwell Young, *Mind, Brain and Adaptation in the Nineteenth Century* (1970). Aram Vartanian, *La Mettrie's L'homme machine: A Study in the Origins of an Idea* (1980). Stanley Finger, *Origins of Neuroscience: A History of Explorations into the Brain* (1994). Roger Smith, *The Fontana History of the Human Sciences* (1997).

W. F. BYNUM

MESMERISM AND ANIMAL MAGNETISM. Mesmerism was the creation of the Viennese physician Franz Mesmer, who regarded his technique of "animal magnetism" as an application of Newtonian principles to physiology. His doctoral dissertation at the University of Vienna concerned the influence of the Moon and planets on the human body, the idea being that the movement of the universal ether had an effect on health. He argued that the body required a certain quantity and rhythm of ethereal motion; disorders arose through an imbalance or incorrect type of motion.

Mesmer tried out this "tidal" theory of physiology in his medical practice in 1773–1774. He found that the application of magnets established an "artificial tide" in a patient.

She initially reacted with pain, then improved dramatically. After this first therapeutic success Mesmer extended the practice to patients of all kinds. He could treat several at once by using a structure called a "bacquet," a circular tub studded round the edges with metal rods. Patients held fast to the rods, from which the magnetic influence supposedly flowed.

Mesmer's hometown of Vienna proved less receptive to his ideas than Paris. There his practice drew crowds of patients from the nobility. Magnetic clinics sprung up throughout France, although orthodox doctors and natural philosophers remained skeptical. Several of his pupils became prominent magnetists in their own right—Charles d'Eslon, Nicolas Bergasse, and Guillaume Kornmann among them—and mesmeric societies sprang up across the country. The success of Mesmer and his protégés drew the attention of the Paris Faculty of Medicine and the Royal Academy of Sciences, each of which appointed a commission to investigate. The commissioners attended magnetic clinics and tried the bacquet. Some of them experienced strong effects; others, nothing. Some of the commissioners suspected that imagination might play a role in the phenomena. To test this, they persuaded a number of patients to believe they were being magnetized when in fact they were not (and vice versa). The magnetized patients exhibited no effects, while the unmagnetized patients experienced the expected "crisis." The commissioners concluded that although the phenomena were real, no physical agency caused them. Since the imagination was not then considered a legitimate cause of natural phenomena, attributing to it the mesmeric phenomena amounted to a form of dismissal. Mesmer lost control of the movement after this point, and several historians have suggested that he went mad.

Initiative passed to the marquis de Puysegur, who transformed mesmeric phenomena in one crucial respect. Before his time, mesmeric effects had been brief and violent, followed by a transformation in the patient's physical comfort. Puysegur produced a different phenomenon: an altered state of mind. In this lucid trance the patient said and acted in ways that seemed to indicate the presence of new mental powers. Subjects displayed signs of clairvoyance, foretold the future, diagnosed their own and other peoples' diseases, and spoke languages they had never learned. This new form of the mesmeric state could be used as a flexible tool for psychic and medical experimentation.

Mesmerism spread and diversified throughout Europe in the early nineteenth century, and many mesmeric societies and clinics were founded. In France it again became a serious contender for scientific and medical respect, based on the work of Puysegur and of other major advocates such as Joseph Philippe François Deleuze and, from the 1830s, Charles Dupotet. In the 1820s another royal commission investigated the practice, with mixed results—validating some effects, failing to reach consensus on others. Meanwhile mesmerism became well established in Germany and Switzerland; its more influential advocates included Johann Kaspar Lavater, Christoph Wilhelm Hufeland, and Carl Kluge. It took hold in Britain in the 1830s, a little later than on the Continent. Throughout the 1840s and 1850s it remained prominent in British public and medical discussion, and in 1842 its advocates could claim that they had used the mesmeric trance to create the first widespread surgical anesthetic (the first operation using ether anesthesia took place in 1846).

Medical mesmerism declined as an organized and prominent force in the second half of the nineteenth century, partly because of the proliferation of related sciences of mind. Hypnotism, for instance, was created as an alternative to mesmerism in 1842 and the spiritualist movement, which began in the 1850s, included many mesmeric elements. By the twentieth century the practice had been entirely subsumed within hypnotism, leaving only its terminology in popular usage. At that stage the terminology came to refer to fascination, charisma, and sex appeal rather than to a formal state of altered consciousness.

See also MENTAL SCIENCES.

Robert Darnton, *Mesmerism and the End of the Enlightenment in France* (1969). Henri F. Ellenberger, *Discovery of the Unconscious: The History and Evolution of Dynamic Psychiatry* (1970). Alan Gauld, *A History of Hypnotism* (1992). Adam Crabtree, *From Mesmer to Freud: Magnetic Sleep and the Roots of Psychological Healing* (1993). Alison Winter, *Mesmerized: Powers of Mind in Victorian Britain* (1998).

ALISON WINTER

METABOLISM designates the totality of the chemical changes that take place among the constituents of living organisms.

The Hippocratic physicians called the conversion of food to the humors of the body "coction" by analogy with cooking. The analogy served as the basis for imagining the underlying process until the seventeenth century. As Greek anatomists from *Aristotle to *Galen described the internal organs of the body in increasing detail, they came to see nutrition as a sequence of similar transformations taking place from the mouth, through the stomach and intestines, and into the blood vessels, progressively changing into blood, the final nutritive fluid. The old anatomists also knew that the body continuously lost matter, not only through excretions, but also through "invisible transpiration." One of the earliest recorded experiments was an effort to measure the loss by comparing the quantity of food taken in by a bird with that accounted for in its solid excretions.

The systematic study of these exchanges goes back to the experiments of the Venetian physician Santorio Santorio, who, in the early seventeenth century, weighed himself on a large scale together with his food and excretions. From the differences measured daily over a very long time, he determined the quantity of matter lost through his skin and lungs in "insensible perspiration" and its variation according to internal and external factors.

During the seventeenth and eighteenth century, the nutritional exchanges of the body began to be explained by analogy with acid-alkali reactions, or fermentations, studied by the emerging science of chemistry. Often such transformations were referred to as composing the "animal economy," a phrase derived originally from analogy with the management of a household. A wholly new understanding of the overall significance of these exchanges came with the theory of Antoine-Laurent *Lavoisier that respiration is a slow combustion yielding heat and work. Lavoisier thus explained the continuous food supply as a replenishment of carbon and hydrogen continuously consumed.

The rapid development of "plant" and "animal" chemistry in the early nineteenth century, and their fusion into the emerging field of organic chemistry, provided more robust foundations for investigating the chemical changes in living organisms. By the 1840s, the term "Stoffwechsel" had emerged as the preferred designation of these processes in German literature; its most common English translation, "metamorphosis," had previously been used both in German and English to describe the multiple chemical reactions, mainly partial decompositions, that organic compounds can undergo in the laboratory.

In his well-known treatise on the *cell theory, Theodor Schwann introduced the adjective *metabolische* to signify the phenomena of chemical change that cells can produce either on molecules within themselves or in the surrounding fluids. French texts began in the 1860s to translate *Stoffwechsel* as *le métabolisme.* Michael Foster adopted the word "metabolism" in his textbook of physiology in 1878, and it soon afterward became the standard term in English.

Throughout most of the nineteenth century, efforts to define the intermediate chemical reactions that connect the foodstuffs entering the body with final decomposition products amounted to conjectures based on knowledge of the chemical properties of the substances involved. The experimental study of the *Stoffwechsel* consisted of quantitative measurements of the intake of carbohydrates, fats, proteins, and oxygen, and of the excretions in the form of urea, carbon dioxide, and water, according to different dietary regimens in health and disease. In 1904, Franz Knoop fed dogs a diet of phenyl-substituted fatty acids. The animals metabolized these compounds only with difficulty, with the result that they excreted some intermediate breakdown products. In this way, Knoop established that fatty acids are decomposed by a succession of "β-oxidations," each shortening the carbon chain by two atoms. This was the first experimentally established sequence of intermediate reactions; unfortunately, Knoop's method did not work for other reactions. During the next three decades, Knoop insistently advocated the view that the goal of *biochemistry was to establish the complete sequences of decomposition reactions connecting foodstuffs to final end-products and the synthetic reactions producing body constituents.

During the first decade of the twentieth century, investigators found more direct methods of studying the chemical changes that take place between the beginning and end points of metabolism, and a new subspecialty, increasingly called "intermediary metabolism," began to take form. Federico Batelli and Linda Stern in Italy, and Thorsten Thunberg in Sweden, independently devised manometric methods to examine the phenomena in isolated tissues by testing the effects on their rate of respiration of adding substances suspected of being intermediates. Gustav Embden in Germany and others studied similar problems by means of perfused isolated organs. They identified a number of substances, such as pyruvic acid, acetic acid, and several dicarboxylic acids, that must take part, but could not connect them and other suspected intermediates into well-demonstrated reaction sequences. Meanwhile, the achievement of cell-free fermentation by Eduard Buchner stimulated the study of the intermediate phenomena in yeast, which were more accessible to direct examination than animal tissues. A series of partially speculative fermentation schemes culminated in a sequence proposed by Carl Neuberg in 1913 in which hexose sugars broke down into methylglyoxal. The proposal guided investigators for more than a decade, until they showed that methylglyoxal played no part at all. By 1935, a modified sequence of reactions, resulting in particular from the work of Embden and Otto Meyerhof, defined the pathway of anaerobic carbohydrate metabolism.

In 1930, a young physician named Hans Adolf *Krebs entered the field of intermediate metabolism, bringing with him methods that he had learned from Otto Warburg. Krebs discovered a cyclic process in 1932 through which urea is synthesized in mammals. He went on to discover several other pathways, including most notably the citric-acid cycle of carbohydrate metabolism in 1937. By then, individual pathways were beginning to link up in extended networks. The introduction of radioactive isotope tracers after 1940 accelerated progress in their identification. By mid-century, "metabolic maps" were complicated enough to fill large charts on biochemistry laboratory walls, and the main routes in the decomposition and synthesis of the amino acids, sugars, fatty acids, nucleic acids, and their many derivatives seemed nearly complete. Investigators began turning their attention to the question of the regulation of the pathways to maintain steady states while responding to the changing energy and other requirements of the organism, a quest that occupies metabolic biochemists to the present day.

Franklin C. Bing, "The History of the Word 'Metabolism,'" *Journal of the History of Medicine*, 26 (1971): 158–180. Frederic L. Holmes. *Hans Krebs* (2 vols., 1991–1993). Joseph S. Fruton, *Proteins, Enzymes, Genes* (1999).

FREDERIC LAWRENCE HOLMES

METALLURGY may be defined as the extraction of metals from their ores, their working and processing, the study of their structure and the development of alloys.

Although metals had been extracted and shaped for thousands of years, it was not until the sixteenth century that Georgius Agricola, a physician working in a German mining district, codified known metallurgical processes. His posthumous publication, *De re metallica* (1556), became an important text for later metallurgists, although it was based on theory and observation rather than on practical experience. Agricola's contemporary, Vannoccio Biringuccio, who was the chief armorer of Siena, produced a practical guide to metallurgical techniques, *Pirotechnia* (1540), based on his own practical experience.

René de Réaumur published an account of the French iron and steel industry in the 1720s based on his own observations and experimental work. His memoirs on the art of converting iron into steel had a mixed reception. Some critics believed that he had little understanding of metallurgy, while others viewed him as one of its pioneers.

During the eighteenth century, major developments in iron and steel manufacture occurred in Britain. In 1709, Abraham Darby successfully used coke instead of charcoal to smelt iron ore. This made possible the production of cast iron in great quantities and the construction of major cast-iron structures such as the first iron bridge in the world, a 100-foot span across the River Severn (1779). Benjamin Huntsman invented the crucible process for making steel in 1740 and Henry Cort developed the puddling process for the conversion of pig iron into wrought iron in 1784.

Sir Henry Bessemer's patented process of 1856 made possible the bulk production of cheap steel. A decade later, the German engineer Carl Wilhelm Siemens introduced the open-hearth furnace into Britain, while Emile Martin and his son Pierre did the same in France. This process could melt scrap steel, leading to the production of large quantities of high-quality structural steel, fundamental to the world-wide expansion of the railways and mass transportation.

In 1854, Henri-Étienne Sainte-Claire Deville introduced a method for the commercial extraction of aluminum from bauxite. At first, aluminum was a luxury metal, and Napoléon III had a banqueting set made from it. In 1886, the American Charles Hall and the Frenchman Paul Héroult, working independently, found that aluminum oxide could be dissolved in molten cryolite and the aluminum extracted by electrolysis. This method is still in use today.

Metallography, the study of metal surfaces, has proved particularly important to metallurgy's development. Between 1863 and 1865, Henry Clifton Sorby examined the surfaces of polished and etched specimens of steel under the microscope, the results being referred to as microstructures.

Steel is a complex alloy, based on iron with varying amounts of carbon. Steel has a number of phases which are visible in the various microstructures and which possess different physical and mechanical properties. Sorby's micrographic work on the structure of steel was confirmed by *x-ray diffraction. An analytical technique that provides clues to the internal structure of a material, it helped to confirm the atomic structure of metals and other materials.

The American physical chemist Willard *Gibbs proposed the phase rule in 1876 in a maze of mathematics and using a terse style difficult for most of his readers to understand. This rule defined and classified phase changes in metals and metal mixtures using the then new techniques of *thermodynamics. A phase may be defined as a distinct homogenous material that can be physically separated from other parts of the system. This work prompted the development of phase diagrams that visually illustrate these changes.

By using metallography and phase diagrams, metallurgists could predict the phases present in alloys at different temperatures. This led to the development and understanding of the heat treatment of alloys—for example, the quenching and tempering of steels and the age-hardening of aluminium alloys, introduced by the German Alfred Wilm in 1906.

Subsequent advances in the theory of solids gave rise to the concept, associated with Geoffrey Ingram Taylor, that deformation in metals takes place by the movement of dislocations (defects in the atomic arrangement) through the crystalline matrix. This concept led in turn to an understanding of failure through creep, fatigue, and brittle fracture, and to the development of strengthening mechanisms in modern alloys. The movement of dislocations was later confirmed with the aid of the electron microscope. Subsequently, more powerful field-ion microscopes, and high-resolution electron and atomic-force microscopes, have enabled metallurgists to specify the position of individual atoms.

Willliam Hume-Rothery carried out definitive work on intermetallic phases, electron compounds, and metallic structures, and proposed rules regarding the solid solubility of metals. The development of *solid state physics, *quantum theory, and the theory of the *periodic table have increased our understanding of the atomic structure of metals and their behavior.

In the late twentieth century, the development of the jet engine led to the demand for higher performance metals. A variety of light alloys, such as those based on aluminum and titanium, as well as superalloys (complex alloys that perform well at high temperatures), have been created for use in aerospace. Nanotechnology, the study and manufacture of materials and structures on the nanometer scale, and metal matrix composites are emerging as significant areas of metallurgical development in the twenty-first century.

Cyril Stanley Smith, *A History of Metallography: The Development of Ideas on the Structure of Metals Before 1890* (1960, repr. 1988). Kenneth C. Barraclough, *Steelmaking Before Bessemer*, 2 vols. (1984). Kenneth C. Barraclough, *Steelmaking: 1850–1900* (1990). Fathi Habashi, ed., *A History of Metallurgy* (1994). Robert W. Cahn, *The Coming of Materials Science* (2001).

SUSAN T. I. MOSSMAN

METALS. Seven metals, distinguished from other substances by their malleability and ductility, have been known since antiquity. Astrologers associated them with the seven planets: gold with the Sun, silver with the Moon, iron with Mars, quicksilver with Mercury, tin with Jupiter, copper with Venus, and lead with Saturn. The "perfect" metals, gold and silver, retained their metallic properties when strongly heated; the others turned into powdery substances known as calxes. Quicksilver, uniquely, was liquid; because it shared the other defining properties of metals, it was regarded as a metal in a permanently molten state.

Since early modern chemists thought that all substances were composed of a few simple principles, they inferred that one metal could be changed into another; the transmutation of lead into gold thus fell within the processes that even ordinary chemists, who generally dissociated themselves from alchemy in the late seventeenth century, considered possible.

All of the metals, including gold, could be dissolved in one or more of the common mineral acids. (Gold required a combination of nitrous and vitriolic acid known as *acqua regia* or *eau régale*.) Evaporation of the solution created crystalline precipitates that chemists gradually recognized as a class of neutral salts. During the seventeenth century they also recognized that one metal could replace another in solution with an acid, causing the other to precipitate in metallic form. For example, copper dissolved in a solution of silver in concentrated nitrous acid, turning the solution blue and precipitating the silver; the copper could be recovered by adding iron. Chemists learned further that all metals could be precipitated from their solutions in acids by adding fixed alkalis. In the early eighteenth century Étienne-François Geoffroy included these displacements prominently in his *Table des rapports*, which summarized the accumulated knowledge of the order in which substances replace one another in combination with a third substance.

During the seventeenth century chemists identified substances associated with and resembling metals, for example, bismuth, obtained from tin-bearing marcassites. By the mid-eighteenth century bismuth-like substances had emerged as a class with the properties of imperfect metals except ductility. These semi-metals included antimony, bismuth, zinc, cobalt, and arsenic.

During the 1740s a third perfect metal, platinum, was discovered. Not only could it not be destroyed by heating, it could not even be melted by the strongest heat available in the furnaces of the time. Pierre Joseph Macquer managed to fuse small amounts of platinum with a burning lens.

During the era of the *phlogiston theory, chemists considered metals to be compounds of calxes that resulted from heating them and phlogiston. This interpretation agreed with the common way of obtaining metals from their ores—reducing them by heating with charcoal—since charcoal provided an abundant source of phlogiston. Phlogiston chemists debated whether the other constituent of metals was an earth common to all of them or a distinct earth specific to each. On this unsolved question rested the possibility or impossibility of transmuting one metal into another.

The chemical revolution led by Antoine-Laurent *Lavoisier settled this question by redefining metals as elements and their calxes as combinations of the metals with *oxygen. Humphry Davy's decomposition of the alkalis by electrolysis early in the nineteenth century showed that they consisted, as Lavoisier had suspected, of a base combined with *oxygen. The alkali metals, potassium and sodium, did not, however, possess the same properties of ductility and malleability that had originally defined metals. With the advent of the dualistic theory of chemical combination (see *BERZELIUS), metallicity became a consequence of the electropositivity of metal atoms.

Nicolas Lemery, *Cours de chymie*, 4th ed. (1681). Pierre Joseph Macquer, *Dictionnaire de chymie* (1766). James R. Partington, *A History of Chemistry*, 4 vols. (1961–1970). Antoine Lavoisier, *Elements of Chemistry* (1965). Mary Elvira Weeks, *Discovery of the Elements*, 7th ed. (1968).

FREDERIC LAWRENCE HOLMES

METAMORPHOSIS, the Greek word used in ancient mythology to describe the process by which a given object, animal, or person changes form or substance. Since the final stage of the transformation—for example, of a human being into a beast or a star—often bore no resemblance to its initial one, the entire process was held to be either miraculous or magical. Both Aristotle and Pliny knew that frogs and butterflies undergo a series of mutations from their larval state to the adult form. During the seventeenth century, metamorphosis referred to the sudden transformation of one creature into another, as appears from Johannes Goedaert's *Metamorphosis et historia naturalis insectorum* (1667). But Jan Swammerdam challenged this idea on the ground that it favored a conception of nature regulated by chance rather than by laws. By dissecting larvae and chrysalides, he showed the rudiments of their limbs growing under their skin and maintained that these limbs changed as gradually and invisibly as those of a chick in a hen's egg (*Historia insectorum generalis*, 1669).

The notion of metamorphosis was also used in botany during the seventeenth and eighteenth centuries. Carl *Linnaeus linked it to his theory of sexual reproduction, and described the emergence and transformation of two anatomical structures: the cortex into stamens and the medulla into pistils (*Metamorphoses plantarum*, 1755). Johann Wolfgang von Goethe turned the notion into an analytical tool of plant morphology by adopting a powerful alchemical imagery in his *Versuch die Metamorphose der Pflanzen zu Erklären* (1790). Restricting his analysis to annual flowering plants, Goethe interpreted their development from germination to blossoming as the metamorphosis of a single form, the leaf, ruled by a polarity of successive expansions and contractions. He conceived each stage of the process (foliage leaf, corolla, petals, stamen, and pistils) as intensifications (*Steigerungen*) of an underlying form driven by purified fluids ascending up the axis of the plant. Extending his views to osteology, and specifically to the vertebrate skeleton, Goethe considered the skull as the transformation of a modified vertebral column. Lorenz Oken put forth a similar vertebral theory of the skull in *Über die Bedeutung der Schädelknochen* (1807).

Most scholars of the Romantic age held metamorphosis to be a general law of nature. It entered the vocabulary of plant and animal morphology, teratology, physiology, *chemistry, and *geology. By the middle of the nineteenth century, opposition to metamorphosis as a general law arose because of its speculative nature. During the twentieth century, with the decline of idealist morphology, it remained alive within a small group of professional botanists including Wilhelm Troll, while its narrower usage was confined to zoology, where it denoted a set of irreversible morphological and structural modifications of an individual in its larval stage, and a change of its mode of life. Most experimental work of the twentieth century sought to show how hormones regulate metamorphosis.

Agnes Arber, *The Natural History of Plant Form* (1950). Abraham Schierbeek, *Jan Swammerdam (12 February 1637–17 February 1680): His Life and Works* (1967). Adolf Portmann, "Goethe and the Concept of Metamorphosis," in *Goethe and the Sciences: A Re-Appraisal*, F. Amrine, F. J. Zucker, and H. Wheeler, eds., (1987): 133–145.

RENATO G. MAZZOLINI

METEOROLOGICAL STATION. Meteorological stations—fixed locations for weather observations—existed before the seventeenth century, but the invention of meteorological instruments in the first half of the century, and the new approach to nature adopted during the *Scientific Revolution, motivated significant numbers of observers and their organization into cooperative networks. Blaise Pascal's brother-in-law Florin Périer, who

A collection of meteorological instruments (barometers, thermometers, wind gauges) at the U.S. Weather Bureau's station at National Airport in Washington in 1943. A bobby-soxer reads the output.

set up stations on the Puy de Dôme for his famous measurements in 1649 (*see* BAROMETER), organized an early network of a few observers; around 1659 Robert *Boyle and Robert *Hooke began keeping regular observations at Oxford.

The new scientific societies sponsored networks of observers. Ferdinand II, grand duke of Tuscany and founder of the Accademia del Cimento, dispatched instruments and a form for recording observations to stations in ten European towns; Hooke worked out a "method for making a history of the weather" for the Royal Society of London, which he distributed with thermometers to English observers. Beginning in 1723 James Jurin, the Society's secretary, solicited and published observations in the Society's *Philosophical Transactions*. Unreliable instruments, a multiplicity of instrument scales, and lack of discipline among observers hobbled these and similar efforts. Significant progress was made during the late Enlightenment with the development of precise instruments and stronger bureaucracies. The Société Royale de Médicine, a product of the reforming ministry of Turgot, recruited physician observers across France; the Societas Meteorologica Palatina collected contributions from European and American scientific institutions; the Royal Society of London set up its own station, while interested observers in England, British travelers abroad, and agents of the British East India and Hudson's Bay Companies contributed long-term records. The age regarded measurement and calculation of this sort as a high form of intellectual activity, and placed considerable hopes on the ability of weather observation to improve the lot of mankind.

The Napoleonic wars interrupted these activities. In the first half of the nineteenth century networks of stations arose at the sub-national level or, in the case of the German states, as meteorological unions among several states. The *telegraph made practical the organization of national networks in the second half-century, often motivated by the need for storm warnings and organized under a maritime agency. These networks revolutionized meteorology, furnishing immediate weather information on a continental scope (*see* METEOROLOGY). The last quarter-century saw renewed emphasis on precision and *standardization and the establishment of national weather services.

In the 1890s kites, balloons, and mountain stations offered early glimpses into the three-dimensional structure of the atmosphere. The airplane increased this knowledge while multiplying requirements for upper-air forecasting. The much denser networks of both ground and air stations established during World War I contributed significantly to the invention and development of the polar-front model of cyclones and the air-mass method of analysis. Immediately after World War II rockets began to gather weather data from above the atmosphere and in 1960 the first weather satellite, *TIROS* (television infrared observational satellite), went into orbit. The *computer, applied to meteorology around the same time, has made possible the management of the tremendous flow of data from satellite stations.

Aleksandr Khristoforovich Khrgian, *Meteorology: A Historical Survey*, vol. 1, 2d ed., ed. Kh. P. Pogosyan (1959). Robert Marc Friedman, *Appropriating the Weather: Vilhelm Bjerknes and the Construction of a Modern Meteorology* (1989). Theodore S. Feldman, "Late Enlightenment Meteorology," in *The Quantifying Spirit in the 18th Century*, Tore Frängsmyr, J. L. Heilbron, and Robin E. Rider, eds. (1990): 143–178.

THEODORE S. FELDMAN

METEOROLOGY. The history of modern meteorology begins with the *Scientific Revolution. The late sixteenth and first half of the seventeenth century saw the invention of the meteorological instruments—*thermometer, *barometer, hygrometer, wind and rain gauges—and around 1650 natural philosophers began using them to record weather observations (*see* METEOROLOGICAL STATION). They immediately understood the importance of coordinating observations over as wide a space as possible. Scientific academies solicited weather diaries and organized observational networks: Leopold de' Medici, Grand Duke of Tuscany, founder of the Accademia del Cimento, and Robert *Hooke of the Royal Society of London both sponsored networks of observers in the 1650s and 1660s. They were motivated in part by theories deriving from the Hippocratic treatise *Airs, Waters, and Places* that related the weather to disease; for two and a half centuries meteorologists attempted to correlate weather patterns with epidemic outbreaks and climate with public health. The application of meteorology to agriculture provided further motivation. In addition, Enlightenment meteorology attempted to rationalize traditional weather lore, including astrological meteorology, searching through recorded observations for patterns confirming traditional wisdom.

Among the few attempts at a theoretical understanding of weather phenomena were explanations of the trade winds by Edmond Halley and George Hadley. According to them, the rising mass of heated equatorial air is replaced by an inflow of cooler air from higher latitudes. This north-south circulation is deflected, according to Halley, by the movement of the subsolar point with the earth's diurnal motion, or, in Hadley's theory, by the acceleration we now call Coriolis (*see* MECHANICS). A flow of warm air at high altitude from equator to poles completes these early pictures of the general circulation. They illustrate the role of oceangoing

commerce both as a source of data and incentive for meteorology.

Early modern meteorologists were frustrated by observers' lack of discipline and by the poor quality of instruments, which rendered observations nearly useless. The late Enlightenment resolved these problems. Emerging modern states organized large networks of disciplined observers, instrument makers developed precise instruments of all types, and natural philosophers devised methods of systematic measurement (*see* MATHEMATIZATION OR QUANTIFICATION). By the end of the eighteenth century meteorologists had access to large quantities of reliable weather data for the first time.

Enlightenment meteorologists, seeking weather patterns and correlations with agricultural harvests or outbreaks of disease, lacked a sense of the geographical expanse of weather events and of their development over time. Romantic natural philosophers worked out geographical and temporal syntheses. Alexander von *Humboldt's famous isothermal lines synthesized temperature observations over the globe; Humboldt integrated all the factors of climate into a unified science of the earth that he called "physique générale." Heinrich Wilhelm Brandes drew (or perhaps proposed to draw) synoptic maps of the weather over Europe for every day of 1783, tracing the progress of temperature changes across the Continent, uncovering the geographic distribution of barometric pressure, and relating wind direction to barometric differences. In the 1830s meteorologists took up the kinetics of storms. Heinrich Wilhelm Dove's "Law of Gyration" described the veering of storm winds resulting, he argued, from the conflict of equatorial and polar air currents; William C. Redfield insisted on the rotary motion of storms. James Espy introduced thermodynamic considerations, pointing to the adiabatic cooling of rising moist air and the energy of latent heat released in precipitation as the "motive" force of tropical storms.

The advent of the *telegraph around midcentury made possible the nearly immediate collection of meteorological data on a continental basis; at the same time the growing importance of meteorology for agriculture and oceangoing commerce led governments to establish national weather services to coordinate observation, particularly for storm warning. The resulting inflow of data fed the systematic production of synoptic weather charts, which became important research tools. A community of meteorologists evolved, its members more consistently trained in physics and mathematics, while the discipline acquired journals and professional societies. These factors, along with the emergence of *thermodynamics after midcentury, led to quantitative treatment of Espy's supposition. A consistent body of work emerged, known as the "thermal" or "convective" theory of cyclones, that derived the kinetic energy of storms from the release of latent heat and the adiabatic cooling of rising currents of air. William Ferrel applied hydrodynamics to the process, showing that air movement caused by any chance pressure gradient will be bent into a spiral by the earth's motion, generating a barometric low and the beginnings of a storm system. Hermann von *Helmholtz and Vilhelm Bjerknes were the best known among scientists applying hydrodynamics to meteorology.

Around the turn of the twentieth century balloons, kites, and airplanes made available observations of the upper atmosphere, while aviation generated demand for detailed forecasts in three dimensions. World War I sharpened these requirements. Discrepancies in the temperature distribution above storms had led meteorologists around the beginning of the century to consider the role of air masses of differing temperatures and geographic origin in the formation of storms. The polar front theory, developed immediately after the war by Bjerknes and his Bergen (Norway) school of meteorologists, demonstrated the origin of cyclones in the encounter of cool, polar air masses with warmer air. In the 1920s the Bergen school extended the air-mass approach to weather not associated with storms.

Around the same time Lewis Richardson succeeded in computing (after the fact) a six-hour advance in the weather using numerical algorithms. The effort consumed six weeks, generated disappointing results, and convinced meteorologists of the uselessness of a computational approach. The advent of the electronic *computer during World War II encouraged a new attempt at computational forecasting. John von Neumann, who selected meteorology to demonstrate the computer's usefulness, had by 1956 shown that it could generate accurate forecasts. The computer has enabled meteorologists to exploit the immense quantities of data arriving from weather satellites and a greatly increased number of observational sources in the atmosphere and at the earth's surface. Computational models of the atmosphere have since blurred the distinctions among observation, experiment, and theory.

Gisela Kutzbach, *The Thermal Theory of Cyclones: A History of Meteorological Thought in the Nineteenth Century* (1979). Robert Marc Friedman, *Appropriating the Weather: Vilhelm Bjerknes and the Construction of a Modern Meteorology* (1989). Theodore S. Feldman, "Late Enlightenment Meteorology," in *The Quantifying Spirit in the 18th Century*, Tore Frängsmyr, J. L. Heilbron, and Robin E. Rider, eds. (1990): 143–178. James Rodger Fleming, *Meteorology in America, 1800–1870* (1990). Frederik Nebeker, *Calculating the Weather* (1995).

THEODORE S. FELDMAN

METEORS. See COMETS AND METEORS.

METHOD, SCIENTIFIC. In the seventeenth century, most mathematicians, astronomers, natural philosophers, and natural historians shared an education in the classics. They knew and respected the definition of true knowledge—the kind of knowledge they thought exemplified by Euclidean geometry—that Aristotle had laid down in the *Posterior Analytics*: certain because demonstrable from self-evidently true first principles. They also knew that most of the theories about the world that they admired, such as the astronomy of Nicholas *Copernicus, Johannes *Kepler and Galileo *Galilei, had not been reached by a method of derivation from self-evidently true first principles.

Thus natural philosophers tried to find methods of discovering knowledge or science that began not from first principles but from observations and experiments. Because the knowledge they sought always went beyond the available data, they asked what methods allowed the discovery of theories. Seventeenth-century thinkers outlined five possible methods of discovery: hypothesis, enumerative induction, eliminative induction, analogy, and true causes. For the next two centuries natural philosophers debated their merits.

René *Descartes in his *Principles of Philosophy* (1644) pursued the method of hypothesis, postulating indiscernibly small corpuscles or atoms, deriving consequences, and testing them. Because it was known to be logically fallacious, the method of hypothesis was regarded with much suspicion until the twentieth century (*see* HYPOTHESIS). Francis Bacon had already advocated the method of induction in the *Novum organum* (1620). He did not recommend enumerative induction, the heaping up of data, because like everyone else he realized that from such a method general theories would never arise. He argued for eliminative induction in which every hypothesis that might explain a given phenomenon was assembled and then all but the correct one eliminated. Many subsequent natural philosophers and scientists shared his enthusiasm and hoped to prove theories true by the crucial experiments that decided between these hypotheses (*see* PROOF).

The two remaining possibilities, true causes and analogy, attempted to mediate between hypothesis and induction. Isaac *Newton proposed allowing only "true causes" in his "Rules of Reasoning" in the *Principia* (1687). Philosophers could postulate unseen causes (gravity, for example) provided that independent evidence could be adduced for their existence and that they sufficed to explain the phenomena. In the nineteenth century, John Herschel advocated this method in his *Preliminary Discourse on the Study of Natural Philosophy* (1830), as did Charles *Lyell in his *Principles of Geology* (1830–1833); Charles *Darwin used it to great effect, though not always to the liking of his critics, in the *Origin of Species* (1859). The method of analogy was essentially the use of models (*see* UNIFORMITARIANISM; MODEL).

For 250 years natural philosophers and methodologists tried to show that these methods of discovery cranked out certain truths about the world. The scientists of the later nineteenth and the twentieth centuries abandoned that dream. Instead they conceived of methods as ways of justifying the acceptance of theories, regardless of how those theories had been discovered. Throughout the twentieth century, the method of hypothesis was the most widely accepted of these methods among both scientists and philosophers of science, though scientists, no longer trained in philosophy, did not make their use of methods explicit. Meanwhile some historians of science, perhaps reading back into the past the more cavalier twentieth-century attitudes about method, argued that earlier methodological claims were simply rhetorical.

Ralph M. Blake, Curt J. Ducasse, and Edward H. Madden, *Theories of Scientific Method* (1960). David Hull, *Darwin and His Critics* (1973). Larry Laudan, *Science and Hypothesis* (1981). John A. Schuster and Richard R. Yeo, eds., *The Politics and Rhetoric of Scientific Method* (1991).

RACHEL LAUDAN

METROLOGY is the mundane backbone of science, the routine calibration of instruments with the aim of rendering comparable the measurements performed in different places. The best known example concerns weights and measures. Balances are compared against a standard in order that weighing of the same object on each of them will yield the same result.

Metrology has ancient roots and can be found in virtually every society. It has developed significantly with the growth of institutions, typically government ones, dedicated especially to routine calibration. Metrology loomed large in many ancient cultures. Taxation offered a prime locus of metrological activity because trustworthy measurement appeases the taxpayer. The Old Testament testifies to the ethical import of standard weights and measures: "Ye shall do no unrighteousness in judgment, in meteyard, in weight, or in measure. Just balances, just weights, a just ephah, and a just hin, shall ye have."

These injunctions may not have been effective. During the Middle Ages, metrological legislation was promulgated virtually unchanged in regular intervals, suggesting that standard weights and measures were not enforced and may not have been enforceable.

During the early modern period weights and measures differed from region to region, sometimes from town to town. There were different yardsticks, weights, and units of volume for every kind of commodity. Coinage was the first medium subjected to a metrological regime. With their expensive assaying equipment, regulated work routines, detailed notebooks, accountancy, and control testing mints had much in common with modern laboratories. Isaac *Newton served as master of the Royal London Mint and worked precisely to provide a metrological infrastructure, the mundane backbone of science and of a sound economy.

The increase of trade in the eighteenth century raised demand for intercomparability of local weights and measures; gradually metrological infrastructures similar to those in coinage came into being for grain and alcohol. Simultaneously, new military technologies and modes of organization, such as the standing army, forced European states to become fiscally efficient. Fiscal efficiency required metrological infrastructures.

These infrastructures rested on a primary standard produced by a trusted natural philosopher and kept in the metropolis. Secondary standards calibrated against the primary were transported to provincial towns and used to control tertiaries employed in the wider society, at ports, town gates, turnpikes, mills, breweries, and public houses. Tertiary and secondary standards were recalibrated at regular intervals; official measurers (surveyors and tax officials) were better trained and supervised. Tampering with standards was punished harshly. The insignia of the highest authority in the land, usually the king, stamped calibrated instruments. Tampering with these insignia was considered lèse-majesté and punishable by death. A metrological infrastructure of such complexity could only be realized within a civil service, which explains why the change from a haphazard system of unenforced decrees to the nascent modern metrological system took place during the eighteenth century, a period of immense civil service growth.

Enlightenment thinkers provided additional support for this development by associating rational thinking and metrological order. The metric system, first promulgated during the French Revolution, was a flower of rational thinking. The practical dimension of, and the main popular objection to, the metric system lay in its rigorous use of the decimal system, which reduced labor by greatly simplifying calculation. Growth in trade, particularly in large-scale trade, provided a groundswell for the metric system. However, the promulgation of the new rational system required numerous local changes, prompting resistance. For example, the duodecimal system had a practical advantage in weighing because halves, quarters, and eighths are easier to balance than tenths. In France, characteristically, the *cahiers de doléance* brought by the delegates to the Estates General in 1789 reveal widespread discontent with the abuse of weights and measures, but the introduction of an imperfect system with novel and learned names did not seem at first to be a clear improvement. Although the metric system became the legal standard in France before 1800, it suffered various transient alterations to please the public until the new units became obligatory in 1840.

The metric system spread first through the force of French arms and more peacefully via postal systems and international trade. It spread through continental Europe and with colonialism to most of the globe. Its association with Jacobin France fuelled resistance to it in nineteenth-century Britain. Both the British empire and the United States developed metrological infrastructures in parallel with continental Europe, but with nonmetrical units. These large infrastructures are now obstacles against the adoption of the metric system. Nonetheless, global trade drives piecemeal adoption of the metric system.

Metrology also expanded beyond weights and measures into electrical quantities, energy, radiation, hardness, heat, and so forth. New instruments were developed for the purpose, and the competition between candidate instruments has at times been fierce. The definition of the standard ohm in the 1880s engaged industrial and colonial interests through its importance for submarine telegraphy. James Clerk *Maxwell, Hermann von *Helmholtz, and Ernst Werner von Siemens championed the rival British and German ways of defining the standard. Metrological research and high-status science still stood close.

Metrology enjoyed considerable institutional growth during the twentieth century. Most large industrialized nations set up standards bureaus, for example the U.S. National Institute of Science and Technology (formerly National Bureau of Standards), which followed the Physikalisch-Technische Bundesanstalt (formerly Reichsanstalt) created in Germany during the 1880s. In the United States, trade associations oversee many of the standards relevant to their business, whereas in Europe and Japan responsibility for standards falls to the state or, more recently, international bodies such as the European Union. In the late twentieth century, many international standards bodies grew in significance, notably the International Organization for Standardization (ISO).

Karl Marx defined the commodity as a piece of merchandise characterized primarily (or only) by its price. The commodity is contrasted with merchandise exchanged in local barter where participants had themselves invested time and labor in their merchandise and consequently could not reduce its meaning to a price tag. Metrology participates in the capitalist process of commodification by undergirding stable values. A more favorable view sees metrology as a force generating accountability. The image of the eighteenth-century tax official subject to monitoring is a potent symbol of the kind of fairness imposed upon civil society by a metrological regime. A more recent such symbol would be the water analyst downstream from a factory making measurements to enforce environmental responsibility.

See also BUREAUS OF STANDARDS; STANDARDIZATION.

Rexmond Cochrane, *Measures for Progress: A History of the National Bureau of Standards* (1966). Witold Kula, *Measures and Men* (1986). Bruno Latour, *Science in Action* (1987). David Cahan, *An Institute for an Empire—The Physikalisch-Technische Reichsanstalt 1871–1918* (1989). Simon Schaffer, "Late Victorian Metrology and Its Instrumentation: A Manufactory of Ohms," in *Invisible Connections—Instruments, Institutions, and Science*, Robert Bud and Susan Cozzens, eds. (1992). M. Norton Wise, ed., *The Values of Precision* (1995). Arne Hessenbruch, "The Spread of Precision Measurement in Scandinavia 1660–1800", in *The Sciences in the European Periphery During the Enlightenment*, Kostas Gavroglu, ed., (1999): 179–224. Arne Hessenbruch, "Calibration and Work in the X-ray Economy, 1896–1928," *Social Studies of Science* 30 (2000): 397–420.

ARNE HESSENBRUCH

MICHELSON, Albert Abraham (1852–1931), physicist.

A. A. Michelson was a master of precision optical measurement. His determinations of the speed of *light and the lengths of light waves were the best of his day, and his 1887 attempt with Edward Morley to detect the motion of the earth through the *ether, the hypothetical medium of light, helped set the stage for Albert *Einstein's theory of *relativity. In 1907 Michelson became the first American to receive a Nobel Prize in the sciences.

Born in Strelno, Prussia (now Poland), Michelson emigrated to America with his family while still a child. He grew up in gold rush towns in California and Nevada. In 1869 he entered the

U.S. Naval Academy at Annapolis and in 1875 became a physics instructor there.

While teaching optics at Annapolis, Michelson set out to repeat Jean Bernard Léon Foucault's 1850 measurement of the speed of light. He improved Foucault's rotating mirror method and published his first results in 1878. Two years later, while pursuing advanced work in Europe, he first confronted an enduring puzzle: the effect of motion on light. Many physicists believed that as the earth moved through the stationary ether, it was swept by a continual "ether wind" blowing at about 30 km/sec (18 mi/sec). This wind should speed up or slow down light waves moving with or against it. Calculations indicated that the effect would be very small, perhaps too small to detect. In 1880 Michelson hit on a way to measure it. In a "Michelson interferometer," a beam of light is split into two parts moving at right angles to one another; when reflected back and recombined, they produce interference fringes that provide an exquisitely sensitive gauge of any changes in the speeds or path lengths of the beams. Michelson tried the device at Potsdam in 1881 but found no shift of the fringes when he turned the interferometer: the ether wind seemed to have no effect. The result was not yet conclusive, however, for the expected shift lay at the limit of the sensitivity of his apparatus.

Michelson resigned from the Navy in 1881 and the next year began teaching at the Case School of Applied Science in Cleveland. There he and Morley, of neighboring Western Reserve University, carried out two important investigations, first confirming Armand-Hippolyte-Louis Fizeau's 1859 demonstration that moving water drags along light waves passing through it, and then in 1887 performing their famous repetition of Michelson's Potsdam experiment. Using a larger and more sensitive interferometer set on a block of sandstone 1.5 m (5 ft) square and floating in a trough of mercury, they found no shift of the fringes and no sign of an ether wind.

Michelson and Morley's null result seemed impossible to reconcile with the known facts of optics. George Francis FitzGerald in 1889 and Hendrik Antoon Lorentz in 1892 independently proposed a striking solution: perhaps motion through the ether slightly alters the forces between molecules, causing Michelson and Morley's sandstone block to shrink by just enough to nullify the effect they had been seeking. The "FitzGerald-Lorentz contraction" later became an important part of relativity theory. Although scholars have often exaggerated the influence of Michelson and Morley's experiment on Einstein's thinking, Einstein knew at least indirectly of their result and it certainly loomed large in later

discussions of his ideas. Michelson himself did not welcome the rise of relativity theory; he remained a firm believer in the ether.

In 1889 Michelson left Case for Clark University in Massachusetts. Disappointed by the negative result of his ether-drift experiments, he turned his interferometer to other uses, traveling to Paris in 1892 to measure the standard meter in terms of light waves. In 1893 he moved to the University of Chicago, where he headed the physics department until 1929. There he continued his optical work, making precision diffraction gratings and inventing the echelon spectroscope. His Lowell lectures of 1899 were published in 1903 as *Light Waves and Their Uses*; his *Studies in Optics* appeared in 1927.

Michelson spent much of the 1920s at Mount Wilson Observatory in California applying interferometric methods to astronomical problems and refining his measurements of the speed of light. In 1926 he bounced light between mountain peaks 32 km (20 mi) apart, pinning the speed of light down to within 4 km/sec (2.5 mi/sec). After Dayton Miller announced in 1926 that he had found positive evidence of ether drift, Michelson repeated a refined version of his old experiment, but again found no shift of the fringes and Miller's results came to be regarded as erroneous.

Loyd S. Swenson, Jr., *The Ethereal Aether: A History of the Michelson-Morley-Miller Aether-Drift Experiments, 1880–1930* (1972). Dorothy Michelson Livingston, *The Master of Light: A Biography of Albert A. Michelson* (1973). Stanley Goldberg and Roger H. Stuewer, eds., *The Michelson Era in American Science, 1870–1930* (1988).

BRUCE J. HUNT

MICROBIOLOGY. See BACTERIOLOGY AND MICROBIOLOGY.

MICROCHIP. Microchips are latter-day versions of the integrated circuits (ICs) first developed in the late 1950s independently at the Texas Instruments Corporation and Fairchild Semiconductor Laboratories. At Texas Instruments, Jack Kilby devised an IC that formed all of its components, including transistors, diodes, and resistors, out of a single wafer of germanium, a semiconducting material. Kilby's IC was difficult to produce because each component had to be connected laboriously by hand. At Fairchild, Robert Noyce constructed an IC that, like Kilby's, made its components from a semiconducting wafer (in this case silicon). But it capitalized on a recent innovation by Fairchild engineers called the planar process that photo-etched the components and the connections between them onto the silicon. By removing the need to connect the components by hand, Noyce's method made the IC commercially practical. Noyce, Kilby, and their companies battled for years over professional and commercial credit for inventing the IC until they agreed to share the honors.

Although ICs were not developed with military patronage, the defense department and NASA provided most of the early market for them. One Minuteman II missile used two thousand; the Apollo guidance system, five thousand. The missile production schedule alone required four thousand ICs a week, necessitating high-volume production methods for the chips. In 1964 Gordon Moore, a Fairchild engineer, noted that the number of circuits on a single IC doubled every year, a trend that came to be known as "Moore's law." The emergence of metal-oxide semiconductors (MOS) in the late 1960s promised to extend the doubling rate well into the future and raised the possibility that a single chip could contain logic circuits equivalent to those in a mainframe computer of the 1950s. By then, more than half the chip market comprised civilian producers of electronic *calculators, video games, and test and control equipment.

In 1968 Noyce and Moore left Fairchild to found the Intel Corporation. Like other chip manufacturers, Intel first concentrated on custom chips for the burgeoning civilian market. But in 1969, Marcian E. Hoff, an Intel engineer, designed a general-purpose circuit that could then be programmed to suit the needs of specific customers; software stored in memory chips would take the place of the hard-wired logic functions of an integrated circuit. By 1971, Intel was selling an IC array that it described as the first "microprogrammable *computer on a chip" at a cost of $1,000. In 1974, Intel announced the 8080, a chip capable of processing 8-bit binary signals, for $360. Other manufacturers soon produced comparable chips, dropping the price of some of them to $100. Such chips enabled the creation of personal computers, which became commercially available in the late 1970s.

At the end of the twentieth century, Moore's law still operated. Microchips that contained more than a thousand transistors on a square wafer of silicon just a few millimeters on a side and operated at speeds many times faster than those of the first ICs had become commonplace. Some carried more than a million transistors. The chips are indispensable elements in laptop computers, cell phones, pagers, and the rest of the vast menu of consumer electronic goods.

Martin Campbell-Kelly and William Aspray, *Computer: A History of the Information Machine* (1996). Paul E. Ceruzzi, *A History of Modern Computing* (1998).

DANIEL J. KEVLES

MICROMETER. Micrometers, which are used to measure small distances or angles, come in three basic types: filar, ocular, and scale-division. The earliest of them, conceived by Lucas Brunn in Dresden and made by Christof Treschler in 1609, employed two precise screws, and could measure small distances. The micrometer did not come into wide use until William Gascoigne's filar device for measuring the diameters of astronomical objects (1638–1639), which used screws driving fiducial knife-edges or wires placed at a telescope's focal plane. The micrometer's various mechanical problems were overcome by Ole Rømer, John Rowley, and Jesse Ramsden. Heliometers, first made by Servington Savery (1743) and Bouguer (1748), employed a *telescope objective lens cut in half. Each half traversed along the cut using a screw, and so produced a double image.

Robert *Hooke (1674) first successfully used a screw to subdivide the scale divisions on the limb of an astronomical quadrant. The scale micrometers conceived by the Chevalier de Louville (1714) incorporated improved features and were more precisely made, while those used on zenith sectors made by George Graham advanced the design and accuracy even further. Jesse Ramsden applied the same technique to sextants around 1780.

Ocular micrometers, such as those made by Philip de La Hire (1700) and Georg Brander (around 1760), used glass elements ruled with fine lines using a diamond and placed in the focal plane of an eyepiece. Precision of the line separation in Brander's element depended on the master screw on a ruling machine. The ultimate expressions in this form were Alfred Nobert's mid-nineteenth century optical test plates used for measuring small features viewed in microscopes. However, screw micrometers had been applied to microscopes by Stephen Gray (to measure the height of mercury in a barometer, 1698) and C. G. Hertel (the first to employ screw and net micrometers, around 1716). The Duc de Chaulnes's use of microscope micrometers such as the points of a compass (around 1760) to divide the scale of his circular dividing engine was an important innovation later also successfully by Edward Troughton.

Micrometers made the greatest scientific impact in astronomy. During the seventeenth century, Richard Towneley, Robert Hooke, Johannes Hevelius, Adrien Auzout, and John Flamsteed all turned their micrometer-equipped telescopes to the problem of measuring stellar parallax—in vain. Success awaited Friedrich Wilhelm Bessel's deployment in 1838 of a heliometer made by Joseph Fraunhofer. During the long wait, James Bradley detected *aberration and nutation with a Graham sector equipped with a micrometer. In the more practical application of filar micrometers to surveying and navigational instruments, Ramsden made important innovations.

Micrometers suffer from periodic and random errors arising from the accuracy of the screw threads. In the seventeenth and early eighteenth centuries, errors associated with the calibration method escaped recognition until La Hire analyzed the problem in 1717. Instrumental errors were not well understood until J. B. Chabert's research in 1753. Circular scales with several micrometers, as implemented by Edward Troughton, permitted simple averaging of errors; their statistical treatment of errors did not become common before the 1830s. J. G. Repsold's impersonal micrometers employing traveling wires (1889–1895) brought the art to its highest point. With the advent of astrophysical research, the importance of the micrometer rapidly waned.

See also PRECISION AND ACCURACY; ERROR AND THE PERSONAL EQUATION.

Randall C. Brooks, "The Development of Micrometers in the Seventeenth, Eighteenth and Nineteenth Centuries," *Journal for the History of Astronomy*, 22 (1991): 127–173.
RANDALL C. BROOKS

MICROSCOPE. The microscope is an instrument for obtaining an enlarged image of an object, which may be viewed directly, photographed, or recorded electronically. The number of times the investigated object appears enlarged determines the magnification of the instrument, and the resolving power specifies its capacity to distinguish clearly between two points.

The earliest simple light microscope consisted of a single lens framed by a ring, plate, or cylinder, combined with a device for holding the object and a focusing mechanism. The compound microscope consisted of a movable tube containing two lenses or lens systems, the objective forming an enlarged image of the object and the ocular or eyepiece magnifying it; a stand; and a specimen stage. Illumination was provided either by a mirror placed under the stage to reflect light into the instrument or by a lamp built into the stand.

The microscope probably was invented during the second decade of the seventeenth century since the earliest printed descriptions and illustrations of a microscopically observed object, the bee, appeared in 1625, in the *Melissographia* and the *Apiarium* published by members of the Accademia dei Lincei. Robert *Hooke's enthusiastically received *Micrographia* of 1665 presented a lavishly illustrated survey of "minute bodies" ranging from the point of a needle and the pores of cork to bookworms and fleas. It also contained an analysis of methods of microscopy and a description of Hooke's

compound microscope, an instrument with a single-pillar stand on a solid base and a tiltable tube. Other seventeenth-century microscopists, notably Antoni van Leeuwenhoek, Marcello Malpighi, and Jan Swammerdam, used simple microscopes. Leeuwenhoek made more than three hundred of them, small plates of metal encasing a lens with provisions for focusing, the majority magnifying between 75 and 150 times. Communicating his findings largely through letters to the Royal Society of London, he became known particularly for his observations of "animalcules" in sperm.

In the eighteenth century, investigators throughout Europe continued to use both simple and compound microscopes. Spherical and chromatic aberrations, however, impaired the performance of these instruments. Objects appeared blurred and surrounded by colored fringes, because a spherical lens surface brings the light rays passing through different parts of the lens to a focus at different points, and light rays of different colors are refracted differently. Reducing the aperture with a diaphragm could lessen the spherical aberration, but at the cost of dimming the image. The aberrations affect the compound microscope more strongly than the simple microscope.

Achromatic telescope objectives consisting of lenses combining crown and flint glass were first made in the later eighteenth century (*see* OPTICS AND VISION). Although manufacturing the smaller lenses of microscopes proved much more difficult, Joseph Fraunhofer and other instrument makers produced achromatic microscope objectives of low magnification by the early nineteenth century. After Joseph Jackson Lister showed in 1830 how to construct aplanatic objectives (achromats with minimal spherical aberration), achromatic instruments of higher magnification became widely available. Together with advancing preparation techniques, they greatly enlarged the scope of microscopy. Microscopists examining plant and animal tissue established the cell as the unit of life, thus transforming histology and pathology and massively stimulating microscopical investigation. Facilitated by the expansion of laboratory and academic science, microscopy became an integral part of medical education, diagnosis, and research, as well as a tool for organic chemistry and the physical sciences.

In 1873, Ernst Abbe published a theory of image formation that enabled the design of optically improved instruments. In the early 1880s, Abbe and Otto Friedrich Schott produced apochromatic objectives, which eliminated the residual secondary spectrum of the achromat and almost completely corrected both chromatic and spherical aberration. At that time, novel techniques of staining and cultivation allowed the identification of pathogenic bacteria, which offered a potent new explanation of diseases.

In the late 1930s the development of phase contrast microscopy, which utilizes the principles of diffraction to convert variations in optical paths into intensity variation, allowed the direct study of unstained transparent biological specimens. The illumination system improved with the incorporation of electric lamps in the 1930s, and again with the introduction of low-voltage tungsten-halogen lamps in the 1960s, which are now the norm. The light microscope has become ubiquitous. Current applications include routine medical tests and examinations of the surface quality of materials.

The twentieth century introduced new kinds of microscopes that utilize rays other than light. The electron microscope contains a source supplying a high-voltage *electron beam, an evacuated column with electromagnetic fields along its length that act as a lens, a specimen stage, and an imaging system. In the transmission electron microscope (TEM), an electron beam passes through the object. In the scanning electron microscope (SEM), a comparatively small electron beam scans the object to produce an image of its surface.

The electronic engineers Max Knoll and Ernst Ruska constructed the first electron microscope in the early 1930s, utilizing contemporary *cathode-ray technology, improved electronic tubes, and vacuum technology, as well as the quantum mechanical principle that the electron can be regarded as a wave. Ruska developed an instrument for commercial manufacture in association with a group of engineers at the electrical firm of Siemens in Germany. Their TEM entered the market in 1939, one year earlier than the TEM constructed by the physicist James Hillier and his group at the Radio Corporation of America (RCA). In 1948, Charles William Oatley at the University of Cambridge launched a research project to construct an SEM, which led to commercial production in 1965.

Both the biomedical and physical sciences use the TEM and SEM. The TEM requires elaborate preparation techniques. To maintain the vacuum, the specimens must be dehydrated; since electrons interfere strongly with matter, the sections must be extremely thin; and chemical fixation is needed to prevent alterations under the electron beam. The introduction of the ultramicrotome and plastic embedding material in the late 1940s facilitated preparation of delicate biological specimens, and rapid freezing has supplemented chemical fixation since the late 1960s. Preparation for the SEM, which can accommodate large specimens, is easier. Biological specimens are rendered conductive by metal coating; other specimens can frequently be investigated without preparation.

Current applications of the electron microscope include virus biophysics, polymer morphology, and materials characterization and inspection in semiconductor technology.

Savile Bradbury, *The Evolution of the Microscope* (1967). Peter Hawkes, ed., *The Beginnings of Electron Microscopy* (1985). Gerard L'E. Turner, *The Great Age of the Microscope* (1989). Marian Fournier, *The Fabric of Life: Microscopy in the Seventeenth Century* (1996). Nicolas Rasmussen, *Picture Control: The Electron Microscope and the Transformation of Biology in America, 1940–1960* (1997).

JUTTA SCHICKORE

MILITARY-INDUSTRIAL-SCIENTIFIC COMPLEX. "If any merchant, selling spears or shields, would fain have battles to improve his trade, may he be seized by thieves and eat raw barley!"—Aristophanes.

In his Farewell Address of 17 January 1961, President Dwight D. Eisenhower warned against "the acquisition of unwarranted influence . . . by the military-industrial complex" (MIC): a "conjunction of an immense military establishment and a large arms industry." Since the days of Aristophanes, critics have accused arms makers of wholesale profiteering, of fomenting war scares to increase sales, of interference in political and military decisions, and of heating up arms races by selling to both sides of a conflict. Eisenhower warned also against the influence of a "scientific-technological elite." The scientific and technological aspects of the MIC are examined here. Contrary to Eisenhower's conviction that the MIC was "new in the American experience," the phenomenon appears early in the nation's history.

The threat of war with France in 1798 brought a crowd of incompetent and fraudulent weapons contractors to the doors of the U.S. Treasury. Among them was Eli Whitney, whose famous demonstration before Congress of interchangeable musket lock parts in 1801 was an early example of a rigged weapons test. Whitney used his influence to advocate machine manufacture of interchangeable parts while delaying fulfillment of his contract for ten years. The combination of technological innovation and contract overruns recurred in the late twentieth-century MIC; in Whitney's time the U.S. Army simply lacked the resources to manage weapons procurement.

The true father of interchangeable parts was the U.S. Army's Bureau of Ordnance, which, after developing interchangeability and standardization at its own armories between 1815 and 1845,

A symbol of the complex: Leslie R. Groves, commanding officer of the *Manhattan Project, pins the Medal for Merit on Ernest O. *Lawrence. Robert Gordon Sproul, the President of the University of California, which ran (and runs) the Los Alamos weapons laboratory, looks on.

encouraged the spread of innovation throughout the American arms industry and to other economic sectors. These advances enabled American firms to expand into international markets after 1850. European arms makers eagerly adopted the American innovations, and a global armaments business emerged. Nascent MICs appeared in several European nations and Japan. In England, William Armstrong used his position as Engineer for Rifled Ordnance to channel contracts to his gunmaking firm. The influence of the Krupp firm in late nineteenth-century Germany may have rested more on Friedrich Krupp's personal friendship with the Kaiser than on relations with the Army, which preferred brass to the superior Krupp steel cannon. Against the advice of the War Department, the Kaiser granted Krupp a monopoly on German arms purchases, while permitting him to sell to other nations, some of whom used Krupp munitions against Germany in World War I. Krupp became the largest industrial firm in the world, earning profits of 100 percent. In the United States, the Navy acted similarly to the Army's Bureau of Ordnance earlier in the century, as technical stimulus and demanding advisor, prodding a reluctant steel industry into the manufacture of armor plate. The industry gained instant profits and new markets; several firms owed their survival to government orders. The nation gained a modernized fleet and a steel industry and labor force capable of producing steel shapes and plates of a quality hitherto unknown in the country.

World War I mobilized entire national economies for war. In the United States, business leaders organized industry for war production in such a way that the largest corporations reaped the greatest profits from munitions contracts and the existing business structure was greatly strengthened. These leaders rated the war "the greatest business proposition since time began." In a parallel fashion, civilian scientific leaders saw in the war "the greatest chance we ever had to advance research in America." Perhaps because the science sector was less mature than the business sector, the war served less to strengthen its existing structures than to open up new opportunities.

The science sector's great success in areas such as submarine detection, range finding, chemical weapons, and radio signaling won it influence after the war. In an age of philanthropy, money flowed from foundations into university science departments. Departments developed close ties with business, as faculty started corporations to exploit new discoveries, while local technology companies loaned equipment, funded scholarships, sponsored research, and hired graduates. These developments helped build such institutions as the California Institute of Technology and MIT into great schools. The military connection weakened during the interwar period. Only the National Advisory Committee on Aeronautics (NACA) maintained significant ties among the military, academic science, and industry, letting contracts to academic and industrial laboratories for research and development on problems of military and civil aviation.

The U.S. scientific community parlayed its newfound strength into civilian control of scientific war work during World War II. Modeled after the NACA, the Office of Scientific Research and Development (OSRD) established successful cooperation among academic scientists, industry, and the military, contracting war research and development to universities and technical schools. As did the largest corporations in both wars, so now the largest universities received most of the contracts; and their scientists greatly expanded working relationships begun with business in the interwar years. The war engendered a military-industrial-academic complex.

The OSRD enjoyed tremendous success, and deserves much credit for winning the war. The nation understood that science was indispensable to its security. Scientists at the top universities and technical schools had already positioned their departments for a great influx of federal money after the war. The military did not disappoint. By late 1951, the Department of Defense and the Atomic Energy Commission were funding 40 percent of all industrial and academic research; defense research occupied two-thirds of the nation's scientists and engineers.

War and defense work shaped postwar science, particularly physics. A few top universities and defense contractors continued to benefit most from the military interest. Fields that answered military needs—microwave electronics, particle physics, and automated systems—dominated. The fertile wartime mix of physics and engineering continued at the most successful departments. Military officers populated graduate programs, professors taught and wrote on military-related topics, and even undergraduate programs were colored by professors' military orientation. Scientists participated in strategic decision-making bodies at the highest levels, as they had during the war.

Although the strategic needs of the military are supposed to drive weapons development, often scientists and engineers at the design level, generating ideas for new technology, spurred the arms race in directions unforeseen by military planners. The "R&D establishment pushes against the technological frontiers without waiting to be asked." "The most exuberant and persuasive of our technologists . . . promote ideas and sell hardware that often take us far beyond the point that mere prudence requires."

Other nations, lacking a superpower's compulsion always to be in the forefront of weapons advancement, have not developed MICs like that of the United States. Only very recently has industry in England, for example, taken over a significant fraction of defense-related R&D from the military, and British universities have never been deeply involved in military-related research. Nevertheless, about the same percentage of the GNP in Britain goes to defense R&D as in the United States, and 10 percent of all British manufacturing employees are associated with defense production.

Carroll W. Pursell, Jr., ed., *The Military-Industrial Complex* (1972). Benjamin Franklin Cooling, *Gray Steel and Blue Water Navy: The Formative Years of America's Military-Industrial Complex* (1979). Paul A. Koistinen, *The Military-Industrial Complex: A Historical Perspective* (1980). Merritt Roe Smith, ed., *Military Enterprise and Technological Change: Perspectives on the American Experience* (1985). Everett Mendelsohn, Merritt Roe Smith, and Peter Weingart, eds., *Science, Technology, and the Military* (1988). Stuart W. Leslie, *The Cold War and American Science: The Military-Industrial-Academic Complex at MIT and Stanford* (1993).

THEODORE S. FELDMAN

MILITARY INSTITUTIONS. The contributions of science to warfare, especially to the wars of the twentieth century, are familiar to students of history. Not so well understood is the equal and opposite action: the influence of military institutions on science and technology. Although interest in the subject has mushroomed over the last decade, historians have not yet forged an overall synthesis. Recent research has focused on U.S. science during the Cold War, but older judgments regarding the two world wars need revisiting. Because the line separating science from technology has remained shifting and indistinct—thanks in part to military influence—"science" here stands for science and technology.

The early modern constellation of the sciences reflects a military influence. Early modern mathematics included disciplines directly applicable to military affairs: fortification or military architecture, naval architecture, ballistics, cartography, and navigation. Early modern natural philosophy, on the other hand, included a broad range of nonquantitative subjects, from physics to botany and psychology; these pursuits offered little to the practice of warfare. *Galileo Galilei consulted for the Venetian Arsenal, a naval institution, as a mathematician and courted the Medici family with a military compass. But when the *telescope brought in his own ship he abandoned his career as a mathematician for the higher-status occupation of court philosopher.

Because mathematics was essential to the education of its military officers, the early modern state became an important patron of the subject. In France during the Ancien Régime military and naval academies employed the top

Instruction of German soldiers in gas warfare around 1916. The instructor is Fritz
*Haber (pointing, second from the left), one of the leading chemists of Germany and
the prime mover of its initiatives with poison gas.

mathematicians of the age, including Gaspard
Monge, who trained Charles Augustin Coulomb
and Lazare Carnot (see CAVENDISH, HENRY, AND
CHARLES AUGUSTIN COULOMB). Carnot's technical
contributions to French arms earned him the
title "father of victories." *Natural philosophy
also received support as part of the cadets'
general education. The Ecole Polytechnique,
revolutionary successor of these institutions,
produced much of the most important *physics,
mathematics, and engineering of the early
nineteenth century. In the United States the
Ecole's sister school, Thomas Jefferson's Military
Academy at West Point, served as the only source
of trained civil engineers for several decades
and graduated military men who achieved
international stature as scientists. In an era when
military officers were broadly educated and
mathematics included so much military content,
mathematicians and engineers naturally found a
home in military institutions.

Military support, intensifying after the Seven
Years War, helped transform mathematics and
natural philosophy. In France and England,
national cartographic projects originated in
military campaigns. These projects developed
advanced methods of triangulation and measurement; along with other military and naval requirements, they generated intense demand for precise
*instruments (see GEODESY). Profits from the sale of
these instruments to armies and navies supported
instrument makers' research and development
and helped industrialize the scientific instrument
trade; Jesse Ramsden's revolutionary instruments

and his innovations in the organization and
technology of instrument manufacture depended
on his sale of sextants to the British Navy. At
the same time, military engineers like Coulomb
and England's William Roy crossed over into
natural philosophy, applying mathematical techniques to fields like *electricity, *magnetism, and
the study of gases (see PNEUMATICS), contributing
to their quantification. Military needs and military engineers thus played an important role in
the emergence of precise scientific instrumentation and in the quantification of physics in the
late Enlightenment.

The early modern state also experimented with
the mobilization of civilian scientists for military
research. In France the monarchy employed
members of the Royal Academy of Sciences in
the reform of the manufacturing sector. Antoine-
Laurent *Lavoisier took on the gunpowder
industry. Setting up a laboratory at the Arsenal in
Paris—where he also carried out his more famous
chemical researches—he developed methods of
gunpowder manufacture that made France an
exporter of high-quality gunpowder and laid
the foundations of industrial chemistry. The
Revolutionary government employed a number
of Lavoisier's colleagues at a weapons laboratory
at Meudon, where Claude Louis Berthollet and
others developed incendiary weapons, explosive
shells, and new types of gunpowder. The
rediscovery of these explosive shells in the late
1830s revolutionized naval warfare.

In the United States the army revolutionized
manufacturing in the first half of the nineteenth

century, developing at its national armories the so-called "American System": machine manufacture of weapons with interchangeable parts. Besides introducing modern methods of gauging and pattern replication, the army developed advanced business and accounting practices to manage its inventory and distribution requirements. The American System eventually took over much of American manufacturing and was eagerly received in Europe. Both the British Navy—in the case of the scientific instrument trade—and the American Army provided markets without which it would have been more difficult, if not impossible, to develop these technologies. No private arms manufacturer of the early nineteenth century would have undertaken as risky and pointless a venture as attempting to make weapons with interchangeable parts.

Closely related to military *cartography was exploration, also often a military venture. James Cook earned his reputation charting the Saint Lawrence River for the Battle of Quebec. On his Pacific voyages he applied the most advanced techniques of land-based cartography to hydrography and contributed to the solution of the problem of *longitude, which had long preoccupied the British Navy. Cook's voyages, part of a series of naval expeditions to the Pacific, set a precedent for the union of scientific and naval operations. The published goal of Cook's first voyage was the observation of the Transit of Venus—a purely scientific activity. But secret instructions charged Cook with establishing a foothold on the reputed Southern Continent and with surveying resources for naval stations and colonization. Cook carried with him civilian naturalists and artists who made important contributions to geography, *botany, ethnography, and other sciences. Charles *Darwin's Beagle voyage belongs to this same series of imperial hydrographical expeditions. In the United States the army, beginning with the expedition of Meriwether Lewis and George Rogers Clark (1803–1806), played a leading role in the exploration and conquest of North America. The expeditions of its topographical engineers, from the Revolution until their dissolution in 1853, ranged broadly through cartography, *anthropology, *geology, survey of natural resources, and diplomacy.

Notwithstanding its nurturing of manufacturing technology, the nineteenth-century U.S. military generally failed to appreciate the larger potential of science and technology for warfare. The navy dragged its feet on steam power and propellers in the first half of the nineteenth century; the army notoriously resisted new technology during the Civil War. Nor were rank-and-file

scientists, jealous of their independence, anxious for close relations with the military. But the scientific leadership grasped the technological character of modern war and perceived that solutions to technological problems required not only engineering but also basic scientific research. During the Civil War they urgently lobbied for an agency to direct military research and development. The navy responded with a permanent commission that enjoyed civilian status, no funding, no research capabilities, no power, and little meaningful activity. The National Academy of Sciences (NAS), born in 1863, likewise contributed little to the war effort.

Much seemed the same when World War I began. In Europe and the United States the military misapprehended the character of technological warfare, but the stalemate rendered the military more receptive to new ideas. Fritz *Haber convinced a skeptical German headquarters in 1915 to experiment with poison gas. A gas arms race ensued, which mobilized scientists on both sides: England's Porton Down occupied 7,000 acres, with a breeding colony for animal and human experimentation. Its scientists were better paid than those in other scientific occupations. In the United States, Edgewood Arsenal in Maryland, employing 2,000 people and turning out 200,000 bombs and shells per day, had no rival in size until the *Manhattan Project.

The U.S. scientific leadership tried several approaches to mobilizing science for war. The Naval Consulting Board, Thomas Edison's child, applied methods of invention and engineering to problems of military technology and made little use of academic scientists. The National Research Council (NRC, a creation of the NAS) took a different approach, organizing basic research for war and uniting both academic and nonacademic scientists with engineers. Its leaders recognized not only what science could do for the war, but also what war could do for science.

By war's end, the NRC's success had demonstrated the need for fundamental scientific research in modern warfare. Academic science won greatly increased status. "Those who shared in the consciousness of the University's power and resourcefulness," remarked one dean, "can never be fully content to return to the old routine of the days before the war." Private foundations poured money into research projects. Industry, impressed with scientists' practical and managerial skills, welcomed them. The war introduced team-based research to American science, and, as it crippled European science, the war made the United States a world scientific power. But the military-scientific alliance was not perfect. The NRC could only coordinate projects; it had little money and could not contract out research. The military feared that academics would exploit its sponsorship for their own purposes. The National Advisory Committee on Aeronautics (NACA) offered a third, more promising approach to war-related research. Well funded and with authority to maintain its own laboratories, NACA forged strong ties among academics, government scientists, and the military. This model would shape the organization of science in the next war.

In 1941 the scientific rank and file again showed little interest in military work. Scientists who had served in World War I favored preparedness. Vannevar Bush had worked at the NRC's submarine detection facility and had served as chair of the NACA, as well as dean of engineering and vice president at the Massachusetts Institute for Technology (MIT), uniting, in his person, the historical precedents for a scientific-military alliance. Under his leadership the Office for Scientific Research and Development (OSRD) organized most scientific work for the war. Patterned after the NACA, the OSRD contracted out research and development to universities and industry, allowing scientists to work on military projects in civilian and academic settings. Huge projects at the nation's top universities and engineering schools—MIT's Radiation Laboratory employed 4,500, for instance—made essential contributions to the war effort. The Manhattan Project, an exception to this pattern, fell to the army under conditions that academic scientists were glad to escape at war's end.

World War II taught the nation that its security depended on technological superiority; that new military technology demanded fundamental research not tied to immediate needs; that big projects and lots of money produced great results; that the nation required scientists, who alone understood the new weapons, at the highest levels of strategic planning; and that an unfailing supply of them must be guaranteed through the cultivation of science education. These lessons shaped the institutional structure and the content of Cold War science. National laboratories at Argonne, Berkeley, Oak Ridge, Brookhaven, and later Livermore took up where the Manhattan Project left off, while universities and engineering schools carried on the approach of the OSRD. Wartime projects had built Stanford and MIT into great schools. Scientists from these and other select institutions favored by the OSRD had foreseen and jockeyed for greatly increased postwar support. Interdisciplinary departments in military-related fields like microwave electronics proved highly successful, as did quasi-independent laboratories managed by academic institutions, such as Johns Hopkins University's Applied Physics

Laboratory and Charles Stark Draper's Instrumentation Laboratory at MIT. Besides microwave electronics, crucial to radar and communications, favored subject areas included materials science for semiconductor research, automated systems, and nuclear science.

Through these institutions military relations helped to direct the orientation and shape the content of science. Military officers populated graduate programs, which in the immediate postwar years might include classified courses among their offerings. Graduate students wrote theses on military topics, then filled the ranks of the next cohort of professors. In subjects like *quantum electrodynamics—the applications of which include lasers and semiconductors—theoretical approaches and results might respond to military requirements. Semi-military agencies like the *RAND Corporation, applying the social sciences to military purposes, fostered pervasive quantification in these fields. The Korean War intensified the military influence in science. By late 1951 the Department of Defense and the Atomic Energy Commission (AEC) funded 40 percent of all academic and industrial research, while defense-related research occupied two-thirds of the nation's scientists and engineers. Through panels and summer study groups scientists advised the government on issues of technology and strategy. The General Advisory Committee to the AEC, for example, contributed to the decision to create the hydrogen bomb; it directly succeeded the Manhattan Project's Scientific Panel, which had advised the War Department on targeting and deployment of the atomic bomb. In all these ways American scientists helped the United States win the arms races of *World War II and the Cold War.

Merritt Roe Smith, *Harper's Ferry Armory and The New Technology* (1977). L. F. Haber, *The Poisonous Cloud: Chemical Warfare in the First World War* (1986). Paul Forman, "Behind Quantum Electronics: National Security as the Basis for Physical Research in the United States, 1940–1960," *Historical Studies in the Physical and Biological Sciences* 18 (1987): 149–229. Daniel Kevles, "Cold War and Hot Physics: Science, Security, and the American State, 1945–56," *Historical Studies in the Physical and Biological Sciences* 20 (1990): 239–264. Charles Coulston Gillispie, "Science and Secret Weapons Development in Revolutionary France, 1792–1804: A Documentary History," *Historical Studies in the Physical and Biological Sciences* 23 (1992): 35–152. Daniel J. Kevles, *The Physicists: The History of a Scientific Community in Modern America* (1995). David Hounshell, "The Cold War, RAND, and the Generation of Knowledge, 1946–1962," *Historical Studies in the Physical and Biological Sciences* 27 (1997): 237–269.

THEODORE S. FELDMAN

MILKY WAY. The Copernican revolution rendered obsolete the Aristotelian universe, but the seventeenth century saw little advance in understanding the Milky Way, a dense band of stars the color of milk across the sky, beyond *Galileo's confirmation of the ancient opinion that it was a congeries of innumerable stars. William Whiston, Isaac *Newton's successor at Cambridge in 1703, argued that the system of the stars, the work of the Creator, had a beautiful proportion, even if frail man could not know it. Whiston demonstrated his frailty by failing to propose an order for the Milky Way. The self-taught English astronomer Thomas Wright did better in 1750. Five years later, Immanuel Kant, inspired by an incorrect summary of Wright's book, explained the Milky Way as a disk-shaped system containing the earth. Kant conjectured that the disk structure arose in the same manner that the planets came to orbit almost in the same plane around the Sun. And the same cause that gave the planets their centrifugal force and directed their orbits into a plane could also have given the power of revolving to the stars and have brought their orbits into a plane.

Kant's manuscript perished in his printer's bankruptcy. A condensed version of his hypothesis appeared in 1763, hidden in the appendix of another book. Meanwhile (in 1761), the polymath Johann Heinrich Lambert published a similar theory. Kant had emphasized Newtonian dynamics and the process by which the world achieved its current shape. Lambert, while acknowledging the existence of Newton and gravitation, emphasized God and the harmonious order He had given the world. Unaware of Wright's, Kant's, and Lambert's ideas, the English astronomer William Herschel began in the 1780s his own investigation of the construction of the heavens. He observed stars in a stratum seemingly running to great lengths and identified the Milky Way as the appearance of the stars as seen from the earth. To determine the position of the Sun in the sidereal stratum, Herschel counted stars in different directions. This number, he argued, should be proportional to the length of the stratum in the direction of the count. Herschel was an observer, not a theoretician. His Milky Way was an observed stratum of stars extending different distances in different directions, not a theoretical disk the result of the force of attraction.

Herschel lacked the means to measure distances. Not until early in the twentieth century could the American astronomer Harlow Shapley argue convincingly that the Milky Way was a hundred times larger than previous estimates and that the Sun lies tens of thousands of light years away from the center of the *galaxy.

Stanley L. Jaki, *The Milky Way: An Elusive Road for Science* (1972).

NORRISS S. HETHERINGTON

MIND-BODY PROBLEMS. The question of the relation between mind and body is generally thought to be a philosophical problem more likely to yield to conceptual analysis than to empirical studies. Nonetheless accounts of this relation have had and continue to have profound implications within science, particularly in *psychology, *physiology, and *medicine.

The problem received its modern form from René *Descartes, who overthrew the account of body and soul that Medieval scholasticism inherited from the Aristotelian and medical traditions. Every living thing had a form—also called the soul—that was the principle of all its activities. The human form contained an intellectual part capable of apprehending universals. Intellectual apprehension began with sense experience and then, with the assistance of memory and imagination, abstracted intelligible structures from their material embodiment. Aristotelian psychology made the intellect depend upon, while being ultimately separable from, matter.

To undermine the notion that human knowledge derived from the senses, Descartes used his method of radical doubt. He concluded that ideal and mathematical knowledge did not depend on sensation and that reality is divided into two radically different substances, the thinking thing (mind) and extension (material reality). Because mind and extension had nothing in common, their union in the human person, although familiar in everyday life, was unintelligible.

Descartes believed that although the fundamental questions of why and how mind and body were united were unanswerable, the instrumental question of how acts of the mind are related to the functions of body could be addressed. Conceiving the body as matter in motion, he suggested that the brain's pineal gland provided the locus for the union of mind and body. The gland's motions correlated with perceptions and acts of will that produced bodily motions by controlling animal spirit flows through the nerves to and from the brain.

Few of his successors accepted Descartes's instrumental solution, much less the fundamental unintelligibility of mind-body union. Some postulated a parallelism, harmony, or fundamental unity between the mind and the laws of the physical universe. Thomas Hobbes and George Berkeley opted for more extreme solutions. Berkeley adopted an idealism that denied matter's reality. Hobbes reduced mind to the actions and reactions of matter. After the mid-eighteenth century, reductionism gained in appeal, especially in light of the continuing advance of the sciences.

From the beginning of the nineteenth century, it became widely accepted that the brain was the organ of the mind and since then the mind-body problem has increasingly been treated as the mind-brain problem. Franz Joseph Gall argued that different parts of the cerebral cortex housed different powers and functions of mind, a theory called *phrenology. Although hotly debated at the time, John Hughlings Jackson's experiments stimulating different areas of the cortex during brain surgery indicated that different mental functions occurred in different areas of the brain.

In the twentieth century, with the help of brain-imaging techniques and theories of computation, scientists made rapid progress correlating behavior and thought with the activity of different parts of the brain and in organizing brain functions into lateral and hierarchical structures. The last two decades of the century brought concerted efforts to explain noncognitive phenomena such as emotion and will. The philosophical question of how to relate reductive materialism, particular theories of brain functions, and common sense (and philosophical) descriptions of mental experience continues to be debated.

See also COGNITIVE SCIENCE.

Jonathan Shear, ed., *Explaining Consciousness: The Hard Problem* (1997). David M. Rosenthal, ed., *Materialism and the Mind-Body Problem*, 2d ed. (2000).

DENNIS SEPPER

MINERALOGY AND PETROLOGY. Mineralogy's disciplinary status has undergone three distinct shifts. From the sixteenth through the early nineteenth century, it bridged *chemistry and *natural history. It used the laboratory techniques of the former and the principles of classification of the latter to study the whole of the mineral kingdom. During the nineteenth century, mineralogy lost this commanding position and became a subdiscipline of *geology. The study of minerals (chemicals that occur naturally in the earth's crust) separated from that of rocks (distinctive assemblages of minerals), leading to a distinction between mineralogy and petrology in the latter part of the century. Following World War II, geology came under the *earth sciences, and mineralogy and petrology were transformed by the theory of *plate tectonics and by new instrumentation.

In spite of changes in disciplinary status, from the eighteenth century to the present, mineralogists have concentrated on two problems. The first, classification, has been essential for geological theory and practical applications in mining. It has also been a scientific nightmare. Classifying depends on being able to make clear distinctions at various levels of organization. In the case of animals and plants, individuals can usually be distinguished easily and most of the time species too by the test of reproductive capability. Although in the eighteenth century, Carl *Linnaeus made a gallant attempt to extend these methods to minerals, it was a doomed strategy. Over the centuries, mineralogists have oscillated between using chemical composition and crystal form. Unfortunately, these do not map onto each other. Two minerals of the same composition can have different forms, and vice versa. Adding to the difficulties, the chemical composition of many of the commonest minerals is not fixed but allows a range of variation. Formal classifications have always been supplemented by keys to field identification using a variety of visible characters.

The second key mineralogical problem is how minerals and rocks originated. Since they occur interlocked with one another, mineralogists from the seventeenth century on have assumed that they originated as fluids and subsequently hardened in their present positions. The fluidity could have been caused by heat, water, or some combination of the two. Mineralogists have fought recurrent battles over the relative importance of these factors. In either case, they have always hoped they would be able to use their knowledge of chemical reactions to reconstruct a genetic account of rock history—a geogony based on an invariable sequence of chemical reactions following from some initial state.

Mineralogy differed greatly from the historical geology better known to the public and more thoroughly studied by historians. Mineralogists have always focused their attention on the hard rocks (igneous and metamorphic). The sedimentary rocks so dear to historical geologists because of their embedded *fossils have taken second place, even though mineralogists have attended to clay mineralogy and sedimentary petrology. Mineralogists have always worked closely with chemists and with crystallographers. They have found experiments and microscope work as important as fieldwork. Germans and Scandinavians dominated mineralogy partly because of the abundance of hard rocks in those countries, but also because of their distinguished traditions in chemistry and crystallography. Only in the twentieth century did they begin yielding to Canadians and Americans.

From the Renaissance through the eighteenth century, mineralogists produced one classification of rocks and minerals after another. Among the more important were Georg Bauer, better known as Agricola, the Swedish chemist Johan G. Wallerius, and Abraham Gottlob Werner. All distinguished four major groups with different chemistries: *earths, *metals, *salts and combustibles. Earths resisted heat and water, metals became fluid on heating, salts dissolved in liquids, and the combustible substances (coal, for example) burned. Because chemistry formed the basis for classifying rocks, the students spent as much time in the cabinet or laboratory working with chemicals and blowpipes as they did in the field. Werner drew the pessimistic conclusion that no theoretically sound principles of mineral classification were to be found. Hence he instructed his students to begin dividing up rocks by the time of their formation (*see* STRATIGRAPHY).

Employed by European states as mining inspectors or surveyors, mineralogists left their laboratories to climb mountains and descend mine shafts. By the second half of the eighteenth century, these men—including Lazzaro Moro and Giovanni Arduino in Italy, Johann Lehmann in Germany, and Guillaume-François Rouelle in France—opted for an alternative approach to rock classification. They divided rocks into two main kinds: primary and secondary. Primary rocks were hard, often crystalline and the matrix in which metals and precious minerals were to be found. They made up the core of mountain chains. Secondary rocks were relatively soft and granular, layered or stratified and banked up against the primary rocks that formed the mountain cores. Often, secondary rocks contained fossils, which by then most mineralogists agreed were the indurated remains of animals and plants.

In seeking the causes for this twofold division of rocks, mineralogists found common ground with cosmogony, the study of the development of the globe. Since the seventeenth century, cosmogonists had argued from the earth's globular figure that at some time in the past it had been fluid. Fluidity could have been caused by heat, as a minority of cosmogonists had argued. Mineralogists, though, preferred water, as suggested by the chemist Johann Joachim Becher in his *Physica subterranea* (1669). They believed that a thick, chemical laden ocean had once covered the earth's surface. The primary rocks crystallized out of the ocean as the high mountain chains, a conclusion supported by the chemists' belief that crystals could be deposited only from watery solutions and not from hot melts. As the water became less saturated, and as waves and rivers wore away the mountains, the ocean began depositing the silt that solidified as the secondary, stratified rocks. This theory, *Neptunism, was most fully developed by Werner at Freiberg.

From about 1830 to 1880, mineralogists looked to new developments in chemistry to aid them with mineral classifications. Jöns Jacob *Berzelius, the Swedish chemist, distinguished the silicates and aluminates—the classes of chemicals most abundant in the earth's crust for the first time.

Gustav Rose offered the most comprehensive classification of minerals to date in his *Mineralsystem* (1852). For practical purposes, mineralogists continued to use external features. Friedrich Mohs best known for developing a hardness scale for minerals, developed one of these. James Dwight Dana adapted it for an American audience. His *System of Mineralogy* published in 1837 must be one of the most enduring of textbooks in the history of modern science. In modified form, it was still in use in the 1960s.

In 1860, Henry Clifton Sorby invented the polarizing *microscope, transforming the process of identifying minerals. Thin sections of minerals or rocks were placed on slides that could be rotated beneath polarizing lens. The characteristic color changes that were observed on rotating the slide served to identify the mineral. This new technique allowed mineralogists for the first time to see and identify mineral assemblages formerly invisible to the naked eye. It gave an enormous boost to petrology. Karl Rosenbusch used it to particularly good effect, summarizing the new results in his classic textbook, *Mikroskopische Physiographie der petrographisch wichtigen Mineralien* (1873).

In the nineteenth century, theories of the origin of rocks and minerals also became more sophisticated. Charles *Lyell suggested that besides volcanic, plutonic, and sedimentary rocks, geologists needed a fourth category, which he called metamorphic. These arose through transformation of the other classes by heat and pressure. While agreeing that gneisses, schists, and perhaps granite might be problematic, continental mineralogists continued to believe that water, perhaps under heat and pressure, perhaps containing many strong chemicals, was crucial to petrographic change. Carl Gustav Christoph Bischof summed up the state of the argument in what became the standard geochemical text, *Lehrbuch der chemischen und physikalischen Geologie* (1848). In Canada, Thomas Sterry Hunt made another stab at a chemical geogony. His theory was rejected.

At the end of the nineteenth century, mineralogists and petrologists found that new research in *thermodynamics, particularly that of J. Willard *Gibbs on the phase rule, offered an alternative way to think about mineral and rock origins. They began constructing phase diagrams for certain common rocks to clarify the sequence and manner in which the different crystals had formed. The founding of the Carnegie Institute of Washington in 1902, and the construction of a lavishly equipped laboratory there, aided this program of research. The United States could now compete with Germany. In 1928, Norman Bowen summed up then recent developments in his classic, *The Evolution of the Igneous Rocks* (1928).

The question of the origin of the rocks was pursued in the field as well as in the laboratory by two opposing camps that frequently compared themselves to the Neptunists and Plutonists of a century earlier. The minority camp, the migmatists, believed that migmatites (as they called the puzzling hard rocks of varied composition) were formed in place as circulating fluids converted extant rocks into something completely different. The majority camp, the magmatists, led by the Canadians Norman Bowen at the Carnegie Institute and Reginald Daly at Harvard, argued that they were intruded from reservoirs of molten magma beneath the earth's crust. Within this camp, heated debates raged about whether there was one magma or many, and whether magmas were homogenous or differentiated. There the matter stood at the beginning of World War II. Following the war, novel techniques in the laboratory and the field, including deep sea drilling, suggested new directions for mineralogical research.

See also Crystallography; Stratigraphy.

Karl Zittel, *History of Geology and Paleontology* (1962, for 1901). Rachel Laudan, *From Mineralogy to Geology* (1987). David Oldroyd, *Thinking About the Earth* (1996).

Rachel Laudan

MINING ACADEMY. Mining academies, university-level institutions for training managers of state mining monopolies, were first established in central Europe when the demand for silver increased following the Seven Years War (1756–1763). States constructed the facilities, paid the professors, and supported the local students, to whom they offered jobs on graduation (though foreign students paid fees). The curriculum included *mathematics, *chemistry, *metallurgy, *mineralogy, and mining techniques. Probably the first mining academy to be founded, and certainly the most famous, was the Bergakademie in 1765 in the town of Freiberg in Saxony where silver mining had been a major industry for five hundred years. Its fame derived from Abraham Gottlob Werner, who taught courses in practical mining, mineralogy, and geognosy (what we now call geology) from 1775 to 1817. He attracted students from across Europe and the Americas who made up the "School of Werner" or the "School of Freiberg," a description much more accurate than the term "Neptunism" later used to describe the school's emphasis on the role of water in shaping the earth. The Bergakademie included many of the important geologists of the early nineteenth century, most notably Alexander von *Humboldt and Leopold von Buch. It also attracted figures important in the Romantic movement—for example, the scientist Heinrich Steffens and the poet Friedrich von Hardenberg.

Other mining academies and institutions followed: in the Hapsburg empire, Schemnitz was founded in 1770; in Prussia, Berlin in 1770 and Clausthal in 1775; in Spain, Alamadén in 1771; in Russia, St. Petersburg in 1773; and in the Spanish empire, Potosí in 1780 and Mexico City in 1792. In Sweden and in Italy, special teaching posts were created within the universities, while the French set up an École Royale des Mines in 1783. Many of these institutions still exist, albeit sometimes under different names and auspices. They provided most of the jobs for geologists until the mid-nineteenth century when geological surveys began to offer an alternative.

England went its own way. Apart from tin, most mining there concerned coal and iron ore, widely distributed, fairly easy to find, and not requiring complex extraction, refining, and assaying techniques. Applied geology fell to "practical men" without formal education, while theoretical geology lay in the hands of independently wealthy gentlemen. Only in 1851 did the government found what would be called the Royal School of Mines to work in concert with the Geological Survey and the Museum of Practical Geology. The Royal School quickly turned into a training ground for scientists from all over the British Empire. In the United States, the Colorado School of Mines, its most distinguished mining school, was founded in 1870. Today, schools of mines, mining engineering, and petroleum engineering occur across the globe, together with associated journals, professional organizations, and meetings.

Apart from producing some individual institutional histories, historians of science, particularly outside the German-speaking world, have scarcely begun to study eighteenth-century mining academies as a general phenomenon. That hampers understanding of eighteenth- and nineteenth-century *geology, the Romantic interest in science, *cameralism and science, the history of education in applied science, and the institutionalization of science. The mining academies offer an instructive counter-example to the frequently heard claim that science flourished best in central metropolitan areas. Almost all the mining academies were located in mining areas far from the capital cities that housed leading academies and societies of science.

See also Humboldtian Science.

Rachel Laudan, *From Mineralogy to Geology: The Foundations of a Science* (1994).

Rachel Laudan

MINORITY GROUPS Minority religious or racial groups in a country or region have often experienced greater difficulty than the majority in cultivating natural knowledge and receiving recognition for their work. Sometimes, as with Huguenots in seventeenth-century France, Quakers in eighteenth-century England, Jews in nineteenth-century Germany, and Asians in twentieth-century America, minorities have done well despite, and in different measures because of, prejudice that allowed them to develop strengths favorable for scientific work. Though not a minority in the general population, women have been severely underrepresented in the sciences owing to a spectrum of beliefs now unsustainable in the Western world: that women have no capacity for abstract thought, cannot raise children while performing other exacting jobs, cannot work outside the home without damaging it, and so on (*see* WOMEN IN SCIENCE).

The history of the position and opportunities of minorities in Europe and the United States has not tended uniformly toward improvement. Huguenots, "tolerated" in France until 1685, then lost their civil rights; women figured more prominently in science during the eighteenth century than in the nineteenth, when increased professionalization of the various *disciplines disadvantaged them; and Jews have suffered from eruptions of anti-Semitism of which the Nazi attempt at a "final solution" was the most loathsome manifestation (*see* *NAZI SCIENCE). The ousting of the Jesuits from several European states during the 1760s and the suppression of the order between 1773 and 1814 removed one of the strongest contributors to science and scientific education of the early modern period.

To the difficulties in historical analysis presented by these and similar advances and retreats must be added the always controversial question of who or what constitutes a minority. In the case most frequently discussed, England in the later seventeenth century, analysis hinges on the definition of "Puritan." If defined loosely, "Puritan" permits the generalization that a minority dominated the pursuit of natural knowledge after the Restoration of Charles II. But a different and perhaps more plausible concept of the "Puritans" in question identifies them with the more numerous low-church Anglicans. How should Catholics whose works earned places on the Index of Prohibited Books be counted? Theoretical physics at certain times and places has been almost a Jewish monopoly. Should Gentile theorists then qualify as a minority group? This article ignores these questions to report more obvious inclusions and exclusions. In considering the twentieth century it makes use of a rough, limited, and problematic indicator of relative access to training, laboratories, and resources in science—the statistics of the Nobel Prizes.

Catholics and Jews

From the late seventeenth century until nearly the present, Roman Catholics have been underrepresented in science relative to their numbers in the general population. The studies of Robert K. Merton documented this situation for the early modern period; the work of Alphonse de Candolle did the same for the nineteenth century. In 1887 when Catholics exceeded Protestants by three to one in Europe, Candolle counted more Protestant than Catholic scientists of note. In the United States, until quite recently, the imbalance between Protestants and Catholics in science was larger than Candolle had found in Europe. On the eve of World War I, Catholics in the United States amounted to about 15 percent of total population but only 2 percent of the scientific community. Even at midcentury, the modest numbers of Catholics in science frustrated American Catholic intellectuals, especially as Catholics then made up nearly a quarter of the total population.

Europe produced prominent Catholic scientists even during the anticlerical nineteenth century: Gregor *Mendel, the founder of genetics; Theodor *Schwann, co-originator of the cell theory in biology; Louis *Pasteur. Many leading figures active in the first half of the twentieth century also came from Catholic families: Henri Becquerel, Louis de Broglie, and Peter Debye in physics, Georges Lemaître in astronomy, and many more. All of them worked in environments in which Catholicism, whether practiced or not, was the dominant religion. In the United States the Catholic minority did not do so well. Leaving aside distinguished émigrés, perhaps the most prominent Catholic scientist of American birth and education in this period was Edward Doisy. Born in 1893 and educated at the University of Illinois and Harvard, Doisy became professor of biochemistry at Washington University in Saint Louis, then at Saint Louis University. In 1944 he received the 1943 Nobel Prize for his research on vitamins K1 and K2 and synthesis of K2, which had significant therapeutic value.

The relatively meager performance of the Catholic minority in the United States may be explained by a combination of factors. Perhaps, as Merton argued, Protestants have particular incentives for scientific work arising from Calvinist asceticism, and, as critics of the Roman Catholic church asserted, the episode of *Galileo has continued to frame relations between it and the natural sciences. Certainly many Catholics in the United States have not had either the tradition of or access to the higher education needed for a career in science. Although conditions have changed greatly since World War II, during the nineteenth century and early twentieth centuries, American Catholics were disproportionately immigrants from rural areas in Europe.

Catholics confronted crippling disabilities in the educational systems of Protestant countries well into the nineteenth century. Catholics in Ireland under British rule had far fewer opportunities for education than Protestants well into the nineteenth century; Catholics in Britain could not obtain degrees at Oxford or Cambridge until after mid-century. Sweden did not allow Catholics even to reside in the country until 1860. But even in Catholic countries instruction in natural science and encouragement to pursue it were often inadequate. Santiago Ramón y Cajal, modern Spain's most eminent scientist, a founding figure in histology, and professor of medicine lastly at the University of Madrid (1892), often complained of the climate of "indifference" (sometimes verging on "actual hostility") toward science in his native country. Often condescended to by colleagues in countries where science was better developed, he made his work known through lectures in Germany in 1889. Those who heard him wondered that Spain could produce an investigator of his stature. In 1906 Ramón y Cajal shared the Nobel Prize in physiology or medicine with a colleague from the Catholic periphery, the Italian Camillo Golgi.

In sharp contrast to Catholics, Jews have been statistically overrepresented in science by every reasonable measure: professorships in research universities, memberships in academies of science, Nobel Prizes, and so forth. Albert *Einstein, by broad consensus the world's preeminent scientist of the twentieth century, was Jewish. Other prominent Jewish Nobel Prize winners in physics included James Franck and Gustav Hertz, who received their prize in 1925 for their confirmation of insights of Niels *Bohr (himself half-Jewish) governing collisions between electrons and atoms. After leaving Germany, Franck attracted additional notice for his work in the quite different field of *photosynthesis. There were also Otto Stern, Wolfgang Pauli (a baptized Jew), Felix Bloch, Max Born, Emilio Segrè, Maria Goeppert Mayer, Eugene Wigner, and Hans Albrecht Bethe, all permanent or temporary émigrés to the United States or Great Britain. They do not exhaust the list even in physics.

The usual explanation for the prominence of European and American Jews in science turns on the deep respect for formal education and learning more generally in the Jewish tradition.

With the opening of the ghettoes after the French Revolution and the Napoléonic wars, Jews could pursue careers previously closed to them. They succeeded in science even before obtaining full civil rights. Formal emancipation of Jews in Germany dates from 1869–1871. Yet Heinrich Gustav Magnus became professor of physics at Berlin in 1845 and M. A. Stern professor of mathematics at Göttingen in 1859.

The answer to the question "Who is Jewish?" determines the degree of Jewish overrepresentation. By the Nazi definition—the "blood" of a single grandparent would do—the degree would be immense. By the Jewish definition—having a Jewish mother—the figure would perhaps be too low. Elie Metchnikoff, born in Russia (1845) where full emancipation came only in 1917, had a Jewish mother but was not penalized by the czarist education system on that account. After taking his first degree from the University of Kharkov, Metchnikoff worked in Germany and France where he did the research on the immune system that brought him a Nobel Prize in medicine (with Paul Ehrlich) in 1908. Otto *Warburg (Nobel laureate in physiology or medicine in 1931) had a Jewish father; he nonetheless remained at his laboratory in Berlin throughout Hitler's rule, protected by Hermann Göring who declared that he decided who was Jewish. (Göring expected Warburg to find a cure for cancer.)

Although the United States was by no means free from anti-Semitism ("The United States is a Protestant country," Franklin Roosevelt once declared. "The Catholics and Jews are here on sufferance") the country's institutions absorbed many Jewish refugees from totalitarian regimes during the 1930s. Against the background of current sensibilities, one prefers to think of Roosevelt's statement as descriptive, rather than prescriptive.

African Americans, Asian Americans, and Affirmative Action

In the United States educational barriers and prejudice long handicapped descendants of former slaves who might have made careers in science. In 1876 Edward Bouchet became the first African American to receive a Ph.D.—in physics, from Yale. But as late as 1930 only a dozen other African Americans had received the Ph.D. in science; in 1943 the number was still only 119. Nonetheless, a few African Americans did manage to achieve distinction in science. Ernest Just was the son of a construction worker and a seamstress. After high school he won a scholarship to Dartmouth College and earned a Ph.D. from the University of Chicago in 1916. A marine biologist, Just studied fertilization in marine invertebrates and wrote on the biological basis of embryogenesis as well as on genetic aspects of differentiation in organisms. Denied the faculty appointment at a major research university for which his achievements qualified him, Just spent his teaching career at Howard University in Washington. Apart from early summers at Woods Hole, he did much of his work in Europe where he generally received easier acceptance than he did in the United States.

Just's younger contemporary Percy Julian was a chemist. His father, the son of a slave, was a postal worker deeply committed to his son's education. Julian graduated as valedictorian from DePauw University in 1916; after some delay he went to Harvard where he received a master's degree in 1923. Following a teaching appointment at Howard, Julian earned a Ph.D. at the University of Vienna (1931). The synthesis of naturally occurring, but expensive, chemicals with industrial and medical applications became his career-long focus; his first big success, physostigmine, was a muscle relaxant valued in the treatment of glaucoma. Eventually he set up his own firm, Julian Laboratories, where he managed to synthesize the sex hormones testosterone and progesterone.

During the past thirty years the presence or relative absence of minority groups in the scientific community and the academy has acquired great salience in the United States. *Science* magazine on behalf of the American Association for the Advancement of Science regularly reports on the progress of minority citizens in the sciences and other disciplines. Advertisements for professors stress the hiring institution's commitment to "affirmative action" and "equal opportunity." University presidents and deans proclaim the necessity of hiring and awarding tenure to more women, and to African Americans, Native Americans, and other minority groups. A product of the modern civil rights and feminist movements, affirmative action—the search for and nurturing of talent in minority groups—also derives from widespread popular reaction to the horrors of the Nazi period and World War II.

Many feel that changes to date have been modest. Among the 461 U.S. citizens who received the Ph.D. in mathematics in 1991, there were 10 African Americans, 6 Hispanics, and 2 Native Americans. In science and engineering, just 2.6 percent of Ph.D.s went to African Americans (10 percent of the U.S. work force), 1.8 percent to Hispanics (7 percent of the work force), and 0.4 percent to Native Americans (1 percent of the work force). The composition of university faculties tells the same story. Walter Massey, former director of the National Science Foundation,

professor of physics at Brown University, and an African American, noted that the University of Chicago (where he once taught) had seventeen African American professors in all fields in 1972 and only four more in 1992. During the same period Brown's complement of African American professors increased from fifteen to seventeen. By contrast, Asian Americans—0.2 percent of the work force in 1972—received 5 percent of the doctorates in engineering and science that year.

While students and professors from Asia have been a significant presence in American science and engineering for many years, commentary on their success was unusual until recently. Some question whether Asian Americans should be considered a "minority" in technical fields at all. Since the removal of racist-inspired restrictions on immigration from Asia after World War II, increasing numbers of students from the region have pursued advanced studies in the United States; and many have made significant careers here. Several Asian Americans have received the Nobel Prize.

Women's presence in American science, medicine, and engineering has undergone significant change. In 1978 only 9 percent of scientists and engineers in the United States were women; in 1988, 16 percent. Among university professors, the number of women holding the rank of full professor in natural science and engineering rose 120 percent between 1977 and 1987, compared to a 46 percent increase for men. At the highest levels of international achievement, however, there are still very few women.

Alphonse de Candolle, *Histoire des Sciences et des Savants*, 2nd ed. (1888). John Tracy Ellis, "American Catholics and the Intellectual Life," *Thought: Fordham University Quarterly*, 30:118 (Autumn 1955): 351–388. Santiago Ramón y Cajal, *Recollections of My Life*, trans. E. Horne Craigie (1969). Robert K. Merton, *Science, Technology and Society in Seventeenth-Century England* (1970). James M. Jay, *Negroes in Science: Natural Science Doctorates, 1876–1969* (1971). Margaret W. Rossiter, *Women Scientists in America: Struggles and Strategies to 1940* (1982). Kenneth Manning, *Black Apollo of Science: The Life of Ernest Everett Just* (1984). Willie Pearson, Jr., *Black Scientists, White Society, and Colorless Science: A Study of Universalism in American Science* (1985). Frank N. Magill, ed., *The Nobel Prize Winners: Physics, Chemistry, Medicine*, 9 vols. (1989–1990). National Science Foundation, *Women and Minorities in Science and Engineering* (1990). Robert Hayden, *7 African American Scientists* (1992). "Minorities in Science: The Pipeline Problem," *Science* 258 (13 November 1992): 1176–1232. Yakov Rabkin and Ira Robinson, eds., *The Interaction of Scientific and Jewish Cultures in Modern Times* (1995). Jack Morrell, *Science at Oxford, 1914–1939* (1997).

JAMES BARTHOLOMEW

MISSION. Historians have typically used "mission" in two senses so far as science is concerned.

In one case the word denotes a formalized attempt to obtain desired scientific information or intelligence by sending a delegation of scientists and others to a particular site or location. While missions in this sense have many of the same properties as "expeditions," they are generally more open-ended and less focused than the latter. Nearly all expeditions or missions have typically been sponsored by a large organization—a nation state, science academy, corporation, or church. Historical examples of this sort of mission include the Jesuit presence in China in the seventeenth and early eighteenth centuries, the French inspections of Vietnam in the early twentieth century, and several American scientific expeditions to Japan after World War II.

In the other case, "mission" signifies a narrowly defined, closely directed research project with a predetermined aim. Examples include visits by French scientists to Germany after the Franco-Prussian War to study the organization of university institutes; the dispatch of young Italians around 1930 to study methods and instrumentation in scientifically more advanced countries; and American and British excursions to Germany at the end of World War II to sieze information, apparatus, and individuals thought useful for military and civilian science and technology.

The Jesuit missions to China between 1583 and 1715, led initially by the Italian Matteo Ricci, introduced a substantial body of European science to China and, by some estimations, allowed Western and Eastern science to "flow in the same channel" for the first time. The well-recognized deficiencies of the Chinese calendar gave the Jesuits their opening. Despite imposing scientific achievements in an earlier era, the rules of calendar formation were no longer known in China by the late sixteenth century. Yet Confucian ideology connected heavenly events with political legitimacy, creating strong incentives to predict eclipses and otherwise demonstrate fundamental harmonies between the natural and political worlds. Ricci attracted much support at the Ming court (1368–1644) for the mission when he and his colleagues produced an accurate calendar of eclipses and other conspicuous celestial events. The mission was discontinued as a result of the Rites Controversy, a dispute among the Europeans and between the papacy and Chinese officials over the acceptability from a Christian perspective of ceremonies honoring the memory of Confucius.

European colonial ventures did not always assure timely introduction of modern science to regions not previously familiar with it. The experience of Vietnam during the nineteenth and early twentieth centuries illustrates one pattern. Over several decades, Vietnam came under the competing cultural influence of its traditional guide, China, and its new colonial master (after 1885), France. Because France was a major scientific power, the competition might have ended quickly in the rapid, efficient introduction of modern science to Vietnam. It did not. In 1902 the French established a "permanent" scientific mission to Indochina to inventory the natural wealth of the region. Staffed by three biologists, the purview of this mission extended only to botany and zoology; mining and agriculture were the objects of separate missions launched in 1903 and 1911 respectively. The botanical and zoological mission aimed at generating basic knowledge; the mining and agricultural, at producing practical results. While the Academy of Sciences in Paris had a part in defining the aims of the "permanent" scientific mission, it did not control the budget, and a myopic governor general of the colony abolished the mission in 1908. In 1917 another governor general revived it as the Scientific Institute of Indochina, but a successor abolished it for a second time in 1925.

French use of the "mission" concept in introducing modern science to Vietnam contrasts sharply with the substantially more effective efforts of the Dutch in the nearby East Indies (today's Indonesia). The Dutch rarely, if ever, used the terminology or explicit strategies of missions to promote scientific pedagogy or research in the region; they also relied heavily on private initiatives to promote their aims in science. Dutch scientific projects also benefited from a blatantly exploitative system of agricultural wealth extraction in the Indies; moreover, the Netherlands saved money by remaining neutral in World War I. By contrast, the enormous expenditures of the French in the same conflict impeded positive efforts on behalf of science in the financially troubled decade of the 1920s.

Although American observers had paid little attention to Japanese scientific and technical endeavors before 1940, the war brought a fundamental change of attitude; and several formal efforts to survey and evaluate Japanese science were commissioned after the surrender in August 1945. The most substantial of these was headed by Karl Taylor Compton, president of MIT and a veteran of the *Manhattan Project. Carefully planned and organized in the Philippines for several weeks before the war ended, the Compton mission consisted of five scientists, six military officers, three translators, two secretaries, and a chauffeur. Intended primarily to evaluate the wartime achievements of the Japanese in science and technology, the mission formally excluded nuclear research, military medicine, and aviation from its purview. In practice, the Compton team carried out a very broad inquiry, not wholly omitting the excluded fields reserved for later missions.

After 135 meetings with some 300 scientific informants, the Compton mission reached several important conclusions: the Japanese had received very little assistance from the Germans; possible breakthroughs had been thwarted by extreme lack of cooperation between the Japanese Army and Navy; the military services had made remarkably little use of academic research talent; poor organization had typified wartime research overall. Only in chemical warfare, meteorology, and rocket development did the mission consider that Japanese scientists had made significant gains during the war.

The concept of scientific mission might be extended to include goal-oriented war (or "defense") research done under pressure. Examples from the twentieth century include chemical weapons, advanced wireless communication technology, mathematical tables and calculating devices for improving the accuracy of artillery barrages, sonar, atomic weaponry, and rockets. The refinement of rocket science by the Soviet Union, marked by the launch of *Sputnik* in October 1957, resulted from a concerted push by the USSR to demonstrate its scientific acumen to the Western powers, especially the United States. Stretching the concept further, the migration or removal of Wernher von Braun and several colleagues who had worked on the V-1 and V-2 rocket programs during the war to the United States and the Soviet Union might be considered a mission in reverse. Their work in their new countries literally were missions. Soviet rocket engineer Sergei Korolev used the same V-1 and V-2 based knowledge to build the giant rockets that sent Soviet spacecraft (*Vostok*, *Voshkod*, and *Soyus*) into space. Von Braun followed a similar course in developing the American program, designing and building rockets that were the backbone of the Gemini, Mercury, and Apollo space missions.

See also DIFFUSION; PAPERCLIP OPERATIONS; SCIENCE AND WAR.

Pasquale M. D'Elia, SJ, *Galileo in China*, trans. Rufus Suter and Matthew Sciascia (1960). Lewis Pyenson, *Empire of Reason: Exact Sciences in Indonesia, 1840–1940* (1989). Lewis Pyenson, *Civilizing Mission: Exact Sciences and French Overseas Expansion, 1830–1940* (1993). Shigeru Nakayama, *A Social History of Science and Technology in Contemporary Japan*, vol. 1 (2001).

ROBERT K. STEWART

MODEL. Models are representations of one thing (the structure of an atom, the behavior of an airplane, the decision-making of a human being, the elevation of mountains) by another (the Bohr picture of the atom, a structure in a wind tunnel, a set of mathematical equation, or layers of clay under pressure in the laboratory). In pure science, models aid the prediction of new effects and the testing of theories. In applied science, they help decide whether to employ new structures, new machines, and new drugs.

The classic debate over the reliability and utility of models occurred in the nineteenth century. Physicists, including figures as eminent as James Clerk *Maxwell and William *Thomson (Lord Kelvin) employed mechanical models to understand physical processes such as the behavior of electromagnetic fields and gases. They followed a precedent going back at least to René *Descartes. He and other mechanical philosophers had tried to understand optical, mechanical, and physiological phenomena in terms of simple mechanical systems: billiard balls, for example, or pumps or pulleys (*see* MECHANICAL PHILOSOPHY).

In 1906, the French physicist and historian of science Pierre Duhem launched a major attack on model-making (*La théorie physique: son object, sa structure*). He judged it to be "broad but shallow," oversimplified, unlikely to lead to new discoveries (and also typically English). He believed that in many cases, scientists had to limit themselves to abstract mathematical analysis of physical systems even if in consequence this meant the systems could not easily be visualized.

Duhem's attack on models came at a time when scientists were deeply divided about the physical reality of atoms. By and large, realists, who thought it possible to comprehend the deep structure of the universe, were willing to use models, while instrumentalists, who agreed with Duhem that the best that could be hoped for were laws and classifications, eschewed them.

Duhem and his supporters lost the battle with the model makers. In physics, the effort to understand the subatomic world in the 1920s and 1930s led to the wave and particle models of *quantum physics, both of them inadequate to capture fully what mathematical theories pointed to, yet at the same time useful props for comprehending these abstractions. Biologists began thinking about the genetics of populations in terms of mathematical models. Geologists reasoned about the crumpling and folding of mountain chains via physical models of layers of clay and sand subjected to pressures. Medicines are approved on the assumption that mice and rats make good models for human beings. Computers simulate one natural system after another—from the weather to the most intricate workings of the human body. Scientists are unlikely to abandon such a useful method.

Duhem's warnings, though, retain their force. In terms of scientific method, models depend on analogies. If scientists want to find the unknown cause of some effect (or the unknown effect of some cause), they look for a known situation with a similar cause and effect. From the three knowns, they infer the unknown effect (or cause). How similar must the elements be to support a reliable inference? The more dissimilar the model or analogy from the thing being modeled, the more fallible is the inference. Though essential in science, models do not guarantee certain knowledge.

See also REDUCTIONISM.

Pierre Duhem, *The Aim and Structure of Physical Theory* (1954, for 1906). M. B. Hesse, *Models and Analogies in Science* (1963).

RACHEL LAUDAN

MODERNITY AND POSTMODERNITY historical eras whose definitions rest on different conceptions of science, technology, and reason and their roles in the historical process. The term "modern" first appeared in the sixteenth century as a way of signifying the separation of the present from the past. Although in the seventeenth century experimental natural philosophers distinguished themselves from past practices through the debate between the "ancients" and the "moderns," it was Georg Wilhelm Friedrich Hegel who provided the first formal definition of modernity as the persistent re-creation of the self and the conditions of life. Since Hegel, scholars have generally defined the period of modernity from the French Revolution in 1789 to the end of World War II. In 1947 architects defined a new "postmodern" style and in the later stages of the Cold War philosophers and sociologists articulated the meaning of postmodernity.

After Hegel modernity referred to two "master" narratives: the unfolding of the Enlightenment promise of emancipation through reason, and the unification of all forms of knowledge through the extension of the rational means of science and technology. Jürgen Habermas has maintained that both narratives have been used to legitimate science and technology in the social realm since 1800. Science, for theorists of the modern, is a representational practice, a reproduction of nature, whose features are the result of rational consensus among practitioners who establish objective, scientific truths untainted by social interests. Since consensus implies the absence of strife, in the modernist point of view science is a rational instrument of peace. Some scholars, like Habermas and Bruno Latour, believe that

modernity is an unfinished project still worth pursuing in order to achieve a conflict-free society where rational and clear communication is possible. Projects in the 1960s of the industrial world to modernize developing countries by applying principles of the social sciences and exporting technologies in order to stimulate the growth of capitalism expressed this belief. From the modernist perspective, the history of science and technology falls securely within the teleological narratives of progress, rationalization, secularization, bureaucratization, and the nation-state.

The origins of postmodernity lie in the critique of reason by Friedrich Nietzsche and others at the end of the nineteenth century. But not until after the atrocities of the Third Reich did many come to doubt seriously modernity's project of emancipation and unification through reason. Postmodernism rejects not only the two master narratives of modernity, emancipation and unification through reason, but also the idea that representation (in the arts and the sciences) can be an accurate mirror of objective reality. Postmodernists like Jean-François Lyotard reconceptualize the sciences as activities oriented toward the reproduction of the practices of research rather than the production of results about the world. (In this postmodernists share a perspective similar to Max Weber's *The Protestant Ethic and the Spirit of Capitalism* [1904–1905].) What Lyotard calls "performativity" in the sciences may be seen in the penchant for interdisciplinarity and teamwork in the sciences, both of which result in new ways of doing research. In the incessant search for the "new"—the invention of new vocabularies, practices, and rules of research, consensus, and investigation—postmodern science (such as the science of fractals) is a form of knowledge that does not reproduce the known, but that constantly seeks the unknown and indeterminate. The antirepresentational character of postmodern science means that images (and theories) do not stand for nature, but rather project the political, social, and economic environments in which scientific activity is found. Nature itself disappears in a created world of simulacra and the history of the sciences becomes the history of images and of what is believed embedded in them.

In postmodernity the sciences remain a part of the economy, as in the capitalist phase of modernity, but in a different way. Especially since the 1980s, the industrialized world has become a knowledge-based economy in which the production of scientific and technological information has become more important than the manufacture of material goods. This new capitalism, based on information technologies, places a high priority on industries that can increase the production and circulation of scientific knowledge. Hence greater cooperation between universities and industry has occurred in the industrialized world after World War II. Prominent examples include Silicon Valley, Route 128 in the Boston area, and Measurement Valley in the state of Lower Saxony in Germany.

See also MARXISM; SCIENCE, HISTORIOGRAPHY OF.

Daniel Bell, *The Coming of Post-Industrial Society* (1973). Jürgen Habermas, *Legitimation Crisis* (1975). Jean-François Lyotard, *The Postmodern Condition: A Report on Knowledge* (1984). Jürgen Habermas, *The Philosophical Discourse of Modernity* (1987). Fredric Jameson, *Postmodernism or the Cultural Logic of Late Capitalism* (1991). Bruno Latour, *We Have Never Been Modern* (1993).

KATHRYN OLESKO

MOHOLE PROJECT AND MOHOROVIČIĆ DISCONTINUITY.

The purpose of the Mohole Project (1957–1966) was to drill through the earth's crust to the Mohorovičić discontinuity, the seismic interface between the earth's crust and mantle. This boundary was discovered in 1909 by Yugoslav geophysicist Andrija Mohorovičić who noted that seismic waves returning from depth indicated there was a zone of abrupt change in the speed of seismic waves some kilometers below the earth's surface. Called Moho for short, this zone defines the base of the earth's crust and marks a change in composition. The depth of Moho varies from about 25–40 km (15–25 mi) beneath the continents, to 5–10 km (3–6 mi) beneath the ocean floor. Recently, following the acceptance of plate tectonic theory, geoscientists have decided that changes in deformational behavior are a more significant aspect of the earth's structure than changes in composition. They divide the outer earth into the rigid lithosphere (crust and upper mantle) overlying the more deformable asthenosphere.

Project Mohole was the brainchild of AMSOC, the American Miscellaneous Society, an informal group of geoscientists formed in the 1950s. Prior to this, crustal drilling was primarily undertaken to explore for oil and gas and had been limited to land and shallow water. Drilling for scientific purposes was much less common. It had begun with efforts to determine the structure, composition, and history of coral islands. In 1877 the Royal Society of London sponsored a borehole that went down 350 m (1,140 ft) on Funafuti in the South Pacific. In 1947, pre-bomb-test drilling of Bikini reached 780 m (2,556 ft). In 1952 drilling on Eniwetak finally reached basaltic crust beneath coralline rock at a depth of over 1,200 m (4,000 ft), still well short of the Moho. In

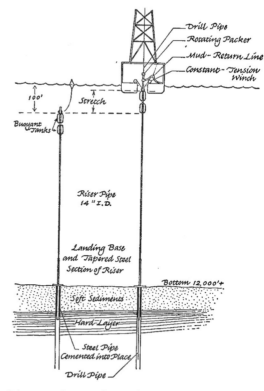

Schematic diagram for producing a hole in the bottom of the sea. The riser pipe floats on the submerged buoys when the drilling rig is not on station.

the 1950s, some countries, including Canada and the Soviet Union, proposed drilling deep holes in continental crust.

Project Mohole, funded by the U.S. National Science Foundation, was designed to drill in the deep sea. The technological challenge was to drill the deep-ocean floor in water depths of thousands of meters. Although Mohole successfully drilled cores in water depths of 950 and 3,560 m (3,111 and 11,672 ft), it did not come close to reaching the Moho. Mohole, commonly seen as one of the first big-science projects in the United States, was terminated by Congress in 1966 and widely considered to have been a failure.

Nonetheless the project demonstrated that ship-based ocean drilling was feasible. It helped spawn the highly successful Deep Sea Drilling Project, begun in 1968, to drill through cover sediments on the ocean floor. Since then scientific deep-ocean drilling has become an international endeavor. Oil companies are reaching deeper and deeper objectives. Moreover, a number of countries and consortia have developed continental deep-drilling projects. None of this drilling has reached the Mohorovičić discontinuity, however.

Even so, scientists now believe that we can observe the Moho on land. By the late nineteenth century, a number of European scientists had recognized that in the Alps uplifted oceanic crust was represented by layered chert (lithified deep-sea sediments) overlying basalt (oceanic crust) overlying ultramafic rocks (high-density rocks rich in iron and magnesium), the so-called Steinmann trinity, after Gustav Steinmann. This package of rocks is called an ophiolite, and the transition from basalt to ultramafic rocks is believed to be the crust-mantle boundary—the Mohorovičić discontinuity.

Willard Bascom, *A Hole in the Bottom of the Sea: The Story of the Mohole Project* (1961). Elizabeth N. Shor, "A Chronology from Mohole to JOIDES," in *Geologists and Ideas: A History of North American Geology*, Ellen T. Drake and William M. Jordan, eds., (1985): 391–399.

JOANNE BOURGEOIS

MOHOROVIČIĆ DISCONTINUITY. See MO-HOLE PROJECT AND MOHOROVIČIĆ DISCONTINUITY.

MOLECULAR BIOLOGY. The second half of the twentieth century witnessed a transformation in our understanding of certain key mechanisms fundamental to life. These mechanisms concern the biological phenomena of *heredity and development and the chemical processes of the synthesis of *proteins and nucleic acids. The techniques developed to investigate these phenomena and the conceptual structure into which they have been fitted constitute the subject known as molecular biology. The former can be likened to a toolkit and the latter to a manual. They have been put to work in numerous areas of biology from phylogeny and population *genetics to *immunology and neuroscience. With advances in these techniques have come applications to *pharmacology (designer drugs), in agriculture (recombinant strains of crop plants), and in forensic science (DNA fingerprinting).

More than a toolkit and manual, molecular biology represents a combination of approaches of the chemist, biochemist, geneticist, and microbiologist, and offers a unifying conceptual structure provided by the mechanism and genetic determinants of protein synthesis. It can hardly be called a discipline, however, because its techniques have been absorbed by biochemists and geneticists, by embryologists, immunologists, and ecologists, and their disciplines have not blended. The practice of molecular biological techniques in so many areas of biology has made it a common currency in the many areas of the science of life.

Since all living things are constituted of molecules, and biochemists have always studied the behavior of molecules, why has molecular biology stirred up so much adverse comment?

As Erwin Chargaff once quipped, a molecular biologist is a biochemist practicing without a license. Molecular biology is interdisciplinary to an extent that biochemistry never was before biochemists faced those who were to call themselves molecular biologists. The approach of the biochemists, dominated by their concern to unravel the nature of *metabolism, its pathways and energetics, rarely included genetics, and paid only occasional attention to structural crystallography. And among chemists, those studying natural products concentrated on the proteins and the carbohydrates, leaving the nucleic acids to a mere handful of researchers. The change in biochemistry came when scientists accepted the notion that the genetic material consists of nucleic acid, not protein, and that the nature of a protein depends on the nature of the nucleic acid in the presence of which it is formed.

But again, had not biophysics a long history before 1950? Biophysicists had been very active in the study of the nerve impulse, the behavior of membranes, and the effects of ionizing radiations, and, in conjunction with physical chemists, they had investigated large molecules by ultracentrifugation and *electrophoresis. But as a group, biophysicists concentrated on technique and did not immerse themselves in the conceptual problems of biology. Thus, when *deoxyribonucleic acid (DNA) rose to prominence as the stuff of the genes, those involved gave their work a new title to distinguish it from that of the biochemists and biophysicists around them. This they achieved through such instruments as the *The Journal of Molecular Biology* published by Academic Press beginning in 1959, the renaming of the Medical Research Council Unit for the study of the Molecular Structure of Biological Systems in Cambridge as the MRC Laboratory of Molecular Biology (1956), and the publication of James D. *Watson's *The Molecular Biology of the Gene* (1965).

The Festschrift to the German physicist-turned-biologist, Max Delbrück, *Phage and the Origins of Molecular Biology* (1967), pointed to the work of the group formed around Delbrück to study the process of replication in bacterial viruses as the source of molecular biology. This book provoked a response from John Kendrew in *Scientific American* pointing out that the phage group was one source, but not the only one. Structural x-ray crystallography of the proteins was the other. He dubbed the former the "Informational School" and the latter the "Structural School," using the term *school in a loose sense to refer to a network of researchers who, though scattered, constitute specialist communities. By the time Watson's enormously influential fragment of autobiography, *The Double Helix*, appeared in

1968, the three elements—phage, structural chemistry, and the addition of physics to biology—had become the distinguishing features of the popular history of molecular biology.

This historiography has been criticized. The salience given to the physicists, it is argued, exemplifies the construction of history by the actors to legitimate their importance. True, the search for a parallel to the physicists' complementarity principle in biology led nowhere, and the solution to the mysteries of the gene—how it duplicates, maintains its constancy, mutates, and is expressed—came from a combination of chemistry, genetics, and cytology, not physics. But indirectly, physics informed the disciplines of x-ray crystallography and physical chemistry by establishing the nature of the chemical bond, and offered a model in Delbrück's career for collaborative research among the phage workers. Also, physicists helped with funding; for example, Sir John Randall used the influential position he had gained for his war work to mastermind the establishment of the first research unit of the UK's MRC given to the biophysical study of the cell. Revisionist histories also complain that too much emphasis has been placed on the influence of the Austrian physicist Erwin *Schrödinger's book, *What Is Life?* (1944) in drawing physicists and chemists to biology; but it is difficult not to accept the testimony of Francis Crick and Maurice Wilkins who were inspired by the eloquent appeal of *What Is Life?* to explore the remarkable properties of the gene, which, like an aperiodic crystal, so faithfully replicates its structure.

The conceptual structure that constitutes molecular biology consists first of the assertion that proteins arise through an interaction between certain proteins and certain nucleic acids. They owe their shape, long held to determine their specificity, to the folding of the long polypeptide chains that compose them. The manner of folding is determined by the specific sequence of the building blocks or "residues" (amino acids) along the chains, not, as formerly suggested, by the presence of another molecule to which they adopt a complementary shape. The sequence does not arise, as many had assumed, from the sequential action of a battery of proteolytic enzymes, but to the sequence of residues or bases on the polynucleotide chain of a nucleic acid (DNA) in the cell's nucleus acting as a template. The sequence represents a code or cipher for the amino acids of the protein being synthesized. It is copied from the DNA of the chromosome in the nucleus onto a "messenger" ribonucleic acid (mRNA). Passing out into the cytoplasm, the mRNA determines the sequence of amino acids in the protein being synthesized. DNA makes RNA make protein.

In 1957, Crick codified this picture in an address to biologists entitled "On Protein Synthesis." He stated two general principles: *the sequence hypothesis*, "the specificity of a piece of nucleic acid is expressed solely by the sequence of its bases ... this sequence is a (simple) code for the amino acid sequence of a particular protein," and the *central dogma*, "once information has passed into protein *it cannot get out again*...the transfer of information from nucleic acid to protein may be possible, but transfer from protein to protein, or from protein to nucleic acid is impossible" (Crick, 1958). Crick defined carefully his use of the term "information": "the *precise* determination of sequence, either of bases in the nucleic acid or of amino acid residues in the protein." This understanding of the term implied a functional meaning in contrast to the purely syntactic meaning found in information theory.

In the decade following Crick's address, molecular biologists addressed the major features of protein synthesis and discovered all parts of the code relating nucleic acid and protein sequences. A sense of triumph marked the completion of this first phase of molecular biology extending from the 1950s to the mid-1960s. The emerging picture was simple. The information in DNA determines the information in the proteins according to a universal code. Since all that remained was filling in the details, many of the leaders among the molecular biologists began to look elsewhere for fresh problems to solve. Those who stayed the course engaged in a surprising and revealing task.

Exceptions were soon reported, both real and apparent, to the universality of the genetic code, to the sequence hypothesis, and to the central dogma. The concept of the *gene, once unambiguous, became a term with multiple meanings as molecular tools dissected it in varied ways. With the introduction of the techniques of recombinant DNA in the 1970s came novel methods to manipulate the genetic material and to reveal the multiple levels of control of gene expression possessed by the cell. The discovery in 1970 of enzymes (reverse transcriptases) that copy RNA base sequences back into DNA upset the complacent assumption that information only flows from DNA to RNA—a direction allowed for in Crick's statement of the *Central Dogma*, but widely assumed not to be possible. The revelation in 1977 that often only parts of the message (mRNA) from the DNA is translated into protein destroyed the simple one-to-one relationship between a DNA sequence and a polypeptide chain. Other discoveries revealed that much of the chromosomal DNA of higher organisms does not carry the information to determine a given polypeptide chain, but has some other function, gene regulation being one.

This second phase of molecular biology lasted from the 1970s to the present. In several respects it is a transformed version of the first phase. In the past, critics accused molecular biologists of adopting a too-reductionist approach, but the multi-level picture of the system of gene expression that has emerged in the second phase of the subject makes this accusation hard to justify. The most significant intellectual outcomes of the current phase of the subject relate to evolution and development. On the one hand, the extent to which the molecular machinery of the cell has been conserved in evolution has surprised everyone. On the other, the attribution of so much of biological diversity to the variety of ways in which the same genetic equipment is expressed has directed increasing attention to the genetic control of development. The recent success in sequencing the human genome, and those of several model organisms, has strikingly emphasized these points.

This more sophisticated molecular biological science has in the last three decades of the twentieth century been put to work in many fields, in none more dramatically than embryology. By 1980, all knowledgeable people recognized that the tools of recombinant DNA technology were opening up methods for the manipulation of genetic material of considerable commercial promise to industry. With these techniques, organisms and their products could be designed for specific purposes. No longer did humans have to wait for nature to turn up the right genetic combination through long, continued breeding. Parasexual processes came to complement or replace sexual processes and to bridge the chasm separating genera. Such power to produce trans-genic organisms has led to accusations that molecular biologists are "playing God" and endangering nature. The development in 1985 of the polymerase chain reaction (PCR) has made possible the amplification of minute quantities of DNA—even a single copy—a technique of great versatility for medicine and science. Thus, from an esoteric exploration of the nature of the gene, its message, and its product, molecular biology has moved into the marketplace, and now plays a significant part in the economic and social culture of the twenty-first century.

Francis H. C. Crick, "On Protein Synthesis," *Symposium of the Society for Experimental Biology*, 12 (1958): 138–163. James D. Watson, *The Double Helix: Text, Commentary, Reviews, Original Papers. A Norton Critical Edition* (1980). Pnina Abir Am, "Themes, Genres and Orders of Legitimation in the Consolidation of the New Scientific Disciplines: Deconstructing the Historiography of Molecular Biology," *History of Science*, 23 (1985): 73–117. Robert Olby, "The Molecular Revolution in Biology," in *Companion to the History of Modern Science*, R. C. Olby, G. N. Cantor, J. R. R. Christie, and M. J. S. Hodge, eds., (1990): 503–520. Robert Olby, *The Path to the Double Helix*, Dover edition (1994). Horace Judson, *The Eighth Day of Creation: Makers of the Revolution in Biology*, expanded edition (1996). Hans-Jörg Rheinberger, *Towards a History of Epistemic Things: Synthesizing Proteins in the Test Tube* (1997). Michel Morange, *A History of Molecular Biology* (1998). Soraya de Chadarevian, *Designs for Life. Molecular Biology after World War II* (2001).

ROBERT OLBY

MOLECULE. See ATOM AND MOLECULE.

MONOD, Jacques Lucien. See JACOB, FRANCOIS, AND JACQUES LUCIEN MONOD.

MONSTERS. From the earliest times, the learned and the unlearned have believed in monsters. For many, they made up a necessary part of natural history, either representing punishment for sin or the fecundity and diversity of creation, and blended seamlessly with religious and mythological beliefs in devils, spirits, ancestors, and legendary beasts. Monstrous births, both human and animal, particularly inspired fear. Curious natural objects such as strangely shaped horns, fossils, skins, and minerals, and human artifacts such as gravel stones from the bladder, were all subjects for allegory and commentary. Folklore, cultural history, medicine, morality stories, and scientific curiosity mingled together, as reflected in works by Albertus Magnus.

With the revival of learning in the West during the Renaissance, and the constant disclosure of new lands and new species, almost anything seemed possible. The principle of plenitude—belief in a world without gaps—endorsed this view. Encyclopedists such as Konrad Gesner took the all-inclusive approach, covering fossils, coins, minerals, animals, plants, and monstrosities. Ulisse Aldrovandi in Bologna published an authoritative thirteen-volume encyclopedia that included emblematic natural history—for example, the lamb as a symbol of Christ. These authors made much of the symbolism of monstrosities as punishments for human sin, or as the result of curses, or as creatures with magical properties. These books were followed by treatises by learned writers such as Athanasius Kircher. Museums and collections of curiosities in which monstrosities took a central, edifying role became popular throughout Europe, often enjoying courtly patronage; Kircher's museum at the Collegio Romano in Rome drew visitors from around Europe. Pygmies, mermaids, crocodiles, mandrakes, deformed fetuses, and other natural marvels were displayed and discussed in many similar places throughout the sixteenth and seventeenth centuries.

During the eighteenth century, and particularly during the Enlightenment, it became common for philosophical naturalists to think of life as a fluid, developmental sequence. In the learned world, monstrosities lost much of their moral symbolism for most people, and came to be regarded as extremes or variations of normal development that could indicate useful facts about morphology. An exception was Carl *Linnaeus, whose belief in the fixity of species did not permit fluidity. For Linnaeus, a deformity violated the natural plan. Monsters, in his view, could not replicate themselves. Étienne Geoffroy Saint-Hilaire coined the term "teratology" in 1822 to denote the experimental study of embryological malformations. He believed that the inducement of abnormalities under controlled conditions would reveal much about normal morphology. Malformations in chicks were further investigated by P. L. Panum in Berlin. Robert Chambers, the anonymous author of the evolutionary tract *Vestiges of the Natural History of Creation* (1844), regarded monstrosities as an important guide to the evolutionary past, tentatively suggesting that developmental surges along particular lines might lead to the birth of new species. Chambers himself had six fingers. Charles *Darwin also interested himself in aberrations and variations, studying pigeon squabs and abortive plants.

Curiosities remained a significant part of popular culture to the middle of the twentieth century. During the nineteenth and twentieth centuries, however, the deliberate inducement of aberrations for research purposes—for example, the creation of strains of special laboratory animals such as the genetically pure oncomouse—was discreetly veiled. Toward the end of the twentieth century, however, with increased freedom of information and public concern over biological manipulations of life, these practices became more commonly known and regulated.

Katherine Park and Lorraine Daston, "Unnatural Conceptions: The Study of Monsters in 16th and 17th Century France and England," *Past and Present* 92 (1981): 20–54. Marie-Helene Huet, *Monstrous Imagination* (1993). David Williams, *Deformed Discourse: The Function of the Monster in Mediaeval Thought and Literature* (1996). Jan Bondeson, *A Cabinet of Medical Curiosities* (1997).

JANET BROWNE

MOON. Since classical Greek times astronomers have been occupied with the orbital characteristics of the Moon, with its proposed epicycles, eccentrics, and, after Johannes *Kepler, ellipses. Concern with lunar dynamics extends to the present. Astronauts placed reflectors on the Moon's surface so that its orbital characteristics could be measured to a factor of inches using laser beams.

The study of the Moon as a physical world began in 1609, when *Galileo Galilei first observed its mountains and "maria" (seas) through a telescope, and announced his findings in *Siderius nuncius* (1610). Since classical times, it had been assumed that the Moon was a perfectly smooth, albeit tarnished, sphere: its tarnish, like its phases, perhaps deriving from its pivotal location at the boundary between the mutable "sublunary," or earthly, realm and the "superlunary" realm of eternal perfection, as in Aristotle's *De caelo*. Galileo interpreted his lunar discoveries (as he also did his planetary and sidereal discoveries) against *Aristotle. For if the Aristotelians had incorrectly understood the surface of the Moon, could they not also be wrong regarding geocentric cosmology? In his measurement (by shadows) of the height of the lunar Apennine Mountains (named after their Tuscan counterparts), and of certain crater formations, Galileo demonstrated the Moon to be a world with topographical features. (Galileo's lunar observations may have been preceded slightly by those of Thomas Harriot and Sir Richard Lower, who also recorded mountains, "seas," and craterlike formations, but neither of them published their findings, nor did they use them to advance a particular scientific argument.)

For almost forty years after Galileo's announcements few further discoveries were made in selenography (from Selene, the Greek lunar goddess), largely owing to the inability of opticians to make better lenses capable of showing additional detail. Meanwhile popular science writers such as John Wilkins, William Godwin, and Cyrano de Bergerac speculated about lunar voyages and moon men.

Johannes Hevelius's *Selenographia* (1647) reopened lunar study. Using a telescope with a 3.5 centimeter (1.5 in) aperture convex object glass, ground to a 3.6-meter (12-ft) radius curvature, Hevelius produced a series of drawings of the lunar surface under changing conditions of illumination, which vastly improved on Galileo's sketches. Developments in glass-making and the development of lens-polishing machines after 1650 opened up new technological potential. Giambattista Riccioli (1651), Gian Domenico Cassini (1679), and others produced maps displaying remarkable and faithful detail. Riccioli introduced the custom of naming lunar craters after eminent scientists.

During most of the eighteenth century, the Moon's surface received little attention. It had been mapped in great detail and seemed to display no change. Instead, eighteenth-century astronomers devoted themselves to the Moon's dynamical behavior (*see* CELESTIAL MECHANICS). Isaac *Newton's gravitational theory of the complex lunar motion, building on the work of Jeremiah Horrocks in the 1630s, became one of the enduring achievements of human intellect. Of special importance was Newton's demonstration of how the combined lunar and solar attraction produced two terrestrial tides per day.

Newton showed that water particles on the moonward and sunward side of the earth at new moon would be more powerfully attracted than the particles on the opposite side. Instead of producing one great water-bulge on one side of the earth, therefore, the most distant, least-attracted particles would produce their own lesser bulge, giving an egg-shaped ocean mass surrounding the earth. The earth's continents would dip into each of these water bulges once a day, to produce high and low tides.

Johann Schröter revived detailed selenographical mapping in Germany around 1790 by using large-aperture reflecting telescopes giving high magnifications. The new fascination with the Moon as a geological body from the 1830s further stimulated interest. Were the craters and "seas" the products of a once geologically active Moon? Did the supposed disappearance of the Crater Linné in 1866 indicate that a central heat might still be present? Those who sought the answers, including Schröter, Wilhelm Beer (who employed the professional Johann von Mädler), Wilhelm Lohrmann, Johann Krieger, James Nasmyth, and the members of the short-lived British Selenographical Society, were "Grand Amateurs." Nasmyth, an iron master by profession, attempted to replicate lunar features in the blast furnace, much as Robert *Hooke had tried to replicate crater formations by blowing air through molten alabaster and dropping bullets into pipe-clay. Nonetheless the nineteenth century thought that craters were *not* meteoritic in origin.

Lunar study in the twentieth century was overwhelmingly dominated by space exploration. In 1959, the Russian spacecraft *Luna 3* succeeded in photographing the "dark" side of the Moon; the pictures revealed a difference in the distribution of crater and "maria" from that seen on the side visible from the earth. Then, both before and after the American Apollo 11 mission's first landing of astronauts on the Moon in 1969, a succession of manned and robotic expeditions yielded a wealth of new information about lunar geology, magnetism, orbital characteristics, and much besides. This recent work has now itself entered the historiographical canon of astronomy. From Galileo's *Siderius nuncius* down to the present day the Moon, in its scientific, poetic, and science-fiction aspects, has probably generated a larger body of literature than any other single astronomical body.

Robert Grant, *History of Physical Astronomy* (1852). James Nasmyth and James Carpenter, *The Moon, Considered as a Planet, a World, and a Satellite* (1874). Ewen A. Whitaker, *Mapping and Naming the Moon: A History of Lunar Cartography and Nomenclature* (1999). Scott L. Montgomery, *The Moon and the Western Imagination* (2000).

ALLAN CHAPMAN

MORGAN, Thomas Hunt (1866–1945), zoologist and geneticist.

Thomas Hunt Morgan was born into a prominent family in Lexington, Kentucky. His father, Charlton Hunt Morgan, served as a U.S. consul at Messina, Italy; his mother, Ellen Key Howard, was the granddaughter of Francis Scott Key, the composer of America's national anthem.

Like many zoologists of his generation, Thomas Hunt Morgan displayed an interest in natural history as a child. He enrolled at the State College (now University) of Kentucky and graduated with a degree in *zoology in 1886. He continued his studies at Johns Hopkins University and at the Marine Biological Laboratory in Woods Hole. He completed his doctoral research in 1890 with a project on sea spiders. From 1890 to 1904 he taught at Bryn Mawr College; from 1904 until 1928, at Columbia University; and thereafter at the California Institute of Technology. He had four children with his wife Lillian Vaughan Sampson, a talented cytologist, who had been his student at Bryn Mawr.

Morgan's early work was in classical morphology, which dominated zoology until the turn of the century. Increasingly dissatisfied with the field's imprecision, Morgan turned to the more experimental studies of biology under the influence of Jacques Loeb, his close associate at Bryn Mawr, who sought understanding of life in terms of strict mechanistic principles. Morgan also collaborated with Hans Driesch and other experimental embryologists at the marine station in Naples, Italy. By 1895 Morgan's work shifted to experimental embryology of marine organisms like the comb jellies, *Ctenophora*. He then took up evolution and the newer science of *heredity, which was flourishing thanks to the rediscovery of the work of Gregor *Mendel. Morgan became sharply critical of evolutionary studies because they lacked experimental rigor, and devoted himself to elucidating the principles of heredity. His first genetic study was the cytological examination of sex determination.

In 1908 Morgan began the work for which he was to become famous: the genetic study of the common fruit fly, *Drosophila melanogaster*. The work began with his discovery of a change in eye color (from red to white) of one male fly. After selectively crossing white-eyed males with red-eyed sisters, Morgan discovered what he called the phenomenon of sex-linkage: only the males displayed the white-eye pattern. The phenomenon of sex linkage confirmed that the material carriers of heredity resided in the chromosomes and that the particular genes for the character of eye color were linked to, or carried on, the X chromosome in flies.

Morgan's project grew in scope as he discovered more small mutations that he could follow through successive generations. To perform this laborious work, he enrolled a talented group of young researchers who would form the classical school of modern *genetics: Calvin Blackman Bridges, Hermann Joseph Muller, Alfred Henry Sturtevant, L. C. Dunn, and Theodosius *Dobzhansky. The work of the group took two primary directions: determining the mutant genes and mapping their positions on the appropriate chromosomes. The group's major accomplishment was the confirmation that chromosomes bore the material carriers of heredity, the genes, in linear fashion in specific locations. Further research into *Drosophila* genetics disclosed the mechanisms responsible for sex determination and the effects of the location of a gene on its expression.

Morgan, Bridges, Sturtevant, and Muller summarized their pathbreaking work in *The Mechanism of Mendelian Heredity* (1915). It presented their evidence that the genes resided linearly on the chromosomes, and that the Mendelian principles could be observed operating at the cytological, and therefore chromosomal, level. It also showed how to determinate the location of genes and how gene changes could lead to mutations.

The year before the publication of *The Mechanism*, Morgan aired his criticism of the methodology of evolutionary studies in *A Critique of the Theory of Evolution*. He revised and republished it as *Evolution and Genetics* (1925) when he realized that his work on eye color supported Darwinian *evolution, which postulated that evolutionary change took place at the level of such small, individual differences. In 1934, Morgan attempted to bring his knowledge of the mechanism of Mendelian *heredity to bear on problems of development or embryology in his *Embryology and Evolution*, but his colleagues did not think his synthesis successful.

Morgan set up a new *Drosophila* group at the California Institute of Technology in 1928. Although he did not move into newer areas emerging in the early 1940s that questioned gene function (and not just location and behavior), Morgan—who had a reputation for helpfulness—was active in supporting the research efforts of geneticists like George Wells Beadle and Max Delbrück.

Morgan's pioneering studies on the fruit fly made him the leading American geneticist of his generation. In introducing experimental and quantitative methods to the study of heredity, embryology, and evolution, he exerted a powerful transformative effect on the whole of biological sciences. In recognition for this work, Morgan received almost every award granted in the life sciences including, in 1933, the Nobel Prize.

Alfred Henry Sturtevant, *Biographical Memoirs: National Academy of Sciences* 33 (1959): 283–325. Ian Shine and Sylvia Wrobel. *Thomas Hunt Morgan: Pioneer of Genetics* (1976). Garland Allen, *Thomas Hunt Morgan: The Man and His Science* (1978). Elof Axel Carlson, *Genes, Radiation, and Society: The Life and Work of H. J. Muller* (1981). Robert E. Kohler, *Lords of the Fly: Drosophila Genetics and the Experimental Life* (1994).

VASSILIKI BETTY SMOCOVITIS

MOVIES. See FILM, TELEVISION, AND SCIENCE.

MULTINATIONAL LABORATORIES. During the Cold War few nations outside the two superpowers had the resources to drive big science. Countries wishing to compete with them had to pool resources. The pooling worked particularly well in Europe and in the physical sciences. Reconstruction requirements, Cold War dilemmas, the threat of Stalinism, competition from the United States, support from international organizations, and an understanding of limitations combined with the politics of European unification to provide the setting for the emergence of European multinational laboratories. The Soviet bloc and other regions also housed multinational projects, which remained, however, under national control.

Several overarching political institutions assisted the development of European multinational laboratories: the Organization for European Economic Cooperation (OEEC) formed in April 1948 to manage the Marshall Plan and transformed in 1960 into the Organization for Economic Cooperation and Development (OECD); the North Atlantic Treaty Organization (NATO), created in April 1949; and the European Coal and Steel Community (ECSC), agreed to by France, the Federal Republic of Germany, Belgium, the Netherlands, Luxembourg, and Italy in April 1951. In 1957, Belgium, France, the Federal Republic of Germany,

Italy, Luxembourg, and the Netherlands signed treaties setting up the European Economic Community (EEC) and European Atomic Energy Community (EURATOM), both of which merged with the ECSC in 1967. The High Authority established to supervise the ECSC treaty became the European Commission. The EEC made explicit its European Community Research, Technology, and Development policy in 1974, the same year that the first elections to the European Parliament were held. The first laboratory created within this broad political context was the European Organization for Nuclear Research (*CERN), which gained broad support.

Louis de Broglie issued the first high-level call for a multinational laboratory as an instrument to revive European science at the Lausanne European Cultural Conference in 1949. Supporters of the call included Raoul Dautry, administrator-general of the French Atomic Energy Commission; Pierre Auger, director of UNESCO's Department of Exact and Natural Sciences from 1948 to 1959; and Edoardo Amaldi, one of the founders of Italy's National Institute for Nuclear Physics. UNESCO provided the institutional framework for CERN. In 1951 Cornelis Jan Bakker and other UNESCO consultants presented their recommendations. They called for a temporary organization, $200,000 in funding, and eighteen months to prepare an administrative, financial, and technical program.

Niels Bohr offered his institute as a home for CERN; the offer was accepted, but only for the theoretical group. The representatives of the eleven countries that set up CERN in February 1952 chose Amaldi as secretary general of the new laboratory, and in October decided to build it near Geneva. The formal founding of CERN by the Federal Republic of Germany, Belgium, Denmark, France, Greece, Italy, Norway, the Netherlands, the United Kingdom, Sweden, Switzerland, and Yugoslavia took place in 1954. In October 1954, Felix Bloch succeeded Amaldi and became CERN's first permanent director general. Though motivated by American competition, the CERN management did not reject Ford Foundation funding. The Soviet inauguration of the Dubna 10 GeV proton-synchrotron in 1957 introduced a second major competitor. CERN's first *accelerator, a 600 MeV proton synchrocyclotron, was inaugurated in 1957, coordinated by Bakker. The CERN proton-synchrotron, whose development was led by Odd Dahl, began operating in November 1959 at 28 GeV.

Nearly half of the particle physicists in the world work at CERN, supported by a staff of three thousand. The size of the operation and the number of researchers complicate relations among project managers. Traditional European disciplinary divisions created more problems than did national differences. In contrast to the American tradition, CERN at first kept science apart from engineering and performed poorly in comparison to the Brookhaven laboratory. European scientists had to learn how to do big science.

Good management and a research focus on reproducible results raised CERN gradually from an institution that missed opportunities in the 1960s and that still had a low reputation in the mid-1970s, to the renowned producer of W and Z boson results and Nobel Prizes.

The failure to establish CERN in Copenhagen—the theoretical group moved out in 1957—motivated the creation in its place of the Nordisk Institut for Theoretisk Atomfysik (Nordita). Bohr and his colleagues recognized that none of the Nordic countries alone commanded sufficient resources to compete in nuclear and elementary particle physics—despite the Norwegians' success in operating JEEP, the first nuclear reactor outside the sphere of the major powers in 1951. Moreover, experience with the UNAEC (*see* INTERNATIONAL ORGANIZATION) revealed some of the barriers that the Cold War introduced into nuclear science.

In January 1953, Bohr from Denmark, Torsten Gustafson from Sweden, and Egil Hylleraas from Norway met at Gothenburg to organize an atomic committee. The timing was politically appropriate: the following month, the Nordic Council had its first meeting in Copenhagen. Finland joined the Nordita project in November 1955 and Iceland in January 1956. Bohr gained the Nordic Council's approval for Nordita in February 1957. Nordita started operations on 1 September 1957, within a month of the departure of CERN's theoretical group from Copenhagen.

In the same year EURATOM set up its Joint Research Center (JRC) at Ispra, near Lago Maggiore in the north of Italy. Centers at Karlsruhe (1960), Geel (1961), and Petten (1961) followed. Until 1973 JRC focused on nuclear energy. After that, it diversified into other fields. During its first decade EURATOM hesitated over its goals, which made the management of the JRC very difficult. Should it be an instrument for industrial or energy policy, or a nuclear research organization? JRC's functions varied from selling nuclear energy to advocating its peaceful uses.

The U.S. *Atoms for Peace policy had motivated the creation of EURATOM; Europeans wanted nuclear independence. The Germans had experienced U.S. restrictions since an American intervention had stopped a joint project with

Brazil in the early 1950s. The French supported EURATOM and the JRC only on condition that they could develop nuclear weapons. Nothing in the EURATOM charter appeared to prevent it. EURATOM also permitted the free movement of nuclear raw materials.

The JRC employed the concept of the five-year research program now a cornerstone of the European Commission's research policy. The first program, 1958–1962, divided the tasks into in-house or direct research and external contracts or indirect research. In 1959, thanks to Francesco Giordani, head of the Italian National Research Council, the nuclear research center at Ispra was transferred to EURATOM. The second program, 1963–1967, continued this trend, putting more weight on direct research.

A crisis ensued in 1968, when it became clear that EURATOM did not synchronize with national programs. A European atomic consensus did not exist. Most of EURATOM's projects stagnated or stopped, notable exceptions being the biological program and the fusion research project, the Joint European Torus (JET). The crisis was linked to tensions between Great Britain and France. The JET lab ended up near London.

Support for the JRC returned once it became clear that a growing scientific gap existed between Europe and America, and that European scientists were emigrating to the United States (*see* BRAIN DRAIN). In 1974, the Europeans decided on a common science and technology policy and chose in 1977 six areas of action for the JRC: energy; raw materials; environment; living and working conditions; the network for the exchange of technical information (Euronet); information science, telecommunications, and transport. The JRC became a research infrastructure for these areas. In December 1982 the European Strategic Program for Research and Development in Information Technology (ESPRIT) was approved. It engaged European information technology industry in JRC projects.

The reorganization of the JRC in 1988 brought further diversification: a headquarters in Brussels, the Institute for Reference Materials and Measurements (IRMM) in Geel, the Institute for Transuranium Elements (ITU) in Karlsruhe, the Institute for Advanced Materials (IAM) in Petten, the Institute for Systems, Informatics and Safety (ISIS), the Environment Institute (EI), the Space Applications Institute (SAI), and the Institute for Health and Consumer Protection (IHCP) in Ispra, and the Institute for Prospective Technological Studies (IPTS) in Seville. The JRC acquired a staff of nearly 2,500, of which 1,600 were researchers. After 1998, the JRC became not only the scientific and technical support for implementation of the European Union's policies but also its reference center for science and technology. Nearly 30 percent of the JRC's total budget of approximately 300 million euros now goes to nuclear activities. The JRC competes increasingly for funding from private as well as public sources.

Another important European multinational science institution is the European Space Agency (ESA). Again a unifying theme facilitated the alliances between scientists. ESA evolved from the European Space Research Organization (ESRO) and the European Launcher Development Organization (ELDO), which had confronted gigantic political and economic difficulties. As at CERN, Auger was a central figure. His years, from 1959 to 1967, were learning years. But they were also years when several European countries tried to develop their own space programs. Under Auger's inspiration, ESRO and ELDO achieved some successes, for instance launching fifty-six sounding rockets, but fell short of expectations.

From 1969 to 1973, under Hermann Bondi, ESRO evolved into ESA. Research fields comprised fundamental physics, plasma physics, high-energy x-ray and gamma ray astrophysics, and special cosmic-ray studies and measurements of solar neutrons and charged particles. The Bondi era began optimistically, but the agency soon confronted the realities of costs and the need to accept cooperation with the Americans and the Soviets. Cooperation collapsed under the strains of the Skylab project, forcing a search for alternatives. The deployment of Ariane in December 1979 gave the Europeans autonomy. Despite substantial scientific achievements, ESA's budget is no larger than ESRO's in real terms. ESA has achieved cooperation (and competition) with the United States on equal terms, which none of the European nations alone could have done.

There is a difference between integrated and interdependent multinational organizations. An integrated laboratory like the JRC has a supranational organization behind it, like the ECSC treaty of 1951 and the EEC treaty of 1957. CERN does not have such a structure and, therefore, its members have an interdependent relationship. That the JRC carried the main burden of integration may have allowed CERN the institutional space to pursue pure science. Historians of science have not yet evaluated the contributions of key JRC actors such as David Wilkinson, Marc Cuypers, and, especially, Hans Jörgen Helms, a former JRC director and one of the key institutional builders of European science. Indeed, the history of European big science in general has yet to receive the attention it deserves. When the overdue examination comes, it will find a clear temporal marker: in a single year, 1957, Nordita started operations, CERN's synchrocyclotron came on line, the JRC was created and, on the other side of the Berlin wall, the first *Sputnik* flew.

Hermann Armin, John Krige, Dominique Pestre et al., *History of CERN*, 3 vols. (1987–1996). Tore Frängsmyr, ed., *Solomon's House Revisited. The Organization and Institutionalization of Science* (1990). Peter Galison and Bruce Hevly, eds., *Big Science: The Growth of Large-Scale Research* (1991). Regis Cabral, *Knowledge Flows Between Scandinavia and Ibero America: A Network Approach* (1992). Arnold Thackray, ed., *Science After '40 [Osiris 7]* (1992). John Krige and Arturo Russo, *Europe in Space 1960–1973* (1994). Luca Guzzetti, *A Brief History of European Union Research Policy* (1995). Martin J. Dedman. *The Origins and Development of the European Union, 1945–95: A History of European Integration* (1996). John Krige and Luca Guzzetti, eds., *History of European Scientific and Technological Cooperation* (1997). JRC, *40 Years of Service for the Commission* (2000).

REGIS CABRAL

MUSEUM. Derived from the name of the celebrated seat of learning established in Hellenistic Alexandria, from the late sixteenth century in Europe "museum" designated a place where a collection of natural and man-made objects afforded amusement and recreation to the studious. Knowledge being broad and recreation many-sided, the objects included in the museum varied greatly depending on the culture, intentions, and means of those promoting the collections.

In Renaissance Italy, wealthy merchant and banking families, responding to a renewed interest in the classical heritage and merging with new forms of patronage of the arts, formed collections of art and antiquities. Similar collections established later by royal families throughout Europe were occasionally opened to the public. *Cabinets and collections focusing on natural objects, curiosities, artifacts, and exotica, as well as antiquities, archeological pieces, or rare manuscripts, usually were organized on the initiative of scholars and polymaths supported by secular or religious institutions, wealthy patrons, and universities. The early modern collections of Ulisse Aldrovandi in Bologna, Ferrante Imperato in Naples, and Athanasius Kircher in Rome, focal points for the exchange of knowledge and cultivated conversation, are celebrated examples.

During the sixteenth and seventeenth centuries similar museums prospered in many European cities, thriving on the ambitions of increasingly entrepreneurial scholarly communities and benefiting from the rich harvest of new natural objects and artifacts that Europe's colonial expansion brought to the metropolises. While often retaining features sanctioned by Aristotelian tradition, reverence for classical antiquity, and Christian belief,

these collections were gradually rearranged and elucidated in conformity with the natural philosophy of the *Scientific Revolution. When in the second half of the seventeenth century the newly established scientific academies decided to make their own repositories of natural objects and scientific instruments, the old Renaissance collections became objects of contempt and were often dismembered to fit into the new maps of knowledge. The revised collections served the needs of the emerging groups of experts attached to scientific academies and universities, emphasizing what would now be called the museums' research missions. They contributed to the development of new forms of expertise and skills and emphasized utility, often in connection with the needs of the technical corps of state administrations. This trend continued into the Enlightenment, spread to the colonies, and inspired the creation of many natural history and science museums still active today.

The Natural History Museum in London, for example, originated from the private collection of Sir Hans Sloane, physician, naturalist, traveler, and president of the Royal Society of London. After Sloane's death his trustees sold his collection to the Crown. It opened to the public in Bloomsbury in 1756 and formed the nucleus of the British Museum until the expansion of the natural history sections made a separate building necessary. This was inaugurated in South Kensington in 1881. In the meantime, through its specialized departments, the direction of naturalists like Richard Owen, and the involvement of widening popular audiences in support of its educational mission, the museum acted as a powerful engine for the development of the life and earth sciences.

The Musée National d'Histoire Naturelle in Paris originated in the eighteenth century's royal botanical gardens, the Jardin des Plantes, and was reestablished under its current name in 1793 as part of the reforms of the Revolutionary period. The museum had a research and teaching program as well as a mission of preservation and display. Its twelve chairs divided the old domain of natural history into new, specialized disciplines. During the early decades of the nineteenth century the collections and the intellectual space provided by the museum nurtured vigorous debates in the life sciences and supported major scientific achievements.

The Paris Conservatoire (now Musée) des Arts et Métiers, regarded as the first museum devoted to technology and industry, dates from 1794. The initiative of a priest loyal to the new republican regime, the conservatory was meant for both preservation and education. It hosted collections ranging from the celebrated automata of Jacques de Vaucanson, once the property of Louis XVI, to prototypes of every machine and instrument newly invented or perfected in France. In order to instruct artisans and "improve national industry," the conservatory had its own demonstrators; from 1819 it had chairs devoted to technological subjects.

Natural history museums proliferated throughout the nineteenth century: in 1900 France boasted 300, Great Britain 250, the United States 250, and Germany 150. In 1910 the *Encyclopaedia Britannica* counted 2,000 scientific museums worldwide. These now increasingly included museums devoted to science and technology like the Science Museum in London, founded in the wake of the Great Exhibition (1851), the Arts and Industries Building of the Smithsonian Institution in Washington (founded 1881), and the Deutsches Museum in Munich, founded in 1903. The science and technology collections of the Smithsonian expanded with the addition of the National Museum of Natural History in 1910, the Museum of History and Technology in 1964 (now the National Museum of American History), and the National Air and Space Museum in 1976.

There has long been tension between the two missions of most scientific museums—research and public education. The zoologist and comparative anatomist William Henry Flower, in charge of the natural history collections of the British Museum from 1884 until 1898, was a powerful advocate both of the research mission of museums and of the parallel development of collections and exhibition techniques aimed especially at the general public. Research demands have supported the establishment and expansion of zoological, botanical, geological, anthropological, and ethnological collections. Institutions such as the Harvard Museum of Comparative Zoology (founded 1859) and the American Museum of Natural History in New York City (founded 1869) have played important roles in the development of the life sciences while succeeding as public education centers at the same time.

Science and technology museums, on the other hand, have focused primarily on public education and preservation. Many of the science centers—now numbering about four hundred in forty-three countries—established after the example set by physicist Frank Oppenheimer with the San Francisco Exploratorium (1969) can claim to be creating new and challenging experiences: the scientists who help conceive the exhibits and the persons using them join in realizing Oppenheimer's vision of "laboratories of learning."

The handful of museums specifically devoted to the history of science established during the twentieth century recognize the problematic relationship between research and public education. Originating mostly in collections of ancient instruments belonging to universities or private collectors, museums of the history of science have been created in Oxford (1924), Florence (1927), Leiden (1928), and Cambridge (1944). Until the 1960s they functioned mainly as centers of preservation and display, emphasizing antiquarian concerns; since that time they have benefited from the history of science's development as an academic discipline and its practitioners' interest in artifacts, practices, instrument makers, and social contexts. It is thus a propitious era for the new Nobel Museum, established in Stockholm in 1998.

Susan Sheets-Pyenson, *Cathedrals of Science* (1988). Ronald Rainger, *An Agenda for Antiquity: Henry Fairfield Osborn and Vertebrate Paleontology at the American Museum of Natural History, 1890–1935* (1991). Mary P. Winsor, *Reading the Shape of Nature: Comparative Zoology at the Agassiz Museum* (1991). John Durant, ed., *Museums and the Public Understanding of Science* (1992). Paula Findlen, *Possessing Nature* (1994). Graham Farmelo and Janet Carding, eds., *Here and Now: Contemporary Science and Technology Museums and Science Centres* (1996). Nicholas Jardine, James A. Secord, and Emma C. Spary, eds., *Cultures of Natural History* (1996). Svante Lindqvist, ed., Marika Hedin and Ulf Larsson, associate eds., *Museums of Modern Science* (2000).

GIULIANO PANCALDI

MUSIC AND SCIENCE. In the early twenty-first century the relationship between science and music is typically understood in terms of particular specialties within the natural and human sciences (e.g., architectural *acoustics, the physics of musical instruments, neurosciences), and what these can reveal about music's physical and cognitive effects. A longer historical perspective, however, shows that the disciplines of both science and music have altered dramatically over the last few hundred years. New technologies have transformed musical as well as scientific practice, and through their use of instruments and shared understanding of measure and time, the two domains have been closely intertwined. Before Isaac *Newton transformed the mathematics of physics, the paradigm was "nature is musical"; thereafter the formula became "music is natural."

Since the Enlightenment music has been classified among the fine and performing arts; consequently, its role in the *Scientific Revolution of the seventeenth century has generally been overlooked. Until around 1700, however, music was considered to be a science as well as an art, that is, a body of systematic theory

Nodal lines of vibrations produced by drawing a bow across thin metal plates; the patterns can be generated by sprinkling sand on the plates.

including both practical and speculative aspects. It was also an academic discipline: music, astronomy, geometry, and arithmetic made up the mathematical sciences of the quadrivium, which, together with the trivium (grammar, rhetoric, and logic), formed the core of the medieval arts degree course. Up to the sixteenth century the most authoritative source on musical science was Boethius's *De musica* (sixth century A.D.), which promoted the Pythagorean harmonic doctrine that audible music is a tangible expression of the underlying mathematical principles (harmonia) governing the relations between the elements of all significant structures in the cosmos.

Some musicians were already declaring this harmonic tradition irrelevant to their art in the early sixteenth century. Practitioners who played instruments with fixed pitches like keyboard and lutes realized that polyphonic music composed for two or more parts did not agree with the Pythagorean scale defined by the ratios of small integers. This conflict between musical practice and theory played a significant part in the emergence of experimental science as the physical properties of musical instruments, especially the vibration of strings, increasingly became the subject of philosophical enquiry. In his *Istitutioni harmoniche* (1558), the Venetian composer and music theorist Gioseffo Zarlino presented a new theory of consonance, which he claimed had been tested experimentally. The lutenist Vincenzio Galilei disputed Zarlino's scenario, and in the process became the first person to check the results of Pythagoras's legendary experiments with hammers, strings, and other sounding bodies. He proved that most of them were wrong. *Galileo Galilei—an accomplished lutenist like his father—further investigated the properties of vibrating strings and announced his findings in the *Discourses Concerning Two New Sciences* (1638).

Although Galileo understood the relationship between pitch and frequency, credit for the discovery of the physical variables governing the pitch of musical strings goes to Marin Mersenne, whose *Harmonie universelle* (1636) contained the first published account of "Mersenne's laws." These boil down to the formula that frequency is proportional to the square root of string tension, and inversely proportional to string length as well as to the square root of the string's thickness.

The universal harmony of Mersenne's title indicates that although musicians rejected Pythagorean intonation on empirical grounds, early modern mathematicians continued to find the tradition of speculative harmonics inspiring. Johannes *Kepler saw his search for universal mathematical laws as an extension of Ptolemy's *Harmonics* (second century A.D.), which assumes that sounding music (*musica instrumentalis*) embodies the harmonic principles governing human and also cosmic bodies (*musica humana et mundana*). Kepler's *Harmonices mundi* (1619) builds on the concept that God created the world in accordance with geometric archetypes reflecting musical consonances. Kepler calculated the "harmony of the spheres" derived from the maximum and minimum angular velocities of the planets measured from the sun. The musical ratios that express these intervals are not Pythagorean, however, but those shown by Zarlino to be the basis of modern polyphonic practice.

The relationship between music and science thus developed in two complementary directions during the seventeenth century. First, musical phenomena continued to be investigated in their own right, most notably in the new field of acoustics. Joseph Sauveur claimed to have established this discipline in 1701, but Francis Bacon had already identified the "Acoustique Art" in his *Advancement of Learning* (1605) as exemplifying his new scientific method. Mersenne's systematic investigation of the properties of musical sound was taken up in the Royal Society of London and the Paris Académie Royale des Sciences. At the same time, however, musical/harmonic models continued to be fruitfully applied to other branches of the physical sciences. Robert *Hooke, for example, postulated a unified theory of matter in which all particles act like musical strings vibrating sympathetically. Mersenne's laws similarly provided Newton with the basis for his analogy between colors and musical tones in the *Opticks* (1704), as well as his wave theory of sounds presented in the *Principia mathematica* (1687).

The success of the Newtonian synthesis brought the study of the relationship between music and science during the eighteenth century almost entirely within the framework of mechanics. In his *Traité de l'harmonie* (1722) the composer and theorist Jean-Philippe Rameau reduced the rules of harmony to a single principle of fundamental bass, which he compared to Newton's gravitational laws. Musical phenomena such as resonance and the overtone series appeared in popular texts of experimental physics, and were a staple of lecture demonstrations of natural philosophy. Meanwhile, leading European mathematicians like Jean Le Rond d'Alembert, Daniel Bernoulli, and Leonhard *Euler disputed the mathematics of vibrating strings, a debate that contributed to the development of a viable theory of sound propagation. Pierre-Simon de *Laplace's employment of the caloric theory of heat (*see* FIRE AND HEAT) to resolve discrepancies between calculated and observed velocities of sound was a centerpiece of the physical science of the early eighteenth century. Enlightenment medical theorists also sought to explain music's effects in terms of vibration. Richard Browne's *Medicina musica; or, A Mechanical Essay on the Effects of Music, Singing and Dancing* (1729) was one of the earliest treatises to address the physiological and psychological mechanisms governing human responses to music.

The nineteenth century marked a new, instrumental phase, which coincided with the shift to laboratory science as the dominant model of scientific practice. Ernst Chaldini's *Die Akustik* (1802) showed that most complex musical phenomena could not be explained by prevailing theory. Charles Wheatstone, Félix Savart, and other nineteenth-century physicists relied on new acoustical instruments (tonometer, siren, resonator, kaleidophone, speaking machine, etc.) in their quest to analyze and synthesize musical sounds and human speech. These acoustical experiments had important consequences for the scientific analysis of music. They provided the essential background for *Die Lehre von den Tonempfindungen als physiologische Grundlage für die Theorie der Musik* (1863), in which Hermann von *Helmoltz established his new discipline of physiological acoustics. They also contributed to the invention of Alexander Graham Bell's telephone (1876) and Thomas Edison's phonograph (1877), technologies where parallel research into acoustical and electromagnetic phenomena converged for the first time. These inventions signaled a long-term shift from mechanical to electrical forms of sound production. In the short term, they had immediate application in the emerging field of psychological acoustics. Wilhelm Wundt and Carl Stumpf investigated the perception and judgment of musical tones, while Lord Rayleigh, whose *Theory of Sound* (1877) set the research agenda for the next fifty years, consolidated Hermann von *Helmholtz's advances in physical acoustics. On the basis of Rayleigh's work and new materials for the absorption of sound the Harvard physicist Wallace Sabine developed a new applied science of architectural acoustics.

During the twentieth century, the ability to record and reproduce sounds electronically had an enormous impact on the production and consumption of music. For the first time (except for barrel organs and other musical automata) it became possible to hear music without requiring live musicians. The music industry expanded in the wake of demand stimulated by the gramophone, radio, and movies. Experiments with electric pianos, guitars, and other amplified instruments in the 1920s and 1930s led to their commercial manufacture in the 1950s and 1960s, and the creation of a distinctive pop and rock musical culture. Electroacoustic and computer-generated music, in which "sound objects" abstracted from their acoustic and notated sources could be manipulated in new ways, made their appearance. Pierre Schaeffer, who invented *musique concrète* in 1948, and Iannis Xenakis, whose stochastic compositions of the 1950s and 1960s were based on the applicability of the kinetic theory of gases to music, exemplify this genre of modern composition stimulated by new electronic and digital technologies.

Laboratories continued to be important sites for exploring and reconfiguring relationships between music and science throughout the late twentieth century, especially in the recording and telecommunications industries. In 1957 the first experiments in digital sound synthesis took place at the Bell Telephone Laboratories. They led not only to new musical applications but also to new research in the psychophysics of music. Since the 1960s advances in computer technologies and research into artificial intelligence have strongly dominated approaches to music analysis and music psychology, which in keeping with the increase in passive listening by mass audiences has concentrated more on studying responses to music than on analyzing the skills necessary for performance. The computational methods used by cognitive psychologists to investigate the processes of hearing, perception, and memory have also been complemented by neurophysiologists' studying music's effects on brain activity by means of PET (positron emission tomography) scans, which permit real-time observations of variations in blood flow across the brain. More recently, scientists have raised questions about a genetic basis for musical ability as the human genome project nears completion, but have found no definitive evidence.

The close relationship between Western science and Western music that persisted into the seventeenth century has continued. Yet once harmonic motion became expressible in a mathematical equation rather than vibrating strings, the importance of musical models for the natural and human sciences diminished. Although the string theory currently fascinating physicists resembles Hooke's concept of universal vibrating matter, modern cosmologists have not come to strings via music. And as the sciences become ever more specialized—acoustics itself being divided into numerous subdisciplines—the connection between music and any single science has fragmented. Nevertheless, some consistent patterns have emerged, especially the role of instruments in generating scientific knowledge about music. New instruments and techniques have been created to investigate the changes effected on human nature by music. These musical and scientific experiments have typically been linked through a shared reliance on mechanical, electrical, and, most recently, digital technologies.

Albert Cohen, *Music in the French Royal Academy of Sciences: A Study in the Evolution of Musical Thought* (1981). H. F. Cohen, *Quantifying Music: The Science of Music at the First Stage of the Scientific Revolution, 1580–1650* (1984). Claude V. Palisca, *Humanism in Italian Renaissance Musical Thought* (1985). Thomas Christensen, *Rameau and Musical Thought in the Enlightenment* (1993). Stephan Vogel, "Sensation of Tone, Perception of Sound, and Empiricism: Helmholtz's Physiological Acoustics," in *Hermann von Helmholtz and the Foundations of Nineteenth-Century Science*, David Cahan, ed., (1993): 259–287. Thomas L. Hankins and Robert J. Silverman, *Instruments and the Imagination* (1995). Jamie C. Kassler, *Inner Music: Hobbes, Hooke and North on Internal Character* (1995). Penelope Gouk, *Music, Science and Natural Magic in Seventeenth-Century England* (1999). Emily Thompson, "Listening to/for Modernity: Architectural Acoustics and the Development of Modern Spaces in America," in *The Architecture of Science*, Peter Galison and Emily Thompson, eds., (1999): 253–280.

PENELOPE GOUK

MUTATION. The term mutation, of ancient vintage, was first adopted for a specific biological purpose by Wilhelm Heinrich Waagen in 1869 to refer to abrupt changes in the fossil record. Subsequently, the Dutch botanist Hugo de Vries used it to refer to the "sudden and spontaneous production of new forms from the old stock." This definition comes from the lectures he delivered in California in 1904, but he had first expounded his mutation theory in German in his great work, *Die Mutationstheorie* (1901–1903, English translation 1909–1910). Despite some skepticism, De Vries's work was very influential. The fall of his theory began in the 1920s. In its place came a concept of mutation that could be synthesized with Darwinian *evolution. Securely founded on the experimental studies of the fruit fly (*Drosophila*), it accounted for the strange results De Vries had reported in his studies of the evening primrose (*Oenothera*).

A specimen of the Harlequin bug (Pentatomidae) found near Three Mile Island after the nuclear accident there. The beautiful drawing by Cornelia Hesse-Honegger (1991) shows the deformed scutellum and asymmetrical patches apparently caused by the leaked radiation.

The inspiration for his theory came in 1886 when De Vries observed among plants of the evening primrose some individuals that departed markedly from the rest of the population. Grown in his experimental garden, the original form (*Oenothera lamarckiana*) gave rise to new and distinctive forms. De Vries classified these examples of abrupt change as *progressive* mutations, since a novel characteristic had appeared that bred true, and hence the plants bearing it could be considered a new *elementary species*. Abrupt changes involving the apparent loss of a character he called *retrogressive mutations*, the character having become latent, whereas those that cease to appear in all individuals, but only occasionally here and there, he termed *digressive* mutations.

From the beginning, botanists suspected that De Vries's mutants were really hybrid. It required more than two decades to establish that evening primroses behaved in an atypical manner, that they bred true because the variants that would normally be found among the hybrid offspring were all nonviable combinations of the genetic material. With the overthrow of De Vries's theory came a reconceptualization of the nature of mutation. Instead of involving necessarily discontinuous alterations—saltations or macromutations—mutation could result in small changes. Furthermore, the magnitude of the changes could be altered by other so-called "modifier" genes, and selection could act in a creative and cumulative fashion upon step-wise variations. De Vries had assumed that species are normally quiescent but occasionally undergo phases of mutation. Research in the 1920s led to the view that mutations are "recurrent" and that their frequency of recurrence can be measured statistically. This reformulation of the concept of mutation played a decisive part in bringing together Mendelian *genetics and Darwinian evolution.

Debate long continued over whether a given mutation arose from a qualitative change in the substance of the gene at a given place—a "point mutation"—or from a rearrangement of the same genetic material. With the continued refinement of genetic analysis, now going down to the molecular level, many of these cases have been resolved. But mutation remains a broad term that includes both changes in the chemical constitution of the gene (base changes in *DNA) and rearrangements in the same genetic material.

Hugo de Vries, *Species and Varieties. Their Origin by Mutation* (1904, Garland Series Reprint 1988). Peter Bowler, *The Eclipse of Darwinism: Anti-Darwinian Evolution Theories in the Decades around 1900* (1983).

ROBERT OLBY

N

NATIONAL ASSOCIATIONS. See ACADEMIES AND LEARNED SOCIETIES; ADVANCEMENT OF SCIENCE, NATIONAL ASSOCIATIONS FOR THE.

NATIONAL CULTURE AND STYLES. "Style" originated as a term in literature. Art historians, notably Heinrich Wölfflin, appropriated it for their own use, and in art the term is firmly lodged. Cultural anthropologists also borrowed the term as an analytical category. In *Man's Rage for Chaos* (1967), Morse Peckham observes that style usually has symbolic, rather than emblematic, meaning. "Thus, the style of each of the arts is an expression, or symbol, of the same orientation, the dominant orientation of an age, which pervades all behavior, at least at the higher cultural levels." This behavior includes or presupposes values. To say that the style produced the values, however, would be as mistaken as to say that the style of dress among university students in the 1960s produced rebellion.

Physicist Polykarp Kusch defines style as "an expression of approbation and admiration" in his *Style and Styles in Research* (1968). "It describes the cohesiveness and internal structure of the work as well as the clarity and effectiveness of the mode of expression of an idea." Kusch contrasts style in the sense of brilliance or flare with "styles," which relate to the "externally imposed circumstances in which the work was done." Styles of research change with each generation; in the twentieth century, styles of research derived largely from levels of funding. Kusch's understanding of style recapitulates the notion of Scottish geologist Hugh Miller (*First Impressions of England and Its People* [1846]), who emphasized that cultural expressions of national character derive from political and social conditions. Scotland had no *Newton or Bacon, men of the first order, because they "belong to ages during which the grinding persecutions of the Stuarts repressed Scottish energy." Kusch's notion of styles is not far from "fashion" as used derogatively by scientists to indicate a topic made popular by the availability of funding. We are far from Herbert Spencer's restriction of fashion to costume and style to literature.

At the beginning of the twentieth century, national scientific styles were discussed in terms of national spirit or intellectual orientation. The French physicist and historian Pierre Duhem contrasted British model-building (suitable for broad and shallow minds) to French mathematical abstraction (appropriate for the deep and narrow). John Theodore Merz, who earned a doctor's degree in philosophy in Germany that served him well in an industrial career in England, characterized the ideal type of German scientist by an attention to thoroughness, an awareness of the larger picture, a desire to create acolytes, and a predilection for philosophy; he rated English scientists as idiosyncratic and practical-minded and the French as analytical and pedagogical. The descriptive device of national characteristics in science has proved remarkably resilient. Quoting Henry Guerlac, historian of science and archetypical Enlightenment scholar: "To deny national variations altogether is quite unrealistic; for there are subtle cultural tendencies which set apart the scientific achievements of every nation." They do not hold universally—the "French animus in favor of pure science is just a tendency; so also is the American scientific genius which manifests itself best in engineering application"—but widely enough that the French are rightly identified with mathematics and the English with experiment and fact-collecting (Guerlac, *Essays and Papers* [1977]). Jonathan Harwood uses the notion of style to indicate "the range of questions that geneticists" in Germany and the United States "took to be central to their discipline." He concludes that stylistic differences are strongest in weakly institutionalized disciplines. Persistence of the theme is found in the way that Nathan Reingold, an early advocate of the integrity of American science as a field of inquiry, expresses his distaste for rigid, ideal types in a collection entitled, precisely, *Science, American Style* (1991): "I will have nothing to do with the whole apparatus of cultural monuments, essences, determinisms, and idealized national styles."

The domain of policy studies has generated systematic inquiry about national styles in government administration. Contrasting consultative policymaking in Sweden with authoritarian policymaking in France, political scientists have postulated criteria for describing how policies are formulated (Gary Freeman [1987] and Jeremy Richardson et al. [1982]). Science-policy writers readily accept the notion of national scientific styles. "We all know that there are national differences in science," wrote Andrew Jamison in 1987. "Science does differ from country to country; there are differences in emphasis, in direction, in orientation, in application." Jamison proposed analyzing national differences according to metaphysical biases, institutional structures, and national scientific and technical interests (that is, emphases on one or another scientific field). Hebe Vessuri has borrowed from Jamison to propose a model for national styles with four levels: metaphysical slant, national scientific interest, institutional structures, and relative congruence among national traditions (*Quipu, 11* [1994]). Relating the social organization of science to broader political practice, notably in international affairs, has engendered stylistic characterizations. There is also a literature on national styles in technological innovation, in which management structures appear.

Alistair C. Crombie has provided the most extended treatment of style in science in his *Styles of Scientific Thinking in the European Tradition* (3 vols., 1994). He identifies "in the history of the classical scientific movement a taxonomy of six styles, distinguished by their objects of inquiry and their methods of argument." The styles are postulation, as in geometry and arithmetic; experimentation; hypothetical modeling; taxonomy; probability and statistics; and historical reasoning. Crombie sees style extending from science to culture generally: "The specific style of a culture is defined by all of its commitments and expectations and methods of procedure"; styles of thinking in all fields of inquiry "nearly always have the marks of a recognizably common provenance"; the *Renaissance virtuoso, for example, was "diagnostic of Western civilization, expressing its characteristic style." *Galileo's enduring achievement is "as the designer of an explicit scientific style." This style, when broadened, allowed the intellectual culture of Europe to engender a scientific movement that could "solve problems and generalize explanations." Crombie avoids a stage-theory of cultural evolution. His view suggests that style is a transcendent quality that resists qualification by the artifact of nation.

Stylistic distinctions across cultures can be made without appeal to particular states or nations, as G. E. R. Lloyd does in his collaboration with Nathan Sivin, which provides a close reading of Greek and Chinese texts from the period of the "Greek miracle" and the 100 schools in the Warring States era (*Adversaries and Authorities* [1996]). Lloyd finds merit in a revision of the notion that the Greeks were adversarial while the Chinese were irenic. Both settings knew schools and discursive traditions, and included a wide range of behavior. The assertive Greek emphasis on the priority of pure science, and the argumentative path to achieving certainty in it, in part reflected the relative lack of influence that philosophers exerted over temporal rulers; Chinese philosophers were more closely tied to political life, and their precepts intensified the connection. Savants in both settings sought to know causes, but Greek schools spoke at cross-purposes, while the Han period managed to "consolidate a comprehensive world-view." Consequences of the difference can be found in astronomy. "The types of astronomical model developed in Greece and China reflect the influence of the styles of intellectual exchange cultivated in each society. The modes of rivalry among astronomers differed, in that in Greece the stakes were those of strict proof. The Chinese demand ... was for accuracy in prediction: there was plenty of competitiveness in delivering that. But deductive certainty, incontrovertibility, were a red herring. The enterprise of demonstrating the movements of the planets by way of geometrical models, if it had been attempted—which it was not—would have been considered irrelevant." Lloyd's study suggests the fruitfulness of elaborating scientific style from the perspective of broad civilizations.

Science in particular settings has been generalized as a national style. *Naturphilosophie and the dialectical and anti-Newtonian science of Goethe, among others, are taken as a style of science characteristic of German-speaking Europe. Niels *Bohr's complementarity has been related to an existential style of Danish philosophy. But the notion of national style has declining relevance in our age, when the nation itself is an amorphous and contentious category.

John Theodore Merz, *A History of European Scientific Thought in the Nineteenth Century* (1904; 1965). Pierre Duhem. *The Aim and Structure of Physical Theory*, trans. Philip P. Wiener (1962). Jeremy Richardson, Gunnel Gustafsson, and Grant Jordan, "The Concept of Policy Style," in Jeremy Richardson, ed., *Policy Styles in Western Europe* (1982). Gary P. Freeman, "National Styles and Policy Sectors: Explaining Structured Variation," *Journal of Public Policy*, 5 (1985): 467–496. Jonathan Harwood, "National Styles in Science: Genetics in Germany and the United States Between the World Wars," *Isis*, 78 (1985): 390–414. Andrew Jamison, "National Styles of Science and Technology: A Comparative

Model," *Sociological Inquiry*, 57 (1987): 144–158. Hebe Vessuri, "Estilos nacionales en ciencia?" *Quipu*, 11 (1994): 103–118. G. E. R. Lloyd, *Adversaries and Authorities: Investigations into Ancient Greek and Chinese Science* (1996). Herbert Spencer, "The Philosophy of Style," in Spencer, *Essays, Scientific, Political, and Speculative* Vol. 2, 333–369 (1999).

LEWIS PYENSON

NATIONAL PARKS AND NATURE RE-SERVES. The establishment of national parks as areas of land set aside for public use originated in the United States during the second half of the nineteenth century. The concept responded to the cultural influence of the nature romantics and transcendentalists, the environmental effects of rapid industrialization, the unveiling of spectacular, unspoiled landscapes in the rapid westward expansion of the United States, the rise of tourism, the negative example of the ruthless commercial exploitation of Niagara Falls, and businesses, especially railroads, interested in promoting travel.

The North American ideal of the national park established in the late 1800s also embraced the egalitarian ideal of preserving nature as a public good. This ideal distinguished national parks from enclosed "parcs" created during the Middle Ages and the Age of Absolutism primarily as hunting reserves for the privileged classes of Europe. These earlier reserves controlled the exploitation of natural resources, such as forests; in contrast, national parks were established to preserve scenery, nature, and wildlife for recreation, study, or admiration.

On 30 June 1864, the U.S. Congress authorized a land grant to the state of California that reserved the Yosemite Valley and Mariposa Big Tree Grove for public use and recreation. This first of all parks did not join the national park system until 1905, however, when it was reacquired by the federal government for Yosemite National Park, founded in 1890. Yellowstone Park, established by Congress on 1 March 1872, was the park under the direction of the secretary of the interior from its inception. In June 1916, Congress created the National Park Service, thus establishing the national park system and rejecting efforts to integrate it with the U.S. Forest Service.

Canada's first national park was established at Banff in 1885. Outside North America, national park movements, as in Africa and Latin America, lagged behind until the 1930s and 1940s. The London Convention for the Protection of African Fauna and Flora, signed in 1933, and the Pan American Convention on Nature Protection in the Western Hemisphere, enacted in 1942, greatly stimulated these efforts.

Wildlife protection, as distinct from the reservation of land, also gained attention within the conservation movement of the late nineteenth and early twentieth centuries. In the United States, the Pelican Island National Wildlife Refuge, created by President Theodore Roosevelt in 1903, became the first national wildlife reserve. In Britain, the passing of the Sea Birds Preservation Act (1869), the Wild Birds Protection Act (1876), and the creation of the first bird reserve at Breydon Water (1888) were important steps on the path to the founding of the National Trust for England and Wales in 1895 and the Society for the Promotion of Nature Reserves in 1912. Legislation enabling the creation of national parks was not enacted until 1949, reversing the order of U.S. steps, and in Britain the mission of these parks was tied more closely to nature conservancy than in the United States.

In the U.S. National Park Service, public access and recreational use often played a greater role in the management of national parks than wildlife preservation or ecological science. The Leopold Committee and National Academy reports of 1963 questioned these priorities, raised ecological concerns, and called for more attention to biological science and natural history. The Wilderness Act passed by Congress in 1964 and the National Environmental Policy Act of 1969 revised Park Service priorities by requiring management for public use in a manner that would not impair permanent preservation of wilderness areas.

Alfred Runte, *National Parks: The American Experience*, 3d ed. (1997). Richard West Sellars, *Preserving Nature in the National Parks: A History* (1997).

HENRY LOWOOD

NATIONALISM. See INTERNATIONALISM AND NATIONALISM.

NATURAL HISTORY. Originally, natural history was the study of the natural world in all its observable diversity, embracing animals, plants, minerals, and aspects of what now fall under archaeology, *anthropology, *geography, *meteorology, and *geology. The objects of study were not always of living origin, as in the observation of weather patterns; nor were they always of natural origin, as in the collection of ancient coins or man-made artifacts such as prehistoric pottery. The term "natural history" perhaps ought to be regarded as a designating technique rather than as a subject area. To conduct a natural historical study was to collect, describe, classify, and from the seventeenth century onward, to perform minor experimental investigations into observable phenomena. At heart, it was descriptive, depending on a large

A stuffed giraffe looking down on other species at the Musée d'Histoire Naturelle, Paris.

variety of examples collected and arranged for edification, inclusive rather than exclusive, and rarely reductive in the form otherwise developed by Western natural philosophers. In this sense, natural history takes its origin in the encyclopedic endeavors of authors such as Pliny the Elder. Pliny's *Natural History* in thirty-seven books dealt with *cosmology, geography, *anthropology, *zoology, *botany, *medicine, chemical recipes, *mineralogy, *magic, and human industry and *art, a miscellany of information for the most part comprising extracts from other authors.

The ancient and medieval scholar described living things by particular properties, uses, or wonders so that others might use or wonder at them. Moral and didactic themes were overwhelmingly important. Manuscripts and early printed books included references to fables, *astrological correspondences, religious symbolism, and what are now considered imaginary beasts. The illustrations that accompanied these descriptions were often partly imaginary and stylized, especially in illuminated manuscripts of the Romanesque tradition. Alexander Neckham discussed the virtues of herbs alongside the theological cause for spots on the moon and insect venoms. Studies of nature typically took the form of catalogs and encyclopedias, like that of Bartolomaeus Anglicus's *On*

the Properties of Things, which became one of the earliest printed works. Bartolomaeus and others displayed great erudition in their sources and shrewd observation by adding their own opinions and commentaries to older sources. They covered herring fisheries, agriculture, mining techniques, falconry, and such like, and gave the first accounts of exotic animals, plants, and medicines brought back from early voyages of exploration. A form of emblematic natural history emerged, based on a complex system of associations and similitudes, mostly astrological and moral, that was also learned and eclectic, full of compilations from earlier texts and oral knowledge.

Highly naturalistic representations of plants and animals were displayed throughout Europe in church decorations and illuminated manuscripts. Manuscript illuminators learned three-dimensional realism from the artists of Italy and Flanders. Both the naturalistic and Romanesque traditions continued without a break into the early printed herbals. The level of popular knowledge of natural history in the medieval and Renaissance periods can be inferred from these artifacts.

After 1650, naturalists abandoned emblematic meanings in order to focus more obviously on description and anatomical investigation. They studied and collected natural objects both for themselves and as part of a broader interest in curiosities. Books such as Robert Plot's *Natural History of Oxfordshire* (1677) discussed all sorts of local features, including country houses of the region. Books based on particular geographical regions became an established genre associated particularly with the recreations of gentlefolk, the "virtuosi" or gentleman-scholars of the period. Collecting rarities, performing experiments, and traveling and discussing their finds in clubs and societies were accepted activities in European aristocratic culture. In describing curiosities, scholars consciously adopted a new style of descriptive language based on observation. Frequently, they reported measurements.

During the late seventeenth and eighteenth centuries, observations of the weather formed an important part of local natural history. Weather-books and garden calendars, in which the owner recorded the weather regularly, were introduced, emerging from the long-standing tradition of almanacs. Small museums and displays, often including antiquities, multiplied, and private menageries became popular among the wealthy. The older universities established physic gardens in which the whole of God's creation might be collected together and *museums or *cabinets in which curiosities from around the world could be shown (*see* CABINETS AND COLLECTIONS).

Beginning in 1620, the Society of Apothecaries, one of the old-established livery companies of London, organized herborizing expeditions; similar expeditions followed in Edinburgh from 1670. These activities were seminal in establishing the field tradition that characterized European natural history thereafter. Since many enthusiasts for natural history were also physicians, collections often ranged widely over the disciplines, and frequently became the subject of inquiry or experiment. Putting together a "natural history of mankind" was a favorite endeavor that variously included reports of new races of humanity in exotic lands, folklore or other accounts of human societies, interesting artifacts, musings on the geographical distribution of mankind, and comparative anatomy. *Natural history came to mean an account of nature based on information acquired by observation and inquiry (including textual and verbal inquiry)—a point of view generally endorsed by forward-looking physicians. Doctors figured prominently among the early Fellows of the Royal Society, and medical societies and colleges formed some of the finest natural history collections still extant. Other enthusiasts typically formed a small club or local society to facilitate expeditions and the exchange of specimens and information. Generally these groups

embraced a wide range of subject matter, and were among the first nonprofessional associations in many provincial towns in Europe. They played a major part in developing the strongly localized basis of British natural history.

During the eighteenth century, the subject's diffuse qualities gave natural history a low position in the hierarchy of the sciences. Countless descriptive papers appeared in the *Philosophical Transactions* of the Royal Society, devoid of the analytical spirit typifying French or Italian scholarly endeavor. The intellectual futility of these articles was lampooned by Jonathan Swift in *Gulliver's Travels*, and Jean Le Rond d'Alembert deprecated mere description in his preliminary discourse to the *Encyclopédie* (1751).

Outside the learned world, natural history was understood to relate mostly to animals, plants, meteorology, and geology. Collecting objects and making observations became the central defining feature, encouraged by the growth of convivial clubs and outings. Techniques for recording data and collecting specimens consolidated. Naming and arranging specimens preoccupied many, both inside and outside museums. Books and catalogs were produced, and specimens sold or exchanged in a rapidly developing market economy.

In Britain, Gilbert White, the chronicler of the *Natural History of Selborne* (1789), became the exemplar of these localized pursuits. An Anglican clergyman, White regarded natural history as an appropriate way to appreciate the wonder and perfect design of God's creation. The Abbé Pluche in France wrote in much the same manner on the *Spectacle of Nature, wherein the wonderful works of Providence in the animal, vegetable and mineral creation are laid open* (1754). Natural history became closely associated with the natural theological tradition (*see* NATURAL THEOLOGY). Not all exponents made obeisance to religion, however. Thomas Pennant barely referred to the Creator in his *British Zoology* (1766), noted for its naturalistic illustrations and keen observation of bird behavior and habitat.

Fashionable interest during the eighteenth century perceived natural history as suitable for cultivation by women and gentlefolk, both literally in the development of lavish gardens and landscapes, greenhouses, menageries, and the breeding of rare livestock, and more passively in providing a responsive market for fine illustrated books, elementary teaching manuals, artworks, and newly fashionable decorated wallpapers and fabrics. Few other sciences have been taken up by consumers in such an obvious manner.

To be sure, Carl *Linnaeus's simplification of classification and nomenclature enabled people who worked with animals and plants to codify their observations. Travelers and collectors could quickly assign unidentified plant specimens into a class or family. Linnaeus trained and sent plant collectors all over the increasingly accessible globe—Japan, South Africa, the Carolinas, Asia, central Spain, and so on. These men, whom he called "apostles," and others using his system, sent numerous collections back to Europe throughout the eighteenth century; Linnaeus consciously served as the hub of a botanical exploring network fanning out from his base at the medical school of Uppsala University. His classification scheme also helped botanical knowledge to move out of the restricted domain of universities, museums, and physic gardens into a broader constituency. Many eighteenth-century figures, including gentlefolk, women, and working men, are all known to have first encountered botany through popularizations of Linnaeus's system. His "sexual system" provided the basis for much cultural satire and parody, ranging from sexual innuendo about plants acting as humans to a scurrilous anti-Catholic classification of monks.

A growing interest in landscape gardening and the cult of the picturesque also stimulated interest in natural history. Plants and animals played a significant role in the life of the landed elite, who often possessed illustrated books and synopses of classification schemes in their libraries. The landed gentry patronized specimen collectors and landscape gardeners. Hothouses and stove-plants became an increasing possibility for the wealthy.

Natural history's greatest impact, however, came through geographical expeditions. These introduced a large number of new species to the West, initiating a rage for choice specimens and providing the foundations of national herbaria and museum collections. Many of these exotics came from government voyages of exploration to the Pacific, Australia, the Cape (of Good Hope), and the Americas. Living exotics could be seen at Kew Gardens (royal property until 1841), Chatsworth, or Syon Park near Chiswick, and the great European gardens such as the Jardin des Plantes in Paris. Enterprising private societies such as the Royal Horticultural Society (founded in 1804) and the Société d'acclimatisation in Paris sponsored collecting trips abroad and ran public gardens and competitions for their members. Animals were displayed in private and commercial menageries as well as in museums. The arrival of living examples of Nubian giraffes in Paris or London for these collections aroused national interest.

European prosperity overseas increasingly depended on the development of the plantation system in which staple crops such as tea or sugar-cane were relocated for colonial purposes. Local observers—frequently European expatriates—helped collectors back home. Naval officers, commercial entrepreneurs, and overseas residents gradually opened up the knowledge base of many different regions. In Britain, the East India Company took the lead in establishing botanical gardens in Saharanpore and Calcutta, and Kew Gardens, under Sir Joseph Banks, became a hub of proto-imperial science. The rich collections in Paris and the Netherlands were already the envy of Europe.

During the eighteenth and nineteenth centuries, natural history underwent a profound intellectual transformation. Interest in producing classification schemes was edged out, though never entirely replaced, by investigations into the inward functions of organisms and the history of living beings over time. Natural history itself slowly separated into subdisciplines such as *paleontology, each with their own methods, agendas, and subject-matter. When the Museum d'Histoire Naturelle was reformed in post-Revolutionary France, the academic work undertaken within its domain was especially significant. Georges Cuvier explored the *fossil world, and restructured classification schemes on the basis of four irreducible "types." His analysis of the elephant genus codified the concept of extinction.

Throughout Europe and America, a Christianized version of Cuvier's explanation of earth history became generally accepted. This invoked a series of creations of living communities of animals and plants periodically subject to catastrophic extinctions. Each successive creation advanced in character beyond its predecessor. The modern epoch, which included mankind, dated only from the Biblical flood, which coincided with the watery, icy period that naturalists recognized in the geologically recent past. This cold period divided the ancient history of the earth and the reign of mankind—and, if necessary, more or less matched the Biblical story. Cuvier's work was extensively popularized and had an impact on general natural history as well as in elite academic fields. The transformist schemes of his contemporary, Jean Baptiste de Lamarck, were less acceptable to the elite, but quickly spread in radically politicized form among progressive medical men in Europe (*see* Georges Cuvier and Jean Baptiste de Lamarck).

Nineteenth-century natural history for the most part served Christianity. The geographical expansion of Europe's political and trading dominions, and the rapid economic development of North America, included a generalized Christianizing and educative mission in which natural history was frequently a significant element. Colonialists thought it important that natural history specimens, once collected overseas, be brought back to developed nations where they would be identified, itemized, and displayed as physical tokens of Western knowledge and, in many cases, colonial possession. In this fashion, butterflies from Surinam (say) served as effective devices to display scientific imperialism. Once named, said the British entomologist William Kirby, a thing becomes a possession. An extensive network of naval officers, surgeons, travelers, and local residents, distributed over the globe, contributed to a supply system—a flow of information mostly from the peripheries to the centers of knowledge in Europe's and America's capital cities. This flow of information to an administrative center, and the human networks on which it depended, mimicked the bureaucracies and administrative structures springing up in colonial countries.

Nineteenth-century natural history, like astronomy, lent itself to an imperialist structure. To a large degree, Charles *Darwin's work was built on just such a structure. He considered himself a naturalist of the widest possible remit, interesting himself in turn in geology, plant life, animal breeding, coral reefs, insects, barnacles, and earthworms, and generally wrote in the pastoral mode created by Gilbert White. His work was nevertheless distinguished from the milieu in which he felt most relaxed by the strongly reductionist laws that he proposed to account for species change in nature. Since his day, natural historians have often justified their work and taken comfort from the fact that Darwin was at heart a naturalist and observer like themselves.

After reaching a peak in the middle years of the nineteenth century, the social and intellectual status of natural history underwent a relative decline with the rise of laboratory biology. During the same period, the *popularization of science developed as a large-scale enterprise, disseminating scientific information of all kinds to the public at large. Natural history remained a favorite topic of books and magazines, lectures, exhibitions, and museums. This commercial success distinguished it from most of the other sciences. Lively, picturesque descriptions became the accepted norm, often featuring accounts of animal behavior.

Women formed an important category of authors, readers, and gardeners. They wrote books, attended public lectures, and encouraged

the discipline in a domestic setting. An exceptionally accessible subject at a time when science was become increasingly abstract and mathematical, natural history seemed to belong to everybody. Anyone could advance it by observing or collecting. Amateur naturalists and volunteers in the field collaborated with newly professional university experts, although relationships could be tense. An indication of the growing distance between professionals and amateurs is the development of *ecology. It did not emerge from field observations or inherit the characteristics of nineteenth-century natural history, even though one of its founders, Charles Elton, wrote that ecology was in essence "scientific natural history." In fact, it emerged in opposition to the natural history tradition in a movement in which professionals rejected the natural historical as amateurish and disorganized, not subject to rigorous controls or experiment.

This aura of accessibility continued to be the defining feature of natural history through the twentieth century, aided immeasurably by the development of cinema, photography, and television. Environmentalism and "green" policies have furthered the public appeal of the subject.

David E. Allen, *The Naturalist in Britain: A Social History* (1976). Susan Sheets-Pyenson, *Cathedrals of Science: The Development of Colonial Natural History Museums During the Late Nineteenth Century* (1988). Pascal Duris, *Linné et la France, 1770–1850* (1993). Roger French, *Ancient Natural History: Histories of Nature* (1994). N. Jardine, J. A. Secord, and E. C. Spary, *Cultures of Natural History* (1996). Greg Mitman, *Reel Nature: America's Romance with Wildlife on Film* (1999). Paul Lawrence Farber, *Finding Order in Nature: The Naturalist Tradition from Linnaeus to E. O. Wilson* (2000).

JANET BROWNE

NATURAL MAGIC. See MAGIC, NATURAL.

NATURAL SELECTION. See DARWINISM; EVOLUTION.

NATURAL THEOLOGY. In Western religions, knowledge of a deity has been supposed to derive from three sources: religious experience, revelation (as in sacred texts), and rational inference. Natural theology is principally concerned with the third category, embracing attempts to establish God's existence and attributes through "natural reason" independently of revelation. It has been used to defend the monotheistic religions, exploiting the argument that the only satisfying explanation for why the universe exists at all must refer to a First Cause or Creator. It has also been used by critics of revelation to construct a rational, "natural" religion opposed to established religious authorities. In eighteenth-century France, for example, Voltaire, although deeply critical of the Catholic Church's intolerance toward dissenters, felt obliged to ascribe the order and harmony of the Newtonian universe to an intelligent designer. Newton's own calculations on the finely balanced forces that kept the planets in closed orbits had persuaded him that the deity, like himself, was highly skilled in mechanics and geometry. Through this and other forms of the argument for design, natural theology found its way into popular scientific discourse.

That God's existence could be demonstrated had been affirmed by the medieval theologian Thomas Aquinas, one of whose arguments shows how a natural theology might complete a philosophy of nature. Aquinas asked why inanimate objects should behave in the orderly way Aristotle had described, apparently governed by final causes or goals inherent in nature itself. Physical bodies lacked knowledge and yet acted in concert to achieve certain ends, contributing to a viable world. For Aquinas, such behavior implied direction by a transcendent being having intelligence and intention.

Later, natural theology became more directly relevant to *natural philosophy. Seventeenth-century philosophers associated the assumption of an ordered and intelligible universe with belief in a Creator who had so designed the human mind that it could grasp hidden harmonies in nature through mathematical analysis. Nicholas *Copernicus considered his heliocentric model of the universe more elegant than that of Ptolemy. Johannes *Kepler spoke of his rapture after uncovering the precise relationship between the period of a planet's orbit and its mean distance from the Sun. Regularities still referred to as natural laws were routinely ascribed by European thinkers to a transcendent legislator. Isaac *Newton particularly emphasized the free will of the Creator in choosing which laws to implement and which outcomes to guarantee through the choice of initial conditions.

Natural theology sometimes provided motivation for the study of nature. A moral obligation to study the Creator's works rested on the claim that we alone can appreciate their beauty and economy. The pursuit of natural history was justified in such terms by the English taxonomist John Ray and his Swedish successor Carl *Linnaeus. They found persuasive evidence for design in the structures of living organisms like the human eye, which could appear wonderfully adapted to their functions. William Paley's *Natural Theology* (1802), which shows how theological ideas could affect perceptions of specific scientific theories, epitomized the argument from design. He made his arguments the more acceptable by adjusting them to the political sensibilities of the time. Following the terror of the French Revolution, Paley rejected the anticonservative evolutionary speculations of Erasmus Darwin.

For Paley and many predecessors, the design argument was rooted in an analogy between the world and finely wrought, but unchanging, clockwork. With the gradual disclosure that the earth had not been fixed at Creation but had a long, convoluted history, religious apologists faced a new challenge. Because the argument for design proved particularly vulnerable to the Darwinian theory of natural selection (*see* *DARWINISM), one might suppose that it has only a negative effect on scientific theorizing. On the contrary, the assumption that organisms were designed as perfectly correlated wholes helped paleontologists in reconstructing past creatures from *fossil fragments. Thus the Oxford clergyman and naturalist William Buckland imitated the French academician Georges Cuvier in developing a science of paleoecology in which the eating habits and habitats of extinct creatures were deduced from their fossil teeth and bones. Cuvier's proofs of *extinction, however, could constitute an embarrassment to natural theology. Some nineteenth-century commentators, including the poet Alfred, Lord Tennyson, found it profoundly disturbing that a beneficent Creator should permit entire species to perish.

Despite this and other difficulties, natural theology continued to be relevant to the promotion of the sciences in the English-speaking world. It provided a vocabulary in which scientific knowledge could be presented as spiritually edifying, destructive of atheism, and above suspicion. Cambridge geologist Adam Sedgwick could still argue in the 1830s that the repeated introduction of new species, apparent from the fossil record, indicated a deity with a continuing interest in the world. Precisely because the sciences were purposed to support natural theology, the shock was all the greater when they provided ammunition for the forces of secularism. The Scottish skeptic David Hume had already identified flaws in the design argument, exposing the limitations of inferences based on analogy. Showing how nature could counterfeit design through the refining operation of natural selection, Charles *Darwin would complete the critique. By stressing that the variations on which natural selection worked appeared randomly and by depicting the evolutionary process as one involving repeated divergence from common ancestors, Darwin made it difficult to assert that the process had been instigated with humans in mind. There were, nevertheless, Christian thinkers who welcomed Darwin's theory, believing that it indicated a deeper unity of nature, that it helped to show how pain and suffering could be concomitants of a creative process, and that, by extending the domain of natural laws, it would reinforce respect for moral laws. Even today there are calls for the revival of natural theology based on the fineness of tuning in the earliest moments of the *Big Bang, seemingly a precondition of the possibility of intelligent life.

Robert Hurlbutt, *Hume, Newton and the Design Argument* (1965). Dov Ospovat, *The Development of Darwin's Theory: Natural History, Natural Theology, and Natural Selection* (1981). John Brooke and Geoffrey Cantor, *Reconstructing Nature: The Engagement of Science and Religion* (1998). Gary Ferngren, ed., *The History of Science and Religion in the Western Tradition: An Encyclopedia* (2000).

JOHN H. BROOKE

NATURALISM AND PHYSICALISM. In the philosophy of science, naturalism is the view that all phenomena belong to *nature and should be studied with the concepts and tools provided by the natural sciences. Physicalism assumes in addition that all phenomena (including the phenomena of mind) can be explained in the language of physics. Physicalism is occasionally identified with materialism (*see* MATERIALISM AND VITALISM; MIND-BODY PROBLEMS; REDUCTIONISM). Recently some authors have called "naturalism" the position of those who, like Thomas S. Kuhn, have conceived the philosophy of science as an activity continuous with the sciences and their history, rather than as the kind of logical enterprise advocated earlier in the century by many Anglo-American philosophers.

The history of naturalism and physicalism is a complex one. In early modern Europe "naturalists" were philosophers who (often covertly because of ecclesiastical censorship) endorsed the materialistic tenets of Epicurus, Lucretius, and their followers in the *Renaissance tradition. During the seventeenth century similar positions were credited to philosophers like Thomas Hobbes and Baruch Spinoza. The latter inspired a literature, circulating mostly underground, that attacked traditional beliefs in spirits, sorcery, the devil, and the supernatural, and was thus regarded as atheistic in its consequences if not in its premises. Meanwhile, milder, more widespread forms of naturalism—associated with the names of John Locke, Robert *Boyle, and Isaac *Newton—claimed that *natural philosophy, by revealing "nature's laws," shed new light on traditional issues such as the laws of reason, moral philosophy, and God's intentions (*see* LAW OF SCIENCE; RELIGION AND SCIENCE).

Appeals to both "nature" and "enlightenment"—the latter conceived as a philosophical and social agenda rooted in the natural sciences—figured prominently in an important section of the literature on science produced during the eighteenth century. Enlightened philosophers appealed to nature in order to bring new, objective support upon moral imperatives that, due to increasing secularization, lacked other, notably religious sanctions (*see* ENLIGHTENMENT AND INDUSTRIAL REVOLUTION). Philosophers like David Hume, however, showed they were aware of the tensions—between "nature" and liberty for example—implicit in naturalism.

Whatever the implications, toward the end of the eighteenth century naturalism—conceived as the exclusion of preternatural causes from scientific discourse—had been introduced even in fields like the earth and life sciences previously impervious to it. Throughout the following century the movement known as positivism capitalized on these developments, and on the related notion that the natural sciences displayed *progress in their history (*see* DARWINISM). Positivists asserted that science was the main if not the only valid form of knowledge, a frequent corollary being that the social sciences had to be modeled on the natural sciences.

After the criticisms leveled against previous naturalism by physicists like Ernst Mach, during the first half of the twentieth century naturalism and physicalism adopted a linguistic and logical orientation, and they were often pursued under the description of logical positivism (*see* POSITIVISM AND SCIENTISM). This earlier, linguistic turn explains why later in the century the renewed interest of philosophers of science for the history of science, and for notions drawn from the biological and the social sciences, was perceived by some as a return to naturalism.

Jeffrey Poland, *Physicalism* (1994). Jonathan I. Israel, *Radical Enlightenment* (2001).

GIULIANO PANCALDI

NATURE. In early modern Europe, religious tradition and classical Greek philosophy, as reinterpreted by medieval and *Renaissance philosophers, molded conceptions of the natural world and its relation to humankind. Since then the ideas of nature circulating in western cultures seem to have oscillated between two poles, expressing different perceptions of the needs of human life and social relations.

At one end of the spectrum stand the conceptions that view humankind as having an undisputed ascendancy—sanctioned by religion, or else by humanistic and social concerns—over the natural world. Within this tradition the needs and goals of human (or divine) actors come first.

At the other end of the spectrum stand the conceptions that view nature as independent of human minds, needs, and concerns. In this view the natural world must be studied in its own right, priority is given to objective knowledge, and humans should refrain from projecting their own needs and goals onto the natural world. This view itself, however, often credits what is found in nature with special, normative powers that refer back to humans.

Somewhere between the two poles stand the instrumentalist views of natural knowledge, which pursue knowledge and practices making only limited claims about the natural world and the role of human actors in it (*see* INSTRUMENTALISM AND REALISM).

Whatever the particular conception of the relationship between nature and humankind, the language used in discourses about nature has often been rich in anthropocentric metaphors, occasionally gender oriented. This was the case even after the Copernican revolution had removed the earth, and presumably humankind, from the center of the universe. The metaphor of nature as a "book," for example, has helped to frame the agendas of natural philosophers since *Galileo. So has the notion of the "laws" of nature (*see* LAW OF SCIENCE). In shaping the concepts of the "order" and "economy" of nature, Enlightenment and Romantic naturalists relied explicitly on notions of social order and equilibrium (*see* CLASSIFICATION IN SCIENCE). Metaphors inspired by human concerns continued to play a role in the natural sciences even after Charles *Darwin had dethroned humankind from its special place in nature, as shown by Darwin's own concept of "natural selection," modeled on breeders' techniques.

Dissatisfaction with previous naturalism and anthropocentrism loomed large among the motives that, around 1900, drove scientists and especially physicists to reassess the conceptual foundations of science (*see* NATURALISM AND PHYSICALISM). One result of the reassessment was the realization that even physicists continued to rely on notions of nature appropriate only to the particular experiences, dimensions, and speed of locomotion of the human species. According to an early twentieth-century commentator, Bertrand Russell, one major achievement of Albert *Einstein's theory of relativity was to force scientists (and, hopefully, the lay public as well) to try to adopt a "point of view" on nature proper to an electron.

The subsequent, further moves away from an anthropocentric view of nature during the twentieth century produced a dramatic, partly unexpected increase in the power that scientists and industrial societies exerted over natural resources, and the other inhabitants of the planet. Predictably, these developments have given renewed vigor to expectations and concerns—about nature and human action—that have accompanied the public perception of science and technology since the industrial revolution.

See also ECOLOGY; ENLIGHTENMENT AND INDUSTRIAL REVOLUTION; SOCIAL RESPONSIBILITY OF SCIENTISTS.

Bertrand Russell, *The ABC of Relativity* (1925; 1997). Keith Thomas, *Man and the Natural World* (1983).
GIULIANO PANCALDI

NATURPHILOSOPHIE. Naturphilosophie was a school of thought characterized by a speculative, idealistic, and holistic approach to the study of nature that had its origins in Germany at the end of the eighteenth century. Since many of its proponents were also associated with Romanticism, Naturphilosophie is often referred to as Romantic science. In much the same way that Romantics reacted against rationalism, adherents of Naturphilosophie reacted against the traditions they saw as deriving from Francis Bacon and Isaac *Newton: the idea that the world was atomistic, that an inductive and empiricist methodology was the best way of exploring it, and that mathematics was the language of nature.

Naturphilosophie combined a version of Neoplatonism with a reading of Immanuel Kant. From the former, which they found in the works of Paracelsus and Jean Baptiste van Helmont, came the Naturphilosophen's belief that all the forces we perceive in the world are manifestations of one basic force. From the latter, they took the idea that in the construction of knowledge, the mind imposes its categories such as space, time, and cause and effect on nature. But whereas Kant emphasized that the mind cannot know the nature of things, the Naturphilosophen reinterpreted the imposition of our mental categories not as veils but as insights. They celebrated human reason for its capacity to participate in the divine reason and thereby to comprehend nature in its entirety.

Friedrich Schelling, who wrote his influential *Ideas for a Philosophy of Nature* in 1797, is regarded as the founder of the movement. For Schelling, nature consisted of opposites or polarities: positive and negative for electrical phenomena, north and south for magnetic, acids and bases for chemical. In each case, the opposing forces resolved and unified in new phenomena and new forces on a higher plane. All manifested a single underlying force and could be converted one into the other in the proper circumstances.

Although he stood apart from Naturphilosophie, Johann Wolfgang von Goethe shared many of its ideas and was held in high regard by many of its leaders. In his seminal work, *On the Metamorphosis of Plants* (1790), Goethe interpreted the organs of the individual flowering plant as modifications, or metamorphoses, of a single basic form, an idealized, primal leaf. As a group, flowering plants were variations of an ideal plant archetype, or Urpflanze. Goethe made his opposition to Newton most explicit in his study of the perception of colors, *Zur Farbenlehre* (1810).

Lorenz Oken, who worked in comparative anatomy, synthesized many of Naturphilosophie's themes in his *Elements of Physio-Philosophy* (1809–1811, English translation 1847). He influenced a generation of students at the University of Jena and abroad. Schelling's student Heinrich Steffens and Johann Wilhelm Ritter also edged Naturphilosophie in the empirical direction with their research on *geology and *electricity, respectively.

During the early nineteenth century, Naturphilosophie spread beyond Germany. It inspired some natural philosophers, particularly those working on problems in electricity, chemistry, magnetism, and anatomy. Naturphilosophie helped Hans Christian Ørsted formulate the questions that led him to the discovery of *electromagnetism (1820). Similarly, it stimulated Thomas Seebeck's research that detected thermal electricity (1822), Humphry Davy's and Michael *Faraday's exploration of electrochemical and *electromagnetic phenomena, and the concept of the *conservation of energy. Schelling himself believed that Faraday's work on electromagnetism confirmed his own theories.

In the life sciences, Naturphilosophie postulated that the succession of higher life forms on earth was the outcome of opposing forces present in lower forms. Although suggestive of modern evolutionary thought, this succession was not evolution in the sense of genetic descent, but rather a process of ascent toward a preordained ideal, comparable to embryological gestation. Such ideas crop up in the transcendental anatomy of Étienne Geoffroy-Saint Hilaire, Henri de Blainville, Robert Knox, Robert Grant, Richard Owen, Edward Forbes, Jr., and Louis Agassiz.

By the mid–nineteenth century, reactions to Naturphilosophie, like its fundamental theory, were sharply polarized. Within German philosophy, its influence continued, thanks largely to Georg Wilhelm Friedrich Hegel's influential *Enzyklopädie der philosophischen Wissenschaften* (1817); it re-emerged in the critique of modern science mounted by Continental philosophy in the twentieth century. In the scientific mainstream, the rise of *positivism, Darwinism, and professional specialization, among other factors, led to Naturphilosophie's demise. Its influence on early nineteenth-century science was largely forgotten until the 1960s. Then, with the turn away from positivism as the dominant historiography of science, historians of science (as well as historians of ideas and literary historians) began uncovering the paths by which Naturphilosophie affected the growth of modern science.

Alexander Gode-von Aesch, *Natural Science in German Romanticism* (1941). L. Pearce Williams, *The Origins of Field Theory* (1966). Trevor Levere, *Poetry Realized in Nature: Samuel Taylor Coleridge and Early Nineteenth-Century Science* (1981). Philip F. Rehbock, *The Philosophical Naturalists: Themes in Early Nineteenth-Century British Biology, Part I* (1983). Andrew Cunningham and Nicholas Jardine, eds., *Romanticism and the Sciences* (1990).

PHILIP F. REHBOCK

NAVIGATION. Seafaring has strongly influenced both natural knowledge and practical mathematics. Voyages have expanded our view of the world and obliged us to accommodate to new natural histories and geographies. The technical problems associated with sailing reliably, safely, and efficiently have profoundly affected applied geometry and *instrument making while helping to shape the programs of *cartography, horology, and *astronomy.

Because seamen had to deal with the world as they found it, not as tradition or authority prescribed, there was a strongly empirical influence at work in the evolution of navigational practice. Early sea charts—the *portolan* charts produced by Italian and Catalan cartographers in the early fourteenth century—present a realistic representation of the European coastline at a time when land maps were more concerned with morality than geography. These charts, with their names of ports crowding the coastlines and sets of radiating compass bearings or "rhumb lines" intersecting across the seas, indicate contemporary navigational technique. This was a mixture of coastal sailing or pilotage, using chart, sounding line, and local knowledge, and sailing farther out to sea, following the compass bearing indicated by the chart and estimating speed while recording the passage of time by a sandglass (*see* MAGNET AND COMPASS).

Bearing and distance relied on "dead reckoning," the plotting of position by recording progress rather than by an independent fix, a technique of limited use in more ambitious voyages in unfamiliar seas. Astronomers offered some help through the relationship between geographical location and the appearance of the heavens. Latitude could be found from the elevation of the Pole Star or from the meridian altitude of any heavenly body of known declination, including the Sun. A quadrant, cross-staff, or simple form of *astrolabe could be used for the measurement; tables gave the annual cycle of change in the Sun's declination, and a "backstaff" allowed measuring solar altitude without staring at the Sun. This technique for measuring latitude led to a new method of navigation by the sixteenth century: in "latitude sailing" a course would be fixed well to the east or west of the target, and the latitude gained and then maintained while sailing west or east until making landfall. Because no complementary method for longitude existed and because the sky might well be covered by cloud, there was still a place for dead reckoning.

Other important developments were the introduction of a projection for sea-charts, used in Gerardus Mercator's world map of 1569, explained by Edward Wright in 1599, and rendered amenable to calculating instruments by Edmund Gunter in 1623. The straight line between two points on a Mercator chart indicated the compass bearing to one from the other. The search for a *longitude solution took many turns in the seventeenth and eighteenth centuries but eventually yielded two practical methods employed in the nineteenth—using a chronometer to keep standard time or finding it by measuring the Moon's position among the stars. More accurate measurement at sea was made possible by the introduction of portable octants, sextants, and circles that relied on using reflection in two mirrors, at least one being mounted on an index arm, to bring two targets (Moon, star, Sun, or horizon) into apparent coincidence by rotating one mirror through a measured angle.

The solution of the longitude problem meant that navigators had the possibility of a complete position fix, but success still depended on the visibility of the sky. Older techniques of recording were refined through the development of mechanical logs for measuring speed and distance, while magnetic compasses were improved and adapted to work in the hostile environment of a metal ship. William *Thomson, Lord Kelvin worked successfully on improving the compass and its housing or binnacle, and also devised a sounding machine that utilized the relationship between underwater depth and pressure. Also in the nineteenth century, new ways of combining the techniques of measurement and timekeeping were devised, going under the general name of "position-line navigation," which freed the navigator of his dependence on a noonday sight for latitude.

In the twentieth century the gyrocompass, depending on the directional properties of a spinning gyroscope and under development in the early decades of the century, came to supplement the magnetic compass, while echo sounding replaced mechanical devices for finding depth. Radio provided the basis of complete position-finding systems, first developed during World War II. From the 1960s there has been sustained development of a satellite-based positioning system with navigation radio beacons transmitting from orbits in space. The Global Positioning System (GPS), under construction since the 1970s, came on line with twenty-four *satellites in 1995.

Eva G. R. Taylor, *The Haven-Finding Art: A History of Navigation from Odysseus to Captain Cook* (1956). Charles H. Cotter, *A History of Nautical Astronomy* (1968). W. E. May, *A History of Marine Navigation* (1973). Jean Randier, *Marine Navigation Instruments* (1977).

JIM BENNETT

NAZI SCIENCE refers in a general sense to all science, its practitioners, and its policy during the Third Reich (1933–1945), and more specifically only to those sciences directly tainted ideologically in content or in application by National Socialist policy. To the occupying powers, all scientists who remained in the Third Reich were "Nazi scientists" for purposes of denazification. For decades after World War II, the historical examination of science under National Socialism was taboo. Since the publication in 1977 of Alan Beyerchen's groundbreaking study of Nazi physics, the extent of Nazi influence has been revealed and the difference between "science in the Third Reich" and "Nazi science" has diminished. Under Hitler, nearly all areas of science, technology, and medicine, and indeed of scholarship in general, were influenced by the Nazi regime in content, practice, policy, or administration. Historians have tried to understand how and why sciences not obviously relevant to National Socialism still became more or less Nazified in content or approach. Nazi science is thus one of the most important historical examples of the relationship between science and ideology.

The most obvious of the Nazi sciences was Aryan science. The purging of Jews from state positions under the Nuremberg Laws of 1933, and the accompanying elimination of "Jewish theories," such as *relativity, from textbooks, led attempts to create ethnically pure varieties of science that embodied National Socialist values. Aryan science included the application of rationality and efficiency to pressing state problems, especially the "Jewish Question" and the infamous Final Solution. It also became manifest in fields as diverse as physics, chemistry, mathematics, and psychology. Once politically objectionable individuals had left Germany, the state or party orchestrated the decline of Aryanized disciplinary science—except for certain forms of the biomedical, engineering, and social sciences useful in attaining Nazi social and political objectives.

Although Nazi racial policy had links both to Social Darwinism in Imperial Germany (1871–1914) and the racial hygiene and *eugenics movements in the Weimar Republic (1919–1932), the execution and administration of those policies were distinctive and tragic occupations of the Third Reich. Already, Weimar Germany used euthanasia in mental institutions and co-opted biology instruction, even at introductory levels, to teach racial purity, civic-mindedness, health, and hygiene. Although some biologists remained unaffected by the Third Reich, biology and the biomedical sciences were mobilized in war-related projects in research institutes or concentration camps on twins, eye color, air pressure, and reproduction. Genocide could only be accomplished with the chemical weapons created during World War I and further developed and deployed in medical settings outside the military in the 1920s. Part of Nazi racial policy involved the resettlement of Germans in eastern Europe. Area studies and spatial planning policy, using a multi-disciplinary approach based on geography, provided the practical details needed to implement the plan. This *Ostforschung*, or research on the east, presumed that the Polish, Jewish, Russian, and Ukrainian populations of eastern Europe would be expelled or murdered. The loss of the war thwarted the project to resettle the east, but the principles of urban and spatial planning developed for it remained useful. Elements of it figured in the reconstruction of West Germany, especially in handling the influx of German citizens forced to leave the eastern sections of the former Reich.

In defiance of the Versailles treaty, the Nazi regime supported military science and technology beginning with the introduction of armaments research in engineering courses at the technical universities in the early 1930s. In 1936–1937, the Reich openly supported military weapons projects; its devastating loss in the *Blitzkrieg* (lightning war) of 1941–1942 intensified its weapons research. In principle, Hitler himself had to approve all major military research and development projects, but bureaucratic conflicts led to inconsistent oversight of even the most strategic of them, including aeronautics, the uranium bomb, and rocketry.

Aeronautical research, based at Ludwig Prandtl's Göttingen institute, effectively circumvented the restrictions of the Versailles treaty, but was stymied occasionally by covert acts of resistance. Nevertheless, *aeronautics became not only a Nazi strength, but later a West German one as well.

The uranium project under Werner *Heisenberg, one of the most enigmatic projects of the Third Reich, failed to build a bomb; Heisenberg nonetheless went on to become the most important public spokesman for science in democratic West Germany and a major figure in promoting West German *nuclear power plants. The guided missile and rocketry project under Wernher von Braun was the most outstanding technological success of the Third Reich. Under the Army's administration at first, the rocketry project expanded when the Air Force took over in 1935, and then again at Peenemünde and eventually in an abandoned mine in the Harz Mountains near Nordhausen, where concentration camp prisoners from Buchenwald supplied the labor for the protection of the vengeance weapon, the V-2, which Hitler believed would win the war. Both the Americans and the Soviets captured nearly all of Hitler's rocket scientists. During the Cold War, the Germanies had hardly any space program, while the superpowers explored space and escalated international tensions with the help of former leaders of Nazi missile projects.

That the Third Reich was not more successful in science and technology has to do in large part with the fractious nature of its internal politics. A state that appeared to be run rationally was weakened by competition among power groups that precluded success in major projects. Where broad definitions and goals guided policy and administration—in matters concerning military technologies, race, gender, *Lebensraum* (living space) in *Mitteleuropa*, and *Gleichschaltung* (social coordination, or alignment with the regime)—science and technology were for the most part effectively co-opted. The most successful of the Nazi policies in science, racial cleansing and genocide, were also the most tragic. The alliance of science, technology, and medicine with the National Socialist technocracy led to a reorganization of scientific administration that survived the Reich and helped to create the big science methods of management found in both Germanies after 1945. These administrative strengths contributed significantly to the rebuilding of the scientific community after 1945, and eventually enabled both Germanies to compete scientifically with the superpowers during the Cold War. In both its short-term and long-term successes, however, Nazi science cannot evade moral judgement.

See also KAISER WILHELM/MAX PLANCK GESELL-SCHAFT; TEXTBOOK.

Alan Beyerchen, *Physicists under Hitler* (1977). Peter Hayes, *Industry and Ideology: IG Farben in the Nazi Era* (1987). Robert N. Proctor, *Racial Hygiene: Medicine under the Nazis* (1988). Mark Walker, *German National Socialism and the Quest for Nuclear Power, 1939–1949* (1989). Michael Burleigh, *Death and Deliverance: 'Euthanasia'* in Germany, 1900–1945 (1994). Michael Neufeld, *The Rocket and the Reich* (1995). Ute Deichmann, *Biologists under Hitler* (1996). Neil Gregor, *Daimler-Benz in the Third Reich* (1998).

KATHRYN OLESKO

NEBULA. With little more to recommend himself than neat handwriting, Charles Messier began his working life as a clerk to a French astronomer, and eventually became the world's foremost hunter of nebulae. M31, the thirty-first object in his 1781 catalog of 103 nebulous-appearing objects that might be mistaken for comets, is now known as the great spiral galaxy in the constellation Andromeda. By 1800, Willam Herschel, with his reflecting telescope, had raised the number of known nebulae to around 2,000.

Opinion swung back and forth over whether nebulae were gaseous or stellar, clouds of luminous fluid or remote star systems. Herschel's big instruments resolved several nebulae into stars. In 1790, however, he encountered a planetary nebula (now known to be an expanding shell of gas ejected from and surrounding a very hot star) that defied resolution. Astronomers came to doubt that any nebulae were composed of stars, or constituted island universes. They came to consider even the Andromeda Nebula as a mass of nebulous matter, definitely not a stellar system. Larger reflecting telescopes constructed around the middle of the nineteenth century resolved more nebulae into groups of stars, and opinion again swung toward the concept of stellar composition of all nebulae. Even the Orion Nebula (now known to be a cloud of glowing gas lit by young stars in the process of formation) was resolved into stars. The director of the Harvard College Observatory exclaimed to the college president, "You will rejoice with me that the great nebula in Orion has yielded to the powers of our incomparable telescopes."

The end of the nineteenth century saw opinion swing yet again. Spectroscopic observations by William Huggins revealed that about a third of some seventy nebulae displayed a gaseous character. Furthermore, Huggins insisted that the nebulae with stellar spectra were composed of gas under special conditions. Astronomers persisted in an "either-or" choice, unable to imagine that both gaseous and stellar nebulae might exist, even as the number of specimens grew. The *New General Catalogue of Nebulae and Clusters of Stars* (1888, 1895, 1908) listed over 13,000 nebulae, and astronomers estimated that 150,000 nebulae were within reach of existing instruments.

The distribution of nebulae was revealing, their avoidance of the plane of the Milky Way seemingly linking them physically to our *galaxy. Later, after the extra-galactic nature of many nebulae was established, the apparent zone of avoidance was attributed to obscuring gas and dust in the plane of our galaxy. As late as 1917, however, the American astronomer Edwin *Hubble could note that "extremely little is known of the nature of nebulae, and no significant classification has yet been suggested; not even a precise definition has been formulated."

The spectroscope revealed the true nature of gaseous nebulae; other nebulae, including the great spirals, are stellar systems. During the 1920s, Hubble proved that the spirals in fact lie outside our galaxy.

Edwin Hubble, *The Realm of the Nebulae* (1936). Norriss S. Hetherington, *Encyclopedia of Cosmology: Historical, Philosophical, and Scientific Foundations of Modern Cosmology* (1993).

NORRISS S. HETHERINGTON

NEOPLATONISM names a philosophical tradition stemming from late Antiquity. Between the third and fifth centuries A.D., Plotinus, Porphyry, and Iamblichus developed a system of ideas that drew on Plato's work while modifying it significantly at the same time. Neoplatonists believed that the world was unified, not dualistic. It depended on a supreme source, variously called the One, the Divine Mind, the Logos, the Demiurge, or the World Soul, from which emanated all the other intelligences and levels of reality, including the one inhabited by humans. As far as knowledge of the natural world was concerned, Neoplatonists drew heavily on the *Timaeus*. In this work, Plato outlined a mythical cosmogony in which the Demiurge, the creating spirit, used the five perfect solids of mathematics—the tetrahedron, the cube, the octahedron, the dodecahedron, and the icosahedron—as templates for the heavens. Musical harmonies went with these solids. Light was the efflorescence of the Demiurge and the means by which humans acquired knowledge.

The Neoplatonic tradition penetrated Islam, Byzantium, and the Christian West. At the beginning of the fifth century, Augustine made much use of it, as did Boethius at the beginning of the sixth century. Unlike Plato's other works, the *Timaeus* was available in Latin translation throughout the Middle Ages.

In the late 1430s, the Byzantine Neoplatonist George Plethon traveled to Florence, uniting Byzantine and Western Neoplatonism. He persuaded Cosimo de Medici to establish a Platonic Academy that although oriented more to the new humanism than to natural science, provided a model for academies of all kinds, some of them devoted to science, that were set up from the mid-sixteenth century on. The Florentine Academy had as members Marsilio Ficino, who translated all of Plato and Plotinus into Latin, and Giovanni Pico de la Mirandola. They were largely responsible for the melding of Neoplatonism with Hermeticism and natural *magic that intrigued intellectuals for the next two centuries. North of the Alps, Nicholas of Cusa also drew on the Neoplatonic tradition when he argued for the certainty of mathematical knowledge and its centrality to natural philosophy, doctrines that influenced Giordano Bruno.

Since the publication of Edwin A. Burtt's *Metaphysical Foundations of Modern Science* in 1932 and Alexandre Koyré's many historical studies, particularly his *Études galiléennes* in 1939, historians of science have debated the extent to which Neoplatonism influenced or inspired *Copernicus, *Kepler, *Galileo and *Newton. The case of Johannes Kepler seems particularly clear. Throughout his astronomical work, culminating in his *Harmonices mundi* (1618), Kepler sought the geometrical structure of the universe, believing it to be based on the five perfect solids and characterized by musical harmonies.

Newton was influenced by another important school of Neoplatonism that emerged at the University of Cambridge in the seventeenth century. The Cambridge Platonists, appalled by the *mechanical philosophy of Thomas Hobbes, created a synthesis of Christianity and Platonism. Although most of the group applied themselves to theology, metaphysics, and ethics, Henry More, a Fellow of the Royal Society, ventured closer to natural philosophy. After an initial flush of enthusiasm, he reacted against *Descartes's identification of space and matter with extension. He made space an attribute of God's, and the means by which God acted on bodies. Newton studied with More, whose concept of space and time as the "sense organs of God" helped shape Newton's concept of absolute space and time.

From the seventeenth century on, Neoplatonism ceased to be an important living philosophical tradition as philosophers concentrated on recovering the true Plato. In science, one important exception occurred in the late eighteenth century. Leaders of *Naturphilosophie, such as Friedrich Schelling and Hegel, owed their belief that all the seemingly separate forces of nature, such as electricity, magnetism, and gravity were expressions of one underlying unified force to their reading of the Neoplatonic tradition.

Ernst Cassirer, *The Platonic Renaissance in England* (trans. 1953, repr. 1970). David Lindberg, "The Genesis of Kepler's Theory of Light: Light Metaphysics from Plotinus to Kepler," *Osiris* 2 (1986), 5–42.

RACHEL LAUDAN

NEPTUNISM AND PLUTONISM. The debate between Neptunists and Plutonists about the origin of the rocks of the earth's crust took place in the late eighteenth and early nineteenth centuries. The Neptunists (named for the Roman god of the ocean) believed that essentially all rocks had been formed in water. According to the most prominent Neptunist, Abraham Gottlob Werner, who taught at the Mining Academy of Freiberg in Saxony, originally the earth was covered by a hot, thick, basic aqueous brew. As this cooled, the rocks that form the core of mountain chains crystallized out. Later, noncrystalline rocks were deposited as layers of strata banked up against the primary, crystalline rocks. Volcanoes were late and largely inconsequent phenomena caused by the burning of plant remains. Underlying his theory was a tradition, stretching back to Johann Joachim Becher and Georg Ernst Stahl in the seventeenth century, of chemical cosmogonies based on the assumption that processes observed in the laboratory could inform theories about mineral formations observed in the field.

The Vulcanists Nicolas Desmarest and Rudolph Eric Raspe argued that basalts, often found interbedded with strata, had actually flowed from volcanoes. Studies of extinct volcanoes in the Massif Central in France confirmed this assertion. The point was quickly accepted by the School of Freiberg, some of whose members, particularly Leopold von Buch and Alexander von *Humboldt, carried out major studies on volcanoes in the first part of the nineteenth century.

James *Hutton, who was influenced by Newtonian traditions, is regarded as the leading representative of the Plutonist theory (named for the Greek god of the underworld). According to Hutton, heat was responsible both for the consolidation of rocks at the bottom of the ocean and for their subsequent elevation to form land. John Playfair, in the *Illustrations of the Huttonian Theory of the Earth* (1802), concentrated on the evidence for the theory and downplayed its natural philosophical and chemical foundations. James Hall carried out dangerous experiments with limestone heated under pressure and lived to report that it did indeed consolidate under sufficient pressure. However, neither Playfair nor Hall succeeded in convincing the geological community that the strata had been consolidated by heat.

By the 1820s, most geologists agreed that strata formed under water and that basalt and certain other igneous rocks were spewed out from volcanoes. The origin of the hard and often crystalline rocks such as granites and gneisses continued to pose a problem. In the late nineteenth century, petrologists fought about the origin of granite in what appeared to many of them as a replay of the Neptunist-Plutonist (or Vulcanist) debates of a hundred years earlier.

Following and amplifying Charles *Lyell's influential historical introduction to the *Principles of Geology* (1830), historians and geologists characterized this debate as a sterile consequence of the intrusion of nonscientific issues into geology. They criticized the Neptunists for taking the idea of a universal ocean from the Bible, and esteemed the Plutonists as excellent field geologists. A more adequate understanding of the debate recognizes that both sides supported their positions with a mixture of empirical data and theoretical arguments, and that in doing so both contributed to the clarification of the basic principles of *geology.

Roy Porter, *The Making of Geology* (1977). Rachel Laudan, *From Mineralogy to Geology: The Foundations of a Science, 1650–1830* (1987).

RACHEL LAUDAN

NETWORK AND VIRTUAL COLLEGE. That scholars belong to an ideal "republic of letters," independent of local power and constraints, has been a frequent claim of savants, and especially natural philosophers and scientists, since the Renaissance. In the 1660s the early scientific societies tried to implement the claim: they presented the establishment of a regular correspondence with distant cultivators of the sciences as a prominent goal of their mission. During the Enlightenment that same claim, and the related notion that natural knowledge increases by being communicated, became widespread tenets, shared by the educated public and many state administrators.

At the beginning of the twentieth century scientists and historians of science like John Theodore Merz and Pierre-Maurice-Marie Duhem emphasized again that intellectual interchange is constitutive of scientific endeavor. Merz noted that during the nineteenth century scientists had built such an impressive system of periodical publications, associations, exchange schemes, and travels that isolation had become almost impossible for anybody engaged in serious scientific activity (*see* INTERNATIONALISM AND NATIONALISM). Botanist Alphonse de Candolle and Merz himself, on the other hand, noted that the centers of scientific excellence had migrated over time from place to place, thus adding an historical dimension to the old notion.

Between 1930 and 1950 sociologists like Robert K. Merton developed the concept of a "scientific community." In response to the challenges coming from the involvement of scientists in two world wars, as well as in totalitarian regimes, the "scientific community" was conceived, again,

as enjoying substantial independence from social and political constraints, and evolving its own norms and communication patterns.

In the flow of studies on communication among scientists that followed, the notion of "invisible colleges" received special attention. Borrowed from the self-ascribed name of an informal group of natural philosophers active during the English Civil War (some of whom later associated with the Royal Society of London), the phrase designated the informal, small groups of scientists who—especially amid the flood of information and opportunities generated by twentieth-century science—managed to control through personal contacts the resources needed to foster their discipline and increase their power.

The diffusion in the 1980s of computer networks and public talk about them offered a trendy new word for renaming the old objects of study passed on by historians and sociologists of science. When used to designate communication and collaborative work among experts, as suggested by Bruno Latour, "network" and its derivatives carried a host of associated metaphors molded on "knots," "nodes," "nets," "links," and "meshes." The jargon favored a critical revision of some by then venerable notions; it also invited occasional abstraction from the concrete attitudes and means that make teamwork in the sciences possible.

In the 1990s, the diffusion of computer mediated communication and of research and educational resources made available online through the Internet led to the introduction of catch-phrases such as "the virtual college" and "the virtual university." In dealing with these, as with earlier fortunate concepts, one must work to detect the real new potentialities within the long-term, often traditional individual and social attitudes that continue to sustain (and constrain) research and education in science and technology.

Bruno Latour, *Science in Action: How to Follow Scientists and Engineers through Society* (1987). John P. Walsh and Todd Bayma, eds., "The Virtual College: Computer-Mediated Communication and Scientific Work," *The Information Society* 12 (1996): 343–363.

GIULIANO PANCALDI

NEUMANN, John (Johann) von (1903–1957), mathematician, mathematical physicist, computer scientist.

Born into a well-to-do liberal Jewish family, John von Neumann displayed his exceptional mathematical gifts early. Following his father's wishes, he studied chemistry at the University of Berlin and later at the Technische Hochschule in Zurich, and received his diploma in chemical engineering in 1926. He managed to obtain a

Ph.D. in mathematics in the same year with a dissertation in set theory from the University of Budapest. This was to be the first scientific domain where von Neumann made a lasting contribution. Using an axiomatic approach (one of the hallmarks of his style), he cast set theory in a new framework; his definition of ordinal numbers (1923) is still used today.

Von Neumann devoted himself to mathematics while a privatdocent in Berlin (1927–1929) and Hamburg (1929–1930). During that time, he collaborated with David Hilbert at Göttingen. This period marked the start of his interest in the development of *quantum physics. He gave the quantum formalism what many regarded as its final form, that of an operator calculus in Hilbert spaces. His axiomatic treatise, *The Mathematical Foundations of Quantum Mechanics* (1932), remains one of the best expositions of the theory. It stimulated discussion of issues still widely discussed today, among them hidden variables and theory of measurement. He claimed to have proved that no theory consistent with quantum mechanics could be expressed by hidden variables (hypothetical quantities yet to be measured) that eluded the uncertainty principle (*see* COMPLEMENTARITY AND UNCERTAINTY; QUANTUM MECHANICS).

Von Neumann continued working on operator theory during the 1930s. His "rings of operators," now called von Neumann algebras, and their important outgrowth, "continuous geometry," cover an important part of present-day mathematics.

Following an invitation to lecture in Princeton in 1930, von Neumann accepted a professorship there in 1931. Two years later, he became one of the first faculty members of the newly created Institute for Advanced Study. He worked on the foundations of statistical mechanics and in 1932 gave a formulation and proof of the ergodic hypothesis. Another highlight of his Princeton years was his solution of an important special case of Hilbert's fifth problem, the case of compact groups.

When World War II broke out, von Neumann, a naturalized American citizen since 1937, served his new country as a consultant to many military and civilian agencies. He worked on ignition methods of nuclear weapons for the *Manhattan Project and later participated in the development of the hydrogen bomb. From 1940 on, his main interests centered on applied research, from statistics and numerical analysis to hydrodynamics, aerodynamics, ballistics, and meteorology. One famous example was the application of his minimax theorem (going back to 1928) to game theory in a book with Oskar Morgenstern, *Theory of Games and Economic Behavior* (1944).

For his various applied projects, von Neumann needed powerful computational tools (*see* CALCULATOR). In the mid-1940s, he made many important contributions to the theoretical and practical aspects of computing. Theory owes him his advocacy of a stored-program machine; technology views him as one of the fathers of the modern electronic computer because of the lineage that he initiated with the machine he built at the Institute for Advanced Study (the ECP, Electronic Computer Project). He was later a pioneer in general automata theory. He studied the curious problems of designing reliable machines out of unreliable components and of constructing automata that can reproduce themselves (*Theory of Self-Reproducing Automata* [1966], with Arthur W. Burks).

Von Neumann's multi-faceted brilliance brought him many honorary doctorates, memberships in many academies of science, and the highest American civil awards: the Distinguished Civilian Service Award (1947), the Medal for Merit (1947), the Medal of Freedom (1956), and the Enrico Fermi Award (1956).

Although indisputably one of the most powerful minds of the twentieth century, von Neumann lived fully in his time. He led an intense and busy life, both scientifically and socially, and is remembered as a spirited and jovial character. His untimely death annihilated a universal intellect at the peak of its creativity.

See also CALCULATOR; COMPLEMENTARITY AND UNCERTAINTY.

Garrett Birkhoff et al., "Memorial Papers on John von Neumann," *Bulletin of the American Mathematical Society* 64 (1958): 3, 2. Steve J. Heims, *John von Neumann and Norbert Wiener: From Mathematics to the Technologies of Life and Death* (1980). William Aspray, *John von Neumann and the Origins of Modern Computing* (1990). James Glimm, John Impagliazzo, and Isadore Singer, eds., The Legacy of John Von Neumann, *Proceedings of Symposia in Pure Mathematics*, vol. 50 (1990). William Poundstone, *Prisoner's Dilemma*, rep. (1993). Norman Macrae, *John von Neumann: The Scientific Genius Who Pioneered the Modern Computer, Game Theory, Nuclear Deterrence, and Much More*, rep. (1999).

JAN LACKI

NEWTON, Isaac (1642–1727), English physicist and mathematician, one of the greatest natural philosophers the world has known.

Newton was born in Woolsthorpe, near Lincoln, on Christmas Day 1642. He was educated at Trinity College, Cambridge, becoming a fellow of the college in 1667. In 1669 he succeeded his teacher, Isaac Barrow, as Lucasian Professor of mathematics.

While the university closed from summer 1665 to April 1667 because of plague, Newton spent most of his time on the family farm in Woolsthorpe. He returned to Cambridge briefly from March to June 1666. Newton made himself master of the latest mathematics and created a wholly new branch, the differential and integral calculus. At about the same time, Gottfried Leibniz independently invented a similar calculus. Newton was the first to produce this new mathematics, but Leibniz was the first to publish it.

During those brilliantly creative months in plague-fearing Cambridge, Newton also set the foundations of modern optics. He analyzed the solar spectrum, revealing the phenomena of dispersion and composition of light and their causes. He invented a new type of telescope, featuring a magnifying mirror instead of a magnifying lens. Newton also devoted time to force and motion, but he did not then—as he later claimed—find that gravity extends to the Moon or discover the law of gravity.

In 1672, he published an account of his new discoveries concerning light and color. This paper gave rise to extensive criticism, to which Newton wrote careful replies. He declared that he so regretted having to reply to critics that he would never again publish his discoveries. For almost twenty years he remained faithful to this vow. He gave the lectures required by the terms of his professorship and busied himself with investigations that may seem far removed from science. They included alchemy, biblical prophecy, the interpretation of Scripture, the chronology of ancient kingdoms, but not astrology, one of the few sorts of ancient knowledge Newton thought invalid. He believed that these investigations were intimately related to his work in optics, *mathematics, rational *mechanics, and celestial dynamics.

In 1679, Newton learned of Robert *Hooke's idea that orbital or curved motion could be explained by a combination of a linear inertial component along the orbit's tangent and a continual falling inward toward the center. Newton wrote that he had never before heard of this "hypothesis." But he perceived a connection between Hooke's suggestion and Johannes *Kepler's law of areas, and showed that they implied that the tendency toward the center in planetary elliptical orbits must vary as the inverse square of the distance from the Sun. He informed no one about this great breakthrough.

In 1684 Newton received a visit from Edmond Halley, who asked for help in solving a problem that had stumped everyone in London: the force that produces planetary elliptical orbits. Newton replied that he had already solved it. He wrote up his solution in a little tract called De motu. While revising and expanding it, he discovered that the same force that keeps the planets in orbit must cause perturbations in the orbital motions of other planets, the key to the great principle and law of universal gravitation.

Encouraged by Halley, Newton now began to develop his work in detail. In 1687 he published the resulting masterpiece, Philosophiae naturalis principia mathematica (Mathematical Principles of Natural Philosophy). Here Newton gave his new concept of mass and the principle of inertia, and his famous three laws of motion, the foundation of the new science of rational mechanics.

The first of the Principia's three "books" sets forth the science of motion; the second, the conditions of fluid resistance and their consequences; and the third, the system of the world, built up from his mechanical principles in the law of universal gravity. In the second and third editions, the Principia concludes with a General Scholium containing Newton's famous slogan, Hypotheses non fingo—"I do not feign hypotheses"—referring to his disinclination (and inability) to declare a mechanical cause of gravitation.

Book three gives explanations of the tides, the motions of the Moon and of the comets, the shape of the earth, the variation of weight with change in terrestrial position, the acceleration of falling bodies, and much more besides.

A decade or so after the publication of the Principia, Newton moved to London, where he became warden and then master of the mint and president of the Royal Society. In 1704, he published his Opticks, an account of his many optical discoveries, which also contained, in the form of "Queries," hints and experiments on all sorts of physical and chemical phenomena. These sometimes contradictory queries, expanded in successive editions of the Opticks, helped to guide the experimental philosophy of the eighteenth century. Newton died in London in 1727 and was buried in Westminster Abbey.

I. Bernard Cohen, Franklin and Newton (1956). Richard S. Westfall, Never at Rest (1983). Derek Gjertson, The Newton Handbook (1986). Newton's major works exist in modern formats: Opticks, ed. E. T. Whittaker (1952); Principia, trans. I. B. Cohen et al., together with a "Guide" by Cohen (1999).

I. BERNARD COHEN

NEWTONIANISM. Isaac Newton joined terrestrial and celestial mechanics together in 1687 with the publication of his Philosophiae naturalis principia mathematica, transforming the two kinds of physics into a single system oriented around the inverse-square law of gravitational attraction. Twenty years after this feat of mathematical synthesis, Newton contributed to framing experimental physics in his *Opticks; or, A Treatise of the Reflections, Refractions, Inflections and Colours of Light*, first published in 1704 and going through four editions by 1730. The *Opticks* laid out an experimentally derived geometry of light-rays, including the use of prisms to analyze white light into the colors of the spectrum and then resynthesize them into white light. The *Opticks* also included a final section of "queries" containing all the unfinished business of Newton's career. By means of this laundry list, Newton set the agenda for his eighteenth-century followers. Most importantly, he proposed that weightless *ethers were the medium and material cause of forces and phenomena including light, heat, *electricity, magnetism, gravitational attraction, and animal sensation.

Over the past three centuries, "Newtonianism" has meant several things. On the model of the *Principia*, it has meant a mathematical, synthetic approach to physics, and more specifically, the confirmation and promulgation of inverse-square laws of force. Important examples are John Mitchell's demonstration of an inverse-square law for magnetic force (1750) and Charles Augustin Coulomb's announcement of an inverse-square law governing electrical attraction and repulsion (1785–1789). Eighteenth- and nineteenth-century Newtonians also developed the field of *astronomy by testing Newton's law of gravitational attraction against new observations and resolving apparent conflicts. Alexis-Claude Clairaut's explanation of the motion of the lunar apogee (1749) and his accurate prediction of the return of Halley's comet in 1759 confirmed and vindicated Newtonian astronomy. These efforts gave rise to an increasingly complex picture of the mutual gravitational influences of celestial bodies. The culmination of eighteenth and early nineteenth century Newtonian astronomy was Pierre-Simon *Laplace's *Traité de mécanique céleste* (1798–1827), in which Laplace used Newton's law of gravitation to develop a complete theory of the solar system, taking into account complexities such as the perturbations in the orbits of the planets and the satellites caused by their mutual attraction.

On the model of the *Opticks*, meanwhile, Newtonianism has meant an inductive, experimental approach to physics. Users of this meaning of the word cite Newton's promise, in the preface to the *Principia*, to "feign no hypotheses." An example of a Newtonian in this sense is Benjamin *Franklin, who presented his electrical science as one founded in experimental tinkering rather than theory. His followers and historians have likened him, on that basis, to the Newton of the *Opticks*, the empirical essayer and querist. The empiricist meaning of Newtonianism has also referred, more specifically, to the use of *analysis and synthesis experiments, for which Newton's investigations of white light served as the paradigm. The leading eighteenth-century example of experimental Newtonianism in this sense is Antoine-Laurent Lavoisier's analysis of water into hydrogen and oxygen and his resynthesis of these elements into water (1785).

The *Opticks* gave rise, finally, to a third meaning of Newtonianism, to describe the eighteenth- and nineteenth-century research program of so-called *imponderable fluids that grew from Newton's hypothesis regarding force-bearing ethers. The hypothesis that an imponderable fluid medium carried each force informed theories of electricity, magnetism, heat, and light well into the nineteenth century. An eighteenth-century example of a fluid theory was Franklin's account of electricity, according to which a weightless electrical fluid, whose particles were mutually repulsive, permeated common matter, balancing the mutual attraction of its particles. Electrical effects accordingly resulted from the depletion (negative charge) or overabundance (positive charge) of the electrical fluid in a body. Another important eighteenth-century example was Joseph Black's understanding of heat. Black noticed that it took a great deal of heat simply to melt ice, without changing its temperature. He gave the name "latent heat" to the thermal fluid that seemed to disappear during phase changes, and he distinguished the quantity of this fluid in a heated object from its density, defining temperature as density of heat. An object's temperature depended, Black reasoned, upon its substance's capacity to contain the thermal fluid. These early notions of negative and positive electricity, of latent heat and of heat capacities, or specific heats, were thus informed by the Newtonian paradigm of imponderable fluids.

In addition to the work set forth in the *Principia* and the *Opticks*, another factor was crucial in shaping the meaning of Newtonianism, particularly during the eighteenth century: the contrast—partly genuine, but also overdrawn by Newton and his followers—between Newton's approach to physics and that of the French mathematician and natural philosopher René *Descartes, to whose example Newton owed the beginnings of many of his ideas. Descartes notoriously allowed his rationalism and his commitment to rigorously mechanical explanations of natural phenomena to get the better of his physics. Based upon the principle that there could be no intelligible difference between matter and space, and on the conviction that physical events must have mechanical causes in the form of pushes between bits of matter, Descartes derived a picture of the universe as a great plenum in which all things were constrained to move in vortices. Newton's followers called Cartesian physics dogmatic, misguided, and arrogant in its claims to completeness. They pointed to Newton's abstention, in the *Principia*, from assigning a mechanical cause for gravitational attraction as the epitome of empiricist open-mindedness and humility. In celebration of what they took to be his epistemological modesty, Newtonians often referred to Newton's remark that he was merely collecting pretty pebbles beside the ocean of truth.

Leaving a gap at the heart of his system of mechanical causation, Newton allowed his disciples to fill in the metaphysics of their choosing. He himself wrote, in the queries to the *Opticks*, that natural phenomena arose not from mechanical causes, but from the will of a divine intelligence. This appeal to a final cause lying beyond the efficient ones pleased Enlightenment eulogists of Newton's mechanical system, who showed a remarkable tendency to cite its breaches. An example is David Hume's satisfaction that although Newton "seemed to draw off the veil from some of the mysteries of nature," he also demonstrated "the imperfections of the *mechanical philosophy," restoring Nature's secrets "to that obscurity in which they ever did and ever will remain" (*The History of England* [1754–1762]). Voltaire, in his *Lettres philosophiques* (1734), popularized for a French audience the contrast between Newton's heroic acceptance, and Descartes's dogmatic refusal, of obscurity.

We now have four meanings of Newtonianism: a mathematical, synthetic approach to natural philosophy (particularly one founded in inverse-square laws of force); an inductive, experimental approach to natural philosophy (particularly one founded in analysis and synthesis experiments); the attribution of forces to weightless, force-bearing ethers or "imponderable fluids"; and the appeal to final causes, manifestations of the will of a divine intelligence, as the ultimate cause of natural phenomena, in contrast with Descartes's and his followers' strict adherence, in their natural philosophy, to mechanical causes.

The promulgation of Newtonianism coincided with an increasing interest in natural knowledge among the literate public. Some of the first people to teach courses of experimental physics were Newton's propagandists: Francis Hauksbee and John Theophilus Desaguliers, demonstrators at the Royal Society of London, and Willem Jacob

'sGravesande, professor of mathematics at the University of Leiden, who was inspired by a meeting with Newton during a visit to London. These lecturers professed to translate Newton's physics from the language of mathematics into the language of experience, using demonstration experiments to make complicated ideas accessible to polite audiences. Popular written expositions of Newton's physics, including Desaguliers's and 'sGravesande's published lectures, emerged during the first third of the eighteenth century. Turning Newton's natural philosophy into a source of philosophical amusement, lecturers and authors established in the minds of their public a particular model of natural knowledge: quantitative and synthetic but also rigorously experimental; materialist and mechanist but also resting upon an underlying assumption that the ultimate causes in nature were final rather than efficient, reasons rather than mechanisms. The same model of knowledge took root in universities, academies, and technical and professional schools during the eighteenth century, beginning with the Royal Society, Cambridge University, and the University of Leiden, and spreading after about 1730 to France, Italy, Russia, and Sweden, where it mixed with continental traditions informed by the work of Descartes, Gottfried Leibniz, and others. Not only mathematicians and philosophers but doctors and engineers studied and taught Newtonian curricula by the end of the eighteenth century. Thus Newtonianism, with its several meanings, permeated the emerging professional and popular cultures of the Enlightenment.

I. Bernard Cohen, *Franklin and Newton* (1956). Gerd Buchdahl, *The Image of Newton and Locke in the Age of Enlightenment* (1961). Henry Guerlac, *Essays and Papers in the History of Modern Science* (1977). Margaret C. Jacob, "Newtonianism and the Origins of the Enlightenment: A Reassessment," *Eighteenth-Century Studies* 11 (1977): 1–25. I. Bernard Cohen, *The Newtonian Revolution* (1980). Betty Jo Teeter Dobbs and Margaret C. Jacob, *Newton and the Culture of Newtonianism* (1995).

JESSICA RISKIN

NOBEL PRIZE. By his will of 1895, Alfred Nobel, inventor of dynamite and smokeless powder, left his estate to a nonexistent corporation, soon established as the Nobel Foundation. He directed that the net income from his endowment be given in equal shares for discoveries or inventions in physics, chemistry, and physiology/medicine made during the previous year that had contributed most to the welfare of mankind; and also for literature of an idealist tendency and for contributions to world peace. The prize in economics in the name of Alfred Nobel was established by the Bank of Sweden in 1968. The Royal Swedish Academy of Sciences awards the prizes in physics,

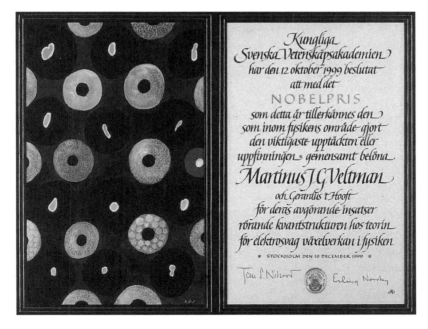

Martinus Veltman's diploma for his physics prize of 1999. The illustration on the left is an artist's impression of Veltman's prize-winning work on the electro-weak interaction between subnuclear particles.

chemistry, and economics, and the Karolinska Institute the prize in physiology/medicine.

The requirement that the prizes be given for current work was not observed in the first awards, made in 1901, and rarely thereafter. Since prizes can be given for old work newly recognized as significant for recent innovations, the lag can be considerable, almost fifty years in the case of one of the physics laureates of 2000. Early winners, like Wilhelm Conrad *Röntgen for *X rays (physics, 1901), Guglielmo Marconi and Ferdinand Braun for wireless telegraphy (physics, 1909), Adolf von Baeyer for contributions to chemical industry (chemistry, 1905), Emil von Behring for serum therapy (physiology/medicine, 1901), Ronald Ross for the etiology of malaria (physiology/medicine, 1902), and many more in physiology/medicine, were the sorts of awardees Nobel had in mind. But the requirement of applicability dissolved along with that of currency. Many of the science prizes, particularly in physics, have gone to academics for discoveries whose applications, if any, were not foreseen at the time of the award.

The distribution of awards has often been scrutinized for signs of bias and indications of national rankings. Some have noticed biases toward Germany, experimental discoveries, and exact measurement in the prizes given during the first two or three decades. But the leading theorists usually were honored if they lived

long enough and the early favoritism of Germans, owing to the closeness of the German and Swedish scientific communities, did not undermine confidence in the merit of the decisions. Since World War II Americans have dominated the science prizes. The first crop of winners, in 1901, were two Germans and a Dutchman working in Germany. The prizes of 1926 (to proceed in 25-year intervals) went to two Germans, an Austrian working in Germany, a Frenchman, a Swede, and a Dane. This increase in awardees does not indicate the arrival of team research but the operation of a rule that played an important part in Nobel politics: a prize committee can hold over a prize for a year, after which it must either be awarded or pocketed by the academy for the research of those who failed to find a winner. The accounting of 1926 includes two prizes postponed in 1925.

Twenty-five years later we reach a new world. The science prizes of 1951 went to two Americans, a South African working in the United States, an Englishman, and an Irishman. (The quantity of winners reflected not belated prizes but small-team collaboration in accelerator physics.) In 1976 Americans made a clean sweep, five of them sharing the three prizes available.

Of greater interest to many winners than the competition among nations is the monetary value of their awards. Although it has varied, it has ended the century as it began, at about six times a top professorial salary. One of Nobel's purposes in making the prizes so valuable was to draw attention to the importance of science and technology. The size of the prize, inflated by the participation of the Swedish royal family in the gala award ceremony and dinner, helped make its message. Science needs no such advertisement today. Nobel had in mind the lone investigator, such as himself, working out ideas that strike only the prepared mind of an individual genius. By insisting that no more than three people can share one science prize, the Nobel Foundation has institutionalized its founder's romantic notion of scientific creativity. In our age of team research, when experiments can involve hundreds of people, Nobel's notion of a single informing creative impulse no longer catches the circumstances of science.

In this predicament, the Nobel Foundation might become more companionable, and permit the award of science prizes to groups as it does the peace prize. Failing that, it might consider enlarging the scope of science to include its history. Most historians still work alone.

Les prix Nobel (annual). The Nobel Foundation, Alfred Nobel: The Man and His Work (1962). Ragnar Sohlman, The Legacy of Alfred Nobel (1983). Elisabeth Crawford, The Beginnings of the Nobel Institution: The Science Prizes, 1901–1915 (1984). Kenne Fant, Alfred Nobel (1993). Elizabeth Crawford, J. L. Heilbron, and Rebeccca Ullrich, The Nobel Population: A Census of Nominees and Nominators for the Prizes in Physics and Chemistry (2001).
J. L. Heilbron

NOBLE GASES. The least reactive of the chemical elements proved the most fruitful guide to their interrelations. Some twenty-five years after Dmitrii *Mendeleev first worked out his *periodic table, Lord Rayleigh (John William Strutt), formerly James Clerk *Maxwell's successor as Cavendish professor of physics at Cambridge, discovered that the nitrogen he drew from the air had a specific weight greater than that of the nitrogen derived from mineral sources. He asked publicly for ways to resolve the discrepancy and then hit on the solution himself. He read a paper, then a century old, in which Henry *Cavendish mentioned an unoxydizable residue of gas he obtained after sparking atmospheric nitrogen with oxygen.

While Rayleigh tried to collect enough of this residue to weigh it, William Ramsay, a leading British chemist alerted to the problem by Rayleigh's request, isolated the residue by more effective chemical means. It weighed enough to account for the discrepancy of one part in two hundred that had started Rayleigh's quest. By 1895 they could announce the discovery of a constituent of the atmosphere they named "argon" (from the Greek for "lazy") because it declined chemical intercourse. Ramsay then looked for other trace gases by examining air and argon liquefied by then-new cryogenic techniques (see Cold). Neon ("novel"), krypton ("hidden"), and xenon ("strange") quickly put in an appearance in the spectroscope and then the balance. So did helium, already named and known as the suppositious source of certain otherwise unattributable lines in the solar spectrum.

The sluggishness and aloofness of the noble gases put them in a class apart. Astonishingly, Mendeleev's chart could accommodate the five newcomers, confirming its importance and renewing its mystery. The only pinch came with argon, whose atomic weight placed it after potassium, but whose nobility placed it before. It took almost twenty years and the invention of the concept of isotope to resolve this problem of precedence. In 1919 Francis Aston at the Cavendish made the first crisp separation of isotopes using neon gas in a *mass spectroscope he had invented. Meanwhile the reversal at argon-potassium (and at cobalt-nickel and iodine-tellurium) helped to alert chemists and physicists that something other than atomic weight regulated the properties of the elements (see Atomic Structure).

The discovery of the noble gases was almost a prerequisite to unraveling the complexities of *radioactivity. Ernest *Rutherford and Frederick Soddy identified the "emanation" from thorium as a new and flighty member of the noble family, now called radon; the occurrence of a decaying nonreactive gas in their experiments provided the clue for working out their theory of the transmutation of atoms. Radium also gives off a radioactive emanation and the two similar (indeed chemically identical) noble gases offered an early example of isotopy. However, the lightest of the noble gases proved the weightiest. Helium is often found with uranium and other active ores. With the spectroscopist Thomas Royds and an apparatus made by the virtuoso glass blower Otto Baumbach, Rutherford demonstrated in 1908 that the alpha particles emitted from radioactive substances turned into helium atoms when they lost their electric charge. In 1910–1911 he showed that alpha particles acted as point charges when fired at metal atoms, and devised the nuclear model of the atom to explain the results of the scattering and to deduce that helium atoms have exactly two electrons. The replacement of atomic weight by atomic number (the charge on the nucleus) as the ordering principle of the periodic table followed. Rayleigh, Ramsay, Aston, and

Rutherford all received Noble Prizes in large measure owing to their work on noble gases.

See also ELEMENT.

Morris T. Travers, *The Discovery of the Rare Gases* (1928; expanded ed., *A Life of Sir William Ramsay* [1956]). Isaac Asimov, *The Noble Gases (1966).* John Robert Strutt, *The Life of John William Strutt, Third Baron Rayleigh*, rev. ed. (1968).

J. L. HEILBRON

NON-EUCLIDEAN GEOMETRY. Much of elementary Euclidean geometry involves definitions and properties of plane figures bounded by straight lines and circles. Even before Euclid's *Elements* were written, around 300 B.C., Greek mathematicians and philosophers knew that these definitions were delicate. Attention centered on the idea that there can be parallel lines—lines that never meet however far they are extended. The idea invokes knowledge of indefinitely large regions, yet without it, very few of the theorems in the *Elements* can be proved. Euclid's own explicit assumption (his famous fifth postulate) amounts to saying that through a given point there is a unique line parallel to a given line. Many Greek, Arab, and later European writers sought to prove the existence of this unique line using only the other, much more intuitive, assumptions of the *Elements*. All such attempts failed. Particularly noteworthy among Western attempts were those of Gerolamo Saccheri and Johann Heinrich Lambert, whose work contained results that could later be interpreted as theorems in a new, non-Euclidean geometry.

In the early nineteenth century, two mathematicians on the fringes of Europe succeeded in the contrary task. They developed a geometry in which infinitely many lines parallel to a given line can be drawn through a given point. Nicolai Ivanovich Lobachevskii in Kazan in Russia and Janos Bolyai in Hungary independently used novel methods analogous to those of spherical trigonometry to give a complete description of the elementary features of this new, non-Euclidean geometry. However, the novelty of their ideas, and the obscure and imperfect way they were published, prevented the acceptance of their work in their lifetimes. Only Carl Friedrich *Gauss, the mathematician who dominated the early nineteenth century and who had come to some of these ideas himself but not published them, acknowledged their merit. Janos Bolyai's father Farkas and Lobachevskii's teacher Martin Bartels were friends of Gauss, but Janos Bolyai and Lobachevskii worked independently of Gauss and of each other. Widespread acceptance of non-Euclidean geometry began with the posthumous discovery of Gauss's opinion, and more importantly with the creation of new, much more general foundations for geometry. Bernhard Riemann vastly enlarged the field of differential geometry so that it embraced Euclidean geometry as a special case, whereas previously differential geometry had been only part of Euclidean geometry. Eugenio Beltrami then showed how non-Euclidean geometry could be expressed rigorously in the new framework, thus eliminating the imperfections of the original accounts.

Non-Euclidean geometry offers an alternative account of physical space, in which familiar figures have novel properties. Although some philosophers, notably Gottlob Frege, found it difficult to accept, almost all mathematicians and scientists, and many members of the public, became interested in the question of whether space was Euclidean or very slightly non-Euclidean. In the early years of the twentieth century, Henri Poincaré proposed that any experimental test involved assumptions about what physically constituted a straight line (for example, a ray of light), and so any result rested on a distinction between physics and geometry that was ultimately arbitrary and a matter of convention. Conventionalism in the philosophy of science became a mainstay of the logical positivists.

Einstein's theory of general *relativity put an end to the simple opposition between Euclidean and non-Euclidean forms through his sophisticated use of differential geometry to formulate a theory of gravity. However, Henri Poincaré had also shown that non-Euclidean geometry is fundamental in mainstream branches of mathematics. It is now important in several domains, including the study of three-dimensional manifolds and the theory of knots.

Jeremy Gray, *Ideas of Space: Euclidean, Non-Euclidean, and Relativistic*, 2d ed. (1989). Jeremy Gray, *Bolyai and Geometry* (2002).

JEREMY GRAY

NON-WESTERN TRADITIONS. Revolutions do not occur in isolation. The *Scientific Revolution of seventeenth-century Europe followed innovations in printing, the development of new markets across the Atlantic, theological reforms, the application of new technology to well-established fields of study, and the cross-cultural exchange of ideas, especially from *Asia. This latter influence, the intellectual contribution of non-Western cultures to the Scientific Revolution in the West, is the subject of this essay. The higher civilizations of India, *Africa, the Americas, and China all developed sophisticated knowledge systems prior to the Scientific Revolution and many of these cultures contributed significantly to it through the exchange of ideas as Greek, Persian, Oriental, and Arabic texts were translated into Latin.

The synthesis of intellectual, social, and technological change, which swept across seventeenth century Europe, made the conditions right for reexamining a number of old ideas, such as the view in non-Western cultures linking science with religion. The Scientific Revolution broke this link. The dominant Christian view in the West at the time of *Copernicus was of God as Creator in contrast to creation as god. This understanding fostered a freedom to examine nature as an object of study rather than an object of worship. Under these circumstances, evidence derived from repeatable observations offered new insights into the natural world and established distinctive areas of inquiry (mathematics, astronomy, physics, religion, magic, logic, and philosophy) where only one or a few had previously existed.

Systematic observations of the sky revealed an apparent correlation between heavenly events and earthly activities, such as the annual appearance of the bright star Sirius and the flooding of the Nile River in Egypt. The development of astrological prognostications demanded careful, systematic, long-term celestial observations. By 300 B.C. Chinese astrologers had compiled an extensive star catalog of transitory heavenly activities such as comets, sunspots, and novae. So precise were the records of Chinese, Japanese, and Korean astrologers of the supernova in A.D. 1054 that from them modern astronomers have determined its present remains as the Crab Nebula. Centuries of careful observations enabled Chinese and Islamic astrologers to observe that points on the celestial sphere where the Moon's orbit crosses the ecliptic (the Sun's apparent annual orbit) drift in a westerly direction at a rate of one complete revolution every 18.6 years. They also noticed that the Sun's equinoxes—the interactions of the Sun's orbit with the plane of the earth's equator—demonstrated a similar but much slower westerly motion.

Islamic cosmology synthesized the works of Greek, Hindu, Chinese, and Persian sky watchers. The translation of Islamic astrological works into Latin proved decisive for Western astronomy. For example, Copernicus may have found a Latin edition of a thirteenth-century Persian astronomical work useful in resolving the perplexing pattern of planetary movements observed from Earth. This information may have helped to guide the thinking that led to his theory of a heliocentric solar system.

Non-Western sky watchers recognized features of the heavens that lent themselves to keeping time on earth—especially the apparent motions of the Sun, Moon, and (as in Mesoamerica) Venus. Owing to the perceived religious significance between heavenly and earthly activities, solar and lunar movements underlay ritualistic calendars and the timing of daily prayer.

Non-Western cultures developed several instruments such as the gnomon and clepsydra to track celestial events more accurately. Hindu and Islamic astronomers developed and applied trigonometrical principles to such instruments with the *astrolabe. The spherical astrolabe appears to be an Islamic development. Chinese expertise in optics precipitated the development of telescope-like instruments coincident with or earlier than their counterpart in the West.

African Natural History

There is more similarity and continuity among ancient African beliefs about nature than might be expected among such a diverse array of indigenous peoples. Typically, they put human beings at the center of events but not as masters of the universe. The earth existed for human use, but humans had to respect natural processes to ensure a healthy relationship between the land and the people. Iron, cattle, and the Sun were thought to be gifts from the gods for use as tools, food, and light respectively. Religious beliefs stood at the core of natural philosophies across the continent. Animated spirits embodied nature. Numerous ceremonies brought human activities into harmony with the natural order.

African agriculture proceeded from a patchwork of natural history principles. Less than 10 percent of Africa is well suited for agriculture. Maximum production focused on making the most of social services, such as the local labor pool, rather than technological innovations or sweeping scientific theories. A long tradition of human migration complicated the picture, but fostered sustainable developments in harmony with the local ecology.

The diversity of local soil types contributed to the level of biodiversity across the African continent, playing an important role in the conservation of genetic resources. Hardy cultivars developed in Africa found use elsewhere, for example, the dry land rice imported by Thomas Jefferson for use on the inland farms of South Carolina and Georgia late in the eighteenth century. He believed that this variety would help reduce health problems associated with rice farming along the southeastern coast of the United States.

African agroforestry illustrates a holistic appreciation and utilization of natural resources. Trees for lumber, shrubs for forage, herbs for pasture, and crops for human consumption were grown and harvested in an intricate balance that sustained human settlement in harmony with natural ecosystems. Africans planted leguminous trees like *Acacia albida* among other crops to enrich the soil through nitrogen fixation. Its pods and leaves were used as forage, its timber was used for building, and various other bits as medicines.

An intriguing level of mystery surrounds astronomical knowledge in sub-Saharan Africa. A distinctive calendar system developed by the Borana people ignored the Sun entirely. It tracked the coincidental rising of new moons with seven star groups: Triangulum, Pleiades, Aldebaran, Bellatrix, Orion's Belt, Saiph, and Sirius. The Borana year began when a new moon appeared in conjunction with Triangulum. Each subsequent month began when a new moon rose with one of the six other star groups. Different phases of the Moon found in alignment with Triangulum divided up the second half of the year. This system of astronomical time keeping was in use around 300 B.C.

Archaeoastronomical sites in Africa include the pyramids of Kush, lunar cave symbols in Tanzania, and the Namoratunga stone pillars. The Kingdom of Kush flourished from 1000 B.C. to A.D. 200. After they conquered Egypt, the Kushitic Pharaohs incorporated Egyptian design concepts into their pyramids including alignment with the rising of the star Sirius. Mary Leakey discovered what may represent a lunar-based timekeeping system in some Tanzanian caves—a series of concentric circular markings consisting of twenty-nine or thirty circles that could coincide with the lunar synodic cycle. Namoratunga resembles a miniature Stonehenge, in which numerous stone pillars bearing petroglyphs seem to be aligned with the seven star groups used in the Borana calendrical system. These square magnetic columns of basalt are reminiscent of local grave markers. Namoratunga means "people of stone" in the local Turkana language. These and innumerable other puzzling structures across the continent were associated with religious ceremonies, agricultural practices, and astronomical observations.

Among the many astronomical myths widespread across sub-Saharan Africa, one told by the Dogon people involved the binary star Sirius. Though the Dogon could not resolve the individual members of this star complex, they associated the smaller star (Sirius B) with the locally grown grain *fonio* because it is small and white. Sirius B turns out to be a million times more dense than a normal star, the exemplar of "white dwarfs." Not until 1844 did Western science confirm what the Dogon believed. In that year Friedrich Wilhelm Bessel deduced from the orbital wobble of the larger member of the pair (Sirius A) that Sirius must be a double star. More powerful telescopes brought Sirius B to light for the first time in 1862. The Dogon celebrated the fifty-year orbit of Sirius B around Sirius A every sixty years and believed that a third star existed in this system. Modern calculations agree with the orbital time period but scientists have yet to verify another star in the complex.

Unlike many fields of study considered distinct in the West, African mathematics figured as part of an integrated whole. Mathematics south of the Sahara was intimately associated with art, cosmology, philosophy, medicine, natural science, technology, and religion. Consequently, mathematical skills were developed widely and early across Africa. One of the oldest mathematical artifacts known, a piece of baboon bone with twenty-nine clearly marked notches found in the Lebombo Mountains between South Africa and Swaziland, dates from around 35,000 B.C. It resembles calendar sticks still in use today in some parts of the African continent.

The mathematics of African games, puzzles, riddles, and other contests included hidden algorithms and a thorough knowledge of probabilities. Sorcerers and wise men who knew the algorithms had a distinct advantage over their opponents, therefore appearing to be endowed with magical and religious powers.

American Natural History and Agriculture

The natural and supernatural graded into one another in the Americas, making for a practical philosophy comprehensible only in light of their religious activities. They regarded plants and animals as mythical associates of the religious realm as well as creatures of the material world. Humans were brothers to the bear, deer, eagle, and jaguar. All had the same mother, Mother Earth. Humans and animals mutually cared for one another. For example, the American bittern expressed its concern for its human brothers and sisters by signaling the arrival of heavy rains and an abundant supply of fish with its nighttime singing.

New World farmers domesticated a number of plants that contributed subsequently to Old World diets—squash (*Cucurbita*), several species of beans (*Phaseolus*), chili peppers (*Capsicum*), cassava (*Manihoc*), sweet potatoes (*Ipomea batatas*), potatoes (*Solanum*), avocado (*Persea americana*), yams (*Dioscorea*), and maize (*Zea mays*). Agriculture played an influential role in political and economic success as well. Cocoa (*Theobroma cacao*) traded widely, especially in Mesoamerica as merchandise, tribute, and ritual ingredient. Tobacco (*Nicotiana*) served primarily in ceremonial activities. Bananas (*Musa*) were a staple in the diet of indigenous people across the Americas. A variety of techniques maximized production, including irrigation, burning, terracing, and fertilization. Some of these practices did severe ecological damage. Biodiversity often suffered, causing unforeseen changes in the landscape.

Star study by pre-European peoples in America served both sacred and practical functions. Harvesting and hunting followed the stars. All New World cultures incorporated the stars into ritual myths. Extensive cultural interaction throughout the Americas led to similar calendars and a shared body of star knowledge. The sky was a realm of power that influenced human lives. It helped to explain human behavior and contributed to social balance and cohesion.

Indigenous American people tracked phases of the Moon, monitored solstices and equinoxes, and assigned stars to constellations. Infrequent celestial events such as comets, eclipses, and meteor showers provoked ritualistic responses. Some cultures distinguished planets from stars. Mesoamericans rated Venus as the most important planet and associated its movements with wars. The Classic Mayans identified Jupiter with several of their rulers. Celestial movements influenced architectural configurations and designs. Buildings in the Puuc region of northern Yucatan, the Templo Mayor in Mexico City, and Alta Vista near Zacatecas on the Tropic of Cancer lie so that solstices and equinoxes could be observed from them with relative ease and precision.

In North America the Hopi Sun Chief established key ceremonial and agricultural cycles based on the solar calendar. Accurately anticipating these events to initiate ceremonial activities amounted to celestial magic, especially important in times of war. Evidence of ancient celestial observations in North America includes petroglyphs in Chaco Canyon that may be associated with the Crab supernova of 1054, which the Chinese reported as visible in daylight for twenty-three days and coincident with a crescent moon. Prehistoric Mimbres ceramics depict a crescent-shaped rabbit and a star with twenty-three rays. Medicine wheels, like Stonehenge, may have been associated with astronomical observations or they may merely have served as geographical markers on the Great Plains. Whatever the detailed significance of these facts, they indicate that the indigenous peoples of North America took close notice of the natural world.

Navigation

Many cultures exploited the stars as reliable reference points during travel. The Caroline Islanders of the central Pacific navigated by means of Polaris, the North Star, which served as a reference for thirty-two points on the horizon where bright stars and constellations rose and set. The Polynesians observed stellar positions with a "sacred calabash," a gourd filled with water to the level of four holes in the neck that were used to measure the altitude of specific stars. With the help of the sacred calabash, they could cross wide expanses of the South Pacific.

Early interest in spatial ordering among the Chinese formed the mathematical foundation of subsequent works on navigation and the development of measuring devices such as the compass and the gnomon (a post of standard height) between 1100 and 300 B.C. Chinese star catalogs date back as far as the Warring States period (481–221 B.C.), from which came a number of star maps.

The magnetic compass directed the voyages of Chinese sailors by the middle of the ninth century A.D. Its use derived from the south pointing lodestone and a traditional ceremony referred to as the sinan ceremony. Sinan can be translated as "south controller" or "south verification." Whatever its construction may have been, apparently it kept people engaged in various trades from getting lost. The Chinese seem to have discovered that the directional properties of the lodestone could be transferred to small pieces of iron by the first century A.D. By the seventh century, magnetic needles appeared in some compasses. Shen Kua in 1086 calculated a specific figure for magnetic declination. Another method of magnetizing iron was described in the *Wujing Zongyao* (*Collection of the Most Important Military Techniques*) in 1044. It recommends taking a piece of red-hot iron, placing it north-south, and quenching it. The process works.

Land maps played an increasingly important role in navigation among the Chinese as economic, political, and military forces fostered innovations and improvements in cartographic accuracy. Initially, maps illustrated rather than aided travel. They utilized a variety of media depending on the designer and available materials. Chinese maps appeared in tombs and caves, on stone, silk, wood, and paper. By the second century B.C. map makers began to draw to scale and a reference framework like that utilized by Chinese astronomers/astrologers gained acceptance. Pei Xiu (A.D. 223–271) formulated mapping principles to improve accuracy, such as consistent use of scale, directional indicators, and adjustments in land measurements to correct for irregularities in the terrain. Chinese navigational tools and techniques helped travelers across Asia for hundreds of years.

Islamic mariners relied heavily upon seasonal winds (monsoon) to provide reliable guidance on transoceanic voyages. Travelers on land or sea also used celestial markers, primarily the North Star (Jah), from which they took their latitude as well as direction. The compass though known was infrequently used through the ninth century.

Islamic maps often came in strips corresponding to a particular route of travel. They were oriented to the south and generally centered on Mecca. Frequently they focused on matters considered more important than the needs of travelers, such as imperial glory, military power, or spiritual journeys. When used for practical ends, they might be accompanied by textual notes, geographical tables, astronomical information, and illustrations, but lacked a standard scale. By the tenth century, Arab map makers were eager to improve the accuracy of their products. Al-Biruni, a notable Muslim astronomer, mathematician, geographer, and historian, helped them determine how to correct for variations in distance owing to the curvature of the earth's surface.

Looking Ahead

Twentieth-century approaches to agriculture and medicine in the West have proven unsustainable and unsatisfactory to a growing constituency of consumers and practitioners. The holistic model of sustainable food production, which underpinned the high levels of biodiversity in the indigenous cultures of Africa and the Americas, offers continuing guidance. Alternative health treatments from Asia, Africa, and the Americas now have advocates among Western physicians. Acupuncture, botanicals, music therapy, prayer, nutrition, and exercise are used worldwide for relief from pain and disease. Frequently, patients adopt these options alongside Western medical treatments. These and other non-Western sciences may foster the next revolution in natural knowledge.

See also Asia; Latin America.

Joseph Needham, *Science and Civilisation in China* (1954). Anthony F. Aveni, ed., *Native American Astronomy* (1977). Peter D. Harrison and B. L. Turner II, eds., *Pre-Hispanic Maya Agriculture* (1978). Anthony F. Aveni, ed., *Skywatchers of Ancient Mexico* (1980). Graham Connah, *Three Thousand Years in Africa: Man and His Environment in the Lake Chad Region of Nigeria* (1981). J. Donald. Hughes, *American Indian Ecology* (1983). Debiprasad Chattopadhyaya, *History of Science and Technology in Ancient India* (1986). David Freidel, Linda Schele, and Joy Parker, *Maya Cosmos* (1993). Paulus Gerdes, "On Mathematics in the History of Sub-Saharan Africa," *Historia Mathematica* 21 (1994): 345–376. Rashid Rushdi, ed., *Encyclopedia of the History of Arabic Science* (1996). Helaine Selin, ed., *Encyclopaedia of the History of Science, Technology, and Medicine in Non-Western Cultures* (1997). Paulus Gerdes. *Women, Art, and Geometry in Southern Africa* (1998). Osman Bakar, *The History and Philosophy of Islamic Science* (1999).
WILLIAM THEODORE JOHNSON

NUCLEAR BOMB. After the discovery of *radioactivity and the relativistic equivalence of matter and energy, scientists realized that vast amounts of energy must be stored within the atom. Most insisted, however, that the enormous potential of atomic energy for industry and also for war could never be exploited in practice. More power would have to be fed into an atomic device than could be extracted; the threat of atomic weapons was consigned to science fiction. Increasing knowledge of the atomic nucleus did not affect this assumption. Once James Chadwick in 1932 had demonstrated the existence of neutrons, Leo Szilard inferred that a nucleus would not repel a neutron as it would a positive particle and might release more neutrons than it captured, leading to a chain reaction. But he was unable to pursue his ideas further.

Meanwhile astrophysicists suggested as early as the 1920s that the energy of stars came from the fusion of hydrogen atoms to form helium. Hans Albrecht Bethe and George Gamow worked out a convincing mechanism for the process in agreement with observation in 1937–1938. Their theory was of purely cosmological interest, since nobody imagined that humans could attain the temperatures required to set the process off. But then came the surprise discovery of fission at the end of 1938 by Otto Hahn and Fritz Strassmann, as interpreted by Lise Meitner and Otto Robert Frisch. Within a month it was widely understood that nuclear fission could release energy far in excess of any chemical combustion or explosion. To achieve the necessary chain reaction required obtaining a critical mass of potentially fissile material (that is, a large enough sample that the neutrons can participate in the chain before being lost to the environment).

Szilard persuaded Albert *Einstein to inform President Franklin Delano Roosevelt about the possibility of a nuclear bomb. Roosevelt set up a scientific committee under the National Bureau of Standards to look into it. Meanwhile, atomic scientists were investigating the question in Germany, France, and Britain. Niels *Bohr and John Wheeler had in 1939 proposed a mechanism for fission from which it followed that uranium 235 (U-235), although only about 0.7 percent of natural uranium, would be more suitable than the common isotope U-238. In March 1940 Frisch, now in England, and another refugee, Rudolf Peierls, completed their memorandum, "On the Properties of a Radioactive Super-Bomb," which showed that the critical mass required for fission on a large scale was much less than had been presumed. The British government set up the MAUD committee to investigate; they reported in the summer of 1941 that a bomb would be feasible. They also suggested methods to enrich the proportion of U-235 in uranium. In Germany Werner *Heisenberg, as head of the Kaiser-Wilhelm Institut für Physik, set out the main tasks of a nuclear energy program. For about a year German research stayed ahead of the British, who at first thought to develop their own nuclear weapon but soon felt obliged to accept the absorption of their program into America's.

Anglo-American research disclosed that the newly discovered plutonium (created by bombarding U-238 with neutrons) would probably also be fissile. The American government, through the National Defense Research Council, began to pump money into the nuclear program even before Pearl Harbor. In December 1941 what was to become the *Manhattan Project began to take shape. In almost exactly a year Enrico *Fermi and his team at Chicago achieved an atomic pile capable of a controlled chain reaction. Their accomplishment indicated that a plutonium bomb might be no more difficult to make than a uranium bomb. Huge factories in remote parts of the United States mushroomed into life—Oak Ridge in the Tennessee Valley (for separation of U-235) and Hanford in Washington state (for production of plutonium).

A third major laboratory, established at Los Alamos, New Mexico, under J. Robert Oppenheimer, assembled many outstanding nuclear scientists, including Bohr, Bethe, and Edward Teller, to design the bombs. They decided that uranium could be detonated by firing (with ordinary explosives) two subcritical masses together, but the plutonium would fizzle if made critical so slowly. Instead, a subcritical mass of plutonium would be made critical by suddenly increasing its density via an imploding charge outside it. War ended in Europe, and the failure of the German program was known, before the test in June 1945 of the plutonium bomb. The uranium bomb fell (without prior test) on Hiroshima and a plutonium bomb on Nagasaki in August. The shock of the invincible weapon forced the military leaders of Japan to capitulate.

The United States resumed testing in summer 1946. The Soviets, who had known about the Manhattan Project, resumed their own project, spurred on after August 1945 by fears that the Americans would use atom bombs to impose their supremacy over all. Helped by espionage (just how much is still debated), the Soviet Union exploded its own bomb in 1949. Shocked by this unwelcome surprise and revelations of past spying, the American government now pressed on to develop a thermonuclear device, the so-called H-bomb (for hydrogen bomb). Some scientists involved in the Manhattan Project, Teller for example, had even then wished to work on such a weapon, nicknamed the "super," since the fission bomb would create the temperatures and pressures to make fusion feasible. Both the United States and the Soviet Union developed hydrogen bombs and tested them in the early 1950s. Meanwhile the British, feeling they had been shut out of a program to which their scientists had originally contributed much, decided in 1947 to make their own nuclear armory. They have been followed by a number of other states.

See also KURCHATOV, IGOR VASILIEVICH, AND J. ROBERT OPPENHEIMER; WORLD WAR II AND COLD WAR.

Alwyn McKay, The Making of the Atomic Age (1984). Richard Rhodes, The Making of the Atomic Bomb (1986). John Newhouse, The Nuclear Age (1989). Lawrence Badash, Scientists and the Development of Nuclear Weapons (1995). Rachel Fermi and Esther Samra, Picturing the Bomb (1995). Richard Rhodes, Dark Sun: The Making of the Hydrogen Bomb (1995).

ALEX KELLER

NUCLEAR CHEMISTRY. See NUCLEAR PHYSICS AND NUCLEAR CHEMISTRY.

NUCLEAR DIPLOMACY. Even before the United States dropped atomic bombs on Hiroshima and Nagasaki in August 1945, high-level American officials recognized that the scientific achievement of nuclear fission and the development of nuclear weapons would have a major impact on international affairs. Secretary of War Henry L. Stimson told President Harry S. Truman in April 1945 that the existence of atomic weapons would become a "primary question" in American diplomacy, and Secretary of State James F. Byrnes hoped that the atomic bomb would make the Soviet Union more tractable. Some scholars have suggested that the main reason that Truman authorized the atomic attacks on Japan was to intimidate the Soviets with a show of American power. Most experts agree however, that Truman used the bomb primarily to end the war as soon as possible.

Soviet Premier Joseph Stalin viewed the bomb as a political weapon that threatened the long-term international position of the Soviet Union. He immediately ordered Soviet scientists to build one as a top priority. A nuclear arms race was inevitable, despite efforts of the newly created United Nations to promote arms control. The race reached new levels of destructive potential when both nations took steps to build a hydrogen bomb, which promised an explosive yield a thousand times greater than the comparatively primitive weapons that the United States dropped on Japan. The Soviets tested their first atomic bomb in August 1949 and immediately accelerated their work on a hydrogen weapon. Truman authorized a crash program to develop a hydrogen bomb in January 1950 without knowing that the Soviet effort was already under way.

During the 1950s the United States and the Soviet Union embarked on an increasingly perilous arms race that began with the development of hydrogen bombs and ended with the deployment of ballistic missiles that could deliver warheads over long distances within minutes. President Dwight D. Eisenhower was deeply troubled by the ever-growing nuclear capabilities, but he failed to slow the arms race with the Soviet Union. Both nations conducted a series of hydrogen bomb tests that alerted the world to another nuclear danger—far-spreading radioactive fallout that might increase the incidence of cancer, birth defects, and other afflictions. The scientific debate over the effects of fallout sharply elevated public fears of radiation and doubts about weapons testing (*see* ANTINUCLEAR MOVEMENT).

The nuclear rivalry between the United States and the Soviet Union produced its most harrowing encounter in 1962 when Soviet Premier Nikita Khrushchev decided to place nuclear missiles in Cuba. The Cuban missile crisis came closer to causing a nuclear exchange than any other event of the Cold War. Once it began, both Khrushchev and President John F. Kennedy, operating under enormous pressure and bewildering uncertainty, sought a negotiated settlement rather than a military confrontation. They eventually agreed that Khrushchev would remove the missiles and Kennedy would pledge not to attack Cuba. Kennedy also promised to withdraw American missiles from Turkey, but carefully kept this part of the bargain from becoming public knowledge. Despite the trauma that it caused, the Cuban missile crisis had the salutary effect of convincing Kennedy and Khrushchev to strive to reduce Cold War hostilities. The most important result, the Limited Test Ban Treaty of 1963, prohibited nuclear testing in the atmosphere and underwater by its signatories, and in that way greatly diminished the problem of radioactive fallout.

The test ban treaty did not end the arms race. Indeed, the Soviet Union undertook strenuous efforts to narrow the lead held by the United States, and by the early 1970s it approached parity. New arms control measures became more desirable for both the major nuclear powers. In 1972 they reached agreement on the Anti-Ballistic Missile (ABM) Treaty, which placed limits on defensive missile systems, and the Strategic Arms Limitation Treaty (SALT), which for the first time established ceilings on the number of offensive weapons. In 1979, the SALT-II treaty placed additional restrictions on offensive capabilities. Although never formally ratified, it slowed the arms race because both the United States and the Soviet Union observed its provisions.

When the superpowers signed the SALT-II treaty, arms control was highly controversial in the United States. Supporters claimed that the agreements reduced the chances of a catastrophic nuclear war. Opponents insisted that SALT-I and SALT-II enabled the Soviet Union to fight and win a nuclear war. This allegation received wide attention when Ronald Reagan forcefully advanced it while running for president in 1980. As president, Reagan acted to improve U.S. military strength by modernizing its nuclear forces with new strategic bombers, missile-launching submarines, and ballistic missiles. He also called for the construction of a defense system that could intercept ballistic missiles launched against the United States, a proposal that became widely known as "Star Wars."

Despite Reagan's anti-Soviet rhetoric and policies, he and his Soviet counterpart, President Mikhail Gorbachev, found ways to curb the arms race. Reagan supporters claimed that the U.S. arms build-up and the prospect of Star Wars bankrupted the Soviets and won the Cold War. Others argued that Gorbachev took the initiative on arms control for reasons that had little to do with Reagan's programs. The two leaders concluded an agreement that virtually eliminated intermediate range nuclear forces in Europe in 1987. President George H. W. Bush went further by signing the Strategic Arms Reduction Treaties (START) with the Soviets in 1991 and 1993; these sharply cut the numbers of strategic warheads of both nations. The START treaties culminated negotiations begun during the Reagan administration.

Gregg Herken, *The Winning Weapon: The Atomic Bomb in the Cold War*, 1945–1950 (1980). McGeorge Bundy, *Danger and Survival: Choices about the Bomb in the First Fifty Years* (1988). Richard G. Hewlett and Jack M. Holl, *Atoms for Peace and War: Eisenhower and the Atomic Energy Commission* (1989). Aleksandr Fursenko and Timothy Naftali, *"One Hell of a Gamble": Khrushchev, Castro, and Kennedy, 1958–1964* (1997). J. Samuel Walker, *Prompt and Utter Destruction: Truman and the Use of Atomic Bombs against Japan* (1997). Frances Fitzgerald, *Way Out There in the Blue: Reagan, Star Wars, and the End of the Cold War* (2000).

J. SAMUEL WALKER

NUCLEAR MAGNETIC RESONANCE. The technique of nuclear magnetic resonance (NMR) emerged as a consequence of *quantum mechanics. The orbital motions of electrons give atoms a magnetic dipole moment, which according to quantum theory can take only particular orientations in a magnetic field. One method to explore the effect involved sending a beam of atoms or molecules from a gas through a magnetic field and measuring the deflection. The so-called molecular-beam method could also measure the magnetic moment of an atomic nucleus, which resulted from quantum-mechanical spin. In 1937 I. I. Rabi, an expert in molecular beams, proposed applying an alternating magnetic field, oscillating at radio frequency, on top of the static magnetic field. The particles in the beam would suddenly switch orientation for particular frequencies of the alternating field, and the frequency depended on the magnetic moment—in particular, when the frequency of precession of the spin axis matched, or resonated with, the frequency of the applied magnetic field. Rabi and his group at Columbia University used the magnetic resonance technique to produce surprising results for the magnetic moment of the proton and the quadrupole moment of the deuteron by 1939.

World War II interrupted research on magnetic resonance, but also fostered the development of new radio frequency electronics that would aid future research. At the end of the war physicists resumed work with resonance. One direction led to atomic clocks, which inverted the approach and used the minute difference between quantum-mechanical energy levels as the basis for a constant frequency. Another direction looked from the gases studied in the 1930s to solid matter. In 1946 a group under Felix Bloch at Stanford University and another under Edward Purcell at Harvard succeeded in detecting nuclear magnetic resonance in condensed matter. Bloch and Purcell shared the Nobel Prize in physics for 1952 for the work; Rabi had won in 1944 for the molecular-beam resonance technique.

Subsequent research brought NMR from the nuclear physics lab into diverse fields of science. In the early 1950s Bloch's group found that the chemical environment of the nucleus—the close presence of nuclei of other chemical elements or molecules—affected the resonance frequency, as did couplings between the spins of nearby nuclei. The so-called chemical shift and spin-spin coupling suggested a powerful tool for chemistry. Organic chemists began using resonance signatures to tease out variations in molecular structure, and increasing use of NMR in the 1950s helped transform organic chemistry from painstaking test-tube analysis of chemical reactions to a quick mechanical process of structure elucidation. Solid-state physicists similarly took up NMR to reveal the internal structure of their samples.

The provision of standardized NMR devices spurred the spread of the technique to other fields. Varian Associates, a firm formed in 1948 near Stanford, led the market. Varian's founders had long experience with microwave electronics and close associations with Stanford physicists and engineers. These ties extended to Bloch's group, and when Bloch and Stanford lagged in pursuing a patent on NMR the company convinced Bloch to file a patent and assign it the license. Varian would also hire several students of Bloch. Varian quickly capitalized when the chemical uses of the machine became apparent; it introduced the first commercial NMR spectrometer in 1952 for $26,000 and improved models in subsequent years. The role of Varian in NMR development demonstrates the indistinct boundary between industrial and academic research, and between basic and applied science.

Yet another use for NMR, perhaps the most widely known, appeared in medicine. In 1972 the chemist Paul Lauterbur proposed to use magnetic field gradients to distinguish NMR signatures from different portions of an object. The resulting scan could identify biochemical properties across a tissue sample, such as water content, blood flow, and cancerous growths. Groups in Aberdeen, Scotland, and Nottingham, England, soon scaled up NMR scanners for human use and commercial firms brought them to the medical marketplace. In the 1980s the technique acquired its current name of magnetic resonance imaging, or MRI, which skirted public fear of things nuclear.

John S. Rigden, *Rabi: Scientist and Citizen* (1987). Timothy Lenoir and Christophe Lécuyer, "Instrument Makers and Discipline Builders: The Case of Nuclear Magnetic Resonance," in *Instituting Science: The Cultural Production of Scientific Disciplines*, ed. Timothy Lenoir (1997): 239–292.

PETER J. WESTWICK

NUCLEAR PHYSICS AND NUCLEAR CHEMISTRY. Research into the atomic nucleus derived from the study of *radioactivity, which flourished starting around 1900 as a program in both chemistry and physics; chemists sorted out the different radioactive elements and their decay products, and physicists elucidated the nature of the emitted rays. The physicist Ernest *Rutherford, who would later deride most science outside physics as stamp-collecting, won a *Nobel Prize in chemistry for his radioactivity research, but his work divided the physical from the chemical in the study of radioactivity. He posited the existence of the atomic nucleus in 1911 (*see* ATOMIC STRUCTURE).

The subsequent atomic model of Niels *Bohr, which combined Rutherford's nuclear atom with the emergent *quantum theory, would establish the nucleus as the seat of radioactivity and encourage physicists to speculate about its contents. In 1919 Rutherford achieved the artificial disintegration of the nucleus by bombarding nitrogen atoms with alpha particles from radioactive substances. Hydrogen nuclei (protons) were knocked out of the nucleus by the impact of the alphas.

Most nuclear models through the 1920s built nuclei out of protons and electrons, since their opposite electric charges would prevent electrical repulsion; also, the apparent expulsion of electrons from the nucleus in beta decay argued for their inclusion. Application of the new quantum mechanics to the nucleus later in the decade, especially by George Gamow, provided a model of a liquid drop held together by surface tension, which nuclear particles could escape through quantum tunneling. But the presence of both protons and electrons defied attempts at detailed descriptions, and the nucleus seemed to reveal deficiencies in quantum mechanics. Bohr believed, characteristically, that the problem required bold steps and suggested that conservation of energy and momentum failed at the nuclear level. Wolfgang Pauli instead proposed a new nuclear particle, later dubbed the neutrino, to save the energy balance in beta decay.

Another new particle solved the problem and reconciled nuclear models with quantum theory. In 1920 Rutherford suggested that protons and electrons could combine to form a neutral particle, a "neutron," which might reside in the nucleus. The hypothetical neutron attracted little attention, but James Chadwick, a protégé of Rutherford's at the Cavendish Laboratory in Cambridge, looked for it through the 1920s. In 1932 Chadwick found it in radiation, studied by Frédéric Joliot and Irène Joliot-Curie, emitted by beryllium when it was exposed to alpha rays. Physicists soon accepted the neutron as a single elementary particle, and new models of the nucleus, notably one proposed by Werner *Heisenberg, gradually accommodated the neutron alongside protons and removed electrons from nuclei. The neutrino and neutron, along with the positron (detected, like the neutrino, in 1932), were followed in the 1930s by other new particles, such as the *cosmic-ray mesons. The floodgates opened to the proliferation of so-called elementary particles that emerged from the heads and machines of physicists in ensuing decades.

The neutron provided a useful projectile with which to probe the nucleus, since its neutral charge experienced no electrical repulsion. Physicists at the time were developing devices to accelerate charged particles to energies high enough to penetrate the electrical barrier of the nucleus. In another corner of the Cavendish, John D. Cockroft and Ernest T. S. Walton had built a high-voltage apparatus, which they used in 1932 to fire a proton into a lithium nucleus and disintegrate it into two alpha particles. Simultaneously, Ernest Lawrence in Berkeley was developing a circular particle *accelerator as an easier route to high energies. Then in 1934 Joliot and Joliot-Curie bombarded aluminum with alpha particles from a polonium source and found that they created an isotope of phosphorous, which decayed radioactively. The process provided a way to produce radioactive isotopes in the lab. A group under Enrico *Fermi in Rome quickly extended the results in a systematic bombardment of elements with neutrons; when they got up to uranium, their results suggested that neutron capture produced new transuranic elements.

The artificial production of radioisotopes, whether using rays from natural radioactivity or accelerated particles, required chemistry to disentangle the decay processes and identify short-lived parent species and their daughter isotopes by comparison to elements with similar chemical behavior. Nuclear research labs of the 1930s restored the sort of collaboration between physicists and chemists that had marked radioactivity research thirty years earlier. Lawrence, who recruited a group of chemists to his cyclotron program, noted the indistinct boundary between the existing disciplines and wondered whether to call the field nuclear physics or nuclear chemistry. An international community emerged among the major centers in nuclear science: the Cavendish under Rutherford; Lawrence's group in Berkeley; Joliot and Joliot-Curie in Paris; Fermi's group in Rome; Bohr's institute in Copenhagen; a group under Igor Vasilyevich *Kurchatov in Leningrad; and the Riken laboratory in Tokyo under Yoshio Nishina.

A momentous collaboration formed in Otto Hahn's chemistry institute in Berlin, where the physicist Lise Meitner worked with chemists Hahn and Fritz Strassmann. The Berlin group began studying the decay modes of the postulated transuranics produced by neutron bombardment of uranium; after several years of analysis, Hahn and Strassmann convinced themselves that the suppositious transuranics behaved chemically like elements much further down the periodic table. The chemical evidence implied that uranium could split into two lighter elements in a nuclear version of cell fission, but the conclusion challenged the results of Fermi's group and seemed to contradict the nuclear theory developed by physicists. As chemists, Hahn and

Strassman hesitated to take such a bold step. They appealed to the physicist Meitner, in exile from the Nazis, who with her nephew and fellow physicist Otto Robert Frisch showed how to reconcile fission with nuclear physics.

Physicists and chemists alike recognized the importance of nuclear fission after its announcement in January 1939, and both disciplines would play central roles in the military and industrial application of nuclear energy in World War II and afterwards. The discovery required the sort of interdisciplinary collaboration possible in Hahn's institute, which helps explain why physicists alone failed to notice the effect earlier. Fission exemplifies the interplay of chemistry and physics within a common field of nuclear science; Rutherford's early Nobel Prize symbolized the difficulties Nobel committees experienced in distinguishing physics from chemistry in the nuclear science of the 1930s.

The contributions of nuclear scientists to the war ensured public prestige and government funding for their postwar programs, which attracted new practitioners and allowed the construction of more and larger particle accelerators. The devices encouraged the separation of a new field of *high-energy physics from nuclear physics, but also supported thriving research. Nuclear physicists recognized the existence of several different mesons and incorporated the various types into the theory of nuclear forces. Based on evidence of the stability of certain isotopes, Maria Goeppert Mayer postulated an alternative nuclear structure based on shells of nucleons instead of the liquid drop; theorists later reconciled the competing models in a rotational scheme combining single-particle states with surface oscillations. Nuclear chemists, especially a group under Glenn Seaborg, meanwhile were using cyclotrons, nuclear reactors, and even nuclear bombs to produce new transuranic elements up to and beyond element 100 in the period table. Both nuclear physics and nuclear chemistry earned state support by continuing as prime contributors to the military and industrial development of nuclear energy in the cold war.

Roger H. Stuewer, ed., *Nuclear Physics in Retrospect* (1979). William R. Shea, ed., *Otto Hahn and the Rise of Nuclear Physics* (1983). Abraham Pais, *Inward Bound: Of Matter and Forces in the Physical World* (1986). Finn Aaserud, *Redirecting Science: Niels Bohr, Philanthropy, and the Rise of Nuclear Physics* (1990). Laurie Brown and Helmut Rechenberg, *The Origin of the Concept of Nuclear Forces* (1996).

PETER J. WESTWICK

NUCLEAR POWER. The dream of obtaining cheap and useful power from the atomic nucleus originated with the discovery in 1896 that

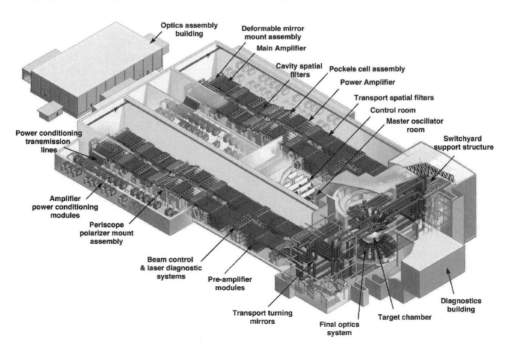

An artist's version of the National Ignition Facility, Livermore National Laboratory. The huge complex focuses the energy of 192 laser beams simultaneously on a speck of fuel in the target chamber at the far right. 1.8 million joules of light energy converge for a billionth of a second to ignite the nuclear reaction.

energy emanated from radioactive atoms. By the beginning of the 1920s, *Einstein's theory of the equivalence of mass and energy and Ernest *Rutherford's experiments in nuclear disintegration together indicated that energy is contained in every atomic mass. Popularizers of the day declared that the energy in just one glass of water could power a giant ship across the Atlantic. But leading physicists at the time agreed with Rutherford in dismissing the prospect of nuclear power as "moonshine." However, nuclear power suddenly became a realistic possibility when Otto Hahn and his collaborators in Berlin in 1938 discovered that uranium is capable of fission upon bombardment with neutrons. Physicists soon learned that in the process of fission, a large quantity of energy is released along with more neutrons, possibly enough to make a chain reaction. While an uncontrolled chain reaction would make a nuclear explosion, a controlled one in a nuclear reactor would provide a source of energy that could be transformed into electrical power.

In Chicago at the end of 1942, as part of the *Manhattan Project to build the atomic bomb, a team led by Enrico *Fermi achieved a controlled chain reaction with an experimental nuclear reactor. The project soon established a reactor research station near Chicago at the site of what would become the Argonne National Laboratory. In 1946, the United States Congress established the Atomic Energy Commission (AEC), a civilian agency with sole control of all nuclear research. At Argonne, which in 1948 became the AEC's center for the development of reactors to produce electrical power, investigators pursued the main elements of reactor design: types of fissionable core, moderators to slow and absorb neutrons, and coolants.

In the late 1940s, as the Cold War set in, the AEC concentrated on nuclear weapons and gave nuclear power a relatively small budget. Admiral Hyman Rickover grew impatient. Rickover, a career navy officer and visionary engineer, was bent on developing nuclear-powered ships. Obtaining authority in 1947 to build a reactor for the navy, he persuaded Congress in 1949 to commit the AEC to the project. At Rickover's urging, the agency contracted with the Westinghouse corporation for a reactor suitable to power a submarine. A prototype version achieved full-power operation in 1953, and the next year a modified version of the reactor was successfully installed in the submarine *Nautilus*.

In 1954, the Republican Congress revised the Atomic Energy Act to permit private ownership of nuclear reactors and the leasing of nuclear materials to industry. The chairman of the AEC told the public that nuclear power would bring "energy too cheap to meter." Both the AEC and partisans of atomic energy in the Congress also spoke of an "atomic power race" with the Russians to bring nuclear power to the Third World. Under AEC sponsorship, Westinghouse constructed a scaled-up version of the *Nautilus* reactor for electrical power generation at Shippingport, Pennsylvania. It went on line in 1957, generating 60 megawatts of power. That year, making conditions for private nuclear power more favorable, Congress limited industry's liability for nuclear accidents and provided hazard insurance at public expense.

More than a dozen nuclear power plants were operating in the United States by 1967. By 1975, in the wake of the Arab oil embargo and the resulting energy shortage, about 225 plants were on order, under construction, or in operation. And by the late 1970s, nuclear power plants generated 13 percent of the nation's electricity. Nuclear power was also under rapid development in France and Japan. Neither had adequate resources of coal or natural gas to fuel its power needs, and the oil embargo prompted both to intensify their commitment to reactor generation of electricity.

Meanwhile, in the United States, environmentalists had been taking aim at the rapidly developing nuclear power industry and its patron, the AEC. Critics raised questions about the disposal of the burgeoning radioactive wastes; about the safety of reactors themselves, particularly the emergency core-cooling systems designed to prevent overheating and meltdown; and about the effect of the heated water that reactors disgorged into rivers, lakes, and streams on aquatic life. In 1970, as the result of a suit brought by environmental groups, the AEC was compelled to take into account all potential environmental hazards in the licensing of nuclear plants. To many observers, the AEC seemed to have a conflict of interest because it was responsible for both the promotion and the regulation of nuclear power. In 1973, Congress broke up the AEC, awarding the agency's promotional and R&D functions to the newly created Energy Research and Development Administration (which in 1977 became the Department of Energy) and its regulatory functions to the new Nuclear Regulatory Commission.

The American nuclear power industry received a severe blow in March 1979 when operator errors at the nuclear power plant at Three Mile Island near Harrisburg, Pennsylvania, caused a failure in the cooling system, a partial meltdown of the nuclear fuel, and the expulsion of some 800,000 gallons of radioactive steam into the air above the surrounding Susquehanna Valley. Utilities canceled orders for more than thirty nuclear plants and no new orders were placed during the rest of the century. The accident at *Chernobyl in 1986 further turned the public in the United States against nuclear power, and stimulated anxious questioning in Europe.

Since the 1950s, several nations have attempted to harness thermonuclear fusion for power generation. Since fusion burns hydrogen, which is abundant, fusion reactors would in principle be cheap to sustain and relatively environmentally clean. However, no nation has come close to making practical fusion reactors. Research focuses on magnetic confinement of fusion fuel, using, for example "Tokomaks," (an acronym from the Russian words for "toroidal magnetic chamber," the standard vessel used to contain the superheated hydrogen fuel) and inertial confinement fusion, which uses a fusion-fuel-containing pellet irradiated by X-rays produced by high-energy lasers to symmetrically implode the pellet. Large laser facilities have been built to study ICF and the world's largest laser system, the stadium-sized 192-beam National Ignition Facility, is under construction in the United States. In the absence of a viable fusion reactor, nuclear power remains indispensable in the industrial economies of France and Japan, for example, where in the mid-1990s it provided 75 percent and 27 percent of their electricity, respectively. In recent years, the specter of global warming has revived interest in nuclear power in the West because nuclear plants do not produce greenhouse gases.

Irvin C. Bupp and Jean-Claude Derian, *Light Water: How the Nuclear Dream Dissolved* (1978). U.S. Office of Technology Assessment, *Starpower: The U.S. and the International Quest for Fusion* (1987). Spencer Weart, *Nuclear Fear: A History of Images* (1988). Brian Balogh, *Chain Reaction: Expert Debate and Public Participation in American Commercial Nuclear Power, 1945–1975* (1991). Richard Rhodes, *Nuclear Renewal: Common Sense About Energy* (1993). Gabrielle Hecht, *Radiant France: Nuclear Power and National Identity after World War II* (1998).

DANIEL J. KEVLES

NUTRITION. French chemists around 1700 characterized animal materials as "alkalescent" because, when kept warm and moist, they putrefied, emitting an alkaline vapor, whereas most vegetable materials were "acescent," turning acid as they fermented. If digestion were a kind of fermentation, how could grain and vegetables fed to a growing calf turn into meat? In 1728, Jacopo Beccari of the University of Bologna reported that white flour, kneaded and washed to remove the starch, left an alkalescent residual pellet (gluten) that had other properties of animal tissues: "this was the *true nutrient* in wheat."

By 1780 the vapor from alkalescent materials had been identified as ammonia containing nitrogen—something absent from starch and fats.

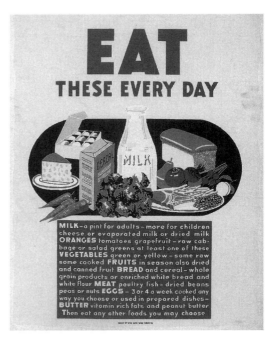

Poster published during World War II to promote the consumption of healthy foods.

Feeding trials with animals conducted between 1815 and 1840 showed that they failed when fed nitrogen-free diets. Animals could not utilize atmospheric nitrogen. In 1842 Justus von *Liebig still believed that "animal substance" (protein) broke down to supply the energy for muscle contraction. He inferred that physical work required a high-protein diet. Adolf Fick and Johannes Wislicenus disproved this conclusion in 1866 by demonstrating that insufficient protein broke down during a mountain climb to supply even the minimum energy needed to balance the work done against gravity, given the heat of combustion of protein and James *Joule's "mechanical equivalent of heat" (*see* CONSERVATION OF ENERGY).

In 1902, Wilbur Atwater and Francis Benedict found that both carbohydrates and fats could be used as muscle fuels with about 25 percent efficiency, based on their heats of combustion, and that an average man needed no more than 70 grams of protein per day regardless of activity.

Study of clinical disease paved another route to the broader understanding of nutritional needs. For example, beriberi, characterized first by numbness and weakness in the legs, and finally by sudden heart failure, had long been a problem in Asia. In 1803 one British army surgeon in Sri Lanka related it to the lack of "some nice chemical combination" in the sufferers' diet. But by 1885 microbial infection had explained so much that people sought the *germ causing beriberi. In Java

NYLON 591

(then Dutch), beriberi was common among native soldiers. The physician Christiaan Eijkman tried infecting chickens with blood from hospitalized patients, and some (both injected and uninjected) developed leg weakness reminiscent of beriberi. But the next batch remained healthy. He found that when leg weakness had appeared the chickens had been fed surplus cooked white rice from the hospital instead of feed-grade unmilled rice. Further studies showed that giving supplementary "polishings" (removed during milling) would keep birds on white rice healthy. Javanese prisons were then classified as to whether they used white rice or brown rice (without the "polishings" removed)—only the former were found to have endemic beriberi, apparently confirming the value of the chicken model.

Painstaking fractionation of rice polishings, followed by bioassays with birds, finally yielded active crystals present at only a few parts per million in the polishings (1926). The molecule, $C_{12}H_{18}N_4SOCl_2$, contained both a thiazole and a pyrimidine ring. In 1936 the structure was identified, synthesized, and named thiamin(e). In thiamin deficiency the pyruvic acid level in the blood goes up. This observation led to the identification of thiamin pyrophosphate as an important coenzyme in carbohydrate metabolism. Thiamin is now added to flour in many countries and available cheaply in pill form. Humans do not instinctively avoid thiamin-deficient diets.

More vitamins came to light in similar ways once an animal model was available: guinea pigs for vitamin C (which prevents scurvy) and dogs for nicotinic acid (which prevents pellagra). Still other vitamins and trace mineral nutrients such as selenium were identified by feeding young rats on purified "protein-energy-mineral" diets and finding what more they needed to thrive. Work on the nutritional needs of bacteria, where assays are rapid and needed less material, also yielded factors needed for humans—particularly folic acid, deficient in the diets of pregnant Indian women with macrocytic anemia.

Since 1950 research has focused on overconsumption in relation to diabetes, heart disease, and other conditions, and the identification of food chemicals that may affect cancer risks for good or ill. In every period advances have come from chemists with broad interests (such as Antoine-Laurent *Lavoisier) and physiologists. Physicians have also been important as have bacteriologists for demonstrating the value of animal models.

Elmer V. McCollum, *A History of Nutrition* (1957). Karl Y. Guggenheim, *Nutrition and Nutritional Disease: The Evolution of Concepts* (1981). Frederic L. Holmes, *Lavoisier and the Chemistry of Life* (1985). Kenneth J. Carpenter, *Protein and Energy: A Study of Changing Ideas in Nutrition* (1994). Harmke Kamminga and Andrew Cunningham, eds., *The Science and Culture of Nutrition, 1840–1940* (1995). Kenneth J. Carpenter, *Beriberi, White Rice and Vitamin B* (2000).

KENNETH J. CARPENTER

NYLON. DuPont Corporation first synthesized and put into commercial production nylon, a polyamide, in 1938. Although not the first synthetic fiber (the artificial silk, viscose or rayon, had been made from cellulose since the 1890s), it was the first commercial fiber to be prepared as a result of basic scientific research into the disputed field of polymer chemistry. Debate over the existence of macromolecules posited by the German chemist Hermann Staudinger in 1922 arose from a tension between two distinct traditions of chemical theory current from the 1880s to the 1930s. A "physicalist tradition" viewed polymers as molecules of low molecular weight held together in aggregates by physical forces that gave rise to their colloidal properties. The "organic structuralist" tradition, to which Staudinger and the American Wallace Hume Carothers belonged, explained the properties of polymers in terms of giant molecules whose repeating submolecules possessed the familiar chemical structures established by August Kekulé and were held together by ordinary covalent bonds. These two approaches, which led their exponents to bitter controversies in the 1930s, were only resolved in the 1940s by their fusion at the hands of Herman Mark and Paul Flory. By then, considerable support for the concept of macromolecules had come from Carothers's industrial research on condensation reactions (for which he developed a new kind of molecular still) at DuPont's central laboratories in Wilmington, Delaware.

In 1928, following a career of academic teaching and research at the universities of South Dakota and Illinois and at Harvard, Carothers went to the DuPont Company to do fundamental research aimed at producing artificial rubber and new textile fibers. Using exacting standards of purification, Carothers brought order to the hitherto empirical production of polymers by showing that the principal methods of generation were by means of addition and condensation reactions. In 1935 Carothers found that a polyamide with a molecular weight of over ten thousand made from the condensation of hexamethylenediamine and adipic acid could be melt-spun into a fine thread that possessed both a high melting point and considerable tensile strength. After devising cheap and reliable methods for producing the raw materials in quantity, and constructing a test plant, DuPont began large-scale manufacture of the new yarn in January 1938. The term "nylon," the generic patented name for linear condensation polymers, was first used in October 1938. The product was initially called "no run." Since stockings made from the new material and first exhibited at the San Francisco exhibition in February 1939 did run or snag, DuPont changed the name to "nylon" to resonate with rayon. A year later, nylon stockings went on sale throughout America. They became a prized luxury during World War II, after which nylon found a huge variety of industrial, commercial, and household applications. Carothers did not live to see this triumph of basic research as a tool of industry. A manic depressive, he committed suicide in Philadelphia on 29 April 1937.

Mathew Hermes, *Enough for One Lifetime: Wallace Carothers, Inventor of Nylon* (1996). Yasu Furukawa, *Inventing Polymer Science. Staudinger, Carothers, and the Emergence of Macromolecular Chemistry* (1998).

WILLIAM H. BROCK

O

OBSERVATION AND EXPERIMENT lie at the heart of the scientific enterprise. The increasing sophistication of observational techniques from simple naked-eye scrutiny to reliance on intricate instrumentation is one of the most striking features of the history of modern science. Another is the progressive extension of the experimentation, the manipulation of aspects of nature in controlled settings, from a few limited areas of natural knowledge around 1700 to a vast range of phenomena from particle *physics and *molecular biology to *medicine and psychology. Since the 1980s, historians and philosophers have given increased attention to the process, as opposed to the results, of observation and experiment.

Scientists began experimenting systematically in the seventeenth century. Scientific academies and societies provided specialized spaces for experimentation. The *mechanical philosophy, with its postulation of corpuscles invisible to the naked eye, encouraged the trend as natural philosophers such as Robert *Boyle and Robert *Hooke used experiments to try to infer more about the behavior of these supposed corpuscles. The *microscope, the *thermometer, the *pendulum, the *telescope, and the *air pump were invented and improved. Natural philosophers of an empiricist bent such as Francis Bacon proposed that crucial experiments, rather than the time-honored strategy of derivation from self-evident first principles, could be used to select or demonstrate scientific theories (*see* BACONIANISM).

By the eighteenth century, makers of specialized instruments had found a profitable niche. By the mid-nineteenth century, teaching laboratories where students could learn bench skills were being established in universities, first in Germany and then elsewhere. Two kinds of experiment attracted special attention. In the early nineteenth century, crucial experiments became yet more prestigious when spectacular theoretical predictions, such as Augustin *Fresnel's diffraction experiment, apparently decisively proved one or another of contending theories. Around the beginning of the twentieth century, the power of experiments to uncover phenomena that could

revolutionize a discipline was driven home by the discoveries of *radioactivity and the *electron.

From the 1930s until the 1960s, observation, according to the dominant positivist philosophy of science, provided the basis for verifying the meaning of terms, and hence for all scientific statements. According to one variant, developed by Percy Bridgman and known as operationalism, terms that referred to unobservables, such as electric force, could be defined and understood as the operations required to measure the force. Historians and scientists, such as J. B. Conant, also stressed the centrality of observation and experiment. His *Harvard Case Studies in Experimental Science* (1957) selected key historical experiments, such as those carried out with the air pump, as instructive case studies of the scientific method.

Such was the prestige of experiment that the term was stretched to cover cases in which the mind alone discovered new things about the world via what the physicist and philosopher Ernst Mach called "thought experiments." Mach believed that we have a store of instinctive knowledge that can be made explicit by thinking about specific cases. As an example he pointed to Simon Stevin's analysis in *De Beghinselen der Weeghconst* (1586) of the problem of a chain draped over a double frictionless plane. By mentally adding links joining the two ends of the chain, we can see that it must have been in static equilibrium; if not, perpetual motion would ensue. Other famous and productive thought experiments were Isaac *Newton's consideration of the way in which water climbed the walls of a spinning bucket and James Clark *Maxwell's fiction of a demon who opened and closed trapdoors to separate fast and slow moving molecules. Later, Thomas Kuhn argued that by helping us reconceptualize the world, thought experiments could create trouble for dominant theories or paradigms just as laboratory experiments could.

In the 1960s and 1970s, observations began to appear more problematic as confirmations of theory. A number of philosophers and historians, important among them Kuhn, Norwood Russell Hanson, Nelson Goodman, and Paul Feyerabend, argued that all observations were theory-laden because scientists bring to any observation all kinds of theoretical expectations that affect what they see. They supported this contention by turning to the thesis enunciated by Pierre Duhem in *La théorie physique: Son object, sa structure* (1906) that no hypothesis could be tested in isolation, and by citing the results of gestalt psychology. Johannes *Kepler and Tycho *Brahe, these philosophers liked to argue, literally saw different things when they looked at the solar system. Suddenly, observations and experimental reports that for centuries had seemed unproblematic, at least by the time they appeared in print, were now open for investigation in the post positivist era.

One of the first to emphasize experiment as a field of inquiry was the philosopher Ian Hacking. Rejecting the notion that observations must be laden with the theory under test, he argued for the relative independence of experimental work. He suggested that historians and philosophers should study how instruments were built, measurements made, and phenomena created in the laboratory. Historians and scientists who had always been more interested in the process of observation and experiment than philosophers took up the challenge. Allan Franklin tried to find formal ways of deciding the validity of the experimental outcome, while Peter Galison explored the complex social and intellectual worlds of the huge experiments of big science.

Their work complemented the laboratory studies pursued by social scientists. As participant observers, Harry Collins, Karin Knorr, Bruno Latour, and Steve Woolgar acted as anthropologists observing the habits of their subjects, the scientists who carried out experiments in the Salk Institute or hunted to find the quark. They drew the dramatic conclusion that scientists did not simply report their observations of the world but that they made or constructed the world. Some historians—for example, Steven Shapin and Simon Schaffer in their study of Boyle's work with the air pump—claimed to find support for this social constructivist position. By the end of the twentieth century, observations and experiments, so long taken to be the foundation of a reliable account of the world, were being used to suggest that science did not describe the world as it really was or appeared to be, but rather as it reflected the presuppositions of the scientists. Many scientists and historians found this an outrageous conclusion. The first shots of the so-called Science Wars followed.

Ian Hacking, *Representing and Intervening* (1983). Andrew Pickering, *Constructing Quarks* (1984). Steven Shapin and Simon Schaffer, *Leviathan and the Air Pump* (1985). Allan Franklin, *The Neglect of Experiment* (1986). Bruno Latour and S. Woolgar, *Laboratory Life* (1986). Peter Galison, *How Experiments End* (1987). D. Gooding, T. Pinch, and S. Schaffer, eds., *The Uses of Experiment* (1989). Paul Gross and Norman Levitt, *Higher Superstition: The Academic Left and its Quarrels with Science* (1998).

See also Proof.

Rachel Laudan

OBSERVATORY. Celestial phenomena have long been observed from buildings or sites specially equipped for the purpose. Notable examples before the early modern period include that at Hamadan (1024), established to remedy tables of planetary positions; at Maragha (1259) in northwest Persia, set up by Genghis Khan's nephew Hulagu, a devotee of astrology, and equipped with renowned astronomers from China to Spain; and at Samarkand (c. 1420), built by Ulugh Beg, grandson of Tamerlane and also a skilled mathematician and astronomer. The main instrument at Samarkand, the largest sextant ever made, consisted of two hinged rods whose free ends roamed over a circular arc 40 meters in radius. Using it, Samarkand's astronomers drew up a catalog of 1,018 stellar positions, the greatest achievement of fifteenth-century observational astronomy. Observatories were yet to be harmonized and integrated into Islamic culture, however. The Istanbul observatory, built in 1577, the year of a famous comet, did not long survive its completion. The faithful suspected that its attempt to pry into the secrets of nature had brought on misfortunes including plague, defeats of Turkish armies, and deaths of important individuals.

The comet of 1577 and an earlier supernova in 1572 undercut Aristotelian astronomical theory in Europe and encouraged the Danish nobleman Tycho *Brahe to build an observatory surpassing everything before it. The Danish king, Frederick, offered him the island of Hven. There Tycho built Uraniborg, his manor and observatory, with large instruments for observing stellar and planetary positions, a paper mill and a printing press, and other necessities for self-sufficiency. Its great mural quadrant was a brass quarter-circle arc 2 meters in radius mounted to a wall oriented precisely north-south. Governmental largesse, equivalent to a great many millions in today's dollars, flowed to Tycho during King Frederick's reign and the regency of his son Christian, but shortly after Christian's coronation in 1596 it ceased.

The seventeenth and eighteenth centuries saw the rise of nation-states in Europe and growing sea trade, increasing navigational demands for accurate astronomical data. Charles II of England issued a royal warrant in 1675 appointing John

Flamsteed "astronomical observator" at a salary of £100 per year to rectify "the tables of the motions of the heavens, and the places of the fixed stars, so as to find out the so much-desired longitude of places for the perfecting the art of navigation." A second warrant established a small observatory in the royal park at Greenwich. There was no provision for instruments, some of which Flamsteed paid for himself, and which his widow removed from Greenwich.

The Paris Observatory, begun in 1667, though more generously funded by the state, combined the pomp of the French court with poor design. Observing through windows rather than a rotating dome limited observations. The Cassini family ruled the observatory for four generations, until driven out during the French Revolution.

The observatory at St. Petersburg also functioned more for show than science until replaced in 1839 by the Pulkovo Observatory south of the city. Equipped with fine instruments, the new astronomical capital of the world figured as an object of high utility and importance to the scientific honor of Russia. The United States Naval Observatory, devoted, like Pulkovo, to positional astronomy, was completed in 1844, though surreptitiously in the face of congressional disapproval.

President John Quincy Adams in his message to Congress of 1825 had lamented that Europe had 130 "light-houses of the skies" and America none. Soon the United States had observatories too, mostly in colleges. Harvard lured William Cranch Bond away from his private observatory to work for no salary and supply his own instruments. They pointed out windows until interest in the great comet of 1843 led to a public subscription to construct and endow the Harvard College Observatory. The municipal Cincinnati Observatory also was funded by public subscription in 1843. Cincinnati purchased an 11.25-inch refractor from the German firm Merz and Mahler, makers of the Pulkovo 15-inch refractor, then the largest and best telescope in the world. Cincinnati's astronomer spent most of his time displaying the heavens to subscribers. Harvard, too, ordered a telescope from Merz and Mahler, and received a twin to the Pulkovo 15-inch. Trying too hard to justify its purchase, Bond mistakenly reported that he had resolved stars in the Orion *nebula.

College and government observatories with precision instruments focused their efforts on positional astronomy. Individuals in small, private observatories were freer to explore new fields. The industrial revolution and British technology facilitated the construction of large metal reflecting telescopes from the 1780s to the

1860s, but steerable mountings and shelter from inclement weather, both important elements of working observatories, lagged in development, leaving the telescopes largely unusable. Application of the new technologies of *photography and *spectroscopy to astronomy also occurred first in private observatories. In 1854 William Huggins disposed of his silk and linen business and moved to Tulse Hill, a suburb of London, where he set up his own telescopes. By the 1870s he was identifying elements in the spectra of stars and nebulae and detecting the motion of stars from shifts in their spectral lines.

Huggins's telescopes limited his investigations to bright, relatively near objects. Meanwhile in the United States, men wealthier than the former draper endowed the observatories with more expensive instruments: the Lick Observatory of the University of California with a 36-inch refractor (1887) and the Yerkes Observatory of the University of Chicago with a 40-inch (1897). The Lick, the first of the mountain observatories, demonstrated the value of good astronomical seeing at high altitude.

Growing interest in astrophysics and in distant stars and nebulae encouraged the development of new observatories with large steerable reflecting telescopes suitable for photography and auxiliary instruments for the analysis of starlight. In 1902 Andrew Carnegie, rich from innovations in the American steel industry, created the Carnegie Institution of Washington to encourage investigation, research, and discovery. George Ellery Hale left the Yerkes Observatory to build, with Carnegie money, the Mount Wilson Observatory on a mountain above Los Angeles. A 60-inch photographic reflecting telescope, completed in 1908, and a 100-inch, completed in 1918, were the largest telescopes in the world and they revolutionized astronomical knowledge. Lights from the expanding city affected astronomical seeing, however, and after World War II an even larger telescope was constructed at Mount Palomar, northeast of San Diego.

The Mount Wilson observatory's relationship with physicists at the nearby California Institute of Technology was also crucial to its dominance of the study of astrophysics during the twentieth century. Many observatory staff at the beginning of the century had acute vision but little scientific education; soon a Ph.D. degree and considerable theoretical understanding were necessary for admittance.

Supposedly only men could withstand the rigors of observing all night in an unheated telescope dome, but the presence of women in observatories kept pace with their progress in university science programs. Women were first employed in an observatory in 1886, at Harvard, for lower wages than men would have received, not to observe but to examine photographs of stellar spectra and catalog the spectral lines.

World War II marked a turning point in the relationship between science and the state. The development of *radar, ballistic missiles, and the atomic bomb relied on and demonstrated the power of state-sponsored and directed research and development. Observatories were a major beneficiary of increased government patronage.

Research on radar during World War II had led to the University of Manchester's Jodrell Bank Research Station and its large steerable radio telescope. Government assistance proved insufficient, however, until the observatory was rescued from financial disaster by its ability to track *Sputnik* in 1958. In 1962 British astronomers discovered a radio source that Australian astronomers identified with a faint star; but the honor of measuring the highest redshift observed up to that time and establishing the first known quasar went to the United States, which had the only optical telescope powerful enough to study the object's spectrum. Commonwealth pride was assuaged by the creation of the Anglo-Australian Observatory in New South Wales in 1974.

The Soviet Union's *Sputnik* challenged American aerospace supremacy. The United States responded with a new institution, the National Aeronautics and Space Administration (NASA). Among its accomplishments are automated mini-observatories launched into space, topped off by the two-billion-dollar Hubble Space Telescope. This instrument, operated by the Space Telescope Science Institute, is managed by a university consortium under contract to NASA. Observatory personnel and much of the auxiliary instrumentation are earthbound. In general, NASA adjudicates questions of scientific priority and supplies the money for space observatories; industry helps build them; and universities or consortiums of universities design and operate them, and analyze the resulting observations.

The Kitt Peak National Observatory on a mountain near Tucson supports the largest collection of big telescopes in the northern hemisphere. Seventeen universities have come together in the Association of Universities for Research in Astronomy (AURA) to manage the observatory. After *Sputnik*, the National Science Foundation supplied many millions of dollars for construction of AURA facilities.

Observatories moved from cities to mountain-tops and then into space. They have also moved south, to observe the sky not visible from the Northern Hemisphere. The British Board of Longitude established the Royal Observatory at the Cape of Good Hope in 1820; the Harvard College Observatory had a southern station in Peru at the end of the nineteenth century; and Mount Wilson established a southern station in Las Campanas, Chile, in 1976. Its dark-sky site is better for extragalactic studies than Mount Wilson, which diminished its activity in the 1980s to free up funds for Las Campanas. Chile also boasts AURA's Cerro Tololo Inter-American Observatory and the European Southern Observatory. Astronomers from any institution may apply for observing time, and automation increasingly allows astronomers to control the telescopes from anywhere in the world.

It remains to be seen whether generous governmental patronage for observatories will long outlive the Cold War. Already, though, immense, technologically complex, and expensive observatories on Earth and in space memorialize our civilization as pyramids do ancient Egypt and Mexico, cathedrals the Middle Ages in Europe, and Stonehenge ancient Britain.

See also ASTRONOMY, RADIO; INSTRUMENTS AND INSTRUMENT MAKING; PHILANTHROPY; TELESCOPE.

Bessie Zaban Jones and Lyle Gifford Boyd, *The Harvard College Observatory: The First Four Directorships, 1839–1919* (1971). *The Royal Observatory at Greenwich and Herstmonceux. 1675–1975* (1975): *Vol. 1: Origins and Early History (1675–1835)*, by Eric G. Forbes; *Vol. 2: Recent History (1836–1975)*, by A. J. Meadows; *Vol. 3: Buildings and Instruments*, by Derek Howse. Kevin Krisciunas, *Astronomical Centers of the World* (1988). Donald E. Osterbrock, John R. Gustafson, and W. J. Shiloh Unruh, *Eye on the Sky: Lick Observatory's First Century* (1988). J. A. Bennett, *Church, State, and Astronomy in Ireland: 200 Years of Armagh Observatory* (1990). S. C. B. Gascoigne, K. M. Proust, and M. O. Robins, *The Creation of the Anglo-Australian Observatory* (1990). Victor E. Thoren, *The Lord of Uraniborg: A Biography of Tycho Brahe* (1990). Frank K. Edmundson, *AURA and its U.S. National Observatories* (1997). Donald E. Osterbrock, *Yerkes Observatory 1892–1950: The Birth, Near Death, and Resurrection of a Scientific Research Institution* (1997).

NORRISS S. HETHERINGTON

OCCULT QUALITY. See SYMPATHY AND OCCULT QUALITY.

OCEANOGRAPHIC INSTITUTIONS. Most modern oceanographic institutions date from the second half of the twentieth century. During this period oceanography came from the periphery of earth science to a more central position. Sea-floor studies played a crucial role in establishing current thinking about the dynamics of continents and oceans. Physical oceanographers improved knowledge of the mechanisms governing ocean circulation and the relationship between ocean and atmosphere, with important consequences for understanding climate change. From the 1980s onwards they joined marine biologists, geologists, and chemists in the study of recently discovered hydrothermal vents and their unique faunas. These developments reflected scientific interests and oceanography's military and other applications in modern industrial society, and took place when funding for large-scale scientific projects became increasingly available. The funding made possible not only new ships and opportunities to explore the oceans on a scale not known before but also a corresponding increase in the number and scale of institutions carrying on marine research.

The first marine laboratories date from the second half of the nineteenth century. By then most nations had organizations such as the United States Coast Survey or the Meteorological Office in Britain, which functioned in the widest sense as oceanographic institutions, and also natural history museums and large aquaria. Zoologists who came to the seaside to be near fresh specimens developed the first dedicated institutions from ad hoc arrangements. A growing awareness of what could be learned about the origins, development, and physiology of living organisms from marine life powered their enthusiasm.

The first permanent institution in Europe was a laboratory dedicated to the study of marine zoology and physiology established by J. J. Coste at Concarneau, France, in 1859. A biological station at Arcachon, maintained by a local scientific society, followed in 1863. By the 1890s marine stations could be found on the coastlines of Europe from the Black Sea to the Scandinavian Arctic. Of these early foundations the most influential was the Stazione Zoologica established at Naples by the German zoologist Anton Dohrn in 1872. Subsidized by the German government but also supporting itself by renting tables to other scientists and institutions and through the sale of specimens, the Stazione Zoologica provided (and provides) scientists of all nationalities with year-round access to research facilities where studies of the anatomy, physiology and development of marine organisms led to important advances in the life sciences.

These institutions varied considerably in their origins, modus operandi, and objectives. Félix-Joseph de Lacaze-Duthiers's stations at Roscoff and Banyuls-sur-Mer were linked to his department at the Sorbonne and welcomed visiting workers on the Naples model. Most laboratories opened only seasonally and a few, such as the station founded by the Zoological Society of the Netherlands in 1876 in its early

days, occupied prefabricated buildings that could be moved from one location to another. Many of these European institutions received at least part of their funding from the state, some governments (notably the British) being stingier than others. In the United Kingdom the Marine Biological Association established the Plymouth Laboratory in 1885. Other early British laboratories were connected with university departments (e.g., Gatty/St. Andrews and Port Erin/Liverpool), local societies, and fishery committees. Until well after 1900 British government funding for marine research was largely restricted to fisheries programs and provided at best an insecure lifeline to the independent laboratories.

Meanwhile in the United States marine biological laboratories were set up at Woods Hole, at Cold Spring Harbor, and elsewhere. The laboratory at Woods Hole, through its predecessors, the Penikese Laboratory established by Louis Agassiz in 1873 and the Annisquam Sea-side Laboratory of 1880, had its roots in teacher training but would later become a leading center for biological research. On the West coast the principal centers were linked to universities—the Hopkins Marine Station of Stanford University at Pacific Grove, founded in 1892, the Scripps Institution for Biological Research (University of California), and the University of Washington's Oceanographic Laboratories.

In 1871 the U.S. Commission for Fisheries under Spencer Fullerton Baird selected Woods Hole for biological investigations, and built the fisheries laboratory there a few years later. Concern over fisheries drove much organized marine research in the late nineteenth century. Norway channeled its marine research effort into specialist fisheries institutions. Scandinavian scientists soon perceived the need for more broadly based and coordinated studies of fisheries and related problems. Their initiative led to the establishment of the International Council for the Exploration of the Sea in 1900. The ICES inspired the creation of national laboratories and a short-lived (1902–1908) Central Laboratory in Christiania (Oslo) that paid special attention to the development of apparatus.

Until 1900 people considered the study of the oceans a shipboard activity. As the collections gathered in expeditions multiplied, however, land facilities for their study were required. The voyage of HMS *Challenger* (1872–1876) resulted in the creation of the Challenger Office in Edinburgh. It shut down after the completion of its report in 1895. Sir John Murray continued its work independently at the Villa Medusa until his death in 1914 but in 1884 he had established a laboratory that he hoped would be more permanent, the Scottish Marine Station for Scientific Research near Granton—the world's first truly oceanographic institution. In spite of Murray's efforts to conform to the prevailing utilitarian ethos by linking the station to long-running research by the Scottish Meteorological Society on the effects of weather on fisheries, the government refused to support it. The Marine Biological Station at Millport, established by the West of Scotland Marine Biological Association in 1897, regarded itself as Murray's successor and its work on productivity in the 1920s and 1930s followed his interdisciplinary vision.

The first permanent oceanographic institutions date from the turn of the twentieth century. The geographer Ferdinand von Richthofen established the Institut für Meereskunde in Berlin in 1900 to cover all aspects of oceanography. It sponsored the influential *Meteor* expedition in the 1920s, which resulted in the first detailed survey of the deep Atlantic circulation. After World War II the Berlin institute's work was transferred to a like-named, preexisting institution attached to the University of Kiel. Kiel was already distinguished in ocean research through the work of Otto Krümmel, whose *Handbuch der Ozeanographie* served as a standard text, and of Victor Hensen and other members of the Kiel school of plankton studies.

One of the foremost leaders of ocean research of the early twentieth century was Prince Albert I of Monaco, an enthusiast for science, education, and international cooperation. His wealth enabled him to mount numerous scientific voyages in Atlantic, Mediterranean, and Arctic waters from the 1880s till 1914. He set up his twin foundations, the Institut Océanographique in Paris and the Musée Océanographique at Monaco, both still in existence, to house his collections and promote the science of the sea through research and teaching.

After World War II Admiralty pressure led to the establishment of a (British) National Institute of Oceanography (1949), now incorporated in the large-scale Southampton Oceanography Centre, established in 1995.

In the United States concern about the state of marine science prompted several initiatives in the interwar period. In 1923 William E. Ritter secured the appointment of Thomas Wayland Vaughan as his successor as director of the renamed Scripps Institution of Oceanography (SIO) of the University of California. However, only after the appointment of the Norwegian physicist Harald Ulrik Sverdrup as Vaughan's successor in 1936 did work expand offshore into the Pacific Ocean. The outbreak of war postponed major development until the 1950s. Since that time SIO has been the premier U.S. West-coast oceanographic institution. On the eastern seaboard, this position is held by the Woods Hole Oceanographic Institution, founded in 1930 when a committee of the National Academy of Sciences enlisted the support of the Rockefeller Foundation for physical oceanography. A key role has also been played by the Lamont-Doherty Geological Observatory of Columbia University, founded by Maurice Ewing in 1948 as a center for seafloor studies.

The diversification of ocean science during recent decades, ranging from the deep-sea research to increasingly comprehensive observations of the oceans from space, has brought many sorts of institutions, including commercial enterprises, into oceanographic work. National and international academic cooperation now maximizes efficiency in what is still an expensive and challenging field of scientific investigation. However, large institutions continue to play an important role as centers of excellence, both in generating and supporting new research projects.

See also FIELD STUDIES; GEOLOGY; METEOROLOGY; PLATE TECTONICS.

Thomas Wayland Vaughan, *International Aspects of Oceanography* (1937). Helen Raitt and Beatrice Moulton, *Scripps Institution of Oceanography. First Fifty Years* (1967). Susan Schlee, *The Edge of an Unfamiliar World. A History of Oceanography* (1976). Elizabeth Noble Shor, *Scripps Institution of Oceanography: Probing the Oceans, 1936 to 1976* (1978). Peter Limburg, *Oceanographic Institutions. Science Studies the Sea* (1979). Jane Maienschein, *One Hundred Years Exploring Life, 1888–1988: The Marine Biological Laboratory at Woods Hole* (1988). Eric L. Mills, *Biological Oceanography: An Early History, 1870–1960* (1989). Margaret Deacon, *Scientists and the Sea, 1650–1900; A Study of Marine Science*, 2d rev. ed. (1997).

MARGARET B. DEACON

OCEANOGRAPHY, as a distinct scientific discipline, and "oceanography," as the standard term for the study of all of the marine sciences, both date from the late nineteenth century, when the first major expeditions were undertaken specifically to explore the physics, chemistry, biology, and geology of the world's oceans. But the prehistory of oceanography begins centuries earlier. Natural philosophers had given attention separately to most of the branches that now constitute oceanography—Aristotle's investigations of marine invertebrates; Robert *Boyle's analysis of the temperature, salinity, and movement of sea-water; and Isaac *Newton's explanation of the forces that cause the tides being among the most famous.

The first text devoted exclusively to marine science was the *Histoire physique de la mer* (1725) of Count Luigi Ferdinando Marsigli, a military man and founder of the Istituto di Bologna (*see* ACADEMIES AND LEARNED SOCIETIES). From studies of the Gulf of Lyons, Marsigli assembled information about water temperature, salinity, specific gravity, tides, waves, currents, depth contours, and marine plants and animals. The diligence required in these efforts convinced Marsigli of the limitations of individual research in marine science; larger scale results required teams of investigators and government support—the hallmarks of oceanography since the nineteenth century.

The second half of the eighteenth century witnessed an acceleration of marine research. Enthusiasm for, and advances in, chemistry underpinned the chemical analysis of sea water by leading chemists, including Antoine-Laurent *Lavoisier and Torbern Bergman. Salinity was also studied during voyages of exploration, such as the Danish expedition to Arabia Felix, and, especially, James Cook's expeditions in the Pacific. Meanwhile, ocean currents and circulation patterns engaged the curiosity not just of sailors but of scientists such as Benjamin *Franklin, chronicler of the Gulf Stream, and Benjamin Thompson, Count Rumford, whose heat experiments led him to attribute ocean circulation to differences in water density, a theory finally accepted after a long delay.

Studies of currents and tides intensified in the early nineteenth century, exemplified by British naval surveyor James Rennell's *An Investigation of the Currents of the Atlantic Ocean* (1832). Also in the 1830s, John Lubbock and William Whewell, Lubbock's mentor at Trinity College, Cambridge, reduced tidal phenomena to mathematical analysis. And the British Admiralty helped to install tidal gauges around the southern coast of England. The limitations of the data for establishing general tidal patterns, and for drawing what Whewell dubbed "cotidal lines," led him to develop an international scheme for the collection of information about tides. The nascent British Association for the Advancement of Science soon took over the program. At the U.S. Naval Observatory, Matthew Fontaine Maury assembled data on winds, currents, and other oceanographic phenomena from ships' captains, and published the results in his textbook, *The Physical Geography of the Sea and Its Meteorology* (8 editions, 1855–1861).

The British Association also played a key role in marine biological enterprises during the middle decades of the nineteenth century. The chief investigator was Edward Forbes, a marine naturalist and paleontologist for the Geological Survey, and a native of the Isle of Man. Excursions of small boats to ascertain the depth and distribution of bottom-dwelling species led to summer-long cruises in British waters, continued aboard Admiralty ships in the late 1860s by naturalists Charles Wyville Thomson and William B. Carpenter. The success of these efforts, and the increasing curiosity about life and conditions in the deep oceans, gave rise to the most ambitious oceanographic endeavor of the era, the expedition of HMS *Challenger* (1872–1876). Wyville Thomson headed a team of five scientists and an artist during this three-and-a-half-year circumnavigation of the globe. They oversaw dredging and trawling at more than 360 stations. International authorities in the various subspecialties analyzed the data and the specimens collected. The resulting *Challenger Reports* (1885–1895), published in fifty volumes, remain the founding benchmark of oceanographic science.

Although the supremacy of its navy allowed England to take a leading role initially, other nations used the precedent to launch important oceanographic enterprises during the late nineteenth and early twentieth centuries. Germany focused initially on the North Sea, but its S. S. *Gazelle* also operated in the Atlantic at the same time as the *Challenger*, and its S. S. *National* (1889) carried out a global Plankton Expedition. Alexander Agassiz headed American cruises in the Atlantic aboard the *Blake*, and in the Pacific aboard the *Albatross*. France, Denmark, Italy, and Russia had also launched projects by the turn of the twentieth century. This period saw the creation of the first marine biological laboratories, the prototype being the Stazione Zoologica created at Naples in 1873 by Anton Dohrn.

The two world wars provided an unprecedented stimulus to physical oceanography at the expense of marine biology. The deployment of submarines and their detection by sonar brought new urgency to studies of bathymetry and the relation of temperature, salinity, and bottom sediments to acoustic transmission. And an intimate knowledge of waves, currents, and surf conditions would become crucial later on to the success of amphibious landings. By World War II, a productive, if sometimes stormy, collaboration had emerged between civilian oceanographers and naval officers. In the United States, the principal centers of this collaboration were the Woods Hole Oceanographic Institution in Massachusetts, headed by Columbus O'Donnell Iselin, and the Scripps Oceanographic Institution in California, under the direction of Harald U. Sverdrup. In the midst of the war, Sverdrup, with co-authors Martin W. Johnson and Richard H. Fleming, published the first modern textbook of oceanography,

The Oceans: Their Physics, Chemistry, and General Biology (1942).

After 1945, the Cold War relentlessly drove the growth of oceanography. As early as 1950, the U.S. Navy defined Soviet submarines patterned after advanced German designs seized at the war's end as the greatest maritime threat to the security of the United States. Understanding the ocean environment became critical to antisubmarine warfare. The Office of Naval Research and the Bureau of Ships quickly and regularly made resources available to address the threat, funding work by hundreds of scientists and institutions around the country. The results quickly outstripped administrative efforts to bring disciplinary coherence and recognition to oceanography, slowing the creation of degree programs and formal technical education at major universities. However, this massive investment made at a dizzying pace resulted in amazingly comprehensive ocean surveys, the understanding and exploitation of the deep sound channel, fundamental advances in sonar, the very rapid development of ocean acoustics as a field of study, the creation of the Navy's ocean surveillance system (SOSUS), the quieting of nuclear and conventional submarines, and the possibility of submerged missile launching. In addition, these circumstances effectively launched the careers of Roger Revelle, Walter Munk, J. Lamar Worzel, Dale Leipper, Waldo Lyon, Henry Stommel, Alan Robinson, and many others. The new sophistication of oceanography made it possible for American nuclear attack submarines to detect and shadow their Soviet missile-carrying counterparts. The deep ocean quickly became the front line in the Cold War.

The Russians followed suit, responding to American determination for the same reasons. While definitely competitive in terms of theoretical understanding and scientific capability, the material resources of the former Soviet Union did not permit it to keep pace. However, its scientists made very significant contributions to understanding antisubmarine acoustics, the study of the Arctic region, ocean surveying, and the construction of advanced oceanographic vessels.

Twentieth-century oceanography achieved a new, mathematically rigorous understanding of the coupling of atmospheric and oceanic phenomena, and of the climatic implications of oceanographic events such as El Niño. But after World War II, the discipline turned increasingly toward questions of marine geology and geophysics. The leading catalyst here was the continental-drift hypothesis proposed by Alfred Wegener in his book *The Origin of Continents and Oceans* (1915, first English translation 1924). Rejecting notions of continental stability and *isostasy, Wegener proposed that the present configuration of the continents, and other phenomena from *stratigraphy, paleontology, and biogeography, could be accounted for by assuming the gradual movement of the continents horizontally over the face of the globe. The theory gained few adherents until the 1960s, by which time new lines of evidence helped bring about the *plate-tectonics revolution. Evidence came from studies of paleomagnetism and polar wandering carried out by P. M. S. Blackett, S. Keith Runcorn, and their colleagues in Britain; heat flow from mid-ocean ridges by British *geophysicist Edward Bullard; seismological activity along mid-ocean ridges by Americans Maurice Ewing and Bruce Heezen; and gravity anomalies, pioneered by the Dutch geophysicist Felix Andries Vening-Meinesz and the American Harry H. Hess. In 1960, Hess proposed the hypothesis, subsequently known as sea-floor spreading, that would explain all of these phenomena. In the mid-1960s, the British geophysicists Frederick Vine and Drummond Matthews confirmed the hypothesis by analyzing patterns of magnetic anomalies around mid-ocean ridges, and the *Glomar Challenger* drilled directly into the Mid-Atlantic Ridge. J. Tuzo Wilson's 1965 concept that the earth's surface consists of several rigid but mobile plates put the finishing touch on the plate tectonics revolution.

See also OCEANOGRAPHIC INSTITUTIONS.

William A. Herdman, *Founders of Oceanography and Their Work* (1923). Margaret Deacon, *Scientists and the Sea 1650–1900: A Study of Marine Science* (1971; 2d ed. 1997). A. Hallam, *A Revolution in the Earth Sciences: From Continental Drift to Plate Tectonics* (1973). Susan Schlee, *The Edge of an Unfamiliar World: A History of Oceanography* (1973). Eric L. Mills, *Biological Oceanography: An Early History, 1870–1960* (1989). Walter Lenz and Margaret Deacon, eds., *Ocean Sciences: Their History and Relation to Man*, Proceedings of the Fourth International Congress on the History of Oceanography, Hamburg, September 1987 (1990). Philip F. Rehbock, *At Sea with the Scientifics: the Challenger Letters of Joseph Matkin* (1993). Gary E. Weir, *An Ocean in Common: American Naval Officers, Scientists, and the Ocean Environment* (2001).

PHILIP F. REHBOCK AND GARY WEIR

ONCOLOGY. See CANCER RESEARCH.

OPERATION PAPERCLIP. See BRAIN DRAINS AND PAPERCLIP OPERATIONS.

OPERATIONS RESEARCH. One of the progenitors of Operations Research (OR), the physicist P. M. S. Blackett, defined it as "social science done in collaboration with and on behalf of executives," but this was both too modest and too limited a definition. Originally forged as a species of interdisciplinary scientific research in the fires of World War II, OR proved to be a major innovation in the management of science and in the application of natural-science models to social phenomena during the postwar period.

OR began as an Anglo-American response to the vast expansion of the funding and organization of science by the military from 1940 onward. Mixed disciplinary units dominated by physicists rationalized the deployment of novel weapons systems such as *radar and mechanized air warfare, and parlayed their scientific expertise into advice for developing a new science of war. OR arose in an Anglo-American context of science organization; the German war effort produced no equivalent, perhaps because German scientists were already well integrated into state bureaucracies. The forerunner of OR, "operational analysis," came into existence in Britain with the Tizard Committee on air defense in 1935; by 1942, several OR groups had become established in military structures in the United States. What in Britain had been a method to optimize the sinking of submarines by altering the color of the attacking airplane, the depth of the torpedo, the angle of attack, and other variables grew in America to encompass the modeling of the targeting of ICBMs as a two-player game with various technological options solved with a Monte Carlo simulation.

Having spread throughout the military in the immediate postwar period, OR subsequently entered business schools as part of the toolkit of academic training for corporate middle management. In 1952, the Operations Research Society of America was formed, providing the professional identity for a haphazard collection of trained natural scientists and mathematicians. Although OR maintained its connections to physics and other natural sciences, recruitment in the second half of the century tended to come increasingly from its own professional base. Thus it became possible to regard OR in the late twentieth century as a social science with an unusually specialized clientele.

OR cannot be reduced to a few discrete doctrines, since its methods tended to diverge between the British and American professions after World War II. Whereas British OR generally maintained its original character as a hands-on project of pragmatic modeling based on thermodynamic metaphors and statistical expertise, American OR, inspired by the contributions of John von *Neumann in the areas of game theory and computers, grew into a mathematically abstract and theoretical discipline including "systems analysis" and "decision theory" at the RAND Corporation and Bell Laboratories. Some of the most illustrious scientists of the postwar

period spent portions of their careers as operations researchers, including P. M. S. Blackett, J. D. Bernal, Conrad Waddington, Philip Morse, George Gamow, William Shockley, John Bardeen, Kenneth Arrow, and Ivan Getting.

The standard curriculum for an operations researcher toward the end of the twentieth century included linear programming, network models, game theory, inventory models, queuing theory, information theory, econometrics, and simulation techniques. OR progressively became allied with "behavioral science," *artificial intelligence, cognitive psychology, organization theory, and neoclassical economics. It became the exemplar of the "cyborg sciences" primarily because of its heavy reliance on the *computer as a source of legitimation and metaphoric inspiration, and its penchant for blurring the conceptual boundaries between men and machines.

Although OR and systems analysis have succeeded as intrinsically interdisciplinary professions, their practitioners experience recurrent bouts of self-doubt about the true or fundamental core competencies of the profession. Initially, the operations researcher regarded himself as a technocratic consultant who reconciled planning with market allocation or military command hierarchy. As planning ambitions receded, the role of the mathematical consultant moved to the fore. Hence, OR has never achieved an altogether stable academic identity, tending to become lodged in schools of management or public policy, but sometimes in departments of mathematics, natural science, or engineering. OR has been predominantly client-driven rather than content-based throughout its half-century of existence; nonetheless, it has played an indispensable role in forging a rapprochement between academic science and the state.

Conrad Waddington, *Operational Research in World War II* (1973). Eric Rau, *Combat Scientists: Operations Research in the U.S. during World War II* (1999). Philip Mirowski, *Machine Dreams* (2001).

PHILIP MIROWSKI

OPPENHEIMER, J. ROBERT. See KURCHATOV, IGOR VASILYEVICH, AND J. ROBERT OPPENHEIMER.

OPTICS AND VISION. During the early seventeenth century Johannes *Kepler, echoing the earlier views of Leonardo da Vinci, likened the eye to a camera obscura, a black box containing a pinhole opening through which an image of external objects is projected on the back wall. Kepler worried that the analogy implied that the eye must invert the images it observes, but his work on light, optics, and the camera obscura forced him to accept the inference. He concluded that light forms images on the back wall of the eye, now called the retina, and not on the crystalline humor, now called the lens, as most scholars of the period thought.

Kepler argued that light as a passive entity followed the laws of geometry. His *Astronomiae pars optica* (1604) and *Dioptrice* (1611) dealt with the refraction of light, whose "law"—a relation between the angle of incidence i and the angle of refraction r—was given in 1621 by Willebrord Snel, professor of mathematics at the University of Leyden, who busied himself with astronomical and triangulation studies. Snel did not print his result; René *Descartes, who may have seen it in manuscript, published it in 1637 as $\sin r = (\sin i)/n$, where n, the index of refraction, is a constant. Christiaan *Huygens elaborated upon the law of refraction in his *Traité de la lumière* of 1690 by assuming that light was undulatory, rather than corpuscular, in nature. Like Kepler, Huygens believed that refraction arises because light moves more slowly in a denser than in a rarer medium.

Descartes had accounted for vision by analogy. According to him, the colors we perceive, like the impulses transmitted by a blind man's stick, result from pressure. He took color to be a secondary quality, not a property of external objects but the mind's interpretation of the pressure registered on the optic nerve. Nicolas Malebranche deepened Descartes's theory of vision in his *De la recherche de la verité* (1688). Sight cannot ascertain the truth of things in themselves; its purpose is to facilitate our navigation through the world. Further to the impugning of vision, George Berkeley (*Essay towards a New Theory of Vision*, 1709) undermined the idea that the concepts the mind creates by working on sense impressions reliably indicate the nature of things.

Isaac *Newton's grand discovery, that sunlight is made up of rays of different refrangibilities, transformed the study of light and vision. Newton showed in 1672 and at length in his *Opticks* of 1704 that sunlight passed through a glass prism yielded a spectrum of rays of different colors refracted at characteristic angles; that the rays of a given color all had the same refrangibility (and so could not be further divided by a prism); and that the differently colored rays, if reunited by a second prism, again produced white light (the so-called experimentum crucis). Newton supposed that the different rays were made up of particles; but to explain why some rays striking glass are refracted and others reflected, as well as to model interference phenomena like the colors of the plates and "Newton's rings," he also supposed that the particles interacted with a pervasive subtle matter or *ether. The emission and motion of the particles set the ether in vibration, and the particles penetrated or rebounded from a surface in accordance with the phase or "fit" of vibration of the ether there.

The different refrangibilities of rays of different colors make single lenses cast colored images. Newton thought, mistakenly, that this evil could not be cured, and he turned his attention to the construction of reflecting telescopes. In the 1750s the instrument maker John Dollond proved Newton wrong by manufacturing achromatic doublets made of crown and flint glass, which compensated one another's dispersion. Drawing upon Dollond's idea of an achromatic doublet, Joseph Fraunhofer, a Bavarian skilled artisan and optician working in the secularized Benedictine monastery of Benediktbeuern, improved upon a complex glass-stirring technique first developed by Pierre Louis Guinand, a Swiss bell pourer. With this method, Fraunhofer was able to manufacture flint glass of unprecedented homogeneity in the 1810s and 1820s. In addition to this crucial technological development, Fraunhofer reckoned that he could use the dark lines of the solar spectrum (later called the Fraunhofer lines), which his superior glass prisms produced, to demarcate precise portions of the spectrum. By altering the ingredients of his glass samples, he adjusted the refractive and dispersive properties in order to produce a second lens that would correct for the chromatic aberration produced by the first. After Fraunhofer's death in 1826, the British firm Chance Brothers of Birmingham and the French company Feil of Paris usurped the world's optical glass market from Bavaria. But during the 1880s, the physicist Ernst Abbe and chemist and glassmaker Otto Friedrich Schott manufactured apochromatic lenses for the Carl Zeiss Company in Jena, Germany, which would monopolize optical glass and equipment production until World War II.

During the eighteenth century prominent natural philosophers like Leonhard *Euler and Benjamin *Franklin challenged Newton's particulate model of light, but their opposition, based on qualitative considerations such as the improbable loss of matter from the sun that the model implied, did not make an effective alternative. The corpuscular theory reached its pinnacle after 1800 in the school of Pierre-Simon de *Laplace, which developed a quantitative theory of optical phenomena based on distance forces between light particles and the particles of ponderable matter. They managed to incorporate the polarization of light, which Huygens had known in the case of birefringent crystals and Étienne Louis Malus had discovered in 1810 in light reflected from glass, into their scheme by supposing that light particles had different properties on different "sides."

The remarkable success and high patronage of the corpuscular theory did not preserve it from an attack launched around 1800 by the English physician Thomas *Young, who drew upon an analogy between sound, which was understood to be a wave motion in air, and light. Young perceived a further, and more useful, analogy between diffraction patterns and the interference of water waves. In 1807 he presented the persuasive demonstration (now famous as the Young double-slit experiment) in which light from a common source passes through two parallel narrow slits in an opaque screen. The superposition of the two transmitted beams on a surface beyond and parallel to the screen produces dark and light stripes rather than clear images of the slits.

Young's initiative was continued by a better mathematician than he, Augustin Fresnel, whose memoirs, composed between 1815 and 1827, challenged his compatriots who championed the corpuscular theory. One of Fresnel's techniques was to treat each point of the wave front as a source of secondary waves, a technique introduced by Huygens. By adding the contributions of those secondary waves in accordance with Young's principle of interference, Fresnel could determine the intensity of light in a diffraction pattern. The Laplacian Simeon-Denis Poisson deduced from Fresnel's equations the paradoxical result that a bright spot should appear at the center of the geometrical shadow of a disk illuminated by a beam of light. Experiment decided for Fresnel.

Around 1862 James Clerk *Maxwell calculated that the speed of propagation of an electromagnetic field is approximately that of the speed of light. He concluded that light must consist of transverse waves of the same medium, which, according to the ideas of Michael *Faraday, gives rise to electric and magnetic phenomena (see ELECTROMAGNETISM). British physicists attempted to ascertain the physical properties of this ubiquitous and omnicompetent ether. To incorporate polarization into the theory, both Young and Fresnel independently suggested that the vibrations constituting light are perpendicular to the direction of travel, not, as in the model proposed by Huygens and, originally, Young, in the direction of travel, as is the case of sound in air. With this addition, the wave theory bettered the corpuscular theory at its strongest point. All efforts to find a mechanical model for the ether failed, and the assumption that one might exist came into increasing conflict with electromagnetic theory. The nature of light and the fate of its medium were resolved—for the twentieth century at least—by the theory of *relativity and *quantum physics.

Young contributed to the study of vision as well as the theory of light by criticizing the custom of separating consideration of the physical and mental states involved in the perception of color. He argued that the eye, being far less complex than the mind, simplifies the information provided by a particular scene and channels it to the brain, which in turn paints the scene. He demonstrated that the most sensitive points of the retina, which are connected directly to the brain, can detect only the three primary colors—blue, red, and yellow. The optic nerve is composed of filaments, portions of which correspond to a primary color. The brain mixes the sensations to create all the possible colors. Young's theory was ignored until Maxwell and Hermann von *Helmholtz elaborated upon his work in the mid-nineteenth century. Individuals suffering from color blindness (a disorder first recognized during Young's lifetime) have an abnormally low number of retinal cones, which detect color.

With the help of the opthalmoscope he invented in 1851, Helmholtz demonstrated that the optic nerve itself is insensitive to light. Sensory nerves apparently merely transmitted stimuli between the end organs and the sensorium. Between 1852 and 1855, Helmholtz studied color mixing and vision. He based his further work on Young's theory of three distinct modes of sensation of the retina, which he had previously rejected. He adopted the interpretation in Maxwell's paper, "Experiments on Colour, as Perceived by the Eye" (1855), which stated as Young's theory the proposition that, although monochromatic light stimulates all three modes of sensation, only one or two color responses will prevail in the resulting mixed color. This reading made one of the foundations for Helmholtz's *Handbuch der Physiologischen Optik* of 1860. In this classic work, Helmholtz developed Young's notion of three distinct sets of nerve fibers, although he observed that three distinct and independent processes might take place in each retinal fiber.

Binocular vision presented another set of physical-psychological problems. Charles Wheatstone's stereoscope, invented in 1838, played a critical role in the debate over how perception transforms two flat, monocular fields into a single, binocular field illustrating objects in relief. One group, led by David Brewster, argued for a theory of projection according to which the mind imagines lines starting at two stimulated points on the retinas and projecting outward to their point of intersection at the observed object. Another theory, proffered by the German physiologist Johannes Peter Müller, argued that retinas possess pairs of corresponding or identical points, each pair providing only one point in the unified field of vision. Wheatstone's stereoscope seemed to disprove Müller's views.

Physiologist Ewald Hering challenged Helmholtz's work on vision, declaring that four primary colors—red, blue, green, and yellow—formed the psychological basis for all color sensation. Hues arrange themselves in opposing pairs: red/green and blue/yellow. Helmholtz characterized the controversy as a clash between opposing epistemologies, nativism (Hering's position) and empiricism (his own view). At a deeper level, Hering opposed reckoning the processes of perception in analogy to functions of the human mind. He viewed vision as immediate impressions on the eyes and so rejected theories of projection in favor of a theory of identity similar to Müller's. Against Wheatstone and Brewster, Hering demanded that corresponding pairs of retinal points determine visual directions and depth perception. Helmholtz argued that the processes governing our spatial perception are psychological in nature and thus conditioned by learning and experience.

The controversy lasted well into the 1920s. Modern research affirms both the Young-Helmholtz three-color theory and Hering's four-color theory. The two theories complement one another. While the trichromatic approach explains how the eye detects and perceives color, Hering's explains how color information is encoded and sent to the brain via the nerve pathways. Research during the 1960s and 1970s confirmed that the eye contains three types of color sensors, the photoreceptors, composed of red, green, and blue cones, so-called because of their absorption of light at those wavelengths.

Lael Wertenbaker, *The Eye: Window to the World* (1981). G. N. Cantor, *Optics after Newton: Theories of Light in Britain and Ireland, 1704–1840* (1983). Jed Z. Buchwald, *The Rise of the Wave Theory of Light: Optical Theory and Experiment in the Early Nineteenth Century* (1989). Margaret Atherton, *Berkeley's Revolution in Vision* (1990). Gary Hatfield, *The Natural and the Normative Theories of Spatial Perception from Kant to Helmholtz* (1990). R. Steven Turner, *In the Eye's Mind: Vision and the Helmholtz-Hering Controversy* (1994). Myles W. Jackson, *Spectrum of Belief: Joseph von Fraunhofer and the Craft of Precision Optics* (2000).

MYLES JACKSON

ORBIT. An orbit is the path of a celestial object around another body, as in planets, comets, and asteroids moving around the sun, and satellites circling their planets. The key historical advances in the study of orbits came when Johannes *Kepler deduced that the orbit of a planet was an ellipse, and Isaac *Newton realized that a planet was bound to its elliptical path by a gravitational *force that was proportional to the product of

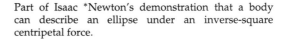

Part of Isaac *Newton's demonstration that a body can describe an ellipse under an inverse-square centripetal force.

the mass of the planet and the mass of the Sun, and inversely proportional to the square of the distance between them.

Prior to Kepler, planetary paths were thought to be a combination of circles and the importance of an "orbit," or single uncompounded curve, had not entered astronomical thought. Kepler used the observations of Mars made by Tycho *Brahe. Fortunately Mars has an orbital eccentricity of 0.093, nearly six times that of the earth. In *Astronomia nova* (1609), Kepler stated his first two laws of planetary motion: (1) planets move in elliptical orbits, the sun being at one of the foci; (2) the line joining the planet to the sun sweeps out equal areas in equal times. Kepler's third law, buried in his *Harmonices mundi* (1619), made the square of the orbital period proportional to the cube of the semi-major axis. The accuracy of the planetary ephemerides published in Kepler's *Rudolphine Tables* (1627) underlined the importance of the three laws, even though their physical foundations were unknown.

Newton's *Principia mathematica* (1687) established the physics of orbits. All orbiting bodies submit to gravity, even comets. Newton described how to calculate their orbits from three accurate observations separated in time by a few weeks.

He used this method to calculate the orbit of the great comet of 1680 (*see* COMETS AND METEORS). Edmond Halley (1656–1742) followed by calculating the orbits of a further twenty-three comets. Both Newton and Halley assumed that the cometary orbits were parabolic. But when Halley found that the comets of 1531, 1607, and 1682 had very similar orbits, he inferred that he was dealing here with a single comet on an elliptical, periodic orbit. Halley understood that the gravitational influence of Jupiter and Saturn perturbed comets and slightly changed their periods.

In 1798, Henry *Cavendish (1731–1810) used a torsion balance to provide the first measurement of the density of the earth. Knowing this quantity, an astronomer could calculate the mass of a large body, provided a much smaller one orbited around it. Previously, only ratios could be deduced—for example, that of the Sun's mass to the earth's from the orbital periods and semi-major axes of the moon and the earth. The production of accurate orbital parameters from diverse and imprecise observations, greatly improved by the work of Carl Friedrich *Gauss, made possible ephemerides of the first asteroids. The greatest success of orbital perturbation analysis came in the 1840s when John Couch Adams and Urbain Jean Joseph Le Verrier independently used observations of the orbit of Uranus to predict the position of a planet beyond the confines of the known solar system. This work led to the discovery of Neptune in 1846. The detailed analysis of the orbits of binary stars about their common centers of mass was carried out by Sir William Herschel between 1782 and 1804. Many other analyses followed, but not until the measurement of stellar parallax—after 1840—would these orbital results be converted into individual star masses.

William Herschel discovered the movement of the solar system as a whole in the direction of the constellation Hercules in 1783. In fact, as uncovered through Harlow Shapley's work of 1917 on the distances to globular clusters, the solar system orbits around the center of the *galaxy. Once again, Kepler's third law could be used. This time, it provided the mass of our galaxy interior to the sun's orbit.

Forest Ray Moulton, *An Introduction to Celestial mechanics*, 2nd rev. ed. (1914). Fred Hoyle, *Astronomy* (1962). Eric M. Rogers, *Astronomy for the Inquiring Mind: The Growth and Use of Theory in Science* (1982). Bruce Stephenson, *Kepler's Physical Astronomy* (1987).

DAVID W. HUGHES

ORGANIC CHEMISTRY. The term "organic" has referred to chemical compounds associated with living creatures only since the late eighteenth century. But many such substances—vinegar; alcohol; sugars; starch; waxes; fats and oils; extracts such as frankincense, camphor, and the volatile oils; and manufactured chemical materials like soap, gelatin, indigo, and tannin—have been known since Antiquity. Many of the processes involved, such as the manufacture of Tyrian purple from certain snails or the preparation of lac-dye and shellac from scale insects, were developed with little or no assistance from philosophical or scientific knowledge. Technology did not wait for science.

During the Middle Ages, more efficient cooling of vapors from stills led to the preparation of nearly pure alcohol, although the place and date of this important innovation are still uncertain. The introduction of powerful mineral acids—likewise a consequence of advances in *distillation apparatus—provided an important new class of reagents for chemical reactions. Common ether (known traditionally as sulfuric ether and in modern times as diethyl ether) was first produced in the sixteenth century by distilling alcohol with sulfuric acid. About the same time, herbalists and apothecaries developed means of preparing distilled or extracted "waters" or "essences" of plant materials for perfumes and pharmaceuticals.

In the seventeenth century, the usual method of analysis of plant and animal materials was dry (destructive) distillation. Often this process produced interesting new substances, such as formic acid from ants; in other cases too great a heat would destroy the delicate constituents sought. Early in the next century, analysis by solvent extraction followed by crystallization and manipulation by chemical reactions gradually came to be preferred to dry distillation. Theoretical understanding framed by ideas about *salts regarded vegetable substances as either acid or neutral, and animal substances as basic or neutral (*see* ACID AND BASE). The Swedish pharmacist Carl Scheele purified various plant acids, such as citric, tartaric, succinic, lactic, and malic acids, and characterized their salts. When Frenchmen Pierre-Joseph Pelletier and Joseph-Bienaimé Caventou identified distinctly basic substances derived from plants in 1818–1820—"alkaloids" such as strychnine and quinine—chemists were greatly surprised.

Antoine-Laurent *Lavoisier's chemical revolution of the 1780s profoundly altered the foundations of organic as well as inorganic chemistry. Lavoisier's work made clear that chemical substances from organisms contain only a very few elements, namely carbon and hydrogen, usually with oxygen as well, and sometimes nitrogen. Lavoisier showed that the carbon in burning organic compounds transformed into carbonic acid (carbon dioxide gas), and the hydrogen in them oxidized to water vapor. By capturing and measuring these two combustion products and knowing their composition, one could easily deduce the percentage of carbon and *oxygen in the original sample. By such means, Lavoisier carried out the first elemental analyses of organic compounds.

Joseph Louis Gay-Lussac and Jöns Jacob *Berzelius developed Lavoisier's combustion method of organic analysis between 1810 and 1815. Thereafter, for the first time, any professional chemist could perform such analyses. In 1831, the German chemist Justus von *Liebig simplified a crucial part of the apparatus. Compared to its predecessor, Liebig's modified method was considerably faster, extraordinarily accurate, and workable by even semiskilled students. When Liebig built up his student clientele at the University of Giessen during the 1830s, he put them to work preparing organic compounds and analyzing the results by his new procedure. Meanwhile in Paris, Liebig's principal rival, Jean-Baptiste-André Dumas, pursued complementary paths. Dumas developed a precise method to determine vapor densities of organic compounds (1826) and an improved way to measure their nitrogen content (1833). Like his German competitor, Dumas developed new theories of organic composition just as fast as the fresh analyses of novel compounds shifted the empirical ground.

Both Liebig and Dumas based their work on the atomic theory developed first by John Dalton and then much more thoroughly by Berzelius. A turning point came in 1830, when Berzelius drew attention to and named the concepts of polymerism and isomerism. Suddenly, new compounds appeared in droves, some of them ("isomers") matching the composition of substances already known. Berzelius had attempted to understand organic compounds by drawing analogies to inorganic ones. He thought that organic hydrocarbon "radicals" play the role that metals do in inorganic chemistry, so that they could be oxidized or exchanged between molecules but never altered. He also regarded electrostatic attraction as the force that held the pieces of molecules together.

At first, both Liebig and Dumas agreed with this scheme. Together with his friend Friedrich Wöhler, Liebig carried out a groundbreaking investigation of the oil of bitter almonds (benzaldehyde), demonstrating the presence of the "benzoyl" radical throughout an entire series of organic substances. However, the "electrochemical radical" theory of organic composition, which seemed so promising in the early 1830s, immediately began to encounter difficulties; these radicals seemed to be not at all

inviolable, and they also seemed not to follow expected electrochemical patterns. In particular, the substitution by chlorine of the hydrogen of organic substances, which Dumas and his students thoroughly explored, led both Liebig and Dumas to have doubts about Berzelius's organic-chemical theories.

Exhausted and exasperated, Liebig gave up attempting to understand organic composition about 1840, and turned to physiological and agricultural chemistry instead. Dumas followed a similar path about the same time. However, their students—Auguste Laurent, Charles Gerhardt, Charles-Adolphe Wurtz, August Wilhelm von Hofmann, Hermann Kolbe, Edward Frankland, and others—pushed further into what Wöhler had called the "primeval forest" of organic chemistry. Laurent and Gerhardt proposed a fundamental reform of atomic weights and molecular formulas that would, they thought, produce a more consistent and fruitful way to think about organic composition. Both died young. In 1850, the English chemist Alexander Williamson provided powerful chemical evidence for their reform, and proposed a quantity regarding oxygen compounds that eventually became known as valence (*see* CHEMICAL BOND AND VALENCE). Wurtz and Hofmann, and eventually also Kolbe and Frankland, adopted Williamson's approach.

The German chemist August *Kekulé, a student of Liebig much influenced also by Williamson and Wurtz, drew these various strands together into a coherent doctrine of organic composition called the theory of chemical structure. According to Kekulé, carbon atoms can bond together to form a carbon "skeleton" that constitutes the framework of the molecule. Trivalent nitrogen, divalent oxygen, and monovalent hydrogen and chlorine schematically add to this skeleton of tetravalent carbon atoms to form the welter of organic compounds by then known: alcohols, organic acids, hydroxyacids, aldehydes, ketones, amines, amino acids, ethers, and esters. Kekulé published this theory in 1857–1858 and expanded its application to "aromatic" compounds (substances derived from benzene) in 1865. Over these same years, Kekulé published an extraordinarily popular textbook of organic chemistry in regular installments, and his influence in the field grew rapidly.

The structure theory sparked a virtual explosion of activity in organic chemistry. Chemists had long since given up a belief in vitalism—Wöhler's celebrated synthesis of urea in 1828 was by no means the first or only reason for this—and Kolbe, Hofmann, and Marcellin Berthelot aggressively pursued the artificial syntheses of novel organic substances. Although most of these syntheses initially had only scientific (i.e., theoretical) motivation and interest, their technological significance became ever clearer. Germany expanded its university science facilities more rapidly than other European countries in the 1860s and 1870s, and many of its increasing number of chemistry professors undertook collaborative ventures with chemical entrepreneurs. The artificial dye industry, born in England and France in the late 1850s and 1860s, transferred to Germany in the 1870s (*see* Dyestuffs). By the 1880s, Germans had established near hegemony in both the pure and applied aspects of organic chemistry.

The structure theory proved to be a master key for unlocking secrets of organic chemistry, but it was neither stable nor uncontroversial. Most structuralists depicted unsaturated hydrocarbons with multiple bonds between carbon atoms. Aromatic molecules were also thought to have multiple bonds, but they had sharply divergent properties; theorists continued to puzzle over the anomaly. More generally, chemists debated whether valence was a constant or variable quantity. A Dutch and a French chemist, Jacobus Henricus van't Hoff and Joseph Le Bel, independently and simultaneously developed similar theories that extended structural formulas into three dimensions. Toward the end of the century, Emil Fischer at Berlin organized an effective investigation of the chemistry of sugars and proteins. Adolf von Baeyer, Liebig's successor at Munich, explored complex natural products such as dye molecules and terpenes.

Twentieth-century organic chemistry continued many of the lines that had already matured, such as *stereochemistry, the development of synthetic methods, structure determinations, syntheses of important natural products, and articulation of theories of composition. In addition, the flourishing of physical chemistry toward the end of the nineteenth century had a powerful effect on the further development of organic chemistry. Physical approaches to the study of organic reaction mechanisms and reaction kinetics were developed. The introduction of electronic theories of the chemical bond by G. N. Lewis and Irving Langmuir led Robert Robinson and Christopher Ingold to a more detailed understanding of organic structures and reactions. Quantum-mechanical approaches to the chemical bond yielded dramatic results in Linus *Pauling's resonance interpretation of the benzene ring, which provided the first truly satisfactory understanding of aromatic chemistry.

World War I alerted the Allied countries to the extent to which they had become dependent on the German chemical industries. As a consequence, Britain, France, and the United States vigorously promoted chemistry during and after the war. The interwar years saw the rise of the United States as a world power in science, including organic chemistry.

The most dramatic change in the practice of organic chemistry during the twentieth century was a revolution in analytical methods that began about 1930 and reached maturity in the postwar years. In 1930, chemists still operated with "wet-chemical" methods that had much in common with procedures of a century earlier. Starting with commercial labs that provided for elementary analysis as an outside source, and pH meters that displaced chemical indicators, organic chemists began increasingly to rely on instruments and specialists rather than their own test-tube methods. Infrared, ultraviolet, and Raman spectroscopy developed in the years around World War II, and mass spectrometers became routinely available to organic-chemical laboratories in the early postwar years (*see* Mass Spectrograph). Chromatography, especially gas chromatography (GC), underwent vigorous development. *Nuclear magnetic resonance (NMR) spectroscopy, which became available in the 1950s, achieved routine use by the 1960s. By the 1970s, GC-NMR combinations provided organic chemists with an exceptional capacity to analyze minute amounts of sample. Many other instrumental methods have been developed for organic-chemical applications, such as electron spin resonance and x-ray diffraction.

Instrumental methods of analysis have had a dramatic and beneficent effect on the practice of organic chemistry, but they have not been unmixed blessings. If not properly understood, machines can be treated too readily as mysterious black boxes, with insufficient appreciation for the subtleties of appropriate application, correct operation, calibration, purification of sample, and so on. Furthermore, they are extremely expensive and often require specialists to operate. It is heartening, therefore, to consider that much of the interesting work being done today in organic chemistry still happens in reaction flasks and crystallization beakers. There is still art in the work.

See also Polymer.

C. Schorlemmer, *Rise and Development of Organic Chemistry* (1894). J. R. Partington, *A History of Chemistry*, vols. 3 and 4 (1962; 1964). A. J. Ihde, *Development of Modern Chemistry* (1964). W. H. Brock, *The Norton [Fontana] History of Chemistry* (1992). A. J. Rocke, *The Quiet Revolution* (1993).

A. J. Rocke

ORGANISM. Toward the end of the eighteenth century, a growing number of naturalists asserted that living beings possess characteristics distinguishing them sharply from inanimate objects. The major distinguishing feature appeared to be the internal "organization" displayed by both animals and plants. Naturalists named them "organisms," a denomination that reshaped the traditional partition between the three "kingdoms" of nature, placing an emphasis on the distance separating living from non-living objects that still survives.

Around 1800, naturalists such as Georges *Cuvier understood by "organization" the internal disposition of organs and parts regarded as necessary to ensure that the whole displayed the fundamental characteristics of *life. According to Cuvier, the comparative anatomy of animals and plants revealed rules or laws governing the mutual dependence of parts as exact as those established within the physical sciences. Such rules exerted a powerful influence well beyond Cuvier's program. The law of the "subordination of characters," for example, asserted that the peculiarities observed in the major organs of a plant or an animal enabled anatomists to predict the structure of the rest of the organism; only some characters could be joined with certain major organs to ensure the "correlation of parts" needed to make life possible. Paleontologists used similar rules throughout the nineteenth century in their efforts to reconstruct the full plan of extinct organisms from surviving, scattered *fossil evidence.

According to Cuvier, the laws of organization, by strictly limiting the range of variable changes, made evolution impossible. For Cuvier's contemporary, Jean-Baptiste de *Lamarck, however, the "organic movements" proper to the "fluids" active inside living beings caused constant changes in their organization. Changes could be handed down by generation, cumulating their effects and producing a tendency toward an ever more complex organization of living beings. The "fluids" had gradually shaped the organisms we know today.

A major challenge for Charles *Darwin in the central decades of the nineteenth century was to reconcile the evolutionary view of life with the orthodox concept of organization set by the anatomical tradition as enriched by the findings of *embryology. *Darwinism substantially met the challenge. Developments within the life sciences such as the affirmation of the *cell theory from the 1830s, the achievements of *biochemistry from mid-century, and the rediscovery of *Mendel's laws in 1900—also modified the notion of organism.

By the 1920s, what made an organism—and, according to some, recommended the adoption of an "organismic," antimechanistic approach to *biology—was a complex, specialized, hierarchical organization, coordinating structures and functions involving organs, tissues, cells, chromosomes, and *proteins regarded as the bearers of the specificity of genus and species. Consistently with a long previous tradition, the metaphor of "organization" throughout the twentieth century continued to encourage conceptual exchanges between the biological and the social sciences based on the notion that in both fields relationships are just as important as constituent elements. After World War II concepts inspired by information technologies began to indicate new ways to exploit the old metaphor.

See also FEEDBACK; HOLISM; NATURE; TELEOLOGY.

François Jacob, *The Logic of Life*, trans. Betty E. Spillmann (1974). Timothy Lenoir, *The Strategy of Life* (1982).

GIULIANO PANCALDI

ORRERY. The widely reproduced painting, *A Philosopher giving that lecture on the Orrery, in which a lamp is put in place of the Sun*, by Joseph Wright of Derby, was first exhibited at the Society of Artists of Great Britain in London in 1766. The painting shows the orrery as an aid to the study of *astronomy, which is portrayed as a social activity in which children as well as adults can participate.

The orrery is a moving model of the motions of the earth, the *Moon, and the Sun. It demonstrates phenomena such as day and night, the seasons, lunar phases, and eclipses. Larger orreries often include additional planets and their accompanying satellites. In contrast with some other planetary models, orreries always depict the sun-centered Copernican cosmology.

The London clock and instrument maker George Graham (1674–1751), working with Thomas Tompion (active 1671–1703), created a number of fine "proto-orreries," astronomical models with more limited scope for representation. The first orrery was made about 1713 by the London instrument maker John Rowley for Charles Boyle, the fourth Earl of Orrery (the eponym of the instrument). The orrery became popular in England (a gentleman's library was incomplete without one) as an introduction to science considered particularly apt to encourage rational thinking and inspire appreciation of the magnificence of the creation. Orreries joined globes and armillaries in the material culture of astronomical education and entertainment produced and marketed during the eighteenth century. A wide variety of books, games, and "souvenirs" of astronomical events (such as eclipses) helped further to disseminate knowledge of astronomy.

As Wright's painting indicates, grand orreries were large pieces of furniture, often expensively and ornately decorated, suitable for permanent display as "conversation pieces" prompting edifying talk. In the 1770s, the lecturer Adam Walker and his sons devised a large transparent orrery, or Eidouranion, about twenty feet in diameter, to provide demonstrations to larger audiences than grand orreries could accommodate. Smaller orreries, including portable models in carrying cases, were used as visual aids by traveling lecturers. Within the English cultural orbit, in the new United States, David Rittenhouse (1732–1796) produced orreries for colleges.

It is sometimes said that Wright aimed to convey an understanding of the Newtonian universe and theory of gravitation. That may be the case. But the orrery itself, as an astronomical model, cannot be described as a specifically Newtonian instrument.

Henry C. King and John R. Millburn, *Geared to the Stars: The Evolution of Planetariums, Orreries, and Astronomical Clocks* (1978).

LIBA TAUB

ORTHOGENESIS involves the evolution of organic forms in a definite direction, determined by intrinsic forces. The expression generally is reserved for relatively long-term nonadaptive evolutionary trends. Jean-Baptiste de *Lamarck and Erasmus Darwin thought that the evolution of life involved an intrinsically driven tendency of organic forms to increase in complexity. Some orthogenetic visions, for example Karl Ernst von Baer's, ground the tendency to form an orderly progression in ontogeny. The term was apparently introduced by the German biologist Theodore Eimer at a meeting of the International Zoological Congress in Leyden (1895). Eimer studied color variation in lizards and insects, finding patterns that he believed had no adaptive significance. What he called "orthogenesis" was meant to explain why *similar* evolutionary sequences occur in *different* lineages. Eimer thought that nonadaptive convergence was widespread.

Paleontologists took a particular interest in orthogenesis, which obtained greater currency through dramatic fossil finds in the late nineteenth century. For example, in the United States, E. D. Cope regarded the evolution of increasing body size and decreasing number of toes in the horse lineage as orthogenetic. At the same time, in Great Britain, St. George Jackson Mivart noticed parallel development of structures and William Carpenter found orthogenetic patterns in the evolution of the *Foraminifera*.

In the United States, there was a diverse group of evolutionary theorists inspired by Cope and Alpheus Hyatt who defended orthogenesis. Cope held that natural selection worked only in the relatively minor character differences between species. His "law of acceleration and retardation" operated independently of the adaptive value of variations and was designed to explain the larger differences among genera rather than species (1868). Hyatt, a student of Louis Agassiz at the Museum of Comparative Zoology, also held a general viewpoint having more affinity with evolutionary views developed on the continent than with Darwinism. Hyatt embraced the law of acceleration on the evidence of his extensive study of fossil cephalopods. For Hyatt, as for Cope, the motive force for evolution is an intrinsic tendency of species to change rather than adaptation. Hyatt described four "epochs" in the history of the tetrabranchiata. Relatively simple forms are followed by successive periods of growth and diversification, ending finally in decline and decay. Like Cope, Hyatt saw a parallel between individual development and phylogenetic changes. This is the idea, in Haeckel's terms, that *ontogeny recapitulates phylogeny*. Hyatt saw the development of species as following a regular and natural trajectory from birth through juvenile forms to senescence. In ontogeny, organisms start out relatively simple, become more complex, and finally degenerate and die; Cope and Hyatt thought that the advance and decline of biological lineages followed a similar pattern.

Orthogenetic views became common in Germany as well as the United States toward the end of the nineteenth century. Henry Fairfield Osborn and William Berryman Scott adopted orthogenesis in order to explain parallelism, ostensibly a view endorsed by Wilhelm Waagen, a German invertebrate paleontologist. Like other advocates of orthogenesis, Osborn thought that internal controls directed the course of variation, and that once phylogenetic orders had been established they evolved stably and linearly (a view he later called "aristogenesis"). Some advocates referred orthogenesis to natural processes; others to vitalistic or theistic processes. Ultimately, orthogenesis was discarded. Most trends have turned out to be illusory, and adaptation explained others.

Stephen Jay Gould, *Ontogeny and Phylogeny* (1977). Ernst W. Mayr, *The Growth of Biological Thought* (1982). Peter J. Bowler, *The Non-Darwinian Revolution* (1988). Matthew H. Nitecki, ed., *Evolutionary Progress* (1988). Stephen Jay Gould, *The Structure of Evolutionary Theory* (2002).

THOMAS KANE AND ROBERT C. RICHARDSON

OSMOSIS, which operates in a great many life processes, occurs whenever two solutions that differ only in the concentration of dissolved substances (solutes) are separated by a semipermeable membrane that allows pure solvent, but not solutes, to move through it. A net flow of solvent from the less concentrated to the more concentrated solution results. Early studies of osmosis used natural membranes such as animal bladders and intestines. If a semipermeable bladder is filled with an aqueous salt solution, sealed, and immersed in pure water it will swell and possibly even burst owing to the osmotic diffusion of pure water into it.

The abbé Jean-Antoine Nollet described osmotic behavior in 1748, but his work attracted little attention. The phenomenon was "rediscovered" in 1812. A generation later (1826), physician and naturalist René-Joachim-Henri Dutrochet began an extended investigation of osmosis in connection with plant movements. In 1828 he constructed a simple osmometer that showed that osmotic pressure could overcome gravity. This observation led Dutrochet to speculate that osmosis might operate in many plant processes, such as the rising of sap. These speculations served ideological as well as scientific ends, because Dutrochet (who coined the verb "osmose") opposed the view that living organisms depend on a unique "vital force" that transcends the boundaries of ordinary chemistry and physics.

One of his successors, Wilhelm Pfeffer, a botanist and fellow antivitalist, published his *Osmotic Investigations* in 1877, ushering in a quarter century of definitive achievements in the understanding of osmosis and its significance. Pfeffer constructed an improved osmometer that employed artificial membranes and collected precise data on the effects of changing solutes, concentrations, and temperatures. Nevertheless, contemporaries ignored Pfeffer's work because his measured osmotic pressures seemed unreasonably high and it seemed unlikely that cells contained the membranes required for osmosis.

Hugo de Vries, a plant physiologist who did appreciate Pfeffer's work, showed that osmosis accounted for turgor, the internal pressure of healthy plant cells. This pressure arises from the resistance of the semirigid cell wall to the osmotic pressure generated by the cell protoplasm, and is essential for cell growth and the transport of nutrients. De Vries's other major contribution to the understanding of osmosis was to bring Pfeffer's work to the attention of Jacobus Henricus van't Hoff, a founder of modern physical chemistry, whose Nobel Prize (1901) recognized his groundbreaking osmotic studies.

Van't Hoff approached the problem by reducing it to its essentials. He calculated osmotic pressures by ignoring the solvent, pretending that the solute had turned into a gas, and applying the simple gas law equation, $p = cRT$ (R a constant). He found a remarkable agreement between the calculated and experimental values, showing that only the concentration (c) and temperature (T) affect the osmotic pressure (p), while the structure of the membrane does not.

The significance of van't Hoff's approach far exceeded its utility for calculations. It gave rise to a precise definition of osmotic pressure for a solution: the counterpressure just necessary to prevent the osmotic solvent from flowing. Van't Hoff's work also made sense of some unexpected results. For example, the osmotic pressure of a sodium chloride solution is twice that of an equally concentrated sugar solution. Following Svante Arrhenius, van't Hoff assumed that sugar molecules remain intact in water while sodium chloride dissociates into Na^+ and Cl^- ions, each of which exerts its own pressure, making the effective concentration of sodium chloride twice that of sugar. At a time when the existence of ions and even molecules was in doubt, these experiments provided convincing evidence of their reality.

The nature of the membrane and solute do not affect the osmotic pressure, but they do strongly influence the rate of osmosis. These factors have been intensively studied because of their significance for various applications such as reverse osmosis, which is used to desalinate water. In this process water diffuses from a concentrated to a dilute solution by imposing a pressure greater than the osmotic pressure. Another application is dialysis, the fundamental technique behind the artificial kidney. It employs a membrane permeable to some but not all solutes, thus mimicking the natural kidney.

See also PHYSICAL CHEMISTRY; VITALISM.

Zvi C. Kornblum, "Osmosis," in *Encyclopedia Americana*, vol. 21 (1995). Juha P. Kokko, "Excretion and Fluid Balance in Vertebrates," in *Encyclopedia of Life Sciences Online* <http://www.els.net> (2001). Floyd H. Meller, "Desalination," in *Grolier Multimedia Encyclopedia* http://gme.grolier.com (2001). Jeremy Pritchard, "Turgor Pressure," in *Encyclopedia of Life Sciences Online* http://www.els.net (2001).

STEPHEN J. WEININGER

OVISM AND ANIMALCULISM. The words ovism and animalculism stand for opposing views on sexual reproduction held during the seventeenth and eighteenth centuries in Europe. Ovists maintained that the offspring of an organism was in some way pre-existent in the female egg, needing only to be activated in some manner in order to produce an embryo. This view was common even before

Regnier de Graaf thought he had observed the mammalian "ovum" in 1672 (he actually saw the ovarian follicle). Animalculists, on the other hand, emphasized the importance of the male element after the observation of "animalcules" by Antonie van Leeuwenhoek in mammalian semen in 1677. Many animalculists believed that the spermatozoa contained in miniature all future generations. However, few were as bold as Nicolaas Hartsoeker, who in 1694 propounded the idea that each animalcule contained within it a "homunculus," a tiny man.

These intellectual positions reflected a reciprocal interaction between contemporary philosophical commitments about the nature of reproduction and the surge in increasingly detailed anatomical and microscopical observations. The years between 1670 and 1690 saw genuinely new and unexpected microscopical observations that needed accommodation within contemporary theories of living beings. Religious questions about the pre-existence and continuation of life played a significant role. Furthermore, since most of the natural philosophers concerned were male and worked within predominantly patriarchal cultures, a number of gender issues also were involved. Male authors seem to have favored views that gave semen a primary role in reproduction, especially when "animalcules" in the semen became an accepted agent in the process. Any contrasting position taken by women of the period is difficult to discern, but some midwifery manuals written by women included information about reproduction that emphasized the female role.

Ovism had roots in a long intellectual tradition that dated from Aristotle and received strong endorsement from William *Harvey. In his *De generatione animalium* (1651) he suggested that all animals were derived from an "ovum." According to Harvey, the ovum produced by the female animal possessed an innate capacity to develop after receiving the influence of the male seminal fluid. He believed that the ovum then developed by the process of epigenesis, by which the embryo took shape gradually, each stage building on the previous one. The male semen contained some vital activating principle or spirit that set the process into motion. This vital principle could be identified with aspects of Christian theology.

Observations on embryonic chick development from 1673 led Marcello *Malpighi and others to propose an alternative to epigenesis. They argued that the organism must exist pre-formed in the germ. Development thus involved an unfolding or gradual revelation of a pre-existing but invisible form. Malpighi asserted that this form lay in the egg. Hence, ovists could believe in

either pre-formation or epigenesis. Nevertheless, despite Harvey's and Malpighi's views, earlier theories about the "two semens," in which both male and female produced fluids during intercourse that mixed to form the embryo, continued in vernacular culture for a long time. In addition, most naturalists assumed that a wide range of organisms arose by spontaneous generation. The reproductive processes of plants remained puzzlingly obscure although providing many analogies for animal processes, such as "scattering seed."

When Leeuwenhoek announced his observations on seminal fluids, and most natural philosophers accepted the existence of animalcules in the semen, the pendulum swung the other way. Many scholars and practical observers now asserted that the pre-formed individual must be contained in the spermatozoan. Leeuwenhoek himself was a committed animalculist. Debate between ovists and animalculists continued through the eighteenth century, given a boost by the comte de Buffon's announcement that spermatozoa were merely parasites in the testes, and its refutation by Lazzaro Spallanzani's demonstration of the animality of "spermatic animals." However, animalculism implied enormous wastage, for each animalcule was a potential human being. Theologians grappled with the consequent philosophical problems. Some scholars, such as Carl *Linnaeus, rejected both points of view and opposed any pre-existence theory. Pierre de Maupertuis objected to pre-formation and revitalized the theory of two semens. Toward the middle of the eighteenth century, ovism and epigenesis gained the upper hand, with Caspar Wolff as the foremost epigenesist. The debate died away before the end of the century when attention turned to understanding the process of fertilization and development.

John Farley, *Gametes and Spores: Ideas about Sexual Reproduction, 1750–1914* (1982). Clara Pinto-Correia, *The Ovary of Eve: Egg and Sperm and Preformation* (1997).
JANET BROWNE

OXYGEN. In a late draft of a paper he intended to publish on the nature of acids, Antoine-Laurent *Lavoisier revised one sentence of his manuscript in February 1780 to read, "I shall henceforth designate dephlogisticated air or *eminently respirable air* in its state of fixity by the name of acidifying principle, or, if one prefers the same signification in a Greek word, by that of *oxygen principle*." The name "dephlogisticated air" had been the choice of Joseph Priestley, the first to notice that the air produced by reducing a calx of mercury without charcoal supported respiration and combustion better than ordinary air. Lavoisier had employed the name "eminently respirable air" to designate his emerging view that the air he had obtained by the same experimental procedure that Priestley had used was a distinct portion of the atmosphere, which, unlike the remainder, supported the respiration of humans and animals. Lavoisier regarded eminently respirable air as a combination of the oxygen principle with the heat necessary to maintain it in an aeriform state. When Lavoisier and his colleagues revised the nomenclature of chemistry more thoroughly in 1787, "eminently respirable air" became "oxygen gas."

At the time he chose these terms, Lavoisier had shown that three of the common mineral acids—nitrous, vitriolic (which he later renamed "sulfuric"), and phosphoric acid, as well as the recently isolated oxalic acid derived from plant matter—contained oxygen. From these results, he induced that oxygen was responsible for acidity in general. Since it had not yet been obtained from a few other acids, most notably "marine acid," which he renamed "muriatic acid," he concluded that the oxygen was so tightly joined to the "muriatic radical" that chemists had not yet separated them. He was proven wrong. A similar guess had a happier result. In 1789, he omitted potash and soda from his list of simple substances because they were "evidently composed, although we are still ignorant of the principles that enter their composition." Using the method of *electrolysis that became available only with the invention of the Voltaic *battery, Humphry Davy showed in 1807 that Lavoisier was right: that these alkalis too contained oxygen, in combination with the newly discovered elements potassium and sodium. The discovery further enhanced the role of oxygen as central to the chemistry of alkalis as well as acids and metals.

The general theory of combustion that Lavoisier had first proposed in 1778 assumed that the burning of combustible bodies such as sulfur and phosphorus, as well as the calcination of metals, involved not the emission of *phlogiston, as previously thought, but the addition of oxygen and the release of "matter of heat," or as he later named it, "caloric." Lavoisier himself observed that some combustible bodies and metals reacted in a similar way with sulfur, forming "sulfides" analogous to the "oxides" formed with oxygen; and "oxidation" came gradually to mean a more general process of which combination with oxygen was only the most prominent example.

In electrolysis experiments, one portion of the dissolved compound moved toward the positively charged electrode while another went to the negative. These observations gave rise to the idea that the forces responsible for attracting atoms together to form compounds, or to cause the components of compounds to separate and rejoin in new combinations, were electrical in nature. The Swedish chemist J. J. *Berzelius developed these ideas into a complete "dualistic system," which ordered all elements, and their first- and second-order compounds, into a series ranging from most positive to most negative. The special properties of oxygen he then explained by the fact that "of all bodies, oxygen is the most electronegative." The fact that sodium and potassium had been so difficult to separate from oxygen merely showed that they were the most electropositive of the elements. In 1869, *Mendeleev incorporated these properties into his periodic table, oxygen and sulfur being placed in the sixth vertical column, now one step removed from the most electronegative of all, the halogens.

J. J. Berzelius, *Theorie des proportions chimiques,* 2d ed. (1835). Antoine Lavoisier, *Elements of Chemistry* (1965). Mary Elvira Weeks, *Discovery of the Elements,* 7th ed. (1968). Karl Hufbauer, *The Formation of the Germany Chemical Community (1720–1795)* (1982). Jean Pierre Poirier, *Antoine Lavoisier: Chemist, Biologist, Economist* (1996).

FREDERIC LAWRENCE HOLMES

OZONE HOLE, a colloquial term for the dramatic seasonal thinning of middle atmospheric ozone (O_3, an allotrope of oxygen) over Antarctica. The controversy over the causes of this phenomenon is a defining episode in the history and politics of anthropogenic global climate change.

In the mid-1920s, the English meteorologist Gordon Dobson made observations of atmospheric ozone employing a global network of spectrophotometers of his own design. The total amount of atmospheric ozone is today expressed in Dobson Units, where 1 DU equals 0.01 mm. If all the ozone in the atmosphere were compressed into a single layer, that layer would be about 300 DU (3 mm) thick.

Between 1929 and 1931, the British geophysicist Sydney Chapman hypothesized that ozone arises by the action of ultraviolet light on molecular oxygen (O_2) in what is sometimes called the Chapman layer at about 45 km, a figure now lowered to a variable altitude between 15 and 30 km. Chapman's ideas about atmospheric ozone dominated the scene until 1964.

By 1970, it was clear that the chemistry of stratospheric ozone involved hydrogen, nitrogen (predicted by the Dutch meteorologist Paul Crutzen), and chlorine. In 1974, the Mexican chemist Mario Molina and the American chemist Sherwood Rowland argued that CFCs (chlorofluorocarbons), inert chemicals widely used for sixty years as aerosol propellants and refrigerants, were finding their way into the stratosphere, where they dissociated under intense ultraviolet radiation and produced chlorine species that combined with photolytically dissociated oxygen, thus preventing the formation of ozone; they predicted a long-term decline in the earth's protective ozone shield. It also appeared that aircraft and spacecraft exhausts (the SST and the Space Shuttle) would therefore also deplete atmospheric ozone.

Although these discoveries led to some regulatory efforts in the 1970s to reduce CFCs especially in the form of aerosol propellants, economic and political interests, as well as legitimate scientific disagreement about the causes of ozone depletion, stymied attempts to control ozone-depleting chemicals. Debate intensified when the British Antarctic Survey reported in 1984 that Antarctic ozone levels had plummeted by a third since 1977 (to less than 200 DU). Americans initially doubted these results, since the TOMS (Total Ozone Mapping Spectrometer) on the *Nimbus-7* satellite had not detected sharply lower levels of ozone. Later analysis of the satellite's programming showed that it had been ordered to disregard such low values as outside the "reasonable range," though they had been regularly observed since 1978.

Regulatory efforts to cut CFCs and other depleting chemicals resumed in the 1980s, including the signing of the Montreal Protocol (1987). Nonetheless, the Antarctic ozone hole has continued to grow. In 1998, it covered 27 million km^2, and in 1999, total atmospheric ozone in October fell below 88 DU. The magnitudes of these changes were not predicted in 1985 by any model of stratospheric composition, even looking fifty and a hundred years ahead. Susan Solomon, the scientist who in 1987 proposed that the ozone hole was a direct consequence of the activation of anthropogenic chlorine on stratospheric ice clouds (later experimentally confirmed many times over), has said that the period of our complacency concerning this ozone hole, and our doubts as to its cause, are over.

Richard Wayne, *Chemistry of Atmospheres*, 3rd ed. (2000). David E. Newton, *The Ozone Dilemma* (1995).

MOTT T. GREENE

P

PALEONTOLOGY. Paleontology, the study of the remains of living beings and their traces in the rocks, emerged as a distinct area of investigation in the late eighteenth and early nineteenth centuries. By then, most mineralogists and naturalists concurred that *fossils were the remains of living beings. The German naturalist Johann Friedrich Blumenbach went further, and argued that in the past many animals and plants had become extinct while new and different ones had been created in their place. Two of his followers, Ernst von Schlotheim and Georges Cuvier, realized this meant that extinct species and genera could be used to identify and correlate stratified rocks. In an important study of the geology of the region around Paris published in 1812, Cuvier, used fossil quadrupeds for the purpose. Schlotheim embarked on the massive task of classifying the less glamorous but much more widely distributed invertebrate fossils, as a preliminary to using them in *stratigraphy. Not long after, in 1816, the English surveyor William Smith published his *Strata Identified by Organized Fossils*. By the 1830s, British geologists, ignoring the Continental tradition, claimed that Smith should be credited with the discovery that fossils could be used to identify strata.

Whatever the merits of this dispute, geologists agreed that fossils offered the best way to determine when a stratum was formed and to correlate strata over large distances. They busied themselves collecting, identifying, cataloguing, and producing monographs on the fossils most useful in stratigraphy. Overwhelmingly these were marine invertebrate fossils, usually shells of one kind or another, but also corals, trilobites, ammonites, belemnites, and so on. Geologists also studied fossil plants, especially abundant in the economically important coal-bearing formations. With the possible exception of fishes, vertebrate fossils were too rarely preserved to be of use.

In the 1870s, paleontology matured as a scientific discipline with its own specialist journals, meetings, and societies. Although university departments employed paleontologists, the new specialty derived the bulk of its institutional support by being an ancillary to stratigraphy, which was of great economic importance first in the mining industry and later in the petroleum industry as well. Jobs were offered in national geological surveys and later in laboratories run by oil companies. Because of the small diameters of the rock cores brought up in drilling for oil, paleontologists turned in earnest to the study of fossils visible only under the microscope, such as ostracods, foraminifera, and pollen grains. In 1917, the American Association of Petroleum Geologists was formed. By the end of the century, it would be the largest geological society in the world.

For stratigraphic and economic purposes, questions about the meaning of fossils—how life had been created and evolved, how dinosaurs had lived, how man had evolved—were irrelevant. Chemists, biologists, enthusiastic amateurs, and anthropologists were as likely to pursue these questions as paleontologists. Most important debates within professional paleontology dealt with matters that bore directly on stratigraphy. Was a single fossil species or genera adequate to identify a formation, or did the paleontologist have to take whole assemblages of fossils into account? Did whole fossil faunas and floras change abruptly and simultaneously, as Cuvier claimed? Or did they change slowly and gradually, as Charles *Lyell contested? Were fossils reliable as indicators of the relative age of rocks? From the early nineteenth century on, geologists had recognized that different extinct species, just like different living species, lived in different environments. If so, was it possible that fossils varied with the environment in which they had been deposited, and not with time? By the late nineteenth century, geologists used the concept of facies to describe these environments, and began to factor environmental factors into their stratigraphic investigations.

In two areas, paleontologists concerned themselves with issues outside stratigraphy. One was paleo-ecology. In the late nineteenth century, German and Russian scientists began using fossils to reconstruct past climates, past environments, and past shorelines. They studied the environments of living marine invertebrates, for example, to shed light on the environment in which trilobites might have lived. They used the distribution of fossil coral reefs to infer what earlier climates were like.

The other area was the Cambrian boundary, below which paleontologists had failed to find any traces of life. Above it, fossils appeared abruptly as rather specialized invertebrate animals. In the mid-twentieth century, paleontologists detected microscopic unicellular animals and sedimentary structures built up of colonies of algae and bacteria (called stromatolites). The reason for the puzzlingly abrupt change, they concluded, was that until the beginning of the Cambrian, almost all life on earth consisted of simple unicellular bacteria, algae, and protozoans.

Following World War II, as funding within universities and new and powerful instruments became more readily available, paleontologists began asking a wider range of questions. Paleoecologists used the ratios of stable isotopes of oxygen in fossils to trace temperature fluctuations in past oceans. Other paleontologists used the trace amounts of nucleic compounds preserved in fossils to shed light on evolutionary affinities and relationships. More traditional paleontology used the scanning electron microscope to subdivide strata from the Mesozoic to the present in quite remarkable detail.

Wilfred Norman Edwards, *The Early History of Palaeontology* (1967). Martin Rudwick, *The Meaning of Fossils* (1972). Eric Buffetaut, *Short History of Vertebrate Palaeontology* (1987). Richard Fortey, *Fossils: The Key to the Past* (1991).

RACHEL LAUDAN

PARADIGM. See SCIENTIFIC DEVELOPMENT, THEORIES OF.

PARALLAX. Parallax is the difference in apparent direction of an object seen from two different places. Stellar parallax, a measure of the distance from the earth to a star, is the angle subtended by the radius of the earth's orbit at its distance from the star. The absence of a measurable parallax embarrassed Copernicans, who insisted that the earth circled the sun. They had to argue that the great distance of the stars made the parallax angle too small to observe. *Galileo proposed a variation on the direct parallax measurement, substituting the easier-to-measure relative positions of stars. Making the assumption that all stars are equal in luminosity, and thus a faint star is more distant than a bright star, he hoped to detect an alteration in the relative positions of two such stars over a period of time. Many optical double stars, however, are also physical double stars, close together in space.

Robert *Hooke, curator of experiments for the Royal Society of London, noted in 1674 that "whether the earth move or stand still hath been a problem, that since Copernicus revived it, hath much exercised the wits of our best modern astronomers and philosophers," and he well appreciated the fame that would accrue to the first person to prove that the earth moved. His reports to the Society intimated success, but attracted little attention, perhaps because no one believed them. In the 1720s, the English astronomer James Bradley, attempting to verify Hooke's claim, discovered stellar *aberration, the apparent displacement of a star in the direction in which the earth is moving.

Improved instrumentation (*see* TELESCOPE) ultimately brought success. Friedrich Bessel at the Königsberg Observatory and Friedrich Struve at the Dorpat Observatory (now Tartu in Estonia) and then at the new Pulkovo Observatory near St. Petersburg, working on different stars, observed parallax in 1838 and 1840, respectively. Meanwhile, the Scottish astronomer Thomas Henderson, luckily having chosen a much nearer star to study, detected a parallax with less accurate measurements than Bessel's and Struve's. Henderson was the first to begin his measurements but the last to report them; Struve published first, but Bessel most convincingly.

In addition to the stellar, or trigonometric, parallax, there is a spectroscopic parallax or distance. An empirical correlation between spectral characteristics and absolute magnitudes of stars (the Hertzsprung-Russell diagram), once established, can subsequently be used to estimate the distances to other stars. Statistical parallax is an estimated distance to a group of stars. Assuming that the group members have random movements, the average of the measured radial velocities (line-of-sight components, corrected for the observer's movement) equals any other component, including that perpendicular to the line of sight. This average perpendicular velocity, combined mathematically with its corresponding average observed angular change (proper motion), yields an average distance.

See also ASTRONOMY; COPERNICUS, NICHOLAS.

Norriss S. Hetherington, *Science and Objectivity: Episodes in the History of Astronomy* (1988). Norriss S. Hetherington, *Encyclopedia of Cosmology: Historical, Philosophical, and Scientific Foundations of Modern Cosmology* (1993).

NORRISS S. HETHERINGTON

PASTEUR, Louis (1822–1895), chemist, microbiologist, advocate of the germ theory of disease.

Born in the small town of Dôle, France, to a family of modest means, Louis Pasteur grew up in nearby Arbois. He carried throughout his adult life a desire for fame and financial security, as well as deep patriotism, acquired at the family hearth.

Contemporary painting of Pasteur at his laboratory in Paris.

Educated at the École Normale Supérieure, Paris (1843–1846), Pasteur served as *préparateur* in chemistry there (1846–1848), which allowed him to complete his doctoral degree based on theses both in physics and in chemistry, in 1847. He was professor of chemistry at Strasbourg University (1849–1854), professor of chemistry and dean of the Faculty of Sciences at Lille (1854–1857), administrator and director of scientific studies at the École Normale (1857–1867), professor of chemistry at the Sorbonne (1867–1874), director of the laboratory of physiological chemistry at the École Normale (1867–1888), and, finally, the director of his own private research facility, the Institut Pasteur in Paris (1888–1895).

Pasteur began his scientific studies with the examination of crystals. In 1848 he made the profound discovery that the "optical activity" (ability to rotate polarized light) of organic substances in solution depended on the structure of the crystalline solid. He showed that tartaric acid, which rotated light to the right, had crystals with only right-handed hemihedral facets. Paratartaric acid, which had no apparent optical activity, consisted of an equal mixture of two mirror-image crystal types. When he separated the two types manually, then redissolved them in separate solutions, the right-handed crystals rotated light to the right (almost identically to tartaric acid), and the left-handed crystals rotated light an equal amount to the left. Pasteur had discovered optical isomers, compounds that differed only by being mirror images of one another.

Pasteur soon became convinced that optically active organic substances could be produced only by living things. He studied the fermentations that produced these substances. Because he always saw specific living microorganisms in association with each *fermentation, he took microbes to be the cause of the process and the producers of the substances he studied (lactic acid 1857, alcohol 1858–1860, butyric acid 1861, and acetic acid 1861–1864). This view directly contradicted the prevailing theory championed by German chemist Justus von *Liebig, that fermentation was a chemical process and the microbes mere concomitants, at most by-products of the process. Pasteur acted as consultant to many French industries seeking the cause of variations in the quality of their products, such as sugar beet alcohol, vinegar, wine, and beer. This work led him in 1866 to patent the process of heating fluids briefly to 60°C to destroy any microbes that might lead to spoilage or disease (pasteurization).

From 1860 to 1864 Pasteur did experiments that undermined the doctrine of spontaneous generation supported by Félix-Archimède Pouchet. Historians differ sharply in their interpretation of the Pasteur-Pouchet controversy. Some have argued that the conservative political and religious climate of the Second Empire and the Roman Catholic church's explicit opposition to spontaneous generation influenced the Paris Academy of Sciences's evaluation of Pasteur's experiments as conclusive while their results were still ambiguous. Others have responded that Pasteur's brilliant experiments alone can explain the outcome. Gerald Geison's examination of Pasteur's private laboratory notebooks demonstrated that Pasteur's private views sometimes conflicted with his public statements, particularly on vaccines for anthrax and rabies, and on spontaneous generation. Some historians have taken a middle ground; all agree that Pasteur shrewdly cultivated patronage from the emperor and empress personally, and that his work is a model of close interaction between academic research and industries that depend on biotechnology. The role of politics and religion in his work, however, remains a contentious issue.

Between 1865 and 1870 Pasteur, at the request of government officials and his scientific patrons, studied the diseases of silkworms that had devastated the French silk industry for twenty years. He found two distinct diseases, each caused by a microorganism, and developed methods to rid the breeding colonies of them. This work suggested that many human diseases had a microbial origin. Convinced the microbes did not arise by spontaneous generation, he first believed only sanitation measures could prevent microbial diseases. He and his student Charles Édouard Chamberland contributed importantly to water filtration procedures.

By 1879 Pasteur found that pathogenic microbes could be rendered less virulent by altering their environment (e.g., exposing them to oxygen) for several generations of growth. He tried injecting weakened strains into animals in an attempt to produce immunity to the virulent strain. In this way he developed vaccines for chicken cholera (1880), anthrax (1881), and rabies (1885), which brought him as much fame, and generated more income, than any of his previous discoveries. With that income and donations and fees for the livestock vaccines, he built the institute of which he had dreamed for much of his working life. By the time the Institut Pasteur opened in 1888, however, Pasteur had been weakened by two strokes, and his personal conduct of experiments was greatly curtailed.

Émile Duclaux, *Pasteur: The History of a Mind* (1920). René Dubos, *Louis Pasteur: Free Lance of Science* (1950). John Farley, *The Spontaneous Generation Controversy from Descartes to Oparin* (1977). Nils Roll-Hansen, "Experimental Method and Spontaneous Generation: The Controversy between Pasteur and Pouchet, 1859–64," *Journal of History of Medicine and Allied Sciences* 34 (1979): 273–292. Gerald Geism, *The Private Science of Louis Pasteur* (1995). Patrice Debré, *Louis Pasteur* (1998).

JAMES E. STRICK

PATENTS. A patent is a grant of governmental authority of a right to exclude others from utilizing an invention except upon agreement with the originator or owner. Allowing an inventor a temporary monopoly encourages investment in technology development by reducing the risk that others will take free advantage. Thus, in recent years biotechnology firms have been able to obtain long-term investments in part due to their secure patent rights, while software firms, where patenting is problematic, have had to rely on short-term capital.

Although patent systems have changed significantly over time, the concept itself is old. Municipal regulation of the Venetian glass industry recognized craft knowledge as a form of intangible property as early as the thirteenth century. By the fifteenth century European cities regularly granted limited monopolies to develop newly introduced crafts or new inventions. The common criterion was an expectation of contribution to economic and social life through the granting of a limited monopoly, rather than solely as a reward for discovery itself. In 1474 the Venetian Senate expanded individual grants of patents into a general patent law, creating by public authority a limited private monopoly in new inventions. An English law of 1624 introduced a new element by restricting patent protection to a specific term, fourteen years. While many consider this enactment to mark the beginning of modern patent systems, formalized procedures for awarding patents became common only in the nineteenth century. In 1836 the United States established the Patent Office in the Department of State. The law of 1836 established a system of examination of patent claims, authorized the appointment of patent examiners, and created an appeals process for unsuccessful applicants. Great Britain began to adopt a similar system in 1850; Japan followed suit in 1883. Later modifications of these nineteenth-century systems have been common. Some countries, especially the United States, employ a "first to invent" standard for patentability; others, including Germany, Britain, and Japan, follow a "first to file" procedure in awarding patents.

Patentability is confined to novel devices or their biological and information analogues; because natural laws and theories cannot be patented, science entered the patent arena relatively late. However, the chemical and electrical industries that arose in the middle and late nineteenth century built on scientific discoveries, and made science increasingly patentable. A British patent issued in 1856 for the first aniline dye (mauve) to William Henry Perkin laid the basis for the aniline dye industry as well as his personal fortune. Perkin's green alizarin dye patent in 1869 added further to his wealth. American inventors rushed behind. Taking advantage of the modernized American patent structure, Thomas Edison obtained his first patent in 1868, for an electrical vote recorder. Later patents for the stock ticker, automatic *telegraph, incandescent bulb, *phonograph, and other devices established his reputation as both a shrewd businessman and an inventive genius. Edison ultimately obtained more than a thousand patents. Of course, he did not accomplish this by himself. From earnings, he was able to fund an organization of technicians and scientists, thereby systematizing the invention process and creating a new social invention, the corporate research laboratory.

When a succession of collaborators, as well as competitors, work to develop a technology to the point of viability, as in the case of the diesel engine, invention as a discrete event and the notion of the individual inventor becomes problematic. Inventions made by railroad employees as part of their job raised this issue during the mid-nineteenth century, and companies began to require their employees to sign over patent rights as part of the employment contract. An amendment to the U.S. Patent and Trademark Act in 1980, the so-called Bayh-Dole Act, partially reversed this practice, at least for employees of universities. By law, faculty, student, or staff inventors must receive a significant share of the income generated by a patent from their invention.

Another salient controversy centers on the possibility of patent protection for life forms, as opposed to various mechanical devices and chemical compounds for which patent laws were devised in the nineteenth century. These issues came to the fore in 1972 when Anada Chakrabarty, a researcher for General Electric, invented a new organism that consumed oil. A product neither of recombinant DNA nor of hybridoma technologies, Chakrabarty's work genetically crossed four microorganisms with appetites for different components of crude oil; the resulting bug digested about two-thirds of the oil, converting it to carbon dioxide and microbial protein. General Electric applied for a patent on the assumption that the organism would have commercial value in cleaning up oil spills. Following precedent, the Patent Office initially denied the claim; but in 1980 a Supreme Court ruling awarded a patent to GE.

Yet another issue is the imbalance between developed and developing countries in the control of patent rights. The latter want inexpensive access to advanced technology. The United States shifted from an anti-patent to a pro-patent position in the nineteenth century during the course of accelerating industrialization. Some developing countries, with R&D capabilities, have gained some redress by threatening to ignore patents if the price of AIDS drugs is not lowered.

Is technological and scientific advance assisted or impeded by patent systems? Proponents argue that the process serves to spur innovation. Patents embody an information system and a legal disclosure process since the invention and its material expression become publicly available upon the issuance of a patent. An alternative to patent protection is secrecy. Opponents hold that patents restrict access to the pool of knowledge that supports future innovation. The free flow of innovation assists technological innovation as well as scientific advance.

The patent is a Janus-faced instrument. Despite a history of use by large firms to inhibit competitors and facilitate cartels, patents are also an integral part of the process of firm formation in new technological fields. A patent obtained through a thorough review process provides a relatively secure base on which to construct a new economic enterprise. Patents encourage economic renewal through the paradox of allowing a temporary monopoly.

By its nature any particular intellectual property is replaceable by new knowledge. Property in knowledge with potential economic value must be exploited promptly and constantly renewed and updated to secure maximal pecuniary value. Marketing and licensing are essential to patents; they promote the use of knowledge under the constraint of cognitive obsolescence. Although a few patents—such as those on vitamin B-12—are extremely valuable, most have modest or no value. Nevertheless, organized in databases and available on the Internet, the collectivity of patents provides an inspiration to inventors seeking innovation from a joint public/private good of increasing value.

See also LAW AND SCIENCE; SECRECY IN SCIENCE.

Elizabeth Antebi and David Fishlock, *Biotechnology: Strategies for Life* (1986). Christine Macleod, *Inventing the Industrial Revolution* (1988). Pamela Long, "Invention, Authorship, 'Intellectual Property,' and the Origin of Patents: Notes Toward a Conceptual History," *Technology and Culture* 32:4 (1991): 846–884. Henry Etzkowitz, *MIT and the Rise of Entrepreneurial Science* (2002).

HENRY ETZKOWITZ

PATRIOTIC AND ECONOMIC SOCIETIES. Hundreds of patriotic, economic, and local scientific societies founded in the eighteenth and early nineteenth centuries accepted the challenge of hitching moral and material improvement to the fleet wagon of scientific advance. No other set of institutions, not even the new universities of this period, better captured the scientific, economic, and technological interests of citizens imbued with the spirit of Enlightenment and determined to improve practical arts and useful sciences. The founding of these many public and private institutions built upon institutional models created to promote the new sciences during the seventeenth century, particularly the closed academies and open societies that encouraged scientific discovery, experimentation, and communication, first in Italy, France, and England, then in Germany. More than a hundred of these scientific academies were established by the end of the eighteenth century. The many hundreds of patriotic and economic societies—whether *gemeinnützige Gesellschaften*, societies of arts, agricultural societies, *sociedades económicas*, or *sociétés d'encouragement*—formed a second wave of institutions founded to promote the material fruits of scientific advance.

Active members of patriotic and economic societies concentrated on an important aspect of the Enlightenment's preoccupation with science: mobilizing voluntary action, self-sacrifice, and civic activism to direct scientific progress toward economic improvement. The first societies were founded in larger urban centers such as Edinburgh, Paris, Zürich, Florence, and London during the second quarter of the eighteenth century, but spread quickly to smaller regional centers during the half-century stretching roughly from the Seven Years War through the Napoleonic Wars. Compared to the enduring scientific academies such as those in Paris, Berlin, St. Petersburg, or London (*see* ACADEMIES AND LEARNED SOCIETIES), many of the patriotic and economic societies were short-lived institutions with only a local impact. They made up for their brief existence by their numbers. The German-speaking parts of central Europe saw the foundation of two hundred of them between 1760 and 1815 (not including the numerous Dutch and Scandinavian societies), an average of four new societies annually, providing thousands of government officials, physicians, academics, pastors, merchants, and others with opportunities to contribute useful works. The remainder of this essay concentrates on these German societies.

The northern center of patriotism was Hamburg. In 1721, a small group of its citizens began a series of informal weekly meetings on useful topics. Their lasting accomplishment was the publication of a moral weekly, *Der Patriot*, between 1724 and 1726. With an ear to the pietists' call for useful education and philanthropy, and moved by the spirit of Christian Wolff's "rational thoughts" on virtually any topic, the Hamburg patriots provided the rhetorical unification of moral philosophy, economic progress, science, and education that inspired subsequent societies throughout central Europe and demonstrated how the medium of serial publication could be used effectively. Following the Hamburg model, most of the patriotic journals and moral weeklies of the early Enlightenment in Germany encouraged the airing of suitable topics rather than direct intervention in or sponsorship of economic activities.

Agriculture was at the center of economic life in Germany, but for enlightened "improvers," it symbolized waste and inefficiency. Weighed down by the momentum of habit and the largest war on German soil since the seventeenth century, agriculture in the 1760s remained locked in an unbroken circle of wasteful crop rotations, inadequate attention to the role and importance of fodder crops, underdeveloped dairy farming, and underutilization of fertilizers. Overarching goals for reform took shape by the middle of the eighteenth century, namely, breaking the hold of ignorance and inefficient practices through progressive techniques based on reason and science. The wave of economic societies founded during the 1750s and 1760s provided agronomists with a new institutional base for action aimed at material improvement; intervention caught up with instruction as a central activity in these societies. Their charters challenged their members to promote useful sciences above all others, not as stages for the accomplishments of individual geniuses, but as springboards for encouraging cooperation, mutual encouragement, and fellowship. The meetings, prize competitions, grants, proceedings, experiments, and other projects of these societies linked sciences such as practical mathematics, natural history, or chemistry with economic activities such as agriculture, *forestry, mining, military affairs, or household and estate management.

The patriotic and economic societies also provided venues for the linkage of useful science and religion, a linkage not usual during the Enlightenment. The relationship between work and faith rested on the notion that improvement was a moral duty, the dominant theme in the ideal sounded by the Hamburg patriots. Drawing upon pietism and patriotism, the German improvers never lost sight of bonds between spiritual and secular profit. They sought to make the state prosperous by raising well-educated and enlightened souls. Pastors joined economic societies in large numbers in part because of the multiple reasons the societies' statutes gave for engaging in useful work: diligent and productive people energetically serve moral causes; industriousness is the best "dam against sin"; agriculture increases knowledge of nature and thus of God; and hard work nourishes love of country. At least one local scientific society regularly applauded the "active Christians" in its ranks. The societies frequently published homilies, utilizing opportunities like the obituaries of recently deceased members published in the transactions of the *Gesellschaft Naturforschender Freunde*.

In keeping with the linked obligations of both agricultural and moral improvement and their connection to scientific study, one of the most popular areas of attention in the early economic societies was the science and economy of bees. The several societies devoted to apiary could not only cite the economic benefits of keeping up with the latest developments in bee-keeping, but also welcomed the art as the exemplar of an economic-patriotic subject. Bee-keeping supplemented agricultural income, perhaps replacing it where farms were unprofitable; even town dwellers could keep bees, and so develop a bond with farmers; beeswax and honey could be exported, an attraction for the public official or estate manager; bees interested the scientifically inclined, whether a natural historian, a careful observer of the insect's habits, or a mathematically precise designer of hives; bee gardens provided gathering points for members and the collection of shared observation, and demonstrated

the utility of their studies to the local community; and, finally, bees were associated with positive moral and religious values.

The wave of enthusiasm associated with agricultural improvement and bee-keeping led by the economic improvers of the 1760s subsided after about a decade. Many societies collapsed during the 1770s, especially in the Hapsburg empire and Switzerland, as either the initial members' enthusiasm or state funding (particularly in Austria) waned. As a whole, the movement represented by the patriotic and economic societies and their members shifted its programs from the direct encouragement of specific economic improvements, the primary activity during the 1750s and 1760s, to the systematization of knowledge and the rationalization of practice. The repeated appearance of words such as "Land" or "Vaterland" in the names of societies founded after 1780 suggests a trend toward the investigation and description of local and regional conditions. In a rough sense, scientific activities took on a larger portion of the agenda of economic and agricultural societies. Typical projects included gathering political statistics, describing topography and mineral resources, charting the weather, and cataloging artifacts of nature or human-made tools. A powerful incentive, along with the expected scientific and economic utility of these activities, was the reawakening of political and cultural history in Germany and Switzerland. Members of the newer societies welcomed work that encouraged patriotic feeling and national pride, and a number of them contributed historical works or sponsored prize competitions on historical topics. By the 1780s, many of the original economic societies had folded, giving way to local scientific societies, "physical-economic" societies, or societies concentrating on varieties of *Landeskunde*.

By 1800, the societies, especially those newly founded, had a hand in rationalizing or systematizing occupations such as forestry, agriculture, and state administration. Their new role centered on the discussion and organization of principles by their members and the establishment of specialized institutes to provide professional training. The change of emphasis in the work of the late Enlightenment societies also reflected their inability to bridge social gaps between town-based improvers and rural practitioners, as well as the demographics of the societies' membership, which drew primarily from intellectual elites with scientific training (physicians, academicians, professors) and government officials. Many of the societies adopted the systems and texts of the new Cameralwissenschaft, rationalizing training and practice, and taking stronger roles in some areas such as forestry, mining, agricultural experimentation, systematization of estate management, and commercial studies than either the universities or learned academies of the day.

See also CAMERALISM; ENLIGHTENMENT AND INDUSTRIAL REVOLUTION.

Hans Hubrig, *Die patriotischen Gesellschaften des 18. Jahrhunderts* (1957). Kenneth Hudson, *Patriotism with Profit: British Agricultural Societies in the Eighteenth and Nineteenth Centuries* (1972). Daniel Roche, *Le siècle des lumières en province: Académie et académiciens provinciaux, 1680–1789* (1978). Henry Lowood, *Patriotism, Profit, and the Promotion of Science in the German Enlightenment: The Economic and Scientific Societies, 1760–1815* (1991).

HENRY LOWOOD

PATRONAGE. See POLITICAL ECONOMY OF SCIENCE.

PAULI, WOLFGANG. See HEISENBERG, WERNER, AND WOLFGANG PAULI.

PAULING, Linus (1901–1994), American chemist, leader in quantum chemistry and molecular biology. The importance and range of Pauling's scientific contributions make him the most remarkable and original chemist of the twentieth century. Best known for his foundational work in quantum chemistry and in *biochemistry and *molecular biology, he did important work in x-ray crystallography, biomolecular phylogeny, and the field he called molecular medicine. By the 1950s his crusading zeal had expanded from science to politics, as he became a leader, with his wife, Ava Helen Pauling, in an international movement to ban nuclear weapons tests. He is the only person to win two Nobel Prizes: the chemistry prize in 1954 and the peace prize in 1962.

Linus Pauling was born in Portland, Oregon. He graduated in chemical engineering in 1922 from Oregon Agricultural College at Corvallis (now Oregon State University) and entered the California Institute of Technology (Caltech) on a three-year graduate scholarship offered by Arthur Amos Noyes. After receiving his Ph.D. in 1925, Pauling used a Guggenheim Fellowship to study in Munich, Copenhagen, Göttingen, and Zurich, returning to a new position at Caltech as assistant professor in theoretical chemistry in 1927. He remained at Caltech until 1963, when he resigned because of colleagues' criticism of his peace activities.

Pauling became acquainted with Walter Heitler and Fritz London in Zurich when they were applying *quantum physics to the *valence bond of the hydrogen molecule. Influenced by their famous paper of 1927, Pauling wrote his first paper on electron-pair exchange and the resulting "resonance energy" in a chemical bond in 1928. He continued this work between 1931 and 1933 and embedded it in his classic text *The Nature of the Chemical Bond* (1939).

In explaining the reactions and stability of chemical molecules, Pauling used quantum-mechanical energy values and atomic-electron orbitals. Among his great successes were explanations of the chemical valency of the carbon tetrahedron and the bond structures of molecules of benzene and other organic and inorganic substances that cannot be represented by a single structural formula. Pauling and, independently, John Slater, created a theoretical chemistry that accustomed chemists to the power of the new quantum methods. To Pauling's consternation, the molecular-orbital methods of Friedrich Hund, Erich Hückel, and Robert Mulliken began to replace the atomic-orbital methods in the 1950s. These chemists analyzed the molecule as a whole rather than treating it as a combination of atoms.

Pauling joined his interest in quantum chemistry with the analysis of molecular structure by x-ray crystallography and three-dimensional model building. His rules for the geometry of ions in crystals, published in 1928, became fundamental to the practices of crystallography. His visit to Herman Mark's I. G. Farben laboratory in 1930 familiarized him with electron-diffraction techniques and with Mark's theories of protein structure and the flexibility of polypeptide chains. Beginning in 1932 Pauling turned to investigation of the structures of proteins, including hemoglobin, and other molecules of medical interest. He paid close attention to George Beadle and Edward Tatum's work at Caltech in the 1930s on mutations produced by X rays in bread mold, and he collaborated in 1940 with the biophysicist Max Delbrück on the investigation of gene replication.

From 1937 to 1951, using amino acid crystallization techniques and x-ray investigations of polypeptide dimensions, Pauling and his co-workers at Caltech discovered the role of hydrogen bonding in proteins and developed the hypothesis of a coiled peptide chain. In 1950 Pauling and Robert Corey reported the alpha-helix conformation for proteins, which was confirmed by John Kendrew in 1960 for myoglobin and by Max Perutz in 1962 for hemoglobin. Despite the ideas he had developed with Delbrück on paired templates and structural complementarity in the gene, Pauling declared in 1953 that DNA had a triple, not a double, helical structure. Francis *Crick and James Watson's subsequent announcement of double-helical DNA owed much to Pauling's methods and work.

The discovery by Pauling's co-worker Harvey Itano that sickle-cell anemia is a genetically based disease in which an individual's hemoglobin has less negative charge than normal hemoglobin led to the notions of molecular disease and molecular medicine. Pauling's work in the 1960s and 1970s on the anti-oxidizing properties of vitamin C as a therapy for the common cold and as a treatment for cancer was both controversial and popular. His fears about the long-term genetic effects of radiations from nuclear fallout and his development, with Emil Zuckerkandl, of a molecular evolutionary clock based on changes in polypeptide chains over time fit with his long-standing interests. After 1973 Pauling worked with colleagues in Palo Alto at his own research institute, which moved to Corvallis, Oregon, after his death in 1994.

Ted Goertzel and Ben Goertzel, *Linus Pauling: A Life in Science and Politics* (1995). Thomas Hager, *Force of Nature: The Life of Linus Pauling* (1995). Ramesh Krishnamurthy et al., eds., *The Pauling Symposium: A Discourse on the Art of Biography* (1996). Stephen F. Mason, "The Science and Humanism of Linus Pauling (1901–1994)," *Chemical Society Reviews* 26 (1997): 29–39.

MARY JO NYE

PAVLOV, Ivan Petrovich (1849–1936), physiologist.

Born in Ryazan, in central Russia, Ivan Pavlov was expected to follow his father into the priesthood. Under the influence of the radical scientism popular among Russian youth in the turbulent years after Tsar Alexander II's emancipation of the serfs in 1861, Pavlov abandoned his seminary education to study *physiology at St. Petersburg University (1870–1875) under the tutelage of I. F. Tsion.

After graduation, Pavlov received his medical training at the Military-Medical Academy (1875–1880), where he also acquired his doctorate (1883) for a thesis on the nerves controlling the heart. He spent the next two years in the laboratories of Carl Ludwig in Leipzig and Rudolf Heidenhain in Breslau, returning in 1886 to his duties as *Privatdozent* in physiology at the Military-Medical Academy and manager of the eminent clinician S. P. Botkin's small lab there. In 1891, Prince A. P. Ol'denburgskii appointed Pavlov chief of the Physiology Division at St. Petersburg's newly founded Imperial Institute of Experimental Medicine. Here, between 1891 and 1904, Pavlov and about one hundred coworkers pursued the investigations of the digestive system that resulted in his *Lectures on the Work of the Main Digestive Glands* (1897), which brought him world renown and a Nobel Prize (1904). As a professor at the Military-Medical Academy (in *pharmacology from 1890 to 1895, and in physiology from 1895 to 1924), Pavlov also regularly delivered lectures to medical students and commanded a second, smaller laboratory

there. Elected to the Academy of Sciences in 1907, Pavlov acquired there a third lab, which expanded substantially in the 1920s.

Pavlov's studies of digestive physiology reflected the scientific-managerial style that characterized him throughout his long career. A proponent of organ physiology and believer in the purposiveness of all physiological processes, he sought both to study the intact, normal animal (through the so-called "chronic experiment") and to arrive at fully determined laws of physiological activity. In his research on the digestive system, these laws took the form of "characteristic secretory curves" describing the precise and purposive response of the digestive glands to ingested foods. These curves—and other important findings such as the role of the psyche and the vagus nerves in digestion—resulted from Pavlov's supervision and interpretation of experiments conducted by his numerous coworkers.

By the year of his Nobel Prize, Pavlov had begun to shift his laboratory to the study of what he had earlier termed "psychic secretion" and would eventually make famous as the "conditional reflex." The well-known phenomenon of a dog salivating upon seeing the person who usually feeds it was transformed into a reliable laboratory phenomenon, and the formation and extinction of conditional reflexes became a method for studying the higher nervous processes that produced many such "psychic phenomena." Using the quantitative patterns of salivation generated during tens of thousands of experiments, Pavlov constructed a complex picture of brain activity that included excitation, inhibition, disinhibition, generalization, irradiation, and concentration. Between 1904 and 1914, he also used the method of conditional reflexes to investigate dogs' ability to distinguish among colors, and to keep track of time.

This research was interrupted by the outbreak of World War I and the departure of Pavlov's coworkers for the front. The Bolshevik seizure of power in October 1917 led Pavlov, who had hoped for an evolutionary development to a constitutional monarchy, to consider emigrating. By 1921, however, he had reached an accommodation with the Bolsheviks, who thereafter funded his scientific enterprise generously while allowing him to continue criticizing their policies in public.

The response of Pavlov's dogs to their near drowning during the Leningrad flood of 1924 encouraged his longstanding interest in "experimental neurosis" and psychiatry. Because experimental animals frequently lived for years in the laboratory, he and his coworkers had long noticed their different personalities and their varying responses to identical experiments. This gave rise to an interest in the various inborn "nervous types" of dogs (and temperaments in humans). Pavlov's research on conditional reflexes reached an international audience with the publication and multiple translations of a collection of his speeches and articles, *Twenty Five Years of Experience in the Objective Study of the Higher Nervous Activity of Animals* (1923), and his only monograph on the subject, *Lectures on the Work of the Large Hemispheres of the Brain* (1927).

In 1929, the Communist Party celebrated Pavlov's eightieth birthday by funding his science village in Koltushi, just outside of Leningrad. At this Institute of the Experimental Genetics of Higher Nervous Activity, Pavlov planned a major investigation (with *eugenic goals) of the relationship between heredity and environment in the determination of nervous types and also studied the apes Roza and Rafael. He played a pivotal role in convincing international physiologists to hold their 1935 congress in Russia, where they hailed him as "The Prince of World Physiology." He died early the following year.

During his lifetime, Pavlov became an international cultural icon. He was renowned not just for his many discoveries and ingenious experimental techniques, but as a symbol of the hope (and fear) that experimental science might enable humanity to understand, and even control, human nature.

Boris Babkin, *Pavlov: A Biography* (1949). Hilaire Cuny, *Ivan Pavlov: The Man and His Theories* (1964). Jeffrey Gray, *Ivan Pavlov* (1979). David Joravsky, *Russian Psychology: A Critical History* (1989). Daniel Todes, "Pavlov and the Bolsheviks," *History and Philosophy of the Life Sciences* 17 (1995): 379–418. Daniel Todes, *Pavlov's Physiology Factory: Experiment, Interpretation, Laboratory Enterprise* (2002).

DANIEL P. TODES

PEER REVIEW. To the public, peer review is the ultimate guarantor of good science. Through its process of expert judgment of experts, peer review preserves science's autonomy while assuring society that the money it devotes to science is well spent. In practice, however, the scope of peer review is not nearly as comprehensive and its norms not as unified or systematic as often suggested. Substantial funding for science, the "earmarked" projects directly funded by the U.S. Congress, escapes the system without peer review; not all scientific journals use external referees; and journal editors (especially in the case of proprietary journals) may abdicate substantial manuscript review. More important, there is a substantial performance gap between ideal and actual peer review. Recent studies have detailed widespread negligence, lack of skill, self-interest, and plagiarism among referees, as well

as conservative bias in the overall system—and have recast peer review as, at best, the least imperfect way of evaluating scientific results and proposals.

Peer review emerged within scientific academies in the seventeenth century. As shown by the practices of manuscript evaluation adopted by the Académie Royale des Sciences of Paris and the Royal Society of London, peer review evolved as a by-product of legal obligations that came with these institutions' book licensing privileges granted by their royal patrons. The primary goal of early peer review was not to reassure readers or taxpayers of the quality of what they were reading or paying for, but to spare institutions from the consequences of publishing controversial claims.

The Académie des Sciences established rules for peer review and book licensing privileges in its statutes of 1699. Similarly, the charter of 1622 granted to the Royal Society by Charles II contained a detailed description of its book printing privileges. In 1663, the Society's governing council addressed the implementation of those privileges and introduced its peer review protocols.

The Society behaved more like a licenser than a modern editor. If a text was deemed to contain "nothing but what is suitable to the design and work of the Society," it was licensed and published in the form submitted. The goal was to filter out books whose topic fell outside of the licensing jurisdiction of the Society (*natural philosophy) or that were "unsuitable" to its philosophical style. As with traditional book licensing, the Society's review practices were designed to ensure that a text did not make unacceptable claims rather than that it made good claims. Given natural philosophy's heterogeneity at the time, standardized peer review guidelines were nearly impossible.

The transfer of peer review protocols from books to journal articles, however, proved complicated for both the Académie des Sciences and the Royal Society, largely because of problems posed by the publication of nonmembers' work. The publication of "foreign" texts in the *Mémoires* (the journal of the Académie des Sciences) strained the limits of the royal privileges, which many saw as allowing publication solely of the work of academy members. The problem was displaced with the introduction in 1750 of the *Mémoires des savants étrangers*, a journal dedicated to the work of nonacademicians. Although the Royal Society could legally publish the work of nonacademicians, it feared that publication of "foreign" works could be read as an endorsement of their claims. The defensive attitude was reflected in the Society's editorial style that eliminated controversial passages and toned down claims.

Not blessed with a productive membership, the Society could not produce sufficient native material to sustain a journal; at the same time, it attracted a surfeit of "foreign" submissions. The solution to the Society's "publish foreign or perish" dilemma came in the shape of an interesting hybrid. In March 1665, the Society mandated the publication of the *Philosophical Transactions*. Subsequently, it licensed all issues of the journal, had them printed by its official printer, and ordered its secretary, Henry Oldenburg, to insert in its issues some of the more interesting papers it received. Despite this direct involvement, the Society had Oldenburg state in the journal's first issue that the *Transactions* were not the business of the Society but the editor's own enterprise.

Eager to publish the journal but worried about the dangers, the Society engaged in a bit of creative taxonomy. It did not treat the *Transactions* as a journal that published individually authored essays. Rather, each issue of the *Transactions* was presented as a book produced by one of its members, Oldenburg (who was therefore treated as an author, not an editor). Once the *Transactions* were framed as an internally produced book, Oldenburg could be construed as the author of each issue just as, say, Robert *Hooke was the author of *Micrographia*, a book on microscopic observations commissioned and then licensed by the Society in 1665. Both books were ordered, reviewed, licensed, and printed by the Society. While the Society's leadership occasionally ordered the inclusion of certain content, the books were presented as the sole responsibility of their designated authors, not the Society.

The meaning of both "peer" and "review" have changed since the seventeenth century. The granting of book licensing privileges to early academies marked a step toward the development of a scientific community and, eventually, its own standards of quality. Today, peer review has become depersonalized (referees act as spokespersons of a community) and delocalized (peer review is ubiquitous, not concentrated in a few sites).

A comparison of seventeenth-century peer review practices with modern ones reveals several significant differences: peer review no longer performs a central legal function (largely because the book licensing systems it once served have disappeared); it is now seen as serving an international scientific community, rather than a specific institution and its sponsors; and today's reviewers generally operate in anonymity, rather than promoting themselves and their imprimatur.

Editorial boards of many contemporary scientific journals are selected to represent a cross-section of professional communities; in some cases, editorship rotates among members of a professional association. The beneficiaries of contemporary peer review are identified as the publication's consumers—readers, universities using peer-reviewed publications in promotion cases, and, ultimately, taxpayers—rather than its producers—the journal, the publisher, and the institutions sponsoring the journal. Today's referees receive little public credit and, in most cases, no financial compensation. Their early modern predecessors, in sharp contrast, often took advantage of their reviewing and licensing powers to advance their own standing in the scientific community. The modern erasure of the referees' identity attempts to improve the objectivity of the review by making it "blind," signifying that referees are not to be regarded as specific, identifiable judges associated with specific institutions but rather as nameless voices of a geographically dispersed and democratic scientific community.

A few journals remain connected to academies today, but most are privately owned and run according to a business model. The priority of these modern journals is not the enhancement or protection of the reputation of a parent academy or royal sponsor but the maximization of their own symbolic capital—a capital that translates into "impact factor," more subscriptions, and greater revenue. In the seventeenth century there was little basis for defining a "peer" except by membership in an academy, a status that had little correlation with disciplinary competence. Today the Ph.D. has become the basic requirement for admission into peerdom, and membership in a professional society (in most cases, simply a matter of paying annual dues) no longer confers an automatic right to publish in its journal. The social system of science has become so complicated and dispersed, in both geographical and disciplinary terms, that peer review (no matter its quality or uniformity in action) has come to be regarded as the law of an otherwise lawless land.

Thomas Birch, *The History of the Royal Society of London*, 4 vols. (1756; 1968). Roger Hahn, *Anatomy of a Scientific Institution* (1971): 1–34. Harriet Zuckerman and Robert Merton, "Patterns of Evaluation in Science: Institutionalization, Structure, and Function of the Referee System," *Minerva* 9 (1971): 66–100. Paul Grendler, "Printing and Censorship," in *The Cambridge History of Renaissance Philosophy*, ed. Quentin Skinner and Eckhard Kessler (1988): 25–53. Sheila Jasanoff, *The Fifth Branch: Science Advisers as Policymakers* (1990): 61–83. Drummond Rennie, ed., *Peer Review in Scientific Publishing* (1991). Adrian Johns, *The Nature of the Book* (1998).

MARIO BIAGIOLI

PENDULUM. Any body that oscillates freely about an axis (usually horizontal) under the influence of gravity can act as a pendulum. *Galileo had observed that the time taken for each oscillation depended on the length of the pendulum and was very nearly isochronous (independent of the angle through which it swung). In its simplest form, consisting of a ball suspended by a thin chain or thread, it was used to measure short time intervals; in 1656 Christiaan *Huygens extended the intervals by using a pendulum to control a clock. In 1672 Jean Richer observed that a clock that had been regulated in Paris lost time when taken to French Guiana. Isaac *Newton attributed this loss to a change in gravity as a result of the Earth's equatorial bulge. The pendulum thus found a use in *geodesy, and the different requirements of these two applications—time and gravity measurements—led to a divergence in pendulum design.

The typical pendulum of an early clock consisted of a suspended rod carrying a mass (the bob) at its lower end, which could be moved vertically to adjust the timekeeping. To achieve long-term stability, changes in the length of the pendulum with temperature must be compensated for. The English clock and instrument maker George Graham accomplished this in 1722 using a bob containing mercury. The English clockmaker John Harrison followed about six years later with his gridiron pendulum, which relied on the difference in thermal expansion of steel and brass. A simpler solution became possible toward the end of the nineteenth century with the invention by Charles-Edouard Guillaume of invar, a nickel-steel alloy whose dimensions were essentially independent of temperature. Huygens had shown that a pendulum bob following a cycloidal rather than a circular path would be truly isochronous, but his method was not adopted, and in practice a small and constant amplitude minimised the circular error.

In his *Horologium oscillatorium* of 1673 Huygens derived the period for a simple pendulum (a mathematical concept in which a mass concentrated at a point is supported by a weightless thread) in terms of its length and a measure of the force of gravity. To obtain an absolute value for gravity the geodecist had to determine this length with great accuracy. The convertible or reversible pendulum that Henry Kater described in 1818 gave the necessary information directly. It had two knife-edges facing inwards, and so could be swung either upright or inverted. The pendulum rod carried weights that the operator adjusted until the pendulum swung in the same time in both

positions. Using a theorem of Huygens, the length of the equivalent simple pendulum was then equal to the separation of the knife-edges. Later versions of the reversible pendulum were symmetrical in shape, and in the twentieth century they were usually made of invar. The determination of absolute values for gravity took time, but once calculated for a particular point, it could be extended to other points by making relative measurements with an invariable, non-reversible, pendulum.

Victor F. Lenzen and Robert P. Multhauf, "Development of Gravity Pendulums in the 19th Century," *United States National Museum Bulletin* 240 (1965): 301–348. A. L. Rawlings, *The Science of Clocks and Watches*, 3rd ed. (1993).

DENYS VAUGHAN

PERIODIC TABLE. Chemists of the early nineteenth century had to rely on indirect methods, analogies, and simplifying assumptions to determine the relative *atomic weights of *elements, and thus did not reach a consensus on them. Gradually, however, methods improved, and at the international Karlsruhe Congress of 1860 Stanislao Cannizzaro advocated what is, in essence, the system still in use today (*see* ATOM AND MOLECULE).

Even before agreement on a single set of atomic weights, several theorists noticed that chemical and physical properties reappeared in a regular, periodic fashion in the various elements. Among these predecessors of the

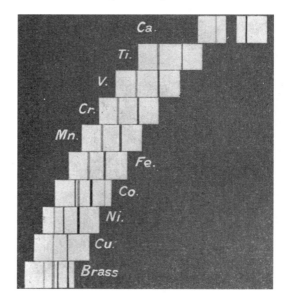

"Moseley's step ladder" showing the progressive displacement of the K lines of the *X-ray emission spectrum towards higher frequencies with increase in atomic number; the irregularity indicates a gap in the sequence of elements investigated.

periodic law were Johann Wolfgang Döbereiner, John Newlands, William Odling, and Jean-Baptiste-André Dumas. Many others proposed partial periodic classifications of the elements.

Dmitrii *Mendeleev reaped the success of these endeavors. (The German chemist Lothar Meyer pursued a closely parallel path independently, and published his similar contribution shortly after Mendeleev's.) Mendeleev arranged the elements horizontally according to increasing atomic weight, and started a new row below the first whenever similar properties in the elements reappeared. The resulting semirectangular table of atomic weights showed many intriguing regularities. The horizontal rows ("periods" or "series") and the vertical columns ("groups" or "families") revealed a "periodic law." For example, Mendeleev's arrangement placed the alkali metals (sodium, potassium, and the then recently discovered rubidium and cesium) in a single vertical group with a marked family resemblance. The set of next-heavier elements to each of these alkali metals formed a second family group, the alkaline earth metals (magnesium, calcium, strontium, and barium).

This example understates the magnitude of the problems Mendeleev faced, for the elements simply failed to order themselves neatly. (Sometimes the greatest genius requires *ignoring* a certain number of anomalies, while discerning the larger pattern hidden within the data.) Lithium, for instance, fell into the alkali metal group by weight order, but it seemed to have more family resemblance to magnesium than to sodium, the element directly below it in the table. By the same token, beryllium seemed to resemble aluminum more than its putative family member below, magnesium. In certain other cases two adjacent elements seemed to come in a different order by weight from that dictated by the family groupings. For the first anomaly, Mendeleev provided an extenuating rationale; for the second, he did not hesitate to violate the weight order and reverse the two elements in the chart, hence preserving the periodicity of properties.

Despite the problems, Mendeleev had sufficient confidence in the periodic system he published in 1869 to hazard predictions based on it. Leaving vacant places in his chart in certain critical instances, he predicted the properties of undiscovered elements. For example, he left a space between calcium and titanium, and two between zinc and arsenic. Here Mendeleev used a strategy that converted yet another anomaly—holes in his periodic table—into potentially powerful evidence for the validity of his discovery. But of course there he risked the danger that his system would fail the test of prediction.

The first of Mendeleev's predicted new elements, discovered by Paul Lecoq de Boisbaudran in 1875, was named "gallium" in honor of France. Gallium's atomic weight of about 70 came close to Mendeleev's prediction of 68, its density of 5.9 grams per cubic centimeter virtually coincided with the prediction, its valence and oxide pattern were as expected, and a long list of observed chemical properties also matched what Mendeleev had forecast. Four years later scandium filled another space, and in another seven years, germanium. Three times Mendeleev had triumphed, not only with the fact of the discoveries, but with the details of physical and especially chemical properties as well.

These confirmed predictions caught the attention of the scientific world. For the first few years after Mendeleev announced his periodic system, it received almost no notice in journals and textbooks, and much less agreement as to its utility. After the discovery of gallium, however, textbook accounts of the periodic law began to appear, and in the 1880s discussions of periodicity of the elements became a common, though not invariable, feature of chemistry textbooks.

Further developments provided both challenges to, and support for, Mendeleev's system. Chemists and then physicists confirmed the weight inversions that Mendeleev had insisted upon to preserve periodicity (*see* ATOMIC STRUCTURE). A dozen or more chemically similar "rare earth elements" discovered in the last quarter of the nineteenth century presented a more worrisome problem in that they did not fit into any periodic system. Eventually chemists grouped them together in an aperiodic category as "lanthanides." In the 1890s William Ramsay and Lord Rayleigh discovered the inert gases argon, helium, neon, krypton, and xenon. They fit perfectly by weight before the alkalis, and soon after the turn of the century chemists decided to create an extra group for them at one end or the other of the periodic chart (*see* NOBLE GASES).

The development of atomic physics in the early twentieth century provided an independent method for assigning positions for elements in the periodic table: the measurement of their "atomic numbers." At the same time the study of radioactivity revealed a number of apparently elementary bodies that did not fit into the table: the solution was to enlarge the concept of element. A group analogous to the lanthanides, the actinides, has been added to accommodate transuranic elements.

See also CHEMICAL BOND AND VALENCE; CHEMISTRY; NUCLEAR PHYSICS AND NUCLEAR CHEMISTRY.

A. J. Ihde, *The Development of Modern Chemistry* (1964). J. W. van Spronsen, *The Periodic System of the Chemical Elements* (1969). E. G. Mazurs, *Graphic Representations of the Periodic System during One Hundred Years*, 2d ed. (1974). W. H. Brock, *The Norton [Fontana] History of Chemistry* (1992). Stephen G. Brush, "The Reception of Mendeleev's Periodic Law," *Isis* 87 (1996): 595–628.

A. J. ROCKE

PERSONAL EQUATION. See ERROR AND THE PERSONAL EQUATION.

PETROLOGY. See MINERALOGY AND PETROLOGY.

PH. Specifying the pH of an aqueous (water-based) solution gives a precise statement of its acidity or alkalinity. Acids and alkalis have been known since ancient times, and their properties and interactions to form *salts have long been basic to chemical operations; a detailed understanding of the mechanism of their interaction, however, dates only from the late nineteenth century, with the development of the theory of solutions and studies of the dynamics of chemical equilibria. In the 1890s, Wilhelm Ostwald and others developed theories relating the color changes of acid-base indicators to chemical reactions that depend on the concentration of hydrogen ions in solution. The concentration of hydrogen ions became a principal measure of acidity. Simultaneously, biochemists and physiologists investigated buffer solutions whose acidity or alkalinity does not change much with the addition of acids or bases. Since living systems maintain acid-base balances within very narrow tolerances, an understanding of buffer solutions was essential to many areas of physiological chemistry.

In 1909, the Danish biochemist Søren Peter Sørensen was studying enzyme activity, working with nearly neutral buffered solutions in which he followed minute changes in the concentration of hydrogen ions by means of a hydrogen electrode. The numerical results were cumbersome, since the concentration of hydrogen ions in pure water at room temperature is only one ten-millionth (10^{-7}) of a mole per liter. To simplify his notation, he found it convenient to express these concentrations in the form of negative base-ten logarithms. He chose the expression p_H to symbolize this convention, apparently taking p to represent the German word *Potenz* (power); H stands for the concentration of hydrogen ions in moles per liter. Thus, $p_H = 7.00$ indicates a neutral solution; $p_H = 5$, an acidic one; $p_H = 9$, an alkaline. That Sørensen's scale reads lower the greater the hydrogen ion concentration and the acidity has been a source of confusion.

At first only Sørensen's circle of specialist peers used the convention, in enzyme chemistry and other investigations into aqueous

biological systems. It was spread by the American physical and biochemist William Mansfield Clark (1884–1964), who adopted it for his book *The Determination of Hydrogen Ions* (1920) and simultaneously altered the relevant symbol to pH. His typesetter must have been grateful. Today, pH has almost the same meaning Sørensen gave it except that it refers to the "activity" or effective concentration of hydrogen ions. The scale ranges from pH = 1, which characterizes the strongest possible solution of hydrochloric acid, through the neutral point of pH = 7, to pH = 14, characteristic of the most powerful base.

See also ACIDS AND BASES; IONIC THEORY.

Ferenc Szabadváry, *History of Analytical Chemistry* (1966). Bruno Jaselskis, Carl Moore, and Alfred von Smolinski, "Development of the pH Meter," in *Electrochemistry, Past and Present*, ed. John Stock and Mary V. Orna (1989): 254–71.

ALAN J. ROCKE

PHARMACOLOGY originally referred to the study of drugs in all their aspects, including their origin, composition, physical and chemical properties, therapeutic uses, preparation, and administration. The word occurs in this sense in the seventeenth century. The science of pharmacology, which involves the study of the interaction of chemicals with living matter, did not emerge as a distinct discipline until the nineteenth century. Pharmacological science uses many of the methods of other biomedical sciences, but is distinguished from them by its focus on the physiological action of drugs and other chemicals. It emerged in many ways as an offshoot of the development of modern *physiology, but it established its niche in the universities by replacing existing chairs in *materia medica* (a term often used synonymously with the older, broader meaning of pharmacology) in schools of medicine.

Although physiological experiments had been carried out much earlier, the modern science of physiology began only with the work of François Magendie in France in the early nineteenth century. Research on the physiological action of drugs and poisons was an important part of Magendie's experimental program. He established that a poison derived from a plant of the *Strychnos* genus (later found to contain the toxic substance strychnine) produced convulsive contractions by acting on the spinal cord of an animal. Attempts to determine the site of action of drugs and poisons helped to mold the emerging science of experimental pharmacology. Magendie's researches also provided strong support for the view that drugs enter the bloodstream and reach the site of action by the circulation.

Magendie's famous pupil Claude *Bernard helped to advance both physiology and pharmacology. Bernard's brilliant studies on curare and carbon monoxide established their specific sites of action and so demonstrated the value of the new science of pharmacology. He showed that the plant poison curare exerted its action on the motor nerves, thus preventing them from stimulating the muscles, and that carbon monoxide exerted its poisonous effect by displacing oxygen in chemical combination with hemoglobin in the blood.

Magendie and Bernard helped to refine many of the experimental techniques used in pharmacology, and these also helped to define the main concerns of the new science. They did not view pharmacology as a separate discipline, however, but as a branch of experimental physiology. The establishment of pharmacology as an independent discipline required individuals who would devote their full time and efforts to it. These individuals worked in German-speaking universities in the second half of the nineteenth century, and there and then pharmacology emerged as a well-defined discipline.

The first person to campaign actively for pharmacology as an independent discipline was Rudolph Buchheim. In 1847 The German-born Buchheim came to the chair of *materia medica*, dietetics, and history and encyclopedia of medicine at the University of Dorpat (later Tartu), an essentially German institution located in Russian-controlled Estonia. The fact that Buchheim taught *materia medica* together with several subjects makes its low position in the medical curriculum clear. Buchheim argued that pharmacology, the science involving the study of the action of drugs, should be pursued by scientists trained as pharmacologists, and not by chemists or pharmacists. He emphasized the need to isolate the active chemical constituents of crude drugs and to determine their physiological action on healthy organisms. Shortly after his appointment at Dorpat, he established the first laboratory and institute of experimental pharmacology.

Buchheim had little influence beyond Dorpat. One of his students, Oswald Schmiedeberg, who succeeded him, played the crucial role in the institutionalization of pharmacology. In 1872, Schemiedeberg moved to the University of Strassburg and began to make his influence felt. Schmiedeberg created the first journal for the new science and published a classic textbook that went through seven editions in several languages. His institute became a mecca for students of pharmacology from all over the world, the training ground for a whole generation of practitioners of the new science. At the time of his death in 1921, his students held more than forty chairs of pharmacology internationally. One of these students, Arthur Cushny, played a key role in the development of the discipline in Britain, and another, John J. Abel, may be considered the founder of American pharmacology.

When Abel completed his studies in Europe in 1891, he took up a position at the University of Michigan's medical school that amounted to the first professorship of modern pharmacology in the United States (although it retained the traditional title of *materia medica* and therapeutics). He brought with him the German tradition of experimental pharmacology as molded by Buchheim and Schmiedeberg. In 1893 Abel moved to the newly opened medical school of The Johns Hopkins University in 1893. From his position there, which he held for four decades, he shaped American pharmacology. His laboratory became the training ground for the first generation of home-grown American pharmacologists. He also established the first American journal (for which he served as the first editor) and the first national society for the discipline.

During the early years of the twentieth century, courses and faculty appointments in pharmacology increasingly supplanted those in *materia medica* in American medical schools, and later in schools of pharmacy as well. Pharmacologists began to find a home in government and industry laboratories as well. Also during the early part of the twentieth century, the dominant theoretical concept of the science, the receptor theory of drug action, began to emerge. Over the course of the century, pharmacologists developed an increasing understanding of the relationship between the chemical structure of drugs and their physiological actions, and of the nature of the receptors in living cells that bind the drugs. Despite this increased focus on the molecular level, however, whole-animal and tissue studies remain an important part of the science. Recently, clinical pharmacology (the study of drugs in humans) has gained in importance as a result of the many advances in pharmaceutical therapy.

See also MOLECULAR BIOLOGY.

B. Holmsted and G. Liljestrand, eds., *Readings in Pharmacology* (1963). David Cowen, "Materia Medica and Pharmacology," in *The Education of American Physicians: Historical Essays*, ed. Ronald Numbers (1980). M. J. Parnham and J. Bruinvels, eds., *Discoveries in Pharmacology*, 3 vols. (1983–1986). John Lesch, *Science and Medicine in France: The Emergence of Experimental Physiology, 1790–1855* (1984). John Swann, *Academic Scientists and the Pharmaceutical Industry: Cooperative Research in Twentieth-Century America* (1988). John Parascandola, *The Development of American Pharmacology: John J. Abel and the Shaping of a Discipline* (1992).

JOHN PARASCANDOLA

PHENOMENOLOGY. "Phenomena," a term derived from the Greek, are things appearing to the senses. The mathematician-astronomer Johann Heinrich Lambert was one of the first to use the term "phenomenology," by which he meant the part of a science that presented basic phenomena relatively free of theory. G. W. F. Hegel in his *Phenomenology of Spirit* (1805) used it to mean the dialectical sequence that begins with simple sense certainty and ends with the total structure of knowledge. These early uses capture the dual goals of later phenomenology: to secure evidence for science and, from this evidence, to develop comprehensive knowledge of ever more complex structures of experience.

The phenomenological tradition originated from Edmund Husserl's philosophy of mathematics. Seeking to avoid psychologism, the notion that mathematical entities (numbers, for instance) result from the peculiar makeup of the mind, Husserl turned to Franz Brentano's theory of intentionality, according to which every act of consciousness is always a consciousness of something. This allowed Husserl to present mathematical entities as the objectively intended, fixed meanings of well-structured acts of consciousness.

Subsequently Husserl used phenomenology as the name of a philosophical movement. Its task was to study, conceive, and describe the objects of consciousness (phenomena) while suspending or "bracketing" the question of the existence of those objects in the "real world." It was neither a descriptive psychology nor a science of conscious experience, but the conceptually articulated foundation that guaranteed the scientific character of all science.

The phenomenological movement evolved rapidly. After an early phase of realism about the intentional objects his method had uncovered, Husserl began to conceive phenomenology as more transcendental, ideal, and intersubjective in character, with the mind's acts "co-constituting" the ideal objects. Max Scheler, Martin Heidegger, and Jean-Paul Sartre developed existential phenomenology, emphasizing consciousness as it is lived in society and in the world, and experienced in feeling and willing (noncognitively) as well as in thinking (cognitively).

Phenomenology's chief influence on the sciences has been in psychology and sociology. Phenomenological psychology emphasized the human subject's co-constitution of a world of meaning and led to Gestalt psychology, which stressed the priority of psychological patterns and structures over the parts into which they can be analyzed. At the end of the twentieth century some cognitive scientists used phenomenology to enrich their sense of the articulated totality of human consciousness and experience. In return they hoped to naturalize some of phenomenology's basic concepts by explaining them in light of cognitive neurobiology and information processing. Alfred Schutz pursued a phenomenological sociology stressing the importance of common-sense beliefs and practices. Many of his ideas were incorporated into the notion of the "social construction of reality," which played an important role in the history and sociology of science in the 1980s and 1990s.

In its basic intentions, phenomenology is holistic, humanistic, and antireductionist, even if it often employs technical philosophical analysis. Husserl expanded his critique of psychologism in psychology to a critique of historicism and vigorously attacked the positivist account of factuality. Intending to counter "the crisis of European sciences" produced by scientific positivism and objectivism, in his last works he developed a historically informed phenomenology of the emergence of the sciences from the world as ordinarily experienced, the "life-world" as he called it. The German philosopher Heidegger took this critique further with his judgment that modern science and technology have fulfilled the nihilistic destiny of Western philosophy. Much of the work on the philosophy of technology in the late twentieth century derives from this tradition.

Joseph J. Kockelmans and Theodore J. Kisiel, eds., *Phenomenology and the Natural Sciences: Essays and Translations* (1970). Jean Petitot, Francisco J. Varela, Bernard Pachoud, and Jean-Michel Roy, eds., *Naturalizing Phenomenology: Issues in Contemporary Phenomenology and Cognitive Science* (1999).

DENNIS SEPPER

PHILANTHROPY. Rooted in the moral imperative of disinterested aid of others, philanthropy has been associated with the support of worthy causes at least since the rise of market capitalism. In the modern era philanthropy increasingly has been directed at institutions and individuals for the increase of knowledge, distinguishing it in some degree from charity (amelioration of human misery) and patronage (support of talented individuals or groups). Substantial philanthropic support has long been a distinctive feature of modern Western science.

Many examples from the history of early modern science illustrate the point. A Medici prince supported the Accademia del Cimento in Florence (1657–1667); seventeenth-century predecessors of the Académie Royale des Sciences in France depended on patrons; and King Frederick II of Denmark financed Tycho *Brahe's installation at Uraniborg. In the eighteenth century, Joseph Priestley had support from the Earl of Shelburne, Jean André Deluc from the queen of England, and many small continental *academies from local magnates.

The growing institutionalization of science in the early nineteenth century is reflected in the philanthropy of Benjamin Thompson, Count Rumford, who created a research prize, helped found the Royal Institution of Great Britain, and left an endowment for a scientific chair at Harvard University. The philanthropic creation of prizes and premiums established a valuable, if post facto, source of financing. Privately endowed university chairs in the sciences helped to give science a permanent role in higher education.

Perhaps the most dramatic instance of philanthropy in science in the nineteenth century was the Englishman James Smithson's bequest of 1826 to found in the city of Washington an institution devoted to the increase and diffusion of knowledge. Finally organized in 1846, the Smithsonian Institution quickly though briefly became the most significant center of scientific research and publication in the United States. American financiers and industrialists soon began to endow chairs and to create science schools (at Harvard and Yale, for example) and entire universities (Chicago, Clark, Johns Hopkins, Stanford). James Lick, a California entrepreneur, greatly enlarged the scale of philanthropy for science by donating $700,000 to build the world's largest *telescope, the 36-inch Lick refractor, which began operation in 1888.

The cumulative effect of small donations could also be substantial. Ormsby Mitchell raised funds for a telescope and observatory in Cincinnati in 1842–1843 by asking local residents for money. A public subscription to publish Louis Agassiz's natural history researches in 1855 was the first nationwide appeal in support of science in the United States; Agassiz personally raised half a million dollars from many sources for his museum at Harvard in the 1850s and 1860s. In 1885 the Pasteur Institute in Paris was founded on broad support from French citizens. Still, in most cases substantial wealth provided the infrastructure for science: Boston's monied class built Harvard's first scientific building, Boylston Hall (1858), and the executor of the Hull fortune funded the biological laboratory at the University of Chicago (1897).

Buildings, university chairs, and astronomical instruments were traditional expressions of philanthropy usually intended to memorialize the donors. Identifying worthy research and researchers required more of the giver as science developed theories and vocabularies beyond the understanding of most educated persons. To pursue leading-edge science demanded tapping various sources for support: Albert A. Michelson and Edward Morley received a grant from the National Academy of Sciences's Bache Fund for their light-wave studies in the 1880s, but their equipment came from local manufacturers who provided it free or below cost. Some philanthropists understood the need: in the 1890s Catherine Wolfe Bruce gave a cumulative $175,000 to researchers around the country to purchase astronomical research instruments, and the Naples Zoological Station in Italy received funds from Alexander H. Davis, a New York businessman, particularly to provide facilities for American researchers.

Although the organized philanthropy of large foundations became the dominant means of nongovernment support of science early in the twentieth century, private philanthropy remained important. In the United States the Cold Spring Harbor Laboratory was established in 1923 on the basis of private gifts, as was the Institute for Advanced Study (founded 1930), a haven for exiled European scientists. In Europe the *Kaiser-Wilhelm-Gesellschaft was created in 1911 with the support of many German bankers and industrialists; the Weizmann Institute of Science in Israel began with gifts that created the Sieff Research Institute in 1934; the Huancayo Laboratory of the Institute of Andean Biology in Peru was built with private funds in 1940; industrialist J. R. D. Tata endowed the Tata Institute of Fundamental Research at Mumbai in India in 1945. Typically, personal philanthropy created new scientific institutions or supported existing institutions, rather than the scientists themselves, and was increasingly shaped by both organized solicitation and tax laws.

The largest private benefactions set up medical foundations, for example, the Wellcome Trust (1936) and the Howard Hughes Medical Institute (1953). Other philanthropy focused on agricultural development, disease, environmental issues, or population control, often directing their programs toward the developing world. As the twenty-first century dawned, Johns Hopkins University received a gift of one hundred million dollars for research on malaria, a scourge of tropical Africa and south Asia.

Philanthropy on a scale that can influence science has derived largely from wealth accumulated in market-capitalist environments. While it has accounted for only a small proportion of science funding and has been limited in the developing world and virtually absent from nations with socialist systems, philanthropy has dramatically affected the conduct of science in many fields and institutions.

Merle Curti, *American Philanthropy Abroad: A History* (1963). Howard S. Miller, *Dollars for Research: Science and Its Patrons in Nineteenth-Century America* (1970). Lewis Pyenson and Susan Sheets-Pyenson, *Servants of Nature: A History of Scientific Institutions, Enterprises, and Sensibilities* (1999).

DARWIN H. STAPLETON

PHILOSOPHY AND SCIENCE. In seventeenth-century Europe, what we would call science was a part of philosophy. By the end of the twentieth century in Europe and America, science had not only become distinct from philosophy (with the qualifications mentioned below) but had grown beyond anything the seventeenth century could have imagined. Most of the rest of the world had accepted Western science more enthusiastically than Western philosophy. The history of this reversal of the intellectual order has yet to be written, not surprising given that even within the different nations of the West the story varies considerably. In contemporary French and German, for example, "science" and "Wissenschaft" continue to mean well-founded knowledge in general, while in English "science" usually signifies knowledge of the natural world.

The word science comes from the Latin *scientia* (itself a translation of the Greek *episteme*), meaning certain knowledge as opposed to mere opinion or *doxa*. The location of the study of the natural within philosophy in the seventeenth and eighteenth centuries may be indicated by the names *Descartes, Leibniz, Locke, d'Alembert, and Kant. Natural philosophy, the study of nature, encompassed much of what we would call science. Those who studied knew the philosophical traditions and described their work as philosophy—Descartes's *Principles of Philosophy* (1644), for example, and Newton's *Mathematical Principles of Natural Philosophy* (1687)—and these works in turn did much to set the agenda for philosophy in general.

In the nineteenth century science and philosophy began to part ways. The word *"scientist," introduced in 1833 to describe this growing breed of professionals, may be taken as a symbol of the parting. The reasons for it included the increasing success of science not only in astronomy and mechanics but in chemistry, physiology, biology, and geology; the creation of new scientific institutions, such as research institutes, university departments, and surveys; the reform of European university systems and the introduction of systematic education in the sciences; the foundation of specialized scientific societies and publications; the increasing inaccessibility of science to those not trained in its practices; and the coupling of science with technology.

Scientists and philosophers alike thus had to rethink their classifications of knowledge. From roughly the mid-nineteenth to the mid-twentieth century, two main schools can be distinguished. The first was the militantly proscience tradition of positivism. Its founder, Auguste Comte (1798–1857), famously demoted philosophy from the umbrella discipline of science to a prior and more primitive form of knowledge, less superstitious and better founded than religion to be sure, but still not as reliable or as free of strange conceptions as science itself (*see* POSITIVISM). Although by the end of the century many scientists had abandoned philosophical worries, others in this tradition, such as Pierre Duhem and Ernst Mach, continued to see a role for philosophy. In works such as the former's *La théorie physique: son object, sa structure* (1906) and the latter's *Knowledge and Error* (1905), they explored the conceptual foundations of science and the methods that they believed had made it successful. By the 1930s, philosophy of science had become an important part of philosophy at least as conceived by the logical positivists and related philosophers such as Karl Popper. Because they saw in science the promise formerly held by philosophy as the prime exemplar of reliable knowledge, and scientific decision-making as the model of rational action, they believed understanding science to be a necessary preliminary to philosophical studies and political life alike.

The other main school of thought about the relations between science and philosophy sheltered a number of different traditions including Kantianism, neo Kantianism, idealism, Hegelianism, and *phenomenology. Some who fell into this eclectic group were interested in science but rejected the positivist analysis of it. William Whewell (1794–1866), for example, thought that positivists had the relation between science and religion absolutely wrong and that they underestimated the formative role of ideas in scientific change. Edmund Husserl (1859–1938) moved from philosophy of mathematics to his phenomenological program. Gaston Bachelard (1884–1962) offered a reading of the history of science alternative to that of the positivists. By and large, though, antipositivists were skeptical about science's claims to epistemic authority and deeply hostile to its growing power in society.

Following World War II, the institutional split between science and philosophy was almost complete. Science rode high and most scientists went through their entire careers without giving a thought to philosophy, now relegated to

Frontispiece to Francis Bacon, *Novum organum* (1620), suggesting the ongoing search for natural knowledge. The Latin signifies "Many will move about, and knowledge will be increased."

one discipline among many in the humanities. Philosophy was itself deeply divided between the descendants of the positivists, now called analytic philosophers, and the descendants of the antipositivists, now known as the continental philosophers (since positivism had moved from the continent of Europe to Britain and America). Academic philosophy, at least in the United States, preferred analytic philosophy, and thus philosophy of science had high status. Philosophers of science (many if not most of whom were trained in the sciences) also found positions in departments of history and philosophy of science, or of science studies.

Philosophers of science continued to study their twin traditional puzzles, the conceptual foundations of the sciences and scientific methodology. Those interested in foundations maintained close ties with the relevant scientific disciplines, talking to physicists about space and time, to biologists about the nature of species or the gene, and to cognitive scientists and psychologists about mental processes, meeting informally within the university structure or more formally at meetings such as those of the American Association for the Advancement of Science. Those concerned with methodology and demarcation found themselves drawn into public affairs. In the case of *McLean vs. The Arkansas Board of Education* in 1981, for example, both philosophers and scientists spoke against the treatment of "creation science" as science (*see* CREATIONISM). Theories of scientific change proliferated, proposed by philosophers such as Dudley Shapere, Imre Lakatos, Larry Laudan, and Thomas Kuhn, whose *Structure of Scientific Revolutions* (1962) was enormously influential not only in philosophy of science but across all academic disciplines. In the 1960s, with the advent of new medical technologies such as in vitro fertilization, biomedical ethics began to take shape as a field. The Hastings Center (founded in 1969) was the first in a series of special research centers and university positions set up to deal with the ethical issues of scientific practices and applications.

Toward the end of the century, continental philosophy gained ground. Sociologists of science in particular began using the language of practices and construction to talk about science and technology. Their generally relativist conclusions that scientific knowledge owed more to human construction than to the natural world challenged a century-old tradition in philosophy of science and horrified scientists. The resultant grandiosely named Science Wars echoed debates about the nature of science carried on by positivists and antipositivists a hundred years earlier.

RACHEL LAUDAN

PHLOGISTON. The German physician and chemist Georg Ernst Stahl introduced the term phlogiston (from the Greek word for "inflammable") in 1697 in a treatise in which he sought to distinguish combustion from fermentation. Seventeenth-century chemists believed that some combustible substances contain an "inflammable principle" driven off during burning. In the paradigmatic example, the burning of wood, flames and smoke, the visible manifestations of the departure of the inflammable substance, escape, leaving an ash that weighs less than the original wood. The inflammable matter was loosely identified with various conspicuously combustible substances, such as sulfur or oil.

Stahl gave these ideas a sharper focus by raising phlogiston to an elementary principle closely related to fire, and unobtainable in isolation. He supposed that sulfur was composed of vitriolic acid + phlogiston since it could be produced by combining vitriolic acid with a substance, like charcoal or oil, rich in phlogiston. Stahl also recognized that the calcination of metals

resembles combustion; since it takes place much more slowly, it does not give off sensible heat. He concluded that metals are composed of their calxes + phlogiston. In later treatises written or published after 1715, Stahl ascribed such properties as the colors and odors of substances to the phlogiston they contain, and claimed that all substances have some phlogiston. In the German states, chemists of the school that Stahl had founded continued to develop his views. In France, Etienne-François Geoffroy redefined the "sulfur principle" in 1709 as a substance that he had found independently to be transferable from combustible substances to acids and metallic calxes to form sulfur and metals. In 1720, Geoffroy recognized the identity of his sulfur principle with Stahl's phlogiston. By 1750, French chemists had come to prefer Stahl's term phlogiston, and gave him the primary credit for introducing a theory of combustion that some of them, in particular the prominent Pierre-Joseph Macquer, believed to have "changed the face" of chemistry.

In his well-known *Dictionnaire de chymie*, Macquer provided a particularly lucid summary of Stahl's phlogiston theory, showing the many chemical operations that could be treated coherently on the premise that phlogiston transfers from one substance to another. Macquer stressed particularly that what had been established for certain cases could be extended to guide further research. By analogy with sulfur, a nitrous sulfur should be formed when phlogiston unites with nitrous acid. The extreme combustibility of nitrous sulfur had, however, prevented chemists from isolating it. Similarly, the marine sulfur that should result from a combination of marine acid with phlogiston had not yet been found, but Macquer encouraged chemists to search for it.

Eighteenth-century chemists knew that some metals gained weight when calcined. Stahl skirted around the difficulty posed by the fact that if phlogiston had weight, its loss in calcination should have caused calxes to weigh less than the metals from which they derived. Others noticed this discrepancy, however, and in 1763, the secretary of the Paris Academy of Sciences called the augmentation of the weight of metals in calcination one of the "true paradoxes of chemistry." When Guyton de Morveau showed by very careful experiments in 1772 that *all* metals gain weight when calcined, the phlogiston theory had entered a period of crisis. Few others accepted Guyton's explanation that phlogiston has negative weight.

At the same time, however, the English natural philosopher Joseph Priestley attached the general idea of phlogiston to his explanations of the relationships between the new airs he had discovered during the preceding years. He gave new life to the concept, even though his own concept of phlogiston had little in common with Stahl's. When Antoine-Laurent *Lavoisier challenged the phlogiston theory with a new theory that combustion and calcination consisted of the combination of combustible bodies and metals with *oxygen, both the adherents of Stahl's theory and the followers of Priestley vigorously resisted the demise of phlogiston.

Modern commentators often express disdain for phlogiston as the exemplar of "wrong" theories that had to be overthrown to make room for modern science. In its time, however, and in the circumstances for which it was intended, phlogiston served chemists well as a means of organizing otherwise disconnected observations into a coherent body of knowledge, and they used it successfully as a guide to further observations.

Hélène Metzger, *Newton, Stahl, Boerhaave et la doctrine chimique* (1969). Karl Hufbauer, *The Formation of the German Chemical Community (1720–1795)* (1982). Sandra Tugnoli Pattaro, *La Teoria del Flogisto* (1983). Jean Pierre Poirier, *Antoine Lavoisier: Chemist, Biologist, Economist* (1996).

FREDERIC LAWRENCE HOLMES

PHONOGRAPH. In 1807, Thomas *Young described a method of recording the vibrations of a tuning fork on the surface of a cylinder coated with lamp-black, but techniques for recording and eventually reproducing the vibratory pattern of sound began in earnest only in the 1840s. Wilhelm Wertheim's method of 1842 resembled Young's, but Jean-Marie-Constant Duhamel's "vibroscope," described in 1843, was a considerable improvement. Duhamel attached a stylus by wax to one prong of a tuning fork; the recording cylinder could move horizontally by means of a screw cut into its axis, which allowed the trace to be extended in a spiral as the cylinder turned. He calculated the sound's frequency by comparing the number of revolutions of the cylinder in a given time with the wave pattern traced in lamp-black. Leon Scott improved this device in 1857. His "phonautograph" was the first device to record the vibrations of a thin membrane, by means of a recording stylus that once again traced the sound waves on a cylinder covered with paper and a thin deposit of lamp-black. Scot's device was the precursor of Edison's phonograph of the 1870s.

William Henry Barlow improved the recording stylus in his "logograph" of 1874, while Clarence Blake in 1876 employed a similar device in which the vibrations were produced by the drumhead of the human ear instead of by an artificial membrane. Other techniques were also developed to record the vibrations of sound

graphically—for example, the manometric flame apparatus of the Parisian acoustic-instrument maker Karl Rudolf König.

König's device showed the delicate movements of a flame vibrated by a bank of Helmholtz resonators by means of a rotating rectangle of mirrors. There were also the aesthetically pleasing figures produced by Jules Antoine Lissajous by reflecting a spot of light from mirrors attached to the prongs of tuning forks. These devices could record the vibrations of sound graphically, but the sound so recorded could not be reproduced. Thomas Alva Edison's "phonograph" of 1877 achieved this feat. Some months before Edison, the French inventor Charles Cros proposed a similar device to Scot's phonautograph except that it recorded the sound trace onto lamp-black deposited onto a transparent disk. The tracing could then be copied onto metal by photoengraving and the sound reproduced by a stylus attached to a membrane riding in the groove of the rotating metal disk. Cros deposited information about his idea in a sealed envelope at the Paris Academy of Sciences in April before he published his invention in October 1877.

In 1876, Edison took out a patent for a *telegraph repeater that recorded ordinary telegraphic signals by a chisel-shaped stylus, indenting a sheet of paper wrapped around a cylinder or a disk along a guide consisting of a groove cut into the metal. This led him to the idea that if the indentation in the paper could reproduce the click of the telegraph instrument, then perhaps sound vibrations recorded in the same way could also be reproduced. The result, his tinfoil phonograph of late 1877, consisted of a brass cylindrical mouthpiece closed by a membrane of parchment or gold-beater's skin using a steel stylus with a sharp chisel-shaped edge at its center. A sheet of tinfoil carrying a helical groove that acted as a guide to the stylus was wrapped round the cylinder. The cylinder was cranked by a handle, its speed regulated by a flywheel. As the cylinder rotated, the membrane vibrated by the sound caused the stylus to impress a series of indentations into the tinfoil. To reproduce these sounds, the stylus traveled over the same indentations in the tinfoil, which caused the membrane to vibrate. A funnel-shaped attachment placed over the membrane amplified the sounds.

Edison's patent of 1878 mentioned the use of a revolving disk rather than the cylinder commonly associated with his device, a plaster-of-Paris process by which copies of an original recording could be made, and other improvements. In the same year, he formed the Edison Speaking Phonograph Company. He hired Jules Levy, a well-known cornetist, to record "Yankee Doodle" and other tunes, and forecast that in time the phonograph would be used in the home for listening to music and novels. At first, however, Edison regarded the phonograph principally as a business aid, for dictating letters and other documents.

Between 1877 and 1888, Edison worked to develop the wax cylinder to replace the tinfoil. In 1885, Chichester Bell (a cousin of Alexander Graham Bell) and Charles Summer Tainter patented the "graphophone," which used wax-coated cardboard cylinders, and in 1887, Emile Berliner, a German inventor living in America, patented the disk gramophone. The original instrument played nursery rhymes and patriotic songs. The sound box had a needle that fitted into the grooves of the disk and a horn amplifier of papier-mâché. Initially, Berliner followed Cros's method of making tracings in lamp-black of which he then made a metal copy by photoengraving, but in 1888 he developed a simpler method. The recording stylus cut through a thin layer of wax on a zinc disk, the exposed metal was etched with acid, and the etched plate became the master from which a number of records could be pressed. In time, the Bell and Berliner patents were merged into the Columbia Gramophone Company, and Edison's interests became the Victor Talking Machine Company. The phonograph, which had started life as a scientific curiosity, had become a key component of the entertainment industry by the early twentieth century.

Oliver Read and Walter L. Welch, *From Tin Foil to Stereo: Evolution of the Phonograph* (1959). Rolan Gelatt, *The Fabulous Phonograph: The Story of the Gramophone from Tin Foil to High Fidelity* (1966; 1956). V. K. Chew, *Talking Machines 1877–1914: Some Aspects of the Early History of the Gramophone* (1967). Paul Israel, *Edison: A Life of Invention* (1998).

W. D. HACKMANN

PHOTOGRAPHY. The technical history of photography comprises three parallel developments: camera negatives, monochrome positive prints, and recording color. The practice of photography was empirically led throughout its first hundred years, frequently outstripping the ability of contemporary science to explain the various phenomena.

Since the late *Renaissance, it had been known that sunlight darkens salts of silver. In 1725, Johann Heinrich Schulze used sunlit stencils to cast images on suspensions of silver salts. Carl Scheele showed in 1777 that the violet rays of the prismatic spectrum were most effective in decomposing silver chloride, the dark product being finely divided silver. During the 1790s,

Photogenic drawing (contact print) from a photographic image (1839). The subject is the supporting frame of William Herschel's telescope during its dismantling; both drawing and image were made by Herschel's scientist son John.

Thomas Wedgwood sun-printed "profiles" of objects onto paper and leather moistened with silver nitrate, but could not prevent their obliteration by daylight. Images in the *camera obscura* were too faint to make any impression, but Wedgwood successfully obtained "copies" of specimens projected by the solar microscope.

The earliest extant camera photograph was not produced in silver, but by *heliography*, a copying process invented by Joseph Nicéphore Niépce in the 1820s. The process places a thin coating of bitumen on a pewter plate, selectively hardens it by sunlight, then dissolves it by oil of lavender to bring out the image. In 1827, Niépce captured the first photograph—now in the Gernsheim Collection of the University of Texas—by a heliographic camera exposure estimated to have taken several days. Heliography was better suited to providing etching-resists for photomechanical printing plates. Attention returned to silver; in 1837, Louis Jacques Mandé Daguerre discovered, fortuitously, that mercury vapor could develop camera images on iodized surfaces of silver-plated copper, the chemical prerequisite, iodine, having been discovered by Bernard Courtois in 1811. The *daguerreotype* process, first publicized in 1839, enjoyed widespread commercial success until photography on paper and glass replaced it in the mid-1850s.

Meanwhile, William Henry Fox Talbot had independently devised *photogenic drawing paper* by 1835. He had noted that the light-sensitivity of silver chloride, precipitated within the paper's fibers, diminished with excess salt, and had thereby discovered the first method for fixing silver images. Talbot's earliest camera negative (1835) is in the National Museum of Photography, Film and Television in Bradford, England. On hearing in 1839 of Talbot's innovation, Sir John Herschel, who originated the terms "photography," "the negative," and "the positive," demonstrated that unchanged silver chloride in a photograph dissolved in a solution of hyposulfite of soda (sodium thiosulfate). Herschel's "hypo-fixing" superseded Talbot's salt-fixing, and remains in use today. In 1839, Talbot also noted the greater sensitivity of silver bromide—now the chief constituent of all modern photographic materials—made possible by Antoine Jerome Balard's isolation of bromine in 1826.

In 1840, Talbot made his third, crucial discovery: that an invisibly weak dormant picture in silver iodide could be brought out by gallic acid, effectively increasing the speed of his camera photography a hundredfold—from hours to minutes. He named this process *calotype*, and patented it in 1841. Although science could not account for these phenomena of latency and development, Talbot had set photography on the path of continuous refinement for the next 150 years. A quest was mounted for shorter camera exposures and higher resolution. The opacity and texture of paper were avoided by suspending the silver halide in organic binders, making "emulsions" (an inaccurate term, but universally employed): hens' egg-white (Claude Félix Abel Niépce de Saint-Victor, 1847), collodion (Frederick Scott Archer, 1851), and finally gelatin (Richard Leach Maddox, 1871). Emulsions were coated on transparent supports ranging from glass plates (Niépce de Saint-Victor, 1847) and waxed paper (Gustave Le Gray, 1851) to the flexible, but dangerously flammable early plastic, celluloid (cellulose nitrate), which permitted the design of roll-film cameras, introduced by George Eastman in 1888. Modern safety films employ polymer bases of cellulose triacetate (1923) or polyethylene terephthalate (1955). Parallel improvements were made in the optical design of camera lenses, notably achromats (Charles and Vincent Chevalier, 1828), large aperture lenses (Josef Petzval, 1841), rapid rectilinear lenses (J. H. Dallmeyer and H. A. Steinheil, 1866), and Zeiss anastigmats (Paul Rudolph and Ernst Abbe, 1890).

Pure silver halides respond only to blue light and the ultraviolet, whose discovery by Johann Wilhelm Ritter in 1801 represents the first contribution of photography to science. To render tonally balanced negatives, emulsions must react to the entire visible spectrum. Sensitizing with dyes, introduced in 1873 by Hermann Wilhelm Vogel, extended the response to green (orthochromatic plates, 1884), and then red wavelengths (panchromatic plates, 1904), reaching the near-infrared by the 1930s. Sensitometry, the photometric study of emulsion response, was originated by Ferdinand Hurter and Vero Charles Driffield in 1890, accompanied by extensive chemical exploration for better developers, such as hydroquinone (William de Wiveleslie Abney, 1880).

The emulsion binder in universal use by 1900, animal gelatin, displayed great variability in speed. In 1926, Samuel Edward Sheppard detected the cause: traces of sulfur-containing substances, arising from the animals' diet, could sensitize silver halides. Modern emulsion technology now uses pure gelatin, with controlled addition of sulfur and gold compounds as sensitizers. Understanding the latent image became possible with the foundation of solid-state physics and *chemistry during the 1920s, especially Yakov Ilyich Frenkel's theory of ionic conductivity and A. H. Wilson's electronic band theory. In 1938, Sir Nevill Francis Mott and R. W. Gurney put forward a mechanism for the formation of the latent image that has found wide acceptance. Because of the granular structure of the developed image, photographic speed is linked inversely to resolution. This trade-off was improved in the 1980s by controlled growth of silver halide crystals with a tabular habit, increasing their surface-to-volume ratio. The speed of modern negative emulsions is approaching the theoretical limit.

As Talbot realized in 1835, printing positives is an essential procedure to rectify the reversed tonality and handedness of camera negatives. Talbot's photogenic drawing paper served at first, but papers coated with albumen emulsion (Louis Désiré Blanquart-Evrard, 1850) displaced Talbot's *salted paper prints*. Silver images suffer from sulfiding, causing them to fade, but gold-toning (1855) mitigated the deterioration, and *albumen prints* remained the chief photographic medium until 1895.

Since speed is not paramount for printing positives, substances less sensitive to light than silver salts were employed in the quest for image permanence. Following the discovery of dichromates (Louis Nicolas Vauquelin, 1798), their light-sensitivity on paper (Mungo Ponton, 1838) led to light-hardening of dichromated gelatin (Talbot, *photoglyphic engraving*, 1852). The addition of artists' pigments to the gelatin matrix, as inert image substances, permitted the development of the *carbon process* (Alphonse Louis Poitevin, 1855; Adolphe Fargier, 1861; Sir Joseph Wilson Swan, 1864).

In 1842, Sir John Herschel discovered that iron (III) citrate was light-sensitive and could yield images in gold, silver, mercury, or Prussian blue—the *cyanotype* or blueprint, the first reprographic process. William Willis patented the analogous platinum process in 1873, and by 1900 his company's *platinotype* paper dominated the market. But demand for platinum as a catalyst in the growing chemical industry brought steep price rises; moreover, the introduction of roll-film cameras, whose small formats required enlargement onto the much faster silver-gelatin development papers, made platinotype commercially unviable by the 1930s. It remains today, together with carbon printing, a minority fine-art practice, yielding images of archival permanence. For the remainder of the twentieth century, silver-gelatin enlarging papers became the commercial norm.

Photography in natural colors was first achieved by James Clerk *Maxwell (1831–1879) in 1861, using optical filters to separate three primary colors and to synthesize them additively. Ducos du Hauron published details of a three-color subtractive process, *heliochrome*, in 1869. Gabriel Lippmann, who received the only *Nobel Prize (physics, 1908) awarded for photographic innovation, devised a unique interference system for recording color photographs in 1891. More successful commercially was the additive *autochrome* process of the brothers Auguste and Louis Lumière (1907), which used a mixture of starch grains dyed with three primary colors to filter the light falling on a panchromatic emulsion. Color photography progressed with research in synthetic organic chemistry of *dyestuffs, notably at the Eastman Kodak Company, which produced Kodachrome in 1935: a triple-layered silver emulsion, sandwiched with subtractive primary dyes.

The scientific value of photography for faithful analogue recording is self-evident; further, by disclosing information imperceptible to the eye, photography permitted new discoveries. Early examples include the "chronophotographic" studies of animal and human locomotion (Eadweard Muybridge and Etienne-Jules Marey, 1870s); recording of shock waves (Ernst Mach, 1884); high-speed photography by stroboscopic flash (Harold Eugene Edgerton, 1932); and time-lapse photography (John Ott, 1940s).

The ability of photography to accumulate a weak optical signal over time makes its enhanced recording sensitivity particularly valuable to *astronomy and optical *spectroscopy. By photographically recording galactic spectra, Edwin Powell *Hubble (1889–1953) discovered in 1929 the "red shift," which implied that the universe was expanding. The response of photographic emulsions to *X rays has found applications since 1895 in medicine and forensic science. The x-ray diffraction patterns for elucidating crystal structure, and electron diffraction patterns from gaseous molecules, were first recorded photographically. Antoine Henri Becquerel owed his discovery of *radioactivity in 1896 to the sensitivity of emulsions to charged particles; the ensuing technique of *autoradiography* finds application in biological studies and metallographic analysis. Emulsions can register the trajectories of *cosmic rays, and particle physics employs photography to record events in its cloud and bubble chambers.

With the advance of modern electronics, photoelectric devices are replacing silver-gelatin emulsions. Digitally-processed electronic imaging has reduced the role of photography in science and the lens-based media. Yet the photograph will endure in the art and archives of humanity.

Josef M. Eder, *History of Photography*, trans. Edward Epstean (1945). Helmut and Alison Gernsheim, *The History of Photography* (1969). T. H. James, ed., *The Theory of the Photographic Process* (1977). William Crawford, *The Keepers of Light: A History & Working Guide to Early Photographic Processes* (1979). Eugene Ostroff, ed., *Pioneers of Photography: Their Achievements in Science and Technology* (1987). Larry J. Schaaf, *Out of the Shadows: Herschel, Talbot, & the Invention of Photography* (1992). Anne Thomas, ed., *Beauty of Another Order: Photography in Science* (1997). Sidney F. Ray, *Scientific Photography and Applied Imaging* (1999).

MIKE WARE

PHOTOSYNTHESIS, the metabolic reactions that provide plants with a source of food, is usually referred to as fixing energy (light) from the sun in the form of carbohydrates. The term dates from 1893, when the major steps in the process were identified. Before then, by analogy with animals, most philosophers regarded plants as breathing organisms deriving their nutrients from water and the air alone. During the seventeenth and eighteenth centuries plant physiology became a branch of *pneumatics. Stephen Hales, who first recorded the exchange of gases, believed that plants "perspired" during the day and "imbibed" at night (*see* RESPIRATION AND TRANSPIRATION). These ideas encouraged the suggestion that plants transformed stale air into fresh or sweet air and led Georg Ernst Stahl and then Joseph Priestley and others to identify the gases involved. Priestley provided evidence that green plants produced dephlogisticated air, the air that Antoine-Laurent *Lavoisier redefined as *oxygen (1777). Jan Ingen-Housz saw this process as a necessary counterpart to the respiration of animals. Air spoiled by the breathing of animals served as a kind of food for plants, and plants in turn supplied animals with purified air. These reciprocal actions constituted a cycle, demonstrating the economy of nature. Ingen-Housz regarded the cycle, which he described accurately, as convincing evidence of the handiwork of the creator. From 1772, Lavoisier developed the theory that combustion arose from or in the fixation or combination of *oxygen in the air with any burning substance. Light and heat appeared as a result. Lavoisier later proposed that the oxidation of the carbon in food produced fixed air (carbon dioxide). The presumed goodness and efficiency of cyclical exchanges of gases, which appealed to religious sentiment as well as to reason, provided a strong basis for future research. In essence, the eighteenth century had demonstrated that plants remove carbon from atmospheric carbon dioxide and release oxygen back into the atmosphere.

With the shift in interest toward cellular activities in the nineteenth century, research began to center on the mechanisms by which plants captured carbon from the air, although only toward the second half of the century was the extent to which plants synthesize the greater part of their substance fully appreciated. Investigators soon recognized the role of light and the significance of the green pigment (chlorophyll). Julius von Sachs established the sites for absorption of gases in leaves and the biochemical pathways leading to the formation of starch and other substances. He demonstrated that the starch usually present in green cells came from the carbon dioxide absorbed and decided that starch was the primary substance produced by photosynthesis. Following up Sachs's theory, researchers discovered that starch makes up only a by-product of the assimilation process.

Subsequent investigation showed that carbon dioxide must combine with chlorophyll before chemical reduction takes place in a number of steps involving oxygen, water, and light. In 1919 Otto *Warburg began work to determine the minimum number of photons required to yield a molecule of oxygen. He found it to be close to four. His initiative opened a period of research into the details of primary light conversion using Chlorella, a unicellular green algae, as the experimental organism of choice. The prevailing view that a plant's output of oxygen derived from carbon dioxide was challenged in the 1930s by Rene Wurmser, and then by Cornelius Bernardus van Niel, at Stanford, who claimed that it must come from water. In 1937 Robin Hill in Cambridge confirmed the water origin and pointed to chloroplasts as the active sites. During the 1940s Melvin Calvin and his group at Berkeley worked out dark reaction sequences thanks in part to the availability of radioactive Carbon 14 and improvements in chromatography. Calvin received the Nobel Prize in chemistry in 1961 for unraveling the path of carbon in photosynthesis.

Another member of Calvin's group, Daniel Arnon, established definitively that isolated chloroplasts could perform all the responses associated with photosynthesis.

In photosynthesis, it is now understood that water acts both as a hydrogen donor and a source of released oxygen. Only part of the process depends on illumination. An important development was the recognition that plants do not create the energy stored in their sugars, but merely fix energy radiated by the sun. Andreas Franz Wilhelm Schimper's discovery that starch stores energy allowed scientists of the early twentieth century to discuss the circulation of energy through the natural world in biochemical terms, as the synthesis and subsequent breakdown of carbohydrates and other vegetable products. From the 1960s, understanding of photochemical action has been enriched by the determination of the detailed molecular processes involved.

Howard S. Reed, "Jan Ingenhousz, Plant Physiologist, with a History of the Discovery of Photosynthesis," *Chronica Botanica* 11, no. 5–6 (1949): 285–396. Robin Hill, "The Growth of Our Knowledge of Photosynthesis," *The Chemistry of Life*, ed. Joseph Needham (1970). Doris T. Zallen, "Redrawing the Boundaries of Molecular Biology: The Case of Photosynthesis," *Journal of the History of Biology* 26 (1993): 65–87. Doris T. Zallen, "The 'Light' Organism for the Job: Green Algae and Photosynthesis Research," *Journal of the History of Biology* 26 (1993): 269–279.

JANET BROWNE

PHRENOLOGY AND PHYSIOGNOMY are two methods of studying cranial anatomy as indicative of intelligence, personality, and temperament. Phrenology arose at the end of the eighteenth century, when physiognomy was in decline. Franz Joseph Gall, founder of phrenology, might have been influenced by the physiognomical tradition, but his methodology was based primarily on his own research.

Physiognomy was endorsed by the *Physiognomica*, which circulated in ancient and medieval times under the authorship (now doubted) of Aristotle. Galen connected Aristotelian physiognomy with the humoral theory of health and illness. Prominent Medieval authors developed the subject, which, however, had played out by the eighteenth century. The term "physiognomy" was in use as late as 1834 but began to be supplanted by "phrenology" in 1819.

Gall received his M.D. in Vienna in 1785 and developed a lucrative medical practice. By 1791 he had published his ideas on brain functions, which he concluded were highly localized. Personality traits and abilities supposedly depended on the size of different parts of the brain. He initiated

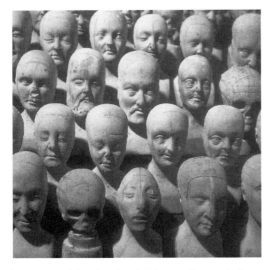

A case containing heads of different sizes and shapes for the teaching and study of phrenology (1831).

studies on the correlation between structures and functions. However, his evidence was anecdotal, and controversy over his claims drove him to seek evidence in dissections and demonstrations. In 1800 he attracted a disciple, Johann Christoph Spurzheim, who assisted in anatomical researches until they parted company in 1813. During their collaboration they published the first two volumes of Gall's *Anatomie et physiologies du systéme nerveux* (5 vols., 1810–1819). These elaborate studies were intended to illustrate rather than test Gall's system, but they did not convince academic and medical authorities in Paris any more than his earlier research and teachings had persuaded the Viennese. A controversy arose over the grounds for this rejection. Did it have to do with substance or divisions of science? Before Gall began his teachings, brain functions had been explained mainly within philosophy rather than physiology.

Spurzheim traveled to London and in 1815 published *The Physiognomical System of Drs. Gall and Spurzheim*. Gall identified twenty-seven faculties of the mind, and Spurzheim thirty-five. Spurzheim won converts in both London and Edinburgh (where he lectured in 1816), but like Gall failed to win over scientific or medical authorities. Spurzheim died during a successful lecture tour in America in 1832. Gall had died in Paris in 1828.

Although Gall and Spurzheim eventually disagreed over details of investigation and interpretation, in general phrenologists used as a reference an anatomical chart of the brain with faculties of the mind labeled on its anatomical parts. The investigator needed to determine the relative size of the different faculties on his subject's brain as visible from the skull. If the subject cooperated, the investigator made careful measurements with calipers and then noted that a given faculty was "very small," "moderate," "full," or "very large." If a subject was uncooperative, an experienced phrenologist could still make a visual analysis reasonably accurately. An expert could do the same with a good portrait or bust.

Spurzheim's most important converts were George Combe, a successful lawyer in Edinburgh, and his younger brother Andrew, the most illustrious British physician to defend phrenology. They organized the Edinburgh Phrenological Society (1820) and founded *The Phrenological Journal* (1823). In 1819 George Combe published his own *Essays on Phrenology*, which he steadily revised and enlarged and eventually retitled as *A System of Phrenology* (5th ed., 1843). He outdid Spurzheim by raising the faculties of the mind to forty-five. Andrew Combe advanced phrenology in a popular textbook, *Principles of Physiology Applied to the Preservation of Health* (1834), which had fifteen editions before his death in 1847. George also wrote a best-seller, *The Constitution of Man* (1828).

George Combe acquired a disciple, Hewett Cottrell Watson, an Englishman who went to Edinburgh in 1828 to study medicine and also phrenology, which was not taught at Edinburgh University or its medical school. Watson did well in his courses but quit medical school a semester before earning a degree having decided that he would never practice medicine. Combe respected Watson's abilities, and Watson began publishing in *The Phrenological Journal* in 1829. After leaving Edinburgh, Watson ceased writing scientific articles on phrenology, but remained a staunch advocate. In 1836 he published *Statistics of Phrenology*, which was not a statistical analysis of phrenological data, but an account of progress made in advancing phrenology. In 1837 he became editor of *The Phrenological Journal*, hoping to raise its scientific standards. He wrote lofty editorials, but his criticism and ridicule of articles submitted only aroused the anger of authors, and in November 1839 he resigned. Watson abandoned phrenology and criticized it in correspondence with Combe until Combe died in 1858. The last serious attempt to defend phrenology was W. Mathieu Williams's *A Vindication of Phrenology* (1894).

Elizabeth C. Evans, "Physiognomics in the Ancient World," *American Philosophical Society Transactions* 59:5 (1969): 1–101. Robert M. Young, *Mind, Brain and Adaptation in the Nineteenth Century: Cerebral Localization and Its Biological Context from Gall to Ferrier* (1970). Madeleine B. Stern, *Heads and Headlines: The*

Phrenological Fowlers (1971). David A. De Guistino, *Conquest of the Mind: Phrenology and Victorian Social Thought* (1975). Roger Cooter, *The Cultural Meaning of Popular Science: Phrenology and the Organization of Consent in Nineteenth-Century Britain* (1984). Frank N. Egerton, "In Quest of a Science: Hewett Watson and Early Victorian Phrenology," *Essays in Arts and Sciences* 24 (1995): 1–20.

FRANK N. EGERTON

PHYSICAL CHEMISTRY. The recognition of physical chemistry as a subfield within the broader field of *chemistry is usually associated with the founding in 1887 of the journal *Zeitschrift für physikalische Chemie* by Svante Arrhenius in Stockholm, Wilhelm Ostwald in Leipzig, and Jacobus Henricus van't Hoff in Amsterdam. In his account of the history of the discipline, Ostwald rooted its foundations in electrochemistry. Somewhat differently, Walther Nernst, in a speech at his newly built physical chemistry institute in Göttingen in 1896, described van't Hoff's work on dissociation in solutions as the path that reunited the sciences of chemistry and physics in a return to Antoine-Laurent *Lavoisier's late eighteenth-century vision of a unified science.

The phrase "physical chemistry" was much in use before the 1880s, for example, by Robert Bunsen, Hans Heinrich Landolt, and Heinrich Rose, and in the title of the *Annalen der Physik und der Physikalischem Chemie* (1819–1823), which became Johann Poggendorff's *Annalen der Physik und Chemie* in 1824. Hermann Kopp received the first independently named university chair in "physical chemistry" at Heidelberg in 1863. Ostwald's predecessor Gustav Heinrich Wiedemann set up the first German instructional laboratory for physical chemistry at Leipzig in 1871.

Increased governmental concern in the 1860s to promote scientific and technological progress made expanding resources for scientific education available first in Germany and then elsewhere in Europe and in the United States. Physical chemistry became a supplement to the curriculum in *organic, agricultural, and medical chemistry that had been the heart of chemical education in universities since the 1830s.

The areas of study that defined physical chemistry at the end of the nineteenth century were *thermodynamics, electrochemistry (*see* ELECTROLYSIS), colloid and surface chemistry, and chemical *spectroscopy. They were enriched during the early decades of the twentieth century by chemical kinetics, electron and quantum chemistry, x-ray crystallography, and new kinds of *spectroscopy.

In the 1850s and 1860s Marcellin Berthelot in Paris and Julius Thomsen in Copenhagen developed theories of chemical affinity based on the assumption that the chemical force (affinity) that holds the parts of a molecule together is directly proportional to the heat evolved during its formation. By the mid-1880s Pierre Duhem had rejected the Berthelot-Thomsen thermochemical approach. Duhem and others adopted the thermodynamical theories of Rudolf Clausius (1854, 1865), J. Willard *Gibbs (1875–1877), and Hermann von *Helmholtz (1882) for chemical purposes.

By the late 1880s a stable chemical system could be defined mathematically as a state of equilibrium in which the *entropy (S) of the system had reached a maximum value and the internal energy (U) a minimum. Entropy measures energy unavailable for work and "free energy" (F) the energy potentially available for work, given values for the mechanical equivalent of heat (J) and temperature (T). Thus, $F = U - JTS$; and chemical affinity is identified with free energy or with work, not with heat.

These results, based on precise experimental work, were mathematical and abstract in character. "Energeticists" like Ostwald and Duhem, who opposed the use of visualizable models in science, lauded this new physical chemistry, but mechanists like Ludwig *Boltzmann in Austria and Marcel Louis Brillouin and Jean Baptiste Perrin in France preferred descriptions linked to particles in motion and the kinetic theory of gases (*see* ENERGETICS).

Drawing on the work of Helmholtz and others, van't Hoff demonstrated in his *Etudes de dynamique chimique* (1884) that the work of chemical affinity could be calculated from vapor pressures, osmotic pressure, or electrical work in a reversible galvanic cell. His starting point in reaction velocities made a contribution to chemical kinetics. The everyday use of thermodynamics in chemical work was greatly simplified by Gilbert N. Lewis and Merle Randall's *Thermodynamics and the Free Energy of Chemical Substances* (1924), which embraced Gibbs's expression for free energy at constant pressure rather than Helmholtz's and van't Hoff's expression for constant volume.

Arrhenius took up van't Hoff's application of the ideal gas law to osmotic pressure and dilute solutions in his doctoral dissertation on electrolytic dissociation (1887). Arrhenius made the surprising argument that at infinite dilution, molecules of electrolytes break up into charged ions even in the absence of electrical current. Arrhenius's notion of independent ions built on Helmholtz's suggestion of 1881 that ions produced in electrolysis carry discrete "atoms of electricity" that might be units of chemical affinity attached to elementary substances. The physicalist electrolytic chemistry of Arrhenius, van't Hoff, and Ostwald became known as "Ionist" chemistry.

Nernst served briefly as Ostwald's assistant in Leipzig. Nernst's practical interest in electrical lamps led to his work of the dissociation of gases at high temperatures in addition to his studies of solutions, solids, and surface chemistry. Irving Langmuir, who took his doctoral degree at Göttingen with Nernst and spent most of his career at the General Electric Company's research laboratory in Schenectady, New York, investigated a similar range of subjects, which preoccupied physical chemists in the early twentieth century.

In late 1905 Nernst stated a third law of thermodynamics, which provided the means for calculating values of specific heat and for predicting the likelihood of chemical reactions from recognition that the values of entropy and specific heats approach zero at very low temperatures. Albert *Einstein's quantum theory of solids, published in 1907, provided theoretical foundations for Nernst's prediction. The utility of the third law persuaded many physical chemists of the usefulness of the statistical mechanics of Gibbs and Boltzmann, as well as of the *quantum theory, in chemical studies.

Optical methods have been consistently fruitful in physical chemistry. Polarimetry resulted from the discovery of Jean-Baptiste Biot and François Arago in 1840 that a monochromatic beam of plane-polarized light is twisted or rotated by solutions of sugar, camphor, or tartaric acid. The new method of study led to the recognition of asymmetries in molecular structure.

Spectroscopy in the chemical context developed from methods used by Robert Bunsen and Gustav Kirchhoff to identify chemical substances by the colors they emit in flames (see BUNSEN BURNER). In 1860 Bunsen and Kirchhoff announced the discovery of cesium, identified by its blue spectral lines; rubidinium followed in 1861, identified by its dark red lines.

In the late nineteenth century, several scientists, including George Johnstone Stoney, Arthur Schuster, and Johann Jakob Balmer, sought quantitative relations among the frequencies of the spectral lines emitted by an element. The first important success, Balmer's formula of 1885, gave the wavelengths of four lines in the visible spectrum of hydrogen in terms of integral whole numbers and a constant. In 1890 Johannes Robert Rydberg revised the Balmer formula by introducing the concept of wave number (reciprocal of the wavelength) in an equation with a constant (Rydberg constant) common to all series and elements.

Max *Planck's early quantum theory resulted in a new understanding of the meaning of spectra. Heinrich Kayser incorporated Planck's radiation formula into the second volume of his *Handbook of Spectroscopy* in 1902. Niels Bjerrum interpreted spectra in terms of quantum theory in 1911, and Niels *Bohr combined spectroscopic data with the quantum hypothesis in his revolutionary paper of 1913 on the hydrogen atom's electron energy.

Gerhard Herzberg's *Atomic Spectra and Atomic Structure* (1936) became one of the fundamental reference books for chemical applications of spectroscopy. By the 1950s laser spectroscopy, nuclear magnetic spectroscopy, and other new techniques came to supplement visible, infrared, and Raman spectroscopy, the latter a technique discovered by C. V. *Raman and K. S. Krishnan in 1928 for studying molecular structure based on the scattering of a beam of monochromatic light.

*Mass spectroscopy, developed by Francis Aston and other physicists beginning in the 1920s, enabled chemists to sort out isotopic and molecular weights. The various spectroscopies not only identified elements and compounds, but also indicated submolecular structures, steric alignments, and bond angles.

By the 1920s x-ray diffraction was employed to study the organization of atoms and submolecular units within molecules. Supplemented by electron diffraction after 1930, it helped to map the structure of metallic alloys, textile fibers, rubber, and large biologically significant molecules like *proteins, nucleic acids, and vitamins. These methods had enormous industrial and commercial application as well as importance in fundamental research.

Paul A. M. *Dirac, Max Born, and other physicists liked to say that quantum mechanics reduced chemistry to physics. Most chemists disagree. Physicists and chemists equally contributed to the development of quantum chemistry. Among the physicists, Walter Heitler, Fritz London, Friedrich Hund, and John Slater made notable contributions. But not until 1931, when Slater and Linus *Pauling developed methods to explain directed chemical valence, did a truly chemical quantum mechanics exist. Their atomic orbital method was largely supplanted in chemical practice after the 1950s by the molecular orbital method developed simultaneously through the work of Hund, Erich Hückel, and, notably, Robert Mulliken.

Chemical kinetics, like quantum chemistry, became an increasingly sophisticated mathematical subject after World War I. Interest in chemical kinetics revived in the 1920s over the question whether activation energy for monomolecular reactions came from radiation or molecular collisions. Farrington Daniels, Hugh S. Taylor, and Cyril Hinshelwood developed evidence and theories for the collision hypothesis in the late 1920s, which replaced the radiation theory defended by Jean Perrin and others at the beginning of the decade. In the 1930s Michael Polányi and Henry Eyring formulated a generalized transition-state theory for elementary chemical processes.

In the study of chemical kinetics in the late twentieth century, physical chemists employed not only the bulk experiments traditional to their field, but also molecular beams in which *lasers excite reactant molecules into desired vibrational and rotational states and chemists identify the states of the products by their fluorescence. Dudley Herschbach and Yuan Tseh Lee pioneered this work.

In the development of physical chemistry, both physicists and chemists played important roles, with their work identified as chemical physics, theoretical chemistry, or theoretical physics as well as physical chemistry. The foundation in 1933 of the *Journal of Chemical Physics* marked a breakthrough in the United States for the application of sophisticated mathematical theories, including quantum mechanics, in chemistry. The journal's first volume included papers from Eyring, Herzberg, Mulliken, Pauling, Slater, and Taylor. What has most distinguished late twentieth-century physical chemistry from its late eighteenth-century counterparts have been innovations in instrumental practice and sophistication in mathematical theory.

See also ION; QUANTUM CHEMISTRY; THERMODYNAMICS.

John W. Servos, *Physical Chemistry from Ostwald to Pauling: The Making of a Science in America* (1990). William H. Brock, *The Norton History of Chemistry* (1992). David Cahan, ed., *Hermann von Helmholtz and the Foundations of Nineteenth-Century Science* (1993). Keith J. Laidler, *The World of Physical Chemistry* (1993). Mary Jo Nye, *From Chemical Philosophy to Theoretical Chemistry: Dynamics of Matter and Dynamics of Disciplines 1800–1950* (1993). Diana Kormos Barkan, *Walther Nernst and the Transition to Modern Physical Science* (1998).

MARY JO NYE

PHYSICALISM. See NATURALISM AND PHYSICALISM.

PHYSICS. The development of physics provides a particularly strong example of the evolution of *natural knowledge into modern physical science. Its traditional meaning of systematized, bookish knowledge about the entire physical world persisted until the eighteenth century; as late as 1798 the Académie Royale des Sciences of Paris could announce, as the subject of a

prize competition in physics, "the nature, form, and uses of the liver in the various classes of animals." To be sure, the assumptions behind the question differed in an essential respect from the implications that would have been drawn a century or so earlier, when "physiologia" was often used as a substitute for "physica." This essential difference was the expectation that competitors would base their answers on experiments, at least some of which they would perform themselves.

Around 1800 "physics" in its old, inclusive sense, modified to imply experiment and some measure of research, lost out to a new division of science bearing the same name. This decisive historical fact is often veiled by the use of "physics" to refer to natural knowledge through the ages. The new physics of 1800 restricted itself to the inorganic world and, within it, to subjects open to investigation by such *instruments as the *air pump, *electrical machine, *balance, *thermometer, and calorimeter. In place of the organic world, physics took on subjects previously the property of *mathematics, especially *mechanics and *optics, and strove to quantify the fields that had distinguished physics during the eighteenth century: *electricity, *magnetism, heat, and *pneumatics, which had garnered attention as particularly suitable for catchy demonstration experiments to large and varied audiences.

The leaders in this reformulation are often grouped together as the school of Pierre-Simon de *Laplace. They applied the mathematical approach developed for the theory of gravity to *imponderable fluids supposed responsible for the phenomena of heat, light, electricity, magnetism, *fire, and flame. The scheme of imponderables was unified by this common approach; but each fluid functioned independently and irreducibly in its theory. Neither the fluids nor the forces they were supposed to carry linked to one another in any fundamental or ontological sense. Rather, the Laplacian school and its fellow travelers deployed the fluids and forces in an *instrumentalist way. It thus promoted the dismemberment of science and the dropping of scruples against instrumentalist mathematical descriptions.

The physics of 1800 differed in scope from later physics by including *meteorology and parts of subjects shared with *chemistry, like atomism, pneumatics, and *thermodynamics. It fell far short of twentieth-century physics in eschewing models of the microworld apart from very general assumptions about the molecular constitution of bodies. A major exception to this generalization was the attachment of *atomic weights to chemical atoms. The French-dominated physics of 1800 also differed from physics in 1900 or even in 1850 by having a vigorous opponent, *Naturphilosophie, which depreciated mathematics and dismemberment and tried to reintegrate physics on general philosophical principles.

The basis of integration proved to be mechanics, which the Naturphilosophen disliked for its mathematics, strictness, and sobriety. This integration, which characterizes the so-called classical physics of the later nineteenth century, began around 1800 with the renewal of a quantitative wave theory of light. By 1840 most physicists (to use a word then just invented) associated light with the vibrations of a world-filling *ether conceived as a medium obeying the laws of mechanics. That substituted a mechanical system for the imponderable fluid of light. As this conception crystallized, the fluid of heat vaporized; by 1860, the mechanical theory of heat and the kinetic theory of gases had replaced the old imponderable caloric. This process brought two fundamentally new concepts and techniques into physics: statistics, in the form of *probability calculations about the herd behavior of molecules, and irreversibility, in the form of the thermodynamic quantity *entropy.

That left the electrical and magnetic fluids, four in all, a positive, a negative, an austral, and a boreal. The discoveries of connections between electricity and magnetism beginning in 1820 led to the representation of magnetism as electricity in motion, a development completed by the demonstration of their identity in the theory of *relativity. Thus, by 1860, the main imponderables of 1800 had either vanished altogether (caloric and the magnetic fluids), metamorphosed into a mechanical system (light), or remained with enhanced properties (electrical fluids). A further diminution occurred with the unexpected discovery that *electromagnetism could be represented as a disturbance in the same ether whose vibrations made up light; or, stated *phenomenologically, that light and other radiations were manifestations of electromagnetism, a representation confirmed by the production of *radio waves in 1887.

Since the ether had been understood as a mechanical medium, physicists, particularly those trained at the University of Cambridge, tried to *model electromagnetic phenomena in mechanical terms. It remained only to bring matter into the system. A way opened with the discovery that whirlpools or vortex rings in a fluid with the sorts of properties usually ascribed to the ether would last forever. These rings could link and separate in accordance with other goings-on in the ether; in short, they could behave in many ways like chemical atoms. Thus arose the grand and remote program of complete mechanical reduction, which few physicists, perhaps, expected to see realized.

The possibility of concocting this program arose from the very substantial progress that mechanics itself had made during the middle third of the eighteenth century. The powerful mathematical theories of the behavior of fluids and solids then developed stood ready for exploitation by theorists of the ether. Also the more general formulations of mechanical principles were extended still further and adapted to electromagnetism and the kinetic theory of heat. The apparent success of the generalized mechanics in describing the phenomena of electromagnetism, light, and heat confirmed the possibility of mechanical reduction without the necessity of exhibiting an explicit model or picture. That was comforting. Of course, the scheme could be turned around and electromagnetism taken as primary. Many distinguished physicists around 1900 played with the idea of electrodynamic reduction (see ELECTRON).

The extraordinary progress of classical physics was supported by new institutional arrangements that made research as well as teaching the business of professors. The physics research of the eighteenth century, insofar as it existed at all, was primarily an easy-going affair associated with academies (see *ACADEMIES AND LEARNED SOCIETIES). Toward the end of the ancien régime and, increasingly after 1830, individual *university professors were conceded the right, and then given the duty, to undertake research, usually at their own expense, as part of their jobs. Their institutions provided space, a collection of teaching apparatus, and perhaps a mechanic to keep it in order. Beginning around 1870, with the foundations of the Cavendish Laboratory in Cambridge and the Physics Institute of the University of Berlin, the idea that a professor's university should furnish him with the instruments he needed and a place to work with collaborators and advanced students steadily gained ground. (Here physics followed chemistry, whose first important university facilities for teaching and research go back to the 1830s.) The instruments employed in the institutes at first were largely home made. By the end of the century, however, most of the important equipment—electrical parts, *air (or *vacuum) pumps, refrigerators (see LOW-TEMPERATURE PHYSICS), *batteries—came from commercial suppliers.

The recognition of the need for institutionally supported research both to advance science and to train students owed much to the demands and opportunities of the new *electrotechnology and the perceived need to give doctors, engineers, and science teachers a grounding in physics. By 1900 every major university and higher school in Europe and the United States had a physics *institute. By far the largest output of papers on physics came from academics. Most of them worked in Britain, France, Germany, or the United States, which then were investing the same proportion of their gross domestic product (GDP), and enlisting the same proportion of their populations, in the physics enterprise. An example of their investment, and an indication of the widening of occupational opportunities beyond the universities, was the establishment of national (or, in the case of France, international) bureaus of standards.

This enterprise reached a climax in the symbolic year 1900, when the world's physicists gathered in Paris for their first and last general conclave. They then awakened, as one participant said, to the news that the keystone of their grand synthesis might have been found. This was the electron, which, it appeared, might be the unique building block of both matter and electricity, the clue to chemical binding, the explanation of the *ion, and the root of *X rays and *radioactivity. These considerations inspired the design of several model atoms based on electromagnetic theory. None worked well beyond the limited range of the phenomena for which it was devised, for example, *magneto-optic effects and the scattering of rays from radioactive substances. Until around 1910 all these models lacked a principal ingredient that would make the theory of *atomic structure into the leading sector of physics in the 1920s. The necessary ingredient was the *quantum, introduced into the theory of black-body *radiation in 1900. Its fundamental importance for physics had only just been recognized when the world's physicists flung themselves into world war.

The performance of physicists during World War I demonstrated their indispensability to modern society. New forms of government, foundation, and philanthropic support (see POLITICAL ECONOMY OF SCIENCE) became available for scholarships, fellowships, and research. The worldwide activities of the Rockefeller Foundation were especially fruitful, since it favored the transfer of *quantum physics from Europe to the United States and of American instrumentation, particularly cyclotrons, abroad. Basic physics, having finished provisionally with radiation and atoms (see QUANTUM ELECTRODYNAMICS), delved into the nucleus; here the United States, with sizable disposable income despite the depression, led the way into *big science. American physics strengthened further as a result of the emigration of European refugees driven from their positions by fascist regimes. *Nazi science policy stultified fundamental physics, in which Jews had been disproportionately prominent, in Germany and its occupied territories. As Germany declined, Japan and the Soviet Union developed their own capacities for training advanced students and set up research institutes often with liberal funding from military or industrial sources. The Soviet capacity centered on the institutes of the Akademiia nauk (Soviet Academy of Sciences) and laboratories supported by the Commissariat for Heavy Industry; the Japanese, on the imperial research institute RIKAN.

As physics spread geographically it also spread intellectually to territory newly conquered by the quantum. The study of *cosmic rays, the behavior of bodies at low temperatures, the *solid state, and the nucleus (see NUCLEAR PHYSICS) advanced prodigiously during the 1930s. The ascendancy of the electron was challenged or shared by the deuteron, meson, neutrino, neutron, positron, and proton (see ELEMENTARY PARTICLES). Molecular physics began to yield to new methods of calculation. Machines and ideas spun off into neighboring and even distant sciences, for example, *astronomy (e.g., stellar energy) and biology (e.g., radioactive tracers). The principles of *complementarity and uncertainty, invented to domesticate a mismatch between ordinary intuitions and the formalism of quantum mechanics, found applications to vitalism, psychology, *philosophy, and theology.

Despite cries for a moratorium on research and movements for *social responsibility among scientists, which gained some strength during the depths of the depression, the public applauded the science-based luxuries it soon found to be necessities: commercial radio, the all-electric kitchen, the long-distance telephone, air mail, air travel, and, above all, the automobile. As this easily extendable list suggests, the public did not (and does not) distinguish between science and technology. The confusion was compounded by the multiplication of industrial research *laboratories, some of which, especially in the United States, permitted their technical staffs considerable leeway in choice and conduct of research. These laboratories helped to keep up demand for physicists so that, except for the loss of a year or two's crop of new graduates, the depression had no effect on recruitment, at least in the United States.

The demand became so great during World War II that physics students who had not completed their degrees were employed on advanced research projects. Once again physicists proved their value: *operations research, *radar, the proximity fuse, cryptography, and the atomic bomb. The great winner in the conflict, the United

States, rewarded its physicists after the war by supporting them in the manner to which they had become accustomed. The infusion of public dollars, intended to keep scientists well disposed toward government, as well as to encourage work related to weaponry, exaggerated a style already characteristic of American science: pragmatic, instrumentalist, democratic, and gigantic. The combination proved potent: countries that wanted to compete had no option but to submit to Americanization, especially in the world's most prestigious science, *high-energy physics.

The United States assisted participation and competition by helping in the recovery of science in Germany (after picking some plums in Operation Paperclip; *see* BRAIN DRAINS AND PAPERCLIP OPERATIONS), in Japan (after throwing its cyclotrons into the sea), and in Europe (by validating *CERN, the Centre Européen pour la Recherche Nucléaire, which brought together nations no longer able to compete alone). The recovery called into existence new mechanisms of national support and international collaboration. As a result, Europe and Japan caught up with, and even outpaced, the United States, which, by canceling the Superconducting Super Collider in the 1990s, showed that it no longer had the will, if it had the means, to dominate the world in high-energy physics.

While the particle physicists enjoyed their limelight, their colleagues working on a smaller scale discovered or invented things of greater importance to the public than *strangeness, parity nonconservation (*see* SYMMETRY), the Eight-Fold Way (*see* THEORY OF EVERYTHING), or the unification of the fundamental forces except for gravity. Radar technologies helped academic laboratories to produce the *laser and airports to handle jet-setters. Research into the solid state returned with silicon chips, the *computer, and high-temperature superconducting *magnets. Color television, direct-dial international telephony, satellite communications, modern banking, the credit card, and so on, demonstrate that physics warmed by cold war can do wonders.

The wonders have included *nuclear power and therewith the pollution that arises from the spent fuel and radioactive parts of reactors, which began to spread worldwide in the late 1950s under the American program, *Atoms for Peace. The pollution, as well as the involvement of physicists and other scientists in advanced weapons projects, fuelled an *antinuclear movement and wider attacks on science (*see* ANTISCIENCE). The end of the cold war redirected some of this sentiment against the diffusion of high-tech industry to developing countries. Nonetheless, the spread of Western science, which began before World War I in the colonies of the European powers and produced important indigenous physics communities in the Soviet Union and Japan before World War II, continued into *Latin America, the Near East, and continental *Asia after 1945. An example of this diffusion of postwar physics, American style, is the synchrotron completed in Brazil in 1996.

The spread of physics to new cultures had a parallel within the old physics-producing countries in the increasing inclusion of groups whose participation had previously been marginal or restricted. Anti-Semitism lost its force in most of the world. *Women gained easier access to training and employment. Opportunities were extended to *minority groups. Acquaintance with physical principles or gadgetry diffused widely in the general culture along with the *computer, high-tech weapons, and space exploration.

On the intellectual and rhetorical side, physicists provided many good accounts of their progress toward a Theory of Everything (*see* POPULARIZATION).

The possibility of an encompassing and final theory arose at the intersection of *cosmology and particle physics. Theories of the largest and smallest structures in the universe combined to make a *Big Bang. Crucial evidence favoring the doctrine came from the characteristic postwar specialty, radioastronomy. The doctrine of the Big Bang and the expectation by some theoretical physicists that a complete and unified theory of the physical world might be achieved during their lifetimes provoked the interest of diverse theologians and the Vatican. In a limited region of vast extent, therefore, physics has returned to the sort of questions with which its predecessor, natural knowledge, had been engaged.

See also HISTORY OF SCIENCE.

Abraham Wolf, *A History of Science, Technology, and Philosophy in the Sixteenth and Seventeenth Centuries* (1950). Abraham Wolf, *A History of Science, Technology, and Philosophy in the Eighteenth Century* (1952). Paul Forman, J. L. Heilbron, and Spencer Weart, "Physics circa 1900," *Historical Studies in the Physical Sciences*, vol. 5 (1971). Richard S. Westfall, *The Construction of Modern Science* (1971). J. L. Heilbron, *Electricity in the Seventeenth and Eighteenth Centuries. A Study in Early Modern Physics* (1979; 1999). Thomas L. Hankins, *Science and the Enlightenment* (1985). Tore Frängsmyr, J. L. Heilbron, and Robin E. Rider, eds., *The Quantitative Spirit in the Eighteenth Century* (1990). Peter Dear, *Discipline and Experience. The Mathematical Way in the Scientific Revolution* (1995). John Krige and Dominique Pestre, eds., *Science in the Twentieth Century* (1997). Robert D. Purrington, *Physics in the Nineteenth Century* (1997). Lisa Jardine, *Ingenious Pursuits* (1999). Helge Kragh, *Quantum Generations. A History of Physics in the Twentieth Century* (1999).

J. L. HEILBRON

PHYSICS, CONDENSED MATTER. See Solid State (Condensed Matter) Physics.

PHYSICS, SOLAR. See Solar Physics.

PHYSICS, SOLID STATE. See Solid State (Condensed Matter) Physics.

PHYSIOGNOMY. See Phrenology and Physiognomy.

PHYSIOLOGY. Originally synonymous with natural philosophy, physics, or the study of all natural bodies, the word "physiology" began to restrict its scope to the structure and composition of the human body during the sixteenth century. The French physician Jean François Fernel then defined physiology as one of the five parts of medicine—the one that "explains the nature of the healthy man, all of his faculties and functions." Fernel contrasted physiology particularly with "pathology," summarized as "the study of the illnesses and affections that can happen to man."

As the study of the structure and functions of the human body, physiology is one of the oldest fields of natural inquiry and links with fewer breaks than almost any other modern science with its roots in classical Antiquity. From Aristotle to Galen, the ancient tradition thoroughly mapped out the internal anatomy of animals closely related to humans and established the principle that the form of the parts must be related to their function. That principle has framed and guided physiological investigation ever since.

Physicians and natural philosophers of the sixteenth and early seventeenth centuries inherited an account of the functional organization of the human body based mainly on Galen. Filtered through twelve centuries and multiple layers of commentary and condensation, the complex, evolving, sometimes contradictory views of the voluminous Galenic corpus has been reduced to an easily remembered system (see Galenism). Despite subtle errors owing to Galen's projection of the anatomy of the apes he had dissected onto the human body he was not permitted to cut, the functional arrangements of the skeleto-muscular system had been thoroughly worked out and entered with only secondary changes into early modern times. The system divided the body into three regions. In the uppermost of these, the head, resided the brain, the "domicile of the human spirit, the seat of thought and reason, and the source and origin of all movement and sensation." Galen had worked out the gross anatomy of the brain, the cranial nerves emanating directly from it, the spinal cord, and the peripheral nerves in considerable and mostly accurate detail, and established regions innervated by various nerves through vivisection experiments not surpassed until the nineteenth century. The heart and lungs dominated the central region, or thorax. The heart was connected with two sets of vessels: on the right side to the veins, through which nourishment, transformed to blood in several stages, most notably in the liver, flowed to all parts of the body; and on the left side to the thick-walled arteries, which also distributed blood, but of a brighter, more spirituous nature, imbued in the left ventricle with vital spirit. The lower region, or venter, was filled by the organs of digestion, through which food passed from the mouth, converted gradually, by a process known as coction, into the uniform milky chyle absorbed through the portal veins into the liver. These arrangements provided a well integrated, comprehensible account of the relation between the observed internal structure of the human body and the primary functions essential to life, and adapted well to explanations of the humoral imbalances associated with disease.

An engineer, however, would have found some difficulties in this account. The Galenic system sometimes required fluids or pneuma to flow both ways in the same vessels, and even to pass the wrong way through valves in the heart generally understood to permit them only to move in the contrary direction. The Italian anatomist Realdo Colombo contended in 1559 that the heart did not attract blood into its ventricles in its expansion, as Galen had said, but pressed it outward in contraction, so that the arteries dilated when the heart constricted; and that the blood did not pass from right to left ventricle through pores in the septum between them, as Galen had said, but by a more indirect route that went through the lungs. More than half a century later the questions Colombo had raised about the motions of the heart and its relation to the motions of the arteries remained controversial.

Through exquisite observations on different classes of animals, especially on the hearts of cold-blooded and dying ones, William *Harvey settled to his satisfaction that Colombo's interpretation of the motions of the heart was correct: that through its active contractions it moved blood from the veins into the right ventricle, from there through the lungs to the left ventricle, and then into the arteries, which it filled, "as my breath in a glove." As soon as Harvey asked himself the further question, how much blood does the heart move in this way, he entered a pathway "so new and unheard of" that he feared "every man almost will become my enemy." No matter what quantity of blood he assumed to be driven out of the heart at each beat, the amount that passed through in an hour or a day was so abundant that it could neither be consumed in nutrition nor replenished from the food. The heart must, therefore, produce a "perpetual circular motion of the blood."

Although he did encounter some fierce resistance, Harvey prevailed over his opponents so quickly that within his own lifetime he was heralded as the founder of a new era in medicine. The discovery of the circulation not only ended centuries of domination by ancient systems of thought, but stimulated observations and experiments that brought further novelties, such as the discovery of the lacteal vessels by Jean Pecquet and a sustained collaborative effort by followers of Harvey informally associated at Oxford to understand the relation between the circulation of the blood and the process of respiration.

Harvey himself had examined physiological questions within an Aristotelian philosophical framework, which his successors soon abandoned for the newly dominant *mechanical philosophy. That transition wrought a fundamental change in the relation between the living and the nonliving world. Where Aristotle had extended to all of nature the purposefulness evident in vital processes and activities, the new mechanical world eschewed *teleology. Henceforth those who attempted to understand the underlying physical phenomena taking place in humans or other living things had the choice among explaining them in terms of the shapes and motions of the ultimate particles of matter (later also of the forces exerted between them), invoking recently discovered chemical properties (such as acidity and alkalinity), and referring vital phenomena to principles unlike those of the physical world.

The chemical and physical tools available in the seventeenth and eighteenth centuries had little capacity to explain physiological phenomena. Explanations of the formation of secretions in the salivary glands, the pancreas, or the kidney tubules remained at the level of crude images of particles of the blood that fit into holes in sieves passing through them into the secretory ducts. The new physical concepts did, however, offer more effective means to investigate the circulation, the one system in the body that seemed clearly accessible to mechanical analysis. The influential "iatromechanist" Giovanni Alfonso Borelli was the first person to attempt to determine the force exerted by the heart. The fantastically large number at which he arrived in 1678 stimulated others, such as Newton's follower James Keill, to seek more direct means to measure the force. The lack of agreement among the measurers prompted the famous measurements of the pressure in the arteries and veins that the Reverend Stephen

Hales carried out early in the eighteenth century by inserting long vertical tubes into the opened blood vessels of horses and other large animals. When the medically trained mathematician Daniel Bernoulli produced a landmark treatise in 1738 that defined a new field of *hydrodynamics, he quickly applied the theoretical principles and methods of calculation that he had introduced there to the study of the heart and movements of the blood.

In 1747 the Swiss professor of anatomy, surgery, and medicine at the University of Göttingen, Albrecht von Haller, published a textbook titled *First Lines of Physiology*, followed between 1759 and 1776 by an eight-volume *Foundations of the Physiology of the Human Body* that summarized and organized all previous work on its structure and functions. These massive works, which Haller intended both to teach medical students the fundamentals of the subject and to open the way to further discoveries, made physiology visible for the first time as a coherent discipline. Also a prolific experimentalist, Haller used the *microscope effectively to study the circulation of the *blood in the smaller arterioles and veins, and established through numerous experiments a distinction between the sensible parts (nerves) and the irritable parts (muscles) of the body. He met challenges by multiplying his own experiments and by gathering a group of followers who extended the scope of his generalization. He and his followers created an investigative field and raised the standards both for performing and reporting physiological experiments.

By the beginning of the nineteenth century, the physical sciences were supplying new, powerful methods for the investigation of physiological problems. The effects of the advances in chemistry can most easily be seen in the study of digestion. The eighteenth-century investigators René-Antoine de Réaumur and Lazzarro Spallanzani had demonstrated through experiments on various animals that foods are softened into a uniform semifluid mass during digestion, and managed to duplicate the process outside the body with gastric juice withdrawn from the stomach. Chemically, however, they could only say that the process resembled the action of an acid on a metal. During the 1820s Friedrich Tiedemann and Leopold Gmelin used a repertoire of reagents and extraction procedures to identify the changes that foodstuffs undergo in their passage through the stomach, intestines, and lacteal vessels. Although the only definite chemical transformation they could identify was the conversion of starch to sugar, their investigation opened a continuous tradition that lasted into the twentieth century.

Between 1800 and 1810 a group of physiologists in Paris began to apply the surgical skills acquired during their medical training to make vivisection central to an emerging field in which experimentation was to be not an occasional, but the central activity in the formation of an independent scientific discipline. François Magendie became the most emphatic and persistent advocate for this viewpoint. His own discovery in 1821 of the sensory and motor roots of the spinal nerves became the starting point for mapping the sensory and motor functions of the entire nervous system. His student Claude *Bernard, following his lead and succeeding to his chair at the Collège de France in 1855, made between 1848 and 1857 a series of brilliant discoveries, including the action of pancreatic juice on fats, the glycogenic function of the liver, the vasomotor nerves, and the modes of action of carbon monoxide and curare, which made him the most eminent figure of the generation in which physiology became established, both intellectually and institutionally, as the first thoroughly experimental life science.

In Germany physiology was in the early nineteenth century combined with anatomy in the same university chair. At the Anatomical Museum of the University of Berlin, Johannes Peter Müller represented and actively pursued research in both physiology and comparative anatomy. His *Handbook of Human Physiology*, first published in 1833 and in successive editions until 1844, mastered the field more fully than any physiology text since Haller's. Müller trained students who followed him into each of the fields in which he worked, but none who could command, as he had, their whole range. By the 1840s advances in both physiology and in comparative anatomy made it imperative to separate them, and in 1847 Müller's student Ernest von Brücke accepted, at the Medical Faculty of the University of Königsberg, one of the first independent chairs of physiology. Thereafter such chairs spread rapidly through the major German universities.

Brücke and two other students of Müller, Emil du Bois-Reymond and Hermann von *Helmholtz, committed themselves about this time, together with Carl Ludwig, to the goal of reducing physiology to physics and chemistry. This they were never able to do, but they did introduce a style of physiological experimentation based on precise instrumentation, accurate measurement, and rigorous analysis. They sought to examine the magnitude of any observed effect as an unknown function of all the conditions that influenced it and to vary one factor at a time while holding other conditions constant. "The dependence of the effect on each condition," du Bois-Reymond wrote, "can now be shown in a curve, whose exact law, to be sure, remains unknown, but whose course can, in general be outlined." In 1847 Ludwig devised a recording instrument in which the changing magnitudes of a physiological quantity such as the pressure of the blood or the length of a muscle were traced on the smoked surface of a revolving drum, soon known as the kymograph. It quickly became the operational symbol of this mode of functional analysis and a standard fixture in the physiological laboratory of the late nineteenth and early twentieth century. The traces recorded by kymographs filled the pages of the journals of physiology.

Meanwhile the advent of the voltaic *battery, galvanic currents, and then electromagnetic phenomena had supplied other new tools for physiological experiments. Using a very sensitive galvanometer of his own design, du Bois-Reymond was able to detect, during the 1840s, delicate changes in electric potential accompanying the transmission of nerve impulses in frogs. In 1850 Helmholtz used a ballistic galvanometer to measure the velocity of the nerve impulse itself, a feat once believed to be all but impossible.

By the last third of the nineteenth century, physiologists in Germany possessed large, well-equipped laboratories in which they could not only perform complicated experiments, but train a steadily increasing number of German and foreign students. It was the golden age of physiology. No sooner, however, did physiology attain its autonomy from anatomy than it began itself to undergo a slow process of subdivision. The very tools of the physical sciences that had given it such power early in the nineteenth century now served as foundations for more specialized investigations of the phenomena of life. The first to devolve was physiological chemistry, which attained only two independent chairs in Germany in the nineteenth century, but gave rise in the early twentieth century to the dominant field of *biochemistry. The study of the effects of drugs on physiological processes, initiated by physiologists, became by the late nineteenth century the domain of the emerging discipline of *pharmacology. What Claude Bernard and others defined in the nineteenth century as "general physiology," or the study of the phenomena of life common to all organisms, became in the twentieth century *cell biology. In this steadily expanding array of new biological disciplines, physiology functioned as a mature, or even old-fashioned, science by the middle of the twentieth century. Nevertheless, it has retained domains of its own, such as the regulation of heat, respiration, and other functions, and continues to be an active field of inquiry.

Karl E. Rothschuh, *History of Physiology*, trans. Guenter B. Risse (1973). Robert G. Frank, *Harvey and the Oxford Physiologists* (1980). Frederic Lawrence Holmes, *Lavoisier and the Chemistry of Life* (1984). John E. Lesch, *Science and Medicine in France: the Emergence of Experimental Physiology* (1984). Frederic Lawrence Holmes, "La Physiologie et la Médecine Expérimentale," in *Histoire de la Pensée Médicale en Occident*, ed. Mirko D. Grmek, vol. 3 (1999): 59–96.

FREDERIC LAWRENCE HOLMES

PLANCK, Max (1858–1947), German physicist and spokesman for science.

Planck came from a family of lawyers and pastors. His marriage to the daughter of a banker placed him comfortably within the professional classes of the rising German empire. His special interest in *physics, *thermodynamics, appealed to his love of order and generalization. In 1885 he joined the university in his home town of Kiel as an assistant professor (Extraordinarius) in theoretical physics, then a small new subdiscipline. Four years later he obtained a similar position at the University of Berlin, where he remained until his retirement in 1928.

Planck's reputation among physicists rests on his solution to a problem about *radiation inside a closed cavity. Thermodynamical arguments indicated that the energy falling to a given frequency (color) in cavity radiation at equilibrium depended only on the color and the temperature of the enclosure's walls, and that the "radiation formula" describing each color's energy must contain two constants of universal significance.

James Clerk *Maxwell's equations described electromagnetic radiation in general, and *entropy fixed the equilibrium state. To obtain a theory that agreed with the rapidly improving measurements (the energy distribution of cavity radiation had the practical interest of serving as a universal standard for the efficiency of electric light bulbs),

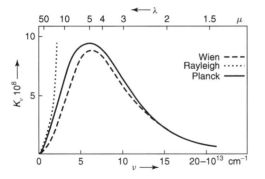

Graph of radiation formulas of Wien, Rayleigh, and Planck.

Planck turned to the definition of entropy that Ludwig *Boltzmann had employed to describe the behavior of a gas. To bring the definition to bear, Planck calculated as if the sum of the energies of the resonators associated with a given frequency f consisted of a very large number of very small elements hf. (The resonators were fictional oscillators that coupled matter to the radiation field.) The key step in the derivation for Planck was thus an *extension* of the concepts of the mechanical world picture to radiation, not a *limitation* on resonator energy.

Planck's scientific reputation and sense of duty soon brought him to public attention. In the mid-1890s he defended a Jewish physicist whom the government wanted to expel for socialist activities. He supported the right of women—though only unusual women, like Lise *Meitner—to study science at the university. He gave unstintingly to his profession, especially as an editor of the *Annalen der Physik*, Germany's leading physics *journal. In this capacity he welcomed and promoted Einstein's theory of *relativity. Planck's famous and severe attack in 1908 on the sensationalist epistemology of Ernst Mach marked a new direction in his efforts to steer the course of physics.

Planck's first important administrative job was to help govern the Berlin Akademie der Wissenschaften. During World War I he prevented the academy from expelling foreign members belonging to enemy countries. His repudiation of his signature of the Manifesto of the Ninety-Three Intellectuals, a declaration in support of the German invasion of Belgium, further exemplifies his wartime balance and courage. He was the only one of the ninety-three to recant publicly. After the war he worked energetically to rebuild German science and took on the presidency of the *Kaiser-Wilhelm-Gesellschaft (KWG), whose many institutes occupied the vanguard of German scientific and technological research.

The Nazi takeover presented Planck with an acute and continuing crisis of conscience; in the end, he decided to retain his positions of influence. He was able to help a few people, but his staying eventually tarnished his reputation. In 1938 he was forced from his remaining posts. Toward the end of World War II, allied bombing destroyed his house and fear of the Russians drove Planck and his wife into the woods, where American soldiers rescued them. Planck still had a role to play. The British decided to revive the KWG under a new name. Planck acted briefly as president of the restructured organization, which took as its name the Max-Planck-Gesellschaft für die Förderung der Wissenschaften. It is now one of the world's premier research networks.

Planck's private life, begun as an idyll, became a tragedy. His first wife, the mother of his two sons and twin daughters, died in 1908. His eldest son was killed in World War I. The twin daughters died two years apart, in 1917 and 1919, both in childbirth. These catastrophes wiped out whatever pleasure might otherwise have come to him from the belated award of the Nobel Prize for physics in 1918. Planck's misery reached its apex when his second son was executed in 1944 for complicity in the plot to assassinate Hitler. Against that the loss of his material possessions in the destruction of Berlin, the last step in the annihilation of the strong, orderly, cultured Germany whose rise to European dominance had been the pride and guide of Planck's early manhood, signified nothing.

T. S. Kuhn, *Black Body Radiation and the Quantum Discontinuity* (1978). J. L. Heilbron, *The Dilemmas of an Upright Man: Max Planck as Spokesman for German Science* (2d edition, 1999).

J. L. HEILBRON

PLANET. The word "planet," derived from the Greek for "wanderer," at first referred to Mercury, Venus, Mars, Jupiter, and Saturn, which wandered amid the fixed stars of the ecliptic plane. The ancients almost universally believed that the planets rotated around the earth, although from the relationship of the orbital retrogrades of Mars, Jupiter, and Saturn to the terrestrial year, Nicholas *Copernicus argued in 1543 that they rotated around the sun. Between 1543 and 1687 the planets became agents of radical intellectual change in astronomy and physics. What was their center of rotation? After the abandonment of Aristotle's crystalline spheres around 1600, what force made the planets move? Were their orbits circular or elliptical, and how did their speed relate to their distance from the center of motion? Planetary motion lay at the heart of the work of Johannes *Kepler and Isaac *Newton; the problems it posed led to the invention of gravitational physics.

Until *Galileo first looked at planets through the telescope and discovered that they subtended disks (indicating that they were spherical worlds), astronomers had envisaged the planets as points of light. The telescope inaugurated physical astronomy. Galileo discovered that Venus showed phases; that Jupiter was slightly compressed at its poles; and that Saturn, as well as being spherical, also displayed peculiar appendages that appeared and disappeared. In 1655, Christiaan *Huygens, using a much more powerful telescope than Galileo's, resolved

these appendages into a ring, and in 1675 Gian Domenico Cassini discovered a division in them.

Following technological breakthroughs that made possible long refracting telescopes with object glasses of diameters up to 15 cm (6 ins), astronomers discovered the Syrtis Major and Polar Caps of Mars (Huygens) and Jupiter's belts and spots (Cassini and Robert *Hooke). Astronomers also obtained accurate timings of the orbital periods of the outer planets using long refracting telescopes, especially in conjunction with Huygens's pendulum clocks after 1658. New planetary satellites came into focus: Saturn's Titan (Huygens, 1655), followed by Lapetus (1671), Rhea (1672), and Tethys and Dione (1684), all discovered by Cassini. In 1668 Cassini produced tables for the motions of Jupiter's moons, which enabled cartographers to fix the respective *longitudes of places on terra firma though not yet of ships at sea.

The first great wave of physical planetary discovery came to an end around 1690 as the long refracting telescope exhausted its research potential. New planetary discoveries awaited the superior resolving power of the relatively large speculum-mirror reflecting telescope, initially in the hands of Sir William Herschel, and, after 1800, the achromatic refractor.

Herschel's discovery of Uranus on the night of 13 March 1781 heralded a new era of planetary astronomy. It suggested that more planets and satellites might be in space awaiting discovery. Herschel himself discovered Uranus's moons Oberon and Titania (1787) and two new Saturnian satellites, Enceladus and Mimas (1789). Then, from the southern latitude of Palermo, Sicily, on 1 January 1801, Giuseppe Piazzi discovered the first asteroid, or minor planet, Ceres, using a Dollond achromatic refractor (*see* TELESCOPE).

During the nineteenth century planetary astronomy made enormous progress. As in the past, advancing instrument technology made it possible. Three new asteroids quickly appeared: Pallas (Heinrich Olbers, 1802), Juno (Karl Ludwig Harding, 1804), and Vesta (Olbers, 1807). Then, after a gap of thirty-eight years, Karl Hencke, a German amateur, discovered Astraea, and, almost in cascade, came hundreds more, 450 in all by 1900. The asteroids fascinated astronomers, eliciting speculation about a former planet between Mars and Jupiter gravitationally destroyed in the early stages of the solar system, and giving support to the empirical Titius-Bode law of planetary distribution (1772). This so-called law, developed by Johann D. Titius and Johann E. Bode in Germany, is an empirical sequence of numbers—four, seven, ten, sixteen, twenty-eight, fifty-two, and one hundred—that correspond to the proportionate distance of the first five planets and the asteroid belt around the Sun. The Titius-Bode law lay at the theoretical heart of the greatest discovery in planetary dynamics of the nineteenth century: the detection of Neptune in 1846. Both John Couch Adams in Cambridge and Urbain J. J. Le Verrier in Paris used the law to model some of the parameters that lay at the heart of their independent predictions of the position of Neptune though, ironically, Bode's number sequence does not hold good for Neptune. This discovery, however, would not have been possible without a combination of fundamental advances in gravitational mathematics and superior tables of the motion of the known planets made at the Greenwich, Königsberg, and other *observatories.

Planetary astronomy was advanced not only by professional scientists in major universities and observatories, but also by self-funded individuals. These "grand amateurs," like the Liverpool brewer William Lassell, who built the most advanced reflecting telescopes of the 1840s to 1860s to discover planetary satellites, and the spectroscopist Sir William Huggins, funded research from their private means. Less wealthy people, like the more modest amateurs in the British Astronomical Association and other bodies worldwide, devoted themselves to monitoring visible changes on the planets.

A succession of interplanetary space probes after the Russian *Venera* flight to Venus (1970) fundamentally transformed our knowledge of the planets (*see* SATELLITE). Mercury, Venus, and Mars turned out to be rocky worlds with widely different sorts of atmospheres. Jupiter, Saturn, Uranus, Neptune, and Pluto (the latter discovered by Clyde Tombaugh in 1930) are made of frozen gases. The American spacecraft *Voyager II* in the late 1980s produced breathtaking fly-past views of the outer planets and discovered several new satellites, rings, and an abundance of surface detail. The future of planetary exploration probably lies in the development of increasingly sophisticated robotics instrumentation operated through space vehicles.

Sir John F. W. Herschel, *Outlines of Astronomy* (1849). Robert Grant, *History of Physical Astronomy* (1852). Michael Hoskin, ed., *Cambridge Illustrated History of Astronomy* (1997).

ALLAN CHAPMAN

PLANETARIUM denotes instruments demonstrating positions and movements of the sun, earth, moon, planets, or stars. Its earliest usage referred to Giovanni de' Dondi's clock and its astronomical dials (around 1360). It later comprised various devices: globe, armillary,

equatorium, *orrery, grand orrery, compound orrery, tellurian, lunarium, cometarium, jovilabe, Copernican sphere, sphère mouvante, and others. These tools have enabled scholars to pursue research and assisted lay audiences to learn basic astronomy. Since 1923, "planetarium" has connoted an optical projector, its architectural facility, and the facility's institution.

Historical planetaria fall into two categories: illustrations of the celestial sphere and associated daily motions, and representations of longer-term motions and principles pertaining to the classical planets. Instruments of the first category, including celestial and terrestrial globes and many armillaries, served as popular instruments for initial astronomical instruction long after the acceptance of Copernicanism, demonstrating earth-based observations in terms of the celestial sphere, its fixed stars, and imaginary geometrical lines. Important globes include the Atlante Farnese, and models by Ptolemy, Islamic and Chinese scholars, Gemma Frisius, Gerhard Mercator, Tycho *Brahe, Jodocus Hondius, Willem Blaeu, Joseph Moxon, Vincenzo Coronelli, and the Cary family. Room-sized hollow spheres by Adam Olearius (Gottorp), Erhard Weigel (Jena), Roger Long (Cambridge), and Wallace Atwood (Chicago) provided experiences of the rotating heavens.

Instruments of the second category featured geared dials, rings, or spheres representing the seven classical planets and, later, the earth and planetary satellites in illustration of Copernican astronomy. Descendants of geared *astrolabes and astronomical clocks, significant examples of these instruments include the first known tellurian, by Blaeu, and models by Ole Rømer, Christiaan *Huygens, George Graham, James Ferguson, Benjamin Martin, David Rittenhouse, George Adams Sr. and Jr., and William and Samuel Jones, as well as Eise Eisenga's room-sized planetarium. Graham's proto-orrery prompted production of elaborate grand orreries for wealthy patrons and numerous inexpensive models for the adult science education market in Hanoverian England.

Instruments addressing both concepts include astronomical clocks with globes and numerous armillaries enveloping planetary models. Twentieth-century Zeiss optical projection systems synthesize the two categories somewhat, but recent computer-based technology offers more fully integrated illustrations of naked-eye observations and astrophysical principles.

Chicago, New York, Los Angeles, London, Stockholm, Calcutta, and other cities around the world feature public planetaria. School programs account for most visits, particularly in Japan, where Goto and Minolta systems are popular, and in the United States, where many Spitz models were built in response to the *Sputnik* crisis.

At times symbols of city, court, wealth, or learning, planetaria originally conveyed creation's divine order and purpose, providentially extended to human affairs. Enlightenment-era lectures and wealthier homes featured orreries illustrating the solar system's mathematical regularity, whereas early Zeiss shows competed with moving pictures for audiences seeking novel theatrical experiences. Contemporary planetaria proclaim the universe's evolutionary history alongside ever-popular presentations detailing the Star of Bethlehem's astronomy. Controversies over rival goals—education or entertainment—have accompanied planetaria for centuries, a natural consequence of their bridging astronomy and popular culture. The mechanisms of the instruments, the techniques, skills, gender, backgrounds, and communication goals of their artisans and operators, and the desires of their audiences have reflected their cultural contexts throughout the history of planetaria, microcosms of their social and scientific macrocosms.

See also CLOCK AND CHRONOMETER; POPULARIZATION.

Planetarian: Quarterly Journal of the International Planetarium Society, 1970 to the present. Henry King, *Geared to the Stars* (1978).

MARVIN BOLT

PLANETARY SCIENCE. The term "planetary science" dates from the 1950s. It applies physics, astronomy, chemistry, geology, biology, atmospheric sciences, and oceanography to discrete bodies in the solar system. Previously the study of the planets had been known as "solar system astronomy" or "solar system science." That might have been a better name, because the field takes the whole solar system, including *comets, meteorites, asteroids, and planetary satellites as its object of study (*see* PLANET).

Like the parallel discipline earth science, planetary science emerged in tandem with the new technologies whose development had been spurred by World War II, rocketry and computers in particular. In 1958, the *International Geophysical Year began, in large measure to take advantage of these technologies. In the same year the Soviet Union launched *Sputnik* and the space race began (*see* SATELLITE). In October 1958, the United States established the National Aeronautics and Space Administration (NASA), to carry on and extend the work formerly done by the National Advisory Committee for Aeronautics (NACA) and other government bodies.

The term "planetary science" first appeared in the journal *Science* in 1959 in a job advertisement put out by the Goddard Space Center. In the same year, the first specialist journal, *Planetary and Space Science*, began publication. In 1962, the American Geophysical Union set up a section on planetary sciences. The journal *Earth and Planetary Science Letters* appeared in 1966. Existing institutions like the Houston Lunar Science Institute, university departments, and the journal *Meteoritics* all added "and Planetary Science" to their names.

*Copernicus, *Galileo, Christiaan *Huygens, Gian Domenico Cassini, and William Herschel had asked questions about the configuration of the solar system (cosmology), its mechanics, and its origin (cosmogony), and little by little discovered smaller and more distant bodies in the solar system. Galileo, William Gilbert, and Thomas Harriot mapped the Moon in the seventeenth century. Michael Florent van Langren published the first large full-Moon map in 1645, though his projected series of maps showing the Moon in its different phases never appeared in print. The first full-Mars map was published in 1840.

View of the surface of Mars taken in 1979 by Viking Lander II. It shows a thin layer of water ice on the Martian landscape.

The advent of large telescopes and photography vastly improved maps of the Moon and Mars in the remaining years of the century.

Meanwhile physicists, geophysicists, and geologists as well as astronomers pursued many aspects of what would now be planetary science. They asked how the Earth differed from neighboring planets and why. In 1801, the Italian astronomer Giuseppe Piazzi detected the first asteroid. By the end of the century, Maximilian Wolf at the University of Heidelberg had invented a technique for discovering new asteroids by the streaks they left on photographic plates. Astronomers thus discovered the asteroid belt. They also discussed the origin of craters on the Moon and other planetary bodies. In 1803, Jean-Baptiste Biot confirmed that certain stones in Normandy really had fallen from the sky, thus establishing the extraterrestrial origin of meteorites.

After a period in which interest in the solar system waned, two American astronomers, Gerard Peter Kuiper and Harold C. Urey, renewed interest in the subject in the 1940s. Kuiper discovered the carbon dioxide atmosphere on Mars and a disk-shaped region of minor planets (now called the Kuiper belt) outside Neptune's orbit, which he proposed as the source of certain types of comets. He pioneered the development of infrared astronomy, helped identify landing sites for the first manned landing on the moon, and edited two influential works, *The Solar System* (1953–1958) and *Stars and Stellar Systems* (1960–1968). Urey synthesized his investigations of the distribution of elements in the solar system in his *The Planets, Their Origin and Development* (1952).

Since 1960, planetary science has developed rapidly with the help of new optical and radio *telescopes. In 1990, the Hubble Space Telescope reached a position high above the distorting effects of the Earth's atmosphere. Project Apollo, which culminated in 1969 with the first human moon landing, the *Pioneer* and *Voyager* spacecraft that explored the Moon and other parts of the solar system, and the *Viking* and *Mars Pathfinder* spacecraft that investigated Mars enabled new kinds of data collection, whether by humans on the moon, by robots on the moon, or by photography and sampling of these and more distant planets. Planetary mapping, aided by radar techniques, proceeded apace. Mercury and Venus, whose surfaces had been difficult to study, Mercury because of its small size and proximity to the Sun, and Venus because of its dense atmosphere, have now been mapped.

New specialties have emerged, such as astrogeology, astrobiology, planetary atmospheres, planetary tectonics, and planetary physics. Topics of active investigation include planetary origins, the structure and composition of planets, vulcanism and tectonic activity, the atmospheres and magnetic fields of planets, and the planets of Jupiter. Public interest in planetary science, though not as high as in the 1960s, is still fueled by dramatic photographs, press coverage, and fascination with perennial puzzles like the canals of Mars and the possibility of life elsewhere in the universe. New discoveries and up-to-date information are posted on the NASA site on the World Wide Web. Planetary scientists, who until the end of the second millennium had been concerned almost exclusively with objects within our own solar system, are beginning to pursue the increasing evidence of planets in other parts of the universe.

See also ASTRONOMY, NON-OPTICAL; EARTH SCIENCES.

Ormsby M. Mitchel, *The Planetary and Stellar Worlds: A Popular Exposition of the Great Discoveries and Theories of Modern Astronomy* (1892; new ed., ed. I. Bernard Cohen, 1980). Dominick A. Pisano and Cathleen S. Lewis, *Air and Space History: An Annotated Bibliography* (1988). Stephen G. Brush, *A History of Modern Planetary Physics*, 3 vols., (1996). Ronald E. Doel, *Solar System Astronomy in America: Communities, Patronage, and Interdisciplinary Science, 1920–1960* (1996). James H. Shirley and Rhodes W. Fairbridge, eds., *Encyclopedia of Planetary Sciences* (1997).

JOANNE BOURGEOIS

PLASMA PHYSICS AND FUSION. A plasma in physics refers to an ionized gas. Scientists first recognized the importance of plasmas in studies of the propagation of radio waves for the nascent radio industry of the early twentieth century. In seeking ways to send radio signals over long distances they realized that waves seemed to bounce off a conducting layer in the earth's atmosphere, allowing signals to travel far beyond the horizon. One of these researchers, Irving Langmuir of the United States, in the 1920s designated the atmospheric matter "plasma" and investigated its properties in gas discharges in the laboratory.

In the late 1920s physicists began to apply new theories of *atomic structure and *quantum mechanics to the energy source of stars. In 1929 Robert Atkinson and Fritz Houtermans predicted that the nuclei of light atoms such as hydrogen, the primary constituent of the sun, could fuse through quantum tunneling, and that the resultant atoms would weigh less than the original constituents. Albert *Einstein's mass-energy relation suggested that fusion would release vast amounts of energy, enough to power the stars. Hans Bethe and others developed the theory of stellar fusion in the 1930s, elucidating the chains of nuclear reactions by which fusion built up heavier chemical elements and calculating the reaction rates and energy release. Astrophysicists were then incorporating plasmas into theories of stellar structure and thus merged the study of plasma with fusion.

Physicists at the time recognized the potential of fusion for a new energy source, but the high temperatures required to produce it seemed out of reach of available technology. Although World War II and the coincident discovery of nuclear fission diverted attention from fusion, they would eventually provide the motivation and means to attain it. Nuclear fission and the subsequent development of *nuclear bombs brought stellar conditions down to earth and offered a way to ignite fusion. During the war scientists working on the atomic bomb project in the United States discussed thermonuclear weapons, or the hydrogen bomb (named after its fuel), with explosive force orders of magnitude beyond fission bombs. Both the United States and the Soviet Union would pursue the hydrogen bomb in the Cold War; in the meantime, work on fission bombs advanced knowledge about plasma behavior, and the complicated hydrodynamic calculations for bomb physics spurred the development of electronic *computers, which would then aid the development of fusion weapons.

Research into controlled fusion revived in 1951 with the help of Juan Perón, the dictator of Argentina. A few years earlier Perón had set up a laboratory for Ronald Richter, an expatriate German with a scheme for controlled fusion power. In 1951 Perón announced Richter's successful production of power from a fusion reactor. The news made headlines in major newspapers, and though American and European scientists quickly discounted Richter's results, they did start thinking more seriously about the problem of fusion reactions.

One physicist so inspired was Lyman Spitzer, Jr., who was familiar with plasmas from his background in astrophysics and who had just joined a group at Princeton University working on the crash program to build the hydrogen bomb in the United States. Spitzer quickly devised a device to contain a plasma at high temperatures and obtained the support of the Atomic Energy Commission for the work. Commission laboratories at Los Alamos, New Mexico; Livermore, California; and Oak Ridge, Tennessee, soon followed suit. Controlled fusion seemed to offer unlimited power from an abundant fuel without the lingering radioactivity of nuclear fission reactors, and also provided a peaceful application of nuclear research to balance the fearful implications of nuclear weapons. It hence did not lack for support: by the late 1950s the United States was spending tens of millions of dollars a year on fusion research.

Other countries joined what became an international race for controlled fusion. British scientists led by George P. Thomson began investigating fusion soon after the war, and the British government sponsored a major fusion effort at its nuclear research laboratory at Harwell. In the Soviet Union, Igor Kurchatov, Igor Tamm, Andrei Sakharov, and other scientists in the nuclear weapons project took up fusion research in the early 1950s (see KURCHATOV, IGOR VASILYEVICH, AND J. ROBERT OPPENHEIMER, and SAKHAROV, ANDREI, AND EDWARD TELLER). The connection to nuclear weapons kept work in each country secret until Kurchatov revealed the Soviet program on a visit to Harwell in 1956; an international conference on atomic energy in Geneva in 1958 opened up the field for good. Japan, France, Italy, and other nations also entered the race, but the high cost of fusion experiments spurred efforts at international collaboration.

Most fusion reactors used various configurations of magnetic fields to bottle up the charged particles of the plasma, which at the high temperatures involved proved difficult to control. Fusion research engaged scientists from diverse fields: astrophysics, cosmic ray physics, accelerator engineering, gas discharges, and weapons physics. But no unified framework emerged from this eclectic background, and the initial optimism of the early 1950s soon gave way to realization of the technical difficulty of the endeavor—skeptics compared it to trying to push all the water to one side of your bathtub with your hands—and lack of knowledge about the basic behavior of plasmas. In the late 1950s fusion researchers instead turned to the underlying theory of magnetohydrodynamics, although work on fusion reactors continued under more empirical techniques.

In the mid-1960s Soviet scientists provided a new impetus with their development of the tokamak, which combined linear and toroidal configurations of previous devices in a single toroidal device. In 1968 a Soviet team under Lev Artsimovich revealed the attainment of temperatures of around ten million degrees and confinement times of about a millisecond in a tokamak. The tokamak thereafter became the preferred device for fusion, but the conditions it produced remained far below those required for fusion. Only after decades of technical refinements did a tokamak at Princeton University provide the first definitive success in late 1993 and 1994, when it confined a plasma of hydrogen isotopes at 300 million degrees Celsius for about a second to produce 10 million watts of power. The Princeton tokamak, however, still consumed more power in heating and confining the plasma than it produced.

The development of lasers in the 1960s suggested another route to fusion. Focusing high-energy lasers on a stationary solid pellet of hydrogen isotopes could compress and heat the pellet enough for fusion. Laser fusion—a form of inertial confinement—offered a way around the difficult problems posed by confining hot moving plasma with magnetic fields, and several nations started laser fusion programs. But laser fusion also presented daunting technical problems, especially the manufacture of laser optics capable of the high energies necessary. Connections with nuclear weapons persisted in laser fusion, since it also offered a way to model miniature nuclear explosions, and secrecy began to return to fusion research. In the 1990s the United States began building the billion-dollar National Ignition Facility at Livermore to substitute laser fusion for full-scale nuclear tests (see NUCLEAR POWER).

Still another route to fusion energy was announced in Utah in March 1989 by B. Stanley Pons and Martin Fleischmann, who claimed to have obtained fusion at room temperature in a cheap and simple electrochemical experiment. The announcement set off a frenzy of popular discussion of limitless energy, but attempts to replicate the experiment and to adjust theory to accommodate the results failed. In addition, the disciplinary background of Pons and Fleischmann in chemistry did not inspire confidence in the physicists who dominated the fusion community, nor did their mode of announcement, in a press conference instead of through peer-reviewed publication. Cold fusion quickly joined N rays, polywater, and other famous nondiscoveries in the history of science.

Joan Lisa Bromberg, *Fusion: Science, Politics, and the Invention of a New Energy Source* (1982). John R. Huizenga, *Cold Fusion: The Scientific Fiasco of the Century* (1992). Gary Taubes, *Bad Science: The Short Life and Weird Times of Cold Fusion* (1993). Richard F. Post, "Plasma Physics in the Twentieth Century," in *Twentieth Century Physics*, eds. Laurie M. Brown, Abraham Pais, and Sir Brian Pippard, vol. 3 (1995): 1617–1690. T. Kenneth Fowler, *The Fusion Quest* (1997).

PETER J. WESTWICK

PLASTICS. In ordinary language "plastics" refers to synthetic *polymers and evokes artificial materials. Technically the term applies to all materials that can be molded and shaped by heat and pressure. Plastics can be artificial or natural. Celluloid, the first commercial plastic, was made from cotton treated with nitric acid mixed with camphor. John Wesley Hyatt developed it

in 1867 as an imitation ivory for billiard balls. Bakelite—a synthetic material made from phenol and formaldehyde, patented by Leo Baekeland in 1907—was marketed as a "protean" material suited for a "thousand uses" rather than as a cheap substitute for natural materials. Vinyl was first manufactured in 1928, followed by urea formaldehyde plastics in 1929. The campaigns launching the synthetic plastics on the market during the depression of the 1930s emphasized their economic benefits—certainty of supply, light weight, easy processing, minimal labor costs of assembly and finishing operations—and vaunted their social virtues as agents of democratization.

The plastics industry emerged without any scientific understanding of the nature of polymers or the mechanism of polymerization; the search for synthetic substitutes for natural rubber, driven by the increasing demand for rubber in the early twentieth century, stimulated academic research. Polymers were first viewed as colloids, aggregates of molecules held together by physical forces. Herman Staudinger from Zurich challenged the aggregate theory in the 1920s. He described polymers as "macromolecules" made of billions of atoms linked by valence bonds. Experimental data did not resolve the controversy. Industrial research helped establish the macromolecular view. The young American organic chemist Wallace Hume Carothers took up the subject in the context of a fundamental research program launched by the DuPont Company in 1927, and focused on the mechanism of polymerization, a topic neglected until then by the Staudinger school. Using step-by-step organic reactions, he synthesized artificial molecules as probes for the existence of macromolecules, although he had to transform them into the commercial products that interested DuPont. After Carother's suicide in 1937, his collaborator Paul Flory continued theoretical studies on polymers in an academic context, and polymer science became a new and autonomous field of research after World War II.

Meanwhile a new generation of plastics changed public perception of their quality. Early plastics were thermoset, formed by polycondensation reactions, and then molded. Thermoplastics, formed by polyaddition or juxtaposition of monomers, polymerize by heating and can be reversibly fluidified by heating. Light and brightly colored molded thermoplastic items proliferated in daily uses, everywhere from kitchens and children's rooms to garbage cans. They encouraged an intensive consumption based on the notions of disposability and impermanence.

By the 1970s plastics were no longer perceived merely as cheap substitution products; artists and architects were finding new uses for them and bringing them new status. Plastic materials created an aesthetics of their own in which artificiality became a supreme value. The French philosopher Roland Barthes has described the mythology surrounding plastics. Plastic, he wrote, "is less a substance than the trace of a movement" (*Mythologies*, 1957).

Technically what is called the "plastics era" emerged in the 1970s when the volume of plastics used in the world first exceeded the volume of steel. The substitution of plastics for metals and wood could not exclusively rely on the properties of polymers. In fact, manufactured plastics were never made out of pure macromolecules. The formulation of a resin included multiple ingredients: agents of elasticity, agents of reticulation, adhesives, coatings. . . .

The strategies of mixture helped extend the applications of plastics. Chemists began to reinforce plastics with glass fibers in the 1940s for military purposes, such as aircraft noses and boats for the U.S. Navy. These laminates of polyester resins molded at low pressures started to be mass produced for civil applications such as electric insulators and tankers in the 1950s, when the difficulty of molding large pieces was overcome. Gradually with the introduction of long fibers, reinforced plastics became composites. Although composite technology eventually became autonomous, it originated as an outgrowth of the plastics industry.

Robert Friedel, *Pioneer Plastic: The Making and Selling of Celluloid* (1983). Susan T. I. Mossman and Peter Morris, eds., *The Development of Plastics* (1994). Jeffrey L. Meikle, *American Plastic. A Cultural History* (1995). Yasu Furakawa, *Inventing Polymer Science: Staudinger, Carothers, and the Emergence of Macromolecular Chemistry* (1998).

BERNADETTE BENSAUDE-VINCENT

PLATE TECTONICS. The theory of plate tectonics, proposed in the 1960s, asserts that the creation, motion, and destruction of a small number of rigid plates, thin in relation to the earth's diameter, shape the earth's surface. Quickly termed a revolution, the switch to plate tectonics was one of the most exciting scientific developments of the mid-twentieth century.

The discoveries that stimulated scientists to propose plate tectonics came from paleomagnetism and *oceanography. At the end of the 1950s, a small but influential group of physicists based at the universities of London and Newcastle and at the Australian National University were studying paleomagnetism. They became convinced that to explain the apparent global wandering of the magnetic pole over geological time, they had to assume that the continents had moved relative to one another. They saw this as new evidence for the theory of continental drift, still widely discussed in Britain and Australia because it had been advocated in 1945 in the *Principles of Physical Geology* by the distinguished geologist Arthur Holmes.

Meanwhile, oceanographers had been surveying the ocean floor and measuring heat flow and gravitational and magnetic anomalies. They discovered a global system of mid-ocean ridges. These enormous mountain chains had some peculiar physical characteristics, such as patterns of magnetic anomalies and a median rift valley with high heat flow. In the early 1960s, Harry Hess of Princeton University suggested that these were tensional cracks through which lava welled up, created new sea floor, and spread. His conjecture of sea floor spreading was quickly corroborated by two confirmed predictions. In 1963, Fred Vine and Drummond Matthews of Cambridge University predicted that magnetic anomalies observed on either side of the mid-ocean ridges recorded global magnetic reversals preserved in the solidified lava. Physicists had dated global magnetic reversals on the continents using radioactivity and so had a magnetic time scale. It was only necessary to find parallel zebra stripes of anomalies on either side of the ocean ridges. In 1965, J. Tuzo Wilson predicted that if sea floor spreading occurred, scientists should be able to detect seismically a new kind of fault that he named "transform." In 1966, scientists at the Lamont Doherty Geological Observatory found evidence supporting both predictions.

If the sea floor was spreading, where was the new material being accommodated? Could it be that the earth was expanding? Scientists gave this possibility serious consideration. It was quickly displaced, however, by the theory of plate tectonics independently conceived by Jason Morgan at Princeton and Dan McKenzie at Cambridge in 1967 and 1968, respectively. They proposed that rigid plates, each perhaps a hundred km thick, covered the earth's surface. They, and not continents and oceans, were the important structural surface features. Created at the mid-ocean ridges, they moved apart until they sank into and were consumed in "subduction zones" signaled by intense earthquake activity and negative gravity anomalies. Abstract mathematical models of plate movements agreed well with field observations. By the early 1970s, almost all earth scientists, except in Russia, had accepted plate tectonics.

Such a rapid shift to an account of the earth so radically different from previous orthodoxy stimulated popular interest. Earth scientists published in the popular scientific press, appeared on television programs, and revised school textbooks.

Once their immediate euphoria waned, many earth scientists suffered a crisis of confidence. Were they wrong to have resisted the theory of continental drift for half a century? And if science proceeded by the patient accumulation of facts, as most of them believed, was it scientific to switch in just a few years from believing that the continents stayed in place to believing that they moved?

Many earth scientists, particularly younger ones, wondered how their predecessors could have rejected continental drift and derided it as pseudoscientific when it had been supported by some evidence from similarities of paleontology and lithology on the two sides of the Atlantic and from the fit of the continents. Their reaction misread history. Continental drift, like other theories put forward when the geological synthesis proposed by Eduard Suess in *Face of the Earth* (1883–1904) collapsed in the early years of the twentieth century, had been given a serious hearing. It was widely accepted in South Africa and viewed with an open mind by geologists in the British Isles and Australia. In the 1950s, some American geologists mocked it in their undergraduate classes largely because they believed its proponents lacked evidence. Plate tectonics, with its confirmed predictions, had much stronger evidential support; moreover, it was a different theory. The introduction of plates made continental movement an incidental theoretical consequence and not the key theoretical claim.

Earth scientists still had to face the fact that the speed with which they accepted plate tectonics did not sit well with their image of science as the gradual accumulation of facts. Casting around for an alternative picture of science, they came across Thomas Kuhn's *Structure of Scientific Revolutions* (1962). By the late 1960s, J. Tuzo Wilson and Allen Cox were describing plate tectonics as a Kuhnian revolution—an attribution still debated by historians and philosophers of science.

Allan Cox, ed., *Plate Tectonics and Geomagnetic Reversals* (1973). Ursula B. Marvin, *Continental Drift: The Evolution of a Concept* (1973). Anthony Hallam, *Great Geological Controversies* (1983). H. W. Menard, *The Ocean of Truth* (1986). Homer LeGrand, *Drifting Continents and Shifting Theories* (1988). John Stewart, *Drifting Continents and Colliding Paradigms* (1990). Naomi Oreskes, *The Rejection of Continental Drift* (1999).

RACHEL LAUDAN

PLUTONISM. See NEPTUNISM AND PLUTONISM.

PNEUMATICS. The discovery of the different gas types during the third quarter of the eighteenth century caused a revolution in physical science. The new field of pneumatics made large demands on experimental technique and

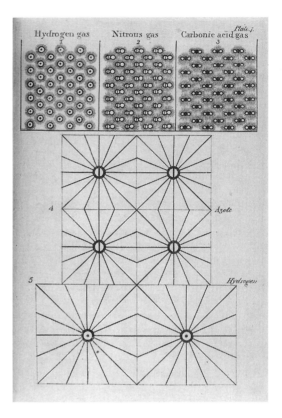

The (static) structure of various gases, indicating the repulsive caloric atmospheres of their particles.

apparatus, and required unusual accuracy in calculating the weights of small quantities of matter. It played an important part in quantifying physical science and in forging fruitful connections among the branches of natural knowledge from anatomy (as in the work of Luigi *Galvani) to chemistry (Joseph Priestley and Antoine-Laurent *Lavoisier), meteorology (Jean André Deluc and John *Dalton), and physics (Alessandro *Volta).

The English clergyman Stephen Hales, who had learned Newtonian experimentation at Cambridge around 1700, pointed the way to pneumatics in his *Vegetable Staticks* of 1727. Hales described many ways of fixing "air" in, and liberating it from, vegetable and other matter. He collected liberated air over water in a "pneumatic trough" of his invention, measured its quantity, and studied its quality; but, although he handled several chemically distinct gases, he regarded them all as the same basic substance. The variety and quantity of substances from which he drew his "air," however, supported his conclusion, which he expressed in the Newtonian style as a query: "may we not with good reason adopt this now fixt, now volatile *Proteus* among the Chymical principles ... notwithstanding it has hitherto been overlooked and rejected by Chymists, as no way entitled to the Denomination?"

Hales studied fixed air while following up his interest in the mechanics (physiology) of plants; Joseph Black came to the problem as a medical student concerned with kidney stones. For his doctoral thesis of 1754 he examined the air (carbon dioxide) released from magnesium alba (magnesium carbonate) when heated or treated with acid. He determined that it differed from common air in its inability to support combustion and respiration, and occurred fixed in the limestone implicated in urinary calculi. Novel airs then began to rise promiscuously. In 1766 Henry Cavendish identified a special "inflammable air" (hydrogen) as a product of metals dissolved in acids. In 1772 Priestley, teacher, divine, and experimental philosopher, inspired by reading Hales, announced the new species "nitrous air" (NO) and hydrochloric acid gas; and in 1774–1775 he introduced "eminently respirable air" (*oxygen), the peculiar portion of ordinary air that maintains life. In 1776 his correspondent Volta discovered a second inflammable air (methane) while gas hunting in a swamp.

In the early 1780s Cavendish, Lavoisier, and the inventor James Watt discovered that inflammable and eminently respirable air made water when sparked together. The rationale for the spark originated in Priestley's test for the respirability of gases: mix nitrous air with a sample under test and determine the contraction of the volume; the greater the diminution, the better the sample. (For oxygen the maximum contraction would be a third: $2NO + O_2 = 2NO_2$.) Volta had substituted inflammable gas for nitrous air and added the spark to speed up the process. He and other devotees of the new pneumatics devised "eudiometers" to test air by sparking. Thus they set up for themselves one of the grandest of all discoveries in physical science, the counterintuitive realization that gases that support combustion or burn freely combine to make the enemy of fire, water, and deprive it of its ancient right to be considered an element.

The discovery of the gas types led to a sweeping reformation of chemistry. It impelled natural philosophers to study the effects of heat on gases, which strengthened the caloric theory (*see* IMPONDERABLES) and supported measurements later important in *thermodynamics. It had practical consequences before the end of the eighteenth century in the application of eudiometry to ventilation, in the craze of ballooning initiated by the Montgolfier brothers, and in the use of laughing gas (nitrous oxide) as an anesthetic.

See also CAVENDISH, HENRY, AND CHARLES-AUGUSTIN COULOMB.

J. G. Crowther, *Scientists of the Industrial Revolution: Joseph Black, Henry Cavendish, Jospeh Priestley, James Watt* (1962). Henry Guerlac, *Essays and Papers in the History of Modern Science* (1977). J. L. Heilbron, *Weighing Imponderables* (1993).

J. L. HEILBRON

POLARIMETER. Although the double refraction of Iceland spar (a transparent and colorless variety of crystallized calcium carbonate) has been known since the seventeenth century, the systematic study of related polarization phenomena did not begin until the nineteenth century. In 1808 Étienne Louis Malus discovered polarization by reflection, and his observations attracted the interest of several leading scientists of the time such as François Arago, David Brewster, and Augustin Fresnel. In 1820, Fresnel, who assumed that light was a transversal wave, elaborated a mathematical theory of polarization (*see* OPTICS AND VISION).

Polarimeters measure the angle of rotation of a linear polarized beam of light passing through optically active substances such as crystals and many liquids. The very first apparatus had nonmetallic adjustable mirrors for polarizing and analyzing light; William Nicol made the first polarizing prism in 1828. In the following decades the apparatus gained better prisms and optical elements for more precise measurement of angles. Jean-Baptiste-François Soleil's polarimeters (with a quartz wedge compensator), half-shadow polarimeters, and polaristrobometers (which measure the rotations by using interference fringes) were among the most successful apparatus. Polarimeters proved to be very useful analytical and testing tools in chemistry, crystallography, biology, medicine, and later astrophysics, as well as in the chemical, pharmaceutical, and food industries. Saccharimeters, a special type of polarimeter, provided a fast method of measuring the sugar concentration of solutions.

In the 1930s the photoelectric detector began to supersede eye observation. At the same time, Edwin Land developed polarizing filters and sheets (Polaroid). Polaroids found many domestic and industrial uses. More recently, nuclear and high-energy physics have made use of special polarimetric apparatuses. Today, thanks to lasers, sophisticated optics and detectors, and computer technology, polarimeters are among the most versatile and useful instruments for scientific research and industrial analysis.

D. Clark and J. F. Grainger, *Polarized Light and Optical Measurement* (1971). Jean Rosmorduc, *La polarisation rotatoire naturelle de la structure de la lumière à celle des molécules* (1983).

PAOLO BRENNI

POLAR SCIENCE. Exploration of the Arctic and Antarctic, regions above 66E°32' North or South latitude, has been one of the most difficult and perilous chapters in the history of scientific investigation. Mortality has often been high, scientific returns often meager. The best known expeditions—those of Cook or Peary to the North Pole, and those of Scott, Amundsen, or Shackleton to the South Pole—are as remarkable for their lack of scientific interest as they are for their human drama. Most scientific information has come from scientists whose names barely register even among historians of science.

Much exploration of the North American and Eurasian Arctic from the sixteenth to the nineteenth centuries was driven by the search for Northwest and Northeast Passages—conjectural open-ocean northern routes to the Pacific. Equally important were pressures to exploit fisheries, whaling, and sealing grounds, and attempts to extend national sovereignty. Even where no obvious economic benefit lay, the enhancement of national prestige stimulated competition. The quest for personal glory and adventure also drove many of the most celebrated explorers.

Polar science includes geography and cartography, the study of flora and fauna, Arctic ethnology, and geology and paleontology; it is in this way like science in other latitudes. Its unique contribution has come in the study of ice and snow, weather and climate (including paleoclimate), and ocean circulation. Sea ice covers 10 million square miles (7 percent of the surface) of the sea; glacier ice covers about 6.5 million square miles (11 percent of the earth's surface), most of it in Greenland and Antarctica; and this mass of ice has a profound effect on the earth's thermal regime and on ocean and atmospheric circulation.

Modern polar science began with the first International Polar Year (1882–1883). It was the brainchild of Lt. Carl Weyprecht of the Austrian navy, who had discovered Franz Josef Land (80E°N, 60E°E) in 1874 as part of the search for the Northeast Passage, a typically nineteenth-century geographic enterprise under aristocratic patronage. Weyprecht argued in 1875 that geographical finds like his would be unimportant unless they enlarged scientific inquiry. He urged a year of coordinated scientific observations (using the same instruments) by manned scientific stations ringing the North and South poles. This idea, supported by the newly created International Meteorological Congress, led to an International Polar Congress in 1879. A plan for eleven Arctic and four Antarctic stations was largely carried out. It involved 700 scientific workers and produced thirty-one quarto volumes of data. Analysis of this and other data by Vilhelm

Bjerknes and his group at Bergen during World War I led to the discovery of the Polar Front and to the theory of "frontal weather."

Danish, Norwegian, and Swedish scientists played the most prominent role in polar science until quite recently. Denmark's sovereignty over Greenland (and Iceland) in the first part of the century occasioned more than thirty major expeditions. Notable among them were studies of the Greenland ice cap by the Swede Nils Nordenskiöld in the 1870s, the Norwegian Fridtjof Nansen in the 1880s, and the German Erich von Drygalski in the 1890s. Nansen's traverse of the inland ice (the first) led him to formulate a program of glaciological and geophysical investigation that guided studies of Greenland's ice (and later of Antarctica's) for almost fifty years. Shortly after, Nansen's specially built ship, the *Fram*, drifted for three years (1893–1896) locked in the Arctic sea ice between Eurasia and the North Pole, and provided data concerning ocean currents and the layering of Arctic waters. Using this, Nansen's coworker, the Swedish scientist Vagn Ekman, worked out a number of important problems in dynamic oceanography. His results, and the theoretical work of Vilhelm Bjerknes, spurred still more investigations by Johan Sandstrom and Bjørn Helland-Hansen on movement of water masses and on geostrophic currents, laying the foundations of physical oceanography.

Polar science was curtailed during World War I and, except for the relentless Danes, remained in limbo into the 1920s. In the early 1930s, interest in transatlantic flight led the United States, France, Britain, and Germany to set up overwintering stations in Greenland. The most important of these scientifically was Alfred Wegener's expedition of 1930–1931. In spite of Wegener's death in 1930, the year-long scientific program at a mid-ice station was completed. Using explosion seismology, it demonstrated that Greenland was a basin weighed down by the ice cap itself and pioneered a technique later used to map Antarctica's geology. The world economic collapse of the later 1920s cut back plans for, and delayed the publication of results of, a second International Polar Year in 1932–1933.

Antarctic science, as opposed to Antarctic exploration, effectively began with the *International Geophysical Year of 1956–1957, which established fifty-five scientific stations in Antarctica. There followed an international treaty to keep the continent open as a scientific laboratory. This also disposed of the disputes over ownership of the Southern continent that emerged in the first half of the century. Interest in Antarctica intensified with the formulation of the geological theory of plate tectonics in the 1970s, since the continent was a key portion of the giant protocontinent Pangaea. Techniques for oxygen isotope dating have made Greenland and Antarctic ice cores an indispensable part of the climate record of the last 200,000 years. In recent years, Antarctica has proven an important resource in astronomy and astrobiology partly because iron meteorites, accumulated over millions of years, were not depleted there by aboriginal hunters seeking raw materials for weapons, and partly because the dry valleys of Antarctica provide extreme environments useful in planning for research on Mars. Investigations of ozone depletion in the atmosphere also have focused on Antarctica.

Jeanette Mirsky, *To the Arctic! The Story of Northern Exploration from Earliest Times to the Present* (1934; 1997). G. E. Fogg, *A History of Antarctic Science* (1992).

MOTT T. GREENE

POLITICAL ECONOMY OF SCIENCE. During the past four centuries, the political economy of science—the nature and sources of its funding and the pertinent institutional arrangements—has changed primarily, though not exclusively, according to the political regime under which science has been pursued. In medieval Europe, the Church had created what Jacques Barzun once called the "House of Intellect"—those interested in scholarly subjects, including the phenomena of nature, pursued their interests in the universities of the time, usually as ordained clergy or as candidates for ordination. There were few divisions between the academic specialties that would later constitute "natural science." The sixteenth century, however, began to see major changes in the political economy of science, setting in motion complex adjustments between science and its environment that continue to the present.

*Galileo's era marks the beginning of modern science not only because of the intellectual and technological advances of his time, but also because of an important shift in science's economic and institutional foundations. As a professor of mathematics at the University of Padua, Galileo belonged to the least admired stratum of academic life, and he multiplied the fault by deriving inspiration from the activities of artisans. When he left Padua in 1611 to join the Medici court in Florence, however, he insisted on the title of Philosopher and Mathematician, a coupling of the more prestigious contemplative work with the slighted practical pursuit. This combination of philosophy (*physics) and *mathematics became the high road of the *Scientific Revolution. That Galileo felt he had to leave the university for a court

to take his first steps on this road indicate the importance of private patronage in the political and intellectual economy of early modern science.

Galileo also enjoyed membership in another extra-university institutional form of first importance in the cultivation of natural knowledge from the mid-seventeenth to the mid-eighteenth centuries. This was the Academy dei Lincei, an academy without walls, a group of correspondents promoted and supported by a prince (Frederico Lesi) to pursue natural knowledge. Lesi's academy did not survive him. Nor did the more substantial group, the Academy del Cimento (1657–1667), composed of Galileo's disciples, outlast the interest of its patrons, the Medicis. But by the 1660s, the institution of the Academy, which brought people together to investigate nature and serve their patrons, had become the most important sites for natural science. The Royal Society of London and the Paris Académie Royale des Sciences date from that decade.

The Royal Society provided its own financing via dues and subscriptions and a large membership made up primarily of gentleman drones. The Paris Academy consisted of a small number of salaried experts who advised the state on technical matters and devoted most of their effort to *astronomy, *geology, *cartography, and other mathematical sciences. The English model was imitated in Sweden, Holland, and the French provinces, the Parisian model in Berlin and St. Petersburg. The academic movement, which reached its height toward the end of the eighteenth century, provided institutional inspiration—though seldom the laboratory space—for the advancement rather than (as was the assignment of the universities) the spread of science (*see* ACADEMIES OF SCIENCE).

The French Revolution inaugurated a new epoch in the political economy of science. The revolutionary government suppressed both the Académie and the universities in 1793. The Académie was to some degree reborn in 1795, as the First Class of the Institut de France, but for almost a decade, scientific and technical training took place exclusively in professional schools that survived the Revolution or in the newly founded grandes école. These institutions—especially the École Polytechnique and the École Normale Supérieur in Paris—had extraordinarily good faculties and adequate facilities. Their graduates enabled France to dominate European science. However, the new regime did not succeed in replacing the old universities and Jesuit colleges with suitable higher schools, especially in the provinces. The University of Paris had "facultés" elsewhere, but these were generally poor, small, and ineffectual.

Many analysts considered the weakness of French higher education (apart from the grandes écoles) to be an important factor in the French defeat in the Franco-Prussian War (1870–1871). A number of initiatives—including financial contributions by local magnates and industrialists—sought to transcend these restraints. In 1896, after years of study, the national government reinstated the universities as comprehensive institutions and built important new science facilities at the Sorbonne. But the patterns of centralized administrative control, concentration of resources at the center, and the resulting flow of talent to Paris still hampered development.

Napoleon's invasions of Germany had aroused powerful nationalistic reactions that prompted sweeping reforms in German-speaking territories. Illustrated by the founding of the University of Berlin in 1810, these reforms, taken together laid the basis for the shift of the leadership of science to Germany. Paradoxically because of its long history of political disunity, Germany—newly united in 1871 after three wars with neighbors—had an important institutional advantage that helped it gain ascendancy in science. Each of its constituent states—Prussia, Bavaria, Saxony, and others—had its own universities; the newly unified Reich possessed twenty-two in all, and shortly after 1900, Frankfurt and Hamburg created their own municipal universities.

These schools, together with German-speaking universities elsewhere in Europe and the better Technische Hochshule, created a cooperative-competitive system that drove its faculties toward increasingly higher performance. Professors also moved between institutions based on their accomplishments, primarily publications and to some extent reputation, and they usually made a condition of their transfer (or of their staying put) the improvement of the facilities, an increase in their research provision, or the acquisition of more technical assistance. This internal elite migration spread innovation and expertise and enlarged the disciplines. A stream of important discoveries, inventions, and publications poured out of the German universities; an impressive number of students from European and other nations flowed into them.

German-speaking higher education retained the old tradition of representing whole fields of knowledge by a single full professor who supervised the work of subordinate junior faculty or graduate students in the same specialty. Under the system, new professorships could only be created at any given university by dividing an established discipline into smaller parts. Expansion could also occur by founding an entirely new university from the ground up. A strong and innovative professor could mobilize and concentrate resources; an indifferent and negligent one could ruin his discipline in his university for decades. And in any case, the low limit on the number of professors the system could absorb eventually sapped the morale of younger scientists. The problem was addressed—though never adequately solved—by various partnerships between government and private donors. In 1887, a large donation from Werner von Siemens had facilitated the creation of the state-run Physikalische-Technische Reichsanstalt (see BUREAU OF STANDARDS). The years before World War I witnessed the founding of the *Kaiser Wilhelm Society, which developed a wide network of specialized laboratories. Instigated by the chemist Emil Fischer, the humanist Adolf von Harnack, and several others, the Society attracted substantial private contributions from bankers, industrialists, and a few landed aristocrats. Their donations, together with funds allocated from the Prussian treasury, made possible the creation of an institution with historical importance for the political economy of science. But it too operated on the Führer-Prinzip: distinguished professors became institute directors with full authority—if they cared to exercise it—over their research staffs.

Invariably citing German developments in the political economy of science, the United States and Britain began to expand their educational and research facilities in the 1870s (see ENGLISH-SPEAKING WORLD). Private capital in Sweden supported the creation of the Stockholm Högskola in 1878; dedicated almost exclusively to the natural sciences in its early years, the Högskola in 1960 became the state-run Stockholm University with a broader array of academic programs. Japan as a reformist state in the same period found both German and French models instructive. German precedents shaped the design of the so-called chair system in which a senior professor managed the budget together with the pedagogical and research activities of a subordinate younger staff. French examples influenced the national organization and important details of the internal structure. Following the Napoleonic model, Japan divided itself into regions, each of which was to have its own so-called imperial university. Each institution would have several chairs per discipline, dependent on enrollment and perceived academic needs. The one-chair rule, though widely debated, was explicitly rejected in Japan. A national system of elite universities, more equal to one another than the parent French institutions, gave rise to a competitive system favorable as such to the needs of modern science. The United States operated similarly, with multiple professorships, competitive appointments, and a unique use of government (state) and private monies (see FOUNDATION; PHILANTHROPY). The Great War of 1914–1918 transformed the political economy of science. Geopolitical rivalries among several nations in the years leading up to the war had already stimulated greater political and financial support for science, and from a broader and more diverse stratum of society. Germany's rapidly paced industrialization between 1870 and 1914 had given it imposing strength in the realm of applied science—as in the large chemical companies and their corporate laboratories—to complement and support the strengths of science based in universities. The United States, and to some extent Britain and France, had also developed strength in company-based science; and although the German industrial research *laboratories may have been more numerous and more generously funded, American corporations such as Westinghouse, AT&T, and DuPont were beginning to invest in long-term research. Germany's strength in applied chemistry enabled it to hold out against the Allied Powers until the spring of 1918. All of the leading military powers, including Japan, which entered the war on the Allied side, tried with mixed success to achieve equality with Germany in science, including its military applications.

Cut off by British blockade from its usual sources of raw materials, Germany soon ran short of strategic materials, notably nitrates for fertilizers and explosives. Fritz *Haber, later director of the Kaiser Wilhelm Society's Institute for Physical Chemistry, had developed a means of producing ammonia synthetically in 1908. This achievement, when scaled up industrially, made Germany independent of imported nitrates. Haber also developed and oversaw the deployment of poison gas. The Allied nations faced the problem of finding substitutes for strategic materials previously obtained from Germany—optical glass, laboratory glassware, dyes, fine chemicals, and magnets. By the war's end, France, Britain, and the United States could make what they needed. Scientists on both sides helped to improve weapons, airplanes, radios, medical treatments, and so on. Agencies set up to mobilize scientific manpower and to provide it with resources—the Department of Scientific and Industrial Research in Britain, the National Research Council in the United States, and equivalents in the other belligerent countries—continued in many cases into the peace.

Imperial Russia was the first casualty of the war, not only because of inadequate military equipment, leadership, and training, but also because of its general backwardness in science. Following its defeat by Britain and France in the Crimean War (1854–1856), Russia had made efforts to catch its European rivals, founding several new universities, freeing the serfs, encouraging foreign investment, and promoting railway construction and industrialization. Significant weaknesses persisted, however. The underdeveloped state of Russian capitalism meant that private philanthropic support of science—increasingly important in Europe, the United States, and even Japan—had scarcely begun in Russia. With the Bolshevik Revolution in 1917, many Russians and their former allies believed that even the gains made in Czarist Russia would be lost. Nonetheless, the new regime preserved nineteenth-century advances and used them as a partial basis for subsequent development.

Several factors contributed to a smoother initial transition from one system of political economy in science to another than many observers expected. The new leadership dominated by V. I. Lenin, however autocratic by basic instinct, was flexible at first. Vigorous attempts by radical Bolsheviks to abolish the Academy of Sciences were defeated, with Lenin's concurrence. The regime desperately needed the services of the well-trained scientists and technical experts in rebuilding the country. Many professors, including leading figures like Ivan *Pavlov, cooperated with the government. Moreover, the more prominent Bolshevik revolutionaries shared the Enlightenment vision of science. For them, science was not only the preeminent but also the only valid form of knowledge and the only reliable template for the reconstruction and maintenance of society.

Bolshevik radicals viewed the relatively tolerant policies of the 1920s as expedient at best, a betrayal of revolutionary egalitarianism at worst, and they were determined to dismantle the Academy of Sciences as a bastion of elitist reaction inherited from a discredited regime. Others charged the Academy with being obsessed with pure science, and insisted that all state agencies based on expert knowledge should have a strongly applied orientation. Many also favored a decentralization of research activities as a means of bringing technical knowledge closer to the clienteles that needed it. Stalin's consolidation of power created a favorable climate in which radicals might realize their vision of "science." They were only partly successful. Members of the Academy admitted a few party activists with minimally appropriate credentials, and opened its doors to engineers for the first time. Equally important in the Academy's survival was Stalin's own preference for centrally controlled institutions, even if they harbored individuals unsympathetic to his goals. The Academy of Sciences not only survived but became the institutional centerpiece and dominant agency in a new, highly centralized political economy of science. With minimal changes, this structure persisted almost to the end of the twentieth century.

Though the most extreme effects of the political economy of science were manifested in Russia, World War I had a large impact on the political economy of science in many countries. Scientists and their work in defeated Germany and Austria suffered intense privation in the first three years after the war, giving the hundreds of Japanese scientists who had earlier studied there the opportunity to repay their academic benefactors with gifts of cash, research equipment, and laboratory animals. Scientific establishments in the United States, Canada, and Britain acquired material resources on an unprecedented scale. Japanese scientists—accustomed to working under conditions of chronic privation—viewed the war as a "blessing from heaven." One consequence of the new largesse available to scientists in Japan was the founding of the Research Institute for Physics and Chemistry (1917–1922). A committee of three scientists proposed an Institute organized precisely in the manner of the Kaiser Wilhelm Society. With the active involvement of the wartime prime minister, Count Shigenobu Okuma, a finance committee raised large private gifts from the country's wealthiest citizens, the nobility, and the Imperial Family. Parliament matched these contributions, as in Germany, helping to create a more robust political economy of science than had existed in Japan previously.

Yet another result of wartime experience had broad implications for science's political economy in many—probably most—of the nations active in research. This was the establishment of formalized systems by which investigators could submit research plans to review committees of peers and request a budget for particular projects. In many countries (Germany being an exception), pre-war university budgets had made no specific provision for research activity, even though research was becoming an essential part of the academic role. Scientists who worked in specialized laboratories—the Pasteur Institute in Paris, any of the Kaiser Wilhelm Institutes, the Carnegie Institute of Washington, the Solvay Institute (Belgium), and others—naturally had budgets dedicated to research. And in many countries, scientists could and did submit ad hoc requests to governments, private foundations, and other potential benefactors. In the wake of the war, governments began to create state agencies established specifically for the purpose of funding research without respect to the particular institutional affiliations of applicants. In some cases—Britain, the United States, and Japan among them—a newly created National Research Council performed this function; Japan created a special program for grant applications by scientists through its Ministry of Education. The Conseil National pour la Récherche Scientifique (CNRS) and Consiglio Nazionale delle Ricerche (CNR) had their beginnings in the interwar period. Germany created a number of special, self-help government and private agencies during the Weimar Republic.

By the 1930s, most features of the political economy of science that would appear in the twentieth century had cropped up in one country or another. Except in the Soviet Union, academies of science had come to function—when they existed at all—mostly as institutions for honoring scientific achievement after the fact, rather than supporting it materially in advance. The fundamental reason for the decline of the academies of Western Europe from the leadership of science in the eighteenth century to purveyors of honors in the twentieth was the professionalization of science in the nineteenth century and the establishment in most countries of the modern research university.

The Japan Society for the Promotion of Science (JSPS), founded in 1932, had as its primary purpose the allocation of financial support, after reviewing proposals, to large research projects considered important to national objectives but too costly for single universities to support (see ADVANCEMENT OF SCIENCE, NATIONAL ASSOCIATIONS FOR THE). Itself a product of the surging militarism that followed Japan's seizure of Manchuria in 1931, the JSPS symbolized the trend toward ever-greater state influence in, and control over, science characteristic of the dictatorial states of the twentieth century. The trend began with World War I and continued with the Stalinist program in the Soviet Union and the Nazi regime in Germany. But apart from certain police interventions, the Nazis used ideological rather than institutional tools to achieve their purpose. The pre-Nazi political economy of science in a formalist sense remained intact even as Jewish academics were ejected or imprisoned (see NAZI SCIENCE). Partly because of the Soviet Union's seemingly rapid development in the 1930s, and partly also owing to their fears of Nazism and Fascism, some scientists—especially in Britain, France, and the United States—began to promote Soviet notions of the political economy of science, or at least to debate the relative merits of democratic and socialistic support systems. Much of what transpired during World War II with respect to the political economy of science repeated the experience of the Great War on an even larger scale. The United States as well as Germany, Japan, and the Soviet Union sought to capitalize on the late 1930s discovery of fission by developing an atomic bomb for wartime use (see MANHATTAN PROJECT). But these projects are only the best-known examples of institutional changes in science's political economy. The earlier development of the cyclotron by E. O. Lawrence after 1929 set in motion a process of institutional growth in high-energy physics that continued almost to the present (see ACCELERATOR). A similar trend later appeared in other research specialties, and continues to the present, where it most recently has arrived in the biological sciences (see BIG SCIENCE; GENOME). The early decades of the Cold War saw a continuation of wartime patterns. Rivalry in almost every sphere of human activity led the United States and the Soviet Union to inexorably greater expenditures on science and technology, exemplified by the *Sputnik program of artificial satellite development and the successive *Apollo* missions of lunar exploration (see SPACE). The Soviet satellites followed the Soviet lead as best they could. Western Europe did the same with the United States as its model, and after Americanizing, have caught up in some respects. The innovative, cooperative research enterprise of *CERN, for example, brought western Europe to parity with the United States in accelerator and particle physics, and it will soon be ahead.

Partly because of Soviet wartime successes and the evident promise of rapid economic and technological development, many developing nations after World War II tried to copy the Soviet model of the political economy of science. Building on a small-scale model developed in the 1930s before the Japanese invasion of 1937, the People's Republic of China under Mao Zedong imported hundreds of Soviet technicians and science advisers. With their help, China in the 1950s created a near-perfect replica of the Stalinist system, based on the central role of the Chinese Academy of Sciences. Attracting émigré scientists from the United States and elsewhere, the PRC managed to develop its own atomic bomb in 1964 and even more advanced weapons in later years. While the Chinese initiatives may seem extreme, the general model of state-led development in science they embodied was more broadly characteristic of the period as a whole.

It appears that the state-led pattern of growth in the political economy of science is sharply decelerating. Countries that once had no private universities, such as Australia and Britain, now have one or more; in others, such as the United States, the ranking private universities are better funded than ever, while the leading public institutions sometimes have to struggle to stay even. Japan, long a bastion of French-style étatism in higher education, has announced a plan to privatize virtually all so-called national and other public universities over the next decade or so. Russia, in the aftermath of the Soviet Union's dissolution, has begun to restore a role for the private sector in science that had disappeared in the Revolution. The government of the United States, at the insistence of Congress, in 1993 terminated the costly Super-Conducting Super Collider in Texas. And in a public-private competition between researchers investigating the human genome, a private firm won a tentative victory over a government-funded team. How far these trends will proceed or when they might reverse cannot be anticipated. The future of the political economy of science is as difficult to predict as the future of nation-states themselves. But the two have developed in tandem, and each is a work in progress.

See also ASIA; LATIN AMERICA; WORLD WAR II AND COLD WAR.

Henry Lyons, *The Royal Society, 1660–1940: A History of its Administration under its Charters* (1944). Joseph Ben-David, *The Scientist's Role in Society: A Comparative Study* (1971). Roger Hahn, *The Anatomy of a Scientific Institution: The Paris Academy of Sciences, 1666–1803* (1971). Alan D. Beyerchen, *Scientists Under Hitler: Politics and the Physics Community in the Third Reich* (1977). Mary Jo Nye, *Science in the Provinces: Scientific Communities and Provincial Leadership in France, 1860–1930* (1986). James R. Bartholomew, *The Formation of Science in Japan: Building a Research Tradition* (1989). Government-University-Industry Research Roundtable, ed., *The Academic Research Enterprise Within the Industrialized Nations: Comparative Perspectives* (1990). Jeffrey Allan Johnson, *The Kaiser's Chemists: Science and Modernization in Imperial Germany* (1990). Loren Graham, *Science in Russia and the Soviet Union* (1993).

JAMES BARTHOLOMEW

POLYMER. The Swedish chemist Jöns Jacob *Berzelius coined the word "polymer" (meaning "multiple parts") in 1830 to refer to substances such as ethylene (C_2H_4) and butylene (C_4H_8) that share the same reduced empirical formula but whose molecular formulas are multiples of it. Natural polymers—starch, cellulose, proteins, and rubber—have been known empirically for centuries, but until the twentieth century they were not considered to belong to Berzelius's category. As late as 1920, chemists regarded them as small molecules loosely associated with each other in a colloidal suspension; the presumed small molecular weights had even been confirmed repeatedly by chemical analyses. The analyses were always problematical, however, since these substances refused to form clean crystals with definite melting points, criteria that had traditionally marked chemical substances as pure and homogeneous.

The Zurich chemist Hermann Staudinger took polymers to be "macromolecules," huge molecular chains made up of a single unit (a "monomer") joined head to tail perhaps by the hundreds, and connected by ordinary chemical bonds. For starch and cellulose the monomer was a simple sugar, glucose; for proteins, amino acids; and for rubber, a molecule called "isoprene." Staudinger supported his viewpoint through the methods of classical organic chemistry, but he also applied new physical methods, such as the ultracentrifuge (*see* CENTRIFUGE AND ULTRACENTRIFUGE), the ultramicroscope, and the diffraction of *X rays. His arguments had convinced most chemists of the reality of macromolecules by the mid-1930s.

Even without a modern understanding of the molecular character, important empirical work on polymers long predated Staudinger's contributions. The nitration of cellulose in the 1840s led to the production of smokeless gunpowder in the following generation. Mixtures of nitrocellulose with other organic compounds led to the first mass-marketed manufactured polymer materials: collodion, xylonite, and celluloid. The last of these was the earliest chemical *"plastic," or shapeable material. Despite its flammability, celluloid found wide application in the late nineteenth century for combs, shirt collars, photographic films (*see* PHOTOGRAPHY), and other articles.

In 1910, Leo Baekeland produced the first wholly synthetic plastic polymer from the reaction of phenol and formaldehyde. "Bakelite" proved much superior to celluloid in many respects. Over the next decades a stream of new plastics emerged from corporate laboratories: polymerized versions of acrylic acid, vinyl chloride, styrene, ethylene, and many others.

Plastics were by no means the only successful commercial polymers. Like celluloid, cellulose-based "rayons" date to the late nineteenth century, and the purely synthetic *"nylon" developed by DuPont Corporation created a commercial revolution during the 1940s. Orlon, dacron, and other synthetic fibers have dramatically supplemented the world of natural textiles. Chemists devoted attention to natural rubber and to developing artificial substitutes for it to make industrial nations independent of natural supplies in the tropics. Already by World War I German chemists had barely adequate substitute materials. The first adequate artificial rubber materials were "neoprene," developed by DuPont in the 1930s, and "buna" rubber, developed in Germany. First produced in the early 1930s, both proved important in World War II.

R. B. Seymour, ed., *Pioneers in Polymer Science* (1989).
Y. Furukawa, *Inventing Polymer Science* (1998).

A. J. ROCKE

POLYP. During three sessions of the Paris Academy of Sciences in March 1741, René-Antoine Ferchault de Réaumur read out two letters he had received from the Genevan naturalist Abraham Trembley. These concerned observations and experiments on small aquatic "organized bodies" whose animal or plant nature was in doubt. Replying to Trembley on 25 March, Réaumur stated that these "polyps," as he called them, were "certainly animals." Trembley thought they resembled the Hydra, the many-headed monster challenged by Heracles. When he cut off one head, two new ones grew in its place. Trembley used the name Hydra to stress the extraordinary ability of freshwater polyps to regenerate amputated parts, a concept present in the Greek myths of the Hydra and of Prometheus's liver, which was devoured by an eagle during the day and regrew at night. In 1758, Carl *Linnaeus used *Hydra* as a general term for polyps without distinguishing the three species described by Trembley in 1744 (*Mémoires, pour servir à l'histoire d'un genre de polypes d'eau douce, à bras en forme de cornes*) and now identified as *Chlorohydra viridissima, Hydra vulgaris,* and *Pelmatohydra oligactis.*

When Trembley's work appeared in 1744, the phenomenon of regeneration was not unknown. In 1686, the Paris Academy had discussed the regeneration of lizard's tails and in 1712 Réaumur described his own experiments confirming the popular belief that crayfish can replace missing appendages such as chelae. But Trembley's experiments showed more than just replacement. By sectioning the polyps transversally or longitudinally, or into many pieces, he demonstrated that a small piece of a polyp could regenerate into a complete animal and that therefore there existed animals that could be multiplied by artificial divisions. One polyp cut into four pieces could generate four complete animals. Trembley also managed to join the anterior half of one specimen to the posterior of another. His astonishing experiments aroused much speculation among naturalists and philosophers during the second half of the eighteenth century on the differences between animal and vegetative life, on the existence of an animal soul responsible for the form of an organism, on the modes and causes of animal reproduction, and on the nature of animal organization. Julien Offroy de La Mettrie and Denis Diderot saw Trembley's polyp as proof that organization was inherent to organic matter and not the result of mechanical laws, whereas supporters of the main competing theories of generation—*epigenesis and pre-formation—adapted their system to include an explanation of the phenomenon of regeneration. In 1768, Lazzaro Spallanzani showed that regenerative phenomena took place in many animal species such as earthworms, slugs, snails, salamanders, tadpoles, and frogs. Trembley's work on polyps is considered to be the origin of experimental research into regeneration.

M. Trembley, ed., *Correspondance inédite entre Réaumur et Abraham Trembley* (1943). Virginia P. Dawson, *Nature's Enigma: The Problem of the Polyp in the Letters of Bonnet, Trembley and Réaumur* (1987). Charles E. Dinsmore, ed., *A History of Regeneration Research* (1991).

RENATO G. MAZZOLINI

POPULARIZATION has many meanings, but generally refers to the creation and diffusion of information for audiences beyond the specialists who create new knowledge. Since knowledge is shaped by the forms in which it is presented, popularizations of science themselves constitute new (and different) knowledge about the natural world.

By publishing their results in Latin, early natural philosophers limited their audience to fellow scholars. They recognized that writing for wider audiences might attract financial and political support—or attack. The trouble aroused by Galileo's publication of his *Dialogue on the Two Chief Systems of the World* (1632) in Italian suggests the danger of broadcasting esoteric knowledge. While *Galileo made his ideas more widely available, he also challenged the Roman Catholic Church in a way it decided not to ignore. Nonetheless, by the end of the seventeenth century, natural philosophers regularly published in their own languages, thus contributing to the rapid spread of new ideas.

In the eighteenth century, the professional and merchant classes began using *natural philosophy as a tool for social advancement. Natural knowledge appeared in *textbooks, *encyclopedias, and other material within a broader discourse, often closely associated with lessons on moral virtues and the revealed beauty of God's creation. As technical knowledge became more complex, these new expositors developed a language and style of their own: simplified, with more dramatic narrative conventions, they recast

knowledge into different forms for different purposes. *The Newtonian System of Philosophy, Adapted to the Capacities of Young Gentlemen and Ladies* (1761, by John Newbery, writing as Tom Telescope, A.M.), for example, presented technical information in the context of instruction in good manners, moral values, and virtuous citizenship.

By the 1800s, concern with moral instruction gave way to the interests of a literate middle-class seeking advancement through education. Later editions of *The Newtonian System of Philosophy* focused on technical issues. A similar separation of philosophy from technical theory marked Jane Marcet's *Conversations on Chemistry* (1806), which had special appeal to women's academies in the United States because it avoided spiritual lessons in favor of concrete instruction in laboratory science. The Royal Institution in London began popular lectures on scientific topics in 1826, emphasizing entertaining spectacles of technical knowledge.

The shift from private knowledge to public presentation also manifested itself in the collections of minerals, plants, and other items brought to Europe from global explorations. Initially housed in private cabinets of curiosities, these collections became sites for scientific investigation. Beginning in the eighteenth century, the great colonial powers established new institutions (the British Museum in 1759, the Muséum National d'Histoire Naturelle in 1793) to showcase their empires; and public exhibitions mixed education, entertainment, and celebration of political power. Local natural history museums became a tool for colonies to express their advancing level of "civilization." The great international expositions and world's fairs of the nineteenth and twentieth centuries similarly provided opportunities for combining explanations of science and technology with claims to international status.

By the mid-nineteenth century, the specialization of science and the growth of the middle classes and industrial technology had changed the demand. Scientists and nonscientists produced books, magazines, dictionaries, juvenile literature, and lectures for those looking for "useful knowledge," as well as for those seeking inspiration in natural history or general science. *Scientific American* (1845), for example, served inventors and mechanics. Camille Flammarion's *Astronomie populaire* (1880) was widely translated, carrying with it his reflections on the probability of extraterrestrial life as well as basic information on astronomy. Public interest in fields such as botany and phrenology also shaped popularizations, as local naturalists and social reformers used them to seek resources for their own activities, and to create their own local knowledge of the natural world. Scientific proselytizers, notably John Tyndall and "Darwin's bulldog" Thomas Huxley, undertook extensive lecture tours, later publishing their lectures in magazines and books. Museum curators shaped their displays both to generate new knowledge and to appeal in entertaining ways to broad publics.

After the emergence of mass markets, stories about scientific findings and technological developments began to appear regularly in newspapers around the turn of the twentieth century. Shaped by the constraints of journalism, these stories shunned philosophical reflections, focusing on simplified reports of new findings. By 1910, motion pictures were featuring nature and the wilderness, providing education mixed with entertainment (*see* FILM AND TELEVISION). In the mid-twentieth century, scientifically trained authors such as Isaac Asimov turned to the development of *science fiction.

Leaders of the new scientific societies and research institutions created at the end of the nineteenth century learned that advocating "public science" could lead to greater financial and political resources. In the 1910s, the new science-based industries of medicine and chemistry built public affairs operations in the United States to defend against what they perceived to be charlatans and to lobby for laws and investment that would serve their interests. After World War II, the growth of government-funded science generated new political needs. Scientists used popularizations to "sell science," to argue that investing in science created practical benefits for society (such as jet engines, radar, and penicillin). Scientists also committed resources to popularization to help educate potential recruits to science. Science and technology news increased in newspapers, books, radio, television, and other media. Worldwide, scientists and governments built hundreds of interactive science centers, a new form of science museum characterized by specially constructed didactic exhibits that relied heavily on entertainment. These new media also carried articles and exhibits questioning the role of science in society. Critics of nuclear power, genetic engineering, oil drilling, and other technologically based elements of industrialized culture combined the tools of popularization and public relations to argue their positions (*see* ANTISCIENCE).

By the beginning of the twenty-first century, popularization had become a network of media and institutions for creating knowledge to serve the interests of various groups. Popularizations are not only simplifications of technical information, but also sites for expressing new ideas about the natural world.

See also ENCYCLOPEDIA; FILM AND TELEVISION; NATIONAL PARKS AND NATURE RESERVES; PLANETARIUM; ZOOLOGICAL GARDEN.

Roger Cooter, *The Cultural Meaning of Popular Science: Phrenology and the Organization of Consent in Nineteenth Century Britain* (1984). Robert W. Rydell, *All the World's a Fair: Visions of Empire at American International Expositions, 1876–1916* (1984). Terry Shinn and Richard Whitley, eds., *Expository Science: Forms and Functions of Popularisation* (1985). John Burnham, *How Superstition Won and Science Lost: Popularizing Science and Health in the United States* (1987). Susan Sheets-Pyenson, *Cathedrals of Science: The Development of Colonial Natural History Museums During the Late Nineteenth Century* (1988). Marcel C. LaFollette, *Making Science Our Own: Public Images of Science, 1910–1955* (1990).

BRUCE V. LEWENSTEIN

POPULATION GENETICS is the science concerned with the inheritance and distribution of gene frequencies in a given population. Evolutionary population genetics, from which this branch of study arose, deals with changes in these distribution and inheritance patterns over time. The roots of the discipline originated in the period following the publication of Charles *Darwin's *On the Origin of Species* in 1859, when a debate raged within biology about the kind of traits upon which natural selection works. Were they small physical changes in an organism, as Darwin, August Weismann, and Alfred Russel Wallace thought, eliciting a slow, gradual *evolution? Or did nature proceed by jumps, fashioning new species in rapid discontinuous leaps, as Thomas Henry Huxley and Francis Galton believed? Galton, Darwin's first cousin, argued that selection of continuous traits would be rendered ineffective by the law of regression—a mathematical formulation of the tendency of offspring characters to regress to the mean of the population. The further down an ancestral path from a new trait, such as a bent nose, the likelier it became that the nose would straighten out. If so, natural selection acting on small, continuous changes could not be responsible for evolution. Galton concluded that only by acting upon discontinuous novelties, such as a completely new nose form, could selection bring about evolution.

With the rediscovery of Gregor *Mendel's laws in 1900, the old debate quickly found new proponents. The experiments of Hugo de Vries on the evening primrose in his Amsterdam garden gave rise in 1903 to the theory that spontaneous, internal, and discontinuous mutations in the hereditary substance were genetic mechanisms capable of explaining evolution. Mendelians, led in England by William Bateson, came to believe that Darwin, who thought of natural selection as a creative force, had it wrong on both counts:

Evolution was neither gradual, nor brought about by selection. It was disruptive, motored by discontinuous, internal changes to the hereditary material unaffected by the environment. The most nature could do was to eliminate those deleterious varieties presented before her and propagate the rest.

A Biometric School, led by Karl Pearson and Walter Frank Raphael Weldon in England, rejected the Mendelian recourse to a heredity based on unseen, theoretical, genetic "factors." They defined inheritance instead in terms of Galton's law of regression, based on the wide range of measurable, perceivable, and continuous physical variations. Pearson argued that Galton had misinterpreted his own law: if the relation between traits of offspring to the mean of their parents, rather than that of the population, were considered, swamping (the gradual erasure of a new trait) would disappear, and Darwin's picture of gradual, continuous evolution by natural selection would be upheld. Fierce debates raged between the biometricians and the Mendelians in England. Mendelian particulate heredity and Darwinian gradual evolution by natural selection appeared to be diametrically opposed.

Darwin's proposed mechanism of natural selection was by now coming under severe attack. In 1902 the Swedish biologist Wilhelm Johannsen had shown that natural selection acting on continuous variation could not overcome Galton's law of regression: it failed to induce evolution within pure lines. Experimental Mendelians took this as strong evidence against Darwinian gradualism. They refused to admit a role for adaptation and selection in the control of those novel variations produced by mutation in their laboratories. At the opposite extreme, field naturalists, who saw variation manifested abundantly, rejected mutationism and endorsed the principle of gradualism, but advocated Lamarckian mechanisms of heredity to buttress, or replace, what they took to be inadequate selection. Embryologists and paleontologists added to the chorus of discontent with natural selection, advocating *Lamarckism and other forms of orthogenetic, or directed inheritance, to explain the adaptations they recognized (see ORTHOGENESIS). The divided biological landscape made imperative the need to establish both the kind of variations upon which selection worked, and the extent to which such a force could be shown to be responsible for evolution.

The first problem was gradually settled in the first two decades of the twentieth century through the work of W. E. Castle, Edward Murray East, H. Nilsson-Ehle, and the so-called Fly Room group of Thomas Hunt *Morgan. Their experimental results showed how continuous hereditary variation could be explained by discontinuous variation produced by small genetic mutations. The Biometric-Mendelian debate had been settled, but entrenched conceptions proved hard to dispel. In 1908 an English mathematician, G. H. Hardy, and a German physician, Wilhelm Weinberg, independently derived a simple yet crucial quantitative rule, expressing the idea that gene frequencies, or proportions, would remain constant unless acted upon by external forces. The Hardy-Weinberg equilibrium, $P^2 + 2PQ + Q^2 = 1$, where P is the frequency of a dominant allele A, and Q the frequency of a recessive allele a, had some very useful properties. (Alleles are alternative genes at the same locus.) In a population with a gene locus with a dominant allele A and a recessive allele a, only two phenotypes are visible: homozygous recessives aa, and a class exhibiting the dominant trait comprising a mix of AA and Aa individuals. The Hardy-Weinberg rule allowed calculation of the proportion of heterozygotes, or carriers, of recessive alleles for human genetic disease in the population, and thus a better understanding of the propagation of disease. Notwithstanding the importance of this basic and useful rule, it remained to be shown whether selection, acting on discontinuous genes and in conjunction with other variables, would explain evolution. R. A. Fisher and J. B. S. Haldane in England, and Sewall Wright in the United States, tackled this problem and founded theoretical population genetics.

Adopting a biometric, mathematical population approach, the three men defined evolution as the differential inheritance and propagation of gene frequencies over time. By assigning adaptive values with respect to fitness to different gene alleles, it became possible to calculate how selection could fashion their respective frequencies in the population. Fisher worked primarily on models with low selection pressures on individual genes in large populations; Haldane on high selection pressures on individual genes in large populations; and Wright on intermediate selection pressures on interactive gene systems in small, partially isolated, interbreeding populations. Although their models emphasized different aspects of evolution and led to disagreements, they all used the gene, not the individual organism, as the functional unit of selection. Together with the quantifiable factors of mutation and migration rate, effective population size, and mating behavior, selection was shown to suffice as a mechanism for evolution. The force it exercised was powerful enough to render Lamarckian and other forms of directed change superfluous. Haldane, for example, estimated the mutation rate of a deleterious gene in a human population, that for hemophilia, and suggested that differential susceptibility to disease might have driven much of human evolution.

Theoretical population genetics, developed in the 1920s and generally adopted after 1932, dispelled the legacy of anti-Darwinian feeling by synthesizing genetic gradualism with Darwin's theory of natural selection. But while providing a correct picture of genetic change in a given local population, what has been called the "bean-bag" genetics of these pioneers failed to explain adequately the interaction between genes and gene systems, macro-evolutionary changes such as adaptation, the origin of higher taxa and novel evolutionary forms, and the multiplication of species. In order to satisfy the organismic phenomena studied by ecologists, systematists, paleontologists, and students of behavior, and to complete what became known as the Evolutionary Synthesis, a further step of translation and modification of the mathematical models to the more complex realities of natural populations was necessary. The work of the Russian school of experimental population genetics led by Sergei Chetverikov, the interpretation of Fisher's work by the ecologist E. B. Ford and of Wright's work by the geneticist Theodosius *Dobzhansky, the labors of the systematist Ernst Mayr, and the popularizations of Huxley's grandson Julian all figured significantly in this endeavor. The synthesis was generally achieved by mid-century.

Population genetics has had important implications for a range of fields from *eugenics to breeding, *ecology, and demography. With the development of techniques in molecular genetics, human population genetics has been employed in studies traditionally dominated by *anthropology, archaeology, and linguistics to help researchers understand the history of global and local migration patterns, language evolution, and cultural and technological diffusion. Medical research has increasingly employed methods from population genetics in the study of disease.

William B. Provine, *The Origins of Theoretical Population Genetics* (1971). Richard C. Lewontin, *The Genetic Basis of Evolutionary Change* (1974). Peter Bowler, *The Eclipse of Darwinism: Anti-Darwinian Evolution Theories in the Decades around 1900* (1983). L. L. Cavalli-Sforza, P. Menozzi, and A. Piazza, *The History and Geography of Human Genes* (1994). Ernst Mayr and William B. Provine, eds., *The Evolutionary Synthesis: Perspectives in the Unification of Biology*, 2d ed. (1998).

OREN SOLOMON HARMAN

POSITIVISM AND SCIENTISM. Positivists believed that empirical knowledge is the only kind of knowledge worth having (apart from logic and mathematics). The best examples of empirical knowledge are the most successful sciences, above all physics. Positivists also had a political vision—science should displace theology and philosophy as the source of social ideology, and the scientific community should lead and be the model for the rest of society. The idea that society should be based on scientific principles is also sometimes referred to as scientism. Positivism reached its height between the 1830s and the 1960s. Although always hotly debated, it has in the last two centuries been the single most influential account of science and its relationship to society.

Positivism was named and first formulated by Auguste Comte, a French philosopher deeply impressed by the sciences. In his vastly influential *Cours de philosophie positive* (1830–1842), he argued that human thought had passed through three stages. In the first, or theological, stage, humans attributed observable phenomena to the actions of supernatural beings. In the second, metaphysical, stage they put phenomena down to mysterious forces or energies. In the third, scientific, stage, which Comte believed the human race was just entering, they would explain the phenomena by theories based on observation, hypothesis, and experiment. This positivistic stage could be further subdivided. Each successive science was founded on a prior one. Mathematics came first, followed by physics, chemistry, biology, and sociology, the last still to become truly scientific. Once sociology revealed the laws governing social interaction, scientists (the priests of the positivist world) would be able to transform society and the political order.

Comtean positivism made its mark across Europe, in the United States, and in *Latin America. Influential French scientists such as Claude *Bernard and Marcellin Berthelot adopted positivism as their philosophy of science, and promoted it as a social philosophy in the 1870s. The English philosopher John Stuart Mill translated the *Cours* into English. Its ideas were disseminated in works as disparate as the novels of George Eliot and the essays of Herbert Spencer. The American diplomat and educator Andrew Dickson White drew on positivist theory in his plans for a nonsectarian college. In 1867, he put his positivist ideas into action when he became first president of Cornell University. He also found time to write the positivist *History of the Warfare of Science with Theology in Christendom* (1896). Political elites across the newly independent countries of Latin America adopted positivism as a guide for social action in nations seeking to modernize and throw off their Catholic past.

Positivism was revived and reinvented in the 1920s in Germany and Austria. At the University of Vienna, the Vienna Circle gathered around

Moritz Schlick. Among its important members were Rudolf Carnap, Kurt Gödel, and Otto Neurath. At the University of Berlin, the Society for Empirical Philosophy, which included Carl Gustav Hempel, assembled around Hans Reichenbach. The Vienna and Berlin groups jointly called themselves logical positivists, or logical empiricists, because they joined traditional positivism and empiricism with the logical concepts of Gottlob Frege and Bertrand Russell. Not officially members of either group, both Ludwig Wittgenstein and Karl Popper nonetheless had close associations with logical positivism. In 1929, the Vienna Circle issued its manifesto, the *Wissenschaftliche Weltauffassung* (the Scientific Conception of the World).

The logical positivists targeted the idealism that had dominated German philosophy in the last half of the nineteenth century, deeming it obscurantist and authoritarian. Instead, they looked to the empiricism of Hermann von *Helmholtz and Ernst Mach, to the mathematics of Henri Poincaré, and to the theories of special and general *relativity of Albert *Einstein. Although they had their differences, as a group they held that all meaningful statements had to be founded on observation (the verification principle), that metaphysical speculation made no sense, that scientific theories had a hypothetico-deductive structure, and that the sciences would one day be unified and reduced to the most basic science, physics. Because they insisted on the careful analysis of language, their kind of philosophy became known as analytic. Otto Neurath, who most fully expressed the social theory of the group, believed that the approach to social problems, as to scientific problems, should be co-operative, piecemeal, and technological.

In 1933, as the threat of National Socialism became more menacing, the movement, with its high proportion of socialists, Marxists, Jews, and atheists, began to scatter. Many moved to the United States, taking positions at such prestigious institutions as Chicago, UCLA, and Princeton. Together they planned and began publishing a series called the *Encyclopedia of the Unified Sciences*. In the 1950s in the United States, when science was appreciated for its part in winning World War II, scientists welcomed positivism, philosophers admired its rigor, and historians of science, many of whose traditions went back to Comtean positivism, found it congenial.

In the 1960s, positivism came under attack from what was later known as the post-positivist movement. Norwood Russell Hanson, Stephen Toulmin, Thomas S. Kuhn, Paul Feyerabend, and Larry Laudan challenged the idea that observations could provide an unproblematic basis for science. They reiterated the point, already made in the nineteenth century by figures such as William Whewell and Pierre Duhem, that observations were theory-laden. Most of the so-called post-positivist philosophers in fact remained sympathetic to many aspects of the positivist program. The publication of Kuhn's *Structure of Scientific Revolutions* (1962) as a volume of the *Encyclopedia of the Unified Sciences* indicates as much. Nonetheless, the critique of positivism dominated science studies in the last third of the twentieth century.

Phillip Frank, *Modern Science and its Philosophy* (1949). Walter Simon, *European Positivism in the Nineteenth Century* (1963). Alberto Coffa, *The Semantic Tradition from Comte to Carnap* (1991). Mary Pickering, *Auguste Comte* (1993).

RACHEL LAUDAN

POSTMODERNITY. See MODERNITY AND POST-MODERNITY.

POWELL, John Wesley (1834–1902), American geologist and ethnologist.

Powell owed his name and his spotty education to his Methodist parents, who spent their lives promoting the Gospel in one small community after another. Local naturalists encouraged his interest in nature and the outdoors, and on coming of age Powell taught in a rural school to earn enough money to attend a few semesters of college. He might have remained a teacher had he not joined the Union forces at the outbreak of the Civil War. He was rapidly promoted and served with distinction in spite of losing an arm at the Battle of Shiloh.

After the war, Powell taught *geology, first at Illinois Wesleyan and then at Illinois State Normal University. After several long solitary expeditions to the West, he brought his wife and some of his students on a trip to Colorado. There he conceived the idea of taking a boat through the unexplored canyons of the Colorado River. Supported by funds from private sources and the Illinois State Natural History Society, he and nine companions, none of them with scientific training, set off in 1869 on a 1,000-mile journey. Four did not make it. The rest succeeded, and Powell became a national hero.

During the early 1870s, the U.S. Congress supported three surveys, well-planned and staffed, to lay out the West for settlement and development: Clarence King's geological exploration of the 40th parallel (the route of the transcontinental railroad); F. V. Hayden's survey of the territories of the United States; and George Wheeler's survey of the lands west of the 100th meridian. In 1870, Congress created a fourth survey, the forerunner of the Geographical and

Geological Survey of the Rocky Mountain Region, and Powell, thanks to his newly won reputation, was placed in charge. In 1871–1872, he made a second trip down the Colorado that was more productive scientifically than the first. It resulted in a topographical map of the region, a rich haul of drawings and photographs, and Powell's monograph, *Exploring the Colorado River of the West* (1875).

Thereafter, Powell worked productively with three distinguished scientists, Grove Karl Gilbert, Clarence Edward Dutton, and William Henry Holmes. In the American West, they found geological structures previously unknown and evidence of geological forces, particularly forces of erosion, on an unprecedented scale. Powell's most important contributions were the recognition of the immense time span represented by the rocks revealed in the Grand Canyon (geologists and physicists then fiercely debated the age of the earth) and his classification of streams with respect to their geological activity.

Powell did not limit himself to geology. In his classic *Introduction to the Study of Indian Languages* (1877), he proposed the first systematic classification of Indian tribes based on their languages, several of which he had learned. Observing that the pattern of homesteading that had worked for the well-watered states to the east of the Rockies would bring nothing but degradation of the soil and personal tragedy if adopted further west, he fought for the conservation of the unique West. His *Report on the Lands of the Arid Region* (1878) proposed that instead of settlements of uniform 160 acre lots, a mix of small irrigated plots and large non-irrigated holdings be adopted. Political maneuvering in Washington delayed consideration of the plan until the Dust Bowl years of the 1930s.

In 1879, Congress established the United States Geological Survey (USGS) with Clarence King as director. Powell became head of the Bureau of Ethnology (later American Ethnology) at the Smithsonian Institution, a post he held for the rest of his life. In 1881, on King's resignation from the USGS, Powell took over that agency too. He tried to further his plans for land use in the West, but, failing once again, resigned after three years. He spent the rest of his life writing about human evolution and the nature of science.

John Wesley Powell has attracted little attention from historians of science who have traditionally been more interested in the scientists who have made major theoretical or experimental discoveries. By contrast, American historians have found his life an attractive subject. Homespun and self taught, a heroic explorer and opener of the West, an exponent of the accessible and useful science of geology, an advocate of Indian rights, a man who could handle the politics of Washington for the public good, a writer who could reach a wide audience, and an early conservationist, Powell fit mainstream themes in the history of the American West.

Wallace Stegner, *Beyond the Hundredth Meridian: John Wesley Powell and the Second Opening of the West* (1954). Donald Worster, *A River Running West: The Life of John Wesley Powell* (2001).

RACHEL LAUDAN

PRECISION AND ACCURACY. The historical study of precision and accuracy has focused primarily on the period from 1700 to about 1930. During that time, the definitions of the two concepts took shape, instrumentation diversified and its construction improved, experimental protocols sharpened, techniques for certifying the reliability of data emerged, procedures for determining metrological standards were greatly enhanced, and the use of gauges and the manufacture of interchangeable parts spread precisely made objects across the landscape of everyday life (*see* STANDARDIZATION). The intensification of quantification of the sciences after 1700 also helped to produce a more positive assessment of precise numbers, which had been regarded either as implausible or as useless in the seventeenth century. In the critique of modernity at the end of the nineteenth century, precision and accuracy lost some of their popularity, at least in science instruction and among the general public. Both nonetheless remained defining features of science, technology, and indeed of modern life.

Precise numerical measurement at first had little currency outside astronomy. The gentlemanly discourse of Robert *Boyle, which otherwise proved important in the shaping of the early modern scientific persona and the discursive standards of scientific proof, worked against the acceptance of precision in measurement. English natural philosophers held exactitude to be possible only in mathematics; physical investigations, which recorded properties of the real world, were thus necessarily imprecise. To claim that experiments could be precise was not only pedantic, but also dishonest in the culture of gentlemanly civility associated with Boyle. In France, critics of Isaac *Newton charged that he had invested the laws of motion with too much "geometrical precision" for application to the physical world. Doubts regarding precision persisted until the end of the eighteenth century, even though by then improved instruments, new graduation and comparative devices, and new techniques (including double weighing) led to new standards that seemed to justify as well as embody claims for the arithmetical precision of measurement. The chemist Antoine-Laurent *Lavoisier gave some of his measurements to eight decimal places, not always persuasively. The British instrument maker and textbook writer William Nicholson doubted Lavoisier's numbers, and limited their reliability to three decimal places. The problem of superfluous significant figures in this arithmetical precision was not resolved until after 1800.

Precision and accuracy now designate different concepts. For a long time, accuracy, which meant both the quality of the observations and their approximation to the true value, was the more common term. Not until after Carl Friedrich *Gauss's publications on the method of least squares in 1809 and 1823 did relatively stable definitions of both terms emerge. In the method of least squares, which related an error of given magnitude to its frequency of occurrence, the degree of agreement among a set of measurements could be numerically specified. "Precision" became a measure of the agreement or reliability of a set of observations after all known sources of error (including the constant and systematic errors in instrumentation and protocol) were computed. "Accuracy" concerned the agreement between the precise measure and its true result, which, of course, was not known. Not all scientists accepted the method of least squares, however. Even when the practice of exact experiment became relatively standardized, scientists still treated different kinds of errors differently. Hence sharp disagreements arose over the "precise" number, as occurred at the end of the nineteenth century in the computation of a standard of electrical resistance. The debates did not raise the possibility of doubts over precise measurements, but rather over what errors afflicted them.

Trust in precision measurements increased as a result of improvements in the construction of standards of measurement, beginning with the French determination of the meter in 1792 and continuing in the reforms of weights and measures nearly everywhere in the nineteenth century. The determination of the comparative precision of different national systems of weights and measures was taken up by astronomers like Gauss, Friedrich Wilhelm Bessel, and Heinrich Schumacher; Gauss believed that the accuracy of an original standard of measure should be 1/1,000,000 (the precision of weighing that could be achieved by mid-century). Large-scale triangulation projects, including the U.S. Coastal Survey, improved the precision of linear measures to about 1/50,000, also by mid-century. The development of interchangeable parts in manufacturing—initially for the military—led to a wide distribution of gauges, tolerances, and precision instruments like vernier calipers that considerably enhanced the degrees of precision found in the objects of daily life. In 1850, for instance, a tolerance of 0.01 inch in machine parts was customary; by 1880 it had narrowed to between 0.001 and 0.0001 inch. The importance of the preservation, maintenance, and production of precise standards of measure to nation-states resulted in the establishment of national *bureaus of standards institutes, the first of which was the German Imperial Institute for Physics and Technology (1887).

Precision of measure improved technically in the twentieth century, but, as in the past, it could still excite political debate. For example, the determination of missile accuracy developed from the conflict and collaboration of scientists, engineers, the military, politicians, and corporate managers. Because of the bias and error that are always present in the calculation of the circular error probability (CEP)—the radius of the circle within which 50 percent of warheads will fall if fired at the same target—even the extraordinary precision of the gyroscopes and accelerometers of guided missiles could not eradicate the social processes at work in the construction of precision and accuracy in deciding what degree of precision makes the claim of a certain accuracy plausible.

See also ERROR AND THE PERSONAL EQUATION; METROLOGY; PROBABILITY AND CHANCE; STANDARDIZATION.

Tore Frängsmyr, J. L. Heilbron, and Robin E. Rider, eds., *The Quantifying Spirit in the Eighteenth Century* (1990). Donald Mackenzie, *Inventing Accuracy* (1990). Kathryn M. Olesko, *Physics as a Calling* (1991). J. L. Heilbron, *Weighing Imponderables and Other Quantitive Science around 1800* (1993). Steven Shapin, *The Social History of Truth* (1994). M. Norton Wise, ed. *The Values of Precision* (1995). Dieter Hoffmann and Harald Wittoff, *Genauigkeit und Präzision in der Geschichte der Wissenschaft und des Alltags* (1996).

KATHRYN OLESKO

PREFORMATION. See EPIGENESIS AND PREFORMATION.

PRIMATOLOGY. Europeans who penetrated the jungles of Asia and Africa during the eighteenth century brought back tales of half-human monsters. Skeptics demanded proof, and the obliging adventurers returned to the jungles to kidnap infant chimpanzees, gorillas, and orangutans. These small animals arrived in Europe orphans—their mothers had been killed to capture them—and sick from weeks at sea and a poor diet. They seldom survived more than a year, leaving their skulls and skins for examination.

Carl *Linnaeus used these relics to classify the primates, today divided into four lineages: apes, Old World monkeys, New World monkeys, and prosimians. Linnaeus lumped the tailless great apes with humans in the family *Hominoidea*. Primatology has focused on *Hominoidea*, where Linnaeus's taxonomy, though fine-tuned during the last two centuries, has persisted with surprisingly few major changes. Today some taxonomers add a fourth species to his three great apes—chimpanzees (*Pan troglodytes*), gorillas (*Gorilla gorilla*), and orangutans (*Pongo pygmaeus*); the fourth candidate, bonobo, is sometimes classified as *Pan paniscus*, a distinct species of chimpanzee, or as *Bonobo paniscus*, a separate genus. Linnaeus's arrangement of the hominids has also expanded to include the discoveries of hominid fossils, which have added extinct species to both ape and human family trees. Radiocarbon dating has shown that our ancestors diverged from the ape ancestors around thirty million years ago. DNA studies have shown the apes to be more closely related to humans than to any other primates. The chimpanzee stands genetically closest of all to *Homo sapiens*.

Charles *Darwin would not have been surprised. In *The Origin of Species* (1859) he suggested that human beings share a common ancestor with other primates. Later, in *The Descent of Man* (1871), he characterized the haunting familiarity between humans and apes as a family resemblance. We share, he demonstrated, emotions and expressions as well as a similar skull and skeleton.

At the end of the nineteenth century students of the new science of psychology began exploring the abilities of the great apes in comparison to each other and to human beings. With the exception of Richard L. Garner, an American zoologist who went to West Africa in the 1890s where he sat in a cage while observing wild chimpanzees, the early primatologists studied only captive apes. In 1927 the German psychologist Wolfgang Köhler described his colony of chimpanzees on the island of Tenerife, near Spain, who stacked boxes to retrieve bananas. At about the same time in the United States Robert Mearns Yerkes studied a pair of chimpanzees at his summer home in New Hampshire and traveled to meet and test gorillas and orangutans wherever he could find them. Eventually he established breeding colonies of apes in Orange Park, Florida. He compared the three great apes anatomically, physiologically, and behaviorally and published his results in 1929 in a massive volume, *The Great Apes: A Study of Anthropoid Life*, coauthored with his wife, Ada. Yerkes was remarkably prescient about ape intelligence and, contrary to the habits of his colleagues, he gave his experimental animals names, rather

than numbers, because he found that their personalities were so distinctive as to make numbers inappropriate. Yerkes acquired his first chimp in 1923 and named him Bill, after William Jennings Bryan, the prosecutor in the Scopes Trial.

As much as he enjoyed keeping them in his laboratory, Yerkes realized the importance of observing apes in their natural habitats and found enough money in the depths of the depression in 1930 to send his student Henry Nissen to West Africa for four months. But the continued economic woes of the 1930s, followed by World War II and revolutions and civil wars in Africa and Asia, prevented the development of field studies until the 1960s. Then, at almost the same time, Dutch, Japanese, British, and American primatologists went individually and in groups to observe wild apes in Africa, Sumatra, and Borneo.

These observers habituated the wild animals to their presence and established field stations to plot the behavior of individual animals as well as groups. With support from the National Geographic Society Jane Goodall, Dian Fossey, and Birute Galdikas set up projects in Tanzania, Rwanda, and Borneo. At the same time Junichero Itani and his Japanese colleagues began studies of chimpanzees and bonobo in central Africa. Most of these projects continued for several decades. They revealed that each ape had an individual personality that determined the nature of each family group, and that each group of apes had a separate culture. Some used tools, some hunted for meat. The primatologists who conducted these studies in the early years of the women's movement consisted of women and men in equal number. Whether gender bias had skewed earlier research is hard to assess, but research in the last decades of the twentieth century showed the powerful role of females in ape societies.

Meanwhile increasing evidence of the genetic nature of apes made them the ideal stand-in for humans in studies ranging from the evolution of the brain to medical models for HIV-*AIDS. NASA's confidence in chimp intelligence led to chimpanzee "pilots" in spacecraft of the early 1960s. At the same time psychologists and linguists in the United States began experiments to test whether apes could communicate with American Sign Language and with symbols. Though their degree of grammatical mastery of language is controversial, some chimpanzees have passed on the skills they learned to two generations of offspring.

The species most similar in behavior to humans is the bonobo, who live in a single area in the Congo. They enjoy a very complicated social life that includes frequent sexual activity, homosexual as well as heterosexual. These exchanges apparently serve as displacement behavior to avert aggression within the group.

Great apes behave differently in the wild and in captivity. Chimpanzees use tools in the wild. Gorillas do not, but in captivity perform very well on tests demanding small muscle coordination. Orangutans occasionally use tools in the wild and are especially clever at escaping from zoo confinement. All reveal a sense of humor in the way they play tricks on each other and on their human guardians.

The population of wild apes dwindled drastically in the last decade of the twentieth century as their habitats became the sites of human wars and ecological exploitation. Wild apes probably will not survive for long without dramatic policy changes. Apes in zoos and wild animal parks may live a life similar to one in the wild. Others may continue to suffer as medical models for human disease.

Philosophers and lawyers now dispute the nature of ape awareness and self-consciousness, and their entitlement to greater protection. Activists, led by Jane Goodall, are trying to have laws interpreted so as to give apes legal protection against experimentation and exploitation. Without increased protection, our fellow *Hominoidea* will soon be extinct, and their branch of primatology will become a historic science like *paleontology.

Bettyann Kevles, *Watching the Wild Apes: The Primate Studies of Galdikas, Goodall and Fossey* (1976). Richard W. Wrangham et al. *Chimpanzee Cultures* (1994). Robert M. Yerkes, "Creating a Chimpanzee Community," *Yale Journal of Biology and Medicine* 73 (2000): 221–234.

BETTYANN HOLTZMANN KEVLES

PRINTING HOUSE. The emergence of printing technology contributed fundamentally to scientific communication and the preservation of knowledge during the *Scientific Revolution. Since the beginnings of European printing in the fifteenth century, various circumstances and negotiations conditioned the appearance of printed scientific works. Printers, publishers, editors, illustrators, block-cutters, and engravers embodied economic interests and craft practices requiring control over the production of printed books. Well into the eighteenth century, authors tolerated the dissipation of control over printed works shared with the printing house, though on occasion they intervened personally or through agents to review proofs, supply images, or even self-publish.

Printers and publishers—roles that were often joined well into the seventeenth century—defined a marketplace for scientific texts in various ways. Early printed books, for example, followed popular works of the manuscript tradition not only in the choice of texts, but also in certain typographic conventions. At the same time, printers adopted innovations that gradually altered scientific communication. Erhard Ratdolt's printing of Euclid's *Elements* (1482), for example, reproduced the appearance of manuscript copies of the text in many ways, but also presented nearly six hundred original diagrams, the first in a printed geometry book. The work of the Aldine press and other scholar-publishers of the Renaissance underpinned the humanist project for the recovery and restoration of ancient learning, including science. Printers also commissioned fresh images drawn from "true life," as in the case of the great sixteenth- and seventeenth-century herbals and other works of natural history, which often proved more useful than the accompanying texts.

Issues of intellectual control over the system of printed scientific communication intensified in the eighteenth century with the rise of scientific academies, increasing technical specialization in the sciences, an emerging legal system of copyright, and the growing cultural prestige of authors, whether literary or scientific. Academies, societies, and other collective groups acquired means, including privileges such as imprimatur and evasion of censorship, to control publication (*see* PEER REVIEW). Notions of intellectual property, such as copyright, stemmed from the need to protect publishers from competition in the form of pirated and unauthorized editions, but by the nineteenth century also guarded creative and economic rights flowing from authorship.

Innovations in nineteenth-century papermaking, printing technology, and techniques of scientific illustration—especially wood engraving, lithography, and photography—combined with changes in publishing practice to produce a markedly different and much larger body of scientific literature. Since the late nineteenth century, increasing specialization, the separation of technical from "popular" science publishing (*see* POPULARIZATION), the rise of conferences and professional societies as sites for the exchange of information, the proliferation of journal literature, and preprint distribution networks have reconfigured patterns of scientific communication.

At the end of the twentieth century, increasingly rapid electronic communication challenged the viability of all print media and accelerated changes in the nature of scientific discourse. The printed monograph, originally central, having already given way first to journals and then to preprints, now has had to yield to e-mail. Scientific disciplines are moving quickly to embrace electronic modes of publication, with important consequences for publishers, readers, and libraries, and the printed volumes that fill them.

Elizabeth Eisenstein, *The Printing Press as an Agent of Change: Communications and Cultural Transformations in Early Modern Europe* (1979). Adrian Johns, *The Nature of the Book: Print and Knowledge in the Making* (1998).

HENRY LOWOOD

PRIORITY. Disputes over the priority of a discovery or an invention are documented throughout the history of science and technology. The pervasiveness of the phenomenon inspired a number of articles on the subject by Robert K. Merton from 1957 and many later investigations of its bearing on the nature of science and its practitioners.

Following a tradition that dates back to Plato's dialogues, part of the teaching carried out in late medieval and early modern universities took the form of the *disputatio*, in which teachers and students defended opposing theses in public. The rituals of the *disputatio* varied considerably, but the issue of the timing and the certification of results occasionally entered into the rules. For example, in the late fifteenth century the University of Bologna decided that theses submitted for public discussion among the lecturers (an event held at least twice yearly, the rector present) had to be deposited eight days in advance. Within fifteen days of the *disputatio* written conclusions would be made public and the reputation of the authors enlarged or diminished as a result.

Early modern mathematicians developed especially precise rules for this kind of contest. One took place in Venice in 1536, revolving around an algebraic solution for cubic equations. It provided for two contenders, thirty questions leveled from each side and kept in sealed envelopes, and two months for submitting the solutions. The whole procedure was certified by a notary. Concerns for priority thus seem to have entered the field quite early and independently of the assertion of the superiority of the moderns over the ancients—an assertion later regarded as the cement of the modern, progressive view of science with its emphasis on priority and competition.

Baroque culture and the courts took special pleasure in the kind of contests just described. The patronage system supporting the leading astronomers and natural philosophers of the late sixteenth and seventeenth centuries brought controversies over priority to new heights and depths. The resultant mud-slinging priority disputes might best be analyzed as duels. Among the best-known such episodes were Tycho *Brahe's controversy with Nicolaus Ursus over priority in the invention of the Tychonic world

system (which prompted Johannes *Kepler's *Defense of Tycho* in 1600); several battles between *Galileo and the Jesuits; Gottfried Leibniz versus Robert *Hooke on the interpolation of series; Leibniz versus Isaac *Newton on the motions of celestial bodies and on the calculus; Christiaan *Huygens versus Hooke over pendulum clocks; and Hooke versus Newton on the principle of universal gravity.

Scientists employed a variety of concealed communication techniques, including anagrams and mirror writing, in the battle for priority. Galileo used ciphers to protect his precedence and give him leisure to follow up his discoveries. His first hint that he had observed Venus showing phases like the *Moon was a cipher sent in letters to four of his correspondents on 11 December 1610; it concealed the sentence, "The mother of love emulates the figures of Cynthia." Sending a cipher alerted privileged correspondents that Galileo would soon be announcing something important. Should his priority later be challenged, disclosure of the cipher would provide Galileo with four witnesses able to certify the date of his claim.

From the 1660s, the Royal Society of London, the Paris Academy of Sciences, and the several academies established on their model in Europe offered certifying procedures to their members eager to establish priority but not ready to publish. Description of the new notion or invention could be sealed in an envelope or box deposited with one of the secretaries. A letter sent to the secretary, to be read at the next meeting of the society, had the similar but distinct purpose of publicly certifying the date of a claim or invention. Such procedures served natural philosophers well: they were unchanged a century later.

The mixture of individual ambition and communality implicit in procedures certifying priority is perhaps best conveyed by the self-justification that Antoine-Laurent *Lavoisier offered in his deposit with the secretary of the Paris Academy of Sciences on 1 November 1772. The envelope contained a description of the crucial observation that sulfur gains weight when burning rather than losing it, as was believed. Lavoisier had taken the step, he declared in his sealed message, because he regarded his observation as one of the most interesting modern discoveries in chemistry and he knew how difficult it was, in conversations with colleagues and friends, not to convey hints that might lead them to the same discovery.

Broad social concerns as well as semilegal procedures inspired the decision within Charles *Darwin's circle of friends about how to deal with the news that the then little-known Alfred Russel Wallace had arrived at conclusions Darwin could certify having reached (but not published) long before. The solution: a joint presentation of papers by Darwin and Wallace at a meeting of the Linnean Society, on 1 July 1858, under the auspices of two of Darwin's colleagues who could testify to Darwin's priority through private letters and personal communications. In this well-known story, however, as in similar ones, retrospective vindication of priority was achieved above all through the production, by the winning party, of an impressive series of outstanding publications.

Some 144 years later—in an entirely different context, involving thousands of researchers, huge amounts of public and private money, nation states, ministries, corporations, legal offices, and possible *Nobel Prizes—the decision reached by two competing research consortia to publish on the same day, 15 February 2001, two different draft outlines of the human genome retains more than a hint (and some of the wisdom) of the old procedures.

In these and many other cases, appropriating a theory, invention, observation, or instrument entailed a mobilization of individual efforts, resources, and social relationships that went well beyond the legal or semilegal procedures believed to settle priority disputes. The history of *patents seems to point in the same direction.

Many analysts argue that our age of large team research, e-mail, electronic journals, and advance online publication articles is reshaping priority claims and assessment. It seems likely, though, that some of the long-employed practices will continue to serve the apparently inextricable mixture of individual ambition and communality that characterizes expert groups.

See also DISCOVERY; PEER REVIEW; PROGRESS.

Robert K. Merton, *The Sociology of Science* (1973). A. Rupert Hall, *Philosophers at War* (1980). Nicholas Jardine, *The Birth of History and Philosophy of Science* (1984). James A. Secord, *Controversy in Victorian Geology* (1986). Mario Biagioli, *Galileo Courtier* (1993). Margaret A. Boden, ed., *Dimensions of Creativity* (1994).
                                    GIULIANO PANCALDI

PROBABILITY AND CHANCE. Civilization has always known games of chance. Throughout history, societies have guessed their future by reading entrails, tea leaves, or the accidental arrangement of other objects, and have made crucial legal decisions by lot. Christians deemed gambling a vice, but new games of chance entered Europe from the Arab world during the Crusades. In the mid–thirteenth century, about the time clocks were introduced into the town square,

a poem called "De vetula" conceived of dice throwing in terms of frequencies. Cards appeared in the fourteenth century, complicating games of chance. Girolamo Cardano's *De ludo aleae* (1663) defined chance events in terms of their frequency of occurrence. During the *Scientific Revolution, when the nature of evidence and causes was debated, Blaise Pascal and Pierre de Fermat corresponded about chance, Christiaan *Huygens wrote a book on it, and Jakob Bernoulli I, in his *Ars conjectandi* (1713), presented the binomial distribution and considered how random events might create a regularity with more than a grain of truth.

Classical probability can be regarded as both empirical (chance as the frequency of events) and epistemic (probability as the mental state of doubt related to degrees of belief, a notion that goes back to medieval times). Probability was thus a state of opinion, perhaps related to authority and testimony, but not as demonstratively true knowledge. The empirical and the epistemic issues in probability shaped discussions on the nature and acceptability of evidence, proof, belief, and truth in early modern Europe. But the subjective and objective sides of probability could not be so cleanly separated in practice. When David Hume showed in 1737 how probability could be used to validate inductive evidence, opinion and knowledge became only differences of degree, not kind. The age of classical probability, which lasted until the mid-nineteenth century, had opened.

The shift to a mathematical theory of probability started with Pierre-Simon de *Laplace's *Analytical Theory of Probabilities* (1812). Chance had begun to look lawful in the eighteenth-century analysis of errors, but now with Laplace's definition of the probability of an event as the ratio of the number of favorable cases to all possible cases, probability theory could be applied to a wide variety of events in the social and physical worlds. Yet probability continued also to refer to incomplete knowledge about matters both esoteric and everyday, about matters that possessed only partial or moral certainty, not absolute certainty. The objective and subjective sides of probability thus collapsed on one another during its classical phase: objective frequencies of events were also subjective degrees of belief. Probability had to do with uncovering the lawful in randomness, and with degrees of ignorance and doubt.

Around the mid-nineteenth century, consideration of large-scale regularities—gambling problems, actuarial computations, demographic patterns, errors as treated in the method of least squares, and the like—shifted the emphasis in probability theory more toward the frequency or ratio of events in the world and less toward assessments of epistemic certainty. Especially in what was then known as "social physics," probability appeared to be more about objectivity than subjectivity. Lambert-Adolphe-Jacques Quetelet, the founder of modern statistical methods, announced in 1844 that the error law used in astronomy could be applied to the distribution of human features, such as height: human variation thus could be understood in the same terms as errors of observation. Quetelet used the bell curve of error analysis to propose the "average man," a new object of investigation that occupied the same position on the error curve as the most probable measurement. The most likely value with the smallest error was an experimental ideal; the average man, a moral one. Mass phenomena or repeatable events became the focus of Wilhelm Lexis's sociology in his *Theory of Mass Phenomena in Human Society* (1877), and the foundation of Richard von Mises's probability calculus in 1919.

In no field was the shift to an objective probability more apparent than physics. The deterministic ideal of classical physics allowed only for an epistemic interpretation of probability (which had been the interpretation of Jakob Bernoulli and Laplace), as was the custom in error theory. Moreover, objects of investigation in physics seemed so well defined that recourse to a composite construction, analogous to Quetelet's average man, was unnecessary. But then the British physicist James Clerk *Maxwell announced in 1860 that the velocities of gas molecules had a bell curve distribution, like errors. Classical physics, which was deterministic and postulated continuous (if not reversible) behavior, appeared unable to describe the intimate behavior of a system of gas molecules. Ludwig *Boltzmann found that in order to understand the second law of thermodynamics—announced by Rudolf Clausius in 1850 as the irreducible increase in *entropy over time—he had to use probability theory applied to molecular behavior. In Boltzmann's interpretation, entropy became a macroscopic measure of the probability of finding a system in a particular state, the most probable state being one of maximum entropy. Finally, discontinuities in matter and energy could only be understood in statistical terms. Using probability theory, Albert Einstein confirmed the existence of discrete atoms in his study of Brownian motion (1905). Discontinuous energy processes—black body radiation, radioactive decay, the photoelectric effect, and atomic and molecular spectra—could also only be understood through probability theory. In his study of black body radiation, Max *Planck introduced what

became known as energy quanta via a statistical theory of heat (1900). In *quantum mechanics, in both Werner Heisenberg's matrix mechanics and Erwin Schrödinger's wave mechanics, the likelihood of measuring a system in a given state became a probability of how likely the state was to occur. Heisenberg's famous uncertainty principle incorporates the notion that at the subatomic level properties could not be measured precisely, but only given a probability distribution. Despite his reliance on the word "uncertainty," Heisenberg did not refer to degree of belief, but to the irreducibly statistical character of subatomic reality.

See also ERROR AND THE PERSONAL EQUATION; PRECISION AND ACCURACY; STATISTICAL MECHANICS.

Ian Hacking, *The Emergence of Probability* (1975). Theodore Porter, *The Rise of Statistical Thinking* (1986). Stephen Stigler, *The History of Statistics* (1986). Lorenz Krüger et al., eds., *The Probabilistic Revolution*, 2 vols. (1987). Lorraine Daston, *Classical Probability in the Enlightenment* (1988). Gerd Gigerenzer et al., eds., *The Empire of Chance* (1989). Ian Hacking, *The Taming of Chance* (1990).

KATHRYN OLESKO

PROFESSIONAL IDENTITY. See SCIENTIST.

PROFESSIONALIZATION. See DISCIPLINE(S); PROFESSIONAL SOCIETY.

PROFESSIONAL SOCIETY. Societies designed to affirm and serve the professional interests of members of disciplines or other scientific groupings came into existence in the late nineteenth century. The fields that first provided professional scientific employment on a large scale were those like chemistry that found a significant role not only in the burgeoning educational system but also in industrial, commercial, and state organizations in the major economies of the world. The sharing and advancement of scientific knowledge had been the chief rationale of *academies and learned societies since the seventeenth century, but professional societies had a wider occupational relationship to scientific knowledge, its possessors, and its diffusion.

Early professional societies, such as the Institute of Chemistry in the United Kingdom (founded in 1877), were designed primarily as qualifying associations. The Institute membership initially overlapped considerably with the disciplinary Chemical Society of London, but had different objectives: to promote the thorough study of chemistry and related sciences in their application to the arts, manufacture, agriculture, and public health (and thereby promote professional standards); to ensure the competence via certified training of consulting or analytical chemists; and to maintain the profession on a satisfactory basis.

Acquiring a Royal Charter in 1885, the Institute gained the right to grant certificates of competence. However, the strong academic membership and the practical and ideological difficulties of representing the membership in salary questions limited its role. The Institute strenuously avoided any suggestion of trade unionism. The British Institute of Physics was established in 1921, when the expansion and diversification of physicists' employment had generated enough concern for coherent representation and promotion of their collective fortunes. The Institute of Biology (1950) met the felt need to give an authoritative voice to British biology as a whole.

The United States followed the same sequence of foundations, though telescoped in time because of the later maturity, but greater volume and more rapid growth, of scientific employment there. Thus the establishment of the American Institute of Chemistry (1923), the American Institute of Physics (1931), and the American Institute of Biological Societies (1947). The membership of the American Physical Society (1899) had been ambivalent about representing the employment and other interests of physicists. Most members gladly ceded this role to the American Institute of Physics (AIP). Apart from catering to the occupational and professional interests of their members, the institutes also sought to secure a sense of common identity among scientists from ever more specialized fields. The AIP consisted of the Physical Society, Optical Society, Acoustical Society, the Society of Rheology, and the American Association of Physics Teachers. More societies joined after the mid-1960s. By the 1990s, the AIP, with a staff of 500, represented over 100,000 physicists. The striving for a common identity among the biological disciplines in the United States resulted first in the American Society of Naturalists and then the American Society of Zoologists. World War II brought home the need for an integrated society in the biological sciences to make the government and the public aware of the coherent body of manpower it represented.

In Continental Europe, patterns of development differed. The Deutsche Physikalische Gesellschaft (DPG) traces its origins to the Physikalische Gesellschaft of Berlin formed in 1845. The DPG, with its range of national and professional functions, did not emerge from these local beginnings until 1899. After World War II, the DPG reformed without the Eastern bloc members and began publication of a general news magazine, *Physikalische Blätter*. By contrast, the Société Française de Physique, formed in 1873, was nationally recognized by decree in 1881 and then grew to coordinate an elaborate structure of local sections, specialist divisions, and publications, including the *Journal de Physique* and *Annales de Physique*.

Professional societies typically have several levels of membership: student (or associate), full, and fellow. Honorary fellows, presidents, and patrons selected from the disciplinary elite symbolized the bridge between the professional institutes and the more traditional learned societies. Sometimes, new institutes stimulated older societies to take on new roles. The American Institute of Chemistry was founded in part because of dissatisfaction with the professional consciousness of the American Chemical Society, which greatly expanded its role in professional affairs beginning in the 1930s. As a result, the society remained the dominant institution for rank-and-file chemists. The American Psychological Association (APA) began as a disciplinary society, saw the foundation of more professionally oriented bodies (notably the American Association for Applied Psychology, founded in 1937), and re-formed after World War II along the institute model.

Publication of research journals, abstracts, and periodicals concerned with the educational, applied, and popular aspects of their field has been a major concern of professional societies. The AIP was formed in part to publish the journals of its member societies in uniform format through a common editorial office. It subsequently acquired or established other journals, including *Physics Today*, like *Physikalische Blätter* a vehicle of news and articles of general interest to physicists. The APA bought up several psychology journals. However, its stable of journals, even as it expanded, represented a decreasing relative proportion of the psychology journals published. Compared with the United States, European and other professional societies have been less active in publication and more reliant on private-sector science publishers.

British and American institutional models have been widely adopted elsewhere in the English-speaking world. Examples include the Royal Australian Chemical Institute, the New Zealand Institute of Physics, and the Canadian Association of Physicists. Cognate institutions exist in most developed societies. More recent foundations for physicists include Nippon Butsuri Gakkai (1946), the Israel Physical Society (1954), the South African Institute of Physics (1955), and the Dansk Fysisk Selskab (1972). The international European Physical Society (founded in 1968) represents over 80,000 physicists through 36 national member societies, for whom it acts as a federation and presents a European perspective in international forums.

The effort continues to coordinate, integrate, and effectively represent to the wider world the knowledge production and professional interests of scientific researchers and workers in the face of ever more specialization and occupational diversity.

See also JOURNAL; NATIONAL ASSOCIATIONS FOR THE ADVANCEMENT OF SCIENCE; POPULARIZATION.

Société Française de Physique, Le Livre du Cinquantenaire de la Société (1925). Colin A. Russell, Noel G. Coley, and Gerrylynn K. Roberts, Chemists by Profession (1977). Spencer R. Weart, "The Physics Business in America, 1991–1940: A Statistical Reconnaissance," in The Sciences in the American Context, ed. Nathan Reingold (1979): 295–358. Toby A. Appel, "Organizing Biology: The American Society of Naturalists and its 'Affiliated Societies', 1883–1923," in The American Development of Biology, eds. Ronald Rainger, Keith R. Benson, and Jane Maienschein (1988): 87–120. James H. Capshew, Psychologists on the March: Science, Practice, and Professional Identity in America, 1929–1969 (1998).

DAVID PHILIP MILLER

PROGRESS. Whether or not it had some distant roots in the Christian concept of providence, as some have claimed, the idea of progress began to acquire its present connotations during the seventeenth century. Francis Bacon and Bernard de Fontenelle first developed the notion that natural knowledge and the mechanical arts advance over time, as evidenced by a comparison of the state of knowledge and the arts among "the moderns" to that of "the ancients."

During the eighteenth century the idea was expanded, debated, and popularized. Voltaire suggested that the progress of society (once people that mattered had been freed from prejudice and ignorance) was a necessary complement to that of the sciences. Economist and polymath Anne-Robert-Jacques Turgot held the more cautious view that humankind, like the sciences, advanced through mistakes, by building on discarded theories. Philosopher and historian David Hume, while inclined toward admitting progress, noted the role of chance in its course. Jean Jacques Rousseau argued that, as the sciences and the arts progressed, the minds and habits of human beings corrupted. He was thus among the first to express a critical attitude toward progress—an attitude that, like the opposite belief in progress, has since obtained intermittent approval and rejection.

After having been used to support explicit political agendas by the Marquise de Condorcet and Henri de Saint-Simon in revolutionary France, and William Godwin among British radical activists, the idea of a necessary progress of the sciences was incorporated into the system of the philosopher, social reformer, and polymath Auguste Comte. In Comte's view the sciences—including the science of society, whose development he urged—passed through three distinct phases: "theological," "metaphysical," and "positive." Comte's ideas inspired important sectors of the intellectual and political elites of the "Age of Progress," and around 1900 played a role in the early establishment of the *history of science as a discipline.

In the meantime, some naturalists had detected what they thought were hints of progress in nature itself, and included them in the burgeoning life sciences of the nineteenth century. Jean Baptiste de *Lamarck signaled a "power" in life that tended to produce a gradual increase in the complexity of organization (see CUVIER, GEORGES, AND JEAN BAPTISTE DE LAMARCK). Half a century later Charles *Darwin, while denying the possibility of establishing the superiority of one living form over another, used arguments that contemporaries found compatible with the idea of progress (natural and social), and with the notion of the superiority of western civilizations proclaimed in those years by Herbert Spencer and political propagandists of many kinds (see DARWINISM).

During the twentieth century, in the context of logical positivist and post-positivist philosophies of science—and amid unprecedented scientific, technological, and military developments—efforts were made to build a critical approach to the notion of progress in the sciences, avoiding the speculations that bound it to controversial, social and/or biological *"proofs." The contributions of Karl Popper, Thomas S. Kuhn, Imre Lakatos, and Larry Laudan are especially noteworthy. Historians of science, however, seem to have derived from their own studies of progress a lesson recommending, above all, a circumstanced, parsimonious use of the very word and concept.

See also ANTISCIENCE; HISTORIOGRAPHY OF SCIENCE; POSITIVISM AND SCIENTISM.

John B. Bury, The Idea of Progress (1932; 1982). Larry Laudan, Progress and Its Problems: Toward a Theory of Scientific Growth (1977).

GIULIANO PANCALDI

PROOF. The prospect of proving scientific theories to be true has been an alluring one. In his influential *Posterior Analytics*, Aristotle stipulated that any genuine knowledge (as opposed to mere opinion) had to be derived from first principles self-evidently true. For centuries, Euclid's geometry seemed to satisfy these conditions, and thus served as the model of proven knowledge.

From the frontispiece of Euclid's *Opera omnia* (1703). The caption to the frontis reads "Aristippus the Socratic philosopher being shipwrecked in Rhodes noticed some diagrams drawn on the beach and said to his companions, 'We can hope for the best for I see the signs of men.'"

In the seventeenth century, some natural philosophers, pre-eminently René *Descartes, tried to achieve this high standard for proven knowledge. Doubts about the standard, though, were setting in. Even mathematicians had to rethink the Euclidean notion of proof in light of the development of the calculus in the seventeenth and eighteenth centuries and alternatives to Euclidean geometry in the nineteenth.

Many natural philosophers had already concluded that they were unlikely to be able to find self-evident principles from which they could derive truths about the natural world. An alternative that many found attractive was the crucial experiment. In 1620 in the *Novum organum*, Francis Bacon introduced the term "instantiae crucis," usually translated as "Instances of the Fingerpost." Investigators had to enumerate all possible paths or theories, and then find a crucial instance that would point the finger at the correct one and eliminate all the others. Later reformulated as the method of eliminative induction, it appealed to many natural philosophers including Isaac *Newton.

In the mid-nineteenth century, many scientists and philosophers believed that the use of crucial experiments had proved its worth. They pointed to the debate between Thomas *Young and Augustin Jean Fresnel, who believed that light had to be understood as a wave motion, and the opposing camp, including David Brewster and the followers of Pierre-Simon de Laplace, who believed that light consisted of a stream of particles, or corpuscles. These two hypotheses

seemed to exhaust the range of possibilities. Fresnel had predicted that if the wave theory were true, then a small opaque disk placed within a suitable beam of light would produce a shadow with a small bright spot in the middle. If the particle theory were true, the whole disk would be dark. The crucial experiment was performed, the bright spot observed, and the wave theory was triumphant.

In the late nineteenth century, scientists and philosophers were becoming increasingly skeptical about what the American pragmatist John Dewey called the "quest for certainty." Then, in 1906, the physical chemist Pierre Duhem in his *La théorie physique: Son object, sa structure*, flatly denied that crucial experiments were possible. No hypothesis can be tested in isolation, he argued, only in conjunction with auxiliary hypotheses (about the instruments used for testing, for example). If a test fails, scientists have no way of telling whether the hypothesis or the auxiliary assumptions are false. Thus, no experiment rules out all hypotheses but the true one.

Although some philosophers of science, notably Karl Popper, continued to hold out hope for some form of crucial experiments in science, many philosophers and sociologists of science took Duhem's argument to heart. Sociologists used it to insist that if scientific theories could not be proved true, then no belief about the world could be known to be better than any other. Philosophers of science responded that even if theories could not be proved true, that did not mean there were no ways of deciding that some offered better accounts of the world than others. The debate continues.

Pierre Duhem, *The Aim and Structure of Physical Theory* (1954, repr. of 1906 ed.). K. R. Popper, *Conjectures and Refutations* (1963). L. J. Cohen, *The Probable and the Provable* (1977).

RACHEL LAUDAN

PROTEIN. The Utrecht chemist Gerardus Johannes Mulder invented the name "protein" (from the Greek for "first") in 1838 to designate albuminous substances he thought to be the primary molecules of life. Mulder regarded proteins as composed of a common nitrogen-containing radical together with small amounts of sulfur and phosphorus. During the nineteenth century, chemists established that nitrogen-containing amino acids, themselves just beginning to be studied, derive from protein hydrolysis.

William Wollaston found what Jöns Jacob *Berzelius identified as an amino acid ("cystine") in urinary calculi in 1810. In 1820 Henri Braconnot obtained a sweet-tasting substance by the hydrolysis of gelatin and incorrectly identified it as a sugar. In 1846 Eben Norton Horsford, working in Justus von *Liebig's laboratory, identified the nitrogen content of Braconnot's sweet substance, and Berzelius named it glycine by analogy to cystine. In 1875, Felix Hoppe-Seyler suggested a classification system for proteins. Some, like egg albumin, cheese casein, and blood plasma proteins, dissolve in water or in aqueous solutions of acids, bases, or salts. Others, like fibroin, collagen, and keratin are insoluble in water.

In 1902 Emil Fischer at Berlin and Franz Hofmeister at Strassburg independently proposed that proteins consist of amino acids joined together through the condensation of the amino group (NH_2) of one with the carboxyl group (COOH) of another, forming amide bonds (−CONH−). Using a method pioneered by Theodor Curtius, Fischer succeeded in forming glycylglycine. By 1907 he had moved from the synthesis of this "dipeptide" to a "polypeptide" of eighteen amino acids.

Opinion differed about how large proteins might be. Fischer believed that natural proteins are smaller than synthetic ones, which might have molecular weights as high as 4,000. Some chemists, including Wilhelm Ostwald's son Carl Wilhelm Wolfgang Ostwald, favored a colloidal theory that made proteins physical aggregates of small molecules.

Herman Mark, a pioneer in the application of x-ray and electron diffraction to proteins, suggested that they might be long flexible chains.

Dorothy Crowfoot Hodgkin produced the first x-ray diffraction photograph of the protein pepsin in 1934. Beginning in 1932, Linus *Pauling, who had visited Mark in 1930, targeted hemoglobin (molecular weight of approximately 65,000) and other proteins for study, working out with Alfred Mirsky a model of proteins as coiled or folded chains of polypeptide units linked together by hydrogen bonds. The fragility of proteins results from the easy breakage of these bonds.

There was some resistance to Pauling's theory, for example from the British mathematician Dorothy M. Wrinch, who proposed a cyclol theory of hexagonal rings. Nor had the aggregate colloidal theory died. Coming back again and again to the problem, Pauling proposed with Robert Corey in 1950 what turned out to be the successful model of protein as a helical structure with a turn in the helix every 3.7 amino acids. In this "alpha-helix" model, the alpha carbon in the carbonyl bond extends parallel to the axis of the helix.

The following year Corey and Pauling proposed an additional model, featuring stacked sheets of polypeptide chains in a pleated structure. John Kendrew confirmed the alpha helix model for myoglobin in 1960 as did Max Perutz for hemoglobin in 1962. The pleated sheet conformation was confirmed for the enzyme lysozyme from eggwhite in 1965 by David Philipps and his group at the Royal Institution. Structural elucidation of protein constituted one of the great triumphs of natural product chemistry assisted by physical methods in the twentieth century.

See also DEOXYRIBONUCLEIC ACID; X RAYS.

Mary Jo Nye, *Before Big Science: The Pursuit of Modern Chemistry and Physics 1800–1940* (1999). Joseph S. Fruton, *Proteins, Enzymes, Genes: The Interplay of Chemistry and Biology* (1999).

MARY JO NYE

PROTYLE. William Prout (1785–1850) was an English physician with a strong interest in medical chemistry and physiology. Like William Wollaston and Humphry Davy, he objected to John *Dalton's hypothesis of chemical elements as indivisible atoms, preferring the view that all the chemical atoms except hydrogen are compound, not simple. In a paper published in Thomas Thomson's journal *Annals of Philosophy* in 1815, Prout noted that gas densities appeared to be exact multiples of the density of hydrogen and suggested that hydrogen is the "protyle" or basic building block of matter. Davy favored Prout's hypothesis, as did Thomson, who attempted to press his experimental values for atomic weights into multiples of hydrogen.

Jöns Jacob Berzelius's atomic weights of 1826 differed markedly from Thomson's. Edward Turner argued in 1833 that English chemists should admit the inaccuracy of Thomson's results in comparison to Berzelius's. Nonetheless, the protyle hypothesis continued to enjoy occasional favor; Jean-Baptiste-André Dumas, for example, supported it in 1840. By 1860, however, chemists had become convinced that atomic weight determinations by Jean-Servais Stas demonstrated that atomic weights could not be taken to be multiples of 1.0 or 0.5 or 0.25 times the weight of hydrogen.

In 1882, John William Strutt (Lord Rayleigh) began measuring the relative densities of hydrogen and oxygen, then of nitrogen, partly influenced by the protyle hypothesis. The work led him to the recognition, with William Ramsay, of the existence of the chemically inert gas argon in atmospheric nitrogen (*see* Noble Gases). Throughout its history, the protyle hypothesis was valuable in stimulating the improvement of analytic techniques and the calculation of more precise values for atomic weights.

Following Francis Aston's determination in the 1920s that all weights of chemical isotopes of elementary atoms depart slightly from whole numbers, a modified protyle hypothesis could be sustained invoking the relativistic mass decrease in the building up of nuclei from proteins and neutrons.

Alan J. Rocke, *Chemical Atomism in the Nineteenth Century: From Dalton to Cannizzaro* (1984). William H. Brock, *From Protyle to Proton* (1985).

MARY JO NYE

PSEUDO-SCIENCE AND QUACKERY. For much of the nineteenth and twentieth centuries, the notion of pseudo-science was taken for granted. In contrast to a shared view of science as factual, objective, and experimentally rigorous, pseudo-science was synonymous with dishonesty, credulity, and fraud. Creationists, scientologists, astrologers, spiritualists, faith healers, naturopaths, and ufologists, no less than alchemists, phrenologists, mesmerists, chiromancers, and nostrum-mongers before them, tended to be regarded as cranks, mountebanks, racketeers, and hoaxers. They were the dangerous ideologues of fanatical ideas and practices, the incompetent opponents of science and scientific medicine.

The eighteenth century had understood quackery as blatant fraud (especially in relation to medicine) but lacked a developed concept of pseudo-science. For the stuff of belief (religion) and the stuff of experiment and analysis (science or natural philosophy) had not yet undergone their rhetorical separation and ranking.

Moreover, people who sought to denigrate certain bodies of would-be natural knowledge as quackery often had reasons for doing so other than the advancement of science. When, for example, Richard Steele disparaged astrologers as fortune-telling tricksters in *The Spectator* in 1712, and associated them with "Wizards, Gypsies and Cuning Men," he was seeking to legitimize representations of truth that did not rely on God for their ultimate authority. For him, discrediting astrology as fraudulent and linking it to supernaturalism challenged the credibility of spiritually legitimated power. In contrast, many natural philosophers, some of whom, like *Newton, had attachments to alchemy and other soon-to-be marginalized practices, did not participate in the rhetorical lambasting of astrology.

By the end of the twentieth century, the concept of pseudo-science had become problematic in new ways. Far from being an objective, timeless, and transcendent descriptive category, "pseudo-science" was a pejorative term, usually deployed defensively by ideologues of science in the interest of demarcating and defending disciplinary boundaries. Race eugenics in Nazi Germany and neo-Lamarckian genetics in the Soviet Union indicate the relativity of scientific legitimacy (and subsequent de-legitimacy). These theories, which had strong opposition from contemporary scientists, and are now entirely rejected, may be contrasted with theories once widely accepted but now either abandoned or limited in application. Examples include *phlogiston theory, caloric explanations for heat, corpuscular descriptions of light, and *imponderable fluids for explaining electricity and magnetism—to choose only some eighteenth-century examples of deposed bodies of natural knowledge. The late-twentieth century studies that focused on the social construction of rejected knowledge (in order to strengthen the case for the social construction of all natural knowledge) reflected the waning faith in science as a necessarily ethical and uniform authority. These studies articulated the view that objective scientific truth was at best partial and relative, at worst a commodity that could be traded like any other.

It does not follow that there can be no history of the subject beyond that of its labeling. Crooks and cultists may be as common in science as in religion. There probably never has been a time since the *Scientific Revolution when there were not those who promoted as "scientific" theories and practices ideas and behavior that they knew were not conformable to, or verifiable by, experimental method. Some of these malefactors acted purely for financial gain, in the manner of vendors of bogus cures for VD, TB, and cancer. Some have deliberately falsified scientific evidence for the success of their theories or the sake of their egos (Cyril Lodowic Burt's fraud on the genetic basis of intelligence), while others have planted false evidence, perhaps simply to expose the pretensions of scientific expertise (the perpetrator of the hoax of "Piltdown Man").

There are many reasons why people subscribed to pseudo-science so labeled by the scientific establishment. In some cases, the label itself has been sufficient cause, the veracity of the knowledge being secondary to the anti-authoritarian principle involved in its support. Very often, as with astrology in the eighteenth century or mesmerism in the nineteenth, the appeal has been fed by anti-materialist and anti-reductionist opposition to orthodox science. The lure of other worlds and other dimensions to secular reality have also played a part in the popularization of heterodox and fundamentalist religious beliefs such as spiritualism and creationism.

Pseudo-sciences have not always been in opposition to science. On the contrary, they have operated as vehicles for the popularization of science. For example, nineteenth-century *phrenology, the hugely popular materialist theory of brain and science of character, extolled a naturalistic theory of evolution that helped pave the way for the reception of *Darwinism. Phrenologists popularized what in effect was a layman's version of the would-be scientific and science-celebrating positivist philosophy of Auguste Comte, the philosophy that proclaimed the secular scientific worldview as the final, inevitable, and altogether superior successor to past eras of theology and metaphysics. From the history of phrenology and other such pseudo-sciences, it is clear there is more to be lost than gained historically by seeking retrospectively to draw sharp distinctions between the "real" and the "pseudo" in science.

Roy Wallis, ed., *On the Margins of Science: The Social Construction of Rejected Knowledge* (1979). Marsha P. Hanen, Margaret Osler, and Robert G. Weyant, eds., *Science, Pseudo-Science and Society* (1980). Arthur Wrobel, ed., *Pseudo-Science and Society in Nineteenth-Century America* (1987). Roger Cooter, *Phrenology in the British Isles: An Annotated Historical Biobibliography and Index* (1989). Roy Porter, *Health for Sale: Quackery in England, 1660–1850* (1989). Roger Cooter and Mary Fissell, "Exploring Natural Knowledge: Science and the Popular in the Eighteenth Century," in Roy Porter, ed., *Cambridge History of Science*, vol. 4: *Science in the Eighteenth Century* (in press).

ROGER COOTER

PULSARS AND QUASARS. Exotic astronomical objects were first detected in the 1960s by their radio emissions, though most quasars (quasi-stellar radio sources) are radio quiet. Pulsars are pulsating radio sources.

In 1960, astronomers identified the radio source 3C48 (number 48 in the Third Cambridge Catalogue) with a star-like object. Three years later, they did the same for 3C 273, which had emission lines at unusual wavelengths. Maarten Schmidt at the Mount Palomar Observatory in Southern California recognized the mysterious spectral lines as lines from common elements shifted far toward the red. One mystery solved led to another. Assuming the red shifts arose from the expansion of the universe, astronomers had to endow quasars with tremendous speeds and distances. To be visible at vast distances, a quasar would have to be enormously bright, a thousand times brighter than all the stars in our galaxy. But a quasar's rapid variation in time would require that its energy be produced in a small volume. No known nuclear process can yield the observed energy output from a small volume.

The need for new physical explanations for new astronomical phenomena expanded further. In 1967, Anthony Hewish and Jocelyn Bell at Cambridge University, looking for rapid variations in the radio brightness of quasars, discovered a rapidly pulsating radio source. The radiation had to be from a source not larger than a planet if the signal could spread so quickly across the object to trigger bursts of radiation. Hewish won a *Nobel Prize for the discovery, though his student Bell made the actual observation.

The period of this pulsar, 1.3 seconds, was so regular that Hewish and Bell briefly thought it might be an interstellar beacon or radio lighthouse built by an alien civilization. Hence the name they gave the source LGM 1—LGM for Little Green Men.

A few months before the discovery, Francis Pacini published a theoretical paper showing that a rapidly rotating neutron star with a strong magnetic field could act as an electric generator and emit radio waves. (Once thermonuclear sources of energy are exhausted, stars of less than 1.4 solar mass shrink until they become white dwarfs; more massive stars continue contracting into even more dense stars composed of neutrons.) Most millisecond pulsars, pulsating many times per second, have white dwarf companions, and their amplified spins may be somehow attributable to the accretion of mass from the companion star.

The cosmological hypothesis for quasars—that their red shifts are associated with the expansion of the universe, and they are at great distances from the earth—is generally accepted. Observations by Halton Arp at Palomar, however, suggested possible physical connections between a few quasars and nearby galaxies. A committee at Palomar judged Arp's controversial research to be without value, and terminated his observing time in 1983. Quasars probably are powered by the energy released when matter falls into a gigantic rotating *black hole. Why so many quasars have red shifts around 2 remains to be explained.

See also ASTRONOMY, NON-OPTICAL; CHANDRA-SEKHAR; COSMOLOGY; RELATIVITY.

F. G. Smith, *Pulsars* (1977). Ajit K. Kembhavi and Jayant V. Narliker, *Quasars and Active Galactic Nuclei: An Introduction* (1999).

NORRISS S. HETHERINGTON

PUTTI IN SCIENCE. Peter Paul Rubens strewed his canvases with putti. These angelic children helped people to heaven, sat on cornices, and carried the instruments of torture to the saints. Their agility, good nature, and cleverness also fitted them for lab assistants. In his drawings for François d'Aguilon's *Opticorum libri sex* (1613), Rubens depicted putti anatomizing the eye and demonstrating the laws of perspective, binocular vision, and projection. The Jesuits, to whom Aguilon belonged, immediately perceived the utility of using angels in experiments. Two reasons, apart from Saint Ignatius of Loyola's taste for cherubs, may explain the frequent use of this unusual iconography. First, the translation of the putti from their familiar and approved haunts to the sites of natural philosophy helped to domesticate the new science of the seventeenth century. If angels thought it acceptable to play with barometers, experimental investigation of the atmosphere could not menace church or state. Second, putti, who are as much alike as electrons, could replace individual Jesuits to represent the Society of Jesus as a whole. In the impressive volume issued by the Jesuits of the Belgian Province to celebrate the centennial of their society, a small, naked winged boy revolves a crank in the heavens that turns the earth via a series of gears. The caption reads, "Societas Jesu convertit mundum." As the angel turns the world, the Jesuits he represents convert it.

Putti appear as lab assistants in many books on mathematics and natural philosophy published

Putti working an air pump.

during the seventeenth and early eighteenth centuries. The famous *Nova experimenta magdeburgica* (1658) by the Jesuit Gaspar Schott was the first general description of the air pump, invented by the Lutheran engineer Otto von Guericke. It shows cherubs working the instrument, simultaneously relieving their human coworkers of hard labor and presiding over an equally unlikely collaboration between a Jesuit and a Protestant.

William R. Shea, ed., *Science and the Visual Image in the Enlightenment* (2000).

J. L. HEILBRON

Q

QED. See Quantum Electrodynamics.

QUACKERY. See Pseudoscience and Quackery.

QUANTIFICATION. See Mathematization and Quantification.

QUANTUM CHEMISTRY focuses on the application of wave functions to electrons in atomic or molecular orbitals. The first successful applications of wave mechanics to a molecule occurred in the late 1920s in the work of the German physicists Walter Heitler, Fritz London, and Friedrich Hund. The first successful applications to a distinctively chemical molecule with spatially directed chemical valences was the work of the Americans Linus *Pauling and John Slater in 1931.

Niels *Bohr had used the quantum hypothesis in 1913 to explain the stability of the electron's orbit about a positively charged nucleus. The connection that Bohr worked out in 1913 between the periodicity of electron-shell configurations in the atom and the periodicity of properties for chemical elements in the *periodic table was improved in a famous paper of 1922. Bohr worked out—but did not calculate—a neat correlation of electron levels containing 2, 8, up to 18, and up to 32 electrons in the so-called K, L, M, and N shells with chemical periodic groups containing 2, 8, 8, 18, 18, and 32 elements. By this time Bohr and others were making use of Niels Bjerrum's formulation of a quantum theory for molecules (1911–1914) that quantified a classical model of a vibrating rotator.

Werner *Heisenberg's formulation of resonance in wave functions (1926) suggested a breakthrough for chemists in the long-standing puzzle of how two electrical particles of like charge can unite to form a stable covalent chemical bond. In 1927 Heitler and London successfully applied Heisenberg's resonance formulation of electron wave functions to the two-atom, two-electron hydrogen molecule. At the same time Hund published a different treatment generalizing the work of the Danish physicist Oyvind Burrau on the hydrogen molecule ion, which produced reliable energy values using elliptical

coordinates for the electron in orbit around two protons. Hund assumed that each electron moves in a potential field that results from all the nuclei and from other electrons present in the molecule. The Heitler-London approach became known as the atomic orbital method (AO) and the Hund approach as the molecular orbital method (MO).

Independently, Slater and Pauling developed ways to explain directed chemical valence by proposing the mixing of the s (spherical) and p (elliptical) energy levels of the four valence electrons in the carbon atom. The mixing, or hybridization, creates wave functions of equal energy with electron distributions oriented toward the corners of a tetrahedron. Pauling and Slater demonstrated that deviation from a 90-degree bond angle could be taken as evidence of mixing, or hybridization, of the electron orbitals.

During 1931–1935 Robert Mulliken used Hund's approach and Douglas Rayner Hartree's group-theoretical methods to develop molecular orbital theory, as did Erich Hückel and Bernhard Eistert in Germany. While Pauling's atomic orbital method prevailed in the United States in the next decade, in England, John Edward Lennard-Jones, Hugh Christopher Longuet-Higgins, and Charles Alfred Coulson took up the molecular orbital method with its greater potential for mathematical application and development. Coulson successfully demonstrated in 1939 how electrons in benzene move over the whole molecule instead of being restricted to the region between two particular atoms. With the development of high-speed digital *computers after the war, the more mathematically difficult MO method came to be preferred by most theoretical chemists despite Pauling's argument that the AO method was more natural to the chemist and its rougher approximations often did the job. Among proponents of MO theory, Coulson persuaded many chemists (although not Pauling) that MO methods complemented AO methods and offered considerably greater capability for solving chemical problems. Coulson's influential textbook *Valence* appeared in 1952.

The development of orbital symmetry rules by Robert Simpson Woodward and Roald Hoffmann in the 1960s, which allowed highly specific predictions of stereochemical details and reaction outcomes, assisted the switch to the MO method. Building on the work of Coulson and Longuet-Higgins, the Kyoto chemist Ken'ichi *Fukui developed the frontier orbital theory of reactions (1950) that showed how the progress of reactions depends upon the geometry and relative energies of the highest recipient molecular orbital of one reactant and the lowest molecular orbital of the other.

See also ATOMIC STRUCTURE; QUANTUM MECHANICS.

Mary Jo Nye, *From Chemical Philosophy to Theoretical Chemistry: Dynamics of Matter and Dynamics of Disciplines 1800–1950* (1993). Ana I. Simoes and Kostas Gavroglu, "The Americans, the Germans, and the Beginnings of Quantum Chemistry: The Confluence of Diverging Traditions," *Historical Studies in the Physical and Biological Sciences* 25 (1994): 47–110. Kostas Gavroglu, *Fritz London: A Scientific Biography* (1995). Stephen G. Brush, "Dynamics of Theory Change in Chemistry," *Studies in the History and Philosophy of Science* 30 (1999): 21–79, 263–302. Buhm Soon Park, "Chemical Translators: Pauling, Wheland, and their Strategies for Teaching the Theory of Resonance," *British Journal for the History of Science* 32 (1999): 21–46. Ana I. Simoes and Kostas Gavroglu, "Quantum Chemistry qua Applied Mathematics: The Contributions of Charles Alfred Coulson (1910–1974)," *Historical Studies in the Physical and Biological Sciences* 29 (1999): 363–406.

MARY JO NYE

QUANTUM ELECTRODYNAMICS (QED) is the *quantum field theory describing the interaction of charged particles with photons. It represents positive and negative electrons by a quantized field satisfying the *Dirac equation in the presence of an electromagnetic (e.m.) field; charged spin 0 particles, such as pi mesons, by quantized field operators satisfying the Klein-Gordon equation; and the electromagnetic field by quantized field operators satisfying *Maxwell's equations. The source terms in these Maxwell equations are the charge-currents arising from the matter field in the presence of the quantized e.m. field. The small dimensionless constant $\alpha = 2\pi e^2/hc = 1/137$, where e is the electronic charge, h Planck's constant, and c the velocity of light, measures the coupling between the charged matter field and the electromagnetic field. Since α is so small, the coupled equations are usually solved "peturbatively," that is, as a power series expansion in α. This perturbative approach has had amazing success in calculating extremely fine details in atomic spectra, accounting for the electromagnetic properties of electrons and muons, and predicting with precision the outcome of collisions between high-energy positive and negative electrons. As Toichiro Kinoshita

and Donald Yennie, two theorists who have carried out some of the most extensive and difficult calculations testing the limits of QED, wrote in 1990, "it is inconceivable that any theory which is conceptually less sophisticated could produce the same results."

Richard *Feynman's great contribution to QED was a technique by which perturbations could be visualized and calculated by straightforward diagrams. The diagrams indicate both why and how certain processes take place in particular systems. In Feynman's approach, as generalized by Freeman Dyson, each quantized field (and associated particle) is characterized by a "propagator" represented in a Feynman diagram by a line, which if internal connects to two vertices, and if external, that is, if corresponding to an incoming or outgoing particle, connects to a single vertex. Each interaction is represented by a vertex characterized by a coupling constant and a factor describing the interaction between the fields. For a given process relatively simple expressions occur in the lowest order of perturbation theory. The diagrams that correspond to higher order contributions contain closed loops and entail integrations over the momenta of the propagators involved in the loops. In almost all cases these integrals diverge because of contributions from large momenta.

The anomalous magnetic moment of a (quasi-free) electron means the deviation from the value predicted by the Dirac equation. According to Dirac's theory, the electron has an intrinsic magnetic moment accompanying its spin, the value of which when expressed in the form $g_e = eh/4\pi mc$ is given by $g_e = 2$. The electron's anomalous magnetic moment is defined as $a_e = (g_e - 2)/2$. The first term in the expansion to calculate the anomalous magnetic moment of the electron (in QED) corresponds to the Feynman diagram reproduced below.

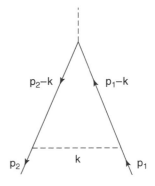

A Feynman diagram illustrating the exchange of a photon (the dashed line) between two electrons, which diverge in consequence.

Julian Schwinger's computation of a_e in 1947 constituted a landmark in the postwar developments of QED. It confirmed the experimental value that had been obtained by Isador Isaac Rabi and his associates and also the ideas of mass and charge renormalization in the low orders of QED. Since that time, both the experiments and the theory have been improved by several orders of magnitude and have provided the most precise and rigorous tests for the validity of QED. To date, the best theoretical and experimental values of the anomalous magnetic moment of the electron agree to ten significant figures—(in units of $eh/4\pi mc$) 1.001 159 652 17 (theoretical) against 1.001 159 652 19 (measured).

Richard Feynman, *QED: The Strange Theory of Light and Matter* (1985). Toichiro Kinoshita, ed., *Quantum Electrodynamics* (1990). Silvan S. Schweber, *QED and the Men Who Made It: Dyson, Feynmen, Schwinger, and Tomonaga* (1994).

SILVAN S. SCHWEBER

QUANTUM FIELD THEORY. When initially formulated *quantum mechanics described non-relativistic systems with a finite number of degrees of freedom. The extension of the formalism to include the interaction of charged particles with the electromagnetic field treated quantum mechanically brought out the difficulties connected with the quantization of systems with an infinite number of degrees of freedom. The effort to make the quantum theory conform with special relativity disclosed further difficulties. To address both sets of problems, Ernst Pascual Jordan, Oskar Klein, Eugene Wigner, Werner *Heisenberg, Wolfgang Pauli, Enrico *Fermi, and others developed quantum field theory (QFT) during the late 1920s. P. A. M. *Dirac had taken the initial step in 1927 with a quantum mechanical description of the interaction of charged particles with the electromagnetic field, which he described as an (infinite) assembly of photons, that is, of massless spin 1 particles. Dirac considered "particles" (whether they had a rest mass or, like photons, had none) to be the "fundamental" substance. In contrast, Jordan insisted that fields constituted the "fundamental" substance.

The history of theoretical elementary particle physics until the mid-1970s can be narrated in terms of oscillations between the particle and field viewpoints epitomized by Dirac and by Jordan. QFT proved richer in potentialities and possibilities than the quantized-particle approach. By the mid-1930s the imposition of special relativity on QFT had produced genuinely novel features: the possibility of particle creation and annihilation, as first encountered in the quantum mechanical description of the emission and absorption of photons by charged particles; the existence of anti-particles; and the complexity of the "vacuum." The latter was now seen to be not a simple substance but the seat of fluctuations in the measured observables, which fluctuations are the larger the smaller the volume probed.

Fermi's theory of beta-decay (1933–1934) was the important landmark of field theoretic developments of the 1930s. It had been recognized since 1915 that the nucleus was the site of all radioactive processes, including β-radioactivity. The process of β-decay—in which a radioactive nucleus emits an electron (β-ray) and increases its electric charge from Z to Z + 1—had been studied extensively during the first decade of the century (*see* RADIOACTIVITY). In 1914 James Chadwick found that the energy of the emitted electrons varied continuously up to some maximum at which conservation held to the accuracy of the measurements in the experiment. By the end of the 1920s no satisfactory explanation of the continuous β-spectrum had been found and some physicists, in particular Niels *Bohr, proposed giving up energy conservation in β-decay processes. In December 1930 Pauli, in a letter addressed to the participants of a conference on radioactivity, countered with "a desperate remedy." He suggested that "there could exist in the nuclei electrically neutral particles that I wish to call neutrons [later renamed neutrinos by Fermi], which have spin 1/2.... The continuous β-spectrum would then become understandable by the assumption that in β-decay a [neutrino] is emitted together with the electron, in such a way that the sum of the energies of the [neutrino] and electron is constant."

Fermi took Pauli's hypothesis seriously when he heard about it for the first time at the Solvay Congress of 1933. Fermi soon formulated a theory of β-decay that marked a change in the concept of "elementary" processes. Fermi supposed that electrons do not exist in nuclei before their emission, but that (to quote his version of 1934) "they, so to say, acquire their existence at the very moment when they are emitted; in the same manner as a quantum of light, emitted by an atom in a quantum jump, can in no way be considered as pre-existing in the atom prior to the emission process. In this theory, then, the total number of the electrons and of the neutrinos (like the total number of light quanta in the theory of radiation) will not necessarily be constant, since there might be processes of creation or destruction of these light particles."

Both Fermi's theory of β-decay and quantum electrodynamics (QED) made clear the power of a quantum field theoretical description. In particular, they indicated that the electromagnetic forces between charged particles could be understood as arising from the exchange of virtual photons between the particles—virtual particles because they do not obey the energy-momentum relation that holds for free photons. When one of the charged particle emits a (virtual) photon of momentum k, it changes its momentum by this amount. When the second charged particle absorbs this virtual photon, it changes its momentum by k. This exchange is the mechanism of the force between the interacting particles. The range of the force generated is inversely proportional to the mass of the virtual quantum exchanged. Zero-mass photons generate electromagnetic forces of infinite range. Spin-zero quanta of mass m generate forces with a range of the order h/mc. This insight led Hideki *Yukawa to postulate that the short-range nuclear forces between nucleons could arise from the exchange of massive spin 0 bosons. Another important lesson learned from QED and Fermi's theory of β-decay was the protean nature of particles. When interacting with one another "particles" can metamorphose their character and number: in a collision between an electron and its anti-particle, the positron, the electron and positron can annihilate and give rise to a number of photons.

By the late 1930s, physicists understood the formalism of quantum field theory and its difficulties. All relativistic QFTs have the mathematical problem that the calculations of the interactions between particles give infinite, that is, nonsensical results. The root cause—fields definable at a point in space-time of these divergences was the assumption of locality, the assumption that the local fields—fields definable at a point in space-time point—whose quanta are the experimentally observed particles interact locally, i.e. at a point in space time.

Local interaction terms implied that in QED photons will couple with (virtual) electron-positron pairs of arbitrarily high momenta, and that electrons and positrons will couple with (virtual) photons of arbitrary high momenta, in both cases giving rise to divergences. The problem impeded progress throughout the 1930s and caused most of the workers in the field to doubt the correctness of QFT. The many proposals to overcome these divergences advanced during the 1930s all ended in failure. The pessimism of the leaders of the discipline—Bohr, Pauli, Heisenberg, Dirac, and J. Robert Oppenheimer—was partly responsible for the lack of progress. They had witnessed the overthrow of the classical concepts of space-time and had themselves rejected the classical concept of determinism in the description of atomic phenomena. They had brought about the quantum mechanical revolution. They were convinced that only further conceptual revolutions would solve the divergence problem in quantum field theory.

The way to circumvent the difficulties was indicated by Hendrik Kramers in the late 1930s and his suggestions were implemented after World War II. These important developments stemmed from the attempt to explain quantitatively the discrepancies between the empirical data and the predictions of the relativistic Dirac equation for the level structure of the hydrogen atom and the value it ascribed to the magnetic moment of the electron. These deviations had been observed in reliable and precise molecular beam experiments carried out by Willis Eugene Lamb, Jr., and by Isidor Isaac Rabi and coworkers at Columbia, and were reported at the Shelter Island Conference in the fall of 1947. Shortly after the conference, Hans Albrecht Bethe showed that the Lamb shift (the deviation of the $2s$ and $2p$ levels of hydrogen from the values given by the Dirac equation) was of quantum electrodynamical origin, and that the effect could be computed by making use of what became known as "mass renormalization," the idea that had been put forward by Hendrik Kramers.

The parameters for the mass m_o and for the charge e_o that appear in the equations defining QED are not the observed charge and mass of an electron. The observed mass m enters the theory through the requirement that the energy of the physical state corresponding to an electron moving with momentum p be equal to $(p^2 + m^2)^{1/2}$. The observed charge e enters through the requirement that the force between two electrons at rest separated by a large distance r satisfy Coulomb's law e^2/r^2. Julian Schwinger and Richard *Feynman showed that the divergences encountered in the low orders of perturbation theoretic calculations could be eliminated by re-expressing the parameters m_o and e_o in terms of the observed values m and e, a procedure that became known as mass and charge renormalization. In 1948 Freeman Dyson working at the Institute for Advanced Study in Princeton proved that these renormalizations could absorb all the divergences arising in scattering processes (the S-matrix) in QED to all orders of perturbation theory. More generally, Dyson demonstrated that only for certain kinds of quantum field theories can all the infinities be removed by a redefinition of a finite number of parameters. He called such theories renormalizable. Renormalizability thereafter became a criterion for theory selection.

The idea of mass and charge renormalization, implemented through a judicious exploitation of the symmetry properties of QED, made it possible to formulate and to give physical justifications for algorithmic rules to eliminate all the ultraviolet divergences that had plagued the theory and to secure unique finite answers. The success of renormalized QED in accounting for the Lamb shift, the anomalous magnetic moment of the electron and of the muon, the scattering of light by light, the radiative corrections to the scattering of photons by electrons, and the radiative corrections to pair production was spectacular.

Perhaps the most important theoretical accomplishment between 1947 and 1952 was providing a firm foundation for believing that local quantum field theory was the framework best suited for the unification of quantum theory and special relativity. Perspicacious theorists, like Murray Gell-Mann also noted the ease with which symmetries—both space-time and internal symmetries—could be incorporated into the framework of local quantum field theory. Gauge invariance became a central feature of the quantum field theoretical description of the electromagnetic field. Subsequently, the weak and the strong forces were similarly described in terms of gauge theories.

See also ELEMENTARY PARTICLES; QUANTUM ELECTRODYNAMICS; QUARK.

Abraham Pais, *Inward Bound* (1986). Olivier Darrigol, *From C-Numbers to Q-Numbers: The Classical Analogy in the History of Quantum Theory* (1992). Silvan S. Schweber, *QED and the Men Who Made It* (1994). Michael E. Peskin and Daniel. V. Schroeder, *An Introduction to Quantum Field Theory* (1995). Helge Kragh, *Quantum Generations: A History of Physics in the Twentieth Century* (1999).

SILVAN S. SCHWEBER

QUANTUM PHYSICS. The proximate origin of the quantum theory was a perplexing paper published by Max *Planck in 1900. In it he showed that the formula he had proposed for the empirically determined spectral density of blackbody radiation could be derived by setting the energy of the collection of charged harmonic "resonators" (which he used to represent atoms capable of emitting and absorbing electromagnetic radiation) of frequency ν equal to an integral multiple of $h\nu$. Here h stood for a new physical constant necessary to fit the empirical spectrum and ν for the frequency of the resonator. The derivation required recourse to Ludwig *Boltzmann's probability calculation for the *entropy of a gas. It appears that, in adapting it to the blackbody problem, Planck did not recognize that he had

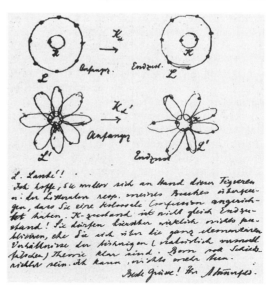

Depiction of the quantum process that, according to Arnold Sommerfeld, gave rise to certain X-ray lines. Sommerfeld drew the diagram to persuade Alfred Landé to abandon a different model, which was also wrong.

made a break with the physics he had used to describe radiation.

In any case, Planck had full confidence in the representation of the electromagnetic field given by James Clerk *Maxwell and Hendrik Antoon Lorentz. The unification of light with electromagnetism, the demonstration of the "reality" of electromagnetic waves by Heinrich Hertz, and the description given by the Maxwell-Lorentz equations of a multitude of wave phenomena was for Planck, and for almost all of his contemporaries, convincing evidence of the continuous nature of radiation. Albert *Einstein entertained doubts. Having scrutinized the statistical mechanical foundations upon which Planck based his derivation of his formula for the spectral density of blackbody, Einstein concluded in 1905 that a few phenomena, like the photo-electric effect, could be explained easily if "the energy of monochromatic light consists of a finite number of energy quanta of magnitude $h\nu$, localized at various points of space [that] can be produced or absorbed only as units." At about the same time, Einstein realized that Planck's radiation theory required a radical discontinuity in the energy content of the individual resonators; with his "heuristic hypothesis" concerning the photoeffect, Einstein extended the discontinuity to the free electromagnetic field, and to the interaction between light and matter.

Einstein's explanation in 1907 of the observed deviation at low temperature of the specific heat of simple solids from their classical value of $3Nk$ (N = the number of molecules in a gram, k = "Boltzmann's constant," a second universal constant from the blackbody formula) corroborated the quantum hypothesis. In Einstein's model of a solid the potential that an atom experiences near its equilibrium position is the same for all the atoms of the solid. Hence for small vibrations near their equilibrium point all the atoms oscillate with the same frequency v. Quantization implies that each oscillator can only have an energy equal to $\epsilon_n = nhv$, and Planck's formula gives, in the limit where hv is small in comparison with kT (T = temperature), the specific heat $3Nk$. At low enough temperatures, where the limit does not hold, characteristic deviations from the classical value occur, which Walther Nernst and others detected around 1910.

In his doctoral thesis on the electron theory of metals (1911), Niels *Bohr concluded that atoms constructed according to the principles of classical physics could not represent the magnetic properties of metals. Working in Ernest *Rutherford's laboratory in Manchester just after Rutherford proposed the nuclear model of the atom, Bohr seized upon it because its radical mechanical instability made it a promising candidate for repair by a quantum hypothesis (see ATOMIC STRUCTURE). Bohr stabilized the Rutherford atom by supposing that it could exist in various "stationary states" constrained by certain quantum rules but otherwise governed by the laws of classical mechanics. However, the laws of mechanics do not hold for the transition of the system between two stationary states during which the atom radiates a quantum of energy hv equal to the difference in energy between the two states. On Bohr's theory, radiation is not emitted (or absorbed) in the continuous way assumed by Maxwell-Lorentz electrodynamics.

Bohr's first postulate, which limited the validity of classical mechanics in the atomic domain, restricted the angular momentum of each atomic electron to an integral multiple of $h/2\pi$. The second postulate, which denied the validity of classical electrodynamics for radiative processes in atoms and made the frequencies of atomic spectral lines different from the orbital frequencies of the electronic motions, required surrendering the classical connection between the frequency v of the emitted radiation and the mechanical frequency of the electron in its orbit.

With the help of these quantum rules Bohr accounted for the phenomenological regularities that had been discerned in the hydrogen spectrum, in particular, the Balmer formula for transitions to the $n = 2$ level, and also, and more dramatically, for the spectrum of ionized helium (1913–14). During World War I, Arnold Sommerfeld generalized Bohr's postulates to elliptical electron orbits and then to motions in three dimensions. He recorded his success in calculating regularities in doublet and triplet spectra, in the Zeeman effect, and in x-ray spectra in a long book, *Atombau und Spektrallinien* (first edition 1919), with which all physicists interested in quantum and atomic physics during the early 1920s began their work.

In the early 1920s Bohr gave a phenomenological explanation of the periodic table based on the occupancy by electrons of Coulomb-like orbits in multi-electron atoms. Thereafter, many theorists tried to justify Bohr's explanation, but, except for Wolfgang Pauli's formulation of the exclusion principle early in 1925, none of their efforts provided a stable foundation for the dynamics of atoms. They were seminal, however, in that they made manifest the problems a more complete quantum mechanics would have to solve.

In 1917 Einstein took what in retrospect was an important step toward this mechanics. Still flirting with the corpuscular nature of radiation, Einstein introduced the concept of the probability for the spontaneous emission of a light quantum by a "molecule" in a transition from one state to another. The concept allowed an easy derivation of Planck's blackbody formula. In 1923, Arthur Holly Compton's experiment on the scattering of *X rays by electrons indicated that the shift in the wave length of the scattered X ray and the recoil energy of the electron could be derived on the assumption that the X rays acted as particles with energy hv and momentum hv/c (c the velocity of light). The positive result of the Compton experiment led Einstein to declare that there are "two theories of light, both indispensable, and . . . without any logical connection." The corpuscular viewpoint accounted for the optical properties of atoms, whereas macroscopic phenomena like diffraction and interference required the wave theory of light. The two theories coexisted without any resolution during the early 1920s.

Another important guide to a more powerful quantum physics was the correspondence principle Bohr refined between 1913 and 1918. It stated that the frequencies calculated by Bohr's second postulate (during "quantum jumps") in the limit where the stationary states have large quantum numbers that differ very little from one another will coincide with the frequencies calculated with the classical theory of radiation from the motion of the system in the stationary states. Bohr's assistant Hendrik Kramers cleverly applied the correspondence idea to compute the intensity and polarization of the light emitted from simple atoms. Kramers and Werner *Heisenberg extended the same idea to the dispersion of light and worked out ways to translate classical quantities involving a single stationary state into quantum mechanical quantities involving two or more states. Max Born, Heisenberg's teacher at the University of Göttingen, called for a "quantum mechanics" for calculating with the quantum mechanical quantities directly. That was in 1924. In less than a year Heisenberg provided him with one. Its guiding principles were satisfaction of Bohr's correspondence principle (in the appropriate limit the theory should yield the classical results); recognition that the troubles of the "old quantum theory" arose primarily from breakdown of the kinematics underlying classical dynamics; and restriction of the theory to relations between observable quantities.

Born, Heisenberg, and a fellow student of Heisenberg's, Pascual Jordan, soon developed the new mechanics into an elaborate mathematical formalism. They built a closed theory that displayed strikingly close analogies with classical mechanics, but at the same time preserved the characteristic features of quantum phenomena. Their work laid the foundations of a consistent quantum theory but at the price of relinquishing the possibility of giving a physical, visualizable picture of the processes it could calculate. Hence the relief felt by Planck, Einstein, and Lorentz when Erwin Schrödinger, who followed a route entirely different from Heisenberg's, began to publish his wave mechanics in 1926. It seemed to avoid the unconventional features of Heisenberg's formulation and rested on more traditional foundations and easier calculations: variational principles, differential equations, and the properties of waves.

Schrödinger had followed up insights and suggestions by Louis de Broglie and Einstein. In 1923 de Broglie published an idea that was the obverse of Einstein's attribution of particle properties to wave radiation—to endow discrete matter with wave properties.

By following sometimes fanciful analogies and the principle of *relativity, de Broglie associated a wave of frequency ν and wavelength λ with a particle of momentum p and energy E according to $\nu = E/h$, $1/\lambda = p/h$. He thus extended the particle-wave duality of radiation to matter. Knowing the wavelength, Schrödinger soon found an appropriate differential equation for a wave of amplitude Ψ. He interpreted the Ψ function as describing a real material wave and considered the electron not a particle but a charge

distribution whose density is given by the square of the wave function. In a short paper dated June 1926, Born rejected Schrödinger's viewpoint and proposed a probabilistic interpretation for the Ψ function. He stipulated that the wave function $\Psi(x, t)$ determines the probability of finding the electron at the position x at time t. In 1927 two different sets of experimentalists—George P. Thomson (the son of Joseph John *Thomson) in Britain and Clinton Davisson and Lester Germer in the United States—detected diffraction patterns from an electron beam.

Several physicists proved in 1926 that wave mechanics gave the same numerical answers as the "matrix mechanics" of Born, Heisenberg, and Jordan. Together they are known as quantum mechanics. In contrast to classical physics, which contained no scale and was assumed to apply both in the micro and macro domain, quantum mechanics asserted that the physical world presented itself hierarchically. Certain constants of nature layered the world. As P. A. M. *Dirac emphasized in the first edition of his *Principles of Quantum Mechanics*, Planck's constant allows the parsing of the world into microscopic and macroscopic realms.

The conquest of the microrealm during the first years after the invention of quantum mechanics stemmed from the confluence of two factors: the apperception of an approximately stable ontology of electrons and nuclei, and the formulation of the dynamical laws governing the motion of electrons and other microscopic particles moving with velocities small compared to the velocity of light. Approximately stable meant that electrons and (non-radioactive) nuclei, the building blocks of atoms, molecules, simple solids, could be treated as ahistoric objects, with physical characteristics seemingly independent of their mode of production and lifetimes effectively infinite. These electrons and nuclei behaved as if they were "elementary," almost point-like objects specified only by their mass, their intrinsic spin, electric charge, and magnetic moment. In addition, the members of each species were indistinguishable: all electrons are identical, as are all protons, and all (stable) nuclei of a given charge and mass when in their ground state. Their indistinguishability implied that an assembly of them obeyed characteristic statistics depending on whether they had integral or half odd integral spin (measured in multiples of $h/2\pi$). Bosons (particles with zero or integral spins) can assemble in any number in a given quantum state. Fermions (particles with half odd integral spins) do not share a quantum state. A one-particle quantum state can be characterized either by the position and the spin state of the particle or by its momentum and its spin state. Thus no two identical Fermions can be at the same position if they have the same spin. More generally, the wave function describing a system of identical bosons remains unchanged under the interchange of any two particles, whereas that describing fermions changes sign under such a transposition.

The quantum mechanical explanation of chemical bonding resulted in a unification of physics and chemistry. In 1929, following the enormous success of nonrelativistic quantum mechanics in explaining atomic and molecular structure and interactions, Dirac, a main contributor to these developments, declared that "the general theory of quantum mechanics is now almost complete." Whatever imperfections still remained were connected with the synthesis of the theory with the special theory of relativity. But these were of no importance in the consideration of atomic and molecular structure and ordinary chemical reactions. "The underlying physical laws necessary for the mathematical theory of a large part of physics and the whole of chemistry are thus completely known, and the difficulty is only that the exact application of these laws leads to equations much too complicated to be soluble." Dirac's assertion may still have the validity it had, but, as emphasized by Phillip Anderson, "the reductionist hypothesis does not by any means imply a 'constructionist' one: The ability to reduce everything to simple fundamental laws does not imply the ability to start from those laws and reconstruct the universe. In fact, the more the elementary particle physicists tell us about the nature of the fundamental laws, the less relevance they seem to have to the very real problems of the rest of science, much less to those of society. The constructionist hypothesis breaks down when confronted with the twin difficulties of scale and complexity." Still, physics can be regarded as more foundational (not fundamental) than chemistry because the laws of physics encompass in principle the phenomena and the laws of chemistry.

See also QUANTUM ELECTRODYNAMICS; QUANTUM FIELD THEORY.

Thomas S. Kuhn, *Blackbody Theory and the Quantum Discontinuity, 1894–1912* (1978). Abraham Pais, *Inward Bound* (1982). Jagdish Mehra and Helmut Rechenberg, *The Historical Development of Quantum Theory* 6 vols. (1986–2001). Olivier Darrigol, *From C-Numbers to Q-Numbers* (1992). Mara Beller, *Quantum Dialogue: The Making of a Scientific Revolution* (1999). Helge Kragh, *Quantum Generations* (1999).

J. L. HEILBRON

QUARK. During the 1950s and 1960s progress in classifying and understanding the phenomenology of the ever increasing number of hadrons

(strongly interacting microscopic particles) came not from fundamental theory but by shunning dynamical assumptions in favor of symmetry and kinematical principles that embodied the essential features of a relativistic quantum mechanics.

In 1961, Murray Gell-Mann and Yuval Ne'eman independently proposed classifying the hadrons into families based on a symmetry later known as the "eightfold way." They realized that the mesons (hadrons with integral spins) grouped naturally into octets; the baryons (heavy hadrons with half integral spins) into octets and decuplets. The "eightfold way" can be represented mathematically in three dimensions, a property that led Gell-Mann, and independently George Zweig, to build hadrons out of three elementary constituents. Gell-Mann called these constituents "quarks" (from a line in James Joyce's *Finnegans Wake*, "Three quarks for Muster Mark!"); Zweig called them aces. The elaboration of the quark scheme is briefly indicated here as an indication of the methods and madness of elementary particle physics.

To account for the observed spectrum of hadrons Gell-Mann and Zweig defined three "flavors" of quarks (generically indicated by q), called up (u), down (d), and strange (s), each with spin $\frac{1}{2}$ but differing in two other quantum numbers (isotopic spin and strangeness) that defined them. Ordinary matter contains only u and/or d quarks. ("Strange" hadrons would contain strange quarks.) The three quarks had two other features: a baryonic mass of $\frac{1}{3}$, and an electrical charge of $\frac{2}{3}$ (for the u) and of $-\frac{1}{3}$ (for the d and s) those of the proton. This last feature startled physicists who had no experimental evidence for any macroscopic object carrying a positive charge smaller than a proton's or a negative charge smaller than the electron's.

Since a relativistic quantum mechanical description implies that for every charged particle there exists an "anti-particle" with the opposite charge, Gell-Mann and Zweig provided for antiquarks (generically denoted by $\bar{q}$) having an electric charge and strangeness opposite to those of the corresponding quarks.

Quarks bind together into hadrons as follows. An up and an antidown quark make a positive meson; two ups (with electrical charge $\frac{4}{3}$) and a down (with electrical charge $-\frac{1}{3}$), a proton. All baryons can be made up of three quarks, all mesons of one quark and one antiquark. That, however, did not provide quite enough possibilities so quarks had to have another attribute, which, in the playful quark nomenclature, was called "color." Quark color comes in three varieties (sometimes taken to be red, yellow, and blue), each of which can be "positive" or "negative." Quarks carry positive color charges and antiquarks carry negative ones. The observed hadrons have no net color charge.

In the late 1960s the Stanford Linear Accelerator (SLAC) could produce electrons of sufficiently high energy to probe the internal structure of protons. If the proton's charge were uniformly distributed, penetrating electrons would tend to go through it without being appreciably deflected. If, on the other hand, the charge was localized on internal constituents, then—in analogy to Ernest *Rutherford's demonstration of the atomic nucleus—an electron that passed close to one of them would be strongly deflected. The SLAC experiments showed this effect, which prompted Richard *Feynman to suggest that the proton contained pointlike particles with spin $\frac{1}{2}$, which he called "partons." The partons soon were assimilated to the quarks, although they (the partons/quarks) appeared to be too light and too mobile to make up protons. These difficulties were eventually resolved.

The discovery in November 1974 of the J/ψ meson gave further support for the quark picture and reason to accept a fourth quark with a new flavor, "charm" (denoted by c), whose existence had been proposed by Sheldon Glashow and others in 1964 and, with greater insistence, in 1970. The J/ψ (its two names resulted from its simultaneous discovery by two different groups) appeared to be a bound state of c and $\bar{c}$. The discovery of November 1974 revolutionized high-energy physics by establishing the representation of hadrons as quark composites. As the number of hadrons grew, however, the scheme had to be extended by the addition of the "bottom" (or "beauty," b) quark in 1977 and the "top" (t) quark in 1994. Each successively discovered quark has a larger mass than its predecessors on a scale in which the u weighs 1, the d weighs 2, the s 36, the c 320, the b 960, and the t 34,800. They all have spin $\frac{1}{2}$; partake in the strong, electromagnetic, and weak interactions; and come in pairs: up and down (u, d), charm and strange (c, s), and top and bottom (t, b). The first member of each pair has electric charge $\frac{2}{3}$ and the second $-\frac{1}{3}$. Each flavor comes in three colors.

If hadrons are made up of fractionally charged quarks, why have fractionally charged particles not been observed? Even if a plausible mechanism could be devised for confining quarks, what reality can be attached to them as constituents of hadrons if they can never be observed empirically? Quantum chromodynamics (QCD), which emerged a decade after the introduction of quarks, explained how quarks could be so strongly bound that they could never escape, while nevertheless behaving as quasifree particles in deep inelastic scattering.

K. Gottfried and V. Weisskopf, *Concepts of Particle Physics* (1986). M. Riordan, *The Hunting of the Quark: A True Story of Modern Physics* (1987).

SILVAN S. SCHWEBER

QUASARS. See PULSARS AND QUASARS.

R

RACE. The term "race" and its equivalent in several languages gained currency in the seventeenth century to describe descendents of the same family or house. The word was also used to refer to a tribe or nation, as in the Germanic races. Only in the nineteenth century did the term take on the taxonomic meaning of a distinctive group or variety within a human or animal species.

During the eighteenth century, naturalists developed comprehensive categories to classify human beings among other animal groups. In the tenth edition (1758) of his *Systema naturae*, Carl Linnaeus placed the genus *Homo* within the order *Primates* (which included monkeys, bats, and sloths) and distinguished two species: *Homo sapiens* and *Homo troglodytes* (anthropoid apes). He divided *Homo sapiens* (wise man) into four varieties: American (copper-colored, choleric, regulated by custom), Asiatic (sooty, melancholic, and governed by opinions), African (black, phlegmatic, and governed by caprice), and European (fair, sanguine, and governed by laws). Linnaeus conceived such differences as expressive of divine intent.

Johann Friedrich Blumenbach, the most influential racial theorist at the turn of the nineteenth century, argued that human beings constituted one species, with five varieties grading into one another. His *De generis humani varietate native liber* (3d ed., 1795) distinguished the Caucasian (originating in Georgia), the Mongolian (including Greenlanders and Eskimos), the Ethiopian (Africans), the American (Indians of North and South America), and the Malayan (including the islanders of the South Pacific). These groups differed in skin color, facial traits, hair texture, stature, and skull shape. Blumenbach speculated that the large penis of his Ethiopian specimen would support tales of sexual prowess, but he did not venture whether the trait characterized the variety. He thought the Caucasian race the most beautiful and inferred that it might constitute the original people, whence the others had descended and altered through the effects of climate.

During the early nineteenth century, cranial capacity became an important marker of racial difference and hierarchy. The anatomists Peter Camper, Samuel Thomas Soemmerring, and Georges Cuvier held that the Ethiopian brain resembled the ape's and displayed comparable intellectual ability. Friedrich Tiedemann, stimulated by debates in the British Parliament over slavery, undertook comprehensive measurement of brains and skulls of the various human groups. In his *Das Hirn des Negers* (1837), he found no significant differences among the groups. He completed his study with accounts of Negroes who had received an education and had made important contributions to the sciences and literature.

In the *Descent of Man* (1871), Charles Darwin concluded that since there was no clear distinction between species and varieties; we could regard human groups as forming either several species of one genus or several varieties of one species. He preferred the latter way of putting it. He recognized physical and intellectual differences among the races, but their common descent made these of small moment. By contrast, Darwin's German disciple Ernst Haeckel, in the later editions of his *Naturliche Schöfungsgeschichte* (1868–1911), identified twelve species that derived from the original ape-man. He maintained that the Mediterraneans—consisting of the Indo-Germans, Caucasians, and the Hamo-Semites—were the most evolved. He believed the hereditary effect of language to be the engine producing the various species. Languages that had the most potential for human thought produced races with brains having the greatest mental capacity.

At the beginning of the twentieth century, immigration into the United States from all quarters of the globe focused the attention of anthropologists and psychologists on characteristic human differences. Madison Grant, in his *Passing of the Great Race* (1916), distinguished three species of men: Caucasian, Mongoloid, and Negroid. He worried that within the Caucasian species, the higher Nordic race was in danger of sacrificing intellectual and psychological superiority by intermarriage with races from the Alpine (central and eastern Europe) and Mediterranean regions. During World War I, considerations of race became closely tied to theories of intelligence. The link was fostered by the development of the U.S. Army intelligence tests, which helped sort men into military ranks. Lewis Madison Terman, Robert Mearns Yerkes, and Carl Campbell Brigham believed that the tests demonstrated the inferiority of the Alpine and Mediterranean races. Terman argued that because of immigration the average mental age of the American public had sunk to that of a fourteen-year old. The journalist Walter Lippmann countered that the observation was ludicrous, since the tests of over one million individuals ought to indicate that the average I.Q. of Americans was average.

In the wake of World War II and in recognition of the Nazis' genocidal acts, UNESCO sponsored two symposia (1950 and 1952) on the scientific status of race. The document coming from the second symposium—formulated by Julian Huxley, Theodosius *Dobzhansky, and J. B. S. Haldane among others—declared that all men derived from a common stock and thus belonged to a single species and that insignificant physical differences gave no support to claims of racial hierarchy. The declaration, however, did not resolve the issue. In his *Origin of Races* (1962), the distinguished physical anthropologist Carleton Coon maintained that *Homo erectus* migrants had come out of Africa and had settled in five locations throughout the world, where they evolved into *Homo sapiens* at different rates. Europeans and Asians, he assumed, left the other races still clinging to more primitive physical, intellectual, and cultural traits.

In the current debate, those holding the regional hypothesis face those arguing the "out-of-Africa" theory. These latter maintain that *Homo erectus* populations went extinct in all regions but Africa, where they evolved into archaic *Homo sapiens*. About 100,000 years ago this group decamped from their homeland and settled in various parts of the world, where only slight differences evolved. Today most geneticists prefer to drop the category of race and to speak rather of human populations whose genetic structures blend into one another at the margins of contact.

John R. Baker, *Race* (1974). Nancy Stepan, *The Idea of Race in Science: Great Britain, 1800–1960* (1982). Daniel J. Kevles, *In the Name of Eugenics: Genetics and the Uses of Human Heredity* (1985). George W. Stocking, Jr., *Victorian Anthropology* (1987).

ROBERT J. RICHARDS

RADAR is today ubiquitous, with central uses in navigation, traffic control of aircraft and ships, military weaponry and defense, mapping, weather forecasting, and scientific experimentation. The basic technique employs pulsed transmission of radio beams that reflect off distant targets, are received, and analyzed. The returned signals allow determination of the location, speed, and nature of the target, even through fog or clouds.

Radio waves (electromagnetic radiation of wavelength longer than 1 cm) were first recognized in 1887 by Heinrich Hertz in Germany, and soon exploited for communications; by 1910 a new industry, led by the British Marconi company, had established many overseas circuits. In the process of developing better receivers and antennas, it was noticed that a passing ship or airplane could affect a radio signal, but early attempts to exploit this phenomenon as a warning device proved impractical. By the 1930s, however, radio techniques had sufficiently advanced that shorter wavelengths could be employed (allowing greater directivity and sensitivity), and political instability in Europe fueled military interest in systems that could reliably detect distant aircraft and ships. Working radar systems existed by 1940 in Great Britain, Germany, and the United States; by far the most advanced in operational terms was the network of "Chain Home" radars lining the English coast (developed by a team led by the Scottish physicist Robert Watson-Watt), a network that provided unprecedented early warning against German bomber attacks in the Battle of Britain.

By the end of the war all of the major combatants had mobilized their best scientific and engineering talent into developing a huge range of radar sets, for use in every type of military fixed installation, vehicle, aircraft, and ship. The leading laboratories were the Telecommunications Research Establishment (England), Telefunken and GEMA (Germany), the Radiophysics Lab (Australia), and the MIT Radiation Lab (United States)—the resources poured into radar research by the United States were second only to those used to develop an atomic bomb.

After the war, the electronics and antenna techniques that had been developed for radar were transferred not only to civilian applications, but also to nascent scientific fields such as radio astronomy, microwave spectroscopy, particle accelerators, semiconductors, and masers. Many wartime radar veterans continued work in related fields; a number went on to win the Nobel Prize in Physics, including Edward Purcell for work on nuclear magnetic resonance, Martin Ryle for the development of radio interferometers for studying cosmic radio sources, and Charles Hard Townes for the invention of the maser amplifier.

In recent years radar has become ever more sophisticated and miniaturized, and can now be used, for example, to study the rings of Saturn at a distance of over one billion kilometers. The history of radar demonstrates the impossibility of decisively answering the question of whether in the twentieth century science led technology, or vice versa; in radar, and in its many associated fields, they were involved in an intricate dance, with neither partner in the lead.

Robert Buderi, *The Invention that Changed the World* (1996). Louis Brown, *A Radar History of World War II: Technical and Military Imperatives* (1999).

WOODRUFF T. SULLIVAN III

RADICAL, CHEMICAL. The word "radical," meaning "root," was introduced into chemistry by L. B. Guyton de Morveau (1785) and adopted by Antoine-Laurent Lavoisier (1789). It designated basic organic material that, when combined with oxygen, became an acid, such as sugar (oxidized to oxalic acid), alcohol (oxidized to acetic acid), or benzoin (oxidized to benzoic acid). Organic *radicals* thus played the same role as the inorganic *elements*, which also could be oxidized to form acids. There could also be inorganic radicals, Lavoisier thought, such as muriatic or fluoric. Consequently, a "radical" was any compound entity that functioned chemically as an element.

As organic chemistry began to grow in importance, additional radicals appeared. Joseph Louis Gay-Lussac investigated the reactions of cyanogen (1815), a compound of carbon and nitrogen that formed a variety of compounds. In 1828, Jean-Baptiste-André Dumas and Polydore Boullay explored the chemistry of olefiant gas ("etherin" or ethylene). Dumas, Justus von Liebig, and others published important work on the compounds of "ethyl" and "methyl." But probably the most dramatic investigation was that of Liebig and Friedrich Wöhler on the oil of bitter almonds (1832). They showed that this substance, now known as benzaldehyde, occurred in an impressive series of derivatives. In 1837, Dumas and Liebig published an important joint paper in which they declared that *organic and *inorganic chemistry differed only in that the latter dealt with the compounds of the elements, while the former treated compounds of radicals.

Serious strains in the theory had already appeared, however, for organic radicals declined to remain inviolable like elements; they could be altered in many ways, especially by substitution of their hydrogen by chlorine. In the 1830s the number of radicals continued to increase (acetyl and cacodyl, for instance), while chemists strove to save the theoretical significance of the concept. Dumas abandoned the idea, proposing a theory of organic "types" that emphasized substitution over the addition mechanism of radical theory. Liebig himself began to have doubts, although the chief theorist, Jöns Jacob Berzelius, held firm.

In the 1840s, Edward Frankland and Hermann Kolbe attempted to salvage the radical theory by isolating radicals. They succeeded, more or less (we now know that they had made dimers—doubled versions—of the radicals). A few years later, Charles Frédéric Gerhardt, Alexander Williamson, and Charles-Adolphe Wurtz, created a new theory of types, which treated organic radicals as merely conventional groupings. They regarded "ethyl," for example, as a group of two carbon and five hydrogen atoms, and the ethyl and other radicals as a group of atoms that could be added, subtracted, altered, or traded between larger molecules.

This modified theory of radicals and types led directly to the formulation of the theories of valence and chemical structure. Once chemists learned how to construct organic molecules schematically from carbon, hydrogen, oxygen, and nitrogen atoms, "radicals" became pieces of molecules or of formulas. Only in the twentieth century did the concept reemerge in the development of "free radicals," neutral molecules with one unpaired electron.

Colin Russell, *The History of Valency* (1971). W. H. Brock, *The Norton [Fontana] History of Chemistry* (1992).

A. J. ROCKE

RADIO AND TELEVISION. Soon after Heinrich Hertz demonstrated the existence of electromagnetic waves by a series of classical experiments between 1886 and 1888, physicists and entrepreneurs turned wireless or radiotelegraphy (the term suggested by John Munro in 1898) into reality. In 1892, William Crookes listed the three essential requirements: reliable transmitters, sensitive tuned receivers, and directional aerials. Édouard Branley invented the coherer (a glass tube containing metal filings used as a receiver) and may well have been the first to use the word "radio" in its modern sense when he suggested the name "radioconducteur" for his device. In 1894, Oliver Joseph Lodge popularized the coherer (his name for this device) and demonstrated the importance of tuning, for which he obtained a patent in 1897.

This work was extended in practical ways by Henry Jackson in Britain, Aleksandr Popov in Russia, Adolf Slaby in Germany, and Augusto Righi in Italy, but none had the commercial success of Guglielmo Marconi, who turned wireless telegraphy into commercial reality through the Marconi Wireless Telegraph and Signal Company, established in 1897. Much of his early apparatus, apart from the detector, derived from telegraphy and telephony. In 1901, he transmitted the letter "s" in Morse code from Poldhu in Cornwall to St. John's, Newfoundland, demonstrating the feasibility of trans-Atlantic radio communication. William du Bois Duddell in Britain and the Danish engineer Valdemar Poulson improved early arc transmitters, while the American engineer Reginald Aubrex Fessenden developed Nikola Tesla's high-frequency alternator for radio use. Radio amateurs followed these technical developments with great interest. The first wireless club in Britain, formed in 1911, was followed by many others in Britain, the United States, and elsewhere.

World War I influenced radio design centered on the thermionic valve and stimulated the growth of radio in other ways as well. The accomplishment of radio communication between airplanes and the ground was particularly significant. Many of the thousands of radio operators returning to civilian life helped to establish public broadcasting. A glut of radio spare parts and firms experienced in manufacturing radio receivers stood ready. In most industrialized nations, public broadcasting started in the early 1920s. In Britain, the privately owned British Broadcasting Company (1922) received a Royal Charter as the British Broadcasting Corporation in 1927, when it became a government-owned monopoly. In the United States, after a chaotic start, previously rival stations were formed into networks, the National Broadcasting Company (NBC) being the first (1926).

The rapid advances in electronics during the inter-war years had an impact on the circuit designs of domestic radio receivers. The manufacturing industry became increasingly active, especially in the United States, where commercial receivers were on the whole more innovative and technically better than elsewhere. In Britain, multi-valve sets proved too expensive, so the market concentrated on receivers with simple tuning circuits with some form of positive feedback to amplify the signal. These TRF (tuned radio frequency) sets remained common in Britain until well into the 1930s.

In the United States, two multi-valve circuits competed for the domestic receiver: the neutrodyne and the superheterodyne. The neutrodyne, invented by Louis Alan Hazeltine in the United States to bypass RCA's feedback patents, was devised independently by John Scott-Taggart in Britain. The superheterodyne ("superhet") also had two independent inventors, Edwin Howard Armstrong and Walter Schottky in 1918, who developed the heterodyne principle patented by Reginald Aubrey Fessenden in 1901. Armstrong

devised his version while serving with the American Expeditionary Force in order to detect enemy aircraft by the high-frequency radiation emitted by the aircraft's ignition system. It was never used for this purpose, but applied (by 1924) to domestic receivers in the United States. Schottky's employers, Siemens in Germany, did not develop his proposals.

The neutrodyne circuit was made obsolete in the late 1920s by the new generation of thermionic valves, the screen-grid tetrode, and the pentode. The first commercially successful thermionic valves with indirectly heated cathode, the KL1 by Marconi-Osram in Britain and the Radiotron UX225 and UY227, appeared in the late 1920s, which made the all-mains domestic receiver possible.

One of the chief problems plaguing radio reception from its earliest days was "static." To overcome it, Armstrong suggested that receiver design should switch from amplitude modulation (AM) to frequency modulation (FM), but FM radio stations and receivers did not become commercially feasible until after World War II. During the 1930s, the radio's appearance changed markedly through the growing use of bakelite and other molded plastics. The commercial introduction of the transistor in the 1950s also had a profound affect both on radio circuitry and on outward appearance.

The idea of "seeing by electricity" goes back to late nineteenth-century electrical technology. Images were transmitted by telegraphy as early as the 1860s, and built up electrochemically. The photo-conductive properties of selenium, discovered in 1873, were exploited to transmit photographs by Arthur Korn in 1907. Two rival television systems arose in the 1920s: one electromechanical, the other electronic. The low-definition electromechanical system of John L. Baird in Britain and C. Jenkins in the United States used the revolving disk scanner invented by Paul Nipkow in 1884.

The rapidly revolving Nipkow disk, with holes or lenses arranged into a spiral, broke up the image into parallel lines of varying intensity and projected them onto a photoelectric cell. The resultant variations in current were transmitted by a radio to a receiver with a similarly arranged Nipkow disk and neon tube whose light varied in intensity with the strength of the radio signal. The lenses in the disk projected the light of the neon tube onto the television screen. Although an ordinary radio receiver could be used, mechanical scanning was too slow and cumbersome for building up high-resolution images. In 1929, Baird achieved a 30-line, 12.5 frames-per-second picture, and at its best, 240 lines. He achieved these higher scanning speeds using Lazare Weiller's revolving mirror-drum scanner invented in 1889.

At the heart of electronic scanning was the *cathode-ray oscilloscope (CRT) invented by Ferdinand Braun in Germany in 1897. This device evolved from the earlier vacuum tubes developed by Crookes and Wilhelm Hittorf. In 1907, Boris Rosin in Russia used a CRT to display images from a mechanical transmitter, while the following year, Alan Archibald Campbell-Swinton in Britain suggested using such an electronic tube for both transmission and display. Vladimir Zworykin, a former student of Rosin's, moved to the United States, where he developed the iconoscope, the first photo-electric television camera tube, in 1928. In 1930, he transferred to RCA, and two years later demonstrated a 120-line all-electronic television. An American, Philo T. Farnsworth, also inspired by Rosin, undertook similar work, and eventually RCA felt the need to take out Farnsworth licenses. The death knell of electromechanical television took place at Alexandra Palace in London in 1936 when, in a series of transmissions of both systems, the BBC adopted the Marconi-EMI electronic system as its standard. During the war years, television went into hibernation, but both the necessary CRT and high-frequency transmission and receiving circuitry were developed for radar. The early 1950s saw a veritable explosion in domestic television. Color television came in the 1950s, first in the United States and then elsewhere, and the industry adopted various dot and line standards.

W. M. Dalton, *The Story of Radio*, 3 vols. (1975). H. J. G. Aiken, *Syntony and Spark: The Origins of Radio* (1976). W. A. Atherton, *From Compass to Computer: A History of Electrical and Electronics Engineering* (1984), pp. 182–215. H. J. G. Aiken, *The Continuous Wave: Technology and American Radio, 1900–1932* (1985). R. W. Burns, *British Television: The Formative Years*, IEE History of Technology Series 7 (1986). Keith Geddes in collaboration with Gordon Bussey, *The Set Makers: A History of the Radio and Television Industry* (1991).

W. D. HACKMANN

RADIOACTIVE TRACER. Marie Sklodowska *Curie first established the principle of using the emissions of a radioactive element to follow a nuclear or chemical process. Between 1896 and 1898 she used the fractional crystallization of barium and bismuth chlorides to separate the less soluble radium and polonium chloride impurities in pitchblende samples. She monitored the progress of the fractionation by measuring the *radioactivity by means of a sensitive plate-condenser electrometer devised by her husband, Pierre Curie.

In 1903, Sir William Crookes invented the spinthariscope, a device for counting the light flashes that rays from radium produced on a zinc sulfide screen. Although widely used by Ernest *Rutherford and others in the early investigation of radioactivity, Crookes's technique tires the eye, and in 1908 Rutherford's student Hans Geiger returned to measuring radiation by the ionizing capacity of the alpha, beta, and gamma particles emitted during radioactive decay. An electrometer registered the passage of each charged particle. Aided by Walther Müller, Geiger greatly improved the sensitivity of this instrument in 1928. The Geiger-Müller counter more or less replaced the spinthariscope until the 1940s, when developments in photoelectricity and of photomultipliers made a return to the scintillation technique possible and accurately quantifiable (*see* GEIGER AND ELECTRONIC COUNTERS).

The Curies' radioactive tracer, or indicator, method was first expressed as a general technique in 1913 by György Hevesy, who pointed out the utility of using radium D (a lead isotope) as a tracer for lead. In 1918 Hevesy's colleague Friedrich A. Paneth used tracers to prove the existence of bismuth hydride by dissolving radio-bismuth (thorium C) in hydrochloric acid and noting the radioactivity of the hydrogen produced. Hevesy then used the technique to study the adsorption of lead by plants; otherwise its use was rather limited. Following the discovery and separation of deuterium by Harold Urey in 1932, however, and the subsequent separation of the radioactive isotopes of oxygen, nitrogen, and carbon, chemists saw that they might be valuable in tracing chemical reaction mechanisms. Thus, in 1935, the Welsh organic chemist Edward Hughes elucidated the phenomenon of the Walden inversion in which optically active substances are transformed into their stereoisomeric derivatives. Urey also encouraged the use of these radioisotopes, which were present naturally in biological systems, to trace compounds in intermediary metabolism. Accordingly, he made radioisotopes available to his Chicago colleague Rudolf Schoenheimer who used them creatively to follow the metabolism of fats and nitrogen. Such natural isotopes had the advantage of being detectable by the mass spectrometer as well as by the Geiger-Müller counter. Scientists also used isotopes produced by cyclotrons, particularly phosphorus 32, to study organic reactions. Following the *Manhattan Project and the end of World War II, a range of artificial radioisotopes (including phosphorus and sulfur) became available as part of the peaceful use of atomic energy. The ready availability of such biologically significant radioisotopes, together with the replacement of the Geiger-Müller counter by the automated liquid scintillation spectrometer developed by Lyle E. Packard in 1953, led to an explosive growth of biochemical knowledge in the 1950s and 1960s. Such research-enabling technology was an important factor in the emergence of *molecular biology.

F. A. Paneth, *Radioelements as Indicators* (1928). R. E. Kohler, "Rudolf Schoenheimer, Isotopes, Tracers, and Biochemistry in the 1930s," *Historical Studies in the Physical Sciences* 8 (1977): 257–298.

WILLIAM H. BROCK

RADIOACTIVITY. Studies in radioactivity produced the scientific research fields of *nuclear physics, *cosmic-ray physics, and *high-energy physics, and also *nuclear chemistry, nuclear medicine, and nuclear engineering. Beginning in 1940 with the isolation of neptunium and plutonium, the creation of short-lived transuranium elements extended the periodic table of the chemical elements into new territory.

That radioactivity has rested in the shared disciplinary terrains of physics and chemistry is indicated by the shared Nobel Prize in physics awarded to its discoverers Henri Becquerel, Marie Sklodowska *Curie, and Pierre Curie, and Ernest *Rutherford's receipt of the Nobel Prize in chemistry in 1908 for his (and Frederick Soddy's) elucidation of the mechanism of the radioactive disintegration of atoms.

Becquerel discovered that uranium salts emit radiation by accident while investigating whether naturally phosphorescent minerals produce the *X rays discovered by Wilhelm Conrad *Röntgen in late 1895. By 1897 Becquerel and others had demonstrated that uranium radiations carry electrical charge, a physical property that led Marie Curie to apply the quartz electrometer

Marie *Curie at work in her laboratory in Paris toward the end of World War I.

A German advertisement for a radioactive toothpaste for cleaning the teeth, whitening the smile, and invigorating the gums.

(then recently invented by her husband and his brother Jacques-Paul Curie) to a variety of minerals in search of the property that she termed "radioactivity." She identified polonium and radium as radioactive elements in 1898 and André-Louis Debierne discovered actinium in 1899. They and others found that thorium, like uranium, is radioactive.

Rutherford distinguished two kinds of radiation, which he called "alpha" (distinguished by its ready absorption) and "beta" (about one hundred times more penetrating). At the Curies' laboratory in 1900, Paul Villard discerned a gamma radiation from radioactive substances that appeared to behave exactly like X rays. By 1900, Becquerel, Rutherford, and others had established that the beta rays consist of negatively charged particles similar if not identical to *electrons. In 1903 Rutherford showed that the alpha rays also carried a charge and so had to be understood as a stream of particles. By 1908 Rutherford and his colleagues had decisive evidence that the "alpha" radiations are helium ions. Rutherford and Soddy recognized in 1903 that thorium continuously produces a new radioactive gas (radon), and a second radioactive substance (chemically identical to radium) from which a longer line of

radioactive substances descends. They settled on the hypothesis that radioactive decay is accompanied by the expulsion of an alpha or beta particle from the decaying atom and that the decay can be expressed in terms of a half life, defined as the time during which half of the mass of a radioactive element is transformed into a new substance. Rutherford's continuing study of alpha particles and their interactions with matter prompted the experiments that led to his invention in 1910–1911 of the nuclear atom.

By 1912 some thirty radioactive elements had been identified. This achievement derived from the work of Rutherford's groups in Montreal and Manchester; Marie Curie's laboratory in Paris; Otto Hahn and Lise Meitner's work in Berlin; and other investigators, including Bertram Borden Boltwood at Yale University in New Haven. They differentiated natural decay series beginning with uranium (238), actinium (227) or uranium (235), and thorium (232). In 1913 Soddy, Alexander Russell, and Kasimir Fajans independently developed a generalized radioactive displacement law. In 1911 Soddy had noted that loss of an alpha particle produced a chemical element two places to the left of the original in the periodic table. Similarly, Russell remarked that beta decays lead to the next element in the periodic table. Soddy coined the word "isotope" for radioactive elements chemically identical to another element while differing in atomic weight. Several studies confirmed in 1914 that atomic weights of lead derived from radioactive ores varied from each other and from the established value of 207.2.

Early methods for detecting radioactivity were both electrical and visual, relying initially on the electrometer and on microscopically observed scintillations of light caused as alpha particles strike zinc sulfide. Electrical methods of detection improved in 1928 when Hans Geiger and Walther Müller succeeded in making a reliable counter using the technique with which Geiger and Rutherford had counted alpha particles before the war. C. T. R. Wilson's cloud chamber, invented in principle in 1899 but not perfected until after World War I, also provided a means of "seeing" nuclear events. Using the older scintillation technique, Rutherford proposed in 1919 that collision of an alpha particle with nitrogen gas resulted in disintegration of the nitrogen atom and expulsion of a long-range hydrogen atom (proton). After automating a cloud chamber, P. M. S. Blackett in 1924 fired alpha particles into nitrogen atoms and obtained dramatic photographs, showing the path of a proton ejected from a recoiling nitrogen nucleus and the capture of the alpha particle by the nitrogen nucleus, creating an isotope of oxygen.

Francis Aston, who began his career as an assistant to Joseph John Thomson, developed a mass spectrograph that provided a photographic record of particles separated by their masses. Aston and others found evidence for isotopes not only among the radioactive and heavy elements, but also among the light elements. The concept of the isotope thus became generalized for all chemical elements.

In 1932, the same year that James Chadwick identified the neutron at the Cavendish Laboratory, John Douglas Cockroft and Ernest T. S. Walton, working with lithium, became the first scientists to split the atom by accelerating protons in a high-voltage, high-tension machine. Ernest O. Lawrence and Milton Stanley Livingston built a circular particle *accelerator that provided a rival model for achieving high energies. However, alpha particles from radium provoked the first radioactivity artificially induced by humans. That was in 1934, when Frédéric Joliot and Irène Joliot-Curie produced a radioactive isotope of phosphorus from aluminum.

During the late 1920s and early 1930s Meitner, Charles D. Ellis, and Enrico *Fermi argued over the existence and interpretation of apparent anomalies in the energy in beta decay. Wolfgang Pauli proposed in 1930 that a nuclear particle with a mass similar to an electron but with no charge might be expelled with an electron in beta decay in order to conserve energy in the reaction. Fermi suggested in 1933 that Pauli's little "neutron" be renamed the "neutrino" following James Chadwick's identification of the proton-sized neutral neutron in 1932. Fermi's theory was adopted, although the neutrino eluded detection until 1956.

Rays from naturally occurring sources continued to serve nuclear physicists and chemists even in the era of the cyclotron. Fermi and his collaborators irradiated every element they could find with neutrons derived from a radon-beryllium source, discovered the efficacy of slow neutrons in inducing nuclear reactions, and, mistakenly, believed that they had made transuranic elements by shining neutrons on uranium. Meitner, Hahn, and Fritz Strassmann also used neutrons from natural sources in the experiments from which, by the end of 1938, Hahn and Strassmann obtained the results whereby Meitner (by then a refugee in Sweden) and her nephew Otto Robert Frisch deduced the existence of nuclear fission.

During the course of the *Manhattan Project for the development of uranium and plutonium bombs, the health hazards of radioactivity increasingly came to the fore. Radiations that had been touted since the early 1900s as a general curative and a specific agent against cancer were demonstrated to cause leukemia and other cancer-related diseases. Facial creams and mineral waters, as well as watch dials and curios containing uranium salts, disappeared from store shelves and health resorts in the 1950s as international movements against the atmospheric testing of nuclear weapons gained force from evidence of the hazards of nuclear debris or "fallout" (see ANTINUCLEAR MOVEMENT). After enthusiasm in the early 1950s for the use of nuclear energy not only as a commercial power source but also for explosives in dam-building and road-building, public suspicion of radioactivity curtailed *nuclear energy projects in countries like the United States and Great Britain, although not in France or the Soviet Union.

In chemotherapy and in medical tests radioactive isotopes continue to serve as tagging or tracer devices for studying the metabolism or pathways of iodine, barium, and other elements in the body. In his first efforts in the 1930s to get large-scale funding for his accelerator program at the University of California at Berkeley, Lawrence emphasized medical applications. Large philanthropies such as the Rockefeller Foundation increasingly turned their funding priorities to medical research in the 1930s. Lawrence and his brother John Lawrence, who served as director of the university's medical physics laboratory, argued for the medical benefits of the production of radioactive isotopes in the accelerator. One result, phosphorus-32, was used in early attempts to treat leukemia.

Radioactive isotopes had uses beyond medicine. Samuel Ruben, W. Z. Hassid, and Martin Kamen at Lawrence's radiation laboratory used carbon-11 to study the chemistry of carbon dioxide in the photosynthesis of barley and the green algae chlorella. In 1941 Ruben and Kamen identified the radioactive isotope carbon-14 produced from nitrogen in the Berkeley accelerator. Melvin Calvin followed up at Berkeley by using a combination of paper *chromatography and radiochemical techniques to unravel in detail the mechanism of photosynthesis.

Willard F. Libby attracted more public attention when he showed in 1946 that living matter contains carbon-14, produced by the collision of cosmic ray neutrons with atmospheric nitrogen, which enters the carbon dioxide and carbon monoxide metabolism. Since the quantity of carbon-14 decays after death, Libby's discovery made a brilliant new means of dating very old organic remains.

Since World War II, particle accelerators and nuclear reactors have entirely superseded natural elements as the sources of radioactive materials for laboratory experiments and medical procedures.

See also ATOMIC STRUCTURE; ELEMENT; PERIODIC TABLE.

Aaron J. Ihde, *The Development of Modern Chemistry* (1964). T. J. Trenn, *The Self-Splitting Atom: A History of the Rutherford-Soddy Collaboration* (1977). Lawrence Badash, *Radioactivity in America: Growth and Decay of a Science* (1979). Roger H. Stuewer, "Artificial Disintegration and the Cambridge-Vienna Controversy," in *Observation, Experiment and Hypothesis in Modern Physical Science*, eds. P. Achinstein and Owen Hannaway (1985): 239–307. J. L. Heilbron and R. W. Seidel, *Lawrence and His Laboratory: A History of the Lawrence Berkeley Laboratory* (1989). Susan Quinn, *Marie Curie: A Life* (1995). Ruth Lewin Sime, *Lise Meitner: A Life in Physics* (1996). S. Boudia and X. Roque, eds., "Science, Medicine and Industry: The Curie and Joliot-Curie Laboratories," special issue of *History and Technology* 13 (1997): 241–354. Jeffrey Hughes, "Radioactivity and Nuclear Physics," in *Modern Physical and Mathematical Sciences*, ed. Mary Jo Nye [*The Cambridge History of Science*, vol. 5] (in press).

MARY JO NYE

RADIOASTRONOMY. See ASTRONOMY, NON-OPTICAL.

RADIUM, named for its spontaneous emission of ionizing radiation, was introduced to the world in 1898 by Marie and Pierre *Curie in Paris. This new element had revealed its presence solely by its radioactivity, the property that enabled the Curies, assisted by the chemist Gustave Bémont, to follow it through a series of chemical separations performed on pitchblende, a uranium ore.

The route to radium's discovery began with Marie Sklodowska Curie's doctoral research. While testing various minerals for radioactivity, Curie noticed that some uranium ore samples emitted more radiation than expected from their uranium content. She deduced that these samples contained a new radioactive element. Pierre Curie joined her in a search for the unknown substance, which yielded two new elements, polonium in July 1898 and radium in December.

These findings met with both astonishment and skepticism: astonishment, because radium appeared so highly radioactive, and because radioactivity had never been used to detect an element; skepticism, for similar reasons. Radioactivity, thought to be of no great importance, had been discovered barely two years earlier. Dubious reports and fanciful speculations about all sorts of invisible radiations circulated widely at the turn of the century.

Most chemists did not regard evidence from the new, insignificant, and poorly understood

The popular reaction to radium as depicted on the cover of the British magazine *Punch*.

phenomenon of radioactivity as convincing. Although the spectroscopist Eugène-Anatole Demarçay had identified a new spectral line in the Curies' purified pitchblende sample, acceptance into the periodic table required establishment of a distinct atomic weight. Marie Curie achieved this feat in 1902, after laborious chemical purifications. Radium took its place in the alkaline earth group above barium. Later it was assigned an atomic number, 88. In 1910 Marie Curie and André-Louis Debierne isolated metallic radium. Several isotopes of radium were identified after isotopy was recognized around 1913.

Radium's strong radioactivity made new experiments possible, which led to further discoveries and insights. Demand increased for this scarce element, which at first could only be obtained from the Curies and from the German chemist Friedrich Giesel, who barely missed discovering it. Radium's prodigious, seemingly endless energy output defied explanation. After scientists adopted the transmutation theory of radioactivity (1903–1906), researchers viewed radium as one of the decay products of uranium.

Eventually the greatest demand for radium came not from scientists but from physicians, who proposed to use the destructive effects radium wrought on living tissues to destroy tumors. The hope of a cure for cancer launched a prospecting fever and fueled a burgeoning radium industry, first in Europe, then in North America, and, from the 1920s, in central Africa. The worldwide search for uranium and its daughter radium led to investigations of environmental radioactivity and cosmic rays.

The promise of a cure for a dreaded disease coupled with the unusual circumstances of radium's discovery and its remarkable powers stoked the public imagination. Radium became a metaphor for the magic elixir. During the 1920s and 1930s the well-meaning and charlatans alike hawked patent medicines and household products purportedly containing radium. Several towns built their economies on "radium water" spas. Radium was also used in phosphorescent paints. These became especially popular for watch dials, but fatal for many dial painters who accidentally ingested the paint. After the health hazards of radioactivity were recognized, most markedly after the first atomic bomb explosions in 1945, radium's image took on a sinister cast. Radium nevertheless continued to find medical and industrial uses.

Alfred Romer, ed., *Radioactivity and the Discovery of Isotopes* (1970). Susan Quinn, *Marie Curie: A Life* (1995).

MARJORIE C. MALLEY

RAINBOW. The diaphanous multicolored arcs in the sky with pots of gold at their ends puzzled humankind from the cloudy era of Noah to the daylight of René *Descartes. From Genesis 9:13 we know that God set the rainbow in a cloud as a covenant that he would not again drown the creatures of the earth. Natural philosophers kept their eyes on the cloud and not the covenant. As early as the time of Aristotle they knew that the Sun, the sinner, and the bow's center all lie on a straight line, and supposed that the colors arose from rays from the Sun (or eye) reflected in the clouds. Aristotle considered the bow in his *Meteorologica*, which contains most of his terrestrial physics; and subsequent learned discussions about it, up through Descartes's *Météores* (1637), typically occurred in or around commentaries on Aristotle's book. When Descartes took up the problem as part of his demonstration of the superiority of his natural philosophy to the physics taught in the schools, the *Aristotelian approach had been refined to place the reflection in individual raindrops rather than in the cloud as a whole, and the origin of the rays in the Sun rather than in the eye.

Descartes demonstrated superiority by calculating. He derived the angle at which an observer

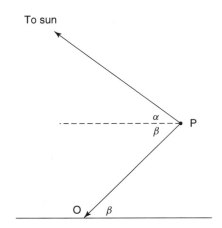

The relation between the altitude of the sun α and of the top of the bow β at P as seen by an observer at O.

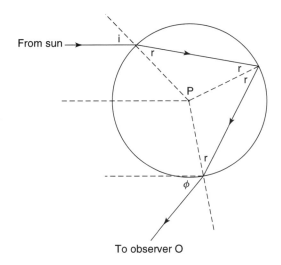

The path of a solar ray through a droplet at P incident at angle i, refracted at angle r, and turned through angle ϕ.

(at O in the first diagram) sees the uppermost part P of the bow. Aristotle could not do it and would have thought it unimportant to try; like Noah, he attended to the qualitative features of phenomena. One quantitative feature, however, was known to the Aristotelian school: the angle ϕ through which the Sun's rays that make the bow are turned at reflection by the raindrops is the sum of the altitudes of the sun (α in Figure 1) and of the bow at P β: $\phi = \alpha + \beta$. The angle ϕ is always around 42°. If God showed Noah a rainbow when the Sun stood higher than 42°, it was a true miracle. For, as Descartes demonstrated, the angle of 42° follows directly from the values of the refractive indexes of air and water.

Willebrord Snel, professor of mathematics at the University of Leyden, established by measurement around 1620 the exact amount by which a ray of light traveling in air bends, or refracts, on entering water (*see* OPTICS AND VISION). With this law, Descartes could trace the path of a light ray reflected by a raindrop, as in the second diagram. He had no room to wiggle: geometry and the number 1.5, the refractive index of water relative to air, fixed ϕ. Since every angle of incidence i gives a different value for ϕ, this geometrical apparatus might not seem an advance. But, by tracing many rays, Descartes showed that for a certain range of values of i (around $60°$), ϕ stays very close to $42°$. In a word, the droplets focus the rays. Explaining the colors and the pale secondary bow, with colors reversed, sometimes seen above the primary, required more words. Descartes ascribed the secondary to rays suffering two reflections in the drop. As for the colors, he suggested that in bouncing around the drop, light rays pick up spin, as if they were tennis balls sliced by a racquet, and that the different spins create the various colors when received by the eye (*see* MECHANICAL PHILOSOPHY). Isaac *Newton replaced this account with his revolutionary doctrine that each of the colors that he thought made up white light had a different index of refraction. On his theory, the drops focus the rays of different colors at slightly different angles ϕ.

Newton gave his theory of light and colors in a volume entitled *Opticks* (1704). Descartes explained the rainbow in his *Principles of Philosophy* (1644) as well as in his *Météores*. The difference in venue suggests the distance between the natural knowledge of the Continent around 1640, when the Aristotelian approach still provided the main competition to innovative natural philosophers, and that of England around 1700, when the claims of the Cartesians were the mark. As the most striking early example of the geometrization of a physical (as opposed to an astronomical) phenomenon, the rainbow as elucidated by Descartes and Newton represented a new covenant: that God would enable humankind to discover the numbers, weights, and measures that were employed in creation.

Carl B. Boyer, *The Rainbow from Myth to Mathematics* (2d edition, 1987).

J. L. HEILBRON

RAMAN, Chandrasekhara Venkata (1888–1970), physicist, Nobel Prize winner.

Raman's ancestral village lay on the banks of the River Kaveri in the Tanjore (Thanjavur) district of Southern India. The art, literature, and especially music that flourished in the region in the late seventeenth century are still palpable. Raman's mother, Parvati Ammal, was the daughter of a distinguished Sanskrit scholar. Raman's father, Chandrasekharan, a lecturer in physics, mathematics, and physical geography, was an avid reader and a connoisseur of *Carnatic* music. Raman developed an early and enduring interest in both music and science.

He graduated with a B.A. degree from the Presidency College of the University of Madras at the age of sixteen, ranked first in the university, and won gold medals in English and physics. In 1907, he obtained an M.A. degree in physics with the highest honors. Colonial contingencies prevented Raman from pursuing an academic career since only Indians who held advanced degrees from British universities were considered for such positions. So he joined the Financial Civil Service in 1907 as assistant accountant general and was posted to Calcutta. Before moving to Calcutta, he married Lokasundari, an accomplished *veena* player.

In Calcutta, Raman came into contact with the Indian Association for the Cultivation of Science (IACS), which had been founded in 1876 by the physician Mahendralal Sircar to foster discussions and public exposition of new scientific developments. The laboratories of IACS provided an unexpected opportunity for Raman to pursue experimental research. He worked in the IACS laboratories before and after his daytime job as an accountant. His researches in optics, *acoustics, and musical instruments, and in particular his studies on the violin, won wide acclaim (*see* OPTICS AND VISION).

In 1917, Raman accepted the Palit Chair in Physics at the University of Calcutta even though the salary levels in the university were considerably less than in the Civil Service. The move to the university brought Raman a group of students and coworkers and greatly expanded the scope of his research. In 1921, on a voyage through the Mediterranean, he was fascinated by the deep blue color of the sea, and conducted experiments on board the ship with the help of a small telescope fitted with polarizers, analyzers, and a diffraction grating. He rejected Rayleigh's explanation that the color of the sea was "simply the blue of the sky seen by reflection," and showed instead that it arose from scattering of light by the water molecules.

On his return to India, Raman and his associates began a systematic study of molecular scattering of light in fluids and solids irradiated with sunlight. By placing complementary light filters in the path of the incident and scattered radiation, they showed the persistence of a weak secondary radiation in the scattered spectrum containing frequencies not present in the incident beam. Raman and K. S. Krishnan further established that the secondary radiation evident in the scattered spectrum of aromatic and aliphatic liquids was strongly polarized. The discovery of the polarized secondary radiation, now known as the "Raman effect," was immediately recognized for its exceptional importance. It provided yet another proof of quantum theory and offered a powerful new tool for investigating the internal structure of molecules and the chemical composition of substances.

The Raman effect can be envisaged as the transfer of energy associated with inelastic collisions between molecules and incident photons. The incident photon can either impart some of its energy to the molecule, raising it to a higher energy state, or collect energy from the molecule, leaving it in a lower energy state. The frequency shifts observed in Raman scattering thus correspond to the vibrational and rotational energies of the irradiated molecules. Raman's researches into the scattering of light won him numerous honors. He was knighted by the British government in 1928 and received the Nobel Prize for physics in 1930.

In 1933, Raman moved to Bangalore to accept the directorship of the Indian Institute of Science. At the Institute, where he also headed the physics department, Raman established a thriving research program. His work with N. S. Nagendra Nath on the scattering of light by ultrasonic waves in a liquid is particularly noteworthy as it explains both the appearance of diffraction bands and the variation in their intensity as a function of the amplitude of the sound wave.

In the early 1940s, Raman took issue with the Born-von Kármán theory, which predicted quasi-continuous second-order Raman spectra of crystals. Experiments in Raman's laboratory, however, showed discrete, line-like structures in the spectra of rock salt and diamond. The resulting controversy between Raman and Max Born spurred new interest in the field. Raman's own laboratory took the lead in exploring different aspects of lattice dynamics, including enumeration and calculation of normal mode frequencies, as well as the absorption, emission, and Raman spectra of a number of crystals. Leon Van Hove resolved the controversy by showing that the quasi-continuum frequency spectrum of a crystal would have singularities (or line-like features) for a subset of the normal mode frequencies.

Raman founded the *Indian Journal of Physics* (1926), The Indian Academy of Sciences (1934) with its monthly *Proceedings*, and the Raman Research Institute (1948). He trained more than 100 physicists, many of whom went on to important positions in universities and research institutes in India. For Raman, science was the exploration of the beautiful and wondrous in nature. Sounds of music, radiance of light, and the vibrancy of colors enthralled him, and he sought to understand their physics. His physics, in turn, was marked profoundly by his aesthetics.

G. Venkataraman, *Journey into Light: Life and Science of C. V. Raman* (1988). Abha Sur, "Aesthetics, Authority, and Control in an Indian Laboratory: The Raman-Born Controversy on Lattice Dynamics," *Isis 90* (1999): 25–49.
ABHA SUR

RAND, a contraction of "Research and Development," is a nonprofit research and development organization set up by the Douglas Aircraft Company after World War II to provide technical analyses for the new U.S. Air Force. Donald Douglas and General Henry Harley Arnold of the air force agreed to create Project RAND largely because the new service lacked the technical infrastructure of the army and navy, but also because operational research had proven to be a significant tool of military planning during the war. Arnold and General Curtis E. LeMay, the air force chief of staff intended to use operation, research to evaluate weapons systems, logistical and strategic planning, and long-range Cold War situations. They first had to override objections against relying on contract research from Wright Patterson's staff. RAND was the forerunner of the nonprofit research corporation now commonly employed by the Department of Defense and other federal agencies to exploit scientific and technical knowledge in planning future acquisitions, reorganization and rationalization of logistical and tactical functions, and effective strategy and tactics.

RAND separated from Douglas Aircraft shortly after its founding. It remained in Santa Monica, California, and acquired other clients in the late 1940s, most notably the Atomic Energy Commission, which, like the air force, had been created out of the army, and which held a monopoly on the development of nuclear energy. It was the first of many federal agencies to join the air force in using the expertise assembled by Douglas and the new RAND Corporation in Southern California.

RAND's early research focused upon problems like space weapons platforms and the disposition of strategic bombers in anticipation of nuclear war with the Soviet Union. New tools were employed in the analyses of these questions, including one of the first of John von *Neumann's computers, the JOHNNIAC, built by RAND's staff to provide numerical capabilities essential to operations analysis and the new field of "systems analysis" that RAND pioneered in the 1950s. This experience in computer programming led the air force to call upon RAND to provide programs and training for the Semi-Automatic Ground Environment (SAGE) system of air defense. RAND psychologists devised experimental air defense stations to test and to train the operators of SAGE, and these functions came to consume most of RAND's research effort until the Systems Development Corporation was created to assume these responsibilities in 1958.

RAND's psychologists were followed by many of its economists who left the organization to staff the Office of the Secretary of Defense under Robert McNamara in the 1960s. Charles Hitch and his colleagues applied the analytical techniques developed at RAND to the reorganization of the Defense Department over the objections of the air force and other services, eroding much of the goodwill that had built up between RAND and its sponsors.

RAND's expansion into civil affairs in 1966 reflected the Johnson administration's "Great Society" programs. Systems analysis was applied to social problems in the so-called War on Poverty, even as RAND shifted its military analysis to support counterinsurgency efforts. This diversification continues as RAND seeks new sources of patronage.

Bruce L. R. Smith, *The Rand Corporation: Case Study of a Nonprofit Advisory Corporation* (1966). David Raymond Jardini, *Out of the Blue Yonder: The Rand Corporation's Diversification into Social Welfare Research, 1946–1968* (Ph.D. Diss., Carnegie-Mellon, 1996).

ROBERT SEIDEL

RATIONAL/IRRATIONAL. An eagerness to emphasize, or else to conflate, the distinction between rational and irrational seems to have been typical of philosophers reflecting on science, rather than of natural philosophers and scientists themselves. So while, for example, few would deny that *Galileo's *Dialogue* (1632) belongs to a rationalist tradition, Galileo himself never appealed to reason in its pages, preferring to use "reason" in association with "reasonableness," and always combining it with other so-called reasonable things, like "experience" and "sound argumentation."

In the history of philosophy the traditional divide between rationalists and irrationalists ascribes René *Descartes, Baruch Spinoza, and Gottfried Leibniz to the former, and authors like Søren Kierkegaard and Friedrich Nietzsche to the latter. The divide, however, has proved to be a mobile one, as in the history of early modern science, with its shifting boundaries between astronomy and astrology, chemistry and alchemy (*see* PSEUDO-SCIENCE AND QUACKERY).

Imre Lakatos made an eloquent plea for the history and philosophy of science to pursue "rational reconstructions" of the past. Because there are always competing conceptions of rationality at stake, Lakatos urged philosophers to adopt canons of rationality able to accommodate most science as rational (this being the burden of the "internal history" of science), while leaving the residues of irrationality to the cures of "external history." Later in the century charges of irrationality could be leveled against Thomas S. Kuhn because of his conception of paradigm changes in the sciences. Programs and episodes like these reinforce the point made by those (like Bruno Latour) who suggest avoiding the divide between rational and irrational altogether. According to them, charges of irrationality signify that a controversy is under way, and will lose most of their import after the controversy ends.

In the wake of the challenges posed by relativist and feminist historiography (*see* HISTORIOGRAPHY OF SCIENCE), and the *science wars, the divide between rationalist and irrationalist approaches to the history of science can perhaps be placed according to the answer given to the following question: Does science possess a distinguishable, epistemic status compared to other forms of knowledge and practice? If your answer is "yes" you may reasonably be called a rationalist; if your answer is "no" others will call you an irrationalist.

Imre Lakatos and Alan Musgrave, eds., *Criticism and the Growth of Knowledge* (1970). Paul Horwich, ed., *World Changes: Thomas Kuhn and the Nature of Science* (1993).

GIULIANO PANCALDI

REACTOR. See NUCLEAR POWER.

REALISM. See INSTRUMENTALISM AND REALISM.

RECORD. See PHONOGRAPH.

REDUCTIONISM is the replacement of facts, objects, concepts, laws, theories, or mental events in a given domain by items from another domain assumed to be more basic. Scientific examples include *Newton's subsumption of *Galileo's law of free fall and *Kepler's laws of planetary motion under his mechanics, or the success of *molecular biologists in understanding *genes in terms of molecules. Because successful reduction seems to open the way to a unified scientific theory, it has always been high on the agenda of positivists, with their belief in the unity of science. Thus it is not surprising that the period of greatest scientific and philosophical interest in reductionism coincided with the apogee of *positivism between the 1930s and 1960s.

Although the term "reductionism" comes from philosophy, the idea has a long history in science going back at least to the seventeenth century. In the late nineteenth century, scientists applauded two great achievements of this kind: the kinetic theory of gases that explained the observable properties of gases by the movement of tiny particles, and James Clerk *Maxwell's demonstration in 1873 that optical, electrical, and magnetic phenomena could all be described by the same basic set of mathematical laws. In the twentieth century, positivists saw the absorption of Newtonian mechanics by relativity theory, and much of *chemistry by quantum mechanics, as further steps on the road to a unified science.

It has been in biology, though, that reductionism has posed the greatest challenge. Scientists have always been deeply divided about whether life can be reduced to the dance of underlying particles. In the nineteenth century, the German physiologists Hermann von *Helmholtz, Ernst Wilhelm von Brücke, and Emil du Bois-Reymond, reacting against the then-predominant vitalism, formed the "1847 school" of physiology. But reductionism came to fruition in Helmholtz's *Handbook of Physiological Optics* (1856–1867), which explained many aspects of vision in terms of physics and chemistry. In the twentieth century, efforts to reduce biology continued. In the 1960s, flushed with the success of deciphering the *double helix structure of DNA, many molecular biologists agreed with Francis *Crick that the appropriate aim was to explain all biology in terms of physics and chemistry.

Reductionism, both in biology and as a general program, has run into a number of criticisms. One persistent argument has been that reductions involve losses in understanding. In 1923, C. Lloyd Morgan coined the term "emergent properties" to refer to characteristics that even if ultimately rooted in physics and chemistry, could only be detected at the macro-level. Another line of attack has been to show that supposed reductions in fact often had to sneak in additional "bridge" assumptions. Furthermore, as realism gained at the expense of positivism in the 1960s, critics of reduction pointed out that supposed reductions worked only if theories were treated as collections of mathematical laws and not as claims about the nature of the world. If the latter claims were taken seriously, then *Newton's theory could not be reduced to *Einstein's because it was based on quite different ideas about space and time. Yet one more criticism of reductionism, popular with postmodernists, who believe that all knowledge is local and restricted, is that reductionism amounts to scarcely disguised disciplinary imperialism on the part of physicists.

By the last third of the twentieth century, historians and philosophers have turned away from a belief in the unity of science. Instead, they have emphasized the ways in which science is disparate, diverse, and disunified.

Robert Olbey, *The Path to the Double Helix* (1968). Robert Causey, *Unity of Science* (1977). Alexander Rosenberg, *Instrumental Biology or the Disunity of Science* (1994). Peter Galison and David Stump, eds., *The Disunity of Science* (1996).

RACHEL LAUDAN

REFLEX. Although a reflex is a nonconscious response to a simple stimulus, this basic act has a complex physiology, discovered over a long history. The search for the neural substrate of the reflex began with René *Descartes, who described it as a unique involuntary function. He went further, however, by proposing a physiological mechanism that could operate without the guidance of the will. This theory suggested that voluntary and involuntary action derived from the fluid in the ventricles of the brain. Thus, for Descartes, the neural pathway of reflex action proceeded from the sensory organs, into the brain, and back to the effectors.

Thomas Willis utilized the same theory to organize his thoughts on reflexive action, but separated voluntary and involuntary functions within the brain. He assigned involuntary actions to the cerebellum and voluntary action to the cerebrum. This influential proposal spawned a host of studies designed to test the relative contribution of each area to these unique movements. Robert Whytt added a series of critical distinctions. He provided the first detailed account of the diversity of reflex responses elicited by a wide array of stimuli. More significantly, his experiments identified the medulla and spinal cord as a vital component of reflex action.

Charles Bell and François Magendie clarified these observations further by demonstrating the functional partition in the spinal cord that we know today—sensory information enters the spinal cord through its dorsal portion and exits as motor commands ventrally. This discovery solidified the spinal cord as in principle the site of reflex processing.

This new physiological understanding combined with an influential theory later known as the reflex chain hypothesis. Alexander Bain stated that the sensory consequences of one action could trigger subsequent actions by reflex response. Thus, a series of automated responses, either inborn or learned, could propagate from beginning to end solely through spinal mechanisms. This proposal strongly influenced the thinking behind William James's theory of habits, the postulates of Ivan Sechenov and Ivan Pavlov, and the entire behaviorist tradition of the early twentieth century. Although scientists later dismissed the hypothesis, a more limited form has recently received support from modern neuroscience research.

Perhaps the most detailed account of the reflex was given by Charles Sherrington, who combined his formidable physiological skill with several converging notions, such as the discovery of inhibition, the neuron doctrine, and the notion that many cortical areas descended to the spinal cord. According to him, the reflex arc interacts with a host of descending commands to result in a final motor command. This model has been modified, but in essence is still held today. The basic spinal reflex is a building block of behavior modified by descending influences coming from midbrain and cortical sources. They form the vital substrate for all action.

Charles Sherrington, *The Integrative Action of the Nervous System* (1906). Franklin Fearing, *Reflex Action: A Study in the History of Physiological Psychology* (1930).

WILLIAM T. CLOWER

RELATIVITY, a theory and program that set the frame of the physical world picture of the twentieth century. The article on Albert *Einstein describes the genesis of the theory in the mind of its inventor; the present article concerns its content and reception. Four stages may be distinguished: the invention of the misnamed "special theory of relativity" (SRT), which covered only phenomena observed in bodies moving among themselves with constant relative velocities, in 1905; the recognition of the equivalence of a body's inertial mass (the measure of its resistance to change of velocity) and gravitational mass (the measure of the pull of gravity on it) in 1907; the removal of the restriction to constant velocities in the general theory of relativity (GRT) in 1915; and the application of GRT to cosmology especially in recent decades.

The infelicity of the term SRT involves both the "S" and the "R." It is not "special" but limited, whence the French term "restricted relativity." Again, what distinguishes SRT from classical *mechanics is not "relativity"—the idea that the laws of motion look the same in all inertial frames (reference frames moving with respect to one another with constant velocity)—but rather the startling concept that all observers measure the same velocity for light in free space. This distinguishing principle is absolute, not relative. It disagrees with the principle of relativity if, as many physicists believed around 1900, radiation is reducible to mechanics. Einstein insisted on both the absolute and the relative—the constancy of the light velocity and the equivalence of inertial frames—and, in consequence, had to surrender common intuitions of space and time. One forgone concept was absolute simultaneity.

In the thought experiment Einstein often described, an observer on an embankment sees a light flash from the middle of a passing railroad carriage equipped with a mirror at either end. According to relativity, an observer seated at the center of the car would see the light returned from the mirrors simultaneously. But the observer on the embankment would see the flash from the forward mirror after that from the rear: the light has farther to go to meet and return from the forward than from the rear mirror, and the speed of light by hypothesis is the same in both directions. On a ballistic theory of light, the return flashes occur simultaneously for both parties: to

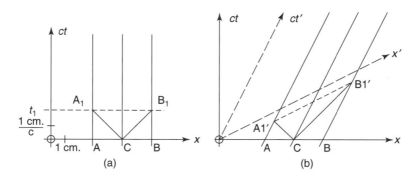

(a)   (b)

The points A and B, at rest in the system x, t, describe "world lines" parallel to the t-(time) axis; a light signal sent from C reaches them simultaneously (at time t_1) along the "light lines" CA_1, CB_1. If A and B move along x at the speed v, they describe world lines inclined to x at an angle $\tan^{-1} = (vt/x)$ and meet the signals from C at A_1', B_1', that is, at different times measured in the x, t system; since relativity requires the signal to be simultaneous in the system x', t' in which A and B are at rest, the x' axis must be parallel to $A_1 B_1$.

the traveling observer, the light has the same speed in both directions; to the stationary one, the speed in the forward direction exceeds that in the opposite direction by twice the speed of the carriage. Further arithmetic shows that holding to the absolute and the relative simultaneously required that meter sticks and synchronized clocks moving at constant velocity with respect to a stationary observer appear to him to be shorter and tick slower than they do to an observer traveling with them.

Einstein showed that the form and magnitude of these odd effects follow from the stipulation that the coordinates of inertial systems are related by a set of equations that he called the Lorentz transformation. Hendrik Antoon Lorentz had introduced them as a mathematical artifice to make the principle of relativity apply to Maxwell's equations of the electromagnetic field (*see* ELECTROMAGNETISM). Einstein now insisted that the Lorentz transformation apply also to mechanics in place of what came to be called the "Galileo transformation" that guaranteed relativity to Newton's laws of motion. The Galilean, which transforms only space, reads, for relative motion at a constant velocity v, $x' = x - vt$, $t' = t$, where the primed letters refer to a coordinate system moving along the x axis of the unprimed system; the Lorentz transformation replaces these relations with $x' = \gamma(x - vt)$, $t' = \gamma(t - vx/c^2)$, $\gamma = (1 - v^2/c^2)^{-1/2}$. When v is negligibly small in comparison with c, the equations have the same form, but not the same meaning, as the Galilean transformation between the same variables. Since the Galilean transformation supports the Newtonian expressions for force, mass, kinetic energy, and so on, replacing it required reworking the formalism of the old mechanics. This labor, in which Max *Planck and his student Max von Laue played leading parts, produced expressions differing from the Newtonian ones by factors of γ. A new form of Newton's second law emerged that satisfied the demand of relativity ("remained invariant") under the Lorentz transformation.

As an afterthought, also in 1905, Einstein argued that energy amounts to ponderable mass and vice versa. The relativistic expressions for the momentum and kinetic energy require that, for conservation of momentum to hold, the mass m of an isolated system of bodies must increase when the system's kinetic energy E decreases. The increase occurs at a rate (change of mass) = (change of energy)/c^2, $\Delta m = \Delta E/c^2$. Einstein thus united the previously distinct principles of the conservation of energy and of mass. From a practical point of view, the equivalence of mass and energy as applied to nuclear power is by far the most important consequence of relativistic mechanics.

Relativity had an enormous appeal to people like Planck, who regarded the surrender of common-sense expectations about space, time, and energy as a major step toward the complete "deanthropomorphizing of the world picture" begun by *Copernicus. The mathematician Hermann Minkowski declared in a famous speech to the Society of German Scientists and Physicians in 1908 that "space by itself, and time by itself, are doomed to fade away into mere shadows, and only a kind of union of the two will preserve an independent reality." He interpreted the Lorentz transformation as a geometrical rotation in his four-dimensional space, whose points represented "world events" and whose lines represented the histories of all the particles in the universe. "The word relativity postulate... seems to me very feeble," he said, "[to express] the postulate of the absolute world," that is, Einstein's theory as geometrized by Minkowski.

Einstein's compulsion to remove the "all too human" from physics and his desire to overcome the limitation of SRT brought him to a more profound generalization than Minkowski's. Taking the equivalence of inertial and gravitational masses as his guide, he worked out by 1911 that gravitational forces should affect electromagnetic fields (for example, by giving radiation potential energy) and deduced that the sun would bend the path of a ray of starlight that passed close to it. But the major conquest wrested from the equivalence of the masses was the elimination of gravity: all freely falling bodies in the same region experience the same acceleration because the presence of large objects distorts the space around them and the bodies have no alternative but to follow the "geodesics"—straight lines in the curved space in which they find themselves. In "flat spacetime," without distorting masses, bodies move in inertial straight Euclidean lines. Falling bodies apparently coerced to rejoin the earth under a gravitational force in Euclidean space in fact move freely along geodesics in the local warp in the absolute four-dimensional space-time occasioned by the earth's presence. By 1915 Einstein had found the mathematical form (tensors) and the field equations describing the local shape of space-time that constituted the backbone of GRT, and had added two more tests: an explanation of a peculiarity in the orbit of Mercury and a calculation of the effect of gravity on the color of light.

By 1910 or 1911 German theorists had accepted SRT and a few physicists elsewhere recognized its importance. Arnold Sommerfeld grafted it onto Niels *Bohr's quantum theory of the atom in 1915–1916, with spectacular results (the explanation of the fine structure of helium). Physicists demobilizing from World War I interested in *atomic structure or *quantum physics perforce had to learn relativity and, ultimately, to find ways of employing SRT systematically in quantum mechanics (*see* DIRAC, PAUL ADRIEN MAURICE; QUANTUM ELECTRODYNAMICS). And then the positive results of the eclipse expedition organized by Arthur Eddington and other English astronomers to test Einstein's prediction of the deflection of starlight engaged the public imagination. It appeared that a lone pacifist had by pure thought bettered Newton while most of the world's scientists had devoted themselves to war. Einstein traveled, quipped, became a favorite of newspaper reporters around the world and the bête-noire of anti-Semites back home. He spent the rest of his life, in Berlin and in Princeton, trying to generalize the general by bringing electromagnetism within GRT. But neither his efforts nor those of the few other theorists who thought the game worth the candle managed to reduce electricity and magnetism to bumps in space.

Application to cosmology at first seemed more promising. A solution to the field equations by Aleksandr Friedmann in 1922 indicated the possibility of a finite expanding universe. The concept agreed with measurements of red shifts in *galaxies (*see* HUBBLE, EDWIN). The notion of the origin of the universe in a compact space, or "cosmic egg," was bruited by Georges Lemaître in 1927. The resultant *Big-Bang universe, with origin in time (initially set at two billion years ago from the accepted value of the Hubble constant), gained some acceptance but little development until the 1960s. Then discoveries in astronomy (*quasars, pulsars, supposititious *black holes, the cosmic background radiation), laboratory demonstrations of the gravitational red shift and tests with rockets and atomic clocks, and advances in particle physics and the mathematics of gravity made GRT fashionable. Relativistic astrophysics now has the panoply of journals, textbooks, meetings, and money that mark a flourishing science. Although gravity remains outside the unified forces of the *Standard Model, the pursuit of the vanishingly small and the ineffably large depend upon GRT for clues to the origin and evolution of the universe.

See also ASTRONOMY, NON-OPTICAL; COSMOLOGY.

Max Born, *Einstein's Theory of Relativity* (1962). Abraham Pais, *"Subtle Is the Lord...": The Science and the Life of Albert Einstein* (1982). Helge Kragh, *Cosmology and Controversy: The Historical Development of Two Theories of the Universe* (1996). Helge Kragh, *Quantum Generations. A History of Physics in the Twentieth Century* (1999).

J. L. HEILBRON

RELIGION AND SCIENCE. Religion and science are both so important and diverse that any attempt to encapsulate their interrelations is bound to fail. Their interaction is also a topic that attracts not only scientists and theologians but also an interested and concerned public. Yet contemporary discussion of religion and science is dominated by partisan writers who promote grand but simplistic theses—some proclaim science to be in perpetual conflict with religion while others posit complete harmony. Moreover, much current discussion draws heavily on such historical episodes as the *Galileo affair. The historical study of religion and science has therefore an important contemporary role since it can illuminate the complexity of religion-science interrelations, place the grand popular theses in context, and inject a dose of critical analysis.

Centrality of Religion to the Scientific Revolution

Most of the major figures associated with the rise of modern science in the sixteenth and seventeenth centuries paid close attention to the religious issues and their bearing on science. Thus Francis Bacon could both advocate the separation of science from religion and stress the religious character of science, since the study of nature displayed "the footprints of the Creator." In many cases, too, religion provided the motivation for pursuing science. Johannes *Kepler felt compelled to discover the harmony, order, and (ultimately) mathematical laws governing the motions of the planets by the conviction that God had created the universe on a mathematical model. Isaac *Newton, who likewise saw the physical world manifesting divine design, adopted a priestly role in determining the laws that God had imposed on matter at the creation. Of the many points of contact between his science and his religion, Newton's innovative concept of "force" possessed strong theological significance since he considered all force to be derived from God. Newton's religious writings were not confined to physicotheology; he also wrote extensively on biblical chronology and prophesy. Nor were his theological views conventional; he tried to conceal his Socinianism (denying the divinity of Christ) in order to avoid the charge of heresy.

The Galileo affair, culminating in his censure by the Roman Catholic church in 1633, has often been portrayed as a conflict between reactionary clerics and the harbinger of scientific progress. Yet, despite the heavy-handed action of a committee of the Holy Office that prohibited Nicholas *Copernicus's book, this is an untenable interpretation since not only did many church members support Galileo, especially in the 1610s and 1620s, but he himself also remained a staunch

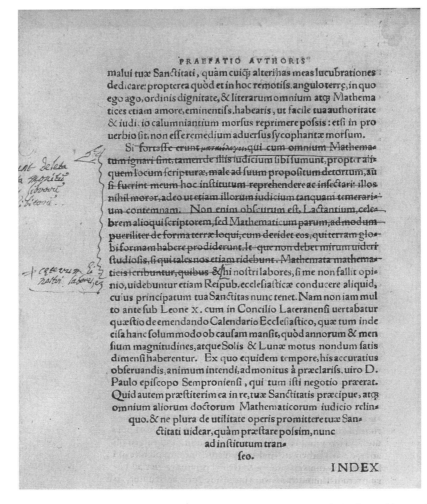

A page from the preface to the second edition of Copernicus's *De revolutionibus orbium coelestium* corrected according to the instructions of the Holy Office. The few emendations required consisted mainly in replacing "theory" with "hypothesis;" here the excision removes a criticism of the cosmology of the patristic writer Lactantius.

Catholic. The affair raises many substantive issues including the meaning of certain scriptural passages that might bear on astronomy, the relation of empirical knowledge to theology, and the strength of Galileo's arguments in favor of Copernicus's theory. (He failed to prove the truth of the Copernican system.) Other factors that bear directly on this episode were Galileo's ability to create enemies and his refusal to heed Cardinal Robert Bellarmine's advice to treat the Copernican system as a mathematical hypothesis and not as physical truth. Galileo can also be seen as a pawn in the vicious political struggles between the Pope, the Dominicans, and the Jesuits. More controversially, some scholars have interpreted the 1633 trial as the final act in Galileo's long-running battle over atomism with his Aristotelian opponents, who believed it challenged the doctrine of the eucharist. From this perspective the theological acceptability of Copernican astronomy was not at issue. The jury is still out, but historians continue to discover new interpretations of the Galileo affair.

Since early modern science was a product of Christians, some have argued that Christian theology provided the necessary conditions for the rise of science. In particular, Christianity specifies a rational creator who constructed a lawlike universe; these laws can, in turn, be decoded by scientists. While this argument might bear on the rise of science, it fails to explain why modern science was delayed a millennium and a half after the rise of Christianity. Likewise, in a classic article first published in 1938 the sociologist Robert K. Merton attributed the values required for the rise of science in the seventeenth century to the "Puritan ethic"—the assurance of salvation among English Protestants, which found expression in good works, science included. Science was therefore the offspring of a religious system. While his argument applies reasonably well to Britain, it makes it difficult to understand how Catholics contributed to science. Yet many eminent seventeenth-century natural philosophers were Catholic, including Galileo, René *Descartes, and Marin Mersenne. Through their schools and colleges, Jesuits were particularly active in teaching modern science. The structure of the Society of Jesus also aided the acquisition of natural knowledge since Jesuits traveled widely and transmitted astronomical data and botanical specimens through their diplomatic channels.

Many of the major intellectual controversies of the seventeenth and eighteenth centuries manifested both religious and scientific dimensions. For example, in the controversy between Gottfried Leibniz and Samuel Clarke (a follower of Newton's), in 1715–1716, Leibniz criticized not only Newton's views about space, time, matter, and force, but also his theology, which emphasized God's will. This is a classic example of a controversy in which scientific and theological issues were thoroughly intermixed. Moreover, "scientists" were not pitted against "theologians"; instead different theologies and different natural philosophies clashed. It is also typical in possessing political and personal connotations, since both Leibniz and Newton were seeking the patronage of the newly enthroned King George I.

Many of those who challenged Newton's science drew attention not only to its scientific difficulties but also to its philosophical and theological defects. For example, Newton considered space to be empty, except for widely dispersed atoms, but his critics seized on emptiness as wasteful and incompatible with a divinely planned universe in which the principle of plenitude applies. Moreover, they objected that Newton's philosophy encouraged pantheism by attributing activity to matter and thereby failing to differentiate the creation from the creator. These critics were not antiscience *per se*, but rather adopted alternative natural philosophies, such as the "scriptural physics" of John Hutchinson, which they considered truly Christian.

Local Contexts

Owing to the diversity of religious traditions, each with its own complex history, it is difficult to draw any global generalizations about science-religion interactions. We should also recognize the existence of local scientific traditions; for example, in France responses to Newton's ideas were affected by both the dominance of Catholicism and by commitments to Cartesian natural philosophy. Much recent research on science and religion has indicated the importance of specific religious and scientific traditions within local contexts. The following examples, which are drawn from English history, illustrate typical loci of science-religion interaction but are not necessarily applicable to other locations. During the eighteenth and nineteenth centuries science proved particularly attractive to religious dissenters, such as Unitarians and Quakers, who emphasized education and promoted science in their schools. For example, the Unitarian Joseph Priestley, best known for his work on gases, was connected with Warrington Academy (county of Cheshire, England) and published a number of educational works, while the Quaker schoolmaster John *Dalton began research in meteorology and subsequently developed his theory of the chemical atom. There are many reasons why dissenters were attracted to science: they were not from the land-owning establishment nor did they have access to some traditional routes for social advancement. Instead they were drawn to new and precarious ventures in science and technology—thus many dissenters became prominent in key industries, such as iron smelting and later railway construction. Since dissenters could not graduate from either Oxford or Cambridge Universities until the repeal of the Test Acts in 1871, many studied abroad or in Scotland, where scientific and medical education was of a far higher standard. Thus in many British cities dissenters promoted the provision of hospitals and dispensaries for the poor. Quakers in particular gravitated to pharmacy where their reputation for honesty and their ability at networking worked to their advantage.

The audience for science increased significantly in eighteenth- and nineteenth-century Britain, with a plethora of itinerant lecturers and publications aimed at the upper, middle, and (increasingly) working classes. Often science was presented in a theological frame, a strategy that not only helped legitimate science but also conveyed its intellectually integrity. Thus signs of design, such as the webbed feet of water birds (enabling them to walk on water) or the appropriateness of the distance of the earth from the Sun (preventing the earth from being fried or frozen), were interpreted as evidence of God's wisdom, power, and goodness. Despite being disparaged by David Hume, design arguments remained very popular, as can be seen from the frequent reprinting of such works as John Ray's *The Wisdom of God Manifested in the Works of Creation* (1691), William Paley's *Natural Theology*

(1802), and the Bridgewater Treatises of the 1830s. Design arguments performed many functions and were capable of many different modalities. They could, for example, perform an irenic function by appealing to all denominations. By demonstrating the harmony and integrity of God's creation they also functioned as a creative aesthetic. Many people who would not otherwise have known much science studied it in popular natural theology texts.

Following the French Revolution fear of a popular uprising was fuelled by the materialism and atheism associated with such leading French scientists as Pierre-Simon de *Laplace, Jean-Baptiste de Lamarck, and Georges-Louis LeClerc, Comte de Buffon. Over the next few decades the "The March of the Mind," as manifested by Mechanics' Institutes, popular science lectures, and cheap books and periodicals, added to this concern. Attitudes to science became increasingly divergent in Britain. Some saw it as a legitimate handmaiden of religion, while others conceived it as materialistic and destructive to the old social and religious order. The success of such "alternative sciences" as *phrenology (often painted as materialist) and spiritualism (antimaterialist), both of which attracted large followings, further challenged the established church. Secular theories in *geology and *natural history likewise questioned not only the biblical narrative but also the notion of a divinely crafted universe. By the early nineteenth century few writers maintained a six thousand year-old Earth, most others being prepared to accept a nonliteral interpretation of Genesis.

The Challenge of Evolution

The publication of Charles *Darwin's On the Origin of Species in 1859 drew fire from other simmering controversies. As with the Galileo affair, responses to Darwin cannot be reduced to a simple conflict between science and religion—although this interpretation was constructed and widely publicized by some of the protagonists. Historians have uncovered diverse responses from writers holding widely different religious commitments, ranging from those who tried to keep *evolution at bay to those who welcomed the new theory as advancing our understanding of how God governs the universe through laws. The short-term impact of evolution on religion is difficult to determine since the "Victorian crisis of faith" had no single cause and the decline in religious observance resulted as much, if not more, from demographic changes and from the impact of such publications as Essays and Reviews (1860), which questioned the basis of revealed religion.

In the long term, however, evolution has been central to discussions of science and religion. The

Origin of Species acted as a rallying point for a number of scientists who not only questioned the truth of Christian belief but also challenged the political power vested in the established church. Thomas Henry Huxley, Joseph Dalton Hooker, John Tyndall, and others—through to Richard Dawkins and Peter Atkins in our own day—have alighted on evolution as the scientific theory with considerable antireligious potential. Following the publication of Darwin's theory it became far more difficult to articulate views of special creation and the argument from design took a severe blow. Nevertheless the polarization that has occurred around evolution is not simply a result of Darwin's theory but also demonstrates the impact of social and political factors. For example, in the southern United States evolution encountered relatively little opposition until the 1920s when it became the focus for many local grievances and the foe of fundamentalists. The 1925 trial of John Scopes, who defied the law and taught evolution in the public schools of Tennessee, has often been used to symbolize the conflict between science and religion. Opponents of "evilution" have more recently developed "creation science" as an alternative to evolutionary science. Thus they have sought to redefine the conflict between science and religion as a disagreement between two scientific theories.

Darwin's theory also fed into a much more protracted philosophical dispute that challenged Christian ontology. For many Christians science is suspect if it implies that all phenomena can be reduced to inert matter and a set of laws. Not only does this *reductionism portray the universe as inhospitable but it removes God from nature. Materialists also deny the need for any form of creator, instead attributing natural phenomena to the properties of matter. Huxley, who coined the word "agnostic" to describe those who refused to accept a creator-God owing to lack of evidence, was an ardent materialist. Yet the most conspicuous site of conflict was the meeting of the British Association for the Advancement of Science in 1874 when Huxley's friend John Tyndall devoted his presidential address to the topic. Tyndall traced the history of materialism from the ancients to contemporary ideas about evolution, showing that the advance of science had been achieved through our increasing understanding of matter. Although Tyndall was no naive materialist or reductionist, he was widely attacked as an atheist and for supporting a thoroughly materialistic view of science.

In contrast to Tyndall, religious writers have repeatedly challenged materialism by affirming that the physical world is dependent on God—as creator and/or conserver of the world. Thus Newton asserted that if God were not in continual interaction with the universe it would simply stop, like a clock whose mechanism has run down. More recently, and in very different theological contexts, quantum theory has proved a boon to those who wish to inject mind or spirit into the universe. Scientists of a religious disposition have often been more attracted to such fields as *cosmology, astronomy, and physics—as opposed to the biological sciences—where considerable scope exists for natural theological speculation.

In the second half of the nineteenth century the Roman Catholic church implemented a number of measures that, in the eyes of its detractors and not a few of its members, made it increasingly rigid and inhospitable to new ideas, especially developments in science. Science was made subservient to theology, while scholastic philosophy and the power of the papacy were enhanced. Among the indirect reactions to this increasing conservatism was the publication in 1874 of John William Draper's *History of the Conflict between Science and Religion*, which portrayed Catholicism as the natural enemy of science. In this classic statement of the conflict thesis, science and Catholicism are two opposing forces that have been played out in history. In response to this interpretation, many Catholic scientists, historians, and theologians have tried to rescue the church by offering less antagonistic accounts of the Galileo affair. Particularly since the Second Vatican Council positions have shifted significantly and in 1979 Pope John Paul II conceded that the church had made a mistake. In 1992, following a report by the Pontifical Commission, the church acknowledged that the Galileo affair resulted from "tragic reciprocal incomprehension."

Beyond Christianity

An interesting twist to science-religion interactions has been the development of "scientism," the view that science can be extended beyond its usual disciplinary boundaries to encompass all other areas, religion included. For example, the late nineteenth-century chemist Carl Wilhelm Wolfgang Ostwald advocated a substitute religion based on contemporary scientific ideas, especially thermodynamics. He believed that his new "religion" would fulfill the psychological functions traditionally performed by Christianity, which he utterly rejected. He envisaged a secular festival that would replace Christmas at which candles would be lit and hymns sung in praise of "energy." The example of Ostwald's new religion and the many subsequent versions of scientism raise the question whether it is a religion or an antireligion.

The literature on science and religion has been dominated by writers seeking to defend or attack Christianity. Little attention has been paid to other religious traditions, which have often experienced far less difficulty in relating to mainstream science. In Judaism, for example, there has been relatively little antagonism to science. Jewish participation in science was however often limited by the civil and educational disabilities imposed by the countries in which Jews lived. Paradoxically, in the late nineteenth and early twentieth centuries Jews tended toward science since other professions were closed to them. Thus Jews made up a significant proportion of scientists, especially in German-speaking countries. Many of them were subsequently forced to leave owing first to the economic recession during the interwar years and later to Hitler's purges. American, British, and Russian science benefited.

China developed independent scientific traditions commensurate with the religious philosophies of Buddhism and Taoism, often emphasizing human harmony with nature rather than nature's exploitation for human ends. Since the seventeenth century there has been considerable but selective interchange of Western science and technology for alternative medical therapies. Probably the most important contemporary encounter is between Western science and Islam. Drawing on both the Koran and its early encounters with Greek science there is a strong Islamic tradition of analyzing the physical world. Yet the contemporary encounter with Western science is also deeply marked by the various politically charged responses to Western culture and capitalism.

Seyyed Hussein Nasr, *Science and Civilisation in Islam* (1968). Frank Miller Turner, *Between Science and Religion: The Reaction of Scientific Naturalism in Late Victorian England* (1974). David C. Lindberg and Ronald L. Numbers, eds., *God and Nature: Historical Essays on the Encounter between Christianity and Science* (1986). Pietro Redondi, *Galileo Heretic* (1987). John Hedley Brooke, *Science and Religion: Some Historical Perspectives* (1991). Christopher Kaiser, *Creation and the History of Science* (1991). Annibale Fantoli, *Galileo: For Copernicanism and for the Church* (1994). David B. Ruderman, *Jewish Thought and Scientific Discovery in Early Modern Europe* (1995). John Brooke and Geoffrey Cantor, *Reconstructing Nature: The Engagement of Science and Religion* (1998).

GEOFFREY CANTOR

RENAISSANCE. The study of nature underwent sweeping transformations in fifteenth- and sixteenth-century Europe. Many of those who taught formal courses about the natural world—for example, professors of natural philosophy—paid little attention to these changes. Textbooks on "the world" were basically as

Aristotelian in 1600 as they had been in 1400. Sweeping plans to transform teaching and scholarship—like Martin Luther's effort to remove Aristotle from the curriculum at his local university in Wittenberg—generally failed. Nonetheless, when *Galileo Galilei and Johannes *Kepler turned *telescopes on the heavens, they drew on the accomplishments of generations of earlier writers, thinkers, and craftsmen.

Change began outside the universities. In the fifteenth century, the literary scholars known as humanists abandoned the traditional prejudice against the study of nature and began to argue that mastery of natural history, medicine, geography, or astronomy required correcting and explicating the works of the Greeks and Romans. By the middle of the fifteenth century, erudite men like George of Trebizond and Johannes Regiomontanus had begun to study the full text of Ptolemy's *Almagest* and *Geography*. Regiomontanus's *Epitome* of the former text, published in 1496, made its full range of data and techniques available to Europeans for the first time in centuries. Angelo Poliziano and other humanists began to study the classics of Greek medicine and botany and to reprove those practicing medicine for their reliance on translations. By the end of the fifteenth century, medical specialists like Nicolò Leoniceno had accepted the challenge of the humanists, learned Greek, and begun to argue back, claiming that only they, not the scholars, could interpret the technical works of Theophrastus, Dioscorides, and Pliny. The study of nature now required not just mastery of Greek and Arabic works translated into Latin, but knowledge of Greek and perhaps Arabic as well.

A second transformation took place in the realm of work. The fifteenth century witnessed rapid technological changes: the rise of effective artillery, the invention of printing, the development of new techniques for mining and construction, and the discovery of one-point perspective. Older techniques like navigation by compass were measurably improved. Leading practitioners of the technical arts, who often called themselves "engineers," showed the new self-confidence these successes engendered by writing elaborate treatises that illustrated and described their devices. Many insisted that they too were learned men, not mere artificers.

By the end of the fifteenth century, bold individuals began to cross the boundaries that had once separated the realm of texts and learning from that of practice. Leonardo da Vinci, for example, read far more widely than his brash remarks suggested. But he also insisted that his practical skills in dissection, perspective, and

shading gave him an advantage over any merely erudite student of the human body. Artists like Leonardo, Piero della Francesca, and Albrecht Dürer tried to give their mastery of nature a rigorous foundation in geometry and optics. And learned people like the botanist Leonhart Fuchs realized that they must collaborate with artists—at least if they hoped to record not only general descriptions of the plants they studied, but their precise images and colors, which could be used in the field.

By the 1540s, finally, the phase of primitive accumulation gave way to one of rapid and radical transformation. Nicholas *Copernicus's *De revolutionibus orbium coelestium* and Andreas *Vesalius's *De humani corporis fabrica*, both of which appeared in 1543, rested on detailed study of the ancient authorities, Ptolemy and Galen respectively. But both works also challenged the ancients on central points, *Copernicus by placing the Sun, rather than the earth, at the center of the universe, and *Vesalius by showing that direct dissection of humans revealed that Galen had gone wrong relying on Barbary apes. Some writers on method in philosophy—above all Peter Ramus—argued that the entire curriculum needed to be reconfigured, perhaps by recovering the ancient learning of the Chaldeans and Egyptians. More than one intelligent reader of these innovative texts agreed with John Donne's later comment: For Robert Recorde and Giordano Bruno, among others, the new philosophy really did call all in doubt, since it suggested that the universe was infinite.

Other writers highlighted traditions that had long flourished on the margins of the world of learning. During the Middle Ages, alchemists had insisted, as natural philosophers did not, that human work and intelligence could emulate natural processes. The sixteenth-century medical man Paracelsus based his new theories of disease and cure on alchemy, denounced the medical establishment, and insisted that the practical knowledge of surgeons and cunning women should take precedence over the erudite traditions of the schools. His pamphlets shook the world of learning and won him many followers. So, more profoundly, did Italian mathematicians like Girolamo Cardano, who provided formal Latin expositions of the algebraic techniques long taught in the vernacular—and by doing so changed the definition of mathematics itself. No school proved more influential than the learned form of medical astrology pioneered by Marsilio Ficino, which penetrated courts and universities across Europe.

At the same time, new sites of scientific work sprang up. Monarchs in Italy, Germany, and Scandinavia supported efforts in cartography and astronomy; some, like Maurice of Hesse-Kassel, built new instruments of great precision. A generation later, Tycho *Brahe's magnificent observatory on Hveen, with its mural quadrants and printing house, became a model for a new notion of the scientific institution. Museums took shape, in courts and cities, where naturalists like Ulisse Aldrovandi gathered large collections of fossils, stones, and plants, as well as masses of images of them on paper, and challenged traditional taxonomies as they did so. Merchant cities like Nuremberg, Antwerp, and London provided easy access to books, instruments, and news. The printing press multiplied copies of modern as well as ancient texts—and gave new, inexplicable phenomena like the nova of 1572 and the comet of 1577 the status of news, widely discussed and simultaneously observed in a way previous celestial phenomena had not been.

At the end of the sixteenth century, most learned men still took their physics from Aristotle, their medicine from Galen, and their astronomy from Ptolemy. When Jean Bodin set out to describe what he called *The Theater of Nature* (1596), he drew his ideas and his information largely from ancient texts. But Kepler and Galileo were far from the only men of learning who knew that they had to collaborate with artisans, who took a passionate interest in the development of new technologies, and who believed the time had come to abandon ancient authorities, painful as that might be, and read the book of nature rather than the books of men.

Marie Boas Hall, *The Scientific Renaissance, 1450–1630* (1962). Alexandre Koyré, *The Astronomical Revolution* (1973). Allen Debus, *Man and Nature in the Renaissance* (1978). J. V. Field and Frank A. J. L. James, eds., *Renaissance and Revolution* (1993). Paula Findlen, *Possessing Nature* (1994). Ann Blair, *The Theater of Nature* (1997).

ANTHONY GRAFTON

RESEARCH SCHOOLS. See SCHOOLS, RESEARCH.

RESPIRATION AND TRANSPIRATION. The ancients generally viewed breathing (respiration) as a process that cooled the heart's innate heat. The Hippocratic corpus further emphasized that living bodies needed air to fan or feed the vital flame. This view received experimental support in William *Harvey's demonstration that blood was vivified by passing through the lungs (1628). From 1662 Robert *Boyle and others at the Royal Society of London confirmed that air was necessary for life (*see* AIR PUMP).

In the 1640s, between the work of Harvey and Boyle, Jean Baptiste van Helmont performed the first controlled experiments into the living functions of plants to throw light on the chemical composition of water. Helmont weighed and planted

a willow tree in a container and allowed it to grow for five years: at the end, he weighed the tree and the annual leaf fall and concluded that the tree had gained some 170 lbs. from water alone. Some eighty years later, influenced by the Royal Society's experimental ethos, the reverend Stephen Hales began experiments on plant physiology, publishing the results in his *Vegetable Staticks* (1727). He used the word "staticks" to emphasize the precise measurement of his experiments, which—unusually for the period—he subjected to controls which typically took the form of growing a plant by conventional means alongside. Taking his cue from animal physiology, Hales investigated the movement of sap and water in the plant. His major achievement was to establish the constant uptake of water by plants and its loss by transpiration ("perspiration," he called it). He showed that these movements took place by root and leaf suction along a one-way system, reluctantly but conclusively demonstrating that sap did not circulate like blood in animals. He also performed experiments on the gas relations of plants that later guided the work of Joseph Black and Joseph Priestley (*see* PNEUMATICS). Investigations into "breathing" in plants played a significant role in the discovery of oxygen—corroborating parallel developments in animal respiration (*see* BIOCHEMISTRY; PHYSIOLOGY).

The mechanism by which sap moved through the plant was the focus of much later research. René-Joachim-Henri Dutrochet identified *osmosis, thereby explaining some of the functions of leaves in controlling the rate of flow. Ludolph Christian Treviranus proposed the Romanticist alternative of a vital force drawing the fluids upwards. Controversy over the methods of uptake and exhalation continued through the nineteenth century and was not fully resolved until Henry Horatio Dixon's compromise between the suction power of the leaf and the tensile strength of very narrow water columns. Scientists consequently abandoned the popular view that plants merely "evaporate" excess water from their leaves. Investigations into the opening and closing of stomata and the permeability of cell membranes became crucial aspects of a new ideology of the plant as a self-regulating, environmentally sensitive organism.

Julius von Sachs, *Lectures on the Physiology of Plants*, trans. H. Marshall Ward (1887). Robert Frank, *Harvey and the Oxford Physiologists: A Study of Scientific Ideas* (1980). A. G. Morton, *History of Botanical Science: An Account of the Development of Botany from Ancient Times to the Present Day* (1981).

JANET BROWNE

RESTORATION. See REVOLUTION, RESTORATION, AND THE ROYAL SOCIETY.

REVOLUTION. See SCIENTIFIC REVOLUTION; SCIENTIFIC REVOLUTIONS.

REVOLUTION, RESTORATION, AND THE ROYAL SOCIETY. During the English Civil War an informal group met weekly at various members' lodgings in Gresham College, London, to consider questions in medicine, geometry, astronomy, chemistry, and mechanics. Another group in London, organized by Samuel Hartlib, pursued the idea, popularized by Francis Bacon in the preceding generation, that knowledge should be put to practical and charitable use. Hartlib envisioned a new educational system to replace traditional studies in divinity, law, and logic with more practical subjects such as glassmaking and metalworking. He brought together such diverse intellectuals as the evangelist John Dury, the poet John Milton, and the natural philosophers Kenelm Digby, William Petty, Frederick Clod, and Robert Boyle.

Meanwhile an "Experimental Philosophy Club" had begun meeting in Oxford. By 1657, its membership included Boyle, Robert *Hooke, John Locke, and Christopher Wren. They had much in common with Hartlib's circle. They advocated

Frontispiece to Thomas Sprat, *The History of the Royal Society* (1667). The president of the Society, William Brouncker, and its inspirer, Francis Bacon, flank the bust of its patron, King Charles II.

charitable works and religious toleration and believed that natural philosophy could help to put an end to sectarian controversy. However they were more politically diverse and had less interest in social engineering.

At the Restoration, London became more hospitable to intellectual pursuits, in part because those who had been exiled during the interregnum brought back new ideas and fashions from their foreign travels. Members of the London and Oxford groups proposed the establishment of a Royal Society of London for the pursuit of natural knowledge to which Charles II gave a charter in 1662. Most of the society's ambitious plans for experimental investigations came to nothing. It had one signal success, however. Its *Philosophical Transactions* quickly became an international journal of news and reviews. In 1672, Isaac Newton made his public debut by publishing his "New Theory about Light and Colours" in the *Transactions*.

Scholars have speculated about how the various social, economic, religious, and political factors of Restoration England influenced the theories and practices of the nascent sciences. One of the earliest and most influential studies was Robert K. Merton's "Science, Technology, and Society in Seventeenth-Century England," which traced the development of experimental sciences and technology to the influence of Puritan ethical and charitable ideals. J. R. Jacob, in *Robert Boyle and the English Revolution* (1977), argued instead that the productive connection involved Anglicans and Royalists. He suggested that the passivity of matter preached by the mechanical philosophers paralleled the passive obedience that High Church Anglicans (and the state) wanted to induce in the masses, thus creating an alliance between natural philosophers and the powerful elite. Steven Shapin and Simon Schaffer, in their *Leviathan and the Air-Pump* (1985), extended Jacob's analysis to experiments. They argued that natural philosophers insisted on civility and gentlemanly deportment in the conduct of experiments as a reaction to the strife and instability of the revolutionary period. All of these arguments have some plausibility and continue to be debated. However because they consider only England, none can be used to explain parallel developments in continental Europe.

See also ACADEMIES AND LEARNED SOCIETIES; MECHANICAL PHILOSOPHY; PEER REVIEW; RELIGION AND SCIENCE.

Christopher Hill, *Change and Continuity in Seventeenth-Century England* (1975). Charles Webster, *The Great Instauration: Science, Medicine, and Reform 1626–1660* (1975). Robert G. Frank, Jr., *Harvey and the Oxford Physiologists* (1980). Michael Hunter, *Science and Society in Restoration England* (1981). Barbara Shapiro, *Probability and Certainty in Seventeenth-Century England* (1983). I. B. Cohen, ed., *Puritanism and the Rise of Modern Science: The Merton Thesis* (1990). Rose-Mary Sargent, *The Diffident Naturalist: Robert Boyle and the Philosophy of Experiment* (1995).

ROSE-MARY SARGENT

REVOLUTIONS IN SCIENCE. See SCIENTIFIC REVOLUTIONS.

RHETORIC IN SCIENCE. Plato introduced the term "rhetoric" to describe the power to persuade others, particularly in political or other public contexts. He contrasted it unfavorably with the knowledge (*scientia*) sought by philosophers. Later classical thinkers presented systematic treatments of rhetoric, the most important being Aristotle's *Rhetoric* in the third century B.C., the works of Cicero, particularly *De Oratore* in the first century A.D., and Quintilian's *Institutio oratoria*, also in the first century A.D. They divided rhetoric into three main classes: deliberative, that sought to persuade people (in legislatures, for example) to approve a matter of public policy; forensic, that sought to persuade people (in trials, for example) of the merits or demerits of an individual's actions; and epideidectic (display) rhetoric, for use on public occasions. None of these dealt with the presentation of *scientia*, which, being derived from self-evident principles, was supposed to stand by itself.

Rhetoric continued to be taught in Medieval universities, and flourished in the Renaissance when humanists such as Peter Ramus promoted it as a living practical art essential in politics, religion, and the law. While occasionally praising eloquence as useful, philosophers, including natural philosophers, continued to contrast it unfavorably with the knowledge that was their goal. In the seventeenth century, the founders of the Royal Society of London, expressing their intention of relying on their own experience and not on the persuasive skills of authorities, took as their motto the Latin phrase, "Nullius in verba," from Horace's, "Nullius addictus in verba magistri," sometimes translated as "not pledged to echo the opinions of any master." They insisted on plain language that eschewed the devious persuasive devices of rhetoric for communications to the Society. To this day, most scientists continue to think of themselves as saying what they mean in the most unadorned manner possible.

From the mid-1970s, however, historians, philosophers, specialists in speech and communication, and literary theorists have poured out books and articles on rhetoric and science, and even an American Association for the Rhetoric of Science and Technology was organized.

At least three shifts in intellectual climate helped create this field of interest. In the history of science, scholars who had dedicated themselves to researching the internal workings of science began shifting their focus to its cultural context. They recognized that well into the nineteenth century (and much longer than that in many countries), natural philosophers and scientists (like other educated people) learned the precepts of Cicero and Quintilian and other rhetoricians as a matter of course. For example, around 1750, undergraduates at the University of Edinburgh, many of whom went on to become physicians or natural philosophers, learned rhetoric, including the rhetorical forms appropriate to the sciences, from Adam Smith.

In the philosophy of science, many academicians, deeply impressed by the arguments of Pierre Duhem (see PROOF), decided that scientific theories could not be proved either true or false by resorting to experience. From this they concluded that given the inevitable inadequacy of evidence, scientists have to have other reasons for making the leap to belief, or have other tactics to persuade others to make the leap. At least some philosophers have explored the possibility that it is rhetoric that bridges this gap.

In the humanities more generally, many intellectuals have taken the so-called "linguistic turn"—coming to believe that language so shapes our vision of the world that we cannot get behind it. They regard as naive scientists' confident belief that language merely denotes facts in the real world. Language makes, constitutes, or creates the world. For these intellectuals, showing that this generalization applies even to scientific language is a splendid, if self-referentially incoherent, project.

Along with these intellectual trends, the institutional exigencies of the American university system pushed professors to teach and write about rhetoric in science. They found it a handy way to make required humanities courses relevant to science majors, to deal with directives to teach writing across the curriculum, and to make arguments for required courses that would increase department enrollments.

By the end of the twentieth century, detailed case studies of rhetoric in science had become too numerous for quick summary. Authors took on many topics: attitudes to language among members of the Royal Society; language as a "literary technology"; the use of eloges of defunct members of the Paris Académie des sciences to create a public image of the scientific character;

rhetoric as self-persuasion in Charles *Darwin's notebooks; the structure of the modern scientific article; disparities between James D. *Watson's accounts of the discovery of the double helix in his original publication (with Francis *Crick) and in his autobiography; and scientific method as rhetoric.

A few of these cases fall into the time-honored traditions of intellectual history. They deal with scientists' co-optation (or rejection) of classical rhetorical theory in persuading a public (both inside and outside their community) of the nature and value of specific theories and experiments or the nascent scientific enterprise as a whole. Most of the cases, though, have little to do with rhetoric as classically understood. Instead, it stakes out positions in the debate about science and relativism (see SCIENCE WARS). Non-relativists are willing to accept the point that persuasion occurs in science just as it does in politics, since scientists are well advised, if they want their fellows to accept their discoveries and theories, to present them as convincingly as possible. They draw the line, though, at asserting that science is mere rhetoric. Relativists, on the other hand, insist that "facts" themselves are consensual, constructed, and established by persuasion. For them, the pursuit of science comes down to a series of rhetorical battles.

See also PROOF.

Owen Hannaway, *The Chemist and the Word* (1975). Charles Bazerman, *Shaping Written Knowledge: The Genre and Activity of the Experimental Article in Science* (1988). Alan Gross, *The Rhetoric of Science* (1990). Greg Myers, *Writing Biology: Texts in the Social Construction of Scientific Knowledge* (1990). Peter Dear, *The Literary Structure of Scientific Argument* (1991). Marcello Pera and William Shea, eds., *Persuading Science: The Art of Scientific Rhetoric* (1991). John A. Schuster and Richard R. Yeo, eds., *The Politics and Rhetoric of Scientific Method* (1991).

RACHEL LAUDAN

RISK ASSESSMENT/SCIENTIFIC EXPERTISE. Risk is the possibility of injury, harm, or other adverse or unwanted effects. Risk assessment is the practice of using observations about what we know to make predictions about what we don't know. The modern concept of risk emerged when people came to believe that their lives were neither completely in the hands of fate nor completely random. Models of the past came to guide decisions about the future based on the theory of probability. The development of probability theory during the eighteenth century, and the influence of Pierre-Simon de *Laplace during the eighteenth and early nineteenth centuries, established the statistical basis for estimating the likelihood of events or outcomes.

During the last century, industrialized countries have routinely analyzed financial risks as part of conducting banking, insurance, and business operations. More recently, risk assessment has been applied to the analysis of health, safety, engineering, natural hazards, and ecological risks.

Before risk assessment became a well-recognized and codified discipline in the United States, risk decisions were guided by expert scientific judgment and a policy of precaution. For example, in 1958 the "Delaney clause" required the Food and Drug Administration to ban outright food and color additives that had been shown to produce tumors in humans or laboratory animals. In 1976, the *Ethyl* decision set a legal basis for a precautionary, non-risk-based policy, when the court upheld an EPA rule against an industry challenge, allowing the EPA to proceed with its plans to ban leaded gasoline even if the science could not demonstrate just what the benefits of removing lead would be. In 1980, however, the *Benzene* decision overturned the precautionary basis of *Ethyl* and substituted a risk-based principle by requiring some form of evaluation as a basis for deciding if a risk is or is not "significant" enough to deserve regulation. A series of executive orders requiring cost-benefit analysis of proposed decisions increased the demand for risk assessment.

Most U.S. law that seeks to ensure safety, or at least mitigate risk, was in place before risk assessment emerged as a discipline. Most of the methodology of risk assessment arose in response to the calls by these laws to define limits on chemical exposures that will "protect the public health with an adequate margin of safety" (Clean Air Act) or "protect the public welfare" (Clean Water Act). In passing these laws, the U.S. Congress called on the regulatory agencies to develop means to assess risks so as to define exposure levels that would achieve the stated goals of health protection. Executive orders and decisions of the Supreme Court enforced the demand that regulatory agencies develop risk assessment to mobilize science in the effort to limit risk.

In other industrialized countries, government regulatory decisions are not subject to judicial challenges in court to nearly the same degree as in the United States. Consequently, other countries do not insist on such elaborate procedures for marshaling and analyzing scientific evidence before a decision can be reached. In Europe, standards limiting exposure to chemicals in the workplace are routinely set based on a consensus of expert judgment. In contrast, U.S. courts have held that expert consensus does not provide a sufficient factual basis for regulation. The countries of the European Union often make decisions about health, safety, and environmental protection on a precautionary basis—that is, based more on the need to take action in the face of uncertain science than on the need to establish a science- and risk-based justification for an action. In Europe, chemicals or products can be banned from commerce on the basis of concern about their inherent toxicological properties without consideration of whether they pose an unacceptable risk.

Scientists participate in policy in the United States by testifying as "expert witnesses" in courts or other proceedings. By law, only a person qualified as an "expert" may express opinions in court (as opposed to ordinary witnesses, who are supposedly limited to testifying about "facts" they have personally observed rather than opinions). The Federal Rules of Evidence adopted for use in federal court trials in 1975 greatly liberalized the requirements for expert testimony, and made it easier for people to qualify as experts. In succeeding decades, a controversy arose over allowing expert testimony based on so-called "junk science," theories not accepted by the mainstream of the scientific community. The U.S. Supreme Court responded in 1993 in a famous decision, *Daubert* v. *Merrell Dow Pharmaceuticals*, which held that the trial judge must serve as a gatekeeper to exclude expert testimony that does not meet minimum standards of scientific reliability. To qualify to be admitted into evidence under the Daubert standard, evidence must be based on a scientific methodology that is "scientifically valid" and "relevant" to the circumstances at issue.

See also LAW AND SCIENCE.

John Adams, *Risk* (1995). Peter L. Bernstein, *Against the Gods: The Remarkable Story of Risk* (1996). National Academy of Sciences, *Understanding Risk* (1996). Richard Wilson and Edmund Crouch, *Risk/Benefit Analysis* (2001).

GAIL CHARNLEY

RÖNTGEN, Wilhelm Conrad (1845–1923), discoverer of X rays.

The son of a German draper and his Dutch wife, Röntgen was born in the Rhineland fifty years before he made his capital discovery. Those who knew him as a young man did not expect to see him become a German professor. His parents left the Rhineland and German citizenship for the Netherlands when Röntgen was three. He did not quite complete his secondary education at a technical school in Utrecht because he was expelled for refusing to identify a schoolmate who had caricatured a teacher. By not finishing high school he did not qualify for enrollment in a German university.

Röntgen went to the then-new Swiss Polytechnic in Zurich, which did not require a high school diploma. He enrolled in 1865 and graduated three years later as a mechanical engineer and protégé of the professor of physics, August Kundt. Since the Polytechnic did not grant doctoral degrees, Röntgen took one under a professor who also taught at the University of Zurich, an itinerary later repeated by Albert *Einstein. When Kundt took a chair at the University of Würzburg, Röntgen went along as his assistant in the hope of habilitating, that is, putting his foot on the first rung of the academic ladder. The university of which he was to be the glory denied his request to habilitate because his Swiss Ph.D. did not compensate for the want of a high school diploma.

Röntgen habilitated at Strasbourg, where Kundt was called to teach in the German university then (in 1872) newly seized from the French, and where the professors determined the qualifications for habilitation within their faculties. Röntgen could thus begin a teaching career that brought him Kundt's old chair at Würzburg in 1888. There he practiced the scrupulous, exact, careful experimental physics he had learned from Kundt. German academic physicists admired him for his work and his colleagues at Würzburg esteemed him for his mandarin qualities. In 1894 they made him their rector.

Röntgen's great discovery may be understood as a reward for his administrative service. When he left the rectorate in 1895, he decided to refresh himself with a new research topic. He began with an experiment designed by Philipp Lenard, who had replaced a bit of the glass wall in a standard discharge tube with a thin aluminum window in order to permit *cathode rays to pass from the tube into the laboratory (*see* ELECTRON). In November 1895, almost certainly using a Lenard tube, Röntgen made the observation that led to his grand discovery: a detecting screen lying upon a table fluoresced far beyond the range of cathode rays studied by Lenard.

After making the fateful observation, Röntgen holed up in his laboratory working out the properties of what he called "a new sort of rays." One was the capacity to photograph the bones in a living human hand. Here the staid Röntgen displayed a bit of showmanship. Among the pictures circulated with the announcement of his discovery was one of his wife's hand. The mystery of the rays and of their discoverer, and their obvious application to medicine, fascinated physicists, physicians, and the general public. No less puzzling to the wider world was Röntgen's

refusal to patent his discovery, to interest himself in the application of his rays, or to improve the method of their production. He received nothing for his gift to humanity except academy memberships, medals, and prizes, including the first Nobel Prize in physics (1901).

Röntgen did not tout himself or relish the many honors he received. But he was proud as well as shy, and felt bitterly the unfounded allegations that not he, but an assistant, had first noticed the effects of *X rays, and that his grand discovery was but an extension of Lenard's work.

Röntgen's status suited him to a more distinguished university than Würzburg. In 1900 he became professor of experimental physics and the director of the physical institute at the University of Munich. Most of his time went to administration. He took no part in the demonstration by Max von Laue and others at Munich that X rays are light of very high frequency.

Ever the loyal civil servant, Röntgen was as shocked by the discovery that his Kaiser and his government had lied systematically about the military situation during World War I as he was by the final collapse. He suffered greatly after the war from political unrest, scarcity of food, rampant inflation, and loneliness. He died in 1923, some say of intestinal cancer, others say of malnutrition. In his will he responded to the innuendoes about his discovery by directing that all his papers relative to the early history of X rays be burned.

Otto Glasser, *Wilhelm Conrad Röntgen and the History of Röntgen Rays* (1933). W. Robert Niske, *The Life of Wilhelm Conrad Röntgen* (1971). Elisabeth Crawford, *The Beginnings of the Nobel Institution: The Science Prizes, 1901–1915* (1984). Albrecht Fölsing, *Wilhelm Conrad Röntgen: Aufbruch ins Innere der Materie* (1995).

J. L. HEILBRON

ROYAL SOCIETY. See REVOLUTION, RESTORATION, AND THE ROYAL SOCIETY.

RUSSIA. See EUROPE AND RUSSIA.

RUTHERFORD, Ernest (1871–1937), the dominant personality in the early investigation of radioactivity and nuclear physics, born the son of a farmer in New Zealand, died an English baron.

Supported by fellowships, Rutherford graduated from Canterbury College, New Zealand, well educated in mathematics and physics, adept at the then-new art of wireless telegraphy (*see* RADIO AND TELEVISION), and armed with an economical style of writing and reasoning. He received an "exposition," a scholarship founded on the proceeds from the Great Exposition of London of

Rutherford's heraldic crest as Baron Rutherford of Nelson. The figures represent Hermes Trismegistus (an allusion to alchemy) and a maori warrior (a reference to Rutherford's native New Zealand); the Latin phrase signifies "to inquire into the elements of things."

1851, to study abroad. Rutherford chose Cambridge, England, where he planned to perfect a wireless detector under the guidance of Joseph John *Thomson at the Cavendish Laboratory. In December 1895, a few months after Rutherford's arrival in England, Wilhelm Conrad *Röntgen discovered *X rays. Thomson invited Rutherford to join him in studying the ionization the rays produced when passing through gases. As they finished this work, news of *radioactivity arrived. Rutherford applied the techniques that he and Thomson had developed to the ionization produced by uranium rays. He found that the rays had two distinct fractions, a short-range, or alpha type, and a long-range, or beta type.

With Thomson's support, Rutherford obtained a professorship in physics at McGill University in Montreal, Canada. With McGill's excellent equipment and the help of several colleagues—R. B. Owens (McGill's professor of electrical engineering), Frederick Soddy (briefly lecturer in chemistry, a chance visitor from England), and Otto Hahn (a determined visitor from Germany)—Rutherford discovered the emanation, radon, and its active deposit (with Owens); invented the disintegration theory of radioactive decay (with Soddy); and monitored the sequence of decay products (with Hahn). Rutherford set

forth the principles of radioactivity in a textbook, first published in 1904, which defined the field for decades.

In 1907 Rutherford left Montreal for the professorship of physics at the University of Manchester, England. There he made discoveries at least equal to those of his McGill years and gathered a research group superior to anything imaginable in Canada. The findings included the demonstration of the identity of alpha particles and ionized helium atoms, a theory of the scattering of alpha particles, and the nuclear model of the atom. In 1908 Rutherford received the Nobel Prize in chemistry, for which he expressed surprise. He had expected to win one in physics. His transformation into a chemist, he said, was the fastest and oddest disintegration he had ever witnessed.

The research group in Manchester included Niels *Bohr, who combined Rutherford's nuclear model into the theory of *atomic structure that guided microphysics for a decade; György Hevesy, who developed the technique of radioactive tracers and helped define the idea of isotopes; and Henry Moseley, whose work on characteristic X rays established the concept and utility of atomic number (*see* ATOMIC STRUCTURE). Both Bohr and Hevesy won Nobel Prizes for their work; Moseley would have had one too if he not died in World War I.

Rutherford helped to mobilize British scientists during the war and investigated acoustical methods of detecting submarines. He led a delegation of British and French scientists to Washington to show what science adapted to war had wrought in Europe and to encourage development in the United States. After the war Rutherford strove to moderate the treatment of German scientists, whom many allied scientists wanted to ostracize from international meetings and research projects.

In 1919 Rutherford succeeded Thomson as Cavendish professor. Again he surrounded himself with a powerful research group: James Chadwick, with whom he pursued his wartime discovery that alpha particles from radioactive substances can transmute nitrogen into oxygen, and who on his own detected (in 1932) the neutral particle (neutron) whose existence Rutherford had predicted a decade earlier; John Douglas Cockroft and Ernest T. S. Walton, who made the first *accelerator that disintegrated an atom with a particle beam (also in 1932); C. T. R. Wilson, a former fellow student from Cambridge, who invented the cloud chamber; P. M. S. Blackett, who used the cloud chamber to discover the positive electron (again, in 1932); Pyotr Leonidovich Kapitsa, who made the world's most powerful magnet; and Francis Aston, Thomson's last collaborator, who demonstrated experimentally the agreement between apparent atomic and true isotopic weights. All of these men won Nobel Prizes.

Rutherford's elevation to life peer in 1931, together with his other honors, scientific accomplishments, and international connections, made him an ideal figurehead for the Academic Assistance Council, a nongovernmental organization set up in 1933 to rescue scholars driven from their positions by the Nazis. During Rutherford's presidency the council found jobs in Britain for over two hundred émigrés and helped broker positions for many more elsewhere.

Rutherford died suddenly of complications arising from a strangulated hernia. He is buried in Westminster Abbey near Isaac Newton, and in the periodic table of elements at number 104.

A. S. Eve, *Rutherford* (1939). Ernest Rutherford, *Collected Papers* (3 vols., 1962–1965). Lawrence Badash, ed., *Rutherford and Boltwood: Letters on Radioactivity* (1969). David Wilson, *Rutherford, Simple Genius* (1983). John Campbell, *Rutherford, Scientist Supreme* (1999). J. L. Heilbron, *Rutherford, A Force of Nature* (2001).

J. L. HEILBRON

S

SAGAN, Carl. See HAWKING, STEPHEN, AND CARL SAGAN.

SAKHAROV, Andrei (1921–1989), theoretical physicist, key figure in the Soviet thermonuclear program, prominent advocate of human rights; and **Edward TELLER** (b. 1908), theoretical physicist, leading figure in the U.S. hydrogen bomb program, staunch proponent of position-of-strength policy toward the Soviet regime.

Although Teller and Sakharov have in common the fathering of the hydrogen bomb and both were prominent political and public figures, their lives on opposite sides of the Iron Curtain exhibited striking differences.

Teller was born in Budapest, Austria-Hungary, to a family of assimilated Jews of the upper middle class. After a course in chemical engineering he went to study physics in Germany (under Werner *Heisenberg) and Denmark (under Niels *Bohr). In 1935 he emigrated to the United States to work at George Washington University. In the 1930s he made a few significant contributions to theoretical physics, including the Jahn-Teller effect, which concerns crystal symmetry arising from interactions between electrons and nuclei, and turned out to be very important for material science.

Teller was involved in the American atomic project from its very beginning before the United States entered World War II. In the late 1940s, preoccupied with an idea of the hydrogen bomb, he helped convince the U.S. government that the weapon was feasible and indispensable for national security in the Cold War. He also made a key contribution to the first successful design of a thermonuclear weapon (tested in 1952). Teller, together with Ernest O. Lawrence, urged the creation of a second nuclear-weapons laboratory, now the Lawrence Livermore Laboratory. As a prominent government adviser on nuclear policy, he opposed the ban of nuclear testing, championed peaceful uses for nuclear explosives, and promoted strategic antiballistic missile defense.

Teller pushed the hydrogen bomb against the opposition of J. Robert *Oppenheimer and the General Advisory Committee of the Atomic Energy Commission that Oppenheimer chaired. During the commission's hearing over its decision to revoke Oppenheimer's security clearance (1954), Teller testified that he regarded Oppenheimer loyal to the United States, but that he "would like to see the vital interests of this country in hands which I understand better, and therefore trust more." The great majority of the American scientific community regarded his testimony as an unacceptable violation of ethics and ostracized Teller for life. Nonetheless, in 1962 the U.S. government awarded Teller the Enrico *Fermi prize "for leadership in research on thermonuclear reactions, and for his efforts to strengthen national security and to insure the peace."

Sakharov was born in Moscow to a family of the Russian intelligentsia. World War II interrupted his study of physics in Moscow University. After two years of work at a munitions factory, in 1945 he went on to graduate study in theoretical physics under Igor Tamm. Among his first results was an idea of muon-catalyzed fusion. In 1948 the government assigned Tamm's group, including Sakharov, to check the feasibility of an H-bomb design that, unknown to Sakharov, had been developed in part through espionage. In a few months he invented a brand-new design realized in the first Soviet thermonuclear bomb (tested in 1953). In 1951 he pioneered a research of controlled thermonuclear fusion that led to the tokamak reactor. His was the main contribution to the full-fledged H-bomb tested in 1955.

In 1958 Sakharov calculated the number of casualties that would result from an atmospheric test of the "cleanest" H-bomb: 6,600 victims for 8,000 years per megaton. "What moral and political conclusion must be drawn from these numbers?" he asked. Sakharov was proud of his contribution to the 1963 Test Ban Treaty, which saved the lives of many people who would have perished had testing continued in the atmosphere. In the 1960s he started his return to pure physics. The most successful consequence was his 1966 explanation of the disparity of matter and antimatter in the universe, or baryon asymmetry.

The major turn in Sakharov's political evolution took place in 1967, when the antiballistic missile defense (ABM), became a key issue in U.S.-Soviet relations. Sakharov wrote the Soviet leadership to argue that the moratorium proposed by the United States on ABM would benefit the Soviet Union and that otherwise the arms race in this new technology would increase the likelihood of nuclear war. The government ignored his memorandum and refused to let him initiate a public discussion of ABM in the Soviet press. Sakharov felt compelled to make his views public in an essay "Reflections on Progress, Peaceful Coexistence and Intellectual Freedom," published in samizdat (underground self-publishing in the Soviet Union) and in the West in the summer of 1968. The secret father of the Soviet H-bomb emerged as an open advocate of peace and human rights.

Sakharov was immediately dismissed from the military-scientific complex. He then concentrated on theoretical physics and human rights activity. The latter brought him the Nobel Prize for Peace in 1975 and internal exile from 1980 until 1986, when the new Soviet leader Mikhail Gorbachev released him. Upon his return from exile Sakharov enjoyed three years of freedom and seven months of professional politics as a member of the Soviet parliament. Those were the last months of his life.

Although Sakharov opposed Teller over nuclear testing and the ABM problem, he considered the attitude of his American colleagues toward Teller to be "unfair and even ignoble." Sakharov had reason to know that Americans who considered Teller's position "anti-Soviet paranoia" did not understand the Soviet regime. Teller had had inside information from two of his physicist friends, pro-socialist Lev Landau and Laszlo Tisza. They had witnessed the great Soviet terror of 1937 that destroyed, among much else, the Kharkov Physics Institute, one of the best scientific centers in the USSR, and killed people dedicated to science and devoted to their country. Teller concluded that "Stalin's Communism was not much better than the Nazi dictatorship of Hitler," and never changed his mind.

Sakharov underwent a conversion. For many years he lived intoxicated by socialist idealism. He later said that he "had subconsciously . . . created an illusory world to justify" himself. Totalitarian control over information enabled Soviet propaganda to brainwash even the best and brightest. Sakharov wanted to make his country strong enough to ensure peace after a horrible war. Experience brought him to a "theory of symmetry": all governments are bad and all nations face common dangers. In his dissident years he realized that the symmetry "between a normal cell and a cancerous one" could not be perfect, although he kept thinking that the theory of symmetry did "contain a measure of truth." That did not close his mind: "We should nevertheless continue to think about these matters and give advice to others that is guided by reason and conscience."

For both theoreticians the statement that "the future is unpredictable" was meaningful far beyond quantum physics. It made them feel personally responsible for the future of humanity.

Herbert York, *The Advisers: Oppenheimer, Teller and the Superbomb* (1976). Andrei Sakharov, *Memoirs*, trans. R. Lourie (1990). Andrei Sakharov, *Moscow and beyond, 1986 to 1989*, trans. A. W. Bouis (1991). *Andrei Sakharov: Facets of a Life* (1991). David Holloway, *Stalin and the Bomb: the Soviet Union and Atomic Energy, 1939–1956* (1994). Edward Teller with Judith Shoolery, *Memoirs: A Twentieth-Century Journey in Science and Politics* (2001). Gennady Gorelik with Antonina W. Bouis, *Andrei Sakharov: Science and Freedom* (forthcoming).

GENNADY GORELIK

SALAM, Abdus (1926–1996), Pakistani physicist who together with Sheldon Glashow and Steven Weinberg won the Nobel Prize in physics in 1979 for work on the unification of the weak and the electromagnetic forces.

Salam was born into a religious family with a long tradition of learning on 29 January 1926 in Jhang Maghiana, a small market town in the then-undivided Punjab province, now in Pakistan. His father, Choudhari Mohammed Hussain, a teacher, rose to head clerk in the Department of Education of the district. A very precocious child, Salam could read and write at four and perform lengthy multiplication and division. An outstanding student, Salam won a scholarship to the University of the Punjab. In 1946, after completing his studies in Lahore he went to St. John's College, Cambridge, again on a scholarship, and emerged with a double first in mathematics and physics (1949), the Smith's Prize for the most outstanding predoctoral contribution to physics (1950), and a Ph.D. in theoretical physics (1951).

Salam's Smith Prize paper offered a resolution of the difficult problem of the overlapping divergences in the proof of the renormalizability of the S-matrix in *quantum electrodynamics and gained him an international reputation. He returned to Pakistan in 1952 and became head of the Mathematics Department of the Punjab University with the intention of founding a school of research. But there were no postgraduate studies at Punjab University, no journals, and no possibility to attend any scientific conferences. He had to make a choice between physics or

Pakistan. In 1954 he returned to Cambridge as a lecturer and committed himself to making it possible for young Ph.D.s to remain active and productive scientists while working in their own communities. In 1960 he conceived of the idea of an International Centre for Theoretical Physics (ICTP). With funds from the international community the ICTP in Trieste started operations in 1964. Salam instituted "associateships" that allowed deserving young physicists to spend their vacations at ICTP in close touch with the leaders in their field of research. They could thus overcome their sense of isolation and return to their own country reinvigorated. Earlier, in 1957, P. M. S. Blackett invited Salam to Imperial College, London, to found and head a Theoretical Physics Group. He remained at Imperial as professor of physics for the rest of his life.

To explain the experiments that demonstrated the violation of parity in weak interactions (*see* QUANTUM PHYSICS), Salam proposed in 1957 that the spin of neutrinos always points in the direction opposite to that of their momentum. The work that won him the 1979 Nobel Prize, which he shared with Sheldon Glashow and Steven Weinberg, advanced a model for the unification of the weak and electromagnetic forces. He had earlier worked with Jeffrey Goldstone and Weinberg on spontaneous symmetry breaking and with John Ward on theories of the weak interactions. In 1971 Jogesh Pati and Salam proposed that the strong (nuclear) forces might also be included in this unification. In 1974 Salam and his lifelong collaborator John Strathdee introduced the idea of superspace, a space with both commuting and anticommuting coordinates, which underlay all research on supersymmetry at the end of the twentieth century.

Salam's scientific achievements reflect only one facet of his personality. He also devoted his life to nurturing international cooperation and bridging the gap between developed and developing nations. He believed that the eradication of this disparity demanded that every country become the master of its own scientific and technological destiny. The first step, the ICTP, has been a major forum for the international scientific community and a model for similar establishments in various countries. Since 1965 over 60,000 scientists from 150 countries have taken part in its activities. He expounded his vision of science and technology in the Third World in his book *Ideals and Realities* (1984).

Salam won the Atoms for Peace Prize (1968), the Einstein Medal (1979), and the Peace Medal (1981) as well as the Nobel Prize; received over forty honorary degrees; and earned a knighthood for his services to British science (1989). He died in Oxford on 21 November 1996 after a long, debilitating illness. He was a devout Muslim, whose religion did not occupy a separate compartment of his life. He wrote, "The Holy Quran enjoins us to reflect on the verities of Allah's created laws of nature; however, that our generation has been privileged to glimpse a part of His design is a bounty and a grace for which I render thanks with a humble heart."

Abdus Salam, *Ideals and Realities* 2d ed. (1987). Abdus Salam, *Science in the Third Word* (1989).

SILVAN S. SCHWEBER

SALONS. See COURTS AND SALONS.

SATELLITE. The first artificial satellite appeared in the serial novel *The Brick Moon*, published by Edward E. Hale in 1869 in *Atlantic Monthly* magazine. The satellite, intended to be used as a navigational aid, was built of bricks to withstand the heat generated during its motion through the atmosphere. The launch system consisted of two gigantic fly wheels. An accident during the launch dispatched the brick moon prematurely, and the construction workers on board with their families organized a pleasant life in their new home planet.

The advent of rockets made the idea of artificial Earth satellites a concrete possibility. In February 1945, with the German V2 missiles still hitting London, science fiction writer and amateur scientist Arthur C. Clarke suggested that V2 rockets could be used to launch communication satellites. He proposed the use of geostationary satellites to achieve a global telecommunications system (*Wireless World*, October 1945). Intercontinental ballistic missiles (ICBM), a part of the V2 legacy in the Cold War framework, would eventually provide the technical basis for satellite launchers.

Strategic reconnaissance was the main objective of satellites from the military point of view. A spy satellite flying over foreign territory, however, violated international law, which guaranteed national sovereignty over air space. Hence the nations able to do so launched nonmilitary, scientific satellites to set a legal precedent of the "freedom of space" for subsequent military space activities. The *International Geophysical Year (IGY) offered a framework to establish the precedent.

The IGY, planned to run from 1 July 1957 to 31 December 1958, eventually had the support of sixty-six nations. One of its scientific objectives was to gain information about upper atmosphere phenomena by the use of balloons and sounding rockets. In October 1954, the IGY Special Committee recommended that governments try to launch Earth satellites for scientific purposes.

Both the United States and the Soviet Union undertook to meet the challenge.

The Soviet Union placed the project within its ICBM military program and developed the future R-7 (Semyorka) missile as a launch vehicle for its satellite. The Dwight D. Eisenhower administration, on the contrary, wanted to stress the scientific image of the venture. It rejected the U.S. Army's Orbiter, based on the Jupiter missile developed by former V2 designer Wernher von Braun, in favor of the Naval Research Laboratory's Vanguard, based on the Viking sounding rocket originally designed for upper atmosphere research.

The Soviet path proved more successful. On 4 October 1957, a modified Semyorka rocket launched *Sputnik 1*, the first artificial "fellow-traveler" of the earth. (Russian astronomers had used "sputnik" to designate any hypothetical small natural satellite of the earth.) It was an aluminum sphere 58 centimeters (22 inches) in diameter weighing 83.6 kilograms (184 pounds) that circled the earth once every 96.3 minutes. Its radio emitters sent its familiar "beep-beep" sound all over the world for 92 days.

One month later, on 3 November, the Soviet Union put up *Sputnik 2* to celebrate the fortieth anniversary of the October Revolution. It weighed more than 500 kilograms (1,100 pounds) and carried the first living being into space, the dog Laika, wired up for medical and biological studies. Space technology had not developed enough to return Laika to Earth alive.

The competition proved too strong for Vanguard. The satellite had been reduced to a small sphere of 1.5 kilograms (3.3 pounds) and the launcher not fully tested when, on 6 December 1957, Vanguard rose four feet off its pad at Cape Canaveral, Florida, and slumped back to Earth in a ball of thunder and flame. American pride was restored on 31 January 1958 when von Braun's Jupiter-C rocket put the *Explorer 1* satellite into orbit. It carried a cosmic-ray counter designed by James A. Van Allen that revealed the radiation belt trapped in the earth's magnetic field eventually named after him. Later in 1958 the Eisenhower administration created the National Aeronautics and Space Administration (NASA).

During the 1960s space was an important field of political confrontation between the two superpowers. A superior space program might indicate a superior ideology, a more efficient political institution, a greater industrial capability, and a stronger armed force. Both countries undertook ambitious programs of manned space flight aimed at capturing the public imagination. In April 1961, Soviet cosmonaut Yury Gagarin became the first human being to visit outer space. In July 1969, the United States succeeded in landing the first astronauts on the Moon.

In the following decade, the Soviet Union undertook to establish staffed space stations in Earth's orbit. Seven Salyut stations went up between 1971 and 1982, followed in 1986 by the large Mir station. The United States' space shuttle program culminated in the first launch of the shuttle Columbia in April 1981.

In the early 1960s, a number of western European nations started civilian space programs. Europe's first satellite, the Italian *San Marco 1*, was carried aloft by an American rocket in December 1964. The French *Astérix*, launched by a French rocket, followed in November 1965. The United Kingdom collaborated with NASA in the Ariel scientific satellite program, whose first launch occurred in 1962. Western Europeans joined together in the European Space Research Organization (ESRO), whose first satellite, *Iris*, flew on an American rocket in 1968. ESRO's success prompted the creation of the European Space Agency (ESA) in 1975. Eastern Europe realized a parallel collaborative effort through the Intercosmos program (1969–1971).

Japan and China joined the list of space nations in 1970. The former launched its first satellite (*Ohsumi*) in February, the latter the *China-1* satellite in April. The first Indian satellite, *Aryabata*, rose from a Soviet range in April 1975.

Artificial Earth satellites opened new frontiers of experimental research in many fields. It became possible to study the upper atmosphere and the ionosphere by in situ measurements. Astronomers could investigate the electromagnetic spectrum from celestial objects in wavelengths absorbed by the atmosphere (infrared, ultraviolet, X- and gamma rays). Satellites and space probes made it possible to study the structure and properties of the magnetosphere and its interaction with the solar wind and the interplanetary plasma. The Moon and other bodies of the solar system became objects of important geophysical studies. The advent of space stations and space laboratories on board the shuttle (Spacelab) opened new opportunities in the life sciences and material sciences. *The Century of Space Science* contains a list of all scientific satellites launched between 1957 and 2001.

Satellites have had important civilian applications, particularly in telecommunications, meteorology, navigation, and Earth observation. Following a number of experimental satellites (*Score, Echo, Telstar*), the era of commercial satellite telecommunications began in April 1965 with Early Bird (*Intelsat 1*), the first communications satellite in geostationary orbit. Four years later, three *Intelsat 3* satellites realized Clarke's vision of global space communications. Several communications satellite systems besides the Intelsat global network were set up, notably the Soviet Union's Molnyia system.

Meteorology from space started with NASA's Tiros satellite program (1960–1965). This was followed by the Environmental Science Services Administration (ESSA), National Oceanic and Atmospheric Administration (NOAA), and Geostationary Operational Environmental Satellite in the United States (GOES); Meteor in the Soviet Union; Geostationary Meteorological Satellites (GMS-Himawari) in Japan; and the European Space Agency's Meteosat. The use of satellites made it possible to establish a world meteorological service under the aegis of the World Meteorological Organization (WMO).

The United States and the Soviet Union developed navigation satellites for military use. In February 1978, the United States launched the first *Navstar* satellite of the Global Positioning System (GPS), capable of providing the position of any moving object with a precision of a few meters. The Soviet Glonass system (1982) had similar capabilities. Both are also used in the civilian sector.

The United States paved the way for remote sensing satellites aimed at surveying Earth and ocean resources. The first *Landsat* satellite flew in July 1972, the seventh and last in 1999. The Soviet Union, France, Japan, India, and the European Space Agency have all had Earth observation programs.

The most ambitious project of the space age's first half century was and is the International Space Station, now under construction 400 kilometers (250 miles) above the earth's surface through a joint effort of the United States, Russia, the European Space Agency, Italy, Japan, and Canada.

Homer E. Newell, *Beyond the Atmosphere: Early Years of Space Science* (1980). Walter A. Mc Dougall, ...*The Heavens and the Earth: A Political History of the Space Age* (1985). Roger D. Launius, *NASA: A History of the U.S. Civil Space Program* (1994). T. A. Heppenheimer, *Countdown: A History of Space Flight* (1997). Asif A. Siddiqi, *Challenge to Apollo: The Soviet Union and the Space Race, 1945–1974* (1998). John Krige, Arturo Russo, and Lorenza Sebesta, *A History of the European Space Agency, 1958–1987* (2000). Johann Bleeker, Johannes Geiss, and Martin Huber, eds., *The Century of Space Science* (2001).

ARTURO RUSSO

SCHOOLS, RESEARCH. A research school is a group that shares ways of thinking, solving problems, and deploying methods and tools of investigation within a discipline. School members

thus follow the same guidelines for making value judgments in scientific practice. Some schools are further defined or differentiated by more refined criteria. Practitioners working on investigations that are all descended from the same ancestor problem may constitute a school. Similarly, scientists espousing a distinctive way of interpreting theory, as in the Copenhagen school of quantum theory, or doctoral students trained under a common mentor, may qualify as members of the mentor's school.

Schools have distinctive social features. Linkage to a master figure can confer a group discipleship. Members of a school understand, appreciate, and can defend the work of fellow members against outside attacks. Schools may control journals, scientific societies, or institutions, especially laboratories. Schools are always subsets within the larger scientific discipline, which often has competing schools. Since schools perpetuate standardized investigative techniques, they influence the direction of research and the evolution of disciplinary knowledge. Because of this standardization, schools are often derided for their indoctrination and conformity.

Not until the nineteenth century, with the advent of distinct disciplines, did schools become a prominent feature in scientific practice. The best-known of the early schools was Justus von *Liebig's at the University of Giessen, where he established a laboratory institute for training pharmacists and chemists in 1826. Not immediately popular, Liebig's institute eventually acquired a reputation for rigorous instruction, productive research, and the successful training not only of research chemists, but also of physicians, apothecaries, industrialists, and agriculturalists. Although the doctoral students worked on problems originating in Liebig's own research, the vast majority of the other members of the laboratory merely completed a set of laboratory exercises that Liebig had designed. The appellation "school" to the Giessen laboratory in its early years refers more to the coherence in the chemical techniques Liebig achieved through instruction than to the research results of his students. With these techniques, almost any student could achieve an accuracy in the elementary analysis of organic compounds comparable to that obtained previously by the best chemists. Through this democratization of research techniques, Liebig's Giessen laboratory literally became a knowledge factory. In addition to Liebig's pedagogical program and his guidance in choosing suitable research topics, students benefited from the personal care and attention he bestowed on them. Affective bonds between mentor and student strengthened the school tie. The standardization of procedures and results, the reduction of *error, and the enhancement of *precision that Liebig's apparatus and methods introduced enlarged the domain over which chemists agreed on the results of experiments and, thus, the effective reach of his school.

The school of physiology around Michael Foster at the University of Cambridge in the late nineteenth century derived more from Foster's impact as a teacher than from his achievements as a researcher. Although the Cambridge School's conceptual unity stemmed largely from the common ancestral problem of the heartbeat as well as from a novel evolutionary approach to physiological phenomena, Foster's most important contribution seems to have been his personal way of encouraging students. He assimilated and generously cited the work of his students in his textbook, and sought contributions from them for the two new journals he inaugurated in 1873 and 1878. Even though schools existed everywhere by the time his emerged, Foster advised his students against conformity. Unlike Liebig's research school, the fate of Foster's did not depend on the productive work of its director.

In the school of physics based on the Königsberg *seminar founded in 1834 by the physicist Franz Ernst Neumann, the defining characteristic was his way of assembling, presenting, and assigning value to the investigative techniques of an exact experimental physics. A minor technique in Neumann's research—the determination of constant errors of an apparatus and the accidental errors of quantitative results—became a dominant element in his problem sets and in the fledgling professional investigations of his students. Especially through a novel presentation of mechanics in which Neumann assigned a central role to Friedrich Wilhelm Bessel's investigations with the seconds *pendulum, his seminar students learned a scientific style that emphasized precision achieved by the method of least squares. Bessel's investigation provided an accessible and practical model for learning how to analyze instruments theoretically. Like Foster and Liebig, Neumann encouraged students to publish research based to a greater or lesser degree on his own findings under their own names. Students considered their passage through the seminar an important ritual in establishing their identities as scientists. Neumann welcomed students into his own home as if they were his sons. Their letters reveal the intensity of their bond and their published works express the gratitude they felt. The hundreds of students who attended his seminar included physicists, secondary school teachers, and precision instrument makers.

Despite Neumann's success, his method of training brought problems that highlight the weaknesses of schools that insist on adherence to their own investigative norms. Neumann's instruction did not convey a sense of when the pursuit of *precision and accuracy should end; students worried that their investigations were woefully incomplete when a residue of error remained. Many members returned again and again to the reduction of error as a means of concluding an investigation, without full success since error-free physics cannot be achieved. The ethos of exactitude cultivated in the seminar demanded an unrealistic goal. This ethos became an ethic that guided professional actions and decisions, defined professional identities, structured investigative strategies, and identified significant problems.

Leaders of schools—Liebig, Foster, and Neumann, and also August Böckh in philology at Berlin, Carl Ludwig in physiology at Leipzig, Joseph John *Thomson at Cambridge, Ernest *Rutherford at Manchester, Wilhelm Wundt in psychology at Leipzig, Arnold Sommerfeld in physics at Munich, and many others—all trained their students in techniques of investigation and guided them toward promising problems. Much of what they taught was not rationalized, codified, standardized, or routine. Through such pedagogy, tacit knowledge becomes explicit, and the ineffable elements of a mentor's enthusiasm and inspiration bring students through the difficult process of becoming a professional.

Even where schools are strong, instabilities inevitably creep in. A school with political control over journals, institutions, financial resources, or societies, but whose modus operandi no longer points toward new and significant results, can retard a discipline. The emotional bonding that takes place in a school can bias the assessment of competitors outside of the school. Originality often entails breaking from the conceptual patterns and practice of the school, and hence leaving it; if the renegade succeeds in establishing a school, the existing one might disappear. A school that was once in a common geographical setting might not survive dispersion. Too great a process can also finish a school. If its methods become universal—for example, through publication of a dominant textbook—it may lose its reason for existence—the Göttingen school of physics did not survive the publication of Friedrich Kohlrausch's *Introduction to Practical Physics* (1870), which codified its approach to research. Kohlrausch's textbook enjoyed immense popularity, and established a uniformity hitherto not seen in the practical exercises of physics.

Finally, schools may simply be ill-suited for dealing with certain problems. In the late twentieth century, the sheer complexity of investigations in areas such as high-energy particle physics and genetics required the creation of other social units, large interdisciplinary teams of international collaborators. Schools may thus be transient historical phenomena, characteristic of one stage in the evolution of scientific communities.

See also DISCIPLINE(s).

Jerome Ravetz, *Scientific Knowledge and Its Social Problems* (1971). J. B. Morrell, "The Chemist Breeders: The Research Schools of Liebig and Thomson," *Ambix* 19 (1972): 1–46. Gerald L. Geison, *Michael Foster and the Cambridge School of Physiology* (1978). Gerald L. Geison, "Scientific Change, Emerging Specialties, and Research Schools," *History of Science* 10 (1981): 20–40. Leo J. Klosterman, "A Research School of Chemistry in the Nineteenth Century: Jean Baptiste Dumas and His Research Students," *Annals of Science* 42 (1985): 1–40. Frederic L. Holmes, "The Complementarity of Teaching and Research in Liebig's Laboratory," in Kathryn M. Olesko, ed., *Science in Germany, Osiris* 5 (1989): 121–164. Joseph S. Fruton, *Contrasts in Scientific Style: Research Groups in the Chemical and Biological Sciences* (1990). Kathryn M. Olesko, *Physics as a Calling: Discipline and Practice in the Königsberg Seminar for Physics* (1991). Gerald L. Geison and Frederic L. Holmes, eds., "Research Schools: Historical Reappraisals," *Osiris* 8 (1993).

KATHRYN OLESKO

SCHRÖDINGER, Erwin (1887–1961), theoretical physicist, discoverer of the equation governing the wave representation of quantum mechanics.

Schrödinger's mother's father, Alexander Bauer, held the principal chair of chemistry at the Vienna Polytechnic. His mother's mother was English and Protestant, and through those relations Erwin spoke English from childhood. Erwin's father, Rudolf Schrödinger, a non-practicing Catholic, one of Bauer's students, operated half-heartedly an inherited linoleum and oilcloth factory and store. His free time he devoted first to painting and etching, and then to microscopic botany, being in his later years a mainstay of the Zoological-Botanical Society of Vienna.

Erwin was a bright and beautiful only child in the most favorable familial circumstances, attached affectionately and intellectually to his father, doted upon by his mother, nurses, maids, and mother's unmarried sisters. Educated at home until he entered the Gymnasium in 1898, where he was then always first in his class, Erwin continued to live in the spacious parental apartment in central Vienna until he married in 1920.

The marriage endured, but almost from the outset Schrödinger sought through extramarital sexual conquests to re-create his childhood conditions of abundant feminine love and to create the religio-mystical experience of mergence with non-self. In his youth nonreligious, and in adulthood disdainful of "official Western creeds," Schrödinger, following Arthur Schopenhauer, had become, by his early thirties at the latest, strongly attached to the Eastern concept of a cosmic intelligence in which every individual soul participates.

Schrödinger entered Vienna University in 1906, receiving his Ph.D. in 1910 with an experimental dissertation on humidity as a source of error in electroscopes. The core of Schrödinger's education was Friedrich Hasenöhrl's extended cycle of lectures over the various fields of theoretical physics transmitting the outlook of Ludwig *Boltzmann. The views of both Boltzmann and Ernst Mach were incorporated in the lectures of the professor of experimental physics, Franz Exner, whose assistant Schrödinger became upon returning in autumn 1911 from his obligatory year of military service. (He would serve throughout World War I, 1914–1918.) His researches now were chiefly theoretical, applying Boltzmann-like statistical-mechanical concepts to magnetic and other properties of bodies. The results were not notably successful—Schrödinger's physical intuition profited little from his familiarity with instruments—but gained him the advanced doctorate (Habilitation) in 1914. By contrast, Schrödinger's spectroscopic studies of human color perception (in particular his own) complemented his development (1918–1920) of a theory, based on the Machist concept of elementary sensations of color, that indeed almost represents the facts. This was Schrödinger's only significant original contribution while he remained in Vienna, where all his work was prompted by the topics and problems of his teachers and friends.

The war ended in a complete economic collapse of Austria, ruining Schrödinger's family and forcing him to pursue his career in the wider German-language world of Central Europe. Between spring 1920 and autumn 1921, Schrödinger took up, successively, positions at the Jena University, the Stuttgart Technical University, the Breslau University, and the Zurich University. Finally safe and secure in Swiss employ, he collapsed after the stresses of the previous two years, which had also seen the deaths of both his parents and grandfather Bauer.

A seven-month rest cure in Arosa in 1922 restored Schrödinger physically and opened the only period in which he contributed importantly to the mainstream of theoretical physics. This creative phase culminated in 1926 with the publication of several lengthy papers introducing his differential equation for the quantum-mechanical treatment of atomic systems, and demonstrating its power by applying it successfully to several standard problems. (In his fifty-year career, Schrödinger's median annual output of research papers amounted to forty pages; in 1926 he published 265.) Schrödinger achieved this "wave" equation, arguably the single most important contribution to theoretical physics in the twentieth century, by a creative union of his Viennese statistical-mechanical concerns with the elaborations of the Bohr theory of the atom then the vogue in Germany. The special advantages of his situation in Switzerland, which brought him in close contact with a mathematician of exceptional power, Hermann Weyl, and which permitted him to consider sympathetically the work of a Frenchman, Louis de Broglie, enabled Schrödinger to seize the opportunity inherent in the latter's attribution of a wave process to material particles.

The wave mechanics of 1926 brought Schrödinger in 1927 the succession to Max *Planck in the chair of theoretical physics at the Berlin University. Schrödinger remained in it until the summer of 1933, accomplishing little. Although he was acceptable to the Nazis, they were not acceptable to him. He resigned to hold a succession of positions at the universities of Oxford, Graz, and Ghent, landing (after hasty, anxious escapes) in October 1939 in Dublin as senior professor at the new Dublin Institute for Advanced Studies. There Schrödinger remained nearly to the end of his career, returning to Vienna in 1956 in poor health and retiring two years later.

At the Dublin Institute Schrödinger devoted himself largely, but unsuccessfully, to the problem then also occupying Einstein, also unsuccessfully, at the similarly named Princeton institute: the creation of a field theory uniting gravity with electromagnetic and nuclear forces. As a contribution to the Dublin Institute's series of public lectures, Schrödinger, who was an engaging speaker, delivered several in February 1943 under the title "What Is Life?" In these popular-scientific lectures Schrödinger, who had only a very slight knowledge of the literature on the physical bases of life, dragged his audience into and then out of a series of blind alleys, leaving them at the end just about where he began. Nonetheless these lectures, printed the following year, achieved an immediate and great reputation with both physicists and biologists, and rank still today as one of the most overrated scientific writings of the twentieth century.

Paul A. Hanle, "The Coming of Age of Erwin Schroedinger: His Quantum Statistics of Ideal Gases," *Archive for History of Exact Sciences* 17 (1977): 165–192. Paul A. Hanle, "Indeterminacy before Heisenberg: The Case of Franz Exner and Erwin Schroedinger," *Historical Studies in the Physical Sciences* 10 (1979): 225–269. Walter J. Moore, *Schroedinger: Life and Thought* (1989), abridged as *A Life of Erwin Schroedinger* (1994). Lily E. Kay, *The Molecular Vision of Life* (1993). Lily E. Kay, *Who Wrote the Book of Life?* (2000).

PAUL FORMAN

SCHWANN, Theodor (1810–1882), physiologist, inventor of the cell theory.

Theodor Schwann was born on 7 December 1810 in Neuss, Palatinate, the son of a goldsmith and printer, and died on 11 January 1882 in Cologne. A Catholic, Schwann studied at the Jesuit Gymnasium in Cologne before entering the University of Bonn, where he received a baccalaureate in philosophy in 1831. He then studied medicine at Würzburg (1831–1833) and Berlin (1833–1834) under the supervision of Johannes Peter Müller, a prominent physiologist. After graduating in 1834 with a doctoral thesis on the function of air in the developing chick embryo, he became Müller's assistant and collaborated in the experimental work for Müller's *Handbuch der Physiologie des Menschen* (1833–1840). Schwann devised a "balance" to measure muscle contractions, connected protein digestion to a pepsin-induced fermentation, and co-discovered in 1837 the reproductive process and role of yeast in alcoholic fermentation. But his main contribution to biological science was the *cell theory he framed in 1838–1839.

Cells had often been described by naturalists since first application of the *microscope to organic structures in the second part of the seventeenth century. But the instrumental limitations of the time made it difficult to distinguish reliable observations. Cells seemed to be only one among several types of microstructures, including globules, vesicles, and fibers. The situation changed dramatically with the fabrication of achromatic microscopes in the early 1830s. They allowed the botanist Robert Brown to establish the recurring presence of nuclei in vegetal cells, and many to observe nuclei inside globule-like structures. At the same time, in his "Beiträge zur Phytogenesis" (1838), the botanist Jacob Mathias Schleiden asserted that all plant structures derive from membrane-circumscribed cells, which form in layerlike fashion around nuclei. Schwann had observed similar cells and nuclei in the cartilage and notochord of animal embryos. The coincidence inspired him to try to demonstrate that all elementary plant and animal structures are nucleated membrane-bound cells or cells variously transformed. His cell model implied microstructures that develop successively, and form inside out. A nucleolus appears first and a nucleus forms around it; as the nucleus develops, a membrane enwraps it as a further organic layer. Schwann believed that the nucleolus and nucleus emerge outside (exogenesis) as well as inside (endogenesis) existing cells from organic fluid, the "cytoblastema." In his *Untersuchungen über die Uebereinstimmung in der Struktur und dem Wachstum der Thiere and Pflanzen* (1839), Schwann unveiled the several morphogenetic processes by which all complex organic structures derive from cells through differentiation. His demonstration was not only anatomical but also physiological, since he considered the cell as an elementary organism that displays organic functions: the "life of the cell," namely, its transformation and action, would depend on "plastic" and "metabolic" properties manifested in physical and chemical processes. He expected that observable morphogenetic sequences would follow from the laws of a special kind of "crystallization": at least, that was the notion Schwann promoted in the speculative part of his treatise, which he termed "Theory of Cells." The descriptive elements in Schwann's histological derivation fostered a major research program of contemporary *biology and resulted in the advent of *cytology as the anatomic and physiological science of organic units. In the mid–1850s, though, the principle of cytoblastemic formation was discarded and replaced, thanks to Robert Remak and Rudolf Virchow, by the principle that cells can only derive from preexisting cells. At about the same time, the cell conceived as a nucleated vesicle underwent systematic redefinition as a lump of nucleated protoplasm with or without a surrounding membrane.

After establishing the cell theory, Schwann, deceived in his hope of securing a physiology chair in Berlin, left Prussia for an academic career in Belgium. In 1839 he was appointed professor of anatomy at the University of Louvain; in 1848 he moved to an equivalent chair at the University of Liège, where he taught until his retirement in 1879. He did not contribute much to experimental physiology during his time in Belgium.

Arthur Hughes, *A History of Cytology* (1959). Marcel Florkin, *Naissance et déviation de la théorie cellulaire dans l'oeuvre de Théodore Schwann* (1960). Theodor Schwann, *Microscopical Researches into the Accordance in the Structure and Growth of Animals and Plants* (1969). François Duchesneau, *Genèse de la théorie cellulaire* (1987). John R. Baker, *The Cell Theory: A Restatement, History, and Critique* (1988).

FRANÇOIS DUCHESNEAU

SCIENCE FICTION. Given the human affinity for storytelling, the rise of modern science inevitably prompted narratives involving science. One of the first writers of science fiction was Johannes *Kepler, whose posthumously published *Somnium* (1634) explained and defended the heliocentric model of the solar system by describing the astronomical observations of a man transported to the Moon by demons. Unmemorable as literature, *Somnium* indicated that scientific discussions in fictional contexts might have special virtues.

Despite sporadic attention to figures like Kepler and Jonathan Swift for his satirical *Gulliver's Travels* (1726), historians of science fiction emphasize four figures of the nineteenth and early twentieth century as central to its origins and development. They accept Mary Shelley, author of *Frankenstein* (1818), as the first science fiction writer. Jules Verne achieved popular success with adventurous novels featuring new means of transportation and journeys into exotic realms. H. G. Wells effected a brilliant synthesis of earlier writers' generic models—including Shelley's Gothic horror, Verne's travel tales, Swift's satire, and Edward Bellamy's utopia *Looking Backward* (1888)—to produce several famous novels, including *The Time Machine* (1895), *The Island of Dr. Moreau* (1896), *The Invisible Man* (1897), *The War of the Worlds* (1898), and *First Men in the Moon* (1901), which established patterns for all later science fiction writers. Hugo Gernsback named the genre "science fiction" and transformed it into a recognized category of literature by editing the first science fiction magazine, *Amazing Stories*, beginning in 1926.

Despite Gernsback's lasting influence on the American variety of science fiction, not all writers and editors shared his devotion to science, and the genre expanded and diversified in magazines of the 1930s and 1940s. Like Gernsback, editor John W. Campbell, Jr., of *Astounding Science-Fiction* favored scrupulously scientific stories, or "hard science fiction," but other magazines like *Planet Stories* emphasized the subgenre of "space opera," exciting interplanetary adventures that might display little awareness of scientific realities. Such stories were also prominent in the science fiction films and television programs of the 1950s and 1960s. Despite the genre's burgeoning presence in these and other media like comic books and (later) video games, science fiction enjoyed little success in theater and radio, with exceptions like the depictions of humanoid robots in the play *R.U.R.* (1920) by Czech writer Karel Čapek and the 1938 radio adaptation of Wells's *The War of the Worlds* (1898).

After World War II, English-language science fiction dominated the world, led by major writers like Isaac Asimov, Arthur C. Clarke, Robert A. Heinlein, and Frank Herbert, all capable of crafting both realistic accounts of near-future developments and expansive visions of humanity's distant future. Science fiction that stressed literary values more than science was produced by Ray Bradbury, Kurt Vonnegut, Jr., Philip K. Dick, Ursula K. Le Guin, J. G. Ballard, Harlan Ellison, Samuel R. Delany, and other writers who sometimes labeled their work "speculative fiction." Non-English science fiction usually dwelt more on satire than on science, though intriguing speculations about alien life and technological breakthroughs came from Russia's Boris and Arkady Strugatsky, Poland's Stanislaw Lem, and Japan's Kobo Abe.

In the 1970s and afterwards, the *Star Trek* television series and *Star Wars* films brought new popularity to space opera, in print and film. New schools of science fiction emerged, including outspoken feminists led by Joanna Russ and Pamela Sargent, and "cyberpunks" like William Gibson, Bruce Sterling, and Neal Stephenson, who were fascinated by technology and committed to literary sophistication. Other distinctive new voices like Kim Stanley Robinson and Octavia E. Butler resisted easy categorization.

While science fiction has grown more variegated to appeal to wider audiences, many still believe that science fiction should have a strong relationship with the scientific community. In his essay "Old Legends" (1995), physicist and writer Gregory Benford reports that many scientists—himself, Freeman Dyson, Stephen Hawking, Edward Teller, Steven Weinberg, and researchers of the *Manhattan Project and at Livermore National Laboratory—often read and discussed science fiction stories. Carl Sagan, who later advocated scientific searches for alien life, stood under the stars as a child and longed to be transported to Mars like John Carter, hero of several novels by Edgar Rice Burroughs. Leo Szilard read Wells's *The World Set Free* (1913), which first predicted atomic bombs, during the 1930s, when he was developing the idea of a chain reaction leading to an explosion. Gerald Feinberg first envisioned faster-than-light tachyons after reading "Beep" (1954) by James Blish. And following Kepler's example, some scientists, including Fred Hoyle, Marvin Minsky, Szilard, and Sagan, have moved from reading to writing science fiction.

Gernsback had expressed hopes of institutionalizing science fiction as a stimulus to scientific advances: Writers would send stories with promising ideas to magazines, experts would review submissions to ensure their scientific plausibility, and scientists reading the stories would be inspired to build new inventions. Although Gernsback's plans (which included allowing science fiction writers to patent their ideas) have been ridiculed, the notion of mining science fiction for potentially useful concepts endured. In

2000, the European Space Agency's Innovative Technologies from Science Fiction project enlisted scholars to compile ideas from futuristic stories that might lead to new scientific initiatives. A possible illustration of the process would be the artificial hands used to manipulate radioactive material, named "waldoes" to acknowledge their first appearance in Heinlein's "Waldo" (1942). The terms "astronaut," "genetic engineering," "robotics," and "terraforming" also originated in science fiction stories.

Campbell extended Gernsback's theories to suggest that development of scientific ideas in narrative form could provide scientists and policymakers with important insights not obtainable from everyday scientific activities. Campbell's suggestion might apply to works like Heinlein's "Solution Unsatisfactory" (1940) and Lester del Rey's "Nerves" (1942), which predicted not only atomic energy but also the technical and political problems it might engender.

Along with its potential power to predict future inventions and their effects (which many would argue is illusory or unimportant), science fiction may also encourage worthwhile exercise in scientific thinking. The spectacular but scientifically plausible new worlds created by masters of science fiction like Hal Clement and Larry Niven enable thoughtful readers to critique or expand upon the scientific logic deployed. When Niven posited a huge artificial ring around a star in *Ringworld* (1970), students analyzing the concept determined that such a structure would not be mechanically stable, which prompted Niven to add "stabilizing rockets" to the construct in *Ringworld Engineers* (1979). Other provocative subjects for informed scrutiny surfaced in Hoyle's *The Black Cloud* (1957), featuring an intelligent cloud traveling through space, and Robert F. Forward's *Dragon's Egg* (1980), describing the evolution of life on the surface of a neutron star. Science fiction might also be lauded for sustaining interest in the possibility of time travel during the decades when working scientists dismissed the idea as fanciful.

In addressing the general public, science fiction has affected attitudes toward science and influenced policy decisions. Gernsback hoped science fiction would educate readers about science and encourage them to support scientific progress. At times it has done so. In the 1950s, novels like Clarke's *Prelude to Space* (1951) and films like *Destination Moon* (1950) portrayed and advocated human flight into space; in the 1970s, the American government exploited the popularity of *Star Trek* to publicize the space shuttle. The National Aeronautics and Space Administration (NASA) named its prototypical shuttle the Enterprise after the ship in *Star Trek* and enlisted cast members for promotional films and appearances. H. Bruce Franklin's *War Stars* (1988) suggests that an American tradition of future-war novels featuring successful superweapons may have inspired Harry S. Truman to use atomic weapons.

Science fiction works have also helped to hinder or prevent scientific developments. The nightmarish scenario of a totalitarian government's employing ubiquitous surveillance technology in George Orwell's *Nineteen Eighty-Four* (1947) sparked determined efforts to prevent that future from occurring. Likening plans for space-based antimissile devices to the death rays in the space battles of *Star Wars*, opponents derided the proposals as "Star Wars" systems. And since science fiction, particularly the celluloid variety, has perpetuated stereotypical images of "mad scientists" who thoughtlessly pursue dangerous projects leading to monsters, mutations, and mayhem, researchers seeking to bioengineer new plants and animals are obliged to proceed carefully, anticipating public fears of new "Frankenstein monsters."

Paul Carter, *The Creation of Tomorrow* (1977). Peter Nicholls, David Langford, and Brian Stableford, *The Science of Science Fiction* (1984). Brian W. Aldiss with David Wingrove, *Trillion Year Spree* (1986). H. Bruce Franklin, *War Stars* (1988). Robert Lambourne, Michael Shallis, and Michael Shortland, *Close Encounters? Science and Science Fiction* (1990). John Clute and Peter Nicholls, eds., *The Encyclopedia of Science Fiction* (1993). Gregory Benford, "Old Legends," in *New Legends*, ed. Greg Bear (1995): 292–306. Gary Westfahl, *Cosmic Engineers* (1996). Gary Westfahl, *The Mechanics of Wonder* (1998).

GARY WESTFAHL

SCIENCE WARS. In the so-called science wars, the sciences square off against the humanities in battles over epistemology and methodology, academic politics, and national policy. A recent flare-up in the wars has brought them to widespread public attention, but intermittent skirmishes have appeared throughout the modern period.

Science wars have generally arisen whenever the two sides competed for institutional support, particularly within the university. The emergence of the modern university itself, however, derived from an alliance of science and humanists. In the reform of German universities in the early nineteenth century, professors of the natural sciences and humanities joined forces to promote the philosophy faculty against the traditional faculties of law, theology, and medicine. In the process humanists and scientists submerged their differences and emphasized their common methodology in exact, empirical research. Despite the programmatic unity, however, the German conception of academia placed natural science subordinate to philosophy and the humanities, which dealt with the more important subjects of culture and the human spirit, and excluded utilitarian science and technology altogether. German university reformers also insisted on the principle of academic freedom, which ensured the rights of the individual researcher to an independent voice even while in the employ of the state.

The subsequent evolution of the German research university reversed the relative standings of the sciences and humanities, in large part because of increasing recognition later in the nineteenth century of the practical relevance of science and the need to train larger numbers of students in technical fields, and because of the resultant willingness of the state to furnish scientists with more appointments and better facilities. When other countries, especially the United States, drew on this later German model for their own systems of higher education, they exhibited similar trends. At Cambridge University during World War I, natural scientists leveraged the newfound importance of science for military and industrial strength to launch a program of education reform, including an assault on the perceived domination of the classics in the curricula and adoption of the Ph.D. on the German model. The scientific reformers succeeded in abolishing requirements for Greek and Latin and bolstered the position of science by obtaining state support of research. The reform movement, however, did not always pit science versus the humanities: instead it arrayed modern, utilitarian fields such as modern languages as well as natural sciences and engineering, against the traditional fields of mathematics and classics.

National governments came to appreciate even more the relevance of science to economic and military strength as the twentieth century unfolded. Although scientific contributions to industrial and military technology—especially poison gas in World War I and the atomic bomb in World War II—led humanists to charge that science had run amok and that humanistic studies were necessary to restore morality amid the technoscientific onslaught, public and political appreciation of the intellectual achievements and technological manifestations of scientific research raised the standing of science and oriented academic curricula toward science and engineering. In 1959 C. P. Snow, a British physicist, published an essay entitled *The Two Cultures* on the growing gulf between the sciences and humanities. Although Snow emphasized the need to bridge the gap from both directions, his audience understood his work as promoting the need for greater understanding of science by cultural and political leaders. The background to Snow's essay betrayed the political implications of the separation between the two groups of intellectuals. Snow served on Britain's Civil Service Commission, which at the time was engaged in debates over the preponderance of civil servants with a classical, liberal education instead of specialized technical or scientific training. Critics of the tendency toward generalists over specialists noted the increasing importance of science to the state and decried the exclusion of scientists from policymaking.

Starting in the 1960s a broad romantic rebellion in Western society in general, and the United States in particular, included attacks on science as another form of authority. The countercultural movement criticized the contributions of science, especially the physical sciences, to military technology and industrial pollution and the failure of science to solve pressing social problems. Public protests led to a decline in political support for science and governmental funding for science leveled off or declined in the late 1960s and 1970s. Scientists would weather the storm with their institutional standing battered but intact, but in the meantime an intellectual challenge had arisen within the university. Anthropologists, historians, philosophers, and sociologists, loosely combined in a field known as science studies, challenged the values and methodology of science. The literature of science studies argued that scientific knowledge is socially constructed; it does not refer to objective truth about nature, but instead reflects the rhetorical strategies, power relationships, and political considerations of the scientists who profess it. Hence scientific knowledge should not enjoy a privileged epistemological position with respect to other forms of knowledge—nor, by extension, should scientists with respect to scholars in other fields (*see* ANTISCIENCE).

In the early 1990s scientists reacted against the growing intellectual and institutional presence of science studies. In 1994 two scientists, Paul Gross and Norman Levitt, fired a broadside entitled *Higher Superstition* against the social constructivism of science studies and the relativism of postmodern theory, which they perceived as its philosophical underpinning. Gross and Levitt feared in particular the consequences for universities, where science studies was making increasing inroads at the expense of science, and was presuming to evaluate the work of scientists from a distant disciplinary perspective. The academic freedom of scholars in other fields to choose their subjects did not, according to Gross and Levitt, include the right to analyze work in the

sciences—or, if it did, scientists should then get to evaluate the work of science studies.

Two years later Alan Sokal, a physicist, wrote a parody of a postmodern analysis of quantum gravity, filled with intentional scientific errors and the latest academic jargon, and submitted it to an academic journal noted for its postmodernist approach. The editors of the journal took it as a serious contribution and published it—in an issue dedicated to the science wars, apparently viewing Sokal as one scientist willing to engage in science studies. When Sokal subsequently revealed the hoax the science wars spilled into the pages of the popular press.

The Sokal affair provoked much laughter at the expense of credulous cultural theorists and talk of tempests in academic teapots. But it came at a time when federal support of science in the United States was dwindling with the end of the Cold War and when a conservative Congress was threatening further cuts in government funding of science as well as the humanities. The cancellation in 1993 of the Superconducting Super Collider in the United States provided the immediate context for the renewed science wars. The continuing disparity between federal funding of sciences and the humanities, and the probable influence of science studies literature on federal funding patterns, did not diminish the sensitivity of scientists to any challenge to their intellectual and institutional standing.

C. P. Snow, *The Two Cultures and the Scientific Revolution* (1959). Joseph Ben-David, *The Scientist's Role in Society: A Comparative Study* (1971). Gerald Holton, *Science and Anti-Science* (1993). Paul R. Gross and Norman Levitt, *Higher Superstition: The Academic Left and Its Quarrels with Science* (1994). Zuoyue Wang, "The First World War, Academic Science, and the 'Two Cultures': Educational Reforms at the University of Cambridge," *Minerva* 33 (1995): 107–127. Bruce Robbins and Andrew Ross, eds., *Science Wars* (1996).

PETER J. WESTWICK

SCIENTIFIC DEVELOPMENT, THEORIES OF. All general theories claim to identify patterns in the development of science, but they differ as to whether the significant patterns are found in the atemporal logical structure of the products of science or in its historical process of development, and in whether development equates with progress toward the truth. They also differ in how to explain the patterns and the apparent progress of science. Is it because of the use of a distinctive scientific method or not, and if so, a fixed method or one that changes over time? Does the theory focus on empirical facts or deep, speculative theory, or both? On factors internal or external to science? On special individuals or the special structure of the scientific community? On the nature of social or individual cognition? On the objective constraint that nature imposes on scientific work or the interpretative flexibility of human cultural communities?

Syllogistic logic and Euclidean geometry determined the structure of science from ancient Greece to the Scientific Revolution. For Aristotle, *epistēmē*, or *scientia*, is the set of syllogisms from evident premises to causal-explanatory conclusions. Scientific knowledge is certain, necessary, causal, and pertains to things beyond human control. Science progresses by adding more syllogisms. Logic and method remain fixed and fall outside of science proper, as do the inferior domains of practical reasoning and productive knowledge. Theory and practice are distinct as are the various fields of science. Yet Aristotle's account possessed a remarkable unity, given his teleological metaphysics, with its general theory of change and his integration of inquiry with politics and ethics.

Nicholas *Copernicus attempted to return astronomy to a genuine causal-explanatory science by rejecting the purely instrumental method of hypothesis frequently employed by mathematical astronomers on the ground that the truth of hypotheses is underdetermined by the facts. To assert that successful predictions prove a hypothesis commits the fallacy of affirming the consequent. But for the Copernicans, an internally consistent, heliocentric astronomy, realistically interpreted, was virtually dictated by the observed facts, properly interpreted. The underdetermination problem was a key issue between *Galileo and the Roman Catholic church—and remains central to this day.

Strong empiricists claim that the discovery of a new, empirical method and a rejection of metaphysics produced the *Scientific Revolution, a sharp break from the past that launched modern science. These and similar claims made by rationalists for the Enlightenment have been largely discredited. Natural philosophy emerged only gradually from its past, and recognizably modern scientific communities did not appear until the nineteenth century.

Francis Bacon was celebrated as the founder of the method of induction: clear the mind of all prejudice, gather all the relevant facts, arrange them into tables, notice positive and negative associations, induce patterns, and thus discover laws (*see* BACONIANISM). Science proceeds bottom-up; fact gathering precedes theory. Sometimes Bacon is portrayed as an eliminative inductivist: from the tables, formulate all possible laws, then seek facts that refute all but one candidate, which must be the true one. Both accounts whiggishly ignore the medieval and *Renaissance

elements in Bacon's thinking. His emphasis on experiment, control of nature, and practical results, however, strongly challenged Aristotle's distinction of genuine science from practical and productive action.

René *Descartes claimed to provide a general deductive method of discovering first principles, based on reason. Experiment fills in details. Thus conservation of motion follows from God's immutability and nondeceitfulness. Science proceeds top-down, both logically and temporally, from high principle to experiment. Descartes never specified the method in usable detail, and in practice he proposed numerous mechanical hypotheses. But his work in algebraic mathematics helped to displace the syllogism as the preferred mode of reasoning and Euclidean geometry as the prime form of mathematics.

Isaac *Newton in turn rejected Cartesian speculation, saying "I do not feign hypotheses," and attempted to return science to a higher standard by his own method of "deduction from the phenomena." But how to achieve causal explanation without hypotheses or metaphysical assumptions? Newton's answer: From Johannes Kepler's laws I can demonstrate the *existence* of universal gravitation, although I am unable to explain the essential nature of gravitation. He thus claimed to resolve most of the tension between the Aristotelian demands for demonstrative certainty and causal explanation, the latter now understood as deep-structural causation, making the tension all the worse. Although Newton employed hypotheses in the process of working out his mechanics and optics, he set out his final theories in Euclidean geometrical form.

The nineteenth century brought a profound logical turn in the dominant methodology of science. Most methodologists gave up the quest for certain foundations and the corresponding cumulative model of scientific progress. They made science out to be fallible and self-correcting, a process of successive refinement. With immediate truth no longer the goal, the method of hypothesis gradually replaced both Baconian inductivism and Newtonian deduction from the phenomena. Facts were now chiefly important as observable logical consequences of hypotheses to be tested rather than as antecedent premises yielding the claim as a conclusion. William Whewell and John Herschel were transitional figures. Whewell combined elements of Bacon and Immanuel Kant, but his main innovation was to derive his philosophy of science from the history of science. Unlike Bacon, he could draw upon two centuries of scientific progress. In 1837 Whewell published the first comprehensive history of science, *History of the*

Inductive Sciences, followed in 1840 by *Philosophy of the Inductive Sciences Founded upon Their History*. In a controversy with John Stuart Mill, Whewell argued that simple empiricism is not enough. The lawful patterns in empirical data are not simply *given* to us in experience but are also a product of how the mind *takes* the experience by bringing it under a conception that "colligates" the facts. Whewell also stressed the epistemic importance of "consilience of inductions," including novel prediction, as when a theory designed for one purpose makes correct predictions in another area or theoretically unites the two.

At the end of the nineteenth century in France, the physicist Pierre Duhem more fully articulated the underdetermination argument and concluded that empirical science could never determine the Aristotelian, metaphysical truth about the universe. The mathematician Henri Poincaré stated that the most basic principles of space, time, and mechanics are conventional, not determined fully by the evidence, and that scientists would always choose to retain classical mechanics and Euclidean geometry for their convenience. Meanwhile, in America, Charles Sanders Peirce defended the long-run convergence upon the truth of self-corrective, fallibilistic science; and he blurred the distinction between method and empirical content, "for each chief step in science has been a lesson in logic." Bacon had previously said that "the art of discovery advances as discoveries advance." Method progresses as science progresses.

In the twentieth century, both philosophy of science and history of science became professional disciplines. The logical positivists—Moritz Schlick, Rudolf Carnap, Otto Neurath, Hans Reichenbach, Carl Gustav Hempel, and others—inspired by the new symbolic logic and the radical empiricism they found in David Hume and Ernst Mach, developed a strongly empiricist, "verifiability theory" of meaning and justification in order to banish metaphysics and to secure a true scientific enlightenment. They aimed to articulate the nonhistorical logical structure of theories, confirmation, explanation, and so on, and they sharply distinguished logic, method, and philosophical analysis from empirical claims. For them, science progresses in a cumulative manner; facts and low-level laws are the primary carriers of knowledge. As positivism mellowed into logical empiricism, theories and theoretical explanation gained in importance.

Karl Popper's approach was similar, although he rejected verifiability as impossible on Humean grounds and championed falsification in place of verification. He readmitted metaphysics as the intellectual root of many deep problems

that later became scientific. Popper defended a fallibilistic, hypothetico-deductive model of science that maximizes criticism and leads to perpetual revolution. Like the positivists, he focused on the logical structure of the products of science and denied that anything methodologically interesting can be said about the creative process of "discovery." Contrary to inductivism, it does not matter how we hit upon hypotheses, only how they are tested.

Thomas S. Kuhn's *Structure of Scientific Revolutions* dismissed these static, logical models of scientific knowledge and highlighted historical but noncumulative patterns of change. For Kuhn, mature normal science under a paradigm is closed, minimally innovative, dogmatic, convergent, tradition-bound, and anti-Enlightenment, even quasimedieval, in character. Eventually, the accumulation of anomalies produces a crisis that may lead to the wrenching discontinuity of a paradigm-shattering scientific revolution. This *revolutionary* model of long periods of normal science occasionally punctuated by discontinuity stands in some tension with Kuhn's *evolutionary* model, according to which science develops and ramifies with identifiable lineages. On neither model does science progress toward a unique truth about the universe waiting to be found.

The apparent relativism of Kuhn's views on the social-communal nature of justification, his emphasis on the underdetermination of decisions by logic plus data and on the tacit knowledge of skilled practice, and his call for empirical study of the process of science all helped to inspire postmodern science studies. Kuhn's account of concept learning and problem solving as pattern matching using an acquired similarity relation signaled a turn from logic to rhetoric as the key to both cognition and theory structure. Thinking is not simply applied logic, and theories are not deductive logical structures but clusters of exemplars or models held together by analogy and similarity. Kuhn's attack on Popper and the positivists led to a debate among several competing "big systems": those of the positivists, Popper, Kuhn, Paul Feyerabend, Stephen Toulmin, Imre Lakatos, and Larry Laudan.

Lakatos's "methodology of scientific research programmes" combined elements of Popper and Kuhn. Initially a Popperian with Hegelian tendencies, Lakatos followed Kuhn in moving to larger, dynamic, historical units of and for analysis—entire research programs instead of individual theories. A research program consists of a series of successively more sophisticated theories governed by a negative heuristic and a positive heuristic. The negative heuristic is a "hard core" of basic principles, for example, Newton's laws, protected from falsification by a belt of auxiliary hypotheses. The positive heuristic, with its metaphysical motivation, guides future planning and constructs the protective belt by anticipating anomalies. Research programs compete with one another to see which can produce the most theoretical and predictive results.

The turn to historical models of development owed much to the maturation of internalist history of science. In turn, the emergence of social history and sociology of science brought new, more externalist, and local models of scientific development such as the "interest" theory of the Edinburgh Strong Programme in sociology of knowledge and the actor-network theory of Bruno Latour and Michel Callon. Derek J. De Solla Price had demonstrated the exponential growth of science until the 1970s and thereby encouraged the formulation of economic and policy-oriented models of science. Today we emphasize the great cultural and topical diversity of scientific work and no longer make physics the model for all science. In short, the day of the big systems has seemingly passed. Gone also are the sharp distinctions between pure science and technology and between empirical content and methods or techniques.

Meanwhile, scientists themselves have proposed interesting models of scientific development. Most of the older models are vaguely positivist. More recently, physicist Steven Weinberg defends the strongly "Copernican" view that physics is moving toward a final, highly reductive, aesthetically pleasing, unitary Theory of Everything, where no internal changes will be possible without inconsistency. Gerald Holton offers a more historical model of scientific change, according to which the scientific work of an age or a generation is shaped by salient members of a set of "themata": conceptual and methodological resources that are imaginative and metaphysical, because untestable, rather than either empirical or analytical. Themata typically come in thema-antithema pairs, like continuous versus discrete, causal-mechanistic versus nonmechanistic, teleological versus nonteleological. Holton holds that there exist only a few dozen basic themata, most inherited from the ancient Greeks. Niels *Bohr's complementarity amounts to a new thema that combines a former, seemingly contradictory pair. Several of these themata manifest themselves in models of scientific change, for example the search for unitary theory versus an antireductive emphasis on diversity.

Imre Lakatos and Alan Musgrave, eds., *Criticism and the Growth of Knowledge* (1970). Frederick Suppe, *The Structure of Scientific Theories*, 2d ed. (1977). Larry Laudan, *Science and Hypothesis* (1981). Helge Kragh, *An Introduction to the Historiography of Science* (1987). Barry Gower, *Scientific Method: An Historical and Philosophical Introduction* (1997). Thomas Nickles, ed., *Thomas Kuhn* (2002).

THOMAS NICKLES

SCIENTIFIC METHOD. See METHOD, SCIENTIFIC.

SCIENTIFIC REVOLUTION. The concept of the Scientific Revolution was introduced in the late 1930s to give analytical unity to a historical period regarded as pivotal in the rise of modern science. Although the period is defined variously, it always includes the seventeenth century. *Galileo Galilei, Johannes *Kepler, and René *Descartes mark an early stage and Isaac *Newton's *Principia* of 1687 brings the revolution to a provisional end.

The constituent element in the Scientific Revolution most clearly recognizable from a present-day point of view is the mathematical treatment of natural phenomena. Around 1600 Kepler subjected planetary trajectories to a unique blend of mathematical-physical-musical analysis yielding, as a by-product, his famous three descriptive laws. Guided likewise by a conviction that our world is at bottom mathematical, Galileo, in employing a new idea of motion retained (inertial motion) rather than spontaneous return to rest, arrived via a maze of reconceptualizations and trial-and-error experiments at his laws of falling and projected bodies.

Both men worked at mechanical concepts to make stronger and more persuasive the hypothesis of a moving Earth proposed half a century earlier by Nicholas *Copernicus. Its long drawn-out victory, owing in good part to Galileo's telescopic observations and to his and others' answers to standing objections, is among the most significant and enduring accomplishments of the Scientific Revolution. Another capital accomplishment was the anchoring of Galileo's highly abstract and counterintuitive conception of motion, heatedly debated subsequently and also extended to topics like flood control and air pressure (*see* VACUUM), in empirical reality through the systematic application of experiment. The connection did not always go smoothly. To take a refined example, Evangelista Torricelli ran into trouble fitting fact to his theorem (derived in analogy to Galileo's law of free fall) that the speed with which a jet of water squirts from a hole in a container is proportional to the square root of the distance between that hole and the water's surface. He had no recourse but to declare the uncontrollable circumstances of his experimental set-up to be irremovable impediments to a satisfactory match between the mathematically ideal case and experience.

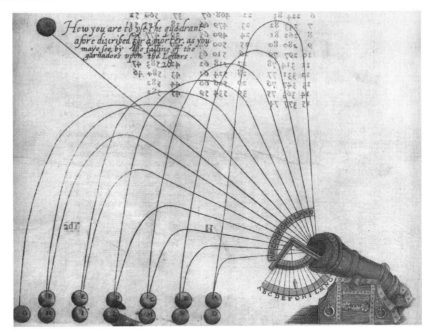

A typical piece of 17th century applied mathematics, one of the main weapons of the new science. It indicates that the greatest range for the same shot and charge come at an angle of 45° and that the short fall at equal angles on either side of 45° is the same.

A different sort of experiment, directed not toward the empirical confirmation of abstract, mathematical regularities but toward discovering hidden properties of nature, also developed during the first decades of the century. This activity usually went forward under the aegis of Francis Bacon's call for a new knowledge of nature based on "natural histories" listing readily observable properties of heat, air, plant life, etc. William Gilbert's study of magnetism and the investigation of chemical reactions by Paracelsists like Jean Baptiste van Helmont supplied models and precedents. These painstaking descriptions of natural phenomena went beyond the traditional goal of understanding. They also aimed at the "improvement of man's estate"—an objective in conformity with *Baconianism and its associations with craft traditions and a spiritual-magical view of the world. A Baconian natural history of sound would include whispering galleries, the making of musical instruments, and the construction of hearing aids for the deaf.

Besides the piecemeal approaches of the mathematicians and natural historians, the early decades of the Scientific Revolution witnessed the creation by Isaac Beeckman, Descartes, Pierre Gassendi, and Thomas Hobbes of philosophies of nature intended to explain the totality of natural phenomena. They reduced natural processes to particles of matter of various shapes and sizes moving in accordance with such laws of motion as were enunciated in most systematic and appealing fashion by Descartes. For example, he explained the fall of bodies by positing a whirlpool of tiny particles of matter surrounding the earth and turning with it. The centrifugal motion of these particles overcame that of gross bodies in the atmosphere and pushed them down.

Whatever divided these innovators, they all shared an urge to find an alternative to the received worldview, that of Aristotle, with its unmediated, commonsense approach to natural phenomena. Although the three novel modes of pursuing natural knowledge spread quickly, sometimes hedged or qualified, most scholars stuck to traditional philosophies. Examples of hedging are Marin Mersenne's ever-skeptical, half-heartedly Galilean experimentation, and the preoccupations of Jesuits with a quantitative variety of *Aristotelianism or a blend of magico-Aristotelian experimentation. Outright rejection came mainly from the arts faculties of Europe's universities. Grounds for rejection, however, stretched far beyond disputes over the best mode of interpreting nature. They touched fundamental worldviews, which, in a religious time, were naturally debated in religious terms. The worldview implied in the Cartesian as well as the Galilean conception of things was a clockwork universe constructed and set in motion by God but left for regulation to the impersonal working of the laws of nature. Not only did the mathematical approach to natural phenomena present an almost incomprehensibly strange and esoteric mode of acquiring knowledge of nature—humankind's view of the cosmos at large and of its own place in it appeared to be at stake. Within decades the harmony attained centuries before between Christian religion and the pursuit of knowledge of nature in the Aristotelian mode went out of tune, as Galileo's trial and controversies over Descartes's philosophy demonstrated spectacularly.

Resolution came from further novelty. Not only had conceptual breakthroughs of the first half of the century been grafted onto Greek traditions and their medieval and Renaissance refinements, but in most other relevant respects continuity also prevailed. Court patronage remained a primary source of institutional support; exchange of views and information still took place largely through travel, books, and letters; and the mathematical, natural-philosophical, and natural-history modes themselves were kept apart, just as their counterparts had in the venerated Greek antiquity that continued to provide the standard of reference for whatever new was taking place in a not yet modern Europe.

These matters changed radically over the second half of the seventeenth century. The printing press, exile owing to religious wars, and the development of absolutist forms of state patronage leading to the establishment of scientific societies together made possible unprecedented rates and modes of interaction that drew the Galilean, Cartesian/Gassendist, and Baconian approaches together. Outside Galileo's Italy the upset balance between religion and what we now call the new science was resolved with the help of the ideology of useful knowledge underlying the Baconian approach. The ongoing movement of renewal of knowledge of nature began to be widely perceived not as sacrilegious but as directed toward beneficial ends sanctioned by Christian faith. This perception too was an act of faith, since the vision of a science-based technology leading to the extension and improvement of human life was not fulfilled for well over a century.

A range of novel tools—*telescope, *microscope, *barometer, *air pump—sped up empirical research and altered it in kind. They made possible the analysis of light rays in telescopes (Christiaan *Huygens, Newton) and made actual the old question whether void spaces can and do exist (Blaise Pascal, Otto von Guericke, Robert *Boyle). The Galilean approach, reinforced by the corpuscularian doctrine taken hypothetically,

not dogmatically, produced in the hands of Huygens and Newton a new understanding of light and also, through increasingly sophisticated mathematical tools, the analysis of vibrational motion, centrifugal motion, and impact. The Baconian approach, reinforced likewise by the corpuscularian doctrine taken hypothetically not dogmatically, produced in the hands of Boyle, Robert *Hooke, and Newton sophisticated patterns of experimentation (Newton's work on colors) guided by subtle mechanisms of particle motion while allowing some room for the working of active principles as well. These various developments came together in the hands of Newton, notably in his *Principia*. There we find the diverse theories of motion established over past decades assembled in a first statement of the discipline of rational mechanics. There we find Galileo's and Kepler's laws jointly derived from a unifying principle—motion on Earth and in the heavens both subjected to a mathematical law of universal gravitation. There we find those as yet barely articulated active principles turned into clearly distinguished, specific forces. There we find the first fruits of the calculus employed in the mathematical treatment of nonuniform motion. In accomplishing all this and much more, the book did what every truly great human achievement has always done. In closing one period, that of the Scientific Revolution, it opened a new one, that of a more truly modern science.

Richard S. Westfall, *The Construction of Modern Science* (1971). David C. Lindberg and Robert S. Westman, eds., *Reappraisals of the Scientific Revolution* (1990). H. Floris Cohen, *The Scientific Revolution: A Historiographical Inquiry* (1994). John Henry, *The Scientific Revolution and the Origins of Modern Science* (1997). Wilbur Applebaum, ed., *Encyclopedia of the Scientific Revolution* (2000). Margaret J. Osler, ed., *Rethinking the Scientific Revolution* (2000).

H. FLORIS COHEN

SCIENTIFIC REVOLUTIONS are abrupt and convulsive stages of scientific advance. The idea that science moves forward in cycles of revolutionary discontinuities seems at odds with the widespread notion that in the *history of science there has been just one major break, known since the late 1930s as the Scientific Revolution of the seventeenth century. The idea of scientific revolutions also runs contrary to the positivist view, which dominated between 1830 and 1970, of scientific advance as a gradual, cumulative enterprise proceeding in a rectilinear, progressive way. Positivists held that new insights arise by heaping bits of knowledge on earlier pieces through the faithful application of sound scientific method. Still, spectacular events with a profound impact, such as the triumph of heliocentrism or the shift from a phlogiston-to an oxygen-based chemistry, seem to many still to deserve celebration as the "Copernican" or the "Chemical" Revolution (*see* POSITIVISM AND SCIENTISM).

During the early eighteenth century, "revolution" came to denote radical progress rather than "cycle." Bound up with political revolutions, the metaphor of revolutions in science has since spread to the point of robbing it of more and more of its content. Much scientific creativity is now regularly celebrated by attaching the label "revolutionary" to it—for example, *relativity and *quantum physics, the structure of DNA, or many an advance in computer science.

A big conceptual step stands between such loose labeling and a comprehensive view of the process of science as essentially revolutionary. One early theorist of specific mechanisms for regular sequences of scientific revolutions was William Whewell, who between 1830 and 1850 organized his comprehensive history and philosophy of all "inductive sciences" around the idea that science advances through the binding together of ever more facts by means of new, sometimes revolutionary, unifying conceptions. By far the best known mechanism held to underlie revolutionary scientific change is that proposed by Thomas S. Kuhn in his *Structure of Scientific Revolutions* of 1962, which also combines in an ingenious way elements of continuity and discontinuity, of gradual and revolutionary advance. Scientific domains are governed by "paradigms," specialized packages of concepts, theories, and practices laid down in textbooks, which express the standing in a community of practitioners and guide their further research, which Kuhn called "puzzle solving." Pre-paradigmatic domains lack consensus (e.g., conceptions of motion in ancient Greece, or the social sciences today). Kuhn dubbed research guided by a paradigm "normal science." Revolutionary change takes place when, as a result of the accretion of more and more anomalies, ongoing puzzle solving appears (at first to a few individuals only) to break down. Out of the crisis a new paradigm arises, usually incompatible with the old one. Science is cumulative only in the framework of a paradigm.

Kuhn anchored his schema on the best existing work in the historiography of science, which he thought was opposed to the then-dominant positivist image of science. Positivist historiography has long since been superseded, owing in part to Kuhn's own example as a historian of science but not to his schema of scientific revolutions. Few historians of science appear to have found it helpful for their own understanding of the development of past science. Among

historians, Kuhn's influence has been much more noticeable in by-products of his schema such as shifting "disciplinary boundaries" between scientific domains and "scientific communities" held together by common practices. Philosophers of science have debated whether the incommensurability between successive paradigms posited by Kuhn is a necessary ingredient of revolutionary science. Among students of science at large, Kuhn's conception of scientific revolutions has been a major landmark among shifting conceptions of what scientists do and what science is. The shifts reflect changing perceptions in society at large of science as the benefactor or nemesis of humankind, or as something in between.

I. Bernard Cohen, *Revolution in Science* (1985). Paul Hoyningen-Huene, *Reconstructing Scientific Revolutions* (1993). H. Floris Cohen, *The Scientific Revolution: A Historiographical Inquiry* (1994).

H. FLORIS COHEN

SCIENTISM. See POSITIVISM AND SCIENTISM.

SCIENTIST. The term "scientist" was coined by the Cambridge don and polymath William Whewell to designate a person interested in promoting natural knowledge. He introduced the term in a review of Mary Somerville's *On the Connexion of the Physical Sciences* in 1834. Many other technical terms still in use come from him: "electrode," "anode," "cathode," and "ion," in response to Michael *Faraday's needs for words to describe his electrolytic experiments; "Pleistocene," "Pliocene," and "Miocene," to help geologists keep track of epochs; and several more. In marked contrast to current practice (*see* TERMINOLOGY IN SCIENCE), he grafted his neologisms on ancient stock. This, however, did not preserve him from posthumous condemnation by the great lexicographer Henry Watson Fowler for perpetuating "regrettable barbarisms."

Whewell concocted "scientist" for the British Association for the Advancement of Science (BA), founded in 1832, so that the participants in its annual meetings and eatings (the banquets being the true "spread of science" according to Charles Dickens) would have a common label. The BA floated "philosopher," but rejected it as too wide and lofty; "savant" (too pretentious and French); and "nature-peeper" or "nature-poker," after the *Gesellschaft der deutschen Naturforscher und Ärzte*, one of the antecedents of the BA (too ridiculous). "Scientist" also met with disapproval. Like "physicist," which Whewell invented at the same time and which Faraday dismissed as unpronounceable, it did not appeal to its beneficiaries. "Un-English, unpleasing, and meaningless," sniffed William Thomson, Lord Kelvin, of "physicist"; as for "scientist," Thomas

Henry Huxley declared it "about as pleasing a word as 'Electrocution.'"

British "men of science" or "scientific men" (the terms they preferred) resisted "scientist" primarily because of its association with practice. It resonated with "dentist," a paid professional in a nasty business, connotations they wanted to avoid. Its miscegenation of Latin and Greek further indicated its low breeding as did its supposed origin in the United States. Not until well into the twentieth century did "scientist" oust "man of science" as the preferred term for professionalized, Americanized, classically uneducated nature pokers.

"Scientist" won out because, despite the wide differences among the sciences, their practitioners came to feel a need to band together to obtain social recognition and financial support. Scientists in non-English–speaking countries felt the same need. The French made a substantive of the old adjective "scientifique" to distinguish practitioners of natural science from savants in general. The first noted occurrence of this usage, which would become standard French for "scientist," dates from 1884. The Germans responded with "Naturwissenschaftler." Further indications of the success of Whewell's coinage may be seen in the eagerness of cultivators of non-natural knowledge to claim it. Today there are political scientists expert in the art of politics, economic scientists eligible for *Nobel Prizes, and Christian Scientists opposed to medicine. French "scientiste" and Italian "scientisto" signify both a practitioner of Christian Science and an exponent of scientism, a doubter and an overly enthusiastic admirer of modern science.

Charles Dickens, "The Mudfog Association for the Advancement of Everything," in *Sketches by Boz* (1836). Sydney Ross, "Scientist: The Story of a Word," *Annals of Science* 18 (1962): 65–85. Jack Morrel and Arnold Thackray, *Gentlemen of Science: Early Years of the British Association for the Advancement of Science* (1981).

J. L. HEILBRON

SECRECY IN SCIENCE. The ideal of open communication entered science in the sixteenth and seventeenth centuries. Previously, philosophical, social, and economic factors limited the free circulation of knowledge about nature. Pythagorean and Aristotelian traditions distinguished esoteric from popular knowledge and restricted the former, which included natural philosophy, to elite disciples. The Hermetic philosophy that flourished in the *Renaissance and embraced alchemy and other mystical systems particularly reserved the secrets of nature for initiates. The guild system discouraged the dissemination of craft techniques beyond the artisan's workshop and the absence

of intellectual property rights deterred engineers and inventors from publishing their work; instead they composed their treatises in cipher or, as Leonardo da Vinci did, in mirror writing, to foil copycats.

Several developments during the *Scientific Revolution combined to overcome secrecy. The technology of the printing press allowed the wide availability of work and encouraged the ethic of publication, a step that would culminate in the current academic doctrine of publish or perish. Italian city-states first implemented patent laws in the fifteenth century, which proved so fruitful that legal systems for patent protection spread throughout Europe in the sixteenth century and eventually extended property rights to intellectual creations through copyrights. Finally, new scientific societies, such as the Royal Society of London and the Académie Royale des Sciences in Paris, both founded in the mid-seventeenth century, provided the practical avenue of publication through their journals and promoted the public ideal of communication.

Secrecy persisted in modern science in three forms. Personal secrecy stemmed from the reward system of the scientific community, which emphasized priority of individual discovery and encouraged scientists to keep unpublished results to themselves; otherwise, a colleague could learn about work in progress and get credit for it by publishing first. Industrial secrecy derived from the rewards of proprietary priority in capitalism. Although the patent system encouraged publication once a patent was secured, until then industrial researchers exercised discretion. Thus, for example, proprietary concerns prevented researchers in the nascent *radio industry of the early twentieth century from publishing all their results and induced AT&T to compartmentalize information within its laboratories. Similar proprietary concerns stifled *peer review and fostered duplication in genomics research at the end of the twentieth century.

Secrecy also stems from the relevance of science to national security. In a precocious example, Antoine-Laurent *Lavoisier and other French chemists worked on new forms of explosives in a secret weapons laboratory they set up during the French Revolution. As this example demonstrates, secrecy is not necessarily forced on scientists by national governments. The excusable ignorance of military and political leaders of the details of the latest scientific and technical developments, combined with the usual technological conservatism of the military, has often left scientists to assume the initiative in developing new military technologies and keeping them secret. To take another example, after the discovery of nuclear fission in 1939 nuclear physicists in the United States imposed self-censorship on further work in their field in order to keep results with military potential away from the Germans. American nuclear scientists eventually consented to work on atomic weapons in another secret lab, Los Alamos. The United States and the other belligerent nations created similar secret labs for other technologies during the war and also classified work done under government auspices outside the labs as secret. The contributions of science to warfare in World War II convinced several governments to continue the support of research labs for national security and thus institutionalized secrecy in science.

Scientists in these secret institutions evolved classified conferences, journals, and textbooks to reproduce elements of the open scientific community. The participation of universities and industrial firms and their scientists in U.S. military research made secrecy a pervasive feature of postwar science and exposed scientists to charges of espionage. Other nations also imposed secrecy on national security research, thus undermining internationalism and isolating scientists in national communities; secret research programs in each country proceeded in ignorance of similar work done elsewhere. Scientists agreed to work under these conditions out of patriotism and pragmatic recognition of the demands of their sponsors. Secrecy for national security survived the end of the Cold War, as national governments continued to value the contributions of science to national military and industrial strength and scientists continued to accept the tradeoff to ensure support for their research.

David Hull, "Openness and Secrecy in Science: Their Origins and Limitations," and Ernan McMullin, "Openness and Secrecy in Science: Some Notes on Early History," *Science, Technology, & Human Values* 10: 2 (1985): 4–13; 14–23. William Eamon, "From the Secrets of Nature to Public Knowledge," in *Reappraisals of the Scientific Revolution*, ed. David C. Lindberg and Robert S. Westman (1990): 333–365. Peter J. Westwick, "Secret Science: A Classified Community in the National Labs," *Minerva* 55 (2001): 363–391.

PETER J. WESTWICK

SECRETION. The most spectacular physiological discovery of the seventeenth century, the circulation of the *blood, reordered functions such as *nutrition and the distribution of the heat and vital spirits thought essential to the maintenance of life. Another set of discoveries following in the next decades, however, connected the circulating blood with a new general class of functions, the secretions. Although since antiquity the kidneys had been known to separate urine from the

blood, and the liver to separate bile, the functions of the pancreas, thyroid, and other glands were unknown. Galen, who strove to attribute a function to every anatomical feature of the body, could think of nothing better for the thyroid and the others than to cushion the structures they surrounded (*see* GALENISM).

In 1643 Johann Georg Wirsung found a duct connecting the pancreas with the duodenum of the small intestine. In 1656 Thomas Wharton published his discovery of a duct of the submaxillary gland, through which, he asserted, saliva is secreted into the mouth. Wharton thought that the blood entering the gland furnished material to the secretion, but that some of it came also from a nervous fluid flowing. Nicolaus Steno added in 1662 the discovery of the ducts of the parotid gland, the sublingual gland, the small buccal gland, and the tear glands, and a more general theory of secretions. The material for any secretion, he believed, comes from the arterial blood entering the beginning of the duct where the blood passes from the arteries into the veins. The nerves supplying the glands do not contribute material, but control the flow of secretions by constricting the veins so as to divert more material into the duct.

Marcello Malpighi, the first person to apply the compound microscope extensively to the study of animal tissue, extended the conception of secretions to structures invisible to the naked eye. The substance of the liver, he observed, contains many microscopic lobules, each of which is thoroughly irrigated with blood vessels supplied by the portal vein. Malpighi inferred that each lobule amounts to a "conglomerate" gland such as the pancreas, and that each must have a secretory duct connected to the bile duct. In the kidney he discovered equivalent bodies (the Malpighian corpuscles), the tuft of blood vessels surrounding them, and the structure of the kidney tubules. Carrying his observations further, Malpighi thought he observed similar glandular structures filling the brain, and assigned to them an analogous secretory function.

The discovery of the general secretory function raised the question how the materials composing the secretions were selectively removed or transformed from the blood. Mechanistically oriented philosophers like René *Descartes and Giovanni Alfonso Borelli imagined sieve-like structures at the entrances to the secretory ducts permitting only particles of appropriate size and shape to pass through. The Newtonian physiologist James Keill proposed in 1717 an explanation based on short-range attractive forces between particles in the blood. None of these explanations was specific enough to explain either the detailed anatomy of the glands or the specific properties of individual secretions, however, and they helped provoke the vitalistic reactions of the later eighteenth century led by the French physician Théophile de Bordeu.

Up until the early nineteenth century physiologists continued to debate whether the special substances comprising the secretions were formed in the blood and separated in its passage through the glands, or prepared by some chemical transformation. The *cell theory transferred the problem. Noting in 1839 that all the mucus membranes that separated the arteries and veins of the secretory organs from their ducts were cellular structures, Jacob Henle inferred that the secretions must be produced in cells that burst to release their contents into the secretory ducts. Meanwhile, new chemical methods helped to characterize the properties of individual secretions, notably in Theodor *Schwann's identification in 1836 of the "ferment" pepsin in the gastric secretions.

For two centuries the definition of a secretory organ included the duct through which the secretion flowed. This conception left mysterious the functions of a number of long-known organs, such as the thyroid, pituitary, and thymus bodies, and the "suprarenal capsule" that were generally similar in structure to the other glands except that they lacked ducts. Late in the nineteenth century the idea spread that these glands secreted some particular substance *into* the blood. The search for such secretions and the identification of their functions accelerated rapidly at the beginning of the twentieth century to become the investigative field of the newly emerging subspecialty of endocrinology.

Michael Foster, *Lectures on the History of Physiology During the Sixteenth, Seventeenth and Eighteenth Centuries* (1924) 83–119. John F. Fulton and Leonard G. Wilson, eds., *Selected Readings in the History of Physiology* (1966) 410–444.

FREDERIC LAWRENCE HOLMES

SEISMOLOGY, a branch of geophysics, examines the behavior and products of elastic (seismic) waves traveling within the earth. Earthquakes are the most significant generators of these waves. Other sources include volcanic eruptions, explosions (including nuclear explosions), and meteorite impacts. Trucks, trains, and thunder produce seismic "noise." The discipline of seismology covers documentation of events and effects (observations, maps, catalogs), instrumentation and analysis, theory and application. Because large earthquakes can be detected worldwide, the science is international.

Since ancient civilizations arose in earthquake-prone country, a long history of observation and speculation about earthquakes exists. Aristotle proposed a classification of earthquakes while the Chinese philosopher Chang Heng is credited with designing an inertial seismoscope in A.D. 132

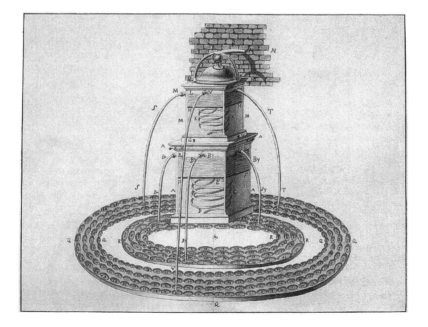

Seismic recorder invented by Atansio Cavalli (1785): shaking the wall causes the mercury-filled basins to oscillate on their springs; mercury, jetting out to a distance determined by the severity of the jolt, is caught in little cups moved around by clockwork.

that could determine the source direction of an earthquake. Many early treatises attributed earthquakes to the movement of air or water vapor within the earth. The role of rock fracturing in earthquake generation was not realized until the mid-nineteenth century.

The great Lisbon earthquake and tsunami (tidal wave) of 1755 mark the beginning of the systematic study of earthquakes in western science. For the next century or so, studies of earthquakes consisted primarily of observations of how earthquakes behaved, their geological effects, catalogs of historical events, and continued speculation about their causes. By the early eighteenth century, inertial seismoscopes in Europe included pendulums and bowls of liquid mercury. Earthquake catalogs and other regular observations of earthquakes were widely compiled. By 1840 enough information had accumulated for Karl Ernst Adolf von Hoff to produce the first global historical catalog of earthquakes.

Around 1850, seismology began to take shape as a separate field of inquiry. Robert Mallet studied the great Italian earthquake of 1857 and wrote the landmark *First Principles of Observational Seismology*. Johann J. Noggerath in 1847 first used isoseismals to map earthquake intensity, Luigi Palmieri produced an electromagnetic seismograph first used in 1856, and M. S. de Rossi and François Alphonse Forel cooperated to publish the first widely used, standardized intensity scale in 1883. By the end of the century, scientists were designing the first reliable seismographs (recording instruments). They sought ways to reduce friction between the recording needle and recording paper, to damp triggered motion in the instrument, and to eliminate local disturbance of the mechanism. In 1892 John Milne in Japan produced the first compact, simple (though still not entirely accurate) seismograph. Boris Golitsyn's galvanometric seismograph, perfected in 1911, suspended the *pendulum in an electromagnetic field, but largely independently, scientists developed the basic theory of elastic wave behavior. Simeon-Denis Poisson worked out a theory for the primary and secondary (P and S) waves; George Gabriel Stokes and Lord Rayleigh made further contributions; and the effort culminated with the publication of theoretical work on surface and other seismic waves by Horace Lamb in 1906 and A. E. H. Love in 1911.

By the early twentieth century, international cooperation and the standardization of observations worldwide, including accurate timing mechanisms and travel-time tables, allowed the development of new standard scales of magnitude and detailed maps of global seismicity. Using Milne's seismograph scientists in the British empire had set up the first uniform, international network. In the 1930s Charles Francis Richter and Beno Gutenberg developed a standard scale to measure the relative sizes of earthquake sources, commonly called the Richter scale. It is one of several magnitude scales in use today. Gutenberg and Richter also wrote textbooks that became standards in the field. Seismology figured among the founding six sections at the first meeting of the International Union of Geodesy and Geophysics (IUGG) in 1922. The International Association of Seismology and Physics of the Earth's Interior (the IASPEI), a branch of the IUGG, continues to coordinate international seismological research.

The twentieth century saw major advances in geophysics as a result of the accumulation of seismic data and analysis. In the first half of the century seismologists and geophysicists such as Gutenberg, Richter, Hugo Benioff, Inge Lehmann, Harold Jeffreys, and Francis Birch delineated the interior, layered structure of the earth. By the end of the century investigators were mapping heterogeneity in the earth's mantle and its boundaries using a method called seismic tomography. A second important spin-off followed from the more accurate location (both geographically and within the earth) of earthquake sources and the differentiation of fault motions, known as focal mechanisms. This information played a major role in the development of the theory of *plate tectonics. A third area of importance was the investigation of the structure of the crust itself. During World War II Maurice Ewing and others developed the technology to make seismic investigations offshore. This allowed the mapping of layered strata beneath the sea floor. The investigation of the earth's crustal structure, at ever greater depths and resolution, has continued to the present day, particularly in the field of seismic stratigraphy.

From the 1920s on, seismology has been put to practical uses like the search for oil and gas in the subsurface by using artificial sources of elastic waves, primarily dynamite explosions. Beginning in the 1950s, seismology has been used for monitoring nuclear testing and for understanding other seismic events caused by humans such as those triggered by pumping fluids such as water or oil into or out of the ground. Volcanic seismology has improved to the point that it can help predict volcanic eruptions. While seismologists still cannot predict earthquakes with accuracy, they have done much to explore their preconditions. As human population has increased, more funding has been devoted to earthquake preparedness, including the design of structures, the education of the public, and hazard analysis. In places where the historical record of earthquakes is too short to analyze earthquake probability, workers in the field of paleoseismology use geological evidence to document prehistoric earthquakes.

*Satellite and digital technology, notably the Internet, have had major effects on seismology. In 1984 a consortium of American universities founded the Incorporated Research Institutions for Seismology (IRIS) to develop, deploy, and support modern digital seismic instrumentation. By the year 2000, IRIS had more than ninety member institutions and four major programs: the Data Management System (DMS), the Global Seismographic Network (GSN), a program for the study of the continental lithosphere (PASSCAL), and an education and outreach program.

Charles Davison, *The Founders of Seismology* (1927). Beno Gutenberg and Charles Francis Richter, *Seismicity of the Earth and Associated Phenomena* (1954). Charles Francis Richter, *Elementary Seismology* (1958). Benjamin F. Howell, Jr., *An Introduction to Seismological Research: History and Development* (1990). Bruce A. Bolt, *Earthquakes and Geological Discovery* (1993). Stephen G. Brush, *Nebulous Earth: The Origin of the Solar System and the Core of the Earth from Laplace to Jeffreys* (1996).

JOANNE BOURGEOIS

SEMINAR. Seminars are instructional and research units and are considered hallmarks of the German university system. They evolved from the early-modern state pedagogical seminars for training secondary school teachers, from private societies in universities, and from the private collegium for informal, small-group professorial instruction. In the early nineteenth century, seminars incorporated the ideals of pure research and self-cultivation.

At the height of the seminar system, between 1825 and 1888, the German states founded some fifteen of them in the natural sciences, *mathematics, or *physics, or some combination of these. Typically they were poorly funded; attempts to establish laboratories in them usually failed, and their research results were meager. Most accommodated beginning students and so had to bridge the school and the university. The attention to pedagogical detail sometimes left little room for intellectual creativity. Yet seminars became important forums for combining instruction in physics and mathematics, providing practical training in the sciences, training secondary school teachers, and laying the foundations for university science laboratories. At some of the more successful seminars, such as the Königsberg mathematico-physical seminar (founded in 1834), research results, although sporadic, were sometimes groundbreaking. At Königsberg in the late 1840s, physicist Gustav Kirchhoff calculated the value of the dielectric constant and laid the

foundation for the rules, which bear his name, for the analysis of electrical circuits.

In Germany after World War II, seminars became larger institutional units dedicated to directing discussions and handling assignments for a hundred or more students. At the end of the twentieth century, larger disciplinary units in the German university system were still organized as seminars; smaller ones, as institutes. Elsewhere, seminars acted as forums for the presentation of research-in-progress and reviews of recent literature. Examples include the Astrophysical Seminar Series at the Los Alamos National Laboratory and the Space Telescope Science Institute Seminars (Baltimore) as well as the weekly seminars in every department and sub-department of research universities. In its migration from Europe, the seminar also served as a foundation for the colloquium format in teaching.

See also DISCIPLINES; INSTITUTE; UNIVERSITY.

William Clark, "On the Dialectical Origins of the Research Seminar," *History of Science* 27 (1989): 112–154. Kathryn M. Olesko, *Physics as a Calling: Discipline and Practice in the Königsberg Seminar for Physics* (1991).

KATHRYN OLESKO

SEX. Societal taboos still significantly affect the dispassionate study of sex. Investigations have been discontinuous and scattered, findings contradictory, and samples small and often not directly comparable. There has been little interdisciplinary discussion.

A necessary precondition for a scientific study of sex was the perception that it had medical and social, not just individual moral, repercussions. These became more apparent as urban spaces enabled semivisible homosexual subcultures and other manifestations of "deviancy," providing many opportunities for sexual gratification away from its "legitimate" locus within marriage. By the late nineteenth century, several factors facilitated the emergence of a sexual science. These were a growing awareness of the serious consequences of venereal diseases and the role of accepted sexual mores (the "double standard") in disseminating them; anthropological reports destabilizing assumptions of one "natural" pattern of sexual behavior; and increasing refusal among homosexuals to accept stigmatization, with a search for validatory models; and the influence of *Darwinism and evolutionary theory in which sexual selection played the central role.

In the 1860s Karl Heinrich Ulrichs, a German lawyer, argued from theories of embryological development that homosexuality was neither a crime nor a disease but an inborn condition whereby one individual might have characteristics of the other gender to the one externally apparent, leading to sexual desire for the same, rather than the other, sex. In the following decades French psychologist Alfred Binet defined "fetishism" and Italian criminologist Cesare Lombroso's writings on the sexually deviant gained wide currency. The German psychiatrist Richard von Krafft-Ebing's *Psychopathia sexualis* (1886) cataloged a vast variety of deviations: recent scholarship locates him as a champion of the homosexual against the notorious Prussian Code. He listened to and learned from the numerous homosexuals of good social position who consulted him. Magnus Hirschfeld, who argued in early-twentieth-century Berlin for the rights of the "third sex," followed Krafft-Ebing's lead in an even more radical direction.

In Britain, works by self-acknowledged "inverts" such as John Addington Symonds and Edward Carpenter drew on a range of contemporary scientific discourses to reinforce their arguments that homosexuality was not a crime, a disease, a sin, or a vice but "natural" and inborn in certain individuals, even benign in its effects. Havelock Ellis, a friend and colleague of Symonds and Carpenter, gave an exhaustive review of human sexuality in his seven-volume *Studies in the Psychology of Sex* (1897–1928), displaying a grasp of several intellectual disciplines and the international literature. Other significant figures in this endeavor were the Swiss Auguste-Henri Forel and the Russian-German Iwan Bloch. Sigmund Freud took a different approach, penetrating the depths of the human psyche rather than scanning cross-cultural differences. Meanwhile biologists began to investigate the "sexual secretions" significantly later than the products of other endocrine glands. By the early 1930s they had discovered that no male or female secretion existed: though differing in their proportions, testosterone and estrogen were found in both sexes. In the United States, a few pioneering social scientists started to survey what people actually did. Alfred Kinsey's huge report of 1948 on the sexual life of the American male revealed the extent of the disjunction between conventional assumptions about sexual life and the doings in the bedrooms of the nation. During the following decade, William Howell Masters and Virginia E. Johnson mapped the orgasm using human subjects in the laboratory.

The advent of the *AIDS-HIV epidemic directed attention to the magnitude of ignorance about human sexuality and behavioral motivation. Claims to have discovered the biological basis for sexual orientation or intellectual and behavioral differences between the sexes routinely arrive, receive widespread publicity, and soon fade. In spite of increasingly meticulous attention to the vexed questions of research methodology and problems of population sampling, surveys by social scientists on individual sexual attitudes and behavior continue to be criticized as unrepresentative, methodologically flawed, or based on deceitful responses, and therefore invalid, even though ever since Krafft-Ebing and Ellis, personal testimonies have demonstrated the mismatch between reality and conventional assumptions. As we enter the twenty-first century, sex remains an area of contested knowledge.

Paul Robinson, *The Modernization of Sex: Havelock Ellis, Alfred Kinsey, William Masters, and Virginia Johnson* (1989). Vern Bullough, *Science in the Bedroom: A History of Sex Research* (1994). Roy Porter and Lesley Hall, *The Facts of Life: The Creation of Sexual Knowledge in Britain* (1995). Vernon Rosario, ed., *Science and Homosexualities* (1997). Harry Oosterhuis, "Medical Science and the Modernization of Sexuality," in *Sexual Cultures in Europe: National Histories*, ed. Franz Eder, Lesley Hall, and Gert Hekma (1999).

Lesley A. Hall

SEXUALLY TRANSMITTED DISEASES. See AIDS; Contraception; Sex; Syphilis.

SHIFT OF HEGEMONY. Hegemony, as understood by Antonio Gramsci early in the twentieth century, implies a compact between the dominator and the dominated. Political authority, in Gramsci's view, could not be maintained merely by force and terror; it required complicity on the part of the governed, a process in which intellectuals were key agents. Hegemony in science, by extension, is also a construction by and between people of learning.

Leading natural philosophers of the *Scientific Revolution, notably Francis Bacon and Johannes *Kepler, spoke about the long sleep of science separating their own time from the fall the Rome, a view held with remarkable consistency by David Hume and Enlightenment writers such as Jean Le Rond d'Alembert, Voltaire, and the Marquis de Condorcet. William Whewell, in his history of science, wrote about the rise of inductive science that followed the stagnation of the Middle Ages. Then, Jacob Burckhardt formulated the notion of the *Renaissance, where Italians delighted in studying nature when "other nations" neglected it. Notwithstanding the recognition by writers such as physicist Pierre Duhem that medieval Europe was by no means barren to natural knowledge, the Renaissance has remained central to writing in the history of science as the herald of reliable scientific method.

Examples from the middle of the twentieth century are instructive. Introducing his survey of modern science written on the eve of the World War II, H. T. Pledge, librarian at the Science Museum in London, retraced the established view that science was born in Egypt and Mesopotamia, "converged on the Mediterranean Sea, reaching a climax in the Near East as regards industrial arts, in Greece as regards cultural quality, in the Roman Empire as regards extent." After the decline of Alexandria, "Greek science lingered in Southern Italy and in Byzantium, and revived and spread, with the fiery religion of Islam, east to Baghdad and west to Spain." H. W. Tyler and Robert Payne Bigelow presented the same scheme to their students at MIT in the 1930s: "The torch of science now passes from the Greeks to the Indians of the far East, to be in turn surrendered to the Mohammedan conquerors of Alexandria, A.D. 641. By them it is kept from extinction until in later ages it once more fanned to ever increasing radiance in western Europe." Sheer density of illustrious thinkers determined hegemony for the MIT historians of science. In their view, during the *Scientific Revolution Nicholas *Copernicus, Tycho *Brahe, Johannes *Kepler, *Galileo, and Isaac *Newton were situated from Poland to England. Internationalism subsequently declined: "As in the century following Newton, France became the great center of mathematical activity, so in the nineteenth century the leadership passed to Germany."

In his history of *mathematics, another historian of science at MIT, Dirk Struik, titles one chapter "The Orient after the Decline of the Greek Society," emphasizing that despite the disappearance of Greek "political hegemony" in the Near East, a Greek tradition survived under Islamic rule. Just as Egypt produced a "most glorious" blending of Oriental and Greek culture under Roman rule, so "with the decline of the Roman Empire the center of mathematical research began to shift to India and later back to Mesopotamia." (But in his treatment of the Scientific Revolution, Struik favored a materialist view of history and approached gradualism, observing the progressive abandonment of Archimedean rigor and "the gradual evolution of the calculus.") In his survey of the history of science, Charles C. Gillispie at Princeton University is unambiguous about the rise of European science: "The creation of modern science in the *Renaissance was at once a rebirth of Greek science and a bursting of its confines. To separate the new from the old in the Renaissance is always difficult, for humanists steeped in classical learning found antique words for new ideas. It is, however, no falsification of a complex situation—it is rather a first approximation toward resolving it—to say that science stirred

into new life under the inspiration of Plato working against the cramping of learning within a fossilized *Aristotelianism."

Historians of science have continued to wrestle with the issue of decline. Attention has focused on the debate surrounding British science in the 1820s and French science at the end of the nineteenth century. Rises and declines have provided subjects for world historians from Oswald Spengler to William Hardy McNeil, and a large literature has developed around the "Needham Question"—the reasons for the absence of a Scientific Revolution in China. This world-encircling field of study has generated grand explanatory systems, invoking political stability, economic prosperity, geographical factors, and so on to explain why natural science prospered under Hellenistic inspiration and almost nowhere else.

A quantitative measure for scientific hegemony has been sought since the late nineteenth century. Anticipating the historical focus of the *Annales* school in Paris, and inspired by long-term economic fluctuations known as Kondratieff waves, T. J. Rainoff proposed sinusoidal patterns for scientific accomplishment over the *longue durée*. Rainoff observed that both Felix Klein and Charles Fabry, in their histories of mathematics and *physics, respectively identified a flowering of exact sciences at the end of the eighteenth century in France and a decline of thermodynamics in the middle of the nineteenth century there. Rainoff counted discoveries in Felix Auerbach's *Geschichtstafeln der Physik* (1910) and assigned them a nationality from indications in Johann Poggendorff's *Biographisch-literarisches Handwörterbuch zur Geschichte der exacten Wissenschaften* (1863 and on). Rainoff concluded that creative productivity followed a similar pattern in England and Germany. Scientific productivity in England corresponded directly with fluctuations in the consumer price index, and in France and Germany inversely with foreign trade. Stimulated by Rainoff, the prominent systems-builder in sociology Pitirim Sorokin and his student Robert K. Merton charted the rise and fall of Islamic science by tabulating, in fifty-year periods, "the *comparative* importance of the contributions of Arabic-writing scientists and men of letters" who received mention in the first two volumes of George Sarton's *Introduction to the History of Science*; peaks and valleys corresponded to "actual movements insofar as these are accurately described by the qualitative judgments of the historian."

In 1962, Derek J. De Solla Price elaborated a quantitative measure of scientific hegemony. He charted the output of scientific papers in

Chemical Abstracts across the twentieth century by nation and concluded that while the British Commonwealth kept its percentage constant and France suffered steady erosion, the USSR, Japan, and "all the minor scientific countries have spectacularly improved their world position, from about 10 percent at the beginning of the century to nearly 50 percent now." The increase was at the expense of Germany and the United States, whose combined total declined from 60 percent to 35 percent. The rise of late starters guaranteed, in Price's view, that "the older scientific countries will necessarily come to their mature state of saturation, and the newly scientific population masses of China, India, Africa, and others will arrive almost simultaneously at the finishing line." "This process is historically inevitable," he concluded, notwithstanding scientific migration: "We suffer the troubles consequent upon a flow from regions of scarcity to regions of plenty, and upon crystallization of the world's supply of the mother liquor of scientific manpower which causes such manpower to aggregate in already overflowing centers."

Price's scientometric observations found detailed extension in a 1975 study by Paul Forman, J. L. Heilbron, and Spencer Weart, who set out to provide a comparative, statistical picture of physics at academic establishments around the world in the year 1900, the eve of the quantum revolution. They compared physical size, budget, and staff of laboratories and institutes, as well as the literature output of national sectors. The information was assembled from a wide variety of published sources, and it drew on the extensive archival record. They concluded that as a fraction of gross national product and of total population, the financial and human resources devoted to academic physics in Britain, France, Germany, and the United States were the same; that, as measured by publications per person, Germans were more productive than the others; and that, between the turn of the century and World War I, Britain and Germany retained their positions, France declined, and the United States rose. The United States gained hegemony in physics during the 1930s and dominated most sciences for most of the second half of the twentieth century.

T. J. Rainoff, "Wave-Like Fluctuations of Creative Productivity in the Development of West-European Physics in the Eighteenth and Nineteenth Centuries," *Isis*, 12 (1929): 287–319. Pitirim A. Sorokin and Robert K. Merton, "The Course of Arabian Intellectual Development, 700–1300 A.D.: A Study in Method," *Isis*, 22 (1935): 516–24. H. W. Tyler and R. P. Bigelow, *A Short History of Science* (1939), revision of earlier edition by W. T. Sedgwick and H. W. Tyler (1917). H. T. Pledge, *Science since 1500: A Short History of Mathematics, Physics, Chemistry, Biology* (1940). Dirk Struik, *A Concise History of Mathematics* (1948, 1967). Charles Coulston Gillispie, *The Edge of Objectivity: An Essay in the History of Scientific Ideas* (1960). Derek J. De Solla Price, *Little Science, Big Science* (1963). Paul Forman, J. L. Heilbron, and Spencer Weart, "Physics ca. 1900: Personnel, Funding, and Productivity of the Academic Establishments," *Historical Studies in the Physical Sciences*, 5 (1975): 1–185. Nathan Sivin, "Why the Scientific Revolution Did Not Take Place in China—or Didn't It?" *Chinese Science*, 5 (1982): 45–66.

LEWIS PYENSON

SIMULATION. Computer simulation is a method for studying complex systems that has had applications in almost every field of scientific study from *quantum chemistry to the analysis of traffic-flow patterns. Its history, as long as that of the digital *computer, begins in America during World War II. When the physicist John Mauchly visited the Ballistic Research Laboratory at Aberdeen and saw an army of women calculating firing tables on mechanical *calculators, he suggested that the laboratory begin work on a digital computer. The Electrical Numerical Integrator and Computer (ENIAC), the first truly programmable digital computer, was born in 1945. John von *Neumann took an immediate interest in it, and at the urging of fellow Hungarian-American physicist Edward Teller, he enlisted the help of Nicholas Metropolis and Stanislaw Ulam to begin work on a computational model of a thermonuclear reaction.

Their effort typified computer simulation techniques. They began with a mathematical model depicting the time-evolution of the system being studied in terms of equations, or rules-of-evolution, for the variables of the model. In their work (as is typical in the physical sciences), they constructed the model from well established theoretical principles. In many other instances of simulation, models rest on speculations, the consequences of which need to be studied. In either case, the model, transformed into a computable algorithm, drives the computer, whose evolution "simulates" that of the system under study.

The outcome of the simulation of a thermonuclear reaction persuaded Teller, von Neumann, and Enrico *Fermi of the feasibility of a hydrogen bomb. It also convinced the military of the practicability of electronic computation. Soon after, meteorology followed weapons research as one of the earliest disciplines to make use of the computer. Von Neumann also identified hydrodynamics as an important candidate for vast computational resources (*see* HYDRODYNAMICS AND HYDRAULICS). He enlisted meteorologists, with the resources at their disposal, as allies. In 1946 he launched the Electronic

Computer Project at the Institute for Advanced Study at Princeton University and chose numerical *meteorology as one of its largest projects. While working on the problem of simulating simplified weather systems, meteorologist and mathematician Edward Lorenz discovered a simple model that displayed characteristics now called "sensitive dependence on initial conditions" and "strange attractors," the hallmarks of a system well described by "*chaos theory," a field he helped to create.

In the last thirty-five years, simulations have proliferated in the sciences. Simulations are classified according to the type of algorithm they employ, rather than the subject matter they study. "Discretization" techniques transform continuous differential equations into step-by-step algebraic expressions. "Monte Carlo" methods use random sampling algorithms the randomness of which need not correspond to an underlying indeterminism in the system. "Cellular automata" assign a discrete state to each node of a network of elements and rules of evolution for each node based on its local environment.

William Aspray, *John von Neumann and the Origins of Modern Computing* (1990). William Kaufmann and Lawrence Smarr, *Supercomputing and the Transformation of Science* (1993).

Eric Winsberg

SKEPTICISM. Among the ancient Greek texts translated during the Renaissance were those by Pyrrho of Ellis, who maintained that humans must suspend judgment about ultimate reality, and the Academic Skeptics, who maintained that humans could not know how things really were. Skepticism about the reliability of knowledge remained influential into the seventeenth century and paved the way for the overthrow of the dominant Aristotelian philosophy.

In his *Meditations* (1641), René *Descartes used a skeptical method to cast doubt upon previously accepted knowledge claims, as was conventional in his time. But he turned the weapon against skeptics by using it to establish what he believed to be the truth about ultimate reality. Pierre Gassendi, on the other hand, tried to revive the skepticism advocated by Sextus Empiricus, who appealed to postulated atomic properties to explain sense experiences while suspending judgment about whether atoms existed.

By the end of the seventeenth century, the *mechanical philosophy had displaced atomistic and Aristotelian explanations. It tried to discover actual properties (primary qualities) of the least parts of matter that caused macroscopic bodies to possess dispositions (secondary qualities) to produce sensible qualities in human observers.

In the eighteenth century, this project too came under skeptical scrutiny. According to David Hume's *Treatise of Human Nature* (1740), all we can ever know is how things appear. He extended this to *causality: although we can see the constant conjunction of two events, we cannot see whether there is some cause that brings about the conjunction. So-called laws of nature are merely statements of previously observed regular occurrences. Hume's version of "mitigated skepticism" helped early twentieth-century logical positivists to support their call for an end to metaphysical speculation within the sciences.

Richard H. Popkin, *The History of Skepticism from Erasmus to Spinoza* (1979). Miles Burnyeat, ed., *The Skeptical Tradition* (1983).

Rose-Mary Sargent

SLEEP. Greek mythology describes sleep (*hypnos*) and death (*thanatos*) as "Twins of the Night" who reside in the underworld. Judaic and Christian morning prayers praise the Lord for restoring the sleeper's soul upon awakening. However, early naturalistic thought perceived sleep as a passive condition created by isolation of the brain from the other parts of the body and the environment. Aristotle, in *De somno et vigilia*, posited that the cooling of the heart caused sleep. Plato and Galen agreed with this mechanical concept, but saw the brain rather than the heart as the source of sensations blocked by cooling. These concepts of sleep survived more than fifteen hundred years. Physicians and philosophers during the Middle Ages and the *Renaissance considered sleep only in regard to the supposed isolation of the brain.

Two main schools of thought can be identified in the eighteenth and nineteenth centuries. One ascribed sleep to "anemia" or lack of blood in the brain, the other to excess blood in the brain. Still, an imaginative researcher in the nineteenth century even thought that isolation of the brain from the body resulted from the swelling of the thyroid gland in the neck or swelling of the lymph glands. Others like Duval, Legendre, and Lépine around the turn of the twentieth century went even further and perceived sleep as a result of a cessation of cerebral activity—a kind of short-circuit caused by physical separation of the nerve cells.

Theories of hypnotoxins have a special place in the history of sleep. These prevailed from the late eighteenth to the middle of the twentieth century, and conceptualized sleep within the framework of homeostatic principles. According to Legendre and Henri Piéron, among others, during sleep, energy or essential brain or bodily ingredients, depleted during waking, were restored. A complementary view posited the accumulation of toxic substances during wakefulness that are detoxified or removed from circulation during sleep. The immediate cause of sleep was the production of hypnotoxins that inhibit brain activities. Piéron's experiments in 1913, demonstrating that injecting cerebral spinal fluid from a sleep-deprived dog into the fourth ventricle of a wide-awake dog induced sleep, greatly enhanced the attraction of the hypnotoxin theory. Constantin von Economo, famous for his discovery of encephalitis lethargica in 1917, developed the hypnotoxin theory to explain sleep-wake periodicity. He supposed that the accumulation and removal of the toxins caused the sleep-wake cycle. He further assumed that a brain center located between the diencephalon and the midbrain functioned as a "sleep center" by being more sensitive to hypnotoxins than the rest of the brain. Its activation induced sleep in other parts of the brain by Pavlovian inhibition, thus preventing widespread brain intoxication.

Eugene Aserinsky and Nathaniel Kleitman's discovery of rapid eye movement (REM) sleep in 1953 revolutionized the theories about sleep. They showed that about one and a half hours after a person falls asleep, characteristic brain wave changes associated with dreaming occur: the appearance of rapid eye movements and disappearance of muscle tonus for ten to fifteen minutes. This phase of sleep reappears periodically every hour throughout sleep and makes up approximately 20 percent of total sleep time in adults.

REM has become the cornerstone of modern sleep research. In the 1960s, researchers found that distinct brain stem mechanisms control the generation of REM sleep. Further studies have shown that the cycle of sleep and wakefulness obeys a biological clock located in the suprachaismatic nucleus of the hypothalamus, and that both the timing and duration of sleep are determined by an interaction between the clock and homeostatic principles.

Nathaniel Kleitman, *Sleep and Wakefulness* (1963). Peretz Lavie, *The Enchanted World of Sleep* (1996).

PERETZ LAVIE

SLIDE RULE. The slide rule served generations of scientists, engineers, and technical professionals as a quick, convenient, and portable calculating device. First devised in the seventeenth century, it achieved its widest use in the nineteenth and twentieth centuries before being rapidly eclipsed by handheld electronic calculators in the early 1970s.

The most basic of slide rules are particularly suitable for multiplication and division. With additional scales, the instrument can tackle trigonometric problems and higher-order tasks such as the extraction of roots. By modifying standard scales, formulas and constants for a myriad of special-purpose calculations can be incorporated into the design of the slide rule.

Slide rules have been made in a variety of formats, with scales arranged on straight rules, circles, cylinders, and spirals. The key ingredient common to all types is the logarithmic scale. First described by the Scottish mathematician John Napier in 1614, logarithms were quickly adopted and reformulated by his contemporaries in London. Arranged in printed tables, logarithms translated lengthy multiplications and divisions into much simpler additions and subtractions.

Edmund Gunter made the first logarithmic instrument by placing logarithmic scales on a ruler and performing calculations with the help of a pair of dividers. Introduced in 1623, the Gunter rule was soon transformed into a self-contained slide rule by the simple expedient of placing two rules side by side. Although surrounded by considerable controversy, the first to devise this arrangement was probably William Oughtred, who also invented the first slide rule in a circular format.

The slide rule stood briefly at the forefront of innovation in practical mathematics, but soon found its niche as a routine tool for particular trades and professions. Early designs for gauging, excise calculations, and carpenter's work were gradually supplemented by new applications in navigation and surveying. During the nineteenth century, the predominantly British slide rule became a genuinely international instrument. As a general-purpose calculator, its design was standardized, while the number of its specialized uses—from chemistry to monetary calculation—continuously increased.

By the middle years of the twentieth century, slide rules could be found in almost any activity requiring calculation, from engineering construction and electronics to meteorology and blackbody radiation. In addition to the professional market, their use in schools created a demand for literally millions of instruments. Traditionally hand-divided on wood, slide rules appeared from the late nineteenth century in celluloid (and other plastics), bamboo, magnesium, and aluminum. New manufacturing methods such as machine dividing and printing increased the accuracy and rate of production.

While efforts were made in the late nineteenth century to produce slide rules operating to four and even five significant figures, precision remained subsidiary to ease of use, low cost, and speed. The introduction of handheld electronic calculators in the 1970s combined these benefits with high precision, and within a very few years the mass market for slide rules collapsed. Some specialist niches survive—slide rules can still be found as circular bezels on watches for aviators and divers—but most instruments are now collector's pieces.

Peter M. Hopp, *Slide Rules: Their History, Models, and Makers* (1999). Dieter von Jezierski, *Slide Rules: A Journey through Three Centuries*, trans. Rodger Shepherd (2000).
STEPHEN JOHNSTON

SLOGANS FROM SCIENCE. "Eppure si muove" ("Still it moves"), quipped Pope John Paul II after hobbling to the podium on a new hip. He was quoting *Galileo Galilei, who is said to have muttered the words after repudiating the Copernican theory under the persuasive pressure of the Inquisition. "Eppure si muove" [1] is an example of "slogans from science," phrases born in or around science and used widely outside it.

It was a "Eureka moment" [2] when "Newton's apple" [3] struck young Isaac. It caused him to associate the force that drew the apple to the earth with the force that retained the moon in its orbit ("essentia non multiplicanda sine necessitate" [4]). Two decades later, *Newton had tied the universe together by gravity, without presuming to know its cause ("hypotheses non fingo" [5]). He did know that God had to intervene now and again to keep the planets going, which God did until Pierre-Simon de *Laplace showed that no intervention was necessary ("Sire, je n'ai pas besoin de cette hypothèse" [6]). No wonder France had decided that "la République n'a pas besoin des savants" [7].

Changing topics in a "quantum leap" [8], we observe that although "Cogito ergo sum" [9] may be a good test for the existence of a philosopher, it was fatal for "Buridan's ass" [10]. Nature is interested in the "survival of the fittest" [11], not of the smartest. That makes no difference in the long run, however, since despite the vast and constant quantity of energy in the universe ("$E = mc^2$," "$E = hv$" [12]), "die Entropie strebt einen Maximum zu" [13]; not even "Maxwell's demon" [14] can save the cosmos from "heat death" [15]. Of course, war or pollution may get us first. "Nuclear winter" [16] would be followed by a very "silent spring" [17], which could also be achieved by baking ("greenhouse effect" [18]) or grilling ("hole in the *ozone layer" [19]) the creatures of the earth. Whether "ontogeny recapitulates phylogeny" [20] then would not even be of academic interest. Since "ex nihilo nihil fit" [21], there would be little comfort in knowing that "ex ovo omnia" [22] and that life was a "double helix" [23].

From absolute gravity to absolute levity or "stark fool;" the horizontal figure represents "good sense."

[1] The Pope's use of the phrase is recorded in George Weigel, *Witness to Hope* (2001).

[2] "I've got it," Archimedes's triumphal cry as he leaped from his bath having invented a hydrostatic balance for testing whether his sovereign's crown was pure gold.

[3] The first record of this most famous anecdote in science occurs in notes from an interview with a very elderly Newton around 1725. It owes its popularization to Voltaire.

[4] "Principles are not to be increased unnecessarily," a formulation of "Ockham's razor," a precept attributed to the scholastic philosopher William of Ockham.

[5] "I do not frame [or feign] hypotheses," Newton's refusal to speculate about the cause of gravity. This has erroneously been seen as the core of his, and the correct, scientific method.

[6] "Sire, I do not need that hypothesis," Laplace's perhaps apochryphal answer to Napoleon's question, why God did not appear in his world system.

[7] "The republic does not need scientists," the answer to efforts to save Antoine-Laurent *Lavoisier from the guillotine.

[8] The much-abused concept taken from Niels *Bohr's atomic theory (1913), where it refers to an abrupt, but energetically minute, transition of an electron from one orbit to another.

[9] "I think, therefore I am," René *Descartes's final ground of knowing.

[10] The overly logical donkey who, in the form attributed to the scholastic philosopher Jean Buridan, starved to death when the principle of sufficient reason prevented him from choosing between two equally close and attractive bales of hay.

[11] The supposed lesson and basis of Charles *Darwin's theory of evolution.

[12] The energy (E) equivalents of mass (m) and radiant frequency (v) in the relativity theory of Albert *Einstein and the quantum theory of Max *Planck, respectively; c, the speed of light in vacuum, and h, the quantum of action, are universal *constants.

[13] Half of Rudolf Clausius's formulation (1865) of the principles of *thermodynamics: "The energy of the universe is constant; the *entropy strives toward a maximum."

[14] The quick-witted and fast-handed microman invented by James Clerk *Maxwell to defeat increase in entropy.

[15] The late Victorian term for the miserable situation when the stars go out.

[16] The consequence of shutting off the earth's surface from sunlight by dust and dirt thrown into the atmosphere by nuclear explosions, a concept introduced by Carl Sagan (*see* HAWKING, STEPHEN, AND CARL SAGAN) in analogy to the event supposed to have wiped out the dinosaurs.

[17] The consequence of the unbridled use of pesticides, a concept introduced by Rachel Carson.

[18] The action of atmospheric gases to prevent solar energy, degraded into heat, from escaping from the earth, in the manner of the glass in a greenhouse.

[19] A rift in the barrier that helps prevent harmful ultraviolet rays from reaching the earth.

[20] The notion that the "development of the embryo is an abstract of the history of the genus," thus formulated by Ernst Haeckel in 1874.

[21] "Nothing comes from nothing," a scholastic slogan.

[22] "Everything comes from the egg."

[23] The DNA model proposed by James Watson and Francis *Crick (*see* MOLECULAR BIOLOGY).

J. L. HEILBRON

SOCIAL RESPONSIBILITY IN SCIENCE requires recognition of the social, political, and cultural context and consequences of scientific activity. These consequences include potential risks to health, safety, and the environment. Social responsibility requires acknowledgement of the fact that the institutions in which scientists work also have agendas that go beyond research for the sake of research (e.g., national laboratories for defense, profit-making corporations). On the epistemological level, social responsibility involves discussion of the extent to which scientists have a privileged view of nature and ought to involve the public in decisions about the direction of research.

Between 1600 and 1750, natural philosophers and other creators or disseminators of natural knowledge did not recognize social responsibility beyond loyalty to the patrons who supported them and to devotion to the topic at hand. The patrons included wealthy individuals from the growing mercantile class like the Medicis of Florence; the state and the church seeking technical advice about draining mines and swamps, canalizing rivers, navigating oceans, mapping lands, managing forests; universities and other higher schools; and *academies of science.

Between 1750 and 1920 social responsibility evolved in concert with *Enlightenment thought and the expanding institutional bases of science. The Enlightenment recommended the study of nature for the advancement of human well-being. The philosophes' engine, the *Encyclopédie* of Denis Diderot and Jean d'Alembert, indicated the responsibility of savants—men of science—to rationalize arts and manufactures. On the eve of World War I, this Enlightenment tradition came together with strong scientific institutions developed in and around the universities. In public health, scientists conducted epidemiological studies, built hospitals, and planned water works and sewer systems to fight epidemics in burgeoning cities. In agriculture, they organized experimental plots and sought new hybrids, fertilizers, and methods of pest control. In the forest and along bodies of water they studied how to manage resources "scientifically" to ensure their availability for present and future generations.

After the war, scientists publicly recognized their responsibility to society. The process centralized in *professional societies founded in the late nineteenth century to lobby the government to secure the privileges of recognition, licenses, and financial reward in exchange for service to the greater interests of society. Participation in the war was such a service but an equivocal one. The improvement of the means of destruction, especially chemical weapons, prompted a reevaluation of the potential human costs of research.

The economic depression of the 1930s, together with the claims of the leaders of the Soviet Union that its scientists served the interests of the masses, rather than the profit motive and the capitalist, contributed to debate about the moral and social compass of scientists. Scientists like John Desmond Bernal in his book *The Social Function of Science* (1939) asked about the impact of science on society, its role in warfare, employment, and social forecasting. The American and the British Associations for the Advancement of Science set up special sections to air these questions.

Three major arenas of scientific activity after World War II forced scientists to engage the notion of responsibility more intimately. The first concerned arms control and disarmament. In the late 1930s physicist Leo Szilard, with the hope of slowing weapons development, urged his counterparts throughout the world not to publish nuclear research. Early in 1945 scientists in the Metallurgical Laboratory in Chicago issued the so-called Franck Report (named after the chairman of the committee that produced this memorandum, James Franck), recommending all steps possible to avoid the use of nuclear devices. In the postwar years, a large number of scientific groups condemned these new weapons as immoral. They called for a moratorium on their development, outlawing of their use, international control of materials, and sharing of knowledge. Those who spoke out against nuclear weapons included the Federation of American Scientists, the One-World-Or-None movement, and such prominent figures as Niels *Bohr, Albert *Einstein, and Bertrand Russell. These three and several others established the Pugwash organization, named after Pugwash, Nova Scotia, where scientists from around the world first met in 1957 to pursue verifiable arms control agreements. On behalf of Pugwash, Joseph Rotblat received the Nobel Peace Price in 1995; another arms control organization, Physicians for Social Responsibility, received the Nobel Peace Prize in 1985 (*see* ANTINUCLEAR MOVEMENT).

The second arena raising new questions of social responsibility concerned *eugenics, medicine, and other biological sciences. Eugenics,

the effort to improve the hereditary quality of society, was a mainstream science. But its notorious manifestations, ranging from antimiscegenation policies and laws that barred the "unfit" from reproducing to the sterilization of tens of thousands of allegedly feeble-minded people throughout the world, and especially the horrors of National Socialist "racial hygiene," led the vast majority of scientists to decry biologically determinist arguments. The extreme determinist argument—humans are entirely the product of their genetic material, programs to improve their circumstances are futile—reached its obscene logical endpoint in the Nazis' Final Solution and horrific experimentation on humans. Another grotesque violation of the sanctity of human life was the so-called Tuskegee experiment initiated during the 1930s. The U.S. Public Health Service observed the path of secondary and tertiary syphilis in four hundred black American males while providing no therapeutic treatment. Although a lead editorial in the *Journal of the American Medical Association* (*JAMA*) condemned the Nazi research in 1948 as totally without merit and morally bankrupt, the Tuskegee experiment continued for nearly forty years. Meanwhile the U.S. Atomic Energy Commission conducted experiments on unwitting subjects in prisons and on Army recruits. An editorial in *JAMA* in 1965 that promulgated a strict definition of informed consent forced many scientists to reconsider their obligations to society. Proper consent took precedence over data generation. This shift in values evolved into institutional commitments such as research advisory committees in government, medical centers, and universities to consider potential risks before the initiation of biomedical research. By the 1990s, the Ethical, Legal and Social Implications program of the *Human Genome Project incorporated the commitment to social responsibility directly in genomic research. Some scientists understood that if they did not regulate research activities from within their disciplines, public officials, responding to public concerns, might impose standards that scientists would find too restrictive.

The third major arena for issues of social responsibility concerned the research and production of new pesticides, herbicides, and fertilizers (much of which grew out of chemical weapons efforts) and the well-intended but too frequently disastrous attempts to produce new drugs and food additives, illustrated by the thalidomide birth defects of the late 1950s and early 1960s. Increasing evidence that overuse of chemicals had significant environmental and public health consequences contributed to stricter controls on the development of various chemical products and gave rise to the environmental movement in which many scientists took part.

In many countries of the postwar world, scientists' notion of their social responsibility also came to include obligations to the state in the name of the public welfare, national security, and international prestige that often superceded notions of human welfare. Many scientists have justified research as "science for its own sake," claiming that scientific activity is apolitical (even, and perhaps especially, when it is defined as patriotic). Scientists who embraced internationalist notions of human welfare frequently came under pressure to conform to national norms. In France, for example, Frédéric Joliot became the first high commissioner for atomic energy and directed the construction of the first French atomic pile. Yet he was relieved of his duties in 1950 because of his communist leanings. Scientists and politicians feared he might interfere with national programs to develop *nuclear power.

In the Soviet Union the notion of social responsibility was tied directly to state programs for economic growth and the production of nuclear weapons. Scientists who embraced a Western view of responsibility, such as Andrei Sakharov, and who attempted to advance the notions of academic freedom, the right of contact with foreigners, and human rights issues, usually were demoted or lost their positions. Sakharov and many others were exiled or sentenced to terms in psychiatric hospitals or labor camps.

In Germany, scientists who had been connected with the Nazi war effort—in V-2 rockets, the atomic bomb, or cruel "medical" research conducted at concentration camps—were either "denazified" or co-opted by the new governments. Denazification penalties were determined on the basis of the severity of the activities committed, the date of entry into the Nazi Party, and functions held in both state and party during the National Socialist period. In some cases this meant prison or loss of job. But many Nazi scientists claimed they had been coerced into research and bore no responsibility for their actions. Many others were useful to German (East and West), Soviet, and American bomb and rocket programs, and willingly declared allegiance to their new states. As a result denazification was ineffective and inconsistent, and many Nazis who were considered "rehabilitated" never addressed publicly notions of social responsibility. By the 1970s these forces had generated significant discussion among many scientists about the need to engage society—the public, government officials, corporate leaders, university and laboratory personnel—in considering the potential risks of unregulated research and development. These scientists and engineers sought to encourage a culture in which scientists need not fear reprisals, reassignment, or dismissal for raising concerns over risky science and technology. They were involved in the creation of public interest science groups and the formation of national and international organizations to spread the cause of social responsibility. These include environmental and consumer product organizations.

Andrei Sakharov, *Progress, Coexistence, and Intellectual Freedom*, trans. Harrison E. Salisbury (1968). Alice Kimball Smith, *A Peril and a Hope: The Scientists Movement in America, 1945–47* (1970). William McGucken, *Scientists, Society, and State: The Social Relations of Science Movement in Great Britain, 1931–1947* (1984). Bernadette Bensaude-Vincent, *Langevin: Science et Vigilance* (1987). Peter Kuznick, *Beyond the Laboratory* (1987).

PAUL R. JOSEPHSON

SOCIETIES. See ACADEMIES AND LEARNED SOCIETIES; ADVANCEMENT OF SCIENCE, NATIONAL ASSOCIATIONS FOR THE; PROFESSIONAL SOCIETY.

SOCIOBIOLOGY. The Harvard biologist Edward O. Wilson, an authority on insect societies, coined the term "sociobiology" in the 1970s to designate a field concerned with "the study of the biological basis of social behavior in every kind of organism, including man." Central to sociobiology was a behavioral trait that *Darwin and succeeding evolutionary biologists had noticed. Self-sacrificial acts seemed to be commonplace among many animals—for example, honeybee workers, which would sting at mortal cost to themselves for the sake of the hive, or certain small birds, which would whistle upon the approach of a hawk, placing themselves in jeopardy in order to warn the rest of the flock. Later called "altruism," the trait suggested to some biologists—for example, the Russian Peter Kropotkin, in *Mutual Aid: A Factor in Evolution* (1902)—that natural selection acts on the group rather than on the individual. Through the mid-twentieth century, mathematical population geneticists argued against the theory of group selection on the grounds that it required conditions unlikely to occur in nature and that in any case, natural selection would favor the genetically advantaged individual at the expense of the group.

Beginning in 1962, the theory of group selection enjoyed a vigorous revival as a result of biologist William D. Hamilton's idea of "kin selection." Hamilton pointed out that every individual animal shares some fraction of its genes with its relatives, and the closer the relative, the greater the fraction, on average, of shared genes. The individual honeybee worker might sacrifice its own genes in defense of the hive, but its act would help to perpetuate the gene pool of its kin, which included some of its own. Thus, natural selection favored kinship groups comprised of at least some individuals with a genetic tendency to surrender themselves to the Darwinian good of the whole.

Wilson summarized much of the research in the field in his *Sociobiology: The New Synthesis* (1975), whose last chapter, devoted to a speculative analysis of human social behavior, won the book great popular attention. The hypothesis of kin selection figured significantly in shaping Wilson's speculative extrapolations to man of the genetic traits that sociobiologists had found in animals. Salient among the traits were aggressiveness (protection of the group) and territoriality (safeguarding its ecological niche). Male dominance over females also derived from genetic considerations since a single dominant male could repeatedly impregnate many different females, while a single female could herself be impregnated only periodically. Wilson used altruism to suggest, among other things, that homosexuality, which was common among most intelligent primates, might be a genetic trait selected because homosexuals could help relatives of the same sex in their tasks. Genetic drives, Wilson argued, might even lie at the emotional source of certain ethical propositions that mankind regarded highly—the obligation of parents to sacrifice for their children, and citizens for the nation.

Although Wilson considered himself a political and social liberal, theories of human sociobiology struck many observers as a revival of social Darwinism and genetic determinism, reactionary political doctrines disguised as science. That aside, Stephen Jay Gould, another Harvard biologist, struck at two arguments central to Wilson's case. One held that if altruistic acts in animals expressed the natural selection of genes for the trait, then by some principle of continuity, altruism in human beings must also be genetically grounded; the other argument insisted that if a given form of social behavior were adaptive in an evolutionary sense, then its origins were genetic. Gould stressed that both claims foundered in principle on the simple proposition that similar behaviors in man and primates could proceed from dissimilar causes—in primates from genes, but in man, whom natural selection had equipped with the potential for a vast range of behavioral patterns, from culture.

Human sociobiology has nevertheless continued to command attention, especially from evolutionary psychologists and the lay public. The sociobiology of non-human animals, however,

escaped such condemnation; indeed, Wilson's synthesis of the field won praise from many of the scientific opponents of human sociobiology. Since the 1970s, studies in animal sociobiology have been intellectually enlarged to include attention to different types of altruism and to the role of female animals in family and group behavior. The field flourishes as a respected branch of evolutionary biology.

Arthur L. Caplan, ed., *The Sociology Debate: Readings an Ethical and Scientific Issues* (1978). Bettyann Kevles, *Females of the Species: Sex and Survival in the Animal Kingdom* (1986). Robert J. Richards, *Darwin and the Emergence of Evolutionary Theories of Mind and Behavior* (1987). Carl N. Degler, *In Search of Human Nature: The Decline and Revival of Darwinism in American Social Thought* (1991). Michael Ruse, *The Evolution Wars: A Guide to the Debates* (2000). Ernst Mayr, *What Evolution Is* (2001).

DANIEL J. KEVLES

SOLAR PHYSICS. As with many topics in the history of astronomy, solar physics started with Isaac *Newton and his *Principia mathematica* (1687). Using his new theory of gravity, the contemporary values for both the Earth-Sun and Earth-Moon distances, and the length of the year and the month, Newton calculated the Sun/Earth mass ratio. Because the Earth-Sun distance was inaccurate, Newton obtained a mass ratio that was eight times too small. By the time of the second edition of *Principia* (1715), however, a better value for the Earth-Sun distance gave the astonishing result that the Sun was about 330,000 times more massive than the earth, and 110 times larger. Doubts about why the earth should orbit the Sun and not vice versa evaporated.

Newton also concerned himself with the source of the solar energy. In his *Opticks* (1704), he often treated the "rays" of light as corpuscular. On this view, the Sun, in emitting light, lost mass. If the loss were not made good, the Sun should dwindle. For Newton, the great comet of 1680 came to the rescue. He calculated that it had passed within 250,000 km of the solar surface. Newton suggested that comets that came closer would fall into the Sun and thus provide it with new fuel and light.

Worries about the solar energy source resurfaced in the nineteenth century with the realization that the earth, and thus the Sun, had to be much older than the 6,000 years allowed by Bishop Ussher, who in the seventeenth century had calculated from Biblical chronology that creation had occurred in the last week of October 4004 B.C. Measurements by Claude-Sevais-Mathias Pouillet in 1837 of the flux of radiation passing the earth indicated that the sun had a luminosity of about 3.8×10^{26} watts. The suggestion that the sun compensated for this continuous energy loss by accreting meteorites was discarded on the realization that a mass equivalent to about 86 percent the mass of the Moon would have to hit the Sun each year, a gain that results in an unobserved annual increase in the length of the year of 1.5 seconds. In 1854, Hermann von *Helmholtz suggested that the sun was contracting and thus converting potential energy into radiated energy (*see* CONSERVATION LAWS). Even though the required rate amounted to only 91 m per year, the accumulated reduction of the Sun's diameter of 50 percent over 5 million years was regarded as untenable. The discovery of *radioactivity in 1895 opened the possibility of another source of solar heating. The solution to the problem came closer with Albert *Einstein's deduction of the energy equivalent of mass (*see* RELATIVITY). The direct conversion of solar mass into energy became more plausible in the early 1920s when mass spectroscopy disclosed that four hydrogen atoms outweighed one helium atom (*see* ELEMENTS). In the mid-1920s, spectroscopic analysis by Cecilia Payne-Gaposkin indicated that about 75 percent of the solar mass was in the form of hydrogen. Astrophysical modeling of the solar interior by Sir Arthur Eddington made the Sun gaseous throughout, and fixed the central temperature high enough, at 15 million degrees, for collisions between nuclei to lead to fusion and energy release. The physicist Hans *Bethe worked out the nuclear chemical equations governing the conversion of hydrogen into helium in 1939. It became clear that the sun could produce energy at its present rate for a further 5,000 million years.

The relationship between solar astronomy and *physics also appears in the problem of surface temperature. The sun's radiant power was known, but not its law of cooling. Father Pietro Angeli Secchi assumed Newton's exponential formula for cooling held, and calculated in 1861 a surface temperature of 10,000,000° K. At the same time, J. M. H. E. Vicaire used the empirical power law proposed by Pierre Louis Dulong and Alexis Thérèse Petit and obtained a value of 1750° K. The introduction of Stephan's law (bodies radiate as the fourth power of their temperature) in 1879 and Wien's law in 1896 pointed to 5,770° K (*see* QUANTUM THEORY).

Pieter Zeeman's observation of the splitting of sunspot spectral lines (*see* MAGNETO-OPTICS) enabled George Ellery Hale to measure the magnetic fields in these temporary refrigerated dark regions on the surface (1910). These fields change polarity every cycle, indicating that solar activity varied with a period of around 22 years. Harold D. Babcock and his son Horace W. Babcock used a magnetograph in the late 1940s to show how the magnetic fields of sunspots related to the general solar magnetic field. By the 1960s, the Babcocks had linked the 11-year sunspot cycle, the 22-year magnetic cycles, the magnitude of the general solar magnetic field, and the way in which the solar spin-rate varied with latitude.

Solar flares in the vicinity of bipolar spot groups lead to the ejection of clouds of plasma from the overlaying solar corona. Richard C. Carrington and Edward Sabine in the mid-nineteenth century had related the occurrences of these solar phenomena to the observations of aurorae and geomagnetic storms on Earth. It takes a day or two for the charged particles to travel between the Sun and the earth. Eugene N. Parker put forward a hydrodynamical model that explained how the continuously expanding high-temperature corona produces a solar wind that expands past the planets into interstellar space (1958).

Robert B. Leighton's high-precision Doppler spectrometry of 1960 has measured the velocity of the solar surface along the line of sight and has revealed that the whole solar surface oscillates vertically. Analysis of the modes of this oscillation allow astrophysicists to determine the conditions inside the sun.

See also CONSERVATION LAWS; ELEMENTS; MAGNETO-OPTICS; QUANTUM THEORY; RELATIVITY.

C. A. Young, *The Sun* (1881). Giorgio Abetti, *The Sun, Its Phenomena and Physical Features* (1938). A. Jack Meadows, *Early Solar Physics* (1970). Peter V. Foukal, *Solar Astrophysics* (1990). Karl Hufbauer, *Exploring the Sun: Solar Science since Galileo* (1991).

DAVID W. HUGHES

SOLID STATE (CONDENSED MATTER) PHYSICS. The publisher's note of Galileo's last work, *Dialogues Concerning the Two New Sciences* (1638), informs the reader that the author considered "the resistance which solid bodies offer to fracture by external forces a subject of great utility, especially in the sciences and mechanical arts, and one also abounding in properties and theorems not hitherto observed." By the 1660s Robert *Hooke had discovered the proportionality of stress and strain. Notwithstanding the amazing volume of empirical data amassed by engineers and craftsmen up to the mid-seventeenth century concerning the elastic properties of materials, only in Galileo's and Hooke's work did characteristic features of what came to be known as solid state physics surface: specific assumptions to simplify calculations, which, nevertheless, express underlying physical processes and mechanisms, together with the introduction of "constants" specific to each substance.

Jakob Bernoulli initiated the mathematical study of elasticity, which his nephew Daniel

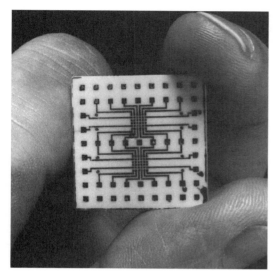

Superconducting wires on a ceramic base. The enclosing fingers give the scale. Computer elements made from superconducting materials work at very high speeds.

Bernoulli and Leonhard *Euler developed in the eighteenth century. By the middle of the nineteenth century the systematic experimental and theoretical investigations of these and other phenomena of the solid state were closely associated with engineering. Slowly the study of solids independently of the problems encountered in their practical use led to the establishment of the first laboratories devoted to work in solid state physics. The main areas of study have been elasticity, crystal structure, strength of materials, thermal and electrical conductivity in metals, thermoelectricity, optical, magnetic, and electric properties of solids, the Hall effect, low and high temperature *superconductivity, incandescent lamps, semiconductors, transistors, and computer chips. The amazingly successful explanations of these phenomena rested on general theories such as *electrodynamics and *quantum mechanics and many new specific concepts. The development of calibration instruments, the needs for *standardization, the extensive use of *x-ray diffraction, and the attainment of temperatures close to absolute zero helped further to consolidate the study of solids. The discovery of the *electron in 1897, the establishment of quantum theory, and, most importantly, the advent of *quantum physics in 1926 provided the framework encompassing almost all of the phenomena of condensed matter physics, as solid state physics was called at the end of the twentieth century.

In 1827 Augustin-Louis Cauchy considered solids as continua and proposed a theory of elasticity involving the use of a large number of disposable parameters that could not readily be determined by experiment. The study of crystals brought about a significant change in the treatment of solids. Louis *Pasteur discovered in 1848 that tartaric acid can have two distinct crystal forms, each polarizing in a different direction. He inferred that crystals acted as an aggregate of "unit cells" rather than of atoms. The shift from a strict atomic viewpoint facilitated the introduction of symmetry techniques and the understanding of a number of properties as deviations from symmetry.

In 1900 Paul Drude, using the methods of the kinetic theory of gases, showed that the quotient of thermal conductivity and electrical conductivity was proportional to the absolute temperature. This derivation of the empirical Wiedemann-Franz law rested on the assumption that the conduction electrons in a metal could be considered as a free gas. Hendrik Antoon Lorentz proceeded to refine the calculations of Drude by taking into consideration the statistical distribution of the electron velocities and their collision with what he considered to be positively charged atomic cores. His results, which differed from Drude's by a factor of $1/2$, also explained other properties such as normal and anomalous dispersion of light, rotation of polarization, and the Zeeman effect (see MAGNETO-OPTICAL EFFECTS).

The experimentally determined dependence of specific heats on temperature at low temperature could not be understood with the model of the free electron gas in solids. It appeared necessary either arbitrarily to reject the gas equation for free electrons or to make the number of free electrons much smaller than the number of atoms in the metals. By 1905 Walther Nernst formulated his heat theorem (see ENTROPY; THERMODYNAMICS) and surmised that as temperature goes to absolute zero, the specific heat of a body must approach a limiting value independent of the nature of the body. Experimental corroboration in 1910 of the behavior predicted by Nernst helped to consolidate Albert *Einstein's treatment in 1907 of specific heats using Max *Planck's radiation law.

Undoubtedly the most decisive developments concerning solid state physics resulted from the advent of quantum mechanics in 1926. The electrical conductivity of metals, paramagnetism, diamagnetism, ferromagnetism, magnetoresistance, the Hall effect, the behavior of semiconductors, and superconductivity found a satisfactory explanation within the framework of quantum mechanics. Theoretical solid state physicists faced two particularly vexing problems in their attempts to solve the wave equation in a periodic crystalline potential: the numerical solution of the Schrödinger equation for the problem and the establishment of the proper expression for the potential used in the equation, which involved approximating interelectron effects.

Long before the extensive use of digital *computers, physicists developed effective numerical solutions. They handled interelectron effects by likening a solid to a periodic array of associated nuclei and their core electrons immersed in a sea of valence or conduction electrons. The treatment of conduction electrons in solids as nearly free particles occupying a series of energy bands that correspond to the electronic shells of atoms turned out to be so successful that the success itself required an explanation. Why did the electrostatic interactions among the electrons not restrict their freedom? Lev Landau showed in 1950 that a system of strongly interacting fermions (particles that, like the electron, have half-integral spin) can be regarded as a collection of "quasi-particles" resembling free fermions. This idea underlay the development of the excitation model of the solid that formally solved the puzzle of how supposedly highly correlated electrons can act as if free.

The understanding of the magnetic properties of matter remained elusive for a long time. Empirical methods and a mass of data about magnets provided sufficient information for the construction and manipulation of magnetic and magnetized materials. By 1903 Joseph John *Thomson had come to consider *magnetism as a property of atoms and to attribute both paramagnetism and diamagnetism to the motion of the atomic electrons under their reciprocal repulsions and the externally applied field. Pierre *Curie's systematic experimental treatment of magnets and his empirical law that the magnetization of a paramagnetic body is proportional to the intensity of the magnetic field divided by the absolute temperature became the background for Paul Langevin's theory of 1905. This theory, based on ideas of Thomson and Lorentz's electrodynamics, derived para- and diamagnetism from magnetic moments arising from motion of the atomic electrons. In the late 1920s Werner *Heisenberg showed that the exchange interaction between electrons might be the key to an understanding of the success of Langevin's approach.

In 1926 Wolfgang Pauli calculated the paramagnetic susceptibility quantum mechanically on the assumption that an electron gas consists of free fermions. In contrast to the prediction of the classical treatment, Pauli found that at low temperatures the susceptibility approached a constant. Only electrons in metals within a certain small range of energy can be aligned by the magnetic field, an effect that dramatically decreases the magnetic susceptibility in metals.

The electron theory of metals was systematically developed after 1928, the year that Felix Bloch defended his doctoral dissertation under Heisenberg at the University of Leipzig. Bloch assumed that the electrons did not act on one another and that they moved freely through a lattice (the metal). A perfect lattice of identical atoms would give an infinite conductivity; electrical resistance resulted from lattice imperfections or ionic motion. Bloch also proved that if the electron Fermi distribution was in equilibrium and at rest with respect to the lattice, and if it had the same temperature as that which the lattice vibrated, then the electrons and the vibrations would be in equilibrium. In this case the motion of electrons did not have any consequences for the thermal motion in a solid. From these considerations Bloch derived a temperature dependence for electrical resistivity.

Hans Bethe's doctoral dissertation with Arnold Sommerfeld was another turning point of the electron theory of metals. He showed that electrons with negative potential energy in a metal have a larger kinetic energy inside it than outside, and concluded that the crystal shortened their wavelengths. This explained the positions of the maxima in electron-diffraction experiments, which did not agree with the predictions of the previous theory. Two additional notions associated with Bethe's close friend Rudolph Peierls—holes and band gaps—became important in understanding the conduction processes in metals. In his attempt to deal with the anomalous Hall effect, Peierls found that the Hall constant in the limit of a slightly filled energy band had the same value as that derived by the classical electron theory. In a nearly full band, it again had the value derived by the classical electron theory, but now for carriers of positive charge equal in number to the unfilled states in the band. These vacancies or holes in an otherwise full band behaved as positively charged bodies. Since, as Peierls showed, electrical conductivity vanished in the case of completely occupied bands, the holes (the negative electrons in Paul *Dirac's *quantum electrodynamics) became an indispensable notion for understanding the behavior of electrical insulators.

In 1934 Eugene Wigner and Frederick Seitz calculated the bands of sodium, a monovalent metal. They modeled the crystal as a network of identical cells surrounding single metallic ions and thus considered the conduction electron in each cell to be influenced only by its "own" ion's field. This calculation initiated other studies of band structures of real materials.

The nearly free electron model could not account for low temperature phenomena such as superconductivity and superfluidity or for phase transitions and critical (or collective) phenomena. Here interactions between the electrons became important. Not until after Wold War II, however, were the necessary many-body methods developed. New concepts like elementary excitations (phonons, spin waves, quasi-particles), macroscopic wave functions (long range order), order parameters and changes of symmetry in phase transitions, collective modes, low-lying excitations above the ground state, Bose-Einstein condensation, pairing, and broken symmetries played an important role in understanding these interactions. The articulation of the notion of macroscopic quantum effects, first formulated by Fritz London in 1936, brought about a deeper understanding of these phenomena.

Electronic conductivity of semiconductors and the mechanical properties of metals are controlled by minute additives of foreign atoms or by irregularities in crystalline structure. Small concentrations of defects in a largely undisturbed lattice have strong effects on macroscopic crystal phenomena such as optical properties and electrical conduction. Semiconductors (flawed crystals) became a separate class of materials and the objects of intensive research, much of it done in industrial laboratories. In one of the largest of them, Bell Laboratories in the United States, researchers invented the *transistor. The contact of two crystals, one with a minute excess of electrons and the other with a predominance of holes, formed the so-called p-n junction, the basis of the transistor. The need for purer materials, the understanding of structure-dependent properties, and the introduction of specific impurities defined the possibilities of semiconductors, and led to the elucidation of their most important property, the rectifying contact that made the transistor practicable. Lasers soon followed (see LASER AND MASER).

Research and development in many universities and *industrial research laboratories, the direct and intense interest of the military, and the profitability of the new materials and inventions based on the understanding of the many properties of the solids played a decisive role in shaping the characteristics of solid state physics. These characteristics derived from many schools of thought and experimental practices from all over the world.

The only person to have been awarded the *Nobel Prize twice for the same science was John Bardeen, who shared the physics prize for his work on semiconductors and again for his work on superconductivity, in 1956 and 1972, respectively. This unique honor reflects solid state physics' idiosyncratic inheritance of diverse phenomena, intriguing problems, and miscellaneous methods.

See also ATOMIC STRUCTURE; ENTROPY; MAGNETO-OPTICAL EFFECTS; STATISTICAL MECHANICS; TEMPERATURE; THERMODYNAMICS.

Marcus Fierz and Victor F. Weiskopf, eds., *Theoretical Physics in the Twentieth Century: A Memorial Volume to Wolfgang Pauli* (1960). Crosbie S. Smith, *A History of Metallography* (1960). John G. Burke, *Origins of the Science of Crystals* (1966). Lillian Hoddeson, George Baym, and Michael Eckert, "The Development of the Quantum Mechanical Electron Theory of Metals, 1926–33," *Reviews of Modern Physics* 59 (1987): 287–327. Lillian Hoddeson, Ernest Braun, Jurgen Teichmann, and Spencer Weart, *Out of the Crystal Maze, Chapters from the History of Solid State Physics* (1992).

KOSTAS GAVROGLU

SOLVAY CONGRESSES AND INSTITUTE. In 1910 Ernest Solvay, an industrial chemist, had a chance encounter with Walther Nernst and learned of the crisis physicists faced in reconciling classical theories with the new quantum theory of radiation. Solvay had made a fortune in the chemical industry through his development of a process to make soda, but he maintained an avid avocational interest in physics. Nernst had already discussed with Max *Planck the possibility of a conference of physicists to discuss the quantum quandary, and he recognized a potential benefactor in Solvay. Solvay agreed to sponsor a conference and Nernst took the initiative in planning it. Unlike general scientific conferences, this one invited only a select group: twenty-one of the physicists most prominently concerned with quantum theory convened for the first Solvay Conference in Physics in the fall of 1911. The intense discussions at what Albert *Einstein, an invitee, called the "witches' Sabbath in Brussels" sharpened the issues and established the seriousness of the challenge of quantum theory to classical physics.

The success of the first conference convinced Solvay to establish the Solvay International Institutes of Physics and of Chemistry, which provided some small grants to individual scientists but mainly supported periodic conferences in Brussels, in which leading figures considered a pressing topic. The second Solvay conference on physics, on the structure of matter, took place in 1913; interrupted by World War I, the conferences resumed in 1921 and then convened every three years. Solvay explicitly intended the conferences to be international, although Belgian outrage at the Germans led to their exclusion after the war;

the only German invited in 1921 and 1924 was Einstein, whose vocal pacifism and Swiss citizenship made him acceptable. Hendrik Antoon Lorentz, a physicist of great scientific, linguistic, and diplomatic skill, presided over the first five Solvay congresses in physics, through 1927, all of which addressed some aspect of quantum theory and atomic structure.

The physics congresses thus provided an important forum for the development of quantum physics and its implications, from the first one through the famous debate between Einstein and Niels *Bohr over the Copenhagen interpretation of quantum effects at the congress of 1927 (see COMPLEMENTARITY AND UNCERTAINTY). There was another hiatus that lasted from 1933 to 1948 owing first to the ill health of organizer Paul Langevin and then to World War II. After the war the conferences began addressing topics beyond quantum physics, in elementary particle physics, *solid state physics, and *astrophysics.

The Solvay congresses on chemistry followed a similar format, with small, elite groups meeting every few years. Although the first and third meetings, in 1922 and 1928, had no announced theme, all three meetings in the 1920s addressed the roles of electrical charge and radiation in molecular dynamics. They also indicated increasing willingness to apply recent physical theories to chemistry, including tentative engagements with quantum theory. Subsequent Solvay congresses in the 1930s covered organic molecules, reactions of oxygen, and vitamins and hormones, and in the 1950s extended to proteins, nucleoproteins, and inorganic chemistry. The published proceedings of the congress offer an unsurpassed insight into the state of leading sectors of physics and chemistry at a given time.

See also COMPLEMENTARITY AND UNCERTAINTY.

Jagdish Mehra, *The Solvay Conferences on Physics: Aspects of the Development of Physics since 1911* (1975). Mary Jo Nye, ''Chemical Explanation and Physical Dynamics: Two Research Schools at the First Solvay Chemistry Conferences, 1922–1928,'' *Annals of Science* 46 (1989): 461–480.

PETER J. WESTWICK

SONAR. From *so*und *na*vigation and *r*anging, sonar describes the use of sound waves to determine the location, velocity, and other characteristics of underwater objects. Sonar systems fall into two categories: passive systems, designed to detect sound emanating from the object under investigation, and active systems, which bounce sound waves off their targets and extract information from the reflected return. In both cases, ''noise,'' extraneous sound in the environment, is a major problem. Noise is dealt with in three ways: by increasing detector sensitivity, by increasing the number of detectors and the distance between them, and by processing and analyzing the returns to factor out extraneous sounds. An operator's trained ear originally discriminated useful data from noise and to some extent still does; in recent decades mathematical analysis using high-speed computers increasingly performs this task.

Sonar emerged from the discovery during the eighteenth century that sound was propagated by vibrations within an elastic medium, and that the velocity of sound was a function of the elasticity, temperature, and density of the medium through which it traveled. Researchers experimentally determined the velocity of sound in water during the 1820s, but without immediate effect. Proposed initially as a means of locating icebergs, sonar was used to detect submarines during World War I. At first the British (who developed the scheme) employed a towed line of microphones. When a submarine was detected, typically through propeller noise, its direction was determined by triangulation based on differences in the time of arrival of sound waves at the various microphones. Later, they used multiple microphones within the hull of the ship in the same manner. Toward the end of the war, Britain and the United States fielded active systems that bounced sound waves off targets, measuring distance and direction by time difference and triangulation. World War II saw such systems widely used, but submariners were often able to negate them by attacking on the surface at night. World War II also saw the development of active sonar homing torpedos. During the Cold War, the efficiency and sensitivity of sonar systems was enhanced by the development of piezoelectric, magnetostrictive, and electrostrictive transducers that directly converted pressure changes into an electrical current and by the development of powerful analytical computers. Perhaps the most impressive Cold War initiative was the U.S. Navy's SOSUS (Sound Surveillance System), which filtered the collective input of a massive array of sonar collectors on the ocean floor through equally massive batteries of analytical computers to track Soviet submarines. Current military trends seek to develop increasingly sophisticated passive systems with active systems held in reserve for tactically critical situations. An important concomitant of sonar technology is that of suppressing the sound signatures of submarines, both active and passive, a field of engineering with close links to optical theory and radar signature suppression.

In addition to military applications, sonar is used for depth location, to find schools of fish, for detailed mapping of the ocean floor, and for underwater archaeological investigation.

Robert J. Urick, *Principles of Underwater Sound* (1983). Willem D. Hackmann, *Seek and Strike: Sonar, Anti-submarine Warfare, and the Royal Navy, 1914–54* (1984).

JOHN GUILMARTIN

SOUND. See ACOUSTICS AND HEARING.

SPACE AND TIME. The evolution of the modern understanding of space and time, which is closely related to the formation and development of physical sciences, can be divided into three stages dominated by Newton's absolute space and time, Minkowski's spacetime, and Einstein's spatial and temporal structures constituted by gravitational fields interacting with material bodies or other physical fields.

Absolutism

The scientific revolutionaries of the seventeenth century (*see* SCIENTIFIC REVOLUTION) rejected the scholastic view of space and time as accidents of substance along with most other fundamental tenets of *Aristotelianism. Against this view, which left no room for a void and assumed that time was the same everywhere at once, the revolutionaries flirted with *atomism and other systems that gave space an independent existence.

Here, René *Descartes took an intermediate position. Although he rejected the notion of space as an immaterial, infinite, immobile container with indistinguishable parts, he did allow it an independent, even a material, existence, by characterizing it by its extension and identifying it with matter. He thus rejected a *vacuum in favor of a plenum, and deduced that motion can be transmitted only by impact, that motion of a body can be measured only relative to other bodies, and that the total motion in the universe is conserved. Conservation of motion suggested to him a principle of inertia, according to which uniform rectilinear motion is equivalent to rest, but rest requires for its definition an immobile frame of reference, which, alas, cannot be supposed within the ceaselessly moving Cartesian universe.

In forming his theory of motion, Isaac *Newton recognized the importance of Descartes's principle of inertia, transformed it radically, and provided it with new conceptual foundations. By conceiving the inertia of a body not as an expression of the conservation of its motion, but as its inertness measured as its mass, Newton purchased ground for taking force, an external mover of inert matter, as a primitive entity existing independently of bodies. To give meaning to the revised notion of inertia and to make the revised law of inertia self-consistent, Newton promoted the atomists' void into a primitive entity.

If motion occurs only relatively to other bodies, immaterial and immobile space as a frame of reference is dispensable; but if absolute motion exists, then the frame of reference has to be taken as a primitive concept (absolute space). Newton adduced the motion of water in a rotating vessel as evidence of a centrifugal force generated by a rotation in absolute space. He described this space eloquently as an entity that, in its own nature, without relation to anything external, remains always similar and immovable. In the same manner, without much argument, Newton defined true time as absolute time, which, of itself and from its own nature, flows equably without relation to anything external. He understood absolute space and absolute time as attributes of God, one expressing divine omnipresence, the other divine eternity.

Newton's absolutist view of space and time came under criticism from Christiaan *Huygens and Gottfried Leibniz. Huygens tried to interpret rotation as a relative motion of the parts of the rotating body, driven to different sides and in different directions, and argued that this relative motion gave the appearance of centrifugal force in Newton's bucket experiment. But this argument failed because in a rotating coordinate system, parts rest but the centrifugal force does not disappear. Leibniz employed his principles of the identity of indiscernibles and sufficient reason to dismiss Newton's absolute spatial and temporal relations and to insist, characteristically, that in Newton's homogeneous absolute space, God would have had no reason to create the world in the way he did rather than in infinitely many other ways. But since Leibniz had to accept rotation as an example of absolute motion, and offered no relational theory to accommodate it, his metaphysical arguments did not carry much force for his contemporaries.

In the two centuries after Newton, natural philosophers accepted absolute space and time as the bedrock of physical theory. The only significant challenge came from Immanuel Kant. In his influential teaching, space and time are imposed by us on the world as the ground or possibility of our intuitions of it. Kant's a priori view of space collapsed with the discovery of *non-Euclidean geometry in the mid-nineteenth century. But his transcendental arguments about space and time as necessary prerequisites for experience was revived by Niels *Bohr and many others concerned to anchor *quantum physics on classical observables in space and time (*see* COMPLEMENTARITY).

Spacetime

Newton defined absolute space in terms of the resting center of gravity of the world. However, for the validity of mechanics, any "inertial system," that is, any body moving uniformly with respect to absolute space, could serve as a reference system. The question of inertial systems in mechanics was entangled with the question of the *ether, the carrier of electromagnetic waves, in the late nineteenth century. Since physicists tended to identify the ether with absolute space, they expected to be able to detect the effect of motion relative to it. The negative result of the Michelson-Morley experiment (1887) posed a puzzle, explained away by the hypothesis, suggested by George Francis FitzGerald and Hendrik Antoon Lorentz, that moving bodies, owing to their interactions with the ether, contract along their line of motion. Lorentz's explanation (1895, 1902), also proposed by Joseph Larmor in 1900, made use of a quantity they called "local time," different for different observers, which they regarded as a mathematical artifice. When local time was taken to be the real time for a moving observer, first by Henri Poincaré in 1902 and then by Albert *Einstein in 1905, the absolutist notion of a single universal time collapsed, and absolute simultaneity could no longer be defined. Times and locations can be defined meaningfully only in accordance with the states of motion of inertial systems. The relation between the space and time coordinates in two inertial systems in relative motion can be obtained mathematically from the principle of *relativity (physical laws take the same form in all inertial systems) and the postulate of the constancy of the speed of light, first suggested by Poincaré in 1902, without resorting to the contraction hypothesis.

In 1905, Poincaré noted that the mathematics (Lorentz transformations) that relate spatial and temporal intervals of inertial systems to another mixed spatial and temporal coordinates, but left the formula for the spacetime intervals between events the same in all coordinate systems. Poincaré observed further that this formula behaved as if it represented a four-dimensional analogue of a line in three-dimensional space, so that the Lorentz transformations could be pictured as four-dimensional analogues of ordinary rotations. These observations suggested a complete change in the ideas of space and time to Hermann Minkowski. He conceived the relative spaces and times of inertial systems as projections of an absolute four-dimensional spacetime manifold, the true and independent stage for physical events to occur, onto the three-dimensional space of the observer.

Absolute spacetime has richer structures than absolute space had. Most important among them is the light cone, defined at each point by the

events that can be causally related to the observer and those that lie absolutely elsewhere and absolutely elsewhen. Minkowskian spacetime, together with its kinematic and causal structures, has replaced Newtonian absolute space and time and played a foundational role in all forms of relativistic dynamical theories, including *quantum mechanics and *quantum field theories, except for the general theory of relativity and its variations.

Dynamical Spatial and Temporal Structures
Non-Euclidean geometries make possible use of intrinsic local variations of curvature to designate positions in space without resorting to a material coordinate system and thus opened the way to a new version of absolute space. But Bernhard Riemann observed in 1854 that since the structure of physical space had to be determined by physical forces, the new notion of absolute space could not be sustained. Einstein vigorously pursued Riemann's idea in developing his general theory of relativity. Here Einstein's work shows the influence of Ernst Mach, whose program of freeing science from metaphysics included ridding the world of the concept of absolute space. In 1883 he rejected Newton's crucial bucket experiment with the argument that the centrifugal forces on the water arose because of its relative motion with respect to the mass of the earth and the other celestial bodies. Mach thus replaced absolute space with the cosmic distribution of matter, which would determine the inertia of bodies and the spatial structures of local inertial systems, and thus provide dynamics with a relationist foundation.

Einstein's general relativity (1915) has a spatial structure (curvature or metric) that varies with the distribution of matter. But an interpretation of general relativity along Mach's lines ran into trouble with the discovery of the "vacuum solution" to the theory's equations, which showed that spatial structures exist in the absence of matter. Further reflection showed that any description of the properties and state of matter necessarily involves a metric as an indispensable ingredient, and thus presupposes the existence of spatial structure. Thus, although Einstein initially liked Mach's idea, which, in 1918, he raised to "Mach's Principle," he later (1953) rejected it. In his final formulation dynamical (gravitational or metric) fields, but not masses, determined the spatiotemporal structures that grounded the dynamic behavior of everything in the world. Spacetime as a quality of the field had no independent existence.

There seems to be unanimity that spatiotemporal structures are not conventional, but specified or constituted by metric fields or their variations. Serious disagreements nevertheless persist over the ontological status of spacetime. Substantivalists ascribe spatial-temporal positions and structures directly to the individual points of a spacetime manifold, and only in a derivative sense to physical entities occupying points of the manifold; relationists claim that the spacetime characteristics of a physical entity belong to it in a primary and underived sense.

Some ardent substantivalists argue that an immovable spacetime substratum as a primitive existence has to be presumed if we wish to ground absolute motions and field theories. The relationist counters that absolute motions can be measured by deviations from geodesic motions and that chirality (right- or left-handedness), as Kant had realized, cannot be understood by reference to the points in absolute space. It is an intrinsic spacetime characteristic of physical entities and belongs to them in a primary sense.

See also COMPLEMENTARITY; SCIENTIFIC REVOLUTION.

Hans Reichenbach, *The Philosophy of Space and Time* (1958). Lawrence Sklar, *Space, Time, and Spacetime* (1974). John Earman, Clark Glymour, and John Starchel, eds., *Foundations of Space-Time Theories* (1977). Michael Friedman, *Foundations of Space-Time Theories* (1983). Robert Torretti, *Relativity and Geometry* (1983). Julian Barbour, *Absolute or Relative Motion? Vol. 1: The Discovery of Dynamics* (1989). John Earman, *World Enough and Space-Time* (1989). Max Jammer, *Concepts of Space*, 3d enlarged ed. (1993). Julian Barbour and Herbert Pfister, eds., *Mach's Principle: From Newton's Bucket to Quantum Gravity* (1995).

TIAN YU CAO

SPACE SCIENCE. The term "space science" came into use in the late 1950s just after the Soviet Union launched *Sputnik I*, but it had antecedents in the 1930s, when astronomers climbed mountains to observe the heavens and meteorologists and physicists sent instruments aboard high-flying balloons to study *cosmic rays.

After World War II, captured German V-2 missiles and their immediate successors, like the Navy's *Viking* or the long line of Soviet missiles based upon V-2 technology, carried probes into near space to examine Earth's upper atmosphere, the nature of cosmic rays, the Sun's high-energy spectrum, and the particles and fields contained within Earth's magnetic system. After *Sputnik*, space science research relied on rockets powerful enough to send nuclear warheads ballistically to another continent or satellites into orbit. Space science identified its programs in terms of the capabilities of specific transport vehicles—balloons, aircraft, sounding rockets, *satellites, and space probes.

In Britain, most of the activity in the 1950s centered on the Gassiot Committee of the Royal Society and on less formal splinter groups at

Clean-room inspection of a probe for Pioneer III (1958). The surgical performance was in vain; Pioneer III, intended to investigate the Van Allen radiation belt, flew only briefly because the rocket that launched it failed.

major university centers. The society's members had established interests in the physics of the upper atmosphere and had followed closely the progress of the rocketry groups in the United States. Starting in 1955, they championed what became known as the *Skylark*, a sounding rocket on the scale of the American *Aerobee*, and eventually gained access to space post-Sputnik with the Ariel series of scientific satellites sent up by American vehicles. Meanwhile other British launchers were being developed—*Blue Streak* (based on preexisting military systems) and *Black Knight* (developed out of research programs)—but they did not survive as scientific launchers once Britain decided to work within an international structure.

In the United States, those who instrumented the V-2s between 1946 and 1951 came from disciplines that traditionally had not inquired into the natural phenomena they now addressed with rockets. They were practical, tool-making, problem-solving physicists and engineers experienced in building and maintaining long-range radio networks, rugged and reliable high-speed optical systems, proximity fuses for artillery shells, and radiation detectors for atomic tests. These were the skills needed to make an instrument perform delicate observations in the violently hostile realm of the rocket.

A second generation of practitioners, typically trained in the university groups that had access to rockets in the 1950s, did postdoctoral work in military laboratories. They tended to identify more with the disciplines they could address with the instruments they built than with the objects of their handiwork. Starting in the 1960s, leading academic scientists like Lyman Spitzer, Jr., at Princeton, Leo Goldberg at Harvard, John Simpson at Chicago, James van Allen at Iowa, Fred Whipple at Harvard, Charles Hard Townes at Berkeley, Joshua Lederberg at Wisconsin, William Dow at Michigan, and Joseph Kaplan at UCLA supported graduate students and assistants on contracts from NASA, the Air Force, and the Navy to develop instruments and techniques to pursue science from space.

Graduate students in astronomy were attracted to groups conducting solar physics from rockets, and became specialists in methods most suited to studying the Sun from space. Many went on to satellite-based research in the 1960s and to manned orbiting platforms in the 1980s and 1990s. As the generational cohorts established stronger and stronger interdisciplinary ties with traditional areas of research, they migrated more freely within their subject matter disciplines; generally they no longer moved from one scientific discipline to another but were attracted

to problems within their discipline where they could exploit their expertise with rocket and satellite technologies.

Space science thus came to lie at the intersection of three elements: a technical capability (the use of rockets and satellites as platforms to make observations of any accessible phenomenon) with a scientific interest (framing problems that can be addressed by observations from rockets and satellites) and a military or commercial need (creating a capability to use and manage space for communication, weapon delivery systems, reconnaissance, and command and control). At the intersection, expensive, government-sponsored technologies made research in space possible.

Scientific satellite development roughly followed the growth of the launch capabilities of the vehicles. The very first American scientific satellites, those typical of the long-lived Explorer series, were single-purpose instrument packages weighing thirty to one hundred pounds. They contained primitive telemetry systems, on-board data storage, and rudimentary temperature stabilization. By the early 1960s, spin-stabilized "observatory" class satellites (the Orbiting Solar and Geophysical Observatories) and the multifunctional but unstabilized Ariel series were flying. These satellite series coexisted throughout the 1960s and offered access to a wide range of electromagnetic phenomena from the Sun, Earth's magnetosphere, and the interplanetary medium. Dedicated sounding rockets and a few experimental high-energy satellites began to detect nonsolar x-ray sources: the first fully dedicated x-ray mapping satellite, *Uhuru*, was launched only in 1970 as the forty-second in the Explorer series.

The first of the OAO series appeared in 1968. With the OAO, in a weight class of thousands rather than hundreds of pounds, lead times crept up during the 1970s from a few to many years and began to slow the pace of training and advancement in the participating disciplines. Astronomers have called OAO-2 the first true observatory in space because the scale and resolution of its instrumentation complemented those available on the ground. It operated from December 1968 to January 1973 and could achieve a pointing accuracy as good as 1 minute of arc with a stability of some seconds of arc providing the capability to secure sustained photometric and spectroscopic data from tiny celestial sources.

In the post-Apollo era, two drivers of the American space program both propelled and severely limited the continuing development of science satellites. The primary driver was NASA's preoccupation with establishing a permanent human presence in space. As the Apollo program wound down in the wake of the successful lunar landings between 1969 and 1972, NASA decided that national goals could best be met in a reusable launch system, the Space Shuttle. Accordingly it concentrated access to space in the Shuttle bay, severely reducing or eliminating altogether suborbital and orbital programs based upon conventional launch vehicles. This had the effect of requiring even the smallest packages to be rated for human space flight, vastly increasing costs and lead times for development and testing.

At the same time, NASA's propensity for mission-based, rather than problem-based, programming drove up the scale of successive satellites. While the OAO's were flying and scientific groups tried to keep alive programs in the Explorer and smaller observatory and planetary probe classes, NASA and a vocal faction within the scientific community set their sights on a "great observatory" class of satellites, instruments that would fill the Shuttle bay and offer truly high resolution and broadband access to the faintest of celestial sources. This older class had been wildly successful beginning in the 1970s. By the end of the century, all the planets save Pluto had been visited, mapped, revisited, and, in the cases of Mars and Venus, investigated by landers. In the 1980s and 1990s, however, the costs of these probes were competing directly with the great observatories like the Hubble Space Telescope, the Compton Gamma Ray Observatory, the Advanced X-Ray Astronomy Observatory (AXAF, renamed *Chandra*), and the Space Infra-Red Telescope Facility (SIRTF). Pressure mounted under competition from satellites of the European Space Agency and Japan.

For these reasons, as well as a lack of public enthusiasm for the continuation of NASA programs at the levels enjoyed in the Apollo era, American promoters of space research increasingly sought out new and more substantial modes of international cooperation at all levels, from Explorer to great observatory class. Few nations have the resources to conduct science from space. The United States from the start entertained an international program, mainly to launch instruments, and in some cases, satellites, for other countries. The Soviet Union soon followed, and all eventually recognized the need for the creation of a new international body, now the Committee on Space Research (COSPAR), that would coordinate international participation in space research. Many countries, including Great Britain, created new organizational structures to communicate with the international body. Thus, COSPAR formed and quickly assumed far broader international responsibilities.

At the same time, the United States announced that it would provide launch vehicles for COSPAR countries as its contribution to international cooperation. Initially western Europe, Great Britain, and the Commonwealth countries dominated COSPAR, with a strong representation from the United States and hardly any from the Soviet Union and Eastern Bloc. This was not satisfactory to member countries of ICSU (*see* INTERNATIONAL ASSOCIATIONS), and eventually the Soviet Union joined COSPAR, which thus became an important forum for international cooperation, eventually as an agency within the United Nations. Its existence, and the American offer of launch vehicles, weakened British resolve to purchase its own systems. The United States sealed the arrangement when it announced at a meeting of COSPAR in March 1959 that it would provide launch systems without charge.

The emergence of the European Space Research Organization (ESRO), and out of that the European Space Agency (ESA) in the 1960s, even with Britain's initial reluctance to join in, marked the completion of the overall structure for space science in Europe and the Americas. Led by France, ESA produced a competitive launch vehicle, *Ariane*, which prompted creation of similar-scaled vehicles in China and Japan. Thus at the beginning of the twenty-first century there exist five competing national—or, in the case of ESA, transnational—sources for placing satellites into orbit. Before the multinational armada of space probes that met Halley's Comet in the spring of 1986, the United States and Soviet Union were the only countries actively supporting interplanetary probes, landers, and orbiters. By 2000 at least four of the five major launching programs were considering new probes, orbiters, and landers.

See also INTERNATIONAL ASSOCIATIONS.

Homer E. Newell, *Beyond the Atmosphere: Early Years of Space Science* (1980). James A. van Allen, *Origins of Magnetospheric Physics* (1983). Richard Hirsh, *Glimpsing an Invisible Universe: The Emergence of X-Ray Astronomy* (1983). Harrie Massey and M. O. Robbins, *History of British Space Science* (1986). Robert W. Smith, *The Space Telescope: A Study of NASA, Science, Technology and Politics* (1989). Joseph N. Tatarewicz, *Space Technology and Planetary Astronomy* (1990). Karl Hufbauer, *Exploring the Sun: Solar Science Since Galileo* (1991). David H. DeVorkin, *Science with a Vengeance: How the Military Created the U.S. Space Sciences After World War II* (1992). Ronald Edmund Doel, *Solar System Astronomy in America: Communities, Patronage, and Interdisciplinary Science, 1920–1960* (1996). *A History of the European Space Agency, 1958–1987. Vol. 1: The Story of ESRO and ELDO, 1958–1973*, ed. John Krige and Arturo Russo; *Vol. 2: The Story of ESA, 1973–1987*, ed. John Krige, Arturo Russo, and Lorenza Sebesta (2000).

DAVID DEVORKIN

SPACE STATION. The space age began in October 1957 when the Russian *Sputnik* satellite orbited the earth, but the first practical proposal for a space station is credited to Wernher von Braun in an article published in *Collier's* magazine in 1954. Von Braun saw space stations as a stepping stone to the moon, Mars, and beyond. His concept of a rotating wheel became fixed in the public consciousness as illustrated in Stanley Kubrick's *2001: A Space Odyssey.*

In order for an object to orbit the earth, it must achieve a sufficient velocity (around 28,000 kph [17,500 mph]) and altitude (160 km [100 miles]) to fall around the earth and counteract the effect of the earth's gravity. Since 1957, four uniquely different manned space stations have orbited the earth. The latest version orbits at an altitude of 400 kilometers (250 miles).

While the Americans were landing on the Moon, the Russians modified their space program to concentrate on the challenges of long-duration space flight. The first of many Salyut space stations was launched in 1971. The program provided almost two decades of continuous human presence in space.

In May 1973, the 78-ton *Skylab* Station became the largest object in earth-orbit. By using the third stage of the Saturn V moon rocket, it became the logical follow-on to the Apollo program, which ended in 1972 with the return of *Apollo 17*. Due to a damaged solar panel, only three manned missions were completed.

The orbital station *Mir* (meaning "peace") went into orbit on 20 February 1986. With two docked spaceships attached, it weighed 136 tons and measured 33 meters (108 ft) across. Ninety-six cosmonauts visited the station over almost thirteen years of continual habitation. Valery Polyvok established the world endurance record of 438 days in space. Starting in 1995, the U.S. Space Shuttle visited the station to gain experience for sending crews or "expeditions" to the upcoming International Space Station. In 1999, the Russian Space Agency controlled the re-entry of the station into the earth's atmosphere.

The fourth version, called the International Space Station, is scheduled to be completed in 2005. It will be as long as a football field and weigh close to 460 tons. It will have over one hundred major component parts from sixteen nations and a pressurized living/working area equal to the volume (9,000 cubic m [46,000 cubic ft]) of a Boeing 747 aircraft. Over 100,000 people in national space agencies and contractor/subcontractor companies will work together in the largest nonmilitary joint effort in history.

ISS, begun as a funded program in 1984 during the Reagan administration, evolved into a foreign-policy initiative under the Clinton administration. From 1994 to 1998, the National Aeronautics and Space Administration (NASA) paid $728 million to Russia for space-station work and joint-training flights to the *Mir* space station. On-orbit construction began in December 1998. In July 2000, a Russian Proton

The International Space Station photographed by a member of the crew just after separation from the Space Shuttle Endeavor (2001).

booster lifted up the key 20-ton *Zvezda* service module to provide living quarters, life support, propulsion, navigation, and communications during the next phases of assembly. NASA's latest cost figure for the project is forty billion dollars, but the General Accounting Office puts it closer to one hundred billion.

The overall goal of the sixteen nations is to build and operate the ISS as a world-class research center in the unique environment of space. This involves doing medical and scientific research, laying the foundation for space-based commerce, encouraging space-related education, and fostering world peace through long-term international cooperation in space. One derived benefit is research in a microgravity environment. The various lab facilities are designed to study the human physiological adaptation to spaceflight, the effects of microgravity on animal and plant cells, and growth of high-quality crystals and protein structure research for pharmaceutical products.

The international space station is the next logical step toward building a true spacefaring civilization. NASA views the ISS as a jumping-off point for missions to Mars. In just a few years, the ISS will be the brightest object in the night sky, and a reminder that the human race has begun the ultimate journey to the stars.

W. David Compton and Charles D. Benson, *Living and Working in Space: A History of Skylab* (1983). Tim Beardsley, "The International Space Station: A Work in Progress," *Scientific American* (May 1999): 20–23. Leonard David, "Special Report: International Space Station," *Aerospace America* (July 1999): S1–S15. Kirsten Roundtree, "The International Space Station a Year Later," *Launchspace* (October/November 1999): 35–39. http://spaceflight.nasa.gov/station/index.html.

VICTOR P. BUDURA, JR.

SPECIES. See BOTANY; CLASSIFICATION IN SCIENCE; DARWINISM.

SPECTOGRAPH, MASS. See MASS SPECTROGRAPH.

SPECTROSCOPY is the science on the borders of chemistry and physics that studies the properties of matter by analyzing, usually prismatically, the light it emits when rendered incandescent. Spectroscopy's progress has depended on the development of the necessary equipment. It uses physical methods to study chemical phenomena. Not until a chemist and a physicist collaborated in working on spectra did spectroscopy begin to yield useful chemical knowledge.

While working to improve optical glass, Joseph von Fraunhofer found that flame spectra were characterized by discrete bright lines (*see* VISION AND OPTICS). He also found a number of dark lines crossing the continuous spectrum of the sun and noted that their positions did not change with intensity. These dark lines, subsequently named after him, still bear the letters he used to designate them.

While spectral lines facilitated the calibration of optical instruments, their meaning eluded satisfactory explanation for many years. The physical interpretation of the lines played a major role in the wave-particle debate over the nature of light that raged, especially in Great Britain, during the 1820s and 1830s. William Henry Fox Talbot suggested in 1826 that spectral lines might be used for chemical analysis. This idea, however, was not pursued, largely because the generally poor quality of glass prisms made it difficult to achieve replicable results, as did the impurities present in chemical substances. Attempts in the 1840s and 1850s to analyze the spectra of electric sparks also came to little, although the theory behind them favored the development of spectroscopy.

From his work on photochemistry in the 1850s, Robert Bunsen, professor of chemistry at the University of Heidelberg, became convinced that the light emitted from flames was uniquely characteristic of the chemical elements present. He pursued this idea with the school's professor of physics, Gustav Kirchhoff. Together they showed conclusively, in 1859, that a chemical element emitted a uniquely characteristic spectrum that could be used for chemical analysis.

In 1860–1861, using what was then known as spectrochemical analysis, Bunsen detected and then chemically isolated two hitherto unknown chemical elements, cesium and rubidium, which occurred in trace quantities in mineral waters. In 1861, William Crookes in London discovered the chemical element thallium using spectrochemical methods. These discoveries placed the new method on a secure evidential basis. Furthermore, they helped popularize knowledge of the new method and to arouse widespread interest. In 1865 August Wilhelm von Hofmann demonstrated spectroscopy to Queen Victoria at Windsor. Thus publicized, the method quickly became established as an invaluable laboratory technique.

During the same period, Bunsen also collaborated with Kirchhoff to show experimentally that the bright yellow lines characteristic of sodium corresponded with Fraunhofer's dark D lines in the solar spectrum. Kirchhoff provided a thermodynamic explanation of the coincidence. This extension of chemical analysis to the sun and stars (entities that the French philosopher Auguste Comte had pointed to in 1835 as examples of things forever unknowable) led to the new science of astrophysics. In the ensuing decades, spectroscopic observations allowed astronomers to develop theories of the evolutionary sequence of stars. Measurements made in the late 1890s on the spectrum of cavity radiation prompted Max *Planck's *quantum theory. Somewhat later, using the measurements and analyses of the distribution of lines emitted by particular elements, physicists began to investigate the internal composition of matter, leading to Niels *Bohr's theories of *atomic structure. The discovery of the diffraction of *X rays in 1911 led to high-frequency spectroscopy with crystals rather than glass prisms as analyzers, and to important information about the deeper reaches of atoms.

The obvious benefit of the spectroscope to a wide range of scientific and technical activities prompted its commercial manufacture by a large number of instrument makers throughout Europe. The instrument was refined and developed during the late nineteenth century by substituting diffraction gratings or hollow prisms filled with carbon bisulphide for the glass prism. Some spectroscopes had prisms so arranged that they appeared to resemble telescopes. The basic principles and uses of the spectroscope did not undergo any fundamental change until the invention of the *mass spectrograph in 1919.

See also OPTICS AND VISION.

Frank A. J. L. James, "The Establishment of Spectro-Chemical Analysis as a Practical Method of Qualitative Analysis, 1854–1861," *Ambix* 30 (1983): 30–53. Frank A. J. L. James, "Of 'Medals and Muddles'. The Context of the Discovery of Thallium: William Crookes's Early Spectro-Chemical Work," *Notes and Records of the Royal Society of London* 39 (1984): 65–90. Frank A. J. L. James, "The Discovery of Line Spectra," *Ambix* 32 (1985): 53–70. Frank A. J. L. James, "The Practical Problems of 'New' Experimental Science: Spectro-Chemistry and the Search for Hitherto Unknown Chemical Elements in Britain 1860–1869," *British Journal for the History of Science* 21 (1988): 181–194. Myles W. Jackson, *Spectrum of Belief: Joseph von Fraunhofer and the Craft of Precision Optics* (2000). Klaus Hentschel, *Mapping the Spectrum: Techniques of Visual Representation in Research and Teaching* (2002).

FRANK A. J. L. JAMES

SPEED OF LIGHT. See LIGHT, SPEED OF.

SPONTANEOUS COMBUSTION. In 1847 the Stadtarzt (municipal physician) of the city of Giessen, Germany, was called to the grisly remains of the Countess of Görlitz, who had been consumed, down to a vertebra or two, in a fire in her private apartment. The windows and doors were locked and the furniture intact. The Stadtarzt declared that the countess had died from spontaneous combustion.

When her butler turned up in France with her jewels, another explanation seemed more likely. A German court found him guilty of murder and arson. It then took the unusual step of putting the original cause of death, spontaneous combustion of the human body (*Selbstverbrennung*), on trial. Although few doctors had attended the final moments of a *Selbstverbrennung*, enough clinical evidence had accumulated to establish it in forensic medicine as the culmination of a disease that afflicted corpulent, alcoholic women. Once ignited, they burned like candles. The countess was neither fat nor alcoholic, however (in fact, abstemious at only a bottle of wine a day), circumstances that helped arouse suspicion that the Stadtarzt had misdiagnosed the case.

The main witnesses called against spontaneous combustion were professors at the University of Giessen, Justus *Liebig and Theodor Ludwig Wilhelm Bischoff. They testified that the human body contains far too much water to sustain a flame, let alone ignite spontaneously; and they exhibited the results of unsavory experiments on cadavers that showed how the butler could have created the conditions discovered in the countess's room. The court found for the professors—perhaps the first time ever that a court preferred the testimony of experimental scientists about a medical matter to the accumulated experience of physicians.

The world would be different from what it is if the decision taken in Giessen had wiped out a disease that made so fitting and uplifting an end to besotted lives. It took some time before spontaneous combustion disappeared from forensic medicine and imaginative literature. A miscreant in Dickens's *Bleak House* who died by spontaneous combustion rekindled the controversy over its possibility. Its leading participants were the positivist philosopher George Lewes, who admired chemistry, and Dickens himself, who believed in doctors. Although now supposedly eradicated, spontaneous combustion still claims victims. A book chronicling hundreds of recent cases was published in 1990.

Michael Harrison, *Fire from Heaven* (1990). J. L. Heilbron, "The Affair of the Countess Görlitz," American Philosophical Society *Proceedings* 138 (1994): 284–316.

J. L. HEILBRON

SPONTANEOUS GENERATION. "Spontaneous" or "equivocal" generation denotes the widespread belief in the production without parents and by chance of living beings either from decomposing living matter (heterogenesis) or from nonliving matter (abiogenesis). This belief, which dates back to ancient times, has often taken the form of sophisticated doctrines related either to the generation of specific living beings like insects, *infusoria, and bacteria, or to the origin of life itself.

Like the ancients, most authors of the seventeenth century believed that flies sprang from rotting meat. By experiments and microscopic observations Francesco Redi demonstrated that the maggots observed in rotting meat came from eggs laid by flies (*Esperienze intorno alla generazione degl'insetti*, 1668). Jan Swammerdam and Marcello *Malpighi, who showed that plant galls also resulted from insect eggs, confirmed Redi's results.

Although the microscopist John Turberville Needham (1713–1781) explicitly rejected the doctrine of equivocal generation, his complex theory of generation by vegetation seemed to most of his contemporaries and successors to support it. He carried out observations on the animalcules arising in infusions and explained their generation as the result of the decomposition of living matter, which released "active" forces able to produce new "vegetations" (*Nouvelles observations microscopiques*, 1750). In 1765 Lazzaro Spallanzani argued against Needham's active forces, proposing instead that the animalcules were true animals generated from eggs present in the air that contaminated the infusions. By repeating Needham's experiments with sealed and heated infusions, Spallanzani showed that if boiled infusions were placed in previously heated hermetic flasks to destroy all existing eggs, no animalcules would be generated. By contrast, if air entered the flasks animalcules proliferated.

Between 1858 and 1864 a fresh controversy took place in France. Félix-Archimède Pouchet claimed that airborne contamination of infusions was unlikely because he had rarely found spores and living particles in air. Therefore only spontaneous generation could account for the microorganisms found in infusions (*Hétérogénie*, 1859). Stimulated by the debate begun by Puchet, Louis Pasteur filtered air with an instrument of his own making and concluded that air did contain microorganisms and that flasks with sterilized sugared yeast water remained sterile if not contaminated by it.

At the same time that the doctrine of spontaneous generation seemed to have been finally superseded by the postulate that life could only be transmitted by preexisting life, some evolutionists posed the question of the origin of life in general from nonliving matter. New terms were introduced to indicate the passage from ordinary to living matter: "abiogenesis" by Thomas Henry Huxley, "archeobiosis" by Henry Charlton Bastian, and "Urzeugung" (primordial generation) by many German authors. According to August Weismann, spontaneous generation was a logical necessity notwithstanding the failures to demonstrate it. In 1924 the Russian biochemist Aleksandr Oparin argued that life did not arise immediately but slowly emerged from a long-term chemical evolution. His ideas, expressed in many widely circulating books, received near-universal acceptance.

See also GENERATION.

John Farley, *The Spontaneous Generation Controversy from Descartes to Oparin* (1977). Renato G. Mazzolini and Shirley A. Roe, *Science against the Unbelievers: The Correspondence of Bonnet and Needham, 1760–1780* (1986).
RENATO G. MAZZOLINI

SPUTNIK. Shortly before midnight on 4 October 1957, the Soviet Union launched the world's first artificial Earth *satellite, a 184-pound sphere about 22 inches in diameter dubbed *Sputnik* (loosely translated as "fellow traveler"). It was made of aluminum alloy and contained two zinc-oxide batteries, a thermal regulation system, and a radio that transmitted temperature and pressure data and signaled its presence to the world with a "beep-beep" sound. A month later, *Sputnik II* went up, weighing 1,121 pounds, packed with a maze of scientific instruments and signaling back the condition of the live dog named Laika it was carrying.

The prime mover behind the satellite program was Sergei Korolev, an imaginative engineer and efficient organizer and the chief designer in the Soviet effort to develop intercontinental ballistic missiles (ICBMs). A longtime enthusiast of space exploration, Korolev urged in 1954 that one of the missile rockets be used to launch an Earth-orbiting satellite. In mid-1955, the United States announced that it intended to launch such a satellite during the *International Geophysical Year, scheduled to begin in mid-1957. Early in 1956, having proposed to Soviet policymakers that the nation beat the Americans to the punch, Korolev and his allies obtained operational authorization for the satellite project. *Sputnik* was launched less than two months after the Soviets first successfully tested an ICBM.

The Soviet achievement stunned the West. The *Sputnik*s demonstrated that the Soviets possessed the rocket and guidance capability for ICBMs, and that by putting a live dog on board, they were well on the way toward putting a man into space. President Dwight Eisenhower refused to panic. He knew from secret intelligence, particularly the information gathered by overflight surveillance of the Soviet Union, that the United States held the lead over the Soviets in intercontinental rocketry. (The United States launched its own satellite soon thereafter.) Eisenhower nevertheless accelerated the nation's program to build and deploy

Laika the space dog of Sputnik II, launched 3 November 1957.

ICBMs and authorized the deployment of short-range missiles in Italy and Turkey. *Sputnik* also prompted him to strengthen American science and its role in policymaking by creating the President's Science Advisory Committee. And it led the government in 1958 to establish the National Aeronautics and Space Administration (NASA) to oversee and coordinate all the nation's nonmilitary activities in space research and development.

Eisenhower thought of the new agency more as a consolidation of existing agencies than as a bold departure, remarking to his cabinet that he did not want to pay to learn "what's on the other side of the Moon." However, Congress, resolute in its sense of Cold War competition, pushed the president hard. "I want to be firstest with the mostest in space," a member of the House of Representatives declared.

Sputnik's tiny "beep-beep" precipitated a space race between the United States and the Soviet Union, and among its achievements were the first exposure of the far side of the *Moon to human observation via the Soviet *Luna 3* in 1959, and the first landing of men on the Moon through the American Apollo program in 1969.

See also SPACE SCIENCE; SPACE STATION.

Robert A. Divine, *The Sputnik Challenge: Eisenhower's Response to the Soviet Satellite* (1993). Asif A. Siddiqi, *Challenge to Apollo: The Soviet Union and the Space Race, 1945–1974* (2000).

DANIEL J. KEVLES

STANDARDIZATION is ancient and ubiquitous, because community requires it, or at least functions better when standards of weights, measures, currency, morals, and the like exist. The nineteenth and twentieth centuries, however, saw a great extension and acceleration of movements toward standardization—in science, technology, industry, and society in general.

In a narrow sense, standardization refers to the establishment of specifications for the measurement, design, and performance of scientific and technological processes and products, enforced either voluntarily or by some authority. In a broader sense, the term "standardization" may be used to characterize wide-ranging historical movements in the nineteenth and twentieth centuries, on par with terms like "nationalism" and "liberalism."

The origins of the modern movements of standardization may be traced to the eighteenth century, when a quantifying and systematizing spirit began to grow among the Enlightenment *philosophes*. They took many of their cues from Isaac *Newton's triumphant system of the world. As he quantified and systematized the solar system in the seventeenth century, so they would extend his techniques to all fields and endeavors, from chemistry to forestry to public health to bureaucracy. Prominent results of this spirit included Linnaeus's system of natural classification, the great *Encyclopédie* of Diderot and d'Alembert, and the beginnings of the metric system in France.

Though the metric system did not always receive unqualified support from French revolutionaries in the 1790s because of the royalist connections of some of its proponents and the usual academic rivalries, the system did have the advantage of suppressing the standards and measures of the old order. Napoleon, in turn, did his part to spread the system by mandating its use in the states he conquered. After Napoleon, metric-based weights and measures reform became a key lever that the new centralized bureaucratic states of Europe used to establish their control over trade and taxes.

The same period saw the beginnings of the revolutions in transportation and communications centered around the railroad and the telegraph, which spurred movements for standardization on a number of levels. As railroads expanded from local to regional to national lines, mismatches in track gauge often occurred. Locals often worked against standard gauges to protect their commercial interests. Eventually, however, the forces encouraging railroad system integration overcame local barriers and inertia. In the United States the increase in east-west trade, the military urgencies of the Civil War, and the first transcontinental railroad forced standardization of gauges.

The expansion of the railroads also precipitated standardization throughout whole economies by encouraging the mass manufacturing and mass marketing of standardized products. The long-distance, high-volume trade in grain, for example, meant that buyers no longer could personally inspect and oversee each purchase. The new impersonality required the establishment of standardized units and grades of grain.

Similar developments during the second half of the nineteenth century may be observed in manufactured products, especially in the United States. Increased speed and other innovations in continuous-process manufacturing enabled the mass production of items such as matches, cigarettes, soap, dressed meat, beer, and canned goods in standard sizes and grades. The celebrated American system of manufacture—the use of standardized, interchangeable parts to enable the mass production of durable goods—had its origins in mid-nineteenth-century small-arms production, from which it spread to sewing machines, bicycles, and typewriters. Overall the implementation of the system was spotty and its full impact came only with the rise of the automobile industry in the twentieth century.

Railroads also acted to standardize timekeeping. For its schedules each railroad usually chose a standard based on the local time of the principal city it served, and thus, by means of its schedules, extended the use of that time to the surrounding communities. The expansion of the railroads initially caused a multiplication of time standards and confusion where railroads met. But lofty proposals from scientists to establish an international system of time, much debated in the 1870s, had little appeal to businesses and railroads. In the United States a mixed system of quasi-official standard time zones and a patchwork of local times persisted until 1918, when Congress mandated a national system.

These developments—the coming of the railroads, the telegraph, mass production, and mass distribution—unintentionally worked to erase local cultures and create national ones in the Western world. More intentionally, nationalistic leaders used the new public education, conscription,

and mass media of the late nineteenth century to turn "peasants into Frenchmen" or whatever nationality was at stake.

Nor did scientific, technological, and cultural standardization stop at national borders. Organizations like the International Telegraphic Bureau (founded 1865), the International Metric Union (founded 1875), and the Universal Postal Union (founded 1878) represented an international cooperative trend that prompted or forced reluctant states to yield sovereignty over what historically had been internal matters. The proliferation of international congresses in the second half of the nineteenth century, bringing together devotees of subjects ranging from the usual academic fields to interests such as world peace, the abolition of tobacco, and liturgical chanting, marked the trend on the non-governmental level.

Scientists were in the vanguard scheduling and promoting such congresses, and standardization was a topic that they often raised and debated. They continued to agitate for the universal acceptance of the metric system and also sought to systematize standards of measurement in other fields, such as electricity. The needs of the rapidly expanding *telegraph industry provided the initial push around 1860 for the development of a system of electrical units and standards. At a series of international electrical congresses between 1881 and 1904 electrical practitioners and scientists, including William *Thomson (Lord Kelvin), Hermann von *Helmholtz, John William Strutt (Lord Rayleigh), and Henry Rowland, worked to construct an international system of such units, based on the ohm, volt, and ampere. In addition to nationalistic rivalries, they struggled with the tension between scientific coherence and practical application. Defining magnetic units that had natural relationships to electrical units and also had convenient magnitudes for engineering work turned out to be impossible. Scientists also were enamored over the increasing capabilities of their instruments for precision measurement and the determination of the next decimal point. Most practitioners, meanwhile, felt it to be wasted effort.

A more subtle impediment was raised by the different legal and scientific cultures of the participants. The Germans and the English not only had varying legal approaches to setting standard weights and measures, they often were at odds on issues as fundamental as what constituted an acceptable measurement and the proper way to calculate, report, and interpret the errors in their results. By their very nature, standards are social creations of networks of people and institutions, and they therefore require shared values and a foundation of trust. Constructing that foundation of trust—in instruments, materials, methods, and measurers—is a delicate and often frustrating iterative process.

In the end the forum of the international electrical congress proved inadequate to the task. The organizations that ultimately and ironically succeeded in establishing an international system of units were the *bureaus of standards of Germany, England, and the United States. Though often rivals, the Physikalisch-Technische Reichsanstalt (founded 1887), National Physical Laboratory (founded 1899), and National Bureau of Standards (founded 1901) had common interests in standard-setting and keeping as well as the authority of their governments behind them. Their cooperative work in the first decade of the twentieth century, which also involved France's Laboratoire Central d'Électricité (founded 1888), led to a well-organized system by 1912.

World War I greatly expanded and accelerated the movements for standardization. The war motivated attempts to rationalize the industrial effort on the home front and eventually forced a transformation of the economies of the combatants to a more efficient, or at least more orderly, command structure. As one result, the war and its aftereffects spawned national standardizing agencies in nineteen countries in Europe, North America, and Asia between 1916 and 1924. Their work and the work of similar organizations ranged widely and deeply. In any given country, thousands of committees and subcommittees worked to develop standards of testing, dimensions, and quality for all products imaginable: from bolts to mattresses to biological and medical substances. The effort extended all the way to the shop floor, where Frederick W. Taylor and his followers in "scientific management" sought to determine and prescribe the most efficient motions of workers.

Nor did the standardizers exempt their own skills and conduct, though they had an ulterior motive. The second half of the nineteenth century saw many attempts to construct new scientific disciplines and engineering professions. A key means used to demarcate professionals from lesser practitioners was the establishment of standards of knowledge, skill, and behavior via university degrees and professional exams and codes of conduct. The creation of such standards requires that knowledge itself be organized and standardized. The standardizers helped to accelerate this ancient and ubiquitous process. Its end points today may be found in textbooks, teaching apparatus, competency exams, and encyclopedic companions.

Norman F. Harriman, *Standards and Standardization* (1928). David A. Hounshell, *From the American System to Mass Production, 1800–1932* (1984). Tore Frängsmyr, J. L. Heilbron, and Robin E. Rider, *The Quantifying Spirit in the 18th Century* (1990). Meinolf Dierkes and Ute Hoffmann, eds., *New Technology at the Outset: Social Forces in the Shaping of Technological Innovations* (1992). Theodore M. Porter, *Trust in Numbers: The Pursuit of Objectivity in Science and Public Life* (1995). M. Norton Wise, ed., *The Values of Precision* (1995). Jed Z. Buchwald, ed., *Scientific Credibility and Technical Standards in Nineteenth and Early Twentieth Century Germany and Britain* (1996).

LARRY R. LAGERSTROM

STANDARD MODEL. During the 1980s, physicists who worked on *elementary particles came to agree that matter consists of three pairs of leptons—very light and even weightless particles—and their antiparticles (of which *electrons and their corresponding neutrinos are the exemplars) and three pairs of *quarks and their antiparticles (of which protons, neutrons, and other heavy particles, or "baryons," are made). So much for the bricks. The mortar that holds the quarks together (the "strong force") comes in eight kinds of "gluons" (*see* TERMINOLOGY); the cement that binds leptons to one another and to quarks (the "electroweak force") consists of the photon (the electrical part of the force) and three adhesives, W^+, W^-, and Z^0 (the weak part). The detection of the W and Z particles and the top quark in 1982–1983 completed the experimental identification of the elements of the standard model. The successes of this model gave impetus to Grand Unified Theories (GUTs), intended to unify the strong and electroweak forces, and to dreams of Theories of Everything (TOEs).

Some particle physicists, notably Steven Weinberg (who won the *Nobel Prize in physics in 1979), have asserted that now that the guts of GUTs are in place, the final TOE will soon follow. A glance at earlier claimants to the status of "standard model" does not give cause for confidence in his prediction. To go back no further than 1800, the system of *imponderables developed by Pierre-Simon de *Laplace and his school seemed capable of describing all the phenomena then known in the same terms (though not the same language): several leptons (the weightless "fluids" of electricity, magnetism, heat, light, and so on), a baryon (the particles of "common matter"), and forces of attraction and repulsion. Many natural philosophers looked forward to a unified theory that would connect the various "fluids" (leptons), a project encouraged by the discoveries of radiant heat and *electromagnetism. But with difficulties in the theory of heat (*see* ENTROPY) and new fashions in science (*see* CONSERVATION LAWS; FIELD), the imponderable fluids evaporated. A new standard model was drawn up, based on the unification of light with electromagnetism, heat with kinetic energy, and magnetism with vortical motion, which strove to manage with one sort of material substrate, or *ether, subject to the laws of mechanics. In the most austere of these GUTs, the "vortex atom," developed especially by William *Thomson (Lord Kelvin), James Clerk *Maxwell, and Joseph John *Thomson, all physical phenomena were to be referred to motions of a single, perfect, incompressible space-filling medium.

The program of mechanical reduction, the standard model of the late nineteenth century, collapsed under experimental discoveries (*see* RADIOACTIVITY; X RAYS) and difficulties in the theories of radiant heat and electrodynamics (*see* QUANTUM THEORY; RELATIVITY). The discovery of the *electron and subsequent speculation about *atomic structure suggested that matter might be built from three ingredients: in today's language, a negative lepton (electron), a positive baryon (proton), and, after the Compton effect, a neutral photon. But the study of the nucleus (*see* NUCLEAR PHYSICS AND NUCLEAR CHEMISTRY) and *cosmic rays between the world wars, and the building of ever more powerful *accelerators after World War II, revealed many more "particles" than three. The gigantic instrumental and theoretical effort to classify and comprehend this cornucopia resulted in the standard model of the 1980s.

Like the decuplet of five pairs of twin brothers who ruled Atlantis, the standard model with its three pairs of leptons, eight gluons, and so on, will slip into the sea. Its place may be taken by uncountable numbers of unimaginably small, wriggling, vibrating bits, as in string theory, the latest candidate for the Theory of Everything.

Yuval Ne'eman and Yoram Kirsh, *The Particle Hunters* (1983).

J. L. HEILBRON

STAR. To pretelescopic astronomers the night sky hemisphere contained about three thousand visible stars, grouped into constellations whose shapes did not vary over thousands of years. This pattern formed a backdrop against which the orbital wanderings of the planets could be traced. The most prominent star was only about a hundred times brighter than the faintest. Hipparchus of Nicaea classified star brightness into six degrees of importance, a division that has formed the basis of today's stellar magnitude system.

From Antiquity to the *Renaissance there had been debate over whether the Sun, a heat-giving, golden disc subtending half a degree at Earth, had anything in common with the cold stellar pinpoints seen at night. To Giordano Bruno, the Sun and stars differed only in their distance from Earth. He also suggested that stars moved. Edmond Halley proved as much in 1718 by finding that Sirius, Aldebaran, Betelgeuse, and Arcturus had changed position since Ptolemy drew up his catalogue fifteen hundred years earlier.

The Ptolemaic cosmos had the stellar sphere just beyond the orbit of Saturn. The Copernican heliocentric hypothesis, post-1542, had Earth moving 300 million kilometers every six months. The implied stellar parallax was eagerly sought. In 1632 *Galileo Galilei suggested that monitoring the separations and angular positions of optical double stars might provide some parallaxes. This approach eventually succeeded around 1838 when Friedrich Wilhelm Bessel, Friedrich Struve, and Thomas Henderson observed the parallax of 61 Cygni (0.293 arc seconds), giving this star a distance of 3.4 parsecs (9.8 light years). Now astronomers could compare the solar luminosity with that of other stars. The number of stars with known distances became an effective aid to astrophysics around 1903 when photographic measurements at the Yerkes and Allegheny Observatories in the United States reached a precision of about 0.01 seconds.

Galileo's telescopic observations of 1609 revealed that the Milky Way consisted of a myriad of faint stars. Isaac *Newton and Edmond Halley convinced themselves (between 1692 and 1720) that the starry realm was infinite, and balanced by gravitation. The discovery of the Doppler principle in 1842 revolutionized the study of stellar motion. Radial velocities could then be combined with proper motions perpendicular to the line of sight. In the second half of the nineteenth century, photography further transformed stellar research. Photographic spectra came in 1872 and photographic cataloguing and sky mapping in 1882.

Earlier, around the turn of the century, William Herschel had telescopically observed optically close stars and discovered that many were gravitationally associated binaries orbiting a common center of mass. Astronomers soon realized that about 50 percent of the stars were single, like the Sun, the majority of the remainder being binaries. In 1912 Henrietta Leavitt found that pulsating giant Cepheid stars have periods that are a function of their luminosity.

Early-nineteenth-century estimates of stellar surface temperatures made using Newton's law of cooling gave temperatures far too high. Correct values were obtained by calculating with the radiation laws of Josef Stefan (1879) and Wilhelm Wien (1896). Stellar surface temperatures came out in the range of 3000° to 40,000°K. Stellar spectral classification blossomed. The astronomical spectroscopist William Huggins showed that stars and Earth were made of similar chemical elements.

Around 1914 Ejnar Hertzsprung and Henry Norris Russell independently plotted logarithmic graphs of stellar luminosity against surface temperature. This Hertzprung-Russell diagram indicated two main types of stars, dwarfs (like the Sun) and giants (some one hundred times larger). The discovery of planet-sized white dwarfs and huge supergiants soon followed. The H-R diagram acted as a basis for the study of stellar evolution.

Before 1840, when most people still regarded the universe as relatively young, few worried about the source of solar (and stellar) energy. Some suggested mass accretion, others solar contraction. By the mid-nineteenth century the accepted age of the universe had increased to 400 million years and the problem of the generation of stellar energy became acute. Radioactive decay (discovered in 1896) appeared to be a possible energy provider (see RADIOACTIVITY). Albert *Einstein's equivalence of mass and energy (see RELATIVITY) became the cornerstone of stellar energy generation when in 1919 Francis Aston, using a mass spectrometer, found that a helium atom weighed less than four hydrogen atoms. Around 1926, Arthur Eddington estimated that the central stellar temperatures of 10^{7}°K were high enough to enable the implied transmutation of hydrogen into helium to occur. Energy generation by the transformation of elements indicated that only a little loss in mass took place during stellar evolution. In 1944 Albrecht Otto Johannes Unsöld found that sun-like stars could "burn" nuclearly for about 10^{10} years, whereas hot, luminous O and B stars would have lifetimes as short as 10^{6} years and would still be very close to where they were born.

In 1921 Meghnad Saha proposed a theory of thermal ionization and excitation that made possible the calculation of stellar chemical composition from spectra. By 1925, Cecilia Payne-Gaposkin realized that hydrogen and helium comprised 99 percent of stellar material. Around 1943 Walter Baade discovered groups of younger stars rich, and groups of older stars poor, in metals. In 1957, Fred Hoyle and William Fowler showed how nuclear fusion synthesized all the other elements.

In 1924 Eddington had discovered that all star masses lay in the range of 0.05 to 100 solar masses and that the luminosity of a Main Sequence star (that is, the diagonal band of stars on the H-R diagram, all of which generate energy by converting hydrogen to helium) was approximately proportional to the fourth power of its mass. Thus stars did not evolve along the Main Sequence; they moved to the right as they ran out of hydrogen. From the early 1950s, observations of this break-off point were related to the age of stellar clusters.

Recent observations of pulsating radio stars have confirmed that stars more massive than the Chandrasekhar mass (suggested by Subrahmanyan Chandrasekhar in 1931 to be about 1.4 solar masses) do not stabilize at the white dwarf stage (by which point the star is about the size of the earth and consists mainly of heavy elements), but continue condensing to become neutron stars, about 20 km across. Stars more massive than about 3.2 solar masses (the Oppenheimer-Volkov mass, calculated in 1939) become black holes.

See also ASTRONOMY; COSMOLOGY; SOLAR PHYSICS.

Agnes M. Clerke, *A Popular History of Astronomy During the Nineteenth Century* (1885). Martin Johnson, *Astronomy of Stellar Astronomy and Decay* (1950). A. Pannekoek, *A History of Astronomy* (1961). Otto Struve and Velta Zebergs, *Astronomy of the 20th Century* (1962). Martin Harwit, *Cosmic Discovery: The Search, Scope, and Heritage of Astronomy* (1981). Dieter B. Herrmann, *The History of Astronomy from Herschel to Hertzsprung* (1984).

DAVID W. HUGHES

STATISTICAL MECHANICS. See THERMODYNAMICS AND STATISTICAL MECHANICS.

STEADY-STATE UNIVERSE. Steady-state theory was conceived at a movie theater in 1947. The British Astronomer Fred Hoyle, accompanied by Hermann Bondi and Thomas Gold, Austrian-born physicists who had worked with Hoyle on *radar during World War II, saw a ghost story that ended the same way it began. That inspired thoughts about a universe unchanging yet dynamic. According to Hoyle, "It did not take us long to see that there would need to be a continuous creation of matter."

A universe unchanging in density holds a philosophical advantage over a *big-bang, expanding universe. If density changes, various physical laws might also change, invalidating extrapolations from the present back to a superdense origin of the universe. Steady-state theory also enjoyed an observational advantage over big-bang theory in 1947. The estimated rate of expansion extrapolated back to an initial big bang yielded an age for the universe that was less than the estimated age of the solar system.

Scientific arguments about the steady-state theory in Great Britain turned on philosophical questions, with little appeal to observation. Religion and politics also played a part. Pope Pius XII suggested in 1952 that big-bang cosmology agreed with the notion of a transcendental creator, and was in harmony with Christian dogma—an extrapolation for which he was later criticized at the Second Vatican Council. Steady-state theory may thus have been too tainted with atheism.

Soviet astronomers rejected both steady-state and big-bang cosmologies as idealistic and unsound. Hoyle associated steady-state theory with personal freedom and anti-Communism. Observational challenges to steady-state theory came from the new science of radio astronomy. Martin Ryle at Cambridge University reported in 1955 that his survey of almost 2,000 radio sources contradicted steady-state theory. His conclusion, however, was premature. Hoyle felt bitterly that Ryle was motivated not by a quest for the truth, but by a desire to destroy steady-state theory. For his part, Ryle did not respect theoretical cosmologists. The final blow against steady-state theory came in 1965 with the discovery of cosmic microwave background radiation.

Hoyle responded that cosmic background radiation could arise from interaction between stellar radiation and interstellar needle-shaped grains of iron. But few found the response persuasive. For most purposes, the big bang had defeated the steady state by 1965.

See also COSMOLOGY; UNIVERSE, AGE AND SIZE OF THE.

Norriss S. Hetherington, ed., *Encyclopedia of Cosmology: Historical, Philosophical, and Scientific Foundations of Modern Cosmology* (1993). Helge Kragh, *Cosmology and Controversy: The Historical Development of Two Theories of the Universe* (1996).

NORRISS S. HETHERINGTON

STELLAR ABERRATION. See ABERRATION, STELLAR.

STEREOCHEMISTRY. Although the founder of the atomic theory, John *Dalton, speculated about the three-dimensional arrangements of atoms within molecules, chemists customarily specified formulas without any indication of the structure of molecules, much less the arrangements of their atoms in space, until the 1850s.

Jean-Baptiste Biot noted in 1811 that quartz crystals were "optically active"; when he directed polarized light through them, the plane of polarization rotated. This asymmetrical effect on light suggested that the atoms of the crystals might be arrayed asymmetrically. Further studies revealed that certain organic materials were also optically active, even in solution. Since a dissolved substance cannot have a crystal array of any sort, the asymmetry in these instances had to inhere in each molecule.

In his first important scientific investigation (1848), Louis *Pasteur studied the crystal structures of the salts of the optically active tartaric acid and of an isomeric substance, the optically inactive "racemic" acid. He discovered that a certain racemate salt consisted of intermixed crystals, all of which appeared to be identical. More careful examination, however, revealed minute differences of form, from which Pasteur could distinguish mirror-image pairs of crystals, half right-handed and half left-handed. Separating them painstakingly by hand, Pasteur showed that the two mirror-image crystals rotated polarized light in equal but opposite directions. It was then clear that the natural racemate was optically inactive only because the two kinds of crystals were normally present in equal numbers, canceling out each other's activities. The artificially separated right-handed racemate was identical in optical activity to the naturally active tartrate.

Pasteur's research gave impetus to the chemical study of the three-dimensional structure of molecules (later called "stereochemistry"), but real progress came only after the formulation of the theory of chemical structure (1858), when chemists first began to have confidence in their ability to discern molecular architecture. The chief architect, August *Kekulé, regarded designing molecules in three dimensions as still too difficult, hence premature. Kekulé proposed that every carbon atom could form bonds to precisely four other atoms—that it was tetravalent. In this way carbon atoms could link up together to form chains, the backbones of organic molecules.

A Dutch student of Kekulé's, Jacobus Henricus van't Hoff, provided the first substantial development of stereochemistry in 1874. (Entirely independently, a Frenchman, Joseph Le Bel, published a substantially similar exposition just two months later. Curiously, the two men had studied together with the noted Parisian chemist Adolphe Wurtz earlier that year, but apparently owed nothing to one another.) Van't Hoff taught that the four bonds of the carbon atom should be positioned as far apart from each other as possible, or (equivalently) that they be symmetrically situated in space. Uniform distribution of four bonds coming from a central point in two dimensions would place them at right angles, but in three dimensions they would have to be arranged tetrahedrally, with angles of about 109 degrees between adjacent bonds.

Van't Hoff demonstrated the empirical power of this purely theoretical idea by referring to the numbers of isomers known for various formulas. If carbon compounds were flat, there would have to be many more isomers than chemists had found to date. If carbon bonds formed symmetrically in three dimensions, on the other hand, the predicted isomer numbers matched those known. A crucial case emerged whenever the four groups attached to a given carbon atom were all different. Van't Hoff showed that in the tetrahedral bonding situation, exactly two different arrangements of the four groups in the form of nonsuperimposable mirror images would be possible. One of these molecules would be right-handed, the other left-handed. An organic molecule with such asymmetry should normally exhibit optical activity.

And so it was, at least to a first approximation. The accepted formula of Pasteur's tartaric acid showed the presence of two asymmetric carbons. The two possible mirror-image formulas centered on one of these atoms suggested the existence of a right- and left-handed version of the acid. Such "optical" isomers became known as "enantiomers," and any equal mixture of two enantiomers became known in the general case as a "racemic mixture." After a decade of hesitation, most chemists accepted the van't Hoff–Le Bel theory, and began to pursue the new field of stereochemistry with enthusiasm.

The earliest supporter of van't Hoff had been the Leipzig chemist Johannes Wislicenus. In 1887 Wislicenus published a pathbreaking study of what he called "geometric" isomerism, a kind of stereoisomerism not related to optical activity. In Heidelberg, Victor Meyer (who coined the term "stereochemistry") demonstrated the chemical effect of large space-filling groups on organic molecules; he called this phenomenon "steric hindrance." In Zurich, Arthur Hantzsch and Alfred Werner investigated the stereochemistry of organic compounds containing nitrogen, and Werner later applied stereochemical precepts to the chemistry of inorganic coordination compounds. Emil Fischer in Berlin used stereochemistry to interpret his fundamental research on carbohydrates.

These examples only highlight the vigorous activity in the new field toward the end of the century. In the next generations, many new ideas and research programs appeared alongside established ones. From early important work by Adolf von Baeyer on the stereochemical analysis of cyclic organic compounds came a substantial research program on "conformational analysis." Other chemists showed that Meyer's phenomenon of steric hindrance gave rise not just to stereoisomers, but in certain circumstances to optical activity. The problems of determining not just the relative, but the absolute configurations of atoms have been solved. The stereochemistry of ever more complex natural products was slowly unraveled. Today, chemists talk about the orientation of invisibly small atoms in space as comfortably as they talk about directions on a street map.

A. J. Ihde, *Development of Modern Chemistry* (1964). J. R. Partington, *A History of Chemistry*, vol. 4 (1964). O. B. Ramsay, ed., *Van't Hoff–LeBel Centennial* (1975). O. B. Ramsay, *Stereochemistry* (1981). W. H. Brock, *The Norton [Fontana] History of Chemistry* (1992).

A. J. ROCKE

STETHOSCOPE invented by R. T. H. Laennec of Paris in late 1816 or early 1817, consisted at first of a rolled paper notebook, one end of which was applied to a patient's chest, the other to the examiner's ear. Laennec soon switched from paper to jointed wooden tubes turned on a lathe. He derived the name from the Greek στηθος (chest) and σκοπειν (to explore). Stethoscopes with two earpieces and flexible tubes were not invented until two decades after his death. When combined with percussion, listening through a stethoscope (or auscultation) made it possible to detect anatomical changes in the chest. By analyzing the sounds made in the hearts, lungs, and voices of both healthy and sick people, and by performing autopsies, Laennec developed a vocabulary of auditory signs linked to lesions found in the organs. For example, an increased voice sound called "pectoriloquy" indicated a lung cavity. With unprecedented accuracy, Laennec could predict anatomical findings before his patients died. His *Traité de l'auscultation médiate* (1819) appeared less than three years after his invention. Because of the prevalence of tuberculosis, physicians adopted the stethoscope widely and quickly, and soon applied it also to pregnancy, fractures, and bladder stones.

The concept of auscultation had an even greater impact on medicine than its embodiment in the stethoscope. It enhanced the relevance of anatomy to medicine by changing disease concepts from constellations of subjective symptoms felt by the patient to anatomical alterations detected objectively by the physician. Laennec was among the first to identify organic lesions, such as early tuberculosis, before the patient suffered any symptoms. Consequently, his stethoscope became a double-edged symbol: on the one hand, of medicine's technological emphasis on physical diagnosis; on the other, of a shift in power from the patient to the doctor.

Stanley Joel Reiser, *Medicine and the Reign of Technology* (1979). Jacalyn Duffin, *To See with a Better Eye: A Life of R. T. H. Laennec* (1998).

JACALYN DUFFIN

STRANGENESS. In the first several years after World War II, physicists found evidence for a flurry of new particles in cosmic rays and the products of accelerators. In addition to the muon and π-meson, the new particles included V particles, named after the forked tracks they left in cloud chambers and nuclear emulsions, and the K meson or kaon. By the early 1950s experimenters had distinguished several different V particles (now known as Λ, Σ and Ξ). The high production rate of kaons and V particles suggested that they were governed by the strong force, but they decayed slowly with lifetimes typical of weakly interacting particles.

Physicists thought this behavior strange and sought a theory for the so-called strange particles. In 1951 Abraham Pais in the United States and a group of Japanese physicists independently proposed that kaons and V particles could be produced only in pairs, although they could still decay individually via weak interactions. The notion of associated production seemed to reconcile the strong production rates with the weak decay modes. In 1953 Murray Gell-Mann and, independently, Kazuhiko Nishijima and Tadao Nakano refined the idea, proposing that the strange particles carried a new quantum number, which Gell-Mann called the strangeness quantum number, S. Pions, muons, and nucleons would have $S = 0$, but the strange particles would have nonzero, integer strangeness: for example, $S = +1$ for kaons and $S = -1$ for the existing V particles. According to the theory, strangeness was conserved in strong interactions, but not in weak ones, thus indicating what interactions were possible. As many more new particles were produced in the 1950s they were assigned strangeness values, and strangeness would prove a useful guide to the particle zoo.

"Strangeness" was a strange term. Physicists had previously appealed to Greek to name new particles, as indicated by the deuteron and mesotron or the atom itself, or to physical concepts such as spin. Strangeness represented a whimsical turn in the argot of physics that would flower in later neologisms such as quarks, flavors, and colors. This represents in part the linguistic interests of Gell-Mann. But terms such as strangeness also represent an approach characteristic of American physics, which spurned classical culture and philosophical ruminations for a more pragmatic engagement with the subject (*see* TERMINOLOGY).

Andrew Pickering, *Constructing Quarks: A Sociological History of Particle Physics* (1984). Laurie M. Brown, Max Dresden, and Lillian Hoddeson, *Pions to Quarks: Particle Physics in the 1950s* (1985).

PETER J. WESTWICK

STRATIGRAPHY AND GEOCHRONOLOGY. The principles of stratigraphy—the study of the earth's strata or layers of sedimentary rock—and of geochronology—the naming and describing, though not necessarily dating, of the periods of earth history—were established rapidly between 1810 and 1840. For the next century, stratigraphers filled in the details of the stratigraphic column with ever-greater precision. Much of this research could be put to good use by the mining industry, and from the 1920s and 1930s by the petroleum industry.

Although stratigraphy flowered in the first half of the nineteenth century, it had its roots in the seventeenth century. The Danish Cartesian Niels Stensen (or Steno), in his *Produmus to a Dissertation on Solids Naturally Contained within Solids* (1669), considered bodies that made up the earth, particularly fossils, crystals, and strata. In any sequence of undisturbed strata, he concluded, the oldest strata would be on the bottom and the youngest on the top. This was an early version of the first of the three major principles of stratigraphy—the principle of superposition.

The second principle—that rock types or lithology usually occur in a predictable sequence—followed from the work of eighteenth-century mineralogists in the German states, Italy, France, the British Isles, and Russia. Independently of one another, they became convinced that the strata of the earth occurred in the same order everywhere. On the small scale, they knew that in an individual mining area, for example, they would find the same rocks in the same sequence in adjoining shafts. On a grander scale, they believed that around the globe, the rocks could be sorted into three main groups that appeared to represent a time sequence: the primary rocks, hard and often crystalline; the secondary rocks, softer, layered, and often fossiliferous; and the tertiary rocks, the topmost and softest rocks. Unfortunately, the principle of superposition failed when strata had been disturbed subsequent to deposition, and the principle of lithological regularity broke down when rocks of the same lithology occurred more than once in the sequence.

By the second decade of the nineteenth century, the third principle of stratigraphy—that fossils can be used to identify and correlate strata—had been established. For the next century and a half, *paleontology was to be chiefly a tool for stratigraphy. Armed with these three principles, geologists between 1820 and 1840 established and named the greater part of the stratigraphic column, an accomplishment that has held up in outline to the present day. In practice, it involved one controversy after another about particular puzzles in the sequence. The British played a large part, perhaps because the strata of England are relatively straightforward.

In 1815, a mineral surveyor, William Smith, had published a map of the strata of England that although not fully correct, made a good start. Charles *Lyell gave names to the epochs of the Tertiary—Pleistocene, Pliocene, Miocene, and Eocene—and distinguished them by their proportion of still extant fossils. Adam Sedgwick renamed the older part of the Secondary formations, the Paleozoic. With Roderick Murchison, he introduced the names Cambrian, Silurian, and Devonian. The Carboniferous and the renaming of the upper part of the older Secondary Mesozoic were English suggestions. The renaming gave birth to Permian (another Murchison coinage), Triassic (a German suggestion), Jurassic (largely French), and Cretaceous (Belgian). The establishment of geochronology was a magnificent achievement. Museum panoramas and book illustrations showing the development of life forms still encapsulate what the general public understands by geology.

More important within professional geology were stratigraphic maps, topographic maps colored to show the strata that outcrop at the surface of the earth. Geological mapping developed with great speed between the 1780s and the 1830s when most of the techniques employed until World War II were at hand. As *cartography progressed and accurate topographic base maps that showed change of elevation by contours became more widely available, the task of making geological maps became easier. Stratigraphers used maps both as a record of their fieldwork and as a way to extract new information. Using them in conjunction with the stratigraphic column—a theoretical reconstruction of the strata arranged according to age—and the section—the vertical arrangement of strata along some line or traverse across the surface of the map—they could construct a mental picture of the three-dimensional structure of the strata and thus predict what would be found beneath any spot on the earth's surface.

Until the 1950s, most geological education gave priority to teaching students to construct and interpret maps, and professional stratigraphers were largely occupied with mapping the earth's surface. During the nineteenth century, they extended their mapping beyond northwestern Europe. They resolved problems about the Cambrian-Silurian boundary by introducing the Ordovician Period. American stratigraphers found that the Carboniferous did not work well for their territory, and replaced it with Mississippian and Pennsylvanian. The Canadians began trying to make sense of the afossiliferous pre-Cambrian rocks that made up much of their country.

Stratigraphers saw themselves as men who traveled widely, scaling mountains and descending mines, hammer, notebook, and map in hand, returning to their bases with packages of fossils and rocks to add to growing collections. Some were independently wealthy, but most found employment in universities and national geological surveys. In 1878, at the first International Geological Congress in Paris, they began the huge task of codifying stratigraphic nomenclature, a task that still continues.

Stratigraphers puzzled about how to reconcile the distinct breaks in the fossil record with the gradual changes predicted by evolutionary theory, writing at length about how elevation and erosion had destroyed part of the record. They worried that fossils might indicate changes in the environment of deposition rather than in the time of deposition. Walking through a gorge with sloping strata might be a walk through time or it might be a walk through space, from the deep ocean to a continental shelf to a brackish delta. With the growth of the petroleum industry, further subdividing the stratigraphic sequence became a necessity. New tools were developed, such as well logs and microscopical examination of microfossils, particularly foraminifera.

By World War II, the intellectual excitement in stratigraphy had evaporated. It revived after the war when stratigraphy was subsumed under the *earth sciences, with their host of new concepts and sophisticated instruments.

Claude C. Albritton, Jr., *The Abyss of Time: Changing Conceptions of the Earth's Antiquity after the Sixteenth Century* (1980). Barbara M. Conkin and James E. Conkin, eds., *Stratigraphy: Foundations and Concepts* (1984). James A. Secord, *A Controversy in Victorian Geology: The Cambrian-Silurian Dispute* (1986). Martin J. S. Rudwick, *The Great Devonian Controversy* (1985). Stephen Jay Gould, *Time's Arrow, Time's Cycle* (1987). Brian W. Harland et al., *A Geologic Time Scale* (1990). David R. Oldroyd, *The Highlands Controversy* (1990). Robert H. Dott, Jr., ed., *Eustasy: The Historical Ups and Downs of a Major Geological Concept* (1992).

RACHEL LAUDAN

SUNDIAL. Today's garden variety, horizontal sundial is an impoverished relic of a once vibrant and challenging mathematical discipline. At its zenith between the sixteenth and eighteenth centuries, dialing attracted a huge investment of ingenuity in the design of different dials, mostly portable. Many accounts of the art and of special dials appeared, either as separate tracts or as dialing sections within more general books on astronomy or practical mathematics. At first the only source of personal timekeeping, portable dials survived the coming of pocket watches because watches still needed to be set by reference to the sun. Many museum and private collections testify to the great quantity and diversity of early-modern sundials, probably the most widely distributed scientific instrument of the period. The quantity and quality suggest that the level of everyday astronomical learning was higher in the sixteenth century than it is today.

Telling time by shadows is recorded in Egypt from about 1450 B.C. Ancient China also practiced the art; Herodotus says that the Greeks learned dialling from the Babylonians. In the Old Testament, King Ahaz (eighth century B.C.) received a sign by God in the contrary movement of the shadow on his dial. The Roman architect Vitruvius mentions thirteen different types of dial in his treatise on architecture in the first century A.D.

In the familiar horizontal or vertical dial, the gnomon, which casts the shadow, points to the pole and the hour lines are projected onto the horizontal or vertical plane. Such a dial works throughout the year without adjustment; it does not depend on solar declination since each daily path of the sun takes place in a plane perpendicular to the gnomon, but it will be accurate in only one latitude. If the hour lines are placed on a plane parallel to the equator and perpendicular to the gnomon, they will be equally spaced and the dial easily adjusted for latitude, which makes the "equatorial" or "equinoctial" dial appropriate to portable dials in various forms. Horizontal dials can be made portable through the addition of a magnetic compass for orienting the gnomon in the meridian, as in the ivory diptych dials with string gnomons from sixteenth- and seventeenth-century Nuremberg. Butterfield dials, mostly from eighteenth-century France, also are portable horizontals. Both types can be made adjustable over a limited range of latitudes.

Some dials work not from the progression of the hour angle or right ascension of the sun, but from the solar altitude. The pillar or shepherd's dial, a simple altitude dial known from the Roman period, was used into the twentieth century. The user reads the time from the length of the shadow cast by a horizontal gnomon along a vertical scale selected for date. Horary quadrants, in a variety of designs, also usually depend on a measurement of solar altitude. The design of Edmund Gunter in 1623 is one of the most common. Other notable mathematicians, such as Johannes Regiomontanus and William Oughtred, made special dials, but the examples mentioned will have to stand for a great variety of designs whose originality, novelty, and diversity were cultivated to engage the interest of customers and exercise the ingenuity of mathematicians.

René R. J. Rohr, *Sundials: History, Theory and Practice* (1996). Hester Higton, *Sundials: An Illustrated History of the Portable Dial* (2001).

JIM BENNETT

SUPERNOVA. Supernovae have been observed on several occasions, recently and spectacularly in 1885, when Ernst Hartwig saw a new star brighten the Andromeda galaxy by 25 percent. Six months later this supernova was ten thousand times fainter.

In 1911 the American astronomer Edward Charles Pickering differentiated between low-energy novae, often seen in the Milky Way galaxy, and novae seen in distant nebulae like Andromeda. By 1919 Knut Lundmark had realized that low-energy novae occurred commonly, whereas the brighter novae, up to tens of thousands times more luminous, occurred rarely. The 1920s saw two theories of these brighter novae (named "supernovae" by Fritz Zwicky and Walter Baade in 1934), one relying on runaway instabilities in stellar interiors, the other (by Alexander William Bickerton) suggesting that collisions had occurred between stars.

Zwicky started the first supernova detection patrol in 1933; J. J. Johnson joined him in 1936. Using the new 45-cm Palomar Schmidt *Telescope they found twelve new supernovae in three years based on 1625 photographs of 175 extragalactic regions. The new 1.2-m Palomar Schmidt came into use after 1958, and by 1977 the supernova tally had reached 450.

In 1981 Gustav Tammann estimated that around three supernovae occurred every century in the Milky Way galaxy. Most go undetected due to obscuring interstellar material. During the last millennium local supernovae were detected only in 1006, 1054, 1572, 1604, and 1667.

The Taurus supernova of 1054 was extensively recorded in the East, being visible in daylight and reaching −5 in the magnitude scale. (The magnitude scale is a logarithmic scale of stellar brightness in which the brightest naked eye star is of magnitude 1, the faintest of magnitude 6. Hence, a negative magnitude denotes a body that is brighter than the brightest naked-eye star.) John Bevis discovered the expanding cloud of material that resulted in 1731. In 1758 Charles Messier labeled the cloud M1, the first entry in his catalogue of *nebulae. By 1937, O II, O III, N II, and S II emission lines (spectral lines emitted by excited atoms as they decay) had been found in the cloud. Owing to the large expansion velocity produced by the stellar explosion, the emission lines were particularly broad. After 1948 astronomers found several supernova radio sources (see ASTRONOMY, NON-OPTICAL).

In 1934, two years after the discovery of the neutron, Baade and Zwicky suggested that supernovae arose when giant stars became neutron stars. Their view became generally accepted after Jocelyn Bell Burnell discovered *pulsars in 1967. The Crab Nebula pulsar came to light in 1969. Over 120 supernova remnants have been discovered in the Milky Way. One type, exemplified by Cassiopeia A and the Veil nebula in Cygnus, has a ring-like structure. Others are irregular with a central brightening, like the Crab.

In the mid-1950s, astronomers recognized two supernova varieties. Type I are binary white dwarfs. Mass accretion, pushing the star beyond the Chandrasekhar limit (see STAR), triggers a wave of nuclear reactions and a flood of neutrinos, either destroying the star completely or leaving behind a neutron star. Type II results from the explosion of a young, massive giant star that has exhausted its nuclear fuel. In February 1987 a Type II supernova exploded nearby, in the Large Magellanic Cloud. The pre-nova star was a supergiant. In exploding, its brightness increased by 10^8 in a few hours. The visible energy release of 10^{44} joules was dwarfed by the 10^{46} joules of high-energy neutrinos, many of which were captured by atomic nuclei thus manufacturing elements heavier than iron. The explosion scattered these elements far and wide throughout the galaxy.

David H. Clark and F. Richard Stephenson, *The Historical Supernovae* (1977). Michael Hoskin, ed., *The Cambridge Concise History of Astronomy* (1999).

DAVID W. HUGHES

SURVEYING INSTRUMENTS. See INSTRUMENTS, SURVEYING.

SYMMETRY AND SYMMETRY BREAKING. The modern scientific notion of symmetry begins with the geometric symmetries of objects, both mathematical and physical. A perfect snowflake rotated through 60° about its center is indistinguishable from its original appearance. Rotation through 90°, however, yields an appearance distinguishable from both. Rotating the snowflake transforms it relative to something external. Symmetry transformations of an object leave the initial and final states indistinguishable (at least with respect to the properties we specify as relevant). This concept of symmetry—indistinguishability under transformations—has blossomed in science over the past 400 years. Here, three developments are fundamental: the extension of the concept to "physical symmetries," the development of group theory and its scientific applications, and the increasing importance of "symmetry-breaking."

In science, the distinction between geometric and physical symmetries is the distinction between symmetries of objects and of laws. An object may fail to possess a given geometric symmetry, and still evolve in accordance with laws that do possess that symmetry. For example, a chair is not rotationally symmetric (turn it through any angle other than 360° and the initial and final positions will be distinguishable), but since the laws of nature are rotationally symmetric in the absence of external influences, the natural behavior of the chair does not change with the direction it faces.

*Galileo made an early and famous application of a physical symmetry in the debate over the system of *Copernicus. Opponents of heliocentrism claimed that if the Earth moved around the Sun, the behavior of terrestrial objects would show it. In his *Dialogue Concerning the Two Chief World Systems* (1632), Galileo claimed that no such observations are possible, and he argued for this using an analogy with a ship. He pointed out that someone shut up in a windowless cabin on a ship would be unable to distinguish by means of any experiments carried out within that cabin whether the ship was at rest or in smooth, uniform motion. This so-called "Galilean relativity" is a symmetry of space and time; it quickly found its way to the heart of seventeenth-century natural philosophy, being used by Christiaan *Huygens in his solution to the problem of colliding bodies, and appearing in *Newton's *Principia* as Corollary V to his laws of motion. The Galilean group of symmetries also includes spatial translations and rotations, and temporal translations. The principle of relativity remains at the heart of modern physics as one of the two postulates of *Einstein's 1905 special theory of *relativity. Here, however, it belongs to a different group of space-time transformations, the Poincaré group. This brings us to the second key development: group theory in mathematics.

The group concept emerged from developments in late eighteenth- and early nineteenth-century mathematics. In the early 1830s, Evariste Galois used discrete groups (groups consisting of a finite number of elements) to characterize polynomial equations via the structural properties of their solutions. In the 1870s, Sophus Lie set about extending Galois's theory from algebraic equations to differential equations, and this led him to the concept of continuous analytic groups (Lie groups). Felix Klein's 1872 "Erlanger Program" used group theory to characterize geometries, putting *non-Euclidean geometry (so important for the general theory of relativity) on an equal footing with Euclidean geometry.

One of the first applications of group theory in science was in *crystallography. René-Just Haüy had used symmetry to characterize and classify crystal structure and formation in his *Traité de minéralogie* (1801). With this application, crystallography emerged as a discipline distinct from mineralogy. From Haüy's work, two strands of development led to the 32 point transformation crystal classes and the 14 Bravais lattices, all of which may be defined in terms of discrete groups. These were combined into the 230 space groups by E. S. Fedorov (1891), Artur Schönflies (1891), and William Barlow (1894). The theory of discrete groups continues to be fundamental in *solid state physics, *chemistry, and *materials science, and in *quantum field theory through the CPT (charge conjugation, parity, and time-reversal) theorem.

Continuous symmetries come in two kinds: global symmetries, such as Galilean translations and rotations, and local symmetries, such as the gauge symmetry of *electromagnetism and the invariance under general coordinate transformations of the field equations of general relativity (1915). The importance of continuous symmetries in physical theories, and the power of symmetries in theory construction, was increased in 1918 when Emmy Noether proved the existence of a general connection between continuous symmetries and conserved quantities, and shed new light on the structure of theories with continuous local symmetries. Group theory and symmetries can provide powerful constraints on theories. For example, in particle physics global symmetries are used to classify particles and to predict the existence of new particles, such as the omega-minus particle (predicted in 1962, detected in 1964) via the SU(3) symmetry classification scheme. In 1918, Hermann Weyl introduced local scale symmetry to construct his unified theory of gravitation and electromagnetism, intended to succeed the general theory of relativity; this theory failed, as did the 1954 proposal of Chen Ning Yang and Robert Mills, today credited as the first "modern" local gauge theory. Following the developments of the 1970s, however, theories with local gauge symmetry have come to dominate fundamental physics.

Symmetry breaking has become as important in modern science as symmetry itself. In 1894, Pierre Curie highlighted the importance of symmetry breaking and asserted the so-called "Curie principle"—that an effect cannot be less symmetric than its cause. This assertion was challenged in the 1950s in two ways. First, the phenomenon of spontaneous symmetry breaking came to the attention of physicists in the context of superconductivity (and was later reapplied in the context of *quantum field theory). In fact, the symmetric solution of a symmetric problem may be unstable; in such cases, the observed stable outcome will be less symmetric than the cause, in apparent violation of Curie's principle. Nevertheless, theoretically there exists a set of equally likely effects (only one of which is observed in any given instance) that together possess the symmetry of the cause, and in this "sophisticated" sense, Curie's principle survives the challenge of spontaneous symmetry breaking. The second challenge is the violation of parity, in which one possible outcome of an experiment dominates its mirror-image. This violation, predicted by Tsung Dao Lee and Chen Ning Yang in 1956, was detected soon afterward experimentally by Chien Shiung Wu and her colleagues. The law governing the weak nuclear interaction breaks the symmetry, and the Curie principle can only be saved by including the law within the cause.

During the latter half of the twentieth century, spontaneous symmetry breaking also became important in biology. Brian Goodwin describes one such application in *How the Leopard Changed Its Spots* (1994). All organisms start off as highly symmetric entities, such as a single spherically symmetric cell. As the organism grows, this highly symmetric state becomes unstable, owing either to internal stresses and strains, or to influences from the environment. Enter spontaneous symmetry breaking: the organism will move to one of a set of possible stable, but less symmetric, states. In this way, the dynamics of stability and spontaneous symmetry breaking constrain the possible general forms that an organism may take during its growth. Which of the possible states the organism moves to at each stage can be controlled internally (by a nudge from the DNA, for example) or by the environment (through temperature or a chemical). All this has radical implications for the theory of *evolution. On the standard Darwinian approach, evolution is free to explore a huge variety of possibilities, constrained only in general terms by the laws of physics and chemistry. This approach leaves us several major puzzles, two of the most important being the emergence of the same general forms in different lineages, and the fact that we don't see evolution exploring all possibilities, but instead a rather limited subset. A response to this is to suggest that the domain of the "biologically possible" is highly constrained by dynamical stability, in which spontaneous symmetry breaking plays a key role.

Hermann Weyl, *Symmetry* (1952). Ian Stewart and Martin Golubitsky, *Fearful Symmetry* (1992). Klaus Mainzer, *Symmetries of Nature* (1996).

KATHERINE A. BRADING

SYMPATHY AND OCCULT QUALITY. Aristotelian physics was strong on classification (four elements, four causes, types of motion, categories of being) but weak on dynamics (generation, corruption, physical interaction) (*see* ARISTOTELIANISM). Bodies acted on one another primarily through the "manifest active qualities" of the elements predominating in their constitutions: hotness, coldness, dryness, and moistness. Thus, to take a complicated example used by Aristotle, the sun melts wax and dries mud, the different consequences of the same manifest quality (hotness) depending on the elementary makeup of the recipient body. Two other widespread attributes of matter, gravity and levity, were often treated as if they were manifest qualities, since they characterized the four elements even though they could not be reduced to the tangible qualities hotness, coldness, and so forth, to which Aristotle gave priority.

The world has many physical properties less widely encountered than gravity and levity but, like them, not easily or obviously explainable in terms of the action of manifest qualities. Later Peripatetic philosophy designated these properties "occult," because, although evident in their consequences, their causes were hidden. The exemplar of an occult quality was *magnetism. The ancients knew it as the ability of a peculiar rock to draw bits of iron to it—but why only iron? The answer lay, according to the natural philosophy taught when *Galileo was in school, in an innate sympathy or harmony between lodestones and iron. This example indicates the level of explanation that, in the seventeenth century, made "occult" a byword for nonsense. Originally an expression that aided the classification of properties whose causes were provisionally unknown, the occult became a trash heap of innate and irreducible qualities. A purge or poison, the deadly glance of the basilisk, astrological influences, the powers of talismans, and the force by which that small pesky fish, the remora, stops big ships—all operated by occult sympathies and antipathies between agent and recipient. Molière neatly satirized the level of explanation afforded by the occult in his *Malade imaginaire* (1673), in which he praises a doctor for ascribing the soporific quality of opium to an occult "dormative virtue."

The *mechanical philosophy, especially in its radical form of René *Descartes's limitation of the affections of matter to extension, shape, and motion, appealed to the scientific revolutionaries of the seventeenth century because it annihilated the complex of qualities taught by the traditional philosophy they opposed. Even manifest qualities had to go: the hot, cold, moist, and dry became secondary effects arising from the interaction of the few primary qualities of extended, moving, material bits with the human sensory apparatus. Explanations in mechanical terms, like Descartes's referral of thunder and lightning, and rains of blood, to the precipitous fall of one cloud on another, might appear no more persuasive than magnetic sympathies; nonetheless, the corpuscular philosophy, by seeking a mechanical account of properties held by its opponents to be innate and irreducible, opened the possibility of further analysis. Robert *Boyle's concept of the "spring of the air," for which he offered several mechanical analogies, and Descartes's representation of magnetism by a vortex of specially shaped particles, suggest the range and limitations of seventeenth-century mechanical models.

Against the rhetorical and explanatory advantages of the corpuscular philosophy, Isaac *Newton's apparent invocation of an occult sympathy—the "universal attraction" of the *Principia* (1687)—seemed retrograde to many natural philosophers enlightened by Descartes. They were both right and wrong. Newton did return to an occult quality, but in its most useful and responsible form: a widespread property of matter, exactly described, whose cause had not yet been found. Newton's famous phrase "hypotheses non fingo" ("I feign no hypotheses") meant that, as far as he was concerned, gravity would remain occult. Until we have a Theory of Everything (*see* STANDARD MODEL), and perhaps even then, scientists necessarily will continue to invoke occult qualities.

Brian Vickers, ed., *Occult and Scientific Mentalities in the Renaissance* (1984). David C. Lindberg and Robert S. Westman, eds., *Reappraisals of the Scientific Revolution* (1990). Dennis Des Chene, *Physiologia: Natural Philosophy in Late Aristotelian and Cartesian Thought* (1996).

J. L. HEILBRON

SYNTHESIS. See ANALYSIS AND SYNTHESIS.

SYPHILIS. Venereal syphilis is a chronic contagious disease caused by a spiral-shaped bacterium (spirochete) of the genus *Treponema* (*T. pallidum*). Although predominantly transmitted by sexual contact, it can be congenital. One of the human treponematoses—along with endemic syphilis, yaws, and pinta—venereal syphilis was the most dreadful and deadly of the sexually transmitted diseases (STDs) until the emergence of *AIDS in the early 1980s. Its natural evolution includes chancres, skin rushes, and lesions involving tissues throughout the entire human body through a clinical course with three stages (primary, secondary, and tertiary syphilis) over many years and usually separated from each other by a latent period with no visible signs of infection. Tertiary syphilis often attacks the cardiovascular and central nervous systems fatally.

The disease took its name from Syphilus, the hero of the Latin poem *Syphilis, sive morbus gallicus* (Verona, 1530) by the physician and poet Girolamo Fracastoro. According to the mythological tale, the Sun God (Apollo) inflicts an ugly ulcer on the body of a young shepherd as a punishment for blasphemy. Ultimately Apollo relents and provides Syphilus with the healing tree, the guaiacum, to cure his terrible disease. As a term for the disease, "syphilis" became widely used only in the late eighteenth century, not fully prevailing before the 1820s. Until the turn of the nineteenth century the usage was vague and presumably applied to many symptoms other than those of the infectious disease currently known as venereal syphilis.

The agent responsible for venereal syphilis, *T. pallidum*, was first isolated in serum from a lesion of secondary syphilis in 1905 and in lesions of tertiary syphilis in the aorta wall (1906) and in brain tissue (1913). The first serological procedure for diagnosing syphilis was invented in 1906. A complement-fixation test (an immunity test depending upon the fact that syphilitic patients develop antibodies to a certain lipid component of many tissues),

it became known as the Wassermann reaction. Between 1909 and 1912 syphilis was attacked with two arsenic salts, Paul Ehrlich's "magic bullet," which became the first successful aetiological drugs ever synthesized against a bacterial disease. During the first decades of the twentieth century, campaigns against venereal disease stimulated by the social hygiene movement all over the Western developed countries and their colonial possessions made syphilis its main target. Finally, during World War II syphilis became one of the infectious diseases successfully treated by penicillin.

Between 1932 and 1972, the United States Public Health Service conducted a study of the effects of untreated syphilis on black men—399 of them having seropositive latent syphilis of three or more years' duration, and 201 more free of the disease chosen to serve as controls—in and around the town of Tuskegee in Macon County, Alabama. The Tuskegee Experiment abruptly ended when it was exposed by the journalist Jean Heller. The revelations concerning this racist, nontherapeutic experiment have played a crucial role in defining contemporary standards for medical experimentation on humans (*see* ETHICS IN SCIENCE).

Jon Arrizabalaga, "Syphilis," in *The Cambridge World History of Human Disease*, ed. Kenneth F. Kiple (1993): 1025–1033. James H. Jones, *Bad Blood: The Tuskegee Syphilis Experiment*, 2d ed. (1993). Jon Arrizabalaga, John Henderson, and Roger French, *The Great Pox: The French Disease in Renaissance Europe* (1997). Susan M. Reverby, ed., *Tuskegee's Truths: Rethinking the Tuskegee Syphilis Study* (2000).

JON ARRIZABALAGA

T

TACIT KNOWLEDGE. The notion that we can occasionally know more than we can tell was first articulated as an epistemological concept with sociological implications by Michael Polányi in the 1950s. It has since proved seminal for the historiography of science and technology.

According to Polányi, just as when perceiving an object we occasionally recognize patterns without being able to specify the particulars, so when knowing we often accept something implicit but very concrete. This implicit core binds together the object of our attention, the knowing person, and a tradition shared with the group to which we belong. Polányi's favorite example at the perception level was our ability to recognize the varied expressions of a human face without being able to tell how we recognize them. He extended the notion to include phenomena such as the perception, on the part of experts, of comprehensive objects—like the characteristics of a wine or the "touch" of a pianist—for which common experience shows that the perception and description of the constituent elements can fail to convey the whole.

When applied to science and technology, Polányi's notion was refreshing. The traditional view of science as disembodied knowledge proved insufficient to accommodate the personal and social dimensions involved in the acquisition, practice, and transmission of tacit knowledge. Polányi's writings circulated a view of science conceived as a craft, in which scientific and social norms intertwine, rather than as the impersonal, abstract enterprise described by logical positivists during the central decades of the twentieth century.

In a similar vein Polányi—himself a chemist and a chemical physicist by training—called attention to the importance of tools (instruments, as well as concepts) and the training needed to master their use. As a result, the master-pupil relation, the process of imitation and interiorization of norms, and the role of research *schools received due attention among the ingredients constitutive of science. Similarly influential were Polányi's contributions to the notion of a "scientific community," which by the 1960s had become a basic conception in the sociology of science.

Among historians of science, Thomas Kuhn, Jerome R. Ravetz, Harry M. Collins, and Donald MacKenzie have acknowledged their debt to Polányi. Through these authors the notion of tacit knowledge contributed to the success of studies on scientific "practice" and to the move away from "theory" in the *historiography of science during the 1980s and 1990s. The philosopher Stephen Turner, however, launched a sharp critique of the concept of practice, also involving the associated notions of tacit knowledge and tradition, in 1994.

See also FACT AND THEORY; POSITIVISM AND SCIENTISM.

Michael Polányi, *The Tacit Dimension* (1967; 1983). Stephen Turner, *The Social Theory of Practices: Tradition, Tacit Knowledge and Presuppositions* (1994).

GIULIANO PANCALDI

TECTONICS. See PLATE TECTONICS.

TELEGRAPH. The term "telegraph" was first applied to a system of mechanically operated semaphores designed by the brothers Claude and Ignace Chappe, erected by the revolutionary government of France beginning in 1794, and expanded under Napoleon and subsequent regimes. A dozen nations maintained Chappe networks in the first half of the nineteenth century; France's remained in operation during the Crimean War. An optical telegraph designed by George Murray, using combinations of shutters, was also widely adopted. The low bandwidth of these telegraphs, their restriction to daylight use, and their dependence on clear weather limited their usefulness.

Telegraphs based on static electricity and using pith balls, sparks, and other means of detection had been developed in the eighteenth century, but line losses prevented their use over significant distances. Devices employing current electricity became possible after Alessandro Volta's invention of the battery in 1800 and Hans Christian Ørsted's discovery of the deflection of a magnetic needle by an electric current in 1820.

British inventors William Cooke and Charles Wheatstone patented a multiple-circuit needle telegraph in 1837.

The single-circuit electric telegraph was developed by Samuel F. B. Morse and his partners Leonard Gale and Alfred Lewis Vail. Morse enjoyed a successful career as a painter, president of the National Academy of Design, and professor of art at the City University of New York (later New York University). At this time artists not infrequently applied their visual and spatial thinking to invention and the design of technological systems. Morse had already produced unsuccessful designs for a steamboat, a fire engine pump, and a marble cutting machine. During a return journey to the United States in 1832 following art study in Europe, he fell into conversation about communication via electric currents, and the conception of a telegraph seized him. Back at the university he recruited Gale, a professor of chemistry, geology, and mineralogy, who contributed scientific knowledge, and Vail, a student whose father owned an ironworks and who introduced mechanical refinements. Morse's first designs used printers' letter type molded with projections corresponding to the dots and dashes of his code. The type was set in a composing stick or "portrule" that an operator cranked past a conductor to create the signal. By 1844 Vail had replaced this device with the familiar "Morse" key. At the separate receiving apparatus a stylus actuated by an electromagnet marked the signals on paper tape.

Development and testing required significant capital, and after extensive lobbying Morse won in 1843 a Congressional grant of thirty thousand dollars. The following year the famous message "What hath God wrought!" flashed across overhead lines from Washington to Baltimore, and the Telegraph Age began. News organizations, financial interests, and lotteries jumped to exploit the invention. The telegraph transformed warfare by affording a central headquarters immediate contact with armies over great distances. It perfectly suited the railroad, whose rapid expansion across the American continent it accompanied so closely that the two were called Siamese twins. Enabling communication and logistical control over great distances, the telegraph has rightly been called the quintessential technology of empire.

Certainly development of undersea cable telegraphy proved indispensable for the maintenance of the British Empire. The first successful undersea cable, laid across the Channel in 1851, was followed by hasty and unsuccessful ventures to span the Irish, Red, and Mediterranean seas and, in 1858, the Atlantic Ocean. Further research resulted in working cables across the Persian Gulf to India and then across the Atlantic in 1866. In the 1870s lines laid throughout the empire permitted the British rapidly to deploy their military forces around the world and provided their merchants with news of colonial and foreign markets. As one French official put it, "England owes her influence in the world perhaps more to her cable communications than to her navy."

In the last third of the nineteenth century the undersea cable enterprise provided the largest market for advanced electrical knowledge and exerted a defining influence on theoretical development. In order to locate breaks in cables it proved necessary to develop standards of electrical resistance as equivalents to lengths of wire. The British Association for the Advancement of Science formed a Committee on Electrical Standards in 1861: James Clerk *Maxwell and Fleeming Jenkin took on the calibration of standard coils of wire, which the association issued in boxes to telegraph engineers. The committee, adapting Wilhelm Weber's "absolute" system of electrical units to telegraphy, established the modern system of electrical units—volt, ampere, and ohm. Telegraph engineers developed voltmeters, ammeters, and other instruments that, migrating to physics laboratories, made precision electrical measurement the characteristic experimental endeavor of late Victorian physics (see STANDARDIZATION).

The emergence of *field theory in England likewise owed much to cable telegraphy. Michael *Faraday had proposed in the 1830s that electric and magnetic phenomena result from stresses in the space surrounding magnets and wires. Poorly received at the time, Faraday's field theory easily explained problems faced by cable telegraph engineers. Underground and undersea lines delayed and spread current pulses transmitted along them, preventing rapid signaling and garbling communication. Faraday, examining the effect in 1854, exclaimed that it provided a "remarkable illustration of some fundamental principles of electricity...which I put forth sixteen years ago." He had argued that a current does not arise instantaneously in a wire but consumes time as a state of strain builds up in the surrounding space. William *Thomson, professor of natural philosophy at Glasgow University, used Faraday's ideas to show that the resulting retardation of telegraph signals increased with the square of the cable's length but decreased in thicker cables. The rejection of these arguments by the backers of the first transatlantic cable contributed, along with hasty manufacture and poor handling, to its early failure; Thomson shepherded the second transatlantic attempt to success. For these and other contributions to telegraphy he was knighted and later raised to the peerage as Lord Kelvin. The field theory's successful explanation of cable-related phenomena contributed to its widespread acceptance in England. In Germany, France, and the United States, on the other hand, telegraph engineers confronted few inductive effects from their overhead, single-wire lines; in those nations electrical theory based on action at a distance reigned.

Bern Dibner, *The Atlantic Cable* (1959). Brook Hindle, *Emulation and Invention* (1981). Crosbie Smith and M. Norton Wise, *Energy and Empire. A Biographical Study of Lord Kelvin* (1989). Paul Israel, *From Machine Shop to Industrial Laboratory. Telegraphy and the Changing Context of American Invention, 1830–1920* (1992). Alexander J. Field, "French Optical Telegraphy, 1793–1855: Hardware, Software, and Administration," *Technology and Culture* 35 (1994): 315–347. Bruce J. Hunt, "Doing Science in a Global Empire: Cable Telegraphy and Electrical Physics in Victorian Britain," in *Victorian Science in Context*, ed. Bernard Lightman (1997): 312–333.

THEODORE S. FELDMAN

TELEOLOGY. Teleology, the study of the purpose (Greek *telos*) perceived in some natural phenomena or events, and regarded as analogous to the purpose observed in objects produced by human design, played an important role in the philosophy of Aristotle. During the later Middle Ages, Aristotelian teleology merged with Christian theology: that some natural events seem to display purpose, and to occur in order to bring about future events rather than being caused by preceding ones, was considered a proof of the existence of an intelligent creator and a divine plan in nature.

With Francis Bacon and other proponents of the "new philosophy" in the seventeenth century, recourse to design in natural philosophy came under attack. According to Bacon, "final causes" could be construed easily out of any sort of evidence, and proved sterile in attempts to understand nature. Baruch Spinoza and his radical followers, on the other hand, denounced final causes as inappropriate to a concept of God that credited him with infinite power, but neither wisdom nor perfection in any anthropomorphic sense.

During the eighteenth century, developments in the life sciences and the concept of *organism invited further reflection on the notion of purpose. Teleology was subjected to systematic treatment in Immanuel Kant's *Critique of Judgment* (1790), a work that had widespread influence on the life sciences during the Romantic era.

Kant distinguished between "extrinsic finality" and the kind of finality displayed by parts inside an organism, which he called "internal." Extrinsic finality had been postulated by naturalists such as Carl *Linnaeus when they argued, for example, that carnivores existed in order to check the number of herbivores, which in turn checked the vegetable kingdom. Man acted as an agent of equilibrium between productive and destructive forces in nature. Kant ruled out this brand of teleology. Instead, he thought that a finality internal to living beings could not be denied. Relying on the work of anatomists such as Johann Friedrich Blumenbach, he argued that in an organism, each part acts both as means and end, according to a system of final causes apparently transmitted from one generation of organisms to the next. Nineteenth-century followers of Kant, such as embryologist Karl Ernst von Baer, pursued this kind of teleological thinking out of a concern for good science.

Charles *Darwin's works represented a turning point in reflections on teleology. While abstaining from explicit attacks against teleological thinking, in books such as *The Various Contrivances by Which Orchids are Fertilised by Insects* (1862), Darwin showed that many features displaying purpose could be explained away by the evolutionary mechanism of natural selection, with its contingent element, and by what we now call the co-evolution of different species in the same habitat. To many this meant, and still means, the demise of teleology.

Throughout the twentieth century, however, biologists' recourse to analogies drawn from *cybernetics, *feedback mechanisms, and *computer technologies led to the introduction (1958) of the notion of "teleonomy" (*telos* plus *nomos*, the Greek for "law"), which might seem to incorporate a new set of anthropomorphic metaphors. Ernst Mayr uses the term "teleonomic" to describe the structures, physiological processes, and behaviors that owe their goal-directedness to a "genetic program." Teleonomic activities thus play an important role in the selection pressure that causes the historical construction of the program itself.

See also DARWINISM; NATURAL THEOLOGY; NATURE; SCIENCE AND RELIGION.

Timothy Lenoir, *The Strategy of Life. Teleology and Mechanics in Nineteenth Century German Biology* (1982). Lily E. Kay, *Who Wrote the Book of Life? A History of the Genetic Code* (2000).

GIULIANO PANCALDI

TELEPHONE. The first practical telephone, invented by Alexander Graham Bell in 1876, did not rely on scientific insight and was initially considered a "speaking *telegraph." The apparatus attracted the attention of many physicists, who studied the theory of the telephone, adapted it for acoustical studies, and applied it as a measuring device for feeble currents. Among the physicists who investigated the telephone in the 1880s were Hermann von *Helmholtz, Ludwig *Boltzmann, and Lord Rayleigh. Friedrich Kohlrausch and others used the telephone as a measuring device in studies of electrical conductivity in electrolytes.

Whereas the telephone apparatus raised no scientific problems, the transmission of telephone currents did. During most of the period from 1880 to 1900 engineers considered telephone currents to propagate like telegraph currents and adapted William *Thomson's theory of telegraph transmission to the new technology. This view, formulated in 1887 by William Henry Preece in England, led to a semiempirical expression for the maximum speaking distance as given by the line's capacitance and resistance. However, it disregarded the inductive effects caused by the rapidly varying telephone currents and disagreed with experiences from long lines in the United States. A scientifically based understanding of telephone currents was first obtained in 1887, independently by Oliver Heaviside in England and Aimé Vaschy in France. By taking into account the effects of self-induction, Heaviside found from *Maxwell's electrodynamics how the attenuation and distortion in a line vary with the electrical parameters. Engineers in the British Post Office ridiculed his theory, which indicated that increased self-induction would result in longer speaking distance.

The Heaviside-Vaschy theory was only turned into an engineering theory, and then into a technology, around 1900. At Bell Telephone Company, George Campbell analysed in 1899 the transmission of telephone currents and suggested inserting inductance coils to increase the speaking distance. Michael I. Pupin, a physicist at Columbia University, presented a similar theory of "loaded" lines (1900). In 1904 the Bell System purchased the rights to use Pupin's approach. The Campbell-Pupin loading method was quickly turned into a practical technology, primarily by the Bell System and, in Europe, by the Siemens and Halske firm. It also led to an advanced theory of telephone transmission developed by, among others, Franz Breisig in Germany, John Carson in the United States, and Henning Pleijel in Sweden.

Researchers at Bell Laboratories, founded in 1925, did much of the work on telephone science and technology. Karl Jansky's discovery in 1932 of extraterrestrial radio waves, as well as Robert Wilson and Arno Penzias's discovery in 1965 of the cosmic microwave background radiation, were by-products of work on wireless telephony. Another important result of the Bell Laboratories was the invention in 1948 of the *transistor, originally conceived as a solid-state amplifier in telephone systems. Also in the area of information science, problems of telephone communication served as important stimuli. Examples are the theories of Harry Nyquist (1928) and Claude Shannon (1948), both of the Bell Laboratories.

The technical developments caused a vast increase in telephone traffic over ever longer lines. The growth in subscribers and line length during the 1920s and 1930s, mainly a result of coil loading, coaxial cables, and vacuum tube amplifiers, was only a prelude to the explosion following World War II. About 1930, most European countries were interconnected by telephone and in 1956 the first transatlantic cable went into operation. Fiber-optical cables and *satellites drastically increased the capacity. For example, by 1977 the eight Intelsat 4 satellites offered forty thousand telephone circuits to six continents, and eleven years later the first transatlantic fiber cable nearly doubled the capacity between Europe and North America.

Neil H. Wasserman, *From Invention to Innovation: Long-Distance Telephone Transmission at the Turn of the Century* (1985). Laszlo Solymar, *Getting the Message: A History of Communications* (1999).

HELGE KRAGH

TELESCOPE. Lenses for reading were available in Italy in the thirteenth century, but not until the seventeenth century did spectacle makers in the Netherlands put together a device "by means of which all things at a very great distance can be seen as if they were nearby." In 1609 *Galileo Galilei heard rumors of spyglasses, made more powerful ones, and pointed them at the heavens. In 1611 Johannes *Kepler explained the path of light rays through lenses and the formation of images. The improved Kepler telescope formed images in its focal plane, where they were viewed by a magnifying lens.

Anything placed in the focal plane of a telescope appears sharply alongside the celestial object, as the Englishman William Gascoigne noticed in about 1640 when a spider spun its web inside his instrument. Astronomers inserted cross hairs, facilitating precise alignment of telescopes on objects, and micrometers, to measure small angular distances and diameters. They also developed, though more slowly, stable, precise mountings and large arcs with precisely divided and marked scales against which the telescope's alignment could be noted when pointed at a celestial object. Still their instruments suffered from chromatic and spherical "aberrations"—fuzziness of

Two of Galileo's telescopes and one of his objective lenses. The lens is mounted in an ivory receptacle as if it were a relic of a saint.

The second half of the nineteenth century saw advances in refracting telescopes, especially by the Boston firm of Alvan Clark and Sons. Their metal tubes were stiffer yet lighter than wooden telescopes. Larger pieces of optical glass were now available, from France and England, and five times the Clarks figured the lens for the world's largest refracting telescope, culminating in a 40-inch lens in 1897. It is yet to be surpassed in size. Larger pieces of optical glass are difficult to cast; heavier lenses flex more; and thicker lenses absorb more light.

Lenses also absorb strongly in the blue region of the spectrum, where *photography is most effective. A new interest in astrophysics and distant stars required a new technology. George Willis Ritchey made the photographic reflecting telescope the basic instrument of astronomical research, constructing at the Mount Wilson Observatory a 60-inch telescope in 1908 and a 100-inch in 1918. Later the Rockefeller Foundation paid for a 200-inch reflecting telescope at nearby Mount Palomar. Corning Glass Works cast the mirror in 1934 as a thin piece of Pyrex glass with a system of ribbing in the back. Grinding the lens removed five tons of material, leaving sixteen tons of curved mirror, which received its reflective coating of aluminum in 1949.

To circumvent the problems of casting and supporting large mirrors, many small mirrors can be assembled into a close array. The Keck Telescope, erected in Hawaii in 1993, has thirty-six 1-meter mirrors mounted together on a tracking structure, and the European Southern Observatory in Chile links four 8-meter and three 1.8-meter mirrors into one very large telescope. Its huge cost is shared among nine countries.

Reflecting telescopes bring only rays from stars in the center of the viewing field to a sharp focus. Given a usable field of view of 15 seconds of arc, approximately a million photographic plates would be required to cover the entire sky. In 1930 Bernhard Schmidt, an Estonian-born optician at the Hamburg Observatory, designed a reflecting telescope with a usable field of view of 15 degrees. The Schmidt telescope has a simple spherical mirror plus a thin correction plate for spherical aberration. Palomar completed a 1.2-meter Schmidt telescope in 1948.

Non-optical telescopes can detect radio and gravitational waves unblocked by the earth's atmosphere. Other non-optical telescopes rise above the earth's atmosphere. There X rays incident on mirrors at small "grazing angles" are reflected into a detector, where their interaction with an inert gas generates countable electrons. The telescope, with several mirror surfaces nested concentrically within it, looks like a funnel.

the image—arising from the fact that different wavelengths or colors of light are refracted by different amounts, and light incident on the periphery of the lens focuses closer to the lens than does light striking near the center. To reduce the aberrations astronomers ground lenses with very long focal lengths, which led to long and unwieldy instruments. In 1757 the Englishman John Dollond perfected the achromat, a combination of glass lenses that brought rays of different colors to the same focus, enabling more precise measurements of positions of faint stars by means of shorter instruments easier to use. Inability to make large pieces of optical-quality glass, however, limited the size of refracting telescopes, in which light passes through transparent lenses.

In 1668 Isaac *Newton, having decided that chromatic aberration in lenses could not be defeated, built the first successful reflecting telescope. It employed a concave mirror to collect light and form the image. William Herschel in England built telescopes with large reflecting metal mirrors in the 1780s and William Parsons in Ireland built the 6-foot "Leviathan of Parsonstown" in 1845. Giant reflectors, though producing spectacular observations, ultimately were disappointing: the mirrors flexed under their immense weights and tarnished quickly.

Telescopes in space also detect infrared emissions and gamma rays.

The Hubble Space Telescope enables traditional, optical astronomers to escape our atmosphere. The telescope's primary mirror is eight feet in diameter. Including recording instruments and guidance system, the telescope weighs twelve tons. It has been called the eighth wonder of the world; critics say it should be, given its cost of two billion dollars. It was as much a political and managerial achievement as a technological one; approval for it came only after a political struggle lasting from 1974 to 1977. In 1990, after overcoming a host of problems, its designers launched it into space, only to discover that an error had occurred in the shaping of the primary mirror. One newspaper reported "Pix Nixed as Hubble Sees Double." Addition of a corrective mirror solved the problem.

Over four centuries the telescope has evolved from two small glass lenses affordable and operable by an untrained individual of no great wealth into an immense political, managerial, and technological undertaking beyond the reach of all but the wealthiest countries. Our understanding of the universe has expanded apace, as ever larger, more expensive, and technologically sophisticated telescopes range over ever more of the electromagnetic spectrum to detect ever more distant objects.

See also INSTRUMENTS AND INSTRUMENT MAKING; OBSERVATORY; RADIOASTRONOMY.

Henry C. King, *The History of the Telescope* (1955). Isaac Asimov, *Eyes on the Universe: A History of the Telescope* (1975). James Cornell and John Carr, eds., *Infinite Vistas: New Tools for Astronomy* (1985). J. A. Bennett, *The Divided Circle: A History of Instruments for Astronomy, Navigation and Surveying* (1987). Donald E. Osterbrock, *Pauper and Prince: Ritchey, Hale, & Big American Telescopes* (1993). Robert W. Smith, *The Space Telescope: A Study of NASA, Science, Technology, and Politics* (1993).

NORRISS S. HETHERINGTON

TELEVISION. See FILM, TELEVISION, AND SCIENCE; RADIO AND TELEVISION.

TELLER, Edward. See SAKHAROV, ANDREI, AND EDWARD TELLER.

TERMINOLOGY. The terms invented to denote new entities and instruments may illuminate the state of science as much as the objects of its investigation. At first, neologisms had unexceptionable derivations from ancient languages: *barometer, *microscope, *telescope, *electricity. The custom continued in the nineteenth century with coinages that may not have satisfied all philologists: electrode, ion, *scientist, physicist, *telegraph, *telephone, cesium, *electron, argon,

helium, *radium. These names not only respected the convention that neologisms be based on the languages that once served the so-called Republic of Letters, but also that they indicate a characteristic feature of the object named. In the second half of the nineteenth century, however, both conventions were shaken by the terminology introduced into mechanics by the English (curl, twist) and by the nationalistic naming of new elements by their discoverers (germanium, gallium, scandium, polonium).

In the twentieth century the decline in the humanistic education of scientists and the jocularity of American physicists produced *strangeness, *quark (in its top, bottom, charmed, and colored varieties), and gluon alongside the old-fashioned quantum, proton, neutron, deuteron, positron, meson, and the playful neutrino. A similar development does not seem to have affected the biological sciences, which have kept to safe items like *gene and genome, owing, perhaps, to closeness to medicine, still stuck in ancient argot.

See also COSMIC RAYS; ELEMENTARY PARTICLES; NOBLE GASES.

J. L. HEILBRON

TERRESTRIAL MAGNETISM. Practical needs, particularly of navigators, have inspired interest in terrestrial magnetism since at least the fifteenth century. Equally important have been conceptual puzzles about how to reconcile terrestrial magnetism with basic physical theory and with theories based on laboratory studies of *magnetism. Consequently interplay between field studies and experimental studies has been a regular feature of the history of geomagnetism. So has a tension between explaining the ultimate causes of geomagnetism, usually in terms of some kind of fluid movement in the interior of the earth, and surveying the spatial and temporal variations of geomagnetic declination, dip, and intensity.

Serious work on geomagnetism began in 1600 with the publication of William Gilbert's *De magnete*. By that date, navigators knew that their needles sometimes pointed at an angle to true north (declination) and that sometimes they inclined from the horizontal (dip). Philosophers generally assumed that the earth's magnetism arose through the occult properties of the mineral magnetite or by some Neoplatonic correspondence between the polestar in the heavens and the magnetic north pole on Earth.

Gilbert, a member of the Royal College of Physicians in London, discussed the five motions associated with magnetism—attraction (he called it coition), orientation, declination, dip, and rotation—as a preliminary to presenting his theory of earth magnetism. Experiments with small magnetic needles on a small spherical lodestone (called a terella) showed that irregularities in the lodestone changed the orientation of the needles. They also demonstrated that needles parallel to the surface of the sphere at the equator gradually dipped to a vertical as they moved to the position at the poles. The earth, he concluded, was a giant lodestone with an immaterial rotating magnetic soul.

Because magnetism, including geomagnetism, seemed an exemplary occult force, mechanical philosophers had to find an alternative explanation in terms of matter in motion. In his *Principles of Philosophy* (1644), René *Descartes traced the earth's *magnetism to circulating streams of corkscrew-shaped particles. From this suggestion arose the tradition, predominant until the 1820s, of attributing the earth's magnetism to subtle active magnetic fluids. Edmond Halley in 1683 and again in 1692 proposed that the earth consisted of an inner sphere and outer shell. They rotated at different speeds and each had a north and south pole. The interactions between these four poles accounted for the variations in declination and dip. Between 1698 and 1700, Halley sailed the Atlantic, measured the variations in declination, and charted them on a pioneering map that appeared in different editions between 1701 and 1703.

Descartes's effluvial theory continued to be important until the early nineteenth century. The alternative, most fully articulated by Charles Augustin Coulomb and Simeon-Denis Poisson—whose theory presented to the Paris Academy in 1826 represented the culmination of the tradition—assumed distance forces resulting from fluids locked in magnetic substances. Other important figures in the debate were Gavin Knight; Leonhard *Euler, who with others won the prize of 1746 offered on the subject by the Paris Académie des Sciences; Franz *Aepinus; and Jean-Baptiste Biot. Many researchers attempted to deal with earth magnetism though the requisite mathematics was dauntingly complex. During the 1820s and 1830s theories (like Poisson's) based on "austral" and "boral" fluids were losing their luster. Christopher Hansteen revived Halley's two-axis–four-pole model in his *Investigations Concerning the Magnetism of the Earth* (1819). To look for poles, defined either as regions of maximum magnetic intensity or of vertical dip, Hansteen traveled to Siberia around 1830 and James Clark Ross went to Canada. Although the two-axis theory did not win acceptance, the reintroduction of poles as an object of investigation, the attempt to mathematize the theory, and expeditions brought fresh ideas and evidence to geomagnetic studies. In the same decades, Hans

Christian Ørsted's discovery in 1820 that electric currents produce magnetic effects, Thomas Seebeck's discovery of thermoelectricity in 1822, and Michael *Faraday's discovery in 1831 that magnetism can produce electric currents gave rise to new questions and possibilities concerning earth magnetism. Alexander von *Humboldt, who had been fascinated by the global variations of magnetism since the 1790s, speculated about the similarities between lines of equal magnetic intensity and isothermal lines and about interconnections between geological, meteorological, and magnetic phenomena. In 1805 he reported that magnetic intensity varied across the earth's surface. To plot these variations, Humboldt encouraged the establishment of a network of magnetic observatories. By 1834 the twenty-three European observatories had detected the phenomenon of magnetic storms. In the fifth volume of his *Cosmos* (1845), Humboldt summed up the state of knowledge of magnetic variation, distribution, and storms.

In the 1830s Carl Friedrich *Gauss and his younger collaborator Wilhelm Weber took over from Humboldt as leaders in geomagnetism, tackling problems from instrumentation to basic theory. Early in the decade Gauss designed the bifilar magnetometer, developed for the first time an absolute measure of magnetic intensity, and launched his own version of the Magnetische Verein (magnetic union) to establish a network of magnetic observatories worldwide. The results from these observatories came out in six volumes, *Resultate aus den Beobachtungen des magnetischen Vereins*, between 1836 and 1841. With new data about variations of magnetic intensity in hand, Gauss could publish his mathematical analysis of the vertical and horizontal components of earth magnetism in 1839. He analyzed the magnetic potential at any point on the earth's surface by an infinite series of spherical functions. Not dependent on a theory about the ultimate causes of geomagnetism, his method of analysis shaped theoretical work on geomagnetism for the rest of the century.

In Britain, a follower of Hansteen, Edward Sabine, fretted that Britain was letting slip the chance of contributing to the growing field of geomagnetism. In 1838 he enlisted the astronomer John Herschel to help him raise support for a British magnetic survey. The publication of James Clerk *Maxwell's *Treatise on Electricity and Magnetism* in 1873 encouraged investigators to speak of the earth's magnetic field, not its magnetic forces, and gave them another set of mathematical tools.

Between 1890 and 1900 geomagnetism began to take on the trappings of a separate discipline. A new generation of mathematically trained physicists, notably Arthur Schuster, continued working on mathematical analyses of the earth's field although they did not propose new comprehensive theories. With the establishment of national surveys and observatories, the amount of data available multiplied. The beginning of submarine warfare accelerated military interest in geomagnetism. International organizations expanded; in 1896 the journal *Terrestrial Magnetism* was founded. Another period of rapid breakthroughs in geomagnetism occurred in the years following World War II. In 1947, following measurements of the magnetic fields of the sun and some stars, the English physicist Patrick Blackett suggested that magnetism (including the earth's magnetism) might be a property common to all rotating bodies. A decade earlier, Göttingen-trained physicist Walter Maurice Elsasser had published a series of papers suggesting that a self-excited magneto hydro dynamo in the earth's core created its field. For a few exciting years, scientists explored the consequences of the two theories in the hope of deciding between them. Then in 1952, after obtaining negative results from an experiment intended to detect the effects of rotation in the laboratory, Blackett himself rejected his own theory. Versions of Elsasser's theory held sway for the rest of the century.

Other major developments, interesting in their own right and for what they contributed to *plate tectonics, occurred in paleomagnetism. Already in the nineteenth century, scientists had detected remanent or fossilized magnetism. They noticed that ferrous minerals in baked clays and cooled lava flows preserved the alignment of the earth's main field as it was when they had cooled. In the late 1950s, physicists in London, Newcastle, and the Australian National University who systematically surveyed remanent magnetism found that the magnetic north pole appeared to have wandered widely over the globe in the past. They proposed various hypotheses to explain this result: their instruments created the effect, the earth's field had not always been dipolar, the continents had moved relative to one another, or the earth's magnetic poles had wandered independently. By the end of the decade, a small but influential group of scientists, Keith Runcorn prominent among them, had convinced themselves that the continents had moved. This served to give the largely discredited theory of continental drift new life.

In the 1920s and 1930s, scientists had discovered another peculiarity about remanent magnetism. In some rocks the magnetism had a polarity opposite to that of the present geomagnetic field. In the 1940s, researchers in the Carnegie Institution of Washington developed a spinning magnetometer capable of detecting weak magnetic fields. From the 1950s through the 1960s, paleomagnetists at the United States Geological Survey and the Australian National University raced to reconstruct the history of these reversals, using radioactive dating to determine their sequence. By the mid-1960s, they had constructed a fairly complete scale. It proved to be a key piece of evidence for the theory of sea floor spreading.

See also IMPONDERABLE; MAGNET AND COMPASS; NAVIGATION; SYMPATHY AND OCCULT QUALITY.

Sydney Chapman and Julius Bartels, *Geomagnetism* (1940). R. W. Home and P. J. Conner, *Aepinus's Essay on the Theory of Electricity and Magnetism* (1979). William Glen, *The Road to Jaramillo* (1982). Christa Jungnickel and Russell McCormick, *Intellectual Mastery of Nature* (1986). Robert P. Multhauf and Gregory Good, *Brief History of Geomagnetism* (1987). David Barraclough, "Geomagnetism: Historical Introduction," in *The Encyclopedia of Solid Earth Geophysics*, ed. David E. James (1989).

RACHEL LAUDAN

TEXTBOOK. Textbooks synthesize canonical mathematical and natural scientific knowledge and organize it hierarchically for instructional purposes. Before 1700, professors usually read directly from classical works by Euclid and Aristotle or scholastic writers; students often wrote down these lectures word for word. A wider distribution of standardized works incorporating new discoveries became possible after the invention of printing, for example, Andreas *Vesalius's *De Fabrica* and Nicholas *Copernicus's *De Revolutionibus*, both from 1543. But almost none adopted the comprehensive and hierarchical nature of the textbook format until after 1700, when didactic works that synthesized and reorganized recent literature began to appear for use by students, practitioners, bureaucrats, and the general public. Antoine-Laurent *Lavoisier, for instance, not only designed his *Elementary Treatise on Chemistry* (1789) for beginning students, but in it he also used his own discoveries to redefine the content and practice of chemistry. Early textbooks also constituted early forms of scientific popularization.

At first textbooks did not include problem sets or other exercises, which were introduced in the nineteenth century when the textbook industry expanded greatly. The expansion was closely linked to the differentiation of scientific disciplines. Textbooks mark the intellectual maturation of a field of scientific knowledge into a discipline with agreed-upon conceptual definitions, standardized conventions of presentation and representation, and investigative methods (observational, mathematical, and experimental). Since the nineteenth century several science textbooks have attained the status of classical works used to train generations of students. One notable example is Friedrich Kohlrausch's *Introduction to Practical Physics* (1870), which in various updated forms was still in use in German university laboratory physics courses in the 1950s.

In 1935 Ludwik Fleck located textbook science, which was for teaching, between journal science and *vademacum* (handbook) science for experts and popular science for the educated public. Repositories of the mental rituals by which a neophyte entered the community of scientific practitioners, textbooks in Fleck's view presented research results as hardened scientific facts while still maintaining contact with common knowledge. Textbooks thus mediated esoteric knowledge and general cultural traits. Thomas Kuhn's influential *Structure of Scientific Revolutions* (1962) considered textbooks as representations of the paradigms governing the normal science of a scientific community. In Kuhn's view textbooks make transparent the rules that govern scientific work through the problems and applications associated with specific theories; from these problems practitioners learn their trade, using paradigmatic examples to guide their thinking and their research protocols. Because paradigms work only as long as consensus reigns about the objects, techniques, and results of investigation, textbooks in Kuhn's schema represent a form of indoctrination without linkages to culture. Since Kuhn, anthropologists of science like Sharon Traweek have reopened the connection between textbooks and culture, demonstrating how textbooks shape professional behaviors in the sciences, including those that erect gender barriers, and create the canonical history of a discipline its leaders desire. Other scholars have examined how textbooks convey ideologies. Depictions of the environment in biology textbooks during industrialization and of the cost to the Nazi state of supporting the physically and mentally disabled suggest ways in which science teaching can shape and has shaped attitudes toward both nature and polity.

See also DISCIPLINES; SCHOOLS.

Thomas S. Kuhn, *The Structure of Scientific Revolutions* (1962). Ludwik Fleck, *Genesis and Development of a Scientific Fact* (1979). Sharon Traweek, *Beamtimes and Lifetimes: The World of High Energy Physicists* (1988). Gunter Lind, *Physik im Lehrbuch 1700–1850* (1992).

KATHRYN OLESKO

THEOLOGY, NATURAL. See NATURAL THEOLOGY.

THEORY. See FACT AND THEORY; HYPOTHESIS.

THEORY OF EVERYTHING. By the 1920s physicists had constructed a standard model of fundamental particles and forces, which consisted of two particles, the electron and proton, interacting through the forces of electromagnetism and gravity. Following the development of general *relativity, Albert *Einstein, Theodor Kaluza, Oskar Klein, and a few other physicists and mathematicians searched in vain for a theory that would combine the forces of gravity and electromagnetism in a unified field theory. Most physicists at the time focused instead on quantum theory and *nuclear physics, but these fields and the subsequent development of high-energy physics introduced new forces (the weak and strong nuclear forces), new particles, and new conditions to physical theory. Physicists in the last few decades of the twentieth century thus embraced the program of unification in order to simplify the scheme of particles and forces, in search of what they called a grand unified theory, or theory of everything. A theory of everything would embrace in a single mathematical structure the different sets of equations needed to describe the actions of the four basic forces.

In the early 1970s theorists succeeded in unifying the electromagnetic and weak forces in the electroweak theory, based on an idea first advanced by Steven Weinberg and Abdus Salam in 1967. Experimental evidence of neutral currents in the early 1970s and the detection of W and Z particles a decade later supported the electroweak unification. Physicists then sought to link the electroweak theory to the recently developed theory, called quantum chromodynamics (QCD), for the strong nuclear force. Promising early attempts by Howard Georgi, Sheldon Glashow, and others accounted for the elementary particles of each theory—leptons for electroweak, quarks for QCD—as well as the carriers of the three forces, but experiments designed to test the theory, in particular its prediction of proton decay, failed to provide convincing evidence.

Physicists recognized that these "grand unified theories" included only three of the four forces. Gravity proved difficult to accommodate. The development in the mid-1980s of string theory, which treated constituents of matter not as particles but as strings, offered a candidate for complete unification. But string theory, while mathematically elegant, increasingly departed from experimentally verifiable predictions and, despite frequent intimations of imminent success, by the end of the century had failed to incorporate gravity with the other three forces.

A theory of everything did not imply an end to scientific research, but rather that the quest for fundamental knowledge had ended and all that remained was to fill in the details. Claims of completeness in physics echoed similar anticipations in the past—for instance, in the late 1920s after the formulation of *quantum physics, or at the end of the nineteenth century after the construction of classical physics. Theories of everything also assumed that elementary particle physics was the foundation for the rest of science, an assumption disputed by *solid-state physicists and chaos theorists, and likewise by biological scientists, for whom *quarks or string theory offered few clues to the meaning of life or consciousness. Some theoretical physicists strayed from science altogether into the realm of theology, and claimed that a theory of everything would give humankind a glimpse of the mind of God.

Steven Weinberg, *Dreams of a Final Theory* (1992). David Lindley, *The End of Physics: The Myth of a Unified Theory* (1993).

PETER J. WESTWICK

THERMIONIC VALVE. See VALVE, THERMIONIC.

THERMODYNAMICS AND STATISTICAL MECHANICS. The development of the theory of heat in the first half of the nineteenth century, which eventually led to thermodynamics, was linked with the technology of steam engines. Their operation was originally analyzed in terms of the caloric theory, which represented heat as a conserved *imponderable fluid. In 1824 the French military engineer Sadi Carnot employed the caloric theory in his analysis of an idealized heat-engine, which aimed at improving the efficiency of real engines. On the basis of an analogy with the production of work by the fall of water in a waterwheel, Carnot assumed that a heat-engine produced work by the "fall" of caloric from a higher to a lower temperature. The analogy suggested that the work produced was proportional to the amount of caloric and the temperature difference of the two bodies between which caloric flowed. Carnot proved that no other engine could surpass his reversible ideal engine in efficiency by showing that the existence of a more efficient engine would imply the possibility of perpetual motion. In 1834 a mining engineer, Benoit-Pierre-Émile Clapeyron, reformulated Carnot's analysis, using calculus and the indicator (pressure-volume) diagram. Carnot's theory was virtually ignored, however, until its discovery in the mid-1840s, via Clapeyron's paper, by William *Thomson (Lord Kelvin), and Hermann von *Helmholtz.

James *Joule's experimental work of the 1840s, which indicated the interconversion of heat and work, undermined the caloric theory. His precise measurements supported the old idea that heat consists in the motion of the microscopic constituents of matter. The interconversion of heat and work, along with other developments spanning several fields (from theoretical mechanics to physiology), led to the formulation of the principle of energy conservation. In the early 1850s all these parallel developments were seen, with the benefit of hindsight, as "simultaneous" discoveries of energy conservation, which became the first law of thermodynamics.

Joule's experiments, however, presented a problem for Carnot's analysis of a reversible heat-engine based on the assumption of conserved heat. In the early 1850s Thomson and the German physicist Rudolf Clausius resolved the problem by introducing a second principle. Carnot's analysis could be retained, despite the rejection of the conservation of heat, because, in fact, it dealt with a quantity—the amount of heat divided by the temperature at which the heat is exchanged—that is conserved in reversible processes. During the operation of Carnot's engine, part of the heat dropped from a higher to a lower temperature and the rest became mechanical work.

In 1847 Thomson diagnosed another problem, also implicit in Carnot's analysis. Carnot had portrayed heat transfer as the cause of the production of work. In processes like conduction, however, heat flows from a warmer to a colder body without doing any work. Since the heat does not spontaneously flow from cold to hot, conduction resulted in the loss of potential for doing work. Both Joule and Thomson agreed that energy cannot perish, or, rather, that only a divine creator could destroy or create it. Thomson resolved the difficulty in 1852 by observing that in processes like conduction, energy is not lost but "dissipated," and by raising the dissipation of energy to a law of nature. "Real"—that is, irreversible—processes continually degrade energy and, in a good long time, will cause the heat-death of the universe. The Scottish engineer William Rankine and Clausius proposed a new concept that represented the same tendency of energy toward dissipation. Initially called "thermodynamic function" (by Rankine) or "disgregation" (by Clausius), it later (in 1865) received the name *"entropy" from Clausius, who grafted onto the Greek root for transformation. Every process (except ideal reversible ones) that takes place in an isolated system increases its entropy. This principle constituted the second fundamental law of thermodynamics, and its interpretation remained the subject of discussion for many years.

The dynamical conception of heat provided a link between mechanics and thermodynamics and led eventually to the introduction of statistical methods in the study of thermal phenomena. In 1857 Clausius correlated explicitly thermodynamic and mechanical concepts by identifying the quantity of heat contained in a gas with the kinetic energy (translational, rotational, and vibrational) of its molecules. He made the simplifying assumption that all the molecules of a gas had the same velocity and calculated its value, which turned out to be of the order of the speed of sound. Clausius's idealized model faced a difficulty, however, as pointed out by the Dutch meteorologist C. H. D. Buys Ballot. On the model, gases should diffuse much faster than actually observed. In 1858, in response to that difficulty, Clausius attributed the slow rate of diffusion to the molecules' collisions with each other and introduced the new concept of "mean free path," the average distance traveled by a molecule before it collides with another one.

In 1859 James Clerk *Maxwell became aware of Clausius's kinetic interpretation of thermodynamics and, in the following years, developed it further by introducing probabilistic methods. In 1860 he developed a theory in which the velocities of the molecules in a gas at equilibrium distribute according to the laws of probability. He inferred from "precarious" assumptions that the distribution followed a bell-shaped curve, the so-called normal distribution, which had been familiar from the theory of errors and the social sciences. Following up these ideas, he published in 1871 an ingenious thought experiment that he had invented four years earlier to suggest that heat need not always flow from a warmer to a colder body. In that case the second law of thermodynamics could have only a statistical validity. A microscopic agent ("Maxwell's demon," as Thomson called it), controlling a diaphragm on a wall separating a hot and a cold gas, could let through either molecules of the cold gas faster than the average speed of the molecules of the hot gas, or molecules of the hot gas slower than the average speed of the molecules of the cold gas. Heat thus would flow from the cold to the hot gas. This thought experiment indicated that the "dissipation" of energy did not lie in nature but in human inability to control microscopic processes.

Ludwig *Boltzmann carried further Maxwell's statistical probing of the foundations of thermodynamics. In 1868 he rederived, in a more general way, the distribution of molecular velocities, taking into account the forces exerted between molecules as well as the influence of external forces like gravity. In 1872 he extended the second law of thermodynamics to systems not in equilibrium by showing that there exists a mathematical function, the negative counterpart of entropy, that decreases as a system approaches thermal equilibrium. This behavior was subsequently called the "H-theorem."

Furthermore, Boltzmann attempted to resolve a severe problem, pointed out by Thomson in 1874 and Joseph Loschmidt in 1876, which undermined the mechanical interpretation of the second law. The law defines a time asymmetry in natural processes: the passage of time results in an irreversible change, the increase of entropy. However, if the laws of mechanics govern the constituents of thermodynamic systems, their evolution should be reversible, since the laws of mechanics run with equal validity toward the past and the future. *Prima facie*, there seems to be no mechanical counterpart to the second law of thermodynamics.

Boltzmann eluded the difficulty in 1877 by construing the second law probabilistically. To each macroscopic state of a system correspond many microstates (particular distributions of energy among the constituents of the system), which Boltzmann ranked as equally probable. He defined the probability of each macroscopic state by the number of microstates corresponding to it and identified the entropy of a system with a simple logarithmic function of the probability of its macroscopic state. On that interpretation of entropy, the second law asserted that thermodynamic systems have a tendency to evolve toward more probable states. The interpretation came at the cost of demoting the law. A decrease of entropy was unlikely, but not impossible.

Maxwell's and Boltzmann's statistical approach to thermodynamics was developed further by J. Willard *Gibbs, who avoided hypotheses concerning the molecular constitution of matter. He formulated statistical mechanics, which analyzed the statistical properties of an ensemble, a collection of mechanical systems. This more general treatment proved to be very useful for the investigation of systems other than those studied by the kinetic theory of gases, like electrons in metals or ions in solutions.

D. S. L. Cardwell, *From Watt to Clausius: The Rise of Thermodynamics in the Early Industrial Age* (1971). S. G. Brush, *The Kind of Motion We Call Heat: A History of the Kinetic Theory of Gases in the 19th Century*, 2 vols. (1976). Lawrence Sklar, *Physics and Chance: Philosophical Issues in the Foundations of Statistical Mechanics* (1993). Crosbie Smith, *The Science of Energy* (1998).

THEODORE ARABATZIS

THERMOMETER. The notion of a scale or degrees of heat and cold dates back at least to the second-century physician Galen, as does the idea of using a standard—such as a mixture of ice and boiling water—as a fixed point for the scale. Ancient philosophers' experiments, such as Hero of Alexandria's "fountain that drips in the sun," demonstrated the expansion of air with heat, and were known among natural philosophers of the sixteenth century. In the second decade of the seventeenth century, *Galileo, Santorio Santorio, and others began to use long-necked glass flasks partially filled with air and inverted in water to measure temperature, applying them to medical and physical experiments and keeping meteorological records. The first sealed liquid-in-glass thermometers, filled with spirit of wine, were constructed for the Accademia del Cimento in Florence in 1654 by the artisan Mariani; though not calibrated from fixed points, his thermometers agreed very closely among themselves.

The succeeding century saw experimentation with thermometric liquids, among which spirit of wine was favored for its quick response and because no cold then known would freeze it. Several natural philosophers, including Robert *Hooke, Christiaan *Huygens, and Edme Mariotte, worked out methods for graduating their instruments from a single fixed point, typically the freezing or boiling point of water. Toward the end of the seventeenth century, Italian investigators began using two fixed points, as did the Dutch instrument maker Daniel Fahrenheit in the first few decades of the eighteenth century. Fahrenheit's excellent thermometers spread his method and his preference for mercury throughout England and the Low Countries, while the dominance of France and the fame of its Académie Royale des Sciences secured the position of academician René-Antoine Ferchault de Réaumur's thermometer on the rest of the Continent.

Réaumur and his contemporaries despaired of precision in their instruments. The inconstant composition of spirit of wine; air dissolved or trapped in the liquid, whether mercury or spirits; the lack of good glass—these and the lack of motivation to precision rendered the thermometer's readings at best qualitative indications of the temperature. After the Seven Years' War, the rational bureaucratic state and industrial manufacturers generated pressure for precise measurement for cartography and navigation, enclosures and canals, and the construction of steam engines and other machinery. In England, instrument making grew from a handicraft to an operation of industrial scale, exploiting advances in glass-making and metallurgy and serving an international clientele. The thermometer played an auxiliary role in the precise measurements of the late Enlightenment, but from about 1760, the Genevan natural philosopher Jean-André Deluc developed methods for rigorously calibrating the instrument and for using it in exhaustive series of systematic measurements. In England, a committee of the Royal Society of London under the chairmanship of Henry *Cavendish worked out methods in 1776 of setting the upper fixed point in a water bath, methods that remain in use today. Late-eighteenth-century thermometers achieved a precision of $1/10°$ F. The chief development of the nineteenth century was the discovery that the glass of new thermometers contracted in time so that their zero point fell; in the 1880s, glasses were developed that did not experience these effects.

W. E. Knowles Middleton, *A History of the Thermometer and Its Use in Meteorology* (1966).

THEODORE S. FELDMAN

THIRD WORLD ACADEMY OF SCIENCES is an autonomous international organization created in 1983 to promote science-based development in the Third World. The establishment of the Academy culminated initiatives of Abdus Salam, winner of the *Nobel Prize for physics in 1979, who was its first president. The most prominent among these initiatives was the foundation, in 1964, of the "Abdus Salam" International Centre for Theoretical Physics (ICTP) at Trieste, which hosts the Academy secretariat. As a consequence of decolonization in *Asia and *Africa, beginning in the late 1940s, small but growing scientific communities in the new nations emerged. In *Latin America an unprecedented institutionalization of science took place in the 1950s and 1960s. Scientists in all of these regions felt isolated both nationally and internationally, due to a lack of local recognition, a concomitant lack of material research support, and a scarcity of international contacts. International collaboration between North and South, to use the terms deployed by the Academy to designate industrialized and developing countries, focused on technology rather than on science, which deepened the sense of frustration of Third World scientists. The ICTP and the Third World Academy of Sciences aim to provide the necessary resources and stimuli to scientists working in or with the Third World.

The original goal of the Academy was to honor the most prominent scientists from developing countries by electing them members and, from 1985, by conferring awards. The Academy distinguishes Third World scientists for contributions to basic and applied sciences, and for promoting science in developing regions. Academy members propose and elect colleagues to two categories of membership: Fellows, who must be citizens of developing countries, and Associate Fellows, citizens of industrialized countries who either were born in developing countries or have contributed to Third World science. The Academy has about six hundred members, 80 percent from developing countries and 20 percent from industrialized countries, which corresponds to the ratio of the corresponding world populations.

The Academy receives funds from the government of Italy, the Swedish Agency for Research Cooperation with Developing Countries, the United Nations Educational, Scientific and Cultural Organization (UNESCO), and the Kuwait Foundation for the Advancement of Science. China, Brazil, India, Kuwait, Mexico, and other developing countries contribute to a special endowment fund to run the secretariat. Since 1991, UNESCO has been in charge of the administration of the Academy.

The Academy takes an active role in supporting research in the Third World through grants in biology, chemistry, mathematics, physics, and basic medical sciences. Since the early 1990s, it has been developing an international program to identify research centers and create networks aimed at addressing problems of critical importance for the South such as conservation of biodiversity and production of food and medicine. The Academy provides funds for South-South scientific exchange and organizes international conferences every two years. In order to foster science-based development, the Academy decided to set up a common space open to Third World scientists and politicians. Hence the Academy's General Conference gathers not only its members, but also ministers of science and technology, presidents of science academies and research councils, and representatives of international organizations.

ALEXIS DE GREIFF

THOMSON, Joseph John (1856–1940), physicist, Nobel Prize winner, often known as the "discoverer of the electron."

Thomson was born on 18 December 1856 at Cheetham Hill near Manchester, England, the son of Joseph James Thomson, a bookseller, and Emma Swindells, who came from a textile manufacturing family. His parents intended him to become an engineer, entering him at Owens College, Manchester, at the age of fourteen. When his father died two years later, Thomson could no longer afford an engineering apprenticeship and was compelled to rely on scholarships, concentrating on mathematics and physics (taught by Thomas Barker and Balfour Stewart, respectively), in which he excelled.

In 1876 Thomson obtained a minor scholarship at Trinity College, Cambridge, to study mathematics. He was coached by Edward Routh, who gave him a thorough grounding in analytical dynamics (the use of Joseph-Louis Lagrange's equations and William Hamilton's principle of least action). This emphasis on physical analogies and a mechanical worldview is evident throughout the rest of his work. In 1880 he graduated as Second Wrangler (second place in mathematics).

Thomson's early work was dominated by his reliance on analytical dynamics to explore James Clerk *Maxwell's electrodynamics, which he had first encountered at Owens College, and then learned from William Niven at Cambridge. In 1881 he was the first to show that the mass of a charged particle increases as it moves, and suggested that the particle dragged some of the ether with it. In 1882 he won Cambridge's Adams Prize for "A Treatise on Vortex Motion," investigating the stability of interlocked vortex rings and developing the then-popular theory that atoms were ethereal vortices to account for the *periodic table. This work laid the foundations for all of his subsequent atomic models.

In 1884 Thomson was elected Cavendish Professor of Experimental Physics at Cambridge, becoming almost overnight a leader of British science and training a high proportion of the next generation of British physicists. He held an increasing number of positions in scientific administration, was on the Board for Invention and Research during World War I, president of the Royal Society from 1915 to 1920, and from 1919 to 1927 was an active member of the Advisory Council to the Department for Scientific and Industrial Research. His position was confirmed by a knighthood in 1908 and the Order of Merit in 1912. In 1890 he had married Rose Paget, daughter of Cambridge's Regius Professor of Physic. They had a son, George, and a daughter, Joan.

As Cavendish Professor, with a secure appointment for life, Thomson was able to take up an unpopular experimental subject, the discharge of electricity through gases. By 1890 he had developed the concept of a discrete electric charge, modeled by the terminus of a vortex tube in the *ether, which guided his later experimental work. The discovery of *X rays in 1895 proved a turning point: X rays ionized the gas in a controllable manner and clearly distinguished the effects of ionization and secondary radiation. Within a year, Thomson and his student Ernest *Rutherford had convincing evidence for Thomson's theory of discharge by ionization of gas molecules. X rays also rekindled interest in the *cathode rays that caused them. With new confidence in his apparatus and theories, Thomson in 1897 showed that all the properties of cathode rays could be explained by assuming that they were subatomic charged particles, which were a universal constituent of matter. He called these "corpuscles," but they soon became known as "electrons." In the following years Thomson unified his ionization and corpuscle theories into a general theory of gaseous discharge of wide applicability, for which he won the Nobel Prize in 1906.

Thomson's sophisticated model of the atom, in which thousands of corpuscles orbited in a sphere of positive electrification, went some way toward explaining the periodic table and chemical bonding. But his discovery in 1906 that the number of corpuscles in the atom was comparable with the atomic weight (i.e., that the atom contained 100s rather than 1,000s of corpuscles) raised problems with the origin of the atom's mass and its stability, to which he sought solutions through experiments on the positive ions in a discharge tube. This work led to recognition of the H_3+ ion and the discovery of the first nonradioactive isotopes, those of neon, in 1913, prompting the invention by Thomson's collaborator Francis Aston of the *mass spectrograph in 1919.

In 1919, following his appointment as Master of Trinity College, a Crown appointment, Thomson resigned the Cavendish Professorship. Under his leadership the Laboratory had become a place of lively debate with a social life of its own, though it suffered from financial stringency. Thomson continued to experiment until a few years before his death, laying the foundations for, among other things, plasma physics. He died on 30 August 1940 and was buried in Westminster Abbey.

Joseph John Thomson, *Recollections and Reflections* (1936; reprinted 1975). Lord Rayleigh, *The Life of Sir J. J. Thomson* (1942; reprinted 1969). E. A. Davis and I. J. Falconer, *J. J. Thomson and the Discovery of the Electron* (1997).

ISOBEL FALCONER

THOMSON, William (Lord Kelvin) (1824–1907), natural philosopher, inventor of electrical and navigational instruments.

Born in Belfast, William Thomson was the fourth child of James and Margaret Thomson. His father taught mathematics in the Belfast Academical Institution (noted for its political and religious radicalism). His mother came from a Glasgow commercial family, but died when William was just six. Encouraged throughout his formative years by his father (mathematics professor at Glasgow University from 1832 until his death in 1849), William moved from the broad philosophical education of Glasgow to the intensive mathematical training at Cambridge University, where he came second in the Mathematics Tripos of 1845. While working in Paris for some weeks to acquire experimental skills, Thomson was elected to the Glasgow Chair

of Natural Philosophy in 1846, a post he held for fifty-three years.

Thomson fashioned for himself a career that took him to the very pinnacle of British imperial science. His capacity to direct his physics toward practical ends placed him among the most eminent of Victorian scientists and engineers. His central position in a network of elite mathematical physicists—including James Clerk *Maxwell, George Gabriel Stokes, Hermann von *Helmholtz, and Peter Guthrie Tait—gave him a leading role in the emergence of physics as a scientific, laboratory-based discipline. And his active part in geological and cosmological controversies following publication of Charles *Darwin's *Origin of Species* (1859) located him in the mainstream of Victorian debates about humanity's place in nature.

Thomson's early scientific papers owed much to Joseph Fourier's *Théorie analytique de la chaleur* (1822). In an original paper written at the age of seventeen, Thomson used Fourier's mathematical treatment of heat flow to reformulate the orthodox theory of electrostatics. Replacing action-at-a-distance forces by continuous flow models, Thomson's radical approach was a principal inspiration for Maxwell's later construction of electromagnetic field theory, exemplified in his famous *Treatise on Electricity and Magnetism* (1873).

Thomson extended Fourier's treatment to the analysis of electric signals transmitted by very long telegraph wires, and advised on the optimum dimensions and operating conditions for transatlantic and imperial cables. He also constructed delicate measuring instruments (notably his "marine mirror galvanometer") for use in telegraphy. In recognition of his services to the Empire, he was created Sir William Thomson by Queen Victoria in 1866, following completion of the first successful Atlantic telegraph.

Through the engineering influence of his older brother James, William became committed in the late 1840s to Sadi Carnot's analogy between the motive power of heat and the fall of water driving a waterwheel. This representation gave Thomson the means of formulating in 1848 an "absolute" temperature scale (later named the Kelvin scale) that correlated temperature difference with work done, independent of the working substance. At the same time, Thomson pondered over the significance of James *Joule's experiments. Committed to Carnot's theory, Thomson could not accept Joule's proposition that work converted into heat could be recovered as useful work. Prompted by the competing investigations of Macquorn Rankine and Rudolf Clausius, Thomson reconciled the theories of Carnot and Joule in 1850–1851. The production of motive power required both the transfer of heat from high temperature to low temperature and the conversion of an amount of heat exactly equivalent to the work done.

Thomson and Rankine introduced the terms "actual" energy (later "kinetic") and "potential" energy. The laws of energy conservation and dissipation became the foundation of a new "science of energy." Thomson and Tait began a monumental project to extend the energy treatment throughout physics, but completed only one volume, on dynamics, of the *Treatise on Natural Philosophy*.

Working from energy principles, Thomson calculated ages for Earth and Sun in the range of 20 to 100 million years, a time scale that contradicted the geological assumptions on which Charles Darwin had built his theory of evolution. Darwin found Thomson's challenge the most difficult he had to counter.

Having built up a university physical laboratory (the first in Britain) for research and teaching, Thomson played a leading role in developing absolute standards of electrical measurement. Closely allied to this work were extensive business interests in the patenting and manufacture of scientific, navigational, and electrical instruments. In 1892, he became the first British scientist to be made a peer, and took the title Kelvin from the tributary of the River Clyde that flowed close to the University.

See also STANDARDIZATION; THERMODYNAMICS AND STATISTICAL MECHANICS.

Crosbie Smith and M. Norton Wise, *Energy and Empire. A Biographical Study of Lord Kelvin* (1989). Crosbie Smith, *The Science of Energy. A Cultural History of Energy Physics in Victorian Britain* (1998).

CROSBIE SMITH

TIDE. See MOON.

TIME. See SPACE AND TIME.

TOMONAGA, Shin'ichirō (1906–1979), Japanese theoretical physicist.

Born in Tokyo, Tomonaga moved with his family to Kyoto in 1913 when his father became professor of philosophy at the imperial university there. His father had studied in Germany and appreciated both the potency and weaknesses of Western culture. He became a close friend of the chief priest of the head temple of the Tendai sect of Buddhism, who invited him to live in a large house on the temple ground. Shin'ichirō grew up in this house, filled with his father's books, and in the temple grounds, communing with nature and Japanese culture. He developed a deep sense of responsibility to the nation that

led him to accept the burdens of academic and governmental administrative posts after 1951.

Tomonaga was a sickly child, frequently absent from school, sensitive, poor at gymnastics, and bullied by classmates. He attended a prestigious senior high school, where he became an outstanding student, and formed a lifelong friendship with his classmate Hideki *Yukawa. By middle school he had decided on a career in biology, but Albert *Einstein changed that. Einstein's visit to Japan in 1922 resulted in extensive popular accounts of the theory of *relativity that Tomonaga found unsatisfactory. He turned to a book by Jun Ishiwara and became fascinated by the four-dimensional world and non-Euclidian geometry.

Yukawa and Tomonaga both entered Kyoto University in 1923 and majored in physics. The theoretical physicist there, Kajûrô Tamaki, worked on relativity and hydrodynamics but had no interest in the old quantum theory. In their senior year, with Tamaki's encouragement, Tomonaga and Yukawa studied *quantum physics. They read the papers of Werner *Heisenberg, P. A. M. *Dirac, Ernst Pascual Jordan, Erwin *Schrödinger, and Wolfgang Pauli that had laid the foundations of the field, and explained them to each other. Shortly after their graduation, they attended the lectures that Dirac and Heisenberg gave at the Institute of Physical and Chemical Research (Riken) during their visit to Japan in 1929.

Tomonaga completed the work for his Rigakushi (bachelor's degree) in physics in 1929. With Japan in the throes of an economic depression and no prospect for a job, he decided to stay at Kyoto for graduate work. In 1931 Yoshio Nishina—who had worked closely with Niels *Bohr and Oskar Klein in Copenhagen in the 1920s and had contributed to the new quantum mechanics—lectured in Kyoto. He had come back to Japan to organize and oversee *cosmic-ray and *nuclear physics at Riken. Nishina was deeply impressed by the acuity and incisiveness of the questions Tomonaga posed after the lecture and invited him to work at Riken. Tomonaga thrived in Nishina's laboratory and assumed the position of house theorist. He, Minoru Kobayasi, and Shoichi Sakata translated the second edition of Dirac's *Quantum Mechanics* into Japanese. In 1937 Tomonaga went to Leipzig to work with Heisenberg. On his return in 1939, he accepted a professorship at Bunrika (Liberal Arts and Science) University in Tokyo. His socratic teaching became legendary. Students said of him that "he was like a magician."

In 1940 Tomonaga married Ryoko Sekiguchi, the daughter of the director of the Tokyo

Metropolitan Observatory. They had two sons and one daughter.

In 1943, while engaged in wartime work on magnetrons and radar devices, Tomonaga recast quantum field theories in a form that explicitly satisfied the requirements of the theory of special relativity. In the ruins of Tokyo after World War II, making use of this formalism, Tomonaga—independently of Hendrik Kramers, Hans Bethe, Julian Schwinger, Richard *Feynman, and Freeman Dyson—formulated the renormalization procedures that made it possible to circumvent the difficulties that all *quantum field theories faced. He was able to isolate and discard in a consistent manner the divergences encountered in perturbation theoretic calculations of *quantum electrodynamics and thus to perform a fully relativistic calculation of the Lamb shift. It had been definitely established experimentally by Willis Lamb in 1947 that the 2s and 2p states of the hydrogen atom were not degenerate—in contradiction to the prediction of the Dirac equation for an electron in a Coulomb field. The energy difference between these two states became known as the Lamb shift. For this work, Tomonaga shared the Nobel Prize with Feynman and Schwinger in 1965.

Nishina died unexpectedly in 1951 and Tomonaga took over many of his duties on governmental committees. He became chair of the Liaison Committee for Nuclear Research of the Science Council responsible for the establishment and oversight of national research institutes in high-energy and nuclear physics. Tomonaga's technical knowledge, sagacity, and even-handedness led to an almost full-time involvement in science policy. Eventually he became president of the Science Council. He practiced a style of arbitrating among competing claims that became known as the Tomonaga method: "waiting for the fruit to ripen and fall." Tomonaga often served as the official representative of Japanese culture and science at international gatherings. As a personal statement and commitment, he actively participated in the Pugwash conferences.

Most of Tomonaga's time after 1951 went to administrative duties. He welcomed his retirement in 1969. It gave him the leisure to give lectures, write essays (collected in a volume entitled *Birds That Come to My Garden*), and nurture his friendships. At the time of his death he had completed a manuscript published posthumously in two volumes as *What Is Physics*. After his death in 1979, his friends constructed a book of testimonials under a threnody from *Man'yoshu*, an anthology of eighth- and ninth-century Japanese poems. It runs, "If I could believe that there were/Two men like you in Japan,/I would never grieve."

S. S. Schweber, *QED and the Men Who Made It* (1994). H. Ezawa, ed., *Sin-intiro Tomonaga—Life of a Japanese Physicist*, trans. C. Fujimoto and T. Sano (1995).

SILVAN S. SCHWEBER

TRANSISTOR. The precursor of the transistor was the crystal detector, a point-contact rectifier. In the 1870s Ferdinand Braun investigated the electrical properties of certain metal sulphides; in 1906 he patented his "psilomelan" detector (made of a hydrated oxide of manganese). This was one of a number of patents for crystal detectors filed in that year, including Louis Winslow Austen's tellurium-silicon detector in Britain, W. Pickard's silicon-metal rectifier in Germany, and General H. H. C. Dunwoody's carborandum and steel detector in America. In Japan, Wichi Torikata investigated some two hundred minerals; in 1908 he patented the "Koseki" detector consisting of crystals of zincite and bornite, found to be as sensitive as the carborandum and steel combination but requiring no polarizing battery.

Many crystal detectors were used during World War I and became popular with amateur radio operators when public broadcasting began in the early 1920s. Standard combinations were galena (lead sulphide) with gold, silver or copper contacts; molybdenite (molybdenum sulphide) and copper, zincite, and tellurium; and carborandum (silicon carbide) and steel. Although crystals could not amplify incoming signals, they required no batteries when rectifying the input signal from the aerial. Crystals also tended to have a much better tonal fidelity than many of the much more expensive thermionic valve sets on the market in the early 1920s.

Solid-state electronics developed further in the 1920s with the commercial exploitation of copper oxide and selenium rectifiers and photocells used as rectifiers, battery chargers, and photographic exposure meters. The application of *quantum physics to semiconductor theory from the mid-1920s, especially the band theory of A. H. Wilson (1931), became an important factor in invention of the transistor after World War II. Another factor was the demand for purer germanium and silicon detectors for shorter radio wavelengths and for *radar. Because of the war, security meant that work on these detectors developed in parallel in various countries, in particular in Germany and America. J. H. Scaff and H. C. Theuerer at Bell Laboratories developed silicon crystals with specific impurities designated as n-type and p-type, including one that turned out to be p-type at one end and n-type at the other. In 1941 another researcher at Bell, R. S. Ohl, discovered that the p-n junction so formed was an excellent rectifier and gave a strong photovoltage. A team led by Karl Lark-Horovitz at Purdue University conducted similar research with germanium.

Marvin Kelly of Bell Laboratories assembled a team of scientists in 1939 to develop a solid-state amplifier. Although unsuccessful, a new team was put together after the war. Headed by William Shockley, a theoretician confident in the possibility of a solid-state analogue of the triode thermionic valve, the team's other members were theoretician John Bardeen and Walter Brattain, a gifted experimenter. They discovered the point-contact transistor in December 1947, for which they received the Nobel Prize in 1956.

Much development had to take place to turn their invention into a reliable transistor that could be mass-produced. The point-contact diode gave way to the p-n junction diode, which, in turn, developed into the grown-junction and alloyed-junction transistors successfully mass-produced from the early 1950s. In 1953 Philco introduced the surface-barrier transistor based on their jet electrolytic etching technique; the next year Texas Instruments introduced the silicon grown-junction transistor. In the mid-1950s doping techniques were developed to control the impurities introduced into the semiconductor to achieve the designed electrical characteristics. These techniques made it possible to manufacture hundreds of transistors on a single slice of germanium or silicon. Two key developments in the early 1960s were an improved diffusion and oxide masking technique known as the planar process patented by the Fairfield Semiconductor Company, and the plastic encapsulation of the transistor introduced by General Electric. Integrated circuits could now be manufactured for transistor radios, television, digital calculators and computers, and other electronic devices. The world's first transistor radio, the Regency TR-1 designed by Texas Instruments and manufactured by I.D.E.A. of Indianapolis, appeared in 1954 in time for Christmas.

V. J. Phillips, *Early Radio Wave Detectors* [IEE History of Technology Series 2] (1980). P. R. Morris, *A History of the World Semiconductor Industry* [IEE History of Technology Series 12] (1990). Michael Brian Schiffer, *The Portable Radio in American Life* (1991). Spencer R. Weart, Lilian Hoddeson, Ernst Braun, and Jurgen Teichman, *Out of the Crystal Maze: Chapters from the History of Solid State Physics* (1992).

W. D. HACKMANN

TRANSPIRATION. See RESPIRATION AND TRANSPIRATION.

TRANSURANIC ELEMENTS. Glenn T. Seaborg popularized the term "transuranic elements"

after the declassification of information on the elements with atomic numbers greater than uranium's.

The Rutherford-Bohr model of the atom, established between 1911 and 1922, explained for the first time why the lanthanides had similar properties—namely that they possessed identical outer electronic configurations and slightly different numbers of electrons in interior subshells. Both Johannes Rydberg in 1913 and Niels *Bohr in 1921 speculated that a similar group of heavier elements might exist, and Vicktor Goldschmidt proposed that they should be named the neptunium group. Following the discovery of nuclear fission in 1939, Edwin M. McMillan and Philip H. Abelson, working at the University of California at Berkeley, showed that neutron bombardment of uranium produced a synthetic radioactive element. Because of its position in the periodic table following uranium (with its association with the planet Uranus), they followed Goldschmidt in naming it neptunium. At the same time, and using the Berkeley 150-cm (60-inch) cyclotron, Seaborg and his colleagues prepared plutonium (element 94) by the beta-decay of neptunium. McMillan and Seaborg received the *Nobel Prize in chemistry in 1951 for this work. During the war, Seaborg found time to pursue further examples of synthetic transuranic elements. Americium and curium were identified between 1944 and 1945, and berkelium, californium, einsteinium, fermium, and mendelevium by 1953. Altogether Seaborg participated in the identification of nine of the fifteen actinides (as chemists called Goldschmidt's neptunides) between 1940 and 1970.

Although the chemistry of neptunium and plutonium proved analogous to uranium's, their successors did not. This anomaly forced a reexamination of their place in the periodic table. As early as 1944 Seaborg noted that elements 95 and 96 should have properties analogous to those of the lanthanides europium and gadolinium. He therefore suggested that elements 90–92 (thorium, protactinium, and uranium) should be moved from the seventh period of the periodic table to form a second rare earth (actinide) family of elements that extended from 89 (actinium) to 102 (nobelium), and subsequently to lawrencium, made in an accelerator in Berkeley in 1961 under the leadership of Seaborg's colleague Albert Ghiorso. This amendment to the periodic table proved the key to discovering the remaining elements. The preparation and confirmation of a few fleeting atoms of synthetic transactinides beyond lawrencium saw rival groups from the Soviet Union and Germany competing with the

Americans. Settling the names of these synthetic elements, rutherfordium, dubnium, seaborgium, bohrium, and meitnerium, proved controversial and protracted. In the fifty years between 1940 and 1990, seventeen new elements were added to the periodic table beyond uranium.

Seaborg also speculated that beyond transuranic elements 113 and 164 there may be islands of stability containing super-heavy elements with long radioactive half-lives that would allow detailed comparisons with natural elements in their group positions within the periodic table. Besides forming an outstanding example of twentieth-century big science, the investigation of the transuranic elements has shown the continuing value and power of Dmitrii *Mendeleev's *periodic table.

See also ATOMIC STRUCTURE.

Glenn T. Seaborg, "Some Recollections of Early Nuclear Age Chemistry," *Journal of Chemical Education* 45 (1968): 278–289. George B. Kauffman, "Beyond Uranium," *Chemical and Engineering News* (19 November 1990): 18–29.

WILLIAM H. BROCK

TURING, Alan (1912–1954), English mathematician, computer pioneer, and cryptanalyst.

Until the 1970s, Turing was renowned for his contributions to mathematical logic and the theoretical basis of computer intelligence. Since then, as an increasing number of government papers have been declassified, he has become famous for his role in cracking German military codes during World War II. Belying his central importance to modern computing, Turing remains an obscure figure not only because of the secrecy surrounding much of his work, but also because of his homosexuality, illegal in Britain until after his death, which affected his career as well as his subsequent reputation.

Because his father was in the Indian Civil Service, Turing was often separated from his parents for long periods of time, and was educated in English male boarding schools. At the age of seventeen, devastated by the sudden death of his closest friend, he started to question the conventional Christian Cartesian separation of spirit and body, an early move toward his materialist convictions and mechanical approach to human thought processes. In 1931, he gained a scholarship to King's College, Cambridge, where he was awarded a fellowship for his dissertation on the Gaussian error function. From 1936 to 1938 he worked at Princeton University, where he published what has now become a classic but difficult paper, a discussion of the decision problem articulated by the German mathematician David Hilbert. To consolidate the foundations of mathematical certainty, Hilbert had proposed that there must be a method for deciding whether a meaningful mathematical statement can be proved true or false. Turing's innovation was to introduce imaginary machines that could generate computable numbers, thus converting the mathematical concept of decidability into the automated behavior of a symbol-manipulating machine that performed the repetitive operations of a human calculator. Through envisaging a theoretical universal computing machine that produced contradictory solutions, Turing was able to overturn Hilbert's suggestion.

When war broke out in 1939, the year after Turing's return to England, he had already been studying cryptanalysis and promptly joined the team of academic code-breakers at Bletchley Park, a large but secret intelligence center between Oxford and Cambridge. German naval authorities believed that messages generated by their Enigma machines, typewriter-like devices incorporating electronic circuitry to convert ordinary text into incomprehensible letter sequences, were totally secure because deciperment involved knowing a daily code specifying just one among a vast number of initial settings for the machines. Using a logical, statistical approach, and exploiting the fallibility of human operators, Turing devised ways of reducing the number of possibilities that had to be searched and helped to construct electronic apparatus to speed up the processes of selection. Because the successful strategies of Turing and his colleagues enabled them to decode many German communications, they played a vital role in the outcome of the war. He was awarded the Order of the British Empire for his cryptanalytical work.

In 1948, frustrated by the slow progress of a group at the National Physics Laboratory building an electronic digital computer, Turing moved to the University of Manchester, where he spent the rest of his life. In addition to further research into mathematical theory, he worked on practical problems of developing computers controlled by stored programs. Continuing his earlier ideas, Turing turned to the topic of machine intelligence, drawing close parallels between the function of human minds and of electronic machines to reinforce his claim that his machines offered a valid approach toward the whole of mathematics. In 1950, the journal *Mind* published his "Computing Intelligence and Machinery," an acclaimed and accessible paper that tackled the question of whether or not machines can think. Reflecting his experiences both of cryptanalysis and of sexual subterfuge, Turing first imagined an interrogator who, linked by a teleprinter to a man and a woman in another room, tried to determine which was which. Then, insisting that machines could successfully imitate human thought, he wondered whether the interrogator would be more successful at distinguishing between a person and a machine. Turing forcefully rejected nine different theological, philosophical, and psychological objections to his premise that machines can think, and his article profoundly influenced explorations of human and computer creativity.

In the intensive intellectual atmosphere at Bletchley Park, Turing had gained a reputation for eccentric behavior and had broken off his engagement to a fellow mathematician. In 1952, because of a robbery Turing himself reported, the police became aware of his homosexuality and, after pleading guilty in a public trial, Turing underwent a year's experimental hormone therapy. In 1954, while actively investigating the mathematical growth patterns of plants, Turing died, almost certainly by suicide. The ever-escalating importance of computers, changed attitudes towards sexuality, and reappraisals of military history presage that Turing's status as a cult hero will undoubtedly increase.

Andrew Hodges, *Alan Turing: The Enigma* (1983). Martin Campbell-Kelly and William Aspray, *Computer: A History of the Information Machine* (1996). Andrew Hodges, *Turing: A Natural Philosopher* (1997). Gordon Welchman, *The Hut Six Story: Breaking the Enigma Codes* (1997). Simon Singh, *The Code Book: The Secret History of Codes and Codebreaking* (1999). Jon Agar, *Turing and the Universal Machine: The Making of the Modern Computer* (2001).

PATRICIA FARA

TV. See FILM, TELEVISION, AND SCIENCE; RADIO AND TELEVISION.

U

ULTRACENTRIFUGE. See CENTRIFUGE AND ULTRA-CENTRIFUGE.

ULTRASONICS. There has been an interest in underwater sound from Antiquity. Both Aristotle and Pliny the Younger speculated on the ability of fish to hear, and many centuries later Leonardo da Vinci heard the movement of a ship by listening through a tube held under water. In the early 1700s Francis Hauksbee the Elder, the Curator of Experiments at the Royal Society in London, demonstrated that sound did not transmit through a vacuum, but did travel through water—a medium much denser than air. An improved version of this experiment published by Willem Jacob 'sGravesande in 1720 aroused much interest. William Arderon reported to the Royal Society in 1747 that a bell could be heard distinctly under water, and Benjamin *Franklin made a similar observation in 1762. The best early measurement of the velocity of underwater sound was made in 1826 on Lake Geneva by Jean-Daniel Colladon and the mathematician Charles-François Sturm; they obtained 1435 metres per second, very close to the value now accepted.

Around 1900 a sudden increase in experimental activity in underwater sound occurred because of its potential use as a navigational aid, brought about by growth of the world's shipping and the increase in the size of ships with steam propulsion. Underwater sound navigation became practical when the telephone receiver was modified into the "hydrophone" to pick up underwater sound from underwater bells warning of hazards. Elisha Gray, who had invented the telephone independently from Alexander Graham Bell, and who, together with A. J. Mundy, developed an underwater bell and hydrophone system, coined the name. The term had originally been used to describe a water bag attached to a stethoscope. The Submarine Signal Company, founded in 1901 to exploit this underwater navigational system, had a great influence on the early development of underwater sound.

In 1910 competition from wireless telegraphy as a navigational aid led to the invention of an underwater analogue, the "Fessenden oscillator," named after the electrical engineer Reginald Aubrey Fessenden. Since this electromagnetic device, which moved a large metal diaphragm, could both transmit Morse code underwater and receive signals, it was the first transducer, although operated at sonic frequencies. Two events now occurred that led to the rapid development of underwater acoustics: the sinking of the *Titanic* in 1912 and the U-boat campaign of World War I.

Hydrophones operating at sonic frequencies were used in great numbers during the war, but led to the destruction of few U-boats. The way forward turned out to be detecting submerged objects by the echoes produced by supersonic, later renamed ultrasonic, waves. The Russian electrical engineer Constantin Chilowsky began developing such an echo-ranging system inspired by the disaster of the *Titanic*; at the start of World War I he emigrated to France and in 1915 persuaded Paul Langevin to apply his idea against U-boats. In 1917 they developed an ultrasonic transducer based on the piezoelectric effect discover many years previously by Pierre *Curie and his brother Jacques. When quartz was compressed it produced a very small electric current and when subjected to a current, expanded slightly. The device thus behaved similarly to the Fessenden transducer but operated at ultrasonic frequencies; when placed in a circuit with the then recently developed valve amplifier, it could pick up extremely weak echoes. In England, Sir Ernest *Rutherford, who advised the Admiralty during the war, assigned the Canadian physicist R. W. Boyle to this problem. The British worked closely with the French. If the war had continued for a few more months after the armistice of November 1918, "asdics," as the system was code-named in Britain, would have gone operational. Asdics signified antisubmarine division (ASD), the Admiralty department that supervised the research. British scientists and technicians improved the system during the interwar years so that it was fully operational at the start of Word War II. During the same time the ultrasonic echo sounder was developed, which from the late 1940s could be used increasingly to locate fish.

The major advances in sonar technology in World War II aimed to combat ever faster and deeper submarines. Higher resolution sonars could detect submarines lying on the seabed and mines and other small objects. From these came, during the 1950s and 1960s, side-scan sonars for mapping the sea floor, such as Gloria (geological long-range inclined asdics), constructed by the British National Institute of Oceanography.

"Sonar" replaced "asdics" in the late 1950s. Invented by the American sonar scientist F. V. Hunt in 1942, the name stood for sound navigation and ranging, the sound equivalent of *"radar" ("radio detection and ranging"). Apart from military and oceanographic application, ultrasound techniques began to be developed for industry and in medicine. Langevin had already observed certain biological effects of ultrasound during his work on echo-ranging. Robert Williams Wood, Alfred L. Loomis, and Edmund Newton Harvey made the first systematic studies in the late 1920s and early 1930s, but concentrated research on medical ultrasound (or ultrasonics) did not start before the late 1940s and 1950s, and clinical devices became widely available only in the 1970s.

F. V. Hunt, *Electroacoustics: The Analysis of Transduction and its Historical Background* (1954). Willem D. Hackmann, *Seek and Strike: Sonar, Anti-Submarine Warfare and the Royal Navy 1914–54* (1984). Stuart Blume, *Insight and Industry: The Dynamics of Technological Change in Medicine* (1992).

WILLEM D. HACKMANN

UNCERTAINTY. See COMPLEMENTARITY AND UNCERTAINTY.

UNIFORMITARIANISM AND CATASTROPHISM. William Whewell, Master of Trinity College, Cambridge, coined the terms uniformitarianism and catastrophism in 1832 in his review of the second volume of Charles *Lyell's *Principles of Geology* (1830–1833) in the *Quarterly Review*. Uniformitarianism referred to Lyell's methodology and theory of the earth, catastrophism to the mainstream geological doctrines that he himself favored. No terms in the history of *geology have created more confusion.

Lyell in his *Principles* had wanted to avoid two methodological extremes—the conservative enumerative induction adopted by the Geological Society of London as its quasi-official methodology, and the method of hypothesis of the seventeenth- and eighteenth-century cosmogonists. The first inhibited theorizing; the second put no empirical checks on the free rein of the imagination. Lyell opted for the vera causa

method (method of true causes) advocated by Isaac *Newton in the "Rules of Reasoning" in his *Principia mathematica* (1687), given canonical form by the Scottish philosopher Thomas Reid, and used by John Playfair in his *Illustrations of the Huttonian Theory of the Earth* (1802). According to the vera causa principle, any postulated cause has to satisfy two conditions: it must be known to exist, and it must be known to be adequate to produce the effect.

Observing both causes and their effects was difficult in geology. Causes frequently acted over long periods or out of sight in the center of the earth, concealing their effects from geologists. Lyell therefore modified the method to accommodate the special problems of his science. Geologists could be sure of the existence of the cause they postulated only by observing one like it in action. In practice, this meant that geologists had to restrict themselves to causes operating at present. The same applied to the adequacy of the cause. Geologists could not postulate causes that acted more forcefully than those they observed in action.

Uniformitarianism, as Lyell understood it, thus consisted of three theses. The first, that the laws of nature had not changed, was uncontroversial. All geologists accepted this constancy except in the case of the creation of new species, which Lyell, like everyone else, believed required divine intervention. The second was that the kinds of causes operating on the surface of the earth had never changed (the existence condition of the vera causa principle), and the third, that the intensity of causes had not changed either (the adequacy condition of the vera causa principle). The highly regarded astronomer John Herschel, the mathematician Charles Babbage, and Charles *Darwin, who made the vera causa method central to the extended argument of his *Origin of Species* (1859), welcomed uniformitarianism.

Almost everyone else rejected Lyell's principles. Accepting them would have meant repudiating the most successful causal theory in the discipline, Élie de Beaumont's theory of the cooling, contracting earth with its spasmodic bursts of mountain-building and species-extinction. If the earth had cooled, then presumably in the past no glaciers had gouged the earth as they now did. And presumably volcanoes would have been more active in the past. Not surprisingly, Lyell's denial that causes had ever varied in intensity seemed both arbitrary and improbable, and therefore almost all geologists rejected uniformitarianism. For the geologists' position, with its commitment to a directional theory of the earth's history, Whewell adapted the Greek word for the denouement of a drama: *catastrophe*.

Uniformitarians and catastrophists also divided along religious lines. Lyell, like Hutton before him, rejected Christianity for deism. He was delighted that his methodology supported the theory of a steady-state earth because, in his opinion, a benevolent deity would not have created an earth that contained the seeds of its own destruction. Most geologists, though, were Christians. Although they did not believe the literal truth of the Biblical accounts of the Creation and the Flood, they did believe that the earth had been created at a specific time in the past, and would come to an end at a specific time in the future. This predisposed them to a belief in a directional Earth history.

After Lyell's death in 1875, uniformitarianism seemed to have lost its luster. Then Sir Archibald Geikie, head of the Geological Survey of Great Britain, praised it in his *Founders of Geology* (1905), and, perhaps because of his endorsement, uniformitarianism became the rallying cry of geologists in the first half of the twentieth century. They had no knowledge of its roots in the old vera causa tradition. They summed it up with the slogan, "The present is key to the past," and used it in a number of not always consistent ways. They argued that the earth had suffered no dramatic upheavals, as catastrophes were now conceived; that geologists should study present processes, though they rarely did; that geologists should not resort to religious explanations; and that they should not invoke extraterrestrial causes.

As geologists came to accept the destruction of crust implied by *plate tectonics, the possibility of non-gradual evolutionary change asserted by proponents of punctuated equilibria, the fact of meteor impacts, and the likelihood that one such impact had killed the dinosaurs, catastrophism (understood as dramatic large-scale events) regained plausibility.

See also EARTH SCIENCES; EARTH, AGE OF.

Reijer Hooykaas, *The Principle of Uniformity in Geology, Biology and Theology* (1963). Rachel Laudan, *From Mineralogy to Geology* (1987). Charles Lyell, *Principles of Geology*, ed. Martin Rudwick (1991, of 1830–1833 ed.).

RACHEL LAUDAN

UNIVERSE, AGE AND SIZE OF THE. At the end of the seventeenth century, Isaac *Newton's concept of an infinite universe created instantaneously by God rendered the question of the size of the universe moot and the question of its age a matter for historical rather than scientific determination. For more than two centuries after the publication of Newton's *Principia*, efforts to measure astronomical distances were confined to our solar system and a few nearby stars. Ever larger *telescopes and the development of *photography and *spectroscopy greatly increased the variety and accuracy of the data available to astronomers, but at the end of the nineteenth century, the size of our own Milky Way *galaxy was still unknown and, despite earlier speculations about the possibility of other systems, few, if any, astronomers believed that anything existed outside our own galaxy in the infinite void of the universe.

Between 1910 and 1930, new techniques for estimating interstellar distances finally enabled astronomers to determine the approximate size and shape of the Milky Way galaxy. During the same period, Vesto M. Slipher and Edwin *Hubble measured the red shift in the spectra of a number of spiral nebula, and determined that almost all of them were moving away from the earth at high radial velocities. By 1929, Hubble had also calculated the distance to several nebula, and for the first time had provided convincing evidence that spiral nebula were clusters of stars (he did not identify them as galaxies) far beyond the borders of our own galaxy. Even more significantly, Hubble had found that the farther away the nebulae were, the greater their radial velocities away from the earth. Hubble's discoveries not only populated Newton's cosmic void with myriads of stars, they also suggested that the universe itself was expanding rather than static. The concept of an expanding universe had already been proposed theoretically by Willem de Sitter (in 1917), Alexander Freedman (1922), and Georges Lemaître (1927) as a way to resolve an anomaly in the solution to *Einstein's theory of general relativity, but it was Hubble's work that caused astronomers to begin seriously to consider the idea of an expanding universe and the implication that it had a calculable age and size.

The key to calculating the age, and hence the size, of an expanding universe is the determination of the intergalactic velocity/distance ratio (which became known as the "Hubble Constant"). In the 1930s, Hubble's original value for the velocity/distance ratio produced an estimate of about 2×10^9 years for the age of the universe. This result briefly created the curious anomaly of a calculated value of the age of the universe that was smaller than radiometric measurements of the age of the earth. The work of Walter Baade in the 1940s and Allan Sandage in the 1950s resulted in substantial revisions in the accepted value of Hubble's Constant, and by the 1960s astronomers agreed that the universe was between 10 and 20×10^9 years old, with a corresponding size of the order of 10^{10} light-years. Also by the 1960s, the success of the "Big Bang" Hypothesis had provided astronomers with a causal physical model of an expanding universe with an instantaneous beginning and a finite age and size. Since the 1950s, the rapid proliferation of new techniques and technologies has enabled astronomers to probe ever closer toward the outer edge of the universe, and in the 1990s, a more precise redetermination of the Hubble Constant was made a priority of the new Hubble Space Telescope. The results of the several independent efforts to recalculate the Hubble Constant were not consistent among themselves, however, and the late 1990s saw renewed controversy over the age and size of the universe. At the century's end, some astronomers believed that the universe's age had been determined to lie within the range of 12 to 15×10^9 years, while their more cautious colleagues remain unwilling to allow for a greater certainty than the earlier 10 to 20×10^9 years. The size of an expanding universe, of course, depends on its age, but recent cosmological theories, such as Alan Guth's inflationary model, suggest the possibility that our observable universe may be only a small part of a much larger structure.

See also SPACE AND TIME.

Kitty Ferguson, *Measuring the Universe* (1999). John Gribbin, *The Birth of Time* (1999). Stephen Webb, *Measuring the Universe, the Cosmological Distance Ladder* (1999).
                                   JOE D. BURCHFIELD

UNIVERSITY. The university is an institution of higher learning that first appeared in late medieval Europe in the form of the *universitas magistrorum et scholarium*, a corporation of masters and students. Signs of the existence of such corporations in Bologna (with an emphasis on law) and Salerno (medicine) date from the eleventh century; but not until the later twelfth century, at Oxford and Paris, can we speak of confidence of a *studium generale*, a university offering instruction in all the faculties and granting degrees to students who had passed certain tests. The degree conferred the *ius ubique docendi*, the right to teach at the level specified (arts, law, theology, or medicine) anywhere. From Bologna, Oxford, and Paris universities spread throughout Europe and the new world, and today are found around the globe. Until the Protestant Reformation of the early sixteenth century, popes or their local representatives, bishops, chartered universities, and they or the municipalities in which the universities resided controlled the content of instruction in the arts and theology faculties. Before the seventeenth century the study of the natural world could be found piecemeal in the medical curriculum or in subjects intrinsic to or associated with the quadrivium (arithmetic, geometry, astronomy, and music). Not until the late eighteenth century and in most cases not until the nineteenth century did universities become the principal institutions for the development as well as for the preservation and transmission of organized forms of knowledge.

Universities have been crucial for the sustenance and evolution of the natural sciences and the social systems that support them. Five aspects of university history are especially important for a historical understanding of the natural sciences: the gradual emancipation from forms of political control over the intellectual content of university teaching, especially the bias against the mechanical arts, and the subsequent integration into university study of laboratories and instruction in experiment; the transformation of the curriculum in the philosophical faculty from the classical seven liberal arts to other forms of organized knowledge; the establishment of a social system for teaching and learning the natural sciences through the inauguration of professorial positions, curricula, and examinations in the natural sciences and of ways of accrediting the next generation of scientists; the addition of research to teaching in the professorial role; and the interaction of the university with state, industrial, and military objectives.

Although not all medieval universities had the full four faculties of medicine, law, theology, and philosophy, all regarded the philosophical faculty as the "lowest" of the four, offering primarily preparatory study for the three "higher" faculties, each of which trained students for particular professions. The faculties maintained the boundaries between branches of knowledge, and all four monitored a common boundary that kept the mechanical arts outside the university. Teaching took place by lecture (the reading of an authorized text by the regent master or professor), commentary, and disputation (an oral debate conducted in Latin according to the rules of Aristotelian syllogistics). The central importance of memorization and repetition helped to bar experimentation and measurement. Teachers were held to high moral principles and were required to "read" (lecture) approximately two hours per day on about 150 days of the year. The close intellectual and affective relationship between masters and pupils led to the formation of locally defined schools, so that registering at a university came to mean studying with a master.

During the thirteenth century the newly recovered works of Aristotle, particularly his logic and books on nature, dominated the arts curriculum. Often the commentaries on these books drew heavily on the work of the Arabic scholars al-Kindi, Avicenna, and Averroës, which took the text more literally than the masters of theology approved. In 1277 the bishop of Paris condemned some 219 theses being taught under the name of Aristotle; the eventual consequence was to drive the arts faculties at Oxford and Paris toward a nominalism favorable to natural science. (Some of the findings of the French school of the fourteenth century, particularly the analyses of motion by Jean Buridan and Nicole Oresme, influenced *Galileo.) In southern Europe, where Arabic texts were especially popular, instrumentation common to the Arabic tradition such as the armillary sphere, *astrolabe, and abacus also became a part of instruction. Following the practice of Muslims, most universities taught astrology along with astronomy, and for some time did not distinguish clearly between the two. Music provided another route within the quadrivium for the study of quantitative relationships, especially in combination with the works of Boethius. Instruction in geometry rested on Euclid's *Elements*, of which several translations circulated in the later Middle Ages, one of which was published in 1482 among the earliest printed books. A very few students who mastered Euclid might study Ptolemy's *Almagest* or a summary of it; but most stopped with Johannes de Sacrobosco's *Sphaera*, an astronomical work of the thirteenth century that saw hundreds of editions up to the seventeenth century. Statics and optics, the latter enhanced by Arab commentaries, also were cultivated in the late medieval university. But the growth in quantitative topics within the university did not keep pace with investigations of the natural world outside the university before the nineteenth century.

The pope's control over universities began to weaken after the Black Death (1348) and the Great Schism (1378). By the fifteenth century the advent of humanistic studies based on the teaching of recovered ancient texts and the concomitant interest in Greek and Hebrew enriched sources on the natural world. A typical syllabus included the *Physics* of Aristotle, as well as mathematics, astronomy, and music, all based on quantitative techniques. Interest in the study of the natural world expanded especially in the medical faculties of the fifteenth and sixteenth centuries, especially with the establishment of a university at Montpellier in 1593 and the botanical gardens of the medical faculties of Bologna, Pisa, and Padua in the mid-sixteenth century. In 1543 Andreas *Vesalius published *De humani corporis fabrica*, which emphasized anatomical dissection as an essential element of medical instruction.

The role of the university in early modern science, especially in the seventeenth century, has recently been downplayed as historians have attended to the role of natural philosophy in absolutist court culture, English arguments over constitutionalism, and the novelty of new royal academies of the 1660s founded in Paris and London. Yet important contributions were made to the *Scientific Revolution by men working in the universities, for example, *Galileo, who as a professor at the University of Padua criticized Aristotelian texts and made use of artisan's knowledge in his manufacture of instruments, and professors at Oxford and Cambridge, who between 1560 and 1640 kept alive mathematical instruction in the general university curriculum.

On the continent the close linkage in the eighteenth century between the absolutist state bureaucracies and universities led to significant modifications in the curriculum, often beyond the philosophical faculty. At new "modern" universities, such as at Göttingen (founded 1737), instruction in several subjects was linked to the administration of resources in the territorial state. Public administration required instruction not only in law and accounting, but also in domestic economy, commercial applications of technologies, agricultural administration, and even applied chemistry, which was essential to mineralogy, metallurgy, glass, pharmacy, and ceramics. Branches of natural philosophy constituted an important part of the science of *cameralism, which took shape as part of state administrative training in the eighteenth century. Utilitarian uses of natural philosophy did not, however, lead easily to independent instruction in the branches of natural philosophy, which foundered until the early nineteenth century. Nor at the end of the eighteenth century was research a required part of professorial obligations. The practical instruction associated with efficient state administration in the eighteenth century gave way in the early nineteenth century to a transformation in university teaching and learning.

Educational reformers, especially in the German states, raised questions not only about the types of knowledge taught, but also about its social functions. Following late Enlightenment critique of utilitarian forms of knowledge like cameralism, administrators nonetheless reaffirmed a commitment to the practical uses of natural knowledge for a liberal, industrializing economy especially after the revolutions of 1848. Baden's science policy shaped the experimental orientation and professorial staffing of the natural sciences in the universities of Freiburg and Heidelberg and at the Karlsruhe Technical High School. In the second half of the century major chemical firms in Baden joined forces with higher education in the training of chemists whose number grew from 350 in 1850 to over 10,000 in 1914.

In Prussia, the largest of the German states, education abetted social change in the creation of an educated middle class. In part the outcome of a succession of changes in the state's certifying examinations for the professions and in part the result of the educational reforms of Wilhelm von Humboldt (1810), who argued for the unity of teaching and research in the university, Prussian emphasis on the purity of knowledge unleashed a series of institutional changes that resulted in the expansion of the natural sciences within the philosophical faculty and the first modern research universities committed to both the transmission and expansion of knowledge. The ideological emphasis upon purity and the presumed autonomy of the university from other institutions, including the economy, in Prussia proved to be chimeras; for the natural sciences even at the Prussian universities had close ties to practical endeavors, including the state reform of weights and measures, surveying, and matters concerning health, agriculture, and industry. But from within the internal dynamic of the Prussian reforms emerged patterns of university institutional practice that have constituted the social infrastructure of the natural sciences since then, including professorial positions and standards of advancement demarcated by levels of research productivity; methods of recruiting, educating, training, and certifying students as the next generation of practitioners; and the patterns of labor associated with the *seminar, *institute, and especially the *laboratory. By the middle of the nineteenth century universities no longer simply transmitted knowledge; they were also prime locations for its production. The German model of the research university spread throughout the Western world; in the United States, the first university to adopt the research ethos was Johns Hopkins University in Baltimore (founded 1876).

Historical examinations of the university require consideration of contextual factors that intersect with what appear superficially to be autonomous institutional developments. Although women's colleges and schools for minorities existed in the West in the nineteenth century, universities generally did not open their doors to women and minorities until the twentieth century, and at many of the most elite institutions, not until well after World War II. The impact of social integration upon the demographic profile of the natural sciences has been mixed. Minorities remain underrepresented in all sciences, while the ratio of women in the social and biological sciences is greater than in the physical and mathematical sciences where barriers persist. The demographic imbalance in the natural sciences is due partly to weaknesses in their social support systems, especially educational and career mentoring.

Large-scale global warfare changed forever the relationship between the university and the military, whether in the granting of federal funds for military projects, as in the United States, or in the excision of overt military objectives in university instruction, as in the Federal Republic of Germany after the Third Reich. After 1945 the ties between the university and industry all over the world became stronger with geographical centers of excellence, such as Silicon Valley in California and Route 128 in the Boston area of the United States or Measurement Valley and Silicon Saxony in the Federal Republic of Germany. The university, initially a European corporation of teachers and students who regarded the natural world as secondary to their concerns, became by the twentieth century an institution responsible not only for the perpetuation and regeneration of the natural sciences, but one through which the natural sciences maintained connections to strategic political, economic, and social endeavors across the globe.

Since World War II, the university's connections to economic development and political change became troublesome issues in regions where Western science was held responsible for the disintegration of traditional ways of life and thinking. In South America universities have been viewed both as seedbeds for labor and civil unrest and, through their cultivation of scientific expertise, as antithetical to participatory politics. The official response has been to restrict the freedoms associated with university teaching, research, and learning. In South Asia and parts of the Middle East, particularly in Pakistan, Afghanistan, Iran, Iraq, and parts of India, governments view science as a political pawn in need of control to maintain domestic stability, rather than as an object of free investigation. In India and Pakistan, where universities were originally established by colonial powers, the link between university administration and government control has become stronger since decolonization, resulting in the retention of outdated systems of learning and teaching and an overall decline in intellectual quality.

In China and Africa, faculties have become particularly sensitive to the homogenization of knowledge systems imposed by Western science. These faculties have become outspoken proponents of an intellectual pluralism that respects and retains indigenous knowledge about medicine, agriculture, and other daily activities. Their positions have been supported by some Western ecologists and environmentalists, who themselves have come to appreciate the value of indigenous knowledge in preserving biodiversity and environmental stability. Yet despite the tensions between local indigenous knowledge and universal scientific knowledge, representatives of non-Western universities nonetheless argue that social justice is not served by global imbalance in the distribution of technical and scientific expertise.

See also ACADEMY; DISCIPLINE; MINORITIES AND OUTSIDERS; SCHOOL; TEXTBOOK; WOMEN IN SCIENCE.

Peter Borscheid, *Naturwissenschaft, Staat, und Industrie in Baden, 1848–1914* (1976). Charles McClelland, *State, Society, and the University in Germany, 1700–1914* (1980). R. Steven Turner, "The Prussian Professoriate and the Research Imperative, 1790–1840," in *Epistemological and Social Problems of the Sciences in the Early Nineteenth Century*, ed. H. N. Janke and M. Otte (1981): 109–121. Mordechai Feingold, *The Mathematicians' Apprenticeship: Science, the Universities, and Society in England, 1560–1640* (1984). Byron K. Marshall, *Academic Freedom and the Japanese Imperial University, 1868–1939* (1992). Hilde de Ridder-Symoens, ed., *A History of the University in Europe, vol. 1: Universities in the Middle Ages* (1992). Stuart W. Leslie, *The Cold War and American Science* (1994). Parker Rossmann, *Emerging Worldwide Electronic University* (1994). Roger L. Geiger, *Research and Relevant Knowledge: American Research Universities Since World War II* (1997). Sohail Inayatullah, ed., *The University in Transformation: Global Perspectives on the Futures of the University* (2000).

KATHRYN OLESKO

UREY, Harold (1893–1981), American physical chemist, discoverer of deuterium, and Nobel Prize winner.

Urey was born in Walkerton, Indiana, the son of a schoolteacher. He read zoology and chemistry at the University of Montana at Missoula, graduating in 1917 just as America entered World War I. Although he retained a passionate interest in biology for the rest of his life, the war forced him to become an industrial research chemist. He returned to Montana as a chemistry instructor at Montana State University before entering graduate school under G. N. Lewis at the University of California at Berkeley in 1921. His doctoral thesis (1923), concerned with the *entropy of gases, introduced him to spectroscopy and whetted his appetite for the new *quantum physics then gaining the attention of physicists and physical chemists. He spent the years 1923–1924 on an American-Scandinavian Fellowship studying with Niels *Bohr and Hendrik Kramers at the Institute for Theoretical Physics in Copenhagen. On his return to America in 1924, he taught chemistry at Johns Hopkins University in Baltimore. He married a bacteriologist, Frieda Daum, in 1926. They had four children.

In 1929 Urey moved to New York as associate professor of chemistry at Columbia University. His interest and expertise in *quantum physics led him to coauthor a pioneering textbook on quantum mechanics with A. E. Ruark (*Atoms, Molecules, and Quanta*, 1930) and to become founding editor of the *Journal of Chemical Physics* in 1933. During World War II he served as director of research of the so-called Substitute Alloys Materials Laboratory at Columbia, a part of the Manhattan Project. In 1945 he became professor of nuclear studies at the University of Chicago, where, in 1952, he published *The Planets: Their Origin and Development*. From 1958 until his retirement in 1970, he was professor of chemistry-at-large at the University of California's San Diego campus.

Urey's early research and publications embraced chemical kinetics, quantum mechanics, and infrared spectroscopy. During the 1920s, Francis Aston of the Cavendish Laboratory in Cambridge, England, had pioneered, and virtually monopolized, the identification of isotopes using the mass spectrometer. Aston's instrument had failed to detect a heavier isotope of hydrogen. In 1931, however, while working with Ferdinand Brickwedde and George Murphy, Urey found evidence for what he named "deuterium" in commercially available hydrogen at room temperature. In the following year he separated deuterium from deuterium oxide ("heavy water") by the electrolysis of ordinary water, and encouraged colleagues like Rudolf Schoenheimer to use deuterium as a radioactive tracer in metabolic studies. Urey received the *Nobel Prize in chemistry in 1934 for this discovery.

Urey's unrivalled expertise in techniques of isotope separation was put to use during World War II when he helped separate uranium isotopes for the atomic bomb. He also had a hand in the separation of tritium (the third isotope of hydrogen), an ingredient of the hydrogen bomb. While willing to serve in what he saw as a just war, Urey was sternly critical of postwar secrecy and the continuation of the manufacture of atomic weapons. Joseph McCarthy's House Un-American Activities Committee accused him of having communist beliefs, which he denied.

In the 1950s, reflections on thermodynamic relations between isotopes led Urey to develop a method of estimating ocean temperatures in past geological eras by using fossil evidence. In 1952 he published *The Planets, Their Origin and Development*, in which he suggested that the earliest atmosphere of the earth must have been a reducing one. His speculation that ultraviolet light and atmospheric electrical discharges could have formed organic molecules that became the basis of life was supported by experiment by his doctoral student, Stanley Miller, in 1953.

The fact that lunar meteoric craters appeared much older than terrestrial ones led him to conclude that the moon had been formed quite separately from the earth by the accretion of interstellar particles. Urey was forced to abandon his lunar hypothesis when the *Apollo* 2 mission (1969) confirmed that the moon had once been a molten mass. Nevertheless, his cosmochemical speculations provided a considerable stimulus to space exploration in the second half of the twentieth century.

H. C. Urey, "Some Thermodynamic Properties of Hydrogen and Deuterium," in *Nobel Lectures, Chemistry, 1922–1941* (1966): 333–356. F. G. Brickwedde, "Harold Urey and the Discovery of Deuterium," *Physics Today* 35 (September 1982): 34–39. S. G. Brush, "Nickel for Your Thoughts: Urey and the Origin of the Moon," *Science* 217 (1982): 891–898. K. P. Cohen et al, "Harold Clayton Urey," in *Biographical Memoirs of Fellows of the Royal Society* (1983): 623–659. J. N. Tatarewicz, "Urey, Harold Clayton," in *Dictionary of Scientific Biography*, ed. F. L. Holmes, 18 (1990): 943–948.

CURTIS WILSON

V

VACCINATION refers to the process, initially developed as a preventative treatment for smallpox, in which an immune state is induced in an animal by exposure to attenuated pathogens or other material immunologically related to the infectious agent. Immunizations with microbial toxins rather than with the infectious agent itself are sometimes called vaccinations, although such immunizing agents are seldom called vaccines, but rather toxoids or antigens.

Between 1773 and 1795 Edward Jenner, a country physician in Gloucestershire, England, tested a local folk belief that someone who had contracted cowpox would not subsequently contract smallpox. Jenner deliberately inoculated patients in his practice with the relatively benign cowpox agent (*Variolae vaccinae*) and then challenged them with smallpox inoculations. He observed that these patients did not contract the disease in several natural outbreaks of smallpox. Jenner documented the procedure's success in his landmark *Inquiry into the Causes and Effects of the Variolae Vaccinae* (1798). By 1800 the procedure was known as vaccination.

Vaccination with cowpox had two significant advantages over the then widespread procedure of "variolation," or "inoculation," for smallpox. Variolation (inoculation through the skin or intranasally with a small amount of material from active smallpox lesions) actually caused the patient to contract smallpox, a mild case it was hoped. This process brought the risks associated with the disease itself as well as the potential spread of smallpox to others. Cowpox inoculation avoided both of these hazards. By the mid-nineteenth century vaccination had replaced variolation almost completely in Europe and North America as protection against smallpox.

Louis *Pasteur extended Jenner's work on protection from disease by exposure to weakened pathogens. Pasteur devised ways to attenuate the causative agent of fowl cholera so as to diminish its virulence. In investigations similar to Jenner's, he showed in 1880 that inoculation of chickens with attenuated fowl cholera microbes protected the animals from challenge with a virulent strain of the pathogen. Pasteur called this procedure "vaccination" and thus generalized the term from immunization based on cowpox to any immunization with attenuated or inactivated microbial products. He subsequently developed vaccines for other infectious diseases including, famously, anthrax and rabies.

Another early attempt at vaccine development was made at the Pasteur Institute by Albert Calmette, who prepared a vaccine aimed at tuberculosis. After many cycles of growth on artificial medium, he obtained an avirulent strain of the tubercle bacillus (bacille Calmette-Guérin, BCG) and used it as a vaccine. Several such strains, known as BCG strains, have been isolated and proven effective in prevention of tuberculosis.

By the middle of the twentieth century the search for a vaccine that protects against poliomyelitis had succeeded. Based on the tissue culture growth of poliovirus, developed by John F. Enders and his colleagues, Jonas Salk prepared a vaccine from killed poliovirus, and Albert Sabin developed a live, attenuated poliovirus vaccine. These vaccines have reduced the incidence of poliomyelitis dramatically, and now the entire Western Hemisphere is free of disease caused by wild-type poliovirus, the very rare cases being accidents associated with vaccine strains.

Anne Marie Moulin, *Le dernier langage de la médecine: Histoire de l'immunologie de Pasteur au Sida* (1991). René Dubos, *Pasteur and Modern Science*, ed. Thomas D. Brock (1998).

WILLIAM C. SUMMERS

VACUUM. Nature abhors a vacuum. So said medieval natural philosophers, following Aristotle. Against the ancient atomists, who held that material atoms move in an infinite void, Aristotle presented several arguments for the impossibility of a vacuum: the lack of resistance would produce infinite velocities; the homogeneity of the void precluded natural motion, which for Aristotle relied on a distinction between up and down; and the void likewise prevented violent motion, which needed an external medium for continued propulsion. The plenum persisted into the seventeenth century, notably in the system of René *Descartes, who identified matter with

The famous experiment of the Magdeburg spheres (1654). A pair of hollow hemispheres easily separable under ordinary conditions could not be pulled apart by two teams of horses when put together, sealed hermetically, and exhausted of air. The experiment demonstrated the power of atmospheric pressure and of the air pump newly invented by Otto von Guericke.

space. But the possibility of the void received some discussion from medieval scholastics, who wondered in particular about the space beyond the stars, where God perhaps resided.

Experimental refutation of the *horror vacui* came in the seventeenth century. Mining pumps of the time operated according to the abhorrence of the vacuum, up to a point—thirty feet, in fact, above which they could not draw water. In 1644 an Italian mathematician, Evangelista Torricelli, explained the limitation by a mechanical equivalence between the weight of atmospheric air and the weight of the column of water, and demonstrated it using a glass tube, closed at one end, inverted in a basin of mercury. The mercury rose to a height one-fourteenth that to which water attained. Torricelli's new device—what we now call a *barometer—figured in a famous experiment four years later by Blaise Pascal, who initially doubted Torricelli's explanation of the barometer and thought it showed only the limits to the force of vacuum. Pascal pursued barometric experiments with a variety of liquids and glass tubes up to 14 m (46 ft) long, for the latter relying on the state-of-the-art products of the glass factory in his hometown of Rouen. In the decisive experiment, Pascal in 1648 sent his brother-in-law up a mountain in France with a barometer; the lower level of the mercury at the peak convinced Pascal and others that the weight of the atmosphere, not the vacuum inside the barometer, was forcing up the mercury in barometers and the water in mining pumps.

Otto von Guericke soon provided an equally famous demonstration using his new air pump, a piston-driven suction pump with valves that could suck the air out of sealed chambers and thus make a vacuum. In 1657 in Magdeburg, where he was the mayor, Guericke worked his air pump on two copper hemispheres stuck together and showed that a team of horses could not pull them apart; the force of the vacuum—or, rather, of the air on the outside of the hemispheres—held them together. Robert *Boyle developed the air pump into a means of easy production of a vacuum. When Boyle placed a barometer inside a glass globe, the level of mercury descended as the pump evacuated the enclosed space until the mercury no longer stood. Boyle's account of the results, *New Experiments Physico-Mechanicall Touching the Spring of the Air* (1660), showed along the way that cats and candles could not survive in a vacuum but that electric and magnetic effects could.

The technology of the vacuum would henceforth be crucial for modern science. The fruitful program of experiment with evacuated *cathode ray tubes in the nineteenth century relied on a new generation of vacuum pumps, the first major advancements over von Guericke's original design; in particular, a pump made by German instrument maker J. H. W. Geissler using a mercury column instead of pistons, which improved residual pressures from one inch to one millimeter of mercury. Rotary pumps made by Wolfgang Gaede in Germany in the early twentieth century proved crucial for the development of vacuum tube technology in the commercial electronics industry. High-energy particle accelerators later in the century required further advances in the production of large empty spaces.

The vacuum of physicists since the early modern period has not been empty. The imponderable fluids of electricity, light, and magnetism in Laplacian physics pervaded it, as did the ether and electromagnetic fields of Maxwellian electromagnetism. Even after Albert *Einstein banished the *ether, his postulated equivalence of energy and mass implied that matter could still intrude in empty space, and fields, electron holes, and ghost particles continue to clog up the vacuum of modern physics.

Edward Grant, *Much Ado about Nothing: Theories of Space and Vacuum from the Middle Ages to the Scientific Revolution* (1981). Steven Shapin and Simon Schaffer, *Leviathan and Air-Pump: Hobbes, Boyle, and the Experimental Life* (1985). Per F. Dahl, *Flash of the Cathode Rays: A History of J. J. Thomson's Electron* (1997).

PETER J. WESTWICK

VACUUM PUMP. See AIR PUMP AND VACUUM PUMP.

VALENCE. See CHEMICAL BOND AND VALENCE.

VALVE, THERMIONIC. The thermionic valve, or electron or vacuum tube, evolved from the "Edison effect," discovered by Thomas Edison in the early 1880s in the course of improving the life of his incandescent lamp. He observed a small current flowing from the heated filament to another wire inserted into the evacuated light bulb. He patented this "Edison Effect" in 1884 as a possible way of regulating and measuring the flow of electric currents, without understanding its mechanism. In 1897 Joseph John *Thomson, through his experiments on cathode rays, explained this effect as thermionic emissions of electrons. At this time pioneers of wireless telegraphy like Oliver Joseph Lodge, Edouard Branly, and Guglielmo Marconi were experimenting with detectors based on the "cohering" properties of different metal filings when subjected to radio waves. John Ambrose Fleming realized that the one-way conductivity or rectifying properties of the Edison-effect lamps could be used to detect radio waves. In October 1904, while acting as a consultant to the Marconi Company, he patented his "oscillation valve": a vacuum diode acting towards oscillatory currents in effect as a nonreturn valve acts in a water pipe. However, Fleming's device played only a small role in the early years of *radio—Fleming faced both a legal wrangle with the American radio pioneer Lee De Forest (eventually settled in Fleming's favor) and the

development of a significant rival to his invention, the crystal detector.

In 1906–1907, De Forest introduced a third electrode, the grid, between the filament and the anode of the diode. This triode, which De Forest called the "audion," not only rectified but also amplified the input signal from the aerial. These early, handmade valves contained a fair amount of residual gas and so varied considerably in their working characteristics. H. J. Round of the Marconi Company and Robert Lieben, Eugen Reisz, and Sigmund Straus at Telefunken developed similar "soft" triodes. In both cases the condition of the vacuum could be altered in the valve by heating, a technique developed for early x-ray tubes. In 1913, Harold Arnold of the American Western Electric Company made the first successful high-vacuum or "hard" valve. It was followed closely by the hard triode developed by Colonel Gustav Ferrié for the French army; widely deployed by the Allies during World War I, it remained in use after the war as the "R" valve of the early broadcasting era. By 1925 thermionic valves had fully replaced crystal detectors for all ordinary radio purposes.

Walter Hans Schottky made an important advance in 1916 when he placed a second grid between the control grid and the anode to improve amplification. Screened-grid valves did not go into commercial production until 1926, and then served only as low-level radio frequency (r. f.) amplifiers to improve the radio's sensitivity. They developed into the five-electrode pentode by adding a suppressor grid between the screened grid and the anode. Future valve advances were coupled to major developments in radio receivers based on the superheterodyne circuits of the 1930s: frequency changer valves with complicated grid arrangements, the beam tetrode, and the "magic-eye" tuning indicator, which combined a triode amplifier valve and a small cathode ray tube. In the 1940s and early 1950s the thermionic valve was miniaturized but gradually gave way to the *transistor, although it continued to be used in power stages of radio and television transmitters and in military equipment.

Gerald F. J. Tyne, *Saga of the Vacuum Tube* (1977). John W. Stokes, *70 Years of Radio Tubes and Valves* (1982). Keith R. Thrower, *History of the Radio Valve to 1940* (1992).

W. D. HACKMANN

VAVILOV, Nikolai Ivanovich (1887–1943), Soviet botanist, geographer, and organizer of agricultural research, and **Sergei Ivanovich VAVILOV** (1891–1951), Soviet physicist and statesman.

The Vavilov brothers were born in Moscow into a one-time peasant family that had risen to become wealthy merchants. Both chose science after graduating, Nikolai from the Agricultural Academy and Sergei from the physico-mathematical department of Moscow University. Nikolai was among the first Russian researchers to take up the new field of *genetics just before World War I. After the war that interrupted their academic pursuits, and the subsequent revolution and civil war that forced their father to emigrate, both brothers stayed in Soviet Russia and continued their scholarly careers. In 1920, Nikolai made a sensational presentation at a congress of Russian breeders, suggesting a new "law of homological series" in hereditary variation of plants. Shortly thereafter, he was appointed director of the Bureau (subsequently Institute) of Applied Botany in Petrograd (Leningrad). In this position, he became the key organizer of the Soviet system of agricultural research, which grew to a countrywide network of several hundred government-sponsored breeding stations and experimental research institutions. In 1929, these institutions were united under the newly created All-Union Lenin Academy of Agricultural Sciences (VASKhNiL), with Nikolai Vavilov as president. The same year, he was elected ordinary member of the USSR Academy of Sciences, where he directed the Institute of Genetics.

While neither he nor his younger brother ever joined the communist party, Nikolai shared many of the social and economic goals of the revolutionary modernizing regime. He endorsed the Stalinist collectivization campaign, hoping that large-scale collective farms would introduce modern scientific practices into backward agriculture. To help breed and introduce better varieties of cultivated plants, Vavilov built up, at his Leningrad institute, a comprehensive worldwide collection of seeds, which grew to include some 300,000 specimens. Many of them came from Vavilov's expeditions to remote areas of Asia, Africa, and Latin America. Another product of these travels was Vavilov's innovative and acclaimed theory of the geographical centers of origin of cultivated plants, which relied on genetic and cytological analysis. The most widely traveled Soviet scientist, Vavilov also served as president of the All-Union Geographical Society. He actively promoted the revolutionary science of genetics, attracting to his institutes some of the best Soviet and foreign researchers. Soviet genetics flourished under his patronage during the 1920s and early 1930s, second only to American genetics.

Geneticists first came under criticism in the early 1930s because of their association with eugenics, denounced by Soviet Marxists as racist.

Stalinist officials often blamed agricultural failures on the inadequate application of scientific research. The charismatic but poorly educated agronomist T. D. Lysenko attracted followers with his rival brand of "Michurinist" genetics (names after a plant breeder), which, he claimed, was more practical and ideologically more sound than "formalist" and "idealistic" Mendelian genetics (*see* LYSENKO AFFAIR). No less dangerous to Vavilov personally were accusations from militants at his own institutes, who challenged the agricultural usefulness of his research program and the wisdom of spending government funds on expensive overseas expeditions. Under mounting pressure, Vavilov stepped down from the VASKhNiL presidency in 1935, but remained its vice president. The more difficult his situation became, the more outspoken his courageous, but losing, defense of genetics against Lysenkoist ideological accusations. Along with many important agricultural officials, Vavilov fell victim to Stalinist purges. He was arrested in 1940, convicted of "wrecking" in the field of agriculture, and died of malnutrition in Saratov prison in 1943. Most key administrative positions in Soviet biological and agricultural research came under the control of Lysenko, whom many in the scientific community held responsible for Vavilov's fate. Having lost its most important spokesman and patron, Soviet genetics fell on hard times.

Sergei Vavilov lacked his elder brother's charisma, rebelliousness, and revolutionary vigor, but excelled in hard work, self-discipline, and manners. His career advanced more slowly than Nikolai's until 1929, when he became professor at Moscow University. His research in experimental optics, on luminescence and the quantum structure of light, required patience and tedious work. This bore important fruit in 1933, when Vavilov asked one of his graduate students, P. A. Cherenkov, to look for luminescence induced by rays from radioactive substances. Cherenkov noticed in the dark a very weak blue glow coming from the liquid. Vavilov recognized that the glow could not be luminescence. It was a new physical phenomenon now known as Cherenkov or Vavilov-Cherenkov radiation. In 1937, Vavilov's colleagues I. E. Tamm and I. M. Frank explained the effect theoretically as caused by *electrons traveling faster than light propagates in the liquid, an electromagnetic analogue of the acoustical shock waves produced in the atmosphere by supersonic projectiles. The discovery later found important application in high-energy physics as the basis of the Cherenkov detectors of elementary particles. In 1958 (after Vavilov's death), Cherenkov, Tamm, and Frank shared the *Nobel Prize in physics.

The discovery did not contribute much to Sergei's rapid career advance during the 1930s, which rested on his administrative talent and unmatched sense of responsibility. In 1932, Vavilov was elected ordinary member of the USSR Academy of Sciences and assumed the scientific directorship of the State Optical Institute in Leningrad, where he coordinated R&D in the national optical industry, both military and civilian. Vavilov's responsibilities included the Academy's own small physical institute (FIAN, where Cherenkov made his discovery), which turned into a major assignment in 1934 with the Academy's move from Leningrad to Moscow. Under Vavilov's directorship, FIAN developed into the nation's largest center for advanced research in fundamental physics and the home for six researchers who would later become Nobel laureates.

After the war's end in 1945, Vavilov became Stalin's choice for the Academy's president. He oversaw the major expansion of the Academy's scientific research in response to the American atomic bomb. As the country's major political representative of science and its public spokesman—in the political climate characterized by Stalin's personality cult, the outbreak of the Cold War, and ideological campaigns in sciences—Vavilov faced difficult compromises. Some were very bitter, such as the Academy's compliance with the ban on Mendelian genetics imposed in 1948 by Lysenko, and others more rhetorical, such as Vavilov's elaborate glorification of Stalin.

Vavilov received a political commission to write about the history of science. He wrote excellent historical works on optics, Isaac *Newton, and eighteenth-century Russian science. In his philosophical publications, Vavilov argued that the twentieth-century revolution in physics (theories of *relativity and the quanta) fully confirmed the predictions of Marxist-Leninist philosophy. This stance, along with Vavilov's subtle political maneuvering, helped deflect the danger of an ideological pogrom in physics. While *genetics struggled to survive underground, Soviet physics continued its spectacular successes throughout the Stalin years. Decades of hard work took a toll on Vavilov's health. He died in 1951: to Stalinist leadership, an ultimately reliable and ideologically loyal scholar; to scientists, especially physicists, an effective protector in dangerous times.

N. I. Vavilov, *The Origin, Variation, Immunity and Breeding of Cultivated Plants: Selected Writings*, trans. from Russian by K. Starr Chester (1951). Andrei Sakharov, foreword to M. A. Popovskii *Vavilov Affair* (1984). Alexei Kojevnikov, "President of Stalin's Academy: The Mask and Responsibility of Sergei Vavilov," *Isis* 87 (1996): 18–50. B. M. Bolotovskii, Yu. N. Vavilov, A. N. Kirkin, "Sergei Ivanovich Vavilov: The Man and the Scientist: A View from the Threshold of the 21st Century," *Physics Uspekhi* 41(5) (1998): 487–504.

ALEXEI KOJEVNIKOV

VAVILOV, SERGEI IVANOVICH. See VAVILOV, NIKOLAI IVANOVICH, AND SERGEI IVANOVICH VAVILOV.

VESALIUS, Andreas (1514–1564), anatomist.

The son of an apothecary to Emperor Charles V, Vesalius had a humanist education at Louvain before studying medicine at Paris in 1533. His professors, notably Jacobus Sylvius and Joannes Guinter, worked in the forefront of the new *Galenism; they denounced medieval obscurities in favor of the pure founts of ancient Greek medicine, particularly Galen and Hippocrates. They were also enthusiasts for anatomy, translating Galen's recently discovered anatomical works and emphasizing the significance of dissection for proper medical practice. Guinter recognized Vesalius's genius with the knife and engaged him to cut up a human corpse to illustrate his anatomical lectures.

War between France and the Holy Roman Empire prompted Vesalius to return to Louvain, where he graduated in 1537. He gained greater experience in anatomy, at autopsies and through private dissections of human corpses, with the connivance of the burgomaster. By late 1537 he was in Padua, where he took his M.D. and accepted a lectureship in surgery and anatomy. His first public dissection began on 6 December with two innovations. Instead of the traditional division of labor between lecturer, demonstrator, and dissector, Vesalius alone held the stage, explaining the significance of his findings as he opened the body. He also introduced drawings as an aid to understanding and memory.

In 1538 he published a revision of Guinter's *Introduction to Anatomy* (without informing its original author) and a series of six *Anatomical Plates*, drawn by Jan van Calcar on the basis of Vesalius's own sketches. Still largely faithful to Galen, these plates, the most detailed yet printed, were a considerable success; plagiaries quickly appeared in many cities of Europe. In 1540–1541, Vesalius revised the Latin versions of Galen's anatomical works for the Giuntine press, working from Greek manuscripts and assisted by his housemate, the Englishman John Caius. But at a dissection he made at Bologna in 1540 he challenged his colleague's belief in Galen's general accuracy. He argued that Galen had made many mistakes about human anatomy because he never dissected human corpses, relying solely on animals. This new message dominated Vesalius's greatest achievement, his

De humani corporis fabrica (*On the Fabric of the Human Body*), published in Basle in 1543. He also published for students a Latin *Epitome*, immediately translated into German.

The *Fabrica* is the most important book in the history of anatomy because of its insistence, and demonstration, that the human body could be understood only through dissection. Its errors and misrepresentations of Galen outraged Vesalius's teachers, who became yet more determined to show that Galen was never wrong. But they made only a vociferous minority. Most contemporaries saw Vesalius as himself a Galenist, putting into practice the program that Galen had advocated but could not follow.

The *Fabrica* is a masterpiece of rhetoric, printing, and art. Its beautiful typography and large format made space for the extensive use of plates, 83 in all, with 420 separate images. Earlier anatomical illustrations had been exiguous, like those of Giacomo Berengario da Carpi. Leonardo da Vinci never fulfilled his plan (if it existed) for an anatomical atlas. The illustrated *De dissectione* by Charles Estienne was delayed by a lawsuit for almost seven years until 1545. Artists like Raphael and Michelangelo had drawn dissected bodies, but links between them and the university anatomists had been slender. Vesalius and his unknown artist (possibly Titian) integrated text and image, solving at a stroke many of the difficult problems of representing a three-dimensional body on a two-dimensional page by using shading and a variety of angles and image sizes. Playful opening initials enhanced the impression of anatomy as a cultured art.

Vesalius made substantial corrections of Galen, particularly in his description of the bones and muscles, which show masterly dissection and careful observation. He denied the existence in man of the *rete mirabile* (a network of veins and arteries at the base of the skull in ruminants) and doubted the permeability of the septum within the heart. He also raised questions about anomalies based on his extensive experience with autopsies.

The *Fabrica* was an instant success. It ensured Vesalius new employment from 1543 as personal physician to Charles V. His university career had lasted hardly six years. He continued to dissect sporadically, bringing out a substantially revised edition of the *Fabrica* and *Epitome* in 1555. Otherwise he largely lived the life of a courtier, marrying in 1544. His surgical skills were called on to treat the wife of William the Silent, Don Carlos of Spain, and King Henry II of France after his fatal wounding in 1559. Still, he was not entirely happy in Spain. In 1564 he went on a pilgrimage to Jerusalem. He died on his way back on the island of Zakynthos. Later rumors that he had gone in penance for having mistakenly cut up a living patient are unlikely to be true, but reflect tensions at the Spanish court.

Andreas Vesalius, *De humani corporis fabrica* (*On the Fabric of the Human Body*), trans. W. F. Richardson and J. B. Carman (1543; 1998). H. Cushing, *A Bio-Bibliography of Andreas Vesalius* (1943). C. D. O'Malley, *Andreas Vesalius of Brussels, 1514–1564* (1964). R. K. French, *Dissection and Vivisection in the European Renaissance* (1999).

VIVIAN NUTTON

VIROLOGY. The term "virus" in its original Latin meaning (poison, morbid principle) was used up until the beginning of the twentieth century to indicate the causative agent in the transmission of a specific disease, as in the virus of cholera or the virus of rabies. In 1892 Dmitri Ivanovsky found that the virus of tobacco mosaic disease passed through a filter believed to trap all known bacteria. This observation gave rise to the concept of "filterable viruses." Martinus Willem Beijerinck conceived of this infectious agent as a "living contagious fluid" on the basis that only true fluids were filterable. Filtration made possible a new classification of infectious agents. Soon clinicians recognized that, in addition to tobacco mosaic virus (TMV), the agents of many well-known diseases such as rabies, hog cholera, influenza, poliomyelitis, herpes simplex, and vaccinia were filterable viruses. Thomas M. Rivers's influential textbook *Filterable Viruses* (1928) was the last major work to retain the term "filterable virus"; by the mid-1930s, "virus" alone would do. From 1900 to about 1930, some authors used the term "ultravirus" to emphasize the invisibility of these filterable agents to light microscopy, but this terminology was not widely adopted.

Although Beijerinck's concept of a contagious fluid would suggest a continuous, nonparticulate nature, except for bacteriophages (bacterial viruses), there seemed to be little challenge to the belief in the particulate nature of viruses. The bacteriological paradigm of tiny organisms seemed to dominate thinking about viruses, filterable and nonfilterable (*see* BACTERIOLOGY). The only controversy concerned the nature of bacteriophages. Their codiscoverer Félix d'Hérelle conceived of them as filterable viruses that infect bacteria. Scientists who believed in the simplicity of bacteria as a class of organisms fundamentally distinct from complex multicellular organisms resisted this concept. They also pointed to experimental results confounded by lysogenic behavior of some phage isolates. Lysogenic bacteria harbor the genetic material of a bacterial virus in a repressed state; under certain conditions, the virus can be induced to enter its regular growth cycle, which often kills the host cell. The analogy between the process of lysogenic virus induction to the then recently discovered autocatalytic activation of proteases was advocated from 1920 to the mid-1940s by two Nobel Prize–winners, Jules Bordet and John H. Northrop. Their authority relegated the bacterial virus model for bacteriophage to a secondary status corrected only when electron microscopy made possible visualization of phages.

The microbial conception of viruses prompted attempts to study their metabolism and their growth in pure culture. It proved impossible to grow viruses in the absence of living host cells, and viruses were reconceptualized as obligate intracellular parasites. Viruses could be obtained only from infected plants, animals, or bacteria. Embryonated chicken eggs, and later animal cells in tissue cultures, became the standard laboratory media for growing animal viruses for diagnosis and study. Bacterial cultures, both in liquid media and on solid agar plates, provided the cells for growing. In a classic study of a virus disease, Edward Jenner confirmed common folk wisdom that prior sickness with a benign disease of cows, cowpox, protected humans from subsequent susceptibility to a lethal disease, smallpox. He exploited this finding to develop the procedure of *vaccination. Louis *Pasteur extended this procedure to several other diseases, both bacterial and viral. The use of attenuated or weakened strains of virus to induce mild or subclinical illness that confers immunity is the basis for vaccines developed against poliovirus through the work of John Franklin Enders and his colleagues Jonas Salk and Albert Sabin.

In the mid-1930s Ernst Ruska and his colleagues turned the electron microscope they invented to the study of viruses. The higher resolution obtained with this instrument revealed the regular structures of virus particles, but also showed that viruses exist in many shapes, from simple rods such as TMV to elaborately tailed and "decorated" structures such as the T-even bacteriophages.

Chemical analysis of viruses became possible in the 1930s because of their purification by sedimentation in the ultracentrifuge (see CENTRIFUGE). Initial analyses of some viruses such as TMV and poliovirus suggested that they were composed entirely of proteins, a finding that fit well with the notion that viruses resembled enzymes in both specificity and catalytic power. Soon, however, it was discovered that purified viruses always contained phosphate in the form of nucleic acids. The conception of viruses as simple nucleoprotein particles raised the conundrum of the living nature of such simple chemical entities.

This paradox became especially acute when Wendell Stanley crystallized TMV and poliovirus in 1935, since crystallization was believed to be the ultimate criterion for chemical purity.

The chemical simplicity and rapidity of growth of viruses and the observation that they had heredity properties (such as the range of host species on which they grow) led investigators in the late 1930s to employ viruses to study the chemistry of the *gene. Max Delbrück, Alfred D. Hershey, and Salvador Luria used bacteriophages as a model for gene replication and heredity transmission; their work helped to set the current directions of molecular biology. Alfred Geirer, Gerhard Schramm, and Heinz Fraenkel-Conrat used biochemical approaches to show that the genetic properties of TMV resided in the RNA component of the virus, not in the protein coat.

As more viruses were isolated from nature and studied, their diversity became apparent. One class, first represented by a virus that causes cancers in chickens (Rous sarcoma virus), was found to copy its RNA genome into DNA and integrate this DNA provirus into the host cell chromosome. This mechanism, first proposed by Howard Temin, seemed implausible because it required a reverse of the normal flow of genetic information. In 1970 both Temin and David Baltimore discovered the enzymatic mechanism for this process, demonstrating that it characterized a large class of RNA viruses, now known as retroviruses. When a mysterious immune deficiency disorder appeared in epidemiologically recognizable groups in 1982, virological studies soon identified the causative agent as a retrovirus, now called the human immunodeficiency virus (HIV) (see AIDS).

Anthony P. Waterson and Lisa Wilkinson, *An Introduction to the History of Virology* (1978). Ton van Helvoort, "What Is a Virus? The Case of Tobacco Mosaic Disease," *Studies in History and Philosophy of Science* 22 (1991): 557–588. Ton van Helvoort, "History of Virus Research in the Twentieth Century: The Problem of Conceptual Continuity," *History of Science* 32 (1994): 185–235. Geoffrey M. Cooper, Rayla Greenberg Temin, and Bill Sugden, eds., *The DNA Provirus: Howard Temin's Scientific Legacy* (1995).

WILLIAM C. SUMMERS

VIRTUAL COLLEGE. See NETWORK AND VIRTUAL COLLEGE.

VIRTUOSI. See CABINETS AND COLLECTIONS.

VISION. See OPTICS AND VISION.

VITALISM. See MATERIALISM AND VITALISM.

VIVISECTION. See ANIMAL CARE AND EXPERIMENTATION.

VOLTA, Alessandro. See GALVANI, LUIGI, AND ALESSANDRO VOLTA.

W

WARBURG, Otto Heinrich (1883–1970), German biochemist.

Warburg discovered the role of iron in the fixation of oxygen, an accomplishment for which he received the *Nobel Prize in medicine for 1931. The oxydase of the last, reversible step of the fixation is the Warburg enzyme. Warburg made several major discoveries after receiving his Nobel Prize, for example, the coenzymes for the hydrogen-fixing part of the respiratory process.

Otto Warburg came from a scientific family. His father, Emil Warburg, professor of physics in Berlin and president of the Physikalisch-Technische Reichsanstalt (the Federal Bureau of Standards) augured that his son would be more successful than he. Perhaps for that reason Otto always set himself high goals and would only work on problems of the greatest significance. Uncertain whether to choose cancer or photosynthesis as his research field, he asked his teacher Walter Nernst for advice; work on cancer, Nernst replied, photosynthesis can get along without you. Warburg devoted himself to both fields. Most modern researchers think that these were just the fields in which he was least successful.

Very early in his career Warburg took up the causes of, and influences on, cell growth, and his dissertation, *Observations on the Oxidation Process in Sea Urchin Eggs* (1908), attracted much attention. The sea urchin was an unusual model organism. The suggestion to use it came from the German-American physiologist Jacques Loeb, with whom Warburg corresponded frequently. Because of his excellent contacts with the Rockefeller Foundation, Loeb later was able to help Warburg build an institute of his own.

In 1923 Warburg published his thesis that tumors, in contrast to normal tissues, metabolize by fermentation rather than respiration. He was so convinced of the truth of this thesis that he clung to it to the end of his life. In fact, although many tumors behave as Warburg thought, not all do.

Warburg saw brief military service in World War I. He was able to curtail his service through the intervention of Albert *Einstein, a good friend of his father, and was made a department leader in the new *Kaiser-Wilhelm-Institut für Biologie. With Otto Meyerhof, who also had an appointment there, he pursued a collaborative study on sea urchins and other biological material at the Naples marine station. As their published correspondence shows, Warburg dominated the collaboration. Although Meyerhof received the Nobel Prize before Warburg did, Warburg referred to Meyerhof with some justification as his student.

In the 1920s Warburg presented his research results in the United States. His ideas on the origin of cancer raised particular interest. The Rockefeller Foundation, which supported German research during this economically difficult time, contributed substantially to the construction of an institute—the Kaiser-Wilhelm-Institut für Zellphysiologie—for him. This unusual building, a copy of a Prussian manor house, opened in 1931. In it Warburg pursued an idiosyncratic personnel policy: he brought in many technicians who would translate his ideas into action and very few scientists. Among the scientists, however, were two—Hans Adolf *Krebs and Hugo Teodor Theorell—who would later win Nobel Prizes for their own work.

The rise of National Socialism began a time of great difficulty for Warburg. His father, although from a Jewish family, had converted to Protestantism; his mother, the daughter of a Prussian officer, was Evangelical. According to the calculations of the Nazis, Warburg was a half-Jew. As a further complication, Warburg lived with a man, Jacob Heiss, whom he had met at the front during the war and who served officially as his secretary. At the time, homosexuality in Germany was a penal offense. Still, the Nazis allowed Warburg to remain as institute director until 1941, when they briefly discharged but soon reappointed him. The reappointment angered many of his foreign friends and former colleagues who had sought refuge abroad. Meyerhof, who had emigrated, wrote him that they stood on different sides of the barricade. It appears that high-ranking Nazis—many of whom, including Hitler, had severe anxieties concerning cancer—wanted Warburg reinstated to continue his research on the disease.

A relative of a friend of Warburg's reported that one Viktor Brack spoke to the Reichschancellor on Warburg's behalf. What was said can be inferred from Warburg's letter of 1945 to the lawyer defending Brack in his trial as a war criminal. Warburg wrote that Brack had not only saved his life, but also enabled him to return to work. It did not help. Brack was found guilty of crimes against humanity and executed.

Hans Krebs, with Roswitha Schmid, *Otto Warburg: Cell Physiologist, Biochemist, and Eccentric*, trans. Hans Krebs and Anne Martin (1981). Petra Werner, *Ein Genie irrt seltener. Otto Heinrich Warburg—ein Lebensbild in Dokumenten* (1990).

PETRA WERNER

WARDIAN CASE, a closed glass compartment for transporting living plants, initially developed to capitalize on the Victorian taste for cultivating indoor plants but soon employed by scientists and collectors for transporting living specimens from place to place. The case was invented by Nathaniel Bagshaw Ward (1791–1868), a medical surgeon of Whitechapel, London, a keen amateur botanist and microscopist, and one of the founders of the Microscopical Society in 1839. He based it on the self-sustaining principle of an aquarium, in the development of which he had had a hand. He noted the preservative power of almost-sealed cases where living plants thrive without any further need for watering, and tested his hypothesis by sending a sealed case of ferns to Australia and back in 1833. The ferns did not need watering once. In essence he had discovered that plants can create their own primitive ecosystem. Ward took his idea to Loddiges, the English firm of nurserymen, who produced cases to his designs commercially. They were discussed at the British Association for the Advancement of Science in 1837 and formed the subject of his book, *On the Growth of Plants in Closely-Glazed Cases* (1842).

Ward's invention transformed the transport of exotic species. Previously, travelling naturalists and collectors had been forced mostly to rely on sending dried or pickled specimens back home. Wooden kegs were customarily used for living plants, for example by Captain William Bligh in 1788 when transporting breadfruit from Tahiti to the West Indies. The system had severe drawbacks: the kegs stayed on deck, open to spray and excess light, and the crew often neglected to collect rainwater for watering. Even though a code of accepted practice evolved through the seventeenth and eighteenth centuries, kegged specimens often arrived in a sorry state. Joseph Banks's (1743–1820) conviction that plants stood a better chance of survival when protected in wooden cabins or portable glasshouses erected on the foredeck was confirmed by Ward's cases. The successful transport of Chinese banana stock from Chatsworth to Samoa in 1840 opened up new possibilities for commercial plantation, realized in Robert Fortune's conveyance of twenty thousand tea plants from Shanghai to the Himalayas. As late as 1940 the same kind of protective travelling case was in common use.

David Elliston Allen, *The Naturalist in Britain: A Social History* (1976). Henry Hobhouse, *Seeds of Change; Five Plants that Transformed Mankind* (1985).

JANET BROWNE

WATSON, James D. See CRICK, FRANCIS, AND JAMES D. WATSON.

WEATHER. See CLIMATE; CLIMATE CHANGE AND GLOBAL WARMING; METEOROLOGICAL STATION; METEOROLOGY.

WILSON, Edward O. (b. 1929), entomologist, investigator of the biological basis of social behavior.

Born in Birmingham, Alabama, Edward Osborne Wilson was a devout Southern Baptist as a child, reading the Bible cover to cover more than once. He left his church as an adult but transferred an almost evangelizing zeal towards disseminating his scientific insights. Both of his parents had forbears who settled in the Mississippi Delta in the eighteenth and nineteenth centuries. They divorced when he was seven years old. Sent to board in Pensacola, Florida, during the proceedings, he accidentally tossed a fishhook into his right eye. Already an incipient naturalist, with an interest in snakes and small mammals, he turned his partial blindness to advantage by focusing especially on insects, which he could see well by holding them close to his left eye.

Wilson grew up in a culture that admired the military and in which none of his relatives had gone to college. At the age of eight he spent a term at a military academy, where he acquired a rigid sense of discipline. He continued exploring wildlife during his high school years and applied to the University of Alabama, determined to become a naturalist. As a freshman there in 1946 he declared his interest in ants to the chairman of the biology department and was sent to an assistant professor of botany, who took him on as a research assistant. Wilson remained at Alabama to earn an M.A. for which he wrote a report on the imported fire ant *Solenopsis Saevissima var. richteri* for the Department of Conservation office in Montgomery, Alabama. This, his first publication, appeared in 1951.

That autumn Wilson entered graduate school at Harvard University, where in 1953 he was selected as a junior fellow in the Society of Fellows. The Society funded him the following year to work on ants in Cuba and Mexico and later in New Guinea. From the theoretical implications he derived from this fieldwork he defined himself as an evolutionary biologist as well as an entomologist. He also became a spokesman for the new field of island biogeography and an environmentalist concerned with evolutionary *ecology and the threat to biodiversity. Wilson received his Ph.D. in 1955 and remained at Harvard as a professor. From 1971 until 1997 he was also curator of entomology at the university's Museum of Comparative Zoology.

During the 1950s and 1960s, Harvard's biology department was rent with dissention between the microbiologists, growing in number and importance with the discovery of *DNA, and "macrobiologists"—zoologists, botanists, entomologists, and others whose disciplines centered on groups of organisms. The department eventually fissioned. All the while Wilson was working simultaneously on comparative studies of social behavior and on detailed accounts of the behavior of ants. These efforts resulted in the publication of *The Insect Societies* (1971) and *The Ants* (1990), coauthored with Bert Holldobler. Wilson published his first comparative study of animal behavior, *Sociobiology: The New Synthesis*, in 1974. This enormous work describes animal behavior across the taxa and incorporates theories that today are widely but not universally accepted. These include William D. Hamilton's theory of "kin selection" as a special kind of natural selection, and Robert Trivers's concept of "reciprocal altruism," which explained mathematically how the behavior of individual animals toward siblings reflects their genetic relationship.

In the last chapter, "From Sociobiology to Society," Wilson went beyond the world of nonhuman animals to argue that genetic components determine much human behavior including aggression and altruism. These assertions triggered outrage within and beyond the scientific community. Wilson was accused of being a racist and a fascist. Undeterred, he continued his efforts to explain human behavior, publishing *On Human Nature* in 1978. He continued to try to integrate different spheres of knowledge in *Consilience: The Unity of Knowledge* (1998), in which he advocated the application of scientific reductionism to unify the sciences, the social sciences, and the humanities.

Wilson has received many recognitions including election to the National Academy of Sciences in the United States and the Royal Society in Great Britain. In 1990 he received the Craaford Prize from the Swedish Academy. A graceful writer, two of his books have won the Pulitzer Prize for nonfiction, *On Human Nature* and *The Ants*.

BETTYANN HOLTZMANN KEVLES

WOMEN IN SCIENCE. Women have been and remain underrepresented in science. During the *Scientific Revolution of the seventeenth century aristocratic women were active as both patrons and interlocutors of natural philosophers. For example, Queen Christina of Sweden patronized many natural philosophers, astrologers, and astronomers, including René *Descartes and Gian Domenico Cassini; Princess Caroline of Wales coordinated Leibniz's activities in England, especially his debate with Isaac *Newton via Samuel Clarke; and Queen Charlotte Sophia of England supported Jean André Deluc.

Some women gained access to the practice of science as wives or relatives of scientists who worked at a home base, whether an observatory or a museum—Caroline Lucretia Herschel, for example, who discovered eight comets, the sister of William Herschel and the aunt of John Herschel, England's leading astronomers for half a century. A few women attained positions of scientific leadership. The marquise du Chatelet, the friend of several eighteenth-century philosophers, notably Voltaire, made the first complete translation of Newton's *Principia* into French. Laura Bassi, professor of physics at the University of Bologna, was a protégée of the city's cultural aristocracy. Most women in science, however, were relegated to the supporting cast—collecting, illustrating, or entertaining.

In the nineteenth century the professionalization of science and its increasing location in the laboratory moved science away from its domestic base. The transition further excluded women from science, because social pressure kept them tied to the home and child rearing. Nevertheless, some women made important contributions: Mary Somerville and Jane Haldimand Marcet in England through their informal textbooks and Maria Mitchell in the United States, who detected comets and taught science to many women at Vassar College. The rise of women's colleges in the last third of the century, as well as the willingness of some universities to allow foreign women to study or even graduate created the first generation of formally educated women scientists. For example, Sonya Kovalevsky, a Russian-born mathematician, obtained a Ph.D. in Germany in 1874, became professor at a university in Sweden, and wrote books and plays advocating women's equality through education. In the twentieth century, the shortage of technical personnel during

An early example of women studying science—in this case chemistry—at the University of Leeds in 1908.

the two world wars temporarily opened further opportunities for women in science.

Women's place in science tended to vary significantly by discipline. They encountered substantial resistance in the laboratory sciences. They were more readily accepted in observational sciences such as botany and classical astronomy, where the need for large-scale collecting and observing encouraged the persistence of an amateur subculture for both men and women. Theoretical sciences—except for mathematics, in which there was an often explicit bias against women—also presented fewer barriers since the equipment and facilities required were minimal. Both observational and theoretical pursuits seemed tolerably compatible with women's

primary familial responsibilities; observation additionally belonged to a socially acceptable tradition of amateur practice.

During the twentieth century an increasing number of women, many of them aided by enlightened male mentors, excelled in the experimental as well as in the theoretical sciences. Among them were Marie Sklodowska *Curie, a codiscoverer of *radioactivity and *radium, who worked in state-funded laboratories in Paris and won two Nobel Prizes; her elder daughter, Irène Joliot-Curie, who shared the 1935 Nobel Prize in chemistry with her husband Frédéric Joliot for discovering artificial radioactivity; Hertha Ayrton, who worked at the interface of physics and engineering with support from the feminist

Langham Place Group and from her physicist husband William Ayrton, and who became the first woman nominated for fellowship in the Royal Society of London; Lise *Meitner, a protégée of Max *Planck, who headed the physics division at the *Kaiser Wilhelm Institute for Physical Chemistry between the wars and figured in the discovery of nuclear fission; and Rita *Levi-Montalcini, who shared a Nobel Prize for her research in neurobiology.

British culture proved relatively conducive to the enrollment of women in science, to some extent because the British Empire provided scope for imagination, independence, and travel for women of the middle upper class. The biologist Dorothy M. Wrinch, born in the British colony in Argentina, distinguished herself in mathematics at Cambridge, became the first woman to receive a doctorate in science from Oxford (1929) and a pioneer of research in *protein structure and *molecular biology.

In Britain during the 1930s, socialist commitments helped others flourish within scientific partnerships such as those of Dorothy and Joseph Needham, Tony and Norman Pirie, and Kathleen Yardley Lonsdale and Thomas Lonsdale. Among the accomplished women scientists from this group was Dorothy Crowfoot Hodgkin, who won a Nobel Prize in chemistry (1964) for solving the structure of key biomedical compounds, most notably *penicillin and vitamin B-12. Still others made their marks by venturing into offbeat areas of their disciplines, a career path exemplified by Barbara McClintock, who in 1983 received the Nobel Prize for her discovery of genomic transposition (jumping genes).

Still women continued to face discrimination in most fields of science, and many found it difficult to reconcile the demands of research with those of home and family. Some women scientists, believing that the demands of science could not be squared with the traditional role of women, remained single. Three-quarters of women scientists who did marry collaborated with their scientist husbands. But while a scientific marriage often facilitated acceptance into a broader scientific community, it often led colleagues to assume that the work of the wife was secondary to that of the husband.

Women scientists often had difficulty isolating the role of gender discrimination in their professional lives. For example, socialist women scientists who believed in the Marxist emphasis upon class as the dominant social relation long resisted the idea of struggling against gender bias because it would detract from the presumably more basic class struggle. Many other women scientists disliked claiming attention as women, as opposed to being acknowledged as just (presumably gender-neutral) scientists.

Beginning in the 1960s the civil rights movement and the women's movement combined to subject gender inequality in science to social, cultural, and legal scrutiny. Overt discrimination became illegal. Shortages in scientific personnel in the late 1980s led to development of a gender-responsive science policy that would enable more women to become scientists and combine their professional careers with diverse family choices. Several policy initiatives by the National Science Foundation, the National Institutes of Health, and other agencies made significant headway on the problem of recruitment. Nevertheless, gender segregation continues to plague fields such as medicine, where most women congregate in pediatrics, obstetrics, and psychiatry. New fields such as biotechnology and informatics also display a consistent pattern of gender hierarchy, inequality, and de facto segregation.

In the late twentieth century, the sociocultural expectation that women are the primary family caregivers remained a major bottleneck in reaching gender parity in science. A combination of science policy and social policy has been suggested as a plausible solution. In the twenty-first century, as innovation in reproductive biology redefines the concept of the family, and as globalization eases national and cultural constraints upon formerly dominated groups such as women, science may become a vanguard of gender equality.

See also ENGLISH-SPEAKING WORLD; GENDER AND SCIENCE.

Anne Sayre, *Rosalind Franklin and DNA* (1975). Margaret W. Rossiter, *Women Scientists in America. Struggles and Strategies to 1940* (1982). Vivian Gornick, *Women in Science: Portraits from a World in Transition* (1983; 1993). Evelyn Fox Keller, *Reflections on Gender in Science* (1985). Pnina G. Abir-Am and Dorinda Outram, eds., *Uneasy Careers and Intimate Lives, Women in Science, 1789–1979* (1987; 1989). Margaret W. Rossiter, *Before Affirmative Action, Women Scientists in America, 1940–1972* (1995). Ruth Sime, *Lise Meitner, A Life* (1996). Caroline L. Herzenberg and Ruth H. Howes, *Their Day in the Sun, Women of the Manhattan Project* (1999). Henry Etzkowitz et al., *Athena Unbound, The Advancement of Women in Science and Technology* (2000).

PNINA G. ABIR-AM

WOODWARD, Robert (1917–1979), chemist.

Raised by his Scots-born mother in Quincy, Massachusetts (his father having died when he was only eighteen months old), Robert Woodward demonstrated precocious talent at school, especially in science. Given a chemistry set, he soon mastered all of the experiments through the level of an introductory university course. He matriculated at MIT at the age of sixteen, but rarely attended classes in order to provide more time for library and laboratory study. Despite a semester's suspension for these unorthodox practices, he still graduated in three years. He then required but one additional year for the Ph.D., which he received at the age of twenty. Hired immediately at Harvard University, he rose rapidly through the ranks and spent the rest of his career there.

Woodward's research productivity was remarkable, a result both of the intensity of his work habits and of his extraordinary skills. His memory of the details of the chemical literature was legendary and he quickly mastered every innovation in his field. In particular, he led in understanding the possibilities of the new instruments of the organic chemical laboratory that made their appearance in the 1930s and 1940s: infrared and ultraviolet *spectrometers, mass spectrometers, and *nuclear magnetic resonance.

Woodward's specialty was what chemists call the "total synthesis" of complex natural organic products: the creation of an organic substance in the laboratory, beginning with the simplest possible starting materials. Successful natural product synthesis requires a combination of rigorous thinking, ingenuity of approach and method, a high level of skill in laboratory manipulations, and an almost intuitive sensibility for how molecules can and will combine.

Woodward was the finest master of this difficult craft. He and his collaborators synthesized alkaloids such as quinine and strychnine, antibiotics such as tetracycline, and the extremely complex molecule *chlorophyll. He received the *Nobel Prize in 1965 for his collective "contributions to the art of organic synthesis." Woodward's highest accomplishment in this field came after his prize: the synthesis of vitamin B-12, possibly the most complicated molecule ever successfully reproduced in the laboratory. This synthesis, completed in 1972, required the collaboration of more than a hundred postdoctoral fellows extending over a decade, and the assistance of the Zurich chemist Albert Eschenmoser.

Natural product synthesis was only one of several areas of Woodward's activity. He undertook structure determinations, for example proving the structures of the molecules strychnine (seven years before he synthesized it), tetrodotoxin, penicillin, aureomycin, and terramycin. Organic synthetic methodology also occupied Woodward's attention. All of these synthetic, structural, and methodological contributions, extending over a period of forty years, had a dramatic cumulative effect on the study of organic chemistry. Indeed, Woodward's influence has been

described as a "revolution"; after Woodward, multistep syntheses of a complexity previously unimagined became commonplace.

Another interest of Woodward's was biogenesis, the ways organisms create the complex substances that organic chemists study. His important contributions in this area include the biogenesis of indole alkaloids and steroids. In his Nobel autobiography, Woodward described his "endless fascination" with natural products, emphasizing not only their intrinsic attractions, but also the fact that they provided "unparalleled opportunities for the discovery, testing, development, and refinement of general principles."

One such general principle emerged from a curious anomaly that Woodward encountered during his work on vitamin B-12. Collaborating on this problem with a young theoretician named Roald Hoffmann (then a Junior Fellow at Harvard, later at Cornell University), the two men formulated "orbital symmetry" rules that seemed to apply to all concerted (i.e., single-step) thermal or photochemical organic reactions. One striking aspect of this episode was the close interplay between empirical experiments and theoretical principles required to develop the idea. Another was the combination of simplicity and precision of the predictions generated by the new orbital symmetry rules, including highly specific stereochemical details of the product of the reaction. The rules had broad applicability and they invariably succeeded in predicting reaction outcomes.

Hoffmann received the Nobel Prize in chemistry in 1981 for his part in this major work. No doubt Woodward would have received a second Nobel if he had not suddenly died of a heart attack in July 1979. Perhaps the greatest organic chemist of the twentieth century passed away at the age of sixty-two.

A. R. Todd and J. W. Cornforth, "R. B. Woodward," *Biographical Memoirs of Fellows of the Royal Society* 27 (1981): 629–695. O. T. Benfey and P. J. T. Morris, eds., *Robert Burns Woodward: Architect and Artist in the World of Molecules* (2001).

ALAN J. ROCKE

WORLDVIEW. The term "worldview" became popular among English-speaking historians of science in the 1950s and 1960s as they searched for words to describe the large conceptual frameworks in which scientists situated specific theories. Its origin, though, like that of the related terms "world picture" (*Weltbild*) and "world conception" (*Weltauffasung*), lay in the German philosophical tradition.

Weltanschaung (worldview), meaning the basic presuppositions that an investigator brought to his work, was probably coined around 1800. The philosopher F. W. J. Schelling adopted the term, as did the theologian F. D. E. Schleiermacher. It found its most influential formulation in the work of Wilhelm Dilthey, for many years professor of the history of philosophy at the University of Berlin, particularly in his *Types of World View* (1911). Dilthey proposed three comprehensive views of the universe and the human situation within it: naturalism, libertarian idealism, and objective idealism, based respectively on intellect, will, and feeling. World picture and world conception, which also referred to basic conceptions, occurred less commonly. The phenomenologist Martin Heidegger favored world picture. The philosophers of science who made up the Vienna Circle, and who were no friends of Dilthey or Heidegger, both critics of modern science, opted for world conception. From the 1880s, English-speaking intellectuals began to pick up this cluster of terms, perhaps hearing them in institutions such as Harvard College and Columbia University, whose professors kept track of Continental thought, perhaps running across them in influential works translated into English such as Karl Mannheim's *Ideology and Utopia* (1936).

The frequent use of *Weltbild* in popular writings by German physicists, notably Max *Planck and Albert *Einstein, to indicate a comprehensive physical theory, helped to familiarize scientists with the concept. Historians of science were particularly influenced by the American philosopher Stephen Pepper's *World Hypotheses*, which argued for four basic ways of viewing the world (formism, mechanism, contextualism, and organicism); by the German physicist Carl von Weizäcker's *Zum Weltbild der Physik*, translated in 1952 by another American philosopher, Marjorie Grene, as *The World View of Physics* (1952); and above all by a major work on seventeenth-century science, *Mechanization of the World Picture* (1961), published in Dutch in 1950 by the historian of science E. J. Dijksterhuis. Dijksterhuis historicized the notion, arguing that the world picture of medieval natural philosophy had been succeeded by the mechanical and mathematical world picture of seventeenth-century *physics. For the next decade, historians of science commonly used terms such as "worldview" and its variants without necessarily, or even usually, accepting the Continental philosophical traditions that engendered them. Struggling to find ways of talking about a range of presuppositions that played a role in science, including religious and philosophical beliefs, that could not be captured by the more limited language of theory, observation and experiment, they found it a useful, if vague, coinage. They experimented with other

locutions too: metaphysical foundations, conceptual schemes, and paradigms. Since the 1960s, worldview has fallen by the wayside, perhaps because of the success of "paradigm," perhaps because of the shift to the social history of science.

RACHEL LAUDAN

WORLD WAR II AND COLD WAR. The blurring of boundaries between scientists, engineers, and the military was crucial to Allied success in World War II and remained a key feature of the Cold War. As a consequence, science was increasingly pulled toward engineering and technology during the latter part of the twentieth century.

Although gas attacks were widely anticipated at the outset of World War II, it became less a chemist's than a physicist's war. Despite use of gases by the Italians in Ethiopia (1935–1936), the Japanese in China (1937–1945), and the Germans in the Final Solution, gas warfare did not reach the European battlefields. The major belligerents stockpiled gases, but did not use them.

J. Robert Oppenheimer (left) and General Leslie R. Groves visiting the site of the Trinity test of the plutonium bomb two months after the trial.

The deterrent effect of stockpiles would have a major influence on Cold War thinking about nuclear deterrence. Meanwhile, chemists on all sides did produce new explosives, materials, and fuels. Allied biologists and medical personnel helped develop and deploy DDT and life-saving penicillin. The appropriation of human beings for pseudo-scientific experiments by the Germans and Japanese played a role in the later development of research safeguards for human subjects.

From 1939 to 1942 the Germans, on the offensive, used *radio to coordinate large tank and aircraft formations as well as far-flung submarine operations. The British defended with *long-range sensing by *radar, developed under such scientists as Henry Tizard, Robert Watson-Watt, John D. Cockroft, and Mark Oliphant and integrated into the Fighter Command and the Royal Navy. The magnetron made radar mobile, radically expanding its utility. P. M. S. Blackett led the development of *operations research, greatly enhancing the efficient allocation of resources, particularly in anti-submarine warfare. Further advances came from code breaking; mathematician Alan *Turing's ingenious ideas and devices laid the foundation for postwar computers. By the time the Russians and Americans had joined the war and the Allies went on the offensive, beacons, signals intelligence, range-finding radar, proximity fuses, and shaped charges offered further evidence of how much physicists and engineers were contributing to the war effort. On the German side, missiles and jet engines came too late to be significant factors.

The atomic bomb was the conclusive weapon of World War II. Under the auspices of Vannevar Bush and the massive American Office of Scientific Research and Development, the *Manhattan Project dwarfed all other technical projects. A feasibility study launched with Albert *Einstein's 1939 letter to President Franklin D. Roosevelt became a race by combined teams of American, British, and refugee scientists to produce a nuclear fission device that would deter the Germans from using such a weapon. The list of distinguished figures involved in the operation was headed by Arthur Holly Compton, Enrico *Fermi, Ernest Lawrence, and J. Robert Oppenheimer. Industrial, electrical, and chemical engineering played essential roles. Under the crush of desperate wartime measures, a much smaller Soviet team under physicist Igor V. Kurchatov worked aided by informants from the Manhattan Project. The small German team led by Werner *Heisenberg lagged behind in both materiel and organization; smaller Japanese teams led by Yoshio Nishina were even further behind.

Following the atomic bombing of Japan in August 1945, the Russians raced to catch up with the Americans. The first phase of the Cold War, from 1947 to the Cuban Missile Crisis in 1962, was marked by Soviet nuclear weapons success, a growing climate of fear and distrust in the United States that produced the Oppenheimer hearing, expanding bomber and submarine forces on both sides, and the Soviet aerospace coups of the first artificial *satellite and first manned spaceflight (see SPUTNIK). During the era, the importance of physicists for national security—and thus funding for them—increased in every major Western country. The Cold War—in large part, a preparation for a nuclear war that on occasion seemed imminent—engaged many scientists as closely as any conventional war.

Yet, as with gas in World War II, the great fear was misplaced. In the end, the Cold War was a battle for information led by operations research, code breaking, miniaturization, cybernetics, digital processing of remotely sensed imagery, behavioral science projects, computers, and electro-optics. In the United States, these developments transformed intelligence from the work of spies to the analysis of massive amounts of machine-acquired data. Along with code breaking, the American Corona and Soviet Zenith satellite programs replaced vulnerable surveillance aircraft and reduced uncertainties that could have led to the outbreak of nuclear war. After 1962, in the era of the space race, mutually assured destruction, and nuclear proliferation, these advanced means of verification stabilized Soviet-American arms agreements. Precision imagery transformed the *earth sciences and engaged scientists in universities and such agencies as the U.S. Geological Survey. Underwater surveys for submarines used vastly expanded scientific knowledge of the oceans (see OCEANOGRAPHY). Government efforts to gain influence with developing countries led to the often-problematic agricultural products of the green revolution, while the behavioral sciences received major funding for topics of political import. *Computers designed to solve nonlinear weapons and telecommunications problems gradually linked up into ARPANET, a military-sponsored network that laid the basis for the Internet. The demand for precise missile guidance and weapons targeting involved not only satellites and development of lasers, but improvements in the basic metrology of time and length. Scientific talent flowed globally into military-related advisory groups, laboratories, and programs. Weapons advisors like Edward Teller had great influence throughout the Cold War.

Many scientists decried the militarization of science, some only after intense participation in defense activities that made them ultimately question their own actions—among the most prominent were Solly Zuckerman in Great Britain, Andrei Sakharov in the USSR, and Hans Bethe, Richard Garwin, and Herbert York in the United States. The Federation of Atomic (later American) Scientists, the Union of Concerned Scientists, and the Pugwash movement gave scientists collaborative opportunities and public venues. Linus *Pauling led the successful effort to ban atmospheric nuclear testing. By the end of the Cold War, scientists raised the nightmare of "nuclear winter" and challenged the wisdom of the Strategic Defense Initiative while scholars debated how Cold War funding and priorities affected the pace, direction, and norms of science itself.

See also BLACKETT, P. M. S., AND ERNEST LAWRENCE; KURCHATOV, IGOR VASILIEVICH, AND J. ROBERT OPPEN-HEIMER; SAKHAROV, ANDREI, AND EDWARD TELLER.

Richard Hewlett et al., *A History of the United States Atomic Energy Commission*, 3 vols. (1962–1989). Solly Zuckerman, *Scientists and War: The Impact of Science on Military and Civil Affairs* (1966). John Wilson Lewis et al., *China Builds the Bomb* (Isis Studies in International Security and Arms Control) (1988). Richard Rhodes, *Dark Sun: The Making of the Hydrogen Bomb* (1995). Paul Forman and Jose-Manuel Sanchez-Ron, eds., *National Military Establishments and the Advance of Science* (1996). Robert Buderi, *The Invention that Changed the World: The Story of Radar from War to Peace* (1997). Stephen I. Schwartz, ed., *Atomic Audit: The Costs and Consequences of U.S. Nuclear Weapons since 1940* (1998). Jeffrey T. Richelson, *The Wizards of Langley: The CIA's Directorate of Science and Technology* (2001). *Social Studies of Science* 31:3, "Science in the Cold War" special issue, ed. Mark Solovey (2001). National Security Archive website at http://www.gwu.edu/~nsarchiv. Carnegie Mellon website at http://www.cmu.edu/coldwar/.

ALAN BEYERCHEN

WU, C. S. See LEE, T. D., C. S. WU, AND C. N. YANG.

X

X RAYS. When the president of the Berlin Physical Society spoke at its jubilee in 1896, he could not manage much enthusiasm about the future of its science. Later he said that, had he known about the discovery of X rays, he would instead have expressed his joy "that the second fifty years in the life of the society had begun as brilliantly as the first." His reaction was representative: from the minute Wilhelm Conrad *Röntgen made known his discovery, physicists recognized it as a tonic to their senescent science. X rays challenged theory, abetted experiments, made a public sensation, and gave doctors a diagnostic tool of unprecedented power. Until the medical profession could provide itself with the necessary apparatus, people who swallowed pins or stopped buckshot appealed to physicists to locate the mischief.

X rays refused easy classification into the available categories. They did not bend in electric or magnetic fields and thus did not belong among charged particles; and since they could not be reflected or refracted, they failed the test for light. Most physicists supposed them to be a peculiar form of electromagnetic radiation. The peculiarities of X rays included behavior unbefitting a wave, however. As the English physicist William Henry Bragg stressed, an X ray could impart to an electron almost as much energy as had gone into the ray's creation. But if a wave, the ray should have spread out from its point of origin, diffusing its energy; how then could the entire original energy reassemble when a small section of the wave front encountered an electron? This difficulty appeared the stronger when in 1912 Max von Laue and his colleagues at the University of Munich, and in 1913 Bragg and his son William Lawrence Bragg, then a student at Cambridge, showed, respectively, that a crystal can both refract and reflect X rays. Röntgen's discovery thus appeared to have properties characteristic of a wave and of a particle.

At first physicists did not worry over the properties that conflicted with the wave model confirmed by the diffraction experiments. The wave model allowed investigations of crystal structure, pioneered by the Braggs. It also made possible determination of the frequencies of the characteristic X rays emitted by the elements. These rays resemble the visible line spectrum but are simpler, depending only on the atomic number, Z, and a "screening constant" σ, indicative of the place within the atom where the electron involved in the emission of a given line ends up. The study of characteristic x-ray spectra thus helped to elucidate *atomic structure. In his influential theory of the constitution of atoms of 1922, Niels *Bohr made systematic use of x-ray data to determine the quantum numbers of atomic electrons. The doublet structure of some x-ray lines helped Samuel Goudsmit and George Uhlenbeck to construct the concept of electron spin in 1925.

Meanwhile, the American physicist Arthur Holly Compton reopened the shelved problem of the nature of X rays by his discovery in 1922 of what was soon called the Compton effect. According to Compton, a high-frequency X ray collides with an electron as if both were billiard balls. From relativity and the quantum theory Compton assigned the X ray a frequency, v; an energy, hv; and a momentum, hv/c. By assuming that energy and momentum are conserved in the collision, he obtained a relation between Δv (the ray's loss in frequency), T (the electron's gain in momentum), and the angles between the velocities of the interacting particles.

Measurement confirmed Compton's equations and lent such support to the material conception of X rays that thereafter atomic theorists felt obliged to work both particle and wave properties into their descriptions. Louis de Broglie moderated the behavior of photons (the word Albert *Einstein introduced in 1905 for hypothetical particles of high-frequency light) by coupling them to unobservable waves; Bohr, Hendrik Kramers, and John Slater abandoned the conservation of energy in considering the relation between electron jumps and emitted light. Erwin Schrödinger followed up de Broglie's lead, and Werner *Heisenberg reworked the Bohr–Kramers–Slater approach and other work of Kramers to arrive at their alternative versions of *quantum physics.

A playful Swedish depiction of a beach holiday recorded by Roentgen photography (ca. 1900).

The completion of the quantum theory of the electronic cloud of the atom did not end the usefulness of characteristic high-frequency radiation in studying the fundamental structure of matter. Nature produces very hard X rays in spontaneous nuclear decay. Beginning in the 1930s, physicists analyzed the energies of these "gamma rays" for data about nuclear transformations and for help in specifying nuclear energy levels. More recently, X rays from stars have given information about stellar processes.

The diagnostic uses of X rays soon gave rise to a new profession, radiology, and, when their effect on lesions and tumors was noticed, to therapeutics against skin diseases and cancers. Demand for more penetrating and more plentiful X rays prompted a rapid development of apparatus, culminating just before World War I in the high-voltage, heated-cathode Coolidge tube. The modest gains of physics from these developments rapidly multiplied after the war, in part because of wartime electrical engineering and surplus electrical equipment. Pioneers in California—at the newly established Caltech in Pasadena and at the University of California at Berkeley—pushed x-ray generators to gigantic sizes to produce radiation that could reach deep-lying tumors. Medically they achieved little, but technically they advanced substantially the art of high-power electrical engineering in the service of science. Some of the techniques and funding for these machines supported the early development of particle *accelerators for *nuclear physics.

The capacity of X rays to peer into previously secret places has had many applications beyond medicine. They have been used to inspect welds, test materials, fit shoes, detect dental cavities, search pyramids for mummies, etch circuits, and so on. The use of characteristic x-ray spectra to analyze the elementary makeup of even minute samples of materials became a staple in chemical assays. The optimism immediately inspired by Röntgen's discovery, and reaffirmed in the presentation to him of the first *Nobel Prize in physics (1901), has been justified repeatedly in the sciences, medicine, and industry, although long exposures and inappropriate therapies have claimed martyrs among physicians and patients.

See also RADIOACTIVITY.

P. P. Ewald, *Fifty Years of X Ray Diffraction* (1962). J. L. Heilbron, *H. G. J. Moseley* (1974). J. L. Heilbron and Robert W. Seidel, *Lawrence and His Laboratory* (1989). Bettyann H. Kevles, *Naked to the Bone* (1997).

J. L. HEILBRON

Y

YANG, C. N. See LEE, T. D., C. S. WU, AND C. N. YANG.

YOUNG, Thomas (1773–1829), natural philosopher, Egyptologist, physician, and man of letters.

The son of a cloth merchant and banker, Young was born at Milverton, Somerset, on 13 June 1773, and died at his home in Park Square, London, on 10 May 1829. His strict Quaker upbringing, which stressed the importance of education, was the major formative factor. Although he attended school for six years Young was largely self-taught and he is often cited as a child prodigy, having mastered several languages by his mid-teens. Encouraged by his kinsman Dr. Richard Brocklesby, Young then decided to pursue a career in medicine. He attended both the medical school in London founded by William Hunter and also St. Bartholomew's Hospital.

Like many dissenters, who were barred from Oxford and Cambridge, he attended the Edinburgh medical school (1794–1795) and then continued his medical training at the University of Göttingen (1795–1796), where he submitted his dissertation, entitled *De corporis humani viribus conservatricibus*, and graduated doctor of physic in July 1796. By this time Young had repudiated his Quaker background and was finally disowned in February 1798. Soon, however, he embraced the Church of England and entered Emmanuel College, Cambridge, intent on gaining a Cambridge degree that would open the way to a Fellowship of the Royal College of Physicians. He gained his Cambridge M.B. in 1803, but the degree of M.D. and the coveted Fellowship eluded him until 1808. Marrying into the minor aristocracy in 1804 further increased his distance from his Quaker upbringing and helped launch his medical career in London.

Brocklesby introduced Young to the London scientific elite, including many of the leading medical men, who were among his supporters in the Royal Society. Elected in June 1794—at the early age of twenty—he had already gained a scientific reputation from having presented a paper to the society. He closely associated himself with the Royal Society, serving as its foreign secretary from 1804 until his death, and he supported Joseph Banks and other traditionalists, who wished to maintain a cozy alliance with the aristocracy, against the reformers, who wanted active scientists to control the society.

In July 1801 he accepted the position of professor of natural philosophy at the recently founded Royal Institution. His main duty during his two-year tenure was to deliver lectures on natural philosophy; these lectures were subsequently revised and published as *A Course of Lectures on Natural Philosophy and the Mechanical Arts* (1807). A number of innovations can be traced to Young's lectures, the most important being his research on the wave theory of light, which was also published in several papers between 1799 and 1804.

Rejecting the dominant corpuscular theory of light—usually attributed to Isaac *Newton—Young developed a wave theory that attributed light to the vibrations of a ubiquitous *ether. His main intellectual sources were Leonhard *Euler and those often-overlooked passages in which Newton had entertained an ether and suggested that light is a periodic vibration. Young's primary innovation was a two-ray version of the principle of interference, which he developed from his work earlier on acoustics. He showed how this principle could explain such phenomena as the colors of thin plates and those seen when a fiber is held close to the eye. Only in the published *Lectures* of 1807 did he apply his principle to the "two-slit" experiment that has become associated with his name. Among his several later optical contributions was the proposal that polarization could be explained by assuming that light is a transverse vibration of the ether particles. Yet despite a number of insights into optical theory, Young's interpretation of the wave theory of light attracted little interest and was subsequently superseded by Augustin Fresnel's mathematically more sophisticated theory.

Young also laid the basis of a three-color theory of color sensation, often called the Young–Helmholtz–Maxwell theory. In rejecting the seven-color theory, often attributed to Newton, Young suggested that the visible spectrum is continuous but that color vision is due to just three types of receptor, according to the three principal colors: red, yellow, and blue. Having first proposed this in his 1801 Bakerian lecture, he subsequently improved it in the published *Lectures* and in an 1817 article on "Chromatics."

Among the other fields in which Young made brief but incisive excursions was hieroglyphics. He helped decipher the inscriptions on the Rosetta stone. However, his writings on this subject were less detailed and not as clearly focused as those of Jean-François Champollion; Young's main excursion was his article "Egypt," which appeared in a supplement to the *Encyclopaedia Britannica* in 1819. Despite his medical practice, which was never very remunerative, Young was principally an essayist who contributed a large number of articles on diverse subjects to the *Britannica* and to numerous medical and periodical publications. Always the gentleman scholar, his contributions show him as well informed, well read, and often brilliantly insightful but lacking in depth and application.

Alexander Wood, *Thomas Young: Natural Philosopher, 1773–1829* (1954). P. D. Sherman, *Colour Vision in the Nineteenth Century: The Young–Helmholtz–Maxwell Theory* (1981). G. N. Cantor, *Optics after Newton: Theories of Light in Britain and Ireland, 1704–1840* (1983). N. S. Kipnis, *History of the Principle of Interference of Light* (1990).

GEOFFREY CANTOR

YUKAWA, Hideki (1907–1981), Japanese theoretical physicist.

Hideki Yukawa, who originated the meson theory of nuclear forces, was born in Tokyo and spent most of his life in Kyoto. The fifth of seven children, Yukawa came from a line of Japanese scholars of samurai origin, including both grandfathers and his geologist father, Takuji Ogawa, professor of geography at Kyoto Imperial University. Hideki's youthful interests centered on literature and philosophy, although he also enjoyed mathematics. Physics took precedence for him in high school, where he had an excellent teacher and a classmate, Shin'ichirō *Tomonaga, a future *Nobel Prize winner. In 1932, Yukawa married Sumi Yukawa, adopting her family name as his own. In the same year he became a lecturer at Kyoto Imperial University (now Kyoto University), where he served as professor from 1939 to 1969.

Yukawa's quantum field theory of nuclear forces of 1934 proposed that the forces between nuclear particles, protons and neutrons, are transmitted through the exchange of "heavy quanta." This mechanism resembles the transmission of electromagnetic forces by the exchange of light quanta (photons). The field theory employing photons is called *quantum electrodynamics

(QED). While light quanta are electrically neutral and massless, Yukawa's proposed quanta carried unit electric charge (positive or negative) and had a mass intermediate between that of the electron and the proton (hence "meson"). Using a simple relation that he discovered between the range of force and the mass of the corresponding quantum, Yukawa estimated that the range of the strong nuclear force, a distance much smaller than the size of the atom, implied a meson mass about two hundred times that of the electron.

In addition to explaining the origin of the strong forces that hold nuclei together, Yukawa's theory also gave an account of beta-decay, a form of weak interaction in which a neutron decays into a proton, electron, and a very light neutral particle, the neutrino. Through its own weak interaction, a meson produced in free space would decay in less than a microsecond. Yukawa argued that, owing to this short lifetime, free mesons should not be found in nature except where, as in *cosmic rays, there is sufficient energy ($E = mc^2$). Later physicists observed mesons produced by beams of the high-energy elementary particle accelerators of the late 1940s.

Although Yukawa published his paper in English in 1934, his theory was ignored until in 1937 American cosmic-ray physicists Carl Anderson and Seth Neddermeyer discovered charged particles of intermediate mass and both signs of charge. Their result led to worldwide recognition of Yukawa's idea. Physicists in the West, and also Yukawa and his students in Japan, extended the scope of his theory. Meanwhile, American nuclear experimentalists showed that the nuclear forces between any two nuclear particles, whether neutron or proton, were the same (a property called charge-independence), suggesting that an additional electrically neutral meson was required to mediate between like particles.

Over the next decade, further cosmic-ray experiments proved that the particles discovered by Anderson and Neddermeyer did not behave as mesons should, but more like heavy electrons. They belong to a family of weakly interacting particles (leptons) and are now called "muons." In 1947 cosmic-ray researchers in Bristol, England, led by Cecil Powell used a new technique to observe the charged mesons ("pions") predicted by Yukawa and their decay into muons. (Particle accelerators in California produced the neutral pi meson in 1950.) In 1949 Yukawa, then a visiting professor at Columbia University in New York City, received the Nobel Prize in physics, the first Japanese citizen to be so honored.

The success of the meson theory established a new paradigm of elementary particle physics. The current *Standard Model of elementary particle interactions, involving *quarks that interact by exchange of particles (gluons), follows the Yukawa pattern, and has replaced meson theory as an underlying fundamental theory of nuclear forces. However, Yukawa's theory is still a useful way of understanding nuclear processes especially at the low and intermediate energies. Physicists now understand mesons, which come in many types, as composites of a quark and an antiquark.

Yukawa's Nobel Prize was a point of pride for the Japanese people at a time when they badly needed one. He received many additional accolades, notably the Research Institute for Fundamental Physics (now Yukawa Institute) founded in his honor at Kyoto University. Yukawa wrote many essays on science and Eastern thought and participated in world movements for peace.

Hideki Yukawa, *Creativity and Intuition*, trans. John Bester (1973). Hideki Yukawa, *"Tabibito" (The Traveler)*, trans. Laurie Brown and Rick Yoshida (1982). Nicholas Kemmer, "Hideki Yukawa," *Biographical Memoirs of Members of the Royal Society* 29 (1983): 661–676. Laurie M. Brown, "Yukawa, Hideki," *Dictionary of Scientific Biography* 18 (supplement II) (1990): 999–1005. Laurie M. Brown and Helmut Rechenberg, *The Origin of the Concept of Nuclear Forces* (1996).

LAURIE M. BROWN

Z

ZOOLOGICAL GARDEN. Live animals taken from the wild amused and entertained aristocratic households and were part of public culture in the ancient world. Perhaps because acquiring and maintaining living animals was more complex than transporting and sustaining living plants, zoological gardens have a history quite different from botanical gardens, although the two sometimes coexisted with each other and with natural history museums. During the nineteenth century zoos in Europe, North America, and areas affected by imperial expansion became distinguishable from earlier menageries as, over the course of the century, they moved deliberately toward public education, conservation measures, and habitat exhibition.

Exotic animal holdings acquired through trade and exploration created the basis for early systematic discussions like Aristotle's *The History of Animals*, which gave some attention to comparative aspects of species from different parts of the world. Animals kept in parks and hunting preserves were occasionally paraded before the public as status symbols of the wealth and power of their owners. Trade and exploration during the Renaissance stimulated European interest in and access to rare and exotic animals, especially from Africa and the Far East. Itinerant showmen took lions, bears, and monkeys as well as birds, snakes, and other exotica to smaller towns and diverse audiences willing to pay for novel entertainment. These individual specimens might also be documented and described by naturalists interested in identifying them according to an increasingly sophisticated taxonomy (*see* CLASSIFICATION IN SCIENCE). Only the very wealthy had resources sufficient to purchase, establish housing, and provide appropriate food and care for multiple species.

Louis XIV established a menagerie in 1665 at Versailles that eventually became part of the post-Revolution Garden of Plants in Paris in 1794. Naturalists at the Museum of Natural History managed and studied this inheritance and also live animals acquired by a French acclimatization society at midcentury. The Schönbrunn menagerie developed under Austrian emperor Franz I included a botanical garden and natural history museum, built largely on donations from other rulers and their diplomats, and unusual in being open to the public.

Zoos became part of urban culture in London (1828), Amsterdam (1839), Berlin (1844), New York (1873), and Philadelphia (1874). These independent and classic zoos emphasized the dissemination of systematic knowledge and provided descriptive labels with scientific names, geographical origin of the species, and observations about physical or behavioral characteristics. Larger zoos experimented with architecture to display their collections, using motifs from Egypt or Thailand to enhance exotic exhibitions. In 1853 the London Zoological Society built an aquarium in response to public interest in marine life, and the short-lived Liverpool Zoological Garden (1832–1863) included an aviary where birds enjoyed the company of squirrels, reptiles, and monkeys.

Wild and domestic animals gained attention as part of economic efforts to find productive sites to adapt flora and fauna to new climates during the era of imperial expansion. Often active acclimatization or zoological societies corresponded across imperial lines, from Paris to Victoria, Australia, for example. The active acclimatizers in Australia exchanged koalas, kangaroos, and swans for llamas, deer, cashmere goats, hare, trout, and other animals. Early enthusiasm waned under the expense of maintaining living animals, the frequency of disease, the failure to reproduce, and a growing recognition that some species that did acclimatize could become pests endangering farm production or indigenous wildlife. While the acclimatization projects continued, however, they stimulated systematic study of the habits of animals, the relationships between environment and distinctive but related species, and the possibilities of cross breeding.

At the end of the century public and scientific concern about the apparent danger of extinction, as well as humanitarian concern about the treatment of caged animals, pushed zoos toward conservation. William Hornaday, who had helped to establish the modest National Zoological Park (1889) attached to the Smithsonian Institution, was attracted to the Bronx Zoo (1899) by its mission to use the facility to conserve endangered species. The following decade, the merchant naturalist Karl Hagenbeck collaborated with an architect to create a dramatic Tierpark near Hamburg where moats rather than bars created safe space between observer and observed. Zookeepers at first resisted his decision to group animals ecologically or geographically, with hoofed stock, carnivores, and birds in proximity, rather than by taxonomic groups. But by the end of the century, zoos had become immensely popular, providing a borderland between wilderness and civilization attractive to middle-class audiences who willingly paid taxes and site admission to maintain them.

In the twentieth century veterinarians (then a relatively new medical field) became managers of zoos and enhanced scientific investigation of captive wild animals. At the Berlin Zoo, Oskar Heinroth, physician and ornithologist, studied his charges carefully; his analyses of animal behavior helped establish the field of ethology. Zoos far from Europe, like the Giza Zoo in Cairo, Egypt, with its significant representation of African species, provided important data to visiting naturalists and artists, whether or not they intended to go further in the field.

The world wars of the twentieth century had a significant impact on European zoos, several of which closed for lack of food, inadequate medical staff, or warfare itself; the interwar years of depression also were difficult. At the same time, zoos with resources increasingly studied nutrition and disease among captive wild animals. After World War II, many zoos in Europe were rebuilt and many in North America expanded their holdings and purposes. A significant expansion of zoos occurred in Asia, India, Japan, and parts of South America. Some of these featured regional species and many adopted the goal of conservation within their region, like the Arizona-Sonora Desert Living Museum. They created attractive, naturalistic settings and established libraries and research facilities led by professional veterinary and scientific staff. Conservation became a fundamental goal as even antagonistic nations shared rare animals in hopes of breeding nearly extinct species like the giant panda from China or the Arabian oryx.

By the end of the twentieth century, zoos were educational facilities with a purposeful and scientific orientation that emphasized environmental studies and conservation as well as recreational values. Many had collaborative arrangements with national parks and wildlife preserves that themselves were becoming metazoos.

See also Botanical Garden; Ecology; National Parks and Nature Reserves; Zoology.

Stanley Smyth Flower, *A List of Zoological Gardens of the World* (1912). Gustave Loisel, *Histoire des menageries de l'antiquite a nos jours*, 3 vols. (1912). Elizabeth Hanson, *A Cultural History of American Zoos* (1996). R. J. Hoage and William A. Deiss, eds., *New Worlds, New Animals: From Menagerie to Zoological Park in the Nineteenth Century* (1996). Vernon N. Kisling, Jr., *Zoo and Aquarium History: Ancient Animal Collections to Zoological Gardens* (2000).

Sally Gregory Kohlstedt

ZOOLOGY. Many early modern medical treatises entitle the section or book devoted to the remedies and drugs obtained from animals "Zoology." The word's current meaning as the general science of animals emerged later, with the splitting of the old domain of *natural history into several branches during the eighteenth century. Already in the sixteenth and seventeenth centuries, however, the study of animals and subfields like ornithology and entomology were the subject of independent studies by Guillaume Rondelet, William Turner, Konrad Gesner, Pierre Belon, Ulisse Aldrovandi, Thomas Moffett, John Jonston, Thomas Willis, John Ray, and Filippo Buonanni.

In 1728, the map of knowledge provided by Ephraim Chambers's *Encyclopaedia* depicted zoology as a branch of natural history, on a par with meteorology, hydrology, mineralogy, and "phytology," that is, *botany. The *Encyclopédie* of Denis Diderot and Jean d'Alembert (1751) placed under "zoologie" a list of subjects ranging from anatomy and medicine to hunting and falconry. The first edition of the *Encyclopaedia Britannica* (1771) devoted the entire article "Natural history" to zoology, reserving separate articles for botany and mineralogy. Its compilers adopted for the animal kingdom the system of classification provided by Carl *Linnaeus, which they presented as the best and the least understood among the many systems that had been invented to introduce some order into the diversity of animal life.

In the meantime, French naturalists like René-Antoine de Réaumur, Mathurin-Jacques Brisson, and George-Louis LeClerc, Comte de Buffon, and their collaborators had devoted studies of unprecedented thoroughness and breadth to whole or parts of the animal kingdom, inspiring frequent imitation. In the second half of the eighteenth century, a growing number of authors used "zoology" in the title of books devoted to the description of the animals living in a certain

A white heron as drawn around 1830 by John James Audubon for his *Birds of America*.

region or country. From the 1770s these local and national zoologies became a successful genre; naturalists throughout Europe and the colonies felt pressed to contribute the zoology of a region to the growing body of knowledge about animals. Similar works later became the source for the new field of biogeography (*see* HUMBOLDTIAN SCIENCE).

Recognition of the independence of the study of animals came in the later 1790s with the creation at the Musée d'Histoire Naturelle in Paris of chairs devoted to specialized areas of zoological research. Jean-Baptiste de Lamarck, Georges Cuvier, and Étienne Geoffroy Saint-Hilaire were among those appointed. While adhering to diverse, often conflicting views on broad, potentially inflammatory issues such as *science and religion, science and politics, *evolution, and the fundamental properties of life, these naturalists and their followers in several countries shared basic notions concerning form, function, development, and classification. They also shared basic working tools offered by the cognate fields of comparative *anatomy, *physiology, *embryology, and *paleontology. During the central decades of the nineteenth century these common notions and tools—together with the widely perceived significance for culture at large of the areas of disagreement within the science of animals—helped zoologists to establish a new, powerful academic field. Among the milestones in its development were the establishment in 1848 of the *Zeitschrift für Wissenschaftliche Zoologie* by Rudolf Albert von Kölliker, the creation in 1826 of the Zoological Society that ran the London zoo (soon imitated by similar societies elsewhere), the

foundation in 1859 of the Museum of Comparative Zoology at Harvard on the initiative of Louis Agassiz, and the creation of specialized zoological departments within the natural history museums set up in several countries during the second half of the century (*see* MUSEUM). Beginning around 1860, universities throughout Europe recognized zoology as a major specialty within their natural science schools; chairs devoted to zoology multiplied in the following decades.

Between the 1870s and the 1890s, institutional development profited from the appeal of evolutionary theories. The fame of well-known supporters of *evolution like Carl Gegenbaur and Ernst Haeckel, and the renown of the teaching in German universities, attracted zoology students from other European countries and North America. The establishment of the Zoological Station in Naples (1872) on the initiative of the German Anton Dohrn, supported by Charles *Darwin, and of the International Commission on Zoological Nomenclature (1895) indicated the international perspective of zoologists of the period.

Mainstream late-nineteenth-century zoology centered on morphology, the determination of phylogenetic relationships among living forms. From the 1890s, however, a new generation of zoologists trained within that same tradition launched a "revolt from morphology," adopting an experimental rather than a descriptive approach. The protagonists of the new trend were the German zoologists Wilhelm Roux, Theodor Boveri, Hans Driesch, and Hans Spemann and the Americans Edmund Beecher Wilson and Thomas Hunt *Morgan. With their work, substantial portions of zoological research moved from natural history into experimental biology, which had developed in the meantime pursuing goals set by experimental physiology and embryology.

In the new context, which characterized much twentieth-century frontier biology, zoology and its specialties provided convenient frameworks for teaching and institutional organization rather than major subjects for research (*see* OCEANOGRAPHY). Yet the fundamental contributions of zoologists to the evolutionary synthesis (*see* DARWINISM); the introduction of new fields of study like *ethology, behavioral *ecology, and mathematical *biology; and the discovery in the 1990s of organisms in environmental niches formerly regarded as impossibly hostile have kept zoology a thriving branch of the life sciences.

Erwin Stresemann, *Ornithology. From Aristotle to the Present*, foreword and epilogue by Ernst Mayr (1975). Mary P. Winsor, *Starfish, Jellyfish, and the Order of Life* (1976). William Coleman, *Biology in the Nineteenth Century* (1977). Garland E. Allen, *Life Science in the 20th Century* (1978). Garland E. Allen, *Thomas Hunt Morgan* (1978). Mary P. Winsor, *Reading the Shape of Nature: Comparative Zoology at the Agassiz Museum* (1991).

GIULIANO PANCALDI

FURTHER READING

Readers who seek more information than entries in this *Companion* can provide will find a rich array of reference sources relevant to history of science. In particular, those in search of biographical information about people who are not given separate entries here will find in the *Dictionary of Scientific Biography* (*DSB*), ed. Charles C. Gillispie (18 vols., including supplements and index, 1970–1990), compact accounts of life and work for figures from antiquity through the middle of the twentieth century, as well as information about major publications and archival sources. Among other specialized biographical sources are *World Who's Who in Science: A Biographical Dictionary of Notable Scientists from Antiquity to the Present*, ed. Allen G. Debus (1968); *The Biographical Dictionary of Scientists*, ed. Roy Porter and Marilyn Ogilvie (3rd ed., 2000), with "historical reviews of the sciences" as well as biographical entries; *Biographical Encyclopedia of Scientists*, ed. Richard Olson (5 vols., 1998); David Millar et al., *The Cambridge Dictionary of Scientists* (1996); and Robert M. Gascoigne, *A Historical Catalogue of Scientists and Scientific Books from the Earliest Times to the Close of the Nineteenth Century* (1984). Many entries in *McGraw-Hill Modern Scientists and Engineers* (3 vols., 1980) are autobiographical.

Of particular relevance to historians of science is a bio-bibliographical project of broad scope and long standing: Johann Christian Poggendorff, *Biographisch-literarisches Handwörterbuch der exakten Naturwissenschaften* (1863–); vols. 1–7:1 are now available on CD-ROM, ed. Sächsische Akademie der Wissenschaften zu Leipzig (2000), facilitating searches across volumes. See http://www.poggendorff.com/.

Those who wish to pursue topics addressed in this *Companion* will also profit from a bibliography of admirable scope and complexity, the annual *Current Bibliography of the History of Science and Its Cultural Influences* (formerly *Critical Bibliography* [etc.]), which appears in the journal *Isis*; various cumulations under the title *Isis Cumulative Bibliography* cover 1913–1965 (ed. Magda Whitrow), 1966–1975, 1976–1985, and 1986–1995 (ed. John Neu). Both annual and cumulative versions cover general topics in history of science, as well as specific scientific subjects and periods. The online *History of Science, Technology, and Medicine* (*HSTM*) bibliographical database made available by the Research Libraries Group (http://www.rlg.org) now integrates four bibliographies: the *Isis Current Bibliography of the History of Science* (as described above), the *Current Bibliography in the History of Technology* (as published in the journal *Technology and Culture*), the *Bibliografia Italiana di Storia della Scienza* and the *Wellcome Library for the History and Understanding of Medicine*. Updated annually, the database covers 1975 to the present. History of science is an international field, as the coverage of the *DSB*, the *Isis Current Bibliography*, and the *HSTM* database make amply clear; this essay, however, emphasizes sources in English, and is, of necessity, highly selective.

GUIDES TO REFERENCE SOURCES FOR HISTORY OF SCIENCE

The field of history of science benefits from numerous lists of specialized reference sources, among them *Reference Books for the Historian of Science: A Handlist*, ed. S. A. Jayawardene (1982), and the section on reference sources in Gordon L. Miller, *The History of Science: An Annotated Bibliography* (1992). The possibility of browsing (print) titles added annually makes Doug Stewart, *History of Science/Science Studies: Reference Sources* (http://www.wsulibs.wsu.edu/hist-of-science/) especially useful. *Starting Points in the Study of Science, Medicine and Technology* (comp. Ed Morman, 1996; rev. Christine Ruggere, 1999, http://www2.h-net.msu.edu/~smt/starting-points.html) includes chronologies, biographical sources, dictionaries and encyclopedias, periodical lists, and manuscript lists and archival guides, among other categories. The *WWW Virtual Library for the History of Science, Technology, & Medicine*, established in 1994, is now available through *echo: Exploring & Collecting History Online* (http://echo.gmu.edu/center/). See also Ronald C. Tobey, *Horus Gets In Gear. A Beginner's Guide to Research in the History of Science* (http://www.horuspublications.com/guide/tp1.html) and "History of Science" at *Teaching History's World Wide Web Links for History Teachers* (http://www.emporia.edu/socsci/journal/links.htm). Sites that are updated frequently are less likely to be compromised by broken links.

DICTIONARIES, THESAURI, ENCYCLOPEDIAS, CHRONOLOGIES, BIBLIOGRAPHIES

The tradition in reference sources represented by *A Guide to the History of Science* (1952) by George Sarton, founder of *Isis*, has continued. Research guides, dictionaries, thesauri, encyclopedias, and chronologies in English include *Companion to the History of Modern Science*, ed. R. C. Olby et al. (1990); *Dictionary of the History of Science*, ed. W. F. Bynum et al. (1981); Anton Sebastian, *A Dictionary of the History of Science* (2001); *Companion Encyclopedia of the History of Medicine*, ed. W. F. Bynum and Roy Porter (2 vols., 1993); *Encyclopaedia of the History of Science, Technology, and Medicine in Non-western Cultures*, ed. Helaine Selin (1997); *Information Sources in the History of Science and Medicine*, ed. Pietro Corsi and Paul Weindling (1983), especially part II on reference materials and sources; *Reader's Guide to the History of Science*, ed. Arne Hessenbruch (2000); and Ellis Mount and Barbara A. List, *Milestones in Science and Technology* (2nd ed., 1994). Such works as Robert M. Gascoigne, *A Chronology of the History of Science, 1450–1900* (1987); Alexander Hellemans and Bryan Bunch, *The Timetables of Science* (1988; 1991), *The History of Science and Technology: A Narrative Chronology* (2 vols., 1988, trans. of *Scienza e tecnica*, 1975); and Claire L. Parkinson, *Breakthroughs: a Chronology of Great Achievements In Science and Mathematics, 1200–1930* (1985) aid the chronologically inclined. For others, see Stewart and Morman/Ruggiere's online guides.

Works of somewhat closer focus still contain much of general interest. Routledge, for example, has published encyclopedias for history of mathematical sciences, of medicine, of technology, and of Arabic science. The series *Garland Encyclopedias in the History of Science* includes works on science and religion, science in the United States, astronomy, and scientific instruments. William E. Burns, *The Scientific Revolution: An Encyclopedia* (2001) addresses a period of central importance; Stephen G. Brush, *The History of Modern Science: A Guide to the Second Scientific Revolution, 1800–1950* (1988) is intended for undergraduate students and those who teach them. Similar sources appear in other major languages. *The Magic Lantern: A Guide to Audiovisual Resources for Teaching the History of Science, Technology, and Medicine*, ed. Robert K. DeKosky et al. (1997), produced under the auspices of the History of Science Society, and Paul S. Cohen and Brenda H. Cohen, *America's Scientific Treasures. A Travel Companion* (1998) serve specialized purposes.

Several series of reference works afford bibliographical guidance for both primary and secondary sources. These include Berkeley Papers in History of Science, Uppsala Studies in History of Science, and Garland Bibliographies of the History of Science and Technology.

Many encyclopedias, research guides, dictionaries, and similar reference works concerning science more generally may also prove useful, especially for studies of recent science. Examples include Rudi Volti, *The Facts on File Encyclopedia of Science, Technology, and Society* (3 vols., 1999); *Science & Technology Encyclopedia* (2000); and *Academic Press Dictionary of Science and Technology*, ed. Christopher Morris (1992). The *Guide to Reference Books*, ed. Robert Balay (11th ed., 1996) lists more.

Historians of science often turn to general encyclopedias in various editions. For entries on scientific subjects, see, for example, the *Encyclopedia Britannica* (first published 1771; current ed. at http://www.eb.com/); Ephraim Chambers, *Cyclopaedia* (first published 1728); John Harris, *Lexicon Technicum* (first published 1704–1710; 1st ed. reprinted 1990, 1997); and the *Encyclopédie* of Diderot and d'Alembert (1751–1765; available online at http://www.lib.uchicago.edu/efts/ARTFL/projects/encyc/).

OTHER BIOGRAPHICAL RESOURCES

Among bibliographies of biographies with special reference to science are E. Scott Barr, *An Index to Biographical Fragments in Unspecialized Scientific Journals* (1973); *Biographical Sources in the Sciences*, comp. Janet Turner et al. (1988); Leslie Howsam, *Scientists since 1660: A Bibliography of Biographies* (1997); *Prominent Scientists: An Index to Collective Biographies*, comp. Paul Pelletier (3rd ed., 1994), supplementing Norma Olin Ireland, *Index to Scientists of the World from Ancient to Modern Times: Biographies and Portraits* (1962); Roger Smith, *Biographies of Scientists: An Annotated Bibliography* (1998). Of broader scope is K. G. Saur Publishing, *World Biographical Index* (1998–; online at http://www.biblio.tu-bs.de/acwww25u/wbi_en/wbi.html and on CD-ROM as *Internationaler biographischer Index*).

Other useful biographical dictionaries take a national tack, as in *American Men and Women of Science* (20th ed., 1998–1999, now included in *SciTech Reference Plus* on CD-ROM, 1989–   ). Biographical and bio-bibliographical undertakings focused on particular fields, time periods, or both—e.g., R. V. and

P. J. Wallis, Biobibliography of British Mathematics and its Applications (1986–   )—note those not necessarily honored by inclusion in Poggendorff or national compilations. Stewart's online guide and *The History of Women and Science, Health, and Technology: A Bibliographic Guide to the Professions and the Disciplines*, ed. Phyllis Holman Weisbard (2nd ed., 1993), now available online through *BiblioLine* as *Women's Resources International* (1989, 1996–   ), are convenient sources for titles concerning women in science.

PUBLISHED SCIENTIFIC WORKS

Compilers of the Royal Society *Catalogue of Scientific Papers, 1800–1900* (19 vols. in 4 series with supplements 1867–1902, 1914–1925; continued as the *International Catalogue of Scientific Literature* covering 1901–1914 in 33 vols., 1902–1921) tried to capture a broad array of scientific literature published in periodicals, society proceedings, and so on. Stewart and Morman/Ruggiere list numerous guides to twentieth-century scientific periodicals, including those in individual fields and subspecialties. Works like Denis Grogan, *Science and Technology: An Introduction to the Literature* (3rd ed., rev. 1976); Saul Herner et al., *A Brief Guide to Sources of Scientific and Technical Information* (2nd ed., 1980); Bernard Houghton, *Scientific Periodicals: Their Historical Development, Characteristics, and Control* (1975); David M. Knight, *Natural Science Books in English, 1600–1900* (1972); H. Robert Malinowsky, *Reference Sources in Science, Engineering, Medicine, and Agriculture* (1994, especially the general descriptions in part I of scientific information practices and products); *Development of Science Publishing in Europe*, ed. A. J. Meadows (1980); Krishna Subramanyam, *Scientific and Technical Information Resources* (1981); and Richard D. Walker and C. D. Hurt, *Scientific and Technical Literature* (1990) examine scientific publishing in the print era. Recent studies such as Carol Tenopir and Donald W. King, *Towards Electronic Journals: Realities for Scientists, Librarians, and Publishers* (2000) and position papers (e.g., Jean-Claude Guédon, "In Oldenburg's Long Shadow: Librarians, Research Scientists, Publishers, and the Control of Scientific Publishing," http://www.arl.org/arl/proceedings/138/guedon.html) consider electronic possibilities. Dramatic growth and pricing policies of a few commercial science publishers have drawn particular fire: see documents concerning the case of Gordon & Breach v. American Institute of Physics and American Physical Society at www.library.yale.edu/barschall/ and http://barschall.stanford.edu/. The Scholarly Publishing and Academic Resources Coalition (http://www.arl.org/sparc/) and HighWire Press (http://highwire.stanford.edu/), among others, offer alternatives.

Those in quest of specific published works have long relied on such national resources as *The [U.S.] National Union Catalog, Pre-1956 Imprints* (754 vols., 1968–1981), the *Union List of Serials in Libraries of the United States and Canada*, ed. Edna Brown Titus (3rd ed., 5 vols., 1965), and the British Library *General Catalogue of Printed Books to 1975* (360 vols., 1979–1987), with their supplements. Online library catalogs such as *WorldCat* (http://www.oclc.org/firstsearch/), the *RLG Union Catalog* (http://www.rlg.org/libres.html), and the *British Library Public Catalogue* (http://www.bl.uk/catalogues/blpc.html) enable creative bibliographic searches; the utility of such databases depends, however, on the extent to which participating libraries have converted paper catalog records to machine-readable ones. Such projects as JSTOR (http://www.jstor.org/) make long runs of periodicals, including the *Philosophical Transactions* of the Royal Society, *Proceedings* of the (U.S.) National Academy of Sciences, and *Science*, available online.

Catalogs of specific library or personal collections often provide detailed information on individual books. For history of science, examples include *The Barchas Collection at Stanford University* (1999); Louis A. Kenney, *Catalogue of the Rare Astronomical Books in the San Diego State University Library* (1988); and *Catalog of the Naval Observatory Library* (6 vols., 1976). Judith Ann Overmier, *Scientific Rare Book Collections in Academic and Research Libraries in Twentieth Century America* (1985) provides an overview. Although William A. Cole, *Chemical Literature, 1700–1860* (1988) grew out of a personal collection impressive in its own right, its coverage is much broader, as the subtitle indicates: *A Bibliography with Annotations, Detailed Descriptions, Comparisons, and Locations*. Book dealer and auction catalogs, current and otherwise, are likewise useful. See such examples as Zeitlin & Ver Brugge, *The Physical World Encompassed* (1979) and *Classics of Science … Including Many Titles from the Robert B. and Marian S. Honeyman Collection* (1981), and Christie's auction catalog for *The Haskell F. Norman Library of Science and Medicine* (3 vols., 1998). The Antiquarian Booksellers Association of America (http://search2.abaa.org/abaa/searchform.php3) permits searches across the catalogs of multiple dealers, including those with a specialty in history of science.

Collectors' guides and exhibit catalogs offer both overview and detail, sometimes ornamented by useful images. *Thornton and Tully's Scientific Books, Libraries, and Collectors: A Study of Bibliography and the Book Trade in Relation to the History of Science*, ed. Andrew Hunter (1999) is a thorough revision of John L. Thornton and R. I. J. Tully, *Scientific Books, Libraries and Collectors* (1954; 1962; 1971). Among relevant exhibit catalogs are William B. Ashworth, *Theories of the Earth 1644–1830* (Linda Hall Library, 1984) and Anthea Waleson, *Nature Disclosed: Books from the Collections of the John Crerar Library Illustrating the History of Science* (1984). Exhibit catalogs now enjoy online counterparts, often containing more illustrations than print versions permit—e.g., "Women and Nature," http://www.library.wisc.edu/libraries/SpecialCollections/womennature/index.html.

The number and use of image compilations continue to grow. The *Album of Science Series*, ed. I. Bernard Cohen (1978-1989) is now supplemented by specialized online image collections like the *Caltech Archives PhotoNet* (http://archives.caltech.edu//photoNet.html), *Emilio Segrè Visual Archives* at the Center for History of Physics (http://www.aip.org/history/esva/), the Edgar Fahs Smith collection at the University of Pennsylvania (http://dewey.library.upenn.edu/sceti/smith/), and online catalogs of instruments compiled at http://www.mhs.ox.ac.uk/links/.

GUIDES TO UNPUBLISHED SOURCES

Historians of science rely heavily on unpublished sources, and have both benefited from and undertaken ambitious projects to collect, describe, and/or reproduce manuscripts and other unpublished resources. Cumbersome searching in multi-volume printed guides such as the *National Union Catalog of Manuscript Collections*, or *NUCMC* (1959/61–1992), has been eased by its availability online and by online library catalogs that contain descriptions of archival holdings. The Library of Congress provides at http://www.loc.gov/coll/nucmc/ a *Gateway to the Archival and Mixed Collections (AMC)* file in the *RLG Union Catalog*. Finding aids in paper to archival collections are sometimes published: e.g., Mary F. I. Smyth, *Catalogue of the Archives of the Royal Observatory, Edinburgh 1764–1937* (1981); and Jeannine Alton and Julia Latham-Jackson, *Report on the Papers and Correspondence of Otto Robert Frisch* (1982). Most online finding aids follow the standards of Encoded Archival Descriptions, as in the Bancroft Library Finding Aids (http://bancroft.berkeley.edu/collections/findingaids.html), part of the Online Archive of California. Such finding aids can be "populated" by digital images of items in the collections thus described. *Archives USA: Integrated Collection and Repository Information* (http://archives.chadwyck.com/) and RLG Archives Resources (http://www.rlg.org/arr/index.html) combine collection descriptions and links to online finding aids.

National, regional, and institutional archival guides such as Peter Harper, *Guide to the Manuscript Papers of British Scientists* (1993); *Guide to the Archives of Science in Australia: Records of Individuals*, comp. and ed. Gavan McCarthy (1991); Ann Mozley Moyal, *A Guide to the Manuscript Records of Australian Science* (1966); Edwin T. Layton, *A Regional Union Catalog of Manuscripts Relating to the History of Science and Technology Located in Indiana, Michigan, and Ohio* (1971); Keith Moore, *A Guide to the Archives and Manuscripts of the Royal Society* (1995); and Grazyna Rosi'nska, *Scientific Writings and Astronomical tables in Cracow: A Census of Manuscript Sources (XIVth–XVIth Centuries)* (1984) point to research possibilities.

A subject approach with chronological boundaries characterizes such projects as *Archival Sources for the History of Biochemistry and Molecular Biology*, ed. David Bearman and John T. Edsall (1980); Bruce Bruemmer, *Resources for the History of Computing: A Guide to U.S. and Canadian Records* (1987); *Understanding Progress as Process: Documentation of the History of Post-war Science and Technology in the United States*, ed. Clark A. Elliott (1983); Thomas S. Kuhn et al., *Sources for History of Quantum Physics: An Inventory and Report* (1967); David C. Lindberg, *A Catalogue of Medieval and Renaissance Optical Manuscripts* (1975); and Bruce R. Wheaton, *Inventory of Sources for History of Twentieth Century Physics* (1992).

Guides like Roger Hahn, *The New Calendar of the Correspondence of Pierre Simon Laplace* (1994) and *The Letters of Georges Cuvier: A Summary Calendar of Manuscript and Printed Materials Preserved in Europe, the United States of America, and Australasia*, ed. Dorinda Outram (1980) and editing projects like the Einstein Papers Project (http://www.einstein.caltech.edu/) and the Darwin Correspondence Project (http://www.lib.cam.ac.uk/Departments/Darwin/) have cast their net widely. SiliconBase (http://www-sul.stanford.edu/siliconbase/) about Silicon Valley's past and present, the Ava Helen and Linus Pauling Papers (http://www.orst.edu/Dept/Special_Collections/subpages/ahp/), the Memory Bank within *echo* (see above), and the History of Recent Science & Technology project (http://hrst.mit.edu/hrs/public/SiteInfo.htm), among others, exploit new technologies to preserve and collect unpublished sources and make them available for teaching and research. The pioneering *arXiv.org E-Print Archive* (http://arXiv.org/), as conceived by Paul Ginsparg in 1991, also deserves note: in this context "e-print" denotes "self-archiving by the author."

While this essay suggests the breadth of reference sources for history of science, it does not provide a comprehensive list of general sources. It only hints at the array of specialized reference works within history of science. It also excludes regions of overlap between science and technology, medicine, and literature, and fields of philosophy and sociology of science and science studies more generally. However, the *HSTM* database, along with other bibliographies cited in this essay, provide valuable guidance to such collateral subjects as well.

ROBIN E. RIDER

WINNERS OF THE NOBEL SCIENCE PRIZES

As detailed in the entry Nobel Prizes, the proceeds of prizes not awarded are retained by the cognizant academy either in its Main Fund (MF, below) or in the Special Funds (SF) of the responsible section (physics or chemistry at the Royal Swedish Academy of Science).

PHYSICS

1901 Wilhelm Röntgen
1902 Hendrik A. Lorentz, Pieter Zeeman
1903 Henri Becquerel, Pierre Curie, Marie Curie
1904 Lord Rayleigh
1905 Philipp Lenard
1906 J. J. Thomson
1907 Albert A. Michelson
1908 Gabriel Lippmann
1909 Guglielmo Marconi, Ferdinand Braun
1910 Johannes Diderik van der Waals
1911 Wilhelm Wien
1912 Gustaf Dalén
1913 Heike Kamerlingh Onnes
1914 Max von Laue
1915 William Bragg, Lawrence Bragg
1916 The prize money went to the SF
1917 Charles Glover Barkla
1918 Max Planck
1919 Johannes Stark
1920 Charles Edouard Guillaume
1921 Albert Einstein
1922 Niels Bohr
1923 Robert A. Millikan
1924 Manne Siegbahn
1925 James Franck, Gustav Hertz
1926 Jean Baptiste Perrin
1927 Arthur H. Compton, C. T. R. Wilson
1928 Owen Willans Richardson
1929 Louis de Broglie
1930 Venkata Raman
1931 The prize money went to the SF
1932 Werner Heisenberg
1933 Erwin Schrödinger, Paul A. M. Dirac
1934 One third of the prize money went to the MF and two thirds to the SF
1935 James Chadwick
1936 Victor F. Hess, Carl D. Anderson
1937 Clinton Davisson, George Paget Thomson
1938 Enrico Fermi
1939 Ernest Lawrence

1940 The prize money was allocated as in 1934
1941 The prize money was allocated as in 1934
1942 The prize money was allocated as in 1934
1943 Otto Stern
1944 Isidor Isaac Rabi
1945 Wolfgang Pauli
1946 Percy W. Bridgman
1947 Edward V. Appleton
1948 Patrick M. S. Blackett
1949 Hideki Yukawa
1950 Cecil Powell
1951 John Cockcroft, Ernest T. S. Walton
1952 Felix Bloch, E. M. Purcell
1953 Frits Zernike
1954 Max Born, Walther Bothe
1955 Willis E. Lamb, Polykarp Kusch
1956 William B. Shockley, John Bardeen, Walter H. Brattain
1957 Chen Ning Yang, Tsung-Dao Lee
1958 Pavel A. Cherenkov, Il'ja M. Frank, Igor Y. Tamm
1959 Emilio Segrè, Owen Chamberlain
1960 Donald A. Glaser
1961 Robert Hofstadter, Rudolf Mössbauer
1962 Lev Landau
1963 Eugene Wigner, Maria Goeppert-Mayer, J. Hans D. Jensen
1964 Charles H. Townes, Nicolay G. Basov, Aleksandr M. Prokhorov
1965 Sin-Itiro Tomonaga, Julian Schwinger, Richard P. Feynman
1966 Alfred Kastler
1967 Hans Bethe
1968 Luis Alvarez
1969 Murray Gell-Mann
1970 Hannes Alfvén, Louis Néel
1971 Dennis Gabor
1972 John Bardeen, Leon N. Cooper, Robert Schrieffer
1973 Leo Esaki, Ivar Giaever, Brian D. Josephson
1974 Martin Ryle, Antony Hewish
1975 Aage N. Bohr, Ben R. Mottelson, James Rainwater
1976 Burton Richter, Samuel C. C. Ting
1977 Philip W. Anderson, Sir Nevill F. Mott, John H. van Vleck
1978 Pyotr Kapitsa, Arno Penzias, Robert Woodrow Wilson

1979 Sheldon Glashow, Abdus Salam, Steven Weinberg
1980 James Cronin, Val Fitch
1981 Nicolaas Bloembergen, Arthur L. Schawlow, Kai M. Siegbahn
1982 Kenneth G. Wilson
1983 Subramanyan Chandrasekhar, William A. Fowler
1984 Carlo Rubbia, Simon van der Meer
1985 Klaus von Klitzing
1986 Ernst Ruska, Gerd Binnig, Heinrich Rohrer
1987 J. Georg Bednorz, K. Alex Müller
1988 Leon M. Lederman, Melvin Schwartz, Jack Steinberger
1989 Norman F. Ramsey, Hans G. Dehmelt, Wolfgang Paul
1990 Jerome I. Friedman, Henry W. Kendall, Richard E. Taylor
1991 Pierre-Gilles de Gennes
1992 Georges Charpak
1993 Russell A. Hulse, Joseph H. Taylor Jr.
1994 Bertram N. Brockhouse, Clifford G. Shull
1995 Martin L. Perl, Frederick Reines
1996 David M. Lee, Douglas D. Osheroff, Robert C. Richardson
1997 Steven Chu, Claude Cohen-Tannoudji, William D. Phillips
1998 Robert B. Laughlin, Horst L. Störmer, Daniel C. Tsui
1999 Gerardus 't Hooft, Martinus J. G. Veltman
2000 Zhores I. Alferov, Herbert Kroemer, Jack S. Kilby
2001 Eric A. Cornell, Wolfgang Ketterle, Carl E. Wieman
2002 Raymond Davis Jr., Masatoshi Koshiba, Riccardo Giacconi

CHEMISTRY

1901 Jacobus Henricus van 't Hoff
1902 Hermann Emil Fischer
1903 Svante August Arrhenius
1904 Sir William Ramsay
1905 Johann Friedrich Wilhelm Adolf von Baeyer
1906 Henri Moissan
1907 Eduard Buchner
1908 Ernest Rutherford
1909 Wilhelm Ostwald
1910 Otto Wallach
1911 Marie Curie
1912 Victor Grignard, Paul Sabatier
1913 Alfred Werner
1914 Theodore William Richards
1915 Richard Martin Willstätter
1916 The prize money went to the SF
1917 The prize money went to the SF
1918 Fritz Haber
1919 The prize money was allocated to the Special Fund of this prize section

1920 Walther Hermann Nernst
1921 Frederick Soddy
1922 Francis William Aston
1923 Fritz Pregl
1924 The prize money was allocated to the Special Fund of this prize section
1925 Richard Adolf Zsigmondy
1926 The (Theodor) Svedberg
1927 Heinrich Otto Wieland
1928 Adolf Otto Reinhold Windaus
1929 Arthur Harden, Hans Karl August Simon von Euler-Chelpin
1930 Hans Fischer
1931 Carl Bosch, Friedrich Bergius
1932 Irving Langmuir
1933 One third of the prize money went to the MF and two thirds to the SF
1934 Harold Clayton Urey
1935 Frédéric Joliot, Irène Joliot-Curie
1936 Petrus (Peter) Josephus Wilhelmus Debye
1937 Walter Norman Haworth, Paul Karrer
1938 Richard Kuhn
1939 Adolf Friedrich Johann Butenandt, Leopold Ruzicka
1940 The prize money was allocated as in 1933
1941 The prize money was allocated as in 1933
1942 The prize money was allocated as in 1933
1943 György Hevesy
1944 Otto Hahn
1945 Artturi Ilmari Virtanen
1946 James Batcheller Sumner, John Howard Northrop, Wendell Meredith Stanley
1947 Sir Robert Robinson
1948 Arne Wilhelm Kaurin Tiselius
1949 William Francis Giauque
1950 Otto Paul Hermann Diels, Kurt Alder
1951 Edwin Mattison McMillan, Glenn Theodore Seaborg
1952 Archer John Porter Martin, Richard Laurence Millington Synge
1953 Hermann Staudinger
1954 Linus Carl Pauling
1955 Vincent du Vigneaud
1956 Sir Cyril Norman Hinshelwood, Nikolay Nikolaevich Semenov
1957 Lord (Alexander R.) Todd
1958 Frederick Sanger
1959 Jaroslav Heyrovsky
1960 Willard Frank Libby
1961 Melvin Calvin
1962 Max Ferdinand Perutz, John Cowdery Kendrew
1963 Karl Ziegler, Giulio Natta
1964 Dorothy Crowfoot Hodgkin
1965 Robert Burns Woodward
1966 Robert S. Mulliken
1967 Manfred Eigen, Ronald George Wreyford Norrish, George Porter

1968 Lars Onsager
1969 Derek H. R. Barton, Odd Hassel
1970 Luis F. Leloir
1971 Gerhard Herzberg
1972 Christian B. Anfinsen, Stanford Moore, William H. Stein
1973 Ernst Otto Fischer, Geoffrey Wilkinson
1974 Paul J. Flory
1975 John Warcup Cornforth, Vladimir Prelog
1976 William N. Lipscomb
1977 Ilya Prigogine
1978 Peter D. Mitchell
1979 Herbert C. Brown, Georg Wittig
1980 Paul Berg, Walter Gilbert, Frederick Sanger
1981 Kenichi Fukui, Roald Hoffmann
1982 Aaron Klug
1983 Henry Taube
1984 Robert Bruce Merrifield
1985 Herbert A. Hauptman, Jerome Karle
1986 Dudley R. Herschbach, Yuan T. Lee, John C. Polanyi
1987 Donald J. Cram, Jean-Marie Lehn, Charles J. Pedersen
1988 Johann Deisenhofer, Robert Huber, Hartmut Michel
1989 Sidney Altman, Thomas R. Cech
1990 Elias James Corey
1991 Richard R. Ernst
1992 Rudolph A. Marcus
1993 Kary B. Mullis, Michael Smith
1994 George A. Olah
1995 Paul J. Crutzen, Mario J. Molina, F. Sherwood Rowland
1996 Robert F. Curl Jr., Sir Harold W. Kroto, Richard E. Smalley
1997 Paul D. Boyer, John E. Walker, Jens C. Skou
1998 Walter Kohn, John A. Pople
1999 Ahmed H. Zewail
2000 Alan J. Heeger, Alan G. MacDiarmid, Hideki Shirakawa
2001 William S. Knowles, Ryoji Noyori, K. Barry Sharpless
2002 John B. Fenn, Koichi Tanaka, Kurt Wüthrich

PHYSIOLOGY/MEDICINE

1901 Emil Adolf von Behring
1902 Ronald Ross
1903 Niels Ryberg Finsen
1904 Ivan Petrovich Pavlov
1905 Robert Koch
1906 Camillo Golgi, Santiago Ramón y Cajal
1907 Charles Louis Alphonse Laveran
1908 Ilya Ilyich Mechnikov, Paul Ehrlich
1909 Emil Theodor Kocher
1910 Albrecht Kossel
1911 Allvar Gullstrand
1912 Alexis Carrel

1913 Charles Robert Richet
1914 Robert Bárány
1915 The prize money went to the SF
1916 The prize money went to the SF
1917 The prize money went to the SF
1918 The prize money went to the SF
1919 Jules Bordet
1920 Schack August Steenberg Krogh
1921 The prize money was allocated to the Special Fund of this prize section
1922 Archibald Vivian Hill, Otto Fritz Meyerhof
1923 Frederick Grant Banting, John James Richard Macleod
1924 Willem Einthoven
1925 The prize money was allocated to the Special Fund of this prize section
1926 Johannes Andreas Grib Fibiger
1927 Julius Wagner-Jauregg
1928 Charles Jules Henri Nicolle
1929 Christiaan Eijkman, Sir Frederick Gowland Hopkins
1930 Karl Landsteiner
1931 Otto Heinrich Warburg
1932 Sir Charles Scott Sherrington, Edgar Douglas Adrian
1933 Thomas Hunt Morgan
1934 George Hoyt Whipple, George Richards Minot, William Parry Murphy
1935 Hans Spemann
1936 Sir Henry Hallett Dale, Otto Loewi
1937 Albert von Szent-Györgyi Nagyrapolt
1938 Corneille Jean François Heymans
1939 Gerhard Domagk
1940 One third of the prize money was allocated to the MF and two thirds to the SF
1941 The prize money was allocated as in 1940
1942 The prize money was allocated as in 1940
1943 Henrik Carl Peter Dam, Edward Adelbert Doisy
1944 Joseph Erlanger, Herbert Spencer Gasser
1945 Sir Alexander Fleming, Ernst Boris Chain, Sir Howard Walter Florey
1946 Hermann Joseph Muller
1947 Carl Ferdinand Cori, Gerty Theresa Cori, née Radnitz, Bernardo Alberto Houssay
1948 Paul Hermann Müller
1949 Walter Rudolf Hess, Antonio Caetano de Abreu Freire Egas Moniz
1950 Edward Calvin Kendall, Tadeus Reichstein, Philip Showalter Hench
1951 Max Theiler
1952 Selman Abraham Waksman
1953 Hans Adolf Krebs, Fritz Albert Lipmann
1954 John Franklin Enders, Thomas Huckle Weller, Frederick Chapman Robbins
1955 Axel Hugo Theodor Theorell
1956 André Frédéric Cournand, Werner Forssmann, Dickinson W. Richards

1957 Daniel Bovet

1958 George Wells Beadle, Edward Lawrie Tatum, Joshua Lederberg

1959 Severo Ochoa, Arthur Kornberg

1960 Sir Frank Macfarlane Burnet, Peter Brian Medawar

1961 Georg von Békésy

1962 Francis Harry Compton Crick, James Dewey Watson, Maurice Hugh Frederick Wilkins

1963 Sir John Carew Eccles, Alan Lloyd Hodgkin, Andrew Fielding Huxley

1964 Konrad Bloch, Feodor Lynen

1965 François Jacob, André Lwoff, Jacques Monod

1966 Peyton Rous, Charles Brenton Huggins

1967 Ragnar Granit, Haldan Keffer Hartline, George Wald

1968 Robert W. Holley, Har Gobind Khorana, Marshall W. Nirenberg

1969 Max Delbrück, Alfred D. Hershey, Salvador E. Luria

1970 Sir Bernard Katz, Ulf von Euler, Julius Axelrod

1971 Earl W. Sutherland, Jr.

1972 Gerald M. Edelman, Rodney R. Porter

1973 Karl von Frisch, Konrad Lorenz, Nikolaas Tinbergen

1974 Albert Claude, Christian de Duve, George E. Palade

1975 David Baltimore, Renato Dulbecco, Howard Martin Temin

1976 Baruch S. Blumberg, D. Carleton Gajdusek

1977 Roger Guillemin, Andrew V. Schally, Rosalyn Yalow

1978 Werner Arber, Daniel Nathans, Hamilton O. Smith

1979 Allan M. Cormack, Godfrey N. Hounsfield

1980 Baruj Benacerraf, Jean Dausset, George D. Snell

1981 Roger W. Sperry, David H. Hubel, Torsten N. Wiesel

1982 Sune K. Bergström, Bengt I. Samuelsson, John R. Vane

1983 Barbara McClintock

1984 Niels K. Jerne, Georges J. F. Köhler, César Milstein

1985 Michael S. Brown, Joseph L. Goldstein

1986 Stanley Cohen, Rita Levi-Montalcini

1987 Susumu Tonegawa

1988 Sir James W. Black, Gertrude B. Elion, George H. Hitchings

1989 J. Michael Bishop, Harold E. Varmus

1990 Joseph E. Murray, E. Donnall Thomas

1991 Erwin Neher, Bert Sakmann

1992 Edmond H. Fischer, Edwin G. Krebs

1993 Richard J. Roberts, Phillip A. Sharp

1994 Alfred G. Gilman, Martin Rodbell

1995 Edward B. Lewis, Christiane Nüsslein-Volhard, Eric F. Wieschaus

1996 Peter C. Doherty, Rolf M. Zinkernagel

1997 Stanley B. Prusiner

1998 Robert F. Furchgott, Louis J. Ignarro, Ferid Murad

1999 Günter Blobel

2000 Arvid Carlsson, Paul Greengard, Eric R. Kandel

2001 Leland H. Hartwell, R. Timothy (Tim) Hunt, Sir Paul M. Nurse

2002 Sydney Brenner, H. Robert Horvitz, John E. Sulston

INDEX

ILLUSTRATION SOURCES AND CREDITS

The publishers wish to thank those who have kindly given permission to reproduce illustrations at the pages identified. While every effort has been made to secure permission, we apologize if in any case we have failed to trace the copyright holder.

325   Giorgio Tabarroni, *Tre disegni inediti delle experienze de Galvani* (Bologna, 1969)

326   Science Photo Library

330   T. S. Painter, "A New Method for the Study of Chromosome Arrangements and the Plotting of Chromosome Maps," *Science*, December 22, 1933: 586. Copyright 2002 American Association for the Advancement of Science

339   The Royal Society

344   Simon Fraser/Science Photo Library

349   National Science Foundation

352   J. H. Lambert, *Pyrometrie* (1779), Pl. VII

362   The New York Public Library

370   Johannes Kepler, *Tabulae rudolphinae* (1627)

376   Science Photo Library

382   Courtesy of Nature

384   Heinrich Berghaus, *Physikalischer Atlas* (Gotha, 1845–48), pt. 7. By Permission of the British Library

389   The New York Public Library

403   Leipzig University

406   Science Museum, London/Heritage-Images

416   Science Photo Library

419   Courtesy of The Bancroft Library, University of California, Berkeley

430   After D. J. de Solla Price, *Little Science, Big Science* (Columbia University Press)

434   Otto Hahn, *Vom Radiothor zur Uranspaltung* (Braunschweig, 1962)

442   By Permission of the British Library

448   Photograph by Mario Belloni, Feb. 6, 2002

454   AKG London

460   J. A. Nollet, *Essai sur l'électricité* (2nd edn, Paris, 1750)

467   Louis Figuier, *Les merveilles de la science* (1867–70), vol. 1

469   Singer-Mendenhall Collection, Annenberg Rare Book and Manuscript Library, University of Pennsylvania

474   Rijksmuseum voor de Geschiedenis der Natuurwetenschappen, Leyden, courtesy AIP Emilio Segrè Visual Archives

480   Francesco Lana-Terzi, *Magia naturalis* (1670): 64. By Permission of the British Library

482   C. F. Gauss and Wilhelm Weber, *Resultate aus der Beobachtungen des magnetischen Vereins im Jahre 1836* (Göttingen, 1837). Courtesy University of Göttingen

486   R. G. Hewlett and O. E. Anderson Jr., *The New World, 1939–46. A History of the United States Atomic Energy Commission*, vol. 1 (Pennsylvania State University, 1962).

486   From David Irving, *The German Atomic Bomb* (Simon and Schuster, 1968)

517   Library of Congress

526   Ernest Orlando Lawrence Berkeley National Laboratory, courtesy AIP Emilio Segrè Visual Archives

528   Archiv zur Geschicte der Max-Planck-Gesellschaft, Berlin-Dahlem

532   Camille Flammarion, *Popular Astronomy*, tr. J. E. Gore (New York, 1907). Courtesy of The New York Public Library

541   The New York Public Library

554   Cornelia Hesse-Honegge (pictures and text), *Heteroptera, the Beautiful and the Other, or Images of a Mutating World* (Scalo Zürich, Berlin, New York)

560   Museum National d'Histoire Naturelle

577   The Nobel Foundation

588   Courtesy of the Lawrence Livermore National Laboratory

590   Library of Congress

604   Library of Congress

617   AKG London

626   H. G. J. Moseley, "The high frequency spectre of the elements," *Philosophical Magazine*, 26 (1913):1024–1034. Courtesy of Taylor and Francis Ltd, http://www.tandf.co.uk/journals.

633   Francis Bacon, *Novum organum* (1620)

636   Museum of the History of Science, University of Oxford

640   Science Museum/Science & Society Picture Library

653   NASA

658   John Dalton, *A New System of Chemical Philosophy*, pt. 1 (Manchester, 1808)

681   Euclid, *Opera omnia* (1703). Courtesy of the Royal Society

685   Gaspar Schott, *Mechanica hydraulica-pneumatica* (1657). Courtesy of Heritage Images

691   J. L. Heilbron, *Isis*, 58 (1967), 451–85

702   Mary Evans Picture Library

703   David Irving, *The German Atomic Bomb* (Simon and Schuster, 1968)

705   Courtesy of *Punch*

711   After Max Born, *Einstein's Theory of Relativity* (London 1924)

714   Copernicus, *De revolutionibus orbium coelestium*

720   Thomas Sprat, *The History of the Royal Society* (1667)

725   Reprinted with the permission of the Cambridge University Press

742   The Beinecke Rare Book and Manuscript Library, Yale University

747   Biblioteca Universitaria di Bologna

755   J. Trusler, *The Works of William Hogarth* (London, 1833), vol. 2

761 Bettmann/CORBIS
767 NASA
770 NASA
774 Science Photo Library
818 Gaspar Schott, *Mechanica hydraulica-pneumatica* (1657)
828 *The Illustrated London News*. From Simon Nowell-Smith, ed., *Edwardian England* (Oxford Univ. Press, 1964)
831 Corbis
836 Jean-Loup Charmet/Science Photo Library
843 The British Library/Heritage-Images

COLOR SECTION, CLOCKWISE FROM TOP LEFT

1 Karl Ernst von Baer, *Über Entwickelungsgeschichte der Thiere* (Koenogsberg 1828)

2 Museum of American History, Smithsonian Institution
   Museum of the History of Science, University of Oxford
   Heine Sadler, *Sonne, Zeit und Ewigkeit* (Harebnburg 1985)
3 Alte Pinakothek, Munich
   Stedelijk Museum de Lakenhal, Leiden
4 The Bancroft Library, University of California, Berkeley
5 CERN
   Fermilab
6 Dr. Seyyed Hossein Nasr
7 Department of Oriental Antiquities, The British Museum
8 *Scientific American*
   Henry E. Roscoe, *Spectrum Analysis* (1885)
   Otto Lehman, *Die elektrishchen Lichterscheinungen* (1898)